BIOLOGY

Second Edition

Robert J. Brooker
University of Minnesota – Minneapolis

Eric P. Widmaier
Boston University

Linda E. Graham
University of Wisconsin – Madison

Peter D. Stiling
University of South Florida

Mc
Graw
Hill

Connect
Learn
Succeed™

BIOLOGY, SECOND EDITION

1 2 3 4 5 6 7 8 9 0 DOW / DOW 1 0 9 8 7 6 5 4 3 2 1 0

ISBN 978–0–07–122180–1
MHID 0–07–122180–8

www.mhhe.com

Brief Contents

About the Authors

Robert J. Brooker

Rob Brooker (Ph.D., Yale University) received his B.A. in biology at Wittenberg University in 1978. At Harvard, he studied the lactose permease, the product of the *lacY* gene of the *lac* operon. He continues working on transporters at the University of Minnesota, where he is a Professor in the Department of Genetics, Cell Biology, and Development and has an active research laboratory. At the University of Minnesota, Dr. Brooker teaches undergraduate courses in biology, genetics, and cell biology. In addition to many other publications, he has written three editions of the undergraduate genetics text *Genetics: Analysis & Principles*, McGraw-Hill, copyright 2009.

Eric P. Widmaier

Eric Widmaier received his Ph.D. in 1984 in endocrinology from the University of California at San Francisco. His research is focused on the control of body mass and metabolism in mammals, the hormonal correlates of obesity, and the effects of high-fat diets on intestinal cell function. Dr. Widmaier is currently Professor of Biology at Boston University, where he recently received the university's highest honor for excellence in teaching. Among other publications, he is a coauthor of *Vander's Human Physiology: The Mechanisms of Body Function*, 11th edition, published by McGraw-Hill, copyright 2008.

Linda E. Graham

Linda Graham received her Ph.D. in botany from the University of Michigan, Ann Arbor. Her research explores the evolutionary origin of land-adapted plants, focusing on their cell and molecular biology as well as ecological interactions. Dr. Graham is now Professor of Botany at the University of Wisconsin–Madison. She teaches undergraduate courses in biology and plant biology. She is the coauthor of, among other publications, *Algae*, copyright 2000, a major's textbook on algal biology, and *Plant Biology*, copyright 2006, both published by Prentice Hall/Pearson.

Left to right: Eric Widmaier, Linda Graham, Peter Stiling, and Rob Brooker

Peter D. Stiling

Peter Stiling obtained his Ph.D. from University College, Cardiff, Wales, in 1979. Subsequently, he became a postdoc at Florida State University and later spent 2 years as a lecturer at the University of the West Indies, Trinidad. During this time, he began photographing and writing about butterflies and other insects, which led to publication of several books on local insects. Dr. Stiling is currently a Professor of Biology at the University of South Florida at Tampa. He teaches graduate and undergraduate courses in ecology and environmental science as well as introductory biology. He has published many scientific papers and is the author of *Ecology: Global Insights and Investigations*, soon to be published by McGraw-Hill. Dr. Stiling's research interests include plant-insect relationships, parasite-host relationships, biological control, restoration ecology, and the effects of elevated carbon dioxide levels on plant herbivore interactions.

*The authors are grateful for the help, support,
and patience of their families, friends, and students,
Deb, Dan, Nate, and Sarah Brooker,
Maria, Rick, and Carrie Widmaier,
Jim, Michael, and Melissa Graham, and
Jacqui, Zoe, Leah, and Jenna Stiling.*

Improving Biology Education: We Listened to You

A Step Ahead

A Step Ahead describes what we set out to accomplish with this second-edition textbook. As authors and educators, we know your goal is to ensure your students are prepared for the future—their future course work, lab experiences, and careers in the sciences. Building a strong foundation in biology will put your students a step ahead on this path.

The illustrations are outstanding and better than in most textbooks. They are clear, eye-catching, and compactly illustrate the important features without the cluttering that so often accompanies diagrams. The essential features can be seen and understood at a glance.

Harold Heatwole, North Carolina State University

Through our classroom experiences and research work, we became inspired by the prospect that the first edition of *Biology* could move biology education forward. We are confident that this new edition of *Biology* is a step ahead because we listened to you. Based on our own experience and our discussions with educators and students, we continue to concentrate our efforts on these crucial areas:

- Experimentation and the process of science
- Modern content
- Evolutionary perspective
- Emphasis on visuals
- Accuracy and consistency
- Critical thinking
- Media—active teaching and learning with technology

Continued feedback from instructors using this textbook has been extremely valuable in refining the presentation of the material. Likewise, we have used the textbook in our own classrooms. This hands-on experience has provided much insight regarding areas for improvement. Our textbook continues to be comprehensive and cutting-edge, featuring an evolutionary focus and an emphasis on scientific inquiry.

The first edition of *Biology* was truly innovative in its visual program, and with the second edition it remains a step ahead. In watching students study as well as in extensive interviews, it is clear that students rely heavily on the artwork as their primary study tool. As you will see when you scan through our book, the illustrations have been crafted with the student's perspective in mind. They are very easy to follow, particularly those that have multiple steps, and have very complete explanations of key concepts. We have taken the approach that students should be able to look at the figures and understand the key concepts, without having to glance back and forth between the text and art. Many figures contain text boxes that explain what the illustration is showing. In those figures with multiple steps, the boxes are numbered and thereby guide the students through biological processes.

A Step Ahead in Serving Teachers and Learners

To accurately and thoroughly cover a course as wide ranging as biology, we felt it was essential that our team reflect the diversity of the field. We saw an opportunity to reach students at an early stage in their education and provide their biology training with a solid and up-to-date foundation. We have worked to balance coverage of classic research with recent discoveries that extend biological concepts in surprising new directions or that forge new concepts. Some new discoveries were selected because they highlight scientific controversies, showing students that we don't have all the answers yet. There is still a lot of work for new generations of biologists. With this in mind, we've also spotlighted discoveries made by diverse people doing research in different countries to illustrate the global nature of modern biological science.

As active teachers and writers, one of the great joys of this process for us is that we have been able to meet many more educators and students during the creation of this textbook. It is humbling to see the level of dedication our peers bring to their teaching. Likewise, it is encouraging to see the energy and enthusiasm so many students bring to their studies. We hope this book and its media package will serve to aid both faculty and students in meeting the challenges of this dynamic and exciting course. For us, this remains a work in progress, and we encourage

This is an excellent textbook for biology majors, and the students should keep the book as a future reference. The thoughts flow very well from one topic to the next.

Gary Walker, Youngstown State University

you to let us know what you think of our efforts and what we can do to serve you better.

Rob Brooker, Eric Widmaier, Linda Graham, Peter Stiling

CHANGES TO THIS EDITION

The author team is dedicated to producing the most engaging and current textbook that is available for undergraduate students who are majoring in biology. We want our students to be inspired by the field of biology and to become critical thinkers. To this end, we have made the following changes throughout the entire book.

- Each chapter in the second edition begins with an interesting story or a set of observations that will capture the students' interests as they begin to read a chapter.
- To help students test their knowledge and critical-thinking skills, we have increased the number of Concept check questions that are associated with the figure legends and revised many of the questions at the end of each chapter so they are at a higher level in Bloom's taxonomy. An answer key for the questions is now provided in an appendix at the end of the book.
- To further help students appreciate the scientific process, the Feature Investigation in each chapter now includes three new elements: a Conclusion, the original journal citation for the experiment, and questions that are directly related to the experiment.
- Many photographs and micrographs have been enlarged or replaced with better images.
- The presentation of the material has been refined by dividing some of the chapters into smaller sections and by the editing of complex sentences.

With regard to the scientific content in the textbook, the author team has worked with hundreds of faculty reviewers to refine the first edition and to update the content so that our students are exposed to the most cutting-edge material. Some of the key changes that have occurred are summarized below.

Chemistry Unit

- **Chapter 2. The Chemical Basis of Life I: Atoms, Molecules, and Water:** This stage-setting chapter now introduces the concepts of matter and energy, chemical equilibrium, condensation/hydrolysis reactions, and expands upon the properties of water (for example, introducing such concepts as specific heat). The nature and importance of radioisotopes in biology and medicine has also been expanded and clarified, along with a new photo of a whole-body PET scan of a person with cancer.
- **Chapter 3. The Chemical Basis of Life II: Organic Molecules:** Enzymes are now defined in this early chapter. A new figure has been added that reinforces and elaborates upon the mechanism and importance of dehydration and hydrolysis reactions, which were first introduced in Chapter 2. This figure includes the principles of polymer formation and breakdown. Carbohydrates, lipids, proteins, and nucleic acids have been reorganized into distinct major headings for sharper focus.

Cell Unit

- **Chapter 4. General Features of Cells:** You will find improved illustrations of the cytoskeleton and new content regarding the origin of peroxisomes. The chapter has a new section on Protein Sorting to Organelles and ends with a new

section called System Biology of Cells: A Summary, which summarizes the content of Chapter 4 from a systems biology perspective.
- **Chapter 5. Membrane Structure, Synthesis, and Transport:** This chapter has a new section on the Synthesis of Membrane Components in Eukaryotic Cells. In this section, you will find a description of how cells make phospholipids, a critical topic that is often neglected.
- **Chapter 6. An Introduction to Energy, Enzymes, and Metabolism:** Based on reviewer comments, this newly created chapter splits the material that was originally in Chapter 7 of the first edition. Chapter 6 provides an introduction to energy, enzymes, and metabolism. It includes added material on ribozymes and a novel section at the end of the chapter that describes the important topic of how cells recycle the building blocks of their organic macromolecules.
- **Chapter 7. Cellular Respiration, Fermentation, and Secondary Metabolism:** In the second edition, Chapter 7 is now divided into three sections: Cellular Respiration in the Presence of Oxygen, Anaerobic Respiration and Fermentation, and Secondary Metabolism.
- **Chapter 8. Photosynthesis:** The discussion of the light-dependent reactions is now divided into two sections: Reactions that Harness Light Energy and Molecular Features of Photosystems.
- **Chapter 9. Cell Communication:** Two sections that were in the first edition on Cellular Receptors and Signal Transduction and the Cellular Response have been streamlined and simplified. A new section called Apoptosis: Programmed Cell Death has been added, which includes a pioneering Feature Investigation that describes how apoptosis was originally discovered.
- **Chapter 10. Multicellularity:** The figures in this chapter have been greatly improved with a greater emphasis on orientation diagrams that help students visualize where an event is occurring in the cell or in a multicellular organism.

Genetics Unit

- **Chapter 11. Nucleic Acid Structure, DNA Replication, and Chromosome Structure:** The section on Chromosome Structure has been moved from Chapter 15 in the first edition to this chapter so that the main molecular features of the genetic material are contained within a single chapter. To help students grasp the major concepts, the topic of DNA replication has been split into two sections: Overview of DNA Replication and Molecular Mechanism of DNA Replication.
- **Chapter 12. Gene Expression at the Molecular Level:** Several topics in this chapter have been streamlined to make it easier for students to grasp the big picture of gene expression.
- **Chapter 13. Gene Regulation:** Topics in gene regulation, such as micro RNAs, have been updated.
- **Chapter 14. Mutation, DNA Repair, and Cancer:** Information regarding the effects of oncogenes has been

modified so that students can appreciate how mutations in particular oncogenes and tumor suppressor genes promote cancer.

- **Chapter 15. The Eukaryotic Cell Cycle, Mitosis, and Meiosis:** This chapter now begins with a section on the eukaryotic cell cycle, which was in Chapter 9 of the first edition. This new organization allows students to connect how the cell cycle is related to mitosis and meiosis. Also, a new Genomes and Proteomes Connection on cytokinesis has been added, which explains new information on how cells divide.
- **Chapter 16. Simple Patterns of Inheritance:** To make the topics stand out better for students, this chapter has been subdivided into six sections: Mendel's Laws of Inheritance, The Chromosome Theory of Inheritance, Pedigree Analysis of Human Traits, Sex Chromosomes and X-Linked Inheritance Patterns, Variations in Inheritance Patterns and Their Molecular Basis, and Genetics and Probability.
- **Chapter 17. Complex Patterns of Inheritance:** The coverage of X inactivation, genomic imprinting, and maternal effect genes has been streamlined to focus on their impacts on phenotypes.
- **Chapter 18. Genetics of Viruses and Bacteria:** In response to reviewers of the first edition, this chapter now begins with viruses. The topics of viroids and prions are set apart in their own section. Also, the information regarding bacterial genetics comes at the end of the chapter and is divided into two sections on Genetic Properties of Bacteria and on Gene Transfer Between Bacteria.
- **Chapter 19. Developmental Genetics:** Invertebrate development has been streamlined to focus on the major themes of development. The topic of stem cells has been updated with new information regarding their importance in development and their potential uses in medicine.
- **Chapter 20. Genetic Technology:** New changes to this chapter include an improved figure on polymerase chain reaction (PCR) and new information regarding the engineering of Bt crops in agriculture.
- **Chapter 21. Genomes, Proteomes, and Bioinformatics:** This chapter has been updated with the newest information regarding genome sequences. Students are introduced to the NCBI website, and a collaborative problem at the end of the chapter asks the students to identify a mystery gene sequence using the BLAST program.

Evolution Unit

- **Chapter 22. The Origin and History of Life:** The topic of fossils has been separated into its own section. The second edition has some new information regarding ideas about how polymers can be formed abiotically in an aqueous setting. The role of oxygen has been expanded.
- **Chapter 23. An Introduction to Evolution:** To help the students make connections between genes and traits, newly discovered examples, such as the role of allelic differences in the *Igf2* gene among dog breeds, have been added.

- **Chapter 24. Population Genetics:** To bring the topics into sharper focus, this chapter is now subdivided into five sections: Genes in Populations, Natural Selection, Sexual Selection, Genetic Drift, and Migration and Nonrandom Mating. The important topic of single nucleotide polymorphisms is highlighted near the beginning of the chapter along with its connection to personalized medicine.
- **Chapter 25. Origin of Species and Macroevolution:** The topic of species concepts has been updated with an emphasis on the general lineage concept. Sympatric speciation has been divided into three subtopics: Polyploidy, Adaptation to Local Environments, and Sexual Selection.
- **Chapter 26. Taxonomy and Systematics:** The chapter begins with a modern description of taxonomy that divides eukaryotes into eight supergroups. To make each topic easier to follow, the chapter is now subdivided into five sections: Taxonomy, Phylogenetic Trees, Cladistics, Molecular Clocks, and Horizontal Gene Transfer.

Diversity Unit

- **Chapter 27. Bacteria and Archaea:** In this chapter featuring bacterial and archeal diversity, several illustrations have been improved. New information has been added to the Feature Investigation highlighting radiation resistance in Deinococcus.
- **Chapter 28. Protists:** In this exploration of protist diversity, recent research findings have been incorporated into chapter organization and phylogenetic trees. The evolutionary and ecological importance of cryptomonads and haptophytes are emphasized more completely. Life-cycle diagrams have been improved for clarity. A new Genomes and Proteomes Connection features genomic studies of the human pathogens trichomonas and giardia.
- **Chapter 29. Plants and the Conquest of Land:** This chapter on seedless plant diversity incorporates new molecular phylogenetic information on relationships. A new Genomes and Proteomes Connection features the model fern genus *Ceratopteris*. Life cycles have been improved for greater clarity.
- **Chapter 30. The Evolution and Diversity of Modern Gymnosperms and Angiosperms:** This chapter, highlighting seed plant diversity, features a new Genomes and Proteomes Connection on the role of whole genome duplication via autopolyploidy and allopolyploidy in the evolution of seed plants.
- **Chapter 31. Fungi:** The fungal diversity chapter's position has been changed to emphasize the close relationship of fungi to animals. There is an increased emphasis upon the role of fungi as pathogens and in other biotic associations. For example, a new Genomes and Proteomes Connection explores the genetic basis of beneficial plant associations with ectomycorrhizal fungi, and a new Feature Investigation features experiments that reveal a partnership between a virus and endophytic fungi that increases heat tolerance in plants. Life cycles of higher fungi have been modified to highlight heterokaryotic phases.

- **Chapter 32: An Introduction to Animal Diversity:** A brief evolutionary history of animal life has been added. A new figure shows the similarity of a sponge to its likely ancestor, a colonial choanoflagellate. The summary characteristics of the major animal phyla have been simplified.

- **Chapter 33: The Invertebrates:** With the huge number of invertebrate species and the medical importance of many, a new Genomes and Proteomes Connection discusses DNA barcoding, which may allow for rapid classification of species. The taxonomy of the annelids, arthropods, and chordates has been updated.

- **Chapter 34: The Vertebrates:** The organization of the section headings now follows the vertebrate cladogram introduced at the start of the chapter. A more modern approach to the taxonomy of vertebrates has been adopted, particularly in the discussion of primates. In addition, there is an extended treatment of human evolution and a new Genomes and Proteomes Connection comparing the human and chimpanzee genetic codes.

Plant Unit

- **Chapter 35. An Introduction to Flowering Plant Form and Function:** This overview of flowering plant structure and function has been revised to better serve as an introduction to Chapters 36–39. A new Genomes and Proteomes Connection features the genetic control of stomatal development and emphasizes the role of asymmetric division in the formation of specialized plant cells. A new Feature Investigation reveals how recent experiments have demonstrated the adaptive value of palmate venation in leaves.

- **Chapter 36. Flowering Plants: Behavior:** In this chapter on plant behavior, the general function of plant hormones in reducing gene repression, thereby allowing gene expression, provides a new unifying theme. As an example, new findings on the stepwise evolution of the interaction between gibberellin and DELLA proteins are highlighted. The Feature Investigation, highlighting classic discoveries concerning auxin's role in phototropism, has been condensed to achieve greater impact.

- **Chapter 37. Flowering Plants: Nutrition:** In this chapter on plant nutrition, a new Genomes and Proteomes Connection features the development of legume-rhizobium symbioses.

- **Chapter 38. Flowering Plants: Transport:** In this chapter on plant transport, the recent use of synthetic tree models has been added to further highlight experimental approaches toward understanding plant structure-function relationships.

- **Chapter 39. Flowering Plants: Reproduction:** In this chapter on flowering plant reproduction, greater attention is paid to the trade-offs of sexual versus asexual reproduction, explaining why both commonly occur and are important in nature and agricultural applications. A new Genomes and Proteomes Connection describes a study of the evolution of plants that reproduce via only asexual means from sexually reproducing ancestral species.

Animal Unit

Key changes to the animal unit include reorganization of the chapters such that animal nervous systems are presented first, an expanded emphasis on comparative features of invertebrate and vertebrate animal biology, and updates to each of the Impact on Public Health sections.

- **Chapter 40. Animal Bodies and Homeostasis:** This opening chapter has numerous new and improved photos and illustrations, such as those associated with an expanded discussion of different types of connective and epithelial tissue. The utility of the Fick diffusion equation has now been explained, and the very important relationship between surface area and volume in animals has been more thoroughly developed.

- **Chapter 41. Neuroscience I: Cells of the Nervous System:** Discussion of animal nervous systems has been moved to the beginning chapters of the animal unit, rather than appearing midway through the unit. This was done to better set the stage for all subsequent chapters. In this way, students will gain an appreciation for how the nervous system regulates the functions of all other organ systems. This concept will be continually reinforced as the students progress through the unit. Specific changes to Chapter 41 include an expanded treatment of equilibrium potential, a new discussion and figure on spatial and temporal summation in neurons, and a false-color SEM image of a synapse.

- **Chapter 42. Neuroscience II: Evolution and Function of the Brain and Nervous Systems:** In this second chapter devoted to nervous systems, the many functions of individual regions of animals' brains have been more extensively described and also summarized for easy reference in a new table. The epithalamus is now included in this discussion, and the structure and function of the autonomic nervous system has received expanded coverage.

- **Chapter 43. Neuroscience III: Sensory Systems:** An expanded, detailed, and step-by-step treatment of visual and auditory signaling mechanisms has been added to this third and concluding chapter on animal nervous systems. A fascinating comparison of the visual fields of predator and prey animals has been added, along with a figure illustrating the differences. A new figure illustrating how people see the world through eyes that are diseased due to glaucoma, macular degeneration, or cataracts is now included.

- **Chapter 44. Muscular-Skeletal System and Locomotion:** The events occurring during cross-bridge cycling in muscle have been newly illustrated and detailed. A new figure showing the histologic appearance of healthy versus osteoporotic bones, and the skeleton of a child with rickets has been added.

- **Chapter 45. Nutrition, Digestion, and Absorption:** An overview figure illustrating the four basic features of energy assimiliation in animals has been added to the beginning of the chapter to set the stage for the later discussions of ingestion, digestion, absorption, and elimination. A more developed comparative emphasis on ingestive and digestive processes in animals has been added, with expanded treatment of adaptations that occur in animals that live in freshwater or marine environments. This is accompanied by newly added photographs of different animals' teeth being used to chew, tear, grasp, and nip food in their native environments.

- **Chapter 46. Control of Energy Balance, Metabolic Rate, and Body Temperature:** The text and artwork in this chapter have been considerably streamlined to emphasize major principles of fat digestion and absorption in animals.
- **Chapter 47. Circulatory Systems:** The local and systemic relationships between pressure, blood flow, and resistance are now distinguished more clearly from each other and described in separate sections to emphasize the differences between them. The organization of major topics has been adjusted to better reflect general principles of circulatory systems that apply across taxa, as well as comparative features of vertebrate circulatory systems.
- **Chapter 48. Respiratory Systems:** This chapter has benefited from a general upgrade in artwork, but particularly that of the human respiratory system, including the addition of a cross section through alveoli to illustrate their cellular structures and associations with capillaries.
- **Chapter 49. Excretory Systems and Salt and Water Balance:** A new photo of proximal tubule cells that reveals their extensive microvilli has been added to reinforce the general principle of surface-area adaptations described in earlier chapters. A major reorganization of the manner in which the anatomy and function of nephrons has been introduced; each part of the nephron has now been separated into multiple figures for easier understanding.
- **Chapter 50. Endocrine Systems:** The layout of many figures has been adjusted to improve readability and flow; this has also been facilitated with new orientation illustrations that reveal where within the body a given endocrine organ is located. Along with the new layouts, several figures have been simplified to better illustrate major concepts of hormone synthesis, action, and function. As part of a unit-wide attempt to increase quantitative descriptions of animal biology, additional data have been added in the form of a bar graph to this chapter's Feature Investigation.
- **Chapter 51. Animal Reproduction:** The major concepts of asexual and sexual reproduction have been pulled together from various sections of the text into a newly organized single section immediately at the start of the chapter. This reorganization and consolidation of material has eliminated some redundancy, but more importantly allows for a direct, integrated comparison of the two major reproductive processes found in animals. In keeping with a unit-wide effort to improve the flow of major illustrations, certain complex, multipart figures have been broken into multiple figures linked with the text.
- **Chapter 52. Animal Development:** To better allow this chapter to be understood on its own, a new introductory section has been added that reinforces basic concepts of cellular

and molecular control of animal development that were first introduced in Chapter 19 (Developmental Genetics). The complex processes occurring during gastrulation have been rendered in a newly simplified and clarified series of illustrations. The treatment of Frzb and Wnt proteins in the Genomes and Proteomes Connection has been removed and replaced with a discussion of Spemann's organizer to better reflect the genetic basis of development across taxa in animals. An amazing series of photographs depicting cleft lip/palate and its surgical reconstruction has also been added to the Impact on Public Health section.
- **Chapter 53. Immune Systems:** A key change in this chapter is the effective use of additional or reformatted text boxes in illustrations of multistep processes. The layout of nearly every figure has been modified for clarity and ease of understanding. The topic of specific immunity has been reorganized such that the cellular and humoral aspects of immunity are clearly defined and distinguished. A new figure illustrating clonal selection has been added.

Ecology Unit

- **Chapter 54: An Introduction to Ecology and Biomes:** A new table summarizes the various abiotic factors and their effects on organisms. New information on greenhouse gases is provided, including their contributions to global warming.
- **Chapter 55: Behavioral Ecology:** Portions of the section on mating systems have been rewritten in this updated chapter on behavior.
- **Chapter 56: Population Ecology:** Additional information on population growth models has been provided by discussing the finite rate of increase, λ, and by discussing growth of black-footed ferret populations in Wyoming, which are recovering after being pushed to the brink of extinction.
- **Chapter 57: Species Interactions:** This new treatment of species interactions has been streamlined, but at the same time, new information is provided on how shark fishing along the eastern seaboard of the United States has disrupted marine food webs.
- **Chapter 58: Community Ecology:** The content of this chapter has been updated and rewritten, and historical information regarding community recovery following volcanic eruption on the island of Krakatau, Indonesia, has been added. The section of species richness has also been reorganized.
- **Chapter 59: Ecosystems Ecology:** New art and text on the pyramid of numbers has been provided in the first section. The carbon cycle has been rewritten, and information on net primary production in different biomes has been updated.
- **Chapter 60: Biodiversity and Conservation Biology:** The link between biodiversity and ecosystem function has been underscored by better explaining Tilman's field experiments. The chapter also provides a new section on climate change as a cause of species extinction and loss of biodiversity. A new discussion of bioremediation has been provided in the restoration ecology section.

A STEP AHEAD IN PREPARING STUDENTS FOR THE FUTURE

I really like the Feature Investigation so students can begin to grasp how scientists come to the conclusions that are simply presented as facts in these introductory texts.

Richard Murray, Hendrix College

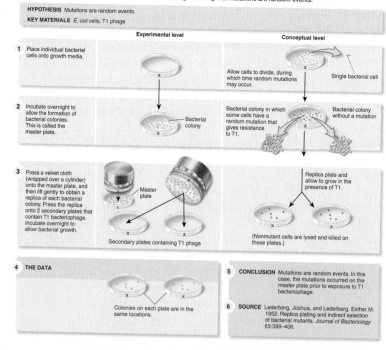

FEATURE INVESTIGATION

The Lederbergs Used Replica Plating to Show That Mutations Are Random Events

Mutations can affect the expression of genes in a variety of ways. Let's now consider the following question: Do mutations that affect the traits of an individual occur as a result of pre-existing circumstances, or are they random events that may happen in any gene of any individual? In the 19th century, French naturalist Jean Baptiste Lamarck proposed that physiological events (such as use or disuse) determine whether traits are passed along to offspring. For example, his hypothesis sug-

gested that an individual who practiced and became adept at a physical activity, such as the long jump, would pass that quality on to his or her offspring. Alternatively, geneticists in the early 1900s suggested that genetic variation occurs as a matter of chance. According to this view, those individuals whose genes happen to contain beneficial mutations are more likely to survive and pass those genes to their offspring.

These opposing views were tested in bacterial studies in the 1940s and 1950s. One such study, by Joshua and Esther Lederberg, focused on the occurrence of mutations in bacteria (Figure 14.2). First, they placed a large number of *E. coli* bac-

Figure 14.2 The experiment performed by the Lederbergs showing that mutations are random events.

EXPERIMENTAL APPROACH

Feature Investigations provide a complete description of experiments, including data analysis, so students can understand how experimentation leads to an understanding of biological concepts. There are two types of *Feature Investigations*. Most describe experiments according to the scientific method. They begin with observations and then progress through the hypothesis, experiment, data, and the interpretation of the data (conclusion). Some *Feature Investigations*

involve discovery-based science, which does not rely on a preconceived hypothesis. The illustrations of the *Feature Investigations* are particularly innovative by having parallel drawings at the experimental and conceptual levels. By comparing the two levels, students will be able to understand how the researchers were able to interpret the data and arrive at their conclusions.

This is one of the best features of these chapters. It is absolutely important to emphasize evolution themes at the molecular level in undergraduate biology courses.

Jorge Busciglio, University of California – Irvine

EVOLUTIONARY PERSPECTIVE

Modern techniques have enabled researchers to study many genes simultaneously, allowing them to explore genomes (all the genes an organism has) and proteomes (all the proteins encoded by those genes). This allows us to understand biology in a more broad way. Beginning in Chapter 3, each chapter has a topic called the *Genomes & Proteomes Connection* that provides an understanding of how genomes and proteomes underlie the inner workings of cells and explains how evolution works at the molecular level. The topics that are covered in the *Genomes & Proteomes Connection* are very useful in preparing students for future careers in biology. The study of genomes and proteomes has revolutionized many careers in biology, including those in medicine, research, biotechnology, and many others.

Conceptual Questions

1. The electron transport chain is so named because electrons are transported from one component to another. Describe the purpose of the electron transport chain.

2. What causes the rotation of the γ subunit of the ATP synthase? How does this rotation promote ATP synthesis?

3. During fermentation, explain why it is important to oxidize NADH to NAD⁺.

CRITICAL THINKING

Students can test their knowledge and critical thinking skills with the *Concept check* questions that are associated with the figure legends. These questions go beyond simple recall of information and ask students to apply or interpret information presented in the illustrations.

Conceptual Questions can be found at the end of each chapter. Again, these questions take students a step ahead in their thought process by asking them to explain, describe, differentiate, distinguish, and so on, key concepts of the chapter.

A VISUAL OUTLINE

Working with a large team of editors, scientific illustrators, photographers, educators, and students, the authors have created an accurate, up-to-date, and visually appealing illustration program that is easy to follow, realistic, and instructive. The artwork and photos serve as a "visual outline" and guide students through complex processes.

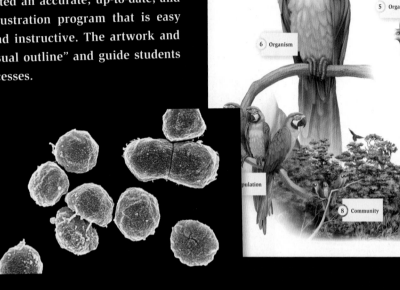

I'm very impressed with the accuracy and quality of the figures. I especially like the explanatory captions within certain figures.

Ernest DuBrul, University of Toledo

The illustrations were very effective in detailing the processes. The drawings were more detailed than our current book, which allowed for a better idea of what the proteins' (or whatever the object) structure was.

Amy Weber, student, Ohio University

COMPANION WEBSITE

Students can enhance their understanding of the concepts with the rich study materials available at www.brookerbiology.com. This open access website provides self-study options with chapter quizzes to assess current understanding, animations that highlight topics students typically struggle with and textbook images that can be used for notetaking and study.

Overall, this is a great chapter where the text, photos, and diagrams come together to make for easy reading and easy understanding of concepts and terminology. Depth of coverage is right on, and bringing in current research results is a winner.

Donald Baud, University of Memphis

Chapter Outline

4.1 Microscopy
4.2 Overview of Cell Structure
4.3 The Cytosol
4.4 The Nucleus and Endomembrane System
4.5 Semiautonomous Organelles
4.6 Protein Sorting to Organelles
4.7 Systems Biology of Cells: A Summary
Summary of Key Concepts
Assess and Discuss

General Features of Cells

4

Emily had a persistent cough ever since she started smoking cigarettes in college. However, at age 35, it seemed to be getting worse, and she was alarmed by the occasional pain in her chest. When she began to lose weight and noticed that she became easily fatigued, Emily decided to see a doctor. The diagnosis was lung cancer. Despite aggressive treatment of the disease with chemotherapy and radiation therapy, she succumbed to lung cancer 14 months after the initial diagnosis. Emily was 36.

Topics such as cancer are within the field of **cell biology**—the study of individual cells and their interactions with each other. Researchers in this field want to understand the basic features of cells and apply their knowledge in the treatment of diseases such as cystic fibrosis, sickle-cell disease, and lung cancer.

The idea that organisms are composed of cells originated in the mid-1800s. German botanist Matthias Schleiden studied plant material under the microscope and was struck by the presence of many similar-looking compartments, each of which contained a dark area. Today we call those compartments cells, and the dark area is the nucleus. In 1838, Schleiden speculated that cells are living entities

A cell from the pituitary gland. The cell in this micro viewed by a technique called transmission electron which is described in this chapter. The micrograph w colored using a computer to enhance the visualizatio cell structures.

experiments provided the first evidence that secreted proteins are synthesized into the rough ER and move through a series of cellular compartments before they are secreted. These findings also caused researchers to wonder how proteins are targeted to particular organelles and how they move from one compartment to another. These topics are described later in Section 4.6.

Experimental Questions

1. Explain the procedure of a pulse-chase experiment. W is the pulse, and what is the chase? What was the pu of the approach?
2. Why were pancreatic cells used for this investigation?
3. What were the key results of the experiment of Figure 4.19? What did the researchers conclude?

Concept check: What is the advantage of having a highly invaginated inner membrane?

Summary of Key Concepts

4.1 Microscopy

- Three important parameters in microscopy are magnification, resolution, and contrast. A light microscope utilizes light for

Assess and Discuss

Test Yourself

1. The cell doctrine states
   a. all living things are composed of cells.
   b. cells are the smallest units of living organisms.
   c. new cells come from pre-existing cells by cell division.

Conceptual Questions

1. Describe two specific ways that protein-protein interactions are involved with cell structure or cell function.
2. Explain how motor proteins and cytoskeletal filaments can interact to promote three different types of movements: movement of a cargo, movement of a filament, and bending of a filament.
3. Describe the functions of the Golgi apparatus.

Collaborative Questions

1. Discuss the roles of the genome and proteome in determining cell structure and function.
2. Discuss and draw the structural relationship between the nucleus, the rough endoplasmic reticulum, and the Golgi apparatus.

Online Resource

www.brookerbiology.com

Stay a step ahead in your studies with animations that bring concepts to life and practice tests to assess your understanding. Your instructor may also recommend the interactive ebook, individualized learning tools, and more.

THE LEARNING SYSTEM

Each chapter starts with a simple outline and engaging story that highlights why the information in the chapter is important and intriguing. Concept checks and the questions with the Feature Investigations throughout the chapter continually ask the student to check their understanding and push a bit further. We end each chapter with a thorough review section that returns to our outline and emphasizes higher-level learning through multiple-question types.

A STEP AHEAD IN PREPARING YOUR COURSE

McGRAW-HILL CONNECT PLUS™ BIOLOGY

Connect™ Biology is a web-based assignment and assessment platform that gives students the means to better connect with their course work, with their instructors, and with the important concepts that they will need to know for success now and in the future.

With Connect Biology, you can deliver assignments, quizzes, and tests online. A robust set of questions and activities are presented and tied to the textbook's learning objectives. As an instructor, you can edit existing questions and author entirely new problems. Track individual student performance—by question, assignment, or in relation to the class overall—with detailed grade reports. Integrate grade reports easily with Learning Management Systems (LMS) such as WebCT and Blackboard. And much more.

ConnectPlus™ Biology provides students with all the advantages of Connect Biology, plus 24/7 access to an eBook—a media-rich version of the book that includes animations, videos, and inline assessments placed appropriately throughout the chapter. ConnectPlus Biology allows students to practice important skills at their own pace and on their own schedule. By purchasing eBooks from McGraw-Hill students can save as much as 50% on selected titles delivered on the most advanced eBook platforms available. Contact your McGraw-Hill sales representative to discuss eBook packaging options.

Powerful Presentation Tools

Everything you need for outstanding presentation in one place!

- FlexArt Image PowerPoints—including every piece of art that has been sized and cropped specifically for superior presentations as well as labels that you can edit, flexible art that can be picked up and moved, tables, and photographs.
- Animation PowerPoints—numerous full-color animations illustrating important processes. Harness the visual

impact of concepts in motion by importing these slides into classroom presentations or online course materials.
- Lecture PowerPoints with animations fully embedded.
- Labeled and unlabeled JPEG images—full-color digital files of all illustrations that can be readily incorporated into presentations, exams, or custom-made classroom materials.

Presentation Center

In addition to the images from your book, this online digital library contains photos, artwork, animations, and other media from an array of McGraw-Hill textbooks that can be used to create customized lectures, visually enhanced tests and quizzes, compelling course websites, or attractive printed support materials.

Quality Test Bank

All questions have been written to fully align with the learning objectives and content of the textbook. Provided within a computerized test bank powered by McGraw-Hill's flexible electronic testing program EZ Test Online, instructors can create paper and online tests or quizzes in this easy-to-use program! A new tagging scheme allows you to sort questions by difficulty level, topic, and section. Now, with EZ Test Online, instructors can select questions from multiple McGraw-Hill test banks or author their own, and then either print the test for paper distribution or give it online.

Active Learning Exercises

Supporting biology faculty in their efforts to make introductory courses more active and student-centered is critical to improving undergraduate biological education. Active learning can broadly be described as strategies and techniques in which students are engaged in their own learning, and is typically characterized by the utilization of higher order critical thinking skills. The use of these techniques is critical to biological education because of their powerful impact on students' learning and development of scientific professional skills.

Active learning strategies are highly valued and have been shown to:

- Help make content relevant
- Be particularly adept at addressing common misconceptions
- Help students to think about their own learning (metacognition)

- Promote meaningful learning of content by emphasizing application
- Foster student interest in science

Guided Activities have been provided for instructors to use in their course for both in-class and out-of-class activities. The Guided Activities make it easy for you to incorporate active learning into your course and are flexible to fit your specific needs.

FLEXIBLE DELIVERY OPTIONS

Brooker et al. *Biology* is available in many formats in addition to the traditional textbook to give instructors and students more choices when deciding on the format of their biology text.

Also available, customized versions for all of your course needs. You're in charge of your course, so why not be in control of the content of your textbook? At McGraw-Hill Custom Publishing, we can help you create the ideal text—the one you've always imagined. Quickly. Easily. With more than 20 years of experience in custom publishing, we're experts. But at McGraw-Hill, we're also innovators, leading the way with new methods and means for creating simplified, value-added custom textbooks.

The options are never-ending when you work with McGraw-Hill. You already know what will work best for you and your students. And here, you can choose it.

Foundations of Life—Chemistry, Cells, and Genetics
ISBN: 007740565X
Units 1, 2, and 3

Evolution, Diversity, and Ecology
ISBN: 0077405889
Units 4, 5, and 8

Plants and Animals
ISBN: 0077405897
Units 6 and 7

LABORATORY MANUALS

Biology Lab Manual, Ninth Edition,
Vodopich/Moore
ISBN 0073383066

This laboratory manual is designed for an introductory majors biology course with a broad survey of basic laboratory techniques. The experiments and procedures are simple, safe, easy to perform, and especially appropriate for large classes. Few experiments require a second class-meeting to complete the procedure. Each exercise includes many photographs, traditional topics, and experiments that help students learn about life. Procedures within each exercise are numerous and discrete so that an exercise can be tailored to the needs of the students, the style of the instructor, and the facilities available.

Biological Investigations,
Ninth Edition,
Dolphin et al.
ISBN 0073383058

This independent laboratory manual can be used for a one- or two-semester majors-level general biology lab and can be used with any majors-level general biology textbook. The labs are investigative and ask students to use more critical thinking and hands-on learning. The author emphasizes investigative, quantitative, and comparative approaches to studying the life sciences.

FOCUS ON EVOLUTION

Understanding Evolution,
Seventh Edition,
by Rosenbaum and Volpe
ISBN 0073383236

As an introduction to the principles of evolution, this paperback textbook is ideally suited as a main textbook for general evolution or as a supplement for general biology, genetics, zoology, botany, anthropology, or any life science course that utilizes evolution as the underlying theme of all life.

A STEP AHEAD IN QUALITY

360° DEVELOPMENT PROCESS

McGraw-Hill's 360° Development Process is an ongoing, never-ending, education-oriented approach to building accurate and innovative print and digital products. It is dedicated to continual large-scale and incremental improvement, driven by multiple user feedback loops and checkpoints. This is initiated during the early planning stages of our new products, intensifies during the development and production stages, then begins again upon publication in anticipation of the next edition.

This process is designed to provide a broad, comprehensive spectrum of feedback for refinement and innovation of our learning tools, for both student and instructor. The 360° Development Process includes market research, content reviews, course- and product-specific symposia, accuracy checks, and art reviews. We appreciate the expertise of the many individuals involved in this process.

Contributing Authors

Photo Consultant - Alvin Telser, *Northwestern University*
Instructors Manual - Mark Hens, *University of North Carolina–Greensboro*
Integrated eBook Study Quizzes - Anita Baines, *University of Wisconsin–LaCrosse;* **Matthew Neatrour,** *North Kentucky University*
Test Bank - Bruce Stallsmith, *University of Alabama–Huntsville*
 Regina Wiggins-Speights, *Houston Community College–Northeast*
 Punnee Soonthornpoct, *Blinn College*
 Sheila Wicks, *Malcom X Junior College*
 James Mickle, *North Carolina State University*
Website - Lisa Burgess, *Broward Community College;* **Marceau Ratard,** *Delgado Community College;* **Amanda Rosenzweig,** *Delgado Community College*
Instructor Media - Sharon Thoma, *University of Wisconsin–Madison;* **Brenda Leady,** *University of Toledo*
Active Learning - Frank Bailey, *Middle Tennessee State University,* **Steve Howard,** *Middle Tennessee State University;* **Michael Rutledge,** *Middle Tennessee State University*

Connect Content Contributors

Russell Borski, *North Carolina State University*
Scott Cooper, *University of Wisconsin—LaCrosse*
Phil Gibson, *Oklahoma University*
Susan Hengeveld, *Indiana University*
Lelsie Jones, *Valdosta State*
Morris Maduro, *University of California– Riverside*
Matt Neatrour, *Northern Kentucky University*
Lynn Preston, *Tarrant County College*
Brian Shmaefsky, *Lone Star College*

Digital Board of Advisors

We are indebted to the valuable advice and direction of an outstanding group of advisors, led by Melissa Michael, University of Illinois at Urbana-Champaign. Other board members include

Russel Borski, *North Carolina State University*
Karen Gerhart, *University of California—Davis*
Jean Heitz, *University of Wisconsin—Madison*
Mark Lyford, *University of Wyoming*
John Merrill, *Michigan State*
Randy Phillis, *University of Massachusetts*
Deb Pires, *University of California— Los Angeles*
Lynn Preston, *Tarrant County College*
Michael Rutledge, *Middle Tennessee State*

David Scicchitano, *New York University*
Bill Wischusen, *Louisiana State University*

General Biology Symposia

Every year McGraw-Hill conducts several General Biology Symposia, which are attended by instructors from across the country. These events are an opportunity for editors from McGraw-Hill to gather information about the needs and challenges of instructors teaching the major's biology course. It also offers a forum for the attendees to exchange ideas and experiences with colleagues they might not have otherwise met. The feedback we have received has been invaluable, and has contributed to the development of Biology and its supplements. A special thank you to recent attendees:

Sylvester Allred, *Northern Arizona University*
Michael Bell, *Richland College*
Arlene Billock, *University of Louisiana–Lafayette*
Stephane Boissinot, *Queens College, the City University of New York*
David Bos, *Purdue University*
Scott Bowling, *Auburn University*
Jacqueline Bowman, *Arkansas Technical University*
Arthur Buikema, *Virginia Polytechnic Institute*
Anne Bullerjahn, *Owens Community College*
Helaine Burstein, *Ohio University*
Raymond Burton, *Germanna Community College*
Peter Busher, *Boston University*
Richard Cardullo, *University of California–Riverside*
Jennifer Ciaccio, *Dixie State College*
Anne Barrett Clark, *Binghamton University*
Allison Cleveland, *University of South Florida–Tampa*
Jennifer Coleman, *University of Massachusetts–Amherst*
Sehoya Cotner, *University of Minnesota*
Mitch Cruzan, *Portland State University*
Laura DiCaprio, *Ohio University*
Kathyrn Dickson, *California State College–Fullerton*
Cathy Donald-Whitney, *Collin County Community College*
Stanley Faeth, *Arizona State University*
Donald French, *Oklahoma State University*
Douglas Gaffin, *University of Oklahoma*
Karen Gerhart, *University of California–Davis*
Cynthia Giffen, *University of Wisconsin–Madison*

William Glider, *University of Nebraska–Lincoln*
Christopher Gregg, *Louisiana State University*
Stan Guffey, *The University of Tennessee*
Bernard Hauser, *University of Florida–Gainesville*
Jean Heitz, *Unversity of Wisconsin–Madison*
Mark Hens, *University of North Carolina–Greensboro*
Albert Herrera, *University of Southern California*
Ralph James Hickey, *Miami University of Ohio–Oxford*
Brad Hyman, *University of California–Riverside*
Kyoungtae Kim, *Missouri State University*
Sherry Krayesky, *University of Louisiana–Lafayette*
Jerry Kudenov, *University of Alaska–Anchorage*
Josephine Kurdziel, *University of Michigan*
Ellen Lamb, *University of North Carolina–Greensboro*
Brenda Leady, *University of Toledo*
Graeme Lindbeck, *Valencia Community College*
Susan Meiers, *Western Illinois University*
Michael Meighan, *University of California–Berkeley*
John Mersfelder, *Sinclair Community College*
Melissa Michael, *University of Illinois–Urbana-Champaign*
Leonore Neary, *Joliet Junior College*
Shawn Nordell, *Saint Louis University*
John Osterman, *University of Nebraska–Lincoln*
Stephanie Pandolfi, *Michigan State University*
C.O. Patterson, *Texas A&M University*
Nancy Pencoe, *University of West Georgia*
Roger Persell, *Hunter College*
Marius Pfeiffer, *Tarrant County College NE*
Steve Phelps, *University of Florida*
Debra Pires, *University of California–Los Angeles*
Lynn Preston, *Tarrant County College*
Rajinder Ranu, *Colorado State University*
Marceau Ratard, *Delgado Community College–City Park*
Melanie Rathburn, *Boston University*
Robin Richardson, *Winona State University*
Amanda Rosenzweig, *Delgado Community College–City Park*
Connie Russell, *Angelo State University*
Laurie Russell, *St. Louis University*
David Scicchitano, *New York University*
Timothy Shannon, *Francis Marion University*
Brian Shmaefsky, *Lone Star College–Kingwood*
Richard Showman, *University of South Carolina*
Robert Simons, *University of California–Los Angeles*
Steve Skarda, *Linn Benton Community College*
Steven D. Skopik, *University of Delaware*

Phillip Sokolove, *University of Maryland—Baltimore County*
Brad Swanson, *Cental Michigan University*
David Thompson, *Northern Kentucky University*
Maureen Tubbiola, *St. Cloud State University*
Ashok Upadhyaya, *University of South Florida–Tampa*
Anthony Uzwiak, *Rutgers University*
Rani Vajravelu, *University of Central Florida*
Gary Walker, *Appalachian State University*
Pat Walsh, *University of Delaware*
Elizabeth Weiss-Kuziel, *University of Texas–Austin*
David Williams, *Valencia Community College, East Campus*
Holly Williams, *Seminole Community College*
Michael Windelspecht, *Appalachian State University*
Mary Wisgirda, *San Jacinto College, South Campus*
Jay Zimmerman, *St. John's University*

Second Edition Reviewers

Eyualem Abebe, *Elizabeth City State University*
Nihal Ahmad, *University of Wisconsin-Madison*
John Alcock, *Arizona State University*
Myriam Alhadefl-Feldman, *Lake Washington Technical College*
Dennis Arvidson, *Michigan State University*
David K. Asch, *Youngstown State University*
Tami Asplin, *North Dakota State University*
Amir Assad-Rad, *Delta College*
Karl Aufderheide, *Texas A&M University*
Idelisa Ayala, *Broward Community College*
Anita Davelos Baines, *University of Texas–Pan American*
Adebiyi Banjoko, *Chandler-Gilbert Community College*
Gerry Barclay, *Highline Community College*
Susan Barrett, *Massasoit Community College*
Donald Reon Baud, *University of Memphis*
Vernon Bauer, *Francis Marion University*
Chris Bazinet, *St. John's University*
Michael C. Bell, *Richland College Giacomo Bernardi University of California-Santa Cruz*
Giacomo Bernardi, *University of California-Santa Cruz*
Deborah Bielser, *University of Illinois–Urbana–Champaign*
Arlene G. Billock, *University of Louisiana–Lafayette*
Eric Blackwell, *Delta State University*
Andrew R. Blaustein, *Oregon State University*
Harriette Howard-Lee Block, *Prairie View A&M University*
Steve Blumenshine, *California State University-Fresno*
Jason Bond, *East Carolina University*
Russel Borski, *North Carolina State University*
James Bottesch, *Brevard Community College/Cocoa Campus*
Scott Bowling, *Auburn University*
Robert S. Boyd, *Auburn University*
Robert Brewer, *Cleveland State Community College*
Randy Brewton, *University of Tennessee*
George Briggs, *State Univerity College–Geneseo*
Mirjana M. Brockett, *Georgia Institute of Technology*
W. Randy Brooks, *Florida Atlantic University*
Jack Brown, *Paris Junior College*
Rodolfo Buiser, *University of Wisconsin-Eau Claire*
Anne Bullerjahn, *Owens Community College*
Carolyn J.W. Bunde, *Idaho State University*
Scott Burt, *Truman State University*
Stephen R. Burton, *Grand Valley State University*
Jorge Busciglio, *University of California*
Thomas Bushart, *University of Texas-Austin*
Malcolm Butler, *North Dakota State University*
David Byres, *Florida Community College South Campus*
Jennifer Campbell, *North Carolina State University*

Jeff Carmichael, *University of North Dakota*
Timothy H. Carter, *St. John's University*
Domenic Castignetti, *Loyola University of Chicago*
Deborah A. Cato, *Wheaton College*
Tien-Hsien Chang, *Ohio State University*
Estella Chen, *Kennesaw State University*
Sixue Chen, *University of Florida*
Brenda Chinnery-Allgeier, *University of Texas–Austin*
Young Cho, *Eastern New Mexico University*
Genevieve Chung, *Broward Community College-Central*
Philip Clampitt, *Oakland University*
T. Denise Clark, *Mesa Community College*
Allison Cleveland Roberts, *University of South Florida–Tampa*
Randy W. Cohen, *California State University-Northridge*
Patricia Colberg, *University of Wyoming*
Joanne Conover, *University of Connecticut*
John Cooley, *Yale University*
Craig Coleman, *Brigham Young University–Provo*
Ronald H. Cooper, *University of California–Los Angeles*
Vicki Corbin, *University of Kansas–Lawrence*
Anthony Cornett, *Valencia Community College*
Will Crampton, *University of Central Florida*
Charles Creutz, *University of Toledo*
Karen Curto, *University of Pittsburgh*
Kenneth A. Cutler, *North Carolina Central University*
Cara Davies, *Ohio Northern University*
Donald H. Dean, *The Ohio State University*
James Dearworth, *Lafayette College*
John Dennehy, *Queens College*
William Dentler, *University of Kansas*
Smruti A. Desai, *Lonestar College–Cy Fair*
Donald Deters, *Bowling Green State University*
Laura DiCaprio, *Ohio University*
David S. Domozych, *Skidmore College*
Kristiann M. Dougherty, *Valencia Community College*
Kari M.H. Doyle, *San Jacinto College*
John Drummond, *Lafayette College*
Ernest Dubrul, *University of Toledo*
James N. Dumond, *Texas Southern University*
Tod Duncan, *University of Colorado–Denver*
Richard Duhrkopf, *Baylor University*
Susan Dunford, *University of Cincinnati*
Roland Dute, *Auburn University*
Ralph P. Eckerlin, *Northern Virginia Community College*
Jose L. Egremy, *Northwest Vista College*
David W. Eldridge, *Baylor University*
Lisa K. Elfring, *University of Arizona*
Kurt J. Elliot, *Northwest Vista College*
Seema Endley, *Blinn College*
Bill Ensign, *Kennesaw State University*
David S. Epstein, *J. Sergeant Reynolds Community College*
Gary N. Ervin, *Mississippi State University*
Frederick Essig, *University of Southern Florida*
Brent E. Ewers, *University of Wyoming*
Susan Fahrbach, *Wake Forest University*
Peter Fajer, *Florida State University*
Zen Faulkes, *University of Texas–Pan American*
Miriam Ferzli, *North Carolina State University*
Fleur Ferro, *Community College of Denver*
Jose Fierro, *Florida State College–Jacksonville*
Melanie Fierro, *Florida State College–Jacksonville*
Teresa G. Fischer, *Indian River College*
David Fitch, *New York University*
Sandra Fraley, *Dutchess Community College*
Steven N. Francoeur, *Eastern Michigan University*
Barbara Frase, *Bradley University*
Robert Friedman, *University of South Carolina*
Bernard L. Frye, *University of Texas–Arlington*
Caitlin Gabor, *Texas State University–San Marcos*
Mike Ganger, *Gannon University*
Deborah Garrity, *Colorado State University*
Shannon Gerry, *Wellesley College*
Cindee Giffen, *University of Wisconsin–Madison*

Chris Gissendanner, *University of Louisiana at Monroe*
Florence K. Gleason, *University of Minnesota*
Elmer Godeny, *Baton Rough Community College*
Elizabeth Godrick, *Boston University*
Robert Gorham, *Northern Virginia Community College*
Brian Grafton, *Kent State University*
John Graham, *Bowling Green State University*
Christopher Gregg, *Louisiana State University*
John Griffis, *Joliet Junior College*
LeeAnn Griggs, *Massasoit Community College*
Tim Grogan, *Valencia Community College–Osceola*
Richard S. Groover, *J. Sergeant Reynolds Community College*
Gretel Guest, *Durham Technical Community College*
Cameron Gundersen, *University of California*
Patricia Halpin, *UCLA*
George Hale, *University of West Georgia*
William Hanna, *Massasoit Community College*
David T. Hanson, *University of New Mexico*
Christopher J. Harendza, *Montgomery County Community College*
Sally E. Harmych, *University of Toledo*
Betsy Harris, *Appalachian State University*
M.C. Hart, *Minnesota State University–Mankato*
Barbara Harvey, *Kirkwood Community College*
Mary Beth Hawkins, *North Carolina State University*
Harold Heatwole, *North Carolina State University*
Cheryl Heinz, *Benedictine University*
Jutta B. Heller, *Loyola University–Chicago*
Susan Hengeveld, *Indiana University–Bloomington*
Mark Hens, *University of North Carolina–Greensboro*
Steven K. Herbert, *University of Wyoming–Laramie*
Edgar Javier Hernandez, *University of Missouri–St. Louis*
Albert A. Herrera, *University of Southern California*
David S. Hibbert, *Clark University*
R. James Hickey, *Miami University of Ohio–Oxford*
Terri Hildebrand, *Southern Utah University*
Juliana Hinton, *McNeese State University*
Anne Hitt, *Oakland University*
Robert D. Hollister, *Grand Valley State University*
Richard G Holloway, *Northern Arizona University*
Dianella Howarth, *St. John's University*
Carrie Hughes, *San Jacinto College*
Kelly Howe, *University of New Mexico*
Barbara Hunnicutt, *Seminole Community College*
Bradley Hyman, *University of California–Riverside*
Vicki J. Isola, *Hope College*
Joseph J. Jacquot, *Grand Valley State University*
Desirée Jackson, *Texas Southern University*
John Jaenike, *University of Rochester*
Ashok Jain, *Albany State University*
Eric Jellen, *Brigham Young University*
Elizabeth A. Jordan, *Moorpark College*
Robyn Jordan, *University of Louisiana at Monroe*
Susan Jorstad, *University of Arizona*
Nick Kaplinsky, *Swarthmore College*
Vesna Karaman, *University of Texas at El Paso*
Istvan Karsai, *East Tennessee State University*
Nancy Kaufmann, *University of Pittsburgh*
Stephen R. Kelso, *University of Illinois–Chicago*
Denice D. King, *Cleveland State Community College*
Bridgette Kirkpatrick, *Collin County Community College*
Ted Klenk, *Valencia Community College–West*
Anna Koshy, *Houston Community College–NW*
David Krauss, *Borough of Manhattan Community College*
William Kroll, *Loyola University–Chicago*
Pramod Kumar, *University of Texas–San Antonio*
Allen Kurta, *Eastern Michigan University*
William Lamberts, *College of St. Benedict/Saint John's University*
David Lampe, *Duquesne University*
John C. Law, *Community College of Allegheny County*
Jonathan N. Lawson, *Collin County Community College*
Brenda Leady, *University of Toledo*

David Leaf, *Western Washington University*
John Lepri, *University of North Carolina–Greensboro*
Hugh Lefcort, *Gonzaga University*
Army Lester, *Kennesaw State University*
Q Quinn Li, *Miami University, Ohio*
Nardos Lijam, *Columbus State Community College*
Yusheng Liu, *East Tennessee State University*
Robert Locy, *Auburn University*
Albert R. Loeblich III, *University of Houston*
Thomas A. Lonergan, *University of New Orleans*
Craig Longtine, *North Hennepin Community College*
Donald Lovett, *The College of New Jersey*
Paul T. Magee, *University of Minnesota–Minneapolis*
Jay Mager, *Ohio Northern University*
Charles H. Mallery, *University of Miami*
Nilo Marin, *Broward College*
Joe Matanoski, *Stevenson University*
Patricia Matthews, *Grand Valley State University*
Barbara May, *College of St. Benedict/St. John's University*
Kamau Mbuthia, *Bowling Green State University*
Norah McCabe, *Washington State University*
Chuck McClaugherty, *Mount Union College*
Regina S. McClinton, *Grand Valley State University*
Mark A. McGinley, *Texas Tech University*
Kerry McKenna, *Lord Fairfax College*
Carrie McMahon Hughes, *San Jacinto College–Central Campus*
Joseph McPhee, *LaGuardia Community College*
Judith Megaw, *Indian River Community College*
Mona C. Mehdy, *University of Texas–Austin*
Brad Mehrtens, *University of Illinois–Urbana–Champaign*
Susan Meiers, *Western Illinois University*
Michael Meighan, *University of California–Berkeley*
Catherine Merovich, *West Virginia University*
Richard Merritt, *Houston Community College*
Jennifer Metzler, *Ball State University*
James Mickle, *North Carolina State University*
Brian T. Miller, *Middle Tennessee State University*
Manuel Miranda-Arango, *University of Texas at El Paso*
Michael Misamore, *Texas Christian University*
Jasleen Mishra, *Houston Community College–Southwest*
Alan Molumby, *University of Illinois, Chicago*
W. Linn Montgomery, *Northern Arizona University*
Daniel Moon, *University of North Florida*
Jennifer Moon, *University of Texas–Austin*
Richard C. Moore, *Miami University*
David Morgan, *University of West Georgia*
Roderick M. Morgan, *Grand Valley State University*
Ann C. Morris, *Florida State University*
Christa P.H. Mulder, *University of Alaska–Fairbanks*
Mike Muller, *University of Illinois–Chicago*
Darrel C. Murray, *University of Illinois–Chicago*
Richard J. Murray, *Hendrix College*
Jennifer Nauen, *University of Delaware*
Raymond Neubauer, *University of Texas–Austin*
Jacalyn Newman, *University of Pittsburgh*
Robert Newman, *University of North Dakota*
Laila Nimri, *Seminole Community College*
Shawn E. Nordell, *St. Louis University*
Olumide Ogunmosin, *Texas Southern University*
Wan Ooi, *Houston Community College–Central*
John C. Osterman, *University of Nebraska–Lincoln*
Ravishankar Palanivelu, *University of Arizona*
Peter Pappas, *Community College of Morris*
Lisa Parks, *North Carolina State University*
David Pennock, *Miami University*
Beverly Perry, *Houston Community College*
John S. Peters, *College of Charleston*
David K. Peyton, *Morehead State University*
Marius Pfeiffer, *Tarrant County College NE*

Jerry Phillips, *University of Colorado–Colorado Springs*
Susan Phillips, *Brevard Community College*
Paul Pilliterri, *Southern Utah University*
Debra B. Pires, *University of California–Los Angeles*
Terry Platt, *University of Rochester*
Peggy E. Pollack, *Northern Arizona University*
Uwe Pott, *University of Wisconsin-Green Bay*
Linda F. Potts, *University of North Carolina–Wilmington*
Jessica Poulin, *University at Buffalo, SUNY*
Kumkum Prabhakar, *Nassau Community College*
Joelle Presson, *University of Maryland*
Gregory Pryor, *Francis Marion University*
Penny L. Ragland, *Auburn University*
Rajinder S. Ranu, *Colorado State University*
Marceau Ratard, *Delgado Community College*
Melanie K. Rathbun, *Boston University*
Flona Redway, *Barry University*
Melissa Murray Reedy, *University of Illinois Urbana-Champaign*
Stuart Reichler, *University of Texas–Austin*
Jill D. Reid, *Virginia Commonwealth University*
Anne E. Reilly, *Florida Atlantic University*
Kim Risley, *Mount Union College*
Elisa Rivera-Boyles, *Valencia Community College*
Laurel B. Roberts, *University of Pittsburgh*
James V. Robinson, *The University of Texas–Arlington*
Kenneth R. Robinson, *Purdue University*
Luis A. Rodriguez, *San Antonio College*
Chris Romero, *Front Range Community College–Larimer Campus*
Doug Rouse, *University of Wisconsin–Madison*
Ann E. Rushing, *Baylor University*
Laurie K. Russell, *St. Louis University*
Sheridan Samano, *Community College of Aurora*
Hildegarde Sanders, *Stevenson University*
David K. Saunders, *Augusta State University*
H. Jochen Schenk, *California State University–Fullerton*
Deemah Schirf, *University of Texas–San Antonio*
Chris Schneider, *Boston University*
Susan Schreier, *Towson University*
David Schwartz, *Houston Community College–Southwest*
David A. Scicchitano, *New York University*
Erik Scully, *Towson University*
Robin Searles-Adenegan, *University of Maryland*
Pat Selelyo, *College of Southern Idaho*
Pramila Sen, *Houston Community College–Central*
Tim Shannon, *Francis Marion University*
Jonathan Shaver, *North Hennepin Community College*
Brandon Sheafor, *Mount Union College*
Ellen Shepherd Lamb, *The University of North Carolina–Greensboro*
Mark Sheridan, *North Dakota State University*
Dennis Shevlin, *The College of New Jersey*
Patty Shields, *University of Maryland*
Cara Shillington, *Eastern Michigan University*
Richard M. Showman, *University of South Carolina*
Scott Siechen, *University of Illinois–Urbana-Champaign*
Anne Simon, *University of Maryland*
Sue Simon Westendorf, *Ohio University–Athens*
John B. Skillman, *California State University–San Bernadino*
Lee Smee, *Texas A&M University*
Dianne Snyder, *Augusta State University*
Nancy Solomon, *Miami University*
Sally Sommers Smith, *Boston University*
Punnee Soonthornpoct, *Blinn College*
Vladimir Spiegelman, *University of Wisconsin–Madison*
Bryan Spohn, *Florida State College at Jacksonville*
Bruce Stallsmith, *University of Alabama–Huntsville*
Richard Stalter, *St. John's University*
Susan J. Stamler, *College of Dupage*
William Stein, *Binghampton University*
Mark E. Stephansky, *Massasoit Community College*

Dean Stetler, *University of Kansas–Lawrence*
Brian Stout, *Northwest Vista College*
Mark Sturtevant, *Oakland University*
C.B. Subrahmanvam, *Florida A&M University*
Mark Sutherland, *Hendrix College*
Brook Swanson, *Gonzaga University*
Debbie Swarthout, *Hope College*
Judy Taylor, *Motlow State Community College*
Randall G. Terry, *Lamar University*
Sharon Thoma, *University of Wisconsin*
Carol Thornber, *University of Rhode Island*
Patrick A. Thorpe, *Grand Valley State University*
Scott Tiegs, *Oakland University*
Kristina Timmerman, *St. John's University*
Paul Trombley, *Florida State University*
John R. True, *Stony Brook University*
Encarni Trueba, *Community College of Baltimore County Essex*
Cathy Tugmon, *Augusta State University*
Marshall Turell, *Houston Community College*
Ashok Upadhyaya, *University of South Florida–Tampa*
Anthony J. Uzwiak, *Rutgers University*
William Velhagen, *New York University*
Wendy Vermillion, *Columbus State Community College*
Sara Via, *University of Maryland*
Thomas V. Vogel, *Western Illinois University*
R. Steven Wagner, *Central Washington University–Ellensburg*
John Waldman, *Queens College-CUNY*
Randall Walikonis, *University of Connecticut*
Gary R. Walker, *Youngstown State University*
Sean E. Walker, *California State University–Fullerton*
Delon E. Washo-Krupps, *Arizona State University*
Fred Wasserman, *Boston University*
R. Douglas Watson, *University of Alabama–Birmingham*
Doug Wendell, *Oakland University*
Jennifer Wiatrowski, *Pasco-Hernando Community College*
Sheila Wicks, *Malcolm X College*
Donna Wiersema, *Houston Community College*
Regina Wiggins-Speights, *Houston Community College–Northeast*
David H. Williams, *Valencia Community College*
Lawrence R. Williams, *University of Houston*
Ned Williams, *Minnesota State University–Mankato*
E. Gay Williamson, *Mississippi State University*
Mark S. Wilson, *Humboldt State University*
Bob Winning, *Eastern Michigan University*
David Wood, *California State University–Chico*
Bruce Wunder, *Colorado State University*
Mark Wygoda, *McNeese State University*
Joanna Wysocka-Diller, *Auburn University*
Marlena Yost, *Mississippi State University*
Robert Yost, *Indiana University—Purdue*
Kelly Young, *California State University–Long Beach*
Linda Young, *Ohio Northern University*
Ted Zerucha, *Appalachian State University*

First Edition Reviewers and Contributors

James K. Adams, *Dalton State College*
Sylvester Allred, *Northern Arizona University*
Jonathan W. Armbruster, *Auburn University*
Joseph E. Armstrong, *Illinois State University*
David K. Asch, *Youngstown State University*
Amir M. Assadi-Rad, *Delta College*
Karl J. Aufderheide, *Texas A&M University*
Anita Davelos Baines, *University of Texas–Pan American*
Lisa M. Baird, *University of San Diego*
Diane Bassham, *Iowa State University*
Donald Baud, *University of Memphis*
Vernon W. Bauer, *Francis Marion University*
Ruth E. Beattie, *University of Kentucky*
Michael C. Bell, *Richland College*

Steve Berg, *Winona State University*
Arlene G. Billock, *University of Louisiana–Lafayette*
Kristopher A. Blee, *California State University–Chico*
Heidi B. Borgeas, *University of Tampa*
Russell Borski, *North Carolina State University*
Scott A. Bowling, *Auburn University*
Robert Boyd, *Auburn University*
Eldon J. Braun, *University of Arizona*
Michael Breed, *University of Colorado–Boulder*
Randy Brewton, *University of Tennessee–Knoxville*
Peggy Brickman, *University of Georgia*
Cheryl Briggs, *University of California–Berkeley*
Peter S. Brown, *Mesa Community College*
Mark Browning, *Purdue University*
Cedric O. Buckley, *Jackson State University*
Don Buckley, *Quinnipiac University*
Arthur L. Buikema, Jr., *Virginia Tech University*
Anne Bullerjahn, *Owens Community College*
Ray D. Burkett, *Southeast Tennessee Community College*
Stephen P. Bush, *Coastal Carolina University*
Peter E. Busher, *Boston University*
Jeff Carmichael, *University of North Dakota*
Clint E. Carter, *Vanderbilt University*
Patrick A. Carter, *Washington State University*
Merri Lynn Casem, *California State University–Fullerton*
Domenic Castignetti, *Loyola University–Chicago*
Maria V. Cattell
David T. Champlin, *University of Southern Maine*
Jung H. Choi, *Georgia Institute of Technology*
Curtis Clark, *Cal Poly–Pomona*
Allison Cleveland, *University of South Florida*
Janice J. Clymer, *San Diego Mesa College*
Linda T. Collins, *University of Tennessee–Chattanooga*
William Collins, *Stony Brook University*
Jay L. Comeaux, *Louisiana State University*
Bob Connor II, *Owens Community College*
Daniel Costa, *University of California–Santa Cruz*
Sehoya Cotner, *University of Minnesota*
Mack E. Crayton III, *Xavier University of Louisiana*
Louis Crescitelli, *Bergen Community College*
Charles Creutz, *University of Toledo*
Karen A. Curto, *University of Pittsburgh*
Mark A. Davis, *Macalester College*
Mark D. Decker, *University of Minnesota*
Jeffery P. Demuth, *Indiana University*
Phil Denette, *Delgado Community College*
Donald W. Deters, *Bowling Green State University*
Hudson R. DeYoe, *University of Texas–Pan American*
Laura DiCaprio, *Ohio University*
Randy DiDomenico, *University of Colorado–Boulder*
Robert S. Dill, *Bergen Community College*
Kevin Dixon, *University of Illinois–Urbana–Champaign*
John S. Doctor, *Duquesne University*
Warren D. Dolphin, *Iowa State University*
Cathy A. Donald-Whitney, *Collin County Community College*
Robert P. Donaldson, *George Washington University*
Kristiann Dougherty, *Valencia Community College*
Marjorie Doyle, *University of Wisconsin–Madison*
Ernest F. Dubrul, *University of Toledo*
Jeffry L. Dudycha, *William Patterson University of New Jersey*
Charles Duggins, Jr., *University of South Carolina*
Roland R. Dute, *Auburn University*
William D. Eldred, *Boston University*
Johnny El-Rady, *University of South Florida*
Dave Eldridge, *Baylor University*
Inge Eley, *Hudson Valley Community College*
Frederick B. Essig, *University of South Florida*
Sharon Eversman, *Montana State University*
Stan Faeth, *Arizona State University*
Peter Fajer, *Florida State University*
Paul Farnsworth, *University of Texas–San Antonio*

Paul D. Ferguson, *University of Illinois–Urbana–Champaign*
Margaret F. Field, *Saint Mary's College of California*
Jorge A. Flores, *West Virginia University*
Irwin Forseth, *University of Maryland*
David Foster, *North Idaho College*
Paul Fox, *Danville Community College*
Pete Franco, *University of Minnesota*
Wayne D. Frasch, *Arizona State University*
Barbara Frase, *Bradley University*
Adam J. Fry, *University of Connecticut*
Caitlin R. Gabor, *Texas State University–San Marcos*
Anne M. Galbraith, *University of Wisconsin–La Crosse*
John R. Geiser, *Western Michigan University*
Nicholas R. Geist, *Sonoma State University*
Patricia A. Geppert, *University of Texas–San Antonio*
Frank S. Gilliam, *Marshall University*
Chris R. Gissendanner, *University of Louisiana–Monroe*
Jon Glase, *Cornell University*
Florence K. Gleason, *University of Minnesota*
Elizabeth Godrick, *Boston University*
James M. Grady, *University of New Orleans*
John S. Graham, *Bowling Green State University*
Barbara E. Graham-Evans, *Jackson State University*
Christine E. Gray, *Blinn College*
Stan Guffey, *University of Tennessee*
Rodney D. Hagley, *University of North Carolina–Wilmington*
Gary L. Hannan, *Eastern Michigan University*
Kyle E. Harms, *Louisiana State University*
M. C. Hart, *Minnesota State University–Mankato*
Carla Ann Hass, *The Pennsylvania State University*
Brian T. Hazlett, *Briar Cliff University*
Harold Heatwole, *North Carolina State University*
Mark D. Hens, *University of North Carolina*
Stephen K. Herbert, *University of Wyoming*
Albert A. Herrera, *University of Southern California*
David L. Herrin, *University of Texas–Austin*
Helen Hess, *College of the Atlantic*
R. James Hickey, *Miami University*
Tracey E. Hickox, *University of Illinois–Urbana–Champaign*
Mark A. Holbrook, *University of Iowa*
Ella Ingram, *Rose-Hulman Institute of Technology*
Jeffrey Jack, *University of Louisville*
Judy Jernstedt, *University of California–Davis*
Lee Johnson, *Ohio State University*
Robyn Jordan, *University of Louisiana–Monroe*
Walter S. Judd, *University of Florida*
David Julian, *University of Florida*
Stephen R. Kelso, *University of Illinois–Chicago*
Heather R. Ketchum, *Blinn College*
Eunsoo Kim, *University of Wisconsin–Madison*
Stephen J. King, *University of Missouri–Kansas City*
John Z. Kiss, *Miami University*
Ted Klenk, *Valencia Community College*
David M. Kohl, *University of California–Santa Barbara*
Anna Koshy, *Houston Community College System*
Sherry Krayesky, *University of Louisiana–Lafayette*
John Krenetsky, *Metropolitan State College–Denver*
Karin E. Krieger, *University of Wisconsin–Green Bay*
Paul Kugrens, *Colorado State University*
Josephine Kurdziel, *University of Michigan*
David T. Kurjiaka, *Ohio University*
Allen Kurta, *Eastern Michigan University*
Paul K. Lago, *University of Mississippi*
Ellen Shepherd Lamb, *University of North Carolina–Greensboro*
Pamela Lanford, *University of Maryland*
Marianne M. Laporte, *Eastern Michigan University*

Arlen T. Larson, *University of Colorado–Denver*
John Latto, *University of California–Berkeley*
Brenda Leady, *University of Toledo*
Shannon Erickson Lee, *California State University–Northridge*
Tali D. Lee, *University of Wisconsin–Eau Claire*
Michael Lentz, *University of North Florida*
Jennifer J. Lewis, *San Juan College*
Pauline A. Lizotte, *Valencia Community College*
Jason L. Locklin, *Temple College*
Robert Locy, *Auburn University*
James A. Long, *Boise State University*
David Lonzarich, *University of Wisconsin–Eau Claire*
James B. Ludden, *College of DuPage*
Albert MacKrell, *Bradley University*
P. T. Magee, *University of Minnesota*
Christi Magrath, *Troy University*
Richard Malkin, *University of California–Berkeley*
Charles H. Mallery, *University of Miami*
Kathleen A. Marrs, *IUPUI–Indianapolis*
Diane L. Marshall, *University of New Mexico*
Peter J. Martinat, *Xavier University of Louisiana*
Joel Maruniak, *University of Missouri*
Kamau Mbuthia, *Bowling Green State University*
Greg McCormac, *American River College*
Andrew McCubbin, *Washington State University*
David L. McCulloch, *Collin County Community College*
Tanya K. McKinney, *Xavier University of Louisiana*
Brad Mehrtens, *University of Illinois–Urbana–Champaign*
Michael Meighan, *University of California–Berkeley*
Douglas Meikle, *Miami University*
Allen F. Mensinger, *University of Minnesota–Duluth*
John Merrill, *Michigan State University*
Richard Merritt, *Houston Community College*
Melissa Michael, *University of Illinois–Urbana–Champaign*
Brian T. Miller, *Middle Tennessee State University*
Hugh A. Miller III, *East Tennessee State University*
Thomas E. Miller, *Florida State University*
Sarah L. Milton, *Florida Atlantic University*
Dennis J. Minchella, *Purdue University*
Subhash C. Minocha, *University of New Hampshire*
Patricia Mire, *University of Louisiana–Lafayette*
Daniela S. Monk, *Washington State University*
Daniel C. Moon, *University of North Florida*
Janice Moore, *Colorado State University*
Mathew D. Moran, *Hendrix College*
Jorge A. Moreno, *University of Colorado–Boulder*
Roderick M. Morgan, *Grand Valley State University*
James V. Moroney, *Louisiana State University*
Molly R. Morris, *Ohio University*
Michael Muller, *University of Illinois–Chicago*
Michelle Mynlieff, *Marquette University*
Allan D. Nelson, *Tarleton State University*
Raymond L. Neubauer, *University of Texas–Austin*
Jacalyn S. Newman, *University of Pittsburgh*
Colleen J. Nolan, *St. Mary's University*
Shawn E. Nordell, *St. Louis University*
Margaret Nsofor, *Southern Illinois University–Carbondale*
Dennis W. Nyberg, *University of Illinois–Chicago*
Nicole S. Obert, *University of Illinois–Urbana–Champaign*
David G. Oppenheimer, *University of Florida*
John C. Osterman, *University of Nebraska–Lincoln*
Brian Palestis, *Wagner College*
Julie M. Palmer, *University of Texas–Austin*
C. O. Patterson, *Texas A&M University*
Ronald J. Patterson, *Michigan State University*
Linda M. Peck, *University of Findlay*
David Pennock, *Miami University*
Shelley W. Penrod, *North Harris College*
Beverly J. Perry, *Houston Community College System*
Chris Petersen, *College of the Atlantic*
Jay Phelan, *UCLA*

ACKNOWLEDGMENTS xix

Randall Phillis, *University of Massachusetts–Amherst*
Eric R. Pianka, *The University of Texas–Austin*
Thomas Pitzer, *Florida International University*
Peggy E. Pollak, *Northern Arizona University*
Mitch Price, *Pennsylvania State University*
Richard B. Primack, *Boston University*
Lynda Randa, *College of Dupage*
Marceau Ratard, *Delgado Community College*
Robert S. Rawding, *Gannon University*
Jennifer Regan, *University of Southern Mississippi*
Stuart Reichler, *University of Texas–Austin*
Jill D. Reid, *Virginia Commonwealth University*
Anne E. Reilly, *Florida Atlantic University*
Linda R. Richardson, *Blinn College*
Laurel Roberts, *University of Pittsburgh*
Kenneth R. Robinson, *Purdue University*
Chris Ross, *Kansas State University*
Anthony M. Rossi, *University of North Florida*
Kenneth H. Roux, *Florida State University*
Ann E. Rushing, *Baylor University*
Scott Russell, *University of Oklahoma*
Christina T. Russin, *Northwestern University*
Charles L. Rutherford, *Virginia Tech University*
Margaret Saha, *College of William and Mary*
Kanagasabapathi Sathasivan, *University of Texas–Austin*
Stephen G. Saupe, *College of St. Benedict*
Jon B. Scales, *Midwestern State University*
Daniel C. Scheirer, *Northeastern University*
H. Jochen Schenk, *California State University–Fullerton*
John Schiefelbein, *University of Michigan*
Deemah N. Schirf, *University of Texas–San Antonio*
Mark Schlueter, *College of Saint Mary*
Scott Schuette, *Southern Illinois University–Carbondale*
Dean D. Schwartz, *Auburn University*
Timothy E. Shannon, *Francis Marion University*
Richard M. Showman, *University of South Carolina*
Michele Shuster, *New Mexico State University*
Martin Silberberg, *McGraw-Hill chemistry author*
Robert Simons, *UCLA*
J. Henry Slone, *Francis Marion University*
Phillip Snider, Jr., *Gadsden State Community College*
Nancy G. Solomon, *Miami University*
Lekha Sreedhar, *University of Missouri–Kansas City*
Bruce Stallsmith, *University of Alabama–Huntsville*

Susan J. Stamler, *College of Dupage*
Mark P. Staves, *Grand Valley State University*
William Stein, *Binghamton University*
Philip J. Stephens, *Villanova University*
Kevin Strang, *University of Wisconsin–Madison*
Antony Stretton, *University of Wisconsin–Madison*
Gregory W. Stunz, *Texas A&M University–Corpus Christi*
Julie Sutherland, *College of Dupage*
David Tam, *University of North Texas*
Roy A. Tassava, *Ohio State University*
Sharon Thoma, *University of Wisconsin–Madison*
Shawn A. Thomas, *College of St. Benedict/St. John's University*
Daniel B. Tinker, *University of Wyoming*
Marty Tracey, *Florida International University*
Marsha Turell, *Houston Community College*
J. M. Turbeville, *Virginia Commonwealth University*
Rani Vajravelu, *University of Central Florida*
Neal J. Voelz, *St. Cloud State University*
Samuel E. Wages, *South Plains College*
Jyoti R. Wagle, *Houston Community College System–Central*
Charles Walcott, *Cornell University*
Randall Walikonis, *University of Connecticut*
Jeffrey A. Walker, *University of Southern Maine*
Delon E. Washo-Krupps, *Arizona State University*
Frederick Wasserman, *Boston University*
Steven A. Wasserman, *University of California–San Diego*
R. Douglas Watson, *University of Alabama–Birmingham*
Cindy Martinez Wedig, *University of Texas–Pan American*
Arthur E. Weis, *University of California–Irvine*
Sue Simon Westendorf, *Ohio University*
Howard Whiteman, *Murray State University*
Susan Whittemore, *Keene State College*
David L. Wilson, *University of Miami*
Robert Winning, *Eastern Michigan University*
Jane E. Wissinger, *University of Minnesota*
Michelle D. Withers, *Louisiana State University*
Clarence C. Wolfe, *Northern Virginia Community College*
Gene K. Wong, *Quinnipiac University*
Richard P. Wunderlin, *University of South Florida*
Joanna Wysocka-Diller, *Auburn University*
H. Randall Yoder, *Lamar University*
Marilyn Yoder, *University of Missouri–Kansas City*
Scott D. Zimmerman, *Southwest Missouri State University*

International Reviewers

Dr. Alyaa Ragaei, *Future University, Cairo*
Heather Addy, *University of Calgary*
Mari L. Acevedo, *University of Puerto Rico at Arecibo*
Heather E. Allison, *University of Liverpool, UK*
David Backhouse, *University of New England*
Andrew Bendall, *University of Guelph*
Marinda Bloom, *Stellenbosch University, South Africa*
Tony Bradshaw, *Oxford-Brookes University, UK*
Alison Campbell, *University of Waikato*
Bruce Campbell, *Okanagan College*
Clara E. Carrasco, Ph.D., *University of Puerto Rico–Ponce Campus*
Keith Charnley, *University of Bath, UK*
Ian Cock, *Griffith University*
Margaret Cooley, *University of NSW*
R. S. Currah, *University of Alberta*
Logan Donaldson, *York University*
Theo Elzenga, *Rijks Universiteit Groningen, Netherlands*
Neil C. Haave, *University of Alberta*
Tom Haffie, *University of Western Ontario*
Louise M. Hafner, *Queensland University of Technology*
Annika F. M. Haywood, *Memorial University of Newfoundland*
William Huddleston, *University of Calgary*
Shin-Sung Kang, *KyungBuk University*
Wendy J. Keenleyside, *University of Guelph*
Christopher J. Kennedy, *Simon Fraser University*
Bob Lauder, *Lancaster University*
Richard C. Leegood, *Sheffield University, UK*
Thomas H. MacRae, *Dalhousie University*
R. Ian Menz, *Flinders University*
Kirsten Poling, *University of Windsor*
Jim Provan, *Queens University, Belfast, UK*
Richard Roy, *McGill University*
Han A.B. Wösten, *Utrecht University, Netherlands*

A NOTE FROM THE AUTHORS

The lives of most science-textbook authors do not revolve around an analysis of writing techniques. Instead, we are people who understand science and are inspired by it, and we want to communicate that information to our students. Simply put, we need a lot of help to get it right.

Editors are a key component that help the authors modify the content of their book so it is logical, easy to read, and inspiring. The editorial team for this *Biology* textbook has been a catalyst that kept this project rolling. The members played various roles in the editorial process. Lisa Bruflodt, Senior Developmental Editor, has been the master organizer. Coordinating the efforts of dozens of people and keeping them on schedule is not always fun. Lisa's success at keeping us on schedule has been truly amazing.

Our Freelance Developmental Editors worked directly with the authors to greatly improve the presentation of the textbook's content. Suzanne Olivier and Joni Frasier did an outstanding job in editing chapters and advising the authors on improvements for the second edition. Deborah Brooker painstakingly analyzed all of the illustrations in the textbook to make sure they are accurate, consistent, and student-friendly. She also took the lead role in the editing of the glossary. We would also like to acknowledge our copy editor, Jane DeShaw, for keeping our grammar on track.

Another important aspect of the editorial process is the actual design, presentation, and layout of materials. It's confusing if the text and art aren't on the same page or if a figure is too large or too small. We are indebted to the tireless efforts of Peggy Selle, Lead Project Manager, and David Hash, Senior Designer of McGraw-Hill. Their artistic talents, ability to size and arrange figures, and attention to the consistency of the figures have been remarkable.

We would like to acknowledge the ongoing efforts of the superb marketing staff at McGraw-Hill. Kent Peterson, Vice President, Director of Marketing, oversees a talented staff of people who work tirelessly to promote our book. Special thanks to Michelle Watnick, Marketing Director, and Chris Loewenberg, Marketing Manager, for their ideas and enthusiasm for this book.

Finally, other staff members at McGraw-Hill Higher Education have ensured that the authors and editors were provided with adequate resources to achieve the goal of producing a superior textbook. These include Kurt Strand, President, Science, Engineering, and Math; Marty Lange, Vice President, Editor-in-Chief; and Janice Roerig-Blong, Publisher.

Contents

UNIT I   Chemistry

UNIT II   Cell

UNIT III   Genetics

UNIT V Diversity

UNIT VI Plants

UNIT VII   Animals

UNIT VIII   Ecology

An Introduction to Biology

1

Biology is the study of life. The diverse forms of life found on Earth provide biologists with an amazing array of organisms to study. In many cases, the investigation of living things leads to unforeseen discoveries that no one would have imagined. For example, researchers determined that the venom from certain poisonous snakes contains a chemical that lowers blood pressure in humans. By analyzing that chemical, drugs were later developed to treat high blood pressure (**Figure 1.1**). Certain ancient civilizations, such as the Greeks, Romans, and Egyptians, discovered that the bark of the white willow tree can be used to fight fever. Modern chemists determined that willow bark contains a substance called salicylic acid, which led to the development of the related compound acetylsalicylic acid, more commonly known as aspirin (**Figure 1.2**). In the last century, biologists studied soil bacteria that naturally produce "chemical weapons" to kill competing bacteria in their native environment. These chemicals have been characterized and used to develop antibiotics such as streptomycin to treat bacterial infections (**Figure 1.3**). Finally, for many decades, biologists have known that the Pacific yew tree produces a toxin in its bark and needles that kills insects.

The crystal jelly (*Aequorea victoria*), which produces a green fluorescent protein (GFP).   The gene that encodes GFP has been widely used by researchers to study gene expression and to determine the locations of proteins in cells.

ACE inhibitor (Lotensin®)

Figure 1.1   **The Brazilian arrowhead viper and inhibitors of high blood pressure.**   Derivatives of a chemical found in the venom of the Brazilian arrowhead viper, called angiotensin-converting enzyme (ACE) inhibitors, are commonly used to treat high blood pressure.

Aspirin

Figure 1.2   **The white willow and aspirin.**   Modern aspirin, acetylsalicylic acid, was developed after analyzing a chemical found in the bark of the white willow tree.

Since the 1990s, this toxin, known by the drug name Taxol, has been used to treat patients with ovarian and breast cancer (**Figure 1.4**). These are but a few of the many discoveries that make biology an intriguing discipline. The study of life not only reveals the fascinating characteristics of living species but also leads to the development of drugs and research tools that benefit the lives of people.

To make new discoveries, biologists view life from many different perspectives. What is the composition of living things? How is life organized? How do organisms reproduce? Sometimes the questions posed by biologists are fundamental and even philosophical in nature. How did living organisms originate? Can we live forever? What is the physical basis for memory? Can we save endangered species? Can we understand intriguing changes in body function, such as the green light given off by certain jellyfish?

Future biologists will continue to make important advances. Biologists are scientific explorers looking for answers to some of the world's most enduring mysteries. Unraveling these mysteries presents an exciting challenge to the best and brightest minds. The rewards of a career in biology include the excitement of forging into uncharted territory, the thrill of making discoveries that can improve the health and lives of people, and the impact of biology on the preservation of the environment and endangered species. For these and many other compelling reasons, students seeking challenging and rewarding careers may wish to choose biology as a lifelong pursuit.

In this chapter, we will begin our survey of biology by examining the basic features that are common to all living organisms. We will consider how evolution has led to the development of modern genomes—the entire genetic compositions of living organisms—which can explain the unity and diversity that we observe among modern species. Finally, we will explore the general approaches that scientists follow when making new discoveries.

Figure 1.3 The soil bacterium *Streptomyces griseus*, which naturally produces streptomycin that kills competing bacteria in the soil. Doctors administer streptomycin to people to treat bacterial infections.

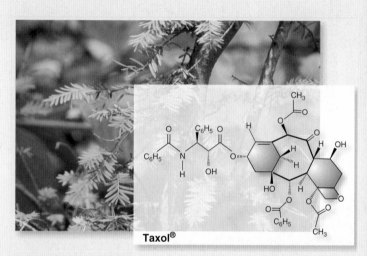

Figure 1.4 **The Pacific yew and Taxol.** A toxin called Taxol, found in the Pacific yew tree, is effective in the treatment of certain cancers.

Concept check: How does biology—the study of life—benefit humans?

1.1 The Properties of Life

A good way to begin a biology textbook is to distinguish living organisms from nonliving objects. At first, the distinction might seem intuitively obvious. A person is alive, but a rock is not. However, the distinction between living and nonliving may seem less obvious when we consider microscopic entities. Is a bacterium alive? What about a virus or a chromosome? In this section, we will examine the characteristics that are common to all forms of life and consider the levels of organization that biologists study.

A Set of Characteristics Is Common to All Forms of Modern Life

Living organisms have consistent features that set them apart from nonliving things. Biologists have determined that all living organisms display seven common characteristics, as described next.

Cells and Organization The concept of organization is so fundamental to biology that the term **organism** can be applied to all living species. Organisms maintain an internal order that is separated from the environment (**Figure 1.5a**). The simplest unit of such organization is the **cell**, which we will examine in Unit II. The **cell theory** states that (1) all organisms are made of cells, (2) cells are the smallest units of life, and (3) cells come from pre-existing cells via cell division. Unicellular organisms are composed of one cell, whereas multicellular organisms such as plants and animals contain many cells. In plants and animals, each cell has internal order, and the cells within the body have specific arrangements and functions.

Energy Use and Metabolism The maintenance of organization requires energy. Therefore, all living organisms acquire

(a) Cells and organization:
Organisms maintain an internal order. The simplest unit of organization is the cell. Yeast cells are shown here.

(b) Energy use and metabolism:
Organisms need energy to maintain internal order. These algae harness light energy via photosynthesis. Energy is used in chemical reactions collectively known as metabolism.

(c) Response to environmental changes:
Organisms react to environmental changes that promote their survival.

(d) Regulation and homeostasis:
Organisms regulate their cells and bodies, maintaining relatively stable internal conditions, a process called homeostasis.

(e) Growth and development:
Growth produces more or larger cells, whereas development produces organisms with a defined set of characteristics.

(f) Reproduction:
To sustain life over many generations, organisms must reproduce. Due to the transmission of genetic material, offspring tend to have traits like their parents.

(g) Biological evolution:
Populations of organisms change over the course of many generations. Evolution results in traits that promote survival and reproductive success.

Figure 1.5   Seven characteristics common to life.

energy from the environment and use that energy to maintain their internal order. Cells carry out a variety of chemical reactions that are responsible for the breakdown of nutrients. Such reactions often release energy in a process called **respiration**. The energy may be used to synthesize the components that make up individual cells and living organisms. Chemical reactions involved with the breakdown and synthesis of cellular molecules are collectively known as **metabolism**. Plants, algae, and certain bacteria can directly harness light energy to produce their own nutrients in the process of **photosynthesis** (Figure 1.5b). They are primary producers of food on Earth. In contrast, some organisms, such as animals and fungi, are consumers—they must use other organisms as food to obtain energy.

Response to Environmental Changes   To survive, living organisms must be able to respond to environmental changes. For example, bacterial cells have mechanisms to detect that certain nutrients in the environment are in short supply while others are readily available. In the winter, many species of mammals develop a thicker coat of fur that protects them from the cold temperatures. Also, plants can respond to changes in the angle of the sun. If you place a plant in a window, it will grow toward the light (Figure 1.5c). The response shown in Figure 1.5c is a short-term response. As discussed later, biological evolution over the course of many generations can lead to more permanent adaptations of a species to its environment.

Regulation and Homeostasis   As we have just seen, one way that organisms can respond to environmental variation is to change themselves. The growth of thick fur in the wintertime is an example. Although life is a dynamic process, living cells and organisms regulate their cells and bodies to maintain relatively stable internal conditions, a process called **homeostasis** (from the Greek, meaning to stay the same). The degree to which homeostasis is achieved varies among different organisms. For example, most mammals and birds maintain a relatively stable body temperature in spite of changing environmental temperatures (Figure 1.5d), whereas reptiles and amphibians tolerate a wider fluctuation in body temperature. By comparison, all organisms continually regulate their cellular metabolism so that nutrient molecules are used at an appropriate rate and new cellular components are synthesized when they are needed.

Growth and Development   All living things grow and develop. **Growth** produces more or larger cells, whereas **development** is a series of changes in the state of a cell, tissue, organ, or organism. The process of development produces organisms with a defined set of characteristics. Among unicellular organisms such as bacteria, new cells are relatively small, and they increase in volume by the synthesis of additional cellular components. Multicellular organisms, such as plants and animals, begin life at a single-cell stage (for example, a fertilized egg) and then undergo multiple cell divisions to develop into a complete organism with many cells (Figure 1.5e).

Reproduction All living organisms have a finite life span. To sustain life over many generations, organisms must **reproduce** (**Figure 1.5f**). A key feature of reproduction is that offspring tend to have characteristics that greatly resemble those of their parent(s). How is this possible? All living organisms contain genetic material composed of **DNA** (**deoxyribonucleic acid**), which provides a blueprint for the organization, development, and function of living things. During reproduction, a copy of this blueprint is transmitted from parent to offspring. As discussed in Unit III, **genes**, which are segments of DNA, govern the characteristics, or traits, of organisms. Most genes are transcribed into a type of **RNA** (**ribonucleic acid**) molecule called messenger RNA (mRNA) that is then translated into a **polypeptide** with a specific amino acid sequence. A **protein** is composed of one or more polypeptides. The structures and functions of proteins are largely responsible for the traits of living organisms.

Biological Evolution The first six characteristics of life, which we have just considered, apply to individual organisms over the short run. Over the long run, another universal characteristic of life is **biological evolution**, which refers to the phenomenon that populations of organisms change from generation to generation. As a result of evolution, organisms may become more successful at survival and reproduction. Populations become better adapted to the environment in which they live. For example, the long snout of an anteater is an adaptation that enhances its ability to obtain food, namely ants, from hard-to-reach places (**Figure 1.5g**). Over the course of many generations, the long snout occurred via biological evolution in which modern anteaters evolved from populations of organisms that did not have such long snouts. Unit IV is devoted to the topic of evolution, and Unit V surveys the evolutionary diversity among different forms of life.

Living Organisms Can Be Viewed at Different Levels of Organization

As we have just learned, life exhibits a set of characteristics, beginning with the concept of organization. The organization of living organisms can be analyzed at different levels of complexity, starting with the tiniest level of organization and progressing to levels that are physically much larger and more complex. **Figure 1.6** depicts a scientist's view of biological organization at different levels.

1. **Atoms:** An **atom** is the smallest unit of an element that has the chemical properties of the element. All matter is composed of atoms.
2. **Molecules and macromolecules:** As discussed in Unit I, atoms bond with each other to form **molecules**. Many molecules bonded together to form a polymer such as a polypeptide is called a **macromolecule**. Carbohydrates, proteins, and nucleic acids (for example, DNA and RNA) are important macromolecules found in living organisms.

3. **Cells:** Molecules and macromolecules associate with each other to form larger structures such as membranes. A **cell** is formed from the association of these larger structures.
4. **Tissues:** In the case of multicellular organisms such as plants and animals, many cells of the same type associate with each other to form **tissues**. An example is muscle tissue (Figure 1.6).
5. **Organs:** In complex multicellular organisms, an **organ** is composed of two or more types of tissue. For example, the heart is composed of several types of tissues, including muscle, nervous, and connective tissue.
6. **Organism:** All living things can be called **organisms**. A single organism possesses the set of characteristics that define life. Biologists classify organisms as belonging to a particular **species**, which is a related group of organisms that share a distinctive form and set of attributes in nature. The members of the same species are closely related genetically. In Units VI and VII, we will examine plants and animals at the level of cells, tissues, organs, and complete organisms.
7. **Population:** A group of organisms of the same species that occupy the same environment is called a **population**.
8. **Community:** A biological **community** is an assemblage of populations of different species. The types of species found in a community are determined by the environment and by the interactions of species with each other.
9. **Ecosystem:** Researchers may extend their work beyond living organisms and also study the environment. Ecologists analyze **ecosystems**, which are formed by interactions of a community of organisms with their physical environment. Unit VIII considers biology from populations to ecosystems.
10. **Biosphere:** The **biosphere** includes all of the places on the Earth where living organisms exist. Life is found in the air, in bodies of water, on the land, and in the soil.

1.2 The Unity and Diversity of Life

Unity and diversity are two words that often are used to describe the living world. As we have seen, all modern forms of life display a common set of characteristics that distinguish them from nonliving objects. In this section, we will explore how this unity of common traits is rooted in the phenomenon of biological evolution. As you will learn, life on Earth is united by an evolutionary past in which modern organisms have evolved from pre-existing organisms.

Evolutionary unity does not mean that organisms are exactly alike. The Earth has many different types of environments, ranging from tropical rain forests to salty oceans, hot and dry deserts, and cold mountaintops. Diverse forms of life have evolved in ways that help them prosper in the diverse environments the Earth has to offer. In this section, we will begin to examine the diversity that exists within the biological world.

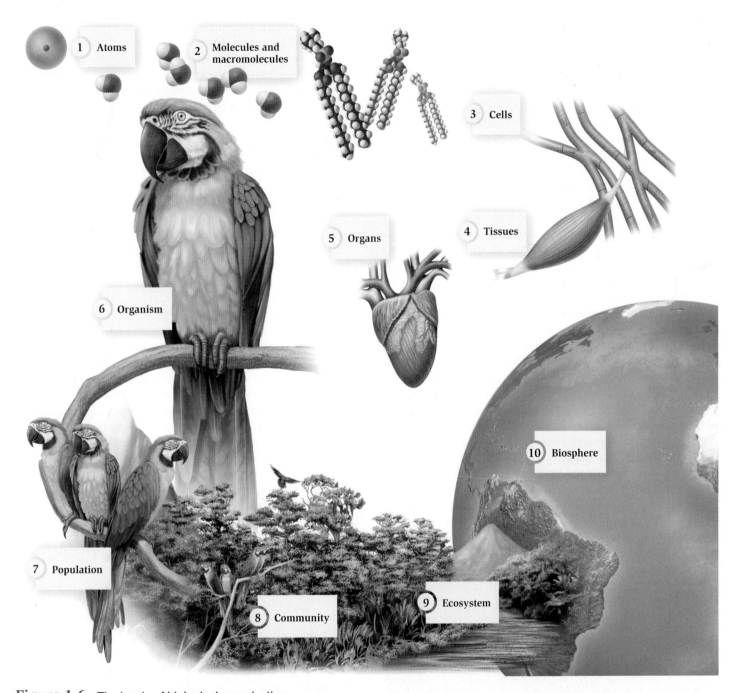

Figure 1.6   The levels of biological organization.

Concept check: At which level of biological organization would you place a herd of buffalo?

Modern Forms of Life Are Connected by an Evolutionary History

Life began on Earth as primitive cells about 3.5 to 4 billion years ago. Since that time, those primitive cells underwent evolutionary changes that ultimately gave rise to the species we see today. Understanding the evolutionary history of species often provides key insights into the structure and function of an organism's body. As a way to help you appreciate this idea, **Figure 1.7** shows a photograph of a bird using a milk carton

in which to build a nest. If we did not already know that the milk carton had served an earlier purpose—namely, to contain milk—we might wonder why the bird had made a nesting site with this shape. Obviously, we do not need to wonder about this because we immediately grasp that the milk carton had a previous history and that it has been modified by a person to serve a new purpose—a nesting site for a bird. Understanding history allows us to make sense out of this nest.

Likewise, evolutionary change involves modifications of characteristics in pre-existing populations. Over long periods of time, populations may change such that structures with a particular function may become modified to serve a new function. For example, the wing of a bat is used for flying, and the flipper of a dolphin is used for swimming. Both structures were modified from a limb that was used for walking in a pre-existing ancestor (**Figure 1.8**).

Evolutionary change occurs by two mechanisms: vertical descent with mutation and horizontal gene transfer. Let's take a brief look at each of these mechanisms.

Vertical Descent with Mutation   The traditional way to view evolution is in a vertical manner, which involves a progression of changes in a series of ancestors. Such a series is called a **lineage**. **Figure 1.9** shows a portion of the lineage that gave rise to modern horses. This type of evolution is called **vertical evolution** because it occurs in a lineage. Biologists have traditionally depicted such evolutionary change in a diagram like the one shown in Figure 1.9. In this mechanism of evolution, new species evolve from pre-existing species by the accumulation of **mutations**, which are random changes in the genetic material of organisms. But why would some mutations accumulate in a population and eventually change the characteristics of an entire species? One reason is that a mutation may alter the traits of organisms in a way that increases their chances of survival and reproduction. When a mutation causes such a beneficial change, the frequency of the mutation may increase in a population from one generation to the next, a process called **natural selection**. This process is discussed in Units IV and V. Evolution also involves the accumulation of neutral changes that do not either benefit or harm a species, and sometimes even involves rare changes that may be harmful.

With regard to the horses shown in Figure 1.9, the fossil record has revealed adaptive changes in various traits such as size and tooth morphology. The first horses were the size of dogs, whereas modern horses typically weigh more than a half ton. The teeth of *Hyracotherium* were relatively small compared to those of modern horses. Over the course of millions of years, horse teeth have increased in size, and a complex pattern of ridges has developed on the molars. How do evolutionary biologists explain these changes in horse characteristics? They can be attributed to natural selection producing adaptations to changing global climates. Over North America, where much of horse evolution occurred, large areas changed from dense forests to grasslands. The horses' increase in size allowed them to escape predators and travel great distances in search of food. The changes seen in horses' teeth are consistent with a dietary shift from eating more tender leaves to eating grasses and other vegetation that are more abrasive and require more chewing.

Horizontal Gene Transfer   The most common way for genes to be transferred is in a vertical manner. This can involve the transfer of genetic material from a mother cell to daughter cells, or it can occur via gametes—sperm and egg—that unite to form a new organism. However, as discussed in later chapters, genes are sometimes transferred between organisms by other mechanisms. These other mechanisms are collectively known as **horizontal gene transfer**. In some cases, horizontal gene transfer can occur between members of different species. For example, you may have heard in the news media that resistance to antibiotics among bacteria is a growing medical problem. As discussed in Chapter 18, genes that confer antibiotic resistance are sometimes transferred between different bacterial species (**Figure 1.10**).

Figure 1.7   **An example of modification of a structure for a new function.**   The bird shown here has used a modified milk carton in which to build its nest. By analogy, evolution also involves the modification of pre-existing structures for a new function.

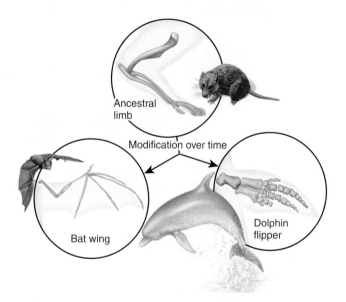

Figure 1.8   **An example showing a modification that has occurred as a result of biological evolution.**   The wing of a bat and the flipper of a dolphin were modified from a limb that was used for walking in a pre-existing ancestor.

Concept check: *Among mammals, give two examples of how the tail has been modified for different purposes.*

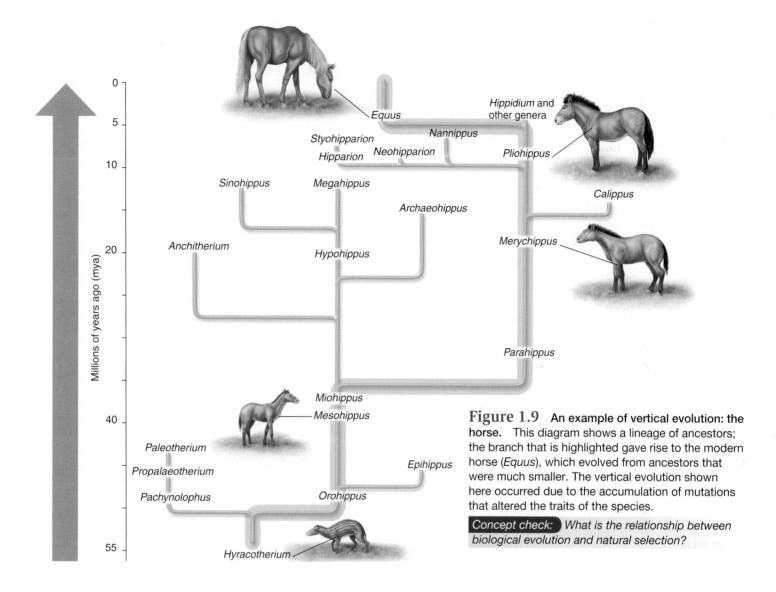

Figure 1.9   **An example of vertical evolution: the horse.**   This diagram shows a lineage of ancestors; the branch that is highlighted gave rise to the modern horse (*Equus*), which evolved from ancestors that were much smaller. The vertical evolution shown here occurred due to the accumulation of mutations that altered the traits of the species.

Concept check: *What is the relationship between biological evolution and natural selection?*

In a lineage in which the time scale is depicted on a vertical axis, horizontal gene transfer between different species is shown as a horizontal line (**Figure 1.11**). Genes transferred horizontally may be subjected to natural selection and promote changes in an entire species. This has been an important mechanism of evolutionary change, particularly among bacterial species. In addition, during the early stages of evolution, which occurred a few billion years ago, horizontal gene transfer was an important part of the process that gave rise to all modern species.

Traditionally, biologists have described evolution using diagrams that depict the vertical evolution of species on a long time scale. This is the type of evolutionary tree that was shown in Figure 1.9. For many decades, the simplistic view held that all living organisms evolved from a common ancestor, resulting in a "tree of life" that could describe the vertical evolution that gave rise to all modern species. Now that we understand the great importance of horizontal gene transfer in the evolution of life on Earth, biologists have needed to re-evaluate the concept of evolution as it occurs over time. Rather than a tree of life, a more appropriate way to view the unity of living organisms is

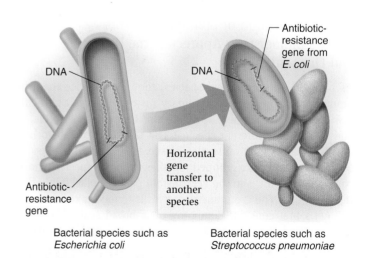

Figure 1.10   **An example of horizontal gene transfer: antibiotic resistance.**   One bacterial species may transfer a gene to a different bacterial species, such as a gene that confers resistance to an antibiotic.

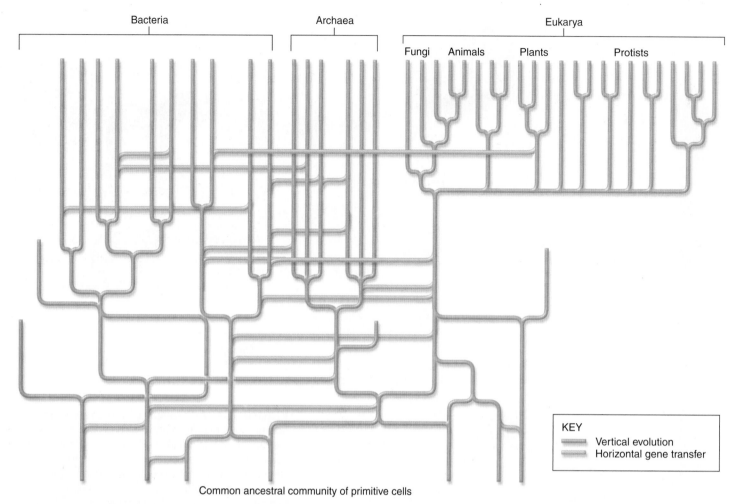

Figure 1.11   **The web of life, showing both vertical evolution and horizontal gene transfer.**   This diagram of evolution includes both of these important mechanisms in the evolution of life on Earth. Note: Archaea are unicellular species.

Concept check: *How does the concept of a tree of life differ from that of a web of life?*

to describe it as a "web of life," which accounts for both vertical evolution and horizontal gene transfer (Figure 1.11).

The Classification of Living Organisms Allows Biologists to Appreciate the Unity and Diversity of Life

As biologists discover new species, they try to place them in groups based on their evolutionary history. This is a difficult task because researchers estimate the Earth has between 10 and 100 million different species! The rationale for categorization is usually based on vertical descent. Species with a recent common ancestor are grouped together, whereas species whose common ancestor is in the very distant past are placed into different groups. The grouping of species is termed **taxonomy**.

Let's first consider taxonomy on a broad scale. You may have noticed that Figure 1.11 showed three main groups of organisms. All forms of life can be placed into three large categories, or domains, called **Bacteria**, **Archaea**, and **Eukarya** (**Figure 1.12**). Bacteria and Archaea are microorganisms that are also termed **prokaryotic** because their cell structure is

relatively simple. At the molecular level, bacterial and archaeal cells show significant differences in their compositions. By comparison, organisms in domain Eukarya are **eukaryotic** and have larger cells with internal compartments that serve various functions. A defining distinction between prokaryotic and eukaryotic cells is that eukaryotic cells have a **cell nucleus** in which the genetic material is surrounded by a membrane. The organisms in domain Eukarya had once been subdivided into four major categories, or kingdoms, called Protista (protists), Plantae (plants), Fungi, and Animalia (animals). However, as discussed in Chapter 26 and Unit V, this traditional view has become invalid as biologists have gathered new information regarding the evolutionary relationships of these organisms. We now know that the protists do not form a single kingdom but instead can be divided into seven broad groups.

Taxonomy involves multiple levels in which particular species are placed into progressively smaller and smaller groups of organisms that are more closely related to each other evolutionarily. Such an approach emphasizes the unity and diversity of different species. As an example, let's consider the clownfish, which is a common saltwater aquarium fish (**Figure 1.13**).

(a) Domain Bacteria: Mostly unicellular prokaryotes that inhabit many diverse environments on Earth.

(b) Domain Archaea: Unicellular prokaryotes that often live in extreme environments, such as hot springs.

Protists: Unicellular and small multicellular organisms that are now subdivided into seven broad groups based on their evolutionary relationships.

Plants: Multicellular organisms that can carry out photosynthesis.

Fungi: Unicellular and multicellular organisms that have a cell wall but cannot carry out photosynthesis. Fungi usually survive on decaying organic material.

Animals: Multicellular organisms that usually have a nervous system and are capable of locomotion. They must eat other organisms or the products of other organisms to live.

(c) Domain Eukarya: Unicellular and multicellular organisms having cells with internal compartments that serve various functions.

Figure 1.12   **The three domains of life.**   Two of these domains, **(a)** Bacteria and **(b)** Archaea, are prokaryotes. The third domain, **(c)** Eukarya, comprises species that are eukaryotes.

Taxonomic group	Clown anemonefish is found in	Approximate time when the common ancestor for this group arose	Approximate number of modern species in this group	Examples
Domain	Eukarya	2,000 mya	> 5,000,000	
Kingdom	Animalia	600 mya	> 1,000,000	
Phylum	Chordata	525 mya	50,000	
Class	Actinopterygii	420 mya	30,000	
Order	Perciformes	80 mya	7,000	
Family	Pomacentridae	~ 40 mya	360	
Genus	*Amphiprion*	~ 9 mya	28	
Species	*ocellaris*	< 3 mya	1	

Figure 1.13   Taxonomic and evolutionary groupings leading to the clownfish.

Concept check: *Why is it useful to place organisms into taxonomic groupings?*

Several species of clownfish, also called clown anemonefish, have been identified. One species of clownfish, which is orange with white stripes, has several common names, including Ocellaris clownfish, false clownfish, and false-clown anemonefish. The broadest grouping for this clownfish is the domain, namely, Eukarya, followed by progressively smaller divisions, from kingdom (Animalia) to species. In the animal kingdom, clownfish are part of a phylum, Chordata, the chordates, which is subdivided into classes. Clownfish are in a class called Actinopterygii, which includes all ray-finned fishes. The common ancestor that gave rise to ray-finned fishes arose about 420 million years ago (mya). Actinopterygii is subdivided into several smaller orders. The clownfish are in the order Perciformes (bony fish). The order is, in turn, divided into families; the clownfish belong to the family of marine fish called Pomacentridae, which are often brightly colored. Families are divided into

genera (singular, genus). The genus *Amphiprion* is composed of 28 different species; these are various types of clownfish. Therefore, the genus contains species that are very similar to each other in form and have evolved from a common (extinct) ancestor that lived relatively recently on an evolutionary time scale.

Biologists use a two-part description, called **binomial nomenclature**, to provide each species with a unique scientific name. The scientific name of the Ocellaris clownfish is *Amphiprion ocellaris*. The first part is the genus, and the second part is the specific epithet or species descriptor. By convention, the genus name is capitalized, whereas the specific epithet is not. Both names are italicized. Scientific names are usually Latinized, which means they are made similar in appearance to Latin words. The origins of scientific names are typically Latin or Greek, but they can come from a variety of sources, such as a person's name.

Genomes & Proteomes Connection

The Study of Genomes and Proteomes Provides an Evolutionary Foundation for Our Understanding of Biology

The unifying concept in biology is evolution. We can understand the unity of modern organisms by realizing that all living species evolved from an interrelated group of ancestors. However, from an experimental perspective, this realization presents a dilemma—we cannot take a time machine back over the course of 4 billion years to carefully study the characteristics of extinct organisms and fully appreciate the series of changes that have led to modern species. Fortunately though, evolution has given biologists some wonderful puzzles to study, including the fossil record and, more recently, the genomes of modern species. As mentioned, the term **genome** refers to the complete genetic composition of an organism (**Figure 1.14a**). The genome is critical to life because it performs these functions:

- *Stores information in a stable form:* The genome of every organism stores information that provides a blueprint to create its characteristics.

- *Provides continuity from generation to generation:* The genome is copied and transmitted from generation to generation.

- *Acts as an instrument of evolutionary change:* Every now and then, the genome undergoes a mutation that may alter the characteristics of an organism. In addition, a genome may acquire new genes by horizontal gene transfer. The accumulation of such changes from generation to generation produces the evolutionary changes that alter species and produce new species.

The evolutionary history and relatedness of all living organisms can be illuminated by genome analysis. The genome of every organism carries the results and the evidence of millions of years of evolution. The genomes of prokaryotes usually contain a few thousand genes, whereas those of eukaryotes may contain tens of thousands. An exciting advance in biology over the past couple of decades has been the ability to analyze the DNA sequence of genomes, a technology called **genomics**. For instance, we can compare the genomes of a frog, a giraffe, and a petunia and discover intriguing similarities and differences. These comparisons help us to understand how new traits evolved. For example, all three types of organisms have the same kinds of genes needed for the breakdown of nutrients such as sugars. In contrast, only the petunia has genes that allow it to carry out photosynthesis.

An extension of genome analysis is the study of **proteomes**, which refers to all of the proteins that a cell or organism can make. The function of most genes is to encode polypeptides that become units in proteins. As shown in **Figure 1.14b**, these include proteins that form a cytoskeleton, proteins that function in cell organization and as enzymes, transport proteins, cell signaling proteins, and extracellular proteins. The genome of each species carries the information to make its proteome, the hundreds or thousands of proteins that each cell of that species makes. Proteins are largely responsible for the structures and functions of cells and organisms. The technical approach called **proteomics** involves the analysis of the proteome of a single species and the comparison of the proteomes of different species. Proteomics helps us understand how the various levels of biology are related to one another, from the molecular level—at the level of protein molecules—to the higher levels, such as how the functioning of proteins produces the characteristics of cells and organisms and affects the ability of populations of organisms to survive in their natural environments.

As a concrete way to understand the unifying concept of evolution in biology, a recurring theme in the chapters that follow is a brief topic called "Genomes & Proteomes Connection" that will allow you to appreciate how evolution produced the characteristics of modern species. These topics explore how the genomes of different species are similar to each other and how they are different. You will learn how genome changes affect the proteome and thereby control the traits of modern species. Ultimately, these concepts provide you with a way to relate information at the molecular level to the traits of organisms and their survival within ecosystems.

The Textbook Cover Provides an Example of How Genomes and Proteomes Are Fundamental to an Organism's Characteristics

As shown on the cover of your textbook, the crystal jelly (*Aequorea victoria*) is a bioluminescent jellyfish found off the west coast of North America. What is **bioluminescence**? The term refers to the ability of some living organisms, such as jellyfish, to produce and emit light due to reactions in which chemical energy is converted to light energy. Biologists currently do not know the function of bioluminescence in this species. Possible roles could be defense against predators or attracting prey.

In the case of the crystal jelly, most of the organism is transparent and not bioluminescent. The bioluminescence is largely restricted to a ring of discrete spots around the bell margin (**Figure 1.15a**). The spots occasionally give off flashes of green light, which is due to a protein the jellyfish makes, called green fluorescent protein (GFP). From the perspective of genomes and proteomes, biologists would say that the GFP gene is found in the genome of this jellyfish, but the green fluorescent protein is expressed only in the proteome of the cells that form these spots around the bell margin.

Researchers interested in bioluminescence have studied how it occurs at the molecular level. The crystal jelly produces light in a two-step process. First, the release of Ca^{2+} in a cell interacts with a protein called aequorin, which produces a blue light. Why don't the jellyfish glow blue? The answer is that, in a second step, the blue light is absorbed by GFP, which then emits a green light.

Because GFP is easily activated by UV or blue light and then specifically gives off green light, researchers have also adapted and used GFP as a visualization tool in medicine, research,

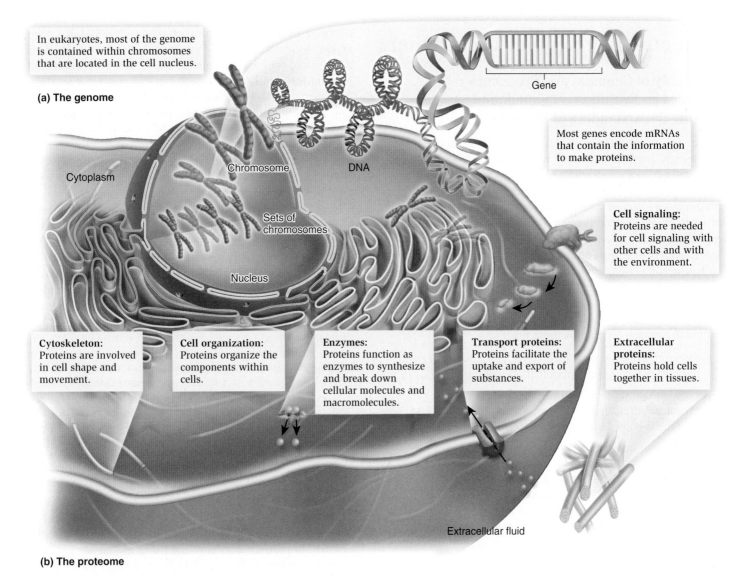

In eukaryotes, most of the genome is contained within chromosomes that are located in the cell nucleus.

(a) The genome

Cytoplasm

Chromosome

DNA

Sets of chromosomes

Nucleus

Gene

Most genes encode mRNAs that contain the information to make proteins.

Cell signaling: Proteins are needed for cell signaling with other cells and with the environment.

Cytoskeleton: Proteins are involved in cell shape and movement.

Cell organization: Proteins organize the components within cells.

Enzymes: Proteins function as enzymes to synthesize and break down cellular molecules and macromolecules.

Transport proteins: Proteins facilitate the uptake and export of substances.

Extracellular proteins: Proteins hold cells together in tissues.

Extracellular fluid

(b) The proteome

Figure 1.14 **Genomes and proteomes.** **(a)** The genome, which is composed of DNA, is the entire genetic composition of an organism. Most of the genetic material in eukaryotic cells is found in the cell nucleus. Its primary function is to encode the proteome. **(b)** The proteome is the entire protein complement of a cell or organism. Six general categories of proteins are illustrated. Proteins are largely responsible for the structure and function of cells and complete organisms.

Concept check: *Biologists sometimes say the genome is a storage unit, whereas the proteome is largely the functional unit of life. Explain this statement.*

and biotechnology. With the aid of GFP, researchers can "see" where genes are expressed in a multicellular organism and where in a cell a particular protein is located. How is this possible? As mentioned, the gene for GFP is found in the genome of the crystal jelly. Using molecular techniques, copies of the GFP gene have been made from this species and placed into the cells of other species. Researchers can create hybrid genes in which a gene from a species of interest is fused with the GFP gene. For example, **Figure 1.15b** shows the results of an experiment where researchers created a hybrid gene by fusing a gene that encodes a protein called tubulin to the GFP gene. Tubulin is a component of microtubules that form a spindle in dividing cells. This hybrid gene encodes a protein in which tubulin is linked to GFP. When this hybrid protein is made in dividing cells and the

cells are exposed to UV light, the spindle glows green, enabling researchers to visualize its location. These results confirm that tubulin is a component of the spindle.

The discovery of GFP and its development as a molecular tool has involved the efforts of several scientists. In the 1960s, Osamu Shimomura was the first researcher to identify and purify GFP from *Aequorea victoria*. Over 20 years later, Martin Chalfie and colleagues obtained the GFP gene from Douglas Prasher, who was also interested in GFP as a molecular tool. Chalfie's work demonstrated that GFP could be used as a colored tag in both bacteria and animals. In addition, Roger Tsien studied the molecular properties of GFP, enabling biologists to understand how GFP gives off light and leading to the development of altered forms of GFP that glow in different colors

(a) Bioluminescence in *Aequorea victoria*

(b) Using GFP to label a spindle in a dividing cell

Figure 1.15 Expression of green fluorescent protein (GFP) in the crystal jelly and its use as a molecular tool. (a) This jellyfish is mostly transparent. GFP is naturally expressed in spots along the bell margin. (b) When GFP is linked to tubulin, the spindle (described in Chapter 15) glows green.

such as cyan, yellow, and red. In 2008, Shimomura, Chalfie, and Tsien received the Nobel Prize for the discovery and the development of GFP, which has become a widely used tool in biology.

1.3 Biology as a Scientific Discipline

What is science? Surprisingly, the definition of science is not easy to state. Most people have an idea of what science is, but actually articulating that idea proves difficult. In biology, we might define **science** as the observation, identification, experimental investigation, and theoretical explanation of natural phenomena.

Science is conducted in different ways and at different levels. Some biologists study the molecules that compose life, while others try to understand how organisms survive in their natural environments. In some cases, experiments are designed to test the validity of ideas suggested by researchers. In this section, we will examine how biologists follow a standard approach, called the **scientific method**, to test their ideas. We will learn that scientific insight is not based solely on intuition. Instead, scientific knowledge makes predictions that can be experimentally tested.

Even so, not all discoveries are the result of researchers following the scientific method. Some discoveries are simply made by gathering new information. As described earlier in Figures 1.1 to 1.4, the characterization of many plants and animals has led to the development of many important medicines and research tools. In this section, we will also consider how researchers often set out on "fact-finding missions" that are aimed at uncovering new information that may eventually lead to modern discoveries in biology.

Biologists Investigate Life at Different Levels of Organization

Earlier, in Figure 1.6, we examined the various levels of biological organization. The study of these different levels depends not only on the scientific interests of biologists but also on the tools available to them. The study of organisms in their natural environments is a branch of biology called **ecology** (Figure 1.16a). In addition, researchers examine the structures and functions of plants and animals, which are disciplines called **anatomy** and **physiology** (Figure 1.16b). With the advent of microscopy, **cell biology**, which is the study of cells, became an important branch of biology in the early 1900s and remains so today (Figure 1.16c). In the 1970s, genetic tools became available to study single genes and the proteins they encode. This genetic technology enabled researchers to study individual molecules, such as proteins, in living cells and thereby spawned the field of **molecular biology**. Together with chemists and biochemists, molecular biologists focus their efforts on the structure and function of the molecules of life (Figure 1.16d). Such researchers want to understand how biology works at the molecular and even atomic levels. Overall, the 20th century saw a progressive increase in the number of biologists who used a reductionist approach to understanding biology. **Reductionism** involves reducing complex systems to simpler components as a way to understand how the system works. In biology, reductionists study the parts of a cell or organism as individual units.

In the 1980s, the pendulum began to swing in the other direction. Scientists have invented new tools that allow us to study groups of genes (genomic techniques) and groups of proteins (proteomic techniques). Biologists now use the term **systems biology** to describe research aimed at understanding how the properties of life arise by complex interactions. This term is often applied to the study of cells. In this context, systems biology may involve the investigation of groups of proteins with a common purpose (Figure 1.16e). For example, a systems biologist may conduct experiments that try to characterize an entire cellular process, which is driven by dozens of different proteins.

Ecologists study species in their native environments.

(a) Ecology—population/ community/ecosystem levels

Anatomists and physiologists study how the structures of organisms are related to their functions.

(b) Anatomy and physiology— tissue/organ/organism levels

Cell biologists often use microscopes to learn how cells function.

(c) Cell biology—cellular levels

Molecular biologists and biochemists study the molecules and macromolecules that make up cells.

(d) Molecular biology— atomic/molecular levels

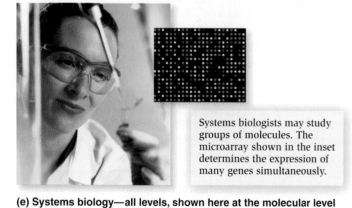

Systems biologists may study groups of molecules. The microarray shown in the inset determines the expression of many genes simultaneously.

(e) Systems biology—all levels, shown here at the molecular level

Figure 1.16 Biological investigation at different levels.

However, systems biology is not new. Animal and plant physiologists have been studying the functions of complex organ systems for centuries. Likewise, ecologists have been characterizing ecosystems for a very long time. The novelty and excitement of systems biology in recent years have been the result of new experimental tools that allow us to study complex interactions at the molecular level. As described throughout this textbook, the investigation of genomes and proteomes has provided important insights regarding many interesting topics in systems biology.

A Hypothesis Is a Proposed Idea, Whereas a Theory Is a Broad Explanation Backed by Extensive Evidence

Let's now consider the process of science. In biology, a **hypothesis** is a proposed explanation for a natural phenomenon. It is a proposition based on previous observations or experimental studies. For example, with knowledge of seasonal changes, you might hypothesize that maple trees drop their leaves in the autumn because of the shortened amount of daylight. An alternative hypothesis might be that the trees drop their leaves because of colder temperatures. In biology, a hypothesis requires more work by researchers to evaluate its validity.

A useful hypothesis must make **predictions**—expected outcomes that can be shown to be correct or incorrect. In other words, a useful hypothesis is testable. If a hypothesis is incorrect, it should be **falsifiable**, which means that it can be shown to be incorrect by additional observations or experimentation. Alternatively, a hypothesis may be correct, so further work will not disprove it. In such cases, we would say that the researcher(s) has failed to reject the hypothesis. Even so, a hypothesis is never really proven but rather always remains provisional. Researchers accept the possibility that perhaps they have not yet conceived of the correct hypothesis. After many experiments, biologists may conclude that their hypothesis is consistent with known data, but they should never say the hypothesis is proven.

By comparison, the term **theory**, as it is used in biology, is a broad explanation of some aspect of the natural world that is substantiated by a large body of evidence. Biological theories incorporate observations, hypothesis testing, and the laws of other disciplines such as chemistry and physics. The power of theories is they allow us to make many predictions regarding the properties of living organisms. As an example, let's consider the theory that DNA is the genetic material and that it is organized into units called genes. An overwhelming body of evidence has substantiated this theory. Thousands of living species have been analyzed at the molecular level. All of them have been found to use DNA as their genetic material and to express genes that produce the proteins that lead to their characteristics. This theory makes many valid predictions. For example, certain types of mutations in genes are expected to affect the traits of organisms. This prediction has been confirmed experimentally. Similarly, this theory predicts that genetic material is copied and transmitted from parents to offspring. By comparing the DNA of parents and offspring, this prediction has also been confirmed. Furthermore, the theory explains the observation that offspring resemble their parents. Overall, two key attributes of a scientific theory are (1) consistency with a vast amount of known data and (2) the ability to make many correct predictions. Two other important biological theories we have touched on in this chapter are the cell theory and the theory of evolution by natural selection.

The meaning of the term theory is sometimes muddled because it is used in different situations. In everyday language, a "theory" is often viewed as little more than a guess or a hypothesis. For example, a person might say, "My theory is that Professor Simpson did not come to class today because he went to the beach." However, in biology, a theory is much more than a guess. A theory is an established set of ideas that explains a vast amount of data and offers valid predictions that can be tested. Like a hypothesis, a theory can never be proven to be true. Scientists acknowledge that they do not know everything. Even so, biologists would say that theories are extremely likely to be true, based on all known information. In this regard, theories are viewed as **knowledge**, which is the awareness and understanding of information.

Discovery-Based Science and Hypothesis Testing Are Scientific Approaches That Help Us Understand Biology

The path that leads to an important discovery is rarely a straight line. Rather, scientists ask questions, make observations, ask modified questions, and may eventually conduct experiments to test their hypotheses. The first attempts at experimentation may fail, and new experimental approaches may be needed. To suggest that scientists follow a rigid scientific method is an oversimplification of the process of science. Scientific advances often occur as scientists dig deeper and deeper into a topic that interests them. Curiosity is the key phenomenon that sparks scientific inquiry. How is biology actually conducted? As discussed next, researchers typically follow two general types of approaches—discovery-based science and hypothesis testing.

Discovery-Based Science The collection and analysis of data without the need for a preconceived hypothesis is called **discovery-based science**, or simply **discovery science**. Why is discovery-based science carried out? The information gained from discovery-based science may lead to the formation of new hypotheses, and, in the long run, may have practical applications that benefit people. Drug companies, for example, may test hundreds or even thousands of compounds to determine if any of them are useful in the treatment of disease (**Figure 1.17a**). Once a drug has been discovered that is effective in disease treatment, researchers may dig deeper and try to understand how the drug exerts its effects. In this way, discovery-based science may help us learn about basic concepts in medicine and biology. Another example involves the study of

genomes (**Figure 1.17b**). Over the past few decades, researchers have identified and begun to investigate newly discovered genes within the human genome without already knowing the function of the gene they are studying. The goal is to gather additional clues that may eventually allow them to propose a hypothesis that explains the gene's function. Discovery-based science often leads to hypothesis testing.

Drug companies may screen hundreds or thousands of different compounds, trying to discover ones that may prove effective in the treatment of a particular disease.

(a) Drug discovery

Genetic researchers search through the genomes of humans and other species, trying to discover new genes. Such discoveries may help us understand molecular biology and provide insight into the causes of inherited diseases in people.

(b) Discovery of genes

Figure 1.17 Discovery-based science.

Concept check: *How is discovery-based science different from hypothesis testing?*

Hypothesis Testing In biological science, the scientific method, also known as **hypothesis testing**, is usually followed to test the validity of a hypothesis. This strategy may be described as a five-stage process:

1. Observations are made regarding natural phenomena.
2. These observations lead to a hypothesis that tries to explain the phenomena. A useful hypothesis is one that is testable because it makes specific predictions.
3. Experimentation is conducted to determine if the predictions are correct.
4. The data from the experiment are analyzed.
5. The hypothesis is considered to be consistent with the data, or it is rejected.

The scientific method is intended to be an objective way to gather knowledge. As an example, let's return to our scenario of maple trees dropping their leaves in autumn. By observing the length of daylight throughout the year and comparing that data with the time of the year when leaves fall, one hypothesis might be that shorter daylight causes the leaves to fall (**Figure 1.18**). This hypothesis makes a prediction—exposure of maple trees to shorter daylight will cause their leaves to fall. To test this prediction, researchers would design and conduct an experiment.

How is hypothesis testing conducted? Although hypothesis testing may follow many paths, certain experimental features are common to this approach. First, data are often collected in two parallel manners. One set of experiments is done on the **control group**, while another set is conducted on the **experimental group**. In an ideal experiment, the control and experimental groups differ by only one factor. For example, an experiment could be conducted in which two groups of trees would be observed and the only difference between their environments would be the length of light each day. To conduct such an experiment, researchers would grow small trees in a greenhouse where they could keep factors such as temperature and water the same between the control and experimental groups, while providing them with different amounts of daylight. In the control group, the number of hours of light provided by lightbulbs would be kept constant each day, while in the experimental group, the amount of light each day would become progressively shorter to mimic seasonal light changes. The researchers would then record the number of leaves dropped by the two groups of trees over a certain period of time.

Another key feature of hypothesis testing is data analysis. The result of experimentation is a set of data from which a biologist tries to draw conclusions. Biology is a quantitative science. When experimentation involves control and experimental groups, a common form of analysis is to determine if

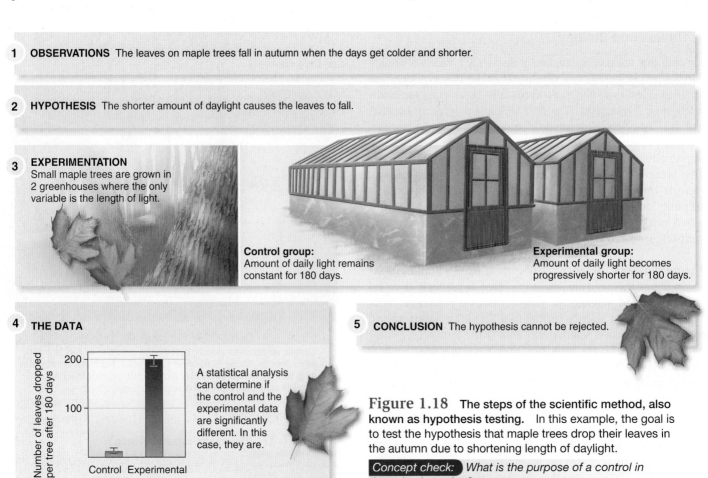

1 **OBSERVATIONS** The leaves on maple trees fall in autumn when the days get colder and shorter.

2 **HYPOTHESIS** The shorter amount of daylight causes the leaves to fall.

3 **EXPERIMENTATION** Small maple trees are grown in 2 greenhouses where the only variable is the length of light.

Control group: Amount of daily light remains constant for 180 days.

Experimental group: Amount of daily light becomes progressively shorter for 180 days.

4 **THE DATA**

A statistical analysis can determine if the control and the experimental data are significantly different. In this case, they are.

5 **CONCLUSION** The hypothesis cannot be rejected.

Figure 1.18 The steps of the scientific method, also known as hypothesis testing. In this example, the goal is to test the hypothesis that maple trees drop their leaves in the autumn due to shortening length of daylight.

Concept check: *What is the purpose of a control in hypothesis testing?*

the data collected from the two groups are truly different from each other. Biologists apply statistical analyses to their data to determine if the control and experimental groups are likely to be different from each other because of the single variable that is different between the two groups. When they are statistically significant, this means that the differences between the control and experimental data are not likely to have occurred as a matter of random chance. In our tree example shown in Figure 1.18, the trees in the control group dropped far fewer leaves than did those in the experimental group. A statistical analysis could determine if the data collected from the two greenhouses are significantly different from each other. If the two sets of data are found not to be significantly different, the hypothesis would be rejected. Alternatively, if the differences between the two sets of data are significant, as shown in Figure 1.18, biologists would conclude that the hypothesis is consistent with the data, though it is not proven. These results may cause researchers to ask further questions. For example, they may want to understand how decreases in the length of daylight promote cellular changes that cause the leaves to fall.

As described next, discovery-based science and hypothesis testing are often used together to learn more about a particular scientific topic. As an example, let's look at how both approaches have led to successes in the study of the disease called cystic fibrosis.

The Study of Cystic Fibrosis Provides Examples of Both Discovery-Based Science and Hypothesis Testing

Let's consider how biologists made discoveries related to the disease cystic fibrosis (CF), which affects about 1 in every 3,500 Americans. Persons with CF produce abnormally thick and sticky mucus that obstructs the lungs and leads to life-threatening lung infections. The thick mucus also blocks the pancreas, which prevents the digestive enzymes this organ produces from reaching the intestine. For this reason, CF patients tend to have excessive appetites but poor weight gain. Persons with this disease may also experience liver damage because the thick mucus can obstruct the liver. The average life span for people with CF is currently in their mid- to late 30s. Fortunately, as more advances have been made in treatment, this number has steadily increased.

Because of its medical significance, many scientists are interested in cystic fibrosis and have conducted studies aimed at gaining greater information regarding its underlying cause. The hope is that a better understanding of the disease may lead to improved treatment options, and perhaps even a cure. As described next, discovery-based science and hypothesis testing have been critical to gaining a better understanding of this disease.

The CF Gene and Discovery-Based Science   In 1945, Dorothy Anderson determined that cystic fibrosis is a genetic disorder. Persons with CF have inherited two faulty *CF* genes, one from each parent. Over 40 years later, researchers used discovery-based science to identify this gene. Their search

for the *CF* gene did not require any preconceived hypothesis regarding the function of the gene. Rather, they used genetic strategies similar to those described in Chapter 20. In 1989, research groups headed by Lap-Chi Tsui, Francis Collins, and John Riordan identified the *CF* gene.

The discovery of the gene made it possible to devise diagnostic testing methods to determine if a person carries a faulty *CF* gene. In addition, the characterization of the *CF* gene provided important clues regarding its function. Researchers observed striking similarities between the *CF* gene and other genes that were already known to encode proteins called transporters, which function in the transport of substances across membranes. Based on this observation, as well as other kinds of data, the researchers hypothesized that the function of the normal *CF* gene is to encode a transporter. In this way, the identification of the *CF* gene led researchers to conduct experiments aimed at testing a hypothesis of its function.

The CF Gene and Hypothesis Testing   Researchers considered the characterization of the *CF* gene along with other studies showing that patients with the disorder have an abnormal regulation of salt balance across their plasma membranes. They hypothesized that the normal *CF* gene encodes a transporter that functions in the transport of chloride ions (Cl^-), a component of common table salt (NaCl), across the membranes of cells (**Figure 1.19**). This hypothesis led to experimentation in which researchers tested normal cells and cells from CF patients

Lung cell with normal *CF* gene   Lung cell with faulty *CF* gene

Figure 1.19   A hypothesis that suggests an explanation for the function of the gene that is defective in patients with cystic fibrosis.   The normal *CF* gene, which does not carry a mutation, encodes a transporter that transports chloride ions (Cl^-) across the plasma membrane to the outside of the cell. In persons with CF, this transporter is defective due to a mutation in the *CF* gene.

Concept check: *Explain how discovery-based science helped researchers to hypothesize that the CF gene encodes a transporter.*

for their ability to transport Cl⁻. The CF cells were found to be defective in chloride transport. In 1990, scientists successfully transferred the normal *CF* gene into CF cells in the laboratory. The introduction of the normal gene corrected the cells' defect in chloride transport. Overall, the results showed that the *CF* gene encodes a transporter that transports Cl⁻ across the plasma membrane. A mutation in this gene causes it to encode a defective transporter, leading to a salt imbalance that affects water levels outside the cell, which explains the thick and sticky mucus in CF patients. In this example, hypothesis testing has provided a way to evaluate a hypothesis regarding how a disease is caused by a genetic change.

FEATURE INVESTIGATION

Observation and Experimentation Form the Core of Biology

Because biology is the study of life, a biology textbook that focuses only on a description of living organisms would miss the main point. Biology is largely about the process of discovery. Therefore, a recurring theme of this textbook is discovery-based science and hypothesis testing. While each chapter contains many examples of data collection and experiments, a consistent element is a "Feature Investigation"—an actual study by current or past researchers. Some of these involve discovery-based science in which biologists collect and analyze data in an attempt to make discoveries that are not hypothesis driven. Alternatively, most Feature Investigations involve hypothesis testing in which a hypothesis is stated and the experiment and resulting data are presented. See Figure 1.18 to see the form of these Feature Investigations.

The Feature Investigations allow you to appreciate the connection between science and scientific theories. We hope you will find this a more interesting and rewarding way to learn about biology. As you read a Feature Investigation, you may find yourself thinking about different approaches and alternative hypotheses. Different people can view the same data and arrive at very different conclusions. As you progress through the experiments in this textbook, you will enjoy biology far more if you try to develop your own skills at formulating hypotheses, designing experiments, and interpreting data.

Experimental Questions

1. Discuss the difference between discovery-based science and hypothesis testing.

2. What are the steps in the scientific method, also called hypothesis testing?

3. When conducting an experiment, explain how a control group and an experimental group differ from each other.

Science as a Social Discipline

Finally, it is worthwhile to point out that science is a social discipline. After performing observations and experiments, biologists communicate their results in different ways. Most importantly, papers are submitted to scientific journals. Following submission, most papers undergo a **peer-review process** in which other scientists, who are experts in the area, evaluate the paper and make suggestions regarding its quality. Following peer review, a paper is either accepted for publication, rejected, or the authors of the paper may be given suggestions for how to revise the work or conduct additional experiments before it will be acceptable for publication.

Another social aspect of research is that biologists often attend meetings where they report their most recent work to the scientific community (**Figure 1.20**). They comment on each other's ideas and work, eventually shaping together the information that builds into scientific theories over many years. As you develop your skills at scrutinizing experiments, it is satisfying to discuss your ideas with other people, including fellow students and faculty members. Importantly, you do not need to "know all the answers" before you enter into a scientific discussion. Instead, a more rewarding way to view science is as an ongoing and never-ending series of questions.

Figure 1.20 The social aspects of science. At scientific meetings, researchers gather to discuss new data and discoveries. Research conducted by professors, students, lab technicians, and industrial participants is sometimes hotly debated.

Summary of Key Concepts

- Biology is the study of life. Discoveries in biology help us understand how life exists, and they also have many practical applications, such as the development of drugs to treat human diseases. (Figures 1.1, 1.2, 1.3, 1.4)

1.1 The Properties of Life

- Seven characteristics are common to all forms of life. All living things (1) are composed of cells; (2) use energy; (3) respond to environmental changes; (4) regulate their internal conditions (homeostasis); (5) grow and develop; (6) reproduce; and (7) evolve over the course of many generations. (Figure 1.5)

- Living organisms can be viewed at different levels of complexity: atoms, molecules and macromolecules, cells, tissues, organs, organisms, populations, communities, ecosystems, and the biosphere. (Figure 1.6)

1.2 The Unity and Diversity of Life

- Changes in species often occur as a result of modification of pre-existing structures. (Figures 1.7, 1.8)

- Vertical evolution involves mutations in a lineage that alter the characteristics of species over many generations. During this process, natural selection results in the survival of individuals with greater reproductive success. Over the long run, this process alters species and may produce new species. In addition, evolution involves the accumulation of neutral changes. (Figure 1.9)

- Horizontal gene transfer may involve the transfer of genes between different species. Along with vertical evolution, it is an important force in biological evolution, producing a web of life. (Figures 1.10, 1.11)

- Taxonomy involves the grouping of species according to their evolutionary relatedness to other species. Going from broad to narrow, each species is placed into a domain, kingdom, phylum, class, order, family, and genus. (Figures 1.12, 1.13)

- The genome is the genetic composition of a species. It provides a blueprint for the traits of an organism, is transmitted from parents to offspring, and acts as an instrument for evolutionary change. The proteome is the collection of proteins that a cell or organism can make. Beginning with Chapter 3, each chapter in this textbook has a brief discussion called "Genomes & Proteomes Connection." (Figure 1.14)

- An analysis of genomes and proteomes helps us to understand the characteristics of individuals and how they survive in their native environments. (Figure 1.15)

1.3 Biology as a Scientific Discipline

- Biological science involves the observation, identification, experimental investigation, and theoretical explanation of natural phenomena.

- Biologists study life at different levels, ranging from ecosystems to molecular components in cells. (Figure 1.16)

- A hypothesis is a proposal to explain a natural phenomenon. A useful hypothesis makes a testable prediction. A biological theory is a broad explanation based on vast amounts of data and makes many valid predictions.

- Discovery-based science is an approach in which researchers conduct experiments without a preconceived hypothesis. It is a fact-finding mission. (Figure 1.17)

- The scientific method, also called hypothesis testing, is a series of steps to test the validity of a hypothesis. The experimentation often involves a comparison between control and experimental groups. (Figure 1.18)

- The study of cystic fibrosis is an interesting example in which both discovery-based science and hypothesis testing have provided key insights regarding the nature of the disease. (Figure 1.19)

- Each chapter in this textbook has a "Feature Investigation" to help you appreciate how science has led to key discoveries in biology.

- To be published, scientific papers are usually subjected to peer review. Advances in science often occur when scientists gather and discuss their data. (Figure 1.20)

Assess and Discuss

Test Yourself

1. The process where living organisms maintain a relatively stable internal condition is
   a. adaptation.   d. homeostasis.
   b. evolution.   e. development.
   c. metabolism.

2. Populations of organisms change over the course of many generations. Many of these changes result in increased survival and reproduction. This phenomenon is
   a. evolution.   d. genetics.
   b. homeostasis.   e. metabolism.
   c. development.

3. All of the places on Earth where living organisms are found is
   a. the ecosystem.   d. a viable land mass.
   b. a community.   e. a population.
   c. the biosphere.

4. Which of the following would be an example of horizontal gene transfer?
   a. the transmission of an eye color gene from father to daughter
   b. the transmission of a mutant gene causing cystic fibrosis from father to daughter
   c. the transmission of a gene conferring pathogenicity (the ability to cause disease) from one bacterial species to another
   d. the transmission of a gene conferring antibiotic resistance from a mother cell to its two daughter cells
   e. all of the above

5. The scientific name for humans is *Homo sapiens*. The name *Homo* is the _____ to which humans are classified.
   a. kingdom   d. genus
   b. phylum   e. species
   c. order

6. The complete genetic makeup of an organism is called
   a. the genus.
   b. the genome.
   c. the proteome.
   d. the genotype.
   e. the phenotype.

7. A proposed explanation for a natural phenomenon is
   a. a theory.
   b. a law.
   c. a prediction.
   d. a hypothesis.
   e. an assay.

8. In science, a theory should
   a. be equated with knowledge.
   b. be supported by a substantial body of evidence.
   c. provide the ability to make many correct predictions.
   d. all of the above.
   e. b and c only.

9. Conducting research without a preconceived hypothesis is called
   a. discovery-based science.
   b. the scientific method.
   c. hypothesis testing.
   d. a control experiment.
   e. none of the above.

10. What is the purpose of using a control in scientific experiments?
    a. A control allows the researcher to practice the experiment first before actually conducting it.
    b. A researcher can compare the results in the experimental group and control group to determine if a single variable is causing a particular outcome in the experimental group.
    c. A control provides the framework for the entire experiment so the researcher can recall the procedures that should be conducted.
    d. A control allows the researcher to conduct other experimental changes without disturbing the original experiment.
    e. All of the above are correct.

Conceptual Questions

1. What are the seven characteristics of life? Explain a little about each.

2. Explain how it is possible for evolution to result in unity among different species yet also create amazing diversity.

3. Which two taxonomic groups are very diverse? Which two are the least diverse (see Figure 1.13)?

Collaborative Questions

1. Discuss whether or not you think that theories in biology are true. Outside of biology, how do you decide if something is true?

2. In certain animals, such as alligators, sex is determined by temperature. When alligator eggs are exposed to low temperatures, most alligator embryos develop into females. Discuss how this phenomenon is related to genomes and proteomes.

Online Resource

www.brookerbiology.com

Stay a step ahead in your studies with animations that bring concepts to life and practice tests to assess your understanding. Your instructor may also recommend the interactive ebook, individualized learning tools, and more.

The Chemical Basis of Life I:
Atoms, Molecules, and Water

2

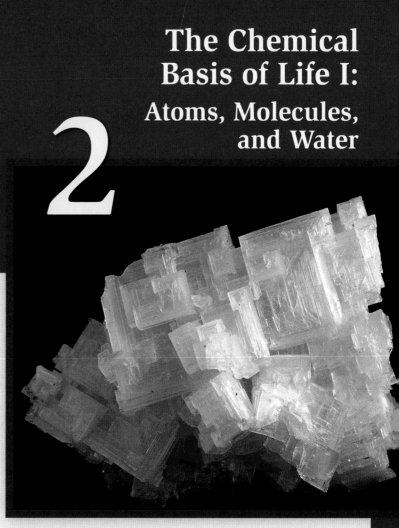

Biology—the study of life—is founded on the principles of chemistry and physics. All living organisms are a collection of atoms and molecules bound together and interacting with each other through the forces of nature. Throughout this textbook, we will see how chemistry can be applied to living organisms as we discuss the components of cells, the functions of proteins, the flow of nutrients in plants and animals, and the evolution of new genes. This chapter lays the groundwork for understanding these and other concepts. We begin with an overview of **inorganic chemistry**—that is, the nature of atoms and molecules, with the exception of those that contain rings or chains of carbon. Such carbon-containing molecules form the basis of **organic chemistry** and are covered in Chapter 3.

Crystals of sodium chloride (NaCl), a molecule composed of two elements.

2.1 Atoms

All life-forms are composed of **matter**, which is defined as anything that contains mass and occupies space. In living organisms, matter may exist in any of three states: solid, liquid, or gas. All matter is composed of **atoms**, which are the smallest functional units of matter that form all chemical substances and ultimately all organisms; they cannot be further broken down into other substances by ordinary chemical or physical means. Atoms, in turn, are composed of smaller, subatomic components collectively referred to as particles. Chemists are interested in the properties of atoms and **molecules**, which are two or more atoms bonded together. A major role of the physicist, by contrast, is to uncover the properties of subatomic particles. Chemistry and physics merge when one attempts to understand the mechanisms by which atoms and molecules interact. When atoms and molecules are studied in the context of a living organism, the science of biochemistry emerges. No living creature is immortal, but atoms never "die." Instead, they exist *ad infinitum* as solitary atoms or as components of a single molecule, or they shuttle between countless molecules over vast eons of time. (An exception to this are unstable atoms called radioisotopes, described later.) In this section, we explore the physical properties of atoms so we can understand how atoms combine to form molecules of biological importance.

Atoms Are Comprised of Subatomic Particles

There are many types of atoms in living organisms. The simplest atom, hydrogen, is approximately 0.1 nanometers (1×10^{-10} meters) in diameter, roughly one-millionth the diameter of a human hair. Each specific type of atom—nitrogen, hydrogen, oxygen, and so on—is called an **element** (or chemical element), which is defined as a pure substance of only one kind of atom.

Three subatomic particles—**protons**, **neutrons**, and **electrons**—are found within atoms. The protons and neutrons are confined to a very small volume at the center of an atom, the **atomic nucleus**, whereas the electrons are found in regions at various distances from the nucleus. With the exception of ions—atoms that have gained or lost one or more electrons (described later in this chapter)—the numbers of protons and electrons in a given type of atom are identical, but the number of neutrons may vary. Each of the subatomic particles has a different electric charge. Protons have one unit of positive charge, electrons have one unit of negative charge, and neutrons are electrically neutral. Like charges always repel each other, and opposite charges always attract each other. It is the opposite charges of the protons and electrons that create an atom—the positive charges in the nucleus attract the negatively charged electrons.

Because the protons are located in the atomic nucleus, the nucleus has a net positive charge equal to the number of protons it contains. The entire atom has no net electric charge, however, because the number of negatively charged electrons around the nucleus is equal to the number of positively charged protons in the nucleus.

The basic structure of the atom was discovered by Ernest Rutherford in a landmark experiment conducted during the years 1909–1911, as described next.

FEATURE INVESTIGATION

Rutherford Determined the Modern Model of the Atom

Nobel laureate Ernest Rutherford was born in 1871 in New Zealand, but he did his greatest work at McGill University in Montreal, Canada, and later at the University of Manchester in England. At that time, scientists knew that atoms contained charged particles but had no idea how those particles were arranged. Neutrons had not yet been discovered, and many scientists, including Rutherford, hypothesized that the positive charge and the mass of an atom were evenly dispersed throughout the atom.

In a now-classic experiment, Rutherford aimed a fine beam of positively charged α (alpha) particles at an extremely thin sheet of gold foil only 400 atoms thick (Figure 2.1). Alpha particles are the two protons and two neutrons that comprise the nuclei of helium atoms; you can think of them as helium atoms without their electrons. Surrounding the gold foil were zinc sulfide screens that registered any α particles passing through or bouncing off the foil, much like film in a camera detects light. Rutherford hypothesized that if the positive charges of the gold atoms were uniformly distributed, many of the positively charged α particles would be slightly deflected, because one of the most important features of electric charge is that like charges repel each other. Due to their much smaller mass, he did not expect electrons in the gold atoms to have any impact on the ability of an α particle to move through the metal foil.

Surprisingly, Rutherford discovered that more than 98% of the α particles passed right through as if the foil was not there and only a small percent were slightly deflected; a few even bounced back at a sharp angle! To explain the 98% that passed right through, Rutherford concluded that most of the volume of an atom is empty space. To explain the few α particles that bounced back at a sharp angle, he postulated that most of the

Figure 2.1   Rutherford's gold foil experiment, demonstrating that most of the volume of an atom is empty space.

HYPOTHESIS Atoms in gold foil are composed of diffuse, evenly distributed positive charges that should usually cause α particles to be slightly deflected as they pass through.

KEY MATERIALS Thin sheet of gold foil, α particle emitter, zinc sulfide detection screen.

Experimental level   Conceptual level

1   Emit beam of α particles.
    α particle emitter
    α particle

2   Pass beam through gold foil.
    Zinc sulfide detection screens   Gold foil
    Gold atom   Gold foil   Positive charges of the gold atom
    α particle
    Undeflected α particles
    Slightly deflected α particle
    α particle that bounced back

α particle that bounced back

α particle that was undeflected

3 Detect α particles on zinc sulfide screens after they pass through foil or bounce back. Record number of α particles detected on zinc sulfide screens and their locations.

α particle that was slightly deflected

Detection of α particles

4 **THE DATA**

% of α particles detected on zinc sulfide screens	Location
98%	Undeflected
<2%	Slightly deflected
0.01%	Bounced back

5 **CONCLUSION** Most of the volume of an atom is empty space, with the positive charges concentrated in a small volume.

6 **SOURCE** Rutherford, E. 1911. The scattering of α and β particles by matter and the structure of the atom. *Philosophical Magazine* 21:669–688.

atom's positive charge was localized in a highly compact area at the center of the atom. The existence of this small, dense region of highly concentrated positive charge—which today we call the atomic nucleus—explains how some α particles could be so strongly deflected by the gold foil. The α particles would bounce back on the rare occasion when they directly collided with an atomic nucleus. Therefore, based on these results, Rutherford rejected his original hypothesis that atoms are composed of diffuse, evenly distributed positive charges.

From this experiment, Rutherford proposed a transitional model of an atom, with its small, positively charged nucleus surrounded at relatively great distances by negatively charged electrons. Today we know that more than 99.99% of an atom's

volume is outside the nucleus. Indeed, the nucleus accounts for only about 1/10,000 of an atom's diameter—most of an atom is empty space!

Experimental Questions

1. Before the experiment conducted by Ernest Rutherford, how did many scientists envision the structure of an atom?

2. What was the hypothesis tested by Rutherford?

3. What were the results of the experiment? How did Rutherford interpret the results?

Electrons Occupy Orbitals Around an Atom's Nucleus

At one time, scientists visualized an atom as a mini–solar system, with the nucleus being the sun and the electrons traveling in clearly defined orbits around it. **Figure 2.2** shows a diagram of the two simplest atoms, hydrogen and helium, which have the smallest numbers of protons. This model of the atom is now known to be an oversimplification, because as described shortly, electrons do not actually orbit the nucleus in a defined path like planets around the sun. However, this depiction of an atom remains a convenient way to diagram atoms in two dimensions.

Electrons move at terrific speeds. Some estimates suggest that the electron in a typical hydrogen atom could circle the Earth in less than 20 seconds! Partly for this reason, it is difficult to precisely predict the exact location of a given electron. In fact, we can only describe the region of space surrounding the nucleus in which there is a high probability of finding that electron. These regions are called **orbitals**. A better model of an atom, therefore, is a central nucleus surrounded by cloud-like orbitals. The cloud represents the region in which a given electron is most likely to be found. Some orbitals are spherical, called *s* orbitals, whereas others assume a shape that is often described as similar to a propeller or dumbbell and are called *p* orbitals (**Figure 2.3**). An orbital can contain a maximum of two

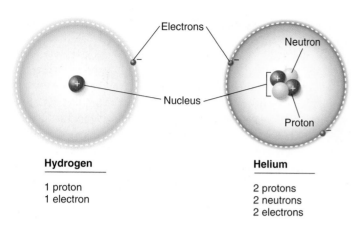

Figure 2.2   Diagrams of two simple atoms.   This is a model of the two simplest atoms, hydrogen and helium. Note: In all figures of atoms, the sizes and distances are not to scale.

Orbital name	1s	2s	2p
	Nucleus		
Number of electrons per energy shell	2	2 per orbital; 8 total	
Orbital shape	Spherical	First orbital: spherical	Second to fourth orbital: dumbbell-shaped

Figure 2.3   Diagrams of individual electron orbitals. Electrons are found outside the nucleus in orbitals that may resemble spherical or dumbbell-shaped clouds. The orbital cloud represents a region in which the probability is high of locating a particular electron. For this illustration, only two shells are shown; the heaviest elements contain a total of seven shells.

electrons. Consequently, any atom with more than two electrons must contain additional orbitals.

Orbitals occupy so-called **energy shells**, or energy levels. **Energy** can be defined as the capacity to do work or effect a change. In biology, we often refer to various types of energy, such as light energy, mechanical energy, and chemical energy. Electrons orbiting a nucleus have kinetic energy, that is, the energy of moving matter. Atoms with progressively more electrons have orbitals within energy shells that are at greater and greater distances from the nucleus. These shells are numbered, with shell number 1 closest to the nucleus. Different energy shells may contain one or more orbitals, each orbital with up to two electrons. The innermost energy shell of all atoms has room for only two electrons, which spin in opposite directions within a spherical *s* orbital (1*s*). The second energy shell is composed of one spherical *s* orbital (2*s*) and three dumbbell-shaped *p*

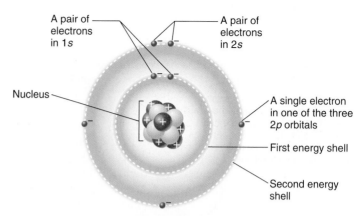

(a) Simplified depiction of a nitrogen atom (7 electrons; 2 electrons in first energy shell, 5 in second energy shell)

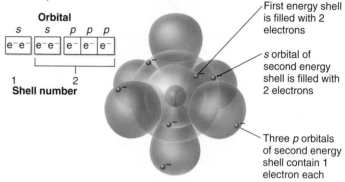

(b) Nitrogen atom showing electrons in orbitals

Figure 2.4   Diagrams showing the multiple energy shells and orbitals of a nitrogen atom.   The nitrogen atom is shown (a) simplified and (b) with all of its orbitals and shells. An atom's shells fill up one by one. In shells containing more than one orbital, the orbital with lowest energy fills first. Subsequent orbitals gain one electron at a time, shown schematically in boxes, where *e* represents an electron. Heavier elements contain additional shells and orbitals.

Concept check: Explain the difference between an energy shell and an orbital.

orbitals (2*p*). Therefore, the second shell can hold up to four pairs of electrons, or eight electrons altogether (Figure 2.3).

Electrons vary in the amount of energy they have. The shell closest to the nucleus fills up with the lowest energy electrons first, and then each subsequent shell fills with higher and higher energy electrons, one shell at a time. Within a given shell, the energy of electrons can also vary among different orbitals. In the second shell, for example, the *s* orbital has lower energy, while the three *p* orbitals have slightly higher and roughly equal energies. In that case, therefore, two electrons fill the *s* orbital first. Any additional electrons fill the *p* orbitals one electron at a time.

Although electrons are actually found in orbitals of varying shapes, as shown in Figure 2.3, chemists often use more simplified diagrams when depicting the energy shells of electrons. **Figure 2.4a** illustrates an example involving the element nitrogen. An atom of this element has seven protons and seven electrons. Two electrons fill the first shell, and five electrons are found in the outer shell. Two of these fill the 2*s* orbital and are shown as a pair of electrons in the second shell. The other three electrons in the second shell are found singly in each of the three *p* orbitals. The diagram in Figure 2.4a makes it easy to see whether electrons are paired within the same orbital and whether the outer shell is full. **Figure 2.4b** shows a more scientifically accurate depiction of a nitrogen atom, showing how the electrons occupy orbitals with different shapes.

Most atoms have outer shells that are not completely filled with electrons. Nitrogen, as we just saw, has a first shell filled with two electrons and a second shell with five electrons (Figure 2.4a). Because the second shell can actually hold eight electrons, the outer shell of a nitrogen atom is not full. As discussed later in this chapter, atoms that have unfilled energy shells tend to share, release, or obtain electrons to fill their outer shell. Those electrons in the outermost shell are called the **valence electrons**. As you will learn shortly, in certain cases such electrons allow atoms to form chemical bonds with each other, in which two or more atoms become joined together to create a new substance.

Each Element Has a Unique Number of Protons

Each chemical element has a specific and unique number of protons that distinguishes it from another element. The number of protons in an atom is its **atomic number**. For example, hydrogen, the simplest atom, has an atomic number of 1, corresponding to its single proton. Magnesium has an atomic number of 12, corresponding to its 12 protons. Recall that except for ions, the number of protons and electrons in a given atom are identical. Therefore, the atomic number is also equal to the number of electrons in the atom, resulting in a net charge of zero.

Figure 2.5 shows the first three rows of the periodic table of the elements, which arranges the known elements according to their atomic number and electron shells (see Appendix for the complete periodic table). A one- or two-letter symbol is used as an abbreviation for each element. The rows (known as "periods") indicate the number of energy shells. For example, hydrogen (H) has one shell, lithium (Li) has two shells, and sodium (Na) has three shells. The columns (called "groups"), from left to right, indicate the numbers of electrons in the outer shell. The outer shell of lithium (Li) has one electron, beryllium (Be) has two, boron (B) has three, and so forth. This organization of the periodic table tends to arrange elements based on similar chemical properties. For example, magnesium (Mg) and calcium (Ca) each have two electrons in their outer shell, so these two elements tend to combine with many of the same other elements. The similarities of elements within a group occur because they have the same number of electrons in their outer shells, and therefore, they have similar chemical

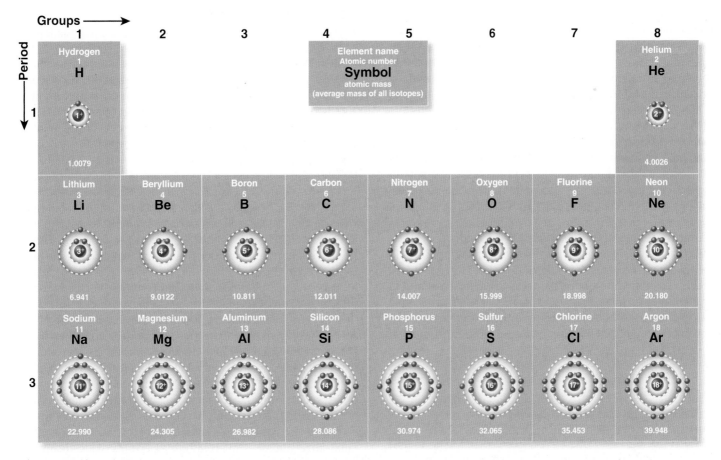

Figure 2.5 **The first three rows of the periodic table of the elements.** The elements are shown in models that depict the energy shells in different colors and the total number of electrons in each shell. The occupancy of orbitals is that of the elements in their pure state. The red sphere represents the nucleus of the atom, and the numerical value with the $^+$ designation represents the number of protons and, therefore, the positive charge of the nucleus. Elements are arranged in groups (columns) and periods (rows). For the complete periodic table, see Appendix.

bonding properties. These properties will be discussed later in this chapter.

Atoms Have a Small but Measurable Mass

Atoms are extremely small and therefore have very little mass. A single hydrogen atom, for example, has a mass of about 1.67×10^{-24} g (grams). Protons and neutrons are nearly equal in mass, and each are more than 1,800 times the mass of an electron (**Table 2.1**). Because of their tiny size relative to protons and neutrons, the mass of the electrons in an atom is ignored in calculations of atomic mass.

The **atomic mass** scale indicates an atom's mass relative to the mass of other atoms. By convention, the most common type of carbon atom, which has six protons and six neutrons, is assigned an atomic mass of exactly 12. On this scale, a hydrogen atom has an atomic mass of 1, indicating that it has 1/12 the mass of a carbon atom. A magnesium atom, with an atomic mass of 24, has twice the mass of a carbon atom.

The term mass is sometimes confused with weight, but these two terms refer to different features of matter. Weight is derived from the gravitational pull on a given mass. For example, a man who weighs 154 pounds on Earth would weigh only 25 pounds if he were standing on the moon, and he would weigh 21 trillion pounds if he could stand on a neutron star. However, his mass is the same in all locations because he has the same amount of matter. Because we are discussing mass on Earth, we can assume that the gravitational tug on all matter is roughly equivalent, and thus the terms become essentially interchangeable for our purpose.

Atomic mass is measured in units called daltons, after the English chemist John Dalton, who, in postulating that matter is composed of tiny indivisible units he called atoms, laid the groundwork for atomic theory. One **Dalton** (**Da**), also known as an atomic mass unit (amu), equals 1/12 the mass of a carbon atom, or about the mass of a proton or a hydrogen atom. Therefore, the most common type of carbon atom has an atomic mass of 12 daltons.

Because atoms such as hydrogen have a small mass, while atoms such as carbon have a larger mass, 1 g of hydrogen would have more atoms than 1 g of carbon. A **mole** of any substance contains the same number of particles as there are atoms in exactly 12 g of carbon. Twelve grams of carbon equals 1 mole of carbon, while 1 g of hydrogen equals 1 mole of hydrogen. As first described by Italian physicist Amedeo Avogadro, 1 mole of

any element contains the same number of atoms—6.022×10^{23}. For example, 12 g of carbon contain 6.022×10^{23} atoms, and 1 g of hydrogen, whose atoms have 1/12 the mass of a carbon atom, also contains 6.022×10^{23} atoms. This number, which is known today as **Avogadro's number**, is large enough to be somewhat mind-boggling, and thus gives us an idea of just how small atoms really are. To visualize the enormity of this number, imagine that people could move through a turnstile at a rate of 1 million people per second. Even at that incredible rate, it would require almost 20 billion years for 6.022×10^{23} people to move through that turnstile!

Isotopes Vary in Their Number of Neutrons

Although the number of neutrons in most biologically relevant atoms is often equal to the number of protons, many elements can exist in multiple forms, called **isotopes**, that differ in the number of neutrons they contain. For example, the most abundant form of the carbon atom, ^{12}C, contains six protons and six neutrons, and thus has an atomic number of 6 and an atomic mass of 12 daltons, as described earlier. The superscript placed to the left of ^{12}C is the sum of the protons and neutrons. The rare carbon isotope ^{14}C, however, contains six protons and eight neutrons. While ^{14}C has an atomic number of 6, it has an atomic mass of 14 Da. Nearly 99% of the carbon in living organisms is ^{12}C. Consequently, the average atomic mass of carbon is very close to, but actually slightly greater than, 12 Da because of the existence of a small amount of heavier isotopes. This explains why the atomic masses given in the periodic table do not add up exactly to the predicted masses based on the atomic number and the number of neutrons of a given atom (see Figure 2.5).

Isotopes of an atom often have similar chemical properties but may have very different physical properties. For example, many isotopes found in nature are inherently unstable; the length of time they persist is measured in half-lives—the time it takes for 50% of the isotope to decay. Some persist for very long times; for example, ^{14}C has a half-life of more than 5,000 years. Such unstable isotopes are called **radioisotopes**, and they lose energy by emitting subatomic particles and/or radiation. At the very low amounts found in nature, radioisotopes usually pose no serious threat to life, but exposure of living organisms to high amounts of radioactivity can result in the disruption of cellular function, cancer, and even death.

Modern medical treatment and diagnosis make use of the special properties of radioactive compounds in many ways. For example, beams of high-energy radiation can be directed onto cancerous parts of the body to kill cancer cells. In another example, one or more atoms in a metabolically important molecule, such as the sugar glucose, can be chemically replaced with a radioactive isotope of fluorine. ^{18}F has a half-life of about 110 minutes. When a solution containing such a modified radioactive glucose is injected into a person's bloodstream, the organs of the body will take it up from the blood just as they would ordinary glucose. Special imaging techniques, such as the PET scan shown in **Figure 2.6**, can detect the amount of the radioactive glucose in the body's organs. In this way, it is

Table 2.1	Characteristics of Major Subatomic Particles		
Particle	Location	Charge	Mass relative to electron
Proton	Nucleus	+1	1,836
Neutron	Nucleus	0	1,839
Electron	Around the nucleus	−1	1

Figure 2.6   Diagnostic image of the human body using radioisotopes.   A procedure called positron-emission tomography (PET) scanning highlights regions of the body that are actively using glucose, the body's major energy source. Radioactivity in this image shows up as a color. The dark patches are regions of extremely intense activity, which were later determined to be cancer in this patient.

possible to visualize whether or not organs such as the heart or brain are functioning normally, or at an increased or decreased rate. For example, a PET scan of the heart that showed reduced uptake of glucose from the blood might indicate the blood vessels of the heart were damaged and thereby depriving the heart of nutrients. PET scans can also reveal the presence of cancer— a disease characterized by uncontrolled cell growth. The scan of the individual shown in Figure 2.6, for example, identified numerous regions of high activity, suggestive of cancer.

The Mass of All Living Organisms Is Largely Composed of Four Elements

Just four elements—oxygen, carbon, hydrogen, and nitrogen— account for the vast majority of atoms in living organisms (**Table 2.2**). These elements typically make up about 95% of the mass of living organisms. Much of the oxygen and hydrogen occur in the form of water, which accounts for approximately 60% of the mass of most animals and up to 95% or more in some plants. Carbon is a major building block of all living matter, and nitrogen is a vital element in all proteins. Note in Table 2.2 that

although hydrogen makes up a small percentage of the mass of the human body, it accounts for about 63% of all the atoms in the body. That is because the atomic mass of hydrogen is so much smaller than that of heavier elements such as oxygen.

Other essential elements in living organisms include the mineral elements. Calcium and phosphorus, for example, are important constituents of the skeletons and shells of animals. Minerals such as potassium and sodium are key regulators of water movement and electrical currents that occur across the surfaces of many cells.

In addition, all living organisms require **trace elements**. These elements are present in extremely small quantities but still are essential for normal growth and function. For example, iron

Table 2.2	Chemical Elements Essential for Life in Most Organisms*		
Element	**Symbol**	**% Human body mass**	**% All atoms in human body**
Most abundant in living organisms (approximately 95% of total mass)			
Oxygen	O	65	25.5
Carbon	C	18	9.5
Hydrogen	H	9	63.0
Nitrogen	N	3	1.4
Mineral elements (less than 1% of total mass)			
Calcium	Ca		
Chlorine	Cl		
Magnesium	Mg		
Phosphorus	P		
Potassium	K		
Sodium	Na		
Sulfur	S		
Trace elements (less than 0.01% of total mass)			
Chromium	Cr		
Cobalt	Co		
Copper	Cu		
Fluorine	F		
Iodine	I		
Iron	Fe		
Manganese	Mn		
Molybdenum	Mo		
Selenium	Se		
Silicon	Si		
Tin	Sn		
Vanadium	V		
Zinc	Zn		

* While these are the most common elements in living organisms, many other trace and mineral elements have reported functions. For example, aluminum is believed to be a cofactor for certain chemical reactions in animals, but it is generally toxic to plants.

plays an important role in how vertebrates store oxygen in their blood, and copper serves a similar role in some invertebrates.

2.2 Chemical Bonds and Molecules

The linkage of atoms with other atoms serves as the basis for life and also gives life its great diversity. Two or more atoms bonded together make up a molecule. Atoms can combine with each other in several ways. For example, two oxygen atoms can combine to form one oxygen molecule, represented as O_2. This representation is called a **molecular formula**. It consists of the chemical symbols for all of the atoms that are present (here, O for oxygen) and a subscript that tells you how many of those atoms are present in the molecule (in this case, two). The term **compound** refers to a molecule composed of two or more different elements. Examples include water (H_2O), with two hydrogen atoms and one oxygen atom, and the sugar glucose ($C_6H_{12}O_6$), which has 6 carbon atoms, 12 hydrogen atoms, and 6 oxygen atoms.

One of the most important features of compounds is their emergent physical properties. This means that the properties of a compound differ greatly from those of its elements. Let's consider sodium as an example. Pure sodium (Na), also called elemental sodium, is a soft, silvery white metal that you can cut with a knife. When sodium forms a compound with chlorine (Cl), table salt (NaCl) is made. NaCl is a white, relatively hard crystal (as seen in the chapter-opening photo) that dissolves in water. Thus, the properties of sodium in a compound can be dramatically different from its properties as a pure element.

The atoms in molecules are held together by chemical bonds. In this section, we will examine the different types of chemical bonds, how these bonds form, and how they determine the structures of molecules.

Covalent Bonds Join Atoms Through the Sharing of Electrons

Covalent bonds, in which atoms share a pair of electrons, can occur between atoms whose outer shells are not full. A fundamental principle of chemistry is that *atoms tend to be most stable when their outer shells are filled with electrons*. **Figure 2.7** shows this principle as it applies to the formation of hydrogen fluoride, a molecule with many important industrial and medical applications such as petroleum refining and fluorocarbon formation. The outer shell of a hydrogen atom is full when it contains two electrons, though a hydrogen atom has only one electron. The outer shell of a fluorine atom is full when it contains eight electrons, though a fluorine atom has only seven electrons in its outer shell. When hydrogen fluoride (HF) is made, the two atoms share a pair of electrons, which spend time orbiting both nuclei. This allows both of the outer shells of those atoms to be full. Covalent bonds are strong chemical bonds, because the shared electrons behave as if they belong to each atom.

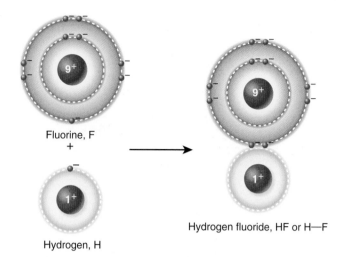

Fluorine, F
+
Hydrogen, H
Hydrogen fluoride, HF or H—F

Figure 2.7 The formation of covalent bonds. In covalent bonds, electrons from the outer shell of two atoms are shared with each other in order to complete the outer shells of both atoms. This simplified illustration shows hydrogen forming a covalent bond with fluorine.

When the structure of a molecule is diagrammed, each covalent bond is represented by a line indicating a pair of shared electrons. For example, hydrogen fluoride is diagrammed as

H—F

A molecule of water (H_2O) can be diagrammed as

H—O—H

The structural formula of water indicates that the oxygen atom is covalently bound to two hydrogen atoms.

Each atom forms a characteristic number of covalent bonds, which depends on the number of electrons required to fill the outer shell. The atoms of some elements important for life, notably carbon, form more than one covalent bond and become linked simultaneously to two or more other atoms. **Figure 2.8** shows the number of covalent bonds formed by several atoms commonly found in the molecules of living cells.

For many types of atoms, their outermost shell is full when they contain eight electrons, an octet. The **octet rule** states that atoms are stable when they have eight electrons in their outermost shell. This rule applies to most atoms found in living organisms, including oxygen, nitrogen, carbon, phosphorus, and sulfur. These atoms form a characteristic number of covalent bonds to make an octet in their outermost shell (Figure 2.8). However, the octet rule does not always apply. For example, hydrogen has an outermost shell that can contain only two electrons, not eight.

In some molecules, a **double bond** occurs when atoms share two pairs of electrons (four electrons) rather than one pair. As shown in **Figure 2.9**, this is the case for an oxygen molecule (O_2), which can be diagrammed as

O=O

Another common example occurs when two carbon atoms form bonds in compounds. They may share one pair of electrons (single bond) or two pairs (double bond), depending on how many other covalent bonds each carbon forms with other

Atom name	Hydrogen	Oxygen	Nitrogen	Carbon
Electron number needed to complete outer shell (typical number of covalent bonds)	1	2	3	4

Figure 2.8 **The number of covalent bonds formed by common essential elements found in living organisms.** These elements form different numbers of covalent bonds due to the electron configurations in their outer shells.

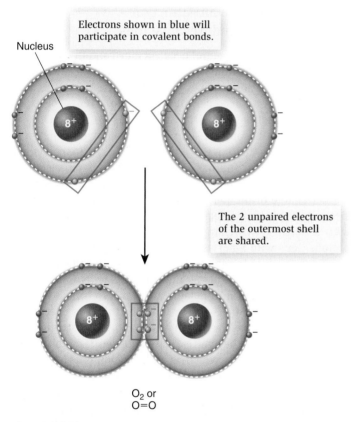

Figure 2.9 A double bond between two oxygen atoms.

Concept check: Explain how an oxygen molecule obeys the octet rule.

atoms. In rare cases, carbon can even form triple bonds, where three pairs of electrons are shared between two atoms.

Electrons Are Not Always Evenly Shared Between Atoms

Some atoms attract shared electrons more readily than do other atoms. The **electronegativity** of an atom is a measure of its ability to attract electrons in a bond with another atom. When

two atoms with different electronegativities form a covalent bond, the shared electrons are more likely to be closer to the nucleus of the atom of higher electronegativity rather than the atom of lower electronegativity. Such bonds are called **polar covalent bonds**, because the distribution of electrons around the nuclei creates a polarity, or difference in electric charge, across the molecule. Water is the classic example of a molecule containing polar covalent bonds. The shared electrons at any moment tend to be closer to the oxygen nucleus rather than to either of the hydrogens. This unequal sharing of electrons gives the molecule a region of partial negative charge and two regions of partial positive charge (**Figure 2.10**).

Atoms with high electronegativity, such as oxygen and nitrogen, have a relatively strong attraction for electrons. These atoms form polar covalent bonds with hydrogen atoms, which have low electronegativity. Examples of polar covalent bonds include O—H and N—H. In contrast, bonds between atoms with similar electronegativities, for example between two carbon atoms (C—C) or between carbon and hydrogen atoms (C—H), are called **nonpolar covalent bonds**. Molecules containing significant numbers of polar bonds are known as **polar molecules**, whereas molecules composed predominantly of nonpolar bonds are called **nonpolar molecules**. A single molecule may have different regions with nonpolar bonds and polar bonds. As we will explore later, the physical characteristics of polar and nonpolar molecules, especially their solubility in water, are quite different.

Hydrogen Bonds Allow Interactions Between and Within Molecules

An important result of certain polar covalent bonds is the ability of one molecule to loosely associate with another molecule through a weak interaction called a **hydrogen bond**. A hydrogen bond forms when a hydrogen atom from one polar molecule becomes electrically attracted to an electronegative atom, such as an oxygen or nitrogen atom, in another polar molecule. Hydrogen bonds, like those between water molecules, are represented in diagrams by dashed or dotted lines to distinguish them from covalent bonds (**Figure 2.11a**). A single hydrogen

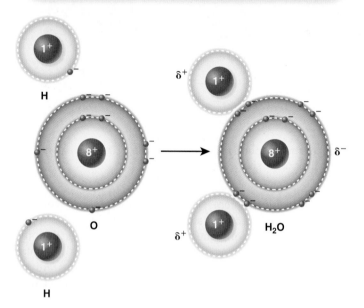

In water, the shared electrons spend more time near the oxygen atom. This gives oxygen a partial negative charge (δ^-) and each hydrogen a partial positive charge (δ^+).

Figure 2.10 Polar covalent bonds in water molecules. In a water molecule, two hydrogen atoms share electrons with an oxygen atom. Because oxygen has a higher electronegativity, the shared electrons spend more time closer to oxygen. This gives oxygen a partial negative charge, designated δ^-, and each hydrogen a partial positive charge, designated δ^+.

bond is very weak. The strength of a hydrogen bond is only a few percent of the strength of the polar covalent bonds linking the hydrogen and oxygen within a water molecule.

Hydrogen bonds can also occur within a single large molecule. Many large molecules may have dozens, hundreds, or more of hydrogen bonds within their structure. Collectively, many hydrogen bonds add up to a strong force that helps maintain the three-dimensional structure of a molecule. This is particularly true in deoxyribonucleic acid (DNA)—the molecule that makes up the genetic material of living organisms. DNA exists as two long, twisting strands of many thousands of atoms. The two strands are held together along their length by hydrogen bonds between different portions of the molecule (**Figure 2.11b**). Due to the large number of hydrogen bonds, it takes considerable energy to separate the strands of DNA.

In contrast to the cumulative strength of many hydrogen bonds, the weakness of individual bonds is also important. When an interaction between two molecules involves relatively few hydrogen bonds, such interactions tend to be weak and may be readily broken. The reversible nature of hydrogen bonds allows molecules to interact and then to become separated again. For example, as discussed in Chapter 7, small molecules may bind to proteins called enzymes via hydrogen bonds. **Enzymes** are molecules found in all cells that facilitate

or catalyze many biologically important chemical reactions. The small molecules are later released after the enzymes have changed their structure.

Hydrogen bonds are similar to a special class of bonds that are collectively known as **van der Waals forces**. In some cases, temporary attractive forces that are even weaker than hydrogen bonds form between molecules. These van der Waals forces arise because electrons orbit atomic nuclei in a random, probabilistic way, as described previously. At any moment, the electrons in the outer shells of the atoms in an electrically neutral molecule may be evenly distributed or unevenly distributed. In the latter case, a short-lived electrical attraction may arise with other nearby molecules. Like hydrogen bonds, the collective strength of temporary attractive forces between molecules can be quite strong.

Ionic Bonds Involve an Attraction Between Positive and Negative Ions

Atoms are electrically neutral because they contain equal numbers of negative electrons and positive protons. If an atom or molecule gains or loses one or more electrons, it acquires a net electric charge and becomes an **ion** (**Figure 2.12a**). For example, when a sodium atom (Na), which has 11 electrons, loses one electron, it becomes a sodium ion (Na^+) with a net positive charge. Ions that have a net positive charge are called **cations**. A sodium ion still has 11 protons, but only 10 electrons. Ions such as Na^+ are depicted with a superscript that indicates the net charge of the ion. A chlorine atom (Cl), which has 17 electrons, can gain an electron and become a chloride ion (Cl^-) with a net negative charge—it has 18 electrons but only 17 protons. Ions with a net negative charge are **anions**.

Table 2.3 lists the ionic forms of several elements. Hydrogen atoms and most mineral and trace elements readily form ions. The ions listed in this table are relatively stable because the outer electron shells of the ions are full. For example, a sodium atom has one electron in its third (outermost) shell. If it loses this electron to become Na^+, it no longer has a third shell, and the second shell, which is full, becomes its outermost shell.

Alternatively, a Cl atom has seven electrons in its outermost shell. If it gains an electron to become a chloride ion (Cl^-), its outer shell becomes full with eight electrons. Some atoms can gain or lose more than one electron. For instance, a calcium atom, which has 20 electrons, loses 2 electrons to become a calcium ion, depicted as Ca^{2+}.

An **ionic bond** occurs when a cation binds to an anion. Figure 2.12a shows an ionic bond between Na^+ and Cl^- to form NaCl. Salt is the general name given to compounds formed from an attraction between a positively charged ion (a cation) and negatively charged ion (an anion). Examples of salts include NaCl, KCl, and $CaCl_2$. Salts may form crystals in which the cations and anions form a regular array. **Figure 2.12b** shows a NaCl crystal, in which the sodium and chloride ions are held together by ionic bonds. Ionic bonds are easily broken in water—the environment of the cell.

Hydrogen bonds

δ^+

δ^+

δ^-

(a) Hydrogen bonds between water molecules

A DNA molecule consists of 2 twisted strands held together along its entire length by millions of hydrogen bonds.

Hydrogen bonds

The hydrogen bond (H bond) is a weak attraction between a partially positive hydrogen and a partially negative atom such as oxygen.

Hydrogen bond

Hydrogen bond

DNA strand

Opposite DNA strand

(b) Hydrogen bonds within a DNA molecule

Figure 2.11   Examples of hydrogen bonds. Hydrogen bonds are important because they allow for interactions between different molecules or interactions of atoms within a molecule. **(a)** This example depicts hydrogen bonds (shown as dashed lines) between water molecules. For simplicity, the partial charges are indicated on only one water molecule. In this diagram, the atoms are depicted as solid spheres, which represent the outer shell. This is called a space-filling model for an atom. **(b)** A DNA molecule is composed of two twisting strands connected to each other by hydrogen bonds (dashed lines). Although each individual bond is weak, the sum of all the hydrogen bonds in a large molecule like DNA imparts considerable stability to the molecule.

Concept check: In Chapter 11, you will learn that the two DNA strands must first separate into two single strands for DNA to be replicated. Do you think the process of strand separation requires energy, or do you think the strands can separate spontaneously?

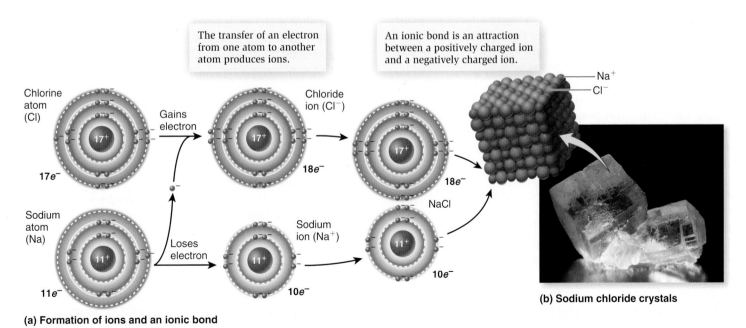

The transfer of an electron from one atom to another atom produces ions.

An ionic bond is an attraction between a positively charged ion and a negatively charged ion.

Chlorine atom (Cl)

17^+

$17e^-$

Gains electron

Chloride ion (Cl$^-$)

17^+

$18e^-$

17^+

$18e^-$

Sodium atom (Na)

11^+

$11e^-$

Loses electron

11^+

$10e^-$

Sodium ion (Na$^+$)

11^+

$10e^-$

NaCl

Na$^+$

Cl$^-$

(a) Formation of ions and an ionic bond

(b) Sodium chloride crystals

Figure 2.12   Ionic bonding in table salt (NaCl).   (a) When an electron is transferred from a sodium atom to a chlorine atom, the resulting ions are attracted to each other via an ionic bond. **(b)** In a salt crystal, a lattice is formed in which the positively charged sodium ions (Na$^+$) are attracted to negatively charged chloride ions (Cl$^-$).

Table 2.3	Ionic Forms of Some Common Elements			
Atom	Chemical symbol	Ion	Ion symbol	Electrons gained or lost
Calcium	Ca	Calcium ion	Ca^{2+}	2 lost
Chlorine	Cl	Chloride ion	Cl^{-}	1 gained
Hydrogen	H	Hydrogen ion	H^{+}	1 lost
Magnesium	Mg	Magnesium ion	Mg^{2+}	2 lost
Potassium	K	Potassium ion	K^{+}	1 lost
Sodium	Na	Sodium ion	Na^{+}	1 lost

Figure 2.13 Shapes of molecules. Molecules may assume different shapes depending on the types of bonds between their atoms. The angles between groups of atoms are well defined. For example, in liquid water at room temperature, the angle formed by the bonds between the two hydrogen atoms and the oxygen atom is approximately 104.5°. This bond angle can vary slightly depending on the temperature and degree of hydrogen bonding between adjacent water molecules.

Molecules May Change Their Shapes

When atoms combine, they can form molecules with various three-dimensional shapes, depending on the arrangements and numbers of bonds between their atoms. As an example, let's consider the arrangements of covalent bonds in a few simple molecules, including water (**Figure 2.13**). These molecules form new orbitals that cause the atoms to have defined angles relative to each other. This gives groups of atoms very specific shapes, as shown in the three examples of Figure 2.13.

Molecules containing covalent bonds are not rigid, inflexible structures. Think of a single covalent bond, for example, as an axle around which the joined atoms can rotate. Within certain limits, the shape of a molecule can change without breaking its covalent bonds. As illustrated in **Figure 2.14a**, a molecule of six carbon atoms bonded together can assume a number of shapes as a result of rotations around various covalent bonds. The three-dimensional, flexible shape of molecules contributes to their biological properties. As shown in **Figure 2.14b**, the binding of one molecule to another may affect the shape of one of the molecules. An animal can taste food, for instance, because food molecules interact with special proteins called receptors on its tongue. When a food molecule encounters a receptor, the two molecules recognize each other by their unique shapes, somewhat like a key fitting into a lock. As molecules in the food interact with the receptor, the shape of the receptor changes. When we look at how an animal's brain receives information from other parts of the body, we will see that the altered shape of the receptor initiates a signal that communicates information about the taste of the food to the animal's brain (see Chapter 43).

Free Radicals Are a Special Class of Highly Reactive Molecules

Recall that an atom or an ion is most stable when each of its orbitals is occupied by its full complement of electrons. A molecule containing an atom with a single, unpaired electron in its outer shell is known as a **free radical**. Free radicals can react with other molecules to "steal" an electron from one of their atoms, thereby filling the orbital in the free radical. In the process, this may create a new free radical in the donor molecule, setting off a chain reaction.

Free radicals can be formed in several ways, including exposure of cells to radiation and toxins. Free radicals can do considerable harm to living cells—for example, by causing a cell to rupture or by damaging the genetic material. Surprisingly, the lethal effect of free radicals is sometimes put to good use. Some cells in animals' bodies create free radicals and use them to kill invading cells such as bacteria. Likewise, people use hydrogen peroxide to kill bacteria, as in a dirty skin wound. Hydrogen peroxide can break down to create free radicals, which can then attack bacteria in the wound.

Despite the exceptional case of fighting off bacteria, though, most free radicals that arise in an organism need to be inactivated so they do not kill healthy cells. Protection from free radicals is afforded by molecules that can donate electrons to the free radicals without becoming highly reactive themselves. Examples of such protective compounds are certain vitamins known as antioxidants (for example, vitamins C and E), found in fruits and vegetables, and the numerous plant compounds

Hydrogen atoms

Rotating this bond creates a new shape.

Subsequent bond rotations may create several additional shapes.

Molecule 1

Molecule 2

Shape changes in molecule 2

(a) Bond rotation in a small molecule

(b) Noncovalent interactions that may alter the shape of molecules

Figure 2.14 **Shape changes in molecules.** A single molecule may assume different three-dimensional shapes without breaking any of the covalent bonds between its atoms, as shown in **(a)** for a six-carbon molecule. Hydrogen atoms above the blue plane are shown in white; those below the blue plane are blue. **(b)** Two molecules are shown schematically as having complementary shapes that permit them to interact. Upon interacting, the flexible nature of the molecules causes molecule 2 to twist sufficiently to assume a new shape. This change in shape is often an important mechanism by which one molecule influences the activity of another.

known as flavonoids. This is one reason why a diet rich in fruits and vegetables is beneficial to our health.

Free radicals are diagrammed with a dot next to the atomic symbol. Examples of biologically important free radicals are superoxide anion, $O_2 \cdot^-$; hydroxyl radical, $\cdot OH$; and nitric oxide, $NO \cdot$. Note that free radicals can be either charged or neutral.

Chemical Reactions Change Elements or Compounds into Different Compounds

A **chemical reaction** occurs when one or more substances are changed into other substances. This can happen when two or more elements or compounds combine with each other to form a new compound, when one compound breaks down into two or more molecules, or when electrons are added to or removed from an atom.

Chemical reactions share many similar properties. First, they all require a source of energy for molecules to encounter each other. The energy required for atoms and molecules to interact is provided partly by heat, or thermal, energy. In the complete absence of any heat (a temperature called absolute zero), atoms and molecules would be totally stationary and unable to interact. Heat energy causes atoms and molecules to vibrate and move, a phenomenon known as Brownian motion. Second, chemical reactions that occur in living organisms often require more than just Brownian motion to proceed at a reasonable rate. Such reactions need to be catalyzed. As discussed in Chapter 6, a catalyst is a substance that speeds up a chemical reaction. As noted earlier, all cells contain many kinds of catalysts called enzymes. Third, chemical reactions tend to proceed in a particular direction but will eventually reach a state of equilibrium unless something happens to prevent equilibrium.

To understand what we mean by "direction" and "equilibrium" in this context, let's consider a chemical reaction between methane (a component found in natural gas) and oxygen. When a single molecule of methane reacts with two molecules of oxygen, one molecule of carbon dioxide and two molecules of water are produced:

$$CH_4 + 2\,O_2 \;\rightleftharpoons\; CO_2 + 2\,H_2O$$
(methane) (oxygen)　　(carbon dioxide) (water)

As it is written here, methane and oxygen are the **reactants**, and carbon dioxide and water are the **products**. The bidirectional arrows indicate that this reaction can proceed in both directions. Whether a chemical reaction is likely to proceed in a forward ("left to right") or reverse ("right to left") direction depends on changes in free energy, which you will learn about in Chapter 6. If we began with only methane and oxygen, the forward reaction would be very favorable. The reaction would produce a large amount of carbon dioxide and water, as well as heat. This is why natural gas is used as a fuel to heat homes. However, all chemical reactions will eventually reach **chemical equilibrium**, in which the rate of the forward reaction is balanced by the rate of the reverse reaction; in other words, there would no longer be a change in the concentrations of products and reactants. In the case of the reaction involving methane and oxygen, this equilibrium would occur when nearly all of the reactants had been converted to products. In biological systems, however, many reactions do not have a chance to reach chemical equilibrium. For example, the products of a reaction may immediately be converted within a cell to a different product through a second reaction, or used by a cell to carry out some function. When a product is removed from a reaction as fast as it is formed, the reactants continue to form new product until all the reactants are used up.

A final feature common to chemical reactions in living organisms is that many reactions occur in watery environments. Such chemical reactions involve reactants and products that are dissolved in water. Next, we will examine the properties of this amazing liquid.

2.3 Properties of Water

It would be difficult to imagine life without water. People can survive for a month or more without food but usually die in less than a week without water. The bodies of all organisms are composed largely of water; most of the cells in an organism's body not only are filled with water, but are surrounded by water. Up to 95% of the weight of certain plants comes from water. In humans, typically 60–70% of body weight is from water. The brain is roughly 70% water, blood is about 80% water, and the lungs are nearly 90% water. Even our bones are about 20% water! In addition, water is an important liquid in the surrounding environments of living organisms. For example, vast numbers of species are aquatic organisms that live in watery environments.

Thus far in this chapter, we have considered the features of atoms and molecules and the nature of bonds and chemical reactions between atoms and molecules. In this section, we will turn our attention to issues related to the liquid properties of living organisms and the environment in which they live. Most of the chemical reactions that occur in nature involve molecules that are dissolved in water, including those reactions that happen inside cells and in the spaces that surround cells of living organisms (**Figure 2.15**).

However, not all molecules dissolve in water. In this section, we will examine the properties of chemicals that influence whether they dissolve in water, and we will consider how biologists measure the amounts of dissolved substances. In addition, we will examine some of the other special properties of water that make it a vital component of living organisms and their environments.

Ions and Polar Molecules Readily Dissolve in Water

Substances dissolved in a liquid are known as **solutes**, and the liquid in which they are dissolved is the **solvent**. In all living organisms, the solvent for chemical reactions is water, which is the most abundant solvent in nature. Solutes dissolve in a solvent to form a **solution**. Solutions made with water are called **aqueous solutions**. To understand why a substance dissolves in water, we need to consider the chemical bonds in the solute molecule and those in water. As discussed earlier, the covalent bonds linking the two hydrogen atoms to the oxygen atom in a water molecule are polar. Therefore, the oxygen in water has a slight negative charge, and each hydrogen has a slight positive charge. To dissolve in water, a substance must be electrically attracted to water molecules. For example, table salt (NaCl) is a solid crystalline substance because of the strong ionic bonds between positive sodium ions (Na^+) and negative chloride ions (Cl^-). When a crystal of sodium chloride is placed in water, the partially negatively charged oxygens of water molecules are attracted to the Na^+, and the partially positively charged hydrogens are attracted to the Cl^- (**Figure 2.16**). Clusters of water molecules surround the ions, allowing the Na^+ and Cl^- to separate from each other and enter the water—that is, to dissolve.

Generally, molecules that contain ionic and/or polar covalent bonds will dissolve in water. Such molecules are said to be **hydrophilic**, which literally means "water-loving." In contrast, molecules composed predominantly of carbon and hydrogen

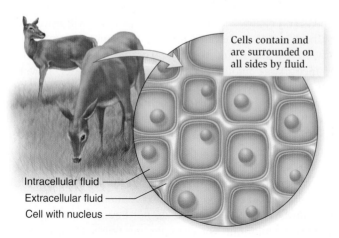

Figure 2.15 Fluids inside and outside of cells.   Aqueous solutions exist in the intracellular fluid and in the extracellular fluid. Chemical reactions are always ongoing in both fluids.

Figure 2.16   NaCl crystals dissolving in water.   The ability of water to dissolve sodium chloride crystals depends on the electrical attraction between the polar water molecules and the charged sodium and chloride ions. Water molecules surround each ion as it becomes dissolved. For simplicity, the partial charges are indicated for only one water molecule.

are relatively insoluble in water, because carbon-carbon and carbon-hydrogen bonds are nonpolar. These molecules do not have partial positive and negative charges and, therefore, are not attracted to water molecules. Such molecules are **hydrophobic**, or "water-fearing." Oils are a familiar example of hydrophobic molecules. Try mixing vegetable oil with water and observe the result. The two liquids separate into an oil phase and water phase. Very little oil dissolves in the water.

Although hydrophobic molecules dissolve poorly in water, they normally dissolve readily in nonpolar solvents. For example, cholesterol is a compound found in the blood and cells of animals. It is a hydrophobic molecule that is barely soluble in water but easily dissolves in nonpolar solvents used in chemical laboratories, such as ether. Biological membranes like those that encase cells are made up in large part of nonpolar compounds. Because of this, cholesterol also inserts into biological membranes, where it helps to maintain the membrane structure.

Molecules that have both polar or ionized regions at one or more sites and nonpolar regions at other sites are called **amphipathic** (or amphiphilic, from the Greek for "both loves"). When mixed with water, long amphipathic molecules may aggregate into spheres called **micelles**, with their polar (hydrophilic) regions at the surface of the micelle, where they are attracted to the surrounding water molecules. The nonpolar (hydrophobic) ends are oriented toward the interior of the micelle (**Figure 2.17**). Such an arrangement minimizes the interaction between water molecules and the nonpolar ends of the amphipathic molecules. Nonpolar molecules can dissolve in the central nonpolar regions of these clusters and thus exist in an aqueous environment in far higher amounts than would otherwise be possible based on their low solubility in water. One familiar example of amphipathic molecules are those in detergents, which can form micelles that help to dissolve oils and nonpolar molecules found in dirt. The detergent molecules found in soap have polar and nonpolar ends. Oils on your skin dissolve in the nonpolar regions of the detergent, and the polar ends help the detergent rinse off in water, taking the oil with it.

In addition to micelles, amphipathic molecules may form structures called bilayers. As you will learn in Chapter 5, lipid bilayers play a key role in cellular membrane structure.

The Amount of a Dissolved Solute per Unit Volume of Liquid Is Its Concentration

Solute **concentration** is defined as the amount of a solute dissolved in a unit volume of solution. For example, if 1 gram (g) of NaCl were dissolved in enough water to make 1 liter of solution, we would say that its solute concentration is 1 g/L.

A comparison of the concentrations of two different substances on the basis of the number of grams per liter of solution does not directly indicate how many molecules of each substance are present. For example, let's compare 10 g each of glucose ($C_6H_{12}O_6$) and sodium chloride (NaCl). Because the individual molecules of glucose have more mass than those of NaCl, 10 g of glucose will contain fewer molecules than 10 g

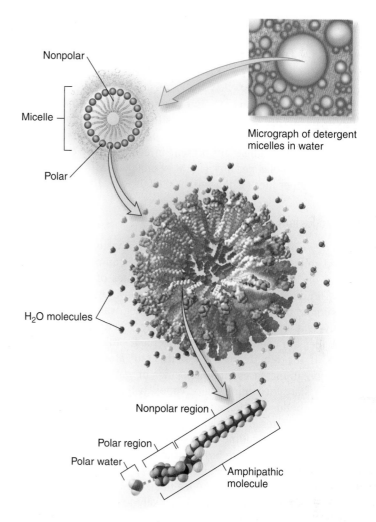

Micrograph of detergent micelles in water

Figure 2.17   **The formation of micelles by amphipathic molecules.**   In water, amphipathic molecules tend to arrange themselves so their nonpolar regions are directed away from water molecules and the polar regions are directed toward the water and can form hydrogen bonds with it.

Concept check: *When oil dissolves in soap, where is the oil found?*

of NaCl. Therefore, another way to describe solute concentration is according to the moles of dissolved solute per volume of solution. To make this calculation, we must know three things: the amount of dissolved solute, the molecular mass of the dissolved solute, and the volume of the solution.

The **molecular mass** of a molecule is equal to the sum of the atomic masses of all the atoms in the molecule. For example, glucose ($C_6H_{12}O_6$) has a molecular mass of 180 ([6 × 12] + [12 × 1] + [6 × 16] = 180). As mentioned earlier, 1 mole (abbreviated mol) of a substance is the amount of the substance in grams equal to its atomic or molecular mass. The **molarity** of a solution is defined as the number of moles of a solute dissolved in 1 L of solution. A solution containing 180 g of glucose (1 mol) dissolved in enough water to make 1 L is a 1 **molar** solution of glucose (1 mol/L). By convention, a

1 mol/L solution is usually written as 1 M, where the capital M stands for molar and is defined as mol/L. If 90 g of glucose (half its molecular mass) were dissolved in enough water to make 1 L, the solution would have a concentration of 0.5 mol/L, or 0.5 M.

The concentrations of solutes dissolved in the fluids of living organisms are usually much less than 1 M. Many have concentrations in the range of millimoles per liter (1 mM = 0.001 M = 10^{-3} M), and others are present in even smaller concentrations—micromoles per liter (1 μM = 0.000001 M = 10^{-6} M) or nanomoles per liter (1 nM = 0.000000001 M = 10^{-9} M).

Water Exists in Three States

Let's now consider some general features of water and how dissolved solutes affect its properties. Water is an abundant compound on Earth that exists in all three states of matter—solid (ice), liquid (water), and gas (water vapor). At the temperatures found over most regions of the planet, water is found primarily as a liquid in which the weak hydrogen bonds between molecules are continuously being formed, broken, and formed again. If the temperature rises, the rate at which hydrogen bonds break increases, and molecules of water escape into the gaseous state, becoming water vapor. If the temperature falls, hydrogen bonds are broken less frequently, so larger and larger clusters of water molecules are formed, until at 0°C water freezes into a crystalline matrix—ice. The water molecules in ice tend to lie in a more orderly and "open" arrangement, that is, with greater intermolecular distances, which makes ice less dense than water. This is why ice floats on water (**Figure 2.18**). Compared to water, ice is also less likely to participate in most types of chemical reactions.

Changes in state, such as changes between the solid, liquid, and gaseous states of water, involve an input or a release of energy. For example, when energy is supplied to make water boil, it changes from the liquid to the gaseous state. This is called vaporization. The heat required to vaporize 1 mole of any substance at its boiling point is called the substance's **heat of vaporization**. For water, this value is very high, because of the high number of hydrogen bonds between the molecules. It takes more than five times as much heat to vaporize water than it does to raise the temperature of water from 0°C to 100°C. In contrast, energy is released when water freezes to form ice. Water also has a high **heat of fusion**, which is the amount of heat energy that must be withdrawn or released from a substance to cause it to change from the liquid to the solid state. These two features, the high heats of vaporization and fusion, mean that water is extremely stable as a liquid. Not surprisingly, therefore, living organisms have evolved to function best within a range of temperatures consistent with the liquid phase of water.

The temperature at which a solution freezes or vaporizes is influenced by the amounts of dissolved solutes. These are examples of **colligative properties**, defined as those properties that depend strictly on the total number of dissolved solutes, not on the specific type of solute. Pure water freezes at 0°C

Ice

Liquid water

Figure 2.18   Structure of liquid water and ice. In its liquid form, the hydrogen bonds between water molecules continually form, break, and re-form, resulting in a changing arrangement of molecules from instant to instant. At temperatures at or below its freezing point, water forms a crystalline matrix called ice. In this solid form, hydrogen bonds are more stable. Ice has a hexagonally shaped crystal structure. The greater space between H_2O molecules in this crystal structure causes ice to have a lower density compared to water. For this reason, ice floats on water.

and vaporizes at 100°C. Addition of solutes to water lowers its freezing point below 0°C and raises its boiling point above 100°C. Adding a small amount of the compound ethylene glycol—antifreeze—to the water in a car's radiator, for instance, lowers the freezing point of the water and consequently prevents it from freezing in cold weather. Similarly, the presence of large amounts of solutes partly explains why the oceans do not freeze when the temperature falls below 0°C. Likewise, the colligative properties of water also account for the remarkable ability of certain ectothermic animals, which are unable to maintain warm body temperatures in cold environments, to nonetheless escape becoming frozen solid. Such animals produce antifreeze molecules that dissolve in their body fluids in very large numbers, thereby lowering the freezing point of the fluids and preventing their blood and cells from freezing in the extreme cold. The emerald rockcod (*Trematomus bernacchii*), found in the waters of Antarctica, for example, manages to live in ocean waters that are at or below 0°C (**Figure 2.19a**). Similarly, many insects, such as the larvae of the parasitic wasp (*Bracon cephi*), also make use of natural antifreeze to stay alive in extreme conditions (**Figure 2.19b**).

Figure 2.19 Antifreeze in living organisms.   Many animals, such as **(a)** the emerald rockcod (*Trematomus bernacchii*) and **(b)** the larvae of the parasitic wasp (*Bracon cephi*), can withstand extremely cold temperatures thanks to natural antifreeze molecules in their body fluids.

(a) Emerald rockcod in the waters of Antarctica

(b) Wasp larvae, which can withstand freezing temperatures

Concept check: *The liquid portion of blood of animals, including humans, is a watery solution containing many dissolved solutes, such as Na^+ and Cl^-. Would you predict that the freezing point of blood is above, below, or the same as that of water?*

Water Performs Many Other Important Tasks in Living Organisms

As discussed earlier, water is the primary solvent in the fluids of all living organisms, from unicellular bacteria to the largest sequoia tree. Water permits atoms and molecules to interact in ways that would be impossible in their nondissolved states. In Unit II, we will consider many ions and molecules that are solutes in living cells. Even so, it is important to recognize that in addition to acting as a solvent, water serves many other remarkable functions that are critical for the survival of living organisms. For example, water molecules themselves take part in many chemical reactions of this general type:

$$R1—R2 + H—O—H \quad \rightarrow \quad R1—OH + H—R2$$

R is a general symbol used in this case to represent a group of atoms. In this equation, R1 and R2 are distinct groups of atoms. On the left side, R1—R2 is a compound in which the groups of atoms are connected by a covalent bond. To be converted to products, a covalent bond is broken in each reactant, R1—R2 and H—O—H, and OH and H (from water) form covalent bonds with R1 and R2, respectively. Reactions of this type are known as **hydrolysis** reactions (from the Greek *hydro*, meaning water, and *lysis*, meaning to break apart), because water is used to break apart another molecule (**Figure 2.20a**). As discussed in Chapter 3 and later chapters, many large molecules are broken down into smaller, biologically important units by hydrolysis.

Alternatively, other chemical reactions in living organisms involve the removal of a water molecule so that a covalent bond can be formed between two separate molecules. For example, let's consider a chemical reaction that is the reverse of our previous hydrolysis reaction:

$$R1—OH + H—R2 \quad \rightarrow \quad R1—R2 + H—O—H$$

Such a reaction involves the formation of a covalent bond between two molecules. Two or more molecules combining to form one larger molecule with the loss of a small molecule is called a **condensation reaction**. In the example shown here, a molecule of water is lost during the reaction; this is a specific type of condensation reaction called a **dehydration reaction**. As discussed in later chapters, this is a common reaction used to build larger molecules in living organisms.

Another feature of water is that it is incompressible—its volume does not significantly decrease when subjected to high pressure. This has biological importance for many organisms that use water to provide force or support (**Figure 2.20b**). For example, water supports the bodies of worms and some other invertebrates, and it provides turgidity (stiffness) and support for plants.

Water is also the means by which unneeded and potentially toxic waste compounds are eliminated from an animal's body (**Figure 2.20c**). In mammals, for example, the kidneys filter out soluble waste products derived from the breakdown of proteins and other compounds. The filtered products remain in solution in a watery fluid, which eventually becomes urine and is excreted.

Recall from our discussion of water's properties that it takes considerable energy in the form of heat to convert water from a liquid to a gas. This feature has great biological significance. Although everyone is familiar with the fact that boiling water is converted to water vapor, water can vaporize into the gaseous state even at ordinary temperatures. This process is known as **evaporation**. The simplest way to understand this is to imagine that in any volume of water at any temperature, some vibrating water molecules will have higher energy than others. Those with the highest energy break their hydrogen bonds and escape into the gaseous state. The important point, however, is that even at ordinary temperatures, it still requires the same energy to change water from liquid to gas. Therefore,

(a) **Water participates in chemical reactions.**

Hydrolysis

(b) **Water provides support.** The plant on the right is wilting due to lack of water.

Blood enters and is purified by kidney cells.

Waste products are carried away in the watery urine.

(c) **Water is used to eliminate soluble wastes.**

(d) **Evaporation helps some animals dissipate body heat.**

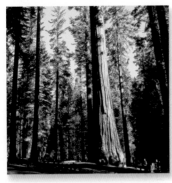

(e) **The cohesive force of water molecules aids in the movement of fluid through vessels in plants.**

(f) **Water in saliva serves as a lubricant during—or as shown here, in anticipation of—feeding.**

(g) **The surface tension of water explains why this water strider doesn't sink.**

Figure 2.20 Some of the amazing roles of water in biology. In addition to acting as a solvent, water serves many crucial functions in nature.

the evaporation of sweat from an animal's skin requires considerable energy in the form of body heat, which is then lost to the environment. Evaporation is an important mechanism by which many animals cool themselves on hot days (**Figure 2.20d**).

Another important feature for living organisms is that water has a very high **specific heat**, defined as the amount of heat energy required to raise the temperature of 1 gram of a substance by 1°C (or conversely, the amount of heat energy that must be lost to lower the temperature by 1°C). This means that it takes considerable heat to raise the temperature of water. A related concept is heat capacity; this refers to the amount of heat energy required to raise the temperature of an entire object or substance. A lake has a greater heat capacity than does a bathtub filled with water, but both have the same specific heat because both are the same substance (ignoring for the moment that a lake is not pure water). These properties of water contribute to the relatively stable temperatures of large bodies of water compared to inland temperatures. Large bodies of water tend to have a moderating effect on the temperature of nearby land masses.

The hydrogen-bonding properties of water affect its ability to form droplets and to adhere to surfaces. The phenomenon of water molecules attracting each other is called **cohesion**. Water exhibits strong cohesion due to hydrogen bonding. Cohesion aids in the movement of water through the vessels of plants (**Figure 2.20e**). A property similar to cohesion is **adhesion**, which refers to the ability of water to be attracted to, and thus adhere to, a surface that is not electrically neutral. Water tends to cling to surfaces to which it can hydrogen bond, such as a paper towel. In organisms, the adhesive properties of water allow it, for example, to coat the surfaces of the digestive tract of animals and act as a lubricant for the passage of food (**Figure 2.20f**).

Surface tension is a measure of the attraction between molecules at the surface of a liquid. In the case of water, the attractive force between hydrogen-bonded water molecules at the interface between water and air is what causes water to form droplets. The surface water molecules attract each other into a configuration (roughly that of a sphere) that reduces the number of water molecules in contact with air. You can see this by slightly overfilling a glass with water; the water will form an oval shape above the rim. Likewise, surface tension allows certain insects, such as water striders, to walk on the surface of a pond without sinking (**Figure 2.20g**) and plays a significant role in the filling of lungs with air in humans and many other animals.

Hydrogen Ion Concentrations Are Changed by Acids and Bases

Pure water has the ability to ionize to a very small extent into hydrogen ions that exist as single protons (H^+) and **hydroxide ions** (OH^-). (In nature or in laboratory conditions, hydrogen atoms may exist as any of several rare types of positively or negatively charged ions; in this text, we will use the term hydrogen ion to refer to the common H^+ form). In pure water, the concentrations of H^+ and OH^- are both 10^{-7} mol/L, or 10^{-7} M. An inherent property of water is that the product of the concentrations of H^+ and OH^- is always 10^{-14} M at 25°C. Therefore, in pure water, $[H^+][OH^-] = [10^{-7}$ M$][10^{-7}$ M$] = 10^{-14}$ M. (The brackets around the symbols for the hydrogen and hydroxide ions indicate concentration.)

When certain substances are dissolved in water, they may release or absorb H^+ or OH^-, thereby altering the relative concentrations of these ions. Substances that release hydrogen ions in solution are called **acids**. Two examples are hydrochloric acid and carbonic acid:

$$HCl \quad \rightarrow \quad H^+ \quad + \quad Cl^-$$

(hydrochloric acid)   (chloride ion)

$$H_2CO_3 \quad \rightleftharpoons \quad H^+ \quad + \quad HCO_3^-$$

(carbonic acid)   (bicarbonate ion)

Hydrochloric acid is called a **strong acid** because it almost completely dissociates into H^+ and Cl^- when added to water. By comparison, carbonic acid is a **weak acid** because some of it will remain in the H_2CO_3 state when dissolved in water.

Compared to an acid, a **base** has the opposite effect when dissolved in water—it absorbs hydrogen ions in solution. This can occur in different ways. Some bases, such as sodium hydroxide (NaOH), release OH^- when dissolved in water:

$$NaOH \quad \rightarrow \quad Na^+ \quad + \quad OH^-$$

(sodium hydroxide)   (sodium ion)

Recall that the product of $[H^+]$ and $[OH^-]$ is always 10^{-14} M. When a base such as NaOH raises the OH^- concentration, some of the hydrogen ions bind to these hydroxide ions to form water. Therefore, increasing the OH^- concentration lowers the H^+ concentration. Alternatively, other bases, such as ammonia, react with water to produce ammonium ion:

$$NH_3 \quad + \quad H_2O \rightleftharpoons NH_4^+ \quad + \quad OH^-$$

(ammonia)   (ammonium ion)

Both NaOH and ammonia have the same effect—they lower the concentration of H^+. NaOH achieves this by directly increasing the OH^- concentration, whereas NH_3 reacts with water to produce OH^-.

The H^+ Concentration of a Solution Determines the Solution's pH

The addition of acids and bases to water can greatly change the H^+ and OH^- concentrations over a very broad range. Therefore, chemists and biologists use a log scale to describe the concentrations of these ions. The H^+ concentration is expressed as the solution's **pH**, which is defined as the negative logarithm to the base 10 of the H^+ concentration.

$$pH = -\log_{10} [H^+]$$

To understand what this equation means, let's consider a few examples. A solution with a H^+ concentration of 10^{-7} M has a pH of 7. A concentration of 10^{-7} M is the same as 0.1 μM. A solution in which $[H^+] = 10^{-6}$ M has a pH of 6. 10^{-6} M is the same as 1.0 μM. A solution at pH 6 is said to be more **acidic**, because the H^+ concentration is 10-fold higher than a solution at pH 7. Note that as the acidity increases, the pH decreases. A solution where the pH is 7 is said to be neutral because $[H^+]$ and $[OH^-]$ are equal. An acidic solution has a pH below 7, and an **alkaline** solution has a pH above 7. **Figure 2.21** considers the pH values of some familiar fluids.

Why is pH of importance to biologists? The answer lies in the observation that H^+ and OH^- can readily bind to many

Figure 2.21   The pH scale and the relative acidities of common substances.

Concept check: What is the OH^- concentration at pH 8?

kinds of ions and molecules. For this reason, the pH of a solution can affect

- the shapes and functions of molecules;
- the rates of many chemical reactions;
- the ability of two molecules to bind to each other;
- the ability of ions or molecules to dissolve in water.

Due to the various effects of pH, many biological processes function best within very narrow ranges of pH, and even small shifts can have a negative effect. In living cells, the pH ranges from about 6.5 to 7.8 and is carefully regulated to avoid major shifts in pH. The blood of the human body has a normal range of about pH 7.35 to 7.45 and is therefore slightly alkaline. Certain diseases, such as kidney disease, or acute illnesses, such as prolonged vomiting (in which stomach acid is vomited) can decrease or increase blood pH by a few tenths of a unit. When this happens, the enzymes in the body that are required for normal metabolism can no longer function optimally, leading to additional illness. As described next, living organisms have molecules called buffers to help prevent such changes in pH.

Buffers Minimize Fluctuations in the pH of Fluids

What factors might alter the pH of an organism's fluids? External factors such as acid rain and other forms of pollution can reduce the pH of water entering the roots of plants. In animals, exercise generates lactic acid, and certain diseases can raise or lower the pH of blood.

Organisms have several ways to cope with changes in pH. Vertebrate animals such as mammals, for example, use structures like the kidney to secrete acidic or alkaline compounds into the bloodstream when the blood pH becomes imbalanced. Similarly, the kidneys can transfer hydrogen ions from the fluids of the body into the urine and adjust the pH of the body's fluids in that way. Another mechanism by which pH balance is regulated in diverse organisms involves the actions of acid-base buffers. A **buffer** is composed of a weak acid and its related base. One such buffer is the bicarbonate pathway, which works to keep the pH of an animal's body fluids within a narrow range.

$$CO_2 + H_2O \rightleftharpoons H_2CO_3 \rightleftharpoons H^+ + HCO_3^-$$

$$\text{(carbonic acid)} \qquad \text{(bicarbonate)}$$

This buffer system can work in both directions. For example, if the pH of an animal's blood were to increase (that is, the H^+ concentration decreased), the bicarbonate pathway would proceed from left to right. Carbon dioxide would combine with water to make carbonic acid, and then the carbonic acid would dissociate into H^+ and HCO_3^-. This would raise the H^+ concentration and thereby lower the pH. Alternatively, when the pH of an animal's blood decreases, this pathway runs in reverse. Bicarbonate combines with H^+ to make H_2CO_3, which then dissociates to CO_2 and H_2O. This process removes H^+ from the blood, restoring it to its normal pH, and the CO_2 is exhaled

from the lungs. Many buffers exist in nature. Buffers found in living organisms are adapted to function most efficiently at the normal range of pH values seen in that organism.

 Summary of Key Concepts

2.1 Atoms

- Atoms are the smallest functional units of matter that form all chemical elements and cannot be further broken down into other substances by ordinary chemical or physical means. Atoms are composed of protons (positive charge), electrons (negative charge), and (except for hydrogen) neutrons (electrically neutral). Electrons are found in orbitals around the nucleus. (Table 2.1, Figures 2.1, 2.2, 2.3, 2.4)

- Each element contains a unique number of protons—its atomic number. The periodic table organizes all known elements by atomic number and energy shells. (Figure 2.5)

- Each atom has a small but measurable mass, measured in daltons. The atomic mass scale indicates an atom's mass relative to the mass of other atoms.

- Many atoms exist as isotopes, which differ in the number of neutrons they contain. Some isotopes are unstable radioisotopes and emit radiation. (Figure 2.6)

- Four elements—oxygen, carbon, hydrogen, and nitrogen—account for the vast majority of atoms in living organisms. In addition, living organisms require mineral and trace elements that are essential for growth and function. (Table 2.2)

2.2 Chemical Bonds and Molecules

- A molecule consists of two or more atoms bonded together. The properties of a molecule are different from the properties of the atoms that combined to form it. A compound is composed of two or more different elements.

- Atoms tend to form bonds that fill their outer shell with electrons.

- Covalent bonds, in which atoms share electrons, are strong chemical bonds. Atoms form two covalent bonds—a double bond—when they share two pairs of electrons. (Figures 2.7, 2.8, 2.9)

- The electronegativity of an atom is a measure of its ability to attract bonded electrons. When two atoms with different electronegativities combine, the atoms form a polar covalent bond because the distribution of electrons around the atoms creates polarity, or difference in electric charge, across the molecule. Polar molecules, such as water, are largely composed of polar bonds, and nonpolar molecules are composed predominantly of nonpolar bonds. (Figure 2.10)

- An important result of polar covalent bonds is the ability of one molecule to loosely associate with another molecule through weak interactions called hydrogen bonds. The van der Waals forces are weak electrical attractions that arise between molecules due to the probabilistic orbiting of electrons in atoms. (Figure 2.11)

- If an atom or molecule gains or loses one or more electrons, it acquires a net electric charge and becomes an ion. The strong attraction between two oppositely charged ions forms an ionic bond. (Table 2.3, Figure 2.12)

- The three-dimensional, flexible shape of molecules allows them to interact and contributes to their biological properties. (Figures 2.13, 2.14)

- A free radical is an unstable molecule that interacts with other molecules by taking away electrons from their atoms.

- A chemical reaction occurs when one or more substances are changed into different substances. All chemical reactions will eventually reach an equilibrium, unless the products of the reaction are continually removed.

2.3 Properties of Water

- Water is the solvent for most chemical reactions in all living organisms, both inside and outside of cells. Atoms and molecules dissolved in water interact in ways that would be impossible in their nondissolved states. All chemical reactions require energy. (Figure 2.15)

- Solutes dissolve in a solvent to form a solution. Solute concentration refers to the amount of a solute dissolved in a unit volume of solution. The molarity of a solution is defined as the number of moles of a solute dissolved in 1 L of solution. (Figure 2.16)

- Polar molecules are hydrophilic, whereas nonpolar molecules, composed predominantly of carbon and hydrogen, are hydrophobic. Amphipathic molecules, such as detergents, have polar and nonpolar regions. (Figure 2.17)

- H_2O exists as ice, liquid water, and water vapor (gas). (Figure 2.18)

- The colligative properties of water depend on the number of dissolved solutes and allow it to function as an antifreeze in certain organisms. (Figure 2.19)

- Water's high heat of vaporization and high heat of fusion make it very stable in liquid form.

- Water molecules participate in many chemical reactions in living organisms. Hydrolysis breaks down large molecules into smaller units, and dehydration reactions combine two smaller molecules into one larger one. In living organisms, water provides support, is used to eliminate wastes, dissipates body heat, aids in the movement of liquid through vessels, and serves as a lubricant. Surface tension allows certain insects to walk on water. (Figure 2.20)

- The pH of a solution refers to its hydrogen ion concentration. The pH of pure water is 7 (a neutral solution). Alkaline solutions have a pH higher than 7, and acidic solutions have a pH lower than 7. (Figure 2.21)

- Buffers are compounds that act to minimize pH fluctuations in the fluids of living organisms. Buffer systems can raise or lower pH as required.

Assess and Discuss

Test Yourself

1. _____ make(s) up the nucleus of an atom.
   a. Protons and electrons
   b. Protons and neutrons
   c. DNA and RNA
   d. Neutrons and electrons
   e. DNA only

2. Living organisms are composed mainly of which atoms?
   a. calcium, hydrogen, nitrogen, and oxygen
   b. carbon, hydrogen, nitrogen, and oxygen
   c. hydrogen, nitrogen, oxygen, and helium
   d. carbon, helium, nitrogen, and oxygen
   e. carbon, calcium, hydrogen, and oxygen

3. The ability of an atom to attract electrons in a bond with another atom is termed its
   a. hydrophobicity.
   b. electronegativity.
   c. solubility.
   d. valence.
   e. both a and b

4. Hydrogen bonds differ from covalent bonds in that
   a. covalent bonds can form between any type of atom and hydrogen bonds form only between H and O.
   b. covalent bonds involve sharing of electrons and hydrogen bonds involve the complete transfer of electrons.
   c. covalent bonds result from equal sharing of electrons but hydrogen bonds involve unequal sharing of electrons.
   d. covalent bonds involve sharing of electrons between atoms but hydrogen bonds are the result of weak attractions between a hydrogen atom of a polar molecule and an electronegative atom of another polar molecule.
   e. covalent bonds are weak bonds that break easily but hydrogen bonds are strong links between atoms that are not easily broken.

5. A free radical
   a. is a positively charged ion.
   b. is an atom with one unpaired electron in its outer shell.
   c. is a stable atom that is not bonded to another atom.
   d. can cause considerable cellular damage.
   e. both b and d

6. Chemical reactions in living organisms
   a. require energy to begin.
   b. usually require a catalyst to speed up the process.
   c. are usually reversible.
   d. occur in liquid environments, such as water.
   e. all of the above

7. Solutes that easily dissolve in water are said to be
   a. hydrophobic.
   b. hydrophilic.
   c. polar molecules.
   d. all of the above.
   e. b and c only.

8. The sum of the atomic masses of all the atoms of a molecule is its
   a. atomic weight.
   b. molarity.
   c. molecular mass.
   d. concentration.
   e. polarity.

9. Reactions that involve water in the breaking apart of other molecules are known as _____ reactions.
   a. hydrophilic
   b. hydrophobic
   c. dehydration
   d. anabolic
   e. hydrolytic

10. A difference between a strong acid and a weak acid is
    a. strong acids have a higher molecular mass than weak acids.
    b. strong acids completely (or almost completely) ionize in solution, but weak acids do not completely ionize in solution.
    c. strong acids give off two hydrogen ions per molecule, but weak acids give off only one hydrogen ion per molecule.
    d. strong acids are water-soluble, but weak acids are not.
    e. strong acids give off hydrogen ions, and weak acids give off hydroxyl groups.

Conceptual Questions

1. Distinguish between the types of bonds commonly found in biological molecules.
2. Distinguish between the terms hydrophobic and hydrophilic.
3. What is the significance of molecular shape, and what may change the shape of molecules?

Collaborative Questions

1. Discuss the properties of the three subatomic particles of atoms.
2. Discuss several properties of water that make it possible for life to exist.

Online Resource

www.brookerbiology.com

Stay a step ahead in your studies with animations that bring concepts to life and practice tests to assess your understanding. Your instructor may also recommend the interactive ebook, individualized learning tools, and more.

The Chemical Basis of Life II:
Organic Molecules

3

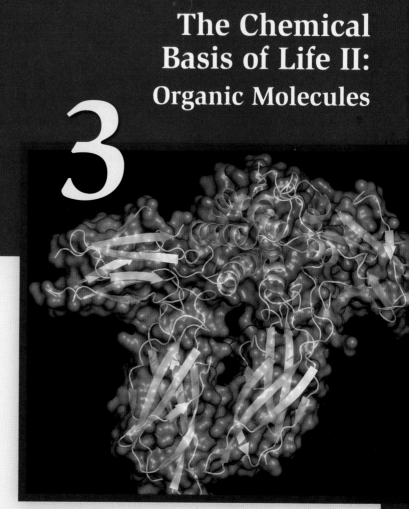

n Chapter 2, we learned that all life is composed of subatomic particles that form atoms, which, in turn, combine to form molecules. Molecules may be simple in atomic composition, as in water (H_2O) or hydrogen gas (H_2), or may bind with other molecules to form larger molecules. Of the countless possible molecules that can be produced from the known elements in nature, certain types contain carbon and are found in all forms of life. These carbon-containing molecules are collectively referred to as **organic molecules**, so named because they were first discovered in living organisms. Among these are lipids and large, complex compounds called **macromolecules**, which include carbohydrates, proteins, and nucleic acids. In this chapter, we will survey the structures of these molecules and examine their chief functions. We begin with the element whose chemical properties are fundamental to the formation of biologically important molecules: carbon. This element provides the atomic scaffold upon which life is built.

A model showing the structure of a protein—a type of organic macromolecule.

3.1 The Carbon Atom and the Study of Organic Molecules

The science of carbon-containing molecules is known as **organic chemistry**. In this section, we will examine the bonding properties of carbon that create groups of atoms with distinct functions and shapes.

Interestingly, the study of organic molecules was long considered a fruitless endeavor because of a concept called vitalism that persisted into the 19th century. Vitalism held that organic molecules were created by, and therefore imparted with, a vital life force that was contained within a plant or an animal's body. Supporters of vitalism argued there was no point in trying to synthesize an organic compound, because such molecules could arise only through the intervention of mysterious qualities associated with life. As described next, this would all change due to the pioneering experiments of Friedrich Wöhler in 1828.

Wöhler's Synthesis of an Organic Compound Transformed Misconceptions About the Molecules of Life

Friedrich Wöhler was a German physician and chemist interested in the properties of inorganic and organic compounds. He spent some time studying urea ($(NH_2)_2CO$), a natural organic product formed from the breakdown of proteins in an animal's body. In mammals, urea accumulates in the urine, which is formed by the kidneys, and then is excreted from the body. During the course of his studies, Wöhler purified urea from the urine of mammals. He noted the color, size, shape, and other characteristics of the crystals that formed when urea was isolated. This experience would serve him well in later years when he quite accidentally helped to put the concept of vitalism to rest.

Figure 3.1 Crystals of urea as viewed with a polarizing microscope (approximately 80x magnification).

Concept check: *How did prior knowledge of urea allow Wöhler to realize he had synthesized urea outside of the body?*

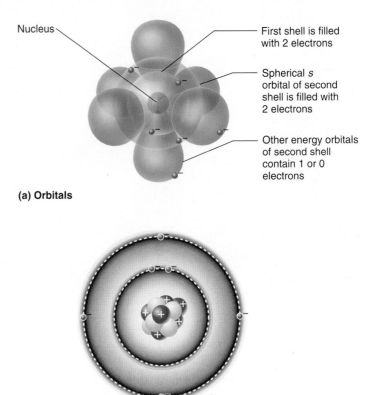

(a) Orbitals

- Nucleus
- First shell is filled with 2 electrons
- Spherical *s* orbital of second shell is filled with 2 electrons
- Other energy orbitals of second shell contain 1 or 0 electrons

(b) Simplified depiction of energy shells

Figure 3.2 Models for the electron orbitals and energy shells of carbon. Carbon atoms have only four electrons in their outer (second) energy shell, which allows carbon to form four covalent bonds. When carbon forms four covalent bonds, the result is four hybrid orbitals of equal energy.

In 1828, while exploring the reactive properties of ammonia and cyanic acid, Wöhler attempted to synthesize an inorganic molecule, ammonium cyanate (NH_4OCN). Instead, to his surprise, Wöhler discovered that ammonia and cyanic acid reacted to produce a third compound, which, when heated, formed familiar-looking crystals (**Figure 3.1**). After careful analysis, he concluded that these crystals were in fact urea. He announced to the scientific community that he had synthesized urea, an organic compound, "without the use of kidneys, either man or dog." In other words, no mysterious life force was required to create this organic molecule. Other scientists, such as Hermann Kolbe, would soon demonstrate that organic compounds such as acetic acid (CH_3COOH) could be synthesized directly from simpler molecules. These studies were a major breakthrough in the way in which scientists viewed life, and so began the field of science now called organic chemistry. From that time to the present, the fields of chemistry and biology have been understood to be intricately related.

Central to Wöhler's and Kolbe's reactions is the carbon atom. Urea and acetic acid, like all organic compounds, contain carbon atoms bound to other atoms. Let's now consider the chemical features that make carbon such an important element in living organisms.

Carbon Forms Four Covalent Bonds with Other Atoms

One of the properties of the carbon atom that makes life possible is its ability to form four covalent bonds with other atoms, including other carbon atoms. This occurs because carbon has four electrons in its outer shell, and it requires four additional electrons for its outer shell to be full (**Figure 3.2**). In living organisms, carbon atoms most commonly form covalent bonds with other carbon atoms and with hydrogen, oxygen, nitrogen, and sulfur atoms. Bonds between two carbon atoms, between carbon and oxygen, or between carbon and nitrogen can be single or double, or in the case of certain C≡C and C≡N bonds, triple. The variation in bonding of carbon with other carbon atoms and with different elements allows a vast number of organic compounds to be formed from only a few chemical elements. This is made all the more impressive because carbon bonds may occur in configurations that are linear, ringlike, or highly branched. Such molecular shapes can produce molecules with a variety of functions.

Carbon and hydrogen have similar electronegativities (see Chapter 2); therefore, carbon-carbon and carbon-hydrogen bonds are nonpolar. As a consequence, molecules with predominantly or entirely hydrogen-carbon bonds, called **hydrocarbons**, tend to be poorly soluble in water. In contrast, when carbon forms polar covalent bonds with more electronegative atoms, such as oxygen or nitrogen, the molecule is much more soluble in water due to the electrical attraction of polar water molecules. The ability of carbon to form both polar and nonpolar bonds (**Figure 3.3**) contributes to its ability to serve as the backbone for an astonishing variety of biologically important molecules.

One last feature of carbon that is important to biology is that carbon bonds are stable in the large range of temperatures associated with life. This property arises in part because the carbon atom is very small compared to most other atoms; therefore, the distance between carbon atoms forming a carbon-carbon bond is quite short. Shorter bonds tend to be stronger and more stable than longer bonds between two large atoms. Thus, carbon bonds are compatible with what we observe about life-forms today; namely, living organisms can inhabit

C—C and C—H bonds are electrically neutral and nonpolar.

Oxygen is more electronegative than carbon; thus, C—O and C=O bonds are polar.

Figure 3.3 **Nonpolar and polar bonds in an organic molecule.** Carbon can form both nonpolar and polar bonds, and single and double bonds, as shown here for the molecule propionic acid, a common food preservative.

environments ranging from the Earth's frigid icy poles to the superheated water of deep-sea vents.

Carbon Atoms Can Bond to Several Biologically Important Functional Groups

Aside from the simplest hydrocarbons, most organic molecules and macromolecules contain **functional groups**—groups of atoms with characteristic chemical features and properties. Each type of functional group exhibits similar chemical properties in all molecules in which it occurs. For example, the amino group (NH_2) acts like a base. At the pH found in living organisms, amino groups readily bind H^+ to become NH_3^+, thereby removing H^+ from an aqueous solution and raising the pH. As discussed later in this chapter, amino groups are widely found in proteins and also in other types of organic molecules. Table 3.1 describes examples of functional groups found in many different types of organic molecules. We will discuss each of these groups at numerous points throughout this textbook.

Carbon-Containing Molecules May Exist in Multiple Forms Called Isomers

When Wöhler did his now-famous experiment, he was surprised to discover that urea and ammonium cyanate apparently contained the exact same ratio of carbon, nitrogen, hydrogen, and oxygen atoms, yet they were different molecules with distinct chemical and biological properties. Two structures with an identical molecular formula but different structures and characteristics are called **isomers**.

Figure 3.4 depicts three ways in which isomers may occur. **Structural isomers** contain the same atoms but in different bonding relationships. Urea and ammonium cyanate fall into this category; a simpler example of a structural isomer is illustrated in Figure 3.4a.

Stereoisomers have identical bonding relationships, but the spatial positioning of the atoms differs in the two isomers. Two types of stereoisomers are *cis-trans* isomers and enantiomers. In **cis-trans isomers**, like those shown in Figure 3.4b, the

Functional group* (with shorthand notation)	Formula	Examples of where they are found
Amino $-NH_2$	$R-N\big\langle{}^H_H$	Amino acids (proteins)
Carbonyl** $-CO$ Ketone	$\underset{R-C-R'}{\overset{O}{\parallel}}$	Steroids, waxes, and proteins
Aldehyde	$\underset{R-C-H}{\overset{O}{\parallel}}$	
Carboxyl $-COOH$	$R-C\big\langle{}^O_{OH}$	Amino acids, fatty acids
Hydroxyl $-OH$	$R-OH$	Steroids, alcohol, carbohydrates, some amino acids
Methyl $-CH_3$	$\underset{\underset{H}{\mid}}{\overset{\overset{H}{\mid}}{R-C-H}}$	May be attached to DNA, proteins, and carbohydrates
Phosphate $-PO_4^{2-}$	$R-O-\underset{\underset{O^-}{\mid}}{\overset{\overset{O}{\parallel}}{P}}-O^-$	Nucleic acids, ATP, attached to amino acids
Sulfate $-SO_4^-$	$R-O-\underset{\overset{\parallel}{O}}{\overset{\overset{O}{\parallel}}{S}}-O^-$	May be attached to carbohydrates, proteins, and lipids
Sulfhydryl $-SH$	$R-SH$	Proteins that contain the amino acid cysteine

Table 3.1 Some Biologically Important Functional Groups That Bond to Carbon

* This list contains many of the functional groups that are important in biology. However, many more functional groups have been identified by biochemists. R and R' represent the remainder of the molecule.

** A carbonyl group is C=O. In a ketone, the carbon forms covalent bonds with two other carbon atoms. In an aldehyde, the carbon is linked to a hydrogen atom.

two hydrogen atoms linked to the two carbons of a C=C double bond may be on the same side of the carbons, in which case the C=C bond is called a *cis* double bond. If the hydrogens are on opposite sides, it is a *trans* double bond. *Cis-trans* isomers may have very different chemical properties from each other, most notably their stability and sensitivity to heat and light. For instance, the light-sensitive region of your eye contains a molecule called retinal, which may exist in either a *cis* or *trans* form because of a pair of double-bonded carbons in its string of carbon atoms. In darkness, the *cis*-retinal form predominates.

Because this –OH group is attached to a different carbon, these 2 molecules are structural isomers.

Isopropyl alcohol   Propyl alcohol

(a) Structural isomers

These 2 hydrogens are *cis* to each other.

These 2 hydrogens are *trans* to each other.

cis-butene   *trans*-butene

Cis–trans isomers

Molecule   Mirror image

Enantiomers

(b) Two types of stereoisomers

Figure 3.4 Types of isomers. Isomers are compounds with the same molecular formula but different structures. The differences in structure, though small, are sufficient to result in very different biological properties. Isomers can be grouped into **(a)** structural isomers and **(b)** stereoisomers.

The energy of sunlight, however, causes retinal to isomerize to the *trans* form. The *trans*-retinal activates the light-capturing cells in the eye.

A second type of stereoisomer, called an **enantiomer**, exists as a pair of molecules that are mirror images. Four different atoms can bind to a single carbon atom in two possible ways, designated a left-handed and a right-handed structure. The resulting structures are not identical, but instead are mirror images of each other (Figure 3.4b). A convenient way to visualize the mirror-image properties of enantiomers is to look at a pair of gloves. No matter which way you turn or hold a left-hand glove, for example, it cannot fit properly on your

right hand. Any given pair of enantiomers shares identical chemical properties, such as solubility and melting point. However, due to the different orientation of atoms in space, their ability to noncovalently bind to other molecules can be strikingly different. For example, as you learned in Chapter 2, **enzymes** are molecules that catalyze, or speed up, the rates of many biologically important chemical reactions. Typically, a given enzyme is very specific in its action, and an enzyme that recognizes one enantiomer of a pair often does not recognize the other. That is because the actions of enzymes depend upon the spatial arrangements of the particular atoms in a molecule.

3.2 Formation of Organic Molecules and Macromolecules

As we have seen, organic molecules have various shapes due to the bonding properties of carbon. During the past two centuries, biochemists have studied many organic molecules found in living organisms and determined their structures at the molecular level. Many of these compounds are relatively small molecules, containing a few or a few dozen atoms. However, some organic molecules are extremely large macromolecules composed of thousands or even millions of atoms. Such large molecules are formed by linking together many smaller molecules called **monomers** (meaning one part) and are thus also known as **polymers** (meaning many parts). The structure of macromolecules depends on the structure of their monomers, the number of monomers linked together, and the three-dimensional way in which the monomers are linked.

As introduced in Chapter 2, the process by which two or more molecules combine into a larger one is called a **condensation reaction**. Such reactions are accompanied by the loss of a small molecule formed as a result of the condensation. When an organic macromolecule is formed, two smaller molecules combine by condensation, producing a larger molecule along with the loss of a molecule of water. This specific type of condensation reaction is called a **dehydration reaction**, because a molecule of water is removed when the monomers combine.

An idealized dehydration reaction is illustrated in **Figure 3.5a**. Notice that the length of a polymer may be extended again and again with additional dehydration reactions. Some polymers can reach great lengths by this mechanism. For example, as you will learn in Chapter 46, nutrients in an animal's food are transported out of the digestive tract into the body fluids as monomers. If more energy-yielding nutrients are consumed than are required for an animal's activities, the excess nutrients may be processed by certain organs into extremely long polymers consisting of tens of thousands of monomers. The polymers are then stored in this convenient form to provide a source of energy when food is not available. An example would be during sleep, when an animal is not eating but nevertheless still requires energy to carry out all the various activities required to maintain cellular function.

A polymer begins as two monomers combine in a dehydration reaction.

Elongation of the polymer continues with additional dehydration reactions.

The final length of a polymer may consist of thousands of monomers.

(a) Polymer formation by dehydration reactions

Polymers are broken down one monomer at a time by hydrolysis reactions.

(b) Breakdown of a polymer by hydrolysis reactions

Figure 3.5 Formation and breakdown of polymers. **(a)** Monomers combine to form polymers in living organisms by dehydration reactions, in which a molecule of water is removed each time a new monomer is added to the growing polymer. **(b)** Polymers can be broken down into their constituent monomers by hydrolysis reactions, in which a molecule of water is added each time a monomer is released.

Polymers, however, are not recognized by the cellular machinery that functions to release the chemical energy stored in the bonds of molecules. Consequently, polymers must first be broken down into their constituent monomers, which then, under the right conditions, can release some of the energy stored in their bonds. The process by which a polymer is broken down into monomers is called a **hydrolysis reaction** (Figure 3.5b) (from the Greek *hydro*, meaning water, and *lysis*, meaning to separate), because a molecule of water is added back each time a monomer is released. Therefore, the formation of polymers in organisms is generally reversible; once formed, a polymer can later be broken down. These processes may repeat themselves over and over again as dictated by changes in the various cellular activities of an organism. Both condensation/dehydration reactions and hydrolysis reactions are catalyzed by enzymes.

By analyzing the cells of many different species, researchers have determined that all forms of life have organic molecules and macromolecules that fall into four broad categories, based on their chemical and biological properties: carbohydrates, lipids, proteins, and nucleic acids. In the next sections, we will survey the structures of these organic compounds and begin to examine their biological functions.

3.3 Carbohydrates

Carbohydrates are composed of carbon, hydrogen, and oxygen atoms in or close to the proportions represented by the general formula $C_n(H_2O)_n$, where n is a whole number. This formula gives carbohydrates their name—carbon-containing compounds that are hydrated (that is, contain water). Most of the carbon atoms in a carbohydrate are linked to a hydrogen atom and a hydroxyl functional group. However, other functional groups, such as amino and carboxyl groups, are also found in certain carbohydrates. As discussed next, sugars are relatively small carbohydrates, whereas polysaccharides are large macromolecules.

Sugars Are Carbohydrate Monomers That May Taste Sweet

Sugars are small carbohydrates that in some, but not all, cases taste sweet. The simplest sugars are the monomers known as **monosaccharides** (from the Greek, meaning single sugars). The most common types are molecules with five carbons, called pentoses, and with six carbons, called hexoses. Important pentoses are ribose ($C_5H_{10}O_5$) and the closely related deoxyribose ($C_5H_{10}O_4$), which are part of RNA and DNA molecules, respectively, which are described later in this chapter. The most common hexose is glucose ($C_6H_{12}O_6$). Like other monosaccharides, glucose is very water-soluble and thus circulates in the blood or fluids of animals, where it can be transported across plasma membranes. Once inside a cell, enzymes can break down glucose into smaller molecules, releasing energy that was stored in the chemical bonds of glucose. This energy is then stored in the bonds of another molecule, called adenosine triphosphate, or ATP (see Chapter 7), which, in turn, powers a variety of cellular processes. In this way, sugar is often used as a source of energy by living organisms.

Figure 3.6a depicts the bonds between atoms in a monosaccharide in both linear and ring forms. The ring structure is a better approximation of the true shape of the molecule as it mostly exists in solution, with the carbon atoms numbered by convention as shown. The ring is made from the linear structure by an oxygen atom, which forms a bond that bridges two carbons. The hydrogen atoms and the hydroxyl groups may lie above or below the plane of the ring structure.

Figure 3.6b compares different types of isomers of glucose. Glucose can exist as D- and L-glucose, which are mirror images of each other, or enantiomers. (The letters D and L are derived from dextrorotatory—rotating to the right—and levorotatory—rotating to the left—which describe the ways in which a beam of polarized light is altered by some molecules.) Other types of isomers are formed by changing the relative positions of the hydrogens and hydroxyl groups along the sugar ring. For example, glucose exists in two interconvertible forms, with the hydroxyl group attached to the number 1 carbon atom lying either above (the β form of glucose, Figure 3.6b) or below (the α form, Figure 3.6a) the plane of the ring. As discussed later, these different isomers of glucose have different biological properties. In another example, if the hydroxyl group on carbon atom number 4 of glucose is switched from below to above the plane of the ring, the sugar called galactose is created (Figure 3.6b).

Monosaccharides can join together by dehydration to form larger carbohydrates. **Disaccharides** (meaning two sugars) are carbohydrates composed of two monosaccharides. A familiar

(a) **Linear and ring structures of D-glucose**

(b) **Isomers of glucose**

Figure 3.6 **Monosaccharide structure.** (a) A comparison of the linear and ring structures of glucose. In solution, such as the fluids of organisms, nearly all glucose is in the ring form. (b) Isomers of glucose. The locations of the C-1 and C-4 hydroxyl groups are emphasized with green and orange boxes, respectively. Glucose exists as stereoisomers designated α- and β-glucose, which differ in the position of the —OH group attached to carbon atom number 1. Glucose and galactose differ in the position of the —OH group attached to carbon atom number 4. Enantiomers of glucose, called D-glucose and L-glucose, are mirror images of each other. D-glucose is the form used by living cells.

Concept check: With regard to their binding to enzymes, why do enantiomers such as D- and L-glucose have different biological properties?

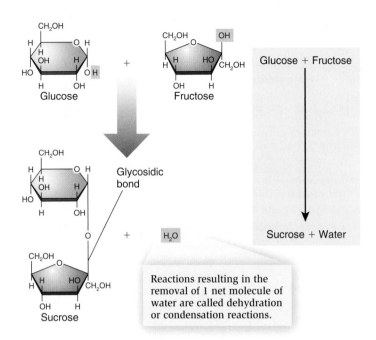

Figure 3.7 **Formation of a disaccharide.** Two monosaccharides can bond to each other to form a disaccharide, such as sucrose, maltose, or lactose, by a dehydration reaction.

Concept check: What type of reaction is the reverse of the ones shown here, in which a disaccharide is broken down into two monosaccharides?

disaccharide is sucrose, or table sugar, which is composed of the monomers glucose and fructose (Figure 3.7). Sucrose is the major transport form of sugar in plants. The linking together of most monosaccharides involves the removal of a hydroxyl group from one monosaccharide and a hydrogen atom from the other, giving rise to a molecule of water and bonding the two sugars together through an oxygen atom. The bond formed between two sugar molecules is called a **glycosidic bond**. Conversely, hydrolysis of a glycosidic bond in a disaccharide breaks the bond by adding back the water, thereby uncoupling the two monosaccharides. Other disaccharides frequently found in nature are maltose, formed in animals during the digestion of large carbohydrates in the intestinal tract, and lactose, present in the milk of mammals. Maltose is α-D-glucose linked to α-D-glucose, and lactose is β-D-galactose linked to β-D-glucose.

Polysaccharides Are Carbohydrate Polymers That Include Starch and Glycogen

When many monosaccharides are linked together to form long polymers, **polysaccharides** (meaning many sugars) are made. **Starch**, found in plant cells, and **glycogen**, present in animal cells and sometimes called animal starch, are examples of polysaccharides (Figure 3.8). Both of these polysaccharides are composed of thousands of α-D-glucose molecules linked together in long, branched chains, differing only in the extent of branching along the chain. The bonds that form in polysaccharides are not random but instead form between specific carbon atoms of each molecule. The carbon atoms are numbered according to convention, as shown in Figure 3.8. The higher degree of branching in glycogen contributes to its solubility in animal tissues, such as muscle. This is because the extensive branching creates a more open structure, in which many hydrophilic hydroxyl (—OH) side groups have access to water and can hydrogen-bond with it. Starch, because it is less branched, is less soluble and contributes to the properties of plant structures (think of a potato or a kernel of corn).

Some polysaccharides, such as starch and glycogen, are used to store energy in cells. Like disaccharides, polysaccharides can be hydrolyzed in the presence of water to yield monosaccharides, which are broken down to provide the energy to make ATP. Starch and glycogen, the polymers of α-glucose, provide efficient means of storing energy for those times when a plant or animal cannot obtain sufficient energy from its environment or diet for its metabolic requirements.

Other polysaccharides provide a structural role, rather than storing energy. The plant polysaccharide **cellulose** is a polymer of β-D-glucose, with a linear arrangement of carbon-carbon bonds and no branching (see Figure 3.8). Each glucose monomer in cellulose is in an opposite orientation from its adjacent monomers ("flipped over"), forming long chains of several thousand glucose monomers. The bond orientations in β-D-glucose prevent cellulose from being hydrolyzed for ATP production in most types of organisms. This is because many enzymes are highly specific for one type of molecule, as noted earlier. The enzymes that break the bonds between monomers of α-D-glucose in starch do not recognize the shape of the polymer made by the bonds between β-D-glucose monomers in cellulose. Therefore, plant cells can break down starch without breaking down cellulose. In this way, cellulose can be used for other functions, notably in the formation of the rigid cell-wall structure characteristic of plants. The linear arrangement of bonds in cellulose provides opportunities for vast numbers of hydrogen bonds between cellulose

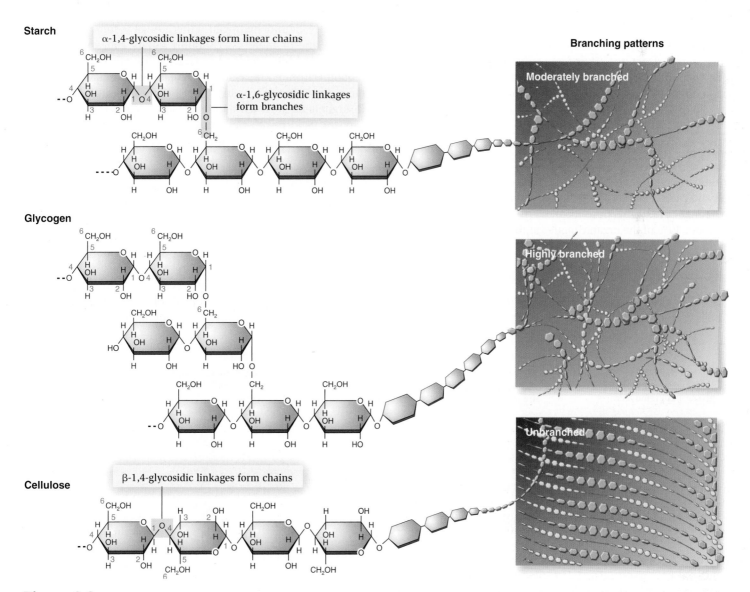

Figure 3.8 **Polysaccharides that are polymers of glucose.** These polysaccharides differ in their arrangement, extent of branching, and type of glucose isomer. Note: In cellulose, the bonding arrangements cause every other glucose to be upside down with respect to its neighbors.

molecules, which stack together in sheets and provide great strength to structures like plant cell walls. Cellulose accounts for up to half of all the carbon contained within a typical plant, making it the most common organic compound on Earth.

Unlike most animals and plants, some organisms do have an enzyme capable of breaking down cellulose. For example, certain bacteria present in the gastrointestinal tracts of grass and wood eaters, such as cows and termites, respectively, can digest cellulose into usable monosaccharides because they contain an enzyme that can hydrolyze the bonds between β-D-glucose monomers. Humans lack this enzyme; therefore, we eliminate in the feces most of the cellulose ingested in our diet. Undigestible plant matter we consume is commonly referred to as fiber.

Other polysaccharides also play structural roles. **Chitin**, a tough, structural polysaccharide, forms the external skeleton of insects and the cell walls of fungi. The sugar monomers within chitin have nitrogen-containing groups attached to them. **Glycosaminoglycans** are large polysaccharides that play a structural role in animals. For example, they are abundantly found in cartilage, the tough, fibrous material found in bone and certain other animal structures. Glycosaminoglycans are also abundant in the extracellular matrix that provides a structural framework surrounding many of the cells in an animal's body (this will be covered in Chapter 10).

3.4 Lipids

Lipids are hydrophobic molecules composed mainly of hydrogen and carbon atoms. The defining feature of lipids is that they are nonpolar and therefore insoluble in water. Lipids account for about 40% of the organic matter in the average human body and include fats, phospholipids, steroids, and waxes.

Fats Are Made from Glycerol and Fatty Acids

Fats, also known as **triglycerides** or **triacylglycerols**, are formed by bonding glycerol to three fatty acids (**Figure 3.9**).

Glycerol is a three-carbon molecule with one hydroxyl group (—OH) bonded to each carbon. A fatty acid is a chain of carbon and hydrogen atoms with a carboxyl group (—COOH) at one end. Each of the hydroxyl groups in glycerol is linked to the carboxyl group of a fatty acid by the removal of a molecule of water by a dehydration reaction. The resulting bond is an example of a type of chemical bond called an ester bond.

The fatty acids found in fats and other lipids may differ with regard to their lengths and the presence of double bonds (**Figure 3.10**). Fatty acids are synthesized by the linking of two-carbon fragments. Therefore, most fatty acids in nature have an even number of carbon atoms, with 16- and 18-carbon fatty acids being the most common in the cells of plants and animals. Fatty acids also differ with regard to the presence of double bonds. When all the carbons in a fatty acid are linked by single covalent bonds, the fatty acid is said to be a **saturated fatty acid**, because all the carbons are saturated with covalently bound hydrogen. Alternatively, some fatty acids contain one or more C=C double bonds and are known as **unsaturated fatty acids**. A fatty acid with one C=C bond is a monounsaturated fatty acid, whereas a fatty acid with two or more C=C bonds constitutes a polyunsaturated fatty acid. In organisms such as mammals, some fatty acids are necessary for good health but cannot be synthesized by the body. Such fatty acids are called essential fatty acids, because they must be obtained in the diet; an example is linoleic acid (Figure 3.10).

Fats (triglycerides) that contain high amounts of saturated fatty acids can pack together tightly, resulting in numerous intermolecular interactions that stabilize the fat. Saturated fats have high melting points and tend to be solid at room temperature. Animal fats generally contain a high proportion of saturated fatty acids. For example, beef fat contains high amounts of stearic acid, a saturated fatty acid with a melting point of 70°C (Figure 3.10). When you heat a hamburger on the stove, the saturated animal fats melt, and liquid grease appears in the

Glycerol   3 Fatty acids   Triglyceride (fat)

Figure 3.9 **The formation of a fat.** The formation of a triglyceride requires three dehydration reactions in which fatty acids are bonded to glycerol. Note in this figure and in Figure 3.10, a common shorthand notation is used for depicting fatty acid chains, in which a portion of the CH_2 groups are illustrated as $(CH_2)_n$, where n may be 2 or greater.

Saturated fatty acid
(Stearic acid)

Double bonds deform
the linear chain and give
the fatty acid a kinked
3-dimensional structure.

Unsaturated fatty acid
(Linoleic acid)

Figure 3.10 **Examples of fatty acids.** Fatty acids are hydrocarbon chains with a carboxyl functional group at one end and either no double-bonded carbons (saturated) or one or more double bonds (unsaturated). Stearic acid, for example, is an abundant saturated fatty acid in animals, whereas linoleic acid is an unsaturated fatty acid found in plants. Note that the presence of two C=C double bonds introduces two kinks into the shape of linoleic acid. As a consequence, saturated fatty acids are able to pack together more tightly than unsaturated fatty acids.

frying pan (**Figure 3.11**). When allowed to cool, however, the liquid grease in the pan returns to its solid form.

As illustrated in Figure 3.10, the presence of an unsaturated bond in a fatty acid introduces a kink into the linear shape of a fatty acid. Because of kinks in their chains, unsaturated fatty acids cannot stack together as tightly as saturated fatty acids. Fats high in unsaturated fatty acids usually have low melting points and are liquids at room temperature. Such fats are called oils. Fats derived from plants generally contain unsaturated fatty acids. Olive oil contains high amounts of oleic acid, a monounsaturated fatty acid with a melting point of 16°C. Fatty acids with additional double bonds have even lower melting points; linoleic acid (see Figure 3.10), found in soybeans and other plants, has two double bonds and melts at −5°C.

Most unsaturated fatty acids, including linoleic acid, exist in nature in the *cis* form (see Figures 3.4 and 3.10). Of particular importance to human health, however, are *trans* fatty acids, which are formed by a synthetic process in which the natural *cis* form is altered to a *trans* configuration. This gives the fats that contain such fatty acids a more linear structure and, therefore, a higher melting point. Although this process has been used for many years to produce fats with a longer shelf-life and with better characteristics for baking, it is now understood that *trans* fats are linked with human disease. Notable among these is coronary artery disease, caused by a narrowing of the blood vessels that supply the muscle cells of the heart with blood.

Like starch and glycogen, fats are important for storing energy. The hydrolysis of triglycerides releases the fatty acids

High temperature converts
solid, saturated fats to liquid.

After cooling, saturated fats
return to their solid form.

Unsaturated fats are oils at
room temperature and below.

(a) Animal fats at high and low temperatures

(b) Vegetable fats at low temperature

Figure 3.11 **Fats at different temperatures.** Saturated fats found in animals tend to have high melting points compared to unsaturated fats found in plants.

Concept check: *Certain types of fats used in baking are called shortenings. They are solid at room temperature. Shortenings are often made from vegetable oils by a process called hydrogenation. What do you think happens to the structure of an oil when it is hydrogenated?*

from glycerol, and these products can then be metabolized to provide energy to make ATP (see Chapter 7). Certain organisms, most notably mammals, have the ability to store large amounts of energy by accumulating fats. As you will learn in Chapter 7, the number of C—H bonds in a molecule of fat or carbohydrate determines in part how much energy the molecule can yield. Fats are primarily long chains of C—H bonds, whereas glucose and other carbohydrates have numerous C—OH bonds. Consequently, 1 gram of fat stores more energy than does 1 gram of starch or glycogen. Fat is therefore an efficient means of energy storage for mobile organisms in which excess body mass may be a disadvantage. In animals, fats can also play a structural role by forming cushions that support organs. In addition, fats provide insulation under the skin that helps protect many terrestrial animals during cold weather and marine mammals in cold water.

Phospholipids Are Amphipathic Lipids

Another class of lipids, **phospholipids**, are similar in structure to triglycerides but with one important difference. The third hydroxyl group of glycerol is linked to a phosphate group instead of a fatty acid. In most phospholipids, a small polar or charged nitrogen-containing molecule is attached to this phosphate (**Figure 3.12a**). The glycerol backbone, phosphate group, and charged molecule constitute a polar hydrophilic region at one end of the phospholipid, whereas the fatty acid chains provide a nonpolar hydrophobic region at the opposite end. Recall from Chapter 2 that molecules with polar and nonpolar regions are called amphipathic molecules.

In water, phospholipids become organized into bilayers, because their hydrophilic polar ends are attracted to the water molecules and their hydrophobic nonpolar ends exclude water. As you will learn in Chapter 5, this bilayer arrangement of phospholipids is critical for determining the structure of cellular membranes, as shown in **Figure 3.12b**.

Steroids Contain Ring Structures

Steroids have a distinctly different chemical structure from that of the other types of lipid molecules discussed thus far. Four fused rings of carbon atoms form the skeleton of all steroids.

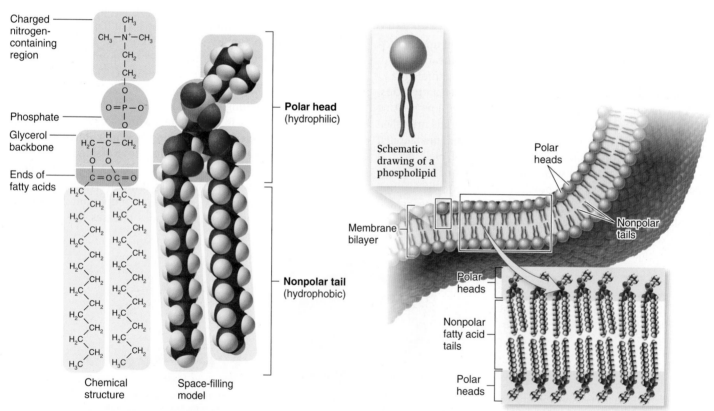

(a) Structure and model of a phospholipid

(b) Arrangement of phospholipids in a bilayer

Figure 3.12 **Structure of phospholipids.** **(a)** Chemical structure and space-filling model of phosphatidylcholine, a common phospholipid found in living organisms. Phospholipids contain both polar and nonpolar regions, making them amphipathic. The fatty-acid tails are the nonpolar region. The rest of the molecule is polar. **(b)** Arrangement of phospholipids in a biological membrane, such as the plasma membrane that encloses cells. The hydrophilic regions of the phospholipid face the watery environments on either side of the membrane, while the hydrophobic regions associate with each other in the interior of the membrane, forming a bilayer.

Concept check: *When water and oil are added to a test tube, the two liquids form two separate layers (think of oil and vinegar in a bottle of salad dressing). If a solution of phospholipids were added to a mixture of water and oil, where would the phospholipids dissolve?*

Figure 3.13 **Structure of cholesterol and steroid hormones derived from cholesterol.** The structure of a steroid has four rings. Steroids include cholesterol and molecules derived from cholesterol, such as steroid hormones. These include the reproductive hormones estrogen and testosterone.

One or more polar hydroxyl groups are attached to this ring structure, but they are not numerous enough to make a steroid highly water-soluble. For example, steroids with a hydroxyl group are known as sterols—one of the most well known being cholesterol (Figure 3.13, top). Cholesterol is found in the blood and plasma membranes of animals. Due to its low solubility in water, at high concentrations cholesterol can contribute to the formation of blockages in major blood vessels.

In steroids, tiny differences in chemical structure can lead to profoundly different biological properties. For example, estrogen is a steroid found in high amounts in female vertebrates. Estrogen differs from testosterone, a steroid found largely in males, by having one less methyl group, a hydroxyl group instead of a ketone group, and additional double bonds in one of its rings (Figure 3.13, bottom). However, these seemingly small differences are sufficient to make these two molecules largely responsible for whether an animal exhibits male or female characteristics, including feather color.

Waxes Are Complex Lipids That Help Prevent Water Loss from Organisms

Many plants and animals produce lipids called waxes that are typically secreted onto their surface, such as the leaves of plants and the cuticles of insects. Although any wax may contain hundreds of different compounds, all waxes contain one or more hydrocarbons and long structures that resemble a fatty acid attached by its carboxyl group to another long hydrocarbon chain. Most waxes are very nonpolar and therefore exclude water, providing a barrier to water loss. They may also be used as structural elements in colonies like those of bees, where beeswax forms the honeycomb of the hive.

3.5 Proteins

Proteins are polymers found in all cells and play critical roles in nearly all life processes (Table 3.2). The word protein comes from the Greek *proteios* (meaning of the first rank), which aptly describes their importance. Proteins account for about 50% of the organic material in a typical animal's body.

Proteins Are Made Up of Amino Acid Monomers

Proteins are composed of carbon, hydrogen, oxygen, nitrogen, and small amounts of other elements, notably sulfur. The building blocks of proteins are **amino acids**, compounds with a structure in which a carbon atom, called the α-carbon, is linked to an amino group (NH_2) and a carboxyl group (COOH). The α-carbon also is linked to a hydrogen atom and a side chain, which is given a general designation R. Proteins are polymers of amino acids.

When dissolved in water at neutral pH, the amino group accepts a hydrogen ion and is positively charged, whereas the carboxyl group loses a hydrogen ion and is negatively charged. The term amino acid is the name given to such molecules because they have an amino group and also a carboxyl group that behaves like an acid.

Table 3.2	Major Categories and Functions of Proteins	
Category	**Functions**	**Examples**
Proteins involved in gene expression and regulation	Make mRNA from a DNA template; synthesize polypeptides from mRNA; regulate genes	RNA polymerase assists in synthesizing RNA from DNA. Transcription factor proteins are involved in gene regulation.
Motor proteins	Initiate movement	Myosin provides the contractile force of muscles. Kinesin is a key protein that helps cells to sort their chromosomes.
Defense proteins	Protect organisms against disease	Antibodies ward off infection due to bacteria or viruses.
Metabolic enzymes	Increase rates of chemical reactions	Hexokinase is an enzyme involved in sugar metabolism.
Cell signaling proteins	Enable cells to communicate with each other and with the environment	Taste receptors in the tongue allow animals to taste molecules in food.
Structural proteins	Support and strengthen structures	Actin provides shape to the cytoplasm of cells, such as plant and animal cells. Collagen gives strength to tendons.
Transporters	Promote movement of solutes across plasma membranes	Glucose transporters move glucose from outside cells to inside cells, where it can be used for energy.

All amino acids except glycine may exist in more than one isomeric form, called the D and L forms, which are enantiomers. Note that glycine cannot exist in D and L forms because there are two hydrogens bound to its α-carbon. Only L-amino acids and glycine are found in proteins. D-isomers are found in the cell walls of certain bacteria, where they may play a protective role against molecules secreted by the host organism in which the bacteria live.

The 20 amino acids found in proteins are distinguished by their side chains (Figure 3.14). The amino acids are categorized as those in which the side chains are nonpolar, or polar and uncharged, or polar and charged. The varying structures of the side chains are critical features of protein structure and function. The arrangement and chemical features of the side chains cause proteins to fold and adopt their three-dimensional shapes. In addition, certain amino acids may be critical in protein function. For example, amino acid side chains found within the active sites of enzymes are important in catalyzing chemical reactions.

Amino acids are joined together by a dehydration reaction that links the carboxyl group of one amino acid to the amino group of another (Figure 3.15a). The covalent bond formed between a carboxyl and amino group is called a **peptide bond**. When many amino acids are joined by peptide bonds, the resulting molecule is called a **polypeptide** (Figure 3.15b). The backbone of the polypeptide in Figure 3.15 is highlighted in yellow. The amino acid side chains project from the backbone. When two or more amino acids are linked together, one end of the resulting molecule has a free amino group. This is the amino end, or N-terminus. The other end of the polypeptide, called the carboxyl end, or C-terminus, has a free carboxyl group. As shown in Figure 3.15c, amino acids within a polypeptide are numbered from the amino end to the carboxyl end.

The term polypeptide refers to a structural unit composed of a single chain of amino acids. In contrast, a protein is a functional unit composed of one or more polypeptides that have been folded and twisted into a precise three-dimensional shape that carries out a particular function. Many proteins also have carbohydrates (glycoproteins) or lipids (lipoproteins) attached at various points along their amino acid chain; these modifications impart unique functions to such proteins.

Proteins Have a Hierarchy of Structure

Scientists view protein structure at four progressive levels: primary, secondary, tertiary, and quaternary, shown schematically in Figure 3.16. Each higher level of structure depends on the preceding levels. For example, changing the primary structure may affect the secondary, tertiary, and quaternary structures. Let's now consider each level separately.

Primary Structure   The **primary structure** (see Figure 3.16) of a polypeptide is its amino acid sequence, from beginning to end. The primary structures of polypeptides are determined by genes. As we will explore in Chapter 12, genes carry the information for the production of polypeptides with a specific amino acid sequence.

Figure 3.17 shows the primary structure of ribonuclease, which functions as an enzyme to degrade ribonucleic acid (RNA) molecules after they are no longer required by a cell. As described later and in Unit III of this textbook, RNA is a key part of the mechanism by which proteins are synthesized. Ribonuclease is composed of a relatively short polypeptide with 124 amino acids. An average polypeptide is about 300–500 amino acids in length, and some genes encode polypeptides that are a few thousand amino acids long.

Secondary Structure   The amino acid sequence of a polypeptide, together with the fundamental constraints of chemistry and physics, cause a polypeptide to fold into a more compact structure. Amino acids can rotate around bonds within a polypeptide. Consequently, polypeptides and proteins are flexible and can fold into a number of shapes, just as a string of beads can be twisted into many configurations. Folding can be irregular or certain regions can have a repeating folding pattern. Such repeating patterns are called **secondary structure**. The two basic types of secondary structure are the α helix and the β pleated sheet.

In an **α helix**, the polypeptide backbone forms a repeating helical structure that is stabilized by hydrogen bonds along the length of the backbone. As shown in Figure 3.16, the hydrogen linked to a nitrogen atom forms a hydrogen bond with an oxygen atom that is double-bonded to a carbon atom. These

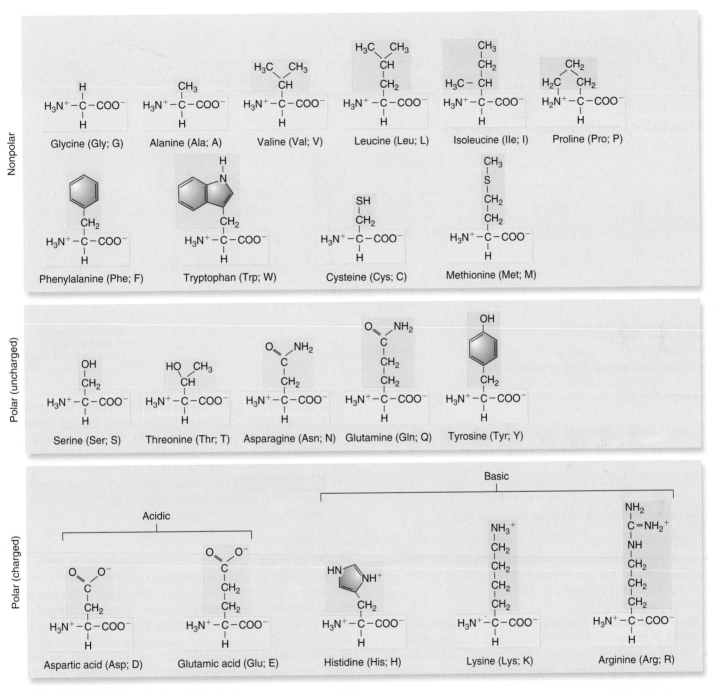

Figure 3.14 **The 20 amino acids found in living organisms.**   The various amino acids have different chemical properties (for example, nonpolar versus polar) due to the nature of their different side chains. These properties contribute to the differences in the three-dimensional shapes and chemical properties of proteins, which, in turn, influence their biological functions. Tyrosine has both polar and nonpolar characteristics and is listed in just one category for simplicity. The common three-letter and one-letter abbreviations for each amino acid are shown in parentheses.

hydrogen bonds occur at regular intervals within the polypeptide backbone and cause the backbone to twist into a helix.

In a **β pleated sheet**, regions of the polypeptide backbone come to lie parallel to each other. Hydrogen bonds between a hydrogen linked to a nitrogen atom and a double-bonded oxygen form between these adjacent, parallel regions. When this occurs, the polypeptide backbone adopts a repeating zigzag—or pleated—shape.

The α helices and β pleated sheets are key determinants of a protein's characteristics. For example, α helices in certain proteins are composed primarily of nonpolar amino acids. Proteins containing many such regions with an α helix structure tend to anchor themselves into a lipid-rich environment, such as a cell's plasma membrane. In this way, a protein whose function is required in a specific location such as a plasma membrane can be retained there. Secondary structure also contributes to

(a) Formation of a peptide bond between 2 amino acids

The amino end of a polypeptide is called the N-terminus.

The backbone of the polypeptide is highlighted in yellow.

The carboxyl end of a polypeptide is called the C-terminus.

(b) Polypeptide—a linear chain of amino acids

This is an octapeptide (8 amino acids).

(c) Numbering system of amino acids in a polypeptide

Figure 3.15   The chemistry of polypeptide formation. Polypeptides are polymers of amino acids. They are formed by linking amino acids via dehydration reactions to make peptide bonds. Every polypeptide has an amino end, or N-terminus, and a carboxyl end, or C-terminus.

Concept check: How many water molecules would be produced in making a polypeptide that is 72 amino acids long by dehydration reactions?

the great strength of certain proteins, including the keratins found in hair and hooves; the proteins that make up the silk webs of spiders; and collagen, the chief component of cartilage in vertebrate animals.

Some regions along a polypeptide chain do not assume an α helix or β pleated sheet conformation and consequently do not have a secondary structure. These regions are sometimes called random coiled regions. However, this term is somewhat misleading because the shapes of random coiled regions are usually very specific and important for the protein's function.

Tertiary Structure   As the secondary structure of a polypeptide chain becomes established due to the particular primary structure, the polypeptide folds and refolds upon itself to assume a complex three-dimensional shape—its **tertiary structure** (see Figure 3.16). The tertiary structure is the three-dimensional shape of a single polypeptide. Tertiary structure includes all secondary structures plus any interactions involving amino acid side chains. For some proteins, such as ribonuclease, the tertiary structure is the final structure of a functional protein. However, as described next, other proteins are composed of two or more polypeptides and adopt a quaternary structure.

Quaternary Structure   Most functional proteins are composed of two or more polypeptides that each adopt a tertiary structure and then assemble with each other (see Figure 3.16). The individual polypeptides are called **protein subunits**. Subunits may be identical polypeptides or they may be different. When proteins consist of more than one polypeptide chain, they are said to have **quaternary structure** and are also known as **multimeric proteins** (meaning multiple parts). Multimeric proteins are widespread in organisms. A common example is the oxygen-binding protein called hemoglobin, found in the red blood cells of vertebrate animals. As you will learn in Chapter 48, four protein subunits combine to form one molecule of hemoglobin. Each subunit can bind a single molecule of oxygen; therefore, each hemoglobin molecule can carry four molecules of oxygen in the blood.

Protein Structure Is Influenced by Several Factors

The amino acid sequences of polypeptides are the defining features that distinguish the structure of one protein from another. As polypeptides are synthesized in a cell, they fold into secondary and tertiary structures, which assemble into quaternary

Figure 3.16 **The hierarchy of protein structure.** The R groups are omitted for simplicity.

Figure 3.17 **The primary structure of ribonuclease.**
The example shown here is ribonuclease from cows.

structures for many proteins. Several factors determine the way that polypeptides adopt their secondary, tertiary, and quaternary structures. As mentioned, the laws of chemistry and physics, together with the amino acid sequence, govern this process.

As shown in **Figure 3.18**, five factors are critical for protein folding and stability:

1. *Hydrogen bonds*—The large number of weak hydrogen bonds within a polypeptide and between polypeptides adds up to a collectively strong force that promotes protein folding and stability. As we have already learned, hydrogen bonding is a critical determinant of protein secondary structure and also is important in tertiary and quaternary structure.

2. *Ionic bonds and other polar interactions*—Some amino acid side chains are positively or negatively charged. Positively charged side chains may bind to negatively charged side chains via ionic bonds. Similarly, uncharged polar side chains in a protein may bind to ionic amino acids. Ionic bonds and polar interactions are particularly important in tertiary and quaternary structure.

3. *Hydrophobic effect*—Some amino acid side chains are nonpolar. These amino acids tend to exclude water. As a protein folds, the hydrophobic amino acids are likely to be found in the center of the protein, minimizing contact with water. As mentioned, some proteins have stretches of nonpolar amino acids that anchor them in the hydrophobic portion of membranes. The hydrophobic effect plays a major role in tertiary and quaternary structures.

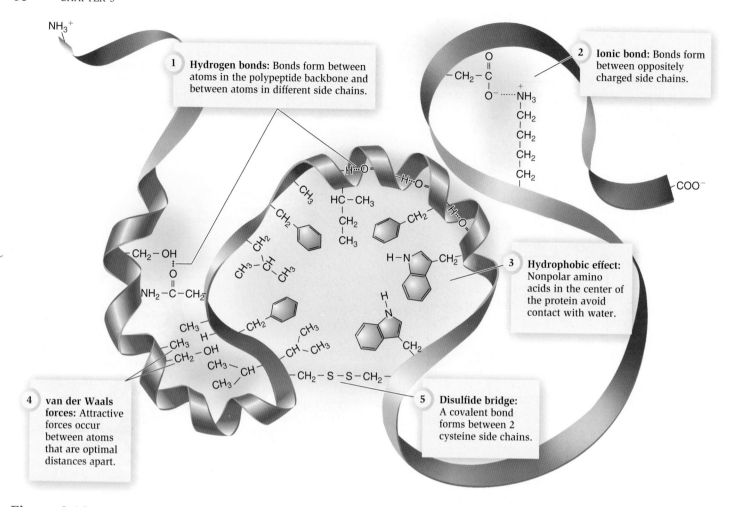

Figure 3.18   Factors that influence protein folding and stability.

4. *van der Waals forces*—Atoms within molecules have weak attractions for each other if they are an optimal distance apart. This optimal distance is called the van der Waals radius, and the weak attraction is the van der Waals force (see Chapter 2). If two atoms are very close together, their electron clouds will repel each other. If they are far apart, the van der Waals force will diminish. The van der Waals forces are particularly important for tertiary structure.

5. *Disulfide bridges*—The side chain of the amino acid cysteine contains a sulfhydryl group (—SH), which can react with a sulfhydryl group in another cysteine side chain. The result is a disulfide bridge or bond, which links the two amino acid side chains together (—S—S—). Disulfide bonds are covalent bonds that can occur within a polypeptide or between different polypeptides. Though other forces are usually more important in protein folding, the covalent nature of disulfide bonds can help to stabilize the tertiary structure of a protein.

The first four factors just described are also important in the ability of different proteins to interact with each other. As discussed throughout Unit II and other parts of this textbook, many cellular processes involve steps in which two or more different proteins interact with each other. For this to occur, the surface of one protein must bind to the surface of the other. Such binding is usually very specific. The surface of one protein precisely fits into the surface of another (**Figure 3.19**). Such **protein-protein interactions** are critically important so that cellular processes can occur in a series of defined steps. In addition, protein-protein interactions are important in building cellular structures that provide shape and organization to cells.

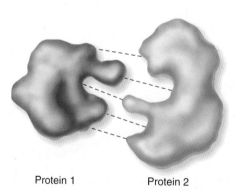

Protein 1   Protein 2

Figure 3.19   **Protein-protein interaction.**   Two different proteins may interact with each other due to hydrogen bonding, ionic bonding, the hydrophobic effect, and van der Waals forces.

Concept check: If the primary structure of Protein 1 in this figure were experimentally altered by the substitution of several incorrect amino acids for the correct ones, would Protein 1 still be able to interact with Protein 2?

FEATURE INVESTIGATION

Anfinsen Showed That the Primary Structure of Ribonuclease Determines Its Three-Dimensional Structure

Prior to the 1960s, the mechanisms by which proteins assume their three-dimensional structures were not understood. Scientists believed either that correct folding required unknown cellular factors or that ribosomes, the site where polypeptides are synthesized, somehow shaped proteins as they were being made. American researcher Christian Anfinsen, however, postulated that proteins contain all the information necessary to fold into their proper conformation without the need for cellular factors or organelles. He hypothesized that proteins spontaneously assume their most stable conformation based on the laws of chemistry and physics (Figure 3.20).

To test this hypothesis, Anfinsen studied ribonuclease, an enzyme that degrades RNA molecules (see Figure 3.17). Biochemists had already determined that ribonuclease has four disulfide bonds between eight cysteine amino acids. Anfinsen began with purified ribonuclease. The key point is that other cellular components were not present, only the purified protein. He exposed ribonuclease to a chemical called β-mercaptoethanol, which broke the S—S bonds, and to urea, which disrupted the hydrogen and ionic bonds. Following this treatment, he measured the ability of the treated enzyme to degrade RNA. The enzyme had lost nearly all of its ability to degrade RNA. Therefore, Anfinsen concluded that when ribonuclease was unfolded or denatured, it was no longer functional.

The key step in this experiment came when Anfinsen removed the urea and β-mercaptoethanol from the solution. Because these molecules are much smaller than ribonuclease, removing them from the solution was accomplished with a technique called size-exclusion chromatography. In size-exclusion chromatography, solutions are layered atop a glass column of beadlike particles and allowed to filter down through the column to an open collection port at the bottom. The particles in the column have microscopic pores that trap small molecules like urea and mercaptoethanol but that permit large molecules such as ribonuclease to pass down the length of the column and out the collection port.

Using size-exclusion chromatography, Anfinsen was able to purify the ribonuclease out of the original solution. He then allowed the purified enzyme to sit in water for up to 20 hours, after which he retested the ribonuclease for its ability to degrade RNA. The result revolutionized our understanding of proteins. The activity of the ribonuclease was almost completely restored! This meant that even in the complete absence of any cellular factors or organelles, an unfolded protein can refold into its correct, functional structure. This was later confirmed by chemical analyses that demonstrated the disulfide bonds had re-formed at the proper locations.

Since Anfinsen's time, we have learned that ribonuclease's ability to refold into its functional structure is not seen in all proteins. Some proteins do require enzymes and other proteins (known as chaperone proteins; see Chapter 4) to assist in their proper folding. Nonetheless, Anfinsen's experiments provided

Figure 3.20 Anfinsen's experiments with ribonuclease, demonstrating that the primary structure of a polypeptide plays a key role in protein folding.

HYPOTHESIS Within their amino acid sequence, proteins contain all the information needed to fold into their correct, 3-dimensional shapes.

KEY MATERIALS Purified ribonuclease, RNA, denaturing chemicals, size-exclusion columns.

Experimental level | **Conceptual level**

1 Incubate purified ribonuclease in test tube with RNA, and measure its ability to degrade RNA.

Purified ribonuclease

Numerous H bonds (not shown) and 4 S—S bonds. Protein is properly folded.

2 Denature ribonuclease by adding β-mercaptoethanol (breaks S—S bonds) and urea (breaks H bonds and ionic bonds). Measure its ability to degrade RNA.

β-mercaptoethanol + Urea

No more H bonds, ionic bonds, or S—S bonds. Protein is unfolded.

Denatured ribonuclease

3 Layer mixture from step 2 atop a chromatography column. Beads in the column allow ribonuclease to escape, while β-mercaptoethanol and urea are retained. Collect ribonuclease in a test tube and measure its ability to degrade RNA.

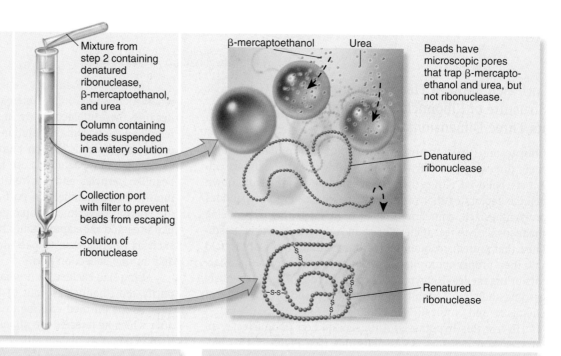

Mixture from step 2 containing denatured ribonuclease, β-mercaptoethanol, and urea

Column containing beads suspended in a watery solution

Collection port with filter to prevent beads from escaping

Solution of ribonuclease

β-mercaptoethanol   Urea

Beads have microscopic pores that trap β-mercapto-ethanol and urea, but not ribonuclease.

Denatured ribonuclease

Renatured ribonuclease

4 **THE DATA**

Activity restored

Ribonuclease function (%)

100

50

0

Purified ribonuclease (step 1)

Denatured ribonuclease (step 2)

Ribonuclease after column chromatography (step 3)

5 **CONCLUSION** Certain proteins, like ribonuclease, can spontaneously fold into their final, functional shapes without assistance from other cellular structures or factors. (However, as described in your text, this is not true of many other proteins.)

6 **SOURCE** Haber, E., and Anfinsen, C.B. 1961. Regeneration of enzyme activity by air oxidation of reduced subtilisin-modified ribonuclease. *Journal of Biological Chemistry* 236:422–424.

compelling evidence that the primary structure of a polypeptide is the key determinant of a protein's tertiary structure, an observation that earned him a Nobel Prize in 1972.

As investigations into the properties of proteins have continued since Anfinsen's classic experiments, it has become clear that most proteins contain within their structure one or more substructures, or domains, each of which is folded into a characteristic shape that imparts special functions to that region of the protein. This knowledge has greatly changed scientists' understanding of the ways in which proteins function and interact, as described next.

Experimental Questions

1. Before the experiments conducted by Anfinsen, what were the common beliefs among scientists about protein folding?

2. Explain the hypothesis tested by Anfinsen.

3. Why did Anfinsen use urea and β-mercaptoethanol in his experiments? Explain the result that was crucial to the discovery that the tertiary structure of ribonuclease may depend entirely on the primary structure.

Genomes & Proteomes Connection

Proteins Contain Functional Domains Within Their Structures

Modern research into the functions of proteins has revealed that many proteins have a modular design. This means that portions within proteins, called modules, motifs, or **domains**, have dis-

tinct structures and functions. These units of amino acid sequences have been duplicated during evolution so that the same kind of domain may be found in several different proteins. When the same domain is found in different proteins, the domain has the same three-dimensional shape and performs a function that is characteristic of that domain.

As an example, **Figure 3.21** shows a member of a family of related proteins that are known to play critical roles in regulating how certain genes are turned on and off in living cells. This

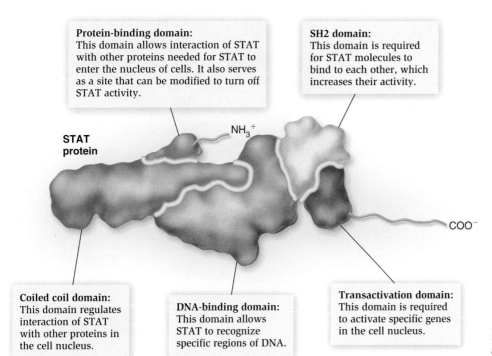

Protein-binding domain:
This domain allows interaction of STAT with other proteins needed for STAT to enter the nucleus of cells. It also serves as a site that can be modified to turn off STAT activity.

SH2 domain:
This domain is required for STAT molecules to bind to each other, which increases their activity.

STAT protein

NH_3^+

COO^-

Coiled coil domain:
This domain regulates interaction of STAT with other proteins in the cell nucleus.

DNA-binding domain:
This domain allows STAT to recognize specific regions of DNA.

Transactivation domain:
This domain is required to activate specific genes in the cell nucleus.

Figure 3.21   The domain structure of a STAT protein.

protein bears the cumbersome name of s̲ignal t̲ransducer and a̲ctivator of t̲ranscription (STAT) protein.

Each domain of this protein is involved in a distinct biological function, a common occurrence in proteins with multiple domains. For example, one of the domains is labeled the SH2 domain (Figure 3.21). Many different proteins contain this domain. It allows such proteins to recognize other proteins in a very specific way. The function of SH2 domains is to bind to tyrosine amino acids to which phosphate groups have been added by cellular enzymes. When an amino acid receives a phosphate group in this way, it is said to be phosphorylated (as is the protein in which the tyrosine exists). As might be predicted, proteins that contain SH2 domains all bind to phosphorylated tyrosines in the proteins they recognize.

As a second example, a STAT protein has another domain called a DNA-binding domain. This portion of the protein has a structure that specifically binds to DNA. Overall, the domain structure of proteins enables them to have multiple, discrete regions, each with its own structure and purpose in the functioning of the protein.

3.6 Nucleic Acids

Nucleic acids account for only about 2% of the weight of animals like ourselves, yet these molecules are extremely important because they are responsible for the storage, expression, and transmission of genetic information. The expression of genetic information in the form of specific proteins determines whether one is a human, a frog, an onion, or a bacterium. Likewise, genetic information determines whether a cell is part of a muscle or a bone, a leaf or a root.

Nucleic Acids Are Polymers Made of Nucleotides

The two classes of nucleic acids are **deoxyribonucleic acid** (**DNA**) and **ribonucleic acid** (**RNA**). DNA molecules store genetic information coded in the sequence of their monomer building blocks. RNA molecules are involved in decoding this information into instructions for linking a specific sequence of amino acids to form a polypeptide chain. The monomers in DNA must be arranged in a precise way so that the correct code can be read. As an analogy, think of the difference in the meanings of the words "marital" and "martial," in which the sequence of two letters is altered.

Like other macromolecules, both types of nucleic acids are polymers and consist of linear sequences of repeating monomers. Each monomer, known as a **nucleotide**, has three components: a phosphate group, a pentose (five-carbon) sugar (either ribose or deoxyribose), and a single or a double ring of carbon and nitrogen atoms known as a **base** (Figure 3.22). Nucleotides in a DNA strand are covalently held together by phosphodiester linkages between adjacent phosphate and sugar molecules, with the bases protruding from the side of the phosphate-sugar backbone (Figure 3.23).

DNA Is Composed of Purines and Pyrimidines

The nucleotides in DNA contain the five-carbon sugar **deoxyribose**. Four different nucleotides are present in DNA, corresponding to the four different bases that can be linked to deoxyribose. The **purine** bases, **adenine** (**A**) and **guanine** (**G**), have double rings of carbon and nitrogen atoms, and the **pyrimidine** bases, **cytosine** (**C**) and **thymine** (**T**), have a single ring (Figure 3.23).

A DNA molecule consists of two strands of nucleotides coiled around each other to form a double helix (Figure 3.24).

Phosphate

Base (uracil)

Sugar (ribose)

OH

Example of a ribonucleotide

Phosphate

Base (cytosine)

Sugar (deoxyribose)

OH

Example of a deoxyribonucleotide

Figure 3.22   **Examples of two nucleotides.**   A nucleotide has a phosphate group, a five-carbon sugar, and a nitrogenous base.

Purines and pyrimidines occur in both strands. The two strands are held together by hydrogen bonds between a purine base in one strand and a pyrimidine base in the opposite strand. The ring structure of each base lies in a flat plane perpendicular to the sugar-phosphate backbone, somewhat like steps on a spiral staircase. This base pairing maintains a constant distance between the sugar-phosphate backbones of the two strands as they coil around each other.

As we will see in Chapter 11, only certain bases can pair with others, due to the location of the hydrogen-bonding groups in the four bases (Figure 3.24). Two hydrogen bonds can be formed between adenine and thymine (A-T pairing), while three hydrogen bonds are formed between guanine and cytosine (G-C pairing). In a DNA molecule, A on one strand is always paired with T on another strand, and G with C. If we know the amount of one type of base in a DNA molecule, we can predict the relative amounts of each of the other three bases. For example, if a DNA molecule were composed of 20% A bases, then there must also be 20% T bases. That leaves 60% of the bases that must be G and C combined. Because the amounts of G and C must be equal, this particular DNA molecule must be composed of 30% each of G and C. This specificity provides the mechanism for duplicating and transferring genetic information (see Chapter 11).

RNA Is Usually Single Stranded and Comes in Several Forms

RNA molecules differ in only a few respects from DNA. Except in some viruses, RNA consists of a single rather than double strand of nucleotides. In RNA, the sugar in each nucleotide is **ribose** rather than deoxyribose. Also, the pyrimidine base

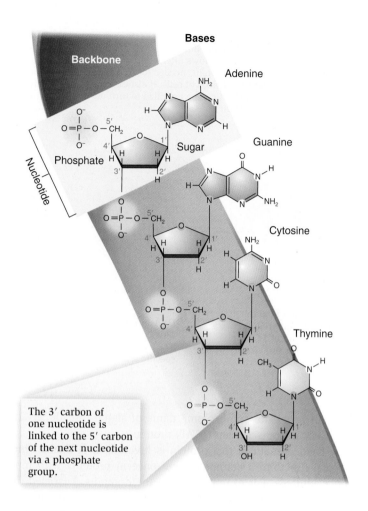

The 3′ carbon of one nucleotide is linked to the 5′ carbon of the next nucleotide via a phosphate group.

Figure 3.23   **Structure of a DNA strand.**   Nucleotides are linked to each other to form a strand of DNA. The four bases found in DNA are shown. A strand of RNA would be similar except the sugar would be ribose, and uracil would be substituted for thymine.

thymine in DNA is replaced in RNA with the pyrimidine base **uracil** (**U**) (see Figure 3.22). The other three bases—adenine, guanine, and cytosine—are found in both DNA and RNA. Certain forms of RNA called messenger RNA (mRNA), ribosomal RNA (rRNA), and transfer RNA (tRNA) are responsible for converting the information contained in DNA into the formation of a new polypeptide. This topic will be discussed in Chapter 12.

▌ Summary of Key Concepts

3.1   The Carbon Atom and the Study of Organic Molecules

- Organic chemistry is the science of studying carbon-containing molecules, which are found in living organisms. (Figure 3.1)

- One property of the carbon atom that makes life possible is its ability to form four covalent bonds (polar or nonpolar) with other atoms. The combination of different elements and different types of bonds allows a vast number of organic

Figure 3.24 **The double-stranded structure of DNA.** DNA consists of two strands coiled together into a double helix. The bases form hydrogen bonds in which A pairs with T, and G pairs with C.

Concept check: *If the sequence of bases in one strand of a DNA double helix is known, can the base sequence of the opposite strand be predicted?*

compounds to be formed from only a few chemical elements. (Figures 3.2, 3.3)

- Organic molecules may occur in various shapes. The structures of molecules determine their functions.

- Carbon bonds are stable at the different temperatures associated with life.

- Organic compounds may contain functional groups. (Table 3.1)

- Carbon-containing molecules can exist as isomers, which have identical molecular composition but different structures and characteristics. Structural isomers contain the same atoms but in different bonding relationships. Stereoisomers have identical bonding relationships but different spatial positioning of their atoms. Two types of stereoisomers are *cis-trans* isomers and enantiomers. (Figure 3.4)

3.2 Formation of Organic Molecules and Macromolecules

- The four major classes of organic molecules are carbohydrates, lipids, proteins, and nucleic acids. Organic molecules exist as monomers or polymers. Polymers are large macromolecules built up by dehydration reactions in which individual monomers combine with each other. Polymers are broken down into monomers by hydrolysis reactions. (Figure 3.5)

3.3 Carbohydrates

- Carbohydrates are composed of carbon, hydrogen, and oxygen atoms. Cells can break down carbohydrates, releasing energy and forming bonds in ATP.

- Carbohydrates include monosaccharides (the simplest sugars), disaccharides, and polysaccharides. The polysaccharides starch (in plant cells) and glycogen (in animal cells) are energy stores. The plant polysaccharide cellulose serves a support or structural function. (Figures 3.6, 3.7, 3.8)

3.4 Lipids

- Lipids, composed predominantly of hydrogen and carbon atoms, are nonpolar and very insoluble in water. Major classes of lipids include fats, phospholipids, steroids, and waxes.

- Fats, also called triglycerides and triacylglycerols, are formed by bonding glycerol with three fatty acids. In a saturated fatty acid, all the carbons are linked by single covalent bonds. Unsaturated fatty acids contain one or more C=C double bonds. Animal fats generally contain a high proportion of saturated fatty acids, and vegetable fats contain more unsaturated fatty acids. (Figures 3.9, 3.10, 3.11)

- Phospholipids are similar in structure to triglycerides, except they are amphipathic because one fatty acid is replaced with a charged polar group that includes a phosphate group. (Figure 3.12)

- Steroids are constructed of four fused rings of carbon atoms. Small differences in steroid structure can lead to profoundly different biological properties, such as the differences between estrogen and testosterone. (Figure 3.13)

- Waxes, another class of lipids, are nonpolar and repel water, and they are often found as protective coatings on the leaves of plants and the outer surfaces of animals' bodies.

3.5 Proteins

- Proteins are composed of carbon, hydrogen, oxygen, nitrogen, and small amounts of other elements, such as sulfur. Proteins are macromolecules that play critical roles in almost all life processes. The proteins of all living organisms are composed of the same set of 20 amino acids, corresponding to 20 different side chains. (Figure 3.14, Table 3.2)

- Amino acids are joined by linking the carboxyl group of one amino acid to the amino group of another, forming a peptide bond. A polypeptide is a structural unit composed of amino acids. A protein is a functional unit composed of one or more polypeptides that have been folded and twisted into precise three-dimensional shapes. (Figure 3.15)

- The four levels of protein structure are primary (its amino acid sequence), secondary (α helices or β pleated sheets), tertiary (folding to assume a three-dimensional shape), and quaternary (multimeric proteins that consist of more than one polypeptide chain). The three-dimensional structure of a protein determines its function—for example, by creating binding sites for other molecules. (Figures 3.16, 3.17, 3.18, 3.19, 3.20, 3.21)

3.6 Nucleic Acids

- Nucleic acids are responsible for the storage, expression, and transmission of genetic information. The two types of nucleic acids are deoxyribonucleic acid (DNA) and ribonucleic acid (RNA). (Figures 3.22, 3.23)

- DNA molecules store genetic information coded in the sequence of their monomers. A DNA molecule consists of two strands of nucleotides coiled around each other to form a double helix, held together by hydrogen bonds between a purine base on one strand and a pyrimidine base on the opposite strand. (Figure 3.24)

- RNA molecules are involved in decoding this information into instructions for linking amino acids in a specific sequence to form a polypeptide chain. RNA consists of a single strand of nucleotides. The sugar in each nucleotide is ribose rather than deoxyribose, and the base uracil replaces thymine.

Assess and Discuss

Test Yourself

1. Molecules that contain the element _____ are considered organic molecules.
   a. hydrogen   d. nitrogen
   b. carbon   e. calcium
   c. oxygen

2. _____ was the first scientist to synthesize an organic molecule. The organic molecule synthesized was _____.
   a. Kolbe, urea   d. Kolbe, acetic acid
   b. Wöhler, urea   e. Wöhler, glucose
   c. Wöhler, acetic acid

3. The versatility of carbon to serve as the backbone for a variety of different molecules is due to
   a. the ability of carbon atoms to form four covalent bonds.
   b. the fact that carbon usually forms ionic bonds with many different atoms.
   c. the abundance of carbon in the environment.
   d. the ability of carbon to form covalent bonds with many different types of atoms.
   e. both a and d.

4. _____ are molecules that have the same molecular composition but differ in structure and/or bonding association.
   a. Isotopes   d. Analogues
   b. Isomers   e. Ions
   c. Free radicals

5. _____ is a storage polysaccharide commonly found in the cells of animals.
   a. Glucose   d. Starch
   b. Sucrose   e. Cellulose
   c. Glycogen

6. In contrast to other fatty acids, essential fatty acids
   a. are always saturated fats.
   b. cannot be synthesized by the organism and are necessary for survival.
   c. can act as building blocks for large, more complex macromolecules.
   d. are the simplest form of lipids found in plant cells.
   e. are structural components of plasma membranes.

7. Phospholipids are amphipathic, which means they
   a. are partially hydrolyzed during cellular metabolism.
   b. are composed of a hydrophilic portion and a hydrophobic portion.
   c. may be poisonous to organisms if in combination with certain other molecules.
   d. are molecules composed of lipids and proteins.
   e. are all of the above.

8. The monomers of proteins are _____, and these are linked by polar covalent bonds commonly referred to as _____ bonds.
   a. nucleotides, peptide   d. amino acids, peptide
   b. amino acids, ester   e. monosaccharides, glycosidic
   c. hydroxyl groups, ester

9. The _____ of a nucleotide determines whether it is a component of DNA or a component of RNA.
   a. phosphate group   d. fatty acid
   b. five-carbon sugar   e. Both b and d are correct.
   c. side chain

10. A _____ is a portion of protein with a particular structure and function.
    a. peptide bond   d. wax
    b. domain   e. monosaccharide
    c. phospholipid

Conceptual Questions

1. Define isomers.

2. List the four classes of organic molecules; give a function of each.

3. Explain the difference between saturated and unsaturated fatty acids.

Collaborative Questions

1. Discuss the differences between different types of carbohydrates.

2. Discuss some of the roles that proteins play in organisms.

Online Resource

www.brookerbiology.com

Stay a step ahead in your studies with animations that bring concepts to life and practice tests to assess your understanding. Your instructor may also recommend the interactive ebook, individualized learning tools, and more.

General Features of Cells

A cell from the pituitary gland. The cell in this micrograph was viewed by a technique called transmission electron microscopy, which is described in this chapter. The micrograph was artificially colored using a computer to enhance the visualization of certain cell structures.

Emily had a persistent cough ever since she started smoking cigarettes in college. However, at age 35, it seemed to be getting worse, and she was alarmed by the occasional pain in her chest. When she began to lose weight and noticed that she became easily fatigued, Emily decided to see a doctor. The diagnosis was lung cancer. Despite aggressive treatment of the disease with chemotherapy and radiation therapy, she succumbed to lung cancer 14 months after the initial diagnosis. Emily was 36.

Topics such as cancer are within the field of **cell biology**—the study of individual cells and their interactions with each other. Researchers in this field want to understand the basic features of cells and apply their knowledge in the treatment of diseases such as cystic fibrosis, sickle-cell disease, and lung cancer.

The idea that organisms are composed of cells originated in the mid-1800s. German botanist Matthias Schleiden studied plant material under the microscope and was struck by the presence of many similar-looking compartments, each of which contained a dark area. Today we call those compartments cells, and the dark area is the nucleus. In 1838, Schleiden speculated that cells are living entities and plants are aggregates of cells arranged according to definite laws.

Schleiden was a good friend of the German physiologist Theodor Schwann. Over dinner one evening, their conversation turned to the nuclei of plant cells, and Schwann remembered having seen similar structures in animal tissue. Schwann conducted additional studies that showed large numbers of nuclei in animal tissue at regular intervals and also located in cell-like compartments. In 1839, Schwann extended Schleiden's hypothesis to animals. About two decades later, German biologist Rudolf Virchow proposed that *omnis cellula e cellula*, or "every cell originates from another cell." This idea arose from his research, which showed that diseased cells divide to produce more diseased cells.

The **cell theory**, or **cell doctrine**, which is credited to both Schleiden and Schwann with contributions from Virchow, has three parts.

1. *All living organisms are composed of one or more cells.*
2. *Cells are the smallest units of life.*
3. *New cells come only from pre-existing cells by cell division.*

Most cells are so small they cannot be seen with the naked eye. However, as cell biologists have begun to unravel cell structure and function at the molecular level, the cell has emerged as a unit of wonderful complexity and adaptability. In this chapter, we will begin our examination of cells with an overview of their structures and functions. Later chapters in this unit will explore certain aspects of cell biology in greater detail. But first, let's look at the tools and techniques that allow us to observe cells.

4.1 Microscopy

The **microscope** is a magnification tool that enables researchers to study the structure and function of cells. A **micrograph** is an image taken with the aid of a microscope. The first compound microscope—a microscope with more than one lens—was invented in 1595 by Zacharias Jansen of Holland. In 1665, an English biologist, Robert Hooke, studied cork under a primitive compound microscope he had made. He actually observed cell

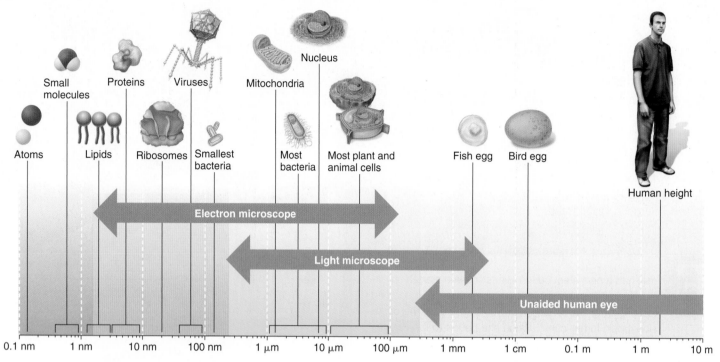

Figure 4.1 A comparison of the sizes of chemical and biological structures, and the resolving power of the naked eye, light microscope, and electron microscope.   The scale at the bottom is logarithmic to accommodate the wide range of sizes in this drawing.

Concept check: *Which type of microscope would you use to observe a virus?*

walls because cork cells are dead and have lost their internal components. Hooke coined the word cell, derived from the Latin word *cellula*, meaning small compartment, to describe the structures he observed. Ten years later, the Dutch merchant Anton van Leeuwenhoek refined techniques of making lenses and was able to observe single-celled microorganisms such as bacteria.

Three important parameters in microscopy are resolution, contrast, and magnification. **Resolution**, a measure of the clarity of an image, is the ability to observe two adjacent objects as distinct from one another. For example, a microscope with good resolution enables a researcher to distinguish two adjacent chromosomes as separate objects, which would appear as a single object under a microscope with poor resolution. The second important parameter in microscopy is **contrast**. The ability to visualize a particular cell structure may depend on how different it looks from an adjacent structure. If the cellular structure of interest can be specifically stained with a colored dye, this makes viewing much easier. The application of stains, which selectively label individual components of the cell, greatly improves contrast. As described later, fluorescent molecules are often used to selectively stain cellular components. However, staining should not be confused with colorization. Many of the micrographs shown in this textbook are colorized to emphasize certain cellular structures (see the chapter opener, for example). In colorization, particular colors are added to micrographs with the aid of a computer. This is done for educational purposes. For example, colorization can help to emphasize different parts of a cell. Finally, **magnification** is the ratio between the size of an image produced by a

microscope and its actual size. For example, if the image size is 100 times larger than its actual size, the magnification is designated 100X. Depending on the quality of the lens and illumination source, every microscope has an optimal range of magnification before objects appear too blurry to be readily observed.

Microscopes are categorized into two groups based on the source of illumination. A **light microscope** utilizes light for illumination, whereas an **electron microscope** uses electrons for illumination. Very good light microscopes can resolve structures that are as close as 0.2 µm (micron, or micrometer) from each other. The resolving power of a microscope depends on several factors, including the wavelength of the source of illumination. Resolution is improved when the illumination source has a shorter wavelength. A major advance in microscopy occurred in 1931 when Max Knoll and Ernst Ruska invented the first electron microscope. Because the wavelength of an electron beam is much shorter than visible light, the resolution of the electron microscope is far better than any light microscope. For biological samples, the resolution limit is typically around 2 nm (nanometers), which is about 100 times better than the light microscope. **Figure 4.1** shows the range of resolving powers of the electron microscope, light microscope, and unaided eye and compares them to various cells and cell structures.

Over the past several decades, enormous technological advances have made light microscopy a powerful research tool. Improvements in lens technology, microscope organization, sample preparation, sample illumination, and computerized image processing have enabled researchers to create different

Standard light microscope (bright field, unstained sample). Light is passed directly through a sample, and the light is focused using glass lenses. Simple, inexpensive, and easy to use but offers little contrast with unstained samples.

Phase contrast microscope. As an alternative to staining, this microscope controls the path of light and amplifies differences in the phase of light transmitted or reflected by a sample. The dense structures appear darker than the background, thereby improving the contrast in different parts of the specimen. Can be used to view living, unstained cells.

Differential-interference-contrast (Nomarski) microscope. Similar to a phase contrast microscope in that it uses optical modifications to improve contrast in unstained specimens. Can be used to visualize the internal structures of cells, and is commonly used to view whole cells or larger cell structures such as nuclei.

(a) Light microscopy on unstained samples

Standard (wide-field) fluorescence microscope. Fluorescent molecules specifically label a particular type of cellular protein or organelle. A fluorescent molecule absorbs light at a particular wavelength and emits light at a longer wavelength. This microscope has filters that illuminate the sample with the wavelength of light that a fluorescent molecule absorbs, and then only the light that is emitted by the fluorescent molecules is allowed to reach the observer. To detect their cellular location, researchers often label specific cellular proteins using fluorescent antibodies that bind specifically to a particular protein.

Confocal fluorescence microscope. Uses lasers that illuminate various points in the sample. These points are processed by a computer to give a very sharp focal plane. In this example, this microscope technique is used in conjunction with fluorescence microscopy to view fluorescent molecules within a cell.

(b) Fluorescence microscopy

Figure 4.2 **Examples of light microscopy.** (a) These micrographs compare three microscopic techniques on the same unstained sample of cells. These cells are endothelial cells that line the interior surface of arteries in the lungs. (b) These two micrographs compare standard (wide-field) fluorescence microscopy with confocal fluorescence microscopy. The sample is a section through a mouse intestine, showing two villi, which are described in Chapter 45. In this sample, the nuclei are stained green, and the actin filaments (discussed later in this chapter) are stained red.

types of light microscopes, each with its own advantages and disadvantages (**Figure 4.2**).

Similarly, improvements in electron microscopy occurred during the 1930s and 1940s, and by the 1950s, the electron microscope was playing a major role in advancing our understanding of cell structure. Two general types of electron microscopy have been developed: transmission electron microscopy and scanning electron microscopy. In **transmission electron**

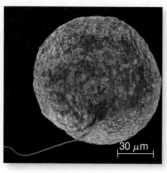

(a) Transmission electron
micrograph

(b) Scanning electron
micrograph

Figure 4.3 A comparison of transmission and scanning electron microscopy. (a) Section through a developing human egg cell, observed by TEM, shortly before it was released from an ovary. (b) An egg cell, with an attached sperm, was coated with heavy metal and observed via SEM. This SEM is colorized.

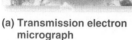 *Concept check:* What is the primary advantage of SEM?

microscopy (**TEM**), a beam of electrons is transmitted through a biological sample. In preparation for TEM, a sample is treated with a chemical that binds to cellular molecules and fixes them in place. The sample is placed in a liquid resin, and the resin polymerizes to form a hardened block. To view cells, the sample embedded within the block is sliced into very thin sections, typically less than 0.2 µm in thickness. To provide contrast, the sample is stained with a heavy metal. During staining, the metal binds to certain cellular structures such as membranes. The thin sections of the sample that have been stained with heavy metal are then adhered to a copper grid and placed in a transmission electron microscope. When the beam of electrons strikes the sample, some of them hit the heavy metal and are scattered, while those that pass through without being scattered are focused to form an image on a photographic plate or screen (**Figure 4.3a**). Because the scattered electrons are lost from the beam, the metal-stained regions of the sample that scatter electrons appear as darker areas, due to reduced electron penetration. TEM provides a cross-sectional view of a cell and its organelles and gives the greatest resolution compared with other forms of microscopy. However, such microscopes are expensive and are not used to view living cells.

Scanning electron microscopy (**SEM**) is used to view the surface of a sample. A biological sample is coated with a thin layer of heavy metal, such as gold or palladium, and then is exposed to an electron beam that scans its surface. Secondary electrons are emitted from the sample, which are detected and create an image of the three-dimensional surface of the sample (**Figure 4.3b**).

4.2   Overview of Cell Structure

Cell structure is primarily determined by four factors: (1) matter, (2) energy, (3) organization, and (4) information. In Chapters 2 and 3, we considered the first factor. The matter found in living organisms is composed of atoms, molecules, and macromolecules. Each type of cell synthesizes a unique set of molecules and macromolecules that contribute to cell structure.

We will discuss the second factor, energy, throughout this unit, particularly in Chapters 6 through 8. Energy is needed to produce molecules and macromolecules and to carry out many cellular functions.

The third phenomenon that underlies cell structure is organization. A cell is not a haphazard bag of components. The molecules and macromolecules that constitute cells have specific sites where they are found. For instance, if we compare the structure of a nerve cell in two different humans, or two nerve cells within the same individual, we would see striking similarities in their overall structures. All living cells have the ability to build and maintain their internal organization. **Protein-protein interactions** are critical to cell structure and function. Proteins often bind to each other in much the same way that building blocks snap together. These types of interactions can build complicated cell structures and also facilitate processes in which proteins interact in a consistent series of steps.

Finally, a fourth critical factor is information. Cell structure requires instructions. These instructions are found in the blueprint of life, namely the genetic material, which is discussed in Unit III. Every species has a distinctive **genome**, which is defined as the entire complement of its genetic material. Likewise, each living cell has a copy of the genome; the **genes** within each species' genome contain the information to create cells with particular structures and functions. This information is passed from cell to cell and from parent to offspring to yield new generations of cells and new generations of life. In this section, we will explore the general structure of cells and examine how the genome contributes to cell structure and function.

Prokaryotic Cells Have a Simple Structure

Based on cell structure, all forms of life can be placed into two categories called prokaryotes and eukaryotes. We will first consider the **prokaryotes**, which have a relatively simple structure. The term comes from the Greek *pro* and *karyon*, which means before a kernel—a reference to the kernel-like appearance of what would later be named the cell nucleus. Prokaryotic cells lack a membrane-enclosed nucleus.

From an evolutionary perspective, the two categories of prokaryotes are **bacteria** and **archaea**. Both types are microorganisms that are relatively small. Bacteria are abundant throughout the world, being found in soil, water, and even our digestive tracts. Most bacterial species are not harmful to humans, and they play vital roles in ecology. However, a few species are pathogenic—they cause disease. Examples of pathogenic bacteria include *Vibrio cholerae*, the source of cholera, and *Bacillus anthracis*, which causes anthrax. Archaea are also widely found throughout the world, though they are less common than bacteria and often occupy extreme environments such as hot springs and deep-sea vents.

Figure 4.4 shows a typical prokaryotic cell. The **plasma membrane**, which is a double layer of phospholipids and

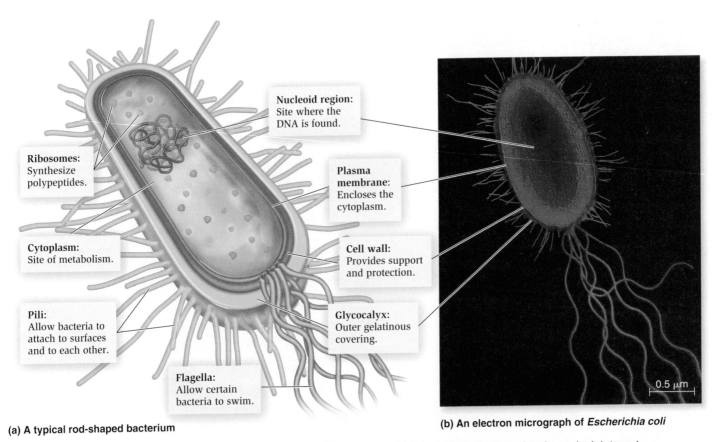

Ribosomes:
Synthesize
polypeptides.

Nucleoid region:
Site where the
DNA is found.

Cytoplasm:
Site of metabolism.

**Plasma
membrane:**
Encloses the
cytoplasm.

Cell wall:
Provides support
and protection.

Pili:
Allow bacteria to
attach to surfaces
and to each other.

Glycocalyx:
Outer gelatinous
covering.

Flagella:
Allow certain
bacteria to swim.

0.5 μm

(a) A typical rod-shaped bacterium

(b) An electron micrograph of _Escherichia coli_

Figure 4.4   Structure of a typical prokaryotic cell.   Prokaryotic cells, which include bacteria and archaea, lack internal compartmentalization.

embedded proteins, forms an important barrier between the cell and its external environment. The **cytoplasm** is the region of the cell contained within the plasma membrane. Certain structures in the bacterial cytoplasm are visible via microscopy. These include the **nucleoid region**, which is where its genetic material (DNA) is located, and **ribosomes**, which are involved in polypeptide synthesis.

Some bacterial structures are located outside the plasma membrane. Nearly all species of prokaryotes have a relatively rigid **cell wall** that supports and protects the plasma membrane and cytoplasm. The cell wall composition varies widely among prokaryotes but commonly contains peptides and carbohydrate. The cell wall, which is relatively porous, allows most nutrients in the environment to reach the plasma membrane. Many bacteria also secrete a **glycocalyx**, an outer viscous covering surrounding the bacterium. The glycocalyx traps water and helps protect bacteria from drying out. Certain strains of bacteria that invade animals' bodies produce a very thick, gelatinous glycocalyx called a **capsule** that may help them avoid being destroyed by the animal's immune (defense) system or may aid in the attachment to cell surfaces. Finally, many prokaryotes have appendages such as **pili** and **flagella**. Pili allow prokaryotes to attach to surfaces and to each other. Flagella provide prokaryotes with a way to move, also called motility.

Eukaryotic Cells Are Compartmentalized by Internal Membranes to Create Organelles

Aside from prokaryotes, all other species are **eukaryotes** (from the Greek, meaning true nucleus), which include protists, fungi, plants, and animals. Paramecia and algae are types of protists; yeasts and molds are types of fungi. **Figure 4.5** describes the morphology of a typical animal cell. Eukaryotic cells possess a true nucleus where most of the DNA is housed. A nucleus is a type of **organelle**—a membrane-bound compartment with its own unique structure and function. In contrast to prokaryotes, eukaryotic cells exhibit **compartmentalization**, which means they have many membrane-bound organelles that separate the cell into different regions. Cellular compartmentalization allows a cell to carry out specialized chemical reactions in different places. For example, protein synthesis and protein breakdown occur in different compartments in the cell.

Some general features of cell organization, such as a nucleus, are found in nearly all eukaryotic cells. Even so, be aware that the shape, size, and organization of cells vary considerably among different species and even among different cell types of the same species. For example, micrographs of a human skin cell and a human nerve cell show that, although these cells contain the same types of organelles, their overall morphologies are quite different (**Figure 4.6**).

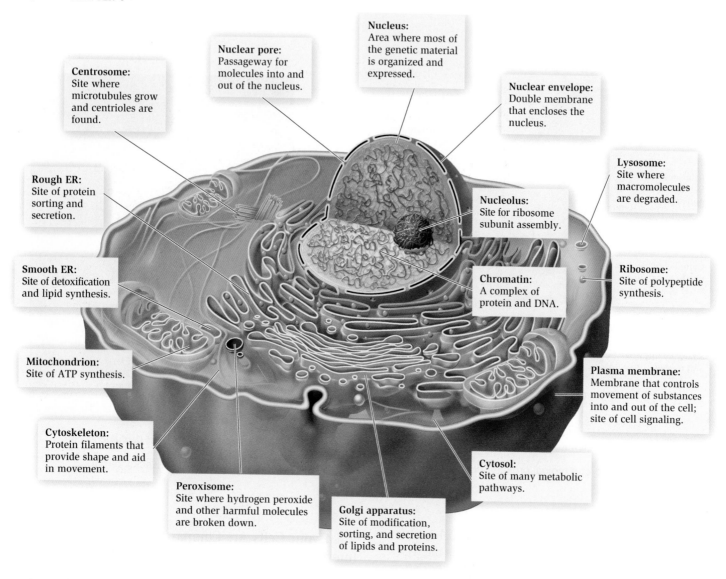

Centrosome: Site where microtubules grow and centrioles are found.

Nuclear pore: Passageway for molecules into and out of the nucleus.

Nucleus: Area where most of the genetic material is organized and expressed.

Nuclear envelope: Double membrane that encloses the nucleus.

Rough ER: Site of protein sorting and secretion.

Nucleolus: Site for ribosome subunit assembly.

Lysosome: Site where macromolecules are degraded.

Smooth ER: Site of detoxification and lipid synthesis.

Chromatin: A complex of protein and DNA.

Ribosome: Site of polypeptide synthesis.

Mitochondrion: Site of ATP synthesis.

Plasma membrane: Membrane that controls movement of substances into and out of the cell; site of cell signaling.

Cytoskeleton: Protein filaments that provide shape and aid in movement.

Cytosol: Site of many metabolic pathways.

Peroxisome: Site where hydrogen peroxide and other harmful molecules are broken down.

Golgi apparatus: Site of modification, sorting, and secretion of lipids and proteins.

Figure 4.5   General structure of an animal cell.

(a) Human skin cell   10 μm

(b) Human nerve cell   46 μm

Figure 4.6   **Variation in morphology of eukaryotic cells.**   Light micrographs of **(a)** a human skin cell and **(b)** a human nerve cell. Although these cells have the same genome and the same types of organelles, note that their general morphologies are quite different.

Concept check: *What is the underlying reason why skin and nerve cells have such different morphologies?*

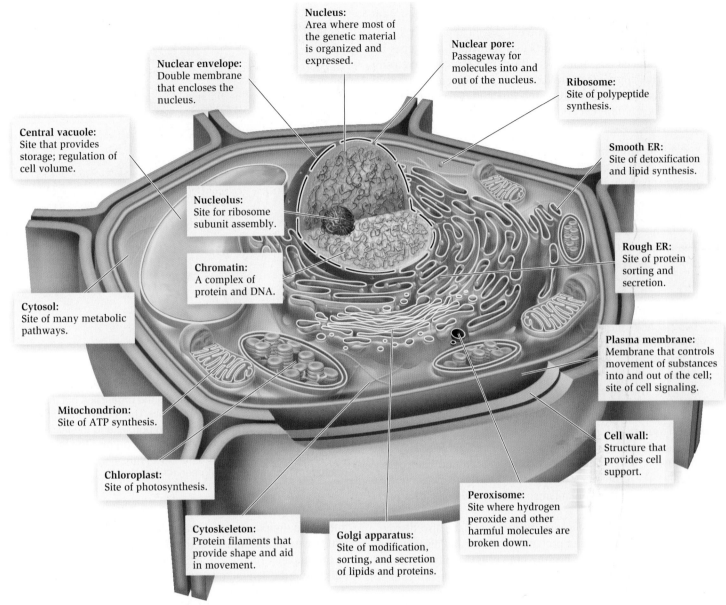

Nucleus:
Area where most of the genetic material is organized and expressed.

Nuclear pore:
Passageway for molecules into and out of the nucleus.

Ribosome:
Site of polypeptide synthesis.

Nuclear envelope:
Double membrane that encloses the nucleus.

Smooth ER:
Site of detoxification and lipid synthesis.

Central vacuole:
Site that provides storage; regulation of cell volume.

Nucleolus:
Site for ribosome subunit assembly.

Rough ER:
Site of protein sorting and secretion.

Chromatin:
A complex of protein and DNA.

Cytosol:
Site of many metabolic pathways.

Plasma membrane:
Membrane that controls movement of substances into and out of the cell; site of cell signaling.

Mitochondrion:
Site of ATP synthesis.

Cell wall:
Structure that provides cell support.

Chloroplast:
Site of photosynthesis.

Peroxisome:
Site where hydrogen peroxide and other harmful molecules are broken down.

Cytoskeleton:
Protein filaments that provide shape and aid in movement.

Golgi apparatus:
Site of modification, sorting, and secretion of lipids and proteins.

Figure 4.7   General structure of a plant cell.   Plant cells lack lysosomes and centrioles. Unlike animal cells, plant cells have an outer cell wall; a large central vacuole that functions in storage, digestion, and cell volume; and chloroplasts, which carry out photosynthesis.

Concept check: *What are the functions of the cell structures and organelles that are not found in both animal and plant cells?*

Plant cells possess a collection of organelles similar to animal cells (**Figure 4.7**). Additional structures found in plant cells but not animal cells include chloroplasts, a central vacuole, and a cell wall.

Genomes & Proteomes Connection

The Proteome Determines the Characteristics of a Cell

Many organisms, such as animals and plants, are multicellular, meaning that a single organism is composed of many cells. However, the cells of a multicellular organism are not all iden-

tical. For example, your body contains skin cells, nerve cells, muscle cells, and many other types. An intriguing question, therefore, is how does a single organism produce different types of cells?

To answer this question, we need to consider the distinction between genomes and proteomes. Recall that the genome constitutes all types of genetic material, namely DNA, that an organism has. Most genes encode the production of polypeptides, which assemble into functional proteins. An emerging theme discussed in this unit is that the structures and functions of proteins are primarily responsible for the structures and functions of cells. The **proteome** is defined as all of the types and relative amounts of proteins that are made in a particular cell at a particular time and under specific conditions. As an example, let's consider skin cells and nerve cells—two cell types that have dramatically different organization and structure (see Figure 4.6). In any particular individual, the genes in a human skin cell are identical to those in a human nerve cell. However, their proteomes are different. The proteome of a cell largely determines its structure and function. Several phenomena underlie the differences observed in the proteomes of different cell types.

1. *Certain proteins found in one cell type may not be produced in another cell type.* This phenomenon is due to differential gene regulation, discussed in Chapter 13.
2. *Two cell types may produce the same protein but in different amounts.* This is also due to gene regulation and to the rates at which a protein is synthesized and degraded.
3. *The amino acid sequences of particular proteins can vary in different cell types.* As discussed in Chapter 13, the mRNA from a single gene can produce two or more polypeptides with slightly different amino acid sequences via a process called alternative splicing.
4. *Two cell types may alter their proteins in different ways.* After a protein is made, its structure may be changed in a variety of ways. These include the covalent attachment of molecules such as phosphate and carbohydrate, and the cleavage of a protein to a smaller size.

These four phenomena enable skin and nerve cells to produce different proteomes and therefore different structures and functions. Likewise, the proteomes of skin and nerve cells differ from those of other cell types such as muscle and liver cells. Ultimately, the proteomes of cells are largely responsible for producing the traits of organisms, such as the color of a person's eyes.

During the last few decades, researchers have also discovered an association between proteome changes and disease. For example, the proteomes of healthy lung cells are different from the proteomes of lung cancer cells. Furthermore, the proteomes of cancer cells change as the disease progresses. One reason for studying cancer-cell proteomes is to improve the early detection of cancer by identifying proteins that are made in the early stages, when the disease is most treatable. In addition, information about the ways that the proteomes of cancer cells change may help researchers uncover new treatment options. A key challenge for biologists is to understand the synthesis and function of proteomes in different cell types and how proteome changes may lead to disease conditions.

4.3   The Cytosol

Thus far, we have focused on the general features of prokaryotic and eukaryotic cells. In the rest of this chapter, we will survey the various compartments of eukaryotic cells with a greater emphasis on structure and function. **Figure 4.8** highlights an animal and plant cell according to four different regions. We will start with the **cytosol** (shown in yellow), the region of a

(a) Animal cell

(b) Plant cell

Figure 4.8   Compartments within (a) animal and (b) plant cells.   The cytosol, which is outside the organelles but inside the plasma membrane, is shown in yellow. The membranes of the endomembrane system are shown in purple, and the fluid-filled interiors are pink. The peroxisome is dark purple. The interior of the nucleus is blue. Semiautonomous organelles are shown in orange (mitochondria) and green (chloroplasts).

eukaryotic cell that is outside the membrane-bound organelles but inside the plasma membrane. The other regions of the cell, which we will examine later in this chapter, include the interior of the nucleus (blue), the endomembrane system (purple and pink), and the semiautonomous organelles (orange and green). As in prokaryotes, the term cytoplasm refers to the region enclosed by the plasma membrane. This includes the cytosol and the organelles.

Though the amount varies among different types of cells, the cytosol is an aqueous environment that typically occupies about 20 to 50% of the total cell volume. In this section, we will consider the primary functions of the cytosol. First, it is the site of many chemical reactions that produce the materials that are necessary for life. Second, we will examine the structure and function of large protein filaments that provide organization to cells and allow cells to move.

Synthesis and Breakdown of Molecules Occur in the Cytosol

Metabolism is defined as the sum of the chemical reactions by which cells produce the materials and utilize the energy that are necessary to sustain life. Although specific steps of metabolism also occur in cell organelles, the cytosol is a central coordinating region for many metabolic activities of eukaryotic cells.

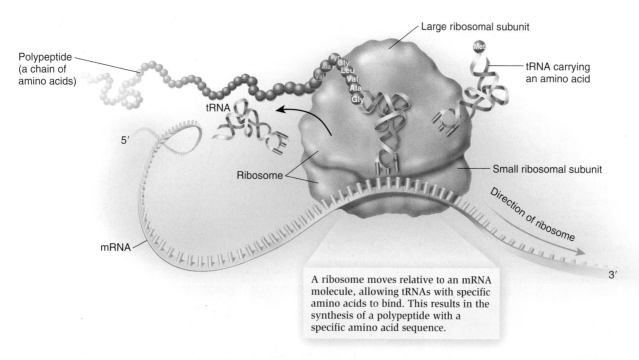

Large ribosomal subunit

Polypeptide (a chain of amino acids)

tRNA carrying an amino acid

tRNA

Met

Ala
Tyr
Gly
Leu
Val
Ala
Gly

5′

Ribosome

Small ribosomal subunit

mRNA

Direction of ribosome

3′

A ribosome moves relative to an mRNA molecule, allowing tRNAs with specific amino acids to bind. This results in the synthesis of a polypeptide with a specific amino acid sequence.

Figure 4.9   **Translation: the process of polypeptide synthesis.**   A ribosome is the site of polypeptide synthesis. It is composed of a small and large subunit. Messenger RNA (mRNA) provides the information for the amino acid sequence of a polypeptide.

Metabolism often involves a series of steps called a metabolic pathway. Each step in a metabolic pathway is catalyzed by a specific **enzyme**—a protein that accelerates the rate of a chemical reaction. In Chapters 6 and 7, we will examine the functional properties of enzymes and consider a few metabolic pathways that occur in the cytosol and cell organelles.

Some pathways involve the breakdown of a molecule into smaller components, a process termed **catabolism**. Such pathways are needed by the cell to utilize energy and also to generate molecules that provide the building blocks to construct cellular macromolecules. Conversely, other pathways are involved in **anabolism**, the synthesis of cellular molecules and macromolecules. For example, polysaccharides are made by linking sugar molecules. To create proteins, amino acids are covalently connected to form a polypeptide. An overview of this process, called translation, is shown in **Figure 4.9**. It is described in greater detail in Chapter 12. Translation occurs on ribosomes, which are found in various locations in the cell. Some ribosomes may float free in the cytosol, others are attached to the endoplasmic reticulum membrane, and still others are found within the mitochondria or chloroplasts.

The Cytoskeleton Provides Cell Shape, Organization, and Movement

The **cytoskeleton** is a network of three different types of protein filaments: **microtubules**, **intermediate filaments**, and **actin filaments** (**Table 4.1**). Each type is constructed from many protein monomers. The cytoskeleton is a striking example of protein-protein interactions. The cytoskeleton is found primarily in the cytosol and also in the nucleus along the inner nuclear membrane. Let's first consider the structure of cytoskeletal filaments and their roles in the construction and organization of cells. Later, we will examine how they are involved in cell movement.

Microtubules   Microtubules are long, hollow, cylindrical structures about 25 nm in diameter composed of subunits called α and β protein tubulin. The assembly of tubulin to form a microtubule results in a polar structure with a plus end and a minus end (Table 4.1). Microtubules grow at the plus end, although they can shorten at either the plus or minus end. A single microtubule can oscillate between growing and shortening phases, a phenomenon termed **dynamic instability**. Dynamic instability is important in many cellular activities, including the sorting of chromosomes during cell division.

The sites where microtubules form within a cell can vary among different types of organisms. Nondividing animal cells contain a single structure near their nucleus called the **centrosome**, or **microtubule-organizing center** (Table 4.1). Within the centrosome are the **centrioles**, a conspicuous pair of structures arranged perpendicular to each other. In animal cells, microtubule growth typically starts at the centrosome in such a way that the minus end is anchored there. In contrast, most plant cells and many protists lack centrosomes and centrioles. Microtubules are created at many sites that are scattered throughout a plant cell. In plants, the nuclear membrane appears to function as a microtubule-organizing center.

Table 4.1	Types of Cytoskeletal Filaments Found in Eukaryotic Cells		
Characteristic	**Microtubules**	**Intermediate filaments**	**Actin filaments**
Diameter	25 nm	10 nm	7 nm
Structure	Hollow tubule	Twisted filament	Spiral filament
Protein composition	Hollow tubule composed of the protein tubulin	Can be composed of different proteins including desmin, keratin, lamin, and others that form twisted filaments	Two intertwined strands composed of the protein actin
Common functions	Cell shape; organization of cell organelles; chromosome sorting in cell division; intracellular movement of cargo; cell motility (cilia and flagella)	Cell shape; provide cells with mechanical strength; anchorage of cell and nuclear membranes	Cell shape; cell strength; muscle contraction; intracellular movement of cargo; cell movement (amoeboid movement); cytokinesis in animal cells

Microtubules are important for cell shape and organization. Organelles such as the Golgi apparatus often are attached to microtubules. In addition, microtubules are involved in the organization and movement of chromosomes during mitosis and in the orientation of cells during cell division. We will examine these events in Chapter 15.

Intermediate Filaments Intermediate filaments are another class of cytoskeletal filament found in the cells of many but not all animal species. Their name is derived from the observation that they are intermediate in diameter between actin filaments and myosin filaments. (Myosin filaments are described in Chapter 44.) Intermediate filament proteins bind to each other in a staggered array to form a twisted, ropelike structure with a diameter of approximately 10 nm (Table 4.1). They function as tension-bearing fibers that help maintain cell shape and rigidity. Intermediate filaments tend to be relatively stable. By comparison, microtubules and actin filaments readily grow by the addition of more protein monomers and shorten by the loss of monomers.

Several types of proteins can assemble into intermediate filaments. Desmins form intermediate filaments in muscle cells and provide mechanical strength. Keratins form intermediate filaments in skin, intestinal, and kidney cells, where they are important for cell shape and mechanical strength. They are also a major constituent of hair and nails. In addition, intermediate filaments are found inside the cell nucleus. As discussed later in this chapter, nuclear lamins form a network of intermediate filaments that line the inner nuclear membrane and provide anchorage points for the nuclear pores.

Actin Filaments   Actin filaments—also known as **microfilaments** because they are the thinnest cytoskeletal filaments—are long, thin fibers approximately 7 nm in diameter (Table 4.1). Like microtubules, actin filaments have plus and minus ends, and they are very dynamic structures in which each strand grows at the plus end by the addition of actin monomers. This assembly process produces a fiber composed of two strands of actin monomers that spiral around each other.

Despite their thinness, actin filaments play a key role in cell shape and strength. Although actin filaments are dispersed throughout the cytosol, they tend to be highly concentrated near the plasma membrane. In many types of cells, actin filaments support the plasma membrane and provide shape and strength to the cell. The sides of actin filaments are often anchored to other proteins near the plasma membrane, which explains why actin filaments are typically found there. The plus ends grow toward the plasma membrane and can play a key role in cell shape and movement.

Motor Proteins Interact with Microtubules or Actin Filaments to Promote Movements

Motor proteins are a category of proteins that use ATP as a source of energy to promote various types of movements. As shown in **Figure 4.10a**, a motor protein consists of three domains called the head, hinge, and tail. The head is the site where ATP binds and is hydrolyzed to ADP and P_i. ATP binding and hydrolysis cause a bend in the hinge, which results in movement. The tail region is attached to other proteins or to other kinds of cellular molecules.

To promote movement, the head region of a motor protein interacts with a cytoskeletal filament (**Figure 4.10b**). When ATP binds and is hydrolyzed, the motor protein attempts to "walk" along the filament. The head of the motor protein is initially attached to a filament. To move forward, the head detaches from the filament, cocks forward, binds to the filament, and cocks backward. To picture how this works, consider the act of walking and imagine that the ground is a cytoskeletal filament, your leg is the head of the motor protein, and your hip is the hinge. To walk, you lift your leg up, you move it forward, you place it on the ground, and then you cock it backward (which propels you forward). This series of events is analogous to how a motor protein moves along a cytoskeletal filament.

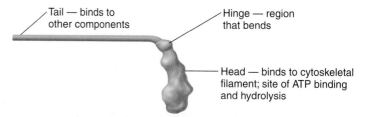

(a) Three-domain structure of a motor protein

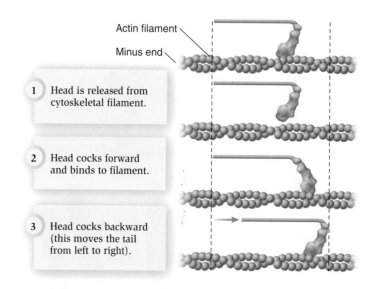

(b) Movement of a motor protein along a cytoskeletal filament

Figure 4.10   **Motor proteins and their interactions with cytoskeletal filaments.**   The example shown here is the motor protein myosin (discussed in Chapter 44), which interacts with actin filaments. **(a)** Three-domain structure of a motor protein. Note: The protein subunits of motor proteins often associate with each other along their tails, such that the motor has two tails, two hinges, and two heads. **(b)** Conformational changes in a motor protein that allow it to "walk" along a cytoskeletal filament.

Interestingly, cells have utilized the actions of motor proteins to promote three different kinds of movements: movement of cargo via the motor protein, movement of the filament, or bending of the filament. In the example shown in **Figure 4.11a**, the tail region of a motor protein called kinesin is attached to a cargo, so the motor protein moves the cargo from one location to another. Alternatively, a motor protein can remain in place and cause the filament to move (**Figure 4.11b**). As discussed in Chapter 44, this occurs during muscle contraction (see Figure 44.7). A third possibility is that both the motor protein and filament are restricted in their movement. In this case, when the motor proteins called dynein attempt to walk toward the minus end, they exert a force that causes microtubules to bend (**Figure 4.11c**). As described next, this occurs during the bending of flagella and cilia.

In certain kinds of cells, microtubules and motor proteins facilitate movement involving cell appendages called **flagella** and **cilia** (singular, flagellum and cilium). Flagella are usually longer than cilia and are found singly or in pairs. Both flagella and cilia cause movement by a bending motion. In flagella, movement occurs by a whiplike motion that is due to the propagation of a bend from the base to the tip. A single flagellum may

Motor proteins "walk" along a microtubule from the minus end to the plus end carrying a cargo.

Cargo

Motor protein (kinesin)

Microtubule

(a) Motor protein moves

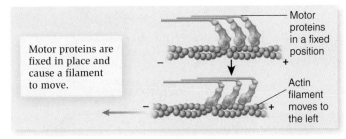

Motor proteins are fixed in place and cause a filament to move.

Motor proteins in a fixed position

Actin filament moves to the left

(b) Filament moves

Both the motor proteins and filaments are fixed in place so the actions of the motor proteins cause the microtubules to bend.

Motor protein (dynein)

Linking protein

(c) Filaments bend

Figure 4.11   Three ways that motor proteins and cytoskeletal filaments cause movement.

propel a cell such as a sperm cell with a whiplike motion (**Figure 4.12a**). Alternatively, a pair of flagella may move in a synchronized manner to pull a microorganism through the water (think of a human swimmer doing the breaststroke). Certain unicellular algae swim in this manner (**Figure 4.12b**). By comparison, cilia are often shorter than flagella and tend to cover all or part of the surface of a cell. Protists such as paramecia may have hundreds of adjacent cilia that beat in a coordinated fashion to propel the organism through the water (**Figure 4.12c**).

Despite their differences in length, flagella and cilia share the same internal structure called the **axoneme**. The axoneme contains microtubules, the motor protein dynein, and linking proteins (**Figure 4.13**). In the cilia and flagella of most eukaryotic organisms, the microtubules form an arrangement called a 9 + 2 array. The outer nine are doublet microtubules, which are composed of a partial microtubule attached to a complete microtubule. Each of the two central microtubules consists of a single microtubule. Radial spokes connect the outer doublet microtubules to the central pair. The microtubules in flagella and cilia emanate from **basal bodies**, which are anchored to the cytoplasmic side of the plasma membrane. At the basal body, the microtubules form a triplet structure. Much like the centrosome of animal cells, the basal bodies provide a site for microtubules to grow.

The movement of both flagella and cilia involves the propagation of a bend, which begins at the base of the structure and proceeds toward the tip (see Figure 4.12a). The bending occurs because dynein is activated to walk toward the minus end of the microtubules. ATP hydrolysis is required for this process. However, the microtubules and dynein are not free to move relative to each other because of linking proteins. Therefore, instead of dyneins freely walking along the microtubules, they exert a force that bends the microtubules (see Figure 4.11c). The dyneins at the base of the structure are activated first, followed by dyneins that are progressively closer to the tip of the appendage. The resulting movement propels the organism.

15 μm

(a) Time-lapse photography of a human sperm moving its flagellum

3 μm

(b) *Chlamydomonas* with 2 flagella

70 μm

(c) *Paramecium* with many cilia

Figure 4.12   Cellular movements due to the actions of flagella and cilia.   (a) Sperm swim by means of a single, long flagellum that moves in a whiplike motion, as shown by this human sperm. (b) The swimming of *Chlamydomonas reinhardtii* also involves a whiplike motion at the base, but the motion is precisely coordinated between two flagella. This results in swimming behavior that resembles a breaststroke. (c) Ciliated protozoa such as this *Paramecium* swim via many shorter cilia.

Concept check: *During the movement of a cilium or flagellum, describe the type of movements that are occurring between the motor proteins and microtubules.*

Figure 4.13 Structure of a eukaryotic cilium. (inset) SEM of a protist, *Tetrahymena themophila*. The core structure consists of a 9 + 2 arrangement of nine outer doublet microtubules and two central microtubules. This structure is anchored to the basal body, which has nine triplet microtubules, in which three microtubules are fused together. Note: The structure of the basal body is very similar to centrioles in animal cells.

4.4 The Nucleus and Endomembrane System

In Chapter 2, we learned that the nucleus of an atom contains protons and neutrons. In cell biology, the term **nucleus** has a different meaning. It is an organelle found in eukaryotic cells that contains most of the cell's genetic material. A small amount of genetic material is also found in mitochondria and chloroplasts.

The membranes that enclose the nucleus are part of a larger network of membranes called the **endomembrane system**. This system includes not only the nuclear envelope, which encloses the nucleus, but also the endoplasmic reticulum, Golgi apparatus, lysosomes, vacuoles, and peroxisomes. The prefix *endo* (from the Greek, meaning inside) originally referred only to these organelles and internal membranes. However, we now know that the plasma membrane is also part of this integrated membrane system (**Figure 4.14**). Some of these membranes, such as the nuclear envelope and the membrane of the

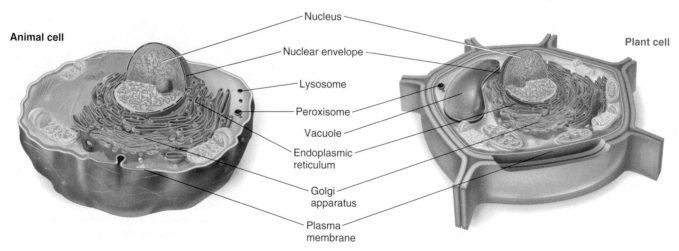

Figure 4.14 The nucleus and endomembrane system. This figure highlights the internal compartment of the nucleus (blue), the membranes of the endomembrane system (purple), and the fluid-filled interiors of the endomembrane system (pink). The nuclear envelope is considered part of the endomembrane system, but the interior of the nucleus is not.

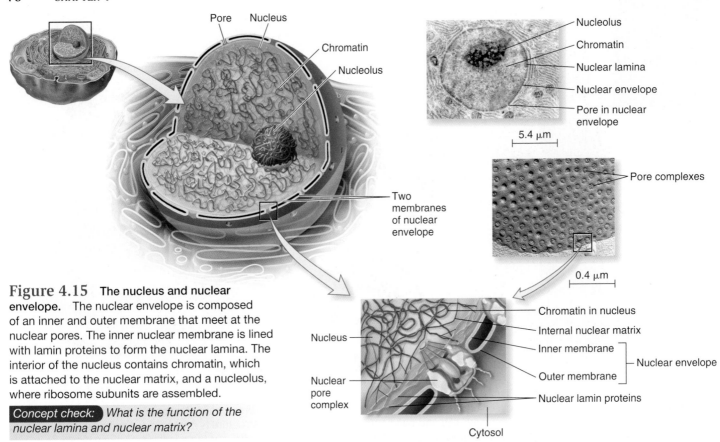

Figure 4.15 **The nucleus and nuclear envelope.** The nuclear envelope is composed of an inner and outer membrane that meet at the nuclear pores. The inner nuclear membrane is lined with lamin proteins to form the nuclear lamina. The interior of the nucleus contains chromatin, which is attached to the nuclear matrix, and a nucleolus, where ribosome subunits are assembled.

Concept check: *What is the function of the nuclear lamina and nuclear matrix?*

endoplasmic reticulum, have direct connections to one another. Other organelles of the endomembrane system pass materials to each other via **vesicles**—small membrane-enclosed spheres (look ahead to Figure 4.18). In this section, we will examine the nucleus and survey the structures and functions of the organelles and membranes of the endomembrane system.

The Eukaryotic Nucleus Contains Chromosomes

The nucleus is the internal compartment that is enclosed by a double-membrane structure termed the **nuclear envelope** (**Figure 4.15**). In most cells, the nucleus is a relatively large organelle that typically occupies 10–20% of the total cell volume. The outer membrane of the nuclear envelope is continuous with the endoplasmic reticulum membrane. **Nuclear pores** are formed where the inner and outer nuclear membranes make contact with each other. The pores provide a passageway for the movement of molecules and macromolecules into and out of the nucleus. Although cell biologists view the nuclear envelope as part of the endomembrane system, the materials within the nucleus are not (Figure 4.15).

Inside the nucleus are the chromosomes and a filamentous network of proteins called the nuclear matrix. Each **chromosome** is composed of genetic material, namely DNA, and many types of proteins that help to compact the chromosome to fit inside the nucleus. The complex formed by DNA and such proteins is termed **chromatin**. The **nuclear matrix** consists of two parts: the nuclear lamina, which is composed of intermediate filaments that line the inner nuclear membrane, and an internal nuclear matrix, which is connected to the lamina and fills

the interior of the nucleus. The nuclear matrix serves to organize the chromosomes within the nucleus. Each chromosome is located in a distinct, nonoverlapping **chromosome territory**, which is visible when cells are exposed to dyes that label specific types of chromosomes (**Figure 4.16**).

The primary function of the nucleus involves the protection, organization, replication, and expression of the genetic material. These topics are discussed in Unit III. Another important

Figure 4.16 **Chromosome territories in the cell nucleus.** Chromosomes from a chicken were labeled with chromosome-specific probes. Seven types of chicken chromosomes are colored with a different dye. Each chromosome occupies its own distinct, nonoverlapping territory within the cell nucleus. Reprinted by permission from Macmillan Publishers Ltd. Cremer, T., and Cremer, C. Chromosome territories, nuclear architecture and gene regulation in mammalian cells. *Nature Reviews/ Genetics*, Vol. 2(4), Figure 2, 292–301, 2001.

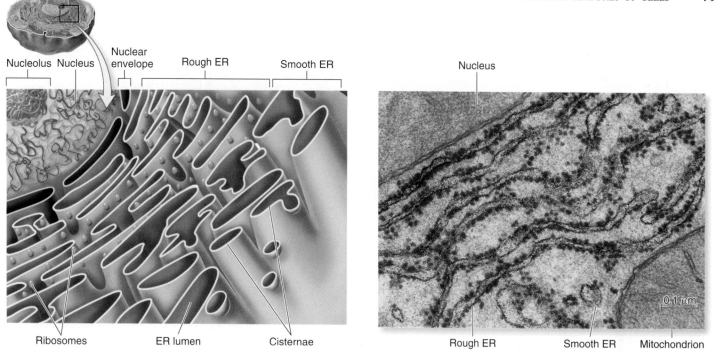

Figure 4.17 **Structure of the endoplasmic reticulum.** (Left side) The ER is composed of a network of flattened tubules called cisternae that enclose a continuous ER lumen. The rough ER is studded with ribosomes, whereas the smooth ER lacks ribosomes. The rough ER is continuous with the outer nuclear membrane. (Right side) A colorized TEM. The lumen of the ER is colored yellow and the ribosomes are red.

function is the assembly of ribosome subunits—cellular structures involved in producing polypeptides during the process of translation. The assembly of ribosome subunits occurs in the **nucleolus** (plural, nucleoli), a prominent region in the nucleus of nondividing cells. A ribosome is composed of two subunits, one small and one large (see Figure 4.9). Each subunit contains one or more RNA molecules and several types of proteins. Most of the RNA molecules that are components of ribosomes are made in the vicinity of the nucleolus. This occurs because the chromosomes that carry the genes that encode most types of ribosomal RNA molecules are located there. By comparison, the ribosomal proteins are produced in the cytosol and then imported into the nucleus through the nuclear pores. The ribosomal proteins and RNA molecules then assemble in the nucleolus to form the ribosomal subunits. Finally, the subunits exit through the nuclear pores into the cytosol, where they are needed for protein synthesis.

The Endoplasmic Reticulum Initiates Protein Sorting and Carries Out Certain Metabolic Functions

The **endoplasmic reticulum** (**ER**) is a network of membranes that form flattened, fluid-filled tubules, or **cisternae** (Figure 4.17). The terms endoplasmic (Greek, for in the cytoplasm) and reticulum (Latin, for little net) refer to the location and shape of this organelle when viewed under a microscope. The term **lumen** describes the internal space of an organelle. The ER membrane encloses a single compartment called the **ER lumen**. In some cells, the ER membrane makes up more than half of the total membrane in the cell. The rough ER has its outer surface studded with ribosomes, giving it a bumpy appearance. Once bound

to the ER membrane, the ribosomes actively synthesize proteins through the ER membrane. The smooth ER lacks ribosomes.

Rough ER The **rough endoplasmic reticulum** (**rough ER**) plays a key role in the sorting of proteins that are destined for the ER, Golgi apparatus, lysosomes, vacuoles, plasma membrane, or outside of the cell. This topic is described later in Section 4.6. In conjunction with protein sorting, a second function of the rough ER is the insertion of certain newly made proteins into the ER membrane. A third important function of the rough ER is the attachment of carbohydrate to proteins and lipids. This process is called **glycosylation**. The topics of membrane protein insertion and protein glycosylation will be discussed in Chapter 5, because they are important features of cell membranes.

Smooth ER The **smooth endoplasmic reticulum** (**smooth ER**), which is continuous with the rough ER, functions in diverse metabolic processes. The extensive network of smooth ER membranes provides an increased surface area for key enzymes that play important metabolic roles. In liver cells, enzymes in the smooth ER detoxify many potentially harmful organic molecules, including barbiturate drugs and ethanol. These enzymes convert hydrophobic toxic molecules into more hydrophilic molecules, which are easily excreted from the body. Chronic alcohol consumption, as in alcoholics, leads to a greater amount of smooth ER in liver cells, which increases the rate of alcohol breakdown. This explains why people who consume alcohol regularly must ingest more alcohol to experience its effects. It also explains why alcoholics often have enlarged livers.

The smooth ER of liver cells also plays a role in carbohydrate metabolism. The liver cells of animals store energy in

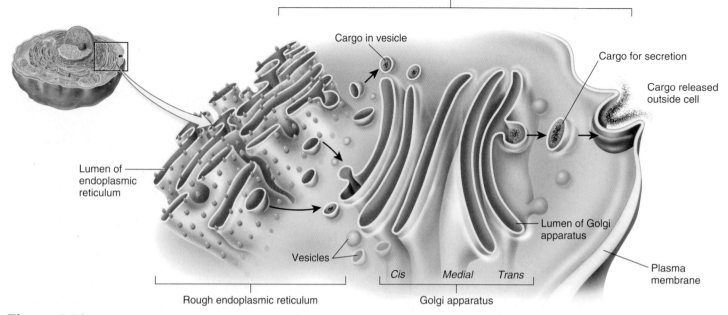

Figure 4.18   **The Golgi apparatus and secretory pathway.**   The Golgi is composed of stacks of membranes that enclose separate compartments. Transport to and from the Golgi compartments occurs via membrane vesicles. Vesicles can bud from the ER and go to the Golgi, and vesicles from the Golgi can fuse with the plasma membrane to release cargo to the outside. The pathway from the ER to the Golgi to the plasma membrane is termed the secretory pathway.

Concept check: *If we consider the Golgi apparatus as three compartments (cis, medial, and trans), describe the compartments that a protein will travel through to be secreted.*

the form of glycogen, which is a polymer of glucose. Glycogen granules, which are in the cytosol, sit very close to the smooth ER membrane. When chemical energy is needed, enzymes are activated that break down the glycogen to glucose-6-phosphate. Then, an enzyme in the smooth ER called glucose-6-phosphatase removes the phosphate group, and glucose is released into the bloodstream.

Another important function of the smooth ER in all eukaryotes is the accumulation of calcium ions. The smooth ER contains calcium pumps that transport Ca^{2+} into the ER lumen. The regulated release of Ca^{2+} into the cytosol is involved in many vital cellular processes, including muscle contraction in animals.

Finally, enzymes in the smooth ER are critical in the synthesis and modification of lipids. For example, steroid hormones such as estrogen and testosterone are derived from the lipid cholesterol. Enzymes in the smooth ER are necessary for certain modifications that are needed to produce these hormones. In addition, the smooth ER is the primary site for the synthesis of phospholipids, which are the main lipid component of eukaryotic cell membranes. This topic is described in Chapter 5.

The Golgi Apparatus Directs the Processing, Sorting, and Secretion of Cellular Molecules

The **Golgi apparatus** (also called the Golgi body, Golgi complex, or simply Golgi) was discovered by the Italian microscopist Camillo Golgi in 1898. It consists of a stack of flattened membranes; each flattened membrane encloses a single compartment. The Golgi stacks are named according to their orientation in the cell. The *cis* Golgi is close to the ER membrane, the *trans* Golgi is near the plasma membrane, and the *medial*

Golgi is found in the middle. Materials are transported between the Golgi stacks via membrane vesicles that bud from one compartment in the Golgi (for example, the *cis* Golgi) and fuse with another compartment (for example, the *medial* Golgi).

The Golgi apparatus performs three overlapping functions: (1) processing, (2) protein sorting, and (3) secretion. We will discuss protein sorting in Section 4.6. Enzymes in the Golgi apparatus modify, or process, certain proteins and lipids. As mentioned earlier, carbohydrates can be attached to proteins and lipids in the endoplasmic reticulum. Glycosylation continues in the Golgi. For this to occur, a protein or lipid is transported via vesicles from the ER to the *cis* Golgi. Most of the glycosylation occurs in the *medial* Golgi.

A second type of processing event is **proteolysis**, whereby enzymes called **proteases** cut proteins into smaller polypeptides. For example, the hormone insulin is first made as a large precursor protein termed proinsulin. In the Golgi apparatus, proinsulin is packaged with proteases into vesicles. The proteases cut out a portion of the proinsulin to create a smaller insulin molecule that is a functional hormone. This happens just prior to secretion, which is described next.

The Golgi apparatus packages different types of materials into **secretory vesicles** that later fuse with the plasma membrane, thereby releasing their contents outside the plasma membrane. Proteins destined for secretion are synthesized into the ER, travel to the Golgi, and then are transported by vesicles to the plasma membrane for secretion. The entire route is called the **secretory pathway** (**Figure 4.18**). The later stage in this process in which vesicles fuse with the plasma membrane is called **exocytosis**. This process can also run in reverse to take substances into the cell; this is called **endocytosis**. This topic is discussed further in Chapter 5.

FEATURE INVESTIGATION

Palade Demonstrated That Secreted Proteins Move Sequentially Through Organelles of the Endomembrane System

As we have seen, one of the key functions of the endomembrane system is protein secretion. The identification of the secretory pathway came from studies of George Palade and his colleagues in the 1960s. He hypothesized that proteins follow a particular intracellular pathway in order to be secreted. Palade's team conducted **pulse-chase experiments**, in which the researchers administered a pulse of radioactive amino acids to cells so they made radioactive proteins. A few minutes later, the cells were given a large amount of nonradioactive amino acids. This is called a "chase" because it chases away the ability of the cells to make any more radioactive proteins. In this way, radioactive proteins were produced only briefly. Because they were labeled with radioactivity, the fate of these proteins could be monitored over time. The goal of a pulse-chase experiment is to determine where the radioactive proteins are produced and the pathway they take as they travel through a cell.

Palade chose to study the cells of the pancreas. This organ secretes enzymes and protein hormones that play a role in digestion and metabolism. Therefore, these cells were chosen because their primary activity is protein secretion. To study the pathway for protein secretion, Palade and colleagues injected a radioactive version of the amino acid leucine into the blood-

stream of male guinea pigs. The radiolabeled leucine would travel in the bloodstream and be quickly taken up by cells of the body, including those in the pancreas. Three minutes later, they injected nonradioactive leucine (**Figure 4.19**). At various times after the second injection, samples of pancreatic cells were removed from the animals. The cells were then prepared for transmission electron microscopy. The sample was stained with osmium tetroxide, a heavy metal that became bound to membranes and showed the locations of the cell organelles. In addition, the sample was coated with a radiation-sensitive emulsion containing silver. When radiation was emitted from radioactive proteins, it interacted with the emulsion in a way that caused the precipitation of silver, which became tightly bound to the sample. In this way, the precipitated silver marked the location of the radiolabeled proteins. Unprecipitated silver in the emulsion was later washed away. Because silver atoms are electron dense, they produce dark spots in a transmission electron micrograph. Therefore, dark spots revealed the locations of radioactive proteins.

The micrograph in the data of Figure 4.19 illustrates the results that were observed 5 minutes after the completion of the pulse-chase injections. Very dark objects, namely radioactive proteins, were observed in the rough ER. As shown schematically to the right of the actual data, later time points indicated that the radioactive proteins moved from the ER to the Golgi, and then to secretory vesicles near the plasma membrane. In this way, Palade followed the intracellular pathway of protein movement. His

Figure 4.19   Palade's use of the pulse-chase method to study protein secretion.

HYPOTHESIS Proteins that are to be secreted follow a particular intracellular pathway.

KEY MATERIALS Male guinea pigs.

Experimental level   **Conceptual level**

1 Inject guinea pigs with a radioactive amino acid, [³H]-leucine. After 3 minutes, inject them with nonlabeled leucine, which is called a chase.

2 At various times after the second injection, remove samples of pancreatic cells.

3 Stain the sample with osmium tetroxide, which is a heavy metal that binds to membranes.

Osmium tetroxide

Sample from pancreas

4 Cut thin sections of the samples, and place a thin layer of radiation-sensitive emulsion over the sample. Allow time for radioactive emission from radiolabeled proteins to precipitate silver atoms in the emulsion.

Thin section

Add radiation-sensitive emulsion

5 Observe the sample under a transmission electron microscope.

6 THE DATA

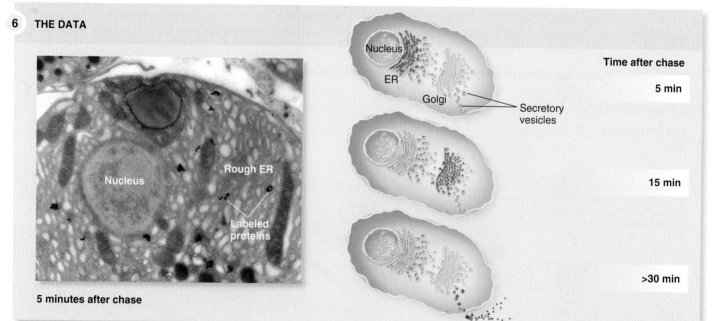

Nucleus

Rough ER

Labeled proteins

5 minutes after chase

Nucleus

ER

Golgi

Secretory vesicles

Time after chase

5 min

15 min

>30 min

7 **CONCLUSION** To be secreted, proteins move from the ER to the Golgi to secretory vesicles and then to the plasma membrane, where they are released to the outside of the cell.

8 **SOURCE** Caro, L.G., and Palade, G.E. 1964. Protein synthesis, storage, and discharge in the pancreatic exocrine cell. An autoradiographic study. *Journal of Cell Biology* 20:473–495.

experiments provided the first evidence that secreted proteins are synthesized into the rough ER and move through a series of cellular compartments before they are secreted. These findings also caused researchers to wonder how proteins are targeted to particular organelles and how they move from one compartment to another. These topics are described later in Section 4.6.

Experimental Questions

1. Explain the procedure of a pulse-chase experiment. What is the pulse, and what is the chase? What was the purpose of the approach?
2. Why were pancreatic cells used for this investigation?
3. What were the key results of the experiment of Figure 4.19? What did the researchers conclude?

Lysosomes Are Involved in the Intracellular Digestion of Macromolecules

We now turn to another organelle of the endomembrane system, **lysosomes**, which are small organelles found in animal cells that are able to lyse macromolecules. Lysosomes contain many **acid hydrolases**, which are hydrolytic enzymes that use a molecule of water to break a covalent bond. This type of chemical reaction is called hydrolysis:

$$R_1-R_2 + H_2O \xrightarrow{\text{Acid hydrolase}} R_1-OH + R_2-H$$

The hydrolases found in a lysosome function optimally at an acidic pH. The fluid-filled interior of a lysosome has a pH of approximately 4.8. If a lysosomal membrane breaks, releasing acid hydrolases into the cytosol, the enzymes are not very active because the cytosolic pH is neutral (approximately pH 7.0) and buffered. This prevents significant damage to the cell from accidental leakage.

Lysosomes contain many different types of acid hydrolases that can break down carbohydrates, proteins, lipids, and nucleic acids. This enzymatic function enables lysosomes to break down complex materials. One function of lysosomes involves the digestion of substances that are taken up from outside the cell via endocytosis. In addition, lysosomes help to break down cellular molecules and macromolecules to recycle their building blocks to make new molecules and macromolecules in a process called autophagy (see Chapter 6).

Vacuoles Are Specialized Compartments That Function in Storage, the Regulation of Cell Volume, and Degradation

The term **vacuole** (Latin, for empty space) came from early microscopic observations of these compartments. We now know that vacuoles are not empty but instead contain fluid and sometimes even solid substances. Most vacuoles are made from the fusion of many smaller membrane vesicles. Vacuoles are prominent organelles in plant cells, fungal cells, and certain protists. In animal cells, vacuoles tend to be smaller and are more commonly used to temporarily store materials or transport substances. In animals, such vacuoles are sometimes called storage vesicles.

The functions of vacuoles are extremely varied, and they differ among cell types and even environmental conditions. The best way to appreciate vacuole function is to consider a few examples. Mature plant cells often have a large **central vacuole** that occupies 80% or more of the cell volume (**Figure 4.20a**). The membrane of this vacuole is called the **tonoplast**. The central vacuole serves two important purposes. First, it stores a large amount of water, enzymes, and inorganic ions such as calcium; it also stores other materials including proteins and pigments. Second, it performs a space-filling function. The large size of the vacuole exerts a pressure on the cell wall, called turgor pressure. If a plant becomes dehydrated and this pressure is lost, a plant will wilt. Turgor pressure is important in maintaining the structure of plant cells and the plant itself, and it helps to drive the expansion of the cell wall, which is necessary for growth.

(a) Central vacuole in a plant cell

(b) Contractile vacuoles in an algal cell

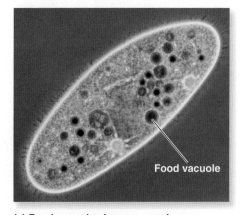

(c) Food vacuoles in a paramecium

Figure 4.20 Examples of vacuoles. These are transmission electron micrographs. Part (c) is colorized.

Certain species of protists also use vacuoles to maintain cell volume. Freshwater organisms such as the alga *Chlamydomonas reinhardtii* have small, water-filled **contractile vacuoles** that expand as water enters the cell (**Figure 4.20b**). Once they reach a certain size, the vacuoles suddenly contract, expelling their contents to the exterior of the cell. This mechanism is necessary to remove the excess water that continually enters the cell by diffusion across the plasma membrane.

Another function of vacuoles is degradation. Some protists engulf their food into large **phagocytic vacuoles**, or **food vacuoles** (**Figure 4.20c**). As in the lysosomes of animal cells, food vacuoles contain digestive enzymes to break down the macromolecules within the food. Macrophages, a type of cell found in animals' immune systems, engulf bacterial cells into phagocytic vacuoles, where the bacteria are destroyed.

Peroxisomes Catalyze Detoxifying Reactions

Peroxisomes, discovered by Christian de Duve in 1965, are relatively small organelles found in all eukaryotic cells. Peroxisomes consist of a single membrane that encloses a fluid-filled lumen. A typical eukaryotic cell contains several hundred of them.

The general function of peroxisomes is to catalyze certain chemical reactions, typically those that break down molecules by removing hydrogen or adding oxygen. In mammals, for example, large numbers of peroxisomes can be found in the cells of the liver, where toxic molecules accumulate and are broken down. A by-product of this type of chemical reaction is hydrogen peroxide, H_2O_2:

$$RH_2 + O_2 \rightarrow R + H_2O_2$$

Hydrogen peroxide has the potential to be highly toxic. In the presence of metals such as iron (Fe^{2+}) that are found naturally in living cells, hydrogen peroxide can be broken down to form a hydroxide ion (OH^-) and a molecule called a hydroxide free-radical ($\cdot OH$):

$$Fe^{2+} + H_2O_2 \rightarrow Fe^{3+} + OH^- + \cdot OH \text{ (hydroxide free-radical)}$$

The hydroxide free-radical is highly reactive and can damage proteins, lipids, and DNA. Therefore, it is beneficial for cells to break down hydrogen peroxide in an alternative manner that does not form a hydroxide free-radical. Peroxisomes contain an enzyme called **catalase** that breaks down hydrogen peroxide to make water and oxygen gas (hence the name peroxisome):

$$2\ H_2O_2 \xrightarrow{\text{Catalase}} 2\ H_2O + O_2$$

Aside from detoxification, peroxisomes usually contain enzymes involved in the metabolism of fats and amino acids. For example, plant seeds contain specialized organelles called **glyoxysomes**, which are similar to peroxisomes. Seeds often store fats instead of carbohydrates. Because fats have higher energy per unit mass, a plant can make seeds that are smaller and less heavy. Glyoxysomes contain enzymes that are needed to convert fats to sugars. These enzymes become active when a seed germinates and the seedling begins to grow.

Peroxisomes were once viewed as semiautonomous because peroxisomal proteins are imported into the peroxisome in a manner that is very similar to the targeting of proteins to the mitochondria and chloroplasts, as described later in this chapter. Another similarity is that new peroxisomes can be produced by the division of pre-existing peroxisomes. However, recent research indicates that peroxisomes are derived from the endomembrane system. A general model for peroxisome formation is shown in **Figure 4.21**, though the details may differ among animal, plant, and fungal cells. To initiate peroxisome

1 Vesicles bud from the ER and fuse with each other to form a premature peroxisome.

2 The import of additional proteins and lipids results in a mature peroxisome.

Premature peroxisome

Mature peroxisome

Division

ER

3 Mature peroxisomes may divide to produce more peroxisomes.

0.25 μm

Figure 4.21   Formation of peroxisomes.   The inset is a TEM of mature peroxisomes.

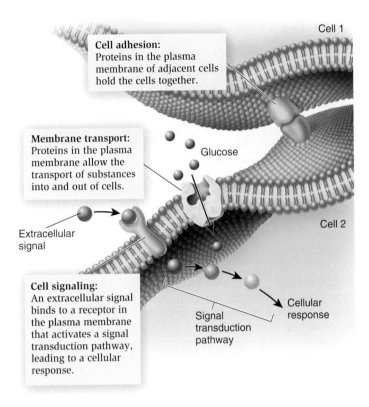

Cell adhesion:
Proteins in the plasma membrane of adjacent cells hold the cells together.

Cell 1

Membrane transport:
Proteins in the plasma membrane allow the transport of substances into and out of cells.

Glucose

Extracellular signal

Cell 2

Cell signaling:
An extracellular signal binds to a receptor in the plasma membrane that activates a signal transduction pathway, leading to a cellular response.

Signal transduction pathway

Cellular response

Figure 4.22   Major functions of the plasma membrane. These include membrane transport, cell signaling, and cell adhesion.

Concept check: *Which of these three functions do you think is the most important for cell metabolism?*

formation, vesicles bud from the ER membrane and form a premature peroxisome. Following the import of additional proteins, the premature peroxisome becomes a mature peroxisome. Once the mature peroxisome has formed, it may then divide to further increase the number of peroxisomes in the cell.

The Plasma Membrane Is the Interface Between a Cell and Its Environment

The cytoplasm of eukaryotic cells is surrounded by a plasma membrane, which is part of the endomembrane system and provides a boundary between a cell and the extracellular environment. Proteins in the plasma membrane perform many important functions that affect the activities inside the cell. First, many plasma membrane proteins are involved in **membrane transport** (Figure 4.22). Some of these proteins function to transport essential nutrients or ions into the cell, and others are involved in the export of substances. Due to the functioning of these transporters, the plasma membrane is selectively permeable; it allows only certain substances in and out. We will examine the structures and functions of a variety of transporters in Chapter 5.

A second vital function of the plasma membrane is **cell signaling**. To survive and adapt to changing conditions, cells must be able to sense changes in their environment. In addition, the cells of a multicellular organism need to communicate with each

other to coordinate their activities. The plasma membrane of all cells contains receptors that recognize signaling molecules—either environmental agents or molecules secreted by other cells. Once signaling molecules bind to a receptor, this elicits a signal transduction pathway—a series of steps that cause the cell to respond to the signal (Figure 4.22). For example, when you eat a meal, the hormone insulin is secreted into your bloodstream. This hormone binds to receptors in the plasma membrane of your cells, which results in a cellular response that allows your cells to increase their uptake of certain molecules found in food, such as glucose. We will explore the details of cell signaling in Chapter 9.

A third important role of the plasma membrane in animal cells is **cell adhesion**. Protein-protein interactions among proteins in the plasma membranes of adjacent cells promote cell-to-cell adhesion (Figure 4.22). This phenomenon is critical for animal cells to properly interact to form a multicellular organism and allows cells to recognize each other. The structures and functions of proteins involved in cell adhesion will be examined in Chapter 10.

4.5 Semiautonomous Organelles

We now turn to those organelles in eukaryotic cells that are considered semiautonomous: mitochondria and chloroplasts. These organelles can grow and divide to reproduce themselves, but they are not completely autonomous because they depend on other parts of the cell for their internal components (**Figure 4.23**). For example, most of the proteins found in mitochondria are imported from the cytosol. In this section, we will survey the structures and functions of the semiautonomous organelles

Animal cell

Mitochondrion

Chloroplast

Plant cell

Figure 4.23   Semiautonomous organelles.   These are the mitochondria and chloroplasts.

Figure 4.24   Structure of a mitochondrion.   This organelle is enclosed in two membranes. The invaginations of the inner membrane are called cristae. The mitochondrial matrix lies inside the inner membrane. The micrograph is a colorized TEM.

Concept check: *What is the advantage of having a highly invaginated inner membrane?*

in eukaryotic cells and consider their evolutionary origins. In Chapters 7 and 8, we will explore the functions of mitochondria and chloroplasts in greater depth.

Mitochondria Supply Cells with Most of Their ATP

Mitochondrion (plural, mitochondria) literally means thread granule, which is what mitochondria look like under a light microscope. They are similar in size to bacteria. Depending on a cell's function, it may contain a few hundred to a few thousand mitochondria. Cells with particularly heavy energy demands, such as muscle cells, have more mitochondria than other cells. Research has shown that regular exercise increases the number and size of mitochondria in human muscle cells to meet the expanded demand for energy.

A mitochondrion has an outer membrane and an inner membrane separated by a region called the intermembrane space (**Figure 4.24**). The inner membrane is highly invaginated (folded) to form projections called **cristae**. These invaginations greatly increase the surface area of the inner membrane, which is the site where ATP is made. The compartment enclosed by the inner membrane is the **mitochondrial matrix**.

The primary role of mitochondria is to make ATP. Even though mitochondria produce most of a cell's ATP, mitochondria do not create energy. Rather, their primary function is to convert chemical energy that is stored within the covalent bonds of organic molecules into a form that can be readily used by cells. Covalent bonds in sugars, fats, and amino acids store a large amount of energy. The breakdown of these molecules into

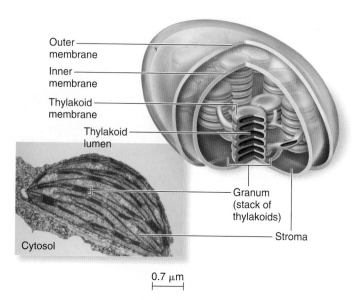

Figure 4.25   Structure of a chloroplast.   Like a mitochondrion, a chloroplast is enclosed in a double membrane. In addition, it has an internal thylakoid membrane system that forms flattened compartments. These compartments stack on each other to form grana. The stroma is located inside the inner membrane but outside the thylakoid membrane. This micrograph is a colorized TEM.

simpler molecules releases energy that is used to make ATP. Many proteins in living cells utilize ATP to carry out their functions, such as muscle contraction, uptake of nutrients, cell division, and many other cellular processes.

Mitochondria perform other functions as well. They are involved in the synthesis, modification, and breakdown of several types of cellular molecules. For example, the synthesis of certain hormones requires enzymes that are found in mitochondria. Another interesting role of mitochondria is to generate heat in specialized fat cells known as brown fat cells. Groups of brown fat cells serve as "heating pads" that help to revive hibernating animals and protect sensitive areas of young animals from the cold.

Chloroplasts Carry Out Photosynthesis

Chloroplasts are organelles that can capture light energy and use some of that energy to synthesize organic molecules such as glucose. This process, called **photosynthesis**, is described in Chapter 8. Chloroplasts are found in nearly all species of plants and algae. **Figure 4.25** shows the structure of a typical chloroplast. Like the mitochondrion, a chloroplast contains an outer and inner membrane. An intermembrane space lies between these two membranes. A third system of membranes, the **thylakoid membrane**, forms many flattened, fluid-filled tubules that enclose a single, convoluted compartment. These tubules tend to stack on top of each other to form a structure called a **granum** (plural, grana). The **stroma** is the compartment of the chloroplast that is enclosed by the inner membrane but outside the thylakoid membrane. The **thylakoid lumen** is enclosed by the thylakoid membrane.

Chloroplasts are a specialized version of plant organelles that are more generally known as **plastids**. All plastids are derived from unspecialized **proplastids**. The various types of plastids are distinguished by their synthetic abilities and the types of pigments they contain. Chloroplasts, which carry out photosynthesis, contain the green pigment chlorophyll. The abundant number of chloroplasts in the leaves of plants gives them their green color. Chromoplasts, a second type of plastid, function in synthesizing and storing the yellow, orange, and red pigments known as carotenoids. Chromoplasts give many fruits and flowers their colors. In autumn, the chromoplasts also give many leaves their yellow, orange, and red colors. A third type of plastid, leucoplasts, typically lacks pigment molecules. An amyloplast is a leucoplast that synthesizes and stores starch. Amyloplasts are common in underground structures such as roots and tubers.

Mitochondria and Chloroplasts Contain Their Own Genetic Material and Divide by Binary Fission

To fully appreciate the structure and organization of mitochondria and chloroplasts, we also need to briefly examine their genetic properties. In 1951, Y. Chiba exposed plant cells to Feulgen, a DNA-specific dye, and discovered that the chloroplasts became stained. Based on this observation, he was the first to suggest that chloroplasts contain their own DNA. Researchers in the 1970s and 1980s isolated DNA from both chloroplasts and mitochondria. These studies revealed that the DNA of these organelles resembled smaller versions of bacterial chromosomes.

The chromosomes found in mitochondria and chloroplasts are referred to as the **mitochondrial genome** and **chloroplast genome**, respectively, and the chromosomes found in the nucleus of the cell constitute the **nuclear genome**. Like bacteria, the genomes of most mitochondria and chloroplasts are composed of a single circular chromosome. Compared to the nuclear genome, they are very small. For example, the amount of DNA in the human nuclear genome (about 3 billion base pairs) is about 200,000 times greater than the mitochondrial genome. In terms of genes, the human genome has approximately 20,000 to 25,000 different genes, whereas the human mitochondrial genome has only a few dozen. Chloroplast genomes tend to be larger than mitochondrial genomes, and they have a correspondingly greater number of genes. Depending on the particular species of plant or algae, a chloroplast genome is about 10 times larger than the mitochondrial genome of human cells.

Just as the genomes of mitochondria and chloroplasts resemble bacterial genomes, the production of new mitochondria and chloroplasts bears a striking resemblance to the division of bacterial cells. Like their bacterial counterparts, mitochondria and chloroplasts increase in number via **binary fission**, or splitting in two. **Figure 4.26** illustrates the process for a mitochondrion. The mitochondrial genome, which is found in a region called the nucleoid, is duplicated, and the organelle divides into two separate organelles. Mitochondrial and chloroplast division are needed to maintain a full complement of these organelles when cell growth occurs following cell

(a) **Binary fission of mitochondria**

(b) **Transmission electron micrographs of the process**

Figure 4.26 Division of mitochondria by binary fission.

division. In addition, environmental conditions may influence the sizes and numbers of these organelles. For example, when plants are exposed to more sunlight, the number of chloroplasts in leaf cells increases.

Mitochondria and Chloroplasts Are Derived from Ancient Symbiotic Relationships

The observation that mitochondria and chloroplasts contain their own genetic material may seem puzzling. Perhaps you might think that it would be simpler for a eukaryotic cell to have all of its genetic material in the nucleus. The distinct genomes of mitochondria and chloroplasts can be traced to their evolutionary origin, which involved an ancient symbiotic association.

A symbiotic relationship occurs when two different species live in direct contact with each other. **Endosymbiosis** describes a symbiotic relationship in which the smaller species—the symbiont—actually lives inside the larger species. In 1883, Andreas Schimper proposed that chloroplasts were descended from an endosymbiotic relationship between cyanobacteria (a bacterium capable of photosynthesis) and eukaryotic cells. In 1922, Ivan Wallin also hypothesized an endosymbiotic origin for mitochondria.

In spite of these interesting ideas, the question of endosymbiosis was largely ignored until the discovery that mitochondria and chloroplasts contain their own genetic material. In 1970, the issue of endosymbiosis as the origin of mitochondria and chloroplasts was revived by Lynn Margulis in her book *Origin of Eukaryotic Cells*. During the 1970s and 1980s, the advent of molecular genetic techniques allowed researchers to analyze genes from mitochondria, chloroplasts, bacteria, and eukaryotic nuclear genomes. Researchers discovered that genes in mitochondria and chloroplasts are very similar to bacterial genes. Likewise, mitochondria and chloroplasts are strikingly similar in size and shape to certain bacterial species. These observations provided strong support for the **endosymbiosis theory**, which proposes that mitochondria and chloroplasts originated from bacteria that took up residence within a primordial eukaryotic cell (**Figure 4.27**). Over the next 2 billion years, the characteristics of these intracellular bacterial cells gradually changed to those of a mitochondrion or chloroplast. A more in-depth discussion of the origin of eukaryotic cells is found in Chapter 22.

Symbiosis occurs because the relationship is beneficial to one or both species. According to the endosymbiosis theory,

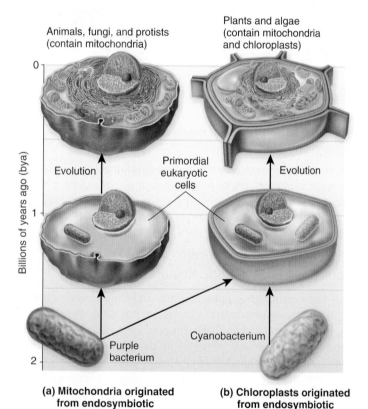

Figure 4.27   A simplified view of the endosymbiosis theory. (a) According to this concept, modern mitochondria were derived from purple bacteria, also called α-proteobacteria. Over the course of evolution, their characteristics changed into those found in mitochondria today. **(b)** A similar phenomenon occurred for chloroplasts, which were derived from cyanobacteria, a bacterium that is capable of photosynthesis.

Concept check: *Discuss the similarities and differences between modern bacteria and mitochondria.*

this relationship provided eukaryotic cells with useful cellular characteristics. Chloroplasts, which were derived from cyanobacteria, have the ability to carry out photosynthesis. This benefits plant cells by giving them the ability to use the energy from sunlight. By comparison, mitochondria are thought to have been derived from a different type of bacteria known as purple bacteria or α-proteobacteria. In this case, the endosymbiotic relationship enabled eukaryotic cells to synthesize greater amounts of ATP. How the relationship would have been beneficial to a cyanobacterium or purple bacterium is less clear, though the cytosol of a eukaryotic cell may have provided a stable environment with an adequate supply of nutrients.

During the evolution of eukaryotic species, many genes that were originally found in the genome of the primordial purple bacteria and cyanobacteria have been transferred from the organelles to the nucleus. This has occurred many times throughout evolution, so modern mitochondria and chloroplasts have lost most of the genes that still exist in present-day purple bacteria and cyanobacteria. Some researchers speculate that the movement of genes into the nucleus makes it easier for the cell to control the structure, function, and division of mitochondria and chloroplasts. In modern cells, hundreds of different proteins that make up these organelles are encoded by genes that have been transferred to the nucleus. These proteins are made in the cytosol and then taken up into mitochondria or chloroplasts. We will discuss this topic next.

4.6   Protein Sorting to Organelles

Thus far, we have considered how eukaryotic cells contain a variety of membrane-bound organelles. Each protein that a cell makes usually functions within one cellular compartment or is secreted from the cell. How does each protein reach its appropriate destination? For example, how does a mitochondrial protein get sent to the mitochondrion rather than to a different organelle such as a lysosome? In eukaryotes, most proteins contain short stretches of amino acid sequences that direct them to their correct cellular location. These sequences are called **sorting signals**, or **traffic signals**. Each sorting signal is recognized by specific cellular components that facilitate the proper routing of that protein to its correct location.

Most eukaryotic proteins begin their synthesis on ribosomes in the cytosol, using messenger RNA (mRNA) that contains the information for polypeptide synthesis (**Figure 4.28**). The cytosol provides amino acids, which are used as building blocks to make these proteins during translation. Cytosolic proteins lack any sorting signal, so they stay there. By comparison, the synthesis of some eukaryotic proteins begins in the cytosol and then halts temporarily until the ribosome has become bound to the ER membrane. After this occurs, translation resumes and the polypeptide is synthesized into the ER lumen or ER membrane. Proteins that are destined for the ER, Golgi, lysosome, vacuole, plasma membrane, or secretion are first directed to the ER. This is called **cotranslational sorting** because the first step in the sorting process begins while translation is occurring.

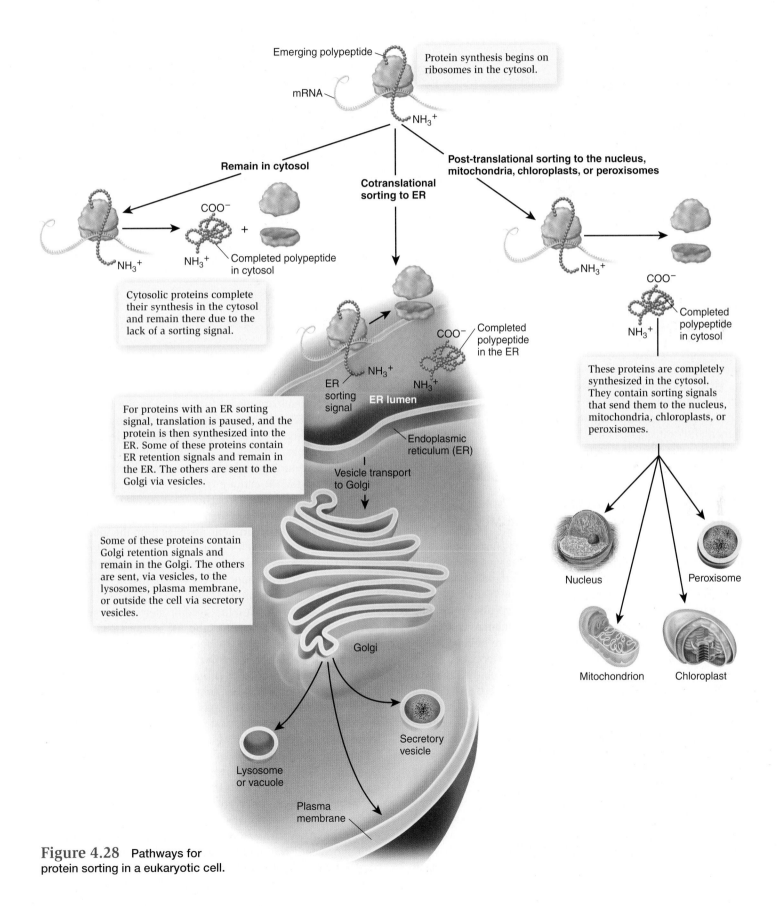

Emerging polypeptide

Protein synthesis begins on ribosomes in the cytosol.

mRNA

NH_3^+

Remain in cytosol

Cotranslational sorting to ER

Post-translational sorting to the nucleus, mitochondria, chloroplasts, or peroxisomes

NH_3^+

COO^-

NH_3^+

Completed polypeptide in cytosol

Cytosolic proteins complete their synthesis in the cytosol and remain there due to the lack of a sorting signal.

NH_3^+

COO^-

Completed polypeptide in the ER

ER sorting signal

ER lumen

NH_3^+

NH_3^+

NH_3^+

COO^-

Completed polypeptide in cytosol

These proteins are completely synthesized in the cytosol. They contain sorting signals that send them to the nucleus, mitochondria, chloroplasts, or peroxisomes.

For proteins with an ER sorting signal, translation is paused, and the protein is then synthesized into the ER. Some of these proteins contain ER retention signals and remain in the ER. The others are sent to the Golgi via vesicles.

Endoplasmic reticulum (ER)

Vesicle transport to Golgi

Some of these proteins contain Golgi retention signals and remain in the Golgi. The others are sent, via vesicles, to the lysosomes, plasma membrane, or outside the cell via secretory vesicles.

Nucleus

Peroxisome

Mitochondrion

Chloroplast

Golgi

Secretory vesicle

Lysosome or vacuole

Plasma membrane

Figure 4.28 Pathways for protein sorting in a eukaryotic cell.

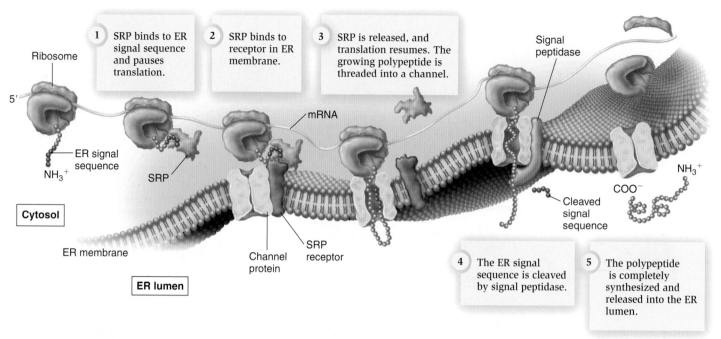

Figure 4.29   First step in cotranslational protein localization: cotranslational sorting.

Concept check: *What prevents an ER protein from being completely synthesized in the cytosol?*

Finally, the uptake of most proteins into the nucleus, mitochondria, chloroplasts, and peroxisomes occurs after the protein is completely made (that is, completely translated). This is called **post-translational sorting** because sorting does not happen until translation is finished. In this section, we will consider how cells carry out cotranslational and post-translational sorting.

The Cotranslational Sorting of Some Proteins Occurs at the Endoplasmic Reticulum Membrane

The concept of sorting signals in proteins was first proposed by Günter Blobel in the 1970s. Blobel and colleagues discovered a sorting signal in proteins that sends them to the ER membrane, which is the first step in cotranslational sorting (**Figure 4.29**). To be directed to the rough ER membrane, a polypeptide must contain a sorting signal called an **ER signal sequence**, which is a sequence of about 6 to 12 amino acids that are predominantly hydrophobic and usually located near the amino terminus. As the ribosome is making the polypeptide in the cytosol, the ER signal sequence emerges from the ribosome and is recognized by a protein/RNA complex called **signal recognition particle (SRP)**. SRP has two functions. First, it recognizes the ER signal sequence and pauses translation. Second, SRP binds to a receptor in the ER membrane, which docks the ribosome over a channel protein. At this stage, SRP is released and translation resumes. The growing polypeptide is threaded through the channel to cross the ER membrane. In most cases, the ER signal sequence is removed by signal peptidase. If the protein is not a membrane protein, it will be released into the lumen of the ER. In 1999, Blobel won the Nobel Prize for his discovery of sorting signals in proteins. The process shown in Figure 4.29 illustrates another important role of

protein-protein interactions—a series of interactions causes the steps of a process to occur in a specific order.

Some proteins are meant to function in the ER. Such proteins contain ER retention signals in addition to the ER signal sequence. Alternatively, other proteins that are destined for the Golgi, lysosomes, vacuoles, plasma membrane, or secretion must be sorted to these other locations (see Figure 4.28). Such proteins leave the ER and are transported to their correct location. This transport process occurs via vesicles that are formed from one compartment and then move through the cytosol and fuse with another compartment. Vesicles from the ER may go to the Golgi, and then vesicles from the Golgi may go to the lysosomes, vacuoles, or plasma membrane. Sorting signals within proteins' amino acid sequences are responsible for directing them to the correct location.

Figure 4.30 describes the second step in cotranslational sorting, vesicle transport from the ER to the Golgi. A cargo, such as protein molecules, is loaded into a developing vesicle by binding to cargo receptors in the ER membrane. Vesicle formation is facilitated by **coat proteins**, which help a vesicle to bud from a given membrane. As a vesicle forms, other proteins called **v-snares** are incorporated into the vesicle membrane (hence the name v-snare). Many types of v-snares are known to exist; the particular v-snare that is found in a vesicle membrane depends on the type of cargo it carries. After a vesicle is released from one compartment such as the ER, the coat is shed. The vesicle then travels through the cytosol. But how does the vesicle know where to go? The answer is that the v-snares in the vesicle membrane are recognized by **t-snares** in a target membrane. After v-snares recognize t-snares, the vesicle fuses with the membrane containing the t-snares. The recognition between v-snares and t-snares ensures that a vesicle carrying a

1. In this example, a cargo of proteins binds to receptors in the ER membrane. The binding of coat proteins helps a vesicle bud from the membrane, and v-snares are incorporated into vesicle.

Coat proteins

Protein cargo

Cargo receptor

v-snare

2. The vesicle pinches off the membrane and is released.

ER membrane

t-snare

Golgi membrane

3. After the vesicle is released, the coat is shed.

4. The vesicle binds to the target membrane by a v-snare/t-snare interaction.

5. The vesicle fuses with the target membrane to deliver the protein cargo to its target destination.

Figure 4.30 Second step in cotranslational protein localization: vesicle transport from the endoplasmic reticulum.

specific cargo moves to the correct target membrane in the cell. Like the sorting of proteins to the ER membrane, the formation and sorting of vesicles also involves a series of protein-protein interactions that cause the steps to occur in a defined manner.

Proteins Are Sorted Post-Translationally to the Nucleus, Peroxisomes, Mitochondria, and Chloroplasts

The organization and function of the nucleus, peroxisomes, and semiautonomous organelles are dependent on the uptake of proteins from the cytosol. Most of their proteins are synthesized

in the cytosol and then taken up into their respective organelles. For example, most proteins involved in ATP synthesis are made in the cytosol and taken up into mitochondria after they have been completely synthesized. For this to occur, a protein must have the appropriate sorting signal as part of its amino acid sequence.

As one example of post-translational sorting, let's consider how a protein is directed to the mitochondrial matrix. Such a protein would have a matrix-targeting sequence as part of its structure, which is a short sequence at the amino terminus with several positively charged amino acids that folds into an α helix. As shown in **Figure 4.31**, the process of protein import into

1. Chaperone proteins keep protein unfolded.

2. Matrix-targeting sequence binds to receptor.

3. Chaperones are released as protein is transferred to a channel in the outer membrane.

4. Protein is transferred to a channel in the inner membrane.

Contact site

Active protein

Chaperone

Matrix-targeting sequence

Cytosol

Outer membrane

Receptor protein

Intermembrane space

Inner membrane

Channel proteins

Mitochondrial matrix

5. Chaperones bind to protein as it enters the matrix.

6. Matrix-targeting sequence is cleaved by an enzyme in the matrix.

7. Protein is completely threaded into the matrix.

8. Chaperones are released, and protein folds into its three-dimensional structure.

Figure 4.31 Post-translational sorting of a protein to the mitochondrial matrix.

Concept check: *What do you think would happen if chaperone proteins did not bind to a mitochondrial matrix protein before it was imported into the mitochondrion?*

the matrix involves a series of intricate protein-protein interactions. A protein destined for the mitochondrial matrix is first made in the cytosol, where proteins called **chaperones** keep it in an unfolded state. A receptor protein in the outer mitochondrial membrane recognizes the matrix-targeting sequence. The protein is released from the chaperone as it is transferred to a channel in the outer mitochondrial membrane. Because it is in an unfolded state, the mitochondrial protein can be threaded through this channel, and then through another channel in the inner mitochondrial membrane. These channels lie close to each other at contact sites between the outer and inner membranes. As the protein emerges in the matrix, other chaperone proteins that were already in the matrix continue to keep it unfolded. Eventually, the matrix-targeting sequence is cleaved, and the entire protein is threaded into the matrix. At this stage, the chaperone proteins are released, and the protein can adopt its three-dimensional active structure.

4.7   Systems Biology of Cells: A Summary

We will conclude this chapter by reviewing cell structure and function from a perspective called systems biology. In **systems biology**, researchers view living organisms in terms of their underlying network structure—groups of structural and functional connections—rather than their individual molecular components. A "system" can be anything from a metabolic pathway to a cell, an organ, or even an entire organism. In this section, we focus on the cell as a system. First, we will compare prokaryotic and eukaryotic cells as systems, and then examine the four interconnected parts that make up the system that is the eukaryotic cell.

Bacterial Cells Are Relatively Simple Systems Compared to Eukaryotic Cells

Bacterial cells are relatively small and lack the extensive internal compartmentalization characteristic of eukaryotic cells (**Table 4.2**). On the outside, bacterial cells are surrounded by a cell wall, and many species have flagella. Animal cells lack a cell wall, and only certain cell types have flagella or cilia. Like bacteria, plant cells also have cell walls but only rarely have flagella.

As stated earlier in this chapter, the cytoplasm is the region of the cell enclosed by the plasma membrane. Ribosomes are found in the cytoplasm of all cell types. In bacteria, the cytoplasm is a single compartment. The bacterial genetic material, usually a single chromosome, is found in the nucleoid region, which is not surrounded by a membrane. By comparison, the cytoplasm of eukaryotic cells is highly compartmentalized. The cytosol is the area that surrounds many different types of membrane-bound organelles. For example, eukaryotic chromosomes are found in the nucleus that is surrounded by a double membrane. In addition, all eukaryotic cells have an endomembrane system and mitochondria, and plant cells also have chloroplasts.

A Eukaryotic Cell Is a System with Four Interacting Parts

We can view a eukaryotic cell as a system of four interacting parts: the interior of the nucleus, the cytosol, the endomembrane system, and the semiautonomous organelles (**Figure 4.32**). These four regions play a role in their own structure and organization, as well as the structure and organization of the entire cell.

Nucleus   The nucleus houses the genome. Earlier in this chapter, we learned how the genome plays a key role in producing the proteome through the process of gene expression. The collection of proteins that a cell makes is primarily responsible for the structure and function of the entire cell. Gene regulation, which largely occurs in the cell nucleus, is very important in creating specific cell types and enabling cells to respond to environmental changes. The nucleus itself is organized by a collection of filamentous proteins called the nuclear matrix.

Cytosol   The cytosol is the region that is enclosed by the plasma membrane but outside of the organelles. It is an important coordination center for cell function and organization. Along with the plasma membrane, the cytosol coordinates responses to the environment. Factors in the environment may stimulate signaling pathways in the cytosol that affect the functions of cellular proteins and the regulation of genes in the cell nucleus.

The cytosol also has a large impact on cell structure because it is the compartment where many small molecules are metabolized in the cell. This region receives molecules that are taken up from the environment. In addition, many pathways for the synthesis and breakdown of cellular molecules are found in the cytosol, and pathways in organelles are often regulated by events there. Most of the proteins that constitute the proteome are made in the cytosol.

A particularly important component of cell organization is the cytoskeleton, which is primarily found in the cytosol. The formation and function of the cytoskeleton is caused by an amazing series of protein-protein interactions. The cytoskeleton provides organization to the cell and facilitates cellular movements. In most cells, the cytoskeleton is a dynamic structure, enabling its composition to respond to environmental and developmental changes.

Endomembrane System   The endomembrane system can be viewed as a smaller system within the confines of a cell. The endomembrane system includes the nuclear envelope, endoplasmic reticulum (ER), Golgi apparatus, lysosomes, vacuoles, peroxisomes, and plasma membrane. This system forms a secretory pathway that is crucial in the movement of larger substances, such as carbohydrates and proteins, out of the cell. The export of carbohydrates and proteins plays a key role in the organization of materials that surround cells.

The endomembrane system also contributes to the overall structure and organization of eukaryotic cells in other ways. The ER and Golgi are involved in protein sorting and in the attachment of carbohydrates to lipids and proteins. In addition,

Table 4.2   A Comparison of Cell Complexity Among Bacterial, Animal, and Plant Cells

Structures	Bacteria	Animal cells	Plant cells
Extracellular structures			
Cell wall*	Present	Absent	Present
Flagella/cilia	Flagella sometimes present	Cilia or flagella present on certain cell types	Rarely present**
Plasma membrane	Present	Present	Present
Interior structures			
Cytoplasm	Usually a single compartment inside the plasma membrane	Composed of membrane-bound organelles that are surrounded by the cytosol	Composed of membrane-bound organelles that are surrounded by the cytosol
Ribosomes	Present	Present	Present
Chromosomes and their location	Typically one circular chromosome per nucleoid region; nucleoid region is not a separate compartment	Multiple linear chromosomes in the nucleus; nucleus is surrounded by a double membrane. Mitochondria also have chromosomes.	Multiple linear chromosomes in the nucleus; nucleus is surrounded by a double membrane. Mitochondria and chloroplasts also have chromosomes.
Endomembrane system	Absent	Present	Present
Mitochondria	Absent	Present	Present
Chloroplasts	Absent	Absent	Present

*The biochemical composition of bacterial cell walls is very different from plant cell walls.
**Some plant species produce sperm cells with flagella, but flowering plants produce sperm within pollen grains that lack flagella.

Nucleus
- Location of most of the genome
- Gene regulation
- Organization and protection of chromosomes via the nuclear matrix

Endomembrane system
1. Nuclear envelope
   - Boundary that surrounds the nucleus
2. Endoplasmic reticulum
   - Protein secretion and sorting
   - Glycosylation
   - Lipid synthesis
   - Metabolic functions and accumulation of Ca^{2+}
3. Golgi apparatus
   - Protein secretion and sorting
   - Glycosylation
4. Lysosome/vacuoles
   - Degradation of organic molecules
   - Storage of organic molecules
   - Accumulation of water (plant vacuoles)
5. Peroxisomes
   - Breakdown of toxic molecules such as H_2O_2
   - Breakdown and synthesis of organic molecules
6. Plasma membrane
   - Uptake and excretion of ions and molecules
   - Cell signaling
   - Cell adhesion

Cytosol
- Coordination of responses to the environment
- Coordination of metabolism
- Synthesis of the proteome
- Organization and movement via a cytoskeleton and motor proteins

Semiautonomous organelles
1. Mitochondria
   - Synthesis of ATP
   - Synthesis and modification of other organic molecules
   - Production of heat
2. Chloroplasts (plants and algae)
   - Photosynthesis

Figure 4.32   **The four interacting parts of eukaryotic cells.**   These include the nucleus, cytosol, endomembrane system, and semiautonomous organelles.

most of a cell's lipids are made in the smooth ER membrane and distributed to other parts of the cell. The smooth ER also plays a role in certain metabolic functions, such as the elimination of alcohol, and is important in the accumulation of Ca^{2+}.

Another important function of the endomembrane system that serves the needs of the entire cell is the breakdown and storage of organic molecules. Lysosomes in animal cells and vacuoles in the cells of other organisms assist in breaking down various types of macromolecules. The building blocks are then recycled back to the cytosol and used to construct new macromolecules. Vacuoles often play a role in the storage of organic molecules such as carbohydrates, proteins, and fats. In plants, vacuoles may store large amounts of water. Finally, peroxisomes are involved in the breakdown and synthesis of organic molecules and can degrade toxic molecules such as hydrogen peroxide.

The plasma membrane is also considered a part of the endomembrane system. It plays an important role as a selective barrier that allows the uptake of nutrients and the excretion of waste products. The plasma membrane also contains different types of receptors that provide a way for a cell to sense changes in its environment and communicate with other cells. Finally, in animals, proteins in the plasma membrane promote the adhesion of adjacent cells.

Semiautonomous Organelles The semiautonomous organelles include the mitochondria and chloroplasts. Regarding organization, these organelles tend to be rather independent. They exist in the cytosol much like a bacterium would grow in a laboratory medium. Whereas a bacterium would take up essential nutrients from the growth medium, the semiautonomous organelles take up molecules from the cytosol. The organelles use these molecules to carry out their functions and maintain their organization. Like bacteria, the semiautonomous organelles divide by binary fission to produce more of themselves.

Although the semiautonomous organelles rely on the rest of the cell for many of their key components, they also give back to the cell in ways that are vital to maintaining cell organization. Mitochondria take up organic molecules from the cytosol and give back ATP, which is used throughout the cell to drive processes that are energetically unfavorable. This energy is crucial for cell organization. Mitochondria also modify certain organic molecules and may produce heat. By comparison, the chloroplasts capture light energy and synthesize organic molecules. These organic molecules also store energy and can be broken down when energy is needed. In addition, organic molecules, such as sugars and amino acids, are used as building blocks to synthesize many different types of cellular molecules, such as carbohydrate polymers and proteins.

Summary of Key Concepts

4.1 Microscopy

- Three important parameters in microscopy are magnification, resolution, and contrast. A light microscope utilizes light for

illumination, whereas an electron microscope uses an electron beam. Transmission electron microscopy (TEM) provides the best resolution of any form of microscopy, and scanning electron microscopy (SEM) produces an image of a three-dimensional surface. (Figures 4.1, 4.2, 4.3)

4.2 Overview of Cell Structure

- Cell structure relies on four factors: matter, energy, organization, and information. Every living organism has a genome. The genes within the genome contain the information to create cells with particular structures and functions.

- We can classify all forms of life into two categories based on cell structure: prokaryotes and eukaryotes.

- The prokaryotes have a relatively simple structure and lack a membrane-enclosed nucleus. The two categories of prokaryotes are bacteria and archaea. Structures in prokaryotic cells include the plasma membrane, cytoplasm, nucleoid region, and ribosomes. Prokaryotes also have a cell wall and many have a glycocalyx. (Figure 4.4)

- Eukaryotic cells are compartmentalized into organelles and contain a nucleus that houses most of their DNA. (Figures 4.5, 4.6, 4.7)

- The proteome of a cell determines its structure and function.

4.3 The Cytosol

- The cytosol is a central coordinating region for many metabolic activities of eukaryotic cells, including polypeptide synthesis. (Figures 4.8, 4.9)

- The cytoskeleton is a network of three different types of protein filaments: microtubules, intermediate filaments, and actin filaments. Microtubules are important for cell shape, organization, and movement. Intermediate filaments help maintain cell shape and rigidity. Actin filaments support the plasma membrane and play a key role in cell strength, shape, and movement. (Table 4.1, Figures 4.10, 4.11, 4.12, 4.13)

4.4 The Nucleus and Endomembrane System

- The primary function of the nucleus involves the organization and expression of the cell's genetic material. A second important function is the assembly of ribosomes in the nucleolus. (Figures 4.14, 4.15, 4.16)

- The endomembrane system includes the nuclear envelope, endoplasmic reticulum, Golgi apparatus, lysosomes, vacuoles, peroxisomes, and plasma membrane. The rough endoplasmic reticulum (rough ER) plays a key role in the initial sorting of proteins. The smooth endoplasmic reticulum (smooth ER) functions in metabolic processes such as detoxification, carbohydrate metabolism, accumulation of calcium ions, and synthesis and modification of lipids. The Golgi apparatus performs three overlapping functions: processing, protein sorting, and secretion. Lysosomes degrade macromolecules and help digest substances taken up from outside the cell (endocytosis) and inside the cell. (Figures 4.17, 4.18)

- Palade's pulse-chase experiments demonstrated that secreted proteins move sequentially through the ER and Golgi apparatus. (Figure 4.19)

- Types and functions of vacuoles include central vacuoles; contractile vacuoles; and phagocytic, or food, vacuoles. (Figure 4.20)

- Peroxisomes catalyze certain chemical reactions, typically those that break down molecules by removing hydrogen or adding oxygen. Peroxisomes usually contain enzymes involved in the metabolism of fats and amino acids. Peroxisomes are made via budding from the ER, followed by maturation and division. (Figure 4.21)

- Proteins in the plasma membrane perform many important roles that affect activities inside the cell, including membrane transport, cell signaling, and cell adhesion. (Figure 4.22)

4.5 Semiautonomous Organelles

- Mitochondria and chloroplasts are considered semiautonomous because they can grow and divide, but they still depend on other parts of the cell for their internal components. (Figure 4.23)

- Mitochondria produce most of a cell's ATP, which is utilized by many proteins to carry out their functions. Other mitochondrial functions include the synthesis, modification, and breakdown of cellular molecules and the generation of heat in specialized fat cells. (Figure 4.24)

- Chloroplasts, which are found in nearly all species of plants and algae, carry out photosynthesis. (Figure 4.25)

- Plastids, such as chloroplasts, chromoplasts, and amyloplasts, differ in their function and the pigments they store.

- Mitochondria and chloroplasts contain their own genetic material and divide by binary fission. (Figure 4.26)

- According to the endosymbiosis theory, mitochondria and chloroplasts have evolved from bacteria that took up residence in early eukaryotic cells. (Figure 4.27)

4.6 Protein Sorting to Organelles

- Eukaryotic proteins are sorted to their correct cellular destination. (Figure 4.28)

- The cotranslational sorting of ER, Golgi, lysosomal, vacuolar, plasma membrane, and secreted proteins involves sorting signals and vesicle transport. (Figures 4.29, 4.30)

- Most proteins are sorted to the nucleus, mitochondria, chloroplasts, and peroxisomes post-translationally. (Figure 4.31)

4.7 Systems Biology of Cells: A Summary

- In systems biology, researchers study living organisms in terms of their structural and functional connections, rather than their individual molecular components.

- Prokaryotic and eukaryotic cells differ in their levels of organization. (Table 4.2)

- In eukaryotic cells, four regions—the nucleus, cytosol, endomembrane system, and semiautonomous organelles—work together to produce dynamic organization. (Figure 4.32)

 # Assess and Discuss

Test Yourself

1. The cell doctrine states
   a. all living things are composed of cells.
   b. cells are the smallest units of living organisms.
   c. new cells come from pre-existing cells by cell division.
   d. all of the above.
   e. a and b only.

2. When using microscopes, the resolution refers to
   a. the ratio between the size of the image produced by the microscope and the actual size of the object.
   b. the degree to which a particular structure looks different from other structures around it.
   c. how well a structure takes up certain dyes.
   d. the ability to observe two adjacent objects as being distinct from each other.
   e. the degree to which the image is magnified.

3. If a motor protein were held in place and a cytoskeletal filament were free to move, what type of motion would occur when the motor protein was active?
   a. The motor protein would "walk" along the filament.
   b. The filament would move.
   c. The filament would bend.
   d. All of the above would happen.
   e. Only b and c would happen.

4. The process of polypeptide synthesis is called
   a. metabolism.   d. hydrolysis.
   b. transcription.   e. both c and d.
   c. translation.

5. Each of the following is part of the endomembrane system except
   a. the nuclear envelope.   d. lysosomes.
   b. the endoplasmic reticulum.   e. mitochondria.
   c. the Golgi apparatus.

6. Vesicle transport occurs between the ER and the Golgi in both directions. Let's suppose a researcher added a drug to cells that inhibited vesicle transport from the Golgi to the ER but did not affect vesicle transport from the ER to the Golgi. If you observed cells microscopically after the drug was added, what would you expect to see happen over the course of 1 hour?
   a. The ER would get smaller, and the Golgi would get larger.
   b. The ER would get larger, and the Golgi would get smaller.
   c. The ER and Golgi would stay the same size.
   d. Both the ER and Golgi would get larger.
   e. Both the ER and Golgi would get smaller.

7. Functions of the smooth endoplasmic reticulum include
   a. detoxification of harmful organic molecules.
   b. metabolism of carbohydrates.
   c. protein sorting.
   d. all of the above.
   e. a and b only.

8. The central vacuole in many plant cells is important for
   a. storage.
   b. photosynthesis.
   c. structural support.
   d. all of the above.
   e. a and c only.

9. Let's suppose an abnormal protein contains three targeting sequences: an ER signal sequence, an ER retention sequence, and a mitochondrial-matrix targeting sequence. The ER retention sequence is supposed to keep proteins within the ER. Where would you expect this abnormal protein to go? Note: Think carefully about the timing of events in protein sorting and which events occur cotranslationally and which occur post-translationally.

   a. It would go to the ER.

   b. It would go the mitochondria.

   c. It would go to both the ER and mitochondria equally.

   d. It would remain in the cytosol.

   e. It would be secreted.

10. Which of the following observations would <u>not</u> be considered evidence for the endosymbiosis theory?

   a. Mitochondria and chloroplasts have genomes that resemble smaller versions of bacterial genomes.

   b. Mitochondria, chloroplasts, and bacteria all divide by binary fission.

   c. Mitochondria, chloroplasts, and bacteria all have ribosomes.

   d. Mitochondria, chloroplasts, and bacteria all have similar sizes and shapes.

   e. all of the above

Conceptual Questions

1. Describe two specific ways that protein-protein interactions are involved with cell structure or cell function.

2. Explain how motor proteins and cytoskeletal filaments can interact to promote three different types of movements: movement of a cargo, movement of a filament, and bending of a filament.

3. Describe the functions of the Golgi apparatus.

Collaborative Questions

1. Discuss the roles of the genome and proteome in determining cell structure and function.

2. Discuss and draw the structural relationship between the nucleus, the rough endoplasmic reticulum, and the Golgi apparatus.

Online Resource

www.brookerbiology.com

Stay a step ahead in your studies with animations that bring concepts to life and practice tests to assess your understanding. Your instructor may also recommend the interactive ebook, individualized learning tools, and more.

Membrane Structure, Synthesis, and Transport

5

When he was 28, Andrew began to develop a combination of symptoms that included fatigue, joint pain, abdominal pain, and a loss of sex drive. His doctor conducted some tests and discovered that Andrew had abnormally high levels of iron in his body. Iron is a mineral found in many foods. Andrew was diagnosed with a genetic disease called hemochromatosis, which caused him to absorb more iron than he needed. This was due to an overactive protein involved in the transport of iron through the membranes of intestinal cells and into the body. Unfortunately, when the human body takes up too much iron, it is stored in body tissues, especially the liver, heart, pancreas, and joints. The extra iron can damage a person's organs. In Andrew's case, the disease was caught relatively early, and treatment—which includes a modification in diet along with medication that inhibits the absorption of iron—prevented more severe symptoms. Without treatment, however, hemochromatosis can cause a person's organs to fail. Later signs and symptoms include arthritis, liver disease, heart failure, and skin discoloration.

The disease hemochromatosis illustrates the importance of membranes in regulating the traffic of ions and molecules into and out of cells. Cellular membranes, also known as biological membranes or biomembranes, are an essential characteristic of all living cells. The **plasma membrane** separates the internal contents of a cell from its external environment. With such a role, you might imagine that the plasma membrane would be thick and rigid. Remarkably, the opposite is true. All cellular membranes, including the plasma membrane, are thin (typically 5–10 nm) and somewhat fluid. It would take 5,000 to 10,000 membranes stacked on top of each other to equal the thickness of the page you are reading! Despite their thinness, membranes are impressively dynamic structures that effectively maintain the separation between a cell and its surroundings. Membranes provide an interface to carry out many vital cellular activities (**Table 5.1**).

In this chapter, we will begin by considering the components that provide the structure of membranes and then explore how they are made. Finally, we will examine one of a membrane's primary functions, membrane transport. Biomembranes regulate the traffic of substances into and out of the cell and its organelles.

A model for the structure of aquaporin. This protein, found in the plasma membrane of many cell types, such as red blood cells and plant cells, allows the rapid movement of water molecules across the membrane.

Table 5.1	Important Functions of Cellular Membranes
Function	
Selective uptake and export of ions and molecules	
Cell compartmentalization	
Protein sorting	
Anchoring of the cytoskeleton	
Production of energy intermediates such as ATP and NADPH	
Cell signaling	
Cell and nuclear division	
Adhesion of cells to each other and to the extracellular matrix	

5.1 Membrane Structure

As we progress through this textbook, a theme that will emerge is that structure determines function. This paradigm is particularly interesting when we consider how the structure of cellular membranes enables them to compartmentalize the cell while selectively importing and exporting vital substances. The two primary components of membranes are phospholipids, which form the basic matrix of a membrane, and proteins, which are embedded in the membrane or loosely attached to its surface. A third component is carbohydrate, which may be attached to membrane lipids and proteins. In this section, we will be mainly concerned with the organization of these components to form a biological membrane and how they are important in the overall function of membranes. We will also consider some interesting experiments that provided insight into the dynamic properties of membranes.

Biological Membranes Are a Mosaic of Lipids, Proteins, and Carbohydrates

Figure 5.1 shows the biochemical organization of cellular membranes, which are similar in composition among all living organisms. The framework of the membrane is the **phospholipid bilayer**, which consists of two layers of phospholipids. The most abundant lipids found in membranes are the phospholipids. Recall from Chapter 3 that phospholipids are **amphipathic** molecules. They have a hydrophobic (water-fearing) or nonpolar region, and also a hydrophilic (water-loving) or polar region. The hydrophobic tails of the lipids, referred to as fatty acyl tails, form the interior of the membrane, and the hydrophilic head groups are on the surface.

Cellular membranes also contain proteins, and most membranes have carbohydrates attached to lipids and proteins. The relative amounts of lipids, proteins, and carbohydrates vary among different membranes. Some membranes, such as the inner mitochondrial membrane, have relatively little

carbohydrate, whereas the plasma membrane of eukaryotic cells can have a large amount. A typical membrane found in cell organelles contains 50% protein by mass; the remainder is mostly lipids. However, the smaller lipid molecules outnumber the proteins by about 50 to 1 because the mass of one lipid molecule is much less than the mass of a protein.

Overall, the membrane is considered a mosaic of lipid, protein, and carbohydrate molecules. The membrane structure illustrated in Figure 5.1 is referred to as the **fluid-mosaic model**, originally proposed by S. Jonathan Singer and Garth Nicolson in 1972. As discussed later, the membrane exhibits properties that resemble a fluid because lipids and proteins can move relative to each other within the membrane. **Table 5.2** summarizes some of the historical experiments that led to the formulation of the fluid-mosaic model.

Half of a phospholipid bilayer is termed a **leaflet**. Each leaflet faces a different region. For example, the plasma membrane contains a cytosolic leaflet and an extracellular leaflet (see Figure 5.1). With regard to lipid composition, the two leaflets of cellular membranes are highly asymmetrical. Certain types of lipids may be more abundant in one leaflet compared to the other. A striking asymmetry occurs with glycolipids—lipids with carbohydrate attached. These are found primarily in the extracellular leaflet such that the carbohydrate portion of a glycolipid protrudes into the extracellular medium.

Membrane Proteins Associate with Membranes in Different Ways

Although the phospholipid bilayer forms the basic foundation of cellular membranes, the protein component carries out many

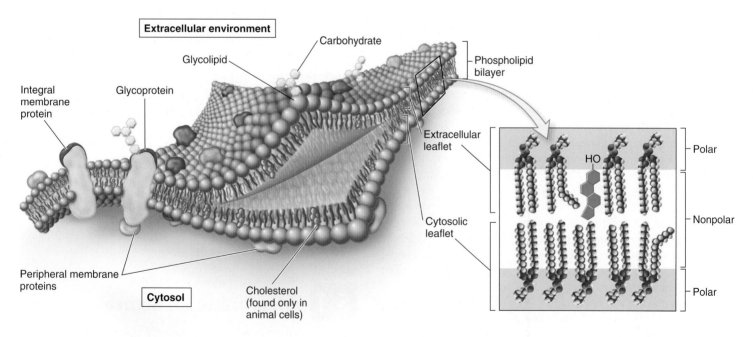

Figure 5.1 **Fluid-mosaic model of membrane structure.** The basic framework of a plasma membrane is a phospholipid bilayer. Proteins may span the membrane and may be bound on the surface to other proteins or to lipids. Proteins and lipids, which have covalently bound carbohydrate, are called glycoproteins and glycolipids, respectively.

Researchers can examine the lateral movements of lipids and transmembrane proteins by a variety of methods. In 1970, Larry Frye and Michael Edidin conducted an experiment that verified the lateral movement of transmembrane proteins (**Figure 5.4**). Mouse and human cells were mixed together and exposed to agents that caused them to fuse with each other. Some cells were cooled to 0°C, while others were incubated at 37°C before being cooled. Both sets of cells were then exposed to fluorescently labeled antibodies that became specifically bound to a mouse transmembrane protein called H-2. The fluorescent label was observed with a fluorescence microscope. If the cells were maintained at 0°C, a temperature that greatly inhibits lateral movement, the fluorescence was seen on only one side of the fused cell. However, if the cells were incubated for several hours at 37°C and then cooled to 0°C, the fluorescence was distributed throughout the plasma membrane of the fused cell. This occurred because the higher temperature allowed the lateral movement of the H-2 protein throughout the fused cell.

Unlike the example shown in Figure 5.4, not all transmembrane proteins are capable of rotational and lateral movement. Depending on the cell type, 10–70% of membrane proteins may be restricted in their movement. Transmembrane proteins may be bound to components of the cytoskeleton, which restricts the proteins from moving (**Figure 5.5**). Also, membrane proteins may be attached to molecules that are outside the cell, such as the interconnected network of proteins that forms the extracellular matrix of animal cells.

Glycosylation of Lipids and Proteins Serves a Variety of Cellular Functions

As mentioned earlier, the third constituent of cellular membranes is carbohydrate. **Glycosylation** refers to the process of covalently attaching a carbohydrate to a lipid or protein. When a carbohydrate is attached to a lipid, this creates a **glycolipid**, whereas attachment to a protein produces a **glycoprotein**.

What is the function of glycosylation? Though the roles of carbohydrate in cell structure and function are not entirely understood, some functional consequences of glycosylation have emerged. The carbohydrates attached to proteins and lipids have well-defined structures that, in some cases, serve as recognition signals for other cellular proteins. For example, proteins destined for the lysosome are glycosylated and have a sugar (mannose-6-phosphate) that is recognized by other proteins in the cell that target the glycosylated protein from the Golgi to the lysosome. Similarly, glycolipids and glycoproteins often play a role in cell surface recognition. When glycolipid and glycoproteins are found in the plasma membrane, the carbohydrate portion is located in the extracellular region. During embryonic development in animals, significant cell movement occurs. Layers of cells slide over each other to create body structures such as the spinal cord and internal organs. The proper migration of individual cells and cell layers relies on the recognition of cell types via the carbohydrates on their cell surfaces.

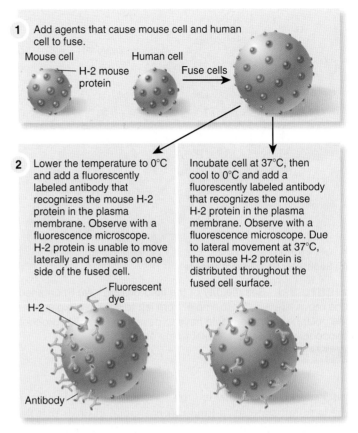

Figure 5.4   A method to measure the lateral movement of membrane proteins.

Concept check: Explain why the H-2 proteins are found on only one side of the cell when the cells were incubated at 0°C.

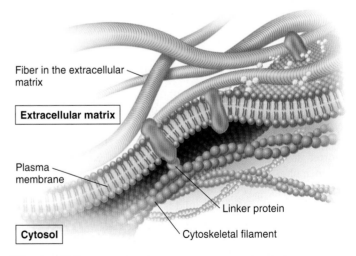

Figure 5.5   Attachment of transmembrane proteins to the cytoskeleton and extracellular matrix of an animal cell.   Some transmembrane proteins have regions that project into the cytosol and are anchored to large cytoskeletal filaments via linker proteins. Being bound to these filaments restricts the movement of these proteins. Similarly, transmembrane proteins may bind to large, immobile components in the extracellular matrix that also restrict the movement of the proteins.

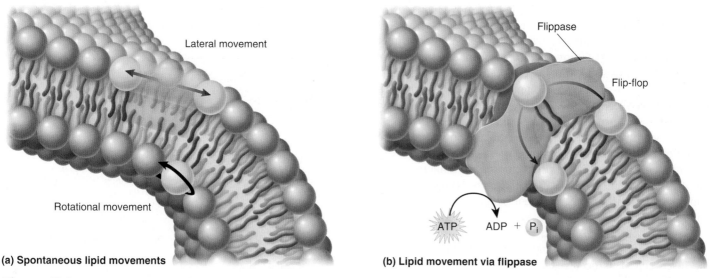

Figure 5.3 **Semifluidity of the lipid bilayer.** **(a)** Spontaneous movements in the bilayer. Lipids can rotate (that is, move 360°) and move laterally (for example, from left to right in the plane of the bilayer). **(b)** Flip-flop does not happen spontaneously, because the polar head group would have to pass through the hydrophobic region of the bilayer. Instead, the enzyme flippase uses ATP to flip phospholipids from one leaflet to the other.

Concept check: *In an animal cell, how can changes in lipid composition affect membrane fluidity?*

associate with each other to form structures called lipid rafts. As the term raft suggests, a **lipid raft** is a group of lipids that float together as a unit within a larger sea of lipids. Lipid rafts have a lipid composition that differs from the surrounding membrane. For example, they usually have a high amount of cholesterol. In addition, lipid rafts may contain unique sets of lipid-anchored proteins and transmembrane proteins. The functional importance of lipid rafts is the subject of a large amount of current research. Lipid rafts may play an important role in endocytosis (discussed later in this chapter) and cell signaling (Chapter 9).

Lipid Composition Affects Membrane Fluidity

The biochemical properties of phospholipids have a profound effect on the fluidity of the phospholipid bilayer. One key factor is the length of fatty acyl tails, which range from 14 to 24 carbon atoms, with 18 to 20 carbons being the most common. Shorter acyl tails are less likely to interact with each other, which makes the membrane more fluid. A second important factor is the presence of double bonds in the acyl tails. When a double bond is present, the lipid is said to be **unsaturated** with respect to the number of hydrogens that can be bound to the carbon atoms (refer back to Figure 3.10). A double bond creates a kink in the fatty acyl tail (see inset to Figure 5.1), making it more difficult for neighboring tails to interact and making the bilayer more fluid. As described in Chapter 3, unsaturated lipids tend to be more liquid compared to saturated lipids that often form solids at room temperature (refer back to Figure 3.11).

A third factor affecting fluidity is the presence of cholesterol, which is a short and rigid planar molecule produced by animal cells (see inset to Figure 5.1). Plant cell membranes contain phytosterols that resemble cholesterol in their chemical structure.

Cholesterol tends to stabilize membranes; its effects depend on temperature. At higher temperatures, such as those observed in mammals that maintain a constant body temperature, cholesterol makes the membrane less fluid. At lower temperatures, such as icy water, cholesterol has the opposite effect. It makes the membrane more fluid and prevents it from freezing.

An optimal level of bilayer fluidity is essential for normal cell function, growth, and division. If a membrane is too fluid, which may occur at higher temperatures, it can become leaky. However, if a membrane becomes too solid, which may occur at lower temperatures, the functioning of membrane proteins will be inhibited. How can organisms cope with changes in temperature? The cells of many species adapt to changes in temperature by altering the lipid composition of their membranes. For example, when the water temperature drops, the cells of certain fish will incorporate more cholesterol in their membranes. If a plant cell is exposed to high temperatures for many hours or days, it will alter its lipid composition to have longer fatty acyl tails and fewer double bonds.

Membrane Proteins May Diffuse in the Plane of the Membrane or Be Restricted in Their Movement

Like lipids, many transmembrane proteins may rotate and laterally move throughout the plane of a membrane. Because transmembrane proteins are larger than lipids, they move within the membrane at a much slower rate. Flip-flop of transmembrane proteins does not occur because the proteins also contain hydrophilic regions that project out from the phospholipid bilayer. It would be energetically unfavorable for the hydrophilic regions of membrane proteins to pass through the hydrophobic portion of the phospholipid bilayer.

Genomes & Proteomes Connection

Approximately 25% of All Genes Encode Transmembrane Proteins

Membrane proteins participate in some of the most important and interesting cellular processes. These include transport, energy transduction, cell signaling, secretion, cell recognition, metabolism, and cell-to-cell contact. Research studies have revealed that cells devote a sizeable fraction of their energy and metabolic machinery to the synthesis of membrane proteins. These proteins are particularly important in human medicine—approximately 70% of all medications exert their effects by binding to membrane proteins. Examples include the drugs aspirin, ibuprofen, and acetaminophen, which are widely used to relieve pain and inflammatory conditions such as arthritis. These drugs bind to cyclooxygenase, a protein in the ER membrane that is necessary for the synthesis of chemicals that play a role in inflammation and pain sensation.

Because membrane proteins are so important biologically and medically, researchers have analyzed the genomes of many species and asked the question, What percentage of genes encodes transmembrane proteins? To answer this question, they have developed tools to predict the likelihood that a gene encodes a transmembrane protein. For example, the occurrence of transmembrane α helices can be predicted from the amino acid sequence of a protein. All 20 amino acids can be ranked according to their tendency to enter a hydrophobic or hydrophilic environment. With these values, the amino acid sequence of a protein can be analyzed using computer software to determine the average hydrophobicity of short amino acid sequences within the protein. A stretch of 18 to 20 amino acids in an α helix is long enough to span the membrane. If such a stretch contains a high percentage of hydrophobic amino acids, it is predicted to be a transmembrane α helix. However, such computer predictions must eventually be verified by experimentation.

Using a computer approach, many research groups have attempted to calculate the percentage of genes that encode transmembrane proteins in various species. **Table 5.3** shows the results of one such study. The estimated percentage of transmembrane proteins is substantial: 20–30% of all genes may encode transmembrane proteins. This trend is found throughout all domains of life, including archaea, bacteria, and eukaryotes. For example, about 30% of human genes encode transmembrane proteins. With a genome size of 20,000 to 25,000 different genes, the total number of genes that encode different transmembrane proteins is estimated at 6,000 to 7,500. The functions of many of them have yet to be determined. Identifying their functions will help researchers gain a better understanding of human biology. Likewise, medical researchers and pharmaceutical companies are interested in the identification of new transmembrane proteins that could be targets for effective new medications

Table 5.3	Estimated Percentage of Genes That Encode Transmembrane Proteins*
Organism	Percentage of genes that encode transmembrane proteins
Archaea	
Archaeoglobus fulgidus	24.2
Methanococcus jannaschii	20.4
Pyrococcus horikoshii	29.9
Bacteria	
Escherichia coli	29.9
Bacillus subtilis	29.2
Haemophilus influenzae	25.3
Eukaryotes	
Homo sapiens	29.7
Drosophila melanogaster	24.9
Arabidopsis thaliana	30.5
Saccharomyces cerevisiae	28.2

*Data from Stevens and Arkin (2000) *Proteins: Structure, Function, and Genetics* 39: 417–420. While the numbers may vary due to different computer programs and estimation techniques, the same general trends have been observed in other similar studies.

Membranes Are Semifluid

Let's now turn our attention to the dynamic properties of membranes. Although a membrane provides a critical interface between a cell and its environment, it is not a solid, rigid structure. Rather, biomembranes exhibit properties of **fluidity**, which means that individual molecules remain in close association yet have the ability to readily move within the membrane. Though membranes are often described as fluid, it is more appropriate to say they are **semifluid**. In a fluid substance, molecules can move in three dimensions. By comparison, most phospholipids can rotate freely around their long axes and move laterally within the membrane leaflet (**Figure 5.3a**). This type of motion is considered two-dimensional, which means it occurs within the plane of the membrane. Because rotational and lateral movements keep the fatty acyl tails within the hydrophobic interior, such movements are energetically favorable. At 37°C, a typical lipid molecule exchanges places with its neighbors about 10^7 times per second, and it can move several micrometers per second. At this rate, a lipid could traverse the length of a bacterial cell (approximately 1 μm) in only 1 second and the length of a typical animal cell in 10–20 seconds.

In contrast to rotational and lateral movements, the "flip-flop" of lipids from one leaflet to the opposite leaflet does not occur spontaneously. Energetically, such movements are unfavorable because the polar head of a phospholipid would have to be transported through the hydrophobic interior of the membrane. How are lipids moved from one leaflet to the other? The transport of lipids between leaflets requires the action of the enzyme flippase, which provides energy from the hydrolysis of ATP (**Figure 5.3b**).

Although most lipids tend to diffuse rotationally and laterally within the plane of the lipid bilayer, researchers have discovered that certain types of lipids in animal cells tend to strongly

Table 5.2	Historical Developments That Led to the Formulation of the Fluid-Mosaic Model
Date	**Description**
1917	Irving Langmuir made artificial membranes experimentally by creating a monolayer of lipids on the surface of water. The polar heads interacted with water, and nonpolar tails projected into the air.
1925	Evert Gorter and F. Grendel proposed that lipids form bilayers around cells. This was based on measurements of lipid content enclosing red blood cells that showed there was just enough lipid to surround the cell with two layers.
1935	Because proteins were also found in membranes, Hugh Davson and James Danielli proposed incorrectly that a phospholipid bilayer was sandwiched between two layers of protein.
1950s	Electron microscopy studies carried out by J.D. Robertson and others revealed that membranes look like a train track—two dark lines separated by a light space. Initially, these results were misinterpreted. Researchers thought the two dark lines were layers of proteins and the light area was the phospholipid bilayer. Later, it was correctly determined that the dark lines in these experiments are the phospholipid heads, which were heavily stained, and the light region is their phospholipid tails.
1966	Using freeze fracture electron microscopy (described later in this chapter), Daniel Branton concluded that membranes are bilayers because the freeze fracture procedure splits membranes in half, thus revealing proteins in the two membrane leaflets.
1972	S. Jonathan Singer and Garth Nicolson proposed the fluid-mosaic model described in Figure 5.1. Their model was consistent with the observation that membrane proteins are globular, and some are known to span the phospholipid bilayer and project from both sides.

Figure 5.2 **Types of membrane proteins.** Integral membrane proteins are of two types: transmembrane proteins and lipid-anchored proteins. Peripheral membrane proteins are noncovalently bound to the hydrophilic regions of integral membrane proteins or to the polar head groups of lipids. Inset: The protein shown in the inset contains seven transmembrane segments in an α helix structure. The transmembrane α helices are depicted as cylinders. This particular protein, bacteriorhodopsin, functions as an ion pump in halophilic (salt-loving) archaea.

other functions. Some of these functions were considered in Chapter 4. For example, we examined how membrane proteins in the smooth ER membrane function as enzymes that break down glycogen. Later in this chapter, we will explore how membrane proteins are involved in transporting ions and molecules across membranes. Other key functions of membrane proteins are examined in later chapters, including ATP synthesis (Chapter 7), photosynthesis (Chapter 8), cell signaling (Chapter 9), and cell-to-cell adhesion (Chapter 10).

Membrane proteins have different ways of associating with a membrane (**Figure 5.2**). An **integral membrane protein**, also called an intrinsic membrane protein, cannot be released from the membrane unless the membrane is dissolved with an organic solvent or detergent—in other words, you would have to disrupt the integrity of the membrane to remove it. The most common type of integral membrane protein is a **transmembrane protein**, which has one or more regions that are physically inserted into the hydrophobic region of the phospholipid bilayer. These regions, the **transmembrane segments**, are stretches of nonpolar amino acids that span or traverse the membrane from one leaflet to the

other. In most transmembrane proteins, each transmembrane segment is folded into an α helix structure. Such a segment is stable in a membrane because the nonpolar amino acids can interact favorably with the hydrophobic fatty acyl tails of the lipid molecules.

A second type of integral membrane protein, known as a **lipid-anchored protein**, has a lipid molecule that is covalently attached to an amino acid side chain within the protein. The fatty acyl tails are inserted into the hydrophobic portion of the membrane and thereby keep the protein firmly attached to the membrane.

Peripheral membrane proteins, also called extrinsic proteins, are another category of membrane protein. They do not interact with the hydrophobic interior of the phospholipid bilayer. Instead, they are noncovalently bound to regions of integral membrane proteins that project out from the membrane, or they are bound to the polar head groups of phospholipids. Peripheral membrane proteins are typically bound to the membrane by hydrogen and/or ionic bonds. For this reason, they usually can be removed from the membrane experimentally by varying the pH or salt concentration.

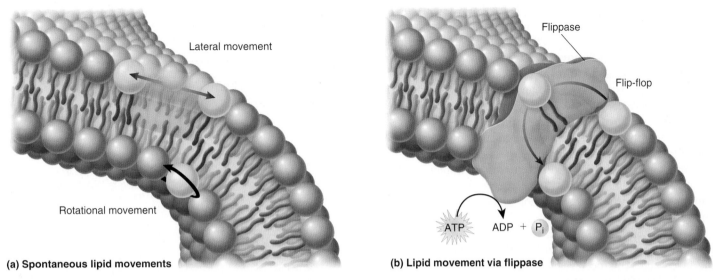

(a) Spontaneous lipid movements

(b) Lipid movement via flippase

Figure 5.3 Semifluidity of the lipid bilayer. (a) Spontaneous movements in the bilayer. Lipids can rotate (that is, move 360°) and move laterally (for example, from left to right in the plane of the bilayer). (b) Flip-flop does not happen spontaneously, because the polar head group would have to pass through the hydrophobic region of the bilayer. Instead, the enzyme flippase uses ATP to flip phospholipids from one leaflet to the other.

Concept check: In an animal cell, how can changes in lipid composition affect membrane fluidity?

associate with each other to form structures called lipid rafts. As the term raft suggests, a **lipid raft** is a group of lipids that float together as a unit within a larger sea of lipids. Lipid rafts have a lipid composition that differs from the surrounding membrane. For example, they usually have a high amount of cholesterol. In addition, lipid rafts may contain unique sets of lipid-anchored proteins and transmembrane proteins. The functional importance of lipid rafts is the subject of a large amount of current research. Lipid rafts may play an important role in endocytosis (discussed later in this chapter) and cell signaling (Chapter 9).

Lipid Composition Affects Membrane Fluidity

The biochemical properties of phospholipids have a profound effect on the fluidity of the phospholipid bilayer. One key factor is the length of fatty acyl tails, which range from 14 to 24 carbon atoms, with 18 to 20 carbons being the most common. Shorter acyl tails are less likely to interact with each other, which makes the membrane more fluid. A second important factor is the presence of double bonds in the acyl tails. When a double bond is present, the lipid is said to be **unsaturated** with respect to the number of hydrogens that can be bound to the carbon atoms (refer back to Figure 3.10). A double bond creates a kink in the fatty acyl tail (see inset to Figure 5.1), making it more difficult for neighboring tails to interact and making the bilayer more fluid. As described in Chapter 3, unsaturated lipids tend to be more liquid compared to saturated lipids that often form solids at room temperature (refer back to Figure 3.11).

A third factor affecting fluidity is the presence of cholesterol, which is a short and rigid planar molecule produced by animal cells (see inset to Figure 5.1). Plant cell membranes contain phytosterols that resemble cholesterol in their chemical structure.

Cholesterol tends to stabilize membranes; its effects depend on temperature. At higher temperatures, such as those observed in mammals that maintain a constant body temperature, cholesterol makes the membrane less fluid. At lower temperatures, such as icy water, cholesterol has the opposite effect. It makes the membrane more fluid and prevents it from freezing.

An optimal level of bilayer fluidity is essential for normal cell function, growth, and division. If a membrane is too fluid, which may occur at higher temperatures, it can become leaky. However, if a membrane becomes too solid, which may occur at lower temperatures, the functioning of membrane proteins will be inhibited. How can organisms cope with changes in temperature? The cells of many species adapt to changes in temperature by altering the lipid composition of their membranes. For example, when the water temperature drops, the cells of certain fish will incorporate more cholesterol in their membranes. If a plant cell is exposed to high temperatures for many hours or days, it will alter its lipid composition to have longer fatty acyl tails and fewer double bonds.

Membrane Proteins May Diffuse in the Plane of the Membrane or Be Restricted in Their Movement

Like lipids, many transmembrane proteins may rotate and laterally move throughout the plane of a membrane. Because transmembrane proteins are larger than lipids, they move within the membrane at a much slower rate. Flip-flop of transmembrane proteins does not occur because the proteins also contain hydrophilic regions that project out from the phospholipid bilayer. It would be energetically unfavorable for the hydrophilic regions of membrane proteins to pass through the hydrophobic portion of the phospholipid bilayer.

Researchers can examine the lateral movements of lipids and transmembrane proteins by a variety of methods. In 1970, Larry Frye and Michael Edidin conducted an experiment that verified the lateral movement of transmembrane proteins (**Figure 5.4**). Mouse and human cells were mixed together and exposed to agents that caused them to fuse with each other. Some cells were cooled to 0°C, while others were incubated at 37°C before being cooled. Both sets of cells were then exposed to fluorescently labeled antibodies that became specifically bound to a mouse transmembrane protein called H-2. The fluorescent label was observed with a fluorescence microscope. If the cells were maintained at 0°C, a temperature that greatly inhibits lateral movement, the fluorescence was seen on only one side of the fused cell. However, if the cells were incubated for several hours at 37°C and then cooled to 0°C, the fluorescence was distributed throughout the plasma membrane of the fused cell. This occurred because the higher temperature allowed the lateral movement of the H-2 protein throughout the fused cell.

Unlike the example shown in Figure 5.4, not all transmembrane proteins are capable of rotational and lateral movement. Depending on the cell type, 10–70% of membrane proteins may be restricted in their movement. Transmembrane proteins may be bound to components of the cytoskeleton, which restricts the proteins from moving (**Figure 5.5**). Also, membrane proteins may be attached to molecules that are outside the cell, such as the interconnected network of proteins that forms the extracellular matrix of animal cells.

Glycosylation of Lipids and Proteins Serves a Variety of Cellular Functions

As mentioned earlier, the third constituent of cellular membranes is carbohydrate. **Glycosylation** refers to the process of covalently attaching a carbohydrate to a lipid or protein. When a carbohydrate is attached to a lipid, this creates a **glycolipid**, whereas attachment to a protein produces a **glycoprotein**.

What is the function of glycosylation? Though the roles of carbohydrate in cell structure and function are not entirely understood, some functional consequences of glycosylation have emerged. The carbohydrates attached to proteins and lipids have well-defined structures that, in some cases, serve as recognition signals for other cellular proteins. For example, proteins destined for the lysosome are glycosylated and have a sugar (mannose-6-phosphate) that is recognized by other proteins in the cell that target the glycosylated protein from the Golgi to the lysosome. Similarly, glycolipids and glycoproteins often play a role in cell surface recognition. When glycolipid and glycoproteins are found in the plasma membrane, the carbohydrate portion is located in the extracellular region. During embryonic development in animals, significant cell movement occurs. Layers of cells slide over each other to create body structures such as the spinal cord and internal organs. The proper migration of individual cells and cell layers relies on the recognition of cell types via the carbohydrates on their cell surfaces.

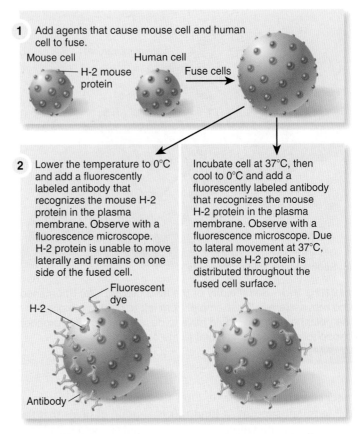

1 Add agents that cause mouse cell and human cell to fuse.
Mouse cell — H-2 mouse protein   Human cell   Fuse cells

2 Lower the temperature to 0°C and add a fluorescently labeled antibody that recognizes the mouse H-2 protein in the plasma membrane. Observe with a fluorescence microscope. H-2 protein is unable to move laterally and remains on one side of the fused cell.

Incubate cell at 37°C, then cool to 0°C and add a fluorescently labeled antibody that recognizes the mouse H-2 protein in the plasma membrane. Observe with a fluorescence microscope. Due to lateral movement at 37°C, the mouse H-2 protein is distributed throughout the fused cell surface.

H-2 — Fluorescent dye
Antibody

Figure 5.4 A method to measure the lateral movement of membrane proteins.

Concept check: Explain why the H-2 proteins are found on only one side of the cell when the cells were incubated at 0°C.

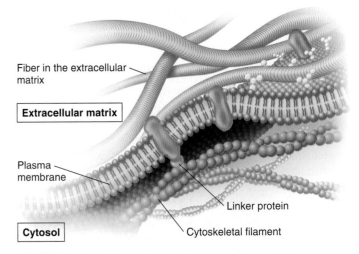

Fiber in the extracellular matrix
Extracellular matrix
Plasma membrane
Cytosol
Linker protein
Cytoskeletal filament

Figure 5.5 Attachment of transmembrane proteins to the cytoskeleton and extracellular matrix of an animal cell. Some transmembrane proteins have regions that project into the cytosol and are anchored to large cytoskeletal filaments via linker proteins. Being bound to these filaments restricts the movement of these proteins. Similarly, transmembrane proteins may bind to large, immobile components in the extracellular matrix that also restrict the movement of the proteins.

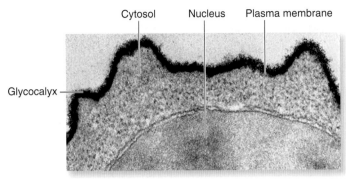

Cytosol   Nucleus   Plasma membrane

Glycocalyx

200 nm

Figure 5.6   **A micrograph of the cell coat, or glycocalyx, of an animal cell.**   This figure shows a lymphocyte—a type of white blood cell—stained with ethidium red, which emphasizes the thick carbohydrate layer that surrounds the cell. Note: The term glycocalyx (from the Greek, meaning sugar coat) is also used to describe other carbohydrate surfaces, such as a carbohydrate layer that surrounds certain strains of bacteria (refer back to Figure 4.4).

Concept check: *What is an important function of the glycocalyx?*

Carbohydrates also play a role in determining blood type, which is described in Chapter 16 (look ahead to Table 16.3).

Carbohydrates can also have a protective effect. The term **cell coat**, or **glycocalyx**, is used to describe the carbohydrate-rich zone on the surface of certain animal cells that shields the cell from mechanical and physical damage (**Figure 5.6**). The carbohydrate portion of glycosylated proteins protects them from the harsh conditions of the extracellular environment and degradation by extracellular proteases, which are enzymes that digest proteins.

Membrane Structure Can Be Viewed with an Electron Microscope

Electron microscopy, discussed in Chapter 4, is a valuable tool to probe membrane structure and function. In transmission electron microscopy (TEM), a biological sample is thin sectioned

Figure 5.7   **Electron micrographs of a cellular membrane.** (a) In the standard form of TEM, a membrane appears as two dark parallel lines. These lines are the lipid head groups, which stain darkly with osmium tetroxide. The fatty acyl tails do not stain well and appear as a light region sandwiched between the dark lines. (b) In the technique of freeze fracture electron microscopy, a sample is frozen in liquid nitrogen and fractured. The sample is then coated with metal and viewed under the electron microscope.

Concept check: *If a heavy metal labeled the hydrophobic tails rather than the polar head groups (as osmium tetroxide does), do you think you would see a bilayer (that is, a railroad track) under TEM?*

and stained with heavy-metal dyes such as osmium tetroxide. This compound binds tightly to the polar head groups of phospholipids, but it does not bind well to the fatty acyl tails. As shown in **Figure 5.7a**, membranes stained with osmium tetroxide resemble a railroad track. Two thin dark lines, which are the stained polar head groups, are separated by a uniform light space about 2 nm thick. This railroad track morphology is seen when cell membranes are subjected to electron microscopy.

Due to the incredibly small size of biological membranes, scientists have not been able to invent instruments small enough to dissect them. However, a specialized form of TEM, freeze fracture electron microscopy (FFEM), can be used to

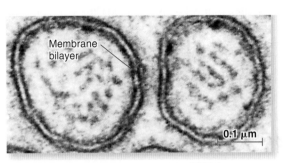

Membrane bilayer

0.1 μm

(a) Transmission electron microscopy (TEM)

Direction of fracture

Transmembrane protein

Lipid bilayer

(b) Freeze fracture electron microscopy (FFEM)

P face exposed

E face exposed

P face   E face

Cytosolic leaflet

Extracellular leaflet

E face

P face

0.4 μm

analyze the interiors of phospholipid bilayers. Russell Steere invented this method in 1957. In FFEM, a sample is frozen in liquid nitrogen and split with a knife (**Figure 5.7b**). The knife does not actually cut through the bilayer, but it fractures the frozen sample. Due to the weakness of the central membrane region, the leaflets separate into a P face (the protoplasmic face that was next to the cytosol) and the E face (the extracellular face). Most transmembrane proteins do not break in half. They remain embedded within one of the leaflets, usually in the P face. The samples, which are under a vacuum, are then sprayed with a heavy metal such as platinum, which coats the sample and reveals architectural features within each leaflet. When viewed with an electron microscope, membrane proteins are visible as bumps that provide significant three-dimensional detail about their form and shape.

5.2 Synthesis of Membrane Components in Eukaryotic Cells

As we have seen, cellular membranes are composed of lipids, proteins, and carbohydrates. Most of the membrane components of eukaryotic cells are made at the endoplasmic reticulum (ER). In this section, we will begin by considering how phospholipids are synthesized at the ER membrane. We will then examine the process by which transmembrane proteins are inserted into the ER membrane and explore how some proteins are glycosylated.

Lipid Synthesis Occurs at the ER Membrane

In eukaryotic cells, the cytosol and endomembrane system work together to synthesize most lipids. This process occurs at the cytosolic leaflet of the smooth ER membrane. **Figure 5.8** shows a simplified pathway for the synthesis of phospholipids. The building blocks for a phospholipid are two fatty acids, each with an acyl tail, one glycerol molecule, one phosphate, and a polar head group. These building blocks are made via enzymes in the cytosol, or they are taken into cells from food. To begin the process of phospholipid synthesis, the fatty acids are activated by attachment to an organic molecule called coenzyme A (CoA). This activation promotes the bonding of the two fatty acids to a glycerol-phosphate molecule, and the resulting molecule is inserted into the cytosolic leaflet of the ER membrane. The phosphate is removed from glycerol, and then a polar molecule already linked to phosphate is attached to glycerol. In the example shown in Figure 5.8, the polar head group contains choline, but many other types are possible. Phospholipids are initially inserted into the cytosolic leaflet. Flippases in the ER membrane transfer some of the newly made lipids to the other leaflet so that similar amounts of lipids are in both leaflets.

The lipids made in the ER membrane can be transferred to other membranes in the cell by a variety of mechan-

| 1 | In the cytosol, fatty acids are activated by the attachment of a CoA molecule. | 2 | The activated fatty acids bond to glycerol-phosphate and are inserted into the cytosolic leaflet of the ER membrane via acyl transferase. | 3 | The phosphate is removed by a phosphatase enzyme. | 4 | A choline already linked to phosphate is attached via choline phosphotransferase. | 5 | Flippases transfer some of the phospholipids to the other leaflet. |

Figure 5.8 A simplified pathway for the synthesis of membrane phospholipids at the ER membrane.   Note: Phosphate is abbreviated P when it is attached to an organic molecule and P_i when it is unattached. The subscript i refers to the inorganic form of phosphate.

Concept check: *How are lipids transferred to the other leaflet of the ER membrane?*

isms. Phospholipids in the ER can diffuse laterally to the nuclear envelope. In addition, lipids can be transported via vesicles to the Golgi, lysosomes, vacuoles, or plasma membrane. A third mode of lipid transfer involves **lipid exchange proteins**, which extract a lipid from one membrane, diffuse through the cell, and insert the lipid into another membrane. Such transfer can occur between any two membranes, even between the endomembrane system and semiautonomous organelles. For example, lipid exchange proteins can transfer lipids between the ER and mitochondria. In addition, chloroplasts and mitochondria can synthesize certain types of lipids that can be transferred from these organelles to other cellular membranes via lipid exchange proteins.

Most Transmembrane Proteins Are First Inserted into the ER Membrane

In Chapter 4 (Section 4.6), we learned that eukaryotic proteins contain sorting signals that direct them to their proper destination. With the exception of proteins destined for semiautonomous organelles, most transmembrane proteins contain an ER signal sequence that directs them to the ER membrane. If a polypeptide also contains a stretch of 20 amino acids that are mostly hydrophobic and form an α helix, this region will become a transmembrane segment. In the example shown in **Figure 5.9**, the polypeptide contains one such sequence. After the ER signal sequence is removed by signal peptidase (refer back to Figure

4.29), a membrane protein with a single transmembrane segment is the result. Other polypeptides may contain more than one transmembrane segment. Each time a polypeptide sequence contains a stretch of 20 hydrophobic amino acids that forms an α helix, an additional transmembrane segment is synthesized into the membrane. From the ER, membrane proteins can be transferred via vesicles to other regions of the cell, such as the Golgi, lysosomes, vacuoles, or plasma membrane.

Glycosylation of Proteins Occurs in the ER and Golgi Apparatus

As mentioned, glycosylation is the attachment of carbohydrate to a lipid or protein, producing a glycolipid or glycoprotein. Two forms of protein glycosylation occur in eukaryotes: N-linked and O-linked. N-linked glycosylation, which also occurs in archaea, involves the attachment of a carbohydrate to the amino acid asparagine in a polypeptide chain. It is called N-linked because the carbohydrate attaches to a nitrogen atom of the asparagine side chain. For this to occur, a group of 14 sugar molecules are built onto a lipid called dolichol, which is found in the ER membrane. This carbohydrate tree is then transferred to an asparagine as a polypeptide is synthesized into the ER lumen through a channel protein (**Figure 5.10**). The carbohydrate

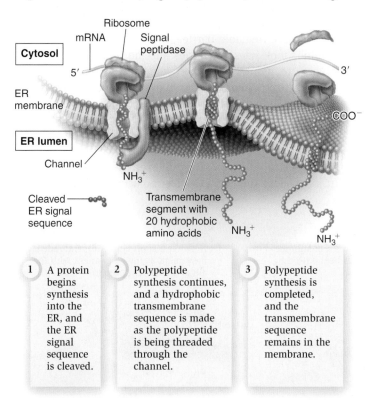

| 1 | A protein begins synthesis into the ER, and the ER signal sequence is cleaved. | 2 | Polypeptide synthesis continues, and a hydrophobic transmembrane sequence is made as the polypeptide is being threaded through the channel. | 3 | Polypeptide synthesis is completed, and the transmembrane sequence remains in the membrane. |

Figure 5.9   Insertion of membrane proteins into the ER membrane.

Concept check: *What structural feature of a protein causes a region to form a transmembrane segment?*

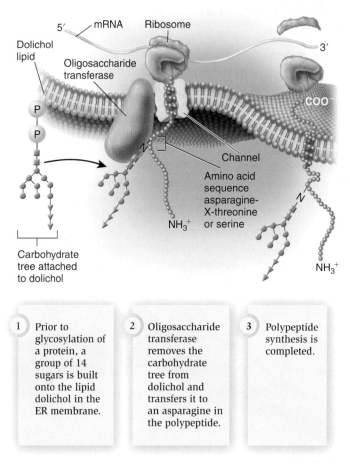

| 1 | Prior to glycosylation of a protein, a group of 14 sugars is built onto the lipid dolichol in the ER membrane. | 2 | Oligosaccharide transferase removes the carbohydrate tree from dolichol and transfers it to an asparagine in the polypeptide. | 3 | Polypeptide synthesis is completed. |

Figure 5.10   N-linked glycosylation in the endoplasmic reticulum.

tree is attached only to asparagines occurring in the sequence asparagine—X—threonine or asparagine—X—serine, where X could be any amino acid except proline. An enzyme in the ER, oligosaccharide transferase, recognizes this sequence and transfers the carbohydrate tree from dolichol to the asparagine. Following this initial glycosylation step, the carbohydrate tree is further modified as other enzymes in the ER attach additional sugars or remove sugars. After a glycosylated protein is transferred to the Golgi by vesicle transport, enzymes in the Golgi usually modify the carbohydrate tree as well. N-linked glycosylation commonly occurs on membrane proteins that are transported to the cell surface.

The second form of glycosylation, O-linked glycosylation, occurs only in the Golgi apparatus. This form involves the addition of a string of sugars to the oxygen atom of serine or threonine side chains in polypeptides. In animals, O-linked glycosylation is important for the production of proteoglycans, which are highly glycosylated proteins that are secreted from cells and help to organize the extracellular matrix that surrounds cells. Proteoglycans are also a component of mucus, a slimy material that coats many cell surfaces and is secreted into fluids such as saliva. High concentrations of carbohydrates give mucus its slimy texture.

5.3 Membrane Transport

We now turn to one of the key functions of membranes, **membrane transport**—the movement of ions and molecules across biological membranes. All cells contain a plasma membrane that is a **selectively permeable** barrier between a cell and its external environment. As a protective envelope, its structure ensures that essential molecules such as glucose and amino acids enter the cell, metabolic intermediates remain in the cell, and waste products exit. The selective permeability of the plasma membrane allows the cell to maintain a favorable internal environment.

Substances can move across a membrane in three general ways (**Figure 5.11**). **Diffusion** occurs when a substance moves from a region of high concentration to a region of lower concentration. Some substances can move directly through a phospholipid bilayer via diffusion. A second way that substances can move across membranes is via **facilitated diffusion**. In this case, a transport protein provides a passageway for the substance to cross the membrane. Both diffusion and facilitated diffusion are examples of **passive transport**—the transport of a substance across a membrane from a region of high concentration to a region of lower concentration. Passive transport does not require an input of energy. In contrast, a third mode of transport, called **active transport**, moves a substance from an area of low concentration to high concentration or against a concentration gradient with the aid of a transport protein. This type of transport requires an input of energy, such as ATP hydrolysis.

In this section, we begin with a discussion of how the phospholipid bilayer presents a barrier to the movement of ions and molecules across membranes, and then consider the concept of gradients across membranes. We will then focus on transport proteins, which carry out facilitated diffusion and active transport. Such proteins play a key role in the selective permeability of biological membranes. Finally, we will examine two mechanisms found in eukaryotic cells for the transport of substances via membrane vesicles.

The Phospholipid Bilayer Is a Barrier to the Diffusion of Hydrophilic Solutes

Because of their hydrophobic interiors, phospholipid bilayers present a formidable barrier to the movement of ions and hydrophilic molecules. Such ions and molecules are called solutes; they are dissolved in water, which is a solvent. The rate of diffusion across a phospholipid bilayer depends on the chemistry of the solute and its concentration. **Figure 5.12** compares the relative permeabilities of various solutes through an artificial phospholipid bilayer that does not contain any proteins or carbohydrates. Gases and a few small, uncharged molecules can passively diffuse across the bilayer. However, the rate of diffusion of ions and larger polar molecules, such as sugars,

(a) Diffusion—passive transport   (b) Facilitated diffusion—passive transport   (c) Active transport

Figure 5.11 Three general types of membrane transport.

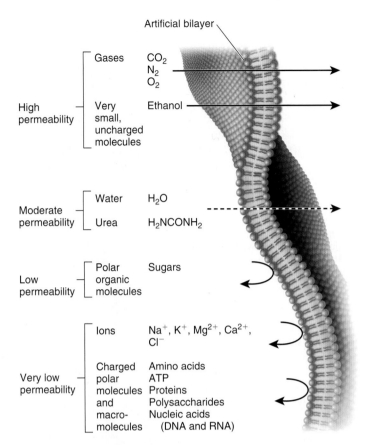

Figure 5.12 Relative permeability of an artificial phospholipid bilayer to a variety of solutes. Solutes that easily penetrate are shown with a straight arrow that passes through the bilayer. The dotted line indicates solutes that have moderate permeability. The remaining solutes shown at the bottom are relatively impermeable.

Concept check: Which amino acid (described in Chapter 3; see Figure 3.14) would you expect to cross an artificial membrane more quickly, leucine or lysine?

$$NH_2-C(=O)-NH_2 \qquad CH_3-CH_2-NH-C(=O)-NH-CH_2-CH_3$$
Urea · · · · · · · · · · · · · · · · · · · Diethylurea

Figure 5.13 Structures of urea and diethylurea.

Concept check: Which molecule would you expect to pass through a phospholipid bilayer more quickly, methanol (CH_3OH) or methane (CH_4)?

is relatively slow. Similarly, macromolecules, such as proteins and polysaccharides, do not readily cross a lipid bilayer.

When we consider the steps of diffusion among different solutes, the greatest variation occurs in the ability of solutes to enter the hydrophobic interior of the bilayer. As an example, let's compare urea and diethylurea. Compared to urea, diethylurea is much more hydrophobic because it contains two nonpolar ethyl groups ($-CH_2CH_3$) (**Figure 5.13**). For this reason, it can more readily pass through the hydrophobic region of the bilayer. The rate of diffusion of diethylurea through a phospholipid bilayer is about 50 times faster than urea.

Cells Maintain Gradients Across Their Membranes

A hallmark of living cells is their ability to maintain a relatively constant internal environment that is distinctively different from their external environment. This involves establishing gradients of solutes across the plasma membrane and organellar membranes. When we speak of a **transmembrane gradient**, we mean the concentration of a solute is higher on one side of a membrane than the other. For example, immediately after you eat a meal containing carbohydrates, a higher concentration of glucose is found outside your cells compared to inside (**Figure 5.14a**). This is an example of a chemical gradient.

Gradients involving ions have two components—electrical and chemical. An **electrochemical gradient** is a dual gradient that has both electrical and chemical components (**Figure 5.14b**). It occurs with solutes that have a net positive or negative charge. For example, let's consider a gradient involving Na^+. An electrical gradient could exist in which the amount of net positive charge outside a cell is greater than inside. In

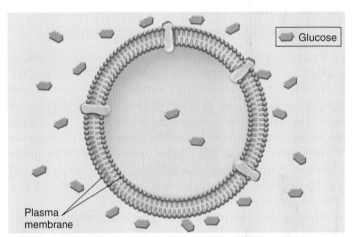

(a) Chemical gradient for glucose—a higher glucose concentration outside the cell

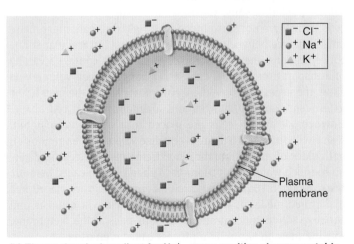

(b) Electrochemical gradient for Na⁺—more positive charges outside the cell and a higher Na⁺ concentration outside the cell

Figure 5.14 Gradients across cell membranes.

Figure 5.14b, an electrical gradient is due to differences in the amounts of different types of ions across the membrane, including Na^+, K^+, and Cl^-. At the same time, a chemical gradient—a difference in Na^+ concentration across the membrane—could exist in which the concentration of Na^+ outside is greater than inside. The Na^+ electrochemical gradient is composed of both an electrical gradient due to charge differences across the membrane along with a chemical gradient for Na^+. Transmembrane gradients of ions and other solutes are a universal feature of all living cells.

One way to view the transport of solutes across membranes is to consider how the transport process affects the pre-existing gradients across membranes. Passive transport tends to dissipate a pre-existing gradient. It is a process that is energetically favorable and does not require an input of energy. As mentioned, passive transport can occur in two ways, via diffusion or facilitated diffusion (see Figure 5.11a,b). By comparison, active transport produces a chemical gradient or electrochemical gradient. The formation of a gradient requires an input of energy.

Osmosis Is the Movement of Water Across Membranes to Balance Solute Concentrations

Let's now turn our attention to how gradients affect the movement of water across membranes. When the solute concentrations on both sides of the plasma membrane are equal, the two solutions are said to be **isotonic** (**Figure 5.15a**). However, we have also seen that transmembrane gradients commonly exist across membranes. When the solute concentration outside the cell is higher, it is said to be **hypertonic** relative to the inside of the cell (**Figure 5.15b**). Alternatively, the outside of the cell could be **hypotonic**—have a lower solute concentration relative to the inside (**Figure 5.15c**).

If solutes cannot readily move across the membrane, water will move and tend to balance the solute concentrations. In this process, called **osmosis**, water diffuses across a membrane from the hypotonic compartment into the hypertonic compartment. Cells generally have a high internal concentration of a variety of solutes, including ions, sugars, amino acids, and so on. Animal cells, which are not surrounded by a rigid cell wall, must maintain a balance between the extracellular and intracellular solute concentrations; they are isotonic. Animal cells contain a variety of transport proteins that can sense changes in cell volume and allow the necessary movements of solutes across the membrane to prevent osmotic changes and maintain normal cell shape. However, if animal cells are placed in a hypotonic medium, water will diffuse into them to equalize solute concentrations on both sides of the membrane. In extreme cases, a cell may take up so much water that it ruptures, a phenomenon called **osmotic lysis** (**Figure 5.16a**). Alternatively, if animal cells are placed in a hypertonic medium, water will exit the cells via osmosis and equalize solute concentrations on both sides of the membrane, causing them to shrink in a process called **crenation**.

How does osmosis affect cells with a rigid cell wall, such as bacteria, fungi, algae, and plant cells? If the extracellular fluid is hypotonic, a plant cell will take up a small amount of water, but the cell wall prevents major changes in cell size (**Figure**

(a) Outside isotonic

The solute concentration outside the cell is **isotonic** to the inside of the cell.

Solute

Cytosol

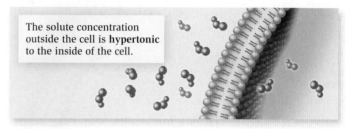

(b) Outside hypertonic

The solute concentration outside the cell is **hypertonic** to the inside of the cell.

(c) Outside hypotonic

The solute concentration outside the cell is **hypotonic** to the inside of the cell.

Figure 5.15   Relative solute concentrations outside and inside cells.

5.16b). Alternatively, if the extracellular fluid surrounding a plant cell is hypertonic, water will exit the cell and the plasma membrane will pull away from the cell wall, a process called **plasmolysis**.

The tendency of water to move into a cell creates an **osmotic pressure**, which is defined as the hydrostatic pressure required to stop the net flow of water across a membrane due to osmosis. In plant cells, osmotic pressure is also called **turgor pressure** or, simply, cell turgor. The turgor pressure pushes the plasma membrane against the rigid cell wall. An appropriate level of turgor is needed for plant cells to maintain their proper structure (**Figure 5.17**). If a plant has insufficient water, the extracellular fluid surrounding plant cells becomes hypertonic. This causes the plasma membrane to pull away from the cell wall, and the turgor pressure drops. Such a loss of turgor pressure is associated with wilting.

Some freshwater microorganisms, such as amoebae and paramecia, can exist in extremely hypotonic environments where the external solute concentration is always much lower than the concentration of solutes in their cytosol. Because of the great tendency for water to move into the cell by osmosis, such organisms contain one or more contractile vacuoles to prevent osmotic lysis. A contractile vacuole takes up water from the cytosol and periodically discharges it by fusing the vacuole with the plasma membrane (**Figure 5.18**).

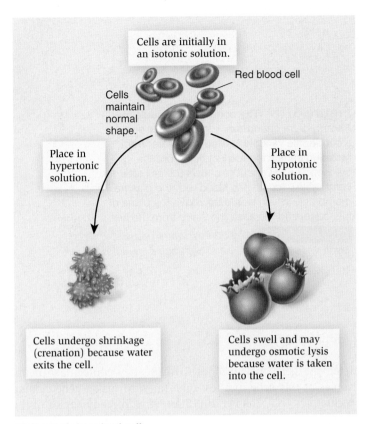

(a) Osmosis in animal cells

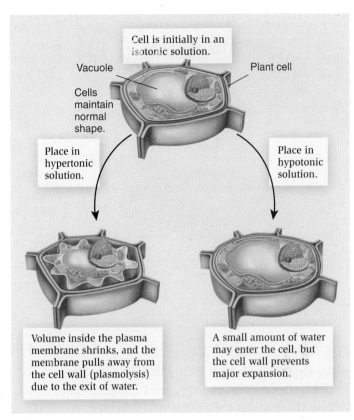

(b) Osmosis in plant cells

Figure 5.16   **The phenomenon of osmosis.**   (a) In cells that lack a cell wall, such as animal cells, osmosis may promote cell swelling or shrinkage (crenation). (b) In cells that have a rigid cell wall, such as plant cells, a hypotonic medium causes only a minor amount of expansion, whereas a hypertonic medium causes the plasma membrane to pull away from the cell wall.

Concept check: *Let's suppose the inside of a cell has a solute concentration of 0.3 M, while the outside is 0.2 M. If the membrane is impermeable to the solutes, which direction will water move?*

(a) Sufficient water

(b) Wilting

Figure 5.17   **Wilting in plants.**   (a) When a plant has plenty of water, the slightly hypotonic surroundings cause the vacuole to store water. The increased size of the vacuole influences the volume of the cytosol, thereby exerting a turgor pressure against the cell wall. (b) Under dry conditions, water is released from the cytosol into the hypertonic extracellular medium. The vacuole also shrinks, because it loses water to the cytosol. Turgor pressure is lost, which causes the plant to wilt.

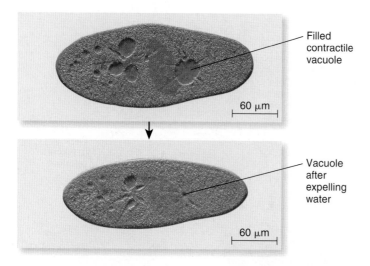

Filled
contractile
vacuole

60 μm

Vacuole
after
expelling
water

60 μm

Figure 5.18   The contractile vacuole in
Paramecium caudatum.   In the upper photo,
a contractile vacuole is filled with water from
radiating canals that collect fluid from the
cytosol. The lower photo shows the cell after the
contractile vacuole has fused with the plasma
membrane (which would be above the plane of
this page) and released the water from the cell.

Concept check: Why do freshwater protists,
such as P. caudatum, need contractile vacuoles?

FEATURE INVESTIGATION

Agre Discovered That Osmosis Occurs More Quickly in Cells with Transport Proteins That Allow the Facilitated Diffusion of Water

In living cells, the flow of water may occur by diffusion through the phospholipid bilayer. However, in the 1980s, researchers also discovered that certain cell types allow water to move across the plasma membrane at a much faster rate than would be predicted by diffusion. For example, water moves very quickly across the membrane of red blood cells, which causes them to shrink and swell in response to changes in extracellular solute concentrations. Likewise, bladder and kidney cells, which play a key role in regulating water balance in the bodies of vertebrates, allow the rapid movement of water across their membranes. Based on these observations, researchers speculated that certain cell types might have proteins in their plasma membranes that permit the rapid movement of water.

One approach to characterize a new protein is to first identify a protein based on its relative abundance in a particular cell type and then attempt to determine the protein's function. This rationale was applied to the discovery of proteins that allow the rapid movement of water across membranes. Peter Agre and his colleagues first identified a protein that was abundant in red blood cells and kidney cells but not found in many other cell types. Though they initially did not know the function of the protein, its physical structure was similar to other proteins that were already known to function as transport proteins. They named this protein CHIP28, which stands for <u>c</u>hannel-forming <u>i</u>ntegral membrane <u>p</u>rotein with a molecular mass of <u>28</u>,000 Da. During the course of their studies, they also identified and isolated the gene that encodes CHIP28.

In 1992, Agre and his colleagues conducted experiments to determine if CHIP28 functions in the transport of water across membranes (**Figure 5.19**). Because they already had iso-

lated the gene that encodes CHIP28, they could make many copies of this gene in a test tube (in vitro) using gene cloning techniques (see Chapter 20). Starting with many copies of the gene in vitro, they added an enzyme to transcribe the gene into mRNA that encodes the CHIP28 protein. This mRNA was then injected into frog oocytes, chosen because frog oocytes are large, easy to inject, and lack pre-existing proteins in their plasma membranes that allow the rapid movement of water. Following injection, the mRNA was expected to be translated into CHIP28 proteins that should be inserted into the plasma membrane of the oocytes. After allowing sufficient time for this to occur, the oocytes were placed in a hypotonic medium. As a control, oocytes that had not been injected with CHIP28 mRNA were also exposed to a hypotonic medium.

As you can see in the data, a striking difference was observed between oocytes that expressed CHIP28 versus the control. Within minutes, oocytes that contained the CHIP28 protein were seen to swell due to the rapid uptake of water. Three to 5 minutes after being placed in a hypotonic medium, they actually burst! By comparison, the control oocytes did not swell as rapidly, and they did not rupture even after 1 hour. Taken together, these results are consistent with the hypothesis that CHIP28 functions as a transport protein that allows the facilitated diffusion of water across the membrane. Many subsequent studies confirmed this observation. Later, CHIP28 was renamed **aquaporin** to indicate its newly identified function of allowing water to diffuse through a pore in the membrane (**Figure 5.20**). More recently, the three-dimensional structure of aquaporin was determined (see chapter-opening photo). Agre was awarded the Nobel Prize in 2003 for this work.

Aquaporin is an example of a transport protein called a channel. Next, we will discuss the characteristics of channels and other types of transport proteins.

Figure 5.19   The discovery of water channels by Agre.

HYPOTHESIS   CHIP28 may function as a water channel.

KEY MATERIALS   Prior to this work, a protein called CHIP28 was identified that is abundant in red blood cells and kidney cells. The gene that encodes this protein was cloned, which means that many copies of the gene were made in a test tube.

Experimental level   Conceptual level

1   Add an enzyme (RNA polymerase) and nucleotides to a test tube that contains many copies of the CHIP28 gene. This results in the synthesis of many copies of CHIP28 mRNA.

Enzymes and nucleotides

CHIP28 mRNA   RNA polymerase

CHIP28 DNA

2   Inject the CHIP28 mRNA into frog oocytes. Wait several hours to allow time for the mRNA to be translated into CHIP28 protein at the ER membrane and then moved via vesicles to the plasma membrane.

Frog oocyte

Nucleus

CHIP28 mRNA

Cytosol

CHIP28 protein is inserted into the plasma membrane.

CHIP28 protein

Ribosome

3   Place oocytes into a hypotonic medium and observe under a light microscope. As a control, also place oocytes that have not been injected with CHIP28 mRNA into a hypotonic medium and observe by microscopy.

Control

4   **THE DATA**

Oocyte

Oocyte rupturing

3–5 minutes

Control   CHIP28

Control   CHIP28

CHIP28 protein

5   **CONCLUSION**   The CHIP28 protein, now called aquaporin, allows the rapid movement of water across the membrane.

6   **SOURCE**   Preston, G.M., Carroll, T.P., Guggino, W.B., and Agre, P. 1992. Appearance of water channels in *Xenopus* oocytes expressing red cell CHIP28 protein. *Science* 256:385–387.

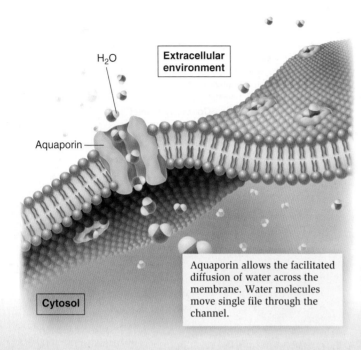

H₂O

Extracellular environment

Aquaporin

Aquaporin allows the facilitated diffusion of water across the membrane. Water molecules move single file through the channel.

Cytosol

Experimental Questions

1. What observations about particular cell types in the human body led to the experimental strategy of Figure 5.19?

2. What were the characteristics of CHIP28 that made Agre and associates speculate that it may transport water? In your own words, briefly explain how they were able to test the hypothesis that CHIP28 may have this function.

3. Explain how the results of the experiment of Figure 5.19 support the proposed hypothesis.

Figure 5.20 Function and structure of aquaporin. Aquaporin is found in the membrane of certain cell types and allows the rapid diffusion of water across the membrane. The chapter-opening photo shows the structure of aquaporin that was determined by X-ray crystallography.

Transport Proteins Alter the Selective Permeability of Biological Membranes

Because the phospholipid bilayer is a physical barrier to the diffusion of ions and most hydrophilic molecules, cells are able to separate their internal contents from their external environment. However, this barrier also poses a severe problem because cells must take up nutrients from the environment and export waste products. How do cells overcome this dilemma? Over the course of millions of years, species have evolved a multitude of **transport proteins**—transmembrane proteins that provide a passageway for the movement of ions and hydrophilic molecules across membranes. Transport proteins play a central role in the selective permeability of biological membranes. We can categorize transport proteins into two classes, channels and transporters, based on the manner in which they move solutes across the membrane.

Channels Transmembrane proteins called **channels** form an open passageway for the facilitated diffusion of ions or molecules across the membrane (**Figure 5.21**). Solutes move directly through a channel to get to the other side. Aquaporin, discussed in the Feature Investigation, is a channel that allows the movement of water across the membrane. When a channel is open, the transmembrane movement of solutes can be extremely rapid, up to 100 million ions or molecules per second!

Most channels are **gated**, which means they can open to allow the diffusion of solutes and close to prohibit diffusion. The phenomenon of gating allows cells to regulate the movement of solutes. For example, gating sometimes involves the direct binding of a molecule to the channel protein itself. One category of channels are ligand-gated channels, which are controlled by the noncovalent binding of small molecules—called

ligands—such as hormones or neurotransmitters. These ligands are often important in the transmission of signals between nerve and muscle cells or between two nerve cells.

Transporters Transmembrane proteins known as **transporters**, or **carriers**, bind their solutes in a hydrophilic pocket and undergo a conformational change that switches the exposure of the pocket from one side of the membrane to the other side (**Figure 5.22**). Transporters tend to be much slower than channels. Their rate of transport is typically 100 to 1,000 ions or molecules per second. Transporters provide the principal pathway for the uptake of organic molecules, such as sugars, amino acids, and nucleotides. In animals, they also allow cells to take

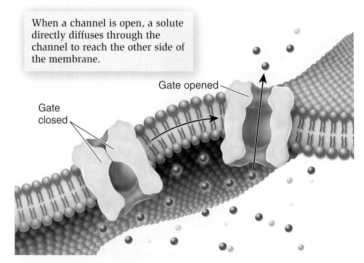

When a channel is open, a solute directly diffuses through the channel to reach the other side of the membrane.

Gate opened

Gate closed

Figure 5.21 Mechanism of transport by a channel protein.

Concept check: What is the purpose of gating?

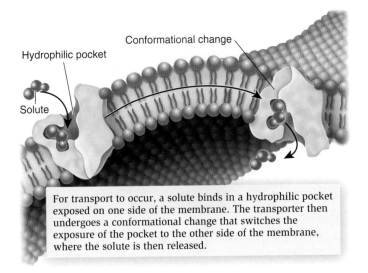

For transport to occur, a solute binds in a hydrophilic pocket exposed on one side of the membrane. The transporter then undergoes a conformational change that switches the exposure of the pocket to the other side of the membrane, where the solute is then released.

Figure 5.22 Mechanism of transport by a transporter, also called a carrier.

up certain hormones and neurotransmitters. In addition, many transporters play a key role in export. Waste products of cellular metabolism must be released from cells before they reach toxic levels. For example, a transporter removes lactic acid, a by-product of muscle cells during exercise. Other transporters, which are involved with ion transport, play an important role in regulating internal pH and controlling cell volume.

Transporters are named according to the number of solutes they bind and the direction in which they transport those solutes (**Figure 5.23**). **Uniporters** bind a single ion or molecule and transport it across the membrane. **Symporters**, or **cotransporters**, bind two or more ions or molecules and transport them in the same direction. **Antiporters** bind two or more ions or molecules and transport them in opposite directions.

Active Transport Is the Movement of Solutes Against a Gradient

As mentioned, active transport is the movement of a solute across a membrane against its gradient—that is, from a region of low concentration to higher concentration. Active transport is energetically unfavorable and requires the input of energy. **Primary active transport** involves the functioning of a **pump**—a type of transporter that directly uses energy to transport a solute against a gradient. **Figure 5.24a** shows a pump that uses ATP to transport H^+ against a gradient. Such a pump can establish a large H^+ electrochemical gradient across a membrane.

Secondary active transport involves the use of a pre-existing gradient to drive the active transport of another solute. For example, a H^+/sucrose symporter can use a H^+ electrochemical gradient, established by an ion pump, to move sucrose against its concentration gradient (**Figure 5.24b**). In this regard, only sucrose is actively transported. Hydrogen ions move down their electrochemical gradient. H^+/solute symporters are more common in bacteria, fungi, algae, and plant cells, because H^+ pumps are found in their plasma membranes. In animal cells, a pump that exports Na^+ maintains a Na^+ gradient across the plasma membrane. Na^+/solute symporters are prevalent in animal cells.

A single solute moves in one direction.

(a) Uniporter

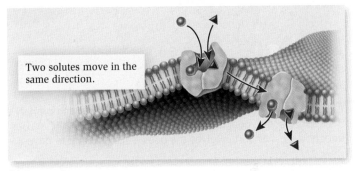

Two solutes move in the same direction.

(b) Symporter

Two solutes move in opposite directions.

(c) Antiporter

Figure 5.23 Types of transporters based on the direction of transport.

Symporters enable cells to actively import nutrients against a gradient. These proteins use the energy stored in the electrochemical gradient of H^+ or Na^+ to power the uphill movement of organic solutes such as sugars, amino acids, and other needed solutes. Therefore, with symporters in their plasma membrane, cells can scavenge nutrients from the extracellular environment and accumulate them to high levels within the cytoplasm.

Different ATP-Driven Ion Pumps Generate Ion Electrochemical Gradients

The phenomenon of active transport was discovered in the 1940s based on the study of ion movements using radioisotopes of Na^+ and K^+. After analyzing the movement of these ions across the plasma membrane of muscle cells, nerve cells, and red blood cells, researchers determined that the export of sodium ions (Na^+) is coupled to the import of potassium ions (K^+). In the late 1950s, Danish biochemist Jens Skou proposed

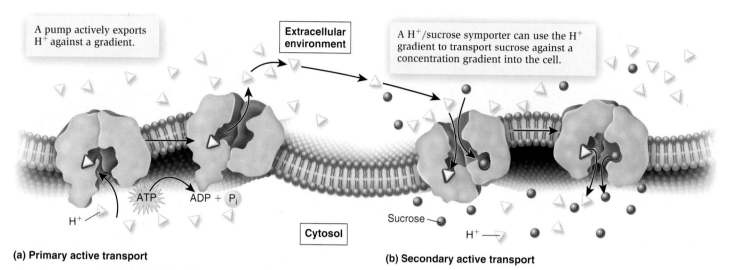

Figure 5.24   **Types of active transport.**   **(a)** During primary active transport, a pump directly uses energy, in this case from ATP, to transport a solute against a gradient. The pump shown here uses ATP to establish a H^+ electrochemical gradient. **(b)** Secondary active transport via symport involves the use of this gradient to drive the active transport of a solute, such as sucrose.

that a single transporter is responsible for this phenomenon. He was the first person to describe an ATP-driven ion pump, which was later named the Na^+/K^+-ATPase. This pump can actively transport Na^+ and K^+ against their gradients by using the energy from ATP hydrolysis (**Figure 5.25a**). The plasma membrane of a typical animal cell contains thousands of Na^+/K^+-ATPase pumps. These pumps establish large gradients in which the concentration of Na^+ is higher outside the cell and the concentration of K^+ is higher inside the cell.

Interestingly, Skou initially had trouble characterizing this pump. He focused his work on the large nerve cells found in the shore crab (*Carcinus maenas*). After isolating membranes from these cells, he was able to identify a transporter that could hydrolyze ATP, but the rate of hydrolysis was too low compared to the level of ATP hydrolysis that was observed in living cells that pump Na^+ and K^+. When he added Na^+ to his membranes, the ATP hydrolysis rate was not greatly affected. Then he tried adding K^+, but the ATP hydrolysis rate still did not

Figure 5.25   **Structure and function of the Na^+/K^+-ATPase.**   **(a)** Active transport by the Na^+/K^+-ATPase. Each time this protein hydrolyzes one ATP molecule, it pumps out three Na^+ and pumps in two K^+. **(b)** Pumping mechanism. The figure illustrates the protein conformational changes between E1 and E2. As this occurs, ATP is hydrolyzed to ADP and phosphate. During the process, phosphate is covalently attached to the protein but is released after two K^+ bind.

Concept check: *If a cell had ATP and Na^+, but K^+ were missing from the extracellular medium, how far through these steps could the Na^+/K^+-ATPase proceed?*

increase. Eventually, he did the critical experiment in which he added both Na^+ and K^+ to his membranes. With both ions present, ATP hydrolysis soared dramatically. This observation led to the identification and purification of the Na^+/K^+-ATPase. Jens Skou was awarded the Nobel Prize in 1997, over 40 years after his original work.

Let's take a closer look at the Na^+/K^+-ATPase that Skou discovered. Every time one ATP is hydrolyzed, the Na^+/K^+-ATPase functions as an antiporter that pumps three Na^+ into the extracellular environment and two K^+ into the cytosol. Because one cycle of pumping results in the net export of one positive charge, the Na^+/K^+-ATPase also produces an electrical gradient across the membrane. For this reason, it is considered an **electrogenic pump**—it generates an electrical gradient.

By studying the interactions of Na^+, K^+, and ATP with the Na^+/K^+-ATPase, researchers have pieced together a molecular road map of the steps that direct the pumping of ions across the membrane (**Figure 5.25b**). The Na^+/K^+-ATPase can alternate between two conformations, designated E1 and E2. In E1, the ion-binding sites are accessible from the cytosol—Na^+ binds tightly to this conformation, whereas K^+ has a low affinity. In E2, the ion-binding sites are accessible from the extracellular environment—Na^+ has a low affinity, and K^+ binds tightly.

To examine the pumping mechanism of the Na^+/K^+-ATPase, let's begin with the E1 conformation. Three Na^+ bind to the Na^+/K^+-ATPase from the cytosol (Figure 5.25b). When this occurs, ATP is hydrolyzed to ADP and phosphate. Temporarily, the phosphate is covalently bound to the pump, an event called phosphorylation. The pump then switches to the E2 conformation. The three Na^+ are released into the extracellular environment because they have a lower affinity for the E2 conformation, and then two K^+ bind from the outside. The binding of two K^+ causes the release of phosphate, which, in turn, causes a switch to E1. Because the E1 conformation has a low affinity for K^+, the two K^+ are released into the cytosol. The Na^+/K^+-ATPase is now ready for another round of pumping.

The Na^+/K^+-ATPase is a critical ion pump in animal cells because it maintains Na^+ and K^+ gradients across the plasma membrane. Many other types of ion pumps are also found in the plasma membrane and in organellar membranes. Ion pumps play the primary role in the formation and maintenance of ion

Table 5.4	Important Functions of Ion Electrochemical Gradients
Function	**Description**
Transport of ions and molecules	Symporters and antiporters use H^+ and Na^+ gradients to take up nutrients and export waste products.
Production of energy intermediates	In the mitochondrion and chloroplast, H^+ gradients are used to synthesize ATP.
Osmotic regulation	Animal cells control their internal volume by regulating ion gradients between the cytosol and extracellular fluid.
Nerve signaling	Na^+ and K^+ gradients are involved in conducting action potentials, the signals transmitted by nerve cells.
Muscle contraction	Ca^{2+} gradients regulate the ability of muscle fibers to contract.
Bacterial swimming	H^+ gradients drive the rotation of bacterial flagella.

gradients that drive many important cellular processes (**Table 5.4**). ATP is commonly the source of energy to drive ion pumps, and cells typically use a substantial portion of their ATP to keep them working. For example, nerve cells use up to 70% of their ATP just to operate ion pumps!

Macromolecules and Large Particles Are Transported via Exocytosis and Endocytosis

We have seen that most small substances are transported via membrane proteins such as channels and transporters, which provide a passageway for the movement of ions and molecules across the membrane. Eukaryotic cells have two other mechanisms, exocytosis and endocytosis, to transport larger molecules such as proteins and polysaccharides, and even very large particles. Both mechanisms involve the packaging of the transported substance, sometimes called the cargo, into a membrane vesicle or vacuole. **Table 5.5** describes some examples.

Exocytosis During **exocytosis**, material inside the cell is packaged into vesicles and then excreted into the extracellular environment (**Figure 5.26**). These vesicles are usually derived from

Table 5.5	Examples of Exocytosis and Endocytosis		
Exocytosis	**Description**	**Endocytosis**	**Description**
Hormones	Certain hormones, such as insulin, are composed of polypeptides. To exert its effect, insulin is secreted via exocytosis into the bloodstream from B cells of the pancreas.	Uptake of vital nutrients	Many important nutrients are insoluble in the bloodstream. Therefore, they are bound to proteins in the blood and then taken into cells via endocytosis. Examples include the uptake of lipids (bound to low-density lipoprotein) and iron (bound to transferrin protein).
Digestive enzymes	Digestive enzymes that function in the lumen of the small intestine are secreted via exocytosis from cells of the pancreas.	Root nodules	Nitrogen-fixing root nodules found in certain species of plants, such as legumes, are formed by the endocytosis of bacteria. After endocytosis, the bacterial cells are contained within a membrane-enclosed compartment in the nitrogen-fixing tissue of root nodules.
Extracellular matrix	Most of the components of the extracellular matrix that surrounds animal cells are secreted via exocytosis.	Immune system	Cells of the immune system, known as macrophages, engulf and destroy bacteria via phagocytosis.

the Golgi apparatus. As the vesicles form, a specific cargo is loaded into their interior. The budding process involves the formation of a protein coat around the emerging vesicle. The assembly of coat proteins on the surface of the Golgi membrane causes the bud to form. Eventually, the bud separates from the membrane to form a vesicle. After the vesicle is released, the coat is shed. Finally, the vesicle fuses with the plasma membrane and releases the cargo into the extracellular environment.

Endocytosis During **endocytosis**, the plasma membrane invaginates, or folds inward, to form a vesicle that brings substances into the cell. A common form of endocytosis is **receptor-mediated endocytosis**, in which a receptor in the plasma membrane is specific for a given cargo (**Figure 5.27**). Cargo molecules binding to their specific receptors stimulate many receptors to aggregate, and then coat proteins bind to the membrane. The protein coat causes the membrane to invaginate and form a vesicle.

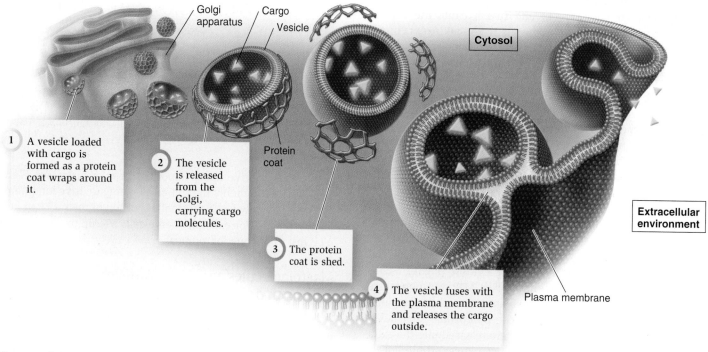

Figure 5.26 Exocytosis.

Concept check: *What is the function of the protein coat?*

Figure 5.27 Receptor-mediated endocytosis.

Once it is released into the cell, the vesicle sheds its coat. In most cases, the vesicle fuses with an internal membrane organelle, such as a lysosome, and the receptor releases its cargo. Depending on the cargo, the lysosome may release it directly into the cytosol or digest it into simpler building blocks before releasing it.

Other specialized forms of endocytosis occur in certain types of cells. **Pinocytosis** (from the Greek, meaning cell drinking) involves the formation of membrane vesicles from the plasma membrane as a way for cells to internalize the extracellular fluid. This allows cells to sample the extracellular solutes. Pinocytosis is particularly important in cells that are actively involved in nutrient absorption, such as cells that line the intestine in animals.

Phagocytosis (from the Greek, meaning cell eating) is an extreme form of endocytosis. It involves the formation of an enormous membrane vesicle called a phagosome, or phagocytic vacuole, that engulfs a large particle such as a bacterium. Only certain kinds of cells can carry out phagocytosis. For example, macrophages, which are cells of the immune system in mammals, kill bacteria via phagocytosis. Once inside the cell, the phagosome fuses with lysosomes, and the digestive enzymes within the lysosomes destroy the bacterium.

 ## Summary of Key Concepts

5.1 Membrane Structure

- Plasma membranes separate a cell from its surroundings, and organellar membranes provide interfaces to carry out vital cellular activities. (Table 5.1)

- The accepted model of membranes is the fluid-mosaic model, and its basic framework is the phospholipid bilayer. Cellular membranes also contain proteins, and most membranes have attached carbohydrates. (Figure 5.1, Table 5.2)

- The three main types of membrane proteins are transmembrane proteins, lipid-anchored proteins, and peripheral membrane proteins. Transmembrane proteins and lipid-anchored proteins are classified as integral membrane proteins. Researchers are working to identify new membrane proteins and their functions because these proteins are important biologically and medically. (Figure 5.2, Table 5.3)

- Bilayer semifluidity is essential for normal cell function, growth, and division. Lipids can move rotationally and laterally, but flip-flop does not occur spontaneously. The chemical properties of phospholipids—such as tail length and the presence of double bonds—and the amount of cholesterol have a profound effect on the fluidity of membranes. (Figures 5.3, 5.4, 5.5)

- Glycosylation, which produces glycolipids or glycoproteins, has a variety of cellular functions. Carbohydrate can serve as a recognition marker or a protective cell coat. (Figure 5.6)

- Electron microscopy is a valuable tool for studying membrane structure and function. Freeze fracture electron microscopy (FFEM) can be used to analyze the interiors of phospholipid bilayers. (Figure 5.7)

5.2 Synthesis of Membrane Components in Eukaryotic Cells

- In eukaryotic cells, most membrane phospholipids are synthesized at the cytosolic leaflet of the smooth ER membrane. Flippases move some phospholipids to the other leaflet. (Figure 5.8)

- Most transmembrane proteins are first inserted into the ER membrane. (Figure 5.9)

- Glycosylation of proteins occurs in the ER and Golgi apparatus. (Figure 5.10)

5.3 Membrane Transport

- Biological membranes are selectively permeable. Diffusion occurs when a solute moves from a region of high concentration to a region of lower concentration. Passive transport of a solute across a membrane can occur via diffusion or facilitated diffusion. Active transport is the movement of a substance against a gradient. (Figure 5.11)

- The lipid bilayers of membranes are relatively impermeable to many substances. (Figures 5.12, 5.13)

- Living cells maintain an internal environment that is separated from their external environment. This involves establishing transmembrane gradients across the plasma membrane and organellar membranes. (Figure 5.14, Table 5.4)

- In the process of osmosis, water diffuses through a membrane from a solution that is hypotonic (lower solute concentration) into a solution that is hypertonic (higher solute concentration). Solutions with identical solute concentrations are isotonic. The tendency of water to move into a cell creates an osmotic (turgor) pressure. (Figures 5.15, 5.16, 5.17, 5.18)

- The two classes of transport proteins are channels and transporters. Channels form an open passageway for the direct diffusion of solutes across the membrane; one example is aquaporin, which allows the movement of water. Most channels are gated, which allows cells to regulate the movement of solutes. (Figures 5.19, 5.20, 5.21)

- Transporters, which tend to be slower than channels, bind their solutes in a hydrophilic pocket and undergo a conformational change that switches the exposure of the pocket to the other side of the membrane. They can be uniporters, symporters, or antiporters. (Figures 5.22, 5.23)

- Primary active transport involves pumps that directly use energy to generate a solute gradient. Secondary active transport uses a pre-existing gradient. (Figure 5.24)

- The Na^+/K^+-ATPase is an electrogenic ATP-driven pump. This protein follows a series of steps that direct the pumping of ions across the membrane. (Figure 5.25, Table 5.4)

- In eukaryotes, exocytosis and endocytosis are used to transport large molecules and particles. Exocytosis is a process in which material inside the cell is packaged into vesicles and excreted into the extracellular environment. During endocytosis, the plasma membrane folds inward to form a vesicle that brings substances into the cell. Forms of endocytosis include receptor-mediated endocytosis, pinocytosis, and phagocytosis. (Figures 5.26, 5.27, Table 5.5)

Assess and Discuss

Test Yourself

1. Which of the following statements best describes the chemical composition of biomembranes?
   a. Biomembranes are bilayers of proteins with associated lipids and carbohydrates.
   b. Biomembranes are composed of two layers—one layer of phospholipids and one layer of proteins.
   c. Biomembranes are bilayers of phospholipids with associated proteins and carbohydrates.
   d. Biomembranes are composed of equal numbers of phospholipids, proteins, and carbohydrates.
   e. Biomembranes are composed of lipids with proteins attached to the outer surface.

2. Which of the following events in a biological membrane would not be energetically favorable and therefore not occur spontaneously?
   a. the rotation of phospholipids
   b. the lateral movement of phospholipids
   c. the flip-flop of phospholipids to the opposite leaflet
   d. the rotation of membrane proteins
   e. the lateral movement of membrane proteins

3. Let's suppose an insect, which doesn't maintain a constant body temperature, was exposed to a shift in temperature from 60°F to 80°F. Which of the following types of cellular changes would be the most beneficial to help this animal cope with the temperature shift?
   a. increase the number of double bonds in the fatty acyl tails of phospholipids
   b. increase the length of the fatty acyl tails of phospholipids
   c. decrease the amount of cholesterol in the membrane
   d. decrease the amount of carbohydrate attached to membrane proteins
   e. decrease the amount of carbohydrate attached to phospholipids

4. Carbohydrates of the plasma membrane
   a. are associated with a protein or lipid.
   b. are located on the outer surface of the plasma membrane.
   c. can function as cell markers for recognition by other cells.
   d. all of the above
   e. a and c only

5. A transmembrane protein in the plasma membrane is glycosylated at two sites in the polypeptide sequence. One site is Asn—Val—Ser and the other site is Asn—Gly—Thr. Where in this protein would you expect these two sites to be found?
   a. in transmembrane segments
   b. in hydrophilic regions that project into the extracellular environment
   c. in hydrophilic regions that project into the cytosol
   d. could be anywhere
   e. b and c only

6. The tendency for Na$^+$ to move into the cell could be due to
   a. the higher numbers of Na$^+$ outside the cell, resulting in a chemical concentration gradient.
   b. the net negative charge inside the cell attracting the positively charged Na$^+$.
   c. the attractive force of K$^+$ inside the cell pulling Na$^+$ into the cell.
   d. all of the above.
   e. a and b only.

7. Let's suppose the solute concentration inside the cells of a plant is 0.3 M and outside is 0.2 M. If we assume that the solutes do not readily cross the membrane, which of the following statements best describes what will happen?
   a. The plant cells will lose water, and the plant will wilt.
   b. The plant cells will lose water, which will result in a higher turgor pressure.
   c. The plant cells will take up a lot of water and undergo osmotic lysis.
   d. The plant cells will take up a little water and have a higher turgor pressure.
   e. Both a and b are correct.

8. What structural features of a membrane are major contributors to its selective permeability?
   a. phospholipid bilayer
   b. transport proteins
   c. glycolipids on the outer surface of the membrane
   d. peripheral membrane proteins on the inside of the membrane
   e. both a and b

9. What is the name given to the process in which solutes are moved across a membrane against their concentration gradient?
   a. diffusion   d. passive diffusion
   b. facilitated diffusion   e. active transport
   c. osmosis

10. Large particles or large volumes of fluid can be brought into the cell by
   a. facilitated diffusion.   d. exocytosis.
   b. active transport.   e. all of the above.
   c. endocytosis.

Conceptual Questions

1. With your textbook closed, draw and describe the fluid-mosaic model of membrane structure.

2. Describe two different ways that integral membrane proteins are anchored to a membrane. How do peripheral membrane proteins associate with a membrane?

3. Solutes can move across membranes via diffusion, facilitated diffusion, active transport, exocytosis, and endocytosis. During which of these five processes would you expect the solute to physically touch the tails of phospholipids in the membrane? For the other processes, describe how the solute avoids an interaction with the phospholipid tails in the membrane.

Collaborative Questions

1. Proteins in the plasma membrane are often the target of medicines. Discuss why you think this is the case. How would you determine experimentally that a specific membrane protein was the target of a drug?

2. With regard to bringing solutes into the cell across the plasma membrane, discuss the advantages and disadvantages of diffusion, facilitated diffusion, active transport, and endocytosis.

Online Resource

Stay a step ahead in your studies with animations that bring concepts to life and practice tests to assess your understanding. Your instructor may also recommend the interactive ebook, individualized learning tools, and more.

An Introduction to Energy, Enzymes, and Metabolism

6

Have you ever taken aspirin or ibuprofen to relieve a headache or reduce a fever? Do you know how it works? If you answered "no" to the second question, you're not alone. Over 2,000 years ago, humans began treating pain with powder from the bark and leaves of the willow tree, which contains a compound called salicylic acid. Modern aspirin is composed of a derivative of salicylic acid called acetylsalicylic acid, which is gentler to the stomach. Only recently, however, have we learned how such drugs work. Aspirin and ibuprofen are examples of drugs that inhibit specific enzymes found in cells. In this case, these drugs inhibit an enzyme called cyclooxygenase. This enzyme is needed to synthesize molecules called prostaglandins, which play a role in inflammation and pain. Aspirin and ibuprofen exert their effects by inhibiting cyclooxygenase, thereby decreasing the levels of prostaglandins.

Enzymes are proteins that act as critical catalysts to speed up thousands of different reactions in cells. As discussed in Chapter 2, a **chemical reaction** is a process in which one or more substances are changed into other substances. Such reactions may involve molecules attaching to each other to form larger molecules, molecules breaking apart to form two or more smaller molecules, rearrangements of atoms within molecules, or the transfer of electrons from one atom to another. Every living cell continuously performs thousands of such chemical reactions to sustain life. The term **metabolism** is used to describe the sum total of all chemical reactions that occur within an organism. The term also refers to a specific set of chemical reactions occurring at the cellular level. For example, biologists may speak of sugar metabolism or fat metabolism. Most types of metabolism involve the breakdown or synthesis of organic molecules. Cells maintain their structure by using organic molecules. Such molecules provide the building blocks to construct cells, and the chemical bonds within organic molecules store energy that can be used to drive cellular processes.

In this chapter, we begin with a general discussion of chemical reactions. We will examine what factors control the direction of a chemical reaction and what determines its rate, paying particular attention to the role of enzymes. We then consider metabolism at the cellular level. First, we will examine some of the general features of chemical reactions that are vital for the energy needs of living cells. We will also explore the variety of ways in which metabolic processes are regulated and how macromolecules are recycled.

Common drugs that are enzyme inhibitors. Drugs such as aspirin and ibuprofen exert their effects by inhibiting an enzyme that speeds up a chemical reaction in the cell.

6.1 Energy and Chemical Reactions

Two general factors govern the fate of a given chemical reaction in a living cell—its direction and rate. To illustrate this point, let's consider a generalized chemical reaction such as

$$a\text{A} + b\text{B} \rightleftharpoons c\text{C} + d\text{D}$$

where A and B are the reactants, C and D are the products, and a, b, c, and d are the number of moles of reactants and products. This reaction is reversible, which means that A + B could be converted to C + D, or C + D could be converted to A + B. The direction of the reaction, whether C + D are made (the forward direction) or A + B are made (the reverse direction), depends on energy and on the concentrations of A, B, C, and D. In this section, we will begin by examining the interplay of energy and the concentration of reactants as they govern the direction of a chemical reaction. You will learn that cells use energy intermediate molecules, such as ATP, to drive chemical reactions in a desired direction.

Energy Exists in Many Forms

To understand why a chemical reaction occurs, we first need to consider **energy**, which we will define as the ability to promote change or do work. Physicists often consider energy in two forms: kinetic energy and potential energy (**Figure 6.1**). **Kinetic energy** is energy associated with movement, such as the movement of a baseball bat from one location to another. By comparison, **potential energy** is the energy that a substance possesses due to its structure or location. The energy contained within covalent bonds in molecules is also a type of potential energy called **chemical energy**. The breakage of those bonds is one way that living cells can harness this energy to perform cellular functions. **Table 6.1** summarizes chemical and other forms of energy important in biological systems.

An important issue in biology is the ability of energy to be converted from one form to another. The study of energy interconversions is called **thermodynamics**. Physicists have determined that two laws govern energy interconversions:

1. **The first law of thermodynamics**—The first law states that energy cannot be created or destroyed; it is also called the law of conservation of energy. However, energy can be transferred from one place to another and can be transformed from one type to another (as when, for example, chemical energy is transformed into heat).

2. **The second law of thermodynamics**—The second law states that the transfer of energy or the transformation of energy from one form to another increases the **entropy**, or degree of disorder of a system (**Figure 6.2**). Entropy is a measure of the randomness of molecules in a system. When a physical system becomes more disordered, the entropy increases. As the energy becomes more evenly distributed, that energy is less able to promote change or do work. When energy is converted from one form to another, some energy may become unusable by living organisms. For example, unusable heat may be released during a chemical reaction.

(a) Kinetic energy   **(b) Potential energy**

Figure 6.1   Examples of energy.   (a) Kinetic energy, such as swinging a bat, is energy associated with motion. **(b)** Potential energy is stored energy, as in a bow that is ready to fire an arrow.

Next, we will see how the two laws of thermodynamics place limits on the ways that living cells can use energy for their own needs.

The Change in Free Energy Determines the Direction of a Chemical Reaction or Any Other Cellular Process

Energy is necessary for living organisms to exist. Energy is required for many cellular processes, including chemical reactions, cellular movements such as those occurring in muscle contraction, and the maintenance of cell organization. To understand how living organisms use energy, we need to distinguish between the energy that can be used to promote change or do work (usable energy) and the energy that cannot (unusable energy).

Total energy = Usable energy + Unusable energy

Why is some energy unusable? The main culprit is entropy. As stated by the second law of thermodynamics, energy transformations involve an increase in entropy, a measure of the disorder that cannot be harnessed in a useful way. The total energy is termed **enthalpy** (**H**), and the usable energy—the

Table 6.1	Types of Energy That Are Important in Biology	
Energy type	**Description**	**Biological example**
Light	Light is a form of electromagnetic radiation. The energy of light is packaged in photons.	During photosynthesis, light energy is captured by pigments (see Chapter 8). Ultimately, this energy is used to reduce carbon and produce organic molecules.
Heat	Heat is the transfer of kinetic energy from one object to another or from an energy source to an object. In biology, heat is often viewed as energy that can be transferred due to a difference in temperature between two objects or locations.	Many organisms, such as humans, maintain their bodies at a constant temperature. This is achieved, in part, by chemical reactions that generate heat.
Mechanical	Mechanical energy is the energy possessed by an object due to its motion or its position relative to other objects.	In animals, mechanical energy is associated with movements due to muscle contraction, such as walking.
Chemical	Chemical energy is stored in the chemical bonds of molecules. When the bonds are broken and rearranged, large amounts of energy can be released.	The covalent bonds in organic molecules, such as glucose and ATP, store large amounts of energy. When bonds are broken in larger molecules to form smaller molecules, the chemical energy that is released can be used to drive cellular processes.
Electrical/ion gradient	The movement of charge or the separation of charge can provide energy. Also, a difference in ion concentration across a membrane constitutes an electrochemical gradient, which is a source of potential energy.	During oxidative phosphorylation (described in Chapter 7), a H^+ gradient provides the energy to drive ATP synthesis.

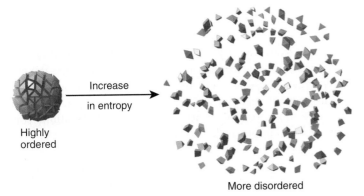

Figure 6.2 Entropy. Entropy is a measure of the disorder of a system. An increase in entropy means an increase in disorder.

Concept check: Which do you think has more entropy, a NaCl crystal at the bottom of a beaker of water or the same beaker of water after the Na⁺ and Cl⁻ in the crystal have dissolved in the water?

amount of available energy that can be used to promote change or do work—is called the **free energy** (*G*). The letter *G* is in recognition of J. Willard Gibbs, who proposed the concept of free energy in 1878. The unusable energy is the system's entropy (*S*). Gibbs proposed that these three factors are related to each other in the following way:

$$H = G + TS$$

where *T* is the absolute temperature in Kelvin (K). Because our focus is on free energy, we can rearrange this equation as

$$G = H - TS$$

A critical issue in biology is whether a process will or will not occur spontaneously. For example, will glucose be broken down into carbon dioxide and water? Another way of framing this question is to ask: Is the breakdown of glucose a spontaneous reaction? A spontaneous reaction or process is one that will occur without being driven by an input of energy. However, a spontaneous reaction does not necessarily proceed quickly. In some cases, the rate of a spontaneous reaction can be quite slow. For example, the breakdown of sugar is a spontaneous reaction, but the rate at which sugar in a sugar bowl would break down into CO_2 and H_2O would be very slow.

The key way to evaluate if a chemical reaction is spontaneous is to determine the free-energy change that occurs as a result of the reaction:

$$\Delta G = \Delta H - T\Delta S$$

where the Δ sign (the Greek letter delta) indicates a change, such as before and after a chemical reaction. If a chemical reaction has a negative free-energy change ($\Delta G < 0$), this means that the products have less free energy than the reactants, and, therefore, free energy is released during product formation. Such a reaction is said to be **exergonic**. Exergonic reactions are spontaneous. Alternatively, if a reaction has a positive free-energy change ($\Delta G > 0$), requiring the addition of free energy from the environment, it is termed **endergonic**. An endergonic reaction is not a spontaneous reaction.

If Δ*G* for a chemical reaction is negative, the reaction favors the formation of products, whereas a reaction with a positive Δ*G* favors the formation of reactants. Chemists have determined

Adenosine triphosphate (ATP)

Adenosine diphosphate (ADP)   Phosphate (Pᵢ)

$$\Delta G = -7.3 \text{ kcal/mol}$$

Figure 6.3 The hydrolysis of ATP to ADP and Pᵢ. As shown in this figure, ATP has a net charge of −4, and ADP and Pᵢ are shown with net charges of −2 each. When these compounds are shown in chemical reactions with other molecules, the net charges will also be indicated. Otherwise, these compounds will simply be designated ATP, ADP, and Pᵢ. At neutral pH, ADP²⁻ will dissociate to ADP³⁻ and H⁺.

Concept check: Because Δ*G* is negative, what does that tell us about the direction of this chemical reaction? What does it tell us about the rate?

free-energy changes for a variety of chemical reactions, which allows them to predict their direction. As an example, let's consider **adenosine triphosphate** (**ATP**), which is a molecule that is a common energy source for all cells. ATP is broken down to adenosine diphosphate (ADP) and inorganic phosphate (Pᵢ). Because water is used to remove a phosphate group, chemists refer to this as the hydrolysis of ATP (**Figure 6.3**). In the reaction of converting 1 mole of ATP to 1 mole of ADP and Pᵢ, Δ*G* equals −7.3 kcal/mole. Because this is a negative value, the reaction strongly favors the formation of products. As discussed later in this chapter, the energy liberated by the hydrolysis of ATP is used to drive a variety of cellular processes.

Chemical Reactions Will Eventually Reach a State of Equilibrium

Even when a chemical reaction is associated with a negative free-energy change, not all of the reactants are converted to products. The reaction reaches a state of **chemical equilibrium** in which the rate of formation of products equals the rate of formation of reactants. Let's consider the generalized reaction

$$a\text{A} + b\text{B} \rightleftharpoons c\text{C} + d\text{D}$$

where again A and B are the reactants, C and D are the products, and a, b, c, and d are the number of moles of reactants and products. An equilibrium occurs, such that:

$$K_{eq} = \frac{[\text{C}]^c[\text{D}]^d}{[\text{A}]^a[\text{B}]^b}$$

where K_{eq} is the equilibrium constant. Each type of chemical reaction will have a specific value for K_{eq}.

Biologists make two simplifying assumptions when determining values for equilibrium constants. First, the concentration of water does not change during the reaction, and the pH remains constant at pH 7. The equilibrium constant under these conditions is designated K_{eq}' (' is the prime symbol). If water is one of the reactants, as in a hydrolysis reaction, it is not included in the chemical equilibrium equation. As an example, let's consider the chemical equilibrium for the hydrolysis of ATP.

$$\text{ATP}^{4-} + \text{H}_2\text{O} \rightleftharpoons \text{ADP}^{2-} + \text{P}_i^{2-}$$

$$K_{eq}' = \frac{[\text{ADP}][\text{P}_i]}{[\text{ATP}]}$$

Experimentally, the value for K_{eq}' for this reaction has been determined and found to be approximately 1,650,000 M. Such a large value indicates that the equilibrium greatly favors the formation of products—ADP and P_i.

Cells Use ATP to Drive Endergonic Reactions

In living organisms, many vital processes require the addition of free energy; that is, they are endergonic and will not occur spontaneously. Fortunately, organisms have a way to overcome this problem. Rather than catalyzing exothermic reactions that release energy in the form of unusable heat, cells often couple exergonic reactions with endergonic reactions. If an exergonic reaction is coupled to an endergonic reaction, the endergonic reaction will proceed spontaneously if the net free-energy change for both processes combined is negative. For example, consider the following reactions:

Glucose + phosphate^{2-} → Glucose-6-phosphate^{2-} + H$_2$O

$$\Delta G = +3.3 \text{ kcal/mole}$$

$$\text{ATP}^{4-} + \text{H}_2\text{O} \rightarrow \text{ADP}^{2-} + \text{P}_i^{2-} \qquad \Delta G = -7.3 \text{ kcal/mole}$$

Coupled reaction:

Glucose + ATP^{4-} → Glucose-6-phosphate^{2-} + ADP^{2-}

$$\Delta G = -4.0 \text{ kcal/mole}$$

The first reaction, in which phosphate is covalently attached to glucose, is endergonic, whereas the second, the hydrolysis

of ATP, is exergonic. By itself, the first reaction would not be spontaneous. If the two reactions are coupled, however, the net free-energy change for both reactions combined is exergonic. In the coupled reaction, a phosphate is directly transferred from ATP to glucose in a process called **phosphorylation**. This coupled reaction proceeds spontaneously because the net free-energy change is negative. Exergonic reactions, such as the breakdown of ATP, are commonly coupled to cellular processes that would otherwise be endergonic or require energy.

6.2 Enzymes and Ribozymes

For most chemical reactions in cells to proceed at a rapid pace, such as the breakdown of sugar, a catalyst is needed. A **catalyst** is an agent that speeds up the rate of a chemical reaction without being permanently changed or consumed. In living cells, the most common catalysts are enzymes. The term was coined in 1876 by a German physiologist, Wilhelm Kühne, who discovered trypsin, an enzyme in pancreatic juice that is needed for digestion of food proteins. In this section, we will explore how enzymes are able to increase the rate of chemical reactions. Interestingly, some biological catalysts are RNA molecules called ribozymes. We will also examine a few examples in which RNA molecules carry out catalytic functions.

Enzymes Increase the Rates of Chemical Reactions

Thus far, we have examined aspects of energy and considered how the laws of thermodynamics are related to the direction of chemical reactions. If a chemical reaction has a negative free-energy change, the reaction will be spontaneous; it will tend to proceed in the direction of reactants to products. Although thermodynamics governs the direction of an energy transformation, it does not control the rate of a chemical reaction. For example, the breakdown of the molecules in gasoline to smaller molecules is a highly exergonic reaction. Even so, we could place gasoline and oxygen in a container and nothing much would happen (provided it wasn't near a flame). If we came back several days later, we would expect to see the gasoline still sitting there. Perhaps if we came back in a few million years, the gasoline would have been broken down. On a time-scale of months or a few years, however, the chemical reaction would proceed very slowly.

In living cells, the rates of enzyme-catalyzed reactions typically occur millions of times faster than the corresponding uncatalyzed reactions. An extreme example is the enzyme catalase, which is found in peroxisomes (see Chapter 4). This enzyme catalyzes the breakdown of hydrogen peroxide (H_2O_2) into water and oxygen. Catalase speeds up this reaction 10^{15}-fold faster than the uncatalyzed reaction!

Why are catalysts necessary to speed up a chemical reaction? When a covalent bond is broken or formed, this process initially involves the straining or stretching of one or more bonds in the starting molecule(s), and/or it may involve the positioning of two molecules so they interact with each other properly. Let's consider the reaction in which ATP is used to phosphorylate glucose.

$$\text{Glucose} + \text{ATP}^{4-} \rightarrow \text{Glucose-phosphate}^{2-} + \text{ADP}^{2-}$$

For a reaction to occur between glucose and ATP, the molecules must collide in the correct orientation and possess enough energy so that chemical bonds can be changed. As glucose and ATP approach each other, their electron clouds cause repulsion. To overcome this repulsion, an initial input of energy, called the **activation energy**, is required (Figure 6.4). Activation energy allows the molecules to get close enough to cause a rearrangement of bonds. With the input of activation energy, glucose and ATP can achieve a **transition state** in which the original bonds have stretched to their limit. Once the reactants have reached the transition state, the chemical reaction can readily proceed to the formation of products, which in this case is glucose-phosphate and ADP.

The activation energy required to achieve the transition state is a barrier to the formation of products. This barrier is the reason why the rate of many chemical reactions is very slow. There are two common ways to overcome this barrier and thereby accelerate a chemical reaction. First, the reactants could be exposed to a large amount of heat. For example, as we noted previously, if gasoline is sitting at room temperature, nothing much happens. However, if the gasoline is exposed to a flame or spark, it breaks down rapidly, perhaps at an explosive rate! Alternatively, a second strategy is to lower the activation energy barrier. Enzymes lower the activation energy to a point where a small amount of available heat can push the reactants to a transition state (Figure 6.4).

How do enzymes lower the activation energy barrier of chemical reactions? Enzymes are generally large proteins that bind relatively small reactants (Figure 6.4). When bound to an enzyme, the bonds in the reactants can be strained, thereby making it easier for them to achieve the transition state. This is one way that enzymes lower the activation energy. In addition, when a chemical reaction involves two or more reactants, the enzyme provides a site in which the reactants are positioned very close to each other in an orientation that facilitates the formation of new covalent bonds. This also lowers the necessary activation energy for a chemical reaction.

Straining the reactants and bringing them close together are two common ways that enzymes lower the activation energy barrier. In addition, enzymes may facilitate a chemical reaction by changing the local environment of the reactants. For example, amino acids in an enzyme may have charges that affect the chemistry of the reactants. In some cases, enzymes lower the activation energy by directly participating in the chemical reaction. For example, certain enzymes that hydrolyze ATP form a covalent bond between phosphate and an amino acid in the enzyme. However, this is a temporary condition. The covalent bond between phosphate and the amino acid is quickly broken, releasing the phosphate and returning the amino acid back to its original condition. An example of such an enzyme is Na^+/K^+-ATPase, described in Chapter 5 (refer back to Figure 5.25).

Figure 6.4 **Activation energy of a chemical reaction.** This figure depicts an exergonic reaction. The activation energy is needed for molecules to achieve a transition state. One way that enzymes lower the activation energy is by straining the reactants so that less energy is required to attain the transition state. A second way is by binding two reactants so they are close to each other and in a favorable orientation.

Concept check: *How does lowering the activation energy affect the rate of a chemical reaction? How does it affect the direction?*

Enzymes Recognize Their Substrates with High Specificity and Undergo Conformational Changes

Thus far, we have considered how enzymes lower the activation energy of a chemical reaction and thereby increase its rate. Let's consider some other features of enzymes that enable them to serve as effective catalysts in chemical reactions. The **active site** is the location in an enzyme where the chemical reaction takes place. The **substrates** for an enzyme are the reactant molecules that bind to an enzyme at the active site and participate in the chemical reaction. For example, hexokinase is an enzyme whose substrates are glucose and ATP (Figure 6.5). The binding between an enzyme and substrate produces an **enzyme-substrate complex**.

A key feature of nearly all enzymes is they bind their substrates with a high degree of **specificity**. For example, hexokinase recognizes glucose but does not recognize other similar sugars very well, such as fructose and galactose. In 1894, the German scientist Emil Fischer proposed that the recognition of a substrate by an enzyme resembles the interaction between a lock and key: only the right-sized key (the substrate) will fit into the keyhole (active site) of the lock (the enzyme). Further research revealed that the interaction between an enzyme and its substrates also involves movements or conformational changes in the enzyme itself. As shown in Figure 6.5, these

1. ATP and glucose bind to enzyme (hexokinase).

2. Enzyme undergoes conformational change that binds the substrates more tightly. This induced fit strains chemical bonds within the substrates and/or brings them closer together.

3. Substrates are converted to products.

4. Products are released. Enzyme is reused.

Figure 6.5 **The steps of an enzyme-catalyzed reaction.** The example shown here involves the enzyme hexokinase, which binds glucose and ATP. The products are glucose-phosphate and ADP, which are released from the enzyme.

Concept check: *During which step is the activation energy lowered?*

conformational changes cause the substrates to bind more tightly to the enzyme, a phenomenon called **induced fit**, which was proposed by American biochemist Daniel Koshland in 1958. Only after this conformational change takes place does the enzyme catalyze the conversion of reactants to products.

Competitive and Noncompetitive Inhibitors Affect Enzyme Function

Molecules or ions may bind to enzymes and inhibit their function. To understand how such inhibitors work, researchers compare the function of enzymes in the absence or presence of inhibitors. Let's first consider enzyme function in the absence of an inhibitor. In the experiment of **Figure 6.6a**, tubes labeled A, B, C, and D each contained one microgram of enzyme. This enzyme recognizes a single type of substrate and converts it to a product. For each data point, the substrate concentration added to each tube was varied from a low to a high level. The samples were incubated for 60 seconds, and then the amount of product in each tube was measured. In this example, the velocity of the chemical reaction is expressed as the amount of product produced per second. As we see in Figure 6.6a, the velocity increases as the substrate concentration increases, but eventually reaches a plateau. Why does the plateau occur? At high substrate concentrations, nearly all of the active sites of the enzyme are occupied with substrate, so increasing the substrate concentration further has a negligible effect. At this point, the enzyme is saturated with substrate, and the velocity of the chemical reaction is near its maximal rate, called its V_{max}.

Figure 6.6a also helps us understand the relationship between substrate concentration and velocity. The K_M is the substrate concentration at which the velocity is half its maximal value. The K_M is also called the Michaelis constant in honor

of the German biochemist Leonor Michaelis, who carried out pioneering work with the Canadian biochemist Maud Menten on the study of enzymes. The K_M is a measure of the substrate concentration required for catalysis to occur. An enzyme with a high K_M requires a higher substrate concentration to achieve a particular reaction velocity compared to an enzyme with a lower K_M.

For an enzyme-catalyzed reaction, we can view the formation of product as occurring in two steps: (1) binding or release of substrate and (2) formation of product.

$$E + S \rightleftharpoons ES \rightarrow E + P$$

where

E is the enzyme

S is the substrate

ES is the enzyme-substrate complex

P is the product

If the second step—the rate of product formation—is much slower than the rate of substrate release, the K_M is inversely related to the **affinity**—degree of attraction—between the enzyme and substrate. For example, let's consider an enzyme that breaks down ATP into ADP and P_i. If the rate of formation of ADP and P_i is much slower than the rate of ATP release, the K_M for such an enzyme is a measure of its affinity for ATP. In such cases, the K_M and affinity show an inverse relationship. Enzymes with a high K_M have a low affinity for their substrates—they bind them more weakly. By comparison, enzymes with a low K_M have a high affinity for their substrates—they bind them more tightly.

Now that we understand the relationship between substrate concentration and the velocity of an enzyme-catalyzed

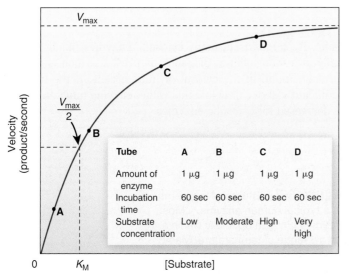

(a) Reaction velocity in the absence of inhibitors

Tube	A	B	C	D
Amount of enzyme	1 μg	1 μg	1 μg	1 μg
Incubation time	60 sec	60 sec	60 sec	60 sec
Substrate concentration	Low	Moderate	High	Very high

(b) Competitive inhibition

(c) Noncompetitive inhibition

Figure 6.6   **The relationship between velocity and substrate concentration in an enzyme-catalyzed reaction, and the effects of inhibitors.**   **(a)** In the absence of an inhibitor, the maximal velocity (V_{max}) is achieved when the substrate concentration is high enough to be saturating. The K_M value is the substrate concentration where the velocity is half the maximal velocity. **(b)** A competitive inhibitor binds to the active site of an enzyme and raises the K_M for the substrate. **(c)** A noncompetitive inhibitor binds outside the active site to an allosteric site and lowers the V_{max} for the reaction.

Concept check: *Enzyme A has a K_M of 0.1 mM, whereas enzyme B has a K_M of 1.0 mM. They both have the same V_{max}. If the substrate concentration was 0.5 mM, which reaction—the one catalyzed by enzyme A or B—would have the higher velocity?*

reaction, we can explore how inhibitors may affect enzyme function. **Competitive inhibitors** are molecules that bind to the active site of an enzyme and inhibit the ability of the substrate to bind. Such inhibitors compete with the substrate for the ability to bind to the enzyme. Competitive inhibitors usually have a structure or a portion of their structure that mimics the structure of the enzyme's substrate. As seen in **Figure 6.6b**, when competitive inhibitors are present, the apparent K_M for the substrate increases—a higher concentration of substrate is needed to achieve the same velocity of the chemical reaction. In this case, the effects of the competitive inhibitor can be overcome by increasing the concentration of the substrate.

By comparison, **Figure 6.6c** illustrates the effects of a **noncompetitive inhibitor**. As seen here, this type of inhibitor lowers the V_{max} for the reaction without affecting the K_M. A noncompetitive inhibitor binds noncovalently to an enzyme at a location outside the active site, called an **allosteric site**, and inhibits the enzyme's function. In this example, a molecule binding to the allosteric site inhibits the enzyme's function, but for other enzymes, such binding can enhance their function.

Additional Factors Influence Enzyme Function

Enzymes, which are composed of protein, sometimes require additional nonprotein molecules or ions to carry out their functions. **Prosthetic groups** are small molecules that are permanently attached to the surface of an enzyme and aid in catalysis. **Cofactors** are usually inorganic ions, such as Fe^{3+} or Zn^{2+}, that temporarily bind to the surface of an enzyme and promote a chemical reaction. Finally, some enzymes use **coenzymes**, organic molecules that temporarily bind to an enzyme and participate in the chemical reaction but are left unchanged after the reaction is completed. Some of these coenzymes can be synthesized by cells, but many of them are taken in as dietary vitamins by animal cells.

The ability of enzymes to increase the rate of a chemical reaction is also affected by the surrounding conditions. In particular, the temperature, pH, and ionic conditions play an important role in the proper functioning of enzymes. Most enzymes function maximally in a narrow range of temperature and pH. For example, many human enzymes work best at 37°C (98.6°F), which is the body's normal temperature. If the temperature was several degrees above or below this value due to infection or environmental causes, the function of many enzymes would be greatly inhibited (**Figure 6.7**). Increasing the temperature may have more severe effects on enzyme function if the protein structure of an enzyme is greatly altered. Very high temperatures may denature a protein—cause it to become unfolded. Denaturing an enzyme is expected to inhibit its function.

Enzyme function is also sensitive to pH. Certain enzymes in the stomach function best at the acidic pH found in this organ. For example, pepsin is a protease—an enzyme that digests proteins—that is released into the stomach. Its function is to degrade food proteins into shorter peptides. The optimal pH for pepsin function is around pH 2.0, which is extremely

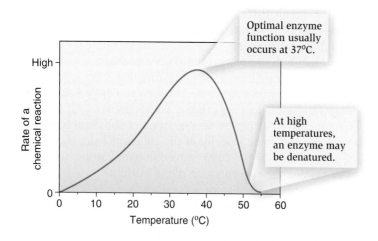

acidic. By comparison, many cytosolic enzymes function optimally at a more neutral pH, such as pH 7.2, which is the pH normally found in the cytosol of human cells. If the pH was significantly above or below this value, enzyme function would be decreased for cytosolic enzymes.

Figure 6.7 **Effects of temperature on a typical human enzyme.** Most enzymes function optimally within a narrow range of temperature. Many human enzymes function best at 37°C, which is body temperature.

FEATURE INVESTIGATION

The Discovery of Ribozymes by Sidney Altman Revealed That RNA Molecules May Also Function as Catalysts

Until the 1980s, scientists thought that all biological catalysts are proteins. One avenue of study that dramatically changed this view came from the analysis of ribonuclease P (RNase P), a catalyst initially found in the bacterium *Escherichia coli* and later identified in all species examined. RNase P is involved in the processing of tRNA molecules—a type of molecule required for protein synthesis. Such tRNA molecules are synthesized as longer precursor molecules called ptRNAs, which have 5′ and 3′ ends. (The 5′ and 3′ directionality of RNA molecules is described in Chapter 11.) RNase P breaks a covalent bond at a specific site in precursor tRNAs, which releases a fragment at the 5′ end and makes them shorter (**Figure 6.8**).

Sidney Altman and his colleagues became interested in the processing of tRNA molecules and turned their attention to RNase P in *E. coli*. During the course of their studies, they purified this enzyme and, to their surprise, discovered it has two subunits—one is an RNA molecule that contains 377 nucleotides, and the other is a small protein with a mass of 14 kDa. A complex between RNA and a protein is called a **ribonucleoprotein**. In 1990, the finding that a catalyst has an RNA subunit was very unexpected. Even so, a second property of RNase P would prove even more exciting.

Altman and colleagues were able to purify RNase P and study its properties in vitro. As mentioned earlier in this chapter, the functioning of enzymes is affected by the surrounding conditions. Cecilia Guerrier-Takada in Altman's laboratory determined that Mg^{2+} had a stimulatory effect on RNase P function. In the experiment described in **Figure 6.9**, the effects of Mg^{2+} were studied in greater detail. The researchers analyzed the effects of low (10 mM $MgCl_2$) and high (100 mM $MgCl_2$) magnesium concentrations on the processing of a ptRNA. At low or high magnesium concentrations, the ptRNA was incubated without RNase P (as a control); with the RNA subunit alone; or with intact RNase P (RNA subunit and protein sub-

Figure 6.8 **The function of RNase P.** A specific bond in a precursor tRNA (ptRNA) is cleaved by RNase P, which releases a small fragment at the 5′ end. This results in the formation of a mature tRNA.

unit). Following incubation, they performed gel electrophoresis on the samples to determine if the ptRNAs had been cleaved into two pieces—the tRNA and a 5′ fragment.

Let's now look at the data. As a control, ptRNAs were incubated with low (lane 1) or high (lane 4) $MgCl_2$ in the absence of RNase P. As expected, no processing to a lower molecular mass tRNA was observed. When the RNA subunit alone was incubated with ptRNA molecules in the presence of low $MgCl_2$ (lane 2), no processing occurred, but it did occur if the protein subunit was also included (lane 3).

The surprising result is shown in lane 5. In this case, the RNA subunit alone was incubated with ptRNAs in the presence of high $MgCl_2$. The RNA subunit by itself was able to cleave the ptRNA to a smaller tRNA and a 5′ fragment! These results indicate that RNA molecules alone can act as catalysts that facilitate the breakage of a covalent bond. In this case, the RNA subunit

Figure 6.9 The discovery that the RNA subunit of RNase P is a catalyst.

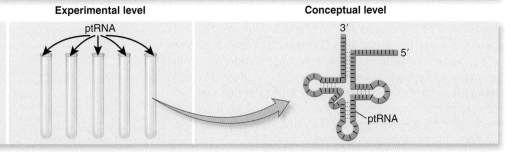

HYPOTHESIS The catalytic function of RNase P could be carried out by its RNA subunit or by its protein subunit.

KEY MATERIALS Purified precursor tRNA (ptRNA) and purified RNA and protein subunits of RNase P from *E. coli*.

Experimental level   **Conceptual level**

1. Into each of five tubes, add ptRNA.

2. In tubes 1–3, add a low concentration of MgCl₂; in tubes 4 and 5, add a high MgCl₂ concentration.

Low MgCl₂ (10 mM)   High MgCl₂ (100 mM)

3. Into tubes 2 and 5, add the RNA subunit of RNase P alone; into tube 3, add both the RNA subunit and the protein subunit of RNAase P. Incubate to allow digestion to occur. Note: Tubes 1 and 4 are controls that have no added subunits of RNase P.

RNA subunit alone   RNA subunit plus protein subunit

RNA subunit alone cuts here   + 5′ fragment

4. Carry out gel electrophoresis on each sample. In this technique, samples are loaded into a well on a gel and exposed to an electric field as described in Chapter 20. The molecules move toward the bottom of the gel and are separated according to their masses: Molecules with higher masses are closer to the top of the gel. The gel is exposed to ethidium bromide, which stains RNA.

Higher mass → Lower mass

ptRNA
tRNA
5′ fragment

Catalytic function will result in the digestion of ptRNA into tRNA and a smaller 5′ fragment.

5. **THE DATA**

ptRNA
tRNA
5′ fragment

6. **CONCLUSION** The RNA subunit alone can catalyze the breakage of a covalent bond in ptRNA at high Mg concentrations. It is a ribozyme.

7. **SOURCE** Altman, S. 1990. Enzymatic cleavage of RNA by RNA. *Bioscience Reports* 10:317–337.

is necessary and sufficient for ptRNA cleavage. Presumably, the high MgCl$_2$ concentration helps to keep the RNA subunit in a conformation that is catalytically active. Alternatively, the protein subunit plays a similar role in a living cell.

Subsequent work confirmed these observations and showed that the RNA subunit of RNase P is a true catalyst—it accelerates the rate of a chemical reaction, and it is not permanently altered. Around the same time, Thomas Cech and colleagues determined that a different RNA molecule found in the protist *Tetrahymena thermophila* also had catalytic activity. The term **ribozyme** is now used to describe an RNA molecule that catalyzes a chemical reaction. In 1989, Altman and Cech received the Nobel Prize in chemistry for their discovery of ribozymes.

Since the pioneering work of Altman and Cech, researchers have discovered that ribozymes play key catalytic roles in cells (**Table 6.2**). They are primarily involved in the processing of RNA molecules from precursor to mature forms. In addition, a ribozyme in the ribosome catalyzes the formation of covalent bonds between adjacent amino acids during polypeptide synthesis.

Experimental Questions

1. Briefly explain why it was necessary to purify the individual subunits of RNase P to show that it is a ribozyme.

2. In the Altman experiment involving RNase P, explain how the researchers experimentally determined if RNase P or

Table 6.2	**Types of Ribozyme**
General function	**Biological examples**
Processing of RNA molecules	1. RNase P: As described in this chapter, RNase P cleaves precursor tRNA molecules (ptRNAs) to a mature form.
	2. Spliceosomal RNA: As described in Chapter 12, eukaryotic pre-mRNAs often have regions called introns that are later removed. These introns are removed by a spliceosome composed of RNA and protein subunits. The RNA within the spliceosome is believed to function as a ribozyme that removes the introns from pre-mRNA.
	3. Certain introns found in mitochondrial, chloroplast, and prokaryotic RNAs are removed by a self-splicing mechanism.
Synthesis of polypeptides	The ribosome has an RNA component that catalyzes the formation of covalent bonds between adjacent amino acids during polypeptide synthesis.

subunits of RNase P were catalytically active or not. Why were two controls—one without protein and one without RNA—needed in this experiment?

3. Describe the critical results that showed RNase P is a ribozyme. How does the concentration of Mg^{2+} affect the function of the RNA in RNase P?

6.3   Overview of Metabolism

In the previous sections, we have examined the underlying factors that govern individual chemical reactions and explored the properties of enzymes and ribozymes. In living cells, chemical reactions are often coordinated with each other and occur in sequences called **metabolic pathways**, each step of which is catalyzed by a specific enzyme (**Figure 6.10**). These pathways are categorized according to whether the reactions lead to the breakdown or synthesis of substances. **Catabolic reactions** result in the breakdown of molecules into smaller molecules. Such reactions are often exergonic. By comparison, **anabolic reactions** involve the synthesis of larger molecules from smaller precursor molecules. This process usually is endergonic and, in living cells, must be coupled to an exergonic reaction. In this

section, we will survey the general features of catabolic and anabolic reactions and explore the ways in which metabolic pathways are controlled.

Catabolic Reactions Recycle Organic Building Blocks and Produce Energy Intermediates Such as ATP and NADH

Catabolic reactions result in the breakdown of larger molecules into smaller ones. One reason for the breakdown of macromolecules is to recycle their building blocks to construct new macromolecules. For example, RNA molecules are composed of building blocks called nucleotides. The breakdown of RNA by enzymes called nucleases produces nucleotides that can be used in the synthesis of new RNA molecules.

Nucleases
RNA →→→→→→→→→→→→ Many individual nucleotides

Polypeptides, which comprise proteins, are composed of a linear sequence of amino acids. When a protein is improperly folded or is no longer needed by a cell, the peptide bonds between amino acids in the protein are broken by enzymes called proteases. This generates amino acids that can be used in the construction of new proteins.

Proteases
Protein →→→→→→→→→→→→ Many individual amino acids

Figure 6.10 **A metabolic pathway.** In this metabolic pathway, a series of different enzymes catalyze the attachment of phosphate groups to various sugars, beginning with a starting substrate and ending with a final product.

The breakdown of macromolecules, such as RNA molecules and proteins that are no longer needed, allows a cell to recycle the building blocks and use them to make new macromolecules. We will consider the mechanisms of recycling later in this chapter.

A second reason for the breakdown of macromolecules and smaller organic molecules is to obtain energy that can be used to drive endergonic processes in the cell. Covalent bonds store a large amount of energy. However, when cells break covalent bonds in organic molecules such as carbohydrates and proteins, they do not directly use the energy released in this process. Instead, the released energy is stored in **energy intermediates**, molecules such as ATP and NADH, that are then directly used to drive endergonic reactions in cells.

As an example, let's consider the breakdown of glucose into two molecules of pyruvate. As discussed in Chapter 7, the breakdown of glucose to pyruvate involves a catabolic pathway called glycolysis. Some of the energy released during the breakage of covalent bonds in glucose is harnessed to synthesize ATP. However, this does not occur in a single step. Rather, glycolysis involves a series of steps in which covalent bonds are broken and rearranged. This process creates molecules that can readily donate a phosphate group to ADP, thereby creating ATP. For example, phosphoenolpyruvate has a phosphate group attached to pyruvate. Due to the arrangement of bonds in phosphoenolpyruvate, this phosphate bond is easily broken. Therefore, the phosphate can be readily transferred to ADP:

$$\text{Phosphoenolpyruvate} + \text{ADP} \rightarrow \text{Pyruvate} + \text{ATP}$$

$$\Delta G = -7.5 \text{ kcal/mole}$$

This is an exergonic reaction and therefore favors the formation of products. In this step of glycolysis, the breakdown of an organic molecule, namely phosphoenolpyruvate, results in the synthesis of an energy intermediate molecule, ATP, which can then be used by a cell to drive endergonic reactions. This way of synthesizing ATP, termed **substrate-level phosphorylation**, occurs when an enzyme directly transfers a phosphate from an organic molecule to ADP, thereby making ATP. In this case, a phosphate is transferred from phosphoenolpyruvate to ADP. Another way to make ATP is via **chemiosmosis**. In this process, energy stored in an ion electrochemical gradient is used to make ATP from ADP and P_i. We will consider this mechanism in Chapter 7.

Redox Reactions Are Important in the Metabolism of Small Organic Molecules

During the breakdown of small organic molecules, **oxidation**— the removal of one or more electrons from an atom or molecule—may occur. This process is called oxidation because oxygen is frequently involved in chemical reactions that remove electrons from other molecules. By comparison, **reduction** is the addition of electrons to an atom or molecule. Reduction is

so named because the addition of a negatively charged electron reduces the net charge of a molecule.

Electrons do not exist freely in solution. When an atom or molecule is oxidized, the electron that is removed must be transferred to another atom or molecule, which becomes reduced. This type of reaction is termed a **redox reaction**, which is short for a <u>red</u>uction-<u>ox</u>idation reaction. As a generalized equation, an electron may be transferred from molecule A to molecule B as follows:

$$\begin{array}{cccccc} Ae^- & + & B & \rightarrow & A & + & Be^- \\ \text{(oxidized)} & & & & & \text{(reduced)} \end{array}$$

As shown in the right side of this reaction, A has been oxidized (that is, had an electron removed), and B has been reduced (that is, had an electron added). In general, a substance that has been oxidized has less energy, whereas a substance that has been reduced has more energy.

During the oxidation of organic molecules such as glucose, the electrons are used to create energy intermediates such as NADH (**Figure 6.11**). In this process, an organic molecule has been oxidized, and **NAD$^+$** (**nicotinamide adenine dinucleotide**) has been reduced to NADH. Cells use NADH in two common ways. First, as we will see in Chapter 7, the oxidation of NADH is a highly exergonic reaction that can be used to make ATP. Second, NADH can donate electrons to other organic molecules and thereby energize them. Such energized molecules can more readily form covalent bonds. Therefore, as described next, NADH is often needed in anabolic reactions that involve the synthesis of larger molecules through the formation of covalent bonds between smaller molecules.

Anabolic Reactions Require an Input of Energy to Make Larger Molecules

Anabolic reactions are also called **biosynthetic reactions**, because they are necessary to make larger molecules and macromolecules. We will examine the synthesis of macromolecules in several chapters of this textbook. For example, RNA and protein biosynthesis are described in Chapter 12. Cells also need to synthesize small organic molecules, such as amino acids and fats, if they are not readily available from food sources. Such molecules are made by the formation of covalent linkages between precursor molecules. For example, glutamate (an amino acid) is made by the covalent linkage between α-ketoglutarate (a product of sugar metabolism) and ammonium.

Figure 6.11   **The reduction of NAD$^+$ to create NADH.**   NAD$^+$ is composed of two nucleotides, one with an adenine base and one with a nicotinamide base. The oxidation of organic molecules releases electrons that can bind to NAD$^+$, and along with a hydrogen ion, result in the formation of NADH. The two electrons and H$^+$ are incorporated into the nicotinamide ring. Note: The actual net charges of NAD$^+$ and NADH are minus one and minus two, respectively. They are designated NAD$^+$ and NADH to emphasize the net charge of the nicotinamide ring, which is involved in oxidation-reduction reactions.

Concept check: *Which is the oxidized form, NAD$^+$ or NADH?*

Subsequently, another amino acid, glutamine, is made from glutamate and ammonium.

$$\text{Glutamate} + NH_4^+ + ATP^{4-} + H_2O \longrightarrow \text{Glutamine} + ADP^{2-} + P_i^{2-}$$

In both reactions, an energy intermediate molecule such as NADH or ATP is needed to drive the reaction forward.

Genomes & Proteomes Connection

Many Proteins Use ATP as a Source of Energy

Over the past several decades, researchers have studied the functions of many types of proteins and discovered numerous examples in which a protein uses the hydrolysis of ATP to drive a cellular process (**Table 6.3**). In humans, a typical cell uses millions of ATP molecules per second. At the same time, the breakdown of food molecules to form smaller molecules releases energy that allows us to make more ATP from ADP and P_i. The turnover of ATP occurs at a remarkable pace. An average person hydrolyzes about 100 pounds of ATP per day, yet at any given time we do not have 100 pounds of ATP in our bodies. For this to happen, each ATP undergoes about

10,000 cycles of hydrolysis and resynthesis during an ordinary day (**Figure 6.12**).

By studying the structures of many proteins that use ATP, biochemists have discovered that particular amino acid sequences within proteins function as ATP-binding sites. This information has allowed researchers to predict whether a newly discovered protein uses ATP or not. When an entire genome sequence of a species has been determined, the genes that encode proteins can be analyzed to find out if the encoded proteins have ATP-binding sites in their amino acid sequences. Using this approach, researchers have been able to analyze

Table 6.3	Examples of Proteins That Use ATP for Energy
Type	**Description**
Metabolic enzymes	Many enzymes use ATP to catalyze endergonic reactions. For example, hexokinase uses ATP to attach phosphate to glucose.
Transporters	Ion pumps, such as the Na$^+$/K$^+$-ATPase, use ATP to pump ions against a gradient (see Chapter 5).
Motor proteins	Motor proteins such as myosin use ATP to facilitate cellular movement, as in muscle contraction (see Chapter 46).
Chaperones	Chaperones are proteins that use ATP to aid in the folding and unfolding of cellular proteins (see Chapter 4).
Protein kinases	Protein kinases are regulatory proteins that use ATP to attach a phosphate to proteins, thereby phosphorylating the protein and affecting its function (see Chapter 9).

The energy to synthesize ATP comes from catabolic reactions that are exergonic.

Energy input (endergonic)

Synthesis

$ADP + P_i$

Hydrolysis

$ATP + H_2O$

Energy release (exergonic)

ATP hydrolysis provides the energy to drive cellular processes that are endergonic.

Figure 6.12   The ATP cycle.   Living cells continuously recycle ATP. The breakdown of food molecules into smaller molecules is used to synthesize ATP from ADP and P_i. The hydrolysis of ATP to ADP and P_i is used to drive many different endergonic reactions and processes that occur in cells.

Concept check: *If a large amount of ADP was broken down in the cell, how would this affect the ATP cycle?*

proteomes—all of the proteins that a given cell can make—and estimate the percentage of proteins that are able to bind ATP. This approach has been applied to the proteomes of bacteria, archaea, and eukaryotes.

On average, over 20% of all proteins bind ATP. However, this number is likely to be an underestimate of the total percentage of ATP-utilizing proteins because we may not have identified all of the types of ATP-binding sites in proteins. In humans, who have an estimated genome size of 20,000 to 25,000 different genes, a minimum of 4,000 to 5,000 of those genes encode proteins that use ATP. From these numbers, we can see the enormous importance of ATP as a source of energy for living cells.

Metabolic Pathways Are Regulated in Three General Ways

The regulation of metabolic pathways is important for a variety of reasons. Catabolic pathways are regulated so that organic molecules are broken down only when they are no longer needed or when the cell requires energy. During anabolic reactions, regulation assures that a cell synthesizes molecules only when they are needed. The regulation of catabolic and anabolic pathways occurs at the genetic, cellular, and biochemical levels.

Gene Regulation   Because enzymes in every metabolic pathway are encoded by genes, one way that cells control chemical reactions is via gene regulation. For example, if a bacterial cell is not exposed to a particular sugar in its environment, it will turn off the genes that encode the enzymes that are needed to break down that sugar. Alternatively, if the sugar becomes available, the genes are switched on. Chapter 13 examines the steps of gene regulation in detail.

Cellular Regulation   Metabolism is also coordinated at the cellular level. Cells integrate signals from their environment and adjust their chemical reactions to adapt to those signals. As discussed in Chapter 9, cell-signaling pathways often lead to the activation of protein kinases—enzymes that covalently attach a phosphate group to target proteins. For example, when people are frightened, they secrete a hormone called epinephrine into their bloodstream. This hormone binds to the surface of muscle cells and stimulates an intracellular pathway that leads to the phosphorylation of several intracellular proteins, including enzymes involved in carbohydrate metabolism. These activated enzymes promote the breakdown of carbohydrates, an event that supplies the frightened individual with more energy. Epinephrine is sometimes called the "fight-or-flight" hormone because the added energy prepares an individual to either stay and fight or run away. After a person is no longer frightened, hormone levels drop, and other enzymes called phosphatases remove the phosphate groups from enzymes, thereby restoring the original level of carbohydrate metabolism.

Another way that cells control metabolic pathways is via compartmentalization. The membrane-bound organelles of eukaryotic cells, such as the endoplasmic reticulum and mitochondria, serve to compartmentalize the cell. As discussed in Chapter 7, this allows specific metabolic pathways to occur in one compartment in the cell but not in others.

Biochemical Regulation   A third and very prominent way that metabolic pathways are controlled is at the biochemical level. In this case, the binding of a molecule to an enzyme directly regulates its function. As discussed earlier, one form of biochemical regulation involves the binding of molecules such as competitive or noncompetitive inhibitors (see Figure 6.6). An example of noncompetitive inhibition is a type of regulation called **feedback inhibition**, in which the product of a metabolic pathway inhibits an enzyme that acts early in the pathway, thus preventing the overaccumulation of the product (**Figure 6.13**).

Many metabolic pathways use feedback inhibition as a form of biochemical regulation. In such cases, the inhibited enzyme has two binding sites. One site is the active site, where the reactants are converted to products. In addition, enzymes controlled by feedback inhibition also have an allosteric site, where a molecule can bind noncovalently and affect the function of the active site. The binding of a molecule to an allosteric site causes a conformational change in the enzyme that inhibits its catalytic function. Allosteric sites are often found in the enzymes that catalyze the early steps in a metabolic pathway. Such allosteric sites typically bind molecules that are the products of the metabolic pathway. When the products bind to these sites, they inhibit the function of these enzymes and thereby prevent the formation of too much product.

Cellular and biochemical regulation are important and rapid ways to control chemical reactions in a cell. For a metabolic pathway composed of several enzymes, which enzyme in a pathway should be controlled? In many cases, a metabolic pathway has a **rate-limiting step**, which is the slowest step in a pathway. If the rate-limiting step is inhibited or enhanced, such

Figure 6.13   Feedback inhibition.   In this process, the product of a metabolic pathway inhibits an enzyme that functions in the pathway, thereby preventing the overaccumulation of the product.

Concept check: What would be the consequences if a mutation had no effect on the active site on enzyme 1 but altered its allosteric site so that it no longer recognized the final product?

changes will have the greatest impact on the formation of the product of the metabolic pathway. Rather than affecting all of the enzymes in a metabolic pathway, cellular and biochemical regulation are often directed at the enzyme that catalyzes the rate-limiting step. This is an efficient and rapid way to control the amount of product of a pathway.

6.4   Recycling of Macromolecules

Except for DNA, which is stably maintained and inherited from cell to cell, other large molecules such as RNA, proteins, lipids, and polysaccharides typically exist for a relatively short period of time. Biologists often speak of the **half-life** of molecules, which is the time it takes for 50% of the molecules to be broken down and recycled. For example, a population of messenger RNA molecules in prokaryotes has an average half-life of about 5 minutes, whereas mRNAs in eukaryotes tend to exist for longer periods of time, on the order of 30 minutes to 24 hours or even several days.

Why is recycling important? To compete effectively in their native environments, all living organisms must efficiently use and recycle the organic molecules that are needed as building blocks to construct larger molecules and macromolecules. Otherwise, they would waste a great deal of energy making such building blocks. For example, organisms conserve an enormous amount of energy by re-using the amino acids that are needed to construct cellular proteins.

As discussed in Chapters 1 and 4, the characteristics of cells are controlled by the genome and the resulting proteome. The genome of every cell contains many genes that are transcribed into RNA. Most of these RNA molecules, called messenger RNA, or mRNA, encode proteins that ultimately determine the structure and function of cells. The expression of the genome is a very dynamic process, allowing cells to

respond to changes in their environment. RNA and proteins are made when they are needed and then broken down when they are not. After they are broken down, the building blocks of RNA and proteins—nucleotides and amino acids—are recycled to make new RNAs and proteins. In this section, we will explore how RNAs and proteins are recycled and consider a mechanism for the recycling of materials found in an entire organelle.

Messenger RNA Molecules in Eukaryotes Are Broken Down by 5′ → 3′ Cleavage or by the Exosome

The degradation of mRNA serves two important functions. First, the proteins that are encoded by particular mRNAs may be needed only under certain conditions. A cell conserves energy by degrading mRNAs when such proteins are no longer necessary. Second, mRNAs may be faulty. For example, mistakes during mRNA synthesis can result in mRNAs that produce aberrant proteins. The degradation of faulty mRNAs is beneficial to the cell to prevent the potentially harmful effects of such aberrant proteins.

As described in Chapter 12, eukaryotic mRNAs contain a cap at their 5′ end. A tail is found at their 3′ end consisting of many adenine bases (look ahead to Figure 12.11). In most cases, degradation of mRNA begins with the removal of nucleotides in the poly A tail at the 3′ end (**Figure 6.14**). After the tail gets shorter, two mechanisms of degradation may occur.

In one mechanism, the 5′ cap is removed, and the mRNA is degraded by an **exonuclease**—an enzyme that cleaves off nucleotides, one at a time, from the end of the RNA. In this case, the exonuclease removes nucleotides starting at the 5′ end and moving toward the 3′. The nucleotides can then be used to make new RNA molecules.

The other mechanism involves mRNA being degraded by an **exosome**, a multiprotein complex discovered in 1997. Exosomes are found in eukaryotic cells and some archaea, whereas in bacteria a simpler complex called the degradosome carries out similar functions. The core of the exosome has a six-membered protein ring to which other proteins are attached (see inset to Figure 6.14). Certain proteins within the exosome are exonucleases that degrade the mRNA starting at the 3′ end and moving toward the 5′ end, thereby releasing nucleotides that can be recycled.

Proteins in Eukaryotes and Archaea Are Broken Down in the Proteasome

Cells continually degrade proteins that are faulty or no longer needed. To be degraded, proteins are recognized by **proteases**—enzymes that cleave the bonds between adjacent amino acids. The primary pathway for protein degradation in archaea and eukaryotic cells is via a protein complex called a **proteasome**. Similar to the exosome that has a central cavity surrounded by a ring of proteins, the core of the proteasome is formed from four stacked rings, each composed of seven protein subunits (**Figure 6.15a**). The proteasomes of eukaryotic cells also contain cap structures at each end that control the entry of proteins into the proteasome.

In eukaryotic cells, unwanted proteins are directed to a proteasome by the covalent attachment of a small protein called **ubiquitin**. **Figure 6.15b** describes the steps of protein degradation via eukaryotic proteasomes. First, a string of ubiquitin proteins are attached to the target protein. This event directs the protein to a proteasome cap, which has binding sites for ubiquitin. The cap also has enzymes that unfold the protein and inject it into the internal cavity of the proteasome core. The ubiquitin proteins are removed during entry and are returned to the cytosol for reuse. Inside the proteasome, proteases degrade the protein into small peptides and amino acids. The process is completed when the peptides and amino acids are recycled back into the cytosol. The amino acids can be used to make new proteins.

Ubiquitin targeting has two advantages. First, the enzymes that attach ubiquitin to its target recognize improperly folded proteins, allowing cells to identify and degrade nonfunctional proteins. Second, changes in cellular conditions may warrant the rapid breakdown of particular proteins. For example, cell division requires a series of stages called the cell cycle, which depends on the degradation of specific proteins. After these proteins perform their functions in the cycle, ubiquitin targeting directs them to the proteasome for degradation.

Autophagy Recycles the Contents of Entire Organelles

As described in Chapter 4, lysosomes contain many different types of acid hydrolases that break down proteins, carbohydrates, nucleic acids, and lipids. This enzymatic function enables lysosomes to break down complex materials. One function of lysosomes involves the digestion of substances that are taken up from outside the cell. This process, called endocytosis, is described in Chapter 5. In addition, lysosomes help digest intracellular materials. In a process known as **autophagy** (from the Greek, meaning eating one's self), cellular material, such as a worn-out organelle, becomes enclosed in a double membrane

(a) 5′ → 3′ degradation by exonuclease

(b) 3′ → 5′ degradation by exosome

Exosome structure

Figure 6.14   Two pathways for mRNA degradation in eukaryotic cells.   Degradation usually begins with a shortening of the poly A tail. After tail shortening, either **(a)** the 5′ cap is removed and the RNA degraded in a 5′ to 3′ direction by an exonuclease, or **(b)** the mRNA is degraded in the 3′ to 5′ direction via an exosome. The reason why cells have two different mechanisms for RNA degradation is not well understood.

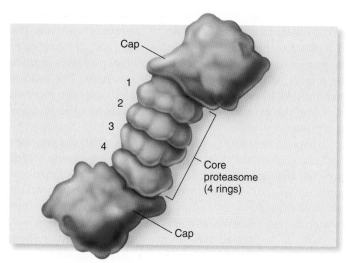

Cap

1

2

3

4

Core
proteasome
(4 rings)

Cap

(a) Structure of the eukaryotic proteasome

Figure 6.15 Protein degradation via the proteasome.

Concept check: *What are advantages of protein degradation?*

(Figure 6.16). This double membrane is formed from a tubule that elongates and eventually wraps around the organelle to form an **autophagosome**. The autophagosome then fuses with a lysosome, and the material inside the autophagosome is digested. The small molecules released from this digestion are recycled back into the cytosol.

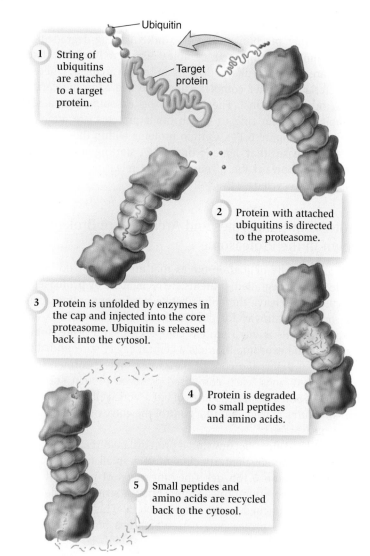

Ubiquitin

1 String of ubiquitins are attached to a target protein.

Target protein

Enzymes

2 Protein with attached ubiquitins is directed to the proteasome.

3 Protein is unfolded by enzymes in the cap and injected into the core proteasome. Ubiquitin is released back into the cytosol.

4 Protein is degraded to small peptides and amino acids.

5 Small peptides and amino acids are recycled back to the cytosol.

(b) Steps of protein degradation in eukaryotic cells

Autophagosome

Outer membrane

Inner membrane

Lysosome

Organelle

1 Membrane tubule begins to enclose an organelle.

2 Double membrane completely encloses an organelle to form an autophagosome.

3 Autophagosome fuses with a lysosome. Contents are degraded and recycled back to the cytosol.

Figure 6.16 Autophagy.

Summary of Key Concepts

6.1 Energy and Chemical Reactions

- The fate of a chemical reaction is determined by its direction and rate.

- Energy, the ability to promote change or do work, exists in many forms. According to the first law of thermodynamics, energy cannot be created or destroyed, but it can be converted from one form to another. The second law of thermodynamics states that energy interconversions involve an increase in entropy. (Figures 6.1, 6.2, Table 6.1)

- Free energy is the amount of available energy that can be used to promote change or do work. Spontaneous reactions, which release free energy, have a negative free-energy change. (Figure 6.3)

- An exergonic reaction has a negative free-energy change, whereas an endergonic reaction has a positive change. Chemical reactions proceed until they reach a state of chemical equilibrium, where the rate of formation of products equals the rate of formation of reactants.

- Exergonic reactions, such as the breakdown of ATP, are commonly coupled to cellular processes that would otherwise be endergonic.

6.2 Enzymes and Ribozymes

- Proteins that speed up the rate of a chemical reaction are called enzymes. They lower the activation energy that is needed to achieve a transition state. (Figure 6.4)

- Enzymes recognize reactants, also called substrates, with a high specificity. Conformational changes lower the activation energy for a chemical reaction. (Figure 6.5)

- Each enzyme-catalyzed reaction exhibits a maximal velocity (V_{max}). The K_M is the substrate concentration at which the velocity of the chemical reaction is half of the V_{max}. Competitive inhibitors raise the apparent K_M for the substrate, whereas noncompetitive inhibitors lower the V_{max}. (Figure 6.6)

- Enzyme function may be affected by a variety of other factors, including prosthetic groups, cofactors, coenzymes, temperature, and pH. (Figure 6.7)

- Altman and colleagues discovered that RNase P is a ribozyme—the RNA molecule within RNase P is a catalyst. Other ribozymes also play key roles in the cell. (Figures 6.8, 6.9, Table 6.2)

6.3 Overview of Metabolism

- Metabolism is the sum of the chemical reactions in a living organism. Enzymes often function in pathways that lead to the formation of a particular product. (Figure 6.10)

- Catabolic reactions involve the breakdown of larger molecules into smaller ones. These reactions regenerate small molecules that are used as building blocks to make new molecules. The small molecules are also broken down to make energy intermediates such as ATP and NADH. Such reactions are often redox reactions in which electrons are transferred from one molecule to another. (Figure 6.11)

- Anabolic reactions involve the synthesis of larger molecules and macromolecules.

- Cells continuously synthesize ATP from ADP and P_i and then hydrolyze it to drive endergonic reactions. Estimates from genome analysis indicate that over 20% of a cell's proteins use ATP. (Table 6.3, Figure 6.12)

- Metabolic pathways are controlled by gene regulation, cell signaling, compartmentalization, and feedback inhibition. (Figure 6.13)

6.4 Recycling of Macromolecules

- Large molecules in cells have a finite half-life.

- Recycling of macromolecules is important because it saves a great deal of energy for living organisms.

- Messenger RNAs in eukaryotes are degraded by 5′ to 3′ exonucleases or by the exosome. (Figure 6.14)

- Proteins in eukaryotes and archaea are degraded by the proteasome. (Figure 6.15)

- During autophagy in eukaryotes, an entire organelle is surrounded by a double membrane and then fuses with a lysosome. The internal contents are degraded, and the smaller building blocks are recycled to the cytosol. (Figure 6.16)

Assess and Discuss

Test Yourself

1. According to the second law of thermodynamics,
   a. energy cannot be created or destroyed.
   b. each energy transfer decreases the disorder of a system.
   c. energy is constant in the universe.
   d. each energy transfer increases the level of disorder in a system.
   e. chemical energy is a form of potential energy.

2. Reactions that release free energy are
   a. exergonic.
   b. spontaneous.
   c. endergonic.
   d. endothermic.
   e. both a and b.

3. Enzymes speed up reactions by
   a. providing chemical energy to fuel a reaction.
   b. lowering the activation energy necessary to initiate the reaction.
   c. causing an endergonic reaction to become an exergonic reaction.
   d. substituting for one of the reactants necessary for the reaction.
   e. none of the above.

4. Which of the following factors may alter the function of an enzyme?
   a. pH   d. all of the above
   b. temperature   e. b and c only
   c. cofactors

5. In biological systems, ATP functions by
   a. providing the energy to drive endergonic reactions.
   b. acting as an enzyme and lowering the activation energy of certain reactions.
   c. adjusting the pH of solutions to maintain optimal conditions for enzyme activity.
   d. regulating the speed at which endergonic reactions proceed.
   e. interacting with enzymes as a cofactor to stimulate chemical reactions.

6. In a chemical reaction, NADH is converted to $NAD^+ + H^+$. We would say that NADH has been
   a. reduced.
   b. phosphorylated.
   c. oxidized.
   d. decarboxylated.
   e. methylated.

7. Currently, scientists are identifying proteins that use ATP as an energy source by
   a. determining whether those proteins function in anabolic or catabolic reactions.
   b. determining if the protein has a known ATP-binding site.
   c. predicting the free energy necessary for the protein to function.
   d. determining if the protein has an ATP synthase subunit.
   e. all of the above.

8. With regard to its effects on an enzyme-catalyzed reaction, a competitive inhibitor
   a. lowers the K_M only.
   b. lowers the K_M and lowers the V_{max}.
   c. raises the K_M only.
   d. raises the K_M and lowers the V_{max}.
   e. raises the K_M and raises the V_{max}.

9. In eukaryotes, mRNAs may be degraded by
   a. a 5′ to 3′ exonuclease.   d. all of the above.
   b. the exosome.   e. a and b only.
   c. the proteasome.

10. Autophagy provides a way for cells to
    a. degrade entire organelles and recycle their components.
    b. automatically control the level of ATP.
    c. engulf bacterial cells.
    d. export unwanted organelles out of the cell.
    e. inhibit the first enzyme in a metabolic pathway.

Conceptual Questions

1. With regard to rate and direction, discuss the differences between endergonic and exergonic reactions.

2. Describe the mechanism and purpose of feedback inhibition in a metabolic pathway.

3. Why is recycling of amino acids and nucleotides an important metabolic function of cells? Explain how eukaryotic cells recycle amino acids found in worn-out proteins.

Collaborative Questions

1. Living cells are highly ordered units, yet the universe is heading toward higher entropy. Discuss how life can maintain its order in spite of the second law of thermodynamics. Are we defying this law?

2. What is the advantage of using ATP as a common energy source? Another way of asking this question is, Why is ATP an advantage over using a bunch of different food molecules? For example, instead of just having a Na^+/K^+-ATPase in a cell, why not have a bunch of different ion pumps each driven by a different food molecule, like a Na^+/K^+-glucosase (a pump that uses glucose), a Na^+/K^+-sucrase (a pump that uses sucrose), a Na^+/K^+-fatty acidase (a pump that uses fatty acids), and so on?

Online Resource

www.brookerbiology.com

Stay a step ahead in your studies with animations that bring concepts to life and practice tests to assess your understanding. Your instructor may also recommend the interactive ebook, individualized learning tools, and more.

Cellular Respiration, Fermentation, and Secondary Metabolism

7

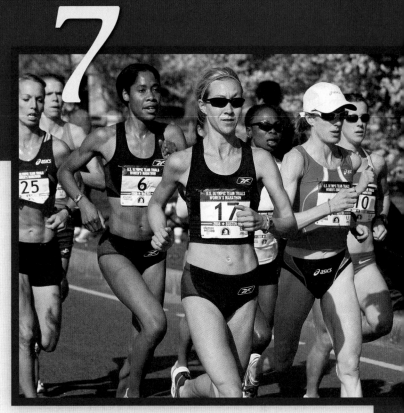

Physical endurance. Conditioned athletes, like these marathon runners, have very efficient metabolism of organic molecules such as glucose.

C armen became inspired while watching the 2008 Summer Olympics and set a personal goal to run a marathon. Although she was active in volleyball and downhill skiing in high school, she had never attempted distance running. At first, running an entire mile was pure torture. She was out of breath, overheated, and unhappy, to say the least. However, she became committed to endurance training and within a few weeks discovered that running a mile was a "piece of cake." Two years later, she participated in her first marathon (42.2 kilometers or 26.2 miles) and finished with a time of 4 hours and 11 minutes—not bad for someone who had previously struggled to run a single mile!

How had Carmen's training allowed her to achieve this goal? Perhaps the biggest factor is that the training had altered the metabolism in her leg muscles. For example, the network of small blood vessels supplying oxygen to her leg muscles became more extensive, allowing the more efficient delivery of oxygen and removal of wastes. Second, her muscle cells developed more mitochondria. With these changes, Carmen's leg muscles were better able to break down organic molecules in her food and use them to make ATP.

The cells in Carmen's leg muscles had become more efficient at **cellular respiration**, which refers to the metabolic reactions that a cell uses to get energy from food molecules and release waste products. When we eat food, we are using much of that food for energy. People often speak of "burning calories." While metabolism does generate some heat, the chemical reactions that take place in the cells of living organisms are uniquely different from those that occur, say, in a fire. When wood is burned, the reaction produces enormous amounts of heat in a short period of time—the reaction lacks control. In contrast, the metabolism that occurs in living cells is extremely controlled. The food molecules from which we harvest energy give up that energy in a very restrained manner rather than all at once, as in a fire. An underlying theme in metabolism is the remarkable control that cells possess when they coordinate chemical reactions. A key emphasis of this chapter is how cells use energy that is stored within the chemical bonds of organic molecules.

We will begin by surveying a group of chemical reactions that involves the breakdown of carbohydrates, namely, the sugar glucose. As you will learn, cells carry out an intricate series of reactions so that glucose can be "burned" in a very controlled fashion when

oxygen is available. We will then examine how cells can use organic molecules in the absence of oxygen via processes known as anaerobic respiration and fermentation. Finally, we will consider secondary metabolism, which is not vital for cell survival, but produces organic molecules that serve unique and important functions.

7.1 Cellular Respiration in the Presence of Oxygen

As mentioned, cellular respiration is a process by which living cells obtain energy from organic molecules and release waste products. A primary aim of cellular respiration is to make ATP. When oxygen (O_2) is used, this process is termed **aerobic respiration**. During aerobic respiration, O_2 is used, and CO_2 is released via the oxidation of organic molecules. When we breathe, we inhale the oxygen needed for aerobic respiration and exhale the CO_2, a by-product of the process. For this reason, the term respiration has a second meaning, which is the act of breathing.

Different types of organic molecules, such as carbohydrates, proteins, and fats, can be used as energy sources to drive aerobic respiration. In this section, we will largely focus on the use of glucose as an energy source for cellular respiration.

$$C_6H_{12}O_6 + 6\ O_2 \rightarrow 6\ CO_2 + 6\ H_2O + \text{Energy intermediates} + \text{Heat}$$
Glucose

$$\Delta G = -686\ \text{kcal/mole}$$

We will examine the metabolic pathways in which glucose is broken down into carbon dioxide and water, thereby releasing a large amount of energy that is used to make many ATP molecules. In so doing, we will focus on four pathways: (1) glycolysis, (2) the breakdown of pyruvate, (3) the citric acid cycle, and (4) oxidative phosphorylation.

Distinct Metabolic Pathways Are Involved in the Breakdown of Glucose to CO$_2$

Let's begin our discussion of cellular respiration with an overview of the entire process. We will focus on the breakdown of glucose in a eukaryotic cell in the presence of oxygen. Certain covalent bonds within glucose store a large amount of chemical bond energy. When glucose is broken down via oxidation, ultimately to CO$_2$ and water, the energy within those bonds is released and used to make three types of energy intermediates: ATP, NADH, and FADH$_2$. The following is an overview of the stages that occur during the breakdown of glucose (**Figure 7.1**):

1. **Glycolysis:** In glycolysis, glucose (a compound with six carbon atoms) is broken down to two pyruvate molecules (with three carbons each), producing a net gain of two ATP molecules and two NADH molecules. The two ATP are made via **substrate-level phosphorylation**, which occurs when an enzyme directly transfers a phosphate from an organic molecule to ADP. In eukaryotes, glycolysis occurs in the cytosol.

2. **Breakdown of pyruvate to an acetyl group:** The two pyruvate molecules enter the mitochondrial matrix, where

Figure 7.1   An overview of glucose metabolism.

Concept check: *The breakdown of glucose produces a lot of NADH. What is this NADH mostly used for?*

each one is broken down to an acetyl group (with two carbons each) and one CO_2 molecule. For each pyruvate broken down via oxidation, one NADH molecule is made by the reduction of NAD^+.

3. **Citric acid cycle:** Each acetyl group is incorporated into an organic molecule, which is later oxidized to liberate two CO_2 molecules. One ATP, three NADH, and one $FADH_2$ are made in this process. Because there are two acetyl groups (one from each pyruvate), the total yield is four CO_2, two ATP via substrate-level phosphorylation, six NADH, and two $FADH_2$. This process occurs in the mitochondrial matrix.

4. **Oxidative phosphorylation:** The NADH and $FADH_2$ made in the three previous stages contain high-energy electrons that can be readily transferred in a redox reaction to other molecules. Once removed from NADH or $FADH_2$ via oxidation, these high-energy electrons release some energy, and that energy is harnessed to produce a H^+ electrochemical gradient. In the process of **chemiosmosis**, energy stored in the H^+ electrochemical gradient is used to synthesize ATP from ADP and P_i. This process is called phosphorylation because ADP has become phosphorylated. Approximately 30 to 34 ATP molecules are made via chemiosmosis. As discussed later, oxidative phosphorylation is accomplished by two components: the electron transport chain and ATP synthase.

In eukaryotes, oxidation phosphorylation occurs along the cristae, which are invaginations of the inner mitochondrial membrane. The invaginations greatly increase the surface area of the inner membrane and thereby increase the amount of ATP that can be made. In prokaryotes, oxidative phosphorylation occurs along the plasma membrane.

Now, let's examine in detail the chemical changes that take place in each of these four stages.

Stage 1: Glycolysis Is a Metabolic Pathway That Breaks Down Glucose to Pyruvate

Glycolysis (from the Greek *glykos*, meaning sweet, and *lysis*, meaning splitting) involves the breakdown of glucose, a simple sugar. This process can occur in the presence or absence of oxygen, that is, under aerobic or anaerobic conditions. During the 1930s, the efforts of several German biochemists, including Gustav Embden, Otto Meyerhof, and Jacob Parnas, determined that glycolysis involves 10 steps, each one catalyzed by a different enzyme. The elucidation of these steps was a major achievement in the field of **biochemistry**—the study of the chemistry of living organisms. Researchers have since discovered that glycolysis is the common pathway for glucose breakdown in bacteria, archaea, and eukaryotes. Remarkably, the steps of glycolysis are virtually identical in nearly all living species, suggesting that glycolysis arose very early in the evolution of life on our planet.

The 10 steps of glycolysis can be grouped into three phases (**Figure 7.2**). The first phase (steps 1–3) involves an energy investment. Two ATP molecules are hydrolyzed, and the phosphates from those ATP molecules are attached to glucose, which is converted to fructose-1,6-bisphosphate. The energy investment phase raises the free energy of glucose and thereby allows later reactions to be exergonic. The cleavage phase (steps 4–5) breaks this six-carbon molecule into two molecules of glyceraldehyde-3-phosphate, which are three-carbon molecules. The third phase (steps 6–10) liberates energy to produce energy intermediates. In step 6, two molecules of NADH are made when two molecules of glyceraldehyde-3-phosphate

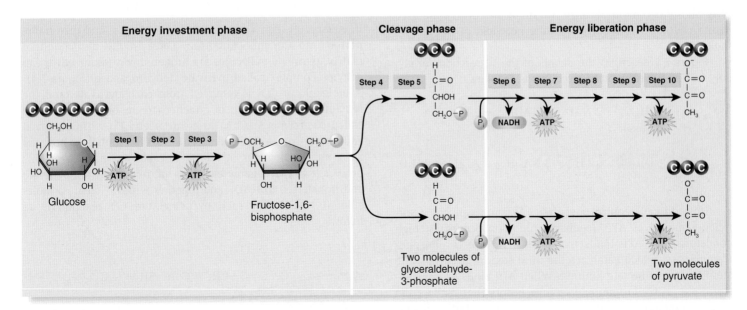

Figure 7.2 Overview of glycolysis.

Concept check: Explain why the three phases are named the energy investment phase, the cleavage phase, and the energy liberation phase.

1 Glucose is phosphorylated by ATP. Glucose-6-phosphate is more easily trapped in the cell compared to glucose.

2 The structure of glucose-6-phosphate is rearranged to make fructose-6-phosphate.

3 Fructose-6-phosphate is phosphorylated to make fructose-1,6-bisphosphate.

4 Fructose-1,6-bisphosphate is cleaved into dihydroxyacetone phosphate and glyceraldehyde-3-phosphate.

5 Dihydroxyacetone phosphate is isomerized to glyceraldehyde-3-phosphate.

Figure 7.3 **A detailed look at the steps of glycolysis.** The pathway begins with a 6-carbon molecule (glucose) that is eventually broken down into 2 molecules that contain 3 carbons each. The notation **x 2** in the figure indicates that 2 of these 3-carbon molecules are produced from each glucose molecule.

Concept check: Which organic molecules donate a phosphate group to ADP during substrate-level phosphorylation?

are oxidized to two molecules of 1,3 bisphosphoglycerate. In steps 7 and 10, four molecules of ATP are made via substrate-level phosphorylation. Because two molecules of ATP are used in the energy investment phase, the net yield of ATP is two molecules.

Figure 7.3 describes the details of the 10 reactions of glycolysis. The net reaction of glycolysis is as follows:

$$\underset{\text{Glucose}}{C_6H_{12}O_6} + 2\ NAD^+ + 2\ ADP^{2-} + 2\ P_i^{2-} \rightarrow$$

$$\underset{\text{Pyruvate}}{2\ CH_3(C{=}O)COO^-} + 2\ H^+ + 2\ NADH + 2\ ATP^{4-} + 2\ H_2O$$

How do cells control glycolysis? When a cell has a sufficient amount of ATP, feedback inhibition occurs. At high concentrations, ATP binds to an allosteric site in phosphofructokinase, which catalyzes the third step in glycolysis, the step thought to be rate limiting. When ATP binds to this allosteric site, a conformational change occurs that renders the enzyme functionally inactive. This prevents the further breakdown of glucose and thereby inhibits the overproduction of ATP. (Allosteric sites and rate-limiting steps are discussed in Chapter 6.)

Stage 2: Pyruvate Enters the Mitochondrion and Is Broken Down to an Acetyl Group and CO_2

In eukaryotes, pyruvate is made in the cytosol and then transported into the mitochondrion. Once in the mitochondrial matrix, pyruvate molecules are broken down (oxidized) by an enzyme complex called pyruvate dehydrogenase (**Figure 7.4**). A molecule of CO_2 is removed from each pyruvate, and the remaining acetyl group is attached to an organic molecule called coenzyme A (CoA) to create acetyl CoA. (In chemical equations, CoA is depicted as CoA—SH to emphasize how the SH group participates in the chemical reaction.) During this process, two high-energy electrons are removed from pyruvate and transferred to NAD^+ and together with H^+ create a molecule of NADH. For each pyruvate, the net reaction is as follows:

$$\underset{\text{Pyruvate}}{\overset{\overset{\displaystyle O \quad O}{\overset{\|\quad\|}{}}}{^-O{-}C{-}C{-}CH_3}} + \underset{\text{CoA}}{CoA{-}SH} + NAD^+ \rightarrow$$

$$\underset{\text{Acetyl CoA}}{CoA{-}S{-}\overset{\overset{\displaystyle O}{\|}}{C}{-}CH_3} + CO_2 + NADH$$

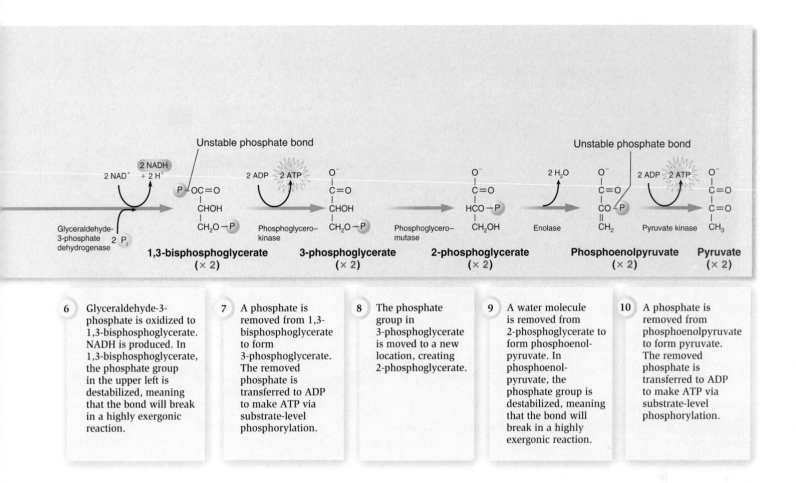

6. Glyceraldehyde-3-phosphate is oxidized to 1,3-bisphosphoglycerate. NADH is produced. In 1,3-bisphosphoglycerate, the phosphate group in the upper left is destabilized, meaning that the bond will break in a highly exergonic reaction.

7. A phosphate is removed from 1,3-bisphosphoglycerate to form 3-phosphoglycerate. The removed phosphate is transferred to ADP to make ATP via substrate-level phosphorylation.

8. The phosphate group in 3-phosphoglycerate is moved to a new location, creating 2-phosphoglycerate.

9. A water molecule is removed from 2-phosphoglycerate to form phosphoenolpyruvate. In phosphoenolpyruvate, the phosphate group is destabilized, meaning that the bond will break in a highly exergonic reaction.

10. A phosphate is removed from phosphoenolpyruvate to form pyruvate. The removed phosphate is transferred to ADP to make ATP via substrate-level phosphorylation.

Pyruvate is made in the cytosol by glycolysis. It travels through a channel in the outer membrane and an H^+/pyruvate symporter in the inner membrane to reach the mitochondrial matrix.

Pyruvate is oxidized via pyruvate dehydrogenase to an acetyl group and CO_2. NADH is made. During this process, the acetyl group is transferred to coenzyme A and will later be removed and enter the citric acid cycle.

Figure 7.4 Breakdown of pyruvate and the attachment of an acetyl group to CoA.

The acetyl group is attached to CoA via a covalent bond to a sulfur atom. The hydrolysis of this bond releases a large amount of free energy, making it possible for the acetyl group to be transferred to other organic molecules. As described next, the acetyl group attached to CoA enters the citric acid cycle.

Stage 3: During the Citric Acid Cycle, an Acetyl Group Is Oxidized to Yield Two CO_2 Molecules

The third stage of sugar metabolism introduces a new concept, that of a **metabolic cycle**. During a metabolic cycle, particular molecules enter the cycle while others leave. The process is cyclical because it involves a series of organic molecules that are regenerated with each turn of the cycle. The idea of a metabolic cycle was first proposed in the early 1930s by German biochemist Hans Krebs. While studying carbohydrate metabolism in England, he analyzed cell extracts from pigeon muscle and determined that citric acid and other organic molecules participated in a cycle that resulted in the breakdown of carbohydrates to carbon dioxide. This cycle is called the **citric acid cycle**, or the Krebs cycle, in honor of Krebs, who was awarded the Nobel Prize in 1953.

An overview of the citric acid cycle is shown in **Figure 7.5**. In the first step of the cycle, the acetyl group (with two carbons) is removed from acetyl CoA and attached to oxaloacetate (with four carbons) to form citrate (with six carbons), also called

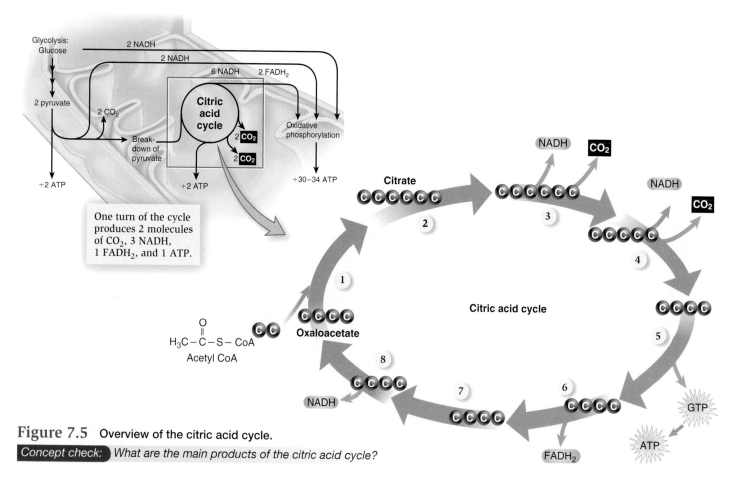

Figure 7.5 Overview of the citric acid cycle.

Concept check: *What are the main products of the citric acid cycle?*

citric acid. Then in a series of several steps, two CO_2 molecules are released. As this occurs, three molecules of NADH, one molecule of $FADH_2$, and one molecule of GTP are made. The GTP, which is made via substrate-level phosphorylation, is used to make ATP. After eight steps, oxaloacetate is regenerated so the cycle can begin again, provided acetyl CoA is available. **Figure 7.6** shows a more detailed view of the citric acid cycle. For each acetyl group attached to CoA, the net reaction of the citric acid cycle is as follows:

$$Acetyl\text{-}CoA + 2\ H_2O + 3\ NAD^+ + FAD + GDP^{2-} + P_i^{2-} \rightarrow$$
$$CoA\text{—}SH + 2\ CO_2 + 3\ NADH + FADH_2 + GTP^{4-} + 3\ H^+$$

How is the citric acid cycle controlled? One way is competitive inhibition. Oxaloacetate is a competitive inhibitor of succinate dehydrogenase, the enzyme that catalyzes step 6 of the cycle (Figure 7.6). When the oxaloacetate level becomes too high, succinate dehydrogenase is inhibited, and the citric acid cycle slows down.

Stage 4: During Oxidative Phosphorylation, NADH and $FADH_2$ Are Oxidized to Power ATP Production

Up to this point, the oxidation of glucose has yielded 6 molecules of CO_2, 4 molecules of ATP, 10 molecules of NADH, and 2 molecules of $FADH_2$. Let's now consider how high-energy electrons are removed from NADH and $FADH_2$ to make more ATP. This process is called **oxidative phosphorylation**. The

term refers to the observation that NADH and $FADH_2$ have had electrons removed and have thus become <u>oxidized</u>, and ATP is made by the <u>phosphorylation</u> of ADP (**Figure 7.7**). As described next, the oxidative process involves the electron transport chain, whereas the phosphorylation occurs via ATP synthase.

Oxidation: The Role of the Electron Transport Chain in Establishing an Electrochemical Gradient The **electron transport chain** (**ETC**) consists of a group of protein complexes and small organic molecules embedded in the inner mitochondrial membrane. These components are referred to as an electron transport chain because electrons are passed from one component to the next in a series of redox reactions (Figure 7.7). Most of the members of the ETC are protein complexes (designated I to IV) that have prosthetic groups, which are small molecules permanently attached to the surface of proteins that aid in their function. For example, cytochrome oxidase contains two prosthetic groups, each with an iron atom. The iron in each prosthetic group can readily accept and release an electron. One of the members of the electron transport chain, ubiquinone (Q), is not a protein. Rather, ubiquinone is a small organic molecule that can accept and release an electron. It is a nonpolar molecule that can diffuse through the lipid bilayer.

The red line in Figure 7.7 shows the path of electron flow. The electrons, which are originally located on NADH or $FADH_2$, are transferred to components of the ETC. The electron path is a series of redox reactions in which electrons are transferred to components with increasingly higher electronegativity. As discussed in Chapter 2, electronegativity is the ability to attract electrons. At the end of the chain is oxygen, which is the most electronegative and the final electron acceptor. The electron transport chain is also called the **respiratory chain** because the oxygen we breathe is used in this process.

NADH and $FADH_2$ donate their electrons at different points in the ETC. Two high-energy electrons from NADH are first transferred one at a time to NADH dehydrogenase (complex I). They are then transferred to ubiquinone (Q, also called coenzyme Q), cytochrome b-c_1 (complex III), cytochrome c, and cytochrome oxidase (complex IV). The final electron acceptor is O_2. By comparison, $FADH_2$ transfers electrons to succinate reductase (complex II), then to ubiquinone, and the rest of the chain.

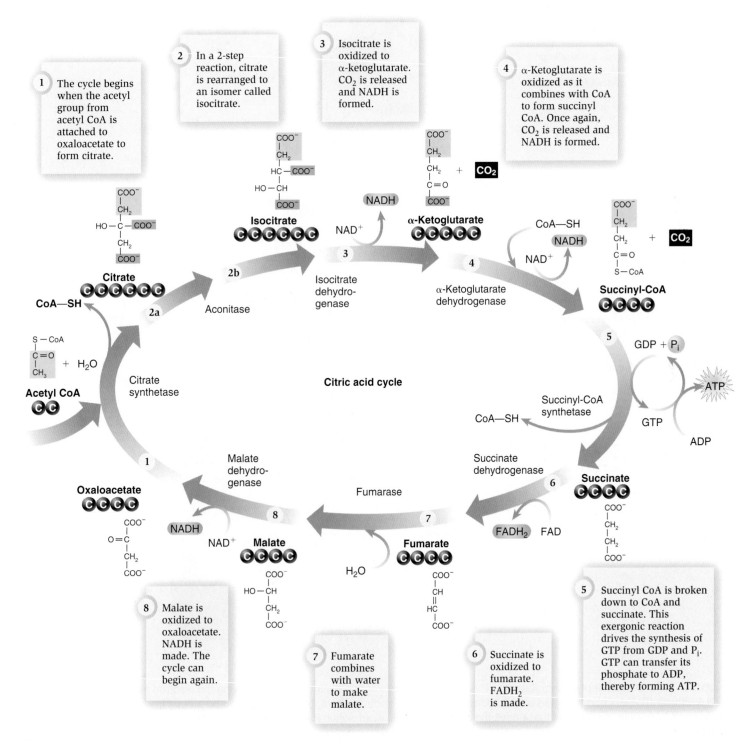

Figure 7.6   A detailed look at the steps of the citric acid cycle. The blue boxes indicate the location of the acetyl group, which is oxidized at step 6. (It is oxidized again in step 8.) The green boxes indicate the locations where CO_2 molecules are removed.

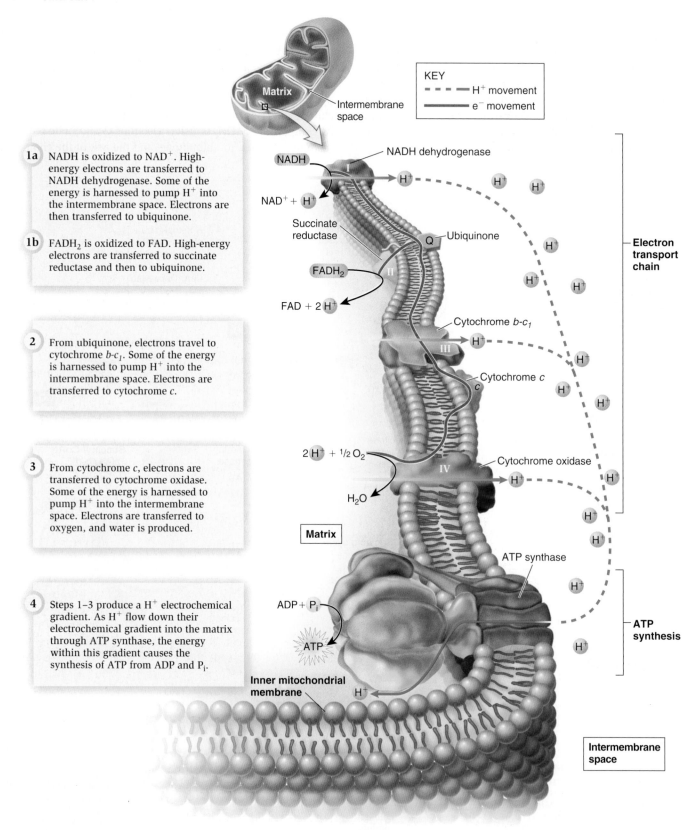

KEY
- - - - H^+ movement
——— e^- movement

1a NADH is oxidized to NAD^+. High-energy electrons are transferred to NADH dehydrogenase. Some of the energy is harnessed to pump H^+ into the intermembrane space. Electrons are then transferred to ubiquinone.

1b $FADH_2$ is oxidized to FAD. High-energy electrons are transferred to succinate reductase and then to ubiquinone.

2 From ubiquinone, electrons travel to cytochrome b-c_1. Some of the energy is harnessed to pump H^+ into the intermembrane space. Electrons are transferred to cytochrome c.

3 From cytochrome c, electrons are transferred to cytochrome oxidase. Some of the energy is harnessed to pump H^+ into the intermembrane space. Electrons are transferred to oxygen, and water is produced.

4 Steps 1–3 produce a H^+ electrochemical gradient. As H^+ flow down their electrochemical gradient into the matrix through ATP synthase, the energy within this gradient causes the synthesis of ATP from ADP and P_i.

Matrix

Intermembrane space

NADH dehydrogenase

NADH

$NAD^+ + H^+$

Succinate reductase

$FADH_2$

$FAD + 2 H^+$

Q Ubiquinone

Cytochrome b-c_1

Cytochrome c

$2 H^+ + \frac{1}{2} O_2$

H_2O

Cytochrome oxidase

Matrix

ATP synthase

$ADP + P_i$

ATP

Inner mitochondrial membrane

H^+

Electron transport chain

ATP synthesis

Intermembrane space

H^+

Figure 7.7 **Oxidative phosphorylation.** This process consists of two distinct events involving the electron transport chain and ATP synthase. The electron transport chain oxidizes, or removes electrons from, NADH or $FADH_2$ and pumps H^+ across the inner mitochondrial membrane. ATP synthase uses the energy in this H^+ electrochemical gradient to phosphorylate ADP and thereby synthesize ATP.

Concept check: *Can you explain the name of cytochrome oxidase? Can you think of another appropriate name?*

As shown in Figure 7.7, some of the energy from this movement of electrons is used to pump H^+ across the inner mitochondrial membrane from the matrix and into the intermembrane space. This active transport establishes a large **H^+ electrochemical gradient** in which the concentration of H^+ is higher outside of the matrix than inside and an excess of positive charge exists outside the matrix. Because hydrogen ions consist of protons, the H^+ electrochemical gradient is also called the **proton-motive force**. NADH dehydrogenase, cytochrome b-c_1, and cytochrome oxidase are H^+ pumps. While traveling along the electron transport chain, electrons release free energy, and some of this energy is captured by these proteins to actively transport H^+ out of the matrix into the intermembrane space against the H^+ electrochemical gradient. Because the electrons from $FADH_2$ enter the chain at an intermediate step, they release less energy and so result in fewer hydrogen ions being pumped out of the matrix than do electrons from NADH.

Why do electrons travel from NADH or $FADH_2$ to the ETC and then to O_2? As you might expect, the answer lies in free-energy changes. The electrons found on the energy intermediates have a high amount of potential energy. As they travel along the electron transport chain, free energy is released (**Figure 7.8**). The movement of one electron from NADH to O_2 results in a very negative free-energy change of approximately −25 kcal/mole. That is why the process is spontaneous and proceeds in the forward direction. Because it is a highly exergonic reaction, some of the free energy can be harnessed to do cellular work. In this case, some energy is used to pump H^+ across the inner mitochondrial membrane and establish a H^+ electrochemical gradient that is then used to power ATP synthesis.

Chemicals that inhibit the flow of electrons along the ETC can have lethal effects. For example, one component of the electron transport chain, cytochrome oxidase, can be inhibited by cyanide. The deadly effects of cyanide occur because the electron transport chain is shut down, preventing cells from making enough ATP for survival.

Phosphorylation: The Role of ATP Synthase in Making ATP via Chemiosmosis

The second event of oxidative phosphorylation is the synthesis of ATP by an enzyme called **ATP synthase**. The H^+ electrochemical gradient across the inner mitochondrial membrane is a source of potential energy. How is this energy used? The passive flow of H^+ back into the matrix is an exergonic process. The lipid bilayer is relatively impermeable to H^+. However, H^+ can pass through the membrane-embedded portion of ATP synthase. This enzyme harnesses some of the free energy that is released as the ions flow through its membrane-embedded region to synthesize ATP from ADP and P_i (see Figure 7.7). This is an example of an energy conversion: Energy in the form of a H^+ gradient is converted to chemical bond energy in ATP. The synthesis of ATP that occurs as a result of pushing H^+ across a membrane is called chemiosmosis (from the Greek *osmos*, meaning to push). The theory behind it was proposed by Peter Mitchell, a British biochemist who was awarded the Nobel Prize in chemistry in 1978.

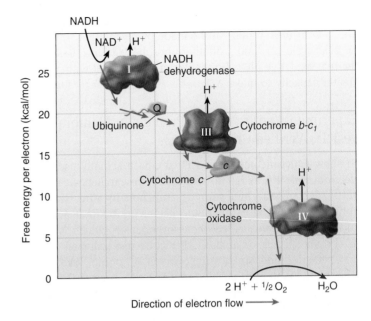

Figure 7.8 **The relationship between free energy and electron movement along the electron transport chain.** As electrons hop from one site to another along the electron transport chain, they release energy. Some of this energy is harnessed to pump H^+ across the inner mitochondrial membrane. The total energy released by a single electron is approximately −25 kcal/mole.

The Relationship Between NADH Oxidation and Amount of ATP Synthesis

For each molecule of NADH that is oxidized and each molecule of ATP that is made, the two chemical reactions of oxidative phosphorylation can be represented as follows:

$$NADH + H^+ + 1/2\ O_2 \rightarrow NAD^+ + H_2O$$

$$ADP^{2-} + P_i^{2-} \rightarrow ATP^{4-} + H_2O$$

The oxidation of NADH to NAD^+ results in a H^+ electrochemical gradient in which more hydrogen ions are in the intermembrane space than are in the matrix. The synthesis of one ATP molecule is thought to require the movement of three to four ions into the matrix, down their H^+ electrochemical gradient.

When we add up the maximal amount of ATP that can be made by oxidative phosphorylation, most researchers agree it is in the range of 30 to 34 ATP molecules for each glucose molecule that is broken down to CO_2 and water. However, the maximum amount of ATP is rarely achieved for two reasons. First, although 10 NADH and 2 $FADH_2$ are available to create the H^+ electrochemical gradient across the inner mitochondrial membrane, a cell may use some of these molecules for anabolic pathways. For example, NADH is used in the synthesis of organic molecules such as glycerol (a component of phospholipids) and lactate (which is secreted from muscle cells during strenuous exercise). Second, the mitochondrion may use some of the H^+ electrochemical gradient for other purposes. For example, the gradient is used for the uptake of pyruvate into the matrix via a H^+/pyruvate symporter (see Figure 7.4). Therefore, the actual amount of ATP synthesis is usually a little less than the maximum number of 30 to 34. Even so, when we compare the amount of ATP that can be made by glycolysis (2), the citric acid cycle (2), and oxidative phosphorylation (30–34), we see that oxidative phosphorylation provides a cell with a much greater capacity to make ATP.

Experiments with Purified Proteins in Membrane Vesicles Verified Chemiosmosis

To show experimentally that ATP synthase actually uses a H^+ electrochemical gradient to make ATP, researchers needed to purify the enzyme and study its function in vitro. In 1974, Ephraim Racker and Walther Stoeckenius purified ATP synthase and another protein called bacteriorhodopsin, which is found in certain species of archaea. Previous research had shown that bacteriorhodopsin is a light-driven H^+ pump. Racker and Stoeckenius took both purified proteins and inserted them into membrane vesicles (**Figure 7.9**). ATP synthase was oriented so its ATP synthesizing region was on the outside of the vesicles.

Bacteriorhodopsin was oriented so it would pump H^+ into the vesicles. They added ADP and P_i on the outside of the vesicles. In the dark, no ATP was made. However, when they shone light on the vesicles, a substantial amount of ATP was made. Because bacteriorhodopsin was already known to be a light-driven H^+ pump, these results convinced researchers that ATP synthase uses a H^+ electrochemical gradient as an energy source to make ATP.

ATP Synthase Is a Rotary Machine That Makes ATP as It Spins

The structure and function of ATP synthase are particularly intriguing and have received much attention over the past few decades (**Figure 7.10**). ATP synthase is a rotary machine. The membrane-embedded region is composed of three types of subunits called a, b, and c. Approximately 9 to 12 c subunits form a ring in the membrane. Each c subunit is a H^+ channel. One a subunit is bound to this ring, and two b subunits are attached to the a subunit and protrude from the membrane. The non-membrane-embedded subunits are designated with Greek letters. One ε and one γ subunit bind to the ring of c subunits. The γ subunit forms a long stalk that pokes into the center of another ring of three α and three β subunits. Each β subunit contains a catalytic site where ATP is made. Finally, the δ subunit forms a connection between the ring of α and β subunits and the two b subunits.

When hydrogen ions pass through a c subunit, a conformational change causes the γ subunit to turn clockwise (when viewed from the intermembrane space). Each time the γ subunit turns 120°, it changes its contacts with the three β subunits,

1 ATP synthase and bacteriorhodopsin were incorporated into membrane vesicles.

ATP synthase

Vesicle

Bacteriorhodopsin (light-driven H^+ pump)

2 ADP and P_i were added on the outside of the vesicles.

ADP

P_i

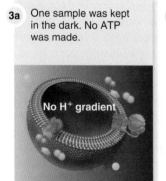

3a One sample was kept in the dark. No ATP was made.

No H^+ gradient

3b One sample was exposed to light. ATP was made.

Light rays

H^+ gradient

ATP

Figure 7.9 The Racker and Stoeckenius experiment showing that a H^+ electrochemical gradient drives ATP synthesis via ATP synthase.

Concept check: Is the functioning of the electron transport chain always needed to make ATP via ATP synthase?

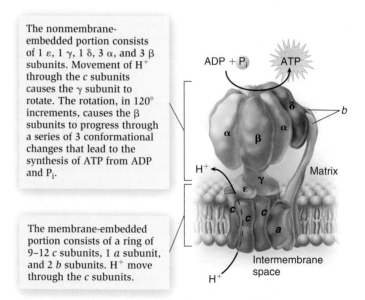

The nonmembrane-embedded portion consists of 1 ε, 1 γ, 1 δ, 3 α, and 3 β subunits. Movement of H^+ through the c subunits causes the γ subunit to rotate. The rotation, in 120° increments, causes the β subunits to progress through a series of 3 conformational changes that lead to the synthesis of ATP from ADP and P_i.

ADP + P_i ATP

δ b

α β α

H^+ Matrix

γ

ε

c c c a

The membrane-embedded portion consists of a ring of 9–12 c subunits, 1 a subunit, and 2 b subunits. H^+ move through the c subunits.

Intermembrane space

H^+

Figure 7.10 The subunit structure and function of ATP synthase.

Concept check: If the β subunit in the front center of this figure is in conformation 2, what are the conformations of the β subunit on the left and the β subunit on the back right?

which, in turn, causes the β subunits to change their conformations. How do these conformational changes promote ATP synthesis? The answer is that the conformational changes occur in a way that favors ATP synthesis and release. The conformational changes in the β subunits happen in the following order:

- Conformation 1: ADP and P_i bind with good affinity.
- Conformation 2: ADP and P_i bind so tightly that ATP is made.
- Conformation 3: ATP (and ADP and P_i) bind very weakly, and ATP is released.

Each time the γ subunit turns 120°, it causes a β subunit to change to the next conformation. After conformation 3, a 120° turn by the γ subunit returns a β subunit back to conformation 1, and the cycle of ATP synthesis can begin again. Because ATP synthase has three β subunits, each subunit is in a different conformation at any given time.

Paul Boyer proposed the concept of a rotary machine in the late 1970s. In his model, the three β subunits alternate between three conformations, as described previously. Boyer's original idea was met with great skepticism, because the concept that part of an enzyme could spin was very novel, to say the least. In 1994, John Walker and colleagues were able to determine the three-dimensional structure of the nonmembrane-embedded portion of the ATP synthase. The structure revealed that each of the three β subunits had a different conformation—one with ADP bound, one with ATP bound, and one without any nucleotide bound. This result supported Boyer's model. In 1997, Boyer and Walker shared the Nobel Prize in chemistry for their work on ATP synthase. As described next in the Feature Investigation, other researchers subsequently visualized the rotation of the γ subunit.

FEATURE INVESTIGATION

Yoshida and Kinosita Demonstrated That the γ Subunit of the ATP Synthase Spins

In 1997, Masasuke Yoshida, Kazuhiko Kinosita, and colleagues set out to experimentally visualize the rotary nature of ATP synthase (**Figure 7.11**). The membrane-embedded region of ATP synthase can be separated from the rest of the protein by treatment of mitochondrial membranes with a high concentration of salt, releasing the portion of the protein containing one γ, three α, and three β subunits. The researchers adhered the $\gamma\alpha_3\beta_3$ complex to a glass slide so the γ subunit was protruding upwards. Because the γ subunit is too small to be seen with a light microscope, the rotation of the γ subunit cannot be visualized directly. To circumvent this problem, the researchers attached a large, fluorescently labeled actin filament to the γ subunit via a linker protein. The fluorescently labeled actin filament is very long compared to the γ subunit and can be readily seen with a fluorescence microscope.

Because the membrane-embedded portion of the protein is missing, you may be wondering how the researchers could get the γ subunit to rotate. The answer is they added ATP. Although the normal function of the ATP synthase is to make ATP, it can also run backwards. In other words, ATP synthase can hydrolyze ATP. As shown in the data for Figure 7.11, when the researchers added ATP, they observed that the fluorescently labeled actin filament rotated in a counterclockwise direction, which is opposite to the direction that the γ subunit rotates when ATP is synthesized. Actin filaments were observed to rotate for more than 100 revolutions in the presence of ATP. These results convinced the scientific community that the ATP synthase is a rotary machine.

Experimental Questions

1. The components of ATP synthase are too small to be visualized by light microscopy. For the experiment of Figure 7.11, how did the researchers observe the movement of ATP synthase?

2. In the experiment of Figure 7.11, what observation did the researchers make that indicated ATP synthase is a rotary machine? What was the control of this experiment? What did it indicate?

3. Were the rotations seen by the researchers in the data of Figure 7.11 in the same direction as expected in the mitochondria during ATP synthesis? Why or why not?

Figure 7.11 Evidence that ATP synthase is a rotary machine.

HYPOTHESIS ATP synthase is a rotary machine.

KEY MATERIALS Purified complex containing 1 γ, 3 α, and 3 β subunits.

Experimental level   Conceptual level

1 Adhere the purified $\gamma\alpha_3\beta_3$ complex to a glass slide so the base of the γ subunit is protruding upwards.

Add purified complex.

γ

$\gamma\alpha_3\beta_3$ complex

α β α

Slide

2 Add linker proteins and fluorescently labeled actin filaments. The linker protein recognizes sites on both the γ subunit and the actin filament.

Add linker proteins and fluorescent actin filaments.

Linker proteins

Fluorescent actin filament

γ

α   β   α

3 Add ATP. As a control, do not add ATP.

Add ATP

Control: No ATP

4 Observe under a fluorescence microscope. The method of fluorescence microscopy is described in Chapter 4.

Fluorescence microscope

+ ATP: counterclockwise rotation

γ

α   β   α

5 THE DATA

Results from step 4:

ATP	Rotation
No ATP added	No rotation observed.
ATP added	Rotation was observed as shown below. This is a time-lapse view of the rotation in action.

Row 1

Row 2

6 CONCLUSION The γ subunit rotates counterclockwise when ATP is hydrolyzed. It would be expected to rotate clockwise when ATP is synthesized.

7 SOURCE Reprinted by permission from Macmillan Publishers Ltd. Noji, H., Yasuda, R.,Yoshida, M., and Kinosita, K. 1997. Direct observation of the rotation of F_1-ATPase. *Nature* 386:299–303.

Genomes & Proteomes Connection

Cancer Cells Usually Favor Glycolysis Over Oxidative Phosphorylation

Thus far, we have examined how eukaryotic cells metabolize glucose under aerobic conditions to produce CO_2 and a large amount of ATP. This occurs in four stages, beginning with glycolysis and ending with oxidative phosphorylation. Our understanding of carbohydrate metabolism has far-reaching medical implications. Many disease conditions, including common disorders such as cancer and diabetes, are associated with alterations in carbohydrate metabolism.

In 1931, the German physiologist Otto Warburg discovered that certain cancer cells preferentially use glycolysis for ATP production while decreasing the level of oxidative phosphorylation. This phenomenon, termed the Warburg effect, is very common among different types of tumors. The Warburg effect is used to clinically diagnose cancer via a procedure called positron emission tomography (PET scan, see Chapter 3). In this technique, patients are given a radiolabeled glucose analogue called [^{18}F]-fluorodeoxyglucose (FDG). The scanner detects regions of the body that metabolize FDG rapidly, which are visualized as bright spots on the PET scan. **Figure 7.12** shows a PET scan of a patient with lung cancer. The bright regions next to the arrows are tumors that show abnormally high levels of glycolysis.

In the past few decades, cancer biologists have analyzed the levels of proteins involved in glycolysis—the glycolytic enzymes described earlier in Figure 7.3. Glycolytic enzymes are overexpressed in approximately 80% of all types of cancer. These include lung, skin, colon, liver, pancreatic, breast, ovarian, and prostate cancer. The three enzymes of glycolysis whose overexpression is most commonly associated with cancer are glyceraldehyde-3-phosphate dehydrogenase, enolase, and pyruvate kinase (see Figure 7.3). In many cancers, all 10 glycolytic enzymes are overexpressed!

What factors cause glycolytic enzymes to be overexpressed? Both genetic and physiological factors are known to play a role. As discussed in Chapter 14, cancer is caused by mutations—changes in the DNA that affect the expression of genes. Mutations that cause cancer are generally not found in the genes that encode glycolytic enzymes themselves. Rather, cancer-causing mutations commonly occur in genes that encode regulatory proteins that control the expression of other genes. As an example, mutations in a human gene called *VHL* are associated with a disorder called von Hippel-Lindau syndrome, which is characterized by different tumor types throughout the body. The *VHL* gene mutations alter the function of the VHL regulatory protein, which then leads to an overexpression of the genes that encode glycolytic enzymes. In addition to mutations, the second factor that affects gene expression is the physiological conditions within a tumor. As a tumor grows, the internal regions of the tumor tend to become deficient in oxygen, a condition called hypoxia. The hypoxic state inside

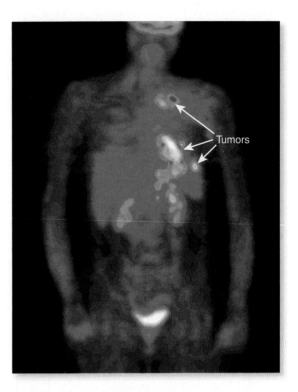

Figure 7.12 **A PET scan of a patient with lung cancer.** The bright regions in the lungs are tumors (see arrows). Organs such as the brain, which are not cancerous, appear bright because they perform high levels of glucose metabolism. Also, the kidneys and bladder appear bright because they filter and accumulate FDG. (Note: FDG is taken up by cells and converted to FDG-phosphate by hexokinase, the first enzyme in glycolysis. However, because FDG lacks an —OH group, it is not metabolized further. Therefore, FDG-phosphate accumulates in metabolically active cells.)

Concept check: *How might a higher level of glycolysis allow tumors to grow faster?*

a tumor may also cause the overexpression of glycolytic genes and thereby lead to a higher level of glycolytic enzymes within the cancer cells. This favors glycolysis as a means to make ATP, which does not require oxygen.

How do changes in the overexpression of glycolytic enzymes affect tumor growth? While the genetic changes associated with tumor growth are complex, researchers have speculated that an increase in glycolysis may favor the growth of the tumor as it becomes hypoxic. This would provide an advantage to the cancer cells, which would otherwise have trouble making ATP via oxidative phosphorylation. Based on these findings, some current research is aimed at discovering drugs to inhibit glycolysis in cancer cells as a way to prevent their growth.

Metabolic Pathways for Carbohydrate Metabolism Are Interconnected to Pathways for Amino Acid and Fat Metabolism

Before we end our discussion of cellular respiration in the presence of oxygen, let's consider the metabolism of other organic

Figure 7.13 Integration of protein, carbohydrate, and fat metabolism. Breakdown products of amino acids and fats can enter the same pathway that is used to break down carbohydrates.

Concept check: What is a cellular advantage of integrating protein, carbohydrate, and fat metabolism?

molecules, namely proteins and fats. When you eat a meal, it usually contains not only carbohydrates (including glucose) but also proteins and fats. These molecules are broken down by some of the same enzymes involved with glucose metabolism.

As shown in **Figure 7.13**, proteins and fats can enter into glycolysis or the citric acid cycle at different points. Proteins are first acted upon by enzymes, either in digestive juices or within cells, that cleave the bonds connecting individual amino acids. Because the 20 amino acids differ in their side chains, amino acids and their breakdown products can enter at different points in the pathway. Breakdown products of amino acids can enter at later steps of glycolysis, or an acetyl group can be removed from certain amino acids and become attached to CoA. Other amino acids can be modified and enter the citric acid cycle. Similarly, fats can be broken down to glycerol and fatty acids. Glycerol can be modified to glyceraldehyde-3-phosphate and enter glycolysis at step 5 (see Figure 7.3). Fatty acyl tails can have two carbon acetyl units removed, which bind to CoA and then enter the citric acid cycle. By using the same pathways for the breakdown of sugars, amino acids, and fats, cellular metabolism is more efficient because the same enzymes can be used for the breakdown of different starting molecules.

Likewise, carbohydrate metabolism is connected to the metabolism of other cellular components at the anabolic level. Cells may use carbohydrates to manufacture parts of amino acids, fats, and nucleotides. For example, the glucose-6-phosphate of glycolysis is used to construct the sugar and phosphate portion of nucleotides, while the oxaloacetate of the citric acid cycle can be used as a precursor for the biosynthesis of purine and pyrimidine bases. Portions of amino acids can be made from products of glycolysis (for example, pyruvate) and components of the citric acid cycle (oxaloacetate). In addition, several other catabolic and anabolic pathways are found in living cells that connect the metabolism of carbohydrates, proteins, fats, and nucleic acids.

7.2 Anaerobic Respiration and Fermentation

Thus far, we have surveyed catabolic pathways that result in the complete breakdown of glucose in the presence of oxygen. Cells also commonly metabolize organic molecules in the absence of oxygen. The term **anaerobic** is used to describe an environment that lacks oxygen. Many bacteria and archaea and some fungi exist in anaerobic environments but still have to oxidize organic molecules to obtain sufficient amounts of energy. Examples include microbes living in your intestinal tract and those living deep in the soil. Similarly, when a person exercises strenuously, the rate of oxygen consumption by muscle cells may greatly exceed the rate of oxygen delivery. Under these conditions, the muscle cells become anaerobic and must obtain sufficient energy in the absence of oxygen to maintain their level of activity.

Organisms have evolved two different strategies to metabolize organic molecules in the absence of oxygen. One mechanism is to use a substance other than O_2 as the final electron acceptor of an electron transport chain, a process called **anaerobic respiration**. A second approach is to produce ATP only via substrate-level phosphorylation. In this section, we will consider examples of both strategies.

Some Microorganisms Carry Out Anaerobic Respiration

At the end of the electron transport chain discussed earlier in Figure 7.7, cytochrome oxidase recognizes O_2 and catalyzes its reduction to H_2O. The final electron acceptor of the chain is O_2. Many species of bacteria that live under anaerobic conditions have evolved enzymes that function similarly to cytochrome oxidase but recognize molecules other than O_2 and use them as the final electron acceptor. For example, *Escherichia coli*, which is a bacterial species found in your intestinal tract, produces an enzyme called nitrate reductase under anaerobic conditions. This enzyme recognizes nitrate (NO_3^-), which is used as the final electron acceptor of an electron transport chain.

Figure 7.14 shows a simplified electron transport chain in *E. coli* in which nitrate is the final electron acceptor. In *E. coli* and other bacterial species, the electron transport chain is in the plasma membrane that surrounds the cytoplasm. Electrons travel from NADH to NADH dehydrogenase to ubiquinone (Q) to cytochrome *b* and then to nitrate reductase. At the end of

the chain, nitrate is converted to nitrite (NO_2^-). This process generates a H^+ electrochemical gradient in three ways. First, NADH dehydrogenase pumps H^+ out of the cytoplasm. Second, ubiquinone picks up H^+ in the cytoplasm and carries it to the other side of the membrane. Third, the reduction of nitrate to nitrite consumes H^+ in the cytoplasm. The generation of a H^+ gradient via these three processes allows *E. coli* cells to make ATP via chemiosmosis under anaerobic conditions.

Fermentation Is the Breakdown of Organic Molecules Without Net Oxidation

Many organisms, including animals and yeast, can use only O_2 as the final electron acceptor of their electron transport chains. When confronted with anaerobic conditions, these organisms must have a different way of producing sufficient ATP. One strategy is to make ATP via glycolysis, which can occur under anaerobic or aerobic conditions. Under anaerobic conditions, the cells do not use the citric acid cycle or the electron transport chain, but make ATP only via glycolysis.

A key issue is that glycolysis requires NAD^+ and generates NADH. Under aerobic conditions, oxygen acts as a final electron acceptor, and the high-energy electrons from NADH can be used to make more ATP. To make ATP, NADH is oxidized to NAD^+. However, this cannot occur under anaerobic conditions in yeast and animals, and, as a result, NADH builds up and NAD^+ decreases. This is a potential problem for two reasons. First, at high concentrations, NADH will haphazardly donate its electrons to other molecules and promote the formation of free radicals, highly reactive chemicals that can damage DNA and cellular proteins. For this reason, yeast and animal cells exposed to anaerobic conditions must have a way to remove the excess NADH generated from the breakdown of glucose. The second problem is the decrease in NAD^+. Cells need to regenerate NAD^+ to keep glycolysis running and make ATP via substrate-level phosphorylation.

How do muscle cells overcome these two problems? When a muscle is working strenuously and becomes anaerobic, the pyruvate from glycolysis is reduced to make lactate. (The uncharged [protonated] form is called lactic acid.) The electrons to reduce pyruvate are derived from NADH, which is oxidized to NAD^+ (Figure 7.15a). Therefore, this process decreases NADH and reduces its potentially harmful effects. It also increases the level of NAD^+, thereby allowing glycolysis to continue. The lactate is secreted from muscle cells. Once sufficient oxygen is restored, the lactate produced during strenuous exercise can be taken up by cells, converted back to pyruvate, and used for energy, or it may be used to make glucose by the liver and other tissues.

Yeast cells cope with anaerobic conditions differently. During wine making, a yeast cell metabolizes sugar under anaerobic conditions. The pyruvate is broken down to CO_2 and a two-carbon molecule called acetaldehyde. The acetaldehyde is then reduced to make ethanol while NADH is oxidized to NAD^+ (Figure 7.15b). Similar to lactate production in muscle cells, this

Figure 7.14 **An example of anaerobic respiration in *E. coli*.** When oxygen is absent, *E. coli* can use nitrate instead of oxygen as the final electron acceptor in an electron transport chain. This generates a H^+ electrochemical gradient that is used to make ATP via chemiosmosis. Note: As shown in this figure, ubiquinone (Q) picks up H^+ on one side of the membrane and deposits it on the other side. A similar event happens during aerobic respiration in mitochondria (described in Figure 7.7), except that ubiquinone transfers H^+ to cytochrome *b*-c_1, which pumps it into the intermembrane space.

Glucose is oxidized to 2 pyruvate molecules. Two pyruvates are reduced to 2 lactate molecules.

Glucose is oxidized to 2 pyruvate molecules. Two acetaldehyde molecules are reduced to 2 ethanol molecules.

(a) Production of lactic acid

(b) Production of ethanol

Figure 7.15 **Examples of fermentation.** In these examples, NADH is produced by the oxidation of an organic molecule, and then the NADH is used up by donating electrons to a different organic molecule such as pyruvate **(a)** or acetaldehyde **(b)**.

decreases NADH and increases NAD^+, thereby preventing the harmful effects of NADH and allowing glycolysis to continue.

The term **fermentation** is used to describe the breakdown of organic molecules to harness energy without any net oxidation (that is, without any removal of electrons). The breakdown of glucose to lactate or ethanol are examples of fermentation. Although electrons are removed from an organic molecule such as glucose to make pyruvate and NADH, the electrons are donated back to an organic molecule in the production of lactate or ethanol. Therefore, there is no net removal of electrons from an organic molecule. Compared with oxidative phosphorylation, fermentation produces far less ATP for two reasons. First, glucose is not oxidized completely to CO_2 and water. Second, the NADH made during glycolysis cannot be used to make more ATP. Overall, the complete breakdown of glucose in the presence of oxygen yields 34 to 38 ATP molecules. By comparison, the anaerobic breakdown of glucose to lactate or ethanol yields only two ATP molecules.

7.3   Secondary Metabolism

Primary metabolism is the synthesis and breakdown of molecules and macromolecules that are found in all forms of life and are essential for cell structure and function. These include compounds such as sugars, amino acids, lipids, and

nucleotides, and the macromolecules that are derived from them. Cellular respiration, which we considered earlier in this chapter, is an example of primary metabolism. By comparison, **secondary metabolism** involves the synthesis of molecules—**secondary metabolites**—that are not essential for cell structure and growth. Secondary metabolites, also called secondary compounds, are commonly made in plants, bacteria, and fungi. Any given secondary metabolite is unique to one species or group of species and is not usually required for survival.

Secondary metabolites perform diverse functions for the species that produce them, often enhancing their chances of survival and reproduction. For example, many secondary metabolites taste bad. When produced in a plant, for example, such a molecule may prevent an animal from eating the plant. In some cases, secondary metabolites are toxic. Such molecules may act as a chemical weapon that inhibits the growth of nearby organisms. In addition, many secondary metabolites produce a strong smell or bright color that attracts or repels other organisms. For example, the scent from a rose is due to secondary metabolites. The scent attracts insects that aid in pollination.

Biologists have discovered thousands of different secondary metabolites, though any given species tends to produce only one or a few types. Plants are particularly diverse in the types of secondary metabolites they produce, perhaps because they have evolved defenses that are effective in stationary organisms. Bacteria and fungi also produce a large array of these compounds, whereas animals tend to produce relatively few. As you will learn, humans have put many of these compounds to practical use, from the spices we use in cooking to the antibiotics we use to treat diseases. In this section, we will survey four categories of secondary metabolites: phenolics, alkaloids, terpenoids, and polyketides.

Phenolic Compounds Are Antioxidants That Defend or Attract with Intense Flavors and Bright Colors

The **phenolic** compounds all contain a cyclic ring of carbon with three double bonds, known as a benzene ring, within their structure. When a benzene ring is covalently linked to a single hydroxyl group, the compound is known as phenol.

Phenol is the simplest of the phenolic compounds, though free phenol is not significantly accumulated in living organisms. However, more complex molecules that are derived from phenol are made in cells. Such phenolic compounds are synthesized using the side groups of the amino acids phenylalanine (which has a benzene ring) or tyrosine (which has a phenol ring). Common categories of phenolics are the flavonoids, tannins, and lignins.

Flavonoids are produced by many plant species and create a variety of flavors and smells. These can play a role as deterrents to eating a plant or as attractants that promote pollination. The flavors of chocolate and vanilla come largely from a mixture of flavonoid molecules. Vanilla is produced by several species of perennial vines of the genus *Vanilla*, native to Mexico and tropical America (**Figure 7.16a**). The primary source of commercial vanilla comes from *V. planifolia*. Vanilla extract is obtained from the seed capsules. Another role of flavonoids is pigmentation. Anthocyanins (from the Greek *anthos*, meaning flower, and *kyanos*, meaning blue) produce the red, blue, and purple colors of many flowers, fruits, and vegetables (**Figure 7.16b**).

Biochemists have discovered that flavonoids have remarkable antioxidant properties that prevent the formation of damaging free radicals. In plants, flavonoids are thought to act as powerful antioxidants, helping to protect plants from ultraviolet (UV) damage. In recent times, nutritionists have advocated the consumption of fruits and vegetables that have high amounts of flavonoids, such as blueberries, broccoli, and spinach. Dark chocolate is also rich in these antioxidants!

Tannins are large polymeric molecules composed of many phenolic units. They are named tannins because they combine with the protein of animal skins in the making of leather. This process, known as tanning, also imparts a tan color to the skins. Tannins are found in many plant species and typically act as a deterrent to animals, either because of a bitter taste or due to toxic effects. If consumed in large amounts, they also can

(a) Flavonoids in vanilla provide flavor

(b) Anthocyanins such as pelargonidin give red color

Figure 7.16   Phenolic compounds as secondary metabolites.   The two examples shown here are flavonoids, which are a type of phenolic compound. **(a)** The flavor of vanilla is largely produced by flavonoids, an example of which is vanillin produced by this *Vanilla planifolia* vine. **(b)** Another group of flavonoids that causes red, blue, or purple color are anthocyanins. The red color of strawberries is caused by pelargonidin, an anthocyanin.

Concept check: *Besides fruits, what other parts of plants may contain anthocyanins?*

inhibit the enzymes found in the digestive tracts of animals. Tannins are found abundantly in grape skins and play a key role in the flavor of red wine. Aging breaks down tannins, making the wine less bitter.

Lignins are also large phenolic polymers synthesized by plants. Lignins are found in plant cell walls and make up about one-quarter to one-third of the weight of dry wood. The lignins form polymers that bond with other plant wall components such as cellulose. This strengthens plant cells and enables a plant to better withstand the rigors of environmental stress. To make paper, which is much more malleable than wood, the lignins are removed.

Alkaloids Form a Large Group of Bitter-Tasting Molecules That Also Provide Defense Mechanisms

Alkaloids are a group of structurally related molecules that all contain nitrogen and usually have a cyclic, ring-like structure. More than 12,000 different alkaloids have been discovered. Their name is derived from the observation that they are basic or alkaline molecules. Alkaloids are usually synthesized from amino acid precursors. Alkaloids are commonly made in plant species and occasionally in fungi and animals (shellfish). Familiar examples include caffeine, nicotine, atropine, morphine, ergot, and quinine.

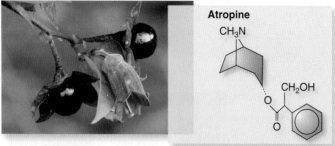

Deadly nightshade

Figure 7.17 **Alkaloids as secondary metabolites.** Atropine is an alkaloid produced by the plant called deadly nightshade (*Hyoscyamus niger*). Atropine is toxic because it interferes with nerve transmission. In humans, atropine causes the heart to speed up to dangerous and possibly fatal rates.

Concept check: *How does the production of atropine provide protection to deadly nightshade?*

Like phenolics, many alkaloids serve a defense function in plants. Alkaloids are bitter-tasting molecules and often have an unpleasant odor. These features may prevent an animal from eating a plant or its fruit. For example, an alkaloid in chile peppers called capsaicin elicits a burning sensation. This molecule is so potent that one-millionth of a drop can be detected by the human tongue. Capsaicin may discourage mammals from eating the peppers. Interestingly, however, birds do not experience the burning sensation of capsaicin and serve to disperse the seeds.

Other alkaloids are poisonous, like atropine, a potent toxin derived from the deadly nightshade plant (**Figure 7.17**). Animals that eat this plant and consequently ingest atropine become very sick and may die. Any animal that eats deadly nightshade and survives would be unlikely to eat it a second time. Atropine acts by interfering with nerve transmission. In humans, for example, atropine causes the heart to speed up to dangerous rates, because the nerve inputs that normally keep a check on heart rate are blocked by atropine. Other alkaloids are not necessarily toxic but can cause an animal that eats them to become overstimulated (caffeine), understimulated (any of the opium alkaloids such as morphine), or simply nauseated because the compound interferes with nerves required for proper functioning of the gastrointestinal system.

Terpenoids Are Molecules with Intense Smells and Color That Have Diverse Functions

A third major class of secondary metabolites are the **terpenoids**, of which over 25,000 have been identified, more than any other family of naturally occurring products. Terpenoids are synthesized from five-carbon isoprene units (shown here) and are also called isoprenoids.

Isoprene units are linked to each other to form larger compounds with multiples of five-carbon atoms. In many cases, the isoprene units form cyclic structures.

Terpenoids have a wide array of functions in plants. Notably, because many terpenoids are volatile (they become gases), they are responsible for the odors emitted by many types of plants, such as menthol produced by mint. The odors of terpenoids may attract pollinators or repel animals that eat plants. In addition, terpenoids often impart an intense flavor to plant tissues. Many of the spices we use in cooking are rich in different types of terpenoids. Examples include cinnamon, fennel, cloves, cumin, caraway, and tarragon. Terpenoids are found in many traditional herbal remedies and are under medical investigation for potential pharmaceutical effects.

Other terpenoids, such as the carotenoids, are responsible for the coloration of many species. An example is β-carotene, which gives carrots their orange color. Carotenoids are also found in leaves, but their color is masked by chlorophyll, which is green. In the autumn, when chlorophyll breaks down, the color of the carotenoids becomes evident. In addition, carotenoids give color to animals such as salmon, goldfish, and flamingos (**Figure 7.18**).

Polyketides Are Often Used as Chemical Weapons to Kill Competing Organisms

Polyketides are a group of secondary metabolites that are produced by bacteria, fungi, plants, insects, dinoflagellates,

Flamingo　　　　　　　　　　　　**β-carotene**

Figure 7.18 **Terpenoids as secondary metabolites.** Carotenoids are a type of terpenoid with bright color. The example shown here is β-carotene, which gives many organisms an orange color. Flamingos (*Phoenicopterus ruber*) receive β-carotene in their diet, primarily from eating shellfish.

Streptomyces griseus, a soil bacterium

Figure 7.19 Polyketides as secondary metabolites. Streptomycin, whose structure is shown here, is an antibiotic produced by *Streptomyces griseus*, a soil bacterium. The scanning electron micrograph shows *S. griseus*.

Concept check: How does the production of streptomycin provide S. griseus with a growth advantage?

mollusks, and sponges. They are synthesized by the polymerization of acetyl (CH_3COOH) and propionyl (CH_3CH_2COOH) groups to create a diverse collection of molecules, often with many ringed structures. During the past several decades, over 10,000 polyketides have been identified and analyzed. Familiar examples include streptomycin, erythromycin, and tetracycline.

Polyketides are usually secreted by the organism that makes them and are often highly toxic to other organisms. For example, the polyketide known as streptomycin is made by the soil bacterium *Streptomyces griseus* (**Figure 7.19**). It is secreted by this bacterium and taken up by other species, where it disrupts protein synthesis and thereby inhibits their growth. In this way, *S. griseus* is able to kill or inhibit the growth of other species in its vicinity.

The toxic effects of polyketides are often very selective, making them valuable medical tools. For example, streptomycin disrupts protein synthesis in many bacterial species, but it does not adversely affect protein synthesis in mammalian cells. Therefore, it has been used as an antibiotic to treat or prevent bacterial infections in humans and other mammals. Similarly, other polyketides inhibit the growth of fungi, parasites, and insects. More recently, researchers have even discovered that certain polyketides inhibit the growth of cancer cells. The production and sale of polyketides to treat and prevent diseases and as pesticides constitute an enormous industry, with annual sales in the U.S. at over $20 billion.

▌ Summary of Key Concepts

7.1 Cellular Respiration in the Presence of Oxygen

- Cells obtain energy via cellular respiration, which involves the breakdown of organic molecules and the export of waste products.

- The breakdown of glucose occurs in four stages: glycolysis, pyruvate breakdown, citric acid cycle, and oxidative phosphorylation. (Figure 7.1)

- Glycolysis is the breakdown of glucose to two pyruvates, producing two net molecules of ATP and two NADH. ATP is made by substrate-level phosphorylation. (Figures 7.2, 7.3)

- Pyruvate is broken down to CO_2 and an acetyl group that becomes attached to CoA. NADH is made during this process. (Figure 7.4)

- During the citric acid cycle, each acetyl group attached to CoA is incorporated into an organic molecule, which is oxidized and releases two CO_2 molecules. Three NADH, one $FADH_2$, and one ATP are made during this process. (Figures 7.5, 7.6)

- Oxidative phosphorylation involves two events. The electron transport chain oxidizes NADH or $FADH_2$ and generates a H^+ electrochemical gradient. This gradient is used by ATP synthase to make ATP via chemiosmosis. (Figures 7.7, 7.8)

- Racker and Stoeckenius showed that ATP synthase uses a H^+ gradient by reconstituting ATP synthase with a light-driven H^+ pump. (Figure 7.9)

- ATP synthase is a rotary machine. The rotation is caused by the movement of H^+ through the c subunits that cause the γ subunit to spin, resulting in conformational changes in the β subunits that promote ATP synthesis. (Figure 7.10)

- Yoshida and Kinosita demonstrated rotation of the γ subunit by attaching a fluorescently labeled actin filament and watching it spin in the presence of ATP. (Figure 7.11)

- Cancer cells preferentially carry out glycolysis due to both genetic changes associated with cancer and physiological changes within the tumor itself. (Figure 7.12)

- Proteins and fats can enter into glycolysis or the citric acid cycle at different points. (Figure 7.13)

7.2 Anaerobic Respiration and Fermentation

- Anaerobic respiration occurs in the absence of oxygen. Certain microorganisms can carry out anaerobic respiration in which the final electron acceptor of the electron transport chain is a substance other than oxygen, such as nitrate. (Figure 7.14)

- During fermentation, organic molecules are broken down without any net oxidation (that is, without any net removal of electrons). Examples include lactate production in muscle cells and ethanol production in yeast. (Figure 7.15)

7.3 Secondary Metabolism

- Secondary metabolites are not usually necessary for cell structure and function, but they provide an advantage to an organism that may involve taste, smell, color, or poison. Four categories of secondary metabolites are phenolic compounds, alkaloids, terpenoids, and polyketides. (Figures 7.16, 7.17, 7.18, 7.19)

Assess and Discuss

Test Yourself

1. Which of the following pathways occurs in the cytosol?
   a. glycolysis
   b. breakdown of pyruvate to an acetyl group
   c. citric acid cycle
   d. oxidative phosphorylation
   e. all of the above

2. To break down glucose to CO_2 and H_2O, which of the following metabolic pathways is <u>not</u> involved?
   a. glycolysis
   b. breakdown of pyruvate to an acetyl group
   c. citric acid cycle
   d. photosynthesis
   e. c and d only

3. The net products of glycolysis are
   a. 6 CO_2, 4 ATP, and 2 NADH.
   b. 2 pyruvate, 2 ATP, and 2 NADH.
   c. 2 pyruvate, 4 ATP, and 2 NADH.
   d. 2 pyruvate, 2 GTP, and 2 CO_2.
   e. 2 CO_2, 2 ATP, and glucose.

4. During glycolysis, ATP is produced by
   a. oxidative phosphorylation.
   b. substrate-level phosphorylation.
   c. redox reactions.
   d. all of the above.
   e. both a and b.

5. Certain drugs act as ionophores that cause the mitochondrial membrane to be highly permeable to H^+. How would such drugs affect oxidative phosphorylation?
   a. Movement of electrons down the electron transport chain would be inhibited.
   b. ATP synthesis would be inhibited.
   c. ATP synthesis would be unaffected.
   d. ATP synthesis would be stimulated.
   e. Both a and b are correct.

6. The source of energy that <u>directly</u> drives the synthesis of ATP during oxidative phosphorylation is
   a. the oxidation of NADH.
   b. the oxidation of glucose.
   c. the oxidation of pyruvate.
   d. the H^+ gradient.
   e. the reduction of O_2.

7. Compared to oxidative phosphorylation in mitochondria, a key difference of anaerobic respiration in bacteria is
   a. more ATP is made.
   b. ATP is made only via substrate-level phosphorylation.
   c. O_2 is converted to H_2O_2 rather than H_2O.
   d. something other than O_2 acts as a final electron acceptor of the electron transport chain.
   e. b and d.

8. When a muscle becomes anaerobic during strenuous exercise, why is it necessary to convert pyruvate to lactate?
   a. to decrease NAD^+ and increase NADH
   b. to decrease NADH and increase NAD^+
   c. to increase NADH and increase NAD^+
   d. to decrease NADH and decrease NAD^+
   e. to keep oxidative phosphorylation running

9. Secondary metabolites
   a. help deter predation of certain organisms by causing the organism to taste bad.
   b. help attract pollinators by producing a pleasant smell.
   c. help organisms compete for resources by acting as a poison to competitors.
   d. provide protection from DNA damage.
   e. do all of the above.

10. Which of the following is an example of a secondary metabolite?
    a. flavonoids found in vanilla
    b. atropine found in deadly nightshade
    c. β-carotene found in carrots and flamingo feathers
    d. streptomycin made by soil bacteria
    e. all of the above

Conceptual Questions

1. The electron transport chain is so named because electrons are transported from one component to another. Describe the purpose of the electron transport chain.

2. What causes the rotation of the γ subunit of the ATP synthase? How does this rotation promote ATP synthesis?

3. During fermentation, explain why it is important to oxidize NADH to NAD^+.

Collaborative Questions

1. Discuss the advantages and disadvantages of aerobic respiration, anaerobic respiration, and fermentation.

2. Discuss the roles of secondary metabolites in biology. Such compounds have a wide variety of practical applications. If you were going to start a biotechnology company that produced secondary metabolites for sale, which type(s) would you focus on? How might you go about discovering new secondary metabolites that could be profitable?

Online Resource

www.brookerbiology.com

Stay a step ahead in your studies with animations that bring concepts to life and practice tests to assess your understanding. Your instructor may also recommend the interactive ebook, individualized learning tools, and more.

Photosynthesis

8

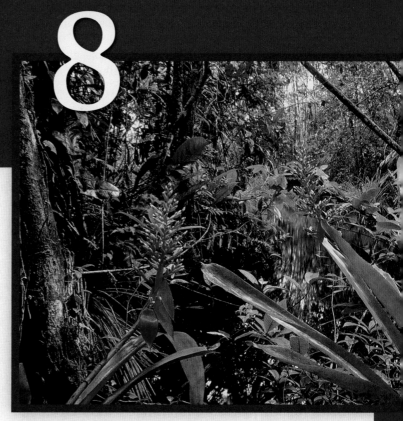

A tropical rain forest in the Amazon. Plant life in tropical rain forests carries out a large amount of the world's photosynthesis and supplies the atmosphere with a sizeable fraction of its oxygen.

T ake a deep breath. Nearly all of the oxygen in every breath you take is made by the abundant plant life, algae, and cyanobacteria on Earth. More than 20% of the world's oxygen is produced in the Amazon rain forest in South America alone (see chapter-opening photo). Biologists are alarmed about the rate at which such forests are being destroyed by human activities. Rain forests once covered 14% of the Earth's land surface but now occupy less than 6%. At their current rate of destruction, rain forests may be nearly eliminated in less than 40 years. Such an event may lower the level of oxygen in the atmosphere and thereby have a harmful impact on living organisms on a global scale.

In rain forests and across all of the Earth, the most visible color on land is green. The green color of plants is due to a pigment called chlorophyll. This pigment provides the starting point for the process of **photosynthesis**, in which the energy from light is captured and used to synthesize carbohydrates. Nearly all living organisms ultimately rely on photosynthesis for their nourishment, either directly or indirectly. Photosynthesis is also responsible for producing the oxygen that makes up a large portion of the Earth's atmosphere. Therefore, all aerobic organisms rely on photosynthesis for cellular respiration.

We begin this chapter with an overview of photosynthesis as it occurs in green plants and algae. We will then explore the two stages of photosynthesis in more detail. In the first stage, called the light reactions, light energy is captured by the chlorophyll pigments and converted to chemical energy in the form of two energy intermediates, ATP and NADPH. During the second stage, known as the Calvin cycle, ATP and NADPH are used to drive the synthesis of carbohydrates. We conclude with a consideration of the variations in photosynthesis that occur in plants existing in hot and dry conditions.

8.1 Overview of Photosynthesis

In the mid-1600s, a Flemish physician, Jan Baptista Van Helmont, conducted an experiment in which he transplanted the shoot of a young willow tree into a bucket of soil and allowed it to grow for 5 years. After this time, the willow tree had added 164 pounds to its original weight, but the soil had lost only 2 ounces. Van Helmont correctly concluded that the willow tree did not get most of its nutrients from the soil. He also hypothesized that the mass of the tree came from the water he had added over the 5 years. This hypothesis was partially correct, but we now know that CO_2 from the air is also a major contributor to the growth and mass of plants.

In the 1770s, Jan Ingenhousz, a Dutch physician, immersed green plants under water and discovered they released bubbles of oxygen. Ingenhousz determined that sunlight was necessary for oxygen production. During this same period, Jean Senebier, a Swiss botanist, found that CO_2 is required for plant growth. With this accumulating information, Julius von Mayer, a German physicist, proposed in 1845 that plants convert light energy from the sun into chemical energy.

For the next several decades, plant biologists studied photosynthesis in plants, algae, and bacteria. Researchers discovered that some photosynthetic bacteria could use hydrogen sulfide (H_2S) instead of water (H_2O) for photosynthesis and these organisms released sulfur instead of oxygen. In the 1930s, based on this information, Dutch-American microbiologist Cornelis van Niel proposed a general equation for photosynthesis that applies to plants, algae, and photosynthetic bacteria alike.

$$CO_2 + 2\,H_2A + \text{Light energy} \rightarrow CH_2O + A_2 + H_2O$$

where A is oxygen (O) or sulfur (S) and CH_2O is the general formula for a carbohydrate. This is a redox reaction in which CO_2 is reduced and H_2A is oxidized.

In green plants, A is oxygen and 2 A is a molecule of oxygen that is designated O_2. Therefore, this equation becomes

$$CO_2 + 2\,H_2O + \text{Light energy} \rightarrow CH_2O + O_2 + H_2O$$

When the carbohydrate produced is glucose ($C_6H_{12}O_6$), we multiply each side of the equation by six to obtain:

$$6\,CO_2 + 12\,H_2O + \text{Light energy} \rightarrow C_6H_{12}O_6 + 6\,O_2 + 6\,H_2O$$

$$\Delta G = +685 \text{ kcal/mole}$$

In this redox reaction, CO_2 is reduced during the formation of glucose, and H_2O is oxidized during the formation of O_2. Notice that the free-energy change required for the production of 1 mole of glucose from carbon dioxide and water is a whopping +685 kcal/mole! As we learned in Chapter 6, endergonic reactions are driven forward by coupling the reaction with an exergonic process that releases free energy. In this case, the energy from sunlight ultimately drives the synthesis of glucose.

In this section, we will survey the general features of photosynthesis as it occurs in green plants and algae. Later sections will examine the various steps in this process.

Photosynthesis Powers the Biosphere

The term **biosphere** describes the regions on the surface of the Earth and in the atmosphere where living organisms exist. Organisms can be categorized as heterotrophs and autotrophs. **Heterotrophs** must consume food—organic molecules from their environment—to sustain life. Heterotrophs include most species of bacteria and protists, as well as all species of fungi and animals. By comparison, **autotrophs** are organisms that make organic molecules from inorganic sources such as CO_2 and H_2O. **Photoautotrophs** are autotrophs that use light as a source of energy to make organic molecules. These include green plants, algae, and some prokaryotic species such as cyanobacteria.

Life in the biosphere is largely driven by the photosynthetic power of green plants and algae. The existence of most species relies on a key energy cycle that involves the interplay between organic molecules (such as glucose) and inorganic molecules, namely, O_2, CO_2, and H_2O (**Figure 8.1**). Photoautotrophs, such as plants, make a large proportion of the Earth's organic molecules via photosynthesis, using light energy, CO_2, and H_2O. During this process, they also produce O_2. To supply their energy needs, both photoautotrophs and heterotrophs metabolize organic molecules via cellular respiration. As described in Chapter 7, cellular respiration generates CO_2 and H_2O and is used to make ATP. The CO_2 is released into the atmosphere and can be re-used by photoautotrophs to make more organic

Figure 8.1   An important energy cycle between photosynthesis and cellular respiration.   Photosynthesis uses light, CO_2, and H_2O to produce O_2 and organic molecules. The organic molecules can be broken down to CO_2 and H_2O via cellular respiration to supply energy in the form of ATP; O_2 is reduced to H_2O.

Concept check: *Which types of organisms carry out cellular respiration? Is it heterotrophs, autotrophs, or both?*

molecules such as glucose. In this way, an energy cycle exists between photosynthesis and cellular respiration that sustains life on our planet.

In Plants and Algae, Photosynthesis Occurs in the Chloroplast

Chloroplasts are organelles found in plant and algal cells that carry out photosynthesis. These organelles contain large quantities of **chlorophyll**, which is a pigment that gives plants their green color. All green parts of a plant contain chloroplasts and can perform photosynthesis, although the majority of photosynthesis occurs in the leaves (**Figure 8.2**). The internal part of the leaf, called the **mesophyll**, contains cells with chloroplasts that carry out the bulk of photosynthesis in plants. For photosynthesis to occur, the mesophyll cells must obtain water and carbon dioxide. The water is taken up by the roots of the plant and is transported to the leaves by small veins. Carbon dioxide gas enters the leaf, and oxygen exits via pores called **stomata** (singular, stoma or stomate; from the Greek, meaning mouth). The anatomy of leaves will be examined further in Chapter 35.

Like the mitochondrion, a chloroplast contains an outer and inner membrane, with an intermembrane space lying between the two. A third membrane, called the **thylakoid membrane**, contains pigment molecules, including chlorophyll. The thylakoid membrane forms many flattened, fluid-filled tubules called the **thylakoids**, which enclose a single, convoluted compartment known as the **thylakoid lumen**. Thylakoids stack on top of each other to form a structure called a **granum** (plural, grana). The **stroma** is the fluid-filled region of the chloroplast between the thylakoid membrane and the inner membrane (Figure 8.2).

Photosynthesis Occurs in Two Stages: Light Reactions and the Calvin Cycle

How does photosynthesis occur? The process of photosynthesis can be divided into two stages called the **light reactions** and the **Calvin cycle**. The term photosynthesis is derived from the association between these two stages: The prefix <u>photo</u> refers to the light reactions that capture the energy from sunlight needed for the <u>synthesis</u> of carbohydrates that occurs in the Calvin cycle. The light reactions take place at the thylakoid membrane, and the Calvin cycle occurs in the stroma (**Figure 8.3**).

The light reactions involve an amazing series of energy conversions, starting with light energy and ending with chemical energy that is stored in the form of covalent bonds. The light reactions produce three chemical products: ATP, NADPH, and O_2. ATP and NADPH are energy intermediates that provide the needed energy and electrons to drive the Calvin cycle. Like NADH, **NADPH (nicotinamide adenine dinucleotide phosphate)** is an electron carrier that can accept two electrons. Its structure differs from NADH by the presence of an additional phosphate group. The structure of NADH is described in Chapter 6 (see Figure 6.11).

Although O_2 is not needed to make carbohydrates, it is still an important product of the light reactions. As described in Chapter 7, O_2 is vital to the process of aerobic cellular respiration. Nearly all of the O_2 in the atmosphere is produced by photosynthesis from green plants and aquatic microorganisms.

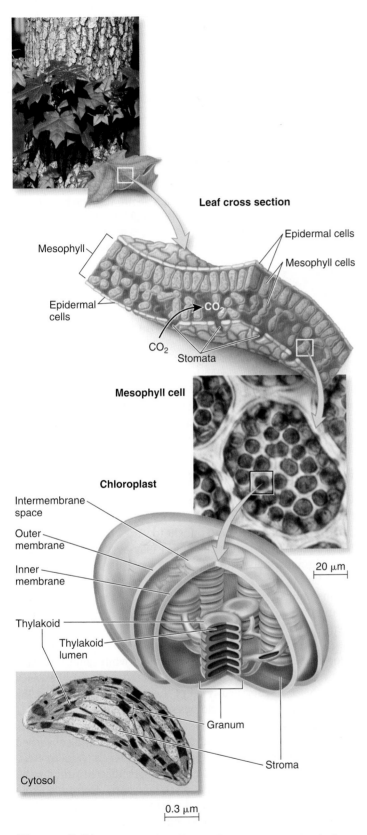

Figure 8.2 Leaf organization.   Leaves are composed of layers of cells. The epidermal cells are on the outer surface, both top and bottom, with mesophyll cells sandwiched in the middle. The mesophyll cells contain chloroplasts and are the primary sites of photosynthesis in most plants.

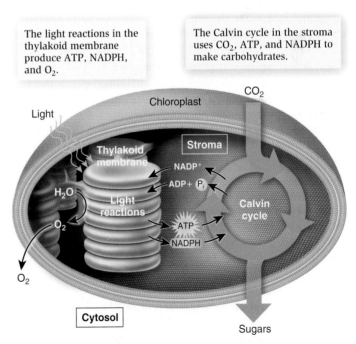

The light reactions in the thylakoid membrane produce ATP, NADPH, and O_2.

The Calvin cycle in the stroma uses CO_2, ATP, and NADPH to make carbohydrates.

Figure 8.3   An overview of the two stages of photosynthesis: light reactions and the Calvin cycle.   The light reactions, through which ATP, NADPH, and O_2 are made, occur at the thylakoid membrane. The Calvin cycle, in which enzymes use ATP and NADPH to incorporate CO_2 into carbohydrate, occurs in the stroma.

Concept check: *Can the Calvin cycle occur in the dark?*

8.2 Reactions That Harness Light Energy

According to the first law of thermodynamics discussed in Chapter 6, energy cannot be created or destroyed, but it can be transferred from one place to another and transformed from one form to another. During photosynthesis, energy in the form of light is transferred from the sun, some 92 million miles away, to a pigment molecule in a photosynthetic organism such as a plant. What follows is an interesting series of energy transformations in which light energy is transformed into electrochemical energy and then into energy stored within chemical bonds.

In this section, we will explore this series of transformations, collectively called the light reactions of photosynthesis. We begin by examining the unique properties of light and then consider the features of chloroplasts that allow them to capture light energy. The rest of this section focuses on how the light reactions of photosynthesis create three important products: ATP, NADPH, and O_2.

Light Energy Is a Form of Electromagnetic Radiation

Light is essential to support life on Earth. Light is a type of electromagnetic radiation, so named because it consists of energy in the form of electric and magnetic fields. Electromagnetic radiation travels as waves caused by the oscillation of the electric and magnetic fields. The **wavelength** is the distance between the peaks in a wave pattern. The **electromagnetic spectrum** encompasses all possible wavelengths of electromagnetic radiation, from relatively short wavelengths (gamma rays) to much longer wavelengths (radio waves) (**Figure 8.4**). Visible light is the range of wavelengths detected by the human eye, commonly between 380–740 nm. As discussed later, it is this visible light that provides the energy to drive photosynthesis.

Physicists have also discovered that light has properties that are characteristic of particles. Albert Einstein formulated the photon theory of light in which he proposed that light is composed of discrete particles called **photons**—massless particles traveling in a wavelike pattern and moving at the speed of light (about 300 million meters/second). Each photon contains a specific amount of energy. An important difference between the various types of electromagnetic radiation, described in Figure 8.4, is the amount of energy found in the photons. Shorter wavelength radiation carries more energy per unit of time than longer wavelength radiation. For example, the photons of gamma rays carry more energy than those of radio waves.

The sun radiates the entire spectrum of electromagnetic radiation, but the atmosphere prevents much of this radiation from reaching the Earth's surface. For example, the ozone layer forms a thin shield in the upper atmosphere, protecting life on Earth from much of the sun's ultraviolet rays. Even so, a substantial amount of electromagnetic radiation does reach the Earth's surface. The effect of light on living organisms is critically dependent on the energy of the photons that reach them.

Figure 8.4 **The electromagnetic spectrum.** The bottom portion of this figure emphasizes visible light, the wavelengths of electromagnetic radiation that are visible to the human eye. Light in the visible portion of the electromagnetic spectrum drives photosynthesis.

Concept check: Which has higher energy, gamma rays or radio waves?

The photons found in gamma rays, X-rays, and UV rays have very high energy. When molecules in cells absorb such energy, the effects can be devastating. Such types of radiation can cause mutations in DNA and even lead to cancer. By comparison, the energy of photons found in visible light is much milder. Molecules can absorb this energy in a way that does not cause permanent harm. Next, we will consider how molecules in living cells absorb the energy within visible light.

Pigments Absorb Light Energy

When light strikes an object, one of three things will happen. First, light may simply pass through the object. Second, the object may change the path of light toward a different direction. A third possibility is that the object may absorb the light. The term **pigment** is used to describe a molecule that can absorb light energy. When light strikes a pigment, some of the wavelengths of light energy are absorbed, while others are reflected. For example, leaves look green to us because they reflect radiant energy of the green wavelength. Various pigments in the leaves absorb the other light energy wavelengths. At the extremes of color reflection are white and black. A white object reflects nearly all of the visible light energy falling on it, whereas a black object absorbs nearly all of the light energy. This is why it's coolest to wear white clothes on a sunny, hot day.

What do we mean when we say that light energy is absorbed? In the visible spectrum, light energy may be absorbed by boosting electrons to higher energy levels (**Figure 8.5**). Recall from Chapter 2 that electrons are located around the nucleus of an atom. The location in which an electron is likely to be found is called its orbital. Electrons in different orbitals possess different amounts of energy. For an electron to absorb light energy and be boosted to an orbital with a higher energy, it must overcome the difference in energy between the orbital it is in and the orbital to which it is going. For this to happen, an electron must absorb a photon that contains precisely that amount of energy. Different pigment molecules contain a

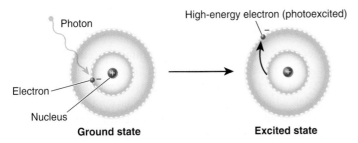

Figure 8.5 Absorption of light energy by an electron. When a photon of light of the correct amount of energy strikes an electron, the electron is boosted from the ground (unexcited) state to a higher energy level (an excited state). When this occurs, the electron occupies an orbital that is farther away from the nucleus of the atom. At this farther distance, the electron is held less firmly and is considered unstable.

Concept check: *For a photoexcited electron to become more stable, describe the three things that could happen.*

variety of electrons that can be shifted to different energy levels. Therefore, the wavelength of light that a pigment absorbs depends on the amount of energy needed to boost an electron to a higher orbital.

After an electron absorbs energy, it is said to be in an excited state. Usually, this is an unstable condition. The electron may release the energy in different ways. First, when an excited electron drops back down to a lower energy level, it may release heat. For example, on a sunny day, the sidewalk heats up because it absorbs light energy that is released as heat. A second way that an electron can release energy is in the form of light. Certain organisms, such as jellyfish, possess molecules that make them glow. This glow is due to the release of light when electrons drop down to lower energy levels, a phenomenon called fluorescence.

In the case of photosynthetic pigments, however, a different event happens that is critical for the process of photosynthesis. Rather than releasing energy, an excited electron in a photosynthetic pigment is removed from that molecule and transferred to another molecule where the electron is more stable. When this occurs, the energy in the electron is said to be "captured," because the electron does not readily drop down to a lower energy level and release heat or light.

Plants Contain Different Types of Photosynthetic Pigments

In plants, different pigment molecules absorb the light energy used to drive photosynthesis. Two types of chlorophyll pigments, termed **chlorophyll a** and **chlorophyll b**, are found in green plants and green algae. Their structure was determined in the 1930s by German chemist Hans Fischer (**Figure 8.6a**). In the chloroplast, both chlorophylls *a* and *b* are bound to integral membrane proteins in the thylakoid membrane.

The chlorophylls contain a porphyrin ring and a phytol tail. A magnesium ion (Mg^{2+}) is bound to the porphyrin ring. An electron in the porphyrin ring can follow a path in which it spends some of its time around several different atoms. Because this electron isn't restricted to a single atom, it is called a delocalized electron. The delocalized electron can absorb light energy.

(a) Chlorophylls *a* and *b*

(b) β-carotene (a carotenoid)

Figure 8.6 Structures of pigment molecules. (a) The structure of chlorophylls *a* and *b*. As indicated, chlorophylls *a* and *b* differ only at a single site, at which chlorophyll *a* has a —CH_3 group and chlorophyll *b* has a —CHO group. (b) The structure of β-carotene, an example of a carotenoid. The dark green and light green areas in parts (a) and (b) are the regions where a delocalized electron can hop from one atom to another.

The phytol tail in chlorophyll is a long hydrocarbon structure that is hydrophobic. Its function is to anchor the pigment to the surface of proteins within the thylakoid membrane.

Carotenoids are another type of pigment found in chloroplasts (**Figure 8.6b**). These pigments impart a color that ranges from yellow to orange to red. Carotenoids are often the major pigments in flowers and fruits. In leaves, the more abundant chlorophylls usually mask the colors of carotenoids. In temperate climates where the leaves change colors, the quantity of chlorophyll in the leaf declines during autumn. The carotenoids become readily visible and produce the yellows and oranges of autumn foliage.

An **absorption spectrum** is a diagram that depicts the wavelengths of electromagnetic radiation that are absorbed by a pigment. Each of the photosynthetic pigments shown in **Figure 8.7a** absorbs light in different regions of the visible spectrum. The absorption spectra of chlorophylls *a* and *b* are slightly different, though both chlorophylls absorb light most strongly in the red and violet parts of the visible spectrum and absorb green light poorly. Green light is reflected, which is why leaves appear green. Carotenoids absorb light in the blue and blue-green regions of the visible spectrum.

Why do plants have different pigments? Having different pigments allows plants to absorb light at many different wavelengths. In this way, plants are more efficient at capturing the energy in sunlight. This phenomenon is highlighted in an **action spectrum**, which shows the rate of photosynthesis plotted as a function of different wavelengths of light (**Figure 8.7b**). The

(a) Absorption spectra

(b) Action spectrum

Figure 8.7 Properties of pigment function: absorption and action spectra. (a) These absorption spectra show the absorption of light by chlorophyll *a*, chlorophyll *b*, and β-carotene. (b) An action spectrum of photosynthesis depicting the relative rate of photosynthesis in green plants at different wavelengths of light.

Concept check: *What is the advantage of having different pigment molecules?*

highest rates of photosynthesis correlate with the wavelengths that are strongly absorbed by the chlorophylls and carotenoids. Photosynthesis is poor in the green region of the spectrum, because these pigments do not readily absorb this wavelength of light.

Photosystems II and I Work Together to Produce ATP and NADPH

Photosynthetic organisms have the unique ability not only to absorb light energy but also to capture that energy in a stable way. Many organic molecules can absorb light energy. For example, on a sunny day, molecules in your skin absorb light energy and release the energy as heat. The heat that is released, however, cannot be harnessed to do useful work. A key feature of photosynthesis is the ability of pigments to capture light energy and transfer it to other molecules that can hold on to

the energy in a stable fashion and ultimately produce energy intermediate molecules that can do cellular work.

Let's now consider how chloroplasts capture light energy. The thylakoid membranes of the chloroplast contain two distinct complexes of proteins and pigment molecules called **photosystem I** (**PSI**) and **photosystem II** (**PSII**) (Figure 8.8). Photosystem I was discovered before photosystem II, but photosystem II is the initial step in photosynthesis. We will consider the structure and function of PSII in greater detail later in this chapter.

As described in steps 1 and 2, light excites electrons in pigment molecules, such as chlorophylls, which are located in regions of PSII and PSI called light-harvesting complexes. Rather than releasing their energy in the form of heat, the excited electrons follow a path shown by the red arrow. Initially, the excited electrons move from a pigment molecule called P680 in PSII to other electron carriers called pheophytin (Pp), Q_A, and Q_B. The excited electrons are moved out of PSII by Q_B. PSII also oxidizes water, which generates O_2 and adds H^+ into the thylakoid lumen. The electrons released from the oxidized water molecules are used to replenish the electrons that leave PSII via Q_B.

After a pair of electrons reaches Q_B, each one enters an **electron transport chain**—a series of electron carriers—located in the thylakoid membrane. The electron transport chain functions similarly to the one found in mitochondria. From Q_B, an electron goes to a cytochrome complex; then to plastocyanin (Pc), a small protein; and then to photosystem I. Along its journey from photosystem II to photosystem I, the electron releases some of its energy at particular steps and is transferred to the next component that has a higher electronegativity. The energy released is harnessed to pump H^+ into the thylakoid lumen. One result of the electron movement is to establish a H^+ electrochemical gradient.

A key role of photosystem I is to make NADPH (Figure 8.8, step 3). When light strikes the light-harvesting complex of photosystem I, this energy is also transferred to a reaction center, where a high-energy electron is removed from a pigment molecule, designated P700, and transferred to a primary electron acceptor. A protein called ferredoxin (Fd) can accept two high-energy electrons, one at a time, from the primary electron acceptor. Fd then transfers the two electrons to the enzyme $NADP^+$ reductase. This enzyme transfers the two electrons to $NADP^+$ and together with a H^+ creates NADPH. The formation of NADPH results in fewer H^+ in the stroma and thereby contributes to the formation of a H^+ electrochemical gradient across the thylakoid membrane.

As described in step 4, the synthesis of ATP in chloroplasts is achieved by a chemiosmotic mechanism similar to that used to make ATP in mitochondria. In chloroplasts, ATP synthesis is driven by the flow of H^+ from the thylakoid lumen into the stroma via ATP synthase (Figure 8.8). A H^+ gradient is generated in three ways: (1) the splitting of water, which places H^+ in the thylakoid lumen; (2) the movement of high-energy electrons from photosystem II to photosystem I, which pumps H^+ into the thylakoid lumen; and (3) the formation of NADPH, which consumes H^+ in the stroma.

A key difference between photosystem II and photosystem I is how the pigment molecules receive electrons. As discussed

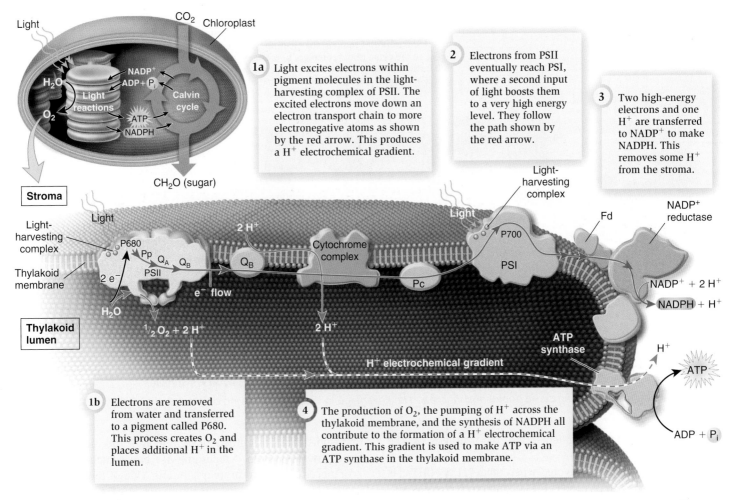

Figure 8.8 The synthesis of ATP, NADPH, and O_2 by the concerted actions of photosystems II and I.

Concept check: *Are ATP, NADPH, and O_2 produced in the stroma or in the thylakoid lumen?*

in more detail later, P680+ receives an electron from water. By comparison, P700+—the oxidized form of P700—receives an electron from Pc. Therefore, photosystem I does not need to split water to reduce P700+ and does not generate oxygen.

In summary, the steps of the light reactions of photosynthesis produce three chemical products:

1. O_2 is produced in the thylakoid lumen by the oxidation of water by photosystem II. Two electrons are removed from water, which creates two H+ and 1/2 O_2. The two electrons are transferred to P680+ molecules.

2. NADPH is produced in the stroma from high-energy electrons that start in photosystem II and are boosted a second time in photosystem I. Two high-energy electrons and one H+ are transferred to NADP+ to create NADPH.

3. ATP is produced in the stroma via ATP synthase that uses a H+ electrochemical gradient.

The combined action of photosystem II and photosystem I is termed **noncyclic electron flow** because the electrons move linearly from PSII to PSI and ultimately reduce NADP+ to NADPH.

Cyclic Electron Flow Produces Only ATP

The mechanism of harvesting light energy described in Figure 8.8 is called noncyclic electron flow because it is a linear process. This electron flow produces ATP and NADPH in roughly equal amounts. However, as we will see later, the Calvin cycle uses more ATP than NADPH. How can plant cells avoid making too much NADPH and not enough ATP? In 1959, Daniel Arnon discovered a pattern of electron flow that is cyclic and generates only ATP (**Figure 8.9**). Arnon termed the process **cyclic photophosphorylation** because (1) the path of electrons is cyclic, (2) light energizes the electrons, and (3) ATP is made via the phosphorylation of ADP. Due to the path of electrons, the mechanism is also called **cyclic electron flow**.

When light strikes photosystem I, high-energy electrons are sent to the primary electron acceptor and then to ferredoxin (Fd). The key difference in cyclic photophosphorylation is that the high-energy electrons are transferred from ferredoxin to Q_B. From Q_B, the electrons then go to the cytochrome complex, then to plastocyanin (Pc), and back to photosystem I. As the electrons travel along this cyclic route, they release energy, and some of this energy is used to transport H+ into the thylakoid

When light strikes photosystem I, electrons are excited and sent to ferredoxin (Fd). From Fd, the electrons are then transferred to Q_B, to the cytochrome complex, to plastocyanin (Pc), and back to photosystem I. This produces a H^+ electrochemical gradient, which is used to make ATP via the ATP synthase.

Figure 8.9 Cyclic photophosphorylation. In this process, an electron follows a cyclic path that is powered by photosystem I. This contributes to the formation of a H^+ electrochemical gradient, which is then used to make ATP by ATP synthase.

Concept check: Why is having cyclic photophosphorylation an advantage to a plant over having only noncyclic electron flow?

lumen. The resulting H^+ gradient drives the synthesis of ATP via ATP synthase.

Cyclic electron flow is favored when the level of $NADP^+$ is low and NADPH is high. Under these conditions, there is sufficient NADPH to run the Calvin cycle, which is described later. Alternatively, when $NADP^+$ is high and NADPH is low, noncyclic electron flow is favored, so more NADPH can be made. Cyclic electron flow is also favored when ATP levels are low.

Genomes & Proteomes Connection

The Cytochrome Complexes of Mitochondria and Chloroplasts Contain Evolutionarily Related Proteins

A recurring theme in cell biology is that evolution has resulted in groups of genes that encode proteins that play similar but specialized roles in cells—descent with modification. When two or more genes are similar because they are derived from the same ancestral gene, they are called **homologous genes**. As discussed in Chapter 23, homologous genes encode proteins that have similar amino acid sequences and may perform similar functions.

A comparison of the electron transport chains of mitochondria and chloroplasts reveals homologous genes. In particular, let's consider the cytochrome complex found in the thylakoid membrane of plants and algae, called cytochrome b_6-f (**Figure 8.10a**) and cytochrome b-c_1, which is found in the electron

transport chain of mitochondria (**Figure 8.10b**; also refer back to Figure 7.7). Both cytochrome b_6-f and cytochrome b-c_1 are composed of several protein subunits. One of those proteins is called cytochrome b_6 in cytochrome b_6-f and cytochrome b in cytochrome b-c_1.

By analyzing the sequences of the genes that encode these proteins, researchers discovered that cytochrome b_6 and cytochrome b are homologous. These proteins carry out similar functions: Both of them accept electrons from a quinone (Q_B or ubiquinone) and both donate an electron to another protein within their respective complexes (cytochrome f or cytochrome c_1). Likewise, both of these proteins function as H^+ pumps that capture some of the energy that is released from electrons to transport H^+ across the membrane. In this way, evolution has produced a family of cytochrome b-type proteins that play similar but specialized roles.

8.3 Molecular Features of Photosystems

Thus far, we have considered how chloroplasts absorb light energy and produce ATP, NADPH, and O_2. Photosystems, namely PSI and PSII, play critical roles in two aspects of photosynthesis. First, both PSI and PSII absorb light energy and capture that energy in the form of excited electrons. Second, PSII is also able to oxidize water and thereby produce O_2. In this section, we will examine how these events occur at the molecular level.

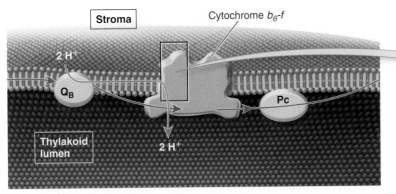

(a) Cytochrome b_6-f in the chloroplast

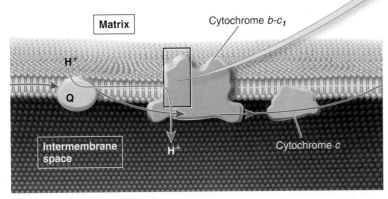

(b) Cytochrome b-c_1 in the mitochondrion

Figure 8.10 Homologous proteins in the electron transport chains of chloroplasts and mitochondria. **(a)** Cytochrome b_6-f is a complex involved in electron and H^+ transport in chloroplasts, and **(b)** cytochrome b-c_1 is a complex involved in electron and H^+ transport in mitochondria. These complexes contain homologous proteins designated cytochrome b_6 in chloroplasts and cytochrome b in mitochondria. The inset shows the three-dimensional structure of cytochrome b, which was determined by X-ray crystallography. It is an integral membrane protein with several transmembrane helices and two heme groups, which are prosthetic groups involved in electron transfer. The structure of cytochrome b_6 has also been determined and found to be very similar.

Concept check: Explain why the three-dimensional structures of cytochrome b and cytochrome b_6 are very similar.

Photosystem II Captures Light Energy and Produces O_2

PSI and PSII have two main components: a light-harvesting complex and a reaction center. **Figure 8.11** shows how these components function in PSII. In 1932, Robert Emerson and an undergraduate student, William Arnold, originally discovered the **light-harvesting complex** in the thylakoid membrane. It is composed of several dozen pigment molecules that are anchored to transmembrane proteins. The role of the complex is to directly absorb photons of light. When a pigment molecule absorbs a photon, an electron is boosted to a higher energy level. As shown in Figure 8.11, the energy (not the electron itself) can be transferred to adjacent pigment molecules by a process called **resonance energy transfer**. The energy may be transferred among multiple pigment molecules until it is eventually transferred to a special pigment molecule designated P680, which is located within the reaction center of PSII. The P680 pigment is so named because it can directly absorb light at a wavelength of 680 nm. However, P680 is more commonly excited by resonance energy transfer from another chlorophyll pigment. In either case, when an electron in P680 is excited, it is designated P680*. The light-harvesting complex is also called the **antenna complex** because it acts like an antenna that absorbs energy from light and funnels that energy to P680 in the reaction center.

A high-energy (photoexcited) electron in a pigment molecule is relatively unstable. It may abruptly release its energy by giving off heat or light. Unlike the pigments in the antenna complex that undergo resonance energy transfer, P680* can actually release its high-energy electron and become P680+.

$$P680^* \rightarrow P680^+ + e^-$$

The role of the reaction center is to quickly remove the high-energy electron from P680* and transfer it to another molecule, where the electron will be more stable. This molecule is called the **primary electron acceptor** (Figure 8.11). The transfer of the electron from P680* to the primary electron acceptor is remarkably fast. It occurs in less than a few picoseconds! (One picosecond equals one-trillionth of a second, also noted as 10^{-12} s.) Because this occurs so quickly, the excited electron does not have much time to release its energy in the form of heat or light.

After the primary electron acceptor has received this high-energy electron, the light energy has been captured and can be used to perform cellular work. As discussed earlier, the work it performs is to synthesize the energy intermediates ATP and NADPH.

Let's now consider what happens to P680+, which has given up its high-energy electron. After P680+ is formed, it is necessary to replace the electron so that P680 can function again. Therefore, another role of the reaction center is to replace the electron that is removed when P680* becomes P680+. This missing electron of P680+ is replaced with a low-energy electron from water (Figure 8.11).

$$H_2O \rightarrow 1/2\ O_2 + 2\ H^+ + 2\ e^-$$

$$2\ P680^+ + 2\ e^- \rightarrow 2\ P680$$
(from water)

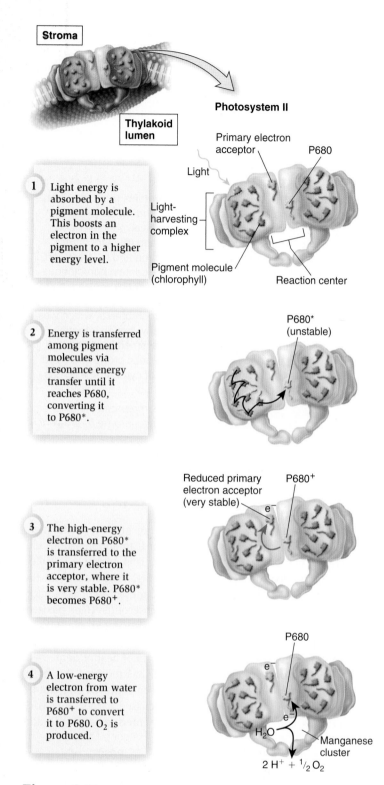

Stroma

Thylakoid lumen

Photosystem II

1. Light energy is absorbed by a pigment molecule. This boosts an electron in the pigment to a higher energy level.

Primary electron acceptor

P680

Light

Light-harvesting complex

Pigment molecule (chlorophyll)

Reaction center

2. Energy is transferred among pigment molecules via resonance energy transfer until it reaches P680, converting it to P680*.

P680* (unstable)

3. The high-energy electron on P680* is transferred to the primary electron acceptor, where it is very stable. P680* becomes P680+.

Reduced primary electron acceptor (very stable)

P680+

P680

e−

4. A low-energy electron from water is transferred to P680+ to convert it to P680. O_2 is produced.

P680

e−

e−

H_2O

Manganese cluster

$2 H^+ + \frac{1}{2} O_2$

Figure 8.11 A closer look at how PSII absorbs light energy and oxidizes water.

The oxidation of water results in the formation of oxygen gas (O_2), which is used by many organisms for cellular respiration. Photosystem II is the only known protein complex that can oxidize water, resulting in the release of O_2 into the atmosphere.

Photosystem II Is an Amazing Redox Machine

All cells rely on redox reactions to store and utilize energy and to form covalent bonds in organic molecules. Photosystem II is a particularly remarkable example of a redox machine. As we have learned, this complex of proteins removes high-energy electrons from a pigment molecule and transfers them to a primary electron acceptor. Perhaps even more remarkable is that photosystem II can remove low-energy electrons from water—a very stable molecule that holds onto its electrons tightly. The removal of electrons is how O_2 is made.

Many approaches have been used to study how photosystem II works. In recent years, much effort has been aimed at determining the biochemical composition of the protein complex and the roles of its individual components. The number of protein subunits varies somewhat from species to species and may vary due to environmental changes. Typically, photosystem II contains around 19 different protein subunits. Two subunits, designated D1 and D2, contain the reaction center that carries out the redox reactions (**Figure 8.12a**). Two other subunits, called CP43 and CP47, bind the pigment molecules that form the light-harvesting complex. Many additional subunits regulate the function of photosystem II and provide structural support.

Figure 8.12a illustrates the pathway of electron movement through photosystem II. The red arrows indicate the movement of a high-energy electron, whereas the black arrows show the path of a low-energy electron. Let's begin with a high-energy electron. When the electron on P680 becomes boosted to a higher energy level, usually by resonance energy transfer, this high-energy electron then moves to the primary electron acceptor, which is a chlorophyll molecule lacking Mg^{2+}, called pheophytin (Pp). Pheophytin is permanently bound to photosystem II and transfers the electron to a plastoquinone molecule, designated Q_A, which is also permanently bound to photosystem II. Next, the electron is transferred to another plastoquinone molecule designated Q_B, which can accept two high-energy electrons and bind two H^+. As shown earlier in Figure 8.8, Q_B can diffuse away from the reaction center.

Let's now consider the path of a low-energy electron. The oxidation of water occurs in a region called the **manganese cluster**. This site is located on the side of D1 that faces the thylakoid lumen. The manganese cluster has four Mn^{2+}, one Ca^{2+}, and one Cl^-. Two water molecules bind to this site. D1 catalyzes the removal of four low-energy electrons from the two water molecules to create four H^+ and O_2. Each low-energy electron is transferred, one at a time, to an amino acid in D1 (a tyrosine, Tyr) and then to P680+ to produce P680.

In 2004, So Iwata, James Barber, and colleagues determined the three-dimensional structure of photosystem II using a technique called **X-ray crystallography**. In this method, researchers must purify a protein or protein complex and expose it to conditions that cause the proteins to associate with each other in an ordered array. In other words, the proteins form a crystal. When a crystal is exposed to X-rays, the resulting pattern can be analyzed mathematically to determine the three-dimensional structure of the crystal's components. Major advances in this technique over the last couple of decades have enabled researchers to

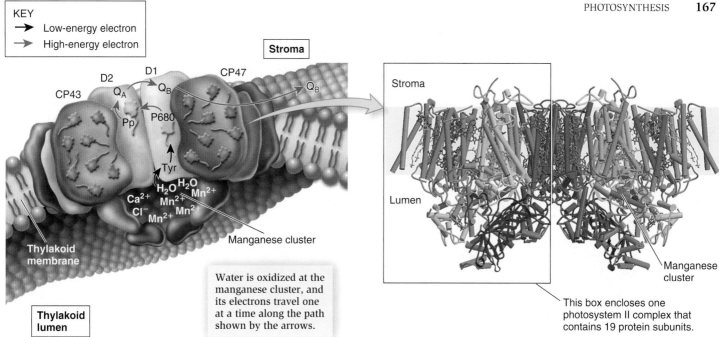

KEY
→ Low-energy electron
⇀ High-energy electron

D2
D1
CP43
CP47
Stroma
Q_A
Q_B
Q_B
Pp
P680
Tyr
H_2O H_2O
Ca^{2+}   Mn^{2+}   Mn^{2+}
Cl^-   Mn^{2+}   Mn^{2+}

Thylakoid
membrane

Thylakoid
lumen

Manganese cluster

Water is oxidized at the manganese cluster, and its electrons travel one at a time along the path shown by the arrows.

Stroma

Lumen

Manganese
cluster

This box encloses one photosystem II complex that contains 19 protein subunits.

(a) The path of electron flow through photosystem II

(b) Three-dimensional structure of photosystem II as determined by X-ray crystallography

Figure 8.12   **The molecular structure of photosystem II.**   (a) Schematic drawing showing the path of electron flow from water to Q_B. The CP43 and CP47 protein subunits wrap around D1 and D2 so that pigments in CP43 and CP47 can transfer energy to P680 by resonance energy transfer. (b) The three-dimensional structure of photosystem II as determined by X-ray crystallography. In the crystal structure, the colors are CP43 (green), D2 (orange), D1 (yellow), and CP47 (red).

Concept check: *According to this figure, how many redox reactions does photosystem II catalyze?*

determine the structures of relatively large macromolecular complexes such as photosystem II (**Figure 8.12b**). The structure shown here is a dimer; it has two PSII complexes, each with 19 protein subunits. As seen in this figure, the intricacy of the structure of photosystem II rivals the complexity of its function.

The Use of Light Flashes of Specific Wavelengths Provided Experimental Evidence for the Existence of PSII and PSI

An experimental technique that uses light flashes at particular wavelengths has been important in helping researchers to understand the function of photosystems. In this method, pioneered by Robert Emerson, a photosynthetic organism is exposed to a particular wavelength of light, after which the rate of photosynthesis is measured by the amount of CO_2 consumed or the amount of O_2 produced. In the 1950s, Emerson performed a particularly intriguing experiment that greatly stimulated photosynthesis research (**Figure 8.13**). He subjected algae to light flashes of different wavelengths and obtained a mysterious result. When he exposed algae to a wavelength of 680 nm, he observed a low rate of photosynthesis. A similarly low rate of photosynthesis occurred when he exposed algae to a wavelength of 700 nm. However, when he exposed the algae to both wavelengths of light simultaneously, the rate of photosynthesis was more than double the rate observed at only one wavelength. This phenomenon was termed the **enhancement effect**.

Simultaneous 680-nm and 700-nm flashes

Enhancement effect

680-nm flash

700-nm flash

Rate of photosynthesis

Time

Figure 8.13   **The enhancement effect observed by Emerson.**   When photosynthetic organisms such as green plants and algae are exposed to 680-nm and 700-nm light simultaneously, the resulting rate of photosynthesis is much more than double the rate produced by each wavelength individually.

Concept check: *Would the enhancement effect be observed if two consecutive flashes of light occurred at 680 nm?*

We know now that it occurs because light of 680-nm wavelength can readily activate the pigment (P680) in the reaction center in photosystem II but is not very efficient at activating pigments in photosystem I. In contrast, light of 700-nm wavelength is optimal at activating the pigments in photosystem I

Figure 8.14   **The Z scheme, showing the energy of an electron moving from photosystem II to NADP$^+$.**   The oxidation of water releases two electrons that travel one at a time from photosystem II to NADP$^+$. As seen here, the input of light boosts the energy of the electron twice. At the end of the pathway, two electrons are used to make NADPH.

Concept check: *During its journey from photosystem II to NADP$^+$, at what point does an electron have the highest amount of energy?*

but not those in photosystem II. When algae are exposed to both wavelengths, however, the pigments in both photosystems are maximally activated.

When researchers began to understand that photosynthesis results in the production of both ATP and NADPH, Robin Hill and Fay Bendall also proposed that photosynthesis involves two photoactivation events. According to their model, known as the **Z scheme**, an electron proceeds through a series of energy changes during photosynthesis. The Z refers to the zigzag shape of this energy curve. Based on our modern understanding of photosynthesis, we now know these events involve increases and decreases in the energy of an electron as it moves from photosystem II through photosystem I to NADP$^+$ (**Figure 8.14**). An electron on a nonexcited pigment molecule in photosystem II has the lowest energy. In photosystem II, light boosts an electron to a much higher energy level. As the electron travels from photosystem II to photosystem I, some of the energy is released. The input of light in photosystem I boosts the electron to an even higher energy than it attained in photosystem II. The electron releases a little energy before it is eventually transferred to NADP$^+$.

8.4 Synthesizing Carbohydrates via the Calvin Cycle

In the previous sections, we learned how the light reactions of photosynthesis produce ATP, NADPH, and O$_2$. We will now turn our attention to the second phase of photosynthesis, the Calvin cycle, in which ATP and NADPH are used to make carbohydrates. The Calvin cycle consists of a series of steps that occur in a metabolic cycle.

The Calvin cycle takes CO$_2$ from the atmosphere and incorporates the carbon into organic molecules, namely, carbohydrates. As mentioned earlier, carbohydrates are critical for two reasons. First, these organic molecules provide the precursors to make the organic molecules and macromolecules of nearly all living cells. The second key reason why the Calvin cycle is important involves the storage of energy. The Calvin cycle produces carbohydrates, which store energy. These carbohydrates are accumulated inside plant cells. When a plant is in the dark and not carrying out photosynthesis, the stored carbohydrates can be used as a source of energy. Similarly, when an animal consumes a plant, it can use the carbohydrates as an energy source.

In this section, we will examine the three phases of the Calvin cycle. We will also explore the experimental approach of Melvin Calvin and his colleagues that enabled them to elucidate the steps of this cycle.

The Calvin Cycle Incorporates CO$_2$ into Carbohydrate

The Calvin cycle, also called the Calvin-Benson cycle, was determined by chemists Melvin Calvin and Andrew Adam Benson and their colleagues in the 1940s and 1950s. This cycle requires a massive input of energy. For every 6 carbon dioxide molecules that are incorporated into a carbohydrate such as glucose (C$_6$H$_{12}$O$_6$), 18 ATP molecules are hydrolyzed and 12 NADPH molecules are oxidized.

$$6\ CO_2 + 12\ H_2O \rightarrow C_6H_{12}O_6 + 6\ O_2 + 6\ H_2O$$

$$18\ ATP + 18\ H_2O \rightarrow 18\ ADP + 18\ P_i$$

$$12\ NADPH \rightarrow 12\ NADP^+ + 12\ H^+ + 24\ e^-$$

Although biologists commonly describe glucose as a product of photosynthesis, glucose is not directly made by the Calvin cycle. Instead, molecules of glyceraldehyde-3-phosphate, which are products of the Calvin cycle, are used as starting materials for the synthesis of glucose and other molecules, including sucrose. After glucose molecules are made, they may be linked together to form a polymer of glucose called starch, which is stored in the chloroplast for later use. Alternatively, the disaccharide sucrose may be made and transported out of the leaf to other parts of the plant.

The Calvin cycle can be divided into three phases. These phases are carbon fixation, reduction and carbohydrate production, and regeneration of RuBP (**Figure 8.15**).

Carbon Fixation (Phase 1)

In **carbon fixation**, CO_2 becomes incorporated into ribulose bisphosphate (RuBP), a five-carbon sugar. The product of the reaction is a six-carbon intermediate

that immediately splits in half to form two molecules of 3-phosphoglycerate (3PG). The enzyme that catalyzes this step is named RuBP carboxylase/oxygenase, or **rubisco**. It is the most abundant protein in chloroplasts and perhaps the most abundant protein on Earth! This observation underscores the massive amount of carbon fixation that happens in the biosphere.

Reduction and Carbohydrate Production (Phase 2)

In the second phase, ATP is used to convert 3PG to 1,3-bisphosphoglycerate. Next, electrons from NADPH reduce 1,3-bisphosphoglycerate to glyceraldehyde-3-phosphate (G3P). G3P is a carbohydrate with three carbon atoms. The key difference between 3PG and G3P is that G3P has a C—H bond, whereas the analogous carbon in 3PG forms a C—O bond (Figure 8.15). The C—H bond can occur because the G3P molecule has been reduced by the addition of two electrons from NADPH. Compared to 3PG, the bonds

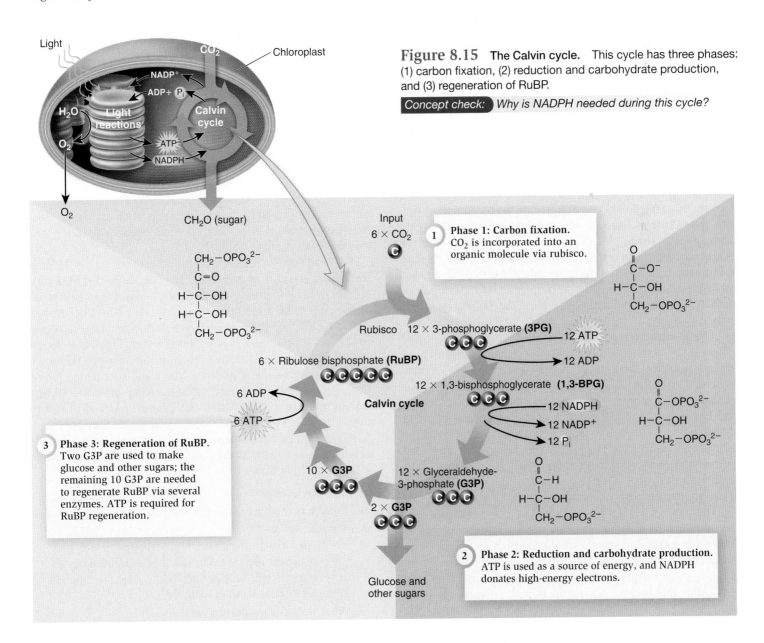

Figure 8.15 **The Calvin cycle.** This cycle has three phases: (1) carbon fixation, (2) reduction and carbohydrate production, and (3) regeneration of RuBP.

Concept check: Why is NADPH needed during this cycle?

in G3P store more energy and enable G3P to readily form larger organic molecules such as glucose.

As shown in Figure 8.15, only some of the G3P molecules are used to make glucose or other carbohydrates. Phase 1 begins with 6 RuBP molecules and 6 CO_2 molecules. Twelve G3P molecules are made at the end of phase 2. Two of these G3P molecules are used in carbohydrate production. As described next, the other 10 G3P molecules are needed to keep the Calvin cycle turning by regenerating RuBP.

Regeneration of RuBP (Phase 3) In the last phase of the Calvin cycle, a series of enzymatic steps converts the 10 G3P molecules into 6 RuBP molecules, using 6 molecules of ATP. After the RuBP molecules are regenerated, they serve as acceptors for CO_2, thereby allowing the cycle to continue.

As we have just seen, the Calvin cycle begins by using carbon from an inorganic source, that is, CO_2, and ends with organic molecules that will be used by the plant to make other compounds. You may be wondering why CO_2 molecules cannot be directly linked to form these larger molecules. The answer lies in the number of electrons that orbit carbon atoms. In CO_2, the carbon atom is considered electron poor. Oxygen is a very electronegative atom that monopolizes the electrons it shares with other atoms. In a covalent bond between carbon and oxygen, the shared electrons are closer to the oxygen atom.

By comparison, in an organic molecule, the carbon atom is electron rich. During the Calvin cycle, ATP provides energy and NADPH donates high-energy electrons, so the carbon originally in CO_2 has been reduced. The Calvin cycle combines less electronegative atoms with carbon atoms so that C—H and C—C bonds are formed. This allows the eventual synthesis of larger organic molecules including glucose, amino acids, and so on. In addition, the covalent bonds within these molecules are capable of storing large amounts of energy.

FEATURE INVESTIGATION

The Calvin Cycle Was Determined by Isotope Labeling Methods

The steps in the Calvin cycle involve the conversion of one type of molecule to another, eventually regenerating the starting material, RuBP. In the 1940s and 1950s, Calvin and his colleagues used ^{14}C, a radioisotope of carbon, to label and trace molecules produced during the cycle (**Figure 8.16**). They injected ^{14}C-labeled CO_2 into cultures of the green algae *Chlorella pyrenoidosa* grown in an apparatus called a "lollipop" (because of its shape). The *Chlorella* cells were given different lengths of time to incorporate the ^{14}C-labeled carbon, ranging from fractions of a second to many minutes. After this incubation period, the cells were abruptly placed into a solution of alcohol to inhibit enzymatic reactions and thereby stop the cycle.

The researchers separated the newly made radiolabeled molecules by a variety of methods. The most commonly used method was two-dimensional paper chromatography. In this approach, a sample containing radiolabeled molecules was spotted onto a corner of the paper at a location called the origin. The edge of the paper was placed in a solvent, such as phenol-water. As the solvent rose through the paper, so did the radiolabeled molecules. The rate at which they rose depended on their structures, which determined how strongly they interacted with the paper. This step separated the mixture of molecules spotted onto the paper at the origin.

The paper was then dried, turned 90°, and then the edge was placed in a different solvent, such as butanol-propionic acid-water. Again, the solvent would rise through the paper (in a second dimension), thereby separating molecules that may not have been adequately separated during the first separation step. After this second separation step, the paper was dried and exposed to X-ray film, a procedure called autoradiography.

Radioactive emission from the ^{14}C-labeled molecules caused dark spots to appear on the film.

The pattern of spots changed depending on the length of time the cells were incubated with ^{14}C-labeled CO_2. When the incubation period was short, only molecules that were made in the first steps of the Calvin cycle were seen. Longer incubations revealed molecules synthesized in later steps. For example, after short incubations, 3-phosphoglycerate (3PG) and 1,3-bisphosphoglycerate (1,3-BPG) were observed, whereas longer incubations also showed glyceraldehyde-3-phosphate (G3P) and ribulose bisphosphate (RuBP).

A challenge for Calvin and his colleagues was to identify the chemical nature of each spot. They achieved this by a variety of chemical methods. For example, a spot could be cut out of the paper, the molecule within the paper could be washed out or eluted, and then the eluted molecule could be subjected to the same procedure that included a radiolabeled molecule whose structure was already known. If the unknown molecule and known molecule migrated to the same spot in the paper, this indicated they were likely to be the same molecule. During the late 1940s and 1950s, Calvin and his coworkers identified all of the ^{14}C-labeled spots and the order in which they appeared. In this way, they were able to determine the series of reactions of what we now know as the Calvin cycle. For this work, Calvin was awarded the Nobel Prize in 1961.

Experimental Questions

1. What was the purpose of the study conducted by Calvin and his colleagues?

2. In Calvin's experiments shown in Figure 8.15, why did the researchers use ^{14}C? Why did they examine samples at several different time periods? How were the different molecules in the samples identified?

3. What were the results of Calvin's study?

Figure 8.16   The determination of the Calvin cycle using CO_2 labeled with ^{14}C and paper chromatography.

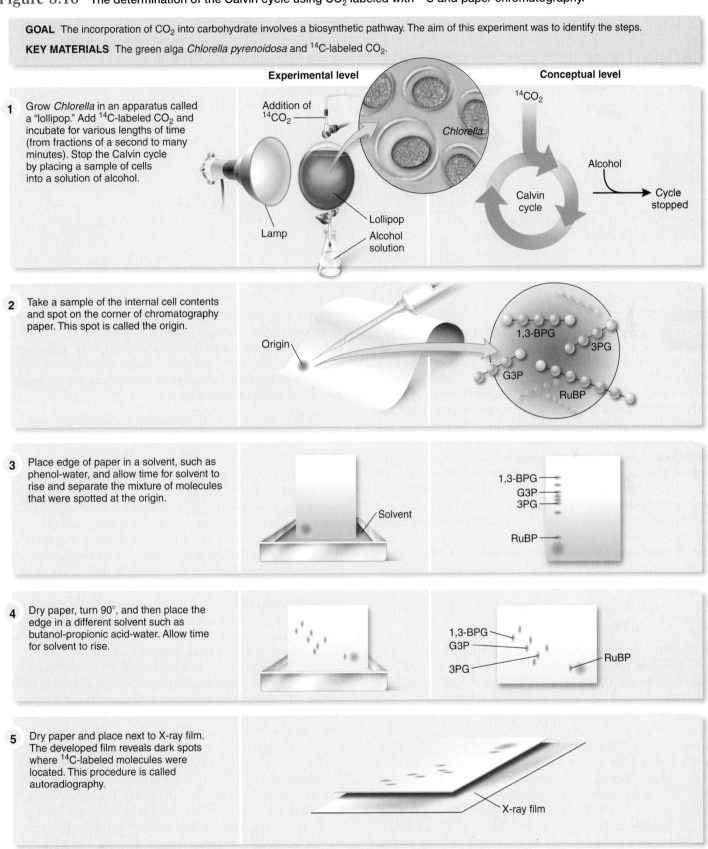

GOAL   The incorporation of CO_2 into carbohydrate involves a biosynthetic pathway. The aim of this experiment was to identify the steps.

KEY MATERIALS   The green alga *Chlorella pyrenoidosa* and ^{14}C-labeled CO_2.

Experimental level   **Conceptual level**

1   Grow *Chlorella* in an apparatus called a "lollipop." Add ^{14}C-labeled CO_2 and incubate for various lengths of time (from fractions of a second to many minutes). Stop the Calvin cycle by placing a sample of cells into a solution of alcohol.

Addition of $^{14}CO_2$

Chlorella

Lollipop

Lamp

Alcohol solution

$^{14}CO_2$

Calvin cycle

Alcohol

Cycle stopped

2   Take a sample of the internal cell contents and spot on the corner of chromatography paper. This spot is called the origin.

Origin

1,3-BPG

3PG

G3P

RuBP

3   Place edge of paper in a solvent, such as phenol-water, and allow time for solvent to rise and separate the mixture of molecules that were spotted at the origin.

Solvent

1,3-BPG
G3P
3PG

RuBP

4   Dry paper, turn 90°, and then place the edge in a different solvent such as butanol-propionic acid-water. Allow time for solvent to rise.

1,3-BPG
G3P

3PG

RuBP

5   Dry paper and place next to X-ray film. The developed film reveals dark spots where ^{14}C-labeled molecules were located. This procedure is called autoradiography.

X-ray film

6 THE DATA*

7 CONCLUSION The identification of the molecules in each spot elucidated the steps of the Calvin cycle.

8 SOURCE Calvin, M. December 11, 1961. The path of carbon in photosynthesis. Nobel Lecture.

*An autoradiograph from one of Calvin's experiments.

8.5 Variations in Photosynthesis

Thus far, we have considered the process of photosynthesis as it occurs in the chloroplasts of green plants and algae. Photosynthesis is a two-stage process in which the light reactions produce ATP, NADPH, and O_2, and the Calvin cycle uses the ATP and NADPH in the synthesis of carbohydrates. This two-stage process is a universal feature of photosynthesis in all green plants, algae, and cyanobacteria. However, certain environmental conditions such as light intensity, temperature, and water availability may influence both the efficiency of photosynthesis and the way in which the Calvin cycle operates. In this section, we begin by examining how hot and dry conditions may reduce the output of photosynthesis. We then explore two adaptations that certain plant species have evolved that conserve water and help to maximize photosynthetic efficiency in such environments.

Photorespiration Decreases the Efficiency of Photosynthesis

In the previous section, we learned that rubisco functions as a carboxylase because it adds a CO_2 molecule to RuBP, an organic molecule, to create two molecules of 3-phosphoglycerate (3PG).

$$RuBP + CO_2 \rightarrow 2 \ 3PG$$

For most species of plants, the incorporation of CO_2 into RuBP is the only way for carbon fixation to occur. Because 3PG is a three-carbon molecule, these plants are called **C_3 plants**. Examples of C_3 plants include wheat and oak trees (**Figure 8.17**). About 90% of the plant species on Earth are C_3 plants.

Researchers have discovered that the active site of rubisco can also function as an oxygenase, although its affinity for CO_2 is over 10-fold better than that for O_2. Even so, when O_2 levels are high and CO_2 levels are low, rubisco adds an O_2 molecule to RuBP. This creates only one molecule of 3-phosphoglycerate and a two-carbon molecule called phosphoglycolate. The phosphoglycolate is then dephosphorylated to glycolate and released from the chloroplast. In a series of several steps, the two-carbon glycolate is eventually oxidized in other organelles to produce an organic molecule plus a molecule of CO_2.

$$RuBP + O_2 \rightarrow 3\text{-phosphoglycerate} + Phosphoglycolate$$

$$Phosphoglycolate \rightarrow Glycolate \rightarrow \rightarrow Organic \ molecule + CO_2$$

This process, called **photorespiration**, uses O_2 and liberates CO_2. Photorespiration is considered wasteful because it reverses the effects of photosynthesis. This reduces the ability of a plant to make carbohydrates and thereby limits plant growth.

Photorespiration is more likely to occur when plants are exposed to a hot and dry environment. To conserve water, the stomata of the leaves close, inhibiting the uptake of CO_2 from the air and trapping the O_2 that is produced by photosynthesis. When the level of CO_2 is low and O_2 is high, photorespiration is favored. If C_3 plants are subjected to hot and dry environmental conditions, as much as 25–50% of their photosynthetic work is reversed by the process of photorespiration.

Why do plants carry out photorespiration? The answer is not entirely clear. Photorespiration undoubtedly has the disadvantage of lowering the efficiency of photosynthesis. One common view is that photorespiration does not offer any advantage and is an evolutionary relic. When rubisco first evolved some 3 billion years ago, the atmospheric oxygen level was low, so photorespiration would not have been a problem. Another view is that photorespiration may have a protective advantage. On hot and dry days when the stomata are closed, CO_2 levels within the leaves will fall, and O_2 levels will rise. Under these conditions, highly toxic oxygen-containing molecules such as free radicals may be produced that could damage the plant. Therefore, plant biologists have speculated that the role of photorespiration may be to protect the plant against the harmful effects of such toxic

(a) Wheat plants

(b) Oak leaves

Figure 8.17 **Examples of C₃ plants.** The structures of **(a)** wheat and **(b)** white oak leaves are similar to that shown in Figure 8.2.

Figure 8.17 **Examples of C$_3$ plants.** The structures of **(a)** wheat and **(b)** white oak leaves are similar to that shown in Figure 8.2.

molecules by consuming O_2 and releasing CO_2. In addition, photorespiration may affect the metabolism of other compounds in plants. Recent research suggests that photorespiration may also help plants to assimilate nitrogen into organic molecules.

C_4 Plants Have Evolved a Mechanism to Minimize Photorespiration

Certain species of plants have developed an interesting way to minimize photorespiration. In the early 1960s, Hugo Kortschak

discovered that the first product of carbon fixation in sugarcane is not 3-phosphoglycerate but instead is a compound with four carbon atoms. Species such as sugarcane are called **C$_4$ plants** because of this four-carbon compound. Later, Marshall Hatch and Roger Slack confirmed this result and identified the compound as oxaloacetate. For this reason, the pathway is sometimes called the Hatch-Slack pathway.

Some C_4 plants employ an interesting cellular organization to avoid photorespiration (**Figure 8.18**). Unlike C_3 plants, an interior layer in the leaves of many C_4 plants has a two-cell

Figure 8.18   **Leaf structure and its relationship to the C$_4$ cycle.**   C$_4$ plants have mesophyll cells, which initially take up CO_2, and bundle-sheath cells, where much of the carbohydrate synthesis occurs. Compare this leaf structure with the structure of C_3 leaves shown in Figure 8.2.

Concept check: *How does this cellular arrangement minimize photorespiration?*

organization composed of mesophyll cells and bundle-sheath cells. CO_2 from the atmosphere enters the mesophyll cells via stomata. Once inside, the enzyme **PEP carboxylase** adds CO_2 to phosphoenolpyruvate (PEP), a three-carbon molecule, to produce oxaloacetate, a four-carbon compound. PEP carboxylase does not recognize O_2. Therefore, unlike rubisco, PEP carboxylase does not promote photorespiration when CO_2 is low and O_2 is high. Instead, PEP carboxylase continues to fix CO_2.

In these types of C_4 plants, a four-carbon compound is transferred between cells. As shown in Figure 8.18, the compound oxaloacetate is converted to the four-carbon compound malate, which is transported into the bundle-sheath cell. Malate is then broken down into pyruvate and CO_2. The pyruvate returns to the mesophyll cell, where it is converted to PEP via ATP, and the cycle in the mesophyll cell can begin again. The main outcome of this C_4 cycle is that the mesophyll cell provides the bundle-sheath cell with CO_2. The Calvin cycle occurs in the chloroplasts of the bundle-sheath cell. Because the mesophyll cell supplies the bundle-sheath cell with a steady supply of CO_2, the concentration of CO_2 remains high in the bundle-sheath cell. Also, the mesophyll cells shield the bundle sheath cells from high levels of O_2. This strategy minimizes photorespiration, which requires low CO_2 and high O_2 levels to proceed.

Which is better—being a C_3 or a C_4 plant? The answer is that it depends on the environment. In warm and dry climates, C_4 plants have an advantage. During the day, they can keep their stomata partially closed to conserve water. Furthermore,

they can avoid photorespiration. C_4 plants are well adapted to habitats with high daytime temperatures and intense sunlight. Examples of C_4 plants are sugarcane, crabgrass, and corn. In cooler climates, C_3 plants have the edge because they use less energy to fix carbon dioxide. The process of carbon fixation that occurs in C_4 plants uses ATP to regenerate PEP from pyruvate (Figure 8.18), which C_3 plants do not have to expend.

CAM Plants Are C_4 Plants That Take Up CO_2 at Night

We have just learned that certain C_4 plants prevent photorespiration by providing CO_2 to the bundle-sheath cells, where the Calvin cycle occurs. This mechanism separates photosynthesis into different cells. Another strategy followed by other C_4 plants, called **CAM plants**, is to separate these processes in time. CAM stands for crassulacean acid metabolism, because the process was first studied in members of the plant family Crassulaceae. CAM plants are water-storing succulents such as cacti, bromeliads (including pineapple), and sedums. To avoid water loss, CAM plants keep their stomata closed during the day and open them at night, when it is cooler and the relative humidity is higher.

How, then, do CAM plants carry out photosynthesis? **Figure 8.19** compares CAM plants with the other type of C_4 plants we considered in Figure 8.18. Photosynthesis in CAM plants occurs entirely within mesophyll cells. During the night, the stomata

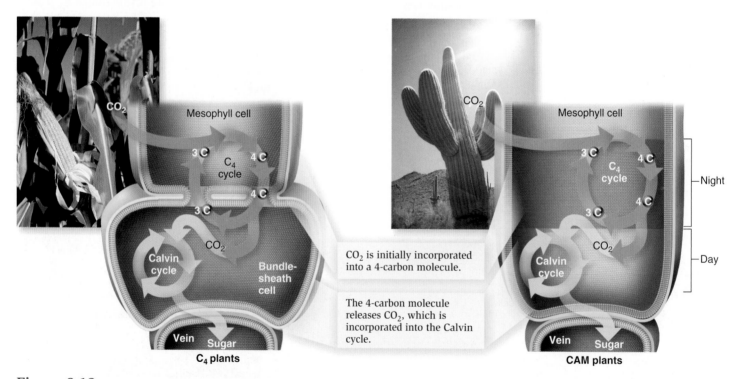

Figure 8.19   A comparison of C_4 and CAM plants.   The name C_4 plant describes those plants in which the first organic product of carbon fixation is a four-carbon compound. Using this definition, CAM plants are a type of C_4 plant. CAM plants, however, do not separate the functions of making a four-carbon molecule and the Calvin cycle into different types of cells. Instead, they make a four-carbon molecule at night and break down that molecule during the day so the CO_2 can be incorporated into the Calvin cycle.

Concept check: *What are the advantages and disadvantages among C_3, C_4, and CAM plants?*

of CAM plants open, thereby allowing the entry of CO_2 into mesophyll cells. CO_2 is joined with PEP to form the four-carbon compound oxaloacetate. This is then converted to malate, which accumulates during the night in the central vacuoles of the cells. In the morning, the stomata close to conserve moisture. The accumulated malate in the mesophyll cells leaves the vacuole and is broken down to release CO_2, which then drives the Calvin cycle during the daytime.

 ## Summary of Key Concepts

- Photosynthesis is the process by which plants, algae, and cyanobacteria capture light energy to synthesize carbohydrates.

8.1 Overview of Photosynthesis

- During photosynthesis, carbon dioxide, water, and energy are used to make carbohydrates and oxygen.

- Heterotrophs must obtain organic molecules in their food, whereas autotrophs can make organic molecules from inorganic sources. Photoautotrophs use the energy from light to make organic molecules.

- An energy cycle occurs in the biosphere in which photosynthesis uses CO_2 and H_2O to make organic molecules, and the organic molecules are broken back down to CO_2 and H_2O via cellular respiration so that organisms can make energy intermediates such as ATP. (Figure 8.1)

- In plants and algae, photosynthesis occurs within chloroplasts, which have an outer membrane, inner membrane, and thylakoid membrane. The stroma is found between the thylakoid membrane and inner membrane. In plants, the leaves are the major site of photosynthesis. (Figure 8.2)

- The light reactions of photosynthesis capture light energy to make ATP, NADPH, and O_2. These reactions occur at the thylakoid membrane. Carbohydrate synthesis via the Calvin cycle happens in the stroma and uses ATP and NADPH from the light reactions. (Figure 8.3)

8.2 Reactions That Harness Light Energy

- Light is a form of electromagnetic radiation that travels in waves and is composed of photons with discrete amounts of energy. (Figure 8.4)

- Electrons can absorb light energy and be boosted to a higher energy level, an excited state. (Figure 8.5)

- Photosynthetic pigments include chlorophylls a and b and carotenoids. These pigments absorb light energy in the visible spectrum. (Figures 8.6, 8.7)

- During noncyclic electron flow, electrons from photosystem II follow a pathway along an electron transport chain in the thylakoid membrane. This pathway generates a H^+ gradient that is used to make ATP. In addition, light energy striking photosystem I boosts electrons to a very high energy level that allows the synthesis of NADPH. (Figure 8.8)

- During cyclic photophosphorylation, electrons are activated in PSI and flow through the electron transport chain back to PSI.

This cyclic electron route produces a H^+ gradient that is used to make ATP. (Figure 8.9)

- Cytochrome b_6 in chloroplasts and cytochrome b in mitochondria are homologous proteins, both of which are involved in electron transport and H^+ pumping. (Figure 8.10)

8.3 Molecular Features of Photosystems

- Pigment molecules in photosystem II absorb light energy, and that energy is transferred to the reaction center via resonance energy transfer. A high-energy electron from P680* is transferred to a primary electron acceptor. An electron from water is then used to replenish the electron that is lost from P680*. (Figures 8.11, 8.12)

- Emerson showed that, compared to single light flashes at 680 nm and 700 nm, light flashes at both wavelengths more than doubled the amount of photosynthesis, a result called the enhancement effect. This occurred because these wavelengths activate pigments in PSII and PSI, respectively. (Figure 8.13)

- Hill and Bendall proposed the Z scheme for electron activation during photosynthesis. According to this scheme, an electron absorbs light energy twice, at both PSII and PSI, and it loses some of that energy as it flows along the electron transport chain in the thylakoid membrane. (Figure 8.14)

8.4 Synthesizing Carbohydrates via the Calvin Cycle

- The Calvin cycle can be divided into three phases: carbon fixation, reduction and carbohydrate production, and regeneration of ribulose bisphosphate (RuBP). During this process, ATP is used as a source of energy, and NADPH is used as a source of high-energy electrons so that CO_2 can be incorporated into carbohydrate. (Figure 8.15)

- Calvin and Benson determined the steps in the Calvin cycle by isotope labeling methods in which products of the Calvin cycle were separated by chromatography. (Figure 8.16)

8.5 Variations in Photosynthesis

- C_3 plants can incorporate CO_2 only into RuBP to make 3PG, a three-carbon molecule. (Figure 8.17)

- Photorespiration can occur when the level of O_2 is high and CO_2 is low, which happens under hot and dry conditions. During this process, some O_2 is used and CO_2 is liberated. Photorespiration is a disadvantage because it reverses the work of photosynthesis.

- Some C_4 plants avoid photorespiration because the CO_2 is first incorporated, via PEP carboxylase, into a four-carbon molecule, which is pumped from mesophyll cells into bundle-sheath cells. This maintains a high concentration of CO_2 in the bundle-sheath cells, where the Calvin cycle occurs. The high CO_2 concentration minimizes photorespiration. (Figure 8.18)

- CAM plants, a type of C_4 plant, prevent photorespiration by fixing CO_2 into a four-carbon molecule at night and then running the Calvin cycle during the day with their stomata closed. (Figure 8.19)

Assess and Discuss

Test Yourself

1. The water necessary for photosynthesis
   a. is split into H_2 and O_2.
   b. is directly involved in the synthesis of carbohydrate.
   c. provides the electrons to replace lost electrons in photosystem II.
   d. provides H^+ needed to synthesize G3P.
   e. does none of the above.

2. The reaction center pigment differs from the other pigment molecules of the light-harvesting complex in that
   a. the reaction center pigment is a carotenoid.
   b. the reaction center pigment absorbs light energy and transfers that energy to other molecules without the transfer of electrons.
   c. the reaction center pigment transfers excited electrons to the primary electron acceptor.
   d. the reaction center pigment does not transfer excited electrons to the primary electron acceptor.
   e. the reaction center acts as an ATP synthase to produce ATP.

3. The cyclic electron flow that occurs via photosystem I produces
   a. NADPH.
   b. oxygen.
   c. ATP.
   d. all of the above.
   e. a and c only.

4. During the light reactions, the high-energy electron from P680*
   a. eventually moves to $NADP^+$.
   b. becomes incorporated in water molecules.
   c. is pumped into the thylakoid space to drive ATP production.
   d. provides the energy necessary to split water molecules.
   e. falls back to the low-energy state in photosystem II.

5. During the first phase of the Calvin cycle, carbon dioxide is incorporated into ribulose bisphosphate by
   a. oxaloacetate.
   b. rubisco.
   c. RuBP.
   d. quinone.
   e. G3P.

6. The NADPH produced during the light reactions is necessary for
   a. the carbon fixation phase, which incorporates carbon dioxide into an organic molecule of the Calvin cycle.
   b. the reduction phase, which produces carbohydrates in the Calvin cycle.
   c. the regeneration of RuBP of the Calvin cycle.
   d. all of the above.
   e. a and b only.

7. The majority of the G3P produced during the reduction and carbohydrate production phase is used to produce
   a. glucose.
   b. ATP.
   c. RuBP to continue the cycle.
   d. rubisco.
   e. all of the above.

8. Photorespiration
   a. is the process where plants use sunlight to make ATP.
   b. is an inefficient way plants can produce organic molecules and in the process use O_2 and release CO_2.
   c. is a process that plants use to convert light energy to NADPH.
   d. occurs in the thylakoid lumen.
   e. is the normal process of carbohydrate production in cool, moist environments.

9. Photorespiration is avoided in C_4 plants because
   a. these plants separate the formation of a four-carbon molecule from the rest of the Calvin cycle in different cells.
   b. these plants carry out only anaerobic respiration.
   c. the enzyme PEP functions to maintain high CO_2 concentrations in the bundle-sheath cells.
   d. all of the above.
   e. a and c only.

10. Plants commonly found in hot and dry environments that carry out carbon fixation at night are
    a. oak trees.
    b. C_3 plants.
    c. CAM plants.
    d. all of the above.
    e. a and b only.

Conceptual Questions

1. What are the two stages of photosynthesis? What are the key products of each stage?

2. What is the function of NADPH in the Calvin cycle?

3. Why is resonance energy transfer an important phenomenon to capture light energy? How do you think photosynthesis would be affected if resonance energy transfer did not occur?

Collaborative Questions

1. Discuss the advantages and disadvantages of being a heterotroph or a photoautotroph.

2. Biotechnologists are trying to genetically modify C_3 plants to convert them to C_4 or CAM plants. Why would this be useful? What genes might you add to C_3 plants to convert them to C_4 or CAM plants?

Online Resource

www.brookerbiology.com

Stay a step ahead in your studies with animations that bring concepts to life and practice tests to assess your understanding. Your instructor may also recommend the interactive ebook, individualized learning tools, and more.

Cell Communication

9

O ver 2 billion cells will die in your body during the next hour. In an adult human body, approximately 50 to 70 billion cells die each day due to programmed cell death— the process in which a cell breaks apart into small fragments (see chapter-opening photo). In a year, your body produces and purposely destroys a mass of cells that is equal to your own body's weight! Though this may seem like a scary process, it's actually keeping you healthy. Programmed cell death, also called apoptosis, ensures that your body maintains a proper number of cells. It also eliminates cells that are worn out or potentially harmful, such as cancer cells. Programmed cell death can occur via signals that intentionally cause particular cells to die, or it can result from a failure of proper cell communication. It may also happen when environmental agents cause damage to a cell. Programmed cell death is one example of a response that involves **cell communication**—the process through which cells can detect and respond to signals in their environment.

In this chapter, we will examine how cells detect environmental signals and also how they produce signals so they can communicate with other cells. As you will learn, cell communication involves an amazing diversity of signaling molecules and cellular proteins that are devoted to this process.

Programmed cell death.   The two cells shown here are breaking apart due to signaling molecules that initiated a pathway that programmed their death.

9.1 General Features of Cell Communication

All living cells, including bacteria, protists, fungi, plant cells, and animal cells, conduct and require cell communication to survive. This phenomenon, also known as cell signaling, involves both incoming and outgoing signals. A **signal** is an agent that can influence the properties of cells. For example, on a sunny day, cells can sense their exposure to ultraviolet (UV) light—a physical signal—and respond accordingly. In humans, UV light acts as an incoming signal to promote the synthesis of melanin, a protective pigment that helps to prevent the harmful effects of UV radiation. In addition, cells can produce outgoing signals that influence the behavior of neighboring cells. Plant cells, for example, produce hormones that influence the pattern of cell elongation so the plant grows toward light. Cells of all

living organisms both respond to incoming signals and produce outgoing signals. Cell communication is a two-way street.

Communication at the cellular level involves not only receiving and sending signals but also their interpretation. For this to occur, a signal must be recognized by a cellular protein called a **receptor**. When a signal and receptor interact, the receptor changes shape, or conformation, thereby changing the way the receptor interacts with cellular factors. These interactions eventually lead to some type of response in the cell. In this section, we begin by considering why cells need to respond to signals. We will then examine various forms of signaling that are based on the distance between the cells that communicate with each other. Finally, we will examine the main steps that occur when a cell is exposed to a signal and elicits a response to it.

Cells Detect and Respond to Signals from Their Environment and from Other Cells

Before getting into the details of cell communication, let's take a general look at why cell communication is necessary. The

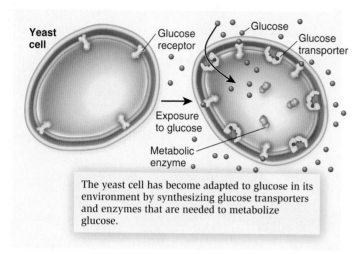

The yeast cell has become adapted to glucose in its environment by synthesizing glucose transporters and enzymes that are needed to metabolize glucose.

Figure 9.1   **Response of a yeast cell to glucose.**   When glucose is absent from the extracellular environment, the cell is not well prepared to take up and metabolize this sugar. However, when glucose is present, some of that glucose binds to receptors in the membrane, which leads to changes in the amounts and properties of intracellular and membrane proteins so the cell can readily use glucose.

Concept check: *What is the signaling molecule in this example?*

first reason is that cells need to respond to a changing environment. Changes in the environment are a persistent feature of life, and living cells are continually faced with alterations in temperature and availability of nutrients and water. A cell may even be exposed to a toxic chemical in its environment. Being able to respond to change, a phenomenon known as **adaptation**, is critical for the survival of all living organisms. Adaptation at the cellular level is called a **cellular response**.

As an example, let's consider the response of a yeast cell to glucose in its environment (**Figure 9.1**). Some of the glucose acts as a signaling molecule that binds to a receptor and causes a cellular response. In this case, the cell responds by increasing

the number of glucose transporters needed to take glucose into the cell and also by increasing the number of metabolic enzymes required to utilize glucose once it is inside. The cellular response has allowed the cell to use glucose efficiently. We could say the cell has become adapted to the presence of glucose in its environment. Note that the term adaptation also refers to more permanent changes in a species as a result of evolutionary changes. We will consider these types of adaptations in Chapter 23.

A second reason for cell signaling is the need for cells to communicate with each other—a type of cell communication also called **cell-to-cell communication**. In one of the earliest experiments demonstrating cell-to-cell communication, Charles Darwin and his son Francis Darwin studied phototropism, the phenomenon in which plants grow toward light (**Figure 9.2**). The Darwins observed that the actual bending occurs in a zone below the growing shoot tip. They concluded that a signal must be transmitted from the growing tip to cells below the tip for this to occur. Later research revealed that the signal is a molecule called auxin, which is transmitted from cell to cell. A higher amount of auxin present on the nonilluminated side of the shoot promotes cell elongation on that side of the shoot only, thereby causing the shoot to bend toward the light source.

Cell-to-Cell Communication Can Occur Between Adjacent Cells and Between Cells That Are Long Distances Apart

Researchers have determined that organisms have a variety of different mechanisms to achieve cell-to-cell communication. The mode of communication depends, in part, on the distance between the cells that need to communicate with each other. Let's first examine the various ways in which signals are transferred between cells. Later in this chapter, we will learn how such signals elicit a cellular response.

One way to categorize cell signaling is by the manner in which the signal is transmitted from one cell to another. Signals are relayed between cells in five common ways, all of

Cells in the growing shoot tip sense light and send a signal (auxin) to cells on the nonilluminated side of the shoot.

Cells located below the growing tip receive this signal and elongate, thereby causing a bend in the shoot. In this way, the tip grows toward the light.

Figure 9.2   **Phototropism in plants.**   This process involves cell-to-cell communication that leads to a shoot bending toward light just beneath its actively growing tip.

Concept check: *Below the shoot tip that is illuminated from one side, where is more auxin located? Does auxin cause cells to elongate or to shorten?*

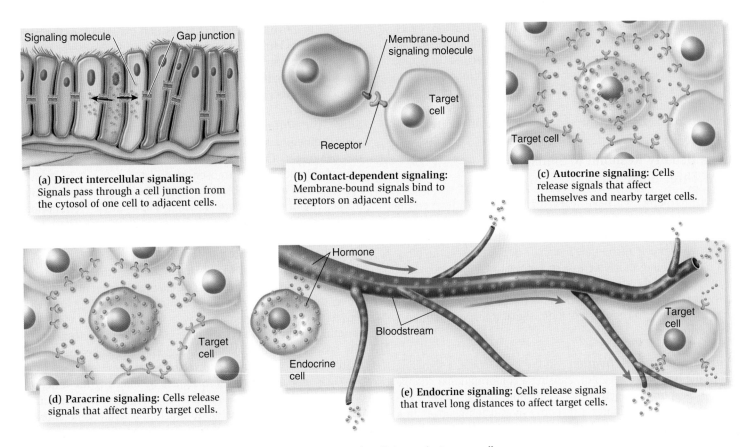

Signaling molecule   Gap junction

(a) Direct intercellular signaling: Signals pass through a cell junction from the cytosol of one cell to adjacent cells.

Membrane-bound signaling molecule

Target cell

Receptor

(b) Contact-dependent signaling: Membrane-bound signals bind to receptors on adjacent cells.

Target cell

(c) Autocrine signaling: Cells release signals that affect themselves and nearby target cells.

Target cell

(d) Paracrine signaling: Cells release signals that affect nearby target cells.

Hormone

Bloodstream

Endocrine cell

Target cell

(e) Endocrine signaling: Cells release signals that travel long distances to affect target cells.

Figure 9.3   Types of cell-to-cell communication based on the distance between cells.

Concept check: *Which type of signal, paracrine or endocrine, is likely to exist for a longer period of time? Explain why this is necessary.*

which involve a cell that produces a signal and a target cell that receives the signal (**Figure 9.3**).

Direct Intercellular Signaling   In a multicellular organism, cells adjacent to each other may have contacts, called cell junctions, that enable them to pass ions, signaling molecules, and other materials between the cytosol of one cell and the cytosol of another (Figure 9.3a). For example, cardiac muscle cells, which cause your heart to beat, have intercellular connections called gap junctions that pass electrical signals needed for the coordinated contraction of cardiac muscle cells. We will examine how gap junctions work in Chapter 10.

Contact-Dependent Signaling   Not all signaling molecules can readily diffuse from one cell to another. Some molecules are bound to the surface of cells and provide a signal to other cells that make contact with the surface of that cell (Figure 9.3b). In this case, one cell has a membrane-bound signaling molecule that is recognized by a receptor on the surface of another cell. This occurs, for example, when portions of nerve cells (neurons) grow and make contact with other neurons. This is important for the formation of the proper connections between neurons.

Autocrine Signaling   In autocrine signaling, a cell secretes signaling molecules that bind to receptors on its own cell surface, stimulating a response (Figure 9.3c). In addition, the

signaling molecule can affect neighboring cells of the same cell type. What is the purpose of autocrine signaling? It is often important for groups of cells to sense cell density. When cell density is high, the concentration of autocrine signals is also high. In some cases, such signals will inhibit further cell growth and thereby prevent the cell density from becoming too high.

Paracrine Signaling   In paracrine signaling, a specific cell secretes a signaling molecule that does not affect the cell secreting the signal but instead influences the behavior of target cells in close proximity (Figure 9.3d). Paracrine signaling is typically of short duration. Usually, the signal is broken down too quickly to be carried to other parts of the body and affect distant cells. A specialized form of paracrine signaling called **synaptic signaling** occurs in the nervous system of animals (see Chapter 41). Neurotransmitters—molecules made in neurons that transmit a signal to an adjacent cell—are released at the end of the neuron and traverse a narrow space called the synapse. The neurotransmitter then binds to a receptor in a target cell.

Endocrine Signaling   In contrast to the previous mechanisms of cell signaling, endocrine signaling occurs over relatively long distances (Figure 9.3e). In both animals and plants, molecules involved in long-distance signaling are called **hormones**. They usually last longer than signaling molecules involved in autocrine and paracrine signaling. In animals, endocrine signaling involves the secretion of hormones into the bloodstream, which may affect virtually all cells of the body, including those that are far from the cells that secrete the signaling molecules. In plants, hormones move through the plant vascular system and can also move through adjacent cells. Some hormones are even gases that diffuse into the air. Ethylene, a gas given off by plants, plays a variety of roles, such as accelerating the ripening of fruit.

Cells Usually Respond to Signals by a Three-Stage Process

Up to this point, we have learned that signals influence the behavior of cells in close proximity or at long distances, interacting with receptors to elicit a cellular response. What events occur when a cell encounters a signal? In most cases, the binding of a signaling molecule to a receptor causes the receptor to activate a signal transduction pathway, which then leads to a cellular response. **Figure 9.4** diagrams the three common stages of cell signaling: receptor activation, signal transduction, and a cellular response.

Stage 1: Receptor Activation   In the initial stage, a signaling molecule binds to a receptor, causing a conformational change in the receptor that activates its function. In most cases, the activated receptor initiates a response by causing changes in a series of proteins that collectively forms a signal transduction pathway, as described next.

Stage 2: Signal Transduction   During signal transduction, the initial signal is converted—or transduced—to a different signal inside the cell. This process is carried out by a group of proteins that form a **signal transduction pathway**. These proteins undergo a series of changes that may result in the production of an intracellular signaling molecule. However, some receptors are intracellular and some do not activate a signal transduction pathway. As discussed later, certain types of intracellular receptors directly cause a cellular response.

Stage 3: Cellular Response   Cells can respond to signals in several different ways. Figure 9.4 shows three common categories of proteins that are controlled by cell signaling: enzymes, structural proteins, and transcription factors.

Many signaling molecules exert their effects by altering the activity of one or more enzymes. For example, certain hormones provide a signal that the body needs energy. These hormones activate enzymes that are required for the breakdown of molecules such as carbohydrates.

Cells also respond to signals by altering the functions of structural proteins in the cell. For example, when animal cells move during embryonic development or when an amoeba moves toward food, signals play a role in the rearrangement of actin filaments, which are components of the cytoskeleton. The coordination of signaling and changes in the cytoskeleton enable cells to move in the correct direction.

Cells may also respond to signals by affecting the function of **transcription factors**—proteins that regulate the transcription of genes. Some transcription factors activate gene expression. For example, when cells are exposed to sex hormones, transcription factors can activate genes that change the properties of cells, which can lead to changes in the sexual characteristics of entire organisms. As discussed in Chapter 51, estrogens and androgens are responsible for the development of secondary sex characteristics in humans, including breast development in females and beard growth in males, respectively.

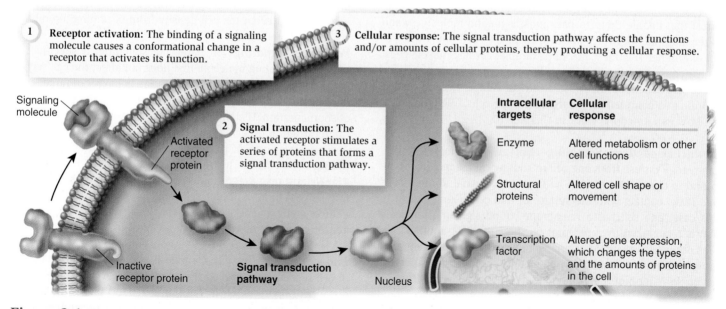

Figure 9.4   The three stages of cell signaling: receptor activation, signal transduction, and a cellular response.

Concept check: *For most signaling molecules, explain why a signal transduction pathway is necessary.*

9.2 Cellular Receptors and Their Activation

In this section, we will take a closer look at receptors and how they interact with signaling molecules. We will compare receptors based on whether they are located on the cell surface or inside the cell. In this chapter, our focus will be on receptors that respond to chemical signaling molecules. Other receptors discussed in Units VI and VII respond to mechanical motion (mechanoreceptors), temperature changes (thermoreceptors), and light (photoreceptors).

Receptors Bind to Specific Signals and Undergo Conformational Changes

The ability of cells to respond to a signal usually requires precise recognition between a signal and its receptor. In many cases, the signal is a molecule, such as a steroid or a protein, that binds to the receptor. A signaling molecule binds to a receptor in much the same way that a substrate binds to the active site of an enzyme, as described in Chapter 6. The signaling molecule, which is called a **ligand**, binds noncovalently to the receptor molecule with a high degree of specificity. The binding occurs when the ligand and receptor happen to collide in the correct orientation with enough energy to form a **ligand • receptor complex**.

$$[\text{Ligand}] + [\text{Receptor}] \underset{k_{off}}{\overset{k_{on}}{\rightleftharpoons}} [\text{Ligand} \bullet \text{Receptor complex}]$$

Brackets [ ] refer to concentration. The value k_{on} is the rate at which binding occurs. After a complex forms between the ligand and its receptor, the noncovalent interaction between a ligand and receptor remains stable for a finite period of time. The term k_{off} is the rate at which the ligand • receptor complex falls apart or dissociates.

In general, the binding and release between a ligand and its receptor are relatively rapid, and therefore an equilibrium is reached when the rate of formation of new ligand • receptor complexes equals the rate at which existing ligand • receptor complexes dissociate:

$$k_{on}[\text{Ligand}][\text{Receptor}] = k_{off}[\text{Ligand} \bullet \text{Receptor complex}]$$

Rearranging,

$$\frac{[\text{Ligand}][\text{Receptor}]}{[\text{Ligand} \bullet \text{Receptor complex}]} = \frac{k_{off}}{k_{on}} = K_d$$

K_d is called the **dissociation constant** between a ligand and its receptor. The K_d value is inversely related to the affinity between the ligand and receptor. Let's look carefully at the left side of this equation and consider what it means. At a ligand concentration where half of the receptors are bound to a ligand, the concentration of the ligand • receptor complex equals the concentration of receptor that doesn't have ligand bound. At this ligand concentration, [Receptor] and [Ligand • Receptor complex] cancel out of the equation because they are equal. Therefore, at a ligand concentration where half of the receptors have bound ligand:

$$K_d = [\text{Ligand}]$$

When the ligand concentration is above the K_d value, most of the receptors are likely to have ligand bound to them. In contrast, if the ligand concentration is substantially below the K_d value, most receptors will not be bound by their ligand. The K_d values for many different ligands and their receptors have been experimentally determined. How is this information useful? It allows researchers to predict when a signaling molecule is likely to cause a cellular response. If the concentration of a signaling molecule is far below the K_d value, a cellular response is not likely because relatively few receptors will form a complex with the signaling molecule.

Unlike enzymes, which convert their substrates into products, receptors do not usually alter the structure of their ligands. Instead, the ligands alter the structure of their receptors, causing a conformational change (**Figure 9.5**). In this case, the binding of the ligand to its receptor changes the receptor in a way that will activate its ability to initiate a cellular response.

Because the binding of a ligand to its receptor is a reversible process, the ligand and receptor will also dissociate. Once the ligand is released, the receptor is no longer activated.

Cells Contain a Variety of Cell Surface Receptors That Respond to Extracellular Signals

Most signaling molecules are either small hydrophilic molecules or large molecules that do not readily pass through the plasma membrane of cells. Such extracellular signals bind to **cell surface receptors**—receptors found in the plasma membrane. A typical cell is expected to contain dozens or even hundreds of different cell surface receptors that enable the cell to respond to different kinds of extracellular signaling molecules. By analyzing the functions of cell surface receptors from many different organisms, researchers have determined that most fall into one

The binding of a ligand to its receptor causes a conformational change in the receptor, resulting in receptor activation.

Figure 9.5   Receptor activation.

of three categories: enzyme-linked receptors, G-protein-coupled receptors, and ligand-gated ion channels, which are described next.

Enzyme-Linked Receptors   Receptors known as **enzyme-linked receptors** are found in all living species. Many human hormones bind to this type of receptor. For example, when insulin binds to an enzyme-linked receptor in muscle cells, it enhances their ability to use glucose. Enzyme-linked receptors typically have two important domains: an extracellular domain, which binds a signaling molecule, and an intracellular domain, which has a catalytic function (**Figure 9.6a**). When a signaling molecule binds to the extracellular domain, a conformational change is transmitted through the membrane-embedded portion of the protein that affects the conformation of the intracellular catalytic domain. In most cases, this conformational change causes the catalytic domain to become functionally active.

Most types of enzyme-linked receptors function as **protein kinases**, enzymes that transfer a phosphate group from ATP to specific amino acids in a protein. For example, tyrosine kinases attach phosphate to the amino acid tyrosine, whereas serine/threonine kinases attach phosphate to the amino acids serine and threonine. In the absence of a signaling molecule, the catalytic domain of the receptor remains inactive (**Figure 9.6b**). However, when a signal binds to the extracellular domain, the catalytic domain is activated. Under these conditions, the cell surface receptor may phosphorylate itself or it may phosphorylate intracellular proteins. The attachment of a negatively charged phosphate changes the structure of a protein and thereby can alter its function. Later in this chapter, we will explore how this event leads to a cellular response, such as the activation of enzymes that affect cell function.

G-Protein-Coupled Receptors   Receptors called **G-protein-coupled receptors** (**GPCRs**) are found in the cells of all eukaryotic species and are particularly common in animals. GPCRs typically contain seven transmembrane segments that wind back and forth through the plasma membrane. The receptors interact with intracellular proteins called **G proteins**, which are so named because of their ability to bind guanosine triphosphate (GTP) and guanosine diphosphate (GDP). GTP is similar in structure to ATP except it has guanine as a base instead of adenine. In the 1970s, the existence of G proteins was first proposed by Martin Rodbell and colleagues, who found that GTP is needed for certain hormone receptors to cause an intracellular response. Later, Alfred Gilman and coworkers used genetic and biochemical techniques to identify and purify a G protein. In 1994, Rodbell and Gilman won the Nobel Prize for their pioneering work.

Figure 9.7 shows how a GPCR and a G protein interact. At the cell surface, a signaling molecule binds to a GPCR, causing a conformational change that activates the receptor. The activated receptor then causes the G protein, which is a lipid-anchored protein, to release GDP and bind GTP instead. GTP binding changes the conformation of the G protein, causing it to dissociate into an α subunit and a β/γ dimer. Later in this chapter, we will examine how the α subunit interacts with other proteins in a signal transduction pathway to elicit a cellular response. The β/γ dimer can also play a role in signal transduction. For example, it can regulate the function of ion channels in the plasma membrane.

When a signaling molecule and GPCR dissociate, the GPCR is no longer activated, and the cellular response will be reversed. For the G protein to return to the inactive state, the α subunit will first hydrolyze its bound GTP to GDP and P_i. After this occurs, the α and β/γ subunits reassociate with each other to form an inactive complex.

(a) **Structure of enzyme-linked receptors**

Signaling molecule
Extracellular signal-binding domain
Extracellular environment
Cytosol
Intracellular catalytic domain

(b) **A receptor that functions as a protein kinase**

Inactive receptor

A signaling molecule binds and activates the catalytic domain of the receptor.

Signaling molecule

Activated receptor

Unphosphorylated protein

ATP   +   ADP

Phosphorylated protein

The receptor then can catalyze the transfer of a phosphate group from ATP to a protein.

Figure 9.6   Enzyme-linked receptors.

Concept check: *Based on your understanding of ATP as an energy intermediate, is the phosphorylation of a protein via a protein kinase an exergonic or endergonic reaction? How is the energy of protein phosphorylation used—what does it accomplish?*

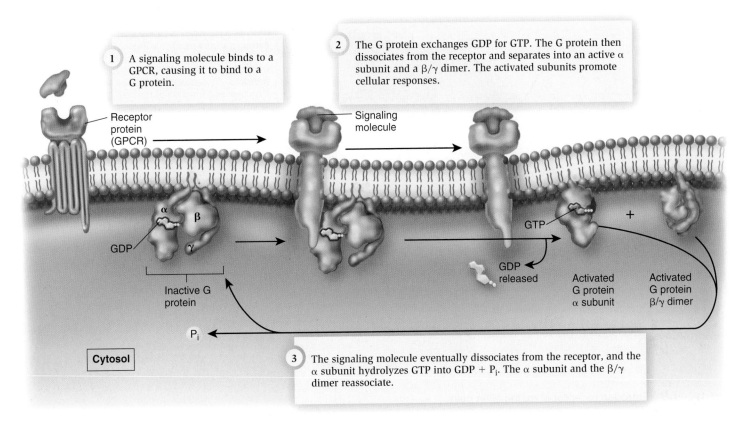

1   A signaling molecule binds to a GPCR, causing it to bind to a G protein.

2   The G protein exchanges GDP for GTP. The G protein then dissociates from the receptor and separates into an active α subunit and a β/γ dimer. The activated subunits promote cellular responses.

Receptor protein (GPCR)

Signaling molecule

α   β

GDP   γ

Inactive G protein

GTP

GDP released

Activated G protein α subunit

Activated G protein β/γ dimer

+

P_i

Cytosol

3   The signaling molecule eventually dissociates from the receptor, and the α subunit hydrolyzes GTP into GDP + P_i. The α subunit and the β/γ dimer reassociate.

Figure 9.7   The activation of G-protein-coupled receptors and G proteins.   Note: The left drawing of the receptor emphasizes that it has seven transmembrane segments.

Concept check: *What has to happen for the α and β/γ subunits of the G protein to reassociate with each other?*

Ligand-Gated Ion Channels   As described in Chapter 5, ion channels are proteins that allow the diffusion of ions across cellular membranes. **Ligand-gated ion channels** are a third type of cell surface receptor found in the plasma membrane of animal, plant, and fungal cells. When signaling molecules (ligands) bind to this type of receptor, the channel opens and allows the flow of ions through the membrane (**Figure 9.8**).

In animals, ligand-gated ion channels are important in the transmission of signals between nerve and muscle cells and between two nerve cells. In addition, ligand-gated ion channels in the plasma membrane allow the influx of Ca^{2+} into the cytosol. As discussed later in this chapter, changes in the cytosolic concentration of Ca^{2+} often play a role in signal transduction.

Cells Also Have Intracellular Receptors Activated by Signaling Molecules That Pass Through the Plasma Membrane

Although most receptors for signaling molecules are located in the plasma membrane, some are found inside the cell. In these cases, an extracellular signaling molecule must pass through the plasma membrane to gain access to its receptor.

In vertebrates, receptors for steroid hormones are intracellular. As discussed in Chapter 51, steroid hormones, such as estrogens and androgens, are secreted into the bloodstream from cells of endocrine glands. The behavior of estrogen is typical of many steroid hormones (**Figure 9.9**). Because estrogen is hydrophobic, it can diffuse through the plasma membrane of a target cell and bind to a receptor in the cell. Some steroids bind to receptors in the cytosol, which then travel into the nucleus. Other steroid hormones, such as estrogen, bind to receptors already in the nucleus. After binding, the estrogen • receptor complex undergoes a conformational change that enables it to form a dimer with another estrogen • receptor complex. The dimer then binds to the DNA and activates the transcription of specific genes. The estrogen receptor is an

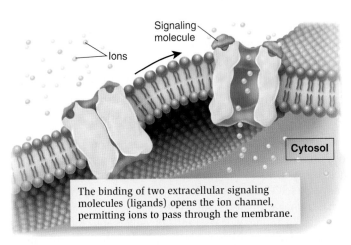

Signaling molecule

Ions

Cytosol

The binding of two extracellular signaling molecules (ligands) opens the ion channel, permitting ions to pass through the membrane.

Figure 9.8   The function of a ligand-gated ion channel.

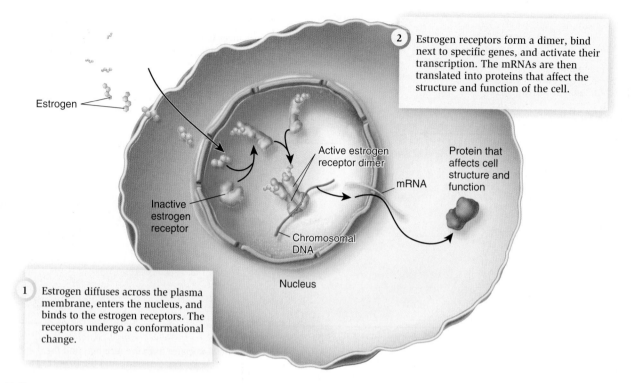

Figure 9.9 Estrogen receptor in mammalian cells.

example of a transcription factor—a protein that regulates the transcription of genes. The expression of specific genes changes cell structure and function in a way that results in a cellular response.

9.3 Signal Transduction and the Cellular Response

We now turn our attention to the intracellular events that enable a cell to respond to a signaling molecule that binds to a cell surface receptor. In most cases, the binding of a signaling molecule to its receptor stimulates a signal transduction pathway. We begin by examining a pathway that is controlled by an enzyme-linked receptor. We will then examine pathways and cellular responses that are controlled by G-protein-coupled receptors. As you will learn, these pathways sometimes involve the production of intracellular signals called second messengers.

Receptor Tyrosine Kinases Activate Signal Transduction Pathways Involving a Protein Kinase Cascade That Alters Gene Transcription

Receptor tyrosine kinases are a category of enzyme-linked receptors that are found in all animals and also in choanoflagellates, which are the protists that are most closely related to animals (see Chapter 32). However, they are not found in bacteria, archaea, or other eukaryotic species. (Bacteria do have receptor histidine kinases, and all eukaryotes have receptor serine/threonine kinases.) The human genome contains about 60 different genes that encode receptor tyrosine kinases that

recognize various types of signaling molecules such as hormones. A type of hormone called a **growth factor** is a protein ligand that acts as a signaling molecule that stimulates cell growth or division.

Figure 9.10 describes a simplified signal transduction pathway for epidermal growth factor (EGF). This protein ligand is secreted from endocrine cells, travels through the bloodstream, and binds to a receptor tyrosine kinase called the EGF receptor. EGF is responsible for stimulating epidermal cells, such as skin cells, to divide. Following receptor activation, the three general parts of the signal transduction pathway are as follows: (1) relay proteins (also called adaptor proteins) activate a protein kinase cascade; (2) the protein kinase cascade phosphorylates proteins in the cell such as transcription factors; and (3) the phosphorylated transcription factors stimulate gene transcription. Next, we will consider the details of this pathway.

EGF Receptor Activation   For receptor activation to occur, two EGF receptor subunits each bind a molecule of EGF. The binding of EGF causes the subunits to dimerize and phosphorylate each other on tyrosines within the receptors themselves, which is why they are named receptor tyrosine kinases. This event is called **autophosphorylation**. Next comes the signal transduction pathway.

Relay Proteins   The phosphorylated form of the EGF receptor is first recognized by a relay protein of the signal transduction pathway called Grb. This interaction changes the conformation of Grb so that it binds to another relay protein in the signal transduction pathway termed Sos, thereby changing the conformation of Sos. The activation of Sos causes a third relay protein

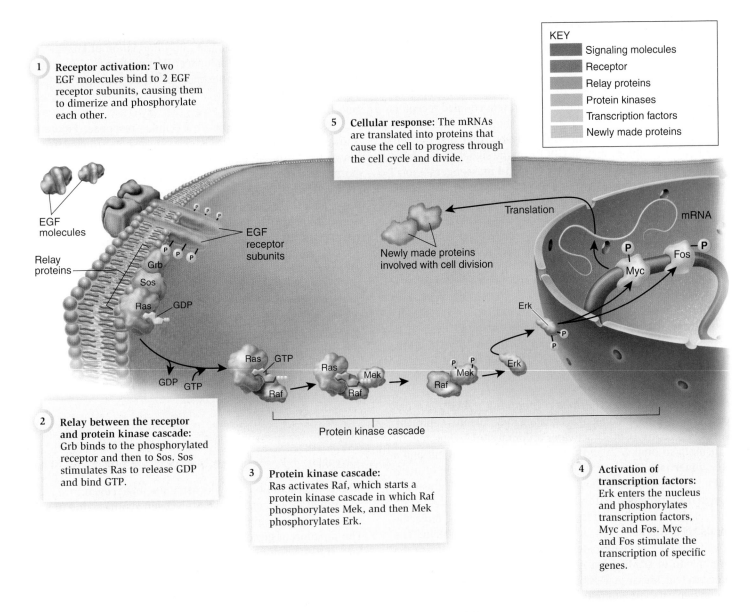

Figure 9.10 The epidermal growth factor (EGF) pathway that promotes cell division.

Concept check: *Certain mutations can alter the structure of the Ras protein so it will not hydrolyze GTP. Such mutations cause cancer. Explain why.*

called Ras to release GDP and bind GTP. The GTP form of Ras is the active form.

Protein Kinase Cascade The function of Grb, Sos, and Ras is to relay a cellular signal to additional proteins in the signal transduction pathway that form a **protein kinase cascade**. This cascade involves the sequential activation of multiple protein kinases. Activated Ras binds to Raf, the first protein kinase in the cascade. Raf then phosphorylates Mek, which becomes active and, in turn, phosphorylates Erk. Raf, Mek, and Erk, the protein kinase cascade, are all examples of **m̲itogen-a̲ctivated p̲rotein k̲inases (MAP-kinases)**. This type of protein kinase was first discovered because it is activated in the presence of mitogens—agents that cause a cell to divide.

Activation of Transcription Factors and the Cellular Response
The phosphorylated form of Erk enters the nucleus and phosphorylates transcription factors such as Myc and Fos, which then activate the transcription of genes involved in cell division. What is the cellular response? Once these transcription factors are phosphorylated, they stimulate the expression of many genes that encode proteins that promote cell division. After these proteins are made, the cell will be stimulated to divide.

Growth factors such as EGF cause a rapid increase in the expression of many genes in mammals, perhaps as many as 100. As we will discuss in Chapter 14, growth factor signaling pathways are often involved in cancer. Mutations that cause proteins in these pathways to become hyperactive result in cells that divide uncontrollably!

Second Messengers Such as Cyclic AMP Are Key Components of Many Signal Transduction Pathways

Let's now turn to examples of signal transduction pathways and cellular responses that involve G-protein-coupled receptors (GPCRs). Cell biologists call signaling molecules that bind to a cell surface receptor the first messengers. After first messengers bind to receptors such as GPCRs, many signal transduction pathways lead to the production of **second messengers**—small molecules or ions that relay signals inside the cell. The signals that result in second messenger production often act quickly, in a matter of seconds or minutes, but their duration is usually short. Therefore, such signaling is typically used when a cell needs a quick and short cellular response.

Production of cAMP   Mammalian and plant cells make several different types of G protein α subunits. One type of α subunit binds to **adenylyl cyclase**, an enzyme in the plasma membrane. This interaction stimulates adenylyl cyclase to synthesize **c**yclic **a**denosine **m**onophosphate (cyclic AMP, or **cAMP**) from ATP (**Figure 9.11**). The molecule cAMP is an example of a second messenger.

Signal Transduction Pathway Involving cAMP   As discussed earlier, the binding of a signaling molecule to a G-protein-coupled receptor (GPCR) activates an intracellular G protein by causing it to bind GTP and dissociate into an α subunit and a β/γ dimer (see Figure 9.7). Let's now follow the role of the α subunit in a signal transduction pathway. **Figure 9.12** illustrates a signal transduction pathway that involves cAMP production and leads to a cellular response. First, a signaling molecule binds to a GPCR, which, in turn, activates a G protein. The α subunit then activates adenylyl cyclase, which catalyzes the production of cAMP from ATP. One effect of cAMP is to activate protein kinase A (PKA), which is composed of four subunits: two catalytic subunits that phosphorylate specific cellular proteins, and two regulatory subunits that inhibit the catalytic subunits when they are bound to each other. Cyclic AMP binds to the regulatory subunits of PKA. The binding of cAMP separates the regulatory and catalytic subunits, which allows each catalytic subunit to be active.

Cellular Response via PKA   How does PKA activation lead to a cellular response? The catalytic subunit of PKA phosphorylates specific cellular proteins such as enzymes, structural proteins, and transcription factors. The phosphorylation of enzymes and structural proteins will influence the structure and function of the cell. Likewise, the phosphorylation of transcription factors leads to the synthesis of new proteins that affect cell structure and function.

As a specific example of a cellular response, **Figure 9.13** shows how a skeletal muscle cell can respond to elevated levels of the hormone epinephrine (also called adrenaline). This hormone is sometimes called the "fight or flight" hormone. Epinephrine is produced when an individual is confronted with a stressful situation and helps the individual deal with that situation. Epinephrine binds to a GPCR, leading to an increase in cAMP, which, in turn, activates PKA. In skeletal muscle cells, PKA phosphorylates two enzymes—phosphorylase kinase and glycogen synthase. Both of these enzymes are involved with the metabolism of glycogen, which is a polymer of glucose used to store energy. When phosphorylase kinase is phosphorylated, it becomes activated. The function of phosphorylase kinase is to phosphorylate another enzyme in the cell called glycogen phosphorylase, which then becomes activated. This enzyme causes glycogen breakdown by phosphorylating glucose units at the ends of a glycogen polymer, which releases individual glucose molecules from glycogen:

$$\text{Glycogen}_n + P_i \xrightarrow{\begin{array}{c}\text{Glycogen}\\ \text{phosphorylase}\end{array}} \text{Glycogen}_{n-1} + \text{Glucose-phosphate}$$

where n is the number of glucose units in glycogen.

When PKA phosphorylates glycogen synthase, the function of this enzyme is inhibited rather than activated (Figure 9.13). The function of glycogen synthase is to make glycogen. Therefore, the effect of cAMP is to prevent glycogen synthesis.

Taken together, the effects of epinephrine in skeletal muscle cells are to stimulate glycogen breakdown and inhibit glycogen synthesis. This provides these cells with more glucose molecules, which they can use for the energy needed for muscle contraction. In this way, the individual is better prepared to fight or flee.

Reversal of the Cellular Response   As mentioned, signaling that involves second messengers is typically of short duration. When the signaling molecule is no longer produced and its level falls, a larger percentage of the receptors are not bound by their

Figure 9.11   The synthesis and breakdown of cyclic AMP.

Figure 9.12 A signal transduction pathway involving cAMP. The pathway leading to the formation of cAMP and subsequent activation of PKA, which is mediated by a G-protein-coupled receptor (GPCR).

Concept check: In this figure, which part is the signal transduction pathway, and which is the cellular response?

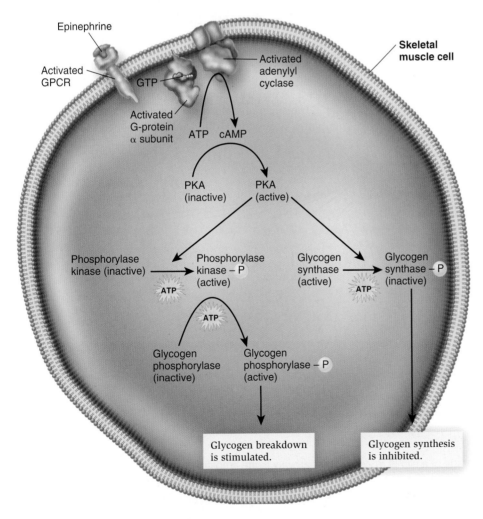

Figure 9.13 The cellular response of a skeletal muscle cell to epinephrine.

Concept check: Explain whether phosphorylation activates or inhibits enzyme function.

ligands. When a ligand dissociates from the GPCR, the GPCR becomes deactivated. Intracellularly, the α subunit hydrolyzes its GTP to GDP, and the α subunit and β/γ dimer reassociate to form an inactive G protein (see step 3, Figure 9.7). The level of cAMP decreases due to the action of an enzyme called **phosphodiesterase**, which converts cAMP to AMP.

As the cAMP level falls, the regulatory subunits of PKA release cAMP, and the regulatory and catalytic subunits reassociate, thereby inhibiting PKA. Finally, enzymes called **protein phosphatases** are responsible for removing phosphate groups from proteins, which reverses the effects of PKA.

The Main Advantages of Second Messengers Are Amplification and Speed

In the 1950s, Earl Sutherland determined that many different hormones cause the formation of cAMP in a variety of cell types. This observation, for which he won the Nobel Prize in 1971, stimulated great interest in the study of signal transduction pathways. Since Sutherland's discovery, the production of second messengers such as cAMP has been found to have two important advantages: amplification and speed.

Amplification of the signal involves the synthesis of many cAMP molecules, which, in turn, activate many PKA proteins (**Figure 9.14**). Likewise, each PKA protein can phosphorylate many target proteins in the cell to promote a cellular response.

A second advantage of second messengers such as cAMP is speed. Because second messengers are relatively small, they can diffuse rapidly through the cytosol. For example, Brian Bacskai and colleagues studied the response of nerve cells to a signaling molecule called serotonin, which is a neurotransmitter that binds to a GPCR. In humans, serotonin is believed to play a role in depression, anxiety, and sexual drive. To monitor cAMP levels, nerve cells grown in a laboratory were injected with a fluorescent protein that changes its fluorescence when cAMP is made. As shown in the right micrograph in **Figure 9.15**, such cells made a substantial amount of cAMP within 20 seconds after the addition of serotonin.

Signal Transduction Pathways May Also Lead to Second Messengers, Such as Diacylglycerol and Inositol Trisphosphate, and Alter Ca²⁺ Levels

Cells use several different types of second messengers, and more than one type may be used at the same time. Let's now consider a second way that an activated G protein can influence a signal transduction pathway and produce second messengers. This pathway produces the second messengers diacylglycerol

Figure 9.14 Signal amplification. An advantage of a signal transduction pathway is the amplification of a signal. In this case, a single signaling molecule can lead to the phosphorylation of many, perhaps hundreds or thousands of, target proteins.

Concept check: In the case of signaling pathways involving hormones, why is signal amplification an advantage?

Add serotonin

+ 20 seconds

Figure 9.15 **The rapid speed of cAMP production.** The micrograph on the left shows a nerve cell prior to its exposure to serotonin; the micrograph on the right shows the same cell 20 seconds after exposure. Blue indicates a low level of cAMP, yellow is an intermediate level, and red/purple is a high level.

(DAG) and inositol trisphosphate (IP$_3$) and ultimately can cause cellular effects by altering the levels of calcium in the cell.

Production of DAG and IP$_3$ To start this pathway, a signaling molecule binds to its GPCR, which, in turn, activates a G protein. However, rather than activating adenylate cyclase as described earlier in Figure 9.12, the α subunit of this G protein activates an enzyme called phospholipase C (**Figure 9.16**). When phospholipase C becomes active, it breaks a covalent bond in a particular plasma membrane phospholipid with an inositol head group, producing the two second messengers DAG and IP$_3$.

Release of Ca^{2+} into the Cytosol Due to active transport via a Ca^{2+}-ATPase, the lumen of the ER contains a very high concentration of Ca^{2+} compared to the cytosol. After IP$_3$ is released into the cytosol, it binds to a ligand-gated Ca^{2+} channel in the ER membrane. The binding of IP$_3$ causes the channel to open, releasing Ca^{2+} into the cytosol. Therefore, this pathway also involves calcium ions, which act as a second messenger. Calcium ions can elicit a cellular response in a variety of ways, two of which are shown in Figure 9.16 and described next.

Cellular Response via Protein Kinase C Ca^{2+} can bind to protein kinase C (PKC), which, in combination with DAG, activates the kinase. Once activated, PKC can phosphorylate specific cellular proteins, thereby altering their function and leading to a cellular response. In smooth muscle cells, for example, protein kinase C phosphorylates proteins that are involved with contraction.

Cellular Response via Calmodulin Ca^{2+} also can bind to a protein called calmodulin, which is a calcium-modulated protein. The Ca^{2+}-calmodulin complex can then interact with specific cellular proteins and alter their functions. For example, calmodulin regulates proteins involved in carbohydrate breakdown in liver cells.

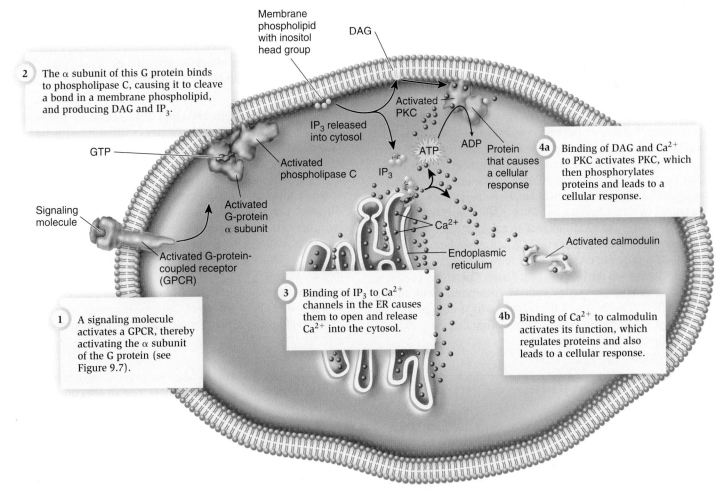

Figure 9.16 A signal transduction pathway involving diacylglycerol (DAG), inositol trisphosphate (IP$_3$), and changing Ca^{2+} levels.

9.4 Hormonal Signaling in Multicellular Organisms

Thus far, we have considered how signaling molecules bind to particular types of receptors, thereby activating a signal transduction pathway that leads to a cellular response. In this section, we will consider the effects of signaling molecules in multicellular organisms that have a variety of cell types. As you will learn, the type of cellular response that is caused by a given signaling molecule depends on the type of cell that is responding to the signal. Each cell type responds to a particular signaling molecule in its own unique way. The variation in a cellular response is determined by the types of proteins, such as receptors and signal transduction proteins, that each cell type makes.

The Cellular Response to a Given Hormone Can Vary Among Different Cell Types

As we have seen, signaling molecules usually exert their effects on cells via signal transduction pathways that control the functions and/or synthesis of specific proteins. In multicellular organisms, one of the amazing effects of hormones is their ability to coordinate cellular activities. One example is epinephrine, which is secreted from endocrine cells. As mentioned, epinephrine is also called the fight-or-flight hormone because it quickly prepares the body for strenuous physical activity. Epinephrine is also secreted into the bloodstream when someone is exercising vigorously.

Epinephrine has different effects throughout the body (**Figure 9.17**). We have already discussed how it promotes the breakdown of glycogen in skeletal muscle cells. In the lungs, it relaxes the airways, allowing a person to take in more oxygen. In the heart, epinephrine stimulates heart muscle cells so the heart beats faster. Interestingly, one of the effects of caffeine can be explained by this mechanism. Caffeine inhibits phosphodiesterase, which converts cAMP to AMP. Phosphodiesterase functions to remove cAMP once a signaling molecule, such as epinephrine, is no longer present. When phosphodiesterase is inhibited by caffeine, cAMP persists for a longer period of time and thereby causes the heart to beat faster. Therefore, even low levels of signaling molecules such as epinephrine will have a greater effect. This is one of the reasons why drinks containing caffeine, including coffee and many energy drinks, provide a feeling of vitality and energy.

Genomes & Proteomes Connection

A Cell's Response to Hormones and Other Signaling Molecules Depends on the Proteins It Makes

As Figure 9.17 shows, a hormone such as epinephrine produces diverse responses throughout the body. How do we explain the

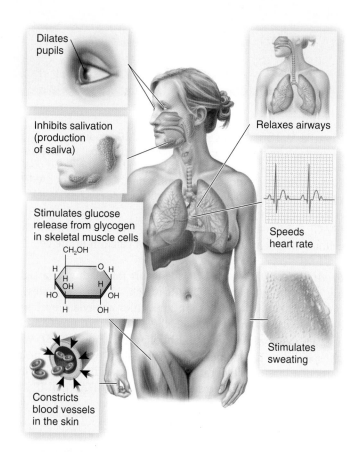

Figure 9.17   The effects of epinephrine in humans. This hormone prepares the body for fight or flight.

observation that various cell types can respond so differently to the same hormone? The answer lies in **differential gene regulation**. As a multicellular organism develops from a fertilized egg, the cells of the body become differentiated into particular types, such as heart and lung cells. The mechanisms that underlie this differentiation process are described in Chapter 19. Although different cell types, such as heart and lung cells, contain the same set of genes—the same genome—they are not expressed in the same pattern. Certain genes that are turned off in heart cells are turned on in lung cells, whereas some genes that are turned on in heart cells are turned off in lung cells. This causes each cell type to have its own distinct proteome. The set of proteins made in any given cell type is critical to a cell's ability to respond to signaling molecules. The following are examples of how differential gene regulation affects the cellular response:

1. *A cell may or may not express a receptor for a particular signaling molecule.* For example, not all cells of the human body express a receptor for epinephrine. These cells are not affected when epinephrine is released into the bloodstream.

2. *Different cell types have different cell surface receptors that recognize the same signaling molecule.* In humans, for example, a signaling molecule called acetylcholine has two different types of receptors. One acetylcholine receptor is a

ligand-gated ion channel that is expressed in skeletal muscle cells. Another acetylcholine receptor is a G-protein-coupled receptor (GPCR) that is expressed in heart muscle cells. Because of this, acetylcholine activates different signal transduction pathways in skeletal and heart muscle cells. Therefore, these cells respond differently to acetylcholine.

3. *Two (or more) receptors may work the same way in different cell types but have different affinities for the same signaling molecule.* For example, two different GPCRs may recognize the same hormone, but the receptor expressed in liver cells may have a higher affinity (that is, a lower K_d) for the hormone than does a receptor expressed in muscle cells. In this case, liver cells will respond to a lower hormone concentration than muscle cells will.

4. *The expression of proteins involved in intracellular signal transduction pathways may vary in different cell types.* For example, one cell type may express the proteins that are needed to activate PKA, while another cell type may not.

5. *The expression of proteins that are controlled by signal transduction pathways may vary in different cell types.* For example, the presence of epinephrine in skeletal muscle

cells leads to the activation of glycogen phosphorylase, an enzyme involved in glycogen breakdown. However, this enzyme is not expressed in all cells of the body. Glycogen breakdown will only be stimulated by epinephrine if glycogen phosphorylase is expressed in that cell.

9.5 Apoptosis: Programmed Cell Death

We will end our discussion of cell communication by considering one of the most dramatic responses that eukaryotic cells exhibit—**apoptosis**, or programmed cell death. During this process, a cell orchestrates its own destruction! The cell first shrinks and forms a rounder shape due to the internal destruction of its nucleus and cytoskeleton (**Figure 9.18**). The plasma membrane then forms irregular extensions that eventually become blebs—small cell fragments that break away from the cell as it destroys itself. In this section, we will examine the pioneering work that led to the discovery of apoptosis and explore its molecular mechanism.

| 1 Cell beginning apoptosis | 2 Condensation of nucleus and cell shrinkage | 3 Multiple extensions of the plasma membrane | 4 Further blebbing |

Figure 9.18 Stages of apoptosis.

FEATURE INVESTIGATION

Kerr, Wyllie, and Currie Found That Hormones May Control Apoptosis

How was this process discovered? One line of evidence involved the microscopic examination of tissues in mammals. In the 1960s, British pathologist John Kerr microscopically examined liver tissue that was deprived of oxygen. Within hours of oxygen deprivation, he observed that some cells underwent a process that involved cell shrinkage. Around this time, similar results had been noted by other researchers, such as Scottish pathologists Andrew Wyllie and Alastair Currie, who had stud-

ied cell death in the adrenal glands. In 1973, Kerr, Wyllie, and Currie joined forces to study this process further.

Prior to their collaboration, other researchers had already established that certain hormones affect the growth of the adrenal glands, which sit atop the kidneys. Adrenocorticotropic hormone (ACTH) was known to increase the number of cells in the adrenal cortex, which is the outer layer of the adrenal glands. By contrast, prednisolone was shown to suppress the synthesis of ACTH and cause a decrease in the number of cells in the cortex. In the experiment described in **Figure 9.19**, Kerr, Wyllie, and Currie wanted to understand how these hormones

Figure 9.19   Discovery of apoptosis in the adrenal cortex by Kerr, Wyllie, and Currie.

HYPOTHESIS Hormones may affect cell number in the adrenal gland by controlling the rate of apoptosis.

KEY MATERIALS Laboratory rats, prednisolone, and ACTH.

Experimental level | Conceptual level

1   Inject 5 rats with saline (control).
Inject 5 rats with prednisolone alone.
Inject 5 rats with prednisolone plus ACTH.
Inject 5 rats with ACTH alone.

Previous studies indicated that prednisolone alone may promote apoptosis by lowering ACTH levels.

2   After 2 days, obtain samples of adrenal tissue from all 20 rats.

Adrenal gland

Cell undergoing apoptosis

3   Observe the samples via light microscopy, described in Chapter 4.

4   **THE DATA**

Micrograph of adrenal tissue showing occasional cells undergoing apoptosis (see arrow)

13.9 μm

Treatment	Number of animals	Glands with enhanced apoptosis*/ Total number of animals
Saline	5	0/10
Prednisolone	5	9/10
Prednisolone + ACTH	5	0/10
ACTH	5	0/10

*Samples from two adrenal glands were removed from each animal. Enhanced apoptosis means that cells undergoing apoptosis were observed in every sample under the light microscope.

5   **CONCLUSION** Prednisolone alone, which lowers ACTH levels, causes some cells to undergo apoptosis. During this process, the cells shrink and form blebs as they kill themselves. Apoptosis is controlled by hormones.

6   **SOURCE** Wyllie, A.H., Kerr, J.F.R., Macaskill, I.A.M., and Currie, A.R. 1973. Adrenocortical cell deletion: the role of ACTH. *Journal of Pathology* 111:85–94.

exert their effects. They subjected rats to four types of treatments. The control rats were injected with saline (salt water). Other rats were injected with prednisolone alone, prednisolone plus ACTH, or ACTH alone. After two days, samples of adrenal cortex were obtained from the rats and observed by light microscopy. Even in control samples, the researchers occasionally observed cell death via apoptosis (see micrograph under The Data). However, in prednisolone-treated rats, the cells in the adrenal cortex were found to undergo a dramatically higher rate of apoptosis. Multiple cells undergoing apoptosis were found in 9 out of every 10 samples observed under the light microscope. Such a high level of apoptosis was not observed in control samples or in samples obtained from rats treated with both prednisolone and ACTH or ACTH alone.

The results of Kerr, Wyllie, and Currie are important for two reasons. First, their results indicated that tissues decrease their cell number via a mechanism that involves cell shrinkage and eventually blebbing. Second, they showed that cell death could follow a program that, in this case, was induced by the presence of prednisolone (which decreases ACTH). They coined the term apoptosis to describe this process.

Experimental Questions

1. In the experiment of Figure 9.19, explain the effects on apoptosis in the control rats (saline injected) versus those injected with prednisolone alone, predinisolone + ACTH, or ACTH alone.

2. Prednisolone inhibits the production of ACTH in rats. Do you think it inhibited the ability of rats to make their own ACTH when they were injected with both prednisolone and ACTH? Explain.

3. Of the four groups—control, prednisolone alone, prednisolone + ACTH, and ACTH alone—which would you expect to have the lowest level of apoptosis? Explain.

Intrinsic and Extrinsic Signal Transduction Pathways Lead to Apoptosis

Since these early studies on apoptosis, cell biologists have discovered that apoptosis plays many important roles. During embryonic development in animals, it is needed to sculpt the tissues and organs. For example, the fingers on a human hand, which are initially webbed, become separated during embryonic development when the cells between the fingers are programmed to die (see Chapter 19, Figure 19.4). Apoptosis is also necessary in adult organisms to maintain the proper cell number in tissues and organs. This process also eliminates cells that have become worn out, infected by viruses or intracellular bacteria, or have the potential to cause cancer. In mammals, apoptosis is also important in the proper functioning of the immune system, which wards off infections. The immune system is composed of a variety of cell types, such as B cells and T cells, that can fight infectious agents and eliminate damaged cells. For this to occur, the immune system creates a large pool of B and T cells and then uses apoptosis to weed out those that are potentially damaging to the body or ineffective at fighting infection.

Apoptosis involves the activation of cell signaling pathways. One pathway, called the intrinsic or mitochondrial pathway, is stimulated by internal signals, such as DNA damage that could cause cancer. Proteins on the surface of the mitochondria play a key role in eliciting the response. Alternatively, extracellular signals can promote apoptosis. This is called the extrinsic or death receptor pathway.

Let's consider how an extracellular signal causes apoptosis. The extrinsic pathway of apoptosis begins with the activation of **death receptors** on the surface of the cell. Death receptors, such as Fas, stimulate a pathway that leads to apoptosis when they become bound to an extracellular ligand. **Figure 9.20** shows a simplified pathway for this process. In this example, the extracellular ligand is a protein composed of three identical subunits—a trimeric protein. Such trimeric ligands that promote cell death are typically produced on the surface of cells of the immune system that recognize abnormal cells and target them for destruction. For example, when a cell is infected with a virus, cells of the immune system may target the infected cell for apoptosis. The trimeric ligand binds to three death receptors, which causes them to aggregate into a trimer. This results in a conformational change that exposes the death domain in the cytosol. Once the death domain is exposed, it binds to an adaptor, such as FADD, which then binds to a procaspase. (FADD is an abbreviation for Fas-associated protein with death domain.) The complex between the death receptors, FADD, and procaspase is called the **death-inducing signaling complex (DISC)**.

Once the procaspase, which is inactive, is part of the death-inducing signaling complex, it is converted by proteolytic cleavage to caspase, which is active. An active **caspase** functions as a protease—an enzyme that digests other proteins. After it is activated, the caspase is then released from the DISC. This caspase is called an initiator caspase because it initiates the activation of many other caspases in the cell. These other caspases are called executioner or effector caspases because they are directly responsible for digesting intracellular proteins and causing the cell to die. The executioner caspases digest a variety of intracellular proteins, including the proteins that constitute the cytoskeleton and nuclear lamina as well as proteins involved with DNA replication and repair. In this way, the executioner caspases cause the cellular changes described earlier in Figure 9.18. The caspases also activate an enzyme called DNase that chops the DNA in the cell into small fragments. This event may be particularly important for eliminating virally infected cells because it will also destroy viral genomes that are composed of DNA.

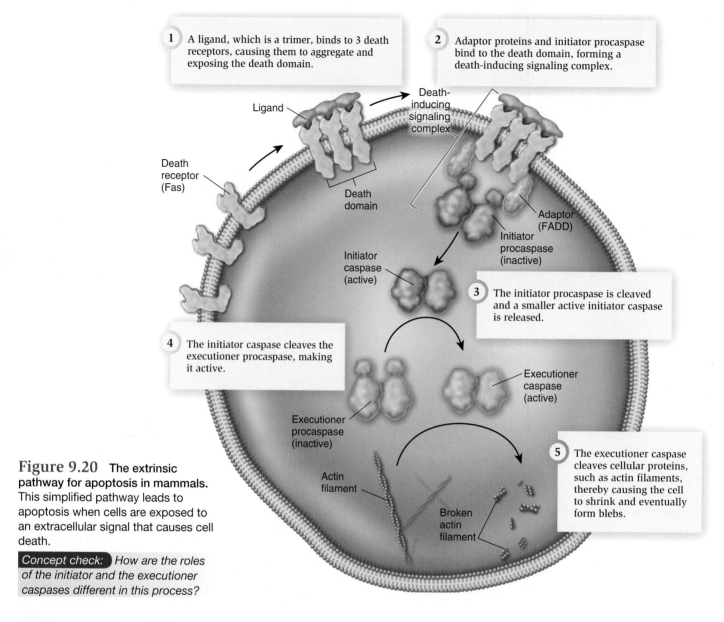

1 A ligand, which is a trimer, binds to 3 death receptors, causing them to aggregate and exposing the death domain.

2 Adaptor proteins and initiator procaspase bind to the death domain, forming a death-inducing signaling complex.

Ligand

Death-inducing signaling complex

Death receptor (Fas)

Death domain

Adaptor (FADD)

Initiator procaspase (inactive)

Initiator caspase (active)

3 The initiator procaspase is cleaved and a smaller active initiator caspase is released.

4 The initiator caspase cleaves the executioner procaspase, making it active.

Executioner caspase (active)

Executioner procaspase (inactive)

Actin filament

Broken actin filament

5 The executioner caspase cleaves cellular proteins, such as actin filaments, thereby causing the cell to shrink and eventually form blebs.

Figure 9.20 **The extrinsic pathway for apoptosis in mammals.** This simplified pathway leads to apoptosis when cells are exposed to an extracellular signal that causes cell death.

Concept check: *How are the roles of the initiator and the executioner caspases different in this process?*

Summary of Key Concepts

9.1 General Features of Cell Communication

- A signal is an agent that can influence the properties of cells. Cell signaling is needed so that cells can sense and respond to environmental changes and communicate with each other.

- When a cell responds to an environmental signal, it has become adapted to its environment. (Figure 9.1)

- Cell-to-cell communication also allows cells to adapt, as when plants grow toward light. (Figure 9.2)

- Cell-to-cell communication can vary in the mechanism and distance that a signal travels. Signals are relayed between cells in five common ways: direct intercellular, contact-dependent, autocrine, paracrine, and endocrine signaling. (Figure 9.3)

- Cell communication is usually a three-stage process involving receptor activation, signal transduction, and a cellular response. A signal transduction pathway is a group of proteins that convert an initial signal to a different signal inside the cell. (Figure 9.4)

9.2 Cellular Receptors and Their Activation

- A signaling molecule, also called a ligand, binds to a receptor with an affinity that is measured as a K_d value. The binding of a ligand to a receptor is usually very specific and alters the conformation of the receptor. (Figure 9.5)

- Most receptors involved in cell signaling are found on the cell surface.

- Enzyme-linked receptors have some type of catalytic function. Many of them are protein kinases that can phosphorylate proteins. (Figure 9.6)

- G-protein-coupled receptors (GPCRs) interact with G proteins to initiate a cellular response. (Figure 9.7)

- Some receptors are ligand-gated ion channels that allow the flow of ions across cellular membranes. (Figure 9.8)

- Some receptors, such as the estrogen receptor, are intracellular receptors. (Figure 9.9)

9.3 Signal Transduction and the Cellular Response

- Signaling pathways influence whether or not a cell will divide. An example is the pathway that is stimulated by epidermal growth factor, which binds to a receptor tyrosine kinase. (Figure 9.10)

- Second messengers, such as cAMP, play a key role in signal transduction pathways, such as those that occur via G-protein-coupled receptors. These pathways are reversible once the signal is degraded. (Figures 9.11, 9.12)

- An example of a pathway that uses cAMP is found in skeletal muscle cells. In these cells, epinephrine enhances the function of enzymes that increase glycogen breakdown and inhibits enzymes that cause glycogen synthesis. (Figure 9.13)

- Second messenger pathways amplify the signal and occur with great speed. (Figures 9.14, 9.15)

- Diacylglycerol (DAG), inositol trisphosphate (IP_3), and Ca^{2+} are other examples of second messengers involved in signal transduction. (Figure 9.16)

9.4 Hormonal Signaling in Multicellular Organisms

- Hormones such as epinephrine exert different effects throughout the body. (Figure 9.17)

- The way in which any particular cell responds to a signaling molecule depends on the types of proteins it makes. These include the types of receptors, proteins involved in signaling transduction pathways, and proteins that carry out the cellular response. The amounts of these proteins are controlled by differential gene regulation.

9.5 Apoptosis: Programmed Cell Death

- Apoptosis is the process of programmed cell death in which the nucleus and cytoskeleton break down, and eventually the cell breaks apart into blebs. (Figure 9.18)

- Microscopy studies of Kerr, Wyllie, and Currie, in which they studied the effects of hormones on the adrenal cortex, were instrumental in the identification of apoptosis. (Figure 9.19)

- Apoptosis plays many important roles in multicellular organisms, including the sculpting of tissues and organs during embryonic development, maintaining the proper cell number in tissues and organs, eliminating cells that have become worn out or have the potential to cause cancer, and the proper functioning of the immune system.

- Apoptosis can occur via intrinsic or extrinsic pathways. (Figure 9.20)

Assess and Discuss

Test Yourself

1. The ability of a cell to respond to changes in its environment is termed
   a. signaling.
   b. apoptosis.
   c. irritability.
   d. adaptation.
   e. stimulation.

2. When a cell secretes a signaling molecule that binds to receptors on neighboring cells as well as the same cell, this is called _____ signaling.
   a. direct intercellular
   b. contact-dependent
   c. autocrine
   d. paracrine
   e. endocrine

3. Which of the following does not describe a typical cellular response to signaling molecules?
   a. activation of enzymes within the cell
   b. change in the function of structural proteins, which determine cell shape
   c. alteration of levels of certain proteins in the cell by changing the level of gene expression
   d. change in a gene sequence that encodes a particular protein
   e. All of the above are examples of cellular responses.

4. A receptor has a K_d for its ligand of 50 nM. This receptor
   a. has a higher affinity for its ligand compared to a receptor with a K_d of 100 nM.
   b. has a higher affinity for its ligand compared to a receptor with a K_d of 10 nM.
   c. will be mostly bound by its ligand when the ligand concentration is 100 nM.
   d. must be an intracellular receptor.
   e. both a and c

5. _____ binds to receptors inside cells.
   a. Estrogen
   b. Epinephrine
   c. Epidermal growth factor
   d. All of the above
   e. None of the above

6. Small molecules, such as cAMP, that relay signals within the cell are called
   a. secondary metabolites.
   b. ligands.
   c. G proteins.
   d. second messengers.
   e. transcription factors.

7. The benefit of second messengers in signal transduction pathways is
   a. an increase in the speed of a cellular response.
   b. duplication of the ligands in the system.
   c. amplification of the signal.
   d. all of the above.
   e. a and c only.

8. All cells of a multicellular organism may not respond in the same way to a particular ligand (signaling molecule) that binds to a cell surface receptor. The difference in response may be due to
   a. the type of receptor for the ligand that the cell expresses.
   b the affinity of the ligand for the receptor in a given cell type.
   c. the type of signal transduction pathways that the cell expresses.
   d. the type of target proteins that the cell expresses.
   e. all of the above.

9. Apoptosis is the process of
   a. cell migration.
   b. cell signaling.
   c. signal transduction.
   d. signal amplification.
   e. programmed cell death.

10. Which statement best describes the extrinsic pathway for apoptosis?
   a. Caspases recognize an environmental signal and expose their death domain.
   b. Death receptors recognize an environmental signal which then leads to the activation of caspases.
   c. Initiator caspases digest the nuclear lamina and cytoskeleton.
   d. Executioner caspases are part of the death-inducing signaling complex (DISC).
   e. all of the above

Conceptual Questions

1. What are the two general reasons that cells need to communicate?

2. What are the three stages of cell signaling? What stage does not occur when the estrogen receptor is activated?

3. What would be some of the harmful consequences if apoptosis did not occur?

Collaborative Questions

1. Discuss and compare several different types of cell-to-cell communication. What are some advantages and disadvantages of each type?

2. How does differential gene regulation enable various cell types to respond differently to the same signaling molecule? Why is this useful to multicellular organisms?

Online Resource

www.brookerbiology.com

Stay a step ahead in your studies with animations that bring concepts to life and practice tests to assess your understanding. Your instructor may also recommend the interactive ebook, individualized learning tools, and more.

Multicellularity

10

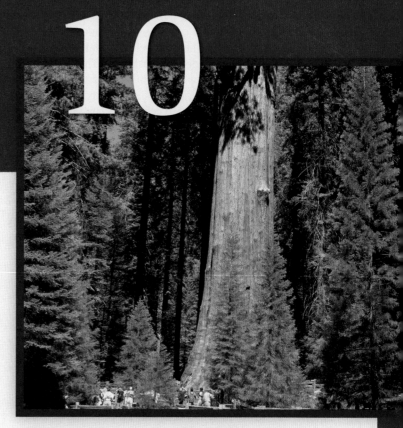

The General Sherman tree in Sequoia National Park, a striking example of the size that multicellular organisms can reach. This tree is thought to be the largest organism (by mass) in the world.

What is the largest living organism on Earth? The size of an organism can be defined by its volume, mass, height, length, or the area it occupies. A giant fungus (*Armillaria ostoyae*), growing in the soil in the Malheur National Forest in Oregon, spans 8.9 km^2, or 2,200 acres, which makes it the largest single organism by area. In the Mediterranean Sea, marine biologists discovered a giant aquatic plant (*Posidonia oceanica*) that is 8 km or 4.3 miles in length, making it the longest organism. With regard to mass, the largest organism is probably a tree named the General Sherman tree that is 83.8 meters tall (275 feet), nearly the length of a football field (see chapter-opening photo). This giant sequoia tree (*Sequoiadendron giganteum*) is estimated to weigh nearly 2 million kg (over 2,000 tons)—equivalent to a herd of 400 elephants!

An organism composed of more than one cell is said to be **multicellular**. The preceding examples illustrate the amazing sizes that certain multicellular organisms have achieved. As we will discuss in Chapter 22, multicellular organisms came into being approximately 1 billion years ago. Some species of protists are multicellular, as are most species of fungi. In this chapter, we will focus on plants and animals, which are always multicellular organisms.

The main benefit of multicellularity arises from the division of labor between different types of cells in an organism. For example, the intestinal cells of animals and the root cells of plants have become specialized for nutrient uptake. Other types of cells in a multicellular organism perform different roles, such as reproduction. In animals, most of the cells of the body—somatic cells—are devoted to the growth, development, and survival of the organism, while specialized cells—gametes—function in sexual reproduction.

Multicellular species usually have much larger genomes than unicellular species. The increase in genome size is associated with an increase in proteome size—multicellular organisms produce a larger array of proteins than do unicellular species. The additional proteins play a role in three general phenomena. First, in a multicellular organism, cell communication is vital for the proper organization and functioning of cells. Many more proteins involved in cell communication are made in multicellular species. Second, both the arrangement of cells within the body and the attachment of cells to each other require a greater variety of proteins in multicellular species than in unicellular species. Finally, additional proteins play a role

in cell specialization because proteins that are needed for the structure and function of one cell type may not be needed in a different cell type, and vice versa. Likewise, additional proteins are needed to regulate the expression of genes so these proteins are expressed in the proper cell types.

In this chapter, we consider characteristics specific to the cell biology of multicellular organisms. We will begin by exploring the material that is produced by animal and plant cells to form an extracellular matrix or cell wall, respectively. This material plays many important roles in the structure, organization, and functioning of cells within multicellular organisms. We will then turn our attention to cell junctions, specialized structures that enable cells to make physical contact with one another. Cells within multicellular organisms form junctions that help to make a cohesive and well-organized body. Finally, we examine the organization and function of tissues, groups of cells that have a similar structure and function. In this chapter, we will survey the general features of tissues from a cellular perspective. Units VI and VII will explore the characteristics of plant and animal tissues in greater detail.

10.1   Extracellular Matrix and Cell Walls

Organisms are not composed solely of cells. A large portion of an animal or plant consists of a network of material that is secreted from cells and forms a complex meshwork outside of cells. In animals, this is called the **extracellular matrix (ECM)**, whereas plant cells are surrounded by a cell wall. The ECM and cell walls are a major component of certain parts of animals and plants, respectively. For example, bones and cartilage in animals and the woody portions of plants are composed largely of ECM and cell walls, respectively. Although the cells within wood eventually die, the cell walls they have produced provide a rigid structure that can support the plant for years or even centuries.

Over the past few decades, cell biologists have examined the synthesis, composition, and function of the ECM in animals and the cell walls in plants. In this section, we will begin by examining the structure and role of the ECM in animals, focusing on the functions of the major ECM components, proteins and polysaccharides. We will then explore the cell wall of plant cells and consider how it differs in structure and function from the ECM of animal cells.

The Extracellular Matrix in Animals Supports and Organizes Cells and Plays a Role in Cell Signaling

Unlike the cells of bacteria, fungi, and plants, the cells of animals are not surrounded by a rigid cell wall that provides structure and support. However, animal cells secrete materials that form an extracellular matrix that also provides support and helps to organize cells. Certain animal cells are completely embedded within an extensive ECM, whereas other cells may adhere to the ECM on only one side. **Figure 10.1** illustrates the general features of the ECM and its relationship to cells. The major macromolecules of the ECM are proteins and polysaccharides. The most abundant proteins are those that form large fibers. The polysaccharides give the ECM surrounding animal cells a gel-like character.

As we will see, the ECM found in animals performs many important roles, including strength, structural support, organization, and cell signaling.

- **Strength:** The ECM is the "tough stuff" of animals' bodies. In the skin of mammals, the strength of the ECM prevents tearing. The ECM found in cartilage resists compression and provides protection to the joints. Similarly, the ECM protects the soft parts of the body, such as the internal organs.

- **Structural support:** The bones of many animals are composed primarily of ECM. Skeletons not only provide structural support but also facilitate movement via the functioning of attached muscles.

- **Organization:** The attachment of cells to the ECM plays a key role in the proper arrangement of cells throughout

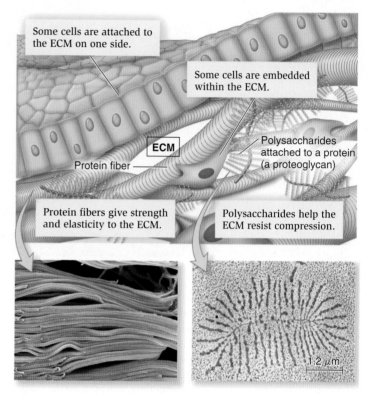

Figure 10.1   The extracellular matrix (ECM) of animal cells. The micrograph (SEM) at the bottom left shows collagen fibers. The micrograph (TEM) at the bottom right shows a proteoglycan.

Concept check: *What are the four functions of the ECM in animals?*

the body. In addition, the ECM binds many body parts together, such as tendons to bones.

- **Cell signaling:** A newly discovered role of the ECM is cell signaling. One way that cells in multicellular organisms sense their environment is via changes in the ECM.

Let's now consider the synthesis and structure of ECM components found in animals.

Adhesive and Structural Proteins Are Major Components of the ECM of Animals

In the 1850s, German biologist Rudolf Virchow suggested that all extracellular materials are made and secreted by cells. Around the same time, biologists realized that gelatin and glue, which are produced by the boiling of animal tissues, must contain a common fibrous substance. This substance was named **collagen** (from the Greek, meaning glue producing). Since that time, the advent of experimental techniques in chemistry, microscopy, and biophysics has enabled scientists to probe the structure of the ECM. We now understand that the ECM contains a mixture of several different components, including proteins such as collagen, that form fibers.

The proteins found in the ECM can be grouped into adhesive proteins, such as fibronectin and laminin, and structural

Table 10.1		Proteins in the ECM of Animals
General type	**Example**	**Function**
Adhesive	Fibronectin	Connects cells to the ECM and helps to organize components in the ECM.
	Laminin	Connects cells to the ECM and helps to organize components in the basal lamina, a specialized ECM found next to epithelial cells (described in Section 10.3).
Structural	Collagen	Forms large fibers and interconnected fibrous networks in the ECM. Provides tensile strength.
	Elastin	Forms elastic fibers in the ECM that can stretch and recoil.

1. Procollagen polypeptides are synthesized into the ER lumen, where they assemble into a triple helix.

Procollagen polypeptide (α chain)

ER lumen

Procollagen triple helix

Extension sequences

Collagen molecule

2. Procollagen is secreted from the cell, and the extension sequences are removed. The protein is now called collagen.

3. The removal of extension sequences allows collagen to assemble into fibrils.

Collagen fibril

4. Collagen fibrils assemble into larger collagen fibers.

Collagen fiber

Figure 10.2 Formation of collagen fibers. Collagen is one type of structural protein found in the ECM of animal cells.

Concept check: *What prevents large collagen fibers from forming intracellularly?*

proteins, such as collagen and elastin (**Table 10.1**). How do adhesive proteins work? Fibronectin and laminin have multiple binding sites that bind to other components in the ECM, such as protein fibers and polysaccharides. These same proteins also have binding sites for receptors on the surfaces of cells. Therefore, adhesive proteins are so named because they adhere ECM components together and to the cell surface. They provide organization to the ECM and facilitate the attachment of cells to the ECM.

Structural proteins, such as collagen and elastin, form large fibers that give the ECM its strength and elasticity. A key function of collagen is to impart tensile strength, which is a measure of how much stretching force a material can bear without tearing apart. Collagen provides high tensile strength to many parts of the animal body. Collagen is the main protein found in bones, cartilage, tendons, and skin and is also found lining blood vessels and internal organs. In mammals, more than 25% of the total protein mass consists of collagen, much more than any other protein. Approximately 75% of the protein in mammalian skin is composed of collagen. Leather is largely a pickled and tanned form of collagen.

Figure 10.2 depicts the synthesis and assembly of collagen. As described in Chapter 4, proteins, such as collagen, that are secreted from eukaryotic cells are first directed to the endoplasmic reticulum (ER), then to the Golgi apparatus, and subsequently are secreted from the cell via vesicles that fuse with the plasma membrane. Individual procollagen polypeptides (called α chains) are synthesized into the lumen (inside) of the ER. Three procollagen polypeptides then associate with each other to form a procollagen triple helix. The amino acid sequences at both ends of the polypeptides, termed extension sequences, promote the formation of procollagen and prevent the formation of a much larger fiber. After procollagen is secreted out of the cell, extracellular enzymes remove the extension sequences. Once this occurs, the protein, now called collagen, can form larger structures. Collagen proteins assemble in a staggered way to form relatively thin collagen fibrils, which then align

and produce large collagen fibers. The many layers of these proteins give collagen fibers their tensile strength.

In addition to tensile strength, elasticity is needed in regions of the body such as the lungs and blood vessels, which regularly expand and return to their original shape. In these places, the ECM contains an abundance of elastic fibers composed primarily of the protein **elastin** (**Figure 10.3**). Elastin proteins form many covalent cross-links to make a fiber with remarkable elastic properties. In the absence of a stretching force, each protein tends to adopt a compact conformation. When subjected to a stretching force, however, the compact proteins become more linear, with the covalent cross-links holding the fiber together. When the stretching force has ended, the proteins naturally return to their compact conformation. In this way, elastic fibers behave much like a rubber band, stretching under tension and snapping back when the tension is released.

Figure 10.3   Structure and function of elastic fibers.
Elastic fibers are made of elastin, one type of structural protein
found in the ECM surrounding animal cells.

Concept check: *Suppose you started with an unstretched
elastic fiber and treated it with a chemical that breaks the cross-
links between adjacent elastin proteins. What would happen
when the fiber is stretched?*

Genomes & Proteomes Connection

Collagens Are a Family of Proteins That Give Animal Cells a Variety of ECM Properties

Researchers have determined that animal cells make many different types of collagen fibers. These are designated as type I, type II, and so on. At least 27 different types of collagens have been identified in humans. To make different types of collagens, the human genome, as well as the genomes of other animals, has many different genes that encode procollagen polypeptides.

Collagens have a common structure, in which three polypeptides wind around each other to form a triple helix (see Figure 10.2). Each polypeptide is an α chain. In some collagens, all three α chains are identical, while in others, the α chains may be encoded by different collagen genes. Nevertheless, the triple helix structure is common to all collagen proteins.

Why are different collagens made? Each of the many different types of collagen polypeptides has a similar yet distinctive amino acid sequence that affects the structure of not only individual collagen proteins but also the resulting collagen fibers. For example, the amino acid sequence may cause the α chains within each collagen protein to bind to each other very tightly, thereby creating rigid proteins that form a relatively stiff fiber. Such collagen fibers are found in bone and cartilage.

The amino acid sequence of the α chains also influences the interactions between the collagen proteins within a fiber. For example, the amino acid sequences of certain chains may promote a looser interaction that produces a more bendable or thin fiber. More flexible collagen fibers support the lining of your lungs and intestines. In addition, domains within the collagen polypeptide may affect the spatial arrangement of col-

lagen proteins. The collagen shown earlier in Figure 10.2 forms fibers in which collagen proteins align themselves in parallel arrays. However, not all collagen proteins form long fibers. For example, type IV collagen proteins interact with each other in a meshwork pattern. This meshwork acts as a filtration unit around capillaries.

Differential gene regulation controls which types of collagens are made throughout the body and in what amounts they are made. Of the 27 types of collagens, **Table 10.2** considers types I to IV, each of which varies with regard to where it is primarily synthesized and its structure and function. Collagen genes are regulated, so the required type of collagen is made in the correct sites in your body. In skin cells, for example, the genes that encode the polypeptides that make up collagen types I, III, and IV are turned on, while the synthesis of type II collagen is minimal.

The regulation of collagen synthesis has received a great deal of attention due to the phenomenon of wrinkling. Many face and skin creams contain collagen as an ingredient! As we age, the amount of collagen that is synthesized in our skin significantly decreases. The underlying network of collagen fibers, which provides scaffolding for the surface of our skin, loosens and unravels. This is one of the factors that causes the skin of older people to sink, sag, and form wrinkles. Various therapeutic and cosmetic agents have been developed to prevent or reverse the appearance of wrinkles, most with limited benefits. One approach is collagen injections, in which small amounts of collagen (from cows) are injected into areas where the body's collagen has weakened, filling the depressions to the level of the surrounding skin. Because collagen is naturally broken down in the skin, the injections are not permanent and last only about 3 to 6 months.

Table 10.2	Examples of Collagen Types	
Type	**Sites of synthesis***	**Structure and function**
I	Tendons, ligaments, bones, and skin	Forms a relatively rigid and thick fiber. Very abundant, provides most of the tensile strength to the ECM.
II	Cartilage, discs between vertebrae	Forms a fairly rigid and thick fiber but is more flexible than type I. Permits smooth movements of joints.
III	Arteries, skin, internal organs, and around muscles	Forms thin fibers, often arranged in a meshwork pattern. Allows for greater elasticity in tissues.
IV	Skin, intestine, and kidneys; also found around capillaries	Does not form long fibers. Instead, the proteins are arranged in a meshwork pattern that provides organization and support to cell layers. Functions as a filter around capillaries.

*The sites of synthesis denote where a large amount of the collagen type is made.

Animal Cells Also Secrete Polysaccharides Into the ECM

In addition to proteins, polysaccharides are the second major component of the extracellular matrix of animals. As discussed in Chapter 3, polysaccharides are polymers of simple sugars. Among vertebrates, the most abundant types of polysaccharides in the ECM are **glycosaminoglycans** (**GAGs**). These molecules are long, unbranched polysaccharides containing a repeating disaccharide unit (**Figure 10.4a**). GAGs are highly negatively charged molecules that tend to attract positively charged ions and water. The majority of GAGs in the ECM are linked to core proteins, forming **proteoglycans** (**Figure 10.4b**).

Providing resistance to compression is the primary function of GAGs and proteoglycans. Once secreted from cells, these macromolecules form a gel-like component in the ECM. How is this gel-like property important? Due to its high water content, the ECM is difficult to compress and thereby serves to protect cells. GAGs and proteoglycans are found abundantly in regions of the body that are subjected to harsh mechanical forces, such as the joints of the human body. Two examples of GAGs are chondroitin sulfate, which is a major component of cartilage, and hyaluronic acid, which is found in the skin, eyes, and joint fluid.

Among many invertebrates, an important ECM component is **chitin**, a nitrogen-containing polysaccharide. Chitin forms

(a) Structure of chondroitin sulfate, a glycosaminoglycan

(b) General structure of a proteoglycan

Figure 10.4 Structures of glycosaminoglycans and proteoglycans. These large molecules are found in the ECM of animal cells. **(a)** Glycosaminoglycans (GAGs) are composed of repeating disaccharide units. They can range in length from several dozen to 25,000 disaccharide units. The GAG shown here is chondroitin sulfate, which is commonly found in cartilage. **(b)** Proteoglycans are composed of a long, linear core protein with many GAGs attached. Note that each GAG is typically 80 disaccharide units long but only a short chain of sugars is shown in this illustration.

Concept check: What structural feature of GAGs and proteoglycans give them a gel-like character?

the hard protective outer covering (called an exoskeleton) of insects, such as crickets and grasshoppers, and shellfish, such as lobsters and shrimp. The chitin exoskeleton is so rigid that as these animals grow, they must periodically shed this outer layer and secrete a new, larger one—a process called molting (look ahead to Figure 32.11).

The Cell Wall of Plants Provides Strength and Resistance to Compression

Let's now turn our attention to the cell walls of plants. Plants cells are surrounded by a **cell wall**, a protective layer that forms outside of the plasma membrane. Like animal cells, the cells of plants are surrounded by material that provides tensile strength and resistance to compression. The cell walls of plants, however, are usually thicker, stronger, and more rigid than the ECM found in animals. Plant cell walls provide rigidity for mechanical support and also play a role in the maintenance of cell shape and the direction of cell growth. As we learned in Chapter 5, the cell wall also prevents expansion when water enters the cell, thereby preventing osmotic lysis.

The cell walls of plants are composed of a primary cell wall and a secondary cell wall (**Figure 10.5**). These walls are named according to the timing of their synthesis—the primary cell wall is made before the secondary cell wall. During cell division, the **primary cell wall** develops between two newly made daughter cells. It is usually very flexible and allows new cells to increase in size. The main macromolecule of the plant cell wall is **cellulose**, a polysaccharide made of repeating molecules of glucose attached end to end. These glucose polymers associate with each other via hydrogen bonding to form microfibrils that provide great tensile strength (**Figure 10.6**).

Cellulose was discovered in 1838 by the French chemist Anselme Payen, who was the first scientist to try to separate wood into its component parts. After treating different types of wood with nitric acid, Payen obtained a fibrous substance that was also found in cotton and other plants. His chemical analysis revealed that the fibers were made of the carbohydrate glucose. Payen called this substance cellulose (from the Latin, meaning consisting of cells). Cellulose is probably the single most abundant organic molecule on Earth. Wood consists mostly of cellulose, and cotton and paper are almost pure cellulose. The mechanism of cellulose synthesis is described in Chapter 30.

In addition to cellulose, other components found in the primary cell wall include hemicellulose, glycans, and pectins (see Figure 10.5). Hemicellulose is another linear polysaccharide, with a structure similar to that of cellulose, but it contains sugars other than glucose in its structure and usually forms thinner microfibrils. Glycans, polysaccharides with branching structures, are also important in cell wall structure. The cross-linking glycans bind to cellulose and provide organization to the cellulose microfibrils. Pectins, which are highly negatively charged polysaccharides, attract water and have a gel-like character that provides the cell wall with the ability to resist compression.

Plasma membrane

Cellulose microfibrils

Cross-linking glycan

Pectin

Secondary cell wall

Hemicellulose

Primary cell wall

50 nm

The primary cell wall is thin and flexible. It contains cellulose microfibrils in a meshwork pattern, along with other components shown to the far right.

The secondary cell wall is made in successive layers. Each layer contains strong cellulose microfibrils in parallel arrays. The direction of cellulose microfibrils in each layer is varied, as shown to the right.

Figure 10.5   Structure of the cell wall of plant cells.   The primary cell wall is relatively thin and flexible. It contains cellulose (tan), hemicellulose (red), cross-linking glycans (blue), and pectin (green). The secondary cell wall, which is produced only by certain plant cells, is made after the primary cell wall and is synthesized in successive layers.

Concept check: *With regard to cell growth, what would happen if the secondary cell wall was made too soon?*

Many polymers associate with each other to form a microfibril.

Microfibril   20 nm

Figure 10.6   Structure of cellulose, the main macromolecule of the primary cell wall.   Cellulose is made of repeating glucose units linked end to end that hydrogen-bond to each other to form microfibrils (SEM).

The **secondary cell wall** is synthesized and deposited between the plasma membrane and the primary cell wall (see Figure 10.5) after a plant cell matures and has stopped increasing in size. It is made in layers by the successive deposition of cellulose microfibrils and other components. While the primary

wall structure is relatively similar in nearly all cell types and species, the structure of the secondary cell wall is more variable. The secondary cell wall often contains components in addition to those found in the primary cell wall. For example, phenolic compounds called lignins are very hard and impart considerable strength to the secondary wall structure. Lignin, a type of secondary metabolite described in Chapter 7, is found in the woody parts of plants.

10.2   Cell Junctions

Thus far, we have learned that the cells of animals and plants create an extracellular matrix or cell wall that provides strength, support, and organization. For an organism to become a multicellular unit, cells within the body must be linked to each other. In animals and plants, this is accomplished by specialized structures called **cell junctions** (Table 10.3). In this section, we will examine different types of cell junctions in animal and plants.

Animal cells, which lack the structural support provided by the cell wall, have a more varied group of junctions than plant cells. In animals, junctions called anchoring junctions play a role in anchoring cells to each other or to the extracellular matrix. In other words, they hold cells in their proper place in the body. Other junctions, termed tight junctions, seal cells together to prevent small molecules from leaking across a layer of cells. Still another type of junction, known as a gap junction, allows cells to communicate directly with each other.

In plants, cellular organization is somewhat different because plant cells are surrounded by a rigid cell wall. Plant cells are connected to each other by a component called the middle lamella, which cements their cell walls together. They also have junctions termed plasmodesmata that allow adjacent cells to communicate with each other. In this section, we will examine these various types of junctions found between the cells of animals and plants.

Table 10.3	Common Types of Cell Junctions
Type	**Description**
Animals	
Anchoring junctions	Cell junctions that hold adjacent cells together or bond cells to the ECM. Anchoring junctions are mechanically strong.
Tight junctions	Junctions between adjacent cells in a layer that prevent the leakage of material between cells.
Gap junctions	Channels that permit the direct exchange of ions and small molecules between the cytosol of adjacent cells.
Plants	
Middle lamella	A polysaccharide layer that cements together the cell walls of adjacent cells.
Plasmodesmata	Passageways between the cell walls of adjacent cells that can be opened or closed. When open, they permit the direct diffusion of ions and molecules between cells.

Anchoring Junctions Link Animal Cells to Each Other and to the ECM

The advent of electron microscopy allowed researchers to explore the types of junctions that occur between cells and within the extracellular matrix. In the 1960s, Marilyn Farquhar, George Palade, and colleagues conducted several studies showing that various types of cellular junctions connect cells to each other. Over the past few decades, researchers have begun to unravel the functions and molecular structures of these types of junctions, collectively called **anchoring junctions**, which attach cells to each other and to the extracellular matrix. Anchoring junctions are particularly common in parts of the body where the cells are tightly connected and form linings. An example is the layer of cells that line the small intestine. Having anchoring junctions keeps intestinal cells tightly adhered to one another, thereby forming a strong barrier between the lumen of the intestine and the blood. A key component of anchoring junctions that form the actual connections are integral membrane proteins called **cell adhesion molecules** (**CAMs**). Two types of CAMs are cadherins and integrins.

Anchoring junctions are grouped into four main categories, according to their functional roles and their connections to cellular components. **Figure 10.7** shows these junctions between cells of the mammalian small intestine.

1. **Adherens junctions** connect cells to each other via cadherins. In many cases, these junctions are organized into bands around cells. In the cytosol, adherens junctions bind to cytoskeletal filaments called actin filaments.
2. **Desmosomes** also connect cells to each other via cadherins. They are spotlike points of intercellular contact that rivet cells together. Desmosomes are connected to cytoskeletal filaments called intermediate filaments.

In adherens junctions, cadherins connect cells to each other and to actin filaments.

In desmosomes, cadherins connect cells to each other and to intermediate filaments.

Ring of actin filaments

Intermediate filaments

Cadherins

Linker proteins

Integrin

Actin filament

ECM

In hemidesmosomes, integrins connect the ECM to intermediate filaments.

In focal adhesions, integrins connect the ECM to actin filaments.

Figure 10.7 Types of anchoring junctions. This figure shows these junctions in three adjacent intestinal cells.

Concept check: Which junctions are cell-to-cell junctions and which are cell-to-ECM junctions?

3. **Hemidesmosomes** connect cells to the extracellular matrix via integrins. Like desmosomes, they interact with intermediate filaments.
4. **Focal adhesions** also connect cells to the extracellular matrix via integrins. In the cytosol, focal adhesions bind to actin filaments.

Let's now consider the molecular components of anchoring junctions. As noted, **cadherins** are CAMs that create cell-to-cell junctions (**Figure 10.8a**). The extracellular domains of two cadherin proteins, each in adjacent cells, bind to each other to promote cell-to-cell adhesion. This binding requires the presence of calcium ions, which change the conformation of the cadherin protein such that cadherins in adjacent cells bind to each other. (This calcium dependence is where cadherin gets its name—Ca^{2+} dependent adhering molecule.) On the inside of the cell, linker proteins connect cadherins to actin or intermediate filaments of the cytoskeleton. This promotes a more stable interaction between two cells because their strong cytoskeletons are connected to each other.

The genomes of vertebrates and invertebrates contain multiple cadherin genes, which encode slightly different cadherin proteins. The expression of particular cadherins allows cells to recognize each other. Dimer formation follows a homophilic, or like-to-like, binding mechanism. To understand the concept of homophilic binding, let's consider an example. One type of cadherin is called E-cadherin, and another is N-cadherin. E-cadherin in one cell will bind to E-cadherin in an adjacent cell to form a homodimer. However, E-cadherin in one cell will not bind to N-cadherin in an adjacent cell to form a heterodimer. Similarly, N-cadherin will bind to N-cadherin but not to E-cadherin in an adjacent cell. Why is such homophilic binding important? By expressing only certain types of cadherins, each cell will bind only to other cells that express the same cadherin types. This phenomenon plays a key role in the proper arrangement of cells throughout the body, particularly during development.

Integrins, a group of cell-surface receptor proteins, are a second type of CAM, one that creates connections between cells and the extracellular matrix. Integrins do not require Ca^{2+} to function. Each integrin protein is composed of two nonidentical subunits. In the example shown in **Figure 10.8b**, an integrin is bound to fibronectin, an adhesive protein in the ECM that binds to other ECM components such as collagen fibers. Like cadherins, integrins also bind to actin or intermediate filaments in the cytosol of the cell, via linker proteins, to promote a strong association between the cytoskeleton of a cell and the extracellular matrix. Thus, integrins have an extracellular domain for the binding of ECM components and an intracellular domain for the binding of cytosolic proteins.

When these CAMs were first discovered, researchers imagined that cadherins and integrins played only a mechanical role. In other words, their functions were described as holding cells together or to the ECM. More recently, however, experiments have shown that cadherins and integrins are also important in cell communication. When cell-to-cell and cell-to-ECM junctions are formed or broken, this affects signal transduction pathways within the cell. Similarly, intracellular signal transduction pathways can affect cadherins and integrins in ways that alter intercellular junctions and the binding of cells to ECM components.

Abnormalities in CAMs such as integrins are often associated with the ability of cancer cells to metastasize, that is, to move to other parts of the body. Cell adhesion molecules are critical for keeping cells in their correct locations. When they become defective due to cancer-causing mutations, cells lose their proper connections with the ECM and adjacent cells and may spread to other parts of the body. This topic is considered in more detail in Chapter 14.

(a) Cadherins

(b) Integrins

Figure 10.8 **Types of cell adhesion molecules (CAMs).** Cadherins and integrins are CAMs that form connections in anchoring junctions. **(a)** A cadherin in one cell binds to a cadherin of an identical type in an adjacent cell. This binding requires Ca^{2+}. In the cytosol, cadherins bind to actin or intermediate filaments of the cytoskeleton via linker proteins. **(b)** Integrins link cells to the extracellular matrix and form intracellular connections to actin or intermediate filaments. Each integrin protein is composed of two nonidentical subunits, a heterodimer.

Tight Junctions Prevent the Leakage of Materials Across Animal Cell Layers

In animals, **tight junctions**, or occluding junctions, are a second type of junction, one that forms a tight seal between adjacent cells and thereby prevents material from leaking between cells. As an example, let's consider the intestine. The cells that line the intestine form a sheet that is one cell thick. One side faces the intestinal lumen, and the other faces the ECM and blood vessels (**Figure 10.9**). Tight junctions between these cells ensure that nutrients pass through the plasma membranes of the intestinal cells before entering the blood, and also prevent the leakage of materials from the blood into the intestine.

Tight junctions are made by membrane proteins, called occludin and claudin, that form interlaced strands in the plasma membrane (see inset to Figure 10.9). These strands of proteins, each in adjacent cells, bind to each other and thereby form a tight seal between cells. Tight junctions are not mechanically strong like anchoring junctions, because they do not have strong connections with the cytoskeleton. Therefore, adjacent cells that have tight junctions also have anchoring junctions to hold the cells in place.

The amazing ability of tight junctions to prevent the leakage of material across cell layers has been demonstrated by dye-injection studies. In 1972, Daniel Friend and Norton Gilula injected lanthanum, a metallic element that is electron dense and can be visualized under the electron microscope, into the bloodstream of a rat. A few minutes later, a sample of a cell layer in the digestive tract was removed and visualized by electron microscopy. As seen in the micrograph in **Figure 10.10**, lanthanum diffused into the region between the cells that faces the blood, but it could not move past the tight junction to the side of the cell layer facing the lumen of the digestive tract.

Gap Junctions in Animal Cells Provide a Passageway for Intercellular Transport

A third type of junction found in animals is called a **gap junction**, because a small gap occurs between the plasma membranes of cells connected by these junctions (**Figure 10.11**). Gap junctions are abundant in tissues and organs where the cells need to communicate with each other. For example, cardiac muscle cells, which cause your heart to beat, are interconnected by many gap junctions. Because gap junctions allow the passage of ions, electrical changes in one cardiac muscle cell are easily transmitted to an adjacent cell that is connected via gap junctions. This is needed for the coordinated contraction of cardiac muscle cells.

In vertebrates, gap junctions are composed of an integral membrane protein called connexin. Invertebrates have a structurally similar protein called innexin. Six connexin proteins in one vertebrate cell align with six connexin proteins in an adjacent cell to form a channel called a **connexon** (see inset to Figure 10.11).

The connexons allow the passage of ions and small molecules with a molecular mass that is less than 1,000 Daltons, including amino acids, sugars, and signaling molecules such as

Figure 10.9 **Tight junctions between adjacent intestinal cells.** In this example, tight junctions form a seal that prevents the movement of material between cells, from the intestinal lumen into the blood, and vice versa. The inset shows the interconnected network of occludin and claudin that forms the tight junction.

Concept check: What do you think is the role of tight junctions in the epidermal layers of your skin?

Figure 10.10 **An experiment demonstrating the function of a tight junction.** When lanthanum was injected into the bloodstream of a rat, it diffused between the cells in the region up to a tight junction but could not diffuse past the junction to the other side of the cell layer.

Concept check: What results would you expect if a rat was fed lanthanum and then a sample of intestinal cells was observed under the EM?

Intercellular gap

Gap junction

Gap junction

Small solute

Connexon

30 nm

Figure 10.11 **Gap junctions between adjacent cells.** Gap junctions form intercellular channels that allow the passage of small solutes with masses less than 1,000 Daltons. A transmembrane channel called a connexon consists of 12 proteins called connexins, 6 in each cell. The micrograph shows a gap junction between intestinal cells.

Ca^{2+}, cAMP, and IP_3. In this way, gap junctions allow adjacent cells to share metabolites and directly signal each other. At the same time, gap-junction channels are too small to allow the passage of RNA, proteins, or polysaccharides. Therefore, cells that communicate via gap junctions still maintain their own distinctive set of macromolecules.

FEATURE INVESTIGATION

Loewenstein and Colleagues Followed the Transfer of Fluorescent Dyes to Determine the Size of Gap-Junction Channels

As mentioned, gap junctions allow the passage of small molecules, those with a mass up to about 1,000 Daltons. This property of gap junctions was determined in experiments involving the transfer of fluorescent dyes. During the 1960s, several research groups began using fluorescent dyes to study cell morphology and function. As discussed in Chapter 4, the location of fluorescent dyes within cells can be seen via fluorescence microscopy. In 1964, Werner Loewenstein and colleagues observed that a fluorescent dye could move from one cell to an adjacent cell, which prompted them to investigate this phenomenon further.

In the experiment shown in **Figure 10.12**, Loewenstein and colleagues grew rat liver cells in the laboratory, where they formed a single layer. The adjacent cells formed gap junctions. Single cells were injected with various dyes composed of fluorescently labeled amino acids or peptide molecules with different masses, and then the cell layers were observed via fluorescence microscopy. As shown in The Data, the researchers observed that dyes with a molecular mass up to 901 Daltons passed from cell to cell. Larger dyes, however, did not move intercellularly. Loewenstein and other researchers subsequently investigated dye transfer in other cell types and species. Though some variation is found when comparing different cell types and species, the researchers generally observed that molecules with a mass greater than 1,000 Daltons do not pass through gap junctions.

Figure 10.12 Use of fluorescent molecules by Loewenstein and colleagues to determine the size of gap-junction channels.

HYPOTHESIS Gap-junction channels allow the passage of ions and molecules, but there is a limit to how large the molecules can be.

KEY MATERIALS Rat liver cells grown in the laboratory, a collection of fluorescent dyes.

Experimental level

Conceptual level

1 Grow rat liver cells in a laboratory on solid growth media until they become a single layer. At this point, adjacent cells have formed gap junctions.

Tissue culture bottle

Rat liver cells

Gap junction

2 Inject 1 cell in the layer with fluorescently labeled amino acids or peptides. Note: Several dyes with different molecular masses were tested.

Rat liver cells

Gap junction channels

3 Incubate for various lengths of time (for example, 40–45 minutes). Observe cell layer under the fluorescence microscope to determine if the dye has moved to adjacent cells.

Note: In this case, the dye was transferred to adjacent cells.

4 **THE DATA**

Mass of dye (in Daltons)	Transfer to adjacent cells*	Mass of dye	Transfer to adjacent cells*
376	++++	851**	–
464	++++	901	+++
536	+++	946	–
559	++++	1004	–
665	+	1158	–
688	++++	1678	–
817	+++	1830	–

*The number of pluses indicates the relative speed of transfer. Four pluses denote fast transfer, whereas one plus is slow transfer. A minus indicates that transfer between cells did not occur. ** In some cases, molecules with less mass did not pass between cells compared to molecules with a higher mass. This may be due to differences in their structures (for example, charges) that influence whether or not they can easily penetrate the channel.

5 **CONCLUSION** Gap junctions allow the intercellular movement of molecules that have a mass of approximately 900 Daltons or less.

6 **SOURCE** Flagg-Newton, J., Simpson, I., and Loewenstein, W.R. 1979. Permeability of the Cell-to-Cell Membrane Channels in Mammalian Cell Junction. *Science* 205:404–407.

Experimental Questions

1. What was the purpose of the study conducted by Loewenstein and colleagues?

2. Explain the experimental procedure used by Loewenstein to determine the size of gap-junction channels.

3. What did the results of Figure 10.12 indicate about the size of gap-junction channels?

The Middle Lamella Cements Adjacent Plant Cell Walls Together

In animals, we have seen that cell-to-cell contact, via anchoring junctions, tight junctions, and gap junctions, involves interactions between membrane proteins in adjacent cells. In plants, cell junctions are quite different. Rather than using membrane proteins to form cell-to-cell connections, plant cells make an additional component called the **middle lamella** (plural, lamellae), which is found between most adjacent plant cells (**Figure 10.13**). When plant cells are dividing, the middle lamella is the first layer formed. The primary cell wall is then made. The middle lamella is rich in pectins, negatively charged polysaccharides that are also found in the primary cell wall (see Figure 10.5). These polymers attract water and make a hydrated gel. Ca^{2+} and Mg^{2+} interact with the negative charges in the

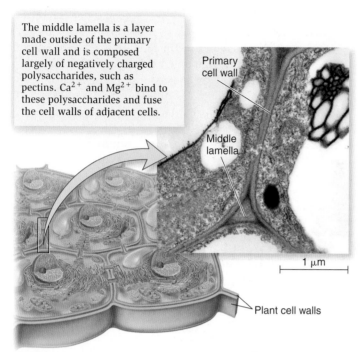

The middle lamella is a layer made outside of the primary cell wall and is composed largely of negatively charged polysaccharides, such as pectins. Ca^{2+} and Mg^{2+} bind to these polysaccharides and fuse the cell walls of adjacent cells.

Primary cell wall

Middle lamella

1 μm

Plant cell walls

Figure 10.13   Plant cell-to-cell junctions known as middle lamellae.

Concept check: How are middle lamellae similar to the anchoring junctions and desmosomes found between animal cells? How are they different?

Plasmodesmata

0.6 μm

Plasma membrane

Cell walls of adjacent plant cells

Smooth endoplasmic reticulum

Desmotubule passing through a plasmodesma

Cytosol
Cell 1

Middle lamella

Cytosol
Cell 2

Figure 10.14   Structure of plasmodesmata.
Plasmodesmata are cell junctions connecting the cytosol of adjacent plant cells, allowing water, ions, and molecules to pass from cell to cell. At these sites, the plasma membrane of one cell is continuous with the plasma membrane of an adjacent cell. In addition, the ER from one cell is connected to that of the adjacent cell via a desmotubule.

polysaccharides and cement the cell walls of adjacent cells together.

The process of fruit ripening illustrates the importance of pectins in holding plant cells together. An unripened fruit, such as a green tomato, is very firm because the rigid cell walls of adjacent cells are firmly attached to each other. During ripening, the cells secrete a group of enzymes called pectinases, which digest pectins in the middle lamella as well as those in the primary cell wall. As this process continues, the attachments between cells are broken, and the cell walls become less rigid. For this reason, a red ripe tomato is much less firm than an unripe tomato.

Plasmodesmata Are Channels Connecting the Cytoplasm of Adjacent Plant Cells

In 1879, Eduard Tangl, a Russian botanist, observed intercellular connections in the seeds of the strychnine tree and hypothesized that the cytoplasm of adjacent cells is connected by ducts in the cell walls. He was the first to propose that direct cell-to-cell communication integrates the functioning of plant cells. The ducts or intercellular channels that Tangl observed are now known as **plasmodesmata** (singular, plasmodesma).

Plasmodesmata are functionally similar to gap junctions in animal cells because they allow the passage of ions and molecules between adjacent plant cells. However, the structure of plasmodesmata is quite different from that of gap junctions. As shown in **Figure 10.14**, plasmodesmata are channels in the cell walls of adjacent cells. At these sites, the plasma membrane of

one cell is continuous with the plasma membrane of the other cell, which permits the diffusion of molecules from the cytosol of one cell to the cytosol of the other. In addition to a cytosolic connection, plasmodesmata also have a central tubule, called a desmotubule, connecting the ER membranes of adjacent cells.

Plasmodesmata can change the size of their opening between closed, open, and dilated states. In the open state, they allow the passage of ions and small molecules, such as sugars and cAMP. In this state, plasmodesmata play a similar role to gap junctions between animal cells. Plasmodesmata tend to close when a large pressure difference occurs between adjacent cells. Why does this happen? One reason is related to cell damage. When a plant is wounded, damaged cells will lose their turgor pressure. (Turgor pressure is described in Chapter 5; refer back to Figure 5.17.) The closure of plasmodesmata between adjacent cells helps to prevent the loss of water and nutrients from the wound site.

Unlike gap junctions between animal cells, researchers have recently discovered that plasmodesmata can dilate to also allow the passage of macromolecules and even viruses between adjacent plant cells. Though the mechanism of dilation is not well understood, the wider opening of plasmodesmata is important for the passage of proteins and mRNA during plant development. It also provides a key mechanism whereby viruses can move from cell to cell.

10.3   Tissues

A **tissue** is a part of an animal or plant consisting of a group of cells having a similar structure and function. In this section, we will view tissues from the perspective of cell biology. Animals and plants contain many different types of cells. Humans, for example, contain over 200 different cell types, each with a specific structure and function. Even so, these cells can be grouped into a few general categories. For example, muscle cells found in your heart (cardiac muscle cells), in your biceps (skeletal muscle cells), and around your arteries (smooth muscle cells) look somewhat different under the microscope and have unique roles in the body. Yet due to structural and functional similarities, all three types can be categorized as muscle tissue. In this section, we will begin by surveying the basic processes that cells undergo to make tissues. Next, we will examine the main categories of animal and plant tissues. We will conclude by taking a more in-depth look at some differences and similarities between selected animal and plant tissues, focusing in particular on the functions of the ECM and cell junctions.

Six Basic Cell Processes Produce Tissues and Organs

A multicellular organism such as a plant or animal contains many cells. For example, an adult human has somewhere between 10 and 100 trillion cells in her or his body. Cells are organized into tissues, and tissues are organized into organs. An **organ** is a collection of two or more tissues that performs a specific function or set of functions. The heart is an organ found in the bodies of complex animals, while a leaf is an organ found in plants. We will examine the structures and functions of organs in Units VI and VII.

How are tissues and organs formed? To form tissues and organs, cells undergo six basic processes that influence their morphology, arrangement, and number: cell division, cell growth, differentiation, migration, apoptosis, and the formation of cell connections.

1. *Cell division:* As discussed in Chapter 15, eukaryotic cells progress through a cell cycle that leads to cell division.
2. *Cell growth:* Following cell division, cells take up nutrients and usually expand in volume. Cell division and cell growth are the primary mechanisms for increasing the size of tissues, organs, and organisms.
3. *Differentiation:* Due to gene regulation, cells differentiate into specialized types of cells. Cell differentiation is described in Chapter 19.
4. *Migration:* During embryonic development in animals, cells migrate to their appropriate positions within the body. Also, adults have cells that can move into regions that have become damaged. Cell migration does not occur during plant development.

5. *Apoptosis:* Cell death, also known as apoptosis (discussed in Chapter 9), is necessary to produce certain morphological features of the body. For example, during development in mammals, the formation of individual fingers and toes requires the removal, by apoptosis, of the skin cells between them.
6. *Cell connections:* In the first section of this chapter, we learned that cells produce an extracellular matrix or cell wall that provides strength and support. In animals, the ECM serves to organize cells within tissues and organs. In plants, the cell wall is largely responsible for the shapes of plant tissues. Different types of cell junctions in both animal and plant cells enable cells to make physical contact and communicate with one another.

Animals Are Composed of Epithelial, Connective, Nervous, and Muscle Tissues

The body of an animal contains four general types of tissue—epithelial, connective, nervous, and muscle—that serve very different purposes (**Figure 10.15**).

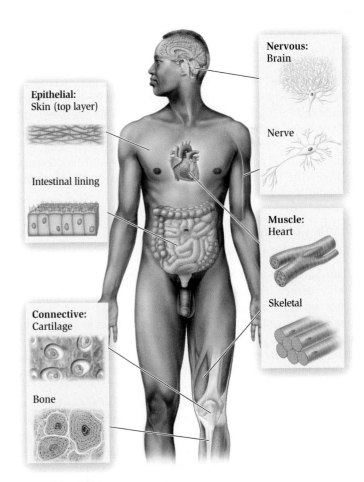

Figure 10.15   Examples of the four general types of tissues—epithelial, connective, nervous, and muscle—found in animals.

Concept check: Which of these four types of tissues would have the most extensive ECM?

Epithelial Tissue   **Epithelial tissue** is composed of cells that are joined together via tight junctions and form continuous sheets. Epithelial tissue covers or forms the lining of all internal and external body surfaces. For example, epithelial tissue lines organs such as the lungs and digestive tract. In addition, epithelial tissue forms skin, a protective surface that shields the body from the outside environment.

Connective Tissue   Most **connective tissue** provides support to the body and/or helps to connect different tissues to each other. Connective tissue is rich in extracellular matrix. In some cases, the tissue contains only a sparse population of cells that are embedded in the ECM. Examples of connective tissue include cartilage, tendons, bone, fat tissue, and the inner layers of the skin. Blood is also considered a form of connective tissue because it provides liquid connections to various regions of the body.

Nervous Tissue   **Nervous tissue** receives, generates, and conducts electrical signals throughout the body. In vertebrates, these electrical signals are integrated by nervous tissue in the brain and transmitted down the spinal cord to the rest of the body. Chapter 41 considers the cellular basis for nerve signals, and Chapters 42 and 43 examine the organization of nervous systems in animals.

Muscle Tissue   **Muscle tissue** can generate a force that facilitates movement. Muscle contraction is needed for bodily movements such as walking and running and also plays a role in the movement of materials throughout the body. For example, contraction of heart muscle propels blood through your body, and smooth muscle contractions move food through the digestive system. The properties of muscle tissue in animals are examined in Chapter 46.

Plants Contain Dermal, Ground, and Vascular Tissues

Plant biologists usually classify tissues as simple or complex. Simple tissues are usually composed of one or possibly two cell types. Complex tissues are composed of two or more cell types but lack an organization that would qualify them as organs. The bodies of most plants contain three general types of tissue—dermal, ground, and vascular—each with a different structure suited to its functions (**Figure 10.16**).

Dermal Tissue   The **dermal tissue** is a complex tissue that forms a covering on various parts of the plant. The **epidermis** refers to the newly made tissue on the surfaces of leaves, stems, and roots. Surfaces of leaves are usually coated with a waxy cuticle to prevent water loss. In addition, leaf epidermis often has hairs, or trichomes, which are specialized types of epidermal cells. Trichomes have diverse functions, including the secretion of oils and leaf protection. Epidermal cells called guard cells form pores in leaves, known as stomata, that permit gas exchange. The function of the root epidermis is the

Figure 10.16   Locations of the three general types of tissues—dermal, ground, and vascular—found in plants.

Concept check: *Which of these three types of tissues would be found on the surfaces of leaves, stems, and roots?*

absorption of water and nutrients. The root epidermis does not have a waxy cuticle because such a cuticle would inhibit water and nutrient absorption.

Ground Tissue   Most of a plant's body is made of **ground tissue**, which has a variety of functions, including photosynthesis, storage of carbohydrates, and support. Ground tissue can be subdivided into three types of simple tissues: parenchyma, collenchyma, and sclerenchyma. Let's look briefly at each of these types of ground tissue.

1. Parenchyma tissue is very active metabolically. The mesophyll, the central part of the leaf that carries out the bulk of photosynthesis, is parenchyma tissue. Parenchyma tissue also functions in the storage of carbohydrates. The cells of parenchyma tissue usually lack a secondary cell wall.

2. Collenchyma tissue provides structural support to the plant body, particularly to growing regions such as the periphery of the stems and leaves. Collenchyma cells tend to have thick, secondary cell walls but do not contain much lignin. Therefore, they provide support but are also able to stretch.

3. Sclerenchyma tissue also provides structural support to the plant body, particularly to those parts that are no longer growing, such as the dense, woody parts of stems. The secondary cell walls of sclerenchyma cells tend to have large amounts of lignin and thereby provide rigid support. In many cases, sclerenchyma cells are dead at maturity, but their cell walls continue to provide structural support during the life of the plant.

Vascular Tissue Most plants living today are vascular plants. In these species, which include ferns and seed plants, the **vascular tissue** is a complex tissue composed of cells that are interconnected and form conducting vessels for water and nutrients. There are two types of vascular tissue called xylem and phloem. The xylem transports water and mineral ions from the root to the rest of the plant, while the phloem distributes the products of photosynthesis and a variety of other nutrients throughout the plant. Some types of modern plants, such as mosses, are nonvascular plants that lack conducting vessels. These plants tend to be small and live in damp, shady places.

Animal and Plant Tissues Have Striking Differences and Similarities

Because plants and animals appear strikingly different, it is not too surprising their cells and tissues show conspicuous differences. For example, the vascular tissue of plants (which transports water and nutrients) does not resemble any one tissue in animals. The blood vessels of animals (which transport blood carrying oxygen and nutrients throughout the body) are hollow tubes that contain both connective and muscle tissue. In addition, animals have two tissue types that are not found in plants: muscle and nervous tissue. Even so, plants are capable of movement and the transmission of signals via action potentials. For example, the Venus flytrap (*Dionaea muscipula*) has two modified leaves that resemble a clamshell. When an insect touches the trichomes on the surface of these leaves, an electrical signal is triggered that causes the leaves to move closer to each other, thereby trapping the unsuspecting insect.

Although cellular differences are prominent between plants and animals, certain tissues show intriguing similarities. Why are there similarities between some animal and plant tissues? The answer is that certain types of cell organization and structure provide functions that are needed by both animals and plants. For example, the epithelial tissue of animals and the dermal tissue of plants both form a protective covering over the organism. Also, the connective tissue of animals and the ground tissue of plants both play a role in structural support. Let's take a closer look at the similarities between these tissues in animals and plants.

A Comparison of Epithelial and Dermal Tissues Both the epithelial tissue in animals (also called an epithelium) and dermal tissue in plants form layers of cells. Let's begin with epithelial tissue. An epithelium can be classified according to its number of layers. Simple epithelium is one cell layer thick,

Apical side—faces air or extracellular region such as the lumen of the intestine

Basal lamina

Basal side—faces inside the body toward the blood

Simple epithelium is composed of a single layer of cells.

Stratified epithelium is composed of multiple layers of cells.

Figure 10.17 Simple and stratified epithelia in animals.

whereas stratified epithelium has several layers (**Figure 10.17**). In both cases, the epithelium has a polarity, which is due to an asymmetry to its organization. The outer, or apical, side of an epithelium is exposed to air or to a watery fluid such as the lumen of the intestine. The inner, or basal, side faces toward the blood. The basal side of the epithelium rests on some type of support, such as another type of tissue or on a form of ECM called the basal lamina.

A hallmark of epithelial cells is they form many connections with each other. For example, in the simple epithelium lining the intestine (**Figure 10.18**), adjacent cells form

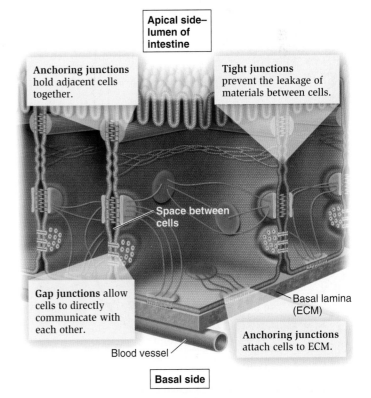

Apical side– lumen of intestine

Anchoring junctions hold adjacent cells together.

Tight junctions prevent the leakage of materials between cells.

Space between cells

Basal lamina (ECM)

Gap junctions allow cells to directly communicate with each other.

Anchoring junctions attach cells to ECM.

Blood vessel

Basal side

Figure 10.18 Connections between cells of a simple epithelium that lines the intestine. This figure emphasizes the three major types of cell junctions that are common in epithelial tissue.

anchoring junctions with each other and with the basal lamina. These anchoring junctions hold the cells firmly in place. Tight junctions, found near the apical surface, prevent the leakage of materials from the lumen of the intestine into the blood. Instead, nutrients are selectively transported from the intestinal lumen into the cytosol of the epithelial cell and then are exported across the basal side of the cell into the blood. This phenomenon, called transepithelial transport, allows the body to take up the nutrients it needs while preventing unwanted materials from getting into the bloodstream. Epithelial cells are also connected via gap junctions, which allow the exchange of nutrients and signaling molecules throughout the epithelium.

In flowering plants, the epidermis covers all of the newly made parts of a plant. For example, the upper and lower sides of leaves are covered by epidermis, which is usually a single layer of closely packed cells (**Figure 10.19a**). Epidermal cells have a thick primary cell wall and are tightly interlocked by their middle lamella. As a consequence, plant epidermal cells are tightly woven together, much like epithelial cell layers in animals.

As a plant ages, the epidermis may be replaced by another dermal tissue called the periderm (**Figure 10.19b**). In woody plants, an example of periderm is the bark on trees. Periderm

protects a plant from pathogens, prevents excessive water loss, and provides insulation. The periderm consists of interconnected cork cells, which may be several layers thick. The cork cells have extremely thick cell walls. When cork cells reach maturity, they die, but the cell walls continue to provide support.

A Comparison of Connective and Ground Tissue   In contrast to epithelial tissue, which is mostly composed of cells, the connective tissue of animals is largely composed of extracellular matrix and has relatively few cells. In animal connective tissue, cell-to-cell contact is somewhat infrequent. Instead, cells are usually adhered to the ECM via integrins, as shown earlier in Figure 10.8b. In some cases, the primary function of cells within connective tissue is to synthesize the components of the ECM. For example, let's consider cartilage, a connective tissue found in joints such as your knees. The cells that synthesize cartilage, known as chondrocytes, actually represent a small proportion of the total volume of cartilage. As shown in **Figure 10.20**, the chondrocytes are found in small cavities within the cartilage called lacunae (singular, lacuna). In some types of cartilage, the chondrocytes represent only 1–2% of the total volume of the tissue! Chondrocytes are the only cells found in cartilage. They are solely responsible for the synthesis of protein fibers, such as collagen, as well as glycosaminoglycans and proteoglycans that are found in cartilage.

Similar to connective tissue in animals, ground tissue in plants provides structural support. **Figure 10.21** shows a scanning electron micrograph of sclerenchyma cells found in *Arabidopsis thaliana*, a model plant studied by plant biologists. At maturity, the cells are dead, but the thick secondary cell walls continue to provide rigid support for the stem. However, not all cells in ground tissue have thick cell walls. For example, mesophyll cells are a type of parenchyma cell in the leaf that carry out photosynthesis. Because a thick cell wall would inhibit the transmission of light, mesophyll cells have relatively thin cell walls.

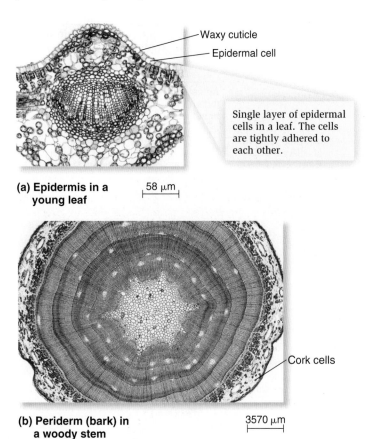

(a) Epidermis in a young leaf   58 μm

Waxy cuticle
Epidermal cell

Single layer of epidermal cells in a leaf. The cells are tightly adhered to each other.

Cork cells

(b) Periderm (bark) in a woody stem   3570 μm

Figure 10.19   Dermal tissues in plants.

Concept check: *Do you notice any parallels between simple epithelium and epidermis, and between stratified epithelium and periderm?*

Lacuna with 2 chondrocytes

ECM

75 μm

Figure 10.20   **An example of connective tissue in animals that is rich in extracellular matrix.**   This micrograph of cartilage shows chondrocytes in the ECM. The chondrocytes, which are responsible for making the components of cartilage, are found in cavities called lacunae.

Secondary
cell wall

2.6 µm

Figure 10.21   **An example of ground tissue in plants.**
This scanning electron micrograph shows sclerenchyma cells
from *Arabidopsis thaliana*. The cells themselves are dead; only
their thick secondary cell walls remain. These cell walls provide
structural support to the plant.

Concept check: *Do you notice any parallels between ground
tissue in plants and connective tissue in animals?*

Summary of Key Concepts

10.1 Extracellular Matrix and Cell Walls

- The extracellular matrix (ECM) is a network of material that
  forms a complex meshwork outside of animal cells. Plant cells
  are surrounded by cell walls.

- In the ECM of animals, proteins and polysaccharides are
  the major constituents. These materials are involved in
  strength, structural support, organization, and cell signaling.
  (Figure 10.1)

- Adhesive proteins, such as fibronectin and laminin, help
  adhere cells to the ECM. Structural proteins form fibers.
  Collagen fibers provide tensile strength, whereas elastic
  fibers allow regions of the body to stretch. (Table 10.1,
  Figures 10.2, 10.3)

- Differential gene regulation controls where in the body
  different types of collagen fibers are made. (Table 10.2)

- Glycosaminoglycans (GAGs) are polysaccharides of repeating
  disaccharide units that give a gel-like character to the ECM of
  animals. Proteoglycans consist of a core protein with attached
  GAGs. (Figure 10.4)

- Plant cells are surrounded by a cell wall. The primary cell wall
  is made first. It is composed largely of cellulose. The secondary
  cell wall is made after the primary cell wall and is often quite
  thick and rigid. (Figures 10.5, 10.6)

10.2 Cell Junctions

- The three common types of cell junctions found in animals are
  anchoring, tight, and gap junctions. Plant junctions include
  middle lamella and plasmodesmata. (Table 10.3)

- Anchoring junctions involve cell adhesion molecules (CAMs),
  which bind cells to each other or to the ECM. The four types

are adherens junctions, desmosomes, hemidesmosomes, and
focal adhesions. (Figure 10.7)

- Two types of CAMs are cadherins and integrins. Cadherins link
  cells to each other, whereas integrins link cells to the ECM.
  In the cytosol, CAMs bind to actin or intermediate filaments.
  (Figure 10.8)

- Tight junctions between cells, composed of occludin and
  claudin, prevent the leakage of materials across a layer of cells.
  (Figures 10.9, 10.10)

- Gap junctions form channels called connexons that permit the
  direct passage of materials between adjacent cells. (Figure 10.11)

- Experiments of Loewenstein and colleagues involving the
  transfer of fluorescent dyes showed that gap junctions permit
  the passage of substances with a molecular mass of less than
  1,000 Daltons. (Figure 10.12)

- The cell walls of adjacent plant cells are cemented together
  via middle lamella. (Figure 10.13)

- Adjacent plant cells usually have direct connections called
  plasmodesmata, which are channels in the cell walls where
  the plasma membranes of adjacent cells are continuous with
  each other. The endoplasmic reticula of adjacent cells are also
  connected via plasmodesmata. (Figure 10.14)

10.3 Tissues

- A tissue is a group of cells that have a similar structure and
  function. An organ is composed of two or more tissues and
  carries out a particular function or functions.

- Six processes—cell division, cell growth, differentiation,
  migration, apoptosis, and the formation of cell connections—
  produce tissues and organs.

- The four general kinds of tissues found in animals are epithelial,
  connective, nervous, and muscle tissues. (Figure 10.15)

- The three general kinds of tissues found in plants are dermal,
  ground, and vascular tissues. (Figure 10.16)

- Epithelial and dermal tissues form layers of cells that are highly
  interconnected. These layers can be one cell thick or several
  cells thick, and they serve as protective coverings for various
  parts of animal and plant bodies. (Figures 10.17, 10.18, 10.19)

- Connective and ground tissues often play a structural role in
  animals and plants. (Figures 10.20, 10.21)

Assess and Discuss

Test Yourself

1. The function of the extracellular matrix (ECM) in animals is
   a. to provide strength.
   b. to provide structural support.
   c. to organize cells and other body parts.
   d. cell signaling.
   e. all of the above.

2. The protein found in the ECM of animals that provides strength
   and resistance to tearing when stretched is
   a. elastin.   c. collagen.   e. fibronectin.
   b. cellulose.   d. laminin.

3. The polysaccharide that forms the hard outer covering of many invertebrates is
   a. collagen.
   b. chitin.
   c. chondroitin sulfate.
   d. pectin.
   e. cellulose.

4. The extension sequence found in procollagen polypeptides
   a. causes procollagen to be synthesized into the ER lumen.
   b. causes procollagen to form a triple helix.
   c. prevents procollagen from forming large collagen fibers.
   d. causes procollagen to be secreted from the cell.
   e. Both b and c are correct.

5. The dilated state of plasmodesmata allows the passage of
   a. water.
   b. ions.
   c. small molecules.
   d. macromolecules and viruses.
   e. all of the above.

6. The gap junctions of animal cells differ from the plasmodesmata of plant cells in that
   a. gap junctions serve as communicating junctions and plasmodesmata serve as adhesion junctions.
   b. gap junctions prevent extracellular material from moving between adjacent cells but the plasmodesmata do not.
   c. gap junctions allow for direct exchange of cellular material between cells but plasmodesmata cannot allow the same type of exchange.
   d. gap junctions are formed by specialized proteins that form channels through the membranes of adjacent cells but plasmodesmata are not formed by specialized proteins.
   e. All of the above are correct.

7. Which of the following is involved in the process of tissue and organ formation in multicellular organisms?
   a. cell division
   b. cell growth
   c. cell differentiation
   d. cell connections
   e. all of the above

8. The tissue type common to animals that functions in the conduction of electrical signals is
   a. epithelial.
   b. dermal.
   c. muscle.
   d. nervous.
   e. ground.

9. A type of tissue that is rich in ECM or has cells with a thick cell wall would be
   a. dermal tissue in plants.
   b. ground tissue in plants.
   c. nervous tissue in animals.
   d. connective tissue in animals.
   e. both b and d.

10. Which of the following is not a correct statement when comparing plant tissues to animal tissues?
    a. Nervous tissue of animals plays the same role as vascular tissue in plants.
    b. The dermal tissue of plants is similar to epithelial tissue of animals in that both provide a covering for the organism.
    c. The epithelial tissue of animals and the dermal tissue of plants have special characteristics that limit the movement of material between cell layers.
    d. The ground tissue of plants and the connective tissue of animals provide structural support for the organism.
    e. All of the above are correct comparisons between animal and plant tissues.

Conceptual Questions

1. What are key differences between the primary cell wall and the secondary cell wall of plant cells?

2. What are similarities and differences in the structures and functions of cadherins and integrins found in animal cells?

3. What are the six basic cell processes required to make tissues and organs?

Collaborative Questions

1. Discuss the similarities and differences between the extracellular matrix of animals and the cell walls of plants.

2. Cell junctions in animals are important in preventing cancer cells from metastasizing—moving to other parts of the body. Certain drugs can bind to CAMs and influence their structure and function. Some of these drugs may help to prevent the spread of cancer cells. What would you hypothesize to be the mechanism by which such drugs work? What might be some harmful side effects?

Online Resource

www.brookerbiology.com

Stay a step ahead in your studies with animations that bring concepts to life and practice tests to assess your understanding. Your instructor may also recommend the interactive ebook, individualized learning tools, and more.

Nucleic Acid Structure, DNA Replication, and Chromosome Structure

11

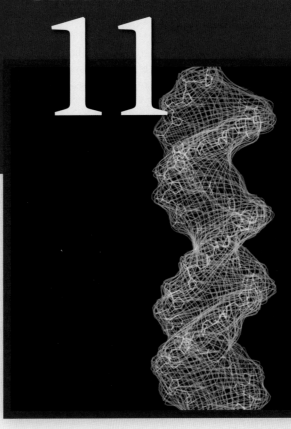

A molecular model for the structure of a DNA double helix.

O n October 17, 2001, Mario K. was set free after serving 16 years in prison. He had been convicted of sexual assault and murder. The charges were dropped because investigators discovered that another person, Edwin M., had actually committed the crime. How was Edwin M. identified as the real murderer? In 2001, he committed another crime, and his DNA was entered into a computer database. Edwin's DNA matched the DNA that had been collected from the victim in 1985, and other evidence was then gathered indicating that Edwin M. was the true murderer. Like Mario K., over 200 other inmates have been exonerated when DNA tests have shown that a different person was responsible for the crime.

Deoxyribonucleic acid, or DNA, is the genetic material that provides the blueprint to produce an individual's traits. Each person's DNA is distinct and unique. Even identical twins show minor differences in their DNA sequences. We begin our survey of genetics by examining DNA at the molecular level. Once we understand how DNA works at this level, it becomes easier to see how the function of DNA controls the properties of cells and ultimately the characteristics of unicellular and multicellular organisms. The past several decades have seen exciting advances in techniques and approaches to investigate and even to alter the genetic material. Not only have these advances greatly expanded our understanding of molecular genetics, such technologies are also widely used in related disciplines such as biochemistry, cell biology, and microbiology. Likewise, genetic techniques have many important applications in biotechnology and are used in criminal justice, including forensics, to provide evidence of guilt or innocence.

To a large extent, our understanding of genetics comes from our knowledge of the molecular structure of DNA. In this chapter, we begin by considering some classic experiments that were consistent with the theory that DNA is the genetic material. We will then survey the molecular features of DNA, which will allow us to appreciate how DNA can store information and be accurately copied. Though this chapter is largely concerned with DNA, we will also consider the components of ribonucleic acid (RNA), which bear some striking similarities to DNA. Lastly, we will examine the molecular composition of chromosomes where the DNA is found.

11.1 Biochemical Identification of the Genetic Material

DNA carries the genetic instructions for the physical characteristics of living organisms. In the case of multicellular organisms such as plants and animals, the information stored in the genetic material enables a fertilized egg to develop into an embryo and eventually into an adult organism. In addition, the genetic material allows organisms to survive in their native environments. For example, an individual's DNA provides the blueprint to produce enzymes that are needed to metabolize nutrients in food. To fulfill its role, the genetic material must meet the following key criteria:

1. **Information:** The genetic material must contain the information necessary to construct an entire organism.
2. **Replication:** The genetic material must be accurately copied.
3. **Transmission:** After it is replicated, the genetic material can be passed from parent to offspring. It also must be passed from cell to cell during the process of cell division.

4. **Variation:** Differences in the genetic material must account for the known variation within each species and among different species.

How was the genetic material discovered? The quest to identify the genetic material really began in the late 1800s, when a few scientists postulated that living organisms possess a blueprint that has a biochemical basis. In 1883, August Weismann and Karl Nägeli championed the idea that a chemical substance exists within living cells that is responsible for the transmission of traits from parents to offspring. During the next 30 years, experimentation along these lines centered on the behavior of **chromosomes**, the cellular structures that we now know contain the genetic material. Taken literally, chromosome is from the Greek words *chromo* and *soma*, meaning colored body, which refers to the observation of early microscopists that the chromosomes are easily stained by colored dyes. By studying the transmission patterns of chromosomes from cell to cell and from parent to offspring, researchers were convinced that chromosomes carry the determinants that control the outcome of traits.

Ironically, the study of chromosomes initially misled researchers regarding the biochemical identity of the genetic material. Chromosomes contain two classes of macromolecules, namely, proteins and DNA. Scientists of this era viewed proteins as being more biochemically complex because they are made from 20 different amino acids. Furthermore, biochemists already knew that proteins perform an amazingly wide range of functions, and complexity seemed an important prerequisite for the blueprint of an organism. By comparison, DNA seemed less complex, because it contains only four types of repeating units, called nucleotides, which will be described later in this chapter. In addition, the functional role of DNA in the nucleus had not been extensively investigated prior to the 1920s. Therefore, from the 1920s to the 1940s, most scientists were expecting that research studies would reveal that proteins are the genetic material. Contrary to this expectation, however, the experiments described in this section were pivotal in showing that DNA carries out this critical role.

Griffith's Bacterial Transformation Experiments Indicated the Existence of a Biochemical Genetic Material

Studies in microbiology were important in developing an experimental strategy to identify the genetic material. In the late 1920s, an English microbiologist, Frederick Griffith, studied a type of bacterium known then as pneumococci and now classified as *Streptococcus pneumoniae*. Some strains of *S. pneumoniae* secrete a polysaccharide capsule, while other strains do not. When streaked on petri plates containing solid growth media, capsule-secreting strains have a smooth colony morphology and therefore look smooth to the naked eye. Those strains unable to secrete a capsule have a colony morphology that looks rough. In mammals, smooth strains of *S. pneumo-*

Treatment	Result	Conclusion

1 Control: Injected living type S bacteria into mouse. — Type S cells are virulent.

2 Control: Injected living type R bacteria into mouse. — Type R cells are benign.

3 Control: Injected heat-killed type S bacteria into mouse. — Heat-killed type S cells are benign.

4 Injected living type R and heat-killed type S bacteria into mouse. — Living type R cells have been transformed into virulent type S cells by a substance from the heat-killed type S cells.

Virulent type S strain in dead mouse's blood

Figure 11.1   Griffith's experiments that showed the transformation of bacteria by a "transformation principle." Note: To determine if a mouse's blood contained live bacteria, a sample of blood was applied to solid growth media to determine if smooth or rough bacterial colonies would form.

Concept check: Let's suppose that the type R strain used by Griffith was resistant to killing by an antibiotic, while the type S strain lacked this trait. For the experiment described in treatment 4, would you expect the living type S bacteria found in the dead mouse's blood to be resistant to the antibiotic?

niae may cause pneumonia and other symptoms. However, in mice, such infections are often fatal.

As shown in **Figure 11.1**, Griffith injected live and/or heat-killed bacteria into mice and then observed whether or not the bacteria caused them to die. He investigated the effects of two strains of *S. pneumoniae*: type S for smooth and type R for rough. When injected into a live mouse, the type S strain killed the mouse (Figure 11.1, step 1). Such a strain is said to be virulent. The capsule present in type S strains prevents the mouse's immune system from killing the bacterial cells. Following the death of the mouse, many type S bacteria were found in the mouse's blood. By comparison, when type R bacteria were injected into a mouse, they did not kill the mouse, and after several days, living bacteria were not found in the live mouse's blood (Figure 11.1, step 2). In a follow-up to these results,

Griffith also heat-killed the smooth bacteria and then injected them into a mouse. As expected, the mouse survived (Figure 11.1, step 3).

A surprising result occurred when Griffith mixed live type R bacteria with heat-killed type S bacteria and then injected them into a mouse—the mouse died (Figure 11.1, step 4). The blood from the dead mouse contained living type S bacteria! How did Griffith explain these results? He postulated that a substance from dead type S bacteria was transforming the type R bacteria into type S bacteria. Griffith called this process **transformation**, and he termed the unidentified material responsible for this phenomenon the "transformation principle."

Now that we have examined Griffith's experiments, it's helpful if we consider what these observations mean with regard to the four criteria for the genetic material that were described previously. According to Griffith's results, the transformed bacteria had acquired the information (criterion 1) to make a capsule from the heat-killed cells. For the transformed bacteria to proliferate and thereby kill the mouse, the substance conferring the ability to make a capsule must be replicated (criterion 2) and then transmitted (criterion 3) from mother to daughter cells during cell division. Finally, Griffith already knew that variation (criterion 4) existed in the ability of his strains to produce a capsule (S strain) or not produce a capsule (R strain). Taken together, these observations are consistent with the idea that the formation of a capsule is governed by genetic material, because it meets the four criteria described at the beginning of this section. The experiment of Figure 11.1, step 4, was consistent with the idea that some genetic material from the heat-killed type S bacteria had been transferred to the living type R bacteria and provided those bacteria with a new trait. At the time of his studies, however, Griffith could not determine the biochemical composition of the transforming substance.

FEATURE INVESTIGATION

Avery, MacLeod, and McCarty Used Purification Methods to Reveal That DNA Is the Genetic Material

Exciting discoveries sometimes occur when researchers recognize that another scientist's experimental approach may be modified and then used to dig deeper into a scientific question. In the 1940s, American physician Oswald Avery and American biologists Colin MacLeod and Maclyn McCarty were also interested in the process of bacterial transformation. During the course of their studies, they realized that Griffith's observations could be used as part of an experimental strategy to biochemically identify the genetic material. They asked the question, What substance is being transferred from the dead type S bacteria to the live type R bacteria?

To answer this question, Avery, MacLeod, and McCarty needed to purify the general categories of substances found in living cells. They used established biochemical procedures to purify classes of macromolecules, such as proteins, DNA, and RNA, from the type S streptococcal strain. Initially, they discovered that only the purified DNA could convert type R bacteria into type S. To further verify that DNA is the genetic material, they performed the investigation outlined in **Figure 11.2**. They purified DNA from the type S bacteria and mixed it with type R bacteria. After allowing time for DNA uptake, they added an antibody that aggregated any nontransformed type R bacteria, which were removed by centrifugation. The remaining bacteria were incubated overnight on petri plates.

When they mixed their S strain DNA extract with type R bacteria, some of the bacteria were converted to type S bacteria (see plate B in step 5 of Figure 11.2). As a control, if no DNA extract was added, no type S bacterial colonies were observed on the petri plates (see plate A in step 5). Though this result was consistent with the idea that DNA is the genetic material, a careful biochemist could argue that the DNA extract might not have been 100% pure. Realistically, any purified extract is likely to contain small traces of other substances. For this reason, the researchers realized that a small amount of contaminating material in the DNA extract could actually be the genetic material. The most likely contaminating substances in this case would be RNA or protein. To address this possibility, Avery, MacLeod, and McCarty treated the DNA extract with enzymes that digest DNA (called **DNase**), RNA (**RNase**), or protein (**protease**) (see step 2). When the DNA extracts were treated with RNase or protease, the type R bacteria were still converted into type S bacteria, suggesting that contaminating RNA or protein in the extract was not acting as the genetic material (see step 5, plates D and E). Moreover, when the extract was treated with DNase, it lost the ability to convert type R bacteria into type S bacteria (see plate C). Taken together, these results were consistent with the idea that DNA is the genetic material.

Experimental Questions

1. Avery, MacLeod, and McCarty worked with two strains of *Streptococcus pneumoniae* to determine the biochemical identity of the genetic material. Explain the characteristics of the *Streptococcus pneumoniae* strains that made them particularly well suited for such an experiment.

2. What is a DNA extract?

3. In the experiment of Avery, MacLeod, and McCarty, what was the purpose of using the protease, RNase, and DNase if only the DNA extract caused transformation?

Figure 11.2   The Avery, MacLeod, and McCarty experiments that identified DNA as Griffith's "transformation principle"—the genetic material.

HYPOTHESIS   A purified macromolecule from type S bacteria, which functions as the genetic material, will be able to convert type R bacteria into type S.

KEY MATERIALS   Type R and type S strains of *Streptococcus pneumoniae*.

Experimental level   Conceptual level

1   Purify DNA from a type S strain. This involves breaking open cells and separating the DNA away from other components by centrifugation.

± DNase
± RNase
± Protease
+ Type R cells

2   Mix the DNA extract with type R bacteria. Allow time for the DNA to be taken up by the type R cells, converting a few of them to type S. Also, carry out the same steps but add the enzymes DNase, RNase, or protease to the DNA extract, which digest DNA, RNA, and proteins, respectively. As a control, don't add any DNA extract to some type R cells.

A   B   C   D   E

Add antibody

Control   + DNA   + DNA + DNase   + DNA + RNase   + DNA + Protease

3   Add an antibody, a protein made by the immune system of mammals, that specifically recognizes type R cells that haven't been transformed. The binding of the antibody causes the type R cells to aggregate.

A   B   C   D   E

4   Remove type R cells by centrifugation. Plate the remaining bacteria (if any) that are in the supernatant onto petri plates. Incubate overnight.

Centrifuge

Type S cells in supernatant

Type R cells in pellet

5   **THE DATA**

A   Control

B   DNA extract

C   DNA extract + DNase

D   DNA extract + RNase

E   DNA extract + protease

6   **CONCLUSION**   DNA is responsible for transforming type R cells into type S cells.

7   **SOURCE**   Avery, O.T., MacLeod, C.M., and McCarty, M. 1944. Studies on the Chemical Nature of the Substance Inducing Transformation of Pneumococcal Types. *Journal of Experimental Medicine* 79:137–156.

Hershey and Chase Determined That DNA Is the Genetic Material of T2 Bacteriophage

In a second avenue of research conducted in 1952, the efforts of Alfred Hershey and Martha Chase centered on the study of a virus named T2. This virus infects bacterial cells, in this case *Escherichia coli*, and is therefore known as a **bacteriophage** or simply a **phage**. A T2 phage has an outer covering called the phage coat that contains a head (capsid), sheath, tail fibers, and base plate (**Figure 11.3**). We now know the phage coat is composed entirely of protein. DNA is found inside the head of T2. From a biochemical perspective, T2 is very simple because it is composed of only proteins and DNA.

The genetic material of T2 provides a blueprint to make new phages. To replicate, all viruses must introduce their genetic material into the cytoplasm of a living host cell. In the case of T2, this involves the attachment of its tail fibers to the bacterial cell wall and the injection of its genetic material into the cytoplasm (Figure 11.3b). However, at the time of Hershey and Chase's work, it was not known if the phage was injecting DNA or protein.

To determine if DNA is the genetic material of T2, Hershey and Chase devised a method to separate the phage coat, which is attached to the outside of the bacterium, from the genetic material, which is injected into the cytoplasm. They reasoned that the attachment of T2 on the surface of the bacterium could be disrupted if the cells were subjected to high shear forces such as those produced by a blender.

They also needed a way to distinguish T2 DNA from T2 proteins. Hershey and Chase used radioisotopes, which are described in Chapter 2, as a way to label these molecules. Sulfur atoms are found in phage proteins but not in DNA, whereas phosphorus atoms are found in DNA but not in phage proteins. They exposed T2-infected bacterial cells to ^{35}S (a radioisotope of sulfur) or to ^{32}P (a radioisotope of phosphorus). These infected cells produced phages that had incorporated ^{35}S into their proteins or ^{32}P into their DNA. The ^{35}S- or ^{32}P-labeled phages were then used in the experiment shown in **Figure 11.4**.

Let's now consider the steps in this experiment. In separate tubes, they took samples of T2 phage, one in which the proteins were labeled with ^{35}S and the other in which the DNA was labeled with ^{32}P, and mixed them with *E. coli* cells for a short period of time. This allowed the phages enough time to inject their genetic material into the bacterial cells. The samples were then subjected to a shearing force, using a blender for up to 8 minutes. This treatment removed the phage coat from the surface of the bacterial cell without causing cell lysis. Each sample was then subjected to centrifugation at a speed that caused the heavier bacterial cells to form a pellet at the bottom of the tube, while the lighter phage coats remained in the supernatant, the solution above the pellet. The amount of radioactivity in the supernatant (emitted from either ^{35}S or ^{32}P) was determined using an instrument called a Geiger counter.

As you can see in the data of Figure 11.4, most of the ^{35}S isotope (80%) was found in the supernatant. Because the shearing force removed only the phage coat, this result indicates that the empty phages contain primarily protein. In contrast, only about 35% of the ^{32}P was found in the supernatant following shearing. This indicates that most of the phage DNA was located within the bacterial cells in the pellet. Taken together, these results suggest that the phage DNA is injected into the bacterial cytoplasm during infection. This is the expected outcome if DNA is the genetic material of T2.

In other experiments, Hershey and Chase measured the amount of radioactivity incorporated into phages that were produced by the infected bacterial cells. Bacteria infected with ^{35}S-labeled phages produced new phages with a negligible amount of radioactivity, whereas bacteria infected with ^{32}P-labeled phages produced new phages with a significant amount of radioactivity. Again, this is the expected result if DNA is the genetic material.

11.2   Nucleic Acid Structure

An important principle in biology is that structure determines function. When biologists want to understand the function of a material at the molecular and cellular level, they focus some of their efforts on the investigation of its biochemical structure. In this regard, an understanding of DNA's structure has proven to be particularly exciting because the structure makes it easier for us to understand how DNA can store information, how it is replicated and then transmitted from cell to cell, and how variation in its structure can occur.

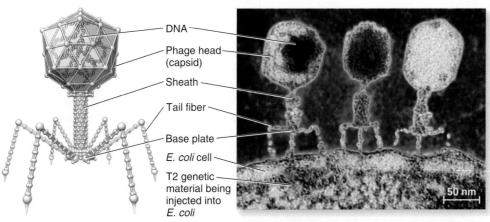

(a) Schematic drawing of T2 bacteriophage

DNA
Phage head (capsid)
Sheath
Tail fiber
Base plate
E. coli cell
T2 genetic material being injected into *E. coli*

(b) An electron micrograph of T2 bacteriophage infecting *E. coli*

50 nm

Figure 11.3   The structure of T2 bacteriophage. The colorized electron micrograph in part **(b)** shows T2 phages attached to an *E. coli* cell and injecting their genetic material into the cell.

Figure 11.4   Hershey and Chase experiment showing that the genetic material of T2 phage is DNA.   Green coloring indicates radiolabeling with either ^{35}S or ^{32}P. Note: The phages and bacteria are not drawn to scale; bacteria are much larger.

Concept check: *In these experiments, what was the purpose of using two different isotopes, ^{35}S and ^{32}P?*

DNA and its molecular cousin, RNA, are known as **nucleic acids**. This term is derived from the discovery of DNA by Friedrich Miescher in 1869. He identified a novel phosphorus-containing substance from the nuclei of white blood cells found in waste surgical bandages. He named this substance nuclein. As the structure of DNA and RNA became better understood, it was found they are acidic molecules, which means they release hydrogen ions (H$^+$) in solution, and have a net negative charge at neutral pH. Thus, the name nucleic acid was coined.

DNA is a very large macromolecule composed of smaller building blocks. We can consider the structural features of DNA at different levels of complexity (**Figure 11.5**):

1. **Nucleotides** are the building blocks of DNA (and RNA).
2. A **strand** of DNA (or RNA) is formed by the covalent linkage of nucleotides in a linear manner.
3. Two strands of DNA can hydrogen-bond with each other to form a **double helix**. In a DNA double helix, two DNA strands are twisted together to form a structure that resembles a spiral staircase.
4. In living cells, DNA is associated with an array of different proteins to form **chromosomes**. The association of proteins with DNA organizes the long strands into a compact structure.
5. A **genome** is the complete complement of an organism's genetic material. For example, the genome of most

bacteria is a single circular chromosome, whereas eukaryotic cells have DNA in their nucleus, mitochondria, and chloroplasts.

The first three levels of complexity will be the focus of this section. Level 4 will be discussed in Section 11.5, and level 5 is examined in Chapter 21.

Nucleotides Contain a Phosphate, a Sugar, and a Base

A nucleotide has three components: a phosphate group, a pentose sugar, and a nitrogenous base. The nucleotides in DNA and RNA contain different sugars: Deoxyribose is found in DNA, and ribose is found in RNA. The base and phosphate group are attached to the sugar molecule at different sites (**Figure 11.6**).

Five different bases are found in nucleotides, although any given nucleotide contains only one base. The five bases are subdivided into two categories, the **purines** and the **pyrimidines**, due to differences in their structures (Figure 11.6). The purine bases, **adenine** (**A**) and **guanine** (**G**), have a double-ring structure; the pyrimidine bases, **thymine** (**T**), **cytosine** (**C**), and **uracil** (**U**), have a single-ring structure. Adenine, guanine, and cytosine are found in both DNA and RNA. Thymine is found only in DNA, whereas uracil is found only in RNA.

A conventional numbering system describes the locations of carbon and nitrogen atoms in the sugars and bases (**Figure 11.7**).

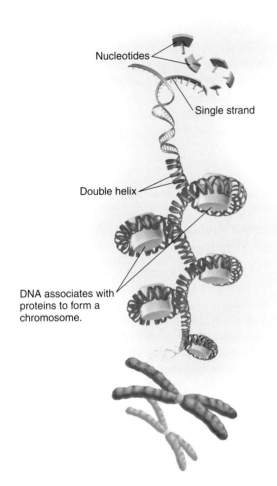

Figure 11.5 Levels of DNA structure to create a chromosome.

(a) DNA nucleotide

(b) RNA nucleotide

Figure 11.6 **Nucleotides and their components.** For simplicity, the carbon atoms in the ring structures are not shown.

Concept check: Which pyrimidine(s) is/are found in both DNA and RNA?

In the sugar ring, carbon atoms are numbered in a clockwise direction starting with the carbon atom to the right of the ring oxygen atom. The fifth carbon is outside the ring. The prime symbol (') is used to distinguish the numbering of carbons in the sugar. The atoms in the ring structures of the bases are not given the prime designation. The sugar carbons are designated 1' (that is, "one prime"), 2', 3', 4', and 5'. A base is attached to the 1' carbon atom, and a phosphate group is attached at the 5' position. Compared to ribose (see Figure 11.6), deoxyribose lacks a single oxygen atom at the 2' position; the prefix deoxy- (meaning without oxygen) refers to this missing atom.

A Strand Is a Linear Linkage of Nucleotides with Directionality

The next level of nucleotide structure is the formation of a strand of DNA or RNA in which nucleotides are covalently attached to each other in a linear fashion. **Figure 11.8** depicts a short strand of DNA with four nucleotides. The linkage is a phosphoester bond (a covalent bond between phosphorus and oxygen) involving a sugar molecule in one nucleotide and a phosphate group in the next nucleotide. Another way of viewing this linkage is to notice that a phosphate group connects

Figure 11.7 **Conventional numbering in a DNA nucleotide.** The carbons in the sugar are given a prime designation, whereas those in the base are not.

Concept check: What is the numbering designation of the carbon atom to which the phosphate is attached?

two sugar molecules. From this perspective, the linkage in DNA and RNA strands is called a **phosphodiester linkage**, which has two phosphoester bonds. The phosphates and sugar molecules form the **backbone** of a DNA or RNA strand, while the bases project from the backbone. The backbone is negatively charged due to the negative charges of the phosphate groups.

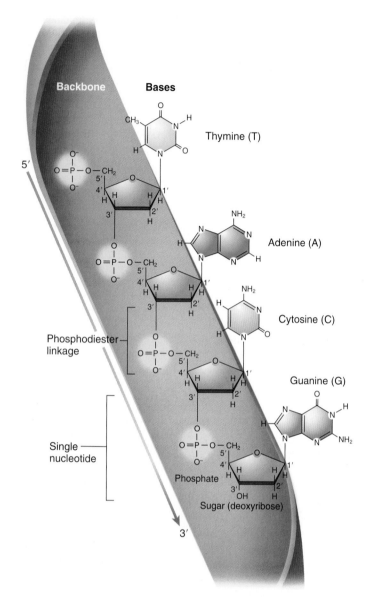

Figure 11.8 **The structure of a DNA strand.** Nucleotides are covalently bonded to each other in a linear manner. Notice the directionality of the strand and that it carries a particular sequence of bases. An RNA strand has a very similar structure, except the sugar is ribose rather than deoxyribose, and uracil is substituted for thymine.

Concept check: *What is the difference between a phosphoester bond versus a phosphodiester linkage?*

An important structural feature of a nucleic acid strand is the orientation of the nucleotides. Each phosphate in a phosphodiester linkage is covalently bonded to the 5′ carbon in one nucleotide and to the 3′ carbon in the other. In a strand, all sugar molecules are oriented in the same direction. For example, in the strand shown in Figure 11.8, all of the 5′ carbons in every sugar molecule are above the 3′ carbons. A strand has a **directionality** based on the orientation of the sugar molecules within that strand. In Figure 11.8, the direction of the strand is said to be 5′ to 3′ when going from top to bottom. The 5′ end

of a DNA strand has a phosphate group, while the 3′ end has an —OH group.

From the perspective of function, a key feature of DNA and RNA structure is that a strand contains a specific sequence of bases. In Figure 11.8, the sequence of bases is thymine—adenine—cytosine—guanine, or TACG. To indicate its directionality, the strand is abbreviated 5′–TACG–3′. Because the nucleotides within a strand are attached to each other by stable covalent bonds, the sequence of bases in a DNA strand will remain the same over time, except in rare cases when mutations occur. The sequence of bases in DNA and RNA is the critical feature that allows them to store and transmit information.

A Few Key Experiments Paved the Way to Solving the Structure of DNA

James Watson and Francis Crick wanted to determine the structure of DNA because they thought this knowledge would provide insights regarding the function of genes. Before we examine the characteristics of the double helix, let's consider the events that led to the discovery of the double helix structure.

In the early 1950s, more information was known about the structure of proteins than that of nucleic acids. Linus Pauling correctly proposed that regions of proteins can fold into a structure known as an α helix. To determine the structure of the α helix, Pauling built large models by linking together simple ball-and-stick units. In this way, he could see if atoms fit together properly in a complicated three-dimensional structure. This approach is still widely used today, except that now researchers construct three-dimensional models on computers. Watson and Crick also used a ball-and-stick approach to solve the structure of the DNA double helix.

What experimental approaches were used to analyze DNA structure? X-ray diffraction was a key experimental tool that led to the discovery of the DNA double helix. When a substance is exposed to X-rays, the atoms in the substance will cause the X-rays to be scattered (**Figure 11.9**). If the substance has a repeating structure, the pattern of scattering, known as the diffraction pattern, is mathematically related to the structural arrangement of the atoms causing the scattering. The diffraction pattern is analyzed using mathematical theory to provide information regarding the three-dimensional structure of the molecule. Rosalind Franklin, working in the 1950s in the same laboratory as Maurice Wilkins, was a gifted experimentalist who made marked advances in X-ray diffraction techniques involving DNA. The diffraction pattern of DNA fibers produced by Franklin suggested a helical structure with a diameter that is relatively uniform and too wide to be a single-stranded helix. In addition, the pattern provided information regarding the number of nucleotides per turn and was consistent with a 2-nm (nanometers) spacing between the strands in which a purine (A or G) bonds with a pyrimidine (T or C). These observations were instrumental in solving the structure of DNA.

Another piece of information that proved to be critical for the determination of the double helix structure came from the studies of Erwin Chargaff. In 1950, Chargaff analyzed the base

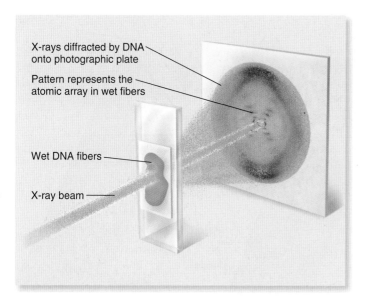

Figure 11.9 Rosalind Franklin's X-ray diffraction of DNA fibers. The exposure of X-rays to DNA wet fibers causes the X-rays to be scattered.

X-rays diffracted by DNA onto photographic plate

Pattern represents the atomic array in wet fibers

Wet DNA fibers

X-ray beam

composition of DNA that was isolated from many different species. His experiments consistently showed that the amount of adenine in each sample was similar to the amount of thymine, and the amount of cytosine was similar to the amount of guanine (**Table 11.1**). As we will see, this observation became crucial evidence that helped Watson and Crick develop the double helix model of DNA.

Watson and Crick Deduced the Double Helix Structure of DNA

Thus far, we have considered the experimental studies that led to the determination of the DNA double helix. These included the biochemical modeling approach of Pauling, the X-ray

Table 11.1	Base Content in the DNA from a Variety of Organisms as Determined by Chargaff			
	% of bases			
Organism	**Adenine**	**Thymine**	**Guanine**	**Cytosine**
Escherichia coli (bacterium)	26.0	23.9	24.9	25.2
Streptococcus pneumoniae (bacterium)	29.8	31.6	20.5	18.0
Saccharomyces cerevisiae (yeast)	31.7	32.6	18.3	17.4
Turtle	28.7	27.9	22.0	21.3
Salmon	29.7	29.1	20.8	20.4
Chicken	28.0	28.4	22.0	21.6
Human	30.3	30.3	19.5	19.9

diffraction work of Franklin, and the base composition studies of Chargaff. Watson and Crick assumed that nucleotides are linked together in a linear fashion and that the chemical linkage between two nucleotides is always the same. Along with Wilkins, they then set out to build ball-and-stick models that incorporated all of the known experimental observations.

Modeling of chemical structures involves trial and error. Watson and Crick initially considered several incorrect models. One model was a double helix in which the bases were on the outside of the helix. In another model, each base formed hydrogen bonds with the identical base in the opposite strand (A to A, T to T, G to G, and C to C). However, model-building revealed that purine-purine pairs were too wide and pyrimidine-pyrimidine pairs were too narrow to fit the uniform diameter of the double helix. Eventually, they realized that the hydrogen bonding of adenine to thymine was structurally similar to that of guanine to cytosine. In both cases, a purine base (A or G) bonds with a pyrimidine base (T or C). With an interaction between A and T and between G and C, the ball-and-stick models showed that the two strands would form a double helix structure in which all atoms would fit together properly.

Watson and Crick proposed the structure of DNA, which was published in the journal *Nature* in 1953. In 1962, Watson, Crick, and Wilkins were awarded the Nobel Prize in Physiology or Medicine. Unfortunately, Rosalind Franklin had died before this time, and the Nobel Prize is awarded only to living recipients.

DNA Has a Repeating, Antiparallel Helical Structure Formed by the Complementary Base Pairing of Nucleotides

The structure that Watson and Crick proposed is a double-stranded, helical structure with the sugar-phosphate backbone on the outside and the bases on the inside (**Figure 11.10a**). This structure is stabilized by hydrogen bonding between the bases in opposite strands to form **base pairs**. A distinguishing feature of base pairing is its specificity. An adenine (A) base in one strand forms two hydrogen bonds with a thymine (T) base in the opposite strand, or a guanine (G) base forms three hydrogen bonds with a cytosine (C) (**Figure 11.10b**). This **AT/GC rule** (also known as Chargaff's rule) is consistent with Chargaff's observation that DNA contains equal amounts of A and T, and equal amounts of G and C. According to the AT/GC rule, purines (A and G) always bond with pyrimidines (T and C) (recall that purines have a double-ring structure, whereas pyrimidines have single rings). This keeps the width of the double helix relatively constant. One complete turn of the double helix is composed of 10 base pairs.

Due to the AT/GC rule, the base sequences of two DNA strands are **complementary** to each other. That is, you can predict the sequence in one DNA strand if you know the sequence in the opposite strand. For example, if one strand has the sequence of 5′–GCGGATTT–3′, the opposite strand must be 3′–CGCCTAAA–5′. With regard to their 5′ and 3′ directionality,

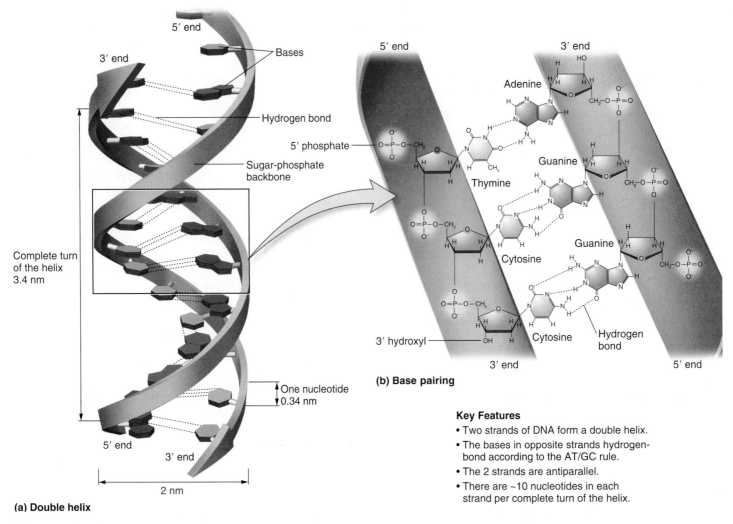

(a) Double helix

(b) Base pairing

Key Features
- Two strands of DNA form a double helix.
- The bases in opposite strands hydrogen-bond according to the AT/GC rule.
- The 2 strands are antiparallel.
- There are ~10 nucleotides in each strand per complete turn of the helix.

Figure 11.10 **Structure of the DNA double helix.** As seen in part **(a)**, DNA is a helix composed of two antiparallel strands. Part **(b)** shows the AT/GC base pairing that holds the strands together via hydrogen bonds.

Concept check: *If one DNA strand is 5′–GATTCGTTC–3′, what is the complementary strand?*

the two strands of a DNA double helix are **antiparallel**. If you look at Figure 11.10, one strand runs in the 5′ to 3′ direction from top to bottom, while the other strand is oriented 3′ to 5′ from top to bottom. Watson and Crick proposed an antiparallel structure in their original DNA model.

The DNA model in Figure 11.10a is called a ribbon model, which clearly shows the components of the DNA molecule. However, other models are also used to visualize DNA. The model for the DNA double helix shown in **Figure 11.11** is a space-filling model in which the atoms are depicted as spheres. Why is this model useful? This type of structural model emphasizes the surface of DNA. As you can see in this model, the sugar-phosphate backbone is on the outermost surface of the double helix; the backbone has the most direct contact with water. The atoms of the bases are more internally located within the double-stranded structure. The indentations where the atoms of the bases make contact with the surrounding water are termed **grooves**. Two grooves, called the **major groove** and the **minor groove**, spiral around the double helix. As discussed

in later chapters, the major groove provides a location where a protein can bind to a particular sequence of bases and affect the expression of a gene.

11.3 An Overview of DNA Replication

In the previous section, we considered the structure of the genetic material. DNA is a double helix that obeys the AT/GC rule. The structure of DNA immediately suggested to Watson and Crick a mechanism by which DNA can be copied. They proposed that during this process, known as **DNA replication**, the original DNA strands are used as templates for the synthesis of new DNA strands. In this section, we will look at an early experiment that helped to determine the mechanism of DNA replication and then examine the structural characteristics that enable a double helix to be faithfully copied.

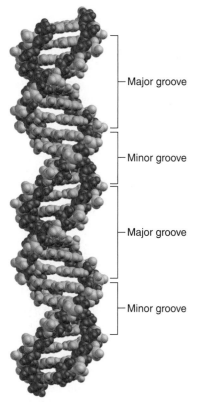

Figure 11.11
A space-filling model of the DNA double helix.
In the sugar-phosphate backbone, sugar molecules are shown in blue, and phosphate groups are yellow. The backbone is on the outermost surface of the double helix. The atoms of the bases, shown in green, are more internally located within the double-stranded structure. Notice the major and minor grooves that are formed by this arrangement.

Major groove

Minor groove

Major groove

Minor groove

Meselson and Stahl Used Density Measurements to Investigate Three Proposed Mechanisms of DNA Replication

Researchers in the late 1950s considered three different models for the mechanism of DNA replication (**Figure 11.12**). In all of these models, the two newly made strands are called the **daughter strands**, and the original strands are the **parental strands**. The first model is a **semiconservative mechanism** (Figure 11.12a). In this model, the double-stranded DNA is half conserved following the replication process such that the new double-stranded DNA contains one parental strand and one daughter strand. This mechanism is consistent with the ideas of Watson and Crick. Even so, other models were possible and had to be ruled out. According to a second model, called a **conservative mechanism**, both parental strands of DNA remain together following DNA replication (Figure 11.12b). The original arrangement of parental strands is completely conserved, while the two newly made daughter strands are also together following replication. Finally, a third possibility, called a **dispersive mechanism**, proposed that segments of parental DNA and newly made DNA are interspersed in both strands following the replication process (Figure 11.12c).

In 1958, Matthew Meselson and Franklin Stahl devised an experimental approach to distinguish among these three mechanisms. An important feature of their research was the use of isotope labeling. Nitrogen, which is found in DNA, occurs in a common light (^{14}N) form and a rare heavy (^{15}N) form. Meselson and Stahl studied DNA replication in the bacterium *Escherichia coli*. They grew *E. coli* cells for many generations in a medium that contained only the ^{15}N form of nitrogen (**Figure 11.13**). This produced a population of bacterial cells in which all of the DNA was heavy labeled. Then they switched the bacteria to

Original double helix	First round of replication	Second round of replication

Parental strand

Daughter strand

(a) Semiconservative mechanism. DNA replication produces DNA molecules with 1 parental strand and 1 newly made daughter strand.

(b) Conservative mechanism. DNA replication produces 1 double helix with both parental strands and the other with 2 new daughter strands.

(c) Dispersive mechanism. DNA replication produces DNA strands in which segments of new DNA are interspersed with the parental DNA.

Figure 11.12 Three proposed mechanisms for DNA replication. The strands of the original double helix are shown in red. Two rounds of replication are illustrated with new strands shown in blue.

a medium that contained only ^{14}N as its nitrogen source. The cells were allowed to divide, and samples were collected after one generation (that is, one round of DNA replication), two generations, and so on. Because the bacteria were doubling in a medium that contained only ^{14}N, all of the newly made DNA strands would be labeled with light nitrogen, while the original strands would remain labeled with the heavy form.

How were the DNA molecules analyzed? Meselson and Stahl used centrifugation to separate DNA molecules based on differences in density. Samples were placed on the top of a solution that contained a salt gradient, in this case, cesium chloride (CsCl). A double helix containing all heavy nitrogen

1. Grow bacteria in ^{15}N media.

^{15}N medium (heavy)

2. Transfer to ^{14}N media and continue growth for <1, 1.0, 2.0, or 3 generations.

^{14}N medium (light)

3. Isolate DNA after each generation. Transfer DNA to CsCl gradient, and centrifuge.

DNA

CsCl gradient

Centrifuge

4. Observe DNA under UV light.

5. **THE DATA**

Approximate generations after transfer to ^{14}N medium.

< 1.0	1.0	2.0	3.0	
				Light
				Half-heavy
				Heavy

Figure 11.13   The Meselson and Stahl experiment showing that DNA replication is semiconservative.

Concept check: *If this experiment were conducted for four rounds of DNA replication (that is, four generations), what would be the expected fractions of light DNA and half-heavy DNA according to the semiconservative model?*

has a higher density and will travel closer to the bottom of the gradient. By comparison, if both DNA strands contained ^{14}N, the DNA would have a light density and remain closer to the top of the gradient. If one strand contained ^{14}N and the other strand contained ^{15}N, the DNA would be half-heavy and have an intermediate density, ending up near the middle of the gradient.

After one cell doubling (that is, one round of DNA replication), all of the DNA exhibited a density that was half-heavy (Figure 11.13, step 5). These results are consistent with both the semiconservative and dispersive models. In contrast, the conservative mechanism predicts two different DNA types: a light type and a heavy type. Because the DNA was found in a single (half-heavy) band after one doubling, the conservative model was disproved. After two cell doublings, both light DNA and half-heavy DNA were observed. This result was also predicted by the semiconservative mechanism of DNA replication, because some DNA molecules should contain all light DNA, while other molecules should be half-heavy (see Figure 11.12a). However, in the dispersive mechanism, all of the DNA strands would have been 1/4 heavy after two generations. This mechanism predicts that the heavy nitrogen would be evenly dispersed among four double helices, each strand containing 1/4 heavy nitrogen and 3/4 light nitrogen (see Figure 11.12c). This result was not obtained. Taken together, the results of the Meselson and Stahl experiment are consistent only with a semiconservative mechanism for DNA replication.

Semiconservative DNA Replication Proceeds According to the AT/GC Rule

As originally proposed by Watson and Crick, semiconservative DNA replication relies on the complementarity of DNA strands according to the AT/GC rule. During the replication process, the two complementary strands of DNA separate and serve as **template strands** (also called parental strands) for the synthesis of new daughter strands of DNA (**Figure 11.14a**). After the double helix has separated, individual nucleotides have access to the template strands in a region called the replication fork. First, individual nucleotides hydrogen-bond to the template strands according to the AT/GC rule. Next, a covalent bond is formed between the phosphate of one nucleotide and the sugar of the previous nucleotide. The end result is that two double helices are made that have the same base sequence as the original DNA molecule (**Figure 11.14b**). This is a critical feature of DNA replication, because it enables the replicated DNA molecules to retain the same information (that is, the same base sequence) as the original molecule. In this way, DNA has the remarkable ability to direct its own duplication.

11.4   Molecular Mechanism of DNA Replication

Thus far, we have considered the general mechanism of DNA replication, known as semiconservative replication, and examined how DNA synthesis obeys the AT/GC rule. In this section,

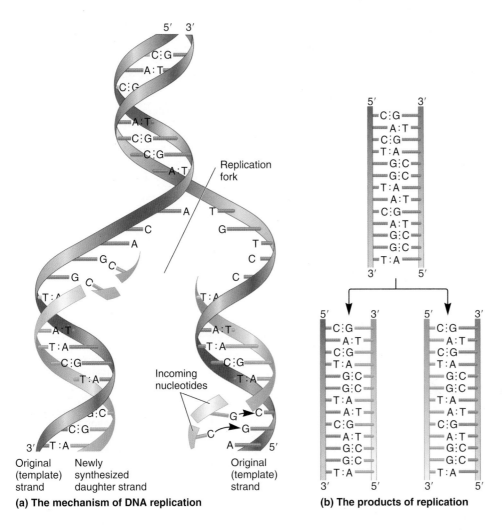

Original (template) strand   Newly synthesized daughter strand   Original (template) strand

(a) The mechanism of DNA replication

(b) The products of replication

Figure 11.14   DNA replication according to the AT/GC rule.   **(a)** The mechanism of DNA replication as originally proposed by Watson and Crick. As we will see in Section 11.4, the synthesis of one newly made strand (the leading strand on the left side) occurs in the direction toward the replication fork, whereas the synthesis of the other newly made strand (the lagging strand on the right side) occurs in small segments away from the fork. **(b)** DNA replication produces two copies of DNA with the same sequence as the original DNA molecule.

we will examine the details of DNA replication as it occurs inside living cells. As you will learn, several cellular proteins are needed to initiate DNA replication and allow it to proceed quickly and accurately.

DNA Replication Begins at an Origin of Replication, Where DNA Replication Forks Are Formed

Where does DNA replication begin? An **origin of replication** is a site within a chromosome that serves as a starting point for DNA replication. At the origin, the two DNA strands unwind (**Figure 11.15a**). DNA replication proceeds outward from two **replication forks**, a process termed **bidirectional replication** (Figure 11.15a). The number of origins of replication varies among different organisms. In bacteria, which have a small circular chromosome, a single origin of replication is found. Bidirectional replication starts at the origin of replication and proceeds until the new strands meet on the opposite side of the chromosome (**Figure 11.15b**). Eukaryotes have larger chromosomes that are linear. They require multiple origins of replication so the DNA can be replicated in a reasonable length of time. The newly made strands from each origin eventually make contact with each other to complete the replication process (**Figure 11.15c**).

DNA Replication Requires the Action of Several Different Proteins

Thus far, we have considered how DNA replication occurs outward from an origin of replication in a region called a DNA replication fork. In all living species, a set of several different proteins is involved in this process. An understanding of the functions of these proteins is critical to explaining the replication process at the molecular level.

Helicase, Topoisomerase, and Single-Strand Binding Proteins: Formation and Movement of the Replication Fork

To act as a template for DNA replication, the strands of a double helix must separate, and the resulting fork must move. As mentioned, an origin of replication serves as a site where this separation initially occurs. The strand separation at each fork then moves outward from the origin via the action of an enzyme called **DNA helicase**. At each fork, DNA helicase binds to one of the DNA strands and travels in the 5′ to 3′ direction toward the fork (**Figure 11.16**). It uses energy from ATP to separate the DNA strands and keeps the fork moving forward. The action of DNA helicase generates additional coiling just ahead of the replication fork that is alleviated by another enzyme called **DNA topoisomerase**.

(a) Bidirectional replication

(b) Single origin of replication in bacteria

(c) Multiple origins of replication in eukaryotes

Figure 11.15 **The bidirectional replication of DNA.** **(a)** DNA replication proceeds in both directions from an origin of replication. **(b)** Bacterial chromosomes have a single origin of replication, whereas **(c)** eukaryotes have multiple origins.

Figure 11.16 **Proteins that facilitate the formation and movement of a replication fork.**

After the two template DNA strands have separated, they must remain that way until the complementary daughter strands have been made. The function of **single-strand binding proteins** is to coat both of the single strands of template DNA and prevent them from re-forming a double helix. In this way, the bases within the template strands are kept exposed so they can act as templates for the synthesis of complementary strands.

DNA Polymerase and Primase: Synthesis of DNA Strands

The enzyme **DNA polymerase** is responsible for covalently linking nucleotides together to form DNA strands. Arthur Kornberg originally identified this enzyme in the 1950s. The structure of DNA polymerase resembles a human hand with the DNA threaded through it (**Figure 11.17a**). As DNA polymerase slides along the DNA, free nucleotides with three phosphate groups, called **deoxynucleoside triphosphates**, hydrogen-bond to the exposed bases in the template strand according to the AT/GC rule. At the catalytic site, DNA polymerase breaks a bond between the first and second phosphate and then attaches the resulting nucleotide with one phosphate group (a deoxynucleoside monophosphate) to the 3′ end of a growing strand via a phosphoester bond. The breakage of the covalent bond that releases pyrophosphate is an exergonic reaction that provides the energy to covalently connect adjacent nucleotides (**Figure 11.17b**). The pyrophosphate is broken down to two phosphates. The rate of synthesis is truly remarkable. In bacteria, DNA polymerase can synthesize DNA at a rate of 500 nucleotides per second, while eukaryotic species can make DNA at a rate of about 50 nucleotides per second.

DNA polymerase has two additional enzymatic features that affect how DNA strands are made. First, DNA polymerase

(a) Action of DNA polymerase

(b) Chemistry of DNA replication

Figure 11.17 Enzymatic synthesis of DNA. **(a)** Incoming deoxynucleoside triphosphates first hydrogen-bond to the template strand according to the AT/GC rule. DNA polymerase recognizes these deoxynucleoside triphosphates and attaches a deoxynucleoside monophosphate to the 3′ end of a growing strand. **(b)** DNA polymerase breaks the bond between the first and second phosphate in a deoxynucleoside triphosphate, causing the release of pyrophosphate. This provides the energy to form a covalent bond between the resulting deoxynucleoside monophosphate and the previous nucleotide in the growing strand. The pyrophosphate is broken down to two phosphates.

Concept check: *Does the oxygen in a new phosphoester bond come from the sugar or from the phosphate?*

is unable to begin DNA synthesis on a bare template strand. However, if a DNA or RNA strand is already attached to a template strand, DNA polymerase can elongate such a pre-existing strand by making DNA. A different enzyme called **DNA primase** is required if the template strand is bare. DNA primase makes a complementary **primer** that is actually a short segment of RNA, typically 10 to 12 nucleotides in length. These short RNA strands start, or prime, the process of DNA replication (**Figure 11.18a**). A second feature of DNA polymerase is that once synthesis has begun, it can synthesize new DNA only in a 5′ to 3′ direction (**Figure 11.18b**).

Leading and Lagging DNA Strands Are Made Differently

Let's now consider how new DNA strands are made at the replication forks. DNA replication occurs near the opening that forms each replication fork (**Figure 11.19**, step 1). The synthesis of a

DNA polymerase is able to covalently link nucleotides together from a primer, which is made by DNA primase.

DNA polymerase can link nucleotides only in the 5′ to 3′ direction.

(a) Need for a primer

(b) 5′ to 3′ direction of synthesis

Figure 11.18 Enzymatic feature of DNA polymerase. **(a)** DNA polymerase needs a primer to begin DNA synthesis, and **(b)** it can synthesize DNA only in the 5′ to 3′ direction.

strand always begins with an RNA primer (depicted in yellow), and the new DNA is made in the 5′ to 3′ direction. The manner in which the two daughter strands are synthesized is strikingly different. One strand, called the **leading strand**, is made in the same direction that the fork is moving. The leading strand is synthesized as one long continuous molecule. By comparison, the other daughter strand, termed the **lagging strand**, is made as a series of small fragments that are subsequently connected to each other to form a continuous strand. The synthesis of these fragments occurs in the direction away from the fork. For example, the lower fragment seen in Figure 11.19, steps 2 and 3, is synthesized from left to right. These DNA fragments are known as **Okazaki fragments**, after Reiji and Tuneko Okazaki, who initially discovered them in the late 1960s. As shown in Figure 11.19, step 4, the RNA primer is eventually removed, and adjacent Okazaki fragments are connected to each other to form a continuous strand of DNA.

Figure 11.20 shows the proteins involved with the synthesis of the leading and lagging strands in *Escherichia coli*. In this bacterium, two different DNA polymerases, called DNA polymerase I and DNA polymerase III, are primarily responsible for DNA replication. In the leading strand, DNA primase makes one RNA primer at the origin, and then DNA polymerase III attaches nucleotides in a 5′ to 3′ direction as it slides toward the opening of the replication fork. DNA polymerase III has a subunit called the clamp protein that allows it to slide along the template strand without falling off. In the lagging strand, DNA is also synthesized in a 5′ to 3′ direction, but this synthesis occurs in the direction away from the replication fork. In the lagging strand, short segments of DNA are made discontinuously as a series of Okazaki fragments, each of which requires its own primer. DNA polymerase III synthesizes the remainder of the fragment.

To complete the synthesis of Okazaki fragments within the lagging strand, three additional events must occur: the removal of the RNA primers, the synthesis of DNA in the area where the primers have been removed, and the covalent joining of adjacent fragments of DNA (Figure 11.20, steps 3 and 4). The RNA primers are removed by DNA polymerase I, which digests the linkages between nucleotides in a 5′ to 3′ direction. After the RNA primer is removed, DNA polymerase I fills in the vacant region with DNA. However, once the DNA has been completely filled in, a covalent bond is missing between the last nucleotide added by DNA polymerase I and the next DNA nucleotide in the adjacent Okazaki fragment. An enzyme known as **DNA ligase** catalyzes the formation of a covalent bond between these two DNA fragments to complete the replication process in the lagging strand (Figure 11.20, step 4). **Table 11.2** provides a summary of the functions of the proteins we have discussed in this section.

DNA Replication Is Very Accurate

Although errors can happen during DNA replication, permanent mistakes are extraordinarily rare. For example, during bacterial DNA replication, only 1 mistake per 100 million nucleotides

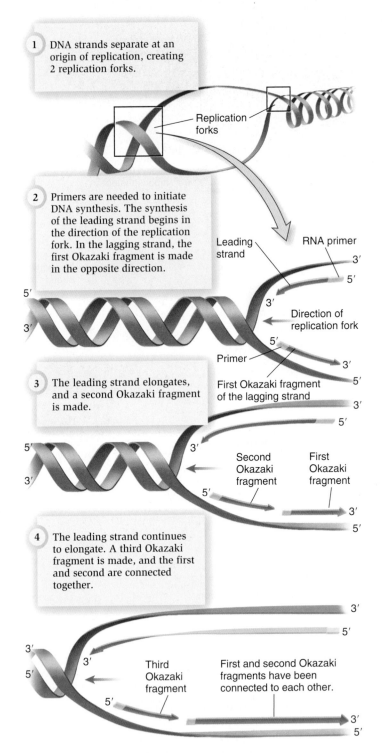

1 DNA strands separate at an origin of replication, creating 2 replication forks.

Replication forks

2 Primers are needed to initiate DNA synthesis. The synthesis of the leading strand begins in the direction of the replication fork. In the lagging strand, the first Okazaki fragment is made in the opposite direction.

Leading strand

RNA primer

3′
5′
3′

Direction of replication fork

5′
Primer
3′
5′

First Okazaki fragment of the lagging strand
3′
5′

3 The leading strand elongates, and a second Okazaki fragment is made.

5′
3′

3′

Second Okazaki fragment

First Okazaki fragment

5′

3′
5′

4 The leading strand continues to elongate. A third Okazaki fragment is made, and the first and second are connected together.

3′
5′

3′
3′
5′

Third Okazaki fragment

First and second Okazaki fragments have been connected to each other.

5′

3′
5′

Figure 11.19 **Synthesis of new DNA strands.** The separation of DNA at the origin of replication produces two replication forks that move in opposite directions. New DNA strands are made near the opening of each fork. The leading strand is made continuously in the same direction the fork is moving. The lagging strand is made as small pieces in the opposite direction. Eventually, these small pieces are connected to each other to form a continuous lagging strand.

Concept check: Which strand, the leading or lagging strand, is made discontinuously in the direction opposite to the movement of the replication fork?

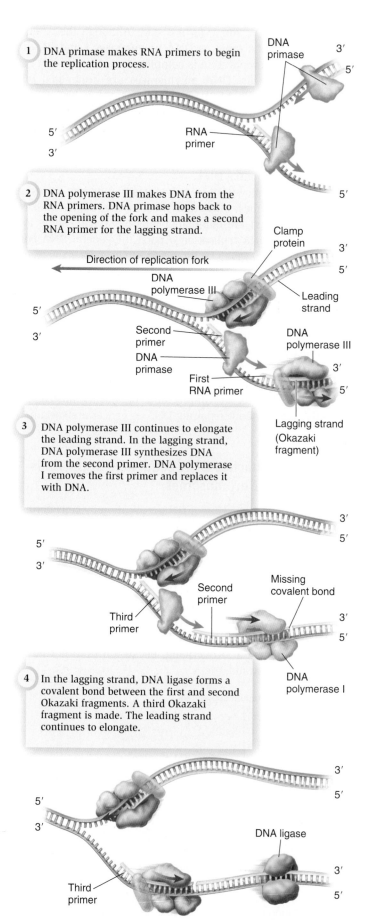

1 DNA primase makes RNA primers to begin the replication process.

2 DNA polymerase III makes DNA from the RNA primers. DNA primase hops back to the opening of the fork and makes a second RNA primer for the lagging strand.

3 DNA polymerase III continues to elongate the leading strand. In the lagging strand, DNA polymerase III synthesizes DNA from the second primer. DNA polymerase I removes the first primer and replaces it with DNA.

4 In the lagging strand, DNA ligase forms a covalent bond between the first and second Okazaki fragments. A third Okazaki fragment is made. The leading strand continues to elongate.

Figure 11.20 Proteins involved with the synthesis of the leading and lagging strands in *E. coli*.

Concept check: Briefly describe the movement of primase in the lagging strand in this figure. In which direction does it move when it is making a primer, from left to right or right to left? Describe how it must move after it is done making a primer and has to start making the next primer at a new location. Does it have to hop from left to right or from right to left?

Table 11.2	Proteins Involved in DNA Replication
Common name	**Function**
DNA helicase	Separates double-stranded DNA into single strands
Single-strand binding protein	Binds to single-stranded DNA and prevents it from re-forming a double helix
Topoisomerase	Removes tightened coils ahead of the replication fork
DNA primase	Synthesizes short RNA primers
DNA polymerase	Synthesizes DNA in the leading and lagging strands, removes RNA primers, and fills in gaps
DNA ligase	Covalently attaches adjacent Okazaki fragments in the lagging strand

is made. Biologists use the term high fidelity to refer to a process that occurs with relatively few mistakes. How can we explain such a remarkably high fidelity for DNA replication? First, hydrogen bonding between A and T or between G and C is more stable than between mismatched pairs. Second, the active site of DNA polymerase is unlikely to catalyze bond formation between adjacent nucleotides if a mismatched base pair is formed. Third, DNA polymerase can identify a mismatched nucleotide and remove it from the daughter strand. This event, called **proofreading**, occurs when DNA polymerase detects a mismatch and then reverses its direction and digests the linkages between nucleotides at the end of a newly made strand in the 3′ to 5′ direction. Once it passes the mismatched base and removes it, DNA polymerase then changes direction again and continues to synthesize DNA in the 5′ to 3′ direction.

Genomes & Proteomes Connection

DNA Polymerases Are a Family of Enzymes with Specialized Functions

Thus far, we have examined the general properties of DNA replication. Three important issues are speed, fidelity, and completeness. DNA replication must proceed quickly and with great accuracy, and gaps should not be left in the newly made strands. To ensure that these three requirements are met, living species produce more than one type of DNA polymerase, each of which may differ with regard to the rate and accuracy of DNA replication and/or the ability to prevent the formation of DNA gaps. Let's first consider how evolution produced these different forms of DNA polymerase and then examine how their functions are finely tuned to the process of DNA replication.

The genomes of living species have multiple DNA polymerase genes, which were produced by random gene duplication events. During evolution, mutations have altered each gene to produce a family of DNA polymerase enzymes with more specialized functions. Natural selection has favored certain mutations that result in DNA polymerase properties that are suited to the organism in which they are found. For comparison, let's consider the bacterium *E. coli* and humans. *E. coli* has five different DNA polymerases, designated I, II, III, IV, and V. In humans, over a dozen different DNA polymerases have been identified (**Table 11.3**). Why does *E. coli* need 5 DNA polymerases, while humans need 12 or more? The answer lies in specialization and the functional needs of each species.

In *E. coli*, DNA polymerase III is responsible for most DNA replication. It is composed of multiple subunits, each with its own functional role. In addition to the catalytic subunit that synthesizes DNA, DNA polymerase III has other subunits that allow it to clamp onto the template DNA and synthesize new DNA very rapidly and with high fidelity. By comparison, DNA polymerase I is composed of a single subunit. Its role during DNA replication is to remove the RNA primers and fill in the short vacant regions with DNA. DNA polymerases II, IV, and V are involved in repairing DNA and in replicating DNA that has been damaged. DNA polymerases I and III become stalled when they encounter DNA damage and may be unable to make a complementary strand at such a site. By comparison, DNA

polymerases II, IV, and V do not stall. Although their rate of synthesis is not as rapid as DNA polymerases I and III, they ensure that DNA replication is complete.

In human cells, DNA polymerases are designated with Greek letters (Table 11.3). DNA polymerase α has its own "built-in" primase subunit. It synthesizes RNA primers followed by short DNA regions. Two other DNA polymerases, δ (delta) and ε (epsilon), then extend the DNA at a faster rate. DNA polymerase γ (gamma) functions in the mitochondria to replicate mitochondrial DNA.

Several additional DNA polymerases function as lesion-replicating enzymes. Although most abnormalities in DNA structure (lesions) are eliminated by DNA repair, some may remain. When DNA replication occurs, the general DNA polymerases (α, δ, or ε) may be unable to replicate over the lesion. If this happens, lesion-replicating polymerases are attracted to the damaged DNA. These polymerases have special properties that enable them to synthesize a complementary strand over the lesion. Each type of lesion-replicating polymerase may be able to replicate over different kinds of DNA damage.

Similarly, other human DNA polymerases play an important role in DNA repair. The need for multiple repair enzymes is rooted in the various ways that DNA can be damaged, as described in Chapter 14. Multicellular organisms must be particularly vigilant about repairing DNA or cancer may occur.

Telomerase Attaches DNA Sequences at the Ends of Eukaryotic Chromosomes

We will end our discussion of DNA replication by considering a specialized form of DNA replication that happens at the ends of eukaryotic chromosomes. This region, called the **telomere**, contains a short nucleotide sequence that is repeated a few dozen to several hundred times in a row (**Figure 11.21**). The repeat sequence shown here, 5′–GGGTTA–3′, is the sequence found in human telomeres. Other organisms have different repeat sequences. For example, the sequence found in the telomeres of maize is 5′–GGGTTTA–3′. A telomere has a region at the 3′ end that is termed a 3′ overhang, because it does not have a complementary strand.

Table 11.3	DNA Polymerases in *E. coli* and Humans
Polymerase types*	**Functions**
E. coli	
III	Replicates most of the DNA during cell division
I	Removes RNA primers and fills in the gaps
II, IV, and V	Repairs damaged DNA and replicates over DNA damage
Humans	
α (alpha)	Makes RNA primers and synthesizes short DNA strands
δ (delta), ε (epsilon)	Displaces DNA polymerase α and then replicates DNA at a rapid rate
γ (gamma)	Replicates the mitochondrial DNA
η (eta), κ (kappa), ι (iota), ζ (zeta)	Replicates over damaged DNA
α, β (beta), δ, ε, σ (sigma), λ (lambda), μ (mu), ϕ (phi), θ (theta)	Repairs DNA or has other functions

*Certain DNA polymerases may have more than one function.

Figure 11.21 Telomere sequences at the end of a human chromosome. The telomere sequence shown here is found in humans and other mammals. The length of the 3′ overhang is variable among different species and cell types.

As discussed previously, DNA polymerase synthesizes DNA only in a 5′ to 3′ direction and requires a primer. For these reasons, DNA polymerase cannot copy the tip of a DNA strand with a 3′ end. Therefore, if this replication problem was not overcome, a linear chromosome would become progressively shorter with each round of DNA replication.

In 1984, Carol Greider and Elizabeth Blackburn discovered an enzyme called **telomerase** that prevents chromosome shortening by attaching many copies of a DNA repeat sequence to the ends of chromosomes (**Figure 11.22**). Telomerase contains both protein and RNA. The RNA part of telomerase is a sequence that is complementary to the DNA repeat sequence. This allows telomerase to bind to the 3′ overhang region of the telomere. Following binding, the RNA sequence beyond the binding site functions as a template, allowing telomerase to synthesize a six-nucleotide sequence at the end of the DNA strand. The enzyme then moves to the new end of this DNA strand and attaches another six nucleotides to the end. This occurs many times and thereby greatly lengthens the 3′ end of the DNA in the telomeric region. This lengthening provides an upstream site for an RNA primer to be made. DNA polymerase then synthesizes the complementary DNA strand. In this way, the progressive shortening of eukaryotic chromosomes is prevented.

Researchers have discovered an interesting connection between telomeres and cellular aging. In humans and other mammals, the cells of the body have a predetermined life span. For example, if a small sample of skin is removed from a person's body and grown in the laboratory, the cells will double a finite number of times. Furthermore, the number of doublings depends on the age of the person from which the sample is taken. If a sample is from an infant, the cells will typically double about 80 times, whereas if a sample is from an older person, the cells will double only 10 to 20 times before division ceases. Cells that have doubled many times and have reached a point where they have lost the capacity to divide any further are termed **senescent**.

The progressive shortening of telomeres is correlated with cellular senescence, though the relationship between the two phenomena is not well understood. The telomerase enzyme is normally present in germ-line cells, which give rise to gametes, and also in many rapidly dividing somatic cells. However, telomerase function is typically reduced as an organism ages. In 1998, Andrea Bodnar and her colleagues inserted a gene that encodes a highly active telomerase into human cells grown in the laboratory, using techniques described in Chapter 20. The results were amazing. The expression of telomerase prevented telomere shortening and cellular senescence. The cells expressing telomerase continued to divide, just like younger, healthy cells!

Telomerase function is also associated with cancer. When cells become cancerous, they continue to divide uncontrollably. In 90% of all types of human cancers, telomerase has been found to be present at high levels in the cancerous cells. This prevents telomere shortening and may play a role in the continued growth of cancer cells. The mechanism whereby cancer cells are able to increase the function of telomerase is not well understood and is a topic of active research.

Figure 11.22   Mechanism of DNA replication by telomerase.

Concept check: *What does telomerase use as a template to make DNA?*

11.5 Molecular Structure of Eukaryotic Chromosomes

We now turn our attention to the structure of eukaryotic chromosomes. A typical eukaryotic chromosome contains a single, linear, double-stranded DNA molecule that may be hundreds of millions of base pairs in length. If the DNA from a single set of human chromosomes were stretched from end to end, the length would be over 1 meter! By comparison, most eukaryotic cells are only 10–100 μm (micrometers) in diameter, and the cell nucleus is only about 2–4 μm in diameter. Therefore, to fit inside the nucleus, the DNA in a eukaryotic cell must be folded and packaged by a staggering amount.

The term chromosome is used to describe a discrete unit of genetic material. For example, a human somatic cell contains 46 chromosomes. By comparison, the term chromatin has a biochemical meaning. **Chromatin** is used to describe the DNA-protein complex that makes up eukaryotic chromosomes. The chromosomes found in the nucleus are composed of chromatin, as are the highly condensed chromosomes found in dividing cells. Chromosomes are very dynamic structures that alternate between tight and loose compaction states. In this section, we will focus our attention on two issues of chromosome structure. First, we will consider how chromosomes are compacted and organized within the cell nucleus. Then, we will examine the additional compaction that is necessary to produce the highly condensed chromosomes that occur during cell division.

DNA Wraps Around Histone Proteins to Form Nucleosomes

The first way that DNA is compacted is by wrapping itself around a group of proteins called **histones**. As shown in **Figure 11.23**, a repeating structural unit of eukaryotic chromatin is the **nucleosome**, which is 11 nanometers (nm) in diameter and composed of double-stranded DNA wrapped around an octamer of histone proteins. Each octamer contains two molecules of four types of histone proteins: H2A, H2B, H3, and H4. Histone proteins are very basic proteins because they contain a large number of positively charged lysine and arginine amino acids. The negative charges found in the phosphate of DNA are attracted to the positive charges on histone proteins. The DNA lies on the surface of the histone octamer and makes 1.65 turns around it. The amount of DNA required to wrap around the histone octamer is 146 or 147 bp (base pairs). The amino terminal tail of each histone protein protrudes from the histone octamer. As discussed in Chapter 13, these tails can be covalently modified and play a key role in gene regulation.

The nucleosomes are connected by linker regions of DNA that vary in length from 20 to 100 bp, depending on the species and cell type. A particular histone named histone H1 is bound to the linker region, as are other types of proteins. The overall structure of connected nucleosomes resembles beads on a string. This structure shortens the length of the DNA molecule about sevenfold.

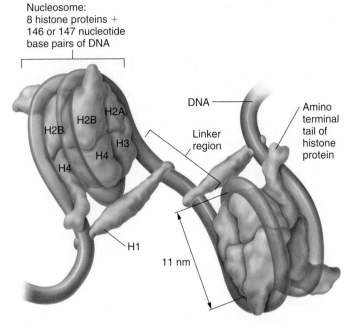

Nucleosome:
8 histone proteins +
146 or 147 nucleotide
base pairs of DNA

H2B H2B H2A DNA Amino terminal tail of histone protein

H4 H3 Linker region

H4

H1

11 nm

Figure 11.23 **Structure of a nucleosome.** A nucleosome is composed of double-stranded DNA wrapped around an octamer of histone proteins. A linker region connects two adjacent nucleosomes. Histone H1 is bound to the linker region, as are other proteins not shown in this figure.

Nucleosomes Form a 30-nm Fiber

Nucleosome units are organized into a more compact structure that is 30 nm in diameter, known as the **30-nm fiber** (**Figure 11.24a**). Histone H1 and other proteins are important in the formation of the 30-nm fiber, which shortens the nucleosome structure another sevenfold. The structure of the 30-nm fiber has proven difficult to determine because the conformation of the DNA may be substantially altered when extracted from living cells. A current model for the 30-nm fiber was proposed by Rachel Horowitz and Christopher Woodcock in the 1990s (**Figure 11.24b**). According to their model, linker regions in the 30-nm structure are variably bent and twisted, with little direct contact observed between nucleosomes. The 30-nm fiber forms an asymmetric, three-dimensional zigzag of nucleosomes. At this level of compaction, the overall picture of chromatin that emerges is an irregular, fluctuating structure with stable nucleosome units connected by bendable linker regions.

Chromatin Loops Are Anchored to the Nuclear Matrix

Thus far, we have examined two mechanisms that compact eukaryotic DNA: the formation of nucleosomes and their arrangement into a 30-nm fiber. Taken together, these two events shorten the folded DNA about 49-fold. A third level of

(a) Micrograph of a 30-nm fiber

30 nm

(b) Three-dimensional zigzag model

Figure 11.24 **The 30-nm fiber.** **(a)** A photomicrograph of the 30-nm fiber. **(b)** In this three-dimensional zigzag model, the linker DNA forms a bendable structure with little contact between adjacent nucleosomes.

Protein fiber inside the nucleus

30-nm fiber

Gene

Radial loop domain

Gene

Gene

Protein that attaches the base of a DNA loop to a protein fiber

Figure 11.25 Attachment of the 30-nm fiber to a protein fiber to form a radial loop domain.

Concept check: *What holds the bottoms of the loops in place?*

compaction involves interactions between the 30-nm fibers and a filamentous network of proteins in the nucleus called the **nuclear matrix**. This matrix consists of the **nuclear lamina**, which is composed of protein fibers that line the inner nuclear membrane (see Chapter 4), and an internal nuclear matrix that is connected to the lamina and fills the interior of the nucleus. The internal nuclear matrix is an intricate network of irregular protein fibers plus many other proteins that bind to these fibers. The nuclear matrix is involved in the compaction of the 30-nm fiber by participating in the formation of **radial loop domains**. These loops, often 25,000 to 200,000 base pairs in size, are anchored to the nuclear matrix (**Figure 11.25**).

How are chromosomes organized within the cell nucleus? Each chromosome in the cell nucleus is located in a discrete and nonoverlapping chromosome territory, which can be experimentally viewed in nondividing cells (refer back to Chapter 4, Figure 4.16). Each chromosome in nondividing cells occupies its own discrete region in the cell nucleus that usually does not overlap with the territory of adjacent chromosomes. In other words, different chromosomes are not substantially intertwined with each other, even when they are in a noncompacted condition.

The compaction level of chromosomes in the cell nucleus is not completely uniform. This variability can be seen with a light microscope and was first observed by the German cytologist E. Heitz in 1928. He used the term **heterochromatin** to describe the highly compacted regions of chromosomes. By comparison, the less condensed regions are known as **euchromatin**. Euchromatin is the form of chromatin in which the 30-nm fiber forms radial loop domains. In heterochromatin, these radial loop domains are compacted even further. In nondividing cells, most chromosomal regions are euchromatic, and some localized regions are heterochromatic.

During Cell Division, Chromosomes Undergo Maximum Compaction

When cells prepare to divide, the chromosomes become even more compacted or condensed. This aids in their proper alignment during metaphase, which is a stage of eukaryotic cell division described in Chapter 15. **Figure 11.26** illustrates the levels of compaction that contribute to the formation of a metaphase chromosome. DNA in the nucleus is always compacted by forming nucleosomes and condensing into a 30-nm fiber (Figure 11.26a,b,c). In euchromatin, the 30-nm fibers are arranged in radial loop domains that are relatively loose, meaning that a fair amount of space is between the 30-nm fibers (Figure 11.26d). The average width of such loops is about 300 nm.

By comparison, heterochromatin involves a much tighter packing of the loops, so little space is between the 30-nm fibers (Figure 11.26e). Heterochromatic regions tend to be wider, in the range of 700 nm. When cells prepare to divide, all of the euchromatin becomes highly compacted. The compaction of euchromatin greatly shortens the chromosomes. In a metaphase chromosome, which contains two copies of the DNA (Figure 11.26f), the width averages about 1,400 nm, but the length of a metaphase chromosome is much shorter than the same chromosome in the nucleus of a nondividing cell.

(a) DNA double helix

(b) Nucleosomes ("beads on a string")

(c) 30-nm fiber

(d) Radial loop domains

(e) Heterochromatin

(f) Metaphase chromosome

DNA double helix

2 nm

11 nm

Histones

Nucleosome

Histone H1

30 nm

300 nm

700 nm

1,400 nm

1 | Wrapping of DNA around histone proteins

2 | Formation of a 3-dimensional zigzag structure via histone H1 and other DNA-binding proteins

3 | Anchoring of radial loop domains to the nuclear matrix

4 | Further compaction of radial loops to form heterochromatin

5 | Metaphase chromosome with 2 copies of the DNA

Figure 11.26   The steps in eukaryotic chromosomal compaction leading to the metaphase chromosome.

Concept check: After they have replicated and become compacted in preparation for cell division, chromosomes are often shaped like an X, as in part (f) of this figure. Which proteins are primarily responsible for this X shape?

Summary of Key Concepts

11.1 Biochemical Identification of the Genetic Material

- The genetic material carries information to produce the traits of organisms. It is replicated and transmitted from cell to cell and generation to generation, and it has differences that explain the variation among different organisms.

- Griffith's work with type S and type R bacteria was consistent with the transfer of genetic material, which he called the transformation principle. (Figure 11.1)

- Avery, MacLeod, and McCarty used biochemical methods to show that DNA is the transformation principle. (Figure 11.2)

- Hershey and Chase labeled T2 phage with ^{35}S and ^{32}P and determined that the ^{32}P-labeled DNA is the genetic material of this phage. (Figures 11.3, 11.4)

11.2 Nucleic Acid Structure

- DNA is composed of nucleotides, which covalently link to form DNA strands. Two DNA strands are held together by hydrogen bonds between the bases to form a double helix. Chromosomes are made of DNA and proteins. (Figure 11.5)

- Nucleotides are composed of a phosphate, sugar, and nitrogenous base. The sugar can be deoxyribose (DNA) or ribose (RNA). The purine bases are adenine and guanine, and the pyrimidine bases are thymine (DNA only), cytosine, and uracil (RNA only). (Figure 11.6)

- The atoms in a nucleotide are numbered in a conventional way. (Figure 11.7)

- In a strand of DNA (or RNA), the sugars are connected by covalent bonds in a 5′ to 3′ direction. (Figure 11.8)

- Watson and Crick used the X-ray diffraction data of Franklin and the biochemical data of Chargaff (that is, A = T, G = C), and constructed ball-and-stick models to reveal the double helix structure of DNA. (Figure 11.9, Table 11.1)

- DNA is a double helix in which the DNA strands are antiparallel and obey the AT/GC rule. (Figures 11.10, 11.11)

11.3 An Overview of DNA Replication

- Meselson and Stahl used ^{15}N- and ^{14}N-isotope labeling methods to show that DNA is replicated by a semiconservative mechanism in which the product of DNA replication is one original strand and one new strand. (Figures 11.12, 11.13)

- New DNA strands are made according to the AT/GC rule in which parental strands serve as templates for the synthesis of new daughter strands. The result of DNA replication is two double helices with the same base sequence. (Figure 11.14)

11.4 Molecular Mechanism of DNA Replication

- DNA synthesis occurs bidirectionally from an origin of replication. The synthesis of new DNA strands happens near each replication fork. (Figure 11.15)

- DNA helicase separates DNA strands, single-strand binding proteins keep them separated, and DNA topoisomerase alleviates coiling ahead of the fork. (Figure 11.16)

- Deoxynucleoside triphosphates bind to the template strands, and DNA polymerase catalyzes the formation of a phosphoester bond between the 3′ end of the strand and a deoxynucleoside monophosphate. (Figure 11.17)

- DNA polymerase requires a primer and can make new DNA strands only in the 5′ to 3′ direction. (Figure 11.18)

- The leading strand is made continuously, in the same direction the fork is moving. The lagging strand is made in the opposite direction as short Okazaki fragments that are connected together. (Figure 11.19)

- DNA primase makes one RNA primer in the leading strand and multiple RNA primers in the lagging strand. In *E. coli*, DNA polymerase III extends these primers with DNA, and DNA polymerase I removes the primers when they are no longer needed. DNA ligase connects adjacent Okazaki fragments in the lagging strand. (Figure 11.20, Table 11.2)

- Living organisms have several different types of DNA polymerases with specialized functions. (Table 11.3)

- The ends of linear, eukaryotic chromosomes have telomeres composed of repeat sequences. Telomerase binds to the telomere repeat sequence and synthesizes a six-nucleotide repeat. This happens many times in a row to lengthen one DNA strand of the telomere. DNA primase, DNA polymerase, and DNA ligase are needed to synthesize the complementary DNA strand. (Figures 11.21, 11.22)

11.5 Molecular Structure of Eukaryotic Chromosomes

- Chromosomes are structures in living cells that carry the genetic material. Chromatin is the name given to the DNA-protein complex that makes up chromosomes.

- In eukaryotic chromosomes, the DNA is wrapped around histone proteins to form nucleosomes. Nucleosomes are further compacted into 30-nm fibers. The linker regions are variably twisted and bent into a zigzag pattern. (Figures 11.23, 11.24)

- A third level of compaction of eukaryotic chromosomes involves the formation of radial loop domains in which the bases of 30-nm fibers are anchored to a network of proteins called the nuclear matrix. This level of compaction is called euchromatin. In heterochromatin, the loops are even more closely packed together. (Figure 11.25)

- Chromosome compaction to produce a metaphase chromosome involves the conversion of all euchromatin into heterochromatin. (Figure 11.26)

▮ Assess and Discuss

Test Yourself

1. Why did researchers initially believe the genetic material was protein?
   a. Proteins are more biochemically complex than DNA.
   b. Proteins are found only in the nucleus, but DNA is found in many areas of the cell.
   c. Proteins are much larger molecules and can store more information than DNA.
   d. all of the above
   e. both a and c

2. Considering the components of a nucleotide, what component is always different when comparing nucleotides in a DNA strand or an RNA strand?
   a. phosphate group   d. both b and c
   b. pentose sugar   e. a, b, and c
   c. nitrogenous base

3. Which of the following equations would be appropriate when considering DNA base composition?
   a. %A + %T = %G + %C
   b. %A = %G
   c. %A = %G = %T = %C
   d. %A + %G = %T + %C

4. If the sequence of a segment of DNA is 5′-CGCAACTAC-3′, what is the appropriate sequence for the opposite strand?
   a. 5′-GCGTTGATG-3′
   b. 3′-ATACCAGCA-5′
   c. 5′-ATACCAGCA-3′
   d. 3′-GCGTTGATG-5′

5. Of the following statements, which is correct when considering the process of DNA replication?
   a. New DNA molecules are composed of two completely new strands.
   b. New DNA molecules are composed of one strand from the old molecule and one new strand.
   c. New DNA molecules are composed of strands that are a mixture of sections from the old molecule and sections that are new.
   d. none of the above

6. Meselson and Stahl were able to demonstrate semiconservative replication in *E. coli* by
   a. using radioactive isotopes of phosphorus to label the old strand and visually determining the relationship of old and new DNA strands.
   b. using different enzymes to eliminate old strands from DNA.
   c. using isotopes of nitrogen to label the DNA and determining the relationship of old and new DNA strands by density differences of the new molecules.
   d. labeling viral DNA before it was incorporated into a bacterial cell and visually determining the location of the DNA after centrifugation.

7. During replication of a DNA molecule, the daughter strands are not produced in exactly the same manner. One strand, the leading strand, is made toward the replication fork, while the lagging strand is made in fragments in the opposite direction. This difference in the synthesis of the two strands is the result of which of the following?
   a. DNA polymerase is not efficient enough to make two "good" strands of DNA.
   b. The two template strands are antiparallel, and DNA polymerase makes DNA only in the 5′ to 3′ direction.
   c. The lagging strand is the result of DNA breakage due to UV light.
   d. The cell does not contain enough nucleotides to make two complete strands.

8. In eukaryotic cells, chromosomes consist of
   a. DNA and RNA.
   b. DNA only.
   c. RNA and proteins.
   d. DNA and proteins.
   e. RNA only.

9. A nucleosome is
   a. a dark-staining body composed of RNA and proteins found in the nucleus.
   b. a protein that helps organize the structure of chromosomes.
   c. another word for a chromosome.
   d. a structure composed of DNA wrapped around eight histones.
   e. the short arm of a chromosome.

10. The conversion of euchromatin into heterochromatin involves
    a. the formation of more nucleosomes.
    b. the formation of less nucleosomes.
    c. a greater compaction of loop domains.
    d. a lesser compaction of loop domains.
    e. both a and c.

Conceptual Questions

1. What are the four key characteristics of the genetic material? What was Frederick Griffith's contribution to the study of DNA, and why was it so important?

2. The Hershey and Chase experiment used radioactive isotopes to track the DNA and protein of phages as they infected bacterial cells. Explain how this procedure allowed them to determine that DNA is the genetic material of this particular virus.

3. Explain or describe the essential features of the Watson and Crick model of the structure of DNA.

Collaborative Questions

1. A trait that some bacterial strains exhibit is resistance to killing by antibiotics. For example, certain strains of bacteria are resistant to tetracycline, whereas other strains are sensitive to this antibiotic. Describe an experiment you would carry out to demonstrate that tetracycline resistance is an inherited trait carried in the DNA of the resistant strain.

2. How might you provide evidence that DNA is the genetic material in mice?

Online Resource

www.brookerbiology.com

Stay a step ahead in your studies with animations that bring concepts to life and practice tests to assess your understanding. Your instructor may also recommend the interactive ebook, individualized learning tools, and more.

Gene Expression at the Molecular Level

12

An electron micrograph of many ribosomes in the act of translating two mRNA molecules into many polypeptides. The short polypeptides are seen emerging from the ribosomes.

M ina, age 21, works part-time in an ice-cream shop and particularly enjoys the double-dark chocolate and chocolate fudge brownie flavors on her breaks. She exercises little and spends most of her time studying and watching television. Mina is effortlessly thin. She never worries about what or how much she eats. By comparison, her close friend, Rezzy, has struggled with her weight as long as she can remember. Compared to Mina, she feels like she must constantly deprive herself of food just to maintain her current weight—a weight she would describe as 30 pounds too much.

How do we explain the differences between Mina and Rezzy? Two fundamental factors are involved. Obesity is strongly influenced by the environment, especially a person's diet, along with social and behavioral factors. The amount and types of food we eat are correlated with weight gain. However, there is little doubt that obesity is also influenced by variation in our genes. Obesity runs in families. A similarity is observed in the degree of obesity between genetically identical twins who have been raised apart. Why has genetic variation resulted in some genes that cause certain people to gain weight? A popular hypothesis is that we have "thrifty genes" as hand-me-downs from our ancestors, who periodically faced famines and food scarcity. Such thrifty genes allow us to store body fat more easily and to use food resources more efficiently when times are lean. The negative side is that when food is abundant, unwanted weight gain can become a serious health problem.

Why do we care about our genes? Let's consider this question with regard to obesity. Researchers have identified several key genes that influence a person's predisposition to becoming obese. Dozens more are likely to play a minor role. By identifying those genes and studying the proteins specified by those genes, researchers may gain a better understanding of how genetic variation can cause certain people to gain weight more easily than others. In addition, this knowledge has led to the development of drugs that are used to combat obesity.

In this chapter, we begin to explore the inner workings of genes. We can broadly define a gene as a unit of heredity. Geneticists view gene function at different biological levels. In Chapter 16, we will examine how genes affect the traits or characteristics of individuals.

For example, we will consider how the transmission of genes from parents to offspring affects the color of the offspring's eyes and their likelihood of becoming bald. In this chapter, we will explore how genes work at the molecular level. You will learn how DNA sequences are organized to form genes and how those genes are used as a template to make RNA copies, ultimately leading to the synthesis of a functional protein. The term **gene expression** can refer to gene function either at the level of traits or at the molecular level. In reality, the two phenomena are intricately woven together. The expression of genes at the molecular level affects the structure and function of cells, which, in turn, determine the traits that an organism expresses.

We begin this chapter by considering how researchers came to realize that most genes store the information to make proteins. Then we will explore the steps of gene expression as they occur at the molecular level. These steps include the use of a gene as a template to make an RNA molecule, the processing of the RNA into a functional molecule, and the use of RNA to direct the formation of a protein.

12.1 Overview of Gene Expression

Even before DNA was known to be the genetic material, scientists had asked the question, How does the functioning of genes create the traits of living organisms? At the molecular level, a similar question can be asked. How do genes affect the composition and/or function of molecules found within living cells? An approach that was successful in answering these questions involved the study of **mutations**, which are changes in the genetic material. Mutations can affect the genetic blueprint by altering gene function. For this reason, research that focused on the effects of mutations proved instrumental in determining the molecular function of genes.

In this section, we will consider two early experiments in which researchers studied the effects of mutations in humans and in a bread mold. Both studies led to the conclusion that the role of some genes is to carry the information to produce enzymes. Then we will examine the general features of gene expression at the molecular level.

The Study of Inborn Errors of Metabolism Suggested That Some Genes Carry the Information to Make Enzymes

In 1908, Archibald Garrod, a British physician, proposed a relationship between genes and the production of enzymes. Prior to his work, biochemists had studied many metabolic pathways that consist of a series of conversions of one molecule to another, each step catalyzed by an enzyme. **Figure 12.1** illustrates part of the metabolic pathway for the breakdown of phenylalanine, an amino acid commonly found in human diets. The enzyme phenylalanine hydroxylase catalyzes the conversion of phenylalanine to tyrosine, another amino acid. A different enzyme, tyrosine aminotransferase, converts tyrosine into the next molecule, called *p*-hydroxyphenylpyruvic acid. In each case, a specific enzyme catalyzes a single chemical reaction.

Much of Garrod's early work centered on the inherited disease alkaptonuria, in which the patient's body accumulates abnormal levels of homogentisic acid (also called alkapton). This compound, which is bluish black, results in discoloration of the skin and cartilage and causes the urine to appear black. Garrod hypothesized that the accumulation of homogentisic acid in these patients is due to a defect in an enzyme, namely, homogentisic acid oxidase (Figure 12.1). Furthermore, he already knew that alkaptonuria is an inherited condition that follows a recessive pattern of inheritance. As discussed in Chapter 16, if a disorder is recessive, an individual with the disease has inherited the mutant (defective) gene that causes the disorder from both parents.

How did Garrod explain these observations? In 1908, he proposed a relationship between the inheritance of a mutant gene and a defect in metabolism. In the case of alkaptonuria, if an individual inherited the mutant gene from both parents,

she or he would not produce any normal enzyme and would be unable to metabolize homogentisic acid. Garrod described alkaptonuria as an **inborn error of metabolism**. An inborn error refers to a mutation in a gene that is inherited from one or both parents. At the turn of the last century, this was a particularly insightful idea because the structure and function of the genetic material were completely unknown.

Figure 12.1   The metabolic pathway that breaks down phenylalanine and its relationship to certain genetic diseases. Each step in the pathway is catalyzed by a different enzyme, shown in the boxes on the right. If one of the enzymes is not functioning, the previous compound builds up, causing the disorders named in the boxes on the left.

Concept check: What disease would occur if a person had inherited two defective copies of the gene that encodes phenylalanine hydroxylase?

Beadle and Tatum Proposed the One Gene–One Enzyme Hypothesis

In early 1940s, George Beadle and Edward Tatum became aware of Garrod's work and were interested in the relationship between genes and enzymes. They focused their studies on *Neurospora crassa*, a common bread mold. *Neurospora* is easily grown in the laboratory and has only a few nutritional requirements: a carbon source (namely, sugar), inorganic salts, and one vitamin known as biotin. Otherwise, *Neurospora* has many different enzymes that synthesize the molecules, such as amino acids and many vitamins, that are essential for growth.

Like Garrod, Beadle and Tatum hypothesized that genes carry the information to make specific enzymes. They reasoned that a mutation, or change in a gene, might cause a defect in an enzyme required for the synthesis of an essential molecule, such as an amino acid or vitamin. A mutant *Neurospora* strain (one that carries such a mutation) would be unable to grow unless the amino acid or vitamin was supplemented in the growth medium. Strains without a mutation are called wild-type. In their original study of 1941, Beadle and Tatum exposed *Neurospora* cells to X-rays, which caused mutations to occur, and studied the resulting cells. By plating the cells on growth media with or without vitamins, they were able to identify mutant strains that required vitamins for growth. In each case, a single mutation resulted in the requirement for a single type of vitamin in the growth media.

This early study by Beadle and Tatum led to additional research by themselves and others to study enzymes involved in the synthesis of other substances, including the amino acid arginine. At that time, the pathway leading to arginine synthesis was known to involve certain precursor molecules, including ornithine and citrulline. A simplified pathway for arginine synthesis is shown in **Figure 12.2a**. Each step is catalyzed by a different enzyme.

Researchers first isolated several different mutants that required arginine for growth. They hypothesized that each mutant strain might be blocked at only a single step in the consecutive series of reactions that lead to arginine synthesis. To test this hypothesis, the mutant strains were examined for their ability to grow in the presence of ornithine, citrulline, or arginine (**Figure 12.2b**). The wild-type strain could grow on minimal growth media that did not contain ornithine, citrulline, or arginine. Based on their growth properties, the mutant strains that had been originally identified as requiring arginine for growth could be placed into three groups, designated 1, 2, and 3. Group 1 mutants were missing enzyme 1, needed for the conversion of a precursor molecule into ornithine. They could grow only if ornithine, citrulline, or arginine was added to the growth medium. Group 2 mutants were missing the second enzyme in this pathway that is needed for the conversion of ornithine into citrulline. The group 2 mutants would not grow if only ornithine was added, but could grow if citrulline or arginine was added. Finally, the group 3 mutants were missing the enzyme needed for the conversion of citrulline into arginine.

Precursor molecule ——(Enzyme 1)——▶ Ornithine ——(Enzyme 2)——▶ Citrulline ——(Enzyme 3)——▶ Arginine

(a) Simplified pathway for arginine synthesis

(b) Growth of strains on minimal and supplemented growth media

Figure 12.2 An experiment that supported Beadle and Tatum's one gene–one enzyme hypothesis. (a) This simplified pathway shows three enzymes that are required for arginine synthesis. (b) Growth of wild-type (WT) and mutant *Neurospora* strains (groups 1, 2, and 3) on minimal plates or in the presence of ornithine, citrulline, or arginine.

Concept check: *What type of enzyme function is missing in group 2 mutants?*

These mutants could grow only if arginine was added. How were these results interpreted? The researchers were able to order the functions of the genes involved in arginine synthesis in the following way:

Group 1   Group 2   Group 3

Precursor ⟶ Ornithine ⟶ Citrulline ⟶ Arginine

From these results and earlier studies, Beadle and Tatum concluded that a single gene controlled the synthesis of a single enzyme. This was referred to as the **one gene–one enzyme hypothesis**.

In later decades, this idea was modified in three ways. First, the information to make all proteins is contained within genes, and many proteins do not function as enzymes. Second, some proteins are composed of two or more different polypeptides. The term **polypeptide** refers to a linear sequence of amino acids; it denotes structure. Most genes carry the information to make a particular polypeptide. By comparison, the term **protein** denotes function. Some proteins are composed of one polypeptide. In such cases, a single gene does contain the information to make a single protein. In other cases, however, a functional protein is composed of two or more different polypeptides. An example is hemoglobin, the protein that carries oxygen in red blood cells, which is composed of two α-globin and two β-globin polypeptides. In this case, the expression of two genes (that is, the α-globin and β-globin genes) is needed to create a functional protein. A third modification to the one gene–one enzyme hypothesis is that some genes encode RNAs that are not used to make polypeptides. For example, as discussed later in this chapter, some genes encode RNA molecules that form part of the structure of ribosomes.

Molecular Gene Expression Involves the Processes of Transcription and Translation

Thus far, we have considered two classic studies that led researchers to conclude that some genes carry the information to make enzymes. Let's now consider the general steps of gene expression at the molecular level. The first step, known as **transcription**, produces an RNA copy of a gene, also called an RNA transcript (**Figure 12.3**). The term transcription literally means the act of making a copy. Most genes, which are termed **structural genes**[1], produce an RNA molecule that contains the information to specify a polypeptide with a particular amino acid sequence. This type of RNA is called **messenger RNA** (abbreviated **mRNA**), because its job is to carry information from the DNA to cellular components called ribosomes. As discussed later, ribosomes play a key role in the synthesis of polypeptides. The process of synthesizing a specific polypeptide on a ribosome is called **translation**. The term translation is used because a nucleotide sequence in mRNA is "translated" into an amino acid sequence of a polypeptide.

Together, the transcription of DNA into mRNA and the translation of mRNA into a polypeptide constitute the **central dogma** of gene expression at the molecular level, which was first proposed by Francis Crick in 1958 (Figure 12.3). The

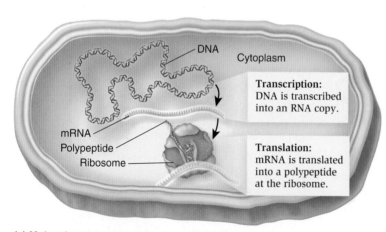

(a) Molecular gene expression in prokaryotes

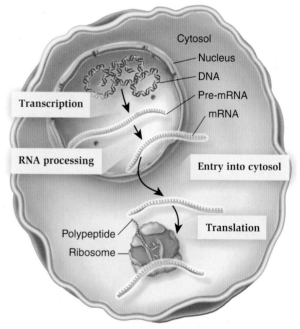

(b) Molecular gene expression in eukaryotes

Figure 12.3   **The central dogma of gene expression at the molecular level.**   **(a)** In bacteria, transcription and translation occur in the cytoplasm. **(b)** In eukaryotes, transcription and RNA processing occur in the nucleus, whereas translation takes place in the cytosol.

Concept check: *What is the direction of flow of genetic information?*

---

[1] Geneticists commonly use the term structural gene to describe all genes that encode polypeptides, which is how it is used in this textbook. Some geneticists, however, distinguish structural genes from regulatory genes—genes that encode proteins regulating the expression of structural genes.

central dogma applies equally to prokaryotes and eukaryotes. However, in eukaryotes, two additional steps occur between transcription and translation. During **RNA processing**, the RNA transcript, termed **pre-mRNA**, is modified in ways that make it a functionally active mRNA (Figure 12.3b). The processing events will be described later in this chapter. The mRNA is then transported into the cytosol. Though the direction of information flow—DNA → RNA → protein—is the most common pathway, exceptions do occur. For example, certain viruses can use RNA as a template to synthesize DNA. Such viruses are described in Chapter 18.

The Protein Products of Genes Determine an Organism's Characteristics

The genes that constitute the genetic material provide a blueprint for the characteristics of every organism. They contain the information necessary to create an organism and allow it to favorably interact with its environment. Each structural gene stores the information for the production of a polypeptide, which then becomes a unit within a functional protein. The activities of proteins determine the structure and function of cells. Furthermore, the characteristics of an organism are rooted in the activities of cellular proteins.

The main purpose of the genetic material is to encode the production of proteins in the correct cell, at the proper time, and in suitable amounts. This is an intricate task, because living cells make thousands of different kinds of proteins. Genetic analyses have shown that a typical bacterium can make a few thousand different proteins, and estimates for eukaryotes range from several thousand in simpler eukaryotes to tens of thousands in more complex eukaryotes like humans.

12.2   Transcription

DNA is an information storage unit. For genes to be expressed, the information in them must be accessed at the molecular level. Rather than accessing the information directly, however, a working copy of the DNA, composed of RNA, is made. This occurs by the process of transcription, in which a DNA sequence is copied into an RNA sequence. Importantly, transcription does not permanently alter the structure of DNA. Therefore, the same DNA can continue to store information even after an RNA copy has been made. In this section, we will examine the steps necessary for genes to act as transcriptional units. We will also consider some differences in these steps between prokaryotes and eukaryotes.

At the Molecular Level, a Gene Can Be Transcribed and Produces a Functional Product

What is a gene? At the molecular level, a **gene** is defined as an organized unit of DNA sequences that enables a segment of DNA to be transcribed into RNA and ultimately results in

the formation of a functional product. When a structural gene is transcribed, an mRNA is made that specifies the amino acid sequence of a polypeptide. After it is made, the polypeptide becomes a functional product. The mRNA is an intermediary in polypeptide synthesis. Among all species, most genes are structural genes. However, for some genes, the functional product is the RNA itself. The RNA from a nonstructural gene is never translated. Two important products of nonstructural genes are transfer RNA and ribosomal RNA. **Transfer RNA (tRNA)** translates the language of mRNA into that of amino acids; **ribosomal RNA (rRNA)** forms part of ribosomes, which provide the site where translation occurs. We'll learn more about these two types of RNA later in this chapter.

A gene is composed of specific base sequences organized in a way that allows the DNA to be transcribed into RNA. Figure 12.4 shows the general organization of sequences in a structural gene. Transcription begins next to a site in the DNA called the **promoter**, whereas the **terminator** specifies the end of transcription. Therefore, transcription occurs between these two boundaries. As shown in Figure 12.4, the DNA is transcribed into mRNA from the end of the promoter through the coding sequence to the terminator. Within this transcribed region is the information that will specify the amino acid sequence of a polypeptide when the mRNA is translated.

Other DNA sequences are involved in the regulation of transcription. These **regulatory sequences** function as sites for genetic regulatory proteins. When a regulatory protein binds to a regulatory sequence, the rate of transcription is affected. Some regulatory proteins enhance the rate of transcription, while others inhibit it.

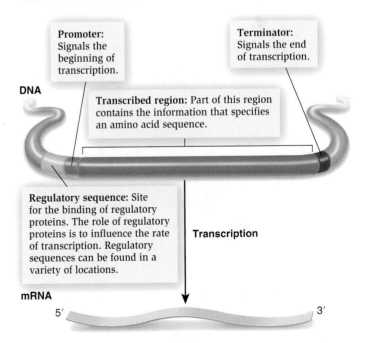

Figure 12.4   A structural gene as a transcriptional unit.

Concept check: If a terminator was removed from a gene, how would this affect transcription? Where would transcription end?

During Transcription, RNA Polymerase Uses a DNA Template to Make RNA

Transcription occurs in three stages, called initiation, elongation, and termination, during which proteins interact with DNA sequences (**Figure 12.5**). The **initiation stage** is a recognition step. In bacteria such as *E. coli*, a protein called **sigma factor** binds to **RNA polymerase**, the enzyme that synthesizes strands of RNA. Sigma factor also recognizes the base sequence of a promoter and binds there. In this way, sigma factor causes RNA polymerase to specifically bind to a promoter sequence. The initiation stage is completed when the DNA strands are separated near the promoter to form an **open complex** that is approximately 10 to 15 base pairs long.

During the **elongation stage**, RNA polymerase synthesizes the RNA transcript. For this to occur, sigma factor is released and RNA polymerase slides along the DNA in a way that maintains an open complex as it goes. The DNA strand that is used as a template for RNA synthesis is called the **template strand**. The opposite DNA strand is called the **coding strand**. The coding strand has the same sequence of bases as the mRNA, except that thymine in the DNA is substituted for uracil in the RNA. The coding strand is so named because, like mRNA, it carries the information that codes for a polypeptide.

During the elongation stage of transcription, nucleotides bind to the template strand and are covalently connected in the 5′ to 3′ direction (see inset of step 2, Figure 12.5). The complementarity rule used in this process is similar to the AT/GC rule of DNA replication, except that uracil (U) substitutes for thymine (T) in RNA. For example, an RNA with a sequence reading 5′–AUGUUACAUCGG–3′ will be transcribed from a DNA template with a sequence of 3′–TACAATGTAGCC–5′. In bacteria, the rate of RNA synthesis is about 40 nucleotides per second! Behind the open complex, the DNA rewinds back into a double helix. Eventually, RNA polymerase reaches a terminator, which causes it and the newly made RNA transcript to dissociate from the DNA. This event constitutes the **termination stage** of transcription.

The catalytic portion of RNA polymerase that is responsible for the synthesis of RNA has a similar structure in all species. The structure of a bacterial RNA polymerase is shown in **Figure 12.6**. RNA polymerase contains a cavity that allows it to slide along the DNA. The DNA strands enter at the side of the protein, and RNA is made in a 5′ to 3′ direction. Both the DNA and the newly made strand of RNA then exit from the top of the protein (see Figure 12.5).

When considering the transcription of multiple genes within a chromosome, the direction of transcription and the

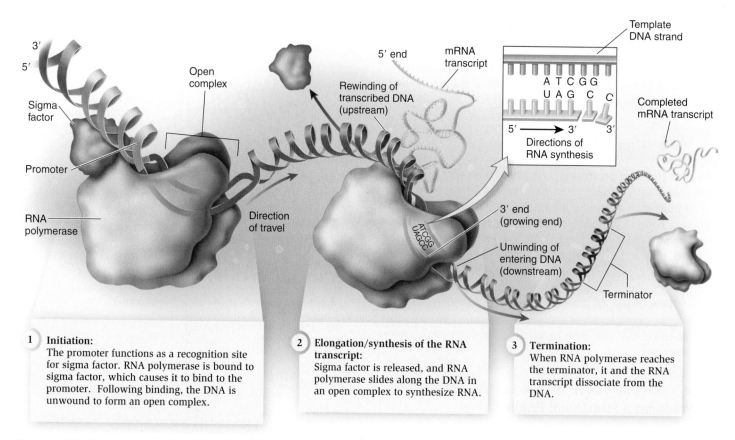

1 Initiation:
The promoter functions as a recognition site for sigma factor. RNA polymerase is bound to sigma factor, which causes it to bind to the promoter. Following binding, the DNA is unwound to form an open complex.

2 Elongation/synthesis of the RNA transcript:
Sigma factor is released, and RNA polymerase slides along the DNA in an open complex to synthesize RNA.

3 Termination:
When RNA polymerase reaches the terminator, it and the RNA transcript dissociate from the DNA.

Figure 12.5 **Stages of transcription.** Transcription can be divided into initiation, elongation, and termination. The inset emphasizes the direction of RNA synthesis and base pairing between the DNA template strand and RNA. Note: Some recent evidence suggests that sigma factor may not always be released during elongation.

Figure 12.6 Three-dimensional structure of a bacterial RNA polymerase.

DNA strand that is used as a template vary among different genes. **Figure 12.7** shows three genes adjacent to each other within a chromosome. Genes *A* and *B* are transcribed from left to right, using the bottom DNA strand as the template strand. By comparison, gene *C* is transcribed from right to left, using the top DNA strand as a template strand. In all three cases, the synthesis of the RNA transcript begins at the promoter and occurs in a 5′ to 3′ direction. The template strand is read in the 3′ to 5′ direction.

Transcription Is Similar in Prokaryotes and Eukaryotes, Except That Eukaryotes Use More Proteins

The basic features of transcription are similar between prokaryotic and eukaryotic organisms. Eukaryotic and prokaryotic genes have promoters, and the transcription process involves the stages of initiation, elongation, and termination.

The transcription of eukaryotic genes tends to involve a greater complexity of protein components. For example, three forms of RNA polymerase are found in eukaryotes, designated I, II, and III. RNA polymerase II is responsible for transcribing the mRNA from eukaryotic structural genes, whereas RNA polymerases I and III transcribe nonstructural genes such as the genes that encode tRNAs and rRNAs. By comparison, bacteria have a single type of RNA polymerase that transcribes all genes.

The initiation stage of transcription in eukaryotes is also more complex. Recall that bacteria such as *E. coli* use a single protein, sigma factor, to recognize the promoter of genes. By comparison, RNA polymerase II of eukaryotes always requires five general transcription factors to initiate transcription. **Transcription factors** are proteins that influence the ability of RNA polymerase to transcribe genes. In addition, the regulation of gene transcription in eukaryotes typically involves the function of several different proteins. The roles of eukaryotic transcription factors are considered in Chapter 13.

12.3 RNA Processing in Eukaryotes

During the 1960s and 1970s, the physical structure of the gene became well established based largely on studies of bacterial genes. Most bacterial mRNAs can be translated into polypeptides as soon as they are made. By comparison, eukaryotic mRNA transcripts undergo RNA processing or modification that is needed for their proper translation. In eukaryotes, transcription initially produces a longer RNA, called **pre-mRNA**, which undergoes certain processing events before it exits the nucleus. The final product is called a **mature mRNA** or simply mRNA.

In the late 1970s, when the experimental tools became available to study eukaryotic genes at the molecular level, the scientific community was astonished by the discovery that the coding sequences within many eukaryotic structural genes are

Figure 12.7 **The transcription of three different genes that are found in the same chromosome.** RNA polymerase synthesizes each RNA transcript in a 5′ to 3′ direction, sliding along a DNA template strand in a 3′ to 5′ direction. However, the use of the template strand can vary from gene to gene. For example, genes *A* and *B* use the bottom strand, while gene *C* uses the top strand.

separated by DNA sequences that are transcribed but not translated into protein. These intervening sequences that are not translated are called **introns**, while coding sequences are found within **exons** contained in the mature mRNA. The exons are <u>ex</u>pressed regions, whereas the introns are <u>inter</u>vening regions that are not expressed because they are removed from the mRNA.

To create a functional mRNA, the pre-mRNA undergoes a process known as **splicing**, in which the introns are removed and the remaining exons are connected to each other (**Figure 12.8**). In addition to splicing, eukaryotic pre-mRNA transcripts are modified in other ways, including the addition of caps and tails to the ends of the mRNA. After these modifications have been completed, the mRNA leaves the nucleus and enters the cytosol, where translation occurs. In this section, we will examine the molecular mechanisms that account for these RNA processing events and consider why they are functionally important.

Splicing Involves the Removal of Introns and the Linkage of Exons

Introns are found in many eukaryotic genes. Splicing is less frequent among unicellular eukaryotic species, such as yeast, but is a widespread phenomenon among more complex eukaryotes. In many animals and flowering plants, most structural genes have one or more introns. For example, an average human gene has about 9 introns. The sizes of introns can vary from a few dozen nucleotides to over one hundred thousand! A few bacterial genes have been found to have introns, but they are rare among all prokaryotic species.

Introns are precisely removed from eukaryotic pre-mRNA by a large complex called a **spliceosome** that is composed

of several different subunits known as snRNPs (pronounced "snurps"). Each snRNP contains <u>s</u>mall <u>n</u>uclear <u>R</u>NA and a set of <u>p</u>roteins. This small nuclear RNA is the product of a nonstructural gene. How are introns identified and removed? Intron RNA is defined by particular sequences within the intron and at the intron-exon boundaries (**Figure 12.9**). These include a 5′ splice site, a branch site, and a 3′ splice site. Spliceosome subunits bind to specific sequences at these three locations. This binding causes the intron to loop outward, which brings the two exons close together. The 5′ splice site is then cut, and the 5′ end of the intron becomes covalently attached to the branch site. In the final step, the 3′ splice site is cut, and then the exons are covalently attached to each other. The intron is released and eventually degraded.

In some cases, the function of the spliceosome can be regulated so the splicing of exons for a given mRNA can occur in two or more ways. This phenomenon, called **alternative splicing**, enables a single gene to encode two or more polypeptides with differences in their amino acid sequences (see Chapter 13).

Although primarily found in mRNAs, introns occasionally occur in rRNA and tRNA molecules of certain species. These introns, however, are not removed by the action of a spliceosome. Instead, such rRNAs and tRNAs are **self-splicing**, which means the RNA itself can catalyze the removal of its own intron. Portions of the RNA act like an enzyme to cleave the covalent bonds at the intron-exon boundaries and connect the exons together. An RNA molecule that catalyzes a chemical reaction is termed a **ribozyme**.

RNA Processing Also Involves Adding a 5′ Cap and a 3′ Poly A Tail to the Ends of Eukaryotic mRNAs

Mature mRNAs of eukaryotes have a modified guanosine covalently attached at the 5′ end, an event known as **capping** (**Figure 12.10a**). Capping occurs while a pre-mRNA is being made by RNA polymerase, usually when the transcript is only 20 to 25 nucleotides in length. What are the functions of the cap? The 7-methylguanosine structure, called a **5′ cap**, is recognized by cap-binding proteins, which are needed for the proper exit of mRNAs from the nucleus. After an mRNA is in the cytosol, the cap structure is recognized by other cap-binding proteins that enable the mRNA to bind to a ribosome for translation.

At the 3′ end, most mature eukaryotic mRNAs have a string of adenine nucleotides, typically 100 to 200 nucleotides in length, referred to as a **poly A tail** (**Figure 12.10b**). A long poly A tail aids in the export of mRNAs from the nucleus. It also causes a eukaryotic mRNA to be more stable and thereby exist for a longer period of time in the cytosol. The poly A tail is not encoded in the gene sequence. Instead, the tail is added enzymatically after a pre-mRNA has been completely transcribed. Interestingly, new research has shown that some bacterial mRNAs also have poly A tails attached to them. However, the poly A tail has an opposite effect in bacteria, where it causes the mRNA to be rapidly degraded. The importance of a poly A tail in bacterial mRNAs is not well understood.

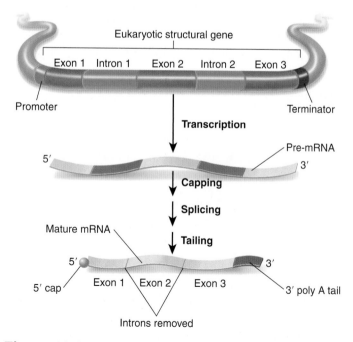

Figure 12.8   Modifications to eukaryotic pre-mRNA that are needed to create a functional (mature) mRNA molecule.

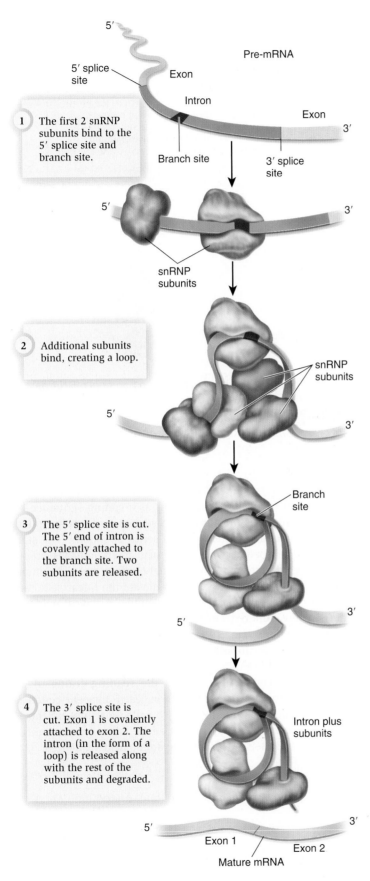

① The first 2 snRNP subunits bind to the 5' splice site and branch site.

② Additional subunits bind, creating a loop.

③ The 5' splice site is cut. The 5' end of intron is covalently attached to the branch site. Two subunits are released.

④ The 3' splice site is cut. Exon 1 is covalently attached to exon 2. The intron (in the form of a loop) is released along with the rest of the subunits and degraded.

Figure 12.9 The splicing of a eukaryotic pre-mRNA by a spliceosome.

(a) Cap structure at the 5' end of eukaryotic mRNA

A poly A tail consisting of 100–200 adenine nucleotides is added after transcription.

(b) Addition of a poly A tail at the 3' end of eukaryotic mRNA

Figure 12.10 Modifications that occur at the ends of mRNA in eukaryotic cells. **(a)** A guanosine cap is attached to the 5' end. This guanosine is modified by having a methyl group attached to it. The linkage between the cap and the mRNA is a 5' to 5' linkage. **(b)** A poly A tail is added to the 3' end.

Concept check: Do the ends of structural genes have a poly T region that provides a template for the synthesis of a poly A tail in mRNA? Explain.

12.4 Translation and the Genetic Code

In the two previous sections, we considered how an RNA transcript is made and how eukaryotes process that transcript. Now we will begin to examine the next process, that of translation, at the molecular level. In 1960, Matthew Meselson and Francois Jacob found that proteins are synthesized on cellular structures known as ribosomes. One year later, Francois Jacob and Jacques Monod made an insightful hypothesis. They proposed that RNA, which is transcribed from DNA, provides the information for protein synthesis via ribosomes. This type of RNA, which they named messenger RNA (mRNA), carries information from the DNA to the ribosome, where polypeptides are made during the process called translation.

Since these early studies, much has been learned about the details of translation. To understand the process of translation, we will first examine the **genetic code**, which specifies the relationship between the sequence of nucleotides in the mRNA and the sequence of amino acids in a polypeptide.

During Translation, the Genetic Code Is Used to Make a Polypeptide with a Specific Amino Acid Sequence

The ability of mRNA to be translated into a polypeptide relies on the genetic code. The code is read in groups of three nucleotide bases known as **codons**. The genetic code consists of 64 different codons (**Table 12.1**). The sequence of three bases in most codons specifies a particular amino acid. For example, the codon CCC specifies the amino acid proline, whereas the codon GGC encodes the amino acid glycine. From the analysis of many different species, including bacteria, protists, fungi, plants, and animals, researchers have found that the genetic code is nearly universal. Only a few rare exceptions to the genetic code have been discovered.

Why are there 64 codons, as shown in Table 12.1? Because there are 20 types of amino acids, at least 20 different codons are needed so that each amino acid can be specified by a codon. With four types of bases in mRNA (U, C, A, and G), a genetic code containing two bases in a codon would not be sufficient, because only 4^2, or 16, different codons would be possible. A three-base system can specify 4^3, or 64, different codons, which is far more than the number of amino acids. The genetic code is said to be **degenerate** because more than one codon can specify the same amino acid (Table 12.1). For example, the codons GGU, GGC, GGA, and GGG all code for the amino acid glycine. In most instances, the third base in the codon is the degenerate or variable base.

Let's look at the organization of a bacterial mRNA to see how translation occurs (**Figure 12.11**). A ribosomal-binding site

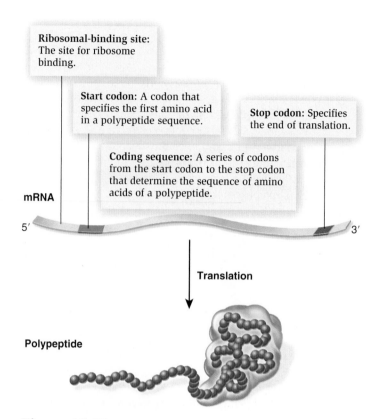

Figure 12.11   The organization of a bacterial mRNA as a translational unit.

Concept check: *If a start codon was missing from a gene, how would that affect transcription, and how would it affect translation?*

is located near the 5′ end of the mRNA. Beyond this site, a large portion of an mRNA functions as a **coding sequence**—a region that specifies the amino acid sequence of a polypeptide. This coding sequence consists of a series of codons. The **start codon**, which specifies the amino acid methionine, is only a few nucleotides from the ribosomal-binding site. The many codons that follow the start codon dictate the linear sequence of amino acids within a given polypeptide. A typical polypeptide is a few hundred amino acids in length. Finally, one of three **stop codons** signals the end of translation. These codons, also known as **termination codons** or **nonsense codons**, are UAA, UAG, and UGA.

The start codon also defines the **reading frame** of an mRNA. Beginning at the start codon, each adjacent codon is read as a group of three bases, also called a **triplet**, in the 5′ to 3′ direction. For example, look at the following two mRNA sequences and their corresponding amino acid sequences.

	Ribosomal-binding site	Start codon

mRNA 5′–<u>AUAAGGAGGU</u>UACG(<u>AUG</u>)(CAG)(CAG)(GGC)(UUU)(ACC)–3′

Polypeptide   Met - Gln - Gln - Gly - Phe - Thr

	Ribosomal-binding site	Start codon

mRNA 5′–<u>AUAAGGAGGU</u>UACG(<u>AUG</u>)(UCA)(GCA)(GGG)(CUU)(UAC)C–3′

Polypeptide   Met - Ser - Ala - Gly - Leu - Tyr

Table 12.1	The Genetic Code*		

		Second position			
		U	**C**	**A**	**G**	
U	UUU } Phe UUC UUA } Leu UUG	UCU UCC } Ser UCA UCG	UAU } Tyr UAC UAA Stop UAG Stop	UGU } Cys UGC UGA Stop UGG Trp	U C A G
C	CUU CUC } Leu CUA CUG	CCU CCC } Pro CCA CCG	CAU } His CAC CAA } Gln CAG	CGU CGC } Arg CGA CGG	U C A G
A	AUU AUC } Ile AUA AUG Met/start	ACU ACC } Thr ACA ACG	AAU } Asn AAC AAA } Lys AAG	AGU } Ser AGC AGA } Arg AGG	U C A G
G	GUU GUC } Val GUA GUG	GCU GCC } Ala GCA GCG	GAU } Asp GAC GAA } Glu GAG	GGU GGC } Gly GGA GGG	U C A G

(First Position on left axis, Third Position on right axis)

*Exceptions to the genetic code are sporadically found among various species. For example, AUA encodes methionine in yeast and mammalian mitochondria.

The first sequence shows how the mRNA codons would be correctly translated into amino acids. In the second sequence, an additional U has been added to the same sequence after the start codon. This shifts the reading frame and thereby changes the codons as they occur in the 5′ to 3′ direction. The polypeptide produced from this series of codons would have a very different sequence of amino acids. From this comparison, we can also see that the reading frame is not overlapping, which means that each base functions within a single codon.

The relationships among the DNA sequence of a gene, the mRNA transcribed from the gene, and the polypeptide sequence are shown schematically in **Figure 12.12**. The coding strand of DNA corresponds to the mRNA strand, except that T in the DNA is substituted for U in the mRNA. The template strand is used to make mRNA. The 5′ end of the mRNA contains an untranslated region as does the 3′ end. The middle portion contains a series of codons that specify the amino acid sequence of a polypeptide.

To translate a nucleotide sequence of mRNA into an amino acid sequence, recognition occurs between mRNA and transfer RNA (tRNA) molecules. Transfer RNA, which is described in Section 12.5, functions as the "translator" or intermediary between an mRNA codon and an amino acid. The **anticodon** is a three-base sequence in a tRNA molecule that is complementary to a codon in mRNA. Due to this complementarity, the anticodon in the tRNA and a codon in an mRNA bind to each other. Furthermore, the anticodon in a tRNA corresponds to the amino acid that it carries. For example, if the anticodon in a tRNA is 3′–AAG–5′, it is complementary to a 5′–UUC–3′ codon. According to the genetic code, a UUC codon specifies phenylalanine (Phe). Therefore, a tRNA with a 3′–AAG–5′ anticodon must carry phenylalanine. As another example, a tRNA with a 3′–GGG–5′ anticodon is complementary to a 5′–CCC–3′ codon, which specifies proline. This tRNA must carry proline (Pro).

As seen at the bottom of Figure 12.12, the direction of polypeptide synthesis parallels the 5′ to 3′ orientation of mRNA. The first amino acid is said to be at the **N-terminus** or **amino terminus** of the polypeptide. The term N-terminus refers to the presence of a nitrogen atom (N) at this end, while amino terminus indicates the presence of an amino group (NH_2). **Peptide bonds** connect the amino acids together. These covalent bonds form between the carboxyl group of the previous amino acid and the amino group of the next amino acid. The last amino acid in a completed polypeptide does not have another amino acid attached to its carboxyl group. This last amino acid is said to be located at the **C-terminus**, or **carboxyl terminus**.

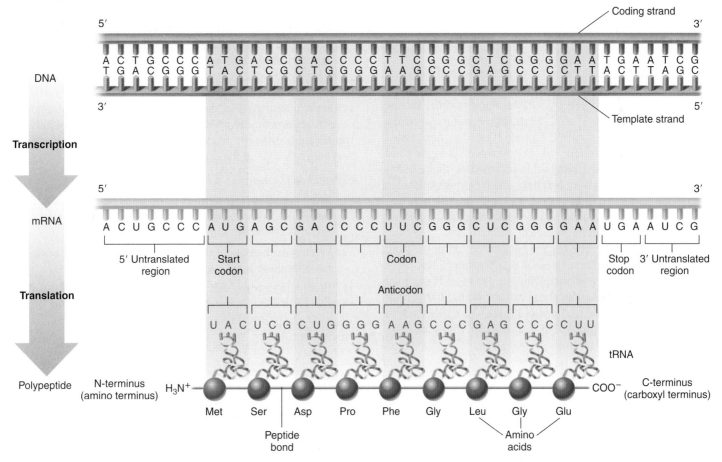

Figure 12.12 Relationships among the coding sequence of a gene, the codon sequence of an mRNA, the anticodons of tRNA, and the amino acid sequence of a polypeptide.

Concept check: *If an anticodon in a tRNA molecule has the sequence 3′–ACC–5′, which amino acid does it carry?*

A carboxyl group (COOH) is always found at this end of the polypeptide chain. Note that at neutral pH, the amino group is positively charged (NH_3^+), whereas the carboxyl group is negatively charged (COO^-).

Synthetic RNA Helped to Decipher the Genetic Code

Now let's look at some early experiments that allowed scientists to decipher the genetic code. During the early 1960s, the genetic code was determined by the collective efforts of several researchers, including Marshall Nirenberg, Severo Ochoa, and Philip Leder. Prior to their studies, other scientists had discovered that bacterial cells can be broken open and components from the cytoplasm can synthesize polypeptides. This is termed an in vitro or cell-free translation system. Nirenberg and Ochoa made synthetic RNA molecules using an enzyme that covalently connects nucleotides together. Using this synthetic mRNA, they then determined which amino acids were incorporated into polypeptides. For example, if an RNA molecule had only adenine-containing nucleotides (for example, 5'-AAAAAAAAAAAAAAAAAAAA-3'), a polypeptide was produced that contained only lysine. This result indicated that the AAA codon specifies lysine.

Another method used to decipher the genetic code involved the chemical synthesis of short RNA molecules. This method is described next in the Feature Investigation.

FEATURE INVESTIGATION

Nirenberg and Leder Found That RNA Triplets Can Promote the Binding of tRNA to Ribosomes

In 1964, Nirenberg and Leder discovered that RNA molecules containing three nucleotides (that is, a triplet) can stimulate ribosomes to bind a tRNA molecule. In other words, an RNA triplet can act like a codon within an mRNA molecule. Ribosomes bind RNA triplets, and then a tRNA with the appropriate anticodon subsequently binds to the ribosome.

To establish the relationship between triplet sequences and specific amino acids, Nirenberg and Leder made triplets with specific base sequences (**Figure 12.13**). For example, in one experiment they studied 5'–CCC–3' triplets. A particular triplet was added to 20 different tubes. To each tube, they next added an in vitro translation system, which contained ribosomes and tRNAs that already had amino acids attached to them. However, each translation system had only one type of radiolabeled amino acid. One translation system had only proline that was radiolabeled, a second translation system had only serine that was radiolabeled, and so on.

As shown in step 2, the triplets became bound to the ribosomes just like the binding of mRNA to a ribosome. The tRNA with an anticodon that was complementary to the added triplet would bind to the triplet, which was already bound to the ribosome. For example, if the triplet was 5'–CCC–3', a tRNA with a 3'–GGG–5' anticodon would bind to the triplet/ribosome complex. This tRNA carries proline.

To determine which tRNA had bound, the contents from each tube were poured through a filter that trapped the large ribosomes but did not trap tRNAs that were not bound to ribosomes (see step 3). If the tRNA carrying the radiolabeled amino acid was bound to the triplet/ribosome complex, radioactivity would be trapped on the filter. Using a scintillation counter, the researchers determined the amount of radioactivity on each filter. Because only one amino acid was radiolabeled in each in vitro translation system, they could determine which triplet corresponded to which amino acid. In the example shown here, CCC corresponds to proline. Therefore, the in vitro translation system containing radiolabeled proline showed a large amount of radioactivity on the filter. As shown in the data, by studying triplets with different sequences, Nirenberg and Leder identified many codons of the genetic code.

Experimental Questions

1. Briefly explain how a triplet mimics the role of an mRNA molecule. How was this observation useful in the study done by Nirenberg and Leder?

2. What was the benefit of using radiolabeled amino acids in the Nirenberg and Leder experiment?

3. Predict the results that Nirenberg and Leder would have found for the following triplets: AUG, UAA, UAG, or UGA.

Figure 12.13   Nirenberg and Leder's use of triplet binding assays to decipher the genetic code.

HYPOTHESIS   An RNA triplet can bind to a ribosome and promote the binding of the tRNA that carries the amino acid that the RNA triplet specifies.

KEY MATERIALS   The researchers made 20 in vitro translation systems, which included ribosomes, tRNAs, and 20 amino acids. The 20 translation systems differed with regard to which amino acid was radiolabeled. For example, in 1 translation system, radiolabeled glycine was added, and the other 19 amino acids were unlabeled. In another system, radiolabeled proline was added, and the other 19 amino acids were unlabeled. The in vitro translation systems also contained the enzymes that attach amino acids to tRNAs.

Experimental level **Conceptual level**

1 Mix together RNA triplets of a specific sequence and 20 in vitro translation systems. In the example shown here, the triplet is 5′–CCC–3′. Each translation system contained a different radiolabeled amino acid. (Note: Only 3 tubes are shown here.)

In vitro translation system with 1 radiolabeled amino acid (for example, proline)

Tubes containing an RNA triplet

Proline

Ribosome

2 Allow time for the RNA triplet to bind to the ribosome and for the appropriate tRNA to bind to the RNA triplet.

Radiolabeled proline
Proline tRNA

RNA triplet that specifies proline

3 Pour each mixture through a filter that allows the passage of unbound tRNA but does not allow the passage of ribosomes.

Ribosomes trapped on filter

Filter

Filter

4 Count radioactivity on the filter.

Scintillation counter

5 THE DATA

Triplet	Radiolabeled amino acid trapped on the filter	Triplet	Radiolabeled amino acid trapped on the filter
5′ – AAA – 3′	Lysine	5′ – GAC – 3′	Aspartic acid
5′ – ACA – 3′	Threonine	5′ – GCC – 3′	Alanine
5′ – ACC – 3′	Threonine	5′ – GGU – 3′	Glycine
5′ – AGA – 3′	Arginine	5′ – GGC – 3′	Glycine
5′ – AUA – 3′	Isoleucine	5′ – GUU – 3′	Valine
5′ – AUU – 3′	Isoleucine	5′ – UAU – 3′	Tyrosine
5′ – CCC – 3′	Proline	5′ – UGU – 3′	Cysteine
5′ – CGC – 3′	Arginine	5′ – UUG – 3′	Leucine
5′ – GAA – 3′	Glutamic acid		

6 CONCLUSION This method enabled the researchers to identify many of the codons of the genetic code.

7 SOURCE Leder, Philip, and Nirenberg, Marshall W. 1964. RNA Codewords and Protein Synthesis, III. On the nucleotide sequence of a cysteine and a leucine RNA codeword. *Proceedings of the National Academy of Sciences* 52:1521–1529.

12.5 The Machinery of Translation

Let's now turn our attention to the components found in living cells that are needed to use the genetic code and translate mRNA into polypeptides. Earlier in this chapter, we considered transcription, the first step in gene expression. To transcribe an RNA molecule, a pre-existing DNA template strand is used to make a complementary RNA strand. A single enzyme, RNA polymerase, can catalyze this reaction. By comparison, translation requires more components because the sequence of codons in an mRNA molecule must be translated into a sequence of amino acids according to the genetic code. A single protein cannot accomplish such a task. Instead, many different proteins and RNA molecules interact in an intricate series of steps to achieve the synthesis of a polypeptide. A cell must make many different components, including mRNAs, tRNAs, ribosomes, and translation factors, so that polypeptides can be made (**Table 12.2**).

Though the estimates vary from cell to cell and from species to species, most cells use a substantial amount of their energy to translate mRNA into polypeptides. In *E. coli*, for example, approximately 90% of the cellular energy is used for this process. This value underscores the complexity and importance of translation in living organisms. In this section, we will focus on the components of the translation machinery. The last section of the chapter will describe the steps of translation as they occur in living cells.

Transfer RNAs Share Common Structural Features

To understand how tRNAs act as carriers of the correct amino acids during translation, researchers have examined their structural characteristics. The tRNAs of both prokaryotes and eukaryotes share common features. As originally proposed by Robert Holley in 1965, the two-dimensional structure of tRNAs exhibits a cloverleaf pattern. The structure has three stem-loops and a fourth stem with a 3′ single-stranded region (**Figure**

Table 12.2	Components of the Translation Machinery
Component	**Function**
mRNA	Contains the information for a polypeptide sequence according to the genetic code.
tRNA	A molecule with two functional sites. One site, termed the anticodon, recognizes a codon in mRNA. A second site has the appropriate amino acid attached to it.
Ribosomes	Composed of many proteins and rRNA molecules. The ribosome provides a location where mRNA and tRNA molecules can properly interact with each other. The ribosome also catalyzes the formation of covalent bonds between adjacent amino acids so that a polypeptide can be made.
Translation factors	Proteins needed for the three stages of translation. Initiation factors are required for the assembly of mRNA, the first tRNA, and ribosomal subunits. Elongation factors are needed to synthesize the polypeptide. Release factors are needed to recognize the stop codon and disassemble the translation machinery. Several translation factors use GTP as an energy source to carry out their functions.

12.14a). The stems are regions where the RNA is double stranded due to complementary base pairing, whereas the loops are regions without base pairing. The anticodon is located in the loop of the second stem-loop region. The 3′ single-stranded region is called the acceptor stem because it accepts the attachment of an amino acid. The three-dimensional structure of tRNA molecules involves additional folding of the secondary structure (**Figure 12.14b**).

The cells of every organism make many different tRNA molecules, each encoded by a different gene. A tRNA is named according to the amino acid it carries. For example, tRNA^{ser} carries a serine. Because the genetic code contains six different serine codons as shown in Table 12.1, a cell produces more than one type of tRNA^{ser}.

Figure 12.14 Structure of tRNA. **(a)** The two-dimensional or secondary structure is that of a cloverleaf. The anticodon is within the middle stem-loop structure. The 3′ single-stranded region (acceptor stem) is where an amino acid can attach. **(b)** The actual three-dimensional structure folds in on itself.

Concept check: *What is the function of the anticodon?*

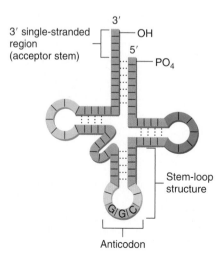

(a) Two-dimensional structure of tRNA

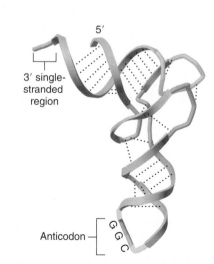

(b) Three-dimensional structure of tRNA

Aminoacyl-tRNA Synthetases Charge tRNAs by Attaching an Appropriate Amino Acid

To perform its role during translation, a tRNA must have the appropriate amino acid attached to its 3' end. The enzymes that catalyze the attachment of amino acids to tRNA molecules are known as **aminoacyl-tRNA synthetases**. Cells make 20 distinct types of aminoacyl-tRNA synthetase enzymes; each type recognizes just one of the 20 different amino acids. Each aminoacyl-tRNA synthetase is named for the specific amino acid it attaches to tRNA. For example, alanyl-tRNA synthetase recognizes alanine and attaches this amino acid to all tRNAs with alanine anticodons.

Aminoacyl-tRNA synthetases catalyze chemical reactions involving an amino acid, a tRNA molecule, and ATP (**Figure 12.15**). First, a specific amino acid and ATP are recognized by the enzyme. Next, the amino acid is activated by the covalent attachment of an AMP molecule, and pyrophosphate is released. In a third step, the activated amino acid is covalently attached to the 3' end of a tRNA molecule, and AMP is released. Finally, the tRNA with its attached amino acid, called a **charged tRNA** or an **aminoacyl tRNA**, is released from the enzyme.

The ability of each aminoacyl-tRNA synthetase to recognize an appropriate tRNA has been called the second genetic code. A precise recognition process is necessary to maintain the fidelity of genetic information. If the wrong amino acid was attached to a tRNA, the amino acid sequence of the translated polypeptide would be incorrect. To prevent this from happening, aminoacyl-tRNA synthetases are amazingly accurate enzymes. The wrong amino acid is attached to a tRNA less than once in 100,000 times! The anticodon region of the tRNA is usually important for recognition by the correct aminoacyl-tRNA synthetase. In addition, the base sequences in other regions may facilitate binding to an aminoacyl-tRNA synthetase.

Ribosomes Are Assembled from rRNA and Proteins

Let's now turn our attention to the **ribosome**, which is often described as a molecular machine. The ribosome is the site where translation takes place. Bacterial cells have one type of ribosome, which translates all mRNAs in the cytoplasm. Because eukaryotic cells are compartmentalized into cellular organelles bounded by membranes, their translation machinery is more complex. Biochemically distinct ribosomes are found in different cellular compartments. The most abundant type of eukaryotic ribosome functions in the cytosol. In addition, mitochondria have ribosomes, and plant and algal cells have ribosomes in their chloroplasts. The compositions of mitochondrial and chloroplast ribosomes are more similar to bacterial ribosomes than they are to eukaryotic cytosolic ribosomes. Unless otherwise noted, the term eukaryotic ribosome refers to ribosomes in the cytosol, not to those found in organelles.

Amino acid
ATP
P P P

tRNA
Charged tRNA

Aminoacyl-tRNA synthetase
P Pi
Pyrophosphate
P
AMP

1. A specific amino acid and ATP bind to aminoacyl-tRNA synthetase.

2. The amino acid is activated by the covalent binding of AMP, and pyrophosphate is released.

3. The correct tRNA binds to the synthetase. The amino acid is covalently attached to the tRNA. AMP is released.

4. The charged tRNA is released.

Figure 12.15 Aminoacyl-tRNA synthetase charging a tRNA.

Concept check: Why is ATP needed to charge a tRNA?

A ribosome is a large complex composed of structures called the large and small subunits. The term subunit is perhaps misleading, because each ribosomal subunit is itself assembled from many different proteins and one or more RNA molecules. In the bacterium *E. coli*, the small ribosomal subunit is called 30S, and the large subunit is 50S (**Table 12.3**). The designations 30S and 50S refer to the rate at which these subunits sediment when subjected to a centrifugal force. This rate is described as a sedimentation coefficient in Svedberg units (S) in honor of Theodor Svedberg, who invented the ultracentrifuge. The 30S subunit is formed from the assembly of 21 different ribosomal proteins and one 16S rRNA molecule. The 50S subunit contains 34 different proteins and two different rRNA molecules, called 5S and 23S. Together, the 30S and 50S subunits form a 70S ribosome. (Svedberg units don't add up linearly, because the sedimentation coefficient is a function of both size and shape.) In bacteria, ribosomal proteins and rRNA molecules are synthesized in the cytoplasm, and the ribosomal subunits are assembled there as well.

Eukaryotic ribosomes consist of subunits that are slightly larger than their bacterial counterparts (Table 12.3). In eukaryotes, 40S and 60S subunits combine to form an 80S ribosome. The 40S subunit is composed of 33 proteins and an 18S rRNA, and the 60S subunit has 49 proteins and 5S, 5.8S, and 28S rRNAs. The synthesis of eukaryotic rRNA occurs in the nucleolus, a region of the nucleus that is specialized for that purpose. The ribosomal proteins are made in the cytosol and imported into the nucleus. The rRNAs and ribosomal proteins are then assembled within the nucleolus to make the 40S and 60S subunits. The 40S

and 60S subunits are exported into the cytosol, where they associate to form an 80S ribosome during translation.

Due to structural differences between bacterial and eukaryotic ribosomes, certain chemicals may bind to bacterial ribosomes but not to eukaryotic ribosomes, and vice versa. Some **antibiotics**, which are chemicals that inhibit the growth of certain microorganisms, bind only to bacterial ribosomes and inhibit translation. Examples include erythromycin and chloramphenicol. Because these chemicals do not inhibit eukaryotic ribosomes, they have been effective drugs for the treatment of bacterial infections in humans and domesticated animals.

Components of Ribosomal Subunits Form Functional Sites for Translation

To understand the structure and function of the ribosome at the molecular level, researchers have determined the locations and functional roles of individual ribosomal proteins and rRNAs. In recent years, a few research groups have succeeded in purifying ribosomes and causing them to crystallize in a test tube. Using the technique of X-ray diffraction, the crystallized ribosomes provide detailed information about ribosome structure. **Figure 12.16a** shows a model of a bacterial ribosome. The overall shape of each subunit is largely determined by the structure of the rRNAs, which constitute most of the mass of the ribosome.

During bacterial translation, the mRNA lies on the surface of the 30S subunit, within a space between the 30S and 50S subunits (**Figure 12.16b**). As a polypeptide is synthesized, it exits through a hole within the 50S subunit. Ribosomes contain discrete sites where tRNAs bind and the polypeptide is synthesized. In 1964, James Watson proposed a two-site model for tRNA binding to the ribosome. These sites are known as the **peptidyl site** (**P site**) and **aminoacyl site** (**A site**). In 1981, Knud Nierhaus and Hans-Jorg Rheinberger expanded this to a three-site model (Figure 12.16b). The third site is known as the **exit site** (**E site**). In Section 12.6, we will examine the roles of these sites in the synthesis of a polypeptide.

Table 12.3	Composition of Bacterial and Eukaryotic Ribosomes		
	Small subunit	Large subunit	Assembled ribosome
Bacterial			
Sedimentation coefficient	30S	50S	70S
Number of proteins	21	34	55
rRNA	16S rRNA	5S rRNA, 23S rRNA	16S rRNA, 5S rRNA, 23S rRNA
Eukaryotic			
Sedimentation coefficient	40S	60S	80S
Number of proteins	33	49	82
rRNA	18S rRNA	5S rRNA, 5.8S rRNA, 28S rRNA	18S rRNA, 5S rRNA, 5.8S rRNA, 28S rRNA

Genomes & Proteomes Connection

Comparisons of Small Subunit rRNAs Among Different Species Provide a Basis for Establishing Evolutionary Relationships

Translation is a fundamental process that is vital for the existence of all living species. Research indicates that the components needed for translation arose very early in the evolution of life on our planet in an ancestor that gave rise to all known living species. For this reason, all organisms have translational components that are evolutionarily related to each other. For example, the rRNA found in the small subunit of ribosomes is similar in all forms of life, though it is slightly larger in eukaryotic species (18S) than in bacterial species (16S). In other words,

(a) Bacterial ribosome model based on X-ray diffraction studies   (b) Schematic model for ribosome structure

Figure 12.16   Ribosome structure.   (a) A model for the structure of a bacterial ribosome based on X-ray diffraction studies, showing the large and small subunits and the major binding sites. The rRNA is shown in gray (large subunit) and turquoise (small subunit), whereas the ribosomal proteins are magenta (large subunit) and purple (small subunit). (b) A schematic model emphasizing functional sites in the ribosome, and showing bound mRNA and tRNA with an attached polypeptide.

the gene for the small subunit rRNA (SSU rRNA) is found in the genomes of all organisms.

How is this observation useful? One way that geneticists explore evolutionary relationships is to compare the sequences of evolutionarily related genes. At the molecular level, gene evolution involves changes in DNA sequences. After two different species have diverged from each other during evolution, the genes of each species have an opportunity to accumulate changes, or mutations, that alter the sequences of those genes. After many generations, evolutionarily related species contain genes that are similar but not identical to each other, because each species will accumulate different mutations. In general, if a very long time has elapsed since two species diverged evolutionarily, their genes tend to be quite different. In contrast, if two species diverged relatively recently on an evolutionary time scale, their genes tend to be more similar.

Figure 12.17 compares a portion of the sequence of the small subunit rRNA gene from three mammalian and three bacterial species. The colors highlight different types of compari-

sons. The bases shaded in yellow are identical in five or six species. Sequences of bases that are very similar or identical in different species are said to be **evolutionarily conserved**. Presumably, these sequences were found in the primordial gene that gave rise to modern species. Perhaps, because these sequences may have some critical function, they have not changed over evolutionary time. Those sequences shaded in green are identical in all three mammals, but differ compared to one or more bacterial species. Actually, if you scan the mammalian species, you may notice that all three sequences are identical to each other in this region. The sequences shaded in red are identical in two or three bacterial species, but differ compared to the mammalian small subunit rRNA genes. The sequences from *Escherichia coli* and *Serratia marcescens* are more similar to each other than the sequence from *Bacillus subtilis* is to either of them. This observation suggests that *E. coli* and *S. marcescens* are more closely related evolutionarily than either of them is to *B. subtilis*.

GATTAAGAGGGACGGCCGGGGGCATTCGTATTGCGCCGCTAGAGGTGAAATTC — Human
GATTAAGAGGGACGGCCGGGGGCATTCGTATTGCGCCGCTAGAGGTGAAATTC — Mouse
GATTAAGAGGGACGGCCGGGGGCATTCGTATTGCGCCGCTAGAGGTGAAATTC — Rat
CAAGCTTGAGTCTCGTAGAGGGGGGTAGAATTCCAGGTGTAGCGGTGAAATGC — E. coli
CAAGCTAGAGTCTCGTAGAGGGGGGTAGAATTCCAGGTGTAGCGGTGAAATGC — S. marcescens
GAGACTTGAGTACAGAAGAAGAGAGTGGAATTCCACGTGTAGCGGTGAAATGC — B. subtilis

Figure 12.17   Comparison of small subunit rRNA gene sequences from three mammalian and three bacterial species.   Note the many similarities (yellow) and differences (green and red) among the sequences.

Concept check: *Based on the gene sequences shown here, pick two species that are closely related evolutionarily and two that are distantly related.*

12.6 | The Stages of Translation

Like transcription, the process of translation occurs in three stages called initiation, elongation, and termination. **Figure 12.18** provides an overview of the process. During initiation, an mRNA, the first tRNA, and the ribosomal subunits assemble into a complex. Next, in the elongation stage, the ribosome moves in the 5′ to 3′ direction from the start codon in the mRNA toward the stop codon, synthesizing a polypeptide according to the sequence of codons in the mRNA. Finally, the process is terminated when the ribosome reaches a stop codon and the complex disassembles, releasing the completed polypeptide. In this section, we will examine the steps in this process as they occur in living cells.

Translation Is Initiated with the Assembly of mRNA, tRNA, and the Ribosomal Subunits

During the **initiation stage**, a complex is formed between an mRNA molecule, the first tRNA, and the ribosomal subunits. In all species, the assembly of this complex requires the help of proteins called **initiation factors** that facilitate the interactions between these components (see Table 12.2). The assembly also requires an input of energy. Guanosine triphosphate

(GTP) is hydrolyzed by certain initiation factors to provide the necessary energy.

In the absence of translation, the small and large ribosomal subunits exist separately. To begin assembly in bacteria, mRNA binds to the small ribosomal subunit (**Figure 12.19**). The binding of mRNA to this subunit is facilitated by a short ribosomal-binding sequence near the 5′ end of the mRNA. This sequence is complementary to a portion of the 16S rRNA within the small ribosomal subunit. For this reason, the mRNA and rRNA hydrogen-bond to each other by base pairing. The start codon is usually just a few nucleotides downstream (that is, toward the 3′ end) from the ribosomal-binding sequence. A specific tRNA, which functions as the **initiator tRNA**, recognizes the start codon in mRNA and binds to it. In eukaryotes, this tRNA carries a methionine, whereas in bacteria it carries a methionine that has been modified by the attachment of a formyl group. To complete the initiation stage, the large ribosomal subunit associates with the small subunit. At the end of this stage, the initiator tRNA is located in the P site of the ribosome.

In eukaryotic species, the initiation phase of translation differs in two ways from the process in bacteria. First, instead of a ribosomal-binding sequence, eukaryotic mRNAs have a guanosine cap at their 5′ end. This 5′ cap is recognized by cap-binding proteins that promote the binding of the mRNA to the small ribosomal subunit. Also, unlike bacteria, in which

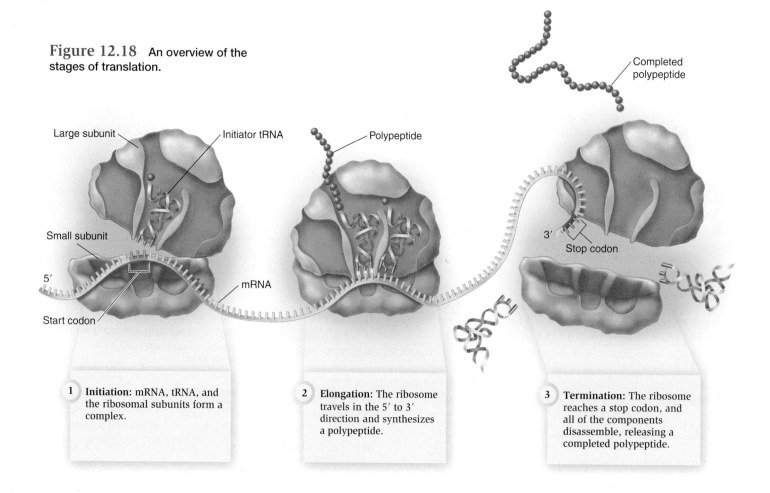

Figure 12.18   **An overview of the stages of translation.**

Completed polypeptide

Large subunit   Initiator tRNA   Polypeptide

Small subunit

5′   mRNA

Start codon

3′   Stop codon

1   **Initiation:** mRNA, tRNA, and the ribosomal subunits form a complex.

2   **Elongation:** The ribosome travels in the 5′ to 3′ direction and synthesizes a polypeptide.

3   **Termination:** The ribosome reaches a stop codon, and all of the components disassemble, releasing a completed polypeptide.

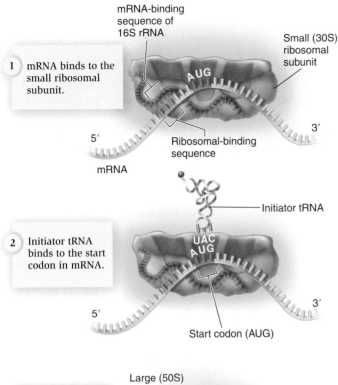

1 mRNA binds to the small ribosomal subunit.

mRNA-binding sequence of 16S rRNA

Small (30S) ribosomal subunit

AUG

5′

Ribosomal-binding sequence

mRNA

3′

2 Initiator tRNA binds to the start codon in mRNA.

Initiator tRNA

UAC
AUG

5′

Start codon (AUG)

3′

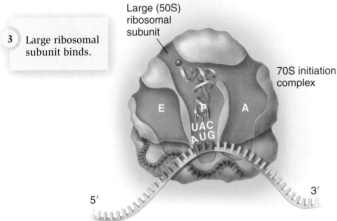

3 Large ribosomal subunit binds.

Large (50S) ribosomal subunit

70S initiation complex

E   P   A

UAC
AUG

5′

3′

Figure 12.19 Initiation stage of translation in bacteria.

Concept check: What promotes the binding between the mRNA and the small ribosomal subunit?

the start codon is very close to a ribosomal-binding sequence, the location of start codons in eukaryotes is more variable. In 1978, Marilyn Kozak proposed that the small ribosomal subunit identifies a start codon by beginning at the 5′ end and then scanning along the mRNA in the 3′ direction in search of an AUG sequence. In many, but not all, cases the first AUG codon is used as a start codon. By analyzing the sequences of many eukaryotic mRNAs, Kozak and her colleagues discovered that the sequence around an AUG codon is important for it to be used as a start codon. The sequence for optimal start codon recognition is shown here:

Upstream of start codon	Start codon	Downstream coding region

. . . G C C (A or G) C C (**A U G**) G . . . .

Aside from an AUG codon itself, a guanine just past the start codon and the sequence of six bases directly upstream from the start codon are important for start codon selection. If the first AUG codon is within a site that deviates markedly from this optimal sequence, the small subunit may skip this codon and instead use another AUG codon farther downstream. Once the small subunit selects a start codon, an initiator tRNA binds to the start codon, and then the large ribosomal subunit associates with the small subunit to complete the assembly process.

Polypeptide Synthesis Occurs During the Elongation Stage

As its name suggests, the **elongation stage** involves the covalent bonding of amino acids to each other, one at a time, to create a polypeptide. Even though this process involves several different components, translation occurs at a remarkable rate. Under normal cellular conditions, the translation machinery can elongate a polypeptide chain at a rate of 15 to 18 amino acids per second in bacteria and 6 amino acids per second in eukaryotes!

To elongate a polypeptide by one amino acid, a tRNA brings a new amino acid to the ribosome, where it is attached to the end of a growing polypeptide chain. In step 1 of **Figure 12.20**, translation has already proceeded to a point where a short polypeptide is attached to the tRNA located in the P site of the ribosome. This is called peptidyl tRNA. In the first step of elongation, a charged tRNA carrying a single amino acid binds to the A site. This binding occurs because the anticodon in the tRNA is complementary to the codon in the mRNA. The hydrolysis of GTP by proteins that function as **elongation factors** provides the energy for the binding of the tRNA to the A site (see Table 12.2). At this stage of translation, a peptidyl tRNA is in the P site and a charged tRNA (an aminoacyl tRNA) is in the A site. This is how the P and A sites came to be named.

In the second step, a peptide bond is formed between the amino acid at the A site and the growing polypeptide chain, thereby lengthening the chain by one amino acid. As this occurs, the polypeptide is removed from the tRNA in the P site and transferred to the amino acid at the A site, an event termed a **peptidyl transfer reaction**. This reaction is catalyzed by a region of the 50S subunit known as the peptidyltransferase center, which is composed of several proteins and rRNA. Thomas Steitz, Peter Moore, and their colleagues proposed that the rRNA is responsible for catalyzing the peptide bond formation between adjacent amino acids. It is now accepted that the ribosome is a ribozyme!

After the peptidyl transfer reaction is complete, the third step involves the movement or translocation of the ribosome toward the 3′ end of the mRNA by exactly one codon. This shifts the tRNAs in the P and A sites to the E and P sites, respectively. Notice that the next codon in the mRNA is now exposed at the unoccupied A site. The uncharged tRNA exits the E site. At this point, the next charged tRNA can enter the empty A site, and the same series of steps will add the next amino acid to the polypeptide chain.

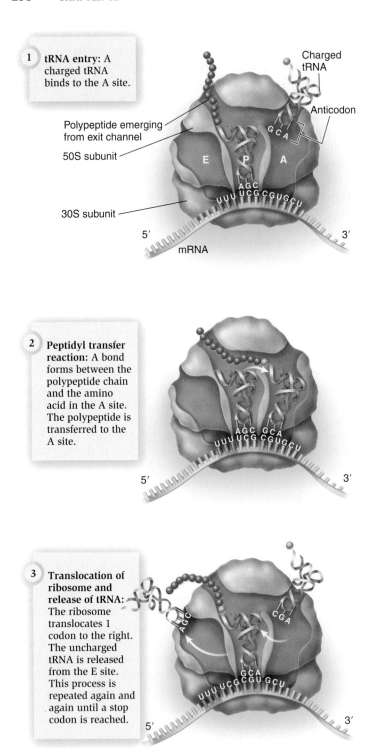

Figure 12.20 Elongation stage of translation in bacteria.

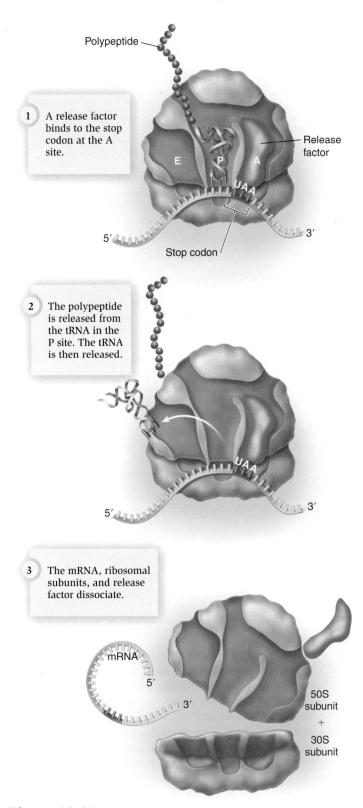

Figure 12.21 Termination stage of translation in bacteria.

Termination Occurs When a Stop Codon Is Reached in the mRNA

Elongation continues until a stop codon moves into the A site of a ribosome. The three stop codons, UAA, UAG, and UGA, are not recognized by a tRNA with a complementary sequence. Instead, they are recognized by a protein known as a **release**

factor. The three-dimensional structure of a release factor protein mimics the structure of tRNAs, which allows it to fit into the A site.

Figure 12.21 illustrates the **termination stage** of translation. In step 1 of this figure, a release factor binds to the stop codon at the A site. The completed polypeptide chain is attached to a tRNA in the P site. In the second step, the bond between

Table 12.4	Comparison of Bacterial and Eukaryotic Translation	
	Bacterial	**Eukaryotic**
Ribosome composition	70S ribosomes:	80S ribosomes:
	30S subunit- 21 proteins + 1 rRNA	40S subunit- 33 proteins + 1 rRNA
	50S subunit- 34 proteins + 2 rRNAs	60S subunit- 49 proteins + 3 rRNAs
Initiator tRNA	tRNA$^{formyl\text{-}methionine}$	tRNA^{methionine}
Initial binding of mRNA	Requires a ribosomal-binding sequence	Requires a 7-methylguanosine cap
Selection of a start codon	Just downstream from the ribosomal-binding sequence	According to Kozak's rules
Termination factors	Two factors: RF1 and RF2	One factor: eRF

the polypeptide and the tRNA is hydrolyzed, causing the polypeptide and tRNA to be released from the ribosome. In the third step, the mRNA, ribosomal subunits, and release factor dissociate. The termination stage of translation is similar in bacteria and eukaryotes except that bacteria have two different termination factors that recognize stop codons (RF1 and RF2), whereas eukaryotes have only one (eRF). **Table 12.4** compares some of the key differences between bacterial and eukaryotic translation.

Summary of Key Concepts

12.1 Overview of Gene Expression

- Based on his studies of inborn errors of metabolism, Garrod hypothesized that some genes encode enzymes. (Figure 12.1)
- By studying the nutritional requirements of bread mold, Beadle and Tatum proposed the one gene–one enzyme hypothesis. (Figure 12.2)
- A polypeptide is a unit of structure. A protein, composed of one or more polypeptides, is a unit of function.
- At the molecular level, the central dogma states that most genes are transcribed into mRNA, and then the mRNA is translated into polypeptides. Eukaryotes modify their RNA transcripts to make them functional. (Figure 12.3)
- The molecular expression of genes is fundamental to the characteristics of an organism's traits.

12.2 Transcription

- The promoter of a gene signals the beginning of transcription whereas the terminator specifies where transcription will end for a given gene. Regulatory sequences control whether a gene is turned on or off. (Figure 12.4)
- In bacteria, sigma factor binds to RNA polymerase and to a promoter, thereby promoting the initiation of transcription. The RNA transcript is made during the elongation stage due to base pairing of nucleotides to the template strand of DNA. RNA

polymerase is released from the DNA at the termination site. (Figures 12.5, 12.6)
- The genes along a chromosome are transcribed in different directions using either DNA strand as a template. RNA is always synthesized in a 5′ to 3′ direction. (Figure 12.7)

12.3 RNA Processing in Eukaryotes

- Eukaryotic mRNA is first made as a pre-mRNA that is capped, spliced, and given a poly A tail. In the process called splicing, introns are removed from eukaryotic pre-mRNA by a spliceosome. The components of a spliceosome first recognize the intron boundaries and the branch site, and then remove the intron and connect the adjacent exons. (Figures 12.8, 12.9, 12.10)

12.4 Translation and the Genetic Code

- The genetic code determines the amino acid sequences of polypeptides. Each of the 64 codons specifies a start codon (methionine), other amino acids, or a stop codon. (Table 12.1, Figure 12.11)
- The template strand of DNA is used to make mRNA with a series of codons. Recognition between mRNA and many tRNA molecules determines the amino acid sequence of a polypeptide. A polypeptide has a directionality in which the first amino acid is at the N-terminus or amino terminus, whereas the last amino acid is at the C-terminus or carboxyl terminus. (Figure 12.12)
- Nirenberg and Leder used the ability of RNA triplets to promote the binding of tRNA to ribosomes as a way to determine many of the codons of the genetic code. (Figure 12.13)

12.5 The Machinery of Translation

- Translation requires mRNA, aminoacyl tRNAs, ribosomes, and many translation factors. (Table 12.2)
- tRNA molecules have a cloverleaf structure. Two important sites are the 3′ single-stranded region, which covalently binds an amino acid, and the anticodon, which base-pairs with a codon in mRNA. (Figure 12.14)
- The enzyme aminoacyl-tRNA synthetase attaches the correct amino acid to a tRNA molecule, creating a charged tRNA. (Figure 12.15)
- Ribosomes are composed of a small and large subunit, each consisting of rRNA molecules and many proteins. Bacterial and eukaryotic ribosomes differ in their molecular composition. (Table 12.3)
- Ribosomes have three sites, the A, P, and E sites, which are locations for the binding and release of tRNA molecules. (Figure 12.16)
- The gene that encodes the small subunit rRNA (SSU rRNA) has been extensively used in the evolutionary comparisons of different species. (Figure 12.17)

12.6 The Stages of Translation

- Translation occurs in three stages, called initiation, elongation, and termination. (Figure 12.18)

- During initiation of translation, the mRNA assembles with the ribosomal subunits and the first tRNA molecule. (Figure 12.19)
- Polypeptide synthesis occurs during the elongation stage, one amino acid at a time. (Figure 12.20)
- During the termination of translation, the binding of a release factor to the stop codon causes the release of the completed polypeptide from the tRNA and the disassembly of the mRNA, ribosomal subunits, and release factor. (Figure 12.21)
- Though translation in bacteria and eukaryotes is strikingly similar, some key differences have been observed. (Table 12.4)

Assess and Discuss

Test Yourself

1. Which of the following best represents the central dogma of gene expression?
   a. During transcription, DNA codes for polypeptides.
   b. During transcription, DNA codes for mRNA, which codes for polypeptides during translation.
   c. During translation, DNA codes for mRNA, which codes for polypeptides during transcription.
   d. none of the above

2. Transcription of a gene begins at a site on DNA called _____ and ends at a site on DNA known as _____.
   a. an initiation codon, the termination codon
   b. a promoter, the termination codon
   c. an initiation codon, the terminator
   d. a promoter, the terminator
   e. an initiator, the terminator

3. The functional product of a structural gene is
   a. tRNA.
   b. mRNA.
   c. rRNA.
   d. a polypeptide.
   e. a, b, and c.

4. During eukaryotic RNA processing, the nontranslated sequences that are removed are called
   a. exons.
   b. introns.
   c. promoters.
   d. codons.
   e. ribozymes.

5. The _____ is the site where the translation process takes place.
   a. mitochondria
   b. nucleus
   c. ribosome
   d. lysosome
   e. ribozyme

6. The small subunit of a ribosome is composed of
   a. a protein.
   b. an rRNA molecule.
   c. many proteins.
   d. many rRNA molecules.
   e. many proteins and one rRNA molecule.

7. The region of the tRNA that is complementary to a codon in mRNA is
   a. the acceptor stem.
   b. the codon.
   c. the peptidyl site.
   d. the anticodon.
   e. the adaptor loop.

8. During the initiation step of translation, the first codon, ____, will enter the _____ and associate with the initiator tRNA.
   a. UAG, A site
   b. AUG, A site
   c. UAG, P site
   d. AUG, P site
   e. AUG, E site

9. The movement of the polypeptide from the tRNA in the P site to the tRNA in the A site is referred to as
   a. peptide bonding.
   b. aminoacyl binding.
   c. translation.
   d. peptidyl transfer reaction.
   e. elongation.

10. The synthesis of a polypeptide occurs during which stage of translation?
    a. initiation
    b. elongation
    c. termination
    d. splicing

Conceptual Questions

1. Describe the one gene–one enzyme hypothesis and the more modern modifications of this hypothesis. Briefly explain how studying the pathway that leads to arginine synthesis allowed Beadle and Tatum to conclude that one gene encodes one enzyme.

2. What is the function of an aminoacyl-tRNA synthetase?

3. A tRNA has an anticodon sequence 3′–GGU–5′. What amino acid does it carry?

Collaborative Questions

1. Why do you think some complexes, such as spliceosomes and ribosomes, have both protein and RNA components?

2. Discuss and make a list of the similarities and differences in the events that occur during the initiation, elongation, and termination stages of transcription and translation.

Online Resource

www.brookerbiology.com

Stay a step ahead in your studies with animations that bring concepts to life and practice tests to assess your understanding. Your instructor may also recommend the interactive ebook, individualized learning tools, and more.

Gene Regulation

13

A model for a protein that binds to DNA and regulates genes. The catabolite activator protein, shown in dark and light blue, is binding to the DNA double helix, shown in orange and white. This protein, described later in Figure 13.10, activates gene transcription.

E milio took a weight-lifting class in college and was surprised by the results. Within a few weeks, he was able to lift substantially more weight. He was inspired by this progress and continued lifting weights after the semester-long course ended. A year later, he was not only much stronger, but he could see physical changes in his body. Certain muscles, such as his biceps and triceps in his upper arms, were noticeably larger. How can we explain the increase in mass of Emilio's muscles? Unknowingly, when he was lifting weights, Emilio was affecting the regulation of his genes. Certain genes in his muscle cells were being "turned on" during his workouts, which then led to the synthesis of proteins that increased the mass of Emilio's muscles.

At the molecular level, **gene expression** is the process by which the information within a gene is made into a functional product, such as a protein or RNA molecule. The majority of genes in all species are regulated so that the proteins they specify are produced at certain times and in specific amounts. The term **gene regulation** refers to the ability of cells to control the expression of their genes. By comparison, some genes have relatively constant levels of expression in all conditions over time. These are called **constitutive genes**. Frequently, constitutive genes encode proteins that are always required for the survival of an organism, such as certain metabolic enzymes.

The importance of gene regulation is underscored by the number of genes devoted to this process in an organism. For example, in *Arabidopsis thaliana*, a plant that is studied as a model organism by plant geneticists, over 5% of its genome is involved with regulating gene transcription. This species has more than 1,500 different genes that encode proteins that regulate the transcription of other genes.

In this chapter, we will begin with an overview that emphasizes the benefits of gene regulation and the general mechanisms that achieve such regulation. Later sections will describe how bacteria regulate gene expression in the face of environmental change and the more complex nature of gene regulation in eukaryotes.

13.1 Overview of Gene Regulation

How do living organisms benefit from gene regulation? One reason is that it conserves energy. Proteins that are encoded by genes will be produced only when they are needed. For example, some proteins function in the metabolism of small molecules, such as sugars, that may or may not be present in the environment. These proteins are required only when the bacterium is exposed to such sugars. Like bacteria, eukaryotic organisms also need to adapt to changes in their environment. For example, all eukaryotic species can respond to environmental stresses such as ultraviolet (UV) radiation by turning on genes that provide protection against this harmful agent. In humans, exposure to UV radiation turns on genes that result in a tanning response.

Gene regulation also ensures that genes are expressed in the appropriate cell types and at the correct stage of development. In multicellular organisms, certain proteins are made only in particular cell types, or their amounts may vary from cell to cell. In humans, for example, some proteins are needed only in muscle cells but not in nerve cells, and vice versa. Similarly, in multicellular organisms that progress through different developmental stages (for example, fertilized egg, embryo, and adult), certain proteins are needed only at particular stages of development. In this section, we will examine a few examples that illustrate the important consequences of gene regulation. We will also survey the major points in the gene expression process at which genes are regulated in prokaryotic and eukaryotic cells.

Bacteria Regulate Genes to Respond to Nutrients in Their Environment

To fully appreciate how gene regulation helps bacteria survive in a changing environment, let's look at an example. The bacterium *Escherichia coli* can use many types of sugars as food sources, thereby increasing its chances of survival. Let's consider the process of how it uses lactose, which is the sugar found in milk. *E. coli* can use lactose because it carries genes that code for proteins that enable it to take up lactose from the environment and metabolize it.

Figure 13.1 illustrates the effects of lactose on the regulation of those genes. In order to utilize lactose, an *E. coli* cell requires a transporter, called lactose permease, that facilitates the uptake of lactose into the cell, and an enzyme, called β-galactosidase, that catalyzes the breakdown of lactose. When lactose is not present in the environment, an *E. coli* cell makes very little of these proteins. However, when lactose becomes available, the bacterium will produce many more copies of these proteins, enabling it to efficiently use lactose from its environment. Eventually, all of the lactose in the environment will be used up. At

this point, the genes encoding these proteins will be shut off, and most of the proteins will be degraded. In the case of lactose utilization, gene regulation ensures that the proteins needed for this process are made only when lactose is present in the environment.

Eukaryotic Gene Regulation Produces Different Cell Types in a Single Organism

One of the most amazing examples of gene regulation is the phenomenon of **cell differentiation**, the process by which cells become specialized into particular types. In humans, for example, cells may differentiate into muscle cells, nerve cells, skin cells, or other types. Figure 13.2 shows micrographs of three types of cells found in humans. As seen here, their morphologies are strikingly different. Likewise, their functions within the body are also quite different. Muscle cells are important in body movements, nerve cells function in cell signaling, and skin cells form a protective outer surface to the body.

Gene regulation is responsible for creating different types of cells within a multicellular organism. The three cell types shown in Figure 13.2 contain the same **genome**, meaning they carry the same set of genes. However, their **proteomes**—the collection of proteins they make—are quite different; this is due to gene regulation. Certain proteins are found in particular cell types but not in others. Alternatively, a protein may be present in all three cell types, but the relative amounts of the protein

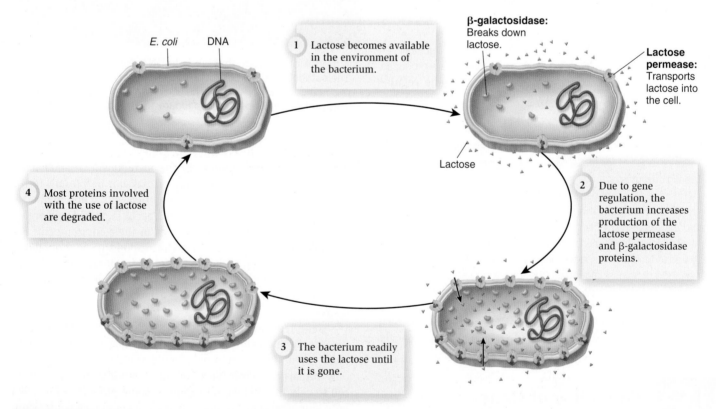

Figure 13.1   Gene regulation of lactose utilization in *E. coli.*

Concept check: *What is the advantage to E. coli of regulating the genes involved with lactose utilization?*

(a) Skeletal muscle cell   118 μm

(b) Nerve cell   179 μm

(c) Skin cell   71 μm

Figure 13.2 **Examples of different cell types in humans.** These cells have the same genetic composition. Their unique morphologies are due to differences in the proteins they make.

Concept check: *How does gene regulation underlie the different morphologies of these cells?*

may be different. The amount of a given protein depends on many factors, including how strongly the corresponding gene is turned on and how much protein is synthesized from mRNA. Gene regulation plays a major role in determining the proteome of each cell type.

Eukaryotic Gene Regulation Enables Multicellular Organisms to Progress Through Developmental Stages

In multicellular organisms that progress through developmental stages, certain genes are expressed at particular stages of development but not others. We'll discuss this topic in greater detail in Chapter 19. Let's consider an example of such developmental gene regulation in mammals. Early stages of development occur in the uterus of female mammals. Following fertilization, an embryo develops inside the uterus. In humans, the embryonic stage lasts from fertilization to 8 weeks. During this stage, major developmental changes produce the various body parts. The fetal stage occurs from 8 weeks to birth (41 weeks). This stage is characterized by continued refinement of body parts and a large increase in size.

Because of this internal development, a system has evolved to provide both the embryonic and fetal cells with the oxygen they require for cellular respiration. The oxygen demands of a rapidly growing embryo and fetus are quite different from the needs of the mother. Gene regulation plays a vital role in ensuring that an embryo and fetus get the proper amount of oxygen. Hemoglobin is the main protein that delivers oxygen to the cells of a mammal's body. The genomes of mammals carry several genes (designated with Greek letters) that encode slightly different globin polypeptides. A hemoglobin protein is composed of four globin polypeptides, two encoded by one globin gene and two encoded by another globin gene (**Figure 13.3**). During the embryonic stage of development, the epsilon (ε)-globin and zeta (ζ)-globin genes are turned on. At the fetal stage, these genes are turned off, and the alpha (α)-globin and gamma (γ)-globin genes are turned on. Finally, at birth, the γ-globin gene is turned off, and the beta (β)-globin gene is turned on.

How do the embryo and fetus acquire oxygen from their mother's bloodstream? The hemoglobin produced during the embryonic and fetal stages has a much higher binding affinity for oxygen than does the hemoglobin produced after birth. Therefore, the embryo and fetus can remove oxygen from the mother's bloodstream and use that oxygen for their own needs. This occurs across the placenta, where the mother's bloodstream is adjacent to the bloodstream of the embryo or fetus. In this way, gene regulation enables mammals to develop internally, even though the embryo and fetus are not breathing on their own. Gene regulation ensures that the correct hemoglobin protein is produced at the right time in development.

	Embryo	Fetus	Adult
Hemoglobin protein	2 ζ-globins 2 ε-globins	2 α-globins 2 γ-globins	2 α-globins 2 β-globins
Oxygen affinity	Highest	High	Moderate
Gene expression α-globin gene β-globin gene γ-globin gene ζ-globin gene ε-globin gene	Off Off Off On On	On Off On Off Off	On On Off Off Off

Figure 13.3 **Developmental regulation of human globin genes.** Note: The delta (δ) globin gene is also expressed in the adult.

Concept check: *How does this form of gene regulation help the embryo and fetus to obtain oxygen?*

Gene Regulation Can Occur at Different Points in the Process from DNA to Protein

Thus far, we have learned that gene regulation can have a dramatic impact on the ability of organisms to respond to environmental changes, differentiate cells, and progress through developmental stages. For structural genes that encode proteins, the regulation of gene expression can occur at any of the steps that are needed to produce a functional protein.

In bacteria, gene regulation most commonly occurs at the level of transcription, which means that bacteria regulate how much mRNA is made from genes (**Figure 13.4a**). When geneticists say a gene is "turned off," they mean that very little or no mRNA is made from that gene, whereas a gene that is "turned on" is transcribed into mRNA. Because transcription is the first step in gene expression, transcriptional regulation is a particularly efficient way to regulate genes because cells avoid wasting energy when the product of the gene is not needed. A second way for bacteria to regulate gene expression is to control the rate at which mRNA is translated into protein. This form of gene regulation is less common in bacteria. Last, gene expression can be regulated at the protein or post-translational level.

In eukaryotes, gene regulation occurs at many levels, including transcription, RNA processing, translation, and after translation is completed (**Figure 13.4b**). Like their bacterial counterparts, transcriptional regulation is a prominent form of gene regulation for eukaryotes. As discussed later in this chapter, eukaryotic genes are transcriptionally regulated in several different ways, some of which are not found in bacteria. As

discussed in Chapter 12, eukaryotes process their mRNA transcripts in ways that do not commonly occur in bacteria (refer back to Figure 12.3). For example, RNA splicing is a widespread phenomenon in eukaryotes. Later in this chapter, we will examine how this process is regulated to create two or more different types of mRNA from a single gene. Eukaryotes can also regulate an mRNA after its modification. The amount of mRNA may be regulated by controlling its degradation. In addition, the translation of mRNA may be regulated by small, inhibitory RNA molecules or by RNA-binding proteins that prevent translation from occurring. As in prokaryotes, eukaryotic proteins can be regulated in a variety of ways, including feedback inhibition, post-translational modification, and protein degradation. These various types of protein regulation are best understood within the context of cell biology, so they were primarily discussed in Unit II. Post-translational modifications are summarized in Chapter 21 (look ahead to Figure 21.10).

13.2 Regulation of Transcription in Bacteria

Due to gene regulation, bacteria can respond to changes in their cellular and environmental conditions. As we have seen, when a bacterium is exposed to a particular nutrient in its environment, such as a sugar, the genes are expressed that encode proteins needed for the uptake and metabolism of that sugar. In addition, bacteria have enzymes that synthesize cellular molecules such as particular amino acids. In such cases, the control of gene expression often occurs at the level of transcription. When we say that a gene is turned on, this refers to a high rate of transcription, whereas a gene that is turned off has a very low rate, perhaps negligible. In this section, we will examine the underlying molecular mechanisms that bring about transcriptional regulation in bacteria.

Transcriptional Regulation Often Involves Regulatory Transcription Factors and Small Effector Molecules

In most cases, transcriptional regulation involves the actions of **regulatory transcription factors**, proteins that bind to DNA in the vicinity of a promoter and affect the rate of transcription of one or more nearby genes. These transcription factors can either decrease or increase the rate of transcription of a gene. **Repressors** are transcription factors that bind to DNA and inhibit transcription, whereas **activators** bind to the DNA and increase the rate of transcription. The term **negative control** refers to transcriptional regulation by repressor proteins; **positive control** refers to regulation by activator proteins (**Figure 13.5a**).

In conjunction with regulatory transcription factors, molecules called **small effector molecules** often play a critical role in transcriptional regulation. A small effector molecule exerts its effects by binding to a regulatory transcription factor and causing a conformational change in the protein. In many cases, the effect of the conformational change determines whether or

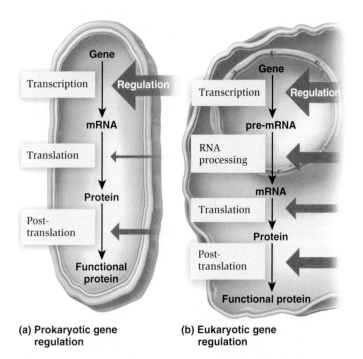

(a) **Prokaryotic gene regulation**

(b) **Eukaryotic gene regulation**

Figure 13.4 Overview of gene regulation in (a) prokaryotes and (b) eukaryotes. The relative width of the red arrows indicates the prominence with which regulation is used to control the production of functional proteins.

Negative control: A repressor inhibits transcription.

Positive control: An activator promotes transcription.

(a) Actions of regulatory transcription factors

The repressor protein is bound to the promoter region when the small effector molecule is not present.

The binding of the small effector molecule causes a conformational change in the repressor protein that prevents it from binding to the DNA.

(b) Action of a small effector molecule on a repressor protein

Figure 13.5 Actions of regulatory transcription factors and small effector molecules. (a) Regulatory transcription factors may exert negative or positive control. (b) One way that a small effector molecule may exert its effects is by preventing a repressor protein from binding to the DNA.

not the protein can bind to the DNA. **Figure 13.5b** illustrates an example. When the small effector molecule is not present in the cytoplasm, the repressor binds to the DNA and inhibits transcription. However, when the small effector molecule is subsequently found in the cytoplasm, it will bind to the repressor and cause a conformational change that inhibits the ability of the protein to bind to the DNA. The gene is turned on because the repressor is not able to bind to the DNA. Regulatory transcription factors that respond to small effector molecules have two functional regions called **domains**. One domain is a site where the protein binds to the DNA, whereas the other is the binding site for the small effector molecule.

The *lac* Operon Contains Genes That Encode Proteins Involved in Lactose Metabolism

In bacteria, structural genes are sometimes clustered together in units that are under the transcriptional control of a single promoter and have a regulatory region called an **operator**. This arrangement is known as an **operon**. The transcription of the structural genes occurs as a single unit and results in the production of a **polycistronic mRNA**, an mRNA that encodes more

than one protein. What advantage is this arrangement? An operon organization allows a bacterium to coordinately regulate a group of genes that encode proteins whose functions are used in a common pathway.

The genome of *E. coli* carries an operon, called the **lac operon**, that contains the genes for the proteins that allow it to metabolize lactose (see Figure 13.1). **Figure 13.6a** shows the organization of this operon as it is found in the *E. coli* chromosome, as well as the polycistronic mRNA that is transcribed from it. The *lac* operon contains a promoter, *lacP*, that is used to transcribe three structural genes: *lacZ*, *lacY*, and *lacA*. *LacZ* encodes β-galactosidase, which is an enzyme that breaks down lactose (**Figure 13.6b**). As a side reaction, β-galactosidase also converts a small percentage of lactose into allolactose, a structurally similar sugar or lactose analogue. As described later, allolactose is important in the regulation of the *lac* operon. The *lacY* gene encodes lactose permease, which is a membrane protein required for the transport of lactose into the cytoplasm of the bacterium. The *lacA* gene encodes galactoside transacetylase, which covalently modifies lactose and lactose analogues by attaching an acetyl group (—COCH$_3$). Although the functional necessity of this enzyme remains unclear, the attachment

(a) Organization of DNA sequences in the *lac* region of the *E. coli* chromosome

(b) Functions of lactose permease and β-galactosidase

Figure 13.6   The *lac* operon.   (a) This diagram depicts a region of the *E. coli* chromosome that contains the *lacI* regulatory gene and the adjacent *lac* operon, as well as the polycistronic mRNA transcribed from the operon. The mRNA is translated into three proteins: lactose permease, β-galactosidase, and galactoside transacetylase. (b) Lactose permease cotransports H⁺ with lactose. Bacteria maintain an H⁺ gradient across their cytoplasmic membrane that drives the active transport of lactose into the cytoplasm. β-galactosidase cleaves lactose into galactose and glucose. As a side reaction, it can also convert lactose into allolactose.

Concept check: *Which genes are under the control of the lac promoter?*

of acetyl groups to nonmetabolizable lactose analogues may prevent their toxic buildup in the cytoplasm.

Near the *lac* promoter are two regulatory sites designated the operator and the CAP site (see Figure 13.6a). The operator, or *lacO* site, is a sequence of nucleotides that provides a binding site for a repressor protein. The **CAP site** is a DNA sequence recognized by an activator protein.

Adjacent to the *lac* operon is the *lacI* gene, which encodes the **lac repressor**. This repressor protein is important for the regulation of the *lac* operon. The *lacI* gene, which is constitutively expressed at fairly low levels, has its own promoter called the *i* promoter. It is called a **regulatory gene** because the sole function of the encoded protein is to regulate the expression of other genes. The *lacI* gene is not considered a part of the *lac* operon. Let's now take a look at how the *lac* operon is regulated by the lac repressor.

The *lac* Operon Is Under Negative Control by a Repressor Protein

In the late 1950s, the first researchers to investigate gene regulation were Francois Jacob and Jacques Monod at the Pasteur Institute in Paris, France. Their focus on gene regulation stemmed from an interest in the phenomenon known as

enzyme adaptation, which had been identified early in the 20th century. Enzyme adaptation refers to the observation that a particular enzyme appears within a living cell only after the cell has been exposed to the substrate for that enzyme. Jacob and Monod studied lactose metabolism in *E. coli* to investigate this phenomenon. When they exposed bacteria to lactose, the levels of lactose-using enzymes in the cells increased by 1,000- to 10,000-fold. After lactose was removed, the synthesis of the enzymes abruptly stopped.

The first mechanism of regulation that Jacob and Monod discovered involved the lac repressor protein, which binds to the sequence of nucleotides found at the *lac* operator site. Once bound, the lac repressor prevents RNA polymerase from transcribing the *lacZ*, *lacY*, and *lacA* genes (**Figure 13.7a**). RNA polymerase can bind to the promoter when the lac repressor is bound to the operator site, but RNA polymerase

When lactose is absent from the environment, allolactose is not made and the lac repressor is free to bind to the *lac* operator. The transcription of the operon is inhibited.

lac operon

lac regulatory gene

RNA polymerase

lac promoter

Operator

lacI   lacP   lacO   lacZ   lacY   lacA

mRNA

Lac repressor binds to the operator and inhibits transcription.

Lac repressor (active)

(a) Lactose absent from the environment

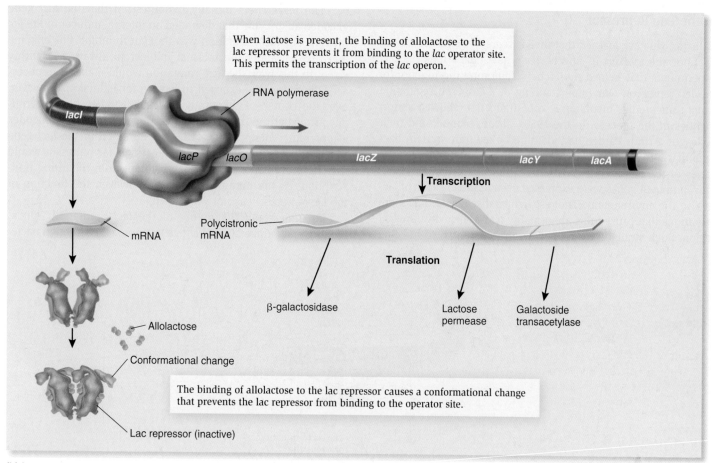

When lactose is present, the binding of allolactose to the lac repressor prevents it from binding to the *lac* operator site. This permits the transcription of the *lac* operon.

RNA polymerase

lacI   lacP   lacO   lacZ   lacY   lacA

Transcription

mRNA

Polycistronic mRNA

Translation

β-galactosidase   Lactose permease   Galactoside transacetylase

Allolactose

Conformational change

The binding of allolactose to the lac repressor causes a conformational change that prevents the lac repressor from binding to the operator site.

Lac repressor (inactive)

(b) Lactose present

Figure 13.7   Negative control of an inducible set of genes: function of the lac repressor in regulating the *lac* operon.

Concept check: With regard to regulatory proteins and small effector molecules, explain the meaning of the terms "negative control" and "inducible."

cannot move past the operator to transcribe the *lacZ*, *lacY*, and *lacA* genes.

Whether or not the lac repressor binds to the operator site depends on allolactose, which is the previously mentioned side product of the β-galactosidase enzyme (see Figure 13.6b). How does allolactose control the lac repressor? Allolactose is an example of a small effector molecule. The lac repressor protein contains four identical subunits, each one recognizing a single allolactose molecule. When four allolactose molecules bind to the lac repressor, a conformational change occurs that prevents the repressor from binding to the operator site. Under these conditions, RNA polymerase is free to transcribe the operon (**Figure 13.7b**). Because transcription has been turned on by the presence of a small effector molecule, this process is called **induction**. Similarly, the *lac* operon is said to be an **inducible operon**.

The regulation of the *lac* operon enables *E. coli* to efficiently respond to changes in the environment. Allolactose is an **inducer**, a small effector molecule that increases the rate of transcription. When the bacterium is not exposed to lactose, no allolactose is available to bind to the lac repressor. Therefore, the lac repressor binds to the operator site and inhibits transcription. In reality, the repressor does not completely inhibit transcription, so very small amounts of β-galactosidase, lactose permease, and galactoside transacetylase are made. Even so, the levels are far too low for the bacterium to readily use lactose. When the bacterium is exposed to lactose, a small amount can be transported into the cytoplasm via lactose permease, and β-galactosidase will convert some of it to allolactose (see Figure 13.6). The cytoplasmic level of allolactose will gradually rise until allolactose binds to the lac repressor, which induces the *lac* operon and promotes a high rate of transcription of the *lacZ*, *lacY*, and *lacA* genes. Translation of the encoded polypeptides will produce the proteins needed for lactose uptake and metabolism as described previously in Figure 13.1.

FEATURE INVESTIGATION

Jacob, Monod, and Pardee Studied a Constitutive Bacterial Mutant to Determine the Function of the Lac Repressor

Thus far, we have learned that the lac repressor binds to the *lac* operator site to exert its effects. Let's now take a look back at experiments that helped researchers determine the function of the lac repressor. Our understanding of *lac* operon regulation came from studies involving *E. coli* strains that showed abnormalities in the process. In the 1950s, Jacob, Monod, and their colleague Arthur Pardee had identified a few rare mutant bacteria that had abnormal lactose use. The mutants expressed the genes of the *lac* operon constitutively, meaning that the *lacZ*, *lacY*, and *lacA* genes were expressed even in the absence of lactose in the environment. The researchers discovered that some mutations that caused this abnormality had occurred in the *lacI* region. Such strains were termed *lacI⁻* (*lacI* minus) to indicate that the *lacI* region was not functioning properly. Normal or wild-type *lacI* strains of *E. coli* are called *lacI⁺* (*lacI* plus).

The researchers initially hypothesized that the *lacI* gene encoded an enzyme that degraded an internal inducer of the *lac* operon. The *lacI⁻* mutation was thought to inhibit this enzyme and thereby allow the internal inducer to always be synthesized. In this way, the *lacI⁻* mutation would make it unnecessary for cells to be exposed to lactose for induction. However, over the course of this study and later studies, they eventually arrived at the correct hypothesis that the *lacI* gene encodes a repressor protein (**Figure 13.8**). A mutation in the *lacI* gene that eliminates the synthesis of a functional lac repressor would thereby prevent the lac repressor protein from inhibiting transcription. At the time of their work, however, the function of the lac repressor was not yet known.

To understand the nature of the *lacI⁻* mutation, Jacob, Monod, and Pardee applied a genetic approach. Although bacte-

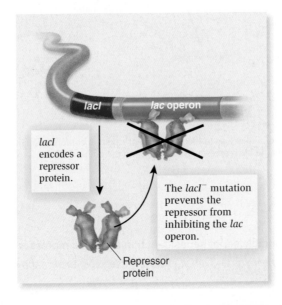

lacI encodes a repressor protein.

The *lacI⁻* mutation prevents the repressor from inhibiting the *lac* operon.

Repressor protein

Figure 13.8
A hypothesis for the function of the *lacI* gene.

rial conjugation is described in Chapter 18, let's briefly examine this process in order to understand this experiment. The earliest studies of Jacob, Monod, and Pardee in 1959 involved matings between recipient cells, termed F⁻ (F minus), and donor cells, which were called Hfr strains. Such Hfr strains were able to transfer a portion of the bacterial chromosome to a recipient cell. Later experiments in 1961 involved the transfer of circular segments of DNA known as F factors. We will consider this later type of experiment here. Sometimes an F factor also carries genes that were originally found within the bacterial chromosome. These types of F factors are called F′ factors (F prime factors). A strain of bacteria containing F′ factor genes is called a **merozygote**, or partial diploid. The production of merozygotes was instrumental in allowing Jacob, Monod, and Pardee to elucidate the function of the *lacI* gene.

As shown in Figure 13.9, these researchers studied the *lac* operon in a bacterial strain carrying a *lacI⁻* mutation that caused constitutive expression of the *lac* operon. In addition, the mutant strain was subjected to mating to create a merozygote that also carried a normal *lac* operon and normal *lacI⁺* gene on an F′ factor. The merozygote contained both *lacI⁺* and *lacI⁻* genes. The constitutive mutant and corresponding merozy-

gote were allowed to grow separately in liquid media and then divided into two tubes each. In half of the tubes, the cells were incubated with lactose to determine if lactose was needed to induce the expression of the operon. In the other tubes, lactose was omitted. To monitor the expression of the *lac* operon, the cells were broken open and then tested for the amount of β-galactosidase they released by measuring the ability of any β-galactosidase present to convert a colorless compound into a yellow product.

The data table of Figure 13.9 summarizes the effects of this constitutive mutation and its analysis in a merozygote. As Jacob, Monod, and Pardee already knew, the *lacI⁻* mutant strain expressed the *lac* operon constitutively, in both the presence and absence of lactose. However, when a normal *lac* operon and *lacI⁺* gene were introduced on an F′ factor into a cell harboring the mutant *lacI⁻* gene on the chromosome, the normal *lacI⁺* gene could regulate both operons. In the absence of lactose, both operons were shut off. How did Jacob, Monod, and Pardee eventually explain these results? This occurred because a single *lacI⁺* gene on the F′ factor can produce enough repressor protein to bind to both operator sites. Furthermore, this protein is diffusible—can spread through the cytoplasm—and can bind to

Figure 13.9   The experiment performed by Jacob, Monod, and Pardee to study a constitutive *lacI⁻* mutant.

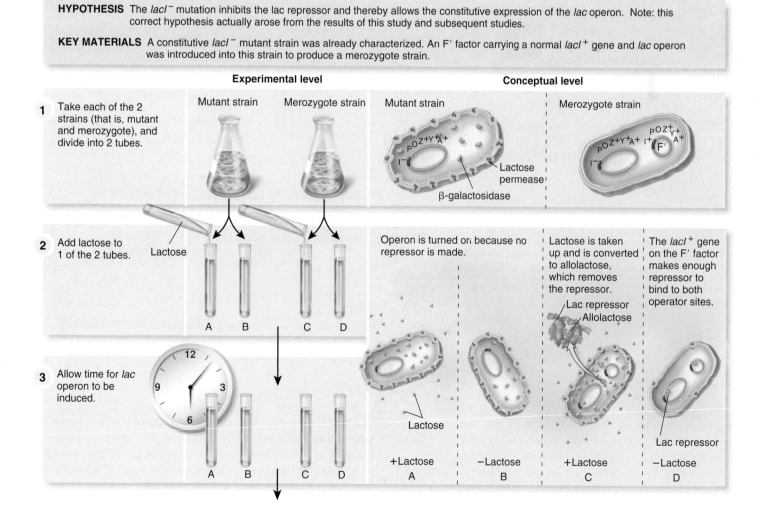

HYPOTHESIS The *lacI⁻* mutation inhibits the lac repressor and thereby allows the constitutive expression of the *lac* operon. Note: this correct hypothesis actually arose from the results of this study and subsequent studies.

KEY MATERIALS A constitutive *lacI⁻* mutant strain was already characterized. An F′ factor carrying a normal *lacI⁺* gene and *lac* operon was introduced into this strain to produce a merozygote strain.

Experimental level   **Conceptual level**

1 Take each of the 2 strains (that is, mutant and merozygote), and divide into 2 tubes.

Mutant strain   Merozygote strain

Mutant strain   Merozygote strain

Lactose permease

β-galactosidase

2 Add lactose to 1 of the 2 tubes.   Lactose

A   B   C   D

Operon is turned on because no repressor is made.

Lactose is taken up and is converted to allolactose, which removes the repressor.

Lac repressor
Allolactose

The *lacI⁺* gene on the F′ factor makes enough repressor to bind to both operator sites.

3 Allow time for *lac* operon to be induced.

A   B   C   D

Lactose

Lac repressor

+Lactose   −Lactose   +Lactose   −Lactose
A   B   C   D

4 Burst the cells with a sonicator, which makes strong sound waves. This releases β-galactosidase.

Sound waves

β-galactosidase
Broken cell

β-galactosidase present
A

β-galactosidase present
B

β-galactosidase present
C

β-galactosidase absent
D

A B C D

5 Measure the function of β-galactosidase. This is done by adding a colorless lactose analogue that β-galactosidase converts to a yellow product. Note: the blue tube is meant to represent colorless liquid.

Add colorless lactose analogue to all 4 tubes.

Lactose analogue

NO₂

Galactose
NO₂
Yellow compound

No change

A B C D
A B C D

6 The amount of yellow color is measured with a spectrophotometer; the deeper the yellow color, the more β-galactosidase was produced.

More β-galactosidase produced

7 THE DATA

Results from step 6:

| | Expression of the *lac* operon | |
	With lactose	Without lactose
Mutant strain	100%	100%
Merozygote strain	220%	<1%

8 CONCLUSION The *lacI* gene encodes a diffusible repressor protein.

9 SOURCE Jacob, Francois, and Monod, Jacques. 1961. Genetic regulatory mechanisms in the synthesis of proteins. *Journal of Molecular Biology* 3:318–356.

lac operons that are on the F′ factor and on the bacterial chromosome. Taken together, the data indicated that the normal *lacI* gene encodes a diffusible protein that represses the *lac* operon.

The interactions between regulatory proteins and DNA sequences illustrated in this experiment have led to the defini-

tion of three genetic terms. In both prokaryotes and eukaryotes, a ***cis*-acting element** is a DNA segment that must be adjacent to the gene(s) that it regulates. The *lac* operator site is an example of a *cis*-acting element. A ***trans*-effect** is a form of gene regulation that can occur even though two DNA segments are not physically adjacent. The action of the lac repressor on the *lac* operon is a *trans*-effect. A **cis-effect** is mediated by a *cis*-acting element that binds regulatory proteins, whereas a *trans*-effect is mediated by genes that encode diffusible regulatory proteins.

Experimental Questions

1. What were the key observations made by Jacob, Monod, and Pardee that led to the development of their hypothesis regarding the *lacI* gene and the regulation of the *lac* operon?

2. What was the eventual hypothesis proposed by the researchers to explain the function of the *lacI* gene and the regulation of the *lac* operon?

3. How did Jacob, Monod, and Pardee test the hypothesis? What were the results of the experiment? How do these results support the idea that the *lacI* gene produces a repressor protein?

The *lac* Operon Is Also Under Positive Control by an Activator Protein

In addition to negative control by a repressor protein, the *lac* operon is also positively regulated by an activator protein called the **catabolite activator protein** (**CAP**). CAP is controlled by a small effector molecule, **cyclic AMP** (**cAMP**), that is produced from ATP via an enzyme known as adenylyl cyclase. Gene regulation involving CAP and cAMP is an example of positive control (**Figure 13.10**). When cAMP binds to CAP, the cAMP-CAP complex binds to the CAP site near the *lac* promoter. This causes a bend in the DNA that enhances the ability of RNA polymerase to bind to the promoter. In this way, the rate of transcription is increased.

The key functional role of CAP is to allow *E. coli* to choose between different sources of sugar. In a process known as **catabolite repression**, transcription of the *lac* operon is inhibited by the presence of glucose, which is a catabolite (it is broken down—**catabolized**—inside the cell). This gene regulation allows *E. coli* to preferentially use glucose compared to other sugars, such as lactose. How does this occur? Glucose inhibits the production of cAMP and thereby prevents the binding of CAP to the DNA. In this way, glucose blocks the activation of the *lac* operon and thereby inhibits transcription. Though it may seem puzzling, the term catabolite repression was coined before the action of the cAMP-CAP complex was understood at the molecular level. Historically, the primary observation of researchers was that glucose (a catabolite) inhibited (repressed) lactose metabolism. Further experimentation revealed that CAP is actually an activator protein.

Figure 13.11 considers the four possible environmental conditions that an *E. coli* bacterium might experience with regard to these two sugars. When both lactose and glucose levels are high, the rate of transcription of the *lac* operon is low, because CAP does not activate transcription. Under these conditions, the bacterium primarily uses glucose rather than lactose. Why is this a benefit to the bacterium? Greater efficiency is achieved if the bacterium uses one type of sugar at a time. If lactose levels are high and glucose is low, the transcription rate of the *lac* operon is very high because CAP is bound to the CAP site and the lac repressor is not bound to the operator site. Under these conditions, the bacterium metabolizes lactose. When lactose levels are low, the lac repressor prevents transcription of the *lac* operon, whether glucose levels are high or low.

Figure 13.10 Positive regulation of the *lac* operon by the catabolite activator protein (CAP). When cAMP is bound to CAP, CAP binds to the DNA and causes it to bend. This bend facilitates the binding of RNA polymerase.

The *trp* Operon Is Also Under Negative Control by a Repressor Protein

So far in this section, we have examined the regulation of the *lac* operon. Let's now consider an example of an operon that encodes enzymes involved in biosynthesis rather than breakdown. Our example is the ***trp* operon** of *E. coli*, which encodes enzymes that are required to make the amino acid tryptophan, a building block of cellular proteins. More specifically, the *trpE*, *trpD*, *trpC*, *trpB*, and *trpA* genes encode enzymes that are involved in a pathway that leads to tryptophan synthesis.

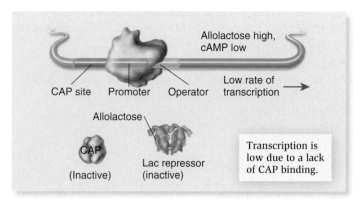

(a) Lactose high, glucose high

(b) Lactose high, glucose low

(c) Lactose low, glucose high

(d) Lactose low, glucose low

Figure 13.11   Effects of lactose and glucose on the expression of the *lac* operon.

Concept check: *What are the advantages of having both an activator and a repressor protein?*

The *trp* operon is regulated by a repressor protein that is encoded by the *trpR* gene. The binding of the repressor to the *trp* operator site inhibits transcription. The ability of the trp repressor to bind to the *trp* operator is controlled by tryptophan, which is the product of the enzymes that are encoded by the operon. When tryptophan levels within the cell are very low, the trp repressor cannot bind to the operator site. Under these conditions, RNA polymerase readily transcribes the operon (**Figure 13.12a**). In this way, the cell expresses the genes that encode enzymes that result in the synthesis of tryptophan, which is in short supply. Alternatively, when the tryptophan levels within the cell are high, tryptophan turns off the *trp* operon. Tryptophan acts as a small effector molecule, or **corepressor**, by binding to the trp repressor protein. This causes a conformational change in the repressor that allows it to bind to the *trp* operator site, inhibiting the ability of RNA polymerase to transcribe the operon (**Figure 13.12b**). Therefore, the bacterium does not waste energy making tryptophan when it is abundant.

When comparing the *lac* and *trp* operons, the actions of their small effector molecules are quite different. The lac repressor binds to its operator in the absence of its small effector molecule, whereas the trp repressor binds to its operator only in the presence of its small effector molecule. The *lac* operon is categorized as inducible because its small effector molecule, namely allolactose, induces transcription. By comparison, the *trp* operon

is considered to be a **repressible operon** because its small effector molecule, namely tryptophan, represses transcription.

Repressible Operons Usually Encode Anabolic Enzymes, and Inducible Operons Encode Catabolic Enzymes

By comparing the mechanisms of regulation among many bacterial operons, geneticists have noticed a general trend. The genes in some operons encode proteins that function in the breakdown, or **catabolism**, of a substance. In such cases, the substance to be broken down (or a related compound) often acts as the inducer. This keeps the genes turned off unless the appropriate substance is available. For example, allolactose, which is a product of lactose metabolism, acts as an inducer of the *lac* operon. An inducible form of regulation allows the bacterium to express the appropriate genes only when they are needed to metabolize lactose.

Other cellular enzymes are important for synthesizing organic molecules, a process termed **anabolism**. Because these molecules are generally needed for the functioning of the cell, the genes that encode these anabolic enzymes tend to be regulated by a repressible mechanism, allowing the genes to be transcribed unless they are turned off. The small effector molecule is commonly a product of the enzymes' biosynthetic activities.

When tryptophan levels are low, the trp repressor is inactive, allowing the entire *trp* operon to be transcribed.

RNA polymerase

trpR

Trp repressor

Inactive trp repressor cannot bind to the operator site.

trpE trpD trpC trpB trpA

Transcription

mRNA

(a) Low tryptophan

When tryptophan levels are high, tryptophan (the corepressor) binds to the trp repressor, enabling it to bind to the operator site, which prevents transcription.

trpE trpD trpC trpB trpA

trpR

Tryptophan

Binding of corepressor causes a conformational change in the trp repressor that allows it to bind to the operator site.

(b) High tryptophan

Figure 13.12 Negative control of a repressible set of genes: function of the trp repressor and corepressor in regulating the *trp* operon.

Concept check: *How are the functions of the lac repressor and trp repressor similar to each other, and how are they different?*

For example, as we learned, tryptophan is produced by the enzymes that are encoded by the *trp* operon. When enough of this amino acid has been made, tryptophan itself acts as a corepressor, turning off the genes required for tryptophan biosynthesis. Therefore, a repressible form of regulation provides the bacterium with a way to prevent the overproduction of the product of a biosynthetic pathway.

13.3 Regulation of Transcription in Eukaryotes

Transcriptional regulation in eukaryotes follows some of the same principles as those found in prokaryotes. For example, activator and repressor proteins are involved in regulating genes by influencing the ability of RNA polymerase to initiate transcription. In addition, many eukaryotic genes are regulated by small effector molecules. However, some important differences also occur. In eukaryotic species, genes are almost always organized individually, not in operons. In addition, eukaryotic gene regulation tends to be more intricate, because eukaryotes are faced with complexities that differ from their prokaryotic counterparts. For example, eukaryotes have more complicated cell structures that contain many more proteins and a variety of cell organelles. Many eukaryotes such as animals and plants are multicellular and contain different cell types. As discussed earlier in this chapter, animal cells may differentiate into nerve cells, muscle cells, and skin cells, among others. Furthermore, animals and plants progress through developmental stages that require changes in gene expression. For these reasons, gene regulation in eukaryotes requires much more coordination and integration.

By studying transcriptional regulation, researchers have discovered that most eukaryotic genes, particularly those found in multicellular species, are regulated by many factors. This phenomenon is called **combinatorial control** because the combination of many factors determines the expression of any given gene. At the level of transcription, common factors that contribute to combinatorial control include the following:

1. One or more activator proteins may stimulate the ability of RNA polymerase to initiate transcription.
2. One or more repressor proteins may inhibit the ability of RNA polymerase to initiate transcription.
3. The function of activators and repressors may be modulated in several ways, which include the binding of small effector molecules, protein–protein interactions, and covalent modifications.
4. Activator proteins are necessary to alter chromatin structure in the region where a gene is located, thereby making it easier for the gene to be recognized and transcribed by RNA polymerase.
5. DNA methylation usually inhibits transcription, either by preventing the binding of an activator protein or by recruiting proteins that inhibit transcription.

All five of these factors may contribute to the regulation of a single gene, or possibly only three or four will play a role. In most cases, transcriptional regulation is aimed at controlling the initiation of transcription at the promoter. In this section, we will survey these basic types of gene regulation in eukaryotic species.

Eukaryotic Structural Genes Have a Core Promoter and Regulatory Elements

To understand gene regulation in eukaryotes, we first need to consider the DNA sequences that are needed to initiate transcription. For eukaryotic structural genes that encode proteins, three features are common among most promoters: a **TATA box**, a **transcriptional start site**, and **regulatory elements** (Figure 13.13).

The TATA box and transcriptional start site form the **core promoter**. The transcriptional start site is the place in the DNA where transcription actually begins. The TATA box, which is a 5′–TATAAA–3′ sequence, is usually about 25 base pairs upstream from a transcriptional start site. The TATA box is important in determining the precise starting point for transcription. If it is missing from the core promoter, transcription may start at a variety of different locations. The core promoter, by itself, results in a low level of transcription that is termed **basal transcription**.

Regulatory elements (or response elements) are DNA segments that regulate eukaryotic genes. As described later, regulatory elements are recognized by regulatory transcription factors that control the ability of RNA polymerase to initiate transcription at the core promoter. Some regulatory elements, known as **enhancers**, play a role in the ability of RNA polymerase to begin transcription and thereby enhance the rate of transcription. When enhancers are not functioning, most eukaryotic genes have very low levels of basal transcription. Other regulatory elements, known as **silencers**, prevent transcription of a given gene when its expression is not needed. When these sequences function, the rate of transcription is decreased.

A common location for regulatory elements is the region that is 50 to 100 base pairs upstream from the transcriptional start site (Figure 13.13). However, the locations of regulatory elements are quite variable among different eukaryotic genes. Regulatory elements can be quite distant from the promoter, even 100,000 base pairs away, yet exert strong effects on the

Figure 13.14   **The preinitiation complex.**   General transcription factors (GTFs) and RNA polymerase II assemble into the preinitiation complex at the core promoter in eukaryotic structural genes.

ability of RNA polymerase to initiate transcription at the core promoter! Regulatory elements were first discovered by Susumu Tonegawa and coworkers in the 1980s. While studying genes that play a role in immunity, they identified a region that was far away from the core promoter but was needed for high levels of transcription to take place.

RNA Polymerase II, General Transcription Factors, and Mediator Are Needed to Transcribe Eukaryotic Structural Genes

As discussed in Chapter 12, eukaryotes have three RNA polymerases designated I, II, and III. RNA polymerase II transcribes structural genes that encode proteins. By studying transcription in a variety of eukaryotic species, researchers have identified three types of proteins that play a role in initiating transcription at the core promoter of structural genes. These are RNA polymerase II, five different proteins called **general transcription factors** (**GTFs**), and a large protein complex called mediator. GTFs are needed for DNA binding at the core promoter and for initiation of transcription. Mediator, which is described later, is also needed for RNA polymerase to proceed to the elongation phase of transcription.

RNA polymerase II and GTFs must come together at the core promoter before transcription can be initiated. A series of interactions occurs between these proteins so that RNA polymerase II can bind to the DNA. **Figure 13.14** shows the structure of the completed assembly of RNA polymerase II and GTFs at the TATA box within the core promoter. This assembly is known as the **preinitiation complex**. In vitro, when researchers mix together RNA polymerase II, GTFs, and a DNA sequence containing a TATA box and a transcriptional start site, the DNA is transcribed into RNA. Therefore, these components are referred to as the **basal transcription apparatus**. In a living cell, however, additional components, such as regulatory transcription factors, control the assembly of RNA polymerase II and GTFs at the core promoter and are responsible for causing transcription to begin at a fast or slow rate.

A third component needed for transcription in eukaryotes is the mediator protein complex. **Mediator** is composed of several proteins that bind to each other to form an elliptical-shaped

Figure 13.13   **A common organization of sequences for the promoter of a eukaryotic structural gene.**   The core promoter has a TATA box and a transcriptional start site. The TATA box sequence is 5′–TATAAA–3′. However, not all structural genes in eukaryotes have a TATA box. The A highlighted in dark blue is the transcriptional start site. This A marks the site of the first A in the RNA transcript. The sequence that flanks the A of the transcriptional start site is two pyrimidines, then C, then five pyrimidines. Py refers to pyrimidine—cytosine or thymine. Regulatory elements, such as enhancers and silencers, are usually found upstream from the core promoter.

complex that partially wraps around RNA polymerase II and the GTFs. Mediator derives its name from the observation that it mediates interactions between the preinitiation complex and regulatory transcription factors such as activators or repressors that bind to enhancers or silencers. The function of mediator is to control the rate at which RNA polymerase can begin to transcribe RNA at the transcriptional start site.

Activators and Repressors May Influence the Function of GTFs or Mediator

In eukaryotes, regulatory transcription factors called activators and repressors bind to enhancers or silencers, respectively, and regulate the rate of transcription of a nearby gene. In some cases, activator proteins interact with **coactivators**—proteins that increase the rate of transcription but do not directly bind to the DNA itself.

Activators and repressors commonly regulate the function of RNA polymerase II by binding to GTFs or mediator. As shown in **Figure 13.15**, some activators bind to an enhancer and then influence the function of GTFs. For example, an activator may improve the ability of a GTF called TFIID to initiate transcription. The function of TFIID is to recognize the TATA box and

begin the assembly process. An activator may recruit TFIID to the TATA box, thereby promoting the assembly of GTFs and RNA polymerase II into the preinitiation complex. In contrast, repressors may inhibit the function of TFIID. Certain repressors exert their effects by preventing the binding of TFIID to the TATA box or by inhibiting the ability of TFIID to assemble other GTFs and RNA polymerase II at the core promoter.

A second way that regulatory transcription factors control RNA polymerase II is via mediator (**Figure 13.16**). In this example, an activator also interacts with a coactivator. The activator/coactivator complex stimulates the function of mediator and thereby causes RNA polymerase II to proceed to the elongation phase of transcription more quickly. Alternatively, repressors have the opposite effect to those seen in Figure 13.16. When a repressor inhibits mediator, RNA polymerase II cannot progress to the elongation stage.

A third way that regulatory transcription factors influence transcription is by recruiting proteins that affect chromatin structure in the promoter region. As described in Chapter 11, the DNA found in eukaryotic chromosomes is wrapped in nucleosomes. The arrangement of nucleosomes also affects the ability of RNA polymerase II and GTFs to initiate transcription. This topic is described next.

1. An activator protein binds to an enhancer.

Activator

Enhancer

2. The activator protein enhances the ability of a GTF called TFIID to bind to the TATA box.

TFIID

TATA box

3. TFIID promotes the assembly of the preinitiation complex.

TFIID

TATA box

Preinitiation complex

Figure 13.15 Effect of an activator via TFIID, a general transcription factor.

1. Mediator binds to the preinitiation complex, but transcriptional initiation does not occur.

Enhancer

Mediator

Preinitiation complex

2. An activator binds to a distant enhancer and a coactivator binds to the activator. A bend in the DNA allows the activator/coactivator complex to interact with mediator. This interaction causes RNA polymerase to proceed to the elongation stage of transcription.

Coactivator

Enhancer   Activator

Figure 13.16 Effect of an activator via mediator.

Concept check: *When an activator interacts with mediator, how does this affect the function of RNA polymerase?*

Figure 13.17 Effects of an activator on chromatin structure.

Concept check: Why are changes in chromatin structure needed for transcription to occur?

Transcription Is Also Controlled by Changes in Chromatin Structure

In eukaryotes, DNA is associated with proteins to form a structure called **chromatin** (see Chapter 11). How does the structure of chromatin affect gene transcription? Depending on the locations and arrangements of nucleosomes, a region containing a gene may be in a **closed conformation**, and transcription may be difficult or impossible. Transcription requires changes in chromatin structure that allow transcription factors to gain access to and bind to the DNA in the promoter region. Such chromatin, said to be in an **open conformation**, is accessible to GTFs and RNA polymerase II so that transcription can take place.

An important role of some activators is to alter the locations and arrangements of nucleosomes where a gene is located. To do this, an activator first binds to an accessible enhancer site (**Figure 13.17**). Next, the binding of the activator recruits proteins to the region that alter the nucleosomes. In some cases, an activator protein attracts **histone acetyltransferase** to the region. This enzyme attaches acetyl groups (—COCH₃) to the amino terminal tails of histone proteins. As described in Chapter 11, histone proteins are critical in the compaction of eukaryotic DNA. When acetylated, histone proteins do not bind as tightly to the DNA. A second effect of an activator protein is to recruit **ATP-dependent chromatin remodeling enzymes** to the site. The overall effect of histone acetyltransferase and ATP-dependent chromatin remodeling enzymes is to alter the locations and arrangements of nucleosomes, sometimes over a fairly long distance such as several hundred or several thousand base pairs of DNA. This change in chromatin structure facilitates the ability of RNA polymerase II to recognize and transcribe a gene.

The Histone Code Controls Chromatin Compaction

As described in Figure 13.17, acetylation of histone proteins can loosen the level of chromatin packing. In recent years, researchers have discovered that many different amino acids

Figure 13.18 Examples of covalent modifications that occur to the amino terminal tails of histone proteins. The amino acids are numbered from the amino terminus. The modifications shown here are m for methylation, p for phosphorylation, and ac for acetylation. Many more modifications can occur to the amino terminal tails. The ones shown here represent common examples.

Concept check: What are the two opposing effects that histone modifications may have with regard to chromatin structure?

in the amino terminal tails of histone proteins are subject to several types of covalent modifications, including acetylation, methylation, and phosphorylation. Over 50 different enzymes have been identified in mammals that selectively modify amino terminal tails. **Figure 13.18** shows examples of sites in the tails of histone proteins H2A, H2B, H3, and H4 that can be modified.

What are the effects of covalent modifications of histones? First, modifications may directly influence interactions between DNA and histone proteins, and between adjacent nucleosomes. Second, histone modifications provide binding sites that are recognized by other proteins. According to the **histone code hypothesis**, proposed by Brian Strahl and David Allis in 2000, the pattern of histone modification is recognized by proteins much like a language or code. For example, one pattern might involve phosphorylation of the serine at the first amino acid in H2A and acetylation of the lysines at the fifth and eighth amino acids in H4. A different pattern could involve acetylation of the fifth amino acid, a lysine, in H2B and methylation of the third amino acid in H4, which is an arginine.

The pattern of covalent modifications of amino terminal tails provides binding sites for proteins that subsequently affect chromatin structure. One pattern of histone modification may attract proteins that cause the chromatin to be less accessible to RNA polymerase and general transcription factors. This would silence the transcription of genes in the region. Alternatively, a different combination of histone modifications may attract proteins, such as chromatin remodeling enzymes, that promote gene transcription. In this way, the histone code plays a key role in accessing the information within the genomes of eukaryotic species.

Steroid Hormones Exert Their Effects by Binding to a Regulatory Transcription Factor and Controlling the Transcription of Nearby Genes

Thus far, we have considered the general ways that regulatory transcription factors control transcription. Let's now turn to a specific example that illustrates how a regulatory transcription factor functions within living cells. Our example involves a transcriptional activator that responds to steroid hormones. This factor is known as a **steroid receptor**, because it binds directly to a steroid hormone. The hormone is an example of a small effector molecule.

As discussed in Chapter 50, steroid hormones are a category of hormones that are synthesized by specialized cells of many organisms, including the endocrine glands of mammals, and then secreted into the bloodstream. The hormones are then taken up by cells that respond to the hormones in different ways. For example, glucocorticoid hormones influence nutrient metabolism in most body cells by promoting the metabolism of glucose, proteins, and fats.

The effect of glucocorticoid hormones is to activate the transcription of specific genes. Glucocorticoids are released from endocrine cells and secreted into the bloodstream when an animal is fasting and needs to regulate its blood levels of glucose, amino acids, and fats. The hormone molecules diffuse across the plasma membrane of target cells and bind to glucocorticoid receptors, which are a type of steroid receptor (**Figure 13.19**). This binding releases proteins called chaperones and thereby exposes an amino acid sequence within the receptor called a nuclear localization signal (NLS). This signal allows

Figure 13.19 Action of the glucocorticoid receptor as a transcriptional activator.

Concept check: If a GRE next to a gene was deleted, how would that affect the regulation of the gene?

the receptor to travel into the nucleus through a nuclear pore. Two glucocorticoid receptors bind to each other noncovalently to form a dimer and then travel through the nuclear pore into the nucleus. The glucocorticoid receptor dimer binds to two adjacent glucocorticoid response elements (GREs) that are next to particular genes. The GREs function as enhancer sequences. The binding of the glucocorticoid receptor dimer to GREs activates the transcription of the adjacent gene, eventually leading to the synthesis of the encoded protein.

How do glucocorticoids affect cell function? Mammalian cells usually have a large number of glucocorticoid receptors within their cytosol. Because GREs are located near several different genes, the uptake of hormone molecules can activate many glucocorticoid receptors and thereby enhance the transcription of several different genes that encode proteins involved with the metabolism of glucose, proteins, and fats. For this reason, glucocorticoid hormones facilitate the coordinated expression of genes that play a role in nutrient metabolism.

DNA Methylation Inhibits Gene Transcription

Let's now turn our attention to a mechanism that usually silences gene expression. DNA structure can be modified by the covalent attachment of methyl groups ($-CH_3$) by an enzyme called **DNA methylase**. This modification, termed **DNA methylation**, is common in some eukaryotic species but not all. For example, yeast and *Drosophila* have little or no detectable methylation of their DNA, whereas DNA methylation in vertebrates and plants is relatively abundant. In mammals, approximately 5% of the

DNA is methylated. Eukaryotic DNA methylation occurs on the cytosine base. The sequence that is methylated is shown here:

$$CH_3$$
$$|$$
$$5'—CG—3'$$
$$3'—GC—5'$$
$$|$$
$$CH_3$$

DNA methylation usually inhibits the transcription of eukaryotic genes, particularly when it occurs in the vicinity of the promoter. In vertebrates and flowering plants, many genes contain sequences called **CpG islands** near their promoters. CpG refers to the nucleotides of <u>C</u> and <u>G</u> in DNA that are connected by a <u>p</u>hosphodiester linkage. A CpG island is a cluster of CpG sites. Unmethylated CpG islands are usually correlated with active genes, whereas repressed genes contain methylated CpG islands. In this way, DNA methylation may play an important role in the silencing of particular genes.

How does DNA methylation inhibit transcription? This can occur in two general ways. First, methylation of CpG islands may prevent an activator from binding to an enhancer element, thus inhibiting the initiation of transcription. A second way that methylation inhibits transcription is by converting chromatin from an open to a closed conformation. Proteins known as **methyl-CpG-binding proteins** bind methylated sequences. Once bound to the DNA, the methyl-CpG-binding protein recruits other proteins to the region that inhibit transcription. A human genetic disorder called Rett syndrome involves a defect in a methyl-CpG-binding protein that is made in nerve cells. The syndrome is primarily found in females and results in a variety of neurodevelopmental problems, including a small head, seizures, and mental impairment. These symptoms are presumably the result of improper gene regulation.

13.4 Regulation of RNA Processing and Translation in Eukaryotes

In the first three sections of this chapter, we have focused on gene regulation at the level of transcription in bacteria and eukaryotes. Eukaryotic gene expression is commonly regulated at the levels of RNA processing and translation. These added levels of regulation provide benefits that are important to eukaryotic species. First, by regulating RNA processing, eukaryotes can produce more than one mRNA transcript from a single gene. This allows a gene to encode two or more polypeptides, thereby increasing the complexity of eukaryotic proteomes. A second issue is timing. Transcriptional regulation in eukaryotes takes a fair amount of time before its effects are observed at the cellular level. During transcriptional regulation, (1) the chromatin must be converted to an open conformation, (2) the gene must be transcribed, (3) the RNA must be processed and exported from the nucleus, and (4) the protein must be made via translation. All four steps take time, on the order of several minutes. One way to achieve faster regulation is to control steps that occur after an RNA transcript is made. In eukaryotes, translational regulation provides a faster way to regulate the levels of gene products, namely, proteins. Translation can be regulated by controlling the stability of an mRNA transcript, causing it to remain in the cytosol for a long time, or causing it to be rapidly degraded. Alternatively, small RNA molecules or RNA-binding proteins can bind to mRNAs and control whether or not a ribosome can translate the mRNA into a polypeptide.

During the past few decades, many critical advances have been made regarding our knowledge of the regulation of RNA processing and translation. Even so, molecular geneticists are still finding new forms of regulation, making this an exciting area of modern research. In this section, we will survey a few of the known mechanisms of RNA processing and translational regulation.

Alternative Splicing of Pre-mRNAs Creates Protein Diversity

In eukaryotes, a pre-mRNA transcript is processed before it becomes a mature mRNA. When a pre-mRNA has multiple introns and exons, splicing may occur in more than one way, resulting in the creation of two of more different polypeptides. Such **alternative splicing** is a form of gene regulation that allows an organism to use the same gene to make different proteins at different stages of development, in different cell types, and/or in response to a change in the environmental conditions. Alternative splicing is an important form of gene regulation in complex eukaryotes such as animals and plants.

As an example of how alternative splicing occurs, let's suppose a human pre-mRNA contains seven exons (**Figure 13.20**). In nerve cells, it is spliced to contain the following pattern of exons: 1-2-3-5-6-7. In muscle cells, it is alternatively spliced to have a different pattern: 1-2-4-5-6-7. In this example, the mRNA from nerve cells contains exon 3, while the mRNA from muscle

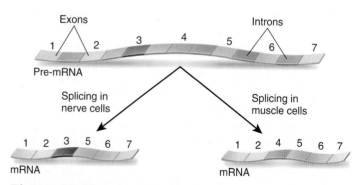

Figure 13.20   Alternative splicing. In this example, the pre-mRNA transcript can be spliced to contain exon 3 (in nerve cells) or exon 4 (in muscle cells), but not both.

Concept check: *What is the biological advantage of alternative splicing?*

cells contains exon 4. When alternative splicing occurs, proteins with significant differences in their amino acid sequences are produced.

What are the consequences of alternative splicing? In most cases, the alternative versions of a protein will have similar functions, because much of their amino acid sequences will be identical to each other. Nevertheless, alternative splicing produces differences in amino acid sequences that will provide each protein with its own unique characteristics. The alternatively spliced versions tend to be expressed in different cell types (for example, nerve versus muscle cells) or at different stages of development (for example, embryonic versus adult). This provides a way for multicellular organisms to fine-tune a given protein to function optimally in a given cell type or stage of development. The advantage of alternative splicing is that two or more different polypeptides can be derived from a single gene, thereby increasing the size of the proteome while minimizing the size of the genome. A small genome size is beneficial because less energy is spent replicating the DNA, and the DNA more easily fits within the nucleus of the cell.

Genomes & Proteomes Connection

Increases in Biological Complexity Are Correlated with Greater Sizes of Genomes and Proteomes

As we have just seen, alternative splicing can increase the proteome size without increasing the total number of genes. For organisms to become more complex, as in plants and animals, evolution has produced more complex proteomes. In the past few decades, many technical advances have improved our ability to analyze the genomes and proteomes of many different species. Researchers have been able to determine the amount of DNA from several species and estimate the total number of genes. In addition, scientists can also estimate the number of polypeptides if information is available concerning the degree of alternative splicing in a given species.

Table 13.1 compares six species: a bacterium (*Escherichia coli*), a eukaryotic single-celled organism (yeast—*Saccharomyces cerevisiae*), a small nematode worm (*Caenorhabditis elegans*), a fruit fly (*Drosophila melanogaster*), a small flowering plant (*Arabidopsis thaliana*), and a human (*Homo sapiens*). One general trend is that less complex organisms tend to have fewer genes. For example, unicellular organisms have only a few thousand genes, whereas multicellular species have tens of thousands. However, the trend is by no means a linear one. If we compare *C. elegans* and *D. melanogaster*, the fly actually has fewer genes even though it is morphologically more complex.

A second trend you can see in Table 13.1 concerns alternative splicing. This phenomenon does not occur in bacteria and is rare in *S. cerevisiae*. The frequency of alternative splicing increases from worms to flies to humans. For example, the level of alternative splicing is 10-fold higher in humans compared to *Drosophila*. This trend can partially explain the increase in complexity among these species. Even though humans have only

Table 13.1	Genome Size and Biological Complexity			
Species	Level of complexity	Genome size (million bp)	Approximate number of genes	Percentage of genes alternatively spliced
Escherichia coli	A unicellular prokaryote	4.2	4,000	0
Saccharomyces cerevisiae	A unicellular eukaryote	12	6,000	<1
Caenorhabditis elegans	A tiny worm (about 1,000 cells)	97	19,000	2
Drosophila melanogaster	An insect	137	14,000	7
Arabidopsis thaliana	A flowering plant	142	26,000	11
Homo sapiens	A complex mammal	3,000	25,000	70

about 25,000 different genes, they can make well over 100,000 different proteins because most genes are alternatively spliced in multiple ways.

RNA Interference May Inhibit mRNA by Translational Repression or mRNA Degradation

Let's now turn our attention to regulatory mechanisms that affect translation. **MicroRNAs (miRNAs)** and **short-interfering RNAs (siRNAs)** are small RNA molecules, typically 22 nucleotides in length, that silence the expression of pre-existing mRNAs. MicroRNAs are partially complementary to certain cellular mRNAs and inhibit their translation, whereas short-interfering RNAs are usually a perfect match to specific mRNAs and cause the mRNAs to be degraded.

In 1993, Victor Ambros and his colleagues, who were interested in the developmental stages that occur in the worm *C. elegans*, determined that the transcription of a particular gene produced a small RNA, now called a microRNA, that does not encode a protein. Instead, this miRNA was found to be partially complementary to an mRNA and inhibit its translation.

Insight into the mechanism of miRNA and siRNA inhibition came from the research of Andrew Fire and Craig Mello, who discovered the mechanism of action of miRNA (Figure 13.21). MiRNAs and siRNAs are first synthesized as a single-stranded molecule that folds back on itself to form a hairpin structure. This double-stranded region is trimmed to a 22-base pair sequence by an enzyme called dicer. One of the strands becomes part of a complex called the **RNA-induced silencing complex (RISC)**, which also includes several proteins. The miRNA or siRNA in the complex then binds to a target mRNA with a complementary sequence. Upon binding, two different

Pre-miRNA or pre-siRNA

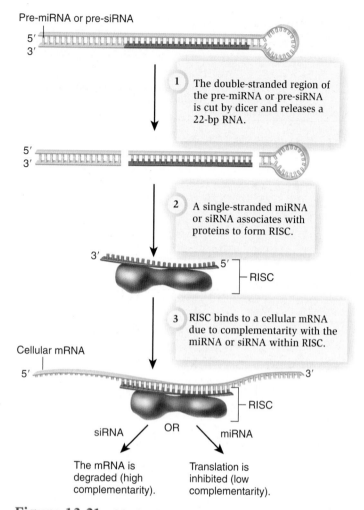

1 The double-stranded region of the pre-miRNA or pre-siRNA is cut by dicer and releases a 22-bp RNA.

2 A single-stranded miRNA or siRNA associates with proteins to form RISC.

RISC

3 RISC binds to a cellular mRNA due to complementarity with the miRNA or siRNA within RISC.

Cellular mRNA

RISC

siRNA   OR   miRNA

The mRNA is degraded (high complementarity).

Translation is inhibited (low complementarity).

Figure 13.21 Mechanism of action of microRNA (miRNA).

things may happen. In some cases, the mRNA is degraded. This occurs when the siRNA and mRNA are a perfect match or highly complementary. Alternatively, the RISC may inhibit translation. This occurs when the miRNA and mRNA are not a perfect match or are only partially complementary. In either case, the expression of the mRNA is silenced. Fire and Mello called this **RNA interference** (**RNAi**), because the miRNA or siRNA interferes with the proper expression of an mRNA.

Since this study, researchers have discovered that genes encoding miRNAs and siRNAs are widely found in animals and plants. In humans, for example, approximately 200 different genes encode miRNAs. MiRNAs and siRNAs represent an important mechanism of gene regulation that results in mRNA silencing. In 2006, Fire and Mello were awarded the Nobel Prize for their studies of RNA interference.

The Prevention of Iron Toxicity in Mammals Involves the Regulation of Translation

Another way to regulate translation involves RNA-binding proteins that directly affect translational initiation. The regulation of iron absorption provides a well-studied example. While iron is a vital cofactor for many cellular enzymes, it is toxic at high

levels. To prevent toxicity, mammalian cells synthesize a protein called ferritin, which forms a hollow, spherical complex that can store excess iron.

The mRNA that encodes ferritin is controlled by an RNA-binding protein known as the **iron regulatory protein** (**IRP**). When iron levels in the cytosol are low and more ferritin is not needed, IRP binds to a regulatory element within the ferritin mRNA known as the **iron regulatory element** (**IRE**). The IRE is located between the 5′-cap, where the ribosome binds, and the start codon where translation begins. This region is called the 5′-untranslated region (5′-UTR). Due to base pairing, it forms a stem-loop structure. The binding of IRP to the IRE inhibits translation of the ferritin mRNA (**Figure 13.22a**). However, when iron is abundant in the cytosol, the iron binds directly to IRP and prevents it from binding to the IRE. Under these conditions, the ferritin mRNA is translated to make more ferritin protein (**Figure 13.22b**).

Why is translational regulation of ferritin mRNA an advantage over transcriptional regulation of the ferritin gene? This mechanism of translational control allows cells to rapidly respond to changes in their environment. When cells are confronted with high levels of iron, they can quickly make more

When iron levels are low, the iron regulatory protein binds IRE and inhibits translation.

Active iron regulatory protein (IRP)

Iron regulatory element (IRE)

Ferritin mRNA

5′   AAAAA–3′

5′ UTR   Start codon   Stop codon

(a) Low iron levels

When iron levels are high, iron regulatory protein binds iron, causing a conformational change that releases it from the IRE; translation proceeds.

Fe^{3+}

Ferritin protein

Inactive iron regulatory protein (IRP)

5′   AAAAA–3′

IRE   Start codon   Stop codon

(b) High iron levels

Figure 13.22 Translational regulation of ferritin mRNA by the iron regulatory protein (IRP).

Concept check: *Poisoning may occur when a young child finds a bottle of vitamins, such as those that taste like candy, and eats a large number of them. One of the toxic effects involves the ingestion of too much iron. How does the IRP protect people from the toxic effects of too much iron?*

ferritin protein to prevent the toxic buildup of iron. This mechanism is faster than transcriptional regulation, which would require the activation of the ferritin gene and the transcription of ferritin mRNA prior to the synthesis of more ferritin protein.

Summary of Key Concepts

13.1 Overview of Gene Regulation

- Most genes are regulated so the levels of gene expression can vary under different conditions. By comparison, constitutive genes are expressed at constant levels.

- Organisms regulate genes so gene products are made only when they are needed. An example is the synthesis of the gene products needed for lactose utilization in bacteria. (Figure 13.1)

- Multicellular eukaryotes regulate genes to produce different cell types, such as muscle, nerve, and skin cells. (Figure 13.2)

- Eukaryotes also regulate genes so the gene products are produced at different developmental stages. An example is the group of globin genes in mammals. (Figure 13.3)

- All organisms regulate gene expression at a variety of levels, including transcription, translation, and post-translation. Eukaryotes also regulate RNA processing. (Figure 13.4)

13.2 Regulation of Transcription in Bacteria

- Repressors and activators are regulatory proteins that bind to the DNA and regulate the transcription of genes. Small effector molecules control the ability of regulatory proteins to bind to DNA. (Figure 13.5)

- The *lac* operon found in *E. coli* is an arrangement of three structural genes controlled by a single promoter. The operon is transcribed into a polycistronic mRNA. The operator and CAP site are involved with gene regulation via the lac repressor and CAP, respectively. (Figure 13.6)

- The lac repressor binds to the operator site and prevents RNA polymerase from transcribing the operon. When allolactose binds to the repressor, a conformational change occurs that prevents the repressor from binding to the operator site so transcription can proceed. (Figure 13.7)

- By constructing a merozygote, Jacob, Monod, and Pardee determined that *lacI* encodes a diffusible protein that represses the *lac* operon. (Figures 13.8, 13.9)

- The catabolite activator protein (CAP) binds to the CAP site in the presence of cAMP. This causes a bend in the DNA, which promotes the binding of RNA polymerase to the promoter. (Figure 13.10)

- Glucose inhibits cAMP production. This inhibits the expression of the *lac* operon because CAP cannot bind to the CAP site. This form of regulation provides bacteria with a more efficient utilization of their resources because the bacteria use one sugar at a time. (Figure 13.11)

- The *trp* operon is repressible. The presence of tryptophan causes the trp repressor to bind to the *trp* operator and stop transcription. This prevents the excessive buildup of tryptophan in the cell, which would be a waste of energy. (Figure 13.12)

13.3 Regulation of Transcription in Eukaryotes

- Eukaryotic genes exhibit combinatorial control, meaning that many factors control the expression of a single gene. (See list on p. 273.)

- Eukaryotic promoters consist of a core promoter and regulatory elements, such as enhancers or silencers, that regulate the rate of transcription. (Figure 13.13)

- General transcription factors (GTFs) are needed for RNA polymerase II to bind to the core promoter. (Figure 13.14)

- Activators and repressors may regulate RNA polymerase II by interacting with GTFs, such as TFIID, or via mediator, a protein complex that wraps around RNA polymerase II. (Figures 13.15, 13.16)

- A change in chromatin structure is needed for eukaryotic genes to be transcribed. (Figure 13.17)

- The pattern of covalent modification of the amino terminal tails of histone proteins, also called the histone code, is one way to control the level of transcription. (Figure 13.18)

- Steroid hormones bind to receptors that function as transcriptional activators. (Figure 13.19)

- DNA methylation, which occurs at CpG islands near promoters, usually inhibits transcription by preventing the binding of activator proteins or by promoting the binding of proteins that inhibit transcription.

13.4 Regulation of RNA Processing and Translation in Eukaryotes

- Alternative splicing occurs when a single type of pre-mRNA can be spliced in more than one way, producing polypeptides with somewhat different sequences. This is a common way for complex eukaryotes to increase the size of their proteomes. (Figure 13.20, Table 13.1)

- MicroRNAs (miRNAs) and short-interfering RNAs (siRNAs) inhibit mRNAs, either by inhibiting translation or by promoting the degradation of mRNAs, respectively. (Figure 13.21)

- RNA-binding proteins can regulate the translation of specific mRNAs. An example is the iron regulatory protein (IRP), which regulates the translation of ferritin mRNA. (Figure 13.22)

Assess and Discuss

Test Yourself

1. Genes that are expressed at all times at relatively constant levels are known as ____ genes.
   a. inducible
   b. repressible
   c. positive
   d. constitutive
   e. structural

2. Which of the following is <u>not</u> considered a common level of gene regulation in prokaryotes?
   a. transcriptional
   b. RNA processing
   c. translational

d. post-translational

e. All of the above are levels at which prokaryotes are able to regulate gene expression.

3. Transcription factors that bind to DNA and stimulate transcription are

a. repressors.

b. small effector molecules.

c. activators.

d. promoters.

e. operators.

4. In prokaryotes, the unit of DNA that contains multiple structural genes under the control of a single promoter is called _____. The mRNA produced from this unit is referred to as _____ mRNA.

a. an operator, a polycistronic

b. a template, a structural

c. an operon, a polycistronic

d. an operon, a monocistronic

e. a template, a monocistronic

5. In the *lac* operon, what would be the expected effects of a mutation in the operator site that prevented the binding of the repressor protein?

a. The operon would always be turned on.

b. The operon would always be turned off.

c. The operon would always be turned on, except when glucose is present.

d. The operon would be turned on only in the presence of lactose.

e. The operon would be turned on only in the presence of lactose and the absence of glucose.

6. The presence of _____ in the medium prevents the CAP from binding to the DNA, resulting in _____ in transcription of the *lac* operon.

a. lactose, an increase

b. glucose, an increase

c. cAMP, a decrease

d. glucose, a decrease

e. lactose, a decrease

7. The *trp* operon is considered _____ operon because the structural genes necessary for tryptophan synthesis are not expressed when the levels of tryptophan in the cell are high.

a. an inducible

b. a positive

c. a repressible

d. a negative

e. Both c and d are correct.

8. Regulatory elements that function to increase transcription levels in eukaryotes are called

a. promoters.

b. silencers.

c. enhancers.

d. transcriptional start sites.

e. activators.

9. DNA methylation in many eukaryotic organisms usually causes

a. increased translation levels.

b. introns that will be removed.

c. regions of DNA that do not contain structural genes.

d. decreased transcription levels.

e. regulatory elements that are not necessary for transcription.

10. _____ refers to the phenomenon where a single type of pre-mRNA may give rise to multiple types of mRNAs due to different patterns of intron and exon removal.

a. Spliceosomes

b. Variable expression

c. Alternative splicing

d. Polycistronic mRNA

e. Induced silencing

Conceptual Questions

1. What is the difference between inducible and repressible operons?

2. Transcriptional regulation often involves a regulatory protein that binds to a segment of DNA and a small effector molecule that binds to the regulatory protein. Do the following terms apply to a regulatory protein, a segment of DNA, or a small effector molecule? a. repressor; b. inducer; c. operator site; d. corepressor; e. activator

3. What are two general ways that DNA methylation inhibits transcription?

Collaborative Questions

1. Discuss the advantages and disadvantages of genetic regulation at the different levels described in Figure 13.4.

2. Let's suppose that a mutation in the glucocorticoid receptor gene does not prevent the binding of the hormone to the mutant glucocorticoid receptor protein but does prevent the ability of the protein to activate transcription. Make a list of all the possible defects that may explain why transcription cannot be activated.

Online Resource

www.brookerbiology.com

Stay a step ahead in your studies with animations that bring concepts to life and practice tests to assess your understanding. Your instructor may also recommend the interactive ebook, individualized learning tools, and more.

Mutation, DNA Repair, and Cancer

14

At a summer camp, the children enjoy ice cream, horseback riding, hay rides, and swimming and learn about the habits of owls. Not such an unusual camp, you might be thinking. However, what makes Camp Sundown unique is that the outdoor fun begins at dusk and runs all night. The children at this camp have inherited a disorder called xeroderma pigmentosum (XP), which makes them highly sensitive to the sun. Their skin will blister or freckle on minimum sun exposure. Of great concern is skin cancer. Persons with XP may have a 1,000-fold greater risk of developing skin cancer, though such a risk is greatly decreased if sun exposure is minimized.

What explains the symptoms of xeroderma pigmentosum? XP individuals are highly susceptible to **mutation**, which is defined as a heritable change in the genetic material. When a mutation occurs, the order of nucleotide bases in a DNA molecule, its base sequence, is changed permanently, and this alteration can be passed from mother to daughter cells during cell division. Mutations that lead to cancer cause particular genes to be expressed in an abnormal way. For example, a mutation could affect the transcription of a gene, or it could alter the functional properties of the polypeptide that is specified by a gene.

Should we be afraid of mutations? Yes and no. On the positive side, mutations are essential to the long-term continuity of life. They supply the variation that enables species to evolve and become better adapted to their environments. Mutations provide the foundation for evolutionary change. On the negative side, however, new mutations are more likely to be harmful than beneficial to the individual. The genes within modern species are the products of billions of years of evolution and have evolved to work properly. Random mutations are more likely to disrupt genes rather than enhance their function. As we will see in this chapter, mutations can cause cancer. In addition, many forms of inherited diseases, such as xeroderma pigmentosum and cystic fibrosis, are caused by gene mutations. For these and many other reasons, understanding the molecular nature of mutations is a compelling area of research.

Because mutations can be harmful, all species have evolved several ways to repair damaged DNA. Such **DNA repair systems** reverse DNA damage before a permanent mutation can occur. DNA repair systems are vital to the survival of all organisms. If these

During the past two decades, over 25% of the beluga whales in Canada's St. Lawrence Seaway have died of cancer. Biologists speculate that these deaths are caused by cancer-causing pollutants, such as polycystic aromatic hydrocarbons (PAHs).

systems did not exist, mutations would be so prevalent that few species, if any, would survive. Persons with XP have an impaired DNA repair system, which is the underlying cause of their disorder. DNA damage from sunlight is normally corrected by DNA repair systems. In people with XP, damaged DNA remains unrepaired, which can lead to cancer. In this chapter, we will examine how such DNA repair systems operate. But first, let's explore the molecular basis of mutation.

14.1 Mutation

How do mutations affect traits? To answer this question at the molecular level, we must understand how changes in DNA structure can ultimately affect DNA function. Most of our understanding of mutation has come from the study of experimental organisms, such as bacteria and *Drosophila*. Researchers can expose these organisms to environmental agents that cause mutations and then study the consequences of the mutations that arise. In addition, because these organisms have a short generation time, researchers can investigate the effects of mutations when they are passed from parent to offspring over many generations.

The structure and amount of genetic material can be altered in a variety of ways. For example, chromosome structure and number can change. We will examine these types of genetic changes in Chapter 15. In this section, we will focus our attention on **gene mutations**, which are relatively small changes in DNA structure that alter a particular gene. We will also consider how the timing of new mutations during an organism's development has important consequences. Finally, we will explore how environmental agents may bring about mutations and examine a testing method that can determine if an agent causes mutations.

Gene Mutations Alter the DNA Sequence of a Gene

Mutations can cause two basic types of changes to a gene: (1) the base sequence within a gene can be changed; and (2) one or more nucleotides can be added to or removed from a gene. A **point mutation** affects only a single base pair within the DNA. For example, the DNA sequence shown here has been altered by a **base substitution** in which a T (in the top strand) has been replaced by a G:

5′–CCCGCTAGATA–3′   ⟶   5′–CCCGCGAGATA–3′
3′–GGGCGATCTAT–5′   3′–GGGCGCTCTAT–5′

A point mutation could also involve the addition or deletion of a single base pair to a DNA sequence. For example, in the following sequence, a single base pair has been added to the DNA:

5′–GGCGCTAGATC–3′   ⟶   5′–GGCAGCTAGATC–3′
3′–CCGCGATCTAG–5′   3′–CCGTCGATCTAG–5′

Though point mutations may seem like small changes to a DNA sequence, they can have important consequences when genes are expressed. This topic is discussed next.

Gene Mutations May Affect the Amino Acid Sequence of a Polypeptide

If a mutation occurs within the region of a structural gene that specifies the amino acid sequence, such a mutation may alter that sequence in a variety of ways. Table 14.1 considers the potential effects of point mutations. **Silent mutations** do not alter the amino acid sequence of the polypeptide, even though the nucleotide sequence has changed. As discussed in Chapter 12, the genetic code is degenerate, that is, more than one codon can specify the same amino acid. Silent mutations can occur in the third base of many codons without changing the type of amino acid it encodes.

A **missense mutation** is a base substitution that changes a single amino acid in a polypeptide sequence. A missense mutation may not alter protein function because it changes only a single amino acid within a polypeptide that is typically hundreds of amino acids in length. A missense mutation that substitutes

Table 14.1	Consequences of Point Mutations Within the Coding Sequence of a Structural Gene

Mutation in the DNA	Effect on polypeptide	Example*
None	None	
Base substitution	Silent— causes no change	
Base substitution	Missense— changes one amino acid	
Base substitution	Nonsense— changes to a stop codon	
Addition (or deletion) of single base	Frameshift— produces a different amino acid sequence	

*DNA sequence in the coding strand. This sequence is the same as the mRNA sequence except that RNA contains uracil (U) instead of thymine (T).

an amino acid with a chemistry similar to the original amino acid is less likely to alter protein function. For example, a missense mutation that substitutes a glutamic acid for an aspartic acid may not alter protein function because both amino acids are negatively charged and have similar side chain structures (refer back to Chapter 3).

Alternatively, some missense mutations have a dramatic effect on protein function. A striking example occurs in the human disease known as **sickle-cell disease**. This disease involves a missense mutation in the β-globin gene, which encodes one of the polypeptide subunits that make up hemoglobin, the oxygen-carrying protein in red blood cells. In the most common form of this disease, a missense mutation alters the polypeptide sequence such that the sixth amino acid is changed from a glutamic acid to a valine (**Figure 14.1**). Because glutamic acid is hydrophilic but valine is hydrophobic, this single amino acid substitution alters the structure and function of the hemoglobin protein. The mutant hemoglobin subunits tend to stick to one another when the oxygen concentration is low. The aggregated proteins form fiber-like structures within red blood cells, which causes the cells to lose their normal disk-shaped morphology and become sickle-shaped. It seems amazing that a single amino acid substitution could have such a profound effect on the structure of cells.

Two other types of point mutations cause more dramatic changes to a polypeptide sequence. A **nonsense mutation** involves a change from a normal codon to a stop or termination codon. This causes translation to be terminated earlier than expected, producing a truncated polypeptide (see Table 14.1). Compared to a normal polypeptide, such a shorter polypeptide is much less likely to function properly. Finally, a **frameshift**

Part of wild-type β-globin gene

A C T C C T G A G G A A

Thr Pro Glu Glu
4   5   6   7

(a) Normal red blood cell

7 μm

Part of mutant β-globin gene

A C T C C T G T G G A A   DNA

Thr Pro Val Glu
4   5   6   7

(b) Sickled red blood cell

7 μm

Hemoglobin fiber

Hemoglobin

Hemoglobin fiber

Mutant β-globin subunit

(c) Fiber-like hemoglobin molecules

Figure 14.1   **A missense mutation that causes red blood cells to sickle in sickle-cell disease.**   Scanning electron micrographs of **(a)** normal red blood cells and **(b)** sickled red blood cells. As shown above the micrographs, a missense mutation in the β-globin gene (which codes for a subunit of hemoglobin) changes the sixth amino acid in the β-globin polypeptide from a glutamic acid to a valine. **(c)** This micrograph shows how this alteration to the structure of β-globin causes the formation of abnormal fiber-like structures. In normal cells, hemoglobin proteins do not form fibers.

Concept check: *Based on the fiber-like structures seen in part (c), what aspect of hemoglobin structure does a glutamic acid at the sixth position in normal β-globin prevent? Speculate on how the charge of this amino acid may play a role.*

mutation involves the addition or deletion of nucleotides that are not in multiples of three nucleotides. For example, a frame-shift mutation could involve the addition or deletion of one, two, four, or five nucleotides. Because the codons are read in multiples of three, these types of insertions or deletions shift the reading frame so that a completely different amino acid sequence occurs downstream from the mutation (see Table 14.1). Such a large change in polypeptide structure is likely to inhibit protein function.

Except for silent mutations, new mutations are more likely to produce polypeptides that have reduced rather than enhanced function. However, mutations can occasionally produce a poly-peptide that has a better ability to function. Changes in protein function may affect the ability of an organism to survive and to reproduce. Such mutations may change in frequency in a population over the course of many generations due to natural selection. This topic is discussed in Chapter 24.

Gene Mutations Can Occur Outside of Coding Sequences and Influence Gene Expression

Thus far, we have focused our attention on mutations in the coding regions of structural genes. In Chapters 12 and 13, we learned how other DNA sequences play important roles during gene expression. A mutation can occur within noncoding sequences and affect gene expression (**Table 14.2**). For example, a mutation may alter the sequence within the promoter of a gene and thereby affect the rate of transcription. A mutation

that improves the ability of transcriptional activators to bind to the DNA may enhance transcription, whereas other mutations may block their binding and thereby inhibit transcription.

Mutations in regulatory elements or operator sites can alter the regulation of gene transcription. For example, in Chapter 13, we considered the roles of regulatory elements such as the *lac* operator site in *E. coli*, which is recognized by the lac repressor protein (refer back to Figure 13.7). Mutations in the *lac* opera-tor site can disrupt the proper regulation of the *lac* operon. An operator mutation may change the DNA sequence so that the lac repressor protein does not bind to it. This mutation would cause the operon to be constitutively expressed.

Table 14.2	Effects of Mutations Outside of the Coding Sequence of a Gene
Sequence	**Effect of mutation**
Promoter	May increase or decrease the rate of transcription
Transcriptional regulatory element/ operator site	May alter the regulation of transcription
Splice junctions	May alter the ability of pre-mRNA to be properly spliced
Translational regulatory element	May alter the ability of mRNA to be translationally regulated
Intergenic region	Not as likely to have an effect on gene expression

FEATURE INVESTIGATION

The Lederbergs Used Replica Plating to Show That Mutations Are Random Events

Mutations can affect the expression of genes in a variety of ways. Let's now consider the following question: Do mutations that affect the traits of an individual occur as a result of pre-existing circumstances, or are they random events that may happen in any gene of any individual? In the 19th century, French naturalist Jean Baptiste Lamarck proposed that physiological events (such as use or disuse) determine whether traits are passed along to offspring. For example, his hypothesis suggested that an individual who practiced and became adept at a physical activity, such as the long jump, would pass that quality on to his or her offspring. Alternatively, geneticists in the early 1900s suggested that genetic variation occurs as a matter of chance. According to this view, those individuals whose genes happen to contain beneficial mutations are more likely to survive and pass those genes to their offspring.

These opposing views were tested in bacterial studies in the 1940s and 1950s. One such study, by Joshua and Esther Lederberg, focused on the occurrence of mutations in bacteria (**Figure 14.2**). First, they placed a large number of *E. coli* bac-

Figure 14.2 The experiment performed by the Lederbergs showing that mutations are random events.

HYPOTHESIS Mutations are random events.

KEY MATERIALS *E. coli* cells, T1 phage

Experimental level / **Conceptual level**

1 Place individual bacterial cells onto growth media.

Allow cells to divide, during which time random mutations may occur.

Single bacterial cell

2 Incubate overnight to allow the formation of bacterial colonies. This is called the master plate.

Bacterial colony

Bacterial colony in which some cells have a random mutation that gives resistance to T1.

Bacterial colony without a mutation

3 Press a velvet cloth (wrapped over a cylinder) onto the master plate, and then lift gently to obtain a replica of each bacterial colony. Press the replica onto 2 secondary plates that contain T1 bacteriophage. Incubate overnight to allow bacterial growth.

Master plate

Secondary plates containing T1 phage

Replica plate and allow to grow in the presence of T1.

(Nonmutant cells are lysed and killed on these plates.)

4 THE DATA

Colonies on each plate are in the same locations.

5 CONCLUSION Mutations are random events. In this case, the mutations occurred on the master plate prior to exposure to T1 bacteriophage.

6 SOURCE Lederberg, Joshua, and Lederberg, Esther M. 1952. Replica plating and indirect selection of bacterial mutants. *Journal of Bacteriology* 63:399–406.

teria onto a master plate that was incubated overnight, so each bacterial cell divided many times to form a bacterial colony composed of millions of cells. Using a technique known as **replica plating**, a sterile piece of velvet cloth was lightly touched to this plate to pick up a few bacterial cells from each colony on the master plate. They then transferred this replica to two secondary plates containing an agent that selected for the growth of bacterial cells with a particular mutation.

In the example shown in Figure 14.2, the secondary plates contained T1 bacteriophages, which are viruses that infect bacteria and cause them to lyse. On these plates, only those rare cells that had acquired a mutation conferring resistance to T1, termed *ton^r*, could grow. All other cells were lysed by the proliferation of bacteriophages in the bacteria. Therefore, only a few colonies were observed on the secondary plates. Strikingly, these colonies occupied the same locations on each plate. How

did the Lederbergs interpret these results? The data indicated that the *ton^r* mutations occurred randomly while the bacterial cells were forming colonies on the nonselective master plate. The presence of T1 bacteriophages in the secondary plates did not cause the mutations to develop. Rather, the T1 bacteriophages simply selected for the growth of *ton^r* mutants that were already in the population. These results supported the idea that mutations are random events.

Experimental Questions

1. Explain the opposing views of mutation prior to the Lederbergs' study.

2. What hypothesis was being tested by the Lederbergs? What were the results of the experiment?

3. How did the results of the Lederbergs support or falsify the hypothesis?

Mutations Can Occur in Germ-Line or Somatic Cells

Let's now consider how the timing of a mutation may have an important impact on its potential effects. Multicellular organisms typically begin their lives as a single fertilized egg cell that divides many times to produce all the cells of an adult organism. A mutation can occur in any cell of the body, either very early in life, such as in a gamete (eggs or sperm) or a fertilized egg, or later in life, such as in the embryonic or adult stages. The number and location of cells with a mutation are critical both to the severity of the genetic effect and to whether the mutation can be passed on to offspring.

Geneticists classify the cells of animals into two types: germ-line and somatic cells. The term **germ line** refers to cells that give rise to gametes, such as egg and sperm cells. A germ-line mutation can occur directly in an egg or sperm cell, or it can occur in a precursor cell that produces the gametes. If a mutant human gamete participates in fertilization, all the cells of the resulting offspring will contain the mutation, as indicated by the red color in **Figure 14.3a**. Likewise, when such an individual produces gametes, the mutation may be transmitted to future generations of offspring. Because humans carry two copies of most genes, a new mutation in a single gene has a 50% chance of being transmitted from parent to offspring.

The **somatic cells** constitute all cells of the body excluding the germ-line cells. Examples include skin cells and muscle cells. Mutations can also occur within somatic cells at early or late stages of development. What are the consequences of a mutation that happens during the embryonic stage? As shown in **Figure 14.3b**, a mutation occurred within a single embryonic cell. This single somatic cell was the precursor for many cells of the adult. Therefore, in the adult, a patch of tissue contains cells that carry the mutation. The size of any patch depends on the timing of a new mutation. In general, the earlier a mutation occurs during development, the larger the patch. An individual

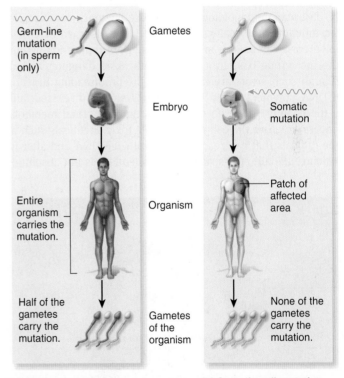

(a) Germ-line mutation   (b) Somatic cell mutation

Figure 14.3 **The effects of germ-line versus somatic cell mutations.**   The red color indicates which cells carry the mutation. (a) If a mutation is passed via gametes, such germ-line mutations occur in every cell of the body. Because humans have two copies of most genes, a germ-line mutation in one of those two copies will be transmitted to only half of the gametes. (b) Somatic mutations affect a limited area of the body and are not transmitted to offspring.

Concept check: *Why are somatic mutations unable to be transmitted to offspring?*

with somatic regions that are genetically different from each other is called a **mosaic**.

Figure 14.4 illustrates a woman who probably had a somatic mutation during an early stage of development. In this case, the woman has a patch of white hair while the rest of her hair is black. Presumably, this individual initially had a single mutation happen in an embryonic cell that ultimately gave rise to the patch that produced the white hair.

Although a change in hair color is not a harmful consequence, mutations during early stages of life can be quite harmful, especially if they disrupt essential developmental processes. Even though it is sensible to avoid environmental agents that cause mutations at any stage of life, the possibility of somatic mutations is a compelling reason to avoid such agents during the early stages of life such as embryonic and fetal development, infancy, and early childhood.

Mutations May Be Spontaneous or Induced

Biologists categorize the causes of mutation as spontaneous or induced. **Spontaneous mutations** result from abnormalities in biological processes (Table 14.3). Spontaneous mutations reflect the observation that biology isn't perfect. Enzymes, for example, can function abnormally. In Chapter 11, we learned that DNA polymerase can make a mistake during DNA replication by putting the wrong base in a newly synthesized daughter strand. Though such errors are rare due to the proofreading function of DNA polymerase and DNA repair systems (discussed later in the chapter), they do occur. In addition, normal metabolic processes within the cell may produce toxic chemicals such as free radicals that can react directly with the DNA and alter its structure. Finally, the structure of nucleotides is not absolutely

Figure 14.4 **Example of a somatic mutation.** This woman has a patch of white hair. This is likely to have occurred because a somatic mutation occurred in a single cell during embryonic development that caused white pigmentation of the hair. This cell continued to divide to produce a patch of white hair.

Concept check: *Can this woman with a patch of white hair transmit this trait to her offspring?*

Table 14.3	Some Common Causes of Gene Mutations
Common causes of mutations	**Description**
Spontaneous:	
Errors in DNA replication	A mistake by DNA polymerase may cause a point mutation.
Toxic metabolic products	The products of normal metabolic processes may be reactive chemicals such as free radicals that can alter the structure of DNA.
Changes in nucleotide structure	On rare occasions, the linkage between purines and deoxyribose can spontaneously break. Changes in base structure (isomerization) may cause mispairing during DNA replication.
Transposons	As discussed in Chapter 21, transposons are small segments of DNA that can insert at various sites in the genome. If they insert into a gene, they may inactivate the gene.
Induced:	
Chemical agents	Chemical substances, such as benzo(a)pyrene, a chemical found in cigarette smoke, may cause changes in the structure of DNA.
Physical agents	Physical agents such as UV (ultraviolet) light and X-rays can damage the DNA.

stable. On occasion, the structure of a base may spontaneously change, and such a change may cause a mutation if it occurs immediately prior to DNA replication.

The rates of spontaneous mutations vary from species to species and from gene to gene. Larger genes are usually more likely to incur a mutation than are smaller genes. A common rate of spontaneous mutation among various species is approximately 1 mutation for every 1 million genes per cell division, which equals 1 in 10^6, or simply 10^{-6}. This is the expected rate of spontaneous mutation, which creates the variation that is the raw material of evolution.

Induced mutations are caused by environmental agents that enter the cell and alter the structure of DNA. They cause the mutation rate to be higher than the spontaneous mutation rate. Agents that cause mutation are called **mutagens**. Mutagenic agents can be categorized as **chemical** or **physical mutagens** (Table 14.4). We will consider their effects next.

Mutagens Alter DNA Structure in Different Ways

Researchers have discovered that an enormous array of agents can act as mutagens. We often hear in the news media that we should avoid these agents in our foods and living environments. We even use products such as sunscreens that help us avoid the mutagenic effects of ultraviolet (UV) light from the sun. The public is often concerned about mutagens for two important reasons. First, mutagenic agents are usually involved in the development of human cancers. Second, because new mutations may be deleterious, people want to avoid mutagens

Table 14.4	Examples of Mutagens
Mutagen	**Effect(s) on DNA structure**
Chemical:	
Nitrous acid	Deaminates bases
5-bromouracil	Acts as a base analogue
2-aminopurine	Acts as a base analogue
Nitrogen mustard	Alkylates bases
Ethyl methanesulfonate (EMS)	Alkylates bases
Benzopyrene	Its metabolic product inserts next to bases in the DNA double helix and causes additions or deletions
Physical:	
X-rays	Causes base deletions, single nicks in DNA strands, cross-linking, and chromosomal breaks
UV light	Promotes pyrimidine dimer formation, which involves covalent bonds between adjacent pyrimidines (C or T)

Figure 14.5 **Deamination and mispairing of modified bases by a chemical mutagen.** Nitrous acid changes cytosine to uracil and adenine to hypoxanthine by replacing NH_2 with an oxygen. During DNA replication, uracil will pair with adenine, and hypoxanthine will pair with cytosine. These incorrect bases will create mutations in the newly replicated strand.

to prevent mutations that may have harmful effects in their future offspring.

How do mutagens affect DNA structure? Some chemical mutagens act by covalently modifying the structure of nucleotides. For example, nitrous acid (HNO_2) deaminates bases by replacing amino groups with keto groups ($-NH_2$ to $=O$). This can change cytosine to uracil, and adenine to a base called hypoxanthine. When this altered DNA replicates, the modified bases do not pair with the appropriate nucleotides in the newly made strand. Instead, uracil pairs with adenine, and hypoxanthine pairs with cytosine (**Figure 14.5**).

Similarly, 5-bromouracil and 2-aminopurine, which are called base analogues, have structures that are similar to particular bases in DNA. When incorporated into DNA, they also cause errors in DNA replication. Other chemical mutagens can disrupt the appropriate pairing between nucleotides by alkylating bases within the DNA. During alkylation, methyl or ethyl groups are covalently attached to the bases. Examples of alkylating agents include nitrogen mustards (used as a chemical weapon during World War I) and ethyl methanesulfonate (EMS).

Some chemical mutagens exert their effects by interfering with DNA replication. For example, benzopyrene, which is found in automobile exhaust and charbroiled food, is metabolized to a compound (benzopyrene diol epoxide) that inserts in between the bases of the double helix, thereby distorting the helical structure. When DNA containing these mutagens is replicated, single nucleotide additions and/or deletions can be incorporated into the newly made strands.

DNA molecules are also sensitive to physical agents such as radiation. In particular, radiation of short wavelength and high energy, known as ionizing radiation, is known to alter DNA structure. Ionizing radiation includes X-rays and gamma rays. This type of radiation can penetrate deeply into biological materials, where it creates free radicals. These molecules can alter the structure of DNA in a variety of ways. Exposure to high doses of ionizing radiation can cause base deletions, breaks in one DNA strand, or even a break in both DNA strands.

Nonionizing radiation, such as UV light, contains less energy, and so it penetrates only the surface of biological materials, such as the skin. Nevertheless, UV light is known to cause DNA mutations. For example, UV light can cause the formation of a **thymine dimer**, which is a site where two adjacent thymine bases become covalently cross-linked to each other (**Figure 14.6**).

Thymine dimers are typically repaired before or during DNA replication. However, if such repair fails to occur, a thymine dimer may cause a mutation when that DNA strand is replicated. When DNA polymerase attempts to replicate over a thymine dimer, proper base pairing does not occur between the template strand and the incoming nucleotides. This mispairing can cause gaps in the newly made strand or the incorporation of incorrect bases. Plants, in particular, must have effective ways to prevent UV damage because they are exposed to sunlight throughout the day.

Testing Methods Can Determine If an Agent Is a Mutagen

Because mutagens are harmful, researchers have developed testing methods to evaluate the ability of a substance to cause mutation. One commonly used test is the **Ames test**, which was developed by Bruce Ames in the 1970s. This test uses a strain of a bacterium, *Salmonella typhimurium*, that cannot

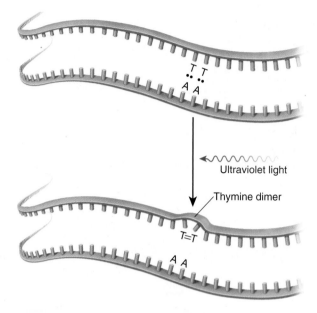

Figure 14.6 Formation and structure of a thymine dimer.

Concept check: *Why is a thymine dimer harmful?*

synthesize the amino acid histidine. This strain contains a point mutation within a gene that encodes an enzyme required for histidine biosynthesis. The mutation renders the enzyme inactive. The bacteria cannot grow unless histidine has been added to the growth medium. However, a second mutation may correct the first mutation and thereby restore the ability to synthesize histidine. The Ames test monitors the rate at which this

second mutation occurs and thereby indicates whether an agent increases the mutation rate above the spontaneous rate.

Figure 14.7 outlines the steps in the Ames test. The suspected mutagen is mixed with a rat liver extract and the bacterial strain of *S. typhimurium* that cannot synthesize histidine. Because some potential mutagens may require activation by cellular enzymes, the rat liver extract provides a mixture of enzymes that may cause such activation. This step improves the ability to identify agents that may cause mutations in mammals. As a control, bacteria that have not been exposed to the mutagen are also tested. After an incubation period in which mutations may occur, a large number of bacteria are plated on a growth medium that does not contain histidine. The *S. typhimurium* strain is not expected to grow on these plates. However, if a mutation has occurred that allows a cell to synthesize histidine, the bacterium harboring this second mutation will proliferate during an overnight incubation period to form a visible bacterial colony.

To estimate the mutation rate, the colonies that grow in the absence of histidine are counted and compared with the total number of bacterial cells that were originally placed on the plate for both the suspected-mutagen sample and the control. The control condition is a measure of the spontaneous mutation rate, whereas the other sample measures the rate of mutation in the presence of the suspected mutagen. As an example, let's suppose that 2 million bacteria were plated from both the suspected-mutagen and control tubes. In the control experiment, 2 bacterial colonies were observed. The spontaneous mutation rate is calculated by dividing 2 (the number of mutants) by

Figure 14.7 The Ames test for mutagenicity. In this example, 2 million bacterial cells were placed on plates lacking histidine. Two colonies were observed in the control sample, whereas 44 were observed in the sample exposed to a suspected mutagen.

Concept check: *Based on the results seen in this figure, what is the rate of mutation that is caused by the suspected mutagen?*

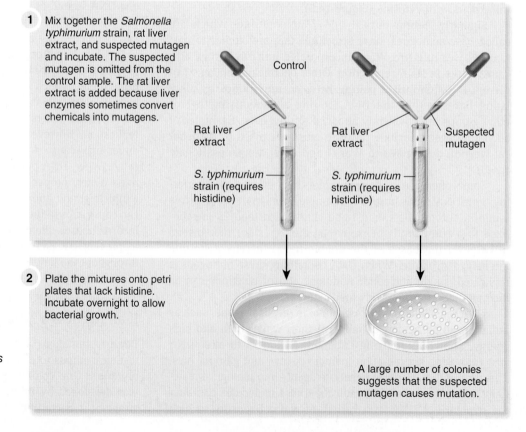

1 Mix together the *Salmonella typhimurium* strain, rat liver extract, and suspected mutagen and incubate. The suspected mutagen is omitted from the control sample. The rat liver extract is added because liver enzymes sometimes convert chemicals into mutagens.

Control

Rat liver extract

S. typhimurium strain (requires histidine)

Rat liver extract

Suspected mutagen

S. typhimurium strain (requires histidine)

2 Plate the mixtures onto petri plates that lack histidine. Incubate overnight to allow bacterial growth.

A large number of colonies suggests that the suspected mutagen causes mutation.

2 million (the number of original cells). This equals 1 in 1 million, or 1×10^{-6}. By comparison, 44 colonies were observed in the suspected-mutagen sample (Figure 14.7). In this case, the mutation rate would be 44 divided by 2 million, which equals 2.2×10^{-5}. The mutation rate in the presence of the mutagen is over 20 times higher than the spontaneous mutation rate.

How do we judge if an agent is a mutagen? Researchers compare the mutation rate in the presence and absence of the suspected mutagen. The experimental approach shown in Figure 14.7 is conducted several times. If statistics reveal that the mutation rate in the suspected-mutagen sample is significantly higher than the control sample, they may tentatively conclude that the agent is a mutagen. Interestingly, many studies have used the Ames test to compare the urine from cigarette smokers to that from nonsmokers. This research has shown that urine from smokers contains much higher levels of mutagens.

14.2 DNA Repair

In the previous section, we considered the causes and consequences of mutation. As we have seen, mutations are random events that often have negative consequences. To minimize mutation, all living organisms must have the ability to repair changes that occur in the structure of DNA. Such DNA repair systems have been studied extensively in many organisms, particularly *E. coli*, yeast, mammals, and plants. The diverse ways of repairing DNA underscore the necessity for the structure of DNA to be maintained properly. The importance of these systems becomes evident when they are missing. For example, as discussed at the beginning of this chapter, persons with xeroderma pigmentosum are highly susceptible to the harmful effects of sunlight because they are missing a single DNA repair system.

How do organisms minimize the occurrence of mutations? Cells contain several DNA repair systems that can fix different types of DNA alterations (**Table 14.5**). Each repair system is composed of one or more proteins that play specific roles in the repair mechanism. DNA repair requires two coordinated events. In the first step, one or more proteins in the repair system detect an irregularity in DNA structure. In the second step, the abnormality is repaired. In some cases, the change in DNA structure can be directly repaired. For example, DNA may be modified by the attachment of an alkyl group, such as $-CH_2CH_3$, to a base. In **direct repair**, an enzyme removes this alkyl group, thereby restoring the structure of the original base. More commonly, however, the altered DNA is removed, and a new segment of DNA is synthesized. In this section, we will examine nucleotide excision repair as an example of how such systems operate. This system, which is found in all species, is an important mechanism of DNA repair.

Nucleotide Excision Repair Removes Segments of Damaged DNA

In **nucleotide excision repair** (**NER**), a region encompassing several nucleotides in the damaged strand is removed from the

Table 14.5	Common Types of DNA Repair Systems
System	**Description**
Direct repair	A repair enzyme recognizes an incorrect structure in the DNA and directly converts it back to a correct structure.
Base excision and nucleotide excision repair	An abnormal base or nucleotide is recognized, and a portion of the strand containing the abnormality is removed. The complementary DNA strand is then used as a template to synthesize a normal DNA strand.
Methyl-directed mismatch repair	Similar to excision repair except that the DNA defect is a base pair mismatch in the DNA, not an abnormal nucleotide. The mismatch is recognized, and a strand of DNA in this region is removed. The complementary strand is used to synthesize a normal strand of DNA.

*Other types of repair systems exist; these are common examples.

DNA, and the intact undamaged strand is used as a template for the resynthesis of a normal complementary strand. NER can fix many different types of DNA damage, including UV-induced damage, chemically modified bases, missing bases, and various types of cross-links (such as thymine dimers). The system is found in all prokaryotes and eukaryotes, although its molecular mechanism is better understood in prokaryotic species.

In *E. coli*, the NER system is composed of four key proteins: UvrA, UvrB, UvrC, and UvrD. They are named Uvr because they are involved in <u>ultraviolet</u> light <u>repair</u> of thymine dimers, although these proteins are also important in repairing chemically damaged DNA. In addition, DNA polymerase and DNA ligase are required to complete the repair process.

How does the NER system work? Two UvrA proteins and one UvrB protein form a complex that tracks along the DNA in search of a damaged site (**Figure 14.8**). Such DNA will have a distorted double helix, which is sensed by the UvrA-UvrB complex. When the complex identifies a damaged site, the two UvrA proteins are released, and UvrC binds to UvrB at the site. The UvrC protein makes incisions in one DNA strand on both sides of the damaged site. After this incision process, UvrC is released. UvrD, which is a helicase, binds to UvrB. UvrD then begins to separate the DNA strands, and UvrB is released. The action of UvrD unravels the DNA, which removes a short DNA strand that contains the damaged region. UvrD is released. After the damaged DNA strand is removed, a gap is left in the double helix. DNA polymerase fills in the gap using the undamaged strand as a template. Finally, DNA ligase makes the final covalent connection between the newly made DNA and the original DNA strand.

Human Genetic Diseases Occur When a Component of the NER System Is Missing

Thus far, we have considered the NER system in *E. coli*. In humans, NER systems were discovered by the analysis of genetic diseases that affect DNA repair. These include xeroderma

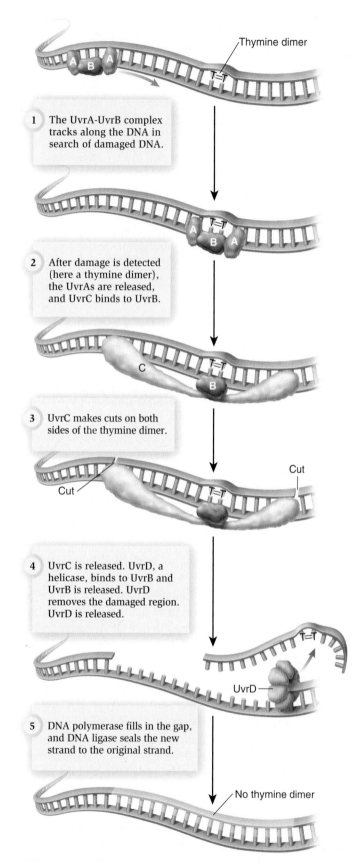

Thymine dimer

1 The UvrA-UvrB complex tracks along the DNA in search of damaged DNA.

2 After damage is detected (here a thymine dimer), the UvrAs are released, and UvrC binds to UvrB.

3 UvrC makes cuts on both sides of the thymine dimer.

Cut

Cut

4 UvrC is released. UvrD, a helicase, binds to UvrB and UvrB is released. UvrD removes the damaged region. UvrD is released.

UvrD

5 DNA polymerase fills in the gap, and DNA ligase seals the new strand to the original strand.

No thymine dimer

Figure 14.8 Nucleotide excision repair in *E. coli.*

Concept check: Which components of NER are responsible for removing the damaged DNA?

Figure 14.9 An individual affected by xeroderma pigmentosum.

Concept check: Why is this person so sensitive to the sun?

pigmentosum (XP), which was discussed at the beginning of this chapter, and also Cockayne's syndrome (CS) and PIBIDS. (PIBIDS is an acronym for a syndrome with symptoms that include photosensitivity [increased sensitivity to sunlight], ichthyosis [a skin abnormality], brittle hair, impaired intelligence, decreased fertility, and short stature.) Photosensitivity is a common characteristic in all three syndromes because of an inability to repair UV-induced lesions. Therefore, people with any of these syndromes must avoid prolonged exposure to sunlight as do the children at Camp Sundown. **Figure 14.9** shows a photograph of a person with XP who has had significant sun exposure. Such individuals may have pigmentation abnormalities, many precancerous lesions, and a high predisposition to developing skin cancer.

14.3 | Cancer

Cancer is a disease of multicellular organisms characterized by uncontrolled cell division. Worldwide, cancer is the second leading cause of death in humans, exceeded only by heart disease. In the United States, approximately 1.5 million people are diagnosed with cancer each year; over 0.5 million will die from the disease. Overall, about one in four Americans will die from cancer.

In about 10% of cancers, a higher predisposition to develop the disease is an inherited trait. Most cancers, though, perhaps 90%, do not involve genetic changes that are passed from parent to offspring. Rather, cancer is usually an acquired condition that typically occurs later in life. At least 80% of all human cancers are related to exposure to **carcinogens**, agents that increase the likelihood of developing cancer. Most carcinogens, such as UV light and certain chemicals in cigarette smoke, are mutagens that promote genetic changes in somatic cells. These DNA alterations can lead to effects on gene expression that

Figure 14.10 Cancer: its progression and effects. **(a)** In a healthy individual, an initial mutation converts a normal cell into a tumor cell. This cell divides to produce a benign tumor. Additional genetic changes in the tumor cells may occur, leading to a malignant tumor. At a later stage in malignancy, the tumor cells will invade surrounding tissues, and some malignant cells may metastasize by traveling through the bloodstream to other parts of the body. **(b)** On the left of the photo is a human lung that was obtained from a healthy nonsmoker. The lung shown on the right has been ravaged by lung cancer. This lung was taken from a person who was a heavy smoker.

ultimately affect cell division and thereby lead to cancer. In this section, we will explore such genetic abnormalities.

How does cancer occur? In most cases, the development of cancer is a multistep process (**Figure 14.10**). Cancers originate from a single cell. This single cell and its lineage of daughter cells undergo a series of mutations that causes the cells to grow abnormally. At an early stage, the cells form a **tumor**, which is an overgrowth of cells. For most types of cancer, a tumor begins as a precancerous or **benign** growth. Such tumors do not invade adjacent tissues and do not spread throughout the body. This may be followed by additional mutations that cause some cells in the tumor to lose their normal growth regulation and become **malignant**. At this stage, the individual has cancer. Cancerous tumors invade healthy tissues and may spread through the bloodstream or surrounding body fluids, a process called **metastasis**. If left untreated, malignant cells will cause the death of the organism.

Over the past few decades, researchers have identified many genes that promote cancer when they are mutant. By comparing the function of each mutant gene with the corresponding nonmutant gene found in healthy cells, these genes have been placed into two categories. In some cases, a mutation causes a gene to be overactive—have an abnormally high level of expression. This overactivity contributes to the uncontrolled cell growth that is observed in cancer cells. This type of mutant gene is called an **oncogene**. Alternatively, when a **tumor-suppressor gene** is normal (that is, not mutant), it encodes a protein that helps to prevents cancer. However, when a mutation eliminates its function, cancer may occur. Thus, the two categories of cancer-causing genes are based on the effects of mutations. Oncogenes are the result of mutations that cause overactivity, whereas cancer-causing mutations in tumor-suppressor genes are due to a loss of activity. In this section, we will begin with a discussion of oncogenes and then consider tumor-suppressor genes.

Oncogenes Cause the Overactivity of Proteins That Promote Cell Division

Over the past four decades, researchers have identified many oncogenes. A large number of oncogenes encode proteins that function in cell growth signaling pathways. Cell division is regulated, in part, by growth factors, which are a type of hormone that regulates cell division. A growth factor binds to a receptor, which results in receptor activation (**Figure 14.11**). This stimulates an intracellular signal transduction pathway that activates

1 Growth factor binds to a receptor, leading to receptor activation.

Growth factor

Receptor

3 The final protein in the signal transduction pathway activates transcription factors in the nucleus. This leads to the transcription of genes that promote cell division.

2 Receptor activation causes a cascade of interactions between intracellular signaling proteins, thereby activating them.

Signal transduction pathway

Figure 14.11 General features of a growth factor signaling pathway that promotes cell division. A detailed description of this pathway is found in Chapter 9 (Figure 9.10).

Concept check: *How does the presence of a growth factor ultimately affect the function of the cell?*

transcription factors. In this way, the transcription of specific genes is activated in response to a growth factor. After they are made, the gene products promote cell division.

Eukaryotic species produce many different growth factors that play a role in cell division. Likewise, cells have several different types of signal transduction pathways that respond to these molecules and promote cell division. Mutations in the genes that produce these signaling proteins can change them into oncogenes (**Table 14.6**). Oncogenes result in an abnormally high level of activity in these proteins, which can include growth factor receptors, intracellular signaling proteins, and transcription factors.

How can an oncogene promote cancer? In some cases, an oncogene may keep the cell division signaling pathway in a permanent "on" position. One way oncogenes can keep cell division turned on is by producing a functionally overactive protein. As a specific example, let's consider how a mutation can alter an intracellular signaling protein called Ras, which is discussed in Chapter 9 (refer back to Figure 9.10). The Ras protein is a GTPase that hydrolyzes GTP to GDP + P_i (**Figure 14.12**). When a signal transduction pathway is activated, the Ras protein exchanges GDP for GTP. When GTP is bound, the activated Ras protein promotes cell division. The Ras protein returns to its inactive state by hydrolyzing its bound GTP, and

Table 14.6	Examples of Genes Encoding Proteins of Growth Factor Signaling Pathways That Can Mutate to Become Oncogenes
Gene*	**Cellular function**
erbB	Growth factor receptor for EGF (epidermal growth factor)
ras	Intracellular signaling protein
raf	Intracellular signaling protein
src	Intracellular signaling protein
fos	Transcription factor
jun	Transcription factor

*The genes described in this table are found in humans as well as other vertebrate species. Most of the genes have been given three-letter names that are abbreviations for the type of cancer the oncogene causes or the type of virus in which the gene was first identified.

GDP-GTP exchange

GDP

GTP

Inactive Ras protein

Active Ras protein

The active Ras protein participates in a signal transduction pathway that promotes cell division.

GDP

GTP

P_i

GTP hydrolysis

Figure 14.12 The function of Ras, a protein that is part of signal transduction pathways. When GTP is bound, the activated Ras protein promotes cell division. When GTP is hydrolyzed to GDP and P_i, Ras is inactivated, and cell division is inhibited.

cell division is inhibited. Mutations that convert the normal *ras* gene into an oncogenic *ras* either decrease the ability of Ras protein to hydrolyze GTP or increase the rate of exchange of bound GDP for GTP. Both of these functional changes result in a greater amount of the active GTP-bound form of the Ras protein. In this way, these mutations keep the signaling pathway turned on when it should not be.

Mutations in Proto-Oncogenes Convert Them to Oncogenes

Thus far, we have examined the functions of proteins that cause cancer when they become overactive. Such overactivity promotes uncontrolled cell division. Let's now consider the common types of genetic changes that create such oncogenes. A **proto-oncogene** is a normal gene that, if mutated, can become an oncogene. Several types of genetic changes may convert a proto-oncogene into an oncogene. **Figure 14.13** describes four common types: missense mutations, gene amplifications, chromosomal translocations, and retroviral insertions.

Missense Mutation   A missense mutation (Figure 14.13a) can alter the function of an encoded protein in a way that promotes cancer. This type of mutation is responsible for the conversion of the *ras* gene into an oncogene. An example is a mutation in the *ras* gene that changes a specific glycine to a valine in the Ras protein. This mutation decreases the ability of the Ras protein to hydrolyze GTP, which promotes cell division (see Figure 14.12). Experimentally, chemical mutagens have been shown to cause this missense mutation and thereby lead to cancer.

Gene Amplification   Another genetic event that occurs in cancer cells is **gene amplification**, in this case an increase in the number of copies of a proto-oncogene (Figure 14.13b). An abnormal increase in the number of genes results in too much of the encoded protein. In 1982, Mark Groudine discovered that the *myc* gene, which encodes a transcription factor, was amplified in a human leukemia. Many human cancers are associated with the amplification of particular proto-oncogenes.

Chromosomal Translocation   A third type of genetic alteration that can lead to cancer is a chromosomal translocation (Figure 14.13c). This occurs when one segment of a chromosome becomes attached to a different chromosome. In 1960, Peter Nowell discovered that a form of leukemia called chronic myelogenous leukemia—a type of cancer involving blood cells—was correlated with the presence of a shortened version of a human chromosome, which he called the Philadelphia chromosome, after the city where it was discovered. This shortened chromosome is the result of a chromosome translocation in which pieces of two different chromosomes, chromosomes 9 and 22, fuse with each other. This activates a proto-oncogene, *abl*, in an unusual way (**Figure 14.14**). In healthy individuals, the *bcr* gene and the *abl* gene are located on different chromosomes. In chronic myelogenous leukemia, these chromosomes break and rejoin in a way that causes the promoter and the first part of *bcr*

A change in the amino acid sequence of a proto-oncogene protein may cause it to function in an abnormal way. For example, missense mutations can convert *ras* genes into oncogenes.

Missense mutation

(a) Missense mutation

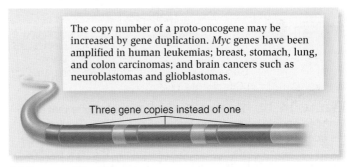

The copy number of a proto-oncogene may be increased by gene duplication. *Myc* genes have been amplified in human leukemias; breast, stomach, lung, and colon carcinomas; and brain cancers such as neuroblastomas and glioblastomas.

Three gene copies instead of one

(b) Gene amplification

A piece of chromosome may be translocated to another chromosome and affect the expression of genes at the breakpoint site. In one form of leukemia, for example, a translocation causes parts of the *bcr* and *abl* genes to fuse, thereby creating a chimeric oncogene.

Fused, chimeric gene

(c) Chromosomal translocation

When a virus integrates into the chromosome, viral regulatory elements may enhance the expression of a nearby proto-oncogene, converting it to an oncogene.

Viral regulatory sequences

Inserted retroviral genome

(d) Retroviral insertion

Figure 14.13   Genetic changes that convert proto-oncogenes to oncogenes.   In addition to these four, other types of genetic changes may also convert proto-oncogenes into oncogenes.

Normal chromosomes   Translocation chromosomes

abl

Breakage and fusion

bcr

bcr
abl

This translocation causes chronic myelogenous leukemia.

Philadelphia chromosome

Figure 14.14   The formation of a chimeric gene that is found in people with certain forms of leukemia.   The fusion of the *bcr* and *abl* genes creates a chimeric gene that encodes an abnormal fusion protein, leading to leukemia. The blue regions are the promoters for the *bcr* and *abl* genes.

Concept check: *The bcr gene is normally expressed in blood cells. Explain how this observation is related to the type of cancer that the translocation causes.*

to fuse with part of *abl*. This abnormal fusion event creates a **chimeric gene** composed of two gene fragments. This chimeric gene acts as an oncogene that encodes an abnormal fusion protein whose functional overactivity leads to leukemia.

Retroviral Insertion   Finally, certain types of viruses can convert proto-oncogenes into oncogenes during the viral replication cycle (see Figure 14.13d). Retroviruses insert their DNA into the chromosomal DNA of the host cell. The viral genome contains promoter and regulatory elements that cause a high level of expression of viral genes. On occasion, the viral DNA may insert into a host chromosome in such a way that a viral promoter and regulatory elements are next to a proto-oncogene. This may result in the overexpression of the proto-oncogene, thereby promoting cancer. This is one way for a virus to cause cancer. Alternatively, a virus may cause cancer because it carries an oncogene in the viral genome. This phenomenon is described next.

Some Types of Cancer Are Caused by Viruses

The great majority of cancers are caused by mutagens that alter the structure and expression of genes that are found in somatic cells. A few viruses, however, are known to cause cancer in plants, animals, and humans (**Table 14.7**).

Table 14.7	Examples of Viruses That Cause Cancer
Virus	**Description**
Rous sarcoma virus	Causes sarcomas in chickens
Simian sarcoma virus	Causes sarcomas in monkeys
Abelson leukemia virus	Causes leukemia in mice
Hardy-Zuckerman-4 feline sarcoma virus	Causes sarcomas in cats
Hepatitis B	Causes liver cancer in several species, including humans
Papillomavirus	Causes benign tumors and malignant carcinomas in several species, including humans; causes cervical cancer in humans
Epstein-Barr virus	Causes Burkitt's lymphoma, which primarily occurs in immunosuppressed individuals such as AIDS patients

In 1911, the first cancer-causing virus to be discovered was isolated from chicken sarcomas by Peyton Rous. A **sarcoma** is a tumor of connective tissue such as bone or cartilage. The virus was named the Rous sarcoma virus (RSV). In the 1970s, research involving RSV led to the identification of a viral gene that acts as an oncogene. Researchers investigated RSV by using it to infect chicken cells grown in the laboratory. This causes the chicken cells to grow like cancer cells, continuously and in an uncontrolled manner. Researchers identified mutant RSV strains that infected and proliferated within chicken cells without transforming them into malignant cells. These RSV strains were missing a gene that is found in the form of the virus that does cause cancer. This gene was called the *src* gene because it causes sarcoma.

Harold Varmus and Michael Bishop, in collaboration with Peter Vogt, later discovered that normal (nonviral-infected) cells also contain a copy of the *src* gene in their chromosomes. It is a proto-oncogene. When the *src* gene is incorporated into a viral genome, it is overexpressed because it is transcribed from a very active viral promoter. This ultimately produces too much of the Src protein in infected cells and promotes uncontrolled cell division.

Tumor-Suppressor Genes Prevent Mutation or Cell Proliferation

Thus far, we have examined one category of genes that can promote cancer, namely oncogenes. We now turn our attention to the second category of genes, those called tumor-suppressor genes. The functioning of a normal (nonmutant) tumor-suppressor gene prevents cancerous growth. The proteins encoded by tumor-suppressor genes usually have one of two functions—maintenance of genome integrity or negative regulation of cell division (**Table 14.8**).

Maintenance of Genome Integrity   Some tumor-suppressor genes encode proteins that maintain the integrity of the genome. Such proteins monitor and/or repair alterations in the genome. The proteins encoded by these genes are vital for the prevention of abnormalities such as gene mutations, DNA breaks, and improperly segregated chromosomes. Therefore, when these

Table 14.8	Functions of Selected Tumor-Suppressor Genes
Gene	**Function of encoded protein**
Maintenance of genome integrity	
p53	p53 is a transcription factor that acts as a sensor of DNA damage. It can promote DNA repair, prevent the progression through the cell cycle, and promote apoptosis.
BRCA-1 *BRCA-2*	BRCA-1 and BRCA-2 proteins are both involved in the cellular defense against DNA damage. They may play a role in sensing DNA damage or act to facilitate DNA repair. These genes are mutant in persons with certain inherited forms of breast cancer.
XPD	This represents one of several different genes whose products function in DNA repair. These genes are defective in patients with xeroderma pigmentosum.
Negative regulation of cell division	
Rb	The Rb protein is a negative regulator that represses the transcription of genes required for DNA replication and cell division.
NF1	The NF1 protein stimulates Ras to hydrolyze its GTP to GDP. Loss of NF1 function causes the Ras protein to be overactive, which promotes cell division.
p16	The p16 protein is a negative regulator of cyclin-dependent protein kinase.

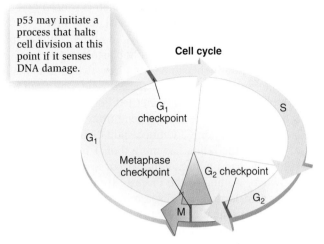

Figure 14.15   The cell cycle and checkpoints.   As discussed in Chapter 15, eukaryotic cells progress through a cell cycle composed of G_1, S, G_2, and M phases (look ahead to Figure 15.2). The red bars indicate common checkpoints that will stop the cell cycle if genetic abnormalities are detected. The p53 protein will stop a cell at the G_1 checkpoint if it senses DNA damage.

Concept check: Why is it an advantage for an organism to have checkpoints that can stop the cell cycle?

proteins are functioning properly, they minimize the chance that a cancer-causing mutation will occur. In some cases, the proteins encoded by tumor-suppressor genes will prevent a cell from progressing through the cell cycle if an abnormality is detected. These are termed **checkpoint proteins** because their role is to <u>check</u> the integrity of the genome and prevent a cell from progressing past a certain <u>point</u> in the cell cycle. Checkpoint proteins are not usually required to regulate normal, healthy cell division, but they can stop cell division if an abnormality is detected.

Proteins called cyclins and cyclin-dependent protein kinases (cdks) are responsible for advancing a cell through the four phases of the cell cycle (see Chapter 15). The formation of activated cyclin/cdk complexes can be stopped by checkpoint proteins. A specific example of a tumor-suppressor gene that encodes a checkpoint protein is *p53*, discovered in 1979 by Arnold Levine. Its name refers to the molecular mass of the p53 protein, which is <u>53</u> kDa (kilodaltons). About 50% of all human cancers are associated with mutations in this gene, including malignant tumors of the lung, breast, esophagus, liver, bladder, and brain, as well as leukemias and lymphomas (cancer of the lymphatic system).

As shown in **Figure 14.15**, p53 is a G_1 checkpoint protein. The expression of the *p53* gene is induced when DNA is damaged. The p53 protein is a regulatory transcription factor that activates several different genes, leading to the synthesis of proteins that stop the cell cycle and other proteins that repair the DNA. When p53 is activated, a cell cannot progress from G_1 to the S, or synthesis, phase of the cell cycle. If the DNA is eventually repaired, a cell may later proceed through the cell cycle.

Alternatively, if the DNA damage is too severe, the p53 protein will also activate other genes that promote programmed cell death. This process, called **apoptosis**, involves cell shrinkage

and DNA degradation. As described in Chapter 9, enzymes known as **caspases** are activated during apoptosis (refer back to Figure 9.20). They function as proteases that are sometimes called the "executioners" of the cell. Caspases digest selected cellular proteins such as microfilaments, which are components of the cytoskeleton. This causes the cell to break down into small vesicles that are eventually phagocytized by cells of the immune system. It is beneficial for a multicellular organism to kill an occasional cell with cancer-causing potential.

When checkpoint genes such as *p53* are rendered inactive by mutation, the division of normal healthy cells may not be adversely affected. For example, mice that are missing the *p53* gene are born healthy. This indicates that checkpoint proteins such as p53 are not necessary for normal cell growth and division. However, these mice are very sensitive to mutagens such as UV light and easily develop cancer. The loss of p53 function makes it more likely that undesirable genetic changes will occur that could cause cancerous growth.

Negative Regulation of Cell Division   A second category of tumor-suppressor genes encodes proteins that are negative regulators or inhibitors of cell division. Their function is necessary to properly halt cell division. If their function is lost, cell division is abnormally accelerated.

An example of such a tumor-suppressor gene is the *Rb* gene. It was the first tumor-suppressor gene to be identified in humans by studying patients with a disease called retinoblastoma, a cancerous tumor that occurs in the retina of the eye. Some people have an inherited predisposition to develop this disease within the first few years of life. By comparison, the noninherited form of retinoblastoma, which is caused by environmental agents, is more likely to occur later in life.

Based on these differences, in 1971, Alfred Knudson proposed a two-hit hypothesis for retinoblastoma. According to this hypothesis, retinoblastoma requires two mutations to occur. People have two copies of the *Rb* gene, one from each parent. Individuals with the inherited form of the disease already have received one mutant gene from one of their parents. They need only one additional mutation to develop the disease. Because the retina has more than 1 million cells, it is relatively likely that a mutation may occur in one of these cells at an early age, leading to the disease. However, people with the noninherited form of the disease must have two *Rb* mutations in the same retinal cell to cause the disease. Because two mutations are less likely than a single mutation, the noninherited form of this disease is expected to occur much later in life, and only rarely. Since Knudson's original work, molecular studies have confirmed the two-hit hypothesis for retinoblastoma.

The Rb protein negatively controls a regulatory transcription factor called E2F that activates genes required for cell cycle progression from G_1 to S phase. The binding of the Rb protein to E2F inhibits its activity and prevents cell division (**Figure 14.16**). When a normal cell is supposed to divide, cyclins bind to cyclin-dependent protein kinases. This binding activates the kinases, which then leads to the phosphorylation of the Rb protein. The phosphorylated form of the Rb protein is released from E2F, thereby allowing E2F to activate genes needed to progress through the cell cycle. By comparison, we can imagine how the cell cycle becomes unregulated without a functional Rb protein. When both copies of Rb are defective, the E2F protein is always active. This explains why uncontrolled cell division occurs in retinoblastoma.

Gene Mutations, Chromosome Loss, and DNA Methylation Can Inhibit the Expression of Tumor-Suppressor Genes

Cancer biologists would also like to understand how tumor-suppressor genes are inactivated, because this knowledge may ultimately help them to prevent or combat cancer. How are tumor-suppressor genes silenced? The function of tumor-suppressor genes is lost in three common ways. First, a mutation can occur within a tumor-suppressor gene to inactivate its function. For example, a mutation could abolish the function of the promoter for a tumor-suppressor gene or introduce an early stop codon in its coding sequence. Either of these would prevent the expression of a functional protein.

Chromosome loss is a second way that the function of a tumor-suppressor gene is lost. Chromosome loss may contribute to the progression of cancer if the missing chromosome carries one or more tumor-suppressor genes.

Recently, researchers have discovered a third way that these genes may be inactivated. Tumor-suppressor genes found in cancer cells are sometimes abnormally methylated. As discussed in Chapter 13, transcription is inhibited when CpG islands near a promoter region are methylated. Such DNA

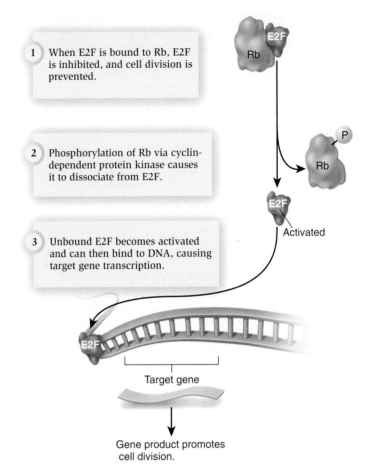

1. When E2F is bound to Rb, E2F is inhibited, and cell division is prevented.

2. Phosphorylation of Rb via cyclin-dependent protein kinase causes it to dissociate from E2F.

3. Unbound E2F becomes activated and can then bind to DNA, causing target gene transcription.

Activated

Target gene

Gene product promotes cell division.

Figure 14.16 Function of the Rb protein. The Rb protein inhibits the function of E2F, which turns on genes that cause a cell to divide. When cells are supposed to divide, Rb is phosphorylated by cyclin-dependent protein kinase, which allows E2F to function. If Rb protein is not properly made due to a mutation, E2F will always be active, and the cell will be stimulated to divide uncontrollably.

Concept check: Would cancer occur if both copies of the Rb gene and both copies of the E2F gene were rendered inactive due to mutations?

methylation near the promoters of tumor-suppressor genes has been found in many types of tumors, suggesting that this form of gene inactivation plays an important role in the formation and/or progression of malignancy.

Most Forms of Cancer Are Caused by a Series of Genetic Changes That Progressively Alter the Growth Properties of Cells

The discovery of oncogenes and tumor-suppressor genes has allowed researchers to study the progression of certain forms of cancer at the molecular level. Cancer usually requires multiple genetic changes to the same cell lineage, perhaps in the range of 10 or more. Many cancers begin with a benign genetic alteration that, over time and with additional mutations, leads to malignancy. Furthermore, a malignancy can continue to accumulate genetic changes that make it even more difficult to treat because the cells divide faster or invade surrounding tissues more readily.

Normal lung epithelium

Hyperplasia

Loss of ciliated cells

Dysplasia (initially precancerous, then cancerous)

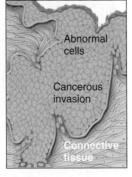

Invasive cancerous cells that can metastasize

(a) Cellular changes

Genes that are commonly mutated in lung cancer include

Oncogenes
erbB – epidermal growth factor receptor
ras – cell signaling
myc – transcription factor
Cyclin D1 – promotes the cell cycle

Tumor-suppressor genes
p53 – checkpoint
XPD – DNA repair
Rb – negative regulator
p16 – negative regulator

(b) Genetic changes

Figure 14.17 **Progression of changes leading to lung cancer.** Lung tissue is largely composed of different types of connective tissue and epithelial cells, including columnar and basal cells. **(a)** A progression of cellular changes in basal cells, caused by the accumulation of mutations, leads to basal cell carcinoma, a common type of lung cancer. **(b)** Mutations in several different genes can contribute to lung cancer.

Lung cancer progresses through different stages of abnormal cell proliferation. Lung cancer is diagnosed in approximately 170,000 men and women each year in the U.S. Worldwide, more than 1.2 million cases are diagnosed. Nearly 90% of these cases are caused by smoking and are thus preventable. Unlike other cancers for which early diagnosis is possible, lung cancer is usually detected only after it has become advanced and is difficult if not impossible to cure. The five-year survival rate for lung cancer patients is approximately 15%.

What is the cellular basis for lung cancer? Most cancers in the lung are **carcinomas**—cancers of epithelial cells (**Figure 14.17a**). Epithelial cells are described in Chapter 10. The top images in this figure show the normal epithelium found in a healthy lung. The rest of the figure shows the progression of cancer that is due to mutations in basal cells, a type of epithelial cell. Keep in mind that cancer occurs due to the accumulation of mutations in a cell lineage, beginning with an initial mutant cell that then divides multiple times to produce a population of many daughter cells (refer back to Figure 14.10). As mutations accumulate in a lineage of basal cells, their numbers increase dramatically. This causes a thickening of the epithelium, a condition called hyperplasia. The proliferation of such basal cells causes the loss of the ciliated, columnar epithelial cells that normally line the airways. As additional mutations accumulate in this cell lineage, the basal cells develop more abnormal morphologies, a condition known as dysplasia. In the early stages of dysplasia, the abnormal basal cells are precancerous. If the source of chronic irritation (usually cigarette smoke) is eliminated, the abnormal cells are likely to disappear. Alternatively, if smoking continues, these abnormal cells may accumulate additional genetic changes and lose the ability to stop dividing. Such cells have become cancerous—the person has basal cell carcinoma.

The basement membrane is a sheetlike layer of extracellular matrix that provides a barrier between the lung cells and the bloodstream. If the cancer cells have not yet penetrated the basement membrane, they will not have metastasized, that is, spread into the blood and to other parts of the body. If the entire tumor is removed at this stage, the patient should be cured. The lower images in Figure 14.17a show a tumor that has broken through the basement membrane. The metastasis of these cells to other parts of the body will likely kill the patient, usually within a year of diagnosis.

The cellular changes that lead to lung cancer are correlated with genetic changes (**Figure 14.17b**). These include the occurrence of mutations that create oncogenes and inhibit tumor-suppressor genes. The order of mutations is not absolute. It takes time for multiple changes to accumulate, so cancer is usually a disease of older people. Reducing your exposure to mutagens such as cigarette smoke throughout your lifetime will minimize the risk of mutations to your genes that could promote cancer.

Genomes & Proteomes Connection

Chromosomal Changes and Mutations in Approximately 300 Human Genes May Promote Cancer

Researchers have identified a large number of genes that are mutated in cancer cells. Though not all of these mutant genes have been directly shown to affect the growth rate of cells, such mutations are likely to be found in tumors because they provide some type of growth advantage for the cell population from which the cancer developed. For example, certain mutations may enable cells to metastasize to neighboring locations. These mutations may not affect growth rate, but they provide the growth advantage that cancer cells are not limited to growing in a particular location. They can migrate to new locations.

How many genes can contribute to cancer when they become mutant? Researchers have estimated that about 300 different genes may play a role in the development of human cancer. With an approximate genome size of 20,000 to 25,000 genes, this observation indicates that over 1% of our genes have the potential to promote cancer if their function is altered by a mutation.

In addition to mutations within specific genes, another common genetic change associated with cancer is abnormalities in chromosome structure and number. Figure 14.18 compares the chromosome composition of a normal male cell and a tumor cell taken from the same person. The normal composition for this person is 22 pairs of chromosomes plus two sex chromosomes (X and Y). By comparison, the chromosome composition of the tumor cell is quite bizarre, including the observation that the tumor cell has two X chromosomes, which is characteristic of females. The tumor cells are also missing several chromosomes. If tumor-suppressor genes were located on these missing chromosomes, their function is lost as well. Figure 14.18 also shows a few cases of extra chromosomes. If these chromosomes contain proto-oncogenes, the expression of those genes may be overactive. Finally, tumor cells often contain chromosomes that have translocations. Such translocations may create chimeric genes (as in the case of the Philadelphia chromosome discussed earlier in this chapter), or they may place a gene next to the regulatory sequences of another gene.

Figure 14.18   A comparison between chromosomes found in a normal human cell (left) and a cancer cell (right) from the same person.   The set found in the cancer cell on the right is highly abnormal, with extra copies of some chromosomes and lost copies of others. Chromosomes made of fused pieces of chromosomes (designated mar in this figure) are also common in cancer cells.

Concept check:) *How might these changes in chromosome structure and number contribute to cancer?*

Summary of Key Concepts

14.1 Mutation

- A mutation is a heritable change in the genetic material.

- Point mutations, which affect a single base pair, can alter the coding sequence of genes in several ways. These include silent, missense, nonsense, and frameshift mutations. (Table 14.1)

- Sickle-cell disease is caused by a missense mutation that changes a single amino acid in β-globin. (Figure 14.1)

- Gene mutations can also alter gene function by changing DNA sequences that are not within the coding region. (Table 14.2)

- The Lederbergs used replica plating to show that mutations conferring resistance to T1 occur randomly. (Figure 14.2)

- Germ-line mutations affect gametes, whereas somatic mutations affect only a part of the body and cannot be passed to offspring. (Figures 14.3, 14.4)

- Spontaneous mutations are the result of errors in natural biological processes. Induced mutations are due to agents in the environment that cause changes in DNA structure. (Table 14.3)

- Mutagens are chemical or physical agents that lead to mutations in DNA. (Table 14.4, Figures 14.5, 14.6)

- Testing methods, such as the Ames test, can determine whether an agent is a mutagen. (Figure 14.7)

14.2 DNA Repair

- DNA repair systems involve proteins that sense DNA damage and repair it before a mutation occurs. (Table 14.5)

- Nucleotide excision repair (NER) systems recognize various types of DNA damage, such as thymine dimers. In this type of system, the damaged strand is excised, and a new strand is made. (Figure 14.8)

- Certain inherited diseases in humans, such as xeroderma pigmentosum, are due to defects in NER. (Figure 14.9)

14.3 Cancer

- Cancer is due to the accumulation of mutations in a lineage of cells that leads to uncontrolled cell growth. (Figure 14.10)

- Mutations in proto-oncogenes that result in overactivity produce cancer-causing genes called oncogenes.

- Oncogenes often encode proteins involved in cell-signaling pathways that promote cell division. (Figures 14.11, 14.12, Table 14.6)

- Four common types of genetic changes, namely, missense mutations, gene amplifications, chromosomal translocations, and retroviral insertions, can change proto-oncogenes into oncogenes. (Figure 14.13)

- A chromosome translocation that fuses parts of the *bcr* gene and the *abl* gene creates an oncogene that causes leukemia. (Figure 14.14)

- Some types of cancer are caused by viruses. (Table 14.7)

- The normal function of tumor-suppressor genes is to prevent cancer. Loss-of-function mutations in such genes can promote cancer. Tumor-suppressor genes often encode proteins that maintain the integrity of the genome or function as negative regulators of cell division. (Table 14.8)

- Checkpoint proteins, such as p53, monitor the integrity of the genome and prevent the cell from progressing through the cell cycle if abnormalities are detected. (Figure 14.15)

- The Rb protein is a negative regulator of cell division because it inhibits E2F, a transcription factor that promotes cell division. (Figure 14.16)

- Gene mutations, chromosome loss, and DNA methylation are common ways that tumor-suppressor genes are inactivated.

- Most forms of cancer, such as lung cancer, involve multiple genetic changes that lead to malignancy. (Figure 14.17)

- Over 300 human genes are known to be associated with cancer when they become mutant. In addition, changes in chromosome number and structure are commonly found in cancer cells. (Figure 14.18)

Assess and Discuss

Test Yourself

1. Point mutations that do not alter the amino acid sequence of the resulting gene product are called _____ mutations.
   a. frameshift
   b. natural
   c. silent
   d. nonsense
   e. missense

2. Some point mutations will lead to an mRNA that produces a shorter polypeptide. This type of mutation is known as a _____ mutation.
   a. neutral
   b. silent
   c. missense
   d. nonsense
   e. chromosomal

3. A mutation in which of the following regions is least likely to affect gene function?
   a. promoter
   b. coding region
   c. splice junction
   d. intergenic region
   e. regulatory site

4. Mutagens can cause mutations by
   a. chemically altering DNA nucleotides.
   b. disrupting DNA replication.
   c. altering the genetic code of an organism.
   d. all of the above.
   e. a and b only.

5. The mutagenic effect of UV light is
   a. the alteration of cytosine bases to adenine bases.
   b. the formation of adenine dimers that interfere with genetic expression.
   c. the breaking of the sugar-phosphate backbone of the DNA molecule.
   d. the formation of thymine dimers that disrupt DNA replication.
   e. the deletion of thymine bases along the DNA molecule.

6. The Ames test
   a. provides a way to determine if any type of cell has experienced a mutation.
   b. provides a way to determine if an agent is a mutagen.
   c. allows researchers to experimentally disrupt gene activity by causing a mutation in a specific gene.
   d. provides a way to repair mutations in bacterial cells.
   e. does all of the above.

7. Xeroderma pigmentosum
   a. is a genetic disorder that results in uncontrolled cell growth.
   b. is a genetic disorder where the NER system is not fully functional.
   c. is a genetic disorder that results in the loss of pigment in certain patches of skin.
   d. results from the lack of DNA polymerase proofreading.
   e. is both b and d.

8. If a mutation eliminated the function of UvrC, which aspect of nucleotide excision repair would not work?
   a. sensing a damaged DNA site
   b. endonuclease cleavage of the damaged strand
   c. removal of the damaged strand
   d. synthesis of a new strand, using the undamaged strand as a template
   e. none of the above

9. Cancer cells are said to be metastatic when they
   a. begin to divide uncontrollably.
   b. invade healthy tissue.
   c. migrate to other parts of the body.
   d. cause mutations in other healthy cells.
   e. do all of the above.

10. Oncogenes can be caused by
    a. missense mutations.
    b. gene amplification.
    c. chromosomal translocation.
    d. retroviral insertion.
    e. all of the above.

Conceptual Questions

1. Is a random mutation more likely to be beneficial or harmful? Explain your answer.

2. Distinguish between spontaneous and induced mutations. Which are more harmful? Which are avoidable?

3. In the treatment of cancer, the basis for many types of chemotherapy and radiation therapy is that mutagens are more effective at killing dividing cells compared to nondividing cells. Explain why. What are possible harmful side effects of chemotherapy and radiation therapy?

Collaborative Questions

1. Discuss the pros and cons of mutation.

2. A large amount of research is aimed at studying mutation. However, there is not an infinite amount of research dollars. Where would you put your money for mutation research?
   a. testing of potential mutagens
   b. investigating molecular effects of mutagens
   c. investigating DNA repair mechanisms
   d. or some other place

Online Resource

www.brookerbiology.com

Stay a step ahead in your studies with animations that bring concepts to life and practice tests to assess your understanding. Your instructor may also recommend the interactive ebook, individualized learning tools, and more.

The Eukaryotic Cell Cycle, Mitosis, and Meiosis

15

A scanning electron micrograph of human chromosomes. These chromosomes are highly compacted and found in a dividing cell.

O ver 10,000,000,000,000! Researchers estimate the adult human body contains somewhere between 10 trillion to 50 trillion cells. It is almost an incomprehensible number. Even more amazing is the accuracy of the process that produces these cells. After a human sperm and egg unite, the fertilized egg goes through a long series of cell divisions to produce an adult with over 10 trillion cells. Let's suppose you randomly removed a cell from your arm and compared it to a cell from your foot. If you examined the chromosomes found in both cells under the microscope, they would look identical. Likewise, the DNA sequences along those chromosomes would also be the same, barring rare mutations. Similar comparisons could be made among the trillions of cells in your body. When you consider how many cell divisions are needed to produce an adult human, the precision of cell division is truly remarkable.

What accounts for this high level of accuracy? As we will examine in this chapter, **cell division**, the reproduction of cells, is a highly regulated process that monitors the integrity of the genetic material. The eukaryotic cell cycle is a series of phases that are needed for cell division. The cells of eukaryotic species may follow one of two different sorting processes so that new daughter cells will receive the correct number and types of chromosomes. The first sorting process we will explore, called mitosis, is needed so that daughter cells will receive the same amount of genetic material as the mother cell that produced them. We will then examine another sorting process, called meiosis, which is needed for sexual reproduction. In meiosis, cells that have two sets of chromosomes produce daughter cells with a single set of chromosomes. Lastly, we will explore variation in the structure and number of chromosomes. As you will see, a variety of mechanisms that alter chromosome structure and number can have important consequences for the organisms that carry them.

15.1 The Eukaryotic Cell Cycle

Life is a continuum in which new living cells are formed by the division of pre-existing cells. The Latin axiom *Omnis cellula e cellula*, meaning "Every cell from a cell," was first proposed in 1858 by a German pathologist, Rudolf Virchow. From an evolutionary perspective, cell division has a very ancient origin. All living organisms, from unicellular bacteria to multicellular plants and animals, have been produced by a series of repeated rounds of cell growth and division extending back to the beginnings of life nearly 4 billion years ago. We now know that cell division is a process that involves remarkable accuracy and precise timing. A cell must be able to sense when conditions are appropriate for division to occur and then orchestrate a series of events that will ensure the production of healthy new cells. A key issue is that the chromosomes must be properly replicated and sorted to new daughter cells. In this section, we will examine the phases of the eukaryotic cell cycle and see how the cell cycle is controlled by proteins that carefully monitor the division process to ensure its accuracy. But first, we need to consider some general features of chromosomes in eukaryotic species.

Eukaryotic Chromosomes Are Inherited in Sets and Occur in Homologous Pairs

To understand the chromosomal composition of cells and the behavior of chromosomes during cell division, scientists observe cells and chromosomes with the use of microscopes.

Cytogenetics is the field of genetics that involves the microscopic examination of chromosomes and cell division. When a cell prepares to divide, the chromosomes become more tightly compacted, a process that decreases their apparent length and increases their diameter. A consequence of this compaction is that distinctive shapes and numbers of chromosomes become visible with a light microscope.

Microscopic Examination of Chromosomes Figure 15.1 shows the general procedure for preparing and viewing chromosomes from a eukaryotic cell. In this example, the cells are obtained from a sample of human blood. Specifically, the chromosomes within leukocytes (white blood cells) are examined. A sample of the blood cells is obtained and treated with drugs that stimulate the cells to divide. The actively dividing cells are centrifuged to concentrate them and then mixed with a hypotonic solution that makes the cells swell. The expansion of the

cell causes the chromosomes to spread out from each other, making it easier to see each individual chromosome.

Next, the cells are concentrated by a second centrifugation and treated with a fixative, which chemically fixes them in place so that the chromosomes will no longer move around. The cells are then exposed to a chemical dye, such as Giemsa, that binds to the chromosomes and stains them. This gives chromosomes a distinctive banding pattern that greatly enhances their contrast and ability to be uniquely identified. The cells are then placed on a slide and viewed with a light microscope. In a cytogenetics laboratory, the microscopes are equipped with an electronic camera to photograph the chromosomes. On a computer screen, the chromosomes can be organized in a standard way, usually from largest to smallest. A photographic representation of the chromosomes, as in the photo in step 5 of Figure 15.1, is called a **karyotype**. A karyotype reveals the number, size, and form of chromosomes found within an actively dividing cell. It

1 A sample of blood is collected and treated with drugs that stimulate cell division. The sample is then subjected to centrifugation.

Supernatant

Blood cells — Pellet

2 The supernatant is discarded, and the cell pellet is suspended in a hypotonic solution. This causes the cells to swell and the chromosomes to spread out from each other.

Hypotonic solution

3 The sample is subjected to centrifugation a second time to concentrate the cells. The cells are suspended in a fixative, stained, and placed on a slide.

Fix   Stain

Blood cells

4 The slide is viewed by a light microscope equipped with a camera; the sample is seen on a computer screen. The chromosomes can be photographed and arranged electronically on the screen.

5

A pair of sister chromatids

G band

Homologues

For a diploid human cell, 2 complete sets of chromosomes from a single cell constitute a karyotype of that cell.

Figure 15.1   **The procedure for making a karyotype.**   In this example, the chromosomes were treated with the dye Giemsa, and the resulting bands are called G bands.

Concept check: *Researchers usually treat cells with drugs that stimulate them to divide prior to the procedure for making a karyotype. Why is this useful?*

should also be noted that the chromosomes that are viewed in actively dividing cells have already replicated to form sister chromatids (see inset to Figure 15.1). We will discuss the formation of sister chromatids later in this section.

Sets of Chromosomes   What type of information is learned from a karyotype? By studying the karyotypes of many species, scientists have discovered that eukaryotic chromosomes occur in sets. Each set is composed of several different types of chromosomes. For example, one set of human chromosomes contains 23 different types of chromosomes (see Figure 15.1). By convention, the chromosomes are numbered according to size, with the largest chromosomes having the smallest numbers. For example, human chromosomes 1, 2, and 3 are relatively large, whereas 21 and 22 are the two smallest. This numbering system does not apply to the **sex chromosomes**, which determine the sex of the individual. Sex chromosomes in humans are designated with the letters X and Y. The chromosomes that are not sex chromosomes are called **autosomes**. Humans have 22 different types of autosomes.

A second feature of many eukaryotic species is that most cells contain two sets of chromosomes. The karyotype shown in Figure 15.1 contains two sets of chromosomes, with 23 different chromosomes in each set. Therefore, this human cell contains a total of 46 chromosomes. A person's cells have 46 chromosomes each because the individual inherited one set from the father and one set from the mother. When the cells of an organism carry two sets of chromosomes, that organism is said to be **diploid**. Geneticists use the letter n to represent a set of chromosomes, so diploid organisms are referred to as $2n$. For example, humans are $2n$, where $n = 23$. Most human cells are diploid. An exception involves **gametes**, namely, sperm and egg cells. Gametes are **haploid**, or $1n$, which means they contain one set of chromosomes.

Homologous Pairs of Chromosomes   When an organism is diploid, the members of a pair of chromosomes are called **homologues** (see inset to Figure 15.1). The term **homology** refers to any similarity that is due to common ancestry. Pairs of homologous chromosomes are evolutionarily derived from the same chromosome. However, homologous chromosomes are not usually identical to each other because over many generations they have accumulated some genetic changes that make them distinct.

How similar are homologous chromosomes to each other? Each of the two chromosomes in a homologous pair is nearly identical in size and contains a very similar composition of genetic material. A particular gene found on one copy of a chromosome is also found on the homologue. Because one homologue is received from each parent, the two homologues may vary with regard to the way that a gene affects an organism's traits. As an example, let's consider a gene in humans called *OCA2*, which plays a major role in determining eye color. The *OCA2* gene is found on chromosome 15. One copy of chromosome 15 might carry the form of this eye color gene that confers brown eyes, whereas the gene on the homologue could confer blue eyes. The topic of how genes affect an organism's traits will be considered in Chapter 16.

The DNA sequences on homologous chromosomes are very similar. In most cases, the sequence of bases on one homologue differs by less than 1% from the sequence on the other homologue. For example, the DNA sequence of chromosome 1 that you inherited from your mother is likely to be more than 99% identical to the DNA sequence of chromosome 1 that you inherited from your father. Nevertheless, keep in mind that the sequences are not identical. The slight differences in DNA sequence provide important variation in gene function. Again, if we use the eye color gene *OCA2* as an example, a minor difference in DNA sequence distinguishes two forms of the gene, brown versus blue.

The striking similarity between homologous chromosomes does not apply to pairs of sex chromosomes (for example, X and Y). These chromosomes differ in size and genetic composition. Certain genes found on the X chromosome are not found on the Y chromosome, and vice versa. The X and Y chromosomes are not considered homologous chromosomes, although they do have short regions of homology.

The Cell Cycle Is a Series of Phases That Lead to Cell Division

Eukaryotic cells that are destined to divide progress through the **cell cycle**, a sequence of growth, replication, and division that produces new cells. **Figure 15.2** provides an overview of the cell cycle. In this diagram, the mother cell has three pairs of chromosomes, for a total of six individual chromosomes. Such a cell is diploid ($2n$) and contains three chromosomes per set ($n = 3$). The paternal set is shown in blue, and the homologous maternal set is shown in red.

The phases of the cell cycle are G_1 (first gap), **S** (synthesis of DNA, the genetic material), G_2 (second gap), and **M phase** (mitosis and cytokinesis). The G_1 and G_2 phases were originally described as gap phases to indicate a pause in activity between DNA synthesis and mitosis. However, we now know these are critical phases of the cell cycle. In actively dividing cells, the G_1, S, and G_2 phases are collectively known as **interphase**. During interphase, the cell grows and copies its chromosomes in preparation for cell division. Alternatively, cells may exit the cell cycle and remain for long periods of time in a phase called G_0 (G zero). The G_0 phase is an alternative to proceeding through G_1. A cell in the G_0 phase has postponed making a decision to divide or, in the case of terminally differentiated cells (such as nerve cells in an adult animal), has made a decision to never divide again. G_0 is a nondividing phase.

G_1 Phase   The G_1 phase is a period in a cell's life when it may become committed to divide. Depending on the environmental conditions and the presence of signaling molecules, a cell in the G_1 phase may accumulate molecular changes that cause it to progress through the rest of the cell cycle. Cell growth typically occurs during the G_1 phase.

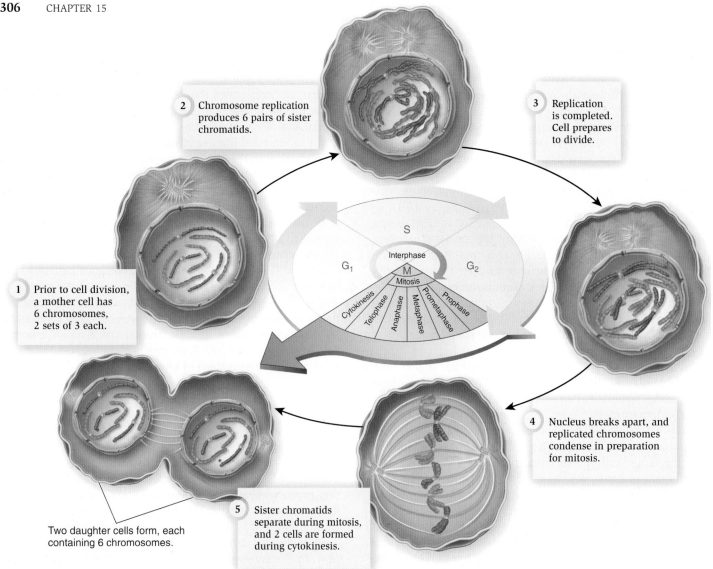

2 Chromosome replication produces 6 pairs of sister chromatids.

3 Replication is completed. Cell prepares to divide.

1 Prior to cell division, a mother cell has 6 chromosomes, 2 sets of 3 each.

4 Nucleus breaks apart, and replicated chromosomes condense in preparation for mitosis.

Two daughter cells form, each containing 6 chromosomes.

5 Sister chromatids separate during mitosis, and 2 cells are formed during cytokinesis.

S

Interphase

G₁   M   G₂

Mitosis

Cytokinesis Telophase Anaphase Metaphase Prometaphase Prophase

Figure 15.2 **The eukaryotic cell cycle.** Dividing cells progress through a series of phases denoted G₁, S, G₂, and M. This diagram shows the progression of a cell through the cell cycle to produce two daughter cells. The original diploid cell had three pairs of chromosomes, for a total of six individual chromosomes. During S phase, these have replicated to yield 12 chromatids. After mitosis is complete, two daughter cells each contain six chromosomes. The width of the phases shown in this figure is not meant to reflect their actual length. G₁ is typically the longest phase of the cell cycle, whereas M phase is relatively short.

Concept check: *Which phases make up interphase?*

S Phase   During the S phase, the chromosomes are replicated, which is discussed in Chapter 11. After replication, the two copies are still joined to each other and referred to as a pair of **sister chromatids** (Figure 15.3). When S phase is completed, a cell actually has twice as many chromatids as the number of chromosomes in the G₁ phase. For example, a human cell in the G₁ phase has 46 distinct chromosomes, whereas the same cell in G₂ would have 46 pairs of sister chromatids, for a total of 92 chromatids.

G₂ Phase   During the G₂ phase, a cell synthesizes proteins that are necessary for chromosome sorting and cell division. This prepares the cell for the last phase of the cell cycle. Some cell growth may occur.

M Phase   The first part of M phase is **mitosis**. The purpose of mitosis is to divide one cell nucleus into two nuclei, distributing the duplicated chromosomes so that each daughter cell will receive the same complement of chromosomes. As noted previously, a human cell in the G₂ phase has 92 chromatids, which are found in 46 pairs. During mitosis, these pairs of chromatids are separated and sorted so that each daughter cell will receive 46 chromosomes. Mitosis is the name given to this sorting process. In most cases, mitosis is followed by **cytokinesis**, which is the division of the cytoplasm to produce two distinct daughter cells.

The length of the cell cycle varies considerably among different cell types, ranging from several minutes in quickly growing embryos to several months in slow-growing adult cells.

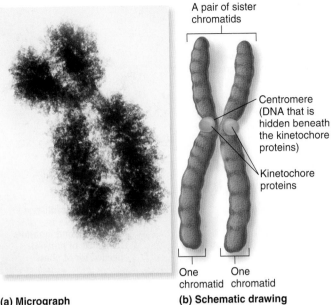

(a) Micrograph

(b) Schematic drawing

A pair of sister chromatids

Centromere (DNA that is hidden beneath the kinetochore proteins)

Kinetochore proteins

One chromatid   One chromatid

Figure 15.3   **Sister chromatids.**   (a) After a chromosome replicates, the two copies remain attached to each other and are called sister chromatids. Sister chromatids are formed during S phase, but they do not become microscopically visible until later, when they become more compact during M phase. This micrograph shows a pair of sister chromatids during metaphase, which is a part of M phase. (b) A schematic drawing of sister chromatids. This structure has two chromatids that lie side-by-side. The two chromatids are held together by cohesin proteins (not shown in this drawing). The kinetochore is a group of proteins that are attached to the centromere and play a role during chromosome sorting.

Concept check: *In a human cell, how many total chromatids would you expect to find during metaphase?*

For fast-dividing mammalian cells in adults, the length of the cycle is typically 24 hours. The various phases within the cell cycle also vary in length. G_1 is often the longest and also the most variable phase, and M phase is the shortest. For a cell that divides in 24 hours, the following lengths of time for each phase are typical:

- G_1 phase: 11 hours
- S phase: 8 hours
- G_2 phase: 4 hours
- M phase: 1 hour

What factors determine whether or not a cell will divide? First, the determination to divide is based on external factors, such as environmental conditions and signaling molecules. The effects of growth factors on cell division are discussed in Chapter 9 (refer back to Figure 9.10). Second, internal controls affect cell division. These include cell cycle control molecules and checkpoints, as we will discuss next.

The Cell Cycle Is Controlled by Checkpoint Proteins

The progression through the cell cycle is a process that is highly regulated to ensure that the nuclear genome is intact and that the conditions are appropriate for a cell to divide. This is necessary to minimize the occurrence of mutations, which could have harmful effects and potentially lead to cancer. Proteins called **cyclins** and **cyclin-dependent kinases** (**cdks**) are responsible for advancing a cell through the phases of the cell cycle. Cyclins are so named because their amount varies throughout the cell cycle. To be active, the kinases controlling the cell cycle must bind to (are dependent on) a cyclin. The number of different types of cyclins and cdks varies from species to species.

Figure 15.4 gives a simplified description of how cyclins and cdks work together to advance a cell through G_1 and mitosis. During G_1, the amount of a particular cyclin termed G_1 cyclin increases. The G_1 cyclin binds to a cdk to form an activated G_1 cyclin/cdk complex. Once activated, cdk functions as a protein kinase that phosphorylates other proteins that are needed to advance the cell to the next phase in the cell cycle. For example, certain proteins involved with DNA synthesis are phosphorylated and activated, thereby allowing the cell to carry on events in S phase. When the cell passes into the S phase, G_1 cyclin is degraded. Similar events advance the cell through other phases of the cell cycle. A different cyclin, called mitotic cyclin, accumulates late in G_2. It binds to a cdk to form an activated mitotic cyclin/cdk complex. This complex phosphorylates proteins that are needed to advance into M phase.

Three critical regulatory points called **checkpoints** are found in the cell cycle of eukaryotic cells (Figure 15.4). At these checkpoints, a variety of proteins, referred to as checkpoint proteins, act as sensors to determine if a cell is in the proper condition to divide. The G_1 checkpoint, also called the **restriction point**, determines if conditions are favorable for cell division. In addition, G_1-checkpoint proteins can sense if the DNA has incurred damage. What happens if DNA damage is detected? The checkpoint proteins will prevent the formation of active cyclin/cdk complexes and thereby stop the progression of the cell cycle.

A second checkpoint exists in G_2. This checkpoint also checks the DNA for damage and ensures that all of the DNA has been replicated. In addition, the G_2 checkpoint monitors the levels of proteins that are needed to progress through M phase. A third checkpoint, called the metaphase checkpoint, senses the integrity of the spindle apparatus. As we will see later, the spindle apparatus is involved in chromosome sorting. Metaphase is a step in mitosis during which all of the chromosomes should be attached to the spindle apparatus. If a chromosome is not correctly attached, the metaphase checkpoint will stop the cell cycle. This checkpoint prevents cells from incorrectly sorting their chromosomes during division.

Checkpoint proteins delay the cell cycle until problems are fixed, or they even prevent cell division when problems cannot be fixed. A primary aim of checkpoint proteins is to prevent the division of a cell that may have incurred DNA damage or harbors abnormalities in chromosome number. As discussed in Chapter 14, when the functions of checkpoint genes are lost due to mutation, this increases the likelihood that undesirable genetic changes will occur that can cause additional mutations and cancerous growth.

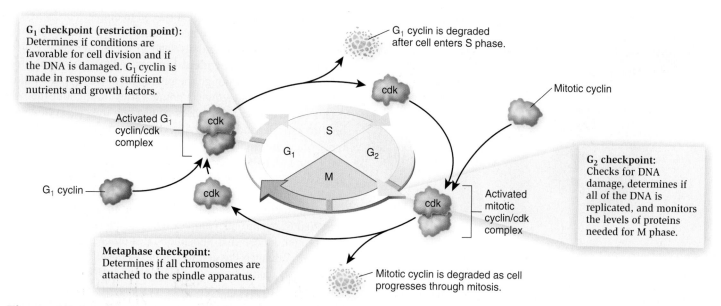

G₁ checkpoint (restriction point): Determines if conditions are favorable for cell division and if the DNA is damaged. G₁ cyclin is made in response to sufficient nutrients and growth factors.

G₁ cyclin is degraded after cell enters S phase.

Mitotic cyclin

Activated G₁ cyclin/cdk complex

G₂ checkpoint: Checks for DNA damage, determines if all of the DNA is replicated, and monitors the levels of proteins needed for M phase.

G₁ cyclin

Activated mitotic cyclin/cdk complex

Metaphase checkpoint: Determines if all chromosomes are attached to the spindle apparatus.

Mitotic cyclin is degraded as cell progresses through mitosis.

Figure 15.4 **Checkpoints in the cell cycle.** This is a general diagram of the eukaryotic cell cycle. Progression through the cell cycle requires the formation of activated cyclin/cdk complexes. Cells make different types of cyclin proteins, which are typically degraded after the cell has progressed to the next phase. The formation of activated cyclin/cdk complexes is regulated by checkpoint proteins.

Concept check: *Why is it beneficial for cells to have checkpoint proteins?*

FEATURE INVESTIGATION

Masui and Markert's Study of Oocyte Maturation Led to the Identification of Cyclins and Cyclin-Dependent Kinases

During the 1960s, researchers were intensely searching for the factors that promote cell division. In 1971, Yoshio Masui and Clement Markert developed a way to test whether a substance causes a cell to progress from one phase of the cell cycle to the next. They chose to study frog oocytes—cells that mature into egg cells. At the time of their work, researchers had already determined that frog oocytes naturally become dormant in the G₂ phase of the cell cycle for up to eight months (Figure 15.5). During mating season, female frogs produce a hormone called progesterone. After progesterone binds to receptors in dormant egg cells, they progress from G₂ to the beginning of M phase, where the chromosomes condense and become visible under the microscope. This phenomenon is called maturation. When a sperm fertilizes the egg, M phase is completed, and the zygote continues to undergo cellular divisions.

Because progesterone is a signaling molecule, Masui and Markert speculated that this hormone affects the functions and/or amounts of proteins that trigger the oocyte to progress through the cell cycle. To test this hypothesis, they developed the procedure described in Figure 15.6, using the oocytes of the leopard frog (*Rana pipiens*). They began by exposing oocytes to progesterone in vitro and then incubating these oocytes for 2 hours or 12 hours. As a control, they also used oocytes that had not been exposed to progesterone. These three types of cells were called the donor oocytes.

Next, they used a micropipette to transfer a small amount of cytosol from the three types of donor oocytes to recipient oocytes that had not been exposed to progesterone. As seen in the data, the recipient oocytes that had been injected with cytosol from the control donor oocytes or from oocytes that had been incubated with progesterone for only 2 hours did not progress to M phase. However, cytosol from donor oocytes that had been incubated with progesterone for 12 hours caused the recipient oocytes to advance to M phase. Masui and Markert concluded that a cytosolic factor, which required more than 2 hours to be synthesized after progesterone treatment, had been

Progesterone

Oocyte dormant in the G₂ phase.

Oocyte advances to the beginning of M phase where chromosomes condense (maturation).

Figure 15.5 Oocyte maturation in certain species of frogs.

Figure 15.6   The experimental approach of Masui and Markert to identify cyclin and cyclin-dependent kinase (cdk).

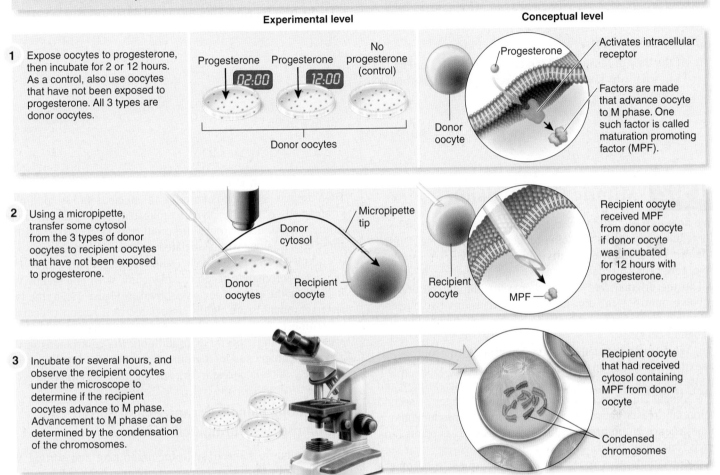

HYPOTHESIS Progesterone induces the synthesis of a factor(s) that advances frog oocytes through the cell cycle from G_2 to M phase.

KEY MATERIALS Oocytes from *Rana pipiens*.

Experimental level | Conceptual level

1 Expose oocytes to progesterone, then incubate for 2 or 12 hours. As a control, also use oocytes that have not been exposed to progesterone. All 3 types are donor oocytes.

Progesterone   Progesterone   No progesterone (control)
02:00   12:00

Donor oocytes

Progesterone — Activates intracellular receptor

Donor oocyte

Factors are made that advance oocyte to M phase. One such factor is called maturation promoting factor (MPF).

2 Using a micropipette, transfer some cytosol from the 3 types of donor oocytes to recipient oocytes that have not been exposed to progesterone.

Donor cytosol — Micropipette tip

Donor oocytes   Recipient oocyte

Recipient oocyte — MPF

Recipient oocyte received MPF from donor oocyte if donor oocyte was incubated for 12 hours with progesterone.

3 Incubate for several hours, and observe the recipient oocytes under the microscope to determine if the recipient oocytes advance to M phase. Advancement to M phase can be determined by the condensation of the chromosomes.

Recipient oocyte that had received cytosol containing MPF from donor oocyte

Condensed chromosomes

4 **THE DATA**

Donor oocytes	Recipient oocytes proceeded to M phase?
Control, no progesterone exposure	No
Progesterone exposure, incubation for 2 hours	No
Progesterone exposure, incubation for 12 hours	Yes

5 **CONCLUSION** Exposure of oocytes to progesterone for 12 hours results in the synthesis of a factor(s) that advances frog oocytes through the cell cycle from G_2 to M phase.

6 **SOURCE** Masui, Y., and Markert, C.L. 1971. Cytoplasmic control of nuclear behavior during meiotic maturation of frog oocytes. *Journal of Experimental Zoology* 177:129–145.

transferred to the recipient oocytes and induced maturation. The factor that caused the oocytes to progress (or mature) from G_2 to M phase was originally called the **maturation promoting factor (MPF)**.

After MPF was discovered in frogs, it was found in all eukaryotic species that researchers studied. MPF is important in the division of all types of cells, not just oocytes. It took another 17 years before Manfred Lohka, Marianne Hayes, and

James Maller were able to purify the components that make up MPF. This was a difficult undertaking because these components are found in very small amounts in the cytosol, and they are easily degraded during purification procedures. We now know that MPF is a complex made of a mitotic cyclin and a cyclin-dependent kinase (cdk), as described in Figure 15.4.

Experimental Questions

1. At the time of Masui and Markert's study shown in Figure 15.6, what was known about the effects of progesterone on oocytes?

2. What hypothesis did Masui and Markert propose to explain the function of progesterone? Explain the procedure used to test the hypothesis.

3. How did the researchers explain the difference between the results using 2-hour-exposed donor oocytes versus 12-hour-exposed donor oocytes?

15.2   Mitotic Cell Division

We now turn our attention to a mechanism of cell division and its relationship to chromosome replication and sorting. During the process of **mitotic cell division**, a cell divides to produce two new cells (the daughter cells) that are genetically identical to the original cell (the mother cell). Mitotic cell division involves mitosis—the division of one nucleus into two nuclei—and then cytokinesis in which the mother cell divides into two daughter cells.

Why is mitotic cell division important? One purpose is **asexual reproduction**. Certain unicellular eukaryotic organisms, such as baker's yeast (*Saccharomyces cerevisiae*) and the amoeba, increase their numbers in this manner. A second important reason for mitotic cell division is the production and maintenance of multicellularity. Organisms such as plants, animals, and most fungi are derived from a single cell that subsequently undergoes repeated cellular divisions to become a multicellular organism. Humans, for example, begin as a single fertilized egg and repeated mitotic cell divisions produce an adult with 10 to 50 trillion cells. As you might imagine, the precise transmission of chromosomes is critical during every cell division so that all cells of the body receive the correct amount of genetic material.

In this section, we will explore how the process of cell division requires the replication, organization, and sorting of chromosomes. We will also examine how a single cell is separated into two distinct cells by cytokinesis.

In Preparation for Cell Division, Eukaryotic Chromosomes Are Replicated and Compacted to Produce Pairs Called Sister Chromatids

As discussed earlier in Figure 15.1, eukaryotic chromosomes are found in sets, and many eukaryotic species are diploid. We will now turn our attention to how those chromosomes are replicated and sorted during cell division. Let's begin with the process of chromosome replication. In Chapter 11, we examined the molecular process of DNA replication. **Figure 15.7** describes the process at the chromosomal level. Prior to DNA replication, the DNA of each eukaryotic chromosome consists of a linear DNA double helix that is found in the nucleus and is not highly compacted. When the DNA is replicated, two identical copies of the original double helix are produced. As discussed earlier, these copies, along with associated proteins, lie side-by-side and are termed sister chromatids. When a cell prepares to divide, the sister chromatids become highly compacted and readily visible under the microscope. As shown in Figure 15.7b, the two sister chromatids are tightly associated at a region called the **centromere**. A protein called cohesin is necessary to hold the sister chromatids together. In addition, the centromere serves as an attachment site for a group of proteins that form the **kinetochore**, which is necessary for sorting each chromosome.

The Mitotic Spindle Organizes and Sorts Chromosomes During Cell Division

What structure is responsible for organizing and sorting the chromosomes during cell division? The answer is the mitotic spindle apparatus, also known simply as the **mitotic spindle** (**Figure 15.8**). It is composed of microtubules—protein fibers that are components of the cytoskeleton (refer back to Table 4.1). In animal cells, microtubule growth and organization starts at two **centrosomes**, regions that are also referred to as microtubule organizing centers (MTOCs). A single centrosome duplicates during interphase. After they separate from each other during mitosis, each centrosome defines a **pole** of the spindle apparatus, one within each of the future daughter cells. The centrosome in animal cells has a pair of **centrioles**. However, centrioles are not found in many other eukaryotic species, such as plants, and are not required for spindle formation.

Each centrosome organizes the construction of the microtubules by rapidly polymerizing tubulin proteins. The three types of spindle microtubules are termed astral, polar, and kinetochore microtubules. The astral microtubules, which extend away from the chromosomes, are important for positioning the spindle apparatus within the cell. The polar microtubules, also called interpolar microtubules, project into the region between the two poles. Polar microtubules that overlap with each other play a role in the separation of the two poles. Finally, the

1 Each chromosome replicates prior to mitosis.

2 At the start of mitosis, the chromosomes become compact.

Sister chromatids

Kinetochore

Centromere (a region of DNA beneath kinetochore proteins)

One chromatid   One chromatid

Pair of sister chromatids

(a) Chromosome replication and compaction

(b) Schematic drawing of a metaphase chromosome

Figure 15.7   Replication and compaction of chromosomes into pairs of sister chromatids.   (a) Chromosomal replication producing a pair of sister chromatids. While the chromosomes are elongated, they are replicated to produce two copies that are connected and lie parallel to each other. This is a pair of sister chromatids. Later, when the cell is preparing to divide, the sister chromatids condense into more compact structures that are easily seen with a light microscope. (b) A schematic drawing of a metaphase chromosome. This structure has two chromatids that lie side-by-side. The two chromatids are held together by cohesin proteins (not shown in this drawing). The kinetochore is a group of proteins that are attached to the centromere and play a role during chromosome sorting.

Concept check: *Look back at the karyotype in Figure 15.1. In this micrograph, is each of the 46 objects a pair of sister chromatids?*

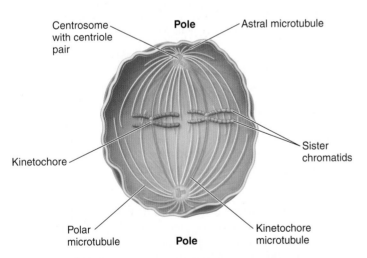

Centrosome with centriole pair

Pole

Astral microtubule

Kinetochore

Sister chromatids

Polar microtubule

Pole

Kinetochore microtubule

Figure 15.8   The structure of the mitotic spindle.   The mitotic spindle in animal cells is formed by the centrosomes, which produce three types of microtubules. The astral microtubules emanate away from the region between the poles. The polar microtubules project into the region between the two poles. The kinetochore microtubules are attached to the kinetochores of sister chromatids. Note: For simplicity, this diagram just shows one pair of homologous chromosomes. Eukaryotic species typically have multiple chromosomes per set.

Concept check: *What are the functions of the three types of microtubules?*

kinetochore microtubules are attached to kinetochores, which are bound to the centromere of each chromosome.

The Transmission of Chromosomes Requires a Sorting Process Known as Mitosis

Mitosis is the sorting process to divide one cell nucleus into two nuclei. The duplicated chromosomes are distributed so that each daughter cell will receive the same complement of chromosomes. Mitosis was first observed microscopically in the 1870s by a German biologist, Walter Flemming, who coined the term mitosis (from the Greek *mitos*, meaning thread). He studied the large, transparent skin cells of salamander larvae as they were dividing and noticed that chromosomes are constructed of "threads" that are doubled in appearance along their length. These double threads divided and moved apart, one going to each of the two daughter nuclei. By this mechanism, Flemming pointed out, the two daughter cells receive an identical group of threads, the same as the number of threads in the mother cell.

Figure 15.9 depicts the process of mitosis in an animal cell, though the process is quite similar in a plant cell. Mitosis occurs as a continuum of phases known as prophase, prometaphase, metaphase, anaphase, and telophase. In the simplified diagrams shown along the bottom of Figure 15.9, the original mother cell contains six chromosomes. One set of chromosomes is depicted in red, whereas the homologous set is blue. These different colors represent maternal and paternal chromosomes.

(a) Interphase   **(b) Prophase**   **(c) Prometaphase**

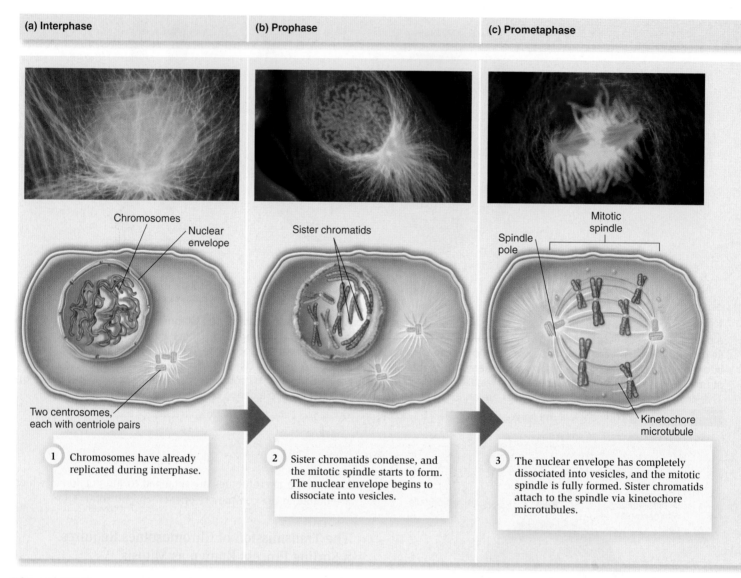

Figure 15.9 **The process of mitosis in an animal cell.** The top panels illustrate the cells of a newt progressing through mitosis. The bottom panels are schematic drawings that emphasize the sorting and separation of the chromosomes in which the diploid mother cell had six chromosomes (three in each set). At the start of mitosis, these have already replicated into 12 chromatids. The final result is two daughter cells, each containing six chromosomes.

Concept check: *With regard to chromosome composition, how does the mother cell compare to the two daughter cells?*

Interphase Prior to mitosis, the cells are in **interphase**. As discussed earlier, interphase includes the G_1, S, and G_2 phases of the cell cycle. During interphase, the chromosomes are decondensed and found in the nucleus (Figure 15.9a).

Prophase At the start of mitosis, in **prophase**, the chromosomes have already replicated to produce 12 chromatids, joined as six pairs of sister chromatids (Figure 15.9b). As prophase proceeds, the nuclear envelope begins to dissociate into small vesicles. At the same time, the chromatids condense into highly compacted structures that are readily visible by light microscopy.

Prometaphase The nuclear envelope completely fragments into small vesicles, and the mitotic spindle is fully formed

during **prometaphase** (Figure 15.9c). As mitosis progresses, the centrosomes move apart and demarcate the two poles. Once the nuclear envelope has dissociated, the spindle fibers can interact with the sister chromatids. How do the sister chromatids become attached to the spindle apparatus? Initially, microtubules are rapidly formed and can be seen under a microscope growing out from the two poles. As it grows, if a microtubule happens to make contact with a kinetochore, it is said to be "captured" and remains firmly attached to the kinetochore. Alternatively, if a microtubule does not collide with a kinetochore, the microtubule will eventually depolymerize and retract to the centrosome. This random process is how sister chromatids become attached to kinetochore microtubules. As the end of prometaphase nears, the two kinetochores on each pair of sister chromatids are attached to kinetochore microtubules

(d) Metaphase

Metaphase plate

4 Sister chromatids align along the metaphase plate.

(e) Anaphase

Individual chromosomes

Polar microtubule

5 Sister chromatids separate, and individual chromosomes move toward the poles as kinetochore microtubules shorten. Polar microtubules lengthen and push the poles apart.

(f) Telophase and cytokinesis

Cleavage furrow

6 Chromosomes decondense, and the nuclear envelope re-forms. Cytokinesis separates the mother cell into two daughter cells, which begins with a cleavage furrow in animal cells.

from opposite poles. As these events are occurring, the sister chromatids are seen under the microscope to undergo jerky movements as they are tugged, back and forth, between the two poles by the kinetochore microtubules.

Metaphase Eventually, the pairs of sister chromatids are aligned in a single row along the **metaphase plate**, a plane halfway between the poles. When this alignment is complete, the cell is in **metaphase** of mitosis (Figure 15.9d). The chromatids can then be equally distributed into two daughter cells.

Anaphase The next step in the sorting process occurs during **anaphase** (Figure 15.9e). At this phase, the connections between the pairs of sister chromatids are broken. Each chromatid, now an individual chromosome, is linked to only one of the two poles by one or more kinetochore microtubules. As anaphase proceeds, the kinetochore microtubules shorten, pulling the chromosomes toward the pole to which they are attached. In addition, the two poles move farther away from each other. This occurs because the overlapping polar micro-

tubules lengthen and push against each other, thereby pushing the poles farther apart.

Telophase During **telophase**, the chromosomes have reached their respective poles and decondense. The nuclear envelope now re-forms to produce two separate nuclei. In Figure 15.9f, two nuclei are being produced that contain six chromosomes each.

Cytokinesis In most cases, mitosis is quickly followed by cytokinesis, in which the two nuclei are segregated into separate daughter cells. While the phases of mitosis are similar between plant and animal cells, the process of cytokinesis is quite different. In animal cells, cytokinesis involves the formation of a **cleavage furrow**, which constricts like a drawstring to separate the cells (Figure 15.10a). In plants, vesicles from the Golgi apparatus move along microtubules to the center of the cell and coalesce to form a **cell plate** (Figure 15.10b), which then forms a cell wall between the two daughter cells.

What are the results of mitosis and cytokinesis? These processes ultimately produce two daughter cells having the same

(a) Cleavage of an animal cell

(b) Formation of a cell plate in a plant cell

Figure 15.10   Micrographs showing cytokinesis in animal and plant cells.

Concept check: *What are the similarities and differences between cytokinesis in animal and plant cells?*

number of chromosomes as the mother cell. Barring rare mutations, the two daughter cells are genetically identical to each other and to the mother cell from which they were derived. Thus, the critical consequence of this sorting process is to ensure genetic consistency from one cell to the next. The development of multicellularity relies on the repeated process of mitosis and cytokinesis. For diploid organisms that are multicellular, most of the somatic cells are diploid and genetically identical to each other.

Genomes & Proteomes Connection

The Genomes of Diverse Animal Species Encode Approximately 20 Proteins That Are Involved in Cytokinesis

To understand how a process works at the molecular and cellular level, researchers often try to identify the genes within a given species that encode proteins necessary for the process. Cytokinesis has been analyzed in this way. By comparing the results from vertebrates, insects, and worms, researchers have

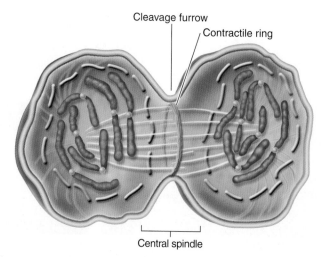

Figure 15.11   A closer look at cytokinesis in animal cells.

identified approximately 20 proteins that are involved with cytokinesis in nearly all animal cells. In any given species, cytokinesis may also involve additional proteins beyond these 20, but these other proteins are not needed among all animal species. Evolutionary biologists would say the 20 proteins that are common to all animals are highly conserved, meaning their structure and function has been retained during the evolution of animals. The 20 conserved proteins are likely to play the most fundamental roles in the process of cytokinesis.

What are the functions of these 20 proteins? To appreciate their functions, we need to take a closer look at cytokinesis in animal cells (**Figure 15.11**). Animal cells produce a contractile ring that is attached to the plasma membrane to create the cleavage furrow. The contractile ring, which encircles a region of the mitotic spindle called the central spindle, is a network of actin (a cytoskeletal protein) and myosin (a motor protein). The motor activity of myosin moves actin filaments in a way that causes the contractile ring to constrict. Once the contractile ring becomes very small, membrane vesicles are inserted into the constricted site to achieve division of the plasma membranes in the two resulting cells.

The 20 conserved proteins perform one of four possible functions.

1. Contractile ring: Seven proteins, including actin, myosin, and other proteins that regulate actin and myosin function, are necessary for the formation of the contractile ring.

2. Central spindle: Eight proteins are known to be components that bind to the central spindle and are necessary for cytokinesis.

3. Cell separation via membrane insertion: Two proteins are needed for the final separation of the two daughter cells.

4. RhoA pathway: Five proteins are components of a cell signal transduction pathway called the RhoA pathway. This pathway initiates the formation of the contractile ring.

The 20 conserved proteins should be considered a minimum estimate. As we gain a deeper understanding of cytokinesis at the molecular level, it is likely that additional proteins may be discovered.

15.3 Meiosis and Sexual Reproduction

We now turn our attention to sexual reproduction. As discussed earlier, a diploid cell contains two homologous sets of chromosomes, whereas a haploid cell contains a single set. For example, a diploid human cell contains 46 chromosomes, but a human gamete—sperm or egg cell—is a haploid cell that contains only 23 chromosomes, one from each of the 23 pairs. **Sexual reproduction** requires a **fertilization** event in which two haploid gametes unite to form a diploid cell called a **zygote**. For multicellular species such as animals and plants, the zygote then grows and divides by mitotic cell divisions into a multicellular organism with many diploid cells.

Meiosis is the process by which haploid cells are produced from a cell that was originally diploid. The term meiosis, which means "to make smaller," refers to the fewer chromosomes found in cells following this process. For this to occur, the chromosomes must be correctly sorted and distributed in a way that reduces the chromosome number to half its original diploid value. In the case of human gametes, for example, each gamete must receive half the total number of chromosomes, but not just any 23 chromosomes will do. A gamete must receive one chromosome from each of the 23 pairs. For this to happen, two rounds of divisions are necessary, termed meiosis I and meiosis II (**Figure 15.12**). When a cell begins meiosis, it contains chromosomes that are found in homologous pairs. When meiosis is completed, a single diploid cell with homologous pairs of chromosomes has produced four haploid cells.

In this section, we will examine the cellular events of meiosis that reduce the chromosome number from diploid to haploid. In addition, we will briefly consider how this process plays a role in the life cycles of animals, plants, fungi, and protists.

Bivalent Formation and Crossing Over Occurs at the Beginning of Meiosis

Like mitosis, meiosis begins after a cell has progressed through the G_1, S, and G_2 phases of the cell cycle. However, two key events occur at the beginning of meiosis that do not occur in mitosis. First, homologous pairs of sister chromatids associate with each other, lying side by side to form a **bivalent**, also called a **tetrad** (**Figure 15.13**). The process of forming a bivalent is termed **synapsis**. In most eukaryotic species, a protein structure called the synaptonemal complex connects homologous chromosomes during a portion of meiosis. However, the synaptonemal complex is not required for the pairing of homologous chromosomes because some species of fungi completely lack such a complex, yet their chromosomes associate with each other correctly. At present, the precise role of the synaptonemal complex is not clearly understood.

The second event that occurs at the beginning of meiosis, but not usually during mitosis, is **crossing over**, which involves a physical exchange between chromosome segments of the bivalent (Figure 15.13). As discussed in Chapter 17, crossing over

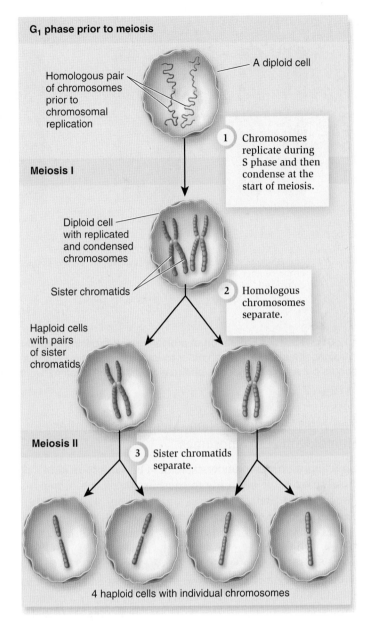

Figure 15.12 How the process of meiosis reduces chromosome number. This simplified diagram emphasizes the reduction in chromosome number as a diploid cell divides by meiosis to produce four haploid cells.

may increase the genetic variation of a species. After crossing over occurs, the arms of the chromosomes tend to separate but remain adhered at a crossover site. This connection is called a **chiasma** (plural, chiasmata), because it physically resembles the Greek letter chi, χ. The number of crossovers is carefully controlled by cells and depends on the size of the chromosome and the species. The range of crossovers for eukaryotic chromosomes is typically one or two to a couple dozen. During the formation of sperm in humans, for example, an average chromosome undergoes slightly more than two crossovers, whereas chromosomes in certain plant species may undergo 20 or more crossovers.

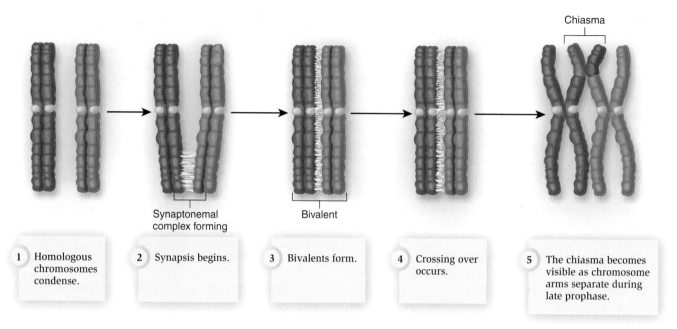

Chiasma

Synaptonemal complex forming

Bivalent

1 Homologous chromosomes condense.

2 Synapsis begins.

3 Bivalents form.

4 Crossing over occurs.

5 The chiasma becomes visible as chromosome arms separate during late prophase.

Figure 15.13 **Formation of a bivalent and crossing over during meiosis I.** At the beginning of meiosis, homologous chromosomes pair with each other to form a bivalent, usually with a synaptonemal complex between them. Crossing over then occurs between homologous chromatids within the bivalent. During this process, homologues exchange segments of chromosomes.

The First Meiotic Division, Meiosis I, Separates Homologous Chromosomes

Now that we have an understanding of bivalent formation and crossing over, we are ready to consider the phases of meiosis (**Figure 15.14**, look two pages ahead to pp. 318 and 319). These simplified diagrams depict a diploid cell (2n) that contains a total of six chromosomes (as in our look at mitosis in Figure 15.9). Prior to meiosis, the chromosomes are replicated in S phase to produce pairs of sister chromatids. This single replication event is then followed by the sequential divisions called meiosis I and II. Like mitosis, each of these is a continuous series of stages called prophase, prometaphase, metaphase, anaphase, and telophase. The sorting that occurs during **meiosis I** separates homologues from each other (Figure 15.14).

Prophase I In prophase I, the replicated chromosomes condense, the homologous chromosomes form bivalents, and crossing over occurs. The nuclear envelope then starts to fragment into small vesicles.

Prometaphase I In prometaphase I, the nuclear envelope is completely broken down into vesicles, and the spindle apparatus is entirely formed. The sister chromatids are attached to kinetochore microtubules. However, a key difference occurs between mitosis and meiosis. In mitosis, a pair of sister chromatids is attached to both poles (see Figure 15.9c). In meiosis, a pair of sister chromatids is attached to just one pole via kinetochore microtubules (Figure 15.14b).

Metaphase I At metaphase I, the bivalents are organized along the metaphase plate. Notice how this pattern of alignment is strikingly different from that observed during mitosis (see Figure 15.9d). In particular, the sister chromatids are aligned in a double row rather than a single row (as in mitosis). Furthermore, the arrangement of sister chromatids within this double row is random with regard to the (red and blue) homologues. In Figure 15.14c, one of the red homologues is to the left of the metaphase plate, and the other two are to the right, while two of the blue homologues are to the left of the metaphase plate and the other one is to the right. In other cells, homologues could be arranged differently along the metaphase plate (for example, three blues to the left and none to the right, or none to the left and three to the right).

Because eukaryotic species typically have many chromosomes per set, homologues can be randomly aligned along the metaphase plate in a variety of ways. For example, consider that humans have 23 chromosomes per set. The possible number of different, random alignments equals 2^n, where n equals the number of chromosomes per set. Thus, in humans, this equals 2^{23}, or over 8 million possibilities. Because the homologues are genetically similar but not identical, we see from this calculation that the random alignment of homologous chromosomes provides a mechanism to promote a vast amount of genetic diversity among the resulting haploid cells. When meiosis is complete, it is very unlikely that any two human gametes will have the same combination of homologous chromosomes.

Anaphase I The segregation of homologues occurs during anaphase I (Figure 15.14d). The connections between bivalents break, but not the connections that hold sister chromatids together. Each joined pair of chromatids migrates to one pole, and the homologous pair of chromatids moves to the opposite pole, both pulled by kinetochore microtubules.

Telophase I   At telophase I, the sister chromatids have reached their respective poles, and they then decondense. The nuclear envelope now re-forms to produce two separate nuclei.

If we consider the end result of meiosis I, we see that two nuclei are produced, each with three pairs of sister chromatids; this is called a reduction division. The original diploid cell had its chromosomes in homologous pairs, whereas the two cells produced as a result of meiosis I and cytokinesis are considered haploid—they do not have pairs of homologous chromosomes.

The Second Meiotic Division, Meiosis II, Separates Sister Chromatids

Meiosis I is followed by cytokinesis and then **meiosis II** (see Figure 15.14). An S phase does not occur between meiosis I and meiosis II. The sorting events of meiosis II are similar to those of mitosis, but the starting point is different. For a diploid cell with six chromosomes, mitosis begins with 12 chromatids that are joined as six pairs of sister chromatids (see Figure 15.9). By comparison, the two cells that begin meiosis II each have six chromatids that are joined as three pairs of sister chromatids. Otherwise, the steps that occur during prophase, prometaphase, metaphase, anaphase, and telophase of meiosis II are analogous to a mitotic division. Sister chromatids are separated during anaphase II, unlike anaphase I in which bivalents are separated.

Changes in a Few Key Steps in Mitosis and Meiosis Account for the Different Outcomes of These Two Processes

How are the outcomes of mitosis and meiosis different from each other? Mitosis produces two diploid daughter cells that are genetically identical. In our example shown in Figure 15.9, the

starting cell had six chromosomes (three homologous pairs of chromosomes), and both daughter cells had copies of the same six chromosomes. By comparison, meiosis reduces the number of sets of chromosomes. In the example shown in Figure 15.14, the starting cell also had six chromosomes, whereas the four daughter cells had only three chromosomes. However, the daughter cells did not contain a random mix of three chromosomes. Each haploid daughter cell contained one complete set of chromosomes, whereas the original diploid mother cell had two complete sets.

How do we explain the different outcomes of mitosis and meiosis? Table 15.1 emphasizes the differences between certain key steps in mitosis and meiosis that account for the different outcomes of these two processes. During prophase of meiosis I, the homologues synapse to form bivalents. This explains why crossing over occurs commonly during meiosis, but rarely during mitosis. During prometaphase of mitosis and meiosis II, pairs of sister chromatids are attached to both poles. In contrast, during meiosis I, each pair of sister chromatids (within a bivalent) is attached to a single pole. This affects their alignment during metaphase. Bivalents align along the metaphase plate during metaphase of meiosis I, whereas sister chromatids align along the metaphase plate during metaphase of mitosis and meiosis II. At anaphase of meiosis I, the homologous chromosomes separate, while the sister chromatids remain together. In contrast, sister chromatid separation occurs during anaphase of mitosis and meiosis II. Taken together, the steps of mitosis produce two diploid cells, whereas the steps of meiosis involve two sequential cell divisions that produce four haploid cells.

Sexually Reproducing Species Produce Haploid and Diploid Cells at Different Times in Their Life Cycles

Let's now turn our attention to the relationship between mitosis, meiosis, and sexual reproduction in animals, plants, fungi, and protists. For any given species, the sequence of events that produces another generation of organisms is known as a **life cycle**. For sexually reproducing organisms, this usually involves

Table 15.1	A Comparison of Mitosis, Meiosis I, and Meiosis II		
Event	**Mitosis**	**Meiosis I**	**Meiosis II**
Synapsis during prophase:	No	Yes, bivalents are formed.	No
Crossing over during prophase:	Rarely	Commonly	Rarely
Attachment to poles at prometaphase:	A pair of sister chromatids is attached to kinetochore microtubules from both poles.	A pair of sister chromatids is attached to kinetochore microtubules from just one pole.	A pair of sister chromatids is attached to kinetochore microtubules from both poles.
Alignment along the metaphase plate:	Sister chromatids align.	Bivalents align.	Sister chromatids align.
Type of separation at anaphase:	Sister chromatids separate. A single chromatid, now called a chromosome, moves to each pole.	Homologous chromosomes separate. A pair of sister chromatids moves to each pole.	Sister chromatids separate. A single chromatid, now called a chromosome, moves to each pole.
End result when the mother cell is diploid:	Two daughter cells that are diploid	—	Four daughter cells that are haploid

Meiosis I

(a) Prophase I

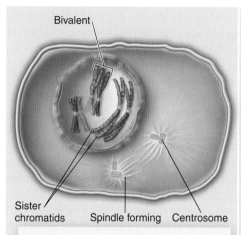

Bivalent

Sister chromatids   Spindle forming   Centrosome

① Homologous chromosomes synapse to form bivalents, and crossing over occurs. Chromosomes condense, and the nuclear envelope begins to dissociate into vesicles.

(b) Prometaphase I

Bivalent

② The nuclear envelope completely dissociates into vesicles, and bivalents become attached to kinetochore microtubules.

(c) Metaphase I

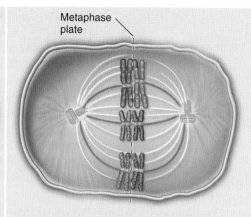

Metaphase plate

③ Bivalents randomly align along the metaphase plate.

Meiosis II

(f) Prophase II

(g) Prometaphase II

(h) Metaphase II

⑥ Sister chromatids condense, and the spindle starts to form. The nuclear envelope begins to dissociate into vesicles.

⑦ The nuclear envelope completely dissociates into vesicles. Sister chromatids attach to the spindle via kinetochore microtubules.

⑧ Sister chromatids align along the metaphase plate.

(d) Anaphase I

(e) Telophase I and cytokinesis

Cleavage furrow

4 Homologous chromosomes separate and move toward opposite poles.

5 The chromosomes decondense, and the nuclear envelope re-forms. The 2 daughter cells are separated by a cleavage furrow.

(i) Anaphase II

(j) Telophase II and cytokinesis

Four haploid cells

9 Sister chromatids separate, and individual chromosomes move toward the poles as kinetochore microtubules shorten. Polar microtubules lengthen and push the poles apart.

10 Chromosomes decondense, and the nuclear envelope re-forms. Cleavage furrows separate the 2 cells into 4 cells.

Figure 15.14 The phases of meiosis in an animal cell.

Concept check:
Relative to the original mother cell, what is the end result of meiosis?

an alternation between haploid cells or organisms and diploid cells or organisms (**Figure 15.15**).

Most species of animals are diploid, and their haploid gametes are considered to be a specialized type of cell. For this reason, animals are viewed as **diploid-dominant species** (Figure 15.15a). Certain diploid cells in the testes or ovaries undergo meiosis to produce haploid sperm or eggs, respectively. During fertilization, sperm and egg unite to form a diploid zygote, which then undergoes repeated mitotic cell divisions to produce a diploid multicellular organism.

By comparison, most fungi and some protists are just the opposite; they are **haploid-dominant species** (Figure 15.15b). In fungi, the multicellular organism is haploid ($1n$). Haploid fungal cells are most commonly produced by mitosis. During sexual reproduction, haploid cells unite to form a diploid zygote, which then immediately proceeds through meiosis to produce four haploid cells called spores. Each spore goes through mitotic cellular divisions to produce a haploid multicellular organism.

Plants and some algae have life cycles that are intermediate between the extreme cases of diploid or haploid dominance. Such

KEY
■ Diploid ($2n$)
■ Haploid ($1n$)

1 Meiosis occurs in cells within testes or ovaries to produce haploid gametes.

2 During fertilization, sperm and egg unite to form a diploid zygote.

Egg ($1n$)
Sperm ($1n$)
Diploid adult ($2n$)
Somatic cells are diploid ($2n$).
Diploid zygote ($2n$)

3 Repeated mitotic cell divisions produce a diploid multicellular organism.

(a) Animal life cycle (mouse)—diploid dominant

4 Repeated mitotic cell divisions produce a haploid multicellular organism.

1 Certain haploid cells act as reproductive cells.

Haploid multicellular organism
Reproductive cells ($1n$)
Somatic cells are haploid ($1n$).

2 Haploid reproductive cells unite to form a diploid zygote.

Diploid zygote ($2n$)
Spore ($1n$)

3 Meiosis of the zygote produces 4 haploid spores.

(b) Fungal life cycle (bread mold)—haploid dominant

1 Certain cells in the diploid sporophyte undergo meiosis to produce haploid spores.

2 Repeated mitotic cell divisions produce a haploid multicellular organism (gametophyte).

Diploid plant sporophyte ($2n$)
Haploid spores ($1n$)
Somatic cells are diploid ($2n$).
Haploid plant gametophyte ($1n$)
Somatic cells are haploid ($1n$).
Diploid zygote ($2n$)
Sperm ($1n$)

5 Repeated mitotic cell divisions produce a diploid multicellular organism (sporophyte).

3 Certain cells within the gametophyte differentiate into gametes.

Egg ($1n$)

4 Two gametes unite during fertilization to form a diploid zygote.

(c) Plant life cycle (fern)—alternation of generations

Figure 15.15   A comparison of three types of sexual life cycles.

Concept check: *What is the main reason for meiosis in animals? What is the main reason for mitosis in animals?*

species exhibit an **alternation of generations** (Figure 15.15c). The species alternate between diploid multicellular organisms called **sporophytes**, and haploid multicellular organisms called **gametophytes**. Meiosis in certain cells within the sporophyte produces haploid spores, which divide by mitosis to produce the gametophyte. Particular cells within the gametophyte differentiate into haploid gametes. Fertilization occurs between two gametes, producing a diploid zygote that then undergoes repeated mitotic cell divisions to produce a sporophyte.

Among different plant species, the relative sizes of the haploid and diploid organisms vary greatly. In mosses, the haploid gametophyte is a visible multicellular organism, whereas the diploid sporophyte is smaller and survives within the haploid organism. In other plants, such as ferns (Figure 15.15c), both the diploid sporophyte and haploid gametophyte can grow independently. The sporophyte is considerably larger and is the organism we commonly think of as a fern. In seed-bearing plants, such as roses and oak trees, the diploid sporophyte is the large multicellular plant, whereas the gametophyte is composed of only a few cells and is formed within the sporophyte.

When comparing animals, plants, and fungi, it's interesting to consider how gametes are made. Animals produce gametes by meiosis. In contrast, plants and fungi produce reproductive cells by mitosis. The gametophyte of plants is a haploid multicellular organism that is created by mitotic cellular divisions of a haploid spore. Within the multicellular gametophyte, certain cells become specialized as gametes.

15.4 Variation in Chromosome Structure and Number

In the previous sections of this chapter, we have examined two important features of chromosomes. First, we considered how chromosomes occur in sets, and second, we explored two sorting processes that determine the chromosome number following cell division. In this section, we will examine how the structures and numbers of chromosomes can vary between different species and within the same species.

Why is the study of chromosomal variation important? First, geneticists have discovered that variations in chromosome structure and number can have major effects on the characteristics of an organism. For example, we now know that several human genetic diseases are caused by such changes. In addition, changes in chromosome structure and number have been an important force in the evolution of new species, which is a topic we will consider in Chapter 25.

Chromosome variation can be viewed in two ways. On relatively rare occasions, the structure or number of chromosomes changes so that an individual is different from most other members of the same species. This is generally viewed as an abnormality. Alternatively, the structure and number of chromosomes among different species tend to show wide variation, which is normal. In this section, we will examine both abnormal and normal types of variation. Let's begin with natural (normal) variation.

Natural Variation Exists in Chromosome Structure and Number

Before we begin to examine chromosome variation, we need to have a reference point for a normal set of chromosomes. To determine what the normal chromosomes of a species look like, a cytogeneticist microscopically examines the chromosomes from several members of the species. Chromosome composition within a given species tends to remain relatively constant. In most cases, normal individuals of the same species will have the same number and types of chromosomes. For example, as mentioned previously, the normal chromosome composition of human cells is two sets of 23 chromosomes, for a total of 46. Other diploid species may have different numbers of chromosomes. The dog has 78 chromosomes (39 per set), the fruit fly has 8 chromosomes (4 per set), and the tomato has 24 chromosomes (12 per set). When comparing distantly related species, such as humans and fruit flies, major differences in chromosomal composition are observed.

The chromosomes of a given species can also vary considerably in size and shape. Cytogeneticists have various ways to classify and identify chromosomes in their metaphase form. The three most commonly used features are size, location of the centromere, and banding patterns that are revealed when the chromosomes are treated with stains. Based on centromere location, each chromosome is classified as **metacentric** (near the middle), **submetacentric** (off center), **acrocentric** (near one end), or **telocentric** (at the end) (**Figure 15.16**). Because the centromere is not exactly in the center of a chromosome, each chromosome has a short arm and a long arm. The short arm is designated with the letter p (for the French *petite*), while the long arm is designated with the letter q. In the case of telocentric chromosomes, the short arm may be nearly nonexistent. When preparing a karyotype, the chromosomes are aligned with the short arms on top and the long arms on the bottom.

Because different chromosomes often have similar sizes and centromeric locations, cytogeneticists must use additional methods to accurately identify each type of chromosome within a karyotype. For detailed identification, chromosomes are treated with stains to produce characteristic banding patterns. Cytogeneticists use several different staining procedures to identify specific chromosomes. An example is Giemsa stain, which produces G banding (see Figure 15.1). The alternating pattern of G bands is unique for each type of chromosome.

Metacentric   Submetacentric   Acrocentric   Telocentric

Figure 15.16 A comparison of centromeric locations among metaphase chromosomes.

The banding pattern of eukaryotic chromosomes is useful in two ways. First, individual chromosomes can be distinguished from each other, even if they have similar sizes and centromeric locations. Also, banding patterns are used to detect changes in chromosome structure that occur as a result of mutation.

Mutations Can Alter Chromosome Structure

Let's now consider how the structures of chromosomes can be modified by a mutation, a heritable change in the genetic material. Chromosomal mutations are categorized as deletions, duplications, inversions, and translocations (**Figure 15.17**).

Deletions and duplications are changes in the total amount of genetic material in a single chromosome. When a **deletion** occurs, a segment of chromosomal material is missing. In other words, the affected chromosome is deficient in a significant amount of genetic material. In a **duplication**, a section of a chromosome occurs two or more times in a row.

What are the consequences of a deletion or duplication? Their possible effects depend on their size and whether they include genes or portions of genes that are vital to the development of the organism. When deletions or duplications have an effect, they are usually detrimental. Larger changes in the amount of genetic material tend to be more harmful because more genes are missing or duplicated.

Inversions and translocations are chromosomal rearrangements. An **inversion** is a change in the direction of the genetic material along a single chromosome. When a segment of one chromosome has been inverted, the order of G bands is opposite to that of a normal chromosome (see Figure 15.17c). A **translocation** occurs when one segment of a chromosome becomes attached to a different chromosome. In a **simple translocation**, a single piece of chromosome is attached to another chromosome. In a **reciprocal translocation**, two different types of chromosomes exchange pieces, thereby producing two abnormal chromosomes carrying translocations (see Figure 15.17d).

Variation Occurs in the Number of Chromosome Sets and the Number of Individual Chromosomes

Variations in chromosome number can be categorized in two ways: variation in the number of sets of chromosomes and variation in the number of particular chromosomes within a set. The suffix -ploid or -ploidy refers to a complete set of chromosomes. Organisms that are **euploid** (the prefix eu- means true) have chromosomes that occur in one or more complete sets. For example, in a species that is diploid, a euploid organism would have two sets of chromosomes in its somatic cells. In *Drosophila melanogaster*, for example, a normal individual has eight chromosomes. The species is diploid, having two sets of four chromosomes each (**Figure 15.18a**). Organisms can vary with regard to the number of sets of chromosomes they have. For example, on rare occasions, an abnormal fruit fly can be produced with 12 chromosomes, containing three sets of 4 chromosomes each (**Figure 15.18b**). Organisms with three or more sets of chromosomes are called **polyploid**. A diploid

(a) Deletion

(b) Duplication

(c) Inversion

(d) Simple translocation

(e) Reciprocal translocation

Figure 15.17 Types of changes in chromosome structure. The letters alongside the chromosomes are placed there as frames of reference.

Concept check: Which types of changes shown here do not affect the total amount of genetic material?

organism is referred to as $2n$, a **triploid** organism as $3n$, a **tetraploid** organism as $4n$, and so forth. All such organisms are euploid.

A second way that chromosome number can vary is a phenomenon called **aneuploidy**. This refers to an alteration in the number of particular chromosomes, so the total number of chromosomes is not an exact multiple of a set. For example, an abnormal fruit fly could contain nine chromosomes instead of eight because it had three copies of chromosome 2 instead of the normal two copies (**Figure 15.18c**). Such an animal is said to have trisomy 2 or to be **trisomic**. Instead of being perfectly diploid, a trisomic animal is $2n + 1$. By comparison, a fruit fly could be lacking a single chromosome, such as chromosome 3, and contain a total of seven chromosomes ($2n - 1$). This animal is **monosomic** and would be described as having monosomy 3.

(a) Normal fruit fly chromosome composition

(b) Polyploidy

(c) Aneuploidy

Figure 15.18   Types of variation in chromosome number.
(a) The normal diploid number of chromosomes in *Drosophila*. The X chromosome is also called chromosome 1. Examples of chromosomes of **(b)** polyploid flies and **(c)** aneuploid flies.

Variations in chromosome number are fairly widespread and have a significant impact on the characteristics of plants and animals. For these reasons, researchers have wanted to understand the mechanisms that cause these variations. In some cases, a change in chromosome number is the result of the abnormal sorting of chromosomes during cell division. The term **nondisjunction** refers to an event in which the chromosomes do not separate properly during cell division. Nondisjunction can occur during meiosis I or meiosis II and produces haploid cells that have too many or too few chromosomes. If such a cell becomes a gamete that fuses with a another gamete during fertilization, the zygote and the resulting organism will have an abnormal number of chromosomes in all of its cells.

Changes in Chromosome Number Have Important Consequences

How do changes in chromosome number affect the characteristics of animals and plants? In many cases, animals do not tolerate deviations from diploidy well. For example, polyploidy in mammals is generally a lethal condition. However, a few cases of naturally occurring variations from diploidy do occur in animals. Male bees, which are called drones, contain a single set of chromosomes and are therefore haploid. They are produced from unfertilized eggs. By comparison, fertilized eggs become female bees, which are diploid. A few examples of vertebrate polyploid animals have been discovered. Interestingly, on rare occasions, animals that are morphologically very similar to each other can be found as a diploid species as well as a separate polyploid species. This situation occurs among certain amphibians and reptiles. **Figure 15.19** shows photographs of a diploid and a tetraploid frog. As you can see, they look very similar to each other. Their differences in chromosome number can be revealed only by a microscopic examination of the chromosomes in the somatic cells of the animals.

In contrast to animals, plants commonly exhibit polyploidy. Among ferns and flowering plants, about 30–35% of species are polyploid. Polyploidy is also important in agriculture. In many instances, polyploid strains of plants display characteristics that are helpful to humans. They are often larger in size and more robust. These traits are clearly advantageous in the production of food. Many of the fruits and grains we eat are produced from polyploid plants. For example, the species of wheat that we use to make bread, *Triticum aestivum*, is a hexaploid (containing six sets of chromosomes) that arose from the union of diploid genomes from three closely related species (**Figure 15.20a**). During the course of its cultivation, two diploid species must have interbred to produce a tetraploid, and then a third species interbred with the tetraploid to produce a hexaploid. Plant polyploids tend to exhibit a greater adaptability, which allows them to withstand harsher environmental conditions. Polyploid ornamental plants commonly produce larger flowers than their diploid counterparts (**Figure 15.20b**).

Although polyploidy is often beneficial in plants, aneuploidy in all eukaryotic species usually has detrimental consequences on the characteristics of an organism. Why is aneuploidy usually

(a) *Hyla chrysoscelis* (diploid)

(b) *Hyla versicolor* (tetraploid)

Figure 15.19 Differences in chromosome number in two closely related frog species. The frog in **(a)** is diploid, whereas the frog in **(b)** is tetraploid. These frogs are in the act of performing their mating calls, which is why the skin under their mouths is protruding as a large bubble.

(a) Wheat, *Triticum aestivum* (hexaploid)

(b) Diploid daylily (left) and tetraploid daylily (right)

Figure 15.20 Examples of polyploid plants. **(a)** Cultivated wheat, *Triticum aestivum*, is a hexaploid. It was derived from three different diploid species of grasses that originally were found in the Middle East and were cultivated by ancient farmers in that region. Modern varieties of wheat have been produced from this hexaploid species. **(b)** Differences in euploidy exist in these two closely related daylily species. The flower stems on the left are diploid, whereas the stems with the larger flowers on the right are tetraploid.

detrimental? To answer this question, we need to consider the relationship between gene expression and chromosome number. For many, but not all genes, the level of gene expression is correlated with the number of genes per cell. Compared to a diploid cell, if a gene is carried on a chromosome that is present in three copies instead of two, approximately 150% of the normal amount of gene product will be made. Alternatively, if only one copy of that gene is present due to a missing chromosome, only 50% of the gene product is usually made. For some genes, producing too much or too little of the gene product may not have adverse effects. However, for other genes, the over- or underexpression may interfere with the proper functioning of cells.

One important reason that geneticists are so interested in aneuploidy is its relationship to certain inherited disorders in humans. Even though most people are born with a normal number of chromosomes, alterations in chromosome number occur at a surprising frequency during gamete formation. About 5–10% of all fertilized human eggs result in an embryo with an abnormality in chromosome number. In most cases, these abnormal embryos do not develop properly and result in a spontaneous abortion very early in pregnancy. Approximately 50% of all spontaneous abortions are due to alterations in chromosome number.

In some cases, an abnormality in chromosome number produces an offspring that can survive. Several human disorders are the result of abnormalities in chromosome number.

The most common are trisomies of chromosomes 21, 18, or 13, or abnormalities in the number of the sex chromosomes (**Table 15.2**). These syndromes are most likely due to nondisjunction. For example, Turner syndrome (XO) may occur when a gamete that is lacking a sex chromosome due to nondisjunction has fused with a gamete carrying an X chromosome. By comparison, Triple X syndrome (XXX) can occur when a gamete carrying two X chromosomes fuses with a gamete carrying a single X chromosome.

Most of the known trisomies involve chromosomes that are relatively small, so they carry fewer genes. Trisomies of the other human chromosomes and most monosomies are presumed to be lethal and have been found in spontaneously aborted embryos and fetuses.

Human abnormalities in chromosome number are influenced by the age of the parents. Older parents are more likely to produce children with abnormalities in chromosome number, because meiotic nondisjunction is more likely to occur in older cells. **Down syndrome**, which was first described by the English physician John Langdon Down in 1866, provides an example. This disorder is caused by the inheritance of three copies of chromosome 21 (Table 15.2). The incidence of Down syndrome rises with the age of either parent. In males, however, the rise occurs relatively late in life, usually past the age when most men have children. By comparison, the likelihood of having a child with Down syndrome rises dramatically during the later reproductive ages of women.

Table 15.2	Aneuploid Conditions in Humans		
Condition	**Frequency (# of live births)**	**Syndrome**	**Characteristics**
Autosomal			
Trisomy 21	1/800	Down	Mental impairment, abnormal pattern of palm creases, slanted eyes, flattened face, short stature
Trisomy 18	1/6,000	Edward	Mental and physical impairment, facial abnormalities, extreme muscle tone, early death
Trisomy 13	1/15,000	Patau	Mental and physical impairment, wide variety of defects in organs, large triangular nose, early death
Sex chromosomal			
XXY	1/1,000 (males)	Klinefelter	Sexual immaturity (no sperm), breast swelling (males)
XYY	1/1,000 (males)	Jacobs	Tall
XXX	1/1,500 (females)	Triple X	Tall and thin, menstrual irregularity
XO	1/5,000 (females)	Turner	Short stature, webbed neck, sexually undeveloped

Summary of Key Concepts

15.1 The Eukaryotic Cell Cycle

- Cytogeneticists examine cells microscopically to determine their chromosome composition. A micrograph that shows the alignment of chromosomes from a given cell is called a karyotype. Eukaryotic chromosomes are inherited in sets. A diploid cell has two sets of chromosomes. The members of each pair are called homologues. (Figure 15.1)

- Haploid cells, such as sperm and egg, have one set of chromosomes.

- The eukaryotic cell cycle consists of four phases called G_1 (first gap), S (synthesis of DNA), G_2 (second gap), and M phase (mitosis and cytokinesis). The G_1, S, and G_2 phases are collectively known as interphase. (Figure 15.2)

- Once a cell passes a restriction point in G_1, it is destined to replicate its DNA and to divide. During S phase, chromosomes are replicated and form pairs of sister chromatids. (Figure 15.3)

- An interaction between cyclin and cyclin-dependent kinase is necessary for cells to progress through the cell cycle. Checkpoint proteins sense the environmental conditions and the integrity of the genome and control whether or not the cell progresses through the cell cycle. (Figure 15.4)

- Masui and Markert studied the maturation of frog oocytes to identify a substance that was necessary for oocytes to progress through the cell cycle. This substance was initially called maturation promoting factor (MPF) and was later identified as a complex of mitotic cyclin and cyclin-dependent kinase. (Figures 15.5, 15.6)

15.2 Mitotic Cell Division

- The process of mitosis involves the sorting of chromosomes to produce two nuclei with the same number and types of chromosomes.

- During S phase, eukaryotic chromosomes are replicated to produce a pair of identical sister chromatids that remain attached to each other. (Figure 15.7)

- The mitotic spindle is composed of astral, kinetochore, and polar microtubules. The spindle organizes the process of cell division and plays a central role in chromosome sorting. (Figure 15.8)

- Mitosis occurs in five phases called prophase, prometaphase, metaphase, anaphase, and telophase. During prophase, the chromosomes condense, and the nuclear envelope begins to dissociate. The spindle apparatus is completely formed by the end of prometaphase. During metaphase, the chromosomes are aligned in a single row along the metaphase plate. At anaphase, the sister chromatids separate from each other and move to opposite poles; the poles themselves also move farther apart. During telophase, the chromosomes decondense, and the nuclear envelope re-forms. (Figure 15.9)

- Cytokinesis, which occurs after mitosis, is the division of the cytoplasm to produce two distinct daughter cells. In animal cells, cytokinesis involves the formation of a cleavage furrow. In plant cells, two separate cells are produced by the formation of a cell plate. Among all animals, 20 different proteins are required for cytokinesis to occur. (Figures 15.10, 15.11)

15.3 Meiosis and Sexual Reproduction

- The process of meiosis begins with a diploid cell and produces four haploid cells with one set of chromosomes each. (Figure 15.12)

- During prophase of meiosis, homologous pairs of sister chromosomes synapse, and crossing over occurs. After crossing over, chiasmata—the site where crossing over occurs—become visible. (Figure 15.13)

- Meiosis consists of two divisions, meiosis I and II, each composed of prophase, prometaphase, metaphase, anaphase, and telophase. During meiosis I, the homologues are separated to different cells, and during meiosis II, the sister chromatids are separated to different cells. (Figure 15.14, Table 15.1)

- The life cycle of animals is diploid dominant, whereas most fungi and some protists show a haploid-dominant life cycle. Plants alternate between diploid and haploid forms. (Figure 15.15)

15.4 Variation in Chromosome Structure and Number

- Chromosomes are named metacentric, submetacentric, acrocentric, and telocentric, according to their centromere location. Each type of chromosome can be uniquely identified by its banding pattern after staining. (Figure 15.16)

- Deletions, duplications, inversions, and translocations are different ways in which mutations alter chromosome structure. (Figure 15.17)

- A euploid organism has chromosomes that occur in complete sets. A polyploid organism has three or more sets of chromosomes. An organism that has one too many (trisomy) or one too few (monosomy) chromosomes is termed aneuploid. (Figure 15.18)

- Polyploid animals are relatively rare, but polyploid plants are common and tend to be more robust than their diploid counterparts. (Figures 15.19, 15.20)

- Aneuploidy in humans is responsible for several types of human genetic diseases, including Down syndrome. (Table 15.2)

Assess and Discuss

Test Yourself

1. In which phase of the cell cycle are chromosomes replicated?
   a. G_1 phase
   b. S phase
   c. M phase
   d. G_2 phase
   e. none of the above

2. If two chromosomes are homologous, they
   a. look similar under the microscope.
   b. have very similar DNA sequences.
   c. carry the same types of genes.
   d. may carry different versions of the same gene.
   e. are all of the above.

3. Checkpoints during the cell cycle are important because they
   a. allow the organelle activity to catch up to cellular demands.
   b. ensure the integrity of the cell's DNA.
   c. allow the cell to generate sufficient ATP for cellular division.
   d. are the only time DNA replication can occur.
   e. do all of the above.

4. Which of the following is a reason for mitotic cell division?
   a. asexual reproduction
   b. gamete formation in animals
   c. multicellularity
   d. all of the above
   e. both a and c

5. A replicated chromosome is composed of
   a. two homologous chromosomes held together at the centromere.
   b. four sister chromatids held together at the centromere.
   c. two sister chromatids held together at the centromere.
   d. four homologous chromosomes held together at the centromere.
   e. one chromosome with a centromere.

6. Which of the following is <u>not</u> an event of anaphase of mitosis?
   a. The nuclear envelope breaks down.
   b. Sister chromatids separate.
   c. Kinetochore microtubules shorten, pulling the chromosomes to the pole.
   d. Polar microtubules push against each other, moving the poles farther apart.
   e. All of the above occur during anaphase.

7. A student is looking at cells under the microscope. The cells are from an organism that has a diploid number of 14. In one particular case, the cell has seven replicated chromosomes (sister chromatids) aligned at the metaphase plate of the cell. Which of the following statements accurately describes this particular cell?

   a. The cell is in metaphase of mitosis.
   b. The cell is in metaphase of meiosis I.
   c. The cell is in metaphase of meiosis II.
   d. All of the above are correct.
   e. Both b and c are correct.

8. Which of the following statements accurately describes a difference between mitosis and meiosis?
   a. Mitosis may produce diploid cells, whereas meiosis produces haploid cells.
   b. Homologous chromosomes synapse during meiosis but do not synapse during mitosis.
   c. Crossing over commonly occurs during meiosis, but it does not commonly occur during mitosis.
   d. All of the above are correct.
   e. Both a and c are correct.

9. During crossing over in meiosis I,
   a. homologous chromosomes are not altered.
   b. homologous chromosomes exchange genetic material.
   c. chromosomal damage occurs.
   d. genetic information is lost.
   e. cytokinesis occurs.

10. Aneuploidy may be the result of
    a. duplication of a region of a chromosome.
    b. inversion of a region of a chromosome.
    c. nondisjunction during meiosis.
    d. interspecies breeding.
    e. all of the above.

Conceptual Questions

1. Distinguish between homologous chromosomes and sister chromatids.

2. The *Oca2* gene, which influences eye color in humans, is found on chromosome 15. How many copies of this gene are found in the karyotype of Figure 15.1? Is it one, two, or four?

3. A diploid cell carries four chromosomes per set. During meiosis I, it undergoes nondisjunction such that one cell receives two copies of chromosome 3 while the other cell receives zero. At the end of meiosis, how many total chromosomes will be found in each of the four resulting cells?

Collaborative Questions

1. Why is it necessary that the chromosomes condense during mitosis and meiosis? What do you think might happen if the chromosomes were not condensed?

2. A diploid eukaryotic cell has 10 chromosomes (five per set). As a group, take turns having one student draw the cell as it would look during a phase of mitosis, meiosis I, or meiosis II; then have the other students guess which phase it is.

Online Resource

www.brookerbiology.com

Stay a step ahead in your studies with animations that bring concepts to life and practice tests to assess your understanding. Your instructor may also recommend the interactive ebook, individualized learning tools, and more.

Simple Patterns of Inheritance

16

An African girl with albinism. This condition results in very light skin and hair color.

N tombi knew she looked different as long as she can remember. Born in Nigeria in 1991, she has accepted her appearance, though she still finds the occasional stare from strangers to be disturbing. Ntombi has albinism, a disorder characterized by a total or a partial lack of pigmentation of the skin, hair, and eyes. As a result, she has very fair skin, blond hair, and blue eyes.[1] In contrast, her parents and three brothers have dark skin, black hair, and brown eyes, as do most of her relatives and most of the people in the city where she lives. Ntombi is very close to her aunt, who also has albinism.

Cases like Ntombi's have intrigued people for many centuries. How do we explain the traits that are found in people, plants, and other organisms? Can we predict what types of offspring that two parents will produce? To answer such questions, researchers have studied the traits among related individuals and tried to make some sense of the data. Their goal is to understand **inheritance**—the acquisition of traits by their transmission from parent to offspring.

The first systematic attempt to understand inheritance was carried out by the plant breeder Joseph Kolreuter between 1761 and 1766. In crosses between two strains of tobacco plants, Kolreuter found that the offspring were usually intermediate in appearance between the two parents. He concluded that parents make equal genetic contributions to their offspring and that their genetic material blends together as it is passed to the next generation. This interpretation was consistent with the concept known as **blending inheritance**, which was widely accepted at that time. In the late 1700s, Jean Baptiste Lamarck, a French naturalist, hypothesized that physiological events (such as use or disuse) could modify traits and such modified traits would be inherited by offspring. For example, an individual who became adept at archery would pass that skill to his or her offspring. Overall, the prevailing view prior to the 1800s was that hereditary traits were rather malleable and could change and blend over the course of one or two generations.

In contrast, microscopic observations of chromosome transmission during mitosis and meiosis in the second half of the 19th century provided compelling evidence for **particulate inheritance**—the idea that the determinants of hereditary traits are transmitted in discrete units or particles from one generation to the next. Remarkably, this idea was first put forward in the 1860s by a researcher who knew nothing about chromosomes (**Figure 16.1**). Gregor Mendel, remembered today as the "father of genetics," used statistical

---

[1] In contrast to popular belief, most people with albinism have blue eyes, not pink eyes. This is particularly the case among Africans with albinism.

Figure 16.1 Gregor Johann Mendel, the father of genetics.

analysis of carefully designed breeding experiments to arrive at the concept of a gene, which is broadly defined as a unit of heredity. Forty years later, through the convergence of Mendel's work and that of cell biologists, this concept became the foundation of the modern science of genetics.

In this chapter, we will consider inheritance patterns and how the transmission of genes is related to the transmission of chromosomes. We will first consider the fundamental genetic patterns known as Mendelian inheritance and the relationship of these patterns to the behavior of chromosomes during meiosis. We will then examine the distinctive inheritance patterns of genes located on the X chromosome, paying special attention to the work of Thomas Hunt Morgan, whose investigation of these patterns confirmed that genes are on chromosomes. Finally, we will discuss the molecular basis of Mendelian inheritance and its variations, and consider how probability calculations can be used to predict the outcome of crosses.

16.1 Mendel's Laws of Inheritance

Gregor Johann Mendel grew up on a small farm in northern Moravia, then a part of the Austrian Empire and now in the Czech Republic. At the age of 21, he entered the Augustinian monastery of St. Thomas in Brno, and was ordained a priest in 1847. Mendel then worked for a short time as a substitute teacher, but to continue teaching he needed a license. Surprisingly, he failed the licensing exam due to poor answers in physics and natural history, so he enrolled at the University of Vienna to expand his knowledge in these two areas. Mendel's training in physics and mathematics taught him to perceive the world as an orderly place, governed by natural laws that could be stated as simple mathematical relationships.

In 1856, Mendel began his historic studies on pea plants. For 8 years, he analyzed thousands of pea plants that he grew on a small plot in his monastery garden. He published his work, entitled "Experiments on Plant Hybrids," in 1866. This paper was largely ignored by scientists at that time, partly because of its title and because it was published in a rather obscure journal (*The Proceedings of the Brünn Society of Natural History*). Also, Mendel was clearly ahead of his time. During this period, biology had not become a quantitative, experimental science. In addition, the behavior of chromosomes during mitosis and meiosis, which provides a framework for understanding inheritance patterns, had yet to be studied. Prior to his death in 1884, Mendel reflected, "My scientific work has brought me a great deal of satisfaction and I am convinced it will be appreciated before long by the whole world." Sixteen years later, in 1900, Mendel's work was independently rediscovered by three biologists with an interest in plant genetics: Hugo de Vries of Holland, Carl Correns of Germany, and Erich von Tschermak of Austria. Within a few years, the impact of Mendel's studies was felt around the world.

In this section, we will examine Mendel's experiments and how they led to the formulation of the basic genetic principles known as Mendel's laws. We will see that these principles apply not only to the pea plants Mendel studied, but also to a wide variety of sexually reproducing organisms, including humans.

Mendel Chose the Garden Pea to Study Inheritance

When two individuals with different characteristics are mated or crossed to each other, this is called a **hybridization** experiment, and the offspring are referred to as hybrids. For example, a hybridization experiment could involve a cross between a purple-flowered plant and a white-flowered plant. Mendel was particularly intrigued by the consistency with which offspring of such crosses showed characteristics of one or the other parent in successive generations. His intellectual foundation in physics and the natural sciences led him to consider that this regularity might be rooted in natural laws that could be expressed mathematically. To uncover these laws, he carried out quantitative experiments in which he carefully analyzed the numbers of offspring carrying specific traits.

Mendel chose the garden pea, *Pisum sativum*, to investigate the natural laws that govern inheritance. Why did he choose this species? Several properties of the garden pea were particularly advantageous for studying inheritance. First, it was available in many varieties that differed in characteristics, such as the appearance of seeds, pods, flowers, and stems. Such general features of an organism are called **characters**. Figure 16.2 illustrates the seven characters that Mendel eventually chose to follow in his breeding experiments. Each of these characters was found in two discrete variants. For example, one character he followed was height, which had the variants known as tall and dwarf. Another was seed color, which had the variants yellow and green. A **trait** is an identifiable characteristic of an organism. The term trait usually refers to a variant.* For example, yellow seed color is a trait.

A second important feature of garden peas is they are normally self-fertilizing. In **self-fertilization**, a female gamete is fertilized by a male gamete from the same plant. Like many flowering plants, peas have male and female sex organs in the same flower (Figure 16.3). Male gametes (sperm cells) are produced within pollen grains, which are formed in structures called stamens. Female gametes (egg cells) are produced in structures called ovules, which form within an organ called an ovary. For fertilization to occur, a pollen grain must land on the receptacle called a stigma, enabling a sperm to migrate to an ovule and fuse with an egg cell. In peas, the stamens and the ovaries are enclosed by a modified petal, an arrangement that greatly favors self-fertilization. Self-fertilization makes it easy to produce plants that breed true for a given trait, meaning the trait does not vary from generation to generation. For example, if a pea plant with yellow seeds breeds true for seed color, all the plants that grow from these seeds will also produce yellow seeds. A variety that continues to exhibit the same trait after several generations of self-fertilization is called a **true-breeding line**. Prior to conducting the studies described in this chapter, Mendel had already established that the seven characters he

---

* Geneticists may also use the term trait to refer to a character.

Figure 16.2 The seven characters that Mendel studied.

Character	Variants	
Flower color	Purple	White
Flower position	Axial	Terminal
Seed color	Yellow	Green
Seed shape	Round	Wrinkled
Pod color	Green	Yellow
Pod shape	Smooth	Constricted
Height	Tall	Dwarf

Concept check: Is having blue eyes a character, a variant, or both?

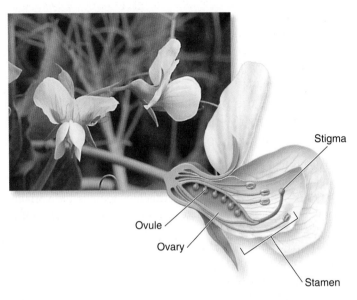

Figure 16.3 **Flower structure in pea plants.** The pea flower produces both male and female gametes. Sperm form in the pollen produced within the stamens; egg cells form in ovules within the ovary. A modified petal encloses the stamens and stigma, encouraging self-fertilization.

chose to study were true-breeding in the strains of pea plants he had obtained.

A third reason for using garden peas in hybridization experiments is the ease of making crosses: The flowers are quite large and easy to manipulate. In some cases, Mendel wanted his pea plants to self-fertilize, but in others, he did not want his plants to self-fertilize. Rather, he wanted to cross plants that differed with respect to some character, a process called hybridization, or **cross-fertilization**. In garden peas, cross-fertilization requires placing pollen from one plant onto the stigma of a flower on a different plant. Mendel's cross-fertilization procedure is shown in Figure 16.4. He would pry open an immature flower and remove the stamens before they produced pollen, so the flower could not self-fertilize. He then used a paintbrush to

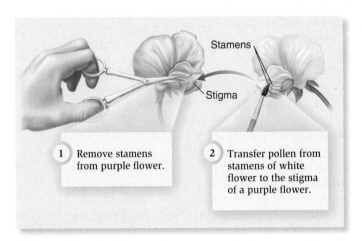

1. Remove stamens from purple flower.
2. Transfer pollen from stamens of white flower to the stigma of a purple flower.

Figure 16.4 A procedure for cross-fertilizing pea plants.

Concept check: Why are the stamens removed from the purple flower?

transfer pollen from another plant to the stigma of the flower that had its stamens removed. In this way, Mendel was able to cross-fertilize any two of his true-breeding pea plants and obtain any type of hybrid he wanted.

By Following the Inheritance Pattern of Single Traits, Mendel's Work Revealed the Law of Segregation

Mendel began his investigations by studying the inheritance patterns of pea plants that differed with regard to a single character. A cross in which an experimenter follows the variants of only one character is called a **monohybrid cross**, or **single-factor cross**. As an example, we will consider a monohybrid cross in which Mendel followed the tall and dwarf variants for height (**Figure 16.5**). The left-hand side of Figure 16.5a shows his experimental approach. The true-breeding parents are termed the **P generation** (parental generation), and a cross of these plants is called a P cross. The first-generation offspring of a P cross constitute the **F_1 generation** (first filial generation, from the Latin *filius*, meaning son). When the true-breeding parents differ with regard to a single character, their F_1 offspring are called single-trait hybrids, or **monohybrids**. When Mendel crossed true-breeding tall and dwarf plants, he observed that all plants of the F_1 generation were tall.

Next, Mendel followed the transmission of this character for a second generation. To do so, he allowed the F_1 monohybrids to self-fertilize, producing a generation called the **F_2 generation** (second filial generation). The dwarf trait reappeared in the F_2 offspring: Three-fourths of the plants were tall and one-fourth were dwarf. Mendel obtained similar results for each of the seven characters he studied, as shown in the data of Figure 16.5b. A quantitative analysis of his data allowed Mendel to postulate three important ideas regarding the properties and transmission of these traits from parents to offspring: (1) traits exist in two forms: dominant and recessive; (2) an individual carries two genes for a given character, and genes have variant forms, which are called alleles; and (3) the two alleles of a gene separate during gamete formation so that each sperm and egg receives only one allele.

Dominant and Recessive Traits Perhaps the most surprising outcome of Mendel's work was that the data argued strongly against the prevailing notion of blending inheritance. In each of the seven cases, the F_1 generation displayed a trait distinctly like one of the two parents rather than an intermediate trait. Using genetic terms that Mendel originated, we describe the alternative traits as dominant and recessive. The term **dominant** describes the displayed trait, whereas the term **recessive** describes a trait that is masked by the presence of a dominant trait. Tall stems and purple flowers are examples of dominant traits; dwarf stems and white flowers are examples of recessive traits. We say that tall is dominant over dwarf, and purple is dominant over white.

Genes and Alleles Mendel's results were consistent with particulate inheritance, in which the determinants of traits

(a) Mendel's protocol for making monohybrid crosses

THE DATA			
P cross	**F_1 generation**	**F_2 generation**	**Ratio**
Purple × white flowers	All purple	705 purple, 224 white	3.15:1
Axial × terminal flowers	All axial	651 axial, 207 terminal	3.14:1
Yellow × green seeds	All yellow	6,022 yellow, 2,001 green	3.01:1
Round × wrinkled seeds	All round	5,474 round, 1,850 wrinkled	2.96:1
Green × yellow pods	All green	428 green, 152 yellow	2.82:1
Smooth × constricted pods	All smooth	882 smooth, 299 constricted	2.95:1
Tall × dwarf stem	All tall	787 tall, 277 dwarf	2.84:1
Total	**All dominant**	**14,949 dominant, 5,010 recessive**	**2.98:1**

(b) Mendel's observed data for all 7 traits

Figure 16.5   Mendel's analyses of monohybrid crosses.

Concept check: Why do offspring of the F_1 generation exhibit only one variant of each character?

are inherited as unchanging, discrete units. In all seven cases, the recessive trait reappeared in the F_2 generation: Some F_2 plants displayed the dominant trait, while a smaller proportion showed the recessive trait. This observation led Mendel to conclude that the genetic determinants of traits are "unit factors" that are passed intact from generation to generation. These unit factors are what we now call **genes** (from the Greek *genos*, meaning birth), a term coined by the Danish botanist Wilhelm Johannsen in 1909. Mendel postulated that every individual carries two genes for a given character and that the gene for each character in his pea plant exists in two variant forms, which we now call **alleles**. For example, the gene controlling height in Mendel's pea plants occurs in two variants, called the tall allele and the dwarf allele. The right-hand side of Figure 16.5a shows Mendel's conclusions, using genetic symbols (letters) that were adopted later. The letters T and t represent the alleles of the gene for plant height. By convention, the uppercase letter represents the dominant allele (in this case, tall), and the same letter in lowercase represents the recessive allele (dwarf).

Segregation of Alleles   When Mendel compared the numbers of F_2 offspring exhibiting dominant and recessive traits, he noticed a recurring pattern. Although some experimental variation occurred, he always observed an approximately 3:1 ratio between the dominant and the recessive trait (Figure 16.5b). How did Mendel interpret this ratio? He concluded that each parent carries two versions (alleles) of a gene and that the two alleles carried by an F_1 plant will **segregate** (separate) from each other during gamete formation, so each sperm or egg carries only one allele. The diagram in Figure 16.6 shows that the segregation of the F_1 alleles should result in equal numbers of gametes carrying the dominant allele (T) and the recessive allele (t). If these gametes combine with one another randomly at fertilization, as shown in the figure, this would account for the 3:1 ratio of the F_2 generation. Note that a Tt individual can be produced by two different combinations of alleles—the T allele can come from the male gamete and the t allele from the female gamete, or vice versa. This accounts for the observation that the Tt genotype is produced twice as often as either TT or tt. The idea that *the two alleles of a gene separate (segregate) during the formation of eggs and sperm so that every gamete receives only one allele* is known today as Mendel's **law of segregation**.

Genotype Describes an Organism's Genetic Makeup, Whereas Phenotype Describes Its Characteristics

To continue our discussion of Mendel's results, we need to introduce a few more genetic terms. The term **genotype** refers to the genetic composition of an individual. In the example shown in Figure 16.5a, TT and tt are the genotypes of the P generation, and Tt is the genotype of the F_1 generation. In the P generation, both parents are true-breeding plants, which means that each have identical copies of the allele of the gene for height. An individual with two identical alleles of a gene is said to be **homozygous** with respect to that gene. In the specific P cross we are considering, the tall plant is homozygous for T, and the dwarf plant is homozygous for t. In contrast, a **heterozygous** individual carries two different alleles of a gene. Plants of the F_1 generation are heterozygous, with the genotype Tt, because every individual carries one copy of the tall allele (T) and one copy of the dwarf allele (t). The F_2 generation includes both homozygous individuals (homozygotes) and heterozygous individuals (heterozygotes).

The term **phenotype** refers to the characteristics of an organism that are the result of the expression of its genes. In

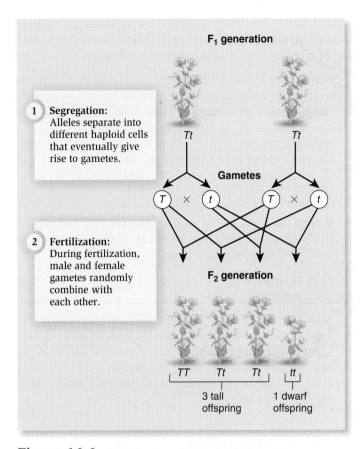

Figure 16.6   **How the law of segregation explains Mendel's observed ratios.**   The segregation of alleles in the F_1 generation gives rise to gametes that carry just one of the two alleles. These gametes combine randomly during fertilization, producing the allele combinations TT, Tt, and tt in the F_2 offspring. The combination Tt occurs twice as often as either of the other two combinations because it can be produced in two different ways. The TT and Tt offspring are tall, whereas the tt offspring are dwarf.

Concept check: *What is ratio of the T allele to the t allele in the F_2 generation? Does this ratio differ from the 3:1 phenotype ratio? If so, explain why.*

the example in Figure 16.5a, one of the parent plants is phenotypically tall, and the other is phenotypically dwarf. Although the F_1 offspring are heterozygous (Tt), they are phenotypically tall because each of them has a copy of the dominant tall allele. In contrast, the F_2 plants display both phenotypes in a ratio of 3:1. Later in the chapter, we will examine the underlying molecular mechanisms that produce phenotypes, but in our discussion of Mendel's results, the term simply refers to a visible characteristic such as flower color or height.

A Punnett Square Can Be Used to Predict the Outcome of Crosses

A common way to predict the outcome of simple genetic crosses is to make a **Punnett square**, a method originally proposed by the British geneticist Reginald Punnett. To construct a

Punnett square, you must know the genotypes of the parents. What follows is a step-by-step description of the Punnett square approach, using a cross of heterozygous tall plants.

Step 1. *Write down the genotypes of both parents.* In this example, a heterozygous tall plant is crossed to another heterozygous tall plant. The plant providing the pollen is considered the male parent and the plant providing the eggs, the female parent. (In self-pollination, a single individual produces both types of gametes.)

Male parent: *Tt*
Female parent: *Tt*

Step 2. *Write down the possible gametes that each parent can make.* Remember the law of segregation tells us that a gamete contains only one copy of each allele.

Male gametes: *T* or *t*
Female gametes: *T* or *t*

Step 3. *Create an empty Punnett square.* The number of columns equals the number of male gametes, and the number of rows equals the number of female gametes. Our example has two rows and two columns. Place the male gametes across the top of the Punnett square and the female gametes along the side.

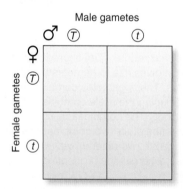

Step 4. *Fill in the possible genotypes of the offspring by combining the alleles of the gametes in the empty boxes.*

Step 5. *Determine the relative proportions of genotypes and phenotypes of the offspring.* The genotypes are obtained directly from the Punnett square. In this example, the genotypic ratios are 1*TT* : 2*Tt* : 1*tt*. To determine the phenotypes,

you must know which allele is dominant. For plant height, *T* (tall) is dominant to *t* (dwarf). The genotypes *TT* and *Tt* are tall, whereas the genotype *tt* is dwarf. Therefore, our Punnett square shows us that the phenotypic ratio is expected to be 3 tall : 1 dwarf. Keep in mind, however, these are predicted ratios for large numbers of offspring. If only a few offspring are produced, the observed ratios could deviate significantly from the predicted ratios. We will examine the topics of sample size and genetic prediction later in this chapter.

A Testcross Can Be Used to Determine an Individual's Genotype

When a character has two variants, one of which is dominant over the other, we know that an individual with a recessive phenotype is homozygous for the recessive allele. A dwarf pea plant, for example, must have the genotype *tt*. But an individual with a dominant phenotype may be either homozygous or heterozygous—a tall pea plant may have the genotype *TT* or *Tt*. How can we distinguish between these two possibilities? Mendel devised a method called a **testcross** to address this question. In a testcross, the researcher crosses the individual of interest to a homozygous recessive individual and observes the phenotypes of the offspring.

Figure 16.7 shows how this procedure can be used to determine the genotype of a tall pea plant. If the testcross produces some dwarf offspring, as shown in the Punnett square on the right side, these offspring must have two copies of the recessive allele, one inherited from each parent. Therefore, the tall parent must be a heterozygote, with the genotype *Tt*. Alternatively, if all of the offspring are tall, as shown in the Punnett square on the left, the tall parent is likely to be a homozygote, with the genotype *TT*.

Analyzing the Inheritance Pattern of Two Characters Demonstrated the Law of Independent Assortment

Mendel's analysis of single-factor crosses suggested that traits are inherited as discrete units and that the alleles for a given gene segregate during the formation of haploid cells. To obtain additional insights into how genes are transmitted from parents to offspring, Mendel conducted crosses in which he simultaneously followed the inheritance of two different characters. A cross of this type is called a **dihybrid cross**, or a **two-factor cross**. We will examine a two-factor cross in which Mendel simultaneously followed the inheritance of seed color and seed shape (Figure 16.8). He began by crossing strains of pea plants that bred true for both characters. The plants of one strain had yellow, round seeds, and plants of the other strain had green, wrinkled seeds. He then allowed the F_1 offspring to self-fertilize and observed the phenotypes of the F_2 generation.

What are the possible patterns of inheritance for two characters? One possibility is that the two genes are linked in some way, so variants that occur together in the parents are always

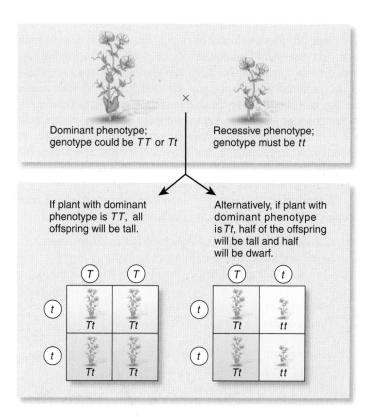

Figure 16.7 A testcross.   The purpose of this experiment is to determine if the organism with the dominant phenotype, in this case a tall pea plant, is a homozygote (*TT*) or a heterozygote (*Tt*).

Concept check: *Let's suppose you had a plant with purple flowers and unknown genotype and conducted a testcross to determine its genotype. You obtained 41 plants, 20 with white flowers and 21 with purple flowers. What was the genotype of the original purple-flowered plant?*

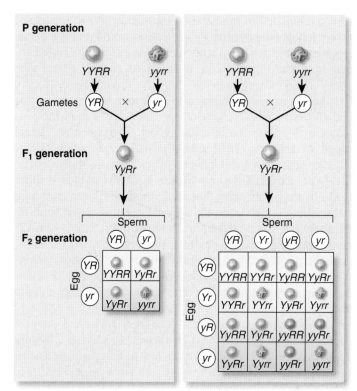

(a) Hypothesis: linked assortment

(b) Hypothesis: independent assortment

P cross	F$_1$ generation	F$_2$ generation
Yellow, round seeds	Yellow, round seeds	315 yellow, round seeds
×		101 yellow, wrinkled seeds
Green, wrinkled seeds		108 green, round seeds
		32 green, wrinkled seeds

(c) The data observed by Mendel

Figure 16.8 Two hypotheses for the assortment of two different genes.   In a cross between two true-breeding pea plants, one with yellow, round seeds and one with green, wrinkled seeds, all of the F$_1$ offspring have yellow, round seeds. When the F$_1$ offspring self-fertilize, the two hypotheses predict different phenotypes in the F$_2$ generation. **(a)** The linkage hypothesis proposes that the parental alleles for the two characters stay associated with each other. In this case, all of the F$_2$ offspring will have either yellow, round seeds or green, wrinkled seeds. **(b)** The independent assortment hypothesis proposes that each allele assorts independently. In this case, the F$_2$ generation will display four different phenotypes. **(c)** Mendel's observations supported the independent assortment hypothesis.

Concept check: *What ratio of offspring phenotypes would have occurred if the linked hypothesis had been correct?*

inherited as a unit. In our example, the allele for yellow seeds (*Y*) would always be inherited with the allele for round seeds (*R*), and the alleles for green seeds (*y*) would always be inherited with the allele for wrinkled seeds (*r*), as shown in Figure 16.8a. A second possibility is that the two genes are independent of one another, so their alleles are randomly distributed into gametes (Figure 16.8b). By following the transmission pattern of two characters simultaneously, Mendel could determine whether the genes that determine seed shape and seed color assort (are distributed) together as a unit or independently of each other.

What experimental results could Mendel predict for each of these two models? The two homozygous plants of the P generation can produce only two kinds of gametes, *YR* and *yr*, so in either case the F$_1$ offspring would be heterozygous for both genes; that is, they would have the genotypes *YyRr*. Because Mendel knew from his earlier experiments that yellow was dominant over green and round over wrinkled, he could predict that all the F$_1$ plants would have yellow, round seeds. In contrast, as shown in Figure 16.8, the ratios he obtained in the F$_2$ generation would depend on whether the alleles of both genes assort together or independently.

If the parental genes are linked, as in Figure 16.8a, the F$_1$ plants could produce gametes that are only *YR* or *yr*. These

gametes would combine to create offspring with the genotypes *YYRR* (yellow, round), *YyRr* (yellow, round), or *yyrr* (green, wrinkled). The ratio of phenotypes would be 3 yellow, round to 1 green, wrinkled. Every F$_2$ plant would be phenotypically like one P-generation parent or the other. None would display a new combination of the parental traits. However, if the alleles assort independently, the F_2 generation would show a wider range of genotypes and phenotypes, as shown by the large Punnett square in Figure 16.8b. In this case, each F_1 parent produces four kinds of gametes—*YR, Yr, yR,* and *yr*—instead of two, so the square is constructed with four rows on each side and shows 16 possible genotypes. The F_2 generation includes plants with yellow, round seeds; yellow, wrinkled seeds; green, round seeds; and green, wrinkled seeds, in a ratio of 9:3:3:1.

The actual results of this two-factor cross are shown in Figure 16.8c. Crossing the true-breeding parents produced **dihybrid** offspring—offspring that are hybrids with respect to both traits. These F_1 dihybrids all had yellow, round seeds, confirming that yellow and round are dominant traits. This result was consistent with either hypothesis. However, the data for the F_2 generation were consistent only with the independent assortment hypothesis. Mendel observed four phenotypically different types of F_2 offspring, in a ratio that was reasonably close to 9:3:3:1.

In his original studies, Mendel reported that he had obtained similar results for every pair of characters he analyzed. His work supported the idea, now called the **law of independent assortment**, that *the alleles of different genes assort independently of each other during gamete formation.* Independent assortment means that a specific allele for one gene may be found in a gamete regardless of which allele for a different gene is found in the same gamete. In our example, the yellow and green alleles assort independently of the round and wrinkled alleles. The union of gametes from F_1 plants carrying these alleles produces the F_2 genotype and phenotype ratios shown in Figure 16.8b.

As we will see in Chapter 17, not all dihybrid crosses exhibit independent assortment. In some cases, the alleles of two genes that are physically located near each other on the same chromosome do not assort independently.

16.2 The Chromosome Theory of Inheritance

Mendel's studies with pea plants led to the concept of a gene, which is the foundation for our understanding of inheritance. However, at the time of Mendel's work, the physical nature and location of genes were a complete mystery. The idea that inheritance has a physical basis was not even addressed until 1883, when the German biologist August Weismann and the Swiss botanist Carl Nägeli championed the idea that a substance in living cells is responsible for the transmission of hereditary traits. This idea challenged other researchers to identify the genetic material. Several scientists, including the German biologists Eduard Strasburger and Walter Flemming, observed dividing cells under the microscope and suggested that the chromosomes are the carriers of the genetic material. As we now know, the genetic material is the DNA within chromosomes.

In the early 1900s, the idea that chromosomes carry the genetic material dramatically unfolded as researchers continued to study the processes of mitosis, meiosis, and fertilization. It became increasingly clear that the characteristics of organisms are rooted in the continuity of cells during the life of an organism and from one generation to the next. Several scientists noted striking parallels between the segregation and assortment of traits noted by Mendel and the behavior of chromosomes during meiosis. Among these scientists were the German biologist Theodor Boveri and the American biologist Walter Sutton, who independently proposed the chromosome theory of inheritance. According to this theory, the inheritance patterns of traits can be explained by the transmission of chromosomes during meiosis and fertilization.

A modern view of the **chromosome theory of inheritance** consists of a few fundamental principles:

1. Chromosomes contain DNA, which is the genetic material. Genes are found in the chromosomes.
2. Chromosomes are replicated and passed from parent to offspring. They are also passed from cell to cell during the development of a multicellular organism.
3. The nucleus of a diploid cell contains two sets of chromosomes, which are found in homologous pairs. The maternal and paternal sets of homologous chromosomes are functionally equivalent; each set carries a full complement of genes.
4. At meiosis, one member of each chromosome pair segregates into one daughter nucleus, and its homologue segregates into the other daughter nucleus. During the formation of haploid cells, the members of different chromosome pairs segregate independently of each other.
5. Gametes are haploid cells that combine to form a diploid cell during fertilization, with each gamete transmitting one set of chromosomes to the offspring.

In this section, we will relate the chromosome theory of inheritance to Mendel's laws of inheritance.

Mendel's Law of Segregation Is Explained by the Segregation of Homologous Chromosomes During Meiosis

Now that you have an understanding of the basic tenets of the chromosome theory of inheritance, let's relate these ideas to Mendel's laws of inheritance. To do so, it will be helpful to introduce another genetic term. The physical location of a gene on a chromosome is called the gene's **locus** (plural, loci). As shown in **Figure 16.9**, each member of a homologous chromosome pair carries an allele of the same gene at the same locus. The individual in this example is heterozygous (*Tt*), so each homologue has a different allele.

How can we relate the chromosome theory of inheritance to Mendel's law of segregation? **Figure 16.10** follows a homologous chromosome pair through the events of meiosis. This example involves a pea plant, heterozygous for height, *Tt*. The top of Figure 16.10 shows the two homologues prior to DNA replication. When a cell prepares to divide, the homologues replicate to produce pairs of sister chromatids. Each chromatid carries a copy of the allele found on the original homologue, either *T* or *t*. During meiosis I, the homologues, each consisting

Gene locus—site on chromosome where a gene is found. A gene can exist as 2 or more different alleles.

Pair of homologous chromosomes

T—Tall allele

t—Dwarf allele

Genotype: *Tt* (heterozygous)

Figure 16.9   **A gene locus.**   The locus (location) of a gene is the same for each member of a homologous pair, whether the individual is homozygous or heterozygous for that gene. This individual is heterozygous (*Tt*) for a gene for plant height in peas.

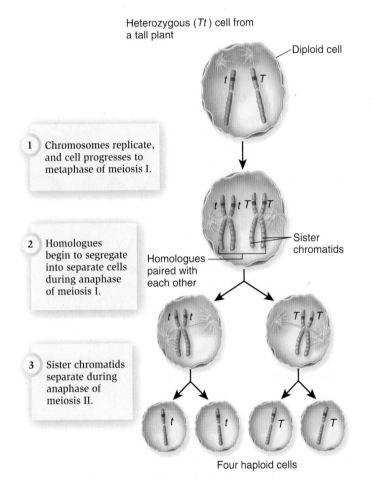

Heterozygous (*Tt*) cell from a tall plant

Diploid cell

1 Chromosomes replicate, and cell progresses to metaphase of meiosis I.

Sister chromatids

Homologues paired with each other

2 Homologues begin to segregate into separate cells during anaphase of meiosis I.

3 Sister chromatids separate during anaphase of meiosis II.

Four haploid cells

Figure 16.10   **The chromosomal basis of allele segregation.** This example shows a pair of homologous chromosomes in a cell of a pea plant. The blue chromosome was inherited from the male parent, and the red chromosome was inherited from the female parent. This individual is heterozygous (*Tt*) for a height gene. The two homologues segregate from each other during meiosis, leading to segregation of the tall allele (*T*) and the dwarf allele (*t*) into different haploid cells. Note: For simplicity, this diagram shows a single pair of homologous chromosomes, though eukaryotic cells typically have several different pairs of homologous chromosomes.

Concept check: *When we say that alleles segregate, what does the word segregate mean? How is this related to meiosis, described in Chapter 15?*

of two sister chromatids, pair up and then segregate into two daughter cells. One of these cells has two copies of the *T* allele, and the other has two copies of the *t* allele. The sister chromatids separate during meiosis II, which produces four haploid cells. The end result of meiosis is that each haploid cell has a copy of just one of the two original homologues. Two of the cells have a chromosome carrying the *T* allele, while the other two have a chromosome carrying the *t* allele at the same locus. If the haploid cells shown at the bottom of Figure 16.10 combine randomly during fertilization, they produce diploid offspring with the genotypic and phenotypic ratios shown earlier in Figure 16.6.

Mendel's Law of Independent Assortment Is Explained by the Independent Alignment of Different Chromosomes During Meiosis

How can we relate the chromosome theory of inheritance to Mendel's law of independent assortment? **Figure 16.11** shows the alignment and segregation of two pairs of chromosomes in a pea plant. One pair carries the gene for seed color: The yellow allele (*Y*) is on one chromosome, and the green allele (*y*) is on its homologue. The other pair of chromosomes carries the gene for seed shape: One member of the pair has the round allele (*R*), whereas its homologue carries the wrinkled allele (*r*). Thus, this individual is heterozygous for both genes, with the genotype *YyRr*.

When meiosis begins, each of the chromosomes has already replicated and consists of two sister chromatids. At metaphase I of meiosis, the two pairs of chromosomes randomly align themselves along the metaphase plate. This alignment can occur in two equally probable ways, shown on the two sides of the figure. On the left, the chromosome carrying the *y* allele is aligned on the same side of the metaphase plate as the chromosome carrying the *R* allele; *Y* is aligned with *r*. On the right, the opposite has occurred: *Y* is aligned with *R*, and *y* is with *r*. In each case, the chromosomes that aligned on the same side of the metaphase plate segregate into the same daughter cell. In this way, the random alignment of chromosome pairs during meiosis I leads to the independent assortment of alleles found on different chromosomes. For two loci found on different chromosomes, each with two variant alleles, meiosis produces four allele combinations in equal numbers, as seen at the bottom of the figure.

If a *YyRr* (dihybrid) plant undergoes self-fertilization, any two gametes can combine randomly during fertilization. Because four kinds of gametes are made, this allows for 16 possible allele combinations in the offspring. These genotypes, in turn, produce four phenotypes in a 9:3:3:1 ratio, as seen earlier in Figure 16.8. This ratio is the expected outcome when a heterozygote for two genes on different chromosomes undergoes self-fertilization.

But what if two different genes are located on the same chromosome? In this case, the transmission pattern may not conform to the law of independent assortment. We will discuss this phenomenon, known as linkage, in Chapter 17.

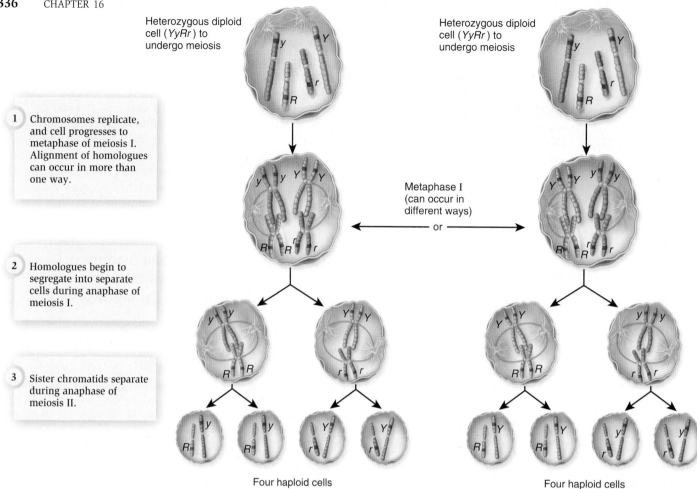

Figure 16.11 **The chromosomal basis of independent assortment.** The genes for seed color (*Y* or *y*) and seed shape (*R* or *r*) in peas are on different chromosomes. During metaphase of meiosis I, different arrangements of the two chromosome pairs can lead to different combinations of the alleles in the resulting haploid cells. On the left, the chromosome carrying the dominant *R* allele has segregated with the chromosome carrying the recessive *y* allele. On the right, the two chromosomes carrying the dominant alleles (*R* and *Y*) have segregated together. Note: For simplicity, this diagram shows only two pairs of homologous chromosomes, though eukaryotic cells typically have several different pairs of homologous chromosomes.

Concept check: Let's suppose that a cell is heterozygous for three different genes (*AaBbCc*) and that each gene is on a different chromosome. How many different ways can these three pairs of homologous chromosomes align themselves during metaphase I, and how many different types of gametes can be produced?

16.3 Pedigree Analysis of Human Traits

As we have seen, Mendel conducted experiments by making selective crosses of pea plants and analyzing large numbers of offspring. Later geneticists also relied on crosses of experimental organisms, especially fruit flies. However, geneticists studying human traits cannot use this approach, for ethical and practical reasons. Instead, human geneticists must rely on information from family trees, or pedigrees. In this approach, called **pedigree analysis**, an inherited trait is analyzed over the course of a few generations in one family. The results of this method may be less definitive than the results of breeding experiments because the small size of human families may lead to large sampling errors. Nevertheless, a pedigree analysis can often provide important clues concerning human inheritance.

Pedigree analysis has been used to understand the inheritance of human genetic diseases that follow simple Mendelian patterns. Many genes that play a role in disease exist in two forms—the normal allele and an abnormal allele that has arisen by mutation. The disease symptoms are associated with the mutant allele. Pedigree analysis allows us to determine whether the mutant allele is dominant or recessive and to predict the likelihood of an individual being affected.

Let's consider a recessive condition to illustrate pedigree analysis. The pedigree in **Figure 16.12** concerns a human genetic disease known as cystic fibrosis (CF). Approximately 3% of Americans of European descent are heterozygous carriers of the recessive *CF* allele. Carriers are usually phenotypically normal. Individuals who are homozygous for the *CF* allele exhibit the disease symptoms, which include abnormalities of the pancreas, intestine, sweat glands, and lungs. A human pedigree, like the one in Figure 16.12, shows the oldest generation

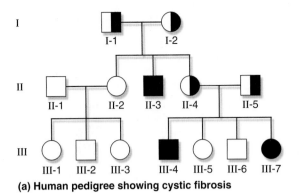

(a) Human pedigree showing cystic fibrosis

(b) Symbols used in a human pedigree

Figure 16.12   **A family pedigree for a recessive trait.**
Some members of the family in this pedigree are affected with cystic fibrosis. Phenotypically normal individuals I-1, I-2, II-4, and II-5 are presumed to be heterozygotes because they have produced affected offspring.

Concept check: *Let's suppose a genetic disease is caused by a mutant allele. If two affected parents produce an unaffected offspring, can the mutant allele be recessive?*

(designated by the Roman numeral I) at the top, with later generations (II and III) below it. A woman (represented by a circle) and a man (represented by a square) who produce offspring are connected by a horizontal line; a vertical line connects parents with their offspring. Siblings (brothers and sisters) are denoted by downward projections from a single horizontal line, from left to right in the order of their birth. For example, individuals I-1 and I-2 are the parents of individuals II-2, II-3, and II-4, who are all siblings. Individuals affected by the disease, such as individual II-3, are depicted by filled symbols.

Why does this pedigree indicate a recessive pattern of inheritance for CF? The answer is that two unaffected individuals can produce an affected offspring. Such individuals are presumed to be heterozygotes (designated by a half-filled symbol). However, the same unaffected parents can also produce unaffected offspring, because an individual must inherit two copies of the mutant allele to exhibit the disease. A recessive mode of inheritance is also characterized by the observation that all of the offspring of two affected individuals will be affected. However, for genetic diseases like CF that limit survival or fertility, there may rarely or never be cases where two affected individuals produce offspring.

Although many of the alleles causing human genetic diseases are recessive, some are known to be dominant. **Figure**

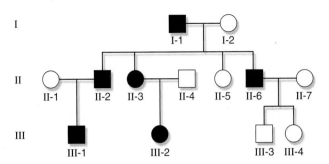

Figure 16.13   **A family pedigree for a dominant trait.**
Huntington disease is caused by a dominant allele. Note that each affected offspring in this pedigree has an affected parent.

Concept check: *What observation in a pedigree suggests a dominant pattern of inheritance?*

16.13 shows a family pedigree involving Huntington disease, a condition that causes the degeneration of brain cells involved in emotions, intellect, and movement. If you examine this pedigree, you will see that every affected individual has one affected parent. This pattern is characteristic of most dominant disorders. However, affected parents do not always produce affected offspring. For example, II-6 is a heterozygote that has passed the normal allele to his offspring and thereby produced unaffected offspring (III-3 and III-4).

The symptoms of Huntington disease, which usually begin to appear when people are 30 to 50 years old, include uncontrollable jerking movements of the limbs, trunk, and face; progressive loss of mental abilities; and the development of psychiatric problems. In 1993, researchers identified the gene involved in this disorder. The gene encodes a protein called huntingtin, which functions in nerve cells. The mutant allele encodes an abnormal form of the protein, which aggregates within nerve cells in the brain. Further research is needed to determine how this aggregation contributes to the disease.

Most human genes are found on the paired chromosomes known as **autosomes**, which are the same in both sexes. Mendelian inheritance patterns involving these autosomal genes are described as autosomal inheritance patterns. Huntington disease is an example of a trait with an autosomal dominant inheritance pattern, whereas cystic fibrosis displays an autosomal recessive pattern. However, some human genes are located on sex chromosomes, which are different in males and females. These genes have their own characteristic inheritance patterns, which we will consider next.

16.4   Sex Chromosomes and X-Linked Inheritance Patterns

In the first part of this chapter, we discussed Mendel's experiments that established the basis for understanding how traits are transmitted from parents to offspring. We also examined the chromosome theory of inheritance, which provided a framework for explaining Mendel's observations. Mendelian patterns of gene transmission are observed for most genes located on autosomes in a wide variety of eukaryotic species.

We will now turn our attention to genes located on **sex chromosomes**. As you learned in Chapter 15, this term refers to a distinctive pair of chromosomes that are different in males and females. Sex chromosomes are found in many but not all species with two sexes. The study of sex chromosomes proved pivotal in confirming the chromosome theory of inheritance. The distinctive transmission patterns of genes on sex chromosomes helped early geneticists show that particular genes are located on particular chromosomes. Later, other researchers became interested in these genes because some of them were found to cause inherited diseases in humans.

In this section, we will consider several mechanisms by which sex chromosomes in various species determines an individual's sex. We will then examine some of the early research involving sex chromosomes that provided convincing evidence for the chromosome theory of inheritance. Finally, we will consider the inheritance patterns of genes on sex chromosomes and why recessive alleles are expressed more frequently in males than in females.

In Many Species, Sex Differences Are Due to the Presence of Sex Chromosomes

Some early evidence supporting the chromosome theory of inheritance involved a consideration of sex determination. In 1901, the American biologist C. E. McClung suggested that the inheritance of particular chromosomes is responsible for determining sex in fruit flies. Following McClung's initial observations, several mechanisms of sex determination were found in different species of animals. Some examples are described in **Figure 16.14**. All of these mechanisms involve chromosomal differences between the sexes, and most involve a difference in a single pair of sex chromosomes.

In the X-Y system of sex determination, which operates in mammals, the somatic cells of males have one X and one Y chromosome, whereas female somatic cells contain two X chromosomes (Figure 16.14a). For example, the 46 chromosomes carried by human cells consist of 22 pairs of autosomes and one pair of sex chromosomes (either XY or XX). Which chromosome, the X or Y, determines sex? In mammals, the presence of the Y chromosome causes maleness. This is known from the analysis of rare individuals who carry chromosomal abnormalities. For example, mistakes that occasionally occur during meiosis may produce an individual who carries two X chromosomes and one Y chromosome. Such an individual develops into a male. A gene called the *SRY* gene located on the Y chromosome of mammals plays a key role in the developmental pathway that leads to maleness.

The X-O system operates in many insects (Figure 16.14b). Unlike the X-Y system in mammals, the presence of the Y chromosome in the X-O system does not determine maleness. Females in this system have a pair of sex chromosomes and are designated XX. In some insect species that follow the X-O system, the male has only one sex chromosome, the X. In other

(a) The X-Y system in mammals

(b) The X-O system in certain insects

(c) The Z-W system in birds

(d) The haplodiploid system in bees

Figure 16.14 **Different mechanisms of sex determination in animals.** The numbers shown in the circles indicate the numbers of autosomes.

Concept check: *If a person is born with only one X chromosome and no Y chromosome, would you expect that person to be a male or a female? Explain your answer.*

X-O insect species, such as *Drosophila melanogaster*, the male has both an X chromosome and a Y chromosome. The insect's sex is determined by the ratio between its X chromosomes and its sets of autosomes. If a fly has one X chromosome and is diploid for the autosomes (2*n*), this ratio is 1/2, or 0.5. This fly will become a male whether or not it receives a Y chromosome. On the other hand, if a diploid fly receives two X chromosomes, the ratio is 2/2, or 1.0, and the fly becomes a female.

Thus far, we have considered examples where females have two similar copies of a sex chromosome, the X. However, in some animal species, such as birds and some fish, the male carries two similar chromosomes (Figure 16.14c). This is called the Z-W system to distinguish it from the X-Y system found in mammals. The male is ZZ, and the female is ZW.

Not all chromosomal mechanisms of sex determination involve a special pair of sex chromosomes. An interesting mechanism known as the haplodiploid system is found in bees (Figure 16.14d). The male bee, or drone, is produced from an unfertilized haploid egg. Thus, male bees are haploid individuals. Females, both worker bees and queen bees, are produced from fertilized eggs and therefore are diploid.

Although sex in many species of animals is determined by chromosomes, other mechanisms are also known. In certain reptiles and fish, sex is controlled by environmental factors such as temperature. For example, in the American alligator (*Alligator mississippiensis*), temperature controls sex development. When eggs of this alligator are incubated at 33°C, nearly all of them produce male individuals. When the eggs are incubated at a temperature significantly below 33°C, they produce nearly all females, whereas at a temperature above 33°C, they produce a mixture of males and females.

Most species of flowering plants, including pea plants, have a single type of diploid plant, or sporophyte, that makes both male and female gametophytes. However, the sporophytes of some species have two sexually distinct types of individuals, one with flowers that produce male gametophytes, and the other with flowers that produce female gametophytes. Examples include hollies, willows, poplars, and date palms. Sex chromosomes, designated X and Y, are responsible for sex determination in many such species. The male plant is XY, whereas the female plant is XX. However, in some plant species with separate sexes, microscopic examination of the chromosomes does not reveal distinct types of sex chromosomes.

In Humans, Recessive X-Linked Traits Are More Likely to Occur in Males

In humans, the X chromosome is rather large and carries over 1,000 genes, whereas the Y chromosome is quite small and has less than 100 genes. Therefore, many genes are found on the X chromosome but not on the Y; these are known as **X-linked genes**. By comparison, fewer genes are known to be Y linked, meaning they are found on the Y chromosome but not on the X. The term **sex linked** refers to genes found on one sex chromosome but not on the other. Because fewer genes are found on the Y chromosome, the term usually refers to X-linked genes. In mammals, a male cannot be described as being homozygous

or heterozygous for an X-linked gene, because these terms apply to genes that are present in two copies. Instead, the term **hemizygous** is used to describe an individual with only one copy of a particular gene. A male mammal is said to be hemizygous for an X-linked gene.

Many recessive X-linked alleles cause diseases in humans, and these diseases occur more frequently in males than in females. As an example, let's consider the X-linked recessive disorder called classical hemophilia (hemophilia A). In individuals with hemophilia, blood does not clot normally, and a minor cut may bleed for a long time. Small bumps can lead to large bruises because broken capillaries may leak blood profusely into surrounding tissues before the capillaries are repaired. Common accidental injuries pose a threat of severe internal or external bleeding for hemophiliacs. Hemophilia A is caused by a recessive X-linked allele that encodes a defective form of a clotting protein. If a mother is a heterozygous carrier of hemophilia A, each of her children has a 50% chance of inheriting the recessive allele. The following Punnett square shows a cross between an unaffected father and a heterozygous mother. X^H designates an X chromosome carrying the normal allele, and X^{h-A} is the X chromosome that carries the recessive allele for hemophilia A.

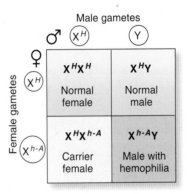

Although each child has a 50% chance of inheriting the hemophilia allele from the mother, only 1/2 of the sons will exhibit the disorder. Because a son inherits only one X chromosome, a son who inherits the abnormal allele from his mother will have hemophilia. However, a daughter inherits an X chromosome from both her mother and her father. In this example, a daughter who inherits the hemophilia allele from her mother will also inherit a normal allele from her father. This daughter will have a normal phenotype, but if she passes the abnormal allele to her sons, they will have hemophilia.

FEATURE INVESTIGATION

Morgan's Experiments Showed a Correlation Between a Genetic Trait and the Inheritance of a Sex Chromosome in *Drosophila*

The distinctive inheritance pattern of X-linked alleles provides a way of demonstrating that a specific gene is on an X chro-

mosome. An X-linked gene was the first gene to be located on a specific chromosome. In 1910, the American geneticist Thomas Hunt Morgan began work on a project in which he reared large populations of fruit flies, *Drosophila melanogaster*, in the dark to determine if their eyes would atrophy from disuse and disappear in future generations. Even after

Figure 16.15 Morgan's crosses of red-eyed and white-eyed *Drosophila*.

GOAL A quantitative analysis of genetic crosses may reveal the pattern of inheritance of a particular gene.

KEY MATERIALS A true-breeding line of red-eyed fruit flies plus one white-eyed male fly that was discovered in the population.

Experimental level | **Conceptual level**

1 Cross the white-eyed male to a red-eyed female.

P generation

$X^w Y$ × $X^{w+} X^{w+}$

2 Record the results of the F₁ generation. This involves noting the eye color and sexes of the offspring.

F₁ generation

$X^{w+} Y$: $X^{w+} X^w$

3 Cross F₁ offspring with each other to obtain F₂ generation of offspring.

$X^{w+} Y$ × $X^{w+} X^w$

4 Record the eye color and sex of F₂ offspring.

F₂ generation

$X^{w+} Y$: $X^w Y$: $X^{w+} X^{w+}$: $X^{w+} X^w$

5 **THE DATA**

Cross	Results
Original white-eyed male to a red-eyed female	F₁ generation All red-eyed flies
F₁ males to F₁ females	F₂ generation 1,011 red-eyed males 782 white-eyed males 2,459 red-eyed females 0 white-eyed females

6 **CONCLUSION** The data are consistent with an inheritance pattern in which an eye-color gene is located on the X chromosome.

7 **SOURCE** Morgan, Thomas H. 1910. Sex limited inheritance in *Drosophila*. *Science* 32:120–122.

many consecutive generations, the flies showed no noticeable changes. After 2 years of looking at many flies, Morgan happened to discover a male fly with white eyes rather than the normal red eyes. The white-eye trait must have arisen from a new mutation that converted a red-eye allele into a white-eye allele.

To study the inheritance of the white-eye trait, Morgan followed an approach similar to Mendel's in which he made crosses and quantitatively analyzed their outcome. In the experiment described in **Figure 16.15**, Morgan crossed his white-eyed male to a red-eyed female. All of the F₁ offspring had red eyes, indicating that red is dominant to white. The F₁ offspring were then mated to each other to obtain an F_2 generation. As seen in the data table, this cross produced 1,011 red-eyed males, 782 white-eyed males, and 2,459 red-eyed females. Surprisingly, no white-eyed females were observed in the F_2 generation.

How did Morgan interpret these results? The results suggested a connection between the alleles for eye color and the sex of the offspring. As shown in the conceptual column of Figure 16.15 and in the Punnett square below, his data were consistent with the idea that the eye-color alleles in *Drosophila* are located on the X chromosome. X^{w+} is the chromosome carrying the normal allele for red eyes, and X^w is the chromosome with the mutant allele for white eyes.

F_1 male is $X^{w+}Y$
F_1 female is $X^{w+}X^w$

Male gametes

♂ X^{w+} Y

♀

Female gametes

X^{w+}

| $X^{w+}X^{w+}$ | $X^{w+}Y$ |
| Red, female | Red, male |

X^w

| $X^{w+}X^w$ | X^wY |
| Red, female | White, male |

The Punnett square predicts that the F_2 generation will not have any white-eyed females. This prediction was confirmed by Morgan's experimental data. However, it should also be pointed out that the experimental ratio of red eyes to white eyes in the F_2 generation is $(2,459 + 1,011):782$, which equals $4.4:1$. This ratio deviates significantly from the ratio of $3:1$ predicted in the Punnett square. The lower than expected number of white-eyed flies is explained by a decreased survival of white-eyed flies.

Following this initial discovery, Morgan carried out many experimental crosses that located specific genes on the *Drosophila* X chromosome. This research provided some of the most persuasive evidence for Mendel's laws and the chromosome theory of inheritance, which are the foundations of modern genetics. In 1933, Morgan became the first geneticist to receive a Nobel Prize.

Experimental Questions

1. Prior to the Feature Investigation, what was the original purpose of Morgan's experiments with *Drosophila*?

2. What results led Morgan to conclude that eye color was associated with the sex of the individual?

3. What crosses between fruit flies could yield female offspring with white eyes?

16.5 Variations in Inheritance Patterns and Their Molecular Basis

The term **Mendelian inheritance** describes the inheritance patterns of genes that segregate and assort independently. In the first section of this chapter, we considered the inheritance pattern of traits affected by a single gene that is found in two variants, one of which is dominant over the other. This pattern is called **simple Mendelian inheritance**, because the phenotypic ratios in the offspring clearly demonstrate Mendel's laws. In the previous section, we examined **X-linked inheritance**, the pattern displayed by pairs of dominant and recessive alleles located on X chromosomes. Early geneticists observed these Mendelian inheritance patterns without knowing why one trait was dominant over another.

In this section, we will discuss the molecular basis of dominant and recessive traits and see how the molecular expression of a gene can have widespread effects on an organism's phenotype. In addition, we will examine the inheritance patterns of genes that segregate and assort independently but do not display a simple dominant/recessive relationship. The transmission of these genes from parents to offspring does not usually produce the ratios of phenotypes we would expect on the basis of Mendel's observations. This does not mean that Mendel was

wrong. Rather, the inheritance patterns of many traits are more intricate and interesting than the simple patterns he chose to study. As described in **Table 16.1**, our understanding of gene function at the molecular level explains both simple Mendelian inheritance and other, more complex, inheritance patterns that conform to Mendel's laws. This modern knowledge also sheds light on the role of the environment in producing an organism's phenotype, which we will discuss at the end of the section.

Protein Function Explains the Phenomenon of Dominance

As we discussed at the beginning of this chapter, Mendel studied seven characters that were found in two variants each (see Figure 16.2). The dominant variants are caused by the common alleles for these traits in pea plants. For any given gene, geneticists refer to a prevalent allele in a population as a **wild-type allele**. In most cases, a wild-type allele encodes a protein that is made in the proper amount and functions normally. By comparison, alleles that have been altered by mutation are called **mutant alleles**; these tend to be rare in natural populations. In the case of Mendel's seven characters, the recessive alleles are due to rare mutations.

How do we explain why one allele is dominant while another allele is recessive? By studying genes and their gene products at the molecular level, researchers have discovered

Table 16.1 Different Types of Mendelian Inheritance Patterns and Their Molecular Basis

Type	Description
Simple Mendelian inheritance	**Inheritance pattern:** Pattern of traits determined by a pair of alleles that display a dominant/recessive relationship and are located on an autosome. The presence of the dominant allele masks the presence of the recessive allele. **Molecular basis:** In many cases, the recessive allele is nonfunctional. Though a heterozygote may produce 50% of the functional protein compared to a dominant homozygote, this is sufficient to produce the dominant trait.
X-linked inheritance	**Inheritance pattern:** Pattern of traits determined by genes that display a dominant/recessive relationship and are located on the X chromosome. In mammals and fruit flies, males are hemizygous for X-linked genes. In these species, X-linked recessive traits occur more frequently in males than in females. **Molecular basis:** In a female with one recessive X-linked allele (a heterozygote), the protein encoded by the dominant allele is sufficient to produce the dominant trait. A male with a recessive X-linked allele (a hemizygote) does not have a dominant allele and does not make any of the functional protein.
Incomplete dominance	**Inheritance pattern:** Pattern that occurs when the heterozygote has a phenotype intermediate to the phenotypes of the homozygotes, as when a cross between red-flowered and white-flowered plants produces pink-flowered offspring. **Molecular basis:** 50% of the protein encoded by the functional (wild-type) allele is not sufficient to produce the normal trait.
Codominance	**Inheritance pattern:** Pattern that occurs when the heterozygote expresses both alleles simultaneously. For example, a human carrying the A and B alleles for the ABO antigens of red blood cells produces both the A and the B antigens (has an AB blood type). **Molecular basis:** The codominant alleles encode proteins that function slightly differently from each other. In a heterozygote, the function of each protein affects the phenotype uniquely.
Sex-influenced inheritance	**Inheritance pattern:** Pattern that occurs when an allele is dominant in one sex and recessive in the other, as in pattern baldness in humans. **Molecular basis:** Sex hormones affect the molecular expression of genes, which can have an impact on the phenotype.

that a recessive allele is often defective in its ability to express a functional protein. In other words, mutations that produce recessive alleles are likely to decrease or eliminate the synthesis or functional activity of a protein. These are called loss-of-function alleles. To understand why many loss-of-function alleles are recessive, we need to take a quantitative look at protein function.

In a simple dominant/recessive relationship, the recessive allele does not affect the phenotype of the heterozygote. In this type of relationship, a single copy of the dominant (wild-type) allele is sufficient to mask the effects of the recessive allele. How do we explain the dominant phenotype of the heterozygote? **Figure 16.16** considers the example of flower color in a pea plant. The gene encodes an enzyme that is needed to make a purple pigment. The *P* allele is dominant because one *P* allele encodes enough of the functional protein—50% of the amount found in a normal homozygote—to provide a normal phenotype. Thus, the *PP* homozygote and the *Pp* heterozygote both make enough of the purple pigment to yield purple flowers. The *pp* homozygote cannot make any of the functional protein required for pigment synthesis, so its flowers are white.

The explanation—50% of the normal protein is enough—is true for many dominant alleles. In such cases, the normal homozygote is making much more of the protein than necessary, so if the amount is reduced to 50%, as it is in the heterozygote, the individual still has plenty of this protein to accomplish whatever cellular function it performs. In other cases, however, an allele may be dominant because the heterozygote actually

Protein P functions as an enzyme. The amount of functional protein P is the molecular connection between the genotype and the phenotype. The normal (dominant) allele encodes a functional enzyme.

Genotype	PP	Pp	pp
Amount of functional protein P produced	100%	50%	0%
Phenotype	**Purple**	**Purple**	**White**
The relationship of the normal (dominant) and mutant (recessive) alleles displays simple Mendelian inheritance.			

Colorless precursor molecule   Protein P   Purple pigment

Figure 16.16 **How genes give rise to traits during simple Mendelian inheritance.** In many cases, the amount of protein encoded by a single dominant allele is sufficient to produce the normal phenotype. In this example, a plant with one or two copies of the normal allele produces enough pigment to produce purple flowers. In a *pp* homozygote, the complete lack of the normal protein results in white flowers.

produces more than 50% of the normal amount of functional protein. This increased production is due to the phenomenon of gene regulation, which is discussed in Chapter 13. The normal gene is "up-regulated" in the heterozygote to compensate for the lack of function of the defective allele.

Genomes & Proteomes Connection

Single-Gene Mutations Cause Many Inherited Diseases and Have Pleiotropic Effects

The idea that recessive alleles usually cause a substantial decrease in the expression of a functional protein is supported by analyses of many human genetic diseases. Keep in mind that many genetic diseases are caused by rare mutant alleles. **Table 16.2** lists several examples of human genetic diseases in which a recessive allele fails to produce a specific cellular protein in its active form.

Over 7,000 human disorders are caused by mutations in single genes. With a human genome size of 20,000 to 25,000 genes, this means that roughly one-third of our genes are known to cause some kind of abnormality when mutations alter the expression or functionality of their gene product. Any particular single-gene disorder is relatively rare. But taken together, about one individual in 100 has a disorder that is due to a single-gene mutation. Such diseases generally have simple inheritance patterns in family pedigrees. Although the majority of these diseases follow a recessive inheritance pattern, some are known to be dominant. We have already discussed Huntington disease as an example of a dominant human disorder (see Figure 16.13). Other examples of diseases caused by dominant alleles include achondroplasia (a form of dwarfism) and osteogenesis imperfecta (brittle bone disease).

Single-gene disorders also illustrate the phenomenon of **pleiotropy**, which means that a mutation in a single gene can have multiple effects on an individual's phenotype. Pleiotropy occurs for several reasons, including the following:

1. The expression of a single gene can affect cell function in more than one way. For example, a defect in a microtubule protein may affect cell division and cell movement.

2. A gene may be expressed in different cell types in a multicellular organism.

3. A gene may be expressed at different stages of development.

In this genetics unit, we tend to discuss genes as they affect a single trait. This educational approach allows us to appreciate how genes function and how they are transmitted from parents to offspring. However, this focus may also obscure how amazing genes really are. In all or nearly all cases, the expression of a gene is pleiotropic with regard to the characteristics of an organism. The expression of any given gene influences the expression of many other genes in the genome, and vice

Table 16.2	Examples of Recessive Human Genetic Diseases	
Disease	**Protein produced by the normal gene***	**Description**
Phenylketonuria	Phenylalanine hydroxylase	Inability to metabolize phenylalanine. Can lead to severe mental retardation and physical degeneration. The disease can be prevented by following a phenylalanine-free diet beginning early in life.
Cystic fibrosis	A chloride-ion transporter	Inability to regulate ion balance in epithelial cells. Leads to a variety of abnormalities, including production of thick lung mucus and chronic lung infections.
Tay-Sachs disease	Hexosaminidase A	Defect in lipid metabolism. Leads to paralysis, blindness, and early death.
Alpha-1 antitrypsin deficiency	Alpha-1 antitrypsin	Inability to prevent the activity of protease enzymes. Leads to a buildup of certain proteins that cause liver damage and emphysema.
Hemophilia A	Coagulation factor VIII	A defect in blood clotting due to a missing clotting factor. An accident may cause excessive bleeding or internal hemorrhaging.

*Individuals who exhibit the disease are homozygous (or hemizygous) for a recessive allele that results in a defect in the amount or function of the normal protein.

versa. Pleiotropy is revealed when researchers study the effects of gene mutations.

As an example of a pleiotropic mutation, let's consider cystic fibrosis (CF), which we discussed earlier as an example of a recessive human disorder (see Figure 16.12). In the late 1980s, the gene for CF was identified. The normal allele encodes a protein called the cystic fibrosis transmembrane conductance regulator (CFTR), which regulates ionic balance by allowing the transport of chloride ions (Cl^-) across epithelial-cell membranes. The mutation that causes CF diminishes the function of this Cl^- transporter, affecting several parts of the body in different ways. Because the movement of Cl^- affects water transport across membranes, the most severe symptom of CF is the production of thick mucus in the lungs, which occurs because of a water imbalance. In sweat glands, the normal Cl^- transporter has the function of recycling salt out of the glands and back into the skin before it can be lost to the outside world. Persons with CF have excessively salty sweat due to their inability to recycle salt back into their skin cells. A common test for CF is the measurement of salt on the skin. Another effect is seen in the reproductive systems of males who are homozygous for the mutant allele. Some males with CF are infertile because the vas deferens, the tubules that transport sperm from the testes, are absent or undeveloped. Presumably, a normally functioning Cl⁻ transporter is needed for the proper development of the vas deferens in the embryo. Taken together, we can see that a defect in CFTR has multiple effects throughout the body.

Incomplete Dominance Results in an Intermediate Phenotype

We will now turn our attention to examples in which the alleles for a given gene do not show a simple dominant/recessive relationship. In some cases, a heterozygote that carries two different alleles exhibits a phenotype that is intermediate between the corresponding homozygous individuals. This phenomenon is known as **incomplete dominance**.

In 1905, Carl Correns discovered this pattern of inheritance for alleles affecting flower color in the four-o'clock plant (*Mirabilis jalapa*). **Figure 16.17** shows a cross between two four-o'clock plants, a red-flowered homozygote and a white-flowered homozygote. The allele for red flower color is designated C^R, and the white allele is C^W. These alleles are designated with superscripts rather than upper- and lowercase letters because neither allele is dominant. The offspring of this cross have pink flowers—they are $C^R C^W$ heterozygotes with an intermediate phenotype. If these F_1 offspring are allowed to self-fertilize, the F_2 generation has 1/4 red-flowered plants, 1/2 pink-flowered plants, and 1/4 white-flowered plants. This is a 1:2:1 phenotypic ratio rather than the 3:1 ratio observed for simple Mendelian inheritance. What is the molecular explanation for this ratio? In this case, the red allele encodes a functional protein needed to produce a red pigment, whereas the white allele is a mutant allele that is nonfunctional. In the $C^R C^W$ heterozygote, 50% of the protein encoded by the C^R allele is not sufficient to produce the red-flower phenotype, but it does provide enough pigment to give pink flowers.

The degree to which we judge an allele to exhibit incomplete dominance may depend on how closely we examine an individual's phenotype. An example is an inherited human disease called phenylketonuria (PKU). This disorder is caused by a rare mutation in a gene that encodes an enzyme called phenylalanine hydroxylase. This enzyme is needed to metabolize the amino acid phenylalanine, which is found in milk, eggs, and other protein-rich foods. If left untreated, phenylalanine builds up, affecting various systems in the body. Homozygotes carrying the mutant allele suffer severe symptoms, including mental retardation, seizures, microcephaly (small head), poor development of tooth enamel, and decreased body growth. By comparison, heterozygotes appear phenotypically normal. For this reason, geneticists consider PKU to be a recessive disorder. However, biochemical analysis of the blood of heterozygotes shows they typically have a phenylalanine blood level double that of an individual carrying two normal copies of the gene.

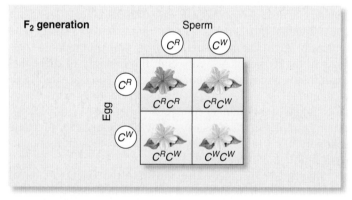

Figure 16.17 Incomplete dominance in the four-o'clock plant. When red-flowered and white-flowered homozygotes ($C^R C^R$ and $C^W C^W$) are crossed, the resulting heterozygote ($C^R C^W$) has an intermediate phenotype of pink flowers.

Individuals with PKU (homozygous recessive) typically have phenylalanine blood levels 30 times higher than normal. Therefore, at this closer level of examination, heterozygotes exhibit an intermediate phenotype in comparison to the homozygous dominant and recessive individuals. At this closer level of inspection, the relationship between the normal and mutant alleles would be defined as incomplete dominance.

ABO Blood Type Is an Example of Multiple Alleles and Codominance

Although diploid individuals have only two copies of most genes, the majority of genes have three or more variants in natural populations. We describe such genes as occurring in **multiple alleles**. Particular phenotypes depend on which two alleles each individual inherits. ABO blood types in humans are an example of phenotypes produced by multiple alleles.

As shown in **Table 16.3**, human red blood cells have structures on their plasma membrane known as surface antigens, which are constructed from several sugar molecules that are connected to form a carbohydrate tree. The carbohydrate tree is attached to lipids or membrane proteins to form glycolipids or glycoproteins, which are described in Chapter 5.

Antigens are substances (in this case, carbohydrates) that may be recognized as foreign material when introduced into the body of an animal. Let's consider two types of surface antigens, known as A and B, which may be found on red blood cells. The synthesis of these antigens is determined by enzymes that are encoded by a gene that exists in three alleles, designated I^A, I^B, and i, respectively. The i allele is recessive to both I^A and I^B. A person who is ii homozygous will not produce surface antigen A or B and will have blood type O. The red blood cells of an $I^A I^A$ homozygous or $I^A i$ heterozygous individual will have surface antigen A (blood type A). Similarly, a homozygous $I^B I^B$ or heterozygous $I^B i$ individual will produce surface antigen B (blood type B). A person who is $I^A I^B$ heterozygous makes both antigens, A and B, on every red blood cell (blood type AB). The phenomenon in which a single individual expresses both alleles is called **codominance**.

What is the molecular explanation for codominance? Biochemists have analyzed the carbohydrate tree produced in people of differing blood types. The differences are shown schematically in Table 16.3. In type O, the carbohydrate tree is smaller than in type A or type B because a sugar has not been attached to a specific site on the tree. People with blood type O have a loss-of-function mutation in the gene that encodes the enzyme that attaches a sugar at this site. This enzyme, called a glycosyl transferase, is inactive in type O individuals. In contrast, the type A and type B antigens have sugars attached to this site, but each of them has a different sugar. This difference occurs because the enzymes encoded by the I^A allele and the I^B allele have slightly different active sites.

As a result, the enzyme encoded by the I^A allele attaches a sugar called N-acetylgalactosamine to the carbohydrate tree, whereas the enzyme encoded by the I^B allele attaches galactose. N-acetylgalactosamine is represented by an orange hexagon in Table 16.3, and galactose by a green triangle.

Blood type is a critical issue for a blood transfusion between a donor and a recipient. Surface antigens A and B have different molecular structures. Such differences allow antibodies, which are produced by the immune system, to recognize and bind very specifically to these antigens. The blood of type A individuals has antibodies that bind to the B antigen. Similarly, type B individuals produce antibodies against the A antigen. Type O individuals produce both kinds of antibodies, and type AB individuals produce neither. (The structure that exists in the absence of added galactose or galactosamine is called the H antigen. Because everyone makes this portion of the carbohydrate tree, no antibodies are formed against it.) When a person receives a blood transfusion, the donor's blood must be an appropriate match with the recipient's blood to avoid a dangerous antigen-antibody reaction. For example, if a person with type O blood is given type A blood, the recipient's anti-A antibodies will react with the donated blood cells and cause them to agglutinate (clump together). This situation is life-threatening because it will cause the blood vessels to clog. Identification of the donor and recipient blood types, called blood typing, is essential for safe transfusions.

The Expression of Certain Traits Is Influenced by the Sex of the Individual

Certain autosomal genes are expressed differently in heterozygous males and females. The term **sex-influenced inheritance** refers to the phenomenon in which an allele is dominant in one sex but recessive in the other. A particular form of baldness, called androgenetic alopecia, or pattern baldness, is an example of a sex-influenced trait in humans. This trait is characterized by a pattern in which hair loss occurs on the front and top but not on the sides (**Figure 16.18**). A male who is heterozygous for the pattern-baldness allele (designated B) will exhibit hair loss,

Table 16.3	The ABO Blood Group			
Blood type:	O	A	B	AB
Genotype:	ii	$I^A I^A$ or $I^A i$	$I^B I^B$ or $I^B i$	$I^A I^B$
Surface antigen:	Neither A nor B	A	B	A and B
Antibodies:	Against A and B	Against B	Against A	Neither

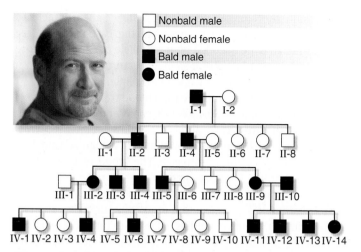

Figure 16.18   A family pedigree showing sex-influenced inheritance.   Pattern baldness, shown in an adult male in the photograph, is an example of sex-influenced inheritance of an autosomal gene. Bald individuals are represented by filled symbols in the pedigree.

Concept check: *Let's suppose two nonbald parents produce a bald son. What are the genotypes of the parents and the son?*

but a heterozygous female will not. In other words, the baldness allele is dominant in males but recessive in females:

Genotype	Phenotype	
	Females	*Males*
BB	bald	bald
Bb	nonbald	bald
bb	nonbald	nonbald

A woman who is homozygous for the baldness allele may develop the trait, although in women it is usually characterized by a significant thinning of the hair that occurs relatively late in life.

As you can see from the pedigree in Figure 16.18, a bald male may have inherited the baldness allele from either parent. Thus, a striking observation is that fathers with pattern baldness can pass this trait to their sons. This could not occur if the trait was X-linked, because fathers transmit only Y chromosomes to their sons.

The sex-influenced nature of pattern baldness is related to the production of the male sex hormone testosterone. Though other genes may influence baldness, the gene that plays a primary role in pattern baldness encodes an enzyme called 5α-reductase, which converts testosterone to 5α-dihydrotestosterone (DHT). This gene is not X-linked. It is located on chromosome 5, which is an autosome. DHT binds to cellular receptors and affects the expression of many genes, including those in the cells of the scalp. The allele that causes pattern baldness results in an overexpression of this enzyme. Because mature males normally make more testosterone than females, this allele has a greater phenotypic impact in males. However, a rare tumor of the adrenal gland can cause the secretion of abnormally large amounts

of testosterone in females. If this occurs in a woman who is heterozygous *Bb*, she will become bald. If the tumor is removed surgically, her hair will return to its normal condition.

The Environment Plays a Vital Role in the Making of a Phenotype

In this chapter, we have been mainly concerned with the effects of genes on phenotypes. In addition, phenotypes are shaped by an organism's environment. An organism cannot exist without its genes or without an environment in which to live. Both are indispensable for life. An organism's genotype provides the plan to create a phenotype, while the environment provides nutrients and energy so that plan can be executed.

The term **norm of reaction** refers to the effects of environmental variation on a phenotype. Specifically, it is the phenotypic range seen in individuals with a particular genotype. To evaluate the norm of reaction, researchers study members of true-breeding strains that have the same genotypes and subject them to different environmental conditions. For example, Figure 16.19 shows the norm of reaction for genetically identical plants raised at different temperatures. As shown in the figure, these plants attain a maximal height when raised at 75°F. At 50°F and 85°F, the plants are substantially shorter. Growth cannot occur below 40°F or above 95°F.

The norm of reaction can be quite dramatic when we consider environmental influences on certain inherited diseases. A striking example is the human genetic disease phenylketonuria (PKU). As we discussed earlier in the chapter, this disorder is caused by a rare mutation in the gene that encodes the enzyme phenylalanine hydroxylase, which is needed to metabolize the amino acid phenylalanine. People with one or two functional

Figure 16.19   The norm of reaction.   The norm of reaction is the range of phenotypes that an organism with a particular genotype exhibits under different environmental conditions. In this example, genetically identical plants were grown at different temperatures in a greenhouse and then measured for height.

Concept check: *Could you study the norm of reaction in a wild population of squirrels?*

Figure 16.20 Environmental influences on the expression of PKU within a single family. All three children in this photo have inherited the alleles that cause PKU. The child in the middle was raised on a phenylalanine-free diet and developed normally. The other two children, born before the benefits of such a diet were known, were raised on diets containing phenylalanine. These two children have symptoms of PKU, including mental impairment.

copies of the gene can eat foods containing the amino acid phenylalanine and metabolize it correctly. However, individuals with two copies of the mutant gene cannot metabolize phenylalanine. When these individuals eat a standard diet containing phenylalanine, this amino acid accumulates within their bodies and becomes highly toxic. Under these conditions, PKU homozygotes manifest a variety of detrimental symptoms, including mental impairment, underdeveloped teeth, and foul-smelling urine. In contrast, when these individuals are identified at birth and given a restricted diet that is free of phenylalanine, they develop normally (**Figure 16.20**). This is a dramatic example of how genes and the environment can interact to determine an individual's phenotype. In the U.S., most newborns are tested for PKU, which occurs in about 1 in 10,000 babies. A newborn who is found to have this disorder can be raised on a phenylalanine-free diet and develop normally.

16.6 Genetics and Probability

As we have seen throughout this chapter, Mendel's laws of inheritance can be used to predict the outcome of genetic crosses. How is this useful? In agriculture, plant and animal breeders use predictions about the types and relative numbers of offspring their crosses will produce in order to develop commercially important crops and livestock. Also, people are often interested in the potential characteristics of their future children. This has particular importance to individuals who may carry alleles that cause inherited diseases. Of course, no one can see into the future and definitively predict what will happen. Nevertheless, genetic counselors can often help couples predict the likelihood of having an affected child. This probability is one factor that may influence a couple's decision about whether to have children.

Earlier in this chapter, we considered how a Punnett square can be used to predict the outcome of simple genetic crosses. In addition to Punnett squares, we can apply the tools of mathematics and probability to solve more complex genetic problems. In this section, we will examine a couple of ways to calculate the outcomes of genetic crosses using these tools.

Genetic Predictions Are Based on the Mathematical Rules of Probability

The chance that an event will have a particular outcome is called the **probability** of that outcome. The probability of a given outcome depends on the number of possible outcomes. For example, if you draw a card at random from a 52-card deck, the probability that you will get the jack of diamonds is 1 in 52, because there are 52 possible outcomes for the draw. In contrast, only two outcomes are possible when you flip a coin, so the probability is one in two (1/2, or 0.5, or 50%) that the heads side will be showing when the coin lands. The general formula for the probability (P) that a random event will have a specific outcome is

$$P = \frac{\text{Number of times an event occurs}}{\text{Total number of possible outcomes}}$$

Thus, for a single coin toss, the chance of getting heads is

$$P_{\text{heads}} = \frac{1 \text{ heads}}{(1 \text{ heads} + 1 \text{ tails})} = \frac{1}{2}$$

Earlier in this chapter, we used Punnett squares to predict the fractions of offspring with a given genotype or phenotype. In a cross between two pea plants that were heterozygous for the height gene (Tt), our Punnett square predicted that one-fourth of the offspring would be dwarf. We can make the same prediction by using a probability calculation.

$$P_{\text{dwarf}} = \frac{1 \, tt}{(1 \, TT + 2 \, Tt + 1 \, tt)} = \frac{1}{4} = 25\%$$

A probability calculation allows us to predict the likelihood that a future event will have a specific outcome. However, the accuracy of this prediction depends to a great extent on the number of events we observe—in other words, on the size of our sample. For example, if we toss a coin six times, the calculation we just presented for P_{heads} suggests we should get heads three times and tails three times. However, each coin toss is an independent event, meaning that every time we toss the coin there is a random chance that it will come up heads or tails, regardless of the outcome of the previous toss. With only six tosses, we would not be too surprised if we got four heads and two tails instead of the expected three heads and three tails. The deviation between the observed and expected outcomes is called the **random sampling error**. With a small sample, the random sampling error may cause the observed data to be quite different from the expected outcome. By comparison, if we flipped a coin 1,000 times, the percentage of heads would be fairly close to the predicted 50%. With a larger sample, we expect the sampling error to be smaller.

The Product Rule Is Used to Predict the Outcome of Independent Events

Punnett squares allow us to predict the likelihood that a genetic cross will produce an offspring with a particular genotype or phenotype. To predict the likelihood of producing multiple offspring with particular genotypes or phenotypes, we can use the **product rule**, which states that *the probability that two or more independent events will occur is equal to the product of their individual probabilities*. As we have already discussed, events are independent if the outcome of one event does not affect the outcome of another. In our previous coin-toss example, each toss is an independent event—if one toss comes up heads, another toss still has an equal chance of coming up either heads or tails. If we toss a coin twice, what is the probability that we will get heads both times? The product rule says that it is equal to the probability of getting heads on the first toss (1/2) times the probability of getting heads on the second toss (1/2), or one in four ($1/2 \times 1/2 = 1/4$).

To see how the product rule can be applied to a genetics problem, let's consider a rare recessive human trait known as congenital analgesia. (Congenital refers to a condition present at birth; analgesia means insensitivity to pain.) People with this trait can distinguish between sensations such as sharp and dull, or hot and cold, but they do not perceive extremes of sensation as painful. The first known case of congenital analgesia, described in 1932, was a man who made his living entertaining the public as a "human pincushion." For a phenotypically normal couple, each heterozygous for the recessive allele causing congenital analgesia, we can ask, What is the probability that their first three offspring will have the disorder? To answer this question, we must first determine the probability of a single offspring having the abnormal phenotype. By using a Punnett square, we would find that the probability of an individual offspring being homozygous recessive is 1/4. Thus, each of this couple's children has a one in four chance of having the disorder.

We can now use the product rule to calculate the probability of this couple having three affected offspring in a row. The phenotypes of the first, second, and third offspring are independent events; that is, the phenotype of the first offspring does not affect the phenotype of the second or third offspring. The product rule tells us that the probability of all three children having the abnormal phenotype is

$$\frac{1}{4} \times \frac{1}{4} \times \frac{1}{4} = \frac{1}{64} = 0.016$$

The probability of the first three offspring having the disorder is 0.016, or 1.6%. In other words, we can say that this couple's chance of having three children in a row with congenital analgesia is very small—only 1.6 out of 100. The phenotypes of the first, second, and third child are independent of each other.

The product rule can also be used to predict the outcome of a cross involving two or more genes. Let's suppose a pea plant with the genotype *TtYy* was crossed to a plant with the genotype *Ttyy*. We could ask the question, What is the probability that an offspring will have the genotype *ttYy*? If the two genes independently assort, the probability of inheriting alleles for one gene is independent of the other gene. Therefore, we can separately calculate the probability of the desired outcome for each gene. By constructing two small Punnett squares, we can determine the probability of genotypes for each gene individually, as shown below.

Cross: *TtYy* × *Ttyy*

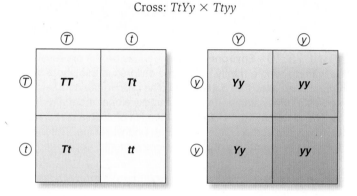

Probability that an offspring will be *tt* is 1/4, or 0.25.

Probability that an offspring will be *Yy* is 1/2, or 0.5

We can now use the product rule to determine the probability that an offspring will be *ttYy*;

$$P = (0.25)(0.5) = 0.125, \text{ or } 12.5\%$$

The Sum Rule Is Used to Predict the Outcome of Mutually Exclusive Events

Let's now consider a second way to predict the outcome of particular crosses. In a cross between two heterozygous (*Tt*) pea plants, we may want to know the probability of a particular offspring being a homozygote. In this case we are asking, What is the chance that this individual will be either homozygous *TT* or homozygous *tt*? To answer an "either/or" question, we use the sum rule, which applies to events with mutually exclusive outcomes. When we say that outcomes are mutually exclusive, we mean they cannot occur at the same time. A pea plant can be tall or dwarf, but not both at the same time. The tall and dwarf phenotypes are mutually exclusive. Similarly, a plant with the genotype *TT* cannot be *Tt* or *tt*. Each of these genotypes is mutually exclusive with the other two. According to the **sum rule**, *the probability that one of two or more mutually exclusive outcomes will occur is the sum of the probabilities of the individual outcomes*.

To find the probability that an offspring will be either homozygous *TT* or homozygous *tt*, we add together the probability that it will be *TT* and the probability that it will be *tt*. If we constructed a Punnett square, we would find that the probability for each of these genotypes is one in four. We can now use the sum rule to determine the probability of an individual having one of these genotypes.

$$\frac{1}{4} \quad + \quad \frac{1}{4} \quad = \quad \frac{1}{2}$$

(probability of *TT*)　　(probability of *tt*)　　(probability of either *TT* or *tt*)

This calculation predicts that in crosses of two *Tt* parents, half of the offspring will be homozygotes—either *TT* or *tt*.

Summary of Key Concepts

16.1 Mendel's Laws of Inheritance

- Mendel studied seven characters found in garden peas that existed in two variants each. (Figures 16.1, 16.2)
- Mendel allowed his peas to self-fertilize, or he carried out cross-fertilization, also known as hybridization. (Figures 16.3, 16.4)
- By following the inheritance pattern of a single character (a monohybrid cross) for two generations, Mendel determined the law of segregation. This law tells us that two alleles of a gene segregate during the formation of eggs and sperm so that every gamete receives only one allele. (Figures 16.5, 16.6)
- The genotype is the genetic makeup of an organism. Alleles are alternative versions of the same gene. Phenotype is a description of the traits that an organism displays.
- A Punnett square can be constructed to predict the outcome of crosses.
- A testcross can be conducted to determine if an individual displaying a dominant trait is a homozygote or a heterozygote. (Figure 16.7)
- By conducting a dihybrid cross, Mendel determined the law of independent assortment, which states that the alleles of different genes assort independently of each other during gamete formation. In a dihybrid cross, this yields a 9:3:3:1 ratio in the F_2 generation. (Figure 16.8)

16.2 The Chromosome Theory of Inheritance

- The chromosome theory of inheritance explains how the steps of meiosis account for Mendel's laws of inheritance. Each gene is located at a particular locus on a chromosome. (Figures 16.9, 16.10, 16.11)

16.3 Pedigree Analysis of Human Traits

- The inheritance patterns in humans are determined from a pedigree analysis. (Figures 16.12, 16.13)

16.4 Sex Chromosomes and X-Linked Inheritance Patterns

- Many species of animals and some species of plants have separate male and female sexes. In many cases, sex is determined by differences in sex chromosomes. (Figure 16.14)
- In mammals, recessive X-linked traits such as hemophilia are more likely to occur in males.

- Morgan's experiments showed that an eye-color gene in *Drosophila* is located on the X chromosome. (Figure 16.15)

16.5 Variations in Inheritance Patterns and Their Molecular Basis

- Several inheritance patterns have been discovered that obey Mendel's laws but yield differing ratios of offspring compared to Mendel's crosses. (Table 16.1)
- Recessive inheritance is often due to a loss-of-function mutation. In many simple dominant/recessive relationships, the heterozygote has a dominant phenotype because 50% of the normal protein is sufficient to produce that phenotype. (Figure 16.16)
- Mutant genes are responsible for many inherited diseases in humans. In many cases, the effects of a mutant gene are pleiotropic, meaning the gene affects several different aspects of bodily structure and function. (Table 16.2)
- Incomplete dominance occurs when a heterozygote has a phenotype that is intermediate between either homozygote. This occurs because 50% of the functional protein is not enough to produce the same phenotype as a homozygote. (Figure 16.17)
- ABO blood type is an example of multiple alleles in which a gene exists in three alleles in a population. The I^A and I^B alleles show codominance, which means that both are expressed in the same individual. These alleles encode enzymes with different specificities for attaching sugar molecules to make antigens. (Table 16.3)
- Pattern baldness in people is a sex-influenced trait that is dominant in males and recessive in females. This pattern occurs because sex hormones influence the expression of certain genes. (Figure 16.18)
- Phenotypes are influenced by an organism's environment as well its genes. The norm of reaction is a description of how a phenotype may change depending on the environmental conditions. (Figures 16.19, 16.20)

16.6 Genetics and Probability

- Probability is the likelihood that an event will occur in the future. Random sampling error is the deviation between observed and expected values.
- The product rule states that the probability of two or more independent events occurring is equal to the product of their individual probabilities. The sum rule states that the probability that two or more mutually exclusive events occurring is the sum of the individual probabilities.

Assess and Discuss

Test Yourself

1. Based on Mendel's experimental crosses, what is the expected F_2 phenotypic ratio of a monohybrid cross?
   - a. 1:2:1
   - b. 2:1
   - c. 3:1
   - d. 9:3:3:1
   - e. 4:1

2. During which phase of cellular division does Mendel's law of segregation physically occur?
   a. mitosis   c. meiosis II   e. b and c only
   b. meiosis I   d. all of the above

3. An individual that has two different alleles of a particular gene is said to be
   a. dihybrid.   d. heterozygous.
   b. recessive.   e. hemizygous.
   c. homozygous.

4. Which of Mendel's laws cannot be observed in a monohybrid cross?
   a. segregation
   b. dominance/recessiveness
   c. independent assortment
   d. codominance
   e. All of the above can be observed in a monohybrid cross.

5. During a _____ cross, an individual with the dominant phenotype and unknown genotype is crossed with a _____ individual to determine the unknown genotype.
   a. monohybrid, homozygous recessive
   b. dihybrid, heterozygous
   c. test, homozygous dominant
   d. monohybrid, homozygous dominant
   e. test, homozygous recessive

6. A woman is heterozygous for an X-linked trait, hemophilia A. If she has a child with a man without hemophilia A, what is the probability that the child will be a male with hemophilia A? (Note: The child could be a male or female.)
   a. 100%   c. 50%   e. 0%
   b. 75%   d. 25%

7. A gene that affects more than one phenotypic trait is said to be
   a. dominant.   d. pleiotropic.
   b. wild type.   e. heterozygous.
   c. dihybrid.

8. A hypothetical flowering plant species produces red, pink, and white flowers. To determine the inheritance pattern, the following crosses were conducted with the results indicated:
   red × red → all red
   white × white → all white
   red × white → all pink
   What type of inheritance pattern does this represent?
   a. dominance/recessiveness   d. incomplete dominance
   b. X-linked   e. pleiotropy
   c. codominance

9. Genes located on a sex chromosome are said to be
   a. X-linked.   c. hemizygous.   e. sex influenced.
   b. dominant.   d. sex linked.

10. A nonbald male has an offspring with a woman who is homozygous for the pattern baldness allele. What is the probability that the offspring will become bald?
    a. 100%   c. 50%   e. 0%
    b. 75%   d. 25%

Conceptual Questions

1. Describe one observation in a human pedigree that would rule out a recessive pattern of inheritance. Describe an observation that would rule out a dominant pattern.

2. A cross is made between individuals of the following genotypes: *AaBbCCDd* and *AabbCcdd*. What is the probability that an offspring will be *AAbbCCDd*? Hint: Don't waste your time making a really large Punnett square. Make four small Punnett squares instead and use the product rule.

3. Explain why recessive X-linked traits in humans are more likely to occur in males.

Collaborative Questions

1. Discuss the principles of the chromosome theory of inheritance. Which principles do you think were deduced via light microscopy, and which were deduced from crosses? What modern techniques could be used to support the chromosome theory of inheritance?

2. When examining a human pedigree, what observations do you look for to distinguish between X-linked recessive inheritance versus autosomal recessive inheritance? How would you distinguish X-linked dominant inheritance from autosomal dominant inheritance from an analysis of a human pedigree?

Online Resource

www.brookerbiology.com

Stay a step ahead in your studies with animations that bring concepts to life and practice tests to assess your understanding. Your instructor may also recommend the interactive ebook, individualized learning tools, and more.

17.1   Gene Interaction

The study of single genes was pivotal in establishing the science of genetics. This focus allowed Mendel to formulate the basic laws of inheritance for traits with a simple dominant/recessive inheritance pattern. Likewise, this approach helped later researchers understand inheritance patterns involving incomplete dominance and codominance, as well as traits that are influenced by an individual's sex. However, all or nearly all traits are influenced by many genes. For example, in both plants and animals, height is affected by genes that encode proteins involved in the production of growth hormones, cell division, the uptake of nutrients, metabolism, and many other functions. Variation in any of the genes involved in these processes is likely to have an impact on an individual's height.

If height is controlled by many genes, how was Mendel able to study the effects of a single gene that produced tall or dwarf pea plants? The answer lies in the genotypes of his strains. Although many genes affect the height of pea plants, Mendel chose true-breeding strains that differed with regard to only one of those genes. As a hypothetical example, let's suppose that pea plants have 10 genes affecting height, which we will call $K, L, M, N, O, P, Q, R, S,$ and T. The genotypes of two hypothetical strains of pea plants may be:

Tall strain: *KK LL MM NN OO PP QQ RR SS TT*

Dwarf strain: *KK LL MM NN OO PP QQ RR SS tt*

In this example, the tall and dwarf strains differ at only a single locus. One strain is *TT* and the other is *tt*, and this accounts for the difference in their height. If we make crosses of tall and dwarf plants, the genotypes of the F_2 offspring will differ with regard to only one gene; the other nine genes will be identical in all of them. This approach allows a researcher to study the effects of a single gene even though many genes may affect a single character.

In this section, we will examine situations in which a single character is controlled by two or more genes, each of which has two or more alleles. This phenomenon is called **gene interaction**. As you will see, allelic variation at two or more loci may affect the outcome of traits in different ways. First we will look at a gene interaction in which an allele of one gene prevents the phenotypic expression of an allele of a different gene. Then we will discuss an interaction in which multiple genes have additive effects on a single character. These additive effects, together with environmental influences, account for the continuous phenotypic variation that we see for most traits.

In an Epistatic Gene Interaction, the Allele of One Gene Masks the Phenotypic Effects of a Different Gene

In some gene interactions, the alleles of one gene mask the expression of the alleles of another gene. This phenomenon is

called **epistasis** (from the Greek *ephistanai*, meaning stopping). An example is the unexpected gene interaction discovered by William Bateson and Reginald Punnett in the early 1900s, when they were studying crosses involving the sweet pea, *Lathyrus odoratus*. A cross between a true-breeding purple-flowered plant and a true-breeding white-flowered plant produced an F_1 generation with all purple-flowered plants and an F_2 generation with a 3:1 ratio of purple- to white-flowered plants. Mendel's laws predicted this result. The surprise came when the researchers crossed two different true-breeding varieties of white-flowered sweet peas (**Figure 17.1**). All of the F_1 generation plants had purple flowers! When these plants were allowed to self-fertilize, the F_2 generation had purple-flowered and white-flowered plants in a 9:7 ratio. From these results, Bateson and Punnett deduced that two different genes were involved. To have purple flowers, a plant must have one or two dominant alleles for each of these genes. The relationships among the alleles are as follows:

C (one allele for purple) is dominant to c (white)

P (an allele of a different gene for purple) is dominant to p (white)

cc masks P, or pp masks C, in either case producing white flowers

A plant that was homozygous for either c or p would have white flowers even if it had a dominant purple-producing allele at the other locus.

How do we explain these results at the molecular and cellular level? Epistatic interactions often arise because two or more different proteins are involved in a single cellular function. For example, two or more proteins may be part of an enzymatic pathway leading to the formation of a single product. This is the case for the formation of a purple pigment in the sweet pea strains we have been discussing:

Colorless precursor $\xrightarrow{\text{Enzyme C}}$ Colorless intermediate $\xrightarrow{\text{Enzyme P}}$ Purple pigment

In this example, a colorless precursor molecule must be acted on by two different enzymes to produce the purple pigment. Gene C encodes a functional protein called enzyme C that converts the colorless precursor into a colorless intermediate. The recessive c allele results in a lack of production of enzyme C in the homozygote. Gene P encodes the functional enzyme P, which converts the colorless intermediate into the purple pigment. Like the c allele, the p allele results in an inability to produce a functional enzyme. A plant homozygous for either of the recessive alleles will not make any functional enzyme C or enzyme P. When either of these enzymes is missing, the plant cannot make the purple pigment and has white flowers. Note that the results observed in Figure 17.1 do not conflict with Mendel's laws of segregation or independent assortment. Mendel investigated the effects of only a single gene on a given

Complex Patterns of Inheritance

17

Photo of a calico cat. This female cat is heterozygous for X-linked alleles that confer orange or black patches of fur.

Darren was invited to a Friday night party at Emily's apartment. Darren took his roommate, Chris, with him. While at the party, Chris noticed that Emily had a cat with a calico coat pattern, in which patches of orange fur are mixed with patches of black fur (see chapter-opening photo). Chris said to Darren, "I'll bet you ten to one that Emily's cat is a female." Since Darren knew that Chris had never met Emily before and certainly had not seen her cat before, Darren said, "You're on." So he walked up to Emily and asked, "What's your cat's name?" "Her name is Patches," Emily replied. "Isn't she cute?" Hmmmm. . . . thought Darren, maybe I shouldn't have made that bet.

How did Chris know that Emily's cat was a female? The answer is that the calico coat pattern almost always occurs in females. It is an example of a complex pattern of inheritance, one that could not have been predicted from Mendel's laws.

In this chapter, we will explore inheritance patterns that would be difficult if not impossible to predict based solely on Mendel's laws of inheritance, which we discussed in Chapter 16. Some of them even violate the law of segregation or the law of independent assortment. Studies of complex inheritance patterns have helped us appreciate more fully how genes influence phenotypes. Such research has revealed an astounding variety of ways that inheritance occurs. The picture that emerges is of a wonderful web of diverse mechanisms by which genes give rise to phenotypes. **Table 17.1** provides a summary of Mendelian inheritance and the types of inheritance patterns we will consider in this chapter.

Table 17.1	Different Types of Inheritance Patterns
Type	**Description**
Mendelian	Inheritance patterns in which a single gene affects a single trait and the alleles obey the law of segregation. These patterns include simple dominant/recessive traits, X-linked traits, incomplete dominance, codominance, and sex-influenced traits (refer back to Table 16.1).
Epistasis	A type of gene interaction in which the alleles of one gene mask the effects of a dominant allele of another gene.
Continuous variation	Inheritance pattern in which the offspring display a continuous range of phenotypes. This pattern is produced by the additive interactions of several genes, along with environmental influences.
Linkage	Inheritance pattern involving two or more genes that are close together on the same chromosome. Linked genes do not assort independently.
Extranuclear inheritance	Inheritance pattern of genes found in the genomes of mitochondria or chloroplasts. Usually these genes are inherited from the mother.
X inactivation	Phenomenon of female mammals in which one X chromosome is inactivated in every somatic cell, producing a mosaic phenotype.
Genomic imprinting	Inheritance pattern in which an allele from one parent is silenced in the somatic cells of the offspring, whereas the allele from the other parent is expressed.
Maternal effect	Inheritance pattern in which the genotype of the mother determines the phenotype of the offspring. This occurs because the mother provides the gene products from maternal effect genes to developing egg cells.

Figure 17.1 **Epistasis in the sweet pea.** The color of the sweet pea flower is controlled by two genes, each with a dominant and a recessive allele. Each of the dominant alleles (*C* and *P*) encodes an enzyme required for the synthesis of purple pigment. A plant that is homozygous recessive for either gene (*cc* or *pp*) cannot synthesize the pigment and will have white flowers.

Concept check: *In a Ccpp individual, which functional enzyme is missing? Is it the enzyme encoded by the C or P gene?*

character. The 9:7 ratio is due to a gene interaction in which two genes affect a single character.

Polygenic Inheritance and Environmental Influences Produce Continuous Phenotypic Variation

Until now, we have discussed the inheritance of characters with clearly defined phenotypic variants, such as red or white

eyes in fruit flies. These are known as **discrete traits**, because the phenotypes do not overlap. For most traits, however, the phenotypes cannot be sorted into discrete categories. Traits that show continuous variation over a range of phenotypes are called **quantitative traits**. In humans, quantitative traits include height, weight, skin color, metabolic rate, and heart size. In the case of domestic animals and plant crops, many of the traits that people consider desirable are quantitative in nature, such as the number of eggs a chicken lays, the amount of milk a cow produces, and the number of apples on an apple tree. Consequently, much of our modern understanding of quantitative traits comes from agricultural research.

Quantitative traits are usually **polygenic**, which means that several or many genes contribute to the outcome of the trait. For many polygenic traits, genes contribute to the phenotype in an additive way. As a hypothetical example, let's suppose that three different genes (*W1*, *W2*, and *W3*) affect weight in turkeys; each gene can occur in heavy (*W*) and light (*w*) alleles. A heavy allele contributes an extra pound to an individual's weight compared to a light allele. A turkey homozygous for all the heavy alleles (*W1W1 W2W2 W3W3*) would weigh 6 pounds more than an individual homozygous for all the light alleles (*w1w1 w2w2 w3w3*). A turkey heterozygous for all three genes (*W1w1 W2w2 W3w3*) would have an intermediate weight that would be 3 pounds lighter than the homozygous turkey carrying all of the heavy alleles, because the heterozygote carries 3 light alleles.

Another important factor regarding quantitative traits is the environment that an individual experiences. As we learned in Chapter 16, the environment plays a vital role in the phenotypic expression of genes. Environmental factors often have a major impact on quantitative traits. For example, an animal's diet affects its weight, and the amount of rain and sunlight that fall on an apple tree affect how many apples it produces.

Because quantitative traits are polygenic and greatly influenced by environmental conditions, the phenotypes among different individuals may vary substantially in any given population. As an example, let's consider skin pigmentation in humans. This character is influenced by several genes that tend to interact in an additive way. As a simplified example, let's consider a population in which skin pigmentation in people is controlled by three genes, which we will designate A, B, and C. Each gene may exist as a dark allele, designated A^D, B^D, or C^D, or a light allele, designated A^L, B^L, or C^L, respectively. All of the alleles encode enzymes that cause the synthesis of skin pigment, but the enzymes encoded by dark alleles cause more pigment synthesis than the enzymes encoded by light alleles. **Figure 17.2** considers a hypothetical case in which people who were heterozygous for all three genes produced a large population of offspring. The bar graph shows the genotypes of the offspring, grouped according to the total number of light and dark alleles. As shown by the shading of the figure, skin pigmentation increases as the number of dark alleles increases. Offspring who have all light alleles or who have all dark alleles—that

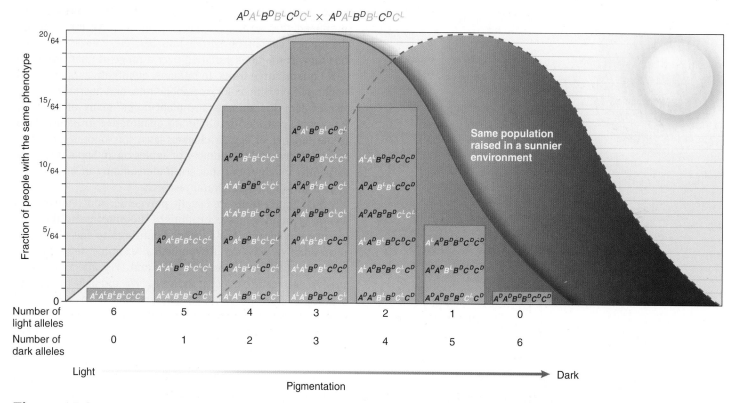

Figure 17.2 **Continuous variation in a polygenic trait.** Skin color is a polygenic character that displays a continuum of phenotypes. The bell curve on the left (solid line) shows the range of skin pigmentation in a hypothetical human population. The bar graphs below the curve show the additive effects of three genes that affect pigment production in this population; each bar shows the fraction of people with a particular number of dark alleles (A^D, B^D, and C^D) and light alleles (A^L, B^L, and C^L). The bell curve on the right (dashed line) represents the expected range of phenotypes if the same population was raised in a sunnier environment.

Concept check: What are the main reasons the pigmentation phenotype displays a continuous distribution?

is, those who are homozygous for all three genes—are fewer in number than those with some combination of light and dark alleles. As seen in the bell-shaped curve above the bar graph, the phenotypes of the offspring fall along a continuum. This continuous phenotypic variation, which is typical of quantitative traits, is produced by genotypic differences together with environmental effects. A second bell-shaped curve (the dashed line) depicts the expected phenotypic range if the same population of offspring had been raised in a sunnier environment, which increases pigment production. This curve illustrates how the environment can also have a significant influence on the range of phenotypes.

In our discussion of genetics, we tend to focus on discrete traits because this makes it easier to relate a specific genotype with a phenotype. This is usually not possible for continuous traits. For example, as depicted in the middle bar of Figure 17.2, seven different genotypes can produce individuals with a medium amount of pigmentation. It is important to emphasize that the majority of traits in all organisms are continuous, not discrete. Most traits are influenced by multiple genes, and the environment has an important impact on the phenotypic outcome.

17.2 Genes on the Same Chromosome: Linkage, Recombination, and Mapping

In the inheritance patterns we have examined so far, the alleles segregate and assort independently as predicted by Mendel's laws. As we have just seen, phenotypes can be influenced by a variety of factors, including gene interactions and environmental effects, which make it difficult to relate a specific genotype to a specific phenotype. Even so, if we understand all of these factors and take them into account, we can see that each of the genes is transmitted according to Mendel's laws.

In this section, we will focus on inheritance patterns that do not conform to the law of independent assortment. We will begin by examining the first experimental cross that demonstrated such a pattern. This pattern was subsequently explained by Thomas Hunt Morgan, who proposed that different genes located close to each other on the same chromosome tend to be inherited together. Finally, we will see how crossing over between such genes provided the first method of mapping genes on chromosomes.

FEATURE INVESTIGATION

Bateson and Punnett's Crosses of Sweet Peas Showed That Genes Do Not Always Assort Independently

In Chapter 16, we learned that the independent assortment of alleles is due to the random alignment of homologous chromosomes during meiosis (refer back to Figure 16.11). But what happens when the alleles of different genes are on the same chromosome? A typical chromosome contains many hundreds or even a few thousand different genes. When two genes are close together on the same chromosome, they tend to be transmitted as a unit, a phenomenon known as **linkage**. A group of genes that usually stay together during meiosis is called a **linkage group**, and the genes in the group are said to be linked. The genes on a single chromosome can be considered to consti-

tute a linkage group. In a two-factor cross, linked genes that are close together on the same chromosome do not follow the law of independent assortment.

The first study showing linkage between two different genes was a cross of sweet peas carried out by William Bateson and Reginald Punnett in 1905. A surprising result occurred when they conducted a two-factor cross involving flower color and pollen shape (**Figure 17.3**). One of the parent plants had purple flowers (*PP*) and long pollen (*LL*); the other had red flowers (*pp*) and round pollen (*ll*). As Bateson and Punnett expected, the F_1 plants all had purple flowers and long pollen (*PpLl*). The unexpected result came in the F_2 generation.

Although the F_2 offspring displayed the four phenotypes predicted by Mendel's laws, the observed numbers of offspring did not conform to the predicted 9:3:3:1 ratio (refer back to

Figure 17.3   A cross of sweet peas showing that independent assortment does not always occur.

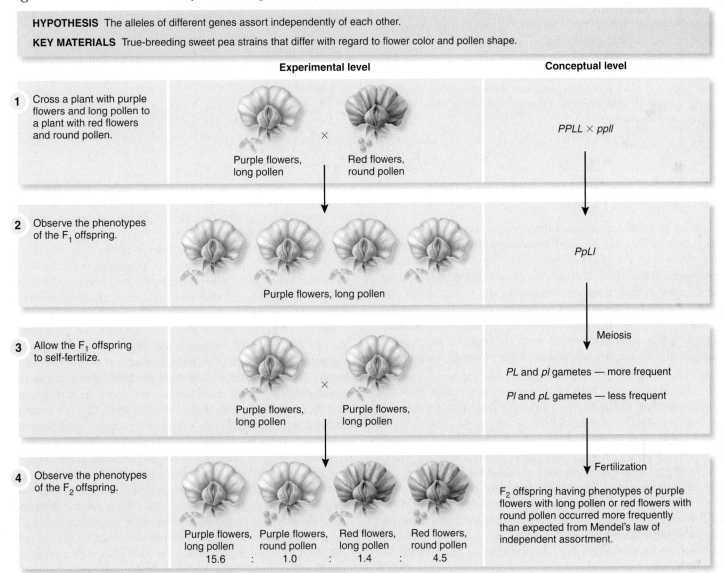

HYPOTHESIS The alleles of different genes assort independently of each other.

KEY MATERIALS True-breeding sweet pea strains that differ with regard to flower color and pollen shape.

	Experimental level	Conceptual level
1 Cross a plant with purple flowers and long pollen to a plant with red flowers and round pollen.	Purple flowers, long pollen × Red flowers, round pollen	$PPLL \times ppll$
2 Observe the phenotypes of the F_1 offspring.	Purple flowers, long pollen	$PpLl$
3 Allow the F_1 offspring to self-fertilize.	Purple flowers, long pollen × Purple flowers, long pollen	Meiosis *PL* and *pl* gametes — more frequent *Pl* and *pL* gametes — less frequent
4 Observe the phenotypes of the F_2 offspring.	Purple flowers, long pollen 15.6 : Purple flowers, round pollen 1.0 : Red flowers, long pollen 1.4 : Red flowers, round pollen 4.5	Fertilization F_2 offspring having phenotypes of purple flowers with long pollen or red flowers with round pollen occurred more frequently than expected from Mendel's law of independent assortment.

5   THE DATA

Phenotypes of F₂ offspring	Observed number	Observed ratio	Expected number	Expected ratio
Purple flowers, long pollen	296	15.6	240	9
Purple flowers, round pollen	19	1.0	80	3
Red flowers, long pollen	27	1.4	80	3
Red flowers, round pollen	85	4.5	27	1

6   CONCLUSION The data are not consistent with the law of independent assortment.

7   SOURCE Bateson, William, and Punnett, Reginald C. 1911. On the inter-relations of genetic factors. *Proceedings of the Royal Society of London, Series B*, 84:3–8.

Figure 16.8). Rather, as seen in the data in Figure 17.3, the F₂ generation had a much higher proportion of the two phenotypes found in the parental generation: purple flowers with long pollen, and red flowers with round pollen. These results did not support the law of independent assortment. How did Bateson and Punnett explain these results? They suggested that the transmission of flower color and pollen shape was somehow coupled, so these traits did not always assort independently. Although the law of independent assortment applies to many other genes, in this example, the hypothesis of independent assortment was rejected.

Experimental Questions

1. What hypothesis were Bateson and Punnett testing when conducting the crosses in the sweet pea?

2. What were the expected results of Bateson and Punnett's cross?

3. How did the observed results differ from the expected results? What did Bateson and Punnett conclude about the results of this particular cross?

Linkage and Crossing Over Produce Parental and Recombinant Types

Bateson and Punnett realized their results did not conform to Mendel's law of independent assortment. However, they did not know why the genes were not assorting independently. A few years later, Thomas Hunt Morgan obtained similar ratios in crosses of fruit flies while studying the transmission pattern of genes in *Drosophila*. Like Bateson and Punnett, Morgan observed many more F₂ offspring with the parental combination of traits than would be predicted on the basis of independent assortment. To explain his data, Morgan proposed three ideas:

1. When different genes are located on the same chromosome, the traits determined by those genes are more likely to be inherited together. This violates the law of independent assortment.

2. Due to crossing over during meiosis, homologous chromosomes can exchange pieces of chromosomes and create new combinations of alleles.

3. The likelihood of a crossover occurring in the region between two genes depends on the distance between the two genes. Crossovers between homologous chromosomes are much more likely to occur between two genes farther apart in the chromosome compared to two genes closer together.

To illustrate the first two ideas, **Figure 17.4** considers a series of crosses involving two genes linked on the same chromosome in *Drosophila*. The two genes are located on an autosome, not on a sex chromosome. The P generation cross is between flies that are homozygous for alleles that affect body color and wing shape. The female is homozygous for the dominant wild-type alleles that produce gray body color (b^+b^+) and straight wings (c^+c^+); the male is homozygous for recessive mutant alleles that produce black body color (bb) and curved wings (cc). The symbols for the genes are based on the name of the mutant allele; the dominant wild-type allele is indicated by a superscript plus sign ($^+$). The chromosomes next to the flies in Figure 17.4 show the arrangement of these alleles. If the two genes are on the same chromosome, we know the arrangement of alleles in the P generation flies because these flies are homozygous for both genes ($b^+b^+c^+c^+$ for one parent and $bbcc$ for the other parent). In the P generation female on the left, b^+ and c^+ are linked, while b and c are linked in the male on the right.

Let's now look at the outcome of the crosses in Figure 17.4. As expected, the F₁ offspring (b^+bc^+c) all had gray bodies and straight wings, confirming that these are the dominant traits. In the next cross, F₁ females were mated to males that were homozygous for both recessive alleles ($bbcc$). Recall from Chapter 16 that a cross in which an individual with a dominant phenotype is mated with a homozygous recessive individual is

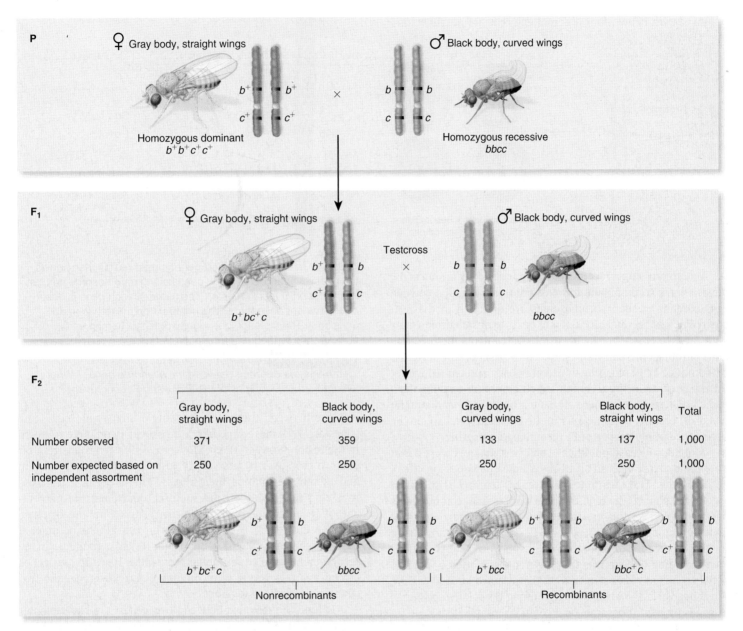

Figure 17.4   **Linkage and recombination of alleles.**   An experimenter crossed $b^+b^+c^+c^+$ and $bbcc$ flies to produce F_1 heterozygotes. F_1 females were then testcrossed to $bbcc$ males. The large number of parental phenotypes in the F_2 generation suggests that the two genes are linked on the same chromosome. F_2 recombinant phenotypes occur because the alleles can be rearranged by crossing over. Note: The b^+ and c^+ alleles are dominant and the b and c alleles are recessive.

Concept check: *In which fly or flies did crossing over occur to produce the recombinant offspring of the F_2 generation?*

called a **testcross**. However, in the crosses we are discussing here, the purpose of the testcross is to determine whether the genes for body color and wing shape are linked. If the genes are on different chromosomes and assort independently, this testcross will produce equal numbers of F_2 offspring with the four possible phenotypes. The observed numbers clearly conflict with this prediction, which is based on independent assortment. The two most abundant phenotypes are those with the combinations of characteristics in the P generation: gray bodies and straight wings or black bodies and curved wings. These offspring are termed **nonrecombinants**, or parental types, because

their combination of traits has not changed from the parental generation. The smaller number of offspring that have a combination of traits not found in the parental generation—gray bodies and curved wings or black bodies and straight wings—are called **recombinants**.

How do we explain the occurrence of recombinants when genes are linked on the same chromosome? As shown beside the flies of the F_2 generation in Figure 17.4, each recombinant individual has a chromosome that is the product of a crossover. The crossover occurred while the F_1 female fly was making egg cells.

As shown below, four different egg cells are possible:

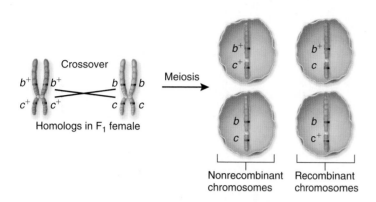

Nonrecombinant chromosomes   Recombinant chromosomes

Due to crossing over, two of the four egg cells produced by meiosis have recombinant chromosomes. What happens when eggs containing such chromosomes are fertilized in the testcross? Each of the male fly's sperm cells carries a chromosome with the two recessive alleles. If the egg contains the recombinant chromosome carrying the b^+ and c alleles, the testcross will produce an F_2 offspring with a gray body and curved wings. If the egg contains the recombinant chromosome carrying the b and c^+ alleles, F_2 offspring will have a black body and straight wings. Therefore, crossing over in the F_1 female can explain the occurrence of both types of F_2 recombinant offspring.

Morgan's ideas about linkage and crossing over were based on similar data, derived from his studies of genes on the X chromosome. The idea that linked genes tend to be inherited together explained the high frequency of parental combinations of traits in certain crosses. The proposal that crossing over produces chromosomes with new allele combinations accounted for the occurrence of recombinant phenotypes. Morgan's third idea regarding linkage was that the frequency of crossing over between linked genes depends on the distance between them. This suggested a method for determining the relative positions of genes on a chromosome, as we will discuss next.

Recombination Frequencies Provide a Method for Mapping Genes Along Chromosomes

The study of the arrangement of genes in a species' genome is called **genetic mapping**. As depicted in Figure 17.5, the linear order of genes that are linked to each other along the same chromosome is shown in a chart known as a **genetic map**. Each gene has its own unique locus at a particular site within a chromosome. For example, the gene for black body color (b) that we discussed earlier is located near the middle of the chromosome, whereas the gene for curved wings (c) is closer to one end. The first genetic map, showing five genes on the *Drosophila* X chromosome, was constructed in 1911 by Alfred Sturtevant, an undergraduate student who studied in Morgan's laboratory.

Genetic mapping allows us to estimate the relative distances between linked genes based on the likelihood that a crossover will occur between them. This likelihood is proportional to the

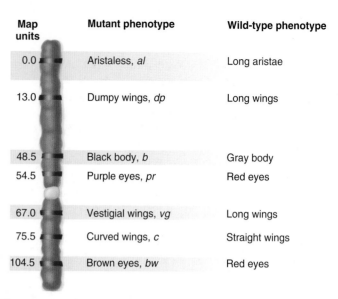

Map units	Mutant phenotype	Wild-type phenotype
0.0	Aristaless, *al*	Long aristae
13.0	Dumpy wings, *dp*	Long wings
48.5	Black body, *b*	Gray body
54.5	Purple eyes, *pr*	Red eyes
67.0	Vestigial wings, *vg*	Long wings
75.5	Curved wings, *c*	Straight wings
104.5	Brown eyes, *bw*	Red eyes

Figure 17.5   **A simplified genetic map.**   This map shows the relative locations of a few genes along chromosome number 2 in *Drosophila melanogaster*. The name of each gene is based on the mutant phenotype. The numbers on the left are map units (mu). The distance between two genes, in map units, corresponds to their recombination frequency in testcrosses.

Concept check: *How would you set up a testcross to determine the distance between the al and dp genes? What would be the genotypes of the P, F_1, and F_2 generations?*

distance between the genes, as Morgan first proposed. If the genes are very close together, a crossover is unlikely to begin in the region between them. However, if the genes are very far apart, a crossover is more likely to be initiated between them and thereby recombine their alleles. Therefore, in a testcross involving two genes on the same chromosome, the percentage of recombinant offspring is correlated with the distance between the genes. This correlation provides the experimental basis for gene mapping. If a two-factor testcross produces many recombinants, the experimenter concludes that the two genes are far apart. If very few recombinants are observed, the two genes must be close together.

To find the distance between two genes, the experimenter must determine the frequency of crossing over between them, called their **recombination frequency**. This is accomplished by conducting a testcross. As an example, let's refer back to the *Drosophila* testcross described in Figure 17.4. As we discussed, the genes for body color and wing shape are on the same chromosome; the recombinants are the result of crossing over during egg formation in the F_1 female. We can use the data from the testcross shown in Figure 17.4 to estimate the distance between these two genes. The **map distance** between two linked genes is defined as the number of recombinants divided by the total number of offspring times 100.

$$\text{Map distance} = \frac{\text{Number of recombinants}}{\text{Total number of offspring}} \times 100$$

$$= \frac{133 + 137}{371 + 359 + 133 + 137} \times 100$$

$$= 27.0 \text{ map units}$$

The units of distance are called **map units** (**mu**), or sometimes **centiMorgans** (**cM**) in honor of Thomas Hunt Morgan. One map unit is equivalent to a 1% recombination frequency. In this example, 270 out of 1,000 offspring are recombinants, so the recombination frequency is 27%, and the two genes are 27.0 mu apart.

Genetic mapping has been useful for analyzing the genes of organisms that are easily crossed and produce many offspring in a short time. It has been used to map the genes of several plant species and of certain species of animals, such as *Drosophila*. However, for most organisms, including humans, genetic mapping via crosses is impractical due to long generation times or the inability to carry out experimental crosses. Fortunately, many alternative methods of gene mapping have been developed in the past few decades that are faster and do not depend on crosses. These newer cytological and molecular approaches, which we will discuss in Chapter 20, are also used to map genes in a wide variety of organisms.

17.3 Extranuclear Inheritance: Organelle Genomes

In the previous section, we examined the inheritance patterns of linked genes that violate the law of independent assortment. In this section, we will explore inheritance patterns that violate the law of segregation. The segregation of genes is explained by the pairing and segregation of homologous chromosomes during meiosis. However, some genes are not found on the chromosomes in the cell nucleus, and these genes do not segregate in the same way. The transmission of genes located outside the cell nucleus is called **extranuclear inheritance**. Two important types of extranuclear inheritance patterns involve genes found in mitochondria and chloroplasts. Extranuclear inheritance is also called **cytoplasmic inheritance** because these organelles are in the cytoplasm of the cell. In this section, we will examine the transmission patterns observed for genes found in the chloroplast and mitochondrial genomes and consider how mutations in these genes may affect an individual's traits.

Genomes & Proteomes Connection

Chloroplast and Mitochondrial Genomes Are Relatively Small, but Contain Genes That Encode Important Proteins

As we discussed in Chapter 4, mitochondria and chloroplasts are found in eukaryotic cells because of an ancient endosymbiotic relationship. They contain their own genetic material, called the mitochondrial genome and chloroplast genome, respectively (**Figure 17.6**). Mitochondrial and chloroplast genomes are composed of a single, circular DNA molecule. The mitochondrial genome of many mammalian species has been analyzed and usually contains a total of 37 genes. Twenty-four genes encode tRNAs and rRNAs, which are needed for translation inside the

(a) An animal cell

(b) A plant cell

Figure 17.6 **The locations of genetic material in animal and plant cells.** The chromosomes in the cell nucleus are collectively known as the nuclear genome. Mitochondria and chloroplasts have small circular chromosomes called the mitochondrial and chloroplast genomes, respectively.

Concept check: *What is the evolutionary origin of mitochondria and chloroplasts in eukaryotic cells?*

mitochondrion, and 13 genes encode proteins that are involved in oxidative phosphorylation. As discussed in Chapter 7, the primary function of the mitochondrion is the synthesis of ATP via oxidative phosphorylation. Among different species of plants, chloroplast genomes typically contain about 110 to 120 genes. Many of these genes encode proteins that are vital to the process of photosynthesis, which we discussed in Chapter 8.

Chloroplast Genomes Are Often Maternally Inherited

One of the first experiments showing an extranuclear inheritance pattern was carried out by German botanist Carl Correns in 1909. Correns discovered that leaf pigmentation in the four-o'clock plant (*Mirabilis jalapa*) follows a pattern of inheritance that does not obey Mendel's law of segregation. Four-o'clock leaves may be green, white, or variegated. Correns observed that the pigmentation of the offspring depended solely on the pigmentation of the female parent, a phenomenon called **maternal inheritance** (**Figure 17.7**). If the female parent had white leaves, all of the offspring had white leaves. Similarly, if the female was green, so were all of the offspring. The offspring of a variegated female parent could be green, white, or variegated.

What accounts for maternal inheritance? At the time, Correns did not understand that chloroplasts contain genetic material. Subsequent research identified DNA present in chloroplasts as responsible for the unusual inheritance pattern observed. We now know that the pigmentation of four-o'clock leaves can be explained by the occurrence of genetically different types of chloroplasts in the leaf cells. As discussed in Chapter 8, chloroplasts are the site of photosynthesis, and their green color is due to the presence of the pigment called chlorophyll. Certain genes required for chlorophyll synthesis are found within the chloroplast DNA. The green phenotype is due to the presence of chloroplasts that have normal genes and synthesize the usual quantity of chlorophyll. The white phenotype is caused by a mutation in a gene within the chloroplast DNA that prevents the synthesis of most of the chlorophyll. (Enough chlorophyll is made for the plant to survive.) The variegated phenotype occurs in leaves that have a mixture of the two types of chloroplasts.

Leaf pigmentation follows a maternal inheritance pattern because the chloroplasts in four-o'clocks are transmitted only through the cytoplasm of the egg (**Figure 17.8**). Recall from Chapter 4 that chloroplasts are derived from proplastids. In four-o'clocks, the egg cell contains several proplastids that are inherited by the offspring. The sperm cell does not contribute any proplastids. For this reason, the phenotype of a four-o'clock plant reflects the types of proplastids it inherits from the maternal parent. If the maternal parent transmits only normal proplastids, all offspring will have green leaves (Figure 17.8a). Alternatively, if the maternal parent transmits only mutant proplastids, all offspring will have white leaves (Figure 17.8b). Because an egg cell contains several proplastids, an offspring from a variegated maternal parent may inherit only normal proplastids, only mutant proplastids, or a mixture of normal and mutant proplastids. Consequently, the offspring of a variegated maternal parent can be green, white, or variegated individuals (Figure 17.8c).

How do we explain the variegated phenotype at the cellular level? This phenotype is due to events that occur after fertilization. As a zygote containing both types of proplastids grows via cellular division to produce a multicellular plant, some cells may receive mostly those that develop into normal chloroplasts. Further division of these cells gives rise to a patch of green tissue. Alternatively, as a matter of chance, other cells may receive all or mostly mutant chloroplasts that are defective in chlorophyll synthesis. The result is a patch of white tissue.

In most species of plants, the egg cell provides most of the zygote's cytoplasm, whereas the much smaller male gamete often provides little more than a nucleus. Therefore,

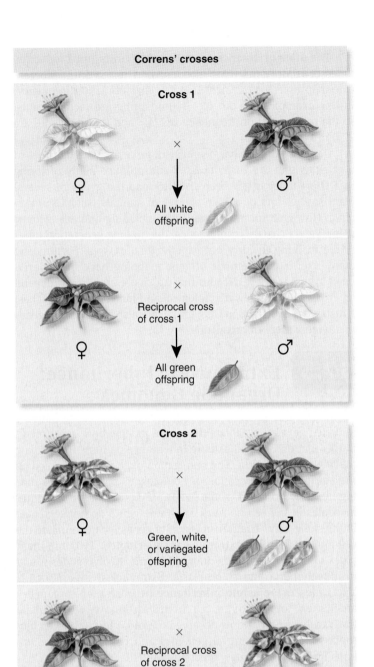

Figure 17.7 **Maternal inheritance in the four-o'clock plant.** Genes for green pigment synthesis in plants are found in the chloroplast genome. The white phenotype in four-o'clocks is due to chloroplasts with a mutant allele that greatly reduces green pigment production. The variegated phenotype is due to a mixture of normal and mutant chloroplasts. In four-o'clocks, the egg contains all of the plastids that are inherited by the offspring, so the phenotype of the offspring is determined by the female parent.

Concept check: *In this example, where is the gene located that causes the green color of four-o'clock leaves? How is this gene transmitted from parent to offspring?*

Normal proplastid will produce chloroplasts with a normal amount of green pigment.

Nucleus

Egg cell

(a) Egg cell from a maternal parent with green leaves

Mutant proplastid will produce chloroplasts with very little pigment.

(b) Egg cell from a maternal parent with white leaves

(c) Possible egg cells from a maternal parent with variegated leaves

Figure 17.8 **Plastid composition of egg cells from green, white, and variegated four-o'clock plants.**   In this drawing of four-o'clock egg cells, normal proplastids are represented as green and mutant proplastids as white. (Note: This drawing is schematic. Proplastids do not differentiate into chloroplasts in egg cells, and they are not actually green.) **(a)** A green plant produces eggs carrying normal proplastids. **(b)** A white plant produces eggs carrying mutant proplastids. **(c)** A variegated plant produces eggs that may contain either or both types of proplastids.

chloroplasts are most often inherited via the egg. In seed-bearing plants, maternal inheritance of chloroplasts is the most common transmission pattern. However, certain species exhibit a pattern called **biparental inheritance,** in which both the pollen and the egg contribute chloroplasts to the offspring. Others exhibit **paternal inheritance,** in which only the pollen contributes these organelles. For example, most types of pine trees show paternal inheritance of chloroplasts.

Mitochondrial Genomes Are Maternally Inherited in Humans and Most Other Species

Mitochondria are found in nearly all eukaryotic species. As with the transmission of chloroplasts in plants, maternal inheritance is the most common pattern of mitochondrial transmission in eukaryotic species, although some species do exhibit biparental or paternal inheritance.

In humans, mitochondria are maternally inherited. Researchers have discovered that mutations in human mitochondrial genes can cause a variety of rare diseases (**Table 17.2**). These

Table 17.2	Examples of Human Mitochondrial Diseases
Disease	**Causes and symptoms**
Leber's hereditary optic neuropathy	A mutation in one of several mitochondrial genes that encode electron transport proteins. The main symptom is loss of vision.
Neurogenic muscle weakness	A mutation in a mitochondrial gene that encodes a subunit of mitochondrial ATP synthase, which is required for ATP synthesis. Symptoms involve abnormalities in the nervous system that affect the muscles and eyes.
Maternal myopathy and cardiomyopathy	A mutation in a mitochondrial gene that encodes a tRNA for leucine. The primary symptoms involve muscle abnormalities, most notably in the heart.
Myoclonic epilepsy and ragged-red muscle fibers	A mutation in a mitochondrial gene that encodes a tRNA for lysine. Symptoms include epilepsy, dementia, blindness, deafness, and heart and kidney malfunctions.

are usually chronic degenerative disorders that affect organs and cells that require high levels of ATP such as the brain, eyes, heart, muscle, kidney, and endocrine glands. For example, Leber's hereditary optic neuropathy (LHON) affects the optic nerve and can lead to the progressive loss of vision in one or both eyes. LHON can be caused by a mutation in one of several different mitochondrial genes.

17.4 X Inactivation, Genomic Imprinting, and Maternal Effect

We will end our discussion of complex inheritance patterns by considering examples in which the timing and control of gene expression create inheritance patterns that are determined by the sex of the individual or by the sex of the parents. The first two patterns, called X inactivation and genomic imprinting, are types of **epigenetic inheritance**. In epigenetic inheritance, modification of a gene or chromosome during egg formation, sperm formation, or early stages of embryo growth alters gene expression in a way that is fixed during an individual's lifetime. Epigenetic changes permanently affect the phenotype of the individual, but they are not permanent over the course of two or more generations, and they do not change the actual DNA sequence. For example, a gene may undergo an epigenetic change that inactivates the gene for an individual's entire life, so it is never expressed in that individual. However, when the same individual produces gametes, the gene may become activated and remain active during the lifetime of an offspring that inherits the gene.

At the end of this section, we will also consider genes that exhibit an intriguing inheritance pattern called the **maternal effect,** in which the genotype of the mother directly determines the phenotype of her offspring. Surprisingly, for maternal effect genes, the genotypes of the father and of the offspring themselves do not affect the offspring's phenotype.

In Female Mammals, One X Chromosome Is Inactivated in Each Somatic Cell

In 1961, the British geneticist Mary Lyon proposed the phenomenon of **X inactivation**, in which one X chromosome in the somatic cells of female mammals is inactivated, meaning that its genes are not expressed. X inactivation is based on two lines of evidence. The first evidence came from microscopic studies of mammalian cells. In 1949, Murray Barr and Ewart Bertram identified a highly condensed structure in the cells of female cats that was not found in the cells of male cats. This structure was named a **Barr body** after one of its discoverers (**Figure 17.9**). In 1960, Susumu Ohno correctly proposed that a Barr body is a highly condensed X chromosome. Lyon's second line of evidence was the inheritance pattern of variegated coat colors in certain female mammals. A classic case is the calico cat, which has randomly distributed patches of black and orange fur (see chapter-opening photo).

How do we explain this patchwork phenotype? According to Lyon's hypothesis, the calico pattern is due to the permanent inactivation of one X chromosome in each cell that forms a patch of the cat's skin, as shown in **Figure 17.10**. The gene involved is an X-linked gene that occurs as an orange allele, X^O, and a black allele, X^B. A female cat heterozygous for this gene will be calico. (The cat's white underside is due to a dominant allele of a different autosomal gene.) At an early stage of embryonic development, one of the two X chromosomes is randomly inactivated in each of the cat's somatic cells, including those that will give rise to the hair-producing skin cells. As the embryo grows and matures, the pattern of X inactivation is maintained during subsequent cell divisions. For example, skin cells derived from a single embryonic cell in which the X^B-carrying chromosome has been inactivated will produce a patch of orange fur, because they express only the X^O allele that is carried on the active chromosome. Alternatively, a group of skin cells in which the chromosome carrying X^O has been inactivated will express only the X^B allele, producing a patch of black fur. The result is an animal with randomly distributed patches of black and orange fur.

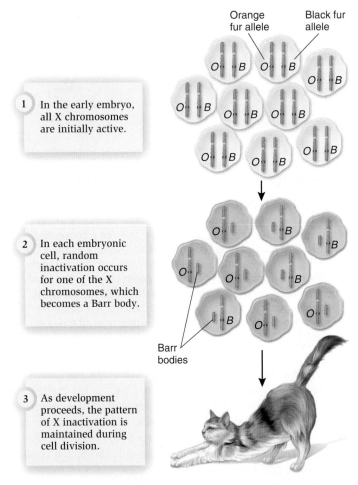

1. In the early embryo, all X chromosomes are initially active.

2. In each embryonic cell, random inactivation occurs for one of the X chromosomes, which becomes a Barr body.

Barr bodies

3. As development proceeds, the pattern of X inactivation is maintained during cell division.

Figure 17.10   Random X-chromosome inactivation in a calico cat.   The calico pattern is due to random X-chromosome inactivation in a female that is heterozygous for an X-linked gene with black and orange alleles. The cells at the top of this figure represent a small mass of cells making up the very early embryo. In these cells, both X chromosomes are active. At an early stage of embryonic development, one X chromosome is randomly inactivated in each cell. The initial inactivation pattern is maintained in the descendents of each cell as the embryo matures into an adult. The pattern of orange and black fur in the adult cat reflects the pattern of X inactivation in the embryo.

Concept check: On rare occasions, a phenotypically male cat is calico. How is this possible?

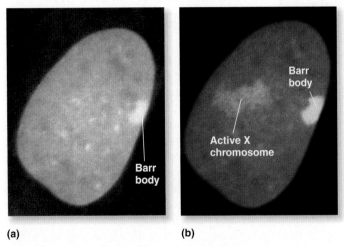

(a)   (b)

Figure 17.9   X-chromosome inactivation in female mammals.   (a) A Barr body is seen on the periphery of a human nucleus (during interphase) after staining with a DNA-specific dye. Because it is compact, the Barr body is the most brightly stained. (b) The same nucleus was labeled using a yellow fluorescent probe that recognizes the X chromosome. The Barr body is more compact compared to the active X chromosome, which is to the left of the Barr body.

Concept check: How is the Barr body different from the other X chromosome in this cell?

In female mammals that are heterozygous for X-linked genes, approximately half of their somatic cells will express one allele, whereas the rest of their somatic cells will express the other allele. These heterozygotes are called **mosaics** because they are composed of two types of cells. The phenomenon of mosaicism is readily apparent in calico cats, in which the alleles affect fur color.

For many X-linked traits in humans, females who are heterozygous for recessive X-linked alleles usually show the dominant trait because the expression of the dominant allele in 50% of their cells is sufficient to produce the dominant phenotype. For example, let's consider the recessive X-linked form of hemophilia that we discussed in Chapter 16. This type of hemophilia is caused by a defect in a gene that encodes a blood-clotting protein, called factor VIII, that is made by cells in the liver and secreted into the bloodstream. In a heterozygous female, approximately half of her liver cells will make and secrete this clotting factor, which is sufficient to prevent hemophilia. Therefore, she will exhibit the dominant trait of normal blood clotting.

On rare occasions, a female who is heterozygous may show mild or even severe disease symptoms. How is this possible? X inactivation in humans occurs when an embryo is 10 days old. At this stage, the liver contains only about a dozen cells. In most females who are heterozygous for the normal and hemophilia alleles, roughly half of their liver cells will express the normal allele. However, on rare occasions, all or most of the dozen embryonic liver cells may inactivate the X chromosome carrying the dominant normal allele. Following growth and development, such a female will have a very low level of factor VIII and as a result will show symptoms of hemophilia.

Why does X inactivation occur? Researchers have proposed that X inactivation achieves **dosage compensation**, a process that equalizes the expression of X-linked genes in male and female mammals. The X chromosome carries many genes, whereas the Y chromosome has relatively few. The inactivation of one X chromosome in the female reduces the number of expressed copies (doses) of X-linked genes from two to one. As a result, the expression of X-linked genes in females and males is roughly equal.

The X Chromosome Has an X Inactivation Center That Controls Compaction into a Barr Body

After Lyon's hypothesis was confirmed, researchers became interested in the genetic control of X inactivation. The cells of humans and other mammals have the ability to count their X chromosomes and allow only one of them to remain active. Additional X chromosomes are converted to Barr bodies. In normal females, two X chromosomes are counted and one is inactivated. In normal males, one X chromosome is counted and none inactivated.

On rare occasions, people are born with abnormalities in the number of their sex chromosomes. In the disorders known as Turner syndrome, triple X syndrome, and Klinefelter syndrome, the cells inactivate the number of X chromosomes necessary to leave a single active chromosome. For example, in triple X syndrome, two X chromosomes convert to Barr bodies.

Phenotype	Chromosome composition	Number of Barr bodies	Number of active X chromosomes
Normal female	XX	1	1
Normal male	XY	0	1
Turner syndrome (female)	XO	0	1
Triple X syndrome (female)	XXX	2	1
Klinefelter syndrome (male)	XXY	1	1

In spite of X inactivation, people with these three syndromes do exhibit some phenotypic abnormalities. The symptoms associated with these disorders may be due to effects that occur prior to X inactivation or because not all of the genes on the Barr body are completely silenced.

Although the genetic control of inactivation is not entirely understood at the molecular level, a short region on the X chromosome called the **X inactivation center** (**Xic**) is known to play a critical role. Eeva Therman and Klaus Patau determined that X inactivation is accomplished by counting the number of Xics and inactivating all X chromosomes except for one. In cells with two X chromosomes, if one of them is missing its Xic due to a chromosome mutation, neither X chromosome will be inactivated, because only one Xic is counted. Having two active X chromosomes is a lethal condition for a human female embryo.

The expression of a specific gene within the X inactivation center is required for compaction of the X chromosome into a Barr body. This gene, discovered in 1991, is named *Xist* (for X inactive specific transcript). The *Xist* gene product is a long RNA molecule that does not encode a protein. Instead, the role of *Xist* RNA is to coat one of the two X chromosomes during the process of X inactivation. After coating, proteins associate with the *Xist* RNA and promote compaction of the chromosome into a Barr body. The *Xist* gene on the Barr body continues to be expressed after other genes on this chromosome have been silenced. The expression of the *Xist* gene also maintains a chromosome as a Barr body during cell division. Whenever a somatic cell divides in a female mammal, the Barr body is replicated to produce two Barr bodies.

The Transcription of an Imprinted Gene Depends on the Sex of the Parent

As we have seen, X inactivation is a type of epigenetic inheritance in which a chromosome is modified in the early embryo, permanently altering gene expression in that individual. Other types of epigenetic inheritance occur in which genes or chromosomes are modified in the gametes of a parent, permanently altering gene expression in the offspring. **Genomic imprinting**, which was discovered in the early 1980s, refers to an inheritance pattern in which a segment of DNA is imprinted or marked so that gene expression occurs only from the genetic material inherited from one parent. It occurs in numerous species, including insects, plants, and mammals.

Genomic imprinting may involve a single gene, a part of a chromosome, an entire chromosome, or even all of the chromosomes inherited from one parent. It is permanent in the somatic cells of a given individual, but the marking of the DNA is altered from generation to generation. Imprinted genes do not follow a Mendelian pattern of inheritance because imprinting causes the offspring to distinguish between maternally and paternally inherited alleles. Depending on how a particular gene is marked by each parent, the offspring will express either the maternal or the paternal allele, but not both.

One of the first imprinted genes to be identified is a gene called *Igf2* that is found in mice and other mammals. This gene encodes a growth hormone called insulin-like growth factor 2 (Igf2) that is needed for proper growth. If a normal copy of this gene is not expressed, a mouse will be dwarf. The *Igf2* gene is known to be located on an autosome, not on a sex chromosome. Because mice are diploid, they have two copies of this gene, one from each parent.

Researchers have discovered mutations in the *Igf2* gene that block the function of the Igf2 hormone. When mice carrying normal or mutant alleles are crossed to each other, a bizarre result is obtained (**Figure 17.11**). If the male parent is homozygous for the normal allele and the female is homozygous for the mutant allele, all the offspring grow to a normal size. In contrast, if the male is homozygous for the mutant allele and the female is homozygous for the normal allele, all the offspring are dwarf. The reason this result is so surprising is that the normal and dwarf offspring have the same genotype but different phenotypes! These phenotypes are not the result of any external influence on the offspring's development. Rather, the allele that is expressed in their somatic cells is dependent upon its parental origin. In mice, the *Igf2* gene is imprinted in such a way that only the paternal gene is expressed, which means it is transcribed into mRNA. The maternal gene is not transcribed.

The baby mice shown on the left side of the photograph of Figure 17.11 are normal because they express a functional paternal gene. In contrast, the baby mice on the right are dwarf because the paternal gene is a mutant allele that results in a nonfunctional hormone.

Why is the maternal gene encoding Igf2 not transcribed into mRNA? To answer this question we need to consider the regulation of gene transcription in eukaryotes. As discussed in Chapter 13, the attachment of methyl ($-CH_3$) groups to the bases of DNA can alter gene transcription. In most genes, DNA methylation silences gene expression by inhibiting the initiation of transcription or by causing the chromatin in a region to become more compact. In contrast, for a few imprinted genes, methylation may enhance gene expression by attracting activator proteins to the promoter or by preventing the binding of repressor proteins. Researchers have discovered that DNA methylation is the marking process that occurs during the imprinting of certain genes, including the *Igf2* gene.

Figure 17.12 shows the imprinting process in which a maternal gene is methylated. The left side of the figure follows the marking process during the life of a female individual; the

	♀	♂		♀	♂
Parents	*Igf2⁻ Igf2⁻* (homozygous dwarf)	× *Igf2 Igf2* (homozygous normal)		*Igf2 Igf2* (homozygous normal)	× *Igf2⁻ Igf2⁻* (homozygous dwarf)
Offspring genotype		*Igf2 Igf2⁻* (heterozygous)			*Igf2 Igf2⁻* (heterozygous)
Allele that is transcribed in offspring		*Igf2* (normal)			*Igf2⁻* (nonfunctional)
Phenotype		Normal			Dwarf

Igf2   Normal allele
Igf2⁻   Mutant allele
▲   Silenced allele (from female parent)
●   Expressed allele (from male parent)

Figure 17.11   **An example of genomic imprinting in the mouse.**   In the cross on the left, a homozygous male with the normal *Igf2* allele is crossed to a homozygous female carrying a defective allele, *Igf2⁻*. Offspring are phenotypically normal because the paternal allele is expressed. In the cross on the right, a homozygous male carrying the defective allele is crossed to a homozygous normal female. In this case, offspring are dwarf because the paternal allele is defective and the maternal allele is not expressed. The photograph shows normal-size (left) and dwarf littermates (right) derived from a cross between a wild-type female and a heterozygous male carrying a loss-of-function *Igf2* allele (courtesy of A. Efstratiadis). The loss-of-function allele was created using methods described in Chapter 20.

Concept check: *If you cross an Igf2 Igf2⁻ male mouse to an Igf2 Igf2 female mouse, what would be the expected results?*

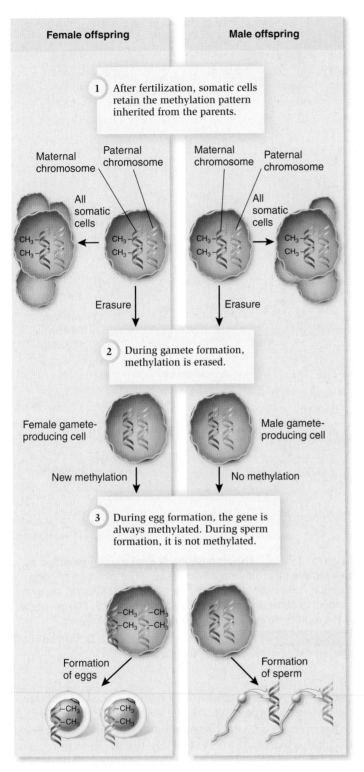

Female offspring   **Male offspring**

1) After fertilization, somatic cells retain the methylation pattern inherited from the parents.

Maternal chromosome   Paternal chromosome   Maternal chromosome   Paternal chromosome

All somatic cells   All somatic cells

Erasure   Erasure

2) During gamete formation, methylation is erased.

Female gamete-producing cell   Male gamete-producing cell

New methylation   No methylation

3) During egg formation, the gene is always methylated. During sperm formation, it is not methylated.

Formation of eggs   Formation of sperm

Figure 17.12   **Genomic imprinting via DNA methylation.** The cells at the top of this figure have a methylated gene inherited from the mother and a nonmethylated version of the same gene inherited from the father. This pattern of methylation is the same in male and female offspring and is maintained in their somatic cells. The methylation is erased during gamete formation, but in females, the gene is methylated again at a later stage in the formation of eggs. Therefore, females always transmit a methylated, transcriptionally silent copy of this gene, whereas males transmit a nonmethylated, transcriptionally active copy.

right side follows the same process in a male. Both individuals received a methylated gene from their mother and a nonmethylated copy of the same gene from their father. Via cell division, the zygote develops into a multicellular organism. Each time a somatic cell divides, enzymes in the cell maintain the methylation of the maternal gene, while the paternal gene remains unmethylated. If methylation inhibits transcription of this gene, only the paternal copy will be expressed in the somatic cells of both the male and female offspring.

The methylation state of an imprinted gene may be altered when individuals make gametes. First, the methylation is erased (Figure 17.12, step 2). Next, the gene may be methylated again, but that depends on whether the individual is a female or male. In females making eggs, both copies of the gene are methylated; in males making sperm, neither copy is methylated. When we consider the effects of methylation over the course of two or more generations, we can see how this phenomenon creates an epigenetic transmission pattern. The male in Figure 17.12 has inherited a methylated gene from his mother that is transcriptionally silenced in his somatic cells. Although he does not express this gene during his lifetime, he can pass on an active, nonmethylated copy of this exact same gene to his offspring.

Genomic imprinting is a recently discovered phenomenon that has been shown to occur for a few genes in mammals. For some genes, such as *Igf2*, the maternal allele is silenced, but for other genes, the paternal allele is silenced. While several hypotheses have been advanced, biologists are still trying to understand the reason for this curious marking process.

For Maternal Effect Genes, the Genotype of the Mother Determines the Phenotype of the Offspring

In epigenetic inheritance, genes are altered in ways that affect their expression in an individual or the individual's offspring. As we have seen, some of these alterations produce unusual inheritance patterns in which organisms with the same genotype have different phenotypes. Another inheritance pattern, with a very different mode of action, involves a category of genes called **maternal effect genes**. Surprisingly, for maternal effect genes, the genotypes of the father and of the offspring themselves do not affect the offspring's phenotype.

Inheritance patterns due to maternal effect genes were first identified in the 1920s by A. E. Boycott, in his studies of the freshwater snail *Lymnaea peregra*. In this species, the shell and internal organs can be arranged in either a right-handed (dextral) or a left-handed (sinistral) direction (**Figure 17.13**). The dextral orientation is dominant to the sinistral orientation. Whether a snail's body curves in a dextral or a sinistral direction depends on the pattern of cell division immediately following fertilization. Boycott crossed true-breeding strains of snails with either a dextral or a sinistral orientation (**Figure 17.14**). When a dextral female (*DD*) was crossed to a sinistral male (*dd*), all of the offspring were dextral. However, crossing a sinistral female (*dd*) to a dextral male (*DD*) produced the opposite result: All of the offspring were sinistral. These seemingly contradictory outcomes could not be explained in terms of Mendelian inheritance.

Figure 17.13 Snail shells (*Lymnaea peregra*) that coil to the right or left. The larger shell on the right is coiling to the right.

Alfred Sturtevant later suggested that snail coiling is due to a maternal effect gene that exists as a dextral (*D*) and a sinistral (*d*) allele. In the cross shown on the upper left in Figure 17.14, the P generation female is dextral (*DD*), and the male is sinistral (*dd*). In the cross on the right, the female is sinistral (*dd*), and the male is dextral (*DD*). All F₁ offspring are *Dd*, but their phenotype depends on their mother's genotype.

When the F₁ individuals from these two crosses are mated to each other, a genotypic ratio of 1 *DD* : 2 *Dd* : 1 *dd* is predicted for the F₂ generation. Because the *D* allele is dominant to the *d* allele, a Mendelian inheritance pattern would produce a 3:1 phenotypic ratio of dextral to sinistral snails. Instead, the snails of the F₂ generation were all dextral. How did Sturtevant explain this result? He proposed that the phenotype of the F₂ offspring depended solely on the genotype of the F₁ mother. Because the F₁ mothers were *Dd* and the *D* allele is dominant, the F₂ offspring were dextral even if their genotype was *dd*!

Sturtevant's hypothesis is also supported by the ratio of phenotypes seen in the F₃ generation. When members of the F₂ generation were crossed, the F₃ generation exhibited a 3:1 ratio of dextral to sinistral snails. These F₃ phenotypes reflect the genotypes of the F₂ mothers. The ratio of genotypes for the F₂ females was 1 *DD* : 2 *Dd* : 1 *dd*. The *DD* and *Dd* females produced dextral offspring, whereas the *dd* females produced sinistral offspring. This is consistent with the 3:1 phenotypic ratio in the F₃ generation.

The peculiar inheritance pattern of maternal effect genes can be explained by the process of egg maturation in female animals (**Figure 17.15**). Maternal cells called nurse cells surround a developing egg cell and provide it with nutrients. Within these diploid nurse cells, both copies of a maternal effect gene are activated to produce their gene products. The gene products are transported into the egg, where they persist for a significant time during embryonic development. The *D*

Figure 17.14 The inheritance of snail-coiling direction as an example of a maternal effect gene. The direction of body coiling is controlled by a single pair of genes. *D* (dextral, or right-handed) is dominant to *d* (sinistral, or left-handed). The genotype of the mother determines the phenotype of the offspring.

Concept check: An offspring has a genotype of *Dd* and coils to the left. What is the genotype of its mother?

and *d* gene products influence the pattern of cell division during the early stages of the snail's embryonic development. If an egg receives only the *D* gene product, the snail will develop a dextral orientation, whereas an egg that receives only the *d* gene product will produce a snail with a sinistral orientation. If an egg receives both *D* and *d* gene products, the snail will be dextral because the *D* gene product is dominant over *d*. In this way, the gene products of nurse cells, which are determined by the mother's genotype, influence the development of the offspring.

Several dozen maternal effect genes have been identified in experimental organisms, such as *Drosophila*. Recently, they have also been found in mice and humans. As we will discuss in Chapter 19, the products of maternal effect genes are critically important in the early stages of animal development.

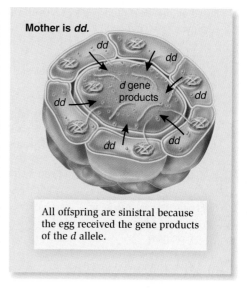

Mother is *DD*.

Nurse cells

DD   *DD*

DD   *D* gene products   *DD*

DD   *DD*

Egg

All offspring are dextral because the egg received the gene products of the *D* allele.

Mother is *Dd*.

Dd   *Dd*

Dd   *D* and *d* gene products   *Dd*

Dd   *Dd*

All offspring are dextral because the egg received the gene products of the *D* and *d* alleles, but the *D* gene products are dominant.

Mother is *dd*.

dd   *dd*

dd   *d* gene products   *dd*

dd   *dd*

All offspring are sinistral because the egg received the gene products of the *d* allele.

Figure 17.15   **The mechanism of maternal effect in snail coiling.**   In this simplified diagram, the mother's diploid nurse cells transfer gene products to the egg as it matures. These gene products persist after fertilization, affecting development of the early embryo.

Concept check: *Can a haploid egg cell that carries a D allele in its genome result in an offspring (following fertilization) that coils to the left? Can a haploid egg that carries a d allele in its genome result in an offspring that coils to the right? Explain your answers.*

Summary of Key Concepts

- A variety of inheritance patterns are more complex than Mendel had realized. Many of these do not obey one or both of his laws of inheritance. (Table 17.1)

17.1 Gene Interaction

- Epistasis is a gene interaction that occurs when the alleles of one gene mask the effects of the alleles of a different gene. (Figure 17.1)

- Quantitative traits such as height and weight are polygenic, which means that several genes govern the trait. Often, the alleles of such genes contribute in an additive way to the phenotype. This produces continuous variation in the trait, which is graphed as a bell-shaped curve. (Figure 17.2)

17.2 Genes on the Same Chromosome: Linkage, Recombination, and Mapping

- When two different genes are on the same chromosome, they are said to be linked. Linked genes tend to be inherited as a unit, unless crossing over separates them. (Figures 17.3, 17.4)

- The percentage of offspring produced in a two-factor testcross can be used to create a genetic map, which shows the relative locations of genes along a chromosome. (Figure 17.5)

17.3 Extranuclear Inheritance: Organelle Genomes

- Mitochondria and chloroplasts carry a small number of genes. The inheritance of such genes is called extranuclear inheritance. (Figure 17.6)

- Chloroplasts in the four-o'clock plant are transmitted via the egg, a pattern called maternal inheritance. (Figures 17.7, 17.8)

- Several human diseases are known to be caused by mutations in mitochondrial genes, which follow a maternal inheritance pattern. (Table 17.2)

17.4 X Inactivation, Genomic Imprinting, and Maternal Effect

- Epigenetic inheritance refers to patterns in which a gene is inactivated during the life of an organism, but not over the course of two or more generations.

- X inactivation in female mammals occurs when one X chromosome in every somatic cell is randomly inactivated. If the female is heterozygous for an X-linked gene, this can lead to a mosaic phenotype, with half of the somatic cells expressing one allele and half expressing the other. (Figures 17.9, 17.10)

- Imprinted genes are inactivated by one parent but not both. The offspring expresses only one of the two alleles. During gamete formation, methylation of a gene from one parent is a mechanism to achieve imprinting. (Figures 17.11, 17.12)

- In inheritance patterns due to maternal effect genes, the genotype of the mother determines the phenotype of the offspring. This is explained by the phenomenon that the mother's nurse cells contribute gene products to egg cells that are needed for early stages of development. (Figures 17.13, 17.14, 17.15)

Assess and Discuss

Test Yourself

1. When two genes are located on the same chromosome they are said to be
   a. homologous.
   b. allelic.
   c. epistatic.
   d. linked.
   e. polygenic.

2. Based on the ideas proposed by Morgan, which of the following statements concerning linkage is <u>not</u> true?
   a. Traits determined by genes located on the same chromosome are likely to be inherited together.
   b. Crossing over between homologous chromosomes can create new gene combinations.
   c. A crossover is more likely to occur in a region between two genes that are close together compared to a region between two genes that are farther apart.
   d. The probability of crossing over depends on the distance between the genes.
   e. Genes that tend to be transmitted together are physically located on the same chromosome.

3. In genetic linkage mapping, 1 map unit is equivalent to
   a. 100 base pairs.
   b. 1 base pair.
   c. 10% recombination frequency.
   d. 1% recombination frequency.
   e. 1% the length of the chromosome.

4. Extranuclear (cytoplasmic) inheritance occurs because
   a. certain genes are found on the X chromosome.
   b. chromosomes in the nucleus may be transferred to the cytoplasm.
   c. some organelles contain DNA.
   d. the nuclear membrane breaks down during cell division.
   e. both a and c.

5. In many organisms, organelles such as the mitochondria are contributed only by the egg. This phenomenon is known as
   a. biparental inheritance.
   b. paternal inheritance.
   c. maternal effect.
   d. maternal inheritance.
   e. both c and d.

6. Modification of a gene during gamete formation or early development that alters the way the gene is expressed during the individual's lifetime is called
   a. maternal inheritance.
   b. epigenetic inheritance.
   c. epistasis.
   d. multiple allelism.
   e. alternative splicing.

7. A male mouse that is homozygous for the normal allele of the *Igf2* gene is mated to a female that is heterozygous, carrying one normal copy and one defective copy of the gene. What would be the expected outcome of this cross?
   a. all normal offspring
   b. 1/2 normal and 1/2 dwarf
   c. all dwarf
   d. 3/4 normal and 1/4 dwarf
   e. none of the above

8. When a gene is inactivated during gamete formation and that gene is maintained in an inactivated state in the somatic cells of offspring, such an inheritance pattern is called
   a. linkage.
   b. X inactivation.
   c. maternal effect.
   d. genomic imprinting.
   e. polygenic inheritance.

9. Calico coat pattern in cats is the result of
   a. X inactivation.
   b. epistasis.
   c. organelle heredity.
   d. genomic imprinting.
   e. maternal inheritance.

10. Maternal effect inheritance can be explained by
    a. gene products that are given to an egg by the nurse cells.
    b. the methylation of genes during gamete formation.
    c. the spreading of X inactivation from the Xic locus.
    d. the inheritance of alleles that contribute additively to a trait.
    e. none of the above.

Conceptual Questions

1. Two genes (called gene *A* and gene *B*) are located on the same chromosome and are 12 map units apart. An *AABB* individual was crossed to an *aabb* individual. The F_1 (*AaBb*) offspring were crossed to *aabb* individuals. What percentage of the F_2 offspring would you expect to be *Aabb*?

2. Certain forms of human color blindness are inherited as X-linked recessive traits. Heterozygous females are not usually color blind, but on rare occasions, a female may exhibit partial color blindness or may be color blind in just one eye. Explain how this could happen.

3. A maternal effect gene in flies affects head size. The dominant allele (*N*) produces a normal head size, whereas a recessive, mutant allele (*n*) causes a smaller head. A female with a small head produces all offspring with normal heads. What are all the possible genotypes of the mother, father, and offspring?

Collaborative Questions

1. As discussed in Chapter 16, Mendel studied seven traits in pea plants, and the garden pea happens to have seven different chromosomes. It has been pointed out that Mendel was very lucky not to have conducted crosses involving two traits that are closely linked on the same chromosome because the results would have confounded his theory of independent assortment. It has even been suggested that Mendel may not have published data involving traits that were linked. An article by Blixt ("Why Didn't Gregor Mendel Find Linkage?" *Nature* 256:206) considers this issue. Look up this article and discuss why Mendel did not find linkage.

2. Discuss the similarities and differences between X inactivation and genomic imprinting.

Online Resource

www.brookerbiology.com

Stay a step ahead in your studies with animations that bring concepts to life and practice tests to assess your understanding. Your instructor may also recommend the interactive ebook, individualized learning tools, and more.

Genetics of Viruses and Bacteria

18

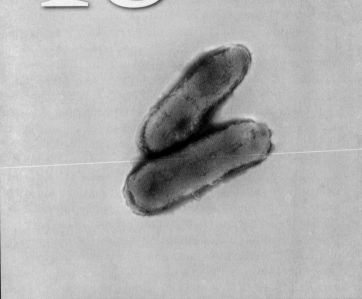

A colorized micrograph of *Haemophilus influenzae,* type b. This bacterium is a common cause of meningitis—a serious infection of the fluid in the spinal cord and the fluid that surrounds the brain.

While studying for his calculus test, Jason was having trouble concentrating due to a severe headache and fever. He thought he must be coming down with a cold. Though he had taken some aspirin, it didn't seem to be working. As he was eating some potato chips, one dropped in his lap. When he tried to look down to see where the chip had fallen, he realized that his neck was extremely stiff; he could barely move his head to look downward. Also, the brightness of his desk light seemed freakishly painful to his eyes. Over the course of that evening, Jason became confused and lethargic, and his roommate urged him to see a doctor. Fortunately, Jason took his advice and went to the college clinic. The diagnosis was bacterial meningitis—an inflammation of the protective membranes that cover the brain and spinal cord, collectively called the meninges. Although a relatively rare disease, bacterial meningitis is up to six times more common among people living in close quarters such as college dormitories. Because Jason sought help early enough, his disease could be treated with antibiotics. Had he not gotten help, the disease could have progressed to the point of causing severe brain damage and even death.

Jason's story highlights a primary reason why biologists are so interested in viruses and bacteria. Infectious diseases caused by viruses and bacteria are a leading cause of human suffering and death, accounting for one-quarter to one-third of deaths worldwide. The spread of infectious diseases results from human behavior, and in recent times, it has been accelerated by changes in land-use patterns, increased trade and travel, and the inappropriate use of antibiotic drugs. Although the incidence of fatal infectious diseases in the U.S. is low compared to the worldwide average, an alarming increase in more deadly strains of viruses and bacteria has occurred over the past few decades. Since 1980, the number of deaths in the U.S. due to infectious diseases has approximately doubled.

In this chapter, we turn our attention to the genetic analyses of viruses and bacteria. We will begin by examining viruses and other nonliving particles that infect living cells. All organisms are susceptible to infection by one or more types of viruses, which use the host's cellular machinery to replicate their own genome. Once a cell is infected, the genetic material of a virus orchestrates a series of events that ultimately leads to the production of new virus particles. We will consider the biological complexity of viruses and explore viral reproductive cycles. We will also examine some of the simplest and smallest infectious agents, called viroids and prions.

In the remaining sections of this chapter, we will examine the bacterial genome and the methods used in its investigation. Like their eukaryotic counterparts, bacteria have genetic differences that affect their cellular traits, and the techniques of modern microbiology make many of these differences, such as sensitivity to antibiotics and differences in nutritional requirements, easy to detect. Although bacteria reproduce asexually by cell division, their genetic variety is enhanced by the phenomenon called gene transfer, in which genes are passed from one bacterial cell to another. Like sexual reproduction in eukaryotes, gene transfer enhances the genetic diversity observed among bacterial species. In this chapter, we will explore three interesting ways that bacteria can transfer genetic material.

18.1 Genetic Properties of Viruses

In earlier chapters, we examined the replication and expression of eukaryotic and prokaryotic genes. Because all living organisms are either eukaryotes or prokaryotes, you may be thinking that we have considered every type of genome. However, certain nonliving things also have genomes. Viruses are nonliving particles with nucleic acid genomes. Why are viruses considered nonliving? They do not exhibit all seven properties associated with living organisms (refer back to Figure 1.5). Viruses are not composed of cells, and by themselves, they do not carry out metabolism, use energy, maintain homeostasis, or even reproduce. A virus or its genetic material must be taken up by a living cell to replicate.

The first virus to be discovered was tobacco mosaic virus (TMV). This virus infects several species of plants and causes mosaic-like patterns in which normal-colored patches are interspersed with light green or yellowish patches on the leaves (Figure 18.1). TMV damages leaves, flowers, and fruit but almost never kills the plant. In 1883, the German scientist Adolf Mayer determined that this disease could be spread by spraying the sap from one plant onto another. By subjecting this sap to filtration, the Russian scientist Dmitri Ivanovski demonstrated that the disease-causing agent was not a bacterium. Sap that had been passed through filters with pores small enough to prevent the passage of bacterial cells was still able to spread the disease. At first, some researchers suggested the agent was a chemical toxin. However, the Dutch botanist Martinus Beijerinck ruled out this possibility by showing that sap could continue to transmit the disease after many plant generations. A toxin would have been diluted after many generations, but Beijerinck's results indicated the disease agent was multiplying in the plant. Around the same time, animal viruses were discovered in connection with a disease of cattle called foot-and-mouth disease. In 1900, the first human virus, the virus that causes yellow fever, was identified.

Since these early studies, microbiologists, geneticists, and molecular biologists have taken a great interest in the structure,

Figure 18.1   A plant infected with tobacco mosaic virus.

genetic composition, and replication of viruses. In this section, we will discuss the structure of viruses and examine viral reproductive cycles in detail, paying particular attention to human immunodeficiency virus (HIV), the virus that causes acquired immunodeficiency syndrome (AIDS) in humans.

Viruses Are Remarkably Varied, Despite Their Simple Structure

A **virus** is a small infectious particle that consists of nucleic acid enclosed in a protein coat. Researchers have identified and studied over 4,000 different types of viruses. Although all viruses share some similarities, such as small size and the reliance on a living cell for replication, they vary greatly in their characteristics, including their host range, structure, and genome composition. Some of the major differences are described next, and characteristics of selected viruses are shown in **Table 18.1**.

Table 18.1	Hosts and Characteristics of Selected Viruses					
Virus or group of viruses	**Host**	**Effect on host**	**Nucleic acid***	**Genome size (kb)†**	**Number of genes†**
Phage fd	E. coli	Slows growth	ssDNA	6.4	10
Phage λ	E. coli	Can exist harmlessly in the host cell or cause lysis	dsDNA	48.5	71
Phage T4	E. coli	Causes lysis	dsDNA	169	288
Phage Qβ	E. coli	Slows growth	ssRNA	4.2	4
Tobacco mosaic virus (TMV)	Many plants	Causes mottling and necrosis of leaves and other plant parts	ssRNA	6.4	6
Baculoviruses	Insects	Most baculoviruses are species specific; they usually kill the insect	dsDNA	133.9	154
Parvovirus	Mammals	Causes respiratory, flulike symptoms	ssDNA	5.0	5
Influenza virus	Mammals	Causes classical "flu," with fever, cough, sore throat, and headache	ssRNA	13.5	11
Epstein-Barr virus	Humans	Causes mononucleosis, with fever, sore throat, and fatigue	dsDNA	172	80
Adenovirus	Humans	Causes respiratory symptoms and diarrhea	dsDNA	34	35
Herpes simplex type II	Humans	Causes blistering sores around the genital region	dsDNA	158.4	77
HIV (type I)	Humans	Causes AIDS, an immunodeficiency syndrome eventually leading to death	ssRNA	9.7	9

*The abbreviations ss and ds refer to single stranded and double stranded, respectively.
†Several of the viruses listed in this table are found in different strains that show variation with regard to genome size and number of genes. The numbers reported in this table are typical values. The abbreviation kb refers to kilobase, which equals 1,000 bases.

Brain and CNS:
Flavivirus—yellow fever
Rhabdovirus—rabies

Skin:
Herpes simplex I—cold sores
Variola virus—smallpox

Respiratory tract:
Influenza virus—flu
Rhinovirus—common cold

Immune system:
Rubella virus—measles
Human immunodeficiency virus—AIDS
Epstein-Barr virus—mononucleosis

Digestive system:
Hepatitis B virus—viral hepatitis
Rotavirus—viral gastroenteritis
Norwalk virus—viral gastroenteritis

Reproductive system:
Herpes simplex II—genital herpes
Papillomavirus—warts, cervical cancer

Blood:
Ebola virus—hemorrhagic fever
Hantavirus—hemorrhagic fever
with renal syndrome

Figure 18.2 **Some viruses that cause human diseases.** Most viruses that cause disease in humans infect cells of specific tissues, as illustrated by the examples in this figure. Note: Herpes simplex I and II infect nerve cells of the peripheral nervous system that are found in the skin and genital region, respectively.

Differences in Host Range A cell that is infected by a virus is called a **host cell**, and a species that can be infected by a specific virus is called a host species for that virus. Viruses differ greatly in their **host range**—the number of species and cell types they can infect. Table 18.1 lists a few examples of viruses with widely different ranges of host species. Tobacco mosaic virus, which we discussed earlier, has a broad host range. TMV is known to infect over 150 different species of plants. By comparison, other viruses have a narrow host range, with some infecting only a single species. Furthermore, a virus may infect only a specific cell type in a host species. **Figure 18.2** shows some viruses that infect particular human cells and cause disease.

Structural Differences Although the existence of viruses was postulated in the 1890s, viruses were not observed until the 1930s, when the electron microscope was invented. Viruses cannot be resolved by even the best light microscope. Most of them are smaller than the wavelength of visible light. Viruses range in size from about 20–400 nm in diameter (1 nanometer = 10^{-9} meters). For comparison, a typical bacterium is

1,000 nm in diameter, and the diameter of most eukaryotic cells is 10 to 1,000 times that of a bacterium. Adenoviruses, which cause infections of the respiratory and gastrointestinal tracts, have an average diameter of 75 nm. Over 50 million adenoviruses could fit into an average-sized human cell.

What are the common structural features of all viruses? As shown in **Figure 18.3**, all viruses have a protein coat called a **capsid** that encloses a genome consisting of one or more molecules of nucleic acid. Capsids are composed of one or several different protein subunits called capsomers. Capsids have a variety of shapes, including helical and polyhedral. Figure 18.3a shows the structure of TMV, which has a helical capsid made of identical capsomers. Figure 18.3b shows an adenovirus, which has a polyhedral capsid. Protein fibers with a terminal knob are located at the corners of the polyhedral capsid. Many viruses that infect animal cells, such as the influenza virus shown in Figure 18.3c, have a **viral envelope** enclosing the capsid. The envelope consists of a lipid bilayer that is derived from the plasma membrane of the host cell and is embedded with virally encoded spike glycoproteins, also called spikes or peplomers.

In addition to encasing and protecting the genetic material, the capsid and envelope enable viruses to infect their hosts. In many viruses, the capsids or envelopes have specialized proteins, including protein fibers with a knob (Figure 18.3b) and spike glycoproteins (Figure 18.3c), that help them bind to the surface of a host cell. Viruses that infect bacteria, called **bacteriophages**, or **phages**, may have more complex protein coats, with accessory structures used for anchoring the virus to a host cell and injecting the viral nucleic acid (Figure 18.3d). As discussed later, the tail fibers of such bacteriophages are needed to attach the virus to the bacterial cell wall.

Genome Differences The genetic material in a virus is called a **viral genome**. The composition of viral genomes varies markedly among different types of viruses, as suggested by the examples in Table 18.1. The nucleic acid of some viruses is DNA, whereas in others it is RNA. These are referred to as DNA viruses and RNA viruses, respectively. It is striking that some viruses use RNA for their genome, whereas all living organisms use DNA. In some viruses, the nucleic acid is single stranded, whereas in others, it is double stranded. The genome can be linear or circular, depending on the type of virus. Some kinds of viruses have more than one copy of the genome.

Viral genomes also vary considerably in size, ranging from a few thousand to more than a hundred thousand nucleotides in length (see Table 18.1). For example, the genomes of some

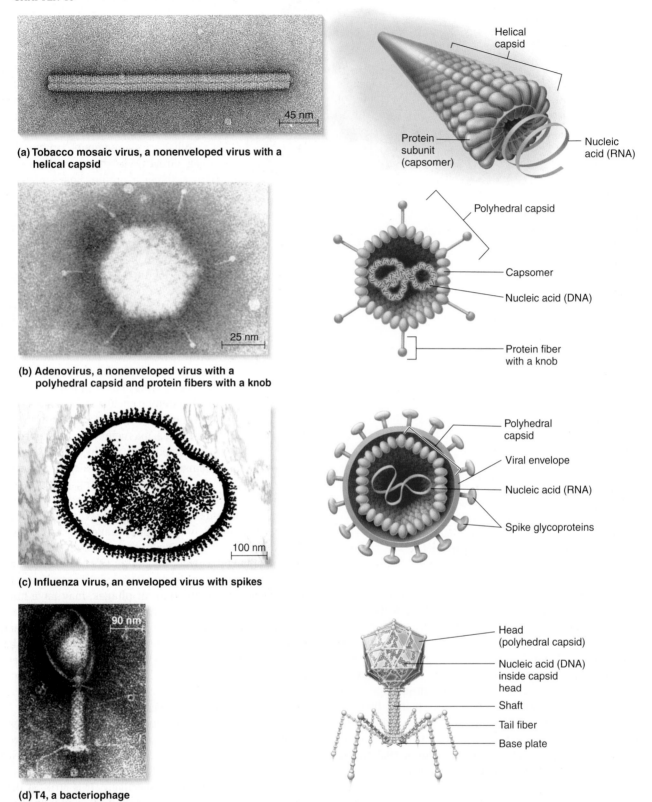

(a) Tobacco mosaic virus, a nonenveloped virus with a helical capsid

Helical capsid

Protein subunit (capsomer)

Nucleic acid (RNA)

(b) Adenovirus, a nonenveloped virus with a polyhedral capsid and protein fibers with a knob

Polyhedral capsid

Capsomer

Nucleic acid (DNA)

Protein fiber with a knob

(c) Influenza virus, an enveloped virus with spikes

Polyhedral capsid

Viral envelope

Nucleic acid (RNA)

Spike glycoproteins

(d) T4, a bacteriophage

Head (polyhedral capsid)

Nucleic acid (DNA) inside capsid head

Shaft

Tail fiber

Base plate

Figure 18.3 **Variations in the structure of viruses, as shown by transmission electron microscopy.** All viruses contain nucleic acid (DNA or RNA) surrounded by a protein capsid. They may or may not have an outer envelope surrounding the capsid. **(a)** Tobacco mosaic virus (TMV) has a capsid made of 2,130 identical protein subunits, helically arranged around a strand of RNA. **(b)** Adenoviruses have polyhedral capsids containing protein fibers with a knob. **(c)** Many animal viruses, including the influenza virus, have an envelope composed of a lipid bilayer and spike glycoproteins. The lipid bilayer is obtained from the host cell when the virus buds from the plasma membrane. **(d)** Some bacteriophages, such as T4, have protein coats with accessory structures that facilitate invasion of a bacterial cell.

Concept check: *What features vary among different types of viruses?*

simple viruses, such as phage Qβ, are only a few thousand nucleotides in length and contain only a few genes. Other viruses, particularly those with a complex structure, such as phage T4, contain many more genes. These extra genes encode many different proteins that are involved in the formation of the elaborate structure shown in Figure 18.3d.

Viruses Reproduce by Mobilizing Their Host Cells to Produce New Viruses

When a virus infects a host cell, the expression of viral genes leads to a series of steps, called a **viral reproductive cycle**, that results in the production of new viruses. The details of the steps may be quite different among various types of viruses, and even the same virus may have the capacity to follow alternative cycles. Even so, by studying the reproductive cycles of hundreds of different viruses, researchers have determined that the viral reproductive cycle consists of five or six basic steps.

To illustrate the general features of viral reproductive cycles, **Figure 18.4** considers these steps for two types of viruses. Figure 18.4a shows the cycle of phage λ (lambda), a bacteriophage with double-stranded DNA as its genome, and Figure 18.4b depicts the cycle of HIV, an enveloped animal virus containing single-stranded RNA. The descriptions that follow compare the reproductive cycles of these two very different viruses.

Step 1: Attachment   In the first step of a viral reproductive cycle, the virus must attach to the surface of a host cell. This attachment is usually specific for one or just a few types of cells because proteins in the virus recognize and bind to specific molecules on the cell surface. In the case of phage λ, the phage tail fibers bind to proteins in the outer bacterial cell membrane of *E. coli* cells. In the case of HIV, spike glycoproteins in the viral envelope bind to protein receptors in the plasma membrane of human blood cells called helper T cells.

Step 2: Entry   After attachment, the viral genome enters the host cell. Attachment of phage λ stimulates a conformational change in the phage coat proteins, so the shaft (also called the sheath) contracts, and the phage injects its DNA into the bacterial cytoplasm. In contrast, the envelope of HIV fuses with the plasma membrane of the host cell, so both the capsid and its contents are released into the cytosol. Some of the HIV capsid proteins are then removed by host cell enzymes, a process called uncoating. This releases two copies of the viral RNA and two molecules of an enzyme called reverse transcriptase into the cytosol. As discussed shortly, reverse transcriptase is needed for step 3.

Once a viral genome has entered the cell, one or several viral genes are expressed immediately due to the action of host cell enzymes and ribosomes. Expression of these key genes leads quickly to either step 3 or step 4 of the reproductive cycle, depending on the specific virus. The genome of some viruses, including both phage λ and HIV, can integrate into a chromosome of the host cell. For such viruses, the cycle may proceed from step 2 to step 3 as described next, delaying the production of new viruses. Alternatively, the cycle may proceed directly from step 2 to step 4 and quickly lead to the production of new viruses.

Step 3: Integration   Viruses capable of integration carry a gene that encodes an enzyme called **integrase**. For integration to occur, this gene is expressed soon after entry so that integrase protein is made. Integrase cuts the host's chromosomal DNA and inserts the viral genome into the chromosome. In the case of phage λ, the double-stranded DNA that entered the cell can be directly integrated into the double-stranded DNA of the chromosome. Once integrated, the phage DNA in a bacterium is called a **prophage**. While it exists as a prophage, this type of viral reproductive cycle is called the **lysogenic cycle**. As discussed later, new phages are not made during the lysogenic cycle, and the host cell is not destroyed. On occasion, a prophage can be excised from the bacterial chromosome and proceed to step 4.

How can an RNA virus integrate its genome into the host cell's DNA? For this to occur, the viral genome must be copied into DNA. HIV accomplishes this by means of a viral enzyme called **reverse transcriptase**, which is carried within the capsid and released into the host cell along with the viral RNA. Reverse transcriptase uses the viral RNA strand to make a complementary copy of DNA, and it then uses the DNA strand as a template to make double-stranded viral DNA. This process is called reverse transcription because it is the reverse of the usual transcription process, in which a DNA strand is used to make a complementary strand of RNA. The viral double-stranded DNA enters the host cell nucleus and is inserted into a host chromosome via integrase. Once integrated, the viral DNA in a eukaryotic cell is called a **provirus**. Viruses that follow this mechanism are called **retroviruses**.

Step 4: Synthesis of Viral Components   The production of new viruses by a host cell involves the replication of the viral genome and the synthesis of viral proteins that make up the protein coat. In the case of a bacteriophage that has been integrated into the host chromosome, the prophage must be excised as described in step 3 before synthesis of new viral components can occur. An enzyme called excisionase is required for this process. Following excision, host cell enzymes make many copies of the phage DNA and transcribe the genes within these copies into mRNA. Host cell ribosomes translate this viral mRNA into viral proteins. The expression of phage genes also leads to the degradation of the host chromosomal DNA.

In the case of HIV, the DNA provirus is not excised from the host chromosome. Instead, it is transcribed in the nucleus to produce many copies of viral RNA. These viral RNA molecules enter the cytosol, where they are used to make viral proteins and serve as the genome for new viral particles.

Step 5: Viral Assembly   After all of the necessary components have been synthesized, they must be assembled into new viruses. Some viruses with a simple structure self-assemble, meaning that viral components spontaneously bind to each other to form a complete virus particle. An example of a self-assembling

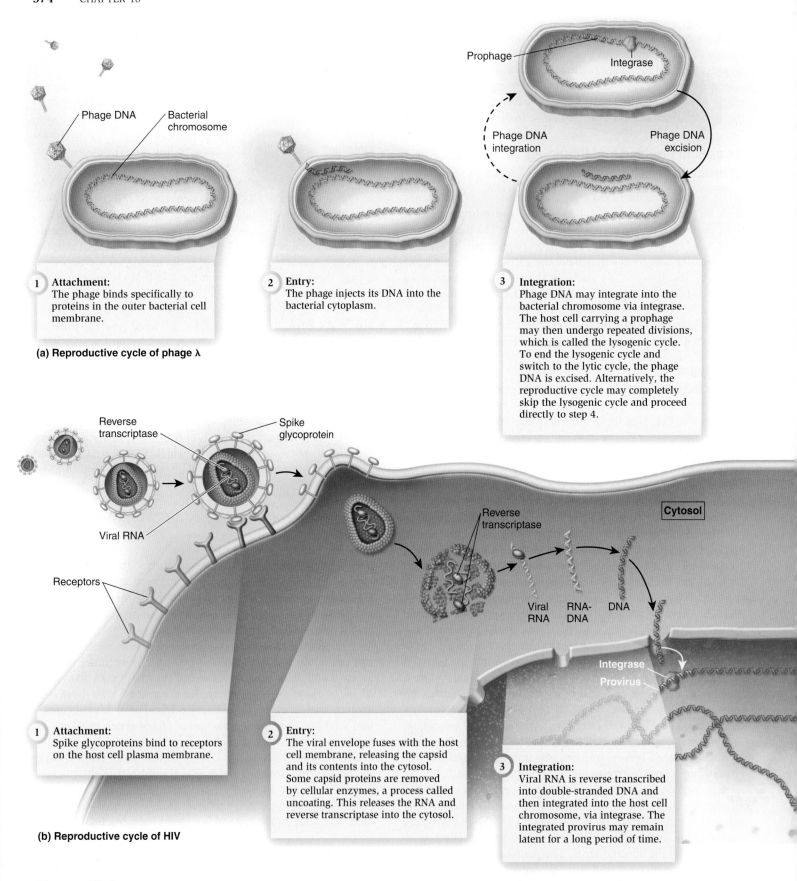

1 **Attachment:**
The phage binds specifically to proteins in the outer bacterial cell membrane.

2 **Entry:**
The phage injects its DNA into the bacterial cytoplasm.

3 **Integration:**
Phage DNA may integrate into the bacterial chromosome via integrase. The host cell carrying a prophage may then undergo repeated divisions, which is called the lysogenic cycle. To end the lysogenic cycle and switch to the lytic cycle, the phage DNA is excised. Alternatively, the reproductive cycle may completely skip the lysogenic cycle and proceed directly to step 4.

(a) Reproductive cycle of phage λ

1 **Attachment:**
Spike glycoproteins bind to receptors on the host cell plasma membrane.

2 **Entry:**
The viral envelope fuses with the host cell membrane, releasing the capsid and its contents into the cytosol. Some capsid proteins are removed by cellular enzymes, a process called uncoating. This releases the RNA and reverse transcriptase into the cytosol.

3 **Integration:**
Viral RNA is reverse transcribed into double-stranded DNA and then integrated into the host cell chromosome, via integrase. The integrated provirus may remain latent for a long period of time.

(b) Reproductive cycle of HIV

Figure 18.4 Comparison of the steps of two viral reproductive cycles. **(a)** The reproductive cycle of phage λ, a bacteriophage with a double-stranded DNA genome. **(b)** The reproductive cycle of HIV, an enveloped animal virus with a single-stranded RNA genome.

4 **Synthesis of viral components:**
In the lytic cycle, phage DNA
directs the synthesis of viral
components. During this process,
the phage DNA circularizes, and the
host chromosomal DNA is degraded.

5 **Viral assembly:**
Phage components are assembled
with the help of noncapsid proteins
to make many new phages.

6 **Release:**
The viral enzyme called lysozyme
causes cell lysis, and new phages
are released from the broken cell.

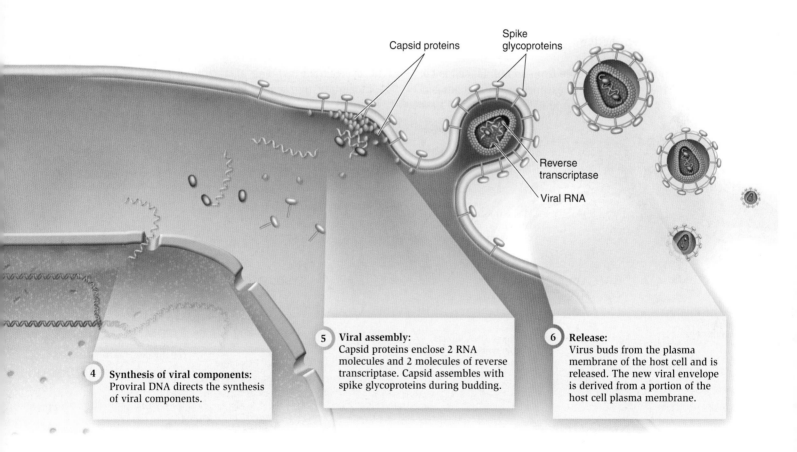

Capsid proteins

Spike
glycoproteins

Reverse
transcriptase

Viral RNA

4 **Synthesis of viral components:**
Proviral DNA directs the synthesis
of viral components.

5 **Viral assembly:**
Capsid proteins enclose 2 RNA
molecules and 2 molecules of reverse
transcriptase. Capsid assembles with
spike glycoproteins during budding.

6 **Release:**
Virus buds from the plasma
membrane of the host cell and is
released. The new viral envelope
is derived from a portion of the
host cell plasma membrane.

virus is TMV, which we examined earlier (see Figure 18.3a). TMV capsid proteins assemble around a TMV RNA molecule, which becomes trapped inside the hollow capsid.

Other viruses, including the two shown in Figure 18.4, do not self-assemble. The correct assembly of phage λ requires the help of noncapsid proteins not found in the completed phage particle. Some of these noncapsid proteins function as enzymes that modify capsid proteins, while others serve as scaffolding for the assembly of the capsid.

The assembly of an HIV virus occurs in two stages. First, capsid proteins assemble around two molecules of HIV RNA and two molecules of reverse transcriptase. Next, the newly formed capsid acquires its outer envelope in a budding process. This second phase of assembly occurs during step 6, as the virus is released from the cell.

Step 6: Release The last step of a viral reproductive cycle is the release of new viruses from the host cell. The release of bacteriophages is a dramatic event. Because bacteria are surrounded by a rigid cell wall, the phages must burst, or lyse, their host cell in order to escape. After the phages have been assembled, a phage-encoded enzyme called lysozyme digests the bacterial cell wall, causing the cell to burst. Lysis releases many new phages into the environment, where they can infect other bacteria and begin the cycle again. Collectively, steps 1, 2, 4, 5, and 6 are called the **lytic cycle** because they lead to cell lysis.

The release of enveloped viruses from an animal cell is far less dramatic. This type of virus escapes by a mechanism called budding that does not lyse the cell. In the case of HIV, a newly assembled virus particle associates with a portion of the plasma membrane containing HIV spike glycoproteins. The membrane enfolds the viral capsid and eventually buds from the surface of the cell. This is how the virus acquires its envelope, which is a piece of host cell membrane studded with viral glycoproteins.

Latency in Bacteriophages As we saw in step 3, viruses can integrate their genomes into a host chromosome. In some cases, the prophage or provirus may remain inactive, or **latent**, for a long time. Most of the viral genes are silent during latency, and the viral reproductive cycle does not progress to step 4.

Latency in bacteriophages is also called lysogeny. When this occurs, both the prophage and its host cell are said to be lysogenic. When a lysogenic bacterium prepares to divide, it copies the prophage DNA along with its own DNA, so each daughter cell inherits a copy of the prophage. A prophage can be replicated repeatedly in this way without killing the host cell or producing new phage particles. As mentioned earlier, this is called the lysogenic cycle.

Many bacteriophages can alternate between lysogenic and lytic cycles (**Figure 18.5**). A bacteriophage that may spend some of its time in the lysogenic cycle is called a **temperate phage**. Phage λ is an example of a temperate phage. Upon infection, it can either enter the lysogenic cycle or proceed directly to the lytic cycle. Other phages, called **virulent phages**, have only lytic cycles. The genome of a virulent phage is not capable of integration into a host chromosome. Phage T2, which we

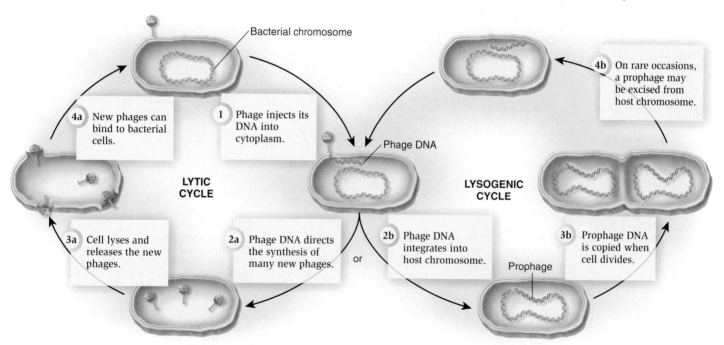

Figure 18.5 **Lytic and lysogenic cycles of bacteriophages.** Some phages, such as phage λ, may follow either a lytic or a lysogenic reproductive cycle. During the lytic cycle, new phages are made, and the bacterial cell is destroyed. During the lysogenic cycle, the integrated phage DNA, or prophage, is replicated along with the DNA of the host cell. Environmental conditions influence how long the phage remains in the lysogenic cycle. Other phages, such as T2, follow only lytic cycles.

Concept check: From the perspective of the virus, what are the primary advantages of the lytic and lysogenic cycles?

examined in Chapter 11, is a virulent phage that infects *E. coli.* Unlike phage λ, which may coexist harmlessly with *E. coli,* T2 always lyses the infected cell.

For phages such as λ that can follow either cycle, environmental conditions influence whether or not viral DNA is integrated into a host chromosome and how long the virus remains in the lysogenic cycle. If nutrients are readily available, phage λ usually proceeds directly to the lytic cycle after its DNA enters the cell. Alternatively, if nutrients are in short supply, the lysogenic cycle is often favored because sufficient material may not be available to make new viruses. If more nutrients become available later, this may cause the prophage to become activated. At this point, the viral reproductive cycle will switch to the lytic cycle, and new viruses will be made and released.

Latency in Human Viruses   Latency among human viruses can occur in two different ways. For HIV, latency occurs because the virus has integrated into the host genome and may remain dormant for long periods of time. In addition, the genomes of other viruses can exist as an **episome**—a genetic element that can replicate independently of the chromosomal DNA but also can occasionally integrate into chromosomal DNA. Examples of viral genomes that can exist as episomes include different types of herpesviruses that cause cold sores (herpes simplex type I), genital herpes (herpes simplex type II), and chickenpox (varicella zoster). A person infected with a given type of herpesvirus may have periodic outbreaks of disease symptoms when the virus switches from the latent, episomal form to the active form that produces new virus particles.

As an example, let's consider the herpesvirus called varicella zoster. The initial infection by this virus causes chickenpox, after which the virus may remain latent for many years as an episome. The disease called shingles occurs when varicella zoster switches from the latent state and starts making new virus particles. Shingles begins as a painful rash that eventually erupts into blisters. The blisters follow the path of the nerve cells that carry the latent varicella zoster virus. The blisters often form a ring around the back of the patient's body, which is why the disease is called shingles—the pattern of blisters line up like the shingles on a house.

Emerging Viruses, Such as HIV, Have Arisen Recently and May Rapidly Spread Through a Population

A primary reason researchers have been interested in viral reproductive cycles is the ability of many viruses to cause diseases in humans and other hosts. Some examples of human disease-causing viruses were presented earlier in Figure 18.2. **Emerging viruses** are viruses that have arisen recently, or are likely to have a greater probability of causing infection. Such viruses often cause public alarm and may lead to a significant loss of human life. New strains of influenza virus arise fairly regularly due to new mutations. An example is the strain H1N1, also called swine flu. In the U.S., despite attempts to minimize influenza deaths by vaccination, over 30,000 people die annually from this disease.

Another emerging virus causes a potentially life-threatening illness called severe acute respiratory syndrome (SARS). This RNA virus was first identified by researchers in Hong Kong, the U.S., and Germany in 2003. The type of virus causing SARS is called a coronavirus. The SARS coronavirus is believed to have originated in bats and acquired the ability to infect humans and livestock.

During the past few decades, the most devastating example of an emerging virus has been **human immunodeficiency virus** (**HIV**), the causative agent of **acquired immune deficiency syndrome** (**AIDS**). AIDS is primarily spread by sexual contact between infected and uninfected individuals, but it can also be spread by the transfusion of HIV-infected blood, by the sharing of needles among drug users, and from infected mother to unborn child. The total number of AIDS deaths between 1981 and the end of 2006 was over 25 million; more than 0.5 million of these deaths occurred in the U.S. During 2008, around 3 million adults and children became infected with HIV. Worldwide, nearly 1 in every 100 adults between ages 15 and 49 is infected. In the U.S., about 55,000 new HIV infections occur each year, 70% of which are in men and 30% in women.

The devastating effects of AIDS result from viral destruction of helper T cells, a type of white blood cell that plays an essential role in the immune system of mammals. Figure 18.6 shows HIV virus particles invading a helper T cell. As described in Chapter 53, helper T cells interact with other cells of the immune system to facilitate the production of antibodies and other molecules that target and kill foreign invaders of the body. When large numbers of helper T cells are destroyed by HIV, the function of the immune system is seriously compromised, and the individual becomes susceptible to infectious diseases called opportunistic infections that would not normally occur in a healthy person. For example, *Pneumocystis jiroveci,* a fungus that causes pneumonia, is easily destroyed by a normal immune system. However, in people with AIDS, infection by this fungus can be fatal.

Figure 18.6   **Micrograph of HIV invading a human helper T cell.**   This is a colorized scanning electron micrograph. The surface of the T cell is purple, and HIV particles are red.

An insidious feature of HIV replication, described earlier in Figure 18.4b, is that reverse transcriptase, the enzyme that copies the RNA genome into DNA, lacks a proofreading function. In Chapter 11, we learned that DNA polymerase can identify and remove mismatched nucleotides in newly synthesized DNA. Because reverse transcriptase lacks this function, it makes more errors and thereby tends to create mutant strains of HIV. This undermines the ability of the body to combat HIV because mutant strains may be resistant to the body's defenses. In addition, mutant strains of HIV may be resistant to antiviral drugs, as described next.

Drugs Have Been Developed to Combat the Proliferation of HIV

A compelling reason to understand the reproductive cycle of HIV and other disease-causing viruses is that such knowledge may be used to develop drugs that stop viral proliferation. For example, in the U.S., the estimated annual number of AIDS-related deaths fell approximately 14% from 1998 to 2002, owing in part to the use of new antiviral drugs. These drugs inhibit viral proliferation, though they cannot eliminate the virus from the body.

One approach to the design of antiviral treatments has been to create drugs that specifically bind to proteins encoded by the viral genome. For example, azidothymidine (AZT) mimics the structure of a normal nucleotide and can bind to reverse transcriptase. In this way, AZT inhibits reverse transcription, thereby inhibiting viral replication. Another way to combat HIV involves the use of antiviral drugs that inhibit proteases, enzymes that are needed during the assembly of the HIV capsid. Certain proteases cut capsid proteins, which makes them smaller and able to assemble into a capsid structure. If the proteases do not function, the capsid will not assemble, and new HIV particles will not be made. Several drugs known as protease inhibitors have been developed that bind to HIV proteases and inhibit their function.

A major challenge in AIDS research is to discover drugs that inhibit viral proteins without also binding to host cell proteins and inhibiting normal cellular functions. A second challenge is to develop drugs to which mutant strains will not become resistant. As mentioned, HIV readily accumulates mutations during viral replication. A current strategy is to treat HIV patients with a cocktail of three or more HIV drugs, making it less likely that any mutant strain will overcome all of the inhibitory effects.

Another approach to fighting HIV and other infectious diseases is vaccination—inoculation with a substance or group of substances that causes the immune system to respond to and destroy infectious agents such as bacteria or viruses. Vaccinations have been successful in the prevention of other viral diseases, such as influenza. In the case of HIV, the ideal vaccine should be both inexpensive and easy to store and administer, and it must confer long-lasting protection against HIV infection by sexual contact or by exposure to infected blood. Importantly, due to the high mutation rate of HIV, the vaccine must protect against exposure to many different strains of the virus.

Several Hypotheses Have Been Proposed to Explain the Origin of Viruses

Because viruses are such small particles, there is no fossil record of their evolution. Researchers must rely on analyses of modern viruses to develop hypotheses about their origin. Viral genomes follow the same rules of gene expression as the genomes of their host cells. Viral genes have promoter sequences similar to those of their host cells, and the translation of viral proteins relies on the genetic code. Viruses depend entirely on host cells for their proliferation. No known virus makes its own ribosomes or generates the energy it requires to make new viruses. Therefore, many biologists have argued that cells must have evolved before viruses.

How did viruses come into existence? A common hypothesis for the origin of viruses is they evolved from macromolecules inside living cells. The precursors of the first viruses may have been plasmids—small, circular DNA molecules that exist independently of chromosomal DNA. (Plasmids are described later in this chapter.) Biologists have hypothesized that such DNA molecules may have acquired genes that code for proteins that facilitate their own replication. Though many biologists favor the idea that viruses originated from primitive plasmids or other chromosomal elements, some have suggested they are an example of regressive evolution—the reduction of a trait or traits over time. This hypothesis proposes that viruses are degenerate cells that have retained the minimal genetic information essential for reproduction.

A new and interesting hypothesis is that viruses did not evolve from living cells but instead evolved in parallel with cellular organisms. As discussed in Chapter 22, the precursors of cellular DNA genomes may have been RNA molecules that could replicate independently of cells. This stage of evolution, termed the RNA world, could have involved the parallel evolution of both viruses and cellular organisms.

18.2   Viroids and Prions

Some nonliving infectious agents are even simpler then viruses. Viroids are composed solely of RNA, and prions are composed solely of protein. In this section, we will begin by examining viroids, infectious agents that cause diseases in plants. Next, we will discuss infectious proteins known as prions, which cause devastating neurological diseases in humans and other mammals. Unlike other agents of infection, prions have no genes and cannot be copied by the replication machinery of a cell. Instead, they increase their numbers by inducing changes in other protein molecules within living cells.

Viroids Are RNA Particles That Infect Plant Cells

In 1971, Theodor Diener discovered that the agent of potato spindle tuber disease is a small RNA molecule devoid of any protein. He coined the term **viroid** for this newly discovered infectious particle. Viroids are composed solely of a single-stranded circular RNA molecule that is a few hundred nucleotides in length.

Viroids infect plant cells, where they depend entirely on host enzymes for their replication. Some viroids are replicated in the host cell nucleus, others in the chloroplast. In contrast to viral genomes, the RNA genomes of viroids do not code for any proteins. How do viroids affect plant cells? The RNA of some viroids is known to possess ribozyme activity, and some researchers think this activity may damage plants by interfering with the function of host cell molecules. However, the mechanism by which viroids induce disease is not well understood.

Since Diener's initial discovery, many more viroids have been characterized as the agents of diseases affecting many economically important plants, including potato, tomato, cucumber, orange, coconut, grape, avocado, peach, apple, pear, and plum. Some viroids have devastating effects, as illustrated by the case of the coconut cadang-cadang viroid, which has killed more than 20 million coconut trees in Southeast Asia and New Guinea (**Figure 18.7**). Other viroids produce less severe damage, causing necrosis on leaves, shortening of stems, bark

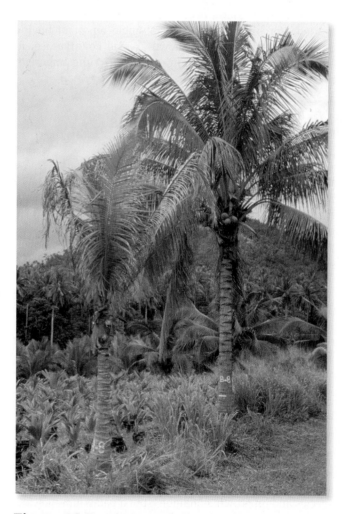

Figure 18.7 Effects of a viroid.   This palm tree in Papua, New Guinea, has been infected with the coconut cadang-cadang viroid.

cracking, and delays in foliation, flowering, and fruit ripening. A few viroids induce mild symptoms or no symptoms at all.

Prions Are Infectious Proteins That Cause Neurodegenerative Diseases

Before we end our discussion of nonliving, infectious particles, let's consider an unusual mechanism in which agents known as **prions** cause a group of rare, fatal brain diseases affecting humans and other mammals. Until the 1980s, biologists thought that any infectious agent, whether living or nonliving, must have genetic material. It seemed logical that genetic material is needed to store the information to create new infectious particles.

In the 1960s, British researchers Tikvah Alper and John Stanley Griffith discovered that preparations from animals with certain neurodegenerative diseases remained infectious even after exposure to radiation that would destroy any DNA or RNA. They suggested that the infectious agent was a protein. In the early 1970s, Stanley Prusiner, moved by the death of a patient from such a neurodegenerative disease, began to search for the causative agent. In 1982, Prusiner isolated a disease-causing particle composed entirely of protein, which he called a prion. The term was based on his characterization of the particle as a proteinaceous infectious agent. In 1997, Prusiner was awarded the Nobel Prize in Physiology or Medicine for his work on prions.

Prion diseases arise from the ability of the prion protein to induce abnormal folding in normal protein molecules (**Figure 18.8**). The prion protein exists in a disease-causing conformation designated PrP^{Sc}. The superscript Sc refers to scrapie, an example of a prion disease. A normal conformation of this same protein, which does not cause disease, is termed PrP^{C}. The superscript C stands for cellular. The normal protein is encoded by an individual's genome, and the protein is expressed at low levels in certain types of nerve cells.

How does someone contract a prion disease? A healthy person may become "infected" with the abnormal protein by eating meat of an animal with the disease. Unlike most other proteins in the diet, the prion protein escapes digestion in the stomach and small intestine and is absorbed into the bloodstream. After being taken up by nerve cells, the prion protein gradually converts the cell's normal proteins to the abnormal conformation. As a prion disease progresses, the PrP^{Sc} proteins are deposited as dense aggregates that form tough fibrils in the cells of the brain and peripheral nervous tissues, causing the disease symptoms. Some of the abnormal prion proteins are also excreted from infected cells, where they travel through the bloodstream. In this way, a prion disease can spread through the body like many viral diseases.

Prions are now known to cause several types of fatal neurodegenerative diseases affecting humans, livestock, and wildlife (**Table 18.2**). Prion diseases are termed transmissible spongiform encephalopathies (TSE). The postmortem examination of the brains of affected individuals reveals a substantial destruction of brain tissue. The brain has a spongy appearance.

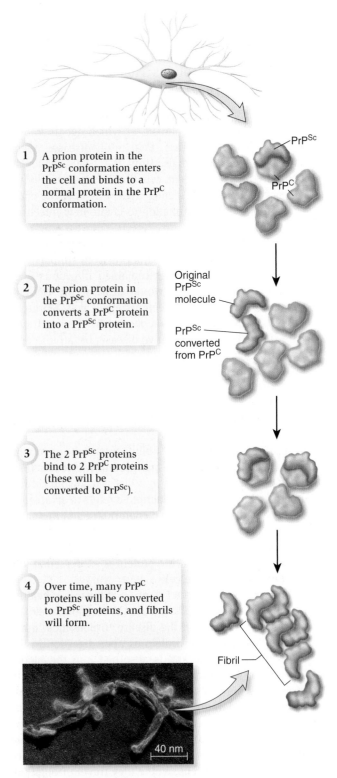

1 A prion protein in the PrP^{Sc} conformation enters the cell and binds to a normal protein in the PrP^C conformation.

PrP^{Sc}

PrP^C

2 The prion protein in the PrP^{Sc} conformation converts a PrP^C protein into a PrP^{Sc} protein.

Original PrP^{Sc} molecule

PrP^{Sc} converted from PrP^C

3 The 2 PrP^{Sc} proteins bind to 2 PrP^C proteins (these will be converted to PrP^{Sc}).

4 Over time, many PrP^C proteins will be converted to PrP^{Sc} proteins, and fibrils will form.

Fibril

40 nm

Figure 18.8   A proposed molecular mechanism of prion diseases.   A healthy neuron normally contains only the PrP^C conformation of the prion protein. The abnormal PrP^{Sc} conformation catalyzes the conversion of PrP^C proteins into PrP^{Sc} proteins, thereby causing the symptoms of prion diseases.

Concept check: Researchers are trying to discover drugs that prevent prion diseases. What are possible effects of a drug that would prevent the spread of the disease?

Table 18.2	Examples of Neurodegenerative Diseases Caused by Infectious Prions
Disease	**Description**
Scrapie	A disease of sheep and pigs characterized by intense itching. The animals tend to scrape themselves against trees or other objects, followed by neurodegeneration.
Mad cow disease	Begins with changes in posture and temperament, followed by loss of coordination and neurodegeneration.
Chronic wasting disease	A disease of deer (genus *Odocoileus*) and Rocky Mountain elk (*Cervus elaphus*). A consistent symptom is weight loss over time. The disease is progressive and fatal.

Most prion diseases progress fairly slowly. Over the course of a few years, symptoms proceed from a loss of motor control to dementia, paralysis, wasting, and eventually death. These symptoms are correlated with an increase in the level of prion protein in the nerve cells of infected individuals. No current treatment can halt the progression of any of the TSEs. For this reason, great public alarm occurs when an outbreak of a TSE is reported. For example, in 2003, a single report of bovine spongiform encephalitis, also known as mad cow disease, in the U.S. prompted several countries to restrict the import of American beef.

18.3   Genetic Properties of Bacteria

Bacteria and archaea, collectively referred to as prokaryotes, are usually unicellular organisms. Individual cells may exist as single units or remain associated with each other after cell division, forming pairs, chains, or clumps. Bacteria are widespread on Earth, and numerous species are known to cause various types of infectious diseases. Many species of archaea are also known. In this chapter, we will focus on bacteria. The genomes of bacteria, archaea, and eukaryotes are compared in Chapter 21.

We begin this section by exploring the structure and replication of the bacterial genome and the organization of DNA sequences along a bacterial chromosome. We then examine how the chromosome is compacted to fit inside a bacterium and how it is transmitted during asexual reproduction.

Bacteria Typically Have Circular Chromosomes That Carry a Few Thousand Genes

The genes of bacteria are found within structures known as bacterial chromosomes. Although a bacterial cell usually has a single type of chromosome, it may have more than one copy of that chromosome. The number of copies depends on the bacterial species and on growth conditions, but a bacterium typically has one to four identical chromosomes. Each bacterial chromosome is tightly packed within a distinct **nucleoid region** of the cell (**Figure 18.9**). Unlike the eukaryotic nucleus, the bacterial

Figure 18.9 Nucleoid regions within the bacterium *Bacillus subtilis.* In the light micrograph shown here, the nucleoid regions are fluorescently labeled and seen as bright, oval-shaped areas within the bacterial cytoplasm. Two or more nucleoid regions are usually found within each cell.

Concept check: How many nucleoid regions are in the bacterial cell to the far right?

nucleoid region is not a separate cellular compartment bounded by a membrane. The DNA in the nucleoid region is in direct contact with the cytoplasm of the cell.

Like eukaryotic chromosomes, bacterial chromosomes contain molecules of double-stranded DNA along with many different proteins. Unlike eukaryotic chromosomes, however, bacterial chromosomes are usually circular and tend to be much shorter, typically only a few million base pairs (bp) long. For example, the chromosome of *Escherichia coli* has approximately 4.6 million bp, and the *Hemophilus influenzae* chromosome has roughly 1.8 million bp. By comparison, an average human chromosome is over 100 million bp in length.

A typical bacterial chromosome contains a few thousand unique genes that are found throughout the chromosome (**Figure 18.10**). Structural gene sequences, nucleotide sequences that encode proteins, account for the largest part of bacterial DNA. Other nucleotide sequences in the chromosome influence DNA replication, gene expression, and chromosome structure. One of these sequences is the origin of replication, which is a few hundred bp long. Bacterial chromosomes have a single origin of replication that functions as an initiation site for the assembly of several proteins that are required for DNA replication (refer back to Figure 11.15b).

The Formation of Chromosomal Loops and DNA Supercoiling Makes the Bacterial Chromosome Compact

Bacterial cells are much smaller than most eukaryotic cells (refer back to Figure 4.1). *E. coli* cells, for example, are approximately 1 μm wide and 2 μm long. To fit within a bacterial cell, the DNA of a typical bacterial chromosome must be compacted about 1,000-fold. How does this occur? The compaction of a bacterial chromosome, shown in **Figure 18.11**, occurs by two processes: the formation of loops and by the supercoiling of the looped DNA.

Key features

- Most, but not all, bacterial species contain circular chromosomal DNA.
- A typical chromosome is a few million base pairs in length.
- Most bacterial species contain a single type of chromosome, but it may be present in multiple copies.
- Several thousand different genes are interspersed throughout the chromosome.
- One origin of replication is required to initiate DNA replication.

Figure 18.10 The organization of nucleotide sequences in bacterial chromosomal DNA.

Unlike eukaryotic DNA, bacterial DNA is not wound around histone proteins to form nucleosomes. However, the binding of proteins to bacterial DNA is important in the formation of **loop domains**—chromosomal segments that are folded into loops. As seen in Figure 18.11, DNA-binding proteins anchor the bases of the loops in place. The number of loops varies according to the size of a bacterial chromosome and the species. The *E. coli* chromosome has 50 to 100 loop domains, each with about 40,000 to 80,000 bp. This looping compacts the circular chromosome about 10-fold. A similar process of loop-domain formation occurs in eukaryotic chromatin compaction, which is described in Chapter 11.

DNA supercoiling is a second important way to compact the bacterial chromosome. Because DNA is a long, thin molecule, twisting can dramatically change its conformation. This compaction is similar to what happens to a rubber band if you twist it in one direction. Because the two strands of DNA already coil around each other, the formation of additional coils due to twisting is referred to as supercoiling. Bacterial enzymes called topoisomerases twist the DNA and control the degree of DNA supercoiling.

Plasmids Are Small Pieces of Extrachromosomal DNA

In addition to chromosomal DNA, bacterial cells commonly contain **plasmids**, small, circular pieces of DNA that exist separate from the bacterial chromosome (**Figure 18.12**). Plasmids

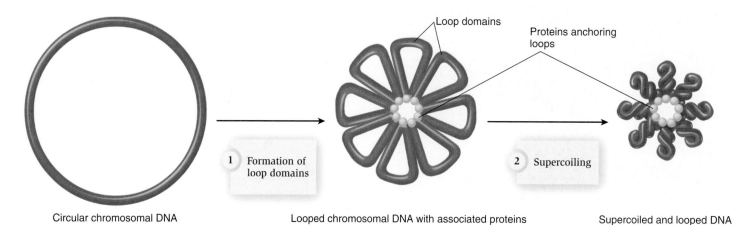

Circular chromosomal DNA

1   Formation of loop domains

Looped chromosomal DNA with associated proteins

2   Supercoiling

Supercoiled and looped DNA

Figure 18.11   **The compaction of a bacterial chromosome.**   As a way to compact the large, circular chromosome, segments are organized into smaller loop domains by binding to proteins at the bases of the loops. These loops are made more compact by DNA supercoiling.

Concept check: *Describe how the loop domains are held in place.*

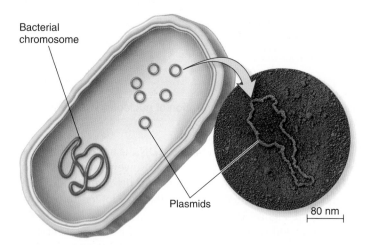

Figure 18.12   **Plasmids in a bacterial cell.**   Plasmids are small, circular DNA molecules that exist independently of the bacterial chromosome.

Concept check: *Describe the similarities and differences between a bacterial chromosome and a plasmid.*

occur naturally in many strains of bacteria and in a few types of eukaryotic cells, such as yeast. The smallest plasmids consist of just a few thousand base pairs and carry only a gene or two. The largest are in the range of 100,000 to 500,000 bp and carry several dozen or even hundreds of genes. A plasmid has its own origin of replication that allows it to be replicated independently of the bacterial chromosome. The DNA sequence of the origin of replication influences how many copies of the plasmid are found within a cell. Some origins are said to be very strong because they result in many copies of the plasmid, perhaps as many as 100 per cell. Other origins of replication have sequences that are much weaker, so the number of copies is relatively low, such as one or two per cell.

Why do bacteria have plasmids? Certain genes within a plasmid usually provide some type of growth advantage to the cell or may aid in survival under certain conditions. By studying plasmids in many different species, researchers have discovered that most plasmids fall into five different categories:

1. Resistance plasmids, also known as R factors, contain genes that confer resistance against antibiotics and other types of toxins.
2. Degradative plasmids carry genes that enable the bacterium to digest and utilize an unusual substance. For example, a degradative plasmid may carry genes that allow a bacterium to digest an organic solvent such as toluene.
3. Col-plasmids contain genes that encode colicins, which are proteins that kill other bacteria.
4. Virulence plasmids carry genes that turn a bacterium into a pathogenic strain.
5. Fertility plasmids, also known as F factors, allow bacteria to mate with each other, a topic described later in this chapter.

On occasion, a plasmid may integrate into the bacterial chromosome. Plasmids that can integrate or remain independent of the chromosome are also termed episomes.

Bacteria Reproduce Asexually by Binary Fission

Thus far, we have considered the genetic material of bacteria and how the bacterial chromosome is compacted to fit inside the cell. Let's now turn our attention to the process of cell division. The capacity of bacteria to divide is really quite astounding. The cells of some species, such as *E. coli*, can divide every 20–30 minutes. When placed on a solid growth medium in a petri dish, an *E. coli* cell and its daughter cells will undergo repeated cellular divisions and form a clone of genetically identical cells called a **bacterial colony** (Figure 18.13). Starting with a single cell that is invisible to the naked eye, a visible bacterial colony containing 10 to 100 million cells will form in less than a day!

Figure 18.13   Growth of a bacterial colony.   Through successive cell divisions, a single bacterial cell of *E. coli* forms a genetically identical group of cells called a bacterial colony.

Concept check: Let's suppose a bacterial strain divides every 30 minutes. If a single cell is placed on a plate, how many cells will be in the colony after 16 hours?

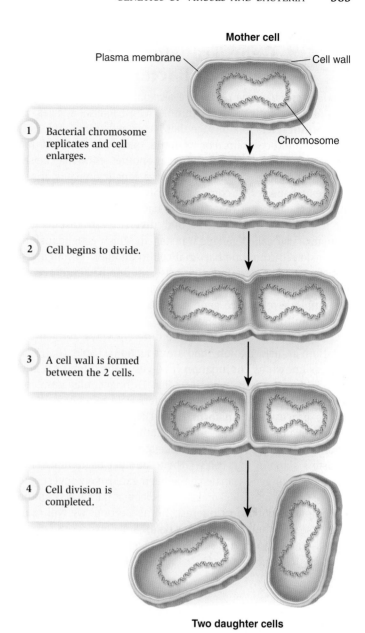

Mother cell

1 Bacterial chromosome replicates and cell enlarges.

2 Cell begins to divide.

3 A cell wall is formed between the 2 cells.

4 Cell division is completed.

Two daughter cells

Figure 18.14   Bacterial cell division.   Bacteria reproduce by a type of cell division called binary fission. Before a bacterium divides, the bacterial chromosome is replicated to produce two identical copies. These two copies segregate from each other during cell division, with one copy going to each daughter cell.

As described in Chapter 15, the division of eukaryotic cells requires a sorting process called mitosis, because eukaryotic chromosomes occur in sets and each daughter cell must receive the correct number and types of chromosomes. By comparison, a bacterial cell usually has only a single type of chromosome. Cell division occurs by a much simpler process called **binary fission**, during which a cell divides into two daughter cells. Figure 18.14 shows this process for a cell with a single chromosome. Before it divides, the cell replicates its DNA. This produces two identical copies of the chromosome. Next, the cell's plasma membrane is drawn inward and deposits new cell-wall

material, separating the two daughter cells. Each daughter cell receives one of the copies of the original chromosome. Therefore, except when a mutation occurs, each daughter cell contains an identical copy of the mother cell's genetic material. Like other types of asexual reproduction, binary fission does not involve genetic contributions from two different parents.

If a bacterial cell contains plasmids, these will replicate independently of the bacterial chromosome. During binary fission, the plasmids are distributed to daughter cells so that each daughter cell usually receives one or more copies of the plasmid.

18.4 Gene Transfer Between Bacteria

Even though bacteria reproduce asexually, they exhibit a great deal of genetic diversity. Within a given bacterial species, the term **strain** refers to a lineage that has genetic differences compared to another strain. For example, one strain of *E. coli* may be resistant to an antibiotic, whereas another strain may be sensitive to the same antibiotic. How does genetic diversity arise in an asexual species? It comes primarily from two sources. First, mutations can occur that alter the bacterial genome and affect the traits of bacterial cells. Second, diversity can arise by **gene transfer**, also called genetic transfer, in which genetic material is transferred from one bacterial cell to another. Through gene transfer, genetic variation that arises in one bacterium can be spread to new strains and even to other species. For example, an antibiotic-resistance gene may be transferred from a resistant strain to a sensitive strain.

Gene transfer occurs in three different ways, termed conjugation, transformation, and transduction (**Table 18.3**). The process known as **conjugation** involves a direct physical interaction between two bacterial cells. During conjugation, one bacterium acts as a donor and transfers DNA to a recipient cell. In the process of **transformation**, DNA that is released into the environment is taken up by another bacterial cell. **Transduction** occurs when a virus infects a bacterial cell and then transfers some of that cell's DNA to another bacterium. These three types of gene transfer have been extensively investigated in research laboratories, and their molecular pathways continue to be studied with great interest. In this section, we will examine these mechanisms in greater detail and consider the experiments that led to their discovery.

Table 18.3	Mechanisms of Gene Transfer Between Bacterial Cells
Mechanism	**Description**
Conjugation: Donor cell Recipient cell	Requires direct contact between a donor and a recipient cell. The donor cell transfers a strand of DNA to the recipient. In the example shown here, DNA from a plasmid is transferred to the recipient cell. Both donor and recipient cells end up with a plasmid.
Transformation: Donor cell Recipient cell (dead)	A fragment of its DNA from a donor cell is released into the environment. This may happen when a bacterial cell dies. This DNA fragment is taken up by a recipient cell, which incorporates the DNA into its chromosome.
Transduction: Donor cell Recipient cell (infected by a virus)	When a virus infects a donor cell, it causes the bacterial chromosome of the donor cell to break up into fragments. A fragment of bacterial chromosomal DNA is incorporated into a newly made virus particle. The virus then transfers this fragment of DNA to a recipient cell.

FEATURE INVESTIGATION

Lederberg and Tatum's Work with *E. coli* Demonstrated Gene Transfer Between Bacteria and Led to the Discovery of Conjugation

In 1946 and 1947, Joshua Lederberg and Edward Tatum carried out the first experiments that clearly demonstrated gene transfer from one bacterial strain to another (**Figure 18.15**). The researchers had been studying strains of *E. coli* that had different nutritional requirements for growth. They designated one strain *met⁻bio⁻thr⁺pro⁺* because its growth required that the amino acid methionine (met) and the vitamin biotin (bio) be added to the growth medium. This strain did not require the amino acids threonine (thr) or proline (pro) for growth. Another strain, designated *met⁺bio⁺thr⁻pro⁻*, had just the opposite requirement. It needed threonine and proline in its growth medium, but not methionine or biotin. These differences in nutritional requirements correspond to allelic differences between the two

strains. The *met⁻bio⁻thr⁺pro⁺* strain had defective genes encoding enzymes necessary for methionine and biotin synthesis, whereas the *met⁺bio⁺thr⁻pro⁻* strain had defective genes for the enzymes required to make threonine and proline.

Figure 18.15 compares the results of mixing the two *E. coli* strains with the results when they were not mixed. The tube shown on the left contained only *met⁻bio⁻thr⁺pro⁺* cells, and the tube on the right had only *met⁺bio⁺thr⁻pro⁻* cells. The middle tube contained a mixture of the two kinds of cells. In each case, the researchers applied about 100 million (10^8) cells to plates containing a growth medium lacking amino acids and the vitamin biotin. When the unmixed strains were applied to these plates, no colonies were observed to grow. This result was expected because the plates did not contain the methionine and biotin that the *met⁻bio⁻thr⁺pro⁺* cells needed for growth or the threonine and proline that the *met⁺bio⁺thr⁻pro⁻* cells required. The striking result occurred when the researchers plated 10^8

Figure 18.15 Experiment of Lederberg and Tatum demonstrating gene transfer in *E. coli*.

HYPOTHESIS Genetic material can be transferred from one bacterial strain to another.

KEY MATERIALS Two bacterial strains, one that was *met⁻ bio⁻ thr⁺ pro⁺* and the other that was *met⁺ bio⁺ thr⁻ pro⁻*.

Experimental level   Conceptual level

1 In 3 separate tubes, add either the *met⁻ bio⁻ thr⁺ pro⁺* strain, the *met⁺ bio⁺ thr⁻ pro⁻* strain, or a mixture of both strains.

   Incubate several hours.

   met⁻ bio⁻ thr⁺ pro⁺   *met⁺ bio⁺ thr⁻ pro⁻*

2 Remove 10⁸ cells from each tube and spread onto plates that lack methionine, biotin, threonine, and proline.

   10⁸   10⁸   10⁸

   Nutrient agar plates lacking amino acids and biotin

   Genetic material was transferred between the two strains.

3 Incubate overnight to allow growth of bacterial colonies.

   No colonies   Bacterial colonies (*met⁺ bio⁺ thr⁺ pro⁺*)   No colonies

4 **THE DATA**

Strain	Number of colonies after overnight growth
met⁻ bio⁻ thr⁺ pro⁺	0
met⁺ bio⁺ thr⁻ pro⁻	0
Both strains together	~10

5 **CONCLUSION** Gene transfer has occurred from one bacterial strain to another.

6 **SOURCE** Lederberg, Joshua, and Tatum, Edward L. 1946. Novel genotypes in mixed cultures of biochemical mutants of bacteria. *Cold Spring Harbor Symposia on Quantitative Biology* 11:113–114.

Tatum, Edward L., and Lederberg, Joshua. 1947. Genetic recombination in the bacterium *Escherichia coli*. *Journal of Bacteriology* 53:673–684.

cells from the tube containing the mixture of the two strains. In this case, approximately 10 cells multiplied and formed visible bacterial colonies on the plates. Because these cells were able to reproduce without supplemental amino acids or vitamins, their genotype must have been *met⁺ bio⁺ thr⁺ pro⁺*. Mutation cannot account for the occurrence of this new genotype because colonies were not observed on the other two plates, which had the same number of cells and also could have incurred mutations.

To explain the results of their experiment, Lederberg and Tatum hypothesized that some genetic material had been transferred between the two strains when they were mixed. This transfer could have occurred in two ways. One possibility is that the genes providing the ability to synthesize threonine and pro-line (*thr⁺ pro⁺*) were transferred to the *met⁺ bio⁺ thr⁻ pro⁻* strain. Alternatively, the genes providing the ability to synthesize methionine and biotin (*met⁺ bio⁺*) may have been transferred to the *met⁻ bio⁻ thr⁺ pro⁺* cells. The experimental results cannot distinguish between these two possibilities, but they provide compelling evidence that at least one of them occurred.

How did the bacteria in Lederberg and Tatum's experiment transfer genes between strains? Two mechanisms seemed plausible. Either genetic material was released from one strain and taken up by the other, or cells of the two different strains made contact with each other and directly transferred genetic material. To distinguish these two scenarios, Bernard Davis conducted experiments using the same two strains of *E. coli*.

The apparatus he used, known as a U-tube, is shown in **Figure 18.16**. The tube had a filter with pores big enough for pieces of DNA to pass through, but too small to permit the passage of bacteria. After filling the tube with a liquid medium, Davis added $met^-bio^-thr^+pro^+$ bacteria on one side of the filter and $met^+bio^+thr^-pro^-$ bacteria on the other. The application of pressure or suction promoted the movement of liquid through the pores. Although the two kinds of bacteria could not mix, any genetic material released by one of them would be available to the other.

After allowing the bacteria to incubate in the U-tube, Davis placed cells from each side of the tube on growth plates lacking methionine, biotin, threonine, and proline. No bacterial colonies grew on these plates. How did Davis interpret these results? He proposed that without physical contact, the two *E. coli* strains could not transfer genetic material from one to the other. The conceptual level of Figure 18.15, step 1, shows the physical connection that explains Lederberg and Tatum's results. Conjugation is the process of gene transfer that requires direct cell-to-cell contact. It has been subsequently observed in other species of bacteria. Many, but not all, species of bacteria can conjugate.

Experimental Questions

1. What was the hypothesis tested by Lederberg and Tatum?

2. During the Lederberg and Tatum experiment, the researchers compared the growth of mutant strains under two scenarios: mixed strains or unmixed strains. When the unmixed strains were plated on the experimental growth medium, why were no colonies observed to grow? When the mixed strains were plated on the experimental growth medium, a number of colonies were seen to grow. What was the significance of the growth of these colonies?

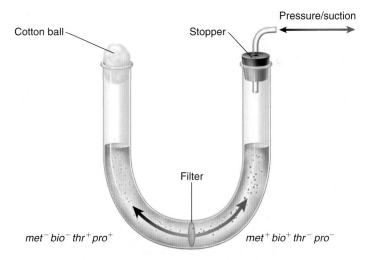

Figure 18.16   **A U-tube apparatus like the one used by Bernard Davis.**   Bacteria of two different strains were suspended in the liquid in the tube and separated by a filter. The liquid was forced through the filter by alternating suction and pressure. The pores in the filter were too small for the passage of bacteria, but they allowed the passage of DNA.

Concept check: Would the results have been different if the pore size was larger and allowed the passage of bacterial cells?

3. The gene transfer seen in the Lederberg and Tatum experiment could have occurred in one of two ways: taking up DNA released into the environment or contact between two bacterial cells allowing for direct transfer. Bernard Davis conducted an experiment to determine the correct process. Explain how his results indicated the correct gene transfer process.

During Conjugation, DNA Is Transferred from a Donor Cell to a Recipient Cell

In the early 1950s, Joshua and Esther Lederberg, William Hayes, and Luca Cavalli-Sforza independently discovered that only certain bacterial strains can donate genetic material during conjugation. For example, only about 5% of *E. coli* strains found in nature can act as donor strains. Further research showed that a strain that is incapable of acting as a donor can acquire this ability after being mixed with a donor strain. Hayes correctly proposed that donor strains contain a type of plasmid called a fertility factor, or **F factor**, that can be transferred to recipient strains. Also, other donor *E. coli* strains were later identified that can transfer portions of the bacterial chromosome at high frequencies. After a segment of the chromosome is transferred, it then inserts, or recombines, into the chromosome of the recipient cell. Such donor strains were named Hfr (for High frequency of recombination). In our discussion, we will focus on donor strains that carry F factors.

The micrograph in **Figure 18.17a** shows two conjugating *E. coli* cells. The cell on the left is designated F^+, meaning that it has an F factor. This donor cell is transferring genetic material to the recipient cell on the right, which lacks an F factor and is designated F^-. F factors carry several genes that are required for conjugation and also may carry genes that confer a growth advantage for the bacterium.

Figure 18.17b describes the events that occur during conjugation in *E. coli*. The process is similar in other bacteria that are capable of conjugating, although the details vary somewhat from one species to another. Contact between donor and recipient cells is often a key step that initiates the conjugation process. Recall from Chapter 4 that many bacteria have appendages called pili that allow them to attach to surfaces and to each other. **Sex pili** are made by F^+ cells that bind specifically to F^- cells. They are so named because conjugation has sometimes been called bacterial mating or bacterial sex. However, these terms are a bit misleading because the process does not involve equal genetic contributions from two gametes and it does not produce offspring. Instead, bacterial mating is a form of gene transfer that alters the genetic composition of the recipient cell. Donor strains have genes responsible for the formation of sex pili. In F^+ strains, the genes are located on the F factor. In *E. coli* and some other species, F^+ cells make very long pili that attempt to make contact with nearby F^- cells. Once contact is

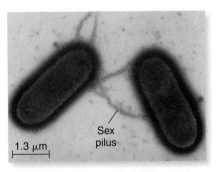

(a) Micrograph of conjugating cells

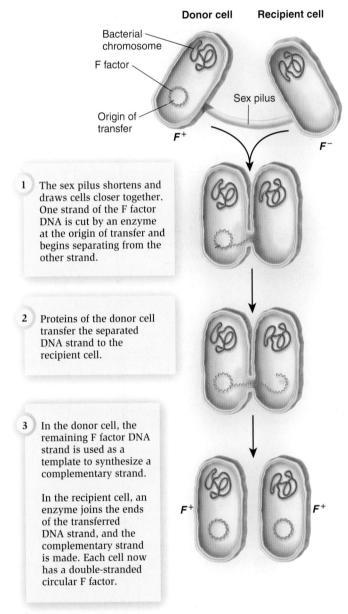

Donor cell   Recipient cell

Bacterial chromosome

F factor

Sex pilus

Origin of transfer

F^+

F^-

1 The sex pilus shortens and draws cells closer together. One strand of the F factor DNA is cut by an enzyme at the origin of transfer and begins separating from the other strand.

2 Proteins of the donor cell transfer the separated DNA strand to the recipient cell.

3 In the donor cell, the remaining F factor DNA strand is used as a template to synthesize a complementary strand.

In the recipient cell, an enzyme joins the ends of the transferred DNA strand, and the complementary strand is made. Each cell now has a double-stranded circular F factor.

F^+ F^+

(b) Transfer of an F factor

Figure 18.17 **Bacterial conjugation.** (a) A micrograph of two *E. coli* cells that are conjugating. The cell on the left, designated F^+, is the donor; the cell on the right, designated F^-, is the recipient. The two cells make contact via sex pili made by the F^+ cell. (b) The transfer of an F factor during conjugation. At the end of conjugation, both the donor cell and the recipient cell are F^+.

Concept check: If a donor cell has only one F factor, explain how the donor and recipient cell both contain one F factor following the transfer of an F factor during conjugation.

made, the pili shorten, drawing the donor and recipient cells closer together.

Successful contact stimulates the donor cell to begin the transfer process. Genes within the F factor encode proteins that promote the transfer of one strand of F factor DNA. This DNA strand is cut at the origin of transfer, and then the strand travels into the recipient cell. The other strand remains in the donor cell, and the complementary strand is synthesized, thereby restoring the F factor DNA to its original double-stranded condition. In the recipient cell, the two ends of the newly acquired F factor DNA strand are joined to form a circular molecule, and its complementary strand is synthesized to produce a double-stranded F factor. The end result of conjugation is that the recipient cell has acquired an F factor, converting it from an F^- to an F^+ cell. The genetic composition of the donor strain has not been changed.

In Transformation, Bacteria Take Up DNA from the Environment

In contrast to conjugation, the process of gene transfer known as bacterial transformation does not require direct contact between bacterial cells. Frederick Griffith first discovered this process in 1928 while working with strains of *Streptococcus pneumoniae*. We discussed early experiments involving transformation in Chapter 11 (refer back to Figures 11.1 and 11.2).

How does a bacterial cell become transformed? First, it imports a strand of DNA from the environment. This DNA strand, which is typically derived from a dead bacterial cell, may then insert or recombine into the bacterial chromosome. The live bacterium is now carrying genes from the dead bacterium—the live bacterium has been transformed. Not all bacterial strains have the ability to take up DNA. Those that do have this ability are described as naturally **competent**, and they have genes that encode proteins called competence factors. Competence factors facilitate the binding of DNA fragments to the bacterial cell surface, the uptake of DNA into the cytoplasm, and the incorporation of the imported DNA into the bacterial chromosome. Temperature, ionic conditions, and the availability of nutrients also affect whether or not a bacterium will be competent to take up genetic material.

In recent years, biologists have unraveled some of the steps that occur when competent bacterial cells are transformed by taking up genetic material from the environment. In the example shown in **Figure 18.18**, the DNA released from a dead bacterium carries a gene, tet^R, that confers resistance to the antibiotic tetracycline. First, a large fragment of the DNA binds to a surface receptor on the outside of a bacterial cell that is sensitive to tetracycline. Enzymes secreted by the bacterium cut this large fragment into fragments small enough to enter the cell. The next step is for a small DNA fragment to begin its entry into the bacterial cytoplasm. One of the two DNA strands of this fragment is degraded. The other strand enters the bacterial cytoplasm via a DNA uptake system that transports the DNA across the plasma membrane. Finally, the imported DNA strand is incorporated into the bacterial chromosome, and the complementary strand is synthesized. Following transformation, the

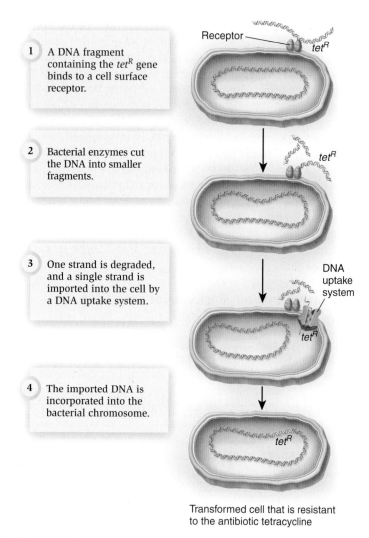

① A DNA fragment containing the *tet^R* gene binds to a cell surface receptor.

② Bacterial enzymes cut the DNA into smaller fragments.

③ One strand is degraded, and a single strand is imported into the cell by a DNA uptake system.

④ The imported DNA is incorporated into the bacterial chromosome.

Transformed cell that is resistant to the antibiotic tetracycline

Figure 18.18 **The steps of bacterial transformation.** This process has transformed a bacterium that was sensitive to the antibiotic tetracycline into one that can grow in the presence of this antibiotic.

recipient cell has been transformed from a tetracycline-sensitive cell to a tetracycline-resistant cell.

In Transduction, Viruses Transfer Genetic Material from One Bacterium to Another

Perhaps the most curious method of gene transfer is transduction, in which bacteriophages transfer bacterial genes from one bacterium to another. As discussed earlier in this chapter, a bacteriophage (or simply phage) is a virus that uses the cellular machinery of a bacterium for its own replication. The new viral particles made in this way usually contain only viral genes. On rare occasions, however, a phage may pick up a piece of DNA from the bacterial chromosome. When a phage carrying a segment of bacterial DNA infects another bacterium, it transfers this segment into the chromosome of its new bacterial host.

Transduction is actually an error in a phage lytic cycle, as shown in **Figure 18.19**. In this example, a phage called P1 infects an *E. coli* cell that has a gene (*his^+*) for histidine

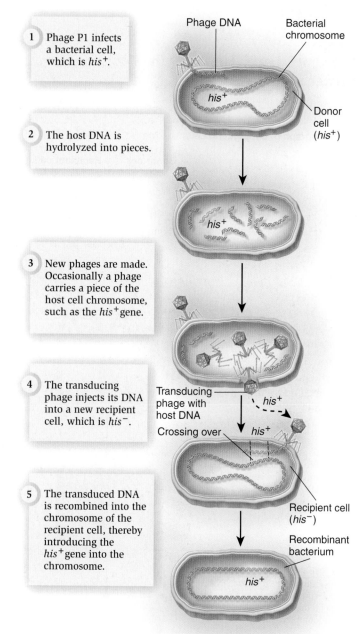

① Phage P1 infects a bacterial cell, which is *his^+*.

② The host DNA is hydrolyzed into pieces.

③ New phages are made. Occasionally a phage carries a piece of the host cell chromosome, such as the *his^+* gene.

④ The transducing phage injects its DNA into a new recipient cell, which is *his^-*.

⑤ The transduced DNA is recombined into the chromosome of the recipient cell, thereby introducing the *his^+* gene into the chromosome.

The recombinant bacterium has a genotype (*his^+*) that is different from the original recipient bacterial cell (*his^-*).

Figure 18.19 Bacterial transduction by P1 phage.

Concept check: Is transduction a normal part of the phage life cycle? Explain.

synthesis. Phage P1 causes the host cell chromosome to degrade into small pieces. When new phages are assembled, coat proteins may enclose a piece of host DNA that carries this gene. This produces a phage carrying bacterial chromosomal DNA. In the example shown in Figure 18.19, this transducing phage is released and binds to an *E. coli* cell that lacks the *his^+* gene. It inserts the bacterial DNA fragment into the recipient cell, which then incorporates this fragment into its own chromosome by recombination. In this case, gene transfer by transduction converts a *his^-* strain of *E. coli* to a *his^+* strain.

Genomes & Proteomes Connection

Horizontal Gene Transfer Is the Transfer of Genes Between the Same or Different Species

So far we have considered gene transfer from one bacterial strain to another strain of the same species. In addition, conjugation, transformation, and transduction occasionally occur between cells of different bacterial species. The term **horizontal gene transfer** refers to a process in which an organism incorporates genetic material from another organism without being the offspring of that organism. Conjugation, transformation, and transduction are examples of horizontal gene transfer. In contrast, vertical gene transfer occurs when genes are passed from one generation to the next—from parents to offspring and from mother cells to daughter cells.

Why is horizontal gene transfer important? In recent years, analyses of bacterial genomes have shown that a sizeable fraction of bacterial genes are derived from horizontal gene transfer. For example, roughly 17% of the genes of *E. coli* and of *Salmonella typhimurium* have been acquired from other species by horizontal transfer during the past 100 million years. Many of these acquired genes are for traits that give cells a selective advantage, including genes that confer antibiotic resistance, the ability to degrade toxic compounds, and the ability to withstand extreme environments. Some horizontally transferred genes confer pathogenicity, turning a harmless bacterial strain into one that can cause disease. Geneticists have suggested that horizontal gene transfer has played a major role in the evolution of different bacterial species. In many cases, the acquisition of new genes allows a bacterium to survive in a new type of environment and can eventually lead to the formation of a new species.

A second reason why horizontal gene transfer is important is its medical relevance. Let's consider the topic of antibiotic resistance. Antibiotics are widely prescribed to treat bacterial infections in humans. They are also used in agriculture to control bacterial diseases. Unfortunately, the widespread use of antibiotics has greatly increased the prevalence of antibiotic-resistant strains of bacteria, strains that have a selective advantage over those that are susceptible to antibiotics. Resistant strains carry genes that counteract the action of antibiotics in various ways. A resistance gene may encode a protein that breaks down the drug, pumps it out of the cell, or prevents it from inhibiting cellular processes.

The term **acquired antibiotic resistance** refers to the common phenomenon of a previously susceptible strain becoming resistant to a specific antibiotic. This change may result from genetic alterations in the bacteria's own genome, but it is often due to the horizontal transfer of resistance genes from a resistant strain. As often mentioned in the news media, antibiotic resistance has increased dramatically worldwide over the past few decades, with resistant strains reported in almost all pathogenic strains of bacteria. For example, the most common cause of pneumonia is infection by *Streptococcus pneumoniae*. In many countries, nearly 50% of all *S. pneumoniae* strains are now penicillin resistant, with resistance to other drugs increasing

as well. Some of the most severe antibiotic resistance problems occur in hospitals. Resistant strains of *Klebsiella pneumoniae* and *Enterococcus faecium* are significant causes of infection and death among critically ill patients in intensive care units.

 ## Summary of Key Concepts

18.1 Genetic Properties of Viruses

- Tobacco mosaic virus (TMV) was the first virus to be discovered. It infects many species of plants. (Figure 18.1)

- Viruses vary with regard to their host range, structure, and genome composition. (Table 18.1, Figures 18.2, 18.3)

- The viral reproductive cycle consists of five or six basic steps, including attachment, entry, integration, synthesis, assembly, and release. (Figure 18.4)

- Some bacteriophages can alternate between two reproductive cycles: the lytic cycle and lysogenic cycle. (Figure 18.5)

- The disease AIDS is caused by a virus called human immunodeficiency virus (HIV). The virus is a retrovirus whose reproductive cycle involves the integration of the viral genome into a chromosome in the host cell. (Figure 18.6)

- Drugs to combat viral proliferation are often created specifically to inhibit viral proteins.

18.2 Viroids and Prions

- Viroids are RNA molecules that infect plant cells. (Figure 18.7)

- Prions are proteins that exist in an abnormal conformation that can cause disease. (Figure 18.8, Table 18.2)

18.3 Genetic Properties of Bacteria

- Bacteria typically have a single type of circular chromosome found in the nucleoid region of the cell. The chromosome contains many genes and one origin of replication. (Figures 18.9, 18.10)

- The bacterial chromosome is made more compact by the formation of loops and by DNA supercoiling. (Figure 18.11)

- Plasmids are small, circular DNA molecules that exist independently of the bacterial chromosome. Examples are resistance, degradative, col-, virulence, and fertility plasmids. (Figure 18.12)

- When placed on solid growth media, a single bacterial cell will divide many times to produce a colony composed of many cells. (Figure 18.13)

- Bacterial cells reproduce by a process called binary fission, during which a cell divides to form two daughter cells. (Figure 18.14)

18.4 Gene Transfer Between Bacteria

- Three common modes of gene transfer among bacteria are conjugation, transformation, and transduction. (Table 18.3)

- Lederberg and Tatum's work demonstrated the transfer of bacterial genes between different strains of *E. coli* by conjugation. Davis showed that direct contact was needed for this type of gene transfer. (Figures 18.15, 18.16)

- During the mechanism of conjugation, a strand of DNA from an F factor is transferred from a donor to a recipient cell. (Figure 18.17)

- Transformation is the process in which a segment of DNA from the environment is taken up by a competent cell and incorporated into the bacterial chromosome. (Figure 18.18)

- Bacterial transduction is a form of gene transfer in which a bacteriophage transfers a segment of bacterial chromosomal DNA to another cell. (Figure 18.19)

- Horizontal gene transfer is a process in which an organism incorporates genetic material from another organism without being the offspring of that organism.

Assess and Discuss

Test Yourself

1. The _____ is the protein coat of a virus that surrounds the genetic material.
   a. host   c. capsid   e. capsule
   b. prion   d. viroid

2. Among the viruses identified, the characteristics of their genomes show many variations. Which of the following does <u>not</u> describe a typical characteristic of viral genomes?
   a. The genetic material may be DNA or RNA.
   b. The nucleic acid may be single stranded or double stranded.
   c. The genome may carry just a few genes or several dozen.
   d. The number of copies of the genome may vary.
   e. All of the above describe typical variation in viral genomes.

3. During viral infection, attachment is usually specific to a particular cell type because
   a. the virus is attracted to the appropriate host cells by proteins secreted into the extracellular fluid.
   b. the virus recognizes and binds to specific molecules in the cytoplasm of the host cell.
   c. the virus recognizes and binds to specific molecules on the surface of the host cell.
   d. the host cell produces channel proteins that provide passageways for viruses to enter the cytoplasm.
   e. the virus releases specific proteins that make holes in the membrane large enough for the virus to enter.

4. HIV, a retrovirus, has a high mutation rate because
   a. the DNA of the viral genome is less stable than other viral genomes.
   b. the viral enzyme reverse transcriptase has a high likelihood of making replication errors.
   c. the viral genome is altered every time it is incorporated into the host genome.
   d. antibodies produced by the host cell mutate the viral genome when infection occurs.
   e. all of the above.

5. A _____ is an infectious agent composed solely of RNA, whereas a _____ is an infectious agent composed solely of protein.
   a. retrovirus, bacteriophage   d. retrovirus, prion
   b. viroid, virus   e. viroid, prion
   c. prion, virus

6. Genetic diversity is maintained in bacterial populations by all of the following except
   a. binary fission.   d. transduction.
   b. mutation.   e. conjugation.
   c. transformation.

7. Bacterial cells divide by a process known as
   a. mitosis.   d. binary fission.
   b. cytokinesis.   e. glycolysis.
   c. meiosis.

8. Gene transfer, whereby a bacterial cell takes up bacterial DNA from the environment, is
   a. conjugation.   d. transformation.
   b. binary fission.   e. transduction.
   c. recombination.

9. A bacterial cell can donate DNA during conjugation when it
   a. produces competence factors.
   b. contains an F factor.
   c. is virulent.
   d. has been infected by a bacteriophage.
   e. all of the above

10. A bacterial species that becomes resistant to certain antibiotics may have acquired the resistance genes from another bacterial species. The phenomenon of acquiring genes from another organism without being the offspring of that organism is known as
    a. hybridization.   d. vertical gene transfer.
    b. integration.   e. competence.
    c. horizontal gene transfer.

Conceptual Questions

1. How are viruses similar to living cells, and how are they different?

2. What are three mechanisms of gene transfer in bacteria? Discuss the evolutionary and medical significance of horizontal gene transfer.

3. If you mix together an equal number of F^+ and F^- cells, how would you expect the proportions to change over time? Do you expect an increase in the relative proportions of F^+ and F^- cells? Explain your answer.

Collaborative Questions

1. Discuss the possible origin of viruses. Which idea(s) do you think is (are) the most likely?

2. Conjugation is sometimes called "bacterial mating." Discuss how conjugation is similar to sexual reproduction in eukaryotes and how it is different.

Online Resource

www.brookerbiology.com

Stay a step ahead in your studies with animations that bring concepts to life and practice tests to assess your understanding. Your instructor may also recommend the interactive ebook, individualized learning tools, and more.

migration does not occur during development in plants. Third, it may cause a cell to differentiate into a specific cell type such as a nerve cell. Finally, positional information may promote cell death. This process, known as **apoptosis**, is a necessary event during normal animal development. Apoptosis is known to play a key role in sculpting the bodies of animals. In plants, programmed cell death is also important during development. For example, cells within xylem tissue undergo programmed cell death to form tracheids that function in water transport (see Chapter 38).

As an example of how the coordination of these four processes is required for pattern formation, Figure 19.4 shows the growth and development of a human arm during the embryonic stage. Cell division with accompanying cell growth increases the size of the limb. Cell migration is also important in this process. For example, embryonic cells that eventually form muscles in the arm and hand must migrate long distances to reach their

correct location within the limb. As development proceeds, cell differentiation produces the various tissues that will eventually be found in the fully developed limb. Some cells will become nerve cells, others will be muscle cells, and still others will form the outer layer of skin. Finally, apoptosis is important in the formation of fingers. If apoptosis did not occur, a human hand would have webbed fingers.

Morphogens and Cell-to-Cell Contacts Convey Positional Information

How does positional information lead to the development of a body plan? Though the details of pattern formation vary widely among different species, two main molecular mechanisms are commonly used to communicate positional information. One of these mechanisms involves molecules called morphogens. **Morphogens** impart positional information and

(a) Limb development in a human embryo

1. Cell division promotes the growth of the limb.

Limb bud

2. Cell migration causes cells to reach their correct location.

3. Cell differentiation produces specialized cells such as muscle cells.

4. Apoptosis eliminates cells in between the fingers.

(b) Four cellular processes that promote limb formation

Figure 19.4   Limb development in humans.   (a) Photographs of limb development in human embryos. The limb begins as a protrusion called a limb bud that eventually forms an arm and hand. **(b)** The development of a human limb from an embryonic limb bud.

Concept check: *How would finger formation be affected if apoptosis did not occur?*

promote developmental changes at the cellular level. A morphogen influences the fate of a cell by promoting cell division, cell migration, cell differentiation, or apoptosis. A key feature of morphogens is they act in a concentration-dependent manner. At a high concentration, a morphogen will restrict a cell into a particular developmental pathway, whereas at a lower concentration, it will not. There is often a critical **threshold concentration** above which the morphogen will exert its effects.

Morphogens typically are distributed asymmetrically along a concentration gradient. Morphogenic gradients may be established in the oocyte (Figure 19.5a). In addition, a morphogenic gradient can be established in the embryo by secretion and diffusion (Figure 19.5b). A certain cell or group of cells may synthesize and secrete a morphogen at a specific stage of development. After secretion, the morphogen may diffuse to neighboring cells, as in Figure 19.5b, or it may be transported to cells that are distant from the cells that secrete the morphogen. The morphogen may then influence the developmental fate of cells exposed to it. The process by which a cell or group of cells governs the developmental fate of other cells is known as **induction**.

Another mechanism used to convey positional information involves **cell adhesion** (Figure 19.5c). Each animal cell makes its own collection of surface receptors that enable it to adhere to other cells and to the extracellular matrix (ECM). Such receptors, known as **cell adhesion molecules** (**CAMs**), are described in Chapter 10 (refer back to Figure 10.8). The positioning of a cell within a multicellular organism is strongly influenced by the combination of contacts it makes with other cells and with the ECM.

The phenomenon of cell adhesion and its role in multicellular development was first recognized by Henry Wilson in 1907. He took multicellular sponges and passed them through a sieve, dissociating them into individual cells. Remarkably, the cells actively migrated until they adhered to one another to form a new sponge, complete with the chambers and canals that characterize a sponge's internal structure! When sponge cells from different species were mixed, they sorted themselves properly, adhering only to cells of the same species. Overall, these results indicate that cells possess specific CAMs, which are critical in cell-to-cell recognition. Cell adhesion plays an important role in governing the position that an animal cell will adopt during development.

(a) Asymmetric distribution of morphogens in the oocyte

(b) Induction: Asymmetric synthesis and extracellular distribution of a morphogen

(c) Cell adhesion: Cell-to-cell contact conveys positional information

Figure 19.5 Molecular mechanisms that convey positional information. Asymmetric distribution of a morphogen in the **(a)** oocyte or **(b)** embryo. **(c)** Positional information may also be obtained by cell-to-cell contact.

Concept check: Why is positional information important during development?

A Hierarchy of Transcription Factors Orchestrates a Program of Development

The formation of a body, in both animals and plants, occurs in a series of organizational phases that overlap with each other. As an overview of this process, let's consider four general phases of pattern formation in an animal (Figure 19.6). The first phase organizes the body along major axes. The anteroposterior axis determines the organization from head to tail, the dorsoventral axis governs the structure from back (dorsal) to front/abdomen (ventral), and the left-right axis provides organization from side to side. During the second phase, the body becomes organized into smaller regions that will eventually contain organs and other structures such as legs. In insects, these regions form well-defined segments. In mammals, some segmentation of the body is apparent during embryonic development, but defined boundaries are lost as the embryo proceeds to the fetal and adult stages. In the third phase, the cells within the segments organize themselves in ways that will produce particular body parts. Finally, during the fourth phase, the cells themselves change their morphologies and become differentiated. This final phase of development produces an organism with many types of tissues, organs, and other body parts with specialized functions.

How does genetics underlie the phases of animal development? Geneticists have discovered a parallel between the expression of specific transcription factors and the four major phases of animal development. As diagrammed in Figure 19.6, a hierarchy of transcription factors control whether or not certain genes are expressed at a specific phase of development in a particular cell type, a phenomenon called **differential gene regulation**.

Many morphogens, particularly those that act at an early phase of development, function as transcription factors. Such transcription factors regulate the expression of genes in a way that controls the formation of the body axes. Next, these early

Hierarchy of transcription factors

Posterior
Right
Dorsal
(ventral is
underneath)
Anterior
Left

1 Phase 1:
Transcription factors determine the formation of the body axes and control the expression of transcription factors of phase 2.

Evidence of segmentation

2 Phase 2:
Transcription factors cause the embryo to become subdivided into regions that have properties of individual segments. They also control transcription factors of phase 3.

Head forming

Limbs forming

3 Phase 3:
Transcription factors cause each segment and groups of segments to develop specific characteristics. They also control transcription factors of phase 4.

4 Phase 4:
Transcription factors cause cells to differentiate into specific cell types such as skin, nerve, and muscle cells.

Figure 19.6 **Pattern formation in a human embryo.** As shown here, pattern formation in animals occurs in four phases that are controlled by a hierarchy of transcription factors. The example shown here involves human development, although research suggests that pattern formation in all complex animals follows a similar plan. The ideas in this scenario are based largely on analogies between pattern formation in *Drosophila* and mammals. Many of the transcription factors that are likely to control the early phases of pattern formation in mammals have yet to be identified.

Concept check: *During which of the four phases of development would you expect cell division and cell migration to be the most prevalent?*

transcription factors cause the expression of other transcription factors that direct the segmented body plan. After the body plan has been segmented, a third category of transcription factors controls what structures will be made within each segment. Finally, a fourth category of transcription factors controls the differentiation of each cell type. Also, note that the phases of development are overlapping. For example, cell differentiation begins to occur as the cells are adopting their correct locations.

19.2 Development in Animals

In this section, we will begin by examining the general stages of *Drosophila* development and then focus our attention on its embryonic stage. During this stage, the overall body plan is determined. We will see how the differential expression of particular genes and the localization of positional information within the embryo control pattern formation. While the roles of genes in the organization of mammalian embryos are not as well understood as they are in *Drosophila*, the analysis of the genomes of mammals and many other species has revealed many interesting parallels in the developmental program of all animals.

This section will end with an examination of cell differentiation. This process is better understood in mammals than in *Drosophila* because researchers have been extensively studying mammalian cells in the laboratory for many decades. To explore cell differentiation, we will consider mammals as our primary example.

Embryonic Development Determines the Pattern of Structures in the Adult: The Development of *Drosophila*

As a way to appreciate the phases of pattern formation in animals, we will largely focus on development in *Drosophila*. However, as described in Chapter 52, animal development is quite varied among different species. **Figure 19.7** illustrates a simplified sequence of events in *Drosophila* development. Let's examine these steps before we consider the differential gene regulation that causes them to happen.

The oocyte is critical to establishing the pattern of development that will ultimately produce an adult organism. It is an elongated cell that contains positional information. As shown in Figure 19.7a, the fertilized oocyte already has anterior and posterior ends that correspond to those found in the adult (compare Figure 19.7a and e).

A key process in *Drosophila* embryonic development is the formation of a segmented body pattern. The embryo is subdivided into visible units. In *Drosophila*, the segments can be grouped into three general areas: the head, the thorax, and the abdomen. Figure 19.7b shows the segmented pattern of a *Drosophila* embryo about 10 hours after fertilization. Later in this section, we will explore how the coordination of gene expression underlies the formation of these segments.

(a) Fertilized oocyte (0 hours)

Dorsal

Anterior — Posterior

Ventral

(b) Embryo (10 hours)

Segments

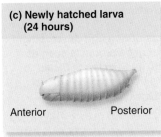

(c) Newly hatched larva (24 hours)

Anterior　　　Posterior

(d) Pupa (5 days)

(e) Adult (10 days)

Figure 19.7
Developmental stages of the fruit fly *Drosophila*.

As an example of one morphogen that plays a role in axis formation, let's consider the product of a gene in *Drosophila* called *bicoid*. Its name is derived from the observation that a mutation that inactivates the gene results in a larva with two posterior ends (**Figure 19.8**). During normal oocyte development, the *bicoid* gene product accumulates in the anterior region of the oocyte. This gene product later acts as a morphogen to cause the development of the anterior end of the embryo.

How does the *bicoid* gene product accumulate in the anterior region of the oocyte? The answer involves specialized nurse cells that are found next to the oocyte, which matures in a follicle within the ovary of a female fly. As discussed in Chapter 17, nurse cells supply the products (for example, mRNA) of maternal effect genes to the developing oocyte. These genes cause an unusual pattern of inheritance called maternal effect (refer back to Figure 17.14). In *Drosophila*, the *bicoid* gene is transcribed in the nurse cells, and *bicoid* mRNA is then transported into the anterior end of the oocyte and trapped there (**Figure 19.9a**). Prior to fertilization, the *bicoid* mRNA is highly concentrated near the anterior end of the oocyte (**Figure 19.9b**). After fertilization, the *bicoid* mRNA is translated, and a gradient of Bicoid protein is established across the zygote (**Figure 19.9c**). This gradient starts a progression of developmental events that will provide the positional information that causes the end of the zygote with a high Bicoid protein concentration to become the anterior region of the embryo.

The Bicoid protein is a morphogen that functions as a transcription factor to activate particular genes at specific times. The ability of Bicoid to activate a given gene depends

A *Drosophila* embryo then develops into a **larva** (Figure 19.7c), which is a free-living organism that is morphologically very different from the adult. Many animal species do not have larval stages. *Drosophila* undergoes three successive larval stages. After the third larval stage, the organism becomes a **pupa** (Figure 19.7d), a transitional stage between the larva and the adult. Through a process known as **metamorphosis**, the organism transforms into a mature adult and emerges from the pupal case (Figure 19.7e). Each segment in the adult develops its own characteristic structures. For example, the wings are on a thoracic segment. From beginning to end, this process takes about 10 days.

Phase 1 Pattern Formation: Maternal Effect Genes Promote the Formation of the Main Body Axes

The first phase in *Drosophila* pattern formation is the establishment of the body axes, which occurs before the embryo becomes segmented. The morphogens necessary to establish these axes are distributed prior to fertilization. In most invertebrates and some vertebrates, certain morphogens, which are important in early developmental stages, are deposited asymmetrically within the egg as it develops. Later, after the egg has been fertilized and development begins, these morphogens will initiate developmental programs that govern the formation of the body axes of the embryo.

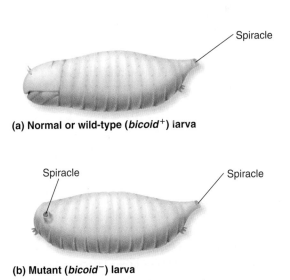

Spiracle

(a) Normal or wild-type (*bicoid*⁺) larva

Spiracle　　　　　　　　Spiracle

(b) Mutant (*bicoid*⁻) larva

Figure 19.8　The bicoid mutation in *Drosophila*. **(a)** A normal *bicoid*⁺ larva. **(b)** A *bicoid*⁻ larva, in which both ends of the larva develop posterior structures. For example, both ends develop a spiracle, a small pore that normally is found only at the posterior end.

> **Concept check:** *What would you expect to be the phenotype of a larva in which the bicoid gene was expressed in both the anterior region and the posterior region of the oocyte?*

(b) Staining of *bicoid* mRNA in an oocyte

(c) Staining of Bicoid protein in an early embryo

Nurse cell / Anterior end / Oocyte
of oocyte

Follicle cell / *bicoid* mRNA

Follicle

(a) Transport of maternal effect gene products (*bicoid* mRNA) into the oocyte

Figure 19.9 Asymmetric localization of gene products during egg development in *Drosophila*. **(a)** The nurse cells transport maternal effect gene products such as *bicoid* mRNA into the anterior end of the developing oocyte. **(b)** Staining of *bicoid* mRNA in an oocyte prior to fertilization. The *bicoid* mRNA is trapped at the anterior region. **(c)** Staining of Bicoid protein after fertilization. The Bicoid protein forms a gradient, with its highest concentration near the anterior end.

Concept check: *What is the function of the Bicoid protein? After fertilization, in which part of the resulting zygote would its function be highest?*

on its concentration. Due to its asymmetric distribution, the Bicoid protein will activate genes only in certain regions of the embryo. For example, a high concentration of Bicoid stimulates the expression of a gene called *hunchback* (that also encodes a transcription factor) in the anterior half of the embryo, but its concentration is too low in the posterior half to activate the *hunchback* gene. The ability of Bicoid to activate genes in certain regions but not others plays a role in the second phase of pattern formation, which is segmentation.

The Study of *Drosophila* Mutants Has Identified Genes That Control the Development of Segments

As described earlier in Figure 19.6, the second phase of pattern formation is the development of segments. The normal *Drosophila* embryo is subdivided into 15 segments: three head segments, three thoracic segments, and nine abdominal segments (**Figure 19.10**). Each segment of the embryo will give rise to unique morphological features in the adult. For example, the second thoracic segment (T2) produces a pair of legs and a pair of wings.

In the 1970s, Christiane Nüsslein-Volhard and Eric Wieschaus undertook a systematic search for *Drosophila* mutants with disrupted development. They focused their search on **segmentation genes**, genes that alter the segmentation pattern of the *Drosophila* embryo and larva. Based on the characteristics of abnormal larva, they identified three classes of segmentation genes: gap genes, pair-rule genes, and segment-polarity genes. When a mutation inactivates a **gap gene**, several adjacent segments are missing in the larva—a gap occurs. A defect in a **pair-rule gene** may cause alternating segments or parts of segments to be absent. Finally, **segment-polarity gene** mutations cause portions of segments to be missing and cause adjacent regions to become mirror images of each other. The role of these segmentation genes during normal *Drosophila* development is described next.

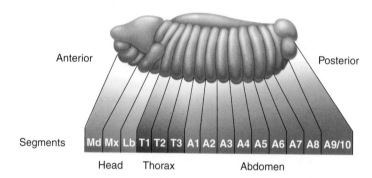

Anterior   Posterior

Segments | Md | Mx | Lb | T1 | T2 | T3 | A1 | A2 | A3 | A4 | A5 | A6 | A7 | A8 | A9/10 |

Head   Thorax   Abdomen

Figure 19.10 The organization of segments in the *Drosophila* embryo.

Phase 2 Pattern Formation: Segmentation Genes Act Sequentially to Divide the *Drosophila* Embryo into Segments

The study of segmentation genes has revealed how segments are formed. To make a segment, particular genes act sequentially to govern the fate of a given region of the body. A simplified scheme of gene expression that leads to a segmented pattern in the *Drosophila* embryo is shown in **Figure 19.11**. Many more genes are actually involved in this process.

In general, the products of maternal effect genes such as *bicoid*, which promote the formation of body axes, activate gap genes. This activation is seen as broad bands of gap proteins in the embryo (Figure 19.11, step 2). Next, products from the gap genes and maternal effect genes function as transcription factors to activate the pair-rule genes in alternating stripes in the embryo (Figure 19.11, step 3). Once the pair-rule genes are activated, their gene products then regulate the segment-polarity genes. As you follow the progression from maternal effect genes to segment-polarity genes, notice that a body pattern is emerging in the embryo that matches the segmentation pattern found in the larva and adult animal. As you can see in step 4 of Figure

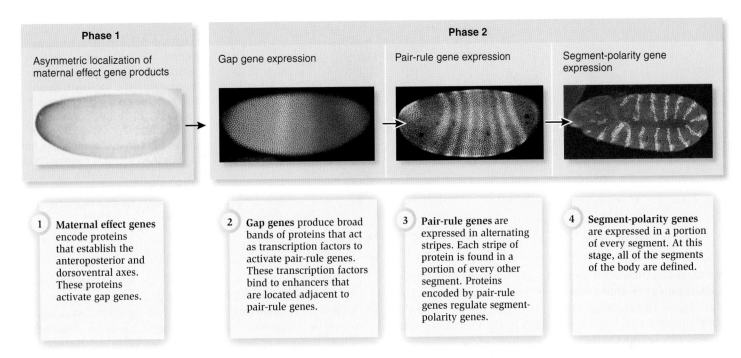

Phase 1	Phase 2		
Asymmetric localization of maternal effect gene products	Gap gene expression	Pair-rule gene expression	Segment-polarity gene expression

1   **Maternal effect genes** encode proteins that establish the anteroposterior and dorsoventral axes. These proteins activate gap genes.

2   **Gap genes** produce broad bands of proteins that act as transcription factors to activate pair-rule genes. These transcription factors bind to enhancers that are located adjacent to pair-rule genes.

3   **Pair-rule genes** are expressed in alternating stripes. Each stripe of protein is found in a portion of every other segment. Proteins encoded by pair-rule genes regulate segment-polarity genes.

4   **Segment-polarity genes** are expressed in a portion of every segment. At this stage, all of the segments of the body are defined.

Figure 19.11   **Overview of segmentation in *Drosophila*.**   The micrographs depict the progression of *Drosophila* development during the first few hours following fertilization. The micrographs also show the expression of protein products of a maternal effect gene (step 1) or segmentation genes (steps 2–4). In step 1, the protein is stained brown and is found in the left side of the early embryo, which is the anterior end. The rest of the embryo is counterstained in yellow. In step 2, one protein encoded by a gap gene is stained in green and another is stained in red. The yellow region is where the two different gap proteins overlap. In step 3, a protein encoded by a pair-rule gene is stained in light blue. In step 4, a protein encoded by a segment-polarity gene is stained pink. When comparing steps 3 and 4, note that the embryo has undergone a 180° turn, folding back on itself.

Concept check: *How many pink stripes can you count in the embryo in step 4? How does this number compare to the number of segments in the embryo in Figure 19.10?*

19.12, the locations of the expression of a segment-polarity gene correspond to portions of segments in the adult fly. To appreciate this phenomenon, notice that the embryo at this stage is curled up and folded back on itself. If you imagine that the embryo was stretched out linearly, the 15 stripes seen in this embryo correspond to portions of the 15 segments of an adult fly.

Phase 3 Pattern Formation: Homeotic Genes Control the Development of Segment Characteristics

Thus far, we have considered how the *Drosophila* embryo becomes organized along axes and then into a segmented body pattern. During the third phase of pattern formation, each segment begins to develop its own unique characteristics (see Figure 19.6). Geneticists use the term **fate** to describe the ultimate morphological features that a cell or group of cells will adopt. For example, the fate of cells in segment T2 in *Drosophila* is to develop into a thoracic segment containing two legs and two wings. In *Drosophila*, the cells in each segment of the body have their fate determined at a very early stage of embryonic development, long before the morphological features become apparent.

Our understanding of developmental fate has been greatly aided by the identification of mutant genes that alter cell fates.

In animals, the first mutant of this type was described by the German entomologist G. Kraatz in 1876. He observed a sawfly (*Cimbex axillaris*) in which part of an antenna was replaced with a leg. During the late 19th century, the English zoologist William Bateson collected many of these types of observations and published them in a book. He coined the term homeotic to describe changes in which one body part is replaced by another. We now know these are caused by mutant alleles of what we call **homeotic genes**. Each homeotic gene specifies the fate of a particular segment or region of the body.

As an example of a homeotic mutation, Figure 19.12 shows a normal fly and one with mutations in a complex of genes called the *bithorax* complex. In a normal fly, two wings are found on the second thoracic segment, and two halteres, which together function as a balancing organ that resembles a pair of miniature wings, are found on the third thoracic segment. In this mutant fly, the third thoracic segment has the characteristics of the second, so the fly has no halteres and four wings. The term *bithorax* refers to the duplicated characteristics of the second thoracic segment. Edward Lewis, a pioneer in the genetic study of development, became interested in the bithorax phenotype and began investigating it in 1946. He discovered that the mutant chromosomal region actually contains a complex of three genes that play a role in the third phase of development.

(a) Normal fly with two wings

(b) Mutant fly with four wings

Figure 19.12 The bithorax mutation in *Drosophila*. **(a)** A normal fly has two wings on the second thoracic segment, and two halteres on the third thoracic segment. **(b)** This fly contains mutations in a complex of genes called the *bithorax* complex. In this fly, the third thoracic segment has the same characteristics as the second thoracic segment, thereby producing a fly with four wings instead of two.

Drosophila has eight homeotic genes that are found in two clusters called the *Antennapedia* complex and the *bithorax* complex (**Figure 19.13**). Both of these complexes are located on the same chromosome, but a long stretch of DNA separates them. As you can see in Figure 19.13, the order of homeotic genes along the chromosome correlates with their expression along the anteroposterior axis of the body. This phenomenon is called the **colinearity rule**. For example, *lab* (for labial) is expressed in the anterior segment and governs the formation of mouth structures. The *Antp* (for antennapedia) gene is expressed in the thoracic region during embryonic development and controls the formation of thoracic structures.

As we have seen in Figure 19.12, the role of homeotic genes in determining the identity of particular segments has been revealed by mutations that alter their function. As a second example, a mutation in the *Antp* gene has been identified in which the gene is incorrectly expressed in an anterior segment (**Figure 19.14**). A fly with this mutation has the bizarre trait in which it develops legs where antennae are normally found!

How do homeotic genes work at the molecular level? Homeotic genes encode homeotic proteins that function as transcription factors. The coding sequence of homeotic genes contains a 180-bp sequence known as a **homeobox** (**Figure 19.15a**). This sequence was first discovered in the *Antp* and *Ubx* genes, and it has since been found in many *Drosophila* homeotic genes. The homeobox is also found in other genes affecting pattern formation. The homeobox encodes a region of the protein called a **homeodomain**, which can bind to DNA (**Figure 19.15b**). The arrangement of α helices in the homeodomain promotes the binding of the protein to the DNA.

The primary function of homeotic proteins is to activate the transcription of specific genes that promote developmental changes in the animal. The homeodomain protein binds to DNA

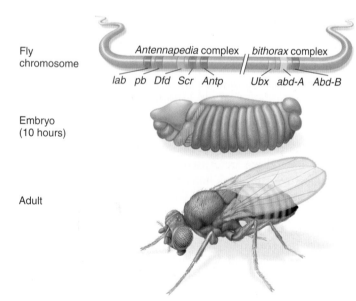

Fly chromosome

Antennapedia complex   *bithorax* complex

lab pb Dfd Scr Antp   *Ubx abd-A Abd-B*

Embryo (10 hours)

Adult

Figure 19.13 Expression pattern of homeotic genes in *Drosophila*. The order of homeotic genes, *labial* (*lab*), *proboscipedia* (*pb*), *deformed* (*Dfd*), *sex combs reduced* (*Scr*), *antennapedia* (*Antp*), *ultrabithorax* (*Ubx*), *abdominal A* (*abd-A*), and *abdominal B* (*Abd-B*), correlates with their spatial order of expression in the embryo. (Note: These genes were discovered and named by different researchers and the capitalization of the names is not consistent.)

(a) Normal fly   **(b) Mutant fly**

Figure 19.14 The Antennapedia mutation in *Drosophila*. **(a)** A normal fly with antennae. **(b)** This fly has a mutation in which the *Antp* gene is expressed in the embryonic segment that normally gives rise to antennae. The abnormal expression of *Antp* causes this region to have legs rather than antennae.

Concept check: *What phenotype would you expect if the Antp gene was expressed where abd-A is supposed to be expressed?*

Region that encodes
the transcriptional
activation domain

DNA

Homeobox (180 bp)

Homeotic gene

(a) Homeotic gene containing homeobox

Transcriptional
activation domain

Homeodomain

(b) Homeodomain binding to DNA

Figure 19.15 Molecular features of homeotic genes and proteins. (a) A homeotic gene (shown mostly in green) contains a 180-bp sequence called the homeobox (shown in blue). (b) Homeotic genes encode proteins that function as transcription factors. The homeobox encodes a region of the protein called a homeodomain, which binds to the DNA at a regulatory site such as an enhancer. The transcriptional activation domain activates RNA polymerase to begin transcription.

sequences called enhancers, which are described in Chapter 13. These enhancers are found in the vicinity of specific genes that control development. Most homeotic proteins also contain a transcriptional activation domain (Figure 19.15b). After the homeodomain binds an enhancer, the transcriptional activation domain of the homeotic protein activates RNA polymerase to begin transcription. Some homeotic proteins also function as repressors of certain genes.

Genomes & Proteomes Connection

A Homologous Group of Homeotic Genes Is Found in All Animals

Homologous genes are evolutionarily derived from the same ancestral gene and have similar DNA sequences. Researchers have found that homeotic genes in vertebrate species are homologous to genes that control development in simpler organisms such as *Drosophila*. For example, in the mouse and other mammals, including humans, homeotic genes are organized into four clusters, designated *HoxA*, *HoxB*, *HoxC*, and *HoxD*. Homeotic genes in vertebrates are called **Hox genes**, an abbreviation for homeobox-containing genes. Thirty-eight genes are found in the four clusters, which represent 13 different gene types. As shown

in **Figure 19.16**, several *Hox* genes in fruit flies and the mouse and other mammals are evolutionarily related. Among the first six types of *Hox* genes in the mouse, five of them are homologous to genes found in the *Antennapedia* complex of *Drosophila*. Among the last seven (genes numbered 7–13), three are homologous to the genes of the *bithorax* complex.

Like the *Antennapedia* and *bithorax* complexes in *Drosophila*, the arrangement of *Hox* genes along the mouse chromosome follows the colinearity rule, reflecting their pattern of expression from the anterior to the posterior end (**Figure 19.17**). Research has shown that the *Hox* genes play a role in determining the fates of regions along the anteroposterior axis. Nevertheless, additional research will be necessary to understand the individual role that each of the 38 *Hox* genes plays during embryonic development.

How widespread are *Hox* genes? They are present in all animals, but in different numbers. Sponges, which are simple animals with no true tissues, have a single *Hox* gene. As noted previously, humans have 38 *Hox* genes organized in four clusters. The study of *Hox* genes in many different animal species has shown that the *Hox* cluster, with its colinear expression and its role of determining the anteroposterior axis, originated very early in the evolution of animals. At the level of genetics, fundamental similarities are observed in the ways that animals, such as worms, fruit flies, and mammals, undergo embryonic development. Researchers have suggested there is a universal body plan for animal development and that a portion of the

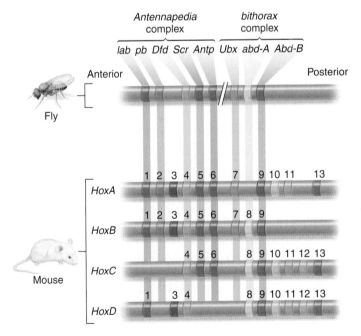

Figure 19.16 A comparison of homeotic genes in *Drosophila* and the mouse. The mouse and other mammals have four gene complexes, *HoxA–D*, that correspond to certain homeotic genes found in *Drosophila*. Thirteen different types of homeotic genes are found in the mouse, although each *Hox* complex does not contain all 13 genes. In this drawing, homologous genes are aligned in columns. For example, *lab* is the homologue to *HoxA-1*, *HoxB-1*, and *HoxD-1*.

Figure 19.17 Expression pattern of *Hox* genes in the mouse.   A schematic illustration of *Hox* gene expression in the embryo and the corresponding regions in the adult. The order of *Hox* gene expression, from anterior to posterior, parallels the order of genes along the chromosome.

genome of all animals is devoted to the execution of this plan. The biological diversity that we see among animals is due to genetic variation from this common plan. The role of *Hox* genes in the evolution of animal body plans is also discussed in Chapter 25 (look ahead to Figure 25.15).

Phase 4 Pattern Formation: Stem Cells Can Divide and Differentiate into Specialized Cell Types

Thus far we have focused our attention on patterns of gene expression that occur during the early stages of development. These genes control the basic body plan of the organism. During the fourth phase of pattern formation, the emphasis shifts to cell differentiation (see Figure 19.6).

Although invertebrates have been instrumental in our understanding of pattern formation in animals, cell differentiation has been studied more extensively in mammals. One reason is because researchers have been able to grow mammalian cells in the laboratory for many decades. The availability of laboratory-grown cells makes it much easier to analyze the process of cell differentiation.

By studying mammalian cells in the laboratory, geneticists have determined that the profound morphological differences between two different types of differentiated cells, such as muscle and nerve cells, arise through gene regulation. Though muscle and nerve cells contain the same set of genes, they regulate the expression of their genes in very different ways. Certain

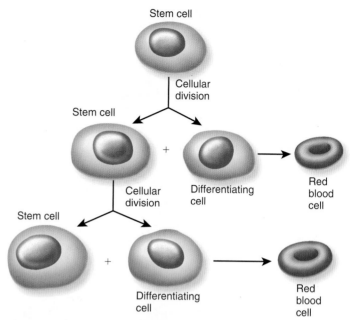

Figure 19.18   Growth pattern of stem cells.   When a stem cell divides, one of the two daughter cells may remain a stem cell, while the other cell can differentiate into a specialized cell type, such as the red blood cells shown here.

Concept check: *What are the two key features of stem cells?*

genes that are transcriptionally active in muscle cells are inactive in nerve cells, and vice versa. Therefore, muscle and nerve cells express different proteins that affect the characteristics of the respective cells in distinct ways. In this manner, differential gene expression underlies cell differentiation.

General Properties of Stem Cells   To understand the process of cell differentiation in a multicellular organism, we need to consider the special properties of **stem cells**, undifferentiated cells that divide and supply the cells that constitute the bodies of all animals and plants. Stem cells have two common characteristics. First, they have the capacity to divide, and second, their daughter cells can differentiate into one or more specialized cell types. The two daughter cells that are produced from the division of a stem cell can have different fates (**Figure 19.18**). One of the cells may remain an undifferentiated stem cell, and the other daughter cell can differentiate into a specialized cell type. With this asymmetric division/differentiation pattern, stem cells can both continue dividing throughout life and generate a population of specialized cells. For example, in mammals, this mechanism is needed to replenish cells that have a finite life span, such as skin cells and red blood cells.

Stem Cells During Development   In mammals, stem cells are commonly categorized according to their developmental stage and their ability to differentiate (**Figure 19.19**). The ultimate stem cell is the fertilized egg, which, via multiple cellular divisions, can give rise to an entire organism. A fertilized egg is considered to be **totipotent** because it can produce all of the cell types in the adult organism. The early mammalian embryo, or blastocyst, contains **embryonic stem cells** (**ES cells**), which are found in the inner cell mass. Embryonic stem cells are **pluripotent**, which means they can also differentiate into every or nearly every cell type of the body. However, a single embryonic stem cell by itself has lost the ability to produce an entire, intact individual. At an early fetal stage of development, the cells that later give rise to sperm or eggs cells, known as the **embryonic germ cells** (**EG cells**), also are pluripotent.

During the embryonic and fetal stages of development, cells lose their ability to differentiate into a wide variety of cell types. Adults have multipotent and unipotent stem cells. A **multipotent** stem cell can differentiate into several cell types, but far fewer than an embryonic stem cell. For example, hematopoietic stem cells (HSCs) found in the bone marrow give rise to multiple blood cell types (**Figure 19.20**). Multipotent HSCs can follow a pathway in which cell division produces a myeloid cell, which then differentiates into various cells of the blood and immune systems. Alternatively, an HSC can follow a path in which it becomes a lymphoid cell that develops into different blood cell types. A **unipotent** stem cell produces daughter cells that differentiate into only one cell type. For example, stem cells in the skin produce daughter cells that develop into skin cells.

Stem Cells in Medicine Why are researchers so interested in stem cells? A compelling medical reason is their potential to treat human diseases or injuries that cause cell and tissue damage. This application has already become a reality in certain cases. For example, bone marrow transplants are used to treat patients with certain forms of cancer, such as leukemia. When bone marrow from a healthy person is injected into the body of a patient who has had her/his immune system wiped out via radiation, the stem cells within the transplanted marrow have the ability to proliferate and differentiate into various types of blood cells within the body of the patient.

Renewed interest in the use of stem cells in the potential treatment of many other diseases has been fostered by studies in 1998 in which researchers obtained ES cells from blastocysts and EG cells from aborted fetuses and successfully propagated them in the laboratory. Because ES and EG cells are pluripotent, they could potentially be used to treat a wide variety of diseases associated with cell and tissue damage (**Table 19.1**). Much progress has been made in testing the use of stem cells in animal models. However, more research will need to be done before the use of stem cells to treat such diseases in humans is realized.

From an ethical perspective, the primary issue that raises debate is the source of stem cells for research and potential treatments. Most ES cells have been derived from human embryos that were produced from in vitro fertilization, a method of assisted conception in which fertilization occurs outside of the mother's body and a limited number of the resulting embryos are transferred to the uterus. Most EG cells are obtained from aborted fetuses, either those that were spontaneously aborted or those in which the decision to abort was not related to donating the fetal tissue to research. Some feel that it is morally wrong to use such tissue in research and/or the treatment of disease. Furthermore, some people fear this technology could lead to intentional abortions for the sole purpose of obtaining fetal tissues for transplantation. Others feel the embryos and fetuses that have been the sources of ES and EG cells were not going to

Figure 19.19 Occurrence of stem cells at different stages of mammalian development.

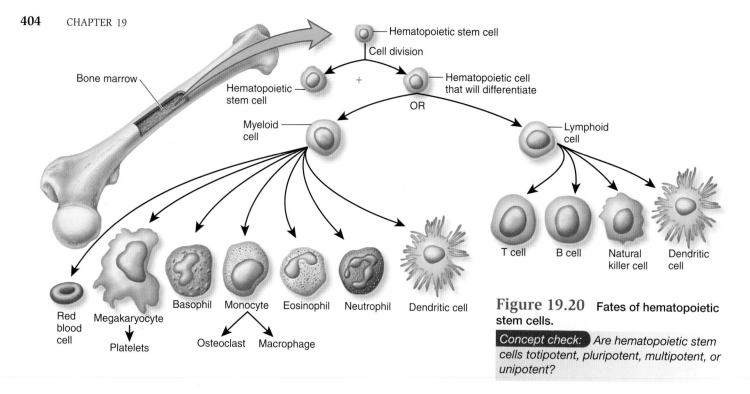

Figure 19.20 Fates of hematopoietic stem cells.

Concept check: *Are hematopoietic stem cells totipotent, pluripotent, multipotent, or unipotent?*

Table 19.1	Some Potential Uses of Stem Cells to Treat Diseases
Cell/tissue type	**Disease treatment**
Nerve	Implantation of cells into the brain to treat Parkinson disease; treatment of spinal cord injuries
Skin	Treatment of burns and skin disorders
Cardiac	Repair of heart damage associated with heart attacks
Cartilage	Repair of joints damaged by injury or arthritis
Bone	Repair or replacement of damaged bone
Liver	Repair or replacement of liver tissue damaged by injury or disease
Skeletal muscle	Repair or replacement of damaged muscle

become living individuals, and therefore it is beneficial to study these cells and to use them in a positive way to treat human diseases and injury. It is not clear whether these two opposing viewpoints can reach a common ground.

If stem cells could be obtained from adult cells and propagated in the laboratory, an ethical dilemma may be avoided because most people do not have serious moral objections to current procedures such as bone marrow transplantation. In 2006, work by Shinya Yamanaka and colleagues showed that adult mouse fibroblasts (a type of connective tissue cell) could become pluripotent by the introduction of four different genes that encode transcription factors. In 2007, Yamanaka's laboratory and two other research groups were able to show that such induced pluripotent stem cells can differentiate into all cell types when injected into mouse blastocysts and grown into baby mice. These results indicate that adult cells can be reprogrammed to become embryonic stem cells.

FEATURE INVESTIGATION

Davis, Weintraub, and Lassar Identified Genes That Promote Muscle Cell Differentiation

A key question regarding the study of stem cells is, What causes a stem cell to differentiate into a particular cell type? Researchers have discovered that certain proteins function as "master transcription factors" that can cause cells to differentiate into a particular cell type. The investigation described here was one of the first studies to reveal this phenomenon.

In 1987, Robert Davis, Harold Weintraub, and Andrew Lassar conducted a study to identify genes that promote skeletal muscle cell differentiation. The initial strategy for their experiments was to identify genes that are expressed only in differ-

entiating skeletal muscle cells, not in nonmuscle cells. Though methods of gene cloning are described in detail in Chapter 20, let's briefly consider these scientists' cloning methods so we can understand their approach. They began with two different laboratory cell lines that could differentiate into muscle cells. From these two cell lines, they cloned and identified about 10,000 different genes that were transcribed into mRNA. Next, they compared the expressed genes in these two muscle cell lines with genes that were expressed in a nonmuscle cell line. Their comparison revealed 26 genes that were expressed only in the two muscle cell lines but not in the nonmuscle cell line. To narrow their search further, they compared these 26 genes with other nonmuscle cell lines they had available. Among the

26, only 3 of them were expressed exclusively in the two muscle cell lines, which they termed *MyoA*, *MyoD*, and *MyoH*.

In the experiment shown in **Figure 19.21**, the scientists' goal was to determine if any of these three genes could cause nonmuscle cells to differentiate into muscle cells. Using techniques described in Chapter 20, the coding sequence of each cloned gene was placed next to an active promoter that caused a high level of transcription, and then the genes were introduced into fibroblasts, which are a type of cell that normally differentiates into osteoblasts (bone cells), chondrocytes (cartilage cells), adipocytes (fat cells), and smooth muscle cells, but never differentiates into skeletal muscle cells in vivo. However, when the cloned *MyoD* gene was expressed in fibroblast cells in a laboratory, the fibroblasts differentiated into skeletal muscle cells. These cells contained large amounts of myosin, which is a protein expressed in muscle cells. The other two cloned genes (*MyoA* and *MyoH*) did not cause muscle cell differentiation or promote myosin production.

Since this initial discovery, researchers have found that *MyoD* belongs to a small group of genes termed myogenic bHLH genes that initiate muscle cell development. Myogenic bHLH genes encode transcription factors that contain two functional regions or domains: a <u>b</u>asic domain and a <u>h</u>elix-<u>l</u>oop-<u>h</u>elix domain (bHLH). They are found in all vertebrates, and they have been identified in several invertebrates, such as *Drosophila* and *C. elegans*. In all cases, myogenic bHLH genes are activated during skeletal muscle cell development.

Experimental Questions

1. What was the goal of the research conducted by Davis, Weintraub, and Lassar?

2. How did Davis, Weintraub, and Lassar's research identify the candidate genes for muscle differentiation?

3. Once the researchers identified the candidate genes for muscle differentiation, how did they test the effect of each gene on cell differentiation? What were the results of the study?

Figure 19.21 Davis, Weintraub, and Lassar and the promotion of skeletal muscle cell differentiation in fibroblasts by the expression of *MyoD*.

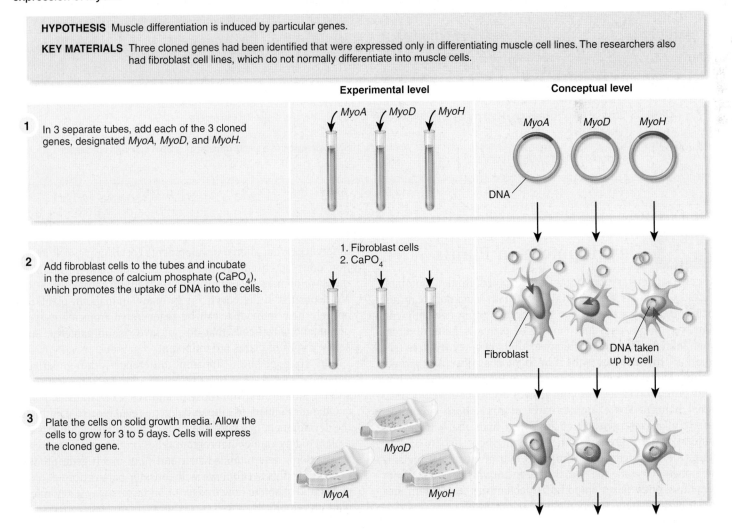

HYPOTHESIS Muscle differentiation is induced by particular genes.

KEY MATERIALS Three cloned genes had been identified that were expressed only in differentiating muscle cell lines. The researchers also had fibroblast cell lines, which do not normally differentiate into muscle cells.

Experimental level | Conceptual level

1 In 3 separate tubes, add each of the 3 cloned genes, designated *MyoA*, *MyoD*, and *MyoH*.

MyoA   MyoD   MyoH

DNA

2 Add fibroblast cells to the tubes and incubate in the presence of calcium phosphate (CaPO₄), which promotes the uptake of DNA into the cells.

1. Fibroblast cells
2. CaPO₄

Fibroblast

DNA taken up by cell

3 Plate the cells on solid growth media. Allow the cells to grow for 3 to 5 days. Cells will express the cloned gene.

MyoD

MyoA   MyoH

4 Examine the cells under a microscope to determine if they have the morphology of differentiating muscle cells.

MyoA   MyoD   MyoH

Now looks like a muscle cell

Still looks like a fibroblast

5 Also, determine if the cells are synthesizing myosin, which is a protein that is abundantly made in muscle cells. This is done by adding a labeled antibody that recognizes myosin and determining the amounts of antibody that bind.

Colony labeled with myosin antibody

MyoD

MyoA   MyoH

Antibodies

6 THE DATA

Results from step 4:

DNA added	Microscopic morphology of cells
MyoA	Fibroblasts
MyoD	Muscle cells
MyoH	Fibroblasts

Results from step 5:

DNA added	Colonies labeled with antibody that binds to myosin?
MyoA	No
MyoD	Yes
MyoH	No

7 CONCLUSION The *MyoD* gene encodes a protein that causes cells to differentiate into skeletal muscle cells.

8 SOURCE Davis, Robert L., Weintraub, Harold, and Lassar, Andrew B. 1987. Expression of a single transfected cDNA converts fibroblasts to myoblasts. *Cell* 51:987–1000.

19.3 Development in Plants

Because all eukaryotic organisms evolved from a common ancestor, animals and plants share many common features, including the types of events that occur during development. However, the general morphology of plants is quite different from animals. Plant morphology exhibits two key features (see Figure 19.2b). The first is the root-shoot axis. Most plant growth occurs via cell division near the tips of the shoots and the bottoms of the roots. Second, this growth occurs in a well-defined radial pattern. For example, early in *Arabidopsis* growth, a rosette of leaves is produced from leaf buds that emanate in a spiral pattern directly from the main shoot (**Figure 19.22**). Later, the shoot generates branches that produce leaf buds as they grow. Overall, the radial pattern in which a plant shoot generates buds is an important mechanism that determines much of the general morphology of the plant.

At the cellular level too, plant development shows some differences from animal development. For example, cell migration does not occur during plant development. In addition, the development of a plant does not rely on morphogens that are deposited in the oocyte, as in many animals. In plants, an entirely new individual can be regenerated from many types of somatic cells—cells that do not give rise to gametes. Such somatic cells of plants are totipotent.

In spite of these apparent differences, the underlying molecular mechanisms of pattern formation in plants still share striking similarities with those in animals. Like animals, plants use the mechanism of differential gene regulation to coordinate the development of a body plan. Like their animal counterparts, a plant's developmental program relies on the use of transcription factors, determining when and how much gene product is made. In this section, we will consider pattern formation in plants and examine how transcription factors play a key role in plant development.

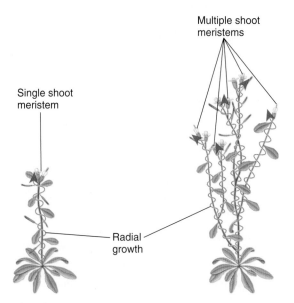

Figure 19.22 **Radial pattern of shoot growth in plants.** Early in development, as shown here in *Arabidopsis*, a single shoot promotes the formation of early leaves on the plant. Later, buds will form from this main shoot that will go on to form branches. The buds that produce the branches and the leaves that form on the branches are also produced in a radial manner.

Plant Development Occurs from Meristems That Are Formed in the Embryo

How does pattern formation occur in plants? **Figure 19.23** illustrates a common sequence of events that takes place in the embryonic development of flowering plants such as *Arabidopsis*. After fertilization, the first cellular division is asymmetrical and produces a smaller apical cell and a larger basal cell (Figure 19.23a). In 2009, Martin Bayer and colleagues conducted experiments indicating that the sperm carries mRNA molecules that are critical for this asymmetric cell division. The apical cell will give rise to most of the embryo and later develop into the shoot of the plant. In *Arabidopsis*, the basal cell will give rise to the root, along with a structure called the suspensor, which will channel nutrients from the parent plant to the embryo (Figure 19.23b).

At the heart stage, which is composed of only about 100 cells, the basic organization of the plant has been established (Figure 19.23c). Plants have organized groups of actively dividing stem cells called **meristems**. As discussed earlier, stem cells retain the ability both to divide and to differentiate into multiple cell types. The meristem produces offshoots of proliferating and differentiating cells. The **root meristem** gives rise only to

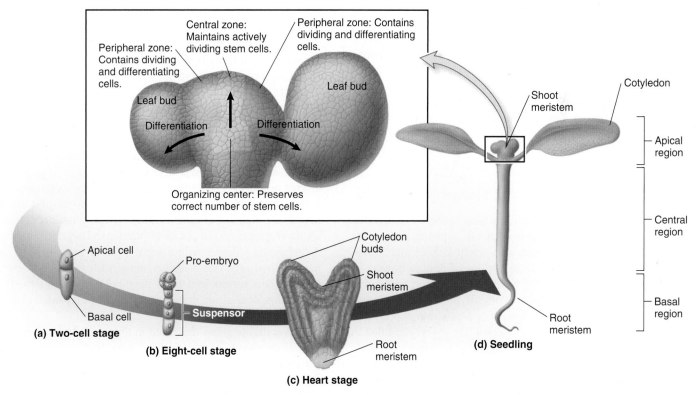

(a) Two-cell stage

(b) Eight-cell stage

(c) Heart stage

(d) Seedling

Figure 19.23 **Developmental steps in the formation of a plant embryo.** **(a)** The two-cell stage consists of the apical cell and basal cell. **(b)** The eight-cell stage consists of a pro-embryo and a suspensor. The suspensor gives rise to extra embryonic tissue that is needed for seed formation. **(c)** At the heart stage, all of the plant tissues have begun to form. The shoot meristem is located between the future cotyledons, and the root meristem is on the opposite side. **(d)** A seedling showing apical, central, and basal regions. The inset shows the organization of the shoot meristem. Note: The steps shown in parts (a), (b), and (c) occur during seed formation, and the embryo would be enclosed within a seed.

Concept check: Where are stem cells found in a growing plant?

the root, whereas the **shoot meristem** produces all aerial parts of the plant, which include the stem as well as lateral structures such as leaves and flowers.

The heart stage then progresses to the formation of a seedling that has two cotyledons, which are embryonic leaves that store nutrients for the developing embryo and seedling. In the seedling shown in Figure 19.23d, you can see three main regions. The **apical region** produces the leaves and flowers of the plant. The **central region** creates the stem. The radial pattern of cells in the central region causes the radial growth observed in plants. Finally, the **basal region** produces the roots. Each of these three regions develops differently, as indicated by their unique cell division patterns and distinct morphologies.

As seen in the inset to Figure 19.23d, the shoot meristem is organized into three areas: the organizing center, the central zone, and the peripheral zone. The **organizing center** ensures the proper organization of the meristem and preserves the correct number of actively dividing stem cells. The **central zone** is an area where undifferentiated stem cells are always maintained. The **peripheral zone** contains dividing cells that will eventually differentiate into plant structures. For example, the peripheral zone may form a bud that will produce a leaf or flower.

By analyzing mutations that disrupt the developmental process, researchers have discovered that the apical, central, and basal regions of a growing plant express different sets of genes. Gerd Jürgens and his colleagues began a search to identify a category of genes, known as **apical-basal-patterning genes**, that are important in early stages of plant development. A few examples are described in **Table 19.2**. Defects in apical-basal-patterning genes cause dramatic effects in one of these three regions. For example, the *Aintegumenta* gene is necessary for apical development. When it is defective, the growth of lateral buds is defective.

Plant Homeotic Genes Control Flower Development

Although William Bateson coined the term homeotic to describe such mutations in animals, the first known homeotic genes were described in plants. Naturalists in ancient Greece and Rome, for example, recorded their observations of double flowers in which stamens were replaced by petals. In current research, geneticists are studying these types of mutations to better understand developmental pathways in plants. Many homeotic mutations affecting flower development have been identified in *Arabidopsis* and also in the snapdragon (*Antirrhinum majus*).

A normal *Arabidopsis* flower is composed of four concentric whorls of structures (**Figure 19.24a**). The first, outer whorl contains four **sepals**, which protect the flower bud before it opens. The second whorl is composed of four **petals**, and the third whorl contains six **stamens**, structures that make the male gametophyte, pollen. Finally, the fourth, innermost whorl contains two carpels that are fused together. The **carpels** produce, enclose, and nurture the female gametophytes.

By analyzing the effects of many different homeotic mutations in *Arabidopsis*, Enrico Coen and Elliot Meyerowitz

Table 19.2	Examples of *Arabidopsis* Apical-Basal-Patterning Genes
Region: *Gene*	**Description**
Apical:	
Aintegumenta	Encodes a transcription factor that is expressed in the peripheral zone. Its expression maintains the growth of lateral buds.
Central:	
Scarecrow	Encodes a transcription factor that plays a role in the asymmetric division that produces the radial pattern of growth in the stem. Note: The Scarecrow protein also affects cell division patterns in roots and plays a role in sensing gravity.
Basal:	
Monopterous	Encodes a transcription factor. When the *monopterous* gene is defective, the plant embryo cannot initiate the formation of root structures, although root structures can be formed post-embryonically. This gene seems to be required for organizing root formation in the embryo.

proposed the ABC model for flower development in 1991. In this model, three classes of genes, called *A*, *B*, and *C*, govern the formation of sepals, petals, stamens, and carpels. More recently, a fourth category of genes, called the *E* genes, was found to also be required for this process. Figure 19.24a illustrates how these genes affect normal flower development in *Arabidopsis*. In whorl 1, gene *A* product is made. This promotes sepal formation. In whorl 2, *A*, *B*, and *E* gene products are made, which promotes petal formation. In whorl 3, the expression of genes *B*, *C*, and *E* causes stamens to be made. Finally, in whorl 4, the products of *C* and *E* genes promote carpel formation.

What happens in certain homeotic mutants that undergo transformations of particular whorls? According to the original ABC model, genes *A* and *C* repress each other's expression, and gene *B* functions independently. In a mutant defective in gene *A* expression, gene *C* will also be expressed in whorls 1 and 2. This produces a carpel-stamen-stamen-carpel arrangement in which the sepals have been transformed into carpels and the petals into stamens (**Figure 19.24b**). When gene *B* is defective, a flower cannot make petals or stamens. Therefore, a gene *B* defect yields a flower with a sepal-sepal-carpel-carpel arrangement. When gene *C* is defective, gene *A* is expressed in all four whorls. This results in a sepal-petal-petal-sepal pattern. If the expression of *E* genes is defective, the flower consists entirely of sepals.

Overall, the genes described in Figure 19.24 promote a pattern of development that leads to sepal, petal, stamen, or carpel structures. But what happens if genes *A*, *B*, and *C* are all defective? This produces a flower composed entirely of leaves (**Figure 19.24c**). These results indicate that the leaf structure is the default pathway and that the *A*, *B*, and *C* genes cause development to deviate from a leaf structure in order to make something else. In this regard, the sepals, petals, stamens, and carpels of plants can be viewed as modified leaves. With astonishing insight, Johann Wolfgang von Goethe, a German poet, novelist, playwright, and natural philosopher, originally proposed this idea over 200 years ago!

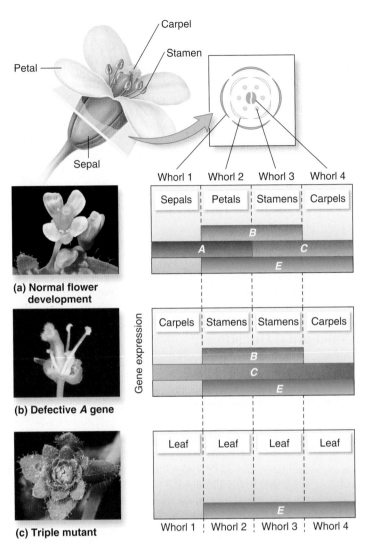

Figure 19.24 Normal and mutant homeotic gene action in *Arabidopsis.* **(a)** A normal flower composed of four concentric whorls of structures: sepals, petals, stamens, and carpels. To the right is the ABC model of homeotic gene action in *Arabidopsis.* This is a revised model based on the recent identification of *E* genes. **(b)** A homeotic mutant defective in gene *A* in which the sepals have been transformed into carpels and the petals have been transformed into stamens. **(c)** A triple mutant defective in the *A*, *B*, and *C* genes, producing a flower with all leaves.

> **Concept check:** *What pattern would you expect if the B gene was expressed in whorls 2, 3, and 4?*

Like the *Drosophila* homeotic genes, plant homeotic genes are part of a hierarchy of gene regulation. All of these plant homeotic genes encode transcription factor proteins that contain a DNA-binding domain, called a MADS domain, and a dimerization domain. However, the *Arabidopsis* homeotic genes do not contain a sequence similar to the homeobox found in animal homeotic genes.

▶ Summary of Key Concepts

- Development refers to a series of changes in the state of a cell, tissue, organ, or organism. Developmental genetics seeks to understand how gene expression controls the development process.

19.1 General Themes in Development

- A cell that is determined has a particular developmental fate. A cell that is differentiated has a specialized morphology and function.
- *Drosophila, C. elegans*, mice, zebrafish, and *Arabidopsis* are model organisms studied by developmental geneticists. (Figure 19.1)
- The process that gives rise to an animal or plant with a particular body structure is called pattern formation. (Figure 19.2)
- Four responses to positional information are cell division, cell migration, cell differentiation, and apoptosis. Cell migration does not occur in plants. (Figures 19.3, 19.4)
- Morphogens and cell adhesion are ways that cells obtain positional information. (Figure 19.5)
- Transcription factors control the program of development in animals and plants. In animals, pattern formation occurs in four general phases. (Figure 19.6)

19.2 Development in Animals

- Embryonic development in *Drosophila* occurs in a series of steps, starting with a fertilized oocyte, then an embryo, larvae, pupa, and an adult. The basic body plan is established in the embryo. (Figure 19.7)
- Maternal effect genes control the formation of body axes, the first phase in *Drosophila* pattern formation. (Figures 19.8, 19.9)
- The second phase of pattern formation is the development of segmentation. The sequential expression of three categories of segmentation genes divides the embryo into segments. Mutations that alter *Drosophila* development have allowed scientists to understand the normal process. (Figures 19.10, 19.11)
- During the third phase of pattern formation, each segment begins to develop its own unique characteristics. Homeotic genes control the development of a particular segment or group of segments. (Figures 19.12, 19.13, 19.14, 19.15)
- Invertebrates, such as *Drosophila*, and vertebrates, such as the mouse, both have a homologous set of homeotic genes. In vertebrates, these are called the *Hox* genes. (Figures 19.16, 19.17)
- Stem cells can divide and then one of the daughter cells can remain a stem cell while the other differentiates. (Figure 19.18)
- In mammals, a fertilized egg is totipotent, certain embryonic and fetal cells are pluripotent, and stem cells in the adult are multipotent or unipotent. (Figures 19.19, 19.20)
- Stem cells have the potential to be used to treat a variety of human disorders. (Table 19.1)
- The fourth phase of pattern formation involves cell differentiation. The differentiation of cell types within certain tissues or organs are controlled by master transcription factors. An example is *MyoD*, a gene that initiates skeletal muscle cell development. (Figure 19.21)

19.3 Development in Plants

- Plants grow in a radial pattern along a root-shoot axis. (Figure 19.22)

- Plant meristems contain dividing cells that promote the development of plant structures such as roots, stems, leaves, and flowers. (Figure 19.23)
- Several types of genes have been identified in plants that influence pattern formation. (Table 19.2)
- Four classes of homeotic genes in plants, *A, B, C,* and *E,* control flower formation and patterning. (Figure 19.24)

Assess and Discuss

Test Yourself

1. The process whereby a cell's morphology and function have changed is called
   a. determination.   d. genetic engineering.
   b. cell fate.   e. both a and c.
   c. differentiation.

2. Pattern formation in plants is along the _____ axis.
   a. dorsoventral   d. root-shoot
   b. anteroposterior   e. All of the above are correct.
   c. left-right

3. Positional information is important in determining the fate of a cell in a multicellular organism. Animal cells respond to positional information by
   a. dividing.   d. undergoing apoptosis.
   b. migrating.   e. all of the above.
   c. differentiating.

4. Morphogens are
   a. molecules that disrupt normal development.
   b. molecules that convey positional information.
   c. mutagenic agents that cause apoptosis.
   d. receptors that allow cells to adhere to the extracellular matrix.
   e. both a and c.

5. What group of proteins plays a key role in controlling the program of developmental changes?
   a. motor proteins   d. restriction endonucleases
   b. transporters   e. cyclins
   c. transcription factors

6. Using the following list of events, determine the proper sequence for the events of animal development:
   1. Tissues, organs, and other body structures in each segment are formed.
   2. Axes of the entire animal are determined.
   3. Cells become differentiated.
   4. The entire animal is divided into segments.
   a. 2, 3, 4, 1   c. 2, 4, 3, 1   e. 2, 4, 1, 3
   b. 1, 2, 4, 3   d. 3, 2, 4, 1

7. The homeotic genes in *Drosophila*
   a. determine the structural and functional characteristics of different segments of the developing fly.
   b. encode motor proteins that transport morphogens throughout the embryo.
   c. are dispersed apparently randomly throughout the genome.
   d. are expressed in similar levels in all parts of the developing embryo.
   e. Both a and c are correct.

8. Which of the following genes do not play a role in the process whereby segments are formed in the fruit fly embryo?
   a. homeotic genes
   b. gap genes
   c. pair-rule genes
   d. segment-polarity genes
   e. All of the above play a role in segmentation.

9. An embryonic stem cell that can give rise to any type of cell of an adult organism but cannot produce an entire, intact individual is called
   a. totipotent.   c. multipotent.   e. antipotent.
   b. pluripotent.   d. unipotent.

10. During plant development, the leaves and the flowers of the plant are derived from
    a. the central region.   d. the apical region.
    b. the basal region.   e. both a and d.
    c. the suspensor.

Conceptual Questions

1. If you observed fruit flies with the following developmental abnormalities, would you guess that a mutation has occurred in a segmentation gene or a homeotic gene? Explain your guess.
   a. Three abdominal segments were missing.
   b. One abdominal segment had legs.

2. The *myoD* gene in mammals plays a role in muscle cell differentiation. The *Hox* genes are homeotic genes that play a role in the differentiation of particular regions of the body. Explain how the functions of these genes are similar and different.

3. What is a meristem? Explain the role of meristems in plant development.

Collaborative Questions

1. It seems that developmental genetics boils down to a complex network of gene regulation. Starting with maternal effect genes and ending with master transcription factors, try to draw/describe how this network is structured for *Drosophila*. How many genes do you think are necessary to describe a complete developmental network for the fruit fly? How many genes do you think are needed for a network to specify one segment?

2. Is it possible for a phenotypically normal female fly to be homozygous for a loss-of-function allele in the *bicoid* gene? What would be the phenotype of the offspring that such a fly would produce if it were mated to a male that was homozygous for the normal *bicoid* allele?

Online Resource

www.brookerbiology.com

Stay a step ahead in your studies with animations that bring concepts to life and practice tests to assess your understanding. Your instructor may also recommend the interactive ebook, individualized learning tools, and more.

Genetic Technology

20

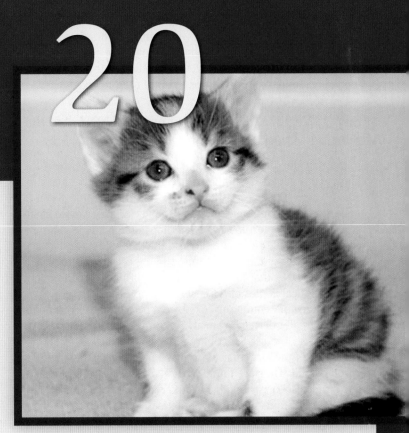

Jacob was diagnosed with type I diabetes, also called insulin-dependent diabetes mellitus, when he was 13 years old. Since then, he's been taking regular injections of insulin, a protein hormone that is normally made by cells in the pancreas. Jacob's pancreas fails to make enough of this hormone. His insulin prescription says Humulin® on the bottle, which stands for <u>hum</u>an in<u>sulin</u>. However, you might be surprised to discover that his insulin is not made by human cells. It's actually made by a laboratory strain of the bacterium *Escherichia coli* that has been genetically modified to synthesize a hormone that is identical in structure to human insulin. This is just one example of how researchers have been able to apply **recombinant DNA technology**—the use of laboratory techniques to isolate and manipulate fragments of DNA—to benefit humans. Such technology produces recombinant DNA, which refers to any DNA molecule that has been manipulated so it contains DNA from two or more sources.

In the early 1970s, the first successes in making recombinant DNA molecules were accomplished independently by two groups at Stanford University: David Jackson, Robert Symons, and Paul Berg; and Peter Lobban and A. Dale Kaiser. Both groups were able to isolate and purify pieces of DNA in a test tube and then covalently link two or more DNA fragments. Shortly thereafter, these recombinant DNA molecules were introduced into living cells. Once inside a host cell, the recombinant molecules are replicated to produce many identical copies. Because such recombinant molecules usually contain a particular gene of interest, this process is known as **gene cloning**. In the first part of the chapter, we will explore recombinant DNA technology and gene cloning, techniques that have enabled geneticists to probe relationships between gene sequences and phenotypic consequences. Such studies have been fundamental to our understanding of gene structure and function.

In the second part of this chapter, we will consider the topic of **genomics**—the molecular analysis of the entire genome of a species. In recent years, molecular techniques have progressed to the point where researchers can study the structure and function of many genes as large integrated networks. For example, the expression of all genes in a genome can be compared between normal and cancerous cells. This information can help us to understand how changes in gene expression can cause uncontrolled cell growth.

The first cloned pet. In 2002, the cat shown here called CC (for carbon copy) or copy cat was produced by reproductive cloning, a procedure described later in this chapter.

In the last section of this chapter, we will explore the topic of **biotechnology**—the use of living organisms or the products of living organisms for human benefit. We will learn that **genetic engineering**, the direct manipulation of genes for practical applications, is playing an ever-increasing role in the production of strains of microorganisms, plants, and animals with characteristics that are useful to people. These include bacteria that make hormones such as human insulin, crops that produce their own insecticides, and farm animals that make human medicines.

20.1 Gene Cloning

As mentioned, the term gene cloning refers to procedures that lead to the formation of many copies of a particular gene. Why is gene cloning useful? **Figure 20.1** provides an overview of the steps and goals of gene cloning. The process is usually done with one of two goals in mind. One goal would be that a researcher or clinician wants many copies of the gene, perhaps to study the DNA directly or to use the DNA as a tool. For example, geneticists may want to obtain the sequence of a gene from a person with a disease to see if the gene carries a mutation. Alternatively, the goal may be to obtain a large amount of the gene product—mRNA or protein. Along these lines, biochemists use gene cloning to obtain large amounts of proteins to study their structure and function. In modern molecular biology, the many uses for gene cloning are remarkable. Gene cloning has provided the foundation for critical technical advances in a variety of disciplines, including molecular biology, genetics, cell biology, biochemistry, and medicine. In this section, we will examine the procedures that are used to copy genes.

Step 1: Vector DNA and Chromosomal DNA Are the Starting Materials to Clone a Gene

In the first step of gene cloning, a key material is a type of DNA known as a **vector** (from the Latin, for one who carries) (Figure 20.1). Vector DNA acts as a carrier of the DNA segment that is to be cloned. In cloning experiments, a vector may carry a small segment of chromosomal DNA, perhaps only a single gene. By comparison, a chromosome carries up to a few thousand genes. When a vector is introduced into a living cell, it can replicate, and so the DNA that it carries is also replicated. This produces many identical copies of the inserted gene.

The vectors commonly used in gene cloning experiments were originally derived from two natural sources, plasmids or viruses. As discussed in Chapter 18, **plasmids** are small circular pieces of DNA that exist independently of the bacterial chromosome. They are found naturally in many strains of bacteria and also occasionally in eukaryotic cells. Commercially available plasmids have been genetically engineered for effective use in cloning experiments. They contain unique sites into which geneticists can easily insert pieces of DNA. An alternative type of vector used in cloning experiments is a **viral vector**. Viruses can infect living cells and propagate themselves by taking control of the host cell's metabolic machinery. When a chromosomal gene is inserted into a viral vector, the gene will be replicated whenever the viral DNA is replicated. Therefore, viruses can be used as vectors to carry other pieces of DNA.

Another material that is needed to clone a gene is the gene itself, which we will call the gene of interest. If a scientist wants to clone a particular gene, the source of the gene is the chromosomal DNA that carries that gene. The preparation of chromosomal DNA involves breaking open cells and extracting and purifying the DNA using biochemical separation techniques such as chromatography and centrifugation.

Step 2: Cutting Chromosomal and Vector DNA into Pieces and Linking Them Together Produces Recombinant Vectors

The second step in a gene cloning experiment is the insertion of chromosomal DNA into a plasmid or viral vector (Figure 20.1). How is this accomplished? DNA molecules must be cut and pasted to produce recombinant vectors. To cut DNA, researchers use enzymes known as **restriction enzymes** or restriction endonucleases. These enzymes, which were discovered

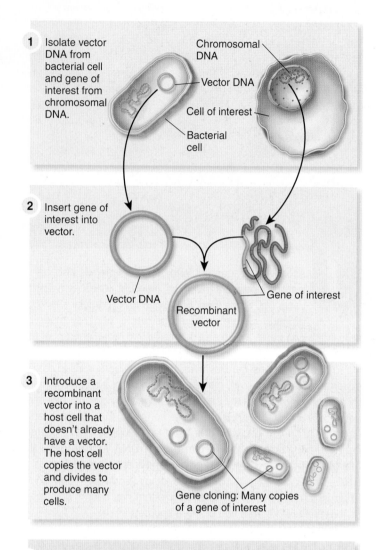

4 **Gene cloning is done to achieve one of two main goals:**

Producing large amounts of DNA of a specific gene	Expressing the cloned gene to produce the encoded protein
Examples	*Examples*
• Cloned genes provide enough DNA for DNA sequencing. The sequence of a gene can help us understand how a gene works and identify mutations that cause diseases.	• Large amounts of the protein can be purified to study its structure and function.
• Cloned DNA can be used as a probe to identify the same gene or similar genes in other organisms.	• Cloned genes can be introduced into bacteria or livestock to make pharmaceutical products such as insulin.
	• Cloned genes can be introduced into plants and animals to alter their traits.
	• Cloned genes can be used to treat diseases—a clinical approach called gene therapy.

Figure 20.1 **Gene cloning.** The process of gene cloning is used to produce large amounts of a gene or its protein product.

by Werner Arber, Hamilton Smith, and Daniel Nathans in the 1960s and 1970s, are made naturally by many different species of bacteria. Restriction enzymes protect bacterial cells from invasion by viruses by degrading the viral DNA into small fragments. Currently, several hundred different restriction enzymes from various bacterial species have been identified and are available commercially to molecular biologists.

The restriction enzymes used in cloning experiments bind to a specific base sequence and then cleave the DNA backbone at two defined locations, one in each strand. The sequences recognized by restriction enzymes are called **restriction sites**. Most restriction enzymes recognize sequences that are palindromic, which means the sequence is identical when read in the opposite direction in the complementary strand (**Table 20.1**). For example, the sequence recognized by the restriction enzyme *Eco*RI is 5′–GAATTC–3′ in the top strand. Read in the opposite direction in the bottom strand, this sequence is also 5′–GAATTC–3′. Certain restriction enzymes are useful in cloning because they digest DNA into fragments with **sticky ends**. Such DNA fragments have single-stranded ends that will hydrogen-bond to other DNA fragments that are cut with the same enzyme due to their complementary sequences.

Figure 20.2 shows the action of a restriction enzyme and the insertion of a gene into a vector. This vector carries the *amp*^R and *lacZ* genes, whose useful functions will be discussed later. The restriction enzyme binds to specific sequences in both the chromosomal and vector DNA. It then cleaves the DNA backbones, producing DNA fragments with sticky ends. The complementary single-stranded ends of fragments from the chromosomal and vector DNA hydrogen-bond with each other (Figure 20.2, step 2). However, this interaction is not stable, because it involves only a few hydrogen bonds between complementary bases. To establish a permanent connection between two DNA fragments, the sugar-phosphate backbones within the DNA strands must be covalently linked together. This linkage is catalyzed by DNA ligase (Figure 20.2, step 3). Recall from Chapter 11 that DNA ligase is an enzyme that

catalyzes the formation of a covalent bond between adjacent DNA fragments.

In some cases, the two ends of the vector will simply ligate back together, restoring it to its original circular structure; this

1 Cut vector and chromosomal DNA with *Eco*RI, a restriction enzyme that recognizes the sequence **GAATTC** and cuts at the arrows. **CTTAAG**

Vector DNA has one *Eco*RI site.

Chromosomal DNA has many *Eco*RI sites.

The restriction enzyme opens up the vector and cuts the chromosomal DNA into many fragments with short single-stranded regions called sticky ends.

2 Allow sticky ends to hydrogen-bond with each other due to complementary sequences.

In this example, a fragment of DNA carrying the gene of interest has hydrogen-bonded to the vector. Four gaps are found where covalent bonds in the DNA backbone are missing.

3 Add DNA ligase to close the gaps by catalyzing the formation of covalent bonds in the DNA backbone.

Figure 20.2 Step 2 of gene cloning: The actions of a restriction enzyme and DNA ligase produce a recombinant vector.

Concept check: In the procedure shown in this figure, has the gene of interest been cloned?

Table 20.1	Examples of Restriction Enzymes Used in Gene Cloning	
Restriction enzyme*	**Bacterial source**	**Sequence recognized†**
*Eco*RI	*Escherichia coli* (strain RY13)	↓ 5′–GAATTC–3′ 3′–CTTAAG–5′ ↑
*Sac*I	*Streptomyces achromogenes*	↓ 5′–GAGCTC–3′ 3′–CTCGAG–5′ ↑

*Restriction enzymes are named according to the species in which they are found. The first three letters are italicized because they indicate the genus and species names. Because a species may produce more than one restriction enzyme, the enzymes are designated I, II, III, etc., to indicate the order in which they were discovered in a given species.

†The arrows show the locations in the upper and lower DNA strands where the restriction enzymes cleave the DNA backbone.

forms what is called a recircularized vector. In other cases, a fragment of chromosomal DNA may become ligated to both ends of the vector. When this happens, a segment of chromosomal DNA has been inserted into the vector. The result is a vector containing a piece of chromosomal DNA, which is called a **recombinant vector** or a hybrid vector. We then have a recombinant vector that is ready to be cloned. A recombinant vector may contain the gene of interest, or it may contain a different piece of chromosomal DNA.

Step 3: Putting Recombinant Vectors into Host Cells and Allowing Those Cells to Propagate Achieves Gene Cloning

The third step in gene cloning is the actual cloning of the gene of interest. In this step, the goal is for the recombinant vector carrying the desired gene to be taken up by bacterial cells treated with agents that render them permeable to DNA molecules (**Figure 20.3**). Some bacterial cells take up a single plasmid, whereas most cells fail to take up a plasmid. The bacteria are then streaked on petri plates containing a bacterial growth medium and ampicillin.

In the experiment shown here, the bacterial cells were originally sensitive to ampicillin. The plasmid carries an antibiotic-resistance gene, called the *amp^R* gene. What is the purpose of this gene in a cloning experiment? Such a gene is called a **selectable marker** because the presence of the antibiotic selects for the growth of cells expressing the *amp^R* gene. The *amp^R* gene encodes an enzyme known as β-lactamase that degrades the antibiotic ampicillin, which normally kills bacteria. Bacteria that have not taken up a plasmid will be killed by the antibiotic. In contrast, any bacterium that has taken up a plasmid carrying the *amp^R* gene will grow and divide many times to form a bacterial colony containing tens of millions of cells. Because each cell in a single colony is derived from the same original cell that took up a single plasmid, all cells within a colony contain the same type of plasmid DNA.

In the experiment shown in Figure 20.3, the experimenter can also distinguish bacterial colonies that contain cells with a recombinant vector from those containing cells with a recircularized vector. In a recombinant vector, a piece of chromosomal DNA has been inserted into a region of the vector that contains the *lacZ* gene, which encodes the enzyme β-galactosidase. The insertion of chromosomal DNA into the vector disrupts the *lacZ* gene and thereby prevents the expression of β-galactosidase. By comparison, a recircularized vector has a functional *lacZ* gene. The functionality of *lacZ* can be determined by providing the growth medium with a colorless compound, X-Gal, which is cleaved by β-galactosidase into a blue dye. Bacteria grown in the presence of X-Gal will form blue colonies if they have a functional β-galactosidase enzyme and will form white colonies if they do not. In this experiment, therefore, bacterial colonies containing recircularized vectors will form blue colonies, whereas colonies containing recombinant vectors carrying a segment of chromosomal DNA will be white.

After a bacterial cell has taken up a recombinant vector, two subsequent events lead to the production of many copies of

1 Mix plasmid DNA with many *E. coli* cells that have been treated with agents that make them permeable to DNA.

In this example, the gene of interest was inserted into a plasmid. This disrupts the *lacZ* gene and renders it nonfunctional. It is also possible for any other chromosomal DNA fragment to be inserted into the plasmid, or the plasmid may recircularize without an insert.

This shows a bacterial cell that has taken up the plasmid carrying the gene of interest. Many bacterial cells fail to take up a plasmid.

2 Plate cells on media containing ampicillin and X-Gal. Incubate overnight. Note: The *ampR* gene allows bacteria to grow in the presence of ampicillin. The *lacZ* gene encodes β-galactosidase that degrades X-Gal to produce a blue color.

Each bacterial colony is derived from a single cell; so all the cells in a colony are genetically identical.

Recircularized vector without an insert—*lacZ* gene is functional and produces blue color.

Recombinant vector with an insert—*lacZ* gene is nonfunctional.

Figure 20.3   Step 3 of gene cloning: introduction of a recombinant vector into a host cell.   For cloning to occur, a recombinant vector is introduced into a host cell, which copies the vector and divides to produce many cells. This produces many copies of the gene of interest.

Concept check: In this cloning experiment, what is the purpose of having the lacZ gene in the vector?

that vector. First, when the vector has a highly active origin of replication, the bacterial host cell produces many copies of the recombinant vector per cell. Second, the bacterial cells divide approximately every 20 minutes. Following overnight growth, a population of many millions of bacteria will be obtained from a single cell. Each of these bacterial cells will contain many copies of the cloned gene. For example, a bacterial colony may comprise 10 million cells, with each cell containing 50 copies of the recombinant vector. Therefore, this bacterial colony would have 500 million copies of the cloned gene!

A DNA Library Is a Collection of Many Different Cloned Genes

In a typical cloning experiment, such as the one described in Figures 20.2 and 20.3, the treatment of chromosomal DNA with restriction enzymes actually yields tens of thousands of different DNA fragments. Therefore, after the DNA fragments are ligated individually to vectors, a researcher has a collection of many recombinant vectors, with each vector containing a particular fragment of chromosomal DNA. This collection of vectors, usually within bacterial cells, is known as a **DNA library** (Figure 20.4). Researchers make DNA libraries using the methods shown in Figures 20.2 and 20.3 and then use those libraries to obtain clones of genes in which they are interested.

Two types of DNA libraries are commonly made. The library is called a **genomic library** when the inserts are derived from chromosomal DNA. Alternatively, researchers can isolate mRNA and use the enzyme reverse transcriptase, which is described in Chapter 18, to make DNA molecules using mRNA as a starting material. Such DNA is called **complementary DNA**, or **cDNA**. A **cDNA library** is a collection of recombinant vectors that have cDNA inserts. From a research perspective, an important advantage of cDNA is that it lacks introns. Because introns can be quite large, it is much simpler for researchers to insert cDNAs into vectors rather than work from chromosomal DNA if they want to focus their attention on the coding sequence of a gene.

Gel Electrophoresis Can Separate Macromolecules, Such as DNA

Gel electrophoresis is a technique that is used to separate macromolecules, such as DNA and proteins, on a gel. This method is often used to evaluate the results of a cloning experiment. For example, gel electrophoresis can be used to determine the sizes of DNA fragments that have been inserted into recombinant vectors.

Gel electrophoresis can separate molecules based on their charge, size/length, and mass. In the example shown in Figure 20.5, gel electrophoresis was used to separate different fragments of chromosomal DNA based on their masses. The gel is a flat semisolid gel with depressions at the top called wells where samples are added. An electric field is applied to the gel, which causes charged molecules, either proteins or nucleic acids, to migrate from the top of the gel toward the bottom—a process called electrophoresis. DNA is negatively charged and moves toward the positive end of the gel, which is at the bottom in this

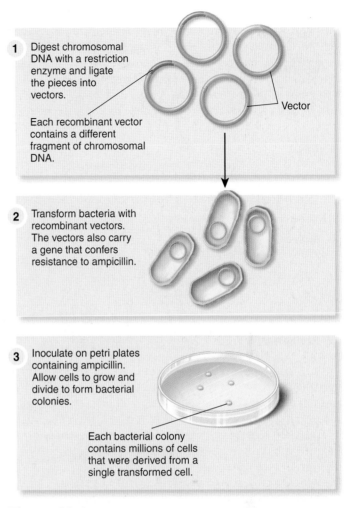

1. Digest chromosomal DNA with a restriction enzyme and ligate the pieces into vectors.

Each recombinant vector contains a different fragment of chromosomal DNA.

Vector

2. Transform bacteria with recombinant vectors. The vectors also carry a gene that confers resistance to ampicillin.

3. Inoculate on petri plates containing ampicillin. Allow cells to grow and divide to form bacterial colonies.

Each bacterial colony contains millions of cells that were derived from a single transformed cell.

Figure 20.4   A DNA library.   Each colony in a DNA library contains a vector with a different piece of chromosomal DNA.

figure. Smaller DNA fragments move more quickly through the gel polymer and therefore are located closer to the bottom of the gel compared to larger ones. As the slab gel runs, the fragments are separated into bands within the gel. The fragments in each band can then be stained with a dye for identification.

Polymerase Chain Reaction (PCR) Can Also Be Used to Make Many Copies of DNA

As we have seen, one method of cloning involves an approach in which the gene of interest is inserted into a vector and then introduced into a host cell. Another technique to copy DNA without the aid of vectors and host cells is **polymerase chain reaction** (**PCR**), which was developed by Kary Mullis in 1985 (Figure 20.6). The goal of PCR is to make many copies of DNA in a defined region, perhaps encompassing a gene or part of a gene. Several reagents are required for the synthesis of DNA. First, two different primers are needed that are complementary to sequences at each end of the DNA region to be amplified. These primers are usually about 20 nucleotides long. One primer is called the forward primer, and the other is the reverse primer. PCR also requires all four deoxynucleoside triphosphates (dNTPs) and a heat-stable form of DNA polymerase called *Taq*

1 Load samples of DNA fragments into wells at the top of the gel.

Samples

Gel

2 Apply an electric field.

⊖

⊕

3 Wait additional time.

Higher-mass molecules

⊖

Each band is a group of DNA fragments with the same mass.

⊕

Lower-mass molecules

Figure 20.5 Separation of molecules by gel electrophoresis. In this example, samples containing many fragments of DNA are loaded into wells at the top of the gel and then subjected to an electric field that causes the fragments to move toward the bottom of the gel. This primarily separates the fragments according to their masses, with the smaller DNA fragments near the bottom of the gel.

Concept check: One DNA fragment contains 600 bp and another has 1,300 bp. Following electrophoresis, which would be closer to the bottom of a gel?

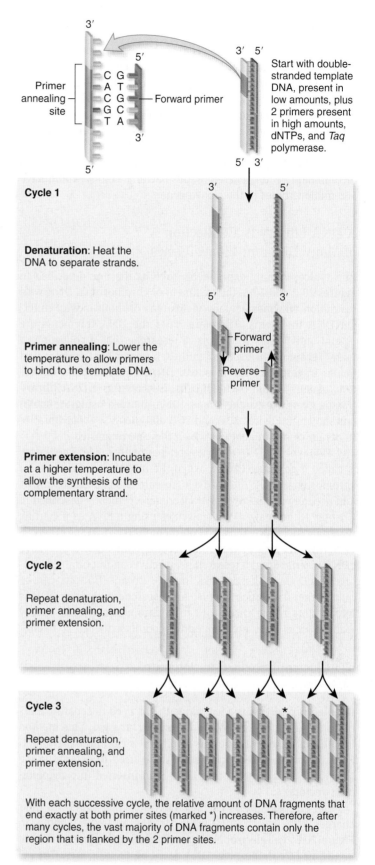

Primer annealing site

Forward primer

Start with double-stranded template DNA, present in low amounts, plus 2 primers present in high amounts, dNTPs, and *Taq* polymerase.

Cycle 1

Denaturation: Heat the DNA to separate strands.

Primer annealing: Lower the temperature to allow primers to bind to the template DNA.

Forward primer

Reverse primer

Primer extension: Incubate at a higher temperature to allow the synthesis of the complementary strand.

Cycle 2

Repeat denaturation, primer annealing, and primer extension.

Cycle 3

Repeat denaturation, primer annealing, and primer extension.

With each successive cycle, the relative amount of DNA fragments that end exactly at both primer sites (marked *) increases. Therefore, after many cycles, the vast majority of DNA fragments contain only the region that is flanked by the 2 primer sites.

Figure 20.6 **Polymerase chain reaction (PCR).** During each PCR cycle, the steps of denaturation, primer annealing, and primer extension take place. The net result of PCR is the synthesis of many copies of DNA in the region that is flanked by the two primers. To conduct this type of PCR experiment, the researcher must have prior knowledge about the base sequence of the template DNA in order to design primers with base sequences that are complementary to the ends of the template DNA.

Concept check: Why do the PCR primers bind specifically to the primer annealing sites (shown in green)?

polymerase. *Taq* polymerase is isolated from the bacterium *Thermus aquaticus*, which lives in hot springs and can tolerate temperatures up to 95°C. A heat-stable form of DNA polymerase is necessary because PCR involves heating steps that would inactivate most other natural forms of DNA polymerase.

To make copies, a sample of chromosomal DNA, called the double-stranded template DNA, is denatured into single-stranded molecules by heat treatment. Then the primers bind to the DNA as the temperature is lowered (Figure 20.6). The binding of the primers to the specific sites in the template DNA is called annealing. After the primers have annealed, the temperature is slightly raised and *Taq* polymerase uses dNTPs to catalyze the synthesis of complementary DNA strands, thereby doubling the amount of DNA in the region that is flanked by the primers. This step is called primer extension because the length of the primers is extended by the synthesis of DNA. The sequential process of denaturation—primer annealing—primer extension is then repeated many times in a row. This method is called a chain reaction because the products of each previous step are used as reactants in subsequent steps. A device that controls the temperature and automates the timing of each step, known as a thermocycler, is used to carry out PCR. The PCR technique can amplify the amount of DNA by a staggering amount. After 30 cycles of denaturation, primer annealing, and primer extension, a DNA sample will increase 2^{30}-fold, which is approximately a billionfold!

20.2 Genomics

As discussed throughout Unit III, the genome is the complete genetic composition of an organism. As genetic technology has progressed over the past few decades, researchers have gained an increasing ability to analyze the composition of genomes as a whole unit. The term genomics refers to the molecular analysis of the entire genome of a species. Segments of chromosomes are cloned and analyzed in progressively smaller pieces, the locations of which are known on the intact chromosomes. This is the mapping phase of genomics. The mapping of a genome ultimately progresses to the determination of the complete DNA sequence, which provides the most detailed description available of an organism's genome at the molecular level. By comparison, **functional genomics** studies the expression of a genome. For example, functional genomics can be used to analyze which genes are turned on or off in normal versus cancer cells. In this section, we will consider a few of the methods that are used in genomics and functional genomics.

BAC Cloning Vectors Are Used to Make Contigs of Chromosomes to Map a Genome

A common goal of genomics research is to clone and analyze the entire genome of a species. For large eukaryotic genomes, which may contain over 1 billion bp (base pairs), cloning an entire genome is much easier when a cloning vector can accept very large chromosomal DNA inserts. In general, most plasmid and viral vectors can accommodate inserts only a few thousand to perhaps tens of thousands of nucleotides in length. If a plasmid or viral vector has a DNA insert that is too large, it will have difficulty with DNA replication and is likely to suffer deletions in the insert. By comparison, a type of cloning vector known as a **bacterial artificial chromosome** (**BAC**) can reliably contain much larger inserted DNA fragments. BACs are derived from large plasmids called F factors (see Chapter 18). They can typically contain inserts up to 500,000 bp. BACs are used in genomic research in the same way as other types of vectors. Similarly, yeast artificial chromosomes (YACs) are used as vectors in yeast. An insert in a YAC can be several hundred thousand to perhaps 2 million bp in length.

The term **mapping** refers to the process of determining the relative locations of genes or other DNA segments along a chromosome. After many large fragments of chromosomal DNA have been inserted into BACs or YACs, the first step of mapping is to determine the relative locations of the inserted chromosomal pieces as they would occur in an intact chromosome. This is called physical mapping. To obtain a complete physical map of a chromosome, researchers need a series of clones that contain overlapping pieces of chromosomal DNA. Such a collection of clones, known as a **contig**, contains a contiguous region of a chromosome that is found as overlapping regions within a group of vectors (**Figure 20.7**). Overlapping regions carry the same genetic material between adjacent clones. For example, clones 7 and 8 both carry gene *K*. These overlapping regions allow researchers to identify the order of the clones along the chromosome.

The Dideoxy Chain-Termination Method Is Used to Determine the Base Sequence of DNA

Once researchers have cloned DNA into vectors and obtained a physical map, the next phase of genomic research is **DNA sequencing**, which is a method to determine the base sequence of DNA. Scientists can learn a great deal of information about the function of a gene if its nucleotide sequence is known. For example, the investigation of genetic sequences has been vital in our understanding of the genetic basis of human diseases.

The most commonly used method for DNA sequencing, developed in 1977 by Frederick Sanger and colleagues, is known as the **dideoxy chain-termination method**, or more simply, dideoxy sequencing. Dideoxy sequencing is based on our knowledge of DNA replication. As described in Chapter 11, DNA polymerase connects adjacent deoxynucleoside triphosphates (dNTPs) by catalyzing a covalent linkage between the 5′–phosphate on one nucleotide and the 3′—OH group on the previous nucleotide. Chemists, however, can synthesize nucleotides, called dideoxynucleoside triphosphates (ddNTPs), that are missing the —OH group at the 3′ position (**Figure 20.8**, step 1). What happens if a dideoxynucleotide is incorporated during DNA replication? Sanger reasoned that if a dideoxynucleotide ddNTP is added to a growing DNA strand, the strand can no longer grow because the 3′—OH group, the site of attachment for the next nucleotide, is missing. This ending of DNA synthesis is called chain termination.

Clone individual pieces into vectors.

A collection of overlapping clones, known as a contig

Figure 20.7 **A contig.** As shown here, a contig is a collection of clones that have overlapping pieces of DNA from a particular chromosome. The numbers denote the order of the members of the contig. The chromosome is labeled with letters that denote the locations of particular genes. The members of the contig have overlapping regions that have the same genes, which allows you to order them.

Concept check: What does it mean when we say that two members of a contig have overlapping regions?

Before describing the steps of this DNA sequencing protocol, let's first consider the DNA segment that is analyzed in a sequencing experiment. The segment of DNA to be sequenced, the target DNA, must be obtained in large amounts by using the cloning techniques that were described earlier in this chapter. In Figure 20.8, the target DNA was inserted into a vector next to a primer-annealing site, the site where a primer will bind. The target DNA is initially double stranded, but Figure 20.8a,

step 1, shows the DNA after it has been denatured into a single strand by heat treatment.

Let's now examine the steps involved in DNA sequencing. Many copies of this single-stranded DNA are placed into four tubes and mixed with primers that bind to the primer-annealing site. DNA polymerase and all four types of regular dNTPs are also added to each tube. Finally, each of the four tubes has a low concentration of just one of the four possible dideoxynucleoside triphosphates: ddGTP, ddATP, ddTTP, or ddCTP. The tubes are then incubated to allow DNA polymerase to make strands that are complementary to the target DNA sequence. However, the ddNTPs will occasionally cause DNA synthesis to terminate early. For example, let's consider the third tube, which contains ddTTP. Synthesis of new DNA strands will occasionally stop at the sixth or thirteenth position after the annealing site if a ddTTP, instead of a dTTP, is incorporated into the growing DNA strand. This means the target DNA has a complementary A at the sixth and thirteenth positions.

Within the four tubes, mixtures of DNA strands of different lengths are made. These DNA strands are separated according to their lengths by subjecting them to gel electrophoresis. The shorter strands move to the bottom of the gel more quickly than the longer strands. To detect the newly made DNA strands, the dNTPs that were added to each reaction are radiolabeled. This enables the strands to be visualized as bands when the gel is exposed to X-ray film. In Figure 20.8a, step 2, the DNA strands in the four tubes were run in separate lanes on a gel. Because we know which ddNTP was added to each tube, we also know which base is at the very end of each DNA strand separated on this gel, because ddNTPs cause chain termination. Therefore, we can determine the DNA sequence by reading which base is at the end of every DNA strand and matching this sequence with the length of the strand.

Dideoxy sequencing can now be done much more quickly with automated sequencing. Instead of having four separate tubes with a single type of ddNTP in each tube, automated sequencing uses one tube containing all four ddNTPs, each of which has a different-colored fluorescent label attached. After incubating the template DNA with DNA polymerase, primers, dNTPs, and the four types of fluorescent ddNTPs, the sample is then loaded into a single lane of a gel, and the fragments are separated by gel electrophoresis. Electrophoresis is continued until each band emerges from the bottom of the gel, where a laser excites the fluorescent dye. A fluorescence detector records the amount of fluorescence emission at four wavelengths, corresponding to the four dyes. An example of a printout from a fluorescence detector is shown in Figure 20.8b. The peaks of fluorescence correspond to the DNA sequence that is complementary to the target DNA. The height of the fluorescent peaks is not always the same because ddNTPs get incorporated at certain sites more readily than at other sites.

Researchers are also developing alternative methods to the dideoxy chain termination in order to sequence DNA. For example, pyrosequencing is a new method of DNA sequencing that is based on the detection of released pyrophosphate (PP_i) during DNA synthesis.

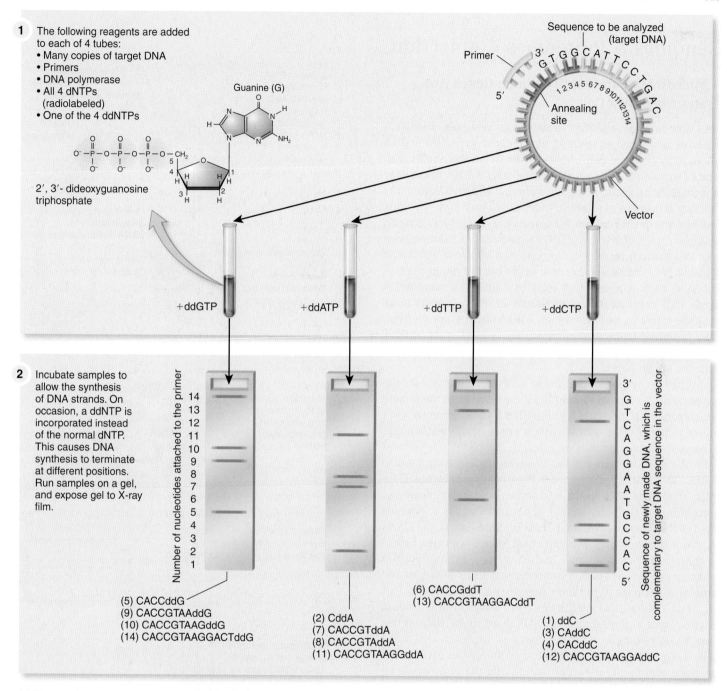

1 The following reagents are added to each of 4 tubes:
- Many copies of target DNA
- Primers
- DNA polymerase
- All 4 dNTPs (radiolabeled)
- One of the 4 ddNTPs

Guanine (G)

2', 3'- dideoxyguanosine triphosphate

Sequence to be analyzed (target DNA)

Primer

3' G T G G C A T T C C T G A C
1 2 3 4 5 6 7 8 9 10 11 12 13 14

5'

Annealing site

Vector

+ddGTP   +ddATP   +ddTTP   +ddCTP

2 Incubate samples to allow the synthesis of DNA strands. On occasion, a ddNTP is incorporated instead of the normal dNTP. This causes DNA synthesis to terminate at different positions. Run samples on a gel, and expose gel to X-ray film.

Number of nucleotides attached to the primer

14
13
12
11
10
9
8
7
6
5
4
3
2
1

3'
G
T
C
A
G
G
A
A
T
G
C
C
A
C
5'

Sequence of newly made DNA, which is complementary to target DNA sequence in the vector

(5) CACCddG
(9) CACCGTAAddG
(10) CACCGTAAGddG
(14) CACCGTAAGGACTddG

(2) CddA
(7) CACCGTddA
(8) CACCGTAddA
(11) CACCGTAAGGddA

(6) CACCGddT
(13) CACCGTAAGGACddT

(1) ddC
(3) CAddC
(4) CACddC
(12) CACCGTAAGGAddC

(a) The procedure used in traditional dideoxy sequencing

C A C C G T A A G G A C T G T C G C C A T C A G G T T G G C C C A A A C G A C A A C C C

(b) Output from automated dideoxy sequencing

Figure 20.8   **DNA sequencing by the dideoxy method.**   (a) Traditional dideoxy sequencing used a mixture of radiolabeled dNTPs and nonlabeled ddNTPs. Following electrophoresis, the bands were detected by autoradiography. Step 1 shows the structure of dideoxyguanosine triphosphate, abbreviated ddGTP. It has a hydrogen, shown in red, instead of a hydroxyl group at the 3' position. The prefix, dideoxy-, means it has two (di) missing (de) oxygens (oxy) compared with ribose, which has —OH groups at both the 2' and 3' positions. (b) Automated sequencing uses a fluorescence detector that measures the four kinds of ddNTPs as they emerge from the gel.

Concept check: *What happens when a ddNTP is incorporated into a growing DNA strand?*

Genomes & Proteomes Connection

A Microarray Can Identify Which Genes Are Transcribed by a Cell

Let's now turn our attention to functional genomics. Researchers have developed an exciting new technology, called a **DNA microarray** (or gene chip), that is used to monitor the expression of thousands of genes simultaneously. A DNA microarray is a small silica, glass, or plastic slide that is dotted with many different sequences of single-stranded DNA, each corresponding to a short sequence within a known gene. Each spot contains multiple copies of a known DNA sequence. For example, one spot in a microarray may correspond to a sequence within the β-globin gene, while another spot might correspond to a different gene, such as a gene that encodes a glucose transporter. A single slide contains tens of thousands of different spots in an area the size of a postage stamp. These microarrays are typically produced using a technology that "prints" spots of DNA sequences onto a slide similar to the way that an inkjet printer deposits ink on paper.

What is the purpose of using a DNA microarray? In the experiment shown in **Figure 20.9**, the goal is to determine which genes are transcribed into mRNA from a particular sample of cells. In other words, which genes in the genome are expressed? To conduct this experiment, the mRNA was isolated from the cells and then used to make fluorescently labeled cDNA. The labeled cDNAs were then incubated with a DNA microarray. The DNA in the microarray is single stranded and corresponds to the sense strand—the strand that has a sequence like mRNA. Those cDNAs that are complementary to the DNAs in the microarray will hybridize and thereby remain bound to the microarray. The array is then washed and placed in a microscope equipped with a computer that scans each spot and generates an image of the spots' relative fluorescence.

If the fluorescence intensity in a spot is high, a large amount of cDNA was in the sample that hybridized to the DNA at this location. For example, if the β-globin gene was expressed in the cells being tested, a large amount of cDNA for this gene would be made, and the fluorescence intensity for that spot would be high. Because the DNA sequence of each spot is already known, a fluorescent spot identifies cDNAs that are complementary to those DNA sequences. Furthermore, because the cDNA was generated from mRNA, this technique identifies genes that have been transcribed in a particular cell type under a given set of conditions. However, the amount of protein encoded by an mRNA may not always correlate with the amount of mRNA due to variation in the rates of mRNA translation and protein degradation.

Thus far, the most common use of microarrays is to study gene expression patterns. In addition, the technology of DNA microarrays has found several other important uses (**Table 20.2**).

Figure 20.9   Identifying transcribed genes within a DNA microarray.   In this simplified example, only three cDNAs specifically hybridize to spots on the microarray. Those genes were expressed in the cells from which the mRNA was isolated. In an actual experiment, there are typically hundreds or thousands of different cDNAs and tens of thousands of different spots on the array.

Concept check: If a fluorescent spot appears on a microarray, what information does this provide regarding gene expression?

Table 20.2	Applications of DNA Microarrays

Application	Description
Cell-specific gene expression	A comparison of microarray data using cDNAs derived from mRNA of different cell types can identify genes that are expressed in a cell-specific manner.
Gene regulation	Because environmental conditions play an important role in gene regulation, a comparison of microarray data using cDNA derived from mRNA from cells exposed to two different environmental conditions may reveal genes that are induced under one set of conditions and repressed under another set.
Elucidation of metabolic pathways	Genes that encode proteins that participate in a common metabolic pathway are often expressed in a parallel manner and can be revealed from a microarray analysis.
Tumor profiling	Different types of cancer cells exhibit striking differences in their profiles of gene expression, which can be revealed by a DNA microarray analysis. This approach is gaining widespread use to classify tumors that are sometimes morphologically indistinguishable.
Genetic variation	A mutant allele may not hybridize to a spot on a microarray as well as a wild-type allele. Therefore, microarrays are gaining widespread use as a tool to detect genetic variation. This application has been used to identify disease-causing alleles in humans and to identify mutations that contribute to quantitative traits in plants and other species.
Microbial strain identification	Microarrays can distinguish between closely related bacterial species and subspecies.

20.3   Biotechnology

Biotechnology is defined as the use of living organisms, or products from living organisms, as a way to benefit humans. Although the term has become associated with molecular genetics, biotechnology is not a new topic. Its use began about 12,000 years ago, when humans began to domesticate animals and plants for the production of food. Since that time, many species of microorganisms, plants, and animals have become routinely used by people. Beginning in the 1970s, genetic engineering has provided new ways to make use of living organisms to benefit humans.

In this section, we will consider the applications of biotechnology that involve the genetic engineering of microorganisms, plants, and animals. We will also examine several topics that you often hear about in the news, such as the cloning of mammals, DNA fingerprinting, and gene therapy.

Important Medicines Are Produced by Recombinant Microorganisms

Several important medicines are now produced by recombinant organisms. These include tissue plasminogen activator (TPA), used to dissolve blood clots in heart attack patients; factor VIII, used to treat people with hemophilia; and insulin, used to treat people with diabetes. In 1982, the U.S. Food and Drug Administration approved the sale of human insulin made by recombinant bacteria. In healthy individuals, insulin is produced by the beta cells of the pancreas. Insulin functions to regulate several physiological processes, including the uptake of glucose into muscle cells. Persons with insulin-dependent diabetes cannot synthesize an adequate amount of insulin due to a defect in their beta cells. Today, people like Jacob, who was introduced at the beginning of the chapter, are usually treated with human insulin that is made by genetically engineered bacteria. Prior to 1982, insulin was isolated from pancreases removed from cattle and pigs. Unfortunately, in some cases, diabetic individuals using cow and pig insulin developed allergic responses. These patients had to use expensive combinations of insulin from human cadavers and other animals. Now, they can use human insulin made by recombinant bacteria.

Insulin is a hormone composed of two polypeptides, the A and B chains, that are held together with disulfide bonds. To make this hormone, the coding sequence of the A or B chains is inserted into a plasmid vector next to the coding sequence of the *E. coli* protein, β-galactosidase (**Figure 20.10**). After such vectors are introduced into bacteria, the cells produce many copies of a fusion protein comprising β-galactosidase and the A or B chain. Why is this step done? The reason is because the A and B chains are rapidly degraded when expressed in bacterial cells by themselves. The fusion proteins, however, are not. These fusion proteins are then extracted from the bacterial cells and treated with a chemical, cyanogen bromide (CNBr), which cleaves a peptide bond at a methionine that is found at the end of the β-galactosidase sequence, thereby separating β-galactosidase from the A or B chain. The A and B chains are purified and mixed together under conditions in which they will fold and associate with each other via disulfide bonds to form a functional insulin hormone molecule.

Microorganisms Can Reduce Pollutants

Bioremediation is the use of living organisms, typically microorganisms or plants, to detoxify pollutants in the environment. As its name suggests, it is a biological remedy for pollution. During microbial bioremediation, enzymes produced by a microorganism modify a toxic pollutant by altering or transforming its structure. In many cases, the toxic pollutant is degraded, yielding less complex, nontoxic metabolites.

Since the early 1900s, microorganisms have been used in sewage treatment plants to degrade sewage. More recently, the field of bioremediation has expanded into the treatment of hazardous and refractory chemical wastes—chemicals that are difficult to degrade and that are usually associated with industrial activity. These pollutants include heavy metals, petroleum hydrocarbons, and halogenated organic compounds such as those with chlorine atoms, as well as pesticides, herbicides, and organic solvents. Many new applications are being tested that use microorganisms to degrade these pollutants. The field of bioremediation has been fostered by better knowledge of how pollutants are degraded by microorganisms, the identification of

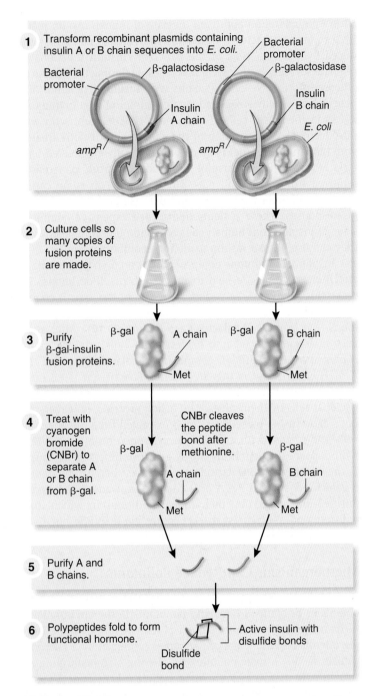

1. Transform recombinant plasmids containing insulin A or B chain sequences into *E. coli.*

Bacterial promoter
β-galactosidase
Bacterial promoter
β-galactosidase
Insulin A chain
Insulin B chain
E. coli
amp^R
amp^R

2. Culture cells so many copies of fusion proteins are made.

3. Purify β-gal-insulin fusion proteins.
β-gal
A chain
β-gal
B chain
Met
Met

4. Treat with cyanogen bromide (CNBr) to separate A or B chain from β-gal.
CNBr cleaves the peptide bond after methionine.
β-gal
A chain
β-gal
B chain
Met
Met

5. Purify A and B chains.

6. Polypeptides fold to form functional hormone.
Active insulin with disulfide bonds
Disulfide bond

Figure 20.10 **The use of bacteria to make human insulin.**

Concept check: *Why are the A and B chains made as fusion proteins with β-galactosidase?*

new and useful strains of microbes, and the ability to enhance bioremediation through genetic engineering.

In 1980, in a landmark case (*Diamond v. Chakrabarty*), the U.S. Supreme Court ruled that a live, recombinant microorganism is patentable as a "manufacture or composition of matter." The first recombinant microorganism to be patented was an "oil-eating" bacterium that contained a laboratory-constructed plasmid. This strain can oxidize the hydrocarbons commonly found in petroleum. It grew faster on crude oil than did any

of the natural strains that were tested. Was it a commercial success? The answer is "No," because this recombinant strain metabolizes only a limited number of toxic compounds. Unfortunately, the strain did not degrade many higher-molecular-weight compounds that tend to persist in the environment.

Bioremediation is a developing industry. Recombinant microorganisms can provide an effective way to decrease the levels of toxic chemicals within our environment. This approach requires careful studies to demonstrate that recombinant organisms are effective at reducing pollutants and safe when released into the environment.

Gene Replacements and Knockouts in Mice Can Be Used to Understand Gene Function and Human Disease

Let's now turn our attention to the genetic engineering of animals. Researchers can introduce a cloned gene into an oocyte, a fertilized egg, or embryonic cells to produce animals that carry the cloned gene. The term **transgenic** is used to describe an organism that carries genes that were introduced using molecular techniques such as gene cloning. Transgenic organisms are also called **genetically modified organisms** (**GMOs**). Such GMOs are typically made in two ways. In some cases, the cloned gene will insert randomly into the genome and result in **gene addition**—the insertion of cloned gene into the genome.

Alternatively, a cloned gene may recombine with the normal gene on a chromosome, a phenomenon called **gene replacement**. For eukaryotic species that are diploid, only one of the two copies is initially replaced. In other words, the initial gene replacement produces a heterozygote carrying one normal copy of the gene and one copy that has been replaced with a cloned gene. Heterozygotes can be crossed to each other to obtain homozygotes, which carry both copies of the cloned gene. If the cloned gene carries a mutation that inactivates the normal gene's function, such a homozygote is said to have undergone a **gene knockout**. The inactive cloned gene has replaced both copies of the normal gene, and the normal gene's function is said to be knocked out. Gene replacements and gene knockouts have become powerful tools for understanding gene function.

A particularly exciting avenue of gene replacement research is its application in the study of human disease. As an example, let's consider the disease cystic fibrosis (CF), which is one of the most common and severe inherited human disorders. In humans, the defective gene that causes CF has been identified. Likewise, the homologous gene in mice was later identified. Using the technique of gene replacement, researchers have produced mice that are homozygous for the same type of mutation that is found in humans with CF. Such mice exhibit disease symptoms resembling those found in humans, namely, respiratory infections and digestive abnormalities. Why is this approach useful? These mice can be used as model organisms to study this human disease. Furthermore, these mice models have been used to test the effects of various therapies in the treatment of the disease.

Biotechnology Holds Promise in Producing Transgenic Livestock

The technology of creating transgenic mice has been extended to other animals, and much research is under way to develop transgenic species of livestock, including fish, sheep, pigs, goats, and cattle. A novel avenue of research involves the production of medically important proteins in the mammary glands of livestock. This approach is sometimes called **molecular pharming**. (The word pharming refers to the use of genetically engineered farm animals or crops to make pharmaceuticals.) Several human proteins have been successfully produced in the milk of domestic livestock such as sheep, goats, and cattle. These include Factor IX to treat a certain type of hemophilia, tissue plasminogen activator (TPA) to dissolve blood clots, and α-1-antitrypsin for the treatment of emphysema.

How are human proteins, such as hormones, made in such a way that they are produced in the milk of livestock? As we learned in Chapter 13, gene regulation may promote the expression of genes in certain cell types. Researchers have identified specific genes that are transcribed only in lactating mammary cells. One example is the gene that encodes β-lactoglobulin, a protein that is found in the milk of sheep and cows. One strategy to produce a human hormone in the milk of livestock is to clone the human hormone gene into a plasmid vector next to the β-lactoglobulin promoter (**Figure 20.11**). The plasmid is then injected into an oocyte, such as a sheep oocyte, where it integrates into the genome. The egg is then fertilized by exposure to sperm and implanted into the uterus of a female sheep. The resulting offspring carries the cloned gene. If the offspring is a female, the protein encoded by the human gene will be expressed within the mammary gland and secreted into the milk. The milk is then obtained from the animal, and the human protein isolated and purified.

Compared with the production of proteins in bacteria, one advantage of molecular pharming is that certain proteins are more likely to function properly when expressed in mammals. This may be due to post-translational modifications of proteins that occur in mammals but not in bacteria. In addition, certain proteins may be degraded rapidly or folded improperly when expressed in bacteria. Furthermore, the yield of recombinant proteins in milk can be quite large. Each dairy cow, for example, produces about 10,000 liters of milk per year. In some cases, a transgenic cow can produce approximately 1 g/L of the transgenic protein in its milk.

Agrobacterium tumefaciens Can Be Used to Make Transgenic Plants

The production of transgenic plants is somewhat easier than producing transgenic animals because many cells from plants are totipotent, which means that an entire organism can be regenerated from somatic cells. Therefore, a transgenic plant can be made by the introduction of cloned genes into somatic tissue, such as the tissue of a leaf. After the cells of a leaf have become transgenic, an entire plant can be regenerated by

1 Clone a human hormone gene into a plasmid vector next to a sheep β-lactoglobulin promoter. This promoter is functional only in mammary cells, so the protein product is secreted into the milk.

Human hormone gene

β-lactoglobulin promoter

Plasmid vector

2 Inject this recombinant plasmid into a sheep oocyte. The plasmid DNA will integrate into the chromosomal DNA, resulting in the addition of the hormone gene into the sheep's genome.

Sheep oocyte

3 The oocyte is fertilized and implanted into a female sheep, which then gives birth to a transgenic sheep offspring.

Transgenic sheep

4 Obtain milk from a female transgenic sheep. The milk contains a human hormone.

Milk containing human hormone

5 Purify the hormone from the milk.

Figure 20.11   Molecular pharming.

treating the leaf with plant growth hormones that cause it to form roots and shoots. The result is a separate plant that is transgenic.

Molecular biologists often use the bacterium *Agrobacterium tumefaciens*, which naturally infects plant cells and causes tumors, to produce transgenic plants. This bacterium contains a plasmid known as the **Ti plasmid**, for tumor-inducing plasmid. The plasmid has a region called the T DNA (for transferred DNA) that is transferred from the bacterium to the plant cell. The T DNA from the Ti plasmid becomes integrated into the chromosomal DNA of the plant cell. After this occurs, genes within the T DNA are expressed that cause uncontrolled plant cell growth. This produces a crown gall tumor, a bulbous growth on the plant.

Researchers have modified the Ti plasmid to use it as a vector to introduce a gene of interest into plant cells. This is achieved by inserting a gene into the T DNA of the Ti plasmid. In the modified Ti plasmid, the genes that cause a crown gall tumor have been removed. In addition, a selectable marker gene is inserted into the T DNA to allow selection of plant cells that have taken up the T DNA. *Kan^R*, a gene that provides resistance to the antibiotic kanamycin, is commonly used as a selectable marker. Last, the Ti plasmid used in cloning experiments is modified to contain unique restriction sites for the convenient insertion of any gene of interest.

Figure 20.12 shows the general strategy for producing transgenic plants via T DNA–mediated gene transfer. A gene of interest is inserted into the T DNA of a genetically engineered Ti plasmid and then introduced into *A. tumefaciens*. Plant cells are then exposed to *A. tumefaciens* carrying the Ti plasmid. After allowing time for T DNA transfer, the plant cells are grown on a solid medium that contains kanamycin and carbenicillin. Kanamycin kills any plant cells that have not taken up the T DNA, and carbenicillin kills *A. tumefaciens*. Therefore, the only surviving cells are those plant cells that have integrated the T DNA into their genome. Because the T DNA also contains the gene of interest, the selected plant cells are expected to have received this gene as well. The cells are then transferred to a medium that contains the plant growth hormones necessary for the regeneration of entire plants.

Many transgenic plants have been approved for human consumption. Their production has become routine practice for several agriculturally important plant species, including alfalfa, corn, cotton, soybean, tobacco, and tomato. Transgenic plants can be given characteristics that are agriculturally useful, such as those that improve plant quality and resistance. In terms of quality, gene additions have been made to improve the nutritional value of some plants, such as making the canola grain produce more oil.

Frequently, transgenic research has sought to produce plant strains that are resistant to insects, disease, and herbicides. A successful example of the use of transgenic plants has involved the introduction of genes from *Bacillus thuringiensis* (Bt). This bacterium produces toxins that kill certain types of caterpillars and beetles and has been widely used as an insecticide for several decades. These toxins are proteins that are encoded in the genome of *B. thuringiensis*. Researchers have succeeded in cloning toxin genes from *B. thuringiensis* and transferring those genes into plants. Such Bt varieties of plants produce the toxins themselves and therefore are resistant to many types of caterpillars and beetles. Examples of commercialized crops include Bt corn (**Figure 20.13a**) and Bt cotton. Since their introduction in 1996, the commercial use of these two Bt crops has steadily increased (**Figure 20.13b**).

The use of transgenic agricultural plants has been strongly opposed by some people. What are the perceived risks? One potential risk is that transgenes in commercial crops could endanger native species. For example, Bt crops may kill pollinators of native species. Another concern is that the planting of transgenic crops could lead to the evolution of resistant insects that would render Bt ineffective. To prevent this from

1 Gene of interest is inserted into the T DNA of the Ti plasmid.

Ti plasmid Gene of interest

T DNA

Kan^R

Site where restriction enzyme cuts

Recombinant Ti plasmid

2 The recombinant Ti plasmid is transformed into *A. tumefaciens*.

A. tumefaciens
Recombinant Ti plasmid

3 Plant cells are exposed to *A. tumefaciens*. The T DNA is transferred and incorporated into the plant cell chromosome.

Inserted T DNA carrying gene of interest

Plant cell

4 The plant cells are placed in a medium containing kanamycin and carbenicillin. Kanamycin kills plant cells that have not taken up T DNA. Carbenicillin kills *A. tumefaciens*. The surviving plant cells are transferred to growth media that has plant hormones necessary for regenerating an entire plant.

Plant with cloned gene

Figure 20.12 Using the Ti plasmid and *Agrobacterium tumefaciens* to create transgenic plants.

Concept check: Which region of the Ti plasmid is transferred to the plant cell?

happening, researchers are producing transgenic strains that carry more than one toxin gene, which makes it more difficult for insect resistance to arise. A third concern is the potential for transgenic plants to elicit allergic reactions. Despite these and other concerns, many farmers are embracing transgenic crops, and their use continues to rise.

(a) A field of Bt corn

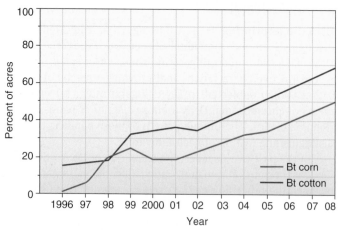

(b) Bt corn and Bt cotton usage since 1996

Figure 20.13 The production of Bt crops. (a) A field of Bt corn. These corn plants carry a toxin gene from *Bacillus thuringiensis* that provides them with resistance to insects such as corn borers, which are a major pest of corn plants. (b) A graph showing the increase in usage of Bt corn and Bt cotton in the U.S. since their commercial introduction in 1996.

Researchers Have Succeeded in Cloning Mammals from Somatic Cells

We now turn our attention to cloning as a way to genetically manipulate plants and animals. The term cloning has several different meanings. At the beginning of this chapter we discussed gene cloning, which involves methods that produce many copies of a gene. The cloning of a multicellular organism, called **reproductive cloning**, is a different matter. By accident, this happens in nature. Identical twins are genetic clones that began from the same fertilized egg. Similarly, researchers can take mammalian embryos at an early stage of development (for example, the two- to eight-cell stage), separate the cells, implant them into the uterus of a female, and obtain multiple births of genetically identical individuals.

As previously noted, the reproductive cloning of new individuals is relatively easy in the case of plants, which can be cloned from somatic cells. Until recently, this approach had not been possible with mammals. Scientists believed that chromosomes within the somatic cells of mammals had incurred irreversible genetic changes that rendered them unsuitable for

reproductive cloning. However, this hypothesis has proven to be incorrect. In 1996, Ian Wilmut and his colleagues created clones of sheep using the genetic material from somatic cells. As you may have heard, they named the first cloned lamb Dolly.

How was Dolly created? As shown in **Figure 20.14**, the researchers removed mammary cells from an adult female

1 Donor sheep's mammary cell is extracted and grown in a tissue culture flask. Another sheep's unfertilized egg is extracted, and the nucleus is removed.

Mammary cell

Unfertilized egg

Nucleus

2 The cells are fused together with electric pulses.

Egg with nucleus removed

Mammary cell

3 The donor nucleus from the mammary cell and the maternal proteins within the egg initiate development of the egg into an embryo.

4 The embryo is transferred into a surrogate sheep.

5 Pregnancy is allowed to proceed.

6 A lamb genetically identical to the donor sheep is then born.

Figure 20.14 Protocol for the successful cloning of sheep. In this procedure, the genetic material from a somatic cell is used to make a cloned mammal, in this case the sheep Dolly.

Concept check: *Did all of Dolly's DNA come from a mammary cell? Explain.*

sheep and grew them in the laboratory. The researchers then extracted the nucleus from a sheep oocyte and fused the diploid mammary cell with the enucleated oocyte cell. Fusion was promoted by electric pulses. After fusion, the zygote was implanted into the uterus of an adult sheep. One hundred and forty-eight days later, Dolly was born.

Dolly was (almost) genetically identical to the sheep that donated the mammary cell. Dolly and the donor sheep were (almost) genetically identical in the same way that identical twins are. They carry the same set of genes and look remarkably similar. The reason that they may not be completely identical is that Dolly and her somatic cell donor may have some minor genetic differences due to possible differences in their mitochondrial DNA and may exhibit some phenotypic differences due to maternal effect genes.

Mammalian reproductive cloning is still at an early stage of development. Nevertheless, creating Dolly was a breakthrough that showed it is technically possible. In recent years, cloning using somatic cells has been achieved in several mammalian species, including sheep, cows, mice, goats, pigs, and cats. In 2002, the first pet was cloned, which was named CC for carbon copy (see chapter-opening photo). The cloning of mammals provides the potential for many practical applications. With regard to livestock, cloning would enable farmers to use the somatic cells from their best animals to create genetically homogeneous herds. This could be advantageous in terms of agricultural yield, although such a genetically homogeneous herd may be more susceptible to certain diseases.

Although reproductive cloning may have practical uses in the field of agriculture, our society has become greatly concerned with the possibility of human cloning. This prospect has raised serious ethical questions. Some people feel it is morally wrong and threatens the basic fabric of parenthood and family. Others feel it is a modern technology that offers a new avenue for reproduction, one that could be offered to infertile couples, for example. Human cloning is a complex subject with many more viewpoints than these two. In the public sector, the sentiment toward human cloning has been generally negative. Many countries have issued a complete ban on human cloning, whereas others permit limited research in this area. In the future, our society will have to wrestle with the legal and ethical aspects of cloning as it applies not only to animals but also to people.

DNA Fingerprinting Is Used for Identification and Relationship Testing

DNA fingerprinting is a technology that can identify and distinguish among individuals based on variations in their DNA. Like the human fingerprint, the DNA of each individual is a distinctive characteristic that provides a means of identification. When subjected to DNA fingerprinting, selected fragments of chromosomal DNA produce a series of bands on a gel (**Figure 20.15a**). The unique pattern of these bands is usually a distinguishing feature of each individual.

In the past two decades, the technique of DNA fingerprinting has become automated, much like the automation that changed the procedure of DNA sequencing described earlier in this chapter. DNA fingerprinting is now done using PCR, which amplifies **short tandem repeat sequences** (**STRs**)—short DNA sequences that are repeated many times in a row. Such tandem repeat sequences, which are noncoding regions of chromosomal DNA, are found at specific locations in the genomes of all species. The number of repeats at each location tends to vary from

(a) Traditional DNA fingerprinting

(b) Automated DNA fingerprinting

Figure 20.15 DNA fingerprinting. (a) Chromosomal DNA from two different individuals was subjected to traditional DNA fingerprinting. Their DNA appears as a series of bands on a gel. The dissimilarities in the patterns of these bands distinguish different individuals, much as the differences in physical fingerprint patterns can be used for identification. DNA evidence at a crime scene (E) is compared to DNA from suspect 1 (S1) and suspect 2 (S2). (b) Automated DNA fingerprinting measures the masses of fluorescently labeled DNA segments called short tandem repeat sequences (STRs) from a selected individual. A printout from the fluorescence detector is shown here. Two individuals are almost always different in the pattern of peaks they exhibit.

Concept check: Which suspect's DNA fingerprint (S1 or S2) matches the DNA evidence collected at a crime scene?

one individual to the next. Using primers that are complementary to DNA sequences that flank each STR, the STRs from a sample of DNA are amplified by PCR and then separated by gel electrophoresis according to their molecular masses. As in automated DNA sequencing, the amplified STR fragments are fluorescently labeled. A laser excites the fluorescent molecule within an STR, and a detector records the amount of fluorescence emission for each STR. The DNA fingerprint yields a series of peaks, each peak having a characteristic molecular mass (**Figure 20.15b**). In this automated approach, the pattern of peaks is an individual's DNA fingerprint.

What are the uses of DNA fingerprinting? First, DNA fingerprinting has gained acceptance as a precise method of identification. In medicine, it is used to identify different species of bacteria and fungi, and it can even distinguish among closely related strains of the same species. This is useful so that clinicians can treat patients with the appropriate antibiotic or fungicide.

A second common use is forensics—providing evidence in a criminal case. DNA fingerprinting can be used as evidence that an individual was at a crime scene. Forensic DNA was first used in the U.S. court system in 1986. When a DNA sample taken from a crime scene matches the DNA fingerprint of an individual, the probability that a match could occur simply by chance can be calculated. Each STR size is given a probability score based on its observed frequency within a reference human population (Caucasian, Asian, and so on). An automated DNA fingerprint contains many peaks, and the probability scores for each peak are multiplied together to arrive at the likelihood that a particular pattern of peaks would be observed. For example, if a DNA fingerprint contains 20 fluorescent peaks, and the probability of an individual having each peak is 1/4, then the likelihood of having that pattern would be $(1/4)^{20}$, or roughly 1 in 1 trillion. Therefore, a match between two samples is rarely a matter of random chance.

Another important use of DNA fingerprinting is to establish paternity and other family relationships. Persons who are related genetically will have some peaks in common. The number they share depends on the closeness of their genetic relationship. For example, offspring are expected to receive half of their peaks from one parent and half from the other. Therefore, DNA fingerprinting can be used as evidence in paternity cases.

FEATURE INVESTIGATION

Blaese and Colleagues Performed the First Gene Therapy to Treat ADA Deficiency

Gene therapy is the introduction of cloned genes into living cells in an attempt to cure disease. Many current research efforts in gene therapy are aimed at alleviating inherited human diseases. More than 4,000 human genetic diseases involve a single gene abnormality. Common examples include cystic fibrosis and sickle-cell disease. Many inherited diseases have been investigated as potential targets for gene therapy. These include metabolic diseases, such as phenylketonuria, and blood disorders, such as hemophilia and severe combined immunodeficiency. Human gene therapy is still at an early stage of development. Relatively few patients have been successfully treated with gene therapy in spite of a large research effort. In addition, experimental gene therapy in humans has been associated with adverse reactions. In 1999, a patient even died from a reaction to a gene therapy treatment.

Adenosine deaminase (ADA) deficiency is an autosomal recessive disorder due to a lack of the enzyme adenosine deaminase. When present, adenosine deaminase deaminates the nucleoside deoxyadenosine. This is an important step in the proper metabolism of nucleosides. If both copies of the *ADA* gene are defective, however, deoxyadenosine will accumulate within the cells of the individual. At high concentration, deoxyadenosine is particularly toxic to lymphocytes in the immune system, namely, T cells and B cells. In individuals with ADA deficiency, the destruction of T and B cells leads to severe combined immunodeficiency disease (SCID). If left untreated, SCID is typically fatal at an early age (generally 1 to 2 years old), because the compromised immune system of these individuals cannot fight infections.

Three approaches are used to treat ADA deficiency. In some cases, a bone marrow transplant is received from a compatible donor. A second method is to treat SCID patients with purified ADA enzyme that is coupled to polyethylene glycol (PEG). This PEG-ADA is taken up by lymphocytes and can correct the ADA deficiency. Unfortunately, these two approaches are not always available and/or successful. A third, more recent approach is to treat ADA patients with gene therapy.

On September 14, 1990, the first human gene therapy was approved for a young girl suffering from ADA deficiency. This work was carried out by a large team of researchers including R. Michael Blaese, Kenneth Culver, W. French Anderson, and several collaborators at the National Institutes of Health (NIH). Prior to this clinical trial, the normal gene for ADA had been cloned into a retroviral vector. The retroviral vector also contained mutations that prevented it from causing a viral disease, yet it still enabled the virus to infect human cells. The general aim of this therapy was to remove lymphocytes from the blood of the girl with SCID, introduce the normal *ADA* gene into the cells via the retrovirus, and then return them to her bloodstream.

Figure 20.16 outlines the protocol for the experimental treatment. The researchers removed lymphocytes from the girl and cultured them in a laboratory. The lymphocytes were then infected with a recombinant retrovirus that contained the normal *ADA* gene. During the reproductive cycle of the retrovirus, the retroviral genetic material was inserted into the host cell's DNA. Therefore, because this retrovirus contained the normal *ADA* gene, this gene also was inserted into the chromosomal

Figure 20.16 The first human gene therapy for adenosine deaminase (ADA) deficiency by Blaese and colleagues.

HYPOTHESIS Infecting lymphocytes with a retrovirus containing the normal *ADA* gene will correct the inherited deficiency of the mutant *ADA* gene in patients with ADA deficiency.

KEY MATERIALS A retrovirus with the normal *ADA* gene.

Experimental level | Conceptual level

1 Remove *ADA*-deficient lymphocytes from the patient with severe combined immunodeficiency disease (SCID).

Lymphocytes

Lymphocyte | Mutant *ADA* gene

2 Culture the cells in a laboratory.

3 Infect the cells with a retrovirus that contains the normal *ADA* gene. As described in Chapter 18, retroviruses insert their DNA into the host cell chromosome as part of their reproductive cycle.

Retrovirus with *ADA* gene | Retrovirus | Lymphocyte

4 Infuse the *ADA*-gene-corrected lymphocytes back into the SCID patient.

Normal *ADA* gene | Retroviral inserted DNA

5 **THE DATA**

6 **CONCLUSION** The introduction of a cloned *ADA* gene into lymphocytes via gene therapy resulted in higher ADA function, even after 4 years.

7 **SOURCE** Blaese, Robert M. et al. 1995. T lymphocyte-directed gene therapy for ADA-SCID: Initial trial results after 4 years. *Science* 270:475–480.

DNA of the girl's lymphocytes, thereby correcting the defect in ADA. After this occurred in the laboratory, the cells were reintroduced back into the patient.

In this clinical trial, two U.S. patients were enrolled, and a third patient was later treated in Japan. The data in Figure 20.16 show the results of this trial for one of the three patients. Over the course of 2 years, this patient was given 11 infusions of lymphocytes that had been corrected with the normal *ADA* gene. Even after 4 years, this patient's lymphocytes were still

making ADA. These results suggest that this first gene therapy trial may offer some benefit. However, the patients in this study were also treated with PEG-ADA as an additional therapy to prevent the adverse symptoms of SCID. Therefore, the researchers could not determine whether gene transfer into T cells was of significant clinical benefit.

Another form of SCID, termed SCID-X1, is inherited as an X-linked trait. SCID-X1 is characterized by a block in T cell growth and differentiation. This block is caused by mutations in the gene encoding the γ_c cytokine receptor, which plays a key role in the recognition of signals that are needed to promote the growth, survival, and differentiation of T cells. A gene therapy trial for SCID-X1 similar to the trial shown in Figure 20.16 was initiated in 2000. A normal γ_c cytokine receptor gene was cloned into a retroviral vector and then introduced into SCID-X1 patients' lymphocytes. The lymphocytes were then reintroduced back into their bodies. At a 10-month follow-up, T cells expressing the normal γ_c cytokine receptor were detected in two patients. Most importantly, the T cell counts in these two patients had risen to levels that were comparable to normal individuals.

This clinical trial was the first clear demonstration that gene therapy can offer clinical benefits, providing in these cases what seemed to be a complete correction of the disease phenotype.

However, in a French study involving 10 SCID-X1 patients, an unexpected and serious side effect occurred. Within 3 years of gene therapy treatment, 3 out of the 10 treated children developed leukemia—a form of cancer involving the proliferation of white blood cells. In these cases, the disease was caused by the integration of the retroviral vector next to an oncogene, a gene that promotes cancer, in the patients' genome. The development of leukemia in these patients has halted many clinical trials involving gene therapy.

Experimental Questions

1. What is gene therapy? What is ADA deficiency?

2. In the investigation of Figure 20.16, how did the researchers treat ADA deficiency?

3. How successful was the gene therapy for ADA deficiency?

Summary of Key Concepts

20.1 Gene Cloning

- Recombinant DNA technology is the use of laboratory techniques to isolate and manipulate fragments of DNA.

- Gene cloning, the process of making multiple copies of a gene, is used to obtain large amounts of the DNA that encodes a particular gene or to obtain large amounts of the gene product. (Figure 20.1)

- Plasmid and viral vectors are used in gene cloning. To obtain recombinant DNA, a vector and chromosomal DNA are cut with restriction enzymes. The DNA fragments bind to each other at their sticky ends, and the pieces are linked together via DNA ligase. (Figure 20.2, Table 20.1)

- When a recombinant vector is introduced into a bacterial cell, the cell replicates the vector and also divides to produce many cells. This achieves gene cloning. (Figure 20.3)

- A collection of recombinant vectors, each with a particular piece of chromosomal DNA, is introduced into bacterial cells to produce a DNA library. If the DNA inserts are derived from cDNA, which is made from mRNA, this is a cDNA library. (Figure 20.4)

- Gel electrophoresis is used to separate macromolecules by using an electric field that causes them to pass through a gel matrix. Gel electrophoresis typically separates molecules according to their charges, sizes, and masses. (Figure 20.5)

- Polymerase chain reaction (PCR) is another common technique to make many copies of a gene. Primers are used that flank the region of DNA to be amplified. (Figure 20.6)

20.2 Genomics

- Genomics is the study of genomes as whole units.

- For genomics, large fragments of chromosomal DNA are cloned into vectors such as BACs. One goal of genomics is to make a contig, which is a collection of clones that cover a contiguous region of a chromosome. This is a type of mapping—determining the relative locations of genes or other DNA segments along a chromosome. (Figure 20.7)

- The dideoxy method of DNA sequencing uses ddNTPs to determine the base sequence of a segment of DNA. (Figure 20.8)

- Functional genomics is aimed at studying the expression of a genome. An important technique used in functional genomics is a DNA microarray that contains a group of spots, each with a specific type of DNA. It is used as a hybridization tool to study which genes are transcribed into mRNA from a particular sample of cells. (Figure 20.9, Table 20.2)

20.3 Biotechnology

- Biotechnology is the use of living organisms, or the products of living organisms, for human benefit. For example, microorganisms can be genetically engineered to produce human products such as insulin. (Figure 20.10)

- Microorganisms are also used to reduce pollutants in the environment, a phenomenon called bioremediation.

- Transgenic organisms, also called genetically modified organisms, are made via gene replacement or gene addition. Transgenic livestock can be genetically engineered to produce human hormones in their milk. (Figure 20.11)

- The Ti plasmid in *Agrobacterium tumefaciens* has been used to produce transgenic plants such as Bt corn, which produces its own insecticide. (Figures 20.12, 20.13)

- The cloning of mammals has been achieved by fusing a somatic cell with an egg that has had its nucleus removed. (Figure 20.14)

- DNA fingerprinting is a method of identification based on the occurrence of segments of DNA in the genomes of all individuals, called STRs, that are variable in length among different individuals. (Figure 20.15)

- Gene therapy is a method to treat human diseases by the introduction of cloned genes into cells. The first gene therapy involved a disease called severe combined immunodeficiency syndrome (SCID). (Figure 20.16)

Assess and Discuss

Test Yourself

1. Vectors used to clone genes were derived originally from
   - a. proteins.
   - b. plasmids.
   - c. viruses.
   - d. all of the above.
   - e. b and c only.

2. Restriction enzymes used in most cloning experiments
   - a. are used to cut DNA into pieces for gene cloning.
   - b. are naturally produced by bacteria cells to prevent viral infection.
   - c. produce sticky ends on DNA fragments.
   - d. All of the above are correct.
   - e. a and c only are correct.

3. DNA ligase is needed in a cloning experiment
   - a. to promote hydrogen bonding between sticky ends.
   - b. to catalyze the covalent attachment of the backbone of DNA strands.
   - c. to digest the chromosomal DNA into small pieces.
   - d. a and b only are correct.
   - e. a, b, and c are correct.

4. Let's suppose you followed the protocols described in Figures 20.2 and 20.3. Which sequence of experiments would you follow to confirm that a white colony really contained a recombinant vector with an insert?
   - a. Pick a bacterial colony and restreak on plates containing X-Gal to confirm that the cells really form white colonies.
   - b. Pick a bacterial colony, isolate plasmid DNA, digest the plasmid DNA with a restriction enzyme, and then run the DNA on a gel.
   - c. Pick a bacterial colony and test it to see if β-galactosidase is functional within the bacterial cells.
   - d. Pick a bacterial colony and retest it on ampicillin-containing plates to double-check that the cells are really ampicillin resistant.
   - e. c and d should both be conducted.

5. Why is *Taq* polymerase used in PCR rather than other DNA polymerases?
   - a. *Taq* polymerase is a synthetic enzyme that produces DNA strands at a faster rate than natural polymerases.
   - b. *Taq* polymerase is a heat-stable form of DNA polymerase that can function after exposure to the high temperatures that are necessary for PCR.
   - c. *Taq* polymerase is easier to isolate than other DNA polymerases.
   - d. *Taq* polymerase is the DNA polymerase commonly produced by most eukaryotic cells.
   - e. All of the above are correct.

6. Let's suppose you want to clone a gene that has never been analyzed before by DNA sequencing. Which of the following statements do you agree with the most?
   - a. Do PCR to clone the gene because it's much faster.
   - b. Do PCR to clone the gene because it is very specific and gives a high yield.
   - c. You can't do PCR because you can't make forward and reverse primers.
   - d. Do cloning by insertion into a vector because it will give you a higher yield.
   - e. Do cloning by insertion into a vector because it is easier than PCR.

7. The method of determining the base sequence of DNA is
   - a. PCR.
   - b. gene cloning.
   - c. DNA fingerprinting.
   - d. DNA sequencing.
   - e. gene mapping.

8. During bioremediation, microorganisms are used to
   - a. clone genes from eukaryotic organisms.
   - b. introduce correct genes into individuals with genetic diseases.
   - c. decrease pollutants in the environment.
   - d. produce useful products such as insulin.
   - e. do all of the above.

9. Organisms that carry genes that were introduced using molecular techniques are called
   - a. transgenics.
   - b. recombinant DNA.
   - c. mutants.
   - d. genetically modified organisms.
   - e. both a and d.

10. DNA fingerprinting is used
    - a. to provide a means of precise identification of an organism, such as the identification of specific strains of bacteria.
    - b. as a forensics tool to provide evidence in a criminal case.
    - c. to determine genetic relationships between individuals.
    - d. to determine the identity of an individual.
    - e. for all of the above.

Conceptual Questions

1. Explain how using one restriction enzyme to cut both a plasmid and a gene of interest will allow the gene to be inserted into the plasmid.

2. Explain and draw the structural feature of a dideoxyribonucleotide that causes chain termination.

3. What is a mouse model for a human disease? Why is such a model useful?

Collaborative Questions

1. Discuss three important advances that have resulted from gene cloning.

2. Discuss the ethical issues associated with genetic engineering, stem cell research, and the cloning of mammals.

Online Resource

www.brookerbiology.com

Stay a step ahead in your studies with animations that bring concepts to life and practice tests to assess your understanding. Your instructor may also recommend the interactive ebook, individualized learning tools, and more.

Genomes, Proteomes, and Bioinformatics

21

I magine a book that is incredibly long and has taken billions of years to write. Such an analogy applies to the human genome, which is written in only four letters—A, T, G, and C—and is over 3 billion base pairs long. If the entire human genome were typed in a textbook like this, with about 3,000 letters per page, it would be over 1 million pages long! Our genome contains many unsolved mysteries that researchers are trying to unravel, such as: What are the functions of every gene in the genome? How do gene mutations cause disease? How are humans evolutionarily related to other species? As you will learn, genomes are full of surprises. For example, did you know that most of your DNA has no known function?

The unifying theme of biology is evolution. The genome of every living species is the product of approximately 4 billion years of evolution. We can understand the unity of modern organisms by realizing that all species evolved from an interrelated group of ancestors. Throughout this textbook, a recurring theme is a series of "Genomes and Proteomes" topics to highlight the evolutionary connections among all forms of life and underscore how the genetic material produces the form and function of living organisms. By now, you may feel familiar with the concept of a **genome**, which is the complete genetic composition of a cell or a species. The genome of each species is critical to its existence in several ways:

- The genome stores information in the form of genes, which provide a genetic blueprint to produce the characteristics of organisms.

- The genome is copied and transmitted from generation to generation.

- The accumulation of genetic changes (mutations) over the course of many generations produces the evolutionary changes that alter species and produce new species.

An extension of genome analysis is the study of proteomes. The term **proteome** refers to the entire complement of proteins that a cell or organism can make. The function of most genes is to encode proteins, which are the key determinants of cell structure and function. Analyzing the proteomes of organisms allows researchers to understand many aspects of biology, including the structure and

Genomes and computer technology. The amount of data derived from the analyses of genomes is so staggering in size and complexity that researchers have turned to the use of computers to unravel the amazing information that genomes contain.

function of cells, the complexity of multicellular organisms, and the interactions between organisms and their environment.

In the first two sections of this chapter, we will consider genome characteristics of prokaryotic and eukaryotic species, respectively. We will then turn our attention to proteomes and examine the roles of the proteins that a species can make. From a molecular perspective, genomes and proteomes contain extensive and complex information, which is studied using computer technology. In the last section of this chapter, we will consider how the field of **bioinformatics**, which employs computers and statistical techniques to analyze biological information, is critical to the study of genomes and proteomes.

21.1 Bacterial and Archaeal Genomes

The past decade has seen remarkable advances in our overall understanding of the entire genome of many species. As genetic technology has progressed, researchers have gained an increasing ability to analyze the composition of genomes as a whole unit. For many species, we now know their complete DNA sequence, which provides the most detailed description available of an organism's genome at the molecular level. In this section, we will survey the sizes and composition of the genomes in selected prokaryotic species, which includes both bacteria and archaea.

Prokaryotic Genomes Often Contain a Circular Chromosome with a Few Thousand Genes

Geneticists have made great progress in the study of bacterial and archaeal genomes. Some of the key features of prokaryotic chromosomes are described in Chapter 18 (refer back to Figure 18.10). Why are researchers interested in the genomes of prokaryotic species? First, bacteria cause many different diseases that affect humans and other animals, as well as plants. Studying the genomes of bacteria reveals important clues about the process of infection and also may help us find ways to combat bacterial infection. A second reason for studying prokaryotic genomes is that the information we learn about these microscopic organisms often applies to larger and more complex organisms. For example, basic genetic mechanisms, such as DNA replication and gene regulation, were first understood in the bacterium *Escherichia coli*. That knowledge provided a critical foundation to understand how these processes work in humans and other eukaryotic species. A third reason is evolution. The origin of the first eukaryotic cell probably involved a union between an archaeal and a bacterial cell, as we will explore in Chapter 22. The study of prokaryotic genomes helps us understand how all living species evolved. Finally, another reason to study the genomes of bacteria is because we use them as tools in research and biotechnology, which was discussed in Chapter 20.

As of 2009, the genomes of over 700 prokaryotic species have been completely sequenced and analyzed. The chromosomes of prokaryotes are usually a few million base pairs in length. Genomic researchers refer to 1 million base pairs as 1 megabase pair, abbreviated Mb. Most prokaryotes contain a single type of chromosome, though multiple copies may be present in a single cell. However, some prokaryotes are known to have different chromosomes. For example, *Vibrio cholerae*, the bacterium that causes the diarrheal disease known as cholera, has two different chromosomes in each cell, one 2.9 Mb and the other 1.1 Mb.

Bacterial chromosomes are often circular. For example, the two chromosomes in *V. cholerae* are circular, as is the single type of chromosome found in *E. coli*. However, linear chromosomes are found in some prokaryotic species, such as *Borrelia burgdorferi*, which is the bacterium that causes Lyme disease, the most common tick-borne disease in the U.S. Certain bacterial species may even contain both linear and circular chromosomes. *Agrobacterium tumefaciens*, which infects plants

and causes a disorder called crown gall disease, has one linear chromosome (2.1 Mb) and one circular chromosome (3.0 Mb).

Table 21.1 compares the sequenced genomes from several prokaryotic species. They range in size from 1.7 to 5.2 Mb. The total number of genes is correlated with the total genome size. Roughly 1,000 genes are found for every Mb of DNA. Compared to eukaryotic genomes, prokaryotic genomes are less complex. Prokaryotic chromosomes lack centromeres and telomeres and have a single origin of replication. Also, prokaryotic chromosomes have relatively little repetitive DNA, a feature typically found in eukaryotic genomes.

In addition to one or more chromosomes, prokaryotes often have plasmids, circular pieces of DNA that exist independently of the bacterial chromosome. Plasmids are typically small, in the range of a few thousand to tens of thousands of base pairs in length, though some can be quite large, even hundreds of thousands of base pairs. The various functions of plasmids are described in Chapter 18, and their use as vectors in genetic engineering is discussed in Chapter 20.

Table 21.1	Examples of Prokaryotic Genomes That Have Been Sequenced*		
Species	**Genome size (Mb)[†]**	**Number of genes[‡]**	**Description**
Methanobacterium thermoautotrophicum	1.7	1,869	An archaeon that produces methane
Haemophilus influenzae	1.8	1,753	One of several different bacterial species that causes respiratory illness and meningitis
Sulfolobus solfataricus	3.0	3,032	An archaeon that metabolizes sulfur
Lactobacillus plantarum	3.3	3,052	A type of lactic acid bacterium used in the production of cheese and yogurt
Mycobacterium tuberculosis	4.4	4,294	The bacterium that causes the respiratory disease tuberculosis
Escherichia coli	4.6	4,289	A naturally occurring intestinal bacterium; certain strains can cause human illness
Bacillus anthracis	5.2	5,439	The bacterium that causes the disease anthrax

*Prokaryotic species often exist in different strains that may differ slightly in their genome size and number of genes. The data are from common strains of the indicated species. The species shown in this table have only one type of chromosome.

[†]Mb equals 1 million base pairs.

[‡]The number of genes is an estimate based on the analysis of genome sequences.

FEATURE INVESTIGATION

Venter, Smith, and Colleagues Sequenced the First Genome in 1995

The first genome to be entirely sequenced was that of the bacterium *Haemophilus influenzae*. This bacterium causes a variety of diseases in humans, including respiratory illnesses and bacterial meningitis. *H. influenzae* has a relatively small genome consisting of approximately 1.8 Mb of DNA in a single circular chromosome.

Scientists can follow different strategies when tackling a genome-sequencing project. One strategy, which has been used for larger eukaryotic genomes, requires extensive mapping. This means that the genome is cut into large pieces of DNA whose locations are known within a given chromosome. These large pieces are then cut into smaller and smaller pieces, whose relative locations are known within the larger pieces. Once small DNA pieces have been mapped throughout the whole genome, their DNA sequences are determined by the dideoxy sequencing method, which is described in Figure 20.8.

An alternative strategy for sequencing an entire genome is called **shotgun DNA sequencing**. In this approach, researchers use the technique of dideoxy sequencing to randomly sequence many DNA fragments from the genome. As a matter of chance, some of the fragments are overlapping—the end of one fragment will contain the same DNA region as the beginning of another fragment. Computers are used to align the overlapping regions and assemble the DNA fragments into a contiguous sequence identical to that found in the intact chromosome. The advantage of shotgun DNA sequencing is that it does not require extensive mapping, a process that can be time-consuming. A disadvantage is that researchers may waste time sequencing the same region of DNA more times than necessary in order to assemble the sequence correctly.

To obtain a complete sequence of a genome with the shotgun approach, how do researchers decide how many fragments to sequence? We can calculate the probability that a base will not be sequenced (P) using this equation:

$$P = e^{-m}$$

where e is the base of the natural logarithm ($e = 2.72$), and m is the number of sequenced bases divided by the total genome size. For example, in the case of *H. influenzae*, with a genome size of 1.8 Mb, if researchers sequenced 9.0 Mb, $m = 5$ (that is, 9.0 Mb divided by 1.8 Mb):

$$P = e^{-m} = e^{-5} = 0.0067, \text{ or } 0.67\%.$$

This means that if we randomly sequence 9.0 Mb, which is five times the length of a single genome, we are likely to miss only 0.67% of the genome. With a genome size of 1.8 Mb, we would miss about 12,000 nucleotides out of approximately 1.8 million. Such missed sequences are typically on small DNA fragments that, as a matter of random chance, did not happen to be sequenced. The missing links in the genome can be sequenced later using mapping methods.

The general protocol conducted by Craig Venter, Hamilton Smith, and colleagues in this discovery-based investigation is described in **Figure 21.1**. This is a shotgun DNA sequencing approach. The researchers isolated chromosomal DNA from *H. influenzae* and used sound waves to break the DNA into small fragments of approximately 2,000 bp in length. These fragments were randomly inserted into vectors, allowing the DNA to be propagated in *E. coli*. Each *E. coli* clone carried a vector with a different piece of DNA from *H. influenzae*. The complete

Figure 21.1 Determination of the complete genome sequence of *Haemophilus influenzae* by Venter, Smith, and colleagues.

GOAL The goal is to obtain the entire genome sequence of *Haemophilus influenzae*. This information will reveal its genome size and also which genes the organism has.

KEY MATERIALS A strain of *H. influenzae*.

Experimental level   Conceptual level

1 Purify DNA from a strain of *H. influenzae*. This involves breaking the cells open by adding phenol and chloroform. Most protein and lipid components go into the phenol-chloroform phase. DNA remains in the aqueous (water) phase.

DNA in aqueous (water) phase
Proteins and lipids in phenol-chloroform phase

H. influenzae chromosomal DNA

2 Sonicate the DNA to break it into small fragments of about 2,000 bp in length.

Sound waves

DNA fragments in aqueous phase

Sound waves

3 Clone the DNA fragments into vectors. The procedures for cloning are described in Chapter 20. This produces a DNA library.

Refer back to Figures 20.2 and 20.3.

Vector DNA

Piece of *H. influenzae* DNA

A DNA library

4 Subject many clones to the procedure of dideoxy DNA sequencing, also described in Chapter 20. A total of 10.8 Mb was sequenced.

Refer back to Figure 20.8.

Produces a large number of overlapping sequences.

5 Use tools of bioinformatics, described in the last section of this chapter, to identify various types of genes in the genome.

Explores the genome sequence and identifies and characterizes genes.

6 THE DATA

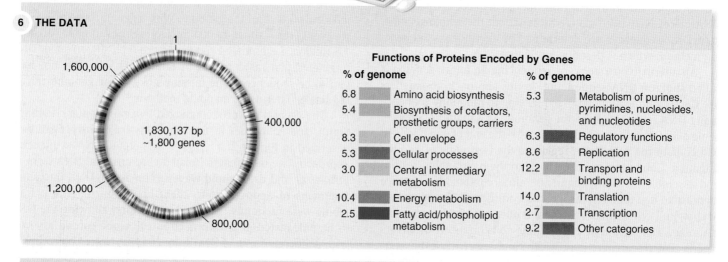

1

1,600,000

400,000

1,830,137 bp
~1,800 genes

1,200,000

800,000

Functions of Proteins Encoded by Genes

% of genome			% of genome	
6.8		Amino acid biosynthesis	5.3	Metabolism of purines, pyrimidines, nucleosides, and nucleotides
5.4		Biosynthesis of cofactors, prosthetic groups, carriers		
8.3		Cell envelope	6.3	Regulatory functions
5.3		Cellular processes	8.6	Replication
3.0		Central intermediary metabolism	12.2	Transport and binding proteins
10.4		Energy metabolism	14.0	Translation
2.5		Fatty acid/phospholipid metabolism	2.7	Transcription
			9.2	Other categories

7 CONCLUSION *H. influenzae* has a genome size of 1.83 Mb with approximately 1,800 genes. The functions of many of those genes could be inferred by comparing them to genes in other species.

8 SOURCE Fleischmann et al. 1995. Whole-genome random sequencing and assembly of *Haemophilus influenzae* Rd. *Science* 269:496–512.

set of vectors, each containing a different fragment of DNA, is called a **DNA library** (refer back to Figure 20.4). The researchers then subjected many of these clones to the procedure of DNA sequencing. They sequenced a total of approximately 10.8 Mb of DNA.

The outcome of this genome-sequencing project was a very long DNA sequence. In 1995, Venter, Smith, and colleagues published the entire DNA sequence of *H. influenzae*. The researchers then analyzed the genome sequence using a computer to obtain information about the properties of the genome. Questions they asked included, How many genes does the genome contain, and what are the likely functions of those genes? Later in this chapter, we will learn how scientists can answer such questions with the use of computers. The data in Figure 21.1 summarize the results that the researchers obtained.

The *H. influenzae* genome is composed of 1,830,137 bp of DNA. The computer analysis predicted 1,743 genes. Based on their similarities to known genes in other species, the researchers also predicted the functions of nearly two-thirds of these genes. The diagram shown in the data of Figure 21.1 places genes in various categories based on their predicted function. These results gave the first comprehensive "genome picture" of a living organism!

Experimental Questions

1. What was the goal of the experiment conducted by Venter, Smith, and their colleagues?

2. How does shotgun DNA sequencing differ from procedures that involve mapping? What is an advantage and a disadvantage of the shotgun DNA sequencing approach?

3. What were the results of the study described in Figure 21.1?

21.2 Eukaryotic Genomes

Thus far, we have examined the size and composition of bacterial and archaeal genomes. In this section, we turn to eukaryotes, which include protists, fungi, animals, and plants. As you will learn, their genomes are larger and more complex than their prokaryotic counterparts. In addition to genes, eukaryotic genomes often have abundant amounts of noncoding sequences. For example, eukaryotic genomes typically have a substantial amount of short repeated sequences called repetitive DNA. We will learn how certain types of repetitive DNA sequences are formed by a process called transposition. We will also examine how the duplication of genes can lead to families of related genes.

The Nuclear Genomes of Eukaryotes Are Sets of Linear Chromosomes That Vary Greatly in Size and Composition Among Different Species

As discussed in Chapter 15, the genome found in the nucleus of eukaryotic species is usually found in sets of linear chromosomes. In humans, for example, one set contains 23 linear chromosomes—22 autosomes and one sex chromosome, X or Y. In addition, certain organelles in eukaryotic cells contain a small amount of their own DNA. These include the mitochondrion, which plays a role in ATP synthesis, and the chloroplast found in plants and algae, which carries out photosynthesis. The genetic material found in these organelles is referred to as the mitochondrial or the chloroplast genome to distinguish it from the nuclear genome, which is found in the cell nucleus. In this chapter, we will focus on the nuclear genome of eukaryotes.

Nuclear Genomes   In the past decade or so, the DNA sequence of entire nuclear genomes has been determined for nearly 100 eukaryotic species, including more than two dozen mammalian genomes. Examples are shown in **Table 21.2**.

Motivation to sequence these genomes comes from four main sources. First, the availability of genome sequences makes it easier for researchers to identify and characterize the genes of model organisms. This has been the impetus for genome projects involving baker's yeast (*Saccharomyces cerevisiae*), the fruit fly (*Drosophila melanogaster*), a nematode worm (*Caenorhabditis elegans*), the simple plant called thale cress (*Arabidopsis thaliana*), and the mouse (*Mus musculus*). A second reason for genome sequencing is to gather more information to identify and treat human diseases, which is an important aim for sequencing the human genome. Researchers hope that knowing the DNA sequence of the human genome will help to identify genes in which mutation plays a role in disease. Third, by sequencing the genomes of agriculturally important species, new strains of livestock and plant species with improved traits can be developed. Fourth, biologists are increasingly relying on genome sequences as a way to establish evolutionary relationships.

Genome Sizes and Repetitive Sequences   Eukaryotic genomes are generally larger than prokaryotic genomes in terms of both

Table 21.2	Examples of Eukaryotic Nuclear Genomes That Have Been Sequenced		
Species	**Nuclear genome size (Mb)**	**Number of genes**	**Description**
Saccharomyces cerevisiae (baker's yeast)	12.1	6,294	One of the simplest eukaryotic species; it has been extensively studied by researchers to understand eukaryotic cell biology and other molecular mechanisms.
Caenorhabditis elegans (nematode worm)	100	~19,000	A model organism used to study animal development.
Drosophila melanogaster (fruit fly)	180	~14,000	A model organism used to study many genetic phenomena, including development.
Arabidopsis thaliana (thale cress)	120	~26,000	A model organism studied by plant biologists.
Oryza sativa (rice)	440	~40,000	A cereal grain with a relatively small genome; it is very important worldwide as a food crop.
Mus musculus (Mouse)	2,500	~20,000–25,000	A model mammalian organism used to study genetics, cell biology, and development.
Homo sapiens (humans)	3,200	~20,000–25,000	Our own genome; the sequencing of the human genome will help to elucidate our understanding of inherited traits and may aid in the identification and treatment of diseases.

Note: The genome size refers to the number of megabase pairs in one set of chromosomes. For species with sex chromosomes, it would include both sex chromosomes.

the number of genes and genome size. The genomes of simpler eukaryotes, such as yeast, carry several thousand different genes, whereas the genomes of more complex eukaryotes contain tens of thousands of genes (Table 21.2). Note that the number of genes is not the same as genome size. When we speak of genome size, this means the total amount of DNA, often measured in megabase pairs. The relative sizes of nuclear genomes vary dramatically among different eukaryotic species (Figure 21.2a). In general, increases in the amount of DNA are correlated with increases in cell size, cell complexity, and body complexity. For example, yeast have smaller genomes than animals. However, major variations in genome sizes are observed among organisms that are similar in form and function. For example, the total amount of DNA found within different species of amphibians can vary over 100-fold.

The DNA content of closely related species can also vary. As an example, let's consider two closely related species of the plant called the globe thistle, *Echinops bannaticus* and *Echinops nanus* (Figure 21.2b,c). These species have similar numbers of chromosomes, but *E. bannaticus* has nearly double the amount of DNA as *E. nanus*. What is the explanation for the larger genome of *E. bannaticus*? The genome of *E. bannaticus* is not likely to contain

(b) Echinops bannaticus

(c) Echinops nanus

(a) Genome size

Figure 21.2 Genome sizes among selected groups of eukaryotes. (a) Genome sizes among various groups of eukaryotes are shown on a log scale. As an example for comparison, two closely related species of globe thistle are pictured. These species have similar characteristics, but *Echinops bannaticus* (b) has nearly double the amount of DNA as *E. nanus* (c) due to the accumulation of repetitive DNA sequences.

Concept check: *What are two reasons why the groups of species shown in (a) have variation in their total amount of DNA?*

twice as many genes. Rather, its genome composition includes many **repetitive sequences**, which are short DNA sequences that are present in many copies throughout the genome. Repetitive sequences are often abundant in eukaryotic species.

Types of Repetitive Sequences The repetitive sequences that are found in genomes fall into two broad categories, moderately and highly repetitive. Sequences found in a few hundred to several thousand copies are called **moderately repetitive sequences**. In some cases, these sequences are multiple copies of the same gene. For example, the genes that encode ribosomal RNA (rRNA) are found in many copies. The cell needs a large amount of rRNA for its cellular ribosomes. This is accomplished by having and expressing multiple copies of the genes that encode rRNA. In addition, other types of functionally important sequences can be moderately repetitive. For example, multiple copies of origins of replication are found in eukaryotic chromosomes. Other moderately repetitive sequences may play a role in the regulation of gene transcription and translation.

Highly repetitive sequences are those that are found in tens of thousands to millions of copies throughout the genome. Each copy of a highly repetitive sequence is relatively short, ranging from a few nucleotides to several hundred nucleotides in length. Most of these sequences have no known function, and whether they benefit the organism is a matter of debate. A widely studied example is the *Alu* family of sequences found in humans and other primates. The *Alu* sequence is approximately 300 bp long. This sequence derives its name from the observation that it contains a site for cleavage by a restriction enzyme known as *Alu*I. It represents about 10% of the total human

DNA and occurs (on average) approximately every 5,000–6,000 bases. Evolutionary studies suggest that the *Alu* sequence arose 65 million years ago from a section of a single ancestral gene known as the 7SL RNA gene. Remarkably, over the course of 65 million years, the *Alu* sequence has been copied and inserted into the human genome so often that it now appears more than 1 million times! The mechanism for the proliferation of *Alu* sequences will be described later.

Some highly repetitive sequences, like the *Alu* family, are interspersed throughout the genome. However, other highly repetitive sequences are clustered together in a tandem array in which a very short nucleotide sequence is repeated many times in a row. In *Drosophila*, for example, 19% of the chromosomal DNA is highly repetitive DNA found in tandem arrays. An example is shown here:

```
AATATAATATAATATAATATAATATAT
TTATATTATATTATATTATATTATATA
```

In this particular tandem array, two related sequences, AATAT and AATATAT, are repeated many times. Highly repetitive sequences, which contain tandem arrays of short sequences, can be quite long, sometimes more than 1 million bp in length!

Figure 21.3 shows the composition of the relative classes of DNA sequences that are found in the nuclear genome of humans. Surprisingly, exons, the coding regions of structural genes, and the genes that give rise to rRNA and tRNA make up only about 2% of our genome! The other 98% is composed of noncoding sequences. Though we often think of genomes as being the repository of sequences that code for proteins, most eukaryotic genomes are largely composed of other types of sequences. Intron DNA comprises the second most common portion of the genome at 24%. Unique noncoding DNA, whose function is largely unknown, constitutes 15%. Repetitive DNA makes up

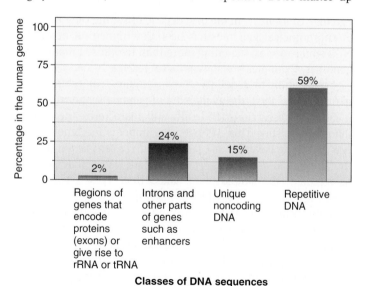

Classes of DNA sequences

Figure 21.3 The composition of DNA sequences that are found in the nuclear genome of humans. Only about 2% of our genome codes for proteins. Most of our genome is made up of repetitive sequences.

Concept check: *Do these results seem surprising to you?*

59% of the DNA in the genome. Much of the repetitive DNA is derived from transposable elements, stretches of DNA that can move from one location to another, which are described next.

Transposable Elements Can Move from One Chromosomal Location to Another

During a process called **transposition**, a short segment of DNA moves from its original site to a new site in the genome. Such segments are known as **transposable elements (TEs)**. They range from a few hundred to several thousand bp in length. TEs have sometimes been referred to as "jumping genes," because they are inherently mobile. Barbara McClintock first identified transposable elements in the late 1940s from her studies with corn plants (**Figure 21.4**). She identified a segment of DNA that could move into and out of a gene that affected the color of corn kernels, producing a speckled appearance. Since that time, biologists have discovered many different types of TEs in nearly all species examined.

Though McClintock identified TEs in corn in the late 1940s, her work was met with great skepticism because many researchers had trouble believing that DNA segments could be mobile. The advent of molecular technology in the 1960s and 1970s allowed scientists to understand more about the characteristics of TEs that enable their movement. Most notably, research involving bacterial TEs eventually progressed to a molecular understanding of the transposition process. In 1983, over 30 years after her initial discovery, McClintock was awarded the Nobel Prize in Physiology or Medicine.

DNA Transposons Researchers have studied TEs from many species, including prokaryotes and eukaryotes. They have found that DNA sequences within transposable elements are organized in several different ways, and they can move by different molecular mechanisms. Let's first begin with transposable elements in which the DNA itself moves from one location

to another. These are also called **DNA transposons** because they move via a DNA molecule. Both ends of DNA transposons usually have inverted repeats (IRs)—DNA sequences that are identical (or very similar) but run in opposite directions (**Figure 21.5a**), such as the following:

5′—CTGACTCTT—3′	5′—AAGAGTCAG—3′
3′—GACTGAGAA—5′	3′—TTCTCAGTC—5′

Depending on the particular transposable element, inverted repeats range from 9 to 40 base pairs in length. In addition, DNA transposons may contain a central region that encodes **transposase**, an enzyme that facilitates transposition.

(a) Organization of a simple transposable element

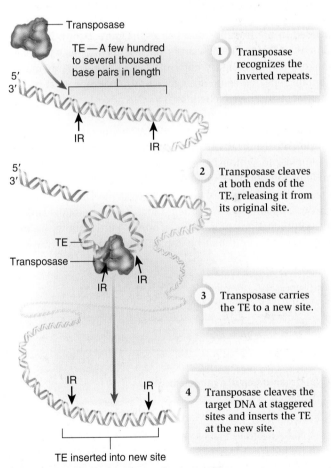

(b) Cut-and-paste mechanism of transposition

Figure 21.5 **DNA transposons and their mechanism of transposition.** (a) DNA transposons contain inverted repeat (IR) sequences at each end and a gene that encodes transposase in the middle. (b) Transposition occurs by a cut-and-paste mechanism.

Concept check: *What is the role of the inverted repeats in the mechanism of transposition?*

(a) Barbara McClintock

(b) Speckled corn kernels caused by transposable elements

Figure 21.4 **Barbara McClintock, who discovered transposable elements.** As shown in part **(b)**, when a transposable element is found within a pigment gene in corn, its frequent movement disrupts the gene, causing the kernel color to be speckled.

As shown in **Figure 21.5b**, transposition of DNA transposons occurs by a cut-and-paste mechanism. Transposase first recognizes the inverted repeats in the TE and then removes the TE from its original site. Next, the transposase/TE complex moves to a new location, where transposase cleaves the target DNA and inserts the TE into the site. Transposition may occur when a cell is in the process of DNA replication. If a transposable element is removed from a site that has already replicated and is inserted into a chromosomal site that has not yet replicated, the TE will increase in number after DNA replication is complete. This is one way for TEs to become more prevalent in a genome.

Retroelements   Another category of transposable elements moves via an RNA intermediate. This form of transposition is very common but is found only in eukaryotic species. These types of elements are known as **retroelements** or retrotransposons. The *Alu* sequence in the human genome is an example of a retroelement. Some retroelements contain genes that encode the enzymes reverse transcriptase and integrase, which are needed in the transposition process (**Figure 21.6a**). Recall from Chapter 18 that reverse transcriptase uses RNA as a template to make a complementary copy of DNA. Retroelements may also contain repeated sequences called terminal repeats at each end that facilitate their recognition.

The mechanism of retroelement movement is shown in **Figure 21.6b**. First, the enzyme RNA polymerase transcribes the retroelement into RNA. Reverse transcriptase uses this RNA as a template to synthesize a double-stranded DNA molecule. The ends of the double-stranded DNA are then recognized by integrase, which catalyzes the insertion of the DNA into the host chromosomal DNA. The integration of retroelements can occur at many locations within the genome. Furthermore, because a single retroelement can be copied into many RNA transcripts, retroelements may accumulate rapidly within a genome. This explains how the *Alu* element in the human genome was able to proliferate and constitute 10% of our genome.

Role of Transposable Elements   What is the biological significance of TEs? The question is not resolved. According to the **selfish DNA hypothesis**, TEs exist solely because they have characteristics that allow them to insert themselves into the host cell DNA. In other words, they resemble parasites in the sense that they inhabit the host without offering any advantage. They can proliferate within the host as long as they do not harm the host to the extent that they significantly disrupt survival. However, TEs can do harm. For example, if they jump into the middle of an important gene and thereby disrupt its function, this may have a negative impact on the phenotype of an organism.

Other biologists have argued that TEs may provide benefits to a given species. For example, bacterial TEs often carry an antibiotic-resistance gene that provides the organism with a survival advantage. In addition, TEs may cause greater genetic variability by promoting chromosomal rearrangements. As discussed next, such rearrangements can cause a misaligned crossover during meiosis and promote the formation of a gene family.

Gene Duplications Provide Additional Material for Genome Evolution, Sometimes Leading to the Formation of Gene Families

Let's now turn our attention to a way that the number of genes in a genome can increase. These gene duplications are important because they provide raw material for the addition of more genes into a species' genome. Such duplications can produce **homologous genes**, two or more genes that are derived from the same ancestral gene (**Figure 21.7a**). Over the course of

(a) Organization of a retroelement

(b) Mechanism of movement of a retroelement

Figure 21.6   Retroelements and their mechanism of transposition.   Retroelements are found only in eukaryotic species. **(a)** Some retroelements contain terminal repeats and genes that encode the enzymes reverse transcriptase and integrase, which are needed in the transposition process. **(b)** The process that adds a copy of a retroelement into a host chromosome.

Concept check: *Based on their mechanism of movement, which type of TEs do you think would proliferate more rapidly in a genome, DNA transposons (see Figure 21.5b) or retroelements?*

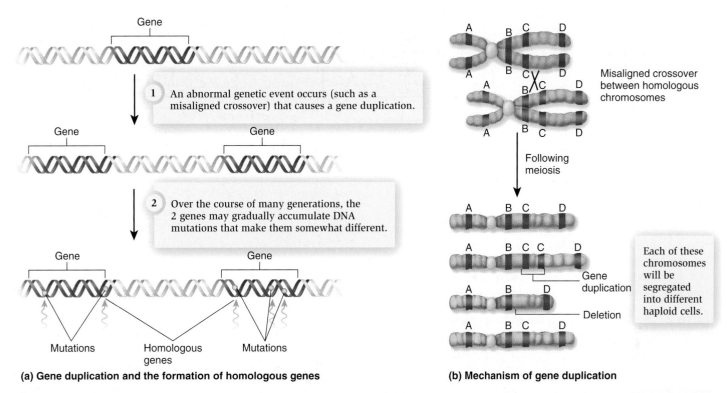

(a) Gene duplication and the formation of homologous genes

(b) Mechanism of gene duplication

Figure 21.7   Gene duplication and the evolution of homologous genes.   (a) A gene duplication produces two copies of the same gene. Over time, these copies accumulate different random mutations, which results in homologous genes with similar but not identical DNA sequences. (b) Mechanism of gene duplication. If two homologous chromosomes misalign during meiosis, a crossover will produce a chromosome with a gene duplication.

many generations, each version of the gene accumulates different mutations, resulting in genes with similar but not identical DNA sequences.

How do gene duplications occur? One mechanism that produces gene duplications is a misaligned crossover (**Figure 21.7b**). In this example, two homologous chromosomes have paired with each other during meiosis, but the homologues are misaligned. If a crossover occurs, this produces one chromosome with a gene duplication, one with a deletion, and two normal chromosomes. Each of these chromosomes will be segregated into different haploid cells. If a haploid cell carrying the chromosome with the gene duplication participates in fertilization with another gamete, an offspring with a gene duplication is produced. In this way, gene duplications can form and be transmitted to future generations. The presence of multiple copies of the same transposable element in a genome can foster this process because the chromosomes may misalign while attempting to align TEs that are at different locations in the same chromosome.

During evolution, gene duplications can occur several times. Two or more homologous genes within a single species are also called **paralogs** or paralogous genes. Multiple gene duplication followed by sequence divergence can result in a **gene family** composed of two or more paralogous genes that carry out related functions. A well-studied example is the globin gene family found in animals. The globin genes encode polypeptides that are subunits of proteins that function in oxygen

binding. Hemoglobin, which is made in red blood cells, carries oxygen throughout the body. In humans, the globin gene family is composed of 14 paralogs that were originally derived from a single ancestral globin gene (**Figure 21.8**). According to an evolutionary analysis, the ancestral globin gene duplicated about 500 million years ago. Since that time, additional duplication events and chromosomal rearrangements have occurred to produce the current number of 14 genes on three different human chromosomes. Four of these are pseudogenes, which are genes that have been produced by gene duplication but have accumulated mutations that make them nonfunctional, so they are not transcribed into RNA.

Gene families have been important in the evolution of complex traits. Even though all of the globin polypeptides are subunits of proteins that play a role in oxygen binding, the accumulation of different mutations in the various family members has produced globins that are more specialized in their function. For example, myoglobin binds and stores oxygen in muscle cells, whereas the hemoglobins bind and transport oxygen via red blood cells. Also, different globin genes are expressed during different stages of human development. The zeta (ζ)-globin and epsilon (ε)-globin genes are expressed very early in embryonic life. During the second trimester of gestation, the alpha (α)-globin and gamma (γ)-globin genes are turned on. Following birth, the γ-globin genes are turned off, and the β-globin gene is turned on. These differences in the expression of the globin genes reflect the differences in the

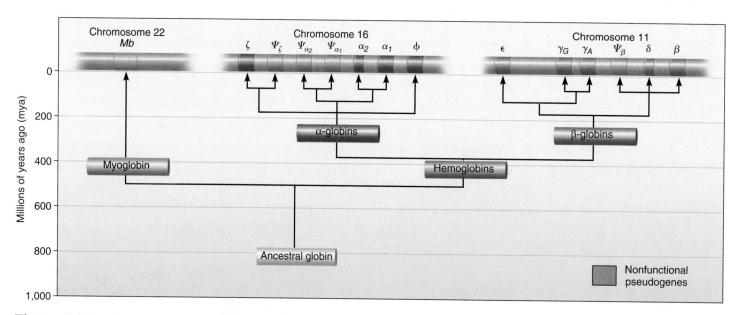

Figure 21.8 **The evolution of the globin gene family in humans.** The globin gene family evolved from a single ancestral globin gene. The first gene duplication produced two genes that accumulated mutations and became the genes encoding myoglobin and the various hemoglobins. The modern myoglobin gene is found on chromosome 22. An ancestral hemoglobin gene duplicated to produce the α- and β-globins. Further duplications of ancestral α-globin and β-globin genes produced several paralogous genes on chromosomes 16 and 11, respectively. The four genes shown in gray are nonfunctional pseudogenes.

Concept check: *What is the biological advantage of a gene family?*

oxygen transport needs of humans during the embryonic, fetal, and postpartum stages of life (refer back to Figure 13.3).

The Human Genome Project Has Stimulated Genomic Research

Before ending our discussion of genomes, let's consider the **Human Genome Project**, a research effort to identify and map all human genes. Scientists had been discussing how to undertake this project since the mid-1980s. In 1988, the National Institutes of Health in Bethesda, Maryland, established an Office of Human Genome Research with James Watson as its first director. The Human Genome Project officially began on October 1, 1990, and was largely finished by the end of 2003. It was an international consortium that included research institutions in the U.S., U.K., France, Germany, Japan, and China. From its outset, the Human Genome Project had the following goals:

1. *To identify all human genes.* This involved mapping the locations of genes throughout the entire genome. The data from the Human Genome Project suggest that humans have about 20,000 to 25,000 different genes.
2. *To obtain the DNA sequence of the entire human genome.* The first draft of a nearly completed DNA sequence was published in February 2001, and a second draft was published in 2003. The entire genome is approximately 3.2 billion base pairs in length.
3. *To develop technology for the generation and management of human genome information.* Some of the efforts of the Human Genome Project have involved improvements in

molecular genetic technology, such as gene cloning, DNA sequencing, and so forth. The Human Genome Project has also developed computer tools to allow scientists to easily access up-to-date information from the project and analytical tools to interpret genomic information.

4. *To analyze the genomes of model organisms.* These include *E. coli, S. cerevisiae, D. melanogaster, C. elegans, A. thaliana,* and *M. musculus.*
5. *To develop programs focused on understanding and addressing the ethical, legal, and social implications of the results obtained from the Human Genome Project.* Who should have access to genetic information? Should employers, insurance companies, law enforcement agencies, and schools have access to our genetic makeup? The answers to these questions are complex and will require discussion among many groups. Another controversial topic is gene patenting. In the U.S., genes can be patented for a variety of reasons. For example, the patenting of genes has been associated with the commercial development of diagnostic tests for genetic diseases. Some argue that patenting fosters greater investment into research and development; others say it can impede basic research and scientific innovation.

Some current and potential applications of the Human Genome Project include the improved diagnosis and treatment of genetic diseases such as cystic fibrosis, Huntington disease, and Duchenne muscular dystrophy. The project may also enable researchers to identify the genetic basis of common disorders such as cancer, diabetes, and heart disease, which involve alterations in several genes.

21.3 Proteomes

Thus far in this chapter, we have considered the genome characteristics of many different species, including humans. Because most genes encode proteins, a logical next step is to examine the functional roles of the proteins that a species can make. As mentioned, the entire collection of proteins that a cell or organism produces is called a proteome. As we move through the 21st century, a key challenge facing molecular biologists is the study of proteomes. Much like the study of genomes, this will require the collective contributions of many scientists, as well as improvements in technologies to investigate the complexities of the proteome. In this section, we will begin by considering the functional categories of proteins, and then examine their relative abundance in the proteome. We also will explore the molecular mechanisms that cause an organism's proteome to be much larger than its genome.

The Proteome Is a Diverse Array of Proteins with Many Kinds of Functions

The genomes of simple, unicellular organisms such as bacteria and yeast contain thousands of structural genes, whereas the genomes of complex, multicellular organisms contain tens of thousands. Such genome sizes can produce proteomes with tens of thousands to hundreds of thousands of different proteins. To bring some order to this large amount of complex information, researchers often organize proteins into different categories based on their functions. **Table 21.3** describes some general categories of protein function and provides examples of each type. Many approaches are used to categorize proteins. Table 21.3 shows just one of the more general ways to categorize protein function. For example, the data of Figure 21.1 describe the functions of proteins encoded by genes in a different, more detailed way.

The relative abundance of proteins can be viewed at two levels. First, we can consider abundance in the genome—the numbers of genes in the genome that encode a particular type or category of protein. For example, if an entire genome encodes 10,000 different types of proteins and 1,500 of these are different types of transporters, we would say that 15% of the genome is composed of transporters. However, such an analysis ignores the phenomenon that genes are expressed at different levels. In other words, various proteins are made in different amounts. Therefore, a second way to view protein abundance is to consider abundance in the cell—the amount of a given protein or protein category that is actually made by a living cell. For example, less than 1% of human genes encode proteins, such as collagen, that are found in the extracellular matrix. Even so, these genes are highly expressed in certain cells, so a large amount of this type of protein is made compared to other types.

Figure 21.9 is a general comparison of protein abundance in two cell types in humans, liver and muscle cells. Liver cells play a key role in metabolism, whereas muscle cells are involved in bodily movements. Both liver and muscle cells have the same

| **Table 21.3** | Categories of Proteins Found in the Proteome |

Function	Examples
Metabolic enzymes—accelerate chemical reactions in the cell.	Hexokinase: phosphorylates glucose during the first step in glycolysis. Glycogen synthetase: uses glucose to synthesize a large carbohydrate known as glycogen.
Structural proteins—provide shape and protection to cells.	Tubulin: forms cytoskeletal structures known as microtubules. Collagen: found abundantly in the extracellular matrix of animals.
Motor proteins—facilitate intracellular movements and the movements of whole cells.	Myosin: involved in muscle cell contraction. Kinesin: involved in the movement of chromosomes during cell division.
Cell-signaling proteins—allow cells to respond to environmental signals and send signals to each other.	Insulin: influences target cell metabolism and growth. Insulin receptor: recognizes insulin and initiates a cellular response.
Transport proteins—involved in the transport of ions and molecules across membranes and throughout the body.	Lactose permease: transports lactose across the bacterial cell membrane. Hemoglobin: found in red blood cells and transports oxygen throughout the body.
Gene expression and regulatory proteins—involved in transcription, mRNA modification, translation, and gene regulation.	Transcription factors: regulate the expression of genes. Ribosomal proteins: components of ribosomes, which are needed for the synthesis of new proteins.
Protective proteins—help cells and organisms to survive environmental stress.	Antibodies: fight viral and bacterial infections in vertebrate species. Heat shock proteins (chaperones): play a role in protein folding and thereby help cells cope with abrupt increases in temperature.

genes. Therefore, at the level of the genome, the percentages of the different protein categories are identical. However, at the cellular level, the relative abundance of certain protein categories is quite different. Liver cells make a large number of different enzymes that play a role in the metabolism of fats, proteins, and carbohydrates. By comparison, their level of structural and motor proteins is relatively small. In contrast, muscle cells

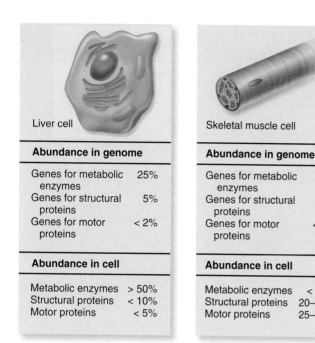

Liver cell		Skeletal muscle cell	
Abundance in genome		**Abundance in genome**	
Genes for metabolic enzymes	25%	Genes for metabolic enzymes	25%
Genes for structural proteins	5%	Genes for structural proteins	5%
Genes for motor proteins	< 2%	Genes for motor proteins	< 2%
Abundance in cell		**Abundance in cell**	
Metabolic enzymes	> 50%	Metabolic enzymes	< 10%
Structural proteins	< 10%	Structural proteins	20–30%
Motor proteins	< 5%	Motor proteins	25–40%

Figure 21.9   A comparison of the proteomes in human liver and skeletal muscle cells.   Because all cells of the human body carry the same genome, the percentages of proteins that are encoded in the genome are the same in each cell type. However, the relative amounts of proteins that are made in different cell types can be vastly different, as is the case between liver and skeletal muscle cells.

Concept check: *What genetic process explains the differences in protein abundance in liver cells versus muscle cells?*

have fairly low levels of enzymes but make a high percentage of structural and motor proteins. These differences in protein composition between liver and muscle cells are largely due to differential gene regulation.

The Number of Different Proteins in a Species' Proteome Is Larger Than the Number of Genes in Its Genome

From the sequencing and analysis of genomes, researchers can identify all or nearly all of the genes that a given species has. For example, the human genome is predicted to contain between 20,000 and 25,000 different genes that encode proteins. Even so, humans can make many more than 25,000 different types of proteins. How is this possible? The larger size of the proteome relative to the genome is primarily due to two types of cellular processes, as described next.

Alternative Splicing   Changes in pre-mRNA structure may ultimately affect the resulting amino acid sequence of a protein. The most important alteration that commonly occurs in eukaryotic species is **alternative splicing**, which is also described in Chapter 13. For many genes, a single pre-mRNA can be spliced in more than one way, resulting in the creation of two or more different proteins (**Figure 21.10a**). The splicing is often cell specific or may be related to environmental conditions. Alternative

(a) Alternative splicing

(b) Post-translational covalent modification

Figure 21.10   Cellular mechanisms that increase protein diversity.   (a) Following alternative splicing, the pattern of exons in the resulting mature mRNA can be different, creating multiple types of transcripts from the same gene. (b) In post-translational covalent modification, after a protein is made, it can be modified in a variety of ways, some of which are permanent and some reversible.

Concept check: *Think back to the Cell unit. What is the advantage of reversible post-translational covalent modifications?*

splicing is widespread, particularly among more complex eukaryotes (refer back to Table 13.1). It can lead to the production of several or perhaps dozens of different polypeptide sequences from the same pre-mRNA. This greatly increases the number of proteins in a species' proteome, while minimizing the size of the genome.

Post-translational Covalent Modification   A second process that greatly diversifies the composition of a proteome is the phenomenon of **post-translational covalent modification**, the modification of the structure of a protein after its translation (**Figure 21.10b**). Such modifications can be permanent or reversible. Permanent modifications are often involved with the assembly and construction of functional proteins. These alterations include proteolytic processing (the cleavage of a polypeptide to a smaller unit), disulfide bond formation, and the attachment of prosthetic groups, sugars, or lipids. In Chapter 6, we also considered the attachment of ubiquitin to a protein, which targets it for degradation. By comparison, reversible modifications, such as phosphorylation, acetylation, and methylation, often transiently affect the function of a protein (Figure 21.10b). Molecules are covalently attached and later removed by cellular enzymes. Because a given type of protein may be subjected to several different types of modifications, this can greatly increase the forms of a particular protein that are found in a cell at any given time.

21.4   Bioinformatics

In the previous sections, we have learned that the number of genes in a genome and the number of proteins that are made by a given cell type are extremely large. In the 1960s and 1970s, when the tools of molecular biology first became available, researchers tended to focus on the study of just one or a few genes and proteins at a time. While this is a useful approach, scientists came to realize that certain properties of life arise by complex interactions involving the expression of many genes and the functioning of many different proteins. Such an awareness challenged researchers to invent new tools to study many genes and proteins at the same time. These tools involved experimental procedures that allowed researchers to simultaneously investigate the various parts of a biological system. Studying such complex interactions is called **systems biology**. To analyze and manage the huge amounts of data produced by these studies, researchers turned to the use of computers.

As a very general definition, **bioinformatics** describes the use of computers, mathematical tools, and statistical techniques to record, store, and analyze biological information. We often think of bioinformatics in the context of examining genetic data, such as DNA sequences. Even so, bioinformatics can also be applied to information from various sources, such as patient statistics and scientific literature. This rapidly developing branch of biology is highly interdisciplinary, incorporating principles from mathematics, statistics, information science, chemistry, and physics.

Why do we need bioinformatics? Simply put, the main issues are size and speed. Earlier in this chapter, we learned

that the human genome has been sequenced and that it is approximately 3.2 billion base pairs long. A single person, or even a group of talented mathematicians, cannot, in a reasonable length of time, analyze such an enormous amount of data. Instead, the data are put into computers, and then scientists devise computational procedures to study and evaluate it.

In this section, we will consider the branch of bioinformatics that focuses on using molecular information to study biology. This area, also called **computational molecular biology**, uses computers to characterize the molecular components of living things. Molecular genetic data, which comes in the form of DNA, RNA, or protein sequences, are particularly amenable to computer analysis. In this section, we will first survey the fundamental concepts that underlie the analysis of genetic sequences. We will then consider how these methods are used to provide knowledge about how biology works at the molecular level.

Sequence Files Are Stored and Analyzed by Computers

The first steps in bioinformatics are to collect and store data in a computer. As an example, let's consider a gene sequence as a type of data. The gene sequence must first be determined experimentally using the technique of DNA sequencing. After the sequence is obtained, the next step is to put that data into a file on a computer. Typically, genetic sequence data are entered into a computer file by laboratory instruments that can read experimental data—such as data from a DNA-sequencing experiment—and enter the sequence directly into a computer.

Genetic sequence data in a computer data file can then be investigated in many different ways, corresponding to the many questions a researcher might ask about the sequence and its functional significance, including the following:

1. Does a sequence contain a gene?
2. Does a gene sequence contain a mutation that might cause a disease?
3. Where are functional sequences, such as promoters, regulatory sites, and splice sites, located within a particular gene?
4. From the sequence of a structural gene, what is the amino acid sequence of the polypeptide encoded by that gene?
5. Is there an evolutionary relationship between two or more genetic sequences?

To answer these and many other questions, computer programs have been written to analyze genetic sequences in particular ways. As an example, let's consider a computer program aimed at translating a DNA sequence into an amino acid sequence. **Figure 21.11** shows a short computer data file of a DNA sequence that is presumed to be part of the coding sequence of a structural gene. In this figure, only the coding strand of DNA is shown. A computer program can analyze this sequence and print out the possible amino acid sequences that this DNA sequence would encode. The program relies on the genetic code (refer back to Table 12.1). In the example shown in Figure 21.11, the computer program shows the results for all three possible reading frames, beginning at nucleotide 1, 2, or 3, respectively. In a newly obtained DNA sequence, a researcher would not know the proper reading frame—the series of codons read in groups of three, which starts with the start codon. Therefore, the computer program provides all three. If you look at the results, reading frames 1 and 2 include several stop codons, whereas reading frame 3 does not. From these results, reading frame 3 is likely to be the correct one. Also, for a new DNA sequence, a researcher may not know which DNA strand is the coding strand. Therefore, the sequence of the other DNA strand, which is not shown in this figure, would also be analyzed by this computer program.

The Scientific Community Has Collected Computer Data Files and Stored Them in Large Computer Databases

Over the past several decades, the amount of genetic information generated by researchers and clinicians has become enormous. The Human Genome Project alone has produced more data than any other undertaking in the history of biology. With these advances, scientists have realized that another critical use of computers is to store the staggering amount of data produced from genetic research.

When a large amount of data is collected, stored in a single location, and organized for rapid search and retrieval, this collection is called a **database**. The files within databases are often annotated, which means they contain a concise description of each gene sequence, the name of the organism from which the sequence was obtained, and the function of the encoded protein, if it is known. The file may also provide a published reference that contains the sequence.

The research community has collected genetic information from thousands of research laboratories and created several large databases. **Table 21.4** describes examples of the major genetic databases in use worldwide, all of which can be accessed online. These databases enable researchers to access and compare genetic sequences that are obtained by many laboratories. Later in this chapter, we will learn how researchers can use databases to analyze genetic sequences.

Many programs are freely available over the Internet to utilize the information within databases. For example, the National Center for Biotechnology Information (NCBI), which is a part of the U.S. National Institutes of Health, manages a website called "Tools for Data Mining," where anyone can run various types of programs that are used to analyze genetic sequences (www.ncbi.nlm.nih.gov/Tools). Like conventional mining, in which a precious mineral is extracted from a large area of land, **data mining** is the extraction of useful information and often previously unknown relationships from sequence files and large databases.

Computer Programs Can Identify Homologous Sequences

Let's now turn our attention to genes that are evolutionarily related. Organisms that are closely related evolutionarily tend to have genes with similar DNA sequences. As an example, let's consider the gene that encodes β-globin. As discussed

Computer DNA sequence file

```
5′ GTGTCCACGC  GGTCCTGGAA  AACCCAGGCT  TGGGCAGGAA
   ACTCTCTGAC  TTTGGACAGG  AAACAAGCTA  TATTGAAGAC
   AACTGCAATC  AAAATGGTGC  CATATCACTG  ATCTTCTCAC
   TCAAAGAAGA  AGTTGGTGCA  TTGGCCAAAG  TATTGCGCTT
   ATTTGAGGAG  AATGATGTAA  ACCTGACCCA  CATTGAATCT
   AGACCTTCTC  GTTTAAAGAA  AGATGAGTAT  GAATTTTTCA
   CCCATTTGGA  TAAACGTAGC  CTGCCTGCTC  TGACAAACAT
   CATCAAGATC  TTGAGGCATG  ACATTGGTGC  CACTGTCCAT
   GAGCTTTCAC  GAGATAAGAA  GAAAGACACA  GTGCCCTGGT
   TTCCCAAG 3′
```

Run a computer program that translates this DNA sequence into an amino acid sequence in all 3 reading frames.

Possible amino acid sequences

5′ ⟶ 3′ Frame 1

Val Ser Thr Arg Ser Trp Lys Thr Gln Ala Trp Ala Gly Asn Ser Leu Thr Leu Asp Arg Lys Gln Ala Ile Leu Lys Thr Thr Ala Ile Lys Met Val Pro Tyr His **STOP** Ser Ser His Ser Lys Lys Lys Leu Val His Trp Pro Lys Tyr Cys Ala Tyr Leu Arg Arg Met Met **STOP** Thr **STOP** Pro Thr Leu Asn Leu Asp Leu Leu Val **STOP** Arg Lys Met Ser Met Asn Phe Ser Pro Ile Trp Ile Asn Val Ala Cys Leu Leu **STOP** Gln Thr Ser Ser Arg Ser **STOP** Gly Met Thr Leu Val Pro Leu Ser Met Ser Phe His Glu Ile Arg Arg Lys Thr Gln Cys Pro Gly Ser Gln

5′ ⟶ 3′ Frame 2

Cys Pro Arg Gly Pro Gly Lys Pro Arg Leu Gly Gln Glu Thr Leu **STOP** Leu Trp Thr Gly Asn Lys Leu Tyr **STOP** Arg Gln Leu Gln Ser Lys Trp Cys His Ile Thr Asp Leu Leu Thr Gln Arg Arg Ser Trp Cys Ile Gly Gln Ser Ile Ala Leu Ile **STOP** Gly Glu **STOP** Cys Lys Pro Asp Pro His **STOP** Ile **STOP** Thr Phe Ser Phe Lys Glu Arg **STOP** Val **STOP** Ile Phe His Pro Phe Gly **STOP** Thr **STOP** Pro Ala Cys Ser Asp Lys His His Gln Asp Leu Glu Ala **STOP** His Trp Cys His Cys Pro **STOP** Ala Phe Thr Arg **STOP** Glu Glu Arg His Ser Ala Leu Val Pro Lys

5′ ⟶ 3′ Frame 3

Val His Ala Val Leu Glu Asn Pro Gly Leu Gly Arg Lys Leu Ser Asp Phe Gly Gln Glu Thr Ser Tyr Ile Glu Asp Asn Cys Asn Gln Asn Gly Ala Ile Ser Leu Ile Phe Ser Leu Lys Glu Glu Val Gly Ala Leu Ala Lys Val Leu Arg Leu Phe Glu Glu Asn Asp Val Asn Leu Thr His Ile Glu Ser Arg Pro Ser Arg Leu Lys Lys Asp Glu Tyr Glu Phe Phe Thr His Leu Asp Lys Arg Ser Leu Pro Ala Leu Thr Asn Ile Ile Lys Ile Leu Arg His Asp Ile Gly Ala Thr Val His Glu Leu Ser Arg Asp Lys Lys Lys Asp Thr Val Pro Trp Phe Pro

Figure 21.11 The use of a computer program to translate a DNA sequence into an amino acid sequence. The top part of this figure shows the sequence of a segment of the coding strand of a structural gene (artificially divided into groups of 10 nucleotide bases for ease of reading). A computer program translates the DNA sequence into an amino acid sequence based on the genetic code. The program produces three different amino acid sequences, as shown at the bottom of the figure. In this example, reading frame 3 is likely to be the correct reading frame because it does not contain any stop codons.

Concept check: Why is it helpful to use a computer program to translate a genetic sequence rather than doing it by hand?

Table 21.4	Examples of Major Computer Databases
Nucleotide sequence	DNA sequence data are collected into three internationally collaborating databases: GenBank (a U.S. database), EMBL (European Molecular Biology Laboratory Nucleotide Sequence Database), and DDBJ (DNA Data Bank of Japan). These databases receive sequence and sequence annotation data from genome projects, sequencing centers, individual scientists, and patent offices. New and updated entries are exchanged daily.
Amino acid sequence	Protein sequence data are collected into a few international databases, including Swiss-Prot (Swiss protein database), PIR (Protein Information Resource), TrEMBL (translated sequences from the EMBL database), and Genpept (translated peptide sequences from the GenBank database).
Three-dimensional structure	PDB (Protein Data Bank) collects the three-dimensional structures of biological macromolecules with an emphasis on protein structure. These are primarily structures that have been determined by X-ray crystallography and nuclear magnetic resonance (NMR), but some models are included in the database.

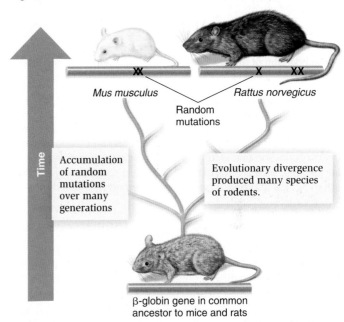

Mouse

GGGCAGGTTGGTATCCAGGTTACAAGGCAGCTCACAAGTAGAAGCTGGGTGCTTGGAGAC

GGGCAGGTTGGTATCCAGGTTACAAGGTAGCTCCTAAGTAGAAGTTTGGTGCTTGGAGAC

Rat

(a) A comparison of one DNA strand of the mouse and rat β-globin genes

Accumulation of random mutations over many generations

Evolutionary divergence produced many species of rodents.

Mus musculus   *Rattus norvegicus*

Random mutations

β-globin gene in common ancestor to mice and rats

(b) The formation of homologous β-globin genes during evolution of mice and rats

Figure 21.12   Structure and formation of the homologous β-globin genes in mice and rats.   **(a)** A comparison of a short region of the gene that encodes β-globin in laboratory mice (*Mus musculus*) and rats (*Rattus norvegicus*). Only one DNA strand is shown. Bases that are identical between the two sequences are connected by a vertical line. **(b)** The formation of these homologous β-globin genes during evolution. An ancestral β-globin gene was found in a rodent species that was a common ancestor to both mice and rats. This ancestral species later diverged into different species, which gave rise to modern rodent species, such as mice and rats. During this process, the β-globin genes accumulated different random mutations, resulting in DNA sequences that are slightly different from each other.

Concept check: *Is it possible for orthologs from two different species to have exactly the same DNA sequence? Explain.*

earlier, β-globin is a polypeptide found in hemoglobin, which carries oxygen in red blood cells. The β-globin gene is found in humans and other vertebrates.

Figure 21.12a compares a short region of this gene from the laboratory mouse (*Mus musculus*) and laboratory rat (*Rattus norvegicus*). As you can see, the gene sequences are similar but not identical. In this 60-nucleotide sequence, five differences are observed. The reason for the sequence similarity is that the genes are derived from the same ancestral gene. This idea is shown schematically in **Figure 21.12b**. An ancestral gene was found in a rodent species that was a common ancestor to both mice and rats. During evolution, this ancestral species diverged into different species, which eventually gave rise to several modern rodent species, including mice and rats. Following divergence, the β-globin genes accumulated distinct mutations that produced somewhat different base sequences for this gene. Therefore, in mice and rats, the β-globin genes have homologous sequences—sequences that are similar because they are derived from the same ancestral gene, but not identical because each species has accumulated a few different random mutations. Homologous genes in different species are also called orthologous genes or **orthologs**. Analyzing genes that are homologous to each other helps biologists understand the evolutionary relationships among modern species, a topic that we will consider in more detail in Units IV and V.

How do researchers, with the aid of computers, determine if two genes are homologous to each other? To evaluate the similarity between two sequences, a matrix can be constructed. **Figure 21.13** illustrates the use of a simplified dot matrix to evaluate two sequences. In Figure 21.13a, the word BIOLOGY is compared with itself. Each point in the grid corresponds to one position of each sequence. The matrix allows all such pairs to

be compared simultaneously. Dots are placed where the same letter occurs at the two corresponding positions. Sequences that are alike produce a diagonal line on the matrix. Figure 21.13b compares two similar but different sequences: BIOLOGY and ECOLOGY. This comparison produces only a partial diagonal line. Overall, the key observation is that regions of similarity are distinguished by the occurrence of many dots along a diagonal line within the matrix. This same concept holds true when homologous gene sequences are compared with each other.

To relate orthologous genes in different species, researchers must compare relatively long DNA sequences. For such long sequences, a dot matrix approach is not adequate. Instead, dynamic computer programming methods are used to identify

(a) Comparison of two identical words

(b) Comparison of two different words

Figure 21.13 **The use of a simple dot matrix.** In these comparisons, a diagonal line indicates sequence similarity. **(a)** The word BIOLOGY is compared with itself. Dots are placed where the same letter occurs at the two corresponding positions. Notice the diagonal line that is formed. **(b)** Two similar but different sequences, BIOLOGY and ECOLOGY, are compared with each other. Notice that only a partial line is formed by this comparison.

similarities between genetic sequences. This approach was first proposed by Saul Needleman and Christian Wunsch in 1970. Dynamic programming methods are theoretically similar to a dot matrix, but they involve mathematical operations that are beyond the scope of this textbook. In their original work, Needleman and Wunsch demonstrated that whale myoglobin and human β-globin genes have similar sequences.

A Database Can Be Searched to Identify Similar Sequences and Thereby Infer Homology and Gene Function

Why is it useful to identify homology between different genes? Because they are derived from the same ancestral gene, homologous genes usually carry out similar or identical functions. In many cases, the first way to identify the function of a newly determined gene sequence is to find a homologous gene whose function is already known. An example is the gene that is altered in cystic fibrosis patients. After this gene was identified in humans, bioinformatic methods revealed that it is homologous to several genes found in other species. A few of the homologous genes were already known to encode proteins that function in the transport of ions and small molecules across the plasma membrane. This observation provided an important clue that cystic fibrosis involves a defect in membrane transport.

The ability of computer programs to identify homology between genetic sequences provides a powerful tool for predicting the function of genetic sequences. In 1990, Stephen Altschul, David Lipman, and their colleagues developed a program called **BLAST** (for basic local alignment search tool). The BLAST program has been described by many biologists as the single most important tool in computational molecular biology. This computer program can start with a particular genetic sequence—either a nucleotide or an amino acid sequence—and then locate homologous sequences within a large database.

As an example of how the BLAST program works, let's consider the human enzyme phenylalanine hydroxylase, which functions in the metabolism of phenylalanine, an amino acid. Recessive mutations in the gene that encodes this enzyme are responsible for the disease called phenylketonuria (PKU). The computational experiment shown in **Table 21.5** started with the amino acid sequence of this protein and used the BLAST program to search the Swiss-Prot database, which contains hundreds of thousands of different protein sequences. The BLAST program can determine which sequences in the Swiss-Prot database are the closest matches to the amino acid sequence of human phenylalanine hydroxylase. Table 21.5 shows some of the results, which includes 10 of the matches to human phenylalanine hydroxylase that were identified by the program. Because this enzyme is found in nearly all eukaryotic species, the program identified phenylalanine hydroxylase from many different species. The column to the right of the match number shows the percentage of amino acids that are identical between the species indicated and the human sequence. Because the human phenylalanine hydroxylase sequence is already in the Swiss-Prot database, the closest match of human phenylalanine hydroxylase is to itself. The next nine sequences are in order of similarity. The next most similar sequence is from the orangutan, a close relative of humans. This is followed by two mammals, the mouse and rat, and then five vertebrates that

Table 21.5	Results from a BLAST Program Comparing Human Phenylalanine Hydroxylase with Database Sequences		
Match	Percentage of identical* amino acids	Species	Function of sequence†
1	100	Human (*Homo sapiens*)	Phenylalanine hydroxylase
2	99	Orangutan (*Pongo pygmaeus*)	Phenylalanine hydroxylase
3	95	Mouse (*Mus musculus*)	Phenylalanine hydroxylase
4	95	Rat (*Rattus norvegicus*)	Phenylalanine hydroxylase
5	89	Chicken (*Gallus gallus*)	Phenylalanine hydroxylase
6	82	Pipid frog (*Xenopus tropicalis*)	Phenylalanine hydroxylase
7	82	Green pufferfish (*Tetraodon nigroviridis*)	Phenylalanine hydroxylase
8	82	Zebrafish (*Danio rerio*)	Phenylalanine hydroxylase
9	80	Japanese pufferfish (*Takifugu rubripes*)	Phenylalanine hydroxylase
10	75	Fruit fly (*Drosophila melanogaster*)	Phenylalanine hydroxylase

*The number indicates the percentage of amino acids that are identical with the amino acid sequence of human phenylalanine hydroxylase. Note: These matches were randomly selected from a long list of matches.

†In some cases, the function of the sequence was determined by biochemical assay. In other cases, the function was inferred due to the high degree of sequence similarity with other species.

are not mammals. The 10th best match is from *Drosophila*, an invertebrate.

As you may have noticed, the order of the matches follows the evolutionary relatedness of the various species to humans. The similarity between any two sequences is related to the time that has passed since they diverged from a common ancestor. Among the species listed in this table, humans are most similar to themselves, followed by the orangutan, other mammals, other vertebrates, and finally invertebrates.

Overall, Table 21.5 is an example of the remarkable computational abilities of current computer technology. In less than a minute, the amino acid sequence of human phenylalanine hydroxylase can be compared with hundreds of thousands of different sequences to yield the data shown in this table! The main power of the BLAST program is its use with newly identified sequences, in which a researcher does not know the function of a gene or an encoded protein. When the BLAST program identifies a match to a sequence whose function is already known, it is likely that the newly identified sequence has an identical or similar function.

Summary of Key Concepts

21.1 Bacterial and Archaeal Genomes

- The genome is the complete genetic makeup of a cell or organism.

- Prokaryotic genomes are typically a single circular chromosome that has a few million base pairs of DNA. Such genomes usually have a few thousand different genes. (Table 21.1)

- Venter, Smith, and colleagues used a whole-genome shotgun sequencing strategy to determine the sequence of a prokaryotic genome, that of *Haemophilus influenzae.* (Figure 21.1)

21.2 Eukaryotic Genomes

- The nuclear genomes of eukaryotic species are composed of sets of linear chromosomes with a total length of several million to billions of base pairs. They typically contain several thousand to tens of thousands of genes. (Table 21.2)

- Genome sizes vary among eukaryotic species. In many cases, this variation is due to the accumulation of noncoding regions of DNA, particularly repetitive DNA sequences. (Figures 21.2, 21.3)

- Much of the repetitive DNA is derived from transposable elements, segments of DNA that can move from one site to another through a process called transposition. (Figure 21.4)

- DNA sequences within transposable elements can move by two different molecular mechanisms. The enzyme transposase mediates the movement of DNA transposons by a cut-and-paste mechanism. Retroelements move to new sites in the genome via RNA intermediates. (Figures 21.5, 21.6)

- Gene duplication may occur by a misaligned crossover during meiosis. This is one mechanism that can create a gene family, two or more homologous genes in a species that have related functions. (Figures 21.7, 21.8)

- The Human Genome Project, an international effort to map and sequence the entire human genome, was completed in 2003.

21.3 Proteomes

- A proteome is the collection of proteins that a given cell or species makes.

- Proteins are often placed into broad categories based on their functions. These include metabolic enzymes, structural proteins, motor proteins, cell-signaling proteins, transport proteins, proteins involved with gene expression and regulation, and those involved with protection. (Table 21.3)

- Protein abundance can refer to the relative abundance of proteins encoded in the genome or produced in the cell. (Figure 21.9)

- Protein diversity can increase via mechanisms such as alternative splicing and post-translational covalent modifications. (Figure 21.10)

21.4 Bioinformatics

- Bioinformatics involves the use of computers, mathematical tools, and statistical techniques to record, store, and analyze biological information, particularly genetic data such as DNA and protein sequences.

- Genetic information is stored in data files that can be analyzed using computer programs. The research community has collected genetic information and created several large databases. (Figure 21.11, Table 21.4)

- Homologous genes are derived from the same ancestral gene and have accumulated random mutations that make their sequences slightly different. (Figure 21.12)

- A simple dot matrix illustrates the approach of identifying regions of similarity between two sequences. (Figure 21.13)

- Computer programs, such as BLAST, can identify homologous genes that are found in a database. (Table 21.5)

Assess and Discuss

Test Yourself

1. The entire collection of proteins produced by a cell or organism is
   a. a genome.
   b. bioinformatics.
   c. a proteome.
   d. a gene family.
   e. a protein family.

2. Important reasons for studying the genomes of prokaryotes include all of the following except
   a. it may provide information that helps us understand how prokaryotes infect other organisms.
   b. it may provide a basic understanding of cellular processes that allow us to determine eukaryotic cellular function.
   c. it may provide the means to understand evolutionary processes.
   d. it will reveal the approximate number of genes that an organism has in its genome.
   e. All of the above are important reasons.

3. The enzyme that allows short segments of DNA to move within a cell from one location in the genome to another is

a. transposase.

b. DNA polymerase.

c. protease.

d. restriction endonuclease.

e. DNA ligase.

4. A gene family includes
   a. one specific gene found in several different species.
   b. all of the genes on the same chromosome.
   c. two or more homologous genes found within a single species.
   d. genes that code for structural proteins.
   e. both a and c.

5. Which of the following was <u>not</u> a goal of the Human Genome Project?
   a. identify all human genes
   b. sequence the entire human genome
   c. address the legal and ethical implications resulting from the project
   d. develop programs to manage the information gathered from the project
   e. clone a human

6. Bioinformatics is
   a. the analysis of DNA by molecular techniques.
   b. the use of computers to analyze and store biological information.
   c. a collection of gene sequences from a single individual.
   d. cloning.
   e. all of the above.

7. Using bioinformatics, evolutionary relationships among species can be characterized by identifying and analyzing
   a. phenotypes of selected organisms.
   b. homologous DNA sequences from different organisms.
   c. fossils of ancestral species.
   d. all of the above.
   e. a and b only.

8. Repetitive sequences
   a. are short DNA sequences that are found many times throughout the genome.
   b. may be multiple copies of the same gene found in the genome.
   c. are more common in eukaryotes.
   d. all of the above
   e. a and c only

9. The BLAST program is a tool for
   a. inserting many DNA fragments into a cell at the same time.
   b. translating a DNA sequence into an amino acid sequence.
   c. identifying homology between a selected sequence and genetic sequences in large databases.
   d. all of the above.
   e. both b and c.

10. Let's suppose you used the BLAST program beginning with a DNA sequence from a *Drosophila* hexokinase gene. (Hexokinase is an enzyme involved with glucose metabolism.) Which of the following choices would you expect to be the closest match?
    a. a *Drosophila* globin gene
    b. a human hexokinase gene
    c. a housefly hexokinase gene
    d. an *Arabidopsis* hexokinase gene
    e. an amoeba hexokinase gene

Conceptual Questions

1. Briefly describe whether or not each of the following could be appropriately described as a genome.
   a. the *E. coli* chromosome
   b. human chromosome 11
   c. a complete set of 10 chromosomes in corn
   d. a copy of the single-stranded RNA packaged into human immunodeficiency virus (HIV)

2. Describe two main reasons why the proteomes of eukaryotes species are usually much larger than their genomes.

3. Why is it useful to search a database to identify sequences that are homologous to a newly determined sequence?

Collaborative Questions

1. Compare and contrast the genomes of prokaryotic and eukaryotic organisms.

2. Below is a DNA sequence from one strand of a gene. Go to the NCBI website (www.ncbi.nlm.nih.gov/Tools) and run the BLAST program to determine which gene it is and in which species it is found.

   gtgaaggctc atggcaagaa agtgctcggt gcctttagtg

   atggcctggc tcacctggac aacctcaagg gcacctttgc

   cacactgagt gagctgcact gtgacaagct gcacgtggat

   cctgagaact tcagggtgag tctatgggac gcttgatgtt

   ttctttcccc ttcttttcta tggttaagtt catgtcatag

   gaaggggata agtaacaggg tacagtttag aatgggaaac

   agacgaatga ttgcatcagt gtggaagtct caggatcgtt

   ttagtttctt ttatttgctg ttcataacaa ttgttttctt

   ttgtttaatt cttgctttct ttttttttct tctccgcaat

Online Resource

www.brookerbiology.com

Stay a step ahead in your studies with animations that bring concepts to life and practice tests to assess your understanding. Your instructor may also recommend the interactive ebook, individualized learning tools, and more.

The Origin and History of Life

22

A fossil fish. This 50-million-year-old fossil of a unicorn fish (*Naso rectifrons*) is an example of the many different kinds of organisms that have existed during the history of life on Earth.

The amazing origin of the universe is difficult to comprehend. Astronomers now think that the universe began with an explosion called the Big Bang about 13.7 billion years ago, when the first clouds of the elements hydrogen and helium were formed. Gravitational forces collapsed these clouds to create stars that converted hydrogen and helium into heavier elements, including carbon, nitrogen, and oxygen, which are the building blocks of life on Earth. These elements were returned to interstellar space by exploding stars called supernovas, forming clouds in which simple molecules such as water, carbon monoxide, and hydrocarbons were formed. The clouds then collapsed to make a new generation of stars and solar systems.

Our solar system began about 4.6 billion years ago after one or more local supernova explosions. According to one widely accepted scenario, hundreds of planetesimals consisting of rocky or icy bodies such as asteroids and comets occupied the region where Venus, Earth, and Mars are now found. The Earth, which is estimated to be 4.55 billion years old, grew from the aggregation of planetesimals over a period of 100–200 million years. For the first half billion years or so after its formation, the Earth was too hot to allow water to accumulate on its surface. By 4 billion years ago, the Earth had cooled enough for the outer layers of the planet to solidify and for oceans to begin to form.

The period between 4.0 and 3.5 billion years ago marked the emergence of life on our planet. The first forms of life that we know about produced well-preserved microscopic fossils, such as those found in western Australia. These fossils, which are 3.5 billion years old, resemble cyanobacteria that are modern photosynthetic bacteria (**Figure 22.1**). Researchers cannot travel back through time and observe how the first life-forms came into being. However, plausible hypotheses regarding how life first arose have emerged from our understanding of modern life.

The first section of this chapter will survey a variety of hypotheses regarding the potential origins of biological molecules and living cells. We will then consider fossils, the preserved remains of organisms that existed in the past. Starting 3.5 billion years ago, the formation of fossils, such as the one shown in the chapter-opening photo, has provided biologists with evidence of the history of life on Earth from its earliest beginnings to the present day. The last section of this chapter surveys a time line for the history of life.

This chapter, the first in the Evolution unit, emphasizes when particular forms of life arose. Later chapters in this unit examine the mechanisms by which populations of organisms change over the course of many generations. This process, termed biological evolution, or simply **evolution**, involves genetic changes that occur from one generation to the next. Such genetic modifications often lead to dramatic changes in traits and the formation of new species and large groups of related species.

(a) Fossil prokaryote

(b) Modern cyanobacteria

Figure 22.1 **Earliest fossils and living cyanobacteria.** (a) A fossilized prokaryote about 3.5 billion years old that is thought to be an early cyanobacterium. (b) A modern cyanobacterium, which has a similar morphology. Cyanobacterial cells are connected to each other to form chains, as shown here.

22.1 Origin of Life on Earth

Living cells are complex collections of molecules and macromolecules. DNA stores genetic information, RNA acts as an intermediary in the process of protein synthesis and plays other important roles, while proteins form the foundation for the structure and activities of living cells. Life as we now know it requires this interplay between DNA, RNA, and proteins for its existence and perpetuation. On modern Earth, every living cell is made from a pre-existing cell.

But how did life get started? As described in Chapter 1, living organisms have several characteristics that distinguish them from nonliving materials. Because DNA, RNA, and proteins are the central players in the enterprise of life, scientists who are interested in the origin of life have focused much of their attention on the formation of these macromolecules and their building blocks, namely, nucleotides and amino acids. To understand the origin of life, we can view the process as occurring in four overlapping stages:

Stage 1: Nucleotides and amino acids were produced prior to the existence of cells.

Stage 2: Nucleotides became polymerized to form RNA and/or DNA, and amino acids become polymerized to form proteins.

Stage 3: Polymers became enclosed in membranes.

Stage 4: Polymers enclosed in membranes acquired cellular properties.

Researchers have followed a variety of experimental approaches to determine how life may have begun, including the synthesis of organic molecules in the laboratory without the presence of living cells or cellular material. This work has led researchers to a variety of hypotheses regarding the origin of life. In this section, we will consider a few scientific viewpoints that wrestle with the question, How did life begin?

Stage 1: Organic Molecules Were Produced Prior to the Existence of Cells

Let's begin our inquiry into the first stage of the origin of life by considering how nucleotides and amino acids may have been made prior to the existence of living cells. In the 1920s, the Russian biochemist Alexander Oparin and the Scottish biologist John Haldane independently proposed that organic molecules such as nucleotides and amino acids arose spontaneously under the conditions that occurred on early Earth. According to this hypothesis, the spontaneous appearance of organic molecules produced what they called a "primordial soup," which eventually gave rise to living cells.

The conditions on early Earth, which were much different than today, may have been more conducive to the spontaneous formation of organic molecules. Current hypotheses suggest that organic molecules, and eventually macromolecules, formed spontaneously. This is termed prebiotic (before life) or abiotic (without life) synthesis. These slowly forming organic molecules accumulated because there was little free oxygen gas, so they were not spontaneously oxidized, and there were as yet no living organisms, so they were also not metabolized. The slow accumulation of these molecules in the early oceans over a long period of time formed what is now called the **prebiotic soup**. The formation of this medium was a key event that preceded the origin of life.

Though most scientists agree that life originated from the assemblage of nonliving matter on early Earth, the mechanism of how and where these molecules originated is widely debated. Many intriguing hypotheses have been proposed, which are not mutually exclusive. A few of the more widely debated ideas are described next.

Reducing Atmosphere Hypothesis Based largely on geological data, many scientists in the 1950s proposed that the atmosphere on early Earth was rich in water vapor (H_2O), hydrogen gas (H_2), methane (CH_4), and ammonia (NH_3). These components, along with a lack of atmospheric oxygen (O_2), produce a reducing atmosphere because methane and ammonia readily give up electrons to other molecules, thereby reducing them. Such oxidation-reduction reactions, or redox reactions, are required for the formation of complex organic molecules from simple inorganic molecules.

In 1953, Stanley Miller, a student in the laboratory of Harold Urey, was the first scientist to use experimentation to test whether the prebiotic synthesis of organic molecules is possible. His experimental apparatus was intended to simulate the conditions on early Earth that were postulated in the 1950s (**Figure 22.2**). Water vapor from a flask of boiling water rose into another chamber containing hydrogen gas (H_2), methane (CH_4), and ammonia (NH_3). Miller inserted two electrodes that sent electrical discharges into the chamber to simulate lightning bolts. A condenser jacket cooled some of the gases from the chamber, causing droplets to form that dropped into a trap. He then took samples from this trap for chemical analysis. In his first experiments, he observed the formation of hydrogen cyanide (HCN) and formaldehyde (CH_2O). Such molecules are precursors of more complex organic molecules. These precursors also combined to make larger molecules such as the amino acid glycine. At the end of 1 week of operation, 10–15% of the carbon had been incorporated into organic compounds. Later experiments by Miller and others demonstrated the formation of sugars, a few types of amino acids, lipids, and nitrogenous bases found in nucleic acids (for example, adenine).

Why were these studies important? The work of Miller and Urey was the first attempt to apply scientific experimentation to our quest to understand the origin of life. Their pioneering strategy showed that the prebiotic synthesis of organic molecules is possible, although it could not prove that it really happened that way. In spite of the importance of these studies, critics of the so-called reducing atmosphere hypothesis have argued that Miller and Urey were wrong about the composition of early Earth's environment.

To vacuum

Electrical discharge

Electrodes

Gases
H_2O
H_2
CH_4
NH_3

Cold water

Condenser

Precipitating droplets

Boiling water

Trap

Sample containing organic molecules such as amino acids

Figure 22.2 Testing the reducing atmosphere hypothesis for the origin of life—the Miller and Urey experiment.

Concept check: With regard to the origin of life, why are biologists interested in the abiotic synthesis of organic molecules?

Since the 1950s, ideas about the atmosphere on early Earth have changed. More recently, many scientists have suggested that the atmosphere on early Earth was not reducing, but instead was a neutral environment composed mostly of carbon monoxide (CO), carbon dioxide (CO_2), nitrogen gas (N_2), and H_2O. These newer ideas are derived from studies of volcanic gas, which has much more CO_2 and N_2 than CH_4 and NH_3, and from the observation that UV radiation destroys CH_4 and NH_3, so these molecules would have been short-lived on early Earth. Nevertheless, since the experiments of Urey, many newer investigations have shown that organic molecules can be made under a variety of conditions. For example, organic molecules can be made prebiotically from a neutral environment composed primarily of CO, CO_2, N_2, and H_2O.

Extraterrestrial Hypothesis Many scientists have argued that sufficient organic carbon would have been present in the materials from asteroids and comets that reached the surface of early Earth in the form of meteorites. A significant proportion of meteorites belong to a class known as carbonaceous chondrites. Such meteorites may contain a substantial amount of organic carbon, including amino acids and nucleic acid bases. Based on this observation, some scientists have postulated that such meteorites could have transported a significant amount of organic molecules to early Earth.

Opponents of this hypothesis argue that most of this material would have been destroyed by the intense heating that accompanies the passage of large bodies through the atmosphere and their subsequent collision with the surface of the Earth. Though some organic molecules are known to reach the Earth via such meteorites, the degree to which heat would have destroyed many of the organic molecules remains a matter of controversy.

Deep-Sea Vent Hypothesis In 1988, the German lawyer and organic chemist Günter Wächtershäuser proposed that key organic molecules may have originated in deep-sea vents, which are cracks in the Earth's surface where superheated water rich in metal ions and hydrogen sulfide (H_2S) mixes abruptly with cold seawater. These vents release hot gaseous substances from the interior of the Earth at temperatures in excess of 300°C (572°F). Supporters of this hypothesis propose that biologically important molecules may have been formed in the temperature gradient between the extremely hot vent water and the cold water that surrounds the vent (**Figure 22.3a**).

Experimentally, the temperatures within this gradient are known to be suitable for the synthesis of molecules that form components of biological molecules. For example, the reaction between iron and hydrogen sulfide (H_2S) yields pyrites and H_2, and has been shown to provide the energy necessary for the reduction of N_2 to NH_3. Nitrogen is an essential component of both nucleic acids and amino acids, the molecular building blocks of life. But N_2, which is found abundantly on Earth, is chemically inert, so it is unlikely to have given rise to life. The conversion of N_2 to NH_3 at deep-sea vents may have led to the production of amino acids and nucleic acids.

Interestingly, complex biological communities are found in the vicinity of modern deep-sea vents. Various types of fish, worms, clams, crabs, shrimp, and bacteria are found in significant abundance in those areas (**Figure 22.3b**). Unlike most other forms of life on our planet, these organisms receive their energy from chemicals in the vent and not from the sun. In 2007, Timothy Kusky and colleagues discovered 1.43 billion-year-old fossils of deep-sea microbes near ancient deep-sea vents. This study provided more evidence that life may have originated on the bottom of the ocean. However, debate continues as to the primary way that organic molecules were made prior to the existence of life on Earth.

Stage 2: Organic Polymers May Have Formed on the Surface of Clay

The preceding three hypotheses provide reasonable mechanisms whereby small organic molecules could have accumulated on early Earth. Scientists hypothesize that the second stage in the origin of life was a period in which simple organic molecules polymerized to form more complex organic polymers such as DNA, RNA, or proteins. Most ideas regarding the origin of life assume that polymers with lengths of at least 30–60 monomers are needed to store enough information to make a viable genetic system. Because hydrolysis competes with polymerization, many scientists have speculated that the synthesis of polymers did not occur in a watery prebiotic soup, but instead took place on a solid surface or in evaporating tidal pools.

(a) Deep-sea vent hypothesis

(b) A deep-sea vent community

Figure 22.3 The deep-sea vent hypothesis for the origin of life. (a) Deep-sea vents are cracks in the Earth's surface that release hot gases such as hydrogen sulfide (H_2S). This heats the water near the vent and creates a gradient between the very hot water adjacent to the vent and the cold water that is farther away from the vent. The synthesis of organic molecules can occur in this gradient. (b) Photograph of a biological community near a deep-sea vent, which includes giant tube worms and crabs.

Concept check: What properties of deep-sea vents made them suitable for the abiotic synthesis of molecules?

In 1951, John Bernal first suggested that the prebiotic synthesis of polymers took place on clay. In his book *The Physical Basis of Life*, he wrote that "clays, muds and inorganic crystals are powerful means to concentrate and polymerize organic molecules." Many clay minerals are known to bind organic molecules such as nucleotides and amino acids. Negative charges within the clay itself attract metal divalent cations, such as magnesium (Mg^{2+}), that can catalyze the chemical reactions that produce polymers. Similarly, minerals such as feldspar, iron oxide, and calcite have properties that attract organic monomers and catalyze chemical reactions.

Experimentally, many research groups have demonstrated the formation of nucleic acid polymers and polypeptides on the surface of clay, given the presence of monomer building blocks. During the prebiotic synthesis of RNA, the purine bases of the nucleotides interact with the silicate surfaces of the clay. Divalent cations, such as Mg^{2+}, bind the nucleotides to the negative surfaces of the clay, thereby positioning the nucleotides in a way that promotes bond formation between the phosphate of one nucleotide and the ribose sugar of an adjacent nucleotide. In this way, polymers such as RNA may have been formed.

Though the formation of polymers on clay remains a reasonable hypothesis, studies by Luke Leman, Leslie Orgel, and M. Reza Ghadiri indicate that polymers can also form in aqueous solutions, which is contrary to popular belief. Their work in 2004 showed that carbonyl sulfide, a simple gas present in volcanic gases and deep-sea vent emissions, can bring about the formation of peptides from amino acids under mild conditions in water. These results indicate that the synthesis of polymers could have taken place in the prebiotic soup.

Stage 3: Cell-Like Structures May Have Originated When Polymers Were Enclosed by a Boundary

The third stage in the origin of living cells is hypothesized to be the formation of a boundary that separated the internal polymers such as RNA from the environment. The term **protobiont** is used to describe an aggregate of prebiotically produced molecules and macromolecules that acquired a boundary, such as a lipid bilayer, that allowed it to maintain an internal chemical environment distinct from that of its surroundings. What are the characteristics that make protobionts possible precursors to living cells? Scientists envision four key features:

1. A boundary, such as a membrane, separated the internal contents of the protobiont from the external environment.
2. Polymers inside the protobiont contained information.
3. Polymers inside the protobiont had enzymatic functions.
4. The protobionts eventually developed the capability of self-replication.

Protobionts were not capable of precise self-reproduction like living cells, but could divide to increase in number. Such protobionts are thought to have exhibited basic metabolic pathways in which the structures of organic molecules were changed. In particular, the polymers inside protobionts must have gained the enzymatic ability to link organic building blocks to create new polymers. This would have been a critical step in the process that eventually provided protobionts with the ability to self-replicate. According to this scenario, metabolic pathways became more complex, and the ability of protobionts to self-replicate became more refined over time. Eventually, these structures exhibited the characteristics that we attribute to living cells. As described next, researchers have hypothesized that protobionts may have exhibited different types of structures, such as coacervates and liposomes.

Russian biologist Alexander Oparin hypothesized in 1924 that living cells evolved from **coacervates**, droplets that form spontaneously from the association of charged polymers such as proteins, carbohydrates, or nucleic acids surrounded by

(a) Coacervates   57 μm

Skin of water

Solid droplet of protein and carbohydrate

Hollow sphere of phospholipid filled with water   Phospholipid bilayer

(b) Liposomes   200 nm

Figure 22.4   Protobionts and their lifelike functions.
Primitive cell-like structures such as these could have given rise
to living cells. **(a)** This micrograph shows coacervates made by
Oparin from a mixture of gelatin (composed primarily of protein)
and gum arabic (composed of protein and carbohydrate). **(b)** An
electron micrograph and illustration of liposomes. Each liposome
is made of a phospholipid bilayer surrounding an aqueous
compartment.

Concept check: Which protobiont seems most similar to real
cells? Explain.

water. Their name derives from the Latin *coacervare*, meaning
to assemble together or cluster. Coacervates measure 1–100 μm
(micrometers) across, possess osmotic properties, and are surrounded by a tight skin of water molecules (**Figure 22.4a**). This
boundary allows the selective absorption of simple molecules
from the surrounding medium.

If enzymes are trapped within coacervates, they can perform primitive metabolic functions. For example, researchers
have made coacervates containing the enzyme glycogen phosphorylase. When glucose-1-phosphate was made available to
the coacervates, it was taken up into them, and starch was
produced. The starch merged with the wall of the coacervates,
which increased in size and eventually divided into two. When
the enzyme amylase was included, the starch was broken down
to maltose, which was released from the coacervates.

As a second possibility, protobionts may have resembled **liposomes**—vesicles surrounded by a lipid bilayer (**Figure 22.4b**). When certain types of lipids are dissolved in water,
they spontaneously form liposomes. As discussed in Chapter 5,

lipid bilayers are selectively permeable (refer back to Figure
5.12) and some liposomes can even store energy in the form of
an electrical gradient. Such liposomes can discharge this energy
in a nerve cell-like fashion, showing rudimentary signs of excit-
ability, which is characteristic of living cells.

In 2003, Martin Hanczyc, Shelly Fujikawa, and Jack Szostak
showed that clay can catalyze the formation of liposomes that
grow and divide, a primitive form of self-replication. Furthermore, if RNA was on the surface of the clay, the researchers
discovered that liposomes were formed that enclosed RNA.
These experiments are exciting because they showed that the
formation of membrane vesicles containing RNA molecules is
a plausible explanation of the emergence of cell-like structures
based on simple physical and chemical forces.

Stage 4: Cellular Characteristics May Have Evolved via Chemical Selection, Beginning with an RNA World

The majority of scientists favor RNA as the first macromolecule
that was found in protobionts. Unlike other polymers, RNA
exhibits three key functions:

1. RNA has the ability to store information in its nucleotide
   sequence.
2. Due to base pairing, its nucleotide sequence has the
   capacity for self-replication.
3. RNA can perform a variety of enzymatic functions. The
   results of many experiments have shown that RNA
   molecules can function as **ribozymes**, acting as catalysts
   for the synthesis of the macromolecules found in living
   cells.

By comparison, DNA and proteins are not as versatile as
RNA. DNA has very limited catalytic activity, and proteins are
not known to undergo self-replication. RNA can perform functions that are characteristic of proteins and, at the same time,
can serve as genetic material with replicative and informational
functions.

How did the RNA molecules that were first made prebiotically evolve into more complex molecules that produced cell-
like characteristics? Researchers propose that a process called
chemical selection was responsible. **Chemical selection** occurs
when a chemical within a mixture has special properties or
advantages that cause it to increase in number compared to
other chemicals in the mixture. In other words, a population
of chemical molecules can change over time to become a new
population with a different composition. (As we will discuss in
Chapter 23, natural selection is a similar process except that it
describes the changing of a population of living organisms over
time due to survival and reproductive advantages.)

Initially, scientists speculate that the special properties
that enabled certain RNA molecules to undergo chemical selection were its ability to self-replicate and to perform other catalytic functions. As a way to understand the concept of chemical
selection, let's consider a hypothetical scenario showing two
steps of chemical selection. Step 1 of **Figure 22.5** shows a group

of protobionts that contain RNA molecules that were made prebiotically. RNA molecules inside these protobionts can be used as templates for the prebiotic synthesis of complementary RNA molecules. Such a process of self-replication, however, would be very slow because it would not be catalyzed by enzymes in the protobiont. In a first step of chemical selection, the sequence of one of the RNA molecules has undergone a mutation that gives it the enzymatic ability to attach nucleotides together, using RNA molecules as a template. This protobiont would have an advantage over the others because it would be capable of faster self-replication of its RNA molecules. Over time, due to its enhanced rate of replication, protobionts carrying such RNA molecules would increase in number compared to the others. Eventually, the group of protobionts shown in the figure contains only this type of enzymatically functional RNA.

In the second step of chemical selection (Figure 22.5, right side), a second mutation in an RNA molecule could produce the enzymatic function that would promote the synthesis of ribonucleotides, the building blocks of RNA. This protobiont would not solely rely on the prebiotic synthesis of ribonucleotides,

which also is a very slow process. Therefore, the protobiont having the ability to both self-replicate and synthesize ribonucleotides would have an advantage over a protobiont that could only self-replicate. Over time, the faster rate of self-replication and ribonucleotide synthesis would cause an increase in the numbers of the protobionts with both functions.

The **RNA world** is a hypothetical period on early Earth when both the information needed for life and the enzymatic activity of living cells were contained solely in RNA molecules. In this scenario, lipid membranes enclosing RNA exhibited the properties of life due to RNA genomes that were copied and maintained through the catalytic function of RNA molecules. Over time, scientists envision that mutations occurred in these RNA molecules, occasionally introducing new functional possibilities. Chemical selection would have eventually produced an increase in complexity in these cells, with RNA molecules accruing activities such as the ability to link amino acids together into proteins and other enzymatic functions.

But is an RNA world a plausible scenario? As described next in the Feature Investigation, chemical selection of RNA molecules can occur experimentally.

First step of chemical selection

1a **Mutation:** A mutation provides an RNA molecule with the enzymatic ability to synthesize new RNA molecules using pre-existing RNA molecules as templates.

RNA

A protobiont with no enzymatic functions

Mutant RNA with enzymatic ability to self-replicate RNA

1b **Chemical selection:** The amount of this mutant RNA with enzymatic function increases because it can self-replicate.

A protobiont with 1 enzymatic function

Second step of chemical selection

2a **Mutation:** A second mutation provides an RNA molecule with the enzymatic ability to synthesize ribonucleotides.

2b **Chemical selection:** The second mutation is also favored, so after many generations, the protobionts have 2 functions—self-replication and ribonucleotide synthesis.

Mutant RNA with the ability to synthesize ribonucleotides

A protobiont with 2 enzymatic functions

Figure 22.5 A hypothetical scenario illustrating the process of chemical selection. This figure shows a two-step scenario. In the first step, RNAs that can self-replicate are selected, and in the second step, RNAs with the ability to synthesize ribonucleotides are selected.

Concept check: *What is meant by the term chemical selection?*

FEATURE INVESTIGATION

Bartel and Szostak Demonstrated Chemical Selection in the Laboratory

Remarkably, scientists have been able to perform experiments in the laboratory that can select for RNA molecules with a particular function. David Bartel and Jack Szostak conducted the first study of this type in 1993 (**Figure 22.6**). Using molecular techniques, they synthesized a mixture of 10^{15} RNA molecules that we will call the long RNA molecules. Each long RNA in this mixture contained two regions. The first region at the 5′ end was a constant region that formed a stem-loop structure. Its sequence was identical among all 10^{15} molecules. The constant region was next to a second region that was 220 nucleotides in length. A key feature of the second region is that its sequence was variable among the 10^{15} molecules. The researchers hypothesized that this variation could occasionally result in a long RNA molecule with the enzymatic ability to catalyze a covalent bond between two adjacent nucleotides.

They also made another type of RNA molecule, which we will call the short RNA, with two important properties. First, the short RNA had a region that was complementary to a site in the constant region of the long RNA molecules. Second, the short RNA had a tag sequence that caused it to bind tightly to column packing material referred to as beads. The short RNAs did not have a variable region; they were all the same.

To begin this experiment, the researchers incubated a large number of the long and short RNA molecules together. During this incubation period, long and short RNA molecules would hydrogen bond to each other due to their complementary regions. Although hydrogen bonding is not permanent, this step allowed the long and short RNAs to recognize each other for a short period of time. The researchers reasoned that a long RNA with the enzymatic ability to form a covalent bond between nucleotides would make this interaction more permanent by catalyzing a bond between the long and short RNA molecules. Following this incubation, the mixture of RNAs was passed through a column with beads that specifically bound the short RNA. The aim of this approach was to select for longer RNA molecules that had covalently bonded to the short RNA molecule (see the Conceptual level of step 2).

The vast majority of long RNAs would not have the enzymatic ability to catalyze a permanent covalent bond between nucleotides. These would pass out of the column at step 2, because hydrogen bonding between the long and short RNAs is not sufficient to hold them together for very long. Such unbound long RNAs would be discarded. Long RNAs with the ability to catalyze a covalent bond to the short RNA would remain bound to the column beads at step 2. These enzymatic RNAs were then flushed out at step 3 to generate a mixture of RNAs termed pool #1. The researchers expected this pool to contain several different long RNA molecules with varying abilities to catalyze a covalent bond between nucleotides.

To further the chemical selection process, the scientists used the first pool of long RNA molecules flushed out at step 3 to make more long RNA molecules. This was accomplished via polymerase chain reaction (PCR). This next batch also had

Figure 22.6 Bartel and Szostak demonstrated chemical selection for RNA molecules that can catalyze the linkage between RNA molecules.

HYPOTHESIS Among a large pool of RNA molecules, some of them may contain the enzymatic ability to catalyze a covalent bond between nucleotides; these can be selected for in the laboratory.

KEY MATERIALS Many copies of short RNA were synthesized that had a tag sequence that binds tightly to column packing material called beads. Also, a population of 10^{15} long RNA molecules was made that contained a constant region with a stem-loop structure and a 220-nucleotide variable region. Note: The variable regions of the long RNAs were made using a PCR step that caused mutations in this region.

Experimental level

Conceptual level

1 Mix together the short RNAs with the 10^{15} different long RNAs. Allow time for covalent connections to form if the long RNA happens to have the enzymatic activity for covalent bond formation.

Short RNAs   Long RNAs (10^{15})

Tag sequence   Short RNA   Stem-loop structure   Long (10^{15}) RNA

Complementary region between long and short RNAs

Constant region

220-nucleotide variable region

Covalent bond

Of the 10^{15} long molecules, the variable region may rarely have the enzymatic ability to covalently connect the 3′ end of the short RNA to the 5′ end of the long RNA.

2 Pass the mixture through a column of beads that binds the tag sequence found on the short RNA. Add additional liquid to flush out long RNAs that are not covalently attached to short RNAs.

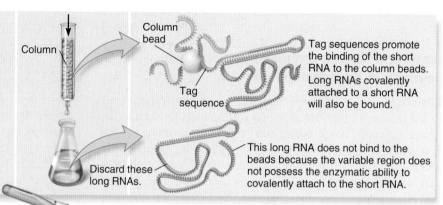

Column

Column bead

Tag sequence

Tag sequences promote the binding of the short RNA to the column beads. Long RNAs covalently attached to a short RNA will also be bound.

Discard these long RNAs.

This long RNA does not bind to the beads because the variable region does not possess the enzymatic ability to covalently attach to the short RNA.

3 Add a low pH solution to prevent the tag sequence from binding to the beads. This causes the tightly bound RNAs to be flushed out of the column.

Low pH wash

4 The flushed-out RNAs are termed pool #1. Use pool #1 to make a second batch of long RNA molecules. This involved a PCR step using reverse transcriptase to make cDNA. The PCR primers recognized the beginning and end of the long RNA sequence and copied only this region. The cDNA was then used as a template to make long RNA via RNA polymerase.

Pool #1

Refer back to Figure 20.6 for a description of PCR.

5 Repeat procedure to generate 10 consecutive pools of RNA molecules.

Pool #1 Pool #2 Pool #3 Pool #4 Pool #5 Pool #6 Pool #7 Pool #8 Pool #9 Pool #10

6 Test a sample of the original population and each of the 10 pools for the enzymatic ability to catalyze a covalent bond between adjacent nucleotides.

Gap

Covalent bond

7 THE DATA

*Original 10^{15} molecules

8 CONCLUSION The increase in covalent bond formation from pool 1 to pool 10 indicates that chemical selection can occur.

9 SOURCE Bartel, David P., and Szostak, Jack W. 1993. Isolation of new ribozymes from a large pool of random sequences. *Science* 261:1411–1418.

the constant and variable regions but did not have the short RNA covalently attached. Because the variable regions of these new RNA molecules were derived from the variable regions of pool #1 RNA molecules, they were expected to have enzymatic activity. The researchers reasoned that additional variation might occasionally produce an RNA molecule with improved enzymatic activity. This second batch of long RNA molecules (pool #2) was subjected to the same steps as was the first batch of 10^{15} molecules. In this case, the group of long molecules flushed out at step 3 was termed pool #2. This protocol was followed eight more times to generate 10 consecutive pools of RNA molecules. During this work, the researchers analyzed the original random collection of 10^{15} RNA molecules and each of the 10 pools for the enzymatic ability to covalently link RNA molecules. As seen in the data, each successive pool became enriched for molecules with higher enzymatic activity. Pool #10 showed enzymatic activity that was approximately 3 million times higher than the original random pool of molecules!

Like the work of Miller and Urey, these experiments showed the feasibility of another phase of the prebiotic process that led to life, in this case, chemical selection. The results showed that chemical selection can improve the functional characteristics of a group of RNA molecules over time by increasing the proportions of those molecules with enhanced function.

Experimental Questions

1. What is chemical selection? What was the hypothesis tested by Bartel and Szostak?

2. In conducting the selection experiment among pools of long RNA molecules with various catalytic abilities, what was the purpose of using the short RNA molecules?

3. What were the results of the experiment conducted by Bartel and Szostak? What impact did this study have on our understanding of the evolution of life on Earth?

The RNA World Was Superseded by the Modern DNA/RNA/Protein World

Assuming that an RNA world was the origin of life, researchers have asked the question, Why and how did the RNA world evolve into the DNA/RNA/protein world we see today? The RNA world may have been superseded by a DNA/RNA world or an RNA/protein world before the emergence of the modern DNA/RNA/protein world. Let's now consider the advantages of a DNA/RNA/protein world as opposed to the simpler RNA world and explore how this modern biological world might have come into being.

Information Storage RNA can store information in its base sequence. If so, why did DNA take over that function, as is the case in modern cells? During the RNA world, RNA had to perform two roles, the storage of information and the catalysis of chemical reactions. Scientists have speculated that the incorporation of DNA into cells would have relieved RNA of its informational role and thereby allowed RNA to perform a greater variety of other functions. For example, if DNA stored the information for the synthesis of RNA molecules, such RNA molecules could bind cofactors, have modified bases, or bind peptides that might enhance their catalytic function. Cells with both DNA and RNA would have had an advantage over those with just RNA, and so they would have been selected. Another advantage of DNA is stability. Compared to RNA, DNA strands are less likely to spontaneously break.

A second issue is how DNA came into being. Scientists have proposed that an ancestral RNA molecule had the ability to make DNA using RNA as a template. This function, known as reverse transcription, is described in Chapter 18. Interestingly, modern eukaryotic cells can use RNA as a template to make DNA. For example, the enzyme telomerase copies the ends of chromosomes by using an RNA template (refer back to Figure 11.22).

Metabolism and Other Cellular Functions Now let's consider the origin of proteins. The emergence of proteins as catalysts may have been a great benefit to early cells. Due to the different chemical properties of the 20 amino acids, proteins have vastly greater catalytic ability than do RNA molecules, again providing a major advantage to cells that had both RNA and proteins. In modern cells, proteins have taken over most, but not all, catalytic functions. In addition, proteins can perform other important tasks. For example, cytoskeletal proteins carry out structural roles, and certain membrane proteins are responsible for the uptake of substances into living cells.

How did proteins come into being from an RNA world? Chemical selection experiments have shown that RNA molecules can catalyze the formation of peptide bonds and even attach amino acids to primitive tRNA molecules. Similarly, modern protein synthesis still involves a central role for RNA in the synthesis of polypeptides. First, mRNA provides the information for a polypeptide sequence. Second, tRNA molecules act as adaptors for the formation of a polypeptide chain. And finally, ribosomes containing rRNA provide a site for polypeptide synthesis. Furthermore, RNA within the ribosome acts as a ribozyme to catalyze peptide bond formation. Taken together, the analysis of translation in modern cells is consistent with an evolutionary history in which RNA molecules were instrumental in the emergence and formation of proteins.

22.2 Fossils

We will now turn our attention to a critical process that has given us a glimpse of the history of life over the past 3.5 billion years. **Fossils** are the preserved remains of past life on Earth. They can take many forms, including bones, shells, and leaves, and the impression of cells or other evidence, such as footprints or burrows. Scientists who study fossils are called

paleontologists (from the Greek *palaios*, meaning ancient). Because our understanding of the history of life is derived primarily from the fossil record, it is important to appreciate how fossils are formed and dated and to understand some of the inherent biases in fossil analyses.

Fossils Are Formed Within Sedimentary Rock

How are fossils usually formed? Many of the rocks observed by paleontologists are sedimentary rocks that were formed from particles of older rocks broken apart by water or wind. These particles, such as gravel, sand, and mud, may settle and bury living and dead organisms at the bottoms of rivers, lakes, and oceans. Over time, more particles pile up, and sediments at the bottom of the pile eventually become rock. Gravel particles form rock called conglomerate, sand becomes sandstone, and mud becomes shale. Most fossils are formed when organisms are buried quickly, and then during the process of sedimentary rock formation, their hard parts are gradually replaced over millions of years by minerals, producing a recognizable representation of the original organism.

The relative ages of fossils can sometimes be revealed by their locations in sedimentary rock formations. Because sedimentary rocks are formed particle by particle and bed by bed, the layers are piled one on top of the other. In a sequence of layered rocks, the lower rock layers are usually older than the upper layers. Paleontologists often study changes in life-forms over time by studying the fossils in various beds from bottom to top (**Figure 22.7**). The more ancient life-forms are found in

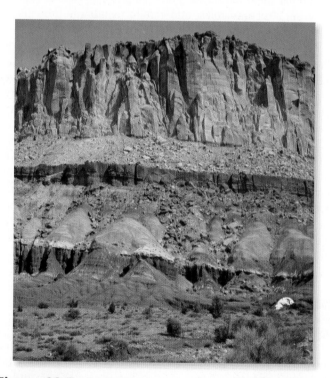

Figure 22.7 An example of layers of sedimentary rock that contain fossils.

Concept check: Which rock layer in this photo is most likely to be the oldest?

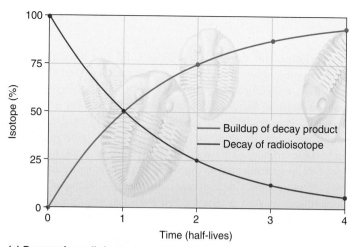

(a) Decay of a radioisotope

Radioisotope	Decay product	Half-life (years)	Useful dating range (years)
Carbon-14	Nitrogen-14	5,730	100–50,000
Potassium-40	Argon-40	1.3 billion	100,000–4.5 billion
Rubidium-87	Strontium-87	47 billion	10 million–4.5 billion
Uranium-235	Lead-207	710 million	10 million–4.5 billion
Uranium-238	Lead-206	4.5 billion	10 million–4.5 billion

(b) Radioisotopes that are useful for geological dating

Figure 22.8 Radioisotope dating of fossils. **(a)** A rock can be dated by measuring the relative amounts of a radioisotope and its decay product within the rock. **(b)** These five isotopes are particularly useful for the dating of fossils.

Concept check: If you suspected a fossil was 50 million years old, which pair of radioisotopes would you choose to analyze?

the lower beds, and newer species are found in the upper beds. However, such an analysis can occasionally be misleading when previous geological processes have scrambled the layers.

The Analysis of Radioisotopes Is Used to Date Fossils

A common way to estimate the age of a fossil is by analyzing the decay of radioisotopes within the accompanying rock, a process called **radioisotope dating**. As discussed in Chapter 2, elements may be found in multiple forms called isotopes that differ in the number of neutrons they contain. A radioisotope is an unstable isotope of an element that decays spontaneously, releasing radiation at a constant rate. The **half-life** is the length of time required for a radioisotope to decay to exactly one-half of its initial quantity. Each radioisotope has its own unique half-life (**Figure 22.8a**). Within a sample of rock, scientists can measure the amount of a given radioisotope as well as the amount of the decay product—the isotope that is produced when the original isotope decays. For dating geological materials, several types of isotope decay patterns are particularly useful: carbon

Table 22.1	Biases That Occur in the Fossil Record
Factor	**Description**
Anatomy	Organisms with hard body parts, such as animals with a skeleton or thick shell, are more likely to be preserved than are organisms composed of soft tissues.
Size	The fossil remains of larger organisms are more likely to be found than those of smaller organisms.
Number	Species that existed in greater numbers or over a larger area are more likely to be preserved within the fossil record than those that existed in smaller numbers or in a smaller area.
Environment	Inland species are less likely to become fossilized than are those that lived in a marine environment or near the edge of water because sedimentary rock is more likely to be formed in or near water.
Time	Organisms that lived relatively recently or existed for a long time are more likely to be found as fossils than organisms that lived very long ago or for a relatively short time.
Geology	Due to the chemistry of fossilization, certain organisms are more likely to be preserved than are other organisms.
Paleontology	Certain types of fossils may be more interesting to paleontologists. In addition, a significant bias exists with regard to the locations where paleontologists search for fossils.

to nitrogen, potassium to argon, rubidium to strontium, and uranium to lead (**Figure 22.8b**).

To determine the age of a rock using radioisotope dating, paleontologists need to have a way to set the clock—extrapolate back to a starting point in which a rock did not have any amount of the decay product. Except for fossils less than 50,000 years old, in which carbon-14 (^{14}C) dating can be employed, fossil dating is not usually conducted on the fossil itself or on the sedimentary rock in which the fossil is found. Most commonly, igneous rock in the vicinity of the sedimentary rock is dated. Why is igneous rock chosen? One reason is that igneous rock derived from an ancient lava flow initially contains uranium-235 (^{235}U) but no lead-207 (^{207}Pb). The decay product of ^{235}U is ^{207}Pb. By comparing the relative proportions of ^{235}U and ^{207}Pb, the age of the igneous rock derived from a lava flow can be accurately determined. Even so, because paleontologists are unlikely to find the fossilized remains of the first member of a particular species, they expect the fossil record to underestimate the actual date that a species came into existence.

The Fossil Record Contains Unintentional Biases

Several factors can bias the fossil record (**Table 22.1**). First, certain organisms are more likely to become fossilized than are others. Organisms with hard shells or bones tend to be overrepresented. Factors such as anatomy, size, number, and the environment and time in which they lived also play important roles in determining the likelihood that organisms will be preserved in the fossil record. In addition, geological processes may create biases because they may favor the fossilization of certain types of organisms. Finally, unintentional biases arise that are related to the efforts of paleontologists. For example, scientific interests may favor searching for and analyzing certain species over others.

The fossil record should not be viewed as a comprehensive and balanced story of the history of life. However, it has provided a wealth of information regarding the types of life that existed in the distant past. The rest of this chapter will survey the emergence of life-forms from 3.5 billion years ago to the present.

22.3   History of Life on Earth

Thus far, we have considered how the first cells may have come into existence and the characteristics of fossils. The first fossils of single-celled organisms were preserved approximately 3.5 billion years ago. In this section, we begin with a brief description of the geological changes on Earth that have affected the emergence of new forms of life and then examine some of the major changes in life that have occurred since life began.

Many Environmental and Biological Changes Have Occurred Since the Origin of the Earth

The **geological time scale** is a time line of the Earth's history and major events from its origin approximately 4.55 billion years ago to the present (**Figure 22.9**). This time line is subdivided into four eons—the Hadean, Archaean, Proterozoic, and Phanerozoic—and then further subdivided into eras. The first three eons are collectively known as the Precambrian because they preceded the Cambrian era, a geological era that saw a rapid increase in the diversity of life. The names of several eons and eras end in *-zoic* (meaning animal life) because we often recognize these time intervals on the basis of animal life. We will examine these time periods later in this chapter.

The changes that have occurred in living organisms over the past 4 billion years are the result of two interactive processes. First, as discussed in the next several chapters, genetic changes in organisms often affect their characteristics. Such changes can influence organisms' abilities to survive in their native environment. Second, the environment on Earth has undergone dramatic changes that have profoundly influenced the types of organisms that have existed during different periods of time. In some cases, a change might allow new types of organisms to come into being. Alternatively, environmental changes can result in **extinction**—the complete loss of a species or group of species. Major types of environmental changes are described next.

Climate/Temperature   During the first 2.5 billion years of its existence, the surface of the Earth gradually cooled. However, during the last 2 billion years, the Earth has undergone major fluctuations in temperature, producing Ice Ages that alternate with warmer periods. Furthermore, the temperature on Earth is not uniform, which produces a range of environments where the temperatures are quite different, such as tropical rain forest and arctic tundra.

Atmosphere   The chemical composition of the gases surrounding the Earth has changed substantially over the past 4 billion years. One notable change involves oxygen. The emergence of organisms that were capable of photosynthesis added oxygen

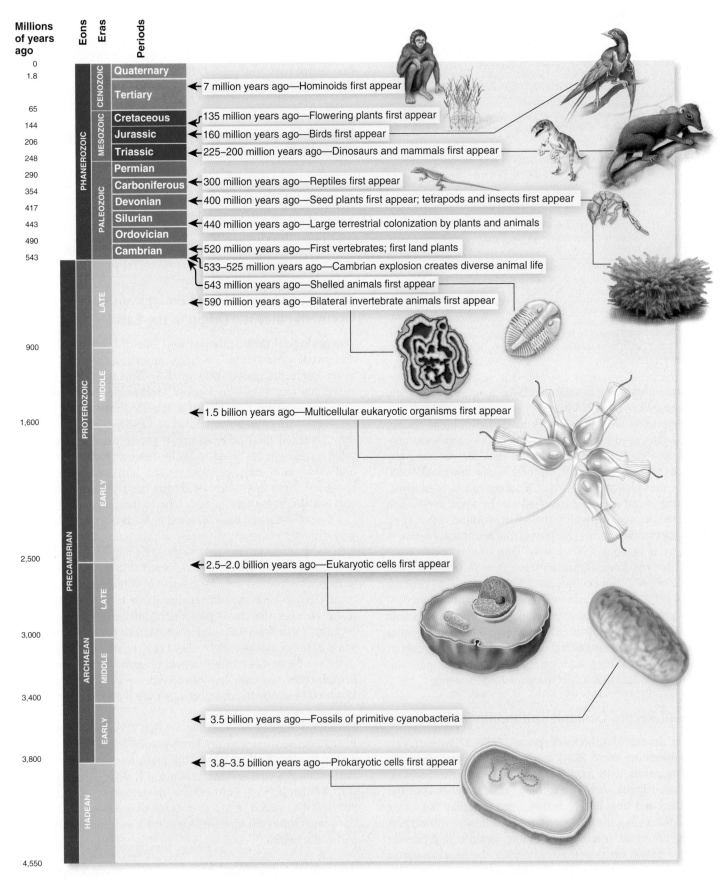

Figure 22.9   The geological time scale and an overview of the history of life on Earth.

to the atmosphere. Prior to 2.4 billion years ago, relatively little oxygen gas was in the atmosphere, but at this time, levels of oxygen in the form of O_2 began to rise significantly. Our current atmosphere contains about 21% O_2. Increased levels of oxygen are thought to have a played a key role in various aspects of the history of life, including the following:

- The origin of animal body plans coincided with a rise in atmospheric O_2.

- The conquest of land by arthropods (about 410 million years ago) and a second conquest by arthropods and vertebrates (about 350 million years ago) occurred during periods in which O_2 levels were high or increasing.

- Increases in animal body sizes are associated with higher O_2 levels.

Higher levels of O_2 could have contributed to these events because higher O_2 levels may enhance the ability of animals to carry out aerobic respiration. These events are also discussed later in this chapter and in more detail in Unit VI.

Landmasses   As the Earth cooled, landmasses formed that were surrounded by bodies of water. This created two different environments, terrestrial and aquatic. Furthermore, over the course of billions of years, the major landmasses, known as the continents, have shifted their positions, changed their shapes, and become separated from each other. This phenomenon, called **continental drift**, is shown in Figure 22.10.

Floods and Glaciation   Catastrophic floods have periodically had major impacts on the organisms in the flooded regions. On a periodic basis, glaciers have moved across continents and altered the composition of species on those landmasses. As an extreme example, in 1992, Joeseph Kirschvink proposed the "Snowball Earth hypothesis," which suggests that the Earth was entirely covered by ice during parts of the period from 790 to 630 million years ago. This hypothesis was developed to explain various types of geological evidence including sedimentary deposits of glacial origin that are found at tropical latitudes. While the prior existence of a completely frozen Earth remains controversial, massive glaciations over our planet have had an important impact on the history of life.

Volcanic Eruptions   The eruptions of volcanoes can harm organisms in the vicinity of the eruption, sometimes causing extinctions. In addition, volcanic eruptions in the oceans can lead to the formation of new islands. Massive eruptions may also spew so much debris into the atmosphere that they affect global temperatures and limit solar radiation, which restricts photosynthetic production.

Meteorite Impacts   During its long history, the Earth has been struck by many meteorites. Large meteorites have had substantial impacts on the Earth's environment.

The effects of one or more of the changes described above have sometimes caused many species to go extinct at the same

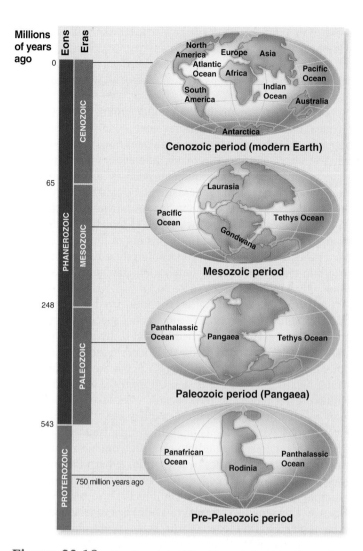

Figure 22.10   Continental drift.   The relative locations of the continents on Earth have changed dramatically over time.

time. Such events are called **mass extinctions**. Five large mass extinctions occurred near the end of the Ordovician, Devonian, Permian, Triassic, and Cretaceous periods. The boundaries between geological time periods are often based on the occurrences of mass extinctions. A recurring pattern seen in the history of life is the extinction of species and the emergence of new species. The rapid extinction of many modern species due to human activities is sometimes referred to as the sixth mass extinction. We will examine mass extinctions and the current biodiversity crisis in more detail in Chapter 60.

Prokaryotic Cells Arose During the Archaean Eon

The Archaean (from the Greek, meaning ancient) was an eon when diverse microbial life flourished in the primordial oceans. As mentioned previously, the first known fossils of living cells were preserved in rocks that are 3.5 billion years old (see Figure 22.1), though scientists postulate that cells arose many millions of years prior to this time. Based on the morphology of fossilized remains, these first cells were prokaryotic. During the more than 1 billion years of the Archaean eon, all life-forms were prokaryotic. Because hardly any free oxygen (O_2) was in the Earth's atmosphere, the single-celled microorganisms of this eon almost certainly used only anaerobic (without oxygen) respiration.

Prokaryotes are divided into two groups: bacteria and archaea. Bacteria are the most prevalent prokaryotic organisms on modern Earth. Many species of archaea have also been identified. Archaea are found in many different environments, and some occupy extreme environments such as hot springs. Both bacteria and archaea share fundamental similarities, indicating they are derived from a common ancestor. Even so, certain differences suggest that these two types of prokaryotes diverged from each other quite early in the history of life. In particular, bacteria and archaea show some interesting differences with regard to metabolism, lipid composition, and genetic pathways (look ahead to Chapter 26, Table 26.1).

An important factor that greatly influenced the emergence of prokaryotic, and eventually eukaryotic, species is the availability of energy. As we learned in Unit II, living cells require energy to survive and reproduce. Organisms may follow two different strategies to obtain energy. Some are **heterotrophs**, which means their energy is derived from the chemical bonds within organic molecules they consume. Because the most common sources of organic molecules today are other organisms, heterotrophs typically consume other organisms or materials from other organisms. Alternatively, many organisms are **autotrophs** that directly harness energy from either inorganic molecules or light. Among modern species, plants are an important example of autotrophs. Plants can directly absorb light energy and use it (via photosynthesis) to synthesize organic molecules such as glucose. On modern Earth, heterotrophs ultimately rely on autotrophs for the production of food.

Were the first forms of life heterotrophs or autotrophs? The answer is not resolved. Some biologists have speculated that autotrophs may have arisen first, such as those living near deep-sea vents. These organisms would have used chemicals that were made near the vents as an energy source to make organic molecules. Alternatively, many scientists have hypothesized that the first living cells were heterotrophs. They reason that it would have been simpler for the first primitive cells to use the organic molecules in the prebiotic soup as a source of energy.

If heterotrophs came first, why were cyanobacteria preserved in the earliest fossils, rather than heterotrophs? One possible reason is related to their manner of growth. Certain cyanobacteria promote the formation of a layered structure called a **stromatolite** (Figure 22.11). The aquatic environment where these cyanobacteria survive is rich in minerals such as calcium. The cyanobacteria grow in large mats that form layers. As they grow, they deplete the carbon dioxide (CO_2) in the surrounding water. This causes calcium carbonate in the water to gradually precipitate over the bacterial cells, calcifying the older cells in the lower layers and also trapping grains of sediment. Newer cells produce a layer on top. Over time, many layers of calcified cells and sediment are formed, thereby producing

(a) Fossil stromatolite

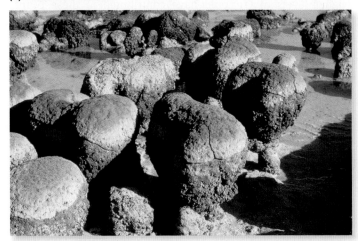

(b) Modern stromatolites

Figure 22.11 Fossil and modern stromatolites: evidence of autotrophic cyanobacteria.   Each stromatolite is a rocklike structure, typically 1 meter in diameter. **(a)** Section of a fossilized stromatolite. These layers are mats of mineralized cyanobacteria, one layer on top of the other. The existence of fossil stromatolites provides evidence of early autotrophic organisms. **(b)** Modern stromatolites that have formed in western Australia.

a stromatolite. This process still occurs today in places such as Shark Bay in western Australia, which is renowned for the stromatolite turfs along its beaches (Figure 22.11).

The emergence and proliferation of ancient cyanobacteria had two critical consequences. First, the autotrophic nature of these bacteria enabled them to produce organic molecules from CO_2. This prevented the depletion of organic foodstuffs that would have been exhausted if only heterotrophs existed. Second, cyanobacteria produce oxygen (O_2) as a waste product of photosynthesis. During the Archaean and Proterozoic eons, the activity of cyanobacteria led to the gradual rise in O_2 that was discussed earlier. The increase in O_2 spelled doom for many prokaryotic groups. Anaerobic species became restricted to a few anoxic (without oxygen) environments, such as deep within the soil. However, O_2 enabled the formation of new prokaryotic species that used aerobic (with oxygen) respiration (see Chapter 7). In addition, aerobic respiration is likely to have played a key role in the emergence and eventual explosion of eukaryotic life-forms, which typically have high energy demands. These eukaryotic life-forms are described next.

Genomes & Proteomes Connection

The Origin of Eukaryotic Cells Involved a Union Between Bacterial and Archaeal Cells

Eukaryotic cells arose during the Proterozoic eon, which began 2.5 billion years ago and ended 543 million years ago (see Figure 22.9). The manner in which the first eukaryotic cell originated is not entirely understood. In modern eukaryotic cells, genetic material is found in three distinct organelles. All eukaryotic cells contain DNA in the nucleus and mitochondria, and plant and algal cells also have DNA in their chloroplasts. To address the issue of the origin of eukaryotic species, scientists have examined the DNA sequences found in these three organelles. From such studies, the nuclear, mitochondrial, and chloroplast genomes appear to be derived from once-separate cells that came together.

Nuclear Genome From a genome perspective, both bacteria and archaea have contributed substantially to modern nuclear genomes. Eukaryotic nuclear genes encoding proteins involved in metabolic pathways and lipid biosynthesis appear to be derived from ancient bacteria, whereas genes involved with transcription and translation appear to be derived from an archaeal ancestor. To explain the origin of the nuclear genome, several hypotheses have been proposed. The most widely accepted involve an association between ancient bacteria and archaea, which could have been symbiotic or endosymbiotic.

A **symbiotic** relationship is one in which two different species live in direct contact with each other. For example, some scientists have postulated an ancient symbiotic relationship in which a bacterium and an archaeal cell (archaeon) formed a close association (**Figure 22.12a**). This eventually led to a fusion event that combined the genetic material of the two organisms. Over time, selection favored the retention of bacterial genes involved in metabolism and lipid biosynthesis and archaeal genes concerned with transcription and translation, and the nuclear genome came into being.

A second possible scenario is an **endosymbiotic** relationship, in which a smaller organism (the symbiont) lives inside a larger organism (the host). According to this idea, one prokaryotic cell engulfed another, which became an endosymbiont (**Figure 22.12b**). For example, an ancient archaeon may have engulfed a bacterium, maintaining the bacterium in its cytoplasm as an endosymbiont. This process may have occurred via endocytosis, which is described in Chapter 5. Over time, genes were transferred to the archaeal host cell, and the resulting genetic material eventually became the nuclear genome.

Mitochondrial and Chloroplast Genomes In 1905, a Russian botanist, Konstantin Mereschkowsky, was the first to suggest that such organelles may have an endosymbiotic origin. However, the question of endosymbiosis was largely ignored until researchers in the 1950s discovered that chloroplasts and mitochondria contain their own genetic material. The issue of endosymbiosis was revived and hotly debated when in 1970, Lynn Margulis published a book presenting evidence in support of this hypothesis entitled *Origin of Eukaryotic Cells*. Analyses of genes from mitochondria, chloroplasts, and prokaryotes are consistent with the endosymbiotic origin of these organelles (refer back to Figure 4.27).

Mitochondria found in eukaryotic cells are likely derived from a bacterial species that resembles modern α-proteobacteria, a diverse group of bacteria that can carry out oxidative phosphorylation to make ATP. One possibility is that an endosymbiotic event involving an ancestor of this bacterial species produced the first eukaryotic cell and that the mitochondrion is a remnant of that event. Alternatively, symbiosis or endosymbiosis may have produced the first eukaryotic cell, and then a subsequent endosymbiosis resulted in mitochondria (Figure 22.12). Over the next few years, the sequencing of more prokaryotic and mitochondrial genomes may help to resolve this controversy. DNA-sequencing data indicate that chloroplasts were derived from an endosymbiotic relationship between a primitive eukaryotic cell and a cyanobacterium.

Curiously, an endosymbiotic relationship involving two different proteobacteria was reported in 2001. In mealybugs, bacteria survive within the cytoplasm of large host cells of a specialized organ called a bacteriome. Recent analyses of these bacteria have shown that the bacteria inside the host cells share their own endosymbiotic relationship. In particular, γ-proteobacteria live endosymbiotically inside β-proteobacteria. Such an observation demonstrates that an endosymbiotic relationship can occur between two prokaryotic species.

Multicellular Eukaryotes and the Earliest Animals Arose During the Proterozoic Eon

Multicellularity was an important evolutionary innovation. The first multicellular eukaryotes are thought to have emerged about 1.5 billion years ago, in the middle of the Proterozoic eon. The oldest fossil evidence for multicellularity is dated at approximately 1.2 billion years old. Simple multicellular organisms are believed to have originated in one of two different ways. One possibility is that several individual cells found each other and aggregated to form a colony. Cellular slime molds, discussed in Chapter 28, are examples of modern organisms in which groups of single-celled organisms can come together to form a small multicellular organism. According to the fossil record, such organisms have remained very simple for hundreds of millions of years.

Alternatively, another way that multicellularity can occur is when a single cell divides and the resulting cells stick together. This pattern occurs in many simple multicellular organisms,

(a) Origin from a symbiotic relationship

1. A symbiotic association occurred between 2 prokaryotic species. The 2 cells fused with each other to make a single cell.

2. Over time, archaeal genes involved with metabolism and lipid biosynthesis were lost. Bacterial genes involved with transcription and translation were also lost.

3. Endosymbiosis of a bacterium resulted in the formation of mitochondria. The endomembrane system created the nucleus and other organelles such as the ER and Golgi.

4. A second endosymbiosis event involving cyanobacteria resulted in chloroplasts.

Eukaryotic cells: Animals, fungi, and some protists

Eukaryotic cells: Plants and algae

(b) Origin from an endosymbiotic relationship

1. A larger archaeal cell engulfed a smaller bacterial cell and established an endosymbiotic relationship.

2. Genes from the endosymbiont were transferred to the host cell.

3. This event may have resulted in mitochondria, or mitochondria may have arisen via a second endosymbiotic event. The endomembrane system created the nucleus and other organelles such as the ER and Golgi.

4. A subsequent endosymbiosis event involving cyanobacteria resulted in chloroplasts.

Eukaryotic cells: Plants and algae

Eukaryotic cells: Animals, fungi, and some protists

Figure 22.12   Possible symbiotic or endosymbiotic relationships that gave rise to the first eukaryotic cells.

Concept check: *What is the fundamental difference between the two scenarios described in this figure?*

such as algae and fungi, as well as in species with more complex body plans, such as plants and animals. Biologists cannot be certain whether the first multicellular organisms arose by an aggregation process or by cell division and adhesion. However, the development of complex, multicellular organisms now occurs by cell division and adhesion.

An interesting example that compares unicellular organisms to more complex multicellular organisms is found among species of volvocine green algae that are related evolutionarily. These algae exist as unicellular species, as small clumps of cells of the same cell type, or as larger groups of cells with two distinct cell types. **Figure 22.13** compares four species of volvocine

(a) *Chlamydomonas reinhardtii*, a unicellular alga

(b) *Gonium pectorale*, composed of 8 identical cells

(c) *Pleodorina californica*, composed of 64 to 128 cells, has 2 cell types, somatic and reproductive

(d) *Volvox aureus*, composed of about 1,000 to 2,000 cells, has 2 cell types, somatic and reproductive

Figure 22.13 Variation in the level of multicellularity among volvocine algae.

Concept check: Describe three differences that occur among these four types of algae.

algae. *Chlamydomonas reinhardtii* is a unicellular alga (Figure 22.13a). It is called a biflagellate because each cell possesses two flagella. *Gonium pectorale* is a multicellular organism composed of eight cells (Figure 22.13b). This simple multicellular organism is formed from a single cell by cell division and adhesion. All of the cells in this species are biflagellate. Other volvocine algae have evolved into larger and more complex organisms. *Pleodorina californica* has 64 to 128 cells (Figure 22.13c), and *Volvox aureus* has about 1,000 to 2,000 cells (Figure 22.13d). A feature of these more complex organisms is they have two cell types—somatic and reproductive cells. The somatic cells are biflagellate cells, whereas the reproductive cells are not. When comparing *P. californica* and *V. aureus*, *V. aureus* has a higher percentage of somatic cells than does *P. californica*.

Overall, an analysis of these four species of algae illustrates three important principles found among complex multicellular species. First, such organisms arise from a single cell that divides to produce daughter cells that adhere to one another. Second, the daughter cells can follow different fates, thereby producing multicellular organisms with different cell types. Third, as organisms get larger, a greater percentage of the cells tend to be somatic cells. The somatic cells carry out the activities that are required for the survival of the multicellular organism, whereas the reproductive cells are specialized for the sole purpose of producing offspring.

Toward the end of the Proterozoic eon, multicellular animals emerged. The first animals were **invertebrates**, which are animals without a backbone. Most modern animals, except for organisms such as sponges and jellyfish, exhibit bilateral symmetry. This is a two-sided body plan with a right and left side that are mirror images of each other. Because each side of the body has appendages such as legs, one advantage of bilateral symmetry is that it facilitates locomotion. Bilateral animals also have anterior and posterior ends, with the mouth at the anterior end, as described in Chapter 19. In southern China in 2004, Jun-Yuan Chen, David Bottjer, and colleagues discovered the

Figure 22.14 Fossil of an early invertebrate animal showing bilateral symmetry.   This fossil of an early animal, *Vernanimalcula guizhouena*, is dated from 580 to 600 million years ago.

Concept check: Name three other species that exhibit bilateral symmetry.

fossil of the earliest known ancestor of animals with bilateral symmetry. This minute creature, with a shape like a flattened helmet, is barely visible to the naked eye (**Figure 22.14**). The fossil is approximately 580–600 million years old.

Phanerozoic Eon: The Paleozoic Era Saw the Diversification of Invertebrates and the Colonization of Land by Plants and Animals

The proliferation of multicellular eukaryotic life has been extensive during the Phanerozoic eon, which started 543 million years ago (mya) and extends to the present day. Phanerozoic means "well-displayed life," referring to the abundance of fossils of plants and animals that have been identified from this eon. As described in Figure 22.9, the Phanerozoic eon is subdivided into three eras—the Paleozoic, Mesozoic, and Cenozoic. Because they are relatively recent and we have many fossils from these eras, each of them is further subdivided into periods. We will consider each era with its associated conditions and prevalent forms of life separately.

The term Paleozoic means ancient animal life. The Paleozoic era covers approximately 300 million years, from 543 to 248 mya, and is subdivided into six periods, known as the Cambrian, Ordovician, Silurian, Devonian, Carboniferous, and Permian. Periods are usually named after regions where rocks and fossils of that age were first discovered.

Cambrian Period (543 to 490 mya)   The climate in the Cambrian period was generally warm and wet, with no evidence of ice at the poles. During this time, there appeared to be a rapid increase (on a geological scale) in the diversity of animal species, an event called the **Cambrian explosion**. However, recent evidence suggests that many types of animal groups that were present during the Cambrian period actually arose prior to this period.

Many fossils from the Cambrian period were found in the Canadian Rockies in a rock bed called the Burgess Shale, which was discovered by Charles Walcott in 1909. At this site, both soft- and hard-bodied (shelled) invertebrates were buried in an underwater mudslide and preserved in water that was so deep and oxygen-free that decomposition was minimal (Figure 22.15a). The excellent preservation of these softer tissues is what makes this deposit unique (Figure 22.15b).

In the middle of the Cambrian period, all of the existing major types of marine invertebrates arose, plus many others that no longer exist. The Cambrian explosion generated over 100 major animal groups with significantly different body plans. Examples that still exist include echinoderms (sea urchins and starfish), arthropods (insects, spiders, and crustaceans), mollusks (clams and snails), and **chordates** (organisms with a dorsal nerve chord). Interestingly, although many new species of animals have arisen since this time, these later species have not shown a major reorganization of body plan, but instead exhibit variations on themes that were established during the Cambrian explosion. Approximately 520 million years ago, the first **vertebrates** (animals with backbones) appeared.

The cause of the Cambrian explosion is not understood. Because it occurred shortly after marine animals evolved shells, some scientists have speculated that the changes observed in animal species may have allowed them to exploit new environments. Alternatively, others have suggested that the increase in diversity may be related to atmospheric oxygen levels. During this period, oxygen levels were increasing, and perhaps more complex body plans became possible only after the atmospheric oxygen surpassed a certain threshold. In addition, as atmospheric oxygen reached its present levels, an ozone (O_3) layer was produced that screens out ultraviolet radiation, thereby allowing complex life to live in shallow water and eventually on land. Another possible contributor to the Cambrian explosion was an "arms race" between predators and their prey. The ability of predators to capture prey and the ability of prey to avoid predators may have been a major factor that resulted in a diversification of animals into many different species.

Ordovician Period (490 to 443 mya)   Like the Cambrian period, the climate of the early and middle parts of the Ordovician period was warm, and the atmosphere was moist. During this period, a diverse group of marine invertebrates, including trilobites and brachiopods, appeared in the fossil record (Figure 22.16). Marine communities consisted of invertebrates, algae, early jawless fishes (a type of early vertebrate), mollusks,

(a) The Burgess Shale

(b) A fossilized arthropod, _Marrella_

Figure 22.15   **The Cambrian explosion and the Burgess Shale.**   (a) This photograph shows the original site in the Canadian Rockies discovered by Charles Walcott. Since its discovery, this site has been made into a quarry for the collection of fossils. (b) A fossil of an extinct arthropod, _Marrella_, that was found at this site.

and corals. Fossil evidence also suggests that early land plants and arthropods may have first invaded the land during this period.

Toward the end of the Ordovician period, the climate changed rather dramatically. Large glaciers formed, which drained the relatively shallow oceans, causing the water levels to drop. This resulted in a mass extinction in which as many as 60% of the existing marine invertebrates became extinct.

Silurian Period (443 to 417 mya) In contrast to the dramatic climate changes observed during the Ordovician period, the climate during the Silurian was relatively stable. The glaciers largely melted, which caused the ocean levels to rise. No new major types of invertebrate animals appeared during this period, but significant changes were observed among existing vertebrate and plant species. Many new types of fishes appeared in the fossil record. In addition, coral reefs made their first appearance during this period.

The Silurian marked a significant colonization by terrestrial plants and animals. For this to occur, certain species evolved adaptations that prevented them from drying out, such as an external cuticle. Ancestral relatives of spiders and centipedes became prevalent. The earliest fossils of **vascular plants**, which

(a) Trilobite

(b) Brachiopod

Figure 22.16 Shelled, invertebrate fossils of the Ordovician period.

can transport water, sugar, and salts throughout the plant body, were observed in this period.

Devonian Period (417 to 354 mya) In the Devonian period, generally dry conditions occurred across much of the northern landmasses. However, the Southern Hemisphere was mostly covered by cool, temperate oceans.

The Devonian saw a major increase in the number of terrestrial species. At first, the vegetation consisted primarily of small plants, only a meter tall or less. Later, ferns, horsetails, and **seed plants**, such as gymnosperms, also emerged. By the end of the Devonian, the first trees and forests were formed. A major expansion of terrestrial animals also occurred. Insects first appeared in the fossil record, and other invertebrates became plentiful. In addition, the first **tetrapods**, vertebrates with four legs, are believed to have evolved in the Devonian. Early tetrapods included amphibians, which lived on land but required water in which to lay their eggs.

In the oceans, many types of invertebrates flourished, including brachiopods, echinoderms, and corals. This period is sometimes called the Age of Fishes, as many new types of fishes emerged. During a period of approximately 20 million years near the end of the Devonian period, a prolonged series of extinctions eliminated many marine species. The cause of this mass extinction is not well understood.

Carboniferous Period (354 to 290 mya) The term Carboniferous refers to the rich coal deposits found in rocks of this age. The Carboniferous had the ideal conditions for the subsequent formation of coal. It was a cooler period, and much of the land was covered by forest swamps. Coal was formed over many millions of years from compressed layers of rotting vegetation.

Plants and animals further diversified during the Carboniferous period. Very large plants and trees became prevalent. For example, tree ferns such as *Psaronius* grew to a height of 15 meters or more (**Figure 22.17**). The first flying insects emerged. Giant dragonflies with a wingspan of over 2 feet inhabited the forest swamps. Terrestrial vertebrates also became more diverse. Amphibians were very prevalent. One innovation that seemed particularly beneficial was the amniotic egg. In reptiles, the amniotic egg was covered with a leathery or hard shell. This prevented the desiccation of the embryo inside. This innovation was critical for the emergence of reptiles during this period.

Permian Period (290 to 248 mya) At the beginning of the Permian, continental drift had brought much of the total land together into a supercontinent known as Pangaea (see Figure 22.10). The interior regions of Pangaea were dry, with great seasonal fluctuations. The forests of fernlike plants were replaced with gymnosperms. Species resembling modern conifers first appeared in the fossil record. Amphibians were prevalent, but reptiles became the dominant vertebrate species.

At the end of the Permian period, the largest known mass extinction in the history of life on Earth occurred; 90–95% of marine species and a large proportion of terrestrial species were eliminated. The cause of the Permian extinction is the subject

Figure 22.17 A giant tree fern, *Psaronius*, from the Carboniferous period. This genus became extinct during the Permian. The illustration is a re-creation based on fossil evidence. The inset shows a fossilized section of the trunk, also known as petrified wood.

Figure 22.18 *Megazostrodon*, the first known mammal of the Triassic period. The illustration is a re-creation based on fossilized skeletons. The *Megazostrodon* was 10 to 12 cm long.

Figure 22.19 A fossil of the first bird-like animal, *Archaeopteryx*, which emerged in the Jurassic period.

of much research and controversy. One possibility is that glaciation destroyed the habitats of terrestrial species and lowered ocean levels, which would have created greater competition among marine species. Another hypothesis is that enormous volcanic eruptions in Siberia produced large ash clouds that abruptly changed the climate on Earth.

Phanerozoic Eon: The Mesozoic Era Saw the Rise and Fall of the Dinosaurs

The Permian extinction marks the division between the Paleozoic and Mesozoic eras. Mesozoic means "middle animals." It was a time period that saw great changes in animal and plant species. This era is sometimes called the Age of Dinosaurs, which flourished during this time. The climate during the Mesozoic era was consistently hot, and terrestrial environments were relatively dry. Little if any ice was found at either pole. The Mesozoic is divided into three periods, the Triassic, Jurassic, and Cretaceous.

Triassic Period (248 to 206 mya) Reptiles were plentiful in this period, including new groups such as crocodiles and turtles. The first dinosaurs emerged during the middle of the Triassic, as did the first true **mammals**, such as the small *Megazostrodon* (Figure 22.18). Gymnosperms were the dominant land plant. Volcanic eruptions near the end of the Triassic are thought to have caused global warming, resulting in mass extinctions that eliminated many marine and terrestrial species.

Jurassic Period (206 to 144 mya) Gymnosperms, such as conifers, continued to be the dominant vegetation. Mammals were not prevalent. Reptiles continued to be the dominant land vertebrate. These included dinosaurs, which were mostly terrestrial reptiles that shared certain anatomical features, such as an erect posture. Some dinosaurs attained enormous sizes, including the massive *Brachiosaurus* that reached a length of 25 m (80 ft) and weighed up to 100 tons! An early bird-like creature, *Archaeopteryx* (Figure 22.19), emerged in the Jurassic period.

Cretaceous Period (144 to 65 mya) On land, dinosaurs continued to be the dominant animals. The earliest **flowering plants**, called angiosperms, which form seeds within a protective chamber, emerged and began to diversify.

The end of the Cretaceous witnessed another mass extinction, which brought an end to many previously successful groups of organisms. The dinosaurs and many other species abruptly died out. As with the Permian extinction, the cause of this mass extinction is still debated. One plausible hypothesis suggests that a large meteorite hit the region that is now the Yucatan

Peninsula of Mexico, lifting massive amounts of debris into the air and thereby blocking the sunlight from reaching the Earth's surface. Such a dense haze could have cooled the Earth's surface by 11–15°C (20–30°F). Evidence also points to strong volcanic eruptions as a contributing factor for this mass extinction.

Phanerozoic Eon: Mammals and Flowering Plants Diversified During the Cenozoic Era

The Cenozoic era spans the most recent 65 million years. It is divided into two periods, the Tertiary and Quaternary. In many parts of the world, tropical conditions were replaced by a colder, drier climate. During this time, mammals became the largest terrestrial animals, which is why the Cenozoic is sometimes called the Age of Mammals. However, the Cenozoic era also saw an amazing diversification of many types of organisms, including birds, fishes, insects, and flowering plants.

Tertiary Period (65 to 1.8 mya)   On land, the mammals that survived from the Cretaceous period began to diversify rapidly during the early part of the Tertiary period. Angiosperms became the dominant land plant, and insects became important for their pollination. Fishes also diversified, and sharks became abundant.

Toward the end of the Tertiary period, about 7 million years ago, hominoids came into existence. **Hominoids** include modern humans, chimpanzees, gorillas, orangutans, and gibbons, plus all of their recent ancestors. The term **hominin** refers to a subset of hominoids, including modern humans, extinct human species (for example, of the *Homo* genus), and our immediate ancestors. In 2002, a fossil of the earliest known hominin, *Sahelanthropus tchadensis*, was discovered in Central Africa. This fossil was dated at between 6 and 7 million years old. Another early hominin genus, called *Australopithecus*, first emerged in Africa about 4 million years ago. Australopithecines walked upright and had a protruding jaw, prominent eyebrow ridges, and a small braincase.

Quaternary Period (1.8 mya to present)   Periodic Ice Ages have been prevalent during the last 1.8 million years, covering much of Europe and North America. This period has witnessed the widespread extinction of many species of mammals, particularly larger species. Certain species of hominins became increasingly more like living humans. Near the beginning of the Quaternary period, fossils were discovered of *Homo habilis*, or handy man, so called because stone tools were found with the fossil remains. Fossils that are classified as *Homo sapiens*—modern humans—first appeared about 170,000 years ago. The evolution of hominins is discussed in more detail in Chapter 34.

▮ Summary of Key Concepts

- Life began on Earth from nonliving material between 3.5 and 4.0 billion years ago. (Figure 22.1)

22.1 Origin of Life on Earth

- The first stage in the formation of life involved the synthesis of organic molecules to form a prebiotic soup. Possible scenarios of how this occurred are the reducing atmosphere, extraterrestrial, and deep-sea vent hypotheses. (Figures 22.2, 22.3)
- The second stage was the bonding of organic molecules to form polymers. This may have occurred on the surface of clay.
- The third stage in the evolution of the first living cells occurred when polymers became enclosed in structures called protobionts that separated them from the external environment. Examples include coacervates and liposomes. (Figure 22.4)
- The fourth stage that led to the first living cells was chemical selection in which molecules with functional properties, such as self-replication and other enzymatic functions, increased in number. (Figure 22.5)
- The precursors of living cells as well as the first living cells themselves are thought to have used RNA for both information storage and catalytic functions. This hypothetical early phase of life is termed the RNA world.
- Bartel and Szostak demonstrated that chemical selection for RNA molecules that can catalyze covalent bond formation is possible experimentally. (Figure 22.6)
- The RNA world was later superseded by the modern DNA/RNA/protein world.

22.2 Fossils

- Fossils, which are preserved remnants of past life-forms, are formed in sedimentary rock. Radioisotope dating is one way to estimate the age of a fossil. Fossils provide an extensive record of the history of life, though the record is incomplete and has biases. (Figures 22.7, 22.8, Table 22.1)

22.3 History of Life on Earth

- The geological time scale, which is divided into four eons and many eras and periods, charts the major events that occurred during the history of life on Earth. (Figure 22.9)
- The formation and extinction of species, as well as mass extinctions, are correlated with changes in temperature, atmosphere, and landmass locations, as well as floods, glaciation, volcanic eruptions, and meteorite impacts. (Figure 22.10)
- During the Archaean eon, bacteria and archaea arose. Organisms such as cyanobacteria became autotrophs and produced oxygen. Cyanobacteria were preserved in structures called stromatolites. (Figure 22.11)
- Eukaryotic cells arose during the Proterozoic eon. This origin involved a union between bacterial and archaeal cells that may have been symbiotic or endosymbiotic or both. The origin of mitochondria and chloroplasts was an endosymbiotic relationship. (Figure 22.12)
- Multicellular eukaryotes evolved during the Proterozoic eon and first emerged about 1.5 billion years ago. Multicellularity now occurs via cell division and the adherence of the resulting cells to each other. A multicellular organism can produce multiple cell types. (Figure 22.13)

- The first bilateral animal emerged toward the end of the Proterozoic eon. (Figure 22.14)
- The Phanerozoic eon is subdivided into the Paleozoic, Mesozoic, and Cenozoic eras. During the Paleozoic era, invertebrates greatly diversified, particularly during the Cambrian explosion, and the land became colonized by plants. Terrestrial vertebrates, including tetrapods, became more diverse. (Figures 22.15, 22.16, 22.17)
- Dinosaurs were prevalent during the Mesozoic era, particularly during the Jurassic period. Mammals and birds also emerged. (Figures 22.18, 22.19)
- During the Cenozoic era, mammals diversified, and flowering plants became the dominant plant species. The first hominoids emerged approximately 7 million years ago. Fossils classified as *Homo sapiens*, our species, appeared about 170,000 years ago.

Assess and Discuss

Test Yourself

1. The prebiotic soup was
   a. the assemblage of unicellular prokaryotes and eukaryotes that existed in the oceans of early Earth.
   b. the accumulation of organic molecules in the oceans of early Earth.
   c. the mixture of organic molecules that was found in the cytoplasm of the earliest cells on Earth.
   d. a pool of nucleic acids that contained the genetic information for the earliest organisms.
   e. none of the above.

2. Which of the following is <u>not</u> a characteristic of protobionts that was necessary for the evolution of living cells?
   a. a membrane-like boundary separating the external environment from an internal environment
   b. polymers capable of functioning in information storage
   c. polymers capable of enzymatic activity
   d. self-replication
   e. compartmentalization of metabolic activity

3. RNA is believed to be the first functional macromolecule in protobionts because it
   a. is easier to synthesize compared to other macromolecules.
   b. has the ability to store information, self-replicate, and perform enzymatic activity.
   c. is the simplest of the macromolecules commonly found in living cells.
   d. All of the above are correct.
   e. a and c only are correct.

4. The movement of landmasses that have changed their positions, shapes, and association with other landmasses is called
   a. glaciation.   d. biogeography.
   b. Pangaea.   e. geological scale.
   c. continental drift.

5. Paleontologists estimate the dates of fossils by
   a. the layer of rock in which the fossils are found.
   b. analysis of radioisotopes found in nearby rock.
   c. the complexity of the body plan of the organism.
   d. all of the above.
   e. a and b only.

6. The fossil record does not give us a complete picture of the history of life because
   a. not all past organisms have become fossilized.
   b. only organisms with hard skeletons can become fossilized.
   c. fossils of very small organisms have not been found.
   d. fossils of early organisms are located too deep in the crust of the Earth to be found.
   e. all of the above.

7. The endosymbiosis hypothesis explaining the evolution of eukaryotic cells is supported by
   a. DNA-sequencing analysis comparing bacterial genomes, mitochondrial genomes, and eukaryotic nuclear genomes.
   b. naturally occurring examples of endosymbiotic relationships between bacterial cells and eukaryotic cells.
   c. the presence of DNA in mitochondria and chloroplasts.
   d. all of the above.
   e. a and b only.

8. Which of the following explanations of the evolution of multicellularity in eukaryotes is seen in the development of complex, multicellular organisms today?
   a. endosymbiosis
   b. aggregation of cells to form a colony
   c. division of cells with the resulting cells sticking together
   d. multiple cell types aggregating to form a complex organism
   e. none of the above

9. The earliest fossils of vascular plants were formed during the _____ period.
   a. Ordovician   c. Devonian   e. Jurassic
   b. Silurian   d. Triassic

10. The appearance of the first hominoids date to the _____ period.
   a. Triassic   c. Cretaceous   e. Quaternary
   b. Jurassic   d. Tertiary

Conceptual Questions

1. What are the four stages that led to the origin of living cells?

2. How are the ages of fossils determined? In your answer, you should discuss which types of rocks are analyzed and explain the concepts of radioisotope dating and half-life.

3. What is meant by the Cambrian explosion? What is the significance of the Cambrian explosion with regard to modern animal species?

Collaborative Questions

1. Discuss possible hypotheses of how organic molecules were first formed.

2. Discuss the key features of a protobiont. What distinguishes a protobiont from a living cell?

Online Resource

www.brookerbiology.com

Stay a step ahead in your studies with animations that bring concepts to life and practice tests to assess your understanding. Your instructor may also recommend the interactive ebook, individualized learning tools, and more.

An Introduction to Evolution

23

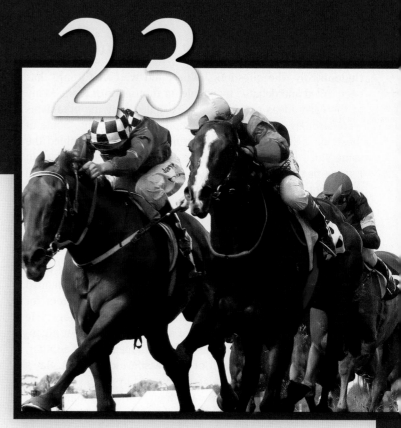

O rganic life beneath the shoreless waves
Was born and nurs'd in Ocean's pearly caves
First forms minute, unseen by spheric glass,
Move on the mud, or pierce the watery mass;
These, as successive generations bloom,
New powers acquire, and larger limbs assume;
Whence countless groups of vegetation spring,
And breathing realms of fin, and feet, and wing.

From The Temple of Nature *by Erasmus Darwin (1731–1802), grandfather of Charles Darwin. Published posthumously in 1803.*

Selective breeding. The horses in this race have been bred for a particular trait, in this case, speed. Such a practice, called selective breeding, can dramatically change the traits of organisms over several generations.

The term **evolution** is used to describe a heritable change in one or more characteristics of a population from one generation to the next. Evolution can be viewed on a small scale (**microevolution**) as it relates to changes in a single gene or allele frequencies in a population over time, or it can be viewed on a larger scale (**macroevolution**) as it relates to the formation of new species or groups of related species.

It is helpful to begin our discussion of evolution with a definition of a species. As a working definition, biologists often define a **species** as a group of related organisms that share a distinctive form. Among species that reproduce sexually, such as plants and animals, members of the same species are capable of interbreeding in nature to produce viable and fertile offspring. The term **population** refers to all members of a species that live in the same area and have the opportunity to interbreed. Some of the emphasis in the study of evolution is on understanding how populations change over the course of many generations to produce new species.

In the first part of this chapter, we will examine the development of evolutionary thought and some of the basic tenets of evolution, particularly those proposed by Charles Darwin in the mid-1800s. Though the theory of evolution has been refined over the past 150 years or so, the fundamental principle of evolution remains unchanged and has provided a cornerstone for our understanding of biology. Theodosius Dobzhansky, an influential evolutionary scientist of the 1900s, once said, "Nothing in biology makes sense except in the light of evolution." The extraordinarily diverse and seemingly bizarre array of species on our planet can be explained within the context of evolution. As is the case with all scientific theories, evolution is called

a theory because it is supported by a substantial body of evidence and because it explains a wide range of observations. The theory of evolution provides answers to many questions related to the diversity of life. In biology, theories such as this are viewed as scientific knowledge.

In the second part of this chapter, we will survey the extensive data that illustrate the processes by which evolution occurs. These data not only support the theory of evolution but also allow us to understand the interrelatedness of different species, whose similarities are often due to descent from a common ancestor. Much of the early evidence supporting evolution came from visual observations and comparisons of living and extinct species. More recently, advances in molecular genetics, particularly those related to DNA sequencing and genomics, have revolutionized the study of evolution. Scientists now have information that allows us to understand how evolution involves changes in the DNA of a given species. These changes affect both a species' genes and the proteins they encode. The term **molecular evolution** refers to the molecular changes in genetic material that underlie the phenotypic changes associated with evolution. A theme of this textbook, namely genomes and proteomes, is rooted in an understanding of these changes. In the last section of this chapter, we consider some of the exciting new information that helps us to appreciate evolutionary change at the molecular level. In the following chapters of this unit, we will examine how such changes are acted upon by evolutionary factors in ways that alter the traits of a given species and may eventually lead to the formation of new species.

23.1 The Theory of Evolution

Undoubtedly, the question, "Where did we come from?" has been asked and debated by people for thousands of years. Many of the early ideas regarding the existence of living organisms were strongly influenced by religion and philosophy. Some of these ideas suggested that all forms of life have remained the same since their creation. In the 1600s, however, scholars in Europe began a revolution that created the basis of empirical and scientific thought. **Empirical thought** relies on observation to form an idea or hypothesis rather than trying to understand life from a nonphysical or spiritual point of view. The shift toward empirical thought encouraged scholars to look for the basic rationale behind a given process or phenomenon.

In the mid- to late-1600s, the first scientist to carry out a thorough study of the living world was an Englishman named John Ray, who developed an early classification system for plants and animals based on anatomy and physiology. He established the modern concept of a species, noting that organisms of one species do not interbreed with members of another, and used it as the basic unit of his classification system. Ray's ideas on classification were later expanded by the Swedish naturalist Carolus Linnaeus. How did their work contribute to the development of evolutionary theory? Neither Ray nor Linnaeus proposed that evolutionary change promotes the formation of new species. However, their systematic classification of plants and animals helped scholars of this period perceive the similarities and differences among living organisms.

Late in the 1700s, a small number of European scientists began to quietly suggest that life-forms are not fixed. A French zoologist, George Buffon, actually proposed that living things change through time. However, Buffon was careful to hide his views in a 44-volume natural history book series. He was a quiet pioneer in asserting that species can change over generations.

Around the same time, a French naturalist named Jean-Baptiste Lamarck suggested an intimate relationship between variation and evolution. By examining fossils, he realized that some species had remained the same over the millennia and others had changed. Lamarck hypothesized that species change over the course of many generations by adapting to new environments. He believed that living things evolved in a continuously upward direction, from dead matter, through simple to more complex forms, toward human "perfection." According to Lamarck, organisms altered their behavior in response to environmental change. He thought that behavioral changes modified traits and hypothesized that such modified traits were inherited by offspring. He called this idea the **inheritance of acquired characteristics**. For example, according to Lamarck's hypothesis, giraffes developed their elongated necks and front legs by feeding on high tree leaves. The exercise of stretching up to the leaves altered the neck and legs, and Lamarck presumed that these acquired characteristics were transmitted to offspring. However, further research has rejected Lamarck's idea that acquired traits can be inherited. Even so, Lamarck's work was important in promoting the idea of evolutionary change.

Interestingly, Erasmus Darwin, the grandfather of Charles Darwin, who was a contemporary of Buffon and Lamarck, was an early advocate of evolutionary change. He was a physician, a plant biologist, and also a poet (see poem at the beginning of the chapter). He was aware that modern species were different from similar types of fossilized organisms and also saw how plant and animal breeders used breeding practices to change the traits of domesticated species (see chapter-opening photo). He knew that offspring inherited features from their parents, and went so far as to say that life on Earth could have descended from a common ancestor.

Overall, Charles Darwin's many scientific predecessors set the stage for the theory of evolution. With this historical introduction, we will now consider Darwin's observations and the tenets that provide the foundation for this theory.

Darwin Suggested That Species Are Derived from Pre-existing Species

Charles Darwin, a British naturalist born in 1809, played a key role in developing the theory that existing species have evolved from pre-existing species. Darwin's unique perspective and his ability to formulate evolutionary theory were shaped by several different fields of study, including ideas of his time about geological and biological processes.

Two main geological hypotheses predominated in the early 19th century. Catastrophism was first proposed by French zoologist and paleontologist Georges Cuvier to explain the age of the Earth. Cuvier suggested that the Earth was just 6,000 years old and that only catastrophic events had changed its geological structure. This idea fit well with religious teachings. Alternatively, uniformitarianism, proposed by Scottish geologist James Hutton and popularized by geologist Charles Lyell, suggested that changes in the Earth are directly caused by recurring events. For example, they suggested that geological processes such as erosion existed in the past and happened at the same gradual rate as they do now. For such slow geological processes to eventually lead to substantial changes in the Earth's characteristics, a great deal of time was required. Hutton and Lyell were the first to propose that the age of the Earth is well beyond 6,000 years. The ideas of Hutton and Lyell helped to shape Darwin's view of the world.

Darwin's thinking was also influenced by a paper published in 1798 called *Essay on the Principle of Population* by Thomas Malthus, an English economist. Malthus asserted that

(a) Charles Darwin

(b) The voyage of the *Beagle*

Figure 23.1 Charles Darwin and the voyage of the *Beagle*, 1831–1836. (a) A portrait of Charles Darwin (1809–1882) while in Ecuador. (b) Darwin's voyage on the *Beagle*, which took almost 5 years to circumnavigate the world.

the population size of humans can, at best, increase linearly due to increased land usage and improvements in agriculture, whereas our reproductive potential is exponential (for example, doubling with each generation). He argued that famine, war, and disease will limit population growth, especially among the poor. The relevant message from Malthus's work was that only a fraction of any population will survive and reproduce.

Darwin's ideas, however, were most influenced by his own experiences and observations. His work aboard the HMS *Beagle*, a survey ship, lasted from 1831 to 1836 and involved a careful examination of many different species (Figure 23.1). The main mission of the *Beagle* was to map the coastline of southern South America and take oceanographic measurements. Darwin's job was to record the weather, geological features, plants, animals, fossils, rocks, minerals, and indigenous people.

Though Darwin made many interesting observations on his journey, he was particularly struck by the distinctive traits of island species. For example, Darwin observed several species of finches found on the Galápagos Islands, a group of volcanic islands 600 miles from the coast of Ecuador. Though it is often assumed that Darwin's personal observations of these finches directly inspired his theory of evolution, this is not the case. Initially, Darwin thought the birds were various species of blackbirds, grosbeaks, and finches. Later, however, the bird specimens from the islands were given to the British ornithologist John Gould, who identified them as several new finch species. Gould's observations helped Darwin in the later formulation of his theory.

As seen in **Table 23.1**, the finches differed widely in the size and shape of their beaks and in their feeding habits. For example, the ground and vegetarian finches have sturdy, crushing beaks they use to crush various sizes of seeds or buds. The tree

finches have grasping beaks they use to pick up insects from trees. The mangrove, woodpecker, warbler, and cactus finches have pointed, probing beaks. They use their beaks to search for insects in crevices. The cactus finches use their probing beaks to open cactus fruits and eat the seeds. One species, the woodpecker finch, even uses twigs or cactus spines to extract insect larvae from holes in dead tree branches. Darwin clearly saw the similarities among these species, yet he noted the differences that provided them with specialized feeding strategies. We now know these finches all evolved from a single species similar to the dull-colored grassquit finch (*Tiaris obscura*), commonly found along the Pacific Coast of South America. Once on the Galápagos Islands, the finches' ability to survive and reproduce in their new habitat depended, in part, on changes in the size and shape of their beaks over many generations. These specializations enabled succeeding generations to better obtain particular types of food.

With an understanding of geology and population growth, and his observations from his voyage on the *Beagle*, Darwin had formulated his theory of evolution by the mid-1840s. He had also catalogued and described all of the species he had collected on his *Beagle* voyage except for one type of barnacle. Some have speculated that Darwin may have felt that he should establish himself as an expert on one species before making generalizations about all of them. Therefore, he spent several additional years studying barnacles. During this time, the geologist Charles Lyell, who had greatly influenced Darwin's thinking, strongly encouraged Darwin to publish his theory of evolution. In 1856, Darwin began to write a long book to explain his ideas. In 1858, however, Alfred Wallace, a naturalist working in the East Indies, sent Darwin an unpublished manuscript to read prior to its publication. In it, Wallace proposed the same ideas

Table 23.1	A Comparison of Beak Type and Diet Among the Galápagos Finches Darwin Studied

Type of finch	Species		Type of beak	Diet
Ground finches	Large ground finch (*Geospiza magnirostris*)		Crushing	Seeds—Ground finches have crushing beaks to crush various sizes of seeds; large beaks can crush large seeds, whereas smaller beaks are better for crushing small seeds.
	Medium ground finch (*G. fortis*)			
	Small ground finch (*G. fuliginosa*)			
	Sharp-billed ground finch (*G. difficilis*)			
Vegetarian finch	Vegetarian finch (*Platyspiza crassirostris*)		Crushing	Buds—Vegetarian finches have crushing beaks to pull buds from branches.
Tree finches	Large tree finch (*Camarhynchus psittacula*)		Grasping	Insects—Tree finches have grasping beaks to pick insects from trees. Those with heavier beaks can also break apart wood in search of insects.
	Medium tree finch (*Camarhynchus pauper*)			
	Small tree finch (*Camarhynchus parvulus*)			
Tree finches	Mangrove finch (*Cactospiza heliobates*)		Probing	Insects—These finches have probing beaks to search for insects in crevices and then to pick them up. The woodpecker finch can also use a cactus spine for probing.
	Woodpecker finch (*Camarhynchus pallidus*)			
Warbler finch	Warbler finch (*Certhidea olivacea*)		Probing	
Cactus finches	Large cactus finch (*G. conirostris*)		Probing	Seeds—Cactus finches have probing beaks to open cactus fruits and take out seeds.
	Cactus finch (*G. scandens*)			

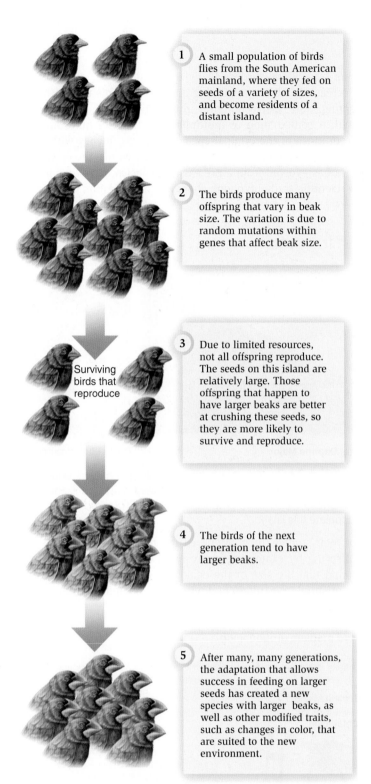

1 A small population of birds flies from the South American mainland, where they fed on seeds of a variety of sizes, and become residents of a distant island.

2 The birds produce many offspring that vary in beak size. The variation is due to random mutations within genes that affect beak size.

3 Due to limited resources, not all offspring reproduce. The seeds on this island are relatively large. Those offspring that happen to have larger beaks are better at crushing these seeds, so they are more likely to survive and reproduce.

Surviving birds that reproduce

4 The birds of the next generation tend to have larger beaks.

5 After many, many generations, the adaptation that allows success in feeding on larger seeds has created a new species with larger beaks, as well as other modified traits, such as changes in color, that are suited to the new environment.

Figure 23.2 **Evolutionary adaptation to a new environment via natural selection.** The example shown here involves a species of finch adapting to a new environment on a distant island. The plants on this island produce larger seeds than do the plants on the mainland from which the birds originated. According to Darwin's theory of evolution by natural selection, the process of adaptation eventually leads to the formation of a new species with larger beaks that are better suited to crushing the large seeds.

Concept check: *The phrase "an organism evolves" is incorrect. Explain why.*

concerning evolution. In response to this, Darwin decided to use some of his own writings on this subject, and two papers, one by Darwin and one by Wallace, were published in the *Proceedings of the Linnaean Society of London*. These papers were not widely recognized. A year later, however, Darwin finished his book *The Origin of Species* (1859), which described his ideas in greater detail and included observational support. This book, which received high praise from many scientists and scorn from others, started a great debate concerning evolution. Although some of his ideas were incomplete because the genetic basis of traits was not understood at that time, Darwin's work remains one of the most important contributions to our understanding of biology.

Natural Selection Can Change Populations from Generation to Generation

As mentioned, evolution is a heritable change of one or more characteristics in a population from one generation to the next. Darwin hypothesized that existing life-forms on our planet result from the modification of pre-existing life-forms. He expressed this concept of evolution as "the theory of descent with modification through variation and natural selection." The term evolution refers to change. What factors bring about evolutionary change? According to Darwin's ideas, evolution may occur from generation to generation due to two interacting factors, genetic variation and natural selection:

1. Variation in traits may occur among individuals of a given species. The heritable traits are then passed from parents to offspring. The genetic basis for variation within a species was not understood at the time Darwin proposed his theory of evolution. We now know that such variation is due to different types of genetic changes such as random mutations in genes. Even though Darwin did not fully appreciate the genetic basis of variation, he and many other people before him observed that offspring resemble their parents more than they do unrelated individuals. Therefore, he assumed that some traits are passed from parent to offspring.

2. In each generation, many more offspring are usually produced than will survive and reproduce. Often times, resources in the environment are limiting for an organism's survival. During the process of **natural selection**, individuals with heritable traits that make them better suited to their native environment tend to flourish and reproduce, whereas other individuals are less likely to survive and reproduce. As a result of natural selection, certain traits that favor reproductive success become more prevalent in a population.

As an example, we can consider a population of finches that migrates from the South American mainland to a distant island (**Figure 23.2**). Variation exists in the beaks sizes among the migrating birds. Let's suppose the seeds produced on the distant island are larger than those produced on the mainland. Those birds with larger beaks would be better able to feed on these larger seeds and therefore would be more likely to survive and pass on that trait to their offspring. What are the consequences of this selection process? In succeeding generations, the population will tend to have a greater proportion of finches with larger beaks. Alternatively, if a trait happens to be detrimental to an individual's ability to survive and reproduce, natural selection is likely to eliminate this type of variation. For example, if a finch in the same environment had a small beak, this bird would be less likely to acquire food, which would decrease its ability to survive and pass on this trait to its offspring. Natural selection may ultimately result in a new species with a combination of multiple traits that are quite different from those of the original species, such as finches with larger beaks and changes in coloration. In other words, the newer species has evolved from a pre-existing species. Now let's look at a scientific study involving change in a population over time.

FEATURE INVESTIGATION

The Grants Have Observed Natural Selection in Galápagos Finches

Since 1973, Peter Grant, Rosemary Grant, and their colleagues have studied natural selection in finches found on the Galápagos Islands. For over 30 years, the Grants have focused much of their work on one of the Galápagos Islands known as Daphne Major (**Figure 23.3a**). This small island (0.34 km²) has a moderate degree of isolation (it is 8 km from the nearest island), an undisturbed habitat, and a resident population of *Geospiza fortis*, the medium ground finch (**Figure 23.3b**).

To study natural selection, the Grants have observed various traits in finches over the course of many years. One example is beak size. The medium ground finch has a relatively small crushing beak, allowing it to more easily feed on small, tender seeds (see Table 23.1). The Grants quantified beak size among the medium ground finches of Daphne Major by carefully measuring beak depth (a measurement of the beak from top to bottom, at its base) on individual birds (**Figure 23.4**). The small size of the island made it possible for them to measure a large percentage of birds and their offspring. During the course

(a) Daphne Major   **(b) Medium ground finch**

Figure 23.3   The Grants' investigation of natural selection in finches.   (a) Daphne Major, one of the Galápagos Islands. (b) One of the medium ground finches (*Geospiza fortis*) that populate this island.

of their studies, they compared the beak sizes of parents and offspring by examining many broods over several years and found that the size of the beak was transmitted from parents to offspring, regardless of environmental conditions, indicating

Figure 23.4 The Grants and natural selection of beak size among the medium ground finch.

HYPOTHESIS   Dry conditions produce larger seeds and may result in larger beaks in succeeding generations of *Geospiza fortis* due to natural selection.

KEY MATERIALS   A population of *G. fortis* on the Galápagos Island called Daphne Major.

	Experimental level	Conceptual level
1 In 1976, measure beak depth in parents and offspring of the species *G. fortis*.	Capture birds and measure beak depth.	This is a way to measure a trait that may be subject to natural selection.
2 Repeat the procedure on offspring that were born in 1978 and had reached mature size. A drought had occurred in 1977 that caused plants on the island to produce mostly large dry seeds and relatively few small seeds.	Capture birds and measure beak depth.	This is a way to measure a trait that may be subject to natural selection.

3 **THE DATA**

4 **CONCLUSION** Because a drought produced larger seeds, birds with larger beaks were more likely to survive and reproduce. The process of natural selection produced postdrought offspring that had larger beaks compared to predrought offspring.

5 **SOURCE** Grant, B. Rosemary, and Grant, Peter R. 2003. What Darwin's Finches Can Teach Us about the Evolutionary Origin and Regulation of Biodiversity. *Bioscience* 53:965–975.

that differences in beak sizes are due to genetic differences in the population. In other words, they found that beak size was a heritable trait.

By measuring many birds every year, the Grants were able to assemble a detailed portrait of natural selection in action. In the study shown in Figure 23.4, they measured beak depth from 1976 to 1978. In the wet year of 1976, the plants of Daphne Major produced an abundance of the small, tender seeds that these finches could easily eat. However, a severe drought occurred in 1977. During this year, the plants on Daphne Major tended to produce few of the smaller seeds, which the finches rapidly consumed. Therefore, the finches resorted to eating larger, drier seeds, which are harder to crush. As a result, birds with larger beaks were more likely to survive and reproduce because they were better at breaking open these large seeds. As shown in the data, the average beak depth of birds in the population increased substantially, from 8.8 mm in predrought

offspring to 9.8 mm in postdrought offspring. How do we explain these results? According to evolutionary theory, birds with larger beaks were more likely to survive and pass this trait on to their offspring. Overall, these results illustrate the power of natural selection to alter the features of a trait—in this case, beak depth—in a given population.

Experimental Questions

1. What features of Daphne Major made it a suitable field site for studying the effects of natural selection?

2. Why is beak depth in finches a good trait for a study of natural selection? What environmental conditions were important to allow the Grants to collect information concerning natural selection?

3. What were the results of the Grants' study following the drought in 1977? What impact did these results have on the theory of evolution?

23.2 Observations of Evolutionary Change

As we have just seen, the Grants provided evidence for evolutionary change in a population of finches. Observations that support the theory of evolution have been gleaned from many sources (**Table 23.2**). Historically, the first evidence of biological evolution came from studies of the fossil record, the distribution of related species on our planet, selective breeding experiments, and the comparison of similar anatomical features in different species. More recently, additional observations that illustrate the process of evolution have been revealed at the molecular level. By comparing DNA sequences from many different species, evolutionary biologists have gained great insight into the relationship between

the evolution of species and the associated changes in the genetic material. In this section, we will survey a variety of observations that show the process of evolutionary change.

Fossils Show Successive Evolutionary Change

As discussed in Chapter 22, the fossil record has provided biologists with evidence of the history of life on Earth. Today, scientists have access to a far more extensive fossil record than was available to Darwin or other scientists of his time. Even though the fossil record is still incomplete, the many fossils that have been discovered often provide detailed information regarding evolutionary change in a series of related organisms. When fossils are compared according to their age, from oldest to youngest, successive evolutionary change becomes apparent.

Table 23.2	Observations of Biological Evolution
Type of observation	**Description**
Fossil record	When fossils are compared according to their age, from oldest to youngest, successive evolutionary change becomes apparent.
Biogeography	Unique species found on islands and other remote areas have arisen because the species in these locations have evolved in isolation from the rest of the world.
Convergent evolution	Two different species from different lineages sometimes become anatomically similar because they occupy similar environments. This indicates that natural selection promotes adaptation to a given environment.
Selective breeding	The traits in domesticated species have been profoundly modified by selective breeding (also called artificial selection) in which breeders choose the parents that have desirable traits.
Homologies:*	
Anatomical	Evolutionarily related species may possess homologous structures that have been modified in ways that allowed them to be used differently by each species. In some cases, such structures are no longer needed and degenerate to vestigial structures.
Developmental	An analysis of embryonic development often reveals similar anatomical features that point to past evolutionary relationships.
Molecular	At the molecular level, certain characteristics are found in all living cells, suggesting that all living species are derived from a common ancestor. In addition, species that are closely related evolutionarily have DNA sequences that are more similar to each other than they are to those in distantly related organisms.

*Homology is a similarity that occurs due to descent from a common ancestor.

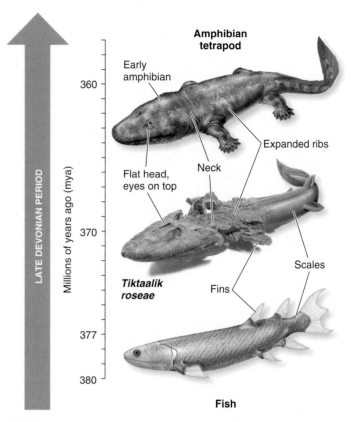

Figure 23.5 **Evolutionary change in the tetrapod lineage, showing a transitional form.** This figure shows two early tetrapod ancestors, a Devonian fish and the transitional form *Tiktaalik roseae*, as well as one of their descendants, an early amphibian. An analysis of the fossils shows that *T. roseae*, also known as a fishapod, had both fish and amphibian characteristics, so it was likely able to survive brief periods out of the water.

Let's consider a couple of examples in which paleontologists have observed evolutionary change. In 2005, fossils of *Tiktaalik roseae*, nicknamed fishapod, were discovered by Ted Daeschler, Neil Shubin, and Farish Jenkins. The discovery of fishapod illuminates one of several steps that led to the evolution of tetrapods, which are animals with four legs. *T. roseae* is called a **transitional form** because it displays an intermediate state between an ancestral form and the form of its descendants (**Figure 23.5**). In this case, the fishapod is a transitional form between fishes, which have fins for locomotion, and tetrapods, which are four-limbed animals. Unlike a true fish, *T. roseae* had a broad skull, a flexible neck, and eyes mounted on the top of its head like a crocodile. Its interlocking rib cage suggests it had primitive lungs. Perhaps the most surprising discovery was that its pectoral fins (those on the side of the body) revealed the beginnings of a primitive wrist and five finger-like bones. These appendages would have allowed *T. roseae* to support its body on shallow river bottoms and lift its head above the water to search for prey, and perhaps even move out of the water for short periods. During the Devonian period, 417–354 million years ago, this could have been an important advantage in the marshy floodplains of large rivers.

One of the best-studied examples of evolutionary change is that of the horse family, modern members of which include horses, zebras, and donkeys. These species, which are large, long-legged animals adapted to living in open grasslands, are the remaining descendants of a long lineage that produced many species that have subsequently become extinct since its origin approximately 55 million years ago. Examination of the horse lineage provides a particularly interesting case of how evolution involves adaptation to changing environments.

The earliest known fossils of the horse family revealed that the animals were small with short legs and broad feet (**Figure 23.6**). Early horses, such as *Hyracotherium*, lived in wooded habitats and are thought to have browsed on leaves and herbs. Between the time of these first members of the horse family and modern horses, the fossil record has revealed changes in size, foot anatomy, and tooth morphology among this group of related species. Early horses were the size of dogs, whereas modern horses typically weigh more than a half ton. *Hyracotherium* had four toes on its front feet and three on its hind feet. Instead of hooves, these toes were encased in fleshy pads. By comparison, the feet of modern horses have a single toe, enclosed in a tough, bony hoof. The fossil record shows an increase in the length of the central toe, the development of a bony hoof, and the loss of the other toes. Finally, the teeth of *Hyracotherium* were relatively small compared to those of modern horses. Over the course of millions of years, horse molars have increased in size and developed a complex pattern of ridges.

How do evolutionary biologists explain these changes in horse characteristics? The changes can be attributed to natural

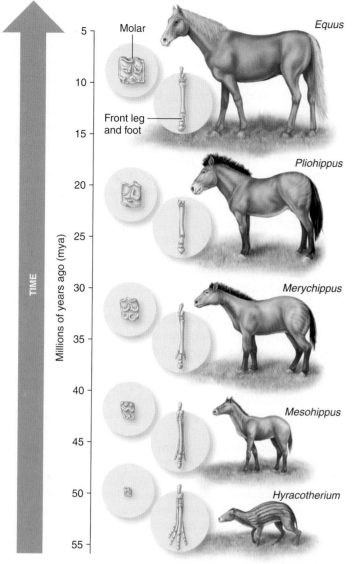

TIME

Millions of years ago (mya)

Molar

Front leg and foot

Equus

Pliohippus

Merychippus

Mesohippus

Hyracotherium

Figure 23.6 Evolutionary changes that led to the modern horse. Some major changes occurred in horses' body size, foot anatomy, and tooth morphology. These changes are hypothesized to be due to natural selection in a changing environment over the last 55 million years. Note: This figure is meant to emphasize changes that led to modern horses. The evolutionary pathway that produced modern horses involves several branches and is described in Chapter 26.

selection, which acted upon existing variation and resulted in adaptations to changes in global climates. Over North America, where much of horse evolution occurred, large areas of dense forests were replaced with grasslands. The increase in size and changes in foot structure enabled horses to escape predators more easily and travel greater distances in search of food. The changes seen in horses' teeth are consistent with a shift from eating more tender leaves to eating grasses and other vegetation that are more abrasive and require more chewing.

Biogeography Indicates That Species in a Given Area Have Evolved from Pre-existing Species

Biogeography is the study of the geographic distribution of extinct and living species. Patterns of past evolution are often found in the natural geographic distribution of related species. From such studies, scientists have discovered that isolated continents and island groups have evolved their own distinct plant and animal communities. As mentioned earlier, Darwin himself observed several species of finches found on the Galápagos Islands that had unique characteristics, such as beak shapes, when compared with similar finches found on the mainland. Scientists now hypothesize that these island species evolved from mainland birds that had migrated to the islands and then became adapted to a variety of new feeding habits (see Figure 23.2).

Around the world, islands, which are isolated from other large landmasses, provide numerous examples in which geography has played a key role in the evolution of new species. Islands often have many species of plants and animals that are **endemic**, which means they are naturally found only in a particular location. Most endemic island species have closely related relatives on nearby islands or the mainland. For example, consider the island fox (*Urocyon littoralis*), which lives on the Channel Islands located off the coast of southern California between Los Angeles and Santa Barbara (**Figure 23.7**). This type of fox is found nowhere else in the world. It weighs about 3–6 pounds and feeds largely on insects, mice, and fruits. The island fox evolved from the mainland gray fox (*Urocyon cinereoargenteus*), which is much larger, usually 7–11 pounds. During the last Ice Age, about 16,000–18,000 years ago, the Santa Barbara channel was frozen and narrow enough for ancestors of the mainland gray fox to cross over to the Channel Islands. When the Ice Age ended, the ice melted and sea levels rose, causing the foxes to be cut off from the mainland. Over the last 16,000–18,000 years, the population of foxes on the Channel Islands evolved into the smaller island fox, which is now considered a different species from the larger gray fox. The gray fox is still found on the mainland. The smaller size of the island fox is an example of island dwarfing, a phenomenon in which the size of large animals isolated on an island shrinks dramatically over many generations. It is an example of natural selection in which smaller size provides a survival and reproductive advantage probably because of limited food.

The evolution of major animal groups is also correlated with known changes in the distribution of landmasses on the Earth. The first mammals arose approximately 200 million years ago, when the area that is now Australia was still connected to the other continents. However, the first placental mammals, which have a long internal gestation and give birth to well-developed offspring, evolved much later after continental drift had separated Australia from the other continents (refer back to Figure 22.10). Except for a few species of bats and rodents that have migrated to Australia more recently, Australia lacks any of the larger, terrestrial placental mammals. How do biologists explain this observation? It is consistent with the idea that placental mammals first arose somewhere other than Australia, and that the barrier of a large ocean prevented most terrestrial placental mammals from migrating there. On the other hand, Australia has more than 100 species of kangaroos, koalas, and other marsupials, most of which are not found on any other continent. Marsupials are a group of mammal species in which young are born in a very immature condition and then develop further in the mother's abdominal pouch, which covers the mammary glands.

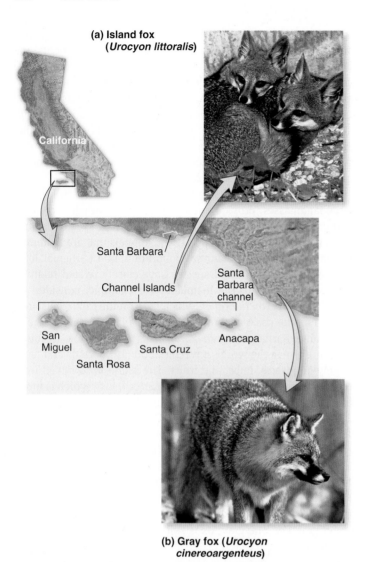

(a) Island fox
(*Urocyon littoralis*)

California

Santa Barbara

Channel Islands

Santa Barbara channel

San Miguel

Santa Cruz

Anacapa

Santa Rosa

(b) Gray fox (*Urocyon cinereoargenteus*)

Figure 23.7 The evolution of an endemic island species from a mainland species. (a) The smaller island fox found on the Channel Islands evolved from (b) the gray fox found on the California mainland.

Concept check: *Explain how geography played a key role in the evolution of the island fox.*

Evolutionary theory is consistent with the idea that the existence of these unique Australian species is due to their having evolved in isolation from the rest of the world for millions of years.

Convergent Evolution Suggests Adaptation to the Environment

The process of natural selection is also evident in the study of plants and animals that have similar characteristics, even though they are not closely related evolutionarily. This similarity is the result of **convergent evolution**, in which two different species from different lineages have independently evolved similar characteristics because they occupy similar environments. For example, both the giant anteater (*Myrmecophaga tridactyla*), found in South America, and the echidna (*Tachyglossus aculeatus*), found

in Australia, have a long snout and tongue. Both species independently evolved these adaptations that enable them to feed on ants (**Figure 23.8a**). The giant anteater is a placental mammal, whereas the echidna is an egg-laying mammal known as a monotreme, so they are not closely related evolutionarily.

Another example of convergent evolution involves aerial rootlets found in vines such as English ivy (*Hedera helix*) and wintercreeper (*Euonymus fortunei*) (**Figure 23.8b**). Based on differences in their structures, these aerial rootlets appear to have developed independently as an effective means to cling to the support on which a vine attaches itself.

A third example of convergent evolution is revealed by the molecular analysis of fishes that live in very cold water. Antifreeze proteins enable certain species of fishes to survive the subfreezing temperatures of Arctic and Antarctic waters by inhibiting the formation of ice crystals in body fluids. Researchers have determined that these fishes are an interesting case of convergent evolution (**Figure 23.8c**). Among different species of fishes, one of five different genes has independently evolved to produce antifreeze proteins. For example, in the sea raven (*Hemitripterus americanus*), the antifreeze protein is rich in the amino acid cysteine, and the secondary structure of the protein is in a β sheet conformation. In contrast, the antifreeze protein in the longhorn sculpin (*Trematomus nicolai*) is encoded by an entirely different gene. The antifreeze protein in this species is rich in the amino acid glutamine, and the secondary structure of the protein is largely composed of α helices.

The similar characteristics in the examples shown in Figure 23.8—for example, the snouts of the anteater and the echidna—are called **analogous structures** or **convergent traits**. They represent cases in which characteristics have arisen independently, two or more times, because different species have occupied similar types of environments on the Earth.

Selective Breeding Is a Human-Driven Form of Selection

The term **selective breeding** refers to programs and procedures designed to modify traits in domesticated species. This practice, also called **artificial selection**, is related to natural selection. In forming his theory of evolution, Charles Darwin was influenced by his observations of selective breeding by pigeon breeders. The primary difference between natural and artificial selection is how the parents are chosen. Natural selection occurs because of natural variation in reproductive success. Organisms that are able to survive and reproduce are more likely to pass their genes on to future generations. Environmental factors often determine which individuals will be successful parents. In artificial selection, the breeder chooses as parents those individuals that possess traits that are desirable from a human perspective.

The underlying phenomenon that makes selective breeding possible is genetic variation. Within a population, variation may exist in a trait of interest. For selective breeding to be successful, the underlying cause of the phenotypic variation is usually related to differences in **alleles**, variant forms of a particular gene, that determine the trait. The breeder will choose

(a) The long snouts and tongues of the giant anteater (left) and the echidna (right), allow them to feed on ants.

(b) The aerial rootlets of English ivy (left) and wintercreeper (right), allow them to climb up supports.

(c) The sea raven (left) and the longhorn sculpin (right) have antifreeze proteins that enable them to survive in frigid waters.

Figure 23.8 Examples of convergent evolution. All three pairs of organisms shown in this figure are not closely related evolutionarily but occupy similar environments, suggesting that natural selection promotes the formation of similar adaptations that are well suited to a particular environment.

Concept check: Can you think of another example in which two species that are not closely related have a similar adaptation?

parents with desirable phenotypic characteristics. For centuries, humans have employed selective breeding to obtain domesticated species with interesting or agriculturally useful characteristics. For example, many common breeds of dog are the result of selective breeding strategies (**Figure 23.9**). All dogs are members of the same species, *Canis lupus*, subspecies *familiaris*, so they can be interbred to produce offspring. Selective breeding can dramatically modify the traits in a species. When you compare certain breeds of dogs (for example, a greyhound and a dachshund), they hardly look like members of the same species! Recent work in 2007 by Nathan Sutter and colleagues indicates that the size of dogs may be determined by alleles in

the *Igf1* gene that encodes a growth hormone called insulin-like growth factor 1. A particular allele of this gene was found to be common to all small breeds of dogs and nearly absent from very large breeds, suggesting that this allele is a major contributor to body size in small breeds of dogs.

Likewise, most of the food we eat—including products such as grains, fruits, vegetables, meat, milk, and juices—is obtained from species that have been profoundly modified by selective breeding strategies. For example, certain characteristics in the wild mustard plant (*Brassica oleracea*) have been modified by selective breeding to create several varieties of domesticated crops, including broccoli, Brussels sprouts, cabbage, and

(a) Bulldog

(b) Greyhound

(c) Dachshund

Figure 23.9   Common breeds of dogs that have been obtained by selective breeding.   By selecting parents carrying the alleles that influence traits desirable to humans, dog breeders have produced breeds with distinctive features. All the dogs in this figure carry the same kinds of genes (for example, genes that affect their size, shape, and fur color). However, the alleles for many of these genes are different among these dogs, thereby allowing humans to select for or against them and produce breeds with strikingly different phenotypes.

cauliflower (**Figure 23.10**). The wild mustard plant is native to Europe and Asia, and plant breeders began to modify its traits approximately 4,000 years ago. As seen here, certain traits in the domestic strains differ dramatically from those of the original wild species. These varieties are all members of the same species. They can interbreed to produce viable offspring. For example, in the grocery store you may have seen broccoflower, which is produced from a cross between broccoli and cauliflower.

As a final example, **Figure 23.11** shows the results of an artificial selection experiment on corn begun at the University of Illinois Agricultural Experiment Station in 1896, several years before the rediscovery of Mendel's laws. This study began with 163 ears of corn with an oil content ranging from 4 to 6%. In each of 80 succeeding generations, corn plants were divided into two separate groups. In one group, members with the highest oil content in the kernels were chosen as parents of the next generation. In the other group, members with the lowest oil content were chosen. After many generations, the oil content in the first group rose to over 18%. In the other group, it dropped to less than 1%. These results show that selective breeding can modify a trait in a very directed manner.

A Comparison of Anatomical, Developmental, and Molecular Homologies Shows Evolution of Related Species from a Common Ancestor

Let's now consider other widespread observations of the process of evolution among living organisms. In biology, the term

Figure 23.10   Crop plants developed by selective breeding of the wild mustard plant.   Although these six agricultural plants look quite different from each other, they carry many of the same alleles as the wild mustard plant. However, they differ from each other in alleles that affect the formation of stems, leaves, and flowers.

Wild mustard plant (*Brassica oleracea*)

Strain	Kohlrabi	Kale	Broccoli	Brussels sprouts	Cabbage	Cauliflower
Modified trait	Stem	Leaves	Flower buds and stem	Lateral leaf buds	Terminal leaf bud	Flower buds

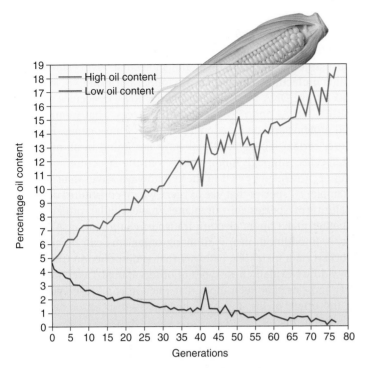

Figure 23.11 Results of selective breeding for oil content in corn plants. In this example, corn plants were selected for breeding based on high or low oil content of the kernels. Over the course of many generations, this had a major impact on the amount of corn oil (an agriculturally important product) made by the two groups of plants.

Concept check: When comparing Figures 23.9, 23.10, and 23.11, what general effects of artificial selection do you observe?

homology refers to a similarity that occurs due to descent from a common ancestor. Two species may have a similar trait because the trait was originally found in a common ancestor. As described next, such homologies may involve anatomical, developmental, or molecular features.

Anatomical Homologies As noted by Theodosius Dobzhansky, many observations regarding the features of living organisms simply cannot be understood in any meaningful scientific way except as a result of evolution. A comparison of vertebrate anatomy is a case in point. An examination of the limbs of modern vertebrate species reveals similarities that indicate the same set of bones has undergone evolutionary changes, becoming modified to perform different functions in different species. As seen in **Figure 23.12**, the forelimbs of vertebrates have a strikingly similar pattern of bone arrangements. These are termed **homologous structures**—structures that are similar to each other because they are derived from a common ancestor. The forearm has developed different functions among various vertebrates, including grasping, walking, flying, swimming, and climbing. The theory of evolution explains how these animals have descended from a common ancestor and how natural selection has resulted in modifications to the original set of bones in ways that ultimately allowed them to be used for several different purposes.

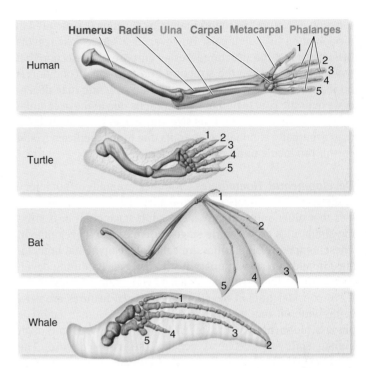

Figure 23.12 An example of anatomical homology: homologous structures found in vertebrates. The same set of bones is found in the human arm, turtle arm, bat wing, and whale flipper, although their relative sizes and shapes have been modified in ways that allow them to perform different functions. This homology suggests that all of these animals evolved from a common ancestor.

Another result of evolution is the phenomenon of **vestigial structures**, which are anatomical features that have no current function but resemble structures of their presumed ancestors (**Table 23.3**). An interesting case is found in humans. People have a complete set of muscles for moving their ears, even though most people are unable to do so. By comparison, many modern mammals can move their ears, and presumably this was an important trait in a distant human ancestor. Why would organisms have structures that are no longer useful? Within the context of evolutionary theory, vestigial structures are evolutionary relics. Organisms having vestigial structures share a common ancestry with organisms in which the structure is

Table 23.3	Examples of Vestigial Structures in Animals
Organism	**Vestigial structure(s)**
Humans	Tail bone and muscles to wiggle ears in adult
Boa constrictors	Skeletal remnants of hip and hind leg bones
Whales	Skeletal remnants of a pelvis
Manatees	Fingernails on the flippers
Hornbills and cuckoos	Fibrous cords that were derived from the common carotid arteries. In certain families of birds, both of the common carotid arteries are nonfunctional, fibrous cords. Their vascular function has been assumed by other vessels.

functional. Natural selection maintains functional structures in a population of individuals. However, if a species changes its lifestyle so that the structure loses its purpose, the selection that would normally keep the structure in a functional condition is no longer present. When this occurs, the structure may degenerate over the course of many generations due to the accumulation of mutations that limit its size and shape. Natural selection may eventually eliminate such traits due to the inefficiency and cost of producing unused structures.

Developmental Homologies   Another example of homology is the way that animals undergo embryonic development. Species that differ substantially at the adult stage often bear striking similarities during early stages of embryonic development. These temporary similarities are called developmental homologies. In addition, evolutionary history is revealed during development in certain organisms, such as vertebrates. For example, if we consider human development, several features are seen in the embryo that are not present at birth. Human embryos possess rudimentary gill ridges like a fish embryo, even though human embryos receive oxygen via the umbilical cord. The presence of gill ridges indicates that humans evolved from an aquatic species that had gill slits. A second observation is that every human embryo has a bony tail. It is difficult to see the advantage of such a structure *in utero*, but easier to understand its presence assuming that an ancestor of the human lineage possessed a tail. These observations, and many others, illustrate that closely related species share similar developmental pathways.

Molecular Homologies   Our last examples of homology due to evolution involve molecular studies. When scientists examine the features of organisms at the molecular level, similarities called **molecular homologies** are found. For example, all living species use DNA to store information. RNA molecules such as mRNA, tRNA, and rRNA are used to access that information, and proteins are the functional products of most genes. Furthermore, certain biochemical pathways are found in all or nearly all species, although minor changes in the structure and function of proteins involved in these pathways have occurred. For example, all species that use oxygen, which constitutes the great majority of species on our planet, have similar proteins that together make up an electron transport chain and an ATP synthase. In addition, nearly all living organisms can break down glucose via a metabolic pathway that is described in Chapter 7. How do we explain these types of observations? Taken together, they indicate that such molecular phenomena arose very early in the origin of life and have been passed to all or nearly all modern forms.

The most compelling observation at the molecular level indicating that modern life-forms are derived from a common ancestor is revealed by analyzing genetic sequences and finding genetic homologies. The same type of gene is often found in diverse organisms. Furthermore, the degree to which a genetic sequence from different species is similar reflects the evolutionary relatedness of those species. As an example, let's consider the *p53* gene, which encodes the p53 protein that plays a role in preventing cancer (see Chapter 14, Figure 14.15). **Figure 23.13** shows a short amino acid sequence that makes up part of

	Short amino acid sequence within the p53 protein	Percentages of amino acids in the whole p53 protein that are identical to human p53
Human (*Homo sapiens*)	Val Pro Ser Gln Lys Thr Tyr Gln Gly Ser Tyr Gly Phe Arg Leu Gly Phe Leu His Ser Gly Thr	100
Rhesus monkey (*Macaca mulatta*)	Val Pro Ser Gln Lys Thr Tyr **His** Gly Ser Tyr Gly Phe Arg Leu Gly Phe Leu His Ser Gly Thr	95
Green monkey (*Cercopithecus aethiops*)	Val Pro Ser Gln Lys Thr Tyr **His** Gly Ser Tyr Gly Phe Arg Leu Gly Phe Leu His Ser Gly Thr	95
Rabbit (*Oryctolagus cuniculus*)	Val Pro Ser Gln Lys Thr Tyr **His** Gly **Asn** Tyr Gly Phe Arg Leu Gly Phe Leu His Ser Gly Thr	86
Dog (*Canis lupus familiaris*)	Val Pro Ser **Pro** Lys Thr Tyr **Pro** Gly **Thr** Tyr Gly Phe Arg Leu Gly Phe Leu His Ser Gly Thr	80
Chicken (*Gallus gallus*)	Val Pro Ser **Thr Glu Asp** Tyr **Gly** Gly **Asp Phe Asp** Phe Arg **Val** Gly Phe **Val Glu Ala** Gly Thr	53
Channel catfish (*Ictalurus punctatus*)	Val Pro **Val Thr Ser Asp** Tyr **Pro** Gly **Leu Leu Asn** Phe **Thr** Leu **His** Phe **Gln Glu** Ser **Ser Gly**	48
European flounder (*Platichthys flesus*)	Val Pro **Val Val Thr Asp** Tyr **Pro** Gly **Glu** Tyr Gly Phe **Gln** Leu **Arg** Phe **Gln Lys** Ser Gly Thr	46
Congo puffer fish (*Tetraodon miurus*)	Val Pro **Val Thr Thr Asp** Tyr **Pro** Gly **Glu** Tyr Gly Phe **Lys** Leu **Arg** Phe **Gln Lys** Ser Gly Thr	41

Figure 23.13   **An example of genetic homology: a comparison of a short amino acid sequence within the p53 protein from nine different animals.**   This figure compares a short region of the p53 protein, which plays a role in preventing cancer. Amino acids are represented by three-letter abbreviations. The orange-colored amino acids in the sequences are identical to those in the human sequence. The numbers in the right column indicate the percentage of amino acids within the whole p53 protein that is identical with the human p53 protein, which is 393 amino acids in length. For example, 95% of the amino acids, or 373 of 393, are identical between the p53 sequence found in humans and in Rhesus monkeys.

Concept check: *In the sequence shown in this figure, how many amino acid differences are there between the following pairs: Rhesus and green monkeys, Congo puffer fish and European flounder, and Rhesus monkey and Congo puffer fish? What do these differences tell you about the evolutionary relationships among these four species?*

the p53 protein from a variety of species, including five mammals, one bird, and three fish. The top sequence is the human p53 sequence, and the right column describes the percentages of amino acids within the entire sequence that are identical to the entire human sequence. Amino acids in the other species that are identical to humans are highlighted in orange. The sequences from the two monkeys are the most similar to humans, followed by the other two mammalian species (rabbit and dog). The three fish sequences are the least similar to the human sequence, but the fish sequences tend to be similar to each other. Taken together, the data shown in Figure 23.13 illustrate two critical points regarding gene evolution. First, specific genes are found in a diverse array of species such as mammals, birds, and fishes. Second, the sequences of closely related species tend to be more similar to each other than they are to distantly related species. The mechanisms for this second observation are discussed in the next section.

23.3 The Molecular Processes That Underlie Evolution

Historically, the study of evolution was based on a comparison of the anatomies of extinct and modern species to identify similarities between related species. However, the advent of molecular approaches for analyzing DNA sequences has revolutionized the field of evolutionary biology. Now we can analyze how changes in the genetic material are associated with changes in phenotype. In this section, we will examine some of the molecular changes in the genetic material that are associated with evolution.

Homologous Genes Are Derived from a Common Ancestral Gene

When two or more genes are derived from the same ancestral gene, they are called **homologous genes**. The analysis of homologous genes reveals details of evolutionary change at the molecular level. How do homologous genes arise? As an example, let's consider a gene in two different species of bacteria that encodes a transport protein involved in the uptake of metal ions into bacterial cells. Homologous genes that are in different species are termed **orthologs**. Millions of years ago, these two species had a common ancestor (**Figure 23.14**). Over time, the common ancestor diverged into additional species, eventually evolving into *Escherichia coli*, *Clostridium acetylbutylicum*, and many other species. Since this divergence, the metal transporter gene has accumulated mutations that altered its sequence, though the similarity between the *E. coli* and the *C. acetylbutylicum* genes remains striking. In this case, the two sequences are similar because they were derived from the same ancestral gene, but they are not identical due to the independent accumulation of different random mutations.

Gene Duplications Create Gene Families

Demonstrations of evolutionary change can also be found within a single species. Two or more homologous genes found

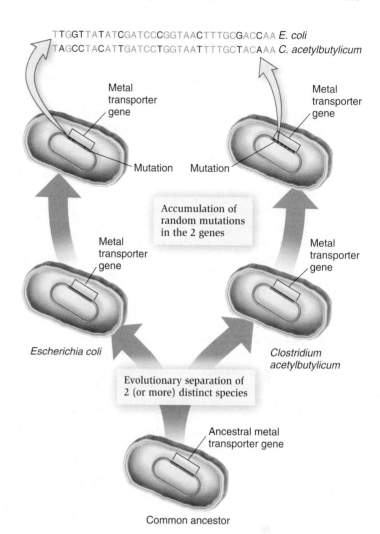

TTGGTTATATCGATCCCGGTAACTTTGCGACCAA *E. coli*
TAGCCTACATTGATCCTGGTAATTTTGCTACAAA *C. acetylbutylicum*

Metal transporter gene — Mutation   Mutation — Metal transporter gene

Accumulation of random mutations in the 2 genes

Metal transporter gene   Metal transporter gene

Escherichia coli   *Clostridium acetylbutylicum*

Evolutionary separation of 2 (or more) distinct species

Ancestral metal transporter gene

Common ancestor

Figure 23.14   The evolution of orthologs, homologous genes from different species.   After the two species diverged from each other, the genes accumulated random mutations that resulted in similar, but not identical, gene sequences called orthologs. These orthologs in *E. coli* and *C. acetylbutylicum* encode metal transporters. Only one of the two DNA strands is shown from each of the genes. Bases that are identical between the two genes are shown in orange.

Concept check: Why do these orthologs have similar gene sequences? Why aren't they identical?

within a single species are termed **paralogs** of each other. Rare gene duplication events can produce multiple copies of a gene and ultimately lead to the formation of a gene family. A **gene family** consists of two or more copies of paralogous genes within the genome of a single species. A well-studied example of a gene family is the globin gene family found in humans and many other animal species. The globin genes encode polypeptides that are subunits of proteins that function in oxygen binding. One such protein is hemoglobin, which is found in red blood cells and carries oxygen throughout the body. The globin gene family is composed of 14 genes that are hypothesized to be derived from a single ancestral globin gene. According to an evolutionary analysis, the ancestral globin gene first duplicated between 500 to 600 million years ago. Since that time, additional duplication events and chromosomal rearrangements have produced the current number of 14 genes on three different human chromosomes (refer back to Figure 21.8).

What is the advantage of a gene family? Even though all of the globin polypeptides are subunits of proteins that play a role in oxygen binding, the accumulation of changes in the various family members has created globins that are more specialized in their function. For example, myoglobin is better at binding and storing oxygen in muscle cells, whereas the hemoglobins are better at binding and transporting oxygen via the red blood cells. Also, different globin genes are expressed during different stages of human development. The epsilon (ε)- and zeta (ζ)-globin genes are expressed very early in embryonic life, while the gamma (γ)-globin genes exhibit maximal expression during the second and third trimesters of gestation. Following birth, the γ-globin gene is turned off and the β-globin gene is turned on. These differences in the expression of the globin genes reflect the differences in the oxygen transport needs of humans during the embryonic, fetal, and postpartum stages of life.

What is the evolutionary significance of the globin gene family regarding adaptation? On land, egg cells and small embryos are very susceptible to drying out if they are not protected in some way. Species such as birds and reptiles lay eggs with a protective shell around them. Most mammals, however, have become adapted to a terrestrial environment by evolving internal gestation. The ability to develop young internally has been an important factor in the survival and proliferation of mammals. The embryonic and fetal forms of hemoglobin allow the embryo and fetus to capture oxygen from the bloodstream of the mother.

Genomes & Proteomes Connection

New Genes in Eukaryotes Have Evolved via Exon Shuffling

Thus far, we have considered how evolutionary change results in the formation of related genes, which are described as orthologs and paralogs. Evolutionary mechanisms are also revealed when exons, the parts of genes that encode protein domains, are compared within a single species. Many proteins, particularly those found in eukaryotic species, have a modular structure composed of two or more domains with different functions. For example, certain transcription factors have discrete domains involved with hormone binding, dimerization, and DNA binding. The glucocorticoid receptor discussed in Chapter 13 has a domain that binds the hormone, a second domain that facilitates protein dimerization, and a third domain that allows the glucocorticoid receptor to bind to glucocorticoid response elements (GREs) next to genes. By comparing the modular structure of eukaryotic proteins with the genes that encode them, geneticists have discovered that each domain tends to be encoded by one exon or by a series of two or more adjacent exons.

During the evolution of eukaryotic species, many new genes have been produced by a type of mutation known as **exon shuffling**. During this process, an exon and part of the flanking noncoding introns from one gene are inserted into another gene, thereby producing a new gene that encodes a protein with an additional domain (**Figure 23.15**). This process may also involve the duplication and rearrangement of exons. Exon shuffling results in novel genes that express proteins with diverse functional modules. Such proteins may alter traits in the organism and therefore be subjected to natural selection.

Exon shuffling may occur by more than one mechanism. One possibility is that a double crossover could promote the insertion of an exon into another gene (Figure 23.15). This is called nonhomologous recombination because the two regions involved in the crossover are not homologous to each other. Alternatively, transposable elements, described in Chapter 21, may promote the movement of exons into other genes.

Horizontal Gene Transfer Also Contributes to the Evolution of Species

At the molecular level, the type of evolutionary change depicted in Figures 23.13 through 23.15 is called **vertical evolution**. In these cases, new species arise from pre-existing species by the accumulation of genetic changes, such as gene mutations, gene duplications, and exon shuffling. Vertical evolution involves genetic changes in a series of ancestors that form a lineage. In addition to vertical evolution, species accumulate genetic changes by **horizontal gene transfer**—a process in which an organism incorporates genetic material from another organism without being the offspring of that organism. Horizontal gene transfer can involve the exchange of genetic material between members of the same species or different species.

How does horizontal gene transfer occur? **Figure 23.16** illustrates one possible mechanism for horizontal gene transfer. In this example, a eukaryotic cell has engulfed a bacterial cell by endocytosis. During the degradation of the bacterium in an endocytic vesicle, a bacterial gene happens to escape to the nucleus of the cell, where it is inserted into one of the chromosomes. In this way, a gene has been transferred from a bacterial species to a eukaryotic species. By analyzing gene sequences among many different species, researchers have discovered that horizontal gene transfer is a common phenomenon. This process can occur from prokaryotes to eukaryotes, from eukaryotes to prokaryotes, between different species of prokaryotes, and between different species of eukaryotes. Therefore, when we view evolution, it is not simply a matter of one species evolving into one or more new species via the accumulation of random mutations. It also involves the horizontal transfer of genes among different species, enabling those species to acquire new traits that foster the evolutionary process.

Gene transfer among bacterial species is relatively widespread. As discussed in Chapter 18, bacterial species may carry out three natural mechanisms of gene transfer known as

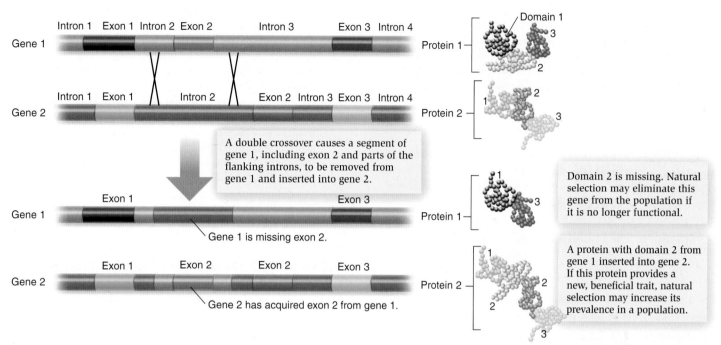

Figure 23.15 **The process of exon shuffling.** In this example, a segment of one gene containing an exon and part of the flanking introns has been inserted into another gene. A rare, abnormal crossing-over event called nonhomologous recombination may cause this to happen. Exon shuffling results in proteins that have new combinations of domains and new combinations of functions.

Concept check: *What is the evolutionary advantage of exon shuffling?*

conjugation, transformation, and transduction. By analyzing the genomes of bacterial species, scientists have determined that many genes within a given bacterial genome are derived from horizontal gene transfer. Genome studies have suggested that as much as 20–30% of the variation in the genetic composition of modern prokaryotic species can be attributed to this process. For example, in *E. coli* and *Salmonella typhimurium*, roughly 17% of

their genes have been acquired via horizontal gene transfer during the past 100 million years. The roles of these acquired genes are quite varied, though they commonly involve functions that are beneficial for survival and reproduction. These include genes that confer antibiotic resistance, the ability to degrade toxic compounds, and pathogenicity (the ability to cause disease).

Evolution at the Genomic Level Involves Changes in Chromosome Structure and Number

Thus far, we have considered several ways that a species might acquire new genetic variation. These include random mutations within pre-existing genes, gene duplications to create gene families, exon shuffling, and horizontal gene transfer. Evolution also occurs at the genomic level, involving changes in chromosome structure and number. When comparing the chromosomes of closely related species, changes in chromosome structure and/or number are common.

As an example, **Figure 23.17** compares the banding patterns of the three largest chromosomes in humans and the corresponding chromosomes in chimpanzees, gorillas, and orangutans. (Refer back to Chapter 15, Figure 15.1 for a description of chromosome banding.) The banding patterns are strikingly similar because these species are closely related evolutionarily. Even so, you can see some interesting differences. Humans have one large chromosome 2, but this chromosome is divided into two separate chromosomes in the other three species. This explains why human cells have 23 pairs of chromosomes, whereas ape

During endocytosis, the bacterial cell is degraded, and the products of degradation are released into the cytosol. However, a gene from the bacterium escapes degradation and becomes inserted into a chromosome in the eukaryotic cell.

Figure 23.16 **Horizontal gene transfer from a bacterium to a eukaryote.** In this example, a bacterium is engulfed by a eukaryotic cell, and a bacterial gene is transferred to one of the eukaryotic chromosomes.

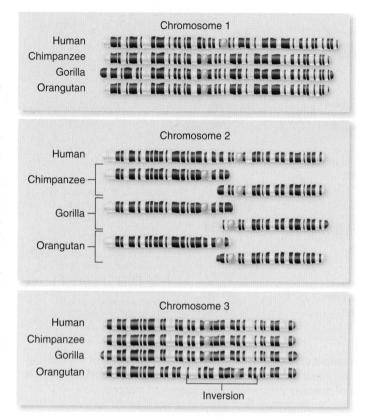

Figure 23.17   **An example of genomic evolution.**
This figure is a comparison of banding patterns among the
three largest human chromosomes and the corresponding
chromosomes in the chimpanzee, gorilla, and orangutan. It is
a schematic drawing of Giemsa-stained chromosomes. The
differences between these chromosomes illustrate the changes
that have occurred during the evolution of these related species.

Concept check: *Describe two changes in chromosome*
structure that have occurred among these chromosomes.

cells have 24. The fusion of the two smaller chromosomes during the development of the human lineage may have caused
this difference in chromosome number. Another interesting
change in chromosome structure is seen in chromosome 3. The
banding patterns among humans, chimpanzees, and gorillas
are very similar, but the orangutan has a large inversion that
flips the arrangement of bands in the centromeric region. As
discussed in Chapter 25, changes in chromosome structure and
number may affect the ability of two organisms to breed with
one another. In this way, such changes have been important in
the establishment of new species.

Summary of Key Concepts

23.1 The Theory of Evolution

- Biological evolution is a heritable change in one or more
  characteristics of a population or species from one generation
  to the next.

- Charles Darwin proposed the theory of evolution based on
  his understanding of geology and population growth and his
  observations of species in their natural settings. His voyage on
  the *Beagle*, during which he studied many species including
  finches on the Galápagos Islands, was particularly influential.
  (Figure 23.1, Table 23.1)

- Darwin expressed his theory of evolution as descent with
  modification through variation and natural selection. As a
  result of natural selection, genetic variation changes from
  generation to generation to produce populations of organisms
  with traits (adaptations) that promote greater reproductive
  success. (Figure 23.2)

- The Grants' research on finches showed how changes in beak
  size in response to changes in the food supply are driven by
  natural selection. (Figures 23.3, 23.4)

23.2 Observations of Evolutionary Change

- Observations of evolution include the fossil record,
  biogeography, convergent evolution, selective breeding,
  and homologies. (Table 23.2)

- Fossils show successive evolutionary change over long periods
  of time. The fossil record often reveals transitional forms that
  link past ancestors to modern species. (Figures 23.5, 23.6)

- Biogeography, the geographic distribution of species, provides
  information on how certain species are evolutionarily related
  to each other. Often, when populations become isolated, they
  evolve into new species. (Figure 23.7)

- Convergent evolution involves independent adaptations that
  result in analogous structures because organisms have evolved
  in similar environments. (Figure 23.8)

- Selective breeding, also known as artificial selection, illustrates
  how changes in genetic variation over the course of many
  generations can dramatically change the traits of organisms.
  (Figures 23.9, 23.10, 23.11)

- Homologous structures are similar to each other because they
  are derived from a common ancestral structure. The set of
  bones in the forearms of vertebrates is one example.
  (Figure 23.12)

- Vestigial structures are found in species because they are
  derived from structures that were once functional, but have
  degenerated because they no longer have use in a modern
  species. (Table 23.3)

- Homologies can also be seen during embryonic development
  and at the molecular level in gene and protein sequences.
  (Figure 23.13)

23.3 The Molecular Processes That Underlie Evolution

- Molecular evolution refers to the molecular changes in
  the genetic material that underlie the phenotypic changes
  associated with evolution.

- Homologous genes are derived from the same ancestral gene.
  They accumulate random mutations that make their sequences
  somewhat different. Orthologs are homologous genes in
  different species. (Figure 23.14)

- Paralogs are homologous genes in the same species that are produced by gene duplication events. Paralogs constitute a gene family. An example is the globin gene family that promoted the evolutionary adaptation of internal gestation.

- Exon shuffling is a form of mutation in which exons are inserted into genes and thereby create proteins with additional functional domains. (Figure 23.15)

- Another mechanism that produces genetic variation is horizontal gene transfer in which genetic material is transferred between different species. Such genetic changes are subject to natural selection. (Figure 23.16)

- Evolution is also associated with changes in chromosome structure and chromosome number. (Figure 23.17)

Assess and Discuss

Test Yourself

1. A change in one or more characteristics of a population that is heritable and occurs from one generation to the next is called
   a. natural selection.
   b. sexual selection.
   c. population genetics.
   d. evolution.
   e. inheritance of acquired characteristics.

2. Lamarck's vision of evolution differed from Darwin's in that Lamarck believed
   a. living things evolved in an upward direction.
   b. behavioral changes modified heritable traits.
   c. genetic differences among individuals in the population allowed for evolution.
   d. a and b only.
   e. none of the above.

3. Which of the following scientists influenced Darwin's views on the nature of population growth?
   a. Cuvier   c. Lyell   e. Wallace
   b. Malthus   d. Hutton

4. An evolutionary change in which an organism's characteristics change in ways that make it better suited to its environment is
   a. natural selection.   d. evolution.
   b. an adaptation.   e. both a and c.
   c. an acquired characteristic.

5. Vestigial structures are anatomical structures
   a. that have more than one function.
   b. that were functional in an ancestor but no longer have a useful function.
   c. that look similar in different species but have different functions.
   d. that have the same function in different species but have very different appearances.
   e. of the body wall.

6. Which of the following is an example of a developmental homology seen in human embryonic development and other vertebrate species that are not mammals?
   a. gill ridges   c. tail   e. all of the above
   b. umbilical cord   d. both a and c

7. Two or more homologous genes found within a particular species are called
   a. homozygous.   c. paralogs.   e. duplicates.
   b. orthologs.   d. alleles.

8. The phenomenon of exon shuffling
   a. creates new gene products by changing the pattern of intron removal in pre-mRNA.
   b. creates new genes by inserting exons and flanking introns into a different gene sequence, thereby introducing a new domain in the gene product.
   c. deletes one or more bases in a single gene.
   d. rearranges the introns in a particular gene, creating new gene products.
   e. both a and d.

9. Horizontal gene transfer
   a. is a process in which an organism incorporates genetic material from another organism without being the offspring of that organism.
   b. can involve the exchange of genetic material among individuals of the same species.
   c. can involve the exchange of genetic material among individuals of different species.
   d. can be all of the above.
   e. can be a and b only.

10. Genetic variation can occur as a result of
    a. random mutations in genes.
    b. exon shuffling.
    c. gene duplication.
    d. horizontal gene transfer.
    e. all of the above.

Conceptual Questions

1. Evolution that results in adaptation is rooted in two phenomena: genetic variation and natural selection. In a very concise way (three sentences or less), describe how genetic variation and natural selection can bring about evolution.

2. What is convergent evolution? How does it support the theory of evolution?

3. What are homologous structures? Explain how the homologous forelimbs of vertebrates support the theory of evolution.

Collaborative Questions

1. The term natural selection is sometimes confused with the term evolution. Discuss the meanings of these two terms. Explain how the terms are different and how they are related to each other.

2. Make a list of the observations made by biologists that support the theory of evolution. Which of the observations on your list do you find the most convincing and the least convincing?

Online Resource

www.brookerbiology.com

Stay a step ahead in your studies with animations that bring concepts to life and practice tests to assess your understanding. Your instructor may also recommend the interactive ebook, individualized learning tools, and more.

Population Genetics

24

Colorful African cichlids. Color is a factor that influences the choice of mates in populations of cichlids.

K imbareta, age 19, lives in the Democratic Republic of Congo (formerly Zaire) with his parents, one brother, and two sisters. Kimbareta has sickle-cell disease, which causes his red blood cells to occasionally form a sickle shape. The sickled cells may block the flow of blood through his vessels. This results in tissue and organ damage along with painful episodes involving his arms, legs, chest, and abdomen. In some people with this disease, stroke may even occur. Sickle-cell disease is an inherited disease with a recessive pattern of inheritance. It is caused by a mutation in a gene that encodes β-globin, a subunit of hemoglobin that carries oxygen in the red blood cells.

Many different recessive diseases have been identified by geneticists. Most of them are very rare. However, in the village where Kimbareta lives, sickle-cell disease is surprisingly common. Nearly 2% of the inhabitants have sickle-cell disease, which is similar to other places in the country. How can we explain such a high occurrence of a serious inherited disease? If natural selection tends to eliminate detrimental genetic variation, as we saw in Chapter 23, why does the sickle-cell allele persist in this population? As we will see later,

biologists have discovered that the effect of the allele in heterozygotes is the underlying factor. Heterozygotes who carry one copy of the sickle-cell allele and one copy of the more common (non-disease-causing) allele have an increased resistance to malaria.

Population genetics is the study of genes and genotypes in a population. A **population** is a group of individuals of the same species that occupy the same environment. For sexually reproducing species, members of a given population can interbreed with one another. The central issue in population genetics is genetic variation—its extent within populations, why it exists, how it is maintained, and how it changes over the course of many generations. Population genetics helps us to understand how underlying genetic variation is related to phenotypic variation. These include issues such as mate preference (see chapter-opening photo) and the prevalence of certain disease-causing alleles, such as the sickle-cell allele, in human populations.

Population genetics emerged as a branch of genetics in the 1920s and 1930s. Its mathematical foundations were developed by theoreticians who extended the principles of Darwin and Mendel by deriving equations to explain the occurrence of genotypes within populations. These foundations can be largely attributed to British evolutionary biologists J. B. S. Haldane and Ronald Fisher, and American geneticist Sewall Wright. As we will see, several researchers who analyzed the genetic composition of natural and experimental populations provided support for their mathematical theories. More recently, population geneticists have used techniques to probe genetic variation at the molecular level. In addition, the staggering improvements in computer technology have aided population geneticists in the analysis of data and the testing of genetic hypotheses.

We will begin this chapter by exploring the extent of genetic variation that occurs in populations and how such variation is subject to change. In many cases, genetic changes are associated with evolutionary adaptations, which are characteristics of a species that have evolved over a long period of time by the process of natural selection. In this chapter, we will examine the various mechanisms that promote genetic change, including natural selection, genetic drift, migration, and nonrandom mating.

24.1   Genes in Populations

Population genetics is an extension of our understanding of Darwin's theory of natural selection, Mendel's laws of inheritance, and newer studies in molecular genetics. All of the alleles for every gene in a given population make up the **gene pool**. Each member of the population receives its genes from its parents, which, in turn, are members of the gene pool. Individuals that reproduce contribute to the gene pool of the next generation. Population geneticists study the genetic variation within the gene pool and how such variation changes from one generation to the next. The emphasis is often on understanding the variation in alleles among members of a population. In this section, we will examine some of the general features of populations and gene pools.

Figure 24.1   An example of polymorphism: the two color variations found in the orchid *Dactylorhiza sambucina*.

Populations Are Dynamic Units

As mentioned, a population is a group of individuals of the same species that occupy the same environment and can interbreed with one another. Certain species occupy a wide geographic range and are divided into discrete populations due to geographic isolation. For example, distinct populations of a given species may be located on different sides of a mountain.

Populations change from one generation to the next. How might populations become different? Populations may change in size and geographic location. As the size and locations of a population change, their genetic composition generally changes as well. Some of the genetic changes involve adaptation, in which a population becomes better suited to its environment, making it more likely to survive and reproduce. For example, a population of mammals may move from a warmer to a colder geographic location. Over the course of many generations, natural selection may change the population such that the fur of the animals is thicker and provides better insulation against the colder temperatures.

Genomes & Proteomes Connection

Genes Are Usually Polymorphic

The term **polymorphism** (from the Greek, meaning many forms) refers to the phenomenon that a trait displays variation within a population. Populations that are polymorphic for a trait have more than one phenotype for that trait. **Figure 24.1** illustrates a striking example of polymorphism in the elder-flowered orchid (*Dactylorhiza sambucina*). Throughout the range of this species in Europe, both yellow- and red-flowered individuals are prevalent.

Polymorphism in a trait is usually due to two or more alleles of a gene that influence the trait. Geneticists also use the term polymorphism to describe the variation in genes; this is sometimes called genetic or gene polymorphism. A gene that commonly exists as two or more alleles in a population is a

polymorphic gene. To be considered polymorphic, a gene must exist in at least two alleles, and each allele must occur at a frequency that is greater than 1%. By comparison, a **monomorphic gene** exists predominantly as a single allele in a population. When 99% or more of the alleles of a given gene are identical in a population, the gene is considered to be monomorphic.

What types of molecular changes cause genes to be polymorphic? A polymorphism may involve various types of changes such as a deletion of a significant region of the gene, a duplication of a region, or a change in a single nucleotide. This last phenomenon is called a **single-nucleotide polymorphism** (**SNP**). SNPs ("snips") are the smallest type of genetic change that can occur within a given gene, and also the most common. For example, the sickle-cell allele discussed at the beginning of the chapter involves a single nucleotide change in a gene that encodes β-globin, a subunit of the oxygen-carrying protein called hemoglobin. The non-disease-causing allele and sickle-cell allele represent a single-nucleotide polymorphism of the β-globin gene:

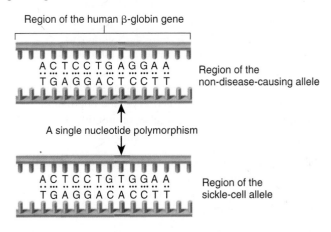

Relative to the non-disease-causing allele, this is a single-nucleotide substitution of an A (in the top strand) to a T (in the sickle-cell allele).

SNPs represent 90% of all the variation in human DNA sequences that occurs among different people. In human populations, a gene that is 2,000–3,000 bp in length will, on average, contain 10 different SNPs. Likewise, SNPs with a frequency of 1% or more are found very frequently among genes of nearly all species. Polymorphism is the norm for relatively large, healthy populations of nearly all species, as evidenced by the occurrence of SNPs within most genes.

Why do we care about SNPs? One reason is their importance in human health. By analyzing SNPs in human genes, researchers have determined that these small variations in DNA sequences can affect the function of the proteins encoded by the genes. These effects on the proteome, in turn, may influence how humans develop diseases, such as heart disease, diabetes, and sickle-cell disease. Variations in SNPs in the human population are also associated with how people respond to viruses, drugs, and vaccines. The analysis of SNPs may be instrumental in the current and future development of **personalized medicine**—a medical practice in which information about a patient's genotype is used to tailor their medical care. For example, an analysis of a person's SNPs may be used to select between different types of medication or customize their dosage. In addition, SNP analysis may reveal that a person has a high predisposition to the development of a particular disease, such as heart disease. Such information may be used to initiate preventative measures to minimize the chances of developing the disease.

Population Genetics Is Concerned with Allele and Genotype Frequencies

To analyze genetic variation in populations, one approach is to consider the frequency of alleles in a quantitative way. Two fundamental calculations are central to population genetics: **allele frequencies** and **genotype frequencies**. Allele and genotype frequencies are defined as follows:

$$\text{Allele frequency} = \frac{\text{Number of copies of a specific allele in a population}}{\text{Total number of all alleles for that gene in a population}}$$

$$\text{Genotype frequency} = \frac{\text{Number of individuals with a particular genotype in a population}}{\text{Total number of individuals in a population}}$$

Although allele and genotype frequencies are related, make sure you keep in mind a clear distinction between them. As an example, let's consider a population of 100 four-o'clock plants (*Mirabilis jalapa*) with the following genotypes:

49 red-flowered plants with the genotype $C^R C^R$

42 pink-flowered plants with the genotype $C^R C^W$

9 white-flowered plants with the genotype $C^W C^W$

When calculating an allele frequency for a diploid species, remember that homozygous individuals have two copies of an allele, whereas heterozygotes have only one. For example, in tallying the C^W allele, each of the 42 heterozygotes has one copy of the C^W allele, and each white-flowered plant has two copies. Therefore, the allele frequency for C^W (the white color allele) equals

$$\text{Frequency of } C^W = \frac{(C^R C^W) + 2(C^W C^W)}{2(C^R C^R) + 2(C^R C^W) + 2(C^W C^W)}$$

$$\text{Frequency of } C^W = \frac{42 + (2)(9)}{(2)(49) + (2)(42) + (2)(9)}$$

$$= \frac{60}{200} = 0.3, \text{ or } 30\%$$

This result tells us that the allele frequency of C^W is 0.3. In other words, 30% of the alleles for this gene in the population are the C^W allele.

Let's now calculate the genotype frequency of $C^W C^W$ homozygotes (white-flowered plants).

$$\text{Frequency of } C^W C^W = \frac{9}{49 + 42 + 9}$$

$$= \frac{9}{100} = 0.09, \text{ or } 9\%$$

We see that 9% of the individuals in this population have white flowers.

Allele and genotype frequencies always sum to less than or equal to one (that is, less than or equal to 100%). If a gene is monomorphic, the allele frequency for the single allele will equal or be close to a value of 1.0. For polymorphic genes, if we add up the frequencies for all the alleles in the population, we should obtain a value of 1.0. In our four-o'clock example, the allele frequency of C^W equals 0.3. Therefore, we can calculate the frequency of the other allele, C^R, as equal to 1.0 − 0.3 = 0.7, because they must add up to 1.0.

The Hardy-Weinberg Equation Relates Allele and Genotype Frequencies in a Population

In 1908, Godfrey Harold Hardy, an English mathematician, and Wilhelm Weinberg, a German physician, independently derived a simple mathematical expression called the Hardy-Weinberg equation that describes the relationship between allele and genotype frequencies when a population is not evolving. Let's examine the Hardy-Weinberg equation using the population of four-o'clock plants that we just considered. If the allele frequency of C^R is denoted by the symbol p and the allele frequency of C^W by q, then

$$p + q = 1$$

For example, if $p = 0.7$, then q must be 0.3. In other words, if the allele frequency of C^R equals 70%, the remaining 30% of alleles must be C^W, because together they equal 100%.

For a gene that exists in two alleles, the **Hardy-Weinberg equation** states that

$$p^2 + 2pq + q^2 = 1$$

If we apply this equation to our flower color gene, then

p^2 = the genotype frequency of $C^R C^R$ homozygotes

$2pq$ = the genotype frequency of $C^R C^W$ heterozygotes

q^2 = the genotype frequency of $C^W C^W$ homozygotes

If $p = 0.7$ and $q = 0.3$, then

Frequency of $C^R C^R = p^2 = (0.7)^2 = 0.49$

Frequency of $C^R C^W = 2pq = 2(0.7)(0.3) = 0.42$

Frequency of $C^W C^W = q^2 = (0.3)^2 = 0.09$

In other words, if the allele frequency of C^R is 70% and the allele frequency of C^W is 30%, the expected genotype frequency of $C^R C^R$ is 49%, $C^R C^W$ is 42%, and $C^W C^W$ is 9%.

To see the relationship between allele frequencies and genotypes in a population, **Figure 24.2** considers the relationship between allele frequencies and the way that gametes combine to produce genotypes. To be valid, the Hardy-Weinberg equation carries the assumption that two gametes combine randomly with each other to produce offspring. In a population, the frequency of a gamete carrying a particular allele is equal to the allele frequency in that population. For example, if the allele frequency of C^R equals 0.7, the frequency of a gamete carrying the C^R allele also equals 0.7. The frequency of producing a $C^R C^R$ homozygote with red flowers is $0.7 \times 0.7 = 0.49$, or 49%. The probability of inheriting both C^W alleles, which produces white flowers, is $0.3 \times 0.3 = 0.09$, or 9%. In our Punnett square, two different gamete combinations can produce heterozygotes with pink flowers (Figure 24.2). An offspring could inherit the C^R allele from the pollen and C^W from the egg, or C^R from the egg and C^W from the pollen. Therefore, the frequency of heterozygotes is $pq + pq$, which equals $2pq$. In our example, this is $2(0.7)(0.3) = 0.42$, or 42%. Note that the frequencies total 100%.

The Hardy-Weinberg equation predicts that allele and genotype frequencies will remain the same, generation after generation, but only when a population is in equilibrium. To be in equilibrium, evolutionary mechanisms that can change allele and genotype frequencies are not acting on a population. For this to occur, the following conditions must be met:

- No new mutations occur.
- No natural selection occurs; that is, no survival or reproductive advantage exists for any of the genotypes.
- The population is so large that allele frequencies do not change due to random chance.
- No migration occurs between different populations.

Frequency of $C^R C^R$ genotype (red flowers) = $(0.7)^2$ = 0.49
Frequency of $C^R C^W$ genotype (pink flowers) = $2(0.7)(0.3)$ = 0.42
Frequency of $C^W C^W$ genotype (white flowers) = $(0.3)^2$ = 0.09
1.00

Figure 24.2 Calculating allele and genotype frequencies with the Hardy-Weinberg equation. A population of four-o'clock plants has allele and gamete frequencies of 0.7 for the C^R allele and 0.3 for the C^W allele. Knowing the allele frequencies allows us to calculate the genotype frequencies in the population.

Concept check: What would be the frequency of pink flowers in a population where the allele frequency of C^R is 0.4 and the population is in Hardy-Weinberg equilibrium? Assume that C^R and C^W are the only two alleles.

- Random mating occurs; that is, the members of the population mate with each other without regard to their phenotypes and genotypes.

In reality, no population satisfies the Hardy-Weinberg equilibrium completely. Nevertheless, in large natural populations with little migration and negligible natural selection, the Hardy-Weinberg equilibrium may be nearly approximated for certain genes. Alternatively, researchers may experimentally examine allele and genotype frequencies for one or more genes in a given species and discover they are not in Hardy-Weinberg equilibrium. In such cases, we would say that the population is in disequilibrium—in other words, evolutionary mechanisms are affecting the population. Disequilibrium indicates that evolution is occurring. Conservation biologists and wildlife managers may wish to identify the reason(s) why disequilibrium has occurred because this may impact the future survival of the species. Next, we will examine the mechanisms that cause such evolutionary change.

Microevolution Involves Changes in Allele Frequencies from One Generation to the Next

The term **microevolution** is used to describe changes in a population's gene pool, such as changes in allele frequencies, from generation to generation. What causes microevolution to happen? Such change is rooted in two related phenomena (**Table 24.1**). First, the introduction of new genetic variation into a population

Table 24.1	Factors That Govern Microevolution
Sources of new genetic variation*	
New mutations within genes that create new alleles	Random mutations within pre-existing genes introduce new alleles into populations, but at a very low rate. New mutations may be beneficial, neutral, or deleterious. For alleles to rise to a significant percentage in a population, evolutionary mechanisms, such as natural selection, genetic drift, and migration, must operate on them.
Gene duplication**	Abnormal crossover events and transposable elements may increase the number of copies of a gene. Over time, the additional copies can accumulate random mutations and create a gene family.
Exon shuffling***	Abnormal crossover events and transposable elements may promote gene rearrangements in which one or more exons from one gene are inserted into another gene. The protein encoded by such a gene may display a novel function that is acted upon by evolutionary mechanisms.
Horizontal gene*** transfer	Genes from one species may be introduced into another species. The transferred gene may be acted upon by evolutionary mechanisms.
Evolutionary mechanisms that alter the frequencies of existing genetic variation	
Natural selection	The process in which individuals that possess certain traits are more likely to survive and reproduce than individuals without those traits. Over the course of many generations, beneficial traits that are heritable become more common and detrimental traits become less common.
Genetic drift	This is a change in genetic variation from generation to generation due to random chance. Allele frequencies may change as a matter of chance from one generation to the next. Genetic drift has a greater impact in a small population.
Migration	Migration can occur between two different populations that have different allele frequencies. The introduction of migrants into a recipient population may change the allele frequencies of that population.
Nonrandom mating	The phenomenon in which individuals select mates based on their phenotypes or genetic lineage. This can alter the relative proportion of homozygotes and heterozygotes that is predicted by the Hardy-Weinberg equation, but it will not change allele frequencies.

* These are examples that affect single genes. Other events, such as crossing over, independent assortment, and changes in chromosome structure and number, may alter the genetic variation among many genes.

** Described in Chapter 21. See Figures 21.7 and 21.8.

*** Described in Chapter 23. See Figures 23.16 and 23.17.

is one essential aspect of microevolution. As discussed in Chapter 23, genetic variation can originate by a variety of molecular mechanisms. New alleles of preexisting genes can arise by random mutation, and new genes can be introduced into a population by gene duplication, exon shuffling, and horizontal gene transfer. Such mutations, albeit rare, provide a continuous source of new variation to populations. In 1926, the Russian geneticist Sergei Chetverikov was the first to suggest that random mutations are the raw material for evolution. However, due to their low rate of occurrence, mutations by themselves do not act as a major factor in dictating the allele frequencies in a population.

The second phenomenon that is required for evolution to happen is one or more mechanisms that alter the prevalence of a given allele or genotype in a population. These mechanisms are natural selection, genetic drift, migration, and nonrandom mating (Table 24.1). Over the course of many generations, these mechanisms may promote widespread genetic changes in a population. In the remainder of this chapter, we will examine how natural selection, genetic drift, migration, and nonrandom mating can affect the type of genetic variation that occurs when a gene exists as two alleles in a population.

24.2 Natural Selection

Recall from Chapter 23 that **natural selection** is the process in which beneficial traits that are heritable become more common in successive generations. In contrast, detrimental traits that are heritable become less common. Natural selection occurs because some individuals possess more favorable phenotypes compared to others. The phenotypes, in turn, are controlled by the individuals' genotypes. Keep in mind that natural selection itself is not evolution. Rather, natural selection is a key mechanism that causes evolution to happen. Over time, natural selection results in **adaptations**—changes in populations of living organisms that promote their survival and reproduction in a particular environment. In this section, we will examine various ways that natural selection produces adaptations.

Natural Selection Favors Individuals with Greater Reproductive Success

Reproductive success is the likelihood of an individual contributing fertile offspring to the next generation. Natural selection occurs because some individuals in a population have greater reproductive success compared to other individuals. Those individuals having heritable traits that favor reproductive success are more likely to pass those traits to their offspring.

Reproductive success is commonly attributed to two categories of traits. First, certain characteristics make organisms better adapted to their environment and more likely to survive to reproductive age. Such organisms have a greater chance to reproduce and pass their favorable traits to offspring of the next generation. Therefore, natural selection favors individuals with characteristics that provide a survival advantage. Second, reproductive success may involve traits that are directly associated with reproduction, such as the ability to find a mate and the ability to produce viable gametes and offspring. Traits that enhance the ability of individuals to reproduce, such as brightly colored plumage in male birds, are often subject to natural selection.

As we discussed in Chapter 23, Charles Darwin and Alfred Wallace independently proposed the theory of evolution by natural selection. A modern description of the principles of natural selection can relate our knowledge of molecular genetics to the process of evolution:

1. Within a population, allelic variation arises from random mutations that cause differences in DNA sequences. A mutation that creates a new allele may alter the amino acid sequence of the encoded protein. This, in turn, may alter the function of the protein.

2. Some alleles encode proteins that enhance an individual's survival or reproductive capability compared to other members of the population. For example, an allele may produce a protein that is more efficient at a higher temperature, conferring on the individual a greater probability of survival in a hot climate.

3. Individuals with beneficial alleles are more likely to survive and contribute their alleles to the gene pool of the next generation.

4. Over the course of many generations, allele frequencies of many different genes may change through natural selection, thereby significantly altering the characteristics of a population. The net result of natural selection is a population that is better adapted to its environment and more successful at reproduction.

Fitness Is a Quantitative Measure of Reproductive Success

As mentioned earlier, Haldane, Fisher, and Wright developed mathematical relationships to explain the phenomenon of natural selection. To begin our quantitative discussion of natural selection, we need to consider the concept of **fitness**, which is the relative likelihood that a genotype will contribute to the gene pool of the next generation as compared with other genotypes. Although this property often correlates with physical fitness, the two ideas should not be confused. Fitness is a measure of reproductive success. An extremely fertile individual may have a higher fitness than a less fertile individual that appears more physically fit.

To examine fitness, let's consider an example of a hypothetical gene existing in *A* and *a* alleles. We can assign fitness values to each of the three possible genotypes according to their relative reproductive success. For example, let's suppose the average reproductive successes of the three genotypes are

AA produces 5 offspring
Aa produces 4 offspring
aa produces 1 offspring

By convention, the genotype with the highest reproductive success is given a fitness value of 1.0. Fitness values are denoted by the variable *w*. The fitness values of the other genotypes are assigned values relative to this 1.0 value.

Fitness of *AA*: $w_{AA} = 1.0$

Fitness of *Aa*: $w_{Aa} = 4/5 = 0.8$

Fitness of *aa*: $w_{aa} = 1/5 = 0.2$

Variation in fitness occurs because certain genotypes result in individuals that have a greater reproductive success compared to other genotypes.

Likewise, the effects of natural selection can be viewed at the level of a population. The average reproductive success of members of a population is called the **mean fitness of the population**. Over many generations, as individuals with higher fitness values become more prevalent, natural selection also increases the mean fitness of the population. In this way, the process of natural selection results in a population of organisms that is well adapted to its native environment and likely to be successful at reproduction.

Natural Selection Can Follow Different Patterns

By studying species in their native environments, population geneticists have discovered that natural selection can occur in several ways. In most of the examples described next, natural selection leads to adaptations such that certain members of a species are more likely to survive to reproductive age.

Directional Selection During **directional selection**, individuals at one extreme of a phenotypic range have greater reproductive success in a particular environment. Different phenomena may initiate the process of directional selection. One way that directional selection may arise is that a new allele may be introduced into a population by mutation, and the new allele may confer a higher fitness in individuals that carry it (**Figure 24.3**). What are the long-term effects of such directional selection? If the homozygote carrying the favored allele has the highest fitness value, directional selection may cause this favored allele to eventually become predominant in the population, perhaps even leading to a monomorphic gene.

Another reason for directional selection is that a population may be exposed to a prolonged change in its living environment. Under the new environmental conditions, the relative fitness values may change to favor one genotype, and this will promote the elimination of other genotypes. As an example, let's suppose a population of finches on a mainland already has genetic variation that affects beak size (refer back to Figure 23.2). A small number of birds migrate to an island where the seeds are generally larger than they are on the mainland. In this new environment, birds with larger beaks would have a higher fitness because they would be better able to crack open the larger seeds and thereby survive to reproduce. Over the course of many generations, directional selection would

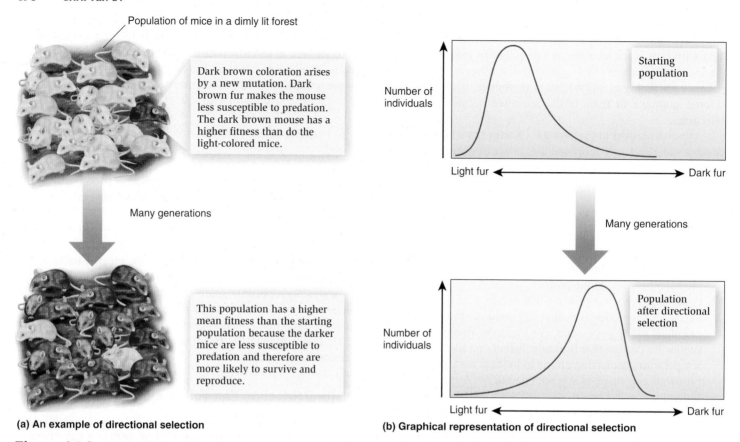

(a) An example of directional selection

Population of mice in a dimly lit forest

Dark brown coloration arises by a new mutation. Dark brown fur makes the mouse less susceptible to predation. The dark brown mouse has a higher fitness than do the light-colored mice.

Many generations

This population has a higher mean fitness than the starting population because the darker mice are less susceptible to predation and therefore are more likely to survive and reproduce.

(b) Graphical representation of directional selection

Number of individuals

Starting population

Light fur ⟷ Dark fur

Many generations

Number of individuals

Population after directional selection

Light fur ⟷ Dark fur

Figure 24.3   Directional selection.   This pattern of natural selection selects for one extreme of a phenotype that confers the highest fitness in the population's environment. **(a)** In this example, a mutation causing darker fur arises in a population of mice. This new genotype confers higher fitness, because mice with dark fur can evade predators and are more likely to survive and reproduce. Over many generations, directional selection will favor the prevalence of darker individuals. **(b)** These graphs show the change in fur color phenotypes in this mouse population before and after directional selection.

Concept check: *Let's suppose the climate on an island abruptly changed such that the average temperature was 10° higher. The climate change is permanent. How would directional selection affect the genetic diversity in a population of mice on the island (1) over the short run and (2) over the long run?*

produce a population of birds carrying alleles that promote larger beak size.

Stabilizing Selection   A type of natural selection called **stabilizing selection** favors the survival of individuals with intermediate phenotypes and selects against those with extreme phenotypes. Stabilizing selection tends to decrease genetic diversity. An example of stabilizing selection involves clutch size (number of eggs laid) in birds, which was first studied by British biologist David Lack in 1947. Under stabilizing selection, birds that lay too many or too few eggs per nest have lower fitness values than do those that lay an intermediate number of eggs. When a bird lays too many eggs, many offspring will die due to inadequate parental care and food. In addition, the strain on the parents themselves may decrease their likelihood of survival and therefore their ability to produce more offspring. Having too few offspring, however, does not contribute many individuals to the next generation. Therefore, the most successful parents are those that produce an intermediate clutch size. In the 1980s, Swedish evolutionary biologist Lars Gustafsson and his colleagues examined the phenomenon of stabilizing

selection in the collared flycatcher (*Ficedula albicollis*) on the island of Gotland south of Sweden. They discovered that Lack's hypothesis concerning an optimal clutch size appears to be true for this species (**Figure 24.4**).

Diversifying Selection   **Diversifying selection** (also known as disruptive selection) favors the survival of two or more different genotypes that produce different phenotypes. In diversifying selection, the fitness values of a particular genotype are higher in one environment and lower in a different environment, whereas the fitness values of the other genotype vary in an opposite manner. Diversifying selection is likely to occur in populations that occupy heterogeneous environments, so some members of the species will be more likely to survive in each type of environmental condition.

An example of diversifying selection involves colonial bentgrass (*Agrostis capillaris*) (**Figure 24.5**). In certain locations where this grass is found, such as South Wales, isolated places occur where the soil is contaminated with high levels of heavy metals due to mining. The relatively recent metal contamination has selected for the proliferation of mutant strains of

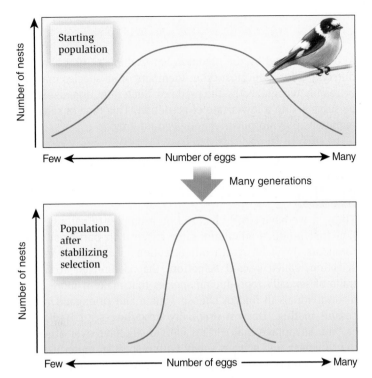

Figure 24.4 **Stabilizing selection.** In this pattern of natural selection, the extremes of a phenotypic distribution are selected against. Those individuals with intermediate traits have the highest fitness. These graphs show the results of stabilizing selection on clutch size in a population of collared flycatchers (*Ficedula albicollis*). This process results in a population with less diversity and more uniform traits.

Concept check: Why does stabilizing selection decrease genetic diversity?

(a) Growth of *Agrostis capillaris* on contaminated soil

Figure 24.5 **Diversifying selection.** This pattern of natural selection selects for two different phenotypes, each of which is most fit in a particular environment. **(a)** In this example, random mutations have resulted in metal-resistant alleles in colonial bentgrass (*Agrostis capillaris*) that allow it to grow on soil contaminated with high levels of heavy metals such as copper. In uncontaminated soils, the grass does not show metal tolerance. Because both metal-resistant and metal-sensitive alleles are maintained in the population, this situation is an example of diversifying selection due to heterogeneous environments. **(b)** These graphs show the change in phenotypes in this bentgrass population before and after diversifying selection.

A. capillaris that are tolerant of the heavy metals (Figure 24.5a). Such genetic changes enable these mutant strains to grow on contaminated soil but tend to inhibit their growth on normal, noncontaminated soil. These metal-resistant plants often grow on contaminated sites that are close to plants that grow on uncontaminated land and do not show metal tolerance.

Balancing Selection Contrary to a popular misconception, natural selection does not always cause the elimination of "weaker" or less-fit alleles. **Balancing selection** is a type of natural selection that maintains genetic diversity in a population. Over many generations, balancing selection can create a situation known as a **balanced polymorphism**, in which two or more alleles are kept in balance and therefore are maintained in a population over many generations.

How can balancing selection maintain a polymorphism? Population geneticists have identified two common ways that balancing selection can occur. First, for genetic variation involving a single gene, balancing selection can favor the heterozygote over either corresponding homozygote. This situation is called **heterozygote advantage**. Heterozygote advantage can sometimes explain the persistence of alleles that are deleterious in a homozygous condition.

A classic example of heterozygote advantage involves the H^S allele of the human β-globin gene. A homozygous $H^S H^S$ individual has sickle-cell disease, such as Kimbareta, discussed at the beginning of the chapter. This disease leads to the sickling of the red blood cells. The $H^S H^S$ homozygote has a lower fitness than a homozygote with two copies of the more

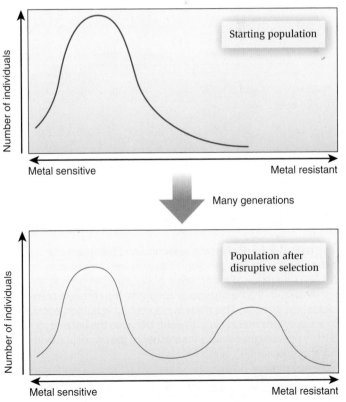

(b) Graphical representation of disruptive selection

common β-globin allele, H^AH^A. Heterozygotes, H^AH^S, do not typically show symptoms of sickle-cell disease but they have an increased resistance to malaria. Compared with H^AH^A homozygotes, heterozygotes have the highest fitness because they have a 10–15% better chance of survival if infected by the malarial parasite *Plasmodium falciparum*. Therefore, the H^S allele is maintained in populations where malaria is prevalent, such as the Democratic Republic of Congo, even though the allele is detrimental in the homozygous state (**Figure 24.6**). This balanced polymorphism results in a higher mean fitness of the population. In areas where malaria is endemic, a population composed of all H^AH^A individuals would have a lower mean fitness.

Negative frequency-dependent selection is a second way that natural selection can produce a balanced polymorphism. In this pattern of natural selection, the fitness of a genotype decreases when its frequency becomes higher. In other words, rare individuals have a higher fitness, and common individuals have a lower fitness. Therefore, rare individuals are more likely to reproduce, whereas common individuals are less likely, thereby producing a balanced polymorphism in which no genotype becomes too rare or too common.

Negative frequency-dependent selection is thought to maintain polymorphisms among species that are preyed upon by predators. Research has shown that certain predators form a mental "search image" for their prey, which is usually based on the common type of prey in an area. A prey that exhibits a rare polymorphism that affects its appearance is less likely

to be recognized by the predator. For example, a prey that is a different color than most other members of its species may not be readily recognized by the predator. Such rarer organisms will be subjected to a lower rate of predation. This type of selection maintains polymorphism among certain prey.

24.3   Sexual Selection

Thus far we have largely focused on examples of natural selection that produce adaptations for survival in particular environments. Now let's turn our attention to a form of natural selection, called **sexual selection**, that is directed at certain traits of sexually reproducing species that make it more likely for individuals to find or choose a mate and/or engage in successful mating. Darwin originally described sexual selection as "the advantage that certain individuals have over others of the same sex and species solely with respect to reproduction." In this section, we will explore how sexual selection can alter traits that play a key role in reproduction.

Sexual Selection Is a Type of Natural Selection That Directly Promotes Reproductive Success

In many species of animals, sexual selection affects male characteristics more intensely than female characteristics. Unlike females, which tend to be fairly uniform in their reproductive success, male success tends to be more variable, with some males mating with many females and others not mating at all. Sexual selection results in the evolution of traits, called secondary sex characteristics, that favor reproductive success. The result of this process is sometimes a significant difference between the appearances of the two sexes within a species, a situation called sexual dimorphism.

Sexual selection operates in one of two ways. In **intrasexual selection**, members of one sex directly compete with each other for the opportunity to mate with individuals of the opposite sex. Examples of traits that result from intrasexual selection in animals include horns in male sheep, antlers in male moose, and the enlarged claw of male fiddler crabs (**Figure 24.7a**). In fiddler crabs (*Uca paradussumieri*), males enter the burrows of females that are ready to mate. If another male attempts to enter the burrow, the male already inside the burrow stands in the burrow shaft and blocks the entrance with his enlarged claw. Males with the largest claws are more likely to be successful at this behavior and more likely to pass their genes to future generations.

Now let's consider an example of **intersexual selection**, namely, mate choice. This type of sexual selection often results in showy characteristics in males. **Figure 24.7b** shows a classic example that involves the Indian peafowl (*Pavo cristatus*), the national bird of India. Male peacocks have long and brightly colored tail feathers, which they fan out as a mating behavior. Females select among males based on feather color and pattern as well as the physical prowess of the display.

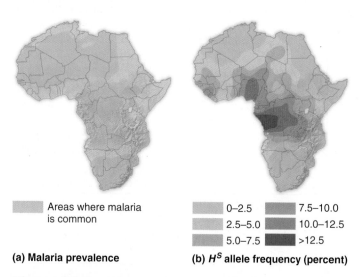

Areas where malaria is common	0–2.5 7.5–10.0
	2.5–5.0 10.0–12.5
	5.0–7.5 >12.5

(a) Malaria prevalence   **(b) H^S allele frequency (percent)**

Figure 24.6   Balancing selection and heterozygote advantage.   (a) The geographic prevalence of malaria in Africa. (b) The frequency of the H^S allele of the β-globin gene in the same area. In the homozygous condition, the H^S allele causes sickle-cell disease. This allele is maintained in human populations in areas where malaria is prevalent, because the heterozygote (H^AH^S) has a higher fitness than either of the corresponding homozygotes (H^AH^A or H^SH^S).

Concept check: If malaria was eradicated, what would you expect to happen to the frequencies of the H^A and H^S alleles over the long run?

(a) Intrasexual selection

(b) Intersexual selection

(c) Sexual selection balanced by predation

Figure 24.7 **Examples of the results of sexual selection, a type of natural selection.**   **(a)** An example of intrasexual selection. The enlarged claw of the male fiddler crab is used in direct male-to-male competition. In this photograph, a male inside a burrow is extending its claw out of the burrow to prevent another male from entering. **(b)** An example of intersexual selection. Female peahens choose male peacocks based on the males' colorful and long tail feathers and the robustness of their display. **(c)** Male guppies (on the right) are brightly colored to attract a female (on the left), but brightly colored males are less common where predation is high.

Concept check: *Male birds of many species have loud and elaborate courtship songs. Is this likely to be the result of intersexual or intrasexual selection? Explain.*

A less obvious type of intersexual selection is cryptic female choice, in which the female reproductive system can influence the relative success of sperm. As an example of cryptic female choice, the female genital tract of certain animals selects for sperm that tend to be genetically unrelated to the female. Sperm from males closely related to the female, such as brothers or cousins, are less successful than are sperm from genetically unrelated males. The selection for sperm may occur over the journey through the reproductive tract. The egg itself may even have mechanisms to prevent fertilization by genetically related sperm. Cryptic female choice occurs in species in which females may mate with more than one male, such as many species of reptiles and ducks. A similar mechanism is found in many plant species in which pollen from genetically related plants, perhaps from the same flower, is unsuccessful at fertilization, whereas pollen from unrelated plants is successful. One possible advantage of cryptic female choice is that it inhibits inbreeding, which is described later in this chapter. At the population level, cryptic female choice may promote genetic diversity by favoring interbreeding among genetically unrelated individuals.

Sexual selection is sometimes a combination of both intrasexual and intersexual selection. During breeding season, male elk (*Cervus elaphus*) become aggressive and bugle loudly to challenge other male elk. Males spar with their antlers, which usually turns into a pushing match to determine which elk is stronger. Female elk then choose the strongest bulls as their mates.

Sexual selection can explain the existence of traits that could decrease an individual's chances of survival but increase their chances of reproducing. For example, the male guppy (*Poecilia reticulata*) is brightly colored compared to the female

(**Figure 24.7c**). In nature, females prefer brightly colored males. However, brightly colored males are more likely to be seen and eaten by predators. In places with few predators, the males tend to be brightly colored. In contrast, where predators are abundant, brightly colored males are less plentiful because they are subject to predation. In this case, the relative abundance of brightly and dully colored males depends on the balance between sexual selection, which favors bright coloring, and escape from predation, which favors dull coloring.

Many animals have secondary sexual characteristics, and evolutionary biologists generally agree that sexual selection is responsible for such traits. But why should males compete, and why should females be choosy? Researchers have proposed various hypotheses to explain the underlying mechanisms. One possible reason is related to the different roles that males and females play in the nurturing of offspring. In some species, the female is the primary caregiver, whereas the male plays a minor role. In such species, mating behavior may influence the fitness of both males and females. Males increase their fitness by mating with multiple females. This increases their likelihood of passing their genes on to the next generation. By comparison, females may produce relatively few offspring, and their reproductive success may not be limited by the number of available males. Females will have higher fitness if they choose males that are good defenders of their territory and have alleles that confer a survival advantage to their offspring. One measure of alleles that confer higher fitness is age. Males that live to an older age are more likely to carry beneficial alleles. Many research studies involving female choice have shown that females tend to select traits that are more likely to be well developed in older males than in immature males. For example, in certain species of birds, females tend to choose males with a larger repertoire of songs, which is more likely to occur in older males.

Sexual selection is governed by the same processes involved in the evolution of traits that are not directly related to sex. The result of sexual selection can be directional, stabilizing, diversifying, or balancing selection. For example, the evolution of the large and brightly colored tail of the male peacock reflects directional selection. Next, we will consider an example in which sexual selection may be diversifying.

FEATURE INVESTIGATION

Seehausen and van Alphen Found That Male Coloration in African Cichlids Is Subject to Female Choice

Cichlids are tropical freshwater fish that are popular among aquarium enthusiasts. The Cichlidae family is composed of more than 3,000 species that vary with regard to body shape, coloration, behavior, and feeding habits, making it one of the largest and most diverse vertebrate families. By far the greatest diversity of these fish is found in Lake Victoria, Lake Malawi, and Lake Tanganyika in East Africa, where more than 1,800 species are found.

Cichlids have complex mating behavior, and females play an important role in choosing males with particular characteristics. In 1998, population geneticists Ole Seehausen and Jacques van Alphen investigated the effect of male coloration of two species of cichlids on female choice. In some locations, *Pundamilia pundamilia* and *Pundamilia nyererei* do not readily interbreed and behave like two distinct biological species, whereas in other places, they behave like a single interbreeding species with two color morphs. They can interbreed to produce viable offspring, and both inhabit Lake Victoria. Males of both species have blackish underparts and blackish vertical bars on their sides (**Figure 24.8a**). *P. pundamilia* males are grayish white on top and on the sides, and they have a metallic blue and red dorsal fin, which is the uppermost fin. *P. nyererei* males are red-orange on top and yellow on their sides.

Seehausen and van Alphen hypothesized that females choose males for mates based, in part, on the males' coloration. The researchers took advantage of the observation that colors are obscured under orange monochromatic light. As seen in **Figure 24.8b**, males of both species look similar under these conditions. In their study, a female of one species was placed in an aquarium that contained one male of each species within an enclosure (**Figure 24.9**). The males were within glass enclosures to avoid direct competition with each other, which would have likely affected female choice. The goal of the experiment was to determine which of the two males a female would prefer. Courtship between a male and female begins when a male swims toward a female and exhibits a lateral display (that is, shows the side of his body to the female). If the female is interested, she will approach the male, and then the male will display a quivering motion. Such courtship behavior was examined under normal light and under orange monochromatic light.

As seen in the data, Seehausen and van Alphen found that the females' preference for males was dramatically different depending on the illumination conditions. Under normal

(a) Males of two species in normal light

(b) Males of two species in artificial light

Figure 24.8 Male coloration in African cichlids.   (a) Two males (*Pundamilia pundamilia*, top, and *Pundamilia nyererei*, bottom) under normal illumination. (b) The same species under orange monochromatic light, which obscures their color differences.

light, *P. pundamilia* females preferred *P. pundamilia* males, and *P. nyererei* females preferred *P. nyererei* males. However, such mating preference was lost when colors were masked by artificial light. If the light conditions in their native habitats are similar to the normal light used in this experiment, female choice would be expected to separate cichlids into two populations, with *P. pundamilia* females mating with *P. pundamilia* males and *P. nyererei* females mating with *P. nyererei* males. In this case, sexual selection appears to have followed a diversifying mechanism in which certain females prefer males with one color pattern, while other females prefer males with a different color pattern. A possible outcome of such sexual selection is that it can separate one large population into smaller populations that selectively breed with each other and eventually become distinct species. We will discuss the topic of species formation in more depth in Chapter 25.

Experimental Questions

1. What hypothesis is tested in the Seehausen and van Alphen experiment?

2. Describe the experimental design for this study, illustrated in Figure 24.9. What was the purpose of conducting the experiment under the two different light conditions?

3. What were the results of the experiment in Figure 24.9?

Figure 24.9 A study by Seehausen and van Alphen involving the effects of male coloration on female choice in African cichlids.

HYPOTHESIS Female African cichlids choose mates based on the males' coloration.

KEY MATERIALS Two species of cichlid, *Pundamilia pundamilia* and *Pundamilia nyererei*, were chosen. The males differ with regard to their coloration. A total of 8 males and 8 females (4 males and 4 females from each species) were tested.

| | Experimental level | Conceptual level |

1. Place 1 female and 2 males in an aquarium. Each male is within a separate glass enclosure. The enclosures contain 1 male from each species.

   This is a method to evaluate sexual selection via female choice in 2 species of cichlid.

2. Observe potential courtship behavior for 1 hour. If a male exhibited lateral display (a courtship invitation) and then the female approached the enclosure that contained the male, this was scored as a positive encounter. This protocol was performed under normal light and under orange monochromatic light.

3. **THE DATA**

Female	Male	Light condition	Percentage of positive encounters*
P. pundamilia	*P. pundamilia*	Normal	16
P. pundamilia	*P. nyererei*	Normal	2
P. nyererei	*P. nyererei*	Normal	16
P. nyererei	*P. pundamilia*	Normal	5
P. pundamilia	*P. pundamilia*	Monochromatic	20
P. pundamilia	*P. nyererei*	Monochromatic	18
P. nyererei	*P. nyererei*	Monochromatic	13
P. nyererei	*P. pundamilia*	Monochromatic	18

*A positive encounter occurred when a male's lateral display was followed by the female approaching the male.

4. **CONCLUSION** Under normal light, where colors can be distinguished, *P. pundamilia* females prefer *P. pundamilia* males, and *P. nyererei* females prefer *P. nyererei* males.

5. **SOURCE** Seehausen, O., and van Alphen, J.J.M. 1998. The effect of male coloration on female mate choice in closely related Lake Victoria cichlids (*Haplochromis nyererei* complex). *Behav. Ecol. Sociobiol.* 42:1–8.

24.4 Genetic Drift

Thus far, we have focused on natural selection as a mechanism that fosters genetic change. Let's now turn our attention to a second important way the gene pool of a population can change. In the 1930s, Sewall Wright played a large role in developing the concept of **genetic drift** (also called random genetic drift), which refers to changes in allele frequencies due to random chance. The term genetic drift is derived from the observation that allele frequencies may "drift" randomly from generation to generation as a matter of chance.

Changes in allele frequencies due to genetic drift happen regardless of the fitness of individuals that carry those alleles. For example, an individual with a high fitness value may, as a matter of bad luck, not encounter a member of the opposite sex. Likewise, random chance can influence which alleles happen to be found in the gametes that fuse with each other in a successful fertilization. In this section, we will examine how genetic drift can alter allele frequencies in populations.

Genetic Drift Has a Greater Impact in Small Populations

What are the effects of genetic drift? Over the long run, genetic drift favors either the elimination or the fixation of an allele, that is, when its frequency reaches 0% or 100% in a population, respectively. However, the number of generations it takes for an allele to be lost or fixed greatly depends on the population size. **Figure 24.10** illustrates the potential consequences of genetic drift in one large (N = 1,000) and two small (N = 10) populations. This simulation involves the frequency of hypothetical B and b alleles of a gene for fur color in a population of mice—B is the black allele, and b is the white allele.

At the beginning of this hypothetical simulation, which runs for 50 generations, all of these populations had identical allele frequencies: B = 0.5 and b = 0.5. In the small populations, the allele frequencies fluctuated substantially from generation to generation. Eventually, in one population, the b allele was eliminated, while in another, it was fixed at 100%. These small populations would then consist of only black mice or white mice, respectively. At this point, the gene has become monomorphic and cannot change any further. By comparison, the frequencies of B and b in the large population fluctuated much less. As discussed in Chapter 16, the relative effect of random chance, also termed random sampling error, is much smaller when the sample size is large. Nevertheless, genetic drift will eventually lead to allele loss or fixation even in large populations, but this will take many more generations to occur than it does in small populations.

In nature, genetic drift may rapidly alter allele frequencies when the population size dramatically decreases. Two common examples of this phenomenon are the bottleneck effect and the founder effect, which are described next.

Bottleneck Effect A population can be reduced dramatically in size by events such as earthquakes, floods, drought, and human destruction of habitat. These occurrences may eliminate most members of the population without regard to their genetic composition. When allele frequencies of the resulting population change due to genetic drift, this is called the **bottleneck effect**. Such changes may happen for two reasons. First, the surviving members may have allele frequencies that differ from those of the original population that was much larger. Second, as we saw in Figure 24.10, genetic drift acts more quickly to reduce genetic variation when the population size is small. In extreme cases, alleles may even be eliminated. Eventually, the bottlenecked population may regain its original size. However, the new population is likely to have less genetic variation than the original one.

A hypothetical example of the bottleneck effect is shown with a population of frogs in **Figure 24.11**. In this example, a starting population of frogs is found in three phenotypes: yellow, dark green, and striped. Due to a bottleneck caused by a drought, the dark green variety is lost from the population.

As a real-life example, the African cheetah has lost nearly all of its genetic variation. This was likely caused by a bottleneck effect. An analysis of the current population by population geneticists has suggested that a severe population bottleneck occurred in this species approximately 10,000–12,000 years ago, reducing it to near extinction. The species eventually rebounded in numbers, but the bottleneck apparently reduced its genetic variation to very low levels. The modern species is monomorphic for many of its genes.

Founder Effect Another common phenomenon in which genetic drift may have a rapid impact is the **founder effect**. This occurs when a small group of individuals separates from a larger population and establishes a colony in a new location. For example, a few individuals may migrate from a large continental population and become the founders of an island

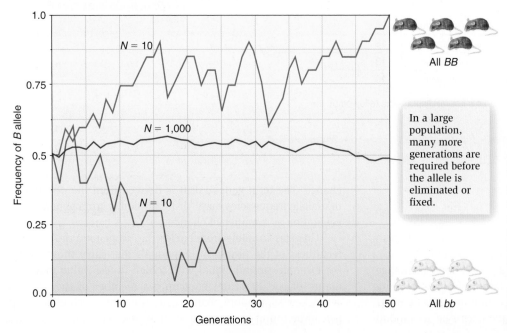

In a large population, many more generations are required before the allele is eliminated or fixed.

Figure 24.10 **Genetic drift and population size.** This graph shows three hypothetical simulations of genetic drift and their effects on small and large populations of black (B allele) and white (b allele) mice. In all cases, the starting allele frequencies are B = 0.5 and b = 0.5. The red lines illustrate two populations of mice in which N = 10; the blue line shows a population in which N = 1,000. Genetic drift eventually leads to either the elimination or fixation of alleles.

1 The starting population includes 3 phenotypes of frogs: yellow, dark green, and striped.

2 A drought causes a bottleneck in which the population size is decreased and the dark green phenotype is lost.

3 The population size recovers, but genetic variation is decreased, and only 2 phenotypes are left.

Figure 24.11 **A hypothetical example of the bottleneck effect.** This example involves a population of frogs in which a drought dramatically reduced population size, resulting in a bottleneck. The bottleneck reduced the genetic diversity in the population.

Concept check: How does the bottleneck effect undermine the efforts of conservation biologists who are trying to save species nearing extinction?

population. The founder effect differs from a bottleneck in that it occurs in a new location, although both effects are related to a reduction in the size of a population. The founder effect has two important consequences. First, the founding population,

which is relatively small, is expected to have less genetic variation than the larger original population from which it was derived. Second, as a matter of chance, the allele frequencies in the founding population may differ markedly from those of the original population.

Population geneticists have studied many examples in which isolated populations were founded via colonization by members of another population. For example, in the 1960s, American geneticist Victor McKusick studied allele frequencies in the Old Order Amish of Lancaster County, Pennsylvania. At that time, this was a group of about 8,000 people, descended from just three couples that immigrated to the U.S. in 1770. Among this population of 8,000, a genetic disease known as the Ellis–van Creveld syndrome (a recessive form of dwarfism) was found at a frequency of 0.07, or 7%. By comparison, this disorder is extremely rare in other human populations, even the population from which the founding members had originated. The high frequency in the Lancaster County population is a chance occurrence due to the founder effect.

The Neutral Theory of Evolution Proposes That Genetic Drift Also Plays an Important Role in Promoting Genetic Change

In 1968, Japanese evolutionary biologist Motoo Kimura proposed that much of the genetic variation seen in natural populations is the result of genetic drift rather than natural selection. Genetic drift is a random process that does not preferentially select for any particular allele—it can alter the frequencies of both beneficial and deleterious alleles. Much of the time, genetic drift promotes **neutral variation**, which does not affect reproductive success. According to Kimura's **neutral theory of evolution**, most genetic variation is due to the accumulation of neutral mutations that have attained high frequencies in a population via genetic drift.

Neutral mutations are not subject to natural selection because they do not affect reproductive success. For example, a new mutation within a structural gene that changes a glycine codon from GGG to GGC would not affect the amino acid sequence of the encoded protein. Both genotypes are equal in fitness. However, such new mutations can spread throughout a population due to genetic drift (**Figure 24.12**). This theory has been called **non-Darwinian evolution** and also "survival of the luckiest." Kimura agreed with Darwin that natural selection is responsible for adaptive changes in a species during evolution. The long neck of the giraffe is the result of natural selection. His main idea is that much of the variation in DNA sequences is explained by neutral variation rather than adaptive variation.

The sequencing of genomes from many species is consistent with the neutral theory of evolution. When we examine changes of the coding sequence within structural genes, we find that nucleotide substitutions are more prevalent in the third base of a codon than in the first or second base. Mutations in the third base are often neutral; that is, they do not change the amino acid sequence of the protein (refer back to Table 12.1). In contrast, random mutations at the first or second base are

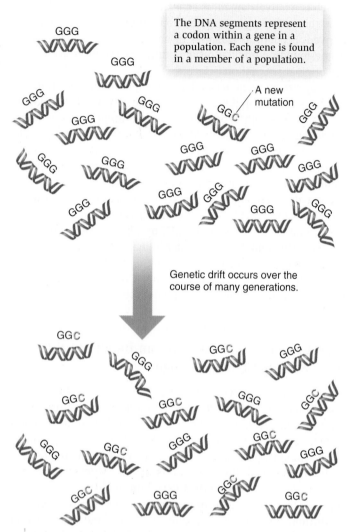

The DNA segments represent a codon within a gene in a population. Each gene is found in a member of a population.

A new mutation

Genetic drift occurs over the course of many generations.

Figure 24.12   **Neutral evolution in a population.**   In this example, a mutation within a gene changes a glycine codon from GGG to GGC, which does not affect the amino acid sequence of the encoded protein. Each gene shown represents a copy of the gene in a member of a population. Over the course of many generations, genetic drift may cause this neutral allele to become prevalent in the population, perhaps even monomorphic.

Concept check: *Describe two other examples of genetic changes that might be neutral.*

more likely to be harmful than beneficial and tend to be eliminated from a population.

24.5   Migration and Nonrandom Mating

Thus far, we have considered how natural selection and genetic drift are key mechanisms that cause evolution to happen. In addition, migration between neighboring populations and nonrandom mating may have an impact on genetic variation and the relative proportions of genotypes. In this section, we will explore how these mechanisms work.

Migration Between Two Populations Tends to Increase Genetic Variation

Earlier in this chapter, we considered how migration to a new location by a relatively small group can result in a founding population with an altered genetic composition due to genetic drift. In addition, migration between two different established populations can alter allele frequencies. As an example, let's consider two populations of a particular species of deer that are separated by a mountain range running north and south (Figure 24.13). On rare occasions, a few deer from the western population may travel through a narrow pass between the mountains and become members of the eastern population. If the two populations are different with regard to genetic variation, this migration will alter the frequencies of certain alleles in the eastern population. Of course, this migration could occur in the opposite direction as well and would then affect the western population. This movement of alleles into or out of a population, called **gene flow**, occurs whenever individuals migrate between populations having different allele frequencies.

In nature, individuals commonly migrate in both directions. What are the consequences of bidirectional migration? First, migration tends to reduce differences in allele frequencies between neighboring populations. Population geneticists can evaluate the extent of migration between two populations by

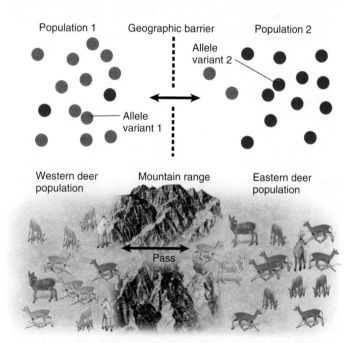

Population 1   Geographic barrier   Population 2

Allele variant 2

Allele variant 1

Western deer population   Mountain range   Eastern deer population

Pass

Figure 24.13   **Migration and gene flow.**   In this example, two populations of a deer species are separated by a mountain range. On rare occasions, a few deer from the one population travel through a narrow pass and become members of the other population. If the two populations differ in regard to genetic variation, this migration will alter the frequencies of alleles in the populations.

Concept check: *How does gene flow affect the genetic compositions of populations?*

analyzing the similarities and differences between their allele frequencies. Populations that frequently mix their gene pools via migration tend to have similar allele frequencies, whereas the allele frequencies of isolated populations are more disparate, due to the effects of natural selection and genetic drift. Second, migration tends to increase genetic diversity within populations. As discussed earlier in this chapter, new mutations are relatively rare events. Therefore, a new mutation may arise in only one population, and migration may then introduce this new allele into a neighboring population.

Nonrandom Mating Affects the Relative Proportion of Homozygotes and Heterozygotes in a Population

As mentioned earlier in this chapter, one of the conditions required to establish the Hardy-Weinberg equilibrium is random mating, which means that individuals choose their mates irrespective of their genotypes or phenotypes. In many species, including human populations, this condition is violated. Such **nonrandom mating** takes different forms. Assortative mating occurs when individuals with similar phenotypes are more likely to mate. If the similar phenotypes are due to similar genotypes, assortative mating tends to increase the proportion of homozygotes and decrease the proportion of heterozygotes in the population. The opposite situation, where dissimilar phenotypes mate preferentially, causes heterozygosity to increase.

Another form of nonrandom mating involves the choice of mates based on their genetic history rather than their phenotypes. Individuals may choose a mate that is part of the same genetic lineage. The mating of two genetically related individuals, such as cousins, is called **inbreeding**. This sometimes occurs in human societies and is more likely to take place in nature when population size becomes very small.

In the absence of other evolutionary factors, nonrandom mating does not affect allele frequencies in a population. However, it will disrupt the balance of genotypes that is predicted by the Hardy-Weinberg equilibrium. As an example, let's consider a human pedigree involving a mating between cousins (**Figure 24.14**). Individuals III-2 and III-3 are cousins and have produced the daughter labeled IV-1. She is said to be inbred, because her parents are genetically related. The parents of an inbred individual have one or more common ancestors. In the pedigree of Figure 24.14, I-2 is the grandfather of both III-2 and III-3.

Inbreeding increases the relative proportions of homozygotes and decreases the likelihood of heterozygotes in a population. Why does this happen? An inbred individual has a higher chance of being homozygous for any given gene because the same allele for that gene could be inherited twice from a common ancestor. For example, let's suppose that individual I-2 is a heterozygote, *Cc*. The *c* allele could pass from I-2 to II-2 to III-2 and finally to IV-1 (see red lines in Figure 24.14). Likewise, the *c* allele could pass from I-2 to II-3 to III-3 and then to IV-1. Therefore, IV-1 has a chance of being homozygous because she inherited both copies of the *c* allele from a common ancestor to both of her parents. Inbreeding does not favor any particular

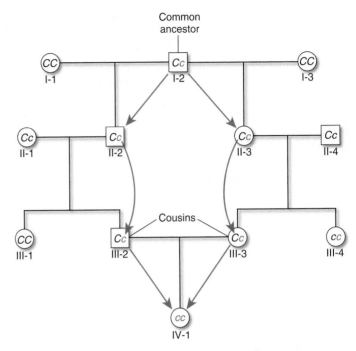

Figure 24.14 A human pedigree containing inbreeding. The parents of individual IV-1 are genetically related (cousins), and therefore, individual IV-1 is inbred. Inbreeding increases the likelihood that an individual will be homozygous for any given gene. The red arrows show how IV-1 could become homozygous by inheriting the same allele (*c*) from the common ancestor (I-2) to both of her parents.

Concept check: *Although inbreeding by itself does not affect allele frequencies, how might inbreeding indirectly affect allele frequencies over the course of many generations if natural selection was also occurring?*

allele—it does not favor *c* over *C*—but it does increase the likelihood that an individual will be homozygous for any given gene.

Although inbreeding by itself does not affect allele frequencies, it may have negative consequences with regard to recessive alleles. Rare recessive alleles that are harmful in the homozygous condition are found in all populations. Such alleles do not usually pose a problem because heterozygotes carrying a rare recessive allele are also rare, making it very unlikely that two such heterozygotes will mate with each other. However, when inbreeding is practiced, homozygous offspring are more likely to be produced. For example, rare recessive diseases in humans are more frequent when inbreeding occurs.

In natural populations, inbreeding will lower the mean fitness of the population if homozygous offspring have lower fitness values. This can be a serious problem as natural populations become smaller due to human habitat destruction. As the population shrinks, inbreeding becomes more likely because individuals have fewer potential mates from which to choose. The inbreeding, in turn, produces homozygotes that are less fit, thereby decreasing the reproductive success of the population. This phenomenon is called **inbreeding depression**. Conservation biologists sometimes try to circumvent this problem by introducing individuals from one population into another. For example, the endangered Florida panther (*Puma concolor coryi*) suffers from inbreeding-related defects, which include poor sperm quality and quantity, and morphological abnormalities. To alleviate these effects, panthers from Texas have been introduced into the Florida population of panthers.

 Summary of Key Concepts

24.1 Genes in Populations

- Population genetics is the study of genes and genotypes in a population, with a focus on understanding genetic variation. A population is a group of individuals of the same species that occupy the same environment and can interbreed with one another. All of the genes in a population constitute a gene pool.

- Polymorphism, which is very common in nearly all populations, refers to a phenotype or genotype that is found in two or more forms in a population. A monomorphic gene exists predominantly (>99%) as a single allele in a population. (Figure 24.1)

- An allele frequency is the number of copies of a specific allele divided by the total number of all alleles in a population. A genotype frequency is the number of individuals with a given genotype divided by the total number of individuals in a population.

- The Hardy-Weinberg equation ($p^2 + 2pq + q^2 = 1$) describes the allele and genotype frequencies in a population that is not evolving. It predicts an equilibrium if no new mutations are formed, no natural selection occurs, the population size is very large, the populations do not migrate, and mating is random. (Figure 24.2)

- Microevolution involves changes in a population's gene pool from one generation to the next.

- Sources of new genetic variation include random gene mutations, gene duplications, exon shuffling, and horizontal gene transfer. Natural selection, genetic drift, migration, and nonrandom mating may alter allele and genotype frequencies and cause a population to evolve. (Table 24.1)

24.2 Natural Selection

- Natural selection favors individuals with greater reproductive success. A measure of reproductive success is fitness, the contribution of a genotype to the genetic composition of the next generation as compared to the contributions of other genotypes.

- Directional selection favors one extreme of a phenotypic distribution. (Figure 24.3)

- Stabilizing selection favors an intermediate phenotype. (Figure 24.4)

- Diversifying selection favors two or more phenotypes. An example is when a population occupies a diverse environment. (Figure 24.5)

- Balancing selection maintains genetic polymorphism in a population. Examples include heterozygote advantage and negative frequency-dependent selection. (Figure 24.6)

24.3 Sexual Selection

- Sexual selection is a form of natural selection in which individuals possessing certain traits are more likely than others to find or choose a mate and/or engage in successful mating. Intrasexual and intersexual selection can lead to traits described as secondary sex characteristics. (Figure 24.7)

- Seehausen and van Alphen discovered that female cichlids' choice of mates is influenced by male coloration. This is an example of sexual selection. (Figures 24.8, 24.9)

24.4 Genetic Drift

- Genetic drift involves changes in allele frequencies due to random chance. It occurs more rapidly in small populations and leads to either the elimination or the fixation of alleles. (Figure 24.10)

- The bottleneck effect is a form of genetic drift in which a population size is dramatically reduced and then rebounds. During the bottleneck, genetic variation may be lost from a population. (Figure 24.11)

- The founder effect occurs when a small population moves to a new geographic location and genetic drift alters the genetic composition of that population.

- The neutral theory of evolution by Kimura indicates that much of the genetic variation observed in populations is due to the accumulation of neutral genetic changes. (Figure 24.12)

24.5 Migration and Nonrandom Mating

- Gene flow occurs when individuals migrate between populations with different allele frequencies. It reduces differences in allele frequencies between neighboring populations and enhances genetic diversity in populations. (Figure 24.13)

- Inbreeding is a type of nonrandom mating in which genetically related individuals mate with each other. This tends to increase the proportion of homozygotes relative to heterozygotes. When homozygotes have lower fitness, this phenomenon is called inbreeding depression. (Figure 24.14)

Assess and Discuss

Test Yourself

1. Population geneticists are interested in the genetic variation in populations. The most common type of genetic change that can cause polymorphism in a population is
   a. a deletion of a gene sequence.
   b. a duplication of a region of a gene.
   c. a rearrangement of a gene sequence.
   d. a single-nucleotide substitution.
   e. an inversion of a segment of a chromosome.

2. The Hardy-Weinberg equation characterizes the genotype frequencies and allele frequencies
   a. of a population that is experiencing selection for mating success.
   b. of a population that is extremely small.

c. of a population that is very large and not evolving.

d. of a community of species that is not evolving.

e. of a community of species that is experiencing selection.

3. Considering the Hardy-Weinberg equation, what portion of the equation would be used to calculate the frequency of individuals that do not exhibit a disease but are carriers of a recessive genetic disorder?

a. q d. q^2

b. p^2 e. both b and d

c. $2pq$

4. By itself, which of the following is not likely to have a major impact on allele frequencies?

a. natural selection   d. inbreeding

b. genetic drift   e. both c and d

c. mutation

5. Which of the following statements is correct regarding mutations?

a. Mutations are not important in evolution.

b. Mutations provide the source for genetic variation, but other evolutionary factors are more important in determining allele frequencies in a population.

c. Mutations occur at such a high rate that they promote major changes in the gene pool from one generation to the next.

d. Mutations are insignificant when considering evolution of a large population.

e. Mutations are of greater importance in larger populations than in smaller populations.

6. In a population of fish, body coloration varies from a light shade, almost white, to a very dark shade of green. If changes in the environment resulted in decreased predation of individuals with the lightest coloration, this would be an example of _____ selection.

a. diversifying   d. sexual

b. stabilizing   e. artificial

c. directional

7. Considering the same population of fish described in question 6, if the stream environment included several areas of sandy, light-colored bottom areas and a lot of dark-colored vegetation, both the light- and dark-colored fish would have selective advantage and increased survival. This type of scenario could explain the occurrence of

a. genetic drift.   d. stabilizing selection.

b. diversifying selection. e. sexual selection.

c. mutation.

8. The microevolutionary factor most sensitive to population size is

a. mutation.   d. genetic drift.

b. migration.   e. all of the above.

c. selection.

9. The neutral theory of evolution differs from Darwinian evolution in that

a. neutral theory states that natural selection does not exist.

b. neutral theory states that most of the genetic variation in a population is due to neutral mutations, which do not affect reproductive success.

c. neutral variation alters survival and reproductive success.

d. neutral mutations are not affected by population size.

e. both b and c.

10. Populations that experience inbreeding may also experience

a. a decrease in fitness due to an increased frequency of recessive genetic diseases.

b. an increase in fitness due to increases in heterozygosity.

c. very little genetic drift.

d. no apparent change.

e. increased mutation rates.

Conceptual Questions

1. The percentage of individuals exhibiting a recessive disease in a population is 0.04, which is 4%. Based on a Hardy-Weinberg equilibrium, what percentage of individuals would be expected to be heterozygous carriers?

2. Describe the four patterns of natural selection that lead to environmental adaptation. You should also discuss sexual selection.

3. Explain how the founder effect is related to genetic drift.

Collaborative Questions

1. Antibiotics are commonly used to combat bacterial and fungal infections. During the past several decades, however, antibiotic-resistant strains of microorganisms have become alarmingly prevalent. This has undermined the ability of physicians to treat many types of infectious disease. Discuss how the following processes that alter allele frequencies may have contributed to the emergence of antibiotic-resistant strains:

a. random mutation

b. genetic drift

c. natural selection

2. Discuss the similarities and differences among directional, disruptive, balancing, and stabilizing selection.

Online Resource

www.brookerbiology.com

Stay a step ahead in your studies with animations that bring concepts to life and practice tests to assess your understanding. Your instructor may also recommend the interactive ebook, individualized learning tools, and more.

Origin of Species and Macroevolution

25

Two different species of zebras. Grey zebra (*Equus grevyi*) is shown on the left and Grant's zebra (*E. quagga* subspecies *boehmi*), which has fewer and thicker stripes, is shown on the right. This chapter will examine how different species come into existence.

T he origin of living organisms has been described by philosophers as the great "mystery of mysteries." Perhaps that is why so many different views have been put forth to explain the existence of living species. At the time of Aristotle (4th century B.C.E.), most people believed that some living organisms could come into being by spontaneous generation, which is the idea that nonliving materials could give rise to living organisms. For example, it was commonly believed that worms and frogs could arise from mud, and mice could come from grain. By comparison, many religious teachings contend that species were divinely made and have remained the same since their creation. In contrast to these ideas, the work of Charles Darwin provided the scientific theory of evolution by descent with modification. This theory helps us to understand the diversity of life, and in particular, it presents a logical explanation for how new species can evolve from pre-existing species.

This chapter provides an exciting way to build on the information that we have considered in previous chapters. In Chapter 22,

we examined how the first primitive cells in an RNA world could have evolved into prokaryotic cells and eventually eukaryotes. Chapter 23 surveyed the tenets on which the theory of evolution is built, and in Chapter 24, we viewed evolution on a small scale as it relates to a single gene. In this chapter, we will consider evolution on a larger scale, as it relates to the formation of new species. **Macroevolution** refers to evolutionary changes that create new species and groups that include many new species. It concerns the diversity of organisms established over long periods of time through the evolution and extinction of many species.

To biologists, the concept of a **species** has come to mean a group of related organisms that maintains a distinctive set of attributes in nature. Members of the same species share an evolutionary history, which makes them more genetically similar to each other than they are to members of a different species. You may already have an intuitive sense of this concept. It is obvious that giraffes and mice are different species. However, as we will learn in the first section of this chapter, the distinction between different, closely related species is often blurred in natural environments. It may not be easy to definitively distinguish two species, as the chapter-opening photo illustrates. Species identification has several practical uses. For example, it allows biologists to plan for the preservation and conservation of endangered species. In addition, it is often important for a physician to identify the correct bacterial species that is causing a disease in a patient so the proper medication can be prescribed.

In this chapter, we will also focus on the mechanisms that promote the formation of new species, a phenomenon called **speciation**. Such macroevolution typically occurs by the accumulation of microevolutionary changes, those that occur in single genes (see Chapter 24). We will also consider how macroevolution can happen at a fast or slow pace and explore how variations in the genes that control development are important in the evolution of new species.

25.1 Identification of Species

How many different species are on Earth? The number is astounding. A study done by biologist E. O. Wilson and colleagues in 1990 estimated the known number of species at approximately 1.4 million. Currently, about 1.75 million species have been identified. However, a vast number of species have yet to be established. This is particularly true among prokaryotic organisms, which are difficult to categorize into distinct species. Common estimates of the total number of species range from 5 to 50 million!

When studying natural populations, evolutionary biologists are often confronted with situations where some differences between two populations are apparent, but it is difficult to decide whether the two populations truly represent separate species. When two or more geographically restricted groups of the same species display one or more traits that are somewhat different but not enough to warrant their placement into different species, biologists sometimes classify such groups as **subspecies**. Similarly, many bacterial species are subdivided into **ecotypes**. Each ecotype is a genetically distinct population adapted to its local environment. In this section, we will consider the different characteristics that biologists examine when deciding if two groups of organisms constitute different species.

Each Species Is Established Using Characteristics and Histories That Distinguish It from Other Species

As mentioned, a species is a group of organisms that maintains a distinctive set of attributes in nature. In the case of sexually reproducing species, members of one species usually cannot successfully interbreed with members of other species. Members of the same species share an evolutionary history that is distinct from other species. While this may seem like a reasonable way to characterize a given species, biologists would agree that the identification of many species is a difficult undertaking. What criteria do we use to distinguish species? How many differences must exist between two populations to classify them as different species? Such questions are often difficult to answer.

The characteristics that a biologist uses to identify a species will depend, in large part, on the species in question. For example, the traits used to distinguish insect species would be quite different from those used to identify different bacterial species. The relatively high level of horizontal gene transfer among bacteria presents special challenges in the grouping of bacterial species. Among bacteria, it is sometimes very difficult and perhaps arbitrary to divide closely related organisms into separate species.

The most commonly used characteristics to identify species are morphological traits, the ability to interbreed, molecular features, ecological factors, and evolutionary relationships. A comparison of these concepts will help you to appreciate the various approaches that biologists use to identify the bewildering array of species on our planet.

Morphological Traits One way to establish that a population constitutes a unique species is based on their physical characteristics. Organisms are classified as the same species if their anatomical traits appear to be very similar. Likewise, microorganisms can be classified according to morphological traits at the cellular level. By comparing many different morphological traits, biologists may be able decide that certain populations constitute a unique species.

Although an analysis of morphological traits is a common way for biologists to establish that a particular group constitutes a species, this approach has drawbacks. First, researchers may have difficulty deciding how many traits to consider. In addition, quantitative traits that vary in a continuous way among members of the same species, such as size and weight, are not easy to analyze. Another drawback is that the degree of dissimilarity that distinguishes different species may not show a simple relationship; the members of the same species sometimes look very different, and conversely, members of different species sometimes look remarkably similar to each other. For example, **Figure 25.1a** shows two different frogs of the species *Dendrobates tinctorius*, commonly called the dyeing poison frog. This species exists in many different-colored morphs, which are individuals of the same species that have noticeably dissimilar appearances. In contrast, **Figure 25.1b** shows two different species of frogs, the Northern leopard frog (*Rana pipiens*) and the Southern leopard frog (*Rana utricularia*), which look fairly similar.

(a) Frogs of the same species

(b) Frogs of different species

Figure 25.1 **Difficulties of using morphological traits to identify species.** In some cases, members of the same species can appear quite different; in other cases, members of different species can look very similar. **(a)** Two frogs of the same species, the dyeing poison frog (*Dendrobates tinctorius*). **(b)** Two different species of frog, the Northern leopard frog (*Rana pipiens*, left) and the Southern leopard frog (*Rana utricularia*, right).

Concept check: *Can you think of another example of two different species that look very similar?*

Reproductive Isolation   Why would biologists describe two species, such as the Northern leopard frog and Southern leopard frog, as being different if they are morphologically similar? One reason is that biologists have discovered that they are unable to breed with each other in nature. Therefore, a second way to identify a species is by the ability to interbreed. In the late 1920s, geneticist Theodosius Dobzhansky proposed that each species is reproductively isolated from other species. Such **reproductive isolation** prevents one species from successfully interbreeding with other species. In 1942, evolutionary biologist Ernst Mayr expanded on the ideas of Dobzhansky to provide a reproductive definition of a species. According to Mayr, a key feature of sexually reproducing species is that, in nature, the members of one species have the potential to interbreed with one another to produce viable, fertile offspring but cannot successfully interbreed with members of other species. As discussed later in this section, reproductive isolation among species of plants and animals can occur by an amazing variety of different mechanisms.

Reproductive isolation has been used to distinguish many plant and animal species, especially those that look alike but do not interbreed. Even so, it suffers from four main problems. First, in nature, it may be difficult to determine if two populations are reproductively isolated, particularly if they are populations with nonoverlapping geographic ranges. Second, biologists have noted many cases in which two different species can interbreed in nature yet consistently maintain themselves as separate species. For example, different species of yucca plants, such as *Yucca pallida* and *Yucca constricta*, do interbreed in nature yet typically maintain populations with distinct characteristics. For this reason, they are viewed as distinct species. A third drawback of reproductive isolation is that it does not apply to asexual species such as bacteria. Likewise, some species of plants and fungi reproduce only asexually. Finally, a fourth drawback is that it cannot be applied to extinct species. For these reasons, reproductive isolation has been primarily used to distinguish closely related species of modern animals and plants that reproduce sexually.

Molecular Features   Molecular features are now commonly used to determine if two different populations are different species. Evolutionary biologists often compare DNA sequences within genes, gene order along chromosomes, chromosome structure, and chromosome number as features to identify similarities and differences among different populations. DNA sequence differences are often used to compare populations. For example, researchers may compare the DNA sequence of the *16S* rRNA gene between different bacterial populations as a way to decide if the two populations represent different species. When the sequences are very similar, such populations would probably be judged as the same species. However, it may be difficult to draw the line when separating groups into different species. Is a 2% difference in their genome sequences sufficient to warrant placement into two different species, or do we need a 5% difference?

Ecological Factors   A variety of factors related to an organism's habitat can be used to distinguish one species from another. For example, certain species of warblers can be distinguished by the habitat in which they forage for food. Some species search the ground for food, others forage in bushes or small trees, and some species primarily forage in tall trees. Such habitat differences can be used to distinguish different species that look morphologically similar.

Many bacterial species have been categorized as distinct species based on ecological factors. Bacterial cells of the same species are likely to use the same types of resources (such as sugars and vitamins) and grow under the same types of conditions (such as temperature and pH). However, a drawback of this approach is that different groups of bacteria sometimes display very similar growth characteristics, and even the same species may show great variation in the growth conditions it will tolerate.

Evolutionary Relationships   In Chapter 26, we will examine the methods that are used to produce evolutionary trees that describe the relationships between ancestral species and modern species. In some cases, such relationships are based on an analysis of the fossil record. For example, in Chapter 26, we will consider how the fossil record was used to construct a tree that describes the ancestors that led to modern horse species. Alternatively, another way to establish evolutionary relationships is by the analysis of DNA sequences. Researchers can obtain samples of cells from different individuals and compare the genes within those cells to see how similar or different they are.

Species Concepts   Thus far, we have considered multiple ways to identify species. The most commonly used characteristics to distinguish species are morphological traits, the ability to interbreed, molecular features, ecological factors, and evolutionary relationships.

A **species concept** is a way to define the concept of a species and/or provide an approach to distinguish one species from another. In 1942, Ernst Mayr proposed one of the first species concepts called the **biological species concept**. According to this idea, a species is a group of individuals whose members have the potential to interbreed with one another in nature to produce viable, fertile offspring but cannot successfully interbreed with members of other species. The biological species concept emphasizes reproductive isolation as the most important criterion for delimiting species.

Since 1942, over 20 different species concepts have been proposed by a variety of evolutionary biologists. Another example is the **evolutionary lineage concept** proposed by American paleontologist George Gaylord Simpson in 1961. A **lineage** is a series of species that forms a line of descent, with each new species the direct result of speciation from an immediate ancestral species. According to Gaylord, species should be defined based on the separate evolution of lineages. A third example is the **ecological species concept**, described by American evolutionary biologist Leigh Van Valen in 1976. According to this viewpoint, each species occupies an ecological niche, which is the unique set of habitat resources that a species requires, as well as its influence on the environment and other species.

Most evolutionary biologists would agree that different methods are needed to distinguish the vast array of species on Earth. Even so, some evolutionary biologists have questioned whether it is valid to have many different species concepts. In

1998, Kevin de Queiroz suggested that there is only a single general species concept, which concurs with Simpson's evolutionary lineage concept, and includes all previous concepts. According to de Queiroz's **general lineage concept**, each species is a population of an independently evolving lineage. Each species has evolved from a specific series of ancestors and, as a consequence, forms a group of organisms with a particular set of characteristics. Multiple criteria are used to determine if a population is part of an independent evolutionary lineage, and thus a species, which is distinct from others. Typically, researchers will use analyses of morphology, reproductive isolation, DNA sequences, and ecology to determine if a population or group of populations is distinct from others. Because of its generality, the general lineage concept has received significant support.

Reproductive Isolating Mechanisms Help to Maintain the Distinctiveness of Each Species

Thus far we have considered various ways to identify different species. In our discussion, you may have realized that the identification of a species is not always a simple matter. With regard to plants and animals, the phenomenon of reproductive isolation has played a major role in the way that biologists study plant and animal species, partly because it identifies a possible mechanism for the process of forming new species. For this reason, much research has been done to try to understand **reproductive isolating mechanisms**, the mechanisms that prevent interbreeding between different species. Why do reproductive isolating mechanisms occur? Populations do not intentionally erect these reproductive barriers. Rather, reproductive isolating mechanisms are a consequence of genetic changes that occur usually because a species becomes adapted to its own particular environment. The view of evolutionary biologists is that reproductive isolation typically evolves as a by-product of genetic divergence. Over time, as a species evolves its own unique characteristics, some of those traits are likely to prevent breeding with other species.

Reproductive isolating mechanisms fall into two categories: **prezygotic isolating mechanisms** prevent the formation of a zygote, whereas **postzygotic isolating mechanisms** block the development of a viable and fertile individual after fertilization has taken place. **Figure 25.2** summarizes some of the more common ways that reproductive isolating mechanisms prevent reproduction between different species. When two species do produce offspring, such an offspring is called an **interspecies hybrid**.

Prezygotic Isolating Mechanisms   We will consider five types of prezygotic isolating mechanisms. One obvious way to prevent interbreeding is for members of different species to never come in contact with each other. This phenomenon, called habitat isolation, may involve a geographic barrier to interbreeding. For example, a large body of water may separate two different plant species that live on nearby islands.

A second prezygotic isolating mechanism is temporal isolation, in which species happen to reproduce at different times of the day or year. In the northeastern U.S., for example, the two most abundant field crickets, *Gryllus veletis* and *Gryllus*

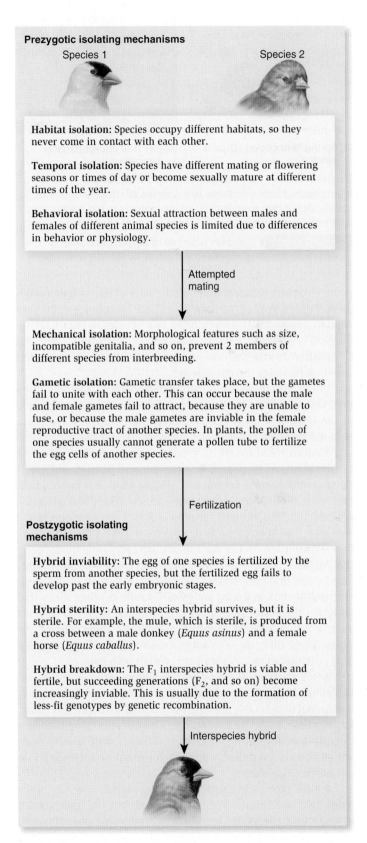

Figure 25.2   Reproductive isolating mechanisms.   These mechanisms prevent successful breeding between different species. They can occur prior to fertilization (prezygotic) or after fertilization (postzygotic).

(a) Spring field cricket (*Gryllus veletis*)

(b) Fall field cricket (*Gryllus pennsylvanicus*)

Figure 25.3 An example of temporal isolation. Interbreeding between these two species of crickets does not usually occur because *Gryllus veletis* matures in the spring, whereas *Gryllus pennsylvanicus* matures in the fall.

Concept check: Is this an example of a prezygotic or a postzygotic isolating mechanism?

pennsylvanicus (spring and fall field crickets, respectively), do not differ in song or habitat and are morphologically very similar (**Figure 25.3**). How do the two species maintain reproductive isolation? *G. veletis* matures in the spring, whereas *G. pennsylvanicus* matures in the fall. This minimizes interbreeding between the two species.

In the case of animals, mating behavior and anatomy often play key roles in promoting reproductive isolation. An example of the third type of prezygotic isolation, behavioral isolation, is found between the western meadowlark (*Sturnella neglecta*) and eastern meadowlark (*Sturnella magna*). Both species are nearly identical in shape, coloration, and habitat, and their ranges overlap in the central U.S. (**Figure 25.4**). For many years, they were thought to be the same species. When biologists discovered that the western meadowlark is a separate species, it was given the species name *S. neglecta* to reflect the long delay in its recognition. In the zone of overlap, very little interspecies mating takes place between western and eastern meadowlarks, largely due to differences in their songs. The song of the western meadowlark is a long series of flutelike gurgling notes that go down the scale. By comparison, the eastern meadowlark's song is a simple series of whistles, typically about four or five notes. These differences in songs enable meadowlarks to recognize potential mates as members of their own species.

A fourth type of prezygotic isolation, called mechanical isolation, occurs when morphological features such as size or incompatible genitalia prevent two species from interbreeding. For example, male dragonflies use a pair of special appendages to grasp females during copulation. When a male tries to mate with a female of a different species, his grasping appendages do not fit her body shape.

A fifth type of prezygotic isolating mechanism can occur when two species attempt to interbreed, but the gametes fail to unite in a successful fertilization event. This phenomenon, called gametic isolation, is widespread among plant and animal species. In aquatic animals that release sperm and egg cells into the water, gametic isolation is important to prevent interspecies hybrids. For example, closely related species of sea urchins

North America

Western meadowlark
Eastern meadowlark
Zone of overlap

(a) Western meadowlark (*Sturnella neglecta*)

(b) Eastern meadowlark (*Sturnella magna*)

Figure 25.4 An example of behavioral isolation. (a) The western meadowlark (*Sturnella neglecta*) and (b) eastern meadowlark (*Sturnella magna*) are very similar in appearance. The red region in this map shows where the two species' ranges overlap. However, very little interspecies mating takes place due to differences in their songs.

may release sperm and eggs into the water at the same time. Researchers have discovered that sea urchin sperm have a protein on their surface called bindin, which mediates sperm–egg attachment and membrane fusion. The structure of bindin is significantly different among different sea urchin species and thereby ensures that fertilization occurs only between sperm and egg cells of the same species.

In flowering plants, gametic isolation is commonly associated with pollination. As discussed in Chapter 39, plant fertilization is initiated when a pollen grain lands on the stigma of a flower and sprouts a pollen tube that ultimately reaches an egg cell (look ahead to Figure 39.4). When pollen is released from a plant, it could be transferred to the stigma of many different plant species. In most cases, when a pollen grain lands on the stigma of a different species, it either fails to generate a pollen tube or the tube does not grow properly and reach the egg cell.

Postzygotic Isolating Mechanisms   Let's now turn to postzygotic mechanisms of reproductive isolation. The first such mechanism is hybrid inviability, in which an egg of one species is fertilized by a sperm from another species, but the fertilized egg cannot develop past the early embryonic stages.

A second mechanism is hybrid sterility, in which an interspecies hybrid may be viable but sterile. A classic example of hybrid sterility is the mule, which is produced by a mating between a male donkey (*Equus asinus*) and a female horse

Male donkey (*Equus asinus*)   ×   **Female horse (*Equus ferus caballus*)**

Mule

Figure 25.5   **An example of hybrid sterility.**   When a male donkey (*Equus asinus*) mates with a female horse (*Equus ferus caballus*), their offspring is a mule, which is usually sterile.

Concept check: *Is this an example of a prezygotic or a postzygotic isolating mechanism?*

(*Equus ferus caballus*) (**Figure 25.5**). All male mules and most female mules are sterile. Why are mules usually sterile? Two reasons explain the sterility. Because the horse has 32 chromosomes per set and a donkey has 31, a mule inherits 63 chromosomes (32 + 31). Due to the uneven number, all of the chromosomes cannot pair evenly. Also, the chromosomes of the horse and donkey have structural differences, which either prevent them from pairing correctly or lead to chromosomal abnormalities if crossing over occurs during meiosis. For these reasons, mules usually produce inviable gametes. Note that the mule has no species name because it is not considered a species due to this sterility.

Finally, interspecies hybrids may be viable and fertile, but the subsequent generation(s) may harbor genetic abnormalities that are detrimental. This third mechanism, called hybrid breakdown, can be caused by changes in chromosome structure. The chromosomes of closely related species may have structural differences from each other, such as inversions. In hybrids, a crossover may occur in the region that is inverted in one species but not the other. This will produce gametes with too little or too much genetic material. Such hybrids would often have offspring with developmental abnormalities.

Postzygotic isolating mechanisms tend to be uncommon in nature compared to prezygotic mechanisms. Why are postzygotic mechanisms rare? One explanation is that they are more costly in terms of energy and resources used. For example, a female mammal would use a large amount of energy to produce an offspring that is sterile. Evolutionary biologists hypothesize that natural selection has favored prezygotic isolating mechanisms because they do not waste a lot of energy.

25.2   Mechanisms of Speciation

Speciation, the formation of a new species, is caused by genetic changes in a particular group that make it different from the species from which it was derived. As discussed in Chapter 24, random mutations in genes can be acted upon by natural selection and other evolutionary mechanisms to alter the genetic composition of a population. New species commonly evolve in this manner. In addition, interspecies matings, changes in chromosome number, and horizontal gene transfer may also cause new species to arise. In all of these cases, the underlying cause of speciation is the accumulation of genetic changes that ultimately promote enough differences so that we judge a population to constitute a unique species.

Even though genetic changes account for the phenotypic differences observed among living organisms, such changes do not fully explain the existence of many distinct species on our planet. Why does life often diversify into the more or less discrete populations that we recognize as species? Two main explanations have been proposed:

1. In some cases, speciation may occur due to abrupt events, such as changes in chromosome number, which can cause reproductive isolation.
2. More commonly, species arise as a consequence of adaptation to different ecological niches. For sexually reproducing organisms, reproductive isolation is typically a by-product of that adaptation.

Depending on the species involved, one or both factors may play a dominant role in the formation of new species. In this section, we will consider how reproductive isolating mechanisms and adaptation to particular environments are critical aspects of the speciation process.

Geographic and Habitat Isolation Can Promote Allopatric Speciation

Cladogenesis is the splitting or diverging of a population into two or more species. In the case of sexually reproducing organisms, the process of cladogenesis requires that gene flow becomes interrupted between two or more populations, limiting or eliminating reproduction between members of different populations. **Allopatric speciation** (from the Greek *allos*, meaning other, and the Latin *patria*, meaning homeland) is the most prevalent way for cladogenesis to occur. This form of speciation occurs when some members of a species occupy a habitat that is isolated from other members. Typically, this isolation may involve a geographic barrier such as a large area of land or a large body of water.

In some cases, geographic separation may be caused by slow, geological events that eventually produce quite large geographic barriers. For example, a mountain range may emerge and split one species that occupies the lowland regions, or a creeping glacier may divide a population. **Figure 25.6** shows an interesting example in which geological separation promoted

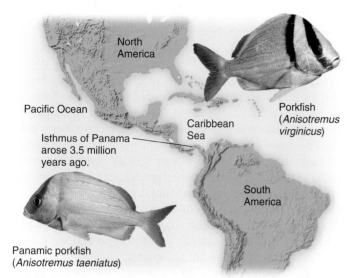

Figure 25.6 **An example of allopatric speciation.** An ancestral fish population was split into two by the formation of the Isthmus of Panama about 3.5 million years ago. Since that time, different genetic changes occurred in the two populations because of their geographic isolation. These changes eventually led to the formation of different species: the Panamic porkfish (*Anisotremus taeniatus*) is found in the Pacific Ocean, whereas the porkfish (*Anisotremus virginicus*) is found in the Caribbean Sea.

speciation. A fish called the Panamic porkfish (*Anisotremus taeniatus*) is found in the Pacific Ocean, whereas the porkfish (*Anisotremus virginicus*) is found in the Caribbean Sea. These two species were derived from an ancestral species whose population was split by the formation of the Isthmus of Panama about 3.5 million years ago. Before that event, the waters of the Pacific Ocean and Caribbean Sea mixed freely. Since the formation of the isthmus, the two populations have been geographically isolated and have evolved into distinct species.

Allopatric speciation can also occur when a small population moves to a new location that is geographically isolated from the main population. For example, a storm may force a small group of birds from a mainland to a distant island. In this case, migration between the island and the mainland population is an infrequent event. In a relatively short period of time, the founding population on the island may evolve into a new species. How does speciation occur rapidly? Because the environment on the island may differ significantly from the mainland environment, natural selection may rapidly alter the genetic composition of the population, leading to adaptation to the new environment. This scenario is thought to be responsible for the speciation of honeycreepers that occurred on the Hawaiian Islands (**Figure 25.7**).

(a) Migration of ancestor to the Hawaiian Islands

Figure 25.7 **An example of adaptive radiation.** (a) The honeycreepers' ancestor is believed to be related to a Eurasian rosefinch that arrived on the Hawaiian Islands approximately 3–7 million years ago. Since that time, at least 54 different species of honeycreepers (*Drepanidinae*) have evolved on the islands. (b) Adaptations to feeding have produced honeycreeper species with notable differences in beak morphology.

Concept check: *Explain why so many different species of honeycreepers evolved on the Hawaiian Islands.*

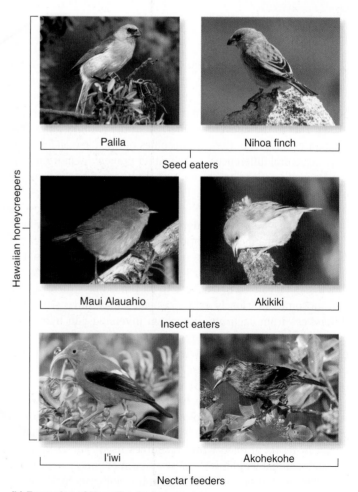

(b) Examples of Hawaiian honeycreepers

The Hawaiian Islands are a showcase of allopatric speciation. The islands' extreme isolation coupled with their phenomenal array of ecological niches has enabled a small number of founding species to evolve into a vast assortment of different species. Biologists have investigated several examples of **adaptive radiation**, in which a single ancestral species has evolved into a wide array of descendant species that differ greatly in their habitat, form, or behavior. For example, approximately 1,000 species of *Drosophila* are found dispersed throughout the Hawaiian Islands. Evolutionary studies suggest that these were derived from a single colonization by one species of fruit fly!

As shown in Figure 25.7, an example of adaptive radiation is seen with a family of birds called honeycreepers (*Drepanidinae*). Researchers estimate that the honeycreepers' ancestor arrived in Hawaii 3–7 million years ago. This ancestor was a single species of finch, possibly a Eurasian rosefinch (genus *Carpodacus*) or, less likely, the North American house finch (*Carpodacus mexicanus*). At least 54 different species of honeycreepers, many of which are now extinct, have evolved from this founding event to fill available niches in the islands' habitats. Seed eaters developed stouter, stronger bills for cracking tough husks. Insect-eating honeycreepers developed thin, warbler-like bills for picking insects from foliage or strong, hooked bills to root out wood-boring insects. Nectar-feeding honeycreepers evolved curved bills for extracting nectar from the flowers of Hawaii's endemic plants.

Before ending our discussion of allopatric speciation, let's consider a common situation in which geographic separation is not complete. The zones where two populations can interbreed are known as **hybrid zones**. Figure 25.8 shows a hybrid zone along a mountain pass that connects two deer populations. For speciation to occur, the amount of gene flow within hybrid zones must become very limited. But how does this happen? As the two populations accumulate different genetic changes, this may decrease the ability of individuals from different populations

Figure 25.8 **Hybrid zones.** Two populations of deer are separated by a mountain range. A hybrid zone exists in a mountain pass where occasional interbreeding may occur.

to mate with each other in the hybrid zone. For example, natural selection in the western deer population may favor an increase in body size that is not favored in the eastern population. Over time, as this size difference between members of the two populations becomes greater, breeding in the hybrid zone may decrease. Larger individuals may not interbreed easily with smaller individuals due to mechanical isolation. Alternatively, larger individuals may prefer larger individuals as mates, while smaller individuals may also prefer each other. Once gene flow through the hybrid zone is greatly diminished, the two populations are reproductively isolated. Over the course of many generations, such populations may evolve into distinct species.

FEATURE INVESTIGATION

Podos Found That an Adaptation to Feeding May Have Promoted Reproductive Isolation in Finches

In 2001, American evolutionary biologist Jeffrey Podos analyzed the songs of Darwin's finches on the Galápagos Islands to determine how environmental adaptation may contribute to reproductive isolation. As in honeycreepers, the differences in beak sizes and shapes among the various species of finches are adaptations to different feeding strategies. Podos hypothesized that changes in beak morphology could also impact the songs that the birds produce, thereby having the potential to affect mate choice. The components of the vocal tract of birds, including the trachea, larynx, and beak, work collectively to produce a bird's song. Birds actively modify the shape of their vocal tracts during singing, and beak movements are normally very rapid and precise.

Podos focused on two aspects of a bird's song. The first feature is the frequency range, which is a measure of the minimum and maximum frequencies in a bird's song. The second feature is the trill rate. A trill is a series of notes or group of notes repeated in succession. Figure 25.9 shows a graphical depiction of the songs of Darwin's finches. As you can see, the song patterns of these finches are quite different from each other.

To quantitatively study the relationship between beak size and song, Podos first captured male finches on one of the Galápagos Islands (Santa Cruz) and measured their beak sizes (Figure 25.10). The birds were banded and then released into the wild. The banding provided a way to identify the birds whose beaks had already been measured. After release, the songs of the banded birds were recorded on a tape recorder, and their range of frequencies and trill rate were analyzed. Podos then compared the data for the Galápagos finches to a large body of data that had been collected on many other bird species. This comparison was used to evaluate whether beak size, in this case beak depth—the measurement of the beak from top to bottom, at its base—constrained either the frequency range and/or the trill rate of the finches.

Figure 25.9 **Differences in the songs of Galápagos finches.** These spectrograms depict the frequency of each bird's song over time, measured in kilohertz (kHz). The songs are produced in a series of trills that have a particular pattern and occur at regular intervals. Notice the differences in frequency and trill rate between different species of birds.

Figure 25.10 Study by Podos investigating the effects of beak depth on song among different species of Galápagos finches.

HYPOTHESIS Changes in beak morphology that are an adaptation to feeding may also affect the songs of Galápagos finches and thereby lead to reproductive isolation between species.

KEY MATERIALS This study was conducted on finch populations of the Galápagos Island of Santa Cruz.

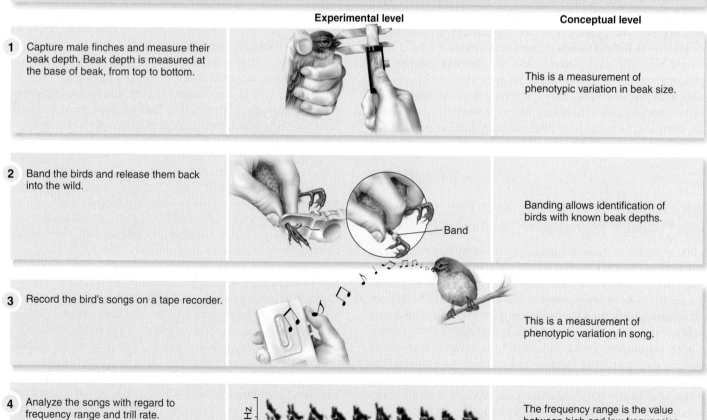

5 THE DATA

The data for the Galápagos finches were compared to a large body of data that had been collected on many other bird species. The relative constraint on vocal performance is higher if a bird has a more narrow frequency range and/or a slower trill rate. These constraints were analyzed with regard to each bird's beak depth.

6 CONCLUSION Larger beak size, which is an adaptation to cracking open large, hard seeds, constrains vocal performance. This may affect mating song patterns and thereby promote reproductive isolation and, in turn, speciation.

7 SOURCE Podos, Jeffrey. 2001. Correlated evolution of morphology and vocal signal structure in Darwin's finches. *Nature* 409:185–188.

The results of this comparison are shown in the data of Figure 25.10. As seen here, the relative constraint on vocal performance became higher as the beak depth became larger. This means that birds with larger beaks had a more narrow frequency range and/or a slower trill rate. Podos proposed that as jaws and beaks become adapted for strength to crack open larger, harder seeds, they will be less able to perform the rapid movements associated with certain types of songs. In contrast, the finches with smaller beaks that were adapted to probe for insects or eat smaller seeds had less constraint on their vocal performance. From the perspective of evolution, the changes observed in song patterns for the Galápagos finches could have played an important role in promoting reproductive isolation, because song pattern is an important factor in mate selection in birds. Therefore, a by-product of beak adaptation for feeding is that it also appears to have affected song pattern, possibly promoting reproductive isolation. In this way, populations of finches would have evolved into distinct species.

Experimental Questions

1. What did Podos hypothesize regarding the effects of beak size on a bird's song? How could changes in beak size and shape lead to reproductive isolation among the finches?

2. How did Podos test the hypothesis that beak morphology caused changes in the birds' songs?

3. Did the results of Podos's study support his original hypothesis? Explain. What is meant by the phrase "by-product of adaptation," and how does it apply to this particular study?

Sympatric Speciation Occurs When Populations Are in Direct Contact

Sympatric speciation (from the Greek *sym*, meaning together) occurs when members of a species that are within the same range diverge into two or more different species even though there are no physical barriers to interbreeding. Although sympatric speciation is believed to be much less common than allopatric speciation, particularly in animals, evolutionary biologists have discovered several ways in which it can occur. These include polyploidy, adaptation to local environments, and sexual selection.

Polyploidy A type of genetic change that can cause immediate reproductive isolation is **polyploidy**, in which an organism has more than two sets of chromosomes. Plants tend to be more tolerant of changes in chromosome number than animals. For example, many crops and decorative species of plants are polyploid. How does polyploidy occur? One mechanism is complete nondisjunction of chromosomes, which can increase the number of chromosome sets in a given species (autopolyploidy). Such changes can result in an abrupt sympatric speciation. For example, nondisjunction could produce a tetraploid plant with four sets of chromosomes from a species that was diploid with two sets. A cross between a tetraploid and a diploid produces a triploid offspring with three sets of chromosomes. Triploid offspring are usually sterile because an odd number of chromosomes cannot be evenly segregated during meiosis. This hybrid sterility causes reproductive isolation between the tetraploid and diploid species.

Another mechanism that leads to polyploidy is interspecies breeding. An **alloploid** organism contains at least one set of chromosomes from two or more different species. This term refers to the occurrence of chromosome sets (ploidy) from the genomes of different (allo-) species. When two different species interbreed, this may produce an allodiploid that has only one set of chromosomes from each species. Species that are close evolutionary relatives are most likely to breed and produce allodiploid offspring. For example, closely related species of grasses may interbreed to produce allodiploids. An organism containing two or more complete sets of chromosomes from two or more different species is called an allopolyploid. An allopolyploid can be the result of interspecies breeding between species that are already polyploid, or it can occur as a result of nondisjunction in an allodiploid organism. For example, complete nondisjunction in an allodiploid could produce an allotetraploid, which is an allopolyploid with two complete sets of chromosomes from two species for a total of four sets.

The formation of an allopolyploid can also abruptly lead to reproductive isolation and thereby promote speciation. As an example, let's consider the origin of a natural species of a plant called the common hemp nettle, *Galeopsis tetrahit*. This species is thought to be an allotetraploid derived from two diploid species, *Galeopsis pubescens* and *Galeopsis speciosa* (Figure 25.11a). These two diploid species contain 16 chromosomes each ($2n = 16$), while *G. tetrahit* contains 32 chromosomes. Though the origin of *G. tetrahit* is not completely certain, research suggests it may have originated from an interspecies cross between *G. pubescens* and *G. speciosa*, which initially

(a) Possible formation of *G. tetrahit*

(b) Outcome of breeding among *G. tetrahit*, *G. pubescens*, and *G. speciosa*

Figure 25.11 **Polyploidy and sympatric speciation.** (a) *Galeopsis tetrahit* may have arisen by an interspecies cross between *G. pubescens* and *G. speciosa*, which was followed by a subsequent nondisjunction event. (b) Polyploidy may have caused reproductive isolation between these three natural species of hemp nettle. If *G. tetrahit* is mated with either of the other two species, the resulting offspring would be monoploid for one chromosome set and diploid for the other set, making them sterile. Therefore, *G. tetrahit* is reproductively isolated from the diploid species, making it a new species.

Concept check: *Suppose that G. tetrahit was crossed to G. pubescens to produce an interspecies hybrid as shown at the bottom of this figure. If this interspecies hybrid was crossed to G. tetrahit, how many chromosomes do you think an offspring would have? The answer you give should be a range, not a single number.*

produced an allodiploid with 16 chromosomes (one set from each species). The allodiploid then underwent complete nondisjunction to become an allotetraploid carrying four sets of chromosomes—two from each species.

How do these genetic changes cause reproductive isolation? The allotetraploid, *G. tetrahit*, is fertile, because all of its chromosomes occur in homologous pairs that can segregate evenly during meiosis. However, a cross between an allotetraploid and a diploid, *G. pubescens* or *G. speciosa*, produces an offspring that is monoploid for one chromosome set and diploid for the other set (**Figure 25.11b**). The chromosomes of the monoploid set cannot be evenly segregated during meiosis. These offspring are expected to be sterile, because they will produce gametes that have incomplete sets of chromosomes. This hybrid sterility causes the allotetraploid to be reproductively isolated from both diploid species. Therefore, this process could have led to the formation of a new species, *G. tetrahit*, by sympatric speciation.

Polyploidy is so frequent in plants that it is a major mechanism of their speciation. In ferns and flowering plants, about 40–70% of the species are polyploid. By comparison, polyploidy can occur in animals, but it is much less common. For example, less than 1% of reptiles and amphibians have been identified that are polyploids derived from diploid ancestors.

Adaptation to Local Environments   In some cases, a geographic area may have variation so that some members of a population may diverge and occupy different local environments that are continuous with each other. An early example of this type of sympatric speciation was described by Jeffrey Feder, Guy Bush, and colleagues. They studied the North American apple maggot fly (*Rhagoletis pomenella*). This fly originally fed on native hawthorn trees. However, the introduction of apple trees approximately 200 years ago provided a new local environment for this species. The apple-feeding populations of this species develop more rapidly because apples mature more quickly than hawthorn fruit. The result is partial temporal isolation, which is an example of prezygotic reproductive isolation. Although the two populations—those that feed on apple trees and those that feed on hawthorn trees—are considered subspecies, evolutionary biologists speculate they may eventually become distinct species due to reproductive isolation and the accumulation of independent mutations in the two populations.

Sara Via and colleagues have studied the beginnings of sympatric speciation in pea aphids (*Acyrthosiphon pisum*), a small, plant-eating insect. Pea aphids in the same geographic area can be found on both alfalfa (*Medicago sativa*) and red clover (*Trifolium pratenae*) (**Figure 25.12**). Although pea aphids on these two host plants look identical, they show significant genetic differences and are highly ecologically specialized. Pea aphids that are found on alfalfa exhibit a lower fitness when transferred to red clover, whereas pea aphids found on red clover exhibit a lower fitness when transferred to alfalfa. The same traits involved in this host specialization cause these two groups of pea aphids to be substantially reproductively isolated. Taken together, the observations of the North American apple maggot fly, pea aphids, and other insect species suggest

Figure 25.12   Pea aphids, a possible example of sympatric speciation in progress.   Some pea aphids prefer alfalfa, whereas others prefer red clover. These two populations may be in the process of sympatric speciation.

Concept check: *How may host preference eventually lead to speciation?*

that diversifying selection (described in Chapter 24) may occur because some members evolve to feed on a different host. This may be an important mechanism of sympatric speciation among insects.

A third example in which sympatric speciation may be occurring due to local environment adaptation involves metal tolerance in colonial bentgrass (*Agrostis capillarus*), which was discussed in Chapter 24 (refer back to Figure 24.5). Along metal mines in south Wales, the soil is contaminated with high levels of heavy metals such as copper. On these sites, natural selection has promoted the spread of plants carrying alleles that confer resistance to the heavy metals. Although these alleles enable *A. capillarus* to grow on contaminated soil, they inhibit growth on normal, uncontaminated soil. Therefore, along metal mines, metal tolerance is advantageous, but in adjacent pastures, it is a disadvantage. Any gene flow between the mine and pasture populations would introduce less favored alleles into either population. In recent years, the metal-tolerant plants are starting to show a change in their flowering season, which will minimize interbreeding. Over time, if this process continues, the metal-tolerant population may no longer breed with the original metal-sensitive species because their flowering periods do not overlap.

Sexual Selection   Another mechanism that may promote sympatric speciation is sexual selection. As discussed in

Chapter 24, one type of sexual selection is mate choice (refer back to Figures 24.8 and 24.9). Ole Seehausen and Jacques van Alphen found that male coloration in African cichlids is subject to female choice. *Pundamilia pundamilia* females preferred *P. pundamilia* males, and *P. nyererei* females preferred *P. nyererei* males. In this case, sexual selection appears to have followed a diversifying mechanism in which certain females prefer males with one color pattern, while other females prefer males with a different color pattern. A possible outcome of such sexual selection is that it can separate one large sympatric population into smaller populations that eventually become distinct species because they selectively breed among themselves.

25.3 The Pace of Speciation

Throughout the history of life on Earth, the rate of evolutionary change and speciation has not been constant. Even Darwin himself suggested that evolution can be fast or slow. Figure 25.13 illustrates contrasting views concerning the rate of evolutionary change. These ideas are not mutually exclusive but represent two different ways to consider the tempo of evolution. The concept of **gradualism** suggests that each new species evolves continuously over long spans of time (Figure 25.13a). The principal idea is that large phenotypic differences that produce new species are due to the accumulation of many small genetic changes. By comparison, the concept of **punctuated equilibria**, advocated in the 1970s by American paleontologists and evolutionary biologists Niles Eldredge and Stephen Jay Gould, suggests that the tempo of evolution is more sporadic (Figure 25.13b). According to this hypothesis, species exist relatively unchanged for many generations. During this equilibrium period, genetic changes are likely to accumulate, particularly neutral changes, but genetic changes that significantly alter phenotype do not substantially change the overall composition of a population. These long periods of equilibrium are punctuated by relatively short periods (that is, on a geological timescale) during which the frequencies of certain phenotypes in a population do substantially change at a far more rapid rate.

A rapid rate of evolution could commonly occur via allopatric speciation in which a small group migrates away from a larger population to a new environment in which different alleles provide better adaptation to the surroundings. By natural selection, the small population may rapidly evolve into a new species. In addition, events such as polyploidy may abruptly create individuals with new phenotypic traits. On an evolutionary timescale, these types of events can be rather rapid, because a few genetic changes can have a major impact on phenotype.

In conjunction with genetic changes, species may also be subjected to sudden environmental shifts that quickly drive the gene pool in a particular direction via natural selection. For example, a species may be subjected to a relatively sudden environmental event that has a major impact on survival. The climate may change, or a new predator may infiltrate the geographic range of the species. Natural selection may lead to a

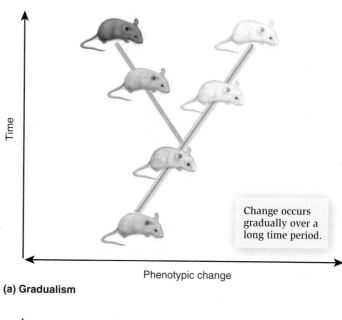

(a) Gradualism

Change occurs gradually over a long time period.

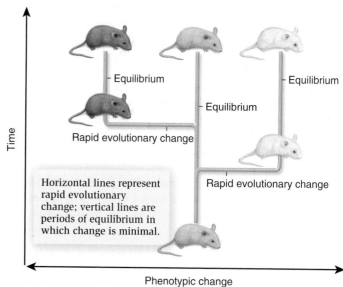

Horizontal lines represent rapid evolutionary change; vertical lines are periods of equilibrium in which change is minimal.

(b) Punctuated equilibrium

Figure 25.13 A comparison of gradualism and punctuated equilibrium. (a) During gradualism, the phenotypic characteristics of a species gradually change due to the accumulation of small genetic changes. (b) During punctuated equilibrium, long periods of equilibrium in which species exist essentially unchanged are punctuated by relatively short periods of evolutionary change during which phenotypic characteristics may change rapidly.

rapid evolution of the gene pool by favoring those alleles that allow members of the population to survive the climatic change or to have phenotypic characteristics that allow them to avoid the predator.

Which viewpoint is correct, punctuated equilibrium or gradualism? Both have merit. The occurrence of punctuated equilibrium is often supported by the fossil record. New species seem to arise rather suddenly in a layer of rocks, persist relatively unchanged for a very long period of time, and then become extinct. In such cases, scientists think that the transition period during which a previous species evolved into a new species was so short that few, if any, of the transitional members were preserved as fossils. Even so, these rapid periods of change were probably followed by long periods of equilibrium that likely involved the additional accumulation of many small genetic changes, consistent with gradualism.

Finally, another issue associated with the speed of speciation is generation time. Species of large animals with long generation times tend to evolve much more slowly than do microbial species with short generations. Many new species of bacteria will come into existence during our lifetime, while new species of large animals tend to arise on a much longer timescale. This is an important consideration for people because bacteria have great environmental impact. They are decomposers of organic materials and pollutants in the environment, and they play a role in many diseases of plants and animals, including humans.

25.4 Evo-Devo: Evolutionary Developmental Biology

As we have learned, the origin of new species involves genetic changes that lead to adaptations to environmental niches and/or to reproductive isolating mechanisms that prevent closely related species from interbreeding. These genetic changes result in morphological and physiological differences that distinguish one species from another. In recent years, many evolutionary biologists have begun to investigate how genetic variation produces species and groups of species with novel shapes and forms. The underlying reasons for such changes are often rooted in the developmental pathways that control an organism's morphology.

Evolutionary developmental biology (referred to as **evo-devo**) is an exciting and relatively new field of biology that compares the development of different organisms in an attempt to understand ancestral relationships between organisms and the developmental mechanisms that bring about evolutionary change. During the past few decades, developmental geneticists have gained a better understanding of biological development at the molecular level. Much of this work has involved the discovery of genes that control development in experimental organisms. As more and more organisms have been analyzed, researchers have become interested in the similarities and differences that occur between closely related and distantly related species. The field of evolutionary developmental biology has arisen in response to this trend.

How do new morphological forms come into being? For example, how does a nonwebbed foot evolve into a webbed foot? How does a new organ, such as an eye, come into existence? As we will learn, such novelty arises through genetic changes, also called genetic innovations. Certain types of genetic innovations have been so advantageous that they have resulted in groups of new species. For example, the innovation

of feathered forearms resulted in the evolution of many different species of birds. In this section, we will learn that proteins that control developmental changes, such as cell-signaling proteins and transcription factors, often play a key role in promoting the morphological changes that occur during evolution.

The Spatial Expression of Genes That Affect Development Can Have a Dramatic Effect on Phenotype

In Chapter 19, we considered the role of genetics in the development of plants and animals. As we learned, genes that play a role in development may influence cell division, cell migration, cell differentiation, and cell death. The interplay among these four processes produces an organism with a specific body pattern, a process called **pattern formation**. As you might imagine, developmental genes are very important to the phenotypes of individuals. They affect traits such as the shape of a bird's beak, the length of a giraffe's neck, and the size of a plant's flower. In recent years, the study of development has indicated that developmental genes are key players in the evolution of many types of traits. Changes in such genes affect traits that can be acted on by natural selection. Furthermore, variation in the expression of such genes may be commonly involved in the acquisition of new traits that promote speciation.

As an example, let's compare the formation of a chicken's foot with that of a duck. Developmental biologists have discovered that the morphological differences between a nonwebbed and a webbed foot are due to the differential expression of two different cell-signaling proteins called bone morphogenetic protein 4 (BMP4) and gremlin. The *BMP4* gene is expressed throughout the developing limb; this is shown in **Figure 25.14a**, in which the BMP4 protein is stained purple. The BMP4 protein causes cells to undergo apoptosis and die. The gremlin protein, which is stained brown in **Figure 25.14b**, inhibits the function of BMP4 and thereby allows cells to survive. In the developing chicken limb, the *Gremlin* gene is expressed throughout the limb, except in the regions between each digit. Therefore, in these regions, the cells die, and a chicken develops a nonwebbed foot (**Figure 25.14c**). By comparison, in the duck, *Gremlin* is expressed throughout the entire limb, including the interdigit regions, and as a result, the duck develops a webbed foot. Interestingly, researchers have been able to introduce gremlin protein into the interdigit regions of developing chicken limbs. This produces a chicken with webbed feet!

How are these observations related to evolution? During the evolution of birds, genetic variation arose such that some individuals expressed the *Gremlin* gene in the regions between each digit, while others did not. This variation determined whether or not a bird's feet were webbed. In terrestrial settings, having nonwebbed feet is an advantage because they enable the individual to hold onto perches, run along the ground, and snatch prey. Therefore, natural selection would favor nonwebbed feet in terrestrial environments. This process explains the occurrence of nonwebbed feet in chickens, hawks, crows, and many

Chicken

Duck

(a) BMP4 protein levels

Future interdigit regions

(b) Gremlin protein levels

(c) Comparison of a chicken foot and a duck foot

Figure 25.14 The role of cell-signaling proteins in the morphology of birds' feet. This figure shows how changes in developmental gene expression can produce certain traits. (a) Expression of the *BMP4* gene in the developing limbs. BMP4 protein is stained purple here and is expressed throughout the limb. (b) Expression of the *Gremlin* gene in the developing limbs. Gremlin protein is stained brown here. Note that *Gremlin* is not expressed in the interdigit regions of the chicken but is expressed in these regions of the duck. Gremlin inhibits BMP4, which causes programmed cell death. (c) Because BMP4 is not inhibited in the interdigit regions in the chicken, the cells in this region die, and the foot is not webbed. By comparison, inhibition of BMP4 in the interdigit regions in the duck results in a webbed foot.

Concept check: What would you expect to happen to the morphology of the feet of ducks if the Gremlin gene was under-expressed?

other terrestrial birds. In aquatic environments, webbed feet are an advantage because they act as paddles for swimming, so genetic variation that produced webbed feet would have been promoted by natural selection. Over time, this gave rise to the webbed feet now found in a wide variety of aquatic birds, including ducks, geese, and penguins.

The *Hox* Genes Have Been Important in the Evolution of a Variety of Body Plans Found in Different Species of Animals

The study of developmental genes has revealed interesting trends among large groups of species. *Hox* genes, which are discussed in Chapter 19, are found in all animals, indicating they have originated very early in animal evolution. Developmental biologists have hypothesized that variation in the *Hox* genes may have spawned the formation of many new body plans. As shown in **Figure 25.15**, the number and arrangement of *Hox* genes varies considerably among different types of animals. Sponges, the simplest of animals, have at least one *Hox* gene, whereas insects typically have nine or more. In most cases, multiple *Hox* genes occur in a cluster in which the genes are close to each other along a chromosome. In mammals, *Hox* gene clusters have been duplicated twice during the course of evolution to form four clusters, all slightly different, containing a total of 38 genes.

Researchers propose that increases in the number of *Hox* genes have been instrumental in the evolution of many animal species with greater complexity in body structure. To understand how, let's first consider *Hox* gene function. All *Hox* genes encode transcription factors that act as master control proteins to direct the formation of particular regions of the body. Each *Hox* gene controls a hierarchy of many regulatory genes that regulate the expression of genes encoding proteins that ultimately affect the morphology of the organism. The evolution of complex body plans is associated with an increase not only in the number of regulatory genes—as evidenced by the increase in *Hox* gene complexity during evolution—but also in genes that encode proteins that directly affect an organism's form and function.

How would an increase in *Hox* genes enable more complex body forms to evolve? Part of the answer lies in the spatial expression of the *Hox* genes. Different *Hox* genes are expressed in different regions of the body along the anteroposterior axis (refer back to Figure 19.16). Therefore, an increase in the number of *Hox* genes allows each of these master control genes to become more specialized in the region that it controls. In fruit flies, one segment in the middle of the body can be controlled by a particular *Hox* gene and form wings and legs, while a segment in the head region can be controlled by a different *Hox* gene and develop antennae. Therefore, research suggests that one way for new, more complex body forms to evolve is by increasing the number of *Hox* genes, thereby making it possible to form many specialized parts of the body that are organized along a body axis.

Three lines of evidence support the idea that *Hox* gene number has been instrumental in the evolution and speciation of animals with different body patterns. First, as discussed in Chapter 19, *Hox* genes are known to control the fate of regions along the anteroposterior axis. Second, as described in Figure 25.15, a general trend is observed in which animals with a more complex body structure tend to have more *Hox* genes and *Hox* clusters in their genomes compared with the genomes of

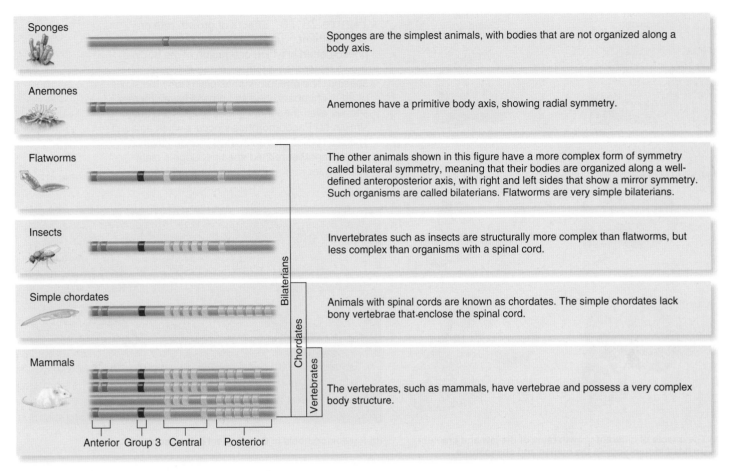

Figure 25.15 *Hox* **gene number and body complexity in different types of animals.** Researchers speculate that the duplication of *Hox* genes and *Hox* gene clusters played a key role in the evolution of more complex body plans in animals. A correlation is observed between increasing numbers of *Hox* genes and increasing complexity of body structure. The *Hox* genes are divided into four groups, called anterior, group 3, central, and posterior, based on their relative similarities. Each group is represented by a different color in this figure.

Concept check: What is the relationship between the total number of Hox genes in an animal species and its morphological complexity?

simpler animals. Third, a comparison of *Hox* gene evolution and animal evolution bears striking parallels. Researchers can analyze *Hox* gene sequences among modern species and make estimates regarding the timing of past events. Using this type of approach, geneticists can estimate when the first *Hox* gene arose by gene innovation. The date is difficult to pinpoint but is well over 600 million years ago. The single *Hox* gene found in the sponge has descended from this primordial *Hox* gene. In addition, gene duplications of this primordial gene produced clusters of *Hox* genes in other species. Clusters such as those found in modern insects were likely to be present approximately 600 million years ago. A duplication of that cluster is estimated to have occurred around 520 million years ago.

Interestingly, these estimates of *Hox* gene origins correlate with major diversification events in the history of animals. As described in Chapter 22, the Cambrian period, which occurred from 543 to 490 million years ago, saw a great diversification of animal species. This diversification occurred after the *Hox* cluster was formed and was possibly undergoing its first duplication

to create two *Hox* clusters. Also, approximately 420 million years ago, a second duplication produced species with four *Hox* clusters. This event preceded the proliferation of tetrapods— vertebrates with four limbs—that occurred during the Devonian period, approximately 417–354 million years ago. Modern tetrapods have four *Hox* clusters. This second duplication may have been a critical event that led to the evolution of complex terrestrial vertebrates with four limbs, such as amphibians, reptiles, and mammals.

Developmental Genes That Affect Growth Rates Can Have a Dramatic Effect on Phenotype and Lead to the Formation of New Species

In Figure 25.14, we saw how differences in the spatial expression of the *Gremlin* gene affected whether a bird species has nonwebbed or webbed feet. Another way that genetic variation can influence morphology is by controlling the relative growth rates of different parts of the body during development. The

Human	Chimpanzee
Fetus	
Infant	
Adult	

(a) The effects of different growth rates of the jaw and cranium on head morphology in the human and chimpanzee

Figure 25.16   **Effects of growth rate on development.**   (a) Heterochrony refers to the phenomenon in which one region of the body grows faster than another among different species. The phenomenon explains why the skulls of adult chimpanzees and humans have different shapes even though their fetal shapes are quite similar. (b) Paedomorphosis occurs when an adult species retains characteristics that are juvenile traits in another related species. Cope's giant salamander reproduces at the tadpole stage.

(b) Paedomorphosis in Cope's giant salamander

term **heterochrony** refers to evolutionary changes in the rate or timing of developmental events. The speeding up or slowing down of growth appears to be a common occurrence in evolution and can lead to different species with striking morphological differences.

As an example, **Figure 25.16a** compares the progressive growth of human and chimpanzee skulls. At the fetal stage, the size and shape of the skulls look fairly similar. However, after this stage, the relative growth rates of certain regions become markedly different, thereby affecting the shape and size of the adult skull. In the chimpanzee, the jaw region grows faster, giving the adult chimpanzee a much larger and longer jaw. In the human, the jaw grows more slowly, and the region of the skull that surrounds the brain, which is called the cranium, grows slightly faster. Therefore, adult humans have smaller jaws but a larger cranium.

Changes in growth rates can also affect the developmental stage at which one species reproduces. This can occur in two ways. One possibility is that the parts of the body associated with reproduction develop faster than the rest of the body. Alternatively, reproduction may occur at the same absolute age, but the development of nonreproductive body parts is slowed down. In either case, the morphological result is the same—reproduction is observed at an earlier stage in one species than it is in another. In such cases, the sexually mature organism may retain traits typical of the juvenile stage of the organism's ancestor, a condition called **paedomorphosis** (from

the Greek *paedo*, meaning young or juvenile, and *morph*, meaning the form of an organism). It is particularly common among salamanders. Even though paedomorphic species have certain juvenile features as adults, they have the ability to reproduce successfully. For example, Cope's giant salamander (*Dicamptodon copei*) becomes mature and reproduces in the aquatic form, without changing into a terrestrial adult as do other salamander species (**Figure 25.16b**). The adult form of Cope's giant salamander has gills and a large paddle-shaped tail, features that resemble those of the larval (tadpole) stage of other salamander species. Such a change in morphology was likely to have been a contributing factor to the formation of this species or an ancestral species to Cope's giant salamander.

Genomes & Proteomes Connection

The Study of the *Pax6* Gene Indicates That Different Types of Eyes Evolved from a Simpler Form

Thus far in this section, we have focused on the roles of particular genes as they influence the development of species with novel shapes and forms. Explaining how a complex organ comes into existence is another major challenge for evolutionary biologists. While it is relatively easy to comprehend how a limb could undergo evolutionary modifications to become a wing, flipper, or arm, it is more difficult to understand how a body structure, such as a limb, comes into being in the first place. In his book *The Origin of Species*, Charles Darwin addressed this question and admitted that the evolution and development of a complex organ such as the eye was difficult to understand. As noted by Darwin, the eyes of vertebrate species are exceedingly complex, being able to adjust focus, let in different amounts of light, and detect a spectrum of colors. Darwin speculated that such complex eyes must have evolved from a simpler structure through the process of descent with modification. With amazing insight, he suggested that a very simple eye would be composed of two cell types, a photoreceptor cell and an adjacent pigment cell. The photoreceptor cell, which is a type of nerve cell, is able to absorb light and respond to it. The function of the pigment cell is to stop the light from reaching one side of the photoreceptor cell. This primitive, two-cell arrangement would allow an organism to sense both light and the direction from which the light comes.

A primitive eye would provide an additional way for an organism to sense its environment, possibly allowing it to avoid predators or locate food. Vision is nearly universal among animals, which indicates a strong selective advantage for better eyesight. Over time, eyes could become more complex by enhancing the ability to absorb different amounts and wavelengths of light and also by refinements in structures such as the addition of lenses that focus the incoming light.

Since the time of Darwin, many evolutionary biologists have wrestled with the question of eye evolution. From an anatomical point of view, researchers have discovered many different types of eyes. For example, the eyes of fruit flies, squid, and humans are quite different from each other. This observation led evolutionary biologists such as Luitfried von Salvini-Plawen and Ernst Mayr to propose that eyes may have independently arisen multiple times during evolution. Based solely on morphology, such a hypothesis seemed reasonable and for many years was accepted by the scientific community.

The situation took a dramatic turn when geneticists began to study eye development. Researchers identified a master control gene, *Pax6*[1]. The protein encoded by the *Pax6* gene is a transcription factor that controls the expression of many other genes, including those involved in the development of the eye in both rodents and humans. In mice and rats, a mutation in the *Pax6* gene results in small eyes. A mutation in the human *Pax6* gene causes an eye disorder called aniridia, in which the iris and other structures of the eye do not develop properly. Similarly, *Drosophila* has a gene named *eyeless* that also causes a defect in eye development when mutant. The *eyeless* and *Pax6* genes are homologous; they were derived from the same ancestral gene.

In 1995, Swiss geneticist Walter Gehring and his colleagues were able to show experimentally that the expression of the *eyeless* gene in parts of *Drosophila* where it is normally inactive

could promote the formation of additional eyes. For example, using genetic engineering techniques, they were able to express the *eyeless* gene in the region where antennae should form. As seen in **Figure 25.17a**, this resulted in the formation of an eye where antennae are normally found! Remarkably, the expression of the mouse *Pax6* gene in *Drosophila* can also cause the formation of eyes in unusual places. For example, **Figure 25.17b** shows the formation of an eye on the leg of *Drosophila*.

Note that when the mouse *Pax6* master control gene switches on eye formation in *Drosophila*, the eye produced is a *Drosophila* eye, not a mouse eye. Why does this occur? It happens because the *Pax6* master control gene activates genes from the *Drosophila* genome. In *Drosophila*, the *Pax6* homolog called *eyeless* switches on a cascade involving several hundred genes required for eye morphogenesis. In more primitive eyes, the *Pax6* gene would be expected to control a cascade of fewer genes.

Since the discovery of the *Pax6* and *eyeless* genes, homologs of this gene have been discovered in many different species. In all cases where it has been tested, this gene is involved with eye development. Gehring and colleagues have hypothesized that the eyes from many different species may have evolved from a common ancestral form consisting of, as proposed by Darwin, one photoreceptor cell and one pigment cell (**Figure 25.18**). As mentioned, such a very simple eye can accomplish a rudimentary form of vision by detecting light and its direction. Eyes such as these are still found in modern species, such as the larvae of certain types of mollusks. Over time, simple eyes evolved into more complex types of eyes by modifications that resulted in the addition of more types of cells, such as lens cells and muscle cells. Alternatively, other researchers argue that

 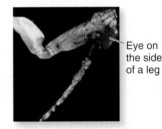

Normal eye — Eye where an antenna is normally found

Eye on the side of a leg

(a) Abnormal expression of the *Drosophila eyeless* gene in the antenna region

(b) Abnormal expression of the mouse *Pax6* gene in a fruit fly leg

Figure 25.17 Formation of additional eyes in *Drosophila* due to the abnormal expression of a master control gene for eye morphogenesis. (a) When the *Drosophila eyeless* gene is expressed in the antenna region, eyes are formed where antennae should be located. (b) When the mouse *Pax6* gene is expressed in the leg region of *Drosophila*, a small eye is formed there.

Concept check: What do you think would happen if the *Drosophila eyeless* gene was expressed at the tip of a mouse's tail?

---

[1] Pax is an acronym for paired box. The protein encoded by this gene contains a domain called a paired box.

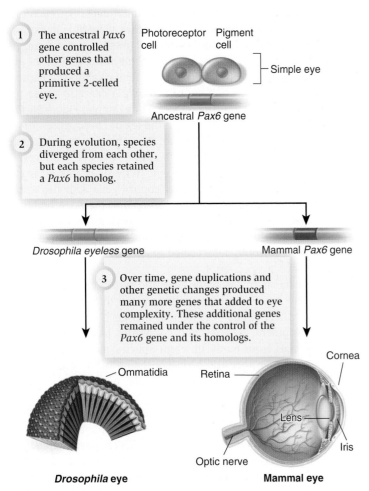

1. The ancestral *Pax6* gene controlled other genes that produced a primitive 2-celled eye.

Photoreceptor cell Pigment cell

Simple eye

Ancestral *Pax6* gene

2. During evolution, species diverged from each other, but each species retained a *Pax6* homolog.

Drosophila eyeless gene   Mammal *Pax6* gene

3. Over time, gene duplications and other genetic changes produced many more genes that added to eye complexity. These additional genes remained under the control of the *Pax6* gene and its homologs.

Ommatidia   Retina   Cornea

Lens

Iris

Optic nerve

Drosophila eye   **Mammal eye**

Figure 25.18 **Genetic control of eye evolution.** In this diagram, genetic changes, under control of the ancestral *Pax6* gene, led to the evolution of different types of eyes.

Pax6 may control only certain features of eye development and that different types of eyes may have evolved independently. Future research will be needed to resolve this controversy.

 Summary of Key Concepts

25.1 Identification of Species

- A species is a group of organisms that maintains a distinctive set of attributes in nature. Speciation is the process whereby new species are formed. Macroevolution refers to the evolutionary changes that create new species and groups of species.

- Different characteristics are used to identify species, including morphological traits, reproductive isolation, molecular features, ecological factors, and evolutionary relationships. (Figure 25.1)

- Researchers have identified many reproductive isolating mechanisms that prevent two different species from breeding with each other. (Figure 25.2)

- Prezygotic isolating mechanisms include habitat isolation, temporal isolation, behavioral isolation, mechanical isolation, and gametic isolation. (Figures 25.3, 25.4)

- Postzygotic isolating mechanisms include hybrid inviability, hybrid sterility, and hybrid breakdown. (Figure 25.5)

25.2 Mechanisms of Speciation

- Allopatric speciation occurs when a population becomes isolated from other populations and evolves into one or more new species. When speciation occurs multiple times due to the migration of populations into new environments, the evolutionary process is called adaptive radiation. If populations are incompletely separated, occasional interbreeding may occur in hybrid zones. (Figures 25.6, 25.7, 27.8)

- Podos discovered that changes in beak depth, associated with evolutionary changes involved with feeding, may also promote reproductive isolation by altering the song pattern of finches. (Figures 25.9, 25.10)

- Sympatric speciation involves the formation of different species that are not geographically isolated from one another. Polyploidy, adaptation to local environments, and sexual selection are mechanisms that can promote sympatric speciation. (Figures 25.11, 25.12)

25.3 The Pace of Speciation

- The pace of evolution may seem relatively constant or it may vary. Gradualism involves gradual evolution due to many small genetic changes, whereas punctuated equilibrium is a pattern of evolution in which new species arise more rapidly and then remain unchanged for long periods of time. (Figure 25.13)

25.4 Evo-Devo: Evolutionary Developmental Biology

- Evolutionary developmental biology compares the development of different species with the goal of understanding ancestral relationships and the mechanisms that bring about evolutionary change. These changes often involve variation in the expression of cell-signaling proteins and transcription factors.

- The spatial expression of genes that affect development can change phenotypes dramatically, as shown by the expression of the *BMP4* and *Gremlin* genes in birds with nonwebbed or webbed feet. (Figure 25.14)

- The evolution and duplication of *Hox* genes played an important role in producing groups of animals with specific body plans that are organized along an anteroposterior axis. (Figure 25.15)

- A difference in the relative growth rates of body parts among different species is called heterochrony. Paedomorphosis occurs when an adult species retains characteristics that are juvenile traits in another related species. (Figure 25.16)

- The *Pax6* gene and its homolog in other species are master control genes that control eye development in animals. (Figures 25.17, 25.18)

Assess and Discuss

Test Yourself

1. Macroevolution refers to the evolutionary changes that
   a. occur in multicellular organisms.
   b. create new species and groups of species.
   c. occur over long periods of time.
   d. cause changes in allele frequencies.
   e. occur in large mammals.

2. The biological species concept classifies a species based on
   a. morphological characteristics.
   b. reproductive isolation.
   c. the niche the organism occupies in the environment.
   d. genetic relationships between an organism and its ancestors.
   e. both a and b.

3. Which of the following would be considered an example of a postzygotic isolating mechanism?
   a. incompatible genitalia
   b. different mating seasons
   c. incompatible gametes
   d. mountain range separating two populations
   e. fertilized egg fails to develop normally

4. Hybrid breakdown occurs when species hybrids
   a. do not develop past the early embryonic stages.
   b. have a reduced life span.
   c. are infertile.
   d. are fertile but produce offspring with reduced viability and fertility.
   e. produce offspring that express the traits of only one of the original species.

5. The evolution of one species into two or more species is
   a. gradualism.
   b. punctuated equilibrium.
   c. cladogenesis.
   d. horizontal gene transfer.
   e. microevolution.

6. A large number of honeycreeper species on the Hawaiian Islands is an example of
   a. adaptive radiation.
   b. genetic drift.
   c. stabilizing selection.
   d. horizontal gene transfer.
   e. microevolution.

7. A major mechanism of speciation in plants but not animals is
   a. adaptation to new environments.
   b. polyploidy.
   c. hybrid breakdown.
   d. genetic changes that alter the organism's niche.
   e. both a and d.

8. The concept of punctuated equilibrium suggests that
   a. the rate of evolution is constant, with short time periods of no evolutionary change.
   b. evolution occurs gradually over time.
   c. small genetic changes accumulate over time to allow for phenotypic change and speciation.
   d. long periods of little evolutionary change are interrupted by short periods of major evolutionary change.
   e. both b and c.

9. Researchers suggest that an increase in the number of *Hox* genes
   a. would lead to reproductive isolation in all cases.
   b. could explain the evolution of color vision.
   c. allows for the evolution of more complex body forms in animals.
   d. results in the decrease in the number of body segments in insects.
   e. does all of the above.

10. The observation that the mammalian *Pax6* gene and the *Drosophila eyeless* gene are homologous genes that promote the formation of different types of eyes suggests that
    a. *Drosophila* eyes are more complex.
    b. mammalian eyes are more complex.
    c. eyes arose once during evolution.
    d. eyes arose at least twice during evolution.
    e. eye development is a simple process.

Conceptual Questions

1. What is the key difference between prezygotic isolating mechanisms and postzygotic isolating mechanisms? Give an example of each. Which type is more costly from the perspective of energy?

2. What are the key differences between gradualism and punctuated equilibrium? How are genetic changes related to these two models?

3. Describe one example in which genes that control development may have played an important role in the evolution of different species.

Collaborative Questions

1. What is a species? Discuss how geographic isolation can lead to speciation. Explain how reproductive isolation plays a role.

2. Discuss the type of speciation (allopatric or sympatric) that is most likely to occur under each of the following conditions:

   a. A pregnant female rat is transported by an ocean liner to a new continent.

   b. A meadow containing several species of grasses is exposed to a pesticide that promotes nondisjunction.

   c. In a very large lake containing several species of fishes, the water level gradually falls over the course of several years. Eventually, the large lake becomes subdivided into smaller lakes, some of which are connected by narrow streams.

Online Resource

www.brookerbiology.com

Stay a step ahead in your studies with animations that bring concepts to life and practice tests to assess your understanding. Your instructor may also recommend the interactive ebook, individualized learning tools, and more.

Taxonomy and Systematics

26

A new species, the African forest elephant. In 2001, biologists decided that this is a unique species of elephant, now named *Loxodonta cyclotis*.

U ntil recently, biologists classified elephants into only two species—the African savanna elephant (*Loxodonta africana*) and the Asian elephant (*Elephas maximus*). However, by analyzing the DNA of African elephants, researchers revised this classification, and proposed a third species, now called the African forest elephant (*Loxodonta cyclotis*) (see chapter-opening photo). How was this new species identified? This surprising finding was made somewhat by accident in 2001. A DNA identification system was set up to trace ivory poachers. By studying the DNA, researchers decided that Africa has two distinctly different *Loxodonta* elephant species. The African forest elephant is found in the forests of central and western Africa. The African savanna elephant, which is larger and has longer tusks, lives on large, dry grasslands. One consequence of this discovery is that it will have an impact on conservation efforts, which had previously been based on a single species of African elephants.

The rules for the classification of newly described species, such as the African forest elephant, are governed by the discipline of taxonomy (from the Greek *taxis*, meaning order, and *nomos*, meaning

law). **Taxonomy** is the science of describing, naming, and classifying **extant** organisms, which still exist today, as well as extinct organisms. Taxonomy results in the ordered division of species into groups based on similarities and dissimilarities in their characteristics. This task has been ongoing for over 300 years. As discussed in Chapter 23, the naturalist John Ray made the first attempt to broadly classify all known forms of life. Ray's ideas were later extended by naturalist Carolus Linnaeus in the mid-1700s, which is considered by some as the official birth of taxonomy.

Systematics is the study of biological diversity and the evolutionary relationships among organisms, both extant and extinct. In the 1950s, German entomologist Willi Hennig began classifying organisms in a new way. Hennig proposed that evolutionary relationships should be inferred from new features shared by descendants of a common ancestor. Since that time, biologists have applied systematics to the field of taxonomy. Researchers now try to place new species into taxonomic groups based on evolutionary relationships with other species. In addition, previously established taxonomic groups are revised as new data shed light on evolutionary relationships. As in any scientific discipline, taxonomy should be viewed as a work in progress.

In this chapter, we will begin with a discussion of taxonomy and the concept of taxonomic groups. We will then examine how biologists use systematics to determine evolutionary relationships among organisms, looking in particular at how these relationships are portrayed in diagrams called **phylogenetic trees**. We will explore how analyses of morphological data and molecular genetic data are used to understand the evolutionary history of life on earth.

26.1 Taxonomy

A hierarchy is a system of organization that involves successive levels. In biology, every species is placed into several different nested groups within a hierarchy. For example, a leopard and a fruit fly are both classified as animals, though they differ in many traits. By comparison, leopards and lions are placed together into a group with a smaller number of species called felines (more formally named Felidae), which are predatory cats. The felines are a subset of the animal group, which has species that share many similar traits. The species that are placed together into small taxonomic groups are likely to share many of the same characteristics. In this section, we will consider how biologists use a hierarchy to group similar species.

Living Organisms Are Now Subdivided into Three Domains of Life

Modern taxonomy places species into progressively smaller hierarchical groups. Each group at any level is called a **taxon** (plural, taxa). The taxonomic group called the **kingdom** was originally the highest and most inclusive. Linnaeus had classified all life into two kingdoms, plants and animals. In 1969, American ecologist Robert Whittaker proposed a five-kingdom system in which all life was classified into the kingdoms Monera, Protista, Fungi, Plantae, and Animalia. (Monera included all prokaryotic organisms.) However, as biologists began to learn more about the evolutionary relationships among these groups, they found that these groupings did not correctly reflect the relationships among them.

In the late 1970s, based on information in the sequences of genes, American biologist Carl Woese proposed the idea of creating a category called a **domain**. Under this system, all forms of life are grouped within three domains: **Bacteria**, **Archaea**, and **Eukarya** (Figure 26.1). The kingdom Monera was split into two domains, Bacteria and Archaea, due to major differences between these two types of prokaryotes. The terms Bacteria and Archaea are capitalized when referring to the domains. Alternatively, these terms can also refer to prokaryotic cells and species, in which case they are not capitalized. A single bacterial cell is called a bacterium, and a single archaeal cell is an archaeon.

The domain Eukarya formerly consisted of four kingdoms called **Protista**, **Fungi**, **Plantae**, and **Animalia**. However, researchers later discovered that Protista is not a separate kingdom but instead is a very broad collection of organisms. Taxonomists now place eukaryotes into seven groups called supergroups. In the taxonomy of eukaryotes, a **supergroup** lies between a domain and a kingdom (Figure 26.1). Kingdom Plantae is within the supergroup called Land plants and algal relatives, and kingdoms Fungi and Animalia are within the supergroup Opisthokonta. **Table 26.1** compares a variety of molecular and cellular characteristics among Bacteria, Archaea, and Eukarya.

Table 26.1	Distinguishing Cellular and Molecular Features of Domains Bacteria, Archaea, and Eukarya*		
Characteristic	**Bacteria**	**Archaea**	**Eukarya**
Chromosomes	Usually circular	Circular	Usually linear
Nucleosome structure	No	No	Yes
Chromosome segregation	Binary fission	Binary fission	Mitosis/meiosis
Introns in genes	Rarely	Rarely	Commonly
Ribosomes	70S	70S	80S
Initiator tRNA	Formylmethionine	Methionine	Methionine
Operons	Yes	Yes	No
Capping of mRNA	No	No	Yes
RNA polymerases	One	Several	Three
Promoters of structural genes	−35 and −10 sequences	TATA box	TATA box
Cell compartmentalization	No	No	Yes
Membrane lipids	Ester-linked	Ether-linked	Ester-linked

*The descriptions in this table are meant to represent the general features of most species in each domain. Some exceptions are observed. For example, certain bacterial species have linear chromosomes.

Figure 26.1 **A classification system for living and extinct organisms.** All organisms can be grouped into three large domains: Bacteria, Archaea, or Eukarya. Eukaryotes are divided into seven supergroups. As discussed in Chapter 28, the division of eukaryotes into supergroups is under current investigation and should be viewed as work in progress.

Every Species Is Placed into a Taxonomic Hierarchy

Why is it important to categorize species into groups? The three domains of life contain millions of different species. Subdividing them into progressively smaller groups makes it easier for biologists to appreciate the relationships among such a large number of species.

As mentioned, the broadest taxon is the domain, which is divided into supergroups among eukaryotes. Below the domain and supergroup is the kingdom, which is divided into **phyla** (singular, phylum). Each phylum is divided into **classes**, then **orders**, **families**, and **genera** (singular, genus). Each of these taxa contains progressively fewer species that are more similar to each other than they are to the members of the taxa above them in the hierarchy. For example, the taxon Animalia, which is at the kingdom level, has a larger number of fairly diverse species than does the class Mammalia, which contains fewer species that are relatively similar to each other.

To further understand taxonomy, let's consider the classification of a species such as the gray wolf (*Canis lupus*) (**Figure 26.2**). The gray wolf is placed in the domain Eukarya, the supergroup Opisthokonta, and then within the kingdom Animalia, which includes over 1 million species of all animals. Next, the gray wolf is classified in the phylum Chordata. The 50,000 species of animals in this group all have four common features at some stage of their development. These are a notochord (a cartilaginous rod that runs along the back of all chordates at some point in their life cycle), a tubular nerve or spinal cord located above the notochord, gill slits or arches, and a post-anal tail. Examples of animals in the phylum Chordata include fishes, reptiles, and mammals.

The gray wolf is in the class Mammalia, which includes all 5,000 species of mammals. Two distinguishing features of animals in this group are hair and mammary glands. The mammary glands produce milk, which nourishes the young, and hair helps insulate the body to maintain a warm, constant body temperature. There are 26 orders of mammals; the order that includes the gray wolf is called Carnivora and has about 270 species. The gray wolf is placed in the family Canidae, which is a relatively small family of 34 species, including different species of wolves, jackals, foxes, wild dogs, and the coyote and domestic dog. All species in the family Canidae are doglike animals. The smallest grouping that contains the gray wolf is the genus *Canis*, which includes four species of jackals, the coyote, two types of wolves, and the domestic dog. The domestic dog (*Canis lupus familiaris*) is a subspecies of the gray wolf.

Binomial Nomenclature Is Used to Name Species

As originally advocated by Linnaeus, **binomial nomenclature** is the standard method for naming species. The scientific name of every species has two names, its genus name and its unique specific epithet. An example is the gray wolf, *Canus lupis*. The

Taxonomic group	Gray wolf found in	Number of species
Domain	Eukarya	~4–10 million
Supergroup	Opisthokonta	>1 million
Kingdom	Animalia	>1 million
Phylum	Chordata	~50,000
Class	Mammalia	~5,000
Order	Carnivora	~270
Family	Canidae	34
Genus	*Canis*	7
Species	*lupus*	1

Figure 26.2 A taxonomic classification of the gray wolf (*Canis lupus*).

Concept check: Which group is broader, a phylum or a family?

genus name is always capitalized, but the specific epithet is not. Both names are italicized. After the first mention, the genus name is abbreviated to a single letter. For example, we would write that *Canis lupus* is the gray wolf, and in subsequent sentences, the species would be referred to as *C. lupus*.

When naming a new species, genus names are always nouns or treated as nouns, whereas species epithets may be either nouns or adjectives. The names often have a Latin or Greek origin and refer to characteristics of the species or to features of its habitat. For example, the genus name of the newly discovered African forest elephant, *Loxodonta*, is from the Greek *loxo*, meaning slanting, and *odonta*, meaning tooth. The species epithet *cyclotis* refers to the observation that the ears of this species are rounder compared to those of *L. africana*. However, sometimes the choice of name for a species can be lighthearted. For example, there is a beetle named *Agra vation* and a spider named *Draculoides bramstokeri*.

The rules for naming animal species, such as *Canis lupus* and *Loxodonta africana*, have been established by the International Commission on Zoological Nomenclature (ICZN). The ICZN provides and regulates a uniform system of nomenclature to ensure that every animal has a unique and universally accepted scientific name. Who is allowed to identify and name a new species? As long as ICZN rules are followed, new animal species can be named by anyone, not only by scientists. The rules for naming plants are described in the International Code of Botanical Nomenclature (ICBN), and the naming of prokaryotes is overseen by the International Committee on Systematics of Prokaryotes (ICSP).

26.2 Phylogenetic Trees

As mentioned, systematics is the study of biological diversity and evolutionary relationships. By studying the similarities and differences among species, biologists can gain information about **phylogeny**, which is the evolutionary history of a species or group of species. To propose a phylogeny, biologists use the tools of systematics. Evolutionary relationships are used to construct taxonomic groups. For example, the classification of the gray wolf described in Figure 26.2 is based on systematics. Therefore, one use of systematics is to place species into taxa and to understand the evolutionary relationships among different taxa.

In this section, we will consider the features of diagrams or trees that describe the evolutionary relationships among various species, both living and extinct. As you will learn, such trees are usually based on morphological or genetic data.

A Phylogenetic Tree Is a Hypothesis That Depicts the Evolutionary Relationships Among Species

A phylogenetic tree is a diagram that describes the evolutionary relationships among various species, based on the information available to and gathered by systematists. Phylogenetic trees should be viewed as hypotheses that are proposed, tested, and later refined as additional data become available. Let's look at what information a phylogenetic tree contains and the form in which it is presented. **Figure 26.3** shows a hypothetical phylogenetic tree of the relationships among various bird species, in which the species are labeled A through K. The vertical axis represents time, with the oldest species at the bottom.

New species can be formed by **anagenesis**, in which a single species evolves into a different species, or more commonly by **cladogenesis**, in which a species diverges into two or more species. The branch points or **nodes** in a phylogenetic tree illustrate times when cladogenesis has occurred. For example, approximately 12 million years ago, species A diverged into species A and species B. Figure 26.3 also shows anagenesis in which species C evolved into species G. The tips of branches may represent species that became extinct in the past, such as species B and E, or living species, such as F, I, G, J, H, and K, which are at the top of the tree. Species A and D are also extinct but gave rise to species that are still in existence.

By studying the branch points of a phylogenetic tree, researchers can group species according to common ancestry. A **clade** consists of a common ancestral species and all of its descendant species. For example, the group highlighted in light green in Figure 26.3 is a clade derived from the common ancestral species labeled D. Likewise, the entire tree forms a clade, with species A as a common ancestor. Therefore, smaller and more recent clades are nested within larger clades that have older common ancestors.

What is the relationship between a clade and taxonomy? The relationship depends on how far back we go to identify a common ancestor. For broader taxa, such as a kingdom, the common ancestor existed a very long time ago, on the order of hundreds of millions or even billions of years ago. For smaller taxa, such as a family or genus, the common ancestor occurred much more recently, on the order of millions or tens of millions of years ago. This concept is shown in a very schematic way in **Figure 26.4**. This small, hypothetical kingdom is a clade that contains 64 living species. (Actual kingdoms are obviously larger and exceedingly more complex.) The diagram emphasizes the taxa that contain the species designated number 43. The common ancestor that gave rise to this kingdom of organisms existed approximately 1 billion years ago. Over time, more recent species arose that subsequently became the common ancestors to the phylum, class, order, family, and genus that contain species number 43.

One goal of modern systematics is to create taxonomic groups that reflect evolutionary relationships. Systematics attempts to organize species into clades, which means that each group includes an ancestral species and all of its descendants. A **monophyletic group** is a taxon that is a clade. Ideally, every taxon, whether it is a kingdom, phylum, class, order, family, or genus, should be a monophyletic group.

How does research in systematics impact taxonomy? As researchers gather new information, they sometimes discover that some of the current taxonomic groups are not monophyletic. **Figure 26.5** compares a monophyletic group with those that are not. A **paraphyletic group** contains a common ancestor

Figure 26.3 **How to read a phylogenetic tree.** This hypothetical tree shows the proposed relationships between various bird species. Species are placed into clades, groups of organisms containing an ancestral organism and all of its descendants.

Concept check: Can two different species have more than one common ancestor?

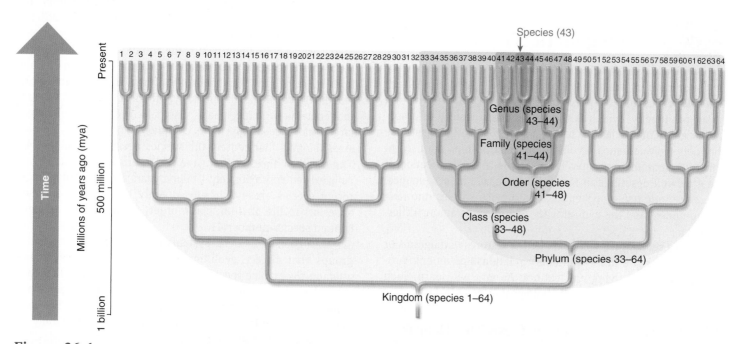

Figure 26.4 **Schematic relationship between a phylogenetic tree and taxonomy, when taxonomy is correctly based on evolutionary relationships.** The shaded areas highlight the kingdom, phylum, class, order, family, and genus for species number 43. All of the taxa are clades. Broader taxa, such as phyla and classes, are derived from more-ancient common ancestors. Smaller taxa, such as families and genera, are derived from more-recent common ancestors. These smaller taxa are subsets of the broader taxa.

Concept check: Which taxon would have a more-recent common ancestor, a phylum or an order?

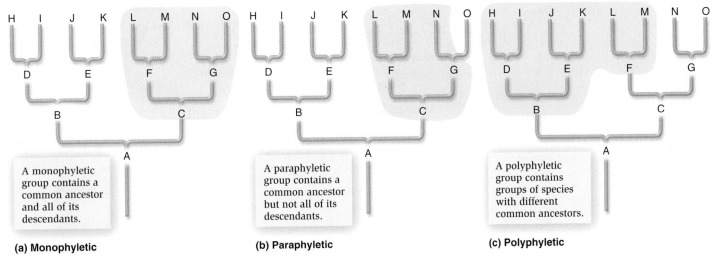

Figure 26.5   A comparison of monophyletic, paraphyletic, and polyphyletic taxonomic groups.

(a) Monophyletic

A monophyletic group contains a common ancestor and all of its descendants.

(b) Paraphyletic

A paraphyletic group contains a common ancestor but not all of its descendants.

(c) Polyphyletic

A polyphyletic group contains groups of species with different common ancestors.

and some, but not all, of its descendants (Figure 26.5b). In contrast, a **polyphyletic group** consists of members of several evolutionary lines and does not include the most recent common ancestor of the included lineages (Figure 26.5c).

Over time, as we learn more about evolutionary relationships, taxonomic groups are being reorganized in an attempt to recognize only monophyletic groups in taxonomy. For example, traditional classification schemes once separated birds and reptiles into separate classes (**Figure 26.6a**). In this scheme, the reptile class (officially named Reptilia) contained orders that included turtles, lizards and snakes, and crocodiles. Research has indicated that the reptile taxon is paraphyletic, because birds were excluded from the group. One way that this group can be made monophyletic is by including birds as a class within the reptile clade and elevating the other groups to a class status (**Figure 26.6b**).

The Study of Systematics Is Usually Based on Morphological or Genetic Homology

As discussed in Chapter 23, the term **homology** refers to similarities among various species that occur because the species are derived from a common ancestor. Such features are said to be homologous. For example, the arm of a human, the wing of a bat, and the flipper of a whale are homologous structures (refer back to Figure 23.12). Similarly, genes found in different species are homologous if they have been derived from the same ancestral gene (refer back to Figure 23.13).

In systematics, researchers identify homologous features that are shared by some species but not by others, which allows them to group species based on their shared similarities. Researchers usually study homology at the level of morphological traits or at the level of genes. In addition, the data they gather are viewed in light of geographic data. Many organisms do not migrate extremely long distances. Species that are closely related evolutionarily are relatively likely to inhabit neighboring or overlapping geographic regions, though many exceptions are known to occur.

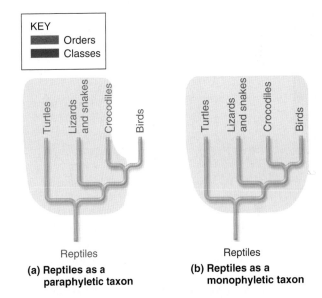

KEY
- Orders
- Classes

(a) Reptiles as a paraphyletic taxon

(b) Reptiles as a monophyletic taxon

Figure 26.6   An example of a taxon that is not monophyletic.   (a) The class of reptiles as a paraphyletic taxon. (b) The group can be made monophyletic if birds and the other orders were classified as classes within the reptile clade.

Morphological Analysis   The first studies in systematics focused on morphological features of extinct and living species. Morphological traits continue to be widely used in systematic studies, particularly in those studies pertaining to extinct species and those involving groups that have not been extensively studied at the molecular level. To establish evolutionary relationships based on morphological homology, many traits have to be analyzed to identify similarities and differences.

By studying morphological features of extinct species in the fossil record, paleontologists can propose phylogenetic trees that chart the evolutionary lineages of species, including those that still exist. In this approach, the trees are based on morphological features that change over the course of many generations. As an example, Figure 26.7 depicts a current hypothesis of the

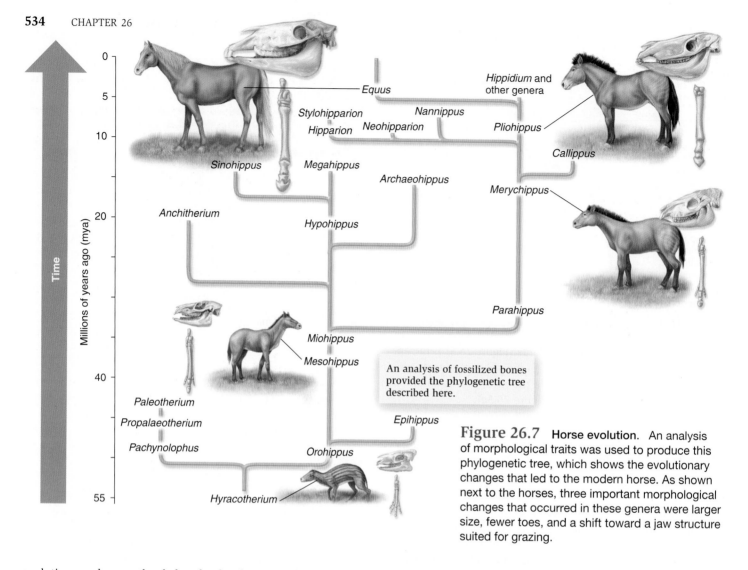

Figure 26.7 **Horse evolution.** An analysis of morphological traits was used to produce this phylogenetic tree, which shows the evolutionary changes that led to the modern horse. As shown next to the horses, three important morphological changes that occurred in these genera were larger size, fewer toes, and a shift toward a jaw structure suited for grazing.

An analysis of fossilized bones provided the phylogenetic tree described here.

evolutionary changes that led to the development of the modern horse. This figure shows representative species from various genera. Many morphological features were used to propose this tree. Because hard parts of the body are more commonly preserved in the fossil record, this tree is largely based on the analysis of skeletal changes in foot structure, lengths and shapes of various leg bones, skull shape and size, and jaw and tooth morphology. Over an evolutionary time scale, the accumulation of many genetic changes has had a dramatic impact on species' characteristics. In the genera depicted in this figure, a variety of morphological changes occurred, such as an increase in size, a reduction in the number of toes, and modifications in the jaw and teeth that were suitable for grazing on fibrous grasses.

Similar morphological features may occasionally confound an evolutionary analysis. As described in Chapter 23, convergent evolution can result in analogous structures, characteristics that arise independently in different lineages because different species have evolved in similar environments. For example, the giant anteater and the echidna have similar structures, such as long snouts and tongues, that enable these animals to feed on ants (refer back to Figure 23.8a). These traits are not derived from a common ancestor. Rather, they arose twice independently during evolution due to adaptation to similar environments. In systematics, the use of analogous structures can

cause errors if a researcher assumes that a particular trait arose only once and that all species having the trait are derived from a common ancestor.

Molecular Systematics The field of **molecular systematics** involves the analysis of genetic data, such as DNA and amino acid sequences, to identify and study genetic homologies and propose phylogenetic trees. In 1963, Austrian biologist Emile Zuckerkandl and American chemist Linus Pauling were the first to suggest that molecular data could be used to establish evolutionary relationships. How can a comparison of genetic sequences help to establish evolutionary relationships? As discussed later in this chapter, DNA sequences change over the course of many generations due to the accumulation of mutations. Therefore, when comparing homologous sequences in different organisms, DNA sequences from closely related organisms are more similar to each other than they are to sequences from distantly related species.

26.3   Cladistics

Cladistics is the classification of species based on evolutionary relationships. A cladistic approach discriminates among

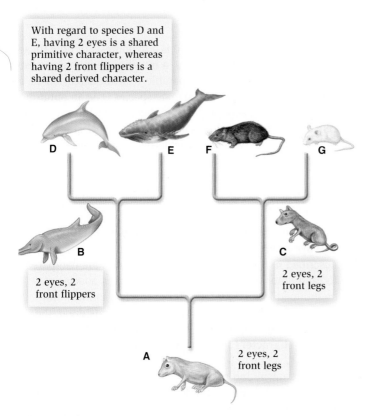

With regard to species D and E, having 2 eyes is a shared primitive character, whereas having 2 front flippers is a shared derived character.

2 eyes, 2 front flippers

2 eyes, 2 front legs

2 eyes, 2 front legs

Figure 26.8   A comparison of shared primitive characters and shared derived characters.

possible phylogenetic trees by considering the possible pathways of evolutionary changes and then choosing the tree based on characteristics that are shared or not shared among various species. Cladistics is a commonly used method for the evaluation of phylogenetic trees, which are known as **cladograms**. In this section, we will consider the cladistic approach and other methods that are used to evaluate phylogenetic trees.

Species Differ with Regard to Primitive and Derived Characters

A cladistic approach compares homologous traits, also called **characters**, which may exist in two or more **character states**. For example, among different species, a front limb, which is a character, may exist in different character states such as a wing, an arm, or a flipper. The various character states are either shared or not shared by different species.

To understand the cladistic approach, let's take a look at a simplified phylogeny (**Figure 26.8**). We can place the extant species that currently exist into two groups: D and E, and F and G. The most recent common ancestor to D and E is B, whereas species C is the most recent common ancestor to F and G. With these ideas in mind, let's focus on the front limbs (flippers versus legs) and eyes. A character that is shared by two or more different taxa and inherited from ancestors older than their last common ancestor is called a **shared primitive character**, or **symplesiomorphy**. Such traits are viewed as being older

traits—ones that occurred earlier in evolution. With regard to species D, E, F, and G, having two eyes is a shared primitive character. It originated prior to species B and C.

By comparison, a **shared derived character**, or **synapomorphy**, is a character that is shared by two or more species or taxa and has originated in their most recent common ancestor. With regard to species D and E, having two front flippers is a shared derived character that originated in species B, their most recent common ancestor (**Figure 26.8**). Compared to shared primitive characters, shared derived characters are more recent traits on an evolutionary timescale. For example, among mammals, only some species have flippers, such as whales and dolphins. In this case, flippers were derived from the two front limbs of an ancestral species. The word "derived" indicates that evolution involves the modification of traits in pre-existing species. In other words, populations of organisms with new traits are derived from changes in pre-existing populations. The basis of the cladistic approach is to analyze many shared derived characters among groups of species to deduce the pathway that gave rise to those species.

Note that the terms primitive and derived do not indicate the complexity of a character. For example, the flippers of a dolphin do not appear more complex than the front limbs of ancestral species A (**Figure 26.8**), which were limbs with individual toes. Derived characters can be similar in complexity, less complex, or more complex than primitive characters.

A Cladistic Approach Is Used to Determine If a Phylogenetic Tree Is a Reasonable Hypothesis

To understand how shared derived characters can be used to propose a phylogenetic tree, **Figure 26.9a** compares several traits among five species of animals. The proposed cladogram shown in **Figure 26.9b** is consistent with the distribution of shared derived characters among these species. A branch point is where two species differ in a character. The oldest common ancestor, which would now be extinct, had a notochord and gave rise to all five species. Vertebrae are a shared derived character of the lamprey, salmon, lizard, and rabbit, but not the lancelet, which is an invertebrate. By comparison, a hinged jaw is a shared derived character of the salmon, lizard, and rabbit, but not of the lamprey or lancelet.

In a cladogram, an **ingroup** is the group whose evolutionary relationships we wish to understand. By comparison, an **outgroup** is a species or group of species that is assumed to have diverged before the species in the ingroup. An outgroup will lack one or more shared derived characters that are found in the ingroup. A designated outgroup can be closely related or more distantly related to the ingroup. In the tree shown in Figure 26.9, if the salmon, lizard, and rabbit are an ingroup, an outgroup would be the lamprey. The lamprey has a notochord and vertebrae but lacks a character shared by the ingroup, namely, a hinged jaw. Thus, for the ingroup, the notochord and vertebrae are shared primitive traits, whereas the hinged jaw is a shared derived trait that is not found in the outgroup.

	Lancelet	Lamprey	Salmon	Lizard	Rabbit
Notochord	Yes	Yes	Yes	Yes	Yes
Vertebrae	No	Yes	Yes	Yes	Yes
Hinged jaw	No	No	Yes	Yes	Yes
Tetrapod	No	No	No	Yes	Yes
Mammary glands	No	No	No	No	Yes

(a) Characteristics among species

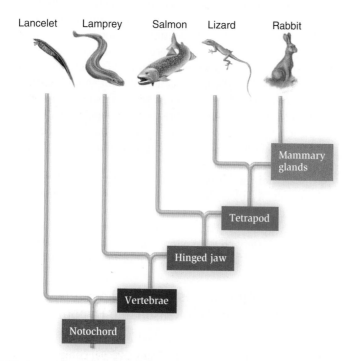

(b) Cladogram based on morphological traits

Figure 26.9   Using shared primitive characters and shared derived characters to propose a phylogenetic tree.   (a) A comparison of characteristics among these species. (b) This phylogenetic tree illustrates both shared primitive and shared derived characters in a cladogram of five animal species.

Concept check: *What shared derived character is common to the salmon, lizard, and rabbit, but not the lamprey?*

Likewise, the concept of shared derived characters can apply to molecular data, such as a gene sequence. Let's consider an example to illustrate this idea. Our example involves molecular data obtained from seven different hypothetical plant species called A–G. In these species, a homologous region of DNA was sequenced as shown here:

```
       12345678910

A:  GATAGTACCC
B:  GATAGTTCCC
C:  GATAGTTCCG
D:  GGTATTACCC
E:  GGTATAACCC
F:  GGTAGTACCA
G:  GGTAGTACCC
```

The cladogram of **Figure 26.10** is a hypothesis of how these DNA sequences arose. In this case, a mutation that changes the sequence of nucleotides is comparable to a modification of a character. For example, let's designate species D as an outgroup and species A, B, C, F, and G as the ingroup. In this case, a G (guanine) at the fifth position is a shared derived character. The genetic sequence carrying this G is derived from an older primitive sequence.

Now that we have an understanding of shared primitive and derived characters, let's consider the steps a researcher would follow to propose a cladogram using a cladistics approach.

1. **Choose the species in whose evolutionary relationships you are interested.** In a simple cladogram, such as those described in this chapter, individual species are compared to each other. In more complex cladograms, species may be grouped into larger taxa (for example, families) and compared with each other. If such grouping is done, the results will not be reliable if the groups are not clades.

2. **Choose characters for comparing the species selected in step 1.** As mentioned, a character is a general feature of an organism and may come in different versions called character states. For example, a front limb is a character in mammals, which could exist in different character states such as a wing, an arm, or a flipper.

3. **Determine the polarity of character states.** In other words, determine if a character state came first and is primitive or came later and is a derived character. This information may be available by examining the fossil record, for example, but is usually done by comparing the ingroup with the outgroup. For a character with two character states, an assumption is made that a character state shared by the outgroup and ingroup is primitive. A character state shared only by members of the ingroup is derived.

4. **Analyze cladograms based on the following principles:**
   - All species (or higher taxa) are placed on tips in a phylogenetic tree, not at branch points.
   - Each cladogram branch point should have a list of one or more shared derived characters that are common to all species above the branch point unless the character is later modified.
   - All shared derived characters appear together only once in a cladogram unless they arose independently during evolution more than once.

5. **Among many possible options, choose the cladogram that provides the simplest explanation for the data.** A common approach is to use a computer program that generates many possible cladograms. Analyzing the data and choosing the cladogram with the fewest number of character changes is a key aspect of this process. As described later, different theoretical approaches, such as the principle of parsimony, can be followed to achieve this goal.

6. **Provide a root to the phylogenetic tree by choosing a noncontroversial outgroup.** In this textbook, most

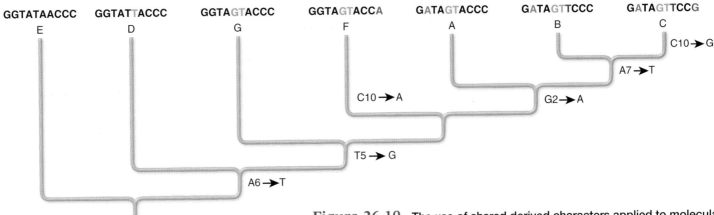

GGTATAACCC
1 2 3 4 5 6 7 8 9 10
Proposed primitive
sequence

Figure 26.10 The use of shared derived characters applied to molecular data such as a sequence of a gene. This phylogenetic tree illustrates a cladogram involving homologous gene sequences found in seven hypothetical plant species. Mutations that alter a primitive DNA sequence are shared among certain species but not others. Note: A, T, G, and C refer to nucleotide bases, and the numbers refer to the position of the base in the nucleotide sequences. For example, A6 refers to an adenine at the sixth position.

Concept check: *What nucleotide change is a shared derived character for species A, B, and C, but not for species G?*

phylogenetic trees are rooted, which means that a single node at the bottom of the tree corresponds to a common ancestor for all of the species or groups of species in the tree. A method for rooting trees is the use of a noncontroversial outgroup. Such an outgroup typically shares morphological traits and/or DNA sequence similarities with the members of the ingroup to allow a comparison between the ingroup and outgroup. Even so, the outgroup must be noncontroversial in that it shows enough distinctive differences with the ingroup to be considered a clear outgroup.

The Principle of Parsimony Can Help Researchers Choose from Among Possible Phylogenetic Trees

One approach for choosing among possible phylogenetic trees is to assume that the best hypothesis is the one that requires the smallest number of evolutionary changes. This concept, called the **principle of parsimony**, states that the preferred hypothesis is the one that is the simplest for all the characters and their states. For example, if two species possess a tail, we would initially assume that a tail arose once during evolution and that both species have descended from a common ancestor with a tail. Such a hypothesis is simpler, and more likely to be correct, than assuming that tails arose twice during evolution and that the tails in the two species are not due to descent from a common ancestor.

The principle of parsimony can also be applied to gene sequence data. Let's consider a hypothetical example involving molecular data from four taxa (A–D), where A is presumed to be the outgroup.

```
12345
```

A: GTACA (outgroup)

B: GACAG

C: GTCAA

D: GACCG

Given that B, C and D are the ingroup, three hypotheses for phylogenetic trees are shown in **Figure 26.11**, although more are possible. Tree 1 requires seven mutations, and tree 2 requires six, whereas tree 3 requires only five. Therefore, tree 3 requires the smallest number of mutations and is considered the most parsimonious. Based on the principle of parsimony, it would be the hypothesis that is the most likely to accurately reflect the evolutionary history of the taxon (or ingroup) in question. In practice, when researchers have multiple sequences that are longer than the ones shown here, computer programs are used to find the most parsimonious tree.

Maximum Likelihood and Bayesian Methods Are Also Used to Discriminate Among Possible Phylogenetic Trees

In addition to the principle of parsimony, evolutionary biologists also apply other approaches, such as maximum likelihood and Bayesian methods, when proposing and evaluating phylogenetic trees. These methods involve the use of an evolutionary model—a set of assumptions about how evolution is likely to happen. For example, mutations affecting the third base in a codon are often neutral because they don't affect the amino acid sequence of the encoded protein and therefore don't affect the fitness of an organism. As discussed in Chapter 24, such neutral

Figure 26.11   Using the principle of parsimony and molecular genetic data to choose a phylogenetic tree.   These are three possible phylogenetic trees for the evolution of a short DNA sequence, but many more are possible. Changes in nucleotide sequence are shown along each tree. For example, T2 → A means that the second base, a T, was changed to an A. According to the principle of parsimony, tree number 3 is the more likely choice because it requires only five mutations.

According to an approach called **maximum likelihood**, researchers ask the question: What is the probability that an evolutionary model and a proposed phylogenetic tree would give rise to the observed data? The rationale is that a phylogenetic tree that gives a higher probability of producing the observed data is preferred to any trees that give a lower probability. By comparison, **Bayesian methods** ask the question: What is the probability that a particular phylogenetic tree is correct given the observed data and a particular evolutionary model? Though the computational strategies of maximum likelihood and Bayesian methods are different (and beyond the scope of this textbook), the goal of both approaches is to identify one or more trees that are most likely to be correct based on an evolutionary model and the available data.

26.4   Molecular Clocks

As we have seen, researchers employ various methods to propose a phylogeny that describes the evolutionary relationships among various species. Researchers are interested not only in the most likely pathway of evolution, but also the timing of evolutionary change. How can researchers determine when different species diverged from each other in the past? As shown in Figure 26.7, the fossil record can sometimes help researchers apply a timescale to a phylogeny. Another way to infer the timing of past events is by analyzing genetic sequences. As discussed in Chapter 24, the neutral theory of evolution proposes that most genetic variation that exists in populations is due to the accumulation of neutral mutations, meaning that such genetic variation is not acted upon by natural selection. The reasoning behind this concept is that favorable mutations are likely to be very rare, and detrimental mutations are likely to be eliminated from a population by natural selection. A large body of evidence supports the idea that much of the genetic variation observed in living species is due to the accumulation of neutral mutations. From an evolutionary point of view, if neutral mutations occur at a relatively constant rate, they can act as a **molecular clock** on which to measure evolutionary time. In this section, we will consider the concept of a molecular clock and its application in phylogenetic trees.

The Timing of Past Events May Be Inferred from Molecular Clock Data

Figure 26.12 illustrates the concept of a molecular clock. The graph's y-axis is a measure of the number of nucleotide differences in a homologous gene between different pairs of species. The x-axis plots the amount of time that has elapsed since each pair of species shared a common ancestor. As discussed in the boxes in this diagram, the number of nucleotide differences is lower when two species shared a common ancestor in the more recent past than it is in pairs that shared a more distant common ancestor. The explanation for this phenomenon is that the gene sequences of various species accumulate independent mutations after they have diverged from each other. A

mutations are more likely to become prevalent in a population compared to mutations in the first or second base. Therefore, one possible assumption of an evolutionary model is that neutral mutations are more likely than nonneutral mutations.

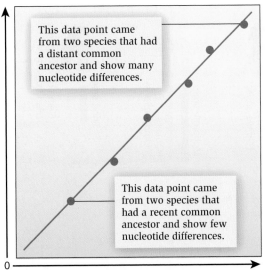

Figure 26.12 **A molecular clock.** According to the concept of a molecular clock, neutral mutations accumulate at a relatively constant rate over evolutionary time. When comparing homologous genes between species, those species that diverged more recently tend to have fewer differences than do those whose common ancestor occurred in the very distant past.

Concept check: *When comparing the sequences of a homologous gene in four different species, one pair of species shows a 1% difference, whereas the other pair shows a 3% difference. Which pair diverged longer ago from their common ancestor?*

longer period of time since their divergence allows for a greater accumulation of mutations, which makes their sequences more different.

Figure 26.12 suggests a linear relationship between the number of nucleotide changes and the time of divergence. For example, a linear relationship predicts that a pair of species that has 20 nucleotide differences in a given gene sequence would have a common ancestor that is roughly twice as old as that of a pair showing 10 nucleotide differences. While actual data sometimes show a relatively linear relationship over a defined time period, evolutionary biologists do not think that molecular clocks are perfectly linear over very long periods of time. Several factors can contribute to nonlinearity of molecular clocks. These include differences in the generation times of the species being analyzed and variation in mutation rates between different species.

To obtain reliable data, researchers must calibrate their molecular clocks. How much time does it take to accumulate a certain percentage of nucleotide changes? To perform such a calibration, researchers must have information regarding the date when two species diverged from a common ancestor. Such information could come from the fossil record, for instance. The genetic differences between those species are then divided by the amount of time since their last common

ancestor to calculate a rate of change. For example, research suggests that humans and chimpanzees diverged from a common ancestor approximately 6 million years ago. The percentage of nucleotide differences between mitochondrial DNA of humans and chimpanzees is 12%. From these data, the molecular clock for changes in mitochondrial DNA sequences of primates is calibrated at roughly 2% nucleotide changes per million years.

Different Genes Are Analyzed to Study Phylogeny

For evolutionary comparisons, the DNA sequences of many genes have been obtained from a wide range of sources. Many different genes have been used to propose phylogenetic trees. For example, the gene that encodes an RNA found in the small ribosomal subunit (SSU rRNA) has been commonly used in evolutionary studies. As noted in Chapter 12, the gene for SSU rRNA is found in the genomes of all living organisms. Therefore, its function must have been established at an early stage in the evolution of life on this planet, and its sequence has changed fairly slowly. Furthermore, SSU rRNA is a rather large molecule, so it contains a large amount of sequence information. This gene has been sequenced from thousands of different species. Slowly changing genes such as the gene that encodes SSU rRNA are useful for evaluating distant evolutionary relationships, such as comparing higher taxa. For example, SSU rRNA data can be used to place eukaryotic species into their proper phyla or orders.

Other genes have changed more rapidly during evolution because of a greater tolerance of neutral mutations. For example, the mitochondrial genome and DNA sequences within introns can more easily incur neutral mutations (compared to the coding sequences of genes), and so their sequences change frequently during evolution. More rapidly changing DNA sequences have been used to study recent evolutionary relationships, particularly among eukaryotic species such as large animals that have long generation times and tend to evolve more slowly. In these cases, slowly evolving genes may not be very useful for establishing evolutionary relationships because two closely related species are likely to have identical or nearly identical DNA sequences for such genes.

Figure 26.13 shows a simplified phylogeny of closely related species of primates. This tree was proposed by comparing DNA sequence changes in the gene for cytochrome oxidase subunit II, a protein located in the mitochondrial inner membrane that is involved in cellular respiration. This gene tends to change fairly rapidly on an evolutionary timescale. The vertical scale represents time, and the branch points that are labeled with letters represent common ancestors. Let's take a look at three branch points (labeled A, D, and E) and relate them to the accumulation of neutral mutations.

> Ancestor A: This ancestor diverged into two species that ultimately gave rise to siamangs and the other five species. Since this divergence, there has been a long time (approximately 23 million years) for the siamang genome to accumulate a relatively high

Figure 26.13 **The use of DNA sequence changes to study primate evolution.** This phylogenetic tree, which shows relationships among closely related species of primates, is based on a comparison of mitochondrial gene sequences encoding the protein cytochrome oxidase subunit II.

Concept check: *Which pair of species would be expected to have fewer genetic differences: orangutans and gorillas or gorillas and humans?*

number of random neutral changes that would be different from the random changes that have occurred in the genomes of the other five species (see the yellow bar in Figure 26.13). Therefore, the gene in the siamangs is fairly different from the genes in the other five species.

Ancestor D: This ancestor diverged into two species that eventually gave rise to humans and chimpanzees. This divergence occurred a moderate time ago, approximately 6 million years ago, as illustrated by the red bar. The differences in gene sequences between humans and chimpanzees are relatively moderate.

Ancestor E: This ancestor diverged into two species of chimpanzees. Since the divergence of species E into two species, approximately 3 million years ago, the time for the molecular clock to "tick" (that is, accumulate random mutations) is relatively short, as depicted by the green bar in Figure 26.13. Therefore, the two existing species of chimpanzees have fewer differences in their gene sequences compared to other primates.

FEATURE INVESTIGATION

Cooper and Colleagues Compared DNA from Extinct Flightless Birds and Existing Species to Propose a New Phylogenetic Tree

Genetic sequence information is primarily used for studying relationships among existing species. Sometimes DNA can be obtained from extinct organisms as well. Starting with small tissue samples from extinct species, scientists have discovered that it is occasionally possible to obtain DNA sequence information. This is called ancient DNA analysis or molecular paleontology. Since the mid-1980s, some researchers have become excited about the information derived from sequencing DNA of extinct specimens. Debate has centered on how long DNA can remain intact after an organism has died. Over time, the structure of DNA is degraded by hydrolysis and the loss of purines. Nevertheless, under certain conditions (cold temperature, low oxygen, and so on), DNA samples may be stable for as long as 50,000–100,000 years. In most studies involving extinct specimens, the ancient DNA is extracted from bone, dried muscle, or preserved skin. In recent years, this approach has been used to study evolutionary relationships between living and extinct species.

As shown in Figure 26.14, Alan Cooper, Cécile Mourer-Chauviré, Geoffrey Chambers, Arndt von Haeseler, Allan Wilson, and Svante Pääbo investigated the evolutionary relationships among some extinct and living species of flightless birds. This is an example of discovery-based science. The researchers gathered data with the goal of investigating the evolutionary relationships among certain species. The kiwis and moas are two groups of flightless birds that existed in New Zealand during the Pleistocene. Species of kiwis still exist, but the moas are now extinct. Eleven known species of moas formerly existed. In this study, the researchers investigated the phylogenetic relationships between four extinct species of moas, which were available as museum samples; three species of New Zealand kiwis; and other living species of flightless birds, including the emu and the cassowary (both found in Australia and/or New Guinea), the ostrich (found in Africa and formerly Asia), and two rheas (found in South America).

Samples from the various species were subjected to PCR to amplify a region of the mitochondrial SSU rRNA gene. This provided enough DNA for sequencing. The data in Figure 26.14 illustrate a comparison of the sequences of a continuous region of the SSU rRNA gene from these species. The first line shows the DNA sequence for one of the four extinct moa species.

Below it are the sequences of several of the other species they analyzed. When the other sequences are identical to the first sequence, a dot is placed in the corresponding position. When the sequences are different, the changed nucleotide base (A, T, G, or C) is placed there. In a few regions, the genes are different lengths. In these cases, a dash is placed to indicate missing nucleotides.

As you can see from the large number of dots, the gene sequences among these flightless birds are very similar, though some differences occur. If you look carefully at the data, you will notice that the sequence from the kiwi (a New Zealand species) is actually more similar to the sequence from the ostrich (an African species) than it is to that of the moa, which was once found in New Zealand. Likewise, the kiwi is more similar to the emu and cassowary (found in Australia and New Guinea) than to the moa. How were these results interpreted? The researchers concluded that the kiwis are more closely related to African and Australian flightless birds than they are to the moas. From these results, they proposed that New Zealand was colonized twice by ancestors of flightless birds. As shown in Figure 26.15, the researchers proposed a new phylogenetic tree that illustrates the revised relationships among these living and extinct species.

Figure 26.14 DNA analysis of phylogenetic relationships among living and extinct flightless birds by Cooper and colleagues.

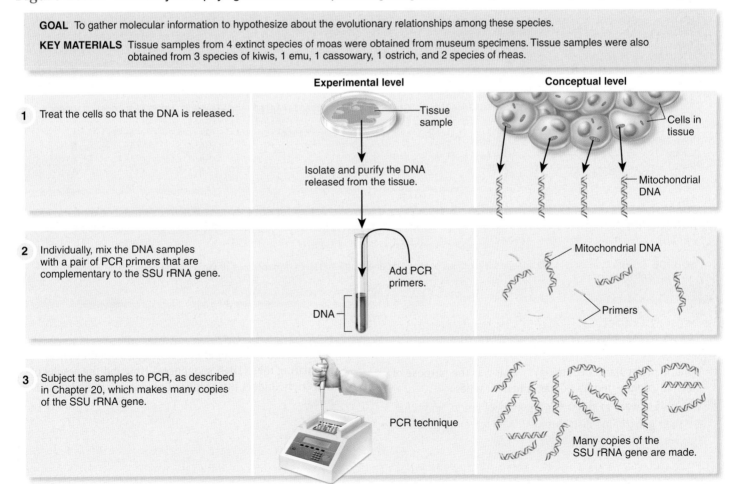

GOAL To gather molecular information to hypothesize about the evolutionary relationships among these species.

KEY MATERIALS Tissue samples from 4 extinct species of moas were obtained from museum specimens. Tissue samples were also obtained from 3 species of kiwis, 1 emu, 1 cassowary, 1 ostrich, and 2 species of rheas.

Experimental level — Conceptual level

1 Treat the cells so that the DNA is released.
Tissue sample — Cells in tissue
Isolate and purify the DNA released from the tissue.
Mitochondrial DNA

2 Individually, mix the DNA samples with a pair of PCR primers that are complementary to the SSU rRNA gene.
Add PCR primers.
DNA
Mitochondrial DNA
Primers

3 Subject the samples to PCR, as described in Chapter 20, which makes many copies of the SSU rRNA gene.
PCR technique
Many copies of the SSU rRNA gene are made.

4 Subject the amplified DNA fragments to DNA sequencing, as described in Chapter 20.

Sequence the amplified DNA.

The amplification of the SSU rRNA gene allows it to be subjected to DNA sequencing.

5 Align the DNA sequences to each other, using computer techniques described in Chapter 21.

Align sequences, using computer programs.

Align sequences to compare the degree of similarity.

6 **THE DATA**

```
Moa 1       GCTTAGCCCTAAATCCAGATACTTACCCTACACAAGTATCCGCCCGAGAACTACGAGCACAAACGCTTAAAACTCTAAGGACTTGGCGGTGCCCCAAACCCA
Kiwi 1      ················T·G····GT··CT····C······································································T·····
Emu         ···············TT····C··T··CAG··C·····T··················································T·····
Cassowary   ···············TT····CG·TA··CTG·······································································T·····
Ostrich     ·······T·····AT·····C··CT···········································································T·····
Rhea 1      ·······T········C··CT·······································································T·····

Moa 1       CCTAGAGGAGCCTGTTCTATAATCGATAATCCACGATACACCCGACCATCCCTCGCCCGT-GCAGCCTACATACCGCCGTCCCCAGCCCGCCT--AATGAAA
Kiwi 1      ················C·····A····T·T··AAC-A···T·······G···T···AA···G·
Emu         ················C·····A····T·T··AA-A·············G··············-·····G·
Cassowary   ················C·····AG···T·T··AA·TA·············G··········--·G··G·
Ostrich     ·················T···A···C···T··A--T············G·············C---··G·
Rhea 1      ················C·················T·T···A···············G···············TA·G·····

Moa 1       G-AACAATAGCGAGCACAACAGCCCTCCCCCGCTAACAAGACAGGTCAAGGTATAGCATATGAGATGGAAGAAATGGGCTACATTTTCTAACATAGAACACC
Kiwi 1      ·-···C···A······TA···-·A··············C·············A····T··T
Emu         ·-········T···AC--TT···········G··························T·T
Cassowary   ·········T···AC--T·············G·························T···
Ostrich     ·-········T···A--·····················GAG··················T·A
Rhea 1      ·-···C··AG···T·T··TA---·················G··············TC······A·

Moa 1       C-------------ACGAAAGAGAAGGTGAAACCCTCCTCAAAAGGCGGATTTAGCAGTAAAATAGAACAAGAATGCCTATTTTAAGCCCGGCCCTGGGGC
Kiwi 1      -·········A·GGT·····T·C··T·G···········C···T···GA·T·········T····A····
Emu         -·········AG·T·····T·AC·T·········C···T···GA·T·····A-·T···T·A····
Cassowary   -·········A·G·T······T·A···T·G·········C···········GA·T·········A----·····A····
Ostrich     -·········G·TA·····T·A···G·········T···GA·T·········-T····T·A····
Rhea 1      -·····G···GGCA·····-AC···CG·········G···G·TC····A···C·C·····--········A····
```

7 **CONCLUSION** This discovery-based investigation led to a hypothesis regarding the evolutionary relationships among these bird species, which is described in Figure 26.15.

8 **SOURCE** Cooper, Alan et al. 1992. Independent origins of New Zealand moas and kiwis. *Proceedings of the National Academy of Sciences* 89:8741–8744.

Experimental Questions

1. What is molecular paleontology? What was the purpose of the study conducted by Cooper and colleagues?

2. What birds were examined in the Cooper study, and what are their geographic distributions? Why were the different species selected for this study?

3. What results did Cooper and colleagues obtain by comparing these DNA sequences? How did the results of this study impact the proposed phylogeny of flightless birds?

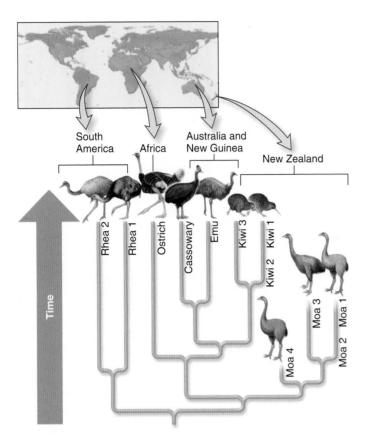

Figure 26.15 **A revised phylogenetic tree of flightless birds.** This tree is based on a comparison of DNA sequences from extinct and living birds as described in Figure 26.14.

Concept check: *With regard to geography, why are the results in this figure surprising?*

26.5 Horizontal Gene Transfer

Thus far, we have considered various ways to propose phylogenetic trees, which describe the relationships between ancestors and their descendents. The type of evolution depicted in previous figures, which involves changes in groups of species due to descent from a common ancestor, is called vertical evolution. Since the time of Darwin, vertical evolution has been the traditional way that biologists view the evolutionary process. However, over the past couple of decades researchers have come to realize that evolution is not so simple. In addition to vertical evolution, horizontal gene transfer has also played a significant role in the phylogeny of living species.

As discussed in Chapter 23, **horizontal gene transfer** is used to describe any process in which an organism incorporates genetic material from another organism without being the offspring of that organism. As discussed next, this phenomenon has reshaped the way that biologists view the evolution of species.

Genomes & Proteomes Connection

Due to Horizontal Gene Transfer, the "Tree of Life" Is Really a "Web of Life"

Horizontal gene transfer has played a major role in the evolution of many species. As discussed in Chapter 18, bacteria can transfer genes via conjugation, transformation, and transduction. The transferred genes may encode proteins that provide a survival advantage such as resistance to antibiotics or the ability to metabolize an organic molecule in the environment. Horizontal gene transfer is also fairly common among certain unicellular eukaryotes. However, its relative frequency and importance in the evolution of multicellular eukaryotes remains difficult to evaluate.

Scientists have debated the role of horizontal gene transfer in the earliest stages of evolution, prior to the emergence of the two prokaryotic domains. The traditional viewpoint was that the three domains of life—Bacteria, Archaea, and Eukarya—arose from a single type of prokaryotic (or pre-prokaryotic) cell called the universal ancestor. However, genomic research has suggested that horizontal gene transfer may have been particularly common during the early stages of evolution on Earth, when all species were unicellular. Horizontal gene transfer may have been so prevalent that the universal ancestor may have actually been an ancestral community of cell lineages that evolved as a whole. If that were the case, the tree of life cannot be traced back to a single prokaryotic ancestor.

Figure 26.16 illustrates a schematic scenario for the evolution of life on Earth that includes the roles of both vertical evolution and horizontal gene transfer. This has been described as a "web of life" rather than a "tree of life." Instead of a universal ancestor, a web of life began with a community of primitive cells that transferred genetic material in a horizontal fashion. Horizontal gene transfer was also prevalent during the early evolution of bacteria and archaea, and when eukaryotes first emerged as unicellular species. In living bacteria and archaea, it remains a prominent way to foster evolutionary change. By comparison, the region of the diagram that contains most eukaryotic species has a more treelike structure. Researchers have speculated that multicellularity and sexual reproduction have presented barriers to horizontal gene transfer in most eukaryotes. For a gene to be transmitted to eukaryotic offspring, it would have to be transferred into a eukaryotic cell that is a gamete or a cell that gives rise to gametes. Horizontal gene transfer has become less common in eukaryotes, though it does occur occasionally.

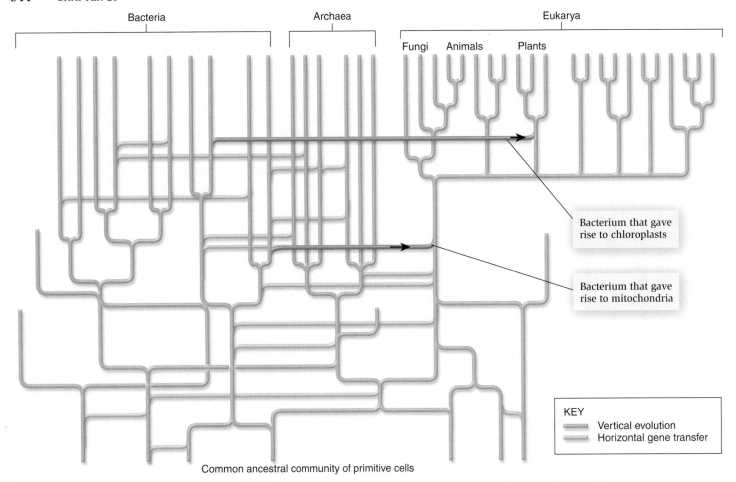

Bacteria

Archaea

Eukarya

Fungi   Animals   Plants

Bacterium that gave
rise to chloroplasts

Bacterium that gave
rise to mitochondria

KEY
Vertical evolution
Horizontal gene transfer

Common ancestral community of primitive cells

Figure 26.16   A web of life.   This phylogenetic tree shows not only the vertical evolution of life on Earth but also the contribution of horizontal gene transfer. In this scenario, horizontal gene transfer was prevalent during the early stages of evolution, when all organisms were unicellular, and continues to be a prominent factor in the speciation of Bacteria and Archaea. Note: This tree is meant to be schematic. Also, while the introduction of chloroplasts into the eukaryotic domain is shown as a single event, such events have occurred multiple times and by different mechanisms, as discussed in Unit V.

Concept check: *How does the phenomenon of horizontal gene transfer muddle the concept of monophyletic groups?*

Summary of Key Concepts

26.1 Taxonomy

- Taxonomy is the field of biology that is concerned with the theory, practice, and rules of classifying living and extinct organisms and viruses.

- Taxonomy places all living organisms into progressively smaller hierarchical groups called taxa. The broadest groups are the three domains, called Bacteria, Archaea, and Eukarya, followed by supergroups, kingdoms, phyla, classes, orders, families, genera, and species. (Figures 26.1, 26.2, Table 26.1)

- Binomial nomenclature provides each species with a scientific name that refers to its genus and species epithet.

26.2 Phylogenetic Trees

- Systematics is the study of biological diversity and evolutionary relationships. The evolutionary history of a species is its phylogeny.

- A phylogenetic tree is a hypothesis that describes the phylogeny of particular species. A monophyletic group or clade includes all of the species that are derived from a common ancestor. (Figure 26.3)

- The hierarchy of taxonomy is related to the process of descent with modification and the timing of common ancestors. Smaller taxa, such as families and genera, are derived from more recent common ancestors than are broader taxa such as kingdoms and phyla. (Figure 26.4)

- Molecular systematics, which involves the analysis of genetic sequences, has led to major revisions in taxonomy. Ideally, all taxa should be monophyletic, consisting of the most recent common ancestor and all of its descendants, though previously established taxa sometimes turn out to be paraphyletic or polyphyletic. (Figures 26.5, 26.6)

- Both morphological and genetic data are used to propose phylogenetic trees. (Figure 26.7)

26.3 Cladistics

- In the cladistic approach to proposing a phylogenetic tree, also called a cladogram, species are grouped together according to shared derived characters. (Figure 26.8)

- An ingroup is the group of interest, whereas an outgroup is a species or group of species that lacks one or more shared derived characters (synapomorphies). A comparison of the ingroup and outgroup is used to determine which character states are derived or primitive. (Figures 26.9, 26.10)

- The cladistic approach can produce many possible cladograms. The most likely phylogenetic tree is chosen by a variety of methods, including the analysis of fossils, the principle of parsimony, maximum likelihood, and Bayesian methods. (Figure 26.11)

26.4 Molecular Clocks

- Assuming that neutral mutations occur at a relatively constant rate, genetic data can act as a molecular clock on which to measure evolutionary time. (Figure 26.12)

- Slowly changing genes are used to analyze broader taxa, whereas rapidly changing genes are used to analyze more closely related species such as eukaryotes with long generation times. (Figure 26.13)

- Cooper and colleagues analyzed DNA sequences from extinct and living flightless birds and proposed a phylogeny in which New Zealand was colonized twice, once by moas and later by kiwis. (Figures 26.14, 26.15)

26.5 Horizontal Gene Transfer

- Due to horizontal gene transfer, the tree of life may more accurately be described as a web of life. (Figure 26.16)

Assess and Discuss

Test Yourself

1. The study of biological diversity based on evolutionary relationships is
   a. paleontology.   c. systematics.   e. both a and b.
   b. evolution.   d. ontogeny.

2. Which of the following is the correct order of the taxa used to classify organisms?
   a. kingdom, domain, phylum, class, order, family, genus, species
   b. domain, kingdom, class, phylum, order, family, genus, species
   c. domain, kingdom, phylum, class, family, order, genus, species
   d. domain, kingdom, phylum, class, order, family, genus, species
   e. kingdom, domain, phylum, order, class, family, species, genus

3. When considering organisms within the same taxon, which level includes organisms with the greatest similarity?
   a. kingdom   c. order   e. genus
   b. class   d. family

4. Which of the following characteristics is not shared by prokaryotes and eukaryotes?
   a. DNA is the genetic material.
   b. Messenger RNA encodes the information to produce proteins.
   c. All cells are surrounded by a plasma membrane.
   d. The cytoplasm is compartmentalized into organelles.
   e. both a and d

5. The branch points or nodes in a phylogenetic tree depict which of the following?
   a. anagenesis   d. a and b only
   b. cladogenesis   e. b and c only
   c. horizontal gene transfer

6. The evolutionary history of a species is its
   a. ontogeny.   c. evolution.   e. embryology.
   b. taxonomy.   d. phylogeny.

7. A taxon composed of all species derived from a common ancestor is referred to as
   a. a phylum.
   b. a monophyletic group or clade.
   c. a genus.
   d. an outgroup.
   e. all of the above.

8. The goal of modern taxonomy is to
   a. classify all organisms based on morphological similarities.
   b. classify all organisms in monophyletic groups.
   c. classify all organisms based solely on genetic similarities.
   d. determine the evolutionary relationships between similar species.
   e. none of the above

9. The concept that the preferred hypothesis is the one that is the simplest is
   a. phenetics.   d. maximum likelihood.
   b. cladistics.   e. both b and d.
   c. the principle of parsimony.

10. Research indicates that horizontal gene transfer is less prevalent in eukaryotes because of
    a. the presence of organelles.   d. all of the above.
    b. multicellularity.   e. b and c only.
    c. sexual reproduction.

Conceptual Questions

1. Explain how species' names follow a binomial nomenclature. Give an example.

2. What are some advantages and potential pitfalls of morphological analysis with regard to phylogenetic trees?

3. What is a molecular clock? How is it used in depicting phylogenetic trees?

Collaborative Questions

1. Discuss how taxonomy is useful. Make a list of some practical applications that are derived from taxonomy.

2. Discuss systematics and how it is used to propose a phylogenetic tree. What are some pros and cons of the principle of parsimony, maximum likelihood, and Bayesian methods for proposing the correct tree?

Online Resource

www.brookerbiology.com

Stay a step ahead in your studies with animations that bring concepts to life and practice tests to assess your understanding. Your instructor may also recommend the interactive ebook, individualized learning tools, and more.

Bacteria and Archaea

27

Cyanobacterial bloom. A visible cyanobacterial bloom gives a pea-soup appearance to this lake water.

A mong Earth's life-forms, the domains Bacteria and Archaea are distinct from domain Eukarya in several ways. These organisms have the simplest cell structure and include the smallest known cells. Bacteria and archaea are also the most abundant organisms on Earth. About half of Earth's total biomass consists of an estimated 10^{30} individuals; just a pinch of garden soil can contain 2 billion prokaryotic cells, and about a million occur in 1 ml of seawater. Bacteria and archaea live in nearly every conceivable habitat, including extremely hot or salty waters that support no other life, and they are also Earth's most ancient organisms, having originated more than 3.5 billion years ago. Their great age and varied habitats have resulted in very high diversity. Today, many millions of species of bacteria and archaea collectively display more diverse metabolic processes than occur in any other group of organisms. Many of these metabolic processes are important on a global scale, influencing Earth's climate, atmosphere, soils, water quality, and human health and technology. In this chapter, we will survey the diversity, structure, reproduction, metabolism, and ecology of bacteria and archaea. This survey will illustrate major principles of diversity, including descent with modification and horizontal gene transfer.

27.1 Diversity and Evolution

As we have noted, one of the prominent features of bacteria and archaea is their astounding diversity. In the past, microbiologists studied diversity by isolating these organisms from nature and growing cultures in the laboratory. Such cultures allowed microbiologists to observe variation in cell structure and metabolism, major features used to classify bacteria and archaea. Today, microbiologists also use molecular techniques to detect diverse bacteria and archaea in nature. By using these new techniques, they have discovered that bacteria and archaea are vastly more diverse than previously realized, and many new species have been discovered.

Though much remains to be learned about the diversity of Earth's microorganisms, extensive molecular analysis has supported the concept that prokaryotic microbes can be classified into two major domains of life: the **Archaea** and **Bacteria** (also called Eubacteria) (**Figure 27.1**). The terms prokaryote and prokaryotic are often used to refer to archaeal and bacterial cells because these organisms lack nuclei with porous envelopes and other cellular features typical of eukaryotes (see Chapter 4). As we noted in Chapter 26, in the 1970s, microbiologist Carl Woese and associates proposed splitting the kingdom Monera, which had included all prokaryotes, into these two domains, based on comparisons of ribosomal RNA sequences from diverse microorganisms. In this section, we will first survey the major kingdoms and phyla of the domains Archaea and Bacteria and then explore how horizontal gene transfer—the transfer of genes between different species—has influenced their evolution.

Domain Archaea Includes Inhabitants of Extremely Harsh Environments

Organisms classified in the domain Archaea, commonly known as archaea, display some unique characteristics. First, archaea possess a number of features in common with the eukaryotic nucleus and cytoplasm, suggesting common ancestry. For example, histone proteins are typically associated with

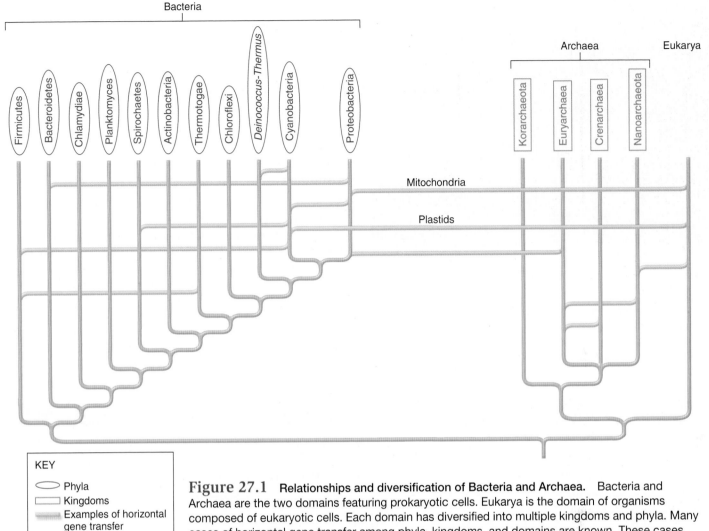

Figure 27.1 **Relationships and diversification of Bacteria and Archaea.** Bacteria and Archaea are the two domains featuring prokaryotic cells. Eukarya is the domain of organisms composed of eukaryotic cells. Each domain has diversified into multiple kingdoms and phyla. Many cases of horizontal gene transfer among phyla, kingdoms, and domains are known. These cases include the acquisition of mitochondria and plastids by eukaryotes. Note: The domain Bacteria includes about 50 phyla, but for simplicity, only 11 are shown in this figure.

the DNA of both archaea and eukaryotes, but they are absent from most bacteria. A distinctive feature of archaea is their membrane lipids, which are formed with ether linkages (in contrast, ester linkages characterize the membrane lipids of bacteria and eukaryotes). Ether-linked membranes are resistant to damage by heat and other extreme conditions, which helps explain why many archaea are able to grow in extremely harsh environments.

Though many archaea occur in soils and surface ocean waters of moderate conditions, diverse archaea occupy habitats of very high salt content, acidity, methane levels, or temperatures that would kill most bacteria and eukaryotes. Organisms that occur primarily in extreme habitats are known as **extremophiles**. One example is the methane producer *Methanopyrus*, which grows best at deep-sea thermal vent sites where the temperature is 98°C. At this temperature, the proteins of most organisms would denature, but those of *Methanopyrus* are resistant to such damage. *Methanopyrus* is so closely adapted to its extremely hot environment that it will not grow when the temperature is less than 84°C. Such archaea are known as **hyperthermophiles**. Some archaea prefer habitats having both

high temperatures and extremely low pH. For example, microbiologist Thomas Brock discovered the archaeal genus *Sulfolobus* in samples taken from sulfur hot springs having a pH of 3 or lower.

Extreme **halophiles** (from the Greek, meaning salt lovers) occupy evaporation ponds used to produce salt from seawater, often growing so abundantly that they color the ponds red (**Figure 27.2**). Halophiles are often red because their plasma membranes contain large amounts of rhodopsins, proteins combined with the light-sensitive red pigment known as retinal. (Similar rhodopsins play important roles in eukaryote light sensing, and they are essential for animal vision.) In bacteria and archaea, rhodopsin functions as a proton pump, a protein that can move protons and other ions across the plasma membrane (see Chapter 5).

The domain Archaea includes several kingdoms: Korarchaeota, Euryarchaeota, Crenarchaeota, and Nanoarchaeota. Korarchaeota is primarily known from DNA sequences found in samples from hot springs. Euryarchaeota includes methane producers, extreme halophiles, and some hyperthermophiles. Crenarchaeota includes organisms that grow in extremely hot

Figure 27.2 Hypersaline waters colored red by numerous halophilic archaea.

Concept check: What material explains the red color of the water in this figure?

or cold habitats and also some that are widespread in aquatic and terrestrial habitats. Nanoarchaeota includes the hyperthermophile *Nanoarchaeum equitans*, which appears to be a parasite of the thermal vent crenarchaeote *Ignicoccus*. Molecular biologist Elizabeth Waters and associates sequenced the exceptionally small genome of *N. equitans* and determined that it represents an early branching archaeal lineage.

Domain Bacteria Includes Proteobacteria, Cyanobacteria, and Many Other Phyla

Molecular studies suggest the existence of 50 or so bacterial phyla. However, the structural and metabolic features of about half of these are unknown. Though some members of domain Bacteria live in extreme environments, many more favor moderate conditions. Many bacteria form symbiotic associations with eukaryotes and are thus of concern in medicine and agriculture. The characteristics of 11 prominent bacterial phyla are briefly summarized in **Table 27.1**. Among these, the Proteobacteria and the Cyanobacteria are particularly diverse and relevant to eukaryotic cell evolution, global ecology, and human affairs.

Table 27.1	Representative Bacterial Phyla
Phyla	**Characteristics**
Firmicutes	Diverse Gram-positive bacteria, some of which produce endospores.
Bacteroidetes	Includes representatives of diverse metabolism types; some are common in the human intestinal tract, and others are primarily aquatic.
Chlamydiae	Notably tiny, obligate intracellular parasites. Some cause eye disease in newborns or sexually transmitted diseases.
Planctomycetes	Reproduce by budding; cell walls lack peptidoglycan; cytoplasm contains nucleus-like bodies.
Spirochaetes	Motile bacteria having distinctive corkscrew shapes, with flagella held close to the body. They include the pathogens *Treponema pallidum*, the agent of syphilis, and *Borrelia burgdorferi*, which causes Lyme disease.
Actinobacteria	Gram-positive bacteria producing branched filaments; many form spores. *Mycobacterium tuberculosis*, the agent of tuberculosis in humans, is an example. Actinobacteria are notable antibiotic producers; over 500 different antibiotics are known from this group. The pharmaceutical industry produces antibiotics from large-scale cultures of the actinobacterium *Streptomyces*. Some fix nitrogen in association with plants.
Thermotogae	Hyperthermophiles.
Chloroflexi	Known as the green nonsulfur bacteria; conduct photosynthesis without releasing oxygen (anoxygenic photosynthesis).
Deinococcus-Thermus	Extremophiles. The genus *Deinococcus* is known for high resistance to ionizing radiation, and the genus *Thermus* inhabits hot springs. *Thermus aquaticus* has been used in commercial production of Taq polymerase enzyme used in the polymerase chain reaction (PCR), an important procedure in molecular biology laboratories.
Cyanobacteria	The oxygen-producing photosynthetic bacteria (some are also capable of anoxygenic photosynthesis). Photosynthetic pigments include chlorophyll *a* and phycobilins, which often give cells a blue-green pigmentation. Occur as unicells, colonies, unbranched filaments, and branched filaments. Many of the filamentous species produce specialized cells: dormant akinetes and heterocysts (heterocytes) in which nitrogen fixation occurs. In waters having excess nutrients, cyanobacteria produce blooms and may release toxins harmful to the health of humans and wild and domesticated animals.
Proteobacteria	A very large group of Gram-negative bacteria, collectively having high metabolic diversity. Includes many species important in medicine, agriculture, and industry. *Myxococcus xanthus* is a Gram-negative bacterium that is able to glide across surfaces, forming swarms of thousands of cells. This behavior aids feeding by concentrating digestive enzymes secreted by the bacteria. When food is scarce, the swarms form tree-shaped structures from which tough spores disperse. By this means, cells move to new, food-rich places.

Proteobacteria Though Proteobacteria share molecular and cell-wall features, this phylum displays amazing diversity of form and metabolism. Genera of this phylum are classified into five major subgroups: alpha (α), beta (β), gamma (γ), delta (δ), and epsilon (ε). As we saw in Chapter 22, the ancestry of mitochondria can be traced to the α-proteobacteria, which also include several genera noted for mutually beneficial relationships with animals and plants. For example, *Rhizobium* and related genera of α-proteobacteria form nutritionally beneficial associations with the roots of legume plants such as beans and peas and are thus agriculturally important (see Chapter 37). Another α-proteobacterium, *Agrobacterium tumifaciens*, causes destructive cancer-like galls to develop on susceptible plants, including grapes and ornamental crops (**Figure 27.3**). *Agrobacterium tumifaciens* induces gall formation by injecting DNA into plant cells, a property that has led to the use of the bacterium in the production of transgenic plants (see Chapter 20).

The genus *Nitrosomonas*, a soil inhabitant important in the global nitrogen cycle, represents the β-proteobacteria. *Neisseria gonorrhoeae*, the agent of the sexually transmitted disease gonorrhea, is a member of the γ-proteobacteria. *Vibrio cholerae*, another γ-proteobacterium, causes cholera epidemics when drinking water becomes contaminated with animal waste during floods and other natural disasters. The γ-proteobacteria *Salmonella enterica* and *Escherichia coli* strain O157:H7 also cause human disease, and food and water are widely tested for their presence. The δ-proteobacteria include the colony-forming myxobacteria and predatory bdellovibrios, which drill their way through the cell walls of other bacteria in order to consume them. *Helicobacter pylori*, which causes stomach ulcers, belongs to the ε-proteobacteria.

Cyanobacteria The phylum Cyanobacteria contains photosynthetic bacteria that are abundant in fresh waters, oceans, and wetlands and on the surfaces of arid soils. Cyanobacteria are named for the typical blue-green (cyan) coloration of their cells. Blue-green pigmentation results from the presence of accessory phycobilin pigments that help chlorophyll absorb light energy. Cyanobacteria are the only prokaryotes that generate oxygen as a product of photosynthesis. Ancient cyanobacteria produced Earth's first oxygen-rich atmosphere, which allowed the rise of eukaryotes. The chloroplasts of eukaryotic algae and plants arose from cyanobacteria.

Cyanobacteria display the greatest structural diversity found among bacterial phyla (**Figure 27.4**). Some occur as single cells called unicells (Figure 27.4a); others form colonies of cells held together by a thick gluey substance called mucilage (Figure 27.4b). Many cyanobacteria form filaments of cells that are attached end-to-end (Figure 27.4c,d). Some of the filamentous cyanobacteria produce specialized cells and display intercellular chemical communication, the hallmarks of multicellular organisms. Many cyanobacteria that grow in conditions of high light intensity produce protective brown sunscreen compounds at their surfaces (Figure 27.4d).

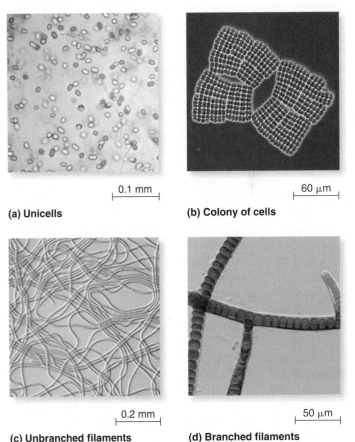

Figure 27.4 Major types of microbial cell aggregations found in the phylum Cyanobacteria. (a) The genus *Chroococcus* occurs as unicells. (b) The genus *Merismopedia* is a flat colony of cells held together by mucilage. (c) The genus *Oscillatoria* is an unbranched filament. (d) The genus *Stigonema* is a branched filament having a mucilage sheath; the brown color is caused by sunscreen compounds that protect the cells from damage by ultraviolet radiation.

Figure 27.3 *Agrobacterium tumifaciens* infection. This bacterium causes cancer-like tumors to grow on plants.

Cyanobacteria play essential ecological roles by producing organic carbon and fixed nitrogen (see Section 27.4). However, several kinds of cyanobacteria, notably the genera *Microcystis*, *Anabaena*, and *Cylindrospermopsis*, form nuisance growths in freshwater lakes during the warm season. Such growths, known as blooms, give the water a pea-soup appearance (see chapter-opening photo). Blooms develop when natural waters receive excess fertilizer from sewage discharges or agricultural runoff. Such blooms are becoming more common every year and are of serious concern because they may produce toxins in amounts sufficient to harm the health of humans and other animals. Consequently, it would be best for people or pets not to swim in or consume water that has a visible cyanobacterial bloom.

Now that we have learned something about the diversity of Archaea and Bacteria, we consider the evolutionary effects of gene exchanges within and between these domains.

Horizontal Gene Transfer Influences the Evolution of Bacteria and Archaea

Horizontal gene transfer, also known as lateral gene transfer, is the process in which an organism receives genetic material from another organism without being the offspring of that organism. This process contrasts with vertical gene transfer from parent to progeny and is increasingly recognized as an important evolutionary mechanism. Horizontal gene transfer affects the evolutionary process by increasing genetic diversity.

Horizontal gene transfer is common among bacteria and archaea, and it can result in large genetic changes. For example, at least 17% of the genes present in the common human gut inhabitant *E. coli* came from other bacteria. Study of nearly 200 genomes has revealed that about 80% of prokaryotic genes have been involved in horizontal transfer at some point in their history. Genes also move among the bacterial, archaeal, and eukaryotic domains. For example, about a third of the genes present in the archaeon *Methanosarcina mazei* originally came from bacteria. Viruses are probably important gene transfer vectors.

Phylogenetic trees displaying our current understanding of relationships (see Figure 27.1) reveal several important concepts. Bacteria and Archaea probably evolved from a common ancestor, and the eukaryotic nucleus and cytoplasm likely arose in an ancient archaeal organism. In addition, mitochondria and plastids originated from proteobacteria and cyanobacteria by endosymbiosis (see Chapter 22). In these cases, endosymbiosis resulted in the transfer of many genes from bacteria to eukaryotes. The next section surveys ways in which bacteria and archaea vary in structure and locomotion.

27.2 Structure and Motility

Bacteria and archaea have several features in common, including small size, rapid growth, and simpler cellular structure than eukaryotes. Some basic features of prokaryotic cell structure are described in Chapter 4 (refer back to Figure 4.4). With few exceptions, bacteria and archaea are 1–5 μm in diameter and

Thylakoids provide a greater surface area for chlorophyll and other molecules involved in photosynthesis.

Thylakoids

Food storage particle

Gas vesicles (cross sections)

Gas vesicles (long sections)

0.6 μm

The gas vesicles buoy this photosynthetic organism to the lighted water surface, where it often forms conspicuous scums.

Figure 27.5 Photosynthetic thylakoid membranes and numerous gas vesicles found in a cell of the cyanobacterial genus *Microcystis*.

Concept check: Do you think this cell would tend to float or sink?

are thus known as microorganisms or microbes. (By contrast, most plant and animal cells are between 10 and 100 μm in diameter.) Small cell size limits the amount of materials that can be stored within cells but allows faster cell division. When nutrients are sufficient, many microorganisms can divide many times within a single day. This explains how bacteria can spoil food rapidly and why bacterial infections can spread quickly within the human body. Despite these common features, prokaryotes differ in many ways. In this section, we explore variation in cell structure and shape, in surface and cell-wall features, and in movement.

Bacteria and Archaea Have More Complex Cellular Structure Than Previously Thought

Prokaryotic cells are much simpler than eukaryotic cells. Even so, many prokaryotes display a surprising complexity of cellular structures resulting from adaptive evolution. For example, cyanobacteria and other photosynthetic bacteria are able to use light energy to produce organic compounds and typically contain large numbers of intracellular tubules known as thylakoids (**Figure 27.5**). The extensive membrane surface of the thylakoids bears large amounts of chlorophyll and other components of the photosynthetic apparatus. Thylakoids therefore enable photosynthetic bacteria and chloroplasts that evolved from them to take maximum advantage of light energy in their environments.

Cyanobacteria and some other bacteria that live in aquatic habitats use cytoplasmic structures known as **gas vesicles** to adjust their buoyancy. This process allows them to move up or down in the water, an advantage in finding nutrients and avoiding harmful conditions. Gas vesicles are hollow cylinders whose water-tight surfaces are made of protein (Figure 27.5). These vesicles do not actually contain gas concentrations different from the rest of the cell, but their density is lower than that of the surrounding cytoplasm, helping bacteria to float. Gas vesicles can repeatedly collapse and reassemble, depending on the cell's solute content. When sugars and other solutes are abundant in the cytoplasm, the high osmotic pressure collapses the vesicles, and heavy particles of stored food are produced within cells. Cells in which the gas vesicles have collapsed and which contain heavy food particles tend to sink. Sinking removes bacteria from damaging radiation at the water's surface and allows them to harvest inorganic minerals from deeper water. When organic solutes and food storages are used up during cell metabolism, the gas vesicles reassemble, providing flotation to the surface and allowing photosynthesis to resume. The presence of gas vesicles explains why cyanobacteria often form buoyant scums at the surfaces of ponds and lakes (see chapter-opening photo).

Earlier described, thylakoids develop in some bacterial cells by ingrowth of the plasma membrane. In other bacteria, plasma membrane ingrowth has generated additional intriguing adaptations—magnetosomes and nucleus-like bodies—that are sometimes described as bacterial organelles. Magnetosomes are tiny crystals of an iron mineral known as magnetite, each surrounded by a membrane. These structures occur in the bacterium *Magnetospirillum* and related genera (**Figure 27.6**). In each cell, about 15 to 20 magnetosomes occur in a row, together acting as a compass needle that responds to the Earth's magnetic field. This helps the bacteria to orient themselves in space and thereby locate the submerged, low-oxygen habitats they prefer. Microbiologists Arash Komeili, Grant Jensen, and their colleagues used rapid freezing techniques and a special type of transmission electron microscope to observe the development of magnetosomes. They found that the process begins with ingrowth of the plasma membrane to form a row of spherical vesicles. If *Magnetospirillum* cells are grown in media having low iron levels, the vesicles remain empty. But if iron is available, a magnetite crystal forms within each vesicle. The investigators also observed that fibrils of an actin-like protein known as MamK keep the magnetosomes aligned in a row.

Figure 27.6 **Magnetosomes found in the spirillum** *Magnetospirillum magnetotacticum.* An internal row of iron-rich magnetite crystals, each enclosed by a membrane, functions like a compass needle, allowing this bacterium to detect the Earth's magnetic field. This feature allows *M. magnetotacticum* to orient itself in space and thereby locate its preferred habitat, low-oxygen subsurface waters. Cells use their flagellum to move from less-favorable to more-attractive locations.

(Recall from Chapter 4 that actin is a major cytoskeletal protein of eukaryotes.) Mutant bacteria lacking a functional form of this protein produce magnetosomes, but they do not remain aligned in a row. Instead, magnetosomes scatter around mutant cells, disrupting their ability to detect a magnetic field.

Plasma membrane invaginations also produce nucleus-like bodies in *Gemmata obscuriglobus* and other members of the bacterial phylum Planktomycetes. In *G. obscuriglobus*, an envelope composed of a double membrane encloses all cellular DNA and some ribosomes. Although this bacterial envelope lacks the nuclear pores characteristic of the eukaryotic nuclear envelope, it likely plays a similar adaptive role in isolating DNA from other cellular influences.

Bacteria and Archaea Vary in Cell Shape and Arrangement

Microbial cells occur in five major shapes (**Figure 27.7**): spheres (**cocci**), rods (**bacilli**), comma-shaped cells (**vibrios**), and spiral-shaped cells that are either flexible (**spirochaetes**) or rigid

1 μm

(a) Sphere-shaped cocci (***Lactococcus lactis***)

11.4 μm

(b) Rod-shaped bacilli (***Lactobacillus plantarum***)

15 μm

(c) Comma-shaped vibrios (***Vibrio cholerae***)

7.5 μm

(d) Spiral-shaped spirochaetes (***Leptospira jaundice***)

Figure 27.7 **Major types of microbial cell shapes.** Scanning electron microscopic views.

(**spirilli**; see Figure 27.6). Cytoskeletal proteins similar to those present in eukaryotic cells control these cell shapes. For example, helical strands of the actin-like protein known as MreB are responsible for the rod shape of bacilli. If MreB is not produced, bacilli become spherical in shape. In addition, these differently shaped cells can occur in several types of arrangements. Some microorganisms occur only as single cells or pairs of cells resulting from recent division (Figure 27.7a). Others occur as cell aggregates or are attached end-to-end to form filaments.

Slimy Mucilage Often Coats Microbial Surfaces

Many microorganisms produce a coat of slimy mucilage, sometimes called a glycocalyx, that is found outside the cell wall and coats the cell and nearby surfaces. Mucilage, which varies in consistency and thickness, is composed of polysaccharides, protein, or both. It is secreted from cells and serves a variety of functions. One example is the mucilage coat (known as a capsule) that helps some disease bacteria evade the defense system of their host. You may recall that Frederick Griffith discovered the transfer of genetic material while experimenting with capsule-producing pathogenic strains and capsuleless non-pathogenic strains of the bacterium *Streptococcus pneumoniae* (refer back to Figure 11.1). The immune system cells of mice are able to destroy this bacterium only if it lacks a capsule.

Mucilage also holds together the cells of colonial microorganisms (see Figure 27.4b), helps aquatic species to float in water, binds mineral nutrients, and repels predators. Slime sheaths (see Figure 27.4d) often coat bacterial filaments, where they may help to prevent drying.

Mucilage is also critical to the formation of biofilms, which are environmentally and medically important. **Biofilms** are aggregations of microorganisms that secrete adhesive mucilage, thereby gluing themselves to surfaces. They help microbes to remain in favorable locations for growth; otherwise body or environmental fluids would wash them away. A process known as **quorum sensing** fosters biofilm formation. During quorum sensing, individual microbes secrete small molecules having the potential to influence the behavior of nearby microbes. If enough individuals are present (a quorum), the concentration of signaling molecules builds to a level that causes collective behavior. In the case of biofilms, populations of microbes respond to chemical signals by moving to a common location and producing mucilage.

From a human standpoint, biofilms have both beneficial and harmful consequences. In aquatic and terrestrial environments, they help to stabilize and enrich sand and soil surfaces. Microbial biofilms can also be important in the formation of mineral deposits. For example, in 2006, microbiologist Frank Reith and colleagues reported that biofilms formed by a bacterium that precipitates gold from solution help to form gold grains and nuggets. Biofilms that form on the surfaces of animal tissues, however, can be harmful. Dental plaque is an example of a harmful biofilm (**Figure 27.8**); if allowed to remain, the bacterial community secretes acids that can damage tooth enamel. Biofilms may also develop in industrial pipelines, where the

Figure 27.8 **A biofilm composed of a community of microorganisms glued by mucilage to a surface.** This SEM shows dental plaque.

attached microbes can contribute to corrosion by secreting enzymes that chemically degrade metal surfaces.

Bacteria and Archaea Differ in Cell-Wall Structure

Whether coated with mucilage or not, most prokaryotes possess a rigid cell wall outside the plasma membrane. Cell walls maintain cell shape and help protect against attack by viruses or predatory bacteria. Cell walls also help microbes avoid lysing in hypotonic conditions, when the solute concentration is higher inside the cell than outside. The structure and composition of cell walls vary in ways that can be important in medical and other contexts.

The cell walls of most archaea and certain bacteria are composed of protein or glycoprotein. In contrast, the polymer known as **peptidoglycan** is lacking from archaea but is an important component of most bacterial cell walls. Peptidoglycan is composed of carbohydrates that are cross-linked by peptides. Bacterial cell walls occur in two major forms that differ in their amount of peptidoglycan, staining properties, and response to antibiotics. These are called Gram-positive and Gram-negative walls, for the Gram stain used to identify them. The Gram-positive cell wall features a relatively thick peptidoglycan layer (**Figure 27.9a**). The Gram-negative cell wall has relatively less wall peptidoglycan and is enclosed by a thin, outer envelope whose outer leaflet is rich in **lipopolysaccharides** (**Figure 27.9b**). This outer layer envelope is a lipid bilayer but is distinct from the plasma membrane.

Lipopolysaccharides and peptidoglycan affect bacterial responses to antibiotics and sometimes disease symptoms. For example, part of the peptidoglycan covering of *Bordetella*

Acidic polysaccharides

Thick peptidoglycan layer

Plasma membrane

(a) Gram-positive: thick peptidoglycan layer, no outer envelope

Lipopolysaccharide-rich outer envelope

Thin peptidoglycan layer

Plasma membrane

(b) Gram-negative: thinner peptidoglycan layer, with outer envelope

Figure 27.9 Cell-wall structure of Gram-positive and Gram-negative bacteria. (a) The structure of the cell wall of Gram-positive bacteria. (b) The structure of the cell wall and lipopolysaccharide envelope typical of Gram-negative bacteria.

pertussis is responsible for the extensive tissue damage associated with whooping cough. Microbiologists use the Gram stain to identify these two major types of bacterial cell walls.

The Gram Stain In the late 1800s, the Danish physician Hans Christian Gram developed the **Gram stain** procedure to more easily detect and distinguish bacteria. The Gram stain remains a useful tool to identify bacteria and predict their responses to antibiotics. To perform a Gram stain, a microbiologist starts by smearing bacteria onto a glass slide and heating it briefly to aid cell attachment. The microbiologist then floods the slide with crystal violet, a purple dye, followed by an iodine solution. The iodine binds the purple dye, forming an insoluble complex. Next, the microbiologist adds alcohol. The alcohol is able to

(a) Gram-positive bacteria   **(b) Gram-negative bacteria**

Figure 27.10 Gram-positive and Gram-negative bacteria. (a) *Streptococcus pneumoniae* stains positive (purple) with the Gram stain. (b) *Escherichia coli* stains negative (pink) when the Gram stain procedure is applied.

Figure 27.11 A commercial fluorescent stain used to distinguish Gram-positive and Gram-negative bacteria. This preparation shows green-fluorescent cells of the Gram-negative bacterium *E. coli* mixed with yellow-fluorescent cells of the Gram-positive bacterium *Staphylococcus aureus*.

Concept check: *For ecological studies, what would be the advantages and disadvantages of this fluorescence staining procedure as compared to the classical Gram stain process?*

remove the purple dye complex from thin peptidoglycan walls, but not dye bound in cell walls with thick peptidoglycan layers. Finally, safranin, a pink stain, is applied. At the end of the procedure, some bacteria will remain purple; these are known as Gram-positive bacteria (**Figure 27.10a**). Other types of bacteria will lose the purple stain at the alcohol step but retain the final pink stain; these are known as Gram-negative bacteria (**Figure 27.10b**). If a fluorescence microscope is available, a single-step fluorescent stain can be used to distinguish Gram-positive from Gram-negative bacteria (**Figure 27.11**). A closer look at Gram-positive and Gram-negative bacteria will reveal how these staining differences are useful.

Gram-Positive Bacteria Gram-positive bacteria occur in the phyla Firmicutes and Actinobacteria (see Figure 27.1 and

Table 27.1). Gram-positive bacteria typically have thick peptidoglycan cell walls lacking a lipopolysaccharide envelope. Such bacteria are typically vulnerable to penicillin and related antibiotics, because these antibiotics interfere with peptidoglycan synthesis. An example of a Gram-positive bacterium is *S. pneumoniae* (see Figure 27.10a). Streptococci of various types cause strep throat, streptococcal pneumonia, streptococcal meningitis (an infection of the spinal fluid), and eye infections. Streptococci also include the infamous "flesh-eating" bacteria that cause necrotizing fasciitis, a disease whose progression is notoriously difficult to control. The entire DNA sequence of *S. pneumoniae* has been determined, revealing the presence of many genes conferring the ability to break down several types of molecules in human tissues for use as food. The presence of these genes explains why streptococci can cause such varying disease symptoms. *Staphylococcus aureus* is another medically important Gram-positive bacterium, which is commonly found on human skin.

Gram-Negative Bacteria Gram-negative bacteria have thin peptidoglycan cell walls enclosed by a lipopolysaccharide envelope. Diverse phyla of bacteria display Gram-negative staining. *E. coli* is an example of a Gram-negative bacterium that normally lives in the human lower intestine (see Figure 27.10b). The outer lipopolysaccharide envelope enables Gram-negative bacteria to resist the effects of penicillin and chemically similar antibiotics. (Infections caused by Gram-negative bacteria would be treated with different antibiotics.) However, the outer envelope impedes the secretion of proteins from bacterial cells into the environment, a process that allows cells to communicate with each other and plays other roles. Gram-negative bacteria have adapted by evolving five or more types of protein systems that function in secretion, known as types I–V secretion systems.

Microbes Display Diverse Types of Motility Structures

Many microorganisms have structures at the cell surface or within cells that enable them to change position in their environment. Motility allows microbes to respond to chemical signals emitted from other cells during quorum sensing and mating and to move to favorable conditions within gradients of light, gases, or nutrients. For example, we have already learned that cyanobacteria use gas buoyancy vesicles for motility (see Figure 27.5). In addition, microbes move by twitching, gliding, or swimming by means of flagella.

Prokaryotic **flagella** (singular, flagellum) differ from eukaryotic flagella in several ways. For example, prokaryotic flagella lack an internal cytoskeleton of microtubules, the motor protein dynein, and a plasma membrane covering, all features that characterize eukaryotic flagella (see Chapter 4). Unlike eukaryotic flagella, prokaryotic flagella do not repeatedly bend and straighten. Instead, prokaryotic flagella spin, propelled by molecular machines composed of a filament, hook, and motor that work together somewhat like a boat's outboard motor and

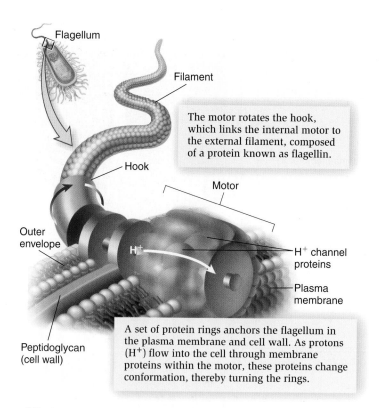

Figure 27.12 Diagram of a prokaryotic flagellum, showing a filament, hook, and motor.

Concept check: Does the filament move more like the arms of a human swimmer or the shaft of a boat propeller?

propeller (**Figure 27.12**). Lying outside the cell, the long, stiff, curved filament, which is composed of the protein flagellin, acts as a propeller. The hook links the filament with the motor that contains a set of protein rings at the cell surface. Hydrogen ions (protons), which have been pumped out of the cytoplasm, usually via the electron transport system, diffuse back into the cell through channel proteins within the motor. This proton movement powers the turning of the hook and filament.

Prokaryotic species differ in the number and location of flagella, which may occur singly or in clumps at one pole of a bacterial cell or may emerge from around the cell (**Figure 27.13**). Differences in flagellar number and location cause microorganisms to exhibit different modes of swimming. For example, spirochaete flagella are located outside the peptidoglycan cell wall but within the confines of an outer membrane that holds them close to the cell. Rotation of these flagella causes spirochaetes to display characteristic bending, flexing, and twirling motions.

Some prokaryotes twitch or glide across surfaces, using threadlike cell surface structures known as **pili** (singular, pilus) (**Figure 27.14**). *Myxococcus xanthus* cells, for example, move by alternately extending and retracting pili from one pole or the other. This process allows directional movement toward food materials. If nutrients are low, cells of these bacteria glide together to form tiny treelike colonies, which are part of a reproductive process. As we will see, pili can also play important roles in bacterial reproduction and disease processes.

(a) Bacteria with a single short flagellum

(b) Bacterium with multiple long flagella

Figure 27.13 **Differences in the number and location of flagella.** Depending on the species, microbial cells can produce one or more flagella at the poles or numerous flagella around the periphery. **(a)** *Vibrio parahaemoliticus*, a bacterium that causes seafood poisoning, has a single short flagellum. **(b)** *Salmonella enterica*, another bacterium that causes food poisoning, has many flagella distributed around the cell periphery.

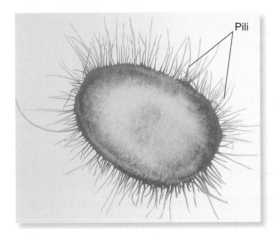

Figure 27.14 **Pili extending from the surface of *Proteus mirabilis*.**

Concept check: *What type of motion does this cell likely use?*

27.3 Reproduction

Bacteria and archaea enlarge their populations by means of cell division. In addition, some microbes produce tough cells that can withstand deleterious conditions for long periods in a dormant condition. Although bacteria and archaea lack meiosis and other features typical of eukaryotic sexual reproduction, they are able to obtain additional genes by a variety of methods.

Populations of Bacteria and Archaea Increase by Binary Fission

The cells of bacteria and archaea divide by splitting in two, a process known as **binary fission** (Figure 27.15a; also refer back to Figure 18.14). Binary fission requires a protein known as FtsZ, which is related to the tubulin protein that makes up eukaryotic microtubules. FtsZ squeezes dividing cells into two progeny cells; if this protein is not functional, bacterial cells cannot complete binary fission, and the cells become very long. When sufficient nutrients are available, an entire population of

(a) Bacterium undergoing binary fission

(b) Colonies developed from single cells

(c) Bacteria stained with fluorescent DNA-binding dye

Figure 27.15 **Binary fission and counting microbes.** **(a)** Division of a bacterial cell as viewed by scanning electron microscopy. **(b)** When samples are spread onto the surfaces of laboratory dishes containing nutrients, single cells of bacteria or archaea may divide repeatedly to form visible colonies, which can be easily counted. The number of colonies is an estimate of the number of culturable cells in the original sample. **(c)** If a fluorescence microscope is available, cells can be counted directly by applying a fluorescent stain that binds to cell DNA. Each cell glows brightly when illuminated with ultraviolet light.

Concept check: *Which procedure would you choose to count bacteria in a sample that is known to include many species that have not as yet been cultured?*

identical cells can be produced from a single parental cell by repeated binary fission. This growth process allows microbes to become very numerous in water, food, or animal tissues, potentially causing harm.

Binary fission is the basis of a widely used method for detecting and counting bacteria in food, water samples, or patient fluids. Microbiologists who study the spread of disease need to quantify bacterial cells in samples taken from the environment. Medical technicians often need to count bacteria in body fluid samples to assess the likelihood of infection. However, because bacterial cells are small and often unpigmented, they are difficult to count directly. One way that microbiologists count bacteria is to place a measured volume of sample into plastic dishes filled with a semisolid nutrient medium. Bacteria in the sample undergo repeated binary fission to form colonies of cells visible to the unaided eye (**Figure 27.15b**). Because each colony represents a single cell that was present in the original sample, the number of colonies in the dish reflects the number of living bacteria in the original sample.

Another way to detect and count bacteria is to treat samples with a stain that binds bacterial DNA, causing cells to glow brightly when illuminated with ultraviolet light. The glowing bacteria can be viewed and counted by the use of a fluorescence microscope (**Figure 27.15c**). The fluorescence method must be used when the microbes of interest cannot be cultured in the laboratory. Microorganisms can also be stained with fluorescent reagents that bind to specific DNA sequences, allowing microbiologists to detect and count bacteria and archaea of precise genetic types. This procedure, known as <u>f</u>luorescence <u>in situ</u> <u>h</u>ybridization (FISH), is a powerful technique for finding particular types of microorganisms in mixed populations.

Some Bacteria Survive Harsh Conditions as Akinetes or Endospores

Some bacteria produce thick-walled cells known as akinetes or endospores that are able to survive unfavorable conditions in a dormant state. Akinetes and endospores develop when bacteria have experienced stress, such as low nutrients or unfavorable temperatures. Such cells are able to germinate into metabolically active cells when conditions improve again. For example, aquatic filamentous cyanobacteria often produce large, food-filled **akinetes** when winter approaches (**Figure 27.16a**). Akinetes are able to survive winter at the bottoms of lakes, and they produce new filaments in spring when they are carried by water currents to the brightly lit surface.

Endospores (**Figure 27.16b**) are cells having tough protein coats that are produced inside bacterial cells and then released when the enclosing cell dies and breaks down. Endospores can remain alive, though in a dormant state, for perhaps hundreds of years.

The ability to produce endospores allows some Gram-positive bacteria to cause serious diseases. For example, *Bacillus anthracis* causes the disease anthrax, a potential agent in bioterrorism and germ warfare. Most cases of human anthrax result when endospores of *B. anthracis* enter wounds, causing

(a) Cyanobacterial akinete   **(b) *Clostridium difficile***

Figure 27.16   Specialized cells capable of dormancy. **(a)** Akinetes are thick-walled, food-filled cells produced by some cyanobacteria. They are able to resist stressful conditions and generate new populations when conditions improve. As discussed later, the heterocyst is a specialized cell in which nitrogen fixation occurs. **(b)** An endospore with a resistant wall develops within the cytoplasm of the pathogen *Clostridium difficile*.

Concept check: *How do endospores influence the ability of some bacteria to cause disease?*

skin infections that are relatively easily cured by antibiotic treatment. But sometimes the endospores are inhaled, or they are consumed in undercooked, contaminated meat, potentially causing more serious illness or death. *Clostridium botulinum* can contaminate improperly canned food that has not been heated to temperatures high enough to destroy its tough endospores. When the endospores germinate and bacterial cells grow in the food, they produce a deadly toxin, as well as NH_3 and CO_2 gas, which causes the can lids to bulge. If humans consume the food, the toxin causes botulism, a severe type of food poisoning that can lead to respiratory and muscular paralysis. The botulism toxin has been recently marketed commercially as Botox, which is injected into the skin, where it paralyzes facial muscles, thereby reducing the appearance of wrinkles. *Clostridium tetani* produces a nerve toxin that causes lockjaw, also known as tetanus, when bacterial cells or endospores from soil enter wounds. The ability of the genera *Bacillus* and *Clostridium* to produce resistant endospores helps to explain their widespread presence in nature and their impact on humans.

Bacteria and Archaea Obtain Genetic Material by Transduction, Transformation, and Conjugation

In Chapter 18, you were introduced to the varied ways in which microorganisms can acquire genetic material from other cells.

In this chapter, we have noted the ecological and evolutionary impacts of horizontal gene transfer. DNA may enter cells by means of viral vectors in the process known as **transduction**. Microbes are able to take up DNA directly from their environments in the process known as **transformation**. Some bacteria transmit DNA during a mating process known as **conjugation**. Pili (see Figure 27.14) help mating cells attach to each other, allowing conjugation to proceed in Gram-negative bacteria. Such gene exchange processes have contributed to the diversity of microbial nutrition and metabolism, our next topic.

27.4 Nutrition and Metabolism

All living cells require energy and a source of carbon to build their organic molecules. Bacteria and archaea use a wide variety of strategies to obtain energy and carbon for growth (**Table 27.2**). Together, bacteria and archaea display more diverse types of metabolism than other groups of organisms. Microbes can be classified according to their energy source, carbon source, response to oxygen, and presence of specialized metabolic processes.

Bacteria and Archaea Display Diverse Types of Nutrition and Responses to Oxygen

Cyanobacteria and some other bacteria are **autotrophs** (from the Greek, meaning self-feeders), organisms that are able to produce all or most of their own organic compounds from inorganic sources. Autotrophic microorganisms fall into two categories: photoautotrophs and chemoautotrophs. **Photoautotrophs** are able to use light as a source of energy for the synthesis of organic compounds from CO_2 and H_2O or H_2S. **Chemoautotrophs** are able to use energy obtained by chemical modifications of inorganic compounds to synthesize organic compounds. Such chemical modifications include nitrification (the conversion of ammonia to nitrate) and the oxidation of sulfur, iron, or hydrogen.

Heterotrophs (from the Greek, meaning other feeders) are organisms that require at least one organic compound, and often more. Some microorganisms are **photoheterotrophs**, meaning that they are able to use light energy to generate ATP,

but they must take in organic compounds from their environment. **Chemoheterotrophs** must obtain organic molecules for both energy and as a carbon source. Among the many types of bacterial chemoheterotrophs is the Gram-positive *Propionibacterium acnes*, which causes acne, affecting up to 80% of adolescents in the U.S. The genome sequence of *P. acnes* has revealed numerous genes that allow it to break down skin cells and consume the products.

Microorganisms differ in their need for oxygen. **Obligate aerobes** require O_2. **Facultative anaerobes** can use O_2 in aerobic respiration, obtain energy via anaerobic fermentation, or use inorganic chemical reactions to obtain energy. One fascinating example of a facultative anaerobe is *Thiomargarita namibiensis*, a giant bacterium about 1 mm in diameter that lives in marine waters off the Namibian coast of Africa. This heterotroph obtains its energy in two ways: by oxidizing sulfide with oxygen when this is available or, when oxygen is low or unavailable, by oxidizing sulfide with nitrate. In either case, the cells convert sulfide to elemental sulfur, which is stored within the cells as large globules.

In contrast to obligate aerobes, **obligate anaerobes**, such as the bacterial genus *Clostridium*, are poisoned by O_2. People suffering from gas gangrene (caused by *Clostridium perfringens* and related species) are usually treated by placement in a chamber having a high oxygen content (called a hyperbaric chamber), which kills the organisms and deactivates the toxins. **Aerotolerant anaerobes** do not use O_2, but they are not poisoned by it either. These organisms obtain their energy by fermentation or anaerobic respiration, which uses electron acceptors other than oxygen in electron transport processes. Anaerobic metabolic processes include denitrification (the conversion of nitrate into N_2 gas) and the reduction of manganese, iron, and sulfate, which are all important in the Earth's mineral cycles.

Bacteria Play Important Roles as Nitrogen Fixers

Many cyanobacteria and some other microbes are known as **diazotrophs** (from the Greek, meaning dinitrogen feeders) because they conduct a specialized metabolic process called **nitrogen fixation**. The removal of nitrogen from the gaseous phase is called fixation. During nitrogen fixation, the enzyme nitrogenase converts inert atmospheric gas (N_2) into ammonia (NH_3). Plants and eukaryotic algae can use ammonia (though not N_2) to produce proteins and other essential nitrogen-containing molecules. As a result, many plants have developed close relationships with diazotrophs, which provide ammonia to the plant partner. For example, nitrogen-fixing cyanobacteria typically live in leaf cavities of water ferns that float in rice paddies. The fern partner and the rice crop both take up ammonia produced by the diazotrophs. This natural fertilizer allows the rice plants to produce more protein. Many types of heterotrophic soil bacteria also fix nitrogen. Examples include *Rhizobium* and its relatives, which live within the roots of legumes (see Chapter 37).

Oxygen binds to nitrogenase, irreversibly disabling it. Thus, most diazotrophs are able to conduct nitrogen fixation only in

Table 27.2	Major Types of Bacteria and Archaea Based on Energy and Carbon Source		
Type	**Energy source**	**Carbon source**	**Example**
Autotroph			
Photoautotroph	Light	CO_2	Cyanobacteria
Chemoautotroph	Inorganic compounds	CO_2	*Sulfolobus*
Heterotroph			
Photoheterotroph	Light	Organic compounds	Chloroflexi
Chemoheterotroph	Organic compounds	Organic compounds	Many

low-oxygen habitats. Many cyanobacteria accomplish nitrogen fixation in specialized cells known as **heterocysts** (or heterocytes) (Figure 27.16a). Heterocysts display many adaptations that reduce nitrogenase exposure to oxygen. These adaptations include thick walls, which reduce inward oxygen diffusion; absence of oxygen-producing photosystems; and increased respiration, which consumes oxygen.

Genomes & Proteomes Connection

Gene Expression Studies Revealed How Cyanobacteria Fix Nitrogen in Hot Springs

The microbial communities of hot springs in Yellowstone National Park and other thermal areas around the world have long fascinated microbiologists interested in the occurrence of life at high temperatures. Thermal pools characteristically display beautiful, multicolored microbial mats (Figure 27.17). Such mats are composed of diverse photoautotrophs, including many types of cyanobacteria, along with chemoautotrophs, photoheterotrophs, and chemoheterotrophs. However, in particular Yellowstone thermal pools where temperatures range from 50 to 70°C, single-celled cyanobacteria of the genus *Synechococcus* are the only photoautotrophic organisms present. Heterocyst-producing cyanobacteria are absent from such hot waters, and few heterotrophic diazotrophs tolerate such temperatures, yet nitrogen fixation occurs in these pools. It was not clear which organisms could fix nitrogen until genomic information provided an essential clue.

Anne-Soisig Steunou, Arthur Grossman, and associates sequenced the genomes of *Synechococcus* cultures obtained from these hot springs and discovered that *nif* (nitrogen fixation) genes were present. They also determined that *Synecho-*

Figure 27.18 Daily changes in photosynthesis and nitrogen fixation gene expression in *Synechococcus* cultures isolated from thermal pools.

Concept check: At what time of the day does the photosynthesis gene psaB reach its lowest level of expression, and when does the nitrogen fixation gene nifK reach its highest expression level?

coccus nif genes were expressed in parts of the microbial mat having temperatures near 60°C. These results indicated that *Synechococcus* conducts nitrogen fixation in the mats, producing fixed nitrogen and other compounds used by other community members. But the investigators wondered how this single-celled bacterium managed to fix nitrogen without the oxygen arising from photosynthesis disabling the functioning of nitrogenase. They tracked the expression of *Synechococcus* genes over a 24-hour period and found that after nightfall the expression of a photosynthesis gene (*psaB*) fell dramatically, whereas expression of nitrogen fixation genes (*nifH, nifD,* and *nifK*) increased (Figure 27.18). They concluded that *Synechococcus* turns on nitrogen fixation at dusk, when oxygen production from photosynthesis drops. But nitrogen fixation is an energy-intensive process, so how does *Synechococcus* fuel it? Steunou and colleagues used differential gene expression analyses to establish that *Synechococcus* uses fermentation, an anaerobic metabolic pathway, to supply the ATP needed for nitrogen fixation. This study not only illustrates important features of microbial metabolism, but also the essential ecological roles played by bacteria and archaea, a topic that we will explore in more detail next.

27.5 Ecological Roles and Biotechnology Applications

Bacteria and archaea play many key ecological roles, including the production and cycling of carbon. Earth's carbon cycle depends on microorganisms that produce and degrade organic compounds, including methane. Bacteria also play fascinating roles as symbionts living in close associations with eukaryotes, as we have seen in the case of nitrogen-fixing bacteria that partner with plants. In this section, we focus on these diverse ecological roles and also provide examples of ways that humans use the metabolic capabilities of bacteria and archaea in biotechnology.

Figure 27.17 A thermal pool in Yellowstone National Park. Brightly colored mats at the pool edge are microbial communities that include thermophilic cyanobacteria.

Bacteria and Archaea Play Important Roles in Earth's Carbon Cycle

The Earth's carbon cycle is the sum of all the chemical changes that occur among compounds that contain carbon. (See Chapter 59 for a detailed discussion of the carbon cycle.) Bacteria and archaea are important in producing and degrading organic compounds. For example, cyanobacteria and other autotrophic bacteria are important **producers**. These bacteria, together with algae and plants, synthesize the organic compounds used by other organisms for food. **Decomposers**, also known as saprobes, include heterotrophic microorganisms (as well as fungi and animals). These organisms break down dead organisms and organic matter, releasing minerals for uptake by living things.

Bacteria and archaea also influence Earth's carbon cycle by producing and consuming methane (CH_4). Methane—the major component of natural gas—is a powerful greenhouse gas, as are CO_2 and H_2O vapor. Atmospheric methane has the potential to alter the Earth's climate, and in recent years, the level of methane has been increasing in Earth's atmosphere as the result of human activities. Several groups of anaerobic archaea known as the **methanogens** convert CO_2, methyl groups, or acetate to methane and release it from their cells. Methanogens live in swampy wetlands, in deep-sea habitats, or in the digestive systems of ungulate animals, such as cattle. Marsh gas produced in wetlands is largely composed of methane, and large quantities of methane produced long ago are trapped in deep-sea and subsurface Arctic deposits. Microbes also play a role in limiting atmospheric methane. The balance of methane in Earth's atmosphere is maintained by the activities of prokaryotes known as **methanotrophs**, which consume methane. In the absence of methanotrophs, Earth's atmosphere would be much richer in the greenhouse gas methane, which would substantially increase global temperatures.

Many Bacteria Live in Symbiotic Associations with Other Prokaryotes or Eukaryotes

An organism that lives in close association with one or more other organisms is said to occur in **symbiosis** (from the Greek, meaning life together with). If symbiotic associations are beneficial to both partners, the association is known as a **mutualism**. If one partner benefits at the expense of the other, the association is known as a **parasitism**. There are numerous examples of mutualistic and parasitic bacteria. Many mutualistic bacteria live in symbioses of two or a few other bacterial species that supply each other with essential nutrients, a type of association known as a **syntrophy**. For example, certain deep-sea archaea are able to metabolize the plentiful methane present in such anaerobic conditions only by partnering with bacteria that reduce sulfate. Other microbes occur in larger communities known as **consortia**, in which metabolite exchange occurs. The marine worm *Olavius algarvensis*, for example, has no mouth, gut, or anus, and depends on a consortium of several types of bacteria that live within the worm, providing it with food and recycling its wastes.

Mutualistic Partnerships with Eukaryotes   Bacteria are involved in many mutually beneficial symbioses in which they provide eukaryotes with minerals or vitamins or other valuable services. For example, the common green seaweed *Ulva* does not display its typical lettuce-leaf shape unless bacterial partners belonging to the phylum Bacteroidetes are present. The bacteria produce a compound that induces normal seaweed development. Bioluminescent bacteria form symbiotic relationships with squid and other marine animals. In deep-sea thermal vent communities, sulfur-oxidizing bacteria live within the tissues of tube worms and mussels, supplying these animals with carbon compounds used as food. Bacteriologist Cameron Curry and his associates have documented a complex land association involving four partners: ants, fungi that the ants cultivate for food, parasitic fungi that attack the food fungi, and mutualistic Actinobacteria, which produce antibiotics. These antibiotics control the growth of the fungal parasite, preventing it from destroying the ants' fungal food supply. The ants rear the useful bacteria in cavities on their body surfaces; glands near these cavities supply the bacteria with nutrients. Humans, too, harbor symbiotic microbes.

The Human Microbiome   On human skin and in our digestive and reproductive systems many types of microbes exist that are known collectively as the human **microbiome**. An estimated 10–100 trillion microbes live in the typical human colon! These microbes provide services using traits that humans do not possess, and the diverse types of metabolism present in the microbiome have co-evolved with human metabolism. For this reason, humans and other multicellular organisms, together with their microbiomes, are considered to function and evolve as superorganisms. Recent research has revealed that human gut microbiome communities contain hundreds of prokaryotic species, dominated by the bacterial phyla Firmicutes and Bacteroidetes, and that extensive horizontal gene transfer has occurred among gut microbial species. Studies also reveal that gut microbiomes differ among healthy people, and between healthy people and those having different types of medical conditions. An ongoing Human Microbiome Project seeks to understand the relationship between human metabolism and the microbiome, with the goal of opening new opportunities to improve human health.

Parasitic and Pathogenic Microbes   **Parasites** are organisms that obtain organic compounds from living hosts. If parasitic microbes cause disease symptoms in their hosts, the microorganisms are known as **pathogens**. Cholera, leprosy, tetanus, pneumonia, whooping cough, diphtheria, Lyme disease, scarlet fever, rheumatic fever, typhoid fever, bacterial dysentery, and tooth decay are among the many examples of human diseases caused by bacterial pathogens. Bacteria also cause many plant diseases of importance in agriculture, including blights, soft rots, and wilts. How do microbiologists determine which bacteria cause these diseases? The pioneering research of the Nobel Prize–winning German physician Robert Koch provides the answer.

In the mid- to late 1800s, Koch established a series of four steps to determine whether a particular organism causes a specific disease. First, the presence of the suspected pathogen must correlate with occurrence of symptoms. Second, the pathogen must be isolated from an infected host and grown in pure culture if possible. Third, cells from the pure culture should cause disease when inoculated into a healthy host. Fourth and finally, one should be able to isolate the same pathogen from the second-infected host. Using these steps, known as **Koch's postulates**, Koch discovered the bacterial causes of anthrax, cholera, and tuberculosis. Subsequent investigators have used Koch's postulates to establish the causes of many other infectious diseases.

How Pathogenic Bacteria Attack Cells Modern research is providing new information about how bacteria attack cells. Such knowledge aids in developing strategies for disease prevention and treatment. Many pathogenic bacteria attack cells by binding to the target cell surfaces and injecting substances that help them utilize cell components. During their evolution, some pathogenic bacteria have developed needle-like systems, made of components also found in flagella, that inject infection proteins into animal or plant cells as part of the infection process. Such structures are known as type III secretion systems, also called injectisomes (**Figure 27.19a**). Examples of bacteria whose injectisomes allow them to attack human cells are *Yersinia pestis* (the agent of bubonic plague), *Salmonella enterica* (which causes salmonellosis), and *Burkholderia pseudomallei* (the cause of melioidosis, a deadly disease of emerging concern in some parts of the world). These bacteria also induce the host cell to form a plasma membrane pocket that encloses the bacterial cell, bringing it into the host cell. Once within a host cell, pathogenic bacteria use the cell's resources to reproduce and spread to nearby tissues.

Some other bacterial pathogens use a type IV secretion system to deliver toxins or to transform DNA into cells (**Figure 27.19b**). Examples of such bacteria that cause human disease include *Helicobacter pylori*, *Legionella pneumophila*, and *Bordetella pertussis*. The plant pathogen *Agrobacterium tumifaciens* uses a type IV secretion system to transfer DNA (T DNA) into plant cells. The bacterial T DNA encodes an enzyme that affects normal plant growth, with the result that cancer-like tumors develop (see Figure 27.3). Type IV systems evolved from pili and other components of bacterial mating. The type IV attack system is an example of descent with modification, the evolutionary process by which organisms acquire new features. Some experts propose that type III and IV secretion systems may be useful in human gene therapy, to deliver DNA to target cells.

Antibiotic-Resistant Pathogens Antibiotics are compounds widely used to treat diseases caused by bacterial pathogens. Such compounds work because they inhibit processes that occur in bacteria, but not eukaryotes. However, the overuse of antibiotics can favor the growth and proliferation of resistant bacteria, a serious health problem. For example, some strains

(a) Type III secretion system

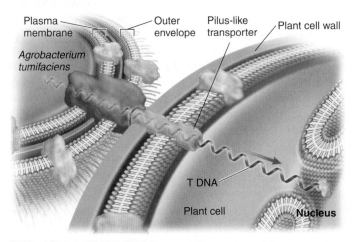

(b) Type IV secretion system

Figure 27.19 Attack systems of pathogenic bacteria. **(a)** The type III secretion system functions like a syringe to inject proteins into host cells, thereby starting a disease process. **(b)** The type IV secretion system forms a channel through which DNA can be transmitted from a pathogen to a host cell, in this case from the bacterium *Agrobacterium tumifaciens* into a plant cell.

Concept check: How does the type IV secretion system illustrate the evolutionary concept of descent with modification?

of *Staphylococcus aureus*, which causes "staph" infections, have developed resistance to methicillin (an antibiotic related to penicillin). Such strains, known as MRSA (methicillin-resistant *Staphylococcus aureus*), have been much in the news. Livestock are often fed antibiotics to reduce the effects of minor bacterial infections or to prevent infections from spreading, thereby increasing the production profit margin. Unfortunately, this activity, as well as the improper use of antibiotics by patients, promotes the evolution of antibiotic-resistant bacterial strains. Antibiotics select against vulnerable strains, allowing resistant strains to thrive, and antibiotic resistance can spread rapidly in bacterial populations by means of horizontal gene transfer. When humans become infected by antibiotic-resistant strains, antibiotic treatments are not effective, and cases have occurred in which patients' infections were resistant to all known antibiotics.

Some Bacteria Are Useful in Industrial and Other Applications

Several industries have harnessed the metabolic capabilities of microbes obtained from nature. The food industry uses bacteria to produce chemical changes in food that improve consistency or flavor—to make dairy products, including cheese and yogurt. Cheese makers add pure cultures of certain bacteria to milk. The bacteria consume milk sugar (lactose) and produce lactic acid, which aids in curdling the milk.

The chemical industry produces materials such as vinegar, amino acids, enzymes, vitamins, insulin, vaccines, antibiotics, and other useful pharmaceuticals by growing particular bacteria in giant vats. For example, bacteria produce the antibiotics streptomycin, tetracycline, kanamycin, gentamycin, bacitracin, polymyxin-B, and neomycin.

The ability of microorganisms to live in harsh environments and break down organic compounds makes them very useful in treating wastewater, industrial discharges, and harmful substances such as explosives, pesticides, and oil spills. This process is known as bioremediation. *Geobacter sulfurreducens*, for example, is used to precipitate metals such as uranium from contaminated water, thereby purifying it. This bacterium uses metal ions and elemental sulfur in an electron transport process to oxidize acetate to CO_2, thereby producing ATP needed for its growth and reproduction. Taking advantage of this observation, engineers drip acetate into the groundwater at contaminated sites, thereby enriching the population of *G. sulfurreducens* to 85% of the microbial community, which more efficiently precipitates uranium.

In addition to food production and waste treatment, bacteria are also used in agriculture. Several species of *Bacillus*, particularly *B. thuringiensis*, produce crystalline proteins known as Bt-toxins. These toxins kill insects that ingest them, but are harmless to many non-insect species that have been tested. Tent caterpillars, potato beetles, gypsy moths, mosquitoes, and black flies are among the pests that can be controlled by Bt-toxin. For this reason, genes involved in Bt-toxin production have been engineered into some crop plants, such as corn, to limit pests, thereby increasing crop yields. *Deinococcus radiodurans* is another example of a bacterial species whose unusual ecological properties may prove useful to humans, as described next.

FEATURE INVESTIGATION

The Daly Experiments Revealed How Mn^{2+} Helps *Deinococcus radiodurans* Avoid Radiation Damage

The bacterial phylum known as *Deinococcus-Thermus* includes *Deinococcus radiodurans*, which is unusually resistant to chemical mutagens and ionizing radiation (see Table 27.1). This bacterium can survive brief radiation doses greater than 10,000 Gray (Gy). (One Gy = the absorption of one joule of radiation energy by one kg of matter.) *Deinococcus radiodurans* can survive continuous radiation levels as high as 50 Gy/hour. (By contrast, a radiation dose of 5 Gy is lethal to humans.) Radiation resistance evolved as an adaptation that aids *Deinococcus* survival in its natural habitats, which include nuclear waste sites, hot springs, and deserts. Like ionizing radiation, drying and solar radiation cause chromosome breakage, and *Deinococcus* has acquired very effective methods for repairing damaged DNA. M. J. Daly and associates were interested in learning more about how *D. radiodurans* survives treatment with high radiation, with the hope that such information might lead to better ways of treating victims of radiation sickness.

As the result of a series of experiments, Daly and colleagues learned that radiation-resistant bacteria tended to have higher levels of manganese—Mn^{2+}—ions than do radiation-sensitive bacteria. From these observations, Daly and associates hypothesized that *D. radiodurans* might protect itself by accumulating Mn^{2+} to high levels. In experiments reported in 2004 and described in **Figure 27.20**, they first grew *D. radiodurans* in basic culture media containing three different levels of manganese ion (50, 100, or 250 nM Mn^{2+}). As they grew, the bacteria took up Mn^{2+} from the media. Researchers then inoculated bacteria from each type of medium into separate sectors of laboratory dishes containing media to which high, medium, or low levels of Mn^{2+} had been added. Next, they exposed some of the dishes to 50 Gy/hour of radiation, leaving the medium concentration as a nonirradiated control. After a period of growth, irradiated bacteria grown in dishes containing high Mn^{2+} levels grew as well as the control, but irradiated bacteria grown in dishes containing low Mn^{2+} levels did not grow as well. Irradiated bacteria that had been grown beforehand with high Mn^{2+} levels fared particularly well.

This experiment revealed that *D. radiodurans* cells that had accumulated high levels of Mn^{2+} were better able to grow in the presence of ionizing radiation. Manganese seems to confer resistance to such damage, possibly by preventing damage to proteins that repair DNA.

Experimental Questions

1. What feature of *Deinococcus radiodurans* attracted the attention of researchers?
2. What hypothesis did Daly and associates develop to explain radiation resistance in *D. radiodurans*?
3. As shown in Figure 27.20, bacterial cells were grown in media having various levels of manganese ion, which they absorbed. Later, some of these bacteria were exposed to high levels of ionizing radiation, whereas control bacteria were not exposed. What results of this experiment support the hypothesis that cellular manganese plays a role in radiation resistance?

Figure 27.20 Manganese helps *Deinococcus radiodurans* avoid radiation damage.

HYPOTHESIS High levels of intracellular Mn²⁺, also called Mn(II), help to protect *Deinococcus radiodurans* from the effects of ionizing radiation.

KEY MATERIALS *Deinococcus radiodurans*.

Experimental level

1 Cultivate *Deinococcus radiodurans* in media having different Mn²⁺ content (50 nM, 100 nM, 250 nM).

50 nM   100 nM   250 nM

2 Inoculate the bacteria onto the surfaces of media differing in Mn²⁺ content.

(1) Control (25 nM)   (2) High Mn medium (2.5 µM)   (3) Low Mn medium (2.5 nM)

50   100   50   100   50   100

250   250   250

3 Expose media plates to 50 Gy/hour of ionizing radiation.

Not irradiated

4 Compare bacterial growth to control plates not irradiated.

(1)   (2)   (3)

50   50   50

100   100   100

250   250   250

5 **THE DATA**

Results from step 4: Even when exposed to ionizing radiation, *D. radiodurans* grew well when high levels of Mn²⁺ were present in its environment (compare plates 2 and 1). The higher the level of Mn²⁺ accumulated by cells, the better the growth after ionizing radiation (plate 2). When environmental levels of Mn²⁺ are low, cells are more readily damaged by ionizing radiation (compare plates 2 and 3).

6 **CONCLUSION** This experiment revealed that *D. radiodurans* cells that had accumulated high levels of Mn²⁺ were better able to grow in the presence of ionizing radiation. Manganese seems to confer resistance to such damage, possibly by preventing damage to proteins that repair DNA.

7 **SOURCE** Daly, M.J., et al. 2004. Accumulation of Mn(II) in *Deinococcus radiodurans* facilitates gamma-radiation resistance. *Science* 306:1025–1028.

Summary of Key Concepts

27.1 Diversity and Evolution

- Domains of life include Bacteria, Archaea, and Eukarya (known informally as bacteria, archaea, and eukaryotes). (Figure 27.1)

- Many representatives of the domain Archaea occur in extremely hot, salty, or acidic habitats. Ether-linked membranes are among the features of archaea that enable their survival in extreme habitats. (Figure 27.2)

- The domain Bacteria includes 50 or more phyla, including Proteobacteria and Cyanobacteria, which are particularly diverse and of great evolutionary and ecological importance. (Table 27.1)

- Mitochondria arose from a proteobacterial ancestor, and many modern proteobacteria occur in close relationships with animals or plants. (Figure 27.3)

- Algal and plant plastids arose from a cyanobacterial ancestor. Modern cyanobacteria play important ecological roles, including the production of harmful blooms in overfertilized waters. (Figure 27.4)

- Widespread horizontal DNA transfer has occurred among bacteria and archaea. Horizontal DNA transfer allows microorganisms to evolve rapidly.

27.2 Structure and Motility

- Bacteria and archaea are composed of prokaryotic cells that are smaller and simpler than those of eukaryotes. Even so, structures such as thylakoids, gas vesicles, magnetosomes, and nucleus-like bodies are examples of bacterial cell structure complexity. (Figures 27.5, 27.6)

- Cells of bacteria and archaea vary in shape. Major cell shape types are spherical cocci, rod-shaped bacilli, comma-shaped vibrios, and coiled spirochaetes and spirilli. Some cyanobacteria display specialized cells and intercellular communication, the hallmarks of multicellular organisms. (Figure 27.7)

- Many microbes occur within a coating of slimy mucilage, which may play a role in diseases or aid in the development of biofilms. Biofilm development is influenced by quorum sensing, a process in which group activity is coordinated by chemical communication. (Figure 27.8)

- Most prokaryotic cells possess a protective cell wall. Archaea and some bacteria have walls composed of proteins, but most bacterial cell walls contain peptidoglycan, which is composed of carbohydrates cross-linked by peptides. (Figure 27.9)

- Gram-positive bacterial cells have walls rich in peptidoglycan, whereas Gram-negative cells have less peptidoglycan in their walls and are enclosed by a lipopolysaccharide envelope. Gram-positive bacteria can be distinguished from Gram-negative bacteria by use of the Gram stain and other staining procedures. (Figures 27.10, 27.11)

- Motility enables microbes to change positions within their environment, which aids in locating favorable conditions for growth. Some microbes adjust their buoyancy in water by means of intracellular gas vesicles, whereas others swim by means of flagella or twitch or glide by the action of threadlike pili. (Figures 27.12, 27.13, 27.14)

27.3 Reproduction

- Populations of bacteria and archaea enlarge by binary fission, a simple type of cell division that provides a means by which culturable microbes can be counted. (Figure 27.15)

- Some bacteria are able to survive harsh conditions as dormant akinetes or endospores. (Figure 27.16)

- Many microorganisms obtain new DNA sequences directly from their environment (transformation), by means of viral vectors (transduction), or by a mating process (conjugation).

27.4 Nutrition and Metabolism

- Bacteria and archaea can be grouped according to nutritional type, response to oxygen, or presence of distinctive metabolic features. Major nutritional types are photoautotrophs, chemoautotrophs, photoheterotrophs, and chemoheterotrophs. (Table 27.2)

- Obligate aerobes require oxygen, whereas facultative aerobes are able to live with or without oxygen by using different processes for obtaining energy. Obligate anaerobes are poisoned by oxygen; aerotolerant anaerobes do not use oxygen but are not poisoned by it. Both obtain their energy by anaerobic respiration.

- Nitrogen fixation is an example of a distinctive metabolism displayed only by certain microorganisms. A number of plants display symbiotic associations with bacteria that fix nitrogen, called diazotrophs; many of these associations are ecologically or agriculturally important. (Figures 27.17, 27.18)

27.5 Ecological Roles and Biotechnology Applications

- Bacteria and archaea play key roles in Earth's carbon cycle as producers, decomposers, symbionts, or pathogens.

- Methane-producing methanogens and methane-using methanotrophs are important in the carbon cycle, and they influence the Earth's climate.

- Parasitic bacteria obtain organic compounds from living hosts, and if disease symptoms result, such bacteria are known as pathogens.

- Bacteria attack eukaryotic cells by means of flagella-like type III secretion systems or type IV secretion systems, which evolved from pili and other mating components. (Figure 27.19)

- Many bacteria and archaea are useful in industrial and other applications; others are used to make food products or antibiotics or to clean up polluted environments. The cellular adaptations of bacteria that are extremely resistant to ionizing radiation may suggest new ways to treat radiation sickness in humans. (Figure 27.20)

Assess and Discuss

Test Yourself

1. Which of the following features is common to prokaryotic cells?
   a. a nucleus, featuring a nuclear envelope with pores
   b. mitochondria
   c. plasma membranes
   d. mitotic spindle
   e. none of the above

2. The bacterial phylum that typically produces oxygen gas as the result of photosynthesis is
   a. the proteobacteria.
   b. the cyanobacteria.
   c. the Gram-positive bacteria.
   d. all of the listed choices.
   e. none of the listed choices.

3. The Gram stain is a procedure that microbiologists use to
   a. determine if a bacterial strain is a pathogen.
   b. determine if a bacterial sample can break down oil.
   c. infer the structure of a bacterial cell wall and bacterial response to antibiotics.
   d. count bacteria in medical or environmental samples.
   e. do all of the above.

4. Place the following steps in the correct order, according to Koch's postulates:
   I. Determine if pure cultures of bacteria cause disease symptoms when introduced to a healthy host.
   II. Determine if disease symptoms correlate with presence of a suspected pathogen.
   III. Isolate the suspected pathogen and grow it in pure culture, free of other possible pathogens.
   IV. Attempt to isolate pathogen from second-infected hosts.
   a. II, III, IV, I
   b. II, IV, III, I
   c. III, II, I, IV
   d. II, III, I, IV
   e. I, II, III, IV

5. Cyanobacteria play what ecological role?
   a. producers
   b. consumers
   c. decomposers
   d. parasites
   e. none of the listed choices

6. Bacterial structures that are produced by pathogenic bacteria for use in attacking host cells include
   a. type III and IV secretion systems.
   b. magnetosomes.
   c. gas vesicles.
   d. thylakoids.
   e. none of the above.

7. The structures that enable some Gram-positive bacteria to remain dormant for extremely long periods of time are known as
   a. akinetes.
   b. endospores.
   c. biofilms.
   d. lipopolysaccharide envelopes.
   e. pili.

8. By means of what process do populations of bacteria or archaea increase their size?
   a. mitosis
   b. meiosis
   c. conjugation
   d. transduction
   e. none of the listed choices

9. By what means do bacterial cells acquire new DNA?
   a. by conjugation, the mating of two cells of the same bacterial species
   b. by transduction, the injection of viral DNA into bacterial cells
   c. by transformation, the uptake of DNA from the environment
   d. all of the above
   e. none of the above

10. How do various types of bacteria move?
    a. by the use of flagella, composed of a filament, hook, and motor
    b. by means of pili, which help cells twitch or glide along a surface
    c. by using gas vesicles to regulate buoyancy in water bodies
    d. all of the above
    e. none of the above

Conceptual Questions

1. Explain why many microbial populations grow more rapidly than do eukaryotes and how bacterial population growth influences the rate of food spoilage or infection.

2. Why does the overuse of antibiotics in medicine and agriculture result in widespread antibiotic resistance?

3. What organisms are responsible for the blue-green blooms that often occur in warm weather on lake surfaces? Think carefully; the answer is not just "cyanobacteria," as you might first guess.

Collaborative Questions

1. How would you go about cataloging the phyla of bacteria and archaea that occur in a particular place?

2. How would you go about developing a bacterial product that could be sold for remediation of a site contaminated with materials that are harmful to humans?

Online Resource

www.brookerbiology.com

Stay a step ahead in your studies with animations that bring concepts to life and practice tests to assess your understanding. Your instructor may also recommend the interactive ebook, individualized learning tools, and more.

Protists

28

Protists are eukaryotes that live in moist habitats and are mostly microscopic in size. Despite their small size, protists have a greater impact on global ecology and human affairs than most people realize. For example, the photosynthetic protists known as algae generate at least half of the oxygen in the Earth's atmosphere and produce organic compounds that feed marine and freshwater animals. The oil that fuels our cars and industry is derived from pressure-cooked algae that accumulated on the ocean floor over millions of years. Today, algae are being engineered into systems for cleaning pollutants from water or air and for producing biofuels.

Protists also include some parasites that cause serious human illnesses. For example, in 1993, the waterborne protist *Cryptosporidium parvum* sickened 400,000 people in Milwaukee, Wisconsin, costing $96 million in medical expenses and lost work time. The related protist *Plasmodium falciparum*, which is carried by mosquitoes in many warm regions of the world, causes the disease malaria. Every year, nearly 500 million people become ill with malaria, and more than 2 million die of this disease. As we will see, sequencing the genomes of these and other protist species has suggested new ways of battling such deadly pathogens.

In this chapter, we will survey protist diversity, including structural, nutritional, and ecological variations. We begin by exploring ways of informally naming protists by their ecological roles, habitat, and motility. We then will focus on the defining features, classification, and evolutionary importance of the major protist phyla. Next, the nutritional modes and defensive adaptations of protists are discussed, and we conclude by looking at the reproductive adaptations that allow protists to exploit and thrive in a variety of environments.

28.1 An Introduction to Protists

Protists are eukaryotes that are not classified in the plant, animal, or fungal kingdoms. The term protist comes from the Greek word *protos*, meaning first, reflecting the observation that protists were Earth's first eukaryotes. Protists display two common characteristics: They are most abundant in moist habitats, and most of them are microscopic in size. Protists are often

Protists such as these green algal cells produce much of the Earth's oxygen. Each of the cells in this population is surrounded by a halo of protective mucilage.

informally labeled according to their ecological roles, habitats, or type of motility.

Protists Can Be Informally Labeled According to Their Diverse Ecological Roles

Protists are often labeled according to their ecological roles, which occur in three major types: algae, protozoa, and fungus-like protists. The term **algae** (from the Latin, meaning seaweeds) applies to protists that are generally photoautotrophic, meaning that most can produce organic compounds by means of photosynthesis. In addition to organic compounds that can be used as food by heterotrophic organisms, photosynthetic algae produce oxygen, which is also needed by most heterotrophs. Despite the general feature of photosynthesis, algae do not form a monophyletic group descended from a single common ancestor.

The term **protozoa** (from the Greek, meaning first life) is commonly used to describe diverse heterotrophic protists. Protozoa feed by absorbing small organic molecules or by ingesting

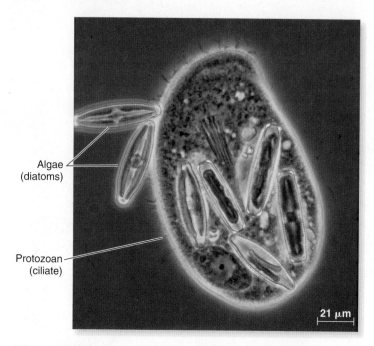

Algae
(diatoms)

Protozoan
(ciliate)

21 μm

Figure 28.1 **A heterotrophic protozoan feeding on photosynthetic algae.** The ciliate shown here has consumed several oil-rich, golden-pigmented, glass-walled algal cells known as diatoms. The algal cells were ingested by the process of phagocytosis, and their organic components will be digested as food. Diatom cells that have avoided capture glide nearby.

Concept check: *Why are diatoms good food for protozoans such as this?*

prey. For example, the protozoa known as ciliates consume smaller cells such as the single-celled photosynthetic algae known as diatoms (**Figure 28.1**). Like the algae, the protozoa do not form a monophyletic group.

Several types of heterotrophic **fungus-like protists** have bodies, nutrition, or reproduction similar to the true fungi. For example, fungus-like protists often have threadlike, filamentous bodies and absorb nutrients from their environment, as do the true fungi (see Chapter 31). However, fungus-like protists are not actually related to fungi; their similar features represent cases of convergent evolution (see Chapter 23). Water molds, some of which cause diseases of fish, and *Phytophthora infestans*, which causes diseases of many crops and wild plants, are examples of fungus-like protists. Various types of slime molds, some of which can be observed on decaying wood in forests, are also fungus-like though not closely related to water molds and *Phytophthora*. These examples illustrate that the terms algae, protozoa, and fungus-like protists, while very useful in describing ecological roles, lack taxonomic or evolutionary meaning.

Protists Can Be Informally Labeled According to Their Diverse Habitats

Although protists occupy nearly every type of moist habitat, they are particularly common and diverse in oceans, lakes, wetlands, and rivers. Even extreme aquatic environments such as Antarctic ice and acidic hot springs serve as habitats for some protists. In such places, protists may swim or float in open water or live attached to surfaces such as rocks or beach sand. These different habitats influence protist structure and size.

Swimming or floating protists are members of an informal group of organisms known as plankton, a group that also includes bacteria, viruses, and small animals. The photosynthetic protists in plankton are called **phytoplankton** (plantlike plankton). Planktonic protists are necessarily quite small in size; otherwise they would readily sink to the bottom. Staying afloat is particularly important for phytoplankton, which need light for photosynthesis. For this reason, planktonic protists occur primarily as single cells, colonies of cells held together with mucilage, or short filaments of cells linked end to end (**Figure 28.2a–c**).

Many protists live within **periphyton**, communities of microorganisms attached by mucilage to underwater surfaces such as rocks, sand, and plants. Because sinking is not a problem for attached protists, these often produce multicellular bodies, such as branched filaments (**Figure 28.2d**). Photosynthetic protists large enough to see with the unaided eye are known as **macroalgae**, or seaweeds. Although the bodies of some macroalgae are very large single cells (**Figure 28.2e**), most macroalgae are multicellular, often producing large and complex bodies. Macroalgae usually grow attached to underwater surfaces such as rocks, sand, docks, ship hulls, or offshore oil platforms. Seaweeds require sunlight and carbon dioxide for photosynthesis and growth, so most of them grow along coastal shorelines, fairly near the water surface.

Protists Can Be Informally Labeled According to Their Type of Motility

Microscopic protists have evolved diverse ways to propel themselves in moist environments. Swimming by means of flagella, cilia, amoeboid movement, and gliding are major types of protist movements.

Many types of photosynthetic and heterotrophic protists are able to swim because they produce one or a few eukaryotic flagella, cellular extensions whose movement is based on interactions between microtubules and the motor protein dynein (refer back to Figure 4.13). Eukaryotic flagella rapidly bend and straighten, thereby pulling or pushing cells through water. Protists that use flagella to move in water are commonly known as **flagellates** (Figure 28.2a). Flagellates are typically composed of one or only a few cells and are small—usually from 2–20 μm long—because flagellar motion is not powerful enough to keep larger bodies from sinking. Some flagellate protists are sedentary, living attached to underwater surfaces. These protists use flagella to collect bacteria and other small particles for food. Macroalgae and other immobile protists often produce small, flagellate reproductive cells that allow these protists to mate and disperse to new habitats.

An alternate type of protist motility relies on **cilia**, tiny hairlike extensions on the outsides of cells, mentioned earlier as occurring on the surfaces of some protozoa. Cilia are structurally similar to eukaryotic flagella but are shorter and more abundant on cells (**Figure 28.3**). Protists that move by means of cilia are **ciliates**. Having many cilia allows ciliates to achieve larger sizes than flagellates yet still remain buoyant in water.

Planktonic protists   **Attached protists**

(a) Single-celled *Chlamydomonas* with flagella

(b) The colonial genus *Pediastrum*
One cell

(c) The filamentous genus *Desmidium*
32 μm

(d) The branched filamentous genus *Cladophora*
25 μm

(e) The seaweed genus *Acetabularia*
15 mm

0.2 mm

Figure 28.2   **The diversity of algal body types reflects their habitats.**   **(a)** The single-celled flagellate genus *Chlamydomonas* occurs in the phytoplankton of lakes. **(b)** The colonial genus *Pediastrum* is composed of several cells arranged in a lacy star shape that helps to keep this alga afloat in water. **(c)** The filamentous genus *Desmidium* occurs as a twisted row of cells. **(d)** The branched filamentous genus *Cladophora* that grows attached to nearshore surfaces is large enough to see with the unaided eye. **(e)** The relatively large seaweed genus *Acetabularia* lives on rocks and coral rubble in shallow tropical oceans.

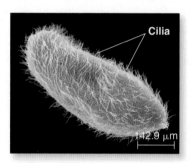

Cilia
42.9 μm

Figure 28.3   A member of the ciliate genus *Paramecium*, showing numerous cilia on the cell surface.

Pseudopod
88 μm

Figure 28.4   A member of the amoebozoan genus *Pelomyxa*, showing pseudopodia.

A third type of motility is amoeboid movement. This kind of motion involves extending protist cytoplasm into lobes, known as pseudopodia (from the Greek, meaning false feet). Once these pseudopodia move toward a food source or other stimulus, the rest of the cytoplasm flows after them, thereby changing the shape of the entire organism as it creeps along.

Protist cells that move by pseudopodia are described as **amoebae** (Figure 28.4).

Finally, many diatoms, the malarial parasite *Plasmodium falciparum*, and some other protists glide along surfaces in a snail-like fashion by secreting protein or carbohydrate slime. With the exception of ciliates, motility classification does not correspond with the phylogenetic classification of protists, our next topic.

28.2   Evolution and Relationships

At one time, protists were classified into a single kingdom. However, modern phylogenetic analyses based on comparative analysis of DNA sequences and cellular features reveal that protists do not form a monophyletic group. The relationships of some protists are uncertain or disputed, and new protist species are continuously being discovered. As a result, concepts of protist evolution and relationships have been changing as new information becomes available.

Even so, molecular and cellular data reveal that many protist phyla can be classified within several eukaryotic **supergroups** that each display distinctive features (Figure 28.5). All of the supergroups shown include phyla of protists; some, in fact, contain only protist phyla. The supergroup Opisthokonta includes the multicellular animal and fungal kingdoms and related protists, while another supergroup includes the multicellular plant kingdom and the protists most closely related to it. The study of such protists helps to reveal how multicellularity originated in animals, fungi, and plants.

In this section, we survey the eukaryotic supergroups, focusing on the defining features and evolutionary importance of the major protist phyla. We will also examine ways in which protists are important ecologically or in human affairs.

A Feeding Groove Characterizes Many Protists Classified in the Excavata

The protist supergroup known as the Excavata originated very early among eukaryotes, so this supergroup is important in understanding the early evolution of eukaryotes. The Exca-

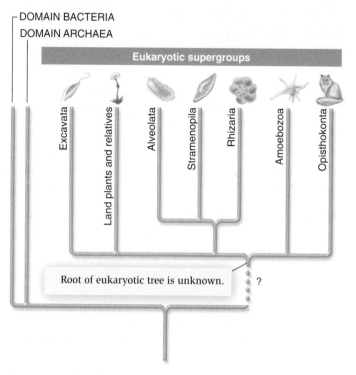

Figure 28.5 A phylogenetic tree showing the major eukaryotic supergroups.   Each of the supergroups shown here includes some phyla of protists. The position of the root of the eukaryotic tree is uncertain, as are the branching patterns of supergroups and many phyla. Many more branches exist than are shown here.

vata is named for a feeding groove "excavated" into the cells of many representatives, such as the genus *Jakoba* (phylum Metamonada) (**Figure 28.6**). The feeding groove is an important adaptation that allows these organisms (informally called excavates) to ingest small particles of food in their aquatic habitats. Once food particles are collected within the feeding groove, they are then taken into cells by a type of endocytosis known as **phagocytosis** (from the Greek, meaning cellular eating). During phagocytosis, a vesicle of plasma membrane surrounds each food particle and pinches off within the cytoplasm. Enzymes within these food vesicles break the food particles down into small molecules that, upon their release into the cytoplasm, can be used for energy. Phagocytosis is also the basis for an important evolutionary process known as endosymbiosis.

Endosymbiosis is a symbiotic association in which a smaller species known as the **endosymbiont** lives within the body of a larger species known as the **host**. Phagocytosis provides a way for protist cells that function as hosts to take in prokaryotic or eukaryotic cells that function as endosymbionts. Such endosymbiotic cells confer valuable traits and are not digested. Endosymbiosis has played a particularly important role in protist evolution. For example, early in protist history, endosymbiotic bacterial cells gave rise to mitochondria, the organelle that is the major site of ATP synthesis in most eukaryotic cells (see Chapter 7). Consequently, most protists possess mitochondria, though these may be highly modified in some species, including certain modern protists that can be linked with Excavata.

Some excavate protists have become parasitic within animals, including human hosts. In addition to feeding by phagocytosis, parasitic species attack host cells and absorb food molecules released from them. For example, *Trichomonas vaginalis* causes a sexually transmitted infection of the human genitourinary tract. In this location, *T. vaginalis* consumes bacteria and host epithelial and red blood cells by phagocytosis, as well

(a) Excavate *Jakoba* with feeding groove

(b) The process of phagocytosis in excavates

Figure 28.6 Feeding groove and phagocytosis displayed by many species of supergroup Excavata.   **(a)** Diagram of *Jakoba libera*, phylum Metamonada, showing flagella emerging from the feeding groove. **(b)** Diagram of phagocytosis, the process by which food particles are consumed at a feeding groove.

Concept check: What happens to ingested particles after they enter feeding cells?

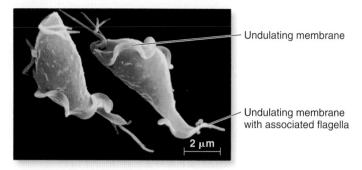

Undulating membrane

Undulating membrane with associated flagella

2 μm

(a) *Trichomonas vaginalis*

Flagella

2.5 μm

(b) *Giardia lamblia*

Figure 28.7   Parasitic members of the supergroup Excavata. **(a)** *Trichomonas vaginalis*. **(b)** *Giardia lamblia*. These specialized flagellates absorb nutrients from living cells of their hosts.

Concept check: How do these two parasitic protists differ in the process of transmission from one human host to another?

as carbohydrates and proteins released from damaged host cells. More than 170 million cases are estimated to occur each year around the globe, and infections can predispose humans to other diseases. *T. vaginalis* has an undulating membrane and flagella that allow it to move over mucus-coated skin (**Figure 28.7a**).

Giardia lamblia, another type of excavate protist, contains two active nuclei and produces eight flagella (**Figure 28.7b**). *G. lamblia* causes giardiasis, an intestinal infection that can result from drinking untreated water or from unsanitary conditions in day-care centers. Flagellate cells attach to the epithelium of the small intestine, causing disease and also producing infectious stages (cysts) that are transmitted in feces. *T. vaginalis* and *G. lamblia* were once thought to lack mitochondria, but they are now known to possess structures that are highly modified mitochondria.

Genomes & Proteomes Connection

Genome Sequences Reveal the Different Evolutionary Pathways of *Trichomonas vaginalis* and *Giardia lamblia*

In 2007, genome sequences were reported for two excavate-related parasites, *Trichomonas vaginalis* and *Giardia lamblia*. A comparison of their genomic features reveals similarities and differences in the evolution of parasitic lifestyles. One common feature is that horizontal gene transfer from bacterial or archaeal

donors has powerfully affected both genomes. About 100 *G. lamblia* genes are likely to have originated via horizontal gene transfer. In *T. vaginalis*, more than 150 cases of likely horizontal gene transfer were identified, with most transferred genes encoding metabolic enzymes such as those involved in carbohydrate or protein metabolism. Another similarity between *T. vaginalis* and *G. lamblia* revealed by comparative genomics is an absence of the cytoskeletal protein myosin, which is present in most eukaryotic cells.

Despite these similarities, the genome sequences of *T. vaginalis* and *G. lamblia* reveal some dramatic differences. The *G. lamblia* genome is quite compact, only 11.7 megabases (MB) in size, with relatively simple metabolic pathways and machinery for DNA replication, transcription, and RNA processing. In contrast, the *T. vaginalis* genome is a surprisingly large 160 MB in size. *T. vaginalis* has a core set of about 60,000 protein-coding genes, one of the greatest coding capacities known among eukaryotes. The additional genes provide an expanded capacity for biochemical degradation. Because most trichomonads inhabit animal intestine, the genomic data suggest that the large genome size of *T. vaginalis* is related to its ecological transition to a new habitat, the urogenital tract.

Euglenoids   The excavate protists known as euglenoids possess unique, interlocking ribbon-like protein strips just beneath their plasma membranes (**Figure 28.8a**). These strips make the surfaces of some euglenoids so flexible that they can crawl through mud. Many euglenoids are colorless and heterotrophic, but *Euglena* and some other genera possess green plastids and are photosynthetic. Plastids are organelles found in plant and algal cells that are distinguished by their synthetic abilities and were acquired via endosymbiosis. Many euglenoids possess a light-sensing system that includes a conspicuous red structure known as an eyespot, or stigma, and light-detecting molecules located in a swollen region at the base of a flagellum. Most euglenoids produce conspicuous storage carbohydrate particles that occur in the cytoplasm. Euglenoids are particularly abundant and ecologically significant in wetlands.

Kinetoplastids   The heterotrophic excavate protists known as kinetoplastids are named for a large mass of DNA known as a kinetoplast that occurs in their single large mitochondrion (**Figure 28.8b**). These protists lack plastids, but they do possess an unusual modified peroxisome that contains glycolytic enzymes; in most eukaryotes, glycolysis occurs in the cytosol. Some kinetoplastids, including *Leishmania* (Figure 28.8b) and *Trypanosoma brucei*, the causative agent of sleeping sickness, are serious pathogens of humans and other animals (**Figure 28.8c**).

Land Plants and Related Algae Share Similar Genetic Features

The supergroup that includes land plants also encompasses several algal phyla (**Figure 28.9**). The land plants, also known as the

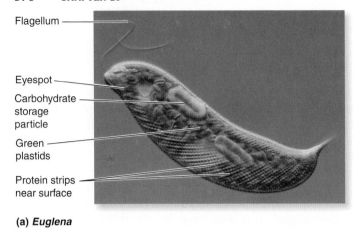

Flagellum

Eyespot

Carbohydrate
storage
particle

Green
plastids

Protein strips
near surface

(a) *Euglena*

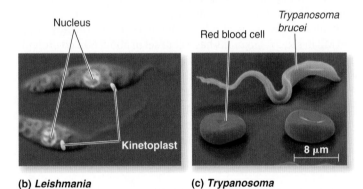

Nucleus

*Trypanosoma
brucei*

Red blood cell

Kinetoplast

8 μm

(b) *Leishmania*   (c) *Trypanosoma*

Figure 28.8   Representative euglenoids and kinetoplastids.
(a) *Euglena* has helical protein ribbons near its surface, internal
green plastids, white storage carbohydrate granules, and a
red eyespot. (b) Fluorescence LM of *Leishmania* showing the
kinetoplast DNA mass typical of kinetoplastid mitochondria.
(c) In this SEM, several undulating kinetoplastids (*Trypanosoma*)
appear near disc-shaped red blood cells.

kingdom Plantae (described more fully in Chapters 29 and 30),
evolved from green algal ancestors. Together, plants and some
closely related green algae, called streptophyte algae, form a
clade, while most green algae are classified in the phylum Chlorophyta. The red algae, classified in the phylum Rhodophyta, are
also regarded as close relatives of green algae and land plants.
Recent molecular sequence data and some similarities in cell
structure link the plants and green and red algae to two additional algal phyla, the Haptophyta and the Cryptophyta.

Green Algae   Diverse structural types of green algae (see Figure 28.2) occur in fresh waters, the ocean, and on land. Most
of the green algae are photosynthetic, and their cells contain
the same types of plastids and photosynthetic pigments as are
present in land plants. Some green algae are responsible for
harmful algal growths, but others are useful as food for aquatic
animals, model laboratory research systems, or even possible
sources of renewable oil supplies. Many green algae possess
flagella or the ability to produce them during the development
of reproductive cells.

Red Algae   Most species of the protists known as red algae
are multicellular marine macroalgae (**Figure 28.10**). The red

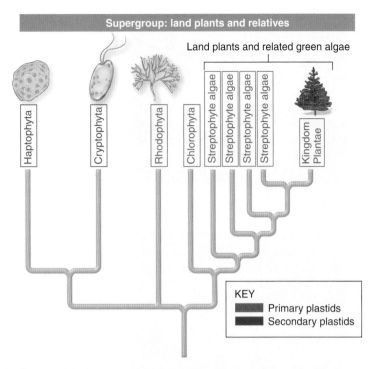

Supergroup: land plants and relatives

Land plants and related green algae

Haptophyta

Cryptophyta

Rhodophyta

Chlorophyta

Streptophyte algae

Streptophyte algae

Streptophyte algae

Streptophyte algae

Kingdom
Plantae

KEY
Primary plastids
Secondary plastids

Figure 28.9   A phylogenetic tree of the supergroup that
includes land plants and their close protist relatives.

appearance of these algae is caused by the presence of distinctive photosynthetic pigments that are absent from green algae
or land plants. Red algae characteristically lack flagella, and
this feature has strongly influenced the evolution of this group,
resulting in unusually complex life cycles (described in Section 28.4). These life cycles are important to humans because
we cultivate red algae in ocean waters for production of food
or industrial and scientific materials. For example, the sushi
wrappers called nori are composed of the sheetlike red algal
genus *Porphyra*, which is grown in ocean farms. Knowledge
of *Porphyra*'s life cycle was critical to the economic development of this important seaweed crop. Red algae are grown or
harvested for other commercially valuable products, such as
carrageenan. Carrageenan is a complex red algal cell-wall polysaccharide that has numerous applications in the food industry;
for example, you may have seen it listed in the ingredients for

(a) *Calliarthron*   (b) *Chondrus crispus*

Figure 28.10   Representative red algae (Rhodophyta).
(a) The genus *Calliarthron* has cell walls that are impregnated with
calcium carbonate. This stony, white material makes the red alga
appear pink. (b) *Chondrus crispus* is an edible red seaweed.

ice cream or chocolate milk, where it is used to keep chocolate particles evenly suspended.

Primary Plastids and Primary Endosymbiosis   The plastids of red algae resemble those of green algae and land plants (and differ from most other algae) in having an enclosing envelope composed of two membranes (**Figure 28.11**). Such plastids, known as **primary plastids**, are thought to have originated via a process known as **primary endosymbiosis**. During primary endosymbiosis, heterotrophic host cells captured cyanobacterial cells via phagocytosis but did not digest them. These endosymbiotic cyanobacteria provided host cells with photosynthetic capacity and other useful biochemical pathways and eventually evolved into primary plastids (**Figure 28.12a**). Endosymbiotic acquisitions of plastids and mitochondria resulted in massive horizontal gene transfer from the endosymbiont to the

Envelope of 2 membranes

Thylakoids

Figure 28.11   A primary plastid, showing an envelope composed of two membranes.   The plastid shown here is red, but primary plastids can also be green or blue-green in color.

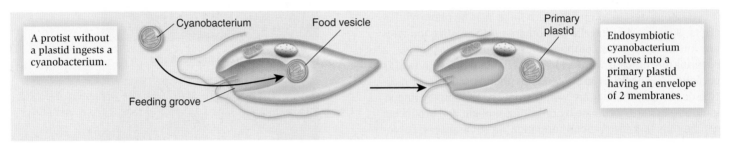

A protist without a plastid ingests a cyanobacterium.

Cyanobacterium

Feeding groove

Food vesicle

Primary plastid

Endosymbiotic cyanobacterium evolves into a primary plastid having an envelope of 2 membranes.

(a) Primary endosymbiosis

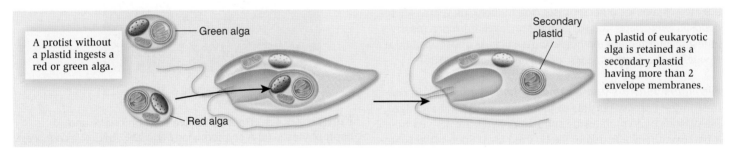

A protist without a plastid ingests a red or green alga.

Green alga

Red alga

Secondary plastid

A plastid of eukaryotic alga is retained as a secondary plastid having more than 2 envelope membranes.

(b) Secondary endosymbiosis

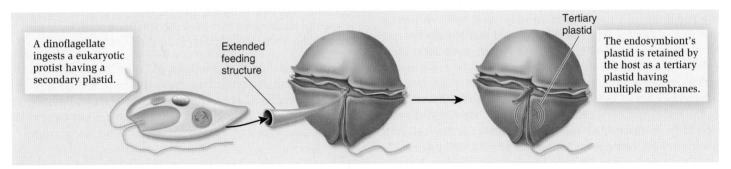

A dinoflagellate ingests a eukaryotic protist having a secondary plastid.

Extended feeding structure

Tertiary plastid

The endosymbiont's plastid is retained by the host as a tertiary plastid having multiple membranes.

(c) Tertiary endosymbiosis

Figure 28.12   Primary, secondary, and tertiary endosymbiosis.   (a) Primary endosymbiosis involves the acquisition of a cyanobacterial endosymbiont by a host cell without a plastid. During the evolution of a primary plastid, the bacterial cell wall is lost, and most endosymbiont genes are transferred to the host nucleus. (b) Secondary endosymbiosis involves the acquisition by a host cell of a eukaryotic endosymbiont that contains one or more primary plastids. During the evolution of a secondary plastid, most components of the endosymbiont cell are lost, but a plastid is often retained within an envelope of endoplasmic reticulum. (c) Tertiary endosymbiosis involves the acquisition by a host cell of a eukaryotic endosymbiont that possesses secondary plastids.

host nucleus. As a result of such gene transfer, many of the proteins needed by plastids and mitochondria are synthesized in the host cytoplasm and then targeted to these organelles. All cells of plants, green algae, and red algae contain one or more plastids, and most of these organisms are photosynthetic. However, some species (or some of the cells within the multicellular bodies of photosynthetic species) are heterotrophic because photosynthetic pigments are not produced in the plastids.

Cryptomonads and Haptophytes Some gene sequence data and cellular features indicate that plants, green algae, and red algae may be closely related to two additional algal groups, the cryptomonads and the haptophytes (see Figure 28.9). Cryptomonads are unicellular flagellates, most of which contain red, blue-green, or brown plastids and are photosynthetic (Figure 28.13a). Occurring in marine and fresh waters, cryptomonads are excellent sources of the fatty-acid-rich food essential to aquatic animals.

Haptophytes are primarily unicellular marine photosynthesizers, some having flagella and others not. Some haptophytes are known as the coccolithophorids because they produce a covering of intricate white calcium carbonate discs known as coccoliths (Figure 28.13b). Coccolithophorids often form massive ocean growths that are visible from space and play important roles in Earth's climate by reflecting sunlight and producing compounds that foster cloud formation. In some places, coccoliths produced by huge populations of ancient coccolithophorids accumulated on the ocean floor, together with the calcium carbonate remains of other protists, for millions of years. These deposits were later raised above sea level, forming massive limestone formations or chalk cliffs such as those visible at Dover, on the southern coast of England (Figure 28.13c).

Secondary Plastids and Secondary Endosymbiosis In contrast to the primary plastids of plants and green and red algae, the plastids of cryptomonads and haptophytes are derived from a photosynthetic eukaryote, likely a red alga. Such plastids are known as **secondary plastids** because they originate by the process of **secondary endosymbiosis** (see Figure 28.12b). Secondary endosymbiosis occurs when a eukaryotic host cell ingests and retains another type of eukaryotic cell that already has one or more primary plastids, such as a red or green alga. Such eukaryotic endosymbionts are often enclosed by endoplasmic reticulum (ER), explaining why secondary plastids typically have envelopes of more than two membranes. Although most of the endosymbiont's cellular components are digested away over time, its plastids are retained, providing the host cell with photosynthetic capacity and other biochemical capabilities.

Cryptomonad plastids are unusual because they retain a highly reduced form of the captured red algal nucleus, known as a nucleomorph (Figure 28.14). Located within the enclosing ER, but outside the two inner plastid membranes, the nucleomorph contains a small amount of DNA that encodes materials needed by the cryptomonad plastid that are not provided by host cells. Evolutionary biologists are intrigued by cryptomonad nucleomorphs because they reflect an intermediate stage in the

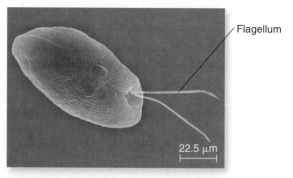

Flagellum

22.5 μm

(a) A cryptomonad

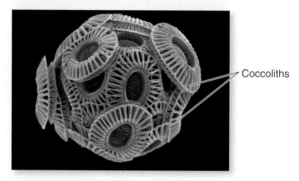

Coccoliths

(b) A haptophyte coccolithophorid

(c) Fossil deposit containing coccolithophorids

Figure 28.13 Representative cryptomonads and haptophytes. (a) A cryptomonad flagellate. (b) *Emiliania huxleyi*, a type of haptophyte known as a coccolithophorid, covered with disc-shaped coccoliths made of calcium carbonate. (c) Fossil carbonate remains of haptophyte algae and protozoan protists known as foraminifera that were deposited over millions of years formed the white cliffs of Dover in England, pictured here.

origin of plastids by secondary endosymbiosis. Haptophyte plastids arose by secondary endosymbiosis involving a red alga, and euglenoid plastids arose by secondary endosymbiosis involving a green alga, but neither haptophytes nor euglenoids have retained the endosymbiont nucleus as a nucleomorph.

Membrane Sacs Lie at the Cell Periphery of Protists Classified in Alveolata

The three supergroups Alveolata, Stramenopila, and Rhizaria seem to form a cluster in recent phylogenetic studies (Figure 28.15). Turning first to Alveolata, we see that it includes three important phyla: (1) the Ciliophora, or ciliates; (2) the Apicomplexa, a medically important group of parasites; and

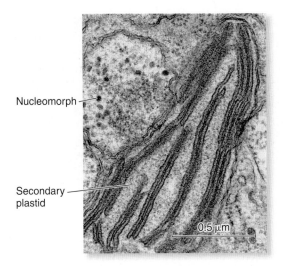

Figure 28.14 A nucleomorph within the secondary plastid of a cryptomonad.

Concept check: What function does the nucleomorph serve?

(3) the Dinozoa, informally known as dinoflagellates. Ciliates are notable for a complex mating behavior known as conjugation (see Section 28.4). Apicomplexans include the malarial agent *Plasmodium falciparum* (see Section 28.4), the related protist *Cryptosporidium parvum*, and other serious pathogens of humans and other animals. Dinoflagellates are recognized for the mutually beneficial relationships that some establish with reef-building corals and the harmful blooms (red tides) that other species produce (see Section 28.3). The Alveolata is named for saclike membranous vesicles known as alveoli that are present at the cell periphery in all of these phyla, illustrated here by dinoflagellates (**Figure 28.16a**).

The alveoli of some dinoflagellates seem empty, so the cell surface appears smooth (**Figure 28.16b**). By contrast, the alveoli

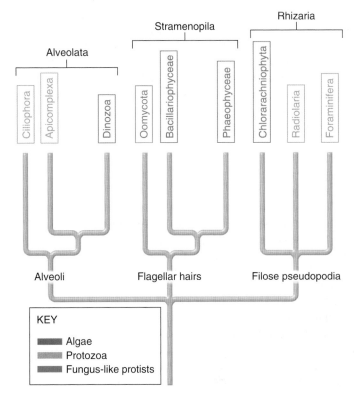

Figure 28.15 A phylogenetic tree illustrating close relationship among the supergroups Alveolata, Stramenopila, and Rhizaria.

of many dinoflagellates contain plates of cellulose, which form an armor-like enclosure (**Figure 28.16c**). These plates are often modified in ways that provide adaptive advantage, such as protection from predators or increased ability to float.

(a) Cross section through characteristic alveoli

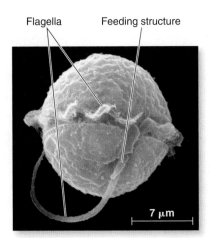

(b) A freshwater dinoflagellate with empty alveoli

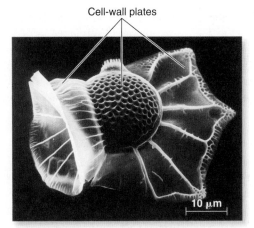

(c) A marine dinoflagellate with cell-wall plate alveoli

Figure 28.16 Dinoflagellates of the supergroup Alveolata and their characteristic alveoli. **(a)** Sac-shaped membranous vesicles known as alveoli lie beneath the plasma membrane of a dinoflagellate, along with defensive projectiles, called trichocysts, ready for discharge. **(b)** The surface of the freshwater dinoflagellate *Peridiniopsis berolinensis* appears smooth because the alveoli seem empty. Dinoflagellates have two types of flagella. One flagellum coils around a cellular groove; as it moves, this flagellum causes the cell to spin. A second flagellum extends from the cell, acting as a rudder to determine the direction of movement. **(c)** The alveoli of the marine dinoflagellate genus *Ornithocercus* contain cellulose cell-wall plates.

Concept check: Why do parts of the cell wall of Ornithocercus resemble sails?

About half of dinoflagellate species are heterotrophic and half possess photosynthetic plastids of diverse types that originated by secondary or even tertiary endosymbiosis; therefore, these are known as secondary or tertiary plastids. **Tertiary plastids** were obtained by **tertiary endosymbiosis**—the acquisition by hosts of plastids from cells that possessed secondary plastids (see Figure 28.12c). Species having tertiary plastids have received genes by horizontal transfer from diverse genomes.

Flagellar Hairs Distinguish Stramenopila

The supergroup Stramenopila (informally known as the stramenopiles) encompasses a wide range of algae, protozoa, and fungus-like protists that usually produce flagellate cells at some point in their lives (see Figure 28.15). The Stramenopila (from the Greek *stramen*, meaning straw, and *pila*, meaning hair) is named for distinctive strawlike hairs that occur on the surfaces of flagella (**Figure 28.17**). These flagellar hairs function something like oars to greatly increase swimming efficiency. Stramenopiles are also informally known as heterokonts (from the Greek, meaning different flagella), because the two flagella often present on swimming cells have slightly different structures.

Stramenopiles include diverse heterotrophic protists as well as many groups of algae having golden- or brown-colored plastids. The plastids of stramenopile algae originated from red algae

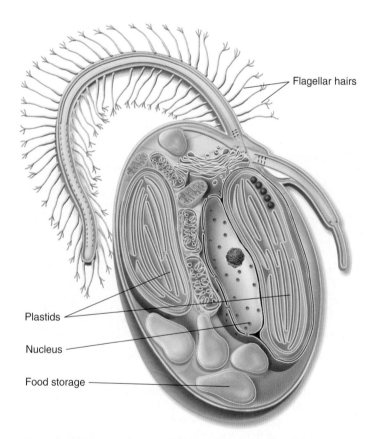

Flagellar hairs

Plastids

Nucleus

Food storage

Figure 28.17   Diagram of a flagellate stramenopile cell, showing characteristic flagellar hairs.

Concept check: *How do the flagellar hairs aid cell motion?*

(a) Diatom   **(b) Kelp forest**

Figure 28.18   Stramenopiles include diatoms and giant kelps.   **(a)** SEM of a diatom cell wall, showing elaborate ornamentation of the silicate structure. **(b)** Forests of giant kelps occur along many ocean shores, providing habitat for diverse organisms.

Concept check: *In what ways are kelp forests economically important?*

by secondary endosymbiosis. Stramenopiles include diatoms (Bacillariophyceae), whose glass cell walls are elaborately ornamented with pores, lines, and other intricate features (**Figure 28.18a**). Recent genome sequencing projects have focused on the processes by which diatoms produce such detailed silicate structures, which may prove useful in industrial microfabrication applications. Vast accumulations of the glass walls of ancient diatoms, known as diatomite or diatomaceous earth, are mined for use in reflective paint and other industrial products.

Diverse brown algae (Phaeophyceae) are sources of industrial products such as polysaccharide emulsifiers known as alginates. The brown algae known as giant kelps are ecologically important because they form extensive forests in cold and temperate coastal oceans (**Figure 28.18b**). Kelp forests are essential nurseries for fish and shellfish. The reproductive processes of kelps are described in Section 28.4.

Spiky Cytoplasmic Extensions Are Present on the Cells of Many Protists Classified in Rhizaria

Several groups of flagellates and amoebae that have thin, hair-like extensions of their cytoplasm, known as filose pseudopodia, are classified into the supergroup Rhizaria (from the Greek *rhiza*, meaning root) (see Figure 28.15). Rhizaria includes the phylum Chlorarachniophyta, whose spider-shaped cells possess secondary plastids obtained from endosymbiotic green algae. Other examples of Rhizaria are the Radiolaria (**Figure 28.19a**) and Foraminifera (**Figure 28.19b**), two phyla of ocean plankton that produce exquisite mineral shells. Fossil shells of foraminiferans are widely used to infer past climatic conditions.

Amoebozoa Includes Many Types of Amoebae with Pseudopodia

The supergroup Amoebozoa includes many types of amoebae that move by extension of pseudopodia (see Figure 28.4). One example is the human parasite *Entamoeba histolytica*, which

Filose pseudopodium

Filose pseudopodium

0.5 mm

(a) Radiolarian

(b) Foraminiferan

Figure 28.19 Representatives of supergroup Rhizaria. **(a)** A radiolarian, *Acanthoplegma* spp., showing long filose pseudopodia. **(b)** A foraminiferan, showing calcium carbonate shell with long filose pseudopodia extending from pores in the shell.

causes a severe form of intestinal illness. Several types of protists known as slime molds are also classified in this supergroup. For example, the phylum Dictyostelia includes the slime mold *Dictyostelium discoideum*, widely used as a model laboratory system for understanding communication among cells.

A Single Flagellum Occurs on Swimming Cells of Opisthokonta

The supergroup Opisthokonta includes the animal and fungal kingdoms and related protists (**Figure 28.20**). This supergroup is named for the presence of a single posterior flagellum on

Supergroup: Opisthokonta

Nuclearia

Kingdom Fungi

Choanomonada

Kingdom Animalia

Multicellularity

Multicellularity

KEY

Critical innovations

Figure 28.20 A phylogenetic tree of the supergroup Opisthokonta. This supergroup includes protist phyla as well as the kingdoms Fungi and Animalia. Notice that a critical innovation of these kingdoms is multicellularity.

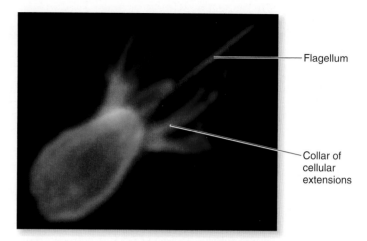

Flagellum

Collar of cellular extensions

Figure 28.21 A choanoflagellate of the supergroup Opisthokonta. This cell has been stained with fluorescent dyes specific for DNA (blue) and the proteins actin (red) and tubulin (green). The single flagellum is stained green because it is rich in tubulin. A collar of cellular extensions that are rich in actin surrounds the flagellum.

Concept check: *What features of the ancient choanoflagellate ancestors of animals were important in the evolution of multicellularity, and what function do such features serve in modern choanoflagellates?*

swimming cells. Closely related to the animal kingdom are the protists known as choanoflagellates (formally, the Choanomonada). These single-celled or colonial protists feature a distinctive collar surrounding the single flagellum (**Figure 28.21**). The collar is made of cytoplasmic extensions that filter bacterial food from water currents generated by flagellar motion.

Evolutionary biologists interested in the origin of animals study choanoflagellates for molecular clues to this important event in our evolutionary history. In 2008, evolutionary biologist Nicole King and associates reported a genome sequence for the choanoflagellate *Monosiga brevicollis* and identified several genes that are present only in choanoflagellates and animals. Some of the shared genes encode cell adhesion and extracellular matrix proteins that help choanoflagellates attach to surfaces and were also essential to the evolution of multicellularity in animals.

The preceding survey of protist diversity, summarized in **Table 28.1**, illustrates the enormous evolutionary and ecological importance of protists. Next, we consider the diverse ways in which protists have become adapted to their environments.

28.3 Nutritional and Defensive Adaptations

Wherever you look in moist places, you will find protists playing diverse and important ecological roles. In this section, we will survey protist nutritional and defensive adaptations, which help to explain their ecological roles.

Table 28.1	Eukaryotic Supergroups and Examples of Constituent Kingdoms, Phyla, Classes, or Species	
Supergroup	**KINGDOMS, Phyla**, classes, or *species*	**Distinguishing features**

EXCAVATA	**Metamonada**
	  Giardia lamblia
	  Trichomonas vaginalis
	
	Kinetoplastea (kinetoplastids)
	  Trypanosoma brucei
	Euglenida (euglenoids)
		Unicellular flagellates, often with feeding groove
		
		Secondary plastids (when present) derived from endosymbiotic green algae

LAND PLANTS AND ALGAL RELATIVES	**Rhodophyta** (red algae) **Chlorophyta** (green algae) KINGDOM PLANTAE and close green algal relatives **Cryptophyta** (cryptomonads) **Haptophyta** (haptophytes, including coccolithophorids)	Land plants, green algae, and red algae have primary plastids derived from cyanobacteria; such plastids have two envelope membranes Haptophytes and most cryptomonads possess secondary plastids derived from red algae; such plastids have more than two envelope membranes. Some experts classify cryptomonads and haptophytes with alveolates, stramenopiles, and rhizaria.

ALVEOLATA	**Ciliophora** (ciliates)
	Apicomplexa (apicomplexans)
	  Plasmodium falciparum
	  Cryptosporidium parvum
	Dinozoa (dinoflagellates)
	  Pfiesteria shumwayae
		Peripheral membrane sacs (alveoli); Apicomplexa sometimes have secondary plastids derived from red algae; some Dinozoa have secondary plastids derived from red algae, some have secondary plastids derived from green algae, and some have tertiary plastids derived from diatoms, haptophytes, or cryptomonads

STRAMENOPILA	Bacillariophyceae (diatoms) Phaeophyceae (brown algae) *Phytophthora infestans* (fungus-like)	Strawlike flagellar hairs; secondary plastids (when present) derived from red algae; fucoxanthin accessory pigment common in autotrophic forms
RHIZARIA	**Chlorarachniophyta** **Radiolaria** **Foraminifera**	Thin, cytoplasmic projections; secondary plastids (when present) derived from endosymbiotic green algae
AMOEBOZOA	*Entamoeba histolytica* **Dictyostelia** (a slime mold phylum) *Dictyostelium discoideum*	Amoeboid movement by pseudopodia
OPISTHOKONTA	*Nuclearia* spp. KINGDOM FUNGI **Choanomonada** (choanoflagellates) KINGDOM ANIMALIA	Swimming cells possess a single posterior flagellum

Protists Display Four Basic Types of Nutrition

Protist nutrition occurs in four basic types: phagotrophy, osmotrophy, photoautotrophy, and mixotrophy. Heterotrophic protists that feed by ingesting particles, or phagocytosis, are known as **phagotrophs**, whereas those relying on osmotrophy (uptake of small organic molecules) are **osmotrophs**. As mentioned earlier, photosynthetic protists are **photoautotrophs**, organisms that can make their own organic nutrients by harvesting light energy. **Mixotrophs** are able to use photoautotrophy as well as phagotrophy or osmotrophy to obtain organic nutrients. The genus *Dinobryon* (**Figure 28.22**), a photosynthetic stramenopile that lives in the phytoplankton of freshwater lakes, is an example of a mixotroph. These protists may switch back and forth between photoautotrophy and heterotrophy, depending on conditions in their environment. If sufficient light, carbon dioxide, and other minerals are available, *Dinobryon* cells produce their own organic food. If any of these resources limits photosynthesis, or organic food is especially

abundant, *Dinobryon* cells can function as heterotrophs, consuming enormous numbers of bacteria. Mixotrophs thus have remarkable nutritional flexibility.

Heterotrophic protists that feed on nonliving organic material function as decomposers, essential in breaking down wastes and releasing minerals for use by other organisms. Heterotrophic

Figure 28.22   A mixotrophic protist. The genus *Dinobryon* is a colonial flagellate that occurs in the phytoplankton of freshwater lakes. The photosynthetic cells have golden photosynthetic plastids and also capture and consume bacterial cells.

protists that feed on the living cells of other organisms are parasites that may cause disease in other organisms. *Trichomonas vaginalis*, *Giardia lamblia*, *Entamoeba histolytica*, and *Phytophthora infestans* are examples of pathogenic protists that have previously been described in this chapter. Humans view such protists as pests when they harm us or our agricultural animals and crops, but pathogenic protists also play important roles in nature by controlling the population growth of other organisms.

Algal Protists Vary in Photosynthetic Pigments and Food Storage Molecules

As we have earlier noted, algae display a surprising diversity of coloration: gold, brown, red, blue-green, and green. Why do so many pigmentation types occur? The answer is related to light availability in these protists' watery environment.

If you dive more than a few feet into lakes or the ocean, your aquatic world will appear intensely blue-green. This occurs because water absorbs the longer red to yellow wavelengths of light to a greater degree than it does the shorter blue and green wavelengths. Little red light penetrates far into natural waters, depriving chlorophyll *a*, the photosynthetic pigment generally present in photosynthetic eukaryotes, of much of the light it would ordinarily absorb. Photosynthetic protists living in water have evolved photosynthetic systems that capture more of the available blue-green light. Macroalgae may be colored red, gold, or brown, in addition to green, because they produce diverse accessory pigments that can absorb more of the light available underwater and transfer the energy to chlorophyll *a* for photosynthesis. For example, the red accessory pigment phycoerythrin is abundant in plastids of red algae, and golden fucoxanthin enriches the color of golden and brown algae (compare Figures 28.10 and 28.18b). Carotene and lutein play similar accessory pigment roles in green algae and were inherited by their land plant descendants, today playing important roles in animal nutrition.

Photosynthetic protists also vary in the types of molecules that serve as food storage. Starch stored in the plastids of green algae, the cytoplasmic polysaccharides of euglenoids, and oil droplets in diatom cells are examples. Algae use such food storage materials for energy when light or minerals are too low to allow photosynthesis. These food storage materials explain why algae are both useful sources of renewable energy and desirable food sources for heterotrophs (see Figure 28.1).

Protists Defend Themselves in Diverse Ways

Protists use a wide variety of defensive adaptations to ward off attack. Major types of defenses are slimy, tough, or spiny cell coverings; sharp projectiles that can be explosively shot from cells; light flashes; and toxic compounds.

Many protists have cell coverings such as slimy mucilages (see chapter-opening photo) or cell walls that provide protection from attack by herbivores or pathogens. Cell walls may also aid in preventing osmotic damage or enhance flotation in water (see Figure 28.16c). Rigid cellulose walls are common in brown and green algae, whereas slimy polysaccharide polymers form a protective matrix around red algal cells. Calcium carbonate forms a stony coat for many protist cells, including foraminifera, haptophytes, and some marine macroalgae (see Figures 28.10a and 28.13b). As noted earlier, ornate glassy coatings of silica protect diatoms, while metallic iron and manganese crystals armor other protists.

Some protist defenses reduce predation by herbivores. Several types of protist cells contain compressed protein structures known as trichocysts (see Figure 28.16a). Upon attack, trichocysts rapidly elongate into spear-shaped projectiles that are shot from these cells, thereby discouraging herbivores from feeding. Some species of ocean dinoflagellates emit flashes of blue light when disturbed, explaining why ocean waters teeming with these protists display bioluminescence. The light flashes may deter herbivores by startling them, but when ingested, the dinoflagellates make the herbivores also glow, revealing them to hungry fishes. Light flashes benefit dinoflagellates by helping to reduce populations of herbivores that consume the algae.

Various protist species produce **toxins**, compounds that inhibit animal physiology and may function to deter small herbivores. Dinoflagellates are probably the most important protist toxin producers; they synthesize several types of toxins that affect humans and other animals. Why does this happen? Under natural conditions, small populations of dinoflagellates produce low amounts of toxin that do not harm large organisms. Dinoflagellate toxins become dangerous to humans when people contaminate natural waters with excess mineral nutrients such as nitrogen and phosphorus from untreated sewage, industrial discharges, or fertilizer that washes from agricultural fields. The excess nutrients fuel the development of harmful algal blooms, which then produce sufficient toxin to affect birds, aquatic mammals, fishes, and humans. Toxins can concentrate in organisms. Humans who ingest shellfish that have accumulated dinoflagellate toxins can suffer poisoning.

In the early 1990s, ecologist JoAnn Burkholder and colleagues reported that the toxic dinoflagellate *Pfiesteria* had caused major fish kills in the overly fertile waters of the Chesapeake Bay. These investigators observed that the dinoflagellate could live by consuming algal cells but also could produce a toxin that damages fish skin, allowing the dinoflagellates to consume fish flesh. These biologists also discovered that *Pfiesteria*-associated toxin caused amnesia and other nervous system conditions in fishers and scientists who were exposed to it, though the cellular basis of the effect on humans was unclear. The discovery of *Pfiesteria* excited the media, which dubbed it a "killer alga" and focused attention on nutrient pollution of Chesapeake Bay, the fundamental cause of *Pfiesteria*'s excessive growth and fish kills. Because of this organism's importance to the fishing industry and human health, teams of aquatic ecologists have continued to study the genus *Pfiesteria* and its toxin, as described next.

FEATURE INVESTIGATION

Burkholder and Colleagues Demonstrated That Strains of the Dinoflagellate Genus *Pfiesteria* Are Toxic to Mammalian Cells

A team of investigators led by JoAnn Burkholder performed an experiment to determine whether or not two strains of *Pfisteria shumwayae* were toxic to mammalian cells (**Figure 28.23**). Although one of these strains (CCMP 1024C) was thought to be toxic to fishes and people, other work suggested that a different strain (CCMP 2089) did not produce toxin, a difference that might have resulted from variation in growth conditions. Neither strain had been tested for its impact on mammalian cells. The experiment was conducted in a biohazard containment facility, for the safety of the investigators.

In the first step of the experiment, the team provided both *Pfiesteria* strains with two forms of food—cryptomonad algal cells or juvenile fish—because the impact of food type on toxicity was unclear. In a second step, the dinoflagellates grown in step 1 were transferred to tanks with fish to elicit maximal toxin production, before treating mammalian cell cultures with the dinoflagellates. Dinoflagellates were not added to a control tank of fish. Toxin was detected, and fish deaths occurred in all of the tanks except the control. In a third step, samples from the step 2 treatments were added to mammalian cell cultures, and investigators determined the relative levels of toxicity to the mammalian cells. They found that both strains of *P. shumwayae* were toxic to mammalian cells and that both types of food supported the growth of toxin-producing dinoflagellates.

Figure 28.23 Burkholder and colleagues demonstrated that some strains of *Pfiesteria shumwayae* are toxic to fish and mammalian cells.

4 THE DATA

Results from step 3:

5 CONCLUSION Both strains of *P. shumwayae* were toxic to mammalian cells, and both types of food supported the growth of toxic dinoflagellates.

6 SOURCE Burkholder, J.M., et al. 2005. Demonstration of toxicity to fish and to mammalian cells by *Pfiesteria* species: Comparison of assay methods and strains. *Proceedings of the National Academy of Sciences of the United States of America* 102:3471–3476.

Experimental Questions

1. Why did the Burkholder team test two different strains of *Pfiesteria shumwayae*?

2. Why did the Burkholder team grow *Pfiesteria shumwayae* with algae or fish as food?

3. Why did the Burkholder team use a biohazard containment system?

28.4 Reproductive Adaptations

Diverse reproductive adaptations allow protists to thrive in an amazing variety of environments, including the bodies of hosts in the cases of parasitic protists. These adaptations include specialized asexual reproductive cells, tough-walled dormant cells that allow protists to survive periods of environmental stress, and several types of sexual life cycles.

Protist Populations Increase by Means of Asexual Reproduction

All protists are able to reproduce asexually by mitotic cell divisions of parental cells to produce progeny. When resources are plentiful, repeated mitotic divisions of single-celled protists will generate large protist populations. By contrast, multicellular protists often generate specialized asexual cells that help disperse the organisms in their environment.

Many protists produce unicellular **cysts** as the result of asexual (and in some cases, sexual) reproduction (**Figure 28.24**). Cysts often have thick, protective walls and can remain dormant through periods of unfavorable climate or low food availability. Dinoflagellates commonly produce cysts that can be transported in ship ballast water from one port to another,

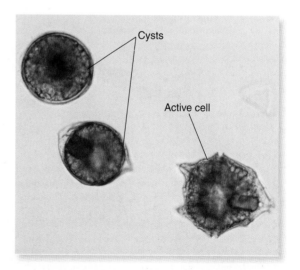

Figure 28.24   Protistan cysts.   The round cells are dormant, tough-walled cysts of the dinoflagellate *Peridinium limbatum*. The pointed cell is an actively growing cell of the same species. As cysts develop, the outer cellulose plates present on actively growing cells are cast off.

Concept check: *How can cysts be involved in the spread of harmful algae and disease-causing parasitic protists?*

a problem that has caused harmful dinoflagellate blooms to appear in harbors around the world. Ship captains can help to prevent such ecological disasters by heating ballast water before it is discharged from ships.

Many protozoan pathogens spread from one host to another via cysts. For example, the apicomplexan pathogen *Cryptosporidium parvum* infects humans via waterborne cysts. *Entamoeba histolytica* is a pathogen that infects people who consume food or water that is contaminated with its cysts. Once inside the human digestive system, *E. histolytica* attacks intestinal cells, causing amoebic dysentery, a worldwide problem.

Sexual Reproduction Provides Multiple Benefits to Protists

Eukaryotic sexual reproduction, featuring gametes, zygotes, and meiosis, first arose among protists. Sexual reproduction has not been observed in some protist phyla but is common in others. Sexual reproduction is generally adaptive because it produces diverse genotypes, thereby increasing the potential for faster evolutionary response to environmental change. Many protists reap additional ecological benefits from sexual reproduction, illustrated by several types of sexual life cycles.

Zygotic Life Cycles Most unicellular protists that reproduce sexually display what is known as a **zygotic life cycle** (**Figure 28.25**). In this type of life cycle, haploid cells develop into gametes. Some protists produce nonmotile eggs and smaller

flagellate sperm. However, many other protists have gametes that look similar to each other structurally but have distinctive biochemical features and hence are known as + and − mating types. Gametes fuse to produce thick-walled diploid zygotes, which give this type of life cycle its name. Such zygotes often have tough cell walls and can survive stressful conditions, much like cysts. When conditions permit, the zygote divides by meiosis to produce haploid cells that increase in number via mitotic cell divisions.

Sporic Life Cycles Many multicellular green and brown seaweeds display a **sporic life cycle**, which is also known as alternation of generations (**Figure 28.26**). Giant kelps and some other protists having sporic life cycles produce two types of multicellular organisms: a haploid gametophyte generation that produces gametes (sperm or eggs) and a diploid sporophyte generation that produces spores by the process of meiosis (Figure 28.26a). This type of life cycle takes its name from the characteristic production of spores as the result of meiosis. Each of the two types of multicellular organisms can adapt to distinct habitats or seasonal conditions, thus allowing protists to occupy more types of environments for longer periods.

Many red seaweeds display a variation of the sporic life cycle that involves alternation of three distinct multicellular generations (Figure 28.26b). This unique type of sexual life cycle has evolved due to the lack of flagella on red algal sperm. Because these sperm are unable to swim to eggs, fertilization occurs only when sperm carried by ocean currents happen to drift close to

Figure 28.25 Zygotic life cycle, illustrated by the unicellular flagellate genus *Chlamydomonas*. In *Chlamydomonas*, most cells are haploid; only the zygote is diploid.

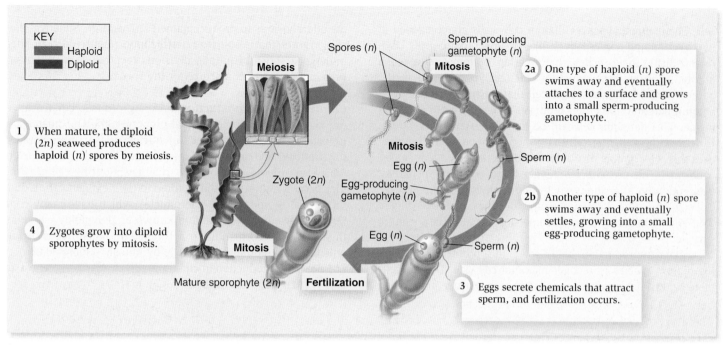

(a) *Laminaria* life cycle—alternation of 2 generations

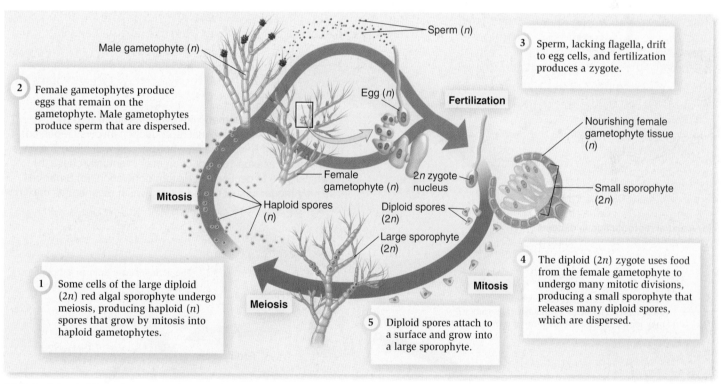

(b) *Polysiphonia* life cycle—alternation of 3 generations

Figure 28.26   Sporic life cycles.   (a) Sporic life cycle with two alternating generations, illustrated by the brown seaweed *Laminaria*. (b) Sporic life cycle involving three alternating generations, illustrated by the red seaweed *Polysiphonia*.

eggs. As a consequence, fertilization can be rare. Many red algae therefore make millions of spores that are produced by two distinct sporophyte generations. A small sporophyte produces diploid spores, while a larger sporophyte produces haploid spores.

Gametic Life Cycle and the Problem of Diatom Size   Diatoms are one of the relatively few types of protists known to

display a **gametic life cycle** (**Figure 28.27**). In gametic life cycles, all cells except the gametes are diploid, and gametes are produced by meiosis. Sexual reproduction in diatoms not only increases their genetic variability, but it also has another major benefit related to cell size.

In many diatoms, one daughter cell arising from mitosis is smaller than the other, and it is also smaller than the parent cell. This happens because diatom cell walls are composed of two overlapping halves, much like two-part round glass laboratory dishes having lids that overlap bottoms. After each mitotic division, each daughter cell receives one-half of the parent cell wall. The daughter cell that inherits a larger, overlapping parental "lid" then produces a new "bottom" that fits inside. This daughter cell will be the same size as its parent. However, the daughter cell that inherits the parental "bottom" uses this wall half as its lid and produces a new, even smaller "bottom." This cell will be smaller than its sibling or parent. Consequently, after many such mitotic divisions, the average cell size of diatom populations often declines over time (Figure 28.27a). If diatom cells become too small, they cannot survive.

Sexual reproduction helps solve this problem by allowing diatom species to recover maximal cell size. Diatom cells mate within a blanket of mucilage, each partner undergoing meiotic divisions to produce gametes. The large, spherical diatom zygotes that result from fertilization (Figure 28.27b) later undergo a series of mitotic divisions to produce new diatom cells having the maximal size for the species.

Ciliate Sexual Reproduction Among protists, ciliates have one of the most complex sexual processes known. Ciliates are unusual in having two types of nuclei: one or more smaller micronuclei and a single large macronucleus. Micronuclei, which are diploid, do not transcribe genes during cell growth; their role is to transmit the genome to the next generation during sexual reproduction. Macronuclei, which have many copies of the genome, a condition known as polyploidy, serve as the source of information for cell function, Macronuclei divide when ciliates reproduce asexually by mitosis (**Figure 28.28a**).

Sexual reproduction in ciliates involves a process known as conjugation (**Figure 28.28b**).

After many cell divisions, some progeny cells are very small.

(a) Asexual reproduction in diatoms

Figure 28.27 **Gametic life cycle, as illustrated by diatoms.** **(a)** Diatom asexual reproduction involves repeated mitotic division. Because a new bottom cell-wall piece is always synthesized, asexual reproduction may eventually cause the mean cell size to decline in a diatom population. **(b)** Small cell size may trigger sexual reproduction, which regenerates maximal cell size.

2 Blanketed by mucilage, each cell produces 1 or more haploid gametes by meiosis. The gametes may look alike or take the form of sperm and eggs.

Meiosis

Haploid (*n*) gametes

3 The gametes fuse to form a diploid zygote that is larger and rounder than a typical diatom cell.

Fertilization

Diploid (2*n*) zygote

KEY
Haploid
Diploid

Mucilage

Mitosis

Lipid food storage

Plastids

1 When diatom cells reach a critical small size or are stimulated by environmental factors, they may begin the process of reproduction.

4 The 2*n* zygote undergoes mitotic divisions to produce diploid cells that have the normal shape and maximum size for that species.

(b) Sexual reproduction in diatoms

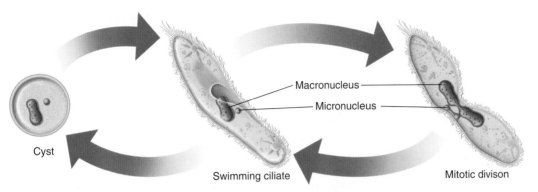

(a) Asexual reproduction by mitosis and cysts

1   Two compatible cells conjugate—line up side by side and partially fuse together.

2   In each cell, the micronucleus undergoes meiosis, producing 4 haploid products, but 3 disintegrate.

3   In each cell, the remaining haploid micronucleus undergoes mitosis.

7   The cell with 8 nuclei undergoes 2 rounds of cytokinesis to produce 4 mature cells that have 1 micronucleus and 1 macronucleus.

6   The diploid nucleus undergoes 3 rounds of mitosis, producing 4 macronuclei and 4 micronuclei. Note: The diagram shows only 1 of the 2 cells from step 5.

5   The paired cells separate. In each cell, the genetically different micronuclei fuse to form a diploid nucleus.

4   The 2 cells exchange a haploid micronucleus, and each cell's macronucleus disintegrates.

(b) Sexual reproduction by conjugation

Figure 28.28   **Ciliate reproduction.**   (a) The asexual reproduction process in ciliates. (b) The sexual reproduction process in the ciliate *Paramecium caudatum*, which involves conjugation.

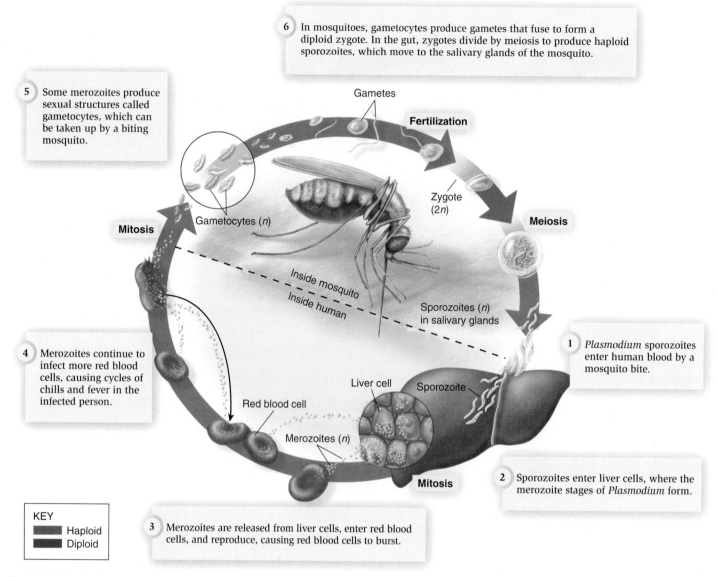

6 In mosquitoes, gametocytes produce gametes that fuse to form a diploid zygote. In the gut, zygotes divide by meiosis to produce haploid sporozoites, which move to the salivary glands of the mosquito.

5 Some merozoites produce sexual structures called gametocytes, which can be taken up by a biting mosquito.

Gametes

Fertilization

Zygote (2n)

Meiosis

Mitosis

Gametocytes (n)

Inside mosquito

Inside human

Sporozoites (n) in salivary glands

4 Merozoites continue to infect more red blood cells, causing cycles of chills and fever in the infected person.

1 *Plasmodium* sporozoites enter human blood by a mosquito bite.

Liver cell   Sporozoite

Red blood cell

Sporozoite

Merozoites (n)

Mitosis

2 Sporozoites enter liver cells, where the merozoite stages of *Plasmodium* form.

KEY
Haploid
Diploid

3 Merozoites are released from liver cells, enter red blood cells, and reproduce, causing red blood cells to burst.

Figure 28.29   Diagram of the life cycle of *Plasmodium falciparum*, the agent of malaria.   This life cycle requires two alternate hosts, humans and *Anopheles* mosquitoes.

Concept check: *In which of the hosts does sexual mating of P. falciparum gametes occur?*

Parasitic Protists May Use Alternate Hosts for Different Life Stages

Parasitic protists are notable for often using more than one host organism, in which different life stages occur. The malarial parasite genus *Plasmodium* is a prominent example. About 40% of humans live in tropical regions of the world where malaria occurs, and as noted earlier, millions of infections and human deaths result each year. Malaria is particularly deadly for young children; in Africa alone, more than 1 million children die each year from malaria. In addition to humans, the malarial parasite's alternate host is the mosquito classified in the genus *Anopheles*. Though insecticides can be used to control mosquito populations and though antimalarial drugs exist, malarial parasites can develop drug resistance. Experts are concerned that cases may double in the next 20 years.

When a mosquito bites a human, *Plasmodium* enters the bloodstream as a life stage known as a sporozoite (**Figure 28.29**). The sporozoites eventually reach the victim's liver and enter the liver cells. Following several cycles of cell division, a life stage known as the merozoite develops and is released from liver cells. Merozoites have protein complexes at their front ends, or apices, that allow them to invade human red blood cells. (The presence of these apical complexes gives rise to the phylum name Apicomplexa.) The merozoites consume the hemoglobin in red blood cells. While living within red blood cells, they form rings, which can be visualized by staining and the use of a microscope, allowing diagnosis. Merozoites reproduce asexually, generating large numbers of new merozoites that synchronously break out of red blood cells at intervals of 48 or 72 hours. These merozoite reproduction cycles correspond to cycles of chills and fever experienced by the infected person.

Some merozoites produce sexual structures—gametocytes—which are transmitted to a female mosquito as she bites.

Within the mosquito's body, the gametocytes produce gametes and fertilization occurs, yielding a zygote, the only diploid cell in *Plasmodium*'s life cycle. Within the mosquito gut, the zygote undergoes meiosis, generating structures filled with many sporozoites, the stage that can be transmitted to a new human host. Sporozoites move to the mosquito's salivary glands, where they remain until injected into a human host when the mosquito feeds.

In recent years, genomic information that has added to our knowledge of these life stages is helping medical scientists to develop new ways to prevent or treat malaria. In the case of *P. falciparum*, genomic data have already highlighted potential new pharmaceutical approaches. About 550 (some 10%) of the nuclear-encoded proteins are likely imported into a nonphotosynthetic plastid known as an apicoplast, where they are needed for fatty-acid metabolism and other processes. *P. falciparum* and some other apicomplexan protists possess plastids because they are descended from algal ancestors that had photosynthetic plastids. Because plastids are not present in mammalian cells, enzymes in apicoplast pathways are possible targets for development of drug therapy. Mammals also lack calcium-dependent protein kinases (CDPKs), enzymes that are essential to *P. falciparum*'s sexual development, offering another potential drug target.

Summary of Key Concepts

28.1 An Introduction to Protists

- Protists are eukaryotes that are abundant in moist habitats, and most are microscopic in size.
- Protists are often informally labeled according to their ecological roles: Algae are mostly photosynthetic protists; protozoa are heterotrophic protists that are often mobile; and fungus-like protists resemble true fungi. (Figure 28.1)
- Protists are particularly diverse in aquatic habitats, occurring as small floating or swimming phytoplankton, attached members of the periphyton, and more complex macroalgae (seaweeds). (Figure 28.2)
- Microscopic protists propel themselves by means of flagella (flagellates), cilia (ciliates), pseudopodia (amoebae), or by gliding across surfaces. (Figures 28.3, 28.4)

28.2 Evolution and Relationships

- Modern phylogenetic analysis has revealed that protists do not form a monophyletic group; instead, many can be classified into several eukaryotic supergroups. (Figure 28.5)
- The supergroup Excavata includes flagellate protists characterized by a feeding groove, the kinetoplastids and euglenoids, some of which are photosynthetic. (Figures 28.6, 28.7, 28.8)
- Land plants are related to green algae and red algae (having primary plastids) and possibly haptophytes and cryptomonads

(featuring secondary plastids). (Figures 28.9, 28.10, 28.11, 28.12, 28.13, 28.14)

- The supergroup Alveolata includes the ciliates, apicomplexans, and dinoflagellates, whose cells feature alveoli. Many dinoflagellates display secondary plastids, and some feature tertiary plastids. (Figures 28.15, 28.16)
- The supergroup Stramenopila includes diverse protists whose flagella have strawlike hairs that aid in swimming. Stramenopiles include diatoms, giant kelps, and other groups of algae, as well as some fungus-like protists. (Figures 28.17, 28.18)
- The supergroup Rhizaria consists of flagellates and amoebae with filose pseudopodia. Three prominent phyla are Chlorarachniophyta, with secondary green plastids; mineral-shelled Radiolaria; and Foraminifera, with calcium carbonate shells. (Figure 28.19)
- The supergroup Amoebozoa is composed of many types of amoebae and includes the parasite *Entamoeba histolytica* and slime molds such as *Dictyostelium discoideum*.
- The supergroup Opisthokonta includes organisms that produce swimming cells having a single posterior flagellum. It includes the fungal and animal kingdoms and choanoflagellate protists related to the ancestry of animals. (Figures 28.20, 28.21, Table 28.1)

28.3 Nutritional and Defensive Adaptations

- Protists display four basic types of nutrition: phagotrophs rely on particle feeding; osmotrophs absorb small organic molecules; photoautotrophs make their own organic food by using light energy; and mixotrophs use both autotrophy and heterotrophy to obtain nutrients. Heterotrophic protists may function as decomposers or parasites. (Figure 28.22)
- Protists possess defensive adaptations such as protective cell coverings, sharp projectiles, light flashes, and toxic compounds. Dinoflagellates are particularly important toxin producers, and aquatic ecologists found that populations of the unicellular dinoflagellate genus *Pfiesteria* kill fishes by using a toxin that can also harm human cells. (Figure 28.23)

28.4 Reproductive Adaptations

- Protist populations grow by means of asexual reproduction involving mitosis, and many persist through unfavorable conditions as tough-walled cysts. (Figure 28.24)
- Protists having zygotic life cycles often use tough-walled zygotes to survive unfavorable conditions. (Figure 28.25)
- Protists displaying sporic life cycles are able to occupy multiple habitats because they produce two or more alternating life stages having differing environmental preferences. (Figure 28.26)
- Diatoms, which have a gametic life cycle, use sexual reproduction to solve a cell size problem that originates from their unique mode of asexual cell division. (Figure 28.27)
- Ciliates display a unique type of sexual reproduction (conjugation) that involves the exchange of genetic material between a mating pair of cells. (Figure 28.28)
- Parasitic protists may have life cycles involving alternate hosts. One example is *Plasmodium*, the malarial agent, whose alternate hosts are humans and mosquitoes. (Figure 28.29)

Assess and Discuss

Test Yourself

1. If you were studying the evolution of animal-specific cell-to-cell signaling systems, from which of the following would you choose representative species to observe?
   a. Rhodophyta
   b. Excavata
   c. Choanomonada
   d. Radiolaria
   e. Chlorophyta

2. If you were studying the origin of land plant traits, which of the following groups would you study?
   a. green algae
   b. radiolarians
   c. choanoflagellates
   d. diatoms
   e. ciliates

3. Which informal ecological group of protists includes photoautotrophs?
   a. protozoa
   b. algae
   c. fungus-like protists
   d. ciliates
   e. all of the choices listed

4. How would you recognize a primary plastid? It would:
   a. have one envelope membrane.
   b. have two envelope membranes.
   c. have more than two envelope membranes.
   d. lack pigments.
   e. be golden brown in color.

5. What organisms have tertiary plastids?
   a. certain stramenopiles
   b. certain euglenoids
   c. certain cryptomonads
   d. certain opisthokonts
   e. certain dinoflagellates

6. What is unusual about mixotrophs?
   a. They have no plastids, but they occur mixed in communities with autotrophs.
   b. They have mixed heterotrophic and autotrophic nutrition.
   c. Their cells contain a mixture of red and green plastids.
   d. Their cells contain a mixture of haploid and diploid nuclei.
   e. They consume a mixed diet of algae.

7. What advantages do diatoms obtain from sexual reproduction?
   a. increased genetic variability
   b. increased ability of populations to respond to environmental change
   c. evolutionary potential
   d. regeneration of maximal cell size for the species
   e. all of the above

8. What are trichocysts?
   a. hairs on flagella
   b. membrane sacs beneath the cell surface
   c. tough-walled asexual cells
   d. spearlike defensive structures shot from cells under attack
   e. special types of survival cysts

9. How do accessory pigments benefit autotrophic protists?
   a. They provide camouflage, so herbivores cannot see algae.
   b. They absorb underwater light and transfer the energy to chlorophyll *a* for use in photosynthesis.
   c. They attract aquatic animals that carry gametes between seaweeds.
   d. They absorb UV light that would harm the photosynthetic apparatus.
   e. All of the above are correct.

10. What are the two alternate hosts of the malarial parasite *Plasmodium falciparum*?
    a. humans and ticks
    b. ticks and mosquitoes
    c. humans and *Anopheles* mosquitoes
    d. humans and all types of mosquitoes
    e. sporophytes and gametophytes

Conceptual Questions

1. Explain why protists are classified into multiple supergroups, rather than a single kingdom or phylum.

2. Why have molecular biologists sequenced the genomes of several parasitic protists?

3. Why are protistan cysts important to epidemiologists, biologists who study the spread of disease?

Collaborative Questions

1. Imagine you are studying an insect species and you discover that the insects are dying of a disease that results in the production of cysts of the type that protists often generate. Thinking that the cysts might have been produced by a parasitic protist that could be used as an insect control agent, how would you go about identifying the disease agent?

2. Imagine you are part of a marine biology team seeking to catalogue the organisms inhabiting a threatened coral reef. The team has found two new seaweeds (macroalgae), each of which occurs during a particular time of the year when the water temperature differs. You suspect that the two seaweeds might be different generations of the same species that have differing optimal temperature conditions. How would you go about testing your hypothesis?

Online Resource

www.brookerbiology.com

Stay a step ahead in your studies with animations that bring concepts to life and practice tests to assess your understanding. Your instructor may also recommend the interactive ebook, individualized learning tools, and more.

Plants and the Conquest of Land

29

A temperate rain forest in Olympic National Park in Washington State, containing diverse plant phyla.

When thinking about plants, people envision lush green lawns, shady street trees, garden flowers, or leafy fields of valuable crops. On a broader scale, they might imagine dense rain forests (see chapter-opening photo), vast grassy plains, or tough desert vegetation. Shopping in the produce section of the local grocery store may remind us that plant photosynthesis is the basic source of our food. Just breathing crisp fresh air might bring to mind the role of plants as oxygen producers—the ultimate air fresheners. Do you start your day with a "wake-up" cup of coffee, tea, or hot chocolate? Then you may appreciate the plants that produce these and many other materials we use in daily life: medicines, cotton, linen, wood, bamboo, cork, and even the paper on which this textbook has been printed.

In addition to their importance to humans and modern ecosystems, plants have played dramatic roles in the Earth's past. Throughout their evolutionary history, diverse plants have influenced Earth's atmospheric chemistry, climate, and soils. Plants have also affected the evolution of many other groups of organisms, including humans. In this chapter, we will survey the diversity of modern plant phyla and their distinctive features. This chapter also explains how early plants adapted to land and how plants have continued to adapt to changing terrestrial environments. During this process, we will gain insight into descent with modification, one of the main principles of evolution and diversity.

29.1 Ancestry and Diversity of Modern Plants

Several hundred thousand modern species are formally classified into the kingdom Plantae, informally known as the plants or land plants (Figure 29.1). Plants are multicellular eukaryotic organisms composed of cells having plastids, and plants primarily live on land. Molecular and other evidence indicates that the plant kingdom evolved from green algal ancestors that lived in aquatic habitats such as lakes or ponds. **Plants** are distinguished from their modern algal relatives by the presence of traits that foster survival in terrestrial conditions, which are drier, sunnier, hotter, colder, and less physically supportive than aquatic habitats. In this section, we will examine the modern

algae that are most closely related to plants and survey the diverse phyla of living land plants. This process reveals how plants gradually acquired diverse structural, biochemical, and reproductive adaptations that foster survival on land.

Modern Green Algae Are Closely Related to the Ancestors of Land Plants

Molecular, biochemical, and structural data indicate that the kingdom Plantae (plant kingdom) originated from a photosynthetic protist ancestor having a relatively complex body, a filament of cells with side branches. Such a body is characteristic of the modern green algal genera *Chara* and *Coleochaete* and their close relatives, the complex **charophyceans** (Figure 29.2a). All charophyceans have features in common with land plants, but complex charophyceans display several derived features shared with land plants. Examples of these shared features are a distinctive type of cytokinesis, intercellular connections known as plasmodesmata (see Chapter 10), and sexual reproduction by means of an egg and smaller sperm (see Figure 29.1). For this reason, *Chara* and *Coleochaete* are good sources of information

Figure 29.1 **Evolutionary relationships of the modern plant phyla.** Land plants gradually acquired diverse structural, biochemical, and reproductive adaptations, allowing them to better survive in terrestrial habitats.

about the ancestors of land plants. By contrast, the simpler charophyceans have less complex reproduction and bodies consisting of single cells or unbranched filaments. Examples include the beautiful desmids such as *Micrasterias radiosa* and the flagellate *Mesostigma viride* (**Figure 29.2b**). Study of both complex and simpler charophyceans helps to illustrate how traits of early plants arose in a sequence of evolutionary events.

By themselves the charophyceans do not form a monophyletic group, but together with land plants, they form a clade known as the streptophytes, which are related to other green algae (see Figure 29.1). Streptophytes and other green algae share distinctive characters such as cellulose-rich cell walls and green, starch-bearing chloroplasts that contain the accessory pigments chlorophyll *b* and β-carotene. Even so, the land plants

(a) Complex charophyceans: *Chara zeylanica* (left) and *Coleochaete pulvinata* (right)

(b) Simple charophyceans: *Micrasterias radiosa* (left) and *Mesostigma viride* (right)

Figure 29.2 Charophycean green algal relatives of the land plants. Charophycean algae inhabit freshwater lakes and ponds, and they display structural, reproductive, biochemical, and molecular features in common with land plants. (a) Complex charophyceans with branched filaments share more traits with land plants than do simpler charophyceans illustrated in (b).

display several common features that distinguish them from green algae, even their closest charophycean relatives.

Distinctive Features of the Land Plants

The features that distinguish land plants represent early adaptations to the land habitat. For example, the bodies of all land plants are primarily composed of three-dimensional tissues, defined as close associations of cells of the same type. Tissues provide land plants with an increased ability to avoid water loss at their surfaces. That's because bodies composed of tissues have lower surface area/volume ratios than do branched filaments. Land plant tissues arise from one or more actively dividing cells that occur at growing tips. Such localized regions of cell division are known as **apical meristems**. The tissue-

producing apical meristems of land plants are able to produce relatively thick, robust bodies able to withstand drought and mechanical stress and to produce tissues and organs having specialized functions.

The land plants also have distinctive reproductive features. These include a life cycle involving alternation between two types of bodies; embryos that depend on maternal tissues during early development; reproductive cells whose tough walls allow dispersal through dry air; and specialized structures that generate, protect, and disperse reproductive cells. The following survey of modern plant phyla and their distinctive features illustrates how these and other fundamental plant features evolved.

Modern Land Plants Can Be Classified into Ten Phyla

Plant systematists use molecular and structural information from living and fossil plants to classify plants into phyla, also known as divisions. In this textbook, 10 phyla of living land plants are described: (1) the plants informally known as **liverworts** (formally called Hepatophyta), (2) **mosses** (Bryophyta), (3) **hornworts** (Anthocerophyta), (4) **lycophytes** (Lycopodiophyta), (5) **pteridophytes** (Pteridophyta), (6) **cycads** (Cycadophyta), (7) **ginkgos** (Ginkgophyta), (8) **conifers** (Coniferophyta), (9) **gnetophytes** (Gnetophyta), and (10) the **flowering plants**, also known as **angiosperms** (Anthophyta) (see Figure 29.1). Fossils reveal that additional plant phyla once lived but are now extinct.

Phylogenetic information suggests that the modern plant phyla arose in a particular sequence (see Figure 29.1). Liverworts diverged first, mosses diverged next, and hornworts seem to be closely related to the vascular plants. The vascular plants are distinguished by internal water and nutrient-conducting tissues that also provide structural support. Lycophytes are sister to all of the other modern vascular plants, pteridophytes arose next, and then seed plants. Our survey of these modern plant phyla reveals how plants acquired more adaptations to life on land.

Liverworts, Mosses, and Hornworts Are the Simplest Land Plants

Liverworts, mosses, and hornworts are Earth's simplest land plants (**Figures 29.3, 29.4,** and **29.5**), and each forms a distinct, monophyletic group. There are about 6,500 species of modern liverworts, 12,000 or more species of mosses, and about 100 species of hornworts. Collectively, liverworts, mosses, and hornworts are known informally as the **bryophytes** (from the Greek *bryon*, meaning moss, and *phyton*, meaning plant). The bryophytes do not form a clade, but the term bryophyte is useful for expressing common structural, reproductive, and ecological features of liverworts, mosses, and hornworts. For example, the bryophytes are all relatively small in stature and are most common and diverse in moist habitats because they lack traits allowing them to grow tall or reproduce in dry places.

Because bryophytes diverged early in the evolutionary history of land plants (see Figure 29.1), they serve as models of the

(a) The common liverwort
Marchantia polymorpha

Asexual gemmae

Upper surface of liverwort

(b) Close-up of common liverwort structures

(c) A species of leafy liverwort

Figure 29.3 **Liverworts.** (a) The common liverwort *Marchantia polymorpha*, with raised, umbrella-shaped structures that bear sexually produced sporophytes on the undersides. Mature sporophytes generate spores, then release them. (b) A close-up of *M. polymorpha* showing surface cups that contain multicellular, frisbee-shaped asexual structures known as gemmae that are dispersed by wind and grow into new liverworts. (c) A species of liverwort having leaflike structures and so known as a leafy liverwort.

Concept check: Why do you think liverworts produce their spores on raised structures?

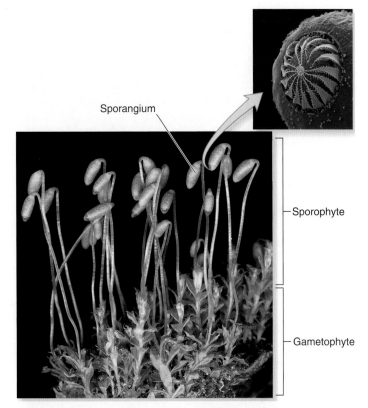

Sporangium

Sporophyte

Gametophyte

Figure 29.4 **Mosses.** The common moss genus *Mnium* has a leafy green gametophyte (multicellular body that generates gametes) and an unbranched, dependent sporophyte that bears a spore-producing sporangium at its tip. Inset: This SEM shows that the tips of moss sporangia are often specialized, so spores are sprinkled into the wind, rather than being released all at once.

Concept check: Why might it be advantageous for a moss sporophyte to release spores gradually?

earliest terrestrial plants. Bryophytes display apical meristems that produce specialized tissues and other features that evolved early in the history of land plants, such as the sporic life cycle. A comparison of the life cycle of aquatic charophyceans with that of bryophytes reveals the bryophyte life cycle's adaptive value on land.

The charophycean algae display a **zygotic life cycle** in which the diploid generation consists of only one cell, the zygote (**Figure 29.6a**). Zygotic life cycles take their name from the observation that the zygote is the only cell that undergoes meiosis (see Figure 28.25). By contrast, sexual reproduction in bryophytes and all other plants follows what is called a **sporic life cycle**, in which there is an alternation of generations (see Figures 15.15c and 28.26). In the sporic life cycle, meiosis results in the formation of spores, which are reproductive cells that allow organisms to disperse in the environment (**Figure 29.6b**). **Alternation of generations** means that land plants produce two types of multicellular bodies that alternate in time. These two types of bodies are known as the diploid, spore-producing **sporophyte** generation and the haploid, gamete-producing **gametophyte** generation. Biologists think the plant life cycle originated by a delay in zygotic meiosis, with the result

Spore dispersal tip

Sporophyte

Gametophyte

Figure 29.5 **Hornworts.** Sporophytes of hornworts generally grow up into the air, whereas the gametophytes grow close to the ground. Hornwort sporophytes become mature and open at the top, dispersing spores.

(a) **Zygotic life cycle of charophyceans**

KEY
- Haploid
- Diploid

(b) **Sporic life cycle of early plants**

Figure 29.6 Origin of the plant life cycle.

a single plant to disperse widely by using meiosis to produce numerous, genetically variable haploid spores. Each spore has the potential to grow into a gametophyte. In bryophytes, the gametophyte generation is generally green and photosynthetic and thus has a major function of organic food production. The more spores that sporophytes produce, the greater the numbers of gametophytes, helping bryophytes to spread in their environments and thereby increase in fitness. We next discuss how gametophytes and sporophytes function together during the life cycle of a bryophyte, starting with gamete production.

Gametophytes   From an evolutionary and reproductive viewpoint, the role of plant gametophytes is to produce haploid gametes. Because the gametophyte cells are already haploid, meiosis is not involved in producing plant gametes. Instead, plant gametes are produced by mitosis; therefore, all gametes produced from a single gametophyte are genetically identical.

The gametophytes of bryophytes and many other land plants produce gametes in specialized structures known as **gametangia** (from the Greek, meaning gamete containers). Certain cells of gametangia develop into gametes, and other cells form an outer jacket of tissue. The gametangial jacket protects delicate gametes from drying out and from microbial attack while they develop. Flask-shaped gametangia that each enclose a single egg cell are known as **archegonia**; spherical or elongate gametangia that each produce many sperm are known as **antheridia** (Figure 29.7).

When the plant sperm are mature, if moist conditions exist, they are released from antheridia into films of water. Under the influence of sex-attractant molecules secreted from archegonia, the sperm swim toward the eggs, twisting their way down the tubular archegonial necks. The sperm then fuse with egg cells in the process of fertilization to form diploid zygotes, which grow into embryos. New sporophytes develop from embryos. Fertilization cannot occur in bryophytes unless water is present because the sperm are flagellate and need water to reach eggs. Conditions of uncertain moisture, common in the land habitat, can thus limit plant reproductive success. As we explain next, plant embryos and sporophytes are adaptive responses to this environmental challenge.

Sporophytes   One reproductive advantage of the plant life cycle is that zygotes remain enclosed within gametophyte tissues, where they are sheltered and fed (a process described in more detail in Section 29.3). This critical innovation, known as **matrotrophy** (from the Latin, meaning mother, and the Greek, meaning food) gives zygotes a good start while they grow into embryos. Because all groups of land plants possess matrotrophic embryos, they are known as **embryophytes**. Sheltering and feeding embryos is particularly important when embryo production is limited by water availability, as is often the case for bryophytes.

Another reproductive advantage to plants of the sporic life cycle is that, when mature, specialized cells within multicellular sporophytes undergo meiosis to produce many genetically diverse spores. Meiosis occurs within enclosures known

that the diploid generation became multicellular before undergoing meiosis. For the earliest land plants, having a multicellular diploid generation provided several advantages in coping with terrestrial conditions. Can you guess what these advantages might be? A closer look at the process of sexual reproduction in bryophytes reveals the answer and also highlights ways in which bryophytes differ from other plants.

Bryophyte Reproduction Illustrates Early Plant Adaptations to Life on Land

As noted, bryophytes and other land plants display alternation of generations (**Figure 29.7**). On land, a multicellular diploid sporophyte generation is advantageous because it allows

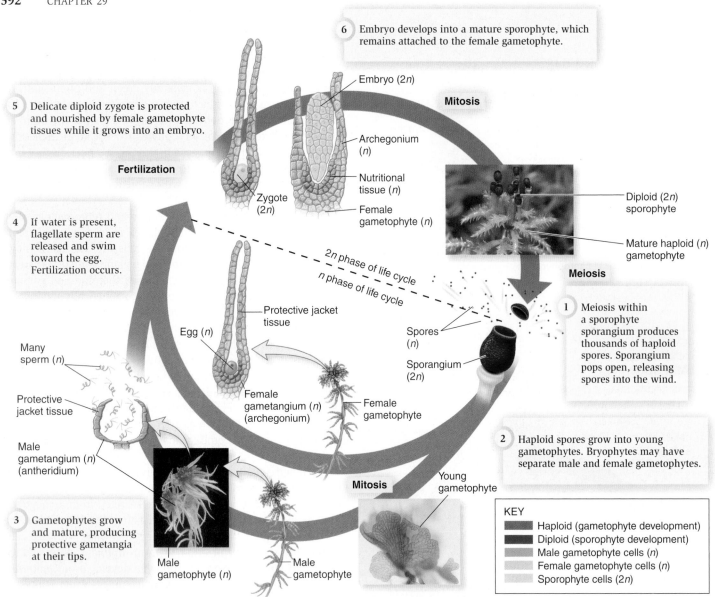

Figure 29.7 The life cycle of the early diverging moss genus *Sphagnum*, illustrating reproductive adaptations that likely helped early plants to reproduce on land.   The life cycle of this organism is illustrated because among modern bryophytes, *Sphagnum* is the single most abundant and ecologically important genus.

as **sporangia** (from the Greek, meaning spore containers), whose tough cell walls protect developing spores from harmful UV radiation and microbial attack. Bryophyte sporangia open in specialized ways that foster dispersal of mature spores into the air, allowing spore transport by wind (see Figures 29.4 and 29.7). Dispersed spores have cell walls containing a tough material, known as **sporopollenin**, that helps to prevent cellular damage during transport in air. If spores reach habitats favorable for growth, their walls crack open, and new gametophytes develop by mitotic divisions, completing the life cycle.

Spore production is a measure of plant fitness, because plants can better disperse progeny throughout the environment when they produce more spores. The larger the diploid generation, the more spores a plant can produce. As a result, during plant evolution, the sporophyte generation has become larger and more complex (**Figure 29.8**).

Bryophytes Display Several Distinguishing Features

As we have seen, bryophytes share several fundamental adaptive traits: alternation of generations, tissue-producing apical meristems, protective gametangia and sporangia, and sporopollenin-walled spores. These same traits are shared with other land plants. Even so, several features distinguish bryophytes from other land plants. First, bryophyte gametophytes are more common in nature, larger, and longer-lived than bryophyte sporophytes. Green patches of moss that you might see in the woods are primarily gametophytes. In order to observe bryophyte sporophytes, you would have to look very closely, because this life stage is quite small and remains attached to gametophytes throughout its short lifetime (see Figures 29.4, 29.5, and 29.7). Plant biologists consider bryophyte gametophytes to be the dominant generation in their life cycle. By contrast, other plants have dominant sporophyte generations (Figure 29.8).

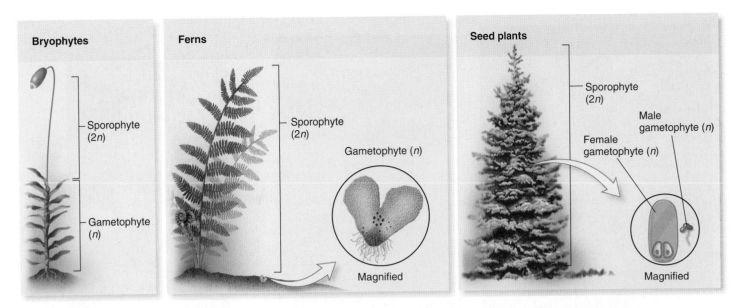

Figure 29.8   Relative sizes of the sporophyte and gametophyte generations of bryophytes, ferns, and seed plants.

Concept check: *What is the advantage to ferns and seed plants of having larger sporophytes than those of bryophytes?*

Bryophyte sporophytes are small; they remain attached to parental gametophytes, are unable to branch, and have short lifetimes. At their tips, bryophyte sporophytes produce only a single sporangium containing a limited number of spores. By contrast, the sporophytes of other land plants become independent, and because they can branch, they can continue to grow and produce sporangia on lateral branches, often for many years (Figure 29.8). In all land plants except bryophytes, the sporophyte generation is the dominant generation, meaning that it is larger, more complex, and longer-lived than the gametophyte.

Yet another distinguishing feature of bryophytes is that they lack tissues that both provide structural support and serve in conduction, known as **vascular tissues**. Although the gametophytes of some bryophytes display simple conducting tissues, these do not provide much structural support. Thus, bryophytes are informally known as **nonvascular plants**. As mentioned, other modern plant phyla are collectively and informally known as **vascular plants** because they produce vascular tissues that function in both conduction and support.

Lycophytes and Pteridophytes Are Vascular Plants That Do Not Produce Seeds

If you take a look outside, most of the plants in view are probably vascular plants. The presence of vascular tissue and the ability to branch allows most of these plants to grow much taller than bryophytes and to produce more spores. As a result, vascular plants are more prominent than bryophytes in most modern plant communities.

Vascular plants have been important to Earth's ecology for several hundreds of millions of years. Fossils tell us that the first vascular plants appeared later than the earliest bryophytes (see Section 29.2) and that several early vascular plant lineages once existed but became extinct. Molecular data indicate that the lycophytes are the oldest phylum of living vascular plants

and that pteridophytes are the next oldest living plant phylum (see Figure 29.1). In the past, lycophytes were very diverse and included tall trees that contributed importantly to coal deposits, but the tree lycophytes became extinct, and now only about 1,000 relatively small species exist (**Figure 29.9**). Pteridophytes have diversified more recently, and there are about 12,000 species of modern pteridophytes, including horsetails, whisk ferns, and other ferns (**Figure 29.10**).

Because the lycophytes and pteridophytes diverged prior to the origin of seeds, they are informally known as seedless vascular plants. Together, lycophytes, pteridophytes, and

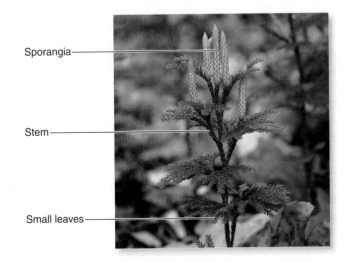

Figure 29.9   An example of a lycophyte (*Lycopodium obscurum*).   The sporophyte stems bear many tiny leaves, and sporangia generally occur in club-shaped clusters. For this reason, lycophytes are informally known as club mosses or spike mosses, though they are not true mosses. The gametophytes of lycophytes are small structures that often occur underground, where they are better protected from drying.

(a) A whisk fern

(b) The giant horsetail

(c) An early-diverging fern

Figure 29.10   Pteridophyte diversity.   (a) The leafless, rootless green stems of the whisk fern (*Psilotum nudum*) branch by forking and bear many clusters of yellow sporangia that disperse spores via wind. The gametophyte of this plant is a tiny pale structure that lives underground in a partnership with fungi. **(b)** The giant horsetail (*Equisetum telmateia*) displays branches in whorls around the green stems. The leaves of this plant are tiny, light brown structures that encircle branches at intervals. This plant produces spores in cone-shaped structures, and the wind-dispersed spores grow into small green gametophytes. **(c)** The early-diverging fern *Botrychium lunaria*, showing a green photosynthetic leaf with leaflets and a modified leaf that bears many round sporangia. **(d)** The later-diverging fern *Blechnum capense*, viewed from above, showing a whorl of young leaves that are in the process of unrolling from the bases to the tips. The leaves have many leaflets. The stem of this fern grows parallel to the ground and thus is not shown. Most ferns produce spores in sporangia on the undersides of leaves.

(d) A later-diverging fern

seed-producing plants are known as the **tracheophytes**. The latter term takes its name from **tracheids**, a type of specialized vascular cell that conducts water and minerals and provides structural support. Vascular tissues occur in the major plant organs: stems, roots, and leaves.

Stems, Roots, and Leaves   **Stems** are branching structures that contain vascular tissue and produce leaves and sporangia. Stems contain the specialized conducting tissues known as **phloem** and **xylem**, the latter of which contains tracheids. Together, such conducting tissues enable vascular plants to conduct organic compounds, water, and minerals throughout the plant body. The xylem also provides structural support, allowing

vascular plants to grow taller than nonvascular plants. This support function arises from the presence of a compression and decay-resistant waterproofing material known as **lignin**, which occurs in the cell walls of tracheids and some other types of plant cells. Most vascular plants also produce **roots**—organs specialized for uptake of water and minerals from the soil—and **leaves**, which generally have a photosynthetic function.

Lycophyte roots and leaves differ from those of pteridophytes. For example, lycophyte roots fork at their tips, whereas roots of pteridophytes branch from the inside like the roots of seed plants (see Chapter 35). Lycophyte leaves are relatively small and possess only one unbranched vein, whereas pteridophyte leaves are larger and have branched veins, as do those of seed plants (compare Figures 29.9 and 29.10d). The evolutionary origins of leaves are discussed in Section 29.4.

Adaptations That Foster Stable Internal Water Content   In relatively dry habitats, lycophytes, pteridophytes, and other vascular plants are able to grow to larger sizes and remain metabolically active for longer periods than can bryophytes. Vascular plants have this advantage because they are better able to maintain stable internal water content by means of several adaptations. Such adaptations include a protective **waxy cuticle**, present on most surfaces of vascular plant sporophytes (**Figure 29.11**). The plant cuticle contains a polyester polymer known as cutin, which helps to prevent attack by pathogens, and wax, which helps to prevent desiccation (drying out). The surface tissue of vascular plant stems and leaves contains many **stomata** (singular, stomate or stoma), pores that are able to open and close (Figure 29.11). Stomata allow plants to take in the carbon dioxide needed for photosynthesis and release oxygen to the air, while conserving water. When the environment is moist, the pores open, allowing photosynthetic gas exchange to occur. But when the environment is very dry, the pores close, which reduces water loss from plants (more information about stomata can be found in Chapters 35 and 38). Though a cuticle and stomata occur in bryophytes, they are not so common as in vascular plants, and bryophytes easily become dry. In vascular plants, conducting tissues, a waxy cuticle, and stomata function together to maintain moisture homeostasis, allowing tracheophytes to exploit a wider spectrum of land habitats.

Life Cycle   Lycophyte and pteridophyte gametophytes are small, delicate, and easily harmed by exposure to heat and drought. This explains why the gametophytes of lycophytes and pteridophytes are restricted to moist places, including underground, or why they have short lifetimes. Lycophyte and pteridophyte sperm are released from antheridia into water films within which they swim to eggs in archegonia (**Figure 29.12**). For this reason, lycophyte and pteridophyte reproduction is limited by dry conditions, as is the case for bryophytes. However, if fertilization occurs, lycophytes and pteridophytes can produce many more spores, because the spore-producing sporophyte generation grows to a much larger size than do bryophyte sporophytes. This fundamental difference has two explanations. First, vascular plant sporophytes are dependent upon maternal gametophytes for only a short time during early embryo development; they eventually become independent by developing a first leaf and roots able to harvest resources needed for photosynthesis (Figure 29.12). Second, the stems of vascular plant sporophytes are able to produce branches, forming relatively large adult plants having many leaves. Roots obtain large amounts of soil water and minerals, supporting the ability of leaves to generate abundant organic compounds by photosynthesis. Lycophytes and pteridophytes use such resources to produce large numbers of sporangia that eject multitudes of spores. You might have seen clusters of sporangia as dark brown dots or lines on the undersides of fern leaves. Dispersed by the wind, spores may land in a suitable place and grow into new gametophytes, completing the life cycle.

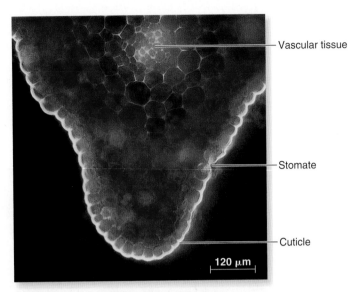

(a) Stem showing tracheophyte adaptations

Vascular tissue

Stomate

Cuticle

120 μm

(b) Close-up of stomata

Stomata

Figure 29.11   A pteridophyte stem with tracheophyte adaptations for transporting and conserving water.   **(a)** This is a cross section through a stem of the pteridophyte *Psilotum nudum*. When viewed with fluorescence microscopy and illuminated with violet light, an internal core of xylem tracheids glows yellow, as does the surface cuticle. **(b)** Surface pores associated with specialized cells—the complexes known as stomata—allow for gas exchange between plant and atmosphere.

Genomes & Proteomes Connection

The Fern *Ceratopteris richardii* Is a Useful Model Genetic System in the Study of Plant Evolution

The fern *Ceratopteris richardii* is useful as a plant model genetic system, as is the flowering plant *Arabidopsis thaliana* (**Figure 29.13**). Comparisons of the genomes and proteomes of plants help to illuminate the genetic changes that occurred during plant evolution.

For example, in 2005, Mari Salmi, Stanley Roux, and associates reported the results of a study of the expression of nearly 4,000 genes during spore germination in *C. richardii*. They

Figure 29.12   The life cycle of a typical fern.

Figure 29.13
A model genetic system, the fern *Ceratopteris richardii*. This fern is a model genetic system for the study of spore germination and other plant features.

identified genes expressed by both fern spores and germinating flowering plant seeds or pollen, a type of spore. One such gene is associated with the role of peroxisomes in making the cellular transition from dormancy to active metabolism, and others are related to cellular use of calcium ions and nitrous oxide (NO) as signaling molecules. In ferns, NO helps the long, thin cells known as rhizoids to perceive gravity, with the result that these elongate cells grow downward at their tips (see Figure 29.12). Such downward growth enables rhizoids to anchor gametophytes in a position that helps to keep gametangia and gametes from drying out. In this way, rhizoid growth fosters the production of embryos, thereby increasing plant fitness. Comparative genomic or proteomic studies suggest how the evolutionary process of descent with modification has generated the reproductive features of the seed plants upon which people greatly depend.

Gymnosperms and Angiosperms Are the Modern Seed Plants

Among the vascular plants, the seed plant phyla dominate most modern landscapes. The modern seed plant phyla commonly known as cycads, ginkgos, conifers, and gnetophytes are collectively known as gymnosperms (**Figure 29.14** shows an example). **Gymnosperms** reproduce using both spores and seeds, as do the flowering plants, the angiosperms (**Figure 29.15**). For this reason, gymnosperms and angiosperms are known informally as the **seed plants**. **Seeds** are complex structures having specialized tissues that protectively enclose embryos and contain stores of carbohydrate, lipid, and protein. Embryos use such food stores to grow and develop. As explained in Section 29.4, the ability to produce seeds helps to free seed plants from the reproductive limitations experienced by the seedless plants, revealing why seed plants are the dominant plants on Earth today.

In addition to the modern seed plant phyla, several additional phyla once existed and left fossils but are now extinct. Collectively, all of the living and fossil seed plant phyla are

formally known as **spermatophytes** (the prefix sperm in this case, from the Greek, meaning seed) (see Figure 29.1). Together with extinct seedless, woody plants, modern and fossil seed plants are known as **lignophytes**, a term that reflects the capacity to produce wood.

Wood is composed of xylem, a tissue whose cellulose-rich cell walls also contain lignin. Functioning something like superglue, lignin cements the fibrils of cellulose together, making wood exceptionally strong. Wood production enables plants to increase in girth and become tall. Though not all modern seed plants produce wood, many are trees or shrubs that produce

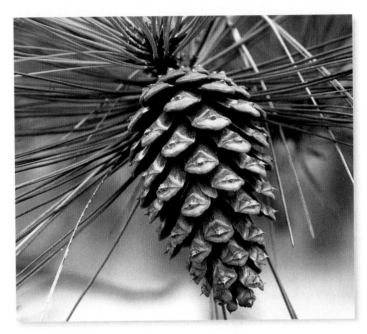

Figure 29.14   An example of a gymnosperm, the pine.

Concept check: *What key characteristics of lignophytes are displayed in this image?*

Figure 29.15   An example of an angiosperm, the bleeding heart plant (genus *Dicentra*).

considerable amounts of wood (described more completely in Chapters 30, 35, and 38).

The angiosperms are distinguished by the presence of flowers, fruits, and a specialized seed tissue known as endosperm. **Flowers** are short stems bearing organs that are specialized in ways that enhance seed production (Figure 29.15). **Fruits** are structures that develop from flower organs, enclose seeds, and foster seed dispersal in the environment. The term angiosperm comes from the Greek, meaning enclosed seeds, reflecting the observation that the flowering plants produce seeds within fruits. **Endosperm** is a nutritive seed tissue that increases the efficiency with which food is stored in the seeds of flowering plants (explained further in Section 29.4). Flowers, fruits, and endosperm are defining features of the angiosperms, and they are integral components of animal nutrition.

Though gymnosperms produce seeds and many are woody plants, they lack flowers, fruits, and endosperm. The term gymnosperm comes from the Greek, meaning naked seeds, reflecting the observation that gymnosperm seeds are not enclosed within fruits. Despite their lack of flowers, fruits, and seed endosperm, the modern gymnosperms are diverse and abundant in many places (see Chapter 30). The plant phyla we have just surveyed played significant roles in the past evolutionary history of plants, our next topic.

29.2 An Evolutionary History of Land Plants

A billion years ago, Earth's terrestrial surface was comparatively bare of life. Green or brown crusts of cyanobacteria most likely grew in moist places, but there would have been very little soil, no plants, and no animal life. The origin of the first land plants was essential to development of the first substantial soils, the evolution of modern plant communities, and the ability of animals to colonize land.

How can we know about events such as the origin and diversification of land plants? One line of information comes from comparing molecular and other features of modern plants. For example, the genome sequence of the moss *Physcomitrella patens*, first reported in 2007, reveals the presence of genes that aid heat and drought tolerance, which are especially useful in the terrestrial habitat. Plant fossils, the preserved remains of plants that lived in earlier times, provide another line of information (**Figure 29.16**). The distinctive plant materials sporopollenin, cutin, and lignin do not readily decay and therefore foster the fossilization of plant parts that contain these tough materials.

The study of fossils and the molecular, structural, and functional features of modern plants has revealed an amazing story—how plants conquered the land. This story can be conceptualized as three dramatic episodes: (1) the first land plants arise from ancestors shared with aquatic charophycean algae and begin to adapt to terrestrial habitats; (2) seedless plants transformed Earth's ecology; and (3) an ancient cataclysm led to the diversification of modern angiosperm lineages.

Figure 29.16 Fossil of *Pseudosalix handleyi*, an angiosperm.

Concept check: What biochemical components of plants favor the formation of fossils?

The First Plants Begin to Adapt to Terrestrial Habitats

Land plants inherited some traits from charophycean algae, as we have noted, but also acquired novel features in response to stresses present on land, but not in the water. For example, plant biologists Zoe Popper and Stephen Fry discovered that the cell walls of all land plants possess xyloglucan carbohydrates that cross-link cellulose microfibrils, but charophycean algae lack this feature. Cell-wall xyloglucans are among the new features that appeared as early plants began to adapt to land, possibly aiding in the development of more complex bodies. As previously mentioned, tissue-producing meristems, a sporic life cycle, tough-walled spores, and the sporophyte generation also appeared very early in plant history.

Seedless Plants Transformed Earth's Ecology

In addition to tough spores, other types of decay-resistant tissues evolved in early seedless plants, likely in response to attack by soil bacteria and fungi. When the plants died, some of their organic constituents were not completely degraded to carbon dioxide, but instead were buried in sediments that were eventually transformed into rock. Such fossil organic carbon can accumulate and remain buried for very long time periods, with the consequence that the amount of carbon dioxide in the atmosphere declined. Carbon dioxide is a greenhouse gas, meaning that an increase in its concentration warms the atmosphere, thereby influencing climate. Very early bryophyte-like land plants and later vascular plants influenced Earth's past climate by reducing the concentration of atmospheric carbon dioxide, and plants continue to do so today.

Ecological Effects of Ancient and Modern Bryophytes   In 2004, plant evolutionary biologists Linda Graham, David Hanson, and colleagues determined the amount of decay-resistant tissue mass produced by several modern, early diverging bryophytes and used these data to estimate the ecological impact of early nonvascular plants. The results suggested that nonvascular plants likely contributed organic substances to early soils, thereby helping to enrich them. The results also indicated that ancient nonvascular plants could have produced organic carbon that was buried before being recycled back to carbon dioxide. By this process, early plants could have thereby influenced climate. The investigators calculated that such effects on soil, atmospheric chemistry, and climate might have been significant because they could have occurred over large geographic areas and for millions of years before vascular plants became dominant.

Modern bryophytes likewise play important roles by storing CO_2 as decay-resistant organic compounds. Plants of the locally abundant modern moss genus *Sphagnum* contain so much decay-resistant body mass that in many places, dead moss has accumulated over thousands of years, forming deep peat deposits. By storing very large amounts of organic carbon for long periods, *Sphagnum* helps to keep Earth's climate steady. Under cooler than normal conditions, *Sphagnum* grows more slowly and thus absorbs less CO_2, allowing atmospheric CO_2 to rise a bit, warming the climate a little. As the climate warms, *Sphagnum* grows faster and sponges up more CO_2, storing it in peat deposits. Such a reduction in atmospheric CO_2 returns the climate to slightly cooler conditions. In this way, peat moss helps to keep the world's climate from changing dramatically.

Ecological Effects of Ancient Vascular Plants   Vascular plants originated from extinct plants called protracheophytes, which had branched sporophytes and produced numerous sporangia. However, protracheophyte water-conducting cells lacked lignin and did not provide structural support. Vascular plant

fossils first appear in rocks deposited 420–429 million years ago and rapidly became diverse and abundant. These fossils reveal that the earliest vascular plants had no leaves or roots, but they did have stems with a central core of lignin-coated water-conducting cells, a tough outer cuticle, and stomata, much like modern vascular plants (see Figure 29.11). These features suggest that early vascular plants had the ability to maintain a stable internal water level. Abundant lignin and cutin also fostered the ability of vascular plant bodies to resist decaying long enough to fossilize, explaining why vascular plants have left a more extensive fossil record than did the earliest bryophyte-like plants.

Fossils tell us that extensive forests dominated by tree-sized lycophytes, pteridophytes, and early lignophytes occurred in widespread swampy regions during the warm, moist Carboniferous period (354–290 million years ago) (**Figure 29.17**). For example, in 2007, paleobotanist William DiMichele and colleagues reported that they had found fossils of an entire coastal forest that was 297 million years old and covered an area of more than 1,000 hectares in what is now Illinois. Large trees related to modern lycophytes dominated this forest, but pteridophytes and representatives of extinct plant phyla were also present. The forest was well preserved because an earthquake had quickly dropped it below sea level into water low in oxygen. Under such conditions, decay microbes could not completely decompose the forest, which was then buried in sediments that later formed coal. Much of today's coal is similarly derived from the abundant remains of ancient plants, explaining why the Carboniferous is commonly known as the Coal Age. Carboniferous plants converted huge amounts of the atmospheric CO_2 into decay-resistant organic materials such as lignin. Long-term burial of these materials, compressed into coal, together with chemical interactions

Giant lycophyte   Giant dragonfly   Giant horsetail (pteridophyte)

Figure 29.17   **Reconstruction of a Carboniferous (Coal Age) forest.**   This ancient forest was dominated by tree-sized lycophytes and pteridophytes, which later contributed to the formation of large coal deposits.

Concept check: *Why did giant dragonflies occur during this time, but not now?*

between soil and the roots of vascular plants, dramatically changed Earth's atmosphere and climate. The removal of large amounts of the greenhouse gas CO_2 from the atmosphere by plants had a cooling effect on the climate, which also became drier because cold air holds less moisture than warm air.

Mathematical models of ancient atmospheric chemistry, supported by measurements of natural carbon isotopes, led paleoclimatologist Robert Berner to propose that the Carboniferous proliferation of vascular plants was correlated with a dramatic decrease in atmospheric carbon dioxide, which reached the lowest known levels about 290 million years ago (**Figure 29.18**). During this period of very low CO_2, atmospheric oxygen levels rose to the highest known levels, because less O_2 was being used to break down organic carbon into CO_2. High atmospheric oxygen content has been invoked to explain the occurrence of giant Carboniferous dragonflies and other huge insects. The great Carboniferous decline in CO_2 level ultimately caused cool, dry conditions to prevail in the late Carboniferous and early Permian periods. As a result of this relatively abrupt global climate change, many of the tall seedless lycophytes and pteridophytes that had dominated earlier Carboniferous forests became extinct, as did organisms such as the giant dragonflies. Cooler, drier conditions favored extensive diversification of the first seed plants, the gymnosperms. Compared to seedless plants, seed plants were better at reproducing in cooler, drier habitats (as we will see in Section 29.4). As a result, seed plants came to dominate Earth's terrestrial communities, as they continue to do.

An Ancient Cataclysm Marked the Rise of Angiosperms

Diverse phyla of gymnosperms dominated Earth's vegetation through the Mesozoic era (248–65 million years ago), which is sometimes called the Age of Dinosaurs. In addition, fossils provide evidence that early flowering plants were present in the early Mesozoic and perhaps before that. Likewise, fossils indicate that early mammals existed in the Mesozoic. Gymnosperms and early angiosperms were probably major sources of food for such early mammals as well as herbivorous dinosaurs. For example, Ruth Stockey and colleagues found many fossils of a new type of angiosperm, named *Cobbania*, which grew in wetlands that are now the Dinosaur Park Formation in Alberta, Canada. These plant fossils occurred with a skeleton of the dinosaur *Ornithomimus* that may have fed on the plant when alive (**Figure 29.19**).

One fateful day about 65 million years ago, disaster struck from the sky, causing a dramatic change in the types of plants and animals that dominated terrestrial ecosystems.

That day, at least one large meteorite crashed into the Earth near the present-day Yucatán Peninsula in Mexico. This episode is known as the **K/T event** because it marks the end of the Cretaceous (sometimes spelled with a K) period and the beginning of the Tertiary (T) period. The impact, together with substantial volcanic activity that also occurred at this time, is thought to have produced huge amounts of ash, smoke, and haze that dimmed the sun's light long enough to kill many of the world's

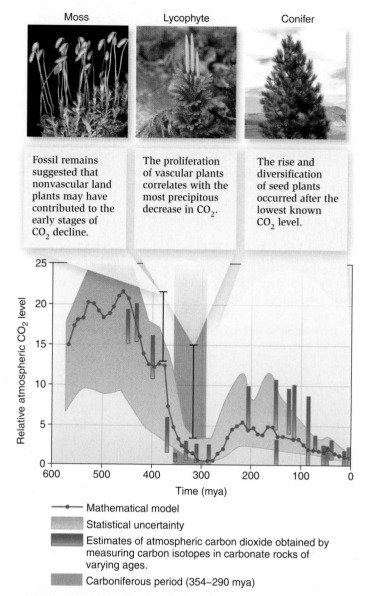

Moss | Lycophyte | Conifer

Fossil remains suggested that nonvascular land plants may have contributed to the early stages of CO_2 decline.

The proliferation of vascular plants correlates with the most precipitous decrease in CO_2.

The rise and diversification of seed plants occurred after the lowest known CO_2 level.

—●— Mathematical model

Statistical uncertainty

Estimates of atmospheric carbon dioxide obtained by measuring carbon isotopes in carbonate rocks of varying ages.

Carboniferous period (354–290 mya)

Figure 29.18 **Changes in Earth's atmospheric carbon dioxide levels over geological time.** Geological evidence indicates that carbon dioxide levels in Earth's atmosphere were once higher than they are now, but that the rise of land plants caused CO_2 to reach the lowest known level about 300 million years ago.

plants. Many groups of plants became extinct, though some survived and persist to the present time. With a severely reduced food supply, most dinosaurs were also doomed, the exceptions being their descendants, the birds. The demise of the dinosaurs left room for birds and mammals to adapt to many kinds of terrestrial habitats.

After the K/T event, ferns dominated long enough to leave huge numbers of fossil spores, and then surviving groups of flowering plants began to diversify into the space left by the extinction of previous plants. The rise of angiosperms fostered the diversification of beetles (see Chapter 33) and other types of insects that associate with plants.

Our brief survey of plant evolutionary history reveals some important diversity principles. While environment certainly influenced the diversification of plants, plant diversification has also changed Earth's environment in ways that affected

Figure 29.19 Early angiosperms, sources of food for large herbivorous dinosaurs of the Mesozoic era. The newly discovered fossil angiosperm *Cobbania corrugata* grew in wetlands that were also inhabited by large dinosaurs such as *Ornithomimus*, whose head is illustrated here.

the evolution of other organisms. In addition, plant evolutionary history serves as essential background for a closer focus on the evolution of **critical innovations**, new features that foster the diversification of phyla. Among the critical innovations that appeared during plant evolutionary history, embryos, leaves, and seeds were particularly important.

29.3 The Origin and Evolutionary Importance of the Plant Embryo

The embryo, absent from charophyceans, was probably one of the first critical innovations acquired by land plants. Recall that plant embryos are young sporophytes that develop from zygotes and are enclosed by maternal tissues that provide sustenance. The presence of an embryo is critical to plant reproduction in terrestrial environments. Drought, heat, ultraviolet light, and microbial attack could kill delicate plant egg cells, zygotes, and embryos if these were not protected and nourished by enclosing maternal tissues. The first embryo-producing plants diversified into hundreds of thousands of diverse modern species, as well as many species that have become extinct. A closer look at embryos reveals why their origin and evolution is so important to all land plants.

Plant Embryos Grow Protected Within the Maternal Plant Body

A plant embryo has several characteristic features, some of which were previously described. First, plant embryos are multicellular and diploid (see Section 29.1). Plant embryos develop by repeated mitosis from a single-celled zygote resulting from

fertilization (see Figure 29.6b). In addition, we have also learned that plant eggs are fertilized while still enclosed by the maternal plant body and embryos begin their development within the protective confines of maternal tissues (see Figure 29.7). Plant biologists say that plants retain their zygotes and embryos. Third, plant embryo development depends on organic and mineral materials supplied by the mother plant. Nutritive tissues composed of specialized **placental transfer tissues** aid in the transfer of nutrients from mother to embryo. Taking a closer look at placental transfer tissues reveals their valuable role.

Placental transfer tissues function similarly to the placenta present in most mammals, which fosters nutrient movement from the mother's bloodstream to the developing fetus. Plant placental transfer tissues often occur in haploid gametophyte tissues that lie closest to embryos and in the diploid tissues of young embryos themselves. Such transfer tissues contain cells that are specialized in ways that promote the movement of solutes from gametophyte to embryo. For example, the cells of placental transport tissues display complex arrays of finger-like cell-wall ingrowths (**Figure 29.20**). Because the plant plasma membrane lines this elaborate plant cell wall, the ingrowths vastly increase the surface area of plasma membrane. This increase provides the space needed for abundant membrane transport proteins, which move solutes into and out of cells. With more transport proteins present, materials can move at a faster rate from one cell to another. Experiments have revealed

Figure 29.20 Placental transfer tissue from a plant in the liverwort genus *Monoclea*. This TEM shows that placental transfer tissues contain specialized cells having extensive finger-shaped cell-wall ingrowths. Such cells help nutrients to move rapidly from parental gametophytes to embryonic sporophytes, a process that fosters plant reproductive success.

Concept check: How does increasing plasma membrane surface area foster rapid movement of nutrients in placental transfer tissue?

that dissolved sugars, amino acids, and minerals first move from maternal cells into the intercellular space between maternal tissues and the embryo. Then, transporter proteins in the membranes of nearby embryo cells efficiently import materials into the embryo. Classic experiments have revealed the adaptive value of this process in land plant reproduction.

FEATURE INVESTIGATION

Browning and Gunning Demonstrated That Placental Transfer Tissues Facilitate the Movement of Organic Molecules from Gametophytes to Sporophytes

In the 1970s, plant cell biologists Adrian Browning and Brian Gunning explored placental transfer tissue function. Using a simple moss experimental system, they investigated the rate at which radioactively labeled carbon moves through placental transfer tissues from green gametophytes into young sporophytes. Recall that embryos are very young, few-celled sporophytes and that in mosses and other bryophytes, all stages of sporophyte development are nutritionally dependent on gametophyte tissues. Browning and Gunning investigated nutrient flow into young sporophytes because these slightly older and larger developmental stages were easier to manipulate in the laboratory than were tiny embryos.

In a first step, the investigators grew many gametophytes of the moss *Funaria hygrometrica* in a greenhouse until young sporophytes developed as the result of sexual reproduction (Figure 29.21).

In a second step, they placed black glass sleeves over young sporophytes as a shade to prevent photosynthesis, enclosed moss gametophytes and their attached sporophytes within transparent jars, and supplied the plants with radioactively labeled carbon dioxide for measured time periods known as pulses. Because the moss gametophytes were not shaded, their photosynthetic cells were able to convert the radioactively labeled carbon dioxide into labeled organic compounds, such as sugars and amino acids. Shading prevented the young sporophytes, which possess some photosynthetic tissue, from using labeled CO_2 to produce organic compounds.

In a third step, the researchers added an excess amount of nonradioactive CO_2 to prevent the further uptake of the radio-

Figure 29.21   Browning and Gunning demonstrated that placental transfer tissues increase plant reproductive success.

HYPOTHESES 1. Placental transfer tissues allow organic nutrients to flow from plant gametophytes to sporophytes faster than such nutrients move through plant tissues lacking transfer cells.
2. The rate of organic nutrient transfer into larger sporophytes is faster than into smaller sporophytes.

KEY MATERIALS Moss *Funaria hygrometrica*, $^{14}CO_2$ (radiolabeled carbon dioxide)

Experimental level   **Conceptual level**

1 Grow moss gametophytes until young sporophytes develop from embryos, and measure sporophyte size.

Young sporophytes receive organic nutrients in the same way as embryos but are easier to handle.

2 Shade young sporophytes from light with blackened glass tubing, and enclose whole plant in clear glass jar. Expose plants to $^{14}CO_2$ for 15 minutes. This is called a pulse.

Clear jar
Sporophyte
Dark tubing
Gametophyte
Threaded lid

$^{14}CO_2$

Photosynthesis, which requires light, will convert $^{14}CO_2$ into ^{14}C-sugar in gametophytes but not sporophytes.

Light
$^{14}CO_2$
$^{14}CO_2$
Labeled sugar

3 Expose plants to a large amount of nonradioactive CO_2. This is called a chase. Incubate up to 8 hours.

Nonradiolabeled CO_2

Addition of excess nonlabeled CO_2 gas is known as a chase because it chases away the ability of the cells to make any more radioactive sugars.

Light
Nonradiolabeled CO_2
$^{14}CO_2$ no longer taken up by plant.

4 Pluck young sporophytes of differing sizes from gametophytes. Assay ^{14}C in both sporophytes and gametophytes using a scintillation counter. This was done immediately following the chase, or 2 or 8 hours after the chase.

Scintillation counter

Determine how much organic carbon flowed into sporophytes during each chase time.

5 THE DATA I

Carbon transfer from gametophyte to sporophyte:

Mean ^{14}C content of 5 gametophytes at 0 chase time	Mean ^{14}C lost from gametophytes after 8-hour chase	Mean ^{14}C gained by sporophytes after 8-hour chase
228 units	145 units	51 units

6 THE DATA II

Sporophyte size effect:

Sporophyte size	Mean ^{14}C content of 8 sporophytes after 2-hour chase
5–7 mm	8.47 ± 4.29 units
11–13 mm	9.93 ± 3.94 units
23–25 mm	24.97 ± 5.30 units

7 CONCLUSION Organic carbon moves from photosynthetic gametophytes into developing sporophytes, facilitated by transfer cell-wall ingrowths. Larger sporophytes absorb more organic carbon than smaller ones.

8 SOURCES Browning, A.J., and Gunning, B.E.S. 1979. Structure and function of transfer cells in the sporophyte haustorium of *Funaria hygrometrica.* Hedw. II. Kinetics of uptake of labelled sugars and localization of absorbed products by freeze-substitution. *Journal of Experimental Botany* 30:1247–1264.

Browning, A.J., and Gunning, B.E.S. 1979. Structure and function of transfer cells in the sporophyte haustorium of *Funaria hygrometrica.* III. Translocation of assimilate into the attached sporophyte and along the seta of attached and excised sporophytes. *Journal of Experimental Botany* 30:1265–1273.

active CO_2 from their experimental system, a process known as a chase. This process stopped the radiolabeling of photosynthetic products because the vast majority of CO_2 taken up was now unlabeled. (Experiments such as these are known as pulse-chase experiments.) In a final step, Browning and Gunning plucked young sporophytes of different sizes (ages) from gametophytes and measured the amount of radioactive organic carbon present in the separated gametophyte and sporophyte tissues at various times following the chase.

From these data, they were able to calculate the relative amount of organic carbon that had moved from the photosynthetic moss gametophytes to their sporophytes. Browning and Gunning discovered that about 22% of the organic carbon produced by gametophyte photosynthesis was transferred to the young sporophytes during an 8-hour chase period. They also calculated the rate of nutrient transfer from gametophyte to sporophyte and compared this rate to the rate (determined in other studies) at which organic carbon moves within several other plant tissues that lack specialized transfer cells. Browning and Gunning discovered that organic carbon moved from moss gametophytes to young sporophytes nine times faster than organic carbon moves within these other plant tissues. These investigators inferred that the increased rate of nutrient

movement could be attributed to placental transfer cell structure, namely, the fact that cell-wall ingrowths enhanced plasma membrane surface area. By comparing the amount of radioactive carbon accumulated by sporophytes of differing sizes, they also learned that larger sporophytes absorbed labeled carbon about three times faster than smaller ones.

These data are consistent with the hypothesis that placental transfer tissues increase plant reproductive success by providing embryos and growing sporophytes with more nutrients than they would otherwise receive. Supplied with these greater amounts of nutrients, sporophytes are able to grow larger than they otherwise would, and eventually they produce more progeny spores.

Experimental Questions

1. What were the goals of the Browning and Gunning investigation?

2. How did Browning and Gunning prevent photosynthesis from occurring in moss sporophytes during the experiment (shown in Figure 29.21), and why did they do this?

3. How did the measurements Browning and Gunning made after adding an excess amount of unlabeled CO_2 lead them to their conclusions?

29.4 The Origin and Evolutionary Importance of Leaves and Seeds

Like plant embryos, leaves and seeds are critical innovations that increased plant fitness and fostered diversification. Unlike the plant embryo, which likely originated just once at the birth of the plant kingdom, leaves and seeds probably evolved several times during plant evolutionary history. Comparative studies of diverse types of leaves and seeds in fossil and living plants suggest how these critical innovations might have originated.

The Large Leaves of Ferns Evolved from Branched Stem Systems

Leaves are the solar panels of the plant world. Their flat structure provides a high surface area that helps leaves to effectively capture sunlight for use in photosynthesis. Among the vascular plants, lycophytes produce the simplest and most ancient type of leaves. Modern lycophytes have tiny leaves, known as lycophylls or microphylls, which typically have only a single unbranched vein (**Figure 29.22a**). Some experts think that these small leaves are modified sporangia.

In contrast, the leaves of ferns and seed plants have extensively branched veins. Such leaves are known as **euphylls** (from the Greek, meaning true leaves) (**Figure 29.22b**), and the clade that includes pteridophytes and seed plants is known as the **euphyllophytes** (see Figure 29.1). The branched veins of euphylls are able to supply relatively large areas of photosynthetic tissue with water and minerals. Thus, euphylls are typically much larger than lycophylls, explaining why euphylls are also known as megaphylls (from the Greek, meaning large leaves). Euphylls provide considerable photosynthetic advantage to ferns and seed plants, because they provide more surface for solar energy capture than do small leaves. Hence, the evolution of relatively large leaves allowed plants to more effectively accomplish photosynthesis, enabling them to grow larger and produce more progeny.

Study of fern fossils indicates that euphylls likely arose from leafless, branched stem systems by a series of steps (**Figure 29.22c**). First, one branch assumed the role of the main

(a) Lycophyll (small leaf)

— Single unbranched leaf vein

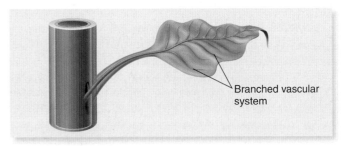

(b) Euphyll (large leaf)

— Branched vascular system

| 1 | Fern ancestors initially had a branched stem system. | 2 | One branch began to dominate the stem system. | 3 | The branch system flattened into a single plane. | 4 | Photosynthetic tissue filled in the spaces between the branches of a system. |

Euphyll

(c) Euphyll evolution process in pteridophytes

Figure 29.22   Lycophylls and euphylls.   (a) Most lycophylls possess only a single unbranched leaf vein having limited conduction capacity, explaining why lycophylls are generally quite small. **(b)** Euphylls possess branched vascular systems having greater conduction capacity, explaining why many euphylls are relatively large. **(c)** Fossil evidence suggests how pteridophyte euphylls might have evolved from branched stem systems.

Concept check: *Imagine that the leaves of some other plant group evolved from stem systems that were more highly branched (that is, there were more branches per unit length of stem) than those of ferns. How do you think the leaves of such plants would differ from those of ferns?*

axis, while the other was reduced in size, became flattened in one plane, and finally, the spaces between the branches of this flattened system became filled with photosynthetic tissue. Such a process would explain why euphylls have branched vascular systems; individual veins apparently originated from the separate branches of an ancestral branched stem. Plant evolutionary biologists suspect that euphylls arose several times, and it is unclear whether the leaves of seed plants originated in the same way as those of ferns.

Seeds Develop from the Interaction of Ovules and Pollen

The seed plants dominate modern ecosystems, suggesting that seeds offer reproductive advantages. Seed plants are also the plants with the greatest importance to humans, as described in Chapter 30. For these reasons, plant biologists are interested in understanding why seeds are so advantageous and how they evolved. To consider these questions, we must first take a closer look at seed structure and development.

Plants produce seeds by reproductive structures unique to seed plants, known as ovules and pollen. An **ovule** is a sporangium that typically contains only a single spore that develops into a very small egg-producing gametophyte, the whole megasporangium enclosed by leaflike structures known as **integuments** (**Figure 29.23a**). You can think of an ovule as being like a nesting doll with four increasingly smaller dolls inside. The smallest doll would correspond to an egg cell; intermediate-sized dolls would represent the gametophyte, spore wall, and megasporangium; and the largest doll would represent the integuments. Fertilization converts such layered ovules into seeds. In seed plants, the sperm needed for fertilization are supplied by **pollen**,

tiny male gametophytes enclosed by walls. A closer look at pollen and ovules will help in understanding how seeds develop.

We have earlier noted that all plants produce spores by meiosis within sporangia, and seed plants are no exception. However, seed plants produce two distinct types of spores in two different types of sporangia, a trait known as heterospory. Small **microspores** develop within microsporangia, and larger **megaspores** develop within megasporangia. Male gametophytes develop from the microspores, and the pollen that develops from these microspores is released from microsporangia. Meanwhile, female gametophytes develop and produce eggs while enclosed by protective megaspore walls. The problem with this process is that female gametophytes are not photosynthetic, so they need help in feeding the embryos that develop from fertilized eggs. Female gametophytes get this help from the previous sporophyte generation by remaining attached to it. This is an advantage because the large sporophyte generation is thereby able to provide gametophytes with the nutrients needed for embryo development.

Embryos develop as the result of fertilization, which cannot occur until after **pollination**, the process by which pollen comes close to ovules. Pollination typically occurs by means of wind or animal transport (see Chapter 30). Fertilization occurs in seed plants when a male gametophyte extends a slender pollen tube that carries two sperm toward an egg. The pollen tube enters through an opening in the integument called the micropyle and releases the sperm. The fertilized egg becomes an embryo, and the ovule's integument develops into a protective, often hard and tough **seed coat** (**Figure 29.23b,c**).

Gymnosperm seeds contain female gametophyte tissue that has accumulated large amounts of protein, lipids, and carbohydrates prior to fertilization. These nutrients are used during both

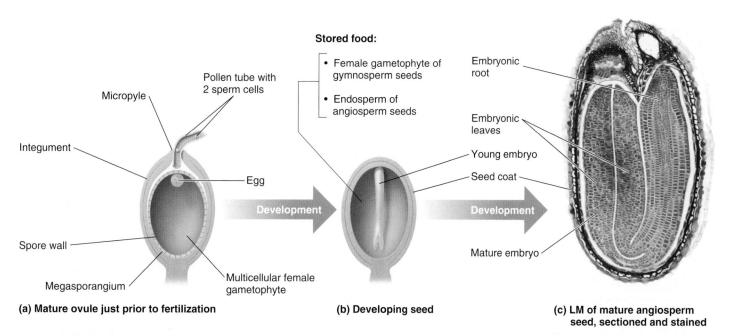

Stored food:
- Female gametophyte of gymnosperm seeds
- Endosperm of angiosperm seeds

(a) Mature ovule just prior to fertilization

Micropyle · Integument · Pollen tube with 2 sperm cells · Egg · Spore wall · Megasporangium · Multicellular female gametophyte

(b) Developing seed

Young embryo

(c) LM of mature angiosperm seed, sectioned and stained

Embryonic root · Embryonic leaves · Seed coat · Mature embryo

Figure 29.23   Structure of an ovule developing into a seed.

Concept check: *Can you hypothesize why this angiosperm seed with its mature embryo does not show obvious endosperm tissue?*

embryo development and seed germination to help support growth of the seedling. Angiosperm seeds also contain this useful food supply, but most angiosperm ovules do not store food materials before fertilization. Instead, angiosperm seeds generally store food only after fertilization occurs, ensuring that the food is not wasted if an embryo does not form. How is this accomplished? The answer is a process known as **double fertilization**. This process produces both a zygote and a food storage tissue known as endosperm, a tissue unique to angiosperms. One of the two sperm delivered by each pollen tube fuses with the egg, producing a diploid zygote, as you might expect. The other sperm nucleus fuses with different gametophyte nuclei to form an unusual cell that has more than the diploid number of chromosomes; this cell generates the endosperm food tissue. More information about endosperm can be found in Chapter 39.

Seeds allow embryos access to food supplied by the previous sporophyte generation, an option not available to seedless plants. The layered structure of ovules explains why seeds are also layered, with a protective seed coat enclosing the embryo and stored food. These seed features improve the chances of embryo and seedling survival, thereby increasing seed plant fitness.

Seeds Confer Important Ecological Advantages

Seeds provide plants with numerous ecological advantages. First, many seeds are able to remain dormant in the soil for long periods, until conditions become favorable for germination and seedling growth. Further, seed coats are often adapted in ways that improve dispersal in diverse habitats. For example, many plants produce winged seeds that are effectively dispersed by wind. Other plants produce seeds with fleshy coverings that attract animals, which consume the seeds, digest their fleshy covering, and eliminate them at some distance from the originating plants.

Another advantage of seeds is that they can store considerable amounts of food, which supports embryo growth and helps plant seedlings grow large enough to compete for light, water, and minerals. This is especially important for seeds that must germinate in shady forests. Finally, the sperm of most seed plants can reach eggs without having to swim through water, because pollen tubes deliver sperm directly to ovules. Consequently, seed plant fertilization is not typically limited by lack of water, in contrast to that of seedless plants. Therefore, seed plants are better able to reproduce in arid and seasonally dry habitats. For these reasons, seeds are considered to be a key adaptation to reproduction in a land habitat.

Ovule and Seed Evolution Illustrate Descent with Modification

As we have seen, seed plants reproduce using both spores and seeds, but note that seed plants have not replaced spores with seeds. Rather, seed plants continue to produce spores, and ovules and seeds have evolved from spore-producing structures by descent with modification. Recall that this evolutionary principle involves changes in pre-existing structures and processes.

Fossils provide some clues about ovule and seed evolution, and other information can be obtained by comparing reproduction in living lycophytes and pteridophytes.

Most modern lycophytes and pteridophytes release one type of spore that develops into one type of gametophyte. Such plants are considered to be homosporous, and their gametophytes live independently and produce both male and female gametangia (see Figure 29.12). However, some lycophytes and pteridophytes produce and release two distinct kinds of spores: relatively small microspores and larger megaspores, which respectively grow into male and female gametophytes. The larger size of megaspores enables them to store food that later supports developing sporophytes. This reproductive process, known as **heterospory** (meaning different spores), also characterizes all seed plants. The gametophytes produced by heterosporous plants also grow within the confines of microspore and megaspore walls and therefore are known as **endosporic gametophytes**.

An advantage of heterospory is that it mandates cross-fertilization. The eggs and sperm that fuse are derived from different gametophytes and hence from different spores and different meiotic events. This makes it likely that the gametes are of distinct genotypes. Cross-fertilization increases the potential for genetic variation, which aids evolutionary flexibility. Endosporic gametophytes receive protection from environmental damage by surrounding spore walls. From these observations, we can infer that heterospory and endosporic gametophytes were probably also features of seed plant ancestors and constitute early steps toward seed evolution (**Figure 29.24**). Fossils and modern plants also illustrate subsequent stages in seed evolution, such as retaining megaspores within sporangia, rather than releasing them.

A further step in seed evolution may have been the production of only one megaspore per sporangium rather than multiple spores per sporangium, which is common in seedless plants. Reduction of megaspore numbers (from those present in seedless plants) would have allowed plants to channel more nutrients into each megaspore. A final step might have been the retention of megasporangia on parental sporophytes by the development of integuments (Figure 29.24). As we have noted, this adaptation would allow food materials to flow from mature photosynthetic sporophytes to their dependent gametophytes and young embryos.

Fossils provide information about when and how the process of seed evolution first occurred. Very early fossil seeds such as *Elkinsia polymorpha* and *Archaeosperma arnoldii* were present 365 million years ago, during the Devonian period, which occurred prior to the Coal Age (see Figure 29.1). Paleobotanists Phillippe Gerrienne, Brigette Meyer-Berthaud, and associates reported that fossil reproductive structures of an extinct plant named *Runcaria heinzelinii* may represent a seed precursor (**Figure 29.25**). These fossil structures had a lacy integument that did not completely enclose the megasporangium.

The evolutionary journey illustrated by the transition from aquatic charophycean algae to bryophytes, to seedless plants, and finally to seed plants reveals how adaptation is related to environmental change, as well as ways in which plants

1 Sporangium containing spores that are similar in size	2a Microsporangium containing many small microspores	2b Megasporangium containing fewer, larger megaspores	3 Reduction to 1 megaspore per megasporangium	4 Enclosure of megasporangium within integuments to form ovule; when fertilized, ovule develops into a seed

Early evolution of heterospory

Microsporangium

+

Megasporangium

Megaspore

Integuments

Ovule

Evolution of megasporangium that led to an ovule

Figure 29.24 Hypothetical stages in the evolution of seeds.   The parallel evolution of heterospory and endosporic gametophytes in some lycophytes and pteridophytes as well as the seed plants suggests that these features were acquired early in the evolution of seeds. Later-occurring events in the origin of seeds included reduction of the number of megaspores to one per megasporangium and enclosure of the megasporangium by protective, leaflike integuments.

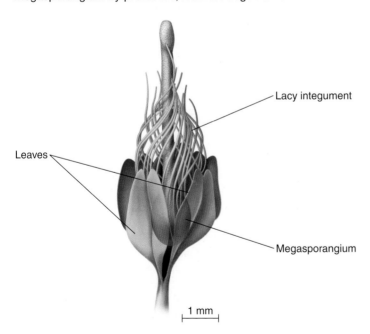

Lacy integument

Leaves

Megasporangium

1 mm

Figure 29.25   The fossil *Runcaria heinzelinii*, a plant with a probable seed precursor.

Concept check: *Based on your knowledge of integument function in modern seed plants, can you hypothesize a function for the lacy integument of Runcaria?*

themselves shaped Earth's ecosystems. As a summary of what we have learned in this chapter, **Table 29.1** provides a list of the distinguishing features of land plants and their charophycean relatives.

Table 29.1	Distinguishing Features of Modern Charophyceans and Land Plants*

Charophyceans

Primarily aquatic habitat; zygotic life cycle; sporangia absent; sporophytes absent; cell-wall xyloglucans absent

LAND PLANTS (EMBRYOPHYTES)

Primarily terrestrial habitat; sporic life cycle consisting of alternation of two multicellular generations—diploid sporophyte and haploid gametophyte; multicellular embryos are nutritionally dependent for at least some time during development; spore-producing sporangia; gamete-producing gametangia; sporopollenin spore walls; cell-wall xyloglucans

Nonvascular plants (Bryophytes) (**liverworts, mosses, hornworts**)

Dominant gametophyte generation; supportive, lignin-containing vascular tissue absent; true roots, stems, leaves absent; sporophytes unbranched and unable to grow independently of gametophytes

VASCULAR PLANTS (TRACHEOPHYTES) (lycophytes, pteridophytes, spermatophytes)

Dominant sporophyte generation; lignified water-conducting tissue—xylem; specialized organic food-conducting tissue—phloem; sporophytes branched; sporophytes eventually become independent of gametophytes

SEEDLESS VASCULAR PLANTS (**lycophytes, pteridophytes**)

Lycophytes Leaves generally small with a single, unbranched vein (lycophylls or microphylls); sporangia borne on sides of stems

EUPHYLLOPHYTES (**pteridophytes,** SPERMATOPHYTES)

Pteridophytes Leaves relatively large with extensively branched vein system (euphylls or megaphylls); sporangia borne on leaves

SEED PLANTS (SPERMATOPHYTES)

Seeds present; leaves are euphylls

Gymnosperms (**cycads, ginkgos, conifers, gnetophytes**)

Flowers and fruits absent; seed food stored before fertilization in female gametophyte, endosperm absent

Angiosperms (**flowering plants**)

Flowers and fruit present; seed food stored after fertilization in endosperm formed by double fertilization

*Key: **Phyla**; LARGER MONOPHYLETIC CLADES (synonyms). All other classification terms are not clades.

Summary of Key Concepts

29.1 Ancestry and Diversity of Modern Plants

- Plants are multicellular eukaryotic organisms composed of cells having plastids; they display many adaptations to life on land. The modern plant kingdom consists of several hundred thousand species classified into 10 phyla, informally known as the liverworts, mosses, hornworts, lycophytes, pteridophytes, cycads, ginkgos, conifers, gnetophytes, and angiosperms. (Figure 29.1)

- The land plants evolved from ancestors that were probably similar to modern, complex charophycean algae of freshwater habitats. (Figure 29.2)

- The monophyletic liverwort, moss, and hornwort phyla are together known informally as the bryophytes. Bryophytes illustrate early-evolved features of land plants, such as a sporic life cycle involving embryos that develop within protective, nourishing gametophytic tissues. (Figures 29.3, 29.4, 29.5, 29.6)

- Bryophytes differ from other plants in having a dominant gametophyte generation and a dependent, nonbranching, short-lived sporophyte generation. Bryophytes also lack supportive vascular tissues in contrast to other modern plant phyla, which are known as the vascular plants or tracheophytes. (Figures 29.7, 29.8)

- Lycophytes, pteridophytes, and other vascular plants generally possess stems, roots, and leaves having vascular tissues composed of phloem and xylem, cuticle, and stomata, but these groups differ in distinctive ways. (Figures 29.9, 29.10, 29.11)

- The fern life cycle illustrates the dominant sporophyte characteristic of vascular plants. The fern *Ceratopteris richardii* is used as a model genetic system. (Figures 29.12, 29.13)

- Cycads, ginkgos, conifers, and gnetophytes are collectively known as gymnosperms. Gymnosperms produce seeds and inherited an ancestral capacity to produce wood. Angiosperms, the flowering plants, produce seeds, and many also produce wood. Flowers, fruits, and seed endosperm are distinctive features of the angiosperms. (Figures 29.14, 29.15)

29.2 An Evolutionary History of Land Plants

- Paleobiologists and plant evolutionary biologists infer the history of land plants by analyzing the molecular features of modern plants and by comparing the structural features of fossil and modern plants. (Figure 29.16)

- Seedless plants transformed Earth's ecology by fostering soil buildup and altering atmospheric chemistry and climate. (Figures 29.17, 29.18, 29.19)

- The K/T meteorite impact event that occurred 65 million years ago helped cause the extinction of previously dominant dinosaurs and many types of gymnosperms, leaving space into which angiosperms, insects, birds, and mammals diversified.

29.3 The Origin and Evolutionary Importance of the Plant Embryo

- Origin of the plant embryo was a critical innovation that fostered diversification of the land plants. Plant embryos are supported by nutrients supplied by female gametophytes with the aid of specialized placental transfer tissues. (Figure 29.20)

- In a classic experiment, Browning and Gunning inferred that placental transfer tissues were responsible for an enhanced flow rate of nutrients from parental gametophytes to embryos. (Figure 29.21)

29.4 The Origin and Evolutionary Importance of Leaves and Seeds

- Leaves are specialized photosynthetic organs that evolved more than once during plant evolutionary history. The lycophylls of lycophytes are relatively small leaves having a single unbranched vein. Euphyllophyte leaves are larger, with an extensively branched vascular system; they are known as euphylls. Fossils indicate that fern euphylls evolved from branched stem systems. (Figure 29.22)

- Seeds develop from ovules, megasporangia enclosed by leaflike integuments. Pollen produces thin cellular tubes that deliver sperm to eggs produced by female gametophytes. Following pollination and fertilization, ovules develop into seeds. Mature seeds contain stored food and an embryonic sporophyte that develops from the zygote. (Figure 29.23)

- Seeds confer many reproductive advantages, including dormancy through unfavorable conditions, greater protection for embryos from mechanical and pathogen damage, seed coat modifications that enhance seed dispersal, and reduction of plant dependence on water for fertilization. Fossil seeds display stages in seed evolution. (Figures 29.24, 29.25)

- The distinctive traits of charophyceans and the different phyla of land plants reveal the occurrence of descent with modification. (Table 29.1)

Assess and Discuss

Test Yourself

1. The simplest and most ancient phylum of modern land plants is probably
   a. the pteridophytes.   d. the angiosperms.
   b. the cycads.   e. none of the listed choices.
   c. the liverworts.

2. An important feature of land plants that originated during the diversification of charophycean algae is
   a. the sporophyte.
   b. spores, which are dispersed in air and coated with sporopollenin.
   c. tracheids.
   d. plasmodesmata.
   e. fruits.

3. A phylum whose members are also known as bryophytes is commonly known as
   a. liverworts.   d. all of the above.
   b. hornworts.   e. none of the listed choices.
   c. mosses.

4. Plants possess a life cycle that involves alternation of two multicellular generations: the gametophyte and
   a. the lycophyte.   c. the pteridophyte.   e. the sporophyte.
   b. the bryophyte.   d. the lignophyte.

5. The seed plants are also known as
   a. bryophytes.
   b. spermatophytes.
   c. pteridophytes.
   d. lycophytes.
   e. euphyllophytes.

6. A waxy cuticle is an adaptation that
   a. helps to prevent water loss from tracheophytes.
   b. helps to prevent water loss from charophyceans.
   c. helps to prevent water loss from bryophytes.
   d. aids in water transport within the bodies of vascular plants.
   e. does all of the above.

7. Plant photosynthesis transformed a very large amount of carbon dioxide into decay-resistant organic compounds, thereby causing a low in atmospheric carbon dioxide levels during the geological period known as
   a. the Cambrian.
   b. the Ordovician.
   c. the Carboniferous.
   d. the Permian.
   e. the Pleistocene.

8. Which phylum among the plants listed is likely to have the largest leaves?
   a. liverworts
   b. hornworts
   c. mosses
   d. lycophytes
   e. pteridophytes

9. Euphylls, also known as megaphylls, probably evolved from
   a. the leaves of mosses.
   b. lycophylls.
   c. branched stem systems.
   d. modified roots.
   e. none of the listed choices.

10. A seed develops from
    a. a spore.
    b. a fertilized ovule.
    c. a microsporangium covered by integuments.
    d. endosperm.
    e. none of the above.

Conceptual Questions

1. List several common traits that lead evolutionary biologists to infer that land plants evolved from ancestors related to modern charophycean algae.

2. Why have bryophytes such as mosses been able to diversify into so many species even though they have relatively small, dependent sporophytes?

3. Explain how several structural features help vascular plants to maintain stable internal water content.

Collaborative Questions

1. Discuss at least one difference in environmental conditions experienced by early land plants and ancestral complex charophycean algae.

2. Discuss as many plant adaptations to land as you can.

Online Resource

www.brookerbiology.com

Stay a step ahead in your studies with animations that bring concepts to life and practice tests to assess your understanding. Your instructor may also recommend the interactive ebook, individualized learning tools, and more.

The Evolution and Diversity of Modern Gymnosperms and Angiosperms

30

The Madagascar periwinkle (*Catharanthus roseus*), one of the many seed plants on which humans depend.

The seed plants, gymnosperms and angiosperms, are particularly important in our everyday lives because they are the sources of many products, including wood, paper, beverages, food, cosmetics, and medicines. Leukemia, for example, is effectively treated with vincristine, a drug extracted from the beautiful flowering plant known as the Madagascar periwinkle (*Catharanthus roseus*), pictured in the chapter-opening photo. Vinblastine—another extract from *C. roseus*—is used to treat lymphatic cancers. Taxol, a compound used in the treatment of breast and ovarian cancers, was first discovered in extracts of the Pacific yew tree, a gymnosperm known as *Taxus brevifolia*. Vincristine, vinblastine, taxol, and many other plant-derived medicines are examples of plant secondary metabolites, which are distinct from the products of primary metabolism (carbohydrates, lipids, proteins, and nucleic acids). Secondary metabolites play essential roles in protecting plants from disease organisms and plant-eating animals, and they also aid plant growth and reproduction. Though all plants produce secondary metabolites, these natural products are exceptionally diverse in gymnosperms and angiosperms.

In this chapter, we will learn how the hundreds of thousands of modern seed plants play many additional important roles in the lives of humans and modern ecosystems. This chapter builds on the introduction to seeds and seed plants provided in Chapter 29, focusing on the diversity of modern lineages of gymnosperms and angiosperms. Coevolutionary interactions among angiosperms and animals are presented as major forces influencing the diversification of these groups. This chapter concludes by considering human impacts on seed plant evolution and the importance of seed plants in modern agriculture.

30.1 Overview of the Seed Plants

Figure 30.1 shows our current understanding of the relationships among modern seed plants. As we discussed in Chapter 29, the first seed plants or spermatophytes evolved from earlier seedless vascular plants similar to modern lycophytes and pteridophytes. Following the early diversification of gymnosperms, one ancestral gymnosperm lineage gave rise to the angiosperms, the flowering plants. Today, humans depend upon modern seed plants, particularly the angiosperms, for many services and products. For example, as noted in the introduction, seed plants are major sources of useful medicines. Modern gymnosperms are important sources of wood and other useful materials, whereas the fruit and grain crops that form the basis for much of modern agriculture are angiosperms.

During their evolution, the seed plants inherited many adaptations to life on land from their ancestors and acquired new traits (Figure 30.1). Major critical innovations shared by all seed plants are pollen, ovules, seeds, and wood. As we learned in Chapter 29, pollen allows seed plants to disperse male gametophytes, ovules provide protection and nutrition to female gametophytes and developing embryos, and seeds allow plants to reproduce in diverse habitats. Wood strengthens plants, allowing them to grow tall and produce many branches, leaves, and seeds. We will learn more about these and other critical innovations in this chapter.

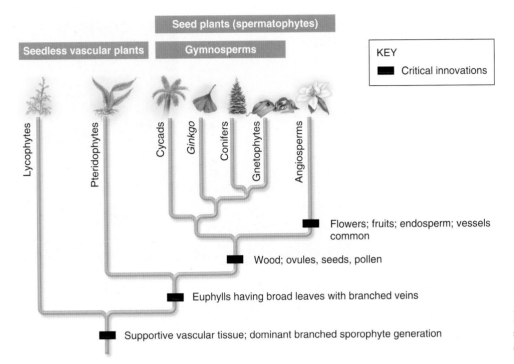

Figure 30.1 Phylogeny of modern seedless and seed plants showing critical innovations.

Labels on figure:
- Seedless vascular plants
- Seed plants (spermatophytes)
- Gymnosperms
- KEY ■ Critical innovations
- Lycophytes
- Pteridophytes
- Cycads
- Ginkgo
- Conifers
- Gnetophytes
- Angiosperms
- Flowers; fruits; endosperm; vessels common
- Wood; ovules, seeds, pollen
- Euphylls having broad leaves with branched veins
- Supportive vascular tissue; dominant branched sporophyte generation

30.2 The Evolution and Diversity of Modern Gymnosperms

Gymnosperms are plants that produce seeds that are exposed rather than enclosed in fruits, as is the case for angiosperms. The word gymnosperm comes from the Greek *gymnos,* meaning naked (referring to the unclothed state of ancient athletes), and *sperma,* meaning seed. Most modern gymnosperms are woody plants that occur as shrubs or trees. Seeds and wood are adaptations that allow gymnosperms to cope with global climate changes and to live in relatively cold and dry habitats. In this section, we will first consider some fossil plants that help to explain important gymnosperm traits. Then we will survey the structure, reproduction, and ecological roles of modern gymnosperm phyla.

Modern Gymnosperms Arose from Woody Ancestors

Modern gymnosperms include the famous giant sequoias (*Sequoiadendron giganteum*) native to the Sierra Nevada of the western U.S. Giant sequoias are among Earth's largest organisms, weighing as much as 6,000 tons and reaching an amazing 100 m in height. The large size of sequoias and other trees is based on the presence of **wood**, a tissue composed of numerous pipelike arrays of empty, water-conducting cells whose walls are strengthened by an exceptionally tough secondary metabolite known as lignin. These properties enable woody tissues to transport water upward for great distances and also to provide the structural support needed for trees to grow tall and produce many branches and leaves. In modern seed plants, a special tissue known as the **vascular cambium** produces both thick

layers of wood and thinner layers of inner bark. The inner bark transports watery solutions of organic compounds. (The structure and function of the vascular cambium, wood, and bark are described in more detail in Chapter 35.) Vascular cambium, wood, and inner bark are critical innovations that helped gymnosperms and other seed plants to compete effectively for light and other resources needed for photosynthesis.

Wood first appeared in a group of plants known as the **progymnosperms** (from the Greek, meaning before gymnosperms). Woody progymnosperms, such as the fossil plant genus *Archaeopteris,* which lived 370 million years ago (mya), were the first trees that had leafy twigs (**Figure 30.2**). Progymnosperms were able to produce a vascular cambium and wood because their vascular tissue was arranged in a ring around a central pith of nonvascular tissue. (In contrast, the vascular tissue of earlier tracheophytes was arranged differently.) This ring of vascular tissue, known as a **eustele**, contained cells that were able to develop into the vascular cambium as seedlings grew into saplings. The vascular cambium then produced wood, allowing saplings to grow into tall trees. Modern seed plants inherited the eustele, explaining why many gymnosperms and angiosperms are also able to produce vascular cambia and wood. Although progymnosperms were woody plants, fossil evidence indicates that they did not produce seeds. This observation reveals that wood originated before seeds evolved.

Diverse early gymnosperms were the major vegetation present during the Mesozoic era, also known as the Age of Dinosaurs. Some groups of gymnosperms became extinct before or as a result of the K/T event at the end of the Cretaceous period 65 mya (see Chapter 29). Only a few gymnosperm phyla have survived to modern times: cycads (the Cycadophyta); *Ginkgo biloba,* the only surviving member of a once-large phylum termed Ginkgophyta; conifers (the Coniferophyta); and

Figure 30.2 The progymnosperm *Archaeopteris*, an early tree. This illustration was reconstructed from fossil data.

gnetophytes (Gnetophyta). These phyla display distinctive reproductive features and play important roles in ecology and human affairs.

Cycads Are Endangered in the Wild but Are Widely Used as Ornamentals

Nearly 300 cycad species occur today, primarily in tropical and subtropical regions. However, many species of cycads are rare, and their tropical forest homes are increasingly threatened by human activities. Many cycads are listed as endangered species, and commercial trade in cycads is regulated by CITES (Convention on International Trade in Endangered Species of Wild Fauna and Flora), a voluntary international agreement.

The structure of cycads is so interesting and attractive that many species are cultivated for use in outdoor plantings or as houseplants. The nonwoody stems of some cycads emerge from the ground much like tree trunks, some reaching 15 m in height, whereas the stems of other cycads are not conspicuous because they are subterranean (**Figure 30.3**). Cycads display spreading, palmlike leaves (*cycad* comes from a Greek word, meaning palm). Mature leaves of the African cycad *Encephalartos laurentianus* can reach an astounding 8.8 m in length!

In addition to underground roots, which provide anchorage and take up water and minerals, many cycads produce coralloid roots. Such roots extend aboveground and have branching shapes resembling corals (**Figure 30.4a**). Coralloid roots harbor light-dependent, photosynthetic cyanobacteria within their

(a) Emergent cycad stem   **(b) Submergent cycad stem**

Figure 30.3   Cycads.   Palmlike foliage and conspicuous seed-producing cones are features of most cycads. **(a)** The stems of some cycads emerge from the ground. **(b)** The stems of other cycads are submerged in the ground, so the leaves emerge at ground level.

Concept check: *Why should people avoid eating cycad seeds or products made from them?*

Root surface

Cyanobacteria

(a) Coralloid roots

(b) Coralloid root cross section

Figure 30.4   Coralloid roots of cycads.   **(a)** Many cycads produce aboveground branching roots that resemble branched corals. **(b)** This magnified cross section of a coralloid root shows a ring of symbiotic blue-green cyanobacteria, which provide the plant with fixed nitrogen.

Concept check: *Why do the coralloid roots grow aboveground?*

tissues. The cyanobacteria, which form a bright blue-green ring beneath root surfaces (**Figure 30.4b**), are nitrogen fixers that provide their plant hosts with nitrogen minerals crucial to their growth (see Chapter 27).

Recent studies have revealed that the cyanobacteria in the cycad *Cycas micronesica* produce an unusual amino acid that is distributed to the host plant's leaves and seeds. The distinctive amino acid, known as BMAA (β-N-methylamino-L-alanine), is harmful to the health of humans who consume flour made from cycad seeds or the meat of bats that have fed on *C. micronesica* seeds. Botanist Paul Alan Cox and associates linked this toxin to the unusually high occurrence of a dementia resembling Alzheimer disease among the Chamorro people of Guam in the Mariana Islands chain. These researchers also found BMAA in the brain tissues of dementia patients who had not consumed foods originating from cycads. In an effort to understand these cases, Cox and associates examined many types of cyanobacteria from diverse habitats. In 2005, these investigators reported that most cyanobacteria produce BMAA in nature, suggesting that the toxic amino acid could be more widely present in the environment than previously thought. Studies of cycad toxicity have thus revealed a potential human health hazard of widespread concern. Because cycads generally produce a variety of toxins that likely deter herbivorous animals, experts recommend that humans should not consume food products made from these plants.

Cycad reproduction is distinctive in several ways. Individual cycad plants produce conspicuous conelike structures that bear either ovules and seeds or pollen (see Figure 30.3). When mature, both types of reproductive structures emit odors that attract beetles. These insects carry pollen to ovules, where the pollen produces tubes that deliver sperm to eggs.

Ginkgo biloba Is the Last Survivor of a Once-Diverse Group

The beautiful tree *Ginkgo biloba* (**Figure 30.5a**) is the single remaining species of a phylum that was much more diverse during the Age of Dinosaurs. *G. biloba* takes its species name from the lobed shape of its leaves, which have unusual forked veins (**Figure 30.5b**). Today, *G. biloba* may be nearly extinct in the wild; widely cultivated modern *Ginkgo* trees are descended from seeds produced by a tree found in a remote Japanese temple garden and brought to Europe by 17th-century explorers.

G. biloba trees are widely planted along city streets because they are ornamental and also tolerate cold, heat, and pollution better than most trees. In addition, these trees are long-lived—individuals can live for more than a thousand years and grow to 30 m in height. Individual trees produce either ovules and seeds or pollen, based on a sex chromosome system much like that of humans. Ovule-producing trees have two X chromosomes; pollen-producing trees have one X and one Y chromosome. Wind disperses pollen to ovules, where pollen grains germinate to produce pollen tubes. These tubes grow through ovule tissues for several months, absorbing nutrients that are used for sperm development. Eventually the pollen tubes burst, delivering flagellate sperm to egg cells. After fertilization, zygotes develop into embryos, and the ovule integument develops into a fleshy, bad-smelling outer seed coat and a hard, inner seed coat (**Figure 30.5c**). For street-side or garden plantings, people usually select pollen-producing trees to avoid the stinky seeds.

Conifers Are the Most Diverse Modern Gymnosperm Lineage

The conifers (**Figure 30.6**) are a lineage of trees named for their seed cones, of which pinecones are familiar examples. Conifers have been important components of terrestrial vegetation since the end of the Carboniferous period some 300 mya. Modern conifer families have existed for about 200 million years, including surviving the K/T event about 65 mya, and today include more than 500 species in 50 genera. Conifers are particularly common in mountain and high-latitude forests and are important sources of wood and paper pulp.

Conifers produce simple pollen cones and more complex ovule-bearing cones (**Figure 30.7**). The pollen cones of conifers

(a) *Ginkgo biloba* **tree**

(b) *Ginkgo biloba* **leaf**

(c) *Ginkgo biloba* **seed**

Figure 30.5 *Ginkgo biloba.* **(a)** A *Ginkgo biloba* tree; **(b)** fan-shaped leaves with forked veins; and **(c)** seeds having foul-smelling, fleshy seed coats.

(a) Pine (*Pinus ponderosa*)   (b) Dawn redwood (*Metasequoia glyptostroboides*)

Figure 30.6   Representative conifers.

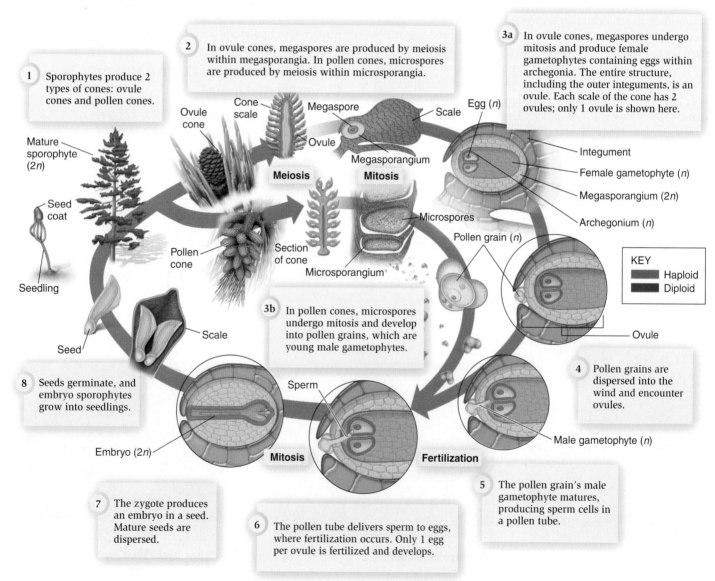

Figure 30.7   The life cycle of the genus *Pinus*.

(a) Pine seed

(b) Yew seeds

(c) Juniper cones with seeds

Figure 30.8 **Conifer seeds.** (a) Winged, wind-dispersed seed of the genus *Pinus*. (b) Fleshy-coated, bird-dispersed seeds of yew (*Taxus baccata*). (c) Fleshy cones of juniper (*Juniperus scopularum*) contain one or more seeds and are dispersed by birds. Juniper seeds are used in the production of gin.

bear many leaflike structures, each bearing a microsporangium in which meiosis occurs and pollen grains develop. The ovule cones, also called seed cones, are composed of many short branch systems that bear ovules. Ovules contain female gametophytes within which eggs develop.

When conifer pollen is mature, it is released into the wind, which transports pollen to ovules. When released from pollen tubes, sperm fuse with eggs, generating zygotes that grow into the embryos within seeds. Altogether, it takes nearly 2 years for pine (the genus *Pinus*) to complete the processes of male and female gamete development, fertilization, and seed development (Figure 30.7). The seeds of pine and some other conifers

develop wings that aid in wind dispersal (**Figure 30.8a**). Other conifers, such as yew and juniper, produce seeds or cones with bright-colored, fleshy coatings that are attractive to birds, which help to disperse the seeds (**Figure 30.8b,c**).

Conifer wood contains many water transport cells known as tracheids that are adapted for efficient conduction even in dry conditions. Like the tracheids of other vascular plants, those of conifers are devoid of cytoplasm and occur in long columns that function like plumbing pipelines (**Figure 30.9a**). Tracheid side and end walls possess many thin-walled, circular **pits** through which water moves both vertically and laterally from one tracheid to another. Conifer pits are unusual in having a porous

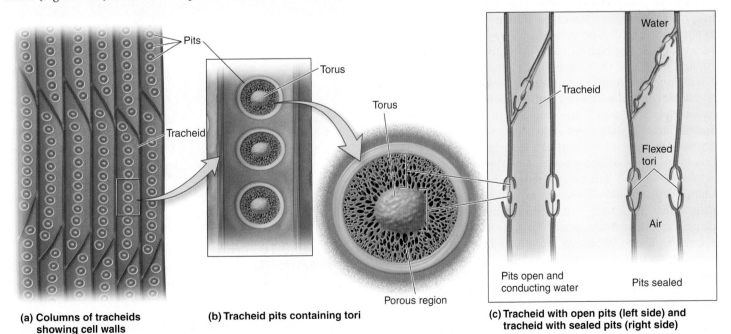

(a) Columns of tracheids showing cell walls

(b) Tracheid pits containing tori

(c) Tracheid with open pits (left side) and tracheid with sealed pits (right side)

Figure 30.9 **Tracheids in conifer wood.** (a) The lignin-rich cell walls of the water-conducting cells called tracheids. (b) Detailed view of a portion of a tracheid that shows the thin-walled areas known as pits, each with a torus. (c) A water-filled tracheid with open pits and an air-filled tracheid with pits sealed by the flexed tori.

Concept check: How does the torus foster conifer survival in arid or cold habitats?

outer region that lets water flow through and a nonporous, flexible central region called the **torus** (plural, tori) that functions like a valve (Figure 30.9b). If conifer tracheids become dry and fill with air, they are no longer able to conduct water. In this case, the torus presses against the pit opening, sealing it (Figure 30.9c). The torus valve thereby prevents air bubbles from spreading to the next tracheid. This conifer adaptation localizes air bubbles, preventing them from stopping water conduction in other tracheids. The presence of tori in their tracheids helps to explain why conifers have been so successful for hundreds of millions of years. Conifer wood (and leaves) may also display conspicuous resin ducts, passageways for the flow of syrup-like resin that helps to prevent attack by pathogens and herbivores. Resin that exudes from tree surfaces may trap insects and other organisms, then harden in the air and fossilize, preserving the inclusions in amber.

Many conifers occur in cold climates and thus display numerous adaptations to such environments. Their conical shapes and flexible branches help conifer trees shed snow, preventing heavy snow accumulations from breaking branches. Conifer leaf shape and structure are adapted to resist damage from drought that occurs in both summer and winter, when liquid water is scarce. Conifer leaves are often scalelike (Figure 30.10a) or needle-shaped (Figure 30.10b); these shapes reduce the area of leaf surface from which water can evaporate. In addition, a thick, waxy cuticle coats conifer leaf surfaces (Figure 30.10c), retarding water loss and attack by disease organisms.

Many conifers are evergreen; that is, their leaves live for more than 1 year before being shed and are not all shed during the same season. Retaining leaves through winter helps conifers start up photosynthesis earlier than deciduous trees, which in spring must replace leaves lost during the previous autumn. Evergreen leaves thus provide an advantage in the short growth season of alpine or high-latitude environments. However, some conifers do lose all their leaves in the autumn. The bald cypress (*Taxodium distichum*) of southern U.S. floodplains, tamarack (*Larix laricina*) of northern bogs, and dawn redwood (*Metasequoia glyptostroboides*) are examples of deciduous conifers. Fossils indicate that dawn redwoods once grew abundantly across wide areas of the Northern Hemisphere until a few million years ago and then disappeared. However, in the 1940s, a forester found a living dawn redwood growing in a remote Chinese village, and subsequent expeditions located forests of the conifers. Since then, dawn redwood trees have been widely planted as ornamentals, prized for their attractive foliage and cones (see Figure 30.6b). As recently as 1994, botanists found a previously unknown conifer species, *Wollemia nobilis*, in an Australian national park. Like the dawn redwood, *Wollemia* is an attractive tree that is likely to become more widely distributed as the result of human cultivation.

Gnetophytes Display Unusual Adaptations

Related to conifers, the modern gnetophytes consist of three genera, *Gnetum*, *Ephedra*, and *Welwitschia*, that feature distinctive adaptations. *Gnetum* is unusual among modern gymnosperms in

(a) Scale-shaped leaves of Eastern red cedar

(b) Needle-shaped leaves of pine

Thick cuticle

Photosynthetic cells

Tracheids

Stomata

(c) Stained cross section of pine needle, showing the thick cuticle

Figure 30.10 Conifer leaves.

Concept check: *In what ways are conifer leaves adapted to resist water loss from their surfaces?*

having broad leaves similar to those of many tropical plants (Figure 30.11a). Such leaves foster light capture in the dim forest habitat. More than 30 species of the genus *Gnetum* occur as vines, shrubs, or trees in tropical Africa or Asia. *Ephedra*, a gnetophyte genus native to arid regions of the southwestern U.S., has tiny brown scalelike leaves and green, photosynthetic stems

(a) Genus *Gnetum*

Reproductive structures

Broad leaf

(b) *Ephedra californica*

Tiny scale-like leaves

Photosynthetic stem

Reproductive structures

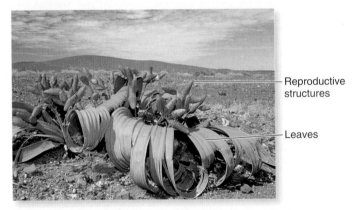

(c) *Welwitschia mirabilis*

Reproductive structures

Leaves

Figure 30.11 **Gnetophytes.** (a) A tropical plant of the genus *Gnetum*, displaying broad leaves and reproductive structures. (b) *Ephedra californica* growing in deserts of North America, showing miniscule brown leaves on green, photosynthetic stems and reproductive structures. (c) *Welwitschia mirabilis* growing in the Namib Desert of southwestern Africa, showing long, wind-shredded leaves and reproductive structures.

(Figure 30.11b). These adaptations help to conserve water by preventing water loss that would otherwise occur from the surfaces of larger leaves. *Ephedra* produces secondary metabolites that aid in plant protection but also affect human physiology. Early settlers of the western U.S. used *Ephedra* to treat colds and other medical conditions. In fact, the modern decongestant drug

pseudoephedrine is based on the chemical structure of ephedrine, which was named for and originally obtained from *Ephedra*. Pseudoephedrine sales are now restricted in many places because this compound can be used as a starting point for the synthesis of illegal drugs. Ephedrine has also been used to enhance sports performance, a practice that has elicited medical concern.

Welwitschia, the third gnetophyte genus, has only one living representative species. *Welwitschia mirabilis* is a strange-looking plant that grows in the coastal Namib Desert of southwestern Africa, one of the driest places on Earth (**Figure 30.11c**). A long taproot anchors a stubby stem that barely emerges from the ground. Two very long leaves grow from the stem but rapidly become wind-shredded into many strips. The plant is thought to obtain most of its water from coastal fog, explaining how *W. mirabilis* can grow and reproduce in such a dry place.

30.3 The Evolution and Diversity of Modern Angiosperms

One ancient gymnosperm group, although it's unclear which one, gave rise to the angiosperms. Angiosperms, the flowering plants, retained many structural and reproductive features from ancestral seed plants. In addition, flowering plants evolved several traits not found or seldom found among other land plants. For example, flowers and fruits are two of the defining features of angiosperms (**Figure 30.12**), because these features do not occur in other modern plants. The term **angiosperm** is from Greek words, meaning enclosed seed, which reflects the presence of seeds within fruits. The seed nutritive material known as endosperm is another defining feature of the flowering plants (see Chapters 29 and 39). Flowers, fruits, and seed endosperm are critical innovations that foster the production and dispersal of seeds and enclosed embryos. In addition, most angiosperms possess distinctive water-conducting cells, known as **vessels**, which are wider than tracheids and therefore increase the efficiency of water flow through plants. Though similar conducting cells occur in gnetophytes and some seedless plants, such vessels are examples of parallel evolution.

Although humans obtain wood, medicines, and other valuable products from gymnosperms, we depend even more on the angiosperms. Our food, beverages, and spices—flavored by an amazing variety of secondary metabolites—primarily come from flowering plants. People surround themselves with ornamental flowering plants and decorative items displaying flowers or fruit. We also commonly use flowers and fruit in ceremonies. In this section, we focus on how flowers, fruits, and secondary metabolites played key roles in angiosperm diversification. We will also learn that features of flowers, fruits, and secondary metabolites are used to classify and identify angiosperm species.

Flower Organs Evolved from Leaflike Structures

Flowers are complex reproductive structures that are specialized for the efficient production of pollen and seeds. The sexual

Figure 30.12 Angiosperm flowers and fruits. Citrus plants display the critical innovations of flowering plants: the flowers and fruits shown here and seed endosperm (not shown).

Concept check: *What other trait occurs widely among angiosperms but rarely among other plants?*

reproduction process of angiosperms depends on flowers. As the flowering plants diversified, flowers of varied types evolved as reproductive adaptations to differing environmental conditions. To understand this process, we can start by considering the basic flower parts and their roles in reproduction.

Flower Parts and Their Reproductive Roles Flowers are produced at stem tips and may contain four types of organs: sepals, petals, pollen-producing stamens, and ovule-producing carpels (**Figure 30.13**). These flower organs are supported by tissue known as a **receptacle**, located at the tip of a flower stalk—a **pedicel**. The functioning of several genes that control flower organ development explains why carpels are the most central flower organs, why stamens surround carpels, and why petals and sepals are the outermost flower organs (refer back to Figure 19.24).

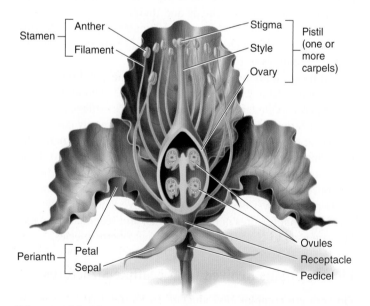

Figure 30.13 Generalized flower structure.

Many flowers produce attractive **petals** that play a role in **pollination**, the transfer of pollen among flowers. **Sepals** of many flowers are green and form the outer layer of flower buds. By contrast, the sepals of other flowers look similar to petals, in which case both sepals and petals are known as **tepals**. All of a flower's sepals and petals (or tepals) are collectively known as the **perianth**. Most flowers produce one or more **stamens**, the structures that produce and disperse pollen. Most flowers also contain **carpels**, structures that produce ovules.

Some flowers lack perianths, stamens, or carpels. Flowers that possess all four types of flower organs are known as **complete flowers**, and flowers lacking one or more organ types are known as **incomplete flowers**. Flowers that contain both stamens and carpels are described as **perfect flowers**, and flowers lacking either stamens or carpels are **imperfect flowers**.

Flowers also differ in the numbers of organs they produce. Some flowers produce only a single carpel, others display several separate carpels, and many possess several carpels that are fused together into a compound structure. Both a single carpel and compound carpels are referred to as a **pistil** (from the Latin *pistillum*, meaning pestle) because it resembles the device people use to grind materials to powder in a mortar (Figure 30.13). Only one pistil is present in flowers that have only one carpel and in flowers with fused carpels. By contrast, flowers possessing several separate carpels display multiple pistils.

Pistil structure can be divided into three regions having distinct functions. A topmost portion of the pistil, known as the **stigma**, receives and recognizes pollen of the appropriate species or genotype. The elongate middle portion of the pistil is called the **style**. The lowermost portion of the pistil is the **ovary**, which encloses and protects ovules.

During the flowering plant life cycle (**Figure 30.14**), the stigma allows pollen of appropriate genetic type to germinate, producing a long pollen tube that grows through the style. The pollen tube thereby delivers sperm cells to ovules and the eggs inside, allowing fertilization. If fertilization occurs, the zygote develops into an embryo, and the ovule develops into a seed. Ovaries (and sometimes additional flower parts) develop into fruits.

Early Flowers Distinctive fossil pollen grains that were deposited about 140 mya are the earliest widely accepted evidence for flowering plants. Fossils of whole plants with flowers and fruits are known from deposits that are about 120 million years old. Flowers were a critical innovation that led to extensive angiosperm diversification. Comparative studies of the structures of modern and fossil flowers suggest how modern stamens and carpels might have arisen. Early fossil flowers and some modern flowers have broad stamens that are leaf-shaped, with elongated, pollen-producing microsporangia on the stamen surface (**Figure 30.15a**). In contrast, the stamens of most modern plants have become narrowed to form **filaments**, or stalks, that elevate **anthers**, clusters of microsporangia that produce pollen and then open to release it (see Figure 30.13). Filaments and anthers are adaptations that foster pollen dispersal.

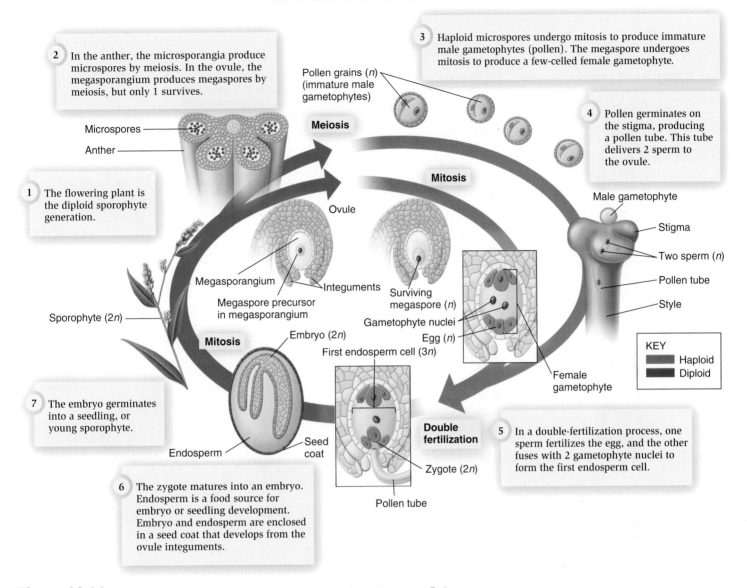

Figure 30.14   The life cycle of a flowering plant, illustrated by the genus *Polygonum*.

Plant biologists hypothesize that carpels evolved from leaf-like structures bearing ovules on their surfaces and that such leaves folded over ovules, protecting them. In support of this hypothesis is the observation that the carpels of some early-diverging modern plants are leaflike structures that fold over ovules, with the carpel edges stuck together by secretions (**Figure 30.15b**). In contrast, most modern flowers produce carpels whose edges have fused together into a tube whose lower portion (ovary) encloses ovules. Plant biologists hypothesize that such evolutionary change increased ovule protection, which would improve plant fitness.

Flowering Plants Diversified into Several Lineages, Including Monocots and Eudicots

Figure 30.16 presents our current understanding of the relationships among angiosperm groups. According to gene-sequencing studies, the earliest-diverging angiosperms are represented by a single modern species called *Amborella trichopoda*, a shrub that lives in cloud forests on the South Pacific island of New Caledonia. The flowers of *A. trichopoda* display hypothesized ancient features. For example, the fairly small flowers have stamens with broad filaments and several separate carpels (**Figure 30.17**). *A. trichopoda* also lacks vessels in the water-conducting tissues. In contrast, typical angiosperm vessels are present in later-diverging groups of angiosperms, including water lilies, the star anise plant, and other close relatives (see Figure 30.16). Magnoliids, represented by the genus *Magnolia*, are the next-diverging group. Magnoliids are closely related to two very large and diverse angiosperm lineages, the **monocots** and the **eudicots**.

Monocots and eudicots are named for differences in the number of embryonic leaves called cotyledons. Monocot embryos possess one cotyledon, whereas eudicots possess two cotyledons. Monocots differ from eudicots in several additional ways (see Chapter 35). For example, monocots typically have flowers with parts numbering three or some multiple of three (**Figure 30.18a**). In contrast, eudicot flower parts often occur in fours, fives, or a multiple of four or five (**Figure 30.18b**).

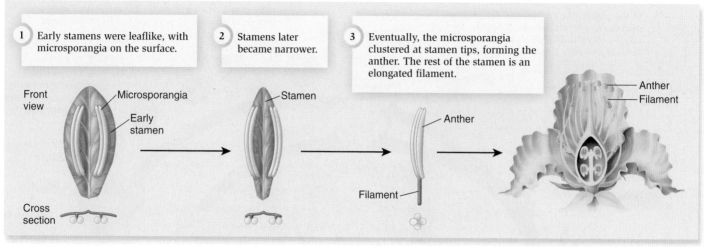

1 Early stamens were leaflike, with microsporangia on the surface.

2 Stamens later became narrower.

3 Eventually, the microsporangia clustered at stamen tips, forming the anther. The rest of the stamen is an elongated filament.

Front view

Microsporangia

Early stamen

Stamen

Anther

Filament

Anther

Filament

Cross section

(a) Stamen evolution

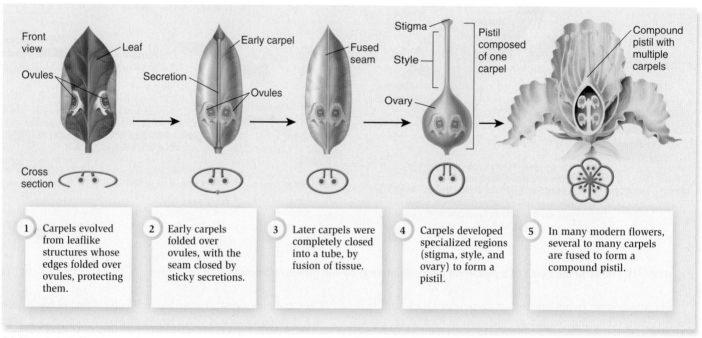

Front view

Leaf

Early carpel

Fused seam

Stigma

Style

Pistil composed of one carpel

Compound pistil with multiple carpels

Ovules

Secretion

Ovules

Ovary

Cross section

1 Carpels evolved from leaflike structures whose edges folded over ovules, protecting them.

2 Early carpels folded over ovules, with the seam closed by sticky secretions.

3 Later carpels were completely closed into a tube, by fusion of tissue.

4 Carpels developed specialized regions (stigma, style, and ovary) to form a pistil.

5 In many modern flowers, several to many carpels are fused to form a compound pistil.

(b) Carpel evolution

Figure 30.15 Hypothetical evolution of stamens, carpels, and pistils.

Genomes & Proteomes Connection

Whole Genome Duplications Influenced Flowering Plant Diversification

Genome doubling, also known as polyploidy, occurs in a wide variety of eukaryotes (see Figure 15.20), and has happened frequently during the evolutionary history of plants. An estimated 40–70% of all plants are polyploid, meaning that the entire diploid genome has been duplicated at least once. An analysis of the grape (*Vitis vinfera*) genome indicates that different patterns of whole genome duplication occurred in monocots and eudicots after their divergence. Whole genome duplication has the potential to affect species' evolutionary pathways because it offers the opportunity for many genes to diverge, forming gene families. For example, Amelie Veron and colleagues showed how whole genome duplications increased the diversity of an important class of plant transcription factors, the MADS-Box proteins, in seed plants.

Two major types of polyploidy occur in plants. Autopolyploidy occurs when homologous chromosome pairs do not separate during meiosis, a process known as nondisjunction. As a result, plants produce diploid spores, gametophytes, and gametes, when these life stages would otherwise be haploid. The mating of two diploid gametes will produce a tetraploid plant,

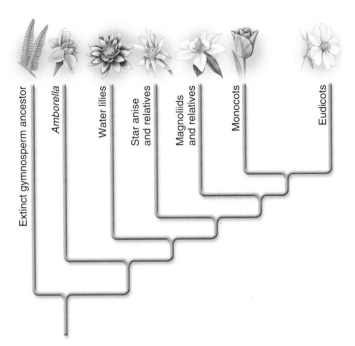

Figure 30.16   A phylogeny showing the major modern angiosperm lineages.

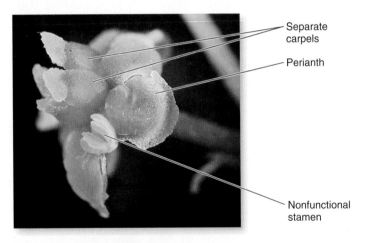

Figure 30.17   *Amborella trichopoda* flower, similar to a hypothesized early flower.   This small flower is only about 3–4 mm in diameter. It displays several central, greenish carpels; nonfunctional stamens; and a pink perianth. This plant species also produces flowers that lack carpels but have many functional stamens.

one having four sets of chromosomes. As the result of a large study of the occurrence of autoploidy in plants, plant evolutionary biologists Douglas and Pamela Soltis and their colleagues concluded that autopolyploidy is an important speciation mechanism in plants. These experts point out that because some autopolyploids have diverged sufficiently from the parental diploid species, they should be given different species names, and that failure to name such polyploids causes natural biodiversity to be underestimated.

(a) A monocot with six tepals

(b) A eudicot with five petals

Figure 30.18   One characteristic difference between monocots and eudicots: flower part number.   (a) Flowers and buds of lily (genus *Lilium*), displaying six tepals. (b) A flower and buds of apple (genus *Malus*), showing five flower petals. Sepals are visible around the pink buds.

A second major type of whole genome duplication in plants involves an initial hybridization between two species having different chromosome numbers. Such a hybrid would not be able to produce viable spores because chromosomes could not pair properly at meiosis. However, if such a hybrid plant then undergoes a whole genome duplication process, producing an allopolyploid, homologous chromosomes are available for pairing during meiosis, and viable spores, gametophytes, and gametes can form (see Chapter 15). Many common crops, including wheat and cotton, are allopolyploids, as are wild plants such as the desert sunflowers *Helianthus anomalus*, *H. deserticola*, and *H. paradoxus*. These sunflower hybrids are better adapted to survive drought conditions than are their parent species, *H. annuis* and *H. petiolaris*.

In addition to allopolyploidy, plants are known to obtain mitochondrial genes from other plant species by means of horizontal gene transfer. Genes have moved from one angiosperm species to another and between angiosperms and nonflowering plants, probably by mitochondrial fusion. For example, of the 31 genes present in the mitochondrial genome of the early-diverging flowering plant *Amborella trichopoda*, at least

20 were obtained from other angiosperms or mosses that grow on *A. trichopoda*'s surfaces. Because plant nuclear genomes commonly take up genes from organelles in the same cell, such foreign mitochondrial genes could end up in the nucleus. Hybridization and mitochondrial transfer are mechanisms by which plant lineages undergo reticulate (network-like) as well as treelike evolution.

Flower Diversification Fosters Efficient Seed Production

During the diversification of flowering plants, flower evolution has involved several types of changes that foster the transfer of pollen from one plant to another. Fusion of flower organs, clustering of flowers into groups, and reducing the perianth are some examples of such changes.

Many flowers have fused petals that form floral tubes. Such tubes tend to accumulate sugar-rich nectar that provides a reward for **pollinators**, animals that transfer pollen among plants. The diameters of floral tubes vary among flowers and are evolutionarily tuned to the feeding structures of diverse animals, which range from the narrow tongues of butterflies to the wider bills of nectar-feeding birds (**Figure 30.19**). Nectar-feeding bats stick their heads into even larger tubular flowers to lap up nectar with their tongues. Orchids provide another example of ways in which flower parts have become fused; stamens and carpels are fused together into a single reproductive column that is surrounded by attractive tepals (**Figure 30.20a**). This arrangement of flower organs fosters orchid pollination by particular insects and is a distinctive feature of the orchid family.

Many plants produce flowers in clusters known as **inflorescences**, which occur in several types. The sunflower family features a type of inflorescence in which many small flowers are clustered into a head (**Figure 30.20b**). The flowers at the center of a sunflower head function in reproduction and lack showy petals, while flowers at the rim have showy petals that attract pollinators. Heads allow pollinators to transfer pollen among a large number of flowers at the same time. The grass family features flowers having few or no perianths, explaining why grass flowers are not showy (**Figure 30.20c**). This adaptation fosters pollination by wind, because petals would only get in the way of such pollen transfer.

Diverse Types of Fruits Function in Seed Dispersal

Fruits are structures that develop from ovary walls in diverse ways that aid the dispersal of enclosed seeds. Seed dispersal helps to prevent seedlings from competing with their larger parents for scarce resources such as water and light. Dispersal of seeds also allows plants to colonize new habitats. Diverse fruit types illustrate the many ways in which plants have become adapted for effective seed dispersal. Like flower types, fruit types are useful in classifying and identifying angiosperms.

Many mature angiosperm fruits, such as cherries, grapes, and lemons, are attractively colored, soft, juicy, and tasty (**Figure 30.21a–c**). Such fruits are adapted to attract animals that consume the fruits, digest the outer portion as food, and eliminate the seeds, thereby dispersing them. Hard seed coats prevent such seeds from being destroyed by the animal's digestive system. Strawberries are aggregate fruits, many fruits that all develop from a single flower having multiple pistils

(a) Zinnia flower and butterfly   **(b) Hibiscus flower and hummingbird**   **(c) Saguaro cactus flower and bat**

Figure 30.19 **Floral tubes and coevolved pollinators.** (a) This zinnia is composed of an outer rim of showy flowers and a central disc of narrow tubular flowers that produce nectar. Butterflies, but not other pollinators, are able to reach the nectar by means of narrow tongues. (b) The hibiscus flower produces nectar in a floral tube whose diameter corresponds to the dimensions of a hummingbird bill. (c) The saguaro cactus (*Carnegiea gigantea*) flower forms a floral tube that is wide enough for nectar-feeding bats to get their heads inside. The cactus flower has been drawn here as if it were transparent, to illustrate bat pollination.

(a) An orchid flower with fused pistil and stamens

(b) A sunflower plant showing inflorescence

(c) Grass flowers lacking showy perianth

Figure 30.20 Evolutionary changes in flower structure.
(a) An orchid of the genus *Cattleya* has fused stamens and pistil, and six tepals, one of which is specialized to form a lower lip.
(b) An inflorescence (head) of sunflower (genus *Helianthus*). This inflorescence includes a rim of flowers with conspicuous petals that attract pollinators and an inner disc of flowers that lack attractive perianths. (c) Grass flowers of the genus *Triticum* lack a showy perianth.

Concept check: What advantage does the nonshowy perianth of grass flowers provide?

Figure 30.21 Representative fruit types. (a–c) The cherry drupe, grape berry, and lemon hesperidium are fleshy fruits adapted to attract animals that consume the fruits and excrete the seeds. (d) Strawberry is an aggregate fruit, consisting of many tiny, single-seeded fruits produced by a single flower. The fruits are embedded in the surface of a fleshy receptacle that is adapted to attract animal seed-dispersal agents. (e) Pineapple is a large multiple fruit formed by the aggregation of smaller fruits, each produced by one of the flowers in an inflorescence. (f) Peas produce legumes, fruits that open on two sides to release seeds. (g) Coconut fruits possess a fibrous husk that aids dispersal in water. (h) Maple trees produce dry fruits with wings adapted for wind dispersal.

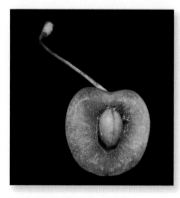

(a) A fleshy fruit (cherry)

(b) A fleshy berry fruit (grape)

(c) A fleshy fruit (lemon)

(d) An aggregate fruit (strawberry)

(e) A multiple fruit (pineapple)

(f) Legumes with dry pods (peas)

(g) Fruit with husk (coconut)

(h) A dry, winged fruit (maple)

(**Figure 30.21d**). The ovaries of these pistils develop into tiny, single-seeded yellow fruits on a strawberry surface; the fleshy, red, sweet portion of a strawberry develops from a flower receptacle. Aggregate fruits allow a single animal consumer, such as a bird, to disperse many seeds at the same time.

Pineapples (**Figure 30.21e**) are juicy multiple fruits that develop when many ovaries of an inflorescence fuse together. Such multiple fruits are larger and attract relatively large animals that have the ability to disperse seeds for long distances. Mulberries and figs are additional examples of multiple fruits that provide similar benefits.

The plant family informally known as **legumes** is named for its distinctive fruits, dry pods that open down both sides when seeds are mature, thereby releasing them (**Figure 30.21f**). Nuts and grains are additional examples of dry fruits. **Grains** are the characteristic single-seeded fruits of cereal grasses such as rice, corn (maize), barley, and wheat. Coconut fruits are adapted for dispersal in ocean currents (**Figure 30.21g**). Maple trees produce dry and thus lightweight fruits having wings, features that foster effective wind dispersal (**Figure 30.21h**). Other plants produce dry fruits with surface burrs that attach to animal fur. These are just a few examples of the diverse mechanisms that flowering plants use to disperse their seeds.

Angiosperms Produce Diverse Secondary Metabolites That Play Important Roles in Structure, Reproduction, and Protection

Secondary metabolism involves the synthesis of molecules that are not essential for cell structure and growth (see Chapter 7). These molecules, called **secondary metabolites**, are produced by various prokaryotes, protists, fungi, some animals, and plants, but they are most diverse in the angiosperms. About 100,000 different types of secondary metabolites are known, most of these produced by flowering plants. Because secondary metabolites play essential roles in plant structure, reproduction, and protection, diversification of these compounds has influenced flowering plant evolution. Three major classes of plant secondary metabolites occur: (1) terpenes and terpenoids; (2) phenolics, which include flavonoids and related compounds; and (3) alkaloids (**Figure 30.22**).

About 25,000 types of plant terpenes and terpenoids are derived from units of the hydrocarbon gas isoprene. Taxol, previously mentioned for its use in the treatment of cancer, is a terpene, as are citronella and a variety of other compounds that repel insects. Rubber, turpentine, rosin, and amber are complex terpenoids that likewise serve important roles in plant biology as well as having useful human applications.

Phenolic compounds are responsible for some flower and fruit colors as well as the distinctive flavors of cinnamon, nutmeg, ginger, cloves, chilies, and vanilla. Phenolics absorb ultraviolet radiation, thereby preventing damage to cellular DNA. They also help to defend plants against insects and disease microbes. Some phenolic compounds found in tea, red wine, grape juice, and blueberries are antioxidants that detoxify free radicals, thereby preventing cellular damage.

(a) Natural rubber produced by *Hevea brasiliensis* **is an example of a complex terpene.**

(b) Capsaicin extracted from capsicum pepper is an example of a phenolic compound.

(c) Caffeine produced by *Coffea arabica* **is an example of an alkaloid.**

Figure 30.22 Major types of plant secondary metabolites.

Alkaloids are nitrogen-containing secondary metabolites that often have potent effects on the animal nervous system. Plants produce at least 12,000 types of alkaloids, and certain species produce many alkaloids. Caffeine, nicotine, morphine, ephedrine, cocaine, and codeine are examples of alkaloids that influence the physiology and behavior of humans and are thus of societal concern. Like flower and fruit structure, secondary metabolites are useful in distinguishing among Earth's hundreds of thousands of flowering plant species.

FEATURE INVESTIGATION

Hillig and Mahlberg Analyzed Secondary Metabolites to Explore Species Diversification in the Genus *Cannabis*

The genus *Cannabis* has long been a source of hemp fiber used for ropes and fabric. People have also used *Cannabis* in traditional medicine and as a hallucinogenic drug. *Cannabis* produces THC (tetrahydrocannabinol), a type of alkaloid called a cannabinoid. THC and other cannabinoids are produced in glandular hairs that cover most of the *Cannabis* plant's surface but are particularly rich in leaves located near the flowers. THC mimics compounds known as endocannabinoids, which are naturally produced and act in the brain and elsewhere in the body. THC affects humans by binding to receptor proteins in plasma membranes in the same way as natural endocannabinoids. Cancer patients sometimes choose to use cannabis to stimulate their appetite, which can decline as a side effect of cancer treatment.

Because humans have subjected cultivated *Cannabis* plants to artificial selection for so long, plant biologists have been uncertain how cultivated *Cannabis* species are related to those in the wild. In the past, plants cultivated for drug production were often identified as *Cannabis indica*, whereas those grown for hemp were typically known as *Cannabis sativa*. However, these species are difficult to distinguish on the basis of structural features, and the relevance of these names to wild cannabis was unknown. At the same time, species identification has become important for biodiversity studies, agriculture, and law enforcement. For these reasons, plant biologists Karl Hillig and Paul Mahlberg investigated the possibility that ratios of THC to another cannabinoid known as CBD (cannabidiol) might aid in defining *Cannabis* species and identifying plant samples at the species level, as shown in **Figure 30.23**.

They began by collecting *Cannabis* fruits from nearly a hundred diverse locations around the world and then growing these plants from seed under uniform conditions in a greenhouse.

Figure 30.23   Hillig and Mahlberg's analysis of secondary metabolites in the genus *Cannabis*.

GOAL   To determine if cannabinoids aid in distinguishing *Cannabis* species.

KEY MATERIALS   *Cannabis* fruits obtained from nearly 100 different worldwide sources.

Experimental level	Conceptual level
1 Grow multiple *Cannabis* plants from seeds under standard conditions in a greenhouse.	Eliminates differential environmental effects on cannabinoid content.
2 Extract cannabinoids from leaves surrounding flowers.	Extracts were made from tissues richest in cannabinoids; this reduces the chance that cannabinoids present in lower levels would be missed.
3 Analyze cannabinoids by gas-liquid chromatography. Determine ratios of THC (tetrahydrocannabinol) to CBD (cannabidiol) in about 200 *Cannabis* plants.	Previous data suggested that ratios of THC to CBD might be different in separate species.

4   THE DATA

Cannabis plants isolated from diverse sources worldwide formed 2 groups—those having relatively high THC to CBD ratios and those having lower THC to CBD ratios.

Plants having low THC to CBD ratios, often used as hemp fiber sources, corresponded to the species *C. sativa*.

Plants having high THC to CBD ratios, often used as drug sources, corresponded to the species *C. indica*.

5   CONCLUSION Differing cannabinoid ratios support a concept of 2 *Cannabis* species.

6   SOURCE Hillig, K.W., and Mahlberg, P.G. 2004. A chemosystematic analysis of cannabinoid variation in *Cannabis* (Cannabaceae). *American Journal of Botany* 91:966–975.

The investigators next extracted cannabinoids, analyzed them by means of gas-liquid chromatography, and determined the ratios of THC to CBD. The results, published in 2004, suggested that the wild and cultivated *Cannabis* samples evaluated in this study could be classified into two species: *C. sativa*, displaying relatively low THC levels, and *C. indica*, having relatively high THC levels. As a result of this work, ecologists, agricultural scientists, and forensic scientists can reliably use ratios of THC to CBD to classify samples. Similar studies of plant secondary metabolites offer the benefit of uncovering potential new medicinal compounds or other applications of significance to humans.

Experimental Questions

1. Why did the investigators in Figure 30.23 obtain nearly a hundred *Cannabis* fruit samples from around the world?

2. Why did Hillig and Mahlberg grow plants in a greenhouse before conducting the cannabinoid analysis?

3. Why did Hillig and Mahlberg collect samples from the leaves growing nearest the flowers?

30.4   The Role of Coevolution in Angiosperm Diversification

In the previous section, we learned that flowering plants are commonly associated with animals in ways that strongly influence plant evolution. Likewise, plants have influenced animal evolution in a diversity-generating process known as **coevolution**, which is the process by which two or more species of organisms influence each other's evolutionary pathway. During the diversification of flowering plants, coevolution with animals has been a major evolutionary force. Coevolution explains the diverse forms of most flowers and many fruits and how plants accomplish effective pollen and seed dispersal. Human attraction to flowers and fruit also reflects coevolutionary associations. This is because human sensory systems are similar to those of various animals that have coevolved with angiosperms.

Pollination Coevolution Influences the Diversification of Flowers and Animals

Animal pollinators transfer pollen from the anthers of one flower to the stigmas of other flowers of the same species. Pollinators thereby foster genetic variability and plant potential for evolutionary change. Insects, birds, bats, and other pollinators learn the characteristics of particular flowers, visiting them preferentially. This animal behavior, known as constancy or fidelity, increases the odds that a flower stigma will receive pollen of the appropriate species. Animal pollinators offer precision of pollen transfer, which reduces the amount of pollen that plants must produce to achieve pollination. By contrast, wind-pollinated plants must produce much larger amounts of pollen because windblown pollen reaches appropriate flowers by chance.

Flowers attract the most appropriate pollinators by means of attractive colors, odors, shapes, and sizes. Secondary metabolites influence the colors and odors of many flowers. Flavonoids, for example, color many blue, purple, or pink flowers. More than 700 types of chemical compounds contribute to floral odors.

Most flowers reward pollinators with appropriate food: sugar-rich nectar, lipid- and protein-rich pollen, or both. In this way, flowering plants provide an important biological service, providing food for many types of pollinator animals. However, some flowers "trick" pollinators into visiting or trap pollinators temporarily, thereby achieving pollination without actually rewarding the pollinator. Examples include flowers that look and smell like dead meat, thereby attracting flies, which are fooled but accomplish pollination anyway.

Table 30.1	Pollination Syndromes
Animal features	**Coevolved flower features**
Bees	
Color vision includes UV, not red	Often blue, purple, yellow, white (not red) colors
Good sense of smell	Fragrant
Require nectar and pollen	Provide nectar and abundant pollen
Butterflies	
Good color vision	Blue, purple, deep pink, orange, red colors
Sense odors with feet	Light floral scent
Need landing place	Provide landing place
Feed with long, tubular tongue	Nectar in deep, narrow floral tubes
Moths	
Active at night	Open at night; white or bright colors
Good sense of smell	Heavy, musky odors
Feed with long, thin tongue	Nectar in deep, narrow floral tubes
Birds	
Color vision, includes red	Often colored red
Often require perch	Strong, damage-resistant structure
Poor sense of smell	No fragrance
Feed in daytime	Open in daytime
High nectar requirement	Copious nectar in floral tubes
Hover (hummingbirds)	Pendulous (dangling) flowers
Bats	
Color blind	Light, reflective colors
Good sense of smell	Strong odors
Active at night	Open at night
High food requirements	Copious nectar and pollen provided
Navigate by echolocation	Pendulous or borne on tree trunks

Although many flowers are pollinated by a variety of animals, others have flowers that have become specialized for particular pollinators, and vice versa. These specializations, which have resulted from coevolution, are known as **pollination syndromes** (**Table 30.1**). For example, odorless red flowers, such as those of hibiscus (see Figure 30.19b), are attractive to birds, which can see the color red but lack a sense of smell. By contrast, bees are not typically attracted to red flowers because bee vision does not extend to the red end of the visible light spectrum. Rather, bees are attracted to blue, purple, yellow, and white flowers having sweet odors. If you are allergic to bee stings or just want to reduce the possibility of being stung, do not dress in bee-attracting flower colors or wear fragrant perfumes when in locales frequented by bees.

Pollination syndromes are also of practical importance in agriculture and in conservation biology. Fruit growers often import colonies of bees to pollinate flowers of fruit crops and so increase crop yields. In recent years, widespread die-offs of bee colonies have become an environmental and agricultural concern. When bee pollinators are not available, growers cannot produce some fruit crops. Some plants have become so specialized to particular pollinators that if the pollinator

becomes extinct, the plant becomes endangered. An example is the Hawaiian cliff-dwelling *Brighamia insignis* (**Figure 30.24**), whose presumed moth pollinator has become extinct. Humans that hand-pollinate *B. insignis* are all that stand between this plant and extinction.

Seed-Dispersal Coevolution Influences the Evolution of Fruits and Particular Animals

As in the case of pollination, coevolution between plants and their animal seed-dispersal agents has influenced both plant fruit characteristics and those of seed-dispersing animals. In addition, flowering plant fruits provide food for animals, an important biological service. For example, many of the plants of temperate forests produce fruits that are attractive to resident birds. Such juicy, sweet fruits have small seeds that readily pass through bird guts. Many plants signal fruit ripeness by undergoing color changes from unripe green fruits to red, orange, yellow, blue, or black (**Figure 30.25**). Because birds have good color vision, they are able to detect the presence of ripe fruits and consume them before the fruits drop from plants and rot. Apples, strawberries, cherries, blueberries, and blackberries are examples of fruits whose seed dispersal adaptations have made them attractive food for humans as well. By contrast, the lipid-rich fruits of Virginia creeper (*Parthenocissus quinquefolia*) and some other autumn-fruiting plants energize migratory birds but are not tasty to humans. The Virginia creeper's leaves often turn fall colors earlier than surrounding plants, thereby signaling the availability of nutritious, ripe fruit to high-flying birds. Such lipid-rich fruits must be consumed promptly because they rot easily, in which case seed dispersal cannot occur.

Figure 30.24 *Brighamia insignis*, a plant endangered by the loss of its pollinator. The pollinator that coevolved with *B. insignis* has become extinct, with the result that the plant is unable to produce seed unless artificially pollinated by humans.

Concept check: What kind of animal likely pollinated *B. insignis*?

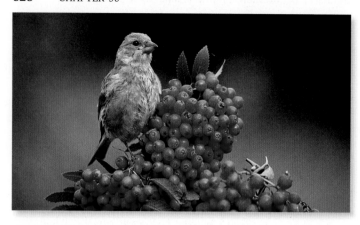

Figure 30.25 Fruits attractive to animal seed-dispersal agents.   Color and odor signals alert coevolved animal species that fruits are ripe, thus favoring the dispersal of mature seeds.

Table 30.2	Critical Innovations of Seed Plant Groups	
Plant group	**Innovation**	**Advantages**
All seed plants	Vascular cambium that makes wood and inner bark	Seed plants have the potential to grow tall and produce many branches and reproductive structures.
	Pollen, ovules, seeds	Pollen allows seed plants to disperse male gametophytes. Ovules provide protection and nutrition to female gametophytes and developing embryos. Seeds allow seed plants to reproduce in arid or shady habitats.
Conifers	Tracheid torus	Fosters water flow in arid or cold conditions
	Scales or needle-shaped leaves	Retard water loss from leaf surface
	Conical shape	Sheds snow, preventing damage
	Resin	Protects against pathogens and herbivores
Angiosperms	Flowers	Foster pollen dispersal, ovule protection, pollination, and seed production
	Fruits	Foster seed dispersal
	Endosperm	An efficient way to provide food to embryo of developing seed
	Vessels	Relatively wide diameter fosters water flow
	Many secondary compounds	Provide flower colors and fragrances and protect against herbivores

To help you remember all of what you've learned by reading this chapter so far, **Table 30.2** provides a summary of the critical innovations of seed plants. These critical innovations are the result of diversity principles at work. Now we turn to how humans have changed seed plants throughout our shared history.

30.5 Human Influences on Angiosperm Diversification

By means of the process known as **domestication**, which involves artificial selection for traits desirable to humans, ancient humans transformed wild plant species into new crop species. More recently, human populations have increased so much that larger areas of natural habitat are being transformed for human use. Deforestation, for example, often results from the conversion of forests to agricultural land. Such habitat destruction is a leading cause of the extinction of plant and other species. How have humans produced new crop species and influenced the loss of wild species?

Between 10,000 and 5,000 years ago, agriculture originated independently in at least 10 different locations around the world. One of the earliest domesticated crops was an African plant commonly known as the bottle gourd (*Lagenaria siceraria*). The bottle gourd was grown for use as containers, musical instruments, and floats for fishing. For planting crops, humans selected seeds from gourds that had thicker rinds because these resisted breakage better than wild gourds. These differences in rind thickness can be detected in fossil remains that are about 10,000 years old. The fossils indicate that bottle gourds were grown as a crop in Asia and from there were transported to the Americas by ancient human colonists.

Cultivated bread wheat (*Triticum aestivum*) was probably among the earliest food crops, having originated more than 8,000 years ago, in what is now southeastern Turkey and northern Syria. Bread wheat originated by a series of steps that included hybridization and whole genome duplication from wild ancestors (*Triticum boeoticum* and *Triticum dicoccoides*). Among the earliest changes that occurred during wheat domestication was the loss of **shattering**, the process by which ears of wild grain crops break apart and disperse their grains. A mutation probably caused the ears of some wheat plants to remain intact, a trait that is disadvantageous in nature but beneficial to humans. Nonshattering ears would have been easier for humans to harvest than normal ears. Early farmers probably selected seed stock from plants having nonshattering ears and other favorable traits such as larger grains. These ancient human selection processes, together with modern breeding efforts, explain why cultivated wheat differs from its wild relatives in shattering and other properties. The accumulation of these trait differences explains why cultivated and wild wheat plants are classified as different species.

About 9,000 years ago, people living in what is now Mexico domesticated a native grass known as teosinte (of the genus *Zea*), producing a new species, *Zea mays*, known as corn or maize. The evidence for this pivotal event includes ancient ears that were larger than wild ones and distinctive fossil pollen. Modern ears of corn are much larger than those of teosinte, corn grains are larger and softer, and modern corn ears do not shatter, as do those of ancestral teosinte (**Figure 30.26**). These and other trait changes reflect artificial selection accomplished by humans. An analysis of the corn genome, reported in 2005 by Stephen Wright, Brandon Gaut, and coworkers, suggests that human selection has influenced about 1,200 genes.

Immature ear of teosinte

Grain

Mature, shattered ear of teosinte

Nonshattering ear of *Z. mays*

Figure 30.26 **Ears and grains of modern corn and its ancestor, teosinte.** Domesticated corn ears are larger than those of the ancestral grass teosinte. In addition, corn fruits are softer and more edible than are grains of teosinte.

Concept check: *In what other way do corn ears differ from those of teosinte?*

Molecular analyses indicate that domesticated rice (*Oryza sativa*) originated from ancestral wild species of grasses (*Oryza nivara* and/or *Oryza rifipogon*). As in the cases of wheat and corn, domestication of rice involved loss of ear shattering. In 2006, Changbao Li, Ailing Zhou, and Tao Sang reported that a key amino acid substitution was primarily responsible for the loss of ear shattering in rice. Ancient humans might have unconsciously selected for this mutation while gathering rice from wild populations, because the mutants would not so easily have shed grains during the harvesting process. Eventually, the nonshattering mutant became a widely planted crop throughout Asia, and today it is the food staple for millions of people.

Although humans generated these and other new plant species, in modern times humans have caused the extinction of plants and other species as the result of habitat destruction. Protecting biodiversity will continue to challenge humans as populations and demands on the Earth's resources increase. Plant biologists are working to identify one or more molecular sequence tools for use in barcoding plants, a process that is also widely used to identify and catalog animals (see Chapter 33). The ability to barcode plants is important to those who monitor international trade in endangered plant species.

▌ Summary of Key Concepts

30.1 Overview of the Seed Plants

- The seed plants inherited features of ancestral seedless plants but display distinctive adaptations. The major phyla of seed plants mainly differ in reproductive features. (Figure 30.1)

30.2 The Evolution and Diversity of Modern Gymnosperms

- Gymnosperms are plants that produce exposed seeds rather than seeds enclosed in fruits. Many gymnosperms produce wood by means of a special tissue called vascular cambium.

- Several phyla of gymnosperms once existed but have become extinct and are known only from fossils. The diversity of modern gymnosperms includes four modern phyla: cycads, *Ginkgo biloba*, the conifers, and the gnetophytes. (Figure 30.2)

- Cycads primarily live in tropical and subtropical regions. Features of cycads include palmlike leaves, nonwoody stems, coralloid roots with cyanobacterial endosymbionts, toxins, and large conelike seed-producing structures. (Figures 30.3, 30.4)

- The tree *Ginkgo biloba* is the last surviving species of a phylum that was diverse during the Age of Dinosaurs. (Figure 30.5)

- Conifers have been widespread and diverse members of plant communities for the past 300 million years, and they are now important sources of wood and paper pulp to humans. Reproduction involves simple pollen cones and complex ovule-producing cones. Conifer wood contains water-conducting tracheids with thin-walled pits with tori that help prevent air bubbles from blocking water flow through the plant. Many conifers display additional adaptations that help them to survive in cold climates. (Figures 30.6, 30.7, 30.8, 30.9, 30.10)

- The gnetophytes display unusual body variations. (Figure 30.11)

30.3 The Evolution and Diversity of Modern Angiosperms

- Angiosperms inherited seeds and other features from ancestors but display distinctive features, such as flowers and fruits. Diversification of flower structure, fruit structure, and secondary metabolites has played important roles in angiosperm evolution. (Figure 30.12)

- Flowers foster seed production and are adapted in various ways that aid pollination in varying circumstances. The major flower organs are sepals, petals, or tepals, stamens, and carpels, but some flowers lack one or more of these organs. Pollination is the transfer of pollen from a stamen to a pistil. Pistils display regions of specialized function: The stigma is a receptive surface for pollen, pollen tubes grow through the style, and ovules develop within the ovary. If pollen tubes successfully deposit sperm near eggs in ovules and fertilization occurs, ovules develop into seeds, and ovaries develop into fruits. Stamens and carpels may have evolved from leaflike structures bearing sporangia. (Figures 30.13, 30.14, 30.15)

- The two largest lineages of flowering plants are the monocots and eudicots. (Figures 30.16, 30.17, 30.18)

- Whole genome duplications arising from autopolyploidy and allopolyploidy and horizontal gene transfer by mitochondrial fusion are mechanisms of reticulate evolution that occur widely in plants.

- Flower diversification involved evolutionary changes such as fusion of parts, changes in symmetry, loss of parts, and aggregation into inflorescences. (Figures 30.19, 30.20)

- Fruits are structures that enclose seeds and aid in their dispersal. Fruits occur in many types that foster seed dispersal in varying circumstances. (Figure 30.21)

- Angiosperms produce three main groups of secondary metabolites: (1) terpenes and terpenoids; (2) phenolics, flavonoids and related compounds; and (3) alkaloids. Secondary metabolites play essential roles in plant structure, reproduction, and defense. (Figure 30.22)

- Hillig and Mahlberg demonstrated the use of particular secondary metabolites in distinguishing species of the genus *Cannabis*. (Figure 30.23)

30.4 The Role of Coevolution in Angiosperm Diversification

- Coevolutionary interactions with animals that serve as pollen- and seed-dispersal agents played a powerful role in the diversification of both flowering plants and animals. (Table 30.1, Figures 30.24, 30.25)

- Human appreciation of flowers and fruits is based on sensory systems similar to those present in the animals with which angiosperms coevolved.

- Seed plants have evolved many critical innovations that have allowed them to be successful for many millions of years. (Table 30.2)

30.5 Human Influences on Angiosperm Diversification

- Humans have produced new crop species by domesticating wild plants. The process of domestication involved artificial selection for traits such as nonshattering ears of wheat, corn, and rice. (Figure 30.26)

- Large human populations are causing habitat destruction, leading to species extinction.

 # Assess and Discuss

Test Yourself

1. What feature must be present for a plant to produce wood?
   a. a type of conducting system in which vascular bundles occur in a ring around pith
   b. a eustele
   c. a vascular cambium
   d. all of the above
   e. none of the above

2. Which sequence of critical adaptations reflects the order of their appearance in time?
   a. embryos, vascular tissue, wood, seeds, flowers
   b. vascular tissue, embryos, wood, flowers, seeds
   c. vascular tissue, wood, seeds, embryos, flowers
   d. wood, seeds, embryos, flowers, vascular tissue
   e. seeds, vascular tissue, wood, embryos, flowers

3. How long have gymnosperms been important members of plant communities?
   a. 10,000 years, since the dawn of agriculture
   b. 100,000 years
   c. 300,000 years
   d. 65 million years, since the K/T event
   e. 300 million years, since the Coal Age

4. What similar features do gymnosperms and angiosperms possess that differ from other modern vascular plants?
   a. Gymnosperms and angiosperms both produce flagellate sperm.
   b. Gymnosperms and angiosperms both produce flowers.
   c. Gymnosperms and angiosperms both produce tracheids, but not vessels, in their vascular tissues.
   d. Gymnosperms and angiosperms both produce fruits.
   e. none of the above

5. Which part of a flower receives pollen from the wind or a pollinating animal?
   a. perianth   c. filament   e. ovary
   b. stigma   d. pedicel

6. The primary function of a fruit is to
   a. provide food for the developing seed.
   b. provide food for the developing seedling.
   c. foster pollen dispersal.
   d. foster seed dispersal.
   e. None of the above is the primary function.

7. What are some ways in which flowers have diversified?
   a. color
   b. number of flower parts
   c. fusion of organs
   d. aggregation into inflorescences
   e. all of the above

8. Flowers of the genus *Fuchsia* produce deep pink to red flowers that dangle from plants, produce nectar in floral tubes, and have no scent. Based on these features, which animal is most likely to be a coevolved pollinator?
   a. bee   c. hummingbird   e. moth
   b. bat   d. butterfly

9. Which type of plant secondary metabolite is best known for the antioxidant properties of human foods such as blueberries, tea, and grape juice?
   a. alkaloids   c. carotenoids   e. terpenoids
   b. cannabinoids   d. phenolics

10. What features of domesticated grain crops might differ from those of wild ancestors?
    a. the degree to which ears shatter, allowing for seed dispersal
    b. grain size
    c. number of grains per ear
    d. softness and edibility of grains
    e. all of the above

Conceptual Questions

1. Explain why humans should not consume food products made from cycads.

2. Explain why fruits such as apples, strawberries, and cherries are attractive and harmless foods for humans.

3. Is a sunflower really a flower?

Collaborative Questions

1. Where in the world would you have to travel to find wild plants representing all of the gymnosperm phyla, including the three types of gnetophytes?

2. How would you go about trying to solve what Darwin called an "abominable mystery," that is, the identity of the seed plant group that was ancestral to the flowering plants?

Online Resource

www.brookerbiology.com

Stay a step ahead in your studies with animations that bring concepts to life and practice tests to assess your understanding. Your instructor may also recommend the interactive ebook, individualized learning tools, and more.

Fungi

31

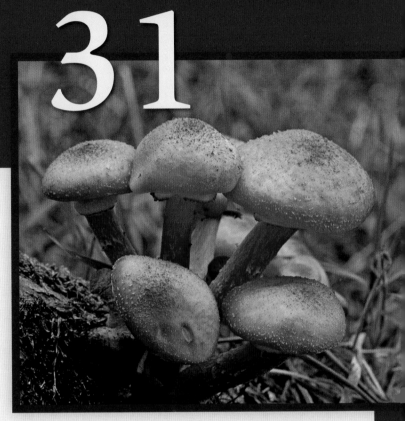

The aboveground reproductive parts of the fungus *Armillaria ostoyae*. Because of the large extent of its underground components, this fungus may be the largest organism in the world.

Y ou might think that the largest organism in the world is a whale or perhaps a giant redwood tree. Amazingly, giant fungi would also be good candidates. For example, an individual of the fungus *Armillaria ostoyae* weighs hundreds of tons, is more than 2,000 years old, and spreads over 2,200 acres of Oregon forest soil! Scientists discovered the extent of this enormous fungus when they found identical DNA sequences in samples taken over this wide area. Other examples of such huge fungi have been found, and mycologists—scientists who study fungi—suspect that they may be fairly common, underfoot yet largely unseen.

Regardless of their size, fungi typically occur within soil or other materials, becoming conspicuous only when reproductive portions such as mushrooms extend above the surface. Even though fungi can be inconspicuous, they play essential roles in the Earth's environment, are associated in diverse ways with other organisms, and have many applications in biotechnology. In this chapter, we will explore the distinctive features of fungal structure, growth, nutrition, reproduction, and diversity. In the process, you will learn how fungi are connected to forest growth, food production and food toxins, sick building syndrome, and other topics of great importance to humans.

31.1 Evolutionary Relationships and Distinctive Features of Fungi

The eukaryotes known as fungi are so distinct from other organisms that they are placed in their own kingdom, the kingdom Fungi (**Figure 31.1**). Together with certain closely related protists, the kingdom Fungi and the animal kingdom (Metazoa) form a eukaryotic supergroup known as Opisthokonta. Several types of slime molds and other fungus-like protists—though often studied with fungi—are classified with protists rather than true fungi (see Chapter 28). The fungi arose from protists closely related to the modern genus *Nuclearia*, an amoeba that feeds by engulfing cells of algae and bacteria. Early-diverging fungi include microsporidia that parasitize animal cells and several aquatic lineages of microscopic species collectively known as chytrids (formally, Chytridiomycota). Zygomycetes (Zygomycota), represented by black bread molds, are another polyphyletic cluster of relatively early-diverging fungi. The arbuscular mycorrhizal fungi (abbreviated AM fungi) are formally called the Glomeromycota. The ascomycetes (also informally known as the sac fungi and formally as Ascomycota) and the basidiomycetes (club fungi, or Basidiomycota) are later-diverging fungal phyla that display many adaptations to life on land. (Note: The suffixes *mycota* and *mycetes* derive from a Greek word meaning fungus.)

Because fungi are closely related to the animal kingdom, fungi and animals display some common features. For example, both are **heterotrophic**, meaning that they cannot produce their own food but must obtain it from the environment. Fungi use an amazing array of organic compounds as food, which is termed their **substrate**. The substrate could be the soil, a rotting log, a piece of bread, a living tissue, or a wide array of other materials. Fungi are also like animals in having **absorptive nutrition**. Both fungi and the cells of animal digestive systems secrete enzymes that break down complex

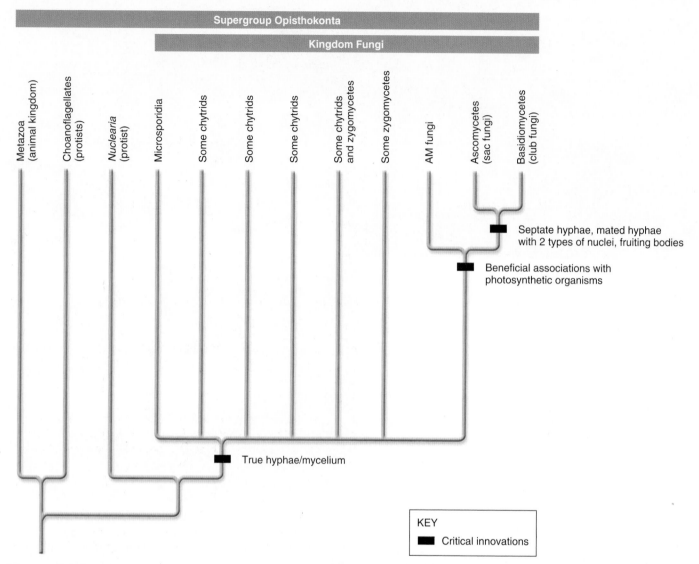

Figure 31.1 **Evolutionary relationships of the fungi.** The kingdom Fungi arose from a protist ancestor similar to the modern genus *Nuclearia*. Five fungal phyla are described in this chapter: the early-diverging chytrids (Chytridiomycota) and zygomycetes (Zygomycota), and the later-diverging AM fungi (Glomeromycota), ascomycetes (Ascomycota), and basidiomycetes (Basidiomycota). Molecular phylogenetic studies indicate that chytrids and zygomycetes are polyphyletic.

organic materials and absorb the resulting small organic food molecules. In addition, both fungi and animals store surplus food as the carbohydrate glycogen in their cells. Despite these nutritional commonalities, fungal structure is quite distinctive.

Fungi Have a Unique Cell-Wall Chemistry and Body Form

Unlike animal cells, which lack rigid cell walls, fungal cells are enclosed by tough cell walls composed of **chitin**, a polysaccharide that contains nitrogen. Chitin, which also forms the exoskeletons of arthropods such as insects, resists bacterial attack, thereby helping to protect fungal cells. Fungal cell walls influence two other features that distinguish fungal cells from animal cells: feeding and motility. Because they have cell walls, fungal cells cannot engulf food particles. In contrast, animal

bodies commonly display food ingestion systems, and some animal cells retain an ancient ability to ingest particles. Additionally, the rigid walls of fungal cells, like those of plant cells, restrict the mobility of nonflagellate cells.

Most fungi have a distinctive body known as a **mycelium** (plural, mycelia), which is composed of microscopic, branched filaments known as **hyphae** (singular, hypha) (**Figure 31.2**). Hyphae and mycelia evolved even before fungi made the transition from aquatic to terrestrial habitats (see Figure 31.1). As mentioned, a fungal mycelium may be very extensive, as in the case of *Armillaria ostoyae* (see chapter-opening photo), but is often inconspicuous because the component hyphae are so tiny and spread out in the substrate. The diffuse form of the fungal mycelium makes sense because most hyphae have the role of gathering food from the substrate. By spreading out, hyphae can absorb food from a large volume of substrate. The gathered

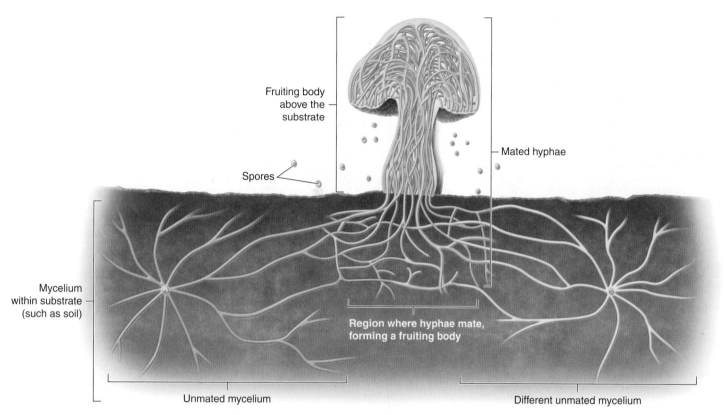

Fruiting body
above the
substrate

Spores

Mated hyphae

Mycelium
within substrate
(such as soil)

**Region where hyphae mate,
forming a fruiting body**

Unmated mycelium

Different unmated mycelium

Figure 31.2   Fungal morphology.   Most of a fungus consists of food-gathering hyphae that grow and branch from a central point to form a diffuse mycelium within a food substrate, such as soil. After a mating process occurs, mated hyphae may aggregate and grow out of the substrate to form fruiting bodies that produce and disperse spores. In suitable sites, spores may germinate, producing new mycelia.

food is used for mycelial growth and for reproduction by means of fruiting bodies, the only parts of fungal bodies that most people see.

Mushrooms and other types of fungal **fruiting bodies** are often noticeable because they are composed of densely packed hyphae that typically emerge from the substrate (Figure 31.2). Fruiting bodies are composed of hyphae that have undergone a mating process and, as a result, differ genetically and biochemically from unmated hyphae. Experts suspect that mated hyphae secrete signaling substances that cause such hyphae to cluster together and grow into the air, where reproductive cells can be more easily dispersed. Amazingly diverse in form, color, and odor, mature fruiting bodies are specialized to produce and disperse reproductive cells known as **spores**. Produced by the process of meiosis and protected by tough walls, spores reflect a major adaptation to the terrestrial habitat. When fungal spores settle in places where conditions are favorable for growth, they produce new mycelia. When the new mycelia undergo sexual reproduction, they produce new fruiting bodies.

Fungi Have Distinctive Growth Processes

If you have ever watched food become increasingly moldy over the course of several days, you have observed fungal growth. When a food source is plentiful, fungal mycelia can grow rapidly, adding as much as a kilometer of new hyphae per day. The mycelia grow at their edges as the fungal hyphae extend their tips through the undigested substrate. The narrow dimensions and extensive branching of hyphae provide a very high surface area for absorption of organic molecules, water, and minerals.

Hyphal Tip Growth   How do hyphae grow? Cytoplasmic streaming and osmosis are important cellular processes in hyphal growth. Osmosis (see Chapter 5) is the diffusion of water through a membrane from an area with a low solute concentration into an area with a high solute concentration. Water enters fungal hyphae by means of osmosis because their cytoplasm is rich in sugars, ions, and other solutes. Water entry swells the hyphal tip, producing the force necessary for tip extension. Masses of tiny vesicles carrying enzymes and cell-wall materials made in the Golgi apparatus collect in the hyphal tip (**Figure 31.3**). The vesicles then fuse with the plasma membrane. Some vesicles release enzymes that digest materials in the environment, releasing small organic molecules that are absorbed as food. Other vesicles deliver cell-wall materials to the hyphal tip, allowing it to extend.

Hyphal Structure and Nuclear Division   The hyphae of most fungi are subdivided into many small cells by cross walls known

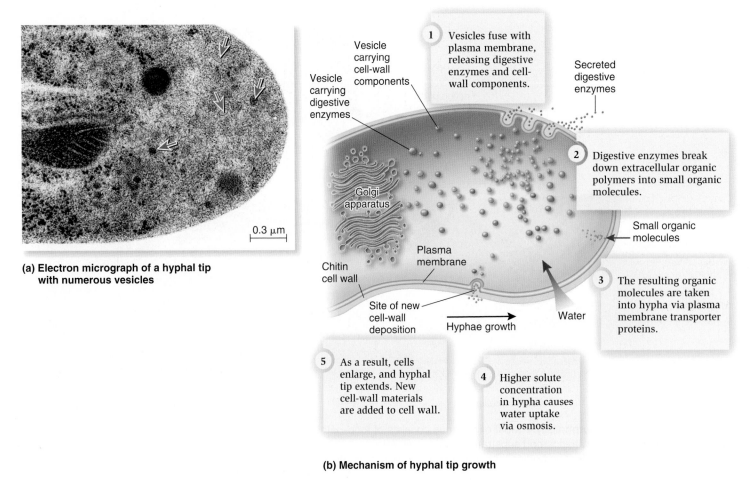

(a) **Electron micrograph of a hyphal tip with numerous vesicles**

1 Vesicles fuse with plasma membrane, releasing digestive enzymes and cell-wall components.

Vesicle carrying cell-wall components

Vesicle carrying digestive enzymes

Secreted digestive enzymes

2 Digestive enzymes break down extracellular organic polymers into small organic molecules.

Golgi apparatus

Small organic molecules

Plasma membrane

Chitin cell wall

3 The resulting organic molecules are taken into hypha via plasma membrane transporter proteins.

Site of new cell-wall deposition

Hyphae growth

Water

5 As a result, cells enlarge, and hyphal tip extends. New cell-wall materials are added to cell wall.

4 Higher solute concentration in hypha causes water uptake via osmosis.

(b) **Mechanism of hyphal tip growth**

Figure 31.3 **Hyphal tip growth and absorptive nutrition.** (a) TEM showing the hyphal tip of *Aspergillus nidulans,* a fungus commonly used as a genetic model system. The tip is filled with membrane-bound vesicles that fuse with the plasma membrane. Purple arrowheads show dark-stained vesicles carrying digestive enzymes, while green arrowheads point out light-stained vesicles carrying cell-wall materials. (b) Diagram of a hyphal tip, with vesicles of the same two types, showing the steps of hyphal tip growth.

Concept check: *What do you think would happen to fungal hyphae that begin to grow into a substrate having higher solute concentration? How might your answer be related to food preservation techniques such as drying or salting?*

as **septa** (singular septum) (**Figure 31.4a**). In such fungi, known as septate fungi, each round of nuclear division is followed by the formation of a septum that is perforated by a small pore. Septate hyphae appeared after the divergence of the AM fungi, but prior to the divergence of ascomycetes from basidiomycetes (see Figure 31.1). The hyphae of early-diverging fungi are not partitioned into smaller cells. Rather, these hyphae are **aseptate**

and multinucleate (**Figure 31.4b**), a condition that results when nuclei repeatedly divide without intervening cytokinesis. Such aseptate hyphae are described as being coenocytic.

When the nuclei of fungi divide, in most cases, a spindle forms within the nuclear envelope, which does not break down. This **intranuclear spindle** distinguishes nuclear division of fungi from that of animals and plants. (By contrast, in land plants and animals, the nuclear envelope vesiculates during prometaphase and then re-forms at telophase.)

Variations in Mycelium Growth Form Fungal hyphae grow rapidly through a substrate from areas where the food has become depleted to food-rich areas. In nature, mycelia may take an irregular shape, depending on the availability of food. A fungal mycelium may extend into food-rich areas for great distances, as noted at the beginning of the chapter. In liquid laboratory media, fungi will grow as a spherical mycelium that resembles a cotton ball floating in water (**Figure 31.5a**). Grown in flat laboratory dishes, the mycelium assumes a more two-dimensional growth form (**Figure 31.5b**).

(a) **Septate hypha**

(b) **Aseptate hypha**

Figure 31.4 **Types of fungal hyphae.**

(a) Mycelium growing in liquid medium

(b) Mycelium growing on flat, solid medium

Figure 31.5 **Fungal shape shifting.** (a) When a mycelium, such as that of this *Rhizoctonia solani*, is surrounded by food substrate in a liquid medium, it will grow into a spherical form. (b) When the food supply is limited to a two-dimensional supply, as shown by *Neotestudina rosatii* in the laboratory dish here, the mycelium will form a disc. Likewise, distribution of the food substrate determines the mycelium shape in nature.

69 μm

Figure 31.6 **Asexual reproductive cells of fungi.** SEM of the asexual spores (conidia) of *Aspergillus versicolor*, which causes skin infections in burn victims and lung infections in AIDS patients. Each of these small cells is able to detach and grow into an individual that is genetically identical to the parent fungus and so is able to grow in similar conditions.

Concept check: *How might you try to protect a burn patient from infection by a conidial fungus?*

31.2 Fungal Asexual and Sexual Reproduction

Many fungi are able to reproduce both asexually and sexually by means of microscopic spores, each of which can grow into a new adult. Asexual reproduction is a natural cloning process; it produces genetically identical organisms. Production of asexual spores allows fungi that are well adapted to a particular environment to disperse to similar, favorable places. Sexual reproduction generates new allele combinations that may allow fungi to colonize new types of habitats. Many fungi reproduce both sexually and asexually, although many reproduce only asexually.

Fungi Reproduce Asexually by Dispersing Specialized Cells

Asexual reproduction is particularly important to fungi, allowing them to spread rapidly by means of asexual spores. To reproduce asexually, fungi do not need to find compatible mates or expend resources on fruiting-body formation and meiosis. More than 17,000 fungal species reproduce primarily or exclusively by asexual means. DNA-sequencing studies have revealed that many types of modern fungi that reproduce only asexually have evolved from ancestors that had both sexual and asexual reproduction.

Many fungi produce asexual spores known as **conidia** (from the Greek *konis*, meaning dust) at the tips of hyphae (**Figure 31.6**). When they land on a favorable substrate, conidia germinate into a new mycelium that produces many more conidia. The green molds that form on citrus fruits are familiar examples of conidial fungi. A single fungus can produce as many as 40 million conidia per hour over a period of 2 days.

Because they can spread so rapidly, asexual fungi are responsible for costly fungal food spoilage, allergies, and diseases. Medically important fungi that reproduce primarily by asexual means include the athlete's foot fungus (*Epidermophyton floccosum*) and the infectious yeast (*Candida albicans*). **Yeasts** are fungi of various lineages that can occur as unicells; they reproduce by asexual budding (**Figure 31.7**).

Fungi Have Distinctive Sexual Reproductive Processes

As is typical for eukaryotes, the fungal sexual reproductive cycle involves the union of gametes, the formation of zygotes, and the process of meiosis. However, many aspects of fungal sexual reproduction are unique, including the function of hyphal branches as gametes and the development of fruiting bodies (**Figure 31.8**).

Fungal Gametes and Mating The gametes of most fungi are cells of hyphal branches. Although most fungi do not have distinguishable male and female gametes, fungal hyphae occur in

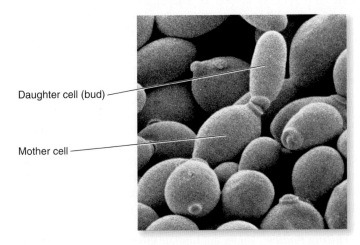

Daughter cell (bud)

Mother cell

Figure 31.7 The budding yeast *Saccharomyces cerevisiae*.

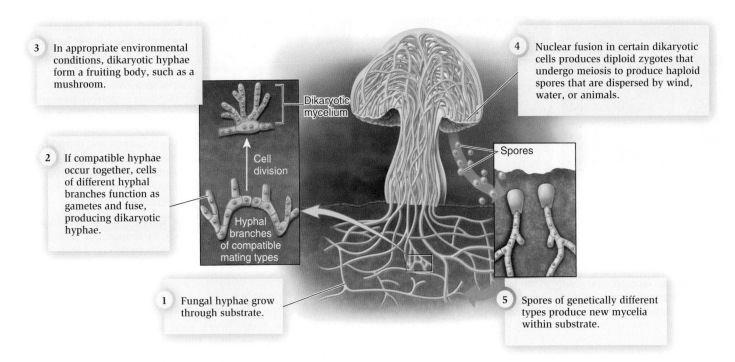

Figure 31.8   The reproductive cycle of many types of fungi.

Concept check: What kinds of fungi would be unlikely to reproduce in the manner shown?

multiple mating types that differ biochemically. The compatibility of these mating types is controlled by particular genes. During sexual reproduction, a hyphal branch of one mycelium fuses with a hyphal branch from a different mycelium of the same species and compatible mating type. Exceptions to this process occur in early-diverging fungi that live in the water, which produce flagellate sperm that swim to nonmotile eggs. During fungal evolution, the ability of gametes to produce flagella has been lost several times as different groups of fungi adapted to life in terrestrial habitats.

The actual mating process in fungi is also remarkable. In most sexual organisms, gametes undergo fusion of their cytoplasms, a process known as **plasmogamy**, and then the nuclei fuse in a process known as **karyogamy**. However, in many fungi, the haploid gamete nuclei remain separate for a long time after plasmogamy occurs. During this time period, the gamete nuclei both divide at each cell division, producing a mycelium that is **dikaryotic** (from the Greek, meaning two nuclei) or heterokaryotic (from the Greek, meaning different nuclei). Each cell of a dikaryotic mycelium possesses two unfused gamete nuclei; these are derived from parental nuclei (Figure 31.8). Some fungi persist as dikaryons, producing clones that can live for hundreds of years. Although the nuclei of dikaryotic mycelia remain haploid, alternate copies of many alleles occur in the separate nuclei. Thus, dikaryotic mycelia are functionally diploid. Eventually, dikaryotic mycelia produce fruiting bodies, the next stage of reproduction.

Fruiting Bodies   Under appropriate environmental conditions, such as seasonal change, a dikaryotic mycelium may produce

a fleshy fruiting body, such as a mushroom, which typically emerges from the substrate (see Figures 31.2, 31.8). All the cells of the fruiting body are dikaryotic. When the fruiting body is mature, the two nuclei in each of many cells at the surface undergo nuclear fusion. This process produces many zygotes, which are the only cells in the fungal life cycle that have a diploid nucleus. In most cases, the fungal zygotes soon undergo meiosis to produce haploid spores. Each spore acquires a tough chitin wall that protects it from drying out and other stresses. Wind, rain, or animals disperse the mature spores, which grow into haploid mycelia. If a haploid mycelium encounters hyphae of an appropriate mating type, hyphal branches will fuse and start the sexual cycle over again.

As we have seen, fruiting bodies usually emerge from the substrate, while most of the fungal mycelium lies inconspicuously within it. Mycelium growth requires organic molecules, minerals, and water provided by the substrate, but in most cases, spores are more easily dispersed if released outside of the substrate. The structures of fruiting bodies vary in ways that reflect different adaptations that foster spore dispersal by wind, rain, or animals. For example, mature puffballs have delicate surfaces upon which just a slight pressure causes the spores to puff out into wind currents (**Figure 31.9a**). Birds' nest fungi form characteristic egg-shaped spore clusters. Raindrops splash on these clusters and disperse the spores. The fruiting bodies of stinkhorn fungi smell and look like rotting meat (**Figure 31.9b**), which attracts carrion flies. The flies land on the fungi to investigate the potential meal and then fly away, in the process dispersing spores that stick to their bodies. The fruiting bodies of fungal truffles are also specialized for spore dispersal

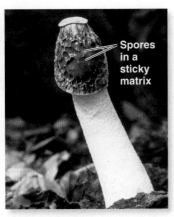

(a) Fruiting bodies adapted for dispersal of spores by wind

(b) Fruiting body adapted for dispersal of spores by insects

Spores in a sticky matrix

Figure 31.9 Fruiting body adaptations that foster spore dispersal. **(a)** When touched by wind gusts or animal movements, spores puff from fruiting bodies of the puffball fungus (*Lycoperdon perlatum*). **(b)** The fruiting bodies of stinkhorn fungi, such as this *Phallus impudicus*, use foul odors and the colors of dung or rotting meat to attract flies, to which the sticky fungal spores attach and are thereby dispersed.

Figure 31.10 Several types of edible, delicious fungi available in the market.

by animals. However, they are unusual in being produced underground. Mature truffles emit odors that attract wild pigs and dogs, which break up the fruiting structures while digging for them, thereby dispersing the spores. When collectors seek to harvest valuable truffles from forests for the market, they use trained leashed pigs or dogs to locate these fungi.

Many fungi such as truffles are edible, and several species of edible fungi are cultivated for market sale (**Figure 31.10**). However, the bodies of many fungi produce toxic substances that may deter animals from consuming them (**Figure 31.11**). For example, several fungi that attack stored grains, fruits, and spices produce **aflatoxins**; these cause liver cancer and are a major health concern worldwide. When people consume the forest mushroom *Amanita virosa*, known as the "destroying angel," they ingest a powerful toxin that may cause liver failure so severe that a transplant may be required. Each year, many people in North America are poisoned when they consume similarly toxic mushrooms gathered in the wild. There is no reliable way for nonexperts to distinguish poisonous from non-toxic fungi; it is essential to receive instruction from an expert before foraging for mushrooms in the woods. Therefore, many experts recommend that it is better to search for mushrooms in the grocery store than in the wild.

Several types of fungal fruiting structures produce hallu-cinogenic or psychoactive substances. As in the case of fungal toxins, fungal hallucinogens may have evolved as antiherbi-vore adaptations. Humans have inadvertently experienced their effects. For example, *Claviceps purpurea*, which causes a dis-ease of rye crops and other grasses known as ergot, produces a psychogenic compound related to LSD (lysergic acid diethyl-amide) (**Figure 31.12**). Some experts speculate that cases of hysteria, convulsions, infertility, and a burning sensation of the skin that occurred in Europe during the Middle Ages resulted

Figure 31.11 Toxic fruiting body of *Amanita muscaria*. Common in conifer forests, *A. muscaria* is both toxic and hallucinogenic. Ancient people used this fungus to induce spiritual visions and to reduce fear during raids. This fungus produces a toxin, amanitin, which specifically inhibits RNA polymerase II of eukaryotes (but not prokaryotes).

Concept check: What effect would the amanitin toxin have on human cells?

Ergot

Figure 31.12 Ergot of rye. The fungus *Claviceps purpurea* infects rye and other grasses, producing hard masses of mycelia known as ergots in place of some of the grains (fruits). Ergots produce alkaloids related to LSD and thus cause psychotic delusions in humans and animals that consume products made with infected rye. Ergots were used in folk medicine to treat migraine or to hasten childbirth.

from ergot-contaminated rye used in foods. Another example of a hallucinogenic fungus is the "magic mushroom" (*Psilocybe*), which is used in traditional rituals in some cultures. Like ergot, the magic mushroom produces a compound similar to LSD, a controlled substance. Consuming hallucinogenic fungi is risky because the amount used to achieve psychoactive effects is dangerously close to a poisonous dose.

31.3 Fungal Ecology and Biotechnology

The ability of fungi to degrade diverse materials is important in ecology, medicine, and human biotechnology applications. Fungal decomposers, also known as saprobes, are able to decompose nonliving organic materials. Recycling of materials in ecosystems depends on the activities of fungal decomposers. Fungal predators that consume tiny soil animals also play an important role in energy and nutrient cycling. Some fungi may attack living animal or plant tissues, in which case they are considered to function as pathogens, or causes of disease. As hyphae of pathogenic fungi grow through the tissues of plants and animals, they absorb food and minerals from the host, causing disease symptoms. Other fungi form beneficial partnerships with photosynthetic organisms that provide the fungi with organic food without experiencing harm. In return, fungi contribute mineral nutrients, water, and other benefits to their photosynthetic partners. In this section, we focus more closely on the ecological roles of decomposer, predatory, pathogenic, and beneficial fungi.

Decomposer and Predatory Fungi Play Important Ecological Roles

Decomposer fungi are essential components of the Earth's ecosystems. Together with bacteria, they decompose dead organisms and organic wastes, preventing litter buildup. For example, only certain bacteria and fungi can break down cellulose, and a few fungi are the major decomposers of lignin, the decay-resistant component of wood. In the absence of decomposers, ecosystems would become clogged with organic debris. Decomposers are needed to break down organic compounds into carbon dioxide, which is used by algae and plants for photosynthesis. Decomposer fungi and bacteria are also Earth's recycling engineers. They release minerals into the soil and water, where plants and algae take up the minerals for growth. In the absence of fungi and bacteria, such minerals would remain forever bound up in dead organisms, unavailable for the growth of forests, coral reefs, and other communities.

More than 200 species of predatory soil fungi use special adhesive or nooselike hyphae to trap tiny soil animals, such as nematodes, and absorb nutrients from their bodies (**Figure 31.13**). Such fungi help to control populations of nematodes, some of which attack plant roots. Other fungi obtain nutrients by attacking insects, and certain of these species are used as biological control agents to kill black field crickets, red-legged earth mites, and other pests.

Figure 31.13 **A predatory fungus.** The fungus *Arthrobotrys anchonia* traps nematode worms in hyphal loops that suddenly swell in response to the animal's presence. Fungal hyphae then grow into the worm's body and digest it.

Pathogenic Fungi Cause Plant and Human Diseases

One of the most important ways in which fungi affect humans is by causing diseases of humans and our crop plants. Five thousand fungal species cause serious crop diseases, and recent results show that new diseases can arise by horizontal gene transfer. Wheat rust is an example of a common crop disease caused by fungi (**Figure 31.14**). Rusts are named for the reddish spores that emerge from the surfaces of infected plants. Many types of plants can be attacked by rust fungi, but rusts are of particular concern when new strains attack crops. For example, in late 2004, agricultural scientists discovered that a devastating rust named *Phakopsora pachyrhizi* had begun to spread in the U.S. soybean crop. This rust kills soybean plants by attacking the leaves, causing complete leaf drop in less than 2 weeks. The disease had apparently spread to U.S. farms by means of spores blown on hurricane winds from South America. To control the spread of fungal diseases, agricultural experts work to

Figure 31.14 **Wheat rust.** The plant pathogenic fungus *Puccinia graminis* grows within the tissues of wheat plants, using plant nutrients to produce rusty streaks of red spores that erupt at the stem and leaf surface where spores can be dispersed. Red spore production is but one stage of a complex life cycle involving several types of spores. Rusts infect many other crops, causing immense economic damage.

identify effective fungicidal chemicals and develop resistant crop varieties.

In addition to being spread by wind, the spores of crop-disease fungi can also be introduced on travelers' clothing and other belongings. To reduce the entry of new crop-disease fungi—as well as crop viruses and insect pests—agricultural customs inspectors closely monitor the entry of plants, soil, foods, and other materials that might harbor these organisms.

Fungi cause several types of disease in humans. For example, athlete's foot and ringworm are common skin diseases caused by fungi that are known as dermatophytes because they colonize the human epidermis. *Pneumocystis carinii* is a fungal pathogen that infects individuals with weakened immune systems, such as AIDS patients, sometimes causing death by pneumonia.

Aspergillus fumigatus is a common conidial mold that can also cause a potentially fatal lung disease. Such harmful conidia-producing molds growing in poorly ventilated, moist places may be the cause of "sick building syndrome," a term used to describe situations in which occupants of a building experience acute health effects that appear to be linked to time spent inside, but where no specific cause can be identified. However, some black molds that also grow on moist building materials are blamed for human illness, even though there is little evidence that they are harmful to adults (**Figure 31.15**).

Another group of pathogenic fungi is made up of the several types of yeast known as the **dimorphic fungi** (from the Greek, meaning two forms). They live as spore-producing hyphae in the soil but transform into pathogenic yeasts when mammals inhale their wind-dispersed spores (**Figure 31.16**). Dimorphic fungi include *Blastomyces dermatitidis*, which causes the disease blastomycosis; *Coccidioides immitis,* the cause of coccidiomycosis; and *Histoplasma capsulatum,* the agent of histoplasmosis. These fungal diseases affect the lungs and may spread to other parts of the body, causing severe illness. Host body temperature triggers the change from hyphal to yeast form. Instead of producing spores, in the mammalian body, these pathogenic yeasts reproduce by forming buds that more effectively stick to lung cells, spread within lung tissue, and move to other organs. In

(a) Soil-dwelling hyphal phase   **(b) Budding yeast phase in host**

Figure 31.16   Dimorphic fungi.   (a) The soil-dwelling hyphal stage reproduces by airborne spores. **(b)** When a mammal inhales the spores, body heat causes the budding yeast phase to develop and attack host tissues.

Concept check: What kind of medicine could be developed to retard the spread of pathogenic yeasts within body tissues without harming the patient?

2006, Julie Nemacek, Marcel Wüthrich, and Bruce Klein identified histidine kinase as the sensor protein of dimorphic yeasts that responds to host body heat by switching on the budding yeast form. Because vertebrates lack histidine kinases, this protein is expected to be a useful target for developing antifungal drugs and vaccines.

Though fungal diseases that attack humans are of medical concern, in nature, fungal pathogens often help to control populations of other organisms, which is an important ecological role. Some fungi play important ecological roles as beneficial partners in associations with other organisms.

Fungi Form Beneficial Associations with Other Species

Fungi form several types of beneficial associations with animals, plants, algae, bacteria, and even viruses. For example, leaf-cutting ants, certain termites and beetles, and the salt marsh snail (*Littoraria irrorata*) cultivate particular fungi for food, much as do human mushroom growers. Other fungi obtain organic food molecules from photosynthetic organisms—plants, green algae, or cyanobacteria—that, in turn, receive benefits from the fungi. Such associations between fungi and photosynthetic organisms are particularly widespread and important in nature and agriculture. We focus next on three types of fungi—mycorrhizal fungi, endophytes, and lichen fungi—that are beneficially associated with photosynthetic organisms.

Mycorrhizae Associations between the hyphae of certain fungi and the roots of most seed plants are known as **mycorrhizae** (from the Greek, meaning fungus roots) (see Chapter 37). Such fungus-root associations are very important in nature and agriculture; more than 80% of terrestrial plants form mycorrhizae. Plants that have mycorrhizal partners receive an increased supply of water and mineral nutrients, primarily

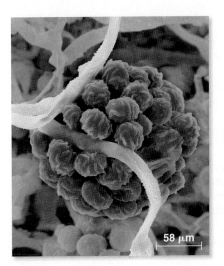

Figure 31.15   An indoor mold fungus. The black mold *Stachybotrys chartarum* is one of several types of conidial fungi that grow on indoor paper and wood that have been wet for an extended period.

phosphate, copper, and zinc. They do so because an extensive fungal mycelium is able to absorb minerals from a much larger volume of soil than roots alone are able to do (**Figure 31.17**). Added together, all the branches of a fungal mycelium in 1 m³ of soil can reach 20,000 km in total length. Experiments have shown that mycorrhizae greatly enhance plant growth in comparison to plants lacking fungal partners. In return, plants provide fungi with organic food molecules, sometimes contributing as much as 20% of their photosynthetic products—a very good investment.

By binding soils, fungal hyphae also reduce water loss and erosion, and they help protect plants against pathogens and toxic wastes. Fungi thereby help plants adapt to and thrive in new sites, thus playing an important role in plant succession (see Chapter 58). For this reason, ecologists increasingly employ mycorrhizal fungi in plant-community restoration projects.

The two most common types of mycorrhizae are endomycorrhizae and ectomycorrhizae. **Endomycorrhizae** (from the Greek *endo*, meaning inside) are partnerships between plants and fungi in which the fungal hyphae penetrate the spaces between root cell walls and plasma membranes and grow along the outer surface of the plasma membrane. In such spaces, endomycorrhizal fungi often form highly branched, bushy arbuscules (from the word "arbor," referring to tree shape). As the arbuscules develop, the root plasma membrane also expands. Consequently, the arbuscules and the root plasma membranes surrounding them have very high surface areas that facilitate rapid and efficient exchange of materials: Minerals flow from fungal hyphae to root cells, and organic food molecules move from root cells to hyphae. These fungus-root associations are known as **arbuscular mycorrhizae**, abbreviated **AM** (**Figure 31.18**). AM fungi are associated with apple trees, coffee shrubs, and many herbaceous plants, including legumes, grasses, tomatoes, strawberries, and peaches.

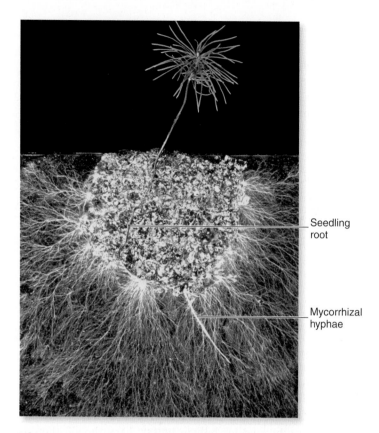

Figure 31.17 **Tree seedling with mycorrhizal fungi.** Hyphae of a mycorrhizal fungus extend farther into the soil than do plant roots, helping plants to obtain mineral nutrients.

Ectomycorrhizae (from the Greek *ecto*, meaning outside) are beneficial interactions between temperate forest trees and soil fungi. The fungi that engage in such associations are known as ectomycorrhizal fungi. The fruiting body shown in **Figure 31.19a** was produced by an ectomycorrhizal fungus. The hyphae of ectomycorrhizal fungi coat tree-root surfaces (**Figure 31.19b**) and grow into the spaces between root cells (**Figure 31.19c**).

(a) Micrograph of arbuscular mycorrhizae

(b) Hyphae growing between cell walls and plasma membranes

Figure 31.18 **Endomycorrhizae.** **(a)** Light micrograph showing black-stained AM fungi within the roots of the forest herb *Asarum canadensis*. Endomycorrhizal fungal hyphae enter plant roots via root hair cells, and then branches grow in the spaces between root cell walls and plasma membranes. **(b)** Diagram showing the position of highly branched arbuscules. Hyphal branches or arbuscules are found on the surface of the plasma membrane, which becomes highly invaginated. The result is that both hyphae and plant membranes have very high surface areas.

Concept check: *What fungal phylum consists entirely of endomycorrhizal fungi that are completely dependent upon plant hosts?*

(a) Ectomycorrhizal fruiting body

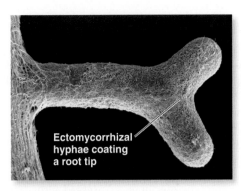

Ectomycorrhizal
hyphae coating
a root tip

(b) SEM of ectomycorrhizal hyphae

Ectomycorrhizal
hyphae

Root
cells

(c) Hyphae invading intercellular spaces

Figure 31.19   Ectomycorrhizae.   (a) The fruiting body of the common forest fungus *Laccaria bicolor*. This is an ectomycorrhizal fungus that is associated with tree roots. **(b)** Ectomycorrhizal fungal hyphae of *L. bicolor* cover the surfaces of young *Pinus resinosa* root tips. **(c)** Diagram showing that the hyphae of ectomycorrhizal fungi do not penetrate root cell walls but grow within intercellular spaces. In this location, fungal hyphae are able to obtain organic food molecules produced by plant photosynthesis.

Concept check: *What benefits do plants obtain from this association?*

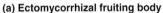

Some species of oak, beech, pine, and spruce trees will not grow unless their ectomycorrhizal partners are also present. Mycorrhizae are thus essential to the success of commercial nursery tree production and reforestation projects. New genetic information has illuminated how mycorrhizal fungi form associations with plants.

Genomes & Proteomes Connection

Gene Expression in Ectomycorrhizal Fungi Explains How They Live Both Independently and in Partnership with Plants

The mushroom-forming fungus *Laccaria bicolor* (Figure 31.19a) has become a model system for experimental studies of ectomycorrhizae, because the mycelium grows well by itself and in symbiosis with tree roots, in both nature and the laboratory. Experimental studies of *L. bicolor* have revealed how the fungus colonizes tree roots by forming a dense coating of hyphae on the root surface, producing hyphae that penetrate between root cells, and spreading hyphae from the root outward into the soil. Radiotracer studies have shown that the hyphae located within roots obtain sugar from the host plant, while the external hyphae harvest soil minerals and transfer them to the plant root.

In 2008, an international group of biologists described the genome of *L. bicolor*, thereby explaining many of its characteristics. Investigators found that *L. bicolor* has a larger genome than nonmycorrhizal fungi, with more and larger gene families, suggesting that symbiotic association has accelerated evolution. *Laccaria* also has an especially high number of genes associated with plasma membrane transport, reflecting the complexity of materials exchanged between the fungus and host roots. They found the fungus has many genes that encode small, secreted proteins; these genes are suspected to be involved in establishing and maintaining the symbiotic relationship. For example,

one secreted protein has been localized to *L. bicolor* hyphae present within roots, but not hyphae of the same species living in soil. This observation indicates that the plant partner influences gene expression in its fungal partner. Further study of gene expression patterns is expected to reveal more about molecular interactions essential to the mycorrhizal symbiosis.

The genome analysis also revealed that unlike most soil fungi, *L. bicolor* lacks the genetic capacity to break down plant cell-wall materials such as cellulose and lignin, which are abundant in soil and litter. As a result, *L. bicolor* cannot use the organic components of these materials as food. The lack of such genes makes sense because a fungus that readily destroys plant cell walls would not likely be able to establish a mutually beneficial relationship with plant roots. When in symbiotic association with roots, *L. bicolor* relies upon organic material supplied by the plant partner. Given this alternative food source, over evolutionary time, the fungus probably lost functional genes required for cellulose and lignin breakdown. How then are nonsymbiotic mycelia of *L. bicolor* able to survive in the soil? The genome sequence provides a likely answer. *Laccaria bicolor*'s genome includes genes encoding enzymes that may allow this fungus to break down dead insects, bacteria, and other protein-rich organic matter in soil. Such examples illustrate the value of genome sequences in explaining how the proteome controls key ecological functions of pivotal organisms such as mycorrhizal fungi.

Fungal Endophytes   Other fungi are known as **endophytes** because they live compatibly within the leaf and stem tissues of various types of plants. The endophytes obtain organic food molecules from plants and, in turn, contribute toxins or antibiotics that deter foraging animals, insect pests, and microbial pathogens. In general, plants that contain nonpathogenic endophytic fungi grow better than plants lacking such fungal partners. Endophytic fungi also help some plants tolerate higher temperatures.

FEATURE INVESTIGATION

Márquez and Associates Discovered That a Three-Partner Association Allows Plants to Cope with Heat Stress

The endophytic fungus *Curvularia protuberata* commonly lives within aboveground tissues of the grass *Dichanthelium lanuginosum*, which is unusual in its ability to grow on very hot soils in thermal areas of Yellowstone National Park. When the soil reaches 38°C, *D. lanuginosum* plants and *C. protuberata* fungi both die unless they live together in a symbiosis. In the symbiotic association, the partners can survive temperatures near 65°C.

In 2007, a team of investigators led by Luis Márquez discovered that a virus is also involved, revealing the occurrence of a three-partner symbiosis (**Figure 31.20**). These biologists were able to isolate the virus from the fungus and named it *Curvularia* thermal tolerance virus (CthTV) to indicate its host and phenotype. The investigators also noticed that some of their fungal cultures contained very little virus, so they were able to use drying and freeze–thaw cycles to cure such cultures of the virus. This procedure allowed them to experimentally determine the relative abilities of virus-infected and virus-free *C. protuberata* fungus to tolerate high temperatures and confer this property to plant partners. They found that plants having viral-infected fungal endophytes tolerated high temperatures much better than plants that lacked fungal endophytes or possessed virus-free fungal endophytes. Márquez and associates also reintroduced virus to their virus-free fungi and found that such fungi acquired the ability to confer heat tolerance to host plants. Finally, the researchers determined that the virus-infected fungus (but not virus-free fungi) could also protect a distantly related crop plant (tomato) from heat stress. These results add to accumulating evidence that multipartner symbioses are more common than previously realized and demonstrated that endophytic fungi may have useful agricultural applications.

Figure 31.20 Márquez and associates discovered that a three-partner symbiosis allows plants to cope with heat stress.

GOAL To determine if a virus is essential to the protective role of endophytic fungi to host plants under heat stress.

KEY MATERIALS *Curvularia* thermal tolerance virus (CthTV), cultures of the endophytic fungus *Curvularia protuberata* infected with CthTV, *C. protuberata* cultures free of CthTV, and *Dichanthelium lanuginosum* plants.

Experimental level		Conceptual level

1 Plant 25 replicate containers with *D. lanuginosum* lacking fungal symbionts (a) or with *C. protuberata* endophytes that either did (b) or did not (c) have virus.

(a) No fungus, no virus (b) Fungus and virus (c) Fungus, no virus

Compare the effects of virus on the ability of the fungus to confer heat stress protection.

2 Expose plants to heat stress treatment (up to 65°C) for 2 weeks in a greenhouse.

Keep environmental conditions constant to reduce experimental error.

3 Count the number of plants that were green (alive), yellow (dying), or brown (dead).

(a) (b) (c)

Assess plant survival in the presence or absence of fungus and/or virus.

4 THE DATA

KEY
- Brown, dead
- Yellow, dying
- Green, healthy

(a) No fungus, no virus
(b) Fungus and virus
(c) Fungus, no virus

Number of plants

5 CONCLUSION A virus enhances the protective role of endophytic fungi in this grass species. The next step will be to try to determine just how the virus changes the fungus so that the fungus is able to protect the plant from heat stress.

6 SOURCE Márquez, Luis M. et al. 2007. A virus in a fungus in a plant: Three-way symbiosis required for thermal tolerance. *Science* 315:513–515.

Experimental Questions

1. Would you expect plants that grow on unusually hot soils to have endophytic fungi or not?

2. How did Márquez and associates demonstrate that a virus was important in the heat tolerance of the *Dichanthelium lanuginosum–Curvularia protuberata* symbiosis?

3. How might the results of the work by Márquez and associates be usefully applied in agriculture?

Lichens Multipartner associations are also represented by **lichens**, which are composed of particular fungi, certain photosynthetic green algae and/or cyanobacteria, and non-photosynthetic bacteria such as actinomycetes. There are at least 25,000 lichen species, but these did not all descend from a common ancestor. DNA-sequencing studies suggest that lichens evolved independently in at least five separate fungal lineages. Molecular studies also show that some fungi have lost their ancestral ability to form lichen associations.

Lichen bodies take one of three major forms: (1) crustose—flat bodies that are tightly adherent to an underlying surface (**Figure 31.21a**); (2) foliose—flat, leaflike bodies (**Figure 31.21b**); or (3) fruticose—bodies that grow upright (**Figure 31.21c**) or hang down from tree branches. The photosynthetic green algae or cyanobacteria typically occur in a distinct layer close to the lichen's surface (**Figure 31.21d**). Lichen structure differs dramatically from that of the fungal component grown separately, demonstrating that the photosynthetic components influence lichen form.

The photosynthetic partner provides lichen fungi with organic food molecules and oxygen, and, in turn, it receives carbon dioxide, water, and minerals from the fungal partner. Lichen fungi also protect their photosynthetic partners from environmental stress. For example, lichens that occupy exposed habitats of high light intensity often produce bright yellow, orange, or red-colored compounds that help prevent damage to the photosynthetic apparatus (see Figure 31.21a). Lichen fungi also produce distinctive organic acids and other compounds that deter animal and microbial attacks.

Many lichens reproduce by both sexual and asexual means, and about one-third of lichen species reproduce only asexually. Asexual reproductive structures include **soredia** (singular, sore-dium), small clumps of hyphae surrounding a few algal cells

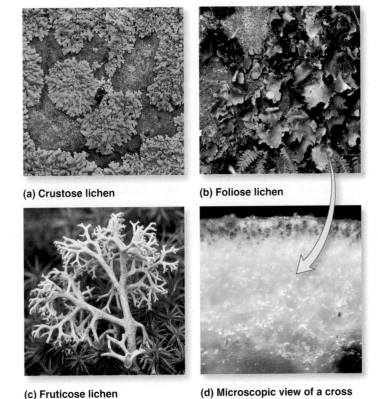

(a) Crustose lichen

(b) Foliose lichen

(c) Fruticose lichen

(d) Microscopic view of a cross section of a lichen

Figure 31.21 Lichen structure. (a) An orange-colored crustose lichen grows tightly pressed to the substrate. (b) The flattened, leaf-shaped genus *Umbilicaria* is a common foliose lichen. (c) The highly branched genus *Cladonia* is a common fruticose lichen. (d) A handmade thin slice of *Umbilicaria* viewed with a light microscope reveals that the photosynthetic algae occur in a thin upper layer. Fungal hyphae make up the rest of the lichen.

Figure 31.22 Lichen asexual reproduction. SEM of a soredium of the lichen *Cladonia coniocraea* that will break off from the parent lichen and disperse in the environment. Soredia are asexual structures that contain algal or cyanobacterial cells wrapped by fungal hyphae.

Concept check: *Do lichens that develop from soredia necessarily contain the same algae that were present in those soredia?*

that can disperse in wind currents (**Figure 31.22**). Soredia are lichen clones. By forming soredia, lichen fungi can disperse along with their photosynthetic partners.

The fungal partners of many lichens can undergo sexual reproduction, producing fruiting bodies and sexual spores much like those of related fungi that do not form lichens. DNA studies have shown that some lichen fungi can self-fertilize, which is advantageous in harsh environments where potential mates may be lacking. To produce new lichens, hyphae that grow from sexual spores must acquire new photosynthetic partners, but only particular green algae or cyanobacteria are suitable. However, lichens do not always contain the same algal partners as their parents because lichen fungi may switch algal partners, "trading up" for better algae. Partner switching allows lichens to adjust to changes in their environments. In 2008, Matt Nelsen and Andrea Gargas reported that lichens often change algal partners even when the fungal components lack sexual reproduction and depend completely upon algae-bearing soredia for dispersal. The partner-switching concept is based upon observations that within the same lichen family, fungal and algal species may display different phylogenies. Such results indicate that lichen partners do not strongly influence each other's evolutionary diversification. Emerging concepts of lichens as multipartner associations and frequent partner switching show that the lichen symbiosis is more complex than previously thought.

Lichens often grow on rocks, buildings, tombstones, tree bark, soil, or other surfaces that easily become dry. When water is not available, the lichens are dormant until moisture returns. Thus, lichens may spend much of their time in an inactive state, and for this reason, they often grow very slowly. However, because they can persist for long periods, lichens can be very old; some are estimated to be more than 4,500 years old. Lichens occur in diverse types of habitats, and a number grow in some of the most extreme, forbidding terrestrial sites on Earth—deserts, mountaintops, and the Arctic and Antarctic—places where most plants cannot survive. In these locations, lichens serve as a food source for reindeer and other hardy organisms. Though unpalatable, lichens are not toxic to humans and have also served as survival foods for aboriginal peoples in times of shortages.

Lichens are recognized for their soil-building activities, which occur over very long periods of time. Lichen acids help to break up the surfaces of rocks, beginning the process of soil development. Lichens having cyanobacterial partners can also increase soil fertility by adding fixed nitrogen. One study showed that such lichens released 20% of the nitrogen they fixed into the environment, where it is available for uptake by plants.

Lichens are useful as air-quality monitors because they are particularly sensitive to air pollutants such as sulfur dioxide. Air pollutants severely injure the photosynthetic components, causing death of the lichens. The disappearance of lichens serves as an early warning sign of air-pollution levels that are also likely to affect humans. Lichens can also be used to monitor atmospheric radiation levels because they accumulate radioactive substances from the air. After the Chernobyl nuclear power plant accident, lichens in nearby countries became so radioactive that reindeer became unfit for use as human food or in milk production.

Fungi Have Many Applications in Biotechnology

The ability of fungi to grow on many types of substrates and produce many types of organic compounds reflects their diverse ecological adaptations. Humans have harnessed fungal biochemistry in many types of biotechnology applications. Fungal biochemistry is a valuable asset to the chemical, food processing, and waste-treatment industries. A variety of industrial processes use fungi to convert inexpensive organic compounds into valuable materials, such as citric acid used in the soft drink industry, glycerol, antibiotics such as penicillin, and cyclosporine, a drug widely used to prevent rejection of organ transplants. In the food industry, fungi are used to produce the distinctive flavors of blue cheese and other cheeses. Other fungi secrete enzymes that are used in the manufacture of protein-rich tempeh and other food products from soybeans. The brewing and winemaking industries find yeasts essential, and the baking industry depends on the yeast *Saccharomyces cerevisiae* for bread production.

S. cerevisiae is also widely used as a model system for fundamental biological studies. Yeasts are useful in the laboratory because they have short life cycles, are easy and safe for lab workers to maintain, and their genomes are similar to those of animals. Some 31% of yeast proteins have human homologs, and nearly 50% of human genes that have been implicated in heritable diseases have homologs in yeast.

Fungi are increasingly being used in industrial processes to replace chemical procedures that generate harmful waste

Hyphae of *Phanerochaete chrysosporium*

Woody tissue; cell walls impregnated with lignin

25 μm

Figure 31.23 Industrial use of fungi in paper production. The forest fungus *Phanerochaete chrysosporium* produces a cobweb-like mycelium on the surfaces of rotting wood. This fungus releases enzymes that break down lignin, thereby making plant cell-wall carbohydrates such as cellulose more accessible.

materials. For example, during the production of paper, wood pulp is chemically bleached to remove lignin, but this bleaching process generates harmful compounds, including dioxin, a carcinogenic compound. The forest fungus *Phanerochaete chrysosporium* produces several enzymes that allow it to break down lignin in wood (**Figure 31.23**). Removal of lignin allows the fungi to gain access to cell-wall polysaccharides, which they break down, thereby releasing sugars that the fungus absorbs as food. Fungal lignin-breaking enzymes such as those produced by *P. chrysosporium* can be used to bleach paper without producing dioxin, thereby reducing a harmful by-product of paper production. In addition, Khadar Valli, Hiroyuki Wariishi,

and Michael Gold demonstrated that *P. chrysosporium* is able to decompose dioxin into nontoxic products, suggesting additional potential uses of the fungus in preventing pollution.

31.4 Diversity of Fungi

As noted earlier, the kingdom Fungi is a monophyletic group that arose from a protist ancestor (see Figure 31.1). The modern fungi have been classified into five phyla, listed in **Table 31.1** by both their common and formal names: chytrids (Chytridiomycota), zygomycetes (Zygomycota), AM fungi (Glomeromycota), ascomycetes (Ascomycota), and basidiomycetes (Basidiomycota). As mentioned, DNA and other data indicate that the chytrids and the zygomycetes are not monophyletic groups. In this section, we will survey the characteristics of the five phyla of true fungi and examine their distinctive reproductive cycles.

Chytrids Primarily Live in Aquatic Environments

Chytrids are the simplest fungi, and molecular evidence indicates that chytrids were among the earliest fungi to appear. Some chytrids occur as single, spherical cells that may produce hyphae (**Figure 31.24**), whereas others exist mainly as branched, aseptate hyphae. Chytrids often produce flagellate cells used for spore or gamete dispersal. The presence of a single, posterior flagellum on chytrid spores or gametes links fungi with the ancestry of choanoflagellate protists and animals (see Chapter 28).

Chytrids live in aquatic habitats or in moist soil. Most chytrids are decomposers, but some are parasites of protists (Figure 31.24) and pathogens of plants or animals. For example, the chytrid *Batrachochytrium dendrobatidis* has been associated with declining harlequin frog populations (look ahead to Figure 54.1).

Table 31.1	Distinguishing Features of Fungal Phyla			
Common name (Formal name)	**Habitat**	**Ecological role**	**Reproduction**	**Examples cited in this chapter**
Chytrids (Chytridiomycota)	Water and soil	Mostly decomposers; some parasites	Flagellate spores or gametes	*Batrachochytrium dendrobatidis*
Zygomycetes (Zygomycota)	Mostly terrestrial	Decomposers and pathogens	Nonflagellate asexual spores produced in sporangia; resistant sexual zygospores	*Rhizopus stolonifer*
AM Fungi (Glomeromycota)	Terrestrial	Form mutually beneficial mycorrhizal associations with plants	Distinctively large, nonflagellate, multinucleate asexual spores	The genus *Glomus*
Ascomycetes (Ascomycota)	Mostly terrestrial	Decomposers; pathogens; many form lichens; some are mycorrhizal	Asexual conidia; nonflagellate sexual spores (ascospores) in sacs (asci) on fruiting bodies (ascocarps)	*Aleuria aurantia, Venturia inaequalis, Saccharomyces cerevisiae*
Basidiomycetes (Basidiomycota)	Terrestrial	Decomposers; many are mycorrhizal; less commonly form lichens	Several types of asexual spores; nonflagellate sexual spores (basidiospores) on club-shaped basidia on fruiting bodies (basidiocarps)	*Coprinus disseminatus, Rhizoctonia solani, Armillaria mellea, Puccinia graminis, Ustilago maydis, Phanerochaete chrysosporium, Laccaria bicolor, Amanita muscaria, Phallus impudicus, Lycoperdon perlatum*

Zygomycetes Produce Distinctive Zygospores

The **zygomycetes** feature a mycelium that is mostly composed of aseptate hyphae (those lacking cross walls) and distinctive reproductive structures. For example, like most zygomycetes, the black bread mold *Rhizopus stolonifer* produces asexual spores in enclosures known as **sporangia** (singular, sporangium) (**Figure 31.25a**). Bread mold sporangia form at hyphal tips in such large numbers that they make moldy bread appear black. Zygomycete sporangia may each release up to 100,000 spores into the air! The great abundance of such spores means that bread molds easily unless the bread contains added retardant chemicals.

Zygomycetes are named for the zygospore, a distinctive feature of their sexual reproduction (**Figure 31.25b**). Zygospore production begins with the development of **gametangia** (from the Greek, meaning gamete bearers). In the zygomycete fungi,

Figure 31.24 **Chytrids growing on a freshwater protist.** The colorless chytrids produce hyphae that penetrate the cellulose cell walls of the dinoflagellate *Ceratium hirundinella*, absorbing organic materials from the alga. Chytrids use these materials to produce spherical flagellate spores that swim away to attack other algal cells.

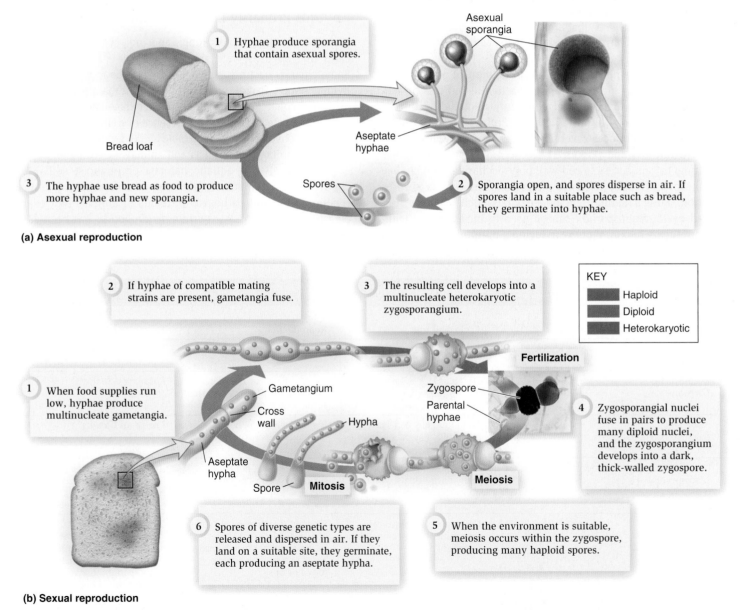

(a) Asexual reproduction

1 Hyphae produce sporangia that contain asexual spores.

2 Sporangia open, and spores disperse in air. If spores land in a suitable place such as bread, they germinate into hyphae.

3 The hyphae use bread as food to produce more hyphae and new sporangia.

Bread loaf

Asexual sporangia

Aseptate hyphae

Spores

(b) Sexual reproduction

1 When food supplies run low, hyphae produce multinucleate gametangia.

2 If hyphae of compatible mating strains are present, gametangia fuse.

3 The resulting cell develops into a multinucleate heterokaryotic zygosporangium.

4 Zygosporangial nuclei fuse in pairs to produce many diploid nuclei, and the zygosporangium develops into a dark, thick-walled zygospore.

5 When the environment is suitable, meiosis occurs within the zygospore, producing many haploid spores.

6 Spores of diverse genetic types are released and dispersed in air. If they land on a suitable site, they germinate, each producing an aseptate hypha.

KEY
Haploid
Diploid
Heterokaryotic

Fertilization

Meiosis

Mitosis

Gametangium
Cross wall
Hypha
Aseptate hypha
Spore
Zygospore
Parental hyphae

Figure 31.25 **The asexual and sexual life cycles of a zygomycete, the black bread mold *Rhizopus stolonifer*.**

Figure 31.26 The genus *Glomus*, an example of an AM fungus. The hyphae of these endomycorrhizal (arbuscular mycorrhizal) fungi are found in roots of many types of plants, aiding them in acquiring water and nutrients. AM fungi produce large, multinucleate spores by asexual processes.

gametangia are hyphal branches whose cytoplasm is isolated from the rest of the mycelium by cross walls. These gametangia enclose gametes that are basically a mass of cytoplasm containing several haploid nuclei. When food supplies run low and if compatible mating strains are present, the gametangia of compatible mating types fuse, as do the gamete cytoplasms. The resulting cell, known as a zygosporangium, contains many haploid nuclei that fuse, producing many diploid nuclei. Eventually, a dark-pigmented, thick-walled **zygospore** matures within the zygosporangium. Each zygospore contains many diploid nuclei and is capable of surviving stressful conditions. When the environment is suitable, the zygospore may undergo meiosis and germinate, dispersing many haploid spores. If the spores land in a suitable place, they germinate to form aseptate hyphae containing many haploid nuclei produced by mitosis. Most zygomycetes live on decaying materials in soil, but some are parasites of plants or animals.

AM Fungi Live with Plant Partners

The phylum Glomeromycota—commonly known as the **AM** (for <u>a</u>rbuscular <u>m</u>ycorrhizal) **fungi**—have aseptate hyphae and reproduce only asexually by means of unusually large spores with many nuclei (**Figure 31.26**). Many vascular plants depend on AM fungi, and these fungi are not known to grow separately from plants or cyanobacterial partners.

Molecular evidence suggests that Glomeromycota originated around 600 million years ago. Fossils having aseptate hyphae and large spores similar to those of modern Glomeromycota are known from the time when land plants first became common and widespread, about 460 million years ago (see Chapter 30). This and other fossil evidence suggests that the ability of early plants to live successfully on land may have depended on help from fungal associates, as is common today.

Ascomycetes Produce Sexual Spores in Saclike Asci

The **ascomycetes** and the basidiomycetes are composed of hyphae subdivided into cells by septa. These septa display distinctive pores, with ascomycete pores being simpler in structure than those of basidiomycetes (**Figure 31.27**). Such pores allow cytoplasmic structures and materials to pass through the hyphae.

The name ascomycetes derives from unique sporangia known as **asci** (from the Greek *asco*, meaning bags or sacs), which produce sexual spores known as **ascospores** (**Figure 31.28**). The asci are produced on fruiting bodies known as **ascocarps**. Although many ascomycetes have lost the ability to reproduce sexually, hyphal septa with simple pores (see Figure 31.27a) and DNA data can be used to identify them as members of this phylum.

Ascomycetes occur in terrestrial and aquatic environments, and they include many decomposers as well as pathogens. Important ascomycete plant pathogens include powdery mildews, chestnut blight (*Cryphonectria parasitica*), Dutch elm

(a) Simple pore—ascomycetes

(b) Complex pore—basidiomycetes

Figure 31.27 Septal pores of ascomycetes and basidiomycetes. (a) The septa of ascomycetes have simple pores at the centers. (b) More complex pores distinguish the septa of most types of basidiomycetes.

disease (the genus *Ophiostoma*), and apple scab (*Venturia inaequalis*) (**Figure 31.29**). Cup fungi (see the ascocarp photo in Figure 31.28) are common examples of ascomycetes. Many yeasts are also ascomycetes. Edible truffles and morels are the fruiting bodies of particular ascomycetes whose mycelia form mycorrhizal partnerships with plants. Ascomycetes are the most common fungal components of lichens.

Basidiomycetes Produce Diverse Fruiting Bodies

DNA-sequencing comparisons indicate that **basidiomycetes**, together with ascomycetes, are the most recently evolved groups of fungi. An estimated 30,000 modern basidiomycete species

Figure 31.29 Apple infected with the ascomycete fungus *Venturia inaequalis*, which causes apple scab disease. This fungus grows on leaves, flowers, and fruits, leaving harmless but unsightly scabs on their surfaces. Growers usually try to control apple scab by spraying trees with fungicides.

Hyphae produce asexual conidia. Conidia grow into new hyphae that are genetically identical to parents.

(a) Asexual reproduction

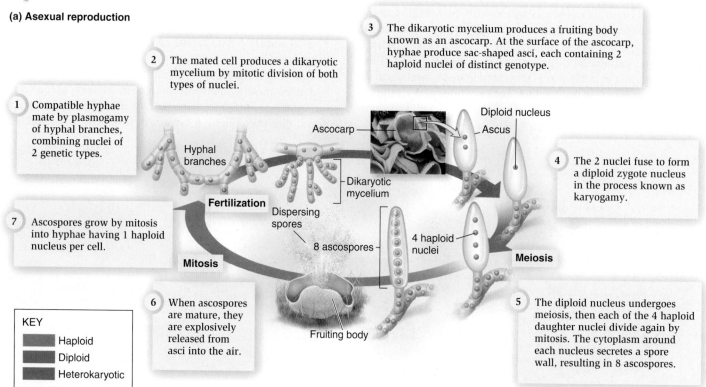

1 Compatible hyphae mate by plasmogamy of hyphal branches, combining nuclei of 2 genetic types.

2 The mated cell produces a dikaryotic mycelium by mitotic division of both types of nuclei.

3 The dikaryotic mycelium produces a fruiting body known as an ascocarp. At the surface of the ascocarp, hyphae produce sac-shaped asci, each containing 2 haploid nuclei of distinct genotype.

4 The 2 nuclei fuse to form a diploid zygote nucleus in the process known as karyogamy.

5 The diploid nucleus undergoes meiosis, then each of the 4 haploid daughter nuclei divide again by mitosis. The cytoplasm around each nucleus secretes a spore wall, resulting in 8 ascospores.

6 When ascospores are mature, they are explosively released from asci into the air.

7 Ascospores grow by mitosis into hyphae having 1 haploid nucleus per cell.

Hyphal branches

Fertilization

Mitosis

Dispersing spores

Dikaryotic mycelium

8 ascospores

Fruiting body

Ascocarp

Diploid nucleus

Ascus

4 haploid nuclei

Meiosis

KEY

Haploid
Diploid
Heterokaryotic

(b) Sexual reproduction of the ascomycete *Aleuria aurantia*

Figure 31.28 The asexual and sexual life cycles of ascomycete fungi.

(a) Corn smut

(b) Shelf fungi

Figure 31.30 **Fruiting bodies of basidiomycetes.** **(a)** Corn smut (*Ustilago maydis*) produces dikaryotic mycelial masses within the kernels (fruits) of infected corn plants. These mycelia produce many dark spores in which karyogamy and meiosis occur. Masses of these dark spores cause the smutty appearance. When the spores germinate, they produce basidiospores that can infect other corn plants. **(b)** Shelf fungi, such as this sulfur shelf fungus, are the fruiting bodies of basidiomycete fungi that have infected trees.

are known. Basidiomycetes are very important as decomposers and mycorrhizal partners in forests, producing diverse fruiting bodies commonly known as mushrooms, puffballs, stinkhorns, shelf fungi, rusts, and smuts (**Figure 31.30**; see also Figures 31.10, 31.11, and 31.12). The fairy rings of mushrooms that people sometimes find in open, grassy areas are ring-shaped arrays of basidiomycete fruiting bodies.

The name given to the basidiomycetes derives from **basidia**, the club-shaped cells that produce sexual spores known as **basidiospores** (**Figure 31.31**). Basidia are typically located on the undersides of fruiting bodies, which are generally known as **basidiocarps**. Though some basidiomycetes have lost sexual reproduction, they can be identified as members of this phylum by unique hyphal structures known as **clamp connections**, which help distribute nuclei during cell division. Basidiomycetes can also be identified by distinctive septa having complex pores (see Figure 31.27b) and by DNA methods. Basidiomycetes reproduce asexually by various types of spores.

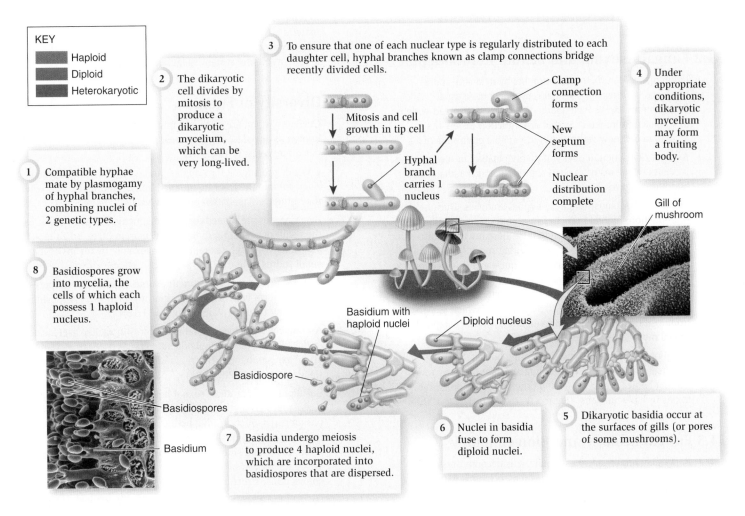

Figure 31.31 The sexual life cycle of the basidiomycete fungus *Coprinus disseminatus*.

Summary of Key Concepts

31.1 Evolutionary Relationships and Distinctive Features of Fungi

- Fungi form a monophyletic kingdom of heterotrophs that are related to the animal kingdom and certain protists. Early-divergent fungi are adapted to aquatic habitats, whereas later-divergent phyla are adapted to life on land. (Figure 31.1)

- Fungal cells possess cell walls composed of chitin, a polysaccharide resistant to microbial attack. Fungal bodies, known as mycelia, are composed of microscopic branched filaments known as hyphae, which appeared early in fungal evolution and grow at their tips. Food-gathering hyphae display absorptive nutrition; they secrete enzymes into food substrates, breaking down complex organic molecules into small organic molecules that are absorbed as food. (Figures 31.2, 31.3)

- The hyphae of later-diverging fungi are subdivided into cells by cross walls known as septa. Early-diverging fungi have aseptate hyphae that are not subdivided into cells. (Figure 31.4)

- Mycelial shape depends on the location of nutrients in the environment, which determines the direction in which cell division and hyphal growth will occur. (Figure 31.5)

31.2 Fungal Asexual and Sexual Reproduction

- Fungi disperse in their environments by means of spores produced by asexual or sexual reproduction. Asexual reproduction allows well-adapted genotypes to spread widely in stable environments. Sexual reproduction fosters the appearance of new traits useful in variable environments.

- Asexual reproduction does not involve mating or meiosis, and it occurs by means of asexual spores such as conidia or by budding. (Figures 31.6, 31.7)

- The gametes of most fungi are hyphal branches. During sexual reproduction, hyphal branches fuse with those of a different mycelium of compatible mating type. In many fungi, dikaryotic hyphae (having two nuclei per cell) that result from mating often persist for long periods before nuclear fusion occurs. Nuclear fusion generates zygotes. Fungal zygotes are the only cells in the life cycle that possess a diploid nucleus. Zygotes undergo meiosis to produce haploid spores, which disperse from fruiting bodies. Spores germinate to produce haploid fungal mycelia. (Figure 31.8)

- Fungi produce diverse types of fruiting bodies that foster spore dispersal by wind, water, or animals. Although many fungal fruiting bodies are edible, many others produce defensive toxins or hallucinogens. (Figures 31.9, 31.10, 31.11, 31.12)

31.3 Fungal Ecology and Biotechnology

- Fungi play important roles in nature as decomposers or predators and by forming beneficial associations with other organisms. Species of pathogenic fungi cause disease. (Figures 31.13, 31.14, 31.15, 31.16)

- Mycorrhizae are common associations between fungi and plant roots. Endomycorrhizae commonly form highly branched arbuscules in the spaces between root cell walls and plasma membranes. Ectomycorrhizae coat root surfaces, extending into root intercellular spaces. (Figures 31.17, 31.18, 31.19)

- A genome sequence newly available for *Laccaria bicolor* helps to explain the molecular basis of key traits of ectomycorrhizal fungi.

- Endophytic fungi often live compatibly within stems and leaves of plants, helping them to resist heat stress and attack by herbivores and disease microbes. (Figure 31.20)

- Lichens are multispecies partnerships between fungi, photosynthetic green algae and/or cyanobacteria, and other bacteria. Lichens can reproduce asexually by means of structures such as soredia, which consist of fungal hyphae wrapped around a few algal cells. When lichen fungi reproduce sexually, the hyphae arising from spore germination must find new algal partners, and lichen fungi often switch algal partners. (Figures 31.21, 31.22)

- Lichens occur in diverse habitats, including harsh environments, and often grow slowly and to great age. Lichens help to build soils and are useful air-quality monitors.

- Fungal biochemistry is useful in the chemical, food processing, and waste-treatment industries, and fungi are increasingly used to replace chemical procedures that generate harmful waste materials. The yeast *Saccharomyces cerevisiae* is important to the brewing and baking industries and is widely used as a laboratory model system. (Figure 31.23)

31.4 Diversity of Fungi

- Currently, fungi are classified into five phyla, commonly known as chytrids, zygomycetes, AM fungi, ascomycetes, and basidiomycetes. (Table 31.1)

- Chytrids are among the simplest and earliest-divergent fungi. They commonly occur in aquatic habitats and moist soil, where they produce flagellate reproductive cells. (Figure 31.24)

- Zygomycetes are named for their distinctive, large zygospores, the result of sexual reproduction. Common black bread mold and other zygomycetes reproduce asexually by means of many small spores. (Figure 31.25)

- The AM fungi produce distinctive large, multinucleate spores, and they form beneficial arbuscular mycorrhizal relationships with many types of plants. (Figure 31.26)

- Ascomycetes produce sexual ascospores in saclike asci located at the surfaces of fruiting bodies known as ascocarps. Many are lichen symbionts. The septa of hyphae have simple pores. (Figures 31.27, 31.28, 31.29)

- Basidiomycetes produce sexual basidiospores on club-shaped basidia located on the surfaces of fruiting bodies known as basidiocarps. Such fruiting bodies take a wide variety of forms, including mushrooms, puffballs, stinkhorns, shelf fungi, rusts, and smuts. The hyphae of basidiomycete fungi are characterized by complex pores in septa and clamp connections, structures that aid in distributing nuclei of two types after cell division occurs in dikaryotic hyphae. (Figures 31.30, 31.31)

Assess and Discuss

Test Yourself

1. Fungal cells differ from animal cells in that fungal cells
   a. lack ribosomes, though these are present in animal cells.
   b. lack mitochondria, though these occur in animal cells.
   c. have cell walls, whereas animal cells lack rigid walls.
   d. lack cell walls, whereas animal cells possess walls.
   e. none of the above

2. Conidia are
   a. cells produced by some fungi as the result of sexual reproduction.
   b. fungal asexual reproductive cells produced by the process of mitosis.
   c. structures that occur in septal pores.
   d. the unspecialized gametes of fungi.
   e. none of the above.

3. What are mycorrhizae?
   a. the bodies of fungi, composed of hyphae
   b. fungi that attack plant roots, causing disease
   c. fungal hyphae that are massed together into stringlike structures
   d. fungi that have symbiotic partnerships with algae or cyanobacteria
   e. mutually beneficial associations of particular fungi and plant roots

4. Where could you find diploid nuclei in an ascomycete or basidiomycete fungus?
   a. in spores
   b. in cells at the surfaces of fruiting bodies
   c. in conidia
   d. in soredia
   e. all of the above

5. Which fungi are examples of hallucinogen producers?
   a. *Claviceps* and *Psilocybe*
   b. *Epidermophyton* and *Candida*
   c. *Pneumocystis carinii* and *Histoplasma capsulatum*
   d. *Saccharomyces cerevisiae* and *Phanerochaete chrysosporium*
   e. *Cryphoenectria parasitica* and *Ventura inaequalis*

6. What role do fungal endophytes play in nature?
   a. They are decomposers.
   b. They are human pathogens that cause skin diseases.
   c. They are plant pathogens that cause serious crop diseases.
   d. They live within the tissues of grasses and other plants, helping to protect plants from herbivores, pathogens, and heat stress.
   e. All of the above are correct.

7. What forms do lichens take?
   a. crusts, flat bodies
   b. foliose, leaf-shaped bodies
   c. fruticose, erect or dangling bodies
   d. single cells
   e. a, b, and c

8. Lichens consist of a partnership between fungi and what other organisms?
   a. red algae and brown algae
   b. green algae, cyanobacteria, and heterotrophic bacteria
   c. the roots of vascular plants
   d. choanoflagellates and *Nuclearia*
   e. none of the above

9. How can ascomycetes be distinguished from basidiomycetes?
   a. Ascomycete hyphae have simple pores in their septa and lack clamp connections, whereas basidiomycete hyphae display complex septal pores and clamp connections.
   b. Ascomycetes produce sexual spores in sacs, whereas basidiomycetes produce sexual spores on the surfaces of club-shaped structures.
   c. Ascomycetes are commonly found in lichens, whereas basidiomycetes are less commonly partners in lichen associations.
   d. Ascomycetes are not commonly mycorrhizal partners, but basidiomycetes are commonly present in mycorrhizal associations.
   e. All of the above are correct.

10. Which group of organisms listed is most closely related to the kingdom Fungi?
    a. the animal kingdom
    b. the green algae
    c. the land plants
    d. the bacteria
    e. the archaea

Conceptual Questions

1. Explain three ways that fungi are like animals and two ways in which fungi resemble plants.

2. Explain why some fungi produce toxic or hallucinogenic compounds.

3. Explain three ways in which fungi function as beneficial partners with autotrophs and what benefit the fungi receive from the partnerships.

Collaborative Questions

1. Thinking about the natural habitats closest to you, where could you find fungi, and what roles do these fungi play?

2. Imagine that you are helping to restore the natural vegetation on a piece of land that had long been used to grow crops. You are placed in charge of planting pine seedlings (*Pinus resinosa*) and fostering their growth. In what way could you consider using fungi?

Online Resource

www.brookerbiology.com

Stay a step ahead in your studies with animations that bring concepts to life and practice tests to assess your understanding. Your instructor may also recommend the interactive ebook, individualized learning tools, and more.

An Introduction to Animal Diversity

32

Two of the staggering variety of animal species on Earth.
These species include thousands of vertebrates and millions of invertebrates.

A nimals constitute a very species-rich kingdom. Well over a million species have been found and described by biologists, with many more species likely awaiting discovery and classification. Beyond being members of this kingdom ourselves, humans are dependent on animals. We eat many different kinds of animals, use a diverse array of animal products, and employ animals such as horses and oxen as a source of labor. We enjoy many animal species as companions and are dependent on other species to test lifesaving drugs. However, we are in competition with animals such as insects that threaten our food supply and are parasitized by others. With such a huge number and diversity of existing animals and with animals featuring so prominently in our lives, understanding animal diversity is of paramount importance. Researchers have spent great effort in determining the unique characteristics of different taxonomic groups and identifying their evolutionary relationships.

Since the time of Carl Linnaeus in the 1700s, scientists have classified animals based on their morphology, that is, on their physical structure. Then, as now, a lively debate has surrounded the question of what constitutes the "correct" animal phylogeny. In the 1990s, animal classifications based on similarities in DNA and rRNA became more common. Quite often, classifications based on morphology and those based on molecular data were similar, but some important differences arose. In this chapter, we will begin by defining the key characteristics of animals and then take a look at the major features of the animal body plans that form the basis of the traditional view of animal classification. We will explore how new molecular evidence has made significant alterations to this categorization of the animal kingdom and examine some of the similarities and differences between the morphological and molecular-based phylogenies. As more molecular-based evidence becomes available, systematists will likely continue to redraw the tree of animal life. Thus, as you read this chapter, keep in mind that animal classification is now, and will continue to be, a work in progress.

32.1 Characteristics of Animals

The Earth contains over a million known animal species, living in environments from the deep sea to the desert and exhibiting an amazing array of characteristics. Most animals move and eat multicellular prey, and therefore, they are loosely differentiated from species in other kingdoms. However, coming up with a firm definition of an animal is tricky, because animals are so diverse that biologists can find exceptions to nearly any given characteristic. Even so, a number of key features exist that can help us broadly characterize the group we call animals (Table 32.1).

In brief, **animals** are multicellular heterotrophs whose cells lack cell walls. Most animals have nerves, muscles, the capacity to move at some point in their life cycle, and the ability to reproduce sexually, with sperm fusing directly with eggs. Unlike plants, animals cannot synthesize their essential organic molecules from inorganic sources and so must feed on other organisms. Many, if not most, animals are capable of some type of movement or locomotion in order to acquire food or escape predators. This ability has led to the development of specialized systems of sensory structures and a nervous system to coordinate movement and prey capture. Sessile species, such as barnacles, use bristled appendages to obtain food. In many such

Table 32.1	Common Characteristics of Animals
Characteristic	**Example**
Multicellularity	Even relatively simple types of animals such as sponges are multicellular, in contrast to the single-celled eukaryotic microorganisms called protists (see Chapter 28).
Heterotrophs	Animals obtain their food by eating other organisms or their products. This contrasts with plants and algae, which are autotrophs and essentially make their own food.
No cell walls	While plant, fungal, and bacterial cells are rigid because they possess a cell wall, animal cells lack a cell wall and are quite flexible.
Nervous tissue	The presence of a nervous system in most animals enables them to respond rapidly to environmental stimuli.
Movement	Most animals have a muscle system, which, combined with a nervous system, allows them to move in their environment.
Sexual reproduction	Most animals reproduce sexually, with small, mobile sperm uniting with a much larger egg to form a fertilized egg, or zygote.
Extracellular matrix	Proteins such as collagen bind animal cells together to give them added support and strength.
Characteristic cell junctions	Animals have characteristic cell junctions, called anchoring, tight, and gap junctions.
Special clusters of *Hox* genes	All animals possess *Hox* genes, which function in patterning the body axis (see Chapter 19).
Similar rRNA	Animals have very similar genes that encode for RNA of the small ribosomal subunit (SSU) rRNA.

sessile species, although adults are immobile, the larvae can swim. The lack of a rigid cell wall and the existence of nerves and muscles facilitate movement but reduce structural support. Instead, animal cells exist in an extensive extracellular matrix that forms strong fibers outside of the cell (see Chapter 10). Additionally, a group of unique cell junctions, called anchoring, tight, and gap junctions, plays an important role in holding animal cells in place and allowing cell communication.

The history of animal life spans over 590 million years, starting at the end of the Proterozoic era, when marine invertebrates appeared (refer back to Figure 22.9). A profusion of animal phyla appeared during the Cambrian explosion, 533–525 million years ago, including sponges, jellyfish, corals, flatworms, mollusks, annelid worms, the first arthropods, and echinoderms, plus many phyla that no longer exist today. The causes of this sudden increase in animal life at this time are not fully understood, but three reasons have been proposed. First, species proliferation may have been related to a favorable environment, which was warm and wet with no evidence of ice at the poles. At the same time, atmospheric oxygen levels

were high, permitting increased metabolic rates, and the ozone layer had developed, blocking out harmful ultraviolet radiation and allowing complex life to thrive in shallow water and, eventually, land. Second, the evolution of the *Hox* gene complex may have permitted much variation in morphology. Third, as new types of predator evolved, prey developed new methods of predator avoidance. This evolutionary arms race may have caused a proliferation of predator and prey types. These hypotheses are not mutually exclusive and may well have operated at the same time.

Five hundred and twenty million years ago, the first vertebrates, fishes, appeared at roughly the same time as the first land plants. The appearance of land plants represented a viable food source for any animals that could utilize them. However, terrestrial organisms faced the problem of drying out, or desiccation. This problem necessitated the development of features that enabled animals to venture onto land, in much the same way as seeds and other adaptations permitted plants to colonize terrestrial habitats (see Chapter 29). Such features include internal fertilization and the development of a tough, protective shell around the eggs of many species. The amniotic egg appeared during the Carboniferous period, about 300 million years ago, and was responsible for the success of the reptiles, which appeared during this period. Reptiles were to dominate the Earth for many millions of years during the rise and fall of the dinosaurs. Mammals also appeared at the same time as dinosaurs and were to inherit the Earth as the dinosaurs disappeared in a changing climate about 65 million years ago.

32.2 Traditional Classification of Animals

Although animals constitute an extremely diverse kingdom, most biologists agree that the kingdom is monophyletic, meaning that all taxa have evolved from a single common ancestor. Today, scientists recognize about 35 animal phyla. At first glance, many of these phyla seem so distantly related to one another (for example, chordates and jellyfish) that making sense of this diversity with a classification scheme seems very challenging. However, over the course of centuries, scientists have come to some basic conclusions about the evolutionary relationships among animals. In this section, we explore the major features of animal body plans and development that are the basis of the traditional view of animal phylogeny. But first, let's look at where this diversity started.

Animals Evolved from a Choanoflagellate Ancestor

With the monophyletic nature of the animal kingdom in mind, scientists have attempted to characterize the organism from which animals most likely evolved. According to research, the most likely ancestor is a colonial flagellated protist that is probably related to the present-day protists known as choanoflagellates. Choanoflagellates are tiny, single-celled organisms, each with a single flagellum surrounded by a collar composed of cytoplasmic tentacles (**Figure 32.1a**). A number of species are colonial, usually taking the form of a cluster of undifferentiated cells on a single stalk. Scientists think that some of these cells may have gradually taken on specialized functions—for

example, movement or nutrition—while still maintaining coordination with other cells and cell types.

As shown in **Figure 32.1b**, colonial choanoflagellate cells bear a striking similarity to cell types found in sponges, which are the simplest animals. As discussed later, evolutionary changes to this simple body plan resulted in critical innovations that led to more complex body plans found in other animals. Molecular data also point to choanoflagellates as the common animal ancestor.

The Traditional Classification of Animals Is Based on Body Plans

Without the availability of modern molecular techniques, biologists traditionally classified animal diversity in terms of these four main morphological and developmental features of animal body plans:

1. presence or absence of different tissue types;
2. type of body symmetry;
3. presence or absence of a true body cavity;
4. specific features of embryonic development.

We will discuss each of these major features of animal body plans next.

Tissues   Collectively, animals are known as **Metazoa**. Animals can be divided into two subgroups based on whether or not they have specialized types of tissues, that is, stable associations of cells that have a similar structure and function. The **Parazoa** (from the Greek, meaning alongside animals) are not generally thought to possess specialized tissue types or organs, although they may have several distinct types of cells. Those cells can change their shape and location, making any associations temporary. The Parazoa consist of a single phylum, Porifera (sponges) (**Figure 32.2a**). In contrast, the **Eumetazoa** (from the Greek, meaning true animals) have more than one type of tissue and, for the most part, have different types of organs.

Choanoflagellate cell   **Sponge cell (choanocyte)**

(a) Colonial choanoflagellate   **(b) Sponge**

Figure 32.1   **Early animal characteristics: A comparison of a colonial choanoflagellate and a sponge.**   Both types of organisms have very similar types of cells, which are used for feeding. The structure of sponges is described in Chapter 33 (look ahead to Figure 33.2).

Concept check: *Why are sponges considered animals but simple choanoflagellates are not?*

(a) Parazoa: no tissue types

(b) Eumetazoa: two tissue types
Radiata: radial symmetry

(c) Eumetazoa: three tissue types
Bilateria: bilateral symmetry

Figure 32.2   **Early divisions in the animal phylogeny.**   Animals can be categorized based on **(a)** the absence of different tissue types (Parazoa; the sponges) or **(b,c)** the presence of tissues (Eumetazoa; all other animals). Further categorization is based on the presence of **(b)** radial symmetry (Radiata; the cnidarians and ctenophores) or **(c)** bilateral symmetry (Bilateria; all other animals).

Symmetry   The Eumetazoa are divided according to their type of symmetry. Symmetry refers to the existence of balanced proportions of the body on either side of a median plane. Radially symmetric animals, the **Radiata**, can be divided equally by any longitudinal plane passing through the central axis (**Figure 32.2b**). Such animals are often circular or tubular in shape, with a mouth at one end, and include the animals called cnidarians and ctenophores (jellyfish and related species).

Bilaterally symmetric animals, the **Bilateria**, can be divided along a vertical plane at the midline to create two halves (**Figure 32.2c**). Thus, a bilateral animal has a left side and a right side, which are mirror images, as well as a **dorsal** (upper) and a **ventral** (lower) side, which are not identical, and an **anterior** (head) and a **posterior** (tail) end. Bilateral symmetry is strongly correlated with both the ability to move through the environment and **cephalization**, the localization of sensory structures at the anterior end of the body. Such abilities allow animals to encounter their environment initially with their head, which is best equipped to detect and consume prey and, in turn, detect and respond to predators and other dangers.

Another key difference between the Radiata and Bilateria is that radial animals have two embryonic cell layers, called **germ layers**, whereas bilateral animals have three germ layers. In all animals except the sponges, the growing embryo develops different layers of cells through a process known as **gastrulation** (**Figure 32.3**).

Fertilization of an egg by a sperm creates a diploid zygote. The zygote then undergoes **cleavage**, a succession of rapid cell divisions with no significant growth that produces a hollow sphere of cells called a **blastula**. In gastrulation, an area in the blastula folds inward, or invaginates, creating in the process a structure called a gastrula with the primary germ layers. The inner layer of cells becomes the **endoderm**, which lines the archenteron, or primitive digestive tract. The outer layer, or **ectoderm**, covers the surface of the embryo and differentiates into the epidermis and nervous system.

The Bilateria develop a third layer of cells, termed the **mesoderm**, between the ectoderm and endoderm. Mesoderm forms the muscles and most other organs between the digestive tract and the ectoderm. Because the Bilateria have these three distinct germ layers, they are often referred to as **triploblastic**, whereas the Radiata, which have only ectoderm and endoderm, are termed **diploblastic**.

Body Cavity   The next three major divisions in the classification of animals concern the development in bilaterally symmetrical animals of a fluid-filled body cavity called a **coelom**. In many animals, the body cavity is completely lined with mesoderm and is called a true coelom. Animals with a true coelom are termed **coelomates** (**Figure 32.4a**). If the fluid-filled cavity is not completely lined by tissue derived from mesoderm, it is known as a pseudocoelom (**Figure 32.4b**). Animals with a pseudocoelom, including rotifers and roundworms, are termed **pseudocoelomates**. Some animals, such as flatworms, lack a fluid-filled body cavity and are termed **acoelomates** (**Figure 32.4c**). Instead of fluid, this region contains mesenchyme tissue.

A body cavity has many important functions, perhaps the most important being that its fluid is relatively incompressible and therefore cushions internal organs such as the heart and intestinal tract, helping to prevent injury from external forces. A body cavity also enables internal organs to move and grow independently of the outer body wall. Furthermore, in some soft-bodied invertebrates, such as earthworms, the coelom functions as a **hydrostatic skeleton**, a fluid-filled body cavity surrounded by muscles that gives support and shape to the

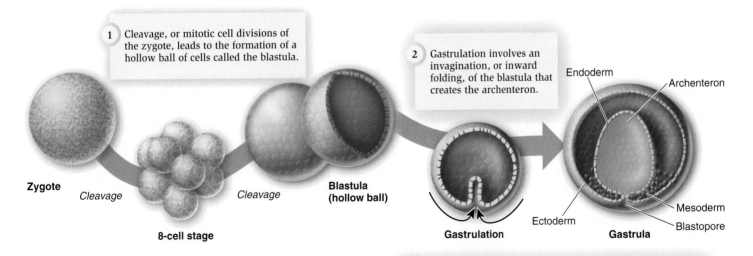

1 Cleavage, or mitotic cell divisions of the zygote, leads to the formation of a hollow ball of cells called the blastula.

2 Gastrulation involves an invagination, or inward folding, of the blastula that creates the archenteron.

Endoderm

Archenteron

Zygote

Cleavage   **8-cell stage**   *Cleavage*   **Blastula (hollow ball)**

Ectoderm

Mesoderm

Blastopore

Gastrulation   **Gastrula**

3 The layer of cells lining the archenteron becomes the endoderm. The cells on the outside of the blastula form the ectoderm. In the Bilateria, a middle layer termed the mesoderm develops between the ectoderm and endoderm.

Figure 32.3   **Formation of germ layers.**   Note: Radially symmetrical animals (Radiata) do not form mesoderm.

Body covering
(from ectoderm)

Muscle layer
(from mesoderm)

Coelom
(fluid-filled
space)

Digestive tract
(from endoderm)

Tissue layer
suspending organs
(from mesoderm)

Earthworm

(a) Coelomate

Body covering
(from ectoderm)

Muscle layer
(from mesoderm)

Pseudocoelom
(fluid-filled
space)

Digestive tract (from endoderm)

Nematode

(b) Pseudocoelomate

Body covering
(from ectoderm)

Muscle layer
(from mesoderm)

Mesenchyme
(from mesoderm)

Digestive tract
(from endoderm)

Flatworm

(c) Acoelomate

Figure 32.4   **The three basic body plans of bilaterally symmetrical animals.**   Cross sections of each animal are shown on the right.

Concept check: *What advantages does a coelom confer for movement?*

body of organisms. Muscle contractions at one part of the body push this fluid toward another part of the body. This type of movement can best be observed in an earthworm (look ahead to Figure 44.1a). Finally, in some organisms, the fluid in the body cavity also acts as a simple circulatory system.

Specific Features of Embryonic Development   In the developing zygote, cleavage may occur by two mechanisms (**Figure 32.5a**). In **spiral cleavage**, the planes of cell cleavage are oblique to the vertical axis of the embryo, resulting in an arrangement in which newly formed upper cells lie centered between the underlying cells. Animals that exhibit spiral cleavage are called **protostomes** and include mollusks, annelid worms, and arthropods. In **radial cleavage**, the cleavage planes are either parallel or perpendicular to the vertical axis of the egg. This results in tiers of cells, one directly above the other. Animals exhibiting radial cleavage are called **deuterostomes** and include echinoderms and chordates.

Protostome development is also characterized by so-called **determinate cleavage**, in which the fate of each embryonic cell is determined very early (**Figure 32.5b**). If one of the cells is removed from a four-cell mollusk embryo, neither the single cell nor the remaining three-cell mass can form viable embryos, and development is halted. In contrast, deuterostome development is characterized by **indeterminate cleavage**, in which each cell produced by early cleavage retains the ability to develop into a complete embryo. For example, when one cell is excised from a four-cell sea urchin embryo, both the single cell and the remaining three can go on to form viable embryos. Other embryonic cells compensate for the missing cells. In human embryos, if individual embryonic cells separate from one another early in development, identical twins can result.

In addition to differences in cleavage patterns, protostomes and deuterostomes differ in other embryonic features. The most fundamental of these concerns the development of a mouth and anus (**Figure 32.5c**). In gastrulation, the endoderm forms an indentation, the **blastopore**, which is the opening of the archenteron to the outside. In protostomes (from the Greek *protos*, meaning first, and *stoma*, meaning mouth), the blastopore becomes the mouth. If an anus is formed in a protostome, it develops from a secondary opening. In contrast, in the deuterostomes (from the Greek *deuteros*, meaning second), the blastopore becomes the anus, and the mouth is formed from the secondary opening.

Further Methods of Classification   In the traditional phylogenetic tree of animal life, more recent branches are based on features such as the possession of an exoskeleton (arthropods) or the development of a notochord (chordates). One other key feature of the animal body plan is the presence or absence of segmentation. In **segmentation**, the body is divided into regions called segments. It is most obvious in the annelids, or segmented worms, but it is also evident in arthropods and chordates (**Figure 32.6**). In annelids, each segment contains the same set of blood vessels, nerves, and muscles. Some segments may differ, such as those containing the brain or the sex organs, but many segments are very similar. In chordates, we can see segmentation in the backbone and muscles.

The advantage of segmentation is that it allows specialization of body regions. For example, as we will see in Chapter 33, arthropods exhibit a vast degree of specialization of their segments. Many insects have wings and only three pairs of legs, whereas centipedes have no wings and many legs. Crabs, lobsters, and shrimp have highly specialized thoracic appendages that aid in feeding. Let's now take a look at how we are

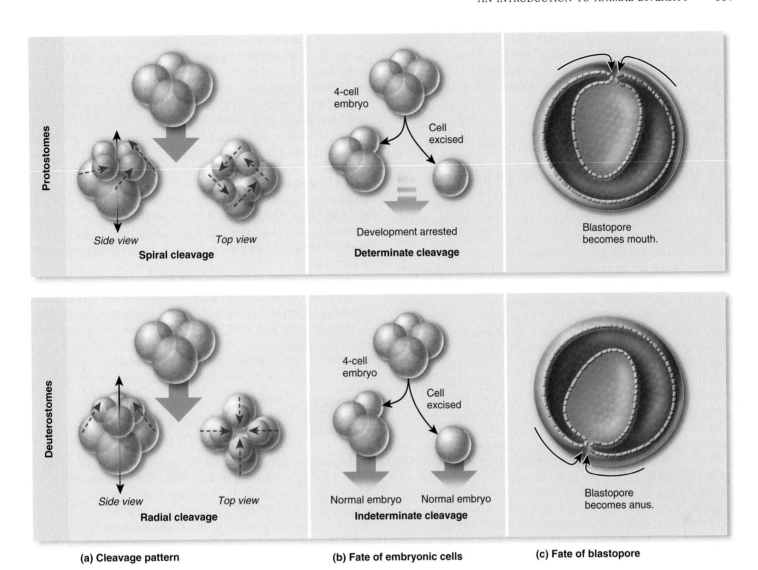

(a) Cleavage pattern   **(b) Fate of embryonic cells**   **(c) Fate of blastopore**

Figure 32.5 **Differences in embryonic development between protostomes and deuterostomes.** (a) Many protostomes have spiral cleavage, and most deuterostomes have radial cleavage. The dashed arrows indicate the direction of cleavage. (b) Protostomes have determinate cleavage, whereas deuterostomes have indeterminate cleavage. (c) In protostomes, the blastopore becomes the mouth. In deuterostomes, the blastopore becomes the anus.

Figure 32.6 **Segmentation.** Annelids, arthropods, and chordates all exhibit segmentation.

beginning to understand the genetic basis for some of these morphological traits. Recent studies have shown that changes in specialization among arthropod body segments can be traced to relatively simple changes in homeotic, or *Hox*, genes.

Genomes & Proteomes Connection

Changes in *Hox* Gene Expression Control Body Segment Specialization

Hox genes, which are present in all animals, are genes that are involved in pattern formation in early embryos. In the 1990s, Michalis Averof and coworkers showed how relatively simple changes in the expression patterns of these genes can account for the large variation in arthropod appendage types. As described in Chapter 19, animals have several *Hox* genes that are expressed in particular regions of the body. Some are expressed in anterior segments; others are expressed in posterior segments (refer back to Figure 19.17). The *Hox* genes are designated with numbers 1 through 13.

Shifts in patterns of expression of *Hox* genes in the embryo along the anteroposterior axis are prominent in evolution. In vertebrates, the transition from one type of vertebra to another, for example, from cervical (neck) to thoracic (chest) vertebrae, is also controlled by particular *Hox* genes (Figure 32.7). The site of the cervical/thoracic boundary appears to be influenced by the *HoxC-6* gene. Differences in its relative position of expression, which occurs prior to vertebrae development, control neck length in vertebrates. In mice, which have a relatively short neck, the expression of *HoxC-6* begins between vertebrae numbers 7 and 8. In chickens and geese, which have longer necks, the expression begins farther back, between vertebrae 14 and

15, or 17 and 18, respectively. The forelimbs also arise at this boundary in all vertebrates. Interestingly, snakes, which essentially have no neck or forelimbs, do not exhibit this boundary, and *HoxC-6* expression occurs immediately behind their heads. This in effect means that snakes got longer by losing their neck and lengthening their chest.

Evolutionary and development biologist Sean Carroll has remarked that it is very satisfying to find that the evolution of body forms and novel structures in two of the most successful and diverse animal groups, arthropods and vertebrates, is shaped by the shifting of *Hox* genes. It also reminds us of one of the basic tenets of diversity—that modern organisms illustrate Darwin's concept of descent with modification. Much of the diversity in animal phyla can be seen as variations on a common theme.

A traditional view of animal phylogeny based on body plans is illustrated in Figure 32.8. This phylogeny was accepted by most biologists for over a century. However, in addition to providing us with new information about the genetic basis of morphological traits, molecular methods are giving us new insights into the relationships between animals.

32.3 Molecular Views of Animal Diversity

Biologists are using molecular techniques to classify animals by comparing similarities in their DNA and ribosomal RNA. An advantage of the molecular approach is that genetic sequences among different species are easier to quantify and compare than are morphological data. The DNA sequence contains four easily identified and mutually exclusive characters: the bases A, T, G, and C (RNA has A, U, G, and C). Contrast this with

Figure 32.7 Relationship between *HoxC-6* gene expression and neck length. In vertebrates, the transition between neck and trunk vertebrae is controlled by the position of the *HoxC-6* gene. In snakes, the expression of this gene is shifted so far forward that a neck does not develop.

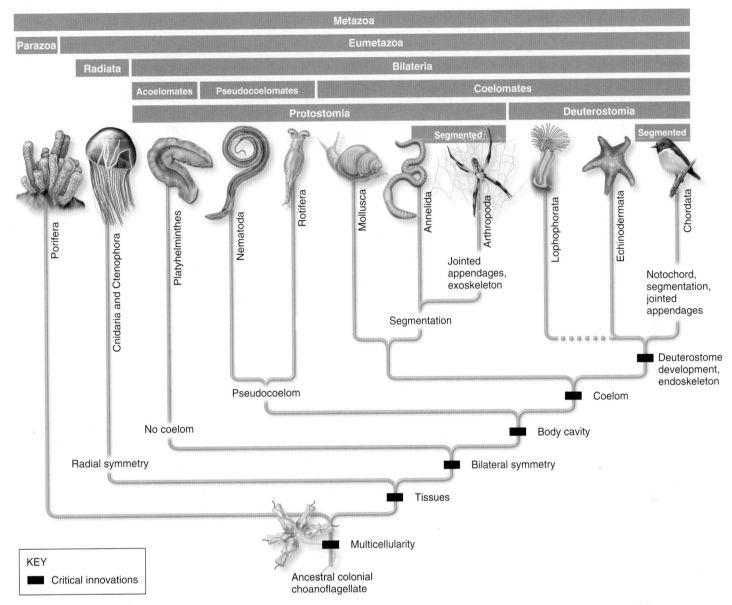

Figure 32.8 **An animal phylogeny based on body plans.** Though there are about 35 different animal phyla, we will focus our discussions here and in the next two chapters on the 11 groups with the greatest numbers of species. The dotted line represents the uncertainty of including the lophophorates with the deuterostomes.

morphological and embryological data, where characters are scored more subjectively, often based on the qualitative assessment of many traits. Traditional analysis left us with many unanswered questions about relationships between animals; we can now use molecular data to try and make sense of these earlier data. Scientists are combining the two types of information to try to arrive at the interpretation that agrees most closely with the available evidence.

To perform molecular analyses, scientists have often focused on comparing sequences of nucleotides in the gene that encodes RNA of the small ribosomal subunit (SSU rRNA) (see Chapter 26). SSU rRNA is a large molecule that contains a lot of genetic information. Furthermore, the molecule is universal in all organisms, and its base pair sequence has changed very

slowly over large periods of time. In addition to the SSU rRNA gene, researchers have also studied *Hox* genes. These genes are important because many branches in the traditional phylogeny are based on early developmental differences, so examination of the genes that regulate these differences should provide insight into the evolution of animal development and to understanding how, when, and why animal body plans diversified.

Phylogenies based on SSU rRNA and *Hox* genes are similar and, in many cases, agree with the structure of the traditional phylogenetic tree. For example, recent analyses of genes have strengthened the view that animals form a single clade. In particular, studies of both small subunit (SSU) rRNA and *Hox* genes provide supportive evidence. We can appreciate this by comparing a portion of the sequence of the SSU rRNA genes

Figure 32.9 Comparison of small subunit (SSU) rRNA gene sequences from three animals and a protist. Note the similarities between the animals, even though they are very different species, and the differences with the protist. This and other comparative studies of gene sequences underscore the likelihood that animals share a common ancestor.

of a sponge, flatworm, and seagull (Figure 32.9), in much the same way as we did in Chapter 12 (refer back to Figure 12.17). The three animal sequences are very similar compared to that of the paramecium (a protist) shown in the figure.

In addition to the clade Metazoa being monophyletic, molecular phylogeny is in agreement with traditional phylogeny on the following features of the animal kingdom.

1. At the earliest stages of evolution, a split occurred between Parazoa and Eumetazoa.
2. There was also an early split between Radiata and Bilateria, with most animal phyla belonging to the Bilateria.

3. Both the echinoderms and the chordates belong to a clade called the Deuterostomia.

However, phylogenies based on molecular data contain some important differences from those based on assessment of body plans. In this section, we examine the similarities and differences between the new molecular phylogeny and the traditional phylogeny. We begin with one of the most influential of the molecular studies, a paper by Anna Marie Aguinaldo and colleagues, which established evidence for a new clade of molting animals, the Ecdysozoa. This study represented a scientific breakthrough because it underscored the value of molecular phylogenies.

FEATURE INVESTIGATION

Aguinaldo and Colleagues Used SSU rRNA to Analyze the Taxonomic Relationships of Arthropods to Other Taxa

In 1997, Anna Marie Aguinaldo, James Lake, and colleagues analyzed the relationships of arthropods to other taxa by sequencing the complete gene that encodes SSU rRNA from a variety of representative taxa (Figure 32.10). Total genomic DNA was isolated using standard techniques and amplified by the polymerase chain reaction (PCR). PCR fragments were then subjected to DNA sequencing, a technique described in Chapter 20. Using approaches that we have described in Chapter 26, the evolutionary relationships among 50 species were determined. The resulting data indicated the existence of a monophyletic clade—the **Ecdysozoa**—containing the nematodes and arthropods.

The hypothesis that nematodes are more closely related to arthropods than previously thought has important ramifications. First, it implies that two well-researched model organisms, *Caenorhabditis elegans* (a nematode) and the fruit fly *Drosophila melanogaster* (an arthropod), are more closely related than had been believed (compare their positions in Figures 32.8 and 32.11). Second, traditional classification assumed that arthropods and annelids were closely related to each other. The new molecular phylogeny suggests otherwise.

Figure 32.10   A molecular animal phylogeny based on sequencing of SSU rRNA.

Concept check: *What are the major differences between molecular and traditional phylogenies?*

GOAL   To determine the evolutionary relationships among many animal species.

KEY MATERIALS   Cellular samples from about 50 animals in different taxa.

	Experimental level	Conceptual level

1   Isolate DNA from animals and subject the DNA to polymerase chain reaction (PCR) to obtain enough material for DNA sequencing. PCR is described in Chapter 20.

For more detail, refer back to Figure 20.6.

The goal of PCR is to amplify a region in the SSU rRNA gene.

2   Sequence the amplified DNA by dideoxy sequencing, also described in Chapter 20.

For more detail, refer back to Figure 20.8.

 CACCGTA

Dideoxy sequencing, in which DNA strands are separated according to their lengths by subjecting them to gel electrophoresis, is used to determine the base sequence of DNA.

3   Compare the DNA sequences and infer phylogenetic relationships using the cladistic approach described in Chapter 26.

The cladistic approach compares traits that are either shared or not shared by different species.

4   **THE DATA**

This process resulted in a large group of DNA sequences that were then analyzed with the use of computer programs. This analysis, along with other studies, suggested a revised phylogeny, shown in Figure 32.11.

5   **CONCLUSION**   The molecular phylogeny agrees with the body plan phylogeny in most areas. It resolves the position of the lophophorates and establishes the Lophotrochozoa and Ecdysozoa as sister groups within the Protostomia.

6   **SOURCE**   Aguinaldo, A.M. et al. 1997. Evidence for a clade of nematodes, arthropods, and other moulting animals. *Nature* 387 (6632):489–493.

Experimental Questions

1. What was the purpose of the study conducted by Aguinaldo and colleagues?

2. What was the major finding of this particular study?

3. What impact does the new view of nematode and arthropod phylogeny have on other areas of research?

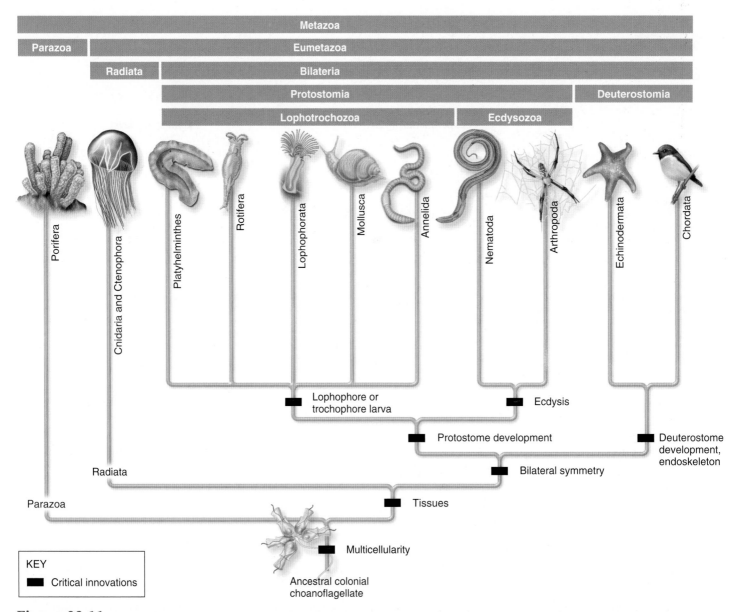

Figure 32.11 A revised animal phylogeny based on molecular data. Compared to Figure 32.8, the main differences are the formation of the Lophotrochozoa and Ecdysozoa clades.

Concept check: *What is the sister group to the deuterostomes in this figure? How is this different from the phylogeny based on the body plan?*

Molecular Phylogeny Has Some Important Differences from Traditional Phylogeny

Recent molecular analyses of evolutionary relationships among animals, including the analysis done by Aguinaldo and colleagues, have revealed two additional key differences between traditional models and newer models (**Figure 32.11**). The two differences concern the division of the protostomes into two separate clades and the division of animals into groups by the presence or absence of a body cavity.

Protostomes: Ecdysozoa and Lophotrochozoa The most important difference between molecular and traditional phylogenies involves relationships among the Bilateria. In the

traditional view of animal phylogeny, the bilaterally symmetrical animals are split into two clades, the Protostomia and Deuterostomia, reflecting two basic modes of embryonic development. If you look back at Figure 32.8, you will see that the various protostome taxa were partitioned into different groups. You will also find that the classification of a group called the Lophophorata under the deuterostomes has been uncertain.

However, recent molecular studies, from James Lake and others, suggest a different grouping. The deuterostomes are still separate, but the protostomes are divided into two major clades: the Ecdysozoa, primarily the nematodes and arthropods, and the **Lophotrochozoa**, which encompasses the mollusks, annelids, and several other phyla (Figure 32.11). When some morphologists looked at their data given this new information,

Figure 32.12 **Ecdysis.** The Ecdysozoa are a clade of animals exhibiting ecdysis, the periodic shedding (molting) and re-formation of the exoskeleton.

Concept check: What are the main members of the Ecdysozoa?

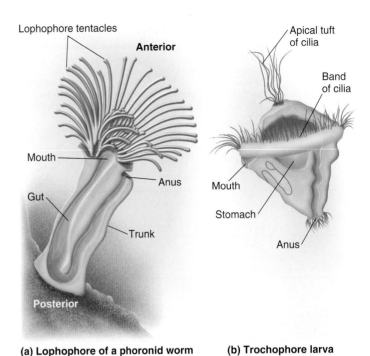

(a) Lophophore of a phoronid worm   **(b) Trochophore larva**

Figure 32.13 **Characteristics of the Lophotrochozoa.** **(a)** A lophophore, a crown of ciliated tentacles, generates a current to bring food particles into the mouth. **(b)** The trochophore (Greek, for wheel-bearer) larval form is found in several animal lineages.

they found there was also morphological support for these new groupings. Let's look at what morphological features make each of these groups unique.

The Ecdysozoa is so named because all of its members secrete a nonliving cuticle, typically an external skeleton (exoskeleton); think of the hard shell of a dragonfly or that of a crab. As these animals grow, the exoskeleton becomes too small, and the animal molts, or breaks out of its old exoskeleton, and secretes a newer, larger one (**Figure 32.12**). This molting process is called ecdysis; hence the name Ecdysozoa. While this group was named for this morphological characteristic, it is strongly supported as a separate clade by molecular evidence such as similarities in DNA.

Although the Lophotrochozoa clade was organized primarily through analysis of molecular data, its name stems from two morphological features seen in organisms of this clade. The "lopho" part is derived from the **lophophore**, a horseshoe-shaped crown of tentacles used for feeding that is present on some members of the group (**Figure 32.13a**). The "trocho" part refers to the **trochophore larva**, a distinct larval stage characterized by having a band of cilia around its middle (**Figure 32.13b**). Trochophore larvae are found in many of the Lophotrochozoa phyla, such as polychaete worms (which are annelids) and mollusks, indicating their similar ancestry.

In traditional phylogeny, much debate surrounded the classification of the Lophophorata, comprised of three minor phyla that possess lophophores—the Bryozoa, Phoronida, and Brachiopoda. Although these taxa exhibited some characteristics of protostomes and some of deuterostomes, as mentioned, they were often classified under Deuterostomia (see Figure 32.8). However,

molecular data support their inclusion within the Lophotrochozoa, along with mollusks and annelids (see Figure 32.11).

Body Cavity   The second important difference between molecular and traditional phylogenies involves the presence or absence of a body cavity. In the traditional view of animal phylogeny, the bilaterally symmetrical animals are divided into those lacking a coelom (acoelomates), those with a pseudocoelom (pseudocoelomates), and those possessing a coelom (coelomates) (see Figure 32.8). The Platyhelminthes, or flatworms, are classified as acoelomate and are thus seen as separate from coelomate phyla. Molecular evidence, however, now suggests that the flatworms should be included with the Lophotrochozoa. In this view, flatworms are not early-diverging acoelomate animals. Rather, they evolved from an ancestor that possessed a coelom but lost it during evolutionary modification. Similarly, molecular-based phylogeny places the rotifers and nematodes, the pseudocoelomate phyla, within the Lophotrochozoa and Ecdysozoa, respectively. Thus, molecular data suggest that the presence or absence of a coelom or pseudocoelom, a distinction traditionally used in the construction of animal phylogenies, may not be a useful way to classify animals.

While molecular analyses have provided key insights regarding animal evolution, a comparison of the molecular and traditional phylogenies reveals more similarities than differences. Most of the major branch points in the phylogenies are in agreement. As a reference, **Table 32.2** summarizes the basic characteristics of the major animal phyla.

In the following two chapters, the discussion of animal phylogeny is based primarily on findings of molecular data.

Table 32.2	Summary of the Basic Characteristics of the Major Animal Phyla											
Feature	Porifera (sponges)	Cnidaria and Ctenophora (hydra, anemones, jellyfish)	Platyhelminthes (flatworms)	Rotifera (rotifers)	Lophophorata (bryozoans and others)	Mollusca (snails, clams, squids)	Annelida (segmented worms)	Nematoda (roundworms)	Arthropoda (insects, arachnids, crustaceans)	Echinodermata (sea stars, sea urchins)	Chordata (vertebrates and others)
Estimated number of species	8,000	11,000	20,000	2,000	4,000+	110,000	15,000	20,000	1,000,000+	7,000+	52,000+
Level of organization	Cellular; lack tissues and organs	Tissue; lack organs	Organs	Organs	Organs	Organs	Organs	Organs	Organs	Organs	Organs
Symmetry	Absent	Radial	Bilateral	Bilateral	Bilateral	Bilateral	Bilateral	Bilateral	Bilateral	Bilateral larvae, radial adults	Bilateral
Cephalization	Absent	Absent	Present	Present	Reduced	Present	Present	Present	Present	Absent	Present
Germ layers	Absent	Two	Three	Three	Three	Three	Three	Three	Three	Three	Three
Body cavity	Absent	Absent	Absent	Pseudocoelom	Coelom	Reduced coelom	Coelom	Pseudocoelom	Reduced coelom	Coelom	Coelom
Segmentation	Absent	Absent	Absent	Absent	Absent	Absent	Present	Absent	Present	Absent	Present (but reduced)

In Chapter 33, we will discuss the diversity of all the phyla in the Parazoa, Radiata, Lophotrochozoa, Ecdysozoa and Deuterostomia, the latter including the phylum Echinodermata and invertebrate members of the phylum Chordata; these are all the animals without a backbone. In Chapter 34, we will turn our attention to the backboned members of the phylum Chordata, including fishes, amphibians, reptiles, birds, and mammals.

 ## Summary of Key Concepts

32.1 Characteristics of Animals

- Animals constitute a very species-rich kingdom. They share a number of key characteristics, including multicellularity, heterotrophic feeding, the possession of nervous and muscle tissues, and sexual reproduction. (Table 32.1)

32.2 Traditional Classification of Animals

- The animal kingdom is monophyletic, meaning that all taxa have evolved from a single common ancestor.
- The traditional classification of animals is based on five morphological and developmental features of animal body plans.
- Biologists hypothesize that animals evolved from a colonial choanoflagellate. (Figure 32.1)
- Animals can be categorized according to the absence of different types of tissues (the Parazoa or sponges) and the presence of tissues (Eumetazoa or all other animals). The Eumetazoa can also be divided according to their type of symmetry, whether radial (Radiata, the cnidarians and ctenophores) or bilateral (Bilateria, all other animals). (Figure 32.2)

- The Radiata have two embryonic cell layers (germ layers) called the endoderm and the ectoderm. The Bilateria develop a third germ layer termed the mesoderm, which develops between the endoderm and the ectoderm. (Figure 32.3)
- Animals can be classified according to the presence or absence of a coelom, or true body cavity. Animals with a coelom are termed coelomates. Animals that possess a pseudocoelom, or coelom that is not completely lined by tissue derived from mesoderm, are called pseudocoelomates. Those animals lacking a fluid-filled body cavity are termed acoelomates. (Figure 32.4)
- Animals are also classified according to patterns of embryonic development. Animals with spiral cleavage are called protostomes, and those exhibiting radial cleavage are considered deuterostomes. In protostomes, the blastopore, or opening of the gut to the outside, becomes the mouth; in deuterostomes, the blastopore becomes the anus. (Figure 32.5)
- Segmentation, the division of the body into identical subunits called segments, is another key feature of the animal body plan. (Figure 32.6)
- Shifts in the pattern of expression of *Hox* genes are prominent in evolution. In vertebrates, the transition from one type of vertebra to another is controlled by certain *Hox* genes. (Figure 32.7)
- Traditional methods of classification based on body plan produce a particular animal phylogeny. (Figure 32.8)

32.3 Molecular Views of Animal Diversity

- New molecular techniques that compare similarities in DNA and ribosomal RNA of animals support the view that all animals share a common ancestor. (Figure 32.9)

- In many cases, phylogenies based on molecular techniques are similar to those of traditional approaches; however, some important differences exist. Recent molecular studies propose a division of the protostomes into two major clades: the Ecdysozoa and the Lophotrochozoa. (Figures 32.10, 32.11)

- The Ecdysozoa is so named because its members secrete a nonliving cuticle, typically an exoskeleton, or external skeleton. Ecdysis is the periodic shedding and re-formation of the exoskeleton. (Figure 32.12)

- The Lophotrochozoa are grouped primarily through analysis of molecular data, but they are distinguished by two morphological features—the lophophore, a crown of tentacles used for feeding, and the trochophore larva, a distinct larval stage. The lophophorates were often classified as deuterostomes, but molecular data support their inclusion within the Lophotrochozoa. (Figure 32.13)

- Molecular evidence suggests that the acoelomate flatworms evolved from a coelomate ancestor and should therefore be classified under the Lophotrochozoa.

- Each animal phylum shows a distinctive set of general characteristics. (Table 32.2)

 ## Assess and Discuss

Test Yourself

1. Which of the following is not a distinguishing characteristic of animals?
   a. the capacity to move at some point in their life cycle
   b. possession of cell walls
   c. multicellularity
   d. heterotrophy
   e. All of the above are characteristics of animals.

2. Which is the correct hierarchy of divisions in the animal kingdom, from most inclusive to least inclusive?
   a. Eumetazoa, Metazoa, Protostomia, Ecdysozoa
   b. Parazoa, Radiata, Lophotrochozoa, Deuterostomia
   c. Metazoa, Eumetazoa, Bilateria, Protostomia
   d. Radiata, Eumetazoa, Deuterostomia, Ecdysozoa
   e. none of the above

3. Bilateral symmetry is strongly correlated with
   a. the ability to move through the environment.
   b. cephalization.
   c. the ability to detect prey.
   d. a and b.
   e. a, b, and c.

4. In triploblastic animals, the inner lining of the digestive tract is derived from
   a. the ectoderm.   d. the pseudocoelom.
   b. the mesoderm.   e. the coelom.
   c. the endoderm.

5. Pseudocoelomates
   a. lack a fluid-filled cavity.
   b. have a fluid-filled cavity that is completely lined with mesoderm.
   c. have a fluid-filled cavity that is partially lined with mesoderm.
   d. have a fluid-filled cavity that is not lined with mesoderm.
   e. have an air-filled cavity that is partially lined with mesoderm.

6. Protostomes and deuterostomes can be classified based on
   a. cleavage pattern.
   b. destiny of the blastopore.
   c. whether the fate of the embryonic cells is fixed early during development.
   d. how the coelom is formed.
   e. all of the above.

7. Indeterminate cleavage is found in
   a. annelids.   d. vertebrates.
   b. mollusks.   e. all of the above.
   c. nematodes.

8. Naturally occurring identical twins are possible only in animals that
   a. have spiral cleavage.
   b. have determinate cleavage.
   c. are protostomes.
   d. have indeterminate cleavage.
   e. a, b, and c

9. Genes involved in the patterning of the body axis, that is, in determining characteristics such as neck length and appendage formation, are called
   a. small subunit (SSU) rRNA genes.
   b. *Hox* genes.
   c. metameric genes.
   d. determinate genes.
   e. none of the above.

10. A major difference between the molecular phylogeny of animals and traditional phylogeny of animals is that
    a. the presence or absence of the mesoderm is not important in molecular phylogeny.
    b. molecular phylogeny suggests that all animals do not share a single common ancestor.
    c. body symmetry, whether radial or bilateral, is not an important determinant in molecular phylogeny.
    d. molecular phylogeny does not include the echinoderms in the deuterostome clade.
    e. molecular phylogeny suggests that the presence or absence of a coelom is not important for classification.

Conceptual Questions

1. The traditional classification is based on what four features of animal body plans?

2. Distinguish between radial and bilateral symmetry.

3. Why might sea urchins be valuable in developmental studies?

Collaborative Questions

1. Discuss the many ways that animals can impact humans, both positively and negatively.

2. Discuss the similarities and differences between the molecular and morphological phylogenies.

Online Resource

www.brookerbiology.com

Stay a step ahead in your studies with animations that bring concepts to life and practice tests to assess your understanding. Your instructor may also recommend the interactive ebook, individualized learning tools, and more.

The Invertebrates

33

Over the next two chapters, we will survey the wondrous array of animal life on Earth (see chapter-opening photo). In this chapter, we examine the **invertebrates**, or animals without a backbone, a category that makes up more than 95% of all animal species. We begin by exploring some of the earliest animal lineages, the Parazoa and Radiata. We then turn to the Lophotrochozoa and Ecdysozoa, the two sister groups of protostomes introduced in Chapter 32. Finally, we will examine the deuterostomes, focusing here on the echinoderms and the invertebrate members of the phylum Chordata.

Figure 33.1 gives a summary of our current understanding of the relationships between these groups. Although the more modern molecular classification outlined in Chapter 32 serves as the basis for this phylogeny and for our discussion of animal lineages, we will not ignore the concept of body plans, because it still can provide clues about how different phyla have evolved. The newer molecular phylogeny is still in its infancy, and many refinements will undoubtedly be made as increasing numbers of genes from more species are sequenced and compared. For this reason, many biologists are not yet ready to totally set aside the older body-plan-based phylogeny.

What is this organism, and how does it feed?

I f you thought the organism shown in the opening photograph was an underwater plant, complete with long leaflike structures and roots, you'd be wrong. This organism is a type of echinoderm, called a feather star, and is related to sea stars. Its long arms catch food particles floating in the ocean current, and tiny tube feet pass these particles into special food gutters that run along the center of each arm and empty into the mouth. The number of arms varies between species and may reach 200. Feather stars can creep along the ocean floor by means of rootlike projections called cirri. While there are about 550 species of feather stars in existence today, some fossil formations are packed with feather star fragments, showing how successful the group was in the past.

The history of animal life on Earth has evolved over hundreds of millions of years. Some scientists suggest that changing environmental conditions, such as a buildup of dissolved oxygen and minerals in the ocean or an increase in atmospheric oxygen, eventually permitted higher metabolic rates and increased the activity of a wide range of animals. Others suggest that with the development of sophisticated locomotor skills, a wide range of predators and prey evolved, leading to an evolutionary arms race in which predators evolved powerful weapons and prey evolved more powerful defenses against them. Such adaptations and counteradaptations would have led to a proliferation of different lifestyles and taxa.

Figure 33.1 **An animal phylogeny.** This phylogenetic tree summarizes our current understanding about the relationships between animal groups.

33.1 Parazoa: Sponges, the First Multicellular Animals

Parazoa

The Parazoa consist of one phylum, Porifera (from the Latin, meaning pore bearers), whose members are commonly referred to as sponges. Sponges are loosely organized and lack true tissues, groups of cells that have a similar structure and function. However, sponges are multicellular and possess several types of cells that perform different functions. Biologists have identified approximately 8,000 species of sponges, the vast majority of which are marine. Sponges range in size from only a few millimeters across to more than 2 m in diameter. The smaller sponges may be radially symmetrical, but most have no apparent symmetry. Some sponges have a low encrusting growth form, whereas others grow tall and erect (**Figure 33.2a**). Although adult sponges are sessile, that is, anchored in place, the larvae are free-swimming.

The body of a sponge looks similar to a vase pierced with small holes or pores (**Figure 33.2b**). Water is drawn through these pores, or ostia (singular, ostium), into a central cavity, the **spongocoel**, and flows out through the large opening at the top called the osculum. The water enters the pores by the beating action of the flagella of the **choanocytes**, or collar cells, that line the spongocoel (**Figure 33.2c**). In the process, the choanocytes trap and eat small particulate matter and tiny plankton. As we noted in Chapter 32, because of striking morphological and molecular similarities between choanocytes and choanoflagellates, a group of modern protists having a single flagellum, scientists think that sponges originated from a common choanoflagellate ancestor.

A layer of flattened epithelial cells similar to those making up the outer layer of other phyla protects the sponge body. In between the choanocytes and the epithelial cells lies a gelatinous, protein-rich matrix called the **mesohyl**. Within this matrix are mobile cells called **amoebocytes** that absorb food from choanocytes, digest it, and carry the nutrients to other cells. Thus, considerable cell-to-cell contact and communication exists in sponges. Sponges are unique among the major animal phyla in using the phagocytic mode of feeding, in common with some protists.

Some amoebocytes can also form tough skeletal fibers that support the body. In many sponges, this skeleton consists of sharp **spicules** formed of calcium carbonate or silica. For

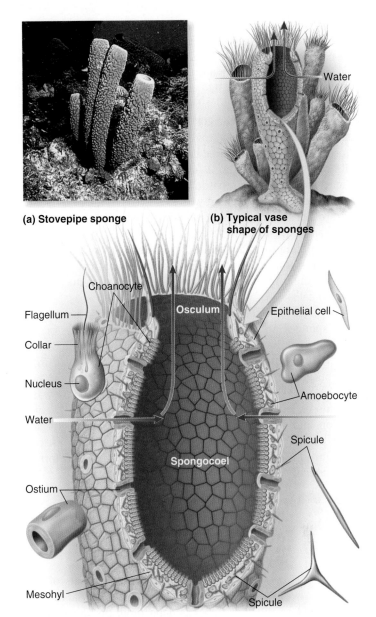

(a) Stovepipe sponge

(b) Typical vase shape of sponges

(c) Cross section of sponge morphology

Figure 33.2 Sponge morphology. **(a)** The stovepipe sponge (*Aplysina archeri*) is a common sponge found on Caribbean reefs. **(b)** Many sponges have a vaselike shape. **(c)** A cross section reveals that sponges are truly multicellular animals, having various cell types but no distinct tissues.

Concept check: If sponges are soft and sessile, why aren't they eaten by other organisms?

example, some deep-ocean species, called glass sponges, are distinguished by needle-like silica spicules that form elaborate lattice-like skeletons. The presence of such tough spicules may help explain why there is not much predation of sponges. Sponge spicules come in a diverse array of shapes and sizes, and they are valuable taxonomic tools by which to distinguish different types of sponges. In a small family of carnivorous sponges, the spicules are sticky and capture small crustaceans. In these sponges, other cells migrate around the immobilized crustaceans and digest them extracellularly. Not all sponges have spicules, however. Others have fibers of a tough protein called **spongin** that lend skeletal support. Spongin skeletons are still commercially harvested and sold as bath sponges. Many species produce toxic defensive chemicals, some of which are thought to have possible antibiotic and anti-inflammatory effects in humans.

Around the turn of the 20th century, biologist Henry V. Wilson made the incredible discovery that if a sponge is dissociated into its individual cells after being passed through a sieve, its cells can reaggregate into a functional sponge within a short time. Wilson concluded that in order to do this, individual cells recognized and reaggregated with other cells of their own kind. Researchers have since discovered that the cells of other multicellular organisms also recognize cells of their own kind and tend to adhere to each other when mixed with other types of cells. For example, in mammals, liver cells recognize and stick better to other liver cells, and brain cells recognize and adhere to brain cells.

Sponges reproduce through both sexual and asexual means. Most sponges are **hermaphrodites** (from the Greek, for the Greek god Hermes and the goddess Aphrodite), individuals that can produce both sperm and eggs. Gametes are derived from amoebocytes or choanocytes. The eggs remain in the mesohyl, and the sperm are released into the water and carried by water currents to fertilize the eggs of neighboring sponges. Zygotes develop into flagellated swimming larvae that eventually settle on a suitable substrate to become sessile adults. In asexual reproduction, a small fragment or bud may detach and form a new sponge.

33.2 Radiata: Jellyfish and Other Radially Symmetrical Animals

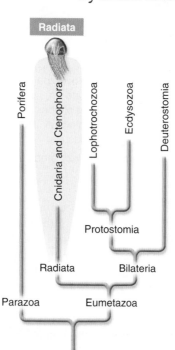

The Radiata consists of two closely related phyla: the Cnidaria (from the Greek *knide*, meaning nettle, and *aria*, meaning related to; pronounced nid-air'-e-ah) and the Ctenophora (from the Greek *ktenos*, meaning comb, and *phora*, meaning bearing; pronounced teen-o-for'-ah). Members of the Radiata phyla, or radiates, are mostly found in marine environments, although a few are freshwater species, such as Hydra. The Cnidaria includes hydra, jellyfish, box jellies, sea anemones, and corals, and the Ctenophora consists of the comb jellies. The diploblastic Radiata have only two embryonic germ layers:

the ectoderm and the endoderm, which give rise to the epidermis and the gastrodermis, respectively. A gelatinous substance called the **mesoglea** connects the two layers. In jellyfish, the mesoglea is enlarged and forms the buoyant, transparent jelly, whereas in coral, the mesoglea is very thin. The Radiata is the first clade with true tissues.

Both cnidarians and ctenophores possess a **gastrovascular cavity**, where extracellular digestion takes place. This feature allows the ingestion of larger food particles and represents a major advance over the sponges, which use only intracellular digestion. Most radiates have tentacles around the mouth that aid in food detection and capture. Radiates also have true nerve cells arranged as a **nerve net** consisting of interconnected neurons with no central control organ. In nerve nets, nerve impulses pass in either direction along a given neuron. In this section, we will provide an overview of the biology and diversity of the cnidarians and ctenophores.

The Cnidarians Have Specialized Stinging Cells

Most cnidarians exist as two different body forms and associated lifestyles: the sessile **polyp** or the motile **medusa** (**Figure 33.3**). For example, jellyfish exist predominantly in the medusa form, and corals exhibit only the polyp form. Many cnidarians, such as *Obelia*, have a life cycle that prominently features both polyp and medusa stages (**Figure 33.4**).

The polyp form has a tubular body with an opening at the oral (top) end that is surrounded by tentacles and functions as both mouth and anus (see Figure 33.3a). The aboral (bottom) end is attached to the substrate. In the 18th century, the Swiss naturalist Abraham Trembley discovered that when a freshwater hydra was cut in two, each part not only survived but could also regenerate the missing half. Polyps exist colonially, as they do in corals, or alone, as in sea anemones.

Corals take dissolved calcium and carbonate ions from seawater and precipitate them as limestone underneath their bodies. In some species, this leads to a buildup of limestone deposits. As each successive generation of polyps dies, the limestone remains in place, and new polyps grow on top. Thus, huge underwater limestone deposits called coral reefs are formed (look ahead to Figure 54.27b). The largest of these is Australia's Great Barrier Reef, which stretches over 2,300 km. Many other extensive coral reefs are known, including the reef system along the Florida Keys, all of which occur in warm water, generally between 20°C and 30°C.

The free-swimming medusa form has an umbrella-shaped body with an opening that serves as both mouth and anus on the concave underside that is surrounded by tentacles (see Figure 33.3b). More mobile medusae possess simple sense organs near the bell margin, including organs of equilibrium called **statocysts** and photosensitive organs known as **ocelli**. When one side of the bell tips upward, the statocysts on that side are stimulated, and muscle contraction is initiated to right the medusa. The ocelli allow medusae to position themselves in particular light levels.

One of the unique and characteristic features of the cnidarians is the existence of stinging cells called **cnidocytes**, which

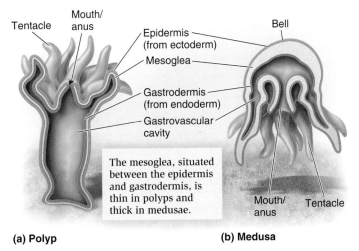

Tentacle
Mouth/anus

Epidermis (from ectoderm)
Mesoglea
Gastrodermis (from endoderm)
Gastrovascular cavity

Bell

The mesoglea, situated between the epidermis and gastrodermis, is thin in polyps and thick in medusae.

Mouth/anus   Tentacle

(a) Polyp   **(b) Medusa**

Figure 33.3 **Polyp and medusa forms of cnidarians.** Both **(a)** polyp and **(b)** medusa forms have two layers of cells, an outer epidermis (from ectoderm) and an inner layer of gastrodermis (from endoderm). In between is a layer of mesoglea, which is thin in polyps, such as corals, and thick in medusae, such as most jellyfish.

function in defense or the capture of prey (**Figure 33.5a**). Cnidocytes contain **nematocysts**, powerful capsules with an inverted coiled and barbed thread. Each cnidocyte has a hairlike trigger called a **cnidocil** on its surface. When the cnidocil is touched or a chemical stimulus is detected, the nematocyst is discharged, and its filament penetrates the prey and injects a small amount of toxin. Small prey are immobilized and passed into the mouth by the tentacles. Alternatively, some nematocyst filaments can be sticky rather than stinging. After discharge, the cnidocyte is absorbed, and a new one grows to replace it. The nematocysts of most cnidarians are not harmful to humans, but those on the

tentacles of the larger jellyfish and the Portuguese man-of-war (**Figure 33.5b**) can be extremely painful and even fatal. Tentacles of the largest jellyfish, *Cyanea arctica*, may be over 40 m long.

Muscles and nerves exist in their simplest forms in cnidarians. Contractile muscle fibers are found in both the epidermis and gastrodermis. Although not true muscles, which only arise from the mesoderm and therefore do not appear in diploblastic animals, these muscle fibers can contract to change the shape of the animal. For example, in the presence of a predator, an anemone can expel water very quickly through its open mouth and shrink down to a very small body form. The muscle fibers work against the fluid contained in the body, which thus acts as a hydrostatic skeleton. The nerve net that conducts signals from sensory nerves to muscle cells allows coordination of simple movements and shape changes.

The phylum Cnidaria consists of four classes—Hydrozoa (small jellyfish, *Obelia* and the Portuguese man-of-war), Scyphozoa (large jellyfish), Anthozoa (sea anemones and corals), and Cubozoa (box jellies). The distinguishing characteristics of these classes are shown in **Table 33.1**.

The Ctenophores Have a Complete Gut

Ctenophores, also known as comb jellies, are a small phylum of fewer than 100 species, all of which are marine and look very much like jellyfish (**Figure 33.6**). They have eight rows of cilia on their surface that resemble combs. The coordinated beating of the cilia, rather than muscular contractions, propels the ctenophores. Averaging about 1–10 cm in length, comb jellies are probably the largest animals to use cilia for locomotion. There are even a few ribbon-like species up to 1 m long.

Comb jellies possess two long tentacles but lack stinging cells. Instead, they have colloblasts, cells on the tentacles that

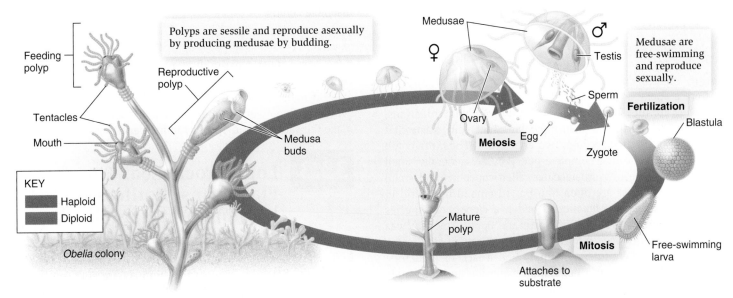

Figure 33.4 **Life cycle of *Obelia*, a colonial cnidarian.** This species exhibits both polyp and medusa stages.

Concept check: What are the dominant life stages of the following types of cnidarians: jellyfish, sea anemone, and Portuguese man-of-war?

(a) Cnidocytes

When triggered, the cnidocyte discharges the nematocyst, which penetrates the prey.

Gastrodermis Mesoglea Epidermis

Sensory cell

Undischarged nematocyst

Cnidocil

Discharged nematocyst

Filament

Stinging cell (cnidocyte)

(b) Portugese man-of-war

Figure 33.5 Specialized stinging cells of cnidarians, called cnidocytes. **(a)** Cnidocytes, which contain stinging capsules called nematocysts, are situated in the tentacles. **(b)** The Portuguese man-of-war (*Physalia physalis*) employs cnidocytes that can be lethal to humans.

Table 33.1	Main Classes and Characteristics of the Cnidaria	
	Class and examples (est. # of species)	**Class characteristics**
	Hydrozoa: *Obelia*, Portuguese man-of-war, *Hydra*, some corals (2,700)	Mostly marine; most have both polyp and medusa stages, with polyp stage often colonial
	Scyphozoa: jellyfish (200)	All marine; medusa stage dominant and large (up to 2 m); reduced polyp stage
	Anthozoa: sea anemones, sea fans, most corals (6,000)	All marine; polyp stage dominant; medusa stage absent; many are colonial
	Cubozoa: box jellies, sea wasps (20)	All marine; medusa stage dominant; box-shaped

Figure 33.6 A ctenophore. Ctenophores are called comb jellies because of the eight rows of cilia on their surfaces that resemble combs.

secrete a sticky substance onto which small prey adhere. The tentacles are then drawn over the mouth. As with cnidarians, digestion occurs in the gastrovascular cavity, but waste and water are eliminated through two anal pores. Thus, the comb jellies possess the first complete gut. Prey are generally small and may include tiny crustaceans called copepods and small fishes. Comb jellies are often transported around the world in ships' ballast water. *Mnemiopsis leidyi*, a ctenophore species native to the Atlantic coast of North and South America, was accidentally introduced into the Black and Caspian seas in the 1980s. With a plentiful food supply and a lack of predators, *Mnemiopsis* underwent a population explosion and ultimately devastated the local fishing industries.

All ctenophores are hermaphroditic, possessing both ovaries and testes, and gametes are shed into the water to eventually form a free-swimming larva that grows into an adult. There

is no polyp stage. Nearly all ctenophores exhibit **bioluminescence**, a phenomenon that results from chemical reactions that give off light rather than heat. Individuals can be particularly evident at night, and ctenophores that wash up onshore can make the sand or mud appear luminescent.

33.3 Lophotrochozoa: The Flatworms, Rotifers, Lophophorates, Mollusks, and Annelids

In the traditional view of animal phylogeny (refer back to Figure 32.8), the bilaterally symmetrical animals are split into those with no coelom (the platyhelminthes), those with a pseudocoelom (the nematodes and rotifers), and those with a coelom (the remaining phyla). However, as we explored in Chapter 32, molecular data suggest a different grouping in which

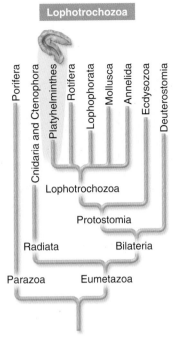

Lophotrochozoa

the deuterostomes are separate, and the protostomes are divided into two major lineages: the Lophotrochozoa and the Ecdysozoa (refer back to Figure 32.11). The Lophotrochozoa are a diverse group that includes taxa that possess either a lophophore (a crown of ciliated tentacles) or a distinct larval stage called a trochophore. In this grouping are seven major Lophotrochozoan phyla: the Platyhelminthes (flatworms), Rotifera (rotifers), Lophophorata (the lophophorates, a group of three phyla), Mollusca (mollusks), and Annelida (segmented worms). In this section, we explore the distinguishing characteristics of these phyla, beginning with some of the simplest lophotrochozoans, the Platyhelminthes.

The Phylum Platyhelminthes Consists of Flatworms with No Coelom

Platyhelminthes (from the Greek *platy*, meaning flat, and *helminth*, meaning worm), or flatworms, lack a specialized respiratory or circulatory system to transport gases. They must obtain oxygen by diffusion, which makes a flattened shape necessary in that no cell can be too far from the surface. Flatworms were among the first animals to develop an active predatory lifestyle. Platyhelminthes, and indeed most animals, are bilaterally symmetrical, with a head bearing sensory appendages (**Figure 33.7**).

The flatworms are also believed to be the first animals to develop three embryonic germ layers—ectoderm, endoderm, and mesoderm—with mesoderm replacing the simpler gelatinous mesoglea of cnidarians. As such, they are said to be triploblastic. The muscles in flatworms, which are derived from mesoderm, are well developed. The development of mesoderm was therefore a critical evolutionary innovation in animals, because it also led to the development of more sophisticated organs. Flatworms are sometimes regarded as the first animals to reach the organ-system level of organization. Because mesoderm fills the body spaces around the gastrovascular cavity, the flatworms are acoelomate—they lack a fluid-filled body cavity in which the gut is suspended.

The digestive system of flatworms is incomplete, with only one opening, which serves as both mouth and anus, as in cnidarians. Most flatworms possess a muscular pharynx that may be extended through the mouth. The pharynx opens to a gastrovascular cavity, where food is digested. In large flatworms, the gastrovascular cavity is branched enough to distribute nutrients to all parts of the body. Any undigested material is egested back out through the pharynx. The incomplete digestive system

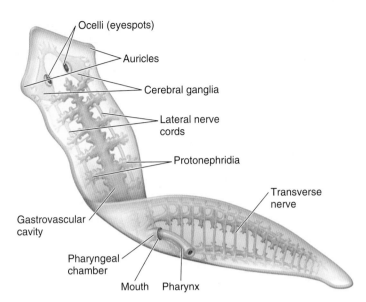

Figure 33.7 Body plan of a flatworm. Flatworm morphology as represented by a planarian, a member of the class Turbellaria.

Concept check: How do flatworms breathe?

of flatworms prevents continuous feeding. Some flatworms are predators, but many species invade other animals as parasites.

Flatworms have a distinct excretory system, consisting of **protonephridia**, two lateral canals with branches capped by **flame cells**. Protonephridia are dead-end tubules lacking internal openings. The flame cells, which are ciliated and waft water through the lateral canals to the outside (look ahead to Figure 49.7), primarily function in maintaining osmotic balance between the flatworm's body and the surrounding fluids. Simple though this system is, its development was key to permitting the movement of animals into freshwater habitats and even moist terrestrial areas.

At the anterior end of some free-living flatworms are light-sensitive eyespots, or ocelli, as well as chemoreceptive and sensory cells that are concentrated in organs called auricles. A pair of **cerebral ganglia** receives input from photoreceptors in eyespots and sensory cells. From the ganglia, a pair of lateral nerve cords running the length of the body allows rapid movement of information from anterior to posterior. In addition, transverse nerves form a nerve net on the ventral surface similar to that of cnidarians. Thus, flatworms retain the cnidarian-style nervous system, while possessing the beginnings of the more centralized type of nervous system (with one-way fast conduction nerves) seen throughout much of the rest of the animal kingdom.

In all the Platyhelminthes, reproduction is either sexual or asexual. Most species are hermaphroditic but do not fertilize their own eggs. Flatworms can also reproduce asexually by splitting into two parts, with each half regenerating the missing fragment.

The four classes of flatworms are the Turbellaria, Monogenea, Trematoda (flukes), and Cestoda (tapeworms) (**Table 33.2**). Turbellarians are the only free-living class of flatworms and are widespread in lakes, ponds, and marine environments. Monogeneans are relatively simple external parasites of fish

Table 33.2	Main Classes and Characteristics of Platyhelminthes	
	Class and examples (est. # of species)	**Class characteristics**
	Turbellaria: planarian (3,000)	Mostly marine; free-living flatworms; predatory or scavengers
	Monogenea: fish flukes (1,000)	Marine and freshwater; usually external parasites of fish; simple life cycle (no intermediate host)
	Trematoda: flukes (11,000)	Internal parasites of vertebrates; complex life cycle with several intermediate hosts
	Cestoda: tapeworms (5,000)	Internal parasites of vertebrates; complex life cycle, usually with one intermediate host; no digestive system; nutrients absorbed across epidermis

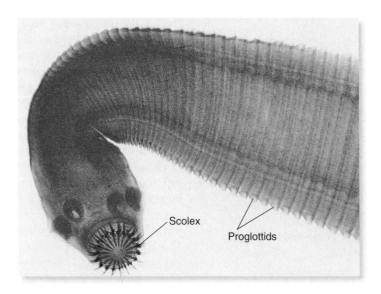

Figure 33.8 A tapeworm, *Taenia pisiformis*, a member of the class Cestoda. Note the tiny hooks and suckers that make up the scolex. Each segment is a proglottid, replete with eggs.

with just one species of host. Both trematodes and cestodes are internally parasitic and therefore are of great medical and veterinary importance. They possess a variety of organs of attachment, such as hooks and suckers, that enable them to remain embedded within their hosts. For example, cestodes attach to their host by means of an organ at the head end called a scolex (**Figure 33.8**). They have no mouth or gastrovascular cavity and absorb nutrients across the body surface. Behind the scolex is a long ribbon of similar segments called proglottids. These are essentially segments of sex organs that develop thousands of eggs. The proglottids are continually shed in the host's feces. Cestodes often require two separate vertebrate host species, such as pigs or cattle, to begin their life cycle and humans to complete their development. Many tapeworms can live inside humans who consume undercooked, infected meat—hence the value of thoroughly cooking meat.

The life history of trematodes is even more complex than that of cestodes, involving multiple hosts. The first host, called the intermediate host, is usually a mollusk, and the final host, or definitive host, is usually a vertebrate, but often a second or even a third intermediate host is involved. In the case of the Chinese liver fluke (*Clonorchis sinensis*) (**Figure 33.9**), (1) the adult parasite lives and reproduces in the definitive host, a human. (2) The resultant embryos are called miracidia. They are encapsulated to form eggs and pass from the host via the feces. (3) An intermediate host, such as a snail, eats the eggs. The miracidia are released and transform into sporocysts. (4) The sporocysts asexually produce more sporocysts, which are called rediae. (5) The rediae reproduce asexually to produce cercariae. Cercariae bore their way out of the snail and (6) infect their second intermediate host, fishes, by entering via the gills. (At this stage, some trematodes infect their definitive hosts directly by boring into their feet when in water.) Here, the cercariae develop into metacercarial cysts (juvenile flukes) and

lodge in fish muscle, which the definitive host will eat, allowing the cycle to continue. In the definitive host, the cyst protects the metacercaria from the host's gastric juices. From the small intestine, the metacercariae travel to the liver and grow into adult flukes, and the life cycle begins anew. The life cycle of a trematode can involve at least seven stages: adult, egg (encapsulated miracidium), miracidium, sporocyst, rediae, cercaria, and metacercaria. Because of the low probability of each larva reaching a suitable host, trematodes must produce large numbers of offspring to ensure that some survive.

Blood flukes, genus *Schistosoma*, are the most common parasitic trematodes infecting humans; they cause the disease known as schistosomiasis. Over 200 million people worldwide, primarily in tropical Asia, Africa, and South America, are infected with schistosomiasis, also called Bilharzia. The inch-long adult flukes can live for years in human hosts, and the release of eggs may cause chronic inflammation and blockage in many organs. Untreated schistosomiasis can lead to severe damage to the liver, intestines, and lungs and can eventually lead to death. Sewage treatment and access to clean water can greatly reduce infection rates.

Members of the Phylum Rotifera Have a Pseudocoelom and a Ciliated Crown

Members of the phylum Rotifera (from the Latin *rota*, meaning wheel, and *fera*, meaning to bear) get their name from their ciliated crown, or **corona**, which, when beating, looks similar to a rotating wheel (**Figure 33.10**). Most rotifers are microscopic animals, usually less than 1 mm long, and some have beautiful colors. There are about 2,000 species of rotifers, most of which inhabit fresh water, with a few marine and terrestrial species. Most often they are bottom-dwelling organisms, living on the pond floor or along lakeside vegetation.

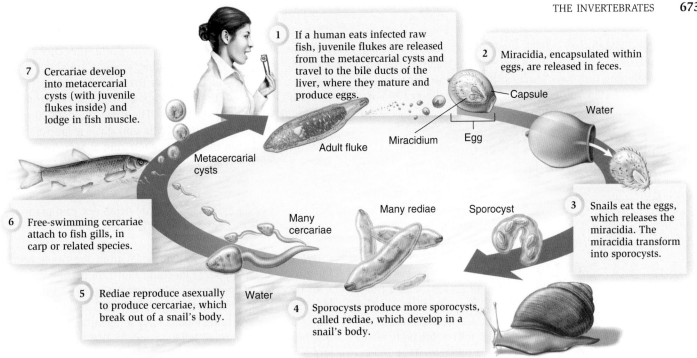

Figure 33.9 **The complete life cycle of a trematode.** As an example, this figure shows the life cycle of the Chinese liver fluke (*Clonorchis sinensis*).

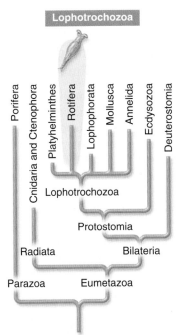

Rotifers have an alimentary canal, a digestive tract with a separate mouth and anus, which means they can feed continuously. The corona creates water currents that propel the animal through the water and that waft small planktonic organisms or decomposing organic material toward the mouth. The mouth opens into a circular, muscular pharynx called a **mastax**, which has jaws for grasping and chewing. The mastax, which in some species can protrude through the mouth to seize small prey, is a structure unique to rotifers. The body of the rotifer bears a jointed foot with one to four toes. **Pedal glands** in the foot secrete a sticky substance that aids in attachment to the substrate. The internal organs lie within a pseudocoelom, a fluid-filled body cavity that is not completely lined with mesoderm. The pseudocoelom serves as a hydrostatic skeleton and as a medium for the internal transport of nutrients and wastes. Rotifers also have a pair of protonephridia with flame bulbs that collect excretory and digestive waste and drain into a cloacal bladder, which passes waste to the anus. The nervous system consists of nerves that extend from the sensory organs, especially the eyespots and some bristles on the corona, to the brain.

Figure 33.10 **Body plan of a common rotifer, *Philodina* genus.**

Reproduction in rotifers is unique. In some species, unfertilized diploid eggs that have not undergone meiotic division develop into females through a process known as **parthenogenesis**. In other species, some unfertilized eggs develop into females, whereas others develop into degenerate males that cannot feed and live only long enough to produce and release sperm that fertilize the females. The resultant fertilized eggs form zygotes, which have a thick shell and can survive for long periods of harsh conditions, such as if a water supply dries up, before developing into new females. Because the tiny zygotes are easily transported, rotifers show up in the smallest of aquatic environments, such as roof gutters or birdbaths.

The Lophophorata Includes Three Closely Related Phyla: Phoronida, Bryozoa, and Brachiopoda

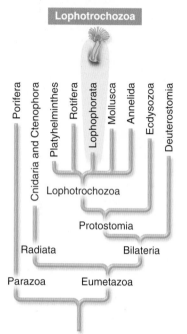

The lophophorates consist of three distinct phyla: the Phoronida, the Bryozoa, and the Brachiopoda. They all possess a lophophore, a ciliary feeding device (refer back to Figure 32.13a), and a true coelom (refer back to Figure 32.4a). The lophophore is a circular fold of the body wall bearing tentacles that draw water toward the mouth. Because a thin extension of the coelom penetrates each tentacle, the tentacles also serve as a respiratory device. Gases diffuse across the tentacles and into or out of the coelomic fluid and are carried throughout the body. All three phyla have a U-shaped alimentary canal, with the anus located near the mouth but outside of the lophophore.

Phylum Phoronida Members of the Phoronida (from the Greek *phoros*, meaning bearer, and the Latin *nidus*, meaning nest) are elongated, tube-dwelling marine worms ranging in size from 1 mm to 50 cm in length. They live in a tough, leather-like chitinous tube that they secrete; this tube is often buried in the ground so only the lophophore sticks out (**Figure 33.11a**). The lophophore can be retracted quickly in the presence of danger. Only about 15 species of phoronids are found worldwide.

Phylum Bryozoa The bryozoans (from the Greek *bryon*, meaning moss, and *zoon*, meaning animal) are small colonial animals, most of which are less than 0.5 mm long, that can be found encrusted on rocks in shallow aquatic environments. They look very much like plants. There are about 4,000 species, many of which encrust boat hulls and have to be scraped off periodically. Within the colony, each animal secretes and lives inside a nonliving case called a zoecium (**Figure 33.11b**). The walls of the zoecium may be composed of chitin or calcium carbonate. For this reason, bryozoans have been important reef-builders. They date back to the Ordovician era, and many fossil forms have been discovered and identified.

Phylum Brachiopoda Brachiopods (from the Greek *brachio*, meaning arm, and *podos*, meaning foot) are marine organisms with two shell halves, much like clams (**Figure 33.11c**). Unlike bivalve mollusks, however, which have a left and right valve (side) of the shell, brachiopods have a dorsal and ventral valve. Brachiopods are bottom-dwelling species that attach to the substrate via a muscular pedicle. Although they are now a relatively small group, with about 300 living species, brachiopods flourished in the Paleozoic and Mesozoic eras—about 30,000 fossil species have been identified. Some of these fossil forms tell of organisms that reached 30 cm in length, although their modern relatives are only 0.5–8.0 cm long.

The Mollusca Is a Large Phylum Containing Snails, Slugs, Clams, Oysters, Octopuses, and Squids

Mollusks (from the Latin *mollis*, meaning soft) constitute a very large phylum, with over 100,000 living species, including organisms as diverse as snails, clams and oysters, octopuses and squid, and chitons. They are an ancient group, as evidenced by the classification of about 35,000 fossil species. Mollusks have a considerable economic, aesthetic, and ecological importance to humans. Many serve as sources of food, including scallops, oysters, clams, and squids. A significant industry involves the farming of oysters to produce cultured pearls, and rare and beautiful mollusk shells are extremely valuable to collectors. Snails and slugs can damage vegetables and ornamental plants, and boring mollusks can penetrate wooden ships and wharfs. Mollusks are

(a) A phoronid worm

(b) A bryozoan

(c) A brachiopod, the northern lamp shell

Figure 33.11 **Lophophorates.** **(a)** A phoronid worm (*Phoronis californica*) buried in the sand with the lophophore extended. **(b)** Bryozoans are colonial lophophorates that reside on a nonliving case called a zoecium. **(c)** Brachiopods, including this northern lamp shell (*Terebratulina septentrionalis*), have dorsal and ventral shells.

Concept check: *What are the two main functions of the lophophore?*

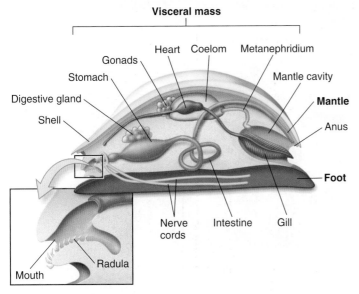

Figure 33.12 The mollusk body plan. The generalized body plan of a mollusk includes the characteristic foot, mantle, and visceral mass.

Concept check: Do molluscan hearts pump blood?

intermediate hosts to many parasites, and several exotic species have become serious pests. For example, populations of the zebra mussel (*Dreissena polymorpha*) were introduced into North America from Asia via ballast water from transoceanic ships. Since their introduction, they have spread rapidly throughout the Great Lakes and an increasing number of inland waterways, adversely impacting native organisms and clogging water intake valves to municipal water-treatment plants around the lakes.

One common feature of the mollusks is their soft body, which exists, in many species, under a protective external shell. Most mollusks are marine, although some have colonized fresh water. Many snails and slugs have even moved onto land, but they survive only in humid areas and where the calcium necessary for shell formation is abundant in the soil. The ability to colonize freshwater and terrestrial habitats has led to a diversification of mollusk body plans. Thus, we again see how organismal diversity is related to environmental diversity.

While great variation in morphology occurs between classes, mollusks have a basic body plan consisting of three parts (**Figure 33.12**). A muscular **foot** is usually used for movement, and a **visceral mass** containing the internal organs rests atop the foot. The **mantle**, a fold of skin draped over the visceral mass, secretes a shell in those species that form shells. The mantle often extends beyond the visceral mass, creating a chamber called the **mantle cavity**, which houses delicate **gills**, filamentous organs that are specialized for gas exchange. A continuous current of water, often induced by cilia present on the gills or by muscular pumping, flushes out the wastes from the mantle cavity and brings in new oxygen-rich water.

Mollusks are coelomate organisms, but the coelom is confined to a small area around the heart. The mollusks' organs are supplied with oxygen and nutrients via a circulatory system. Mollusks have an **open circulatory system** with a heart that pumps body fluid called hemolymph through vessels and into sinuses. Sinuses are the open, fluid-filled cavities between their internal organs. The organs and tissues are therefore continually bathed in hemolymph. From these sinuses, the hemolymph drains into vessels that take it to the gills and then back to the heart. Organs called **metanephridia** extract nitrogenous and other wastes. Metanephridia have ciliated funnel-like openings inside the coelom connected to ducts that lead to the exterior mantle cavity. The pores from the metanephridia discharge wastes into this cavity. The anus also opens into the mantle cavity. The metanephridial ducts may also serve to discharge sperm or eggs from the gonads. The nervous system varies from simple ganglia and nerve chords in most species to much larger brains and sophisticated organs of touch, smell, taste, and vision in octopuses.

The mollusk's mouth may contain a **radula**, a unique, protrusible, tonguelike organ that has many teeth and is used to eat plants, scrape food particles off rocks, or, if the mollusk is predatory, bore into shells of other species and tear flesh. In the cone shells (genus *Conus*), the radula is reduced to a few poison-injecting teeth on the end of a long proboscis that is cast about in search of prey, such as a worm or even a fish. Some Indo-Pacific cone shell species produce a neuromuscular toxin that can kill humans. Other mollusks, particularly bivalves, have lost their radula and are suspension feeders that filter water brought in by ciliary currents.

Most shells are complex three-layered structures secreted by the mantle that continue to grow as the mollusk grows. Shell growth is often seasonal, resulting in distinct growth lines on the shell, much the same as tree rings (**Figure 33.13a**). Using shell growth patterns, biologists have discovered some bivalves that are over 100 years old. The innermost layer of the shells of oysters, mussels, abalone, and other mollusks is a smooth, iridescent lining called nacre, which is commonly known as mother-of-pearl and is often collected from abalone shells for jewelry. Actual pearl production in mollusks, primarily oysters, occurs when a foreign object, such as a grain of sand, becomes lodged between the shell and the mantle, and layers of nacre are laid down around it to reduce the irritation.

Most mollusks have separate sexes, although some are hermaphroditic. Gametes are usually released into the water, where they mix and fertilization occurs. In some snails, however, fertilization is internal, with the male inserting sperm directly into the female. Internal fertilization was a key evolutionary development, enabling some snails to colonize land, and can be considered a critical innovation that fostered extensive adaptive radiation. In many species, reproduction involves the production of a trochophore larva that develops into a **veliger**, a free-swimming larva that has a rudimentary foot, shell, and mantle.

Of the eight molluscan classes, the four most common are the Polyplacophora (chitons), Gastropoda (snails and slugs), Bivalvia (clams and mussels), and Cephalopoda (octopuses, squids, and nautiluses) (**Table 33.3**). Chitons are marine mollusks with a shell composed of eight separate plates (**Figure 33.13b**). Chitons are common in the intertidal, and they creep along when covered by the tide. Feeding occurs by scraping algae off rock surfaces. When the tide recedes, the muscular foot holds the chiton tight to the rock surface, preventing desiccation. The class Gastropoda (from the Greek *gaster*, meaning stomach, and *podos*, meaning foot) is the largest group of mollusks and encompasses about 75,000 living species, including snails, periwinkles, limpets, and other shelled members. The class also includes species such as slugs and nudibranchs, whose shells have been greatly reduced or completely lost during their evolution (**Figure 33.13c**). Most gastropods are marine

Table 33.3 — Main Classes and Characteristics of Mollusks

	Class and examples (est. # of species)	Class characteristics
	Polyplacophora: chitons (860)	Marine; eight-plated shell
	Gastropoda: snails, slugs, nudibranchs (75,000)	Marine, freshwater, or terrestrial; most with coiled shell, but shell absent in slugs and nudibranchs; radula present
	Bivalvia: clams, mussels, oysters (30,000)	Marine or freshwater; shell with two halves or valves; primarily filter feeders with siphons
	Cephalopoda: octopuses, squids, nautiluses (780)	Marine; predatory, with tentacles around mouth, often with suckers; shell often absent or reduced; closed circulatory system; jet propulsion via siphon

(a) A Quahog clam, class Bivalvia

(b) A chiton, class Polyplacophora

Figure 33.13 Mollusks. **(a)** A bivalve shell, class Bivalvia, with growth rings. Quahog clams (*Mercenaria mercenaria*) can live over 20 years. **(b)** A chiton (*Tonicella lineata*), a polyplacophoran with a shell made up of eight separate plates. **(c)** A nudibranch (*Phyllidia ocellata*). The nudibranchs are a gastropod subclass whose members have lost their shell altogether. **(d)** The highly poisonous blue-ringed octopus (*Hapalochlaena lunulata*), a cephalopod.

(c) A sea slug, class Gastropoda

(d) An octopus, class Cephalopoda

or freshwater species, but some species, including snails and slugs, have also colonized land. Most gastropods are slow-moving animals that are weighed down by their shell. Unlike bivalves, gastropods have a one-piece shell, into which the animal can withdraw to escape predators.

The 780 species of Cephalopoda (from the Greek *kephale*, meaning head, and *podos*, meaning foot) are the most morphologically complex of the mollusks and indeed among the most complex of all invertebrates. Most are fast-swimming marine predators that range from organisms just a few centimeters in size to the colossal squid (*Mesonychoteuthis hamiltoni*), which is known to reach over 13 m in length and 495 kg (1,091 lbs) in weight. A cephalopod's mouth is surrounded by many long arms commonly armed with suckers. Octopuses have 8 arms with suckers, and squids and cuttlefish have 10 arms—8 with suckers and 2 long tentacles with suckers limited to their ends. Nautiluses have from 60 to 90 tentacles around the mouth. All cephalopods have a beaklike jaw that allows them to bite their prey, and some, such as the blue-ringed octopus (*Hapalochlaena lunulata*), deliver a deadly poison through their saliva (**Figure 33.13d**). Only one group, the nautiluses, has retained its external shell. In octopuses, the shell is not present, and in squid and cuttlefish, it is greatly reduced and internal. However, the fossil record is full of shelled cephalopods, called ammonites, some of which were as big as truck tires (**Figure 33.14**). They became extinct at the end of the Cretaceous period, although the reasons for this are not well understood.

The foot of some cephalopods has become modified into a muscular siphon. Water drawn into the mantle cavity is quickly expelled through the siphon, propelling the organisms forward or backward in a kind of jet propulsion. Such vigorous movement requires powerful muscles and a very efficient circulatory system to deliver oxygen and nutrients to the muscles. Cephalopods are the only mollusks with a **closed circulatory system**, in which blood flows throughout an animal entirely within a series of vessels. One of the advantages of this type of system is that the heart can pump blood through the tissues rapidly. The blood of cephalopods contains the copper-rich protein hemocyanin for transporting oxygen. Less efficient than the iron-rich hemoglobin of vertebrates, hemocyanin gives the blood a blue color.

The nautiluses are impeded by a coiled, chambered shell and so do not move as fast as the jet-propelled squids and octopuses (**Figure 33.15**). As it grows, the nautilus secretes a new chamber and seals off the old one with a **septum** (plural, septa). The older chambers are gas filled and act as buoyancy chambers. A thin strip of living tissue called the siphuncle removes liquid from the old chamber and replaces it with gas. The gas pressure within the chambers is only 1 atmosphere, despite the fact that nautiluses may be swimming at 400 m depths at a pressure of about 40 atmospheres. The shell's structure is strong enough to withstand this amount of pressure differential.

Cephalopods have a well-developed nervous system and brain that support their active lifestyle. Their sense organs, especially their eyes, are also very well developed. Many cephalopods (with the exception of nautiluses) have an ink sac

Figure 33.14 **A fossil ammonite.** These shelled cephalopods were abundant in the Cretaceous period.

that contains the pigment melanin; the sac can be emptied to provide a "smokescreen" to confuse predators. In many species, melanin is also distributed in special pigment cells in the skin, which allows for color changes. Octopuses often change color when alarmed or during courtship, and they can rapidly change color to blend in with their background and escape detection. The central nervous system of the octopus is among the most complex in the invertebrate world. Behavioral biologists have demonstrated that octopuses can behave in sophisticated ways, and scientists are currently debating to what degree they are capable of learning by observation.

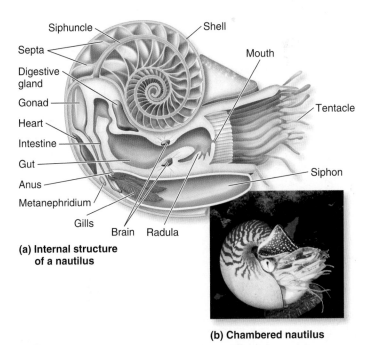

(a) Internal structure of a nautilus

(b) Chambered nautilus

Figure 33.15 **The nautilus.** (a) A longitudinal section of a nautilus, showing the coiled shell with many chambers. The animal secretes a new chamber each year and lives only in the new one. (b) The chambered nautilus (*Nautilus pompilius*).

FEATURE INVESTIGATION

Fiorito and Scotto's Experiments Showed Invertebrates Can Exhibit Sophisticated Observational Learning Behavior

We tend to think of the ability to learn from others as a vertebrate phenomenon, especially among species that live in social groups. However, in 1992, Italian researchers Graziano Fiorito and Pietro Scotto demonstrated that octopuses can learn by observing the behavior of other octopuses (Figure 33.16). This was a surprising finding, in part because *Octopus vulgaris*, the species they studied, lives a solitary existence for most of its life.

In their experiments, octopuses were trained to attack either a red ball or a white ball by use of a reward (a small piece of fish placed behind the ball that it could not see) and a punishment (a small electric shock for choosing the wrong ball). This type of learning is called classical conditioning (see Chapter 55). Because octopuses are color blind, they must distinguish between the relative brightness of the balls. Octopuses were considered to be trained when they made no mistakes in five trials. Observer octopuses in adjacent tanks were then allowed to watch the trained octopuses attacking the balls. In the third part of the experiment, the observer octopuses were themselves tested. In these cases, observers nearly always attacked the same color ball as they had observed the demonstrators attacking. In addition, learning by observation was achieved more quickly than the original training. This remarkable behavior is considered by some to be the precursor to more complex forms of learning, including problem solving.

Figure 33.16 Observational learning in octopuses.

HYPOTHESIS Octopuses can learn by observing another's behavior.

STUDY LOCATION Laboratory setting with *Octopus vulgaris* collected from the Bay of Naples, Italy.

	Experimental level	Conceptual level
1 Train 2 groups of octopuses, one to attack white balls, one to attack red. These are called the demonstrator octopuses.	Reward choice of correct ball (with fish) and punish choice of incorrect ball (with electric shock). Training is complete when octopus makes no "mistakes" in 5 trials.	Conditions a demonstrator octopus to attack a particular color of ball.
2 In an adjacent tank, allow observer octopus to watch trained demonstrator octopus.		Observer octopus may be learning the correct ball to attack by watching the demonstrator octopus.
3 Drop balls into the tank of the observer octopus. Test the observer octopus to see if it makes the same decisions as the demonstrator octopus.		If the observer octopus is learning from the demonstrator octopus, the observer octopus should attack the ball of the same color as the demonstrator octopus was trained to attack.

4 THE DATA

Participant	Color of ball chosen in 5 trials*	
	Red	**White**
Observers (watched demonstrator attack red)	4.31	0.31
Observers (watched demonstrator attack white)	0.40	4.10
Untrained (did not watch demonstrations)	2.11	1.94

*Average of 5 trials; data do not always sum to 5, because some trials resulted in no balls being chosen.

5 CONCLUSION Invertebrate animals are capable of learning from watching other individuals behave, in much the same way as vertebrate species learn from watching others.

6 SOURCE Fiorito, G., and Scotto, P. 1992. Observational learning in *Octopus vulgaris. Science* 256:545–547.

Experimental Questions

1. What was the hypothesis tested by Fiorito and Scotto?

2. What were the results of the experiment? Did these results support the hypothesis?

3. What is the significance of performing the experiment on both trained and untrained octopuses?

The Phylum Annelida Consists of the Segmented Worms

If you look at an earthworm, you will see little rings all down its body. Indeed, the phylum name Annelida is derived from the Latin *annulus*, meaning little ring. Each ring is a distinct segment of the annelid's body, with each segment separated from the one in front and the one behind by septa (**Figure 33.17**). Segmentation, the division of the body into compartments that are often very similar to each other, is a critical evolutionary innovation in the annelids and confers at least three major advantages.

First, many components of the body are repeated in each segment, including blood vessels, nerves, and excretory and reproductive organs. Excretion is accomplished by metanephridia, paired excretory organs in every segment that extract waste from the blood and coelomic fluid, emptying it to the exterior via pores in the skin (look ahead to Figure 49.8). If the excretory organs in one segment fail, the organs of another segment will still function.

Second, annelids possess a fluid-filled coelom that acts as a hydrostatic skeleton. In unsegmented coelomate animals, muscle contractions can distort the entire body during movement. However, such distortion is minimized in segmented animals, which allows for more effective locomotion over solid surfaces. In an earthworm, when the circular muscles around a segment contract against the hydrostatic skeleton, that segment becomes elongated. When the longitudinal muscles contract, the segment becomes compact. Waves of muscular contraction ripple down the segments, which elongate or contract independently.

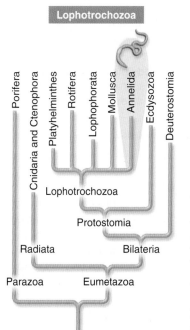

Third, segmentation also permits specialization of some segments, especially at the annelid's anterior end. Therefore, although segments are often similar, they are not identical. This contrasts with tapeworms where most segments are identical. As a rule, as we move from more simple to more complex animals, segments become more specialized. Annelids have a relatively sophisticated nervous system involving a pair of cerebral ganglia that connect to a subpharyngeal ganglion (Figure 33.17). From there, a large ventral nerve cord runs down the entire length of the body. The ventral nerve chord is unusual because it contains a few very large nerve cells called **giant axons** that facilitate high-speed nerve conduction and rapid responses to stimuli. Such axons are found in other invertebrates, the best known of which is the squid.

Annelids have a double transport system. Both the circulatory system and the coelomic fluid carry nutrients, wastes, and respiratory gases, to some degree. The circulatory system

Figure 33.17 **The segmented body plan of an annelid, as illustrated by an earthworm.** The segmented nature of the worm is apparent internally as well as externally. Individual segments are separated by septa.

Concept check: *What are some of the advantages of segmentation?*

is usually closed, with dorsal and ventral vessels connected by five pairs of pumping vessels that serve as muscular hearts. The blood of most annelid species contains the respiratory pigment hemoglobin. Respiration occurs directly through the permeable skin surface, which restricts annelids to moist environments. The digestive system is complete and unsegmented, with many specialized regions: mouth, pharynx, esophagus, crop, gizzard, intestine, and anus. Sexual reproduction involves two individuals, often of separate sexes, but sometimes hermaphrodites, which exchange sperm via internal fertilization. In some species, asexual reproduction by fission occurs, in which the posterior part of the body breaks off and forms a new individual.

Annelids are a large phylum with about 15,000 described species. Its members include marine polychaete worms, the familiar earthworm, and leeches. They range in size from less than 1 mm to enormous Australian earthworms that can reach a size of 3 m. All annelids except the leeches have chitinous bristles, called **setae**, on each segment. In one class, the

polychaetes, these are situated on fleshy, footlike **parapodia** (from the Greek, meaning almost feet) that are pushed into the substrate to provide traction during movement. Many annelid species burrow into soil or into muddy marine sediments and extract nutrients from ingested soil or mud. Some annelids also feed on dead or living vegetation, whereas others are predatory or parasitic.

The phylum Annelida consists of two main classes: the Polychaeta and Clitellata (**Table 33.4**). Members of the Polychaeta have a pair of parapodia on every segment. Members of the Clitellata do not have parapodia but possess a clitellum, a glandular region of the body that has a role in reproduction. There are two main subclasses within Clitellata: the Oligochaeta, the earthworms, and the Hirudinea, the leeches.

Class Polychaeta With over 10,000 species, the Polychaeta is the most species-rich class of the annelids. Many polychaetes are brightly colored, and all have many long setae bristling out of their body (polychaete is from the Greek, meaning many bristles). Most of them are marine organisms, living in burrows in the mud or sand, or in rock crevices, and they are often abundant in the intertidal mudflats. They are important prey for predators such as fishes and crustaceans. The polychaete head is well developed and, in predatory species, may exhibit powerful jaws. Some species are filter feeders and have a crown of tentacles that sticks up out of the mud while the bulk of the worm remains hidden.

Class Clitellata, Subclass Oligochaeta The Oligochaeta (from the Greek, meaning few bristles) includes the common earthworms and many species of freshwater worms. Earthworms play a unique and beneficial role in conditioning the soil, primarily due to the effects of their burrows and castings. Earthworms ingest soil and leaf tissue to extract nutrients and in the process create burrows in the Earth. As plant material and soil pass through the earthworm's digestive system, it is finely ground in

Table 33.4	Main Classes and Characteristics of Annelids	
	Class and examples (est. # of species)	**Class characteristics**
	Polychaeta: marine worms (10,000)	Well-developed head; usually free-living; parapodia present; many setae
	Clitellata, subclass Oligochaeta: terrestrial and freshwater worms such as earthworms (3,500)	Less-developed head; fewer setae; no parapodia
	Clitellata, subclass Hirudinea: leeches (630)	Mostly ectoparasites; suckers present at both ends; flattened body; reduced coelom; no setae

the gizzard into smaller fragments. Once excreted, this material—called castings—enriches the soil. Because a worm can eat its own weight in soil every day, worm castings on the soil surface can be extensive. The biologist Charles Darwin was interested in earthworm activity, and his last work, *The Formation of Vegetable Mould, through the Actions of Worms, with Observations on Their Habits,* was the first detailed study of earthworm ecology. In it, he wrote, "All the fertile areas of this planet have at least once passed through the bodies of earthworms."

Class Clitellata, Subclass Hirudinea Leeches are primarily found in freshwater environments, but there are also some marine species as well as terrestrial species that inhabit warm, moist areas such as tropical forests. Leeches have a fixed number of segments, usually 34, though the septa have disappeared in most species. Most leeches feed as blood-sucking parasites of vertebrates. They have powerful suckers at both ends of the body, and the anterior sucker is equipped with razor-sharp jaws that can bore or slice into the host's tissues. The salivary secretion of leeches (hirudin) acts as an anticoagulant to stop blood clotting. Leeches can suck up to several times their own weight in blood. They were once used in the medical field in the practice of bloodletting, the withdrawal of often considerable quantities of blood from a patient in the erroneous belief that this would prevent or cure illness and disease. Even today, leeches may be used after surgeries, particularly those involving the reattachment of digits (Figure 33.18). In these cases, the blood vessels are not fully reconnected and much excess blood accumulates, causing a swelling called a hematoma. This excess blood switches off the delivery of new blood and stops the formation of new vessels. If leeches remove the accumulated blood, new capillaries will be more likely to form, and the tissues will become healthy.

Unlike cestode and trematode flatworms, which are internally parasitic and quite host specific, leeches are generally external parasites that feed on a broad range of hosts, including fishes, amphibians, and mammals. However, there are always exceptions. *Placobdelloides jaegerskioeldi* is a parasitic leech that lives only in the rectum of hippopotamuses.

Figure 33.18 **A leech, a member of the class Hirudinea.** This species, *Hirudo medicinalis*, is sucking blood from a hematoma, a swelling of blood that can occur after surgery.

33.4 Ecdysozoa: The Nematodes and Arthropods

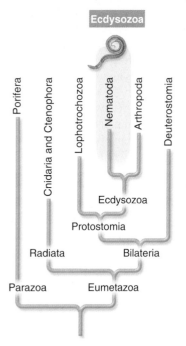

The Ecdysozoa is the sister group to the Lophotrochozoa. While the separation is supported by molecular evidence, the Ecdysozoa is named for a morphological characteristic, the physical phenomenon of ecdysis, or molting (refer back to Figure 32.12). All ecdysozoans possess a **cuticle**, a nonliving cover that serves to both support and protect the animal. Once formed, however, the cuticle typically cannot increase in size, which restricts the growth of the animal inside. The solution for growth is the formation of a new, softer cuticle under the old one. The old one then splits open and is sloughed off, allowing the new, soft cuticle to expand to a bigger size before it hardens. Where the cuticle is thick, as in arthropods, it impedes the diffusion of oxygen across the skin. Such species acquire oxygen by lungs, gills, or a set of branching, air-filled tubes called tracheae. A variety of appendages specialized for locomotion evolved in many species, including legs for walking or swimming and wings for flying.

The ability to shed the cuticle opened up developmental options for the ecdysozoans. For example, many species undergo a complete metamorphosis, changing from a wormlike larva into a winged adult. Animals with internal skeletons cannot do this because growth occurs only by adding more minerals to the existing skeleton. Another significant adaptation is the development of internal fertilization. This trait evolved independently in the vertebrates.

Because of these innovations, ecdysozoans are an incredibly successful group. Of the eight ecdysozoan phyla, we will consider the two most common: the nematodes and arthropods. The grouping of nematodes and arthropods is a relatively new idea and implies that the process of molting arose only once in animal evolution. In support of this, certain hormones that stimulate molting have been discovered to exist in both nematodes and arthropods. Furthermore, in 2007, Julie Dunning Hotopp and colleagues demonstrated the existence of lateral gene transfer (the movement of genes between distantly related organisms) between these groups. Elements of the bacterial genome *Wolbachia pipientis* were found in four insect and four nematode species. This both provides a mechanism for the acquisition of new genes by these eukaryotes and further underscores the idea that nematodes and arthropods are related.

The Phylum Nematoda Consists of Small Pseudocoelomate Worms Covered by a Tough Cuticle

The nematodes (from the Greek *nematos*, meaning thread), also called roundworms, are small, thin worms that range from less than 1 mm to about 5 cm (**Figure 33.19**), although some parasitic species measuring 1 m or more have been found in the placenta of sperm whales. Nematodes are ubiquitous organisms that exist in nearly all habitats, from the poles to the tropics. They are found in the soil, in both freshwater and marine environments, and inside plants and animals as parasites. A shovelful of soil may contain a million nematodes. Over 20,000 species are known, but there are probably at least five times as many undiscovered species.

Nematodes have several distinguishing characteristics. A tough cuticle covers the body. The cuticle is secreted by the epidermis and is made primarily of **collagen**, a structural protein also present in vertebrates. The cuticle is shed periodically as the nematode grows. Beneath the epidermis are longitudinal muscles but no circular muscles, which means that muscle contraction results in more thrashing of the body than smoother wormlike movement. The pseudocoelom functions as both a hydrostatic skeleton and a circulatory system. Diffusion of gases occurs through the cuticle. Roundworms have a complete digestive tract composed of a mouth, pharynx, intestine, and anus. The mouth often contains sharp, piercing organs called **stylets**, and the muscular pharynx functions to suck in food. Excretion of metabolic waste occurs via two simple tubules that have no cilia or flame cells.

Nematode reproduction is usually sexual, with separate males and females, and fertilization takes place internally. Females are generally larger than males and can produce prodigious numbers of eggs, in some cases, over 100,000 per day. Development in some nematodes is easily observed because the organism is transparent and the generation time is short. For these reasons, the small, free-living nematode *Caenorhabditis elegans* has become a model organism for researchers to study (refer back to Figure 19.1b). In 2002, the Nobel Prize in Medicine or Physiology was shared by Sydney Brenner, Robert Horvitz, and John Sulston for their studies of the genetic regulation of development and programmed cell death in *C. elegans*. This nematode has 1,090 somatic cells, but 131 die, leaving exactly 959 cells. The cells die via a genetically controlled cell death. Many diseases in humans, including acquired immunodeficiency syndrome (AIDS), cause extensive cell death, whereas others, such as cancer and autoimmune diseases, reduce cell death so that cells that should die do not. Researchers are studying the process of programmed cell death in *C. elegans* in the hope of finding treatments for these and other human diseases.

A large number of nematodes are parasitic in humans and other vertebrates. The large roundworm *Ascaris lumbricoides* is a parasite of the small intestine that can reach up to 30 cm in length. Over a billion people worldwide carry this parasite. Although infections are most prevalent in tropical or developing countries, the prevalence of *A. lumbricoides* is relatively high in rural areas of the southeastern U.S. Eggs pass out in feces and can remain viable in the soil for years. Eggs require ingestion before hatching into an infective stage. Hookworms (*Necator americanus*), so named because their anterior end curves dorsally like a hook, are also parasites of the human intestine. The eggs pass out in feces, and recently hatched hookworms can penetrate the skin of a host's foot to establish a new infection. In areas with modern plumbing, these diseases are uncommon.

Pinworms (*Enterobius vermicularis*), while a nuisance, have relatively benign effects on their hosts. The rate of infection in the U.S., however, is staggering: 30% of children and 16% of adults are believed to be hosts. Adult pinworms live in the large intestine and migrate to the anal region at night to lay their eggs, which causes intense itching. The resultant scratching spreads the eggs. In the tropics, some 250 million people are infected with *Wuchereria bancrofti*, a fairly large (100 mm) worm that lives in the lymphatic system, blocking the flow of lymph, and, in extreme cases, causing elephantiasis, or extreme swelling of the legs and other body parts (**Figure 33.20**). Females release tiny, live young called microfilariae, which are transmitted to new hosts via mosquitoes.

The Phylum Arthropoda Contains the Spiders, Millipedes and Centipedes, Insects, and Crustaceans, Species with Jointed Appendages

The arthropods (from the Greek *arthron*, meaning joint, and *podos*, meaning foot) constitute perhaps the most successful phylum on Earth. About three-quarters of all described living species are arthropods, and scientists have estimated they are also numerically common, with an estimated 10^{18}, or a billion billion, individual organisms present on Earth. The huge success of the arthropods, in terms of their sheer numbers and diversity, is related to a body plan that permits these animals to live in all the major biomes on Earth, from the poles to the tropics, and from marine and freshwater habitats to dry land.

Figure 33.19   Scanning electron micrograph of a nematode within a plant leaf.

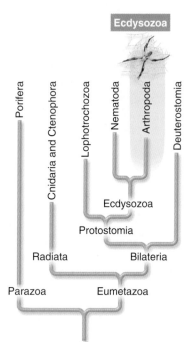

Figure 33.20 **Elephantiasis in a human leg.** The disease is caused by the nematode parasite *Wuchereria bancrofti*, which lives in the lymphatic system and blocks the flow of lymph.

Concept check: What other nematodes are parasitic in humans?

(a) External anatomy

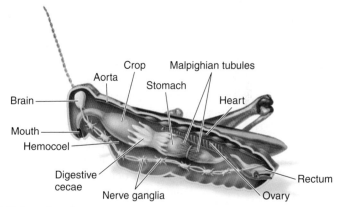

(b) Internal anatomy

Figure 33.21 **Body plan of an arthropod, as represented by a grasshopper.**

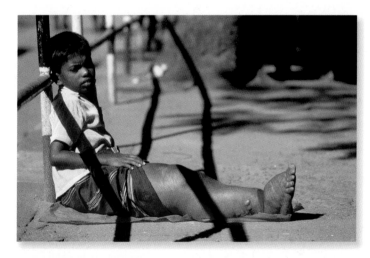

The body of a typical arthropod is covered by a hard cuticle, an **exoskeleton** (external skeleton), made of layers of chitin and protein. The cuticle can be extremely tough in some parts, as in the shells of crabs, lobsters, and even beetles, yet be soft and flexible in other parts, between body segments and segments of appendages, to allow for movement. In the class of arthropods called crustaceans, the exoskeleton is reinforced with calcium carbonate to make it extra hard. The exoskeleton provides protection and also a point of attachment for muscles, all of which are internal. It is also relatively impermeable to water, a feature that may have enabled many arthropods to conserve water and colonize land, in much the same way as a tough seed coat allowed plants to colonize land (see Chapter 29). From this point of view, the development of a hard cuticle was a critical innovation. It also reminds us that the ability to adapt to diverse environmental conditions can itself lead to increased organismal diversity.

Arthropods are segmented, and many of the segments bear appendages for locomotion, food handling, or reproduction. In many orders, the body segments have become fused into functional units, or **tagmata**, such as the head, thorax, and abdomen of an insect (**Figure 33.21**). Cephalization is extensive, and arthropods have well-developed sensory organs, including organs of sight, touch, smell, hearing, and balance. Arthropods have compound eyes composed of many independent visual units called **ommatidia** (singular, ommatidium) (look ahead to Figure 43.14). Together, these lenses render a mosaic-like image of the environment. Some species, particularly some insects, possess additional simple eyes, or ocelli, that are probably only capable of distinguishing light from dark.

The arthropod brain is quite sophisticated, consisting of two or three ganglia connected to several smaller ventral nerve ganglia. Like most mollusks, arthropods have an open circulatory system (look ahead to Figure 47.2), in which hemolymph is pumped from the heart into the aorta or short arteries and then into the open sinuses that coalesce to form the hemocoel. From the hemocoel, gases and nutrients diffuse into tissues. The hemolymph flows back into the heart via pores, called ostia, that are equipped with valves.

Because the cuticle impedes the diffusion of gases through the body surface, arthropods possess special organs that permit gas exchange. In aquatic arthropods, these consist of feathery gills that have an extensive surface area in contact with the surrounding water. Terrestrial species have a highly developed

tracheal system (look ahead to Figure 48.7). On the body surface, pores called **spiracles** provide openings to a series of finely branched air tubes within the body called trachea. The tracheal system delivers oxygen directly to tissues and cells. Some spiders have book lungs, consisting of a series of sheetlike structures, like the pages of a book, extending into a hemolymph-filled chamber on the underside of the abdomen. Gases also diffuse across thin areas of the cuticle.

The digestive system is complex and often includes a mouth, crop, stomach, intestine, and rectum. The stomach has glands called digestive cecae that secrete digestive enzymes. Excretion is accomplished by specialized metanephridia or, in insects and some other taxa, by **Malpighian tubules**, delicate projections from the digestive tract that protrude into the hemolymph (look ahead to Figure 49.9). Nitrogenous wastes are absorbed by the tubules and emptied into the gut. The intestine and rectum reabsorb water and salts. This excretory system, allowing the retention of water, was another critical innovation that permitted the colonization of land by arthropods.

Ancestral
arthropod

The history of arthropod classification is extensive. One scheme proposed two groups: the mandibulates, such as insects, which possessed jaws, and the chelicerates, such as spiders, which instead possessed pincers. Another scheme classified taxa according to whether they had one pair of unbranched antennae (called uniramous) versus one pair of branched antennae (called biramous) or none, the chelicerates. Such differences in classification led many to question whether the group was even monophyletic, or derived from a common ancestor. Three separate ancestral phyla, corresponding to the uniramia (insects and relatives), biramia (crustaceans), and chelicerates (spiders and scorpions) were proposed. However, a 2002 molecular study by Jeffrey Boore and colleagues, together with morphological data, suggest a different scheme with five main subphyla: one now-extinct subphyla, Trilobita (trilobites), and four living subphyla, Chelicerata (spiders and scorpions), Myriapoda (millipedes and centipedes), Hexapoda (insects), and Crustacea (crabs and relatives) (Table 33.5). In Boore's scheme, the Trilobita were among the earliest-diverging arthropods. The lineage then split into two groups. One, often referred to as the Pancrustacea, contained the insects and crustaceans. The other, with no overarching name, contained the Myriapods and Chelicerates. Molecular evidence suggests insects are more closely related to crustaceans than they are to spiders or millipedes and centipedes.

Subphylum Trilobita: Extinct Early Arthropods The trilobites were among the earliest arthropods, flourishing in shallow seas of the Paleozoic era, some 500 million years ago, and

Table 33.5	Main Subphyla and Characteristics of Arthropods	
	Subphyla and examples (est. # of species)	**Class characteristics**
	Chelicerata: spiders, scorpions, mites, ticks, horseshoe crabs, and sea spiders (74,000)	Body usually with cephalothorax and abdomen only; six pairs of appendages, including four pairs of legs, one pair of fangs, and one pair of pedipalps; terrestrial; predatory or parasitic
	Myriapoda: millipedes and centipedes (13,000)	Body with head and highly segmented trunk. In millipedes, each segment with two pairs of walking legs; terrestrial; herbivorous. In centipedes, each segment with one pair of walking legs; terrestrial; predatory, poison jaws
	Hexapoda: insects such as beetles, butterflies, flies, fleas, grasshoppers, ants, bees, wasps, termites and springtails (>1 million)	Body with head, thorax, and abdomen; mouthparts modified for biting, chewing, sucking, or lapping; usually with two pairs of wings and three pairs of legs; mostly terrestrial, some freshwater; herbivorous, parasitic, or predatory
	Crustacea: crabs, lobsters, shrimp (45,000)	Body of two to three parts; three or more pairs of legs; chewing mouthparts; usually marine

dying out about 250 million years ago. Most trilobites were bottom feeders and were generally 3–10 cm in size, although some reached almost 1 m in length (Figure 33.22). Like many arthropods, they had three main tagmata: the head, thorax, and abdomen. Trilobites also had two dorsal grooves that divided the body longitudinally into three lobes—an axial lobe and two pleural lobes—a structural characteristic giving the class its name. Most of the body segments showed little specialization. In contrast, later-diverging arthropods developed specialized appendages on many segments, including appendages for grasping, walking, and swimming.

Subphylum Chelicerata: The Spiders, Scorpions, Mites, and Ticks The Chelicerata consists mainly of the class Arachnida, which contains predatory spiders and scorpions as well as the ticks and mites, some of which are blood-sucking parasites that feed on vertebrates. The two other living classes are the Merostomata, the horseshoe crabs (four species), and the Pycnogonida, the sea spiders (1,000 species), both of which are marine, reflecting the groups' marine ancestry. All species have a body consisting of two tagmata: a fused head and thorax, called a **cephalothorax**, and an abdomen (Figure 33.23a). All

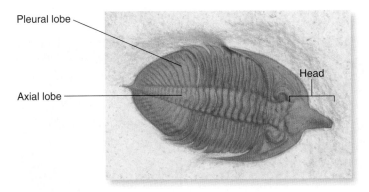

Figure 33.22 A fossil trilobite. About 4,000 fossil species of these early arthropods, including *Huntonia huntonesis* shown here, at about 20 cm long, have been described.

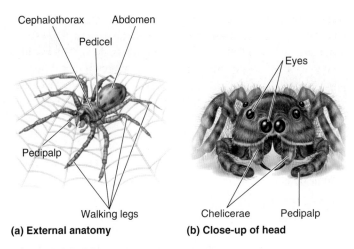

(a) External anatomy (b) Close-up of head

Figure 33.23 Spider morphology.

species also possess six pairs of appendages: the chelicerae, or fangs; a pair of **pedipalps**, which have various sensory, predatory, or reproductive functions; and four pairs of walking legs (Figure 33.23b).

In spiders (order Araneae), the cephalothorax and abdomen are joined by a **pedicel**, a narrow, waistlike point of attachment. The fangs, or chelicerae, are supplied with venom from poison glands. Most spider bites are harmless to humans, although they are very effective in immobilizing and/or killing their insect prey. Venom from some species, including the black widow (*Latrodectus mactans*) and the brown recluse (*Loxosceles reclusa*), are potentially, although rarely, fatal to humans. The toxin of the black widow is a neurotoxin, which interferes with the functioning of the nervous system, whereas that of the brown recluse is hemolytic, meaning it destroys tissue around the bite. After the spider has subdued its prey, it pumps digestive fluid into the tissues via the fangs and sucks out the partially digested meal.

Spiders have abdominal silk glands, called spinnerets, and many spin webs to catch prey (Figure 33.24a). The silk is a protein that stiffens after extrusion from the body because the mechanical shearing causes a change in the organization of the amino acids. Silk is stronger than steel of the same diameter but is more elastic than Kevlar, the material used in bulletproof vests. Each family constructs a characteristic size and style of

web and can do it perfectly on its first attempt, indicating that web spinning is an innate (inherited) behavior (see Chapter 55). Spiders also use silk to wrap up prey and to construct egg sacs. Interestingly, spiders that are fed drugged food (flies) spin their webs differently than undrugged spiders (Figure 33.24b,c). Some scientists have suggested that web-spinning spiders be used to test substances for the presence of drugs or even to indicate environmental contamination. Not all spiders use silk extensively. Some spiders, including the wolf spider, actively pursue their prey (Figure 33.25a).

Scorpions (order Scorpionida) are generally tropical or subtropical animals that feed primarily on insects, though they may eat spiders and other arthropods as well as smaller reptiles and mice. Their pedipalps are modified into large claws, and their abdomen tapers into a stinger, which is used to inject venom. While the venom of most North American species is generally not fatal to humans, that of the *Centruroides* genus from deserts in the U.S. Southwest and Mexico can be deadly. Fatal species are also found in India, Africa, and other countries. Unlike spiders, which lay eggs, scorpions bear live young that the mother then carries around on her back until they have their first molt (Figure 33.25b).

(a) Normal web

(b) Web spun by spider fed with prey containing caffeine

(c) Web spun by spider fed with prey containing marijuana

Figure 33.24 Spider-web construction by normal and drugged spiders.

(a) Wolf spider   **(b) Scorpion with young**   **(c) Chigger mite**   **(d) Bont ticks**

Figure 33.25   Common arachnids.   (a) This wolf spider (*Lycosa tarantula*) does not spin a web but instead runs after its prey. Note the pedipalps, which look like short legs. **(b)** The Cape thick-tailed scorpion (*Parabuthus capensis*) is highly venomous and carries its white young on its back. **(c)** SEM of a chigger mite (*Trombicula alfreddugesi*) that can cause irritation to human skin and spread disease. **(d)** These South African bont ticks (*Amblyomma hebraeum*) are feeding on a white rhinoceros.

Concept check: *What is one of the main characteristics distinguishing arachnids from insects?*

In mites and ticks (order Acari), the two main body segments (cephalothorax and abdomen) are fused and appear as one large segment. Many mite species are free-living scavengers that feed on dead plant or animal material. Other mites are serious pests on crops, and some, like chiggers (*Trombicula alfreddugesi*), are parasites of humans that can spread diseases such as typhus (**Figure 33.25c**). Chiggers are parasites only in their larval stage. Chiggers do not bore into the skin; it is their bite and salivary secretions that cause skin irritation. *Demodex brevis* is a hair-follicle mite that is common in animals and humans. The mite is estimated to be present in over 90% of adult humans. Although the mite causes no irritation in most humans, *Demodex canis* causes the skin disease known as mange in domestic animals, particularly dogs.

Ticks are larger organisms than mites, and all are ectoparasitic, feeding on the body surface, on vertebrates. Their life cycle includes attachment to a host, sucking blood until they are replete, and dropping off the host to molt (**Figure 33.25d**). Ticks can carry a huge variety of viral and bacterial diseases, including Lyme disease, a bacterial disease so named because it was first found in the town of Lyme, Connecticut, in the 1970s.

Subphylum Myriapoda: The Millipedes and Centipedes   Myriapods have one pair of antennae on the head and three pairs of appendages that are modified as mouth parts, including mandibles that act like jaws. The millipedes and centipedes, both wormlike arthropods with legs, are among the earliest terrestrial animal phyla known. Millipedes (class Diplopoda) have two pairs of legs per segment, as their name denotes (from the Latin *diplo*, meaning two, and *podos*, meaning feet), not 1,000 legs, as their common name suggests (**Figure 33.26a**). They are slow-moving herbivorous creatures that eat decaying leaves and other plant material. When threatened, the millipede's response is to roll up into a protective coil. Many millipede species also have glands on their underside that can eject a variety of toxic, repellent secretions. Some millipedes are brightly colored, warning potential predators that they can protect themselves.

Class Chilopoda (from the Latin *chilo*, meaning lip, and *podos*, meaning feet), or centipedes, are fast-moving carnivores

that have one pair of walking legs per segment (**Figure 33.26b**). The head has many sensory appendages, including a pair of antennae and three pairs of appendages modified as mouthparts, including powerful claws connected to poison glands. The toxin from venom of some of the larger species, such as *Scolopendra heros*, is powerful enough to cause pain in humans. Most species do not have a waxy waterproofing layer on their cuticle and so are restricted to moist environments under leaf litter or in decaying logs, usually coming out at night to actively hunt their prey.

Subphylum Hexapoda: A Diverse Array of Insects and Close Relatives   Hexapods are six-legged arthropods. Most are insects, but there are a few earlier-diverging noninsect hexapods, including soil-dwelling groups such as collembolans, that molecular studies have shown represent a separate but related lineage. Insects are in a class by themselves (Insecta), literally and figuratively. There are more species of insects than all other species of animal life combined. One million species of insects have been described, and, according to best estimates, 9 million more species await description. At least 90,000 species of insects have been identified in the U.S. and Canada alone. Genetic barcoding can help resolve many taxonomic dilemmas between closely related species.

(a) Two millipedes   **(b) A centipede**

Figure 33.26   Millipedes and centipedes.   (a) Millipedes have two pairs of legs per segment. **(b)** The venom of the giant centipede (*Scolopendra heros*) is known to produce significant swelling and pain in humans.

Insects are the subject of an entire field of scientific study, **entomology**. They are studied in large part because of their significance as pests of the world's agricultural crops and carriers of some of the world's most deadly diseases. Insects live in all terrestrial habitats, and virtually all species of plants are fed upon by at least one, usually tens, and sometimes, in the case of large trees, hundreds of insect species. Because insects eat approximately one-quarter of the world's crops, we are constantly trying to find ways to reduce insect pest densities. Pest reduction often involves chemical control (the use of pesticides) or biological control (the use of living organisms) to reduce pest populations. Many species of insects are also important pests or parasites of humans and livestock, both by their own actions and as vectors of diseases such as malaria and sleeping sickness.

In contrast, insects also provide us with many types of essential biological services. We depend on insects such as honeybees to pollinate our crops. Bees also produce honey, and silkworms are the source of silk fiber. Despite the revulsion they provoke in us, fly larvae (maggots) are important in the decomposition process of both dead plants and animals. In addition, we use insect parasites and predators in biological control.

Of paramount importance to the success of insects was the development of wings, a feature possessed by no other arthropod and indeed no other living animal except birds and bats. Unlike vertebrate wings, however, insect wings are actually outgrowths of the body wall cuticle and are not true segmental appendages. This means that insects still have all their walking legs. Insects are thus like the mythological horse Pegasus, which sprouted wings out of its back while retaining all four legs. In contrast, birds and bats have one pair of appendages (arms) modified for flight, which leaves them considerably less agile on the ground.

The great diversity of insects is illustrated by the fact that there are 35 different orders, some of which have over 100,000 species. The most common of the orders are discussed in **Table 33.6**. Different orders of insects have slightly different wing structures, and many of the orders are based on wing type (their names often include the root *pter-*, from the Greek *pteron*, meaning wing). In beetles (Coleoptera), only the back pair of wings is functional, as the front wings have been hardened into protective shell-like coverings under which the back pair folds when not in use. Wasps and bees (Hymenoptera) have two pairs of wings hooked together that move as one wing. Flies (Diptera) possess only one pair of wings (the front pair); the back pair has been modified into a small pair of balancing organs, called halteres, that act like miniature gyroscopes. Butterflies (Lepidoptera) have wings that are covered in scales (from the Greek *lepido*, meaning scale); other insects generally have clear, membranous wings. In ant and termite colonies, the queen and the drones (males) retain their wings, whereas female individuals called workers have lost theirs. Other species, such as fleas and lice, are completely wingless.

Insects in different orders have also evolved a variety of mouthparts in which the constituent parts, the mandibles and maxillae, are modified for different functions (**Figure 33.27**). Grasshoppers, beetles, dragonflies, and many others have mouthparts adapted for chewing. Mosquitoes and many plant pests have mouthparts adapted for piercing and sucking. Butterflies and moths have a coiled tongue (**proboscis**) that can be uncoiled, enabling them to drink nectar from flowers. Finally, some flies have lapping, spongelike mouthparts that sop up liquid food. Their varied mouthparts allow insects to specialize their feeding on virtually anything: plant matter, decaying organic matter, and other living animals. The biological diversity of insects is therefore related to environmental diversity, in this case, the variety of foods that insects eat. Parasitic insects attach themselves to other species, and there are even some insect parasites (called hyperparasites) that feed on other parasites, proving, as the 18th-century English poet Jonathan Swift noted

> Big fleas have little fleas
>
> upon their backs to bite 'em;
>
> little fleas have smaller fleas
>
> and so *ad infinitum.*

All insects have separate sexes, and fertilization is internal. During development, the majority (approximately 85%) of insects undergo a change in body form known as **complete metamorphosis** (from the Greek *meta*, meaning change, and *morph*, meaning form) (**Figure 33.28a**). Complete metamorphosis has four types of stages: egg, larva, pupa, and adult. In these species, the larval stage is often spent in an entirely different habitat from that of the adult, and larval and adult forms use different food sources. Consequently, they do not compete directly for the same resources. The dramatic body transformation from larva to adult occurs in the pupa stage. The remaining insects undergo **incomplete metamorphosis**, in which change is more gradual (**Figure 33.28b**). Incomplete metamorphosis has only three types of stages: egg, nymph, and adult. Young insects, called nymphs, look like miniature adults when they hatch from their eggs. As they grow and feed, they shed their skin several times, each time entering a new instar, or stage of growth.

Finally, some insects, such as bees, wasps, ants, and termites, have developed complex social behavior and live cooperatively in underground or aboveground nests. Such colonies exhibit a division of labor, in that some individuals forage for food and care for the brood (workers), others protect the nest (soldiers), and some only reproduce (the queen and drones) (**Figure 33.29**).

Genomes & Proteomes Connection

Barcoding: A New Tool for Classification

As mentioned, insects are a very species-rich taxa. For example, there are at least 3,500 species of mosquitoes, many of them hard to tell apart. Some mosquitoes transmit deadly diseases such as malaria and yellow fever and are subject to stringent control measures in many countries. Other mosquito species are relatively benign. Distinguishing mosquito species in the field is not easy. The mosquito barcoding initiative aims to catalog each species by using DNA and building up a mosquito

Table 33.6	Main Orders and Characteristics of Insects

Order and examples (est. # of species)		Order characteristics
Coleoptera: beetles, weevils (500,000)		Two pairs of wings (front pair thick and leathery, acting as wing cases, back pair membranous); armored exoskeleton; biting and chewing mouthparts; complete metamorphosis; largest order of insects
Hymenoptera: ants, bees, wasps (190,000)		Two pairs of membranous wings; chewing or sucking mouthparts; many have posterior stinging organ on females; complete metamorphosis; many species social; important pollinators
Diptera: flies, mosquitoes (190,000)		One pair of wings with hind wings modified into halteres (balancing organs); sucking, piercing, or lapping mouthparts; complete metamorphosis; larvae are grublike maggots in various food sources; some adults are disease vectors
Lepidoptera: butterflies, moths (180,000)		Two pairs of colorful wings covered with tiny scales; long tubelike tongue for sucking; complete metamorphosis; larvae are plant-feeding caterpillars; adults are important pollinators
Hemiptera: true bugs; assassin bug, bedbug, chinch bug, cicada (100,000)		Two pairs of membranous wings; piercing or sucking mouthparts; incomplete metamorphosis; many plant feeders; some predatory or blood feeders; vectors of plant diseases
Orthoptera: crickets, grasshoppers (30,000)		Two pairs of wings (front pair leathery, back pair membranous); chewing mouthparts; mostly herbivorous; incomplete metamorphosis; powerful hind legs for jumping
Odonata: damselflies, dragonflies (6,500)		Two pairs of long, membranous wings; chewing mouthparts; large eyes; predatory on other insects; incomplete metamorphosis; larvae aquatic; considered early-diverging insects
Phthiraptera: sucking lice (2,400)		Wingless ectoparasites; sucking mouthparts; flattened body; reduced eyes; legs with clawlike tarsi for clinging to skin; incomplete metamorphosis; very host specific; vectors of typhus
Siphonaptera: fleas (2,600)		Wingless, laterally flattened; piercing and sucking mouthparts; adults are bloodsuckers on birds and mammals; jumping legs; complete metamorphosis; vectors of plague
Isoptera: termites (2,000)		Two pairs of membranous wings when present; some stages wingless; chewing mouthparts; social species; incomplete metamorphosis

DNA catalog. Field researchers will be able to quickly analyze the DNA of a particular species and identify it based on existing bar codes. Appropriate control measures can then be instigated if it is a disease carrier.

The barcoding project was developed in 2003 by Dr. Paul Herbert of the University of Guelph, Ontario. He made the analogy that the large diversity of products in a grocery store can each be distinguished with a relatively small barcode. Though the diversity of the world's animal species is considerably larger, Dr. Herbert reasoned that all species could be distinguished using their DNA. The complete genome would be too large to analyze rapidly, so Dr. Herbert suggested analyzing a small piece of DNA of all species. The DNA sequence he proposed is the first 684 base pairs of a gene called *CO1*, for cyto-

(a) Chewing (grasshopper)

(b) Piercing and blood sucking (mosquito)

Proboscis

(c) Nectar sucking (butterfly)

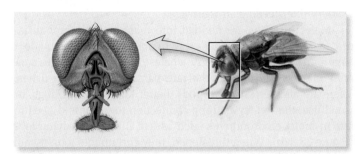

(d) Sponging liquid (housefly)

Figure 33.27 A variety of insect mouthparts. Insect mouthparts have become modified in ways that allow insects to feed by a variety of methods, including **(a)** chewing (Orthoptera, Coleoptera, and others), **(b)** piercing and blood sucking (Diptera), **(c)** nectar sucking (Lepidoptera), and **(d)** sponging up liquid (Diptera).

Concept check: Insects have a variety of mouthparts. Name two other key insect adaptations.

quickly. Aviation groups are also interested in bird barcoding. Following airplane–bird collisions, they would like to remove bird tissue from planes, identify the species, and then possibly be able to avoid its known migration routes. Barcodes for plants, based on a DNA sequence of the *matK* gene, were proposed in 2008. This simple barcode might be insufficient for plant identification in only a few instances, where hybrids are involved. Many scientists anticipate the day when handheld field barcoding identification devices appear. At the moment, barcoding involves a laboratory analysis taking about an hour and costing $2.00 per sample. Nevertheless, Dr. Herbert hopes to have half-a-million barcodes online by 2012.

Subphylum Crustacea: Crabs, Lobsters, Barnacles, and Shrimp The crustaceans are common inhabitants of marine environments, although some species live in fresh water and a few are terrestrial. Many are economically important food items for humans, including crabs, lobsters, crayfish, and shrimp, and smaller species are important food sources for other predators.

The crustaceans are unique among the arthropods in that they possess two pairs of antennae at the anterior end of the body—the antennule (first pair) and antenna (second pair) (**Figure 33.30**). In addition, they have three or more sensory and feeding appendages that are modified mouthparts: the mandibles, maxillae, and maxillipeds. These are followed by walking legs and, often, additional abdominal appendages, called swimmerets, and a powerful tail consisting of a telson and uropod. In some orders, the first pair of walking legs, or chelipeds, is modified to form powerful claws. A lost crustacean appendage can regrow. The head and thorax are often fused together, forming the cephalothorax. In many species, the cuticle covering the head extends over most of the cephalothorax, forming a hard protective fold called the **carapace**. For growth to occur, a crustacean must shed the entire exoskeleton.

Many crustaceans are predators, but others are scavengers, and some, such as barnacles, are suspension feeders. Gas exchange typically occurs via gills, and crustaceans, like

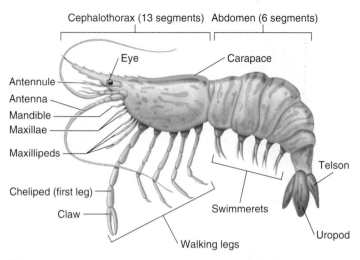

Figure 33.30 Body plan of a crustacean, as represented by a shrimp.

other arthropods, have an open circulatory system. Crustaceans possess two excretory organs: antennal glands and maxillary glands, both modified metanephridia, which open at the bases of the antennae and maxillae, respectively. Reproduction usually involves separate sexes, and fertilization is internal. Most species carry their eggs in brood pouches under the female's body. Eggs of most species produce larvae that must go through many different molts prior to assuming adult form. The first of these larval stages, called a **nauplius**, is very different in appearance from the adult crustacean (**Figure 33.31**).

While there are many crustacean clades, most are small and obscure, although many feature prominently in marine food chains, a series of organisms in which each member of the chain feeds on and derives energy from the member below it. These include the Ostracoda, Copepoda, Cirripedia, and Malacostraca. Ostracods are tiny creatures that superficially resemble clams, and copepods are tiny and abundant planktonic crustaceans, both of which are a food source for filter-feeding organisms and small fish. The clade Cirripedia is composed of the barnacles, crustaceans whose carapace forms calcified plates that cover most of the body (**Figure 33.32a**). Their legs are modified into feathery filter-feeding structures.

Malacostracans are divided into many orders. Euphausiacea are shrimplike krill that grow to about 3 cm and provide a large part of the diet of many whales. The order Isopoda contains many small species that are parasitic on marine fishes. There are also terrestrial isopods, better known as pill bugs, or wood lice, that retain a strong connection to water and need to live in moist environments such as leaf litter or decaying logs (**Figure 33.32b**). When threatened, they curl up into a tight ball, making it difficult for predators to get a grip on them.

The most famous Malacostracan order, however, is the Decapoda, which includes the crabs and lobsters, the largest crustacean species (**Figure 33.32c**). As their name suggests, these decapods have 10 walking legs (five pairs), although the first pair is invariably modified to support large claws. Most decapods are marine, but there are many freshwater species, such as crayfish, and in hot, moist tropical areas, even some terrestrial species called land crabs. The larvae of many larger crustaceans

(a) Worker and soldier ants   **(b) Queen ant**

Figure 33.29 The division of labor in insect societies. Individuals from the same insect colony may appear very different. Among these army ants (*Eciton burchelli*) from Paraguay, there are **(a)** workers that forage for the colony, soldiers that protect the colony from predators, and **(b)** the queen, which reproduces and lays eggs.

Figure 33.31 **Crustacean larva.** The nauplius is a distinct larval type possessed by most crustaceans, which molts several times before reaching maturity.

are planktonic and grow to about 3 cm. These are abundant in some oceans and are a staple food source for many species.

33.5 Deuterostomia: The Echinoderms and Chordates

As we explored in Chapter 32, the deuterostomes are grouped together because they share similarities in patterns of development (refer back to Figure 32.5). Molecular evidence also supports a deuterostome clade. All animals in the phylum Chordata, which includes the vertebrates, are deuterostomes. Interestingly, so is one invertebrate group, the phylum Echinodermata, which includes the sea stars, sea urchins, and sea cucumbers. While there are far fewer phyla and species of deuterostomes than protostomes, the species are generally much more familiar to us. After all, we humans are deuterostomes. While the majority of the deuterostome clade will be discussed in Chapter 34, we will conclude our discussion of invertebrate biology by turning our attention to the invertebrate deuterostomes. In this section, we will explore the phylum Echinodermata and then introduce the phylum Chordata, looking in particular at its distinguishing characteristics and at its two invertebrate subphyla: the cephalochordates, commonly referred to as the lancelets, and the urochordates, also known as the tunicates.

The Phylum Echinodermata Includes Sea Stars and Sea Urchins, Species with a Water Vascular System

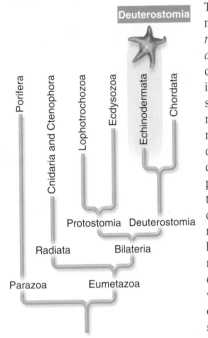

The phylum Echinodermata (from the Greek *echinos*, meaning spiny, and *derma*, meaning skin) consists of a unique grouping of deuterostomes. A striking feature of all echinoderms is their modified radial symmetry. The body of most species can be divided into five parts pointing out from the center. As a consequence, cephalization is absent in most classes. There is no brain and only a simple nervous system consisting of a central nerve ring from which radial branches to each limb arise. The radial symmetry of echinoderms is secondary, however, because the free-swimming larvae have bilateral symmetry and metamorphose into the radially symmetrical adult form.

Most echinoderms have an **endoskeleton** (internal skeleton) composed of calcareous plates overlaid by a thin skin (**Figure 33.33**). The skeleton is covered with spines and jawlike pincers called pedicellariae, the primary purpose of which is to deter settling of animals such as barnacles. These structures can also possess poison glands.

A portion of the coelom has been adapted to serve as a unique **water vascular system**, a network of canals that branch into tiny **tube feet** that function in movement, gas exchange, feeding, and excretion (see inset to Figure 33.33). The water vascular system is powered by hydraulic power, that is, by water pressure generated by the contraction of muscles that enables the extension and contraction of the tube feet, allowing echinoderms to move only very slowly.

(a) Goose barnacles—order Cirripedia

(b) Pill bug—order Isopoda

(c) Coral crab—order Decapoda

Figure 33.32 **Common crustaceans.** (a) Goose barnacles (*Lepas anatifera*). (b) Pill bug, or wood louse (*Armadillium vulgare*). (c) Coral crab (*Carpilius maculates*).

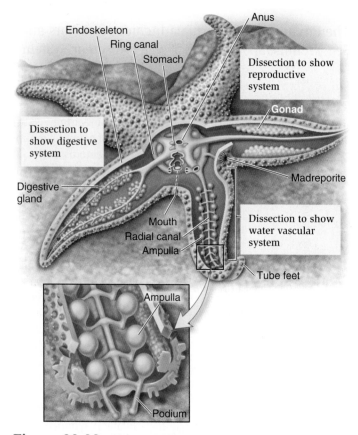

Figure 33.33 **Body plan of an echinoderm, as represented by a sea star.** The arms of this sea star have been dissected to different degrees to show the echinoderm's various organs. The inset shows a close-up view of the tube feet, part of the water vascular system characteristic of echinoderms.

Concept check: Echinoderms and chordates are both deuterostomes. What are three defining features of deuterostomes?

Water enters the water vascular system through the **madreporite**, a sievelike plate on the animal's surface. From there it flows into a **ring canal** in the central disc, into five radial canals, and into the tube feet. At the base of each tube foot is a muscular sac called an **ampulla**, which stores water. Contractions of the ampullae force water into the tube feet, causing them to straighten and extend. When the foot contacts a solid surface, muscles in the foot contract, forcing water back into the ampulla. Sea stars also use their tube feet in feeding, where they can exert a constant and strong pressure on bivalves, whose adductor muscles eventually tire, allowing the shell to open slightly. At this stage, the sea star everts its stomach and inserts it into the opening. It then digests its prey, using juices secreted from extensive digestive glands. Sea stars also feed on sea urchins, brittle stars, and sand dollars, prey that cannot easily escape them.

Echinoderms cannot osmoregulate, so no species have entered freshwater environments. No excretory organs are present. For some species, both respiration and excretion of nitrogenous waste take place by diffusion across their tube feet. Coelomic fluid circulates around the body.

Most echinoderms exhibit **autotomy**, the ability to intentionally detach a body part, such as a limb, that will later regenerate. In some species, a broken limb can even regenerate into a whole animal. Some sea stars regularly reproduce by breaking in two. Most echinoderms reproduce sexually and have separate sexes. Fertilization is usually external, with gametes shed into the water. Fertilized eggs develop into free-swimming larvae, which become sedentary adults.

While over 20 classes of echinoderms have been described from the fossil record, only 5 main classes of echinoderms exist today: the Asteroidea (sea stars), Ophiuroidea (brittle stars), Echinoidea (sea urchins and sand dollars), Crinoidea (sea lilies and feather stars), and Holothuroidea (sea cucumbers). The key features of the echinoderms and their classes are listed in **Table 33.7**.

The most unusual of the echinoderms are members of the class Holothuroidea, the sea cucumbers. These animals really do look like a cucumber rather than a sea star or sea urchin (Figure 33.34). The hard plates of the endoskeleton are less extensive, so the animal appears and feels fleshy. Sea cucumbers possess specialized respiratory structures called respiratory trees that pump water in and out of the anus. They are typically deposit feeders, ingesting sediment and extracting nutrients.

When threatened by a predator, a few tropical species of sea cucumber can eject sticky, toxic substances from their anus. If these do not deter the predator, a member of these species can undergo the process of evisceration—ejecting its digestive

Table 33.7	Main Classes and Characteristics of Echinoderms	
	Class and examples (est. # of species)	**Class characteristics**
	Asteroidea: sea stars (1,600)	Five arms; tube feet; predatory on bivalves and other echinoderms; eversible stomach
	Ophiuroidea: brittle stars (2,000)	Five long, slender arms; tube feet not used for locomotion; no pedicellariae; browse on sea bottom or filter feed
	Echinoidea: sea urchins, sand dollars (1,900)	Spherical (sea urchins) or disc-shaped (sand dollars); no arms; tube feet and moveable spines; pedicellariae present; many feed on seaweeds
	Crinoidea: sea lilies and feather stars (700)	Cup-shaped; often attached to substrate via stalk; arms feathery and used in filter feeding; very abundant in fossil record
	Holothuroidea: sea cucumbers (1,200)	Cucumber-shaped; no arms; spines absent; endoskeleton reduced; tube feet; browse on sea bottom

Figure 33.34 The edible sea cucumber, *Holothuria edulis*, on the Great Barrier Reef, Australia.

Concept check: What are two unique features of an echinoderm?

tract, respiratory structures, and gonads from the anus. If the sea cucumber survives, it can regenerate its organs later. The pearl fish has a particularly intriguing relationship with sea cucumbers. It takes advantage of the sea cucumber's defenses by darting inside its anus for protection. At first, the sea cucumber resists the pearl fish, but because it breathes through its anus, it is only a matter of time before another opportunity is afforded to the fish.

The Phylum Chordata Includes All the Vertebrates and Some Invertebrates

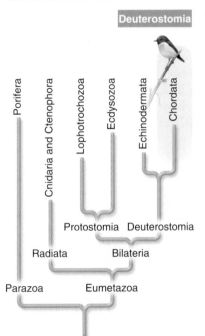

The deuterostomes consist of two major phyla: the echinoderms and the chordates (from the Greek *chorde*, meaning string). As deuterostomes, both phyla share similar developmental traits. In addition, both have an endoskeleton, consisting in the echinoderms of calcareous plates and in chordates, for the most part, of bone. However, the echinoderm endoskeleton functions in much the same way as the arthropod exoskeleton, in that its primary function is providing protection. The chordate endoskeleton serves a very different purpose. In early-divergent chordates, the endoskeleton is composed of a single flexible rod situated dorsally, deep inside the body. Muscles move this rod, and their

contractions cause the back and tail end to move from side to side, permitting a swimming motion in water. The endoskeleton becomes more complex in different lineages that develop limbs, as we will see in Chapter 34, but it is always internal, with muscles attached. This arrangement permits the possibility of complex movements, including the ability to move on land.

Let's take a look at the four critical innovations in the body design of chordates that distinguish them from all other animal life (Figure 33.35):

1. **Notochord.** Chordates are named for the **notochord**, a single flexible rod that lies between the digestive tract and the nerve cord. Composed of fibrous tissue encasing fluid-filled cells, the notochord is stiff yet flexible and provides skeletal support for all early-diverging chordates. In most chordates, such as vertebrates, a more complex jointed backbone usually replaces the notochord; its remnants exist only as the soft material within the discs of vertebrae.

2. **Dorsal hollow nerve cord.** Many animals have a long nerve cord, but in nonchordate invertebrates, it is a solid tube that lies ventral to the alimentary canal. In contrast, the nerve cord in chordates is a hollow tube that develops dorsal to the alimentary canal. In 1822, the French naturalist Geoffroy Saint-Hilaire argued that this difference suggested that the ventral side of nonchordate invertebrates (as exemplified by the lobster) was homologous to the dorsal side of vertebrates. Some recent molecular work suggests that, as Hilaire first proposed, an inversion of the dorsoventral axis occurred during animal evolution. In vertebrates, the dorsal hollow nerve cord develops into the brain and spinal cord.

3. **Pharyngeal slits.** Chordates, like many animals, have a complete gut, from mouth to anus. However, in chordates, slits develop in the pharyngeal region, close to the mouth, that open to the outside. This permits water to enter through the mouth and exit via the slits, without having to go through the digestive tract. In early-divergent chordates, **pharyngeal slits** function as a filter-feeding

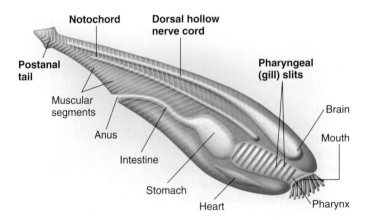

Figure 33.35 **Chordate characteristics.** The generalized chordate body plan has four main features: notochord, dorsal hollow nerve cord, pharyngeal slits, and postanal tail.

device, whereas in later-divergent chordates, they develop into gills for gas exchange. In terrestrial chordates, the slits do not fully form, and they become modified for other purposes, such as the auditory (Eustachian) tubes in ears.

4. **Postanal tail.** Chordates possess a postanal tail of variable length that extends posterior to the anal opening. In aquatic chordates such as fishes, the tail is used in locomotion. In terrestrial chordates, the tail may be used in a variety of functions or may be absent, as in humans. In virtually all other nonchordate phyla, the anus is at the end of the body.

Although few chordates apart from fishes possess all of these characteristics in their adult life, they all exhibit them at some time during development. For example, in adult humans, the notochord becomes the spinal column, and the dorsal hollow nerve cord becomes the central nervous system. However, humans exhibit pharyngeal slits and a postanal tail only during early embryonic development. All the pharyngeal slits, except one, which forms the auditory tubes in the ear, are eventually lost, and the postanal tail regresses to form the tailbone (the coccyx).

The phylum Chordata consists of the invertebrate chordates—the subphylum Cephalochordata (lancelets) and the subphylum Urochordata (tunicates)—along with the subphylum Craniata, which includes Vertebrata. Although the Craniata is by far the largest of these subphyla, biologists have focused on the Cephalochordata and Urochordata for clues as to how the chordate phylum may have evolved. Comparisons of gene sequences coding for 18S rRNA show that these two subphyla are our closest invertebrate relatives (**Figure 33.36**).

Subphylum Cephalochordata: The Lancelets

The cephalochordates (from the Greek *cephalo*, meaning head) look a lot more chordate-like than do tunicates. They are commonly referred to as lancelets, in reference to their bladelike shape and size, about 5–7 cm in length (**Figure 33.37a**). Lancelets are a small subphylum of 26 species, all marine filter feeders, with 4 species occurring in North American waters. Most of them belong to the genus *Branchiostoma*.

(a) Lancelet in the sand

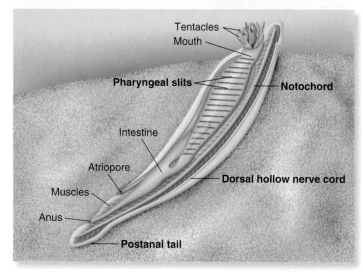

(b) Body plan of the lancelet

Figure 33.37   Lancelets.   (a) A bladelike lancelet. **(b)** The body plan of the lancelet clearly displays the four characteristic chordate features.

The lancelets live mostly buried in sand, with only the anterior end protruding into the water so that they can filter feed through the mouth. Lancelets have the four distinguishing chordate characteristics: a clearly discernible notochord (extending well into the head), dorsal hollow nerve cord, pharyngeal slits, and postanal tail (**Figure 33.37b**). Water enters the mouth and

Figure 33.36   Comparison of small subunit rRNA gene sequences of chordate and nonchordate species.   The similarities between the invertebrate chordates (represented by the lancelet) and the vertebrates (represented by a human) suggest they are indeed our closest invertebrate relatives.

CGGCCCCGCCGGGGTCGGCCCACGGCCTTGGCGGAG-GCCT   Human (vertebrate chordate)

CGGCCCCGCCGGGGTCGGCCCACGGCCCTGGCGGAG-CGCT   Lancelet (invertebrate chordate)

TGGTCTCGGCCCGCCTTTCA-CCGGGCGGCGCCGTT-GGTC   Mollusk (invertebrate)

GTGAACATTTGCTAGTCCCTCGGGATTACATTTGAATCGCT   Mosquito (invertebrate)

(a) Adult tunicate

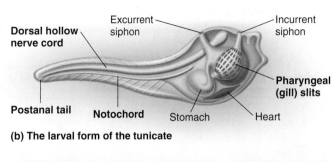

(b) The larval form of the tunicate

(c) Typical tunicate

Figure 33.38   **Tunicates.**   (a) Body plan of the sessile, filter-feeding adult tunicate. (b) The larval form, which shows the four characteristic chordate features, has been proposed as a possible ancestor of modern vertebrates. (c) The blue tunicate, *Rhopalaea crassa.*

moves into the pharynx, where it is filtered through the pharyngeal slits. A mucous net across the pharyngeal slits traps food particles, and ciliary action takes the food into the digestive tract, while water exits via the atriopore. Gas exchange generally takes place across the body surface. Although the lancelet is usually sessile, it can leave its sandy burrow and swim to a new spot, using a sequence of serially arranged muscles that appear like chevrons (<<<<) along its sides. These muscles reflect the segmented nature of the lancelet body.

Subphylum Urochordata: The Tunicates   The urochordates (from the Greek *oura*, meaning tail) are a group of 3,000 marine species also known as tunicates. Looking at an adult tunicate, you might never guess it is a relative of modern vertebrates. The only one of the four distinguishing chordate characteristics that it possesses is pharyngeal slits (**Figure 33.38a**). The larval tunicate, in contrast, looks like a tadpole and exhibits all four chordate hallmarks (**Figure 33.38b**). The larval tadpole swims for only a few days, usually without feeding. Larvae settle on and attach to a rock surface via rootlike extensions called stolons. Here the larvae metamorphose into adult tunicates and in the process lose most of their chordate characteristics. In 1928, the marine biologist Walter Garstang suggested modern vertebrates arose from a larval tunicate form that had somehow acquired the ability to reproduce. Recent analyses of molecular data have led Frédéric Delsuc and colleagues to propose that tunicates are the closest living relatives of vertebrates. These authors also

group the cephalochordates most closely with the echinoderms. This means the common ancestor of living deuterostomes was a free-living, bilaterally symmetrical animal with gill slits, a segmented body, and a sophisticated brain. This ancestral line split into two groups, the echinoderm–cephalochordate group and the tunicate–vertebrate group. Echinoderms lost most of their ancestral features, but cephalochordates did not. In this view, tunicates lost their segmentation and most became sedentary, whereas vertebrates did not.

Adult tunicates are marine animals, some of which are colonial and others solitary, that superficially resemble sponges or cnidarians. Tunicates are filter feeders that draw water through the mouth through an **incurrent siphon**, using a ciliated pharynx, and filter it through extensive pharyngeal slits. The food is trapped on a mucous sheet; passes via ciliary action to the stomach, intestine, and anus; and exits through the excurrent siphon. The whole animal is enclosed in a nonliving **tunic** that it secretes, made of a protein and a cellulose-like material called tunicin. Tunicates are also known as sea squirts for their ability to squirt out water from the excurrent siphon when disturbed. There is a rudimentary circulatory system with a heart and a simple nervous system of relatively few nerves connected to sensory tentacles around the incurrent siphon. The animals are mostly hermaphroditic.

As a reference, **Table 33.8** describes the common body characteristics of the various animal phyla that we have considered in this chapter.

Table 33.8 Summary of the Physical Characteristics of the Major Animal Phyla

Feature	Porifera (sponges)	Cnidaria and Ctenophora (hydra, anemones, jellyfish)	Platyhelminthes (flatworms)	Rotifera (rotifers)	Lophophorata (bryozoans and others)	Mollusca (snails, clams, squid)	Annelida (segmented worms)	Nematoda (roundworms)	Arthropoda (insects, arachnids, crustaceans)	Echinodermata (sea stars, sea urchins)	Chordata (vertebrates and others)
Digestive system	Absent	Gastrovascular cavity; ctenophores have complete gut	Gastrovascular cavity	Complete gut (usually)	Complete gut	Complete gut	Complete gut	Complete gut	Complete gut	Usually complete gut	Complete gut
Circulatory system	Absent	Absent	Absent	Absent	Absent; open; or closed	Open; closed in cephalopods	Closed	Absent	Open	Absent	Closed
Respiratory system	Absent	Absent	Absent	Absent	Absent	Gills	Absent	Absent	Trachae; gills; or book lungs	Tube feet; respiratory tree	Gills; lungs
Excretory system	Absent	Absent	Protonephridia with flame cells	Protonephridia	Metanephridia	Metanephridia	Metanephridia	Excretory tubules	Excretory glands resembling metanephridia	Absent	Kidneys
Nervous system	Absent	Nerve net	Brain; cerebral ganglia; lateral nerve chords; nerve net	Brain; nerve cords	No brain; nerve ring	Ganglia; nerve cords	Brain; ventral nerve cord	Brain; nerve cords	Brain; ventral nerve cord	No brain; nerve ring and radial nerves	Well-developed brain; dorsal hollow nerve cord
Reproduction	Sexual; asexual (budding)	Sexual; asexual (budding)	Sexual (most hermaphroditic); asexual (body splits)	Mostly parthenogenetic; males appear only rarely	Sexual (some hermaphroditic); asexual (budding)	Sexual (some hermaphroditic)	Sexual (some hermaphroditic)	Sexual (some hermaphroditic)	Usually sexual (some hermaphroditic)	Sexual (some hermaphroditic); parthenogenetic; asexual by regeneration (rare)	Sexual; rarely parthenogenetic
Support	Endoskeleton of spicules and collagen	Hydrostatic skeleton	Hydrostatic skeleton	Hydrostatic skeleton	Exoskeleton	Hydrostatic skeleton and shell	Hydrostatic skeleton	Hydrostatic skeleton	Exoskeleton	Endoskeleton of plates beneath outer skin	Endoskeleton of cartilage or bone

Summary of Key Concepts

33.1 Parazoa: Sponges, the First Multicellular Animals

- Invertebrates, or animals without a backbone, make up more than 95% of all animal species. An early lineage—the Parazoa—consists of one phylum, the Porifera, or sponges. Although sponges lack true tissues, they are multicellular animals possessing several types of cells. (Figures 33.1, 33.2)

33.2 Radiata: Jellyfish and Other Radially Symmetrical Animals

- The Radiata consists of two phyla: the Cnidaria (hydra, jellyfish, box jellies, sea anemones, and corals) and the Ctenophora (comb jellies). Radiata have only two embryonic germ layers: the ectoderm and endoderm, with a gelatinous substance (mesoglea) connecting the two layers.

- Cnidarians exist as two forms: polyp and medusa. A characteristic feature of cnidarians is their stinging cells, or cnidocytes, which function in defense and prey capture.

Ctenophores possess the first complete gut, and nearly all exhibit bioluminescence. (Figures 33.3, 33.4, 33.5, 33.6, Table 33.1)

33.3 Lophotrochozoa: The Flatworms, Rotifers, Lophophorates, Mollusks, and Annelids

- The Lophotrochozoa include taxa that possess either a lophophore or trochophore larva. Platyhelminthes, or flatworms, are regarded as the first animals to have the organ-system level of organization. (Figure 33.7, Table 33.2)

- The four classes of flatworms are the Turbellaria, Monogenea, Trematoda (flukes), and Cestoda (tapeworms). Flukes and tapeworms are internally parasitic, with complex life cycles. (Figures 33.8, 33.9)

- Rotifers are microscopic animals that have a complete digestive tract with separate mouth and anus; the mastax, a muscular pharynx, is a structure unique to the rotifers. (Figure 33.10)

- The lophophorates consist of the phoronids, bryozoa, and brachiopods, all of which possess a lophophore, a ciliary feeding structure. (Figure 33.11)

- The mollusks, which constitute a large phylum with over 100,000 diverse living species, have a basic body plan with three parts—a foot, a visceral mass, and a mantle—and an open circulatory system. (Figures 33.12, 33.13)

- The four most common mollusk classes are the polyplacophora (chitons), gastropoda (snails and slugs), bivalvia (clams and mussels), and cephalopoda (octopuses, squids, and nautiluses). (Table 33.3)

- Cephalopods are among the most complex of all invertebrates. They are the only mollusks with a closed circulatory system; they have a well-developed nervous system and brain and are believed to exhibit learning by observation. (Figures 33.14, 33.15, 33.16)

- Segmentation, in which the body is divided into compartments, is a critical evolutionary innovation in the annelids, although specialization of segments is only minimally present at the anterior end. (Figure 33.17)

- Annelids are a large phylum with two main classes: Polychaeta (the most species-rich class) and Clitellata (which includes the earthworms and leeches). (Table 33.4, Figure 33.18)

33.4 Ecdysozoa: The Nematodes and Arthropods

- The ecdysozoans are so named for their ability to shed their cuticle, a nonliving cover providing support and protection. The two most common ecdysozoan phyla are the nematodes and the arthropods.

- Nematodes, which exist in nearly all habitats, have a cuticle made of collagen, a structural protein. The small, free-living nematode *Caenorhabditis elegans* is a model organism. Many nematodes are parasitic in humans, including *Wuchereria bancrofti*, which causes elephantiasis. (Figures 33.19, 33.20)

- Arthropods are perhaps the most successful phylum on Earth. The arthropod body is covered by a cuticle made of layers of chitin and protein, and it is segmented, with segments fused into functional units called tagmata. (Figure 33.21)

- The five main subphyla of arthropods are Trilobita (trilobites; now extinct), Chelicerata (spiders, scorpions, and relatives), Myriapoda (millipedes and centipedes), Hexapoda (insects), and Crustacea (crabs and relatives). (Table 33.5, Figures 33.22, 33.23, 33.24, 33.25, 33.26)

- More insect species are known than all other animal species combined. The barcoding project developed in 2003 by Paul Herbert aims to distinguish all species by a small piece of their DNA; this may be useful in identifying some of the millions of insect species. (Table 33.6)

- The development of a variety of wing structures and mouthparts was a key to the success of insects. (Figure 33.27)

- Insects undergo a change in body form during development, either complete metamorphosis or incomplete metamorphosis, and have developed complex social behaviors. (Figures 33.28, 33.29)

- Most crustacean orders are small and feature prominently in marine food chains. The most well-known order of crustaceans is the Decapoda, which includes the crabs, lobsters, and shrimp. (Figures 33.30, 33.31, 33.32)

33.5 Deuterostomia: The Echinoderms and Chordates

- The Deuterostomia includes the phyla Echinodermata and Chordata. A striking feature of the echinoderms is their radial symmetry, which is secondary; the free-swimming larvae are bilaterally symmetrical. Echinoderms possess a unique water vascular system. (Figure 33.33)

- Five main classes of echinoderms exist today: the Asteroidea (sea stars), Ophiuroidea (brittle stars), Echinoidea (sea urchins and sand dollars), Crinoidea (sea lilies and feather stars), and Holothuroidea (sea cucumbers). (Table 33.7, Figure 33.34)

- The phylum Chordata is distinguished by four critical innovations: the notochord, dorsal hollow nerve chord, pharyngeal slits, and postanal tail. (Figure 33.35)

- The subphylum Cephalochordata (lancelets) and subphylum Urochordata (tunicates) are invertebrate chordates. Genetic studies have shown that tunicates are the closest invertebrate relatives of the vertebrate chordates (subphylum Craniata). (Figures 33.36, 33.37, 33.38, Table 33.8)

Assess and Discuss

Test Yourself

1. Choanocytes are
   a. a group of protists that are believed to have given rise to animals.
   b. specialized cells of sponges that function to trap and eat small particles.
   c. cells that make up the gelatinous layer in sponges.
   d. cells of sponges that function to transfer nutrients to other cells.
   e. cells that form spicules in sponges.

2. Why aren't sponges eaten more by predators?
   a. They are protected by silica spicules.
   b. They are protected by toxic defensive chemicals.
   c. They are eaten; it's just that the leftover cells reaggregate into new, smaller sponges.
   d. Both a and b are correct.
   e. a, b, and c are correct.

3. Which of the following organisms can produce female offspring through parthenogenesis?
   a. cnidarians   c. choanocytes   e. annelids
   b. flukes   d. rotifers

4. What organisms can survive without a mouth, digestive system, or anus?
   a. cnidarians   c. echinoderms   e. nematodes
   b. rotifers   d. cestodes

5. Which phylum does not have at least some members with a closed circulatory system?
   a. Lophophorata
   b. Arthopoda
   c. Annelida
   d. Mollusca
   e. All of the above phyla have some members with a closed circulatory system.

6. A defining feature of the Ecdysozoa is
   a. a segmented body.
   b. a closed circulatory system.
   c. a cuticle.
   d. a complete gut.
   e. a lophophore.

7. In arthropods, the tracheal system is
   a. a unique set of structures that function in ingestion and digestion of food.
   b. a series of branching tubes extending into the body that allow for gas exchange.
   c. a series of tubules that allow waste products in the blood to be released into the digestive tract.
   d. the series of ommatidia that form the compound eye.
   e. none of the above.

8. Characteristics of the class Arachnida include
   a. two tagmata.   d. a lobed body.
   b. six walking legs.   e. both b and d.
   c. an aquatic lifestyle.

9. Incomplete metamorphosis
   a. is characterized by distinct larval and adult stages that do not compete for resources.
   b. is typically seen in arachnids.
   c. involves gradual changes in life stages where young resemble the adult stage.
   d. is characteristic of the majority of insects.
   e. always includes a pupal stage.

10. Echinodermata are <u>not</u> a member of which clade?
    a. protostomia
    b. bilateria
    c. eumetazoa
    d. metazoa
    e. They are a member of all the above clades.

Conceptual Questions

1. Compare and contrast the five main feeding types discussed in the chapter.

2. Define nematocyst, and explain its function.

3. Explain the difference between complete metamorphosis and incomplete metamorphosis.

Collaborative Questions

1. What animals exhibit complete metamorphosis?

2. Discuss the four defining characteristics of chordates.

Online Resource

www.brookerbiology.com

Stay a step ahead in your studies with animations that bring concepts to life and practice tests to assess your understanding. Your instructor may also recommend the interactive ebook, individualized learning tools, and more.

The Vertebrates

34

Vertebrates range in size from tiny fishes weighing 0.1 g to huge whales of over 100,000 kg. They occupy nearly all of Earth's habitats, from the deepest depths of the oceans to mountaintops and the sky beyond. Throughout history, humans have depended on many vertebrate species for their welfare. They have domesticated species such as horses, cattle, pigs, sheep, and chickens and have kept countless species, including cats and dogs, as pets. Many other vertebrate species are the subjects of conservation efforts, as we will see in Chapter 60 (also see chapter-opening photo). Some vertebrate species that require large areas for their survival are considered to be umbrella species. Their preservation will help ensure the conservation of others.

In Chapter 33, we discussed two chordate subphyla, the Cephalochordates (lancelets) and Urochordates (tunicates). The third subphylum of chordates, the Craniata, with about 53,000 species, is by far the largest and most dominant group of chordates. In this chapter, we begin by outlining the characteristics of **craniates**, chordates that have a brain encased in a skull. We will take a brief look at the hagfish, a craniate that has some, but not all, vertebrate characteristics. In the remainder of the chapter, we will explore the characteristics of vertebrates, discussing the evolutionary development of the major vertebrate classes, including fishes, amphibians, reptiles, and mammals.

Our current understanding of the relationships between the vertebrate groups is shown in **Figure 34.1**. Nested within the vertebrates are various clades based on morphological characteristics. For example, most vertebrates have jaws and are collectively known as gnathostomes. Many gnathostomes have four limbs for movement and are known as tetrapods. Next, we explain these sequential divisions in the vertebrate lineage.

34.1 The Craniates: Chordates with a Head

Although all vertebrates are craniates, not all craniates are vertebrates. There is one type of animal, the eel-like hagfish, whose characteristics place it within the craniates but not within the vertebrates. Hagfish do not have a vertebral column—one of the vertebrate characteristics—but they do have a skull, in this

Giant pandas (*Ailuropoda melanoleuca*). The species, shown here in Sichuan, China, is one of many charismatic vertebrate species. Because the giant panda is an endangered species, conservation efforts are underway to try to protect the remaining populations.

case, one made of cartilage and not bone. (Therefore, despite its name, the hagfish cannot be considered a true fish.) Paleontologists in China have recently discovered similar fishlike fossils that date to around 530 million years ago, the time of the Cambrian explosion. These are believed to be the earliest known craniates. What makes an animal a craniate?

Craniates Are Distinguished by a Cranium and Neural Crest

Craniates have two defining characteristics that distinguish them from invertebrate chordates:

1. **Cranium.** In craniates, the anterior end of the nerve cord elaborates to form a more developed brain that is encased in a protective bony or cartilaginous housing called the **cranium**. This continues the trend of cephalization—the development of the head end in animals.

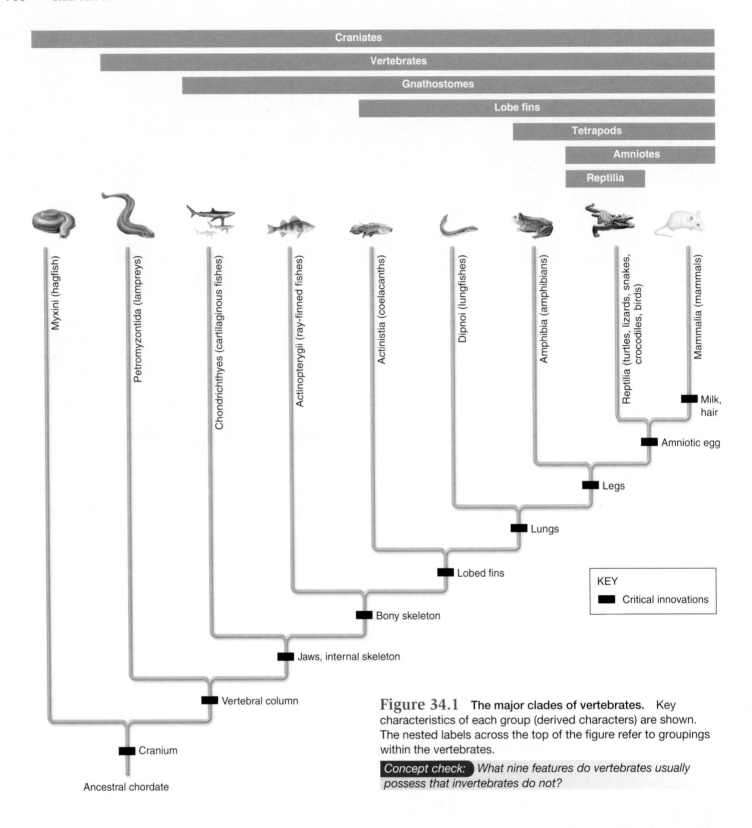

Figure 34.1 **The major clades of vertebrates.** Key characteristics of each group (derived characters) are shown. The nested labels across the top of the figure refer to groupings within the vertebrates.

Concept check: What nine features do vertebrates usually possess that invertebrates do not?

2. **Neural crest.** The **neural crest** is a group of embryonic cells found on either side of the neural tube as it develops. The cells disperse throughout the embryo, where they contribute to the development of the skeleton, especially the cranium and other structures, including jaws and teeth.

Although these are the main distinguishing characteristics of craniates, there are others. For example, craniates have at least two clusters of *Hox* genes, compared to the single cluster of *Hox* genes in tunicates and lancelets. This additional gene cluster is believed to have permitted increasingly complex morphologies than those possessed by invertebrate chordates.

The Hagfish, Class Myxini, Are the Simplest Living Craniates

Ancestral chordate

The hagfish are entirely marine craniates that lack eyes, jaws, and fins as well as vertebrae (**Figure 34.2**). The hagfish skeleton consists largely of a notochord and a cartilaginous skull that encloses the brain. The lack of a vertebral column leads to extensive flexibility. Hagfish live in the cold waters of northern oceans, close to the muddy bottom. Essentially blind, hagfish have a very keen sense of smell and are attracted to dead and dying fish, which they attach themselves to via toothed plates on the mouth. The powerful tongue then rasps off pieces of tissue. However, most of their diet consists of dead or disabled marine worms and other invertebrates. Though the hagfish cannot see approaching predators, they have special glands that produce copious amounts of slime, which may deter predatory attacks. When provoked, the hagfish's slime production increases dramatically, enough to potentially distract predators or coat their gills and interfere with breathing. Hagfish can sneeze to free their nostrils of their own slime.

Figure 34.2 **The hagfish: a craniate that is not a vertebrate.** The hagfish possesses craniate characteristics (a skull and neural crest) but is not a true fish or vertebrate.

Concept check: Why isn't the hagfish a true fish?

34.2 Vertebrates: Craniates with a Backbone

The **vertebrates** (from the Latin *vertebratus*, meaning back boned) retain all chordate characteristics we outlined in Chapter 33 and all craniate characteristics noted previously, as well as possessing several additional traits, including the following:

1. **Vertebral column.** During development in vertebrates, the notochord is replaced by a bony or cartilaginous column of interlocking **vertebrae** that provides support and also protects the nerve cord, which lies within its tubelike structure.

2. **Endoskeleton of cartilage or bone.** The cranium and vertebral column are parts of the endoskeleton, the living skeleton of vertebrates that forms within the animal's body. Most vertebrates also have two pairs of appendages, whether fins, legs, or arms. The endoskeleton is composed of either bone or cartilage, both of which are very strong materials, yet they are more flexible than the chitin found in insects and other arthropods. The endoskeleton also contains living cells that secrete the skeleton, which grows with the animal, unlike the nonliving exoskeleton of arthropods.

3. **Diversity of internal organs.** Vertebrates possess a great diversity of internal organs, including a liver, kidneys, endocrine glands, and a heart with at least two chambers. The liver is unique to vertebrates, and the vertebrate kidneys, endocrine system, and heart are more complex than are analogous structures in other taxa.

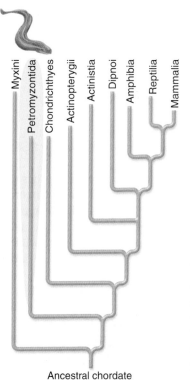

Ancestral chordate

Although these features are exhibited in all vertebrate classes, some classes developed critical innovations that helped them succeed in specific environments such as on land or in the air. For example, birds developed feathers and wings, structures that enable most species to fly. In fact, each of the vertebrate classes is distinctly different from one another, as outlined in **Table 34.1**. One of the earliest innovations was the development of jaws. All vertebrates except some early-diverging

Table 34.1   The Main Classes and Characteristics of Living Vertebrates

Class	Examples (approx. # of species)	Main characteristics
Petromyzontida	Lampreys (41)	Early-diverging fishes with jawless sucking mouth; no appendages, that is, fins; parasitic on fishes
Chondrichthyes	Sharks, skates, rays (850)	Fishes with cartilaginous skeleton; teeth not fused to jaw; no swim bladder; well-developed fins; internal fertilization; single blood circulation
Actinopterygii	Ray-finned fishes, most bony fish (24,600)	Fishes with ossified skeleton; single gill opening covered by operculum; fins supported by rays, fin muscles within body; swim bladder often present; mucous glands in skin
Actinistia	Lobe-finned fishes, of which coelacanths are the only living members (2)	Fishes with ossified skeleton; bony extensions, together with muscles, project into pectoral and pelvic fins; swim bladder filled with oil
Dipnoi	Lungfishes (6)	Fishes with ossified skeleton; rudimentary lungs allow fishes to come to the surface to gulp air; limblike appendages
Amphibia	Frogs, toads, salamanders (4,000)	Tetrapods; adults able to live on land; fresh water needed for reproduction; development usually involving metamorphosis from tadpoles; adults with lungs and double blood circulation; moist skin; shell-less eggs
Testudines	Turtles (330)	Body encased in hard shell; no teeth; head and neck retractable into shell; eggs laid on land
Lepidosauria	Lizards, snakes (7,800)	Lower jaw not attached to skull; skin covered in scales
Crocodilia	Crocodiles, alligators (23)	Four-chambered heart; large aquatic predators; parental care of young
Aves	Birds (9,600)	Feathers; hollow bones; air sacs; reduced internal organs; endothermic; four-chambered heart
Mammalia	Mammals (5,500)	Mammary glands; hair; specialized teeth; enlarged skull; external ears; endothermic; highly developed brains; diversity of body forms

fishes possessed jaws. Today, the only jawless vertebrates are lampreys.

The lampreys, class Petromyzontida, are eel-like fishes that lack jaws. Lampreys are unlike members of other vertebrate clades because they lack both a hinged jaw and true appendages. However, lampreys do possess a notochord surrounded by a cartilaginous rod that represents a rudimentary vertebral column; therefore, they represent one of the earliest-diverging groups of vertebrates.

Lampreys can be found in both marine and freshwater environments. Marine lampreys are parasitic as adults. They grasp other fish with their circular mouth (**Figure 34.3a**) and rasp a hole in the fish's side, sucking blood, tissue, and fluids until they are full (**Figure 34.3b**). Reproduction of all species

is similar, whether they live in marine or freshwater environments. Males and females spawn in freshwater streams, and the resultant larval lampreys bury into the sand or mud, much like lancelets (refer back to Figure 33.37a), emerging to feed on small invertebrates or detritus at night. This stage can last for 3 to 7 years, at which time the larvae metamorphose into adults. In most freshwater species, the adults do not feed at all but quickly mate and die. Marine species migrate from fresh water back to the ocean, until they return to fresh water to spawn and then die. Although many other species of jawless fishes would evolve and thrive for over 300 million years, they all became extinct by the end of the Devonian period. Jawed fishes, which had appeared in the mid-Ordovician period, about 470 million years ago, radiated in both fresh and salt water.

(a) Jawless mouth of a sea lamprey   **(b) A sea lamprey feeding**

Figure 34.3 **The lamprey, a modern jawless fish.** **(a)** The sea lamprey (*Petromyzon marinus*) has a circular, jawless mouth. **(b)** A sea lamprey feeding on a fish.

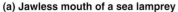

34.3 Gnathostomes: Jawed Vertebrates

All vertebrate species that possess jaws are called **gnathostomes** (from the Greek, meaning jaw mouth) (see Figure 34.1). In contrast, the jawless species are termed agnathans (meaning without jaws). Gnathostomes are a diverse clade of vertebrates that include fishes, amphibians, reptiles, and mammals. The earliest-diverging gnathostomes were fishes. Biologists have identified about 25,000 species of living fishes, more than all other species of vertebrates combined. Most are aquatic, gill-breathing species that usually possess fins and a scaly skin. There are four separate classes of jawed fishes, each of which has distinguishing characteristics (see Table 34.1): the Chondrichthyes (cartilaginous fishes), Actinopterygii (ray-finned fishes), Actinistia (coelacanths), and Dipnoi (lungfishes).

The jawed mouth was a significant evolutionary development. It enabled an animal to grip its prey more firmly, which may have increased its rate of capture, and to attack larger prey species, thus increasing its potential food supply. Accompanying the jawed mouth was the development of more sophisticated head and body structures, including two pairs of appendages called fins. Gnathostomes also possess two additional *Hox* gene clusters (bringing their total to four), which permitted increased morphological complexity.

The hinged jaw developed from the pharyngeal arches, cartilaginous or bony rods that help to support the respiratory tissue. Primitive jawless fishes had nine gill arches surrounding the eight gill slits (**Figure 34.4a**). During the late Silurian period, about 417 million years ago, some of these gill arches became modified in jawless fishes. The first and second gill arches were lost, while the third and fourth pairs became modified to form the jaws (**Figure 34.4b,c**). This is how evolution usually works; body features do not appear *de novo*, but instead, existing features become modified to serve other functions.

By the mid-Devonian period, two classes of jawed fishes, the Acanthodii (spiny fishes) and Placodermi (armored fishes) were common. Some of the placoderms were huge individuals,

(a) Primitive jawless fishes

(b) Early jawed fishes (placoderms)

(c) Modern jawed fishes (cartilaginous and bony fishes)

Figure 34.4 **The evolution of the vertebrate jaw.** **(a)** Primitive fishes and extant jawless fishes such as lampreys have nine cartilaginous gill arches that support eight gill slits. **(b)** In early jawed fishes such as the placoderms, the first two pairs of gill arches were lost, and the third pair became modified to form a hinged jaw. This left six gill arches (4–9) to support the remaining five gill slits, which were still used in breathing. **(c)** In modern jawed fishes, the fourth gill arch also contributes to jaw support, allowing stronger, more powerful bites to be delivered.

over 9 m long. Both classes died out by the end of the Devonian as part of one of several mass extinctions that occurred in the Earth's geological and biological history. The reasons for this extinction are not well understood, but other types of jawed fishes present at the same time—the cartilaginous and bony fishes—did not go extinct. We will discuss the fishes in these classes next.

Chondrichthyans: The Cartilaginous Fishes

Members of the class Chondrichthyes (the **chondrichthyans**)—sharks, skates, and rays—are also called cartilaginous fishes because their skeleton is composed of flexible cartilage rather than bone. The cartilaginous skeleton is not considered an ancestral character but rather a derived character. This means that the ancestors of the chondrichthyans had bony skeletons but that members of this class subsequently lost this feature. This hypothesis is reinforced by the observation that during development, the skeleton of most vertebrates is cartilaginous, and then it becomes bony (ossified) as a hard calcium-phosphate matrix replaces the softer cartilage. A change in the developmental sequence of the cartilaginous fishes is believed to prevent the ossification process.

In the Carboniferous period, 354–290 million years ago, sharks were the great predators of the ocean. Aided by fins, sharks became fast, extremely efficient swimmers (**Figure 34.5a**). Perhaps the most important fin for propulsion is the large and powerful caudal fin, or tail fin, which, when swept from side to side, thrusts the fish forward at great speed. For

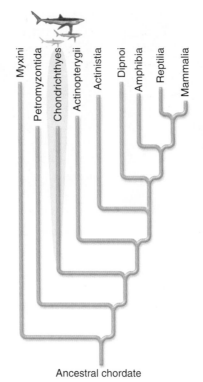
Ancestral chordate

example, great white sharks (*Carcharodon carcharias*) can swim at over 40 km per hour, and Mako sharks (*Isurus oxyrinchus*) have been clocked at nearly 50 km per hour. The paired pelvic fins (at the back) and pectoral fins (at the front) act like flaps on airplane wings, allowing the shark to dive deeper or rise to the surface. They also aid in steering. In addition, the dorsal fin (on the back) acts as a stabilizer to prevent the shark from rolling in the water as the tail fin pushes it forward.

Sharks were among the earliest fishes to develop teeth. Their teeth evolved from rough scales on the skin that also contain dentin and enamel. Although shark's teeth are very sharp and hard, they are not set into the jaw, as are human teeth, so they break off easily. To offset this, the teeth are continually replaced, row by row (**Figure 34.5b**). Sharks may have 20 rows of teeth, with the front pair in active use and the ones behind

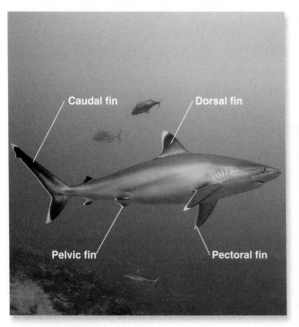
(a) Silvertip shark

(b) Rows of shark teeth

(c) Shark egg pouch

(d) Stingray

Figure 34.5 Cartilaginous fishes. (a) The silvertip shark (*Carcharhinus albimarginatus*) is one of the ocean's most powerful predators. **(b)** Close-up of the mouth of a sand tiger shark (*Carcharias taurus*), showing rows of teeth. **(c)** This mermaid's purse (egg pouch) of a dogfish shark (*Scyliorhinus canicula*) is entwined in vegetation to keep it stationary. **(d)** Stingrays are essentially flattened sharks with very large pectoral fins.

ready to grow in as replacements when needed. Tooth replacement time varies from 9 days in the cookie-cutter shark (named for its characteristic of biting round plugs of flesh from its prey) to 242 days in great whites. Some experts estimate that certain sharks can use up to 20,000 teeth in a lifetime.

All chondrichthyans are denser than water, which theoretically means that they would sink if they stopped swimming. Many sharks never stop swimming and maintain buoyancy via the use of their fins and a large oil-filled liver. Another advantage of swimming is that water continually enters the mouth and is forced over the gills, allowing sharks to extract oxygen and breathe. How then do skates and rays breathe when they rest on the ocean floor? These species, and a few sharks such as the nurse shark, use a muscular pharynx and jaw muscles to pump water over the gills. In these and indeed all species of fishes, the heart consists of two chambers, an atrium and a ventricle, that contract in sequence. They employ what is known as a single circulation, in which blood is pumped from the heart to capillaries in the gills to collect oxygen, and then it flows through arteries to the tissues of the body, before returning to the heart (look ahead to Figure 47.4a).

Many active predators possess a variety of acute senses, and sharks are no exception. They have a powerful sense of smell, facilitated by sense organs in the nostrils (sharks and other fishes do not use nostrils for breathing). They can see well but cannot distinguish colors. Although sharks have no eardrum, they can detect pressure waves generated by moving objects. All jawed fishes have a row of microscopic organs in the skin, arranged in a line that runs laterally down each side of the body, that can detect movements in the surrounding water. This system of sense organs, known as the **lateral line**, senses pressure waves and sends nervous signals to the inner ear and then on to the brain.

Fertilization is internal in chondrichthyans, with the male transferring sperm to the female via a pair of **claspers**, modifications of the pelvic fins. Some shark species are **oviparous**, that is, they lay eggs, often inside a protective pouch called a mermaid's purse (**Figure 34.5c**). The eggs, which are not guarded by either parent, then hatch into tiny sharks. In **ovoviparous** species, the eggs are retained within the female's body, but there is no placenta to nourish the young. A few species are **viviparous**; the eggs develop within the uterus, receiving nourishment from the mother via a placenta. Both ovoviparous and viviparous sharks give birth to live young.

The sharks have been a very successful vertebrate group, with many species identified in the fossil record. Although many species died out in the mass extinction at the end of the Permian period (290–248 million years ago), the survivors underwent a period of further speciation in the Mesozoic era, when most of the 375 modern species appeared.

Skates and rays are essentially flattened sharks that cruise along the ocean floor by using hugely expanded pectoral fins. In addition, their thin and whiplike tails are often equipped with a venomous barb used in defense (**Figure 34.5d**). Most of the 475 or so species of skates and rays feed on bottom-dwelling crustaceans and mollusks.

The Bony Fishes

Bony fishes are the most numerous of all types of fishes, with more individuals and more species (about 24,600) than any other. Most authorities now recognize three living classes: the Actinopterygii (ray-finned fishes), the Actinistia (coelacanths), and the Dipnoi (lungfishes). Fishes in all three classes possess a bony skeleton and scale-covered skin. The skin of bony fishes, unlike the rough skin of sharks, is slippery and slimy because the scales are covered by a thin epidermal layer containing glands that produce mucus, an adaptation that reduces drag during swimming. Just as in the cartilaginous fishes, water is drawn over the gills for breathing, but in bony fishes, a protective flap called an **operculum** covers the gills (**Figure 34.6**). Muscle contractions around the gills and operculum draw water across the gills so that bony fishes do not need to swim continuously to breathe. Some early bony fishes lived in shallow, oxygen-poor waters and developed lungs as an embryological offshoot of the pharynx. These fish could rise to the water surface and gulp air. As we will see, modern lungfishes

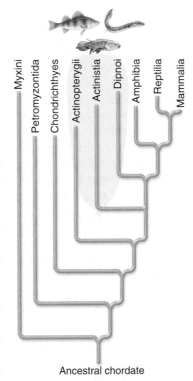

Ancestral chordate

operate in much the same fashion. In most bony fishes, these lungs evolved into a **swim bladder**, a gas-filled, balloon-like structure that helps the fish remain buoyant in the water even when it is completely stationary. In less-derived fishes, the gut and swim bladder are connected via a duct, and the fishes can fill their swim bladder by gulping air. In more-derived species, the swim bladder is connected to the circulatory system, and gases are transported in and out of the blood, allowing the fishes to change the volume of the swim bladder and so to rise and sink. Therefore, unlike the sharks, many bony fishes can remain motionless and use a "sit-and-wait" ambush style. These three features—bony skeleton, operculum, and swim bladder—distinguish bony fishes from cartilaginous fishes.

Reproductive strategies of bony fishes vary tremendously, but most species reproduce via external fertilization, with the female shedding her eggs and the male depositing sperm on top of them. While adult bony fishes can maintain their buoyancy, their eggs tend to sink. This is why many species spawn in shallow, more oxygen- and food-rich waters and why coastal areas are such important fish nurseries.

Bony fishes have colonized nearly all aquatic habitats. Following the cooling of the newly formed planet Earth, rain

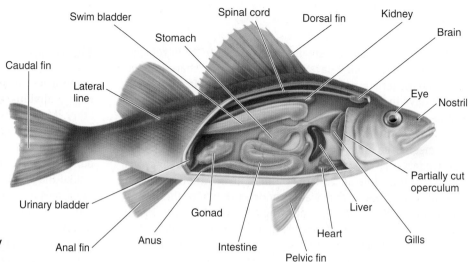

Figure 34.6 Generalized body plan of a bony fish.

fell continuously, and the oceans were filled with fresh water. Later, as water evaporated from the oceans and sodium, potassium, and calcium were added via runoff from the land, the oceans became salty. Therefore, most fishes probably evolved in freshwater habitats and secondarily became adapted to marine environments. This, of course, required the development of physiological adaptations to the different osmotic problems seawater presents compared to fresh water (see Chapter 49).

The most species-rich class of bony fishes is the Actinopterygii, or **ray-finned fishes**, which includes all bony fishes except the coelacanths and lungfishes. In Actinopterygii, the fins are supported by thin, bony, flexible rays and are moved by muscles on the interior of the body. The class has a diversity of forms, from lionfish and large predatory moray eels to delicate sea dragons (**Figure 34.7**). Whole fisheries are built around the harvest of species such as cod, anchovies, and salmon.

The Actinistia (coelacanths) and Dipnoi (lungfishes) are both considered Sarcopterygii, or **lobe fins**. The name Sarcopterygii used to refer solely to the lobe-finned fishes, but since it has become clear that terrestrial vertebrates (tetrapods) evolved from such fishes, the definition of the group has been expanded to include both lobe-finned fishes and tetrapods (see Figure 34.1). In the **lobe-finned fishes**, the fins are supported by skeletal extensions of the pectoral and pelvic areas that are moved by muscles within the fins.

The fossil record revealed that the Actinistia, or coelacanths, were a very successful group in the Devonian period, but all fishes of the class were believed to have died off at the end of the Mesozoic era, some 65 million years ago. You can therefore imagine the scientific excitement when in 1938, a modern coelacanth was discovered as part of the catch of a boat fishing near the Chalumna River in South Africa (**Figure 34.8**). Intensive searches in the area revealed that coelacanths were living in deep waters off the southern African coast and especially off a group of islands near the coast of Madagascar called the Comoros Islands. Another species was found more recently in Indonesian waters.

Early-diverging lobe-finned fishes probably evolved in fresh water and had lungs, but the coelacanth lost them and returned to the sea. One distinctive feature of this group is a special joint in the skull that allows the jaws to open extremely wide and gives the coelacanth a powerful bite. As further evidence of the coelacanth's unusual body plan, its swim bladder is filled with oil rather than gas, although it serves a similar purpose—to increase buoyancy.

The Dipnoi, or **lungfishes**, like the coelacanths, are also not currently a very species-rich class, having just three genera

(a) Lionfish (*Pterois volitans*)

(b) Whitemouth moray eel (*Gymnothorax meleagris*)

(c) Leafy sea dragon (*Phycodurus eques*)

Figure 34.7 The diversity of ray-finned fishes.

Concept check: What features distinguish ray-finned fishes from sharks?

Figure 34.8   A lobe-finned fish, the coelacanth (*Latimeria chalumnae*).

Figure 34.9   An Australian lungfish (*Neoceratodus forsteri*).

Concept check: *How are lungfishes similar to coelacanths?*

and six species (**Figure 34.9**). Lungfishes live in oxygen-poor freshwater swamps and ponds. They have both gills and lungs, the latter of which enable them to come to the surface and gulp air. In fact, lungfish will drown if they are unable to breathe air. When ponds dry out, some species of lungfish can dig a burrow and survive in it until the next rain. Because they also have muscular lobe fins, they are often able to successfully traverse quite long distances over shallow-bottomed lakes that may be drying out.

The morphological features of coelacanths, lungfishes, and primitive terrestrial vertebrates, together with the similarity of their nuclear genes, suggest to many scientists that lobe-fin ancestors gave rise to three lineages: the coelacanths, the lungfishes, and the tetrapods. In the next section we will examine the biology of tetrapods in more detail.

34.4 Tetrapods: Gnathostomes with Four Limbs

During the Devonian period, from about 417 to 354 million years ago, a diversity of plants colonized the land. The presence of plants served as both a source of oxygen and an extensive food source for animals that ventured out of the aquatic environment. Terrestrial arthropods, especially insects, evolved to feed on these plants and provided an additional food source for any vertebrate that could colonize the land.

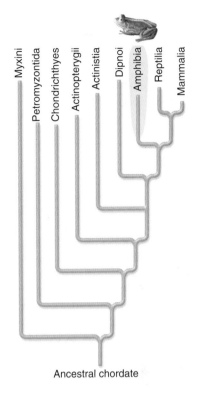

The transition to life on land involved a large number of adaptations. Paramount among these were adaptations preventing desiccation and making locomotion and reproduction on land possible. We have seen that the lungfishes evolved the ability to breathe air. In this section, we begin by outlining the development of the **tetrapods**, vertebrate animals having four legs or leglike appendages. We will discuss the first terrestrial vertebrates and their immediate descendants, the amphibians. We will then explore the characteristic features and diversity of modern amphibians.

The Origin of Tetrapods Involved the Development of Four Limbs

Over the Devonian period, fossils record the evolution of sturdy lobe-finned fishes that became fishes with four limbs. The abundance of light and nutrients in shallow waters encouraged a profusion of plant life and the invertebrates that fed on them. The development of lungs enabled lungfishes to colonize these productive yet often oxygen-poor waters. Here, the ability to move in shallow water chock full of plants and debris was more vital than the ability to swim swiftly through open water and may have favored the progressive development of sturdy limbs. As an animal's weight began to be borne more by the limbs, the vertebral column strengthened, and hip bones and shoulder bones were braced against the backbone for added strength. Such modifications, while seemingly dramatic, are believed to be caused by relatively simple changes in the expression of genes, especially *Hox* genes (see Feature Investigation). In particular, *Hox* genes 9–13 work together to specify limb formation from the proximal to the distal direction, meaning from close to the point of attachment to the body to the terminal end of the limb (**Figure 34.10**).

One of the transitional forms between fish and tetrapods was *Tiktaalik rosae*, nicknamed a fishapod, which we discussed back in Figure 23.5. Fishapods had broad skulls with eyes mounted on the top, and lungs. They also possessed pectoral fins with five finger-like bones. This is an important species, for it may represent a so-called stem species, a common tetrapod

Hox genes

| 9 | 10 | 11 | 12 | 13 |

Scapula   Humerus   Radius, ulna,   Carpals   Metacarpals
                        some carpals   and phalanges

Proximal ◀━━━━━━━━━━━━━━━━━▶ Distal

Figure 34.10 The roles of *Hox* genes 9–13 in specifying limb formation from the proximal to distal direction. The axis of limb development in mice is shown, together with the associated genes.

ancestor located at the beginning of the tetrapod branch on the tree of life. Eventually, species more like modern amphibians evolved, species that were still tied to water for reproduction but increasingly fed on land. In these species, the vertebral column, hip bones, and shoulder bones grew even sturdier. Such changes were needed as the animal's weight was no longer supported by water but was borne entirely on the limbs. The evolution of a rib cage provided protection for the internal organs, especially the lungs and heart.

By the middle of the Carboniferous period, about 320 million years ago, species similar to modern amphibians had become common in the terrestrial environment. For example, *Cacops* was a large amphibian, as big as a pony (**Figure 34.11**). Its skin was heavy and tough, an adaptation that helped prevent water loss; its breathing was accomplished more by lungs than by skin; and it possessed **pentadactyl limbs** (limbs ending

in five digits). With a bonanza of terrestrial arthropods to feast on, the amphibians became very numerous and species rich, and the mid-Permian period, some 260 million years ago, is sometimes known as the Age of Amphibians. However, most of the large amphibians became extinct at the end of the Permian period, coincident with the radiation of the reptiles, though whether the rise of reptiles was responsible for the extinction of amphibians is not known. This was the largest extinction event in history, and many other organisms went extinct as well. Most surviving amphibians were smaller organisms resembling modern species.

Figure 34.11 A primitive tetrapod. *Cacops* was a large, early amphibian of the Permian period.

Concept check: What were the advantages to animals of moving on to land?

FEATURE INVESTIGATION

Davis, Capecchi, and Colleagues Provide a Genetic-Developmental Explanation for Limb Length in Tetrapods

The development of limbs in tetrapods was a vital step that allowed animals to colonize land. The diversity of vertebrate limb types is amazing, from fins in fish and marine mammals to legs and arms in primates to different wing types in bats and birds. Early in vertebrate evolution, an ancestral gene complex was duplicated twice to give rise to four groups of genes, called *Hox A, B, C,* and *D,* which controlled limb development. In 1995, Allen Davis, Mario Capecchi, and colleagues analyzed the effects of mutations in specific *Hox* genes that are responsible for determining limb formation in mice. The vertebrate forelimb is divided into three zones, humerus (upper arm), radius and ulna (forearm), and carpals, metacarpals, and phalanges (digits). The authors had no specific hypothesis in mind; their goal was to understand the role of *Hox* genes in limb formation. As described in **Figure 34.12**, they began with strains of mice carrying loss-of-function mutations in *HoxA-11* or *HoxD-11*

that, on their own, did not cause dramatic changes in limb formation. They bred the mice to obtain offspring carrying one, two, three, or four loss-of-function mutations. Double mutants exhibited dramatically different phenotypes not seen in mice homozygous for the individual mutations. The radius and ulna were almost entirely eliminated.

As seen in the data, the mutations affected the formation of limbs. For example, the wrist contains seven bones: three proximal carpals—called navicular lunate (NL), triangular (T), and pisiform (P)—and four distal carpals (d1–d4). In mice with the genotypes *aaDD* and *AAdd*, the proximal carpal bones are usually fused together. Individual heterozygotes (*AADd* and *AaDD*) do not show this defect, but compound heterozygotes (*AaDd*) often do. Therefore, any two mutant alleles (either from both *HoxA-11* and *HoxD-11* or one from each locus) will cause carpal fusions. Deformities became even more severe with three mutant alleles (*Aadd* or *aaDd*) or four mutant alleles (*aadd*) (data not shown in the figure). Thus, scientists have shown that relatively simple mutations can control relatively large changes in limb development.

Figure 34.12   Relatively simple changes in *Hox* genes control limb formation in tetrapods.

GOAL To determine what role *Hox* genes have in limb development in mice.

KEY MATERIALS Mice with individual mutations in *HoxA-11* and *HoxD-11* genes.

Experimental level	Conceptual level

1 Breed mice with individual mutations in *HoxA-11* and *HoxD-11* genes. (The *A* and *D* refer to wild-type alleles; *a* and *d* are mutant alleles.)

AaDd mice

The mice bred were heterozygous for both genes (*AaDd*).

Based on previous studies, researchers expect mutant mice to produce viable offspring, perhaps with altered limb morphologies.

2 Using molecular techniques described in Chapter 20, obtain DNA from the tail and determine the genotypes of offspring.

The resulting genotypes would occur in Mendelian ratios, generating mice with different combinations of wild-type and mutant alleles.

	(AD)	(Ad)	(aD)	(ad)
(AD)	*AADD*	*AADd*	*AaDD*	*AaDd*
(Ad)	*AADd*	*AAdd*	*AaDd*	*Aadd*
(aD)	*AaDD*	*AaDd*	*aaDD*	*aaDd*
(ad)	*AaDd*	*Aadd*	*aaDd*	*aadd*	← Double mutant

9:3:3:1 phenotypic ratio

3 Stain the skeletons and compare the limb characteristics of the wild-type mice (*AADD*) to those of strains carrying mutant alleles in one or both genes.

Mutant mice may have altered bone morphologies.

aadd

AADD

4 **THE DATA**

Genotype	Carpal bone fusions (% of mice showing the fusion)			
	Normal (none fused)	NL fused to T	T fused to P	NL fused to T and P
AADD	100	0	0	0
AaDD	100	0	0	0
aaDD	33	17	50	0
AADd	100	0	0	0
AAdd	0	17	17	67
AaDd	17	17	33	33

5   CONCLUSION Relatively simple mutations involving two genes can cause large changes in limb development.

6   SOURCE Davis, P. A. et al. 1995. Absence of radius and ulna in mice lacking *Hoxa-11* and *Hoxd-11*. *Nature* 375:791–795.

Experimental Questions

1. What was the purpose of the study conducted by A. P. Davis and colleagues?

2. How were the researchers able to study the effects of individual genes?

3. Explain the results of the experiment shown in Figure 34.12 and how this relates to limb development in vertebrates.

Amphibian Lungs and Limbs Are Adaptations to a Semiterrestrial Lifestyle

Amphibians (from the Greek, meaning two lives) live in two worlds: They have successfully invaded the land but most must return to the water to reproduce. One of the first challenges terrestrial animals had to overcome was breathing air when on land. Amphibians can use the same technique as lungfishes: They open their mouths to let in air. Alternatively, they may take in air through their nostrils. They then close and raise the floor of the mouth, creating a positive pressure that pumps air into the lungs. This method of breathing is called **buccal pumping**. In addition, the skin of amphibians is much thinner than that of fishes, and amphibians absorb oxygen from the air directly through their outer moist skin or through the skin lining of the inside of the mouth or pharynx.

Amphibians have a three-chambered heart, with two atria and one ventricle. One atrium receives blood from the body, and the other receives blood from the lungs. Both atria pump blood into the single ventricle, which pumps some blood to the lungs and some to the rest of the body (look ahead to Figure 47.4b). This form of circulation allows the tissues to receive well-oxygenated blood at a higher pressure than is possible via single circulation, because some of the blood that returns to the heart is directly pumped to the tissues without being slowed down by passage through the lung capillaries. This development enhances the delivery of nutrients and oxygen to the tissues, but it is still not maximally efficient because some oxygenated and deoxygenated blood are mixed in the ventricle.

Because the skin of amphibians is so thin, the animals face the problem of desiccation, or drying out. As a consequence, even amphibian adults are more abundant in damp habitats, such as swamps or rain forests, than in dry areas. Also, most amphibians cannot venture too far from water because their larval stages are still aquatic. In frogs and toads, fertilization is generally external, with males shedding sperm over the gelatinous egg masses laid by the females in water (**Figure 34.13a**). The fertilized eggs lack a shell and would quickly dry out if exposed to the air. They soon hatch into tadpoles (**Figure 34.13b**), small fishlike animals that lack limbs and breathe through gills. As the tadpole nears the adult stage, the tail and gills are resorbed, and limbs and lungs appear (**Figure 34.13c**). Such a dramatic change in body form is known as metamorphosis, a process regulated by hormones from the thyroid gland. There are a few species of amphibians that do not require water to reproduce. These species are ovoviparous or viviparous, retaining the eggs in the reproductive tract and giving birth to live young.

Modern Amphibians Include a Variety of Frogs, Toads, Salamanders, and Caecilians

Approximately 4,800 living amphibian species are known, and the vast majority of these, nearly 90%, are frogs and toads of

(a) Gelatinous mass of amphibian eggs

(b) Tadpole

(c) Tadpole undergoing metamorphosis

Figure 34.13   Amphibian development in the wood frog (*Rana sylvatica*).   (a) Amphibian eggs are laid in gelatinous masses in water. (b) The eggs develop into tadpoles, aquatic herbivores with a fishlike tail that breathe through gills. (c) During metamorphosis, the tadpole loses its gills and tail and develops limbs and lungs.

(a) Tree frog

(b) A caecilian

(c) Mud salamander

Figure 34.14   Amphibians.   (a) Most amphibians are frogs and toads of the order Anura, including this red-eyed tree frog (*Agalychnis callidryas*). **(b)** The order Gymnophiona includes wormlike caecilians such as this species from Colombia, *Caecilia nigricans*. **(c)** The order Caudata includes species such as this mud salamander (*Pseudotriton montanus*).

> *Concept check:* Do all amphibians produce tadpoles?

the order Anura (from the Greek, meaning without tail) (**Figure 34.14a**). The other two orders are the Gymnophiona (from the Greek, meaning naked snake), the wormlike caecilians, and the Caudata (from the Latin, meaning tail visible), the salamanders. Global warming is currently threatening many anurans with extinction (see Chapter 54).

Adult anurans are carnivores, eating a variety of invertebrates by catching them on a long, sticky tongue. In contrast, the aquatic larvae (tadpoles) are primarily herbivores. Frogs generally have smooth, moist skin and long hind legs, making them excellent jumpers and swimmers. In addition to secreting mucus, which keeps their skin moist, some frogs can also secrete poisonous chemicals that deter would-be predators. Some amphibians advertise the poisonous nature of their skin with warning coloration (look ahead to Figure 57.10b). Others use camouflage as a way of avoiding detection by predators. Toads have a drier, bumpier skin and shorter legs than frogs. They are less impressive leapers than frogs, but toads can better tolerate drier conditions.

Caecilians (order Gymnophiona) are a small order of about 160 species of legless, nearly blind amphibians (**Figure 34.14b**). Most are tropical and burrow in forest soils, but a few live in ponds and streams. They are secondarily legless, which means they evolved from legged ancestors. Caecilians eat worms and other soil invertebrates and have tiny jaws equipped with teeth. In this order, fertilization is internal, and females usually bear live young. The young are nourished inside the mother's body by a thick, creamy secretion known as uterine milk. In most

caecilian species, the young grow into adults about 30 cm long, though species up to 1.3 m in length are known.

The salamanders (order Caudata) possess a tail and have a more elongate body than anurans (**Figure 34.14c**). During locomotion, they seem to sway from side to side, perhaps reminiscent of how the earliest tetrapods may have walked. Like frogs, salamanders often have colorful skin patterns that advertise their distastefulness to predators. Salamanders retain their moist skin by living in damp areas under leaves or logs or beneath lush vegetation. They generally range in size from 10 to 30 cm. Fertilization is usually internal, with females using their cloaca to pick up sperm packets deposited by males. A very few salamander species do not undergo metamorphosis, and the newly hatched young resemble tiny adults. However, some species, such as the axolotl, retain the gills and tail fins characteristic of the larval stage into adulthood, a phenomenon known as paedomorphosis (refer back to Figure 25.16b).

34.5 Amniotes: Tetrapods with a Desiccation-Resistant Egg

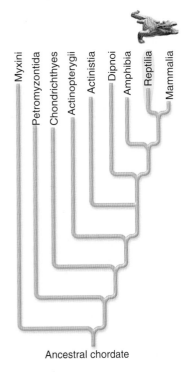

Myxini
Petromyzontida
Chondrichthyes
Actinopterygii
Actinistia
Dipnoi
Amphibia
Reptilia
Mammalia

Ancestral chordate

As mentioned, although amphibians live successfully in a terrestrial environment, they must lay their eggs in water or in a very moist place, so their shell-less eggs do not dry out on exposure to air. Thus, a critical innovation in animal evolution was the development of a shelled egg that sheltered the embryo from desiccating conditions on land. A shelled egg containing fluids was like a personal enclosed pond for each developing individual. Such an egg evolved in the common ancestor of turtles, lizards, snakes, crocodiles, birds, and mammals—a group of tetrapods collectively known as the **amniotes**. The amniotic egg permitted animals to lay their eggs in a dry place so that reproduction was no longer tied to water. It was truly a critical innovation. The amniotic egg untethered animals from water in much the same way as the development of seeds liberated plants from water (see Chapter 29).

In time, the amniotes came to dominate the Earth. Mammals are considered amniotes, too, because even though most of them do not lay eggs, they retain other features of amniotic reproduction. In this section, we begin by discussing in detail the morphology of the amniotic egg and other adaptations that permitted animal species to become fully terrestrial. We then discuss the biology of the reptiles, the first group of vertebrates to fully exploit land.

The Amniotic Egg and Other Innovations Permitted Life on Land

The **amniotic egg** (Figure 34.15) contains the developing embryo and the four separate extraembryonic membranes that it produces:

1. The innermost membrane is the **amnion**, which protects the developing embryo in a fluid-filled sac called the amniotic cavity.
2. The **yolk sac** encloses a stockpile of nutrients, in the form of yolk, for the developing embryo.
3. The **allantois** functions as a disposal sac for metabolic wastes.
4. The **chorion**, along with the allantois, provides gas exchange between the embryo and the surrounding air.

Surrounding the chorion is the albumin, or egg white, which also stores nutrients. The **shell** provides a tough, protective covering that is not very permeable to water and prevents the embryo from drying out. However, the shell remains permeable to oxygen and carbon dioxide, so the embryo can breathe. In birds, this shell is hard and calcareous, whereas in reptiles and early-diverging mammals such as the platypus and echidna, it is soft and leathery. In most mammals, however, the embryos embed into the wall of the uterus and receive their nutrients directly from the mother.

Along with the amniotic egg, other critical innovations that enabled the conquest of land include the following:

- **Desiccation-resistant skin.** Whereas the skin of amphibians is moist and aids in respiration, the skin of amniotes is thicker and water resistant and contains keratin, a tough protein. This requires that most gas exchange takes place through the lungs.
- **Thoracic breathing.** Amphibians use buccal pumping to breathe, contracting the mouth to force air into the lungs. In contrast, amniotes use **thoracic breathing**, in which coordinated contractions of muscles expand the rib cage, creating a negative pressure to suck air in and then forcing it out later. This results in a greater volume of air being displaced with each breath than with buccal pumping.
- **Water-conserving kidneys.** The ability to concentrate wastes prior to elimination and thus conserve water is an important role of the amniotic kidneys.
- **Internal fertilization.** Because sperm cannot penetrate a shelled egg, fertilization occurs internally, within the female's body before the shell is secreted. In this process, the male of the species often uses a copulatory organ (penis) to transfer sperm into the female reproductive tract. However, birds usually transfer sperm from cloaca to cloaca.

Reptiles Include Turtles, Lizards, Snakes, Crocodilians, Dinosaurs, and Birds

Early amniote ancestors gave rise to all modern amniotes we know today, from lizards and snakes to birds and mammals.

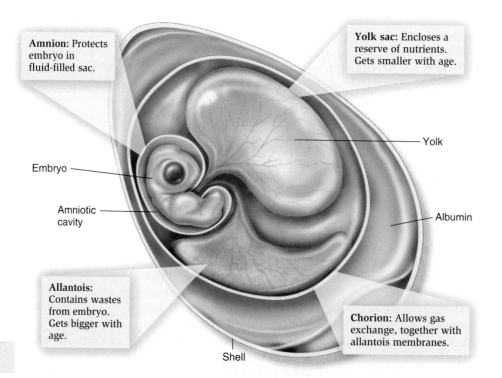

Figure 34.15 **The amniotic egg.**

Concept check: What are some of the other critical innovations of amniotes?

Amnion: Protects embryo in fluid-filled sac.

Yolk sac: Encloses a reserve of nutrients. Gets smaller with age.

Embryo

Amniotic cavity

Yolk

Albumin

Allantois: Contains wastes from embryo. Gets bigger with age.

Chorion: Allows gas exchange, together with allantois membranes.

Shell

(a) Green turtle

(b) Florida worm lizard

(c) Gila monster

Figure 34.16 **A variety of reptiles.** **(a)** A green turtle (*Chelonia mydas*) laying eggs in the sand in Malaysia. **(b)** The Florida worm lizard (*Rhineura floridana*) is an amphisbaenian, a type of legless soil-burrowing lizard. **(c)** The Gila monster (*Heloderma suspectum*), one of only two venomous lizards, is an inhabitant of the desert Southwest of the U.S. and of Mexico.

The traditional view of amniotes involved three living classes: the reptiles (turtles, lizards, snakes, and crocodilians), birds, and mammals. As we will see later in the chapter, modern systematists have argued that enough similarities exist between birds and the classic reptiles that birds should be considered part of the reptilian lineage. This is the classification scheme that we will follow in this chapter. The fossil record includes other reptilian classes, all of which are extinct, including two classes of dinosaurs (ornithischian and saurischian dinosaurs), flying reptiles (pterosaurs), and two classes of ancient aquatic reptiles (icthyosaurs and plesiosaurs).

Class Testudines: The Turtles Turtles is an umbrella term for terrestrial species, also called tortoises, and aquatic species, sometimes known as terrapins. The turtle lineage is ancient and has remained virtually unchanged for 200 million years. The major distinguishing characteristic of the turtle is a hard protective shell into which the animal can withdraw its head and limbs. In most species, the vertebrae and ribs are fused to form this shell. All turtles lack teeth but have sharp beaks for biting.

Most turtles are aquatic and have webbed feet. The forelimbs of marine species have evolved to become large flippers. All turtles, even the aquatic species, lay their eggs on land, usually in soft sand. The gender of hatchlings is temperature dependent, with high temperatures producing more females (see also Chapter 16). Marine species often make long migrations to sandy beaches to lay their eggs (**Figure 34.16**). Most land tortoises are quite slow movers, possibly due to a low metabolic rate and a heavy shell. However, they are very long-lived species, often surviving for 120 years or more. Furthermore, turtles do not appear to show reproductive senescence or aging, reproducing continually throughout their lifetime. Most organs such as the liver, lungs, and kidneys of a centenarian turtle function as effectively as do organs in young individuals, prompting genetic researchers to examine the turtle genome for longevity genes. Many turtle species are in danger of extinction,

due to egg hunting, destruction of habitat and nesting sites, and death from entanglement in fishing nets.

Class Lepidosauria: Lizards and Snakes The class Lepidosauria is the largest class within the traditional reptiles, with about 4,800 species of lizards (order Sauria) and 3,000 species of snakes (order Serpentes). Many species have an elongated body form. One of the defining characteristics of the orders is a **kinetic skull**, in which the joints between various parts of the skull are extremely mobile. The lower jaw does not join directly to the skull but rather is connected by a multijointed hinge, and the upper jaw is hinged and movable from the rest of the head. This allows the jaws to open relatively wider than other vertebrate jaws, with the result that lizards, and especially snakes, can swallow large prey (**Figure 34.17**). Nearly all species are carnivores.

Figure 34.17 **The kinetic skull.** In snakes and lizards, both the top and bottom of the jaw is hinged on the skull, thereby permitting large prey to be swallowed. This Halloween snake (*Pliocercus euryzonus*) is swallowing a Costa Rican rain frog.

A main difference between lizards and snakes is that lizards generally have limbs, whereas snakes do not. Leglessness is a derived condition, meaning snake ancestors possessed legs but later lost them. Also, snakes may be venomous, whereas lizards usually are not. However, there are exceptions to these general rules. Many legless lizard species exist (see Figure 34.16b), and two lizards are venomous: the Gila monster (*Heloderma suspectum*) of the U.S. Southwest (see Figure 34.16c) and the Mexican beaded lizard (*Heloderma horridum*). A more reliable distinguishing characteristic is that lizards have movable eyelids and external ears (at least ear canals), whereas snakes do not.

Class Crocodilia: The Crocodiles and Alligators   The Crocodilia is a small class of large, carnivorous, aquatic animals that have remained essentially unchanged for nearly 200 million years (**Figure 34.18**). Indeed, these animals existed at the same time as the dinosaurs. Most of the 23 recognized species live in tropical or subtropical regions. There are only two extant species of alligators: one living in the southeastern U.S. and one found in China.

While the class is small, it is evolutionarily very important. Crocodiles have a four-chambered heart, a feature they share with birds and mammals (look ahead to Figure 47.4c). In this regard, crocodiles are more closely related to birds than to any other living reptile class. Their teeth are set in sockets, a feature typical of the dinosaurs and the earliest birds. Similarly, crocodiles care for their young, another trait they have in common with birds. These and other features suggest that crocodiles and birds are more closely related than crocodiles and lizards. As with turtles, nest temperature influences the sex ratio of offspring. With crocodiles, however, when temperatures are warm, more males are produced. Biologists are concerned that global warming may reduce the number of breeding females, leading to extinction of some species.

Classes Ornithischia and Saurischia: The Dinosaurs   In 1841, the English paleontologist Richard Owen coined the term **dinosaur** (from the Greek, meaning terrible lizard) to describe some of the wondrous fossil animals discovered in the 19th century. About 215 million years ago, dinosaurs were the dominant tetrapods on Earth and remained so for 150 million years, far longer than any other vertebrate. The two main classes were the ornithischian, or bird-hipped dinosaurs, which were herbivores such as *Stegosaurus*; and the saurischian, or lizard-hipped dinosaurs, which were fast, bipedal carnivores such as *Tyrannosaurus* (**Figure 34.19**). In contrast to the limbs of lizards, amphibians, and crocodiles, which splay out to the side, the legs of dinosaurs were positioned directly under the body, like pillars, a position that could help support their heavy body. Because less energy was devoted to lifting the body from the ground, some dinosaurs are believed to have been fast runners. Members of different but closely related classes—the pterosaurs (the first vertebrates to fly) and ichthyosaurs and plesiosaurs (marine reptiles)—were also common at this time.

Dinosaurs were the biggest animals ever to walk on the planet, with some animals weighing up to 50 tonnes (metric tons) or over 100,000 pounds. The variety of the thousands of dinosaur species found in fossil form around the world is staggering. However, perhaps not surprisingly for such long-extinct species, scientists are still hotly debating the details of their lives. For example, an issue still unresolved is whether some dinosaur species were **endothermic**, that is, capable of generating and retaining body heat through their metabolism, just as birds and mammals are. Another issue is whether dinosaurs exhibited parental care of their young.

(a) American alligator

(b) American crocodile

Figure 34.18   Crocodilians.   The Crocodilia is an ancient class that has existed unchanged for millions of years. **(a)** Alligators, such as this American alligator (*Alligator mississippiensis*), have a broad snout, and the lower jaw teeth close on the inside of the upper jaw (and thus are almost completely hidden when the mouth is closed). **(b)** Crocodiles, including this American crocodile (*Crocodylus acutus*), have a longer, thinner snout, and the lower jaw teeth close on the outside of the upper jaw (and thus are visible when the mouth is closed).

Concept check: *In what ways are crocodilians similar to birds?*

(a) Ornithischian (*Stegosaurus*) **(b) Saurischian (*Tyrannosaurus*)**

Figure 34.19 Classes of dinosaurs. (a) Herbivorous ornithischians included *Stegosaurus*, and (b) carnivorous saurischians included bipedal species such as *Tyrannosaurus rex*.

All dinosaurs, and many other animals, went extinct quite abruptly at the end of the Cretaceous period, about 65 million years ago. Although widely attributed to climatic change brought about by the impact of a meteorite or comet, scientists continue to debate the cause of this mass extinction. We do not yet know why all the dinosaurs died out, while many other animals, including small mammals, survived.

Class Aves: The Birds The defining characteristics of birds (class Aves, plural of the Latin *avis*, meaning bird) are that they have feathers and nearly all species can fly. As we will see, the ability to fly has shaped nearly every feature of the bird body. The other vertebrates that have evolved the ability to fly, the bats and the now-extinct pterosaurs, used skin stretched tight over elongated limbs to fly. Such a surface can be irreparably

damaged, though some holes may heal remarkably quickly. In contrast, birds use feathers, epidermal outgrowths that can be replaced if damaged. Recent studies in cell signaling show that the developmental changes necessary for animals to grow feathers instead of scales may not be that complex at the molecular level.

In the rest of this section, we will discuss the likely evolution of birds from dinosaur ancestors, outline the key characteristics of birds, and provide a brief overview of the various bird orders.

Modern Birds Probably Evolved from Small, Feather-Covered Dinosaurs

To trace the evolution of birds, we must look for transitional forms, the earliest type of animals that had feathers. One of the first known fossils exhibiting the faint impression of feathers was *Archaeopteryx lithographica* (from the Greek, meaning ancient wings and stone picture), found in a limestone quarry in Germany in 1861. The fossil was dated at 150 million years old. Except for the presence of feathers, *Archaeopteryx* appears to have had features similar to those of dinosaurs (**Figure 34.20a**). First, the fossil had an impression of a long tail with many vertebrae, a dinosaur feature. Some modern birds have long tails, but they are made of feathers, with the actual tailbone being much reduced. Second, the wings had claws halfway down the leading edge, another dinosaur-like character. Among modern birds, only the hoatzin, a South American swamp-inhabiting bird, has claws on its wings, which enable the chicks to climb back into the nest if they fall out. A third dinosaur-like feature

  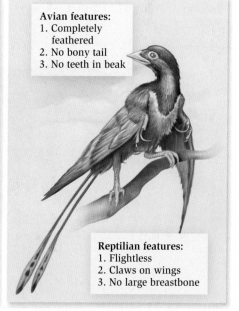

Avian features:
1. Feathered wings and tail

Avian features:
1. Feathered wings and tail

Avian features:
1. Completely feathered
2. No bony tail
3. No teeth in beak

Reptilian features:
1. Long, bony tail
2. Claws on wings
3. Toothed beak
4. No large breastbone

Reptilian features:
1. Short, bony tail
2. Claws on wings
3. Toothed beak
4. No large breastbone

Reptilian features:
1. Flightless
2. Claws on wings
3. No large breastbone

(a) *Archaeopteryx lithographica* **(b) *Caudipteryx zoui*** **(c) *Confuciusornis sanctus***

Figure 34.20 Transitional forms between dinosaurs and birds. (a) *Archaeopteryx lithographica* was a Jurassic animal with dinosaur-like features as well as wings and feathers. (b) *Caudipteryx zoui* was a dinosaur with feathers on its tail and wings. (c) *Confuciusornis sanctus* was a birdlike animal with a horny, toothless beak.

is *Archaeopteryx*'s toothed beak. Fourth, the fossils show that *Archaeopteryx* lacked an enlarged breastbone, a feature that modern birds possess to anchor their large flight muscles, so it likely could not fly.

Similarities between the structure of the skull, feet, and hind leg bones have led scientists to conclude that *Archaeopteryx* is closely related to **theropods**, a group of bipedal saurischian dinosaurs. (Note that despite their name, the ornithiscian, or bird-hipped dinosaurs, were not ancestral to modern birds.) The wings and feathers of *Archaeopteryx* may have enabled it to glide from tree to tree, helped to it keep warm, or cut out the glare when folded over its head when hunting, in much the same way as some herons fold their wings over their heads when they are fishing. Later, the wings and feathers may have taken on functions of flight. We often find that evolution proceeds when previously evolved structures are co-opted into different uses, a diversity principle known as descent with modification.

In 1997, paleontologists unearthed fossils of about the same age as *Archaeopteryx* in China that similarly suggest a close kinship between dinosaurs and modern birds. *Caudipteryx zoui* was a dinosaur-like animal with feathers on its wings and tail and a toothed beak (**Figure 34.20b**). *Confuciusornis sanctus* was a small, flightless but completely feathered dinosaur lacking the long, bony tail and toothed jaw found in other theropod dinosaurs. Its large tail feathers may have functioned in courtship displays (**Figure 34.20c**).

These three species—*Archaeopteryx*, *Caudipteryx*, and *Confuciusornis*—help trace a lineage from dinosaurs to birds. By the early Cretaceous period, and only a relatively short period after *Archaeopteryx* evolved, the fossil record shows the existence of a huge array of bird types resembling modern species. These were to share the skies with pterosaurs for 70 million years, before eventually having the airways to themselves.

Birds Have Feathers, a Lightweight Skeleton, Air Sacs, and Reduced Organs

Modern birds possess many characteristics that reveal their reptilian ancestry. For example, they have scales on their feet and legs, and they lay shelled eggs. In addition, however, among living animals they have four features unique to birds, all of which are associated with flight.

1. **Feathers.** Feathers are modified scales that keep birds warm and enable flight (**Figure 34.21a**). Soft, downy feathers maintain heat, while stiffer contour feathers, supported on a modified forelimb, give the wing the airfoil shape it needs to generate lift. Each contour feather develops from a follicle, a tiny pit in the skin. If a feather is lost, a new one can be regrown. The contour feathers consist of many paired barbs, each of which supports barbules that contain hooks that interlock with barbules from neighboring barbs to give the feather its shape (**Figure 34.21b**).

2. **Air sacs.** Flight requires a great deal of energy generated from an active metabolism that requires abundant oxygen. Birds have nine air sacs—large, hollow sacs that may extend into the bones (look ahead to Figure 48.13). On inhalation, air passes into the posterior air sacs. On exhalation, this air is forced into the lungs, where oxygen is extracted. On the next inhalation, the air moves from the lungs into the anterior air sacs. On the next exhalation,

(a) Wing **(b) Feather structure** **(c) Pelican bone (*Pelicanus occidentalis*)**

Figure 34.21 Features of the bird wing and feather. **(a)** The wing is supported by an elongated and modified forelimb with extended fingers. **(b)** Each feather has a hollow shaft that supports many barbs, which, in turn, support barbules that interlock with hooks to give the feather its form. **(c)** The bones of a pelican (*Pelicanus occidentalis*) are hollow but crisscrossed with a honeycomb structure that provides added strength.

Concept check: *What adaptations in birds help reduce their body weight to enable flight?*

the air leaves the anterior air sacs. Air is therefore being constantly moved across the lungs during inhalation and exhalation. While making bird breathing very efficient, this process also makes birds especially susceptible to airborne toxins (hence, the utility of the canary in the coal mine; the bird's death signaled the presence of harmful carbon dioxide or methane gas that was otherwise unnoticed by miners).

3. **Reduction of organs.** To decrease the total mass the bird must carry, some organs are reduced in size or are lacking altogether. For example, birds have only one ovary and can carry relatively few eggs. As a result, they lay fewer eggs than most other reptile species. In fact, the gonads of both males and females are reduced, except during the breeding season, when they increase in size. Most birds also lack a urinary bladder. In addition, teeth loss has reduced weight at the head end.

4. **Lightweight bones.** Most bird bones are thin and hollow and are crisscrossed internally by tiny pieces of bone to give them a honeycomb structure (Figure 34.21c). However, because the limb bones are large, bird skeletons weigh about the same as the skeleton of a comparably sized mammal. An enlarged breastbone, or **sternum**, provides an anchor on which a bird's powerful flight muscles attach. These muscles may contribute up to 30% of the bird's body weight. Birds' skulls are lighter than skulls from mammals of approximately the same weight. The bird skull lacks teeth, but the keratin of bird beaks is tough and malleable, and it takes very different shapes in different species, depending on the function of the beak (Figure 34.22).

Birds also have other distinct features, though mammals also possess some of these. For example, birds have a warm body temperature, which ensures rapid metabolism and the quick production of ATP that these active organisms need to fuel flight and other activities. In fact, birds' body temperatures are generally 40–42°C, considerably warmer than the human body's average of 37°C. Birds have a double circulation and a four-chambered heart to ensure rapid blood circulation. Rapid flight also requires acute vision, and bird vision is among the best in the vertebrate world. Generally, plant material is not energy-rich enough to supply the dietary needs of birds, so most birds are carnivores, eating insects and other invertebrates. However, some birds, such as parrots, eat just the more-nutrient-rich fruits and seeds. Bird eggs also have to be kept warm for successful development, which entails brooding by an adult bird. Often, the males and females take turns brooding so that one parent can feed and maintain its strength. Picking successful partners is therefore an important task, and birds often engage in complex courtship rituals.

There Are Many Orders of Birds, All with the Same Body Plan

Birds are the most species-rich class of terrestrial vertebrates, with 28 orders, 166 families, and about 9,600 species (Table 34.2). Despite this diversity, birds lack the variety of body

(a) Cracking beak

(b) Scooping beak

(c) Tearing beak

(d) Probing beak

(e) Nectar-feeding beak

(f) Sieving beak

Figure 34.22   A variety of bird beaks. Birds have evolved a variety of beak shapes used in different types of food gathering. (a) Hyacinthe macaw (*Anodorhynchus hyacinthinus*)—cracking. (b) White pelican (*Pelecanus onocrotalus*)—scooping. (c) Verreaux's eagle (*Aquila verreauxii*)—tearing. (d) American avocet (*Recurvirostra americana*)—probing. (e) Lucifer hummingbird (*Calothorax lucifer*)—nectar feeding. (f) Roseate spoonbill (*Ajaia ajaia*)—sieving.

shapes that exist in the other endothermic class of vertebrates, the mammals, where some species swim, others fly, others walk on four legs, and yet others walk only on two legs. Most birds fly, and therefore, most have the same general body shape. The biggest departures from this body shape are the flightless birds, including the cassowaries, emus, and ostriches. These birds have smaller wing bones, and the keel on the breastbone is greatly reduced or absent. Penguins are also flightless birds whose upper limbs are modified as flippers used in swimming.

Table 34.2 The Main Orders of Birds, in Order of Species Richness

Order		Examples (approx. # of species)	Main characteristics
Passeriformes		Robins, starlings, sparrows, warblers (5,300)	Perching birds with perching feet; songbirds
Apodiformes		Hummingbirds, swifts (430)	Fast fliers with rapidly beating wings; small bodies
Piciformes		Woodpeckers, toucans (380)	Large with specialized beaks; two toes pointing forward and two backward
Psittaciformes		Parrots, cockatoos (340)	Large, powerful beaks
Chadradriiformes		Seagulls, wading birds (330)	Shorebirds
Columbiformes		Doves, pigeons (300)	Round bodies; short legs
Falconiformes		Eagles, hawks, kestrels, vultures (290)	Diurnal carnivores; birds of prey; powerful talons; strong beaks
Galliformes		Chickens, pheasants, quail (270)	Often large birds; weak flyers; ground nesters
Coraciiformes		Hornbills, kingfishers (200)	Large beaks; cavity nesters
Anseriformes		Ducks, swans, geese (150)	Able to swim; webbed feet; broad bills
Strigiformes		Owls (150)	Nocturnal carnivores; powerful talons; strong beaks
Pelecaniformes		Pelicans, frigate birds, cormorants (55)	Large, colonial fish eaters; often tropical
Sphenisciformes		Penguins (18)	Flightless; wings modified into flippers for swimming; marine; Southern Hemisphere
Casuariformes		Cassowaries, emus (3)	Large, flightless; Australia and New Guinea
Struthioniformes		Ostrich (1)	Large, flightless; only two toes; Africa only

34.6 Mammals: Milk-Producing Amniotes

Mammals evolved from amniote ancestors earlier than birds. About 225 million years ago, the first mammals appeared in the mid-Triassic period. They evolved from small mammal-like reptiles that went extinct about 170 million years ago. Mammals survived, although most are believed to have been small, insect-eating species that lived in the shadows of dinosaurs. However, in January 2005, two fossils of a 130-million-year-old mammalian genus called *Repenomamus* were discovered that challenge the notion of mammals as small insect eaters. One fossil was of an animal estimated to weigh about 13 kg (30 lb), and the other had the remains of a baby dinosaur in its stomach.

After the extinction of the dinosaurs in the Cretaceous period, some 65 million years ago, mammals flourished. Today, biologists have identified about 5,500 species of mammals with a diverse array of lifestyles, from fishlike dolphins to bird-like bats, and from small insectivores such as shrews to large

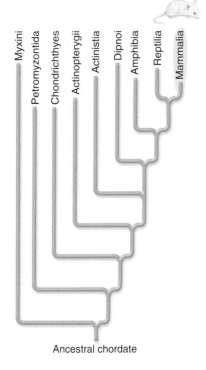

Myxini
Petromyzontida
Chondrichthyes
Actinopterygii
Actinistia
Dipnoi
Amphibia
Reptilia
Mammalia

Ancestral chordate

herbivores such as giraffes and elephants. The range of sizes and body forms of mammals is unmatched by any other group, and mammals are prime illustrations of the concept that organismal diversity is related to environmental diversity. In this section, we will outline the features that distinguish mammals from other taxa. We will also examine the diversity of mammals that exists on Earth and will end by turning our attention to primates and, in particular, to the evolution of humans.

Mammals Have Mammary Glands, Hair, Specialized Teeth, and an Enlarged Skull

The distinguishing characteristics of mammals are the possession of mammary glands, hair, specialized teeth, and an enlarged skull.

- **Mammary glands.** Mammals, or the class Mammalia (from the Latin *mamma*, meaning breast), are named after the female's distinctive **mammary glands**, which secrete milk. Milk is a fluid rich in fat, sugar, protein, and vital minerals, especially calcium. Newborn mammals suckle this fluid, which helps promote rapid growth.

- **Hair.** All mammals have hair, although some have more than others. Whales have hair in utero, but adults are hairless or retain only a few hairs on their snout. Humans

are relatively hairless. In some animals, the hair is dense and is referred to as fur. In some aquatic species such as beavers, the fur is so dense it cannot be thoroughly wetted, so the hair underneath remains dry. Mammals are endothermic, and their fur is an efficient insulator. Hair can also take on functions other than insulation. Many mammals, including cats, dogs, walruses, and whales, have sensory hairs called vibrissae (**Figure 34.23a**). Hair can be of many colors, to allow the mammals to blend into their background (**Figure 34.23b**). In some cases, as in porcupines and hedgehogs, the hairs become long, stiffened, and sharp (quills) and serve as a defense mechanism (**Figure 34.23c**).

- **Specialized teeth.** Mammals are the only vertebrates with highly differentiated teeth—incisors, canines, premolars, and molars—that are adapted for different types of diets (**Figure 34.24**). Although teeth are generally present in all species, different teeth are larger, smaller, lost, or reduced, depending on diet. Of particular importance to carnivores such as wolves are the piercing canine teeth, whereas herbivorous species such as deer depend on their chisel-like incisors to snip off vegetation and on their many molars to grind plant material. Only mammals chew their food in this fashion. Rodent incisors grow continuously throughout life, and species such as beavers wear them down by gnawing tough plant material such as wood. Mammals that have different types of teeth are called heterodont; others, such as dolphins, where the teeth are of uniform size and shape, are called homodont.

- **Enlarged skull.** The mammalian skull differs from other amniote skulls in several ways. First, the brain is enlarged and is contained within a relatively large skull. Second, mammals have a single lower jawbone, unlike reptiles, whose lower jaw is composed of multiple bones. Third, mammals have three bones in the middle ear, as opposed to reptiles, which have one bone in the middle ear. Fourth, most mammals, except some seals, have external ears.

Figure 34.23   **Mammalian hair.**   **(a)** The sensory hairs (vibrissae) of the walrus (*Odobenus rosmarus*). **(b)** The camouflaged coat of a bobcat (*Lynx rufus*). **(c)** The defensive quills of the crested porcupine (*Hystrix africaeaustralis*).

(a) Sensory hairs

(b) Camouflaged coat

(c) Defensive quills

(a) Biting teeth

(b) Grinding teeth

(c) Gnawing teeth

(d) Tusks

(e) Grasping teeth

Figure 34.24 **Mammalian teeth.** Mammals have different types of teeth, according to their diet. **(a)** The wolf has long canine teeth for biting its prey. **(b)** The deer has a long row of molars for grinding plant material. **(c)** The beaver, a rodent, has long, continually growing incisors used to gnaw wood. **(d)** The elephant's incisors are modified into tusks. **(e)** Dolphins and other fishes or plankton feeders have numerous small teeth for grasping prey.

In addition to those uniquely mammalian characteristics, some, but not all, mammals possess these additional features:

- **The ability to digest plants.** Apart from tortoises and marine iguanas, certain species of mammals are the only large vertebrates alive today that can exist on a steady diet of grasses or tree leaves; indeed, most large mammals are herbivores. Though mammals cannot digest cellulose, the principal constituent of the cell wall of many plants, some species have a large four-chambered stomach containing cellulose-digesting bacteria. These bacteria can break down the cellulose and make the plant cell contents available to the animal. Others have an extensive cecum or large intestine where digestion occurs.

- **Horns and antlers.** Mammals are the only living class of vertebrates to possess horns or antlers. Many mammals, especially antelopes, cattle, and sheep, have horns, typically consisting of a bony core that is a permanent outgrowth of the skull surrounded by a hairlike keratin sheath, as shown in the large antelope called a kudu (**Figure 34.25a**). Rhinoceros horns are outgrowths of the epidermis, consisting of very tightly matted hair (**Figure 34.25b**). In contrast, deer antlers are made entirely of bone (**Figure 34.25c**). Deer grow a new set of antlers each year and shed them after the mating season. Hooves are also made of keratin and protect an animal's toes from the impact of its feet striking the ground.

Figure 34.25 **Horns and antlers in mammals.** Mammals have a variety of outgrowths that are used for defense or by males as weapons in contests over females. **(a)** The horns of this male kudu (*Tragelaphus strepsiceros*) are bony outgrowths of the skull covered in a keratin sheath. **(b)** The horns of the black rhinoceros (*Diceros bicornis*) are outgrowths of the epidermis, made of tightly matted hair. **(c)** The antlers of the caribou (*Rangifer tarandus*), also known as reindeer, are made entirely of bone and are grown and shed each year.

(a) Skull outgrowths

(b) Epidermal outgrowths

(c) Bony antlers

Mammals Are the Most Diverse Group of Vertebrates Living on Earth

Modern mammals are incredibly diverse (**Table 34.3**). They vary in size from tiny insect-eating bats, weighing in at only 2 g, to leviathans such as the blue whale, the largest animal ever known, which tips the scales at 100 tonnes (over 200,000 lbs). The 26 different mammalian orders are divided into two distinct subclasses. The subclass Prototheria contains only the order Monotremata, or **monotremes**, which are found in Australia and New Guinea. There are only five species: the duck-billed platypus (**Figure 34.26a**) and four species of echidna, a spiny animal resembling a hedgehog. Monotremes are early-diverging mammals that lay eggs rather than bear live young,

lack a placenta, and have mammary glands with poorly developed nipples. The mothers incubate the eggs, and upon hatching, the young simply lap up the milk as it oozes onto the fur.

The subclass Theria contains all remaining live-bearing mammals. The Theria are divided into two clades, the Metatheria and the Eutheria. The clade Metatheria, or the **marsupials**, is a group of seven orders, with about 280 species, including the rock wallaby pictured in **Figure 34.26b**. Once widespread, members of this order are now largely confined to Australia, although some marsupials exist in South America, and one species—the opossum—is found in North America. Fertilization is internal, and reproduction is viviparous in marsupials. Marsupials have a placenta that nourishes the embryo. Unlike other mammals, however, marsupials are extremely small when they

Table 34.3	The Main Orders of Mammals, in Order of Species Richness		
Order		**Examples (approx. # of species)**	**Main characteristics**
Rodentia		Mice, rats, squirrels, beavers, porcupines (2,277)	Plant eating; gnawing habit, with two pairs of continually growing incisor teeth
Chiroptera		Bats (1,116)	Insect or fruit eating; small; have ability to fly; navigate by sonar; nocturnal
Soricomorpha		Shrews, moles (428)	Insect eaters; primitive placental mammals
Primates		Monkeys, apes, humans (404)	Opposable thumb; binocular vision; large brains
Carnivora		Cats, dogs, weasels, bears, seals, sea lions (286)	Flesh-eating mammals; canine teeth
Artiodactyla		Deer, antelopes, cattle, sheep, goats, camels, pigs (240)	Herbivorous hoofed mammals, usually with two toes, hippopotamus and others with four toes; many with horns or antlers
Diprotodontia		Kangaroos, koalas, opossums, wombats (143)	Pouched mammals mainly found in Australia
Lagomorpha		Rabbits, hares (92)	Powerful hind legs; rodent-like teeth
Cetacea		Whales, dolphins (84)	Marine fishes or plankton feeders; front limbs modified into flippers; no hind limbs; little hair except on snout
Erinaceomorpha		Hedgehogs, moonrats (24)	Insect eaters; nocturnal; hedgehogs with stiff spines
Perissodactyla		Horses, zebras, tapirs, rhinoceroses (18)	Hoofed herbivorous mammals with odd number of toes, one (horses) or three (rhinoceroses)
Monotremata		Duck-billed platypuses, echidna (5)	Egg-laying mammals found only in Australia and New Guinea
Proboscidea		Elephants (3)	Long trunk; large, upper incisors modified as tusks

(a) Prototherian (duck-billed platypus)

(c) Eutherian (orangutan)

(b) Metatherian (rock wallaby)

Figure 34.26 **Diversity among mammals.** **(a)** Prototherians, such as this duck-billed platypus (*Ornithorhynchus anatinus*), lay eggs, lack a placenta, and possess mammary glands with poorly developed nipples. **(b)** Metatherians, or marsupials, such as this rock wallaby (*Petrogale assimilis*), feed and carry their developing young, or "joeys," in a ventral pouch. **(c)** Gestation lasts longer in eutherians, and their young are more developed at birth, as illustrated by this young orangutan (*Pongo pygmaeus*).

are born (often only 1 or 2 cm) and make their way to a ventral pouch called a marsupium for further development.

All the other mammalian orders are members of the clade Eutheria and are considered **eutherians**, or placental mammals, such as the orangutans shown in **Figure 34.26c**. Eutherians have a long-lived and complex placenta, compared to that of marsupials. In eutherians, fertilization is internal, and reproduction is viviparous, but the developmental period, or gestation, of the young is prolonged.

Of all the world's animals, the diversity of vertebrates, and especially mammals, is the most threatened by human activities. Many species are hunted for food. Others, such as wild cats and whales, are hunted for their products (fur and oil, respectively), and still others, such as the oryx, have simply been shot for sport.

Primates Are Mammals with Opposable Thumbs and a Large Brain

Of the mammalian orders, the primates, and specifically humans, have had the greatest impact on the world. Primates are primarily tree-dwelling species that are believed to have evolved from a group of small, arboreal insect-eating mammals about 85 million years ago, before dinosaurs went extinct. Primates have several defining characteristics, mostly relating to their tree-dwelling nature:

- **Grasping hands.** All primates have grasping hands, a characteristic that enables them to hold onto branches (see Figure 34.28a,b). Most primate species also possess an **opposable thumb**, a thumb that can be placed opposite

the fingers of the same hand, which gives them a precision grip and enables the manipulation of small objects. All primates except humans also have an opposable big toe.

- **Large brain.** Acute vision and other senses enhancing the ability to move quickly through the trees require the efficient processing of large amounts of information. As a result, the primate brain is large and well developed. In turn, this has facilitated complex social behaviors.

- **At least some digits with flat nails instead of claws.** This feature is believed to aid in the manipulation of objects.

- **Binocular vision.** Primates have forward-facing eyes that are positioned close together on a flattened face, though some other mammals share this characteristic. Jumping from branch to branch requires accurate judgment of distances. This is facilitated by binocular vision, in which the field of vision for both eyes overlaps, producing a single image.

- **Complex social behavior and well-developed parental care.**

Some of these characteristics are possessed by other animals. For example, binocular vision occurs in owls, grasping hands are found in raccoons, and relatively large brains occur in marine mammals. Primates are defined by possessing the whole suite of these characteristics together. Primates may be classified in several ways. Taxonomists often divide them into two groups: the strepsirrhini and the haplorrhini (**Figure 34.27**). The **strepsirrhini** contain the smaller species such as bush babies, lemurs, and pottos. These are generally nocturnal and smaller-brained primates with eyes positioned a little more

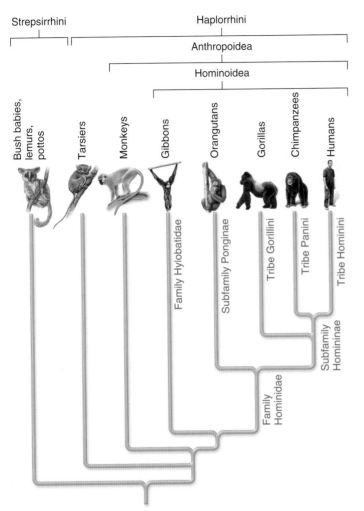

Figure 34.27   Evolutionary tree of the primates.

toward the side of their heads (**Figure 34.28a**). The strepsir-rhini are named for their wet noses with no fur at the tip. The **haplorrhini** have dry noses with a fully furred nose tip and fully forward-facing eyes. This group consists of the larger-brained and diurnal **anthropoidea**: the monkeys (**Figure 34.28b**) and the **hominoidea** (gibbons, orangutans, gorillas, chimpanzees, and humans) (**Figure 34.28c**). The tarsiers also belong in the haplorrhini, despite their small size, based on their forward-facing eyes and DNA similarities to the monkeys and apes.

What differentiates monkeys from hominoids? Most mon-keys have tails, whereas hominoids do not. In addition, apes have more mobile shoulder joints, broader rib cages, and a shorter spine. These features aid in brachiation, a swinging movement in trees. Apes also possess relatively long limbs and short legs and, with the exception of gibbons, are much larger than monkeys. The 20 species of hominoids are split into two groups: the lesser apes (family Hylobatidae), or the gib-bons; and the greater apes (family Hominidae), or the orang-utans, gorillas, chimpanzees, and humans (**Figure 34.29**). The lesser apes are strictly arboreal, whereas the greater apes often descend to the ground to feed.

Figure 34.28   Primate classification.   Many authorities divide the primates into two groups: **(a)** the strepsirrhini (smaller, nocturnal species such as this bush baby), and the haplorrhini (larger diurnal species). Haplorrhini comprise **(b)** the monkeys and tarsiers, such as this Capuchin monkey, *Cebus capucinus*, and **(c)** the hominoids, species such as this white-handed gibbon (*Hylobates lar*).

Concept check: *What are the defining features of primates?*

(a) Strepsirrhini (lesser bush baby)

(b) Anthropoidea (capuchin monkey)

(c) Hominoidea (white-handed gibbon)

(a) Gorilla (*Gorilla gorilla*)

(b) Chimpanzee (*Pan troglodytes*)

(c) Human (*Homo sapiens*)

Figure 34.29 **Members of the family Hominidae.** The orangutan is also a member of this group. **(a)** Gorillas, the largest of the living primates, are ground-dwelling herbivores that inhabit the forests of Africa. **(b)** Chimpanzees are smaller, omnivorous primates that also live in Africa. The chimpanzees are the closest living relatives of modern humans. **(c)** Humans are also members of the family Hominidae.

Although humans are closely related to chimpanzees and gorillas, they did not evolve directly from them. Rather, all hominoid species shared a common ancestor. Recent molecular studies show that gorillas, chimpanzees, and humans are more closely related to one another than they are to gibbons and orangutans, so scientists have split the family Hominidae into groups, including the subfamily Ponginae (orangutans) and the subfamily Homininae (gorillas, chimpanzees, and humans and their ancestors). In turn, the Homininae are split into three tribes: the Gorillini (gorillas), the Panini (chimpanzees), and the Hominini (humans and their ancestors). The sequencing of the chimpanzee genome by the Chimpanzee Sequencing and Analysis Consortium in 2005 allowed detailed comparisons to be made with the human genome.

Genomes & Proteomes Connection

Comparing the Human and Chimpanzee Genetic Codes

A male chimp called Clint who lived at a primate research center in Atlanta provided the DNA used to sequence the chimp genome. The 2005 publication of the chimpanzee genome followed the 2003 publication of the human genome (see Chapter 21) and allowed scientists to make detailed comparisons between the two species. These comparisons revealed that the sequence of base pairs making up both species' genomes differ by only 1.23%, 10 times less than the difference between the mouse and rat genomes. Compared to mice and rats, chimps and humans both carry what appear to be high levels of potentially harmful repetitive sequences. Such sequences may have permitted adaptation to environmental change but also have made primates more prone to genetic diseases. The chimpanzee genome in particular shows that it has been attacked frequently by retroviral elements similar to those present in the HIV virus. Comparisons of human and chimpanzee proteomes revealed

that 29% of all proteins are identical, but most others differ by only one or two amino acid substitutions.

Many of the genetic differences between chimps and humans result from chromosome inversions and duplications. Geneticists have found over 1,500 inversions between the chimp and human genomes. Although many inversions occur in the noncoding regions of the genome, the DNA in these regions may regulate the expression of the genes in the coding regions. Duplications and deletions are also common. For example, one gene that codes for a subunit of a protein found in areas of the brain occurs in multiple copies in a wide range of primates, but humans have the most copies. However, humans appear to have lost a gene called *caspase-12*, which in other primates may protect against Alzheimer disease.

Some interesting genetic differences were apparent between chimps and humans even before their entire genomes were sequenced. In 1998, Ajit Varki and colleagues investigated a molecule called sialic acid that occurs on cell surfaces and acts as a locking site for pathogens such as malaria and influenza. They found an altered form of the molecule in humans, coded for by a single damaged gene, which may explain why humans are more susceptible to these diseases than are chimpanzees. A little later, molecular geneticist Svante Pääbo discovered differences between humans and chimps in a gene called *FOXP2* that plays a role in speech development. Proteins coded for by this gene differ in just two locations of a 715-amino-acid sequence. Humans with a defective *FOXP2* gene have difficulty articulating words, underscoring its importance.

Finally, another team led by David Reich discovered that the human X chromosome diverged from the chimpanzee X chromosome about 1.2 million years later than the other chromosomes. This may indicate that the human and chimp lineages split apart, then later interbred before diverging again. This would explain why many fossils appear to exhibit traits of both humans and chimps, because they may actually have been hybrids.

Humans Evolved from Ancestral Primates

About 6 million years ago in Africa, a lineage that led to humans began to separate from other primate lineages. The evolution of humans should not be viewed as a neat, stepwise progression from one species to another. Rather, human evolution, like the evolution of most species, can be visualized more like a tree, with one or two **hominin** species—members of the Homininae tribe—likely coexisting at the same point in time, with some eventually going extinct and some giving rise to other species (**Figure 34.30**).

The key characteristic differentiating hominins from other apes is that hominins walk on two feet, that is, they are **bipedal**. At about the time when hominins diverged from other ape lineages, the Earth's climate had cooled, and the forests of Africa had given way to grassy savannas. A bipedal method of locomotion and upright stance may have been advantageous in allowing hominins to peer over the tall grass of the savanna to see predators or even prey.

Bipedalism is correlated with many anatomical changes in hominins. First, the opening of the skull where the spinal cord enters shifted forward, allowing the spine to be more directly underneath the head. Second, the hominin pelvis became broader to support the additional weight. And third, the lower limbs, used for walking, became relatively larger than those in other apes. These are the types of anatomical changes paleontologists look for in the fossil record to help determine whether fossil remains are hominin. The earliest group of hominins included several species of a smaller-brained genus, *Australopithecus*. *A. afarensis* is generally regarded as the common ancestor of most of these species. From there, the evolution of different species becomes somewhat hazy. It is generally agreed that two genera evolved from *Australopithecus*, the robust *Paranthropus* and the more gracile *Homo*. The early stages of the evolution of *Homo* species, and their differentiation from at

least two possible *Australopithecus*, have not yet been determined with any certainty. However, the later divergence of various *Homo* species is a little better understood.

Australopithecines Since 1924, when the first fossil australopithecine (from the Latin *austral*, meaning southern, and the Greek *pithecus*, meaning ape) was found in South Africa, hundreds of fossils of this group have been unearthed all over southern and eastern Africa, the areas where fossil deposits are best exposed to paleontologists. This was a widespread group, with at least six species. In 1974, paleontologist Donald Johanson unearthed the skeleton of a female *Australopithecus afarensis* in the Afar region of Ethiopia and dubbed her Lucy. The Beatles' song "Lucy in the Sky with Diamonds" was playing in the camp the night when Johanson was sorting the unearthed bones. Over 40% of the skeleton had been preserved, enough to provide a good idea of the physical appearance of australopithecines. Compared to modern humans, all were relatively small, about 1–1.5 m in height and around 18 kg in weight (**Figure 34.31**). Females were much smaller than males, a condition known as sexual dimorphism. Examination of the bones revealed that *A. afarensis* walked on two legs. They possessed a facial structure and brain size (about 500 cubic centimeters [cc]) similar to those of a chimp.

In the 1930s, the remains of bigger-boned hominids were found. Two of the larger species now considered to be a separate genus, *Paranthropus*, weighed about 40 kg and lived contemporaneously with australopithecines and members of *Homo* species. *Paranthropus* were vegetarians with enormous jaws used for grinding up tough roots and tubers. Both *Paranthropus* species died out rather suddenly about 1.5–2.0 million years ago. Although *Australopithecus africanus* was thought to have evolved slightly later than *A. afarensis*, its bones had been found much earlier than those of *A. afarensis*. In the 1920s, Raymond Dart described *A. africanus* from infant bones

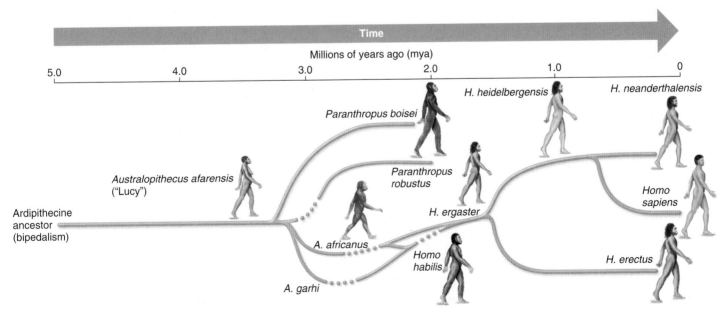

Figure 34.30 A possible scenario for human evolution.   In this human family tree (based on the ongoing work of Donald Johanson), several hominin species lived contemporaneously with one another, but only one lineage gave rise to modern humans (*Homo sapiens*).

Figure 34.31 **A modern woman compared to an australopithecine.** Compared to modern humans, Australopithecines, as illustrated by this reconstruction of the famous fossil "Lucy," were much smaller and lighter.

discovered in a cave in Taing, South Africa. The type specimen was called Taing child. The well-preserved skull was small but was well rounded, quite unlike those of chimpanzees and gorillas. Also, the positioning of the head on the vertebral column suggested bipedalism. These facts suggested to Dart that he had found a transitional form between apes and humans. However, it would take another 20 years and the discovery of more fossils to convince the scientific world to support Dart's view. In 1996, remains of another species, *A. garhi*, were also found in the Afar region. They were somewhat of a surprise in that the dentition suggested similarities with *Paranthropus boisei*. "Garhi" means surprise in the local Afar language. The position of both *A. garhi* and *A. africanus* as ancestors of modern humans has been the subject of much debate. Both have been viewed as dead-end cousins or the ancestors of the first members of the genus *Homo*.

The Genus Homo and Modern Humans   In the 1960s, paleontologist Louis Leakey found hominin fossils estimated to be about 2 million years old in Olduvai Gorge, Tanzania. Two particularly interesting observations stand out about these fossils. First, reconstruction of the skull showed a brain size of about 680 cc, larger than that of *Australopithecus*. Second, the fossils were found with a wealth of stone tools. As a result, Leakey assigned the fossils to a new species, *Homo habilis*, from the Latin, meaning handy man. The discovery of several more *Homo* fossils followed, but there have been no extensive finds, as there were with Lucy. This makes it difficult to determine which *Australopithecus* lineage gave rise to the *Homo* lineage (see Figure 34.30), and scientists remain divided on this point.

Homo habilis lived alongside *Paranthropus* but had much smaller jaws and teeth, indicating that it probably ate large quantities of meat. The smaller jaw provided more space in

the skull for brain development. *Homo habilis* probably scavenged most of its meat from the kills of large predators. A meatier diet is easier to digest and is rich in nutrients and energy. The human brain uses a lot of energy, 20% of the body's total energy production. The meat-eating habit helped propel the increasing brain size in humans. Stone tools also allowed *H. habilis* to smash open bones and extract protein-rich bone marrow. Cut marks on animal bones of the period are testament to this behavior.

Although we are not clear exactly how, we think *H. habilis* probably gave rise to one of the most important species of *Homo*, *Homo ergaster*. *Homo ergaster* was a hominin that evolved in Africa; it had a human-looking face and skull, with downward-facing nostrils. *Homo ergaster* was also a tool user, and now the tools, such as hand axes, were larger and more sophisticated. *Homo ergaster* evolved in a period of global cooling and drying that reduced tropical forests even more and promoted savanna conditions. Hairlessness and sweating may also have evolved at this time as an adaptation to the sunny environment. A leaner body shape was evident. We know this from so-called Turkana boy, a fossil teenage boy found in Kenya in 1984. Though only 13 years old, scientists predict he would have been about 185 cm (6 ft 1 in) when adult, much the same height as the Masai tribesman that inhabit the area today. A dark skin probably protected *H. ergaster* from the sun's rays. The hips had narrowed, and the size of the head increased, which may have produced more difficulty in childbirth and eventually led to human babies being born earlier. Earlier birth leads to prolonged childhood care compared to that in other apes. Prolonged childcare required well-nourished mothers, who would have benefited from the support of their male partner. Some anthropologists have suggested this was the beginning of family groups.

H. ergaster is thought to have given rise to many species, including *Homo erectus*, *Homo heidelbergensis*, *Homo neanderthalensis*, and *Homo sapiens*. A possible time line and geographic location for these species are given in **Figure 34.32**. *Homo ergaster* probably was the first type of human to leave Africa, as similar bones have been found in the Eurasian country of Georgia. *H. ergaster* is believed to be a direct ancestor of modern humans, with *Homo heidelbergensis* viewed as an intermediary step. Living contemporaneously with *H. heidelbergensis* was another descendent of *H. ergaster*, *Homo erectus*.

H. erectus was a large hominin, as large as a modern human but with heavier bones and a smaller brain capacity of between 750 and 1,225 cc (modern brain size is about 1,350 cc). Fossil evidence shows that *H. erectus* was a social species that used tools, hunted animals, and cooked over fires. The meat-eating habit may have sparked a migration in *H. erectus*, because carnivores had larger ranges than similar-sized herbivores, their prey being scarcer per unit area. *H. erectus* spread out of Africa soon after the species appeared, over a million years ago, and fossils have been found as far away as China and Indonesia. The first fossil was found by Dutch physician Eugene Dubois in 1891 on the Indonesian island of Java. Stone tools are rarely found in these Asian sites, suggesting *H. erectus* based their technology on other materials, for example, bamboo, which

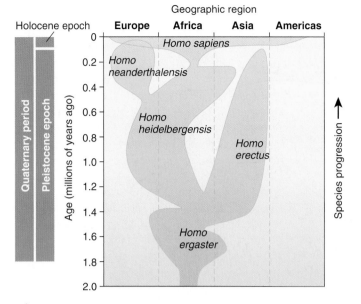

Figure 34.32 One view of the temporal and geographic evolution of hominid populations.

was abundant at that time. Bamboo is strong yet lightweight and could have been used to make spears. These people may even have used rafts to take to the seas. *H. erectus* went extinct about 100,000 years ago, for reasons that are unclear but may be related to the spread of *Homo sapiens* into its range.

Homo heidelbergensis was similar in body form to modern humans. Large caches of their bones were found in Spain, at the bottom of a 14 m (45 ft) shaft known as La Sima de Los Huesos (the pit of bones). Similar remains were also found at Boxgrove in England. Shinbones recovered from Boxgrove suggest males stood around 180 cm (6 ft) and weighed 88 kg (196 pounds). Skulls were large, with brain volumes from 1,100 to 1,400 cc, similar to modern humans. Animal bones from these sites showed cut marks from stone blades beneath tooth marks from carnivores. This showed humans were killing large prey before scavengers arrived. Horses, giant deer, and rhinoceroses were common prey and would have required much skill and cooperation to hunt.

Homo heidelbergensis gave rise to two species, *H. neanderthalensis* and *H. sapiens*. *H. neanderthalensis* was named for the Neander Valley of Germany, where the first fossils of its type were found. In the Pleistocene epoch (Figure 34.32), the ice ages were locked in a cycle of advance and retreat, and the European landscape often turned into a snowy tundra. The more gracile body form of *H. heidelbergensis* evolved into a shorter, stockier body form that was better equipped to conserve heat; we now call this type of human Neanderthal. Neanderthals also possessed a more massive skull and larger brain size than modern humans, about 1,450 cc, perhaps associated with their bulk. Males were about 168 cm (5 ft 6 in) in height and would have been very strong by modern standards. They had a large face with a prominent bridge over the eyebrows, a large nose, and no chin. They lived predominantly in Europe, with a range extending to the Middle East. Their muscular physique was well suited to the rigors of hunting prey. Paleontologists have found a high rate of head and neck injuries in Neanderthal bones, similar to that seen in present-day rodeo

riders. This suggests that close encounters with large prey often resulted in blows that knocked the hunters off their feet. The hyoid bone, which holds the voice box in place, was well developed, suggesting speech was used. However, about 30,000 years ago, this species was replaced by another hominin species, *H. sapiens* (from the Latin, meaning wise man), our own species. *H. sapiens* was a taller, lighter-weight species with a slightly smaller brain capacity than that of the Neanderthals.

Paleontologists remain divided as to whether *H. sapiens* evolved in Africa and spread from there to other areas of the world, or whether premodern humans such as *H. ergaster* that migrated from Africa evolved to become modern humans in different parts of the world. The first model, the Out of Africa hypothesis, suggests that after the evolution of hominins in Africa, they migrated to other continents three times, once for *H. ergaster*, once for *H. erectus*, and once for *H. sapiens*. Then *H. sapiens* would have gradually replaced species such as *H. erectus* and *H. neanderthalensis* in other parts of the world. Some scientists find this difficult to accept, however, and suggest that human groups have evolved from *H. ergaster* populations in a number of different parts of the world, a model known as the multiregional hypothesis. According to this hypothesis, gene flow between neighboring populations prevented the formation of several different species.

Studies of human mitochondrial DNA (mtDNA), which occurs only in the cellular organelles called mitochondria, show that all modern people share a common ancestor, dubbed "mitochondrial Eve," dating to about 170,000 years ago. This evidence is consistent with the Out of Africa model, because the common ancestor would have to be much older than that to support the multiregional hypothesis. Furthermore, 2006 analyses of DNA from Neanderthal bones show it to be distinct from the DNA of *H. sapiens*, even though Neanderthals and humans share 99.5% of their genome. This also suggests that there was no interbreeding between Neanderthals and the *H. sapiens* who migrated into Europe.

Evidence overall appears to support the Out of Africa hypothesis. In this scenario, *H. ergaster* evolved in Africa and spread to Asia and Europe. Later, *H. erectus* evolved in Africa and spread into Asia, and *H. neanderthalensis* evolved in Europe. Both these species shared the same fate, extinction at the hands of the later-evolving *H. sapiens*. *Homo sapiens* evolved in Africa about 170,000 years ago from *H. heidelbergensis*. The mtDNA data suggest a migration of *Homo sapiens* from eastern Africa to other parts of the globe beginning 170,000–150,000 years ago (**Figure 34.33**). Modern humans spread first into the Middle East and Asia, then later into Europe and Australia, finally crossing the Bering Strait to the Americas.

Much remains to be resolved in human evolution, and new data constantly forces us to rethink our hypotheses. Unexpected finds surface and have to be explained. For example, in 2004, the remains of a small human on the Indonesian island of Flores were discovered and were given the name *Homo florensiensis*, nicknamed "hobbits" by the media. Many species—for example, deer and elephants—develop into small forms in insular situations, so hobbit humans seemed plausible. Since then, many

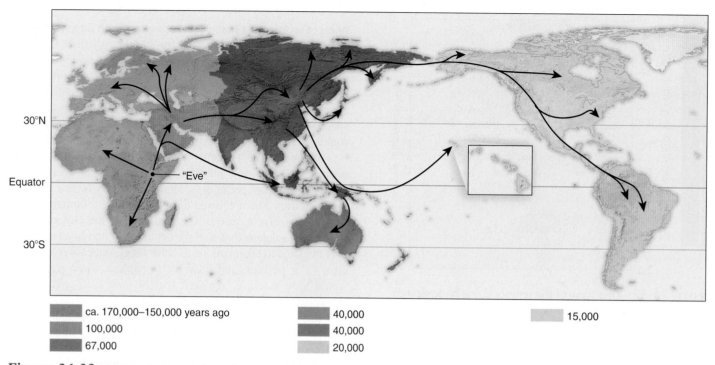

30°N

Equator

"Eve"

30°S

	ca. 170,000–150,000 years ago		40,000		15,000
	100,000		40,000		
	67,000		20,000		

Figure 34.33   **The probable origin and spread of *Homo sapiens* throughout the world.**   This map, based on differences of mtDNA throughout current members of the world's population, suggests *Homo sapiens* originated from "mitochondrial Eve" in east Africa. About 100,000 years ago, the species spread into the Middle East and from there to Asia, Europe, Australia, and the Americas.

researchers have suggested these people were modern humans who were suffering from a genetic disorder. Even modern humans on Flores are pygmies. Pathological dwarfism would have made these people even smaller. Only *H. sapiens* tools have been found at the area where the bones occur, suggesting these individuals were indeed dwarf forms of modern humans.

Summary of Key Concepts

34.1 The Craniates: Chordates with a Head

- The subphylum Vertebrata is the largest and most dominant group of chordates, occupying nearly all of Earth's major habitats. All vertebrates are craniates. (Figure 34.1)

- Craniates have two defining characteristics that distinguish them from invertebrate chordates: a cranium and a neural crest. The hagfish (class Myxini) is considered an early-diverging modern craniate; it is the only living craniate that is not a vertebrate. (Figure 34.2)

34.2 Vertebrates: Craniates with a Backbone

- Vertebrates have several characteristic features, including a vertebral column, endoskeleton of cartilage or bone, and internal organs. Of the 11 different vertebrate classes, 5 are classes of fishes, reflecting the dominance of fishes. (Table 34.1)

- Early diverging vertebrates lacked jaws. Today the only jawless vertebrates are lampreys. (Figure 34.3)

34.3 Gnathostomes: Jawed Vertebrates

- A critical innovation in vertebrate evolution is the hinged jaw, which first developed in fishes. Lampreys, eel-like fish, lack a hinged jaw. The hinged jaw evolved from cartilaginous gill arches. (Figure 34.4)

- The chondrichthyans (sharks, skates, and rays) have a skeleton composed of flexible cartilage and powerful appendages called fins. They are active predators with acute senses and were among the earliest fishes to develop teeth. (Figure 34.5)

- Bony fishes consist of the Actinopterygii (ray-finned fishes, the most species-rich class), Actinistia (coelacanths), and the Dipnoi (lungfishes). In Actinopterygii, the fins are supported by thin, flexible rays and moved by muscles inside the body. (Figures 34.6, 34.7)

- The lobe fins comprise the lobe-finned fishes (Actinistia and Dipnoi) and the tetrapods. In the lobe-finned fishes, the fins contain their own muscles. (Figures 34.8, 34.9)

34.4 Tetrapods: Gnathostomes with Four Limbs

- Fossils record the evolution of lobe-finned fishes to fishes with four limbs. Recent research has shown that relatively simple mutations control large changes in limb development. (Figures 34.10, 34.11, 34.12)

- Amphibians live on land but return to the water to reproduce. The larval stage undergoes metamorphosis, losing gills and tail for lungs and limbs. (Figure 34.13)

- The majority of amphibians belong to the order Anura (frogs and toads). Other orders are the Gymnophiona (caecilians) and Caudata (salamanders). (Figure 34.14)

34.5 Amniotes: Tetrapods with a Desiccation-Resistant Egg

- The amniotic egg permitted animals to become fully terrestrial. Other critical innovations included desiccation-resistant skin, thoracic breathing, water-conserving kidneys, and internal fertilization. (Figure 34.15)

- Living reptilian classes include the Testudines (turtles), Lepidosauria (lizards and snakes), Crocodilia (crocodiles), and Aves (birds). Other reptilian classes, all of which are extinct, include two classes of dinosaurs (Ornithischia and Saurischia). (Figures 34.16, 34.17, 34.18, 34.19)

- Three species—*Archaeopteryx*, *Caudipteryx*, and *Confuciusornis*—help trace a lineage from dinosaurs to birds. (Figure 34.20)

- The four key characteristics of birds are feathers, a lightweight skeleton, air sacs, and reduced organs. Birds are the most species-rich class of terrestrial vertebrates. The diversity of bird beaks reflects the varied methods they use for feeding. (Figures 34.21, 34.22, Table 34.2)

34.6 Mammals: Milk-Producing Amniotes

- The distinguishing characteristics of mammals are the possession of mammary glands, hair, specialized teeth, and an enlarged skull. Other unique characteristics of some mammals are the ability to digest plants and possession of horns or antlers. Mammal tooth shape varies according to diet. (Figures 34.23, 34.24, 34.25)

- Two subclasses of mammals exist: the Prototheria (monotremes) and the Theria (the live-bearing mammals). The live-bearing mammals are, in turn, divided into the Metatheria (marsupials) and Eutheria (placental mammals). (Table 34.3, Figure 34.26)

- Many defining characteristics of primates relate to their tree-dwelling nature and include grasping hands, large brain, nails instead of claws, and binocular vision. (Figures 34.27, 34.28, 34.29)

- About 6 million years ago in Africa, a lineage that led to humans began to separate from other primate lineages. A key characteristic of hominins (extinct and modern humans) is bipedalism.

- Human evolution can be visualized like a tree, with a few hominin species coexisting at the same point in time, some eventually going extinct, and some giving rise to other species. (Figures 34.30, 34.31)

- The Out of Africa hypothesis suggests that the migration of hominins from Africa happened at least three times, with *Homo sapiens* gradually replacing other hominin species in other parts of the world (Figure 34.32). The multiregional hypothesis proposes that human groups evolved in a number of different parts of the world. Most scientists believe the Out of Africa hypothesis is better supported by the data.

- Data from human mitochondrial DNA suggest all humans derive from a "mitochondrial Eve" that originated in east Africa. From there, *Homo sapiens* spread to Asia and then to all other parts of the globe. (Figure 34.33)

▌ Assess and Discuss

Test Yourself

1. Which of the following is not a defining characteristic of craniates?
   a. cranium
   b. neural crest
   c. two clusters of *Hox* genes
   d. protective housing around the brain
   e. cephalization

2. The presence of a bony skeleton, an operculum, and a swim bladder are all defining characteristics of
   a. Myxini.
   b. lampreys.
   c. Chondrichthyes.
   d. bony fishes.
   e. amphibians.

3. Organisms that lay eggs are said to be
   a. oviparous.
   b. ovoviparous.
   c. viviparous.
   d. placental.
   e. none of the above.

4. Which clade does not include frogs?
   a. craniates
   b. gnathostomes
   c. tetrapods
   d. amniotes
   e. lobe fins

5. In some amphibians, the adult retains certain larval characteristics, which is known as
   a. metamorphosis.
   b. parthenogenesis.
   c. cephalization.
   d. paedomorphosis.
   e. hermaphrodism.

6. The membrane of the amniotic egg that serves as a site for waste storage is
   a. the amnion.
   b. the yolk sac.
   c. the allantois.
   d. the chorion.
   e. the albumin.

7. Which characteristic qualifies lizards as gnathostomes?
   a. a cranium
   b. a skeleton of bone or cartilage
   c. a hinged jaw
   d. the possession of limbs
   e. amniotic eggs

8. Which of the following is not a distinguishing characteristic of birds?
   a. amniotic egg
   b. feathers
   c. air sacs
   d. lack of certain organs
   e. lightweight skeletons

9. What is not a derived trait of primates?
   a. opposable thumb
   b. grasping hands
   c. prehensile tail
   d. flat nails
   e. large brain

10. Despite their small size and nocturnal habits, tarsiers are classed with much larger monkeys and apes as *Haplorrhini*. This is based on which of the following characteristics?
    a. dry fully furred noses
    b. forward-facing eyes
    c. DNA similarities
    d. a and b
    e. a, b, and c

Conceptual Questions

1. How is vertebrate movement accomplished in a similar way to arthropod movement, and how is it different?

2. Explain the function of the lateral line and the operculum.

3. List the four extraembryonic membranes in the amniotic egg, and explain the function of each.

Collaborative Questions

1. By what means can vertebrates move?

2. Discuss the three different ways mammals bring their offspring into the world.

Online Resource

www.brookerbiology.com

Stay a step ahead in your studies with animations that bring concepts to life and practice tests to assess your understanding. Your instructor may also recommend the interactive ebook, individualized learning tools, and more.

An Introduction to Flowering Plant Form and Function

35

Representatives of two major groups of flowering plants. These corn and soybean plants display similar architecture common to flowering plants, yet differ in other structural and reproductive features.

Corn, more formally known as *Zea mays*, soybean (*Glycine max*), and the common bean (*Phaseolus vulgaris*) are among the world's most important agricultural crop plants. Farmers grow corn for food and production of ethanol, a gasoline additive. The seeds of beans and related plants are extraordinarily rich sources of protein in the human diet and therefore of particular importance to vegetarians. Because both are flowering plants, corn and beans have many features in common, including stems, roots, leaves, flowers, fruits, and seeds. However, corn and soybean plants also display differences in these features, as you can see in the chapter-opening photo. In fact, the hundreds of thousands of modern flowering plants display an amazing variety of forms that represent adaptations to diverse habitats.

This chapter provides an introduction to the flowering plants, focusing on fundamental principles of structure and function. These principles will help to explain why flowering plants resemble each other in many ways, yet differ in features that influence plant roles in nature and agriculture. We will begin by describing the general life of

a flowering plant—from seed germination to reproductive maturity. A survey of major plant structural features and growth processes will follow. Finally, we will consider diverse stem, leaf, and root adaptations. These topics provide essential background for subsequent chapters in this unit, which focus on flowering plant behavior, nutrition, transport, and reproduction and development.

35.1 From Seed to Seed—The Life of a Flowering Plant

Several major events punctuate the lives of flowering plants, also known as the **angiosperms**. When seeds germinate, dormant embryos wake to metabolic activity and begin the process of seedling development. Seedlings grow and develop into mature plants capable of reproduction. Finally, flowers produce and fruits disperse the next generation of seeds. In this section, we will briefly survey the life of flowering plants, focusing on the basic structural features of each life stage.

Seedlings Develop from Embryos in Seeds

Seeds are reproductive structures produced by flowering plants and other seed plants, usually as the result of sexual reproduction. Seeds contain embryos that develop into young plants—seedlings—when seeds germinate. The embryo is an essential stage in the sexual cycle of a flowering plant. The plant sexual cycle explains how embryos typically arise (**Figure 35.1**).

Sexual reproduction in plants requires two multicellular stages: a gamete-producing **gametophyte** and a spore-producing **sporophyte**. In the life cycle of plants, these two life stages alternate with one another, a process called **alternation of generations**. Flowering plants produce relatively large sporophytes and microscopic gametophytes that grow and develop within flowers. Diploid sporophytes produce haploid spores by the process of meiosis. These spores grow into gametophytes that produce plant gametes—eggs and sperm. Fusion of egg and sperm in the process of fertilization generates a diploid zygote, which undergoes repeated mitotic divisions to form the plant **embryo**.

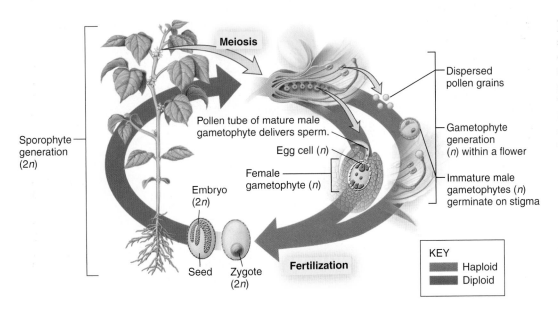

Figure 35.1   **The plant sexual cycle.**   The sexual cycle of flowering plants involves alternation of sporophyte and gametophyte generations. In flowering plants, the sporophyte is the dominant, conspicuous generation, whereas the tiny gametophytes are mostly hidden within flowers. The bean plant shown here produces both male and female gametophytes. Embryos form when eggs produced by female gametophytes are fertilized by sperm produced by male gametophytes. Embryos, which are the young sporophytes of the next generation, are dispersed from plants within seeds.

The plant embryo is a very young sporophyte that lies dormant within seeds, accompanied by a supply of stored food and enclosed by a tough, protective seed coat (**Figure 35.2a**). The seed coat protects the delicate embryo during the dispersal of seeds from parent plants into the environment. Dispersed seeds may remain dormant in the soil—sometimes for long periods— but germinate when temperature, moisture, and light conditions are favorable. Such conditions activate embryo metabolism, causing embryos to enlarge and break the seed coat. Once free of the seed coat, embryos grow into seedlings (**Figure 35.2b**), and if sufficient resources such as water and minerals are available, seedlings develop into mature plants (**Figure 35.2c**). **Growth**

is an increase in weight or size. **Development** is a series of changes in the state of a cell, tissue, organ, or organism.

The plant body is composed of three types of organs: stems, leaves, and roots. **Stems** produce leaves and branches and bear the reproductive structures of mature plants. **Leaves** are flattened structures that emerge from stems and are often specialized in ways that enable photosynthesis. Stems and leaves together make up the plant **shoot** (**Figure 35.2b**). Mature plants often possess multiple stems bearing many leaves, which together form the **shoot system**. **Roots** provide anchorage in the soil and also foster efficient uptake of water and minerals. The aggregate of a plant's roots make up the **root system** (Figure 35.2c).

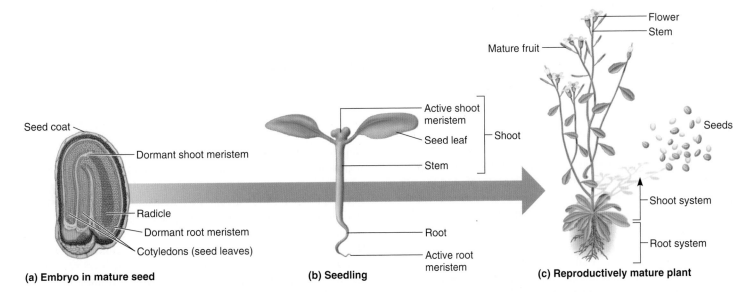

(a) Embryo in mature seed   **(b) Seedling**   **(c) Reproductively mature plant**

Figure 35.2   **The seed-to-seed life of flowering plants.**   This example depicts the life of the eudicot *Arabidopsis thaliana*, widely used as a model organism for understanding the genetics of plant structure and development. **(a)** Seed embryos possess embryonic leaves, known as cotyledons; a dormant shoot meristem; an embryonic root, known as a radicle; and a dormant root meristem. **(b)** When seeds germinate, the shoot and root meristems become active. Meristem activity allows the radicle to produce the seedling root and the young shoot of the seedling to grow and produce leaves. **(c)** Reproductively mature plants have branched shoot and root systems and bear flowers and fruits that disperse seeds.

Stems, leaves and roots are organs so important to plants that they are present even in seed embryos and seedlings. A short, stemlike region of the embryo, known as a **hypocotyl**, produces embryonic leaves known as **cotyledons**, or seed leaves. These structures often store food that supplies energy for seedling development. Peas and peanuts are examples of cotyledons that are rich in nutrients, explaining their food value to humans. The embryonic root, known as the **radicle**, is the first organ to emerge from a germinating seed and allows the uptake of the water and minerals needed for seedling growth (Figure 35.2b).

The process of body and organ development involves the differentiation of specialized cells having distinctive structure and function. Seedlings and mature plants produce new tissues in areas called meristems. A **meristem** (from the Greek *merizein*, meaning to divide) is a region of undifferentiated cells that produces new tissues by cell division. A dormant meristem occurs at the shoot and root tips of seed embryos, and these meristems become active in seedlings (Figure 35.2a,b). In mature plants, active meristems occur at each stem and root tip. Such meristems are known as shoot and root **apical meristems** because they occur at shoot and root tips, also known as apices.

Mature Sporophytes Develop from Seedlings

As seedlings develop into mature sporophytes, the above-ground shoot typically becomes green and photosynthetic and thus able to produce organic food. Photosynthesis powers the transformation of seedlings into mature plants. The development of mature plants encompasses both **vegetative growth**, a process that increases the size of the shoot and root systems, and reproductive development. Vegetative growth and reproductive development involve **organ systems**, structures that are composed of more than one organ. Branches, buds, flowers, seeds, and fruits are organ systems. The hierarchy of structure in a mature plant, ranging from specialized cells, tissues, organs, and organ systems to root and shoot systems, is shown in Figure 35.3.

Vegetative Growth   During their growth, plant shoots produce **buds**, miniature shoots each having a dormant shoot apical meristem. Scaly modified leaves protect bud contents. Under favorable conditions, the bud scales fall off, and the vegetative buds open. Newly opened buds display young leaves on a short shoot. The shoot apical meristem then becomes active, producing new stem tissue and leaves. In this way, buds generate leafy branches. A bud is an example of an organ system because it contains more than one organ.

Shoots often display **indeterminate growth**, meaning that apical meristems continuously produce new stem tissues and leaves as long as conditions remain favorable. This process explains how very large plants can develop from seedlings. However, plant size is also under genetic control, so some plants remain small even when they are mature. The tiny floating plants of *Lemna* species, commonly known as duckweeds,

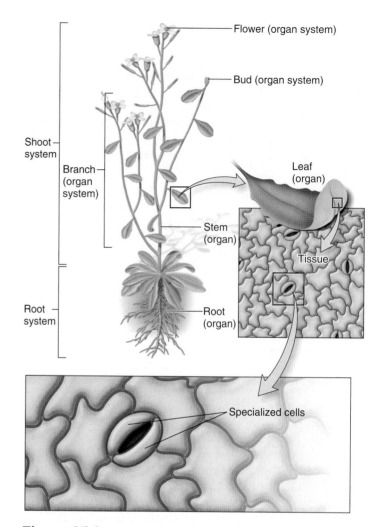

Figure 35.3   Levels of biological organization in a plant. Flowering plant sporophyte bodies consist of a root system and a shoot system. Shoot systems produce organ systems such as buds, flowers, fruits, and seeds, which are composed of organs, tissues, and specialized cells. Root systems are composed of organs, tissues, and specialized cells.

Concept check: *Do roots have organ systems?*

which sometimes cover the surfaces of ponds in summer, are examples of plants whose small size is genetically determined.

Reproductive Development   Under favorable conditions, mature plants produce reproductive structures: flowers, seeds, and fruits. **Flowers** and floral buds are reproductive shoots that develop when shoot apical meristems produce flower parts instead of new stem tissues and leaves. Flower development occurs under the control of several genes whose roles are well understood (refer back to Figure 19.24). In contrast to shoots (which often show indeterminate growth), flowers are produced by **determinate growth**, which is growth of limited duration. A floral shoot no longer produces new stem growth or leaves. Therefore, vegetative growth and reproductive development are alternative processes. In order to flower, a plant must give up some of its potential to continue vegetative growth.

Flower tissues enclose and protect tiny male and female gametophytes during their growth and development (see

Figure 35.1). Female gametophytes produce eggs within structures known as ovules, produced in the ovary of a flower pistil. Male gametophytes begin their development within pollen grains produced in the anthers of a flower stamen. Pollen is dispersed to the flower pistil, whereupon pollen grains may germinate, producing a tube that delivers sperm to eggs. Fertilization generates zygotes, which develop into embryos, and also triggers the process by which ovules develop into seeds and flower parts develop into fruits. **Fruits** thus enclose seeds and function in seed dispersal. Flower buds, flowers, fruits, and seeds are organ systems because they consist of more than one organ. For example, flowers typically contain several leafy organs, including sepals and petals, as well as stamens and pistils, which evolved from leaves (refer back to Chapter 30).

Flowering Plants Vary in the Structure of Organs and Organ Systems

With some exceptions, flowering plants occur in two groups, informally known as the **eudicots** and the **monocots**. These groups take their names from the number of seed leaves (cotyledons) that are present on seed embryos. For example, bean plants and relatives, which possess two (*di*) seed leaves, are examples of eudicots. Corn, which has only one (*mono*) seed leaf, is an example of a monocot. Eudicots and monocots also vary in the structure of other organs and organ systems. For example, eudicot flowers typically have petals and other parts numbering four, five, or a multiple of those numbers, whereas monocot flower parts usually occur in threes or a multiple of three. Stems, roots, leaves, and pollen of eudicots and monocots also vary in distinctive ways, as shown in **Table 35.1**.

Flowering Plants Vary in Seed-to-Seed Lifetime

The seed-to-seed lifetime of a flowering plant can vary from a few weeks to many years. Plants that die after producing seed during their first year of life are known as **annuals**. Corn and the common bean are examples of annual crops whose nutrient-rich seeds are harvested within a few months after planting, and must be replanted at the beginning of each new growing season. Plants that do not reproduce during the first year of life but may reproduce within the following year are known as **biennials**. Such plants often store food in fleshy roots during the first year of growth, and this food fuels reproduction during the second or later year of life. Humans use some of these fleshy roots for food, including carrots, parsnips, and sugar beets. Trees are examples of **perennials**, plants that live for more than 2 years, often producing seed each year after they reach reproductive maturity. Many flowering plants use environmental signals to time flowering in ways that enhance seed production. Temperature and day length are examples of environmental factors that determine flowering time. Plant seed-to-seed lifetimes are also influenced by the longevity of their seeds. Seeds of some plants are able to germinate after more than a thousand years of dormancy, whereas other plant seeds are unable to remain alive for long periods.

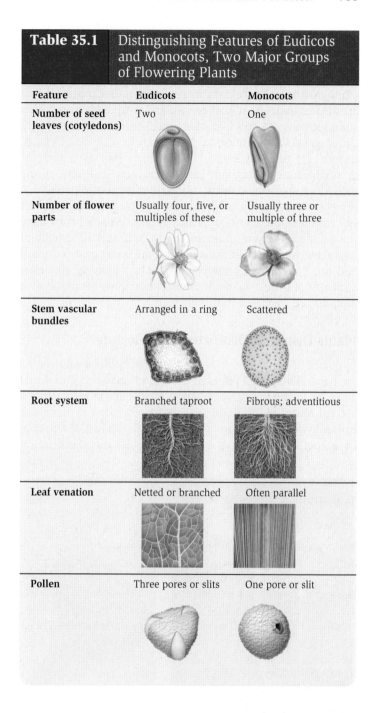

Table 35.1	Distinguishing Features of Eudicots and Monocots, Two Major Groups of Flowering Plants	
Feature	**Eudicots**	**Monocots**
Number of seed leaves (cotyledons)	Two	One
Number of flower parts	Usually four, five, or multiples of these	Usually three or multiple of three
Stem vascular bundles	Arranged in a ring	Scattered
Root system	Branched taproot	Fibrous; adventitious
Leaf venation	Netted or branched	Often parallel
Pollen	Three pores or slits	One pore or slit

In this section, we have seen that the life of a flowering plant includes seed germination, seedling development into mature plants, and reproduction. We have also learned that sexual reproduction requires two stages that alternate with each other, a process known as alternation of generations. During the plant sexual cycle, a parental sporophyte produces many gametophytes in flowers, from which develop the next generation of young sporophytes—embryos in seeds. In this way, a single plant can produce many progeny. We have also surveyed the basic components of embryo, seedling, and adult plant structure. These topics provide a valuable background as we focus more closely on principles of plant structure, growth, and development.

35.2 How Plants Grow and Develop

Plant growth and development depends on four processes that are also essential to animal growth and development: cell division, growth, cell specialization, and programmed cell death. Other general principles of plant growth and development include (1) development and maintenance of a characteristic architecture throughout life, (2) increase in length by the activity of apical meristems, (3) maintenance of a population of youthful stem cells in meristems, and (4) expansion of cells in controlled directions, by water uptake. Woody plants have the additional feature of increasing in width by the activity of lateral meristems. All of these features are under genetic control and can be studied by the use of mutants showing abnormal structure and development. Several examples of informative mutants are discussed in this chapter.

Plants Display a Distinctive Architecture

In plant biology, the term apical has two distinct meanings. As we have seen, apical refers to the tips or apices of shoots and roots, as in shoot apical meristems or root apical meristems. A second meaning for apical is the part of a plant that projects upward, which is the top of the shoot. By contrast, the bottom of a root is termed the basal region. So the shoot apical

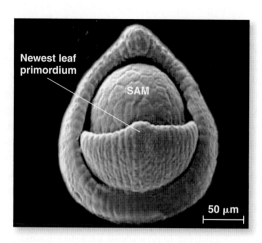

Figure 35.5 **Plant radial symmetry.** This top-down view of a shoot apical meristem reveals the radial symmetry of the shoot. Any line drawn from one edge of the surface to another that also passes through the center will divide the shoot into equal pieces. Leaf primordia are produced in circles or spirals around the shoot axis.

Concept check: *How could you determine that roots have radial symmetry?*

meristem occurs at the apical pole, and the root apical meristem occurs at the basal pole. This property, known as **apical-basal polarity**, explains why plants produce shoots at their tops and roots at their lower regions. Apical-basal polarity originates during embryo development. As seedlings and maturing plants grow in length by the activity of shoot and root meristems, apical-basal polarity is maintained (**Figure 35.4a**). Apical-basal polarity is under the control of genes such as *GNOM*; mutations in such genes result in plant embryos that are cone shaped or spherical and thus lack features of apical-basal polarity (**Figure 35.4b**).

A second architectural feature common to plants is **radial symmetry**. Plant embryos normally display a cylindrical shape, also known as an axis, which is retained in the stems and roots of seedlings and mature plants. A thin slice or cross section of an embryo, stem, or root is typically circular in shape. Another component of radial symmetry is that most plants produce new leaves or flower parts in circular whorls, or spirals, around shoot tips (**Figure 35.5**). Buds and branches likewise emerge from stems in radial patterns, as do lateral roots from a central root axis. Together, apical-basal polarity and radial symmetry explain why diverse plant species have a fundamentally similar architecture.

Primary Meristems Increase Plant Length and Produce Plant Organs

We have previously noted that plant embryos grow into seedlings by adding new cells from only two growth points, the **shoot apical meristem** (commonly abbreviated as **SAM**) and the **root apical meristem** (**RAM**). During plant development, the SAM and RAM of the embryo give rise to many apical meristems located in the buds of shoots and at the tips of roots.

(a) Normal seedling   **(b) Abnormal *GNOM* mutants**

Figure 35.4 **Plant apical-basal polarity.** (a) Normal plants exhibit apical-basal polarity, as shown by this seedling. Such growth occurs at two meristems, one at the shoot (SAM) and one at the root (RAM). (b) *GNOM* mutants of *Arabidopsis thaliana* lack apical-basal polarity and thus produce abnormal embryos and seedlings.

As a SAM grows, it leaves behind meristematic tissues that increase plant length and produce new organs. Such meristems are known as the **primary meristems**, and in a process known as **primary growth**, they ultimately produce primary tissues and organs of diverse types (**Table 35.2**). Tissues differ in their cellular complexity. Simple primary tissues are those composed of only one or two cell types; complex primary tissues are made of more cell types. As described in Section 35.3, the primary meristems of woody plants also give rise to secondary or lateral meristems. In a process known as **secondary growth**, the secondary meristems increase the girth of plant stems and roots by producing secondary tissues (Table 35.2).

Shoot apical meristems produce three primary meristems called protoderm, procambium, and ground meristem, which are present in young stems and leaves (**Figure 35.6**). The **protoderm** generates the outermost **dermal tissue**. The **procambium** produces **vascular tissues**, which make up a vascular system that conducts materials within the plant body and also provide support. The **ground meristem** gives rise to **ground tissues** defined by their locations—either between dermal and vascular tissues or at the center of mature stems, where it is called pith. Even so, plant biologists have discovered that plant

Table 35.2	Examples of Tissues and Specialized Cells Found in Flowering Plants*
Simple primary tissues (composed of one or two cell types)	**Plant cell types found in those tissues**
Parenchyma	Parenchyma cells
Collenchyma	Collenchyma cells
Sclerenchyma	Fibers and sclereids
Root endodermis	Endodermal cells
Root pericycle	Pericycle cells
Complex primary tissues (composed of at least two cell types)	**Plant cell types found in those tissues**
Leaf or stem epidermis	Flattened epidermal cells, trichomes, stomatal guard cells
Root epidermis	Flattened epidermal cells, root hairs
Leaf mesophyll	Spongy parenchyma cells, palisade parenchyma cells
Leaf, stem, or root xylem	Tracheids, vessel elements, fibers, parenchyma cells
Leaf, stem, or root phloem	Sieve-tube elements, companion cells, fibers, parenchyma cells
Simple and complex secondary tissues	**Plant cell types found in those tissues**
Secondary xylem (wood)	Tracheids, vessel elements, fibers, parenchyma
Secondary phloem (inner bark)	Sieve-tube elements, companion cells, fibers, parenchyma
Outer bark	Cork cells

*This list does not include all of the tissues and cell types found in flowering plants. Some of these examples will be described later in this chapter.

Figure 35.6 **Shoot apical meristems and the primary tissues they produce.** Shoot apical meristems generate three primary meristems: Protoderm produces the dermal tissues (D) (light brown), ground meristem produces ground tissues (G) (green or brown), and procambium yields vascular tissues (V) (blue). The vascular systems of leaves, stem, and root are connected.

cell specialization and tissue development do not depend much on the lineage (the parentage) of a cell or tissue. Chemical influences are much more important in determining the type of specialized tissue that will be produced by unspecialized plant cells.

Primary Stem Structure and Development New primary stem tissues arise by the cell division activities of primary meristems located near the bases of SAMs. A layer of dermal tissue known as the **epidermis** develops at the stem surface. The epidermis produces a waxy surface coating known as the **cuticle**, which helps to reduce water loss from the plant surface. The epidermis and cuticle also help to protect plants from damage by insects and disease microorganisms.

Beneath the epidermis lies the stem **cortex**, which is largely composed of a ground tissue known as **parenchyma tissue** (**Figure 35.7a**). This tissue is composed of only one cell type, thin-walled cells known as **parenchyma cells**. These cells often store starch in plastids and therefore serve as an organic food reserve. Stem parenchyma also has the ability to undergo cell division (meristematic capacity), which aids wound healing when stems are damaged. The cell division capability of stem parenchyma also explains how people are able to grow new plants from stem cuttings. Stems also contain **collenchyma tissue** (Figure 35.7b), composed of flexible **collenchyma cells**; and rigid **sclerenchyma tissue** (Figure 35.7c), composed of two types of tough-walled sclerenchyma cells termed **fibers** and **sclereids**.

New water- and food-conducting tissues develop at the core of a young shoot. These conducting tissues are known as **primary vascular tissues** because they develop from new cells produced by the SAM. Vascular tissues occur in two forms, **xylem** and **phloem**, which are each composed of several types of specialized tissues and cells (see Section 35.3 and Table 35.2). Newly formed stem xylem and phloem connect with older conducting tissues that extend throughout the stem system and are linked with vascular tissues of the root system. Xylem and phloem conduct water, minerals, and organic food resources needed for continued cell division in shoot meristems, and for stem growth and development. Primary vascular tissues are typically arranged in elongate clusters known as **vascular bundles** that appear round or oval when cross-cut (Figure 35.7d). In the primary stems of beans and other eudicots, the vascular bundles are arranged in a ring, which is easily seen in thin slices made across a stem (see Table 35.1). In contrast, in the stems of corn and other monocots, the vascular bundles are scattered.

(a) Parenchyma (b) Collenchyma

(c) Sclerenchyma (d) Vascular bundle

Figure 35.7 Examples of tissues produced by primary shoot meristems. (a) Parenchyma, (b) collenchyma, (c) sclerenchyma, and (d) a vascular bundle composed of complex xylem and phloem tissues.

Leaf Structure and Development Young leaves are produced at the sides of a SAM as small bumps known as **leaf primordia** (see Figure 35.5). As young leaves develop, they acquire vascular tissue that is connected to the stem xylem and phloem (Figure 35.8a) and become flattened, a process that expands the area of leaf surface available for light collection during photosynthesis. In the cases of some leaves, thinness is an adaptation that helps them to shed excess heat. Leaves also become bilaterally symmetrical, meaning that they can be divided into two equal halves in only one direction, from the leaf tip to its base (Figure 35.8b).

In addition, upper (adaxial—upper surface) and lower (abaxial—lower surface) leaf tissues develop differently in several ways that foster photosynthesis (Figure 35.8c). For example, the more shaded abaxial leaf epidermis usually displays larger numbers of pores, known as **stomata** (from the Greek word *stoma*, meaning mouth), than the sunnier adaxial leaf surface. When open, stomata allow CO_2 to enter and water vapor and O_2 to escape leaf tissues. Closure of stomata helps to prevent excess water loss from plant surfaces. Many leaves also show another example of differentiation between the adaxial and abaxial regions in the positions of different types of photosynthetic ground tissue. The adaxial **palisade parenchyma** consists of closely packed, elongated cells adapted to absorb sunlight efficiently, whereas abaxial **spongy parenchyma** contains rounder cells separated by abundant air spaces. These air spaces, located near stomata, foster CO_2 absorption and O_2 release by leaves. Together, the palisade and spongy parenchyma are known as the leaf **mesophyll**.

Leaf veins composed of vascular tissue commonly occur at the junction of palisade and spongy parenchyma, or within the spongy parenchyma (Figure 35.8c). Leaf parenchyma tissues are green and active in photosynthesis, a process that requires water, carbon dioxide, and dissolved minerals. Parenchyma cells can only take up carbon dioxide that has first dissolved into water, a feature inherited from ancient aquatic algal ancestors. For this reason, in order to perform photosynthesis, leaf parenchyma cells must be bathed in water. The xylem tissues of veins conduct water and minerals throughout leaf tissues, fostering photosynthesis. Phloem tissues of leaf veins carry the sugar products of photosynthesis from leaf cells to stem vascular tissues. In this way, sugar produced in leaves can be exported to other parts of the plant.

Root System Structure and Development In beans and most other eudicots, a main root develops from the embryonic root and then produces branch roots, also known as lateral roots. Such a root system of eudicots is known as a **taproot system**; this kind of root system has one main root with many branch roots (see Figure 35.6). In contrast, the embryonic root of most monocots dies soon after seed germination, and it is replaced by a **fibrous root system** consisting of multiple roots that grow from the stem base (see Table 35.1). Fibrous roots are examples of **adventitious roots**, structures that are produced on the surfaces of stems (and sometimes leaves) of both monocots and eudicots. Roots that develop at the bases of stem cuttings are also adventitious roots.

(a) SAM and young leaves   **(b) Mature leaf**   **(c) Leaf internal structure**

Figure 35.8 Leaf development and structure.   (a) Young leaves develop at the sides of SAMs, as shown in this thinly sliced, stained shoot tip. Note that the darkly stained vascular tissues of young leaves are connected to those of the stem. (b) Mature leaves are typically thin and flat and show bilateral symmetry. (c) An internal view of a thinly sliced and stained leaf reveals top to bottom (adaxial to abaxial) tissue differentiation. A layer of palisade parenchyma lies just beneath the upper epidermis capped with a waxy cuticle. Veins of conducting tissue (xylem and phloem) are embedded in the photosynthetic mesophyll. Spongy parenchyma lies above the lower epidermis, which displays stomatal pores. These structural features of mature foliage leaves facilitate photosynthesis.

Concept check: What advantage do plant leaves obtain by having stomatal pores on the lower epidermal surface?

As we have earlier noted, the tips of roots and their branches each possess an apical meristem that adds new cells. Expansion of these new cells allows roots to grow into the soil. As they lengthen, roots produce branches but not from buds, as is the case for stems. Instead, branch roots develop from meristematic tissues located within the root (see Section 35.4). The root system both anchors plants in the soil and plays an essential role in harvesting water and mineral nutrients. Root tissues are usually not green and photosynthetic. They must rely on organic compounds transported from the shoot. The plant root system and shoot system therefore depend on each other.

Plant Meristems Contain Youthful Stem Cells

Plant meristems include undifferentiated, youthful cells earlier known as initials and more recently referred to as **stem cells**. In the late 19th century, the biologist Alexander Maximow coined the term *Stammzelle*, which is derived from the German words *stamm*, meaning stem, such as a plant stem, and *zelle*, meaning cell. Maximow used the term "stem cell" to describe animal cells that remain undifferentiated but are also able to generate specialized tissues. Animal stem cells are currently much in the news because of their potential for use in the treatment of damaged human tissues. The term stem cell is now also widely used for cells located within the plant meristem that likewise remain undifferentiated but also can produce new tissues. In the context of plant development, the term does not mean any cell located in a plant stem, only the undifferentiated cells located within the meristems of the shoot and root.

When plant stem cells divide, they produce two cells, one that remains young and unspecialized plus another cell. This second cell may differentiate into various types of specialized cells, but it often retains the ability to divide. As a result of these properties, stem cell numbers influence the size of a meristem, which, in turn, affects plant growth.

Plant Cells Expand in a Controlled Way by Absorbing Water

As we have observed, meristem production of new cells is an important component of plant growth. In addition, plant growth involves cell expansion. The diameters of newly formed stem and root cells are usually equal in all dimensions, but many soon begin to extend lengthwise, thereby helping shoots and roots to grow longer. Cell extension occurs when water enters the central vacuole by osmosis (**Figure 35.9**). As the central vacuole expands, the cell wall also expands and increases the cell's volume. By taking up water, plant cells can enlarge quickly, allowing rapid plant growth. Bamboo, for instance, can grow taller by 2 m within a week and can grow up to 30 m in less than three months! The importance of water uptake in cell expansion helps to explain why plant growth is so dependent on water supply.

Plant cell walls contain cellulose microfibrils that are held together by cross-linking polysaccharides. When plant cells and their vacuoles absorb water, pressure builds on cell walls. In response to this pressure and under acidic conditions, proteins unique to plants—known as expansins—are produced. **Expansins** unzip cross-linking cell-wall polysaccharides from cellulose microfibrils so that the cell wall can stretch (**Figure 35.10**). As a result, cells enlarge, often by elongating in a particular direction, which is important to plant form. Some plant cells are able to elongate up to 20 times their original length.

The direction in which a plant cell expands depends on the arrangement of cellulose microfibrils in its cell wall, which is determined by the orientation of cytoplasmic microtubules.

Figure 35.9 Plant cells expand by taking up water into their vacuoles.

1 Before a cell begins to expand, proton pumps increase cell-wall acidity.

2 Acidic conditions activate expansin proteins, which unzip cross-linking polysaccharides from cellulose microfibrils.

3 The cellulose microfibrils are free to glide apart. As the cell takes up water, the cytoplasm exerts pressure on the cell wall, causing it to expand.

Figure 35.10 A hypothetical model of the process of cell-wall expansion.

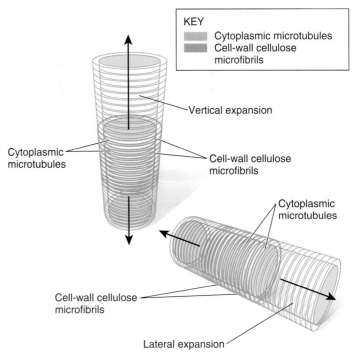

Figure 35.11 Control of the direction of plant cell expansion by microfibrils and microtubules. Plant cells enlarge in the direction perpendicular to encircling cell-wall cellulose microfibrils, which run parallel to the orientation of underlying cytoplasmic microtubules.

These microtubules are thought to influence the positions of cellulose-synthesizing protein complexes located in the plant plasma membrane. The protein complexes connect sugars to form cellulose polymers, spinning cellulose microfibrils onto the cell surface to form the cell wall. As a result, cell-wall cellulose microfibrils encircle cells in the same orientation as underlying cytoplasmic microtubules (**Figure 35.11**). Because cellulose microfibrils do not extend lengthwise, plant cell walls expand more easily in a direction perpendicular to them. To visualize this process, imagine encircling a spherical balloon with several parallel bands of tape before it has been completely inflated. The tape bands, which do not extend lengthwise, represent encircling cellulose microfibrils. As you add more air, the balloon will tend to extend in the direction perpendicular to the bands of tape.

Microtubules control not only the direction of cell expansion but also the plane of cell division, which is also critical to plant form. The *FASS* gene mutation in the model plant *Arabidopsis* illustrates the importance of microtubule orientation to plant structure. In cells of these mutants, microtubules are randomly arranged, resulting in cell division planes that are abnormal, cells that do not elongate, and mature plants with abnormal, stubby organs.

35.3 The Shoot System: Stem and Leaf Adaptations

As we have seen, the shoot system includes all of a plant's stems, branches, leaves, and buds. It also produces flowers and fruits when the plant has reached reproductive maturity. Thus, the shoot system is essential to plant growth, photosynthesis, and reproduction. In this section, we examine stem and leaf structure and development in more detail. We will observe that features of shoot stems and leaves vary among plants in ways that explain plant ecological function in nature and that these features are very useful to us in distinguishing plant species.

Shoot Systems Have a Modular Structure

More than 200 years ago, the German author, politician, and scientist Johann Wolfgang von Goethe realized that plants are modular organisms, composed of repeated units. Shoots are notably modular (**Figure 35.12**). Each shoot module, often known as a phytomere, consists of four parts: a stem node, an internode, a leaf, and an axillary meristem or bud. A **node** is the stem region from which one or more leaves emerge. An **internode** is the region of stem between adjacent nodes. Differences in numbers and lengths of internodes help to explain why plants differ in height. Each time a young leaf is produced at the SAM, a new meristem develops in the upper angle formed where the leaf emerges from the stem. This angle is known as an axil (from the Greek *axilla*, meaning armpit), and the meristem formed there is called an **axillary meristem**. Such axillary meristems generate **axillary buds**, which can produce flowers or branches known as lateral shoots. Such new branches bear a SAM at their tips. SAMs located at the apices of both main and lateral shoots produce new leaves.

What causes new leaves to arise? The answer involves chemical messengers known as hormones.

Hormones and Differential Gene Expression Influence Leaf Development

As we have noted, leaf primordia are surface bumps of tissue that develop at the sides of a SAM (see Figure 35.5). Production of such leaf primordia is under the control of a hormone known as **auxin**. In general, **hormones** are signaling molecules that exert their effects at a site distant from the place where such compounds are produced. Plant hormones are important in coordinating both plant development and plant responses to environmental conditions and are discussed in more detail in Chapter 36.

The outermost epidermal layer of cells at shoot tips produces auxin, which moves from cell to cell by means of specific membrane transport proteins. Auxin accumulates in particular locations because cells of the shoot apex differ in their ability to import and export auxin. When auxin accumulates in a particular apical region, the hormone causes expansin gene expression to increase. When expansin loosens their cell walls (see Figure 35.10), cells expand by taking up water, thereby forming a tissue bulge—a leaf primordium. The development of leaf primordia depletes auxin from nearby tissue, with the result that the next leaf primordium will develop in a different place on the shoot apex, where the auxin level is higher. Such changes in auxin concentration on the surface of the shoot explain why

(a) Modular structure of herbaceous shoot

(b) Modular structure of woody shoot in winter

Figure 35.12   **The modular organization of plant shoots.**   (a) The top end of an herbaceous stem showing the shoot modules. Each module consists of a node with its associated leaf and axillary meristem or bud and an internode. (b) The modular organization shown by a woody stem as it appears during winter. Axillary buds lie above the scars left by leaf fall. Regions between successive sets of bud scale scars mark each year's growth.

Concept check: *If a twig has five sets of bud scale scars, how old is the twig likely to be?*

leaf (and flower) primordia develop in spiral or whorled patterns around the shoot tip. The youngest leaf primordia occur closest to the shoot tip, and successively older leaf primordia occur on the sides of the shoot tip (see Figure 35.5).

Hormones also influence the transformation of primordia into leaves. The cells of leaf primordia do not produce a protein known as KNOX, which is produced by other shoot meristematic cells. KNOX is a transcription factor, a protein that regulates gene transcription. The absence of KNOX proteins induces leaf primordia to produce a plant hormone known as **gibberellic acid**. This hormone stimulates both cell division and cell enlargement, causing young leaves to grow.

Other molecules produced by a SAM direct leaf flattening and differentiation of adaxial and abaxial leaf tissues (see Figure 35.8c). These signaling effects were first demonstrated in 1955 by plant developmental biologist Ian Sussex, who made surgical cuts around leaf primordia, thereby isolating them from chemical substances produced by the meristem. The isolated primordia retained their initial cylindrical, radially symmetrical structure rather than becoming flat and bilaterally symmetrical, as they should. Later, investigators discovered that a particular set of genes expressed in young primordia causes leaf adaxial tissues to develop and that a different set of genes specifies abaxial tissue development. What causes this differential gene expression? In 2004, plant biologists Catherine Kidner and Robert Martienssen reported that a specific type of microRNA (miRNA; described in Chapter 13) accumulates in the abaxial regions of a developing leaf. In this location, the miRNA interrupts expression of the set of genes that specify adaxial tissue development and allows expression of the genes specifying abaxial fate, helping to explain why leaf adaxial and abaxial tissues differ. Other types of miRNA molecules can influence leaf shape, which helps explain the diversity of leaf shapes produced by different plant species.

Leaf Shape and Surface Features Reflect Adaptation to Environmental Stress

As we have noted, leaf flatness facilitates solar energy collection, and thinness helps leaves to avoid overheating. Leaf shape and surface features also reflect adaptation to stressful environmental conditions.

Leaf Form The flattened portion of a leaf is known as the leaf **blade**. In beans and most other eudicots, blades are attached to the stem by means of a stalk known as a **petiole**, and an axillary bud occurs at the junction of stem and petiole (**Figure 35.13a**). In contrast, corn and other monocots have leaf blades that grow directly from the stem, encircling it to form a leaf sheath (**Figure 35.13b**).

Leaf shape can be simple or compound, each having particular advantages. Simple leaves have only one blade, though the edges may be smooth, toothed, or lobed. Simple leaves are advantageous in shady environments because they provide maximal light absorption surface, but they can overheat in sunny environments. As an evolutionary response to the heating stress, the blades of some leaves have become highly dissected into leaflets. Such leaves are known as compound leaves (**Figure 35.13c**). You can distinguish leaflets from leaves because leaflets lack axillary buds at their bases. Compound leaves are common in hot environments because leaflets foster heat dissipation. During the development of at least some compound leaves, the transcription factor KNOX becomes active shortly after the leaf primordia form, causing these primordia to produce multiple growth points that generate the leaflets. In contrast, during simple leaf development, KNOX is not active, because the expression of other proteins suppresses such *KNOX* gene expression.

Leaf Vein Patterns Leaf vein patterns are known as venation. Eudicot leaves occur in two major venation forms. They may have a single main vein from which smaller lateral veins diverge in a feather-like pattern known as **pinnate** venation (Figure 35.13a). Alternatively, several main veins may spread from a common point on the petiole like the fingers of your hand, a pattern known as **palmate** venation (**Figure 35.13d**). In both cases, small veins connect in a netted pattern (see Table 35.1). Most monocot leaves have a distinctive parallel venation (Figure 35.13b).

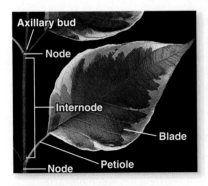

(a) Eudicot stem with simple leaf and pinnate venation

(b) Monocot stem and leaf

(c) A compound leaf

(d) A simple leaf and palmate venation

Figure 35.13 **Examples of variation in leaf form.** (a) A simple eudicot leaf, showing blade, petiole, and axillary bud. This leaf has a pinnate venation pattern. (b) The leaf of a monocot, showing parallel veins. The base of the leaf encircles the stem. (c) A compound leaf divided into leaflets. (d) A simple leaf having palmate venation.

FEATURE INVESTIGATION

Lawren Sack and Colleagues Showed That Palmate Venation Confers Tolerance of Leaf Vein Breakage

In 2008, Lawren Sack and associates studied the adaptive value of leaf venation patterns by comparing water conduction after injury in pinnately and palmately veined leaves (Figure 35.14). They hypothesized that vascular redundancy would confer tolerance of vein breakage of the type that would occur during mechanical injury or insect damage. To test the hypothesis, the investigators experimentally cut a main vein in the leaves of several plants belonging to seven different plant species: four having pinnately veined leaves and three having palmately veined leaves. They conducted the experiments in vivo, that is, in the plant's natural forest environment. After the wounds had healed, the investigators measured the extent of water flow within the leaves at two or three places on each leaf. They found that across all species examined, palmately veined leaves tolerated the disruption in water flow better than pinnately veined leaves. Leaves with palmate venation have several main veins rather than just one, as in the case of leaves with pinnate venation, which could reduce the impact of disruption by providing transport around injured veins. Although palmate venation provides redundancy in case a main vein becomes damaged, it is more costly in terms of materials needed to construct the additional main veins.

Figure 35.14   Sack and colleagues investigated the function of palmate venation.

HYPOTHESIS   Palmate venation provides vascular redundancy, which allows leaves to tolerate vein breakage.

KEY MATERIALS   Seven species of trees or shrubs at Harvard Forest, Petersham, MA.

	Experimental level	Conceptual level
1 Identify 7 species, 4 with pinnately veined leaves and 3 with palmately veined leaves. For each species, the researchers analyzed 10 leaves on 3 different plants.	Single primary vein connecting directly to petiole — Pinnately veined leaves — Petiole — *Quercus rubra*, *Betula alleghaniensis*, *Viburnum cassanoides*, *Kalmia latifolia* — More than one primary vein connecting directly to petiole — Palmately veined leaves — *Viburnum acerifolium*, *Acer saccharum*, *Acer rubrum*	Locate pinnate and palmate leaves for comparison.
2 With the leaf still attached to the plant, use a scalpel to cut across 1 primary vein in 5 experimental leaves but not 5 control leaves.		This procedure initially cuts off water supply via 1 primary vein.
3 Cover the cuts with medical tape on both top and bottom of leaves. Tape same area of uncut control leaves.	Tape	The tape prevents infection. Controls: Tape is applied to controls for experimental consistency.

4 Fold cardboard over base of cut leaves, forming a splint. Splint the same area of uncut control leaves.

Splint

The cardboard prevents leaf collapse.
Control: Cardboard is applied to controls for experimental consistency.

5 Measure water conduction 2–9 weeks after treatment, in 2 regions (A, B) of each pinnate leaf and in 3 regions (A, B, C) of each palmate leaf.

Pinnate leaf

Palmate leaf

After 2–9 weeks, the cuts had healed. Measurement of water conduction will determine the effect of the vein cut.

6 **THE DATA**

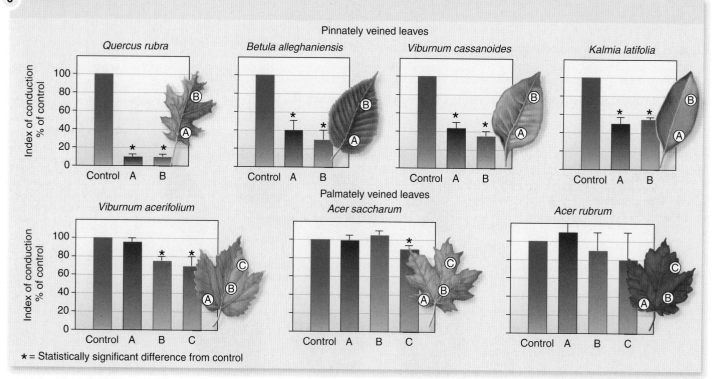

★ = Statistically significant difference from control

7 **CONCLUSION** Palmately veined leaves did not suffer as much conduction loss from a primary vein cut as did pinnately veined leaves.

8 **SOURCE** Sack, L. et al. 2008. Leaf palmate venation and vascular redundancy confer tolerance of hydraulic disruption. *Proceedings of the National Academy of Sciences of the U.S.* 105:1567–1572.

Experimental Questions

1. Why did Sack and associates conduct their studies of palmate venation on plants growing in a forest rather than in a greenhouse?

2. Why did Sack and colleagues splint pinnately veined leaves?

3. Why did Sack and associates measure leaf water conduction at two or more places on each leaf?

Figure 35.15 Leaf surface features. These features, viewed by SEM, include cuticular wax, trichomes, and stomatal pores with guard cells.

Leaf Surface Features Leaf surfaces also show adaptive features. As we have previously noted, a layer of epidermal tissue occurs at upper and lower leaf surfaces (see Figure 35.8c). These epidermal cells secrete a cuticle composed of protective wax and polyester compounds. The cuticle helps plants to avoid drying in the same way that enclosure in waxed paper keeps food moist. Plants that grow in very arid climates often have thick cuticles, whereas plants native to moist habitats typically have thinner cuticles. Cuticles also filter damaging ultraviolet (UV) radiation, reduce attack by microbes and animals, and foster a self-cleaning process by which water droplets and debris wash from the leaf surface.

Some leaf epidermal cells may differentiate into spiky or hairlike projections known as **trichomes** (**Figure 35.15**). Blankets of trichomes offer protection from excessive light, UV radiation, extreme air temperature, excess water loss, or attack by herbivores—animals that consume plant tissues. Broken trichomes of the stinging nettle, for example, release a caustic substance that irritates animals' skin, causing them to avoid these plants. Leaf epidermal cells include pairs of specialized guard cells located on either side of stomatal pores (Figure 35.15). These **guard cells** allow stomatal pores to be open

during moist conditions and to close when conditions are dry, thereby preventing plants from losing too much water. The genetic basis of stomatal guard cell development is becoming increasingly well understood.

Genomes & Proteomes Connection

Genetic Control of Stomatal Development

The flowering plant *Arabidopsis thaliana* is a model plant that is widely used to explore the genetic basis for plant structure and development. Several features increase *A. thaliana*'s utility for such studies: It is small in size, it has a fast seed-to-seed life cycle, it produces a relatively large number of seeds, and the genome has been sequenced. Mutants of this plant have been used to identify the genes controlling many aspects of plant structure and development, including the development of specialized stomatal guard cells.

Guard-cell development begins with an unspecialized protodermal cell that divides unequally (**Figure 35.16**). The larger of the two progeny cells eventually becomes a flat, puzzle piece–shaped epidermal cell, while the smaller is called a meristemoid because it functions like a stem cell. Meristemoids undergo one or more unequal cell divisions, producing more puzzle piece–shaped epidermal cells before finally dividing equally to produce a pair of guard cells. Genetic studies of *A. thaliana* have revealed that the meristemoid secretes a protein that inhibits division by adjacent cells but does not affect cells farther away. This process distributes stomata evenly and prevents too many of them from forming, which could increase the loss of water from plant surfaces.

In 2007, two teams, led by Lynn Pillitteri and Cora MacAlister, respectively, independently reported experiments with *A. thaliana* showing how three closely related genes control guard-cell development. A gene called *SPEECHLESS* starts the process by establishing the first unequal cell divisions of meristemoids. A protein encoded by the *MUTE* gene then causes meristemoids to stop dividing unequally so that equal divisions can produce the two guard cells (Figure 35.16). Disabling mutations of these two genes cause the plant epidermis to completely

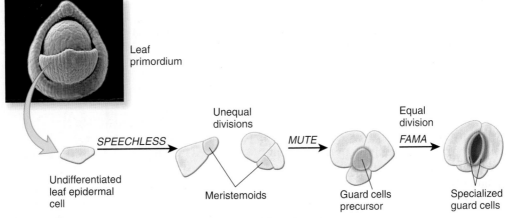

Figure 35.16 The development of stomatal guard cells, controlled by three genes.

(a) Tendrils **(b) Bud scales** **(c) Bracts** **(d) Spines**

Figure 35.17 **Examples of modified leaves.** (a) The tendrils of an American vetch plant are modified leaves that help the plant attach to a trellis. (b) Bud scales, such as those on this sycamore bud, are modified leaves that protect buds from winter damage. (c) The attractive red bracts of poinsettia are modified leaves that function like flower petals to attract pollinator insects to the small flowers. (d) Cactus spines, such as these on this giant saguaro, are modified leaves that function in defense.

Concept check: *Because cactus leaves are so highly modified for defense that they cannot effectively accomplish photosynthesis, how do cacti obtain organic compounds?*

lack stomatal pores (and so lack epidermal "mouths" and be speechless or mute). Finally, the gene *FAMA* directs guard-cell specialization. The proteins encoded by these plant genes are members of a type known as basic helix-loop-helix (bHLH) proteins (see Chapter 13) (similar proteins control the development of muscle and nerve cells in animals).

Modified Leaves Perform Diverse Functions

Though most leaves function primarily as photosynthetic organs, some plants produce leaves that are modified in ways that allow them to play other roles. For example, threadlike tendrils that help some plants attach to a supporting structure are modified leaves or leaflets (**Figure 35.17a**). The tough scales that protect buds on plants such as the sycamore from winter damage are modified leaves (**Figure 35.17b**). Poinsettia "petals" are actually modified leaves known as bracts, which are larger and more brightly colored than the flowers they surround and help attract pollinators (**Figure 35.17c**). Cactus spines are actually modified leaves that have taken on a defensive role, leaving photosynthesis to the cactus stem (**Figure 35.17d**).

Stems May Contain Primary and Secondary Vascular Systems

Stems, leaves, roots, buds, flowers, and fruits all contain vascular systems composed of xylem and phloem tissues that conduct water, minerals, and organic compounds. **Herbaceous plants** such as corn and bean produce mostly primary vascular tissues. In contrast, **woody plants** produce both primary and secondary vascular tissues. A comparison of primary and secondary vascular tissues will aid in understanding their roles.

Primary Vascular Tissues **Primary vascular tissues** are composed of primary xylem and phloem. Primary xylem is a complex tissue containing several cell types (see Table 35.2). These include unspecialized parenchyma cells; stiff fibers that provide structural support; and two types of cells that have differentiated in ways that facilitate water transport: tracheids and vessel elements. Arranged in pipeline-like arrays, **tracheids** and **vessel elements** conduct water, along with dissolved minerals and certain organic compounds (**Figure 35.18**). Mature tracheids and vessel elements are no longer living cells, and the absence

Tracheid

Vessel element

50 µm

Figure 35.18 **Water-conducting cells of the xylem.** In this thinly sliced portion of a stem, the stained, lignin-impregnated walls of narrow tracheids and wider vessel elements can be distinguished.

of cytoplasm facilitates water flow. During development, these cells lose their cytoplasm by the process of programmed cell death. Why then don't the cell walls of tracheids and vessel elements break down or collapse? They are impregnated with a tough polymer known as lignin. The rigid cell walls of tracheids and vessel elements not only foster water conduction but also help support the plant body.

In contrast to xylem, living phloem tissue transports organic compounds such as sugars and certain minerals in a watery solution. Phloem tissue includes **sieve-tube elements**, thin-walled living cells that are arranged end to end to form pipelines (**Figure 35.19**). Pores in the end walls of sieve-tube elements allow solutions to move from one cell to another. Phloem tissue also includes companion cells that aid sieve-tube element metabolism, supportive fibers, and parenchyma cells (see Table 35.2). Phloem fibers are tough-walled sclerenchyma cells that are surprisingly long, 20–50 mm, and valued for their high strength. The phloem fibers of hemp (*Cannabis sativa*), flax (*Linum usitatissimum*), jute (*Corchorus capsularis*), kenaf (*Hibiscus cannabinus*), and ramie (*Boehmeria nivea*) are commercially important in the production of rope, textiles, and paper.

Secondary Vascular Tissues Woody plants begin life as herbaceous seedlings that possess only primary vascular systems. But as these plants mature, they produce secondary vascular tissues and bark. Secondary vascular tissues are composed of secondary xylem and secondary phloem. **Secondary xylem** is also known as **wood**, and **secondary phloem** is the **inner bark**. **Outer bark** is protective layers of mostly dead cork cells that cover the outside of woody stems and roots. Therefore, bark includes both inner bark (secondary phloem) and outer bark (cork). Woody plants produce secondary vascular tissues by means of **secondary meristems**, also known as **lateral meristems**, which form rings of actively dividing cells that encircle the stem. The two types of secondary meristems are vascular cambium and cork cambium, which are derived from primary meristems (**Figure 35.20**). In particular, the vascular cambium is derived from procambium, and cork cambium is derived from ground tissue (**Figure 35.21**).

The secondary meristem known as **vascular cambium** is a ring of dividing cells that produces secondary xylem to its interior and secondary phloem to its exterior (**Figure 35.22**). Secondary xylem conducts most of a woody plant's water and minerals. Cell divisions that occur in secondary meristems increase the girth of woody stems. During each new growing season, the vascular cambium produces new cylinders of secondary xylem and secondary phloem. In temperate trees, each year's addition of new secondary xylem forms growth rings that can be observed on the cut stem surfaces. If environmental conditions favor plant growth, the growth rings formed at that time will be wider than those formed during stressful conditions. Climatologists use growth ring widths in samples of old

Figure 35.19 Food-conducting cells of the phloem. This thinly sliced portion of a stem shows stained, thin-walled sieve-tube elements that conduct watery solutions of organic compounds such as sugar and certain minerals.

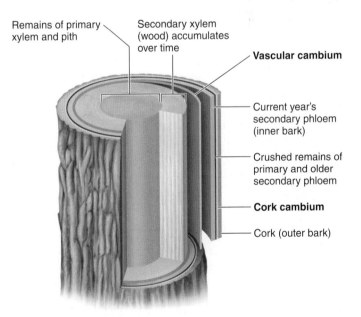

Figure 35.20 Formation of wood and bark by secondary (lateral) meristems. The vascular cambium is a thin cylinder of tissue that produces a thick cylinder of wood (secondary xylem) toward the inside of the stem and a thinner cylinder of inner bark (secondary phloem) toward the outside of the stem. The cork cambium forms an outer coating of protective cork (outer bark).

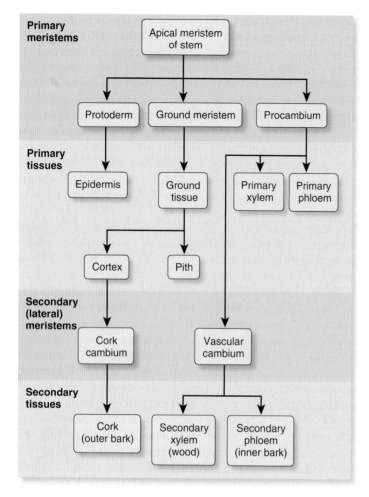

Figure 35.21 An overview of the process of primary and secondary growth in a woody stem.

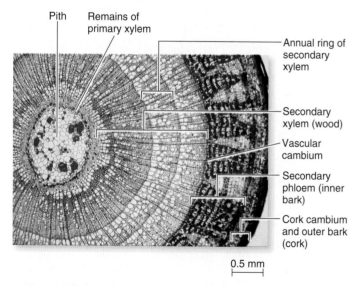

Pith
Remains of primary xylem
Annual ring of secondary xylem
Secondary xylem (wood)
Vascular cambium
Secondary phloem (inner bark)
Cork cambium and outer bark (cork)

0.5 mm

Figure 35.22 The anatomy of a tree trunk. Each year, a new cylinder of wood is produced; this yearly wood production appears as annual rings on the cut surface of a woody stem.

Concept check: Why do tree trunks have a thicker layer of wood (secondary xylem) than of inner bark (secondary phloem)?

wood to deduce past climatic conditions, and archeologists use growth ring data to determine the age of wood constructions and artifacts left by ancient cultures.

Secondary xylem may transport water for several years, but usually only the current year's production of secondary phloem is active in food transport. This is because thin-walled sieve elements typically live for only a year. Thus, only a thin layer of phloem, the **inner bark**, is responsible for most of the sugar transport in a large tree. Deep abrasion of tree bark may damage this thin phloem layer, disrupting a tree's food transport. If a groove is cut all the way around a tree trunk—a process known as girdling—the tree will die because all of its functional phloem transport routes will have been interrupted.

As a young woody stem begins to increase in diameter, its thin epidermis eventually ruptures and is replaced by outer bark, which is composed of protective cork tissues. Cork is produced by a secondary meristem called the **cork cambium**, another ring of actively dividing cells. The cork cambium surrounds the secondary phloem (see Figures 35.20 and 35.22). Cork cells are dead when mature, and their walls are layered with suberin, a material that helps to prevent both attack by microbial pathogens and water loss from the stem surface. Cork tissues also produce tannins, compounds that protect against pathogens by inactivating their proteins. The cracked surfaces of tree trunks are dead cork tissues of the outer bark. Commercial cork is sustainably harvested from the cork oak tree (*Quercus suber*) for production of flooring material, bottle stoppers, and other items. Additional information about the structure and function of primary and secondary xylem and phloem can be found in Chapter 38.

Modified Stems Display Diverse Forms and Functions

Stems mostly grow upright because light is required for photosynthesis. But some stems, known as rhizomes, occur underground and grow horizontally. For example, potato tubers are the swollen, food-storing tips of rhizomes. Grass stems also grow horizontally, as either rhizomes just beneath the soil surface or stolons, which grow along the soil surface. The leaves and reproductive shoots of grasses grow upward from the point where they are attached to these horizontal stems. Grass blades continue to elongate from their bases even if you cut their tips off, explaining why lawns must be mowed repeatedly during the growing season. The horizontal stems of grasses are adaptations that help to protect vulnerable shoot apical meristems against natural hazards such as fire and grazing animals.

35.4 Root System Adaptations

Roots play the essential roles of absorbing water and minerals, anchoring plants in soil, and storing nutrients. The external form of roots varies among flowering plants, reflecting adaptation to particular life spans or habitats. In contrast, root internal structure is more uniform. In this section, we first consider variation in root external structure and then focus on root internal structure and development.

Modified Roots Display Diverse External Forms and Functions

As we have observed, the common bean and other eudicots display an underground taproot system, whereas corn and other monocots have a fibrous root system (see Table 35.1). Plants produce several other types of roots that provide adaptive advantages. For example, corn and many other plants produce supportive prop roots from the lower portions of their stems. Many tropical trees grow in such thin soils that the trees are vulnerable to being blown down in windstorms. Such trees often produce dramatic aboveground buttress roots that help keep trees upright (**Figure 35.23a**). Many mangrove trees that grow along tropical coasts produce pneumatophores (Greek meaning breath bearers), roots that grow upward into the air (**Figure 35.23b**). Functioning like snorkels, pneumatophores absorb oxygen-rich air, which diffuses to submerged roots growing in oxygen-poor sediments. This is necessary because all roots require a supply of oxygen in order to produce ATP, which is needed to power the uptake of mineral nutrients (see Chapter 37).

(a) Buttress roots

—Pneumatophores

(b) Pneumatophores

Figure 35.23 **Modified aboveground roots.** **(a)** Buttress roots help to keep tropical trees such as this *Pterocarpus hayesii* from toppling in windstorms. **(b)** Pneumatophores produced by mangroves are roots that extend upward into the air. These roots take up air and then transmit it to underwater roots that grow in oxygen-poor sediments.

Root Internal Growth and Tissue Specialization Occur in Distinct Zones

We now turn to root internal structure and development. In their study of gene expression in *Arabidopsis* roots, plant molecular biologists Kenneth Birnbaum and associates identified 15 distinct cell types. Such studies reveal that roots are amazingly complex in structure. However, for our purposes, a simpler microscopic examination of root internal structure reveals three major zones: (1) a root apical meristem (RAM), (2) a zone of root elongation, and (3) a zone of maturation in which specialized cells can be observed (**Figure 35.24**).

Root Apical Meristem and Root Cap As we have discussed earlier, an apical meristem occurs at the tips of roots and their branches. Like the SAM, the RAM contains stem cells, but these are organized differently in root apices. Root stem cells surround a tiny region of cells that rarely divide, known as the quiescent center. Signals emanating from the quiescent center keep nearby stem cells in an undifferentiated state. Root stem cells farther away from the quiescent center produce new cells in multiple directions. Toward the root tip, stem cells produce columella cells that sense gravity and touch, which helps roots extend downward into the soil and around obstacles such as rocks. At the sides of the quiescent center, stem cells produce a protective root cap and epidermal cells. Root tip epidermal cells secrete a sticky substance called mucigel that lubricates root growth through the soil and has other beneficial functions. Toward the shoot, root stem cells generate cells that become ground and vascular tissues.

Zones of Elongation and Maturation Above the RAM lies the **zone of elongation**, in which cells extend by water uptake, thereby dramatically increasing root length (Figure 35.24). Above and overlapping with the zone of elongation is the **zone of maturation**, where most root cell differentiation and tissue specialization occur. Specialized root tissues include mature vascular tissues at the root core, an enclosing cylinder of cells known as the pericycle, another cell cylinder called the endodermis (meaning inside skin), and epidermal cells at the root surface. Relatively unspecialized parenchyma cells form a cortex that lies between the endodermis and the epidermis. Starting with the epidermis and moving inward, we will take a closer look at these root tissues and factors that control their development.

The zone of maturation can be identified by the presence of numerous microscopic hairs that emerge from the root epidermis. **Root hairs** are specialized epidermal cells that can be as long as 1.3 cm, about the width of your little finger, but are only 10 μm in diameter. Their small diameter allows root hairs to obtain water and minerals from soil pores that are too narrow for even the smallest roots to enter. Root hair plasma membranes are rich in transport proteins that use ATP to selectively absorb materials from the soil (see Chapter 37).

The production of hairs from root epidermal cells is controlled by the activity of the gene *GLABRA-2* (from the Latin

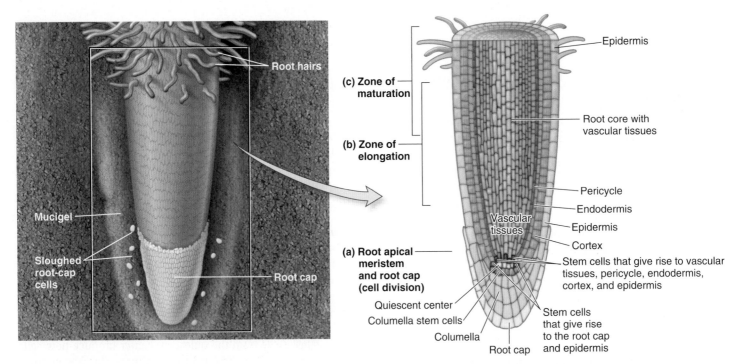

Figure 35.24 **Three zones of root growth.** A longitudinal view of a typical root reveals three major zones: **(a)** a root apical meristem region that includes stem cells, a quiescent zone, columella, and a root cap; **(b)** a zone of elongation; and **(c)** a zone of maturation, characterized by specialized cells and tissues including epidermal root hairs, cylinders of endodermis and pericycle tissue, and a core of vascular tissue.

glaber, meaning bald) (**Figure 35.25**). Whether this gene is expressed or not depends on the position of an epidermal cell. If an epidermal cell contacts two cortical cells, *GLABRA-2* is repressed and other genes are expressed, causing a root hair to develop. In contrast, no hair will develop if an epidermal cell lies in contact with only one cortical cell. A mutation that disables *GLABRA-2* in every epidermal cell causes all of them to produce root hairs.

Root hairs are so delicate that they are easily damaged by abrasion as roots grow through the soil, and they live for only 4

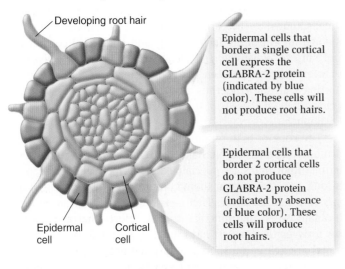

Figure 35.25 **A cross section of a root showing the genetic control of root-hair development.** The *Arabidopsis* gene *GLABRA-2* affects root-hair development.

or 5 days. As a result, root hairs are absent from older regions above the zone of maturation. To compensate, roots must continually produce new root hairs. The average rate of root hair production has been estimated at more than 100 million per day for some plants. One reason that gardeners use care when transplanting seedlings is to prevent extensive damage to the root hair zone.

The epidermis of mature roots encloses a region of ground parenchyma known as the root cortex (**Figure 35.26**). Much like the stem cortex, root cortex cells are often rich in starch and therefore serve as a food storage site for plants. The root cortex of some plants contains intercellular air spaces that arise from programmed cell death and provide routes for oxygen diffusion within the root. Water and dissolved minerals also diffuse from the environment into roots through spaces between cortex cells, stopping only when they reach a specialized tissue known as **endodermis**, an important component of the mechanism by which roots absorb selected minerals (see Chapter 37).

A cylinder of tissue having cell division (meristematic) capacity, known as the **pericycle**, encloses the root vascular tissue (Figure 35.26). The pericycle produces lateral (branch) roots that force their way through the cortex to the surface. This process differs from the way that stems produce branches by means of buds. In the model plant *Arabidopsis thaliana*, the expression of a gene known as *ARABIDILLO* is known to promote lateral root development. In some roots, the pericycle generates a vascular cambium that produces wood—secondary xylem. Such woody roots also possess a cork cambium that makes a protective covering of suberin-coated cork tissue. Like woody stems, woody roots produce primary vascular tissues

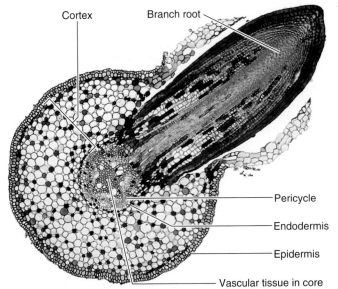

Cortex   Branch root

Pericycle

Endodermis

Epidermis

Vascular tissue in core

Figure 35.26   **Cross section of a mature root.**   This stained light micrograph shows the epidermis and cortex of a root surrounding a central core of vascular tissue. An inner cortex layer is the endodermis, which surrounds a cylinder of meristematic pericycle tissue. The pericycle has produced a young branch (lateral) root that has grown through the cortex and the epidermis.

 Concept check: *Why must lateral roots be produced in this way?*

in their youth and secondary vascular tissues at maturity. The woody roots of trees are sometimes visible aboveground.

Finally, the primary vascular system in a mature root includes xylem and phloem; in dicot roots, strands of phloem enclose a core of xylem. The xylem exports water and minerals upward to shoots, and the phloem imports a watery solution of organic compounds from the shoots.

Summary of Key Concepts

35.1 From Seed to Seed—The Life of a Flowering Plant

- Seed embryos, seedlings, and mature plants are components of the sporophyte generation in the plant sexual cycle; tiny gametophytes develop and grow within flowers. (Figure 35.1)

- Plant organs are composed of tissues that contain specialized cells. The basic plant organs are roots, stems, and leaves. Shoot systems include stems and stem branches, and stems produce leaves, buds, flowers, and fruits. Root systems include one or more main roots with branches. Buds, flowers, fruits, and seeds are organ systems, composed of more than one organ. (Figures 35.2, 35.3)

- The two major groups of flowering plants, eudicots and monocots, differ in the structure of their seed embryos, flowers, stems, roots, leaves, and pollen. (Table 35.1)

35.2 How Plants Grow and Develop

- The principles of plant growth and development include the presence of a fundamental architecture featuring apical-basal

polarity and radial symmetry throughout the life of a plant. (Figures 35.4, 35.5)

- Plants grow by producing new cells at meristems and controlled cell enlargement involving water uptake.

- Shoot apical meristems produce primary meristems that increase plant length and produce organs. Lateral meristems give rise to secondary tissues in many species. (Table 35.2, Figure 35.6)

- The simple plant tissues, containing one or two cell types, include parenchyma, collenchyma, and sclerenchyma. Complex plant tissues include the vascular tissues known as xylem and phloem, and the primary vascular tissues occur in vascular bundles. (Figure 35.7)

- Leaves develop from primordia at shoot apices. Foliage leaves have internal and external structure that is adapted for photosynthetic functions. (Figure 35.8)

- Meristems include youthful stem cells, whose numbers influence plant growth and structure and are thus genetically controlled.

- Plant cells are able to expand under conditions that result in loosening of cell-wall components, and by water uptake into vacuoles. The direction in which plant cells expand is determined by the arrangement of wall cellulose microfibrils, which is influenced by the orientation of microtubules in the nearby cytoplasm. (Figures 35.9, 35.10, 35.11)

35.3 The Shoot System: Stem and Leaf Adaptations

- Shoots are modular systems; each module includes a node, internode, leaf, and axillary meristem or bud. An axillary bud develops in leaf axils; such buds may grow into new branches. (Figure 35.12)

- Variations in leaf structure reflect adaptations that aid photosynthesis or protect against stress. For example, Sack and colleagues demonstrated that palmate leaves provide conducting system redundancy useful in coping with vein damage. (Figures 35.13, 35.14, 35.15)

- Stomatal guard cell differentiation is controlled by several genes. (Figure 35.16)

- Leaves function not only in photosynthesis but they also play other roles, including attachment, attraction, and protection. (Figure 35.17)

- Herbaceous plants are those whose stems produce little or no wood and are mostly composed of primary vascular tissues. The primary vascular tissues are primary xylem and primary phloem. (Figures 35.18, 35.19)

- In addition to primary tissues, woody plants—trees, shrubs, and lianas—possess secondary meristems that produce wood and bark. The vascular cambium produces secondary xylem (wood) and secondary phloem (inner bark). The cork cambium produces cork tissues that form outer bark. (Figures 35.20, 35.21, 35.22)

- Stems occur in diverse forms that reflect adaptation to environmental conditions. Examples include grass rhizomes, which grow horizontally underground and are therefore better protected from fire and grazing animals (as well as lawnmowers).

35.4 Root System Adaptations

- Roots occur in multiple forms that reflect adaptation to environmental conditions. Examples of aboveground roots include prop roots, buttress roots, and pneumatophores. (Figure 35.23)

- The internal organization of roots is comparatively uniform, and three major zones can be recognized with the use of a microscope: the root apical meristem and root cap, a zone of cell and root elongation, and a zone of tissue maturation. Features of the mature root include epidermal root hairs that aid nutrient uptake, a food-storing cortex, an endodermis that functions in mineral selection, a pericycle that produces lateral (branch) roots (and vascular cambium in the case of woody roots), and an inner core of vascular tissue. (Figures 35.24, 35.25, 35.26)

Assess and Discuss

Test Yourself

1. Where would you look to find the gametophyte generation of a flowering plant?
   a. at the shoot apical meristem
   b. at the root apical meristem
   c. in seeds
   d. in flower parts
   e. Flowering plants lack a gametophyte generation.

2. What is a radicle?
   a. an embryonic leaf
   b. an embryonic stem
   c. an embryonic root
   d. a mature root system of a monocot
   e. an organism that has extreme political views

3. Which type of plant is most likely to have food-rich roots that are useful as human food?
   a. an annual
   b. a biennial
   c. a perennial
   d. a centennial
   e. plants that grow along coastal shorelines

4. Which of the following terms best describes the distinctive architecture of plants?
   a. radial symmetry and apical-basal polarity
   b. bilateral symmetry and apical-basal polarity
   c. radial symmetry and absence of apical-basal polarity
   d. bilateral symmetry and absence of apical-basal polarity
   e. absence of symmetry and absence of apical-basal polarity

5. Which is the most accurate description of how plants grow?
   a. by the addition of new cells at meristems that include stem cells
   b. by cell enlargement as the result of water uptake
   c. by both the addition of new cells and cell expansion
   d. by addition of fat cells
   e. all of the above

6. Where would you look for leaf primordia?
   a. at a vegetative shoot tip
   b. at the root apical meristem
   c. at the vascular cambium
   d. at the cork cambium
   e. in a floral bud

7. Which leaf tissues display the greatest amount of air space?
   a. the adaxial epidermis
   b. the abaxial epidermis
   c. the palisade parenchyma
   d. the spongy parenchyma
   e. the vascular tissues

8. What are adventitious roots?
   a. roots that develop on plant cuttings that have been placed in water
   b. buttress roots that grow from tree trunks
   c. the only kinds of roots produced by monocots, because their embryonic root dies soon after seed germination
   d. any root that is produced by stem (or sometimes leaf) tissue, rather than developing directly from the embryonic root
   e. all of the above

9. During its development, a tracheid will elongate in a direction parallel to the shoot or root axis. Based on this information, what can you say about the orientation of cellulose cell-wall microfibrils and cytoplasmic microtubules in this developing tracheid?
   a. The microfibrils will be oriented perpendicularly (at right angles) to the long axis of the developing tracheid, encircling it, but the cytoplasmic microtubules will be oriented parallel to the direction in which the tracheid is elongating.
   b. Microfibrils and microtubules will both be oriented perpendicularly (at right angles) to the elongating axis of the tracheid.
   c. Microfibrils and microtubules will both be oriented parallel to the direction of tracheid elongation.
   d. Microfibrils will be oriented parallel to the direction of tracheid elongation, but microtubules will be perpendicular (at right angles) to both the microfibrils and the elongating tracheid.
   e. None of the above is correct.

10. What are examples of woody plants?
    a. trees
    b. shrubs
    c. woody vines, known as lianas
    d. all of the above
    e. none of the above

Conceptual Questions

1. What would be the consequences if overall plant architecture were bilaterally symmetrical?

2. What would be the consequences if leaves were radially symmetrical (shaped like spheres or cylinders)?

3. Why are most tall plants also woody, rather than herbaceous?

Collaborative Questions

1. Find a tree stump or a large limb that has recently been cut from a tree (or imagine doing so). Which of the following features could you locate with the unaided eye: the outer bark, the inner bark, the secondary xylem, the vascular cambium, annual rings?

2. Which physical factors would you expect influence shoot growth most strongly? Which physical factors would you expect most strongly influence underground root growth?

Online Resource

www.brookerbiology.com

Stay a step ahead in your studies with animations that bring concepts to life and practice tests to assess your understanding. Your instructor may also recommend the interactive ebook, individualized learning tools, and more.

Flowering Plants: Behavior

36

Behavior of the snow buttercup. The snow buttercup (*Ranunculus adoneus*) holds its flowers above the surface of the snow. The flowers move so they always face the sun during the day, a behavior known as sun tracking.

T he snow buttercup (*Ranunculus adoneus*) grows in deep snowbanks in the high Rocky Mountains, with flower stems protruding above the snow surface toward the sun, as shown in the chapter-opening photo. Amazingly, snow buttercup flowers change their position so they face the sun throughout the day, a process known as sun tracking. Experiments have demonstrated that sun tracking warms snow buttercup flowers, thereby favoring pollen development and germination. Thus, sun tracking is an adaptation that increases snow buttercup reproductive fitness. Leaves of alfalfa, lupine, soybean, common bean, cotton, and other plants may also track the sun, a process that aids in photosynthesis. Sun tracking is but one example of the many ways in which plants display behavior—that is, responses to stimuli. Although sun tracking and some other plant responses are relatively rapid, plants often display behavior such as changes in growth and development that occur more slowly.

We begin this chapter with an overview of the diverse types of stimuli that induce plant behavior and consider how plant cells perceive and respond to stimuli by means of signal transduction pathways. Next, we will learn about plant hormones, internal mobile molecules that influence plant behavior. Finally, we consider how plant responses to environmental stimuli such as light foster survival and reproduction.

36.1 Overview of Plant Behavioral Responses

As mentioned, **behavior** is defined as a response to a stimulus. Therefore, plants display behavior because they exhibit many types of responses to stimuli. Examples include plant movements, some types of which were described in 1880 by Charles Darwin and his son Francis in their book *The Power of Movement in Plants*. Modern time-lapse photography reveals that most plants are constantly in motion, bending, twisting, or rotating in dancelike movements known as nutation (**Figure 36.1a**). Some plants display quite rapid movements, illustrated by the sensitive plant (*Mimosa pudica*), whose leaves quickly fold when touched, then open more slowly (**Figure 36.1b**). Plants also display behavior when they grow and develop in

response to stimuli, though such responses are slower. For example, plant shoots typically grow toward light and against the pull of gravity, and most roots grow toward water and in the same direction as the gravitational force. Seeds germinate when they detect the presence of sufficient light and moisture for successful seedling growth. Producing flowers, fruit, and seeds only at the season most favorable for reproductive success is a behavioral response to environmental change. Plants also sense attack by disease microbes or hungry animals and take protective actions.

One reason why humans are often unaware of plant behavior is that it typically occurs on a longer timescale than our sensory systems are adapted to notice. Consequently, time-lapse photography, as shown in Figure 36.1a, is often needed to reveal slow plant movements. Although we typically are more aware of animal behavior than that of plants, plant behavior is important to humans because it influences agricultural productivity and plant roles in nature. To gain a more complete understanding of plant behavior, we will begin by surveying the types of stimuli that cause plant responses.

(a) Nutation movements   **(b) Leaf folding**

Figure 36.1 **Examples of plant movements.** (a) Sixteen superimposed photographs of a shoot of the honeysuckle vine *Lonicera japonica*, taken over a period of 2 hours, reveal the circular movement known as nutation. (b) Photographs of the sensitive plant (*Mimosa pudica*) made before and shortly after a touch reveal the rapid process of leaf folding. Even if only one leaflet is touched, electrical signals travel throughout the complex leaf, causing the entire organ to fold. The leaves will eventually unfold.

Plant Behavior Involves Responses to Internal and External Stimuli

Most people are aware that both internal chemical hormones and environmental factors influence animal behavior. Bird nesting behavior in spring, for example, involves hormonal changes triggered by seasonal conditions. Plants likewise respond to both internal and environmental influences (**Figure 36.2**).

Internal Stimuli Plants respond to mobile internal chemical signals that are produced within the body and move from one location to another, and act at very low concentrations. These chemical signals are small compounds generally known as **hormones**, phytohormones, or plant growth substances. Hormones often interact with each other and with external stimuli to maintain a stable internal environment (homeostasis) and enable progress through life stages.

Environmental Stimuli Plants sense and respond to many types of external physical and biological stimuli (Figure 36.2). Physical stimuli in natural plant environments include light, atmospheric gases such as CO_2 and water vapor, temperature, touch, wind, gravity, soil water, rocks and other barriers to root growth, and soil minerals. Biological stimuli include herbivores (animals that consume plant parts), airborne pathogens (disease-causing microbes), organic chemicals emitted from neighboring plants, and beneficial or harmful soil organisms. Crop plants also respond to applications of agricultural chemicals, which may include hormones. That plants have evolved

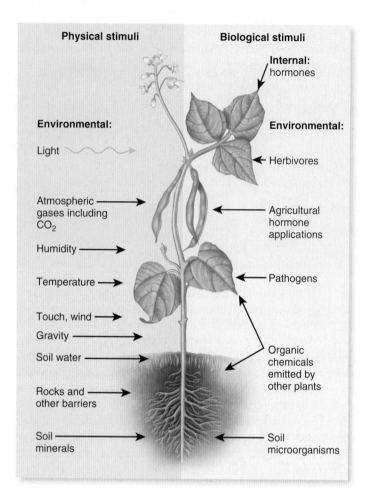

Figure 36.2 **Types of plant stimuli.** Plants respond to both physical and biological stimuli. Stimuli may be internal to the plant or come from the environment.

such a broad array of sensory capacity is not surprising, because all of the listed environmental influences affect plant survival and reproduction.

Plant Responses to Environmental Stimuli Though plants lack the specialized sense organs typical of animals, receptor molecules located in plant cells sense stimuli and cause responses. When many cells of a tissue receive and respond to the same chemical or physical signal, entire organs or plant bodies display behavior. For example, houseplants tend to grow toward a light source such as a window. This process, known as positive **phototropism**, involves both a cellular perception of light and a growth response of stem tissue to an internal chemical signal. In general, a **tropism** is a growth response that is dependent on a stimulus that occurs in a particular direction.

How can an environmental signal such as directional light cause stem movements? In the case of phototropism, the plant senses the direction of light and responds by changing the location of a plant hormone known as auxin. The auxin moves into cells where it influences gene expression. We will next consider how plant cells more generally receive signals and transmit them intracellularly, a process known as signal transduction.

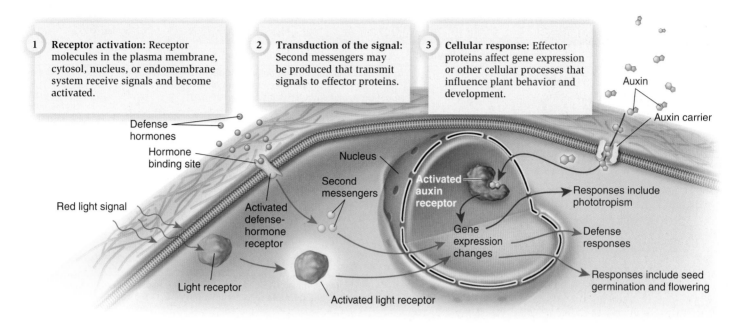

1 **Receptor activation:** Receptor molecules in the plasma membrane, cytosol, nucleus, or endomembrane system receive signals and become activated.

2 **Transduction of the signal:** Second messengers may be produced that transmit signals to effector proteins.

3 **Cellular response:** Effector proteins affect gene expression or other cellular processes that influence plant behavior and development.

Auxin

Auxin carrier

Defense hormones

Hormone binding site

Nucleus

Red light signal

Second messengers

Activated auxin receptor

Responses include phototropism

Activated defense-hormone receptor

Gene expression changes

Defense responses

Light receptor

Responses include seed germination and flowering

Activated light receptor

Figure 36.3   **An overview of plant signal transduction.**   Plant cells respond to hormonal signals produced within the plant body, as well as to environmental stimuli. Three different signal transduction processes are shown here: those started by light, a defense hormone having a plasma membrane receptor, and auxin having a receptor located in the nucleus.

Plant Signal Transduction Involves Receptors, Messengers, and Effectors

Signal transduction is the process in which a cell perceives a physical or chemical signal, thereby switching on an intracellular pathway that leads to a cellular response (**Figure 36.3**). The process of signal transduction occurs in three stages: receptor activation, transduction of the signal via second messengers, and a cellular response (refer back to Figure 9.4).

Receptors (also known as sensors) are proteins that become activated when they receive a specific type of signal (Figure 36.3). Receptors occur in diverse cellular locations. Whereas some plant defense-hormone receptors are located within the plasma membrane, light receptors may occur in the cytosol, and auxin receptors are located in the nucleus. Some activated receptors directly generate a response, such as an increased flow of ions across a membrane. In contrast, many activated receptors first bind to signaling molecules that initiate an intracellular signaling pathway. For example, the binding of brassinosteroid defense hormones to plasma membrane receptors results in the intracellular production of signaling molecules called second messengers.

Second messengers transmit messages from many types of activated receptors. Cyclic AMP, inositol trisphosphate (IP_3), and calcium ions are major types of second messengers in animal and plant cells. Calcium ions are particularly common messengers in plant cells. Touch and various other stimuli cause calcium ions to flow from storage sites in the ER lumen into the cytosol. Calmodulin or other calcium-binding proteins then bind the calcium ions. Calcium binding alters the structure of such proteins, causing them to interact with other cell proteins or to alter enzymatic function.

Effectors are molecules that directly influence cellular responses. In plants, calcium-dependent protein kinases (CDPKs) are particularly important effector molecules. Signal transduction ends when an effector causes a cellular response, such as opening or closing an ion channel or switching the transcription of particular genes on or off. A single activated receptor can dispatch many second messenger molecules, which, in turn, can activate scores of effectors, leading to many molecular responses within a single cell.

Now that we have considered how plant cells perceive and respond to stimuli, we are prepared to focus more closely on the roles of plant hormones.

36.2 Plant Hormones

As we have seen, plant hormones are chemical signals transported within the plant body that bind to cellular receptors, thereby causing responses. Plant hormones include about a dozen types of small molecules that are synthesized in metabolic pathways that also make amino acids, nucleotides, sterols, or secondary metabolites. Auxins, cytokinins, gibberellins, ethylene, abscisic acid, and brassinosteroids are considered to be the major plant hormones (**Table 36.1**). Individual plant hormones often have multiple effects, and different concentrations or combinations of hormones can produce distinct growth or developmental responses. Several plant hormones are known to act by causing the removal of gene repressors, thereby allowing gene expression to occur. This general mechanism allows hormones to cause relatively rapid responses. A closer look at the major types of plant hormones reveals their multifaceted roles.

Table 36.1	The Major Types of Plant Hormones	
Type of plant hormone	**Chemical structure of an example**	**Functions (Note: this is a partial, not complete list)**
Auxins	Indoleacetic acid (IAA)	Establish apical-basal polarity, induce vascular tissue development, mediate phototropism, promote formation of adventitious roots, inhibit leaf and fruit drop, and stimulate fruit development
Cytokinins	Zeatin	Promote cell division, influence cell specialization and plant aging, activate secondary meristem development, promote adventitious root growth, and promote shoot development on callus
Gibberellins	Gibberellic acid	Stimulate cell division and cell elongation, stimulate stem elongation and flowering, and promote seed germination
Ethylene	Ethylene $H_2C = CH_2$	Promotes seedling growth, induces fruit ripening, plays a role in leaf and petal aging and drop, coordinates defenses against osmotic stress and pathogen attack
Abscisic acid	Abscisic acid	Slows or stops metabolism during environmental stress, induces bud and seed dormancy, prevents seed germination in unfavorable conditions, and promotes stomatal closing
Brassinosteroids	Brassinolide	Promote cell expansion, stimulate shoot elongation, retard leaf drop, stimulate xylem development, and promote stress responses

Auxins Are the Master Plant Hormones

Plants produce several types of **auxins**, which are considered to be the master plant hormones because they influence plant structure, development, and behavior in many ways, often working with other hormones. Indoleacetic acid (IAA) is one plant auxin (Table 36.1), but other natural and artificial compounds have similar structures and effects. In this section, we will refer to this family of related compounds simply as auxin.

Auxin exerts so many effects because it promotes the expression of diverse genes, together known as **auxin-response genes**. Under low auxin conditions, proteins called Aux/IAA repressors prevent plant cells from expressing these genes. The repressor proteins prevent gene expression by binding to activator proteins at gene promoters. When the auxin concentration

is high enough, auxin molecules glue repressors onto a protein complex called TIR1, which causes the breakdown of the repressors. Free of the repressors, the activator proteins enhance the expression of auxin-response genes.

Auxin Transport The way in which auxin is transported into and out of cells is integral to its effects in plants. Auxin is produced in apical shoot tips and young leaves, and it is directionally transported from one living parenchyma cell to another. Auxin in an uncharged form (IAAH) may enter cells from intercellular spaces by means of diffusion; however, the negatively charged form (IAA⁻) requires the aid of AUX1, a plasma membrane protein known as the **auxin influx carrier**. Several types of proteins, called PIN proteins, transport auxin out of cells. They are named for the pin-shaped shoot apices of plants having mutations in *PIN* genes. Because they transport auxin out of cells, PIN proteins are called **auxin efflux carriers**. They are necessary because in the cytoplasm, auxin occurs as a charged ion that does not readily diffuse out of cells.

In shoots, AUX1 is located at the apical ends of cells, whereas PIN proteins often occur at the basal ends (**Figure 36.4a**). This polar distribution of auxin carriers explains why auxin primarily flows downward in shoots and into roots, a process called **polar transport** (**Figure 36.4b**). However, the locations of auxin carriers can also change within cell plasma membranes, allowing lateral or upward transport of auxin. Differences in the presence and positions of auxin carrier proteins explain variations in auxin concentration within plants. By measuring the local auxin concentration, plant cells determine their position within the plant body and respond by dividing, expanding, or specializing.

Auxin Effects In nature, auxin influences plants throughout their lifetimes. Auxin establishes the apical-basal polarity of seed embryos, induces vascular tissue to differentiate, mediates phototropism, promotes formation of adventitious roots, and stimulates fruit development. Many of auxin's effects are also of practical importance to humans. Auxin is used to produce some types of seedless fruit, retard premature fruit drop in orchards,

and stimulate root development on stem cuttings. Auxin produced by intact shoot tips inhibits lateral bud growth, a process known as apical dominance. Gardeners know that if they remove the topmost portion of a plant shoot, nearby lateral buds will begin to grow and produce new branches, allowing plants to become bushier. Such decapitation disrupts the flow of auxin from the shoot apex. Recent research suggests that a class of newly discovered hormones known as strigolactones also influence shoot branching. These carotenoid-derived terpenes, synthesized in roots and shoots, are transported to buds where they are thought to work with auxin. Although there is still much to learn about auxin function, auxin's role in phototropism has been elucidated by a series of experiments, as described next.

The Role of Auxin in Phototropism In the 1880s, Charles Darwin and his son Francis were the first to publish results of experiments on plant phototropism. The Darwins performed their experiments on cereal seedlings, whose tips are protected by a sheath of tissue called a coleoptile. In a simple but elegant experiment, the Darwins covered either the tips or lower portions of coleoptiles with shading materials such as blackened glass tubes, left other seedlings uncovered, and removed the tips of some seedlings. They then compared how those seedlings responded to illumination from the side. The seedlings

(a) Cellular mechanism of auxin transport   **(b) Auxin transport throughout a plant**

Figure 36.4 **Auxin transport.** (a) Polar and lateral auxin transport is controlled by the distribution of auxin efflux carriers located in the plasma membrane. When efflux carriers primarily occur at the basal ends of cells, auxin will flow downward. Auxin may flow laterally when auxin efflux carriers occur at the sides of cells. (b) In a whole plant, auxin primarily flows downward from shoot tips to root tips, where it then flows upward for a short distance.

Concept check: How could auxin carriers be organized to allow auxin to move upward in roots?

whose tips were left uncovered grew toward the light, whereas seedlings whose tips were covered or removed did not. The Darwins concluded that seedling tips transmit some "influence" to lower portions, causing them to bend toward the light. You can probably guess what this influence was, but technology available at the time did not allow the Darwins to determine this.

Three decades later, in the 1910s, Danish botanist Peter Boysen-Jensen confirmed the Darwins' results and demonstrated that the influence was a chemical substance that diffused from the tips of the seedlings to other parts. To do this,

Boysen-Jensen cut off the tips of oat seedlings and placed either a porous layer of gelatin or a nonporous material such as a sheet of the mineral mica on the cut surface. Then he replaced the tips. Oat seedlings layered with porous gelatin displayed a normal phototropic response, bending toward the light, but those layered with nonporous mica did not. Boysen-Jensen's experiment demonstrated that the phototropic substance was a diffusible chemical, but exactly which one, and how it worked, remained unknown. A series of additional experiments provided some answers.

FEATURE INVESTIGATION

Experiments Performed by Went and Briggs Revealed the Role of Auxin in Phototropism

In the 1920s, the Dutch plant physiologist Frits Went named the substance discovered by Boysen-Jensen auxin (from the Greek word *auxein*, meaning to increase). Although the chemical structure of auxin was not determined until 1934, Went performed experiments that helped explain how auxin works, as shown in Figure 36.5a. In a first step, Went cut the tips off of oat seedlings and placed these tips onto agar blocks. Agar, a complex polysaccharide derived from red algae, forms a mesh capable of holding considerable water and dissolved compounds. Agar's permeability to auxin is similar to that of the protein gelatin used by Boysen-Jensen, but agar is much more stable at room temperature and more resistant to microbial breakdown and therefore is easier to use in laboratory experiments. In Went's experiment, the auxin diffused from cut seedling tips into these agar blocks. In the next steps, he treated decapitated seedlings in one of four ways: (A) placed auxin-laden agar blocks off-center on some, (B) placed auxin-laden blocks evenly on others, (C) placed plain agar blocks off-center on some, and

(D) left some uncapped. All seedlings were then kept in darkness throughout the experiment. Only seedlings that were capped off-center with an auxin-laden block grew in the direction away from the agar block. This experiment demonstrated that auxin application could substitute for the directional light stimulus, and it suggested that asymmetric auxin distribution is the mechanism by which light causes plants to bend.

Subsequently, Went and N. O. Cholodny independently proposed that light causes auxin to move to the unlit side of seedling tips, causing cells on that side to elongate more, which results in bending. But other scientists argued that bending could result if light destroys auxin on the illuminated side of a seedling. In the 1950s, American plant biologist Winslow Briggs designed two experiments to test these alternate hypotheses.

To test the hypothesis that auxin might be destroyed by light, Briggs first grew corn seedlings in the dark. Then he cut off their tips, put the tips on agar blocks, and exposed some to darkness and others to directional light. During this process, auxin from tips diffused evenly into the agar blocks. If auxin were destroyed by light, agar blocks under lighted tips should receive less auxin than blocks under tips kept in the dark.

Figure 36.5   Went and Briggs demonstrated the relationship between light perception and auxin function.   (a) The Went experiment. (b) The Briggs experiment.

(a) Went experiment

> **HYPOTHESIS** A chemical substance produced by shoot tips causes the shoots to bend.
> **KEY MATERIALS** Oat seedlings.

Experimental level   **Conceptual level**

1   Cut seedling tips and put onto agar blocks. Note: Once auxin enters the agar block, it can diffuse throughout the whole block.
   Scalpel / Seedling — Seedling tip — Agar block
   Diffusible substance in tip of seedling — Diffusible substance in agar block

2   Cut more shoot tips from seedlings. Discard tips.
   Discard tips
   To prepare shoots for experimentation

3 A Cap some shoots off-center with agar blocks from step 1.
   B Cap some shoots evenly with agar blocks from step 1.
   C Cap some shoots off-center with plain agar blocks (control).
   D Leave some shoots uncapped (control).

Diffusion from agar block into shoot

4 Place all shoots in darkness.

Seedling growth will be affected by caps, not by light.

5 THE DATA

Degree of bending among treated shoots:

| Bending | No bending | No bending | No bending |

6 CONCLUSION Uneven distribution of a diffusible tip substance, named auxin, causes bending.

Went's new hypothesis:
Light causes auxin to move to unlit side of shoots, causing cells on that side to elongate more.

7 SOURCE Went, F.W. 1926. On growth-accelerating substances in the coleoptile of *Avena sativa. Proc. Kon. Akad. Wetensch.* Amsterdam 30:10–19.

(b) Briggs experiment 2

HYPOTHESIS Directional light causes auxin to move to the shaded side of shoot tips.

KEY MATERIALS Corn seedlings.

Experimental level   **Conceptual level**

1 Place shoot tips on agar blocks.

2 Divide some tip/block combinations completely with a mica sheet, which prevents diffusion between the 2 halves of the tip and agar block. Divide some tip/block combinations only partially with a mica sheet. This allows auxin diffusion across the tip, but not across the agar block. Expose both to directional light.

Mica sheet   Mica sheet

If directional light causes auxins to move to shaded side of shoot tips, agar block in B will contain more auxin on right side.

3 Remove agar block halves from tips. Place agar halves onto right sides of shoots, which have their tips removed.

If directional light causes auxins to move laterally, the block half beneath the left side of the partially divided tip shown in B should cause the least shoot bending, whereas the block half beneath the right side of B should cause the greatest amount of bending.

4 THE DATA

5 CONCLUSION In A, the mica sheets prevented auxin from moving to the shaded side of tips. In B, auxin was able to move in response to directional light. Agar block pieces from the shaded side of tips contained more auxin and therefore caused greater shoot bending. Hypothesis is correct.

6 SOURCE Briggs, W.R. 1963. Mediation of phototropic responses of corn coleoptiles by lateral transport of auxin. *Plant Physiology* 38(3):237–247.

The auxin destruction hypothesis also predicts that when agar blocks from lighted tips are placed on one side of decapitated seedlings, they should cause less bending than would blocks from tips kept in the dark. However, Briggs discovered that both types of agar blocks caused the same amount of shoot bending. This result is not consistent with the hypothesis that light destroys auxin.

In a second experiment, Briggs tested the Cholodny–Went hypothesis that light causes auxin to move to the shaded side of seedlings (**Figure 36.5b**). Briggs set shoot tips onto agar blocks (step 1) and then used a mica sheet (which is impervious to auxin) to completely divide some tips and blocks into halves (step 2A). In other cases, he divided blocks completely but left tips incompletely divided, allowing auxin to diffuse across tips but not the block halves (step 2B). Then Briggs exposed all sets of tips and blocks to directional light. He predicted that auxin would not be able to move across tips having complete mica barricades but that auxin would be able to move across tips that had been only partially divided. Auxin diffused from tips into blocks, but it could not diffuse evenly across blocks divided by mica sheets. When Briggs later placed the agar block halves on decapitated shoots (step 3), those receiving auxin from completely divided tips were bent by the same amount. By contrast, agar block halves from the lit side of partially divided tips induced less bending, while halves from the unlit side of partially divided tips caused the most bending (see the data in Figure 36.5b). These experimental results support the hypothesis that unidirectional light causes auxin to accumulate on the shaded side. Modern plant scientists would explain such auxin movement as the result of lateral transport involving PIN proteins.

How might auxin accumulation cause phototropic bending? One widely held hypothesis is that auxin accumulation on the shaded side of a plant shoot causes plasma membrane proton pumps located there to work at a faster rate. In response, the cell wall becomes more acidic, which activates expansins, proteins that break cross-links between cellulose microfibrils and allow cells to elongate (refer back to Figure 35.10). This process might explain how auxin accumulation in cells located on the shaded side of shoot tips causes them to elongate more than do cells on the sunny side, causing the tip to bend toward the light.

Experimental Questions

1. Use the text discussion to draw a diagram illustrating the experiment that Briggs did in order to determine whether or not light destroys auxin.

2. What is the current hypothesized mechanism by which auxin accumulation causes shoot bending in response to directional light?

3. Figure 36.5 illustrates experiments performed with four seedling tips. Would this number really be enough to allow conclusions to be made about how such seedling tips would generally respond?

Cytokinins Stimulate Cell Division

Like auxins, the plant hormones known as **cytokinins** play varied and important roles throughout the lives of plants. The name of these hormones reflects their major effect—an increase in the rate of plant cytokinesis, or cell division. Root tips are major sites of cytokinin production, but shoots and seeds also make this plant hormone. Transported in the xylem to meristems and other plants parts, cytokinins bind to receptors in the plasma membrane. At shoot and root tips, cytokinins influence meristem size, stem cell activity, and vascular tissue development. Cytokinins are also involved in root and shoot growth and branching, the production of flowers and seeds, and leaf senescence (aging).

Plant Tissue Culture In the laboratory, cytokinin and auxin are essential to cloning plants. This involves a process, known as **plant tissue culture**, which is used commercially to produce thousands of identical plants having the same desirable characteristics. Plant tissue culture begins with pieces of stem, leaf, or root that have been removed from a plant, and their surfaces are sterilized to prevent growth of microbes (**Figure 36.6**, step 1). The cleaned plant pieces are then placed into dishes containing nutrients (minerals, vitamins, and sugar) and various proportions of auxin and cytokinin. If the proportions of auxin and cytokinin are about the same (1:1), plant cells undergo division, forming a mass of white tissue known as a callus (step 2). If the callus is then transferred to a new dish containing the same nutrients, with auxin-to-cytokinin proportions greater than 10:1, the callus will form roots (step 3). Auxin-to-cytokinin proportions of less than 10:1 will cause the callus to develop green shoots (step 4). Thus, by altering the ratios of auxin and cytokinin, entire plants can be regenerated from a callus. A single callus can be divided into many pieces and each piece treated with these hormones, thereby producing many hundreds of identical new plants.

① A block of tissue is removed from a plant, and the surfaces are sterilized.

② Tissue is cultivated in dishes on nutrient media. Treatment with equal proportions of auxin and cytokinin causes formation of an undifferentiated callus.

③ Treatment with auxin-to-cytokinin ratios greater than 10:1 causes root development on many replicate plantlets.

④ Treatment with auxin-to-cytokinin ratios less than 10:1 induces shoot development on many replicate plantlets.

Plant tissue

Callus

1 : 1
auxin : cytokinin

>10 : 1
auxin : cytokinin

<10 : 1
auxin : cytokinin

Figure 36.6 **The process of plant tissue culture.** Plant tissue culture illustrates the impact of different proportions of auxin and cytokinin on plant organ development.

Concept check: *How do commercial growers use this process to produce many identical plants?*

Gibberellins Stimulate Cell Division and Elongation

The **gibberellins** (also known as gibberellic acids, or GA; see Table 36.1) are another family of plant hormones. Gibberellins are produced in apical buds, roots, young leaves, and seed embryos. In addition to promoting shoot development on laboratory calluses, gibberellins interact with light and other hormones to foster seed germination and enhance stem elongation and flowering in nature. Gibberellins also retard leaf and fruit aging. These multiple effects largely arise from gibberellin's stimulatory effects on cell division and elongation.

More than a hundred different forms of gibberellin have been found. Many kinds of dwarf plants are short because they produce less gibberellin than taller varieties of the same species. The dwarf strain of pea plants Mendel used in some of his breeding experiments is an example. When dwarf varieties of plants are experimentally sprayed with gibberellin, their stems grow to normal heights. However, dwarf wheat and rice crops are valued in agriculture because they can be more productive and less vulnerable to storm damage than taller varieties. Since the discovery of gibberellin, plant scientists have discovered how this plant hormone works at the molecular level and how gibberellin regulation of plant growth evolved, our next topics.

Genomes & Proteomes Connection

Gibberellin Function Arose in a Series of Stages During Plant Evolution

In flowering plants, gibberellin works by helping to liberate repressed transcription factors. In the absence of gibberellin, DELLA proteins bind particular transcription factors needed for the expression of gibberellin-responsive genes (Figure 36.7a). In this way, DELLAs function as brakes that restrain cell division and expansion. But when sufficient gibberellin is present, it binds receptor proteins called GID1 (Figure 36.7b). Gibberellin-binding increases the ability of GID1 proteins to interact with DELLA proteins, starting a process that leads to the destruction of DELLAs. In the absence of DELLA proteins, transcription factors are able to bind the promoter regions of gibberellin-responsive genes, allowing their expression. As a result, cell division and expansion occur, leading to growth. This efficient system integrates the effects of many signals that affect plant growth: the hormones auxin, ethylene, and abscisic acid, as well as light and environmental stress.

In 2007, Yuki Yasumura, Nicholas Harberd, and their colleagues reported that the gibberellin-DELLA mechanism for regulating the growth of flowering plants arose in a stepwise fashion. They discovered this by comparing the DELLA and GID1 proteins of flowering plants with homologous proteins of a bryophyte and a lycophyte. The seedless lycophytes, the oldest living phylum of vascular plants, first appeared millions of years before the first flowering plants, and the seedless, nonvascular bryophytes arose millions of years before the first vascular plants (see Figure 30.1). The investigators studied the interactions of DELLA proteins with GID1 and evaluated the extent to which gibberellin enhanced DELLA-GID1 binding and growth in these different groups of plants.

The results indicated that the bryophyte possesses DELLA and GID1 proteins, but these proteins don't interact, and bryophyte DELLA does not repress growth. In contrast, differences in the protein structure of lycophyte DELLAs enable them to interact with GID1 and gibberellin, but without detectably influencing growth. Neither the bryophyte nor the lycophyte showed detectable growth responses to gibberellin, though their DELLAs were able to repress growth when expressed in a flowering plant. This pattern led the biologists to propose that

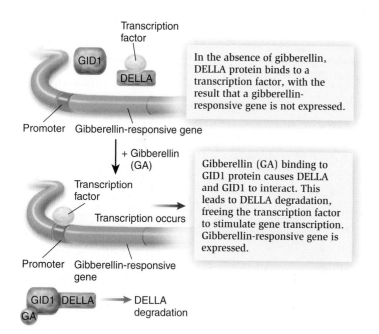

Figure 36.7 **Gibberellin works by releasing trapped transcription factors.** (a) In the absence of gibberellin, DELLA binds transcription factors, and gibberellin-response genes are not expressed. (b) When gibberellin binds to the protein GID1, GID1 can bind DELLA proteins. Such binding causes DELLA proteins to be degraded. As a result, transcription factors that had been bound to DELLA proteins are released and can bind to gene promoters, thereby inducing gene expression.

DELLA-mediated repression of plant growth evolved after the divergence of lycophytes but prior to the appearance of the first flowering plants (Figure 36.8). Though the necessary components (DELLAs and GID1 proteins) were earlier present, they did not assemble into a growth regulation system until later in plant evolutionary history.

Ethylene Influences Cell Expansion

The plant hormone **ethylene** is particularly important in coordinating plant developmental and stress responses. Ethylene is a simple hydrocarbon gas produced during seedling growth, flower development, and fruit ripening (see Table 36.1). In the root tip, ethylene determines how many stem cells remain inactive in the quiescent center and how many cells undergo divisions. This hormone also plays important roles in defense against osmotic stress and pathogen attack, and leaf and petal aging and drop. As a gas, ethylene is able to diffuse through the plasma membrane and cytosol to bind to ethylene receptors localized in the endoplasmic reticulum. When activated by ethylene binding, these receptors inactivate a protein kinase known as CTR1. This action ultimately enables transcription factors to induce the transcription of various genes.

People first noticed the effects of ethylene gas on plants in the 1800s when they observed that street-side trees exposed to leaking street lanterns unexpectedly lost their leaves. A 17-year-old student in St. Petersburg, Russia, Dimitry Neljubov,

Figure 36.8 **Evolution of the gibberellin-DELLA system.** The gibberellin-DELLA system that controls growth of flowering plants evolved in a step-by-step fashion. Although DELLA and GID1 proteins are present in a bryophyte, they do not interact, nor does gibberellin bind to GID1. In lycophytes, GID1 interacts with DELLA, but the system does not influence growth. The growth responses of gibberellin apparently evolved after the divergence of lycophytes.

performed the first experiments to explore the effects of illumination gas on plants. He exposed pea seedlings grown in the laboratory to illumination gas and noticed that the pea seedlings grew sideways rather than upward. Then he tested the individual chemical components of illumination gas for the same effect. After conducting many experiments, in 1901, Neljubov reported that ethylene was the only component of illumination gas that caused the seedlings to grow horizontally and that ethylene was effective in very low concentrations (as low as 0.06 parts per million in air). Later, scientists established that ethylene influences cell expansion, often in association with auxin. Ethylene does this by increasing the disorder of microtubules within cells, thereby causing random orientation of cell-wall microfibrils. As a result, cells exposed to ethylene tend to expand in all directions rather than elongating.

Ethylene's effects on cell expansion explain this hormone's important role in dicot seedling growth. The tender apical meristems of seedlings could be easily damaged during their growth through crusty soil. Ethylene helps seedlings avoid such damage by inducing what is known as the triple response (Figure 36.9). First, ethylene prevents the seedling stem and root from elongating. Second, the hormone induces the stem and root to swell radially, thereby increasing in thickness. Together, these responses strengthen the seedling stem and root. Third,

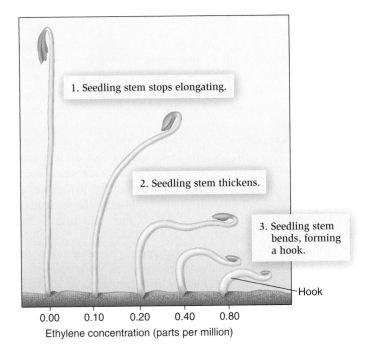

1. Seedling stem stops elongating.

2. Seedling stem thickens.

3. Seedling stem bends, forming a hook.

Hook

0.00   0.10   0.20   0.40   0.80
Ethylene concentration (parts per million)

Figure 36.9   **Seedling growth showing the triple response to ethylene.**   When applied at levels above a particular concentration (0.80 ppm), ethylene causes seedlings to cease elongation, swell radially, and bend to form a hook that can push upward through the soil. Ethylene produced naturally within seedlings causes the same response.

Concept check: *What adaptive advantage does this seedling behavior provide?*

the seedling stem bends so that embryonic leaves and the delicate meristem grow horizontally rather than vertically; this is the sideways growth response that Neljubov first observed. The bent portion of the stem, known as a hook, then pushes up through the soil. The hook forms as the result of an imbalance of auxin across the stem axis, which causes cells on one side of the stem to elongate faster than cells on the other side. Ethylene drives this auxin imbalance.

Knowledge of the effects of ethylene on fruit has been very useful commercially. Ripe fruit can be easily damaged during transit, but tomatoes and apples can be picked before they ripen for transport with minimal damage. At their destination, such fruit can be ripened by treatment with ethylene. However, fruit that becomes overripe may exude ethylene, which hastens ripening in nearby, unripe fruit. For this reason, fruit that must be stored for extended periods is kept in ethylene-free environments.

Several Hormones Help Plants Cope with Environmental Stresses

Several plant hormones share the property of helping plants respond to environmental stresses such as flooding, drought, high salinity, cold, heat, and attack by disease microorganisms and animal herbivores. These protective hormones include the major plant hormones known as **abscisic acid** and **brassinosteroids** (see Table 36.1). Additional protective hormones are salicylic acid (SA), whose chemical structure is similar to that

of aspirin, and a peptide known as systemin. The fragrant compound jasmonic acid, whose structure is similar to that of mammalian prostaglandins, and the gas nitric oxide (NO) are also protective plant hormones.

Abscisic Acid   Abscisic acid, abbreviated as ABA, was named at a time when plant biologists thought that it played a role in leaf or fruit drop, also known as abscission. Later, they discovered that ethylene actually causes leaf and fruit abscission, whereas abscisic acid slows or stops plant metabolism when growing conditions are poor. For example, ABA may induce bud and seed dormancy. Dormant buds and seeds resume growth only when specific environmental signals reveal the onset of conditions suitable for survival. In preparation for winter, ABA stimulates the formation of tough, protective scales around the buds of perennial plants. Seed coats of apple, cherry, and other plants also accumulate ABA, which prevents seeds from germinating unless temperature and moisture conditions are favorable for seedling growth. Water-stressed roots also produce ABA, which is then transported to shoots, where (together with ABA produced by water-stressed leaf mesophyll) it helps to prevent water loss from leaf surfaces by inducing leaf pores (stomata) to close. ABA receptors are G-protein-coupled receptors (GPCRs) located in the plasma membrane (see Chapter 9). ABA-binding causes the release of G protein, which begins the process of signal transduction.

Brassinosteroids   Brassinosteroids are named after the cruciferous plant genus *Brassica* (which includes cabbage and broccoli), from which they were first identified. However, seeds, fruit, shoots, leaves, and flower buds of all types of plants contain brassinosteroids. These plant hormones induce vacuole water uptake and influence enzymes that alter cell-wall carbohydrates, thereby fostering cell expansion. Mutations that affect brassinosteroid synthesis cause plants to exhibit dwarfism. Such plants have small, dark green cells because their tissues are unable to expand. Brassinosteroids also impede leaf drop, help grass leaves to unroll, and stimulate xylem development. They can be applied to crops to help protect plants from heat, cold, high salinity, and herbicide injury.

Brassinosteroids are chemically related to animal steroid hormones, such as human sex hormones. However, unlike animal steroid hormones, which bind to receptors in the nucleus or cytosol, brassinosteroids bind to receptors in the plasma membrane. When they bind brassinosteroids, the membrane receptors inhibit a protein that would otherwise target certain transcription factors for destruction. These transcription factors accumulate and influence gene expression, particularly genes that affect time of flowering. Brassinosteroid action thereby illustrates the general principle that plant hormones often work by removing the brakes from gene expression.

In this section, we have surveyed the roles of major plant hormones produced within plants. While plant hormones coordinate many genetically determined developmental processes, they also help plants to respond to environmental stimuli, our next topic.

36.3 Plant Responses to Environmental Stimuli

Plants encounter many types of environmental challenges and behave accordingly. Take seed germination, for example. If buried seeds were to germinate beneath soil layers too deep for light to penetrate, or beneath a cover of established plants, seedlings would not be able to obtain sufficient light for photosynthesis and would die. A related reproductive challenge for plants is to flower at times of the year that are most beneficial for achieving pollination or seed dispersal. How do plants determine if there is enough light for seeds to germinate and for seedlings to grow? How do plants determine when to flower?

The answer is that plants possess cellular systems for measuring light and determining the seasonal time of year. Using these systems, plant seeds germinate only when there is sufficient light for seedling growth, and flowering occurs during the most advantageous season. Therefore, plants can sense and respond to their light environments. Plants are also able to respond to other physical and biological stimuli. In this section, we survey plant responses to these external stimuli, beginning with light.

Plants Detect Light and Measure Day Length

A plant's ability to measure and respond to light amounts and day length, a process called **photoperiodism**, is based on the presence of light receptors within cells. Such light sensors, known as **photoreceptors**, are distinct from the light-absorbing pigments that function in photosynthesis. Each type of photoreceptor has a light-absorbing component as well as other regions that respond to light absorption by switching on signal transduction pathways. Responses by many cells in a tissue or organ cumulatively result in behaviors such as sun tracking, phototropism, flowering, and seed germination.

Blue-Light Receptors Cryptochromes and phototropins are two types of blue-light receptors—molecules that absorb and respond to blue light. Experiments suggest that receptors called **cryptochromes** help young seedlings determine if their light environment is bright enough to allow photosynthesis. If not, seedlings continue to elongate through the soil, toward the light.

Phototropin is the main blue-light sensor involved in phototropism. The light-activated form of this sensor has two components: a protein that has a kinase domain and a flavin pigment that can absorb blue light. In the dark, the flavin is not covalently bound to the protein. However, when the flavin absorbs blue light, it changes conformation and becomes able to covalently bind to the protein. Flavin binding, in turn, changes the conformation of the phototropin protein, allowing it to phosphorylate itself by means of the protein kinase domain. When a plant organ is exposed to directional blue light, phototropin becomes phosphorylated. In this way, a light signal is converted into a chemical signal. However, the events that connect phosphorylation to changes in auxin movement in phototropism, discussed previously, are so far unknown.

Phytochrome, the Red- and Far-Red-Light Receptor Many plant growth and developmental processes are influenced by **phytochrome**, a red- and far-red-light receptor. Phytochrome operates much like a light switch, flipping back and forth between two conformations (**Figure 36.10**). When red light is

1. P_r, the inactive conformation of phytochrome, occurs in the cytosol and is a receptor for red light.

2. Red light activates phytochrome, converting it to P_{fr}, a receptor for far-red light.

3. Activated P_{fr} moves into the nucleus, where it interacts with specific proteins, thereby regulating genes and causing responses such as seed germination.

Protein

Light-sensitive molecule

Inactive P_r (in cytosol)

Active P_{fr} (in cytosol)

Active P_{fr} (in nucleus)

Red light

Far-red light

Figure 36.10 How phytochrome acts as a molecular light switch.

Concept check: *What kind of light does the active conformation of phytochrome absorb, and what kind of change does such absorption cause?*

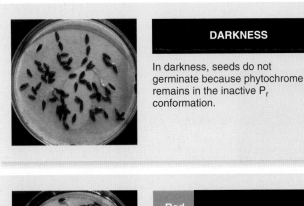

DARKNESS

In darkness, seeds do not germinate because phytochrome remains in the inactive P_r conformation.

Red

Even a brief exposure to red light generates the active P_{fr} conformation of phytochrome, allowing seeds to germinate.

Red | Far red

Exposure to far-red light after red-light exposure converts active P_{fr} to inactive P_r, so seeds do not germinate.

Red | Far red | Red

Exposure to red light after far-red light switches phytochrome back to the active P_{fr} conformation, so seeds germinate.

Red | Far red | Red | Far red

The most recent light exposure determines whether phytochrome occurs in the active P_{fr} or in the inactive P_r conformation. If in the latter, most seeds do not germinate.

Figure 36.11   How phytochrome influences seed germination.

Concept check: *Describe the change in phytochrome that would occur if a deeply buried seed were uncovered enough to receive sunlight.*

abundant, as in full sunlight, phytochrome absorbs red light and changes to a conformation that absorbs only far-red light (light having a wavelength longer than that of red light). This form of phytochrome, known as P_{fr}, activates cellular responses—such as seed germination—which are described later. When left in the dark for a long period, P_{fr} slowly transforms into the inactive red light–absorbing conformation, known as P_r. Far-red light is more abundant than red light in situations such as deep within a canopy of leaves, because chlorophyll absorbs most of the available red light. In this environment, P_{fr} rapidly switches to P_r, which can only absorb red light and does not activate cellular responses.

The role of phytochrome as a plant "light switch" has been shown experimentally in studies of lettuce-seed germination. Researchers have found that water-soaked lettuce seeds will not germinate in darkness, but they will germinate if exposed to as little as 1 minute of red light (**Figure 36.11**). This amount of light exposure is sufficient to transform a critical amount of P_r to the active P_{fr} conformation, which stimulates germination. However, if this brief red-light treatment is followed by a few minutes of treatment with far-red light, the lettuce seeds will not germinate. This short period of far-red illumination is enough to convert seed P_{fr} back to the inactive P_r conformation.

The most recent light exposure determines whether the phytochrome occurs in the active or inactive conformation. In nature, if seeds are close enough to the surface that their phytochrome is switched on by red light, the seeds will germinate. But if seeds are buried too deeply for red light to penetrate, they will not germinate. In this way, seeds can sense if they are close enough to the surface to begin the germination process.

Most of our understanding of phytochrome's effect on gene expression comes from the study of the model plant *Arabidopsis*. This plant has five phytochrome genes (*PHYA* to *PHYE*). Each of the five types of phytochrome is composed of two proteins, each having a light-sensitive chromophore (see Figure 36.10). In the dark, phytochrome molecules in the P_r state reside in the cytosol. After exposure to red light, activated phytochrome (P_{fr}) molecules typically move from the cytosol to the nucleus. Within the nucleus, P_{fr} interacts with a transcription factor protein known as PIF3 (phytochrome interacting factor 3). PIF3 binds to the regulatory elements of several phytochrome-responsive genes, functioning as a positive regulator of some genes and as a negative regulator of others.

Photoperiodism   Phytochromes also play a critical role in photoperiodism, the response to relative lengths of darkness that influence the timing of dormancy and flowering. Flowering plants can be classified as long-day, short-day, or day-neutral plants. When scientists named these groups, they thought that plants measured the amount of daylight. Researchers later discovered that plants actually measure night length.

Lettuce, spinach, radish, beet, clover, gladiolus, and iris are examples of **long-day plants** because they flower in spring or early summer, when the night period is shorter (and thus the day length is longer) than a defined period (**Figure 36.12**). In contrast, asters, strawberries, dahlias, poinsettias, potatoes, soybeans, and goldenrods are examples of **short-day plants**

Figure 36.12 **Flowering and photoperiodism.** Iris is a long-day plant that flowers in response to the short nights of late spring and early summer, whereas goldenrod is a short-day plant that flowers in response to the longer nights of autumn. The length of night is the critical factor, as shown by the effects of light flashes.

Concept check: What would happen if you gave these plants a brief exposure to darkness in the middle of the daytime?

because they flower only when the night length is longer than a defined period. Such night lengths occur in late summer, fall, or winter, when days are short. As shown in Figure 36.12, when plants are given an experimental light flash in the middle of a long dark period, the long-day plants flower while the short-day plants do not. These results indicate that both types of plants measure night length. Roses, snapdragons, cotton, carnations, dandelions, sunflowers, tomatoes, and cucumbers flower regardless of the night length, as long as day length meets the minimal requirements for plant growth, and are thus known as **day-neutral plants**.

Ornamental plant growers manipulate night length to produce flowers for market during seasons when they are not naturally available. For example, chrysanthemums are short-day plants that usually flower in the fall, but growers use light-blocking shades to increase night length in order to produce flowering plants at any season.

Shading Responses Phytochrome also mediates plant responses to shading. These responses include the extension of leaves from shady portions of a dense tree canopy into the light, and growth that allows plants to avoid being shaded by neighboring plants. These growth responses occur by the elongation of branch internodes. Leaves detect shade as an increased proportion of far-red light to red light. This means that more of the

phytochrome in shaded leaves is in the inactive (P_r) state than is the case for leaves in the sun. Activated phytochrome (P_{fr}) inhibits the growth of shoot internodes, but phytochrome in the inactivated state does not, so branches bearing shaded leaves extend toward sunlight.

Plants Respond to Gravity and Touch

Have you ever wondered what causes plant stems to generally grow upward and roots downward? The upward growth of shoots and the downward growth of roots are behaviors known as **gravitropism**, growth in response to the force of gravity. Shoots are said to be negatively gravitropic because they often grow in the direction opposite to gravitational force. If a potted plant is turned over on its side, the shoot will eventually bend and begin to grow vertically again (**Figure 36.13**). Most roots are said to be positively gravitropic because they grow in the same direction as the gravitational force.

Both roots and shoots detect gravity by means of starch-heavy plastids known as **statoliths**, which are located in specialized gravity-sensing cells called statocytes. In shoots, statocytes are located in a tissue known as the endodermis, which forms a sheath around vascular tissues. In roots, gravity-sensing cells primarily occur in the center of the root cap, in a region called the columella.

Figure 36.13 Negative gravitropism in a shoot.   This tomato shoot system has resumed upward growth after being placed on its side. Upward growth started about 4 hours after the plant was turned sideways; this photo was taken 20 hours later. Shoots sense gravity by means of starchy statoliths present in stem tissue near the central vascular tissue.

Gravity causes the relatively heavy statoliths to sink, which causes changes in calcium ion messengers that affect the direction of auxin transport. This process induces changes in the direction of shoot or root growth. For example, in a root that becomes oriented horizontally, the statoliths are pulled by gravity to the lower sides of statocytes (**Figure 36.14**). The change in statolith position causes auxin to move to cells on the lower sides of roots. In roots, auxin inhibits cell elongation (in contrast to its action in shoots). Therefore, root growth slows on the lower side, while cell elongation continues normally on the upper side. This process causes the root to bend, so that it eventually grows downward again.

Recent studies suggest that gravity responses are related to touch responses, known as **thigmotropism** (from the Greek *thigma*, meaning touch). For example, when roots encounter rocks or other barriers to their downward growth in the soil, they display a touch response that temporarily supersedes their response to gravity. Such roots grow horizontally until they get around the barrier, whereupon downward growth in response to gravity resumes. Plant shoots also respond to touch; examples include vines with tendrils that wind around or clasp supporting structures. Wind also induces touch responses. In very windy places, trees tend to be shorter than normal, giving them the advantage of being less likely to blow over than are taller trees. In the laboratory, plant scientists have simulated natural touch responses by rubbing plant stems and found that this treatment can result in shorter plants. Touch causes the release of calcium ion messengers that influence gene expression.

More rapid responses to touch, such as leaf folding by the sensitive plant (see Figure 36.1b), are based on changes in the water content of cells within a structure known as a pulvinus (plural, pulvini), a swelling located at the base of attachment of each pair of leaflets in complex leaves. A pulvinus consists of a thick layer of parenchyma cells that surrounds a core of vascular tissue (**Figure 36.15**). When the leaflet of a sensitive plant is touched, an action potential opens ion channels in parenchyma cells near the lower surfaces of the pulvini. These cells lose potassium and chloride ions, causing water to flow out and the cells to become flattened. This bends the leaflets together, starting the leaflet-folding process. The action potential generated at the touch site also flows through the leaflet, causing many or all of the leaflets to also bend, with the result that the entire leaf folds. Reversal of this process allows the leaf to unfold. This type of movement is known as a nastic response rather than a tropism, because the direction of the stimulus does not control the response.

The action of pulvini also explains some plant movements that are unrelated to touch, including sleep movements and sun tracking. Sleep movements are changes in leaf position that occur in response to day–night cycles. Sun tracking, as mentioned at the beginning of this chapter, is the movement of leaves or flowers in response to the sun's position.

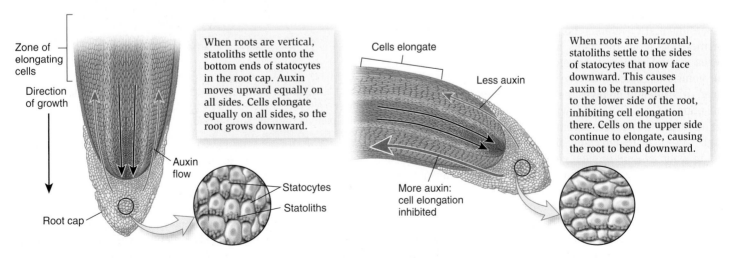

Zone of elongating cells

Direction of growth

When roots are vertical, statoliths settle onto the bottom ends of statocytes in the root cap. Auxin moves upward equally on all sides. Cells elongate equally on all sides, so the root grows downward.

Auxin flow

Root cap

Statocytes

Statoliths

Cells elongate

Less auxin

More auxin: cell elongation inhibited

When roots are horizontal, statoliths settle to the sides of statocytes that now face downward. This causes auxin to be transported to the lower side of the root, inhibiting cell elongation there. Cells on the upper side continue to elongate, causing the root to bend downward.

Figure 36.14 Positive gravitropism in a root.   Root-tip cells sense gravity by means of starchy statoliths present in cells at the center of the root cap.

Concept check: *Is there any other environmental signal that roots could use to achieve downward growth?*

Petiole

Leaf of sensitive plant

Leaflet

Touch

Wind

Action potential

Vascular tissue

Parenchyma cells

Upper (adaxial) leaflet surface

Pulvinus

Lower (abaxial) leaflet surface

K^+ and Cl^- exit from parenchyma cells near the lower leaflet surface, causing H_2O to also exit. These cells flatten, causing the leaflets to fold upward.

Flattened parenchyma cells

Figure 36.15 **Leaf folding in the sensitive plant: how pulvini change the positions of leaflets.** The electrical signals known as action potentials result from the rapid flow of ions through membrane ion channels. Electrical signals spread from one cell to another through plasmodesmata. Cells near the lower surface of the pulvinus respond to ion flow by losing water, causing them to flatten.

Plants Respond to Physical Stresses Such as Flooding and Drought

Plants display many types of adaptations that help them cope with unfavorable growth conditions, such as flooding and drought. These responses are often mediated by hormones.

The major harmful effect of flooding is that too much water makes roots unable to obtain sufficient oxygen to fuel respiratory processes. Without oxygen from the air, roots cannot produce the ATP needed to absorb minerals from soil. Many plants reduce the effects of flooding by producing **aerenchyma**, a tissue containing large, snorkel-like airways that allow more oxygen to flow from shoots to the submerged roots (**Figure 36.16**). In some plants, aerenchyma formation is developmentally programmed, and in others, it is a response to change in environmental conditions. For example, aerenchyma develops in the roots of many plants native to wetland habitats even when the soil is not wet. Aerenchyma can also form in the roots of plants such as corn as a response to flooding. This mechanism of formation involves the action of ethylene, which leads to controlled cell death, followed by cell collapse.

Drought is related to several other environmental stresses—high salinity, heat, and cold—all of which reduce the amount of liquid water present in plant cells. Most plants that lose half or more of their water are unable to recover. Thus, plants possess diverse adaptations that reduce water loss, and the hormone abscisic acid often coordinates these responses.

One way in which plants cope with drought stress is by regulating aquaporins, proteins that form water channels in plasma membranes. Aquaporins allow plant cells to take up or lose water much faster than could be accomplished by diffusion alone. Because water uptake is such an important component of plant growth responses and movements, plants regulate

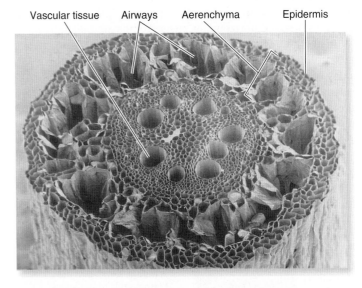

Vascular tissue   Airways   Aerenchyma   Epidermis

Figure 36.16 **A plant response to flooding.** A slice of a root is shown in this micrograph. Air channels in aerenchyma tissue allow air to readily flow to roots from shoots even when the plant is partially submerged.

Concept check: What are two ways in which aerenchyma tissue is formed?

water channel opening and closing. In 2006, Susanna Törnroth-Horsefield, Per Kjellbom, and coworkers described the X-ray crystallographic structure of spinach plasma membrane aquaporin from cells experiencing different levels of water stress. These investigators found that under drought conditions, particular aquaporin amino acids lose phosphate groups, causing a loop of the protein to block the membrane channel. This helps to prevent plant cells from losing water via these pores. When water becomes available again, phosphorylation of aquaporin causes displacement of the loop, thereby opening the pore for water passage. Drought-stressed plants also close their stomata. Stomatal closure and other plant responses to water stress are described in Chapter 38.

Plants Respond to Biological Stresses Such as Herbivore and Pathogen Attack

Plants are vulnerable to attack by animal herbivores and pathogens—disease-causing viruses, bacteria, and fungi. Structural barriers such as cuticles, epidermal trichomes, and outer bark help to reduce infection and herbivore attack (see Chapter 35). Like other organisms, plants use microRNAs and nucleic acid–degrading enzymes as defenses against attacking viruses. Plants also use many types of chemical defenses to deter herbivores and respond to pathogens. Such defenses explain why remarkably little natural vegetation is lost to herbivore or pathogen attack. However, agricultural crops can be more vulnerable to attack than their wild counterparts. This is because some protective adaptations have been lost during crop domestication as the result of genetic changes that increase edibility. For this reason, crop scientists are particularly interested in understanding plant defense behavior, with the goal of being able to breed or genetically engineer crop plants that are better protected from pests.

Plant Responses to Herbivores Plants use a wide variety of chemical defenses against herbivores. Defense compounds include the secondary metabolites: alkaloids, terpenes and terpenoids, phenolics, and phytoalexins. Some of these substances act directly on herbivores, making plants taste bad so that herbivores learn to avoid them. Other chemical compounds function indirectly. For example, when attacked by insect caterpillars, cruciferous plants release terpenoids that attract the bodyguard wasp (*Cotesia rubecula*), which attacks the caterpillars.

Mouth secretions and damage caused by attacking insects mobilize a variety of plant defenses. Self-defenses against herbivores may involve the action of plant hormones (**Figure 36.17**). For example, when insects wound tomato plants, damaged cells release the peptide hormone systemin. This protective hormone induces undamaged cells to produce the hormone jasmonic acid (JA), which functions as an alarm system. JA travels in the phloem or through the air to undamaged parts of the plant, inducing widespread production of defensive compounds (see Figure 36.17, step 2a). Airborne compounds released from herbivore-damaged plants may also attract the enemies of insect attackers.

In addition, nearby plants may also detect these airborne signals and respond by similarly arming themselves against attack (see Figure 36.17, step 2b). JA functions in plant cells by binding to the protein inhibitors of transcription factors. Such JA-binding leads to inhibitor destruction, freeing the

1a Mouth secretions from herbivores and cell damage induce plants to produce the defense hormones systemin and jasmonic acid.

2b Volatile compounds from the damaged plant signal to undamaged neighbor plants, which may lead to the production of defensive compounds or attract enemies of the attackers.

No herbivores allowed!

Volatile compounds

2a Jasmonic acid travels in phloem or through the air to undamaged tissues, inducing defenses throughout the plant.

Figure 36.17   **Plant responses to herbivore attack.**   **(1a)** Leaf damage induces local defensive responses, including production of compounds that travel through the plant and cause defense responses elsewhere. In addition, insect mouth secretions and tissue damage may cause the attacked plant to release volatile compounds. **(2a)** These volatile compounds foster the development of defenses in undamaged parts of the same plant. **(2b)** Such volatile compounds may also induce defenses in neighboring plants, which then become less vulnerable to herbivores. Volatile compounds may also attract predators that feed on the herbivore attacker.

Concept check: *What advantage would a predator obtain by responding to the volatile signals emitted from herbivore-damaged plants? What adaptive advantage could a plant obtain by attracting the enemies of its attackers?*

Figure 36.18   Pathogen/plant interactions and the hypersensitive response to pathogen attack.

transcription factors to start gene expression. In this way, the alarm system is activated only when necessary.

Plant Responses to Pathogen Attack   Bacterial and fungal pathogens produce many types of compounds known as **elicitors**. Such elicitors promote **virulence**, the infection of cells and tissues. A peptide produced by the plant pathogenic bacterium *Pseudomonas syringae* and the fungal cell-wall compound chitin are examples of elicitors. Despite their name, **avirulence genes** (*Avr* genes) encode virulence-enhancing elicitors (**Figure 36.18**, step 1). Some bacteria inject elicitors into plant cells by means of syringe-like systems that are also used to attack animal cells. Fungal pathogens often deliver elicitors by means of structures known as haustoria. These haustoria (from a Latin word, meaning to drink) are also the means by which fungi penetrate the host cell walls and absorb nutrients.

In response to pathogen attack, plants have evolved defense systems based on about 20 types of **resistance genes** (*R* genes) (**Figure 36.18**, step 2). Many alleles of *R* genes occur in plant populations, providing plants with a large capacity to cope with different types of pathogens. Most resistance genes encode proteins that function as receptors for elicitors. Some elicitor receptors occur in plasma membranes, allowing early detection of pathogens. Receptors in the cytosol recognize elicitors that have been injected into cells.

In 1955, based on studies of rust fungi and flax plants (*Linum usitatissimum*), plant pathologist H. H. Flor proposed the gene-for-gene hypothesis to explain how the interaction of *Avr* and *R* gene products influences disease. If a plant is genetically unable to produce a receptor that can recognize a pathogen's elicitor, disease will result. By contrast, plants successfully resist disease when the product of a dominant *R* gene (a receptor) recognizes a pathogen's dominant *Avr* gene product (elicitor). Receptor-elicitor binding stimulates signal transduction pathways that induce plant defensive responses. As described next, the hypersensitive response is a local defensive response, whereas systemic acquired resistance occurs throughout the plant body. These responses are of great importance to agricultural scientists as they endeavor to find new ways to protect crop plants from attack by herbivores and pathogens.

The Hypersensitive Response to Pathogen Attack   The plant **hypersensitive response** (**HR**) occurs when a plant recognizes the elicitors released by a pathogen and responds in such a way that the disease symptoms are limited (Figure 36.18, steps 3 and 4). This plant defensive response has several components. One of these is increased production of hydrogen peroxide (H_2O_2), which can kill infectious agents and helps strengthen the cell wall by cross-binding polymers. The hormone NO is also produced as part of the hypersensitive response. Together with H_2O_2, it stimulates the synthesis of hydrolytic enzymes, defensive secondary metabolites, the hormone salicylic acid (SA), and tough lignin in cell walls of nearby tissues. NO also induces cell death, which deprives pathogens of food and helps

Defensive responses

Defensive responses

Methyl salicylate

Defensive responses

Jasmonic acid

Salicylic acid

Systemin

Sites of pathogen attack

Figure 36.19 Systemic acquired resistance to pathogen attack.

Concept check: *In what way is systemic acquired resistance to pathogens similar to plant responses to herbivore attack?*

to prevent their spread. Necrotic spots are brown patches on plant organs that reveal where disease pathogens have attacked plants and where plant tissues have battled back. SA or the related compound methyl salicylate signals other parts of the plant, which respond by preparing defenses.

Systemic Acquired Resistance A localized hypersensitive response can result in the production of alarm signals that travel to noninfected regions of a plant and induce widespread resistance to diverse pathogens. This response of the whole plant, known as **systemic acquired resistance** (**SAR**), is a type of plant immune system (**Figure 36.19**). The SAR system uses the same type of long-distance signaling process that is associated with herbivore attack (see Figure 36.17). At or near a wound, systemin induces the production of jasmonic acid in nearby vascular tissues. Jasmonic acid is then transported throughout the plant. In addition, the hormone salicylic acid can be converted to volatile methyl salicylate, which diffuses into the air surrounding a plant, inducing resistance in noninfected tissues. In response, plant tissues may produce defensive enzymes, which can break down pathogen cell walls, or they may generate tannins, which are toxic to microorganisms. Experimental infection of *Arabidopsis* leaves with the bacterial pathogen *Pseudomonas syringae* led to induction of SAR within 6 hours.

Summary of Key Concepts

36.1 Overview of Plant Behavioral Responses

- Plants sense and respond to diverse internal stimuli and external signals and thus display behavior. (Figures 36.1, 36.2)

- During the process of signal transduction, cellular receptors respond to environmental stimuli as well as to internal hormonal signals. The process involves receptor activation, transduction of the signal by messengers, and cellular response due to effectors. (Figure 36.3)

- Cellular responses include changes in gene expression and ion channels that influence plant growth, development, reproduction, chemistry, and movements such as sun tracking and leaf folding.

36.2 Plant Hormones

- Plant hormones interact with environmental stimuli to control plant development, growth, and behavior. (Table 36.1)

- Auxin plays an important role in many aspects of plant behavior, including phototropism—as demonstrated by the classic experiments of the Darwins, Went, Briggs, and others. Auxin can be transported downward or upward (polar transport) and sideways (lateral transport) in the plant. The position of auxin influx and efflux carriers determines the direction of auxin transport. (Figures 36.4, 36.5)

- Other major plant hormones are the cytokinins, gibberellins, ethylene, abscisic acid, and brassinosteroids. Auxin and cytokinin are used in plant tissue culture. (Figure 36.6)

- Gibberellin function illustrates the general principle that plant hormones often act to release cellular brakes on gene expression. The components of a system by which gibberellin and cell proteins interact to influence plant growth evolved in a step-by-step fashion. (Figures 36.7, 36.8)

- The gaseous hormone ethylene plays an important role in seed germination. (Figure 36.9)

36.3 Plant Responses to Environmental Stimuli

- Light-sensitive pigments such as cryptochrome, phototropin, and phytochrome allow plants to respond to light stimuli and influence sun tracking, seed germination, and photoperiodic control of flowering. (Figures 36.10, 36.11, 36.12)

- Plant shoots and roots respond to gravity (in the process of gravitropism) by means of starch-heavy statoliths located within statocytes. Sun tracking and rapid touch responses (thigmotropism), such as leaf folding in sensitive plants, depend on changes in the water content of cells in structures known as pulvini. (Figures 36.13, 36.14, 36.15)

- Plants cope with physical stresses such as flooding and drought, often through the production of special tissue (aerenchyma), the involvement of hormones, and the regulation of plasma membrane proteins called aquaporins. (Figure 36.16)

- Plants cope with biological stresses such as herbivore and pathogen attack by means of structural and chemical adaptations. Injured plant parts produce volatile hormones that signal other parts of the same plant and nearby plants to produce defensive responses. Plant pathogens begin their attack by producing elicitor compounds encoded by avirulence (*Avr*) genes. If an attacked plant can produce a compound that interferes with that elicitor, encoded by an *R* (resistance) gene, then the plant can resist infection by that pathogen. The hypersensitive response (HR) is a local defensive response to pathogen attack, whereas systemic acquired resistance (SAR) is a whole-plant defensive response. (Figures 36.17, 36.18, 36.19)

Assess and Discuss

Test Yourself

1. The major types of plant hormones include
   a. cyclic AMP, IP$_3$, and calcium ions.
   b. calcium, CDPKs, and DELLA proteins.
   c. auxin, cytokinin, and gibberellin.
   d. cryptochrome, phototropin, and phytochrome.
   e. statoliths, pulvini, and aerenchyma.

2. Phototropism is
   a. the production of flowers in response to a particular day length.
   b. the production of flowers in response to a particular night length.
   c. the growth response of a plant, organ system, or organ to directional light.
   d. the growth response of a plant, organ system, or organ to gravity.
   e. the growth response of a plant, organ system, or organ to touch.

3. What is the most accurate order of events during signal transduction?
   a. first, receptor activation; then, messenger signaling; and last, an effector response
   b. first, an effector response; then, messenger signaling; and last, receptor activation
   c. first, messenger signaling; then, receptor activation; and last, an effector response
   d. first, an effector response; then, receptor activation; and last, messenger signaling
   e. none of the above

4. Which of the plant hormones is known as the "master hormone," and why?
   a. cytokinin, because many plant functions require cell division
   b. gibberellins, because growth is essential to many plant responses
   c. abscisic acid, because it is necessary for leaf and fruit drop
   d. brassinosteroids, because water uptake is so fundamental to plant growth
   e. auxin, because there are many different auxin-response genes

5. Gaseous hormones are able to enter cells without requiring special membrane transporter systems. Which of the major plant hormones is a diffusible gas?
   a. auxin   c. cytokinin   e. abscisic acid
   b. gibberellin   d. ethylene

6. Photoreceptor molecules allow plant cells to detect light of particular wavelengths. Which of these molecules is considered to be a plant photoreceptor?
   a. cryptochrome   d. a, b, and c are all correct
   b. phototropin   e. no earlier answer listed is
   c. phytochrome   correct

7. Thigmotropism is a plant response to
   a. light.   c. touch.   e. drought.
   b. cold.   d. gravity.

8. Which response is an adaptation to flooding?
   a. geotropism   d. production of aerenchyma
   b. stomatal closure   e. opening aquaporins
   c. photoperiodism

9. What are avirulence genes?
   a. plant genes that encode proteins that prevent infection (virulence)
   b. plant genes that cause infection when the proteins they encode bind to pathogen elicitors
   c. pathogen genes that prevent the pathogens from causing plant disease
   d. pathogen genes that encode elicitors that foster disease in plants
   e. none of the above

10. How do plants defend themselves against pathogens?
    a. Plants produce resistance molecules (usually proteins) that bind pathogen elicitors, thereby preventing disease.
    b. Plants display a hypersensitive response that limits the ability of pathogens to survive and spread.
    c. Plants display systemic acquired resistance, whereby an infection induces immunity to diverse pathogens in other parts of a plant.
    d. All of the above are correct.
    e. None of the above.

Conceptual Questions

1. Why can plants be said to display behavior?

2. Why do plants produce so many types of resistance (*R*) genes?

3. Because diverse plants exude volatile compounds in response to herbivore or pathogen attack, some experts have written about "talking trees." Is there any such thing?

Collaborative Questions

1. Why are most wild plants distasteful, and some even poisonous, to people?

2. How could you increase the resistance of a particular crop plant species to particular types of herbivores?

Online Resource

www.brookerbiology.com

Stay a step ahead in your studies with animations that bring concepts to life and practice tests to assess your understanding. Your instructor may also recommend the interactive ebook, individualized learning tools, and more.

Flowering Plants: Nutrition

37

M any types of fascinating carnivorous (meat-eating) plants grow abundantly in wetlands around the world, even though the soils in these places are infertile. How is this possible? Like most plants, carnivorous plants are photosynthetic and thus produce their own organic food from carbon dioxide and water, using sunlight as an energy source. These resources are abundant in wetlands, but wetland soils are low in other nutrients such as nitrogen that are needed for plant growth. Carnivorous plants, such as the sundews pictured in the chapter-opening photo, have adapted by obtaining nutrients from the bodies of trapped insects and other small animals. Carnivorous plants lure animals with enticing fragrances, brightly colored leaves, or glistening sugar-rich drops of nectar. The unsuspecting prey fall into deep, water-filled pitchers; become ensnared by gluelike mucilage; or are trapped within the walls of leafy jails whose doors suddenly snap shut. Decomposition of the animal bodies releases nutrients that plant leaves quickly absorb. Other wild and cultivated plants face similar nutritional challenges and likewise display adaptations that help them acquire sufficient resources for growth and reproduction.

This chapter focuses on plant nutrition, the processes by which plants obtain essential resources. We will begin by describing the resources needed by plants for completion of their seed-to-seed life cycle in good health. Next, we will explore the role of soil as an essential resource for plants. Last, we will examine the biological sources of plant nutrients, focusing on nutritional associations between plants and microorganisms, sources of nutrients for carnivorous plants, and how some plants obtain nutrients from other plants. An understanding of these topics is crucial for those who seek ways to grow more plant-derived food for humans without causing harmful mineral pollution of Earth's waters. Plant nutritional information is also useful to people who tend gardens or houseplants.

The leaves of the sundews (*Drosera rotundifolia* and *D. intermedia*), shown with a trapped fly, are a plant adaptation for the acquisition of nutrients.

37.1 Plant Nutritional Requirements

Let's begin our consideration of plant nutritional requirements by defining nutrients more specifically. A **nutrient** is a substance that is metabolized by or incorporated into an organism. Deficiency symptoms develop in plants that receive too little of these substances, and the scarcity of nutrients selects for adaptations that help plants to acquire them.

Like other organisms, plants have a specific set of nutritional requirements. **Essential nutrients** are defined as those substances needed by plants in order to complete their reproductive cycle, while avoiding the symptoms of nutrient deficiency. For healthy growth and reproduction, green plants require light and certain minerals, defined here as inorganic substances with specific chemical compositions that are formed by geological processes (**Figure 37.1**). Carbon dioxide (CO_2) is primarily absorbed from air, while water and more than a dozen elements—such as potassium, nitrogen, and calcium—are primarily taken up from soil in the form of dissolved ions.

Elements required by plants play many roles in plant metabolism, often functioning as enzyme cofactors (**Table 37.1**). Elements that are generally required in amounts of at least 1 g/kg of plant dry matter are known as **macronutrients**. In contrast, elements that are needed in amounts at or less than 0.1 g/kg of plant dry mass are known as **micronutrients**, or trace elements. Because insufficient amounts of light, carbon dioxide, water, and other mineral nutrients can limit the extent of green

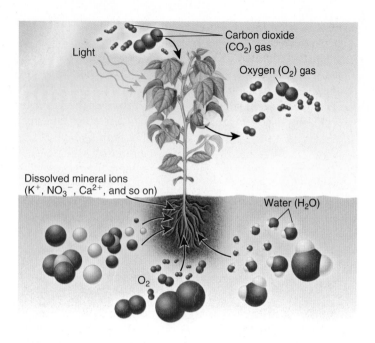

Figure 37.1 The major types of plant nutrients and their sources.

plant growth, these resources are known as **limiting factors**. In this section, we will take a closer look at the factors that can foster or limit plant growth, beginning with light energy.

Light Is an Essential Resource for the Growth of Green Plants

All photosynthetic plants require light for the formation of the covalent bonds of organic compounds that make up the plant body. Green plants' use of light energy as an essential resource parallels animals' nutritional requirements for organic food as a source of chemical energy. For the several hundred species of plants that have lost their photosynthetic capacity, such as the Indian pipe (*Monotropa uniflora*), essential nutrients include organic compounds that replace light as a source of energy (Figure 37.2).

Light-deprived plants display deficiency symptoms, as do plants that receive too little carbon dioxide, water, or other minerals. For example, if houseplants receive too little light, they must use stored carbohydrates for cellular respiration. As a result, the leaves may die and fall off, a symptom of light deficiency. When reserves are exhausted, the plant will die. Even if a plant is able to obtain enough light to survive, it may not absorb enough light to reproduce.

In nature, plants must adapt to environments with varying amounts of light and shade. For example, in forests, light availability limits the growth of tree seedlings and other small plants that are shaded by the leafy tree canopy overhead. But if a canopy tree dies from disease or is blown over by wind, creating a hole in the canopy called a light gap, the plants growing beneath it receive more sunlight and may be able to grow taller. Conversely, plants growing in deserts or on mountains often

Table 37.1	Plant Essential Nutrients			
Element (chemical symbol)	**Percent of plant dry mass**	**Major source**	**Form taken up by plants**	**Function(s)**
Macronutrients				
Carbon (C)	45	Air	CO_2	Component of all organic molecules
Oxygen (O)	45	Air, soil, water	CO_2, O_2, H_2O	Component of all organic molecules
Hydrogen (H)	6	Water	H_2O	Component of all organic molecules; protons used in chemiosmosis and cotransport
Nitrogen (N)	1.5	Soil	NO_3^-, NH_4^+	Component of proteins, nucleic acids, chlorophyll, coenzymes, and alkaloids
Potassium (K)	1.0	Soil	K^+	Has essential role in cell ionic balance
Calcium (Ca)	0.5	Soil	Ca^{2+}	Component of cell walls; messenger in signal transduction
Magnesium (Mg)	0.2	Soil	Mg^{2+}	Component of chlorophyll; activates some enzymes
Phosphorus (P)	0.2	Soil	HPO_4^{2-}	Component of nucleic acids, ATP, phospholipids, and some coenzymes
Sulfur (S)	0.1	Soil	SO_4^{2-}	Component of proteins, some coenzymes, and defense compounds
Micronutrients				
Chlorine (Cl)	0.01	Soil	Cl^-	Required for water splitting in photosystem cell ion balance
Iron (Fe)	0.01	Soil	Fe^{3+}, Fe^{2+}	Enzyme cofactor; component of cytochromes
Manganese (Mn)	0.005	Soil	Mn^{2+}	Enzyme cofactor
Boron (B)	0.002	Soil	$B(OH)_3$	Enzyme cofactor; component of cell walls
Zinc (Zn)	0.002	Soil	Zn^{2+}	Enzyme cofactor
Sodium (Na)	0.001	Soil	Na^+	Required to generate PEP in C4 and CAM plants
Copper (Cu)	0.0006	Soil	Cu^+, Cu^{2+}	Enzyme cofactor
Molybdenum (Mo)	0.00001	Soil	MoO_4^{2-}	Enzyme cofactor
Nickel (Ni)	0.000005	Soil	Ni^{2+}	Enzyme cofactor

experience light that is so intense that it can damage the photosynthetic components. We now consider some of the adaptations that help plants cope with environments that have too little or too much light.

Figure 37.2   The heterotrophic flowering plant, Indian pipe (*Monotropa uniflora*). This plant lacks photosynthetic capacity and therefore must absorb organic compounds for use as an energy source. In contrast, nearby autotrophic green plants use sunlight as a source of energy.

(a) Shade leaf

(b) Sun leaf

Figure 37.3   Shade and sun leaves. These scanning electron micrographs of cut leaves reveal the thinner mesophyll and greater amount of air spaces in (a) shade leaves as compared to (b) sun leaves.

Adaptations to Shade   One way in which plants adapt to shading is by producing thin leaves that allow some light to pass through to other leaves. In a shady environment, a plant or leaf may also produce more total chlorophyll, thereby maximizing the amount of light absorbed. In addition, many plants produce distinctive sun and shade leaves (**Figure 37.3**). Sun leaves have a thicker layer of chlorophyll-containing mesophyll and are able to harvest more of the bright sunlight that penetrates

deeply into the leaf. In contrast, shade leaves have a thinner mesophyll layer, with more air spaces than do sun leaves.

Most epidermal cells at leaf surfaces are transparent because their plastids lack chlorophyll, allowing light to penetrate to green mesophyll. In some plants, these epidermal cells are shaped like lenses that focus more light into the green tissue. The arrangement of leaves on plants and stem branching patterns can also reflect adaptations that reduce shading. For example, many tropical forest trees that must compete for light with closely crowded neighbors are extremely tall, and they produce branches and leaves only at their very tops. Shorter plants native to the shady interiors of tropical rain forests are so well adapted to these moist but dim conditions that they make excellent houseplants (**Figure 37.4**). One example of an adaptation by shade-tolerant plants is that their epidermal cells may contain green, photosynthetically active plastids that aid in collecting light.

Adaptations to Excessive Light   Too much light can damage the photosynthetic machinery of plants by destroying an essential photosynthetic protein, called the D1 protein. As an adaptation to excess light, chloroplasts can change position so that they

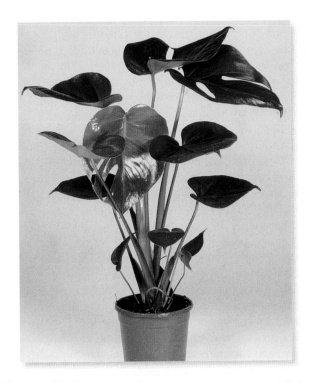

Figure 37.4   The tropical houseplant *Monstera deliciosa*. Houseplants can survive indoors because they have evolved in dimly lit natural habitats. *M. deliciosa* makes an attractive houseplant because its large, deep green leaves are adapted to the moist and shady conditions present in the interiors of tropical forests.

Concept check: *How do the epidermal cells of plants adapted to deep shade sometimes differ from epidermal cells of plants adapted to sunnier habitats?*

are less likely to be damaged. In addition, specific xanthophyll pigments in the chloroplast absorb some of the light energy and dissipate it as harmless heat. Xanthophylls are carotenoids that contain oxygen. These protective pigments can be produced from a precursor xanthophyll within minutes of a plant's exposure to excess light. Other adaptations prevent UV damage. Harmful amounts of UV radiation are absorbed by the plant's surface cuticle, as well as by carotenoid and flavonoid compounds located within leaf cells. These protective compounds are often brightly colored and, when present in large amounts, explain the attractive red or purple colors of some leaves.

Carbon Dioxide Concentration Influences Plant Growth

Although light provides the energy for plant photosynthesis, most plant dry mass originates from carbon dioxide (CO_2). Most plants obtain CO_2 gas from the atmosphere, by absorption through stomatal pores. As CO_2 dissolves in water on the surfaces of leaf mesophyll cells, it combines with water to form carbonic acid, which can dissociate into bicarbonate ion and H^+.

$$H_2O + CO_2 \longleftrightarrow H_2CO_3 \longleftrightarrow HCO_3^- + H^+$$

Carbonic acid   Bicarbonate ion

Leaf mesophyll cells then absorb the carbonic acid and bicarbonate ions. Under experimental conditions, plant photosynthetic rates increase with CO_2 concentration until the Calvin cycle enzyme rubisco has become fully supplied, that is, saturated with CO_2. (To review the role of rubisco and the Calvin cycle in photosynthesis, see Chapter 8.)

CO_2 Limitation   In nature, plants are often unable to obtain enough CO_2 for maximal photosynthesis. As a result, CO_2 limits agricultural crop productivity. This is partly because the modern atmospheric concentration of CO_2 is only 350 μL/L (0.035% of atmospheric content), whereas 1,000 μL/L of CO_2 would be required to saturate photosynthesis in most plants. In addition, oxygen can compete with carbon dioxide for the active site in rubisco, causing a reduction in photosynthetic productivity known as photorespiration (see Chapter 8).

Evidence that plant photosynthesis can be limited by CO_2 availability is provided by studies of crops grown in greenhouses where atmospheric gas content can be controlled. When supplied with air enriched with CO_2, tomatoes, cucumbers, leafy vegetables, and some other crops can double their growth rate.

In nature, plants that experience hot, dry conditions are particularly vulnerable to low CO_2 levels when their stomata close to conserve water. When stomata are closed, plants cannot absorb CO_2, which limits photosynthesis. Many plants possess structural and biochemical adaptations that help them cope with CO_2 limitation by improving CO_2 absorption. Such adaptations may be helpful in genetic engineering efforts to improve crop productivity.

Engineering CO_2-Absorption Adaptations into Crop Plants
Many plants that evolved in arid or hot environments display an adaptation known as C_4 photosynthesis, which improves productivity by aiding CO_2 absorption (see Chapter 8). About 30,000 plant species utilize C_4 photosynthesis, which is thought to have evolved on as many as 70 separate occasions. In addition, 7,500 other plant species possess a variation of C_4 photosynthesis known as CAM (crassulacean acid metabolism; see Chapter 8). However, many other plants that grow in hot environments lack C_4 photosynthesis. Rice, a staple crop for much of the world's population, is a prominent example. Agricultural scientists envision using genetic engineering techniques to endow rice with C_4 photosynthesis with the goal of increasing this crop's productivity.

To understand how this valuable crop modification might be accomplished, a review of C_4 photosynthesis may be useful. Both C_4 and CAM photosynthesis rely on the leaf mesophyll enzyme phosphoenolpyruvate (PEP) carboxylase. PEP carboxylase binds CO_2 more readily than does rubisco, thereby effectively trapping scarce CO_2 within four-carbon compounds. You may recall that this process is the basis for the term C_4 plants. In the leaves of most C_4 plants, such as maize (corn), the four-carbon compounds move through intracellular connections from mesophyll cells into specialized bundle-sheath cells that surround leaf veins. Within the bundle-sheath cells, enzymes release CO_2 from the C_4 compounds, and the CO_2 is incorporated into organic carbon during the Calvin cycle (see Chapter 8).

Molecular biologists have succeeded in transferring maize genes encoding PEP carboxylase and other C_4 enzymes into the rice genome. However, this procedure primarily improved plant stress tolerance rather than photosynthetic performance, which was the primary goal. Because bundle-sheath cells are often central to C_4 photosynthesis, an alternative strategy is to transform rice with the genes that control development of these specialized cells.

Some other plants accomplish C_4 photosynthesis without specialized bundle-sheath cells, offering additional strategies for crop genetic engineering. For example, PEP carboxylase and rubisco occupy the same mesophyll cells in the aquatic plant *Hydrilla verticillata*. In this case, PEP carboxylase is localized within the cytoplasm, while rubisco occurs within plastids. *Borszczowia aralocaspica*, a plant native to salty, near-desert regions of central Asia, and *Bienertia cycloptera*, native to Iranian deserts, are likewise remarkable for single-cell C_4 photosynthesis. Plant biologists Elena Voznesenskaya, Gerald Edwards, and colleagues discovered that *B. aralocaspica* and *B. cycloptera* achieve C_4 photosynthesis within a single photosynthetic cell by producing two types of plastids, one type containing PEP carboxylase and the other containing rubisco (**Figure 37.5**). These and other recent discoveries reveal that wild plants have acquired diverse adaptations that improve CO_2 absorption and show that both basic and applied research are essential to efforts to improve food crops.

Water Is an Essential Plant Resource

Water is essential to plants for several reasons. As a nutrient, water is the source of most of the hydrogen atoms and some of the oxygen atoms in organic compounds (see Table 37.1). You

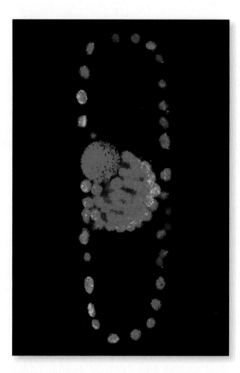

Figure 37.5 A CO$_2$-acquisition adaptation. In contrast to most C$_4$ plants, *Bienertia cycloptera* is able to conduct C$_4$ photosynthesis within the confines of a single cell. In this fluorescence photograph of a *B. cycloptera* cell, the red plastids at the cell periphery function like the mesophyll cells of most C$_4$ plants. In contrast, the red plastids clustered near the green nucleus function much like the bundle-sheath cells of most C$_4$ plants. The plastids appear red because chlorophyll emits red light when it fluoresces.

Concept check: *Which plastids should have more rubisco enzyme in them, plastids at the cell periphery or those clustered near the nucleus?*

may recall that hydrogen is incorporated as CO$_2$ is reduced during the process of photosynthesis, and oxygen is incorporated during hydrolysis reactions. Water is also the solvent for other mineral nutrients and is the main transport medium in plants, allowing movement of minerals and other solutes throughout the plant body via the vascular tissues. Cytoplasmic and vacuolar water also help to support plants by maintaining hydrostatic pressure on the cell wall.

Though plants vary in water content, water typically makes up about 90% of the weight of living plants. Most plants die when their water content falls below half of the amount normal for that particular species. While many types of desiccation-resistant plants display adaptations that allow them to survive for extended periods in nearly dry conditions, all plants require an adequate supply of water for active metabolism and growth.

Soil Water Provides Additional Plant Nutrients

In addition to light, CO$_2$, and water, plants require additional elements. Macronutrients—the elements required in relatively large amounts—include nitrogen, potassium, calcium, magnesium, phosphorus, and sulfur (see Table 37.1). Nitrogen and phosphorus play many important roles in plants, and potassium helps to maintain or alter plant cell water content. Calcium is

used in the construction of cell walls and as a cellular signal, and magnesium is an essential component of chlorophyll. Sulfur occurs in the amino acids cysteine and methionine, as well as other cell constituents. In addition, the distinctive flavors of onions, garlic, broccoli, mustard, and horseradish result from plant defense compounds that contain sulfur. Micronutrients—essential elements that plants require in relatively small (trace) amounts—include chlorine, iron, manganese, boron, zinc, sodium, copper, molybdenum, and nickel (see Table 37.1). Terrestrial plants obtain both macronutrients and micronutrients in the form of ions dissolved in soil water (see Figure 37.1).

Plant biologists have quantitatively analyzed the mineral ion requirements of plants by growing them hydroponically, that is, by bathing plant roots in a water solution to which minerals are added in various combinations and amounts. Hydroponic studies reveal that when plants lack an adequate supply of an essential mineral nutrient, they display characteristic deficiency symptoms. Such symptoms include failure to reproduce, tissue death, and changes in leaf color. Yellowing of leaves, known as **chlorosis**, is a common mineral deficiency symptom, because many nutrients are needed for chlorophyll production (**Figure 37.6**). Deficiency symptoms are often specific to particular plants. For example, zinc deficiency causes chlorosis in corn, but causes smaller than normal leaves in fruit trees. Such deficiency symptoms provide clues that crop, garden, or houseplant soils are deficient in one or more essential minerals. Farmers commonly have their soil analyzed for mineral content, and if the soil is deficient, they may amend it with **fertilizers**, soil additions that enhance plant growth by providing essential elements. Fertilizers are also sold for use in enriching garden soil and growing houseplants.

Figure 37.6 Chlorosis as a symptom of mineral deficiency. This camellia plant is suffering from an iron deficiency, as revealed by the yellow leaves, a symptom known as chlorosis.

Concept check: *Does chlorosis always indicate iron deficiency?*

37.2 The Role of Soil in Plant Nutrition

Soil is an essential resource for most wild and cultivated plants, providing water and other essential nutrients. For this reason, extensive loss of soil by wind and water erosion is of wide concern. Soils vary greatly in fertility, that is, their ability to support plant growth. Thus, plants of many types have had to adapt to the challenges of obtaining nutrients from poor soils. In this section, we explore soil structure and chemistry from the perspective of plant growth and examine how plants take up nutrients from the soil.

The Physical Structure of Soils Affects Their Aeration, Water-Holding Capacity, and Fertility

Natural soils display layers, known as **soil horizons** (Figure 37.7a). The remains of dead plants and other organisms form a layer of litter above the topsoil, the topmost layer of soil. Many of the inorganic minerals and organic materials that enrich high-quality topsoil arise from the activities of microorganisms that decompose the litter. In this way, the minerals contained in living plants are eventually recycled to subsequent generations. Beneath the topsoil lie layers called subsoil and soil base, which are largely composed of mineral materials. Bedrock is the bottom layer that supports the soil horizons. Plant roots play an important role in conveying deep-lying minerals to the surface, thereby helping to enrich topsoil.

Soil horizons vary in composition and thickness, depending on various factors—including climate, vegetation, and bedrock type—and the history of human impact. For example, natural grasslands produce deep, rich topsoil that is used for cropland in many regions of the world (Figure 37.7b). In dramatic contrast, tropical rain forests often have only thin layers of topsoil; their soils are composed mostly of inorganic materials that are not useful to plants (Figure 37.7c). However, certain Amazonian soils, known as "terra preta," are rich and black with organic materials, because ancient inhabitants fertilized extensive gardens with large amounts of decayed plant material dredged from nearby riverbanks. In other parts of the world, however, human activities have increased topsoil loss by wind or water erosion, a worldwide agricultural concern.

These examples illustrate that soils are composed of both organic materials and inorganic minerals. The proportions of organic to inorganic materials and the sizes of inorganic particles are used to classify soils into different types. Soils also display variation in amount of aeration, water-holding capacity, pH, and mineral content. All of these soil properties affect plant growth. For this reason, we will take a closer look at the organic and inorganic constituents of soils and the role of fertilizers.

Organic Soil Constituents   The organic constituents of soils are collectively known as **humus**. Humus is largely derived from plant detritus, the dead and decaying remains of plants, although animal wastes and decayed animal bodies also contribute to the organic content of soils. Organic soil constituents derived from plants include chemically stable lignin and cutin, as well as organic acids containing phenolic groups that are known as humic acids. These organic compounds give many soils their characteristic brown to black coloration. Soils that contain less than 1% organic materials are said to be humus-poor, whereas soils containing more than 8% are considered to be humus-rich. Though some plants are adapted to humus-poor soils, optimal growth generally requires soil that has an organic content close to 8%, because humus helps soils hold more water and ions. Humus particles accomplish this by electrostatically binding positively charged ions, cations such as K^+, to negative charges at their surfaces. In this way, humus helps to prevent ions from washing out of soils, thereby maintaining soil fertility. Humus also gives most soils their soft consistency, a property that fosters root growth.

Horizon A
topsoil

Horizon B
subsoil

Horizon C
soil base

Bedrock

(a) Soil structure   **(b) Thick layer of topsoil in cropland**   **(c) Thin layer of topsoil in rain forest**

Figure 37.7   Soil horizons, the structural layers of soil.   (a) Diagram showing general soil structure. (b) A vertical view of an agricultural soil, showing a relatively deep layer of dark topsoil. (c) A vertical view of a tropical rain forest soil, showing a thin layer of dark topsoil.

Figure 37.8 Composting. Gardeners produce compost by layering small amounts of soil with vegetable waste from the kitchen and yard waste and periodically turning the pile to introduce the oxygen needed for decomposition. Compost can be used to increase the humus and mineral content of garden soils.

The benefits of humus explain why farmers often plow manure and plant residues left after crop harvesting into agricultural soils. Garden soils that are low in organic materials can be amended with peat moss or with composted (partially decomposed) manure, which is available at garden stores. Gardeners often produce their own humus-rich compost by layering vegetable waste from the kitchen and yard waste with soil, and turning the pile occasionally to introduce the oxygen necessary for decomposition (**Figure 37.8**).

Inorganic Soil Constituents Inorganic materials in soil are derived from the physical and chemical breakdown of rock, a process known as **weathering**. Rock, which is an aggregate of two or more minerals, is physically weathered by changes such as cycles of freezing and thawing. Lichens and plant roots may produce organic acids that contribute to chemical weathering of rocks. During chemical weathering, soluble salts are washed out, and minerals are hydrolyzed or oxidized. **Leaching** is the dissolution and removal of inorganic ions as water percolates through materials. Heavy rainfall can reduce the fertility of soils by leaching large amounts of nutrients from them. The leached minerals often end up in natural bodies of water, where they can foster the growth of cyanobacteria, algae, and aquatic plants.

Inorganic soil materials occur as particles that can be categorized according to their size as gravel, coarse sand, fine sand, silt, or clay (**Figure 37.9**). Coarse sand grains range from 200 to 2,000 μm (micrometers) in diameter, whereas fine sand particles can be as small as 20 μm. Particles of silt range from 2 to 20 μm in diameter, and clay particles are less than 2 μm in size. Soils can be classified according to their relative content of coarse and fine materials. For example, soils that contain 45% or more sand and 35% or less clay are classified as sandy soils. The other main types of soil are silty, clay, and loam soils. Loam soil contains a mixture of sand, silt, and clay.

Because of their size differences, sand, silt, and clay particles confer different properties on soils. The relatively large size and irregular shapes of sand particles allow air and water to

Figure 37.9 The relative sizes of inorganic soil components.

move rapidly through sandy soils, which are said to be porous (**Figure 37.10**). Sandy soils are well aerated, which is favorable to the growth of plant roots, because they require oxygen for cellular respiration. However, sandy soils hold less water than the same volume of clay, and rapid percolation of water through sandy soils both reduces the amount of water available to the roots and leaches minerals from the soil.

Silt and clay particles fit closely together, so soils containing larger amounts of these materials are less porous than sandy

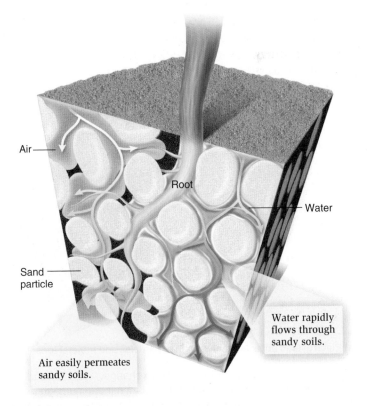

Figure 37.10 The movement of air and water in sandy soil.

Concept check: In which type of soil is nutrient leaching a more common problem, sandy or clay soils?

soils. Water percolates less easily through silty and clay soils, which therefore retain more ionic mineral nutrients than do sandy soils. Like humus, clay particles have negative charges on their surfaces that electrostatically bind cations (such as NH_4^+, Ca^{2+}, and Fe^{2+}) (**Figure 37.11a**). Cations having higher valence numbers (such as Fe^{2+}) are bound more tightly than ions having lower valence numbers (NH_4^+). Hydrogen ions (H^+, protons) are able to replace mineral cations on the surfaces of humus or clay particles, a process known as **cation exchange** (**Figure 37.11b**). Cation exchange releases cations to soil water, making them more available for uptake by plant roots. However, free ions are also more easily washed out of soil. If the H^+ concentration becomes too high, large numbers of mineral ions are released and can be leached from soil by heavy rainfall. Such leached minerals may include heavy metals such as aluminum that would otherwise be bound in soil. Cation exchange is the mechanism by which acid rain, which adds H^+ to soil, causes loss of soil fertility and the pollution of streams with toxic substances, such as aluminum, that can harm human health.

Despite their water- and mineral-retention features, silt- and clay-rich soils may be poorly aerated and therefore unfavorable to root growth. Gardeners often mix organic materials and sand into silt or clay-rich soils to improve their aeration properties. A loam soil is the preferred soil for agriculture because it combines the aeration provided by sand with the mineral and water retention capacity of silt and clay.

The Role of Fertilizers   The addition of fertilizer to soils can compensate for deficiencies in soil humus or mineral content and thus improve soil fertility. Fertilizers occur in organic and inorganic forms. Organic fertilizers are those in which most of the minerals are bound to organic molecules and are thus released relatively slowly. They play an important role in **organic farming**, the production of crops without the use of inorganic fertilizers, growth substances, and pesticides. Manure and compost are examples of organic fertilizers.

In contrast, inorganic fertilizers consist of inorganic minerals, which are immediately useful to plants but are more easily leached from soils by heavy rainfall. Nitrogen (N), phosphorus (P), and potassium (K) are the mineral nutrients that most frequently limit crop growth. For this reason, these minerals are the main components of the most common type of commercial inorganic fertilizers, known as NPK fertilizers. Such fertilizers are available in different ratios of minerals, which are optimal for different types of plants.

Excessive application of fertilizers to farm fields and lawns is undesirable because minerals not taken up by plant roots are easily washed by rain into waterways. In aquatic habitats, excess mineral nutrients fuel large growths of cyanobacteria, algae, and aquatic plants that can harm other aquatic life-forms. For example, large areas of the Gulf of Mexico and other coastal regions are known as "dead zones" because microbial decomposition of large algal populations has depleted oxygen from the water, suffocating the animal life. These large populations of algae are fostered by fertilizers that wash from farm soils into rivers, such as the Mississippi River, that drain into coastal oceans. More careful application of fertilizers and planting stream and river edges with vegetation that helps to absorb mineral nutrients are actions that can reduce or prevent dead zones.

Plants Require Fixed Nitrogen

Nitrogen is frequently limiting to plant growth in nature and in crop fields, because large amounts of it are required for plant

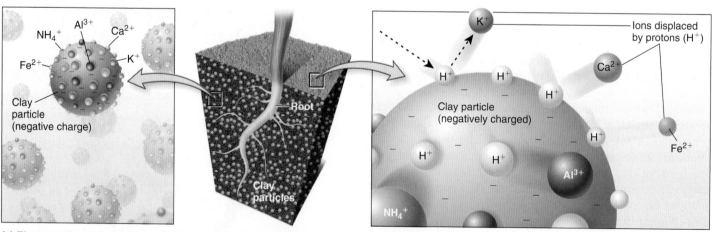

(a) Electrostatic attraction between clay particles and mineral ions

(b) Cation exchange

Figure 37.11   Cation binding and exchange.   (a) Clay and humus particles display negative electrostatic surface charges that bind cations. Bound cations include not only plant mineral nutrients such as NH_4^+ but also cations that are not plant nutrients, such as Al^{3+}. **(b)** Cation exchange occurs when protons (H^+) displace other cations, releasing them to soil water. This process makes cations more available for uptake by plant roots, but it also increases the potential for cations to leach away during rains or floods. Protons in acid soils or arising from acid rain displace so many cations from soil that its fertility decreases, and toxic ions such as Al^{3+} more readily damage plants.

Concept check: *In what parts of the world do acidic soils tend to occur?*

synthesis of amino acids, nucleotides, and alkaloids, among many other cellular constituents. Nitrogen is the largest component of plants by mass after carbon, oxygen, and hydrogen. Although the Earth's atmosphere is 78% nitrogen gas (N_2), plants cannot utilize nitrogen in this form.

To be of use to plants, soil nitrogen must occur in a combined form, such as ammonia (NH_3), ammonium ion (NH_4^+), or nitrate ion (NO_3^-). Combined forms of nitrogen are also known as **fixed nitrogen**. Ammonia and its dissolved form—ammonium ion—can be used directly for amino acid production by plants, explaining why ammonia is often applied as a fertilizer to farm fields in springtime. However, in oxygen-rich soils, microorganisms oxidize much of the ammonium ion to nitrate, so nitrate may be the form in which fixed nitrogen enters most plants.

Nitrate is imported into root cells by means of plasma membrane transporter proteins (**Figure 37.12**). Once inside the plants, the nitrate can be transported in the xylem for use elsewhere, stored for later use, or assimilated for immediate use. Cells store nitrate by transporting it into vacuoles. To assimilate nitrate, plant cells must reduce it to ammonium ion in a two-step process. First, the cytosolic enzyme nitrate reductase reduces nitrate (NO_3^-) to nitrite (NO_2^-). Second, in plastids, nitrite is reduced to ammonium ion (NH_4^+) by nitrite reductase. Another plastid enzyme, glutamine synthetase, uses ammonium ion to produce the amino acid glutamine.

Atmospheric Nitrogen Is Fixed by Natural and Industrial Processes

Much of the fixed nitrogen in soils has been recycled from compounds previously utilized by other organisms. New fixed nitrogen can be added to soils by the action of lightning, fire, and air pollution, as well as biological and industrial nitrogen fixation. **Nitrogen fixation** is the process by which atmospheric nitrogen gas is combined with hydrogen to produce ammonia. Most of the fixed nitrogen in soils is produced by **biological nitrogen fixation**, which is performed in nature only by certain prokaryotes. Nitrogen fertilizers applied to crops are produced by **industrial nitrogen fixation**, a human activity.

Biological Nitrogen Fixation by Bacteria   Nitrogen-fixing prokaryotes include many types of cyanobacteria, which are

1 Uptake:
Protons (H^+) pumped from plant cells provide the driving force for the cotransport of protons (H^+) and nitrate (NO_3^-) into the cytosol.

2 Storage:
Nitrate can be transported into a vacuole for storage.

OR

3a Assimilation:
Cytoplasmic nitrate reductase converts nitrate (NO_3^-) to nitrite (NO_2^-), which enters plastids.

3b In plastids, nitrite reductase converts nitrite to ammonium ion (NH_4^+). Glutamine synthetase uses NH_4^+ to synthesize the amino acid glutamine.

Figure 37.12   How plant cells take up, store, and assimilate nitrate.

Concept check: *List at least five organic compounds essential to photosynthesis whose biosynthesis requires a supply of fixed nitrogen such as nitrate.*

Figure 37.13 A soil surface crust that includes nitrogen-fixing, soil-enriching cyanobacteria. Such crusts are widespread in grasslands and other arid regions.

Concept check: *How might soil crusts influence the ecology and economy of regions in which grazing is important?*

photosynthetic organisms that occur in oceans, lakes, and other aquatic systems, as well as in surface soil crusts (**Figure 37.13**). Various types of nonphotosynthetic bacteria living in water and soil are also able to fix nitrogen; actinobacteria and the genera *Clostridium*, *Klebsiella*, and *Azotobacter* are examples. Nitrogen-fixing prokaryotes often excrete a substantial amount of fixed nitrogen, and their death makes still more fixed nitrogen available to plants. Many plants have nitrogen-fixing, prokaryotic symbionts that transfer fixed nitrogen directly to plant cells. Nitrogen-fixation symbioses are so

important in nature and in agriculture that they are discussed in more detail in Section 37.3.

All nitrogen-fixing prokaryotes utilize relatively large amounts of ATP and an enzyme known as **nitrogenase** to fix nitrogen (**Figure 37.14**). This process occurs in three steps. In the first step, a molecule of nitrogen gas (N_2) binds to nitrogenase. In the second step, the bound nitrogen is reduced by the addition of two hydrogen atoms (2 H), a reaction powered by the breakdown of ATP. Such a reduction occurs three times, with the addition of a total of three hydrogen atoms to each nitrogen atom. In a third and final step, two molecules of ammonia (NH_3) are released and dissolve in cell water to form ammonium ions. The nitrogenase enzyme is then free to bind more nitrogen gas.

Because the O_2 molecule resembles N_2, oxygen can bind to the active site of nitrogenase. Oxygen-binding disables nitrogenase, thereby stopping nitrogen fixation. Many of the genes involved in prokaryotic nitrogen fixation are known, and crop scientists are working to genetically engineer nitrogen fixation capacity into crop plants such as rice and maize (corn). However, the vulnerability of nitrogenase to oxygen means that this enzyme may need to be altered so that it binds oxygen less readily, or plants must be engineered with some mechanism that protects nitrogenase from oxygen.

Industrial Nitrogen Fixation Worldwide, farmers apply more than 80 million metric tons of nitrogen fertilizer per year. The fixed nitrogen found in fertilizer is produced industrially from nitrogen gas by means of a procedure invented by German chemists Fritz Haber and Carl Bosch in 1909. The reduction of N_2 gas to NH_3 is energetically favorable at room temperature, but the activation energy is very high, so the reaction occurs extremely slowly. Using an iron catalyst, temperatures

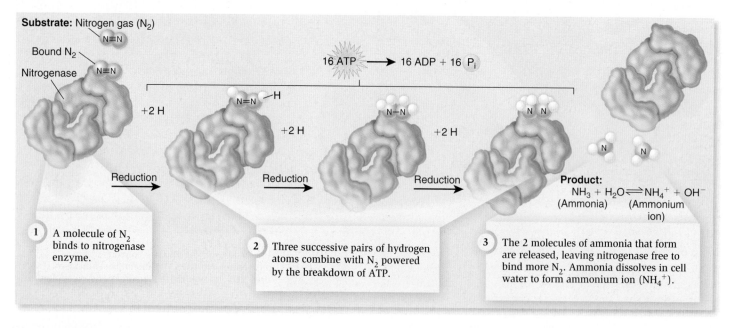

Figure 37.14 The biological process of nitrogen fixation.

Concept check: *What common substance inactivates nitrogenase enzyme by binding to its active site?*

of 400–650°C (752–1,202°F), and high pressures (150–400 atmospheres), the Haber–Bosch process generates NH_3 rapidly. However, because of its high energy requirements, industrial nitrogen fixation can be costly from the perspective of many of the world's farmers. The high cost of fertilizers helps to explain why agricultural scientists are so interested in the possibility of genetically engineering nitrogen fixation into crop plants.

Plants Display Adaptations for Acquiring Phosphorus

Phosphorus (P) is another soil mineral that often limits plant growth. Plants obtain phosphorus from the ion known as phosphate (PO_4^{3-}), which occurs in the soil in three dissolved forms: H_3PO_4, $H_2PO_4^-$, and HPO_4^{2-}. HPO_4^{2-} (called hydrogen phosphate ion) is the form most commonly absorbed by plants. Phosphate uptake involves ATP-requiring proton cotransport, as is the case for other anions such as nitrate and sulfate. The uptake of minerals at the root hair surface is discussed further in Chapter 38.

Although phosphorus is abundant in soil, it is often unavailable to plants. One reason is that phosphate forms tightly bound complexes with clay, iron and aluminum oxides, and calcium carbonate in soils. In addition, soil microbes convert phosphate ions into organic compounds that are not taken up by plants.

Because plants need large supplies of phosphorus for a variety of cell processes, they have evolved various adaptations that increase their ability to obtain phosphate from soil. A common adaptation for acquiring phosphate is the symbiotic association of plant roots with various types of fungi (see Section 37.3). In addition, plants that grow in soils having low phosphorous content may produce more highly branched roots and more and longer root hairs. Plant roots also secrete protons and organic acids such as citrate and malate into the soil, which help release phosphorus from inorganic complexes. For example, the plant *Lupinus alba* releases as much as 25% of its total photosynthetic carbon into the soil as organic acids—a high price to pay but apparently one that is essential for the plant to obtain sufficient phosphorus. Plants may also secrete phosphatase enzymes from roots. These enzymes release phosphorus from organic compounds in the soil. The general process by which phosphorus, nitrogen, CO_2, and other minerals are released from organic compounds is called **mineralization**.

Farmers and gardeners apply phosphate-rich fertilizers to crop fields and gardens as a way of preventing phosphorous deficiencies, which reduce yields. Phosphorous fertilizers are obtained from phosphate-rich mineral deposits, but experts have warned that inexpensive sources of phosphate will be exhausted within the next 90 years. Consequently, there is much interest in devising ways to maximize the efficiency by which plants are able to take up and use phosphorus. Genetic engineering to produce "smart plants" that can sense the levels of nutrients in the soil may offer some options.

FEATURE INVESTIGATION

Hammond and Colleagues Engineered Smart Plants That Can Communicate Their Phosphate Needs

If farmers could apply fertilizer to crops in the precise amounts needed by plants, not only would they save money, but also less fertilizer would run off fields into aquatic habitats, where it can lead to harmful ecological effects. Plant biologists have used genetic engineering to produce smart plants that signal impending nutrient deficiency via a visible marker. Such plants could serve as sentinels, warning farmers of the conditions of an entire field. With this information, farmers could apply just enough mineral nutrients to prevent deficiency, thereby avoiding overapplication of fertilizers.

In 2003, working with the model plant *Arabidopsis*, John Hammond, Philip White, and their associates grew plants hydroponically, which means their roots were in a water solution rather than soil. They then used microarray technology (see Chapter 20) to identify some of the genes whose expression changes when plants are transferred from nutrient solutions containing sufficient phosphorus to solutions lacking phosphorus. They found that some genes were turned on quickly after phosphorus removal, while others took much longer, such as 100 hours or more. This timing is important because genes expressed between 24 and 72 hours after phosphate removal are considered useful as phosphorous monitors. During this window of time, plant tissue levels of phosphate decreased but had not yet affected plant growth. One gene that met this timing criterion was *SQD1*, which is required for the synthesis of sulfur-containing lipids. Expression of *SQD1* allows plants to respond to low phosphorous levels by replacing plastid phospholipids with sulfur-containing lipids, thereby reducing their phosphorous requirement. This evidence suggested that smart plants could be engineered to communicate impending phosphate deficiency when they express *SQD1*.

To make smart plants, the researchers first placed the reporter gene *GUS* under the control of the *SQD1* promoter and transformed this gene into *Arabidopsis* plants (**Figure 37.15**). The researchers then grew these genetically engineered plants for various time periods in hydroponic solutions of differing phosphate levels. After different time periods, they removed leaves and chemically treated them with a compound that produces a blue color when the *GUS* gene is expressed. Some leaves were removed before transfer to the phosphate-deficient solution and served as controls. Because the *GUS* gene was under the control of the *SQD1* promoter, leaves from plants that were developing phosphate deficiency turned blue! These smart plants were able to communicate impending phosphorous deficiency in time for a farmer to apply fertilizer.

In a later and more extensive study of gene expression in *Arabidopsis*, molecular biologists Julie Misson, Marie-Christine Thibaud, and associates discovered that phosphate induces 612 genes and represses 254 genes. Some of these genes may

Figure 37.15 The experiment of Hammond and colleagues showed that plants can be engineered to communicate changes in the level of nutrients.

GOAL To determine if plants can be engineered to signal impending phosphate deficiency.

KEY MATERIALS *Arabidopsis* plants, mineral nutrient solutions.

	Experimental level	Conceptual level
1 Construct DNA with promoter of *SQD1* linked to *GUS* (blue color–producing gene), using methods described in Chapter 20.	Promoter of *SQD1* gene Coding region of *GUS* gene	Plants are genetically engineered with a reporter system (*GUS*).
2 Transform new DNA into *Arabidopsis* (see Chapter 20).		
3 Remove some leaves 20 hours before transfer to phosphate-deficient media, and 24, 100, and 220 hours after transfer.	20 hours before transfer 24 hours 100 hours 220 hours after transfer	Leaves removed 20 hours before transfer to phosphate-deficient media serve as negative (untreated) controls.
4 After each time point, add a reagent that produces blue color when the *GUS* (and *SQD1*) gene is expressed from the *SQD1* promoter. (See the data.)		Leaves removed at increasingly longer times after transfer to phosphate-deficient media reveal how quickly the reporter system works.

5 **THE DATA**

Control   24 hours   100 hours   220 hours

6 **CONCLUSION** Plants can be genetically engineered to express color signals in time for farmers to apply fertilizer sufficient to prevent nutrient deficiency.

7 **SOURCE** Hammond, John P. et al. June 2003. Changes in gene expression in *Arabidopsis* shoots during phosphate starvation and the potential for developing smart plants. *Plant Physiology* 132:578–596.

encode proteins useful in monitoring plant phosphorous status. If smart plant technology can be developed for crop plants, farmers may be able to monitor and fertilize fields with much greater precision.

Experimental Questions

1. Why did Hammond and colleagues seek to identify genes whose expression changed between 24 and 72 hours after plants experience phosphorous limitation?

2. What advantage do plants obtain when the *SQD1* gene is expressed?

3. How were the investigators able to identify potential sentinel plants that were starting to experience phosphorous deficiency?

Some Common Soil Minerals Can Be Toxic to Plants

Some soil minerals, including certain micronutrients, inhibit plant growth when present at high concentrations and therefore are considered toxic to plants. Minerals that may occur in toxic concentrations in soils include aluminum, lead, cadmium, chromium, mercury, and uranium and the micronutrients boron, copper, and nickel. These minerals enter plant roots through water-soaked cell walls and intercellular spaces. Boron is an example of an element that functions as a micronutrient when present in low levels but that is toxic at higher concentrations. Aluminum is not known to be an essential element, but has important ecological effects on plants even in very low concentrations.

Boron Toxicity   At concentrations less than 0.01 g/kg of plant tissue, boron (B) is an essential micronutrient, required for cross-linking polysaccharides in plant cell walls (see Table 37.1). However, boron is toxic to wheat plants, for example, in concentrations of 0.01 g/kg or higher. Boron is easily taken up by plant tissues because this mineral occurs as uncharged boric acid, $B(OH)_3$, which readily crosses membranes. When applying fertilizer to relieve boron deficiency, farmers must take care not to add too much.

Plant scientists Junpei Takano, Toru Fujiwara, and their associates reported in 2005 that the model flowering plant *Arabidopsis* has cellular adaptations that regulate the concentration of boron. These investigators demonstrated that when boron is present in small amounts, a boron transporter protein (BOR1) is expressed in the plasma membranes of root pericycle tissue. In this case, BOR1 moves boric acid into root xylem for transport to shoots, where boron can be used in producing new cell walls. However, when potentially harmful amounts of boric acid are applied to roots, endocytotic vesicles move the BOR1 protein to the cell vacuole for degradation. This process prevents too much boric acid from moving into shoots. Together, these processes protect plants from boron deficiency and toxicity.

Aluminum Toxicity   Aluminum, in the form of Al^{3+}, is one of the most common minerals in soil, so small amounts may seep into plant tissues. However, aluminum is not known to be an essential element and is toxic to plants even at micromolar concentrations. Aluminum binds to cellular carbohydrates, phosphates such as ATP, and sulfates, thereby inhibiting root elongation and uptake of minerals and water. Aluminum toxicity is primarily a problem for plants rooted in acidic soils, which are common in the tropics and subtropics, including at least 12% of cultivated land. In addition, aluminum toxicity can occur in temperate regions that receive acid rain. In the cation exchange process, the abundant protons in acidic soils or acid rain can replace aluminum ions bound to soil particles (see Figure 37.11) and allow the released aluminum ions to enter plant roots more easily.

Some plants defend themselves against aluminum toxicity by secreting organic acids such as citrate, oxalate, and malate from their root tips. In the soil, these acids form complexes with Al^{3+}, which helps to prevent aluminum from entering

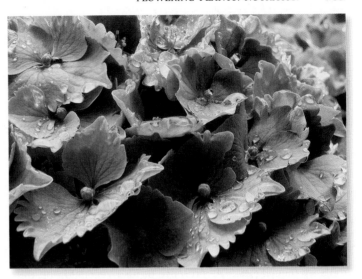

Figure 37.16   **An adaptation protecting against aluminum toxicity.**   The showy sepals of hydrangea (*Hydrangea macrophylla*) flowers turn from pink to blue when soil conditions become more acidic. The sepals shown here are in the process of undergoing the transition from pink to blue.

Concept check: *Explain the processes by which the color change occurs in hydrangea sepals.*

roots. Within plant cells, certain proteins also bind aluminum and other metals, thereby protecting cell components. Some plants accumulate aluminum or other toxic metals in their tissues, often sequestering the metals in cell vacuoles, where they do little harm. For example, when hydrangea plants are grown in acid soil, aluminum accumulations in vacuoles cause the showy sepals of the flowers to change from pink to blue (**Figure 37.16**). Many other kinds of plants protect themselves from toxic metals by storing them in plant tissues.

Hyperaccumulators and Phytoremediation   Some plants are known as hyperaccumulators because they accumulate and safely bind high concentrations of toxic metal minerals within their tissues, an adaptation that allows such plants to grow on metal-rich soils. Some 400 different plant species are hyperaccumulators. Examples include the alpine pennycress (*Thlaspi caerulescens*), which accumulates zinc and cadmium, and the brake fern (*Pteris vittata*), which accumulates arsenic. These plant adaptations can be usefully applied to remove pollutants such as harmful metals from soils, a process known as **phytoremediation**. In this process, hyperaccumulator plants are grown on metal-contaminated soils, harvested, and burned to ashes for disposal and/or metal recovery.

37.3 Biological Sources of Plant Nutrients

This section focuses on several fascinating ways in which plants use other organisms as sources of nutrients. Biological sources of plant nutrients include symbiotic fungi or bacteria, the animal prey of carnivorous plants, and green plants that serve as hosts for nonphotosynthetic plant parasites.

Mycorrhizal Associations Help Most Plants Obtain Mineral Nutrients

About 90% of seed plants have symbiotic associations with fungi that live within the tissues of plant roots or that envelop root surfaces (see Chapter 31). These associations are termed **mycorrhizae**; the prefix *myco* refers to fungi, and *rhiza* means root, so the term literally means "fungus root."

In mycorrhizal associations, soil fungi obtain organic food from the roots of a photosynthetic plant host, while the fungal partner supplies the plant with water and mineral nutrients. Due to the extensive mycelia that fungi produce within the soil, these fungal-root associations provide an exceptionally efficient way for plants to harvest water and minerals, especially phosphate, from a much larger volume of soil than is available to roots by themselves. The presence of lush vegetation on thin, infertile tropical rain forest soils is largely due to the ability of mycorrhizae to rapidly absorb mineral nutrients released by decaying organisms and transmit the nutrients directly to plant roots (**Figure 37.17**). In many tropical rain forests, mineral nutrients occur within the bodies of living organisms, rather than accumulating in the soil where they could easily be leached away by heavy, frequent rains.

Various species of ghostly pale plants have lost their photosynthetic pigments (see Figure 37.2) and have thus become dependent on organic compounds supplied by fungi that form mycorrhizal associations with a photosynthetic host, such as a nearby tree. In this process, known as mycoheterotrophy, the fungus serves as an underground conduit for the flow of organic nutrients from a green, photosynthetic plant to a heterotrophic plant. Many plant seedlings that grow in the shade of taller plants also use mycoheterotrophy to survive until they are able to obtain enough light for photosynthesis.

Plant–Prokaryote Symbioses Provide Some Plants with Fixed Nitrogen

Some kinds of plants have symbiotic relationships with bacteria that provide them with fixed nitrogen. Though many nitrogen-fixing bacteria live freely within the soil, some form nitrogen-fixing partnerships with plants, actually living within plant cells or tissues. Such symbioses are advantageous to both partners. The plants provide organic nutrients to the bacteria, while the bacteria supply the plants with a much higher supply of fixed nitrogen than the plants could obtain from most soils. Representatives of three types of prokaryotes—cyanobacteria, actinobacteria, and proteobacteria—are symbiotically associated with specific types of plants. (For more information about the characteristics of these bacterial groups, see Chapter 27.)

Plant–Cyanobacteria Symbioses   Although cyanobacteria are themselves photosynthetic, organic compounds supplied by plant partners subsidize the high energy costs of nitrogen fixation. This allows the cyanobacteria to fix more nitrogen than they require, secreting the excess to plant partners. Nitrogen-fixing cyanobacteria form symbioses with some bryophytes,

Figure 37.17   Nutrient acquisition via mycorrhizae.   In all forests, but particularly those of tropical regions, mycorrhizal fungi rapidly collect soil minerals released from decaying organisms and transport them directly to plant roots. Such efficient nutrient cycling bypasses the soil, from which mineral ions can be easily leached by heavy rainfall. This process explains how lush forests can grow on thin, infertile soils.

ferns, and gymnosperms, as well as the flowering plant *Gunnera*. This plant, commonly known as the giant rhubarb or prickly rhubarb, can produce leaves almost 3 m across (**Figure 37.18**). Nitrogen-fixing symbionts are advantageous to *Gunnera* because this large plant grows in nitrogen-poor habitats, such as volcanic slopes in Hawaii. *Gunnera* harbors cyanobacteria within stems and leaf petioles. In these locations, the cyanobacteria can use cyclic electron flow to transform light energy into ATP, needed to produce fixed nitrogen. The presence of nitrogen-fixing cyanobacteria helps to explain why *Gunnera* can grow to dramatic size on poor soils.

Woody Plant–Actinobacteria Symbioses   In contrast to cyanobacteria, actinobacteria are heterotrophic, nitrogen-fixing bacteria. Actinobacteria known as *Frankia* occur in nodules formed on the underground roots of certain shrubs or trees, such as alder (*Alnus*) and myrtle (*Myrica*). These plants receive fixed nitrogen from their bacterial partners, which, in turn, obtain organic nutrients. Woody plants, such as *Ceanothus* shrubs, that have *Frankia* symbionts are able to grow abundantly even in places where soil nitrogen is low. This symbiosis helps to explain why *Ceanothus* covers extensive areas in mountainous chaparral regions of the western U.S.

Legume–Rhizobia Symbioses   The nitrogen-fixation symbioses most important in nature and to agriculture involve certain proteobacteria that are collectively known as **rhizobia** (from the Greek *rhiza*, meaning root). Rhizobia live within root cells of wild and cultivated legumes, forming legume–rhizobia symbioses. In nature, legume plants are important sources of fixed nitrogen for other plants. When legumes die, they generate humus that is enriched with fixed nitrogen. Consequently,

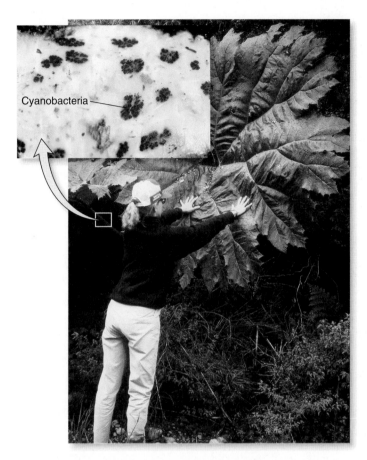

Figure 37.18 *Gunnera* growing on nitrogen-poor soil. Nitrogen-fixing cyanobacteria that live within cavities in this plant's leaf petioles provide the plant with fixed nitrogen, which explains how such a large plant can grow on infertile soils.

wild legumes are regarded as particularly valuable members of natural plant communities. For example, plant conservationists are concerned about the survival of a rare Appalachian legume known as running buffalo clover (*Trifolium stoloniferum*), because it is a valuable source of environmental fixed nitrogen.

Important legume crops include soybeans, peas, beans, peanuts, clover, and alfalfa. Foods produced from soybeans, peas, beans, and peanuts are valued for their high protein content. Clover and alfalfa are used for animal food and to enrich fields with the fixed nitrogen needed by subsequent food crops. The value of these crops arises from their fixed-nitrogen content. The amount of ammonia produced by legume–rhizobia symbioses nearly equals the world's entire industrial production.

Genomes & Proteomes Connection

Development of Legume–Rhizobia Symbioses

Rhizobia can live independently in the soil, but they fix nitrogen only when they occur within lumpy **nodules** that form on legume roots (**Figure 37.19**). Different species of rhizobia

preferentially form symbioses with particular plant species. Because of their agricultural importance, these legume–rhizobia symbioses have been extensively studied, and a great deal is now known about the molecular basis of their development. This information is potentially useful in efforts to genetically engineer nitrogen-fixation capacity into nonlegume crops.

Nodule development involves a series of chemical signals sent back and forth between rhizobia and their host plants (**Figure 37.20**). Legumes start this exchange by secreting particular flavonoid compounds from their roots. Recall that flavonoids are phenolic secondary metabolites that play essential roles in plant structure, reproduction, and protection (see Chapter 30). These flavonoids bind to receptors in the plasma membranes of compatible soil rhizobia (Figure 37.20, step 1). In response, the rhizobia typically secrete **Nod factors** (nodulation factors). Each rhizobial species produces Nod factors with distinctive structural variations that can be recognized by the preferred host species. These Nod factors function something like keys that unlock doors, allowing bacteria to enter roots via root hairs. These factors bind to receptors in the membranes of root hair cells in the host plant (step 2).

Within minutes after its membrane receptors bind Nod factors, the root hair plasma membrane allows an influx of calcium ions, and a few minutes later, root hair calcium concentrations start oscillating rapidly. Root hairs respond to these calcium changes by swelling at their tips and curling around the rhizobia (step 3). The rhizobia then inject infection proteins into root hairs. In response, the cell wall at the root hair tip changes in a way that allows bacterial enzymes to erode a small hole in the wall, allowing bacterial cells to enter. The plasma membrane forms a tubular infection thread through which rhizobia

Figure 37.19 Legume root nodules. The cells of nodules on the roots of this soybean plant (*Glycine max*) and other legumes contain nitrogen-fixing bacteria known as rhizobia.

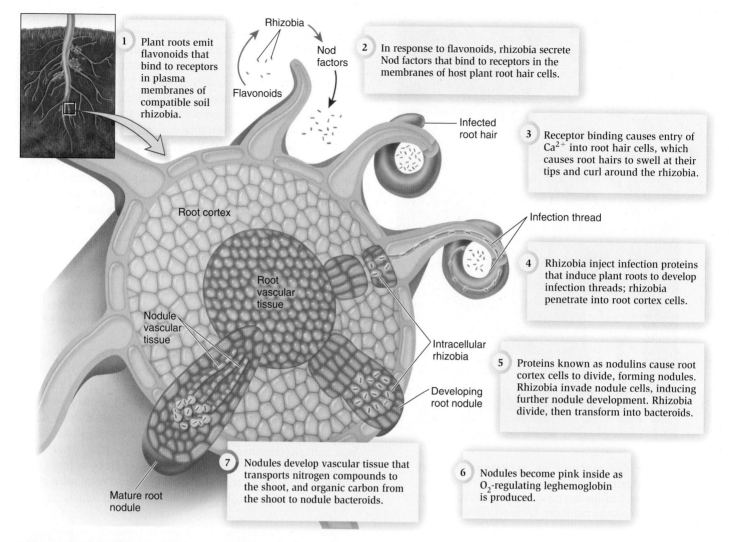

1 Plant roots emit flavonoids that bind to receptors in plasma membranes of compatible soil rhizobia.

Rhizobia

Nod factors

Flavonoids

2 In response to flavonoids, rhizobia secrete Nod factors that bind to receptors in the membranes of host plant root hair cells.

Infected root hair

3 Receptor binding causes entry of Ca^{2+} into root hair cells, which causes root hairs to swell at their tips and curl around the rhizobia.

Root cortex

Infection thread

4 Rhizobia inject infection proteins that induce plant roots to develop infection threads; rhizobia penetrate into root cortex cells.

Root vascular tissue

Nodule vascular tissue

Intracellular rhizobia

Developing root nodule

5 Proteins known as nodulins cause root cortex cells to divide, forming nodules. Rhizobia invade nodule cells, inducing further nodule development. Rhizobia divide, then transform into bacteroids.

7 Nodules develop vascular tissue that transports nitrogen compounds to the shoot, and organic carbon from the shoot to nodule bacteroids.

6 Nodules become pink inside as O_2-regulating leghemoglobin is produced.

Mature root nodule

Figure 37.20   Root nodule development.

move into the root cortex. The tip of the infection thread fuses with the plasma membrane of a cortex cell, then the rhizobia are released into the cortex cell cytoplasm, each bacterial cell enclosed by the host membrane (step 4).

Meanwhile, plants produce proteins known as **nodulins** that foster nodule development. Within 18–30 hours after the initial infection, root cortex cells start to divide to form root nodules. Environmental conditions in developing nodules cause rhizobia to undergo changes in their structure and gene expression patterns. The modified rhizobia present in mature nodules are known as **bacteroids** (step 5). Legume nodules typically produce **leghemoglobin** (legume hemoglobin), a pink protein that helps regulate local oxygen concentrations (step 6). By so doing, leghemoglobin prevents oxygen from poisoning nitrogenase, while still allowing bacteroids to accomplish aerobic respiration. Bacteroid respiration provides the large amounts of ATP that are necessary for nitrogen fixation. Mature nodules also produce vascular tissue that moves nitrogen fixed by bacteroids to the root vascular system for transport throughout the

plant. These nodule vascular tissues also supply organic food produced by the legume to their bacteroid partners (step 7).

In 2004, plant molecular biologists Melanie Barnett, Sharon Long, and their colleagues produced a type of a DNA microarray chip (see Chapter 20) called a symbiosis chip that contains the genomes of both a legume plant host, *Medicago truncatula*, and its rhizobial symbiont, *Sinorhizobium meliloti*. This chip allows investigators to study gene expression in both partners simultaneously, revealing that nearly 5,000 gene expression changes are associated with the legume–rhizobia symbiosis.

Carnivorous Plants Are Autotrophs That Obtain Mineral Nutrients from Animals

About 600 species of flowering plants have adapted to low-nitrogen environments by evolving mechanisms for trapping and digesting animals and are therefore known as carnivorous plants. Their leaves are modified in ways that allow them to capture animal prey, primarily insects, though larger animals are sometimes snared as well. (Despite the popular play and movie *Little Shop of Horrors*, there are no wild or cultivated carnivorous plants that, like Audrey II, are large enough to consume humans!) Carnivorous plants are photosynthetic autotrophs that supply their own organic compounds; prey animals are primarily sources of nitrogen. The experimental use of radioactively labeled prey insects has revealed that carnivorous plants obtain as much as 87% of their nitrogen from animals.

The trapping mechanisms used by carnivorous plants are classified as passive or active. Plants with passive trapping mechanisms depend on the prey to fall or wander into the trap. For example, tropical pitcher plants (genus *Nepenthes*) have leaves that are folded and partially fused to form tubes that collect rainwater (**Figure 37.21a**). The interior walls of these pitchers are slippery and have downward-pointing hairs. Insects and other small animals such as lizards and frogs that fall into the pitchers are unable to climb out. Eventually, the trapped animals drown and are digested by microbes living within the pitchers.

Plants with active mechanisms, such as Venus flytraps and sundews, have traps that are stimulated by touch. Charles Darwin, who was fascinated by carnivorous plants, was one of the first to study the trapping mechanisms of Venus flytraps and sundews. The Venus flytrap (*Dionaea muscipula*) has an active trap formed by two-lobed leaves that are edged with lance-shaped teeth (**Figure 37.21b**). The leaf surface has glands that secrete carbohydrates, which lure prey, as well as glands that secrete digestive enzymes after prey has been trapped. Also present on leaf surfaces are modified hairs, usually three per leaf lobe. If a single hair is touched—perhaps by wind or rain, or debris—and another touch does not occur soon thereafter, nothing happens. But when a fly or similar prey lands on the leaf and brushes against the same hair twice, or touches a second hair within 20–40 seconds, the leaf lobes snap shut around it. Experimental studies indicate that an action potential develops in the stimulated hairs and then travels along plasma membranes, via plasmodesmata, at about 10 cm per second.

This signal causes leaf cells to take up ions and water so that the leaf enlarges and changes shape, springing the trap. Digestion of the prey is typically finished within 10 days, whereupon the trap may reopen. Trap leaves can go through three or four digestive cycles during their lifetime.

Sundews (such as *Drosera rotundifolia*) have leaves bearing glandular hairs whose sticky tips glisten in the sunlight. Insects that land on sundew leaves get mired in the sticky mucilage exuded by these hairs, as shown in the chapter-opening photo. As the insects struggle to get away, they become covered with more mucilage and eventually smother as their breathing pores become clogged. Darwin discovered that sundew leaves bend after being touched and that glandular hairs not originally in contact with the insects also bend, folding over the prey as you would fold your fingers over an object in your palm. Later, investigators discovered that this bending involves the plant hormone auxin. In response to touch, auxin accumulates in sundew leaf tips and then flows downward, stimulating the cell expansion that causes the leaf bending. The glandular hairs also produce enzymes that digest the prey.

Parasitic Plants Obtain Nutrients from Photosynthetic Plants

More than 4,500 species of plants live as complete or partial **parasites**, organisms that obtain all or much of their water, minerals, and organic compounds from another organism. Dodder and witchweed are prominent examples of plants that are completely parasitic.

Dodder (*Cuscuta pentagona*) lacks roots and does not grow from the soil. Instead, all of the 150 species of this parasite live aboveground (**Figure 37.22**). These parasites twine their yellow or orange stems around green plant hosts, into which they sink peg-shaped, absorptive structures known as haustoria. These haustoria tap into the host plant's vascular system, stealing water, minerals, and sugar, which the parasite uses for growth and reproduction. The long, flexible stems of dodder often loop from one plant to another, such that an individual dodder plant

(a) This pitcher plant (genus *Nepenthes*) passively captures animals that accidentally fall into its water-filled pitcher.

(b) The Venus flytrap (*Dionaea muscipula*) has an active trap that is stimulated by the touch of its prey, in this case, a fly.

Figure 37.21 Carnivorous plants.

Figure 37.22   **A parasitic plant.**   Dodder (*Cuscuta pentagona*) is an example of a parasitic plant that obtains all of its water, minerals, and organic compounds from one or more green plant hosts.

can tap into many different plants at the same time. Dodder reproduces very rapidly by means of broken-off stem fragments and seeds. A single dodder plant can produce more than 16,000 seeds. In consequence, dodder is a widespread agricultural pest that attacks citrus, tomatoes, and many other fruit, vegetable, forage, and flower crops.

Another group of parasitic plants, the witchweeds (genus *Striga*), are serious problems for African agriculture, because these parasites attack the major cereal crops: corn, sorghum, and millet. Witchweed seeds lie dormant in soil until secretions from host plant roots stimulate their germination. Genetic engineers are working to find ways to protect crops from the debilitating effects of these crop parasites.

Summary of Key Concepts

37.1 Plant Nutritional Requirements

- In addition to light, green plants require CO_2, absorbed from the air, water, and several types of minerals absorbed from water in the soil. (Figure 37.1, Table 37.1)
- Green plants are autotrophs that use light as the energy source for producing organic compounds. Flowering plants that have lost photosynthetic capacity must obtain energy by metabolizing organic compounds absorbed from their environment. (Figure 37.2)
- Plants display many adaptations that allow them to cope with insufficient or excess amounts of light and inadequate CO_2.

Some of these adaptations may be usefully engineered into crop plants as a way of improving food production. (Figures 37.3, 37.4, 37.5)
- Nutrients can limit plant growth and cause nutrient-deficiency symptoms. (Figure 37.6)

37.2 The Role of Soil in Plant Nutrition

- Soil structure affects aeration, water-holding capacity, and fertility, factors that influence root and plant growth. Soil structure depends largely on particle sizes of inorganic soil components (sand, silt, and clay) and the amount of organic material that is present. (Figures 37.7, 37.8, 37.9, 37.10)
- Organic soil components, collectively known as humus, help soil hold water and mineral nutrients and provide the soil's soft consistency.
- Inorganic soil components include mineral nutrients dissolved in soil water. Clay particles electrostatically bind cations, which can be released by ion exchange. (Figure 37.11)
- The addition of organic fertilizers, which release minerals relatively slowly, can compensate for humus and mineral deficiencies in the soil. Inorganic fertilizers are washed more rapidly from soils and can contaminate natural waters when excessively applied. In aquatic habitats, excess mineral nutrients washed from agricultural fields can cause excessive growths of cyanobacteria, algae, and plants that harm other organisms.
- Plants take up fixed nitrogen in the form of ammonium ion or nitrate. To utilize nitrate, plant cells must convert it to ammonium ion. (Figure 37.12)
- Biological or industrial processes can convert atmospheric nitrogen gas into fixed nitrogen. Biological nitrogen fixation can be performed only by certain prokaryotes. (Figures 37.13, 37.14)
- Plants display several types of adaptations to cope with phosphate deficiency. Genetically modified smart plants can signal impending phosphate deficiency. (Figure 37.15)
- Some soil minerals inhibit plant growth, and some plants, called hyperaccumulators, are adapted to cope with high concentrations of toxic soil minerals. Such plants can be used to remove harmful minerals from soil in the process known as phytoremediation. (Figure 37.16)

37.3 Biological Sources of Plant Nutrients

- Mycorrhizal fungi, which are associated with the roots of most plants, provide plants with water, phosphorus, and other minerals. (Figure 37.17)
- Nitrogen-fixing prokaryotes living within the tissues of some plants provide them with fixed nitrogen. Legume–rhizobia associations are particularly important in nature and in agriculture. (Figures 37.18, 37.19, 37.20)
- Carnivorous plants obtain mineral nutrients from the digested bodies of trapped animals. (Figure 37.21)
- Parasitic plants obtain water, mineral ions, and organic compounds from green plant hosts. (Figure 37.22)

Assess and Discuss

Test Yourself

1. Which of the following substances can limit plant growth in nature?
   a. sunlight
   b. water
   c. carbon dioxide
   d. fixed nitrogen
   e. all of the above

2. In what form do plants take up most soil minerals?
   a. as ions dissolved in water
   b. as neutral salts
   c. as mineral–clay complexes
   d. linked to humus particles
   e. none of the above

3. Why do plants need sulfur?
   a. for the construction of cell walls
   b. as an essential component of chlorophyll
   c. to produce the amino acids cysteine and methionine
   d. all of the above
   e. none of the above

4. Humus is
   a. a food produced from garbanzo beans.
   b. the organic constituents of soils.
   c. the inorganic constituents of soils.
   d. the bedrock layer of soils.
   e. none of the above.

5. Which environments are conducive to heavy leaching of minerals from soils?
   a. those having soils that are composed primarily of sand particles
   b. those having acidic soils
   c. those impacted by acid rain
   d. regions characterized by heavy rainfall
   e. all of the above

6. Which property is not characteristic of clay-rich soils?
   a. high mineral nutrient retention
   b. high water retention
   c. high aeration
   d. lower amounts of sand than clay
   e. all of the above

7. Which of the substances listed is toxic to plants?
   a. carbon dioxide
   b. oxygen
   c. atmospheric nitrogen gas
   d. aluminum
   e. none of the above

8. What kinds of organisms occur in nitrogen-fixing symbioses with plants?
   a. cyanobacteria
   b. actinobacteria
   c. rhizobia bacteria
   d. all of the above
   e. none of the above

9. How do legume roots attract rhizobia?
   a. They secrete flavonoids.
   b. They secrete carotenoids.
   c. They secrete alkaloids.
   d. They secrete Nod factors.
   e. none of the above

10. Which plant uses a passive trap to obtain animal prey as a source of mineral nutrients?
    a. the Indian pipe (*Monotropa uniflora*)
    b. the tropical pitcher plant (*Nepenthes* spp.)
    c. the Venus flytrap (*Dionaea muscipula*)
    d. dodder (*Cuscuta* spp.)
    e. all of the above

Conceptual Questions

1. Why are agricultural experts and ecologists alike concerned about overfertilization of crop fields?

2. Describe how plant roots acquire enough boron to serve essential needs without transporting toxic amounts of this mineral to shoots.

3. Draw a diagram showing how rhizobia and legume roots communicate chemically during nodule formation.

Collaborative Questions

1. Imagine that you have bought a farm and want to start growing a crop to sell at a local market. How could you go about determining if the soil needs to be fertilized and with what mineral nutrients?

2. Imagine that you own a large farm with a trout stream running through it. How would you protect the water quality of the stream?

Online Resource

www.brookerbiology.com

Stay a step ahead in your studies with animations that bring concepts to life and practice tests to assess your understanding. Your instructor may also recommend the interactive ebook, individualized learning tools, and more.

Flowering Plants: Transport

38

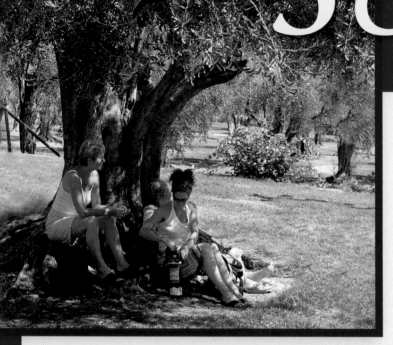

A shade tree. The evaporation of water from plant leaves cools them and us, and even affects local and global climate.

O n hot days, people naturally gravitate to the cool shade beneath trees, as shown in the chapter-opening photo. But most people do not realize that trees are not only sun umbrellas. Plants actually cool the air around them as water evaporates from their surfaces. That's why grass feels cool when you walk barefoot on it, even on a hot day.

Plants benefit from this evaporation process—known as transpiration—because it cools their surfaces and enables the movement of a continuous stream of water from the soil, through roots and stems, to leaves. This evaporation process not only helps to distribute water throughout the plant body, it also aids the movement of dissolved minerals and organic compounds such as sugars, hormones, and other organic materials over long distances within plants. Transport is therefore crucial for the functions of plant growth, behavior, and nutrition, which were described in the preceding three chapters.

In addition, plant transport plays a critical role in Earth's global climate. On a worldwide basis, plant transpiration moves huge amounts of water from the soil into the atmosphere as water vapor. Together with water evaporated from the surfaces of oceans and freshwater bodies, plant-produced water vapor is a source of rain. Along with other atmospheric gases (including carbon dioxide and methane), water vapor also works as a greenhouse gas that helps to warm Earth's climate by absorbing the sun's heat. Plant transport processes are also relevant to agriculture, as humans seek to improve crop productivity and the efficiency of water and nutrient use.

To comprehend plant transport more fully, in this chapter, we will first survey the materials that move through plants and the general directions of material movements. Next, we will focus more closely on water and solute uptake by plant cells. We will then examine how these materials are moved within plants over short and long distances and explore some of the plant adaptations that allow such transport to be as efficient as possible in a variety of environments. In the process, we will learn why plants are so cool!

38.1 Overview of Plant Transport

Our previous studies of angiosperm plant structure and nutrition described the interdependence of plant root and shoot systems (see Chapters 35 and 37). We have observed that in most plants, the root system absorbs water and dissolved minerals from the soil and that the shoot system takes up carbon dioxide (CO_2) from the atmosphere (**Figure 38.1**). Carbon dioxide enters plants via stomata, pores that occur in the surfaces of leaves and other aboveground structures. Photosynthetic cells use these materials to produce sugar and other organic compounds needed for overall plant growth and reproduction. Nonphotosynthetic plant cells, such as those of roots and flowers, depend upon organic food produced by green tissues. Therefore, plants must transport water and minerals upward from roots to shoots and transport organic food from photosynthetic to nonphotosynthetic parts. Since plants can grow to sizeable heights, the tallest trees being over 100 meters tall, transport of materials often occurs over long distances.

The long-distance transport of water, dissolved minerals, and sugar throughout the plant body occurs within a continuous system of conducting tissues. Recall that the complex tissues of vascular plants that primarily conduct water and dissolved minerals are known as the xylem, and those that conduct organic substances in a watery sap are termed the phloem. These conducting tissues are key to the ability of vascular plants to thrive in terrestrial habitats, which can sometimes be quite arid. To

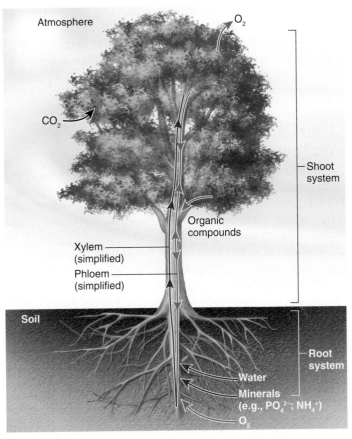

Figure 38.1 Overview of material uptake and long-distance transport processes in plants.

fully understand how plants accomplish long-distance transport, we will begin by reviewing the processes by which minerals, organic compounds, and water are taken up and move at the cellular level.

38.2 Uptake and Movement of Materials at the Cellular Level

Chapter 5 described how cells use both passive and active processes to import or export materials. Here, we briefly review these processes, describing and contrasting them and illustrating how they work in flowering plants.

Passive Transport Does Not Require the Input of Energy

Recall from your earlier studies of plasma membranes that water, gases, and certain small, uncharged compounds can diffuse across membranes in the direction of their concentration gradients. **Passive transport** is the movement of materials into or out of cells down a concentration gradient without the expenditure of energy in the form of ATP. Passive transport across plasma membranes occurs in two ways: by passive diffusion or facilitated diffusion. **Passive diffusion** into or out of cells is the movement of molecules through a phospholipid bilayer down a concentration gradient. **Facilitated diffusion** is the transport of molecules across plasma membranes down a concentration gradient with the aid of membrane transport proteins (Figure 38.2a; see also Chapter 5).

(a) **Passive transport: facilitated diffusion**

(b) **Active transport**

Figure 38.2 Passive and active transport.

The two main types of membrane transport proteins that function in facilitated diffusion are channels and transporters. **Channels** are membrane pores formed by proteins that allow movement of ions and molecules across membranes (refer back to Figure 5.21). **Transporters** are proteins that transport molecules by binding them on one side of the membrane and then changing conformation so that the molecule is released to the other side of the membrane (refer back to Figure 5.22). Transporters increase the rate at which specific mineral ions and organic molecules are able to enter or leave plant cells and vacuoles.

Recall that osmosis is the diffusion of water across a selectively permeable membrane in response to differences in solute concentrations. In the case of plants, water moves from a solution that has a lower solute concentration (soil) to one of higher solute concentration (root cells). Osmotic water uptake into living plant cells is essential to photosynthesis, as well as to cell expansion and structural support. However, the passive diffusion of water does not occur rapidly enough to supply the water needs of rapidly expanding plant cells. In this case, facilitated diffusion of water occurs through protein channels known as **aquaporins**, which occur widely in living things (refer back to Figure 5.20). Thirty-five distinct aquaporin genes have been identified in the genome of the model plant *Arabidopsis*. Aquaporins increase the rate at which water flows into expanding plant cells and their vacuoles. In the same way, many other types of plasma membrane protein channels and transporters facilitate the diffusion of specific mineral ions and organic molecules into and out of plant cells and vacuoles. If a substance must be transported across a plasma membrane against its concentration gradient, however, work must be performed in the process known as active transport.

ATP Hydrolysis Powers Active Transport

During **active transport**, membrane transporter proteins use energy to move substances against their concentration gradients. An example is the H^+-ATPase proton pump, found in the plasma membranes of plant cells, which uses ATP to pump H^+, which are protons, against a gradient (**Figure 38.2b**). This proton gradient generates an electrical difference across the membrane, which is known as a **membrane potential**. Energy is released when protons pass back across the plasma membrane, in the direction of their electrochemical gradient. This energy can then be used to power other active transport processes. For example, it might be used to open or close ion channels or in the functioning of a proton **symporter** (a protein that transports two substances in the same direction across a membrane) that is needed for the uptake of organic materials such as sugars, amino acids, and nucleotide bases.

Proton pumps occur in vast numbers in the vacuolar membrane as well as in the plasma membrane. The large vacuole present in many plant cells serves as a storage site for ions and other substances that move in or out via active transport. For example, vacuoles store calcium ions until they are needed in signal transduction processes, and they store reserves of nitrate

(see Chapter 37). Active transport proteins are particularly abundant in root cell membranes (**Figure 38.3**), allowing root cells to concentrate dissolved mineral nutrients to more than 75 times their abundance in soil. As a result, soil water flows into root cells by osmosis. We next take a closer look at osmotic water movement into and out of plant cells.

Cellular Water Content Is Influenced by Solute Content and Turgor Pressure

The water content of plant cells depends on osmosis, and osmosis depends on two factors: solute content and turgor pressure. **Turgor pressure** is the hydrostatic pressure that increases as water enters plant cells, because their cell walls restrict the extent to which the cells can swell.

A plant cell whose cytosol is so full of water that the plasma membrane presses right up against the cell wall is said to be **turgid** (**Figure 38.4a**). The pressure relationship between the cytosol and cell wall recalls the way that a soccer ball's leather skin presses inward upon the air within, while at the same time the internal air presses on the ball's cover. If you add more air to a limp ball, the ball will stiffen. In the same way, if a nonturgid cell absorbs more water, it will become more rigid as the water exerts pressure on the cell wall. By contrast, a plasmolyzed cell is one that has lost so much water by osmosis

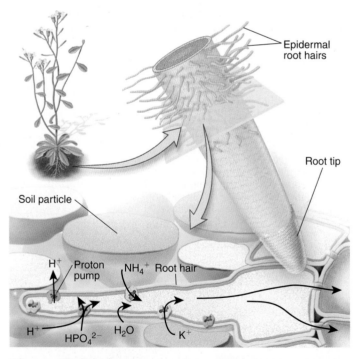

Figure 38.3 Ion uptake at root-hair membranes. When soil mineral ion concentrations are lower than those within cells, root-hair plasma membranes take up nutrient ions by active transport. The H^+-ATPase establishes an electrochemical gradient that drives the active uptake of solutes. The resulting increase in intracellular solute concentration also drives the osmotic diffusion of water into the cell.

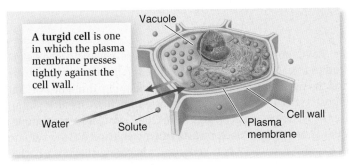

(a) Turgid cell in a hypotonic solution

A turgid cell is one in which the plasma membrane presses tightly against the cell wall.

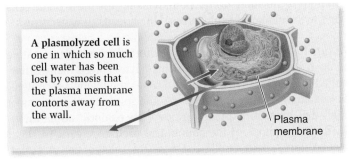

(b) Plasmolyzed cell in a hypertonic solution

A plasmolyzed cell is one in which so much cell water has been lost by osmosis that the plasma membrane contorts away from the wall.

(c) Flaccid cell in an isotonic solution

A flaccid cell is one in which the plasma membrane does not press tightly against the cell wall.

Figure 38.4   Turgid, plasmolyzed, and flaccid plant cells. (a) When the concentration of solutes inside a cell is higher than that outside (the cell is surrounded by a hypotonic solution), more water will enter the cell than will leave it. As a result, a plant cell will become swollen, or turgid. (b) When the concentration of solutes outside a cell is greater than within it (the cell is surrounded by a hypertonic solution), more water will leave the cell than will enter it. As a result, a plant cell will become plasmolyzed. (c) When the concentration of solutes is the same outside and inside a plant cell (the cell is bathed in an isotonic solution), it will be flaccid.

that turgor pressure has also been lost. **Plasmolysis** is the condition in which the plasma membrane no longer presses on the cell wall (**Figure 38.4b**). A cell having a water content between these two extremes is termed a flaccid cell (**Figure 38.4c**).

Together, solute concentration and turgor influence an important plant cell property known as water potential. What is water potential? Recall that potential energy is the stored energy of a material or system, with the capacity to perform work. **Water potential** is the potential energy of water. A waterfall analogy illustrates that water moves from a region of higher water potential to a region of lower water potential, in this case, as the result of gravity. Pressure also influences water potential;

a waterfall would flow upward if pressure greater than the force of gravity were applied. Solutes and some other factors also affect water potential. Water potential is measured in pressure units known as megapascals (MPa) (a pascal is equal to 1 newton per square meter). One MPa is equal to 10 times the average air pressure at sea level. As another reference point, 1 MPa is several times the pressure in typical home plumbing pipes, which you experience when turning on a water faucet.

In the study of plants, the concept of water potential is used in two ways: to understand the movement of water into and out of cells (cellular water potential) and to understand the water status of entire plants or organs (relative water content). We begin by considering the water potential of cells.

Cellular Water Potential   A water potential equation can be used to predict the direction of cellular water movement, given information about the solute concentrations inside and outside of plant cells, and a measure of pressure at the cell wall–membrane interface (**Figure 38.5**). In this equation, cellular water potential is symbolized by the Greek letter psi (ψ) with the subscript W for water: ψ_W. In its simplest form, total

| Water potential (ψ_W) | = | Solute potential (ψ_S) | + | Pressure potential (ψ_P) |

Examples of ψ_W calculations:

Turgid cell	Plasmolyzed cell	Flaccid cell
$\psi_S = -0.1$	$\psi_S = -1.0$	$\psi_S = -0.5$
$\psi_P = +0.1$	$\psi_P = 0.0$	$\psi_P = 0.0$
$\psi_W = 0.0$ MPa	$\psi_W = -1.0$ MPa	$\psi_W = -0.5$ MPa

Figure 38.5   Plant-cell water potential.   The water potential of a plant cell, which predicts the direction of water movement into or out of a cell, can be simplified as the sum of the solute potential and the pressure potential resulting from pressure exerted by the cell wall–plasma membrane complex. Examples of water potential calculations are shown for a turgid cell, a plasmolyzed cell, and a flaccid cell.

Concept check: Make a drawing that shows the direction of water movement when a cell of each of these types is placed into a solution of pure water (whose water potential, ψ_W, is defined as 0 MPa).

ψ_W is calculated as the sum of the **solute potential** (ψ_S), also known as the osmotic potential, and the **pressure potential** (ψ_P). A simple form of the water potential equation is

$$\psi_W = \psi_S + \psi_P$$

Solute potential is the component of water potential due to the presence of solute molecules. As you might expect, solute potential is proportional to the concentration of solutes in a solution. The solute potential of pure water open to the air, at sea level and room temperature, is defined as zero. When solutes are added, they interact with water molecules, thereby diluting the water and affecting its disorder. As a result, fewer free water molecules are present, which reduces the potential energy of water. Thus, in the absence of a pressure potential, water that contains solutes always has a negative solute potential. The higher the concentration of dissolved solutes, the lower (more negative) the solute potential.

Pressure potential is the component of water potential due to hydrostatic pressure. In plant cells, the hydrostatic pressure is determined in part by the resistance provided by the cell wall. Because of this resistance, the value for pressure potential can be either positive or negative. For example, a turgid cell has a positive pressure potential, which typically measures about 1 MPa. This high pressure inside turgid plant cells is a testimony to the strength of their cellulose-rich plant cell walls. In contrast to turgid cells, both flaccid and plasmolyzed cells have a pressure potential of zero. Plants or plant organs having many cells with low turgor pressure will appear wilted. If your houseplants become wilted, watering will enable the cells to increase their pressure potential, restoring cell turgor and normal plant appearance. Therefore, the water content of an entire organ or plant is influenced by the water potential of its component cells.

Relative Water Content

The property known as **relative water content** (**RWC**) is often used to gauge the water content of a plant organ or entire plant. RWC integrates the water potential of all cells within an organ or plant and is thus a measure of relative turgidity. Measurements of RWC can be used to predict a plant's ability to recover from the wilted condition. An RWC of less than 50% spells death for most plants, but some plants can tolerate lower water content for substantial time periods.

A standard method for determining RWC was developed by plant biologists H. D. Barrs and Paul E. Weatherley in 1962. This process involves three simple weight measurements: fresh weight, turgid weight, and dry weight. Sample tissue taken from a plant under a given set of conditions is first weighed to obtain the fresh weight. Then it is completely hydrated in water within an enclosed, lighted chamber until constant turgid weight is achieved. Finally, the sample is dried to a constant dry weight. Researchers use these measurements to calculate RWC using the equation

$$\text{RWC} = \frac{(\text{fresh weight} - \text{dry weight})}{(\text{turgid weight} - \text{dry weight})} \times 100$$

RWC measurements have been very useful in ecological studies of natural plant adaptation to cold, drought, or salt stress and in agricultural research for developing drought-tolerant crops. Agricultural research reported by R. Chandra Babu, Henry Nguyen, and colleagues in 2004 provides an example. These investigators used measurements of leaf RWC to evaluate rice crops that had been genetically engineered with a barley gene that confers dehydration tolerance. Developing new crops that are better able to withstand water stress requires an understanding of not only water potential but also how plant cells cope with cellular osmotic stress, which leads to water stress.

Plant Adaptations to Cellular Osmotic Stress

Plants native to cold, dry, or saline environments have evolved many different adaptations that allow them to cope with low water content. For example, plants often increase the solute concentrations of their cell cytosol, a process known as **osmotic adjustment**. Increased amounts of the amino acid proline, sugars, or sugar alcohols such as mannitol decrease the cells' water potential, thus drawing water into cells. By increasing the concentration of solutes inside cells, cold-resistant plants prevent water from moving out of their cells when ice crystal formation in intercellular spaces lowers the water potential outside cells. The additional solutes also lower the freezing point of the cytosol, for the same reason that adding antifreeze to a car's radiator in winter keeps the radiator fluid from freezing.

Plants of arid lands often possess adaptations that help them survive water stress. Many can survive in a nearly dry state for as much as 10 months of the year, growing and reproducing only after the rains come. The cytosol of such desiccation-tolerant plants is typically rich in sugars that bind to phospholipids to form a glasslike structure. This helps to stabilize the cellular membranes, preventing them from becoming damaged during plasmolysis (see Figure 38.4b). Plant cells under water stress may also increase the number of plasma membrane aquaporins. These additional protein channels increase the rate of water uptake, allowing cells to recover turgor more quickly when water becomes available.

Halophytes are plants that are able to grow in salty habitats, including coastal salt marshes. Plant cells cannot readily absorb salty water because of its highly negative water potential. Halophytes have therefore acquired adaptations to help them cope with osmotic stress. Some halophytes accumulate inorganic salts in their vacuoles and organic solutes in their cytosol. Such storages balance solute concentrations inside and outside cells, preventing osmotic water loss. Halophytes also display adaptations that help rid their bodies of excess salt. Some halophytes excrete salt from root surfaces, while others transport excess salt to their leaf epidermal cells, where it forms crystals (look ahead to Figure 54.16). Some halophytes accumulate salt within multicellular surface glands that break off or burst harmlessly. The water potential adaptations of desiccation-tolerant plants and halophytes are of special interest to agricultural scientists seeking to develop crops that are able to grow in arid lands or in places where the soil or water used for irrigation is salty.

38.3 Tissue-Level Transport

Now that we have learned more about how water, dissolved minerals, and organic compounds enter or leave plant cells, we are prepared to consider short-distance transport within and among nearby tissues. Tissue-level transport occurs in three forms: transmembrane transport, symplastic transport, and apoplastic transport (**Figure 38.6**).

Transmembrane transport involves the export of a material from one cell, followed by import of the same substance by an adjacent cell (Figure 38.6a). One prominent example of transmembrane transport is the movement of the plant hormone auxin downward in shoots. Auxin travels from one phloem parenchyma cell to another in a linear series (Chapter 36). This process explains how auxin produced in one part of the plant body can influence more distant tissues.

Symplastic transport is the movement of a substance from the cytosol of one cell to the cytosol of an adjacent cell via membrane-lined channels called plasmodesmata (Figure 38.6b). Plasmodesmata also have a central tubule called a desmotubule that connects endoplasmic reticulum of adjoining cells (refer back to Figure 10.14). Plasmodesmata are large enough in diameter to allow transport of proteins and nucleic acids, as well as smaller molecules. These molecules primarily move by diffusion, though in some cases, special movement proteins facilitate transport through plasmodesmata. Together, all of a plant's protoplasts (the cell contents without the cell walls) and plasmodesmata form the **symplast**. Symplastic transport has

the potential to move molecules widely among the cells and tissues of the plant body.

In contrast to the symplast, the **apoplast** refers to the continuum of water-soaked cell walls and intercellular spaces (Figure 38.6c). **Apoplastic transport** is the movement of solutes along cell walls and the spaces between cells. Water and dissolved minerals often move through plant tissues for short distances by apoplastic transport.

Both symplastic and apoplastic transport play important roles in mineral nutrient transport through the outer tissues of roots (**Figure 38.7**). As we have noted, the plasma membrane of epidermal root hair cells is rich in channels and transporters that selectively absorb essential mineral ions from soil water. Absorbed ions can move symplastically from the cytosol of root hairs, cortex, and endodermis directly to xylem parenchyma cells. Plasmodesmata make such cell-to-cell, tissue-level transport possible.

Apoplastic transport can also move water and dissolved minerals into root epidermal and cortex tissues. However, apoplastic movement of water and minerals stops at the **endodermis**, a term meaning "inside skin." In roots, the endodermis is a thin cylinder of tissue whose close-fitting cells and specialized cell walls form a barrier between the cortex and the central core of vascular tissue. Materials in the root apoplast cannot penetrate farther into the root unless endodermal cells transport them into their cytosol, a process that requires specific transporter proteins.

Root endodermal cell walls possess ribbon-like strips of waterproof suberin, composed of wax and phenolic polymers. These suberin ribbons, known as **Casparian strips**, prevent apoplastic transport through endodermal cell walls and into the root vascular tissues (**Figure 38.8**). The root endodermis prevents harmful solutes (such as toxic metal ions) from moving through the apoplast to vascular tissues and being transported to the shoot. For example, recall that aluminum ions (Al^{3+}) are commonly dissolved in soil water, but they are not plant nutrients and are highly toxic to plants (see Chapter 37). Aluminum ions can penetrate the root epidermis and cortex by moving through the apoplast, but they stop at the endodermis because they are unable to enter the cytosol of endodermal cells. Keep in mind, however, that while the endodermis is effective in reducing the entry of toxic metal ions, these substances can enter roots at root tips, which lack a mature endodermis, and in places where a branch root has broken through the endodermis (refer back to Figure 35.26).

Endodermal plasma membranes possess specific channels and transporters for essential mineral nutrients (such as K^+), which are thereby able to enter the cytosol of root endodermal cells. By moving through endodermal cytosol, symplastically transported essential minerals are able to bypass the endodermal barrier that limits apoplastic transport. Therefore, the root endodermis functions as a molecular filter that allows the passage of beneficial solutes that have entered from the symplast or have been specifically transported into endodermal cytosol. Once solutes have passed the endodermal barrier, they are transported out of xylem parenchyma cells and into

Figure 38.6 Three routes of tissue-level transport in plants: (a) transmembrane, (b) symplastic, and (c) apoplastic.

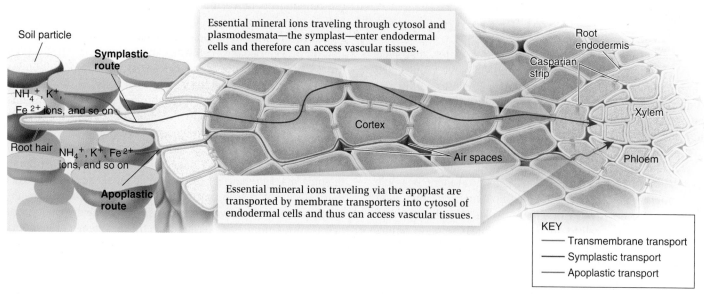

Figure 38.7   Symplastic and apoplastic transport of mineral ions in roots.

the apoplast of the vascular system, which includes conducting cells of the xylem (Figure 38.8). The endodermis prevents solutes from returning to outer root tissues or the soil, so the solute concentrations of xylem parenchyma cells rise, decreasing their cellular water potential. As a result, water flows into

vascular tissues from outer root tissues and the soil. In the process known as **xylem loading**, large amounts of water and dissolved solutes enter the long-distance conducting cells of the xylem. In the next section, we will see how water and solutes are transported for long distances through the plant.

Figure 38.8   Ion transport pathways across the root endodermis.   Casparian strips in endodermal cell walls prevent apoplastic transport across the root endodermis, limiting entry of harmful soil minerals such as Al^{3+} and exit of useful minerals. Mineral nutrients that are transported into the cytosol of endodermal cells are able to pass through the endodermal barrier to xylem parenchyma cells via plasmodesmata. Once past the endodermis, nutrient ions such as K^+ are moved across plasma membranes to the apoplast of the vascular tissue and are thus able to enter xylem. Inset shows a transmission electron micrograph (TEM) of a Casparian strip in the wall of an endodermal cell.

Concept check: *Where in most plants would you expect to find the highest concentration of Al^{3+}?*

38.4   Long-Distance Transport

Plants rely on long-distance transport to move water and dissolved materials from roots to shoots and among organs. Tall trees are able to transport water and minerals to astounding heights, more than 110 m in some cases. This is possible because plants possess an extensive, branched, long-distance vascular system composed of xylem and phloem tissues. Watery solutions move through these tissues by **bulk flow**, the mass movement of liquid caused by pressure, gravity, or both. Plant-conducting tissues are specialized in ways that foster bulk flow and aid plants in adapting to water stress. The evaporation of water at plant surfaces, the cohesion of water molecules to each other, the adhesion of water to the walls of conducting cells, and the tension that pulls water through those cells all are important physical forces that influence long-distance water transport. Gravity, though of negligible effect for shorter plants, must also be considered in the case of water transport in tall trees. In this section, we will take a closer look at bulk flow and the major factors involved in long-distance transport by this process.

Bulk Flow Is Water Movement Under the Influence of Pressure and Gravity

Bulk flow (also known as mass flow) occurs when molecules of liquid all move together from one place to another as the result of differences in pressure. One example of bulk flow is leaching, the movement of water and dissolved minerals downward through soil layers as the result of gravity. Bulk flow is one way in which mineral ions can move through soil toward plant roots. Likewise, once inside the plant, minerals and other dissolved solutes can move through xylem and phloem conducting tissues via bulk flow, which is much faster than diffusion.

Phloem sap, which contains sugars and other dissolved solutes, moves by bulk flow up to 1 m per hour. Bulk flow within xylem moves water from the roots to the uppermost leaves of 100-m-tall trees.

In plants, water and dissolved materials move by bulk flow through xylem tissues as the result of differences in root and shoot water pressure. The water pressure is lower in upper parts of plants than in lower parts because water diffuses through stomata into the atmosphere (Figure 38.9). In the phloem, bulk flow occurs because a high pressure builds up in leaves, which have relatively high solute concentrations. The pressure causes the water to flow to other regions having lower solute concentrations, such as developing fruit. Bulk flow is impeded by obstructions such as cytoplasm. This explains why xylem-conducting cells are devoid of cytoplasm and why phloem-conducting cells have reduced amounts of cytoplasm. Xylem and phloem share additional features related to bulk flow, which we will discuss later in this chapter.

Although xylem serves as the primary transport system for water and minerals, and phloem for organic compounds dissolved in water, the transport functions of xylem and phloem

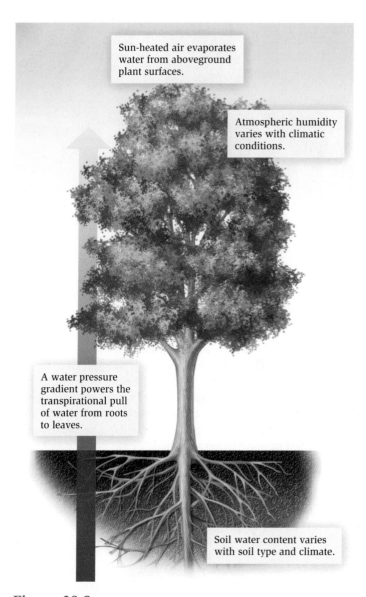

Figure 38.9   **Upward transport of water in xylem.**   Water pressure differences between moist soil and drier air drive the upward movement of water in plants.

overlap somewhat. Phloem can aid in the distribution of certain minerals, and xylem sometimes transports organic compounds. For example, in early spring, trees convert starch stored in stem parenchyma cells into sugars that are used during bud expansion and flower development. These sugars are transported in xylem sap. Maple trees produce such a copious flow of sugar-rich xylem sap that people have long tapped them for making maple syrup (Figure 38.10).

Long-Distance Transport of Water and Minerals Occurs in the Xylem

Xylem structure plays an essential role in its transport function. The xylem of flowering plants contains several types of specialized cells, some of which remain alive at maturity, and some

Figure 38.10   Maple-sugar tapping.   In early spring, xylem transports sugar from storage sites to the shoot buds of woody plants such as this sugar maple. We use the xylem sap of sugar maples to produce maple syrup.

of which are dead when they are fully functional. Xylem parenchyma cells are alive, while thick-walled supportive fibers may be alive or dead at maturity. Two types of specialized water-conducting cells are always dead and empty of cytosol when mature: tracheids and vessel elements. Together, tracheids and vessel elements are known as **tracheary elements**. During the development of tracheary elements, a secondary wall is deposited in patterns on the inside of the primary cell wall. This secondary wall is rich in a plastic-like polymer known as lignin. Because lignin is resistant to compression, microbial decay, and water infiltration, it confers strength, durability, and waterproofing. Like the plumbing pipes of a building, tracheary elements do not readily collapse as water moves through them

under tension. These characteristics explain why tracheary elements contribute to structural support of the plant body as well as transport.

Tracheids   Long and narrow in shape, **tracheids** typically have slanted end walls that fit together to form long tubes (**Figure 38.11a**). The end walls of tracheids are not lignified, nor are large areas of the side walls of tracheids that occur in plant tissues that are still growing. Such tracheids are extensible because they have rings or spirals of lignin that allow tracheids to continue elongating (**Figure 38.11b**). In contrast, tracheids that develop in tissues that have already expanded have more lignin, which makes them rigid and unable to elongate any more. Tracheid walls that are extensively lignified display numerous small, lignin-free cell-wall regions known as **pits**. At such pits, the thin primary wall of the tracheid remains readily permeable to water. Water moves from one tracheid to another both vertically and laterally through pits.

Vessels and Vessel Elements   Mature **vessel elements** are a second type of water-conducting cell present in xylem tissue. Vessel elements are aligned in pipeline-like files known as **vessels** (**Figure 38.12a**). Flowering plants are distinguished from other plant groups by the abundance of vessels; nonflowering plants primarily rely on tracheids for water conduction. Vessel elements are larger in diameter than tracheids, conferring greater capacity for bulk flow and therefore represent one of the many ways in which flowering plants are particularly well adapted to life on land.

Development of vessel elements resembles that of tracheids in some ways. For example, lignified secondary walls are deposited in spirals or sheets on the inside of the primary cell

End wall

Pits

(a) Tracheids   **(b) Extensible tracheids**

8.5 µm

Figure 38.11   Tracheid cells in xylem tissue.   (a) Tracheids are long, tubular cells with slanted end walls. Water and ions move from cell to cell through the pits. (b) Light micrograph of extensible tracheids from the xylem of pumpkin.

Concept check: *If you applied a stain specific for lignin to tracheids present in a longitudinal slice of a plant stem that is still growing in length, then observed the cells with a light microscope, what portions of the tracheids would be stained, and what parts would not be stained?*

Large end-wall perforations

Pits

Vessel element

End-wall perforation

Pit

|———| 43 μm

|———| 6 μm

(a) Vessels made up of vessel elements

(b) Vessel elements in a walnut tree

(c) Perforations in a vessel element end wall

Figure 38.12 **Vessels composed of vessel elements in xylem tissue.** **(a)** This illustration shows the wide diameter of vessel elements with many pits and end-wall perforations. **(b)** SEM of vessels in the wood of the walnut tree (genus *Juglans*). **(c)** SEM of a perforated vessel element end wall from the tulip tree (*Liriodendron tulipifera*).

Concept check: Which structural features of vessel elements explain the vulnerability of vessels to embolism, that is, blockage by air bubbles?

wall. Vessel elements also have numerous pits in their side walls. In contrast to tracheids, the end walls and some side walls of vessel elements are extensively perforated, meaning that all cell-wall material is removed from some areas (**Figure 38.12b,c**). This allows water to flow faster from one vessel element to another than it can flow from one tracheid to another.

Because the perforated end walls and large diameter of vessels allow them to transport more water at a faster rate than tracheids, you might wonder why vessels have not completely replaced tracheids in flowering plants. The answer is that vessels are more vulnerable than tracheids to embolism, meaning blockage by air bubbles. Once an embolism forms in a vessel element, it can extend through the large end-wall perforations into many elements, thereby blocking a vessel. Just as air bubbles can cause disruption of blood circulation in people, sometimes leading to death, such bubbles also disrupt water transport in plants, sometimes severely. An embolism can form within a vessel as the result of physical damage, drought, or repeated cycles of freezing and thawing. Air bubbles form frequently during winter, because air does not dissolve in ice. By the end of a cold winter, the functional vessels of many woody plants have become almost completely blocked by air. Blocked vessels in trees often cease to function in water transport and must be replaced by new growth in the spring. Fortunately, even if vessels become blocked, water conduction can still occur via tracheids. This is because tracheid pits are so small that they do not allow air bubbles to move to other tracheids. Thus, an air bubble is confined to the single tracheid in which it first formed, and water continues to flow through nearby tracheids. Tracheids thereby provide a fail-safe conduction route when vessels have become disabled by embolisms.

Figure 38.13 Guttation, the result of root pressure.

Concept check: What functions can root pressure serve in plants?

Some plants are able to refill embolized vessels by means of a process known as **root pressure**. At night, the xylem of roots may accumulate high concentrations of ions that are not immediately transported upward to shoots. In this case, the root acts much like a cell rich in solutes, with the result that water gushes in so rapidly that it pushes upward to leaves. Evidence of this process can be observed in the early morning as droplets of water at the edges of leaves, a phenomenon known as **guttation** (**Figure 38.13**). As the water rushes upward, it can dislodge air bubbles or dissolve them, thereby reversing an embolism. Root pressure refilling has been observed to occur in nonwoody plants such as corn (*Zea mays*) and in some woody plants, including the sugar maple (*Acer saccharum*).

FEATURE INVESTIGATION

Holbrook and Associates Revealed the Dynamic Role of Xylem in Transport

Plant biologists once thought that xylem sap moved from one conducting cell to another at a uniform rate, unless blocked by an embolism. After all, botanists hypothesized, the cells are dead and thus lack cytoplasmic components that might influence sap flow rates. However, recent experiments have revealed that xylem sap actually moves more readily from one vessel to another through pits when the solute concentration of xylem sap is relatively high. These studies suggest that this variation results from dynamic changes in pit-wall microstructure.

More than 20 years ago, plant biologists noticed that tap water moved more rapidly through cut pieces of stem than did pure water. This observation suggested that ions present in the tap water might have influenced bulk flow rates. However, this hypothesis remained untested until 2001, when N. Michelle Holbrook and colleagues conducted experiments designed to explore this phenomenon, as shown in Figure 38.14. These investigators compared the flow rate of pure water with that of artificial sap, pure water to which they had added various levels of potassium chloride (KCl) or other salts that normally occur in xylem sap. The artificial sap and pure water were introduced into the same experimental tobacco plant through

Figure 38.14   The Holbrook team's dynamic xylem experiment.

6 **THE DATA**

(a) Results of steps 1 and 2:

Flow rate in experimental flap alternately receiving pure water or artificial sap varies by a factor of 2.

Flow rate in control flap receiving pure water is constant.

(b) Results of step 3*—Effect of various solutes on xylem flow rate in stem segments:

Notice that salts increased flow rate, but organic molecules did not.

*Similar results were observed for steps 4 and 5.

7 **CONCLUSION** Artificial sap and other salts increase the rate of sap flow compared to pure water, likely by shrinking cell-wall pectins, thereby opening micropores in the thin pit walls between adjacent vessels.

8 **SOURCE** Zwieniecki, Maciej A., Melcher, Peter J., and Holbrook, N. Michele. 2001. Hydrogel control of xylem hydraulic resistance in plants. *Science* 291:1059–1062.

separate flaps cut into the stem. (Using different flaps on the same plant helps to remove experimental variation that might have resulted if investigators had compared results from separate plants.)

These researchers observed that the xylem flow rate was higher when the flap received artificial sap than when it was receiving pure water. Then the investigators supplied pure water to both stem flaps, with the result that the flow rate equalized (see step 2 in Figure 38.14 and also The Data, part a). The investigators then provided artificial sap containing various types of solutes to one flap, along with alternating additions of pure water. The flow rate increased for artificial sap containing dissolved salts, but not for artificial sap containing the organic compounds sucrose or ethanol (see The Data, part b).

Next, the scientists performed experiments designed to determine if living cells in the xylem tissue were causing these effects. They flushed the stems with boiling water or froze the stems in liquid nitrogen before treating them with artificial sap or water as in step 3. These treatments kill living cells. They found that flow rates through the boiled or frozen stems yielded the same results: The flow rate was higher when the flap received sap containing dissolved salts rather than pure water or organic solutes. These results indicated that living cells did not cause the effect.

The investigators suspected that ions change vessel element structure in some way that allows greater flow through them. In later experiments not shown here, the scientists devised an apparatus for measuring the flow of sap through individual xylem vessels versus sap flow through groups of vessels. They found that flow through an isolated vessel did not respond to changes in sap ion content, whereas flow between adjacent vessels responded. The researchers interpreted these results as evidence that small pores called micropores in the thin pit walls between adjacent vessel elements play a role in the change in flow rates. These micropores occur within a mesh of cellulose microfibrils and pectins. The pectins, which are water-rich gels known as hydrogels, shrink in response to sap ions and swell in response to pure water. As a result, micropores widen in response to sap ions, increasing flow rates, but narrow in response to pure water (Figure 38.14, step 1, Conceptual level). Researchers speculated that this process hastens the delivery of ion-rich sap to transpiring tissues. Investigators also proposed that this process, which increases sap flow, may help plants compensate for partial loss of xylem sap flow when an air embolism occurs.

Experimental Questions

1. Why did the Holbrook team use the same plant to examine the effect of a pure-water control and an experimental, artificial xylem sap?

2. What caused water to flow from the source of artificial sap (or pure water) into the xylem of the plants used in the experiment conducted by Holbrook and associates?

3. Why were sap flow rates the same in living and dead plants in the experiment conducted by the Holbrook team?

Cohesion-Tension Theory Explains the Role of Transpiration in Long-Distance Water Transport

In warm, dry air, water evaporates from plant surfaces. This evaporation process is known as **transpiration** (from the French *transpirer*, meaning to perspire) (**Figure 38.15a**). Transpiration is capable of pulling water by bulk flow up to the tops of the tallest trees and is the primary way in which water is transported for long distances in plants. Plants expend no energy to transport water and minerals by transpiration. Rather, the sun's energy indirectly powers this process by generating a water pressure difference between moist soil and drier air (see Figure 38.9).

(a) **Transpiration occurs when leaf water is exposed to drier air.**

(b) **Cohesion in xylem causes water to form a continuous stream.**

(c) **When water evaporates, the surface tension increases in the intercellular spaces of cells, pulling on the water stream in xylem.**

Figure 38.15 The roles of transpiration, cohesion, adhesion, and tension in long-distance water transport.

How does evaporation at plant surfaces influence long-distance water transport in the xylem? To answer this question, we must consider the unique physical properties of water. Liquid water molecules are linked by hydrogen bonds (see Chapter 2). As a result, liquid water is amazingly cohesive, explaining why water tends to form continuous streams (**Figure 38.15b**). Consequently, when water molecules evaporate from plants, water films present in leaves display high surface tension. This tension causes a curved water surface known as a meniscus to form (**Figure 38.15c**) that pulls on neighboring liquid water molecules and eventually on water in the nearest vein, which is connected to the plant's entire water supply. As the result of water's cohesion and the tension exerted on water at the plant's surface, a continuous stream of water can be pulled up through the plant body from the soil, into roots, through stems, and into leaves. This explanation for long-distance water movement in plants is known as the **cohesion-tension theory**. (Recall that a scientific theory is a well-established concept, not just a hypothesis.) Considerable experimental data support the cohesion-tension concept. For example, in 2008, Tobias Wheeler and Abraham Stroock constructed a synthetic tree and watched the movement of water through it; their observations were consistent with the cohesion-tension theory. Even so, some experts think that additional factors might be important in long-distance water transport. These factors include adhesion of water (its tendency to stick) to the lignified walls of xylem conducting cells (Figure 38.15b) and a squeezing effect exerted by nonconducting tissues on conducting tissues.

Plant transpiration moves huge amounts of water from the soil to the atmosphere. About 99% of the water that enters plants via roots is generally lost as water vapor during transpiration. Each crop season, a single corn plant (*Zea mays*) loses more than 200 L of water, which is more than 100 times the corn plant's mass. A typical tree loses 400 L of water per day! On a regional and global basis, plant transpiration has enormous climate effects. For example, an estimated one-half to three-quarters of rainfall received by the Amazon tropical rain forest actually originates from plant-transpired water vapor, often visible as mist (**Figure 38.16**). Furthermore, about half of the solar heat received by Amazonian plants is dispersed to the atmosphere during transpiration. This heat dispersal has a cooling effect on regional ground temperature, which would be much higher in the absence of plant transpiration. Such cooling effects result from water's unusually high heat of vaporization, the amount of heat needed to isolate water molecules from the liquid phase and move them to the vapor phase. Most of this energy is needed to break the large numbers of hydrogen bonds that occur in liquid water. The evaporation of large amounts of water from plant surfaces effectively dissipates heat, explaining how plants cool themselves and their environments.

Although evaporation of water from plant surfaces plays an essential role in bulk flow through xylem, if plants lose too much water, they will die. Plant surfaces, including those of leaves, typically produce a cuticle, a wax-containing layer that retards water loss. Only about 5% of water evaporated from plant surfaces emerges through the cuticle. More than 90% of the water that evaporates from plants is lost through stomata, surface pores that can be closed to retain water or opened to allow the entry of CO_2 needed for photosynthesis. When the stomata are open, oxygen also exits the plant, as does water vapor when the atmospheric humidity is relatively low. Stomata are often abundantly located on the lower surfaces of leaves. Tobacco leaves, for example, possess an estimated 12,000 stomata per square centimeter of leaf surface! Plants face a constant dilemma: whether to open their stomata for CO_2 intake and suffer the impact of reduced water content or to close stomata to retain water, thereby preventing CO_2 uptake. For this reason, plants have acquired many adaptations to cope with transpirational water loss.

Plant Adaptations Help to Reduce Transpirational Water Loss

Under some conditions, almost all plants experience water stress, which is an inadequate amount of water. Water stress is common for plants of the world's arid regions, and their growth is often limited by water availability. Even plants of moist, forested regions of the world experience water stress during drier or colder seasons or under windy conditions. The leaves at the tops of tall trees are generally under considerable water stress because gravity has a substantial impact on their water potential. Earlier, we considered examples of plant cellular adaptations to deal with osmotic stress. Plants have evolved two additional ways to prevent excessive loss of water by transpiration: regulation of stomata opening and leaf drop.

Stomatal Opening and Closing   Plant stomata close to conserve water under conditions of water stress and open when the stress has been relieved, allowing air exchange with the leaf's spongy mesophyll. Stomata are bordered by a pair of **guard cells**, which are sausage-shaped chloroplast-containing cells attached at their ends (**Figure 38.17a**). The distinctive structural features of guard cells explain how they are able to open and close a pore. As guard cells become fully turgid, their volume expands by 40–100%. This expansion does not occur evenly, however, because the innermost cell walls are thicker and less extensible than are other parts of the guard cell walls. In addition, the cells expand primarily in the lengthwise direction because bands of radially oriented cellulose microfibrils prevent the guard cells from expanding laterally (**Figure 38.17b**). Thus, when guard cells are turgid, a stomatal pore opens between them, allowing air exchange with the leaf's spongy mesophyll. Conversely, when the guard cells lose their turgor, their volume decreases, and the stomatal pore closes.

What causes the change in turgor? Stomata often open early in the morning, in response to sunlight. This response makes sense, given that light, water, and carbon dioxide are all required for photosynthesis. Blue light stimulates H^+-ATPase proton pumps, leading to guard cell uptake of ions, especially potassium (K^+), and other solutes. As a result of increases in solute concentrations inside guard cells, osmotic water uptake

Figure 38.16   Plant-transpired water vapor mist rising from a tropical rain forest.   This mist visually illustrates the enormous amount of water that is transpired from the surfaces of plants into the atmosphere. Water vapor derived from plant transpiration is an important source of rainfall, and the process of evaporation cools plant surfaces as well as affecting the local and global climate.

Concept check: *Why does evaporation of water have such a powerful cooling effect?*

occurs via plasma membrane aquaporins, resulting in cell expansion and stomatal opening (**Figure 38.18a**).

At night, the reverse process closes stomata. Potassium and other solutes are pumped out of guard cells, causing water to exit and deflate guard cells, resulting in pore closure. Plants also close stomata during the daytime under conditions of water stress, a process mediated by the stress hormone abscisic acid (ABA). Water stress causes a 50-fold increase in ABA, which is transported in the xylem sap to guard cells. ABA then binds to a receptor, which elicits a Ca^{2+} second messenger, causing the guard cells to lose solutes and deflate (**Figure 38.18b**).

Leaf Abscission   Angiosperm trees and shrubs of seasonally cold habitats experience water stress every winter, when evaporation from plant surfaces occurs, yet soil water is frozen and therefore unavailable for uptake by roots. Desert plants experience water stress conditions at less predictable times and for much of the year. Both types of plants are adapted to cope with water stress by dropping their leaves, a process known as **leaf abscission**. Dropping leaves lets these plants avoid very low leaf water potentials and the consequent danger of xylem embolism. Leaf abscission also reduces the amount of root mass that plants must produce to obtain water under arid conditions. The ocotillo (*Fouquieria splendens*) of North American deserts can produce leaves after sporadic rains and then drop all of its

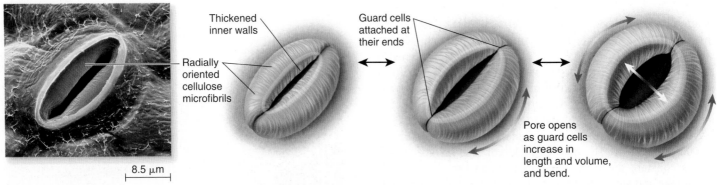

Thickened inner walls

Guard cells attached at their ends

Radially oriented cellulose microfibrils

Pore opens as guard cells increase in length and volume, and bend.

8.5 μm

(a) Stomatal guard cells

(b) The roles of radial orientation of cellulose microfibrils and thickened inner walls in opening or closing guard cells

Figure 38.17 **The structure of stomatal guard cells.** When flaccid, guard cells close stomatal pores. Turgid guard cells produce a stomatal opening. **(a)** An SEM of a stomate in a rose leaf, showing the two guard cells bordering a partly open pore. **(b)** Thickened inner cell walls and radial orientation of cellulose microfibrils in the guard cell walls explain why they separate when turgid, forming a pore.

Concept check: *How could you make a physical model that would illustrate how guard cell structure affects its function?*

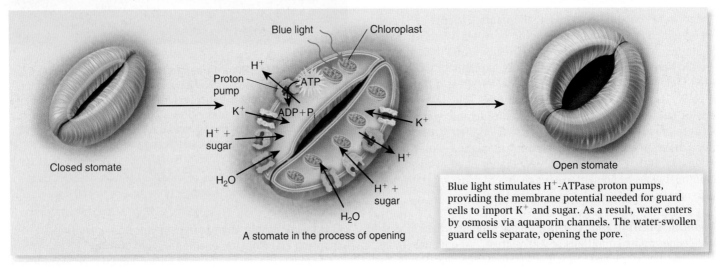

Blue light   Chloroplast

H^+
Proton pump   ATP
K^+ ADP+P_i K^+
H^+ + sugar H^+
H_2O H^+ + sugar
H_2O

Closed stomate

A stomate in the process of opening

Open stomate

Blue light stimulates H^+-ATPase proton pumps, providing the membrane potential needed for guard cells to import K^+ and sugar. As a result, water enters by osmosis via aquaporin channels. The water-swollen guard cells separate, opening the pore.

(a) The process of stomate opening

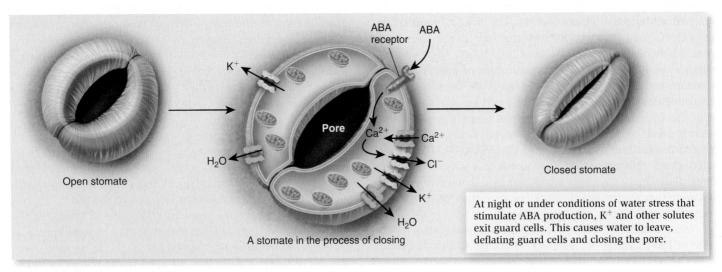

ABA receptor   ABA

K^+
Pore Ca^{2+} Ca^{2+}
H_2O Cl^-
K^+
H_2O

Open stomate

A stomate in the process of closing

Closed stomate

At night or under conditions of water stress that stimulate ABA production, K^+ and other solutes exit guard cells. This causes water to leave, deflating guard cells and closing the pore.

(b) The process of stomate closing

Figure 38.18 **How stomatal pores open and close.** **(a)** Stomata usually open in response to the blue light of sunlight. **(b)** Stomata usually close in response to lack of sunlight. They can also close during the day under conditions of water stress, which induces plants to produce more of the hormone abscisic acid (ABA). Stomatal guard cell plasma membranes possess ABA receptors, which receive the drought signal.

(a) Ocotillo with leaves   **(b) Ocotillo without leaves**

Figure 38.19 Leaf abscission as a drought adaptation. The ocotillo (*Fouquieria splendens*), a plant native to North American deserts, is known for its ability to respond to intermittent rain and drought by producing and dropping leaves multiple times within a year.

Concept check: *Why does the ocotillo not drop its leaves at a single predictable time each year, as do temperate angiosperm trees and shrubs?*

leaves as a direct response to drought as many as six times a year (**Figure 38.19**).

The sugar maple (*Acer saccharum*) is an example of the many types of temperate forest trees or shrubs that drop their leaves each autumn and are thus known as deciduous plants. Deciduous plants contrast with evergreen conifers, whose needle- or scale-shaped leaves are adapted to help these gymnosperms cope with water stress during the cold season (Chapter 30). By contrast, the broader, thinner leaves produced by many angiosperms are well adapted for efficient light-capture, but more vulnerable to the stresses caused by cold. During their evolution, temperate zone angiosperm trees and shrubs have acquired the genetic capacity to predict the onset of cold, dry winter conditions and respond with preemptive leaf abscission. In contrast to the case of ocotillo, autumn leaf drop in temperate angiosperms is not directly induced by drought.

Leaf abscission is a highly coordinated developmental process. The hormone ethylene stimulates an abscission zone to develop at the bases of leaf petioles (**Figure 38.20a**). The abscission zone contains two types of tissues: a separation layer of short, thin-walled cells and an underlying protective layer of suberin-impregnated cork cells (**Figure 38.20b**). Suberin contains both waterproofing wax and phenolic polymers that retard microbial attack. As the abscission layer develops across the vein linking the petiole with the stem, it eventually cuts off the water supply to the leaf. Chlorophyll in the leaves degrades, revealing colorful orange and yellow carotenoid and xanthophyll pigments that were hidden beneath. In addition, some plants synthesize red and reddish blue pigments in response to changing environmental conditions. The presence of these pigments explains colorful autumn vegetation in temperate zones.

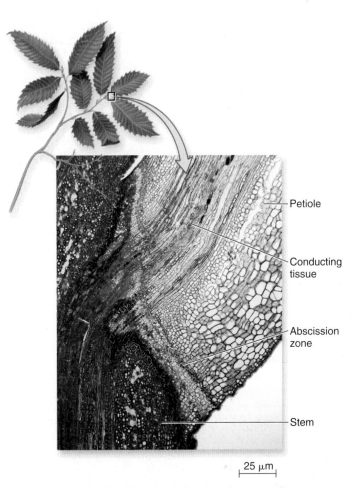

(a) LM of leaf abscission zone at the junction of a petiole and stem, stained with dyes

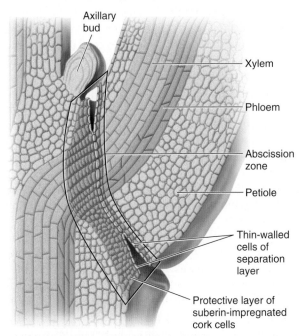

(b) Tissues involved in leaf abscission

Figure 38.20 Leaf abscission.

Enzymes eventually break down the cell-wall components of the separation layer, causing the petiole to break off the stem. The underlying protective layer forms a leaf scar that seals the wound, helping to protect the plant stem from water loss and pathogen attack.

Long-Distance Transport of Organic Molecules Occurs in the Phloem

Our examination of xylem structure and function provides the background needed to understand phloem structure and function. As we have noted, phloem plays an essential role in long-distance transport of organic molecules and some minerals in the plant body. Phloem often transports sugars from where they are produced to other sites where they are also used. Recall that primary phloem occurs in the vascular bundles of herbaceous plants and secondary phloem occurs as the inner portion of bark of woody plants. In contrast to xylem, whose transport tissues are dead and empty of cytoplasm at maturity, mature phloem tissues remain alive and retain at least some cytoplasmic components. A closer look at phloem structure and function will help to illuminate these differences.

Phloem Structure   Phloem tissues of flowering plants include supporting fibers, parenchyma cells, sieve-tube elements, and adjacent companion cells. **Sieve-tube elements** are arranged end to end to form transport pipes (**Figure 38.21**), analogous to the way that the xylem's vessel elements are aligned to form longitudinal vessels. Together, the sieve-tube elements and companion cells form a transport system whose structure explains its function.

Each sieve-tube element and companion cell pair has a common origin. They are produced by an unequal division of a single precursor cell and are therefore linked by plasmodesmata formed at cytokinesis. The smaller of the two cells develops into a companion cell, whose name reflects its life-support function, and the larger of the two cells develops into a sieve-tube element. The sieve-tube element loses its nucleus and most of its cytoplasm as an adaptation that reduces obstruction to bulk flow. Mature sieve-tube elements retain only a thin film of peripheral cytoplasm that includes some endoplasmic reticulum, plastids, and mitochondria. The end walls of developing sieve-tube elements become perforated by the action of wall-digesting enzymes that enlarge existing plasmodesmata. The perforated end walls of mature sieve-tube elements are known as **sieve plates**, and the numerous perforations are known as **sieve plate pores**. Phloem sap passes through these plates from one sieve-tube element to another.

Mature sieve-tube elements are not dead. However, because they lack a nucleus, they are dependent on their neighboring companion cell for messenger RNA (mRNA) and proteins, which are supplied via plasmodesmata. For example, when a plant's conducting system is damaged, a short-term wound response occurs that involves a protein known as **P protein** (for phloem protein). Masses of this protein accumulate along sieve plates, preventing loss of phloem sap (**Figure**

(a) **Sieve-tube elements and companion cells**

Sieve plate pore
Sieve plate
Narrow rim of cytoplasm remaining in sieve-tube element
Sieve-tube element
Companion cell

20 μm

(b) **Light micrograph of phloem stained with blue dye, showing sieve-tube elements**

Figure 38.21   Sieve-tube elements and companion cells.

38.22). This mass functions much like a clot that helps reduce blood loss from wounded animals. P protein also binds to the cell walls of pathogens, thereby helping to prevent infection at wounds. However, sieve-tube elements cannot produce P protein by themselves. Their companion cells provide either P protein mRNA or the protein itself to sieve-tube elements. In the longer term, plants deposit the carbohydrate callose at the wound site.

Phloem Loading   Companion cells also play an essential role in moving sugars into sieve-tube elements for long-distance transport, a process known as **phloem loading**. Although glucose and some other sugars can occur in phloem, the disaccharide sucrose is the main form in which most plants transport

Sieve plate
P protein
Companion cell

5 μm

Figure 38.22   **Phloem wound response.**   When phloem is damaged, the cytoplasm of a sieve-tube element surges toward the sieve plate, depositing P protein, stained red in this light micrograph. In this location, P protein helps to prevent infection and leakage of solutes.

sugar over long distances. Plant biologists think that sucrose is less vulnerable to metabolic breakdown en route than are monosaccharides.

Two types of phloem loading occur, symplastic and partly apoplastic. Many woody plants transport sucrose from sugar-producing cells of the leaf mesophyll to companion cells and then to sieve-tube elements via plasmodesmata, a process known as symplastic phloem loading (**Figure 38.23a**). The advantage of symplastic loading is that it does not require ATP; by moving through plasmodesmata, sugar does not have to cross plasma membranes.

In contrast, most herbaceous plants, including important crop plants and the model plant *Arabidopsis*, load sugar into sieve-tube elements or companion cells from intercellular spaces, often against a concentration gradient. ATP must be used to move the sugar across a plasma membrane into a companion cell or sieve-tube element (**Figure 38.23b**). Therefore, this second type of phloem loading is partly apoplastic and partly a transmembrane process.

The Pressure-Flow Hypothesis Explains Transport in Phloem Tissues

We have learned that transpiration, driven by the sun's energy, moves water in plant xylem by the processes of cohesion and tension. Once sugar has been loaded into phloem sieve-tube elements, how does it move within the plant? The most common explanation is that phloem transport is driven by differences in turgor pressure that occur between cells of a **sugar source**, where sugar is produced, and those of a **sugar sink**, where sugar is consumed. Photosynthetic leaf mesophyll is the main sugar source. Roots and developing leaves, seeds, and fruits are examples of sugar sinks. In the process known as **translocation**, phloem transports substances from source to sink. The direction of phloem movement can be horizontal as well as vertical, depending on the relative positions of the sources and sinks.

Because sieve-tube elements near source tissues have higher solute contents than surrounding tissues, water tends to rush into them from nearby xylem, thereby building turgor pressure. In contrast, sieve-tube elements near sink tissues have lower solute concentration. The resulting water pressure difference drives the bulk flow of phloem sap from source to sink tissues. This explanation for phloem transport is known as the **pressure-flow hypothesis** (**Figure 38.24**). German plant physiologist Ernst Münch first proposed this hypothesis in 1927, though later investigators have modified it. The pressure-flow hypothesis is supported by studies that show relatively high solute concentrations in phloem samples taken from tree leaves (source) and relatively low solute concentrations in samples taken from a tree trunk (sink). The production of turgor pressure requires an intact plasma membrane. This explains why mature phloem sieve-tube elements must be alive in order to function, in contrast to xylem tracheary elements.

At sink tissues, sugar is typically unloaded through plasmodesmata (Figure 38.24). Because plasmodesmata are very narrow in comparison to sieve-tube elements, they slow the flow of phloem sap from sieve-tube elements into sink tissues. This reduction in flow rate helps to equalize the distribution of phloem sap, preventing delivery of too much sap to any single sink. When the solute concentration of phloem sap has been sufficiently reduced, water flows from the phloem back into the xylem, where upward transport occurs. The reliance of phloem bulk flow on water supplied by the xylem explains the close proximity of phloem and xylem tissues in vascular bundles and woody stems.

(a) Symplastic phloem loading

(b) Partly apoplastic phloem loading

Figure 38.23   Symplastic and partly apoplastic phloem loading.

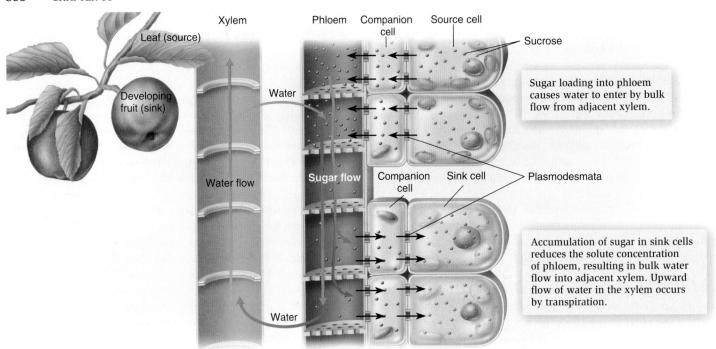

Figure 38.24   Pressure-flow hypothesis for phloem transport.

Genomes & Proteomes Connection

Microarray Studies of Gene Transcription Reveal Xylem- and Phloem-Specific Genes

Xylem and phloem tissues have important economic, agricultural, and ecological roles. For example, secondary xylem produced by the vascular cambium, otherwise known as wood, forms the basis of our forest products industry. The ability of plants to respond to environmental factors also relies on vascular tissues to transport hormones. Seeking increased fundamental understanding and enhanced ability to engineer plants, molecular biologists have tried to identify the genes involved in xylem and phloem development. The use of microarrays, also known as gene chips, is providing a genomewide view of gene expression related to xylem and phloem.

Chengsong Zhao, Eric Beers, and their coworkers recently compared gene expression in immature xylem (X), phloem plus vascular cambium (PC), and nonvascular (NV) tissues dissected from young *Arabidopsis* plants. Recall that the vascular cambium is the meristematic tissue that produces secondary phloem (inner bark) and secondary xylem (wood; see Chapter 35). The investigators used a dissecting microscope to view 1-cm-long pieces of tissue and a razor blade to make a longitudinal cut into nonvascular tissues and phloem, isolating the immature xylem. Then they peeled the nonvascular tissue away from the phloem. They ensured that their tissue samples were pure and authentic by checking for the expression of marker genes already known to be tissue specific. The investigators ascertained that their X and NV tissue samples had very low expression of certain PC genes and that their PC samples had very low expression of certain X and NV genes. They then extracted RNA from the three separate tissues and used it to make labeled nucleic acid

for hybridization to gene microarrays containing about 90% of the *Arabidopsis* genome (refer back to Figure 20.9).

Binding of labeled nucleic acid revealed which genes were expressed. By comparing the binding patterns on different chips hybridized with nucleic acid from distinct tissues, the researchers were able to determine the genes that were preferentially expressed in X, PC, and/or NV tissues. These differences are revealed by a "triangle plot" that reflects the proportions of genes expressed in the three tested tissues (**Figure 38.25**). Genes expressed primarily in only one tissue type cluster at the trian-

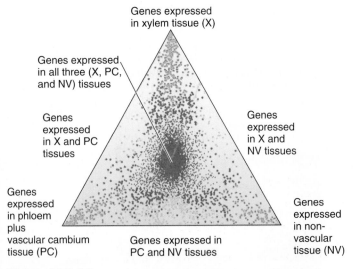

Figure 38.25   Triangle plot showing gene expression related to immature xylem, phloem, and nonvascular tissue in *Arabidopsis*.

Concept check: Suppose you are interested in identifying proteins that are involved in the development of both xylem and phloem. How might results illustrated here help you?

gle corners, and genes expressed in two tissues but not the third are plotted at the triangle edges. Genes expressed in all three tissues lie in the center. The researchers identified 319 X-biased genes, 211 PC-biased genes, and 154 NV-biased genes. Not surprisingly, 17% of the X-expressed genes were related to cell-wall biosynthesis, including lignin production. The investigators then confirmed, via microscopy, that several proteins predicted by microarray data to be located in X, PC, or NV tissues were actually made in these tissues. This study is an essential first step toward determining the function, timing of expression, and cellular locations of xylem- and phloem-specific proteins.

Summary of Key Concepts

38.1 Overview of Plant Transport

- The root system takes up water and minerals, and the shoot system absorbs carbon dioxide from the air. Photosynthetic cells use these materials to produce organic compounds. The vascular system xylem transports water and minerals from root to shoot, while the phloem transports organic compounds to roots and other nonphotosynthetic tissues. (Figure 38.1)

38.2 Uptake and Movement of Materials at the Cellular Level

- Passive transport of substances along a concentration gradient does not require ATP. Simple diffusion of water, gases, and other small, uncharged molecules across plasma membranes may occur in the absence of transport proteins. Facilitated diffusion involves plasma membrane and vacuolar protein channels and transporters. (Figure 38.2)

- Transport of materials across plasma membranes against concentration gradients is known as active transport and usually requires ATP. Plasma membrane proton pumps use the energy released by ATP hydrolysis to move protons from the cytosol into the intercellular space, thereby generating a membrane potential. Potential energy released by the flow of protons back into the cell can be coupled to the transport of ions and organic compounds into and out of cells, a process known as cotransport. (Figure 38.3)

- Turgor pressure arises when the cell wall restricts the extent to which plant cells can swell when water enters as the result of osmosis. A cell that is so full of water that the plasma membrane presses closely against the cell wall is turgid. By contrast, a cell that is not swollen with water is flaccid, and a cell that contains so little water that the plasma membrane pulls away from the cell wall is plasmolyzed. (Figure 38.4)

- Solute potential and pressure potential arising from the presence of a cell wall are major factors affecting cellular water potential. The relative water content (RWC) is a measure of relative turgidity that integrates the water potential of all cells in a plant or an organ. (Figure 38.5)

- Plants display a variety of adaptations that help them cope with osmotic stress. In osmotic adjustment, high solute concentrations accumulate in the cytosol, a process that helps cells retain their water. Halophytes are plants adapted to very saline conditions, which impose extreme osmotic stress.

38.3 Tissue-Level Transport

- Transmembrane transport involves the movement of materials from one cell to another from intercellular spaces, across plasma membranes, and into the cytosol. The symplast is the continuum of all plant cytosolic compartments linked by plasmodesmata. Symplastic transport allows materials to move from one cell to another without crossing plasma membranes. In apoplastic transport, water and solutes move through the apoplast, the water-filled cell walls and intercellular spaces of tissues. (Figures 38.6, 38.7)

- In roots, waxy Casparian strips on cell walls of endodermal tissue act as filters that reduce the movement of toxic ions and concentrate useful minerals into the plant vascular system. (Figure 38.8)

38.4 Long-Distance Transport

- Water and solutes move for long distances by bulk flow within the xylem and phloem. Plant vascular tissues are adapted in ways that reduce resistance to bulk flow. Bulk flow of water upward in xylem is powered by the water pressure difference between moist soil and drier air, the latter resulting from solar heating. (Figure 38.9)

- Xylem is the main conduit for water and dissolved mineral nutrients, but it may also transport certain organic compounds. At maturity, tracheids and vessel elements, together known as tracheary elements, are dead and empty of cytoplasm, and have lignified cell walls. Pits in tracheary element walls allow water entry and exit. Experiments reveal that pits narrow or constrict in response to xylem sap solute content. Narrow tracheids fit together in long cell files. Vessel elements are wider but are more vulnerable than tracheids to blockage by air bubbles or embolisms. Root pressure, the effects of which include the morning water drops on leaf tips known as guttation, helps some plants to refill embolized vessels. (Figures 38.10, 38.11, 38.12, 38.13, 38.14)

- Transpiration is the evaporative loss of water from plant surfaces into sun-heated air. Plant transpiration has extensive local, regional, and global effects on climate. The cohesion-tension theory explains long-distance water transport as the combined effect of the cohesive forces of water and evaporative tension. (Figures 38.15, 38.16)

- Expansion of guard cells causes stomatal pores to open, allowing CO_2 intake. Guard cell deflation causes pores to close, an adaptation that helps to prevent excess water loss. Plants under existing or predicted water stress often drop their leaves in a process known as abscission, an adaptive response that lets plants avoid very low water potentials and the consequent threat of embolism. (Figures 38.17, 38.18, 38.19, 38.20)

- Organic solutes and minerals are transported in phloem sap as the result of osmosis. Phloem sap moves within sieve-tube elements that are living when mature, but lack a nucleus and are thus dependent on companion cells. Phloem loading occurs by symplastic or partly apoplastic transport. The pressure-flow hypothesis explains sugar translocation as a process driven by differences in turgor pressure that occur between a sugar source (for example, leaves) and a sugar sink (for example, developing fruit). (Figures 38.21, 38.22, 38.23, 38.24, 38.25)

Assess and Discuss

Test Yourself

1. An aquaporin is
   a. a channel protein that allows the influx of K$^+$ into cells, causing water to also flow in between the phospholipids of a plasma membrane.
   b. a type of blue-colored pore in the epidermal surfaces of plants.
   c. a protein channel in plasma membranes that facilitates the diffusion of water.
   d. a protein transporter in plasma membranes that uses protons to cotransport water.
   e. none of the above.

2. Why is turgor pressure a property of plant cells?
   a. Plant cells possess the necessary chloroplasts.
   b. Plant cells possess a cell wall, necessary for formation of turgor.
   c. Plant cells possess mitochondria, which provide the ATP needed for turgor.
   d. all of the above
   e. none of the above

3. How might plant cells avoid losing too much water in very cold, dry, or saline habitats?
   a. They may balance the osmotic condition of their cytosol with that of the environment.
   b. Their epidermal cells may be coated with waxy cuticle.
   c. They may stabilize their membranes with sugars.
   d. They may produce more aquaporin water channels to take maximum advantage of available moisture.
   e. All of the above are possible.

4. What are ways in which plants accomplish tissue-level transport?
   a. transmembrane transport of solutes from one cell to another
   b. symplastic transport of materials from one cell to another via plasmodesmata
   c. apoplastic transport of water and dissolved solutes through cell walls and intercellular spaces
   d. all of the above
   e. none of the above

5. A root endodermis is
   a. an innermost layer of cortex cells that each display characteristic Casparian strips.
   b. a layer of cells just inside the epidermis of a root.
   c. a layer of cells just outside the epidermis of a root.
   d. a group of specialized cells that occur within the root epidermis.
   e. none of the above.

6. Xylem loading is
   a. the process by which water from the air enters vascular tissues of the leaf.
   b. the process by which sugar is transported via plasmodesmata directly into vessel elements.
   c. the process by which sugar is transported directly into sieve-tube elements.
   d. the process in which ions are transported across the membranes of root xylem parenchyma into the xylem apoplast, followed by water.
   e. none of the above.

7. What features of water explain how it can be drawn up a tall tree from roots to leaves?
   a. cohesion, the result of extensive hydrogen bonding
   b. adhesion, water's tendency to stick to surfaces such as the inner walls of tracheid and vessels
   c. high surface tension that develops when water evaporates from intercellular leaf spaces
   d. all of the above
   e. none of the above

8. What feature of vascular plants contributes to their ability to maintain relatively stable internal water content?
   a. a waxy surface cuticle
   b. an extensive root system that mines water from soil
   c. specialized water-conducting tracheary elements composed of dead cells
   d. epidermal pores that open and close
   e. all of the above

9. What structural features of stomatal guard cells foster their ability to form an open pore in plant epidermal surfaces?
   a. thickened inner cell walls and radially oriented microfibrils
   b. thickened outer cell walls and radially oriented microfibrils
   c. thickened inner cell walls and longitudinal microfibrils
   d. thickened outer cell walls and longitudinal microfibrils
   e. uniform thickness of cell walls and randomly arranged microfibrils

10. What substances plug wounded sieve-tube elements, thereby preventing the leakage of phloem sap?
    a. X protein and callose
    b. C protein and callose
    c. P protein and callose
    d. P protein and sucrose
    e. none of the previously listed choices

Conceptual Questions

1. Why is it a bad idea to overfertilize your houseplants? If the amount recommended on the package is good, wouldn't more be better?

2. Why is it a bad idea for subsistence farmers (those barely able to grow enough crops to feed themselves) to allow livestock to graze natural vegetation to the point that it disappears?

3. In the desert southwestern U.S., the ocotillo plant is often used as a landscaping plant, thanks to its interesting shape and beautiful floral displays. Imagine that you are responsible for property landscaped with ocotillo but find the plants bare of leaves. Can you assume that the plants are dead and need to be replaced?

Collaborative Questions

1. Imagine that you are part of a team assigned to determine what environmental conditions best suit a new crop so that the crop can be recommended to farmers in appropriate climate regions. What features of the crop plants might you investigate?

2. Take a look outside or imagine a forest or grassland. What can you deduce about the availability of soil water from the types of plants that occur?

Online Resource

www.brookerbiology.com

Stay a step ahead in your studies with animations that bring concepts to life and practice tests to assess your understanding. Your instructor may also recommend the interactive ebook, individualized learning tools, and more.

Flowering Plants: Reproduction

39

The reproductive success of dandelions.

Dandelions sometimes seem to be taking over the world, growing abundantly in open, sunny areas. Cheerful, bright yellow flower heads, as shown in the chapter-opening photo, are one of the secrets of dandelions' success. If you pull a dandelion flower head apart, you can see that it is actually a bouquet of 200 or so small flowers. Each flower produces a tiny, one-seeded fruit equipped with a "parachute" for effective long-distance dispersal by wind. Each dandelion plant can produce up to 5,000 fruits during its lifetime, which explains how dandelions can spread so rapidly across a landscape.

Though most flowering plants produce seeds by means of sexual reproduction, dandelions and some other plants are able to produce seeds by a type of asexual reproduction, a process that does not involve meiosis and fusion of gametes. As a result, the traits of asexually reproducing dandelion parents and their progeny are uniform. Asexual reproduction could be very usefully applied in agriculture, because most crops reproduce only sexually. Each year many U.S. farmers buy and plant hybrid seeds that develop into mature plants having uniform and desirable trait combinations. Such farmers typically do not use seed from one year's crop to plant the next, because sexual reproduction mixes genes into diverse combinations present among the resulting seeds. For this reason, plants that grow from sexually produced seeds do not uniformly express the desirable trait combinations present in their hybrid parents. But if hybrid crop plants could be engineered to produce seed asexually, as dandelions do, farmers might be able to use seed from one crop to plant the next and still harvest uniformly desirable crops. Because reproduction is so important to agriculture, plant biologists aim to better understand both sexual and asexual reproduction in flowering plants.

Flowers, fruits, and seeds are not just important in agriculture but are essential reproductive features of the diverse types of flowering plants found in nature. This chapter begins with an overview of the reproductive cycle of flowering plants that describes how flowers, fruits, and seeds function. This overview provides essential background for a closer look at flower structure and development and some of the genes that control flower production and appearance. We will also examine the sexual reproductive processes by which plants produce gametes and accomplish fertilization, thereby producing zygotes and embryos. We will explore how seeds, fruits, and seedlings develop. Finally, we will take a closer look at the ways in which dandelions and some other plants reproduce without using the sexual process.

39.1 An Overview of Flowering Plant Reproduction

Most flowering plants display **sexual reproduction**, the process by which two gametes fuse to produce offspring that have unique combinations of genes. Flowering plants, also known as angiosperms, inherited their sexual life cycle, known as alternation of generations, from ancestors extending back to the earliest land plants (as we explored in Chapters 29 and 30). Though all plants share the same basic life cycle, flowering plants display unique reproductive features. In this section, we first review the general features of alternation of generations and then consider more specific features of the angiosperm life cycle.

Flowering Plants Display Alternation of Generations

All groups of land plants produce two multicellular life cycle stages, in essence, two distinct plants. These two life cycle stages

Figure 39.1 Alternation of generations, the plant life cycle.

(a) Gametophyte-dominant bryophyte (moss)

(b) Sporophyte-dominant flowering plant (oak)

Figure 39.2 Evolutionary shift in plant life cycle stage dominance. (a) In mosses, the gametophyte is the dominant life cycle stage, and the sporophyte is dependent on the gametophyte for resources. (b) In flowering plants such as oak trees, the sporophyte life cycle stage is dominant. Microscopic flowering plant gametophytes develop and grow within sporophytic flower tissues and are completely dependent on sporophytes.

Concept check: What advantages do flowering plants obtain by having such small and dependent gametophytes?

are the diploid, spore-producing **sporophyte** and the haploid, gamete-producing **gametophyte**. In all groups of plants, haploid spores are typically produced by diploid sporophytes as the result of meiosis. These spores undergo mitotic cell divisions to produce multicellular gametophytes. Certain cells within the gametophytes differentiate into gametes. Thus, meiosis does not directly generate the gametes of plants. The processes of meiosis and fertilization form the transitions between the sporophyte and gametophyte life stages and link them in a cycle (**Figure 39.1**). The land plant life cycle is known as **alternation of generations** because it involves the cycling between distinct sporophyte and gametophyte generations.

During the evolutionary diversification of land plants, the sporophyte generation has become larger and more complex, while the gametophyte generation has become smaller and less complex. To illustrate this, let's compare the life cycle stages of mosses to those of angiosperms. Mosses diverged early in the history of land plants, whereas angiosperms appeared much later. During the intervening time, the relative sizes and dependence of the sporophyte and gametophyte generations changed dramatically. Moss sporophytes are small structures that always grow attached to larger, photosynthetic gametophytes, because moss sporophytes are incapable of independent life (**Figure 39.2a**). Moss sporophytes are dependent on gametophytes, because the gametophytes supply the sporophytes with essential nutrients.

In contrast, flowering plant sporophytes are notably larger and more complex than gametophytes. A tall oak tree, for example, is a single sporophyte. However, oak gametophytes are few-celled, microscopic structures that develop and grow within flowers (**Figure 39.2b**). In addition, photosynthetic oak seedlings and trees grow independently, but nonphotosynthetic oak gametophytes are completely dependent on the sporophyte generation for their nutrition. A closer look at flower structure

will help us to gain a more complete view of angiosperm gametophyte structure and function.

Flowers Produce and Nurture Male and Female Gametophytes

The literary wit Gertrude Stein famously paraphrased Shakespeare when she wrote, "A rose is a rose is a rose." Everyone knows that a rose is a flower, but what, exactly, is a flower? A **flower** is defined as a reproductive shoot, a stem branch that produces reproductive organs instead of leaves. Flower organs are thought to have evolved from leaflike structures by descent with modification (refer back to Figure 30.15). Flower organs are produced by shoot apical meristems much like those that generate leaves.

A flower shoot generally produces four types of organs: sepals, petals, stamens, and carpels (**Figure 39.3**). **Sepals** often function to protect the unopened flower bud. **Petals** usually serve to attract insects or other animals for pollen transport (refer back to Figures 30.19 and 30.20). **Stamens** and **carpels**

(a) **Flower parts**

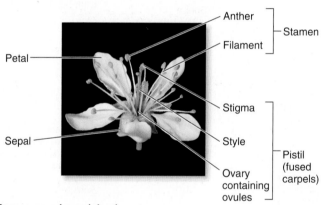

(b) *Prunus americana* (plum)

Figure 39.3   The structure of a typical flower.

Concept check: *Do all flowers have all of the structures illustrated here?*

each produce distinctive types of spores by the process of meiosis. From these spores, multicellular gametophytes develop, and certain gametophytic cells become specialized gametes. Nutritive tissues of stamens and carpels channel organic food, mineral ions, and water from the adult sporophyte into dependent gametophytes.

Stamens   Stamens produce **male gametophytes** and foster their early development. Most stamens display an elongate stalk, known as a **filament**, which is topped by an anther (Figure 39.3). Filaments contain vascular tissue that delivers nutrients from the parental sporophyte to the anthers. Each **anther** is a group of four sporangia, structures in which spores are produced. Within the anther's sporangia, many diploid cells undergo meiosis, each producing four tiny, haploid spores. Because they are so small, generally 25–50 μm in diameter, the spores produced within anthers are known as **microspores**.

Immature male gametophytes, known as **pollen grains**, develop from microspores. The term pollen comes from a Latin word meaning "fine flour," reflecting the small size of pollen grains. Pollen grains are eventually dispersed through pores or slits in the anthers. At the time of dispersal, the pollen grain is a two- or three-celled immature male gametophyte produced by mitotic division. During a later phase of development, a mature male gametophyte produces **sperm cells**.

Carpels   Carpels are vase-shaped structures that produce, enclose, and nurture **female gametophytes**. Carpels contain veins of vascular tissue that deliver nutrients from the parent sporophyte to the developing gametophytes. The term **pistil** (named for its resemblance to the pestle used to grind materials to a powder) refers to a single carpel or several fused carpels (Figure 39.3). The topmost portion of a pistil, known as a **stigma** (Greek, meaning mark), receives pollen grains. The **style** is the middle portion of the pistil, and an ovary is at the bottom of the pistil. As described later in Section 39.3, the **ovary** produces and nourishes one or more ovules. An **ovule** consists of a spore-producing structure (a sporangium) and enclosing tissues consisting of modified leaves known as **integument**. Within an ovule, a diploid cell produces four haploid **megaspores** by meiosis, three of which die. The surviving megaspore generates a female gametophyte by mitosis. The female gametophytes of flowering plants typically consist of seven cells, one of which is the female gamete, the **egg cell**. This basic information about male and female gametophytes will help us to understand how they function to produce a young sporophyte within a seed.

Fertilization Triggers the Development of Embryonic Sporophytes, Seeds, and Fruits

In flowering plants, fertilization leads to the production of a young sporophyte that lies within a seed, completing the life cycle (**Figure 39.4**). Prior to fertilization, pollen grains released from anthers first find their way to the stigma of a compatible flower, a process known as **pollination**. Some plants display **self-pollination**, in which pollen from the anthers of a flower is transferred to the stigma of the same flower or between flowers of the same plant. **Cross-pollination**, which occurs when a stigma receives pollen from a different plant of the same species, is also common. Many flowers are attractive to insects or other animals that transport pollen, but oak flowers and those of some other angiosperms are adapted for pollen transport by wind. A few plants move pollen by means of water currents. The diverse ways that flowers are adapted for pollination are described and illustrated in Chapter 30.

Development of a Mature Male Gametophyte   When pollen grains land on the stigma, the stigma functions as a gatekeeper, allowing only pollen of appropriate genotype to germinate. During germination, a pollen grain produces a long, thin **pollen tube** that contains two sperm cells. The pollen tube grows through the style toward the ovary. Upon reaching the ovules, the pollen tube grows through the **micropyle**, an opening in the

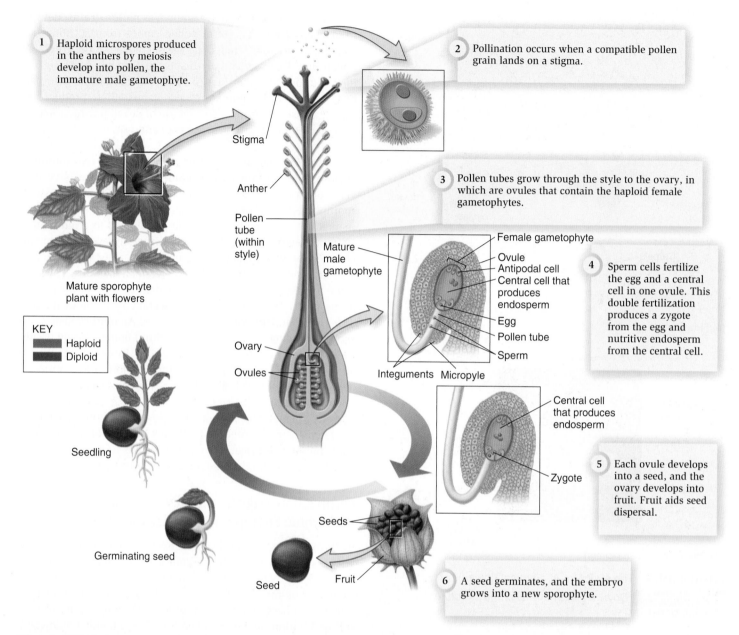

1 Haploid microspores produced in the anthers by meiosis develop into pollen, the immature male gametophyte.

2 Pollination occurs when a compatible pollen grain lands on a stigma.

3 Pollen tubes grow through the style to the ovary, in which are ovules that contain the haploid female gametophytes.

Stigma

Anther

Pollen tube (within style)

Mature sporophyte plant with flowers

KEY
Haploid
Diploid

Mature male gametophyte

Female gametophyte
Ovule
Antipodal cell
Central cell that produces endosperm
Egg
Pollen tube
Sperm

Integuments   Micropyle

4 Sperm cells fertilize the egg and a central cell in one ovule. This double fertilization produces a zygote from the egg and nutritive endosperm from the central cell.

Ovary

Ovules

Central cell that produces endosperm

Zygote

5 Each ovule develops into a seed, and the ovary develops into fruit. Fruit aids seed dispersal.

Seedling

Germinating seed

Seeds

Fruit

Seed

6 A seed germinates, and the embryo grows into a new sporophyte.

Figure 39.4   **The life cycle of a flowering plant.**   The plant reproductive cycle is illustrated here by hibiscus.

Concept check: *What advantage does the hibiscus flower gain by clustering its stamens around the pistil?*

ovule, and delivers sperm to the female gametophyte (Figure 39.4). These sperm unite with haploid cells of the female gametophyte in the process of **fertilization**. Note that pollination and fertilization are distinct processes in flowering plants.

Double Fertilization   Angiosperms display an amazing phenomenon known as **double fertilization**. In this process, two different fertilization events occur. One of the two sperm cells delivered by a pollen tube fertilizes the egg cell, thereby forming a diploid **zygote**. This zygote may develop by mitotic division into a young sporophyte, known as an **embryo**. Fertilization thus begins a new cycle of alternation between sporophyte and gametophyte generations. The other sperm delivered

by the same pollen tube fuses with the two nuclei of the central cell of the female gametophyte. The cell formed by this second fertilization undergoes mitosis, eventually producing a nutritive tissue known as the **endosperm**. The embryo and the endosperm are essential parts of maturing seeds. Fertilization not only starts the development of zygotes into embryos but also triggers the transformation of ovules into seeds and ovaries into fruits. Embryo, seed, and fruit development occur at the same time.

Embryos and Seeds   An embryo is a young, multicellular, diploid sporophyte that develops from a single-celled zygote by mitosis. Because they are not yet capable of photosynthesis,

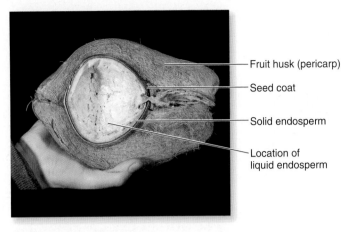

(a) Coconut fruit and seed

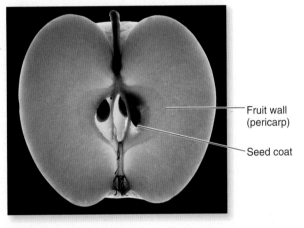

(b) Apple fruit and seed

Figure 39.5   **The structure of fruits as adaptations for seed dispersal.**   (a) The coconut fruit's outer wall, called a husk, allows the fruit to float and thus disperse its seed among tropical shores. (b) The apple is a juicy, sweet fruit that attracts animals to consume it, thereby helping to disperse its seed.

embryos depend on organic food and other materials supplied by sporophytes. Therefore, embryo development occurs within developing seeds located in a flower ovary. Seeds develop from fertilized ovules. Each developing seed contains an embryo and nutritive endosperm tissue, enclosed and protected by a **seed coat** that develops from the ovule integuments. When embryos and the seed coat have fully matured, they undergo drying, and the seed enters a phase of metabolic slowdown known as **dormancy**. Fully mature, dormant seeds are ready to be dispersed.

Fruit and Seed Dispersal   A **fruit** is a structure that encloses and helps to disperse seeds (Figure 39.4). Seed dispersal benefits plants by reducing competition for resources among seedlings and parental plants, and it allows plants to colonize new sites.

Fruits develop from the flower's ovary and sometimes include other flower parts. Young fruits bearing immature seeds are typically small and green. During the time that embryos and seeds are developing, the fruit also matures. The ovary wall changes into a fruit wall known as a **pericarp** (from the Greek, meaning surrounding the fruit). Mature fruits vary greatly

among plant species in size, shape, color, and water content. These variations represent adaptations for seed dispersal in different ways. For example, single-seeded dandelion fruits are dry and lightweight and bear a fluffy "parachute" derived from the flower's sepals (see the chapter-opening photo). These features foster dispersal by wind. In contrast, coconut fruits feature an airy husk (the pericarp) that keeps them afloat in ocean currents that carry coconuts from one tropical shore to another. Inside the coconut fruit is a single, large seed loaded with liquid and solid endosperm, which people use as coconut milk and coconut meat, respectively (**Figure 39.5a**). These large amounts of endosperm provide nutrients that sustain coconut seedling growth on infertile, sandy shores. Fruit variation is also extremely important to wild animals and in human agriculture. For example, most fruit crops are juicy and sweet, with relatively small seeds (**Figure 39.5b**). In nature, these features foster dispersal by birds and other animals that feed on such fruits.

Seed Germination and Seedlings   If a dispersed seed encounters favorable conditions, including sufficient sunlight and water, it will undergo seed **germination**. During seed germination, the embryo absorbs water, becomes metabolically active, and grows out of the seed coat, producing a seedling. If the seedling obtains sufficient nutrients from the environment, it grows into a mature sporophyte capable of producing flowers. In the next section, we focus on flower production, structure, and development.

39.2   Flower Production, Structure, and Development

Flowers are essential sources of food for many animals that help to disperse pollen. As the result of coevolutionary relationships with such animals, flowers occur in a spectacular array of colors and forms that attract particular pollinators (refer back to Table 30.1). Flowers also attract humans because we possess sensory systems much like those of animal pollinators. We give bouquets to show love and appreciation; decorate homes, workplaces, and objects with flowers; display flower arrangements on ceremonial occasions; and make perfume from flowers. Consequently, many types of flowers are grown for the florist and perfume industries. Humans also associate flowers with the eventual presence of fruits, which are a source of our food. Flowers are necessary for the production of grain and other crops. Flower development is controlled by both environmental signals and changes in gene expression.

Environmental Signals Interact with Genes to Control Flower Production

Flowering time is controlled by the integration of environmental information such as temperature and day length (photoperiod) with hormonal influences (see Chapter 36). These stimuli are perceived and integrated by leaves, which then signal shoot meristems to produce flowers. Experiments performed in the 1970s revealed that leaves transmit a flowering signal to the shoot apical meristem via phloem. Based on these results, plant scientists hypothesized the existence of a flowering hormone, which they named **florigen**, meaning "flower generator." However, the chemical nature of florigen remained unknown until recently. In 2007, several groups of plant researchers demonstrated that a protein known as FT (flowering time), produced in leaves, is transported to the shoot tip. In this location, FT protein binds to a transcription factor called FD, and together these proteins stimulate another protein known as LEAFY, which activates many floral-identity genes. As the result, a leaf-producing apical meristem transforms into a reproductive meristem. FT protein is probably the long-sought florigen molecule.

Developmental Genes Control Flower Structure

Organ identity genes specify the four basic flower organs—sepals, petals, stamens, and carpels. Other genes determine flower shape, color, odor, or grouping into bunches known as inflorescences.

The Genetic Basis of Flower Organ Identity Sepals, petals, stamens, and carpels occur in four concentric rings known as **whorls**. Sepals (collectively known as the **calyx**) form the outermost whorl, and petals (together known as the **corolla**) form an adjacent whorl type. Stamens (the **androecium**) create a third type of whorl, and carpels (the **gynoecium**) form the innermost whorl (**Figure 39.6**). The **perianth** consists of the calyx plus the corolla.

You may recall that *A*, *B*, *C*, and *E* genes encode transcription factors that control the production and arrangement of these whorls (refer back to Figure 19.24). Class *A* genes control the development of sepals and petals, class *B* genes specify petal and stamen formation, and class *C* genes determine the identities of the stamen and carpels. *E* gene expression distinguishes

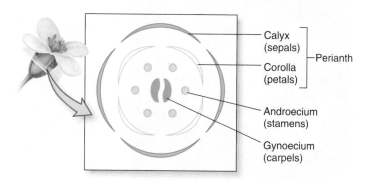

Figure 39.6 The occurrence of flower parts in whorls.

the sepals from the other whorls. However, not all flowers produce all four types of organs.

Variation in Number of Whorls Flowers that possess all four types of flower whorls—calyx, corolla, androecium, and gynoecium—are known as **complete flowers**. In contrast, flowers that lack one or more flower whorls are described as **incomplete flowers**. Flowers having both stamens and carpels are said to be **perfect**, whereas flowers lacking stamens or carpels are described as **imperfect**. An imperfect flower that produces only carpels is known as a carpellate flower (or pistillate flower). Imperfect flowers that produce only stamens are described as staminate flowers.

Corn is a plant that produces both imperfect staminate and carpellate flowers (**Figure 39.7**). The flowers of corn start to develop as perfect flowers, but in carpellate flowers, the stamens stop developing. After pollination and fertilization, each carpellate flower produces one of the kernels on a cob of corn. In contrast, staminate flowers of corn, which are found in corn tassels, produce the pollen. Corn is termed **monoecious** (meaning "one house") because it produces staminate and carpellate

Figure 39.7 Imperfect flowers of corn, a monoecious plant. **(a)** Staminate flowers lack carpels, and **(b)** carpellate flowers lack stamens, but both types of flowers occur on a single corn plant. In contrast, dioecious plants produce staminate and carpellate flowers on separate plants.

Concept check: What inference can you make from the observation that corn flowers lack showy petals?

(a) Staminate flowers of *Zea mays* (corn)

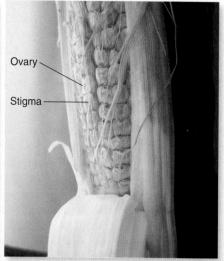

(b) Carpellate flowers of *Zea mays* (corn)

flowers on the same plant. Holly and willow also produce staminate and carpellate flowers, though on separate plants, and are thus described as **dioecious** (meaning "two houses").

Variation in Flower Organ Number   In addition to variation in whorls, flowers vary in number of organs. You may recall that flowering plants occur in two major groups differing in flower structure and other features: the eudicots and the monocots (see Chapter 35). Eudicot flower organs often occur in fours or fives or a multiple of these numbers. By contrast, monocot flower organs often occur in threes or a multiple of three. Many plants sold for use in gardens have been bred so that the flowers produce multiple organs. Garden roses, for example, typically have many more whorls of petals than do wild roses. This change results from a mutation that causes organs that would have become stamens to instead develop into additional petals. Other flowers possess relatively few organs. For example, the miniscule flowers of the tiny aquatic flowering plant *Lemna gibba* produce no perianths and only two stamens.

Variation in Flower Color   Different flower parts may vary in color. The calyx and corolla of monocot flowers such as tulip are often similar in appearance and attractive function. In contrast, eudicot flowers tend to have green, leaflike sepals that are quite distinct from petals, which are often colorful and fragrant. For example, the sepals of petunia (*Petunia hybrida*) resemble leaves in having numerous stomata on both epidermal surfaces, and internal parenchyma tissue rich in chloroplasts. Petunia petal epidermis lacks stomata, and the parenchyma tissue usually lacks chlorophyll. Instead, the cell vacuoles of petals are often colored by **flavonoids**, a type of secondary metabolite (see Chapter 30). Among the pigments responsible for pink, red, blue, violet, purple, or yellow colors of petals and other flower parts, flavonoids are the most important. Color variations arise from differences in gene action that influence pigment biosynthesis pathways. Petunia has become an important model system for understanding the evolution of genes and proteins that control petal color.

Interestingly, research involving flower color has led to important discoveries in genetics. In 1990, plant molecular biologist Rich Jorgensen and colleagues reported one of the first cases of gene silencing in plants. You may recall that gene silencing is one of the ways in which gene expression is controlled (see Chapter 13). In an attempt to produce flowers of deeper color, the researchers introduced an extra copy of a pigment-producing gene into petunia plants. To their surprise, the extra copy of the gene sometimes produced flowers whose petals had white patches or were completely white (**Figure 39.8**). Adding the extra gene caused the plants to produce siRNA that not only silenced the expression of the extra gene, but also the natural pigment-producing genes, causing white patches on the petals.

Variation in Flower Fragrance   The fragrances of flowers result from secondary metabolites that diffuse into the air from petals and other flower organs. One example is the terpene known as geraniol, which is alluded to in Shakespeare's famous phrase, "A rose by any other name would smell as sweet." Diverse fragrances function to attract particular types of animal pollinators. Because humans possess sensory systems similar to those of many pollinators, we are also attracted to many of these fragrances. Genetic variation in the synthesis of different types of secondary metabolites is responsible for the many types of flower scents.

Flower Shape Variation Resulting from Organ Fusion
During their development, many flowers undergo genetically controlled fusion between whorls or fusion of the organs within a whorl. Stamen filaments often partially fuse with the carpel or form a tube surrounding the pistil, a feature displayed by hibiscus flowers (see Figure 39.4). Each small dandelion flower has five petals that are fused at their sides to form a single

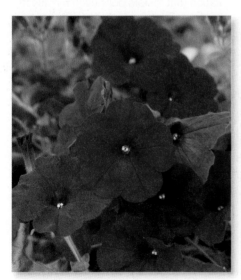

(a) Normal purple petunia flowers

(b) Petunia flower affected by siRNA

Figure 39.8   Gene silencing and flower color in petunias.   **(a)** Normal purple petunia (*Petunia hybrida*) flowers produced by the expression of all genes involved in flavonoid synthesis. **(b)** Flower of a genetically engineered plant displays petal tissues in which one of the genes needed for production of purple pigment production has been silenced (suppressed) by siRNA.

Figure 39.9   Genetic control of flower symmetry. (a) Normal snapdragon flowers, with functioning *CYCLOIDEA* genes, are bilaterally symmetrical. (b) Snapdragon plants carrying mutations in the *CYCLOIDEA* gene produce flowers that have radial symmetry.

(a) Normal snapdragon flower

(b) Snapdragon flower with *CYCLOIDEA* mutation

strap-shaped structure. Some flowers have petals that are fused together to form a tube that holds nectar consumed by animal pollinators (refer back to Figure 30.20).

As we have seen, pistils are often composed of two or more fused carpels. The rosy periwinkle (*Cantharanthus roseus*), the original source of widely used leukemia drugs, provides an example. This plant's flowers each have two carpels, whose surfaces fuse together during flower development. In 1995, plant developmental biologist Judy Verbeke and colleagues reported that when the two young carpels first come into contact with each other, epidermal cells change their pattern of differentiation. This re-differentiation is necessary for carpel fusion. When the investigators experimentally removed one of the carpels or inserted a nonporous barrier between the two carpels, cells did not redifferentiate and carpels did not fuse. If they inserted a porous barrier, however, normal differentiation and carpel fusion occurred. These experiments indicate that carpels communicate with each other by means of diffusible chemical signals that foster the cellular changes necessary for fusion.

Variations in Flower Symmetry and Aggregation Flower shape variation can also result from changes in symmetry. Flowers that possess radial symmetry are described as regular, actinomorphic, or polysymmetric flowers. Flowers having radial symmetry can be divided into two equal parts by more than one plane inserted through the center of the flower. In contrast, flowers that display bilateral symmetry are known as irregular, zygomorphic, or monosymmetric flowers. Flowers having bilateral symmetry can be divided into two equal parts by only a single plane inserted through the center. The evolution of bilateral symmetry in flowers can often be linked to insect pollination.

Symmetry, like other flower features, is under genetic control. The production of flowers having bilateral symmetry is controlled by transcription factors such those encoded by the *CYCLOIDEA* gene. For example, snapdragon (*Antirrhinum majus*) flowers are normally bilaterally symmetrical, but a

Figure 39.10   An inflorescence. The clustering of flowers into a type of inflorescence known as a head, displayed here by a Gerbera daisy, is an adaptation that fosters pollination by insects.

Concept check: *List several ways in which the flowers at the rim of the Gerbera inflorescence differ from those at the center.*

loss-of-function mutation in the *CYCLOIDEA* gene causes these flowers to display radial symmetry (**Figure 39.9**). Another interesting example is the cut-flower crop plant *Gerbera hybrida*, whose "flowers" are actually **inflorescences**, structures composed of many flowers. While many flowers occur singly, many others occur in inflorescences of diverse types that are adapted to particular pollination circumstances. As in the case of sunflowers (refer back to Figure 30.20b), *Gerbera* flowers occur in more than one structural type having distinctive functions. During inflorescence development, a *CYCLOIDEA*-like transcription factor is expressed more strongly at the margins so flowers in that position become bilaterally symmetrical, while the center-most flowers of *Gerbera* are more radially symmetrical (**Figure 39.10**). These central flowers produce stamens and therefore function to supply pollen. In contrast, stamens do not develop in the marginal flowers, and their petals fuse to form larger colorful structures that attract pollinators.

(a) A cut pollen grain showing immature male gametophyte

39.3 Male and Female Gametophytes and Double Fertilization

Mature male gametophytes are pollen tubes that deliver sperm cells, the male gametes. Mature female gametophytes are located within ovules and produce egg cells, the female gametes. In this section, we will begin by examining male and female gametophyte development and how stamens and carpels aid this process. Gametes participate in fertilization, a process we consider next. We will also take an in-depth look at how scientists first achieved the technique of in vitro fertilization, which has applications in producing transgenic crops and allowing plant biologists to gain a better understanding of plant fertilization processes.

Pollen Grains Are Immature Male Gametophytes

As we have earlier noted, pollen grains are immature male gametophytes that are adapted for transport through air from one flower to another. Pollen grains develop within the sporangia of anthers. Inside these sporangia, diploid cells undergo meiosis and cytokinesis. As a result, each diploid cell gives rise to a cluster of four haploid microspores, each having a thin cellulose cell wall. The development of microspores into pollen grains involves two processes that occur at the same time: (1) the microspore divides within its cell wall, producing a two- or three-celled young male gametophyte, and (2) each male gametophyte develops a tough pollen wall that protects the gametophyte during pollen transport. Both of these processes are completed before anthers release pollen.

Early Male Gametophyte Development Each microspore nucleus undergoes one or two mitotic divisions to form a young male gametophyte. The first division gives rise to two specialized cells: a tube cell and a generative cell suspended within the tube cell (**Figure 39.11a**). The **generative cell** divides to produce two sperm cells, either before or (more commonly) after pollination. The **tube cell** produces the pollen tube, which delivers sperm to the female gametophyte.

Pollen Wall Development A mature pollen grain has a tough wall, and each plant species produces pollen whose wall has a distinctive sculptural shape (**Figure 39.11b**). The **pollen wall**, which surrounds the plasma membrane of the tube cell, is composed largely of a nearly indestructible polymer known as **sporopollenin**. Sporopollenin is so chemically inert that its biochemical composition is still uncertain. Named for its presence on the surfaces of mature spores and pollen, sporopollenin confers physical strength, chemical inertness, and resistance to

(b) SEM of whole pollen grains showing distinctive pollen wall ornamentation

189 µm

Figure 39.11 Pollen grains. (a) Diagram of cut pollen grain. (b) SEM of whole pollen grains of different species.

Concept check: What is the maximum number of cells in a mature male gametophyte of a flowering plant?

microbial attack. Sporopollenin protects spores and pollen from damage. It also is responsible for the occurrence of intact fossil pollen grains in deposits that are millions of years old.

Development of the pollen wall starts with deposition of a blanket of the carbohydrate **callose** around each cluster of four microspores after they form by meiosis. The callose blanket seals microspores off from the influences of adjacent sporophyte tissues, thereby aiding pollen differentiation. Callose also provides a surface pattern for sporopollenin deposition and holds microspores together until an anther enzyme degrades the callose, freeing developing pollen grains from each other.

As pollen grains develop, anther cells secrete a **pollen coat**, a layer of material that covers the sporopollenin-rich pollen wall. Coat materials include additional sporopollenin and pigments that give pollen its typically yellow, orange, or brown coloration, and lipids and proteins that aid in pollen attachment to carpels. Certain of these pollen coat compounds are responsible for allergic reactions in people exposed to particular types of airborne pollen. It is no wonder that pollen allergies are common, because about 10% of flowering plants are wind-pollinated, and such plants produce copious amounts of pollen. For example, ragweed plants (genus *Ambrosia*), which are commonly associated with allergies, each produce an estimated 1 billion pollen grains during a year.

The production and dispersal of pollen grains is just the first phase of male gametophyte development. We will learn more about the second phase of male gametophyte development, the production of a pollen tube, after focusing on female gameto-phyte development.

A Female Gametophyte Develops Within Each Ovule

Each ovule produces a single female gametophyte (Figure 39.12). Although flowering plant female gametophytes vary somewhat in size and structure, many possess seven cells and eight nuclei. One of these cells is an egg cell, which lies wedged between two cells known as **synergids**. These synergids have characteristic cell-wall ingrowths that increase the area of the plasma membrane, thereby helping to move nutrients from the larger sporophyte to the nonphotosynthetic female gameto-phyte. The other four cells of the angiosperm female gameto-phyte consist of three antipodal cells, whose functions are not well understood, and a large **central cell**. The central cell contains two nuclei, which accounts for the extra nucleus in the female gametophyte.

After Pollination, the Pistil Controls Pollen Germination

Pollen grains adhere to the surfaces of the stigma, which pro-vides a receptive surface. The stigma and the style determine whether or not pollen grains germinate and pollen tubes grow. In some plants, pistils allow pollen from the same plant to ger-minate, but in many cases, pistils prevent the germination of

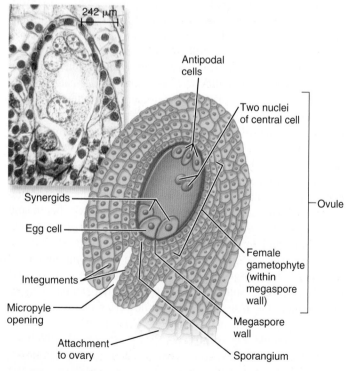

Figure 39.12 Female gametophyte within an ovule.

Concept check: *How do female gametophytes obtain nutrients?*

pollen from different species and pollen that is genetically too similar, such as pollen from the same flower.

Rejection of pollen that is genetically too similar to the pis-til is a phenomenon known as **self-incompatibility** (**SI**). This feature helps to decrease the likelihood of recessive disorders in offspring. In the absence of self-incompatibility, self-fertilization and matings between closely related parent plants are more likely to produce progeny that are homozygous for recessive deleterious alleles, which would then be expressed. How then do pistils recognize compatible pollen grains of the appropri-ate species that are genetically distinct? Recognition involves interactions between proteins of pollen and pistil cells. These interactions influence the ability of pollen to take up water, a process known as rehydration. Pollen is nearly dry when it reaches the stigma, and incompatible pollen will not rehydrate and will eventually die. In contrast, compatible pollen rehy-drates, a process that activates pollen metabolism and allows it to germinate. Pollen–stigma interactions thus form an impor-tant part of a flowering plant's mate recognition and selection system.

A closer look at SI systems explains how they work. In many plants, SI involves the *S* gene locus, which encodes S proteins. Each locus contains two genes, one that determines pollen compatibility traits and another that determines pistil compatibility traits. Multiple *S* alleles for both genes occur in plant populations. Two major types of SI are known—game-tophytic SI and sporophytic SI. Gametophytic SI occurs when an *S* gene within a pollen grain itself determines compatibility. Because pollen is a haploid gametophyte, the pollen genome contains only one pollen-compatibility allele. This allele encodes a specific S protein that is located in the pollen cyto-sol. When the gametophyte controls compatibility, tubes may start to grow from incompatible pollen, but the same specific S protein encoded by pistil cells enters the tubes and destroys the pollen tube RNA. This halts tube growth. In contrast, S protein within genetically compatible pollen binds the pistil-produced protein, thereby preventing destruction of pollen tube RNA and allowing tube growth to continue (Figure 39.13a).

Sporophytic SI occurs when pollen compatibility is deter-mined by the sporophyte that produces the pollen. This control is exerted when anthers deposit proteins into the pollen coat. Because anther cells belong to the sporophyte generation, they possess two *S* genes, and therefore, heterozygotes will produce two different pollen coat S proteins. When the sporophyte con-trols compatibility, pollen cannot germinate when S proteins in the plasma membranes of stigma cells recognize (bind) the S pro-teins of incompatible pollen. In this case, protein binding leads to signal transduction processes that prevent pollen germination. However, pollen can germinate if stigma proteins are unable to bind genetically distinct S proteins in its coat (Figure 39.13b).

If the SI system becomes inactivated by mutation, a plant species can become self-pollinating. This has occurred dur-ing the evolution of self-pollinating lines of the model plant *Arabidopsis* from an ancestor that possessed SI. An estimated 20–50% of all modern plant species, including many crop plants, display self-pollination. The ability to self-pollinate

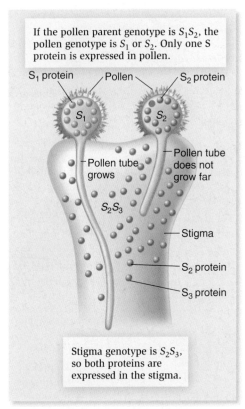

If the pollen parent genotype is S_1S_2, the pollen genotype is S_1 or S_2. Only one S protein is expressed in pollen.

S₁ protein — Pollen — S₂ protein

Pollen tube grows — Pollen tube does not grow far

S_2S_3

Stigma

S₂ protein

S₃ protein

Stigma genotype is S_2S_3, so both proteins are expressed in the stigma.

(a) Gametophytic SI: If pollen *S* allele does not match either stigma allele, pollen will germinate.

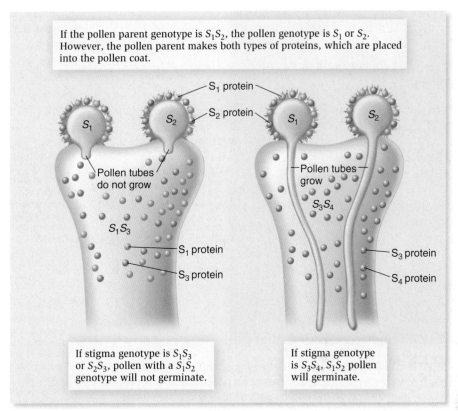

If the pollen parent genotype is S_1S_2, the pollen genotype is S_1 or S_2. However, the pollen parent makes both types of proteins, which are placed into the pollen coat.

S₁ protein

S₂ protein

Pollen tubes do not grow

Pollen tubes grow

S_1S_3

S_3S_4

S₁ protein

S₃ protein

S₃ protein

S₄ protein

If stigma genotype is S_1S_3 or S_2S_3, pollen with a S_1S_2 genotype will not germinate.

If stigma genotype is S_3S_4, S_1S_2 pollen will germinate.

(b) Sporophytic SI: If pollen coat S proteins do not match either stigma S protein, pollen tubes will grow.

Figure 39.13 Self-incompatibility. Self-incompatibility helps plants to avoid the combination of gametes that are too genetically similar. It is controlled by interactions between proteins of pollen and the pistil. **(a)** In gametophytic self-incompatibility, compatibility between pollen and pistil is determined by the haploid genotype of the pollen. In this case, S protein is located in the pollen cytosol. **(b)** In sporophytic self-incompatibility, compatibility between pollen and pistil is determined by the sporophyte that produced the pollen and contributed S proteins to its coat.

can be advantageous to plants as they colonize new habitats, and it is a common characteristic of weedy and invasive plant species. Self-pollination is also useful to scientists performing genetic experiments with *Arabidopsis*. Pea plants are also self-pollinating, a feature that was useful to Mendel in performing his classic experiments on inheritance (see Chapter 16).

During Pollen Germination, the Pollen Tube Grows, and Double Fertilization Occurs

As we have seen, a pollen grain germinates by taking up water, and when this occurs, the pollen generative nucleus usually divides by mitosis to produce two sperm cells. Upon rehydration, it takes a few minutes to an hour or so for the tube cell to produce a pollen tube. Pollen tubes extend from a thin region in the grain's wall into the spaces between cells of the style. To deliver sperm to egg cells, the tube must grow from the stigma, through the style, to reach the ovule (**Figure 39.14**).

The Role of the Style   In some flowers, the style plays a role in mate selection by nourishing or inhibiting pollen tube growth. The style also guides pollen tube growth. For example, in 2005,

Elizabeth Lord and associates reported that the *Arabidopsis* style produces a blue copper-containing protein known as plantacyanin. These investigators also noted that the concentration of plantacyanin is normally low at the stigma but gradually increases in the style, with highest levels occurring near the ovary. Their observation suggested that plantacyanin might guide pollen tube growth toward ovules. Lord and associates tested this hypothesis by producing transgenic plants that expressed abnormally high levels of plantacyanin in the stigma. This change disrupted normal pollen tube behavior, causing pollen tubes to grow in circles around stigma cells or, in rare cases, up into the air, away from the style! This experimental evidence supports the concept that a gradient of plantacyanin in the style likely guides *Arabidopsis* pollen tubes toward ovules.

Tip Growth and Sperm Delivery   A pollen tube transports two sperm cells to a female gametophyte in an ovule. A pollen tube does this by a process called tip growth, which is controlled by the tube cell nucleus. During tip growth, new cytoplasm and cell-wall material are added to the tip of an elongating cell (**Figure 39.15**). Golgi vesicles continuously deliver the carbohydrate

Figure 39.14   Pollen tubes delivering sperm to ovules.
This fluorescence microscopic view shows pollen grains (the pale
objects) germinating on the stigma surface (SG) and pollen tubes
(PT) growing through the style (ST) toward ovules.

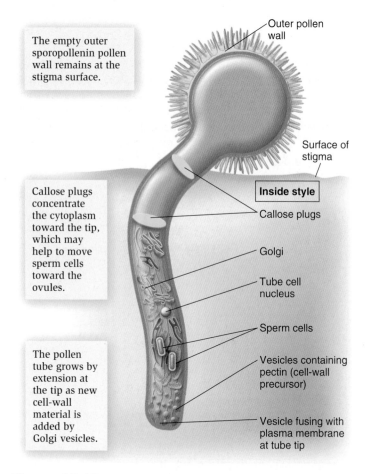

The empty outer sporopollenin pollen wall remains at the stigma surface.

Outer pollen wall

Surface of stigma

Inside style

Callose plugs concentrate the cytoplasm toward the tip, which may help to move sperm cells toward the ovules.

Callose plugs

Golgi

Tube cell nucleus

Sperm cells

The pollen tube grows by extension at the tip as new cell-wall material is added by Golgi vesicles.

Vesicles containing pectin (cell-wall precursor)

Vesicle fusing with plasma membrane at tube tip

Figure 39.15   Tip growth by a pollen tube.

pectin to the tube tip, whose thin cell wall is largely composed of pectin. As a pollen tube grows, callose plugs are commonly deposited in the older parts of the tubes. These plugs concentrate the tube cytoplasm at the tip, which may help to maintain the turgor pressure necessary for continued tip growth.

In some plants, pollen tubes have been observed to grow toward ovules at about 0.5 mm per hour, commonly taking from 1 hour to 2 days to reach their destination. When a pollen tube encounters an ovule, it enters through the micropyle. Attracted by secretions from a synergid, the pollen tube penetrates the synergid. The thin tube tip wall then bursts, releasing the sperm into the female gametophyte. Plant biologists used to think that these two sperm cells were identical, but now it is known that in at least some plants, the sperm differ in structure and role. For example, plant reproductive biologist Scott Russell reported

in 1985 that pollen tubes of leadwort (*Plumbago zeylanica*) contain a smaller sperm cell that is destined to fuse with the egg and a larger sperm that participates in endosperm formation.

Double Fertilization   During the process of double fertilization, one sperm nucleus fuses with the egg cell to produce a zygote, the first cell of a new sporophyte generation (see steps 4 and 5 in Figure 39.4). The other sperm fuses with the two nuclei of the central cell to form the first endosperm cell. As the zygote develops into an embryo, the endosperm develops into a nutritive tissue that is usually triploid. We will next see how plant biologists first achieved plant fertilization in the laboratory using isolated gametes, a process known as in vitro fertilization.

FEATURE INVESTIGATION

Kranz and Lörz First Achieved Plant in Vitro Fertilization

In vitro fertilization (IVF) is a widely practiced technology in human medicine and in agriculture. The term in vitro is Latin, meaning "in glass," referring to lab dishes. In past decades, plant biologists had only been able to perform in vitro pollination. To do this, they removed ovules from plant carpels and applied pollen to them in laboratory dishes. Pollen tubes were

able to deliver sperm cells to female gametophytes, resulting in fertilization and seed development. This process was useful in overcoming mating barriers to hybridization that occur because stigmas and styles usually prevent interspecies matings. Therefore, in vitro pollination allowed agricultural scientists to produce new interspecies hybrids that would otherwise not be possible.

Crop scientists realized that IVF, the direct union between egg and sperm cell, would allow plant biologists to gain a better understanding of plant fertilization processes and might ultimately be useful in producing hybrid crops. However, plant gametes are so deeply buried within gametophytic and sporophytic tissues that scientists found it difficult to isolate and manipulate them. As a result, plant IVF was not accomplished until 1993. In that year, German researchers Erhard Kranz and Horst Lörz reported that they had used isolated gametes to produce mature hybrid corn (maize) plants, as shown in Figure 39.16.

The development of a method for plant IVF required several research steps. First, by conducting many experiments, Kranz and Lörz developed procedures for isolating and handling single eggs and sperm. They also designed a process for using electrical current to induce gamete fusion. Further, they defined the optimal chemical and physical conditions necessary for in vitro development of zygotes into mature plants. For example, the researchers discovered that "nurse" cells grown from excised natural embryos were critical to the survival of

Figure 39.16   Kranz and Lörz first achieved plant in vitro fertilization.

GOAL   To demonstrate in vitro plant fertilization using isolated gametes.

KEY MATERIALS   Maize egg cell donor plants with green stigmas and hairs, and maize sperm cell donor plants with red stigmas and hairs.

Experimental level | **Conceptual level**

1 Isolate sperm cells from pollen ruptured by osmotic shock. Isolate egg cells from ovules by microdissection of female gametophytes with a glass needle, using an enzyme solution that degrades cell walls.

Sperm   Egg

2 Under a light microscope, use a micromanipulator to line up egg and sperm on electrode surface.

Micromanipulator

Sperm and egg are aligned.

3 Fuse gametes with electrical pulses.

Pt (platinum) electrode   Electrical pulse

4 Culture zygotes on porous membrane over a suspension of maize "nurse" cells in a solution of minerals and glucose at pH 5.5, on a rotary shaker, at 25°C, with 16-hour light/day.

Liquid medium   Embryo   Porous membrane   "Nurse" cells

Maize "nurse" cells allow a sustained growth of zygotes into embryos.

5 Transfer embryos to solid media containing various concentrations of sucrose and plant hormones.

Embryo   Solid medium

Embryos develop into young plants.

6 Transplant young plants to soil. When plants flower, examine plants to see if their stigmas and stigma hairs were those expected of a hybrid.

7 **THE DATA**

Plants donating the egg cells (step 1) had green stigma and hairs.

Plants donating the sperm cells (step 1) had red stigma and hairs.

Plants produced by in vitro fertilization (step 6) had green stigma, red hairs.

8 **CONCLUSION** Under the appropriate conditions, plant in vitro fertilization can occur.

9 **SOURCE** Kranz, E., and Lörz, H. 1993. In vitro fertilization with isolated, single gametes results in zygotic embryogenesis and fertile maize plants. *Plant Cell* 5:739–746.

zygotes and embryos produced in vitro. The IVF corn embryos displayed normal development and grew into normal seedlings and reproductive adults. These investigators demonstrated their accomplishment by showing the hybrid origin of traits in progeny plants (see The Data in Figure 39.16). The hybrid progeny plants had flower pistils with green stigmas coated by red hairs. These traits were inherited from different parents: The green stigmas were inherited from the female parent, and the red hairs were inherited from the male parent. This work provided a foundation for additional IVF work with plants and also enabled more detailed investigations of plant embryo development, our next subject.

Experimental Questions

1. Why had in vitro plant fertilization been so difficult for plant biologists to accomplish before the work of Kranz and Lörz?

2. What procedure did Kranz and Lörz use to accomplish fertilization of isolated eggs by sperm cells?

3. How did Kranz and Lörz demonstrate that their plants were actually hybrids arising from in vitro fertilization involving specific genetic parents?

39.4 Embryo, Seed, Fruit, and Seedling Development

Seeds and fruits are major components of plant reproduction. Seeds contain dormant plant embryos that may develop into seedlings under favorable conditions. As noted earlier, fruits aid seed dispersal, which allows plants to colonize new sites. Embryos, seeds, and fruits mature simultaneously, and their development is coordinated by hormonal signals. Seedling development is also hormonally regulated.

Endosperm Provides Nutrients for Developing Embryos Within Seeds

Endosperm is a tissue present in developing angiosperm seeds and also occurs in the mature seeds of many plants. Rich in protein, lipid, carbohydrate, vitamins, and minerals, endosperm supplies the nutritional needs of developing embryos and often seedlings as well. We have already noted how humans use the coconut's liquid and solid endosperm as food (see Figure 39.5a). In fact, a large percentage of human and animal food protein comes from seed endosperm of grain crops. Corn, wheat, rice, and other grain crops generate more than 380 billion pounds of endosperm per year in the U.S. alone.

How does this valuable tissue develop? Recall that the first cell of the endosperm tissue arises by the fusion of a sperm

nucleus with the two nuclei of the female gametophyte's central cell. The resulting nucleus is typically triploid, combining one set of chromosomes from the male parent with two sets from the female parent. Mitotic division of the triploid cell generates endosperm tissue. The food stored in endosperm comes from the parent sporophyte by diffusing through moist cell walls and intercellular spaces.

As the embryo develops, it uses food stored in the endosperm, but food storage differs in eudicots and monocots. By the time that most eudicot seeds are mature, they contain little or no endosperm because such embryos store organic food in two embryonic leaves, the cotyledons. The seeds of legumes such as beans, peas, and peanuts are valued as food because of their protein-rich cotyledons. In contrast, many mature monocot seeds retain considerable endosperm, explaining the food value of grain crops.

Embryos Develop from Zygotes Within Seeds

Embryos develop from single-celled zygotes by mitotic divisions, in the process known as **embryogenesis**. This process begins with an unequal cell division and proceeds through several distinctive stages, as illustrated by the model eudicot plant *Arabidopsis* (Figure 39.17).

Early Embryogenesis Sometime within a period of days to several weeks following fertilization, a zygote begins to divide.

At this point, a zygote is blanketed with a layer of callose, which helps to seal it off from the environment, thereby fostering embryo-specific gene expression. The zygote's first cell division is unequal, producing a smaller cell and a larger cell (Figure 39.17, step 1). These two cells differ in cytoplasmic contents, which Kranz and Lörz observed in IVF corn embryos. This unequal division helps to establish the apical-basal (top-bottom) polarity of the embryo, which persists through the life of the plant. The smaller cell develops into the embryo, whose radial symmetry is established at this point and continues in adult plants. The larger cell develops into a **suspensor**, a short chain of cells anchored near the micropyle at the ovule entrance (Figure 39.17, step 2). The suspensor channels nutrients and hormones from the parent sporophyte into the young embryo, which absorbs them at its surfaces.

Researchers have found that the functional loss of a gene called *TWN* (twin) transforms the suspensor into a second, twin embryo. This indicates that when TWN protein is active, it prevents the suspensor from developing into an embryo, thereby maintaining normal suspensor attachment and nutrient transport functions. The suspensor usually disappears by controlled cell death during embryo development. Older embryos rely on nutrients supplied by the endosperm.

Later Embryogenesis   Young eudicot embryos are spherical, but they soon become heart-shaped as the seedling leaves, called **cotyledons**, start to develop. At this point, auxin and cytokinins are involved in establishing the young shoot and root at the apical and basal poles, respectively. This process is also influenced by the TOPLESS protein, which helps to repress auxin-response genes that would otherwise promote root development at the apical pole. (Mutants that have lost normal *TOPLESS* function have roots at both apical and basal poles, but no shoots, and are thus topless.) Eudicot embryos such as *Arabidopsis* then become torpedo-shaped, and as the cotyledons grow, they often curl to fit within the developing seed (Figure 39.17, step 4). In contrast, mature monocot embryos are cylindrical, with a single cotyledon and a side notch where the apical meristem forms. Mature embryos then become dormant as seeds mature.

Mature Seeds Contain Dormant Embryos

The structure of mature monocot and eudicot seeds differs. Within eudicot seeds, mature embryos often display an **epicotyl**, the portion of an embryonic stem with two tiny leaves in a first bud that is located above the point of attachment of the cotyledons (**Figure 39.18a**). The **hypocotyl** is the portion of an embryonic stem located below the point of attachment of the cotyledons. An embryonic root, the **radicle**, extends from the hypocotyl. Much of the endosperm has been absorbed into the large cotyledons.

In contrast, mature monocot embryos, such as those of corn, feature an epicotyl with a first bud enclosed in a protective sheath known as the **coleoptile**. The young monocot root is enclosed within a protective envelope known as the **coleorhiza** (**Figure 39.18b**).

As seeds mature, they undergo changes leading to dormancy, an adaptation that prevents them from germinating when environmental conditions are not suitable for seedling

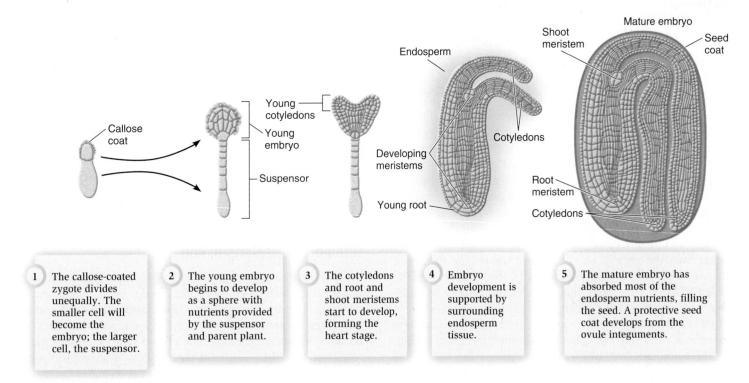

1 The callose-coated zygote divides unequally. The smaller cell will become the embryo; the larger cell, the suspensor.	**2** The young embryo begins to develop as a sphere with nutrients provided by the suspensor and parent plant.	**3** The cotyledons and root and shoot meristems start to develop, forming the heart stage.	**4** Embryo development is supported by surrounding endosperm tissue.	**5** The mature embryo has absorbed most of the endosperm nutrients, filling the seed. A protective seed coat develops from the ovule integuments.

Figure 39.17   Embryogenesis in the eudicot *Arabidopsis*.

Concept check: How would this embryo differ if its TOPLESS genes were nonfunctional?

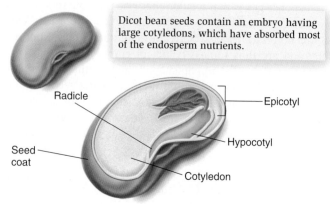

Dicot bean seeds contain an embryo having large cotyledons, which have absorbed most of the endosperm nutrients.

Radicle

Epicotyl

Hypocotyl

Seed coat

Cotyledon

(a) Eudicot bean seed, showing embryo with epicotyl, hypocotyl, and radicle

Monocot corn kernels are single-seeded fruits. The seed embryo has one cotyledon that presses against the nutritive endosperm. The embryonic shoot tip is protected by a tissue sheath—the coleoptile. The embryonic root is protected by a tissue sheath—the coleorhiza.

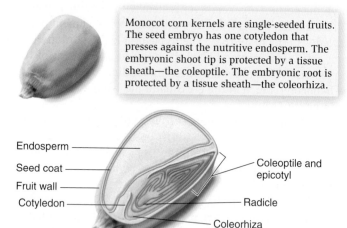

Endosperm

Seed coat

Fruit wall

Cotyledon

Coleoptile and epicotyl

Radicle

Coleorhiza

(b) Monocot corn seed, showing an embryo protected by coleoptile and coleorhiza

Figure 39.18 Structure of mature seeds and embryos.

Concept check: Why do mature seeds of eudicots lack extensive amounts of endosperm?

growth. During this process, embryos become dry and thus able to survive in the absence of water. Seed maturation includes transformation of the ovule's integuments into a tough seed coat (see Figure 39.17, step 5). The seed coat restrains seedlings from growing and prevents the entry of water and oxygen, which maintains low seed metabolism. In addition, the coats of some seeds are darkly colored with pigments that may help to prevent damage by UV radiation or microbial attack. Another change leading to seed dormancy is gradual, controlled loss of water from the embryo and other tissues. As the result, the water content of dispersed seeds is only 5–15%. Abscisic acid (ABA) is a hormone that induces the activity of genes that help embryo tissues to survive the drying process. Some of these desiccation-tolerance genes encode proteins that form loose coils enclosing cell contents, thereby preventing damage as the cytoplasm becomes almost completely dry. When the seeds of flowering plants are dry and ready for dispersal, they are released from the plant while enclosed in a fruit or released when the fruit breaks open.

Fruits Develop from Ovaries and Other Flower Parts

All fruits develop from ovaries and sometimes other flower parts. They occur in diverse forms that aid seed dispersal. Some fruits are dry, whereas others are moist and juicy; some open to release seeds, and others do not. Fruits also display a wide variety of sizes, colors, and fragrances. These variations result from differences in the process of fruit development. Plant hormones, including auxin, gibberellic acid, and cytokinins, control this transformation. Abscisic acid stimulates cell expansion, and ethylene influences fruit ripening. For instance, ethylene helps to ripen nuts, a type of dry fruit, by inducing plasma membranes to rupture, causing water loss. Under the influence of plant hormones, the pericarp (ripened ovary wall) of peaches, plums, and related fruits swells and softens, while orange or red chromoplasts replace green chloroplasts. As fruits mature, the outer protective cuticle often becomes very thick, contributing to peel toughness, which helps to prevent microbe attack. In addition, many maturing fruits increase their sugar and acid content, which produces the distinctive tastes of ripe fruit. Many fruits also produce fragrant volatile compounds.

Differences in the shape, color, fragrance, and moisture content of wild fruits reflect evolutionary adaptation for effective seed dispersal. Though many fruits and seeds are dispersed by wind or water or by attaching to animal fur, others are consumed by fruit-eating animals that are attracted by fruit color and fragrance. Blackberries provide a good example of fruits adapted for animal dispersal. Blackberry flowers produce many separate pistils, each containing a single ovule (**Figure 39.19a**). Following pollination and fertilization, the ovary of each pistil develops into a sweet, juicy fruitlet containing a single seed. As the individual fruitlets develop, they fuse together at the sides. Consequently, the many fruitlets produced by a single blackberry flower are dispersed together, in a structure known as an aggregate fruit (**Figure 39.19b**). Attracted by the color, birds consume the whole aggregate and excrete the seeds, thereby dispersing many at a time. Many other types of fruits occur and these likewise represent adaptations that foster seed dispersal (refer back to Figure 30.21). Although a fruit is usually defined as a mature ovary containing seeds, commercial seedless fruits such as watermelon are produced by genetic modification or treatment with artificial auxin.

Environmental and Internal Factors Influence Seed Germination

Seeds vary greatly in their ability to germinate after dispersal. Small seeds such as those of dandelions and lettuces germinate quickly if light is available. Other seeds require a period of dormancy before germination occurs. Some seeds can remain dormant for amazingly long time periods. For example, a lotus (*Nelumbo nucifera*) seed collected from a lake bed in China germinated at the age of 1,300 years, as determined by radiocarbon dating. In 2005, plant scientists germinated a 2,000-year-old date seed found in Israel.

(a) *Rubus allegheniensis* **(common blackberry) flower**

(b) Blackberry fruit

Figure 39.19   Blackberry flower and fruit.   **(a)** Each of the many separate pistils in a blackberry flower is able to produce a single one-seed fruit (called a fruitlet) if fertilization occurs. **(b)** Together, the individual fruitlets of the blackberry compose an aggregate fruit. A shriveled style with a stigma is attached to each fruitlet.

Water is generally required to rehydrate seeds so that embryos can resume their metabolic activity. Water absorption also swells seeds, helping to break the seed coat and allowing embryonic organs to emerge. In some cases, rainfall of sufficient duration to leach germination-inhibiting compounds out of seeds is required. The optimal temperature for germination of most seeds lies between 25 and 30.25°C (77 and 86.25°F). This explains why gardeners wait until the soil is warm before planting seeds outdoors in spring. However, some seeds need a period of cold treatment or seed coat abrasion before they will germinate.

Such physical stimuli induce the activity of more than 2,000 genes associated with seed germination. In *Arabidopsis*, several phases of cell division occur after dry seeds have been moistened. Cell division in the radicle, the embryonic root, occurs first. As a result, the radicle is the first organ to emerge from a germinating seed, which allows seedlings to take up water even more rapidly. Next, cell division rates rise in the cotyledons and then in the shoot apical meristem, which increases the size of the embryonic shoot.

When grass seeds rehydrate, the young shoot secretes the hormone gibberellic acid from the seed cotyledon into the outermost endosperm layer, known as the aleurone. In response, the aleurone secretes digestive enzymes into the central endosperm, releasing sugars from stored starch (**Figure 39.20**). The seedling uses these sugars for growth. This highly coordinated process allows grass seeds to quickly germinate when it rains, an advantage in arid grassland habitats.

Humans also use this basic process to make beer. In the process known as malting, beer brewers apply gibberellic acid to barley seeds to induce them to germinate simultaneously. The barley seeds are then baked at a high temperature to stop germination, a process that produces malt. Baking also caramelizes sugars, contributing to the distinctive flavors and color of beer. To make beer, brewers treat malt with water and heat, add the dried flowers of the hop plant (the genus *Humulus*), and add yeasts to ferment the plant sugars to alcohol.

1   Grass seeds rehydrate, activating embryo growth and metabolism.

2   The young shoot secretes gibberellic acid from the cotyledon into the outermost endosperm, the aleurone.

3   In response, the aleurone secretes digestive enzymes into the central endosperm. The enzymes break down stored starch into sugars.

4   Sugars are used for seedling growth.

Figure 39.20   Germination of grass seeds.

Once seeds have germinated, plants vary in the process by which the embryonic shoot emerges. When bean and onion seeds germinate, the hypocotyl forms a hook that first breaches the soil surface and then straightens, thereby pulling the rest of the seedling and cotyledons aboveground (**Figure 39.21a,b**). In contrast, when pea seeds germinate, the epicotyl forms a hook that pulls the shoot tip out of the ground, leaving the

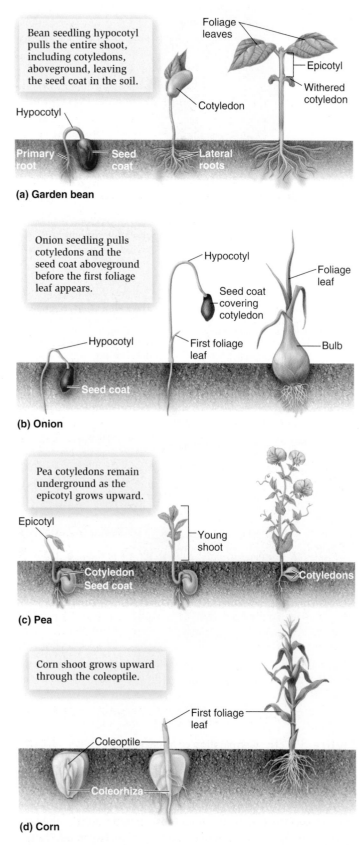

(a) Garden bean

Bean seedling hypocotyl pulls the entire shoot, including cotyledons, aboveground, leaving the seed coat in the soil.

Foliage leaves

Epicotyl

Withered cotyledon

Cotyledon

Hypocotyl

Primary root   Seed coat   Lateral roots

(b) Onion

Onion seedling pulls cotyledons and the seed coat aboveground before the first foliage leaf appears.

Hypocotyl

Seed coat covering cotyledon

Foliage leaf

Hypocotyl

First foliage leaf

Bulb

Seed coat

(c) Pea

Pea cotyledons remain underground as the epicotyl grows upward.

Epicotyl

Young shoot

Cotyledon

Seed coat

Cotyledons

(d) Corn

Corn shoot grows upward through the coleoptile.

First foliage leaf

Coleoptile

Coleorhiza

Figure 39.21   Variations in seed germination and seedling growth patterns.

cotyledons beneath the soil surface (**Figure 39.21c**). In both cases, the tough hook cells bear the brunt of passage through hard surface soil crusts, thereby protecting the delicate shoot tips. The plant hormone ethylene controls seedling hook formation, as described in Chapter 36. However, not all seedlings form hooks. For example, as they grow through the soil, the shoot tips of corn seedlings and those of other grasses are protected by the coleoptile, which functions as a protective tube for the first foliage leaves (**Figure 39.21d**).

39.5   Asexual Reproduction in Flowering Plants

Many plants rely on sexual reproduction. However, a wide variety of other angiosperms reproduce primarily by asexual means, and other plants commonly utilize both sexual and asexual reproduction. **Asexual reproduction** is the production of new individuals from a single parent without the occurrence of fertilization. Although sexual reproduction provides beneficial genetic variation, asexual reproduction can be advantageous in other ways. For example, asexual reproduction maintains favorable gene combinations that allow faster population growth in stable environments. Asexual reproduction is also advantageous in stressful habitats where pollinators or mates can be rare, because it allows a single individual to start a new population. Finally, asexual reproduction allows some plants to persist for very long periods of time. Among the oldest known plants are creosote bushes that are asexual clones of a parent that grew from a seed about 12,000 years ago!

The artificial vegetative propagation of plants from cuttings is a form of asexual reproduction that is widely used commercially and by home gardeners. In this section, we will explore the three main mechanisms of plant asexual reproduction: specialized reproductive structures, apomixis, and somatic embryogenesis.

Vegetative Asexual Reproduction Generates Plant Clones from Organs

Roots, stems, and leaves are vegetative plant organs that can function as asexual reproductive structures. For example, root sprouts, such as those produced by aspens, can generate entire groves of genetically identical trees. Sucker shoots, such as those appearing at the bases of banana plants and date palms, and the pieces of tuber-bearing "eyes" (which are buds) that develop into potato plants are examples of vegetative plant organs that have agricultural importance. Attractive horticultural varieties of African violets and other plants can be propagated from leaf cuttings. Such asexual offspring grow into adult plants that have the same valued properties as their parent plant. In contrast, these crops grown from seed would produce diverse progeny, not all of which would have economically prized properties.

Somatic embryogenesis is the production of plant embryos from body (somatic) cells. Embryos can develop from many

Figure 39.22 Asexual reproduction via vegetative structures. The leaves of this *Kalanchoë* plant bear small plantlets around the edges. When mature, these plantlets drop off and under the right conditions, grow into new plants.

types of plant cells. Somatic embryos develop normally to the torpedo stage but do not dehydrate and become dormant, as is normal for zygotic embryos. Rather, somatic embryos produce root and shoot systems and develop into mature plants. Somatic embryogenesis occurs naturally in citrus, mango, onion, and tobacco plants, but agricultural scientists have made use of it as well.

In the 1950s, plant biologist F. C. Steward and associates were the first researchers to successfully clone a complex organism, carrot plants, by means of somatic embryogenesis. They used differentiated cells from carrot roots, grew the cells in conditions that caused some cells to lose their specialized properties and develop into embryos, and then cultivated each embryo in conditions that favored development into a mature carrot plant. Many types of plants are now cloned by somatic embryogenesis, which allows commercial growers to produce large numbers of genetically identical individuals.

The common houseplant *Kalanchoë daigremontiana* has leaves bearing many tiny plantlets at their edges. When these detach, they are able to take root and grow into new individuals (**Figure 39.22**). Recent molecular studies have revealed that embryogenesis genes are involved in this process and its evolution.

Genomes & Proteomes Connection

The Evolution of Plantlet Production in *Kalanchoë*

Kalanchoë daigremontiana is informally known as mother of thousands because it produces many plantlets at the edges of leaves. Helena Garces, Neelima Sinha, and their colleagues investigated the evolution of this type of asexual reproduction by studying genes that are involved in the development of organs and embryos in four species of *Kalanchoë*. In addition to *K. daigremontiana*, they investigated *K. marmorata*, which does not produce plantlets; *K. pinnata*, which produces plantlets only under certain stressful conditions; and *K. gastonis-bonnieri*, which produces plantlets both normally and when stressed. In leaves of these species, the biologists looked for expression of the gene *STM*, which encodes a key regulator of leaf production

at the shoot meristem. They found that *STM* is expressed in cells at the leaf margins of all *Kalanchoë* species that produce leaf plantlets, but not the species that do not produce plantlets. Then the investigators checked for the expression in leaves of two genes that are involved in embryo development (*LEC1* and *FUS3*). They discovered that these embryo-linked genes were expressed only in the leaf margins of species that normally form plantlets, not the species in which plantlet formation is induced by stress.

In a survey of a larger number of species, these investigators also discovered that *Kalanchoë* species that normally produce plantlets have an altered LEC1 protein. The normal form of LEC1 protein is essential to the process by which embryos become dry and thus tolerant to arid conditions. LEC1 is therefore necessary for the production of viable seeds, those able to germinate. *Kalanchoë* species that produce plantlets only under stressful conditions and those that do not produce plantlets at all were able to produce viable seeds. By contrast, the seeds of plantlet-producing species were not viable.

Together, these data allowed the investigators to infer that the evolution of plantlet formation began when certain leaf cells of some species gained the ability to function like a shoot meristem, thereby producing structures resembling small shoots. In some of the descendents of these species, normal LEC1 function in seeds was lost, as was the ability to produce viable seeds. Some species adapted by expressing the embryo-development process in leaf margin cells, a process that allowed them to produce plantlets. Such plantlets are not affected by loss of LEC1 function because, unlike seeds, they do not undergo a drying process during development.

Apomixis Is Seed Production Without Fertilization

Apomixis (from the Greek, meaning away from mixing, that is, genetic mixing) is a natural asexual reproductive process in which fruits and seeds are produced in the absence of fertilization. More than 300 species of flowering plants, including hawkweeds, dandelions, and some types of citrus, are able to reproduce asexually by apomixis. Dandelions and some other apomictic plants require pollination to stimulate seed development, but others do not. Some weedy plant species are apomictic; this process allows them to reproduce quickly, a characteristic of weeds. In addition to their concern about weeds, agricultural scientists are interested in apomixis as a potential method for producing genetically uniform seeds, propagating hybrids, and removing the need for fertilization in crop plants.

Most studies of apomixis have been carried out with dandelions, because they are widespread and locally abundant. Dandelions are herbaceous perennials that flower in their first year. Although some European dandelion populations contain individuals that reproduce sexually, many dandelion populations are composed mainly of individuals that reproduce by apomixis. In apomictic plants, meiosis produces diploid megaspores because homologous chromosomes do not pair and meiosis II does not occur. These diploid megaspores produce diploid female gametophytes, which generate diploid eggs. These eggs develop into normal embryos without fertilization, a process known as parthenogenesis. In apomictic dandelions, the endosperm develops from diploid central cells. Though such plants produce microspores via meiosis, most pollen grains have abnormal numbers of chromosomes and can produce pollen tubes, but not sperm cells. Hence, fertilization does not occur. However, pollination stimulates dandelions to produce their single-seeded fruits. Fruit that develops in the absence of fertilization is known as parthenocarpic fruit.

Summary of Key Concepts

39.1 An Overview of Flowering Plant Reproduction

- Flowering plants display a sexual life cycle known as alternation of generations, as do all plant groups. The gamete-producing male and female gametophytes of flowering plants are very small and entirely dependent on nurturing sporophytic tissues. Plant gametes arise by the process of mitosis. (Figures 39.1, 39.2)

- Flowers are reproductive shoots that develop from a shoot apical meristem. The role of flowers is to promote seed production. A flower shoot generally produces four types of organs: sepals, petals, stamens, and carpels. Stamens produce pollen, otherwise known as immature male gametophytes. Carpels produce ovules, which contain female gametophytes. (Figure 39.3)

- Mature male gametophytes, or pollen tubes, each produce two sperm and deliver them to ovules. Flowering plants display double fertilization: One of the two sperm released from a pollen tube combines with an egg cell to form a zygote, while the other fuses with two nuclei located in a central cell of the female gametophyte, producing the first cell of endosperm tissue. Endosperm is a nutritive tissue that supports development of an embryonic sporophyte. (Figure 39.4)

- Seeds are reproductive structures that contain a dormant embryo enclosed by a protective seed coat that develops from ovule integuments. Fruits are structures that contain seeds and foster seed dispersal. Like flowers and endosperm, fruits are unique features of flowering plants. (Figure 39.5)

39.2 Flower Production, Structure, and Development

- Plants flower in response to environmental stimuli, such as temperature and day length, by the conversion of a leaf-producing shoot into a flowering shoot.

- Flowers vary in the type of whorls present, the number of flower organs, color, fragrance, organ fusion, symmetry, and arrangement (whether single or in inflorescences). These variations are related to pollination mechanisms and are genetically controlled. (Figures 39.6, 39.7, 39.8, 39.9, 39.10)

39.3 Male and Female Gametophytes and Double Fertilization

- Pollen grains are immature male gametophytes protected by a tough sporopollenin wall. Female gametophyte development occurs within an ovule. Mature female gametophytes include an egg, a central cell with two nuclei, and a few additional cells. (Figures 39.11, 39.12)

- After pollination, interactions between proteins of pistil cells and those of pollen determine pollen germination. (Figure 39.13)

- Germinated pollen delivers two sperm to female gametophytes by means of a long pollen tube. The style plays a role in the guidance, nutrition, and fate of the pollen tube. Pollen tubes that enter female gametophytes burst at their tip, releasing sperm. One sperm nucleus fuses with the egg to produce a zygote, the first cell of a new sporophyte generation; the other sperm nucleus fuses with the two nuclei of the central cell, generating the first cell of the nutritive endosperm tissue. (Figures 39.14, 39.15, 39.16)

39.4 Embryo, Seed, Fruit, and Seedling Development

- Unequal division of a zygote leads to development of a nutritive suspensor and an embryo. Young eudicot embryos are heart-shaped as two embryonic leaves (cotyledons) develop, and the embryo assumes a torpedo shape as the embryonic root forms. Monocot embryos produce only a single cotyledon. (Figures 39.17, 39.18)

- Mature seeds contain embryos that become dry and are protected by desiccation-resistance proteins and a tough seed coat. These adaptations enable seeds to withstand long periods of dormancy, germinating only when conditions are favorable for seedling survival. Mature fruits develop from ovaries and aid in seed dispersal. (Figure 39.19)

- Seed germination is influenced by environmental and internal factors. The embryonic root (radicle) is the first organ to emerge, an adaptation that allows rapid water uptake, essential for seedling development. (Figures 39.20, 39.21)

39.5 Asexual Reproduction in Flowering Plants

- Asexual reproduction is the production of new individuals from a single parent without the occurrence of fertilization. Vegetative reproduction is the development of whole plants from nonreproductive organs. Somatic embryogenesis is the production of embryos from individual body cells. Apomixis is a mechanism by which some plants produce seeds from flowers without fertilization. (Figure 39.22)

Assess and Discuss

Test Yourself

1. Where do the pollen grains of flowering plants develop?
   a. in the anthers of a flower
   b. in the carpels of a flower
   c. while being dispersed by wind, water, or animals
   d. within ovules
   e. within pistils

2. Where do mature male gametophytes of flowering plants primarily develop?
   a. in the anthers of a flower
   b. in the carpels of a flower

c. while being dispersed in wind, water, or by animals
d. within ovules
e. on the surfaces of leaves

3. Where would you find female gametophytes of a flowering plant?
   a. in the anthers of a flower
   b. at the stigma of a pistil
   c. in the style
   d. within ovules in a flower's ovary
   e. in structures that are dispersed by wind, water, or animals

4. How does double fertilization occur in flowering plants?
   a. The two sperm in a pollen tube fertilize the two egg cells present in each female gametophyte.
   b. One of the two sperm in a pollen tube fertilizes the single egg in a female gametophyte, while the other fuses with the two nuclei present in the central cell.
   c. Two sperm, one contributed by each of two different pollen tubes, fertilize the two egg cells in a single female gametophyte.
   d. Two sperm contributed by separate pollen tubes enter a single female gametophyte; one of the sperm fertilizes the egg cell, while the other fertilizes the central cell.
   e. None of the above is correct.

5. A seed is
   a. an embryo produced by the fertilization of an egg, which is protected by a seed coat.
   b. a structure that germinates to form a seedling under the right conditions.
   c. an embryo produced by parthenogenesis that is enclosed by a seed coat.
   d. all of the above.
   e. none of the above.

6. What is the likely chemical composition of florigen, the long-sought chemical stimulus of flowering?
   a. the hormone auxin   d. the mineral ion K^+
   b. the protein STM   e. none of the listed choices
   c. the carbohydrate callose

7. How many whorls of organs occur in complete flowers?
   a. two   c. six   e. ten
   b. four   d. eight

8. If an ovary contains eight ovules, how many seeds could potentially result if pollen tubes reach all eight ovules?
   a. one
   b. four
   c. eight
   d. more than 20
   e. none of the listed choices

9. What function(s) does the polysaccharide callose have in the reproduction of flowering plants?
   a. Callose forms a coat that isolates young embryos during their early development.
   b. Callose forms a coat that isolates groups of four microspores during their early development into pollen grains.
   c. Callose helps to pattern the sculptured sporopollenin walls of pollen grains.
   d. all of the above
   e. none of the above

10. From what structure does a fruit pericarp primarily develop?
    a. the style
    b. a stamen filament
    c. the ovary wall
    d. a group of fused sepals
    e. the stigma

Conceptual Questions

1. Why are pollen grain walls composed of sporopollenin?
2. Why are seed coats often tough?
3. Why do flowers occur in such a diversity of shapes and colors?

Collaborative Questions

1. Observe or think of orchid flowers. Are these flowers bilaterally symmetrical or radially symmetrical? Are these flowers more likely to be wind-pollinated or pollinated by animals? What gene might be involved in the production of orchid flower shape?
2. How do plants prevent the production of many offspring expressing deleterious recessive traits?

Online Resource

www.brookerbiology.com

Stay a step ahead in your studies with animations that bring concepts to life and practice tests to assess your understanding. Your instructor may also recommend the interactive ebook, individualized learning tools, and more.

Animal Bodies and Homeostasis

40

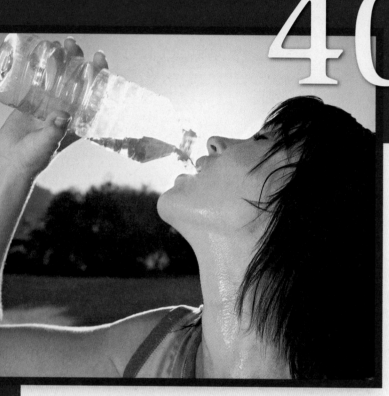

Perspiring and drinking water are both mechanisms that help achieve homeostasis, a stable internal body environment.

I n Chapters 32–34, we considered the diversity of animal life. Yet, despite the obvious differences between a worm and a human, or an insect and a fish, fundamental similarities link the millions of animal species together. We will introduce some of these similarities in this chapter and explore them throughout this unit. For example, all animals must obtain energy to survive, and they must have the ability to cope with or adapt to changes in their environment. Furthermore, all animals are composed of similar types of cells. In this chapter, we will look at some of the characteristics common to all animals and the link between structure and function. We will also see how animals maintain internal conditions within normal limits (homeostasis) despite fluctuations in their external environments. **Anatomy** is the study of an animal's structure, and **physiology** is the study of how those structures function. Collectively, anatomy and physiology form the foundation of animal biology. We begin, therefore, with an examination of the physical structure and organization of animal bodies.

40.1 Organization of Animal Bodies

All animal cells share similarities in the ways in which they exchange materials with their surroundings, obtain energy from organic nutrients, synthesize complex molecules, reproduce themselves, and detect and respond to signals in their immediate environment. Animals usually begin life as a single cell—most commonly a fertilized egg—which divides to create two cells, each of which divides in turn, resulting in four cells, and so on. If cell multiplication were the only event occurring, the end result would be a spherical mass of identical cells. As we will see in Chapter 52, however, cells become specialized during development to perform a particular function (that is, they differentiate). Examples of differentiated cells are muscle and blood cells. Cells also migrate to new locations within the developing organism and form clusters with other cells. In this way, the cells of an animal's body are arranged in various combinations to form organized, multicellular structures (**Figure 40.1**). Cells with similar properties group to form tissues (for example, muscle tissue), which combine with other types of tissues to form organs (for example, a urinary bladder). Organs are anatomically or functionally linked to form organ systems (for example, the urinary system).

Specialized Cells Are Organized into Tissues

Specialized cells of a given type often cluster together to form **tissues**. The tissues in a typical animal's body can be classified into four categories, according to their locations and the types of functions they perform: muscle, nervous, epithelial, and connective tissues. Most structures in an animal's body contain at least two types of tissues; many of these structures contain all four. Within each of these functional categories, subtypes of tissues perform variations of that function, as described next for the three types of muscle tissue.

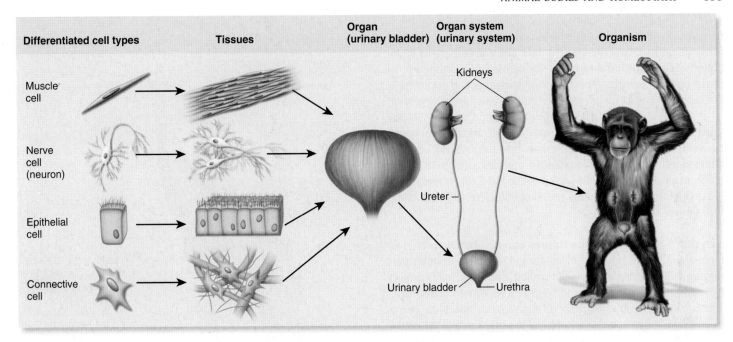

Figure 40.1 The internal organization of cells, tissues, organs, and organ systems in a mammal.   Most animals share the same four tissue types.

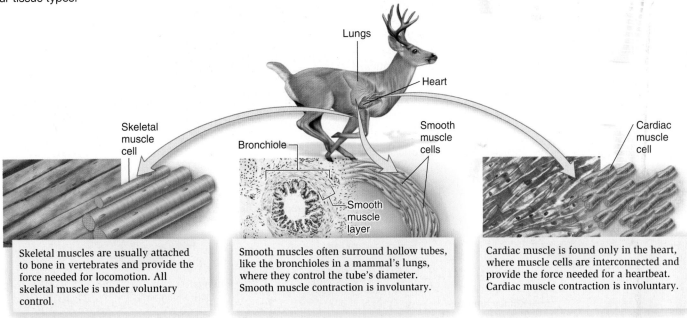

Skeletal muscles are usually attached to bone in vertebrates and provide the force needed for locomotion. All skeletal muscle is under voluntary control.

Smooth muscles often surround hollow tubes, like the bronchioles in a mammal's lungs, where they control the tube's diameter. Smooth muscle contraction is involuntary.

Cardiac muscle is found only in the heart, where muscle cells are interconnected and provide the force needed for a heartbeat. Cardiac muscle contraction is involuntary.

Figure 40.2 Three types of muscle tissue: skeletal, smooth, and cardiac.   All three types produce force, but they differ in their appearance and in their locations within animals' bodies. (Right inset: © Dr. Richard Kessel & Dr. Gene Shih/Visuals Unlimited)

Concept check: *All muscles produce movement, but only skeletal muscle produces locomotion. What is meant by this statement?*

Muscle Tissues **Muscle tissues** consist of cells specialized to shorten, or contract, generating the mechanical forces that produce body movement, decrease the diameter of a tube, or exert pressure on a fluid-filled cavity. Three types of muscle tissue may be found in animals: skeletal, smooth, and cardiac (**Figure 40.2**). **Skeletal muscles** are generally linked to bones in vertebrates via bundles of collagen fibers called tendons, and to the exoskeleton of invertebrates. When skeletal muscles are stimulated by signals from the nervous system, they generate force that leads to the contraction of the muscle (see Chapter 44). Contraction of these muscles may be under voluntary control and can produce the types of movements required for locomotion, such as extending limbs or flapping wings. Skeletal muscles may also attach to skin, such as the muscles producing facial expressions. **Smooth muscles** surround hollow tubes and cavities inside the body's organs, such that their contraction can propel the contents of those organs. For example, the contraction of smooth muscle in the stomach wall propels partially digested food into the intestines, where it can be digested fully. Smooth muscle also surrounds and forms part of small blood vessels and airway tubes (bronchioles). Contraction in those regions reduces blood flow or movement of air, respectively. Contraction of all smooth muscle is involuntary—that is, it occurs automatically without conscious control. In the third type, **cardiac muscle**, physical and electrical connections between individual cells enable many of the cells to contract almost simultaneously. Like smooth muscle, cardiac muscle is involuntary. It is found only in the heart, however, where it provides the force that generates pressure sufficient to pump blood through an animal's body.

Nervous Tissues   **Nervous tissues** initiate and conduct electrical signals from one part of an animal's body to another part (**Figure 40.3**). A single nerve cell, called a **neuron**, may connect two or more other neurons and be only a few micrometers long. Such neurons are found widely throughout the vertebrate brain. In contrast, other neurons located in the brain may send extensions along the length of the spinal cord; in a large animal like a giraffe, this distance may extend over 2 meters! Depending on where it is generated in an animal's body, an electrical signal produced in one neuron may stimulate or inhibit other neurons to initiate new electrical signals, stimulate muscle tissue to contract, or stimulate glandular cells to release chemicals into the animal's body fluids. Thus, nervous tissue provides a critical means of controlling many diverse activities of body cells.

Many neurons, as in a giraffe's brain, are short and connect to other neurons.

Spinal cord

Other neurons, such as those that extend from the brain down the spinal cord, are very long.

Figure 40.3   Variation in shape and length of neurons.

Nasal passage:
Pseudostratified ciliated columnar

Basil lamina

Kidney tubules:
Simple cuboidal

Urinary system:
Transitional

Esophagus lining:
Stratified squamous

Lungs:
Simple squamous

Intestines:
Simple columnar

Figure 40.4   **Examples of epithelial tissue.**   There are several types of epithelial tissue, distinguished by their appearance. Epithelial tissue is used to construct body coverings and the protective sheets that line and cover hollow tubes and cavities.

Blood is composed of red and white blood cells and cell fragments called platelets, all three of which are suspended in a watery fluid called plasma that is rich in electrolytes, proteins, and other solutes.

Adipose tissue is composed of fat-filled cells, which provide a layer of protection and insulation around internal organs and under the skin. Adipose tissue is also a major energy store.

Adipose tissue

Blood

Bone

Bone is composed of bone-forming cells that secrete the protein collagen. The collagen is embedded in a hard casing composed of calcium and phosphorus, which gives bone its inflexible, tough characteristics suitable for support and protection.

Dense connective tissue has tightly packed layers of collagen fibers in parallel arrays, giving the tissue great strength but very little flexibility, as in tendons and ligaments.

192 μm

Cartilage

Dense connective tissue   Loose connective tissue

Loose connective tissue is abundant throughout animals' bodies, where it holds internal organs in place and provides much of the internal framework of the body. It is composed of loosely arranged collagen fibers mixed with elastin fibers, which allows it to be flexible.

160 μm

Cartilage is formed by collagen-secreting cells. Cartilage is not mineralized and is therefore softer and more flexible than bone, providing flexibility of movement and cushioning of joints in animals with bony skeletons.

Figure 40.5 **Examples of connective tissue in mammals.**   Connective tissue connects, surrounds, anchors, and supports other tissues and may exist as isolated cells (blood), clumps of cells (fat), or tough, rigid material (bone and cartilage). The samples have been stained or the micrographs have been computer-colorized to reveal connective tissue.

Epithelial Tissues **Epithelial tissues** are sheets of densely packed cells that cover the body or individual organs or line the walls of various cavities inside the body. Epithelial cells (see Chapter 10) are specialized to protect structures and to secrete and absorb ions and organic molecules. For example, epithelial tissue can invaginate (fold inward) to form sweat glands that secrete water and ions onto the surface of an animal's skin. Epithelial cells come in a variety of shapes, such as cuboidal (cube-shaped), squamous (flattened), and columnar (elongated), and are arranged in epithelial tissues as simple (one layer), stratified (multiple layers), or pseudostratified (one layer, but with nuclei located in such a way that it appears stratified) (**Figure 40.4**). Regardless of their location or function, all epithelial cells are asymmetrical, or polarized. This means that one side of such a cell is anchored to an extracellular matrix called the basal lamina, or basement membrane (see Chapter 10). The other side faces the

internal or external environment of the animal. Thus, epithelial cells form boundaries between different body compartments, as discussed later in this chapter. In this way, epithelial tissues can function as selective barriers that regulate the exchange of molecules between compartments. For example, epithelial tissues in an animal's skin help to form a barrier that prevents most substances in the external environment from entering the body.

Connective Tissues As their name implies, **connective tissues** connect, surround, anchor, and support the structures of an animal's body. Connective tissues include blood, adipose (fat-storing) tissue, bone, cartilage, loose connective tissue, and dense connective tissue (**Figure 40.5**).

An important function of some types of connective tissue cells is to form part of the extracellular matrix around cells by secreting a mixture of fibrous proteins and carbohydrates, such as glycosaminoglycans. These carbohydrates may covalently attach to proteins to form proteoglycans. In some cases, the extracellular matrix is rich in minerals. The final characteristics of any type of connective tissue are determined in part by the relative proportions and types of proteins, proteoglycans, and minerals secreted into the extracellular matrix. The matrix serves several general functions, which include (1) providing a scaffold to which cells attach, (2) protecting and cushioning parts of the body, (3) providing mechanical strength, and (4) transmitting information to the cells that helps regulate their activity, migration, growth, and differentiation.

The proteins of the extracellular matrix of a tissue consist mainly of two types. The first type is insoluble fiber-like proteins such as **collagen** and the rubber band-like protein **elastin**; these proteins are often referred to as fibers. A second category is adhesive proteins (fibronectin and laminin) that serve to organize the protein and carbohydrate components of the extracellular matrix (refer back to Tables 10.1 and 10.2, and Figures 10.1–10.4).

Different Tissue Types Combine to Form Organs and Organ Systems

An **organ** is composed of two or more kinds of tissues arranged in various proportions and patterns, such as sheets, tubes, layers, bundles, or strips. For example, the vertebrate stomach (**Figure 40.6**) consists of the following layers:

- an outer covering of simple squamous epithelial tissue;
- connective tissue layers covering and cementing the organ together;
- layers of smooth muscle tissue whose contraction propels food through the stomach;

- nervous tissue that comes in close contact with the smooth muscle tissues and helps regulate their activity;
- an inner lining of simple columnar epithelial tissue that secretes hormones (long-distance signaling molecules), enzymes and acid (important in the digestive process), and protective mucus.

In an **organ system**, different organs work together to perform an overall function. In the example just described, the stomach is part of the digestive system, along with other structures, such as the mouth, esophagus, small and large intestines, and anus. In another familiar example, the kidneys, the urinary bladder, the tubes leading from the kidneys to the bladder, and the tube leading from the bladder to the exterior of the body constitute the urinary system in mammals (see Figure 40.1). This system helps regulate the composition of body fluids and removes waste products from the blood, which are then excreted in the urine. The organ systems found in animals are listed in **Table 40.1**. Some organ systems may be rudimentary in some animals (such as the simple nervous systems of worms and other early-diverging groups of animals).

Organ systems frequently work together. For example, signals from the nervous system and endocrine system strongly influence how much water the mammalian kidney retains as it forms urine, an adaptation that can be lifesaving under certain circumstances. Likewise, nerve signals from the urinary bladder relay information to the cells of the brain when the bladder is full and needs to be emptied.

The spatial arrangement of organs into organ systems is part of the overall body plan of animals. Organ systems develop at specific times and locations within the body and along the anteroposterior body axis, as do other structures, such as

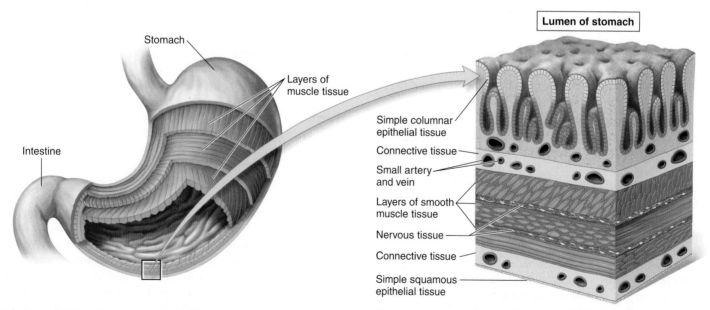

Figure 40.6 The vertebrate stomach as an example of an organ composed of all four tissue types. In this illustration, the thickness and appearance of the layers of nervous tissue have been considerably exaggerated for visual clarity.

Concept check: Do all animal organs contain all four tissue types? Can you think of an example that does not?

Table 40.1   Organ Systems Found in Animals

Organ system	Major components*	Major functions
Circulatory	Contractile element (heart or vessel); distribution network (blood vessels); blood or hemolymph	Transports and distributes solutes (nutrients, gases, wastes, and so on) to and from all parts of an animal's body
Digestive	Ingestion structures (mouth or mouthparts); storage structures (crop, stomach); digestive and absorptive structures (stomach, intestines); elimination structures (rectum, anus); accessory structures (pancreas, gallbladder)	Breaks complex foods into smaller, absorbable units; absorbs organic nutrients, salts, and water; eliminates solid wastes
Endocrine	All glands, organs, or tissues that secrete hormones; examples include the pituitary and thyroid glands	Regulates and coordinates processes such as growth, development, metabolism, mineral balance, and reproduction
Excretory	Filtration system (kidneys or comparable structures); storage sites for soluble wastes (bladder); tubes connecting kidneys and bladder, and bladder to external environment	Eliminates soluble metabolic wastes; regulates body fluid volume and solute concentrations
Immune and lymphatic	Circulating white blood cells; lymph vessels and nodes	Defends against pathogens
Integumentary	Body surfaces (skin)	Protects from dehydration and injury; defends against pathogens; regulates body temperature; in some animals, transports gases and fluids
Muscular-skeletal	Force-producing structures (muscles); support structures (bones, cartilage, exoskeleton); connective structures (tendons, ligaments)	Produces locomotion; generates force; propels materials through body organs; supports body
Nervous	Processing (brain); signal delivery (spinal cord, peripheral nerves and ganglia, sense organs)	Regulates and coordinates many body activities, such as movement, sensation, and learning
Reproductive	Gonads and associated structures	Produces gametes (sperm and egg); in some animals, provides nutritive environment for embryo and fetus
Respiratory	Gas exchange sites (gills, skin, trachea, lungs)	Exchanges oxygen and carbon dioxide with environment; regulates blood pH

*Selected examples only; these do not necessarily pertain to all animals.

limbs, tentacles, antennae, and other animal appendages. Scientists have long wondered how the layout of animal bodies is determined during the period when an embryo is developing. Remarkably, the body plan of animals appears to be under the control of a highly conserved family of genes with homologs in all animals, as described next.

Genomes & Proteomes Connection

Organ Development and Function Are Controlled by Homeotic Genes

In previous chapters, you have learned about a family of ancient, highly conserved genes called homeotic genes that are found in all animals. These genes determine the timing and spatial patterning of the anteroposterior body axis during development. For example, we saw in Chapter 19 how homeotic genes determine the number and position of legs and wings in *Drosophila*. In vertebrates, the homeotic genes are known as *Hox* genes, and they play a similar role in determining the spatial patterning of the vertebrate body and appendages. Recently, scientists have begun exploring the role of *Hox* genes in the development and spatial patterning of the organs that make up animals' organ systems.

By generating mutant mice that fail to express one or more *Hox* genes (the genes are said to have been knocked out; see Chapter 20), researchers have discovered the important role these genes play in determining where within the vertebrate body particular organs form. Recall from Chapter 19 (refer back to Figures 19.16 and 19.17) that mouse *Hox* genes are arranged in four clusters, designated A–D, with multiple genes per cluster. Because these homologous genes (for example, *HoxA-3*, *HoxB-3*, and *HoxD-3*) are found within a single species, they are considered paralogous genes. Such paralogous genes typically act in concert to regulate similar developmental processes. For example, *HoxA-3* is important for development of anterior parts of the body, including the neck. When this gene is knocked out, mouse embryos show defects in neck structure, such as abnormal blood vessels. Also, the organs within the neck—including the thymus, thyroid, and parathyroid glands—do not develop normally. When two or more paralogs of this group are knocked out, certain neck organs fail to form at all. Experimental deletion of *Hox* genes associated with other body segments does not affect the development and function of neck glands and organs. Likewise, investigators have uncovered vital roles of different *Hox* genes in lung development within the thorax and the proper positioning and development of the vertebrate kidneys in the abdomen.

Of particular interest is the discovery that *Hox* genes are important in vertebrates not only for spatial patterning of organs but also for their growth, development, and function. Genes in the *Hox* 1 and 3 groups, for instance, help determine the final branching patterns of the airways of the lungs, the final size of the lungs, and the ability of the lungs to produce secretions that are important for breathing air after birth. Other *Hox* genes have been shown to control cell proliferation, shape changes, apoptosis, cell migration, and cell-cell adhesion within various organs. This is also true in invertebrates, such as the leech, where homeotic genes are first expressed during organ formation, and in *Drosophila*, where the final shape and size of the heart are partly controlled by homeotic genes.

Chapter 52 will describe in greater detail how organs form. *Hox* genes in vertebrates and their homologs in invertebrates are important in organ positioning, development, and function. This explains in part why the thyroid gland forms in the neck region of vertebrates and not in the legs or tail, for example. Interestingly, *Hox* genes continue to be expressed long after embryonic development has ceased, and scientists believe that they play roles in organ function even in adult animals. Researchers have hypothesized that the expression of abnormal *Hox* genes in a mature organ is a contributing factor in organ disease, such as lung disorders and leukemia (cancer of the white blood cells).

Body Fluids Are Distributed into Compartments

All animal bodies are composed primarily of water. Indeed, even terrestrial animals can be thought of as "living in water," because all the cells of an animal's tissues and organs are filled with and surrounded by water.

Most of the water in an animal's body is contained inside its cells and therefore is called **intracellular fluid** (from the Latin *intra*, meaning inside of). The rest of the water in its body exists outside of the cells and is therefore called **extracellular fluid** (from the Latin *extra*, meaning outside of). Extracellular fluid is composed of the fluid part of blood, called **plasma**, and the fluid-filled spaces that surround cells, called **interstitial fluid** (from the Latin *inter*, meaning between) (Figure 40.7). In vertebrates and some invertebrates, plasma and interstitial fluid are kept separate within a closed circulatory system. In many invertebrates with open circulatory systems (see Chapter 47), however, plasma and interstitial fluid are intermingled into a single fluid called hemolymph.

The fluids depicted in Figure 40.7 are said to be enclosed in compartments. In a typical vertebrate, the total water volume in the three compartments (intracellular fluid, plasma, and interstitial fluid) accounts for about two-thirds of body weight, with solids comprising the rest. Of the total body water, up to two-thirds is intracellular and one-third extracellular, the majority of which is located in the interstitial compartment.

Plasma membranes separate the intracellular fluid from the extracellular fluid. The two components of extracellular fluid in vertebrates and some invertebrates—the interstitial fluid and the plasma—are separated by the walls of the blood vessels (for example, arteries, capillaries, and veins).

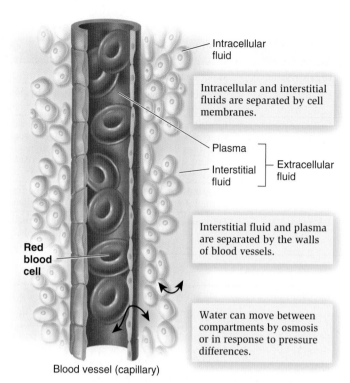

Figure 40.7   **Fluid compartments in a typical vertebrate.** Most of the fluid within an animal's body exists within cells (intracellular fluid). Extracellular fluid is that portion of the body's fluid that lies outside cells (interstitial fluid) and within blood vessels (plasma), such as the capillary shown here. Arrows indicate directions of water movement between adjacent compartments.

Concept check: *What would happen to the distribution of water in the fluid compartments of an animal's body if a blood vessel were damaged such that it leaked its contents?*

The solute composition of the extracellular fluid is very different from that of the intracellular fluid. Maintaining differences in fluid composition across the plasma membrane is an important way in which animal cells regulate their own activity. For example, many different proteins that are important in regulating cellular events such as mitosis, cytokinesis, and metabolism are confined to intracellular fluid.

Movement of Solutes Between Compartments   Solutes must move between body fluid compartments in order for cells in an animal's body to maintain concentrations of ions, nutrients, and gases such as oxygen within their normal ranges. Barriers separating adjacent fluid compartments determine which solutes can move between them. Solute movement, in turn, accounts for the differences in composition of the different compartments. We discussed the mechanisms by which solutes move in Chapter 5. Let's summarize those mechanisms, which apply to all animal cells.

Simple diffusion is the movement of a solute down its concentration gradient without the aid of a transport protein or hydrolysis of ATP. Simple diffusion is one way in which cells gain and lose solutes. Molecules that can cross lipid bilayers are able to passively diffuse into or out of a cell. Examples include nonpolar solutes such as lipids and CO_2, and extremely small polar solutes such as ethyl alcohol. Most polar molecules and ions, however, can diffuse through a plasma membrane only if the membrane has pores or channels of some type that permit the molecule to pass through the bilayer. The rate of diffusion of any solute depends on several factors, notably the concentration gradient of the solute and the area across which it is diffusing. The rate of diffusion of a solute across a membrane of given thickness can be calculated by the Fick diffusion equation, shown here in simplified form:

$$J = KA \, (C_1 - C_2)$$

where J is the rate of diffusion, K is a constant that includes temperature, A is the cross-sectional area of the barrier across which diffusion is occurring, and C_1 and C_2 are the concentrations of the solute at two locations (for example, inside and outside a cell). This equation makes it possible to determine how changes in solute concentrations, temperature, or area can influence the rate at which a substance moves across a plasma membrane. For example, breathing a gas mixture from a tank that is enriched in oxygen will increase the amount of oxygen entering the blood in a mountain climber at high altitude, where oxygen is limited. According to the Fick equation, the difference between C_1 (inhaled oxygen) and C_2 (oxygen in the blood) will be increased by this procedure. Therefore, we can predict that J, the rate of diffusion of oxygen into the blood, will also be increased, an important survival mechanism at very high altitudes.

The movement of most solutes between compartments or across plasma membranes is mediated by transport proteins via facilitated diffusion or active transport (see Chapter 5). Each of these processes is critical to regulation of intracellular and extracellular fluid composition in animal cells. In Chapter 45, we will discuss one example, the mechanism by which cells obtain their most important energy source, glucose. Because glucose is a relatively large and polar molecule, it cannot diffuse through lipid bilayers. Instead, it is transported from the interstitial fluid into cells by membrane-bound proteins.

Movement of Water Between Compartments   Water can readily move between adjacent compartments in an animal's body, because barriers such as plasma membranes tend to be highly permeable to water. This movement depends on pressure differences in the fluids of each compartment and on osmosis (see Chapter 5), in which water moves from a compartment of lower solute concentration to one of higher solute concentration. For cells to function properly, they require a relatively stable internal composition, including ion and protein concentrations, cellular volume, and pH. A decrease in solute concentration outside a cell, for example, would cause water to move by osmosis from outside the cell to inside. In this case, osmosis redistributes fluid from the interstitial to the intracellular compartment. This would cause a cell to become deformed as it swells due to the influx of water. In contrast, an increase in extracellular solute concentration would lead to osmosis of water from inside the cell to outside, causing the cell to shrink. In either case, a swollen or shrunken animal cell generally is more fragile than a normal cell and will die if its membrane ruptures. **Figure 40.8** shows examples of mammalian red blood cells in which intracellular fluid levels have been altered. This could occur, for example, if the blood cells were exposed to extracellular fluids that were either more dilute (hypoosmotic or hypotonic; see Chapter 5) or more concentrated (hyperosmotic or hypertonic) than the fluid inside the blood cell. When red blood cells swell, they may burst, a phenomenon called hemolysis; shrinkage of red blood cells is called crenation (Figure 40.8, middle panel).

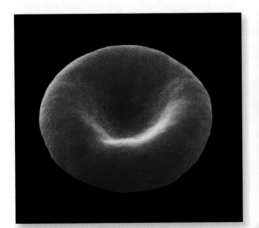

Red blood cell in a solution of normal osmolarity (an isotonic solution)

Red blood cell that has lost intracellular fluid when placed in a solution of higher than normal osmolarity (a hypertonic solution)

Red blood cell that has gained intracellular fluid when placed in a solution of lower than normal osmolarity (a hypotonic solution)

Figure 40.8   **Changes in cell shape due to alterations in intracellular fluid volume.**   Alterations in intracellular fluid volume can have drastic effects on cell shape, as shown by these scanning electron micrographs of red blood cells. Each cell is approximately 7 μm in diameter. Dramatic changes in shape like these are usually lethal for cells.

Concept check: *What effect would changes in intracellular fluid volume have on intracellular solute concentration?*

40.2 The Relationship Between Form and Function

A key theme throughout this unit is that form (structure) and function are closely related. The appearance or structure of an animal's tissues and organs can often help us predict the function of those structures. For example, let's compare the respiratory systems of an insect and a mammal (**Figure 40.9**). The respiratory systems of animals exchange oxygen from the environment with carbon dioxide generated by the body. Although many important differences exist between the respiratory systems of insects and mammals, notably the absence of lungs in insects, certain structural similarities suggest that both systems serve similar functions. In both cases, for example, a series of internal branching tubes composed of epithelial and connective tissues arises from one or more openings that connect with the outside environment (the mouth and nose in the mammal, and the spiracles in the insect). These tubes become smaller and smaller as they continue to branch, eventually terminating in narrow structures that are only one cell thick.

Without knowing anything else about the respiratory systems of these two animals, we can surmise that in both cases these branching tubes serve as conduits for air to flow back and forth between the environment and the internal spaces of the animal. In the insect, the ends of the branching tubes called tracheoles are where oxygen diffuses from the air to the fluid around individual cells (and from there to intracellular fluid) (Figure 40.9a). In the mammal, the ends of the tubes form saclike structures called alveoli across which oxygen diffuses into the bloodstream.

If we examine the mammalian lung in greater detail (Figure 40.9b), we see that the alveoli are composed of extremely thin, squamous epithelial cells. The shape of the cells provides a clue to their function. Their thinness permits diffusion of gases across the cells as rapidly and efficiently as possible. Imagine the resistance to oxygen diffusion if the cells were thick or scaly, like the cells of the body surface of many animals, for example. Therefore, both the gross and microscopic anatomy of the gas-exchange surfaces of respiratory systems facilitates their functions.

An additional structural similarity found in essentially all respiratory surfaces, including gills, is an extensive surface area. In fact, we can expand our discussion to include all cells, tissues, and organs that mediate diffusion or absorption of a solute from one compartment to another, or which require extensive cell-to-cell contacts. Consider, for instance, the finger-like projections of the small intestine of a human, the skin folds of some high-altitude frogs, the cellular extensions on the surface of neurons of a mouse, and the feathery antennae of a moth (**Figure 40.10**). What do these structures have in common? They all display a large surface area, which maximizes the ability of a tissue or organ to absorb solutes (intestine), obtain oxygen from the environment (frog skin), communicate with other cells (neurons), or sense airborne molecules (moth antennae). Increasing surface area of a structure, however, comes at the expense of greatly increasing volume if the shape of the structure is not changed. This is because as an object enlarges, its volume grows relatively more than its surface area; surface area increases by the power of two, while volume increases by the power of three as an object enlarges. For example, if an animal's height, length, and width are increased by a factor of 10, its surface area is

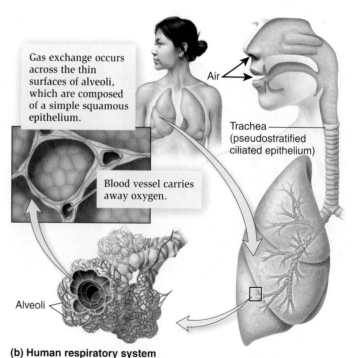

(a) Insect respiratory system

Spiracle
Air
Body surface
Tracheae
Strip of skeletal muscle
Gas exchange occurs across thin, narrow tracheoles.

(b) Human respiratory system

Gas exchange occurs across the thin surfaces of alveoli, which are composed of a simple squamous epithelium.
Air
Trachea (pseudostratified ciliated epithelium)
Blood vessel carries away oxygen.
Alveoli

Figure 40.9 Comparison of the branching air tubes in (a) an insect and (b) a mammal. Note the similar features of highly branching, internalized hollow tubules that connect to the outside air, suggesting that these systems perform similar functions.

(a) Human intestine

(b) Frog skin

(c) Mouse brain neuron stained with a fluorescent marker

(d) Moth antennae

Figure 40.10 Examples of structures in which extensive surface area is important for function. A large surface area allows **(a)** high rates of transport of digested foods across the intestine of a human, **(b)** increased diffusion of oxygen across the folds of skin of a frog living at high altitude, where O_2 is less available, **(c)** extensive communication between neurons in a mouse's brain, and **(d)** detection of airborne chemicals by moth antennae.

Concept check: Is surface area important only for animals, or could it also provide advantages to other living organisms?

increased 100 times, but its volume is increased 1,000 times. This relationship is also true when considering an animal's organs and other structures, and could create certain disadvantages. For example, the ability to obtain sufficient oxygen from water requires a great amount of surface area within the fine structure of a fish's gills. Were the gills to increase their volume in the expected proportion just described, they would become far too unwieldy to allow the normal behaviors required for survival of a fish. The challenge of packaging an extensive surface area into a confined space is overcome by changes in shape (see, for example, the way the inner surface of the intestines in Figure 40.10a folds inward to form finger-like extensions). In a fish gill, numerous thin, flat, platelike structures are packed together, one on another, forming a dense array of surfaces available for oxygen diffusion. The ratio between a structure's surface area and the volume in which the structure is contained is called the **surface area/volume (SA/V) ratio**. This concept will appear throughout this unit as we explore the ways in which animals obtain energy, regulate their metabolism and body temperature, obtain oxygen, and eliminate wastes.

When all of an animal's organ systems operate correctly and body fluid levels and solute composition are maintained within normal limits, the animal is generally considered to be in a healthy condition. In the rest of this chapter, we will examine the process of achieving and maintaining this condition, which is known as homeostasis.

40.3 Homeostasis

The environmental conditions in which organisms—including all animals—live are rarely, if ever, constant. Animals are exposed to fluctuations in air and water temperatures, food and water supplies, pH, and, in some cases, oxygen availability. Any one of these environmental changes could be harmful or even fatal if an organism is unable to respond appropriately. However, as you might expect from the incredible diversity of environments in which they exist, animals can adjust in many ways to their surroundings and thrive.

The process of maintaining a relatively stable internal environment despite changes in the external surroundings is known as **homeostasis** (from the Greek *homoios*, meaning similar, and *stasis*, meaning to stand still). The term was coined in the 20th century by the American physician and physiologist Walter Cannon, but the concept itself originated in the 19th century with the French physician and physiologist Claude Bernard, who postulated that a constant *milieu interieur* (internal environment) was a prerequisite for good health. The ability to maintain homeostasis can impart a selection advantage for animals (as well as other organisms), because many of the fundamental processes of life—for example, enzyme activity—operate most efficiently within narrow limits of such variables as temperature and intracellular pH.

Some Animals Conform to External Environments; Others Regulate Their Internal Environments

Generally speaking, animals use two mechanisms to maintain homeostasis: conforming and regulating. Some animals conform to their environments so that some feature of their internal body composition matches their external surroundings. A marine crab, for example, has about the same solute concentration in its body fluids as is found in seawater. Likewise, the body temperatures of many fishes and aquatic invertebrates match the temperatures of the surrounding waters. Energetically speaking, conforming is a cheap strategy for survival. It would take a great deal of energy for a small fish to maintain its body temperature at, say, 37°C when swimming in the waters of Antarctica. However, because they do not actively adjust their internal body composition, conformers are generally restricted to living in environments that are relatively stable.

Other animals regulate the internal composition of their fluids and solutes at levels that are different from those of the external environment. Most vertebrates are regulators, but as just noted for fishes, exceptions exist. A single animal may be both a conformer and a regulator with respect to different physiological variables. A fish may conform its body temperature to the environment but regulate its internal solute concentrations at levels different from that of fresh or salt water. Regulating the internal environment requires considerable energy in the form of ATP. For this reason, homeostasis for regulators comes with a high price tag. However, the energetic price paid by regulators makes it possible for them to exploit environments that fluctuate significantly, something that is difficult for conformers.

Vertebrates Maintain Most Physiological Variables Within a Narrow Range

In vertebrates, the common physiological variables—levels of bloodborne factors such as minerals, glucose, and oxygen, for example—are usually maintained within a certain range despite fluctuating external environmental conditions (**Table 40.2**). At first glance, homeostasis may appear to be a state of stable balance of physiological variables. However, this simple idea cannot capture the scope of homeostasis. For example, no physiological function is constant for very long, which is why we call them variables. Some variables may fluctuate around an average value during the course of a single day yet still be considered in balance. Homeostasis is a dynamic process, not a static one.

Consider an example in your own body. Normally, blood sugar (glucose) remains at fairly steady and predictable levels in any healthy individual. After a meal, however, the level of glucose in your blood can increase quickly, especially if you have just eaten something sweet. Conversely, if you skip a few meals, your blood sugar level may drop slightly (**Figure 40.11**). Such fluctuations above and below the normal value might suggest that blood glucose levels are not homeostatic, but this is incorrect. Once blood glucose increases or decreases, homeostatic mechanisms restore glucose levels back toward normal. In the case of glucose, the endocrine system is primarily responsible for this quick adjustment, but in other examples, a wide variety of control systems may be initiated. In later chapters, we will see how every organ and tissue of an animal's body contributes to homeostasis, sometimes in multiple ways, and usually in concert with each other.

Homeostasis, then, does not imply that a given physiological function is rigidly constant. Instead, homeostasis means

Table 40.2	Selected Examples of Homeostatic Variables in Animals	
Variable	**Factors that influence homeostasis**	**Examples of functions**
Minerals	Eating food; excreting wastes	
Na^+ and K^+		Establish resting membrane potentials across plasma membranes in all cells and are responsible for transmitting electrical signals in excitable tissues (muscles and nervous tissue).
Ca^{2+}		Important for muscle contraction; neuron function; skeleton and shell formation
Fe^{2+}		Binds and transports oxygen in blood or body fluids (some invertebrates use copper instead of iron)
Fuel sources	Eating food; expending energy	
Glucose		Broken down to provide energy for use by all cells, especially brain cells
Fat		Provides an alternate source of energy, particularly for cells not in the nervous system; major component of plasma membranes
ATP		Provides energy to drive most chemical reactions and body functions; modifies function of many proteins by transferring its terminal phosphate group to proteins
Body temperature	Changing rates of energy expenditure; environmental temperature; behavioral mechanisms (see Chapter 46)	Determines the rate of chemical reactions in an animal's body
pH of body fluids	Hydrogen ion pumps in cells; buffers in body fluids; rates of energy expenditure; breathing rate	Affects enzymatic activity in all cells
Other variables		
Oxygen and carbon dioxide	Movement of air or water across respiratory surfaces (for example, lungs and gills); rates of energy expenditure	Oxygen circulates in body fluids and enters cells, where it is used during the production of ATP; carbon dioxide is a waste product that is eliminated to the environment, but it is also a key factor that regulates the rate of breathing air or water
Water	Drinking, eating, excretion of wastes, perspiration, osmosis across body surface (for example, skin or gills)	Numerous biological functions including participating in chemical reactions, helping to regulate body temperature, and acting as a solvent for biologically important molecules (see Chapter 2)

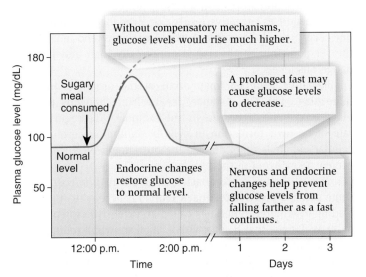

Figure 40.11 **An example of a homeostatically controlled variable, glucose levels in human blood.** Note that glucose levels in the plasma may rise or fall depending on whether an animal has recently eaten. However, even after a sugary meal or a prolonged fast, homeostatic mechanisms either return glucose levels to normal or enable those levels to remain within the range required for survival. Traditional units for glucose used in the U.S. are given on the y-axis; as a reference, a value of 100 mg/dL is equal to 5.5 mM.

that a variable fluctuates within a certain range and that once disturbed from that range, compensatory mechanisms restore the variable toward normal levels.

Homeostatic Control Systems Maintain the Internal Environment

The activities of cells, tissues, and organs must be regulated and coordinated with each other so that any change in the extracellular fluid—the internal environment—initiates a response to correct the change. These compensating regulatory responses are performed by homeostatic control systems. A **homeostatic control system** must have several components, including:

- a **set point**—the normal value for a controlled variable;
- a **sensor**, which monitors the level or activity of a particular variable;
- an **integrator**, which compares signals from the sensor to the set point; and
- an **effector**, which compensates for any deviation between the actual value and the set point.

Figure 40.12 shows an example of a homeostatic control system that regulates body temperature in mammals. This system is somewhat analogous to the heating system of a home. In that case, a sensor and integrator within the thermostat compare the actual room temperature to the set point temperature that was determined by setting the thermostat to a given tem-

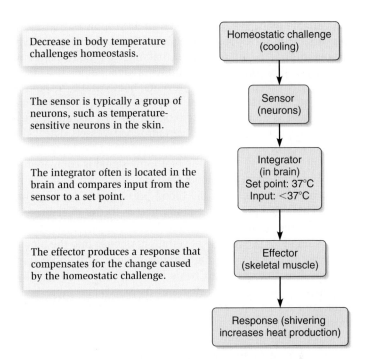

Figure 40.12 **An example of a homeostatic control system.** The sensor and effector for responding to a decrease in body temperature are shown. Different homeostatic control systems have different sensors and effectors.

Concept check: *Would you expect all animals to have similar set points for a given homeostatic variable?*

perature. If the room temperature becomes cooler than the thermostat setting, the effector (furnace) is activated and adds heat to the room. In a mammal, the sensors are temperature-sensitive neurons in the skin and brain, whereas the integrator is within the brain. Signals from the brain are sent along nerves to the effectors, which include skeletal muscles. The muscles contract vigorously in response to these signals, resulting in shivering—a key way in which mammals' bodies generate heat. We will discuss other heat-conserving and heat-generating mechanisms that contribute to this homeostatic control system in Chapter 46.

There are instances where animals make excursions far outside of their homeostatic ranges. For example, the high body temperature associated with a fever or the low body temperature associated with hibernation are temporary resettings of physiological set points.

When a homeostatic control system operates normally, changes in a physiological variable are kept to a minimum. These changes are limited by the process of feedback, as we will now see.

Feedback Is a Key Feature of Homeostasis

Feedback is a fundamental feature of homeostasis and a major way in which disturbances to a physiological variable are minimized. The temperature regulation system just described is an example of a **negative feedback loop**, in which a change in the variable being regulated brings about responses that move the variable in the opposite direction. Thus, a decrease in body temperature leads to responses that increase body temperature—that is, move it back toward its original value.

Negative feedback also prevents homeostatic responses from overcompensating. When the blood pressure of a bleeding animal falls as more and more blood is lost, for example, pressure sensors in the heart and certain blood vessels detect the change in pressure and send the information to the integrator—the brain (**Figure 40.13**). In the brain, the signal is compared to the normal set point for blood pressure. The brain responds to this sharp deviation from the blood pressure set point in two ways. First, signals are sent along nerves to the effectors—in this case, the kidneys, heart, and blood vessels. Second, the brain stimulates the release of certain hormones into the blood; these hormones act with the nervous system on the effectors. The result is that the heart beats more rapidly and forcefully, the kidneys produce less urine and thereby retain more body fluid, and the blood vessels preferentially direct blood to the most vital organs such as the brain. These responses raise the animal's blood pressure back toward the set point. Restoring the blood pressure removes the stimulus from the sensor, and

this, in turn, shuts off further production of the hormonal and neural responses (negative feedback). If feedback inhibition did not occur, the blood pressure would not only rebound back to the set point but might rise to abnormally high and possibly dangerous levels.

Negative feedback may occur at the organ, cellular, or molecular level. For instance, feedback mechanisms regulate many enzymatic processes. In one example, ATP regulates the rate of its own formation in cells by inhibiting certain intracellular enzymes that catalyze the breakdown of glucose molecules, a key event in the production of ATP.

Not all forms of feedback contribute to homeostasis. In some cases, a **positive feedback loop** may accelerate a process, leading to what is sometimes called an explosive system (think of an avalanche that begins with a small snowball rolling down a steep hill). This is contrary to the principle of homeostasis, in which large fluctuations in a variable are minimized and reversed. Not surprisingly, perhaps, positive feedback is far less common than negative feedback in animals. Nonetheless, positive feedback is crucial to some processes in animal biology. One example is the process of birth in mammals (**Figure 40.14**). Birth is triggered by a positive feedback loop between nerve signals arising from smooth muscle cells of the cervix (the connection between the vagina and the uterus) and the mother's brain and pituitary gland (a component of the endocrine system). As the uterus pushes the baby's head against the cervix in late pregnancy, nerve signals from the cervix send information

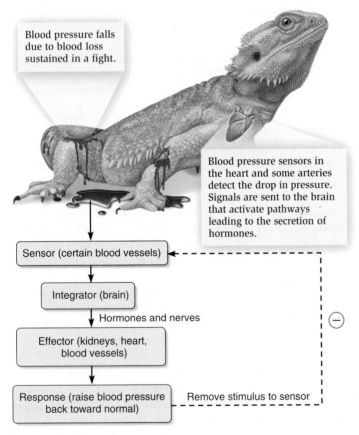

Figure 40.13 **Negative feedback as a mechanism by which homeostatic control systems operate.** In this example, loss of blood results in a drop in blood pressure, which could be life-threatening if not corrected. Effectors such as the kidneys, heart, and blood vessels help restore blood pressure toward normal. They do not raise blood pressure above normal, however, because of negative feedback (as denoted by the minus sign).

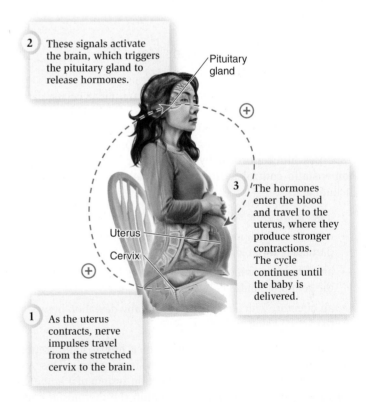

Figure 40.14 **The human birth process as an example of positive feedback.**

to the brain, which triggers the pituitary gland to release certain hormones. These hormones stimulate the uterus to contract with more force, which in turn sends additional nerve signals to the brain, and so on. Eventually, the uterus contracts strongly enough to deliver the baby, which stops the signals from the cervix to the brain and allows the uterus to gradually relax.

Feedforward Regulation Prepares for an Upcoming Challenge to Homeostasis

Built into the homeostatic mechanisms of many animals, particularly those with well-developed nervous systems, is another feature designed to minimize large swings in physiological variables. In **feedforward regulation**, an animal's body begins preparing for a change in some variable (for example, blood glucose levels) before it even occurs. Consider the anticipatory changes that occur when a hungry dog smells or sees food; indeed, the same phenomenon happens in humans. First, the animal starts to salivate, and its stomach begins to churn. Salivation and movements of the stomach are important components of the digestive process, yet at this stage, the animal has not actually eaten any food. Instead, its digestive system is already preparing for the arrival of food in order to maximize digestive efficiency, speed the flow of nutrients into the blood, and minimize the time required for active cells to replenish energy stores. Therefore, feedforward regulation speeds up the body's homeostatic responses and minimizes fluctuations in the variable being regulated—that is, it reduces the amount of deviation from the set point.

In the preceding example, feedforward control uses sensory detectors that recognize odors and sights. Many examples of feedforward control, however, result from, or are modified by, the phenomenon called learning. The result of this process is that the nervous system learns to anticipate a homeostatic challenge. Familiar examples are the increased heart rate and breathing rate that occur just before an athletic competition—for

Figure 40.15 Feedforward control of breathing rate in an animal trained for athletic exercise. Feedforward processes prepare the body for an ensuing challenge or event, such as the race shown here.

Concept check: *What kind of similar feedforward response might occur in nature?*

example, in trained racehorses before the start of a race (**Figure 40.15**). The process of training, in which the horses' bodies learn to prepare for the exertion of the race, allows there to be no delay between the start of exercise and the flow of blood and nutrients to skeletal muscle. The most famous example of feedforward control was first demonstrated in the early 20th century by Russian physiologist Ivan Pavlov, as described next.

FEATURE INVESTIGATION

Pavlov Demonstrated the Relationship Between Learning and Feedforward Processes

Ivan Pavlov made numerous contributions to our understanding of the digestive processes in mammals, for which he earned a Nobel Prize in Physiology and Medicine in 1904. Today, however, Pavlov is best remembered for work he did later in his career, when he demonstrated that feedforward processes associated with digestion could be conditioned to an irrelevant stimulus, that is, one that normally is not associated with digestive processes.

Pavlov was interested in the factors that increase production of saliva in a hungry animal. Saliva is an important secretion made by glands in the mouth, because among other things it aids in swallowing and contains enzymes that kill patho-

gens and begin the process of digesting starches in food. Pavlov discovered that dogs accustomed to being fed by the same researcher each day would begin to salivate any time they saw the researcher approaching, even without receiving, seeing, or smelling any food. In other words, the dogs had become conditioned to associate the researcher with food. This did not happen in dogs that were not yet accustomed to the regular feeding schedule of the laboratory. Pavlov hypothesized that any stimulus could elicit the feedforward process of salivation, even an irrelevant one that is not normally associated with feeding, so long as that stimulus was somehow paired with feeding.

To test this hypothesis, Pavlov presented two groups of dogs with either food (the control group) or food plus an irrelevant stimulus (the experimental group) (**Figure 40.16**). Each dog was isolated in a room so that it could not see, hear, or smell the researcher or sense any other cues that might interfere with the experiment. Contrary to popular belief, the first stimulus Pavlov used was not the ringing of a bell, but the ticking of a metronome. Pavlov reasoned that after some period of time, the experimental dogs would become conditioned to the auditory stimulus and would learn to associate it with the arrival of food. If this was correct, the metronome by itself should eventually be sufficient to elicit the feedforward process of salivation.

To measure salivary production rates, Pavlov surgically altered the ducts leading from the salivary glands located under the dogs' tongues such that the ducts opened up outside the dog's chin. A glass funnel was then glued beneath the chin. In Pavlov's initial experiments, the amount of saliva was quantified simply by counting the number of drops coming from a tube leading from the funnel. This was achieved by allowing the saliva to settle onto a mechanical device attached to a rocker arm that caused a deflection on a rotating electrical recorder. Each time a drop was collected, the arm moved and registered

a hatch mark on the recorder. The number of hatch marks was equal to the total number of drops of saliva collected in any given period. Pavlov's hypothesis was confirmed when, after a conditioning period of several days, the conditioned stimulus by itself elicited increased salivation in the conditioned dogs in under 10 seconds. The control dogs responded slightly sooner to the appearance of food than the conditioned dogs did to the metronome, but the amount of saliva produced was roughly comparable in both groups of dogs.

In later experiments, Pavlov demonstrated that other stimuli, such as touch, produced the same phenomenon he observed with sound. He also showed that the conditioned response was not permanent. If a dog that was already conditioned to the sound of a metronome was then repeatedly exposed to that sound without the simultaneous presentation of food, the salivary response to the sound would eventually cease. These experiments demonstrated that certain feedforward processes can be modulated by experience and learning and suggested

Figure 40.16 Feedforward events can be conditioned to external stimuli.

3 **THE DATA**

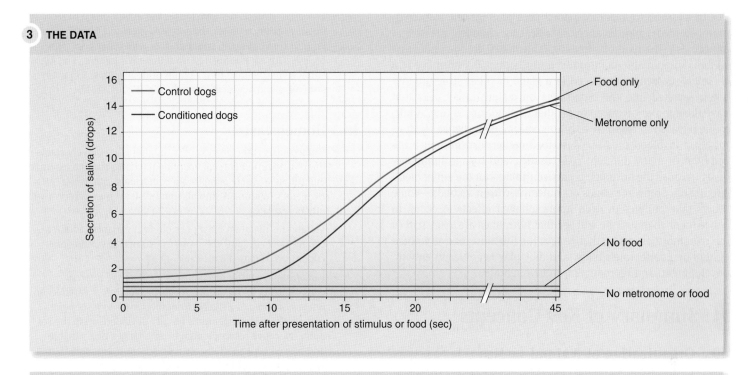

4 **CONCLUSION** The feedforward response of salivation can be modulated by experience and learning, demonstrating the adaptability of animals' physiological responses.

5 **SOURCE** These are idealized data compiled from several of the original published experiments performed by Pavlov and coworkers. The original experiments were reported in Pawlow, J. 1903. Sur la sécrétion psychique des glandes salivares. *Archives Internationales Physiologie* 1:119–135. (Note the alternate spelling of Pavlov's name.)

that animals have a much greater ability to recognize and adapt to changes in their environment than previously thought.

Experimental Questions

1. What process did Pavlov study in his experimental animals, and what was his hypothesis?

2. How did Pavlov control for the possibility that other stimuli, such as the smell of the investigator, were somehow causing the feedforward response?

3. What did Pavlov measure, and what was his major finding?

Local and Long-Distance Chemical Signals Coordinate Homeostatic Responses

A common thread that links all homeostatic processes together is communication between cells, whether cells are close to each other or in different parts of an animal's body. Some homeostatic responses may be highly localized, occurring only in the area of a disturbance. For example, damage to an area of skin causes cells in the injured area to release molecules that help contain the injury, prevent infections, and promote tissue repair in the immediate vicinity (see Chapter 53). Local responses provide areas of the body with mechanisms for local self-regulation. It would be of no benefit to an animal to promote tissue repair in regions of the body that are not injured. This type of cellular communication—in which molecules are released into the interstitial fluid and act on nearby cells—is called **paracrine** signaling (refer back to Figure 9.3).

Another example of extremely localized paracrine signaling occurs between neurons. Most neurons communicate through the release of **neurotransmitters**, small signaling molecules that are synthesized and stored in neurons. When a neuron releases these neurotransmitters, they diffuse to combine with receptor proteins on a single, adjacent neuron (or in some cases a muscle or gland cell), altering the activity of that cell. This type of cell-to-cell communication is typically very rapid, finishing within milliseconds. Consequently, neurotransmitter responses can make immediate homeostatic adjustments, like those associated with reflexes. These are just two of the many types of paracrine signaling that occur in animals' bodies and that will be described in subsequent chapters in this unit.

In addition to localized paracrine signaling, cells can communicate over long distances by releasing chemical messenger molecules. This type of signaling is mediated by the endocrine systems of animals. As introduced in Chapter 9, a hormone is a chemical messenger produced in a gland or other structure that when secreted into the blood acts on distant target cells that have the appropriate receptor for that particular hormone. A hormone released in response to a homeostatic disturbance, such as the drop in blood pressure described earlier, can influence the activities of many different cells, tissues, and organs simultaneously because the hormone is carried throughout the entire blood circulation. Some hormones act quickly—within seconds—while others take minutes or even hours for their effects to occur. In subsequent chapters, we will see that hormones are a key part of the regulatory processes that govern the functions of every organ system in a vertebrate's body, and they play key roles in growth, development, and reproduction in invertebrates.

Summary of Key Concepts

40.1 Organization of Animal Bodies

- In an animal's body, differentiated cells with similar properties group to form tissues, which combine with other types of tissues to form organs. Organs are functionally linked to form organ systems. (Figure 40.1)

- Muscle tissues consist of cells specialized to contract, generating the mechanical forces that produce movement, decrease the diameter of a tube, or exert pressure on a fluid-filled cavity. There are three categories of muscle tissue: skeletal, smooth, and cardiac. Nervous tissues initiate and conduct electrical signals from one part of an animal's body to another part. Epithelial tissues are specialized to protect structures and to secrete and absorb ions and organic molecules. Connective tissues connect, surround, anchor, and support the structures of an animal's body. Connective tissues include blood, adipose (fat-storing) tissue, bone, cartilage, loose connective tissue, and dense connective tissue. (Figures 40.2, 40.3, 40.4, 40.5)

- An organ is composed of two or more kinds of tissues. In an organ system, different organs work together to perform an overall function. (Figure 40.6, Table 40.1)

- The development, spatial positioning, and function of many body organs are under the control of highly conserved homeotic genes, such as *Hox* genes in vertebrates.

- An animal's body fluids are distributed into three compartments (intracellular fluid, plasma, and interstitial fluid). The properties of the barriers between the compartments determine which substances can move between them. (Figures 40.7, 40.8)

40.2 The Relationship Between Form and Function

- Similarities in structure are often associated with similarities in function. (Figure 40.9)

- Extensive surface area maximizes the ability of a tissue or organ to absorb solutes, exchange oxygen and carbon dioxide with the environment, communicate with other cells, and receive sensory information from the environment. The high surface area/volume (SA/V) ratio principle applies to many structures and across all animal taxa. (Figure 40.10)

40.3 Homeostasis

- Homeostasis is the process of maintaining a relatively stable internal environment and adapting to changes in the external environment. Some animals are conformers; others are regulators. Within an animal, some systems may conform while others regulate.

- Vertebrates maintain most physiological variables within a certain range despite variations in external environmental conditions. (Figure 40.11, Table 40.2)

- Homeostatic control mechanisms regulate the internal environments of organisms. A sensor monitors a variable, an integrator compares signals from the sensor to a set point, and an effector compensates for any deviations from the set point. Negative feedback minimizes changes in a variable and prevents homeostatic responses from overcompensating. (Figures 40.12, 40.13)

- Negative feedback is far more common in animals than positive feedback. Feedforward regulation prepares the body for an upcoming challenge to homeostasis. (Figures 40.14, 40.15)

- Pavlov's experiments with dogs demonstrated that certain feedforward responses such as salivation could be conditioned to irrelevant stimuli, such as the ticking of a metronome. (Figure 40.16)

- Chemical communication between cells is essential to homeostasis. Types of chemical messengers include paracrine signals, such as neurotransmitters, and long-distance signals, such as hormones.

Assess and Discuss

Test Yourself

1. Tissue that is specialized to conduct electrical signals from one structure in the body to another structure is _____ tissue.
   a. epithelial
   b. connective
   c. nervous
   d. muscle

2. Structures composed of two or more tissue types arranged in various proportions and patterns are
   a. cells.
   b. tissues.
   c. organs.
   d. organ systems.
   e. organisms.

3. The extracellular matrix of connective tissue
   a. contains different proteins that provide structural support to cells.

b. provides a scaffolding for the cells of the tissue.
c. plays a role in cellular communication.
d. does all of the above.
e. does a and b only.

4. When examining the structure of many animal organs,
   a. it is apparent that all organ systems are fully functional in all animals.
   b. the function can be predicted based on the structural adaptations.
   c. the function of the structure is difficult to determine because of the lack of variation among the different structures of the body.
   d. all four tissues are equally represented in all organs.
   e. None of the above are correct.

5. Most of the water in an animal's body
   a. lacks any type of ions or other solutes.
   b. is found in the spaces between cells.
   c. is contained inside the cells.
   d. is located in the extracellular fluid.
   e. is unable to move between body compartments.

6. The folds, convolutions, or extensions found in many structures of animals results in
   a. decreased level of activity in that particular structure.
   b. interruption in the normal functioning of the structure.
   c. increased surface area for absorption, communication, or exchange.
   d. higher susceptibility to infection.
   e. none of the above.

7. Adapting to changes in the external environment and maintaining internal variables within physiological ranges is
   a. natural selection.
   b. evolution.
   c. positive feedback.
   d. homeostasis.
   e. both c and d.

8. Which of the following statements regarding negative feedback is not correct?
   a. It helps regulate variables such as body temperature and blood pressure.
   b. It is the mechanism by which birth occurs in mammals.
   c. It is a major feature of homeostatic control systems.
   d. It prevents homeostatic responses from overcompensating.
   e. It may occur at the organ, cellular, or molecular level.

9. The ability of an animal's body to prepare for a change in some variable before the change occurs is called
   a. positive feedback.
   b. negative feedback.
   c. homeostasis.
   d. feedforward regulation.
   e. autoregulation.

10. A hormone differs from a neurotransmitter in that
    a. hormones act extracellularly; neurotransmitters act within the cell that synthesized them.
    b. hormones are released only by neurons; neurotransmitters are released by many different types of cells.
    c. hormones only cause fast responses (seconds or less) to stimuli; neurotransmitters cause slow responses (minutes to hours) to stimuli.
    d. hormones affect only epithelial cells; neurotransmitters affect only muscle cells.
    e. hormones are released into the bloodstream and can activate many cells in many parts of the body; neurotransmitters are released by neurons and affect adjacent cells.

Conceptual Questions

1. Define anatomy and physiology.

2. Describe the relationship between structure and function in animals, and describe how surface area and volume are related.

3. Define homeostasis, and give examples of homeostatic variables.

Collaborative Questions

1. Describe the terms negative feedback loop and positive feedback loop, and describe how they differ.

2. Discuss the organization of animal bodies from the cellular to the organ system level.

Online Resource

www.brookerbiology.com

Stay a step ahead in your studies with animations that bring concepts to life and practice tests to assess your understanding. Your instructor may also recommend the interactive ebook, individualized learning tools, and more.

Neuroscience I: Cells of the Nervous System

41

A fluorescently stained human neuron. These types of cells are responsible for sending signals throughout the nervous systems of animals.

A s you begin reading, stop and think. Can you describe everything you are doing right now? Your eyes are sensing light reflected off this page, and your brain is interpreting the meanings of the words you are reading. You may be hearing sounds and, in some cases, choosing to ignore them as you concentrate on reading. Your digestive system may be sending signals about hunger, or perhaps if you've just eaten, you feel full and are digesting what you ate. You are breathing, perspiring, feeling the chair on which you're sitting, and your heart is beating. All of these processes and many others are under the control of the **nervous system**, coordinated circuits of cells that sense internal and environmental changes and transmit signals that enable us and other animals to respond in an appropriate way. Nervous systems help us to exert control over our bodies. They also allow animals such as ourselves to sense what is going on in the outside world, initiate actions that influence events and respond to demands, and regulate internal processes—all while maintaining homeostasis (see Chapter 40). You are conscious of some of these functions, such as reading this text or feeling hungry. Many others, however, such as

maintaining your body temperature and controlling your heart rate, occur without your awareness.

Neuroscience is the scientific study of nervous systems. Neuroscientists are interested in topics such as the structure and function of the brain and the biological basis of consciousness, memory, learning, and behavior. Neuroscience is now experiencing an unprecedented level of new discoveries and rapid growth. It interfaces with other disciplines such as cell and molecular biology, psychology, and behavioral biology.

In this chapter, our focus on neuroscience will be at the cellular level. Nervous systems are composed of circuits of **neurons**, highly specialized cells that communicate with each other and with other types of cells by electrical or chemical signals (see chapter-opening photo). In many animals, such as ourselves, neurons become organized into **nuclei** (singular, nucleus), clusters of cells that carry out a particular function. Nuclei typically form in a central processing area of the nervous system called a **brain**. The brain sends commands to various parts of the body through **nerves**—bundles of neuronal cell extensions projecting to and from various tissues and organs.

In this chapter, we will investigate the special features of neurons that make them suited for rapid communication between cells. Chapters 42 and 43 will explore how nervous systems are organized, how the brain functions, and how animals use their nervous systems to sense the world around them.

41.1 Cellular Components of Nervous Systems

The organization of nervous systems permits extremely rapid responses to changes in an animal's external or internal environment. In complex animals, the **central nervous system** (**CNS**) consists of a brain and a nerve cord, which in vertebrates extends from the brain through the vertebral column and is called the **spinal cord**. The **peripheral nervous system** (**PNS**) consists of all neurons and projections of their plasma membranes that are outside of but connect with the CNS, such as projections that end on muscle and gland cells. In certain invertebrates with simple nervous systems, the distinction

1 **Peripheral nervous system:** Neurons in the nose and mouth detect stimuli (odor and taste) and send signals to the brain.

2 **Central nervous system:** Neurons in the brain interpret the signals as food.

Odor Taste

3 **Peripheral nervous system:** The brain sends a signal to neurons in the PNS that stimulate the salivation response.

Saliva

Figure 41.1 Roles of the central and peripheral nervous systems. In this example, a hungry hyena senses a smell and taste, which the brain interprets as a potential food source. This initiates a biological response (salivation) that prepares the hyena for eating.

Concept check: As you learned in Chapter 40, what type of response is demonstrated by the production of saliva in this figure? Note that saliva is being produced before the animal has begun eating.

between central and peripheral nervous systems is less clear or not present.

The evolution of nervous systems has allowed animals to receive information about the environment via their PNS, transmit that information along nerves to a CNS, interpret that information in a CNS, and, if necessary, initiate a behavioral response via their PNS (**Figure 41.1**). For example, if a hungry hyena receives stimuli such as the smell and taste of food, odor-sensing cells in the nose and taste-sensing cells in the tongue act as receptors for the stimuli and then send signals along nerves to the brain. There, the signals are interpreted and recognized. The brain then sends a signal via nerves that stimulate gland cells in the mouth, which respond by producing saliva in preparation for the arrival of food. In this section, we will survey the general properties of the cells of the CNS and PNS.

Cells of the Nervous System Are Specialized to Transfer Signals

Nervous systems transfer signals from one part of the body to another and direct the activities of cells, tissues, organs, and glands. Although these are complex tasks, nervous systems have only two unique classes of cells: neurons and glia.

Neurons As mentioned, neurons are cells that send and receive electrical and chemical signals to and from other neurons or other cells throughout the body. All animals except sponges have neurons. The number of neurons in the nervous systems of different species varies widely, partly as a function of the size of an animal's head and brain, but also as a function of the complexity of its behavior. As a comparison, the tiny, short-lived nematode *C. elegans* has 302 neurons in its nervous system, compared with several thousand in a wasp, several hundred thousand in a salamander, 300 million in an octopus, and over 100 billion in a human!

Regardless of the total number, neurons in one animal species look and act much like neurons from any other species. A neuron is composed of a **cell body**, or **soma**, which contains the cell nucleus and other organelles (**Figure 41.2a,b**). There are two types of extensions or projections that arise from the cell body: the dendrites and the axon. **Dendrites** (from the Greek word *dendron*, meaning tree) may be single projections of the plasma membrane but more commonly are elaborate treelike structures with numerous branching extensions. Electrical and chemical messages from other neurons are received by the dendrites, and electrical signals move toward the cell body (**Figure 41.2c**). The cell body processes these signals along with some signals that are received directly by the cell body and produces an outgoing signal to a structure called an axon. An **axon** is an extension of the plasma membrane that is involved in sending signals from a neuron to neighboring cells. An axon may be only a few micrometers long or as long as 2 m, such as in a giraffe's neck. A typical neuron has a single axon, which may have branches. The part of the axon closest to the cell body is named the **axon hillock**, also known as the initial segment. As we will see later, the axon hillock is important in the generation of the electrical signals that travel along an axon. At the other end of the axon are **axon terminals** that convey electrical or chemical messages to other cells, such as neurons or muscle cells.

Within an animal's body, many axons tend to run in parallel bundles to form nerves, which are covered by a protective layer of connective tissue. Nerves enter and leave the CNS and transmit signals between the PNS and the CNS. Along the way, the axon terminals of axons communicate with particular cells of the body.

Glia Surrounding the neurons are cells called the **glia** (from the Greek, meaning glue). Glial cells perform various functions and are many times more numerous than neurons, depending on the species. In the human brain, for example, glial cells outnumber neurons by about 50 to 1. Glial cells perform numerous important roles. One type of glia, called astrocytes, provides metabolic support for neurons and also is involved in forming the blood-brain barrier, which is a physical barrier between blood vessels and most parts of the central nervous system. This barrier protects the CNS by preventing the passage of toxins and other damaging chemicals from the blood into the

Figure 41.2 Structure and basic function of a typical vertebrate neuron and associated glial cells. (a) A stained neuron seen at high magnification (confocal microscopy). (b) A diagrammatic representation of a peripheral neuron with glial cells—in this instance, a type of glia called Schwann cells. The Schwann cells wrap their membranes around the axon at regular intervals, creating a myelin sheath that is interrupted by nodes of Ranvier. (c) The structures involved in the processing of information by a neuron; information flows in the direction shown.

CNS. Astrocytes also help to maintain a constant concentration of ions in the extracellular fluid. Other glia, called microglia, remove cellular debris produced by damaged or dying cells. In developing embryos, glia form tracks along which neurons migrate to form the nervous system. In addition, some glia function as stem cells to produce more glial cells and neurons.

In vertebrates, specialized glial cells wrap around the axons at regular intervals to form an insulating layer called a **myelin sheath** (Figure 41.2). The sheath is periodically interrupted by noninsulated regions called **nodes of Ranvier**. In the human brain and spinal cord, the myelin-producing glial cells are called **oligodendrocytes**. **Schwann cells** are the glial cells that form myelin on axons that travel outside the brain and spinal cord. As we will see later, myelin and the nodes of Ranvier increase the speed with which electrical signals pass down the axon.

Sensory and Motor Neurons as Well as Interneurons Form Pathways in a Nervous System

Neurons can be categorized into three main types: sensory neurons, motor neurons, and interneurons. The structures of each type reflect their specialized functions (**Figure 41.3**).

Sensory Neurons As their name suggests, **sensory neurons** detect or sense information from the outside world, such as light, odors, touch, or heat. In addition, sensory neurons detect internal body conditions such as blood pressure or body temperature. Sensory neurons are also called afferent (from the Latin, meaning to bring toward) neurons because they transmit information to the CNS. Many sensory neurons have a long, single axon that branches into a peripheral process and a central process, with the cell body in between (Figure 41.3a). This arrangement allows for the rapid transmission of a sensory signal to the CNS.

Motor Neurons Motor neurons send signals away from the CNS and elicit some type of response. They are so named because one type of response they cause is movement. In addition, motor neurons may cause other effects such as the secretion of hormones from endocrine glands. Because they send signals away from the CNS, motor neurons are also called efferent (from the Latin, meaning to carry from) neurons. Like sensory neurons, motor neurons tend to have long axons (Figure 41.3b), but do not branch into two main processes.

Interneurons A third type of neuron, called the **interneuron** (or association neuron), forms interconnections between other

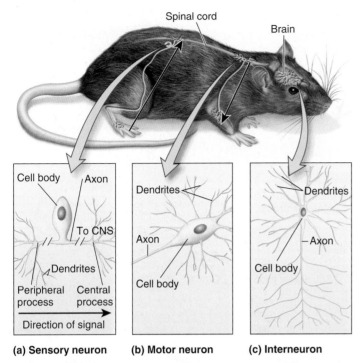

Figure 41.3 Types of neurons. (a) The vertebrate sensory neurons are afferent neurons with an axon that bypasses the cell body and projects to the CNS. (b) Motor neurons are efferent neurons that leave the CNS and usually have long axons that enable them to act on distant cells. (c) Interneurons are usually short neurons that connect two or more other neurons within the CNS. Although short, the axons and dendrites may have extensive branches, allowing them to receive many inputs and transmit signals to many neurons.

neurons in the CNS. The greatest complexity of nervous systems occurs among interneurons, such as those found in the brain. The signals sent between interneurons are critical in the interpretation of information that the CNS receives, as well as the response that it may elicit. Interneurons tend to have many dendrites, and their axons are typically short and highly branched (Figure 41.3c). This arrangement allows interneurons to form complex connections with many other cells.

Reflex Circuits As a way to understand the interplay between sensory neurons, interneurons, and motor neurons, let's consider a simple example in which these types of neurons form interconnections with each other. Neurons transmit information to each other through a series of connections that form a circuit. An example of a simple circuit is a **reflex arc**, which allows an organism to respond rapidly to inputs from sensory neurons and consists of only a few neurons (**Figure 41.4**). The stimulus from sensory neurons is sent to the CNS, but there is little or no interpretation of the signal; typically there are very few, if any, interneurons involved, as in the example shown in Figure 41.4. The signal is then transmitted to motor neurons, which elicit a response, such as a knee jerk. Such a response is very quick and automatic.

Reflexes are among the evolutionarily oldest and most important features of nervous systems, because they allow animals to respond quickly to potentially dangerous events. For instance, many vertebrates will immediately cringe, jump, leap, or take flight in response to a loud noise, which could represent sudden danger. Some animals that live in the water

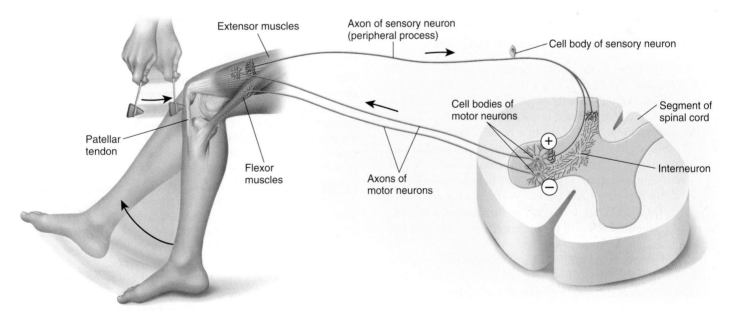

Figure 41.4 A reflex arc. The knee-jerk response is an example of a reflex arc. A tap below the kneecap (also known as the patella) stretches the patellar tendon, which acts as a stimulus. This stimulus initiates a reflex arc that activates (+) a motor neuron that causes the extensor muscle on top of the thigh to contract. At the same time, an interneuron inhibits (−) the motor neuron of the flexor muscle, causing it to relax.

Concept check: Animals have many types of reflexes. Once initiated, must all reflexes occur to completion, or do you think that in some cases they may be overridden or partially suppressed?

will reflexively dive deeper in response to a shadow overhead, which could represent a passing shark or other predator. Infant primates have strong grasping reflexes that help them hold onto their mothers as they move about. Countless examples of useful reflexes are found in animals, and their importance is evident from the observation that they arose early in evolution and exist in nearly all animals.

41.2 Electrical Properties of Neurons

In the late 18th century, Italian scientists Luigi Galvani and Alessandro Volta experimented with ways to stimulate the contraction of frog leg muscles that had been dissected and placed in saline (NaCl) solutions. The saline solutions approximated the ion concentrations in plasma, kept the muscles alive, and also conducted electricity. They discovered that the stimulation of the nerve that led to the muscle or the stimulation of the muscle itself with any source of electric current caused the muscle to contract. Eventually, Galvani postulated that electric current could somehow be generated by the tissue itself, something he called "animal electricity."

Today, we know that Galvani's animal electricity comes from neurons, which use electrical signals to communicate with other neurons, muscles, or glands. These signals, or nerve impulses, involve changes in the amount of electric charge across a cell's plasma membrane. In this section, we first examine the electrical and chemical gradients across the plasma membrane of neurons. Later, we explore how such gradients provide a way for neurons to conduct signals.

Neurons Establish Differences in Ion Concentration and Electric Charge Across Their Membranes

Like all cell membranes, the plasma membrane of a neuron acts as a barrier that separates charges. Ion concentrations differ between the interior and exterior of the cell. Such differences in charge act as an electrical force measured in **volts** (V), named after Alessandro Volta. Analogous to a battery, there are negative and positive poles, but these are the inside and outside surfaces of the plasma membrane. For this reason, a neuron is said to be electrically **polarized**. The difference between the electric charges along the inside and outside surfaces of a cell membrane is called a potential difference, or **membrane potential**. The **resting potential** refers to the membrane potential of a cell that is not sending nerve impulses.

How do scientists measure electrical changes in a structure as tiny as a neuron? Several invertebrates, such as squids, lobsters, and earthworms, have large neurons that have been used successfully to measure membrane potentials. The squid giant axon, for example, has a diameter up to 1 mm. This makes it relatively easy to impale the axon with a microelectrode, which is a recording instrument constructed of a glass pipette with an extremely thin tip of less than 1 μm in diameter. Within the microelectrode is a salt solution that can conduct electric charge (ions). The squid giant axon has been used extensively

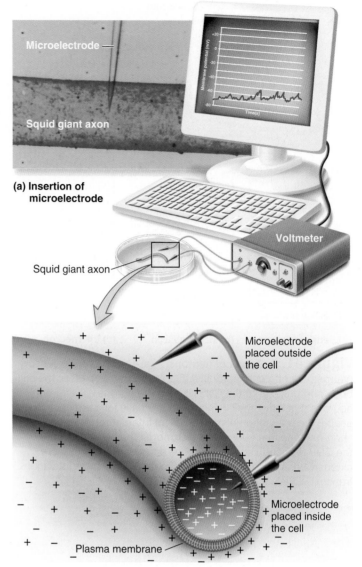

(a) Insertion of microelectrode

(b) Measurement of membrane potential

Figure 41.5 **Recording the membrane potential of neurons.** **(a)** A microelectrode, seen here under a microscope, impales the giant axon from a squid. **(b)** The potential difference across the membrane of the axon is recorded by comparing the electric charge inside and outside the axon. The membrane potential (measured in millivolts, mV) is steady at times and fluctuates at other times, depending on experimental conditions.

Concept check: *What obvious difference (besides its size) can you see between the giant axon of the squid, as illustrated here, and the illustration of the vertebrate axon depicted in Figure 41.2?*

since the mid-20th century to determine the mechanisms by which neurons generate electrical signals (**Figure 41.5**). A neuron is dissected from the squid and placed in a solution that has ionic concentrations similar to that of normal extracellular fluid. A microelectrode is pushed through the axon membrane to record from inside the cell, and another microelectrode is placed within the solution bathing the neuron outside the cell. A voltmeter records the voltage difference between the two microelectrodes, which is a measure of the membrane potential.

This measurement is made as a function of time and displayed on a computer screen.

Let's begin our discussion of electrical signaling by examining how the resting potential is established and maintained. The plasma membrane is not very permeable to cations and anions, so it separates charge by keeping different ions largely inside or outside the cell. When investigators first measured the resting potential of a squid giant axon, they registered a voltage that read about −70 millivolts (mV) inside the cell with respect to the outside bathing medium. This means that the interior of the cell had a more negative charge than the exterior, which turns out to be true of all animal cells in their resting state. A resting potential of −70 mV is tiny compared to the voltages used to provide electric current in a home (approximately 120 V), or even that of a small 1.5-V battery. Nonetheless, this tiny difference in charge across the membrane of a neuron is sufficient to generate a nerve impulse, or action potential, that can travel from one end of a neuron to the other, as we will see later in this chapter.

The resting potential is determined by the ions located along the inner and outer surfaces of the plasma membrane (**Figure 41.6a**). Ions of opposite charges align on either side of the membrane because they are drawn to each other due to electrical forces. Negative ions within the cell are drawn to the positive ions arrayed on the outer surface of the plasma membrane. Although there are more positive charges along the outside surface of a neuron and more negative charges inside, the actual number of ions that contribute to the resting membrane potential is extremely small compared to the total number of ions inside and outside the cell. **Table 41.1** lists the ions that are important in maintaining the resting potential and their intracellular and extracellular concentrations. The ions that are most critical for establishing the resting potential are Na$^+$ and K$^+$ and, to a lesser extent, Cl$^-$, and intracellular anions such as negatively charged proteins.

Three factors are primarily responsible for the resting membrane potential (**Figure 41.6b**). First, the sodium-potassium pump (Na$^+$/K$^+$-ATPase) within the plasma membrane continually moves sodium ions out of the cell and potassium ions into the cytosol (see Chapter 5). The sodium-potassium pump uses the energy of ATP to transport three Na$^+$ out of the cell for every two K$^+$ it moves into the cell. Consequently, the pump contributes modestly to a charge difference across the plasma membrane; more importantly, it establishes concentration gradients for Na$^+$ and K$^+$ by continually transporting them across the membrane in one direction. Second, the plasma membrane contains ion-specific channels that affect the permeability of Na$^+$ and K$^+$ across the membrane. Ungated channels that are specific for Na$^+$ or K$^+$ influence the resting potential by allowing the passive movement of these ions. An ungated channel is one that is open at rest and that does not respond to other stimuli, such as voltage, ligand binding, or mechanical stimulation. Ungated ion channels are sometimes referred to as "leak" channels. In most neurons, there are about 50 times more ungated K$^+$ channels than ungated Na$^+$ channels, so at rest, the membrane is more permeable to K$^+$ than to Na$^+$. As you will learn shortly, Na$^+$ tends to diffuse into cells, and K$^+$ diffuses

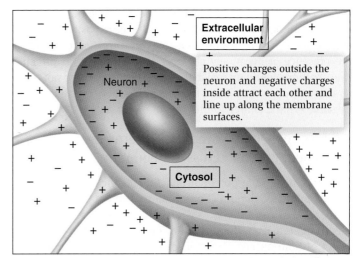

(a) Distribution of charges across the neuronal plasma membrane

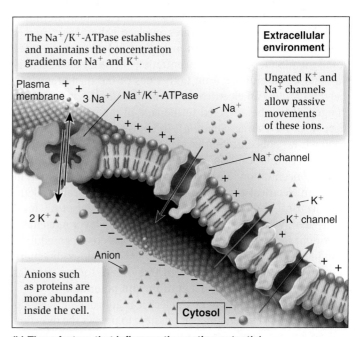

(b) Three factors that influence the resting potential

Figure 41.6   The resting membrane potential.

Table 41.1	Extracellular and Intracellular Concentrations of Ions for a Typical Mammalian Neuron	
	Concentration (mM)	
Ion	**Extracellular**	**Intracellular**
Na$^+$	145	15
K$^+$	5	150
Cl$^-$	110	7
Anions found in macromolecules	<1	65

out of cells when the cell is at rest. The greater number of leak channels for K⁺, therefore, means that excess positive charges exit the cell. Finally, a third contribution to resting potential is the presence of negatively charged molecules such as proteins that are more abundant inside the cell. These anions do not readily move through the plasma membrane, so they contribute some negative electric charge to the interior of the cell.

An Electrochemical Gradient Governs the Movement of Ions Across a Membrane

Both the membrane potential and the chemical concentration of ions influence the direction of ion movement across a membrane. The direction that an ion will move depends on the **electrochemical gradient** for that ion, which is the combined effect of both an electrical and chemical gradient. **Figure 41.7** considers the concept of an electrochemical gradient for K⁺. This hypothetical drawing illustrates two compartments separated by a semipermeable membrane that permits the diffusion of K⁺. Figure 41.7a illustrates an electrical gradient. In this case, the chemical concentration of K⁺ is equal on both sides of the membrane, but the concentrations of other ions such as Na⁺ and Cl⁻ are unequal on both sides of the membrane and thereby produce an electrical gradient. Because K⁺ is positively charged, it will be attracted to the side of the membrane with more negative charge. Figure 41.7b shows a chemical gradient in which K⁺ concentration is higher on one side compared to the other. In this scenario, K⁺ will diffuse from a region of high to low concentration. Finally, Figure 41.7c shows a situation in which an electrical gradient is balanced by a chemical gradient. The electrical gradient would favor the movement of K⁺ from left to right, while the chemical gradient would favor movement from right to left. These opposing forces can create an equilibrium in which there is no net diffusion of K⁺ in either direction. The membrane potential at which this occurs for a particular ion is referred to as that ion's **equilibrium potential**.

With two different forces—electrical and chemical gradients—acting on a given ion, is it possible to predict the direction that an ion will move across a membrane at any concentration gradient? In other words, can we compare the relative strengths of the electrical and chemical gradients and predict their net effect? By measuring the membrane potential of isolated neurons in the presence of changing concentrations of extracellular ions, scientists have deduced a mathematical formula that relates electrical and chemical gradients to each other. This formula, named the **Nernst equation** after the Nobel laureate Walther Nernst, gives the equilibrium potential for an ion at any given concentration gradient. For monovalent cations such as Na⁺ and K⁺, the Nernst equation can be expressed as

$$E = 60 \text{ mV } \log_{10} ([X_{extracellular}]/[X_{intracellular}])$$

where E is the equilibrium potential; [X] is the concentration of an ion, outside or inside the cell; and 60 mV is a value that depends on temperature, valence of the ion, and other factors. (For anions, the value would be −60 mV.)

The Nernst equation allows neuroscientists to predict when an ion is or is not in equilibrium. To understand the usefulness

(a) Electrical gradient, no chemical gradient for K⁺

(b) Chemical gradient for K⁺, no electrical gradient

(c) K⁺ equilibrium, electrical gradient is balanced by the chemical gradient for K⁺

Figure 41.7 Electrical and chemical gradients. This hypothetical example depicts two chambers separated by a membrane that is permeable to K⁺. **(a)** In this example, the compartments initially contain equal concentrations of K⁺, but an electrical gradient exists due to an unequal distribution of Na⁺ and Cl⁻. Potassium ions are attracted to the higher amount of negative charge on the right side of the membrane. **(b)** In this case, there is initially no electrical gradient across the membrane, and the left compartment contains a lower concentration of KCl compared to the right compartment. In water, the KCl dissociates to K⁺ and Cl⁻. Under these conditions, K⁺ diffuses down its chemical concentration gradient from right to left. **(c)** This example illustrates opposing electrical and chemical gradients. The right compartment contains a higher chemical concentration of K⁺, while the left side has a higher net amount of positive charge. These gradients balance each other, so no net movement of K⁺ occurs.

Concept check: In part (b), does the diffusion of K⁺ down its chemical gradient result in an electrical gradient? What will eventually stop the net diffusion of K⁺?

of this equation, consider two examples. First, let's suppose the membrane potential is −88.6 mV and the K^+ concentration is 5 mM outside and 150 mM inside a neuron. If we plug these chemical concentrations into the Nernst equation,

$$E = 60 \text{ mV} \log_{10} (5/150) = 60 \text{ mV} (-1.48)$$

$$E = -88.6 \text{ mV}$$

Under these conditions, where the K^+ equilibrium potential equals the membrane potential, K^+ is in equilibrium, and no net diffusion of K^+ will occur, even when many K^+ channels are open.

As a second example, let's suppose that the membrane potential is at the typical resting potential of −70 mV and the Na^+ concentration is 100 mM outside and 10 mM inside. If we plug these chemical concentrations into the Nernst equation,

$$E = 60 \text{ mV} \log_{10} (100/10) = 60 \text{ mV} (1)$$

$$E = 60 \text{ mV}$$

At a resting potential of –70 mV, the value of +60 mV tells us that Na^+ is not in equilibrium. When the equilibrium potential for a given ion—calculated by the Nernst equation—and the resting membrane potential do not match, there will be a driving force for that ion to diffuse across the membrane. In this example, Na^+ will diffuse into the cell, because both the electrical and chemical gradients favor an inward flow. However, if the membrane potential was +60 mV instead of −70 mV, Na^+ would be in equilibrium at these concentrations, and therefore, there would be no net diffusion of Na^+ in any direction.

By establishing electrochemical gradients and maintaining them with the sodium-potassium pump, neurons have the ability to suddenly allow ions to move across the plasma membrane by opening additional channels that were previously closed. The movement of a charged ion down its gradient results in an electric current. This current is different from the type of current that runs through electrical wires and is much smaller in amplitude. Even so, this small current provides the electrical signal that neurons use to communicate with one another, as we will see next.

41.3 Communication Between Neurons

Communication between neurons begins when one cell receives a stimulus and sends an electrical signal along its plasma membrane via currents generated by ion movements. This signal will then influence the next neuron in a circuit. Each signal is brief (only a few milliseconds), but a neuron may receive and transmit millions of signals in its lifetime. In this section, we will survey the ability of neurons to send brief signals with amazing speed.

Signaling by a Neuron Occurs Through Changes in the Membrane Potential

Recall that a cell is polarized because of the separation of charge across its membrane. The resting potential of the membrane is more negative inside than outside. Changes in the membrane

potential, therefore, are changes in the degree of polarization. **Depolarization** occurs when the cell membrane becomes less polarized, that is, less negative inside the cell relative to the surrounding fluid. As described later, when a neuron is stimulated by an electrical or chemical stimulus, one or more types of gated membrane channels open, and sodium ions may move into the cell, bringing with them their positive charge. This changes the membrane potential to a level somewhat less negative than the resting membrane potential. Consequently, the membrane potential is said to be depolarized. In contrast, **hyperpolarization** occurs when the cell membrane becomes more polarized, that is, more negative relative to the extracellular fluid. For example, increased movement of K^+ out of the cell would make the internal charge of the cell membrane more negative than the resting potential.

Though all cells in an animal's body have a membrane potential, neurons and muscle cells are called **excitable cells** because they can generate electrical signals. For a cell to communicate using electrical signals, it must be able to change its membrane potential very rapidly. This is accomplished by gated ion channels, so called because they open and close in a manner analogous to a gate in a fence (**Figure 41.8**; also refer back to Chapter 5). **Voltage-gated ion channels** open and close in

(a) Voltage-gated ion channel

(b) Ligand-gated ion channel

Figure 41.8 Examples of gated ion channels. (a) A voltage-gated ion channel allows ions to diffuse into the cell. These channels open or close depending on changes in charge (voltage) across the membrane. (b) A ligand-gated ion channel opens or closes in response to ligand binding. In the example here, the binding of a neurotransmitter opens the channel.

Figure 41.9 Graded membrane potentials. Within a limited range, the change in the membrane potential occurs in proportion to the intensity of the stimulus. If the membrane potential becomes less negative, the resulting change is called a depolarization. Hyperpolarization results from the membrane potential becoming more negative following a stimulus.

response to changes in voltage across the membrane. **Ligand-gated ion channels**, also known as chemically-gated ion channels, open or close when ligands, such as neurotransmitters, bind to them. The opening and closing of ligand-gated and voltage-gated channels are responsible for two types of changes in the neuron's membrane potential, called graded potentials and action potentials.

Graded Potentials A **graded potential** is a depolarization or hyperpolarization that varies depending on the strength of the stimulus. A large change in membrane potential will occur when a strong stimulus opens many channels, whereas a weak stimulus causes a small change because only a few channels are opened (**Figure 41.9**).

Graded potentials occur locally on a particular area of the plasma membrane, such as dendrites or the cell body, where an electrical or chemical stimulus opens ion channels. From this area, a graded potential spreads a small distance across a region of the plasma membrane. In a short time, the membrane potential returns to the resting potential because ion pumps restore the ion concentration gradients, and the ion channels close again.

Graded potentials occur on all neurons and are particularly important for the function of sensory neurons, which must distinguish between strong and weak stimuli coming into the organism from the environment. In addition, though, graded potentials can act as triggers for the second type of electrical signal, the action potential.

Action Potentials **Action potentials** are the nerve impulses that carry an electrical signal along an axon. In contrast to a graded potential, an action potential is always a large depolarization; all action potentials in a given neuron have very similar

amplitudes, which is the degree to which an action potential changes the membrane potential away from its resting state. Once an action potential has been triggered, it occurs in an all-or-none fashion. In other words, it cannot be graded. Unlike a graded potential, an action potential is actively propagated along the axon, regenerating itself as it travels. Action potentials travel rapidly down the axon to the axon terminal, where they initiate a response at the junction with the next cell.

Figure 41.10 shows the electrical changes that happen in a localized region of an axon when an action potential is occurring. Voltage-gated Na^+ channels and K^+ channels are present in very high numbers from the axon hillock to the axon terminus. An action potential begins when a graded potential is large enough to spread to the axon hillock and depolarizes the membrane to a value called the threshold potential. The **threshold potential** is the membrane potential, typically around -50 mV, which is sufficient to open voltage-gated Na^+ channels and trigger an action potential.

Voltage-gated Na^+ channels open rapidly when the membrane potential changes from the resting potential (-70 mV) to the threshold potential (-50 mV). The opening of Na^+ channels involves a change in the conformation of the membrane-spanning region of the channel (Figure 41.10). When the protein changes shape, the central pore opens, and Na^+ rapidly diffuses into the cell down its electrochemical gradient. This further depolarizes the cell, causing even more voltage-gated Na^+ channels to open, resulting in the spike in membrane potential that characterizes the action potential. This positive feedback process is so rapid that the membrane potential reaches its peak positive value in less than 1 millisecond (msec)!

When the membrane potential becomes sufficiently positively polarized, a conformational change in the Na^+ channel blocks the continued flow of Na^+ into the cell. The conformational change involves the **inactivation gate**, a string of amino acids that juts out from the channel protein into the cytosol (Figure 41.10). The inactivation gate swings into the channel, thereby preventing the movement of Na^+ into the cell. Under these conditions, the Na^+ channel is said to be inactivated. The inactivation gate does not swing out of the channel until the membrane potential returns to a value that is close to the resting potential.

Voltage-gated K^+ channels are also triggered to open by the change in voltage to the threshold potential, but they open about 1 msec later than Na^+ channels. When K^+ channels open, some potassium ions leave the cell down their electrochemical gradient, and the membrane potential becomes more negative again. The membrane shows a brief period of hyperpolarization because the K^+ channels remain open and the electrical gradient approaches the equilibrium potential for K^+, which is slightly more negative than the resting potential. When the voltage-gated K^+ channels close, the membrane returns to the resting potential. At this stage, both the Na^+ and K^+ voltage-gated channels are closed, and they have the ability to reopen if the resting potential increases again to the threshold potential.

The evolution of K^+ channels with a slightly slower opening time than Na^+ channels was a key event that led to the formation of nervous systems. Imagine what would happen if the voltage-gated Na$^+$ and K$^+$ channels opened simultaneously: As Na$^+$ entered the cell down its electrochemical gradient, K$^+$ would leave the cell down its electrochemical gradient, and they would negate each other's effects on membrane potential.

While the inactivation gate of the Na$^+$ channel is closed, the membrane is in its **absolute refractory period** (Figure 41.10), during which time that portion of membrane is unresponsive to another stimulus. A change in voltage will not open the Na$^+$ channels while they are inactivated. Once the voltage-gated Na$^+$ channels change to the closed state, but while the voltage-gated K$^+$ channels are still open, the membrane enters a brief **relative refractory period** (Figure 41.10). During this time, when the membrane is hyperpolarized, a new action potential may be generated but only in response to an unusually large stimulus. Therefore, the refractory periods place limits on the frequency with which a neuron can generate and transmit action potentials. As we will see, this also ensures that the action potential does not "retrace its steps" by moving backward toward the cell body.

Many natural toxins work by blocking the actions of Na$^+$ and K$^+$ channels. Some animals use these toxins as defensive mechanisms or for capturing prey, and they can be deadly. For instance, tetrodotoxin, a chemical produced by pufferfish, blocks voltage-gated Na$^+$ channels and, therefore, nerve activity. An animal that eats a pufferfish dies because the poison paralyzes muscles, including those that depend on nerve stimulation to control breathing and movement. Animals that avoid pufferfish, therefore, have a selection advantage. Eating even a small amount can paralyze or kill a human. Other toxins from wasps, bees, and scorpions can block the action of voltage-gated K$^+$ channels, resulting in abnormal action potentials and loss of normal muscle responses.

Our present understanding of action potentials began with experiments carried out in the 1940s by two British physiologists, Alan Hodgkin and Andrew Huxley. They noted that Na$^+$ channels are involved in the initiation and early phase of the action potential, while K$^+$ channels control the duration and termination of the action potential. Since those early studies, we have learned a great deal about the structure and function of voltage-gated Na$^+$ and K$^+$ channels. Modern studies have largely involved selectively mutating amino acids within these proteins and observing whether changes occur in the function of the channels. The Na$^+$ channel consists of one long protein that crosses or spans the membrane 24 times. The K$^+$ channel consists of four separate protein subunits that each span the membrane six times. The pores of the Na$^+$ and K$^+$ channels are composed of different amino acids and have different properties,

Figure 41.10 Changes that occur during an action potential. The electrical gradient across the plasma membrane of an axon first depolarizes and then repolarizes the cell. The values given for membrane potential are representative and not necessarily the same in all neurons. These changes in the membrane potential are caused by the opening and closing of voltage-gated Na$^+$ and K$^+$ channels.

Concept check: How would the curve in part (a) be affected if the Na$^+$ channels were missing their inactivation gate?

which results in each ion being able to specifically pass through its own channel. Researchers have identified 10 different Na⁺ channel genes in mammals alone and as many as 100 types of K⁺ channel genes, each encoding channels with slightly different properties, such as how quickly they open or close.

Action Potentials Are Conducted Down the Axon with Great Speed

Thus far we have considered the electrical changes that happen when an action potential is initiated. We will now consider how such potentials are conducted down an axon from the axon hillock to the axon terminal (Figure 41.11). Let's begin at the axon hillock. When the neuron receives stimuli from other cells, this causes a graded potential that reaches the axon hillock. If the change in membrane potential is sufficient to reach threshold, an action potential will be triggered. The action potential starts with the abrupt opening of several voltage-gated Na⁺ channels just beyond the axon hillock, where many Na⁺ channels are found. This action potential, in turn, triggers the opening of nearby Na⁺ channels farther along the axon, which allows even more Na⁺ to flow into the neuron and depolarize a region closer to the axon terminal, leading to another action potential. In this way, the sequential opening of Na⁺ channels along the axon membrane conducts a wave of depolarization from the axon hillock to the axon terminal.

Why doesn't the action potential move backward from the terminal to the hillock? Experimentally, if an axon is stimulated in its middle, action potentials can travel in both directions, toward the cell body and the terminal. The key reason why this doesn't ordinarily happen is the inactivation state of the Na⁺ channel, which contributes to the absolute refractory period. Again, let's begin at the hillock. When these voltage-gated Na⁺ channels open for 1 msec, they allow Na⁺ to rapidly enter the cell, and then they become inactivated. This inactivation prevents the action potential from moving backward (Figure 41.11). While the Na⁺ channels are inactivated, K⁺ channels open, and the resting potential is restored. At the resting potential, the Na⁺ channels switch from the inactivated to the closed state. The opening of K⁺ channels also travels from the axon hillock to the axon terminal, but it causes a wave of hyperpolarization that helps to reestablish the resting potential.

An action potential can be conducted down the axon as fast as 100 m/sec or as slow as a centimeter or two per second. The speed is determined by two factors: the axon diameter and the presence or absence of myelin.

The axon diameter influences the rate at which incoming ions can spread along the inner surface of the plasma membrane. The flow of ions meets less resistance in a wide axon than it does in a thin axon, just as water moves more easily through a wide hose than a narrow one. Therefore, in a wider axon the action potential will move faster. The large axons of the squid or lobster conduct action potentials very rapidly, allowing the animals to move quickly when threatened.

Myelination also influences the speed of an action potential. Myelinated axons conduct action potentials at a faster rate than unmyelinated axons. Invertebrate neurons lack myelin, whereas

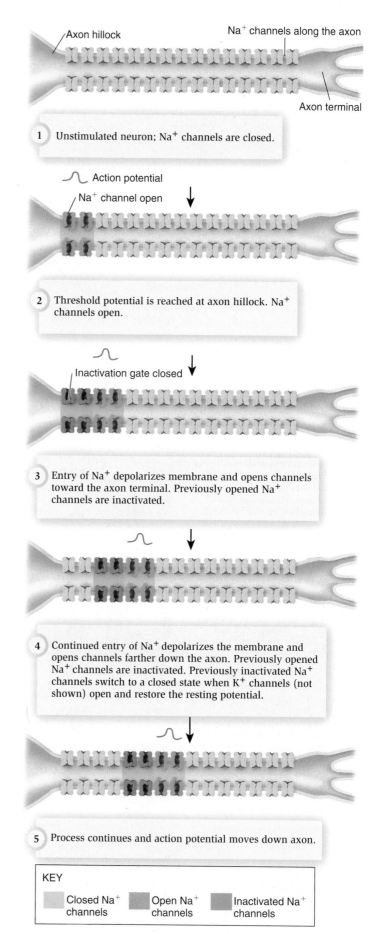

1 Unstimulated neuron; Na⁺ channels are closed.

2 Threshold potential is reached at axon hillock. Na⁺ channels open.

3 Entry of Na⁺ depolarizes membrane and opens channels toward the axon terminal. Previously opened Na⁺ channels are inactivated.

4 Continued entry of Na⁺ depolarizes the membrane and opens channels farther down the axon. Previously opened Na⁺ channels are inactivated. Previously inactivated Na⁺ channels switch to a closed state when K⁺ channels (not shown) open and restore the resting potential.

5 Process continues and action potential moves down axon.

KEY

Closed Na⁺ channels | Open Na⁺ channels | Inactivated Na⁺ channels

Figure 41.11 Conduction of the action potential along an axon. All Na⁺ channels shown refer to voltage-gated channels.

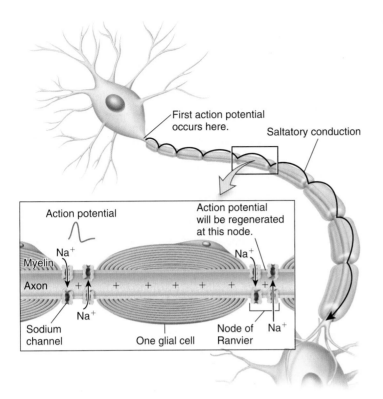

Figure 41.12 Saltatory conduction along a myelinated axon. Action potentials are generated only at the nodes of Ranvier, which lack a surrounding sheath of myelin.

Concept check: What is the main advantage of saltatory conduction?

vertebrate neurons may be myelinated or unmyelinated. Recall that glial cells (oligodendrocytes and Schwann cells) wrap around vertebrate axons to form an insulating sheath of membrane. The insulating layer of myelin reduces charge leakage across the membrane of the axon. However, this myelin sheath is not continuous (Figure 41.12). As described earlier (see Figure 41.2), the axons of myelinated neurons have exposed areas known as the nodes of Ranvier; these nonmyelinated regions are characterized by having many voltage-gated Na$^+$ channels. The nodes of Ranvier are the only areas of myelinated axons that have enough Na$^+$ channels to elicit an action potential. When sodium ions diffuse into the cell at one node, the charge spreads through the cytosol and causes the opening of Na$^+$ channels at the next node, where an action potential is regenerated. This type of conduction is called **saltatory conduction** (from the Latin *saltare*, meaning to leap) because the action potential seems to "jump" from one node to the next. In reality, action potentials do not jump from place to place; saltatory conduction speeds up the conduction process because it takes less time for changes in membrane potential to travel from node to node, eliciting action potentials only at the nodes, than it would if each tiny strip of membrane generated action potentials all along the length of the axon.

Neurons Communicate Electrically or Chemically at Synapses

Neurons communicate with other cells at a **synapse**, which is a junction where an axon terminal meets a target neuron, muscle cell, or gland. At a synapse, an electrical or chemical

signal passes from the axon terminal to the next cell. A synapse includes the axon terminal of the neuron that is sending the message, the nearby plasma membrane of the receiving cell, and the **synaptic cleft**, or extracellular space between the two cells. The **presynaptic cell** sends the signal, and the **postsynaptic cell** receives it (Figure 41.13).

By studying neurons from both invertebrates and vertebrates, researchers have identified two types of synapses, electrical and chemical. The first type, the **electrical synapse**, directly passes electric current from the presynaptic to the postsynaptic cell. The electrical signal passes through this type of synapse extremely rapidly, because the plasma membranes of adjacent cells are connected by gap junctions that can move electric charge freely from one cell to the other (see Chapter 10). Recent research indicates that electrical synapses are more widespread among taxa than originally thought, including vertebrates. The most well-studied examples, however, occur in some aquatic invertebrates such as leeches. In those animals, electrical synapses occur where a group of neurons must fire rapidly and synchronously, such as when an animal must coordinate a number of muscles to swim or escape danger. The second type of synapse is a **chemical synapse**, in which a chemical called a **neurotransmitter** is released from the axon terminal and acts as a signal from the presynaptic to the postsynaptic cell. Chemical synapses are more common, particularly in vertebrates. Some neurons release only one type of neurotransmitter, and some neurons can release two or more different ones.

Figure 41.14 describes the steps that occur when two cells communicate via a chemical synapse. The presynaptic axon terminal of a chemical synapse contains vesicles—small,

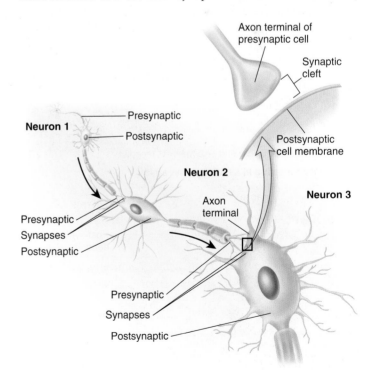

Figure 41.13 Presynaptic and postsynaptic cells. The arrows show the direction of signal transmission from one neuron to the next. Note that neuron 2 is postsynaptic with respect to neuron 1 and presynaptic with respect to neuron 3.

membrane-enclosed packets, each containing thousands of molecules of neurotransmitter. The membranes of axon terminals contain another type of ion channel localized primarily to the terminal, namely, voltage-gated Ca^{2+} channels. When an action potential arrives at the axon terminal, these channels open, allowing calcium ions to diffuse down their electrochemical gradient into the cell. Calcium binds to a protein associated with the vesicle membrane. This triggers exocytosis, in which the vesicle fuses with the presynaptic membrane, thereby releasing its neurotransmitter molecules into the synaptic cleft. The neurotransmitter molecules diffuse across the 10–20-nm-wide synaptic cleft and bind to receptors which are ion channels or other receptor proteins in the postsynaptic cell membrane.

The binding of neurotransmitter molecules opens or closes ion channels and thereby changes the membrane potential of the postsynaptic cell. In some cases, neurotransmitters directly bind to ion channels and cause the channels to open or close. Others bind to receptor proteins on the neuronal membrane and induce the formation of second messengers in the cell (see Chapter 9), which, in turn, open or close ion channels by any of several mechanisms. Some neurotransmitters are called excitatory, because they depolarize the postsynaptic membrane; the response is an **excitatory postsynaptic potential** (**EPSP**). This response is called excitatory because the depolarization of the membrane brings the membrane potential closer to the threshold potential that would trigger an action potential. An EPSP can be caused by the opening of Na^+ channels in a neuronal membrane or by the closing of K^+ channels. In both cases, positive charges would accumulate inside the cell. Conversely, an inhibitory neurotransmitter usually hyperpolarizes the postsynaptic

Presynaptic cell

Action potential — **1** In a presynaptic cell, an action potential opens voltage-gated Ca^{2+} channels. Ca^{2+} enters the cytosol.

Ca^{2+} channel

Vesicle

Ca^{2+} binds to vesicle

Ca^{2+}

2 Intracellular Ca^{2+} binds to vesicles and causes them to fuse with the presynaptic cell membrane, releasing neurotransmitter into the synaptic cleft via exocytosis.

Exocytosis of neurotransmitter

Reuptake of neurotransmitter

Synaptic cleft

Neurotransmitter

3 Neurotransmitter molecules diffuse across the synaptic cleft and bind to receptors in the postsynaptic cell membrane. Some neurotransmitter molecules are taken back up into the presynaptic cell or are broken down by degrading enzymes.

Cation

Receptors with bound neurotransmitter are open

4 In this example, the receptor is a ligand-gated ion channel that opens in response to neurotransmitters and allows the movement of cations into the postsynaptic cell. This depolarizes the membrane.

Degrading enzymes

Postsynaptic neurotransmitter receptor

Postsynaptic cell

(a) Events occurring at a chemical synapse

Axon terminal   Synapse

Post-synaptic dendrite

Synaptic cleft

Synaptic vesicles containing neurotransmitter

Mitochondrion

0.1 μm

(b) False-colored electron micrograph of a chemical synapse

Figure 41.14   Structure and function of a chemical synapse. (a) In response to an action potential, Ca^{2+} enters the presynaptic neuron axon terminal. This results in vesicle fusion with the plasma membrane, which releases neurotransmitter molecules into the synaptic cleft. The neurotransmitter molecules then bind to receptors in the plasma membrane of the postsynaptic cell. This causes ion channels to open (in this example) or close, which, in turn, changes the membrane potential of the postsynaptic cell. (b) Transmission electron micrograph of a chemical synapse from rat brain (magnification 170,000x).

Concept check: What is the benefit of having synaptic enzymes break down neurotransmitter molecules?

membrane, producing an **inhibitory postsynaptic potential** (**IPSP**), which reduces the likelihood of an action potential. An IPSP can be caused by, for example, the opening of Cl^- channels. The equilibrium potential for Cl^- is typically between the resting potential and that of K^+. Therefore, when Cl^- channels are opened, Cl^- moves into cells, bringing them closer to the equilibrium potential for Cl^-.

To end the synaptic signal, neurotransmitter molecules in the synaptic cleft are broken down by enzymes or transported back into the presynaptic terminal and repackaged into vesicles for reuse. The latter event is called reuptake and is an efficient mechanism for recapturing unused neurotransmitters that were released into the synaptic cleft. As we will see at the end of this chapter, drugs that prevent the reuptake process can sometimes be used to treat people with disorders of neurotransmission.

We can appreciate the power of neurotransmitters when we consider the effects of toxins that are known to affect neurotransmitter release. For example, the venom of black widow spiders (genus *Latrodectus*), α-latrotoxin, acts on presynaptic proteins involved in exocytosis to cause massive release of neurotransmitter from motor neurons that act on skeletal muscle. Correspondingly, the symptoms of black widow spider bites in humans are painful muscle cramping and rigidity due to overstimulation of the muscle. Researchers have also demonstrated the function of several proteins involved in vesicle membrane fusion and exocytosis by experimenting with certain toxins that decrease neurotransmitter release. For instance, the toxin secreted by the *Clostridium botulinum* bacterium prevents exocytosis of motor neuron vesicles containing neurotransmitter. If ingested in food, the toxin causes botulism, a sometimes fatal form of food poisoning characterized by weak muscles, trouble focusing the eyes, and difficulty swallowing. As you may know, researchers have discovered several cosmetic and medical uses for this toxin. For example, when injected into a chronically stimulated muscle, it blocks contraction of the muscle by preventing the motor neurons from releasing neurotransmitter, resulting in muscle relaxation. Clinicians may also administer a synthetic form of the toxin (called BOTOX) to patients who wish to reduce facial wrinkles. The effect is to reduce the contractions of the facial muscles that cause frown lines. The procedure is not without risk, however, as the toxin may diffuse from the site of the injection and affect other targets.

Neurons Respond to Multiple Synaptic Inputs

A neuron may receive inputs from many synapses occurring on its dendrites or cell body (**Figure 41.15a**), and some of these synapses may release their neurotransmitter onto the neuron at the same or nearly the same time. Certain neurotransmitters are excitatory, while others are inhibitory. When do these different inputs lead to an action potential? The effect of a single synapse is usually far too weak to elicit an action potential in the postsynaptic neuron (though a single synapse can cause a muscle cell to contract). However, when two or more EPSPs are generated at one time along different regions of the dendrites and cell body, their depolarizations sum together. The resulting larger depolarization may bring the membrane potential at the axon hillock to the threshold potential, initiating an action potential. This is called **spatial summation** and is a type of neuronal processing known as synaptic integration (**Figure 41.15b**). In another example of synaptic integration, two or more EPSPs may arrive at the same location in quick succession, such that the first EPSP has not yet decayed away when the next EPSP arrives. In that case, the depolarizations sum and may reach

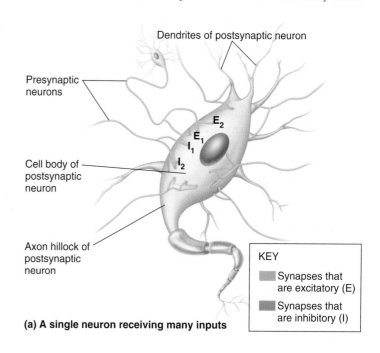

(a) A single neuron receiving many inputs

KEY

■ Synapses that are excitatory (E)

■ Synapses that are inhibitory (I)

(b) Spatial summation of EPSPs

(c) Temporal summation of EPSPs

(d) Cancellation of EPSP and IPSP

(e) Spatial summation of IPSPs

Figure 41.15   Integration of synaptic inputs.

threshold upon arrival at the axon hillock. This is called **temporal summation** (Figure 41.15c). Finally, if EPSPs and IPSPs arrive together at a postsynaptic cell, the two types of signals may cancel each other out, and no action potential is elicited (Figure 41.15d). From this discussion, you might deduce that when two or more IPSPs arrive together, their hyperpolarizations might sum, too. That is exactly what happens. In that case, the membrane potential of the cell receiving multiple IPSPs moves farther away from threshold (Figure 41.15e).

In addition to the number of synapses that stimulate the postsynaptic membrane, the location of the synapses is also important. Synapses initially cause a graded potential in the postsynaptic membrane that spreads only a short distance. If two excitatory synapses are close together and activated at the same time, the resulting depolarization will be larger and will spread farther. This spread increases the chances that the depolarization will reach the axon hillock, where the high concentration of Na^+ channels can trigger an action potential. Synapses that occur far from the axon hillock are less effective than synapses on the cell body nearer the axon hillock.

Several Classes of Neurotransmitters Elicit Responses in Postsynaptic Cells

Neuroscientists have identified more than 100 different neurotransmitters in animals. Generally, neurotransmitters are categorized by size or structure (Table 41.2). The changing balance between excitatory and inhibitory neurotransmission controls the state of nervous system circuits at any one time. To understand how neurotransmitters work, imagine driving a car with one foot on the accelerator and one foot on the brake. To speed up, you could press down on the accelerator, ease up on the brake, or both, whereas to slow down, you could do the opposite. All nervous systems operate in this way, with combined excitatory and inhibitory actions of neurotransmitters.

Next, we highlight major features of the different chemical classes of neurotransmitters found in animals, which include acetylcholine, biogenic amines, amino acids, neuropeptides, and gaseous neurotransmitters. With the exception of acetylcholine, all of these classes contain several different neurotransmitters that are similar in chemical structure but may have different functions in nervous systems.

Acetylcholine Acetylcholine, one of the most widespread neurotransmitters in animals, is released at the synapses of **neuromuscular junctions** in vertebrates, where a neuron contacts skeletal or cardiac muscle. It is also released at synapses within the brain and elsewhere. Acetylcholine acts as an excitatory neurotransmitter in the brain and on skeletal muscle cells, but it is inhibitory when released from neurons that control cardiac muscle contraction. Therefore, it stimulates skeletal muscle but inhibits cardiac muscle. As we will see later, the same neurotransmitter can exert both excitatory and inhibitory effects depending on the type of receptor to which it binds.

Biogenic Amines The biogenic amines are compounds containing amine groups that are formed from amino acids or

their precursors. Common biogenic amines include the catecholamines—dopamine, norepinephrine, and epinephrine—and serotonin and histamine. The catecholamines are formed from the amino acid tyrosine. Dopamine and epinephrine are primarily active in the brain, while norepinephrine acts in the brain and at synapses in the peripheral nervous system. Serotonin is formed from the amino acid tryptophan. In addition to widespread physiological effects such as control of heart and lung function, catecholamines and serotonin are psychoactive; that is, they affect mood, attention, behavior, and learning. In humans, for example, abnormally high or low levels of the biogenic amines have been associated with a variety of mental illnesses, including schizophrenia and depression. Histamine is formed from the amino acid histidine and has several functions in the body. As a neurotransmitter, it is important in modulating sleep. In the brain, neurons that produce histamine are most active during waking and are nearly inactive during sleep. This explains why certain antihistamines (drugs that block the ability of histamine to bind to its receptor, thereby inhibiting its action) used to treat colds and allergies also induce drowsiness.

Table 41.2 Classes of Neurotransmitters and Some Representative Examples and Functions

Transmitter	Some major functions
Acetylcholine	CNS: Stimulates the brain; important in memory, motor control, and many other functions
	PNS: Stimulates skeletal muscle at neuromuscular junctions; inhibits cardiac muscle; promotes digestion
Biogenic amines	
Catecholamines: dopamine, norepinephrine, epinephrine	CNS: Regulate mood, attention, learning, behavior
Serotonin	PNS: Stimulates cardiac muscle; improves lung function; helps animals respond to stressful situations
Histamine	CNS: Helps to maintain awake state
Amino acids	
Excitatory amino acids: glutamate, aspartate	CNS: Widespread mediators of activity in all areas of CNS; the major "on" signal of the CNS
Inhibitory amino acids: gamma-aminobutyric acid (GABA), glycine	CNS: Hyperpolarize neurons; act as a "brake" on the nervous system
Neuropeptides	
Opiate peptides: endorphin	CNS: Modulate postsynaptic cell response to neurotransmitters; play a role in mood, behavior, appetite, pain perception, and many other areas
Gases	
Nitric oxide, carbon monoxide	CNS: Possible role in memory and odor sensation
	PNS: Relax smooth muscle, especially in blood vessels

Amino Acids   In addition to their role in the formation of proteins, the amino acids glutamate, aspartate, and glycine function as neurotransmitters. Glutamate is the most widespread excitatory neurotransmitter found in animal nervous systems. Though not involved in protein formation, another amino acid, gamma-aminobutyric acid (GABA), is the most common inhibitory neurotransmitter. GABA hyperpolarizes the postsynaptic membrane by opening chloride channels, allowing chloride ions to move into the cell (see Table 41.1), and acting as the major "brake" on the central nervous system.

Neuropeptides   Neuropeptides are short chains of 2 to about 15 amino acids. They remained largely unknown until the 1970s, when it became clear that perhaps 100 or so peptides can act as neurotransmitters. Like the other neurotransmitters discussed, neuropeptides can be excitatory or inhibitory. They are produced within the neuron cell body, packaged in vesicles in the Golgi apparatus, and transported to the axon terminal, from where they are released. Neuropeptides are often called **neuromodulators**, because they can alter or modulate the response of the postsynaptic neuron to other neurotransmitters. For example, a cell exposed to a neuropeptide may increase the number of receptors for another neurotransmitter, which makes the cell more responsive to that neurotransmitter. One group of neuropeptides is called the opiate peptides because opium-like drugs, such as morphine, bind to their receptors. Opiate peptides include the endorphins, a group of peptides that decrease pain and cause natural feelings of euphoria (extreme happiness and feeling of invulnerability). Endorphins may also be important in helping animals deal with stress.

Gaseous Neurotransmitters   Certain gaseous molecules such as nitric oxide (NO) and carbon monoxide (CO) act locally in many tissues and sometimes function as neurotransmitters. Unlike other neurotransmitters, they are not sequestered into vesicles and are produced locally as required. Gaseous neurotransmitters are short-acting and influence other cells by diffusion. In humans, NO is responsible for relaxing the smooth muscle surrounding blood vessels, including those in the penis. When a male becomes sexually aroused, NO levels increase in this tissue, dilating the vessels and increasing blood flow into the penis, producing an erection. Several drugs used for male sexual dysfunction enhance erections by increasing or mimicking the action of NO on smooth muscle. The functions of CO are still uncertain, but scientists think it may act as a neurotransmitter in the sense of smell in some animals, such as the terrestrial mollusk *Limax maximus*.

The discovery that the actions of neurons are mediated in large part by neurotransmitters was one of the most significant discoveries in the history of neuroscience. That discovery laid the foundation for our understanding of the nervous system and of many human diseases. Amazingly, the existence of neurotransmitters was demonstrated in a remarkably simple experiment that arose from a dream, as we see next.

FEATURE INVESTIGATION

Otto Loewi Discovered Acetylcholine

German physiologist Otto Loewi was interested in how neurons communicate with skeletal muscle. He already knew that the electrical stimulation of a nerve in a frog's leg would result in muscle contraction, so it appeared that neurons communicate with the muscle by electrical signals. In 1921, he turned his studies to another type of muscle, the heart. As we will see in Chapter 47, all vertebrate and some invertebrate hearts receive both excitatory and inhibitory signals from different nerves that regulate the rhythm and intensity of the heartbeat. Loewi hypothesized that because different nerves produced opposite effects on the heart, the effects of the nerves could not be a direct electrical action on heart muscle, because there would be no way for the heart muscle to discern between the same type of signal from different nerves. Instead, perhaps the neurons in each nerve released different chemicals of some type, and it was these chemicals that exerted opposite actions on the heart.

As shown in **Figure 41.16**, Loewi removed the hearts from two frogs and placed the hearts in baths containing saline (an isotonic NaCl solution). As long as a frog's heart is kept in this solution, it will continue to beat for several hours before eventually dying.

Initially, Loewi began by examining the major inhibitory nerve of the vertebrate heart, called the vagus nerve. When Loewi dissected the hearts from the two frogs, he left the vagus nerve intact in one heart, but removed it from the other. Next, Loewi used an electrode to electrically stimulate the vagus nerve attached to the first frog's heart. As expected, this resulted in a decrease in the rate at which the first heart contracted. He then removed some of the saline solution from within and around the first heart and transferred it to the solution that was bathing the second (unstimulated) heart. The rate and force of beating of the second heart quickly decreased, even though it had no vagus nerve and was not exposed to any electrical stimulation. Loewi concluded that a chemical substance was released from the nerve of the first heart into the surrounding fluid and that when this chemical was added to the second heart, it reproduced the effects of electrical stimulation that were observed on the first heart.

Loewi initially named this substance vagusstoff (vagus substance), after the vagus nerve he stimulated. It was later renamed acetylcholine when its chemical nature (acetic acid bonded to choline) was determined. Acetylcholine was the first neurotransmitter discovered. Loewi's research opened the door for what we now know about chemical transmission at synapses, and the enormous pharmaceutical industry, which builds on this knowledge to treat neurological disorders (diseases of the nervous system).

Interestingly, as Loewi described later, the idea for his experiment, which would be largely responsible for his earning a share of a Nobel Prize, came to him in a dream. He woke in the middle of the night, scribbled down his idea, and returned

Figure 41.16   Loewi's experimental discovery of chemical neurotransmission.

HYPOTHESIS Neurons release chemical substances that influence the activity of the heart.

KEY MATERIALS Two frog hearts, saline solution, and stimulating and recording electrodes.

	Experimental level	Conceptual level

1 Dissect hearts from 2 frogs and place in chambers with a saline solution. Heart 1 still has its vagus nerve attached.

Vagus nerve
Saline solution
Heart 1 in chamber 1 with vagus nerve intact
Heart 2 in chamber 2 with vagus nerve removed

It was known that the vagus nerve has an inhibitory effect on heart activity.

2 Electrically stimulate vagus nerve of heart 1.

Heart 1
Stimulating electrode
Action potential
Branches of vagus nerve
Action potentials travel along vagus nerve to heart 1.

3 Record strength and number of beats in heart 1 before and after electrical stimulation of vagus nerve. Next, remove a sample of the saline solution in and around heart 1, and transfer to heart 2. Record activity of heart 2. This was done using mercury manometers that were connected to each heart. The manometers measure pressure, which is due to the contractile force of the heart beating.

If stimulation of vagus nerve resulted in the release of chemicals onto heart 1, then these same chemicals (some of which may diffuse into the saline solution) should have an identical effect on heart 2.

4   THE DATA

Stimulation of vagus nerve of heart 1
Heart 1 response
Contractile force
Time (sec)

Addition of saline solution from chamber 1
Heart 2 response
Contractile force
Time (sec)

5   CONCLUSION Electrical stimulation causes the vagus nerve to secrete chemicals that decrease heart contractions.

6   SOURCE Loewi, O. 1921. On humoral transmission of the action of heart nerves. Pflügers archives. *European Journal of Physiology* 189:239–242.

to sleep. The next morning, he despaired to find that he couldn't read his sleepy scribblings! Incredibly, he had the dream again the following night and, not taking any chances, got up and went directly to his laboratory.

Experimental Questions

1. What observations led Loewi to develop his hypothesis of how nerves stimulate or inhibit heart muscle contractions?

2. Describe Loewi's experimental design to test his hypothesis.

3. What were the results of Loewi's experiment? Did the results support his hypothesis?

Postsynaptic Receptors Determine the Response to Neurotransmitters

As we have seen, in some cases the same neurotransmitter can have both excitatory and inhibitory effects. The response of the postsynaptic neuron depends on the type of receptor present in the postsynaptic membrane. The two major types of postsynaptic receptors are ionotropic and metabotropic, and many neurotransmitters, such as acetylcholine, act on both (**Figure 41.17**). Neurotransmitter molecules bind to the extracellular portion of these receptors.

(a) Ionotropic receptor

(b) Metabotropic receptor

Figure 41.17   **The two major categories of postsynaptic receptors.**   (a) Ionotropic receptors have several subunits. Neurotransmitters bind to ionotropic receptors and directly open ion channels in the membrane. (b) Metabotropic receptors are G-protein-coupled receptors, which are discussed in Chapter 9. Neurotransmitters bind to metabotropic receptors and initiate a signaling pathway that typically opens or closes ion channels.

Ionotropic receptors are ligand-gated ion channels that open in response to neurotransmitter binding (Figure 41.17a). When neurotransmitter molecules bind to these receptors, ions flow through the channels to cause an EPSP or IPSP. Acetylcholine and amino acids bind to ionotropic receptors. Ionotropic receptors are composed of multiple subunits that associate in a ring to form the receptor's channel.

Metabotropic receptors are G-protein-coupled receptors (GPCRs) (described in Chapter 9). They do not form a channel but instead are coupled to an intracellular signaling pathway that initiates changes in the postsynaptic cell (Figure 41.17b). A common type of response is the phosphorylation of ion channels for sodium, potassium, or calcium ions, which are present in the plasma membrane. In Chapter 43, we will see that metabotropic receptors are important in activating sensory cells that respond to visual and other stimuli.

As mentioned earlier, many neurotransmitters bind to more than one type of receptor. In addition, receptors that are composed of subunits may exist in multiple forms made up of different combinations of subunits. What benefit is there to an organism to express such a variety of receptor types for a given neurotransmitter? One helpful way of understanding this diversity and its importance is illustrated by the amazing complexity of one well-studied receptor family, that of the widespread amino acid neurotransmitter GABA, as we see next.

Genomes & Proteomes Connection

Varied Subunit Compositions of Neurotransmitter Receptors Allow Precise Control of Neuronal Regulation

As mentioned earlier, gamma-aminobutyric acid (GABA) is an inhibitory neurotransmitter that opens Cl^- channels. Though cells can possess different types of GABA receptors, we will focus here on one type, which functions as a ligand-gated ion channel. This ionotropic receptor, which we will simply refer to as the GABA-A receptor, binds GABA and thereby opens the channel. This event allows Cl^- to diffuse into the cell, causing a hyperpolarization of the plasma membrane and shifting the membrane potential toward the equilibrium potential for Cl^- (usually between about -70 to -90 mV) (**Figure 41.18**). In this way, GABA binding to this receptor decreases the likelihood that a neuron will generate an action potential.

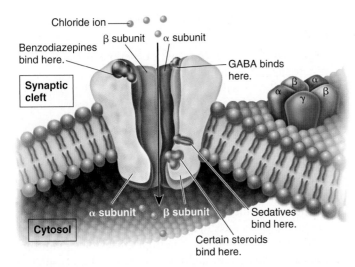

Figure 41.18   **The structure and function of the GABA-A receptor.**   Each type of GABA-A receptor has five subunits that form an anion channel allowing the passage of chloride ions. When GABA binds to the receptor, chloride ions move through the open channel and hyperpolarize the cell. Various other molecules bind to different sites on the receptor. These include molecules that naturally occur in animals' bodies, such as certain steroid hormones, and drugs, such as sedatives and benzodiazepines, used to treat anxiety and other disorders.

The GABA-A receptor is a good example of how receptor subunits can influence a postsynaptic response to a neurotransmitter. GABA-A receptor proteins are usually composed of five subunits (designated α, β, and so on). The genomes of humans and other mammals have a group of homologous genes that encode at least 19 different GABA-A receptor subunits. In addition, subunit variation can be further increased by alternative splicing (see Chapter 13). This amazing variety in subunits allows cells to express dozens of different kinds of GABA-A receptors.

What are the benefits of having such a variety of different GABA-A receptor subunits? Though the answer is not entirely understood, each type of subunit has its own unique properties that can fine-tune the function of the GABA-A receptor so that it works optimally in the neuron in which it is expressed. The various subunits may differ in their affinity for GABA and the rate of Cl^- movement through the channel.

In addition, neuroscientists have been particularly interested in whether the various subunits differ in their ability to recognize molecules other than GABA (Figure 41.18). This work has shown that the subunits of the GABA receptor bind a variety of other molecules, including naturally occurring ones such as certain steroid hormones. Presumably, the binding of these molecules enhances or reduces the effectiveness of GABA in activating the receptor. This knowledge has proven beneficial in understanding how certain drugs exert their actions. For example, ethanol—found in alcoholic drinks—binds to one of the GABA-A receptor subunits expressed in brain and motor neurons and enhances the actions of GABA. This may explain in part why alcohol depresses the activity of the brain and

impairs motor coordination, among other effects. Other subunits of the GABA-A receptor bind drugs, such as benzodiazepines (for example, Xanax), which are used to treat chronic or severe anxiety. Apparently, the inhibitory effects of GABA are part of the mechanism for achieving a balance between alertness, anxiety, and calmness. The ability of the receptor to bind numerous ligands, and the many different combinations of subunits in the receptor, provide an enormous degree of control over precisely how this neurotransmitter system regulates the activity of the brain.

41.4   Impact on Public Health

When neurons fail to develop properly or their function is impaired, the consequences can be devastating, affecting mood, behavior, and even the ability to think or move. Over 100 neurological disorders have been identified in humans, and drugs to treat them are among the most widely prescribed medicines today (Table 41.3). All recreational drugs also exert their effects by altering neurotransmission. The use of these drugs, therefore, can result in symptoms similar to those of neurological disorders.

Disorders of Neurotransmission Can Impact Mood

Several neurological disorders result from disrupted neurotransmission between cells. Genetic processes involved in the production of neurotransmitters or malfunction of synaptic events can increase or decrease activity at synapses, which, in turn, affects emotions and behavior. One mood disorder is **bipolar disorder**, previously called manic depression, in which individuals experience shifts between periods of euphoria and despair.

The most common mood disorder is **major depressive disorder**, also called unipolar depression or, simply, depression. This illness results in prolonged periods of sadness, despair, and lack of interest in daily activities without alternating episodes of euphoria. Depression affects 5–12% of men and 10–25% of women at some point in their lifetime. This condition is thought to result from decreased activity of synapses that release biogenic amines, such as serotonin, which changes neuronal activity within specific areas in the brain involved in processing emotion. Drugs used to treat major depression include the **selective serotonin reuptake inhibitors**, such as Prozac, Zoloft, and Paxil, which reduce the reuptake of serotonin into the presynaptic terminal after it is released. This allows serotonin to accumulate in the synaptic cleft, counteracting the deficit that causes the alteration in mood.

In many respects, the use of such drugs to treat depression reflects a profound change in the public's attitude toward mental illness. Historical attitudes toward mood disorders held that individuals who were depressed lacked the ability to cope with stressful events in their lives. Only recently has it become accepted that mood disorders can be caused by changes in the balance of neurotransmitters in the brain. Perhaps the best

Table 41.3	Representative Effects of Common Therapeutic or Recreational Drugs on Neurotransmitter Action and Mood		
Name of drug	**Actions on neurotransmission**	**Effects on mood**	**Effects of abuse or overdose**
Alcohol (ethanol)	Enhances inhibitory GABA transmission; increases dopamine transmission; inhibits glutamate transmission	Relaxation; euphoria; sleepiness	Liver damage; brain damage
Amphetamines *"uppers," "crystal meth," "pep pills"*	Stimulate the release of dopamine and norepinephrine	Euphoria; increased activity	High blood pressure; psychosis
Antidepressants *Tricyclic antidepressants*; for example, *Elavil, Anafranil*	Block the reuptake of norepinephrine from synapses	Relieve depression and obsessive-compulsive disorder	Drowsiness; confusion
Selective serotonin reuptake inhibitors; for example, *Prozac, Zoloft, Paxil, Lexapro*	Block the reuptake of serotonin from synapses	Relieve depression and obsessive-compulsive disorder	Insomnia; anxiety; headache
Monoamine oxidase inhibitors; for example, *Parnate, Nardil*	Block the breakdown of biogenic amine neurotransmitters	Relieve depression	Liver damage; hyperexcitability
Antianxiety drugs *Benzodiazepines*; for example, *Xanax, Valium, Librium, Rohypnol* ("date rape drug," "roofies")	Bind to GABA receptors and increase inhibitory neurotransmission	Relieve anxiety; cause sleepiness and in some cases amnesia	Drowsiness; memory loss in some cases
Antipsychotic drugs *Phenothiazines*; for example, *Thorazine, Mellaril, Stelazine, Abilify, Risperdal*	Block dopamine receptors	Ease schizophrenic symptoms	Decreased control of movement
Cocaine	Blocks norepinephrine and dopamine reuptake	Intense euphoria followed by depression	Convulsions; hallucinations; death from overdose
LSD (lysergic acid diethylamine)	Binds to serotonin receptors	Hallucinations; sensory distortions	Unpredictable and irrational behavior
Marijuana (tetrahydrocannibinol)	Binds to receptors for natural cannabinoids	Increased sense of well-being; decreased short-term memory; decreased goal-directed behavior; increased appetite	Delusions; paranoia; confusion
Narcotics *Heroin, morphine, Demerol, codeine*	Bind to opiate receptors	Pain relief; euphoria; sedation	Slowed breathing; death from overdose
Nicotine	Initially stimulates but then depresses activity in adrenal medulla and neurons in the peripheral nervous system; increases dopamine in brain	Increased attention; decreased irritability	Heart disease and lung disease
PCP (phencyclidine) *"angel dust," "ozone"*	Blocks channel for excitatory amino acid neurotransmitters; increases dopamine activity	Violent behavior; feelings of power; numbness; disorganized thoughts	Psychosis; convulsions; coma; death

evidence of a physiological basis for depression is that mood disorders occur more frequently in certain families, suggesting a genetic component. Drugs can be very effective in treating these disorders. Patients taking selective serotonin reuptake inhibitors often report decreased sadness, increased energy, and a greater interest in daily activities.

Recreational Drugs Disrupt Normal Neurotransmission

Most so-called recreational drugs work at the synapse to either enhance or interfere with the normal mechanisms of neurotransmission (Table 41.3). In the presynaptic terminal, such drugs can decrease neurotransmitter release by reducing Ca^{2+} entry into the cell or by preventing the exocytosis of vesicles containing stored neurotransmitters. In the synaptic cleft, drugs can slow the rate at which the neurotransmitter is broken down into an inactive form or taken back up into the presynaptic neuron, thereby prolonging the action of the neurotransmitter in the synaptic cleft. Some substances act on the postsynaptic membrane by either preventing the neurotransmitter from binding to its receptor or by acting as a substitute for the neurotransmitter by stimulating the receptor.

In effect, these drugs produce changes or imbalances in neurotransmission similar to those observed in some neurological disorders. These substances can induce euphoria, increase activity, alter mood, and produce hallucinations. They can also have potentially life-threatening effects and may be highly addictive.

Some drugs, such as cocaine, block the removal of dopamine and norepinephrine from the synaptic cleft by preventing their reuptake into the presynaptic terminal. Morphine and marijuana mimic the actions of biological substances already in the brain, binding to receptors on the postsynaptic membrane. With these drugs, the resulting effects are much stronger than are the effects of natural neurotransmitters. It is no surprise that recreational drugs are mind altering. They do, after all, change the ways in which neurons communicate with each other.

Disorders of Conduction May Result in Motor Problems and Abnormal Neuronal Development

Some human diseases are caused by the inability of the axon to properly conduct an action potential. This occurs most commonly because an axon fails to become myelinated or because a myelinated axon becomes demyelinated.

In **congenital hypothyroidism** (formerly called cretinism), axons fail to become wrapped with myelin during fetal development, which leads to slow conduction speeds and abnormal connections between brain neurons. This results in profound mental defects that cannot be reversed unless treatment begins immediately after birth. Congenital hypothyroidism is caused by insufficient levels of thyroid hormones in the fetus. Among their many actions, thyroid hormones stimulate the formation of myelin during fetal development. However, thyroid hormones cannot be synthesized without the element iodine, which is part of the structure of the hormones. The iodine in the fetus comes from the mother's diet. If a mother's dietary intake of iodine is too low, the fetus will not have enough iodine to make its own thyroid hormones, and therefore, the fetus will not be able to make normal amounts of myelin. Although iodine is not an abundant element in most diets, congenital hypothyroidism is rare in the U.S. and many other countries since the advent of iodized table salt; however, it is not uncommon in many parts of the world.

Unlike congenital hypothyroidism, **multiple sclerosis (MS)** usually begins between the ages of 20 and 50 in individuals with apparently healthy nervous systems. With MS, the patient's own immune system, for reasons unknown, attacks and destroys myelin as if it were a foreign substance. Eventually, these repeated attacks leave multiple scarred (sclerotic) areas of tissue in the nervous system and impair the function of myelinated neurons that control movement, speech, memory, and emotion. Multiple sclerosis is a serious and unpredictable disease, characterized by flare-ups followed by periods of remission in which symptoms are reduced or absent. No cure is currently available, but certain drugs may slow its progression and reduce the severity of symptoms. This disease affects roughly 2.5 million people worldwide, 60–75% percent of them women.

Summary of Key Concepts

41.1 Cellular Components of Nervous Systems

- The central nervous system (CNS) is composed of a brain and a nerve cord. The peripheral nervous system (PNS) consists of all neurons and their projections that are outside of and connect with the CNS. Nerves transmit signals between the PNS and CNS. (Figure 41.1)
- The two major classes of cells in nervous systems are neurons and glia. In neurons, information flows from dendrites to the cell body and then to the axon and axon terminal. (Figure 41.2)
- Types of neurons include sensory neurons, motor neurons, and interneurons. A neuron's structure is a reflection of its function. (Figure 41.3)
- The most basic circuit is a reflex arc, which occurs rapidly in response to inputs from sensory neurons and consists of only one or a few afferent and efferent cells. (Figure 41.4)

41.2 Electrical Properties of Neurons

- Neuronal membranes are electrically polarized. The membrane potential is determined by the differential distribution and differential permeability of ions across the plasma membrane. The resting potential is the membrane potential of a cell that is not sending electrical signals. (Figures 41.5, 41.6, Table 41.1)
- Neurons use electrical signals to communicate with other neurons, muscles, or glands. These signals involve changes in the amount of electric charge across a cell's plasma membrane.
- Diffusion of ions through membrane channels occurs as a result of the concentration gradient of an ion across the membrane and the electric charge across the membrane. Ions move in response to an electrochemical gradient. (Figure 41.7)
- The Nernst equation gives the equilibrium potential for an ion at any given concentration gradient.

41.3 Communication Between Neurons

- Gated ion channels enable a cell to communicate by changing its membrane potential rapidly. The opening and closing of voltage-gated and ligand-gated ion channels cause two types of changes in the neuron's membrane potential—graded potentials and action potentials. (Figures 41.8, 41.9)
- When a graded potential is large enough to spread to the axon hillock and depolarize the membrane to its threshold potential, this causes an action potential that carries an electrical signal from the axon hillock to the axon terminal. (Figures 41.10, 41.11)
- Axon diameter and myelination influence the speed of an action potential. (Figure 41.12)
- Electrical synapses are connected via gap junctions. Most vertebrate neurons communicate via chemical synapses, in which a neurotransmitter carries the signal from the presynaptic to the postsynaptic cell. Many EPSPs generated at one time can sum together and bring the membrane potential to the threshold potential, initiating an action potential. (Figures 41.13, 41.14, 41.15)
- Chemical classes of neurotransmitters found in animals include acetylcholine, biogenic amines, amino acids, neuropeptides, and gaseous neurotransmitters. (Table 41.2)
- The discovery that the actions of neurons are mediated in large part by neurotransmitters was one of the most significant discoveries in the history of neuroscience. (Figure 41.16)

- The receptors of the postsynaptic neuron determines the types of signals that pass from one neuron to the other. The two major types of postsynaptic receptors are ionotropic and metabotropic. (Figures 41.17, 41.18)

41.4 Impact on Public Health

- Most neurological conditions can be classified as disorders of either neurotransmission or conduction. Mood disorders caused by disrupted neurotransmission include bipolar disorder and major depressive disorder. Drugs used in the treatment of neurological disorders and recreational drugs usually alter neurotransmission. (Table 41.3)

- Some neurological conditions are caused by the inability of the axon to conduct an action potential. This occurs most commonly because axons fail to become myelinated (congenital hypothyroidism) or because myelinated axons become demyelinated (multiple sclerosis).

Assess and Discuss

Test Yourself

1. In vertebrates, the brain and the spinal cord are parts of
   a. the peripheral nervous system.
   b. the enteric nervous system.
   c. the central nervous system.
   d. the autonomic nervous system.
   e. the endocrine system.

2. The structures of a neuron that function mainly in receiving signals from other neurons are
   a. the myelin sheaths.   d. the dendrites.
   b. the axons.   e. the K^+ channels.
   c. the axon terminals.

3. The glial cells that form the myelin sheath in the peripheral nervous system are called
   a. astrocytes.   d. neurons.
   b. oligodendrocytes.   e. Schwann cells.
   c. microglia.

4. Neurons that function mainly in connecting other neurons in the central nervous system are
   a. sensory neurons.   d. afferent neurons.
   b. efferent neurons.   e. interneurons.
   c. motor neurons.

5. The difference in charges across the plasma membrane of an unstimulated neuron is called
   a. the membrane potential.
   b. the resting membrane potential.
   c. homeostasis.
   d. the graded potential.
   e. the action potential.

6. Which of the following contributes to the resting membrane potential?
   a. negatively charged ions inside and outside of the cell
   b. active transport of ions across the membrane
   c. concentration of Na^+ and K^+ inside and outside of the cell
   d. all of the above
   e. b and c only

7. A neuron has reached a threshold potential when it has depolarized to the point where
   a. most voltage-gated K^+ channels open.
   b. sufficient numbers of voltage-gated Na^+ channels open to initiate a positive feedback cycle, contributing to further depolarization.
   c. voltage-gated K^+ channels close.
   d. voltage-gated Na^+ channels close.
   e. both b and c occur.

8. The speed of transmission of an action potential along an axon is influenced by
   a. the presence of myelin.
   b. an increased concentration of Ca^{2+}.
   c. the diameter of the axon.
   d. all of the above.
   e. a and c only.

9. Gap junctions are characteristic of
   a. electrical synapses.
   b. chemical synapses.
   c. acetylcholine synapses.
   d. GABA synapses.
   e. synapses between motor neurons and muscle cells.

10. The response of the postsynaptic cell is determined by
    a. the type of neurotransmitter released at the synapse.
    b. the type of receptors the postsynaptic cell has.
    c. the number of Na^+ channels in the postsynaptic membrane.
    d. the number of K^+ channels in the postsynaptic membrane.
    e. all of the above.

Conceptual Questions

1. Distinguish between neurons and glial cells, and give an example of a glial cell.

2. Describe the difference between graded and action potentials.

3. In certain diseases, such as kidney failure, the Na^+ concentration in the body's extracellular fluid can become altered. What effect would a high extracellular Na^+ concentration have on neurons?

Collaborative Questions

1. Discuss how nervous systems are organized into central and peripheral nervous systems in animals.

2. Name the parts of a neuron, and give a brief description of their major characteristics.

Online Resource

www.brookerbiology.com

Stay a step ahead in your studies with animations that bring concepts to life and practice tests to assess your understanding. Your instructor may also recommend the interactive ebook, individualized learning tools, and more.

Neuroscience II: Evolution and Function of the Brain and Nervous Systems

42

Three-dimensional reconstruction of the brain of a fruit fly. The brains of animals are organized into anatomic structures with specialized functions.

I t will take you approximately 2–3 seconds to read this sentence. During that time, many of the 100 billion or so neurons of your brain will have fired off millions of action potentials. Some of those signals will help process the visual information reaching your eyes as you scan the page. Others will activate centers of learning and memory to help you understand the meanings of the words you've read. Still other signals will help filter out extraneous inputs—such as background noise—that might distract you from your task. The complexity of the seemingly simple task of reading a single sentence emphasizes the enormous level of activity that goes on continually in the brain, even at rest.

The beauty of the brain lies in its incredible complexity. The human brain, for example, has several thousand miles of interconnected neurons and hundreds of trillions of synapses, resulting in a total surface area that if spread out, would cover more than four soccer fields. The brain allows us to move, think, and experience sensation and emotion. Groups of neurons also coordinate homeostatic functions such as breathing, blood circulation, and body temperature. When we examine the way that groups of neurons

communicate, we begin to understand the complex mental functions of nervous systems, including learning, memory, and motivation.

Neuroscience—the study of nervous systems—is an area of intense research activity worldwide. For example, at one annual scientific meeting, more than 30,000 neuroscientists from dozens of countries meet to present their research and learn about the latest developments in this dynamic field. The challenge of neuroscience is to transform the astounding complexity of nervous systems into manageable proportions. For this reason, most neuroscientists do not study the human nervous system, because of its great complexity and the difficulties inherent in doing research on humans. Many animals, however—including the nematode *Caenorhabditis elegans*, the sea slug *Aplysia californica*, the fruit fly *Drosophila melanogaster* (see chapter-opening photo), the zebrafish *Danio rerio*, and the mouse *Mus musculus*—provide excellent opportunities to study how neurons work and how groups of neurons cooperate to produce animal behavior. The genomes of most of these animals have been sequenced, and neuroscientists are identifying genes that are critical for the structure and function of nervous systems. Many of the relevant genes are homologous to human genes, so the study of the molecular control of the nervous system in these model animals has the potential to reveal new treatments for many genetically inherited neurological diseases in humans.

In this chapter, we will first survey a variety of nervous systems, which allow animals to sense and respond to environmental changes. We will then examine the complex nervous system of humans. However, keep in mind that we still have much to learn about the organization, connectivity, and functions of nervous system structures. Our nervous system is fascinating and mysterious, and the study of how it functions will ultimately tell us much about what makes us human.

42.1 The Evolution and Development of Nervous Systems

The nervous systems of animals are the products of hundreds of millions of years of evolution. They provide advantages to animals that promote reproductive success. For example, nervous systems allow animals to keenly sense their environment and respond to changes in an appropriate way. In addition, nervous systems form connections with muscle systems and facilitate movement, which has allowed animals to travel across distances to obtain food. Likewise, nervous systems help animals to avoid predation and other environmental dangers.

Studying the evolution and development of nervous systems helps us understand how particular nervous systems are adapted to different functions. At the structural level, the organization of nervous systems ranges from a relatively simple network of a few cells to the marvelous complexity of the human brain. The characteristics of an animal's nervous system determine the behaviors that it displays. In this section, we will survey the nervous systems of invertebrates and vertebrates and also examine the brains of vertebrates in greater detail.

Nervous Systems Evolved to Sense and Respond to Changes in the Environment

Precisely when nervous systems first arose and whether or not the nervous systems of most or all animals can be traced back to a common ancestor are questions of active investigation by neuroscientists. For example, recent genetic studies have uncovered remarkable similarities in the expression and activation of genes coding for proteins that regulate neuronal development across taxa in bilaterians. Those studies suggest that the patterning of nervous system development in such animals may be traced to a common ancestor that lived more than 500 million years ago!

Today, all animals except sponges have a nervous system. Interestingly, though, researchers have recently discovered that sponges express about 25 genes with similarity to genes expressed in human neurons, specifically those that code for proteins that regulate synaptic function. The functions of the sponge genes are uncertain, but the proteins coded for by the genes interact in ways that are reminiscent of human synaptic proteins. Thus, the origin of nervous systems almost certainly can be traced to genes of evolutionarily ancient organisms; as animals evolved, these genes were modified and formed the basis of all future nervous systems. The simplest nervous system is the **nerve net** of the radially symmetrical cnidarians (jellyfish, hydras, and anemones) (Figure 42.1a). In these organisms, no single group of neurons controls all the others the way a brain does in other animals, as we'll see shortly. Instead, the neurons are arranged in a network of connections between the inner and outer body layers of the animals. A characteristic feature of nerve nets is that activation of neurons in any one region leads to activation of most or all other neurons, with the excitation spreading in all directions at once. Many of these neurons stimulate contractile cells to contract. This allows the organism to move large areas of its body simultaneously, thereby coordinating simple movements such as swimming. Recent research has identified regions of specialized function in the nerve nets of some cnidarians, such as local sensory neurons in the outer body wall. Some cnidarians, such as the jellyfish, have two nerve nets, one for moving tentacles and one for swimming. By contrast, sea stars and other echinoderms have a slightly more

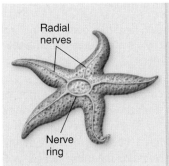

(a) Cnidarian

(b) Echinoderm

(c) Flatworm

(d) Annelid

(e) Arthropod

(f) Mollusk

(g) Chordate

Figure 42.1   Representative nervous systems throughout the animal kingdom.

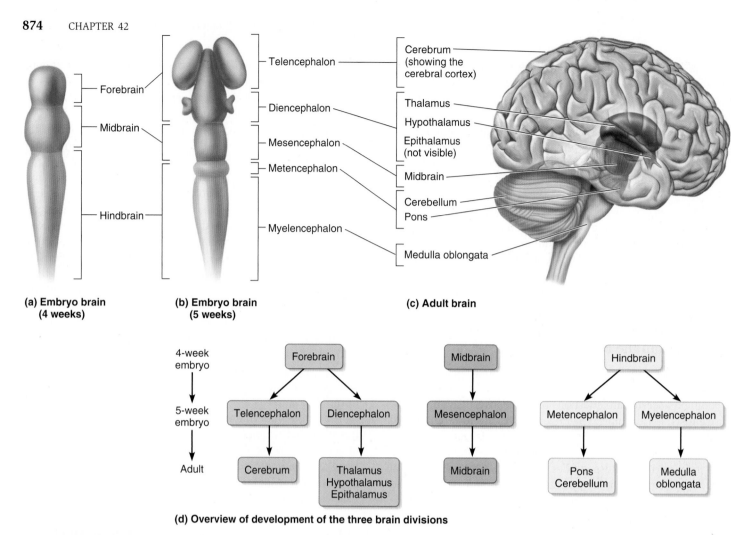

**(a) Embryo brain
(4 weeks)**

**(b) Embryo brain
(5 weeks)**

(c) Adult brain

(d) Overview of development of the three brain divisions

Figure 42.2 **Development of the human brain.** The structures shown here that occur during embryonic development at **(a)** 4 weeks and **(b)** 5 weeks are compared to **(c)** how they appear in the adult. Most structures beneath the cerebrum are not shown. **(d)** This flowchart gives an overview of the development of the three brain divisions.

complex nervous system with a nerve ring around the mouth that is connected to larger radial nerves extending into the arms (**Figure 42.1b**). This arrangement allows the mouth and arms to operate independently.

More complex invertebrate nervous systems usually consist of one or two **nerve cords** that extend from the anterior end to the posterior end. In the flatworm (genus *Planaria*), collections of neurons in the animal's head form **cerebral ganglia** (singular, ganglion). These are groups of neuronal cell bodies that perform the basic functions of integrating inputs from sense organs, such as the eyes, and controlling motor outputs such as those involved in swimming (**Figure 42.1c**). Two lateral nerve cords extend along the ventral surface of the animal into the tail and are connected to each other by transverse nerves. In annelids (segmented worms), the basic structure is similar, except that more neurons are present and the single ventral nerve cord has ganglia located in each body segment (**Figure 42.1d**). In the head, cerebral ganglia form an integrative center, which functions as a rudimentary brain to control body movements. The other ventral ganglia along the nerve cord receive sensory information from a particular body segment and control local movements. In the simpler types of mollusks, such as the snail, the nervous system is very similar to that of the annelids. The head

contains a pair of anterior ganglia; paired nerve cords extend from these ganglia to the eyes, muscular foot, and gut.

During the evolution of animals, more complex body types have been associated with **cephalization** (from the Greek *cephalo*, meaning head), which is the formation in the head of an increasingly complex **brain** that controls sensory and motor functions of the entire body. Within the brain, neuronal pathways provide the integrative functions necessary for an animal to make more sophisticated responses to its environment. Brains are found in all vertebrates and most invertebrates, and they are usually composed of more than one anatomical and functional region. For instance, in *Drosophila* (**Figure 42.1e** and the chapter-opening photo), the brain has several subdivisions with separate functions, such as a region devoted to learning and memory. Some complex mollusks, such as the squid and octopus, have brains with well-developed subdivisions that allow these animals to coordinate the complex visual and motor behaviors necessary for their predatory lifestyle (**Figure 42.1f**). In all chordates, the brain is connected to a dorsally located **spinal cord**; together, the brain and spinal cord constitute the **central nervous system** (**Figure 42.1g**). As discussed later, nerves from the **peripheral nervous system** route information into and out of the central nervous system at separate regions along the spinal cord. The organization of the vertebrate spinal cord and bony vertebrae shares similarities with that of the segmented nervous system of invertebrates.

Brains of Vertebrates Have Three Basic Divisions

Development of the vertebrate brain begins with the formation of a central fold in the embryo called the **neural tube**. This hollow tube is the structure from which the entire nervous system develops. Increased cell proliferation leads to bending and folding of the neural tube during embryonic development, resulting in bulges that become separate divisions of the nervous system. The anterior end develops into the brain, while the posterior portion becomes the spinal cord.

In vertebrates, the brain has three major divisions, the **hindbrain**, **midbrain**, and **forebrain** (**Figure 42.2**). Fossils of agnathan fishes that lived 400 million years ago show that their brains were already organized into the three basic divisions that have been retained in all modern vertebrates.

Let's look at the development of the human brain, which goes through these forms before it reaches its adult structure. At 4 weeks, the human embryo exhibits just hindbrain, midbrain, and forebrain (Figure 42.2a). Just a week later, the hindbrain and forebrain have each formed two separate divisions (Figure 42.2b). The hindbrain subdivides into the metencephalon and the myelencephalon. The forebrain subdivides into the telencephalon and the diencephalon. The midbrain, by contrast, does not subdivide and is termed the mesencephalon. By the time the human brain is fully developed, some of these structures have further divided and specialized (Figure 42.2c). The metencephalon, from the hindbrain, subdivides into the **pons** and **cerebellum**. Also from the hindbrain, the myelencephalon develops into the **medulla oblongata**. In the forebrain, the telencephalon develops into the **cerebrum**, and the diencephalon develops into the **thalamus**, **hypothalamus**, and **epithalamus**. The development of brain subdivisions provides additional specialization and increases the capacity of the brain to perform complex functions.

Hindbrain   The hindbrain includes the medulla oblongata, pons, and cerebellum (Figure 42.2c). The medulla oblongata coordinates many basic reflexes and bodily functions, such as breathing, that maintain the normal homeostatic processes of the animal. The pons and cerebellum are responsible for monitoring and coordinating body movements.

Midbrain   The midbrain processes several types of sensory inputs, including vision, olfaction (smell), and audition (hearing). It controls sophisticated tasks such as coordinating eye movement with visual inspection of the environment and also plays a role in alertness.

Forebrain   The forebrain initiates motor functions and processes sensory inputs. In humans and other mammals, it consists of a group of structures that are responsible for the higher functions of conscious thought, planning, and emotion. Many of these functions are attributed to the cerebrum. The surface layer of the cerebrum, which is only a few millimeters thick, is called the **cerebral cortex**. As we will learn later, this thin area is critical to thought, learning, and movement, among other functions. Other structures of the forebrain are beneath the cerebrum. These include the thalamus, hypothalamus, and epithalamus, which we will discuss later in this chapter.

The relative sizes of the brain divisions vary in different animals. Usually, a larger area plays a greater role in an animal's life. In addition, the functions performed by these divisions have changed over the course of evolution. For instance, the midbrain in fishes is relatively large compared to the other two areas, and it plays a key role in vision. By comparison, the forebrain is the largest division in birds and mammals and is the primary area responsible for vision.

Increased Brain Complexity Involves a Larger, Highly Folded Cerebrum

As evolution produced animals with more complex nervous systems, the size of the cerebrum also increased, making up a greater proportion of the brain. As mentioned, many of the important functions of the cerebrum are carried out by neurons along its outer surface in the cerebral cortex. Therefore, increased complexity of the brain is also correlated with an increased surface area of the cerebral cortex. During the evolution of mammals, this increase in surface area occurred more rapidly than an expansion in the size of the skull. How could this occur? The answer is that the external surface of the cerebrum in animals with increasingly complex brains is highly convoluted, forming many folds called gyri (singular, gyrus), separated by grooves called sulci (singular, sulcus). Compare the relatively smooth-looking surface of the cerebrum of a rat with the highly folded one of a human in **Figure 42.3**.

As body size increases across the animal kingdom, you might expect that brain mass would increase proportionately—that the brain of an elephant would be larger than that of a bat, for instance. That is generally the case, with a few important exceptions (**Figure 42.4**). In particular, the masses of the human

| Rat | Rabbit | Cat | Sheep | Dolphin | Chimpanzee | Human |

Figure 42.3   The degree of cerebral cortex folding in different mammalian species.   The brains are not shown to scale.

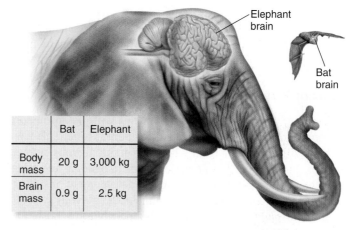

	Bat	Elephant
Body mass	20 g	3,000 kg
Brain mass	0.9 g	2.5 kg

Figure 42.4   **Brain mass as a function of body mass in mammals.**   For most mammals, brain weight is in proportion to body mass. However, humans and dolphins have a much higher brain mass relative to their body mass than do other mammals.

Concept check: *Scientists have determined that the brain mass of Homo neanderthalensis (Neanderthal man) was greater than that of our own. Does this mean that Neanderthals were more intelligent and capable of more complex behaviors than we are?*

and dolphin brains are greater than would be expected on the basis of body mass.

Brain mass and the amount of folding are correlated with more complex behaviors. Why is this so? As we've indicated, the outer surface of the brain, the cerebral cortex, plays a key role in conscious thought, reasoning, and learning. Greater size and folding provide a larger number of neurons and synapses and more surface area, which allows greater processing and interpretation of information. Even so, evidence does not suggest that people with small differences in brain size differ in intelligence. Also, it would be wrong to assume that an animal with a small brain is profoundly limited in its behavioral repertoire. A bat with a 0.9-gram brain and an elephant with a 2,500-gram brain can both perform a great variety of interesting and complex behaviors, such as navigating across great distances and interacting with fellow members of their species.

42.2   Structure and Function of the Human Nervous System

The nervous system of the human is wonderfully complex—the brain alone has over 100 billion neurons and many more billions of glia cells. Moreover, complexity is defined by more than just numbers of cells. Within the human brain, for example, are enormous numbers of connections between neurons—a single neuron in the cerebellum may have as many as 100,000 synapses with other cells! In this section, we will examine the human nervous system, with an emphasis on the functions of the major parts of the brain and spinal cord.

The Nervous System Is Composed of the Central and Peripheral Nervous Systems

The brain and spinal cord constitute the central nervous system (CNS) of vertebrates (**Figure 42.5**). The peripheral nervous system (PNS) is composed of neurons and axons of neurons that are outside the CNS and communicate with it. The CNS

Figure 42.5   **Organization of the human nervous system.** The central nervous system consists of the brain and spinal cord, both of which are encased in bone (not shown). The peripheral nervous system includes cranial nerves, ganglia, and spinal nerves, which carry information to and from the CNS, and many other neurons throughout the body.

and PNS are connected anatomically and functionally. The CNS receives information about the internal or external environment from the PNS. The CNS interprets that information and may initiate a response that is then carried out by the PNS. For example, suppose you accidentally lean against a newly painted fence. Neuronal endings in your skin, which are part of the PNS, would transmit tactile (touch) information through axons that bring information directly into the spinal cord. From there, the information travels to your brain, where the sensation is

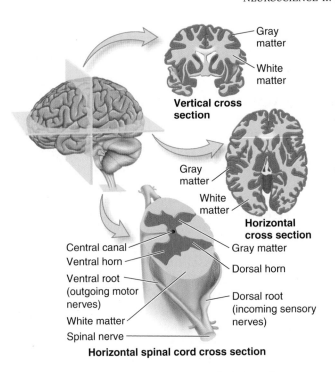

Vertical cross section

Gray matter

White matter

Gray matter

White matter

Horizontal cross section

Gray matter

Dorsal horn

Dorsal root (incoming sensory nerves)

Central canal

Ventral horn

Ventral root (outgoing motor nerves)

White matter

Spinal nerve

Horizontal spinal cord cross section

(a) Gray and white matter in the brain and spinal cord

Gray matter

White matter

(b) Light microscope image of a vertical cross section of the human brain

Figure 42.6 Gray matter and white matter in the CNS. **(a)** The gray matter is composed of groups of cell bodies, dendrites, and unmyelinated axons. The white matter consists of tracks of myelinated axons. **(b)** Photograph of a vertical cross section through an adult human brain.

Concept check: *Is a spinal nerve composed of axons from afferent or efferent neurons, or both?*

analyzed and identified as something sticky. Signals are sent from your brain, down your spinal cord, and through the neurons of the PNS to your muscles, causing you to move away.

Within the nervous system, groups of neurons may associate with each other and perform a particular function. In the CNS, the cell bodies of neurons that are involved in a similar function may be grouped into a structure called a **nucleus** (plural, nuclei), which may include thousands of cells. For instance, cell bodies that regulate body temperature and those that recognize visual information are located in separate nuclei in the brain. In the context of the vertebrate nervous system, the term **ganglion** is used to refer to a group of neuronal cell bodies located in the PNS (Figure 42.5).

Within the vertebrate nervous system, many myelinated axons may run in parallel bundles. (Myelination is described in Chapter 41.) Such a structure is called a **tract** when it is found in the CNS. Tracts convey information from region to region within the brain and between the brain and the spinal cord. Bundles of myelinated axons also form outside the CNS, in the PNS, in which case they are called **nerves**. The cell bodies that give rise to the axons of nerves may be within the PNS or the CNS; in other words, nerves may carry information from outside the CNS into the CNS, from the CNS to structures outside the CNS, or even both. Connections between the PNS and the CNS occur at the brain or spinal cord. **Cranial nerves** are directly connected to the brain, primarily to sites within the hindbrain and midbrain. By comparison, **spinal nerves** are connections between the PNS and spinal cord (Figure 42.5).

One of the most obvious characteristics of the CNS is that some parts look white, and others appear gray (**Figure 42.6**). The **white matter** gets its color from myelin; it consists of myelinated axons bundled together in large numbers to form tracts. The

gray matter consists of neuronal cell bodies, dendrites, and some unmyelinated axons. The cerebral cortex is composed of gray matter that sits on top of a large collection of white matter pathways. In the spinal cord, the gray matter is located in the center and forms two dorsal horns and two ventral horns (Figure 42.6a). Each dorsal horn connects to a dorsal root, which is part of a spinal nerve. Dorsal roots receive incoming information from sensory (afferent) nerves of the PNS. The ventral horn connects to the ventral root, which is also part of a spinal nerve that transmits outgoing information to motor (efferent) nerves. A central canal runs through the spinal cord, carrying a nutritive and protective fluid called cerebrospinal fluid, as described shortly.

Unlike the PNS, the CNS is encased in protective structures including bone (the skull and backbone) and three layers of sheathlike membranes called **meninges** (**Figure 42.7**). The outermost membrane, the dura mater (from the Latin, meaning hard mother), is a thick protective layer that lies just inside the skull and vertebrae. The middle membrane is called the arachnoid mater (from the Latin, meaning spidery mother) because it has numerous weblike tissue connections to the innermost membrane, the pia mater (from the Latin, meaning thin mother). The pia mater is a very thin membrane that lies on the surface of the brain and spinal cord, folding with the brain's surface.

Between the arachnoid mater and pia mater is the subarachnoid space. This space is filled with **cerebrospinal fluid**, which surrounds the exterior of the brain and spinal cord and absorbs physical shocks to the brain resulting from sudden movements or blows to the head. The cerebrospinal fluid contains nutrients, hormones, and other substances that are taken up by cells of the brain. The fluid is also a reservoir for metabolic waste products that are then carried away by the circulatory system. In addition to the subarachnoid space, the cerebrospinal fluid also fills a

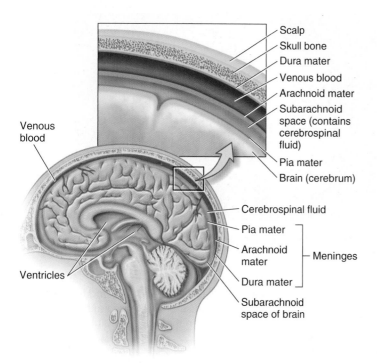

Figure 42.7   The meninges and ventricles of the CNS.
The thickness of the meninges are exaggerated for illustration
purposes. Note that the cerebrospinal fluid encases the entire
CNS and also fills the ventricles.

Concept check: *In a procedure known as a lumbar puncture
(commonly referred to as a spinal tap), physicians use a needle
to withdraw a small portion of cerebrospinal fluid from the
bottom of the spine to help diagnose specific illnesses. What
effects might this procedure have on a patient?*

series of connected cavities called the ventricles that lie deep
within the brain and connect to the central canal that extends the
length of the spinal cord (see Figure 42.6). These fluid-filled structures provide a cushion of support and protection for the CNS.

The PNS Carries Information to and from the CNS

The PNS of vertebrates is subdivided into two major functional
and anatomical components: the somatic nervous system and
the autonomic nervous system. Both divisions have sensory
(afferent) nerves and motor (efferent) nerves.

Somatic Nervous System   The primary function of the **somatic
nervous system** is to sense the external environment and control skeletal muscles. The sensory neurons of the somatic nervous system receive stimuli, such as heat, light, odors, chemicals
(in food), sounds, and touch, and transmit signals to the CNS.
The motor neurons of the somatic nervous system control skeletal muscles. The cell bodies of these motor neurons are actually located within the CNS. The axons from these cells leave
the spinal cord and project directly onto skeletal muscle without any intermediary synapses along the way.

The somatic nervous system is said to be voluntary because
many of the responses can be controlled consciously. For example, we use our somatic nervous system to walk and hold a

pencil. However, not all responses are voluntary. An example is
a reflex arc, such as the knee-jerk response, which is automatic
(refer back to Figure 41.4).

Autonomic Nervous System   The **autonomic nervous system** regulates homeostasis and organ function. For example, it
is involved in regulating the rate of heart contractions, blood
pressure, and the amount of stomach acid secreted. Though the
autonomic nervous system is predominantly composed of motor
neurons, it also has sensory neurons that detect internal body
conditions. For example, baroreceptors are sensory neurons that
detect blood pressure. For the most part, the autonomic nervous
system is not subject to voluntary control. We usually cannot
consciously change our heart rate or blood pressure.

The efferent pathways of the autonomic nervous system
involve two motor neurons. The cell body of the first neuron
is within the CNS and synapses on a second neuron in ganglia outside the spinal cord; these ganglia, therefore, are part
of the PNS. This second neuron sends its axon to an effector
cell, where it alters that cell's function. These neurons control
smooth muscles, cardiac muscle, and glands.

The efferent nerves of the autonomic system are subdivided into the sympathetic and parasympathetic divisions (**Figure 42.8**). Both divisions of the autonomic system act on the
same organs and usually have opposing actions. The **sympathetic division** is responsible for rapidly activating systems that
prepare the body for danger or stress. Imagine, for example,
the physiological responses that would occur if a person was
hiking and came upon a grizzly bear. This is the **fight-or-flight**
response, which is characterized by increased heart rate, stronger pumping action of the heart, relaxed (opened) airways and
faster breathing, inhibition of digestive activity, increased blood
flow to skeletal muscles, and increased secretion of energy-
supplying substances such as glucose and fats into the blood
by the liver and adipose tissue. These features prepare us to
confront (fight) or avoid (flight) a perceived threat.

The **parasympathetic division** of the autonomic nervous
system is involved in maintaining and restoring body functions.
It is active during restful periods or after a meal, which is why
it is sometimes said to mediate the **rest-or-digest** response.
Neurons of the parasympathetic division promote digestion
and absorption of food from the gut, slow the heart rate, and
decrease the amount of fuel supplied to the blood from the liver
and adipose tissue. A summary of these and other major functions of the two divisions of the autonomic nervous system can
be found in Figure 42.8.

The Hindbrain Is Important for Homeostasis and Coordination of Bodily Functions

Let's now turn our attention to the structure and function of the
human brain (**Figure 42.9**). We will begin with the evolutionarily oldest structures of the brain, some of which are located in
the hindbrain and control the basic processes that sustain life.

Medulla Oblongata   The medulla oblongata is located
between the pons and the anterior part of the spinal cord. It

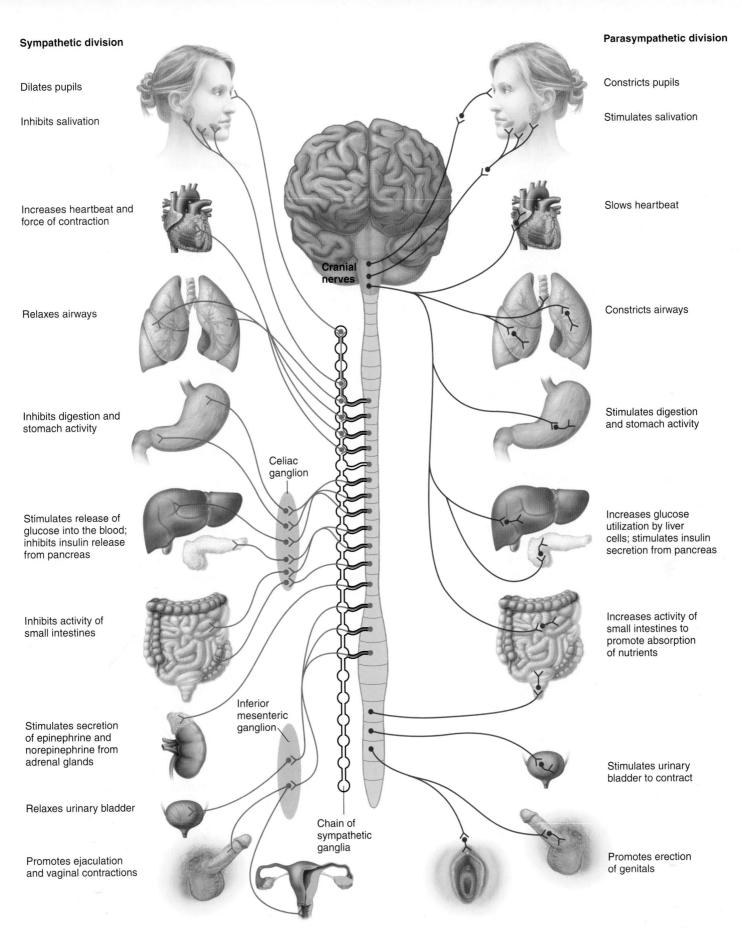

Sympathetic division

Dilates pupils

Inhibits salivation

Increases heartbeat and force of contraction

Relaxes airways

Inhibits digestion and stomach activity

Stimulates release of glucose into the blood; inhibits insulin release from pancreas

Inhibits activity of small intestines

Stimulates secretion of epinephrine and norepinephrine from adrenal glands

Relaxes urinary bladder

Promotes ejaculation and vaginal contractions

Parasympathetic division

Constricts pupils

Stimulates salivation

Slows heartbeat

Constricts airways

Stimulates digestion and stomach activity

Increases glucose utilization by liver cells; stimulates insulin secretion from pancreas

Increases activity of small intestines to promote absorption of nutrients

Stimulates urinary bladder to contract

Promotes erection of genitals

Cranial nerves

Celiac ganglion

Inferior mesenteric ganglion

Chain of sympathetic ganglia

Figure 42.8 **Sympathetic and parasympathetic divisions of the autonomic nervous system.** For simplicity, only some of the major functions of each division are shown in this figure. The sympathetic and parasympathetic systems tend to have opposite effects, and most parts of the body receive inputs from both divisions. Nerves from the sympathetic division make connections with a chain of ganglia, most but not all of which are alongside the spinal cord. Nerves from the parasympathetic division make connections in ganglia near or in their targets (for clarity, the ganglia are only shown near the targets).

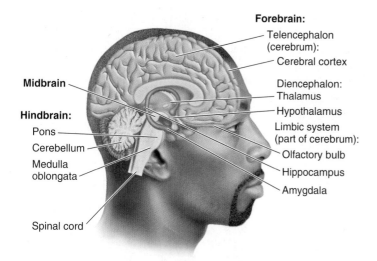

Forebrain:
— Telencephalon (cerebrum):
— Cerebral cortex

Diencephalon:
— Thalamus
— Hypothalamus

Limbic system (part of cerebrum):
— Olfactory bulb
— Hippocampus
— Amygdala

Midbrain

Hindbrain:
Pons
Cerebellum
Medulla oblongata

Spinal cord

Figure 42.9 Major structures of the human brain. An overview of the brain, showing many internal structures (not all structures are visible in this plane, such as the basal nuclei). The limbic system consists of the olfactory bulbs, amygdala, and hippocampus, which are part of the telencephalon (many neuroscientists consider parts of the thalamus and hypothalamus as part of the limbic system). The midbrain, pons, and medulla oblongata collectively comprise the brainstem.

Concept check: If a person received an injury to the cerebellum, how would that affect his or her behavior?

coordinates many basic reflexes and bodily functions that maintain the normal homeostatic processes of a person. It is involved in the control of heart rate, breathing, cardiovascular function, digestion, swallowing, and vomiting, and gives rise to several of the cranial nerves.

Cerebellum and Pons The cerebellum, a large structure that sits dorsal to the medulla oblongata, receives sensory inputs from the cerebral cortex and the auditory and visual areas of the brain. It also receives inputs from the spinal cord that convey information about the position of joints and the contraction or relaxation of muscles. The overall function of the cerebellum is to maintain balance and coordinate hand-eye movements. The cerebellum controls the use of multiple muscles at one time and synchronizes motor activities such as typing, making a jump shot in basketball, or touching the fingers to the tip of the nose with your eyes closed. When the cerebellum is damaged, such as in an accident, people find it difficult to maintain balance and fine-tune motor functions. Reaching for a glass filled with water, for example, might result in spilling some of the water while bringing the glass to the mouth. Although historically scientists have thought that the cerebellum does not play a role in learning, memory, and conscious thought, recent evidence has strongly suggested that the cerebellum may indeed have significant cognitive functions, the full extent of which remains to be discovered.

The pons sits anterior to the medulla oblongata and beneath the cerebellum. Major tracts pass through the pons into and out of the cerebellum, so the pons serves as a relay between the cerebellum and other areas of the brain. In addition to this integrative motor function, the pons contains nuclei that play a very important role in regulating the rate and depth of breathing. The pons also gives rise to some of the cranial nerves.

The Midbrain Processes Sensory Inputs

The midbrain lies anterior to the pons. It processes several types of sensory inputs, including vision, olfaction, and audition. It has tracts that pass this information to other parts of the brain for further processing and interpretation. As one example, the midbrain is responsible for activating neural pathways that change the diameter of the pupil of the eye in response to a change in the amount of ambient light. If the midbrain were damaged, this pupilary reflex would not occur normally or at all.

The medulla oblongata, the pons, and the midbrain collectively comprise the **brainstem**. In addition to the functions just described, all three major parts of the brainstem contain additional nuclei that together form the **reticular formation**. This is a network-like pathway that extends throughout much of the brainstem and that maintains and controls alertness and sleep, plus essential functions such as regulation of the respiratory and cardiovascular systems. Because of the importance of the brainstem's functions, damage to it is catastrophic and may result in coma or death.

The Forebrain Is Responsible for Movement, Sensory Function, and Higher Functions of Thought, Learning, and Emotion

As mentioned, the diencephalon and the telencephalon (cerebrum) comprise the forebrain (**Figure 42.10a**). The diencephalon is made up of the thalamus, hypothalamus, and epithalamus. The cerebrum consists of the basal nuclei, limbic system, and cerebral cortex.

Diencephalon In vertebrates, the thalamus plays a major role in relaying sensory information to appropriate parts of the cerebrum and, in turn, sending outputs from the cerebrum to other parts of the brain. It receives input from all sensory systems except olfaction. The thalamus is organized according to the type of sensation. For example, certain parts of the thalamus process visual inputs, while other parts process sounds. One type of processing performed by neurons in the thalamus is that of filtering out sensory information in a way that allows us to pay attention to important cues while temporarily ignoring less important ones. A good example of this occurs when you focus on what someone is saying to you in a crowded room, with background sounds, sights, and activities that could otherwise be distracting. The thalamus also directs outgoing motor instructions that it receives from the cerebral cortex, sending directions to the spinal cord to generate voluntary movements, and is involved in the perception of pain and the degree of mental arousal in the cerebral cortex.

The hypothalamus, located below the thalamus at the floor of the forebrain, controls functions of the gastrointestinal and reproductive systems, body temperature (thermoregulation), and many basic behaviors such as eating and drinking. This area has

(a) Forebrain hierarchy

Parietal lobe

Occipital lobe

Frontal lobe

Temporal lobe

(b) Lobes of the cerebral cortex

Figure 42.10 **The human forebrain.** (a) Relationships of the various parts of the forebrain. (b) The four lobes of the cerebral cortex as seen on the right hemisphere.

great importance for homeostasis of the body and the control of behavior. Though small in size, it is composed of many nuclei, each with its own vital function. A major role of the hypothalamus is the production of hormones, which travel to the pituitary gland located just beneath the brain. The pituitary gland, in turn, regulates hormone secretion from other glands in the body, including the thyroid, gonads, and adrenal glands. In addition to producing hormones, the hypothalamus is sensitive to their actions. For example, certain hormones produced by cells in the stomach, intestine, adipose tissue, gonads, and elsewhere act within the hypothalamus to facilitate the expression of feeding, drinking, sexual, and aggressive behaviors. Finally, a small pair of hypothalamic nuclei called the suprachiasmatic nuclei acts as the "master clock" of the CNS, establishing circadian rhythms, which control the expression of behavioral, physiological, and hormonal rhythms over the 24-hour day.

The epithalamus is a collection of structures that have various roles in the production of cerebrospinal fluid, control of food and water intake, and rhythmic and seasonal behaviors in some vertebrates. One of these structures, the pineal gland, is located in the center of the brain and secretes a hormone called **melatonin** into the blood. Production of melatonin is regulated by the length of the light period in each day, and, partly for this reason, melatonin is believed to be a key factor in determining

seasonal reproductive behaviors in certain mammals (for example, sheep and hamsters). The function of melatonin in humans in still debated but has been suggested to play a role in daily cycles such as our sleep/wake rhythm.

Telencephalon (Cerebrum) As mentioned, the cerebrum consists of the basal nuclei, limbic system, and cerebral cortex. For historical reasons, the **basal nuclei** are also called the basal ganglia, though the term ganglia in vertebrates normally refers to structures in the PNS. The basal nuclei are a group of nuclei that surround the thalamus and lie beneath the cerebral cortex. Like the cerebellum, the basal nuclei are involved in planning, learning, and fine-tuning movements. They also function via a complex circuitry to initiate or inhibit movements.

Parkinson disease (formerly called Parkinson's disease), a common neurological disorder that affects the basal nuclei, provides a good illustration of the function of these nuclei. People with Parkinson disease have trouble initiating movement, such as beginning to move their legs when they wish to walk. They are capable of walking once movement has begun, but they move slowly with muscle tremors and a shuffling, jerky gait. These symptoms result from the gradual deterioration of dopamine-releasing neurons in an area of the midbrain called the substantia nigra, the neurons of which send axons to the basal nuclei. People in the early stages of Parkinson disease can be treated with L-dopa, a molecule that enters the blood and travels to the basal nuclei. There, axon terminals from cells originating in the substantia nigra take up the L-dopa and convert it into dopamine, which is then released onto cells of the basal nuclei. L-dopa, therefore, substitutes for dopamine and reduces the Parkinson symptoms.

The **limbic system** refers to a collection of evolutionarily older structures that form an inner layer at the base of the forebrain. These include structures such as the **olfactory bulbs** (which process information about smells), **amygdala**, and **hippocampus**. Many neuroscientists also consider parts of the diencephalon as part of the limbic system, because of the extensive connections between these regions. The limbic system is primarily involved in the formation and expression of emotions, and it plays an important role in learning, memory, and the perception of smells. The expression of emotions occurs early in childhood before the more advanced functions of the cerebral cortex are evident. Thus, even very young babies can express fear, distress, and anger as well as bond emotionally with their parents.

Deep within the ventral part of the brain, the amygdala is one of the areas critical for understanding and remembering emotional situations. This area also is involved in the ability to recognize emotional expression in others. Emotions are not unique to humans, however, and are clearly present in other primates and mammals. Being able to express and detect emotions imparts a selective advantage by enabling animals to establish and maintain relationships. Emotions such as fear help an animal defend itself against danger by avoiding conflict. Likewise, anger is associated with aggression, a key behavior by which many animals defend themselves or their territories.

Adjacent to the amygdala and forming a loop within the medial regions within the brain, the hippocampus is composed of several layers of cells that are connected together in a circuit. Its main function appears to be establishing memories for spatial locations, facts, and sequences of events. Damage to certain parts of the hippocampus in humans results in an inability to form new memories, a devastating condition that prevents recognition of other people or even an awareness of daily events. Experiments with laboratory animals have also demonstrated the importance of the hippocampus for memory and learning in other mammals. In a particularly well-studied example, rats are placed into a pool of milky water containing a hidden platform. The animals swim until they find the platform, on which they can safely stand. The time it takes to find the platform in subsequent trials is shorter as they learn and remember its whereabouts. This type of spatial learning depends on activity in the hippocampus. Rats with parts of their hippocampus destroyed fail to improve their times with repeated trials. The hippocampus also receives extensive inputs from the olfactory bulbs, which may explain why smells are such potent triggers of memory in humans and why many animals use their sense of smell as a major way to learn and remember aspects of their environments.

As mentioned earlier, the cerebral cortex is the surface layer of gray matter that covers the cerebrum (see Figure 42.6). The **neocortex** (from the Greek, meaning new cortex), the layer that evolved most recently in mammals, has six layers of cells organized in rows and columns. Neurons within and between layers are connected to form circuits, which integrate information from other nervous system structures and create outgoing signals that reflect that integration. Evolutionarily older parts of the cerebral cortex, including areas involved in processing the sense of smell, have fewer layers. Although the cerebral cortex is only a few millimeters thick, it contains about 10% of all the neurons in the human brain.

The brain has two halves, or **hemispheres**. The cerebral cortex is divided into four lobes in each hemisphere: the frontal, parietal, occipital, and temporal lobes (**Figure 42.10b**). Each lobe has a number of functions, many of which are still being actively investigated by researchers. Among other things, the **frontal lobe** is important for voluntary initiation of movement, decision making, controlling impulses, making plans, and exhibiting judgment and for conscious thought and social awareness. The **parietal lobe** receives and interprets sensory input from visual pathways and somatic pathways, including those from the surface of the body. In addition, the parietal lobe plays an important role in spatial awareness, that is, our ability to use visual cues to orient ourselves in space. The **occipital lobe** controls many aspects of vision and color recognition. The **temporal lobe** is necessary for language, hearing, and some types of memory.

One particularly interesting aspect of temporal lobe activity that illustrates the subtle functions of the various lobes is the recognition of objects and people. One area of the temporal lobe contains neurons known as face cells that are activated specifically in response to seeing faces. Electrical activity of individual cells within the temporal lobe of rhesus monkeys increases when the monkey views human and monkey faces, though not when looking at other images. Humans who have damage to that area are unable to recognize other people by their faces even if they knew them before the brain damage

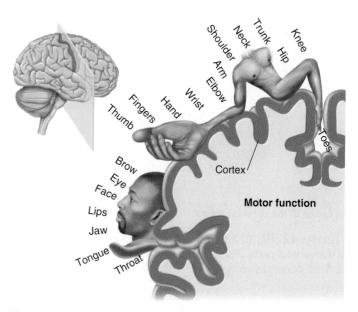

Figure 42.11 Homunculus maps of human body parts along the cerebral cortex.   These maps represent how the cortex interprets sensory information from these body parts and controls body movements of these parts (motor function). The relative sizes of body parts reflect the relative amount of cortex devoted to them.

occurred. Fortunately, these patients can recognize other traits such as gait or tone of voice, which they use to identify the people in their lives. Scientists do not know whether face cells are active in nonprimate mammalian species.

An amazing finding is that sensory inputs enter and motor outputs exit the cerebral cortex in a pattern that forms a sort of map of the body, often depicted as a homunculus (from the Latin, meaning little man) (**Figure 42.11**). The amount of space in the cerebral cortex assigned to a particular part of the body is proportional to either the degree of sensory sensitivity of that body part or the number of muscles required for its movement. For instance, a larger part of the cerebral cortex is devoted to the lips than to other areas of the face. The lips have more nerve endings and are more sensitive to touch than these other areas. Other cortical functions are also mapped in this way; for example, a map that reflects different sound frequencies (the pitch of

Table 42.1	Major Functions of Brain Regions in Humans	
Region		**Major Functions**
Hindbrain		
	Medulla oblongata and pons	Coordinate homeostatic functions such as breathing, heart rate, digestion; form part of reticular formation that controls sleep and alertness; give rise to cranial nerves
	Cerebellum	Fine tuning of complex body movements; maintenance of balance
Midbrain		Processes visual, auditory, and olfactory sensory inputs; forms part of reticular formation
Forebrain		
	Thalamus	Routes sensory information (except olfaction) to discrete parts of cerebrum; filters irrelevant sensory information; directs outgoing motor information from cerebral cortex to spinal cord; involved in pain perception and mental arousal
	Hypothalamus	Regulates activities of gastrointestinal and reproductive systems; controls function of pituitary gland; regulates body temperature, appetite, thirst, aggressive behavior, sexual behavior, and body rhythms
	Epithalamus	Produces cerebrospinal fluid; plays a role in food and water intake; contains the pineal gland, which may regulate sleep/wake and body rhythms
	Basal nuclei	Planning, fine-tuning, initiating, inhibiting and learning movements
	Limbic system	Formation and expression of emotions; perception of odors; learning and memory
	Cerebral cortex	Voluntary motor control; perception of sensory inputs; attention; integration of sensory and motor information; generation of speech; decision-making, impulse control, judgment, and planning; conscious thought; learning, memory, and emotion

sound) exists in the temporal lobes. The organization of the cerebral cortex may not be permanent, however, because the map may change depending on the amount of use or disuse of a given part of the body, as discussed in the Feature Investigation later in this chapter. Some of the major functions of the hindbrain, midbrain, and forebrain are summarized in **Table 42.1**.

Cerebral Hemispheres   As mentioned, one of the most recognizable features of the surface of the cerebrum is its division into two hemispheres. Each hemisphere is connected to the other by a major tract called the **corpus callosum** (Figure 42.12a). In the 1950s, American neuroscientists Roger Sperry and Ronald Meyers examined the separate functions of the hemispheres in laboratory animals by performing split-brain surgeries in which they severed the corpus callosum. The animals that underwent such

surgery maintained their overall health and functioning; therefore, the surgery was considered safe for humans. This became important in 1961, when the procedure was used for the first time to treat patients with severe epilepsy, a disorder characterized by uncontrolled electrical activity that begins in one place in the brain and can spread via the corpus callosum to the other side. Cutting the connection between the hemispheres reduced the severity of epileptic seizures.

Split-brain surgery also provided an opportunity for the researchers to make critical observations that could not be made in laboratory animals. Split-brain humans generally show normal behavior and intellectual function, because both hemispheres can function fairly independently. However, psychological tests revealed that the two sides of the brain process different types of information. One study demonstrated that the

(a) Cross section of brain showing the corpus callosum

(b) Testing of split-brain patient

Figure 42.12   **The hemispheres of the human brain.**
(a) The cerebral hemispheres and their connection by the corpus callosum. (Note: The left hemisphere controls the right side of the body, and the right hemisphere controls the left side.) **(b)** Split-brain patient being tested for hemispheric dominance. By using this apparatus, Roger Sperry and his collaborators showed that the left and right cerebral hemispheres have different capabilities. When a split-brain patient held an object in his right hand but could not see it or touch it with his left hand, he could give it a name (for example, an apple). When he held another object in his left hand, he could describe it (for example, smooth), but could not name it.

Concept check: *With her eyes closed, a split-brain patient was given a rock to hold, and she described it as a rock. Which hand was it in?*

left hemisphere produces a descriptive word for an object but does not identify certain characteristics of that object, such as its shape and texture (**Figure 42.12b**). The right hemisphere, in contrast, cannot use words to name the object but can identify other qualities. The studies of Sperry, Meyers, and other neuroscientists have concluded that the left hemisphere is involved in understanding language and producing speech in most people. Therefore, the left hemisphere is said to be dominant for those functions. The right hemisphere is dominant for nonverbal memories, recognizing faces, and interpreting emotions. In 1981, Sperry received the Nobel Prize for his insight regarding specialization in each hemisphere of the brain.

Although not unique to humans, the cerebral cortex is one of the defining features of the human brain, because it is responsible for much of what we call our individual personalities. Researchers are now beginning to understand the molecular mechanisms by which this important brain structure develops, as we see next.

Genomes & Proteomes Connection

Several Genes Have Been Important in the Evolution of the Cerebral Cortex

A number of genes are known to be involved in the development of the cerebral cortex. Some have been identified by examining genetic mutations in developmentally disabled individuals, and others by comparing human genes with genes known to be involved in brain development in other species such as *Drosophila*. Researchers have also compared these genes in many species that show notable differences in cerebral structure. This last approach can determine whether a relationship exists between the expression of a particular gene and the organization of the cerebral cortex.

One inherited disorder that involves abnormal development of the cerebral cortex is polymicrogyria (from the Greek, meaning many small folds), which results in mental impairment as well as disrupted gait and language production. In people with this disorder, the cerebral cortex is characterized by multiple abnormal surface folds and fewer layers of cells. One type of polymicrogyria is a recessively inherited condition for which eight different mutations of a single gene are known. This gene, called *GPR56*, encodes a G-protein-coupled receptor (described in Chapter 9), which has large extracellular loops. All eight mutations that produce polymicrogyria alter these extracellular loops of the receptor, and scientists think that this alters the ability of the G-protein-coupled receptor to bind its ligand.

Two other genes, called microcephalin or *MCPH1* (from the Greek, meaning small head) and *ASPM* (*abnormal spindle-like microcephaly-associated gene*), have been shown to be determinants of brain size and thus are possibly involved in the evolution of brain size. For example, mutations of these genes in the human population produce individuals with much smaller frontal lobes. Interestingly, the sequences of these genes in several primates, including humans, as well as in other mammals such as dogs and sheep, have shown that the proteins produced by the normal *MCPH1* and *ASPM* genes have undergone greater changes in humans and great apes than in other species. Therefore, these genes may have been under greater selection pressure in animals with larger cerebral cortexes, suggesting that the genes play a key role in cerebral cortex development.

42.3 Cellular Basis of Learning and Memory

In the past few decades, an exciting advance in neuroscience has occurred—researchers have begun to understand complex behaviors, such as learning and memory, at the cellular level. Though it is difficult to separate the two concepts, **learning** can be defined as the process by which new information is acquired. Learning is an evolutionary adaptation that allows past experiences to affect ongoing and future behavior. **Memory** is the ability to retain, retrieve, and use information that was previously learned. Memory connects our experiences throughout life. Our behavior is largely controlled by what we have learned and remember from past experiences. Neuroscientists want to understand how the brain learns and how it captures memories. In this section, we will examine some current ideas about how this may be achieved at the cellular level and consider experimental approaches that neuroscientists follow when investigating such complicated phenomena.

Learning and Memory Occur via Changes Within Neurons and Their Connections with Each Other

Beginning in the 1960s, research along two fronts led to key insights regarding the cellular basis of memory. Norwegian neuroscientist Terje Lomo and British researcher Timothy Bliss focused their efforts on the hippocampus. As described earlier, this is a key region of the brain involved with learning and memory. Lomo and Bliss conducted experiments on anesthetized rabbits in which they monitored signal transmission across particular regions of the hippocampus. Their key discovery involved the effects of multiple stimuli. Experimentally, a series of short, electrical stimulations to a neuron was shown to strengthen, or potentiate, its communication at a synapse with an adjacent cell for minutes or hours. Such multiple stimuli caused neurons to communicate more readily; responses were stronger and more prolonged. This phenomenon was termed **long-term potentiation** (LTP). LTP is the long-lasting strengthening of the connection between neurons. Later work showed that LTP occurs naturally in the hippocampus and can last from hours to days, and even years.

Austrian neuroscientist Eric Kandel also was interested in learning and memory, and some of his early studies involved the hippocampus. In the 1960s, however, he took a different approach by studying learning and memory in a simpler organism called the California sea slug or sea hare (*Aplysia californica*). He chose this organism for several key reasons. First, it has only about 20,000 neurons, making it easier to identify pathways that are involved in specific types of behavior. Second, some of the neurons in this organism are extremely large, allowing the study of action potentials via microelectrodes (as described in Chapter 41). In addition, the large size of the neurons made it possible to inject substances into them and study their effects. Finally, another major advantage is that Kandel and colleagues could isolate proteins and mRNA from these large neurons and identify the biochemical and genetic changes that occur when the animal learns about its outside world and retains memory.

Much of Kandel's work focused on one type of learning involving the gill-withdrawal reflex, which is thought to involve less than 100 neurons in the CNS. The gill and siphon are organs involved in respiration, located in the animal's mantle cavity and protected by muscular appendages called parapodia (**Figure 42.13a**). When the siphon is gently touched with a fine probe, the sea slug closes the siphon and retracts its gills into the mantle cavity for protection (**Figure 42.13b**). Though called a reflex, this behavior is subject to learning. For example, if the touching of the siphon is accompanied with a brief electrical shock to the tail, the sea slug can learn to withdraw its gill in response to a subsequent shock without the siphon being touched. This is similar in some ways to the famous conditioning experiments of Ivan Pavlov, described in Chapter 40. Interestingly, a single tail shock paired with a touch of the siphon will result in conditioning that lasts for a few minutes. Amazingly, though, multiple trials over several days result in a lasting memory—a shock given three weeks later (without siphon touch) still results in the gill-withdrawal reflex!

Over the course of many years, the work of Kandel and colleagues revealed many clues regarding the cellular basis of learning and memory. As in vertebrates, these processes in the sea slug occur in two phases, short-term and then long-term memory (**Figure 42.14**). Short-term memory lasts for minutes or hours. This type of memory is typically caused by a single stimulus. Kandel found that short-term memory does not require the synthesis of new proteins. Rather, a single stimulus activates intracellular second messenger pathways that make it easier for neurons involved in a particular behavior to communicate with each other. For example, as shown in Figure 42.14a, a single stimulus may lead to the activation of protein kinases such as protein kinase A in the presynaptic (sensory) cell. Protein kinase A, in turn, can phosphorylate proteins such as ion channels and proteins involved with the release of neurotransmitter. These changes enhance the transmission of a signal between the presynaptic and postsynaptic cells.

Kandel and colleagues also discovered that repeated stimuli result in long-term memory, which lasts days or weeks. Such repeated stimuli require the synthesis of new proteins (Figure 42.14b). Long-term memory involves the activation of genes in the presynaptic cell, which leads to the synthesis of mRNA and the translation of the encoded proteins. Once made, such proteins cause the formation of additional synaptic connections. These connections also allow the presynaptic and postsynaptic cells to communicate with each other more readily. Such a change in synapses, which occurs as a result of learning, is termed **synaptic plasticity**.

Kandel's work provides a foundation for our ability to understand how learning and memory may occur at the cellular

(a) Sea slug

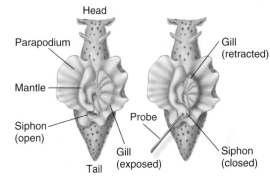

(b) Gill-withdrawal reflex

Figure 42.13 The gill-withdrawal reflex in the sea slug. **(a)** Photo of *Aplysia californica* in its natural habitat; note the parapodium and tail. **(b)** When touched with a probe, the siphon closes and the gill retracts. In this drawing, the parapodia are moved apart for a better view of the gill.

level. Short-term memory may involve changes in pre-existing cellular proteins that make it easier for neurons to communicate. Long-term memory results in protein synthesis that causes physical changes in the synapse itself, also affecting communication. Later studies by Kandel and others showed that such changes also occur in vertebrates such as the mouse. For his work on learning and memory, Kandel was awarded the Nobel Prize in Physiology or Medicine in 2000.

Neurogenesis May Also Contribute to Learning and Memory

Until fairly recently, neuroscientists had thought that the adult brain of vertebrates was incapable of **neurogenesis**—the production of new neurons by cell division. In 1983, a study showing neurogenesis in an adult vertebrate was carried out by Fernando Nottebohm and colleagues. Their research revealed an increase in the number of neurons in certain brain areas of the canary arose during mating season. In the late 1990s, evidence also showed that the primate and human CNS, like other parts of the body, contain stem cells (see Chapter 19). American researchers Elizabeth Gould and Bruce McEwen have demonstrated the appearance of new neurons in the hippocampus and olfactory bulbs of adult marmosets and rhesus monkeys. In 1998, American researcher Fred Gage and Swedish physician Peter Eriksson made a key discovery: They found new hippocampal cells in the brains of recently deceased adult cancer patients. Those patients

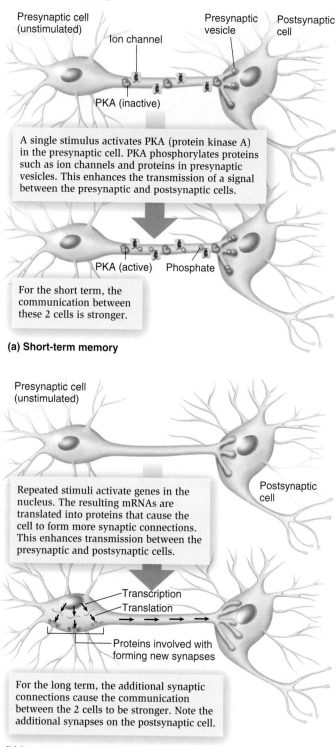

Presynaptic cell (unstimulated)

Ion channel

Presynaptic vesicle

Postsynaptic cell

PKA (inactive)

A single stimulus activates PKA (protein kinase A) in the presynaptic cell. PKA phosphorylates proteins such as ion channels and proteins in presynaptic vesicles. This enhances the transmission of a signal between the presynaptic and postsynaptic cells.

PKA (active)   Phosphate

For the short term, the communication between these 2 cells is stronger.

(a) Short-term memory

Presynaptic cell (unstimulated)

Repeated stimuli activate genes in the nucleus. The resulting mRNAs are translated into proteins that cause the cell to form more synaptic connections. This enhances transmission between the presynaptic and postsynaptic cells.

Postsynaptic cell

Transcription

Translation

Proteins involved with forming new synapses

For the long term, the additional synaptic connections cause the communication between the 2 cells to be stronger. Note the additional synapses on the postsynaptic cell.

(b) Long-term memory

Figure 42.14 Cellular changes associated with short-term and long-term memory in the sea slug.

had previously been treated with bromodeoxyuridine (BrdU) to combat their cancer. BrdU is also taken up by noncancerous cells, but only those that are actively dividing. Its presence in cells can be detected with special stains on sections of brain and other tissue. Gage conducted such staining procedures on the brains of the deceased patients and observed the presence of BrdU in the hippocampus, demonstrating recent neurogenesis.

A key question is whether the neurogenesis observed in adult brains is involved in learning and memory. This question is hotly debated and not resolved. However, some evidence suggests that it could play a role. For example, McEwen and coworkers have shown that the hippocampus of adult rhesus monkeys grows new neurons when the animals are placed in enriching environments, and the formation of these neurons slows when animals are chronically stressed. Also, studies in the rat by Gould and colleagues suggested that new neurons are retained in the hippocampus in response to training in particular tasks that require hippocampus function.

Brain Images Are Used to Assess Brain Structure and Function

Because of the enormous number of neurons and connections between neurons in the vertebrate brain, a key challenge in neuroscience is to understand how such complexity results in sophisticated forms of learning, memory, and responses to environmental conditions. Several imaging techniques allow doctors and researchers to examine the structure and activity level of the brain without anesthesia or surgery. The earliest technique to be developed was computerized tomography (from the Greek *tomas*, meaning a cutting). A **CT scan** involves the use of an X-ray beam and a series of detectors that rotate around the head, producing slices of images that are reconstructed into three-dimensional images based on differences in the density of brain tissue. CT scans can easily visualize the ventricles and differences between white and gray matter, but they cannot examine the brain in great detail.

A more sensitive method, called **magnetic resonance imaging** (**MRI**), was developed in the 1980s. The patient's head is placed in a device that contains a magnet powerful enough to generate a magnetic field many thousands of times greater than that of the Earth. This stabilizes the spinning, or resonance, of atomic nuclei (usually hydrogen atoms in water molecules) so that most of the nuclei align with the magnetic field. When body tissue is stimulated with a beam of radio waves, its atoms absorb the energy of the waves and the resonance of their nuclei changes, thereby altering their alignment with the magnetic field. When the radio wave pulse stops, the atoms release their energy, which is recorded by a detector. This information is analyzed by a computer, and an image is produced. MRI images allow detection of structures as small as one-tenth of a millimeter. For example, they can provide information about abnormal tissue, such as brain tumors, which respond to magnetic and radio frequency pulses differently than normal tissue. MRIs are widely used in medicine to check for injured tissue, cancers, and other abnormalities throughout the body. In 2003, Paul Lauterbur and Peter Mansfield received the Nobel Prize in Medicine for their work in developing this method.

With certain modifications, MRI can be used to assess the functional activity of areas within the brain. This technique, which is widely used by neuroscientists, is called **functional MRI** (**fMRI**). It takes advantage of the observation that blood flow, and therefore oxygen delivery, increases to areas where neurons are more active. This increase in oxygenation is detected via

(a) Brain activity of a person thinking about a task that requires finger movements

(b) Brain activity when the same person is performing this task

Figure 42.15 Exploring the functional activity of brain regions using fMRI scans.   Red color indicates higher O$_2$ use.

Concept check: *What can we conclude from part (a) about the metabolic cost of thinking?*

fMRI. In this way, fMRI determines which neurons in particular areas of the brain are active when an individual performs certain intellectual or motor tasks (**Figure 42.15**). The principle is similar to the standard MRI, except that the higher oxygen content of active tissue alters the resonance of local hydrogen atoms.

The use of fMRI has revealed many fascinating aspects of the activities of different brain regions, notably in people who have suffered brain damage or loss of sensory inputs. For example, individuals who are blind from birth might be predicted to have occipital lobes that are less functional or active than are those in sighted persons (recall that the occipital lobes play the major role in visual processing). However, the work of American researcher Harold Burton and others has revealed with fMRI that the

occipital lobes of blind persons are active but have become adapted to other sensory functions such as tactile signals from the fingers, including those arising from Braille reading. Amazingly, this reassignment of occipital function occurs to some extent even in individuals who have lost their vision later in life. Most likely, this does not represent a brand new function of the occipital lobes, but rather an expanded ability of an existing function that remains relatively minor in sighted persons.

The plasticity of the brain revealed by the work of Burton and coworkers is not restricted to clinical situations, as just described. MRI and fMRI are also revealing differences in brain structure and function in individuals due to the types of activities in which they engage, as described next.

FEATURE INVESTIGATION

Gaser and Schlaug Showed That the Sizes of Certain Brain Structures Differ Between Musicians and Nonmusicians

MRI and fMRI have been extremely useful in revealing which brain areas are involved in a particular function. They have also shown that the human brain is surprisingly adaptable. A number of studies have been carried out on musicians, because they practice extensively throughout their lives, enabling researchers to study the effects of repeated use on brain function.

American neuroscientist Christian Gaser and German neuroscientist Gottfried Schlaug used MRI to examine the sizes of

brain structures in three groups of people—professional musicians, amateur musicians, and nonmusicians (**Figure 42.16**). Individuals were assigned to each group based on their reported history of musical training: professional musicians with over 2 hours of musical practice time each day, amateur musicians who played a musical instrument regularly but not professionally (practicing about 1 hour/day), and those who never played a musical instrument regularly. The researchers hypothesized that repeated exposure to musical training would increase the size of brain areas associated with visual, motor, and auditory skills, because each of these activities is used to read, make, and interpret music. The results showed that brain areas

Figure 42.16 Gaser and Schlaug's study of the size of visual, motor, and auditory nuclei in the brains of musicians and nonmusicians.

HYPOTHESIS Musical training is associated with structural differences in the brain.

KEY MATERIALS Volunteer subjects with different degrees of musical training.

	Experimental level	Conceptual level
1 Establish 3 groups of subjects with different musical backgrounds.	Interview subjects for musical history and assign to 1 of 3 groups.	**Controls (nonmusicians)** No musical training. **Amateur musicians** Play an instrument about 1 hr/day and are not employed as musicians. **Professional musicians** Employed as musicians and practice their instrument > 2 hr/day.

2 Control for possible factors that might affect results.

Make group assignments in a way that matches subjects' age and other characteristics.

	Controls	Amateurs	Professionals
Age (average)	23	26	27
Type of instrument	None	Keyboard	Keyboard
Mental IQ	119	123	118
Sex	Male	Male	Male

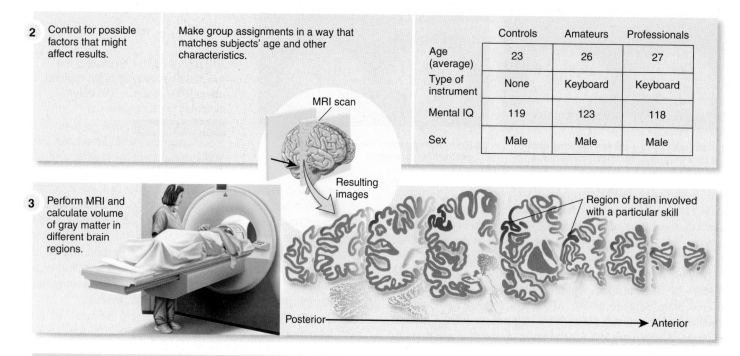

MRI scan

Resulting images

3 Perform MRI and calculate volume of gray matter in different brain regions.

Region of brain involved with a particular skill

Posterior ————————————————→ Anterior

4 THE DATA

Results from step 3:

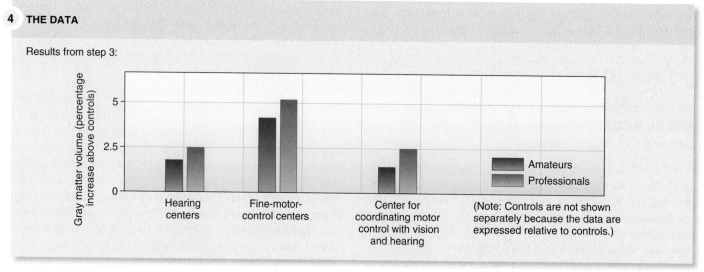

Gray matter volume (percentage increase above controls)

- Hearing centers
- Fine-motor-control centers
- Center for coordinating motor control with vision and hearing

Amateurs
Professionals

(Note: Controls are not shown separately because the data are expressed relative to controls.)

5   CONCLUSION Musical training is associated with increased volumes of brain regions involved in hearing, fine-motor control, and the coordination of motor and sensory information.

6   SOURCE Gaser, C., and Schlaug, G. 2003. Brain structures differ between musicians and non-musicians. *Journal of Neuroscience* 23:9240–9245.

involved in hearing, moving the fingers, and coordinating movements with vision and hearing were larger in professionals than in amateur musicians, and larger in amateurs than in nonmusicians. The region of the brain that controls finger movements was particularly well developed in the professional musicians, an interesting finding because all of the musicians in this study played keyboard instruments such as the piano.

In another study, American researchers Vincent Schmithorst and Scott Holland used fMRI to determine if musicians' brains were activated differently than nonmusicians' brains when they heard music. They found that one area of the cerebral cortex was selectively activated by melodies only in musicians. This study differed from that of Gaser and Schlaug, because Schmithorst and Holland examined the activity of brain areas as well as their sizes. Their results showed that listening to music activates certain neurons and pathways in the brains of musicians but not in nonmusicians.

The human studies of Gaser, Schlaug, Schmithorst, and Holland have not determined the underlying reason(s) for increased brain size. One possibility is that people with increased brain size in these regions are more likely to become musicians. Alternatively, musical training may actually cause certain regions of the brain to grow larger and alter their neuronal pathways. In other research studies involving experimental animals, groups of animals have been randomly separated into those learning a task versus controls, which do not learn the task. Such experiments have shown increases in the size of brain regions that are associated with learning and memory. The increased size may result from formation of new synapses, growth of blood vessels to the region, and/or production of more glial cells.

Experimental Questions

1. What was the hypothesis proposed by Gaser and Schlaug?

2. How did Gaser and Schlaug test this hypothesis? What were the results of their experiment?

3. How did the research of Schmithorst and Holland impact the findings of Gaser and Schlaug? How did their experiments differ from that of Gaser and Schlaug?

42.4 | Impact on Public Health

Most neurological disorders can be classified into several broad groups (**Table 42.2**). These disorders collectively affect hundreds of millions of people around the world. In addition to the conditions already discussed in this chapter and Chapter 41, we will consider two of these disorders, meningitis and Alzheimer disease, that result from very different causes and impact millions of individuals worldwide.

Meningitis Is an Infectious Disease That Attacks the Meninges

An essential nonspecific response to infection is inflammation. This response increases the permeability of blood vessels in infected areas, allowing immune cells to be delivered to the site of an infection. When that infection occurs within the meninges (causing **meningitis**), fluid accumulates in the subarachnoid space. This accumulation compresses the underlying brain tissue, interrupting oxygen flow to the neurons of the cerebral cortex. If not corrected, the resulting loss of oxygen (and nutrients) cause neuronal death and the loss of function of brain regions associated with those neurons.

The initial symptoms of meningitis include severe headaches, fever, or seizures. Many patients with meningitis also develop a stiff neck because the inflammation proceeds down the spinal cord. If the infection progresses untreated, it may lead to unconsciousness and even death within hours.

Several different viruses or bacterial species can cause meningitis. It usually results from an untreated infection in neighboring regions, such as the sinuses behind the eyes, nose, or ears. Meningitis can be confirmed by using a long needle to sample the cerebrospinal fluid in the spinal cord and analyzing the pressure and contents of the fluid. Large numbers of white blood cells, which are the body's chief infection-fighting cells, indicate infection in the cerebrospinal fluid and meninges. If the infection is the result of bacterial invasion, meningitis can be treated with bacterial-killing agents such as antibiotics. Antibiotics do not kill viruses, but fortunately the viral form of meningitis is usually less serious than the bacterial form and runs its course after several days or weeks.

Meningitis strikes roughly 25,000 people a year in the U.S. and can affect people of any age. Its incidence in children has greatly declined since the widespread use of a vaccine against

Table 42.2	Categories of Diseases Affecting the Human Central Nervous System
Category	**Examples**
Infectious diseases	Meningitis (discussed in this section), encephalitis
Neurodegenerative disorders	Alzheimer disease (discussed in this section)
Movement disorders	Parkinson disease (described in Section 42.2)
Seizure disorders	Epilepsy (described in Section 42.2)
Sleep disorders	Sleep apnea (brain fails to regulate breathing during sleep)
Tumors	Glioma (a tumor arising from glial cells)
Headache disorders	Migraines
Mood disorders	Major depressive disorder; bipolar depression (see Chapter 41)
Demyelinating disorders	Multiple sclerosis (see Chapter 41)
Injury-related disorders	Brain and spinal cord injuries due to accidents

the bacterium *Haemophilus influenzae* began in the U.S. in the early 1990s. Despite the vaccine, meningitis is still a dangerous and prevalent disease worldwide, and it tends to occur in individuals living in close quarters, where infections may spread rapidly. Occasionally, meningitis can become epidemic. For example, 250,000 people in sub-Saharan Africa were infected and 25,000 died in 1996, and nearly 75,000 people in Southeast Asia died of meningitis in 2004.

Alzheimer Disease Impairs Memory, Thought, and Language

Alzheimer disease (**AD**) (formerly called Alzheimer's disease) is the leading cause of dementia worldwide. It is characterized by a loss of memory and intellectual and emotional function. AD is a progressive disease that begins with small memory lapses, leading in later stages to problems with language and abstract thinking, and finally to loss of normal motor control. The disease usually appears after age 65, though some inherited forms can strike people in their 30s and 40s.

Although psychological testing can help to diagnose AD, historically a definitive diagnosis is possible only after death when the brain is examined microscopically. Brains of Alzheimer patients show two noticeable changes: senile plaques and neurofibrillary tangles (**Figure 42.17**). Senile plaques are extracellular deposits of an abnormal protein, β-amyloid, that forms large, sticky aggregates. These plaques were first noted in 1906 by German physician Alois Alzheimer, after whom the disease was named. Neurofibrillary tangles are intracellular, twisted accumulations of cytoskeletal fibers. Scientists are unsure how these changes influence intellectual function and memory. AD is also associated with the degeneration and death of neurons, particularly in the hippocampus and parietal lobes, which is why it is considered a neurodegenerative disease.

Researchers have identified variation in a few genes whose products are associated with the likelihood of developing AD later in life, but the underlying changes that result in the expression of these and other possible AD-related genes are still the subject of considerable research. Although genetics undoubtedly plays a role in AD, it is not the only possible cause. For example, when one identical twin develops AD, the other appears to be at increased risk but does not always develop the disease, even if he or she survives to very old age. Moreover, evidence suggests that severe head injuries, metabolic diseases such as diabetes, and heart and blood vessel disease may predispose a person to AD in later life.

Currently, AD cannot be prevented or cured. However, three major clinical approaches are currently being tested to prevent or slow down its progression. These approaches are designed to (1) induce a person's immune system to destroy β-amyloid as soon as it is formed, (2) prevent the formation of β-amyloid with drugs that block its synthesis, or (3) prevent the accumulation of β-amyloid into large aggregates using antiaggregation drugs. Each of these approaches holds great promise but is still unproven.

Until a cure for AD is found, its impact on public health remains enormous. Currently, about 4–5 million Americans have AD, and this number is expected to grow to nearly 16 million by 2050. The prevalence of the disease is about 3% for people between the ages of 65 and 74, and 25–50% for people over 85. Estimated costs associated with providing health care and housing for AD patients (30% of whom live in nursing homes), as well as lost productivity in the workplace, total a staggering $100 billion in the U.S. per year. This number will rise substantially as the population spike known as the baby boomer generation approaches age 65.

Summary of Key Concepts

42.1 The Evolution and Development of Nervous Systems

- All multicellular animals except sponges have a nervous system. Simpler nervous systems include the nerve net of cnidarians and cerebral ganglia, which integrate inputs from sense organs. As animal bodies become more complex, the integrative center in the head becomes a brain that is larger and capable of more functions. (Figure 42.1)

- In all vertebrates, the three major divisions of the brain are the hindbrain, midbrain, and forebrain. Human embryos develop these divisions by 4 weeks. (Figure 42.2)

Figure 42.17 Cellular section from the brain of a person who died from Alzheimer disease. The section has been stained by immunohistochemistry for visualization of proteins found in plaques and neurofibrillary tangles.

- Additional folding of the brain and increased mass allow for expansion of regions associated with conscious thought, reasoning, and learning. (Figures 42.3, 42.4)

42.2 Structure and Function of the Human Nervous System

- In humans and other vertebrates, the brain and spinal cord are the central nervous system (CNS). The neurons and all axons outside the CNS, including the cranial and spinal nerves, constitute the peripheral nervous system (PNS). The CNS relies on the PNS for sensory input, and the PNS relies on commands from the CNS. (Figure 42.5)

- The gray matter of the CNS is composed of dendrites, cell bodies, and unmyelinated axons. The white matter consists of tracts of myelinated axons. (Figure 42.6)

- The meninges are protective coverings of the CNS. Cerebrospinal fluid fills the subarachnoid space and ventricles. (Figure 42.7)

- The PNS can be subdivided into the somatic and autonomic nervous systems. The somatic nervous system senses external environmental conditions and controls skeletal muscles and skin. The autonomic nervous system senses internal body conditions and controls homeostasis.

- The efferent part of the autonomic nervous system is divided into two components: sympathetic (fight or flight) and parasympathetic (rest or digest). (Figure 42.8)

- The evolutionarily oldest structures of the brain, some of which are located in the hindbrain, control the basic processes that sustain life. These structures include the medulla oblongata, cerebellum, and pons. (Figure 42.9)

- In addition to forming part of the reticular formation, the midbrain processes several types of sensory inputs, including vision, olfaction, and audition. (Figure 42.9)

- The forebrain is made of the thalamus, hypothalamus, and epithalamus (diencephalon) and the cerebrum (telencephalon), which consists of the basal nuclei, limbic system, and cerebral cortex. Each side of the human cerebral cortex is divided into four lobes, each of which has a number of functions. (Figures 42.9, 42.10, 42.11, Table 42.1)

- The cerebrum is divided into two hemispheres. Each hemisphere is specialized to perform certain aspects of behavior and can operate independently. (Figure 42.12)

42.3 Cellular Basis of Learning and Memory

- Learning is the process by which new information is acquired; memory involves the retention of that information.

- Repeated stimuli result in long-term potentiation, in which the connections between adjacent neurons become stronger.

- Studies of the sea slug indicate that short-term memory is caused by a single stimulus that activates second messenger pathways. Long-term memory is caused by repeated stimuli that activate genes, which results in more synaptic connections, a phenomenon called synaptic plasticity. (Figures 42.13, 42.14)

- Imaging techniques such as CT scans, MRI, and fMRI allow us to examine the structure and activity of the brain. (Figures 42.15, 42.16)

42.4 Impact on Public Health

- Disorders of the human central nervous system can be placed into several broad categories. (Table 42.2)

- Meningitis is a potentially life-threatening infectious disease in which the meninges become inflamed.

- Alzheimer disease, the leading cause of dementia worldwide, is a progressive disorder characterized by the formation of senile plaques and neurofibrillary tangles in brain tissue. (Figure 42.17)

Assess and Discuss

Test Yourself

1. A nerve net consists of
   a. bilateral neurons that extend from the head of the animal to the tail.
   b. a group of neurons that are interconnected and are activated all at once.
   c. a single nerve cord with ganglia in each body segment.
   d. a central nervous system with peripheral nerves associated with different body structures.
   e. none of the above.

2. The division of the vertebrate brain that includes the cerebellum is
   a. the hindbrain.   d. the forebrain.
   b. the telencephalon.   e. the diencephalon.
   c. the midbrain.

3. In general, the brains of more complex vertebrates
   a. are larger.
   b. have fewer neurons.
   c. have more folds in the cerebral cortex.
   d. use less oxygen.
   e. both a and c

4. The white matter of the CNS is composed of
   a. dendrites.   d. cell bodies.
   b. unmyelinated axons.   e. a and b only.
   c. myelinated axons.

5. The division of the nervous system that controls voluntary muscle movement is
   a. the autonomic nervous system.
   b. the sensory division.
   c. the somatic nervous system.
   d. the parasympathetic division.
   e. the sympathetic division.

6. Which of the following is <u>not</u> a response to activation of the sympathetic division of the autonomic nervous system?
   a. increased breathing rate
   b. decreased heart rate
   c. increased blood flow to the skeletal muscles
   d. increased blood glucose levels
   e. All of the above are characteristic responses to activation of the sympathetic division of the autonomic nervous system.

7. The ___ acts as a relay for the cerebrum.
   a. medulla   d. midbrain
   b. pons   e. thalamus
   c. hypothalamus

8. The _____ is a portion of the limbic system that is important for memory formation.
   a. amygdala   d. thalamus
   b. hippocampus   e. mesencephalon
   c. pons

9. In humans, the _____ hemisphere of the cerebrum is dominant in nonverbal processing.
   a. right
   b. left
   c. both a and b

10. _____ is a progressive disease that causes a loss of memory and intellectual and emotional function.
    a. Meningitis   d. Alzheimer disease
    b. Parkinson disease   e. Stroke
    c. Amnesia

Conceptual Questions

1. One of the most important and fundamental features of all nervous systems is the reflex. Describe why reflexes are adaptive.

2. Explain the differences between white matter and gray matter.

3. What are the two subdivisions of the efferent pathways of the autonomic nervous system, and what are their functions?

Collaborative Questions

1. Describe the basic features of two different types of nervous systems found in animals.

2. List the three major divisions of the brain of vertebrates and briefly describe the function of each in humans.

Online Resource

www.brookerbiology.com

Stay a step ahead in your studies with animations that bring concepts to life and practice tests to assess your understanding. Your instructor may also recommend the interactive ebook, individualized learning tools, and more.

Neuroscience III: Sensory Systems

43

Vision in the dark. As with many animals that are active at night, a cat's eye contains a reflective layer of tissue called a tapetum lucidum. This increases their ability to see in the dark.

A s complex as our sensory environment seems, there exists, in other animals, a world of sensory abilities that differ radically from our own. We do not see the color patterns of flowers produced by the reflection of ultraviolet light the way honeybees do, or hear the very low frequency sounds (such as those produced by earthquakes) that elephants, whales, and alligators hear. We cannot use echoes of our own sounds to locate flying insects the way some bats do. We cannot smell the odor given off by one female Emperor moth from a distance of hundreds of meters the way a male moth can, nor can we detect the presence of chemicals (food) using our entire body surface, the way an earthworm can. We have a sensitive ability to detect touch, but we cannot detect the electric field generated by the muscles and hearts of marine animals, although many sharks and catfish can.

What does the world look, sound, smell, and feel like to other animals? We can never know exactly how an animal perceives its environments. Biologists can perform experiments, however, to determine the capabilities of an animal's sensory systems. For instance, we can examine the structure of a cat's eye, conduct behavioral studies to determine if cats can discriminate between different colors, and measure the electrical responses of neurons in the feline visual system to different visual stimuli. From such research, we know that cats sense color very poorly, if at all, but can see exceptionally well even in the dark (see the chapter-opening photo).

Animals with nervous systems often have acute senses, such as the ability to detect light, touch, and sound. Senses allow living organisms to perceive their environments. In neuroscience, a broad definition of a **sense** is a system that consists of specialized cells that respond to a specific type of chemical, energy, or physical stimulus (also known as a modality) and send signals to the central nervous system, where the signals are received and interpreted. The senses allow animals to detect subtle and complex aspects of their environments. They are the windows through which animals experience the world around them. The nervous systems of most animals also have the ability to sense signals arising from within an animal's body, such as hunger or pain. However, neuroscientists do not always agree on the actual number of different senses. Common examples include sight, smell, taste, touch, hearing, balance, and the ability to sense heat, cold, and pain.

In this chapter, we will examine how nervous systems collect incoming sensory information and how membrane potentials of specialized neurons change in response to sensory inputs. We will learn that other structures of the nervous system may modify or enhance this neural activity before sending it to the brain, where it is interpreted. Finally, we will discuss how problems with sensory systems can affect human health.

43.1 An Introduction to Sensory Receptors

Sensory systems convert chemical or physical stimuli from an animal's body or the external environment into a signal that causes a change in the membrane potential of sensory neurons. **Sensory transduction** is the process by which incoming stimuli are converted into neural signals. Sensory transduction involves cellular changes, such as opening of ion channels, which cause either graded potentials or action potentials in neurons.

Perception is an awareness of the sensations that are experienced. For instance, touching a hot object generates a thermal sensation, which initiates a neuronal response, giving us the perception that this stimulus is hot. Not all sensations are consciously perceived by an organism. Most of the time, for example, we are not aware of the touch of our clothing. The brain also processes sensory information in areas that do not generate conscious thought. For instance, certain neurons constantly monitor blood pressure and the blood levels of oxygen, glucose, and other substances, but we are not aware that this is occurring.

We will begin our study of sensory systems by examining the specialized cells—called sensory receptors—that receive sensory inputs. A **sensory receptor** recognizes an internal or external (environmental) stimulus and initiates sensory transduction by creating graded potentials (described in Chapter 41) in itself or an adjacent cell. Sensory receptor is a term that neuroscientists use to describe certain types of cells, which are either neurons or specialized epithelial cells that respond to internal or environmental stimuli (Figure 43.1). When a response is strong enough, sensory receptors initiate electrical responses to stimuli, such as chemicals, light, heat, and sound, which lead to action potentials that are sent to the CNS.

An Intense Stimulus Generates More Frequent Action Potentials

How do sensory receptors pass along the intensity of a stimulus? Let's consider an example involving weak and strong stimuli to the sense of touch (Figure 43.2). Sensory transduction begins when the specialized endings of a sensory receptor, such as dendrites, respond to a stimulus such as the touch of a glass rod. Such a stimulus opens ion channels that allow sodium ions to diffuse down their electrochemical gradient into the cell, depolarizing the sensory receptor. The amount of depolarization is directly related to the intensity of the stimulus, because a stronger stimulus will open more ion channels.

The first response of a sensory receptor is usually a graded change in the membrane potential of the cell body that is proportional to the intensity of the stimulus (Figure 43.2). The membrane potential, known as the **receptor potential** in these cells, becomes more and more positive as the strength of the stimulus increases. When a stimulus is strong enough, it will depolarize the membrane to the threshold potential at the axon hillock and produce an action potential in a sensory neuron (see Chapter 41).

Recall from Chapter 41 that action potentials proceed in an all-or-none fashion, regardless of the nature of the stimulus that elicits them. How, then, can action potentials provide information about the intensity of a stimulus? The answer is that the strength of the stimulus is indicated by the frequency of action potentials generated. A particularly strong stimulus will generate many action potentials in a short period of time. As a result, the frequency of action potentials is higher when the stimulus is strong than when it is weak. The action potentials are transmitted into the CNS and carried to the brain for interpretation.

(a) A neuron as a sensory receptor

(b) A specialized epithelial cell as a sensory receptor

Figure 43.1   **Sensory receptors.**   **(a)** Many sensory receptors are neurons that directly sense stimuli. **(b)** Others are specialized epithelial cells that sense stimuli and secrete neurotransmitter that stimulates nearby sensory neurons. In both cases, when stimulated, the neurons send action potentials to the CNS, where the signals are interpreted.

Concept check: *What is the difference between the sensory receptor described in this chapter and the membrane receptors described in Chapter 9?*

The brain interprets a higher frequency of action potentials as a more intense stimulus.

The CNS Processes Each Sense Within Its Own Pathway

Different stimuli produce different sensations because they activate specific neural pathways that are dedicated to processing only that type of stimulus. We know that we are seeing light because the signals generated by visual sensory receptors in the eye are transmitted along a neural pathway that sends action potentials into areas of the brain that are devoted to processing vision. For this reason, the brain interprets such signals as

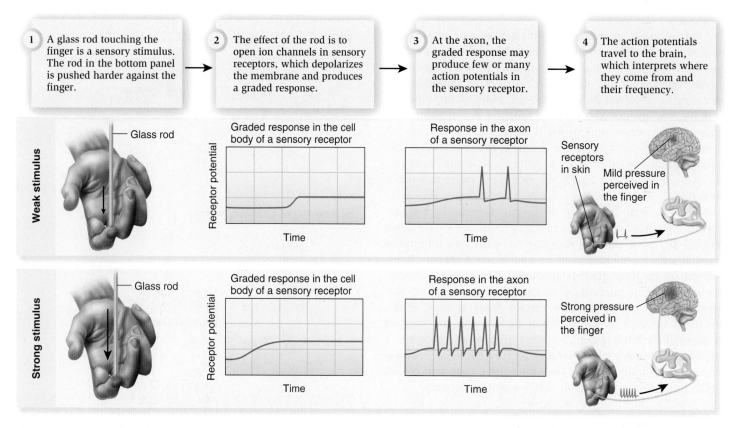

1 A glass rod touching the finger is a sensory stimulus. The rod in the bottom panel is pushed harder against the finger.

2 The effect of the rod is to open ion channels in sensory receptors, which depolarizes the membrane and produces a graded response.

3 At the axon, the graded response may produce few or many action potentials in the sensory receptor.

4 The action potentials travel to the brain, which interprets where they come from and their frequency.

Figure 43.2   Transduction of a sensory stimulus of two different intensities.   In this example, the sensory receptor is a neuron. Note the faster and larger graded response following the stronger stimulus.

visual stimuli. The brain can separate and identify each sense because each one uses its own dedicated pathway.

We can divide sensory receptors into general classes, based on the type of stimulus or modality to which they respond. Each type uses a different mechanism to detect stimuli and to transmit the information to different regions of the CNS. **Mechanoreceptors** transduce mechanical energy such as touch, pressure, stretch, movement, and sound. **Thermoreceptors** detect cold and heat. **Nociceptors**, or pain receptors, detect extreme heat, cold, and pressure, as well as certain molecules such as acids. **Electromagnetic receptors** sense radiation within a wide range of the electromagnetic spectrum, including visible, ultraviolet, and infrared light, as well as electrical and magnetic fields in some animals. **Photoreceptors** are electromagnetic receptors that detect visible light energy. **Chemoreceptors** recognize specific chemical compounds such as odor molecules or blood-borne compounds. Most of the remaining sections of this chapter will examine the structures and functions of these types of sensory receptors and the organs in which they are found.

43.2   Mechanoreception

As mentioned, mechanoreceptors are cells that detect physical stimuli such as touch, pressure, stretch, movement, and sound. Physically touching or deforming a mechanoreceptor cell opens ion channels in its plasma membrane. As discussed in this section, some mechanoreceptors are neurons that send action potentials to the CNS in response to physical stimuli. Other mechanoreceptors are specialized epithelial cells that contain hairlike structures that bend in response to mechanical forces.

Skin Receptors Detect Touch and Pressure

Several types of receptors in the skin detect touch, deep pressure, or the bending of hairs on the skin. Some of these specialized receptors consist of neuronal dendrites covered in dense connective tissue. In mammals, these receptors are located at different depths below the surface of the skin, which makes them suitable for responding to different types of stimuli (**Figure 43.3**). For example, **Meissner's corpuscles** (also called tactile corpuscles) sense touch and light pressure and lie just beneath the skin surface. They are found throughout the skin but are concentrated in areas sensitive to light touch, such as the fingertips, lips, eyelids, and genitals. In contrast, **Pacinian corpuscles** (also called lamellated corpuscles) and **Ruffini corpuscles** are located much deeper beneath the surface, particularly in the soles of the feet and the palms of the hands. These corpuscles respond best to deep pressure or vibration. All the skin corpuscles contain sensory receptor neurons that generate action potentials when the structure of the corpuscle is deformed. Other skin mechanoreceptors located in the hair follicles respond to movements of hairs and whiskers.

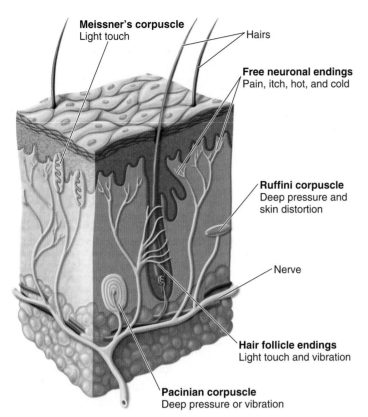

Figure 43.3 Examples of sensory receptors in the skin of mammals.

Concept check: *There are several types of touch receptors in the skin that respond to different stimuli. What different qualities of touch are you aware of?*

Stretch Receptors Detect Expansion

Mechanoreceptors called **stretch receptors** are neuron endings commonly found in the walls of organs that can be distended. They occur in organs such as the stomach and urinary bladder and also in skeletal muscles. Although stretch receptors are probably found throughout the animal kingdom, they have been best studied in crustaceans and mammals. In decapod crustaceans, for example, stretch receptors in muscles of the tail, abdomen, and thorax relay signals to the brain regarding the position in space of the different body parts. This information allows the animal to perform complex motor functions, such as walking backward or sideways. In another example, when the mammalian stomach stretches after a meal, the stretch receptors in the stomach are deformed, causing them to become depolarized and send action potentials to the brain. The brain interprets the signals as fullness, which reduces appetite.

Hair Cells Are Mechanoreceptors with Deformable Stereocilia

Thus far, we have considered skin and stretch receptors, which are neurons that detect physical stimuli. Other mechanoreceptors are specialized epithelial cells called **hair cells**, which have deformable projections called **stereocilia** that resemble hairs under the microscope. The stereocilia, which are different from true cilia (see Chapter 4), are bent by movements of fluid or other physical stimuli (**Figure 43.4**).

Many types of hair cells contain ion channels that open or close when the stereocilia bend and thereby change the cell's membrane potential. When the plasma membrane depolarizes, this opens voltage-gated Ca^{2+} channels and results in the release of neurotransmitter molecules from the hair cells (refer back to

At rest

At rest, a small amount of neurotransmitter is released at all times, resulting in a steady number of action potentials being generated in the sensory neuron.

Excited

Fluid moving in one direction causes the release of more neurotransmitter, which results in more action potentials in the sensory neuron.

Inhibited

Fluid moving in the opposite direction inhibits the release of neurotransmitter, which results in fewer action potentials in the sensory neuron.

Figure 43.4 The response of hair cells to mechanical stimulation.   Note: The stereocilia inside these hair cells are hairlike projections of the plasma membrane that contain actin filaments. They are structurally different from cilia, which are described in Chapter 4.

Figure 41.14). The neurotransmitter then binds to protein receptors in adjacent sensory neurons and can result in action potentials being sent to the CNS. Even when unstimulated, hair cells usually release a small amount of neurotransmitter onto nearby sensory neurons, resulting in a resting level of action potentials in the sensory neurons. In the example shown in Figure 43.4, bending of the stereocilia in one direction in response to fluid movement increases the release of neurotransmitter from the hair cell, whereas bending in the other direction decreases the release of the same neurotransmitter. The result is an increase or decrease, respectively, in the number of action potentials produced in the sensory neurons.

Hair cells are found in the ear and equilibrium (balance) organs of many invertebrates and vertebrates, where they detect sound or changes in head position. They are also found along the body surface of fishes and some amphibians, where they detect external water currents, as described next.

Mechanoreceptors in the Lateral Line System Detect Movements in Water

Fishes and some toads detect changes in their environment through a **lateral line system** (Figure 43.5). This sensory system has hair cells that detect changes in water currents brought about by waves, nearby moving objects, and low-frequency sounds traveling through the water. The lateral line organ runs along both sides of the body and the head of the animal. Small pores let water enter into a lateral line canal. The stereocilia of hair cells protrude into a gelatinous structure called a **cupula** within the lateral line organ. When the cupula is moved by the water, the stereocilia bend, causing the release of neurotransmitter from the hair cell. This stimulates a response in sensory neurons at the base of the hair cells. The response provides information to the brain about changes in water movement, such as the approach of a predator.

Hearing Involves the Reception of Sound Waves

The sense of hearing, called **audition**, is the ability to detect and interpret sound waves. This sense is critical for the survival of many types of animals. For example, a mother seal locates her pup by hearing its calls, and a male bird sings an elaborate song to attract a mate. Hearing is also important for detecting the approach of danger—a predator, a thunderstorm, an automobile—and locating its source.

Sound travels through air or water in waves. The distance from the peak of one wave to the next is a **wavelength**. The number of complete wavelengths that occur in 1 second is called the **frequency** of the sound, measured in number of waves per second, or Hertz (Hz), after the German physicist and pioneer of radio wave research, Heinrich Hertz. The length and frequency of sound waves impart certain characteristics to the stimulus. Short wavelengths have high frequencies that are perceived as a high **pitch** or tone, while long wavelengths have lower frequencies and a lower pitch.

The sense of hearing is present in vertebrates and arthropods, but not in other phyla. Arthropods do not appear to have more than a general sensitivity to sound, although some exceptions exist. For example, some species of moths have sound-sensitive membranes that detect the high frequencies emitted by their chief predators, bats. The sense of hearing, however, is especially well developed in vertebrates (notably birds and mammals); we turn now to a detailed discussion of the mammalian ear and the mechanism by which it detects sound.

Structure of the Mammalian Ear The mammalian ear has three main compartments: the outer, middle, and inner ears (**Figure 43.6**). The **outer ear** consists of the external ear, or pinna (plural pinnae), and the auditory canal. Mammals have a wide variety of shapes and sizes of the external ear, which reflects their different abilities to capture sound waves. The outer ear is separated from the **middle ear** by the tympanic

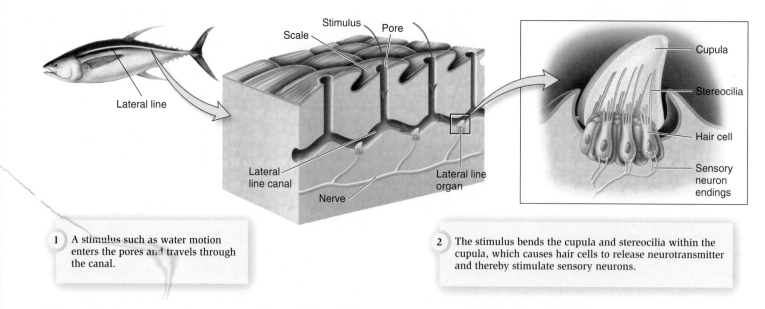

1 A stimulus such as water motion enters the pores and travels through the canal.

2 The stimulus bends the cupula and stereocilia within the cupula, which causes hair cells to release neurotransmitter and thereby stimulate sensory neurons.

Figure 43.5 Mechanoreceptors in the lateral line system that detect changes in water movement.

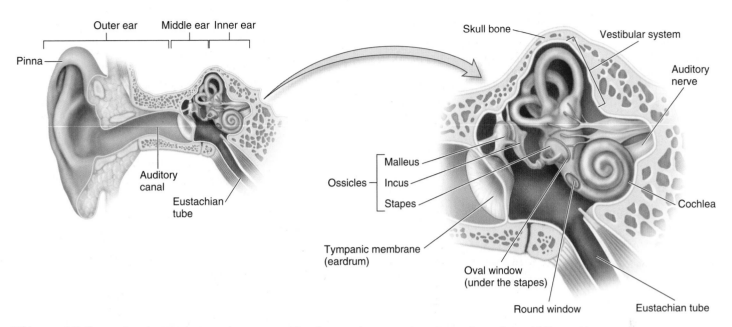

Figure 43.6   The structure of the human ear.   The three main compartments are the outer, middle, and inner ear.

membrane (eardrum). The middle ear contains three small bones called ossicles (named the malleus, incus, and stapes) that link movements of the eardrum with the oval window. The oval window is another membrane similar to the eardrum that separates the middle ear from the **inner ear**. The inner ear is composed of the **cochlea** (from the Latin, meaning snail)—a coiled chamber of bone containing the hair cells and the membrane-like round window—and the vestibular system, which plays a role in balance, as described later. These structures in the inner ear generate the signals that travel via the auditory nerve to the brain.

Both the tympanic membrane and the oval window must be able to vibrate when sound waves meet the eardrum. For this to occur properly, the air pressure in the outer and middle ear compartments must be equal. The pressure in the outer ear is the atmospheric pressure. The **Eustachian** (or **Auditory**) **tube**, which connects the middle ear to the pharynx, maintains atmospheric pressure in the middle ear. If you change altitude quickly, as during a plane takeoff, the pressure in your outer ear changes because atmospheric pressure decreases with altitude. After awhile, the pressures in the outer and middle ear equalize. However, if the pressure in your outer ear becomes greater than the pressure in your middle ear, which would occur when the plane descends, your eardrum will be pushed or bowed inward. The eardrum is less able to vibrate in response to sounds when it is deformed under pressure, and thus your hearing becomes muffled. Swallowing or yawning will open the Eustachian tube and equalize the pressure in the middle ear with atmospheric pressure. The popping sound you hear when the plane ascends or descends is the sudden return of the eardrum to its normal position.

Generation of Electrical Signals in the Mammalian Ear
To understand how we hear, let's first consider how mechanical

forces move through the ear. Sound waves entering the outer ear cause the tympanic membrane to vibrate back and forth (**Figure 43.7**). The ossicles transfer the vibration of the tympanic membrane to the oval window, causing it to vibrate against the cochlea. This sends pressure waves, which also travel in a back-and-forth manner, through a fluid called perilymph. Perilymph is found within two narrow passages in the cochlea called the vestibular and tympanic canals. The waves travel from the vestibular canal to the tympanic canal and eventually strike the round window, where they dissipate. Along the way, the waves cause the vibration of a membrane called the **basilar membrane**, which as you will see shortly is a key part of the mechanism by which different frequency sounds are sensed. Sounds of very low frequency create pressure waves that take the complete route through the vestibular and tympanic canals (see green arrows in Figure 43.7). Sounds of higher frequency produce pressure waves that follow a different route, passing from the vestibular canal through a tube between the canals called the cochlear duct (as shown by the blue arrows in Figure 43.7). They then pass through the basilar membrane, before reaching the tympanic canal.

Mechanical vibrations are changed, or transduced, into electrical signals within the cochlea. This happens in a structure called the organ of Corti (named after the Italian anatomist Alfonso Corti), which rests on top of the basilar membrane. To understand how this works, we need to look at a cross section through the cochlea (**Figure 43.8**). The organ of Corti contains supporting cells and rows of hair cells. The stereocilia of the hair cells are embedded in a gelatinous tectorial membrane. The back-and-forth vibration of the basilar membrane bends the stereocilia in one direction and then the other. When bent in one direction, the hair cells depolarize and release neurotransmitter, which activates adjacent sensory neurons that then send action potentials to the CNS via the auditory nerve. When bent in the

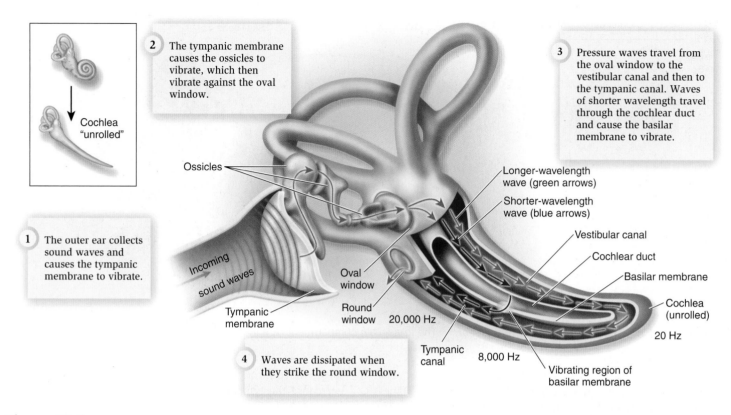

2 The tympanic membrane causes the ossicles to vibrate, which then vibrate against the oval window.

3 Pressure waves travel from the oval window to the vestibular canal and then to the tympanic canal. Waves of shorter wavelength travel through the cochlear duct and cause the basilar membrane to vibrate.

Cochlea "unrolled"

1 The outer ear collects sound waves and causes the tympanic membrane to vibrate.

Ossicles

Longer-wavelength wave (green arrows)

Shorter-wavelength wave (blue arrows)

Vestibular canal

Cochlear duct

Basilar membrane

Incoming sound waves

Oval window

Round window   20,000 Hz

Cochlea (unrolled)

20 Hz

Tympanic membrane

Tympanic canal   8,000 Hz

Vibrating region of basilar membrane

4 Waves are dissipated when they strike the round window.

Figure 43.7   Movement of sound waves through the human ear.

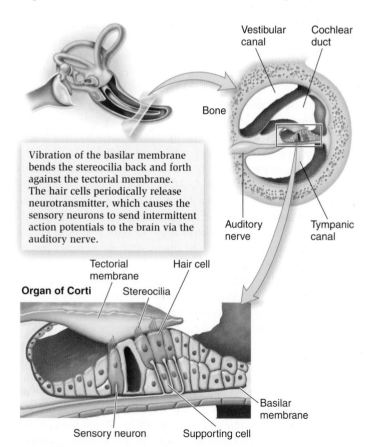

Vestibular canal   Cochlear duct

Bone

Vibration of the basilar membrane bends the stereocilia back and forth against the tectorial membrane. The hair cells periodically release neurotransmitter, which causes the sensory neurons to send intermittent action potentials to the brain via the auditory nerve.

Auditory nerve

Tympanic canal

Tectorial membrane   Hair cell

Organ of Corti   Stereocilia

Basilar membrane

Sensory neuron   Supporting cell

Figure 43.8   Transduction of mechanical vibrations to electrical signals in the organ of Corti.

other direction, the hair cells hyperpolarize and shut off the release of neurotransmitter. In this way, the frequency of action potentials generated by the sensory neurons is determined by the up-and-down vibration of the basilar membrane.

The basilar membrane is lined with protein fibers that span its width. These fibers function much like the strings of a guitar. The fibers near the oval and round windows at the base of the cochlea are short and rigid, and they vibrate in response to high-frequency waves. Longer and more resilient fibers are near the other end of the cochlea and vibrate to lower-frequency waves. For this reason, hair cells closer to the oval and round windows respond to high-pitched sounds, whereas those at the opposite end are triggered by lower-pitched sounds. When we hear a great number of sound frequencies at once, such as at a musical concert, the waves traveling through the cochlea activate hair cells all along the basilar membrane in a physical representation of the music! These cells stimulate sensory neurons, which send multiple action potentials to the auditory areas of the brain for processing. The most incredible feature of this process, however, is that the mammalian ear and brain can "tune in" to all of these frequencies simultaneously.

Adaptations for Hearing Allow Animals to Live in a Wide Range of Habitats

Different animals are capable of hearing sounds of different pitches. For example, humans can hear between 20 and 20,000 Hz (conversation averages 90–300 Hz). Insectivorous bats, toothed whales, and some species of moths may have the highest-frequency range, while baleen whales and elephants may have the lowest-frequency range. These adaptations increase the animals' ability to communicate and survive. For instance, high-pitched sounds are useful to animals that need to locate small prey such as flying insects, while low-pitched sounds carry great distances through water or air and are especially useful for animals with large territories.

Locating a Sound A vital feature of hearing is the ability to locate the origin of a sound. For example, this ability makes the difference between a successful and an unsuccessful hunter. How does an animal locate a sound? Under most circumstances, sound does not arrive at both ears simultaneously. Sound waves coming from the right, for example, may excite the sensory receptors in the right ear first and the left ear some milliseconds later, and therefore, the brain receives action potentials from the auditory nerves of each ear at slightly different times. The brain interprets the time difference to determine the origin of the sound.

Animals such as owls that rely on hearing to locate prey tend to be extremely good at localizing the source of a sound. An interesting experiment demonstrated this by outfitting owls in a dark room with small headphones. Just as in a human hearing test, sounds could be sent to either headphone or to both. If the investigator sent a high-pitched noise that mimicked the sounds of a mouse first to the left headphone, and then a single millisecond later to the right headphone, the owl turned its head to the left, because the owl's brain perceived the sound to be coming from that direction. If the noises reached both headphones simultaneously, the owl interpreted the signal as coming from directly in front of its head.

Hearing in Air and Water Amphibians have the special challenge of hearing both on land and underwater, their two natural environments. The ears of amphibians have several interesting specializations that are adapted to these requirements. First, they do not have external ears. The tympanic membrane is located on the outer surface of the head behind the eye (**Figure 43.9**). This arrangement allows them to swim through the water without being impeded by pinnae and to receive sound waves from any direction. Amphibians also have unusually wide Eustachian tubes. When they are on land, sound waves can pass from one ear through the Eustachian tube to the pharynx and then travel up the Eustachian tube of the other ear, where they hit the backside of the tympanic membrane. Air from the lungs also can pass into each ear in this manner. When a male bullfrog emits its loud call, pressure coming through the Eustachian tubes and from the outside of the head prevents his own eardrum from vibrating and being injured by the loud sound.

Echolocation Bats in the air, whales and dolphins in the sea, and shrews in underground tunnels generate high-frequency sound waves to determine the location of an object. This phenomenon is called **echolocation**, because the sound waves

Tympanic membrane

Figure 43.9 The tympanic membrane of a frog.

bounce off a distant object, like an echo, and return to the animal. The time it takes for the sound to return indicates the distance of the object. Echolocation can be useful in situations where vision is limited, such as in the dark. Although the sound of an echo may be familiar to you, it is hard to imagine how other animals can use echoes to locate objects with such precision. To get a feel for how this works, imagine shouting "hello" across a canyon. After a few seconds, you would hear the echo return to you. Now imagine that you knew the speed of sound in air (approximately 344 m/sec at sea level) and had a stopwatch that measured the length of time it took for the echo to return. From this, you could calculate the distance across the canyon (accounting for the fact that sound first traveled in one direction, then back again, so you would have to divide your result by two). Bats and other echolocating animals perform this feat instantly, without the aid of mathematics or calculators, and can do it while they and the object they are tracking are both in motion!

The Sense of Balance Is Mediated by Statocysts in Invertebrates and the Vestibular System in Vertebrates

Let's now turn our attention to the sense of balance, also called equilibrium. Balance is part of a broader sense called proprioception, which is the ability to sense the position, orientation, and movement of the body. Being able to sense body position is vital for the survival of animals. This is how a lobster, for example, rights itself when flipped over by a predator or how a bird maintains its balance while flying rapidly through the air.

Many aquatic invertebrates have sensory organs called **statocysts** that send information to the brain about the position of the animal in space (**Figure 43.10**). Statocysts are small round structures made of an outer sphere of hair cells and **statoliths**, which are tiny granules of sand or other dense objects. When the animal moves, gravity alters the statoliths' position. If the animal turns on its left side, for example, the movement of statoliths stimulates a new set of hair cells to release neurotransmitter, generating action potentials in sensory neurons that inform the brain of the change in body position.

Several experiments have demonstrated the importance of statoliths. In one particularly dramatic example, researchers replaced the statoliths of crayfish with iron filings. Moving a magnet to different positions around the animal displaced the filings, causing the animal to change its position, and even to swim upside down when the magnet was placed directly above its head.

The organ of balance in vertebrates, known as the **vestibular system**, is located in the inner ear next to the cochlea (**Figure 43.11**). The vestibular system is composed of a series of fluid-filled sacs and tubules, which provide information about either linear or rotational movements. The utricle and saccule, the two sacs nearest the cochlea, detect linear movements of the head (Figure 43.11a), such as those that occur when an animal runs and jumps or changes from lying down to sitting up. The hair cells within these structures are embedded in a gelatinous substance that contains granules of calcium carbonate called **otoliths** (from the Latin, meaning ear stones), which are analogous to statoliths. When the head moves forward, the heavy otoliths are temporarily "left behind" as they are dragged forward more slowly, and the weight of the otoliths bends the

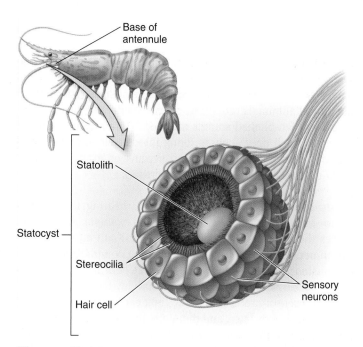

Figure 43.10　**Sensing of balance in aquatic invertebrates.** Statocysts located near the antennae consist of a sphere of sensory hair cells surrounding a stony statolith. When the animal moves, gravity shifts the statolith and stimulates the hair cells beneath it.

Figure 43.11　**The vertebrate vestibular system.**

Concept check: Note the orientation of the three semicircular canals with respect to each other. Why are the canals oriented in three different planes?

stereocilia of the hair cells in the direction opposite that of the linear movement. This changes the membrane potential of the hair cells and alters the electrical responses of nearby sensory neurons. These signals are sent to the brain, which uses them to interpret how the head has moved. (Interestingly, gravity detection by statolith-containing organs is not unique to animals; as described in Chapter 36, the upward growth of many types of plants also depends on the presence of statoliths!)

Three **semicircular canals** connect to the utricle at bulbous regions called the **ampullae**. The function of the semicircular canals is to detect rotational motions of the head (Figure 43.11b). The hair cells in the semicircular canals are embedded in the ampullae in a gelatinous cone called the cupula (similar to the cupula of the fish lateral line). When the head moves, the fluid in the canal shifts in the opposite direction. This movement of fluid pushes on the cupula and bends the stereocilia of the hair cells in the direction of the fluid flow, which is opposite that of the motion. The three canals are oriented at right angles to each other, and each canal is maximally sensitive to motion in its own plane. For example, the canal that is oriented horizontally would respond most to rotations such as shaking the head "no," whereas the other canals respond to "yes" motions and to tipping the ear to the shoulder. Overall, by comparing the signals from the three canals, the brain can interpret the motion of the head in three dimensions.

The vestibular system of vertebrates provides conscious information about body position and movement. It also supplies unconscious information for reflexes that maintain normal posture, control head and eye movements, and assist in locomotion. Motion sickness in humans results when the vestibular system has not adapted to unfamiliar patterns of movement, such as spinning around or sailing aboard a ship. In these cases, the sense from your vestibular system conflicts with information coming from your vision. Vertigo, an illusion of movement or spinning, can result from malfunction of the vestibular system.

43.3 Thermoreception and Nociception

The perception of temperature and pain enables animals to respond effectively to their environments. As described in this section, these sensory stimuli are related in that their receptors are located in some of the same areas (skin), share similar physical features, and under certain conditions result in similar perceptions.

Thermoreceptors Detect Temperature

Sensing the outside temperature is important for animals because their body temperature is affected by the external temperature, particularly in ectotherms. Animals can survive at body temperatures only within certain limits, because the proteins in their cells function optimally within a particular temperature range. Thermoreceptors respond to cold or hot temperatures by activating or inhibiting enzymes within their plasma membranes, which alters membrane channels. There are two types of thermoreceptors: those that respond to hot and those that respond to cold. Both of these types of receptors are free neuronal endings. Thermoreceptors are often linked with reflexive behaviors, such as when an animal steps on a hot surface and pulls its foot away.

In addition to skin receptors that sense the outside temperature, thermoreceptors in the brain also detect changes in core body temperature. Activation of skin or brain thermoreceptors triggers physiological and behavioral adjustments that help maintain body temperature. These changes, described in Chapter 46, include changes in blood flow, shivering, and behaviors such as seeking shade or sunlight.

Nociceptors Warn of Pain

The sense of pain is one of the most important of all the senses. It tells an animal whether it has been injured and triggers behavioral responses that protect it from further danger. Although in many cases we cannot know whether or how animals perceive pain, nociceptors have been identified in mammals and birds. Their existence and physiological significance in other phyla are likely but are still being investigated.

Nociceptors, the receptors for pain, are free neuronal endings in the skin and internal organs (see Figure 43.3). They respond to tissue damage or to stimuli that are about to cause tissue damage. Nociceptors are unusual because they can respond not only to external stimuli, such as extreme temperatures, but also to internal stimuli, such as molecules released into the extracellular space from injured cells. Damaged cells release a number of substances, including acids and prostaglandins, that cause inflammation and make nociceptors more sensitive to painful stimuli. Aspirin and ibuprofen reduce pain by preventing the production of prostaglandins.

Signals arising from nociceptors travel to the CNS and reach the cerebrum, where the type or cause of the pain is interpreted. The signals are also sent to the limbic system, which holds memories and emotions associated with pain, and to the reticular formation, which increases alertness and arousal—an important response to a painful stimulus.

43.4 Electromagnetic Sensing

As mentioned earlier, electromagnetic receptors detect radiation within a wide range of the electromagnetic spectrum, including those wavelengths that correspond to visible light, ultraviolet light, and infrared light, as well as electrical and magnetic stimuli. Photoreceptors are specialized electromagnetic receptors that respond to light and are covered in Section 43.5. Here, we will examine the ability of some animals to sense electrical and magnetic fields and also heat in the form of infrared radiation.

The ability to detect the presence of nearby prey or predators is an important adaptation in certain animals that inhabit low-light environments. The more ways that an animal has to detect other animals, the better it can avoid danger or obtain a meal. Many fishes living in dark waters can detect weak electrical signals given off by other fishes, and sharks and rays can even detect the tiny electrical signals generated by the heartbeats of their prey. Similarly, the platypus, which lives in the murky waters of streams and ponds, has receptors on the skin of its bill that can detect very small electrical currents produced by its prey.

Homing pigeons use one type of electromagnetic sensing to return to their starting points from as far away as 1,500 km. This navigational feat is made possible by small particles of magnetite (iron oxide) in their beaks that indicate direction by acting as a compass. The magnetic particles respond to the Earth's magnetic field and alter the activity of neurons that project to the brain. In one experiment, pigeons were placed individually in large tubes and trained that food was present in only one end of the tube. When the tube was placed in a changeable magnetic field, pigeons readily learned which end contained food based solely on the magnetic polarity of the tube. In another experiment, the pigeons lost this ability when their beaks were anesthetized or cooled down, procedures that block action potentials from being sent to the brain. This demonstrates that their magnetic sensing ability is located in the beak and communicated by nerves to the brain.

Magnetic field sensing is not unique to birds. Magnetite has also been found in the heads of migratory fishes such as rainbow trout. However, this probably does not entirely explain the extraordinary ability of migratory animals to navigate great distances, because other cues, such as smell and visual recognition of landmarks, also appear to play roles in this process.

Venomous snakes known as pit vipers (a group that includes copperheads and rattlesnakes) can localize prey in the dark with detectors that sense the heat emitted from animals as infrared radiation. These detectors are located in pits on each side of the head between the eyes and nostrils (**Figure 43.12**). Within the pit, a thin, nerve-rich, temperature-sensitive membrane becomes activated in response to infrared waves emitted by live animals. When the snake detects the heat of the animal, it localizes its prey by moving its head back and forth until both pits detect the same intensity of radiation. This indicates that the prey is centered in front of the snake, and then the snake is ready to strike.

Electrical, magnetic, and infrared sensing are adaptations for long-distance migration or low-light environments. When light is available, however, photoreception becomes a dominant sensory ability in many animals, as described next.

43.5 Photoreception

Although it is a form of electromagnetic reception, photoreception is such an important and widespread sense that we will cover it separately here. Visual systems employ specialized neurons called **photoreceptors**, which detect photons of light arriving from the sun or other light sources, or reflecting off an object. A photon is the fundamental unit of electromagnetic radiation and has the properties of both a particle and a wave. The properties of light are described in Chapter 8. In this section, we will examine the organs found in animals, usually

Figure 43.12 **Infrared sensing.** Sensory pits enable a white-lipped pit viper to detect the heat given off by its prey.

called **eyes**, that detect light and send signals to the brain. The amazing features of these organs reflect the importance of vision in the animal world.

Eyecups and Compound Eyes Are Found in Certain Invertebrates

One of the simplest visual organs is found in the flatworm (*Planaria*); it is a concave structure called an **eyecup**, or eyespot (**Figure 43.13**). The eyecup contains the endings of photoreceptor cells and a layer of pigment cells that shields the

Figure 43.13 **The eyecup of a flatworm.** The orientation of the eyecup allows light to stimulate photoreceptors from only one direction. This type of eye senses only the presence or absence of light and does not form visual images.

photoreceptors from one side. The left and right eyecups receive light from different directions. This allows the eyecups to detect the direction as well as the presence or absence of light. The nervous system compares the amount of light detected by each eyecup, and the flatworm moves toward darkness, a behavior that protects it from predators. This type of photoreceptor does not form visual images of the environment.

In contrast, arthropods and some annelids have image-forming **compound eyes** (Figure 43.14a), which consist of several hundred to more than 10,000 light detectors called ommatidia (**Figure 43.14b**). Each ommatidium makes up one facet of the eye. Within the ommatidium, a **lens** composed of the cornea and the crystalline cone focuses light onto a long central structure called a rhabdom (**Figure 43.14c**). The rhabdom is a column of light-sensitive microvilli that project from the cell membranes of the photoreceptor (retinula) cells of the ommatidium (**Figure 43.14d**). The light-sensitive molecules required for vision are located in the microvilli; the extensive surface area imparted by the microvilli provides the animal with increased sensitivity to light. Pigmented cells surrounding the photoreceptor cells absorb excess light and thereby isolate each ommatidium from its neighbors.

Each ommatidium senses the intensity and color of light. Combined with the different inputs from neighboring ommatidia, the compound eye forms an image that the brain interprets. Animals such as bees and fruit flies, with large numbers of ommatidia, presumably have sharper vision than those with fewer sensory cells, such as grasshoppers.

As anyone who has tried to swat a fly knows, the compound eye is extremely sensitive to movement and helps flying insects evade birds and other predators. Behavioral studies have shown, however, that the resolving power of even the best compound eye is considerably less than that of the single-lens eye, which we will consider next.

Vertebrates and Some Invertebrates Have a More Complex Single-Lens Eye

Single-lens eyes are found in vertebrates and also in some mollusks, such as squid, octopus, and some snails, and in some annelids. In such eyes, different patterns of light emitted from images in the animal's field of view are transmitted through a small opening, or **pupil**, through the lens, to a sheetlike layer of photoreceptors called the **retina** at the back of the eye (**Figure 43.15a**). This forms a visual image of the environment on the retina. The activation of these photoreceptors triggers electrical changes in neurons that pass out of the eye through the **optic nerves**, carrying the signals to the brain. The brain then interprets the visual image that was transmitted.

As illustrated in Figure 43.15a, the vertebrate eye has a strong outer sheath called the **sclera** (the white of the eye). Between the sclera and the retina is a layer of blood vessels called the choroid. At the front of the eye, the sclera is continuous with a thin, clear layer known as the **cornea**. Within the eye are two cavities, the anterior and posterior cavities. The anterior cavity is the part of the eye between the lens and the cornea. It is subdivided into the anterior chamber and posterior chamber. The anterior chamber is located between the cornea and the **iris**—the circle of pigmented smooth muscle responsible for eye color. This chamber is filled with a thin liquid called the **aqueous humor**. The posterior chamber is the space between the single lens and the iris. The larger posterior cavity between the lens and the retina contains the thicker **vitreous humor**, which helps maintain the shape of the eye.

(a) Compound eyes of Drosophila

(b) Multiple ommatidia in a compound eye

Ommatidia

Axons

Ommatidium

Rhabdom · Pigment cell · Photoreceptor (retinula) cells · Crystalline cone · Cornea

(c) Structure of a single ommatidium

Lens

Microvilli

(d) A single photo-receptor cell

Figure 43.14 The compound eye of insects. (a) Close-up of the eyes of a fruit fly (*Drosophila melanogaster*). (b) Each eye has approximately 1,000 ommatidia, which form a sheet on the surface of the eye. (c) Each ommatidium has a lens that directs light to the photosensitive rhabdom. (d) Extending from each photoreceptor cell are many light-sensitive microvilli.

(a) Human eye structure

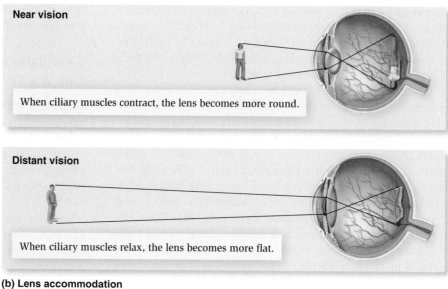

Near vision

When ciliary muscles contract, the lens becomes more round.

Distant vision

When ciliary muscles relax, the lens becomes more flat.

(b) Lens accommodation

Figure 43.15 **The vertebrate single-lens eye.** **(a)** The structure of the human eye. **(b)** Changes in lens shape during accommodation. When an object is near, the ciliary muscles contract and the lens becomes rounder, causing light to bend more. When the object is far away, the ciliary muscles relax and the lens flattens. **(c)** Demonstration of the blind spot. First, hold this picture up in front of your face, or place the book on a table and stand over it. Next, close your left eye and stare at the black spot with your right eye while you move the picture toward and away from your face. At some point, light reflecting off the plus (+) sign will fall directly on your blind spot, and it will seem to disappear.

(c) Demonstration of blind spot

The hole in the center of the iris is the pupil. The size of the pupil changes when the muscles of the iris reflexively relax or contract to allow more or less light to enter the eye.

Because light radiates in all directions from a light source, light must be bent (refracted) inward toward the photoreceptors at the back of the eye. This is accomplished by the cornea and the lens. Whenever light passes from one medium to another medium of a different density, light waves will bend (try looking at a pencil in a glass partly filled with water). The cornea, which is at the interface between the air and the aqueous humor, initially refracts the light. The light then passes through the thick lens, where it is refracted again and focused onto the layer of photoreceptors, the retina, at the back of the posterior chamber. The bending of the incoming light results in an upside-down and laterally inverted image on the retina, but the brain adjusts for this, and the image is perceived correctly.

The lens is adjusted to focus light that comes from different distances. In fishes and amphibians, the lens is moved forward or backward. In the avian and mammalian eye, the lens remains stationary but changes shape to become more or less round. When the lens is stretched, it flattens, and light passing through it bends less than when it is round. Contraction and relaxation of the ciliary muscles adjust the lens according to the angle at which light enters the eye, a process called **accommodation** (Figure 43.15b).

The single-lens eye is often compared to a film camera. Like the aperture mechanism of a camera, the iris regulates the amount of light that enters the eye. Just as the lens of a camera bends light and directs it toward the light-sensitive film, the lens of the eye bends and directs light toward the photoreceptor cells on the surface of the retina. Both the retina and a piece of film can be thought of as sheetlike structures across which different intensities and colors of light imprint an image.

How does the retina form such sharp images? The region on the retina directly in line with the pupil and lens is called the macula. The center of the macula, the **fovea**, contains the highest density of photoreceptors for color. The fovea is responsible for the sharpness with which we and many other animals see in daylight. However, the retina also has limitations in forming images. In the eye, as mentioned, the image initiates signals that travel from the retina to the brain through neurons that exit the eye in the optic nerve. In vertebrates, the point on the retina where the optic nerve leaves the eye is called the **optic disc**. The optic disc does not have any photoreceptors, forming a "blind spot" where light does not activate a response (**Figure 43.15c**). Invertebrates with single-lens eyes do not have a blind spot, because the photoreceptors in their eyes are at the front of the retina; consequently, the optic nerve does not pass through the layer of photoreceptors before leaving the eye. How do photoreceptors respond to light to form these images?

Rods and Cones Are Photoreceptor Cells

The two types of photoreceptors have names that are derived from their shapes: rods and cones (Figure 43.16). **Rods** are very sensitive to low-intensity light and can respond to as little as one photon, but they do not discriminate different colors. Rods are used mostly at night, and they send signals to the brain that generate a black-and-white visual image. **Cones** are less sensitive to low levels of light but, unlike rods, can detect color. Cones are used in daylight by most diurnal (active by day) vertebrate species and by some insects such as the honeybee, which can detect the yellow color of pollen. Compared to rods, the human retina has fewer cones, which are clustered in and around the fovea. Cones provide sharp images because of their density at the fovea. Although they are less sensitive to light than rods, this is less critical in daylight because the amount of light reaching the eyes at this time far exceeds what is needed to stimulate any photoreceptor cell. Because the two types of photoreceptor are specialized for either night or day vision, neither rods nor cones function at peak efficiency at twilight. This accounts for our relatively poor vision at this time.

Rods and cones are cells with three functional parts: the outer segment, inner segment, and synaptic terminal. The **outer segment** of the cell contains folds of membranes that form stacks like discs (Figure 43.17). These discs contain the pigment molecules that absorb light. The **inner segment** of the cell contains the cell nucleus and other cytoplasmic organelles. Rods and cones do not have axons but have synaptic terminals with neurotransmitter-containing vesicles, which synapse with other neurons within the retina.

Nocturnal animals (those active at night) rely predominantly on rod vision, though some have limited color vision, too. In diurnal animals with both rods and cones, such as humans, the rods are located around the periphery of the retina away from the fovea. Therefore, it is easiest to see low-intensity light if it comes into the eye at an angle. You can easily test this. In early evening, before many stars are visible, look at the sky until you notice a star out of the corner of your eye. Now shift your gaze to where you thought you saw the star. You will probably not be able to locate it anymore. When you look away again so that light from the dim star enters your eye at an angle, it will reappear. This demonstrates that under low-light conditions, your vision is better when the light is directed to the part of the retina that contains only rods.

Rods and Cones Contain Visual Pigments That Detect Light

Visual pigments are molecules that absorb light; they are found embedded in the disc membranes of the outer segment of rods and cones. In the mid-20th century, American biologist and Nobel Prize winner George Wald discovered that these pigments consist of two components bonded together. The first is **retinal**, a derivative of vitamin A that is capable of absorbing light energy. The discovery of retinal in the visual pigment explains the need for vitamin A in the diet and its importance in vision. The second component of visual pigments is a protein called **opsin**, of which there are several types. Opsins are examples of G-protein-coupled receptors (see Chapter 9), which trigger a biochemical cascade that changes the permeability of membrane channels to ions.

Rods and cones have visual pigments containing different types of opsin protein. These pigments are named according to the type of opsin they contain. In rods, the visual pigment is

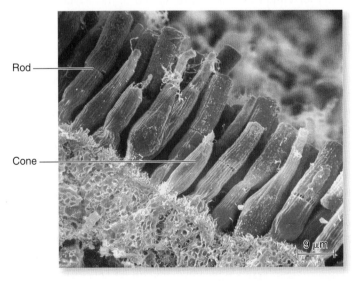

Figure 43.16 Rod and cone photoreceptors. Rods are shown as green and cones as blue in this false-color image (SEM).

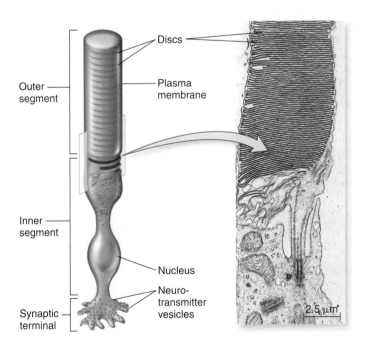

Figure 43.17 Structure of a photoreceptor. The illustration shows the structure of a rod photoreceptor and its appearance in a transmission electron micrograph. Note the multiple stacks of membranous discs in the outer segment of the cell.

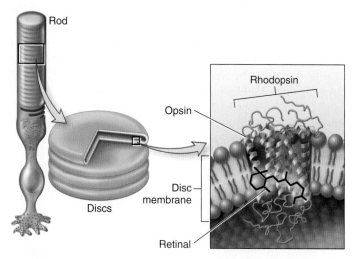

Figure 43.18   A visual pigment.   The visual pigment rhodopsin is found within the membrane of the rod photoreceptor discs. It is composed of a transmembrane protein, opsin, that is bonded to a molecule of retinal, a derivative of vitamin A that is capable of absorbing light energy.

Concept check: *What is the adaptive value of the convolutions and discs of photoreceptors?*

named **rhodopsin** (**Figure 43.18**). Cones contain several types of visual pigments called **cone pigments**, or photopsins.

In humans, cone pigments are composed of retinal plus one of three possible opsin proteins. Each type of opsin protein determines the wavelength of light that the retinal in a cone can absorb. For this reason, each cone pigment can respond to either red, green, or blue light. Any given cone cell makes only one type of cone pigment. Many different shades of these colors can be perceived, however, because the brain uses information about the proportion of each type of cone that was stimulated to generate all other colors. Red, green, and blue opsins are not present in all species, and many species have only one or two opsins. Presumably, the more types of cones an animal has, the more shades of color it perceives. It is intriguing, therefore, to imagine how birds see the world, because they have up to five cones of overlapping wavelength sensitivity. What happens, however, if some cones are defective?

Genomes & Proteomes Connection

Mutations in Cone Pigments Cause Color Blindness

In daylight, about 92% of human males and over 99% of females have normal color vision. However, problems in color vision may result from defects in the cone pigments arising from mutations in the opsin genes. The most common is red-green color blindness, which occurs predominantly in men (1 in 12 males compared with 1 in 200 females). Individuals with red-green color blindness either lack the red or green cone pig-

○ Unaffected female

● Color-blind female

◑ Female carrier of recessive allele

☐ Unaffected male

■ Color-blind male

(a) Pedigree for color blindness

(b) Test for red-green color blindness

Figure 43.19   Color blindness.   (a) A pedigree for red-green color blindness showing all possible offspring. (b) A standard eye test to screen for red-green color blindness. People with red-green color blindness will not see the number 74 hidden in this picture.

Concept check: *Why is red-green color blindness rare in females?*

ments entirely or, more commonly, have one or both of them in an abnormal form. In one form, for example, an abnormal green pigment responds to red light as well as green, making it difficult to discriminate between the two colors.

Color blindness was first described in the scientific literature in 1794 by John Dalton, the chemist mentioned in Chapter 2 after whom units of molecular mass are named. Dalton was himself color blind and willed that his eyes be preserved so that careful examination of them might help to determine the cause of the defect. He hypothesized that it arose because his aqueous humor was filled with a blue-colored medium of some type, but this was proven wrong after his death.

Instead, discovering the cause of color blindness had to wait until the development of genetics. As geneticists determined, color blindness results from a recessive mutation in one or more genes encoding the opsins. Genes encoding the red and green opsins are located very close to each other on the X chromosome, while the gene encoding the blue opsin is located on a different chromosome. In males, the presence of only one X chromosome means that a single recessive allele from the mother will result in red-green color blindness, even though the mother herself is not color blind (**Figure 43.19**). In 1994, DNA testing of John Dalton's retina confirmed that he had classic red-green color blindness.

Dark

Na+ channels in the membrane of the outer segment are open, which depolarizes the cell. Glutamate is released from the cell.

Light

Na+ channels close, which hyperpolarizes the cell. Glutamate is not released from the cell.

Glutamate released

Postsynaptic cell

No glutamate released

Figure 43.20 Membrane potential response of photoreceptors to dark and light.

Photons Change Photoreceptor Activity by Altering the Conformation of Visual Pigments

Photoreceptors differ from other sensory receptor cells because at rest in the dark their membrane potential is slightly depolarized, whereas in response to a light stimulus, it is hyperpolarized rather than depolarized (**Figure 43.20**). In the dark, the cell membranes of the outer segments of resting cells are highly permeable to sodium ions. Sodium ions flow into the cytosol of the cell through open Na+ channels in this membrane, depolarizing the cell. This depolarization results in a continuous release of the neurotransmitter glutamate from the synaptic terminal of the photoreceptor. The photoreceptor synapses with a postsynaptic cell that is the next neuron in the visual pathway. This initiates a series of events within the retina that is interpreted by the brain as an absence of light. In contrast, when exposed to light, the Na+ channels in the outer segment membranes of the photoreceptor close. The resulting reduction in sodium concentrations leads to a hyperpolarization of the cell. In response, the release of glutamate is stopped. This results in a series of cellular activations within the retina and brain that is interpreted as a visual image.

Let's take a more detailed look at the mechanism that allows a photoreceptor to respond to light (**Figure 43.21**). When the photoreceptor is exposed to light, the retinal within the visual pigment absorbs a photon. The energy of the photon alters the retinal from *cis*-retinal to *trans*-retinal, an isomer with a slightly different conformation due to a rotation at one of the molecule's double bonds (Figure 43.21). This change results in retinal briefly dissociating from the opsin protein, causing the opsin to change its three-dimensional shape and activate a G protein called transducin, located in the disc membrane. The activated transducin, in turn, activates another disc protein, the enzyme phosphodiesterase.

The action of phosphodiesterase results in the closure of Na+ channels in the outer segment membrane. Remember that

cis-retinal

Light

trans-retinal

cGMP

Cytosol

Na+

1 In the disc membrane, the *cis*-retinal in the visual pigment absorbs a photon of light, converting it to *trans*-retinal.

Disc

Disc membrane

Phosphodiesterase

Opsin   Transducin

cGMP   GMP

Outer segment membrane

Na+

2 The change to *trans*-retinal causes a conformational change in opsin that activates transducin, a G protein also in the disc membrane.

3 Transducin activates a phosphodiesterase, which converts cGMP into GMP.

4 The levels of cGMP in the cytosol drop, causing cGMP to dissociate from Na+ channels in the outer segment membrane. The Na+ channels then close, resulting in a hyperpolarization of the cell.

Figure 43.21 Signal transduction in photoreceptor (rod) cells in response to light.

in the dark, these channels are open. Sodium ions flow in and are pumped back out of the cell by Na⁺/K⁺-ATPase pumps in the cell membrane of the inner segment. The Na⁺ channels are gated by intracellular cGMP. In the dark, cytosolic concentrations of cGMP are high, keeping sodium channels open. However, when phosphodiesterase is activated, it reduces the amount of cytosolic cGMP by converting it to GMP. GMP molecules cannot hold the sodium channels of the outer segment membrane open, so the channels close, and sodium ions stop moving into the cell. The membrane potential of the cell becomes less positive than it was in the dark. Therefore, the response of the cell is a hyperpolarization that is proportional to the intensity of the light. The final result is a decrease in glutamate release from the photoreceptor (see Figure 43.20), ultimately leading to a visual image. The sequential activation of enzymes following activation of a single photoreceptor results in an amplification of the original signal (see Chapter 9 for a description of signal amplification). Because of this property, we are able to detect extremely low levels of light.

The Visual Image Is Refined in the Retina

Thus far, we have considered the structure of the vertebrate eye and how photoreceptors transduce light. We will now turn our attention to the neural pathways through which the visual signal travels to reach the brain. To do so, we must consider the cellular organization of the retina. The vertebrate retina has three layers of cells (Figure 43.22). The rods and cones form the deepest layer, closest to the inside of the sclera. Immediately behind the photoreceptors is a pigmented epithelium that absorbs light that missed the photoreceptors; this prevents scattering of light within the retina, which could degrade the sharpness of vision. Because the photoreceptors are positioned at the back of the retina, light must pass through two transparent layers of cells before it reaches them. The middle layer contains **bipolar cells**, so named because one end (or "pole") of the cell synapses with the photoreceptors, and the other end relays responses to a top layer of cells, the **ganglion cells**. The ganglion cells send their axons out of the eye into the optic nerve. In addition to these three layers, two other types of cells, horizontal and amacrine cells, are interspersed across the retina.

The pathway for light reception begins at the photoreceptor cells (rods and cones). These photoreceptor cells release neurotransmitter molecules that affect the membrane potential of bipolar cells. The membrane potential of the bipolar cells determines the amount of neurotransmitter that they release, which, in turn, controls the membrane potential of ganglion cells. When a threshold potential is reached in ganglion cells, action potentials are sent out of the eye via the optic nerve to the brain. These signals travel along pathways that include the thalamus, brainstem, cerebellum, and the visual cortex (the occipital lobe of the cerebral cortex). Visual information is further refined and interpreted within the visual cortex. The visual cortex responds to such characteristics of the visual scene as whether something is moving, how far away it is, how one color compares to another, and the nature of the image (for

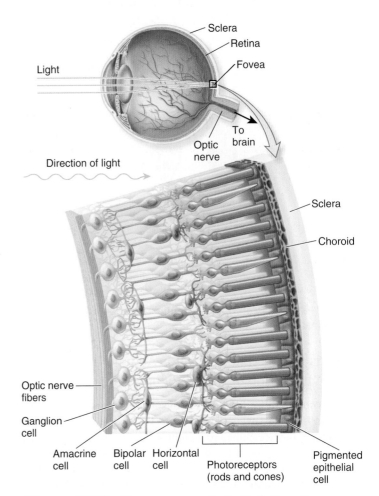

Figure 43.22 **The arrangement of cells in the retina.** Light passes through two layers of cells before it reaches the photoreceptors. Amacrine and horizontal cells integrate the responses of the bipolar and ganglion cells. The ganglion cells generate action potentials to carry the information to the brain.

example, a face). The cortex does not form a picture in the brain, but forms a spatial and temporal pattern of electrical activity that is perceived as an image.

Horizontal and amacrine cells modify electrical signals as they pass from the photoreceptors to the ganglion cells. These cells adjust the signal significantly, enhancing an animal's ability to visualize a scene by emphasizing the differences between images. Horizontal cells make connections between photoreceptors and help to define the boundaries of an image. Amacrine cells are important in adjusting the eye to different light intensities and increasing the sensitivity of the eye to moving images. The ability of the retina to refine the image is especially well developed in birds and reptiles. These animals have complex retinas that process the image extensively before it is interpreted in the brain.

Vertebrate Eyes Are Adapted to Environmental Conditions and Life Histories

Many vertebrates show unique modifications of their visual systems that are the result of evolutionary adaptations to environmental conditions. Other adaptations have occurred as a result of behavioral requirements for obtaining food or attracting a mate.

Variation in Light Intensity   Over the course of a day, the amount of light that stimulates the eye varies widely. The vertebrate eye can adjust to the differences in illumination in part by adjusting the diameter of the pupil. In addition, the eye can adapt to light differences by changing the relative amounts of *cis*- and *trans*-retinal. Such changes can alter the light sensitivity of the eye by as much as a million-fold!

In a dark environment, a large amount of *cis*-retinal is available for light absorption. When light stimulates *cis*-retinal to change into *trans*-retinal, the visual pigment no longer responds to additional light and is said to be bleached. Bleaching makes the eye less sensitive to low levels of light. When you enter a dark movie theater on a bright day, for example, most of your visual pigments have been bleached by the sunlight and are temporarily unavailable for use. For a few minutes, you are unable to detect low levels of light. This process of adjusting to dim light, called dark adaptation, involves the gradual reconversion of *trans*-retinal to *cis*-retinal.

Conversely, when you move from a dark room into bright sunlight, as when you leave the theater after the matinee, the large amount of *cis*-retinal makes your eyes extremely sensitive to light and temporarily overwhelms your ability to see colors and shapes. In light adaptation, the bright light stimulates bleaching of the visual pigments, converting *cis*-retinal back to *trans*-retinal, which reduces the response to bright light.

Differences in Eye Placement   Except for some of the ray-finned fishes, blunt-headed cetaceans, and most amphibians, vertebrate animals have some degree of **binocular** (or stereoscopic) **vision**. Animals with both eyes located at or near the front of the head, such as primates and raptors, have greater binocular vision, because the overlapping images coming into both eyes are processed together in the brain to form one perception (**Figure 43.23**). Binocular vision provides excellent depth perception because the images come into each eye from slightly different angles. The brain processes those tiny differences to determine where an object is relative to other objects in its environment. Predators benefit from binocular vision because it helps them judge distance and determine the location of their prey. Binocular vision is present in predatory birds such as the snowy owl (Figure 43.23a) and mammals (as well as in predatory insects such as Mantids), and also in arboreal animals that must judge distances between tree limbs.

In contrast, animals with eyes on the sides of the head, such as most fishes, blunt-headed cetaceans, amphibians, herbivorous mammals, and insects, have strictly or primarily **monocular vision**. Monocular vision allows an animal to see a wide area at one time, at the cost of reduced depth perception. Many prey species have monocular vision, perhaps because it helps them scan for predators across a wide field of vision. The placement of the eyes in the American woodcock (*Scolopax minor*) actually permits a field of vision of 360°, most of which is monocular (Figure 43.23b); in other words, these and similar birds can see directly behind themselves, even when digging in the dirt for earthworms!

Vision in the Deep Sea   Fishes and other deep-sea vertebrates have color vision that is limited primarily to the color blue. Light with longer wavelengths that would be seen as red or orange does not usually penetrate more than 6 m into the water, whereas the higher-energy, shorter-wavelength light, which is seen as blue, can penetrate to greater depths. Aquatic animals that live in the deep sea are usually capable of seeing only blue, because they generally have only one opsin, which

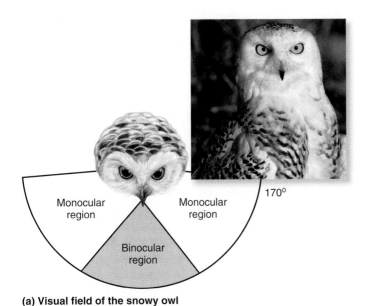

(a) Visual field of the snowy owl

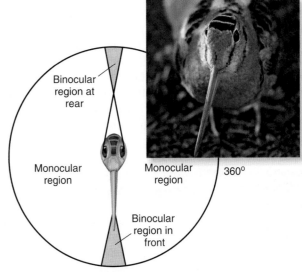

(b) Visual field of the American woodcock

Figure 43.23   **Examples of binocular and monocular fields of vision.**   Visual fields are shown for **(a)** the snowy owl (*Bubo scandiacus*) and **(b)** the American woodcock (*Scolopax minor*). Monocular regions are white; binocular regions are shaded.

is responsive to blue light. Deep-dwelling fishes tend to be drab colors, because most wavelengths of light do not penetrate that far and therefore could not reflect off the surface of the fishes. In those deep-dwelling fishes with more than one type of opsin, the additional visual pigments detect the bioluminescence (self-generated light, like that of a firefly) produced by their own or other species.

By contrast, fishes that live near the water's surface sometimes have four or five different opsins, giving them excellent color vision. Not surprisingly, shallow water and surface-dwelling fishes are often very colorful, because light of all wavelengths penetrates shallow water. These fishes have adapted by using coloration for protection (camouflage) or for identification.

Vision with Speed Peregrine falcons, like other raptors, have exceptionally good eyesight. They dive at speeds up to 150 to 200 miles per hour to capture prey, and their visual system is specialized to maintain great sensitivity during the dive. The retina of a falcon has two foveas, one deep and one shallow. The deep fovea has a central pit and provides the highest visual accuracy when objects are far away and to the side. However, when the falcon uses the deep fovea, it does not have good depth perception. The shallow fovea provides the best images of nearby objects and good depth perception. When the falcon dives, it first uses the deep fovea to locate the prey and steer toward it. When the falcon gets closer, it switches to the shallow fovea. Some details of the image are sacrificed, but the improved depth perception allows it to gauge when it will reach its prey.

Vision in the Dark: The Tapetum Lucidum Did you ever wonder why cats' eyes seem to glow in photographs, as in the chapter-opening photo? What you see is light being reflected off the tapetum lucidum, a reflective layer of tissue located beneath the photoreceptors at the back of the eye in some animals. When light enters the retina, some of it passes into the layer containing the photoreceptors, but some light also passes through this layer. Normally, the pigmented epithelium at the back of the eye absorbs this light. However, in animals with a tapetum lucidum, the light reflects off this structure and back to the photoreceptors. This gives the photoreceptors a "second chance" to capture light, making it easier to detect low levels of light. Many nocturnal animals, both vertebrate and invertebrate, have a tapetum lucidum, which increases their ability to see in the dark.

43.6 Chemoreception

Chemoreception includes the senses of smell (**olfaction**) and taste (**gustation**), both of which involve detecting chemicals in air, water, or food. These chemicals bind to chemoreceptors, which, in turn, initiate electrical responses in other neurons that pass into the brain. Amazingly, the binding of a single molecule to a receptor cell can sometimes be perceived as an odor! Airborne molecules that bind to olfactory receptors must be small

enough to be carried in the air and into the nose. Taste molecules can be heavier because they are conveyed in food and liquid.

Taste and smell are closely related. In fact, the distinction is largely meaningless for aquatic animals, because for them all chemoreception comes through the water. Even in terrestrial animals, about 80% of the perception of taste is actually due to activation of olfactory receptors. (This is why food loses its flavor when the sense of smell is impaired, such as when you have a cold.) In this section, we explore chemoreception in insects and then in mammals.

Olfaction and Taste in Insects Involve Chemoreceptors in Sensory Hairs

Insects are highly dependent on odor and taste for finding food and mates. In insects, chemoreceptors are neurons that are located on sensory hairs on the proboscis, legs, feet, and antennae. Each sensory hair on the proboscis and feet has a pore at the tip through which the substance passes. As the example in **Figure 43.24a** shows, the blowfly has four separate chemoreceptors within each hair, and each of these neurons responds to different molecules. Dendrites of the chemoreceptor cells inside the pore bind to the molecules and initiate a sensory transduction pathway that opens ion channels in the membrane. This depolarizes the plasma membrane of the chemoreceptor cell and generates action potentials, which are sent to the brain for interpretation.

In certain moths, males have elaborate antennae that can sense pheromones, extremely potent signaling molecules given off by a female. The female secretes a sexual-attractant pheromone into the air from an abdominal gland (**Figure 43.24b**), and the chemosensory hairs on the male's antennae (**Figure 43.24c**) can detect extremely low concentrations of it from several kilometers away. This highly sensitive detection system enables the male to locate the female in the dark.

Mammalian Olfactory Receptors Respond to the Binding of Odor Molecules

The olfactory sensitivity of mammals varies widely depending on their supply of olfactory receptor cells, which ranges from 5 or 6 million in humans to 100 million in rabbits and 220 million in dogs. Olfactory receptors are neurons that are located in the epithelial tissue at the upper part of the nasal cavity in mammals (**Figure 43.25a**). These cells are surrounded by two additional cell types: supporting cells and basal cells. Supporting cells are located between the receptor cells and provide physical support for the olfactory receptors. The basal cells differentiate into new olfactory receptors every 30–60 days, replacing those that have died after prolonged exposure of their cell endings.

Olfactory receptors have dendrites from which long, thin extensions called cilia emerge into a mucous layer that covers the epithelium. Despite the similarity in structure, these cells do not function like the mechanoreceptor hair cells of the auditory and vestibular systems; these cilia are different from stereocilia that bend. Instead, olfactory receptor cells have receptor proteins within the plasma membranes of the cilia (**Figure 43.25b**).

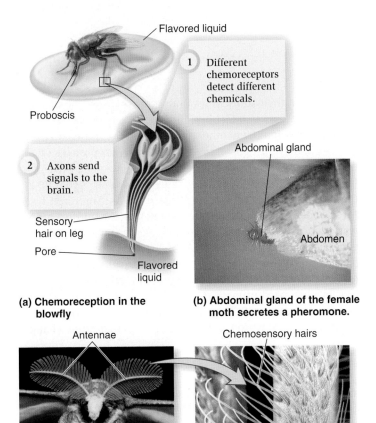

(a) Chemoreception in the blowfly

1. Different chemoreceptors detect different chemicals.

2. Axons send signals to the brain.

Flavored liquid
Proboscis
Sensory hair on leg
Pore
Flavored liquid

(b) Abdominal gland of the female moth secretes a pheromone.

Abdominal gland
Abdomen

(c) Chemosensory hairs on antennae of the male moth detect the pheromone.

Antennae
Chemosensory hairs

Figure 43.24 Chemoreception in insects.
(a) Chemosensory cells of the blowfly located on the proboscis, legs, and feet sense different chemicals. (b) Female moths secrete a pheromone from abdominal glands, which is detected by (c) chemosensory hairs in a male's antennae (shown enlarged in this scanning electron micrograph) that bind odor molecules.

Airborne molecules dissolve in the mucus and bind to these olfactory receptor proteins. When an odor molecule binds to its receptor protein, it initiates a signal transduction pathway that ultimately opens sodium channels in the plasma membrane. The subsequent depolarization results in action potentials being transmitted to the next series of cells in the olfactory bulbs, which are located at the base of the brain, one on each side. The olfactory bulbs are a collection of neurons that act as an initial processing center of olfactory information and as a relay center to the cerebral cortex and limbic system for further processing and interpretation.

Figure 43.25 Olfactory structures in the human nose.
Odor molecules dissolve in a layer of mucus that coats the olfactory receptor cells. The molecules bind to protein receptors in the membrane of cilia that extend from the olfactory receptor cells. Action potentials in the olfactory receptor cells are conducted to cells in the olfactory bulb, and from there to the brain for interpretation. Basal cells periodically replace dead or damaged olfactory receptor cells.

The relative size of the olfactory bulbs indicates the importance of olfaction to an animal. In humans, the olfactory bulbs make up only about 5% of the weight of the brain, whereas in nocturnal animals like rats and mice, they can comprise as much as 20%. Even with their relatively limited olfactory sensitivity, however, humans have the capacity to detect up to 10,000 different odors, and other mammals are thought to respond to many more. Exactly how this is possible remained a mystery until 1991, when two scientists uncovered the molecular basis of olfaction.

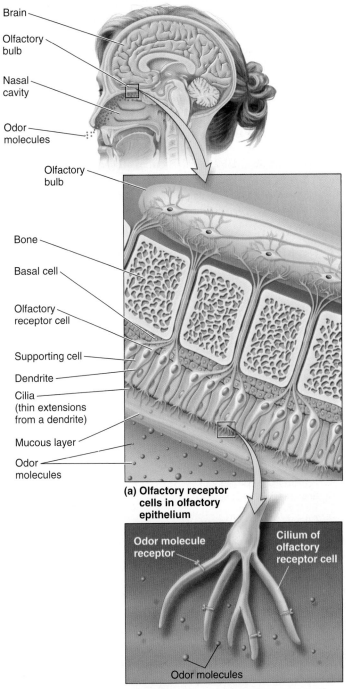

Brain
Olfactory bulb
Nasal cavity
Odor molecules

Olfactory bulb
Bone
Basal cell
Olfactory receptor cell
Supporting cell
Dendrite
Cilia (thin extensions from a dendrite)
Mucous layer
Odor molecules

(a) Olfactory receptor cells in olfactory epithelium

Odor molecule receptor
Cilium of olfactory receptor cell
Odor molecules

(b) Cilia from dendrite with odor molecule receptors

FEATURE INVESTIGATION

Buck and Axel Discovered a Family of Olfactory Receptor Proteins That Bind Specific Odor Molecules

How does the olfactory system discriminate among thousands of different odors? American neuroscientists Linda Buck and Richard Axel set out to study this question. When they began, two hypotheses were proposed to explain this phenomenon. One possibility was that many different types of odor molecules might bind to one or just a few types of receptor proteins, with the brain responding differently depending on the number or distribution of the activated receptors. The second hypothesis was that olfactory receptor cells can make many different types of receptor proteins, each type binding a particular odor molecule or group of odor molecules.

To begin their study, Buck and Axel assumed that olfactory receptor proteins would be highly expressed in the olfactory receptor cells, but not in other parts of the body. Based on previous work, they also postulated that the receptor proteins would be members of the large family of G-protein-coupled receptors (GPCRs), described in Chapter 9. As shown in Figure 43.26, they isolated olfactory receptor cells from rats and then lysed (broke open) the cells to release the mRNA. The purified mRNA was then used to make cDNA via reverse transcriptase. This generated a large pool of cDNAs, representing all of the genes that were expressed in olfactory receptor cells at the time of mRNA collection. To determine if any of these cDNAs encoded GPCRs, they used primers that recognized regions within previously identified GPCR genes that are highly conserved. A highly conserved region is a DNA sequence that rarely changes

Figure 43.26 Buck and Axel identified olfactory receptor proteins in olfactory receptor cells.

HYPOTHESES 1. Many different types of odor molecules bind to just a few types of receptor proteins. 2. Odor molecules are detected by many specific olfactory receptor proteins belonging to the family of G-protein-coupled receptors (GPCRs).

KEY MATERIALS Laboratory rats (*Rattus norvegicus*), PCR reagents, DNA-sequencing gels.

Experimental level   **Conceptual level**

1 Dissect and homogenize olfactory epithelium from laboratory rats. Extract mRNA.

Euthanize rats.   Homogenizer   Blade   Epithelium

mRNA   DNA fragment   Cell fragment   Cell nucleus

2 Purify mRNA. Make cDNA (described in Chapter 20) from the mRNA, using reverse transcriptase.

Add mRNA and reverse transcriptase.

Many double-stranded cDNAs

3 Add primers that bind specifically to genes that encode GPCRs. Subject to PCR as described in Chapter 20.

Add specific primers for GPCR genes.   PCR machine

Primers will hybridize only with cDNA that codes for proteins in the GPCR family and amplify those genes. Many different PCR products are obtained, each corresponding to a different gene.

4 Subject each PCR product to DNA sequencing, also described in Chapter 20.

DNA-sequencing gel

A T G C

Different GPCRs will have slightly different DNA sequences.

5   THE DATA

The DNA sequencing revealed 18 different GPCRs that were specifically expressed in the receptor cells of the olfactory epithelium.

6   CONCLUSION   Olfactory receptor cells express many different receptor proteins that account for an animal's ability to detect a wide variety of odors.

7   SOURCE   Buck, L., and Axel, R. 1991. A novel multigene family may encode odorant receptors: a molecular basis for odor recognition. *Cell* 65:175–187.

among different members of a gene family (see Chapter 21). The primers were used in the PCR technique to amplify cDNAs that encoded GPCRs. This produced many PCR products that were then subjected to DNA sequencing. As shown in The Data in Figure 43.26, Buck and Axel initially identified 18 different genes, each encoding a GPCR with a slightly different amino acid sequence. Further research showed that these 18 genes were expressed in nasal epithelia, but not in other parts of the rat's body. These results were consistent with the second hypothesis stated earlier, namely that organisms make a large number of receptor proteins, each type binding a particular odor molecule or group of odor molecules.

Since these studies, researchers have determined that this family of olfactory genes in mammals is surprisingly large. In humans, over 600 genes that encode olfactory receptor proteins have been identified, though about half of these are pseudogenes that are no longer functional. This value underscores the importance of olfaction even in humans. Each olfactory receptor cell is thought to express only one type of GPCR that recognizes its own specific odor molecule or group of molecules. Most odors are due to multiple chemicals that activate many different types of odor receptors at the same time. We perceive odors based on the combination of receptors that become activated, and then send signals to the brain. This ability to combine different inputs allows us to perceive several thousand different odors despite having only several hundred different olfactory receptor proteins.

The research of Buck and Axel explained, in part, how animals detect a myriad of odors. In 2004, they received the Nobel Prize for this pioneering work. Even so, many intriguing questions remain. We do not yet understand the pattern of gene regulation in the receptor cells or how the brain actually interprets the signals it receives from those cells.

Experimental Questions

1. What were the two major hypotheses to explain how animals discriminate between different odors? How did Buck and Axel test the hypothesis of multiple olfactory receptor proteins?

2. What were the results of Buck and Axel's study?

3. Considering the two hypotheses explaining how animals discriminate between different odors (see Question 1), which one was supported by the results of this experiment? With the evidence presented by Buck and Axel, what is the current hypothesis explaining the discrimination of odors in animals?

Taste Buds Detect Food Molecules

Chemical senses are present in all animal phyla, although not all animals have taste-sensing organs. Many animals use their taste (gustation) sense to select appropriate foods. For example, butterflies select nectar based on the sugars found in a particular flower, and carnivorous animals detect the taste of different meats based on the combination of amino acids, fats, and sugars that are present. Some freshwater and marine animals, such as catfish and lobsters, have exceptionally sensitive chemoreceptors for specific amino acids that are particularly important as neurotransmitters in the nervous system. Taste can also help an animal seek out necessary nutrients, such as salt, and avoid poisonous chemicals. Toxic substances are often perceived as bitter or distasteful, which may cause an animal to stop eating any such substance immediately.

Taste buds are structures containing chemosensory cells that detect particular molecules in food. Humans have about 9,000 taste buds, while other mammals such as pigs and rabbits have approximately 15,000 to 18,000. The bumps that you see on your tongue are not the taste buds but the papillae, elevated structures on the tongue that collect food molecules in depressions called taste pores. These pores contain many taste buds with sensory receptor cells (**Figure 43.27**). The sensory receptor cells form a complex with several supporting cells that is organized like the wedges of an orange. The tips of the sensory receptor cells have microvilli that extend into the taste pore. Here, molecules in food that have dissolved in saliva bind to

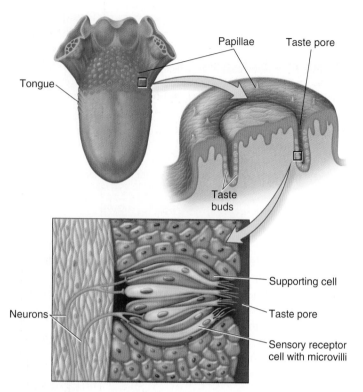

Figure 43.27 Structures involved in the sense of taste. This sense occurs in taste buds, which contain the sensory receptor cells that respond to dissolved food molecules.

Concept check: *Why have many animals, including humans, evolved the ability to sense salty, sweet, sour, and bitter?*

receptor proteins. This triggers intracellular signals that alter ion permeability and membrane potentials. The sensory receptor cells then release neurotransmitters onto underlying sensory neurons. Action potentials travel from these neurons to the thalamus and other regions of the forebrain, where the taste is perceived and recognized.

There are a number of different types of taste cells, and they are distributed on the tongue in different areas, but these areas overlap considerably. Each type of cell has a specific transduction mechanism that allows it to detect specific chemicals present in the foods and fluids we ingest. The subjective experience of their activation includes the perception of sweet (sugar), sour (acidic foods like citrus fruits), salty, and bitter. In addition, a newly recognized taste perception called umami (after the Japanese word for "delicious") is associated with the presence in ingested food of glutamate and other similar amino acids, and is usually described as imparting flavorfulness to food. It may account for the widely recognized effects of monosodium glutamate (MSG) in enhancing the flavor of food.

The senses of taste and smell are enhanced when we are hungry, a phenomenon that most likely occurs in other animals as well. Once we have eaten, we are less aware of the smell and taste of food. The importance of this for survival is clear: A hungry animal needs to eat. An improved sense of smell aids in locating food, and a heightened sense of taste encourages

an animal to eat. Afterward, these senses become temporarily dampened so as not to distract the animal from its other needs. The mechanism by which these changes occur is uncertain, but it may involve a temporary alteration in the number of smell and taste receptors, or in their ability to bind ligands.

43.7   Impact on Public Health

Sensory disorders are among the most common neurological problems found in humans and range from mild (needing eyeglasses) to severe (blindness). In this section, we present an overview of a few representative sensory disorders that have a major impact on the human population.

Visual Disorders Include Glaucoma, Macular Degeneration, and Cataracts

Visual disorders affect an enormous fraction of the world's population. In the U.S. alone, more than 10 million people have severe eye problems that cannot be corrected by eyeglasses, and more than 1 million people are blind (42 million worldwide). The costs to the U.S. government associated with blindness and severe visual loss amount to nearly $4 billion yearly. Although vision loss has many causes, three disorders account for over half of all cases: glaucoma, macular degeneration, and cataracts.

Normally, the fluid that makes up the aqueous humor in the eye is produced and reabsorbed (drained) in a circulation that keeps the fluid levels constant. In **glaucoma**, drainage of aqueous humor becomes blocked, and the pressure inside the eye increases as the fluid level rises. If untreated, this eventually damages cells in the retina and leads to irreversible loss of vision (**Figure 43.28a**). The cause of glaucoma is not always known, but in some cases it is due to trauma to the eye, chronic use of certain medicines, or diseases such as diabetes. Some forms of glaucoma may have a genetic component. Glaucoma can be treated with eye drops that reduce fluid production in the eye or with laser surgery to reshape the structures in the eye. Incidentally, rubbing your eyes increases the pressure in the eyes, and habitual rubbing can also damage vision. The American Foundation for the Blind estimates that 400,000 new cases of glaucoma are diagnosed each year in the U.S., and up to 10 million individuals have elevated eye pressure. Glaucoma accounts for roughly 10% of all cases of blindness in the U.S.

In **macular degeneration**, photoreceptor cells in and around the macula (which contains the fovea) of the retina are lost. Because this is the region where cones are densely packed, this condition is associated with loss of sharpness and color vision (**Figure 43.28b**). It does not usually occur before age 60, but in some cases it is hereditary and can occur at any age. Macular degeneration is the leading cause of blindness in the U.S., accounting for roughly 25% of all cases, but its causes remain obscure.

Cataracts, the accumulation of protein in the lens, cloud the lens and cause blurring, poor night vision, and difficulty

(a) How the world is seen by a person with glaucoma

(b) How the world is seen by a person with macular degeneration

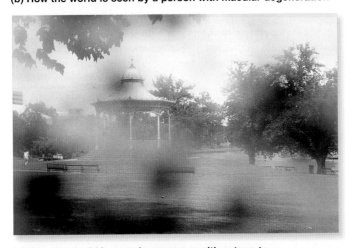

(c) How the world is seen by a person with cataracts

Figure 43.28 Appearance of the visual field in a person with (a) glaucoma, (b) macular degeneration, and (c) cataracts.

focusing on nearby objects (**Figure 43.28c**). By age 65, as many as 50% of individuals have one or more cataracts in either eye, and this jumps to 70% by age 75. Many cataracts are small enough not to affect vision. In fact, many people do not even realize they have cataracts until they undergo an eye exam.

The causes of cataracts are not all known but include trauma, medicinal drugs, diabetes, and heredity. The treatment, when needed, is to wear a powerful contact lens to help do the job formerly done by the person's own lens or to have the affected lens surgically removed and replaced with a plastic lens. Nearly 1.4 million such surgeries are performed each year in the U.S. Without the lens, which helps protect the retina by absorbing some of the high-energy ultraviolet light from the sun, the retina must be protected during the day by wearing dark sunglasses.

Damaged Hair Cells Within the Cochlea Can Cause Deafness

Deafness (hearing loss) is usually caused by damage to the hair cells within the cochlea, although some cases result from functional problems in brain areas that process sound or in the nerves that carry information from the hair cells to the brain. When the hair cells are damaged, noises have to be louder to be detected, and an affected person may require the use of hearing aids to amplify incoming sounds.

Hearing loss may be mild or severe and may result from many causes, including injury to the ear or head, hereditary defects of the inner ear, and exposure to certain diseases (for example, rubella) or toxins during fetal life. By far, however, the most significant cause of hearing loss is repeated, long-term exposure to loud noise. When the amplitude of a sound wave (that is, the distance between the peak and the trough of the wave) is high, the loudness of the sound is great. The loudness of sound is measured in decibels. The decibel (dB) scale is logarithmic; every increase of 10 dB is actually 10 times louder than the previous level. Normal conversation is about 60 dB, a chainsaw is 108 dB, and a jet plane taking off can reach 150 dB. Estimates of noise-induced hearing loss, which usually results from job-related activities, range from 7 to 21% of all cases of hearing loss worldwide; this makes it the leading occupational disorder in the U.S. It is, therefore, one of the most significant disabilities in the U.S. in terms of numbers of people affected and costs to society. Scientists estimate that nearly 40 million people are exposed daily to dangerous noise levels in the U.S. as a result of their occupation.

Recent research has led to a better understanding of the mechanism by which noise impairs hearing. Chronic exposure to loud sounds appears to produce a state of metabolic exhaustion in the hair cells of the cochlea. As a result, the cells become fatigued and are unable to maintain normal biochemical processes. One consequence of this is a buildup of free radicals. These compounds oxidize lipids in cellular membranes, damaging the membranes in the process. Mitochondrial membranes appear to be particularly susceptible to these free radicals. Once mitochondria are destroyed, a cell's ability to produce the ATP needed to fulfill its energy demands is compromised, and the cell dies. As hair cells die, the ear becomes less sensitive to sound.

Researchers are investigating drugs that might prevent the formation of free radicals in the cells of the ear, but such drugs are not yet available. If a cochlea is severely damaged, it can be surgically replaced with artificial cochlear implants. These devices generate electrical signals in response to sound waves that can stimulate the auditory nerve, which communicates with the brain. Cochlear implants cannot restore hearing to normal, but they can make it possible to hear conversations.

 ## Summary of Key Concepts

43.1 An Introduction to Sensory Receptors

- Sensory transduction is the process by which incoming stimuli are converted to neural signals. Perception is an awareness of a sense.

- Sensory receptors are either neurons or specialized epithelial cells that respond to stimuli and begin the process of sensory transduction. (Figure 43.1)

- A sensory receptor often responds to a stimulus by eliciting a graded response that is proportional to the intensity of the stimulus. (Figure 43.2)

43.2 Mechanoreception

- Mechanoreceptors respond to physical stimuli such as touch, pressure, stretch, movement, and sound.

- Mechanoreceptors in the skin are neurons that sense stimuli such as touch and pressure. (Figure 43.3)

- Hair cells have stereocilia that respond to specific types of stimuli and then release neurotransmitter that may result in action potentials in an adjacent sensory neuron. (Figure 43.4)

- The lateral line system of fishes detects water movements. (Figure 43.5)

- Hearing is the ability to sense sound waves. Sound waves that travel at a high frequency are perceived as having a high pitch; those at low frequency have a low pitch.

- The ear has three main compartments called the outer, middle, and inner ear. (Figure 43.6)

- As pressure waves move through the cochlea of the ear, they may cause the basilar membrane to vibrate. This causes the stereocilia of hair cells to bend back and forth, which results in intermittent action potentials being sent to the brain, leading to the perception of sound. (Figures 43.7, 43.8)

- Many adaptations for hearing improve an animal's ability to sense sound. (Figure 43.9)

- Statocysts in certain invertebrates allow an animal to sense its body position or equilibrium. (Figure 43.10)

- The vestibular system in vertebrates allows animals to sense linear and rotational movement. (Figure 43.11)

43.3 Thermoreception and Nociception

- Thermoreceptors in the skin and brain allow an animal to sense external and internal temperatures, respectively.

- Nociceptors in the skin and internal organs sense pain, an important survival adaptation in animals.

43.4 Electromagnetic Sensing

- Some animals can detect electrical and magnetic stimuli; in some cases, this ability is used to locate prey. (Figure 43.12)

43.5 Photoreception

- The eyecup in flatworms is a simple eye that detects light and its direction but does not form an image. (Figure 43.13)

- The compound eye found in many invertebrates consists of many ommatidia that focus light. (Figure 43.14)

- The eye found in vertebrates and certain invertebrates is a single-lens eye. Stretching and flattening of the lens in mammals aids in focusing on objects at varying distances. (Figure 43.15)

- Rods and cones are photoreceptors found in the vertebrate eye. The visual pigment consists of retinal and a protein called opsin. (Figures 43.16, 43.17, 43.18)

- Red-green color blindness is due to a defect in a type of opsin found in cone pigments. (Figure 43.19)

- Photoreceptors respond to the dark by releasing the neurotransmitter glutamate, and to light by stopping its release. (Figures 43.20, 43.21)

- The retina is composed of layers of cells that receive light input and send electrical signals to the brain via the optic nerve. (Figure 43.22)

- The vertebrate eye has several adaptations that aid in proper vision for particular animals. (Figure 43.23)

43.6 Chemoreception

- Invertebrates have taste receptors on their proboscis, legs, feet, and antennae. Some insects can detect pheromones. (Figure 43.24)

- Olfactory receptors in vertebrates detect odor. They contain cilia with protein receptors that bind specific odor molecules. Buck and Axel discovered that olfactory receptors have many different types of protein receptors for odor molecules. (Figures 43.25, 43.26)

- Taste buds contain chemoreceptors that sense molecules in food. (Figure 43.27)

43.7 Impact on Public Health

- Common visual disorders include glaucoma, macular degeneration, and cataracts. (Figure 43.28)

- Prolonged exposure to intense sound can damage hair cells within the cochlea and lead to deafness.

Assess and Discuss

Test Yourself

1. The process whereby incoming sensory stimulation is converted to electrical signals is known as
   a. an action potential.
   b. a threshold potential.
   c. perception.
   d. sensory transduction.
   e. reception.

2. Photoreceptors are examples of
   a. nociceptors.
   b. baroreceptors.
   c. mechanoreceptors.
   d. electromagnetic receptors.
   e. thermoreceptors.

3. The sensory receptors for audition (hearing) are located in
   a. the organ of Corti.
   b. the Eustachian tube.
   c. the vestibular system.
   d. the tympanic membrane.
   e. a, b, and c.

4. Statocysts are sensory organs for
   a. hearing found in many invertebrates.
   b. equilibrium found in mammals.
   c. equilibrium found in many invertebrates.
   d. water current changes found in fish.
   e. hearing found in amphibians.

5. The eyecups of planaria can
   a. focus light to form an image.
   b. detect light.
   c. detect the direction of light.
   d. all of the above.
   e. b and c only.

6. The light detectors of a compound eye are called
   a. retinas.
   b. opsins.
   c. cones.
   d. ommatidia.
   e. rods.

7. In the mammalian eye, light from near or far objects is focused on the retina when
   a. the lens moves forward or backward.
   b. the lens changes shape.
   c. the eyeball changes shape.
   d. the cornea changes shape.
   e. none of the above

8. The level of glutamate release from photoreceptors of the vertebrate eye would be highest when
   a. a person is standing in full sunlight.
   b. a person is in a completely dark room.
   c. a person is in a dimly lit room.
   d. Na^+ channels of the photoreceptor are closed.
   e. both a and d

9. Cone pigments detect different wavelengths of light due to
   a. their location in the retina.
   b. the amount of light they absorb.
   c. the type of retinal they have.
   d. the type of opsin protein they have.
   e. interactions with bipolar cells.

10. The stimulation for olfaction involves odorant molecules
    a. bending the cilia of olfactory receptor cells.
    b. binding to protein receptors of olfactory receptor cells.
    c. entering the cytoplasm of olfactory receptor cells.
    d. opening K^+ channels of olfactory receptor cells.
    e. all of the above

Conceptual Questions

1. Distinguish between sensory transduction and perception.

2. Explain how the mammalian ear is adapted to distinguish sounds of different frequencies.

3. Explain how eye placement affects binocular vision.

Collaborative Questions

1. Discuss two different types of mechanoreceptors.

2. Discuss different types of eyes found in animals.

Online Resource

www.brookerbiology.com

Stay a step ahead in your studies with animations that bring concepts to life and practice tests to assess your understanding. Your instructor may also recommend the interactive ebook, individualized learning tools, and more.

Muscular-Skeletal Systems and Locomotion

44

European tree frog (*Hyla arborea*) using its muscular-skeletal system to leap from a leaf.

Most of us are amazed at the athletic feats performed by humans with great jumping ability, such as long-jumpers or professional basketball players. Although it is impressive to see someone long-jump 8 or so meters or dunk a basketball through a rim that is more than 3 meters high, such feats pale by comparison to those routinely achieved by other animals. For example, some tree frogs, like the one shown in the chapter-opening photo, can leap distances up to 1.4 meters, yet the frog is only 4.5 cm long and weighs only 8 grams! Indeed, the jumping ability of frogs makes them model organisms for the study of muscle function in vertebrates.

Muscles are composed of highly specialized cells that have the ability to contract in response to stimuli. The three types of muscle—skeletal, smooth, and cardiac—were introduced in Chapter 40. In this chapter, you will learn about the structure and function of skeletal muscle and how skeletal muscle controls movements such as chewing, postural changes, and **locomotion**, the movement of an animal from place to place.

For skeletal muscles to produce movement, they must exert a force on an animal's skeleton. We begin the chapter, therefore, with a description of the types of animal skeletons. Then we examine the structure and function of skeletal muscle. Next, we consider various modes of locomotion in animals. We conclude with a few important bone and muscle diseases in humans.

44.1 Types of Animal Skeletons

When we think of the word skeleton, an image of the vertebrate system of bones usually comes to mind. However, invertebrates possess a skeleton as well, but their skeletons are not made of bone. Therefore, a broader definition of a **skeleton** is a structure or structures that serve one or more functions related to support, protection, and locomotion. Using this definition, the three types of skeletons found in animals are hydrostatic skeletons, exoskeletons, and endoskeletons. The first two are found in invertebrates, while endoskeletons are found in some sponges and all echinoderms and vertebrates.

Hydrostatic Skeletons Consist of Fluid-Filled Body Compartments

Many soft-bodied invertebrates use hydrostatic pressure to support their bodies and generate movements. Hydrostatic pressure is the pressure of a fluid at rest in a contained space. The combination of muscles and a water-based fluid in the body constitutes what is known as a **hydrostatic skeleton**. Water is nearly incompressible, which means that if an animal exerts a force on the fluid that fills its body cavities, it can use the resulting hydrostatic pressure to move the body, much like a balloon partly filled with water can be deformed by squeezing it along its length. In cnidarians such as hydras, for example, contractile cells in the body wall can exert a pressure on the fluid in the gastrovascular cavity. Depending on the direction of the force, a hydra's body and tentacles will either be lengthened or shortened by the movement of water into and out of different regions of the cavity.

Annelids or segmented worms, such as earthworms, move forward by passing a wave of muscular contractions along the

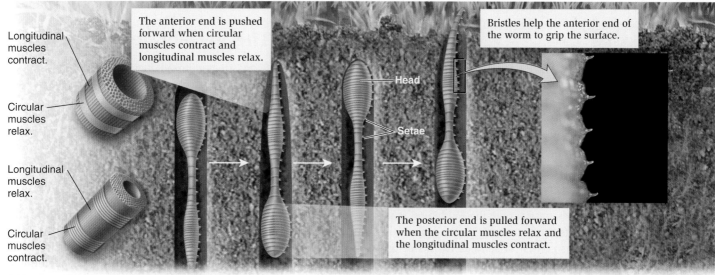

Longitudinal muscles contract.

Circular muscles relax.

Longitudinal muscles relax.

Circular muscles contract.

The anterior end is pushed forward when circular muscles contract and longitudinal muscles relax.

Bristles help the anterior end of the worm to grip the surface.

Head

Setae

The posterior end is pulled forward when the circular muscles relax and the longitudinal muscles contract.

(a) Hydrostatic skeleton in an annelid (earthworm)

(b) Exoskeleton in an arthropod (spiny lobster)

(c) Echinoderm (sea star) and (right) SEM of its endoskeleton

Figure 44.1   **Types of skeletons.**   **(a)** Hydrostatic skeleton. By alternately contracting and relaxing circular and longitudinal muscles, earthworms use differences in hydrostatic pressure to achieve locomotion. Clinging setae (bristles) along the body surface (shown in top-right inset) help prevent backsliding. **(b)** Exoskeleton. An arthropod's skeleton covers and protects its body, but it must be periodically shed and replaced to allow growth, as shown in this photo. **(c)** Endoskeleton. Echinoderms such as the sea star (starfish) have endoskeletons of bony plates made of calcium carbonate (shown in bottom-right inset).

Concept check: Endoskeletons do not provide protection for the body surface of animals. Can the lack of an external skeleton also be an advantage for animals?

length of their body, segment by segment (**Figure 44.1a**). The muscles are arranged in two orientations: circular and longitudinal (lengthwise). Contraction of circular muscles squeezes and lengthens the body, whereas contraction of the longitudinal muscles shortens and widens it. The worm is "squeezed" forward by differences in hydrostatic pressure created along its body by the muscles. Bristles known as setae along its body surface help the animal grip the ground to prevent backsliding.

Exoskeletons Are on the Outside of an Animal's Body

Arthropods have an **exoskeleton**, an external skeleton that surrounds and protects most of the body surface (**Figure 44.1b**). Exoskeletons provide support for the body, protection from the environment and predators, and protection for internal organs. The arthropod skeleton is made of a polysaccharide called chitin, and in crustaceans such as lobsters and shrimp, it is sometimes strengthened with calcium and other minerals.

Exoskeletons are often tough, durable, and segmented to allow movement. However, to allow growth they must be periodically shed, regrown, and strengthened again, a process called ecdysis, or molting. A disadvantage of exoskeletons is that when an animal is molting, its new exoskeleton is temporarily soft, making it more vulnerable to predators and the environment.

Exoskeletons vary enormously in their complexity, thickness, and durability. The differences in exoskeletons are usually adaptations that enhance an animal's survival. Think, for example, of the difference between the wing of a butterfly and the outer skeleton of a lobster. A butterfly's exoskeleton must be light enough for the animal to fly, while the thick, tough exoskeleton of the lobster provides a very effective defense against predators in the sea. Exoskeletons may seem primitive compared to the endoskeletons of vertebrates (discussed next), particularly because of the requirement for molting. Even so, insects are among the most successful of all classes of animals living today, having survived for hundreds of millions of years and inhabiting nearly every possible ecological niche on the planet. Clearly, exoskeletons have been sufficient for one of the planet's greatest success stories.

Endoskeletons Are Internal Support Structures

Like exoskeletons, **endoskeletons** provide support and protection. Unlike exoskeletons, however, endoskeletons are internal structures and do not protect the body surface. Some endoskeletons do, however, protect internal organs such as those in the thorax of vertebrates.

Endoskeletons are found in some species of sponges and all echinoderms and vertebrates. Minerals such as calcium, magnesium, phosphate, and carbonate supply the hardening material that gives the skeleton its firm structure. The endoskeletons of sponges and echinoderms consist of spiky networks of proteins and minerals, and mineralized platelike structures, respectively. Beneath the body surface of echinoderms, for example, arrays of mineralized plates made largely of calcium carbonate extend into the spines and arms that radiate from the main body (**Figure 44.1c**).

Vertebrate skeletons, by contrast, are composed of either cartilage or bone or both. Next we consider the structure and function of the vertebrate skeleton in greater detail.

44.2 The Vertebrate Skeleton

Vertebrate skeletons are composed of either cartilage (see Chapter 40)—as in cartilaginous fishes (sharks, rays, and skates)—or both cartilage and bone, as in bony fishes, amphibians, reptiles, birds, and mammals. First we consider the skeleton as a whole before examining bones in greater detail.

The Vertebrate Skeleton Performs Several Important Functions

In the vertebrate skeleton, bones are connected in ways that allow for support, protection, and movement. The vertebrate endoskeleton is often considered in two parts: the axial and appendicular skeletons (**Figure 44.2**). The axial skeleton is composed of the bones that form the main longitudinal axis of an animal's body. The appendicular skeleton consists of the limb or fin bones and the bones that connect them to the axial skeleton. A **joint** is formed where two or more bones come together. Some joints permit free movement (for example, the shoulders), whereas others allow no movement (fused joints like those interlocking the skull bones) or only limited

Figure 44.2 **The adult human skeleton.** This diagram shows the axial (tan) and appendicular (green) parts of the skeleton in an adult human, an animal with an endoskeleton. The adult human skeleton consists of 206 separate bones. Three examples of movable joints, one of each type, are shown.

movement (such as those of the vertebral column). Figure 44.2 illustrates three types of joints that allow different types of movements: pivot, hinge, and ball-and-socket joints.

The skeleton of vertebrates serves several functions in addition to support, protection, and movement. For example, blood cells and platelets, the latter of which help blood to clot (see Chapter 47), are formed within the marrow of the ilia, vertebrae, the ends of the femurs, and other bones. In addition, calcium and phosphate homeostasis is achieved in large part through exchanges of these ions between bone and blood. For example, if dietary intake of calcium is low, calcium is removed from bone and added to the blood, so that all of the vital cellular activities that depend on calcium, such as nerve and muscle function, can continue to operate normally. When dietary calcium is restored to normal, calcium is redeposited in bone. This calcium cycling is under the control of hormones, such as parathyroid hormone produced by the parathyroid glands (see Chapter 50). About 99% of all the calcium in a typical vertebrate's body exists in bone. This represents a huge reservoir of calcium ions for the blood.

Bone Consists of a Mixture of Organic and Mineral Components

Bone is a living, dynamic tissue with both organic and mineral components. Organic materials include cells that form bone—called osteoblasts—and cells that break it down—called osteoclasts. The organic part of bone is secreted by osteoblasts and consists of the protein collagen, which has a unique triple helical structure that gives bone both strength and flexibility. The mineral component is composed of a crystalline mixture of primarily Ca^{2+} and PO_4^{2-}, and other ions that provide bone its rigidity. These ions must be obtained in an animal's diet, absorbed into the blood, and deposited in bone.

A proper ratio of organic and mineral components is required for normal bone function. Bone lacking sufficient mineral, for example, is easily fractured. Bone is formed at high rates during an animal's growth periods, but even in adulthood, bone is continuously formed, broken down, and re-formed. The skeleton is continually changing—the one in your body right now is completely remodeled from the one that was in your body a few years ago. Similarly, the skeleton you will have a few years from now is not identical to the one you have today.

Bones cannot move by themselves but instead provide the scaffold upon which skeletal muscles act to cause body movement. We turn now to a discussion of skeletal muscle and the mechanism by which it generates force.

44.3 Skeletal Muscle Structure and the Mechanism of Force Generation

As mentioned, vertebrates have three types of muscle that are classified according to their structure, function, and control mechanisms. These are cardiac muscle, smooth muscle, and skeletal muscle. **Cardiac muscle** is found only in the heart and provides the force required for the heart to pump blood. Cardiac muscle will be discussed in Chapter 47. **Smooth muscle** surrounds and forms part of the lining of hollow organs and tubes, including those of the gastrointestinal system, urinary bladder, uterus, blood vessels, and airways. Contraction of the smooth muscle in hollow organs may propel the contents forward or churn them up, as when the stomach contracts after a meal. In other cases, smooth muscle regulates the flow of substances by changing the tube diameter, as in the widening or narrowing of small blood vessels. Smooth muscle contraction is not under voluntary control. Instead, it is controlled by the autonomic nervous system, hormones, and local chemical signals. Some smooth muscles, such as those found in regions of the digestive tract, have the ability to contract on their own, even in the absence of such signals.

Skeletal muscle is found throughout the body and is directly involved in locomotion. In vertebrates, but not invertebrates, skeletal muscle is electrically excitable—it can generate action potentials in response to a stimulus (invertebrate skeletal muscle cells have graded membrane potentials but do not have action potentials). The action potentials of vertebrate skeletal muscle cells result in increased levels of cytosolic calcium ions, which trigger force generation. Before understanding how this is possible, however, let's begin with an overview of skeletal muscle structure and function.

A Skeletal Muscle Is a Contractile Organ That Supports and Moves Bones

A **muscle** is a grouping of cells—called **muscle fibers**—bound together in structures called fascicles by a succession of connective tissue layers (**Figure 44.3**). Skeletal muscles are usually linked to bones by bundles of collagen fibers known as tendons. The transmission of force from contracting muscle to bone can be likened to a number of people pulling on a rope attached to a heavy object. Each person corresponds to a single muscle fiber, the rope corresponds to the tendons, and the bone is the heavy object.

Some tendons are very long, with the site of tendon attachment to bone far removed from the end of the muscle. For example, some of the muscles that move the fingers are in the forearm. You can wiggle your fingers and feel the movement of the muscles in your lower arm. These muscles are connected to the finger bones by long tendons.

Skeletal muscle fibers increase in size during growth from infancy to adulthood, but few new fibers are formed during that time. As described later, enlargement of skeletal muscles in adults, as from weightlifting, is also primarily a function of enlarged fibers, not the formation of new ones.

Muscle Fibers Contain Myofibrils Composed of Arrays of Filaments

The most striking microscopic feature of skeletal muscle is a series of light and dark bands perpendicular to the muscle's long axis. Because of this characteristic striped pattern, it is also

Figure 44.3 **Skeletal muscle structure.** Skeletal muscles attach to bone by tendons, which are bundles of collagen fibers. Each muscle consists of bundles of muscle fibers (cells) bound together by connective tissue in structures called fascicles.

Concept check: *What would happen to the ability of a muscle to move a bone if its tendon were torn, for example, due to injury?*

known as **striated muscle**. This pattern is the result of muscle fibers containing cylindrical bundles known as **myofibrils** (Figure 44.4). Most of the cytosol of a muscle fiber is filled with myofibrils, each of which extends from one end of the fiber to the other and is linked to the tendons at the ends of the fiber.

Each myofibril contains numerous functional structures called thick and thin filaments. These filaments are arranged in a repeating pattern running the length of the myofibril. One complete unit of this repeating pattern is known as a **sarcomere** (from the Greek *sarco*, meaning muscle, and *mer*, meaning part). The **thick filaments** are composed almost entirely of the motor protein **myosin**. The **thin filaments**, which are about half the diameter of the thick filaments, contain the cytoskeletal protein **actin**, as well as other proteins that we will discuss later.

The thick filaments are located in the middle of each sarcomere, where their orderly parallel arrangement produces a wide, dark band known as the **A band** (Figure 44.4). Each sarcomere contains two sets of thin filaments, each of which is anchored to a network of proteins at the **Z line**. The thin filaments overlap a portion of the thick filaments. Two successive Z lines define the boundaries of one sarcomere.

A light band known as the **I band** lies between the A bands of two adjacent sarcomeres. Each I band contains those portions of the thin filaments that do not overlap the thick filaments, and each I band is bisected by a Z line.

Two additional features are present in the A band of each sarcomere. The **H zone** is a narrow, light region in the center of the A band. It corresponds to the space between the two sets of thin filaments in each sarcomere. A narrow, dark band in the center of the H zone, known as the **M line**, corresponds to proteins that link the central regions of adjacent thick filaments.

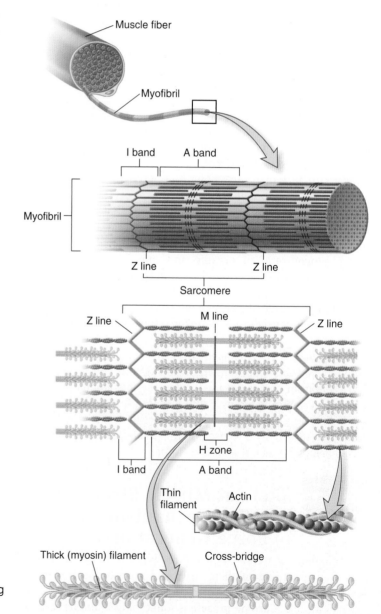

Figure 44.4 **Myofibril structure.** Each muscle fiber consists of numerous myofibrils containing thick and thin filaments. Their arrangement produces the striated banding pattern. Each repeating unit of the arrayed filaments constitutes a sarcomere.

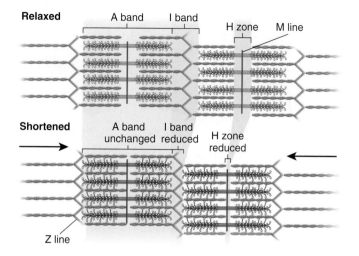

Relaxed

A band | I band | H zone | M line

Shortened

A band unchanged | I band reduced | H zone reduced

Z line

Figure 44.5 Sliding filament mechanism of muscle contraction. The sliding of thin filaments past the overlapping thick filaments shortens the sarcomere but does not change the lengths of the filaments themselves.

The spaces between overlapping thick and thin filaments are bridged by projections known as **cross-bridges**, which are regions of myosin molecules that extend from the surface of the thick filaments toward the thin filaments (Figure 44.4). As we see next, during muscle contraction, the cross-bridges make contact with the thin filaments and exert force on them.

Skeletal Muscle Shortens When Thin Filaments Slide over Thick Filaments

Movement requires shortening of muscles to pull against the attached tendons and bones. However, a muscle can generate

force, or contract, without producing movement. Holding a heavy weight at a constant position, for example, requires muscle contraction, but not muscle shortening. Therefore, as used in muscle physiology, the term contraction refers to activation of the cross-bridges within muscle fibers, which initiates the generation of force. When these mechanisms are turned off, contractions end, allowing the muscle fiber to relax.

How a muscle fiber actually shortens is known as the **sliding filament mechanism** of muscle contraction (**Figure 44.5**). In this mechanism, the sarcomeres shorten, but neither the thick nor the thin filaments change in length. Instead, the thick filaments remain stationary while the thin filaments slide, pulling on the Z lines and shortening the sarcomere.

Let's look at how the structure of the thin and thick filaments allows them to produce this sliding movement (**Figure 44.6**). Each actin molecule of a thin filament contains a binding site for myosin. Actin molecules form polymers that are arranged into two intertwined helical chains (Figure 44.6a). These chains are closely associated with two proteins called tropomyosin and troponin that play important roles in regulating contraction (which we will discuss a bit later).

In the thick filament, myosin is composed of six protein subunits: two large heavy chains and four smaller light chains. These polypeptides combine to form a protein that consists of two heads and a long tail formed by the two intertwined heavy chains. The tail of each myosin molecule lies along the axis of the thick filament, and the two globular heads extend out to the sides, forming the cross-bridges. The myosin proteins at the two ends of each thick filament are oriented in opposite directions, such that their tails are directed toward the center of the filament while the

Thin filament

Actin — Tropomyosin

Troponin

Thick filament — Cross-bridge

Actin-binding site
ATP-binding site
Heavy chains — Light chains
Myosin tails — Myosin heads

(a) Structure of thin and thick filaments

Thin filament | Thick filament | H zone

Relaxed

Z line

M line

Shortened | H zone

Successive movements of cross-bridges pull actin toward the H zone.

(b) Cross-bridge mechanism

Figure 44.6 Structure and function of the thin and thick filaments. (a) Thin filaments are composed of two intertwined actin molecules and associated proteins tropomyosin and troponin. Thick filaments are made of the protein myosin, which has two intertwined heavy chains, each with two associated light chains, an actin-binding site, and an ATP-binding site. The end of a myosin molecule that contains the actin-binding sites is bent at an angle to form the globular heads known as cross-bridges. (b) When the cross-bridges on myosin molecules bind to actin filaments, the thin filaments are pulled toward each other, shortening the sarcomere.

heads are at the ends. Each head contains two binding sites, one for actin and one for ATP. The hydrolysis of ATP provides the energy for the cross-bridge (that is, the myosin head) to move.

The sliding filament movement is propelled by the myosin cross-bridges (Figure 44.6b). During shortening, each cross-bridge attaches to an actin molecule in a thin filament and moves in a motion much like an oar on a boat. Because of the opposing orientation of the thick filaments, the movement of a cross-bridge forces the thin filaments toward the center of the sarcomere, thereby shortening it. One stroke of a cross-bridge produces only a very small movement of a thin filament relative to a thick filament. As long as a muscle fiber is stimulated to contract, however, each cross-bridge repeats its motion many times, resulting in continued sliding of the thin filaments. Thus, the ability of a muscle fiber to generate force and movement depends on the amount of interaction of actin and myosin.

The protein myosin was first discovered in extracts of frog leg muscle in the 1860s by the German physiologist Willi Kühne. We now know that eukaryotic genomes encode a family of related myosin proteins, as described next.

Genomes & Proteomes Connection

Did an Ancient Mutation in Myosin Play a Role in the Development of the Human Brain?

Myosins are among the most ancient of eukaryotic proteins and also among the most highly conserved in amino acid sequence. Nonetheless, small differences in the sequences of genes that arose from a single primordial myosin gene have led to tissue-specific expression of various myosin proteins, each with a characteristic ability to bind actin and hydrolyze ATP. One recently discovered member of this gene family, called *MYH16*, codes for a heavy chain of a myosin molecule that is expressed mainly in the muscles of the jaw in primates. In 2004, H. Stedman and colleagues discovered that this gene, although present in humans, did not code for a functional protein because of a mutation that deleted two base pairs from it. This mutation was present in 100% of people tested worldwide but was not found in the eight nonhuman primate species tested, which included chimpanzees.

Based on genetic comparisons and estimates of genetic divergence among species, the researchers estimated that the mutation occurred approximately 2.4 million years ago. Significantly, the genus *Homo*, with its smaller jaw and larger braincase, is also thought to have first appeared about 2.4 million years ago, leading Stedman to suggest that the loss of the MYH16 protein led to the smaller, less-muscular jaws characteristic of humans (Figure 44.7). Because jaw muscles are attached to skull bones, massive jaw muscles may have placed a mechanical constraint on skull growth in early hominins (as well as in modern nonhuman primates). Smaller, less-muscular jaws would place less strain on the forming skull and allow the brain to grow larger, expanding the braincase to the larger size seen in modern humans.

Macaque   **Gorilla**   **Human**

Figure 44.7   Comparison of jaw size and area of muscle attachment (shown in orange) in modern primates.   Note: Skulls are not to scale. (Reprinted by permission from Macmillan Publishers Ltd: Stedman, Hansell H. et al. Myosin gene mutation correlates with anatomical changes in the human lineage, *Nature*, Volume 428, Issue Number 6981, pages 415–418, 2004.)

Stedman's hypothesis is compelling, given that the myosin mutation occurs in jaw muscles and is present only in humans. Recently, though, another research group, using sophisticated statistical analyses of larger regions of gene sequences including the deletion in *MYH16*, concluded that the mutation may have arisen earlier—perhaps 4.0 to 5.4 million years ago, before the appearance of small-jawed species of *Homo*. Stedman's hypothesis will require confirmation by future research. Scientists are certain that a human-specific mutation in *MYH16* arose at some point in the evolution of hominins. Whether the mutation facilitated expansion of the cranium, which, in turn, permitted enlargement of the brain is unknown. The hypothesis does provide an intriguing piece of the puzzle of how modern humans may have evolved.

The Cross-Bridge Cycle Requires ATP and Calcium Ions

Let's now turn our attention to how actin and myosin interact to promote contraction and shortening. The sequence of events that occurs between the time when a cross-bridge binds to a thin filament and when it is set to repeat the process is known as a **cross-bridge cycle**.

Cross-bridge cycling begins whenever the Ca^{2+} level increases inside the cell. This usually occurs when neural input results in the release of Ca^{2+} from intracellular storage sites (described in detail shortly). In other words, the contraction of skeletal muscle fibers is under nervous control, a subject that we will return to in depth later.

The chemical and physical events that occur during the four steps of the cross-bridge cycle are illustrated in **Figure 44.8**. These steps are cross-bridge binding, power stroke (moving), detaching, and resetting. As the cycle begins, the myosin cross-bridges are in an energized state, which is produced by the hydrolysis of their bound ATP. In this state, the hydrolysis products, ADP and inorganic phosphate, remain bound to myosin. The sequence of storage and release of energy by myosin is analogous to the operation of a mousetrap: Energy is first stored in the trap by cocking the spring (ATP hydrolysis) and is then released by the springing of the trap (head binds to actin and moves in power stroke). Let's look at each step of the process more closely.

Figure 44.8 The four stages of cross-bridge cycling in skeletal muscle.

1 Binding: When Ca^{2+} levels are high, cross-bridge can bind to actin. (ADP + P_i are already bound to the cross-bridge.)

2 Power stroke: Release of P_i causes cross-bridge to move toward the H zone of the sarcomere, which would be to the left of this drawing. This power stroke moves the actin filament toward the H zone. ADP is then released.

3 Detaching: ATP binds to myosin, causing the cross-bridge to detach from the actin filament.

4 Resetting: Hydrolysis of ATP to ADP + P_i provides energy, which causes the cross-bridge to move away from the H zone. ADP and P_i remain bound to the cross-bridge. Cycle can begin again.

Concept check: Rigor mortis is a condition of contracted skeletal muscle that occurs shortly after death. Why would the muscle remain contracted after death, and why does such contraction eventually stop?

Step 1: Ca^{2+} levels rise, triggering the cross-bridge to bind to actin.

When Ca^{2+} levels rise, this triggers the first step in the cross-bridge cycle. In this step, an energized myosin cross-bridge, with its associated ADP and P_i, binds to an actin molecule on a thin filament.

$$\text{Actin} + [\text{Energized myosin} \cdot \text{ADP} \cdot P_i] \longrightarrow$$
$$[\text{Actin} \cdot \text{Myosin} \cdot \text{ADP} \cdot P_i \text{ complex}]$$

Step 2: Release of phosphate (P_i) fuels the power stroke: The cross-bridge and thin filaments move.

The binding of an energized myosin cross-bridge to actin in step 1 triggers the release of P_i. This release causes a conformational change in the shape of the myosin molecule. This causes the cross-bridge to rotate forward toward the M line at the center of the sarcomere. This process is known as the **power stroke**, which moves the actin filament. At the same time, ADP is released.

$$[\text{Actin} \cdot \text{Myosin} \cdot \text{ADP} \cdot P_i \text{ complex}] \longrightarrow$$
$$[\text{Actin} \cdot \text{Myosin complex}] + \text{ADP} + P_i$$

Step 3: ATP binds to myosin, causing the cross-bridge to detach.

During the power stroke, myosin is bound very firmly to actin; this linkage must be broken to allow the cross-bridge to be reenergized and repeat the cycle. The binding of a new molecule of ATP to the myosin cross-bridge breaks the link between actin and myosin.

$$[\text{Actin} \cdot \text{Myosin complex}] \ \text{ATP} \longrightarrow$$
$$\text{Actin} + [\text{Myosin} \cdot \text{ATP}]$$

The dissociation of actin and myosin by ATP is an example of allosteric regulation of protein activity (see Chapter 7). When ATP binds at one site on myosin, myosin's affinity for actin bound at another site decreases. ATP is not hydrolyzed in this step. Instead, ATP acts here only as an allosteric modulator of the myosin head, weakening the binding of myosin to actin and leading to their dissociation.

Step 4: ATP hydrolysis reenergizes and resets the cross-bridge.

After actin and myosin dissociate, the ATP bound to myosin is hydrolyzed. This hydrolysis reenergizes myosin, causing it to reset to the position that allows actin binding.

$$\text{Actin} + [\text{Myosin} \cdot \text{ATP}] \longrightarrow$$
$$\text{Actin} \ [\text{Energized myosin} \cdot \text{ADP} \cdot P_i]$$

If Ca^{2+} is still present at this time, the cross-bridge can reattach to a new actin molecule in the thin filament, and the cross-bridge cycle repeats, causing the muscle fiber to further shorten.

As we have seen, ATP performs two different roles in the cross-bridge cycle. First, the energy released from ATP hydrolysis provides the energy for cross-bridge movement. Second, ATP binding breaks the link formed between actin and myosin during the cycle, allowing the cycle to be repeated. Although the precise mechanisms may differ slightly between vertebrates and invertebrates, biologists think that all skeletal muscle functions according to steps similar to those just described. As mentioned, all skeletal muscle function requires the actions of calcium ions, which we describe next.

The Regulation of Muscle Contraction by Calcium Ions Is Mediated by Tropomyosin and Troponin

How does the presence of Ca^{2+} in the muscle fiber cytosol regulate the cycling of cross-bridges? The answer requires a closer look at the two additional thin filament proteins mentioned earlier, tropomyosin and troponin.

Tropomyosin is a rod-shaped molecule composed of two intertwined proteins, with a length equal to that of about seven actin molecules (**Figure 44.9a**). Chains of tropomyosin molecules are arranged end to end along the actin thin filament. In the absence of Ca^{2+}, they partially cover the myosin-binding site on each actin molecule, thereby preventing cross-bridges from making contact with actin. Each tropomyosin molecule is held in this blocking position by **troponin**, a smaller, globular-shaped protein that is bound to both tropomyosin and actin. One molecule of troponin binds to each molecule of tropomyosin. In this way, troponin and tropomyosin cooperate to block access to myosin-binding sites on actin molecules in the relaxed muscle fiber.

Troponin is capable of binding Ca^{2+}. Such binding produces a change in the shape of troponin, which—through troponin's linkage to tropomyosin—allows tropomyosin to move away from the myosin-binding site on each actin molecule (**Figure 44.9b**). This permits cross-bridge cycling to occur. Conversely, release of Ca^{2+} from troponin reverses the process, turning off contractile activity.

Thus, the concentration of cytosolic Ca^{2+} determines whether or not Ca^{2+} is bound to troponin molecules, which, in turn, determines the number of actin sites available for cross-bridge binding. The cytosolic concentration of Ca^{2+}, however, is very low in resting muscle. How is the Ca^{2+} concentration increased so that contraction can occur? This topic is discussed next.

Contraction of Skeletal Muscle Is Coupled with Electrical Excitation

Like neurons, vertebrate skeletal muscle cells are capable of generating and propagating action potentials in response to an appropriate stimulus. This causes a rise in the concentration of cytosolic Ca^{2+}, which triggers contraction of a muscle fiber. The sequence of events by which an action potential in the plasma membrane of a muscle fiber leads to cross-bridge activity by the mechanisms just described is called **excitation-contraction coupling**. The electrical activity in the plasma membrane does not act directly on the contractile proteins but instead acts as a stimulus to increase cytosolic Ca^{2+} concentration. These increased levels continue to activate the contractile apparatus long after electrical activity in the membrane has ceased.

The source of the increased cytosolic Ca^{2+} that occurs following a muscle action potential is the fiber's **sarcoplasmic reticulum**, which acts as a Ca^{2+} reservoir. The sarcoplasmic reticulum, which is a specialized form of the endoplasmic reticulum found in most cells, is composed of interconnected sleevelike compartments around each myofibril (**Figure 44.10**). Separate tubular structures, the **transverse tubules** (**T-tubules**), are invaginations of the plasma membrane that open to the extracellular fluid. The T-tubules run around each myofibril and conduct action potentials from the outer surface of the muscle fiber inside to the myofibrils. An action potential causes the opening of calcium channels in the sarcoplasmic reticulum, which allows Ca^{2+} to flow into the cytosol and bind to troponin, initiating muscle contraction.

A contraction continues until Ca^{2+} is removed from troponin. This is achieved by ATP-dependent ion pumps in the

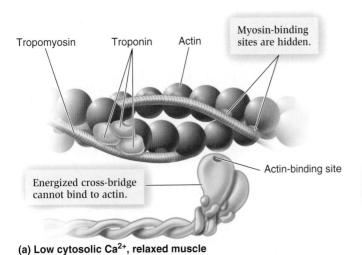

(a) Low cytosolic Ca^{2+}, relaxed muscle

Tropomyosin   Troponin   Actin

Myosin-binding sites are hidden.

Actin-binding site

Energized cross-bridge cannot bind to actin.

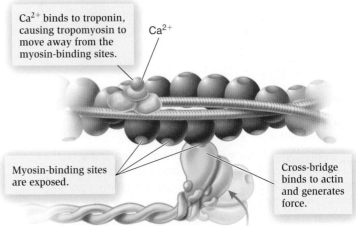

(b) High cytosolic Ca^{2+}, activated muscle

Ca^{2+} binds to troponin, causing tropomyosin to move away from the myosin-binding sites.

Ca^{2+}

Myosin-binding sites are exposed.

Cross-bridge binds to actin and generates force.

Figure 44.9   Role of Ca^{2+}, tropomyosin, and troponin in cross-bridge cycling.   **(a)** When cytosolic Ca^{2+} is low, the myosin-binding sites on actin are blocked by tropomyosin. **(b)** When cytosolic Ca^{2+} increases, Ca^{2+} binds to troponin, which, in turn, allows tropomyosin to move away from the myosin-binding sites on actin.

sarcoplasmic reticulum that lower the Ca^{2+} concentration in the cytosol back to its resting level.

Electrical Stimulation of Skeletal Muscle Occurs at the Neuromuscular Junction

We have seen that an action potential in the plasma membrane of a skeletal muscle fiber is the signal that triggers contraction. How are these action potentials initiated? The mechanism by which action potentials are initiated in a skeletal muscle involves stimulation by a motor neuron. These are neurons located in the central nervous system (CNS) that project their axons outside the CNS and directly or indirectly control muscles.

The junction of a motor neuron's axon and the muscle fiber is known as a **neuromuscular junction** (Figure 44.11). Near the surface of the muscle fiber, the axon divides into several short processes, or terminals, containing synaptic vesicles filled with the neurotransmitter acetylcholine (ACh) (see Chapter 41). The region of the muscle fiber plasma membrane that lies directly under the axon terminal is called the **motor end plate**;

it is folded into what are known as junctional folds, which contain many ACh receptors. These folds therefore increase the total surface area available for the membrane to respond to ACh. The extracellular space between the axon terminal and the motor end plate is called the synaptic cleft.

When an action potential in a motor neuron arrives at the axon terminal, it releases its stored ACh, which crosses the synaptic cleft and binds to receptors in the junctional folds of the muscle fiber. The ACh receptor is a ligand-gated ion channel (see Chapter 5). The binding of ACh causes an influx of Na^+ into the muscle fiber through a channel housed in the ACh receptor. This influx of Na^+ causes the muscle fiber to depolarize, resulting in an action potential that spreads along the membrane of the muscle fiber and through the T tubules. Most neuromuscular junctions are located near the middle of a muscle fiber, and newly generated muscle action potentials propagate from this region toward the ends of the fiber and throughout the T-tubule network. Overstimulation of a muscle fiber is prevented by the action of **acetylcholinesterase**. This enzyme breaks down excess ACh in the synaptic cleft to inactive forms that cannot bind the ACh receptor (Figure 44.11).

Figure 44.10 Structure and function of the sarcoplasmic reticulum, transverse tubules, and myofibrils in a skeletal muscle fiber.

(a) SEM of neuromuscular junction

(b) Structures of, and events at, the neuromuscular junction (only part of the motor neuron is shown)

Figure 44.11 **The neuromuscular junction.** The structure of a neuromuscular junction **(a)** as seen in a scanning electron micrograph and **(b)** as depicted schematically. Action potentials in the motor neuron cause exocytosis of ACh-containing synaptic vesicles. ACh binds to receptors in the plasma membrane of the junctional folds of the skeletal muscle cell. This initiates Na⁺ entry and, consequently, an action potential in the muscle cell. Excess ACh is removed by the enzyme acetylcholinesterase.

Concept check: Why does Na⁺ enter the muscle cell after the ACh receptor is activated?

Now that we have learned about the structural characteristics of skeletal muscle and the events that initiate and produce force generation, we turn to a discussion of how skeletal muscle is adapted to meet the varied functional demands of vertebrates.

44.4 Skeletal Muscle Function

Animals use skeletal muscle for a wide variety of activities, such as locomotion, stretching, chewing, breathing, and maintaining balance, to name a few. Therefore, it is not surprising that not all skeletal muscle fibers share the same mechanical

and metabolic characteristics. In this section, we will consider how different types of fibers can be classified on the basis of their rates of shortening (as either fast or slow) and the way in which they produce the ATP needed for contraction (as oxidative or glycolytic).

Skeletal Muscle Fibers Are Adapted for Different Types of Movement

Different muscle fibers contain forms of myosin that differ in the maximal rates at which they can hydrolyze ATP. This, in turn, determines the maximal rates of cross-bridge cycling and muscle shortening. Fibers containing myosin with low ATPase activity are called **slow fibers**. Those containing myosin with higher ATPase activity are classified as **fast fibers**. Although the rate of cross-bridge cycling is about four times faster in fast fibers than in slow fibers, the maximal force produced by both types of cross-bridges is approximately the same.

The second means of classifying skeletal muscle fibers is based on the type of metabolic pathways available for synthesizing ATP. Fibers that contain numerous mitochondria and have a high capacity for oxidative phosphorylation are classified as **oxidative fibers**. Most of the ATP production by such fibers depends on blood flow to deliver oxygen and nutrients to the muscle. Not surprisingly, therefore, these fibers are surrounded by many small blood vessels. They also contain large amounts of the oxygen-binding protein **myoglobin**, which increases the availability of oxygen in the fiber by providing an intracellular reservoir of oxygen. The large amounts of myoglobin present in oxidative fibers give these fibers a dark-red color. Oxidative fibers are often referred to as red muscle fibers. The benefit of red muscle fibers is they can maintain sustained action over a long period of time without fatigue.

By contrast, **glycolytic fibers** have few mitochondria but possess both a high concentration of glycolytic enzymes (see Chapter 7 for a discussion of glycolysis) and large stores of glycogen, the storage form of glucose. Corresponding to their limited use of oxygen, these fibers are surrounded by relatively few blood vessels and contain little myoglobin. The lack of myoglobin is responsible for the pale color of glycolytic fibers and their designation as white muscle fibers.

On the basis of these two characteristics, three major types of skeletal muscle fibers have been distinguished:

1. **Slow-oxidative fibers** have low rates of myosin ATPase activity but have the ability to make large amounts of ATP. These fibers are used for prolonged, regular types of movement, such as steady, long flight; long-distance swimming; or the maintenance of posture. These muscles, for example, are what give the red color to the dark meat of ducks, which use the muscles for flight. Long-distance runners have a high proportion of these fibers in their leg muscles. These types of activities require muscles that do not fatigue easily.

2. **Fast-oxidative fibers** have high myosin ATPase activity and can make large amounts of ATP. Like slow-oxidative fibers, these fibers do not fatigue quickly and can be used

for long-term activities. They are also particularly suited for rapid actions, such as the rapid trilling sounds made by the throat muscles in songbirds or the clicking sounds generated by the shaking of a rattlesnake's tail.

3. **Fast-glycolytic fibers** have high myosin ATPase activity but cannot make as much ATP as oxidative fibers, because their source of ATP is glycolysis. These fibers are best suited for rapid, intense actions, such as a runner's short sprint at maximum speed or a cat pouncing on its prey. Fast-glycolytic fibers fatigue more rapidly than oxidative fibers. The breast meat of chickens, for example, appears white because, unlike ducks, chickens do not fly except for very short distances and therefore do not require oxidative pectoral muscles. The fast-glycolytic muscles of chickens, however, are ideal for short flights in the air. These flights are useful for quickly escaping predators and breaking scent trails that would otherwise help predators in their chase. When they land, chickens use slow-oxidative fibers in their leg muscles to run long distances as they continue to elude a predator.

Different muscle groups within the animal body have different proportions of each fiber type interspersed with one another; many activities require the action of all three types of fibers at once. This is important when you consider the wide range of animal activities related to locomotion alone, including walking, climbing, running, swimming, flying, crawling, crouching, jumping, and maintaining balance and posture. Depending on the needs of an animal at any given moment, the motor nerve inputs can be adjusted to stimulate different ratios of fiber types. When you lift a heavy weight for a brief time, the motor units containing fast-glycolytic fibers in your arm muscles are activated in large numbers. When a crab uses its pincers to grab prey, fast-glycolytic muscles snap the claws closed quickly, but then slow-oxidative fibers maintain a tight grip for as long as needed. The characteristics of the three types of skeletal muscle fibers are summarized in **Table 44.1**.

Muscles Adapt to Exercise

The regularity with which a muscle is used, as well as the duration and intensity of the activity and whether it includes resistance, affects the properties of the muscle. For example, increased amounts of resistance exercise—such as weightlifting—results in hypertrophy (increase in size) of muscle fibers. Because the number of fibers in a muscle remains essentially constant throughout adult life, the increases in muscle size that occur with resistance exercise do not result from increases in

Table 44.1	Characteristics of the Three Types of Skeletal Muscle Fibers		
	Slow-oxidative	Fast-oxidative	Fast-glycolytic
Primary source of ATP production	Oxidative phosphorylation	Oxidative phosphorylation	Glycolysis
Mitochondria	Many	Many	Few
Blood supply	High	High	Moderate
Myoglobin content	High (red muscle)	High (red muscle)	Low (white muscle)
Rate of fatigue	Slow	Intermediate	Fast
Myosin-ATPase activity	Low	High	High
Rate of contraction	Slow	Fast	Fast

the number of muscle fibers, but instead from increases in the size of each fiber. These fibers undergo an increase in fiber diameter due to the increased synthesis of actin and myosin filaments, which form more myofibrils.

Exercise of relatively low intensity but long duration—popularly called aerobic exercise, including running and swimming—produces increases in the number of mitochondria in the fibers that are needed in this type of activity. In addition, the number of blood vessels around these fibers increases to supply the greater energy demands of active muscle. All of these changes increase endurance with a minimum of fatigue.

By contrast, short-duration, high-intensity exercise, such as weightlifting, primarily affects fast-glycolytic fibers, which are used during strong contractions. In addition, glycolytic activity is enhanced by elevated synthesis of glycolytic enzymes. The results of such high-intensity exercise are the increased strength and bulging muscles of a conditioned weight lifter. Such muscles, although very powerful, have little capacity for endurance and therefore fatigue rapidly.

A decline or cessation of muscular activity results in the condition called **atrophy**, a reduction in the size of the muscle. Likewise, if the neurons to a skeletal muscle are destroyed or the neuromuscular junctions become nonfunctional, the denervated muscle fibers will become progressively smaller in diameter. This condition is known as denervation atrophy. Even with an intact nerve supply, a muscle can atrophy if it is not used for a long period of time, as when a broken limb is immobilized in a cast.

The mechanism by which changes occur in skeletal muscle during exercise is an active area of research, but a recent discovery has provided an intriguing clue, as described next.

FEATURE INVESTIGATION

Evans and Colleagues Activated a Gene to Produce "Marathon Mice"

In the course of investigating possible ways to reverse or prevent obesity in humans and other mammals, Ron Evans and his

colleagues at the Salk Institute in California discovered one way in which the ratios of oxidative and glycolytic fibers change in skeletal muscle. Evans was interested in a gene that codes for a transcription factor called PPAR-δ. Activation of this protein results in the expression of genes that enable skeletal muscle or other cells to more efficiently burn fat instead of glucose for energy. Evans hypothesized that mice in which PPAR-δ was chronically activated at high levels would lose weight due to increased fat burning, as shown in **Figure 44.12.**

To test this hypothesis, Evans created transgenic mice (see Chapter 20) which contained the *PPAR-δ* gene in a modified form. The modified gene had a skeletal muscle-specific promoter so that it would be expressed only in skeletal muscle cells. The region of the gene that encoded PPAR-δ was linked to a region of another gene that encoded a viral protein domain called VP16. This domain also facilitates gene activation. The researchers expected the combination of PPAR-δ and VP16 to strongly activate genes that enable cells to more efficiently burn fat instead of glucose for energy.

In the first part of the experiment, Evans monitored the body weights of the transgenic mice after they reached adulthood. These mice gained significantly less weight compared to wild-type mice when fed high-fat diets (which normally cause mice—like humans—to gain weight), confirming Evans' hypothesis. As is sometimes the case in scientific discovery, an unexpected finding arose from this study when Evans examined several tissues in these mice. Under the microscope, he observed that the skeletal muscle of the transgenic mice showed a dramatic shift from glycolytic fibers to slow-oxidative fibers. The muscle in transgenic mice appeared redder than it did in wild-type mice. It contained more myoglobin and mitochondria and had higher levels of oxidative enzymes capable of providing the cells with sustained levels of ATP. These changes occurred even though the mice had not been subjected to exercise training.

Based on these observations, Evans also tested the hypothesis that the transgenic mice would have a greater capacity for prolonged exercise than wild-type mice. When the transgenic mice were challenged with an endurance exercise test, Evans discovered that they outperformed age- and weight-matched wild-type mice by a factor of nearly twofold! They could sustain a high level of activity on a miniature treadmill for nearly twice as long as wild-type mice (hence the nickname "marathon mice"). This effect occurred in transgenic mice even without prior exercise training. In other words, simply increasing the ratio of oxidative to glycolytic fibers gave the mice greater ability to sustain aerobic activity.

The results of these experiments indicate that elevating the levels of activated PPAR-δ facilitates an oxidative state in skeletal muscle fibers that somehow signals them to convert to types that are best suited for oxidative metabolism. Therefore, the switch in fiber type that occurs normally in exercise training may not require exercise *per se*, and it may be mediated in part by proteins that activate or induce *PPAR-δ* expression. These results may have important implications for exercise endurance in humans, as well as for possible treatments for various muscle diseases.

Experimental Questions

1. What is the normal function of the PPAR-δ protein in mice?

2. What was the hypothesis proposed by Evans in relation to PPAR-δ and obesity?

3. How did Evans and his colleagues test this hypothesis, and what did they observe?

Figure 44.12   Evans and colleagues' activation of a gene to produce "marathon mice."

HYPOTHESES   1. Increased expression of genes that lead to increased fat oxidation in skeletal muscle cells will prevent obesity in mice.
2. Transgenic mice have a greater capacity for prolonged exercise than do wild-type mice.

KEY MATERIALS   Mice, light and electron microscopes, motorized treadmills.

	Experimental level	Conceptual level
1 Prepare a modified gene containing a skeletal muscle-specific promoter and a coding sequence that links *VP16* and *PPAR-δ*. See Chapter 20 for gene cloning methods.	Skeletal muscle-specific promoter VP16 PPAR-δ	Skeletal muscle-specific promoter ensures gene is turned on only in skeletal muscle. VP16 domain is a domain that always activates transcription. *PPAR-δ* codes for a transcription factor that specifically activates genes that allow cells to efficiently burn fat.
2 Make transgenic mice expressing the *VP16–PPAR-δ* gene.	See Chapter 20 for a discussion of gene addition.	All of the cells will carry this gene, but only skeletal muscle cells will express the gene.

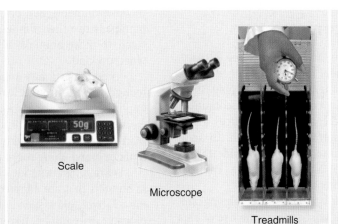

3 Perform the following tests:

(a) Feed wild-type control mice and transgenic mice a normal-fat diet (4%) and then switch to a high-fat diet (35%). Weigh mice weekly.
(b) Examine the muscle fibers in the mice.
(c) Test their endurance on a treadmill.

Scale

Microscope

Treadmills

(a) Eating a high-fat diet is known to cause obesity in mice and other mammals.
(b) The appearance of skeletal muscle can be examined by light and electron microscopy.
(c) The treadmills are motorized to keep mice moving until they become exhausted, at which time the treadmills are stopped.

4 THE DATA

Characteristics of skeletal muscle in transgenic mice:

– redder than wild type
– more myoglobin
– more mitochondria
– more slow-oxidative fibers

(a) Weight gain

(b) Difference in skeletal muscle

(c) Muscle endurance

5 CONCLUSION PPAR-δ contributes to both weight loss and endurance in mice. The fiber-type switching associated with exercise does not require exercise, because increasing fat oxidation in skeletal muscle cells resulted in more oxidative fibers even without exercise training.

6 SOURCE Wang, Y.X. et al. 2004. Gene targeting turns mice into long-distance runners. *Public Library of Science Biology* 2:322.

Skeletal Muscles Can Bend or Straighten a Limb

As mentioned previously, contracting muscle exerts a force on bones through its connecting tendons. When the force is great enough, the bone moves as the muscle shortens. A contracting muscle exerts only a pulling force, so as the muscle shortens, the attached bones are pulled toward or away from each other. Muscles that bend a limb at a joint (that is, reduce the angle between two bones) are called **flexors**, whereas muscles that straighten a limb (increase the angle between two bones) are called **extensors**. Groups of muscles that produce oppositely directed movements at a joint are known as **antagonists**. For example, in Figure 44.13, we can see that contraction of the hamstrings flexes the leg at the knee joint, whereas contraction of the antagonistic muscle, the quadriceps, causes the leg to extend. Both antagonistic muscles exert only a pulling force when they contract.

Figure 44.13 Actions of flexors and extensors. The figure shows how muscles cause flexion or extension of a limb in an animal with an endoskeleton. When the flexor muscle contracts, the extensor relaxes, and vice versa.

44.5   Animal Locomotion

Locomotion is the movement of an animal from place to place. Animal locomotion occurs in several forms, including walking/running, swimming, flying, crawling/sliding, or jumping. In all cases, animals experience certain constraints to locomotion. For example, all animals must overcome frictional forces (drag) generated by the air, water, or surface of the Earth. In addition, all forms of locomotion require energy to provide thrust, defined as the forward motion of an animal in any environment, and/or lift, which is movement against gravity.

Although the precise mechanism may differ among animals, locomotion with few exceptions (such as the use of cilia in ctenophores) results from muscular contractions that exert force on one of the three types of skeletons discussed at the beginning of this chapter. In this section, we will examine the similarities and differences between locomotion in water, on land, and in air.

Aquatic Animals Must Overcome the Resistance of Water

The greatest challenge to locomotion in water is the density of water, which is much greater than that of air. This is apparent when you compare waving your hand through the air and underwater. Water's resistance to movement increases exponentially as the speed of locomotion increases, which is one reason why many fishes swim at relatively slow speeds. Overcoming this resistance requires considerable muscular effort. Most swimming animals, including fishes, amphibians, reptiles, diving birds, and marine mammals, have evolved streamlined bodies that reduce drag and so make swimming more efficient. Other animals that swim only occasionally or spend time on the surface of the water have evolved adaptations, such as the webbed feet of ducks, that assist their muscles in generating greater thrust through the water.

Although the density of water creates challenges to locomotion, it also provides certain benefits. An energetic advantage to swimming is that fishes and other swimmers do not need to provide as much lift to overcome gravity. Because the density of water is similar to that of an animal's body, water provides buoyancy, which helps support the animal's weight.

The mechanism of swimming is similar among many different vertebrates. Most fishes, for example, contract posterior skeletal muscles to move the tail end of the animal from side to side. This pushes water backward and propels the fish forward. Other muscles and fins provide additional thrust and enable changes in direction. Likewise, amphibians and marine reptiles rely predominantly on their hind legs, their tail, or undulations of the posterior parts of the body for propulsion through the water. Cetaceans use up-and-down thrusts of their tail flukes to provide propulsion. Confining most of the swimming muscles to the rear of an animal's body has certain advantages. With the rear end devoted to movement, the front end is free to explore the environment, fight off aggressors, or find food. By contrast, pinnipeds use their flippers to provide thrust and their tails to steer. In these animals, a flexible tail assists with locomotion when on land.

In contrast to swimming vertebrates, many aquatic invertebrates move through water by means other than swimming. Cephalopods such as squids use propulsion provided by the ejection of water to give them brief but rapid bursts of speed. Many other aquatic invertebrates move passively on water currents or crawl along rocks and other underwater surfaces.

Swimming is energetically the cheapest form of locomotion in animals that are adapted to swimming, due to streamlining, the relatively slow speed of most swimmers, and the buoyancy of water. By contrast, terrestrial animals face considerable energetic costs to locomotion, as we examine next.

Walking and Running on Land Are Energetically Costly

In contrast to swimming, locomotion on land is, on average, the most energetically costly means of locomotion (Figure 44.14). Flying may seem more costly, but it is not. Many migratory birds can travel hundreds of miles daily for many days. A recent study by researchers in New Zealand, using satellite transmitters, tracked the nonstop, week-long migration of four bar-tailed godwits (*Limosa lapponica*) over a distance of roughly 10,000 km, from New Zealand to the coast of South Korea. No terrestrial animal could possibly match such a feat by walking or running. (To grasp how astonishing this feat is, imagine running from New York City to Los Angeles and back again in 7 days without any rest or food!)

Whereas gravity is not an important factor for locomotion in swimming animals, terrestrial animals must overcome

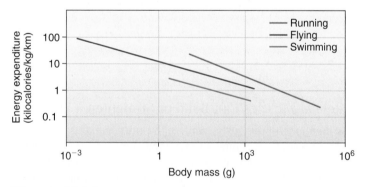

Figure 44.14   Energy costs of locomotion.   The energy costs of different modes of locomotion for animals of different sizes are shown. The y-axis gives the energy costs expressed as kilocalories expended per kilogram (kg) per kilometer (km). Energy costs are highest for runners compared to similarly sized fliers and swimmers. Note: Only a portion of the full range of body sizes of swimmers is shown.

Concept check: *Does this graph indicate that smaller animals expend more energy than larger animals when moving a similar distance?*

gravity each time they take a step. Of even greater importance to walking and running animals, though, is the requirement to accelerate and decelerate the limbs with every step. In essence, each step is like starting a movement from scratch, without the luxury of occasionally gliding through water or air as fishes and birds do. This challenge is even greater when an animal moves uphill or over rough terrain.

Apart from gastropods, which move along the surface of the Earth on a layer of secreted mucus, and snakes, which undulate along the ground on a portion of their ventral body surface, most terrestrial animals limit the amount of contact with the ground while moving, thereby minimizing the amount of friction they encounter. Tetrapods usually have only two feet on the ground at any time when walking, but for brief moments, an animal such as a horse galloping at full speed has all four feet off the ground.

Having fewer legs on the ground surface at any one time helps increase speed but can compromise stability. Arthropods, for example, have at least six legs. This apparently provides excellent stability but reduces maximal speed (although cockroaches can attain rapid speeds by running on only two legs). At the other extreme are animals that move by jumping—fleas, certain spiders, click beetles, grasshoppers, frogs, and kangaroos.

Flying Has Evolved in Four Different Lineages

Flying is a highly successful means of locomotion and is hypothesized to have evolved in four different lineages: pterosaurs (extinct reptiles), insects, birds, and mammals (bats). Flying provides numerous advantages. Animals can escape land-based predators, scan their surroundings over great distances, and inhabit environments such as high cliffs that may be inaccessible to nonflying animals. The mechanics of flying, however, require animals to overcome gravity and air resistance, which makes flying more energetically costly than swimming but still less costly than running on land (see Figure 44.14). As with swimming, resistance to flight is decreased by streamlined bodies. However, earthbound animals have one advantage over their flying cousins—they can grow to much larger sizes than animals that fly. The vast majority of flying animals have a mass between about 1 mg and 1 kg. Only a few large birds have masses exceeding 10 kg. Although this represents an enormous range, it falls far short of the sizes achieved by earthbound or aquatic animals.

In flying vertebrates, lift and thrust are provided by pectoral and other muscles that move the wings. The pectoral muscles are so powerful and massive that they constitute as much as 15–20% of a bird's total body mass and up to 30% in hummingbirds, which use their wings not only to fly but also to hover. The requirement for large, strong pectoral muscles is one reason why the remaining body mass of flying vertebrates is limited. The extinct pterosaurs would seem to be an exception because some species were known to have had wingspans of nearly 10 m. However, scientists think that these large animals were unable to generate the force required to lift their massive bodies off the Earth and instead glided off of trees or cliffs to fly.

In birds and bats, the wings are modifications of the forelimbs. In general, bat wings are far more maneuverable than bird wings, because unlike birds, bats have digits at the end of their forelimbs/wings. This allows bats to precisely alter the shape of their wings and provide for rapid, fine-tuned changes in direction, even at high speeds. Bird wings are more similar to those of a fixed-wing aircraft. The largest birds, such as hawks and eagles, are able to glide because of the great surface area of their large wings. By using a bird's momentum to propel it forward, gliding provides a considerable energy savings. Bats and small birds, however, can glide for only very brief moments.

In this chapter, we have seen how skeletal muscles and (in vertebrates) bones work together to provide animals with protection and to enable them to move around in their environments. Unfortunately, muscles and bone are frequently the sites of significant diseases with great impacts on human health. We conclude the chapter with a discussion of a few of these diseases, each resulting from a different cause.

44.6 Impact on Public Health

A number of diseases can affect bone structure and function in humans. Bone disease may involve defects in either the mineral or organic components of bone. Poor bone formation and structure may result from inadequate nutrition, hormonal imbalances, aging, or skeletal muscle atrophy, to name just a few of the common causes.

In addition, many diseases or disorders directly affect the contraction of skeletal muscle. Some of them are temporary and not serious, such as muscle cramps, whereas others are chronic and severe, such as the disease muscular dystrophy. Also, some diseases result from defects in parts of the nervous system that control contraction of the muscle fibers rather than from defects in the fibers themselves. One example is poliomyelitis (polio), a viral disease in which the destruction of motor neurons leads to skeletal muscle paralysis that may result in death from respiratory failure. Polio, once thought to be nearly eliminated from the human population, is still a threat in parts of the world.

Other diseases that affect skeletal muscle function result when normal processes go awry. For example, the system that normally protects the body from invaders may turn on itself, or faulty genes may produce an abnormal or absent protein. This section looks at a few of these conditions in more detail.

Rickets and Osteoporosis Affect the Bones of Millions of People

Bone diseases are fairly common, particularly among individuals over age 50. Two major abnormalities can occur in bone. The first is improper mineral deposition in bone, usually due to inadequate dietary calcium intake or inadequate absorption of calcium from the small intestine. Without adequate minerals, bone becomes soft and easily deformed, as occurs in the weight-bearing bones of the legs of children with **rickets** (or

osteomalacia, as it is called in adults) (**Figure 44.15a**). These disorders are best prevented or treated with vitamin D, because this vitamin is the most important factor in promoting absorption of calcium from the small intestine.

The second major abnormality we will discuss is a more common disease called **osteoporosis**, in which both the mineral and organic portions of bone are reduced (**Figure 44.15b**). This disease, which affects four times as many women as men, occurs when the normal balance between bone formation and bone breakdown is disrupted.

One cause of osteoporosis is prolonged disuse of muscles. In ways that are not completely clear, the force produced by active skeletal muscle contractions helps maintain bone mass. When muscles are not or cannot be used—due to paralysis or long-term immobilizing illnesses or even during prolonged space flight in which the lack of gravity reduces the need for muscles to work hard—bone mass declines.

More commonly, osteoporosis may result from hormonal imbalances. Some hormones—for example, estrogen—stimulate bone formation. When estrogen levels decline after menopause (the period of time when a woman's reproductive cycles cease), bone density may decline, increasing the risk of bone fractures. By contrast, some hormones—such as parathyroid hormone (see Chapter 50)—act to demineralize bone as part of the way in which the body normally maintains mineral homeostasis in the blood. If such hormones are present in excess, however, they can cause enough demineralization of bone to result in osteoporosis. This may happen in rare cases when the glands that make these hormones malfunction and overproduce the hormones.

Osteoporosis can be minimized with adequate calcium and vitamin D intake and weight-bearing exercise programs. In some cases, postmenopausal women may be given estrogen to replace what their bodies are no longer producing, but this therapy is controversial due to the potential adverse effects of estrogen on cardiovascular health and breast cancer in some

women. Osteoporosis is the most prevalent bone disease in the U.S., affecting up to 15–30 million individuals. It results in annual national expenditures of approximately $15–20 billion in hospital and other medical costs.

Myasthenia Gravis Is an Autoimmune Disease That Affects Skeletal Muscles

Myasthenia gravis is a disease characterized by skeletal muscle fatigue and weakness. It ranges in severity, but at its worst, it can make chewing, swallowing, talking, and even breathing difficult. It affects an estimated 10,000 to 30,000 Americans, usually not becoming apparent until adolescence.

In myasthenia gravis, skeletal muscle function is reduced because the body's immune system produces antibodies—proteins that attack foreign matter—that bind to and inactivate ACh receptors on skeletal muscle cells. Because the destruction of ACh receptors is brought about by the body's own defense mechanisms gone awry, myasthenia gravis is considered an autoimmune disease (see Chapter 53). Thus, although motor nerve signals to the muscles are normal, the muscles cannot fully respond to the ACh released from the nerves because the receptors are not functional.

A number of approaches are currently used to treat the disease. One is to administer inhibitors of acetylcholinesterase, the enzyme that normally breaks down ACh at the neuromuscular junction. The inhibition of this enzyme prolongs the time that acetylcholine is available at the synapse, which to some degree compensates for reduced ACh receptor abundance. Other therapies are directed at suppressing the immune system to prevent destruction of the ACh receptors by antibodies. This is accomplished either by surgically removing certain glands that constitute part of the immune system or with drug therapy, but it has the disadvantage of making the person more susceptible to infections. Another therapy, called plasmapheresis, involves removing the liquid fraction of blood (plasma), which contains antibodies, and reinfusing the blood after the antibodies have been removed. A combination of these treatments has greatly reduced the mortality rate for myasthenia gravis.

Muscular Dystrophy Is a Rare Genetic Disease That Causes Muscle Degeneration

The group of diseases collectively called muscular dystrophy affects 1 of every 3,500 American males; it is much less common in females. **Muscular dystrophy** is associated with the progressive degeneration of skeletal (and cardiac) muscle fibers, weakening the muscles and leading ultimately to death from lung or heart failure. The signs and symptoms become evident at about 2 to 6 years of age, and most affected individuals do not survive beyond the age of 30.

The most common form of muscular dystrophy, called Duchenne muscular dystrophy, is a sex-linked recessive disease resulting from a defective gene on the X chromosome. Females have two X chromosomes, and males only one (plus one Y). Therefore, a heterozygote female with one abnormal

(a) X-ray image of leg bones of a child with rickets

(b) Histologic appearance of normal bone (top) and bone from a person with osteoporosis (bottom)

Figure 44.15 Deformations in human bones.

and one normal gene will not generally develop the disease. In Duchenne muscular dystrophy, the affected gene codes for a protein known as dystrophin, which is absent in patients with the disease. Dystrophin is a large protein that links cytoskeletal proteins to the plasma membrane. Scientists think it is involved in maintaining the structural integrity of the plasma membrane in muscle fibers. In the absence of dystrophin, the plasma membrane of muscle fibers is disrupted, causing extracellular fluid to enter the cell. Eventually, the cell ruptures and dies.

 # Summary of Key Concepts

44.1 Types of Animal Skeletons

- Skeletons are structures that provide support and protection and also function in locomotion.

- Three types of skeletons are found in animals. In hydrostatic skeletons, found in many soft-bodied invertebrates, muscle contractions acting on fluid-filled body cavities create differences in hydrostatic pressure that support the body and generate movements. Exoskeletons, which are found in arthropods, are protective external structures that must be shed to accommodate growth of the animal. Endoskeletons are internal structures that grow with the animal but do not protect its body surface; endoskeletons are found in some species of sponges and all echinoderms and vertebrates. (Figure 44.1)

44.2 The Vertebrate Skeleton

- Vertebrate endoskeletons are made up of the axial and appendicular skeletons. A joint is formed where two or more bones of a vertebrate endoskeleton come together. (Figure 44.2)

- In addition to the functions of support, protection, and locomotion, the vertebrate skeleton produces blood cells and constitutes a reservoir for ions that are crucial to homeostasis.

44.3 Skeletal Muscle Structure and the Mechanism of Force Generation

- A skeletal muscle is a grouping of cells, called muscle fibers, bound together by connective tissue layers. The striations of skeletal muscle result from the presence within muscle fibers of cylindrical bundles known as myofibrils, each of which contains thick and thin filaments arranged in repeating units called sarcomeres. The arrangements of thick and thin filaments within each sarcomere create various light and dark regions designated the A and I bands, respectively; the Z and M lines; and the H zone. Thick filaments are composed almost entirely of the motor protein myosin, whereas thin filaments contain the cytoskeletal protein actin plus two other proteins called troponin and tropomyosin. Portions of the thick filaments are called cross-bridges. (Figures 44.3, 44.4)

- During muscle contraction, the sarcomeres shorten by a process known as the sliding filament mechanism. In muscle contraction, the thick filaments remain stationary while the thin filaments slide past them propelled by action of the cross-bridges. (Figures 44.5, 44.6)

- Mutations in a myosin gene of the jaw muscle may have allowed the human brain to become larger. (Figure 44.7)

- The repeated interaction of cross-bridges on the thick filaments and the protein components of the thin filaments occurs according to a process called the cross-bridge cycle. The steps of the cycle are binding, power stroke, detaching, and resetting. (Figure 44.8)

- Proteins associated with the thin filament called tropomyosin and troponin, along with Ca^{2+}, play a critical role in the regulation of muscle contraction. (Figure 44.9)

- Ca^{2+} also plays an important role in excitation-contraction coupling, the sequence of events by which an action potential in the plasma membrane of a muscle fiber leads to cross-bridge activity.

- The source of the cytosolic Ca^{2+} involved in a muscle fiber action potential is the muscle fiber's sarcoplasmic reticulum. Tubular structures called the transverse tubules (T-tubules) conduct action potentials from the plasma membrane at the outer surface of the muscle fiber to the myofibrils. (Figure 44.10)

- Electrical stimulation of skeletal muscle occurs at a neuromuscular junction, in which a motor neuron's axon and a muscle fiber are in close proximity. (Figure 44.11)

44.4 Skeletal Muscle Function

- Three major types of skeletal muscle fibers have been distinguished. Slow-oxidative fibers have low rates of myosin ATP hydrolysis; they do not fatigue easily and are used for prolonged, regular activities. Fast-oxidative fibers have high myosin activity, do not fatigue quickly, and are particularly suited for rapid actions. Fast-glycolytic fibers have high myosin activity but cannot make as much ATP as oxidative fibers; they are best suited for rapid, intense actions. (Table 44.1)

- Increased expression of *PPAR-δ* in mice results in increased slow-oxidative muscle, greater exercise endurance, and weight loss. (Figure 44.12)

- Groups of muscles that produce oppositely directed movements at a joint are known as antagonists. Muscles that bend a limb at a joint are called flexors, whereas muscles that straighten a limb are called extensors. (Figure 44.13)

- To produce movement, muscles, bones, and joints are arranged in lever systems.

44.5 Animal Locomotion

- Locomotion is the movement of an animal from place to place.

- Due to streamlining, the relatively slow speed of most swimmers, and the buoyancy of water, swimming is energetically the most efficient form of locomotion. Locomotion on land is, on average, the most energetically costly means of locomotion. The energy expenditure required for flight is intermediate between those for swimming and land-based locomotion. (Figure 44.14)

44.6 Impact on Public Health

- Several health conditions affect muscular-skeletal structures. Rickets (osteomalacia in adults) is characterized by soft, deformed bones, usually resulting from insufficient dietary intake or absorption of calcium. In osteoporosis, bone density is reduced when bone formation fails to keep pace with normal bone breakdown. Myasthenia gravis is an autoimmune disease characterized by skeletal muscle fatigue and weakness. Muscular dystrophy is an ultimately fatal genetic disease associated with the progressive degeneration of skeletal and cardiac muscle fibers. (Figure 44.15)

Assess and Discuss

Test Yourself

1. The hydrostatic skeleton common in soft-bodied invertebrates is composed of
   a. muscle and cartilage.
   b. cartilage and a neural net.
   c. muscle and water-based fluid.
   d. cartilage and water-based fluid.
   e. bone and muscle.

2. Which of the following is not a function of the vertebrate skeleton?
   a. structural support
   b. protection of internal organs
   c. calcium reserve
   d. blood cell production
   e. All of the above are functions of the vertebrate skeleton.

3. The protein that provides strength and flexibility to bone is
   a. actin.   c. myoglobin.   e. elastin.
   b. myosin.   d. collagen.

4. In a sarcomere, the _____ contain(s) thin filaments and no thick filaments.
   a. A band   c. I band   e. both a and d
   b. M line   d. H zone

5. The function of ATP during muscle contraction is to
   a. cause an allosteric change in myosin so it detaches from actin.
   b. provide the energy necessary for the movement of the cross-bridge.
   c. expose the myosin-binding sites on the thin filaments.
   d. do all of the above.
   e. do a and b only.

6. The function of calcium ions in skeletal muscle contraction is to
   a. cause an allosteric change in myosin so it detaches from actin.
   b. provide the energy necessary for the movement of the cross-bridge.
   c. expose the myosin-binding sites on the thin filaments.
   d. bind to tropomyosin.
   e. do a and c only.

7. Stimulation of a muscle fiber by a motor neuron occurs at
   a. the neuromuscular junction.
   b. the transverse tubules.
   c. the myofibril.
   d. the sarcoplasmic reticulum.
   e. none of the above.

8. Muscle fibers that have a high number of mitochondria, contain large amounts of myoglobin, and exhibit low rates of ATP hydrolysis are called _____ fibers.
   a. slow-glycolytic   d. fast-oxidative
   b. fast-glycolytic   e. slow-oxidative
   c. intermediate

9. Which of the following statements about movement and locomotion is incorrect?
   a. Terrestrial animals and flying animals expend energy to provide lift.
   b. Swimming animals expend energy to provide thrust but not lift.
   c. Flexors and extensors are examples of muscles called agonists.
   d. Flexors cause bending at a joint.
   e. Extensors cause straightening of a limb.

10. For animals adapted to it, swimming is energetically the cheapest type of locomotion because of
    a. streamlined body forms in aquatic organisms.
    b. slow speed of movement of some swimmers.
    c. buoyancy of water.
    d. a and c only.
    e. a, b, and c.

Conceptual Questions

1. Define locomotion, and list some of the various types that occur in animals.

2. Distinguish between exoskeletons and endoskeletons.

3. List and briefly describe the steps in the cross-bridge cycle.

Collaborative Questions

1. Discuss the structure and function of the three different types of skeletons found in animals.

2. Discuss the three types of muscle tissues found in vertebrates.

Online Resource

www.brookerbiology.com

Stay a step ahead in your studies with animations that bring concepts to life and practice tests to assess your understanding. Your instructor may also recommend the interactive ebook, individualized learning tools, and more.

Nutrition, Digestion, and Absorption

45

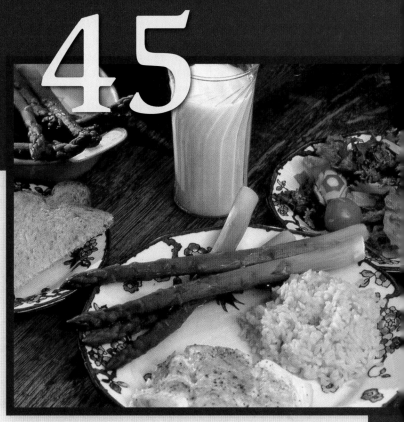

A balanced meal containing many nutrients, including carbohydrates, lipids, proteins, vitamins, minerals, and water.

What would 50,000,000 Calories (or more precisely, kilocalories) worth of food look like if it were assembled all in one place? That is roughly the amount of calories that you will consume in your lifetime. All that energy is required for the trillions of body cells to perform numerous activities such as synthesizing proteins, making cellular organelles, and maintaining concentration gradients of ions across the plasma membrane. These and other activities are necessary for homeostasis and require a lifelong supply of energy and building blocks. These materials must be consumed by animals in the form of nutrients. A **nutrient** is any organic or inorganic substance that is taken in by an organism and is required for survival, growth, development, tissue repair, or reproduction; the process of consuming and using food and nutrients is called nutrition. All organisms require nutrients to survive. Animals receive their nutrients by consuming food, as shown by the nutrionally balanced meal in the chapter-opening photo.

The process of food use in animals occurs in four phases: ingestion, digestion, absorption, and elimination (**Figure 45.1**). **Ingestion** is the act of taking food into the body via a structure such as a mouth. From there, the food moves into a digestive cavity or canal. If the nutrients in food are in a form that cannot be directly used by cells, they must be broken down into smaller molecules, a process known as **digestion**. This is followed by the process of **absorption**, in which ions, water, and small molecules diffuse or are transported from the digestive cavity into an animal's circulatory system or body fluids. **Elimination** is the process of passing undigested material out of the body.

In this chapter, we will look at the nutrients that animals require and the diverse ways in which animals obtain, digest, and absorb those nutrients. We will also discuss how these processes are regulated by the body and some of the common ways in which they may go awry. We begin by examining the types of nutrients animals consume.

Digestive canal

Food

Blood (circulatory system)

1 **Ingestion:** Food is consumed.

2 **Digestion:** Nutrients are broken down into smaller molecules in the digestive canal.

3 **Absorption:** Ions, water, and small molecules diffuse or are transported out of the digestive canal into the body fluids.

4 **Elimination:** Undigested waste products are excreted.

Figure 45.1   An overview of the four phases of food use in animals.

45.1 Animal Nutrition

Animals require both organic and inorganic nutrients. Organic nutrients fall into five categories: carbohydrates, proteins, lipids, nucleic acids, and vitamins. These provide energy and the building blocks of new molecules or serve as cofactors in many enzymatic reactions. Inorganic nutrients include water and minerals such as calcium, copper, and iron. The importance of water to life was covered in Chapter 2. Minerals may function as cofactors in enzymatic reactions and other processes.

Animals have many nutritional requirements in common, because of similarities in the way their bodies work. However, nutritional demands differ depending upon an animal's physiology. The digestive systems of **herbivores**, animals that eat only plants, contain microbes that assist in the digestion of cellulose, so these animals are well adapted to subsist on plants. By contrast, **carnivores** are primarily adapted to consume animal flesh or fluids, and **omnivores**, such as humans, have the ability to survive on both plant and animal products.

The amount of each type of nutrient required by an animal may also differ, depending on its activity level, or metabolic rate, which we will discuss in Chapter 46. Generally, highly active and energetic animals, such as most birds and mammals, require a proportionally greater amount of nutrients each day than do relatively inactive or sedentary animals, such as nonmotile invertebrates. However, exceptions to this general rule exist, as when some mammals enter hibernation and reduce their energy use. In this section, we will examine the various organic and inorganic nutrients consumed by animals and some of the major functions of these molecules.

Animals Require Nutrients for Energy and the Synthesis of New Molecules

Ingested organic macromolecules are used for two general purposes: to provide energy or to make new molecules. As described in Chapter 6, the bonds of organic molecules may be broken down to release energy; this energy can be used in the synthesis of ATP, the energy source of all cells. Indirectly, therefore, organic molecules provide the energy required for most of the chemical reactions that occur in animals' bodies. Alternatively, organic molecules can serve as building blocks to synthesize new cellular molecules. For example, all animals that have muscles use amino acids to make the same specialized proteins that allow their muscles to contract.

Although different species consume a wide variety of foods, all animals require the same fundamental organic macromolecules (**Table 45.1**). Carbohydrates supply energy-yielding glucose and carbon. Proteins supply amino acids that, in addition to building protein, can be used as an energy source. Lipids supply components for membrane-building and thermal insulation and also provide energy. Nucleic acids supply some of the components required for DNA, RNA, and ATP synthesis. The other category of organic nutrients—the vitamins—are required in small quantities and function as coenzymes in various reactions, as described later.

Essential Nutrients Must Be Obtained from the Diet

Certain compounds cannot be synthesized from any ingested or stored precursor molecule. These **essential nutrients** must be obtained in the diet in their complete form. The word essential in this context refers to the fact that they must come from food. It does not mean that other nutrients are less important, because many nutrients are required for an animal's survival. The essential nutrients can be classified into four groups: essential amino acids, essential fatty acids, vitamins, and minerals.

Essential Amino Acids Eight **essential amino acids** are required in the diet of humans and many other animals—isoleucine, leucine, lysine, methionine, phenylalanine, threonine, tryptophan, and valine. These amino acids are required for building proteins but cannot be synthesized by the animal's cells. Also, animal cells do not store amino acids. Therefore, without a recurring supply of these eight amino acids, protein synthesis in each cell in an animal's body would slow down or stop completely. Carnivores and omnivores readily obtain

Table 45.1	Major Nutrients in Animals, Their Dietary Sources, and Some of Their Functions and Symptoms of Deficiency in Vertebrates		
Class of nutrient	**Dietary sources**	**Functions in vertebrates**	**Symptoms of deficiency**
Carbohydrates	All food sources, especially starchy plants	Energy source; component of some proteins; source of carbon	Muscle weakness; weight loss; ketone formation (decreases pH of blood)
Proteins	All food sources, especially meat, legumes, cereals, roots	Provide amino acids to make new proteins; build muscle; some amino acids used as energy source	Weight loss; muscle loss; weakness; weakened immune system; increased likelihood of infections
Lipids	All food sources, especially fatty meats, dairy products, plant oils	Major component of cell membranes; energy source; thermal insulator; building blocks of some hormones	Hair loss; dry skin; weight loss; hormonal and reproductive disorders
Nucleic acids	All food sources	Provide sugars, bases and phosphates that can be used to make DNA, RNA, and ATP	None; components of nucleic acids can be synthesized by cells from amino acids and sugars

all of the essential amino acids, because meat (animal muscle) contains all 20 amino acids. Unlike animal meat, most plants do not contain every essential amino acid in sufficient quantities to supply a human's nutritive requirements. Therefore, people who follow strict vegetarian diets must find ways to balance the protein content of the plant matter they eat. By contrast, some herbivores, such as cows, have evolved the capacity to synthesize the essential amino acids, which allows them to subsist entirely on a plant diet.

Essential Fatty Acids   The **essential fatty acids** are certain polyunsaturated fatty acids, such as linoleic acid, that cannot be synthesized by animal cells. Saturated fatty acids, however, can be synthesized from carbon sources such as glucose (see Chapter 3 for a discussion of the chemistry of fatty acids). Linoleic acid is vital to an animal's health because it is converted in cells to another fatty acid, called arachidonic acid. This fatty acid is the precursor for production of several compounds important in many aspects of animal physiology; such compounds include the prostaglandins, which play roles in pain, blood clotting, and smooth muscle contraction. Some animals—such as felines—cannot synthesize arachidonic acid from linoleic acid, so arachidonic acid is an essential fatty acid in those species. Unsaturated fats are found primarily in plants, which provide a dietary source for both herbivores and omnivores. Strict carnivores such as felines, however, obtain their essential fatty acids from fishes or from the adipose tissue of birds and mammals.

Vitamins   **Vitamins** are important organic nutrients that serve as coenzymes for many metabolic and biosynthetic reactions. The two categories of vitamins are water-soluble and fat-soluble. Water-soluble vitamins, such as vitamin C, are not stored in the body and must be regularly ingested. Fat-soluble vitamins, such as vitamin A, are stored in adipose (fat) tissue. Not all animals require the same vitamins in their diet, however. Among vertebrates, for example, only primates and guinea pigs cannot synthesize their own vitamin C and must therefore consume it in the diet. **Table 45.2** summarizes the vitamins, their dietary sources, some of their important functions, and some health consequences associated with their deficiencies.

Minerals   **Minerals** are inorganic ions required by animals for normal functioning of cells. Minerals such as iron and zinc are required as cofactors or constituents of enzymes and other proteins. Other minerals such as calcium are required for bone, muscle, and nervous system function, while still others—notably potassium and sodium—contribute to electrical differences across plasma membranes and therefore are critical for heart, skeletal muscle, and neuronal activity. **Table 45.3** summarizes some of the most important minerals and their functions. Many minerals are required in only trace amounts, far less than 1 mg/day in a relatively large mammal such as ourselves. Nonetheless, without regular consumption of these small amounts, serious health problems arise.

Some minerals can be stored in an animal's body, reducing the risk of deficiency when the mineral is not available in the diet. Iodine, for example, is stored in large amounts in the vertebrate thyroid gland, where it is required for proper thyroid function. If a vertebrate's diet is deficient in iodine, which is not uncommon in areas where iodine periodically leaches from the soil as a result of considerable rainfall, it usually has enough stored iodine to permit normal thyroid gland function for days or even months. As another example, calcium is stored in huge quantities in bone in vertebrates and in the shells of some invertebrates. If the dietary intake of this mineral falls, calcium concentrations in the body fluids may decrease. Because calcium is important for proper cellular function, under such conditions, some of it is released from these storage sites and made available to the rest of the body.

Not all minerals are used the same way or at the same rate by all animals. For instance, copper helps to bind and transport oxygen in the fluids of some invertebrates such as horseshoe crabs, whereas in all vertebrates and in most other invertebrates, iron serves this function. Factors that affect mineral usage are an animal's species, age, weight, overall health status, and the types of food it eats.

45.2   Ingestion

Several factors determine how and when an animal obtains food, including its energetic demands, whether or not it is capable of locomotion, the local environment, and the structure of its digestive system. In this section, we will see that despite the enormous number of animal species and the highly varied environments in which they live, the major ways in which animals ingest food can be classified into just a few categories.

Animals May Eat Plants, Other Animals, or Both

Unlike plants, which are autotrophs that make their own food, all animals are heterotrophic and must consume their food. As stated earlier, most animals can be grouped into one of three broad dietary categories: herbivores, carnivores, or omnivores.

The gastrointestinal (GI) tracts, or **gut**, of carnivores and omnivores resemble each other much more than either resembles those of herbivores. This reflects differences in the enzymatic processing and the energetic quality of the foods they eat. Generally, the nutritive value of meat is much greater than that of plants. Consequently, carnivores usually need to eat less frequently than herbivores, many of whom eat almost continuously to supply sufficient nutrients for energy and growth.

Although broadly useful, the three dietary categories are limited when describing the diversity of animal species. For example, some animals eat protists, algae, and/or fungi as a major source of food. Also, other animals are almost strictly carnivores at one time of year but herbivores at other times. Many nonmigratory birds, for example, feed on insects and worms during the summer but switch to eating whatever vegetation, buds, or seeds they can find during the winter. Similarly, a coyote prefers to eat only meat but will consume plants and fruits if hungry enough. A hungry fruit-eating bat will eat

Table 45.2   Vitamins Required by Animals

Class of nutrient	Dietary sources	Functions in vertebrates	Symptoms of deficiency
Water-soluble vitamins			
Biotin	Liver; legumes; soybeans; eggs; nuts; mushrooms; some green vegetables	Coenzyme for gluconeogenesis and fatty acid synthesis; required to catabolize certain amino acids	Skin rash; nausea; loss of appetite; mental disorders (depression or hallucinations)
Folic acid	Green vegetables; nuts; legumes; whole grains; organ meats (especially liver; kidney; heart)	Coenzyme required for synthesis of nucleic acids	Anemia; depression; birth defects
Niacin	Legumes; nuts; milk; eggs; meat	Involved in many oxidation/reduction reactions	Skin rashes; diarrhea; mental confusion; memory loss
Pantothenic acid	Nearly all foods	Part of coenzyme A, which is involved in numerous synthetic reactions, including formation of cholesterol	Burning sensation in hands and feet; gastrointestinal symptoms; depression
Vitamin B$_1$ (thiamine)	Meats; legumes; whole grains	Required for certain steps of intermediary metabolism, leading to energy production from some amino acids	Beriberi (muscular weakness, anemia, heart problems, loss of weight)
Vitamin B$_2$ (riboflavin)	Dairy foods; meats; organ meats; cereals; some vegetables	Respiratory coenzyme; required for metabolism of fats, carbohydrates, and proteins	Seborrhea (excessive oil secretion from skin glands resulting in skin lesions)
Vitamin B$_6$ (pyridoxine)	Meats; liver; fish; nuts; whole grains; legumes	Cofactor for over 100 enzymes that participate in amino acid metabolism, lipid metabolism, and heme synthesis	Seborrhea; nerve disorders; depression; confusion; muscle spasms
Vitamin B$_{12}$	Meats; liver; eggs; some shellfish; dairy foods	Required for red blood cell formation	Anemia; nervous system disorders leading to sensory problems; balance and gait problems; loss of bladder and bowel control
Vitamin C (ascorbic acid)	Citrus fruits; green vegetables; tomatoes; potatoes	Antioxidant and free radical scavenger; aids in iron absorption; helps maintain healthy connective tissue and gums	Scurvy (connective tissue disease associated with skin lesions, weakness, poor wound healing, tooth decay); bleeding gums
Fat-soluble vitamins			
Vitamin A (retinol)	Liver; green and yellow vegetables; some fruits in small amounts	Component of visual pigments; regulatory molecule affecting transcription; important for reproduction and immunity	Night blindness due to loss of visual ability; skin lesions; impaired immunity
Vitamin D	Fish oils; fish; egg yolk; liver; synthesized in skin via sunlight	Required for calcium and phosphorus absorption from gut; bone growth	Rickets (weakened, deformed bones) in children; osteomalacia (weak bones) in adults
Vitamin E	Meats; vegetable oils; cereals; small amounts in some fruits and vegetables	Antioxidant; inhibits prostaglandin synthesis	Unknown, possibly skeletal muscle atrophy; peripheral nerve disorders
Vitamin K	Legumes; green vegetables; some fruits; some vegetable oils (olive oil, soybean oil); liver; also synthesized by hindgut bacteria	Component of blood-clotting mechanism	Reduced blood clotting ability

leaves, nectar, or pollen if fruit is not available. Animals like these are said to be opportunistic; they have a strong preference for one type of food but can adjust their diet if the need arises.

An animal's life stage may also influence its diet. Mammals begin life as milk-drinkers and later switch to consuming plants, animals, or both. Caterpillars eat leaves, but after metamorphosis, most species of moths or butterflies are strictly fluid-drinkers, typically consuming the nectar in flowers. Other animals that undergo metamorphosis also change their diet as adults. Tadpoles, for example, are mostly herbivores, while frogs eat insects and therefore are carnivores.

The three major categories also do not indicate what type of plant or animal is consumed. Grasses, cereals, and fruits,

for example, each have different energy and nutrient contents. Some herbivores eat primarily fruits, and some carnivores only drink the blood of other animals. To more fully describe the nutrition of animals, therefore, we need to investigate the strategies by which they obtain food.

Animals Have Evolved Multiple Strategies for Obtaining Food

The ways in which an animal obtains its food are related to its environment. Not surprisingly, for example, a sessile (nonmotile) marine invertebrate and a mobile terrestrial vertebrate face different challenges and opportunities. Although animals have evolved varied ways of obtaining food, there are three main

Table 45.3	Minerals Required by Animals		
Class of nutrient	**Dietary sources**	**Functions in vertebrates**	**Symptoms of deficiency**
Calcium (Ca)	Dairy products; cereals; legumes; whole grains; green leafy vegetables; bones (eaten by some animals)	Bone and tooth formation; exocytosis of stored secretions in nerves and other cells; muscle contraction; blood clotting	Muscular disorders; loss of bone; reduced growth in children
Chlorine (Cl)	Meats; dairy foods; blood; natural deposits of salt	Participates in electrical, acid-base, and osmotic balance across cell membranes, notably nerve and heart cells	Muscular and nerve disorders
Chromium (Cr)	Liver; seafood; some nuts; meats; mushrooms; some vegetables	Required for proper glucose metabolism, possibly by aiding the action of the hormone insulin	Disorders of lipid and glucose balance in blood
Copper (Cu)	Fish; shellfish; nuts; legumes; liver; and other organs	Required for hemoglobin production and melanin synthesis; required for connective tissue formation; serves as oxygen-binding component in some invertebrates	Anemia (abnormally low number of red blood cells); bone changes
Iodine (I)	Seaweed; seafood; milk; iodized salt	Required for formation of thyroid hormones	Inability to make thyroid hormones, resulting in enlarged thyroid gland
Iron (Fe)	Liver and other organs; some meats; eggs; legumes; leafy green vegetables	Oxygen-binding component of hemoglobin; co-factor in some enzymes	Anemia
Magnesium (Mg)	Hay; grasses; whole grains; green leafy vegetables	Cofactor for many enzymes that use ATP as a substrate	Changes in nervous system function
Manganese (Mn) and Molybdenum (Mo)	Nuts; whole grains; legumes; vegetables; liver	Cofactors for many enzymes	Poor growth; abnormal skeletal formation; nervous system disorders (convulsions)
Phosphorus (P)	Dairy foods; grains; legumes; nuts; meats	Bone and tooth formation; component of DNA, RNA, and ATP	Bone loss; muscle weakness
Potassium (K)	Meats; fruits; vegetables; dairy foods; grains	See Chlorine	Muscle weakness; serious heart irregularities; gastrointestinal symptoms
Selenium (Se)	Seafood; eggs; chicken; soybeans; grains	Antioxidant; cofactor for some enzymes	Keshan disease (damage to and loss of heart muscle)
Sulfur (S)	Proteins from any source	Component of two amino acids (methionine and cysteine)	Inability to synthesize many proteins
Sodium (Na)	Many fruits; vegetables; meats; and natural salt deposits	See Chlorine	Muscle cramps; changes in nerve activities
Zinc (Zn)	Widely found in meats; fish; shellfish (oysters); grains	Many functions related to tissue repair; sperm development; cofactor for many metabolic enzymes; required for certain transcription factors to bind to DNA	Stunted growth; loss of certain sensations like taste; impaired immune function; skin lesions

ways that we will focus on here: suspension feeding, bulk feeding, and fluid feeding.

Suspension feeders sift water, filtering out the organic matter and expelling the rest (**Figure 45.2**). Bivalve mollusks filter seawater and capture floating bits of organic material on mucus-covered cilia located in their gills, which move the material into the animal's mouth (Figure 45.2a). Sea squirts have a sticky mucous net in their pharynx that traps suspended food particles. Baleen whales have specialized plates made of keratin suspended like a stiff, frayed comb from the roof of the mouth (Figure 45.2b). The whale engulfs a mouthful of seawater and then uses its tongue to squeeze the water back out through the plates, which act like a sieve, trapping small copepods, protozoa, and other small animals or organic matter.

Carnivores, herbivores, and omnivores can be considered **bulk feeders**, those who eat food in large pieces. Carnivores, which inhabit aquatic and terrestrial environments, are generally **predators** (that kill live prey), such as piranhas and wolves, or **scavengers** (that eat the remains of dead animals), such as vultures. When present, carnivores' teeth show a variety of adaptations for grasping or seizing, biting, slicing, and, in some cases, chewing (**Figure 45.3a,b**). For example, sharp incisors and canine teeth and jagged molars help to slice and tear animal flesh. Carnivores that chew do so mostly to break up food into pieces small enough to swallow. Chewing food also has the advantage of increasing the total surface area of the food that is available for digestive enzymes.

Most carnivores, however, do not extensively chew their food. Some carnivores have mouth parts adapted to tearing off chunks of flesh, which they then swallow. The strategy of consuming entire chunks of food without chewing occurs in species from all the vertebrate classes. Birds of prey, for example, lack teeth and use their sharp beaks to tear away pieces of flesh that can be engulfed all at once. One advantage to swallowing

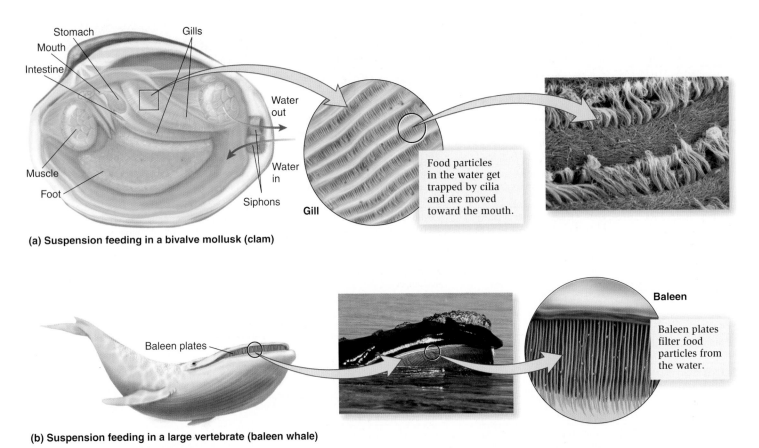

(a) Suspension feeding in a bivalve mollusk (clam)

Food particles in the water get trapped by cilia and are moved toward the mouth.

Baleen

Baleen plates filter food particles from the water.

(b) Suspension feeding in a large vertebrate (baleen whale)

Figure 45.2 **Suspension feeding.** Suspension feeders include species of both invertebrates and vertebrates. **(a)** Bivalves, such as the clam shown here, use a siphon mechanism to move water across their gills. Cilia located along the gills trap and move food particles to the mouth. **(b)** Baleen whales engulf water with their mouth and then use their tongue to force the water back out across the comblike baleen. The baleen filters tiny organisms or bits of organic matter from the water.

whole chunks of food is that it allows some carnivores—particularly those that hunt and eat in packs—to quickly get another bite when competing with other hungry animals. Some carnivores even swallow their prey whole. The reverse-oriented teeth of snakes, for example, seize prey and prevent it from escaping. This does not hinder their ability to digest their food, because meat can still be fully digested if swallowed whole.

Herbivores and algae eaters, however, may have powerful jaw muscles and large, broad molar teeth that are highly adapted for grinding tough, fibrous plants and cell walls (**Figure 45.3c**). Incisors and canine teeth are poorly developed or absent in many herbivores. When present, they are usually for defense (as in hippos and some primates) or for nipping grass (as in horses) or other vegetation. Food is chewed against the molars in a rotary motion, which grinds up the tough plant cell walls and releases the more digestible intracellular contents. In herbivorous fishes, the teeth are often adapted for cropping or scraping plants or algae off the substrate (**Figure 45.3d**). The food is ground up not in the mouth, but in the pharynx or throat, which contains specialized teeth suited for that purpose. In this way, chewing food does not interfere with gill breathing. In some cases, the mouth (or buccal) teeth of herbivorous fishes such as the parrotfish have evolved into a fused beaklike structure that aids in rasping algae off of coral and

other hard surfaces, and even chopping off bits of coral itself (Figure 45.3d).

In omnivores, such as ourselves, the teeth are a mix of the types seen in carnivores and herbivores (**Figure 45.3e**). For example, incisors and canines help cut and slice food, while relatively flat molars assist in chewing. Chewing occurs in a vertical motion, with which food is crushed. This is particularly useful for consuming hard foods such as nuts.

Fluid-feeders lick or suck fluid from plants or animals and so do not need teeth except, perhaps, to puncture an animal's skin. Fluid-feeding has evolved independently in many types of animals, including worms, insects, fishes, birds, and mammals (refer back to Figure 33.27). Many birds and bats drink nectar from plants and have specialized beaks, tongues, or mouths that enable them to reach the nectar of a particular flower (**Figure 45.4a**). When the fluid is obtained from an animal, the fluid-feeder usually has a specialized mouthpart, such as the piercing needle-like extension of a mosquito's mouth or the tiny blade-like jaws of a blood-sucking leech (**Figure 45.4b**). Fluid-feeders that consume the blood of other animals have developed a fascinating set of strategies to ensure a full meal. The European medicinal leech (*Hirudo medicinalis*), for example, secretes a local anesthetic at the site of the bite to dull the pain. An enzyme digests the host's connective tissue to allow the leech

Sharp, backward-facing teeth seize prey and prevent its escape.

(a) Teeth of a carnivore that consumes whole prey

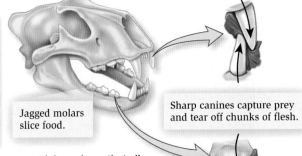

Jagged molars slice food.

Sharp canines capture prey and tear off chunks of flesh.

(b) Canines and molars of a terrestrial carnivore that slices prey into smaller chunks for swallowing

Incisors are used to nip or crop vegetation.

The teeth of parrotfish are fused into a beaklike structure ideal for cropping and scraping food off a hard surface.

Broad, flat molars grind food in a rotary motion.

(c) Teeth of a terrestrial herbivore

(d) Cropping teeth of a marine algae eater

Figure 45.3 **Examples of teeth in bulk feeders.** The shape and size of teeth reflects the way in which animals feed and the types of food they consume. In carnivores, teeth may be used for **(a)** seizing prey that will be swallowed whole or **(b)** slicing food into pieces small enough to be swallowed. The teeth of herbivores and algae eaters may be used for **(c)** nipping and grinding vegetation and **(d)** scraping or cropping food off coral or other hard substrates. **(e)** Omnivores typically have a mix of teeth resembling those found in both carnivores and herbivores. The molars may be adapted for crushing hard foods.

(e) Teeth of a terrestrial omnivore

Incisors and reduced canines help to bite off chunks of food.

Molars are used to crush food by chewing in an up-and-down motion.

to firmly embed its mouth into its host's flesh, and a locally act-ing chemical is secreted to keep the host's blood vessels open. Finally, to ensure that the blood does not clot before the leech has drunk its fill, the saliva of leeches and other blood-drinking animals such as the vampire bat *Desmodus rotundus* contains an anticoagulant. Scientists are studying these anticoagulants to develop anticlotting drugs for people with certain forms of heart and blood vessel disease. Recently, some physicians have applied leeches to surgical sites to prevent blood clots, such as when surgically reattaching a severed finger.

Figure 45.4 **Strategies of fluid-feeders.** **(a)** Hummingbirds extend their tongues through long, thin beaks to consume nectar from flowers. **(b)** Blood-drinking animals use teeth, specialized mouthparts, or, like this leech, bladelike jaws to prick the skin of their host and suck or lap up their blood.

Concept check: To obtain sufficient nutrients, blood-sucking animals such as mosquitoes, leeches, and some bats must consume large amounts of blood with each meal. For mosquitoes and bats, what effect might this have on their ability to take flight after a meal?

Regardless of what an animal eats, the useful parts of the eaten material must be digested into molecules that its cells can absorb. Next, we turn to a discussion of how animals digest their food and absorb the nutrients.

(a) Nectar-feeding hummingbird

(b) Blood-sucking leech

45.3 Principles of Digestion and Absorption of Food

Once food has been ingested, the nutrients from the food must be broken down (digested) so that they can be absorbed by the cells of the digestive tract. In this section, we will examine some of the major principles of digestion and absorption in animals, beginning with where digestion takes place.

Digestion Can Occur Intracellularly or Extracellularly

Food is digested either inside cells (intracellularly) or outside cells (extracellularly). Intracellular digestion occurs only in some very simple invertebrates such as sponges and single-celled organisms. It involves using phagocytosis to bring food particles directly into a cell, where the food is segregated from the rest of the cytoplasm in food vacuoles. Once inside these vacuoles, hydrolytic enzymes digest the food into monomers (the building blocks of polymers), which then are moved out of the vacuole to be used directly by that cell. Intracellular digestion cannot support the metabolic demands of an active animal for long, because only tiny bits of food can be phagocytosed at one time. It also does not provide a mechanism for storing large quantities of food so that an animal can digest it slowly while going about its other activities.

Most animals digest food via extracellular digestion in a cavity of some sort. Extracellular digestion protects the interior of the cells from the actions of hydrolytic enzymes and allows animals to consume large prey or large pieces of plants. Food enters the digestive cavity, where it is stored, slowly digested, and absorbed gradually over long periods of time, ranging from hours (for example, after a human eats a pizza) to weeks (after a python eats a gazelle).

In the simplest form of extracellular digestion—seen in invertebrates such as flatworms and cnidarians—the digestive cavity has one opening that serves as both an entry and an exit port (**Figure 45.5**). The digestive cavity of these animals is called a **gastrovascular cavity**, because not only does digestion occur within it, but fluid movements in the cavity serve as a circulatory—or vascular—system to distribute digested nutrients throughout the animal's body. Food within a gastrovascular cavity is partially digested by enzymes that are secreted into the cavity by the cells lining the cavity. As the food particles become small enough, they are phagocytosed by the lining cells and further digested intracellularly. Undigested material that remains in the gastrovascular cavity is expelled.

Most Digestive Cavities Are Tubes with Specialized Regions and Openings at Opposite Ends

In contrast to the gastrovascular cavities of simple invertebrates, all other animals possess digestive systems that consist of a single elongated tube with an opening at both ends, through which food passes from one end to the other. Along its length,

Figure 45.5 Extracellular digestion in a gastrovascular cavity. In animals with gastrovascular cavities, such as the cnidarian *Hydra* illustrated here, digestion occurs extracellularly within the cavity.

the tube, or **alimentary canal**, contains smooth muscle. When the muscle contracts, it helps churn up the ingested food so that it is mechanically broken into smaller fragments. The canal is lined on its interior surface by a layer of epithelial cells. These cells synthesize and secrete digestive enzymes and other factors into the lumen of the alimentary canal, and they secrete certain hormones into the blood that help regulate digestive processes. The cells are also involved with transporting digested material out of the canal.

The alimentary canal has several specialized regions along its length that vary according to species. Because of these specializations, digestive processes requiring acidic conditions can be segregated from those requiring higher pH, and undigested food can be stored in one region while digestion continues in another area. The ability to store food in the stomach, for example, allows those animals with stomachs to eat less frequently, leaving time for other activities.

Digestion of Food Requires Hydrolytic Enzymes

Digestion is required to convert polymers into smaller units that can be absorbed across plasma membranes. Therefore, digestion requires enzymes that can hydrolyze the chemical bonds in carbohydrates, proteins, lipids, and nucleic acids. These enzymes are named after their substrates: polysaccharidases, proteases, lipases, and nucleases. Not all of the food that is eaten, however, is digested; some substances such as vitamins do not require digestion and are absorbed intact.

Absorption of Food May Be Passive or Active

Once food is digested, the nutrients must be absorbed by the epithelial cells that line specialized portions of the alimentary canal. This occurs in different ways, either by passive or facilitated diffusion or by active transport. Small, hydrophobic molecules such as fatty acids diffuse down concentration gradients across the epithelium. Ions and other molecules are transported by facilitated diffusion or active transport. Minerals are ions and therefore do not readily cross plasma membranes. Instead, they are usually actively transported across the membranes of the epithelial cells of the canal by ATP-driven ion pumps. In other cases, small, hydrophilic organic nutrients are transported by secondary active transport, usually with Na^+.

After nutrients enter the epithelial cells of the alimentary canal, the cells use some of the nutrients for their own requirements. Most of the nutrients, however, are transported out of the epithelial cells and into nearby blood vessels, where they enter the blood and circulate to the other cells of the body. Thus, nutrients enter the alimentary canal in food, are digested within the canal into monomers that can be transported into epithelial cells, and from there are released into the blood, where they can reach all of the body's cells. In the special case of water,

osmotic gradients established by the transport of ions and other nutrients out of the epithelial cells draw water by osmosis from the canal, across the epithelial cells, and from there into the blood.

The mechanisms that activate and control the digestive and absorptive functions of the alimentary canal have been extensively studied in vertebrates and have great importance for human health. In the rest of this chapter, we will explore the structure and function of the digestive systems of vertebrates.

45.4 Overview of Vertebrate Digestive Systems

The vertebrate **digestive system** consists of the alimentary canal—also known as the gastrointestinal, or GI, tract—plus several accessory structures (**Figure 45.6**). As illustrated in the example in Figure 45.6, the human GI tract consists of the oral cavity, pharynx, esophagus, stomach, small and large intestines, rectum, and anus. The accessory structures, not all of which are found in all vertebrates, are the tongue, teeth, salivary glands, liver, gallbladder, and pancreas. The differences between the digestive systems of various vertebrates reveal much about their respective feeding strategies. In this section, we will look at some of the most important differences as we discuss the form

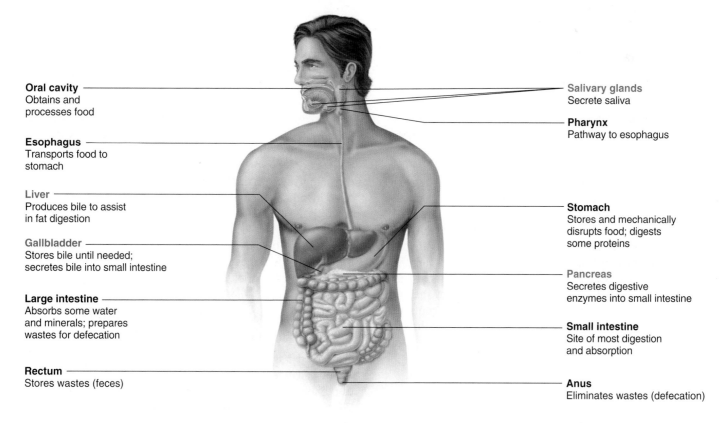

Oral cavity
Obtains and processes food

Esophagus
Transports food to stomach

Liver
Produces bile to assist in fat digestion

Gallbladder
Stores bile until needed; secretes bile into small intestine

Large intestine
Absorbs some water and minerals; prepares wastes for defecation

Rectum
Stores wastes (feces)

Salivary glands
Secrete saliva

Pharynx
Pathway to esophagus

Stomach
Stores and mechanically disrupts food; digests some proteins

Pancreas
Secretes digestive enzymes into small intestine

Small intestine
Site of most digestion and absorption

Anus
Eliminates wastes (defecation)

Figure 45.6 **A vertebrate digestive system, as shown in the human.** This figure shows the organs of the gastrointestinal tract (labeled in black) and the accessory structures (labeled in red). Not all vertebrates share identical features of the digestive system; for example, some fishes lack a stomach, and many birds lack a gallbladder.

and function of each part of the vertebrate digestive system. We begin with an overview of the form, or anatomy, of some typical vertebrate digestive systems.

Alimentary Canals Are Divided into Functional Regions

The alimentary canal is one continuous tube that changes in appearance and function along its length, with three general sections. The first section, at the anterior end, functions primarily in the ingestion of food. It contains the oral cavity, salivary glands, pharynx (throat), and esophagus. The middle portion, which functions in the storage and initial digestion of food, contains one or more food storage or digestive organs, including the crop, gizzard, and stomach(s), depending on species. This section also contains the upper part of the small intestine—where most of the digestion and absorption of food takes place—and accessory structures that connect with the intestine, including the pancreas, liver, and gallbladder. The third section, the posterior part of the canal, functions in final digestion and absorption and the elimination of nondigestible wastes. It consists of the remainder of the small intestine, and in most vertebrates other than fishes, a large intestine. Undigested material is excreted through an opening called an anus, or in many amphibians, reptiles, and birds, a cloaca (a common opening for the digestive and urogenital tracts).

From the mid-esophagus to the anus or cloaca, the gastrointestinal tract has the same general structure, with a hollow cavity called a lumen that is lined by a layer of epithelial cells. Included in the epithelial layer are secretory cells that release a protective coating of mucus into the lumen of the tract, and other cells that release hormones into the blood in response to the presence of food. Connected to the epithelium are secretory glands that release acid, enzymes, water, and ions into the lumen. From the stomach onward, the epithelial cells are linked along the edges of their luminal surfaces by tight junctions that prevent digestive enzymes and undigested food from moving between the cells and out of the canal. In some places, such as the small intestine, the luminal surface is highly convoluted, a feature that increases the surface area available for digestion and absorption.

The epithelial cell layer is surrounded by layers of tissue made up of smooth muscles, neurons, connective tissue, and blood vessels. The neurons are activated by nerves coming from the central nervous system that respond to the sight and smell of food. The gut neurons are also activated directly by the presence of food in the gut. Contraction of the muscles is controlled by these neurons and results in mechanical mixing of the contents within the stomach and intestine. This helps speed up digestion and also brings digested foods into contact with the epithelium to facilitate absorption.

Food Processing and Polysaccharide Digestion Begin in the Mouth

Once food enters the mouth, some animals, as stated earlier, chew it to reduce the size of each piece. Even in those animals

that do not chew their food, however, the initial processing of food begins in the mouth because of the presence of saliva.

Saliva is a watery fluid containing proteins, mucus, and antibacterial agents. In terrestrial vertebrates, salivary glands in and around the mouth, cheeks, tongue, and throat produce a constant flow of saliva, which keeps the mouth moist and clean; fishes, which lack true salivary glands, secrete mucus from specialized cells in their mouth and pharynx. In response to food, chemical and pressure detectors in the walls of the mouth and tongue increase its secretion. Increased secretion of saliva can be produced in terrestrial vertebrates by a large increase in blood flow to the salivary glands, which occurs in response to signals sent from the nervous system. Saliva production in some animals can also be increased simply by the smell or sight of food, a feedforward response discussed in Chapter 40. In mammals, the volume of saliva secreted per gram of tissue is among the largest secretion of any of the glands in the animal.

Saliva has several functions, not all of which pertain to all vertebrates:

1. moisten and lubricate food to facilitate swallowing;
2. dissolve food particles to facilitate the ability of specialized chemical-sensing structures called taste buds to taste food;
3. kill ingested bacteria with a variety of antibacterial compounds, including antibodies; and
4. initiate digestion of polysaccharides through the action of a secreted enzyme called **amylase**.

Digestion is the least important of these functions in most vertebrates, few of which produce salivary amylase. In humans and other primates, salivary amylase is present and accounts for a few percent of total polysaccharide digestion. Depending on how much salivary amylase you produce, you may be able to detect the action of amylase in your own saliva with a simple test. If you chew on a starchy soda cracker and leave it in your mouth for a while, you may notice it begins to taste sweet. That is because some of the polymers in the cracker have been digested to sweet-tasting disaccharides.

The other functions of saliva, however, are very important. For example, imagine trying to swallow unchewed food with a perfectly dry mouth. Also, the antibiotic properties of saliva help cleanse the mouth. In people who have had cancerous salivary glands removed, the teeth and gums often become so diseased that tooth loss may occur.

Peristalsis Moves Swallowed Food Through the Esophagus to a Storage Organ

Once food has been sufficiently processed in the mouth, it is swallowed. The swallowed food moves into the next segments of the alimentary canal, the **pharynx** (throat) and **esophagus**. These structures do not contribute to digestion or absorption but serve as a pathway to storage organs such as the stomach.

Pharynx and Esophagus The muscles in the walls of the pharynx and esophagus contribute to swallowing. In the

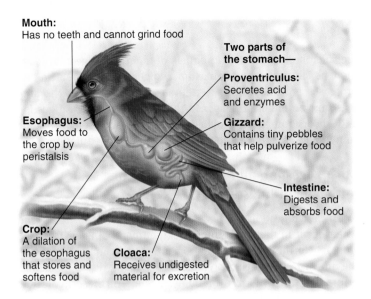

Mouth:
Has no teeth and cannot grind food

Two parts of the stomach—

Proventriculus:
Secretes acid and enzymes

Esophagus:
Moves food to the crop by peristalsis

Gizzard:
Contains tiny pebbles that help pulverize food

Intestine:
Digests and absorbs food

Crop:
A dilation of the esophagus that stores and softens food

Cloaca:
Receives undigested material for excretion

Figure 45.7 **The alimentary canal of birds.** The avian alimentary canal contains specialized regions for storing and softening food (the crop) and pulverizing food (the gizzard). The gizzard and proventriculus constitute the stomach. Undigested material is excreted through the cloaca.

Concept check: *Smooth, polished stones have been found in the stomach region of fossilized skeletons of ancient sauropod dinosaurs. What does this suggest about the alimentary canal of such animals?*

pharynx, swallowing begins as a voluntary action but continues in the esophagus by the process of **peristalsis**—rhythmic, spontaneous waves of muscle contraction that begin near the mouth and end at the stomach. When most vertebrates eat, the mouth and stomach are roughly horizontal with respect to each other. In fact, the head of a terrestrial grazing animal is usually lower than its stomach when eating. The wavelike action of peristalsis ensures that food will be pushed toward the stomach and not sit in the esophagus or even move backward into the mouth if the head is lowered.

The Crop In some animals, food moves directly from the esophagus to a storage organ called the **crop**, which is a dilation of the lower esophagus (**Figure 45.7**). Crops are found in most birds (and many invertebrates, including insects and some worms). Food is stored and softened by watery secretions in the crop, but little or no digestion occurs there. Because they process large amounts of tough food, birds that eat primarily grains and seeds have larger crops than birds that eat insects and worms. The material that birds regurgitate to their young comes from the crop. In some species, such as pigeons and doves, the cells that line the crop wall secrete a lipid-rich watery solution called crop or pigeon milk into the material to be regurgitated.

Food Processing Continues and Protein Digestion Begins in the Stomach

Once food passes through the esophagus and crop, if present, it reaches a storage organ such as a stomach. The **stomach** is a saclike organ that most likely evolved as a means of storing

food. Stomach-like organs are found in all vertebrate classes, but a true stomach (one that produces hydrochloric acid) is absent in many species of herbivorous fishes. In addition to its storage function, the muscular nature of the stomach helps it to mechanically break up large chunks of food into smaller, more easily digestible fragments. Lastly, the stomach partially digests some of the macromolecules in food and regulates the rate at which the contents empty into the small intestine. Glands within the stomach wall secrete hydrochloric acid (HCl) and an inactive molecule called pepsinogen into the stomach lumen. One function of the acid is to convert pepsinogen into the active enzyme **pepsin**, which as a protease begins the digestion of protein. Why do stomach cells secrete pepsinogen instead of pepsin? The answer is that if the cells produced active pepsin, they would digest their own cellular proteins.

Within the stomach lumen, HCl kills many of the microbes that may have been ingested with food. The acid also helps dissolve the particulate matter in food. In addition, the acid environment in the stomach (or gastric) lumen alters the ionization of polar molecules, especially proteins. This disrupts the structural framework of the tissues in food and makes the proteins more accessible to pepsin. The proteins released by the dissolving action of HCl are partially digested in the stomach by pepsin. By contrast, no significant digestion of carbohydrates or lipids occurs in the stomach.

In birds, the stomach is divided into two parts: the proventriculus and the gizzard (Figure 45.7). The **proventriculus** is the glandular portion of the stomach, and it secretes acid and pepsinogen. Partially digested and acidified food then moves to the **gizzard**, a muscular structure with a rough inner lining that grinds food into smaller fragments.

The gizzard contains sand or tiny stones swallowed by the bird. The gritty sand and stones take the place of teeth and help mash and grind ingested food. Scientists think that this function of the gizzard may have evolved partly as an adaptation for flight, because large jaws and associated chewing muscles and teeth would result in a larger, heavier, and less aerodynamic head. Eventually, the pebbles in the gizzard become smaller as they are worn away, and they are excreted. Thus, birds must occasionally restock the gizzard with new grinding stones. Grain-eating birds, particularly chickens and other fowl, many passerines, and pigeons and doves, generally have more muscular gizzards than do insectivorous birds, because of the difficulty in breaking down plant cell walls. In other birds, such as owls, gizzards help compress the nondigestible parts of their meals (bones, teeth, fur, feathers) into a pellet that can be regurgitated. Gizzards, incidentally, are not unique to birds. Certain reptiles that are closely related to birds, such as crocodiles, also contain muscular gizzards. In addition, some species of herbivorous fishes (for example, members of the family Acanthuridae) ingest quantities of inorganic grit with their meals, which help to grind up food in a portion of the stomach that is modified into a strong, muscular grinding organ like a gizzard.

Digestive actions of the stomach reduce food particles to **chyme**, a solution that contains water, salts, molecular fragments of proteins, nucleic acids, polysaccharides, droplets of fat, and various other small molecules. Virtually none of these molecules, except water, can cross the epithelium of the stomach wall; therefore, little or no absorption of organic nutrients occurs in the stomach.

Herbivores Use Microbes to Aid Digestion in the Stomach and Intestine

Herbivores face a special challenge because they must digest cellulose, but they lack the enzyme cellulase required for the job. Instead, herbivores rely on microbes living within their digestive tracts that have the capacity to digest cellulose. The microbes break down the cellulose into monosaccharides that can be absorbed along with other by-products of microbial digestion, such as fatty acids and some vitamins. In this way, bacteria and protists predigest the food for the animal.

Microbial cellulose digestion has been demonstrated in herbivores of all vertebrate classes but has been particularly well studied in mammals. In mammals with simple stomachs, such as horses, the microbes exist in the large intestine and the **cecum**, a blind outpocketing of the gut at the junction of the small and large intestines. Other herbivores, such as **ruminants** (for example, sheep, goats, llamas, and cows), have complex stomachs consisting of several chambers, beginning with three outpouchings of the lower esophagus collectively referred to as the forestomach (**Figure 45.8**). The forestomach is composed of the rumen, reticulum, and omasum, in sequence. The rumen and reticulum contain the microbes that digest cellulose, and the omasum absorbs some of the water and salts released from the chewed and partially digested food. The tough, partially digested food (the cud) is occasionally regurgitated, rechewed, and swallowed again. Eventually, the partially digested food,

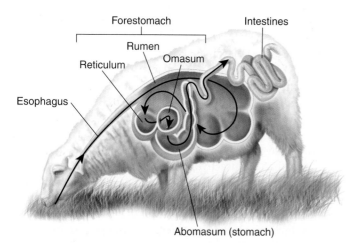

Figure 45.8　Digestive tract of a ruminant. Ruminants have a complex arrangement of three modified pouches, together called the forestomach, arising from the esophagus; these pouches are called the rumen, reticulum, and omasum. The rumen and reticulum act as storage and processing sites (in large ruminants, the rumen may store up to 95 liters of undigested food); the omasum absorbs some water and salts. The forestomach is connected to the true stomach, or abomasum. Digestion by acid and pepsin takes place in the abomasum, which then connects with the intestines.

the microbes, and the by-products of microbial digestion reach the true stomach, the abomasum, which contains the acid and proteases typical of other vertebrate stomachs. From the abomasum, the material passes to the intestines, where digestion and absorption are completed. Some microbes remain in the rumen and quickly multiply to replenish their populations.

Most Digestion and Absorption Occurs in the Small Intestine

Nearly all digestion of food and absorption of nutrients and water occur in the **small intestine**, the tube that leads from the stomach to the large intestine or, in some animals, directly to the anus or cloaca. Hydrolytic enzymes break down molecules of organic nutrients into monomers. Some of these hydrolytic enzymes are on the luminal surface of the intestinal epithelial cells, while others are secreted by the pancreas and enter the intestinal lumen. The products of digestion are absorbed across the epithelial cells and enter the blood. Vitamins and minerals, which do not require enzymatic digestion, are also absorbed in the small intestine. Water is absorbed by osmosis from the small intestine in response to the movement of nutrients across the intestinal epithelium. We will discuss the mechanisms of digestion and absorption in more detail later.

The ability of the small intestine to carry out the bulk of digestion and absorption is aided by mucosal infoldings and specializations along its length, as mentioned previously. Finger-like projections known as **villi** (singular villus) extend into the lumen of the vertebrate small intestine (**Figure 45.9**). The surface of each villus is covered with a layer of epithelial cells whose plasma membranes form small projections called **microvilli**, known collectively as the **brush border**. The combination of folded mucosa, villi, and microvilli increases the small intestine's surface area about 600-fold over that of a flat-surfaced tube having the same length and diameter. The small intestine is small in diameter compared to the large intestine, but it is very long—3 meters in an adult human (the small intestine is almost twice as long if removed from the abdomen, because the muscles relax). This brings the total surface area of the human small intestine to about 300 m^2—roughly the size of a tennis court! This enormous surface area means that the likelihood of an ingested food particle encountering a digestive enzyme and being absorbed across the epithelium is very high, so digestion and absorption proceed rapidly.

The center of each intestinal villus is occupied by a special type of vessel called a **lacteal**, which is part of the lymphatic system, and by capillaries, the smallest blood vessels in the body (Figure 45.9). Most of the fat absorbed in the small intestine exists as bulky protein-bound particles that are too large to enter capillaries. Consequently, absorbed fat enters the larger, wider lacteals. Material absorbed by the lacteals eventually empties into the circulatory system. Other nutrients are absorbed directly into the capillaries and from there into veins.

The length of the small intestine varies among species. Both terrestrial and aquatic herbivores generally have much longer small intestines than do carnivores, which allows added time for plant material to be digested and absorbed. Even within an

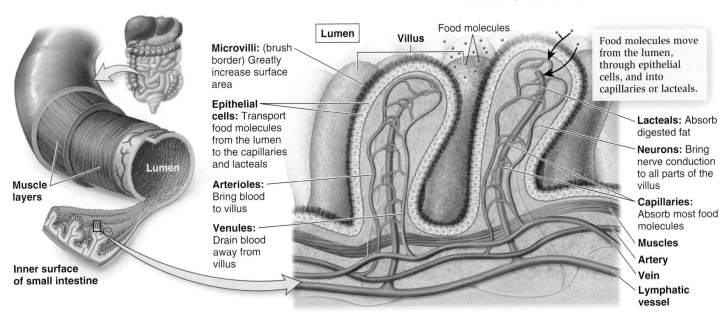

Figure 45.9 **The specialized arrangement of tissues in the small intestine.**   The small intestine is folded into numerous villi, which increase the surface area for digestion and absorption. Within each villus are capillaries and lymphatic vessels (lacteals) into which absorbed nutrients are transported. The epithelial cells of the villi have extensions from their surface called microvilli. The microvilli constitute the brush border of the intestine and greatly add to the total surface area.

individual animal, the length of the small intestine can change. For example, in a bird that switches from eating insects and worms in summer to buds and other nutrient-poor vegetation in winter, the small intestine grows and elongates and increases its total absorptive surface area to meet the digestive challenges associated with an herbivorous diet.

The Pancreas and Liver Secrete Substances That Aid Digestion

During the process of chyme moving through the small intestine, two major organs—the pancreas and liver—secrete substances that flow via ducts into the first portion of the intestine called the duodenum (**Figure 45.10**). The **pancreas**, a complex organ located inferior to the stomach in humans (see Figure 45.6), has several functions, but in this chapter, we will focus on those that are directly involved in digestion. The gland secretes digestive enzymes and a fluid rich in bicarbonate ions (HCO_3^-). The bicarbonate neutralizes the acidity of chyme, which would otherwise inactivate the pancreatic enzymes in the small intestine and could also damage the intestinal epithelium.

The **liver** is the site of bile production. **Bile** contains bicarbonate ions, cholesterol, phospholipids, a number of organic wastes, and a group of substances collectively termed **bile salts**. The bicarbonate ions, like those from the pancreas, help neutralize acid from the stomach, while the bile salts break up dietary fat and increase its accessibility to digestive enzymes.

The liver secretes bile into small ducts that join to form the common hepatic duct (from the Greek *hepar*, meaning liver). Between meals, secreted bile is stored in the **gallbladder**, a small sac underneath the liver. During a meal, the smooth muscles in the gallbladder contract, injecting the bile solution into a connecting duct called the common bile duct (Figure

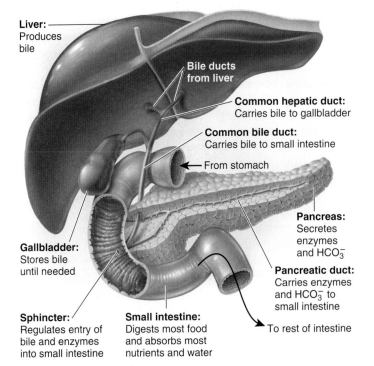

Figure 45.10 **The arrangement and functions of the vertebrate liver, gallbladder, pancreas, and small intestine.** Bile drains from the liver into the gallbladder through the common hepatic duct. During a meal, bile is secreted from the gallbladder and enters the duodenum of the small intestine through the common bile duct. Simultaneously, secretions from the pancreas travel through another duct that joins with that from the gallbladder and empties into the small intestine. A muscular sphincter controls the entrance to the small intestine.

Concept check: *What advantage does an animal gain by having a gallbladder?*

45.10). The opening of a sphincter allows the bile to flow into the lumen of the small intestine. The gallbladder, therefore, is a storage organ that allows the release of large amounts of bile to be precisely timed to the consumption of fats. However, many animals such as horses and doves that secrete bile do not have a gallbladder. In humans, the gallbladder can be surgically removed without impairing bile secretion by the liver or its flow into the intestinal tract. People without a gallbladder can still digest fat but may need to limit the amount of fat they eat at one time because bile secretion can no longer be timed to a meal.

The digested nutrients, along with water, are absorbed across the plasma membranes of the brush border cells. Peristalsis slowly propels the remaining contents through the later portions of the small intestine, called the jejunum and the ileum, where further absorption occurs. Finally, the remaining material enters the large intestine or the anus or cloaca.

The Large Intestine Concentrates Undigested Material, Which Is Then Eliminated via the Anus

The size of the large intestine varies greatly among different vertebrates, and the organ is vestigial or even absent in many animals, notably fishes. In humans, the large intestine is a tube about 6 cm in diameter and 1–1.5 meters long. Its first portion, the cecum, forms a small pouch from which extends the appendix, a finger-like projection having no known essential function but that may contribute to the body's immune defense mechanisms. The next part of the large intestine in humans and other mammals is called the **colon**, which consists of three segments—the ascending, transverse, and descending portions. The terminal portion of the descending colon in humans is S-shaped, forming the sigmoid colon, which empties into the rectum, a short segment of the large intestine that ends at the anus. The characteristic appearance of the mammalian colon is not widely found in other vertebrates, most of whom have a simple, straight large intestine.

Chyme from the small intestine enters the cecum through a sphincter. This sphincter is normally closed, but it periodically relaxes after a meal, allowing chyme to enter. Although the large intestine has a greater diameter than the small intestine, its epithelial surface area is far less—particularly in carnivores and omnivores—because it lacks the convolutions and villi of the small intestine.

The primary functions of the large intestine are to store and concentrate fecal material before defecation and to absorb some of the remaining salts and water that were not absorbed in the small intestine. Because most substances are absorbed in the small intestine, only a small volume of water and salts, along with undigested material, is passed on to the large intestine. The large intestine temporarily stores the undigested material and concentrates it by absorbing additional water and salts. **Defecation** occurs when contractions of the rectum and relaxation of associated sphincter muscles expel the feces through the final portion of the canal, the **anus** (see Figure 45.6). Some terrestrial herbivores—notably certain rabbits, hares, mice, and several families of birds—consume their own feces, a behavior

called **coprophagy**. This is an adaptation to maximize the absorption of water, vitamins, and organic nutrients that were not fully absorbed as food passed through the small intestine and was eventually defecated. The resulting feces are not eaten a second time.

The large intestine also absorbs some of the products formed by bacteria inhabiting this region, including small amounts of vitamins. Although this source of vitamins generally provides only a small part of the normal daily requirement, it may make a significant contribution when dietary vitamin intake is low. Sometimes people develop a vitamin deficiency if treated with antibiotics that inhibit these species of bacteria. Other bacterial products include gas (flatus), which is a mixture of nitrogen and carbon dioxide, with small amounts of hydrogen, methane, and hydrogen sulfide. Bacterial processing of undigested polysaccharides produces these gases, except for nitrogen, which is derived from swallowed air. Certain foods (beans, for example) contain large amounts of carbohydrates that cannot be digested by intestinal enzymes but are readily metabolized by bacteria in the large intestine, producing flatus.

45.5 Mechanisms of Digestion and Absorption in Vertebrates

The preceding sections provided an overview of nutrition and the basic features of digestive systems. We turn now to a more detailed description of how carbohydrates, proteins, and lipids are processed in the vertebrate digestive system and how the end products of digestion are absorbed across intestinal cells.

Carbohydrates Are Digested and Absorbed in the Small Intestine

In a typical omnivore such as ourselves, most of the ingested carbohydrates are the polysaccharides starch and cellulose from plants and glycogen from animals. The remainder consists of simple carbohydrates, such as the monosaccharides fructose and glucose in fruit, and disaccharides, such as lactose in milk. Humans also add the disaccharide sucrose (table sugar) to their food. Other animals consume sucrose in its natural form from sources such as maple sap and sugarcane.

Starch digestion by salivary amylase begins in the mouth, but the acid in the stomach destroys the amylase and prevents further starch digestion there. Starch digestion resumes in the small intestine by amylase secreted into the intestine by the pancreas. The products of starch digestion via amylase are molecules of the disaccharide maltose (**Figure 45.11**). Maltose, along with ingested sucrose and lactose, is broken down into monosaccharides—fructose, glucose, and galactose—by enzymes located on the brush border of the small intestine epithelial cells. The monosaccharides are then absorbed into the epithelial cells. Fructose crosses the apical membrane of the epithelial cells by facilitated diffusion, whereas glucose and galactose undergo secondary active transport coupled to

1 Enzymes on the luminal surface of the small intestine epithelial cells digest disaccharides into monosaccharides.

Digestion — Maltose → 2 glucose

Small intestine

2 Monosaccharides are absorbed into the cells by facilitated diffusion or by secondary active transport with Na$^+$.

Absorption

Lumen of small intestine

Apical membrane

Small intestine epithelial cell

Enzyme (maltase)

Tight junction

Facilitated diffusion

Fructose → Fructose

Secondary active transport

Na$^+$
Glucose → Glucose

Na$^+$
Galactose → Galactose

Interstitial fluid

Basolateral membrane

Capillary

Red blood cells

3 The absorbed monosaccharides leave the epithelial cells by facilitated diffusion and enter the blood. The bloodstream distributes the nutrients throughout the body.

Figure 45.11 Digestion and absorption of carbohydrates in the small intestine. Digestion and absorption occur in the same cells but are shown separately here for clarity. For simplicity, the microvilli (brush border) are not shown.

Concept check: *The absorption of many nutrients requires secondary active transport. What can we conclude from that about the energetic cost of absorption?*

sodium ions. Monosaccharides then leave the epithelial cells by way of facilitated diffusion transporters located in the basolateral membranes of the epithelial cells and enter the blood. The transport of substances from the lumen to the blood is called **transepithelial transport** because it occurs across a layer of epithelial cells. The bloodstream distributes the monosaccharides and other absorbed nutrients to the cells of the body.

Do you feel ill after drinking milk or eating dairy products? If so, you are among the majority of people who cannot adequately digest lactose, the chief disaccharide in milk. A small percentage of humans, however, retain the ability to digest lactose throughout life. Next, we'll examine these phenomena along with their genetic causes.

Genomes & Proteomes Connection

Genetics Explains Lactose Intolerance

With rare exceptions, the milk of all mammals contains lactose. Lactose is digested by the intestinal enzyme lactase, which cleaves lactose into glucose and galactose. Humans with lactose intolerance cannot adequately digest lactose because their lactase is either inactive, absent, or present in small amounts.

Milk is the sole food of most mammals shortly after birth, and the primary food for various lengths of time thereafter until weaning, the transition from consuming mother's milk to eating a diet of solid foods. Once weaned, mammals never again drink milk, except, of course, for humans. The popular notion of adult cats lapping up milk from a bowl is a misconception.

Some visits to the veterinarian are in fact related to gastrointestinal symptoms caused by well-meaning owners who fed milk to their adult pets.

Because the only dietary source for lactose is milk, it is not surprising that older mammals lose the ability to digest this disaccharide. This occurs because the gene that encodes lactase is shut off at the age of weaning or shortly thereafter. The developmental mechanisms that turn off lactase production and activity are not firmly established, but they are known to involve decreased transcription of the lactase gene.

If an adult mammal were to drink milk, the undigested lactose would remain in the gut. As a result, water that normally would be absorbed by osmosis with the digested monosaccharides formed from lactose would also remain in the gut. Farther down the alimentary canal, microbes in the large intestine digest some of the lactose for their own use and, in the process, release by-products such as hydrogen and other gases. The combination of water retention and bacterial action results in gastrointestinal symptoms such as diarrhea, gas, and cramps.

An estimated 90% of the world's human population cannot fully digest lactose after early childhood. In other words, lactose intolerance is not a disorder, it is a normal condition for humans just as it is for all other mammals.

Then why are some people able to consume dairy products without getting ill? Researchers have discovered that the ability of human adults to digest lactose is clearly linked to genetic background. Lactose intolerance is an example of a human polymorphism, a genetic trait that varies among people. Other than individuals who trace their ancestry to northern Europe and a few isolated regions in western Africa, nearly every population of humans shows a considerable degree of lactose intolerance, reaching nearly 100% in most of south and east Asia. One hypothesis for this phenomenon appears to be a behavioral and cultural change that occurred in Neolithic times, when certain populations domesticated cattle and added cow's milk to their diet. Adults who were able to digest lactose—presumably due to some mutation affecting the expression of the lactase gene—enjoyed a selective advantage and tended to thrive and pass on their genes more frequently than those whose digestive systems could not handle milk. In other words, natural selection resulted in certain populations carrying mutations that caused the lactase gene to be expressed after weaning. Eventually, a substantial percentage of people in those regions of the world retained sufficient intestinal lactase to use milk as a staple food.

To understand how this trait was passed on, let's examine the gene that codes for lactase. Surprisingly, the coding regions of the lactase gene and the lactase core promoter are identical whether individuals are lactose tolerant or intolerant. Recently, however, Finnish investigators have uncovered two single nucleotide changes located in presumed regulatory sites that control the expression of the lactase gene. These changes are associated with prolonged lactase expression, allowing it to occur after weaning. People carrying these mutations are lactose tolerant and can consume milk products through adulthood. By comparison, adults who lack these mutations—and are therefore lactose intolerant—can consume dairy products only in small amounts or not at all. However, their ability to do so is greatly improved if the product has been commercially treated with purified lactase to predigest the lactose.

Proteins Are Digested in the Stomach and Small Intestine, and Absorbed in the Small Intestine

Proteins are broken down to peptide fragments in the stomach by pepsin, and in the small intestine by the proteases **trypsin** and **chymotrypsin**. The pancreas secretes the latter two enzymes as inactive precursors, which prevents the active enzymes from digesting the pancreas itself. Once the inactive form of trypsin enters the small intestine, it is cleaved into the active molecule by the enzyme enteropeptidase whose active site is located on the luminal membranes of the intestinal cells. Trypsin then activates the inactive forms of chymotrypsin and trypsin.

The peptide fragments produced by trypsin and chymotrypsin are further digested into individual amino acids by the actions of specific proteases in the small intestine. These proteases, which are called aminopeptidases and carboxypeptidases, are located on the lumen-facing membranes of the epithelial cells or secreted into the lumen by the pancreas. They cleave off one amino acid at a time from the N-terminus and C-terminus of peptides, respectively. Individual amino acids then enter the epithelial cells by secondary active transport coupled to sodium ions. Amino acids leave these cells and enter the blood by facilitated diffusion across the basolateral membrane. As with carbohydrates, protein digestion and absorption are largely completed in the upper portion of the small intestine.

Lipid Digestion and Absorption Occur in the Small Intestine

Most ingested lipid is in the form of triglycerides (fats). Fat digestion occurs almost entirely in the small intestine. The major digestive enzyme in this process is pancreatic **lipase**, secreted by cells of the pancreas into the small intestine. This lipase catalyzes the splitting of bonds in triglycerides, producing two free fatty acids and a monoglyceride.

Triglyceride → 2 fatty acids + Monoglyceride

Fats are poorly soluble in water and aggregate into large lipid droplets, as you can see if you shake a salad dressing made of oil and vinegar. Because pancreatic lipase is a water-soluble enzyme, its digestive action in the small intestine can take place only at the surface layer of a lipid droplet. Therefore, if most of the ingested fat remained in large droplets, lipid digestion would be very slow. The rate of digestion is substantially increased by the process of **emulsification**, which disrupts the large lipid droplets into many tiny droplets, each about 1 mm in diameter, thereby increasing their total surface area and exposure to lipase action. The resulting suspension of small lipid droplets is called an emulsion.

Emulsification of fat requires mechanical disruption of the large fat droplets into smaller droplets and an emulsifying agent, which coats the outer surface of the small droplets and prevents them from recombining back into larger droplets. The muscular contractions of the stomach and small intestine provide mechanical disruption to mix the luminal contents. Bile salts and phospholipids, both secreted in the bile, serve as emulsifying agents.

Although emulsification speeds up digestion of fats, absorption of the poorly soluble products of the lipase reaction is facilitated by a second action of the bile salts, the formation of **micelles**, which are similar in structure to emulsion droplets but much smaller at 4–7 nm in diameter (**Figure 45.12a**). Micelles consist of bile salts, phospholipids, fatty acids, and monoglycerides clustered together. Micelles continuously break down and re-form near the epithelium of the intestine. When a micelle breaks down, small lipid molecules are released into the solution and diffuse across the intestinal epithelium. As the lipids diffuse into epithelial cells, micelles release more lipids into the aqueous phase. Thus, the micelles keep most of the insoluble fat digestion products in small soluble aggregates while gradually releasing very small quantities of lipids to diffuse into the intestinal epithelium. Note that it is not the micelle that is absorbed but rather the individual lipid molecules that are released from the micelle.

During their passage through the epithelial cells, fatty acids and monoglycerides are resynthesized into triglycerides. This occurs in the smooth endoplasmic reticulum (SER), where the enzymes for triglyceride synthesis are located. This process lowers the concentration of cytosolic free fatty acids and monoglycerides in the epithelial cells, and so maintains a diffusion gradient for these molecules from the lumen into the cell. The resynthesized triglycerides aggregate into **chylomicrons**, large droplets coated with proteins that perform an emulsifying

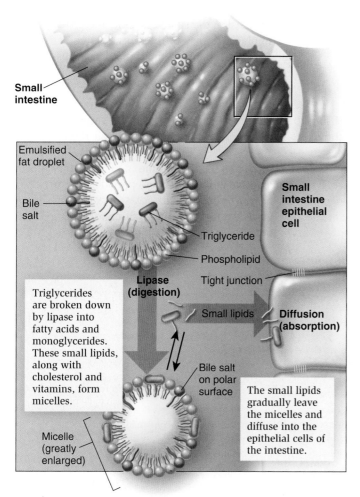

Small intestine

Emulsified fat droplet

Bile salt

Triglyceride

Phospholipid

Small intestine epithelial cell

Lipase (digestion)

Tight junction

Small lipids

Diffusion (absorption)

Triglycerides are broken down by lipase into fatty acids and monoglycerides. These small lipids, along with cholesterol and vitamins, form micelles.

Bile salt on polar surface

The small lipids gradually leave the micelles and diffuse into the epithelial cells of the intestine.

Micelle (greatly enlarged)

(a) Digestion of emulsified fats into small lipids, and absorption into intestinal cells

Figure 45.12 Digestion and absorption of emulsified fat in the small intestine.

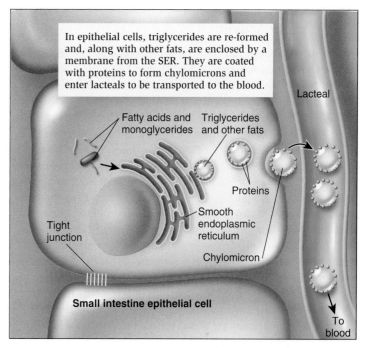

In epithelial cells, triglycerides are re-formed and, along with other fats, are enclosed by a membrane from the SER. They are coated with proteins to form chylomicrons and enter lacteals to be transported to the blood.

Lacteal

Fatty acids and monoglycerides

Triglycerides and other fats

Proteins

Tight junction

Smooth endoplasmic reticulum

Chylomicron

Small intestine epithelial cell

To blood

(b) Synthesis of triglycerides and the formation and release of chylomicrons

function similar to that of bile salts (**Figure 45.12b**). In addition to triglycerides, chylomicrons contain phospholipids, cholesterol, and fat-soluble vitamins that have been absorbed by the same process that led to fatty acid and monoglyceride movement into the epithelial cells of the small intestine.

Chylomicrons are released by exocytosis from the epithelial cells. Chylomicrons are too large to diffuse into blood capillaries. However, lacteals within the intestinal villi have large slit pores in their walls through which the chylomicrons can pass. The fluid from the lacteals eventually empties into a large vein and from there into the general blood circulation.

Vitamins, Minerals, and Water Are Not Digested but Must Be Absorbed

As stated earlier, vitamins, minerals, and water do not require digestion, and they are absorbed in their complete form. Most water-soluble vitamins are absorbed by diffusion or active transport in the small intestine. The fat-soluble vitamins—A, D, E, and K—follow the pathway for lipid absorption described in the previous section. Any interference with the secretion of bile or the action of bile salts in the intestine decreases the absorption of fat-soluble vitamins.

Water is the most abundant substance in chyme. Small amounts of ingested water are absorbed in the stomach, but the stomach has a small surface area available for diffusion and lacks the solute-absorbing mechanisms that create the osmotic gradients necessary for water absorption. The great majority of water absorption occurs in the small intestine. The epithelial membranes of the small intestine are very permeable to water, and water diffuses across the epithelium whenever an osmotic gradient is established by the active absorption of solutes, particularly Na^+, Cl^-, and HCO_3^-. Other minerals present in smaller concentrations, such as potassium, magnesium, and calcium, are also absorbed, as are trace elements such as zinc and iodide. The mechanisms of absorption of these molecules generally involve transport proteins and/or ion pumps.

45.6 Regulation of Digestion

The digestive systems of animals are under complex control regulated in part by other organ systems, especially the nervous and endocrine systems. Neurotransmitters from neurons and hormones from endocrine glands control the volume of saliva produced, the amount of acid produced in the stomach, the timing and amount of secretions from the gallbladder and pancreas, and the rate and strength of muscle contractions along the walls of the alimentary canal. In this section, we examine the major mechanisms by which the nervous and endocrine systems control the activity of the vertebrate digestive system.

The Nervous System Controls Muscle and Secretory Activity

The nervous system can affect the activities of the digestive system in two major ways: (1) local control of muscle and glandular activity by the nerves within the alimentary canal and (2) long-distance regulation by the brain.

Within the walls along the length of the gastrointestinal tract is a highly branched, interconnected collection of neurons that function in local control of the digestive system. These neurons interact with nearby smooth muscles, glands, and epithelial cells located within the tract. Stimulation of neuronal activity at one point along the alimentary canal can lead to impulses that are transmitted up and down the canal. When food enters the small intestine, for example, the intestine is stretched. This directly activates the neurons in the intestinal wall. Impulses are sent from these neurons to the muscles of the stomach, where they decrease the contractions of the stomach. This slows the rate at which chyme moves from the stomach into the small intestine, giving the intestine sufficient time for digestion and absorption. In this way, the alimentary canal can regulate its own function independent of the brain.

However, in long-distance regulation, the brain can communicate with neurons in the walls of the stomach and intestines and thereby influence the movement and secretory activity of the gastrointestinal tract. For example, emotional stress, a brain-related event, can affect digestive processes. Likewise, the sight, smell, and taste of food activate digestive functions even before food reaches the stomach. These stimuli when processed by the nervous system act via nerves from the brain as a feedforward mechanism so that saliva production, stomach activity, and digestion can begin even as the food is first ingested.

Hormones Regulate the Rate of Digestion

Hormones are chemical messengers secreted by specialized cells into the blood, where they travel to all parts of the body and act on various target cells (see Chapter 50). Hormones that control the digestive system are secreted mainly by cells scattered throughout the epithelium of the stomach and small intestine. One surface of each hormone-producing cell is exposed to the lumen of the gastrointestinal tract. At this surface, chemical substances in chyme stimulate the cell to release its hormones into the blood, by which they travel to their target cells. In this process, the presence of food in the stomach induces cells in the stomach epithelium to secrete a hormone called gastrin, which reaches all the parts of the stomach through the bloodstream (Figure 45.13). The presence of gastrin stimulates smooth muscle contraction in the stomach, which helps move chyme into the small intestine. Gastrin also stimulates acid production by

stomach epithelial cells. In the small intestine, the arrival of the acidic chyme stimulates release of the hormones cholecystokinin (CCK) and secretin from intestinal epithelial cells. These hormones travel through the bloodstream to reach their target cells. Cells of the pancreas respond to CCK and secretin by secreting digestive enzymes and acid-neutralizing bicarbonate ions into the small intestine. CCK also has a second function; it stimulates contraction of the gallbladder and therefore bile release.

Secretin holds a special place in the modern history of science. It is a polypeptide whose discovery provided the first clear understanding of how the gut and its accessory structures communicate, as well as initiating the field of endocrinology—the study of hormones.

1. The stomach secretes gastrin into the blood in response to a meal.

2. Gastrin returns to the stomach and stimulates muscle contraction and acid production.

3. The partially digested food, which is called chyme, moves into the small intestine and stimulates the secretion of CCK and secretin into the blood.

4. CCK and secretin stimulate the secretion of digestive enzymes and bicarbonate ions from the pancreas into the small intestine.

5. CCK also stimulates contraction of the gallbladder, which then releases bile into the small intestine.

Figure 45.13 Hormonal regulation of digestion in the stomach and small intestine.

Concept check: *Can you make a prediction about the possible effects, if any, of CCK on the stomach? Do you think CCK might stimulate or inhibit smooth muscle activity and acid production in the stomach? (Hint: Think about feedback principles you learned in Chapter 40.)*

FEATURE INVESTIGATION

Bayliss and Starling Discovered a Mechanism by Which the Small Intestine Communicates with the Pancreas

The dependence of digestive function on the nervous system and the brain was established by the pioneering work of Ivan Pavlov. Pavlov's work in the early 1900s suggested that nerves conveyed signals between the gut and other structures, such as the brain and pancreas, and in this way, the gut could communicate with these structures. Such communication is essential for the synchronized release of pancreatic enzymes and bicarbonate ions with the arrival of food in the intestine. Shortly after Pavlov's work, two English scientists, William Bayliss and Ernest Starling, hypothesized that nerves were not the only structures controlling gut activity.

To test this hypothesis, Bayliss and Starling carefully dissected away and cut the nerves to the small intestine of an anesthetized dog, as shown in **Figure 45.14**. Next, they directly injected a small quantity of acid into the intestinal lumen, because it was known at the time that the acidic contents arriving from the stomach somehow triggered secretion of digestive enzymes and bicarbonate ions from the pancreas. They placed a tube into the duct that carries secretions from the pancreas to the small intestine, so they could collect the pancreatic secretions. Each drop of secretion was recorded as a hatch mark on a chart recorder. They discovered that even in the absence of intact nerves, simply adding acid into the intestinal lumen

resulted in increased pancreatic secretions after a short lag period. Bayliss and Starling postulated that the lag was due to the time required for the intestinal cells to produce and secrete a factor that entered the blood and traveled to the pancreas.

To confirm their results and to counter the suggestion made by Pavlov and others that their first experiment merely failed to sever all the intestinal nerves, Bayliss and Starling performed a follow-up experiment. First, they removed a section of a dog's small intestine, added acid to it, and then ground it up in the acid. The mashed tissue was then filtered to produce an extract of any secretions that may have been produced by the intestine in response to the acid. Next, they injected the filtered extract into the bloodstream of another anesthetized dog whose pancreatic duct was opened as in the first experiment. What they discovered was remarkable: The extract derived from the acid-stimulated intestine of one dog caused the almost immediate secretion of pancreatic enzymes and bicarbonate ions from the second dog. This demonstrated that the small intestine secreted some chemical or chemicals in response to the presence of acid and that these secretions could reach the pancreas—presumably through the bloodstream—and stimulate the gland to release its contents. Bayliss and Starling called these chemicals hormones, from a Greek verb meaning to excite or arouse to action, and they named the hormone that activated the pancreas secretin. Thus was born not only the concept of a hormone but the recognition that these chemical messengers play a vital role in regulating the digestive system.

Figure 45.14   Bayliss and Starling discovered the mechanism by which the small intestine and pancreas work together in digestion.

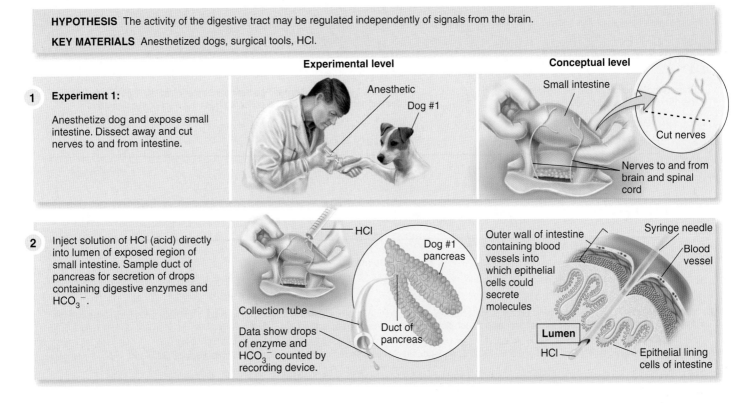

HYPOTHESIS The activity of the digestive tract may be regulated independently of signals from the brain.

KEY MATERIALS Anesthetized dogs, surgical tools, HCl.

Experimental level / Conceptual level

1   Experiment 1:

Anesthetize dog and expose small intestine. Dissect away and cut nerves to and from intestine.

Anesthetic
Dog #1
Small intestine
Cut nerves
Nerves to and from brain and spinal cord

2   Inject solution of HCl (acid) directly into lumen of exposed region of small intestine. Sample duct of pancreas for secretion of drops containing digestive enzymes and HCO_3^-.

HCl
Dog #1 pancreas
Collection tube
Data show drops of enzyme and HCO_3^- counted by recording device.
Duct of pancreas
Outer wall of intestine containing blood vessels into which epithelial cells could secrete molecules
Syringe needle
Blood vessel
Lumen
HCl
Epithelial lining cells of intestine

3 **THE DATA from Experiment 1:**

Time lag between response and injection of acid

Each hatch mark made on a recording device is due to a drop of secretion into the collection tube.

Record of drops

HCl added

Pancreatic secretions increased following HCl injection directly into small intestine.

Time (min)

4 **Experiment 2:**

Cut out a piece of an untreated region of small intestine from dog #1. Expose the piece to acid, grind it up, and filter the ground tissue. Inject filtrate into bloodstream of another dog (dog #2).

Stomach

Small intestine

Cut

HCl and intestine

Filtrate from ground up tissue

Dog #2

5 Test for pancreatic secretions into small intestine of dog #2.

Pancreas of dog #2

Pancreatic secretions present?

If the filtrate from dog #1 contained chemical signals released upon exposure to acid, those signals should stimulate the pancreas of dog #2 to secrete enzymes and HCO_3^-.

6 **THE DATA from Experiment 2:**

Prior to the injection of the filtrate into dog #2, the pancreas did not produce any secretion.

After the injection, pancreatic secretion increased rapidly. Each hatch mark corresponds to a drop of fluid collected from the duct of the pancreas.

Filtrate from dog #1 injected

Time (min)

7 **CONCLUSION** The small intestine contains molecules whose secretion into the blood is stimulated by acid. The molecules travel through the blood to the pancreas, causing secretion of pancreatic enzymes and HCO_3^-.

8 **SOURCE** Bayliss, W.M., and Starling, E.H. 1902. The mechanism of pancreatic secretion. *Journal of Physiology* 28:325–353.

In the rest of this chapter, we will examine the impact of digestive system disorders on human health.

Experimental Questions

1. Explain the first experiment conducted by Bayliss and Starling, which indicated that other factors besides signals from nerves may influence digestive gland secretion.

2. What criticism did Bayliss and Starling need to address to provide more conclusive evidence that the secretion was not due to neural regulation, and how did they address it?

3. What conclusions did the investigators draw from their second experiment?

45.7   Impact on Public Health

As we have seen, the functioning of the vertebrate digestive system is extraordinarily fine-tuned by the brain, nerves, and hormones. When people are well nourished, therefore, you might assume that digestive problems would be rare. However, each year in the U.S. alone, gastrointestinal complaints account for approximately 40 million visits to doctors and 10 million visits to hospitals and emergency rooms. Among the most common GI problems registered by hospitals and physicians are heartburn, ulcers, and diarrhea.

Excess Stomach Acid Production Can Lead to Heartburn

Approximately one in four people in the U.S. suffers at some time from heartburn. The term heartburn is a misnomer, because although the painful burning sensations of this disorder are felt in the vicinity of the heart, they are caused by stomach acid and arise within the esophagus, not the heart. Normally, the stomach contents do not move backward into the esophagus, largely because of the muscular sphincter at the esophageal/stomach juncture. However, under some conditions, the sphincter either does not close entirely or is forced open by the pressure of material in the stomach. When this happens, the acid in the stomach lumen enters the esophagus and irritates nerve endings there.

Many circumstances may contribute to heartburn. Overeating, for example, enlarges the volume of the stomach to the point of forcing its contents through the esophageal sphincter. Lying down after a big meal removes the effect of gravity on food in the stomach and may allow some acid to leak backward. Heartburn is also associated with smoking and consumption of alcohol, citrus fruits (which are acidic), chocolate (which contains caffeine, a known inducer of heartburn), and fatty foods (which take longer to digest than other foods and therefore remain in the stomach longer). One of the most common causes is pregnancy. Toward the middle of pregnancy, the growing fetus pushes up on the abdominal contents, which tends to force material from the stomach into the esophagus.

Common antacids contain carbonate ions, which buffer the acid in the stomach and esophagus. In severe cases, heartburn can damage the walls of the esophagus enough to cause a chronic cough and pain, or even perforate the esophagus. Antacids may not be sufficient to treat these patients, who are instead given drugs that inhibit the stomach's ability to produce acid. These drugs are among the most widely prescribed medications in the U.S.

Erosion of the Walls of the Alimentary Canal Causes an Ulcer

Although heartburn is usually treated before the acid begins eroding away the lining of the esophagus or stomach, occasionally this does occur. Erosion of any portion of the alimentary canal due to any cause is called an ulcer. Most ulcers occur in the stomach, lower esophagus, and the part of the small intestine that connects to the stomach, because these sites have the greatest acid concentrations. Ulcers are typically less than an inch wide; if untreated, luminal contents may leak into the surrounding body cavity where enzymes and acids from the stomach can do considerable damage. As many as 20 million Americans have an ulcer. Each year, 40,000 patients require surgery to repair tissue damaged by ulcers, and ~6,000 die due to complications from the disease.

Acid is essential for ulcer formation, but we now know that it is not usually the primary cause. In fact, many patients with ulcers have perfectly normal rates of acid production. Contrary to popular belief, stress is probably not a major cause of ulcers either. So what is the main cause of ulcers?

In the 1980s, two Australian scientists, Barry Marshall and J. Robin Warren, proposed that most stomach ulcers arise from a bacterial infection. This idea did not gain quick acceptance because many scientists believed that bacteria could not survive the acidic conditions in the stomach. However, Marshall and Warren demonstrated the existence of *Helicobacter pylori* in the stomachs of a majority of patients with ulcers, and killing the bacteria with antibiotics proved to be an effective treatment. Once the bacteria are gone, the normal body repair mechanisms heal the wound created by their secretions. Marshall and Warren received the 2005 Nobel Prize in Medicine for their revolutionary discovery, which has improved the quality of life for millions of people. Currently, a combined treatment of antibiotics and the same drugs used to neutralize or inhibit stomach acid production in heartburn patients is effective in reducing or eliminating the symptoms of ulcers.

Diarrhea Is the Most Common Gastrointestinal Disorder Worldwide

Just in the U.S., there are over 1 billion cases of diarrhea—loose, watery stools occurring at least three times per day—every year. Nearly all episodes of diarrhea run their course in a day or two. Typically they result from infection with a pathogen, such as a virus or bacterium. Other times, however, diarrhea may be caused by food sensitivities (such as lactose intolerance, discussed earlier), reactions to medications, stress-related disorders, or parasites that inhabit the rectum and large intestine.

Most cases of pathogen-related diarrhea in the U.S. result from exposure to one of several related types of bacteria, including those of the genus *Salmonella*, *Shigella*, and *Escherichia*. Often, these infections are handled by the body's immune system within a day or two, but sometimes medical attention is required. One cause of diarrhea stands apart as particularly dangerous, however. Cholera is a disease caused by the bacterium *Vibrio cholerae*, usually ingested by consuming contaminated food or water. According to the World Health Organization, at least 2,000 people die of cholera each year, with another 100,000 people contracting the disease but surviving. Nearly all cases of cholera occur in Africa, parts of Asia (notably China), and India, although scattered outbreaks have occurred nearly everywhere except the U.S., with the most recent worldwide pandemic lasting from 1961 to 1971. *V. cholerae* releases a toxin that alters the permeability of salts in the large intestine, resulting in a massive flow of these ions, followed by water, into the intestinal lumen. As in all cases of diarrhea, the chief concern with cholera is the loss of nutrients and water and the dehydration that ensues. In addition to killing the bacteria with antibiotics, therefore, the major treatment of cholera is the same as for any cause of diarrhea, which includes drinking solutions of salts and water to replace those that were lost in the feces.

■ Summary of Key Concepts

45.1 Animal Nutrition

- The four phases of food use in animals are ingestion, digestion, absorption, and elimination.
- Animals require organic nutrients—carbohydrates, proteins, lipids, and nucleic acids—and inorganic nutrients in the form of water and minerals. Differences in nutritional demands reflect an animal's physiology and environment.
- Essential nutrients (essential amino acids, essential fatty acids, vitamins, and minerals) must be obtained from the diet. There are eight essential amino acids that cannot be synthesized by humans and many other animals; they are required for building proteins. Vitamins are organic nutrients that serve as coenzymes for metabolic and biosynthetic reactions. Minerals are inorganic ions that serve many functions. (Tables 45.1, 45.2, 45.3)

45.2 Ingestion

- Herbivores eat only plants, carnivores consume animal flesh or fluids, and omnivores eat both plant and animal products.
- Animals display a variety of strategies for obtaining food. Three main strategies are suspension, bulk, and fluid-feeding.
- Suspension feeders sift water to filter out organic matter. Bulk feeders include carnivores, herbivores, and omnivores. Carnivores are generally predators or scavengers; they have teeth evolved for biting and tearing. The teeth and jaw muscles of herbivores are adapted for grinding plants. Fluid-feeders do

not need teeth except to puncture an animal's skin. (Figures 45.2, 45.3, 45.4)

45.3 Principles of Digestion and Absorption of Food

- Intracellular digestion occurs in single-celled organisms and simple invertebrates.
- Most animals digest food via extracellular digestion. Flatworms and cnidarians have a gastrovascular cavity with one opening that serves as both entry and exit port. (Figure 45.5)
- All other animals have an alimentary canal, open at the mouth and anus or cloaca and segregated into specialized regions, through which food passes from one end to the other.
- Digestion of food requires hydrolytic enzymes.
- Once food has been digested, the nutrients must be absorbed via passive diffusion, facilitated diffusion, or primary or secondary active transport.

45.4 Overview of Vertebrate Digestive Systems

- The vertebrate digestive system consists of the alimentary canal plus associated structures, not all of which are found in all vertebrates. The anterior portion of the canal contains the oral cavity, salivary glands or mucus-secreting cells, pharynx (throat), and esophagus. The middle portion contains food storage or digestive organs (crop, gizzard, and/or stomach, depending on species), and the upper part of the small intestine. The posterior part of the alimentary canal contains the remainder of the small intestine, the large intestine (except in fishes), the rectum, and the anus or cloaca. (Figure 45.6)
- Saliva helps animals lubricate, chew, and swallow food. Peristalsis moves food through the pharynx and esophagus to the stomach. Some animals have a crop where food is stored and softened. (Figure 45.7)
- The stomach stores food, partially digests proteins, and regulates the rate at which chyme empties into the small intestine.
- Herbivores rely on microbes living within their digestive tracts to digest cellulose. (Figure 45.8)
- Nearly all digestion and absorption occur in the small intestine. The combination of villi and microvilli increases the small intestine's surface area and maximizes the efficiency of digestion and absorption. (Figure 45.9)
- The pancreas secretes digestive enzymes and a bicarbonate-rich fluid that neutralizes the acidic chyme. The liver secretes bile, which is stored in the gallbladder and aids in the digestion of fat. (Figure 45.10)
- The large intestine concentrates undigested material, which is then eliminated from the anus or cloaca.

45.5 Mechanisms of Digestion and Absorption in Vertebrates

- Carbohydrate digestion occurs in the small intestine of most vertebrates. Starch is digested by amylases into molecules of the disaccharide maltose. Maltose and other disaccharides are digested into monomers by enzymes located at the brush border and transported across epithelial cells into the blood. (Figure 45.11)
- Humans with lactose intolerance cannot digest lactose because the intestinal enzyme lactase is not expressed after weaning, a trait that is linked to genetic background.

- Proteins are broken down to peptide fragments in the stomach by pepsin; these fragments are further broken down into amino acids in the small intestine by trypsin and chymotrypsin.

- Lipid digestion occurs entirely in the small intestine by the action of pancreatic lipase. (Figure 45.12)

- Most water-soluble vitamins are absorbed by diffusion or active transport. Fat-soluble vitamins follow the pathway for fat absorption.

- The large intestine absorbs some of the remaining salts and water that were not absorbed in the small intestine.

45.6 Regulation of Digestion

- The digestive systems of animals are regulated in part by other organ systems, especially the nervous and endocrine systems. The nervous system can affect the digestive system in two major ways: (1) local control of muscular and glandular activity by the nerves in the alimentary canal and (2) long-distance regulation by the brain.

- Several hormones work together to regulate the rate of digestion. Bayliss and Starling demonstrated that a hormone from the small intestine triggers secretions from the pancreas into the small intestine. (Figures 45.13, 45.14)

45.7 Impact on Public Health

- Common GI complaints registered by hospitals and physicians include heartburn, ulcers, and diarrhea.

Assess and Discuss

Test Yourself

1. The process of enzymatically breaking down large molecules into smaller molecules that can be used by cells is
   a. absorption.   c. ingestion.
   b. secretion.   d. digestion.

2. The term "essential nutrients" refers to
   a. all the carbohydrates, proteins, and lipids ingested by an organism.
   b. nutrients that an animal cannot manufacture.
   c. nutrients that must be obtained from the diet in their complete form.
   d. all of the above.
   e. b and c only.

3. An animal that has a strong preference for a particular food but can adjust its diet if necessary is said to be
   a. herbivorous.   c. carnivorous.   e. both b and d.
   b. omnivorous.   d. opportunistic.

4. Which of the following statements is not correct?
   a. Fluid-feeders include animals that consume animal or plant fluids.
   b. Suspension feeding occurs only in invertebrates.
   c. Predators are animals that kill and consume live prey.
   d. Herbivory occurs in both aquatic and terrestrial species.
   e. Blood-sucking animals may contain anticoagulants in their saliva.

5. The pancreas connects to which part of the alimentary canal?
   a. esophagus   c. small intestine   e. large intestine
   b. stomach   d. cecum

6. Which of the following statements regarding the vertebrate stomach is not correct?
   a. Its cells secrete the active protease enzyme pepsin.
   b. It is a saclike organ that evolved to store food.
   c. Its cells secrete hydrochloric acid.
   d. It is the initial site of protein digestion.
   e. Little or no absorption of nutrients occurs there.

7. Absorption in the small intestine is increased by
   a. the many villi that are present on the inner surface of the small intestine.
   b. the brush border formed by microvilli on the cells of the villi.
   c. the presence of numerous transporter molecules on the epithelial cells.
   d. all of the above.
   e. a and b only.

8. In birds, the secretion of acid and pepsinogen occurs in
   a. the crop.   d. the cloaca.
   b. the gizzard.   e. the gallbladder.
   c. the proventriculus.

9. Bile is produced by
   a. the liver.   d. the small intestine.
   b. the gallbladder.   e. the cecum.
   c. the pancreas.

10. Which of the following is a function of the large intestine?
    a. It participates in cellulose digestion by microbes that exist in the cecum of herbivores.
    b. It stores and concentrates fecal material.
    c. Its cells absorb salts and water that remain in chyme after it leaves the small intestine.
    d. Its cells absorb certain vitamins produced by bacteria.
    e. All of the above are functions of the large intestine.

Conceptual Questions

1. Distinguish between digestion and absorption.

2. Explain some of the differences between the teeth of herbivores and carnivores and how these differences are adaptive for the animal's diet.

3. Explain the functions of the crop and gizzard in birds. Why don't humans have a crop or gizzard?

Collaborative Questions

1. Describe several strategies that animals use to obtain food.

2. Define "nutrient." On the basis of that definition, do you consider water a nutrient? Briefly discuss the essential nutrients, vitamins, and minerals that animals must obtain through their diet.

Online Resource

www.brookerbiology.com

Stay a step ahead in your studies with animations that bring concepts to life and practice tests to assess your understanding. Your instructor may also recommend the interactive ebook, individualized learning tools, and more.

Control of Energy Balance, Metabolic Rate, and Body Temperature

46

A genetically obese, leptin-deficient mouse and a normal mouse.

In 1997, two young cousins being treated for extreme obesity were brought to the attention of researchers studying a newly discovered hormone called leptin. Leptin had been demonstrated in laboratory rodents to inhibit appetite. The researchers hypothesized that the children—who weighed approximately 30 kg by age 2, and in one case, 86 kg by age 9—were not producing leptin. This hypothesis was confirmed by molecular analyses that identified a mutation in the children's leptin gene. Subsequently, several other individuals with mutations in leptin were identified, leading in each case to extreme childhood obesity. Treatment of these individuals with leptin has proven beneficial in restoring normal body weight. Although leptin deficiency is rare and is only one of many possible causes of obesity in humans, animal models of leptin deficiency, such as the mice shown in the chapter-opening photo, represent a remarkable breakthrough in our understanding of the genetic bases of the control of appetite and metabolism. How nutrients that supply energy are processed, stored, and used by animals' bodies in times of food abundance or fasting is a subject of this chapter.

In Chapter 45, we saw that all animals require nutrients to assemble the macromolecules that make up body tissues. Some of the ingested nutrients, such as carbohydrates, lipids, and proteins, also represent a form of fuel, or energy, that can be used to generate ATP within cells. Most animals, however, do not have a constant supply of nutrients. Insects, cephalopods, and all vertebrates, for instance, sleep or have periods of greatly reduced activity for part of the day, during which time they do not eat. In addition, environmental changes may reduce the food supply, leading to long fasts. As a consequence of the irregular and sometimes unpredictable flow of nutrients into the body, animals have evolved an array of mechanisms to adequately maintain levels of important fuel molecules even during a fast. In the first two sections of this chapter, we will explore how ingested nutrients are stored in the body for such times of need and the mechanisms by which these stores are tapped.

Metabolism refers to all the bodily activities and chemical reactions in an organism that maintain life. **Metabolic rate** describes the rate at which an organism uses energy to power these reactions. Animals may have widely different metabolic rates, which determine the amount of nutrients they need. The greater an animal's metabolic rate, the more heat it generates as a by-product of breaking down nutrients and using the energy of their chemical bonds to synthesize ATP. Some of this heat escapes to the environment, and some is used to warm an animal's body. Metabolism and body temperature are therefore closely related, and we will examine this relationship in the next two sections of the chapter. We will also discuss energy balance, the balance between energy expenditure and consumption. We conclude with the public health impact of human disorders associated with metabolism, including obesity.

46.1 Nutrient Use and Storage

Once nutrients have been ingested, they are used or stored by the body. The use of nutrients can be divided into two alternating phases. The **absorptive state** occurs when ingested nutrients enter the blood from the gastrointestinal tract. The **postabsorptive state** occurs when the gastrointestinal tract is empty of nutrients and the body's own stores must supply energy. During the absorptive period, some ingested nutrients supply the immediate energy requirements of the body. The rest are added to the body's energy stores to be called upon during the next postabsorptive period. An average meal in a human requires about 4 hours for complete absorption. Therefore, our usual three-meal-a-day pattern places us in the postabsorptive state during the late morning and afternoon and part of the night. Total-body energy stores are adequate for the average human to withstand a fast of several weeks. By contrast, some animals can barely survive a single skipped meal—particularly if they have low energy reserves—because their metabolic needs are much higher than our own. In this section, we will focus on nutrient absorption, use, and storage in vertebrates, with a closer look at mammals.

In the Absorptive State, Nutrients Are Absorbed and Used for Energy or Stored

The categories of nutrients that are absorbed during the absorptive state include or are formed from carbohydrates, lipids, proteins, nucleic acids, vitamins, minerals, and water. We will look in depth at the first three of these; **Figure 46.1** gives an overview of what happens to these nutrients. Digested carbohydrates are absorbed as monosaccharides, particularly glucose. Digested lipids (fats) are absorbed after first being resynthesized into triglycerides in intestinal epithelial cells. Proteins are broken down into amino acids and then absorbed.

Absorbed Carbohydrates: Energy or Storage The chief carbohydrate monomers absorbed from the gastrointestinal tract of vertebrates are glucose, galactose, and fructose. In mammals, the latter two sugars are either converted to glucose by the liver or enter the same metabolic pathways as glucose does.

Glucose is one of the body's two major energy sources during the absorptive state (triglycerides being the other). Much of the absorbed glucose enters cells and is enzymatically broken down, resulting in the formation of hydrogen ions, carbon dioxide, and water and, in the process, releasing the energy required to form ATP (Figure 46.1a). Because skeletal muscle makes up a large fraction of body mass in most vertebrates, it is a major consumer of glucose, particularly

when an animal is active. In all vertebrates, skeletal muscle also incorporates some of the glucose into the polymer glycogen (see Chapter 3), which is stored in the muscles to be used later. If more glucose is absorbed than is needed for immediate energy, a portion of the excess is incorporated into glycogen in the liver, and the remainder into triglycerides in fat cells. The structures of glycogen and triglycerides are described in Chapter 3.

Absorbed Triglycerides: Storage Triglycerides are too large to diffuse across the plasma membranes of the intestinal epithelial cells. As described in Chapter 45, they are digested into fatty acids and monoglycerides in the lumen of the small intestine and then resynthesized into triglycerides once they diffuse into the intestinal epithelial cells. The triglycerides and other ingested lipids (for example, cholesterol) are packaged into chylomicrons, which enter lymph and from there the blood circulation. As blood moves through adipose tissue, a blood vessel enzyme called lipoprotein lipase releases the fatty acids from the triglycerides in the chylomicrons. The released fatty acids then enter adipose cells and combine with glycerol to re-form triglycerides (Figure 46.1b). These triglycerides are stored in fat cells until the body requires additional fuel.

As with glucose, some of the ingested fat is not stored but is used by most organs other than the brain during the absorptive state to provide energy. The relative amounts of carbohydrate and fat used for energy during the absorptive period depend largely on the composition of a meal.

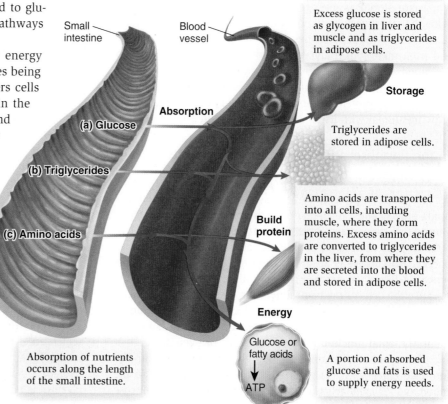

Figure 46.1 **Events of the absorptive state.** The products of digestion are absorbed into the blood along the length of the small intestine. (Note that for simplicity, triglycerides are shown being absorbed directly into blood, not into lacteals; see Chapter 45 for details.) These nutrients are used for immediate energy needs, or they are deposited in cells as energy stores or as macromolecules important in cell function, such as for building proteins.

Excess glucose is stored as glycogen in liver and muscle and as triglycerides in adipose cells.

Storage

Triglycerides are stored in adipose cells.

Amino acids are transported into all cells, including muscle, where they form proteins. Excess amino acids are converted to triglycerides in the liver, from where they are secreted into the blood and stored in adipose cells.

Build protein

Energy

Glucose or fatty acids → ATP

A portion of absorbed glucose and fats is used to supply energy needs.

Absorption of nutrients occurs along the length of the small intestine.

Absorption

(a) Glucose
(b) Triglycerides
(c) Amino acids

Small intestine
Blood vessel

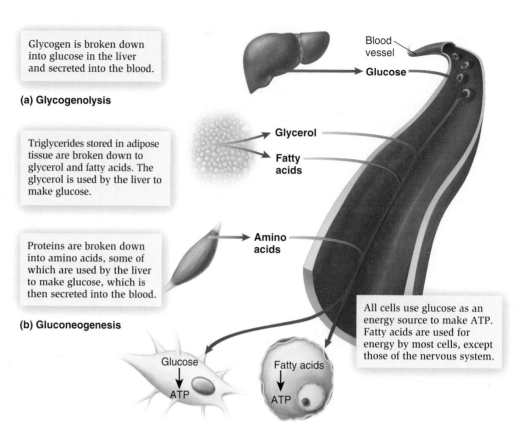

Glycogen is broken down into glucose in the liver and secreted into the blood.

(a) Glycogenolysis

Triglycerides stored in adipose tissue are broken down to glycerol and fatty acids. The glycerol is used by the liver to make glucose.

Proteins are broken down into amino acids, some of which are used by the liver to make glucose, which is then secreted into the blood.

(b) Gluconeogenesis

All cells use glucose as an energy source to make ATP. Fatty acids are used for energy by most cells, except those of the nervous system.

Figure 46.2 Events of the postabsorptive state.   In this state, macromolecules formed during the absorptive state are broken down to supply monomers that can be used for energy. **(a)** This begins with glycogenolysis—the breakdown of glycogen into glucose. **(b)** In gluconeogenesis, the breakdown products of triglycerides and proteins, namely, fatty acids, glycerol, and amino acids, can be used for energy either directly or indirectly by their conversion in the liver to glucose.

Absorbed Amino Acids: Protein Building   Amino acids are taken up by all body cells, where they are used to synthesize proteins (Figure 46.1c). All cells require a constant supply of amino acids, because proteins are constantly being synthesized and degraded. However, unlike excess glucose and fatty acids which are stored as glycogen and triglycerides, excess amino acids that are ingested are not stored as protein. Instead, amino acids are enzymatically converted by liver cells into fatty acids and from there into triglycerides, which then get released into the blood and are taken up and stored in adipose cells. Therefore, eating large amounts of protein does not normally increase stores of body protein. An exception would be a young, rapidly growing animal with its continuous increase in body protein.

In the Postabsorptive State, Stored Nutrients Are Released and Used

As the postabsorptive state begins, synthesis of glycogen and lipids slows, and the breakdown of these substances begins. During this state, macromolecules formed during the absorptive state are broken down to supply monomers that can be used for energy. No glucose is absorbed from the intestines during this time, yet the blood glucose concentration must be maintained because the cells of the central nervous system (CNS) normally use only glucose for energy. A large decrease in blood glucose levels can disrupt CNS functions, ranging from subtle impairment of mental function to seizures, coma, or even death.

The events that maintain blood glucose concentration fall into two categories: (1) reactions that provide glucose to the blood and (2) cellular use of fat for energy, thus sparing glucose for the CNS. Let's look at each of these.

Glucose from Glycogen and Noncarbohydrate Precursors
Vertebrates can increase their blood glucose levels during the postabsorptive period in two major ways: by breaking down glycogen and by making new glucose. First, the glycogen that was formed during the absorptive period can be broken back down into molecules of glucose by hydrolysis, a process known as **glycogenolysis** (Figure 46.2a). This occurs primarily in the liver, following which the glucose is released into the blood, where it can travel to all cells. Muscle glycogen is also broken down into glucose by glycogenolysis, but this glucose is used exclusively by muscle cells and not secreted into the blood.

The amount of liver glycogen available to provide glucose during the postabsorptive period varies among animals, but it is generally sufficient to maintain blood glucose levels for only a brief time, such as an overnight fast in a human. Therefore, a second mechanism for maintaining blood glucose levels is required when the postabsorptive period continues longer. In the process of **gluconeogenesis** (literally, creation of new glucose), enzymes in the liver convert noncarbohydrates into glucose, which is then secreted into the blood (Figure 46.2b). This process occurs in all vertebrates but appears to be especially important in mammals.

A major precursor for gluconeogenesis is glycerol, which is released from triglycerides in adipose tissue by the breakdown process called lipolysis. In lipolysis, enzymes within fat cells hydrolyze triglycerides into fatty acids and glycerol, both of which enter the bloodstream. The fatty acids diffuse into cells, where they are used as an alternate energy source to glucose (except for the CNS, which continues to require glucose). The glycerol is taken up by the liver, where enzymes convert it into glucose, which is then released back into the bloodstream.

Table 46.1	Relative Changes in the Use and Generation of Energy Sources During the Absorptive and Postabsorptive Periods					
	Glucose absorption from gut	**Glucose use by cells**	**Synthesis of triglycerides**	**Use of fatty acids for energy by cells**	**Breakdown of glycogen in muscle and liver**	**Blood levels of glucose**
Absorptive period	High	High	High	Low/moderate	Low	Normal
Postabsorptive period	Absent	Moderate (some glucose is spared for CNS use)	Low	High	High	Normal

When the postabsorptive period continues for an extended period of time—as when an animal fails to find food—protein becomes an important source of blood glucose. Large quantities of protein in muscle and other tissues can be broken down to amino acids without serious tissue damage or loss of function. The amino acids enter the blood and are taken up by the liver. In the liver, the amino group is removed from each amino acid, and the remaining molecule is converted into glucose. This process, however, has limits. Continued protein loss can result in the death of cells throughout the body because they depend on proteins for such vital processes as plasma membrane function, enzymatic activity, and formation of organelles.

Lipid Metabolism by Other Tissues: Glucose Sparing
Another way that glucose is made available to the organs and tissues that need it the most—such as the brain—in the postabsorptive state is by having other organs and tissues reduce their dependence on glucose. They do this by increasing their use of fat as an energy supply during this period. This metabolic adjustment, called **glucose sparing**, reserves (or spares) the glucose produced by the liver by glycogenolysis and gluconeogenesis for use by the nervous system.

The essential step in glucose sparing is lipolysis, the breakdown of adipose tissue triglycerides, which, as stated earlier, liberates fatty acids and glycerol into the blood. In vertebrates, the circulating fatty acids are taken up and used to provide energy by almost all tissues, excluding the nervous system, whose cells do not express the enzymes required to break down fatty acids for energy.

Of the vertebrate body's tissues and organs, the liver is unique in that most of the fatty acids entering it during the postabsorptive state are not used by that organ for energy. Instead they are processed into three small compounds collectively called **ketones**, or ketone bodies. Ketones are released into the blood during prolonged fasting. They provide an important energy source for the many tissues, including the brain, capable of oxidizing ketones through the citric acid cycle.

The use of fatty acids and ketones during fasting provides energy for the body, sparing the available glucose for the brain. Moreover, as just mentioned, the brain can use ketones as energy, and it does so increasingly as ketones build up in the blood during the first few days of a fast. The survival value of this phenomenon is significant. When the brain reduces its glucose requirement by using ketones, much less protein breakdown is required to supply amino acids for gluconeogenesis.

Protein stores last longer, enabling the animal to survive a long fast without serious tissue damage.

The combined effects of glycogenolysis, gluconeogenesis, and glucose sparing are so efficient that, after several days of complete fasting, human blood levels of glucose fall by only a small percentage. Table 46.1 summarizes the relative changes in these and other variables during the absorptive and postabsorptive periods.

46.2 Regulation of the Absorptive and Postabsorptive States

Tight control mechanisms are required to maintain homeostatic levels of energy-providing molecules in the blood. These controls come in two forms: endocrine and nervous. Cells of the endocrine system produce blood-borne signaling molecules called hormones. Several hormones as well as signals arising from cells of the nervous system work together in this coordination. In this section, we will learn that one common function of the endocrine and nervous systems is to regulate the processes of glycogenolysis and gluconeogenesis so that glucose is made available to cells at all times.

Insulin Is a Key Regulator of Metabolism

The blood concentration of insulin, a hormone made by the pancreas, increases during the absorptive state and decreases during the postabsorptive state. **Insulin** regulates metabolism in several ways, primarily by regulating the blood glucose concentration. It does this by promoting the transport of glucose from extracellular fluid into cells, where it can be used for metabolism. Glucose is a relatively large, polar molecule that cannot cross plasma membranes without the aid of a transporter protein. Insulin increases glucose uptake by binding to a cell-surface receptor and stimulating an intracellular signaling pathway. This pathway increases the availability of transport proteins called glucose transporters (GLUTs) in the plasma membrane (Figure 46.3). These GLUTs are located within preformed vesicles that are stored in the cytosol of cells. When these vesicles are stimulated, they fuse with the plasma membrane, making more GLUTs available to transport glucose into the cell. Consequently, insulin lowers the blood glucose level. Insulin also regulates metabolism by inhibiting glycogenolysis and gluconeogenesis in the liver, because these processes are not required during the absorptive state.

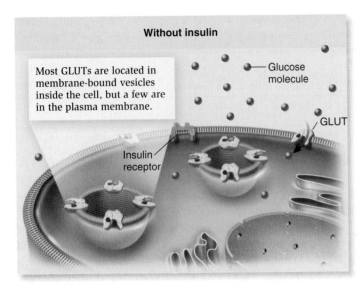

Without insulin

Most GLUTs are located in membrane-bound vesicles inside the cell, but a few are in the plasma membrane.

Glucose molecule

GLUT

Insulin receptor

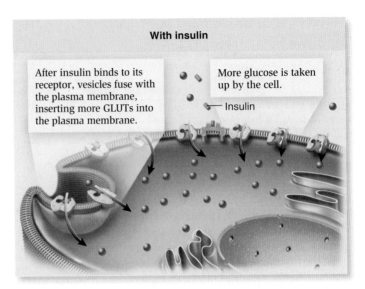

With insulin

After insulin binds to its receptor, vesicles fuse with the plasma membrane, inserting more GLUTs into the plasma membrane.

More glucose is taken up by the cell.

Insulin

Figure 46.3 The effect of insulin on the location of glucose transport (GLUT) proteins.   The presence of insulin results in the fusion of intracellular vesicles containing GLUT proteins with the plasma membrane, where they facilitate glucose uptake into the cell.

Concept check: *What benefit is there to having GLUT proteins premanufactured and stored in intracellular vesicles?*

Insulin exerts its effects mainly on muscle cells and adipose tissue cells, because these cells have insulin receptors in their plasma membranes. However, not all GLUTs are identical, and not all require insulin for their activity, as described next.

Genomes & Proteomes Connection

GLUT Proteins Transport Glucose in Animal Cells

All animal cells require transporters to move glucose across their plasma membranes. GLUTs in mammals make up a family of at least 14 related proteins that share similar structures but are expressed in different tissues. GLUTs typically have 12 membrane-spanning helical domains that anchor the protein in plasma membranes. The structures of GLUTs are very similar across phyla. For example, the coding sequence of one *Drosophila* GLUT gene is nearly 70% identical to that of one of the human GLUTs. In more closely related taxa, such as birds and mammals, GLUTs are even more similar, with up to 95% of their amino acid sequences being the same. These and other considerations indicate that the GLUTs arose by accumulated mutations of a common ancestral gene.

The different GLUTs vary in their ability to bind glucose. For example, some GLUTs have high affinity for glucose, and others have low affinity. High affinity means the protein binds glucose even at very low concentrations of glucose. Let's look at the properties of three mammalian GLUTs, named GLUT1, GLUT3, and GLUT4. Muscle and fat cells express the protein GLUT4, which has a low affinity for glucose, but one that is sufficient for the concentration of glucose normally found in blood. This is also the only GLUT protein whose movement to the plasma membrane requires the cell to be stimulated by insulin. Consequently, as the glucose level rises in the blood

after a meal, insulin recruits more GLUT4 molecules from the cell cytosol to the plasma membrane of muscle and fat cells.

By comparison, GLUT1 and GLUT3 are found predominantly in the brain, where they act in concert to mediate the transport of glucose from blood vessels to the interstitial fluid of the brain, and from there into brain cells. GLUT1 and GLUT3 have much higher affinity for glucose than do other GLUT proteins. This means that neurons of the brain can transport glucose into their cytosol even when the concentration of glucose in the extracellular fluid is very low. In addition, insulin is not required for GLUT1 and GLUT3 to be present in the plasma membrane, unlike the situation for GLUT4. As a result of these properties of GLUT1 and GLUT3, neurons of the brain would still receive adequate energy for survival even if an animal's blood glucose level plummeted, perhaps due to disease or starvation.

Expressing multiple types of the same functional class of protein means that different parts of an animal's body can meet their own particular metabolic demands. Moreover, these needs may change during development. For example, high-affinity GLUTs such as GLUT1 and GLUT3 are present in high numbers in embryonic cells when there is a greater need for glucose but smaller numbers during other stages of life.

Because insulin, through its actions on GLUT4, is the key regulatory molecule that controls blood glucose levels, it is important to understand the regulation of insulin production and release. This is described next.

The Blood Glucose Concentration Is Maintained Within a Normal Range

Like all important variables in an animal's body, blood glucose concentrations are controlled by a system of checks and balances. In the absorptive state, after a meal has been eaten, the blood glucose level increases; in the postabsorptive state, depending on how long it lasts, the level may go too low. What homeostatic mechanisms keep the blood glucose level within a normal range?

Increased Glucose in the Absorptive State   The primary factor controlling the secretion of insulin from the pancreas is the blood glucose level (**Figure 46.4a**). An increase in an animal's blood glucose level directly stimulates the cells of the pancreas to secrete insulin in proportion to the amount of glucose in the blood. Later, after insulin helps the cells of the body take up glucose, the resulting decrease in the blood glucose level removes the signal for insulin secretion. This is an example of negative feedback, as described in Chapter 40.

In addition to blood glucose level, inputs from the nervous system to the pancreas also play a role in the regulation of insulin secretion. During a meal, signals from the parasympathetic nervous system (see "rest-or-digest" response in Chapter 42) stimulate the secretion of insulin into the blood, ultimately reducing circulating glucose levels back to normal.

Decreased Glucose in the Postabsorptive State   Several factors act in concert to prevent blood glucose from decreasing below the normal homeostatic range, even during a short fast. Otherwise, glucose could fall so low—a condition called hypoglycemia—that despite the high-affinity GLUT1 and GLUT3 proteins in brain cells, there would not be enough glucose to keep brain cells alive.

If for any reason—such as a prolonged fast—the blood glucose level decreases below the normal homeostatic range for an animal, neurons that respond to changes in the extracellular concentration of glucose are activated within the hypothalamus (**Figure 46.4b**). Signals from the hypothalamus then stimulate the production of glucose-elevating factors. These include numerous hormones, notably another one produced by the pancreas called glucagon. **Glucagon** is a protein hormone that stimulates the processes of glycogenolysis, gluconeogenesis, and the synthesis of ketones in the liver. In addition, certain other hormones from various endocrine glands, as well as the neurotransmitter norepinephrine released from neurons of the sympathetic nervous system, stimulate adipose tissue to release fatty acids into the blood. The fatty acids diffuse across plasma membranes and provide another source of energy for the synthesis of ATP. The overall effect is to increase the blood concentrations of glucose, fatty acids, and ketones during the postabsorptive period or during a prolonged fast.

More Energy Is Required During Exercise or Stress

We think of exercise as something humans do for fun or fitness, but in its broadest sense, **exercise** can be defined as any physical activity that increases an animal's metabolic rate. Generally, an animal becomes active to seek something, such as food, shelter, or a mate, or to elude something, such as a predator or a storm. What we consider exercise can also be a response to a stressful situation. The type of exercise animals engage in,

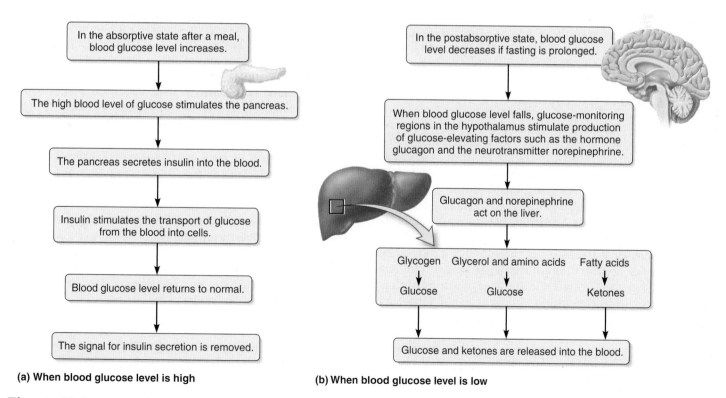

(a) When blood glucose level is high

(b) When blood glucose level is low

Figure 46.4   Maintenance of normal blood glucose levels.

Concept check: *What are the sources of the glycerol, fatty acids, and amino acids used to make glucose?*

therefore, can be quite varied. When a cheetah sprints after a small antelope, for example, the activity of both predator and prey is brief and intense, perhaps lasting only a few seconds. By contrast, a tuna may never stop swimming, and a migrating bird may fly a hundred miles a day or more over a span of weeks.

For any type of exercise, including that which results from an animal's behavioral responses to stress, nutrients must be available to provide the energy required for such things as skeletal muscle contraction, increased heart and lung activity, and increased activity of the nervous system. These energy-providing nutrients include glucose and fatty acids as well as the muscle's own glycogen.

The liver supplies the blood with the glucose used during exercise by breaking down its glycogen stores and by gluconeogenesis. This occurs even in the absorptive state, and thus the blood glucose level increases above normal when an animal exercises at such times. In addition, an increase in adipose tissue lipolysis releases fatty acids into the blood, which provides an additional source of energy for the exercising muscle.

These events are mediated by the same hormones and nerves responsible for the regulation of the postabsorptive state. For example, inputs from the sympathetic nervous system to the pancreas inhibit insulin secretion during exercise or acutely stressful situations. Consequently, glucose transport into muscle and fat cells is decreased, which tends to raise blood glucose levels (this is part of the fight-or-flight response described in Chapter 42). Because the brain does not depend on insulin for glucose transport across neuronal membranes, as just described, more of the body's supply of glucose is available to the brain at such times, while other cells can use fatty acids for energy. Therefore, the body uses all available forms of energy in response to fasting, exercise, and stress.

46.3 Energy Balance

Animals have a wide range of metabolic needs that depend on numerous factors. Active animals, such as migrating birds, burn fuel more rapidly than inactive animals, such as hibernating mammals. Juveniles typically burn fuel more quickly than mature animals. An animal's **energy expenditure** is the same as its metabolic rate, which was already defined as the amount of energy it uses in a given period of time to power all of its metabolic requirements.

A fundamental characteristic of energy is that it can be neither created nor destroyed, but it can be converted from one form to another (see Chapter 6). The breakdown of organic molecules liberates energy locked in their chemical bonds and transfers it to the bonds in ATP. This is the energy that cells use to perform various biological activities such as muscle contraction, active transport, and molecular synthesis. We refer to these functions as work. Not all of the energy is used to do work, however. Some of it appears as heat, which contributes to an animal's body temperature or is dissipated to the environment. In this section, we examine how metabolic rate is measured in animals and how a balance is achieved between energy consumption and expenditure.

Metabolic Rate Can Be Measured by Calorimetry and Compared Among Different Animals

The standard unit of energy is the joule, but biologists have historically quantified the energy of metabolism in calories. A **calorie** (equivalent to 4.187 joules) is the amount of heat required to raise the temperature of 1 gram of water 1 degree Celsius. Most biological activities, however, require much greater amounts of energy than a calorie, and consequently, the more common unit of measurement is the kilocalorie (1,000 calories, or **kcal**). (In food labeling, a Calorie with a capital C is also the same as kcal.) As noted above, the metabolic rate is the total energy expenditure of an animal per unit of time. Biologists often measure and compare the metabolic rates of different animals to learn, for example, how some animals are capable of hibernating, how an animal's body temperature influences its metabolic rate, and how hormones and other factors alter an animal's metabolism.

The most common measure used to compare the metabolic rates of different species is the **basal metabolic rate** (**BMR**). The BMR is called the metabolic cost of living, and most of it can be attributed to the routine functions of the heart, liver, kidneys, and brain. In the basal condition, the animal is at rest in the postabsorptive state and at a standard temperature. For **endotherms**, animals that generate their own internal heat and maintain a relatively narrow range of body temperatures, the standard temperature is within the range that causes the animal to neither gain heat (for example, by shivering) nor lose heat (for example, by perspiration). This temperature range is called an animal's thermoneutral zone. The BMR of **ectotherms**, animals whose body temperature changes with the environmental temperature, must be measured at a standard temperature for each species—one that approximates the average temperature that a species normally encounters. In this case, the term **standard metabolic rate** (**SMR**) is used instead of BMR, because the basal condition in ectotherms is harder to define than for endotherms. Because BMR and SMR apply only to resting, postabsorptive animals at a standard temperature, any animal that has recently eaten or been active has a higher metabolic rate than its basal metabolic rate.

Two methods used to obtain a good estimate of BMR are direct and indirect calorimetry. **Direct calorimetry** was invented by French chemist and biologist Antoine Lavoisier in 1780 (**Figure 46.5**). In Lavoisier's procedure, an animal was placed in an enclosed, insulated chamber surrounded by ice. As the animal metabolized fuel, it generated heat, which dissipated into the chamber and melted the ice. The amount of water collected from an opening at the bottom of the chamber could be used to estimate the amount of heat generated by the animal. Today, direct calorimetry is measured using more sophisticated instrumentation, but the principle remains the same. It provides an accurate measure of metabolism, because energy expenditure and heat production are directly related. However, the method is not very practical, particularly with large animals.

The second and more practical method of measuring BMR, **indirect calorimetry**, is based on the principle that animals require oxygen to metabolize fuel. The more fuel being

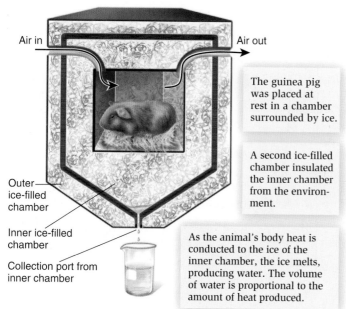

Figure 46.5   Lavoisier's pioneering method of measuring BMR: direct calorimetry.   Direct calorimetry determines an animal's basal metabolic rate, such as that of this guinea pig, by determining how much the heat given off by its body raises the temperature of a surrounding ice-filled chamber.

Air in

Air out

The guinea pig was placed at rest in a chamber surrounded by ice.

A second ice-filled chamber insulated the inner chamber from the environment.

Outer ice-filled chamber

Inner ice-filled chamber

Collection port from inner chamber

As the animal's body heat is conducted to the ice of the inner chamber, the ice melts, producing water. The volume of water is proportional to the amount of heat produced.

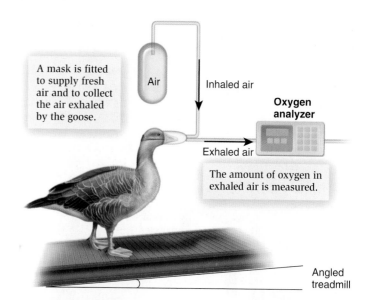

A mask is fitted to supply fresh air and to collect the air exhaled by the goose.

Air

Inhaled air

Oxygen analyzer

Exhaled air

The amount of oxygen in exhaled air is measured.

Angled treadmill

Figure 46.6   Measuring BMR via oxygen consumption with indirect calorimetry.   Many animals, such as this goose, can be trained to walk on a treadmill, which allows scientists to compare metabolism during rest and exercise. Oxygen consumption can be determined by sampling the air exhaled into a tightly fitting mask. One-way valves prevent inhaled and exhaled air from mixing.

Concept check: *If the air had been sampled from the mask just before starting the treadmill while the goose was not yet exercising, would this have been a good estimate of BMR?*

metabolized—that is, the greater the BMR—the more oxygen must be consumed by the animal. By measuring the rate at which an animal uses oxygen, therefore, we can obtain a good estimate of BMR. Indirect calorimetry can also be used to compare the metabolic rates of an animal during rest and activity, when oxygen consumption increases (**Figure 46.6**). One limitation to this method is that a small percentage of fuel is metabolized without oxygen consumption, and thus indirect calorimetry underestimates actual metabolic rate.

Activity, Digestion, and Size Influence Metabolic Rate

BMR and SMR are indicators of total body resting metabolic rates. However, metabolic rate is not always basal. Not all tissues in the body use oxygen and produce heat at the same rate. Some structures, such as skin, consume relatively little oxygen under resting conditions, whereas others, such as the brain, heart, and liver, have high rates of metabolism even when an animal is sleeping. Moreover, the metabolic rates of different tissues can vary independently of each other. For example, the metabolism of the gut increases when food is being digested, and that of skeletal muscle increases during exercise (**Figure 46.7**). You can see from this that many factors have an impact on metabolism, including skeletal muscle activity, whether an animal has recently eaten, the size of an animal, and whether an animal is frightened, sleeping, or hibernating.

The primary factor that increases metabolic rate is altered skeletal muscle activity. Even minimal increases in muscle contraction significantly increase metabolic rate; strenuous activity increases it even more. For example, the total daily expenditure of kilocalories may vary for a healthy adult human from

approximately 1,350 kcal for a small person at rest to more than 7,000 kcal for a cyclist competing in the Tour de France. Changes in muscle activity also affect metabolic rate during sleep due to decreased muscle activity, and during exposure to cold temperatures due to increased muscle activity from shivering.

Eating and digesting food also increase the metabolic rate. Particularly in mammals that eat meat, this may raise metabolic rate (and associated heat production) by 10–50% for a few hours after eating. You may have noticed this **food-induced thermogenesis** after consuming a large meal, such as Thanksgiving dinner. Ingested protein—for example, turkey—produces the greatest effect, whereas carbohydrate and fat produce less. The increased heat is believed to result partly from the processing of the absorbed nutrients by the liver and from the energy expended by the gastrointestinal tract in digestion and absorption. Food-induced thermogenesis is observed in nearly all vertebrates, but it is most notable in certain reptiles that eat large and infrequent meals. Because of food-induced thermogenesis, BMR tests must be performed in the postabsorptive state.

Another factor affecting metabolic rate is body size. In general, a large animal uses greater amounts of energy than does a small animal because the large animal has more mass and more cells, all of which consume fuel and generate heat. The total energy expenditure and heat generation of an elephant is clearly greater than that of a mouse, for instance. However, when the energy expenditures of an elephant and a mouse are scaled to their respective body masses, we find that

Figure 46.7 **Changes in metabolic rate of selected structures in a mammal's body during different activities.** Red areas are regions of high metabolism; blue areas are regions of low metabolism. Note that during exercise, skeletal muscle becomes more active, and areas associated with food digestion and absorption become less active. At all times, however, the heart, liver, and brain are highly active.

Concept check: Why does metabolism decrease in the gut during exercise?

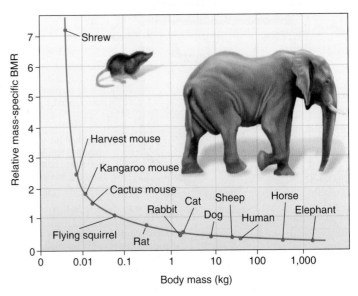

Figure 46.8 **Metabolic rates of animals that differ in size.** Metabolism can be scaled to body mass by measuring oxygen consumption and normalizing it to the animal's body mass (mass-specific BMR). Note that when expressed in this way, the mass-specific BMR of a shrew is higher than that of an elephant, even though the total oxygen consumption and heat output of the elephant would be much greater. The values on the y-axis are relative units of metabolism. Sizes of the animals shown are not to scale.

Concept check: Can the relationship between body size and metabolic rate be used to propose hypotheses about metabolic rates of extinct animals?

the energy expenditure per gram of body mass in a mouse is much higher than the comparable calculation in an elephant. **Mass-specific BMR** is the amount of energy expended per gram of body mass. Mass-specific BMR is a relative term that allows scientists to compare the metabolic rates among animals of different sizes. Research has shown that the relationship between mass-specific BMR and body mass follows an exponential curve (**Figure 46.8**). One possible explanation for this phenomenon is that because the ratio of an animal's surface area to its body mass is greater in smaller animals than in larger animals, smaller animals lose heat more rapidly than larger ones. According to this hypothesis, smaller animals must generate more heat per gram of body mass than larger animals to replace their heat loss. However, while this hypothesis appears to provide an explanation for the relationship between metabolism and body size in endotherms, it does not explain the observation that the same relationship exists in almost all animals, including ectotherms.

The smallest endotherms—hummingbirds, shrews, mice and other rodents, and some bats—face the special challenge of fueling their very high mass-specific metabolic rates. This becomes difficult and even impossible during cold months or any time when food is unavailable. Many animals have evolved a strategy of lowering their internal body temperature to just a few degrees above that of the environment, a process called **torpor**. Torpor may occur on a nightly basis, while the animal

continues to be active during the daylight hours, or it can extend for months, in which case it is called **hibernation**. Torpor-like conditions are not unique to small vertebrates, however, as even some large animals such as bears enter extended periods of rest or torpor during months when food is not available.

A reduction in body temperature reduces the metabolic demands of all body cells. BMR in a small hibernating rodent, for example, may drop to less than 1% of what it would be at the normal body temperature of around 38°C. This allows the animal to conserve energy for remarkably long times. Ground squirrels of the genus *Spermophilus*, for example, may hibernate for up to 8 months.

Total Body Energy Stores Are Balanced Between Consumption and Expenditure

When the daily amount of energy an animal consumes is equal to the amount of energy it expends, the animal's body weight remains stable. Tipping the balance in either direction causes weight gain or loss, that is, the total body mass increases or decreases. Normally, energy is stored in the form of fat in adipose tissue.

Body weight in an adult animal is usually regulated around a predetermined set point that differs among species. Body weight is maintained by adjusting caloric intake and energy expenditure in response to changes in body weight. This mechanism usually works very precisely in those animals in which it has been studied. For example, a mammal that eats less one day will eat enough the next day to compensate for the previous day's deficit. Similarly, if an animal is overfed one day, it may eat less the next day. Scientists currently think that appetite and metabolism change when food intake is more or less than the amount required to maintain the body's set point. This phenomenon explains why some dieters initially lose weight easily and then become stuck at a plateau, because appetite increases and metabolic rate decreases to compensate for the weight loss. Conversely, it also helps explain why some very thin people have difficulty gaining weight.

Hormones and the Nervous System Control Food Intake

Short-term control of feeding generally involves a feeling of **satiety**, that is, satisfaction or fullness. As an animal's stomach and small intestine stretch to accommodate food, nerves send signals from these structures to the hypothalamus. At the same time, the stomach and small intestine release hormones into the blood that reach the hypothalamus and suppress appetite. These **satiety signals** remove the sensation of hunger and set the time period before hunger returns again.

Over the long term, such as weeks, months, or years, an animal's total energy consumption tends to remain fairly stable. The animal may go through a period of fasting and then make up for it when food becomes available. In other words, meals may vary from day to day or during different seasons of the year, but the average amount of food consumed over the long term does not change much. Long-term control of food intake is mediated by many different brain molecules, by hormones, and by emotional state, particularly in humans.

One hormone that has received considerable attention in recent years for its ability to control appetite and metabolic rate is **leptin** (from the Greek *leptos*, meaning thin). Leptin has been identified in all classes of vertebrates but has been most extensively studied in mammals. In these animals, leptin is produced by adipose cells in proportion to fat mass: As more fat is stored in the body, more leptin is secreted into the blood. Leptin acts on the hypothalamus to reduce appetite and increase metabolic rate (**Figure 46.9a**). In this way, the brain is made aware of how much fat is stored in the body at all times, and it can adjust appetite and energy expenditure appropriately if fat stores decline or increase.

If an animal fasts for a period of time, its adipose cells shrink as they release their stored fat into the blood. The decrease in leptin secretion resulting from the decreased fat mass results in a decrease in BMR and an increase in appetite. This may be the true evolutionary significance of leptin, namely that its disappearance from the blood lowers the BMR, consequently prolonging life during periods of starvation (**Figure 46.9b**). Leptin was not discovered until 1994, but its existence was postulated decades before that by the pioneering work of Douglas Coleman, who investigated the nature of mutations in mice that result in obesity.

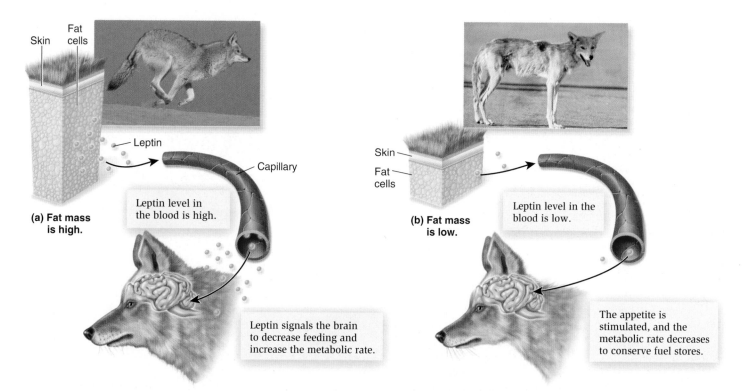

Figure 46.9 **The role of leptin in regulating appetite and metabolic rate.** In animals such as this coyote, changes in the blood leptin level result directly from changes in fat mass. Animals with more fat make more leptin.

FEATURE INVESTIGATION

Coleman Revealed a Satiety Factor in Mammals

For many years, scientists wondered how most animals appeared to regulate their body mass around a predetermined set point, despite fluctuations in the food supply. They postulated that other parts of the body somehow communicated with the brain to signal when energy stores were above or below normal. In the 1970s, Canadian-American researcher Douglas Coleman tested this hypothesis in an experiment involving parabiosis, the surgical connection of the abdominal walls of two animals, such that the blood supply from one animal intermixes with that of the other.

Coleman used two strains of mice called ob and db mice, carrying different mutations that resulted in inherited forms of obesity (an example of an ob mouse is shown in the chapter-opening photo). Coleman first connected a wild-type mouse, one that lacked these mutations, with either an ob mouse or a db mouse, as shown in Figure 46.10. He discovered that when the circulatory system of the ob mouse was in contact with that of the wild-type mouse, the ob mouse ate less and gained less weight than usual. This suggested that the blood of the wild-type mouse contained a circulating factor that signals the brain when an animal has sufficient fat stored in its body and adjusts appetite accordingly. The ob animal apparently lacked this factor, but when exposed to it through the wild-type mouse's cir-

culation, it responded in the appropriate way. The wild-type mouse of the parabiosis pair apparently retained a sufficient amount of the factor in its blood, because it maintained its body weight at a normal level.

Coleman noticed, however, that a db mouse continued to gain weight at an abnormally high rate even when parabiosed with a wild-type mouse. In this case, the wild-type animal actually lost weight while the db animal remained obese. Coleman concluded that the db mouse must produce the same factor as the wild-type mouse, but for some reason, it was unable to respond to it. The wild-type mouse that was parabiosed to the db mouse lost weight because it received the factor from the db mouse, in addition to having its own supply of the factor. Thus, whether the factor was absent as in the ob mouse, or present but unable to function as in the db mouse, the resulting phenotype was the same—obesity.

In 1994, Jeffrey Friedman and coworkers at Rockefeller University identified this circulating factor as the hormone leptin. The ob mice were found to be homozygous for a mutation in the leptin gene, which produced an inactive leptin molecule, whereas db mice produced leptin but did not respond to it. In fact, the db mice were found to produce even greater amounts of leptin than wild-type mice, which explained why the wild-type mouse in Coleman's experiments lost weight when parabiosed with a db mouse. Friedman and others later showed that

Figure 46.10 Coleman's parabiosis experiments revealed a satiety factor in wild-type mice that was absent in genetically obese mice.

HYPOTHESIS Body weight is controlled by a factor that circulates in the blood. This factor is absent in strains of mice that have an inherited form of obesity.

KEY MATERIALS Two different strains of genetically obese mice, normal (wild-type) mice.

Experimental level　　　　　　　　　Conceptual level

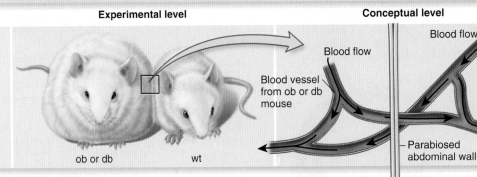

1　Surgically connect the abdominal walls of an obese and normal (wt; wild-type) mouse. After a few days, blood vessels from each mouse cross to the other mouse. Monitor changes in body weight. Note: 2 different strains of obese mice were tested, called ob and db mice.

ob or db　　　wt

Blood flow

Blood vessel from ob or db mouse

Blood flow

Blood vessel from wt mouse

Parabiosed abdominal wall

2　Feed mice a normal diet for several weeks, then visually inspect and weigh each pair.

Weight-reducing factor now appears in circulation of ob mouse.

Ob mouse vessel

Wild-type mouse vessel

Weight-reducing factor

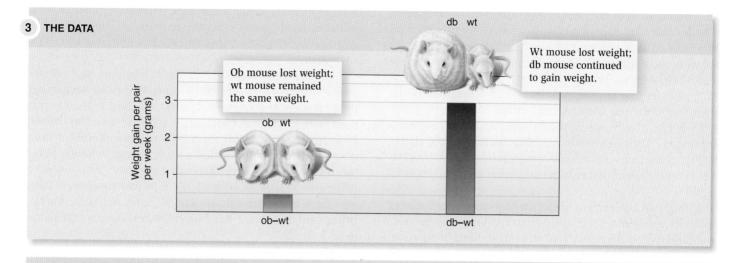

3 **THE DATA**

Ob mouse lost weight; wt mouse remained the same weight.

Wt mouse lost weight; db mouse continued to gain weight.

db wt

ob wt

ob–wt

db–wt

Weight gain per pair per week (grams)

4 **CONCLUSION** Wild-type mice secrete a blood-borne factor that reduces body weight. The factor is absent from ob mice but present in db mice. Ob mice retain the ability to respond to the factor, unlike db mice, which cannot respond to it.

5 **SOURCE** Coleman, D.L. 1973. Effects of parabiosis of obese with diabetes and normal mice. *Diabetologia* 9:294–298.

adipose cells produce leptin in direct proportion to the total fat mass of an animal, as stated earlier. Biologists now know that all classes of vertebrates produce leptin.

At first, the work of Coleman and Friedman generated considerable excitement that leptin might be useful to treat obesity in humans, but this has thus far proven difficult. Why? Recent research has revealed that most obese humans are more like the db mice than the ob mice. That is, they produce leptin but fail to respond adequately to it, and therefore, simply increasing blood levels of leptin may not have a significant effect on body weight. However, other studies have shown that leptin normally acts in nonobese humans in a manner much like it does in wild-type rodents. As noted in the chapter introduction, researchers have identified rare individuals in whom leptin is not produced due to a mutation in the leptin gene. These individuals are extremely obese and respond well to injections of leptin, losing considerable weight. The body weight disorders of such individuals, therefore, are reminiscent of ob mice.

Experimental Questions

1. What observation led to the experiments conducted by Coleman?

2. What was the hypothesis tested by Coleman, and how did he test it?

3. How did the experimental linking of the bloodstreams of the wild-type mice and the mutant mice affect the body weight of both strains?

46.4 Regulation of Body Temperature

As we have seen, metabolic rate is linked to body temperature. Next, we discuss why body temperature is important for the health and survival of all animals, and consider mechanisms by which animals gain or lose heat.

Temperature Affects Chemical Reactions, Protein Function, and Plasma Membrane Structure

Most animals can survive only in a relatively narrow range of temperatures. Temperature has an effect on animals' bodies in three main areas: chemical reactions (and the enzymes that catalyze them), other protein function, and membrane structure. First, chemical reactions depend on temperature. Heat accelerates the motion of molecules, so as an animal's temperature rises, the rates at which the molecules in its body move and contact each other also increase. Consequently, the rate of most chemical reactions in animals doubles or even triples for every 10°C increase in body temperature. In addition, enzymes, which catalyze many reactions in the body, including those involved in metabolism, have an optimal temperature range for their maximal catalytic function. Low temperatures slow down chemical reactions, making it harder for an animal to remain active and carry out internal functions such as digestion, reproduction, and immunity. The latter is particularly important, as many vertebrates become susceptible to disease when their body temperatures are reduced for long periods.

A second effect of unusually high temperature is that it causes many proteins to become denatured; that is, they lose the three-dimensional structure that is crucial to their ability to function properly. This occurs because the bonds that form tertiary and quaternary protein structures result from weak interactions, such as hydrogen bonds, and can be disrupted by heat. Denaturation of enzymes is especially serious because of the major role they play in metabolism. Most animals have an upper limit of body temperature at which they can survive. To start with, most mammals have a resting body temperature of 35–38°C (95–100°F). In humans, a body temperature of 41°C (106°F) causes loss of protein function and breakdown of the nervous system, and a body temperature of 42–43°C (107–109°F) is fatal. Birds, which have slightly higher resting body temperatures than mammals (approximately 40–41°C [104–106°F]), cannot survive at body temperatures above 46–47°C (115–117°F). At environmental temperatures greater than 50°C (122°F), nearly all animals die.

A third effect of temperature is that heat alters the structures of plasma and intracellular membranes. At low temperatures, membranes become less fluid and more rigid. Rigid membranes are less able to perform biological functions, such as transporting ions and fuels and binding extracellular molecules to receptors on the membrane surface. Alternatively, if the temperature becomes too high, membranes can become leaky.

Compared to extreme heat, animals can better tolerate extreme cold. For example, some animals can survive after freezing and thawing. This is normally dangerous because ice crystals form inside cells and rupture membranes. Also, the ice forms from water in the cells' cytosol, which dehydrates the cells. However, many ectotherms, including certain insects such as the woolly caterpillar, a few species of amphibia such as several types of frogs, and a very small number of reptiles such as the painted turtle, can prevent crystal formation in their cells. They do this by responding to ice on their skin surfaces with an enormous outpouring of glucose from the liver. The glucose enters the blood and cells, acting like antifreeze and lowering the freezing point so that the cells do not completely freeze solid. Other regions of the body that are less critical, such as the lumens of the stomach and bladder, do freeze. These animals can have 65% or more of their bodies completely frozen for long periods, only to thaw during warm periods without harmful effects. The glucose is reabsorbed by the liver at that time. An alternative strategy to cope with extremely cold temperatures is seen in certain fishes inhabiting Arctic or Antarctic waters. As described in Chapter 23, such fishes do not freeze like the terrestrial vertebrates just described, but instead contain large amounts of highly specialized antifreeze proteins in their blood that stop ice crystal formation.

Ectotherms and Endotherms May Have Fluctuating or Stable Body Temperatures

In the past, animals were classified as either cold blooded or warm blooded. Cold-blooded animals were thought to require an external heat source such as sunlight to warm themselves. By contrast, warm-blooded animals were thought to use internal heat to maintain their body temperature. These terms are misleading and incorrect, however, because many cold-blooded animals generate considerable heat by exercising their skeletal muscles. Indeed, many have a body temperature during daylight hours that is at least as warm as that of warm-blooded

animals like birds and mammals. Moreover, many so-called warm-blooded animals spend portions of their lives with greatly reduced body temperatures that are close to that of their environment.

Biologists now classify animals according to both their source of heat and their ability to maintain body temperature. Recall that ectotherms depend on external heat sources to warm their bodies, whereas endotherms use their own metabolically generated heat to warm themselves. **Homeotherms** maintain their body temperature within a narrow range, whereas **heterotherms** have body temperatures that vary with the environment (**Figure 46.11**). Most animals fall into two categories. Generally, birds and mammals are endothermic and homeothermic. Other vertebrates and most invertebrates are ectothermic and heterothermic.

Not all animals, however, can be neatly classified into two categories at all times. Hibernating mammals, for example, are endotherms. They are usually homeothermic, but during the winter, their body temperature drops dramatically as their metabolism slows to conserve energy for the winter. Hibernators behave like heterotherms during the transition from fall to winter and again from winter to spring. During the winter, however, they are again homeothermic except for brief periods of arousal, but at a lower body temperature than at other times of year. Similarly, a fish swimming in deep ocean waters is an ectotherm but also homeothermic because the temperature of the water—and therefore of its body—is essentially constant. Fishes that live in waters with fluctuating temperatures, by contrast, are ectothermic and heterothermic.

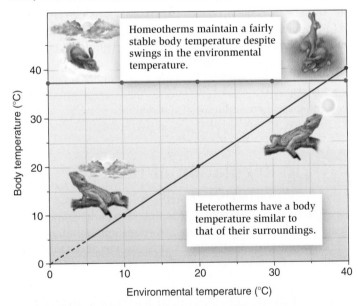

Figure 46.11   Body temperature and environmental temperature in homeotherms and heterotherms. Homeotherms maintain stable body temperatures across a wide range of environmental temperatures. Heterotherms, by contrast, have body temperatures that depend in part on the temperature of their surroundings.

Concept check: *What thermoregulatory categories do humans fit into?*

Even endothermic homeotherms do not have truly constant body temperatures. They have a narrow range of body temperatures that increases or decreases slightly in extreme climates, during exercise, or during sleep. The important feature is that birds and mammals can quickly adjust the body's mechanisms for retaining or releasing heat such that body temperature remains within the required narrow range. This provides the advantage that the body's chemical reactions are at optimal levels even when the environment imposes extreme challenges. The metabolic rate of a resting mammal, for example, is roughly six times greater than that of a comparably sized reptile. A suddenly awakened mammal is instantly capable of intense activity even on a winter day, but an icy-cold reptile could be at the mercy of a predator because of the time required to warm itself in order to flee.

Endothermy does have three major disadvantages, however. First, to produce sufficient heat by metabolic processes, endotherms must consume larger amounts of food to provide the nutrients used by cells in the formation of ATP, during which heat is generated. Small endotherms with high BMRs, such as shrews, must eat almost continuously and may die if deprived of food for as little as a day. By contrast, many ectotherms, such as snakes, can go for weeks without eating. Second, endotherms have a greater risk of hyperthermia, or overheating, during periods of intense activity, even in cold weather. Third, because lowering body temperature requires the evaporation of bodily fluids (as described next), endotherms are often restricted to environments where water is plentiful.

Animals Exchange Heat with the Environment in Four Ways

The surface of an animal's body can lose or gain heat from the external environment via four mechanisms. These are radiation, evaporation, convection, and conduction (Figure 46.12).

Radiation is the emission of electromagnetic waves by the surfaces of objects. The rate of emission is determined by the temperature of the radiating surface. Thus, if the body surface is warmer than the environment, the body loses heat at a rate that depends on the temperature difference. If the outside temperature is warmer than body temperature, the body gains heat, for instance from sunlight. We can observe radiated heat from an animal's body with imaging devices that detect infrared light (Figure 46.13).

Animals can lose body heat through **evaporation** of water from the skin and membranes lining the respiratory tract, including the surface of the tongue. A large amount of energy in the form of heat is required to transform water from the liquid to the gaseous state. Whenever water vaporizes from the body's surface, the heat required to drive the process is conducted from the surface, thereby cooling the animal.

Convection is the transfer of heat by the movement of air or water next to the body. For example, the air close to an endotherm's body is heated by conduction. Because warm air is less dense than cold air, the warm air near the body rises and carries away heat by convection. Convection is aided by

Figure 46.12   Types of heat exchange.   The four ways in which animals exchange heat with the environment are radiation, evaporation, convection, and conduction.

Figure 46.13   Visualization of heat exchange in an ectotherm and an endotherm.   Thermal-imaging cameras can detect heat radiated from an animal's body. Note the warm skin of the endotherm (the human) and the cold skin of the ectotherm (the tarantula), even though both animals are at the same environmental temperature.

creating currents of air around an animal's body. Humans do this by sitting near fans, but other animals can create cooling air currents by other means such as when an elephant waves its ears or a bat flaps its wings.

In **conduction**, the body surface loses or gains heat through direct contact with cooler or warmer substances. The greater the temperature difference, the greater is the rate of heat transfer. Different materials have different abilities to absorb heat, however. As we saw in Chapter 2, water has a higher heat capacity than air, meaning that at any temperature, water will retain greater amounts of heat than will air. Consequently, aquatic animals in water that is 10°C lose considerably more heat in a short time than terrestrial animals lose in air that is 10°C. Indeed, on a hot day, terrestrial animals can lose heat efficiently by immersing themselves in cooler water. An animal's body surface area plays an important role in the rate of heat conduction across its body surface. In some animals, certain body regions are particularly good heat conductors, such as the ears of an elephant or the wings of a bat.

The four processes of heat transfer just described can be regulated in animals, such that heat is retained within the body at some times and lost at other times, as we see next.

Several Mechanisms Can Alter Rates of Heat Gain or Loss in Endotherms

For purposes of temperature control, think of an endotherm's body as a central core surrounded by a shell consisting of skin and subcutaneous (just below the skin) tissue. The temperature of the central core of endotherms is regulated at approximately 37–40°C (97–104°F), but the temperature of the outer surface of the skin varies considerably. If the skin were a perfect insulator, the body would never lose or gain heat by conduction. The skin does not insulate completely, however, so the temperature of its outer surface generally is somewhere between that of the external environment and the core. Only in animals that store large amounts of subcutaneous fat (blubber) does the body surface provide considerable insulation. In endotherms without blubber, the main form of insulation is a covering of hair, fur, or feathers, which traps heat from the body in a layer of warm air near the skin, reducing heat loss. Given these structures, then, let's take a look at mechanisms that different endotherms use to regulate how much heat is gained or lost from their surface. We will cover four such mechanisms here: changes in skin blood flow, countercurrent heat exchange, evaporative heat loss, and behavioral adaptations.

Changes in Skin Blood Flow   Rather than acting as an insulator as in most animals, in some animals the skin functions as a heat exchanger that can be adjusted to increase or decrease heat loss from the body. Skin surface blood vessels dilate (widen to increase blood flow) on hot days to dissipate heat to the environment, and they constrict (get narrower to reduce blood flow) on cold days to retain body heat (**Figure 46.14**). Signals from the nervous system regulate the relaxation or contraction of the smooth muscles that control the opening and closing of these blood vessels. Diving birds and diving mammals are good examples of animals that use this mechanism. Ducks, seals, and walruses dramatically reduce the amount of blood flowing to the skin when they dive in cold waters. This allows them to retain body heat that would otherwise be conducted into the water. In many terrestrial endotherms, certain

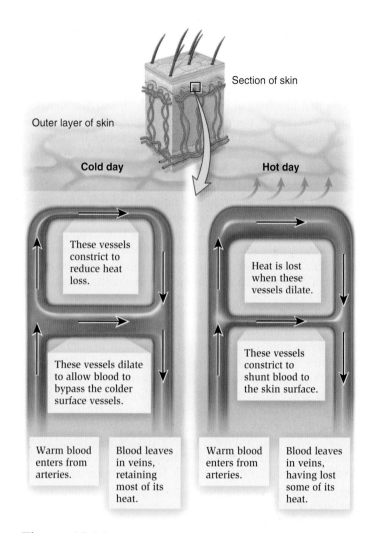

Figure 46.14   Regulation of heat exchange in the skin. As shown in this schematic illustration, the skin functions as a variable heat exchanger. The arrows in the blood vessels indicate direction of blood flow.

areas of skin play a more prominent role in heat exchange than others—recall the elephant ears and bat wings mentioned earlier—so skin temperature varies with its location in the body.

Countercurrent Heat Exchange   Both endotherms and ectotherms regulate heat loss to the environment through **countercurrent heat exchange**, which retains heat by returning it to the body's core and keeping the core much warmer than the extremities. In endotherms, countercurrent heat exchange occurs primarily in the extremities—the flippers of dolphins, for example, or the legs of birds and certain other terrestrial animals (**Figure 46.15a**). As warm blood travels from the core through arteries down a bird's leg, heat moves by conduction from the artery to adjacent veins carrying cooler blood in the other direction (**Figure 46.15b**). By the time the arterial blood reaches the tip of the leg, its temperature has dropped considerably, reducing the amount of heat lost to the environment, and returning the heat to the body's core.

Ectotherms such as many fishes use countercurrent heat exchange to warm their muscles. As the swimming muscles become active, they generate heat from metabolism that warms the blood in the veins leaving the muscles. The veins are in

Warm arterial blood
40°C   38°C

5°C

10°C   5°C

Cold venous blood

Heat is transferred from arteries to nearby veins, thereby reducing heat loss to the water.

(a) Countercurrent heat exchange in the leg of an endotherm

Artery

Veins

Artery

Veins

(b) Cross section and surface view of veins covering an artery

Figure 46.15 Countercurrent heat exchange.
(a) Countercurrent exchange retains heat in the leg of an endotherm such as this bird. Black arrows in vessels indicate direction of blood flow. (b) An SEM of the arrangement of veins surrounding an artery in a bird's leg. The artery is almost completely covered by overlying veins, allowing efficient heat exchange between the vessels.

close contact with nearby arteries bringing fresh, oxygen-rich blood from the gills. In tuna, as in other ectotherms, arterial blood is cold. Blood entering the gills comes in contact with cold seawater, and its temperature rapidly adjusts to that of the seawater. As the cold blood reaches the muscles, however, heat from the warm veins <u>leaving</u> the muscles is conducted from the veins into the nearby arteries <u>entering</u> the muscle. In this way, the heat generated by the muscles is returned to the muscles rather than being lost from the gills. As warm temperatures stimulate the rate of chemical reactions, the muscles are able to operate more efficiently.

Evaporative Heat Loss   Recall that some animals can lose body heat through evaporation of water from the skin and membranes lining the respiratory tract. Heat exchange in some mammals can be regulated by changing the rate of water evaporation through perspiration. Nerves to the sweat glands stimulate the production of sweat, a dilute solution containing sodium chloride. The most important factor determining evaporation rate—and therefore heat loss—is the water vapor concentration, or humidity, of the air. The discomfort you feel on a humid day is due to the failure of evaporation. Your sweat

glands continue to secrete, but the sweat simply remains on your skin with body temperature remaining elevated, especially during exercise.

In endotherms that lack sweat glands, such as birds, or have very few glands, such as dogs and cats, panting (short, rapid breaths with the mouth open) promotes evaporation of water from the tongue surface. Panting has advantages over sweating, because no salt is lost, and panting provides the air current that promotes heat exchange by convection. However, the surface area of the mouth and tongue is relatively small, which limits the rate at which heat can be eliminated. Interestingly, many reptiles, while not endotherms, also pant on hot days, suggesting that panting evolved prior to endothermy.

Animals that neither sweat nor pant can still benefit from evaporative heat loss. The evaporation of fluids deposited on the body surface by licking the skin or splashing the skin with water also draws heat from the body.

Behavioral Adaptations   Behavioral mechanisms can also alter heat loss by radiation and conduction. Two such behaviors involve changing exposed surface area and changing surroundings. On hot days, birds may ruffle their feathers and raise their wings, while many mammals will reduce their activity and spread their limbs; these postural changes increase the surface area available for heat transfer. Terrestrial ectotherms and many endotherms seek shade, partially immerse themselves in water, or burrow into the ground when the sun is high. Pigs, which lack sweat glands, roll in the mud to cool down. Fishes may migrate from hot waters to cooler regions.

Similarly, animals respond to cold temperatures with numerous behavioral adaptations. Huddling in groups, curling up into a ball, hunching the shoulders, burying the head and feet in feathers, and similar maneuvers reduce the surface area exposed to a cold environment and decrease heat loss by radiation and conduction. Changing environments is also a common strategy for coping with cold. Migration from cold to warmer regions occurs in numerous species of fishes, birds, and mammals. Burrowing is not only useful to escape the heat of the sun, as mentioned previously, but is also a useful way to avoid the cold without migrating to a completely new environment. For example, many temperate zone frogs, toads, salamanders, turtles, and snakes burrow as much as a meter beneath the frost line during winter, where they hibernate.

Muscle Activity and Brown Adipose Tissue Metabolism Increase Heat Production

We have discussed how heat is gained or lost to the environment and how heat can be retained by reducing blood flow to the skin on a cold day. Body temperature, however, is a balance between these factors and heat production. Changes in muscle activity constitute the major control of heat production for temperature regulation in endotherms.

When an endotherm is in its thermoneutral zone, no significant adjustments are necessary to maintain core body temperature. When exposed to conditions below the thermoneutral zone, however, core body temperature begins to fall. The primary response to falling temperatures is to reduce the flow of blood to regions that permit conduction of heat. If this does not adequately reduce heat loss, skeletal muscle contraction is increased. This may lead to shivering, which consists of rapid muscle contractions without any locomotion. Virtually all of the energy liberated by the contracting muscles appears as internal heat, a process known as **shivering thermogenesis**. Many birds that remain in cold climates during the winter shiver almost continuously.

In many mammals, chronic cold exposure also induces **nonshivering thermogenesis**, an increase in the metabolic rate and therefore heat production that is not due to increased muscle activity. Nonshivering thermogenesis occurs primarily in **brown adipose tissue** (also called brown fat), a specialized tissue in small mammals such as hibernating bats, small rodents living in cold environments, and many newborn mammals, including humans. Brown adipose tissue is responsive to hormones and signals from the nervous system, which are activated when body temperature falls. The mitochondria in the brown adipose tissue contain uncoupling proteins, a class of proteins that uncouple oxidative phosphorylation, described in Chapter 6. The H^+ gradient becomes uncoupled from ATP synthesis and instead is used to generate heat, which helps maintain body temperature. Recently, researchers have discovered brown adipose tissue in adult humans. Because brown adipose tissue metabolizes fat rather than storing it, researchers are currently investigating ways in which adipose tissue stem cells may be induced to differentiate into brown fat in adult humans. The goal of such research is to increase a person's ability to burn rather than store ingested fat and thereby prevent or reverse obesity.

Endotherms Can Adapt to Chronic Changes in the Temperature of Their Environments

Although many animals—notably deep-sea-dwelling fishes and invertebrates—spend their lives in relatively unchanging environmental temperatures, most others do not. Such animals, therefore, must adapt to environmental changes using the mechanisms we have discussed in this section. In some cases, however, long-term exposure to a challenging environment, either very hot or very cold, results in a fine-tuning of the adaptive mechanisms that persist for as long as the animal lives in that environment. This process of **acclimatization** occurs particularly well in humans, who have more sweat glands than any known mammal. Increases in the amount and rate of sweat production and a decrease in the temperature threshold for initiating sweating, for example, can acclimatize an animal such as ourselves to chronic high temperatures. A person newly arrived in a hot climate initially has trouble coping. Body temperature rises, and too much activity can lead to a breakdown in the body's temperature-regulating systems. After several days to weeks, however, acclimatization occurs. Body temperature stabilizes, and the person finds it much easier to function at a normal activity level. Body temperature does not rise as much

as it did initially upon moving to the hot climate because sweating begins sooner and the volume of sweat produced is greater. Long-term exposure to hot conditions also increases the dilation of skin blood vessels and blood flow to the skin, helping to dissipate heat by conduction.

Cold acclimatization has been less studied than heat acclimatization, but seasonal changes occur in many endotherms that live in variable climates. Birds grow an extra layer of insulating feathers, and mammals may grow additional fur in the winter. The additional insulation reduces heat loss up to 50% in some animals. Such feathers and fur are shed in the summer. Ectotherms, too, can acclimatize to changes in temperature. For example, marine animals that migrate from warm to very cold waters undergo a change in the lipid composition of their cellular membranes. The membranes contain more unsaturated fats and therefore remain fluid even at extremely cold temperatures. In addition, many cold-acclimated ectotherms have cellular enzymes with a wider range of temperature tolerance than those that are found in animals living in mild climates.

46.5 Impact on Public Health

As we have seen, most animals, when provided adequate nutrients, maintain their body mass around a set point that is normal for their species. We rarely observe healthy animals in nature that are overweight. Generally, only domesticated animals become sufficiently sedentary that they gain weight (think of an overweight housecat). Humans, too, are prone to weight gain, particularly when living sedentary lives. Many people maintain a healthy body weight, but, as discussed in this section, an increasing number are unable to meet this goal.

Obesity Is a Global Health Issue

Excess body fat in humans (and in other mammals) increases the risk of many diseases, including high blood pressure, cancer, heart disease, and diabetes. In diabetes, there is either insufficient insulin available from the pancreas to control blood glucose levels, or the insulin that is present is no longer effective. In the U.S. alone, about 8% of the population—nearly 24,000,000 people—have diabetes mellitus. Of all diabetics in the U.S., more than 90% have the form called Type 2 diabetes mellitus that results from insulin being ineffective. Compelling evidence has directly linked the incidence of this type of diabetes with being overweight. At what point does fat accumulation in humans start to pose a health risk? Historically, this question has been evaluated by research studies that investigate possible correlations between disease rates and some measure of body fat.

One of the currently preferred methods for assessing body fat and health risks is the **body mass index** (**BMI**), a ratio of weight compared to height. A person's BMI is calculated by dividing their weight in kilograms by the square of their height in meters. For example, a 70-kg human with a height of 180 cm would have a BMI of 21.6 kg/m^2 (70/1.8^2).

Current National Institutes of Health guidelines categorize BMIs of 25 or more as **overweight**, that is, as having increased health risk because of excess adipose tissue. BMIs of 30 or greater are considered **obese**, with a highly increased health risk. Data compiled by The Centers for Disease Control and Prevention in Atlanta, Georgia, and other U.S. federal agencies indicate that approximately 2/3 of U.S. adults age 20–74 are now overweight or obese. One of the more troubling statistics is that the percentage of adults who are overweight but not obese has remained relatively unchanged since 1960, at roughly 30–35%. However, the percentage of obese adults has risen during that time from about 13% to the current level of 33–34%. Since as recently as the early 1990s, the CDC estimates that the average body weight of Americans has risen by 10 pounds. Even more troubling, the rate of childhood obesity has also risen. In the U.S., the incidence of obesity in children age 6–11 has increased from 2–3% in 1960 to the current estimate of more than 15%.

The recent increase in overweight and obesity categories is a worldwide trend. According to the World Health Organization, more than 1 billion adults globally are overweight and 300 million are obese. One ray of hope is that the incidence of obesity in the U.S. population has leveled off since 2003, suggesting that efforts at educating the public about the health consequences of obesity may be starting to have a positive result.

The increased health risk associated with obesity may be at least partly due to a lack of physical activity, not body fat *per se*. Humans today are more sedentary than were past generations, with many participating in forms of recreation that do not require physical activity, such as television, computer games, and the Internet, and most having careers with reduced physical exertion when compared to past generations. Lack of exercise is associated with fat accumulation because it reduces total daily energy expenditure.

Growing evidence suggests that the location of body fat has important health consequences, too. For reasons that are not yet completely clear, people with mostly abdominal fat are at greater risk for developing diabetes and cardiovascular disease than are people whose fat is mainly in the lower body, on the buttocks, hips, and thighs.

Some studies indicate that genetic factors play an important role in obesity. Identical twins separated soon after birth and raised in different households have strikingly similar body weights as adults. Researchers hypothesize that natural selection favored the evolution of so-called **thrifty genes**, which boosted our ancestors' ability to store fat from each feast in order to sustain them through the next famine. Given today's abundance of high-fat foods in many countries, what was once a survival mechanism may now be a liability.

The methods and goals of treating obesity are undergoing extensive rethinking. An increase in body fat must be due to an excess of energy intake over energy expenditure, and overweight people have traditionally been advised to follow a low-calorie diet. However, such diets alone have limited effectiveness, because over 90% of obese people regain most or all of their lost weight within 5 years. This disturbing phenomenon may be related to the observation that metabolic rate decreases as leptin levels decrease, sometimes falling low enough to prevent further weight loss on as little as 1,000 calories per day.

Research indicates that crash diets are not an effective long-term method for controlling weight. Instead, caloric intake should be set at a realistic level that can be maintained for the rest of one's life. This should lead to a slow, steady weight loss of no more than 1 pound per week until body weight stabilizes at a new, lower level. Most important, any program of weight loss should include increased physical activity. The exercise itself burns calories, but more importantly, it partially offsets the tendency for the metabolic rate to decrease. As a bonus, the combination of exercise and caloric restriction causes the person to lose more fat and less protein than with caloric restriction alone.

The impact of obesity on public health is enormous, accounting for many illnesses requiring hospitalization and chronic drug therapy and well over 100,000 premature deaths per year. Its impact is far-reaching on the economy as well. The economic toll of obesity-related illnesses is felt in the loss of worker-hours in the workplace, in the cost of hospital stays, physician office visits, nursing home care, and medication costs. In fact, current estimates are that nearly 10% of all U.S. health-care expenditures are directly or indirectly related to obesity!

Obesity can impact society in unexpected ways. As an example, the increasing weight load of passengers forces airplanes to burn 350 million additional gallons of fuel each year, compared to just 12 years ago! This translates into nearly 4 million tons of additional pollution released into the atmosphere every year.

Anorexia Nervosa and Bulimia Nervosa Can Have Serious Health Consequences

Being obese can be dangerous, but so can being extremely underweight. Here we will look at two eating disorders that can leave people underweight: anorexia nervosa and bulimia nervosa.

Anorexia nervosa is characterized by weight loss and is found primarily in adolescent girls and young women. People with anorexia nervosa become pathologically obsessed with weight and body image, and they reduce their food intake to the point of starving. Anorexia nervosa may have both biological and psychological causes. Many other abnormalities are associated with the disorder, such as loss of menstrual periods, low blood pressure, low body temperature, and altered secretion of many hormones. These may be simply signs of starvation, although they may also result from malfunction of the parts of the brain that normally control appetite.

Bulimia nervosa (often simply called bulimia) involves recurrent episodes of fasting and overeating. Weight loss may occur because of regular self-induced vomiting and the use of laxatives to stimulate bowel movements or diuretics to stimulate urination, as well as strict dieting, fasting, or vigorous exercise. As with individuals with anorexia nervosa, people with bulimia are obsessed with body weight, although they are usually within 10% of their ideal weight. Successful treatments for anorexia and bulimia rely primarily on counseling, nutritional education, and medications.

Summary of Key Concepts

46.1 Nutrient Use and Storage

- From an energetic point of view, an animal's life can be divided into two phases: the absorptive state, during which ingested nutrients are entering the blood from the gastrointestinal tract, and the postabsorptive state, during which the GI tract is empty of nutrients and the body's own stores must supply energy.

- Glucose and fats are the two major energy sources during the absorptive state. Much of the absorbed glucose immediately enters cells and is enzymatically broken down, providing energy required to synthesize ATP. Most absorbed triglycerides are stored in fat cells until the body requires additional energy. Amino acids are taken up by all body cells and used to synthesize proteins. (Figure 46.1)

- The events that maintain blood glucose concentration in the postabsorptive state fall into two categories: (1) the reactions glycogenolysis and gluconeogenesis, which provide glucose to the blood and (2) cellular use of fat for energy, which spares glucose for use by the nervous system. (Figure 46.2, Table 46.1)

46.2 Regulation of the Absorptive and Postabsorptive States

- Tight control mechanisms, in the form of several hormones and the nervous system, maintain homeostatic levels of fuel in the blood. The hormone insulin acts on cells to facilitate the diffusion of glucose from blood into the cell cytosol via glucose transporters (GLUTs). All animal cells use GLUTs to transport glucose across their plasma membranes. (Figure 46.3)

- In vertebrates, an increase in blood glucose level in the absorptive state stimulates the cells of the pancreas to secrete insulin; a decrease in glucose level removes the signal for secretion. In the postabsorptive state, when the blood glucose level falls, glucose-monitoring regions in the hypothalamus stimulate production of glucose-elevating factors such as glucagon and norepinephrine. (Figure 46.4)

- Exercise is any type of physical activity that increases an animal's metabolic rate. Exercise increases an animal's requirement for nutrients to provide energy. The source of these nutrients includes glucose and fatty acids.

46.3 Energy Balance

- An animal's energy expenditure, or metabolic rate, refers to the amount of energy it uses in a given period of time to power all of its metabolic requirements.

- The most common measure for comparing metabolic rates of different species is to compare the basal metabolic rate (BMR). Most of the basal metabolism is due to the routine functions of the heart, liver, kidneys, and brain. (Figures 46.5, 46.6)

- Many factors have an impact on metabolism, including skeletal muscle activity, whether an animal has recently eaten, and body size. (Figures 46.7, 46.8)

- Torpor and hibernation allow animals to conserve energy during changing environmental conditions, such as reduced environmental temperature and lack of available food.

- When the daily amount of energy consumed equals the amount of energy expended, body weight remains stable. Tipping the balance in either direction causes weight gain or loss by increasing or decreasing total body energy content.

- Short-term control of feeding generally involves satiety signals that remove the sensation of hunger and set the time period before hunger returns again. Experiments by Coleman and Friedman investigated the hormone leptin as a satiety factor in mammals. Leptin has since been found in all classes of vertebrates. (Figures 46.9, 46.10)

46.4 Regulation of Body Temperature

- Most animals can survive only in a relatively narrow temperature range that allows molecules in the body to move and react with one another, maintains the structures of membranes, and avoids denaturing proteins. A few animals, however, have the ability to survive being partially frozen.

- Animals can be classified according to their source of heat and their ability to maintain body temperature. Ectotherms depend on external heat sources to warm their bodies, whereas endotherms use their own metabolically generated heat to warm themselves. Homeotherms maintain their body temperature within a narrow range, whereas heterotherms have body temperatures that vary with environmental conditions. All animals fall into two of these categories. (Figure 46.11)

- The surface of an animal's body can lose or gain heat from the external environment via four mechanisms: radiation, evaporation, convection, and conduction. (Figures 46.12, 46.13)

- The skin can function as a variable heat exchanger; blood vessels near the skin surface dilate to dissipate heat or constrict to retain it. Both endotherms and ectotherms regulate heat loss through countercurrent heat exchange, which retains heat by returning it to the body's core and keeping it warmer than the extremities. Heat exchange can also be regulated by changing the rate of water evaporation via perspiration. Behavioral mechanisms can alter heat loss by radiation, conduction, and convection. (Figures 46.14, 46.15)

- Muscle activity (shivering thermogenesis) and brown adipose tissue metabolism (nonshivering thermogenesis) increase the production of heat.

- Acclimatization can fine-tune an animal's adaptive mechanisms to a changing environment.

46.5 Impact on Public Health

- Excess body fat increases the risk of many diseases. A body mass index (BMI) of 25 kg/m^2 or more is considered overweight; $\geq$ 30 kg/m^2 is considered obese.

- Being underweight is also unhealthy. People with anorexia nervosa become pathologically obsessed with weight and body image and reduce their food intake to the point of starving. Bulimia nervosa involves recurrent episodes of fasting and overeating, plus self-induced vomiting of ingested food. Both eating disorders can be treated with counseling, nutritional education, and medications.

Assess and Discuss

Test Yourself

1. During the absorptive phase, an animal is
   a. fasting.
   b. relying entirely on stored molecules for energy.
   c. absorbing nutrients from a recently ingested meal.
   d. metabolizing lipids stored in adipose tissue to supply ATP to its cells.
   e. both a and b

2. Gluconeogenesis
   a. occurs when the liver synthesizes glucose from noncarbohydrate precursors.
   b. is the process by which glycogen is broken down to glucose.
   c. occurs primarily when an animal is in the absorptive phase.
   d. occurs when triglycerides are being formed and stored in adipose cells.
   e. none of the above

3. In the process of _____, most tissues of the vertebrate body metabolize fat instead of glucose to ensure that _____ tissue has an adequate supply of glucose.
   a. gluconeogenesis, muscle
   b. glucose sparing, epithelial
   c. glycogenolysis, nervous
   d. glucose sparing, nervous
   e. gluconeogenesis, epithelial

4. Ketones are compounds derived from
   a. glucose.   c. fatty acids.   e. proteins.
   b. glycogen.   d. amino acids.

5. Insulin primarily regulates blood glucose levels by
   a. stimulating the recruitment of glucose transporter proteins from the cytosol to the plasma membrane for transport of glucose from extracellular to intracellular fluid.
   b. stimulating gluconeogenesis.
   c. suppressing glucose uptake by muscle tissue.
   d. stimulating the release of glucose from glycogen reserves in the liver.
   e. inhibiting the synthesis of new GLUT proteins.

6. The rate at which an animal uses energy is called
   a. body mass index.
   b. an animal's energy consumption.
   c. metabolic rate.
   d. an animal's energy expenditure.
   e. both c and d.

7. Which factor may increase metabolic rate?
   a. hibernation   d. fasting
   b. reduced muscle activity   e. consumption of a meal
   c. sleeping

8. Which molecule acts on brain centers to reduce appetite in mammals and other vertebrates?
   a. GLUT4   c. leptin   e. ketones
   b. glycogen   d. glucagon

9. Animals that have body temperatures that are maintained within a narrow range are
   a. endotherms.   c. homeotherms.   e. both b and d.
   b. ectotherms.   d. heterotherms.

10. The rate of heat loss in a mammal is regulated by
    a. the degree of blood flow at the surface of the skin.
    b. the amount of perspiration.
    c. behavioral adaptations.
    d. air currents near the animal's body.
    e. all of the above.

Conceptual Questions

1. Explain the functions of insulin. Why do you think a hormone such as insulin is required to carry out these functions?

2. Explain how appetite is controlled by the brain. What is the benefit of having a hormone released from adipose cells in proportion to total fat mass?

3. Explain countercurrent heat exchange.

Collaborative Questions

1. Discuss the differences between being ectothermic and endothermic and between being heterothermic and homeothermic.

2. Discuss four ways animals exchange heat with their environment.

Online Resource

www.brookerbiology.com

Stay a step ahead in your studies with animations that bring concepts to life and practice tests to assess your understanding. Your instructor may also recommend the interactive ebook, individualized learning tools, and more.

Circulatory Systems

Image from cardiac angiography (dye contrast injection to visualize coronary blood vessels).

I n the time it takes you to read this chapter introduction, three people in the United States will have suffered a heart attack, and one of those individuals will likely have died as a result. Diseases of the heart and blood vessels are the number one cause of death in the U.S. and in much of the rest of the world. In the U.S. alone, an estimated 80,000,000 people have one or more diseases that affect the heart and blood vessels. About 1,200,000 of those individuals will have a heart attack in the next year, and nearly 500,000 of those attacks will be fatal.

The heart is a muscular pump that works ceaselessly. Because of this constant activity, heart muscle requires a steady and large amount of blood, bringing with it the nutrients and oxygen required to sustain that muscular effort. The chapter-opening photo shows the extensive network of blood vessels coursing through the human heart. Should any of those vessels become diseased, the regions of the heart supplied by those vessels can die. That is what happens during a heart attack. If a heart attack occurs, the damaged heart may not be able to pump blood forcefully enough to generate the blood pressure required to reach all the cells of the body.

The heart, blood, and blood pressure are all features of the circulatory system. The primary function of **circulatory systems** is to transport necessary materials to all the cells of an animal's body, and to transport waste products away from the cells where they can be released into the environment. In this chapter, we will examine how this occurs. Single-celled organisms are small enough that dissolved substances are able to diffuse into and out of the cell. In larger animals, however, dissolved substances must move greater distances between cells, as well as between the internal and external environments. The time required for osmosis of water and diffusion of solutes throughout the body and between environments would be too great to sustain life in such animals. In the circulatory systems of larger animals, the heart pumps blood through blood vessels throughout the body. The blood carries nutrients and oxygen to all the body's cells, and takes away waste products and carbon dioxide. Most circulatory systems are capable of adjusting their activities to an animal's changing metabolic demands. For instance, we will see how exercise increases the ability of the vertebrate heart to pump blood and directs blood to the skeletal muscles where it is most needed. Understanding the relationships between the heart, blood, and blood pressure is a major concept addressed in this chapter. Finally, we consider some of the causes, symptoms, and treatments of the major types of cardiovascular disease in humans.

47.1 Types of Circulatory Systems

The three basic types of circulatory systems are gastrovascular cavities, open systems, and closed systems. We begin with the simplest form of circulation, that provided by the water in which certain invertebrates live.

Gastrovascular Cavities

In cnidarians (jellyfish, hydras, sea anemones, and corals) and some flatworms, the surrounding water not only contains the nutrients and oxygen needed to sustain life but also functions as a circulatory system. These animals rely on water currents to bring a steady supply of water into contact with their body surfaces. Water enters the **gastrovascular cavity**, a body cavity with a single opening to the outside. As described in Chapter 45, food is digested within the cavity and absorbed by body cells, and wastes are excreted into the cavity. Nutrients obtained from digested food reach all the cells because water circulates throughout the entire cavity. Because the animals are only a few cells thick, all cells in their bodies are located close to the gastrovascular cavity or in slender extensions that branch from it.

In some cnidarians, cilia that line the opening of the gastrovascular cavity propel water through these extensions, increasing the circulation of water. In addition, muscular efforts of the body wall—as a *Hydra* stretches and relaxes, for instance, or a jellyfish propels itself by contracting its bell—help circulate water throughout the cavity (**Figure 47.1**). The harder the muscles work, the more effectively they propel water, thus providing a rudimentary but effective circulation.

Open Circulatory Systems

Arthropods and some mollusks have a circulatory system with three basic components:

1. **hemolymph**, a body fluid containing dissolved solutes;
2. **vessels**, a system of hollow tubes within the body through which hemolymph travels;
3. one or more **hearts**, muscular structures that pump hemolymph through the vessels.

A circulatory system containing these three components is also known as a **cardiovascular system** (from the Greek *kardia*, meaning heart, and Latin *vas*, meaning vessel). In an **open circulatory system**, the vessels open into the animal's body cavity. Therefore, the fluid in the vessels and the interstitial fluid that surrounds cells mingle in one large, mixed compartment, rather than being located in separate body compartments. The mixed fluid is hemolymph. The heart or hearts pump hemolymph through the vessels, where it exits into the animal's body cavity

Figure 47.1 **Circulation of water through the gastrovascular cavity of *Hydra*.** The animal's movements assist the flow of water throughout the cavity, which extends into the tentacles.

Figure 47.2 **An open circulatory system.** In open systems, one or more muscular, tubelike hearts pump hemolymph through open-ended vessels, where it percolates through the body. In arthropods such as this honeybee, hemolymph reenters the heart through ostia. The arrows show the movement of hemolymph.

Concept check: Why is it incorrect to think of open circulatory systems as "primitive" compared to closed systems?

(**Figure 47.2**). Nutrients and wastes are exchanged by diffusion between the hemolymph and body cells, and the hemolymph is recirculated back to the heart. In mollusks with open circulatory systems, hemolymph reenters the heart through vessels, whereas in arthropods, it returns to the heart through small openings called ostia.

In insects, hemolymph primarily transports nutrients and wastes; in mollusks, it also brings oxygen to body cells. Insects have a unique oxygen-delivery system that does not involve the circulatory system (see Chapter 48).

Open circulatory systems are metabolically inexpensive because they do not require much energy to circulate hemolymph. In addition, open systems can adapt to changes in an animal's metabolic demands. As an insect takes flight, for example, its flight muscles contract more forcefully and rapidly, which acts to expand and compress the animal's thorax. This helps propel hemolymph more rapidly throughout the body cavity and into and out of the hearts. In other words, as the animal's activity increases, its circulation becomes more efficient, recharging metabolically active cells with nutrients. As noted earlier, the ability to adjust circulation to meet an animal's requirements is one of the most important features of any circulatory system. Open systems, therefore, are highly effective, particularly for smaller animals.

Despite the phenomenal success of species with open systems, certain limitations are associated with this type of circulation. For instance, because the hemolymph empties in bulk into the general body cavity, it cannot be selectively delivered to body cells whose metabolic activity may have increased compared to other cells. By contrast, during exercise in vertebrates, larger amounts of blood are delivered to leg muscles and away from less metabolically active structures, an adaptation that results in greater endurance. This is possible because these animals have closed circulatory systems, as we see next.

Closed Circulatory Systems

Open circulatory systems are not ideal for larger, more active animals in which higher pressure is required to pump blood to all body cells at a rate sufficient to meet energetic demands. In these groups of animals, a closed circulatory system has evolved.

In a **closed circulatory system**, blood and interstitial fluid are physically separated, and they differ in their components and chemical composition. Closed circulatory systems are found in earthworms, cephalopods (squids and octopuses), and all vertebrates (**Figure 47.3**). Despite differences in structure, closed circulatory systems share certain key features.

- **Blood**—a watery solution containing solutes to be transported throughout the body—is pumped under pressure by one or more contractile, muscular hearts.
- Blood remains within tubelike vessels that distribute it throughout the body.
- The solutes in blood can be exchanged with the environment and the body's cells.
- In most cases, blood contains disease-fighting cells and molecules.
- The activity of the closed circulatory system can be adjusted to match an animal's metabolic demands.
- Closed systems generally can heal themselves when wounded, by forming clots at the site of injury.
- These systems grow in size as an animal grows.

A closed circulatory system offers several advantages. First, animals can grow to a larger size, because blood can be directed to every cell of an animal's body, no matter how large. Nearly all body cells are within one or two cell-widths of a blood vessel. Second, blood flow can be selectively increased or decreased to supply different parts of the body with the precise amount of blood needed at any given moment. After a meal, for example, more blood can be directed to the gut to absorb nutrients, and on a hot day, additional blood can be routed to the skin to dissipate heat. Animals with open systems cannot make these adjustments.

The closed circulatory systems of vertebrates can be divided into two major groups: single and double circulations. In a single circulation, blood is pumped under low pressure from the heart to the respiratory surface (for example, gills), where it picks up oxygen and drops off carbon dioxide. From there, blood circulates to the tissues of the body, where it releases oxygen and picks up carbon dioxide. The blood then circulates back to the heart. In a double circulation, blood is pumped under low pressure from the heart to the lungs, and then back to the heart, to be pumped under high pressure to the tissues and finally returned again to the heart. Single circulations are seen in fishes, and double circulations are seen in crocodiles, birds, and mammals. Amphibians and most reptiles have a type of circulation that combines features of both.

Single Circulation: Fishes   In the single circulation of fishes, the heart has a single filling chamber—an **atrium**—to collect blood from the tissues, and an exit chamber—a **ventricle**—to pump blood out of the heart (**Figure 47.4a**). Blood vessels called **arteries** carry blood away from the heart to the gills, which pick up oxygen from the water in which the fish swims and unload carbon dioxide into the water. The freshly oxygenated blood then circulates via other arteries to the rest of the body. There,

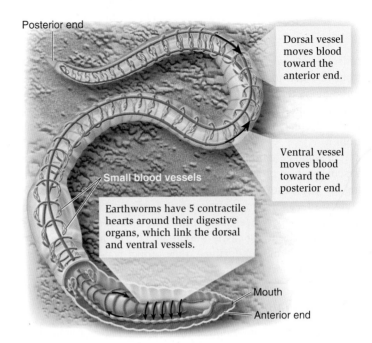

Figure 47.3   **A closed circulatory system.**   In closed systems, such as that of the earthworm, blood remains in vessels and hearts and recirculates without emptying into the body cavity. Arrows indicate direction of blood flow.

oxygen and nutrients are delivered to cells, and carbon dioxide diffuses from cells into the blood. Finally, the deoxygenated blood is returned to the heart via blood vessels called **veins**, where it is pumped back to the gills for another load of oxygen.

An important feature of all respiratory surfaces is that they function best when the blood flowing through them is maintained at a low pressure. The fish heart does not generate high pressure when it pumps blood to the gills. This means that blood leaving the gills will also be under low pressure, therefore limiting the rate at which oxygenated blood can be delivered to the body's cells.

Intermediate Circulatory Features: Amphibians and Most Reptiles   Unlike fishes, most adult amphibians rely on lungs and their permeable skin to obtain oxygen and rid themselves of carbon dioxide. While amphibians are on land, their deoxygenated blood is pumped from the heart to the lungs and, to a lesser extent, the skin. Oxygen diffuses from the air into the blood vessels within the lungs and beneath the skin, and carbon dioxide diffuses in the opposite direction. While amphibians are under water, deoxygenated blood from the heart bypasses the lungs and is directed almost entirely to the skin so that oxygen in the water can diffuse across the skin and into the circulation. Oxygenated blood from the lungs and skin does not travel directly to the rest of the body but instead returns to the heart, where it is pumped out to the tissues of the body.

The amphibian heart, therefore, pumps oxygenated and deoxygenated blood to two separate locations. Blood is routed from the heart through different vessels to either the respiratory surfaces (the **pulmocutaneous circulation**) or the body tissues (the **systemic circulation**). This would not be possible if an amphibian's heart and circulation were the same as that of a

KEY
- O₂-rich blood
- O₂-poor blood
- Mixed blood
- Mostly O₂-rich blood
- Mostly mixed blood
- → Direction of blood flow

(a) Single circulation in fishes

(b) Features of both single and double circulation in most amphibians

(c) Double circulation in birds and mammals

Figure 47.4 **Representative vertebrate circulatory systems.** (a) Fishes have a single circulation in which blood is pumped from the heart to the gills, from where it circulates to the rest of the body tissues, and finally returns to the heart. (b) Amphibians, such as frogs, have a heart with a single ventricle that nevertheless has an internal structure that separates most of the oxygenated and deoxygenated blood entering it from the two atria. Some mixing does occur, however, as represented in this figure by the dashed lines. A significant advantage of this type of circulation is the ability to redirect blood to the skin instead of the lungs when under water. Because the skin is somewhat permeable to oxygen, this partly compensates for the lack of air-breathing. Note that blood returning from the skin has picked up oxygen, but is returned to the right atrium, not the left atrium as is the case with the lungs. (c) Birds and mammals have a double circulation, in which oxygenated blood is pumped under high pressure to the body's tissues, and deoxygenated blood is pumped under low pressure to the lungs. No mixing of the two types of blood occurs.

Concept check: *What is the advantage of completely separating oxygenated and deoxygenated blood in a double circulation?*

fish, where blood travels through a single circulation. Instead, the amphibian heart has two separate atria for collecting blood (**Figure 47.4b**). The right atrium receives blood that has passed through all of the body except the lungs and is therefore depleted of some of its oxygen. However, some oxygen-rich blood returning from the skin also enters the right atrium and mixes with the blood returning from the systemic circulation. This is an important adaptation for amphibians when they are under water, because as stated, the circulation through the

lungs is decreased at such times and, therefore, diffusion across the skin becomes the sole means of obtaining oxygen.

When an amphibian breathes air, oxygen-rich blood from the lungs is delivered to the left atrium. Each atrium of the amphibian heart empties its blood into a single large ventricle. The internal structure of the single ventricle consists of flaps and folds of tissue that keeps the oxygenated and deoxygenated blood from the two atria mostly separate. Some mixing of the two streams of blood, however, does occur in the ventricle.

This would not occur if each atrium emptied into a separate ventricle. Because there is only a single ventricle, however, the slightly mixed blood traveling to the tissues is not fully oxygenated. The more oxygen that is dissolved in an animal's blood, the faster the oxygen is able to diffuse from the blood, through the interstitial fluid, and into cells where it is required. The mixed blood of amphibians, therefore, imposes limits on their metabolic activity.

Like amphibians, all of the reptiles except for crocodiles have hearts with two atria and a single ventricle, although the ventricle is further subdivided into two partially but incompletely separated chambers. These chambers allow more efficient separation of deoxygenated and oxygenated blood entering the ventricle from the two atria; this, in turn, contributes in part to the higher activity levels of reptiles compared to amphibians. Because these reptiles and amphibians have only a single ventricle that must pump blood to both the tissues and respiratory surfaces, blood must be pumped under low or moderate pressure in these animals to minimize the pressure of blood flowing through the delicate lung tissue.

Double Circulation: Crocodiles, Birds, and Mammals In crocodiles, birds, and mammals, oxygenated and deoxygenated blood are completely separated into two distinct circuits, the systemic circulation and the **pulmonary circulation** (Figure 47.4c). This is made possible by a heart that has two complete ventricles. The crocodilian circulatory system has evolved to be somewhat different from that of birds and mammals. It allows crocodiles to spend part of their lives under water and it reflects that they are considerably less active than birds and mammals. When resting or underwater, crocodiles can divert blood between the two circuits in ways that birds and mammals cannot. Nonetheless, all animals with double circulations have a left and right atrium and a left and right ventricle.

In a double circulatory system, the pulmonary circulation delivers oxygenated blood from the lungs to the left atrium, which then passes it on to the left ventricle. The left ventricle pumps blood to all the body's tissues via the systemic circulation. The systemic circulation delivers deoxygenated blood from the tissues to the right atrium, which then sends it to the right ventricle. The right ventricle pumps this blood to the lungs via the pulmonary circulation, and the cycle starts again.

An advantage of a double circulation is that the two ventricles can function as if they were, in effect, two hearts, each with its own ability to pump blood under different pressures. This means that blood from the right ventricle can be pumped under low pressure to the lungs. Once the blood picks up oxygen from the lungs, it can be returned to the left side of the heart. The left ventricle is more muscular than the right ventricle and therefore generates higher pressures when pumping blood. Thus, the oxygenated blood leaving the left side of the heart has sufficiently high pressure to reach all the cells of the animal's body and deliver oxygen and nutrients at a high rate, even to regions above the heart that must contend with gravity. This is no small feat considering the distance blood must travel in some large mammals. The left ventricle of a giraffe, for example, is particularly muscular compared to that of other animals.

47.2 Blood and Blood Components

Blood is the transport medium of many animals' bodies. It moves necessary materials, including nutrients and gases such as oxygen, to all cells and takes away waste products, including carbon dioxide and other breakdown products of metabolism. What components of blood allow it to perform its functions?

Blood Is Composed of Cells and Water with Dissolved Solutes

Blood is a fluid connective tissue consisting of cells and, in mammals, cell fragments suspended in a solution of water containing dissolved nutrients, proteins, gases, and other molecules. If we collect a blood sample and spin it in a centrifuge, the blood separates into three layers (**Figure 47.5**). These layers correspond to three of blood's four components: plasma, leukocytes, and erythrocytes. (A fourth component is the much less abundant platelets or thrombocytes, and is not readily visible.) Let's take a look at each of these.

Plasma The top layer of the centrifuged blood sample is **plasma**, a yellowish solution of water and solutes (Figure 47.5). Plasma typically makes up about 35–60% of the total volume of blood in vertebrates. It contains water and the dissolved organic and inorganic nutrients that were absorbed from the digestive tract or secreted from cells. Plasma also contains dissolved oxygen; waste products of metabolism, such as carbon dioxide; and other molecules released by cells, such as hormones. Plasma also transports cells of the immune system and cells involved in oxygen transport, as well as proteins that serve several important functions, such as forming blood clots, which seal off wounds to blood vessels.

In addition to transporting molecules throughout the body, plasma has other functions. For example, plasma contains buffers that help keep the body's pH within its normal range. It is also important in maintaining the fluid balance of cells. Changes in plasma salt and protein concentrations can affect the movement of water between intracellular and extracellular fluid compartments.

Leukocytes Beneath the plasma in our sample is a narrow white layer of **leukocytes**, also known as **white blood cells** (Figure 47.5). Leukocytes develop from a specialized connective tissue (the marrow) of certain bones in vertebrates. Although there are several types—which we describe further in Chapter 53—all leukocytes perform vital functions that defend the body against infection and disease.

Erythrocytes The bottom visible layer of our blood sample consists of **erythrocytes**, also called **red blood cells** because of their color (Figure 47.5). **Hematocrit** refers to the volume of blood that is composed of red blood cells, usually between 40 and 65% among vertebrates. Red blood cells serve the critical function of transporting oxygen throughout the body. There are approximately a thousand times more red blood cells than white blood cells in the circulation. Like leukocytes, red blood cells are

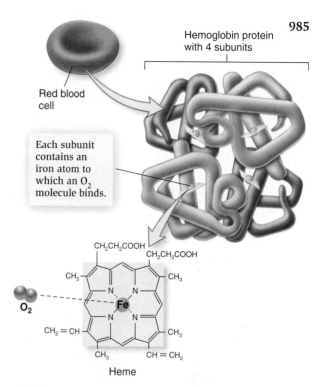

Heme

Figure 47.5 **Components of blood.** When a blood sample that has been prevented from clotting is centrifuged, it forms three visible layers. Leukocytes, shown in the scanning electron micrograph, are the white blood cells that make up part of the immune system. Erythrocytes are red blood cells and function to carry oxygen. An additional component of blood is not visible under these conditions because of small numbers and size. These are the platelets in mammals (or thrombocytes in other vertebrates) that participate in blood clotting reactions. Note: The leukocyte layer is enlarged and not to scale, for illustrative purposes.

Figure 47.6 **Hemoglobin.** Erythrocytes contain large amounts of the protein hemoglobin. Oxygen binds reversibly to iron atoms in the heme portion of each subunit of hemoglobin.

Concept check: *How many <u>atoms</u> of oxygen can be bound to a single molecule of hemoglobin?*

derived from cells in the bone marrow. In most vertebrates, mature red blood cells retain their nuclei and other cellular organelles, but in all mammals (and a few species of fishes and amphibia), the nuclei are lost upon maturation. The lack of a nucleus and many other organelles in the mammalian red blood cell increases the cell's oxygen-carrying capacity and contributes to its characteristic biconcave shape (Figure 47.5). The biconcave shape of the mammalian red blood cell increases its surface area compared to the flattened disc or oval shape seen in most other vertebrates. This is believed to make gas exchange between the red blood cell and the surrounding body fluids more efficient.

Oxygen is poorly soluble in plasma. Consequently, the amount of oxygen that dissolves in plasma usually cannot support a vertebrate's basal metabolic rate, let alone more strenuous activity. The components of red blood cells solve this problem. Within the cytosol of red blood cells are large amounts of the protein **hemoglobin**. Each molecule of hemoglobin contains four protein subunits, each with an atom of iron in a heme group at its core; each iron reversibly binds to a molecule of oxygen (Figure 47.6). Thus, each hemoglobin molecule has four iron atoms and can reversibly bind four oxygen molecules. The oxygen attached to hemoglobin greatly increases the reservoir of oxygen in the blood and enables animals to be more active. Chapter 48 describes the mechanisms by which hemoglobin binds and releases oxygen.

Anemia refers to lower than normal levels of hemoglobin, which reduces the amount of oxygen that can be stored in the blood. Among possible causes are loss of blood, decreased production of hemoglobin, impaired production of erythrocytes, or increased breakdown of erythrocytes, each of which reduces hematocrit.

Platelets Vertebrate blood has a fourth component—called platelets in mammals and thrombocytes in other vertebrates—that would not be visible in our centrifuge sample because of their low numbers. **Platelets** are cell fragments that lack a nucleus, whereas **thrombocytes** are intact cells; they play a crucial role in the formation of blood clots, which limit blood loss after injury. Like leukocytes and erythrocytes, platelets and thrombocytes are formed in the bone marrow.

Although clot formation in mammals is a multistep process, two essential steps involve platelets (**Figure 47.7a**). First, platelets secrete substances that cause platelets to clump together and bind to collagen fibers in the surrounding connective tissue at the wound site; this forms a plug that prevents continued blood loss. Second, other platelet secretions interact with plasma proteins to cause the precipitation from solution of the fibrous protein **fibrin**. Fibrin forms a meshwork of threadlike fibers that wrap around and between platelets and red blood cells, enlarging and thickening the plug to form a clot. Blood clotting begins within seconds and helps prevent injured animals from bleeding to death. Eventually, the body absorbs the clot as the injured vessel heals. The importance of clotting is most evident when it fails to occur normally, as we see next.

Genomes & Proteomes Connection

Hemophilia Is Caused by a Genetic Defect in Clotting Factors

The process of blood clotting is not a simple one—at least a dozen different substances must be present in the right amounts

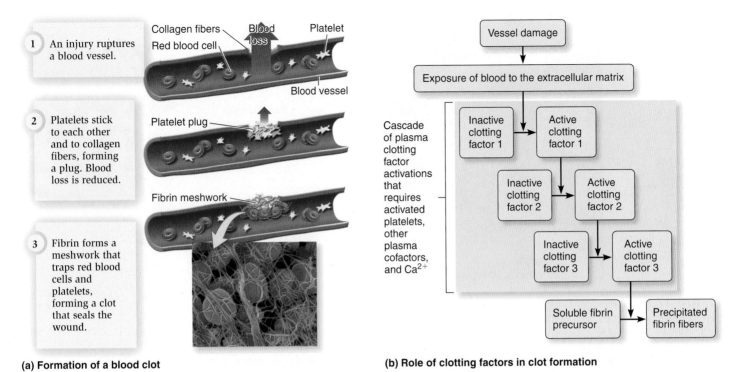

1 An injury ruptures a blood vessel.

Collagen fibers
Red blood cell
Blood loss
Platelet
Blood vessel

2 Platelets stick to each other and to collagen fibers, forming a plug. Blood loss is reduced.

Platelet plug

3 Fibrin forms a meshwork that traps red blood cells and platelets, forming a clot that seals the wound.

Fibrin meshwork

(a) Formation of a blood clot

Vessel damage

Exposure of blood to the extracellular matrix

Cascade of plasma clotting factor activations that requires activated platelets, other plasma cofactors, and Ca^{2+}

Inactive clotting factor 1 → Active clotting factor 1

Inactive clotting factor 2 → Active clotting factor 2

Inactive clotting factor 3 → Active clotting factor 3

Soluble fibrin precursor → Precipitated fibrin fibers

(b) Role of clotting factors in clot formation

Figure 47.7 **Platelets and the process of blood clot formation.** (a) A blood clot forms in two major steps: A platelet plug reduces initial blood loss, and a fibrin clot then seals the wound. An example of a fibrin clot is shown in the scanning electron micrograph. (b) Formation of the clot requires several clotting factors working together. These include enzymes, which catalyze a cascade of reactions that leads to the production of fibrin fibers. Several steps in the process are omitted here for simplicity.

for a clot to form. These substances are known as clotting factors and include proteins and ions such as Ca^{2+}. Platelets secrete some of these substances; others circulate in the plasma in an inactive form until activated by platelet and cellular secretions.

The clotting process is a chain reaction that begins when an injury to a blood vessel exposes blood to the surrounding extracellular matrix (**Figure 47.7b**). When one particular clotting factor in blood interacts with proteins in the matrix, the factor is activated. This initiates a cascade in which clotting factors sequentially activate each other, finally producing fibrin.

In the disease **hemophilia**, an inherited deficiency of a specific clotting factor inhibits the clotting process. The absence of the clotting factor interrupts the process leading to fibrin formation. Consequently, injured blood vessels take longer to heal and blood loss is excessive. Minor cuts are not necessarily serious, but uncontrolled bleeding in internal tissues and joints can be life-threatening.

Hemophilia in humans occurs in three forms, called the A, B, and C forms. In the most common form of hemophilia, the A form, an X-linked recessive mutation results in the inability to make an enzyme that is one of the clotting factors. Because the mutation is X-linked and recessive, it is much more common in males than females. Females who are heterozygous for the mutation are carriers and have a 50% chance of passing the disease genes to their sons.

Treating hemophilia requires transfusions of clotting factors purified from great amounts of human blood or from genetically engineered organisms. These procedures are extremely expensive, and transfusions run the risk of introducing infection or disease from the donor blood. Currently, researchers are

attempting to treat hemophilia with gene therapy by using a viral vector (see Chapter 20) to introduce the normal gene into liver cells, which then secrete the missing factor into the blood. Even if the introduced gene is not extremely efficient at forming new clotting factor, it could still improve patients' symptoms. This is because the missing clotting factor is an enzyme. Like all enzymes, it can act over and over again without being destroyed or chemically altered by the reaction it catalyzes. In addition, this enzyme is known to remain in the circulation for long periods without being degraded or excreted. Recent estimates suggest that restoring the clotting factor to only 2–5% of normal levels should prevent most symptoms of hemophilia. Gene therapy has proven to be effective in curing laboratory rodents and dogs of hemophilia, and testing has recently begun in humans. The results thus far are mixed but hopeful. One problem that needs to be overcome is that human immune systems appear to be more likely than those of experimental animals to destroy cells that incorporate the viral vector.

47.3 The Vertebrate Heart and Its Function

Now let's move on to the heart. As described earlier, all vertebrate hearts have at least one upper or anterior chamber called an atrium and at least one lower or posterior chamber called a ventricle (**Figure 47.8**). In hearts with more than one atrium or ventricle, such as the mammalian heart, which has two of each chamber, the chambers are physically separated by a thick strip

Figure 47.8 **The mammalian heart and the route blood takes through it.** The figure shows the major blood vessels entering and leaving the heart, the locations of the valves, and the direction of blood flow through the chambers of the heart. Oxygenated blood is shown in red, deoxygenated in blue. Note that the pulmonary veins carry oxygenated blood because they return blood from the lungs, while veins from the systemic circulation carry deoxygenated blood.

of connective tissue called a septum. Between each atrium and ventricle there are valves called **atrioventricular (AV) valves** that control the movement of blood between them. Similarly, between each ventricle and the blood vessels it empties into, there are valves called **semilunar valves**. Each atrium is fed by systemic or pulmonary veins; the systemic veins return blood from the body, and the pulmonary veins return blood from the lungs. Each ventricle empties into systemic or pulmonary arteries. The **aorta** leads to the systemic circulation, and the pulmonary arteries lead to the right and left lung.

The heart beats with a steady rhythm; each beat propels blood through the various chambers of the heart. Figure 47.8 also depicts the route blood takes through the mammalian heart. Recall from Figure 47.4c that each ventricle in the mammalian heart functions as a separate pump. The right ventricle pumps deoxygenated blood returning from the body to the lungs, while the left ventricle pumps oxygenated blood coming from the lungs to the body. Blood enters the atria from systemic or pulmonary veins, moves down a pressure gradient through the AV valves into the ventricles, and is pumped out through the semilunar valves into the systemic and pulmonary arteries. Note that the blood can flow only one way through the valves.

Vertebrates Have Myogenic Hearts Capable of Beating on Their Own

What causes the heart to beat so steadily? Animals cannot consciously initiate heart contractions. The beating of the heart is initiated either by nerves or by intrinsic activity of the heart muscle cells themselves. Many arthropods and decapod crustaceans have a **neurogenic heart** that will not beat unless it receives regular electrical impulses from the nervous system. All vertebrates, however, have a **myogenic heart**; that is, the signaling mechanism that initiates contraction resides within the cardiac muscle itself.

Myogenic hearts are electrically excitable and generate their own action potentials. Electrical signals produced by cells of these hearts initiate the cellular events that trigger muscle contraction (see Chapter 44). The rate and forcefulness of the beating of myogenic hearts can, however, be regulated by the nervous system. Nonetheless, myogenic hearts will continue to beat on their own if dissected out of an animal and placed in a nutrient bath, even with no nerves present.

Excitation of the Vertebrate Heart Begins in the SA Node in the Atria and Spreads to the Ventricles via the AV Node

The electrical excitation of the myogenic vertebrate heart has two phases: atrial and ventricular. In atrial excitation, electrical signals are generated at the junction of the veins and single atrium in fishes, and within the wall of the right atrium in other vertebrates, at what is called the **sinoatrial (SA) node**, or **pacemaker** (Figure 47.9). The SA node is a collection of modified cardiac cells that have an inherently unstable resting membrane potential. Ion channels in the membranes of these cells are opened spontaneously and allow the influx of positively charged ions into the cytosol, thereby depolarizing the cell.

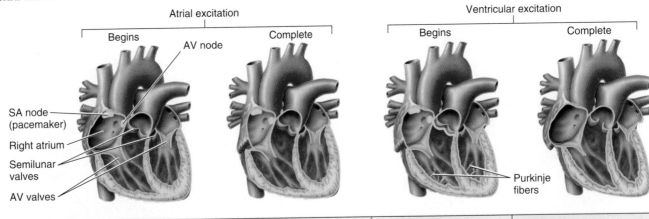

AV valves	OPEN	OPEN	Beginning to close	CLOSED
Semilunar valves	CLOSED	CLOSED	CLOSED	OPEN
Phase of cardiac cycle	Diastole	Diastole	Systole beginning	Systole
Atria	Relaxed	Contracted	Beginning to relax	Relaxed
Ventricles	Relaxed	Relaxed	Beginning to contract	Contracted
Site of highest pressure	Atria	Atria	Ventricles	Ventricles

Figure 47.9 Coupling of electrical and mechanical activity in the mammalian heart. Electrical activity, which is depicted as yellow color, begins in the SA node and quickly spreads through the atria to the AV node. Branches from the AV node transmit electrical activity throughout the ventricles along the Purkinje fibers. The atria and ventricles do not contract until they have become electrically excited. Changes in fluid pressure gradients between the atria and ventricles, and between the ventricles and the aorta and pulmonary arteries, are the forces that open or close the four heart valves. Note: The aortic valve is not visible in this section.

These depolarizations produce action potentials in the SA node cells with a frequency that is characteristic for a given species; this frequency is an animal's heart rate. Once the SA node cells generate action potentials, the potentials quickly spread across one or both atria. This occurs because cardiac muscle forms a structure called a syncytium—a group of cells, all of which are electrically coupled by gap junctions (see Chapter 10). The action potentials trigger an influx of Ca^{2+} into the muscle cell cytosol, which activates contraction. Because the gap junctions between cardiac muscle cells allow electric charge to flow freely from one cell to the next, the impulses spread very rapidly across the atria, so both atria contract almost as if they were one large muscle cell. Atrial contraction pumps blood through the AV valves into the ventricles.

To begin ventricular excitation, action potentials initiated in the SA node first reach another node of specialized cardiac cells, the **atrioventricular (AV) node**. The AV node is located near the junction of the atria and ventricles and conducts the electrical signals from the atria to the ventricles. This is important, because there are no gap junctions connecting atrial and ventricular cells. Like the SA node, the AV node is electrically excitable, but its cells require a longer time to become excited than do the cells of the atria. This allows time for the atria to contract before the ventricles do. Fibers branching from the AV node spread electrical impulses along a conducting system of cardiac muscle cells called Purkinje fibers, which are specialized to rapidly conduct electricity. These fibers branch throughout the muscular walls of the ventricles, ensuring that all of the ventricular muscle gets depolarized quickly and nearly simultaneously (Figure 47.9). In this way, the entire mass of both ventricles contracts in unison in response to depolarization.

The Cardiac Cycle Has Two Phases: Diastole and Systole

Each beat of the vertebrate heart requires the coordinated activities of the atria and ventricles. The events that produce a single heartbeat are known as the **cardiac cycle**, which can be divided into two phases (Figure 47.9). In the first phase, **diastole**, the ventricles fill with blood coming from the atria through the open AV valves. In the second phase, **systole**, the ventricles contract and eject the blood through the open semilunar valves.

The valves open and close as a result of pressure gradients established between the atria and ventricles, and the ventricles and arteries. During diastole, pressure in the ventricles is lower than in the atria and the arteries; therefore, the AV valves are open, but the semilunar valves are closed (**Figure 47.10a**). Once the ventricles are electrically excited, they begin contracting, and the pressure within the ventricles rapidly increases. This marks the beginning of systole. When the pressure in the ventricles exceeds that in the atria, the AV valves are forced closed. Closure of the AV valves makes the "lub" sound of the familiar "lub-dup" heard through a stethoscope. Ventricular pressure continues to rise; when it exceeds the pressure in the arteries, the semilunar valves open (**Figure 47.10b**). Blood is ejected through the open semilunar valves into the arteries.

Because the ventricular walls are thicker and stronger than the atrial walls, ventricular contractions generate much greater pressure. The pressure exerted by the ventricles on the blood within their cavities causes the AV valves to shut, preventing blood from flowing backward into the atria. This is why the electrical delay that is built into the AV node is important. If the atria and ventricles were excited and contracted at the same

(a) Semilunar valves in nearly closed position

(b) Semilunar valves in opened position

Figure 47.10 **Appearance and function of the semilunar valves.** The four heart valves are sheathlike structures with flaps that open and close, as in this illustration of semilunar valves.

time, the AV valves would close too soon, and the ventricles would not receive their normal volume of blood.

The pressure in the arteries increases during systole. As this is happening, the ventricles begin to return to their resting, unexcited state, and the pressure in the ventricles falls below that in the arteries. The higher pressure in the arteries closes the semilunar valves, which prevents blood from flowing back into the heart. Closure of the semilunar valves creates the second heart sound, "dup," heard through a stethoscope. Throughout systole, meanwhile, the atria continue to fill with blood,

which raises the pressure in the atria. Soon the pressure in the atria exceeds that in the relaxed ventricles, and the AV valves open again, bringing a new volume of blood into the ventricles and starting a new diastole.

Blood pressure, which is defined as the force exerted by blood on the walls of blood vessels, is highest in the arteries during systole and lowest during diastole. For this reason, blood pressure is measured with two numbers, the systolic and diastolic pressures. For historical reasons, the units are usually given as mmHg (millimeters of mercury). Blood pressures are highest in mammals and birds and lowest in fishes and invertebrates. For example, a typical blood pressure in humans is around 120/80 mmHg (systolic/diastolic).

An ECG Tracks Electrical Events During the Cardiac Cycle

An **electrocardiogram** (**ECG** or **EKG**) is a medical test used to investigate the function of the heart. An ECG is a record of the electrical potentials between various points on the body. These potentials arise from the electrical signals generated during the cardiac cycle. Sensitive electrodes are placed on the surface of the body to monitor the wave of electricity initiated by the SA node, which travels through the atria, AV node, and ventricles. This procedure works because the body fluids that surround the heart conduct electricity, even the very weak impulses generated by a beating heart.

The trace on an ECG reveals several waves of electrical excitation (**Figure 47.11**). The first is the P wave, which begins when the SA node fires, and ends when the two atria completely depolarize. The next wave of excitation is a cluster of three waves, called the QRS complex. It begins when the branches from the AV node excite the ventricles, and ends when both ventricles depolarize completely. The final wave is the T wave, which results from the repolarization of the ventricles back to their resting state. The T wave is upward on the ECG tracing because the wave of ventricular repolarization proceeds in the opposite direction of depolarization. No wave is visible

Figure 47.11 **An electrocardiogram (ECG).** Electrodes placed on the skin detect electrical impulses occurring in the heart, and an electrocardiogram visualizes them. The resultant waveform is a useful indicator of cardiac health. Note that the ventricular wave (QRS complex) is taller than the atrial wave (P wave). This is because the ventricles are larger than the atria and generate more electrical activity.

Concept check: Why is it possible to detect electrical changes in the heart using electrodes placed on the surface of the skin? *(Hint: Think about body fluids and their ability to conduct electricity.)*

for atrial repolarization because it occurs simultaneously with the large QRS complex.

The ECG monitor displays both the amplitude (strength) of the electrical signal and the direction that the signal is moving in the chest. From this information, physicians can determine whether a person's heart is generating signals with a normal frequency, strength, duration, and direction.

47.4 Blood Vessels

Now that we have discussed the vertebrate heart, let's examine the vessels that transport blood to and from the heart and throughout the body. **Figure 47.12** illustrates the route that blood follows in a closed circulatory system. Blood is pumped by the heart to large arteries and then flows to smaller arteries and eventually to the smallest arteries, called **arterioles**. Arterioles lead the blood to flow through the smallest vessels, which are called **capillaries**. Gas and nutrient exchange occurs between the blood in the capillaries and the cells surrounding the capillaries. The blood then flows back to the heart, leaving the capillaries through the smallest veins, called **venules**, to small veins and, finally, to large veins.

Arteries Distribute Blood to Organs and Tissues

Arteries are thick-walled vessels that consist of layers of smooth muscle and connective tissue wrapped around a single-celled inner layer, the **endothelium**, which forms a smooth lining in

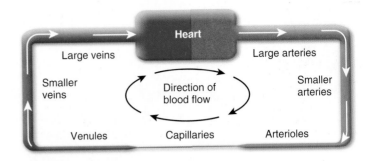

Figure 47.12 Overview of blood flow through vessels in a closed circulatory system.   Regions of gas exchange with the environment have been omitted for clarity.

contact with the blood (**Figure 47.13**). Because of the thick layers of tissue surrounding the endothelium, most dissolved substances cannot diffuse across arteries. Instead, arteries act as conducting tubes that distribute blood leaving the heart to all the organs and tissues of an animal's body.

In vertebrates, the walls of the largest arteries, such as the aorta, also contain one or more layers of proteins that have elastic properties (Figure 47.13). As the aorta stretches to accommodate blood arriving from the heart, the elastic proteins also stretch. The thick layers of tissue in the aorta and other large arteries prevent them from stretching more than a small amount. When the heart relaxes as it readies for another beat, the elastic proteins in the aorta and largest arteries recoil to their original state, something like releasing a stretched rubber band. The recoiling vessels generate a force on the blood within them; this helps prevent blood pressure from decreasing too much while the heart is relaxing during diastole.

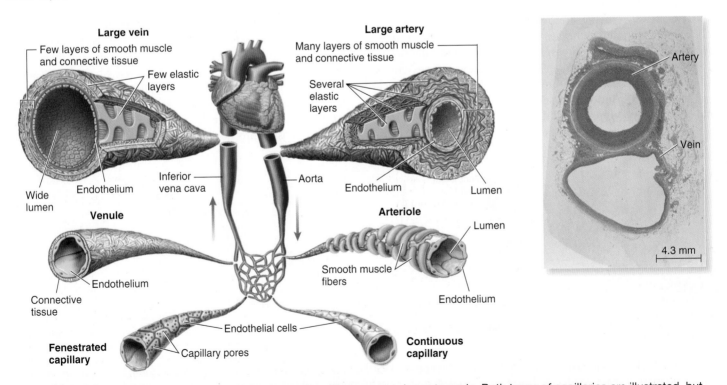

Figure 47.13 Comparative features of blood vessels.   Sizes are not drawn to scale. Both types of capillaries are illustrated, but they do not actually appear together in this way. Fenestrated capillaries are typically found in secretory glands, such as the pancreas, or organs where blood is filtered, such as the kidney. Inset: Light micrograph (enlarged four times) of a medium-sized artery near a vein. Note the difference between the two vessels in wall thickness and lumen diameter.

Figure 47.14 **Arteries and arterioles.** Arteries branch into smaller arteries, and into even smaller arterioles, in order to penetrate to all the cells of tissues and organs.

Concept check: *What would happen to the diameter of the arterioles of the animal's leg muscle when the animal is active?*

Arterioles Distribute Blood to Capillaries

As arteries carry blood away from the heart, they branch repeatedly and become narrower to penetrate to the smallest reaches of an organ or tissue (**Figure 47.14**). Eventually, the vessels are little more than a single-celled layer of endothelium surrounded by one or two layers of smooth muscle and connective tissue (see Figure 47.13). These arterioles deliver blood to the capillaries and distribute blood to regions of the body in proportion to metabolic demands. This is accomplished by changing the diameter of arterioles, such that they widen, or dilate, in areas of high metabolic activity and narrow, or constrict, in inactive regions. Arterioles dilate when the smooth muscle cells around them relax, and they constrict when these cells contract.

Capillaries Are the Site of Gas and Nutrient Exchange

Arterioles branch into the tiny thin-walled capillaries. At the capillaries, materials in the blood are delivered to the other cells of the body, and waste products and secretions from cells are delivered to the blood. Capillaries are tubes composed of a single-celled layer of endothelium resting on an extracellular matrix called a basement membrane.

There are two types of capillaries: continuous and fenestrated. Continuous capillaries have smooth walls and are the most common; fenestrated capillaries contain tiny openings called fenestrations (see Figure 47.13). Because of their fenestrations, such capillaries are much leakier than continuous capillaries. Fenestrated capillaries are found in places such as the vertebrate kidney, where large amounts of plasma are filtered to remove wastes, and in glands such as the pancreas that secrete watery fluids. Capillaries are the narrowest blood vessels in the body; essentially every cell in an animal's body is near one. The diameter of a capillary is about the same as the width of a red blood cell, so erythrocytes move through them in single file (**Figure 47.15**).

Figure 47.15 Erythrocytes (red blood cells; RBCs) moving through a capillary in single file, as seen through a light microscope.

Blood enters capillaries under pressure that is created by the beating of the heart as blood is pumped into the arteries. This pressure forces some of the water in blood out through the fenestrations and other tiny openings in capillary walls into the interstitial fluid (**Figure 47.16**). These openings are wide enough to permit water and small solutes, including O_2, but not red blood cells and most proteins, to leave the capillary. Once in interstitial fluid, the solutes may diffuse into body cells. Conversely, body cells release secretions and metabolic waste

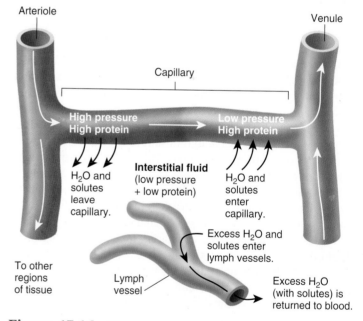

Figure 47.16 **Water movement between capillaries and interstitial fluid.** Water and dissolved solutes (other than proteins) exit capillaries near the arteriolar end because the capillary pressure is much greater than that of the interstitial fluid. As the volume of the water in the capillary decreases, however, the pressure within the capillary also decreases (but remains greater than interstitial fluid). Proteins remaining in the capillary contribute an osmotic force that tends to draw water back into the capillary. The combination of decreased pressure within the capillary and the osmotic force due to the proteins leads to the recapture of much of the water that left the capillary. Lymph vessels drain any excess fluid from the interstitial fluid. For clarity, the system shown here contains a single capillary connecting an arteriole and venule, instead of the typical situation where capillaries divide into numerous branching vessels.

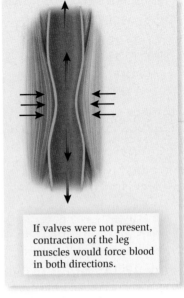

Figure 47.17 One-way valves in veins. Valves are typically present in the limbs, as shown in this dog's leg. **(a)** In some veins, one-way valves assist the return of blood to the heart against the force of gravity. **(b)** Blood moving in a vein without valves would flow in both directions.

Concept check: Unlike most mammals, giraffes have one-way valves in the veins of their long necks. Which way do you think the valves open, toward the heart or toward the head, and why? (Hint: Picture the way a giraffe drinks from a body of water.)

Leg muscle

Blood flows toward heart.

Open valve

Contracting skeletal muscles of leg

When the leg muscle contracts, the lower valve stays closed while the upper valve opens. This causes blood flow toward the heart.

If valves were not present, contraction of the leg muscles would force blood in both directions.

(a) Vein with one-way valves

(b) Vein without valves

products including CO_2 gas into interstitial fluid, where they diffuse into capillaries to be carried away.

If the water that leaves a capillary were to remain in the interstitial fluid, the volume of plasma in the blood would decrease, and the interstitial fluid would swell. Most of the water that leaves at the beginning of a capillary, however, is recaptured at the capillary's end, for two reasons (Figure 47.16). First, near the arteriole, the blood pressure inside the capillary is higher than the fluid pressure of the surrounding interstitial fluid, which creates a force that moves water out of the capillary. However, by the time blood reaches the end of a capillary, its pressure has decreased considerably, partly as a result of the loss of water. Second, proteins are trapped within the capillary because they are too large to diffuse through the capillary walls or to leak out of the tiny openings. This creates an osmotic force that draws water back into the capillary from the interstitial fluid. Near the end of a capillary, the osmotic force drawing water into the capillary is greater than the pressure forcing water out; therefore, water and dissolved solutes reenter the capillary.

Despite the ability of capillaries to recapture their own fluid, the process is not 100% effective. Another set of vessels, those of the **lymphatic system**, collects excess fluid and returns it to the blood. The lymphatic system is part of the defense mechanisms of the animal body (see Chapter 53).

Venules and Veins Return Blood to the Heart

Once blood travels through capillaries, picking up any substances secreted from the cells of the body, it enters the venules, small, thin-walled extensions of capillaries. The venules empty into the larger veins that return blood to the heart. The largest of these veins are the vena cavae. In humans, the superior vena cava drains body tissues above the heart, while the inferior vena cava drains those below the heart. The walls of veins are much thinner, less muscular, and more easily distended or stretched than those of arteries (see Figure 47.13).

By the time blood has traveled through the capillaries and reached the veins, the pressure of the blood is very low. Consequently, veins can fill with considerable volumes of blood,

particularly veins in the lower parts of an animal's body, such as the legs, where gravity tends to cause blood to pool. Several factors assist the blood on its way toward the heart when flowing against gravity. One factor is stimulation of the contraction of smooth muscle in leg veins by the sympathetic nervous system, which compresses the veins and helps force blood back to the heart. Another factor is the activity of skeletal muscles in the limbs that assists the return of venous blood to the heart. For example, each time the leg skeletal muscles contract, they squeeze the veins passing through them (**Figure 47.17**). This alone would not move blood upward toward the heart, though, because each time a vein is squeezed, blood could be forced both up and down. This does not happen because one-way valves inside veins ensure that blood returning from below the heart moves in only one direction, toward the heart. By contrast, veins located above the heart, like those in the necks of bipeds and some quadrupeds, lack valves because gravity pulls blood toward the heart.

Many people can easily observe the effects of gravity on venous blood flow. When the arms are held down by their side, the veins are visible on the backs of their hands. When the arms are raised above the head, the bulging veins quickly lose blood and become less visible. When blood from the veins returns to the heart, it must travel against gravity when your arms are at your side and with gravity when your arms are elevated. Blood drains from veins much more efficiently when gravity works in its favor.

47.5 Relationships Between Blood Pressure, Blood Flow, and Resistance

Blood pressure is responsible for blood flow, or moving blood through the vessels. It is not the same in all regions of an animal's body, however, because of resistance. **Resistance (R)** refers to the tendency of blood vessels to slow down the flow of blood through their lumens. As we saw earlier, the blood pressure in arteries is much higher than in veins. For this reason, we generally speak only of arterial pressure when discussing blood pressure. The relationships between blood pressure, blood flow, and resistance can be considered on two levels, local and systemic, as we see next.

Resistance Determines Local Blood Pressure and Flow

The physical characteristics of a blood vessel determine the amount of blood that flows through it and how fast that blood flows. This affects how much blood reaches particular areas of the body at any given time. For example, the diameter of different blood vessels can be increased or decreased to change the amount of blood flowing through them. As stated earlier, the ability to distribute blood to different parts of the body in amounts proportional to that body part's metabolic requirements is a key adaptation of closed circulatory systems.

Resistance   Resistance is a function of three variables: vessel radius, length, and blood viscosity. Wide, short tubes provide less resistance to flow than narrow, long tubes. Resistance is increased by the blood's viscosity, which is a measure of the blood's hematocrit. A high hematocrit increases the blood's viscosity—it makes the blood more sludgelike—and hinders its smooth flow through vessels.

The relationship between blood pressure, flow, and resistance is stated by Poiseuille's law, which was derived in the 1840s by Jean Marie Louis Poiseuille, a French physician and physiologist. His law is simplified here:

$$\text{Flow (F)} = \Delta\, \text{Pressure (P)}/\text{Resistance (R)}$$

Stated mathematically, blood flow through a blood vessel is directly proportional to the difference (Δ) in pressure of the blood at the beginning and end of the vessel, and inversely proportional to the resistance created by that vessel. Changes in any of these three variables determine how much blood flows through different body regions at any moment. The equation can be rearranged as $\Delta P = F \times R$, which demonstrates that blood pressure depends on both blood flow and resistance.

Poiseuille's law applies to blood flow through a single vessel, an organ, or the entire body.

Vasodilation and Vasocontriction   Changes in arteriolar resistance are the major mechanism for increasing or decreasing blood flow to a region. In the short term, the length of an arteriole and the viscosity of blood do not normally change. Therefore, the radii of arterioles become the most important factor in determining minute-to-minute resistance.

The relationship between arteriolar radius and resistance is not linear. Resistance is inversely proportional to the radius of the vessel raised to the fourth power: $R \propto 1/r^4$, where $\propto$ means "proportional to," and r is the radius of the arteriolar lumen. Let's consider an arteriole with a radius that increases by a factor of two. This would occur if the smooth muscles around the arteriolar wall relax sufficiently to allow the vessel to dilate and double its original radius. Because resistance is inversely proportional to the fourth power of vessel radius, an increase in radius by a factor of two will result in a decrease in resistance of 2^4 ($2 \times 2 \times 2 \times 2$), or 16-fold.

Vasodilation refers to an increase in blood vessel radius, and **vasoconstriction** refers to a decrease in blood vessel radius. The signals that control arteriolar radius come from three sources, including locally produced substances, hormones, and nervous system inputs. Locally, metabolic by-products such as carbon dioxide and lactic acid, as well as potassium ions and other substances secreted by metabolically active tissues, cause nearby arterioles to vasodilate immediately. According to Poiseuille's law, this permits more blood flow to the active region, facilitating oxygen and nutrient delivery and waste removal. Hormones secreted by glands throughout the body can also regulate arteriolar radius. For example, some hormones cause arterioles that deliver blood to the small intestine to vasoconstrict during fight-or-flight responses, routing blood away from the intestine and to areas of more immediate need such as the heart and skeletal muscles. Finally, smooth muscle cells that make up the outer wall of arterioles receive inputs from nerves that can stimulate the muscles to contract or relax. One of the most important regulators of blood vessel diameters, however, is a gas, as we see next.

FEATURE INVESTIGATION

Furchgott Discovered a Vasodilatory Factor Produced by Endothelial Cells

As we have seen, the cells surrounding blood vessels can produce ions, acids, and gases that diffuse to arteriolar smooth muscle cells and cause these cells to relax and vasodilate. Beginning in the 1970s, American biochemist Robert Furchgott provided evidence that an unidentified substance released from within an artery—that is, from the endothelial lining of the vessel—also played a key role in regulating blood vessel radius.

For many years, Furchgott had used flattened strips of rabbit aorta in a culture bath to test whether various compounds stimulated or inhibited contraction of the smooth muscle in the artery. Although large arteries such as the aorta do not show the dra-

matic vasodilation and vasoconstriction that arterioles do, they were more useful in experiments because of their larger size.

One of the compounds he tested was the neurotransmitter acetylcholine (ACh). Scientists knew at the time that when ACh is injected into an animal, it causes vasodilation by relaxing smooth muscle in arterioles. However, in Furchgott's in vitro preparations, ACh caused the flattened strips of artery to contract. To determine if this apparent paradox could be the result of the method in which the tissues were prepared for in vitro tests, Furchgott compared the effects of ACh in different types of preparations. He discovered that ACh produced muscle relaxation when applied to circular rings of aorta, as described in step 1 in **Figure 47.18**.

How did Furchgott explain that the muscle contracted when ACh was applied to the flattened strips of aorta? After examining

Figure 47.18 Furchgott's discovery that endothelial cells produce a vasodilating substance.

HYPOTHESIS The endothelium of arteries produces a vasodilatory substance.

KEY MATERIALS Strips or rings of rabbit aorta, force transducers, acetylcholine.

	Experimental level	Conceptual level

1 Attach a ring of aorta to a pressure gauge. Add acetylcholine (ACh) to bath medium. Record change in force (increased force = increased muscle contraction; decreased force = decreased muscle contraction).

ACh added to medium

Force transducer

Intact aorta with endothelium

In the presence of endothelium, ACh relaxes the muscles of the aorta, causing the vessel to widen.

2 Denude ring of its endothelium. The muscle layer is left behind. Retest effect of ACh.

Wooden stick is used to scrape away endothelium.

Endothelial cells

Muscle layer

The denuded ring is then placed in recording apparatus as before.

In the absence of endothelium, ACh stimulates the muscles to contract, which narrows the vessel.

3 Perform "sandwich" experiment by combining denuded strip of aorta with an intact strip. Record effect of ACh on strip 1 alone, strip 2 alone, and strip 1 while part of a sandwich with strip 2.

Muscle layer only

Endothelium

Muscle layer

Strip 1

Strip 2

Strip 1   Strip 2   Sandwich

+ ACh →

Endothelium produces a vasodilating substance following ACh treatment that diffuses from strip 2 to strip 1, and causes strip 1 to relax.

4 THE DATA

Results of sandwich experiment

Force recorded on force transducer

Contraction ↑   Relaxation ↓

ACh

Time
Strip 1 alone: contraction

ACh

Time
Strip 2 alone: relaxation

ACh

Time
Strip 1 recorded while attached to strip 2: relaxation

5 CONCLUSION Endothelial cells release a factor that diffuses to the surrounding smooth muscle, causing it to relax. Relaxation of the vascular smooth muscle causes dilation of the blood vessel.

6 SOURCE Furchgott, R.F., and Zawadzki, J.V. 1980. The obligatory role of the endothelium in the relaxation of arterial smooth muscle by acetylcholine. *Nature* 288:373–376.

the two preparations of tissue, Furchgott realized that the lining of endothelial cells along the inner wall of the vessel had been scraped away during preparation of the strips, but the endothelial lining was still present in the rings. He hypothesized that the endothelial cells must produce a factor whose secretion was stimulated by ACh, and that not only prevented contraction, but even caused smooth muscles surrounding the blood vessel to relax. If so, this would explain the observation that when ACh was injected into rabbits, it resulted in vasodilation, because the intact vessels of the animals contained their endothelial lining and thus resembled Furchgott's aortic rings.

To test his hypothesis, Furchgott performed two experiments. As shown in step 2 of Figure 47.18, he used a wooden rod to scrape away the endothelium from circular rings of aorta and discovered that the relaxation effect of ACh was lost, as predicted. In step 3, he used a sandwich technique, in which he tested the effects of ACh on muscle contraction in a "denuded" strip of artery with the endothelium removed, and then again on the same strip, which was attached ("sandwiched") to another strip in which the endothelium remained intact. When ACh was added to the water bath before the sandwich was formed, the denuded strip contracted, as expected. When the denuded strip was attached to the intact strip in such a way that its muscle layer was exposed to the intact strip's endothelial surface, the denuded strip relaxed after ACh was added. This was consistent with the hypothesis that the endothelium released a vasodilatory factor that diffused to the denuded strip and relaxed its muscles.

Two other scientists, Americans Louis Ignarro and Ferid Murad, later determined that this factor was the gas nitric oxide (NO). The discovery of this function of NO revealed a new category of signaling molecules. Ignarro and Murad also revealed that nitroglycerin, which had for many years been used to treat patients with cardiovascular disease, acted by generating NO in the blood to produce vasodilation and lower blood pressure. Today NO is considered one of the most potent and important naturally occurring vasodilators in animals. Knowledge of the NO signaling pathway has led to new therapies for treating a wide variety of health disorders associated with blood vessels, including high blood pressure, glaucoma (an eye disease), and erectile dysfunction (see Chapter 51). For their efforts, Furchgott, Ignarro, and Murad shared the 1998 Nobel Prize.

Experimental Questions

1. What observation did Furchgott make when testing the effects of acetylcholine on different treatments of rabbit aorta? How did he explain the results?

2. Based on his observations of the effects of acetylcholine on different preparations of rabbit aorta, what hypothesis did Furchgott propose to explain the relationship of acetylcholine and vasodilation? How did he test this hypothesis?

3. What did Furchgott conclude based on his results?

Cardiac Output and Resistance Determine Systemic Blood Pressure and Flow

Because blood vessels provide resistance to blood flow, the heart must beat forcefully enough to overcome that resistance throughout the whole body. **Cardiac output (CO)** is the amount of blood the heart pumps per unit time, usually expressed in units of L/min. Poiseuille's law can be adapted to the whole body. In this case, pressure refers to arterial blood pressure (BP) recorded using a blood pressure cuff, flow refers to CO, and resistance refers to the sum of all the resistances in all the arterioles (**total peripheral resistance**, or **TPR**). Thus, we have BP = CO × TPR. The CO and TPR determine the pressure the blood exerts in the arteries of a closed circulatory system. Let's examine these variables in more detail.

Cardiac Output   The cardiac output depends on the size of an animal's heart, how often it beats each minute, and how much blood it ejects with each beat. Each beat, or stroke, of the heart ejects an amount of blood known as the **stroke volume (SV)** that is roughly proportional to the size of the heart (**Table 47.1**). Thus, if we know the stroke volume of a heart and can measure the heart rate (HR: the number of beats per minute), we can determine the CO. Simply put, CO = SV × HR.

Naturally, the CO of an elephant is far greater than that of a mouse, a sparrow, or a salmon. Typically, heart size varies in proportion to body mass within a given class of vertebrates, with birds and mammals having larger hearts than do similarly sized

Table 47.1	Comparative Features of Representative Mammalian Hearts			
Animal	Body mass* (kg)	Heart mass* (kg)	Stroke volume* (L)	Heart rate* (bpm)†
Shrew‡	0.0024 (2.4 g)	0.000035 (35 mg)	0.000008 (8 µl)	835
Rat	0.20	0.001	0.0018	360
Dog	23	0.12	0.025	95
Human	75	0.38	0.075	70
Elephant	4,000	25	4.0	25
Blue whale	100,000	600	100	10

*Values are based on average body masses and resting conditions. In some cases, stroke volumes are estimates based on heart size. †bpm = beats per minute. ‡The shrew reported here is the Etruscan shrew, one of the smallest known mammals. Its heart is somewhat larger than would be predicted for its body mass. Note its heart rate; at 835 bpm, the heart beats 14 times per second!

fishes, amphibians, or reptiles. Note in Table 47.1 that heart rate decreases as mammals get larger, but heart mass and stroke volume increase roughly in proportion with body mass. Similar relationships are observed in other vertebrates, notably birds. Smaller animals have smaller hearts and, therefore, smaller stroke volumes. The heart of a typical shrew, for example, is the size of a small pea, whereas the heart of a blue whale is as large as a cow. Their stroke volumes are similarly proportioned. However, small animals have faster heart rates than do large animals. A hummingbird's or shrew's resting heart rate may be more than 800 beats per minute, whereas a blue whale's heart may beat only 10 times per minute (although the amount of blood ejected with each of those beats is enormous!). Recall from Chapter 46 that metabolic rate is relatively greater in smaller animals. The higher heart rates of small animals give them a greater cardiac output than would be predicted for the size of their hearts, which helps them meet the extraordinary oxygen and nutrient demands of such highly metabolic organisms.

Systemic Blood Pressure   The greater the cardiac output and the higher the resistance to blood flow, the higher will be the blood pressure. Imagine that the circulatory system is like a faucet (the heart) connected to a garden hose (the arteries) (**Figure 47.19**). If the faucet is fully open, analogous to maximal cardiac

(a) Maximal cardiac output with low resistance

(b) Moderate cardiac output with low resistance

(c) Moderate cardiac output with high resistance

Figure 47.19   An analogy for how cardiac output, resistance, and blood pressure are related.   A hose analogy shows the way in which cardiac output (the faucet) and resistance (constriction of the hose) impact blood pressure.

output, and the hose is not blocked, analogous to low arteriolar resistance, the amount of water rushing into the hose will be high and so will the water pressure, representing blood pressure. If the faucet is only partially open, the water pressure will be lower. However, now imagine that the faucet is partially open but the end of the hose is constricted, representing a region of high arteriolar resistance. In that case, the pressure of the water in the hose will increase between the faucet and the point where the hose is squeezed, and will decrease beyond the constriction, as will the flow of water (representing blood).

Arterial blood pressure, therefore, is a function of how hard the heart is working and how constricted or dilated the various arterioles are. Blood pressure must be high enough for blood to reach all body tissues even at the farthest extremities but not high enough to damage blood vessels or force excess plasma out of capillaries. Blood pressure is roughly similar among mammals. Although species differences exist, they are not as diverse as, for example, species differences in heart rates. Blood pressure is relatively low in nonmammalian vertebrates and in invertebrates.

47.6   Adaptive Functions of Closed Circulatory Systems

A circulatory system must not only pump blood and return it to the heart, it must also adapt to changes, including sleep, activity, and emergencies such as blood loss or dehydration. In this section, we examine how changes in the circulatory system help maintain cardiovascular activity during exercise and when an animal experiences a change in blood pressure.

Circulatory Function Adapts to Exercise

Nearly all animals have periods of rest and activity. Therefore, the activity of the heart must adjust to fluctuating metabolic demands. For this reason, blood must be sent to different areas in proportion to their requirements for oxygen and nutrients. Exercise provides a dramatic illustration of rapid changes in circulatory function.

We often think of exercise as voluntary. In nature, exercise means an increase in locomotion for any reason. Typically, an animal exercises when it needs something, such as food or shelter, or when it is escaping something, such as a predator. The increased activity may be momentary and moderate, as when a fish briefly swims faster to avoid the tentacles of an anemone, or it may last longer and be more intense, as when a cheetah sprints after a gazelle. Exercise may last for very long periods of time, as when a salmon migrates from the sea to its freshwater spawning grounds or a migratory goose flies thousands of miles over several weeks. The circulatory systems of animals must quickly adapt to these increased metabolic demands and adjust to the intensity, duration, and type of exercise.

We have already noted some adjustments, such as vasodilation and vasoconstriction. These are the result of changes in the radius of arterioles. During exercise, these vessels vasodilate in regions that require more energy and oxygen—for example, skeletal muscles and the heart—and vasoconstrict in areas that require less energy and oxygen, such as the gut.

In addition, cardiac output increases during exercise to supply more blood to the tissues that require it. This happens by increasing stroke volume or heart rate, or both. Fishes typically increase stroke volume more than heart rate, whereas birds and mammals do the opposite. In humans, for example, stroke volume increases from approximately 75 mL (milliliters) at rest to 110 mL during exercise, a change of 1.5-fold, and heart rate can almost triple from about 70 bpm to nearly 200 bpm. Thus, the combined effects of increasing stroke volume and heart rate raise the cardiac output by about 4.5-fold.

Whichever way cardiac output increases, one of the major mechanisms involved is an increase in blood levels of epinephrine, also known as adrenaline. **Epinephrine** is a hormone secreted by the adrenal glands, which are activated by the sympathetic nervous system during exercise. Epinephrine binds to receptors on heart ventricular muscle cells, making them contract more vigorously, thereby increasing stroke volume. In addition, epinephrine binds to receptors in the atrial SA node cells, stimulating them to initiate electrical signals at a faster rate, thereby increasing heart rate. The latter effect of epinephrine is enhanced by the release of the neurotransmitter **norepinephrine** (or noradrenaline) directly from sympathetic nerve endings onto the SA node cells.

At rest, the heart does not pump out its entire content of blood. A reservoir of blood remains in each ventricle at the end of a beat. During exercise, some of this reservoir is tapped when epinephrine stimulates the heart muscles. In addition, the mechanical process of exercise itself enhances the return of blood to the heart, as skeletal muscles squeeze the veins within them and open their one-way valves. The combination of more blood getting returned to the heart and the enhanced emptying of the built-in reservoir of blood from the ventricles provides increased stroke volume during exercise.

Because BP equals CO × TPR, you might imagine that increased cardiac output would raise blood pressure to possibly harmful levels during exercise. Although blood pressure does increase somewhat, it does not rise as much as you might predict. This is because the blood vessels of skeletal muscles dilate (that is, resistance is decreased) to allow more blood to enter; similarly, the vessels in the skin of terrestrial animals dilate as a means of dissipating heat. Therefore, even though CO increases in exercise, the total peripheral resistance in the body decreases, negating some of the effect on blood pressure.

Regular, long-term exercise induces additional adaptive responses in the circulatory system. For example, the thickness of the muscle of the ventricles of the human heart is greater in athletes than in sedentary individuals, which makes the athletic heart a more efficient pump.

Baroreceptors Maintain Blood Pressure

How do vertebrate circulatory systems adapt to changes in blood pressure? Within the walls of certain arteries, notably the aorta and the carotid arteries, are pressure-sensitive regions that contain the endings of neurons. These neurons, known as **baroreceptors** (from the Greek *baros*, weight or pressure), are in constant communication with the brain (**Figure 47.20**).

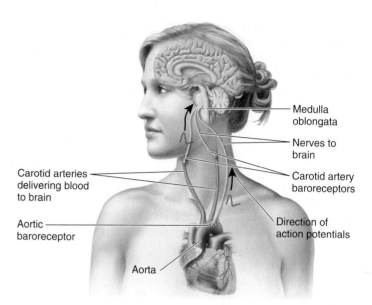

Figure 47.20   Location of the major baroreceptors. Increased blood pressure stretches arteries. This activates baroreceptors in the arteries, which send action potentials to the medulla oblongata in the brain. Nerve signals from the brain then initiate changes in heart function and total peripheral resistance that help restore blood pressure to normal.

Concept check: *In Chapters 41–43, we discussed the different types of ion channels found in neurons and the ways in which such channels open and close. What type of channels would you expect to be activated when baroreceptors are stimulated?*

If blood pressure increases above the normal range, these large arteries stretch more than normal as a result; this makes the endings of the baroreceptors stretch, too. The stretching of the baroreceptors opens ion channels in the cell membrane, resulting in the generation of action potentials. Nerves carry the signals to the brain, which interprets this as a signal that blood pressure is above normal. The response is a decrease in the amount of norepinephrine released from neurons of the autonomic nervous system onto the cardiac SA node, and a decrease in the secretion of epinephrine from the adrenal glands. The result is a decreased heart rate and stroke volume, resulting in lower cardiac output. In addition, nerves throughout the body release less vasoconstricting neurotransmitters from their neurons onto the smooth muscle cells of arterioles, thereby promoting vasodilation and reducing total peripheral resistance. The combination of decreased cardiac output and decreased resistance lowers blood pressure toward normal. Elevations in blood pressure are not common in animals, but when it occurs in humans, it can have serious medical consequences (see Section 47.7 of this chapter).

By contrast, if blood pressure falls below normal, the walls of these arteries would be less stretched than they normally are, and the baroreceptors would send fewer impulses to the brain. This results in a sequence of events that are opposite to those just described. More norepinephrine and epinephrine are secreted, thereby increasing cardiac output and vasoconstriction and consequently increasing blood pressure toward normal. One situation in which blood pressure might decrease is

dehydration, which is a reduction in the amount of water in the body. Because blood is mostly water, dehydration reduces blood volume, and arteries—including those that contain baroreceptors—are filled with less blood than normal. Blood pressure could also drop due to **hemorrhage**, loss of blood from a ruptured blood vessel. Dehydration is primarily a problem for terrestrial animals, but hemorrhage may occur in any animal that has been injured. In a hemorrhage, both plasma and blood cells are lost; therefore, hemorrhage may be more serious than dehydration. Like dehydration, though, hemorrhage results in reduced blood volume and therefore less baroreceptor activity.

Without baroreceptors, the circulatory system would have no means of relaying information about its status to the brain, and the brain would be unable to initiate compensatory actions to maintain normal pressure. Baroreceptors even mediate nearly instantaneous adjustments to blood pressure, such as when you jump up quickly from a reclining position. Upon standing suddenly, gravity causes more blood to pool in the lower legs, temporarily reducing the volume and pressure of blood reaching the brain. This is why you may sometimes feel woozy when you stand up quickly. Baroreceptors immediately detect the change in pressure in your neck arteries, however, and cardiac output is rapidly increased to bring pressure back to normal until the system stabilizes.

47.7 Impact on Public Health

Cardiovascular disease—conditions affecting the heart and blood vessels—accounts for more deaths each year in the U.S. than any other cause, making it one of the most intensely researched areas of human health. Why is cardiovascular disease so devastating? One reason is that damage to these structures often occurs slowly, over many years, and without symptoms until the disease has reached late stages. In this section, we will consider several common cardiovascular disorders and their causes and treatments.

Systemic Hypertension and Atherosclerosis May Cause Heart and Blood Vessel Disease

Systemic hypertension, often called hypertension or high blood pressure, refers to an arterial blood pressure that is chronically above normal. The normal range in humans varies from systolic/diastolic pressures of about 90/60 to 120/80 mmHg. A resting blood pressure above 140/90 mmHg defines hypertension, and values between 140/90 and 120/80 mmHg are considered borderline (sometimes called "prehypertension"). Hypertension can have many causes, including obesity, smoking, aging, kidney disease, excess male hormones, and genetic factors, although in many cases, the cause is unknown. It can often be treated with diet and exercise and with drugs that cause vasodilation, thereby reducing arterial resistance.

Systemic hypertension rarely has any noticeable symptoms. For this reason, it is important to have your blood pressure checked regularly. Systemic hypertension is dangerous because it

Figure 47.21 An atherosclerotic plaque in an artery. Compare this with the normal artery shown in the micrograph in Figure 47.13.

can damage arteries, contributing to the formation of **plaques**—deposits of lipids, fibrous tissue, and smooth muscle cells—inside arterial walls and leading to a condition known as **atherosclerosis** ("hardening of the arteries") (**Figure 47.21**). Large plaques may occlude (block) the lumen of the artery entirely. In addition to hypertension, plaques are known to arise from a variety of factors, including calcium and fat deposits, and are correlated with obesity, high blood cholesterol levels, and smoking.

If plaques form in any artery, the regions of the body supplied with blood by that artery receive less oxygen and nutrients. Although atherosclerosis is dangerous anywhere, it is especially significant if it affects the **coronary arteries**, which carry oxygen and nutrients to the thick heart muscle. **Coronary artery disease** occurs when plaques form in the coronary vessels, and it can be life-threatening. One warning sign of coronary artery disease is **angina pectoris**, chest pain during exertion due to the heart being deprived of oxygen.

Myocardial Infarction Results in Death of Cardiac Muscle Cells

If a region of the heart is deprived of blood for an extended time, the result may be a **myocardial infarction** (**MI**), or **heart attack**. This is usually caused by a blockage of one of the coronary arteries. Some heart attacks are relatively minor; in fact, the discomfort of a small heart attack may not even alarm someone enough to seek medical attention. However, more serious heart attacks can lead to significant damage or death of part of the heart muscle. Dead cardiac muscle tissue does not regenerate, diminishing the heart's ability to pump. As noted earlier, reduced pumping activity of the heart—notably the left ventricle—can result in congestive heart failure. As noted in the chapter introduction, each year in the U.S., more than 1 million people suffer a heart attack, and about 40% of them are fatal.

Preventing a heart attack in the first place is the best way to increase survival rates. Procedures are available that allow physicians to monitor the status of the coronary vessels in people suspected of having heart disease. The coronary vessels can be visualized by injecting a dye into a person's veins and then taking an X-ray image of the chest, allowing a physician to determine if the vessels are narrowed by disease. This procedure is called cardiac angiography.

If problems with the coronary arteries are found, several common treatments can restore blood flow through a blood

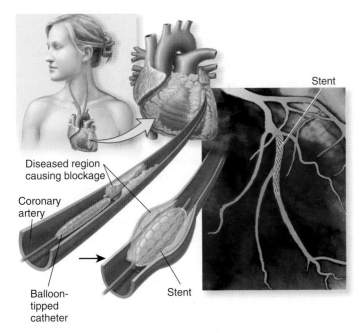

Figure 47.22 A treatment for blocked blood vessels. Balloon angioplasty can widen diseased arteries, followed by insertion of a stent. The inset shows a stent placed in a coronary artery of a human patient.

vessel. One is **balloon angioplasty**, in which a thin tube with a tiny, inflatable balloon at its tip is threaded through the artery to the diseased area. Inflating the balloon compresses the plaque against the arterial wall, widening the lumen. In most cases, a wire-mesh device called a stent is inserted into the diseased artery after angioplasty has expanded it, providing a sort of lattice to hold the artery open (**Figure 47.22**). A treatment for more serious coronary artery disease is a **coronary artery bypass**, in which a small piece of healthy blood vessel is removed from one part of the body and surgically grafted onto the coronary circulation in such a way that blood bypasses the diseased artery.

Summary of Key Concepts

47.1 Types of Circulatory Systems

- Circulatory systems transport necessary materials to all cells of an animal's body and transport waste products away from cells. The three basic types of circulatory systems are gastrovascular cavities, open systems, and closed systems.

- Cnidarians and some flatworms have a gastrovascular cavity that serves two functions: digestion of food and circulation of water containing the digested food. (Figure 47.1)

- An open circulatory system contains three components: hemolymph, vessels, and one or more hearts. The vessels open into the animal's body cavity. (Figure 47.2)

- In a closed circulatory system—found in earthworms, cephalopods, and all vertebrates—blood and interstitial fluid are physically separated by blood vessel walls and differ in their components and chemical composition. (Figure 47.3)

- The closed circulatory systems of vertebrates can be divided into single or double circulations. In the double circulation of crocodiles, birds, and mammals, oxygenated and deoxygenated

blood are completely separated into the systemic circulation and pulmonary circulation. (Figure 47.4)

47.2 Blood and Blood Components

- Blood is a fluid connective tissue consisting of cells and (in mammals) cell fragments suspended in a solution of water containing dissolved nutrients, proteins, gases, and other molecules. Blood has four components: plasma, leukocytes, erythrocytes, and platelets or thrombocytes. (Figures 47.5, 47.6, 47.7)

47.3 The Vertebrate Heart and Its Function

- All vertebrate hearts have at least one atrium and one ventricle. In hearts with more than one atrium or ventricle, connective tissue separates the atria and ventricles. Blood enters the atria from systemic or pulmonary veins, moves down a pressure gradient through the AV valves into the ventricles, and is pumped out through the semilunar valves into the systemic and pulmonary arteries. (Figure 47.8)

- Many arthropods and decapod crustaceans have a neurogenic heart that cannot beat on its own; all vertebrates have a myogenic heart, which beats on its own without electrical stimulation, but which can be regulated by the nervous system.

- The electrical signals of myogenic hearts are generated either at the junction of the veins and the single atrium (in fishes) or at the sinoatrial (SA) node (in other vertebrates). The electrical impulses spread through the atria and ventricles and stimulate the heart muscle to contract. The cardiac cycle has two phases, diastole and systole. (Figures 47.9, 47.10, 47.11)

47.4 Blood Vessels

- Arteries are thick-walled vessels that carry blood away from the heart. Arterioles distribute blood to capillaries, and capillaries are the site of gas and nutrient exchange. (Figures 47.12, 47.13, 47.14, 47.15, 47.16)

- Blood leaving the capillaries empties into venules and then into veins, which return it to the heart. One-way venous valves help to return blood to the heart. (Figure 47.17)

47.5 Relationships Between Blood Pressure, Blood Flow, and Resistance

- Blood pressure is the force exerted by blood on the walls of blood vessels, and it is responsible for moving blood through the vessels. Resistance refers to the tendency of blood vessels to slow down the flow of blood through their lumens.

- Local blood flow through a vessel is directly proportional to the pressure of the blood entering the vessel and inversely proportional to the resistance created by that vessel. Arteriole radius is the major factor that regulates resistance. The endothelium of arteries produces a vasodilatory substance. (Figure 47.18)

- Cardiac output overcomes resistance to generate systemic blood pressure. Cardiac output depends on the size of an animal's heart, how often it beats each minute, and how strongly it contracts with each beat. (Table 47.1)

- Cardiac output and total peripheral resistance determine blood pressure: BP = CO × TPR. Blood pressure must be high enough for blood to adequately reach all body tissues and cells, but not so high as to damage blood vessels. (Figure 47.19)

47.6 Adaptive Functions of Closed Circulatory Systems

- The circulatory system adjusts to fluctuating metabolic requirements through changes in cardiac output, vasodilation, and vasoconstriction. Baroreceptors maintain blood pressure by relaying information about circulatory system status to the brain, which initiates compensatory mechanisms. (Figure 47.20)

47.7 Impact on Public Health

- Cardiovascular disease accounts for more deaths each year in the U.S. than any other cause. Cardiovascular conditions include systemic hypertension, atherosclerosis, coronary artery disease, angina pectoris, and myocardial infarction (MI, or heart attack). (Figure 47.21)

- Cardiovascular diagnostic techniques and treatments include cardiac angiography, balloon angioplasty, and coronary artery bypass. (Figure 47.22)

 Assess and Discuss

Test Yourself

1. Hemolymph differs from blood in that it
   a. does not contain blood cells.
   b. is a mixture of fluid in blood vessels and interstitium.
   c. circulates through closed circulatory systems only.
   d. functions only in defense of the body and not transport.
   e. does not pass through a heart.

2. The heart chamber that receives oxygenated blood from the lungs of animals with a double circulation is
   a. the left atrium.   c. the left ventricle.
   b. the right ventricle.   d. the right atrium.

3. Amphibians are adapted for life in and out of water because
   a. they are the only vertebrates with an open circulatory system.
   b. oxygen diffuses out of their blood into the environment through their skin.
   c. oxygen diffuses into the blood from the lungs or through the skin.
   d. they have a four-chambered heart.
   e. they can direct blood away from the skin when they are submerged under water.

4. A major advantage of a double circulation is that
   a. blood can be pumped to the upper portions of the body by one circuit and to the lower portions of the body by the other circuit.
   b. each circuit can pump blood with differing pressures to optimize the function of each.
   c. the oxygenated blood can mix with the deoxygenated blood before being pumped to the tissues of the body.
   d. less energy is required to provide nutrients and oxygen to the tissues of the body.
   e. all of the above

5. The function of erythrocytes is to
   a. transport oxygen throughout the body.
   b. defend the body against infection and disease.
   c. transport chemical signals throughout the body.
   d. secrete the proteins that form blood clots.
   e. a and d are both correct.

6. The mammalian heart is an example of a myogenic heart, meaning
   a. it is composed of four chambers.
   b. it acts as two pumps for two different circulations.
   c. contraction is regulated solely by the nervous system.
   d. electrical activity is initiated by specialized cells of the heart.
   e. the nervous system plays no role in its function.

7. During systole of the cardiac cycle
   a. the ventricles of the heart are both relaxed.
   b. the ventricles of the heart are both filling.
   c. the ventricles of the heart are both contracting.
   d. the right ventricle is contracting while the left ventricle is at rest.
   e. the left ventricle is contracting while the right ventricle is at rest.

8. Considering blood flow through a closed circulation, which is the correct sequence of vessels beginning at the heart?
   a. arteriole, artery, capillary, vein, venule
   b. artery, capillary, arteriole, venule, vein
   c. vein, venule, capillary, arteriole, artery
   d. artery, arteriole, capillary, venule, vein
   e. artery, arteriole, capillary, vein, venule

9. Gas and nutrient exchange between the circulation and the tissues occurs primarily at
   a. the arteries.   c. the capillaries.   e. the veins.
   b. the arterioles.   d. the venules.

10. Which of the following factors determines arterial blood pressure?
    a. cardiac output   d. a, b, and c
    b. resistance   e. a and b only
    c. arteriole diameter

Conceptual Questions

1. What are the three basic components of a true circulatory system?

2. Discuss the difference between closed and open circulatory systems. What advantages does a closed system provide?

3. Explain the cardiac cycle. Why must heart valves open in only one direction?

Collaborative Questions

1. List and briefly explain the components of blood.

2. Discuss the types of closed circulatory systems found in vertebrates.

Online Resource

www.brookerbiology.com

Stay a step ahead in your studies with animations that bring concepts to life and practice tests to assess your understanding. Your instructor may also recommend the interactive ebook, individualized learning tools, and more.

Respiratory Systems

48

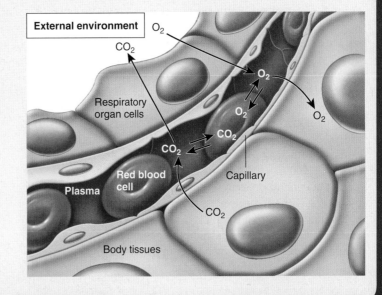

Llama (*Lama glama*) in the Andes Mountains.

E xcept during illnesses such as the flu and during periods of intense exercise, we take our respiratory system for granted. The biology of respiration, however, is filled with many intriguing questions. For example, have you ever wondered why it is easy for an animal like the llama in the chapter-opening photo to live at great altitudes, whereas most humans become ill when they ascend to such heights? What is it about the atmosphere in mountainous regions that makes it hard for us to function normally, and how do some people and animals adapt to those conditions if given sufficient time? As another example, think how limited we are in our ability to control our respiratory system. How long can you hold your breath, for example? Why do you continue breathing while asleep? Why do you not need to consciously adjust your breathing rate to match your level of activity? If our respiratory system did not operate automatically, think how challenging, indeed impossible, life would be.

In Chapter 47, we saw how circulatory systems function to transport blood or hemolymph throughout an animal's body. These fluids also carry the oxygen that animals require. The oxygen enters the body via a respiratory (breathing) structure. It then dissolves into bodily fluids and is distributed to cells throughout the body. As described in Chapter 7, O_2 is used by mitochondria in the formation of ATP. One of the waste products of that reaction is carbon dioxide—CO_2—which diffuses out of cells. CO_2 is returned by the circulatory system to the respiratory organ and is then released into the environment, either the air or water. The process of moving oxygen and carbon dioxide in opposite directions between the environment, bodily fluids, and cells is called **gas exchange** (Figure 48.1). A **respiratory system** includes all the structures of an animal that contribute to gas exchange. This chapter deals with the mechanisms

by which animals obtain and transport oxygen, rid themselves of carbon dioxide, and cope with the challenges imposed by their environments and their changing metabolic needs. We begin with an overview of the properties of gases in air and water.

Figure 48.1 Overview of gas exchange between the environment, blood, and cells. In vertebrates, oxygen diffuses from the environment across the cells of a respiratory organ into the blood (plasma and red blood cells). From there, oxygen diffuses into tissue cells, where it is used during the formation of ATP. Cells generate carbon dioxide as a waste product, which diffuses out of cells and into blood, and then across the respiratory organ into the environment. Note: Interstitial fluid is omitted for simplicity.

48.1   Physical Properties of Gases

Oxygen and carbon dioxide exchange between the environment and the fluids of an animal's body depend on the physical properties of these gases. These properties include such things as the solubility of the gas in water and the rate and mechanism of diffusion of the gas. At one time, it was unknown that air was actually a mixture of gases. In 1774, however, the British physiologist Joseph Priestley demonstrated that heating mercuric oxide produced a pure gas that was "five or six times as good as common air." He based this conclusion on the ability of the pure gas to sustain a burning candle or prevent a mouse from suffocating in an enclosed bell jar. Today, we know that the life-sustaining component of Priestley's experiment was oxygen and that air is composed of about 21% oxygen, 78% nitrogen, and less than 1% carbon dioxide and other gases. From a respiratory standpoint, nitrogen gas can usually be ignored because it plays no role in energy production nor is it created as a waste product of metabolism. In this section, we begin by examining some of the basic properties of the two major gases that are important in respiration.

Gases Exert Pressure Which Depends on Altitude

The gases in air exert pressure on the body surfaces of animals, although the pressure is not perceptible (unless it changes suddenly, like when your ears "pop" in a descending airplane). This pressure is called **atmospheric** (or **barometric**) **pressure**. It can be measured by noting how high a column of mercury is forced upward by the air pressure in a device called a mercury manometer. The traditional unit of gas pressure, therefore, is the same as for blood pressure—mmHg—although there are several other ways of expressing pressure. For example, U.S. weather stations report atmospheric pressure in inches of Hg instead of mmHg, and most researchers today record gas pressures in units called kilopascals (kPa; 1 kPa = 7.5 mmHg).

At sea level, atmospheric pressure is 760 mmHg. It decreases as we ascend to higher elevations because the gravitational effect of the Earth decreases and there are fewer gas molecules in a given volume of air (**Figure 48.2**). To visualize why gas pressure decreases at higher altitudes, think about how hydrostatic pressure decreases from the ocean floor to the ocean surface. Now, imagine that you are standing at the bottom of an "ocean" of air. The closer you get to the "ocean" surface (the top of the atmosphere), the lower the pressure exerted on your body. At an elevation of 1,700 m, as in Denver, the pressure is only about 640 mmHg, while at the highest elevations continuously inhabited by humans—in the Andes Mountains at 5,500 m—the pressure falls to 375 mmHg.

Atmospheric pressure is the sum of the pressures exerted by each gas in air, in exact proportion to their amounts. The individual pressure of each gas is its **partial pressure**, symbolized by a capital P and a subscript depending on the gas. Thus, the partial pressure of oxygen (P_{O_2}) in the air we breathe is 21% of the atmospheric pressure at sea level, or 160 mmHg (0.21

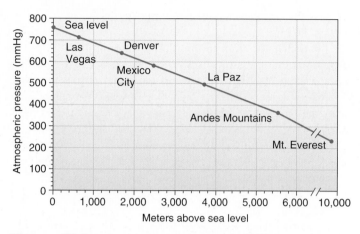

Figure 48.2   **The relationship between atmospheric pressure and altitude.**   Atmospheric pressure decreases as altitude increases.

Concept check: *Is there less percent oxygen in the air over Mt. Everest than there is at sea level?*

× 760 mmHg). The percentage of oxygen and other gases in air remains the same regardless of altitude, but the lower the atmospheric pressure, the lower the partial pressure of oxygen in air.

The partial pressure of oxygen in the environment provides the driving force for its diffusion from air or water across the respiratory surface and into the blood. All gases diffuse along pressure gradients, from regions of high pressure to regions of lower pressure. This is analogous to the diffusion of dissolved solids from regions of high to low concentration, or the way heat moves from hotter to cooler regions. Consequently, the rate of oxygen diffusion into the bloodstream of a terrestrial animal decreases when the animal moves from sea level to a higher altitude, where the partial pressure of oxygen is lower.

Pressure, Temperature, and Other Solutes Influence the Solubility of Gases

Gases dissolve in solution, including fresh water, seawater, and all body fluids. Gases such as oxygen exert their biological effects while in solution. However, most gases dissolve rather poorly in water. There is less oxygen in a given volume of water than in air, for example, which limits how active many aquatic organisms can be. Among the factors that influence the solubility of a gas in water, three are particularly important: the pressure of the gas, the temperature of the water, and the presence of other solutes.

The higher the pressure of a gas that comes into contact with water, the more of that gas that will dissolve, up to a limit that is specific for each gas at a given temperature. Not all gases dissolve equally. Oxygen is less soluble in water than carbon dioxide, but more soluble than nitrogen. The partial pressure of gas in water is given in the same units as atmospheric pressure—mmHg. For example, if a solution of water is in contact with air that contains a P_{O_2} of 160 mmHg, the P_{O_2} of the water is also 160 mmHg.

Solubility also depends on the water's temperature, with more O_2 and CO_2 dissolved in 1 L of cold water than in 1 L of warm water. You can observe this effect of temperature with the following experiment. Open two bottles of a carbonated beverage (that is, one in which carbon dioxide gas has been added) and keep one on ice and the other at room temperature. Note that the "fizz" of the room-temperature beverage disappears sooner, as CO_2 escapes due to its lower solubility at that temperature. Why does this happen? At higher temperatures, gases in solution have more thermal energy and are therefore more likely to escape the liquid. In terms of biological systems, this effect of water temperature means that animals inhabiting very warm waters have less oxygen available to them than do animals living, for example, in the waters off Antarctica. Animals that live in shallow water such as ponds or tide pools, where water temperature can fluctuate widely over a single day, must be able to adapt to large swings in oxygen.

Finally, the presence of other solutes, namely salts, reduces the amount of gas that dissolves in water. Thus, warm salty blood dissolves less oxygen than does cold water. Likewise, less oxygen dissolves in seawater than fresh water at any given temperature and pressure. Animals living in cold freshwater lakes, therefore, typically have more oxygen available to them than do animals living in warm salty seas (although other factors can also influence the amount of dissolved oxygen, such as water depth and the amount of aquatic plant life).

48.2 Types of Respiratory Systems

Despite the wide array of animal shapes, sizes, and habitats, animals obtain oxygen from their surroundings with one of four major types of respiratory structures: the body surface, gills, tracheae, or lungs. In these organs, gas exchange occurs by diffusion of oxygen across the body surface or across specialized respiratory surfaces. **Ventilation** is the process of bringing oxygenated water or air into contact with a gas-exchange surface. In this section, we will examine the mechanisms of ventilation used by animals with different respiratory systems.

The Structures of Respiratory Organs Are Adapted for Gas Exchange

Whether an animal uses specialized respiratory organs or its body surface to obtain oxygen and expel carbon dioxide, all gas exchange structures share certain common features. All of them have moist surfaces in which the gases can dissolve and diffuse. Further, all have adaptations that increase the amount of surface area available for gas exchange. As an example, the total exchange surface of the human lung is roughly equal to that of a tennis court!

The extensive surface area of respiratory organs is coupled with equally extensive blood flow, except in the special case of insects, described later in this chapter. The density of capillaries in gills and lungs, for example, is among the highest found anywhere in an animal's body. The greater the amount of blood flowing to a respiratory surface, the more efficient will be the delivery and removal of gases.

Thick barriers greatly slow the diffusion of a gas. As a result, respiratory organs tend to be thin, delicate structures, which also means they are easily damaged. For instance, the inner structures of fish gills are so thin that they cannot support their own weight out of water, and they collapse when exposed to air.

Water-Breathing and Air-Breathing Animals Face Different Challenges

Exchanging gases in water or in air requires respiratory organs that can deal with the challenges of those different environments. As noted earlier, water-breathing animals generally have less oxygen available to them than do air-breathing animals, which means that an active water-breathing animal requires a very efficient means of extracting oxygen from water. In addition, unlike air-breathing animals, water-breathing animals must adapt to fluctuating oxygen availability when the temperature changes. By contrast, air-breathing animals must cope with the dryness of air. Passing dry air over the lung surfaces runs the risk of drying out the lungs, which would damage them, dehydrate the animal, and reduce the ability of gases to dissolve.

Water-breathing animals bring water, not air, over their respiratory surfaces and therefore do not usually risk dehydration in the way that air-breathing animals do. However, moving water over a gas-exchange surface creates its own challenges. First, because water is denser than air, more energy (muscular work) is required to move water. Second, fresh water or cold seawater moving over the richly vascular gills removes considerable heat from an animal's body, due to the high heat capacity of water. Third, moving fresh or salt water over a gas-exchange surface can create osmotic movement of water across the surface, potentially resulting in water imbalances in an animal's body.

Some Animals Use Their Body Surface to Exchange Gases with the Environment; Some Can Switch Between Water- and Air-Breathing

In those invertebrates that are only one or a few cell layers thick, oxygen and carbon dioxide can diffuse directly across the body surface. In this way, oxygen reaches all the interior cells, in some cases without any specialized transport mechanism or circulatory system.

Even in some large, complex animals, the body surface may be permeable to gases (**Figure 48.3**). Amphibians, and a few species of fishes including eels, have unusually permeable skin. On land, amphibians rely on their lungs and their moist skin to obtain oxygen and release carbon dioxide. Under water, however, the lungs are no longer useful because they are not

Lungless salamander
(*Ensatina eschscholtzii*)

Bullfrog
(*Rana catesbeiana*)

Mudpuppy
(*Necturus maculosus*)

Southern musk turtle
(*Sternotherus minor*)

European eel
(*Anguilla anguilla*)

Boa constrictor
(*Constrictor constrictor*)

Green lizard
(*Lacerta viridis*)

Big brown bat
(*Eptesicus fuscus*)

Brown trout
(*Salmo trutta*)

Human
(*Homo sapiens*)

Lungless salamander

Southern musk turtle

O_2
CO_2

0 10 20 30 40 50 60 70 80 90 100
Percentage of gas exchange through the skin

Figure 48.3 **Gas exchange across body surfaces.** This type of gas exchange occurs readily in amphibians and, to a lesser extent, in marine reptiles and some fishes. Among mammals, it occurs to a negligable extent and primarily in species with body features that have high surface area, such as bat wings. Note: CO_2 determinations were not made in all species.

Concept check: *What would happen to gas exchange in lungless salamanders if the animals were allowed to dry out?*

suited for extracting oxygen from water. Although oxygen diffusion across the skin is less efficient than in the lungs, this secondary ability to exchange gas permits amphibians to spend prolonged times underwater.

Many vertebrates such as seals are amphibious, in that they spend part of their lives on land and part in water, but only amphibians have skin that is highly permeable to gases and can obtain significant amounts of oxygen across their skin. With some exceptions, very little gas diffuses across the body surfaces of fishes, reptiles, birds, and mammals. Some reptiles, however, such as marine iguanas, can obtain oxygen from the water across the highly permeable mucous membranes of the

nose, mouth, and anus. Although this provides less oxygen than does air-breathing, it nonetheless allows them to remain underwater for extended periods, as when foraging for food. Some animals have highly specialized body regions with considerable surface area—like the wings of a bat—that permit a small amount of gas exchange (Figure 48.3). Except for amphibians, however, no vertebrates can rely for long periods of time exclusively on their skin to obtain oxygen and remain active.

Water-Breathing Animals Use Gills for Gas Exchange

Most water-breathing animals use specialized respiratory structures called **gills**. Gills can be either external or internal. External gills are uncovered extensions from the body surface, as occur in many invertebrates and the larval forms of some amphibians. Internal gills, which occur in fishes, are enclosed in a protective cavity.

External Gills External gills vary widely in appearance, but all have a large surface area, often in the form of extensive projections (**Figure 48.4**). External gills may exist in one region of the body or be scattered over a large area. In many cases, they are ventilated by waving back and forth through the water. The ability to move external gills is particularly important for sessile invertebrates, which must otherwise rely on sporadic local water currents or muscular efforts of their bodies to create local currents for ventilation.

Despite the success of marine invertebrates, external gills have several limitations. First, they are unprotected and therefore are susceptible to damage from the environment. Second, because water is much more dense than air, considerable energy is required to continually wave the gills back and forth through the water (think of the difference between waving your hands through air or water). Finally, their appearance and motion may draw the attention of predators.

Internal Gills By contrast, fishes have internal gills, which are covered by a bony plate called an operculum (**Figure 48.5a**). Fish gills have a more uniform appearance than external gills and are confined within the opercular cavity, which protects the gills and helps streamline the fish body.

The internal gills of fishes contain numerous rows of **lamellae**, platelike structures that branch from structures called filaments. The filaments, in turn, arise from the main support structure of the gills, the gill arches (**Figure 48.5b**). Blood vessels run the length of the filaments, with oxygen-poor blood traveling through the afferent vessel along one side of the filament, and oxygen-rich blood returning through the efferent vessel along the other side. Within the lamellae are numerous capillaries, all oriented with blood flowing in the same direction, from the oxygen-poor vessel to the oxygen-rich one.

Water enters a fish's mouth and flows between the lamellae in the opposite direction to blood flowing through the lamellar capillaries. This arrangement of water and blood flow is another example of a **countercurrent exchange mechanism**

Figure 48.4 **Examples of animals with external gills.** The gills of these animals have extensive projections to increase surface area, which facilitates oxygen diffusion from the surrounding water.

(a) Mollusk (nudibranch) with clustered external gill tufts

(b) Larval salamander with external gills

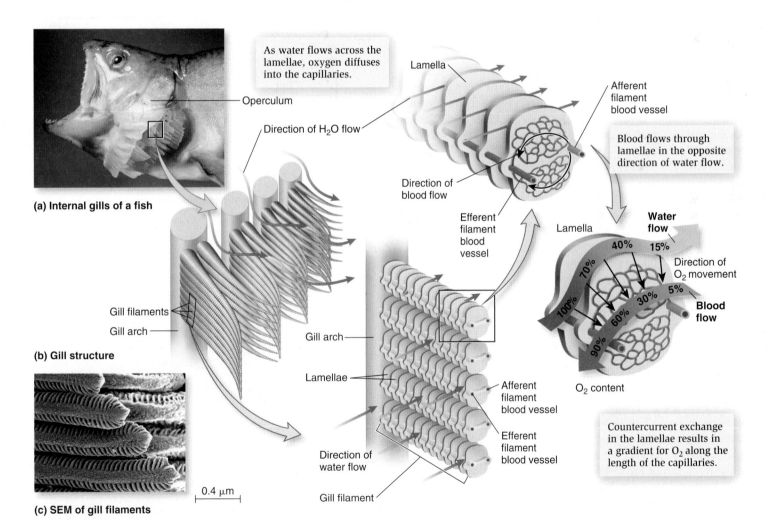

As water flows across the lamellae, oxygen diffuses into the capillaries.

Operculum

Direction of H₂O flow

Lamella

Afferent filament blood vessel

Direction of blood flow

Blood flows through lamellae in the opposite direction of water flow.

Efferent filament blood vessel

(a) Internal gills of a fish

Gill filaments

Gill arch

(b) Gill structure

Gill arch

Lamellae

Afferent filament blood vessel

Efferent filament blood vessel

Direction of water flow

Gill filament

0.4 μm

(c) SEM of gill filaments

Lamella

Water flow

40%   15%

70%

100%   60%   30%   5%

90%

Direction of O₂ movement

Blood flow

O₂ content

Countercurrent exchange in the lamellae results in a gradient for O₂ along the length of the capillaries.

Figure 48.5 **Structure of fish gills.** **(a)** The gills are protected beneath the operculum, which has been lifted open in this photo to reveal the gills underneath. **(b)** The gills are composed of gill arches, from which numerous pairs of filaments arise. Thin, platelike lamellae with large surface areas are arrayed along the filaments. Blood vessels from the filaments form capillaries in the lamellae. Blood flows through the capillaries in the opposite direction of water flowing between lamellae, a process called countercurrent exchange. **(c)** Several filaments with their lamellae as revealed in a scanning electron micrograph.

Concept check: *Why do fishes die when out of water? (Hint: Consider the thin, delicate appearance of the water-covered lamellae.)*

(see Chapters 40 and 46). As oxygenated water encounters the lamellae, it comes into contact with blood in the gill capillaries. Recall that oxygen diffuses along pressure gradients from a region of high oxygen pressure to low oxygen pressure. Thus, oxygen diffuses from the water into the capillaries of the lamellae. As water continues to flow across the lamellar surface, it encounters regions of capillaries that have not yet picked up oxygen—in other words, even as oxygen begins to diffuse from the water into the gill capillaries, a sufficient pressure gradient remains along the lamellae to permit diffusion of more of the remaining oxygen from the water. This is an extremely efficient way to remove as much oxygen from the water as possible before the water passes out of the operculum.

Fishes use one of two mechanisms to ventilate their gills, buccal pumping or ram ventilation. The first method, **buccal pumping** (the buccal cavity refers to the mouth), makes use of the muscles of the mouth and operculum to create a hydrostatic pressure gradient for water to flow in one direction (**Figure 48.6a**). First, the jaw is lowered, which enlarges the buccal cavity and lowers the pressure in the mouth, something like a suction pump. This draws water into the mouth, raising the water pressure again. At roughly the same time, the operculum begins to swing out from the body, enlarging the opercular cavity and lowering the water pressure there. This second suction pump is stronger than the buccal pump, so water flows down its hydrostatic pressure gradient from the outside to the mouth and into the opercular cavity. Next, the mouth closes, and the buccal cavity is compressed. This creates positive pressure that forces water across the gills and out through the operculum, which is now open. A flap of tissue at the back of the mouth helps prevent accidental flow of water down the esophagus. Consequently, fishes do not swallow the water they inhale but instead send it on a one-way journey across their gills. In buccal pumping, a fish can remain stationary and still ventilate even in stagnant water by drawing water into its mouth.

The second way in which some fishes ventilate their gills, **ram ventilation**, consists of swimming with their mouths open, in essence using their large swimming muscles to bring water into their buccal cavity and from there across their gills (**Figure 48.6b**). This method of ventilation is more energy efficient than buccal pumping. Ram ventilation still requires energy expenditure, but in this case, the energy is used for swimming or for using muscles to remain stationary while facing upstream in moving water. Many fishes employ both methods of ventilation, using buccal pumping when swimming slowly or in stagnant water and switching to ram ventilation when swimming quickly or facing upstream. Some fishes—such as tuna—can only ram ventilate and therefore rarely stop moving.

Both buccal pumping and ram ventilation are **flow-through systems**—water moves unidirectionally in such a way that the gills are constantly in contact with fresh, oxygenated water. As we will see, many air-breathing animals use a less efficient method of ventilating their lungs, called tidal ventilation, in which fresh air is breathed in and stale air is breathed out through the same route. Later in this chapter, we will discuss an example of an air-breathing animal that also uses a

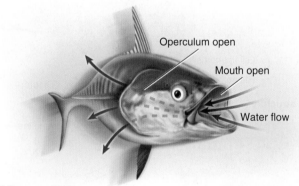

1 Mouth opens and the jaw is lowered to draw water into the buccal cavity. The operculum then pulls away from the body, expanding the opercular cavity.

2 Mouth closes and the lower jaw raises, which squeezes water toward the opercular cavity. Water is forced across the gills and out through the open operculum.

3 Operculum closes and the cycle begins again.

(a) Buccal pumping

(b) Ram ventilation

Figure 48.6 **Mechanisms of gill ventilation.** Some species use both buccal pumping and ram ventilation, while some can only use one. **(a)** Buccal pumping. The expansion of the buccal and opercular cavities acts like a suction pump, and closing these cavities acts like a pressure pump. **(b)** Ram ventilation. Many fishes can ventilate their gills by swimming forward, such as this tuna, or by facing upstream in moving water with their mouths open. This causes the water to move through the mouth and operculum. During both types of ventilation, flaps of tissue in the mouth, throat, and operculum help prevent water from moving in the reverse direction or down the esophagus.

flow-through system. Despite the efficiency of flow-through systems in water for maximizing oxygen extraction, they are energetically costly because of the work required to overcome the density of water. Fishes may devote up to 10–20% of their resting metabolic rate simply to ventilating their gills, whereas a typical air-breather may allocate only 1–2% of its total energy usage to ventilating the lungs at rest.

Insects Use Tracheal Systems to Exchange Gases with the Air

Except for terrestrial isopods such as the pill bug (*Armadillidium vulgare*), which breathe air through very small, moist gills, the delicate nature of gill lamellae makes them unsuitable for gas exchange in air. Air-breathing probably evolved as an adaptation in aquatic animals inhabiting regions that were subject to periodic drought. One of the major mechanisms that animals evolved to breathe air is the tracheal system found in insects.

Running along the surface of both sides of an insect's body are tiny openings to the outside, called spiracles. Arising from the spiracles are **tracheae** (singular, trachea), sturdy tubes that are reinforced with the polysaccharide chitin to keep them open (**Figure 48.7**). Tracheae branch extensively into ever-smaller tubes called tracheoles, which eventually become small enough that their tips contact virtually every cell in the body. At their tips, tracheoles are filled with a small amount of fluid. Air flowing down the tracheoles comes into contact with this fluid. Oxygen from the air dissolves in the fluid, and from there it diffuses across the tracheole wall and into nearby cells. Carbon dioxide diffuses in the opposite direction, from cells into the tracheoles, and from there to the environment.

When an insect's oxygen demands increase, muscular movements of its abdomen and thorax draw air into and out of the tracheae like a bellows, a phenomenon first demonstrated in the early 20th century by Danish physiologist and Nobel Laureate August Krogh, who studied the mechanics of ventilation in locusts. Krogh showed that an insect's muscles and tracheal system match ventilation with the animal's exercise intensity and oxygen requirements. This is particularly important in flying insects, whose metabolic demands are high. The faster an insect flies, the more oxygen it uses. Likewise, the more intensely and rapidly the body wall muscles move, the more air is moved in and out of the tracheae.

As discussed in Chapter 47, the open circulatory system of insects does not participate in gas exchange. Oxygen diffuses directly from air to trachea to tracheoles and finally to body cells. This mechanism of ventilation and oxygen delivery is very efficient. The metabolic rate of insect flight muscles is among the highest known of any tissue in any animal, and the tracheal system supplies enough oxygen to meet those enormous demands.

Air-Breathing Vertebrates Use Lungs to Exchange Gases

Except for some amphibians such as lungless salamanders (family Plethodontidae), all air-breathing terrestrial vertebrates use

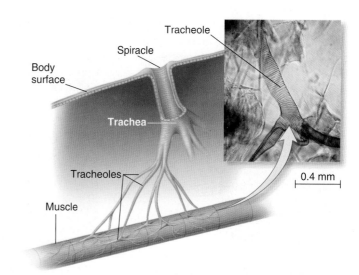

Figure 48.7 **The tracheal system of insects.**   Air enters holes on the body surface called spiracles. Oxygen diffuses directly from the fluid-filled tracheole tips to cells that come into contact with the tips. The circulatory system plays no role in gas exchange. The micrograph illustrates a few branching tracheoles.

Concept check: *How might the structure of the respiratory system of insects be related to why insects tend to be smaller than vertebrates?*

lungs to bring oxygen into the circulatory system and remove carbon dioxide from it. Lungs are internal paired structures that arise from the pharynx (throat). All lungs receive deoxygenated blood from the heart and return oxygenated blood back to the heart. Depending on the class of vertebrate, lungs may be filled using positive or negative pressure, and they may be ventilated by a tidal system or a flow-through system.

Most amphibians have lungs that are simple sacs with relatively little surface area, making them less-effective surfaces for oxygen diffusion than are the lungs of other vertebrates. A 30-g mouse's lung, for example, has nearly 50 times more gas-exchange surface per cubic centimeter of lung tissue than a 30-g frog's lung.

The method by which amphibians ventilate their lungs is similar in some ways to the buccal pumping of fishes. A frog, for example, lowers its jaw, which decreases the air pressure inside the mouth cavity. According to Boyle's law, gas pressure and volume are inversely related (**Figure 48.8**). For example, when the volume in which a gas is contained increases, the pressure of the gas in that container decreases. By expanding its mouth cavity, therefore, a frog creates a pressure gradient for air to move from the atmosphere (higher pressure) into its mouth (lower pressure). The mouth and nostrils then close, and the mouth cavity is compressed by raising the bottom jaw. This raises the air pressure in the mouth, forcing air through a series of valves into the lungs. Thus, **positive pressure filling** means that frogs and most other amphibians gulp air and force it under pressure into the lungs, as if inflating a balloon. They may do this many times in a row before exhaling.

Rest

Pressure gauge

Piston

Increased volume
(analogous to inhaling)

Reduced volume
(analogous to exhaling)

Gas molecules have thermal energy and exert a pressure on the walls of the cylinder.

Fewer collisions of gas molecules occur with cylinder wall (decreased pressure).

Greater number of collisions of gas molecules occur with cylinder wall (increased pressure).

Figure 48.8 **Boyle's law.** At a constant temperature, the volume and pressure of a gas are inversely related. This relationship creates the gas pressure gradients needed to ventilate the vertebrate lungs.

Except for a few species of reptiles—such as members of the genus *Varanus*—that can in some circumstances also employ positive pressure filling, all other terrestrial vertebrates use a different method of ventilation, called negative pressure filling. To understand how negative pressure filling works, we will first examine the anatomy of the respiratory tract in vertebrates, using the mammalian and avian systems as models.

48.3 Structure and Function of the Mammalian and Avian Respiratory Systems

In vertebrates, the respiratory system includes all components of the body that contribute to the exchange of gas between the external environment and the blood. In this section, we begin by examining the structures of the mammalian respiratory system and the mechanisms by which mammals ventilate their lungs. We then compare and contrast this with the highly specialized and unique way in which birds ventilate their lungs. Although both mammals and birds fill their lungs using negative pressure, mammals breathe by tidal ventilation, and birds use a flow-through system.

During Ventilation, Air Follows a Series of Branching Tubes in the Mammalian Respiratory System

In mammals, the respiratory system includes the nose, mouth, airways, lungs, and muscles and connective tissues that encase these structures within the thoracic (chest) cavity (**Figure 48.9a**). When humans and other mammals breathe, air first

enters the nose and mouth, where it is warmed and humidified. These processes protect the lungs from drying out. While in the nose, the air is cleansed as it flows over a coating of sticky mucus in the nasal cavity. The mucus and hairs in the nasal cavity trap some of the larger dust and other particles that are inhaled with air. These are then removed by the body's defense cells or swallowed.

The inhaled air from the mouth and nose converges at the back of the throat, or **pharynx**, a common passageway for air and food. From there, air passes through the **larynx**. Within the larynx are the vocal cords, folds of tissue through which air passes to create sound. Air flows from the larynx down the **trachea**, a tube that leads to the lungs. At the opening to the larynx, a flap of tissue called the epiglottis prevents food from entering the trachea by closing when food is swallowed.

The trachea is ringed by C-shaped cartilage supports that provide rigidity and ensure that the trachea always remains open. The inner wall of the trachea is lined with cilia and cells called goblet cells (**Figure 48.9b**). These cells secrete mucus into the lumen of the trachea, coating the cilia. This mucous layer captures potentially harmful or irritating particles that escaped the nasal cleaning mechanism. The cilia constantly beat toward the mouth and move the mucus and its trapped particles into the mouth, where it can be expelled or swallowed. Without these active cilia, repeated coughing would be needed to force the mucus out of the trachea. This is why people who smoke tobacco develop a chronic "smoker's cough"—the smoke and its components reduce the number of cilia, causing mucus to pool in the trachea.

Inhaled air moves down the trachea as it branches into two smaller tubes, called **bronchi** (singular, bronchus), which lead to each lung. The bronchi branch repeatedly into smaller and

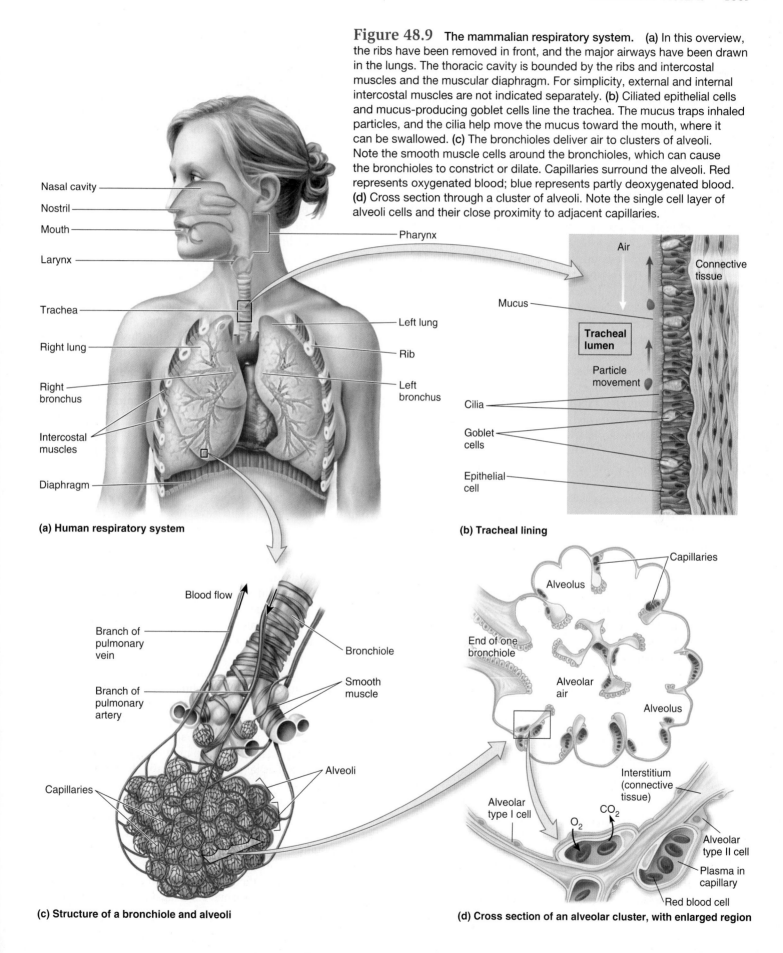

Figure 48.9   The mammalian respiratory system.   (a) In this overview, the ribs have been removed in front, and the major airways have been drawn in the lungs. The thoracic cavity is bounded by the ribs and intercostal muscles and the muscular diaphragm. For simplicity, external and internal intercostal muscles are not indicated separately. **(b)** Ciliated epithelial cells and mucus-producing goblet cells line the trachea. The mucus traps inhaled particles, and the cilia help move the mucus toward the mouth, where it can be swallowed. **(c)** The bronchioles deliver air to clusters of alveoli. Note the smooth muscle cells around the bronchioles, which can cause the bronchioles to constrict or dilate. Capillaries surround the alveoli. Red represents oxygenated blood; blue represents partly deoxygenated blood. **(d)** Cross section through a cluster of alveoli. Note the single cell layer of alveoli cells and their close proximity to adjacent capillaries.

Nasal cavity
Nostril
Mouth
Larynx
Pharynx
Trachea
Right lung
Right bronchus
Intercostal muscles
Diaphragm
Left lung
Rib
Left bronchus

(a) Human respiratory system

Air
Connective tissue
Mucus
Tracheal lumen
Particle movement
Cilia
Goblet cells
Epithelial cell

(b) Tracheal lining

Blood flow
Branch of pulmonary vein
Bronchiole
Branch of pulmonary artery
Smooth muscle
Alveoli
Capillaries

(c) Structure of a bronchiole and alveoli

Capillaries
Alveolus
End of one bronchiole
Alveolar air
Alveolus
Interstitium (connective tissue)
Alveolar type I cell
O_2
CO_2
Alveolar type II cell
Plasma in capillary
Red blood cell

(d) Cross section of an alveolar cluster, with enlarged region

smaller tubes, eventually becoming thin-walled **bronchioles** surrounded by circular rings of smooth muscle (**Figure 48.9c**). Bronchioles can dilate or constrict in a manner analogous to that of arterioles (see Chapter 47). They may partially constrict when a damaging particle—such as a small bit of inhaled pollutant or dust—gets past the mucous layers of the upper airways. Constriction of a bronchiole prevents foreign particles from reaching delicate lung tissue. Once the body's defense cells have removed the particles, the bronchiole reopens to its normal diameter.

The bronchioles empty into the final, saclike regions of the lungs where gas exchange occurs—the **alveoli** (singular, alveolus; Figure 48.9c). Until now, air has flowed through the air tubes without any gas exchange taking place. The alveoli are highly adapted for gas exchange and consist of two major types of cells. Type I cells are those across which gases diffuse, whereas Type II cells are secretory cells (described later). The alveoli are only one cell thick and resemble extremely thin sacs, appearing like bunches of grapes on a stem. Many capillaries containing deoxygenated blood pumped from the right ventricle of the heart surround the alveoli. Oxygen diffuses from the lumen of each alveolus across the alveolar cells, through the interstitial space outside the cells, and into the capillaries (**Figure 48.9d**). Carbon dioxide diffuses in the opposite direction. The oxygenated blood from the lungs then flows to the left atrium of the heart and from there enters the left ventricle, where it is pumped out through the aorta to the rest of the body.

The Pleural Sac Protects the Lungs

The lungs are soft, delicate tissues that could easily be damaged by the surrounding bone, muscle, and connective tissue of the thoracic cavity if not protected. Each lung is encased in a **pleural sac**, a double layer of thin, moist tissue. Between the two layers of tissue is a microscopically thin layer of water that acts as a lubricant and makes the two tissue layers adhere to each other.

In addition to protecting the lungs, the inner pleural sac adheres to its lung, and the outer pleural sac adheres to the chest wall. In this way, movements of the chest wall result in similar movements of the lungs. This is important because the lungs are not muscular and so cannot inflate themselves. Instead, as we will see, the lungs are inflated by the expansion of the thoracic cavity, which results from the contraction of muscles in the thorax.

Negative Pressure Is Used to Fill Both Mammalian and Avian Lungs

The way in which you inflate a balloon, by forcing air from your mouth into the balloon, is called positive pressure filling. It is reminiscent of how amphibians ventilate their lungs. **Negative pressure filling**, by contrast, is the mechanism by which reptiles, birds, and mammals ventilate their lungs. In this process, the volume of the lungs expands, creating a negative pressure that draws air into the lungs. The process differs in some ways among classes of vertebrates, but in mammals, the work

is provided by the **external intercostal muscles**, which surround and connect the ribs in the chest, and a large muscle called the **diaphragm** (Figure 48.9a), which divides the thoracic cavity from the abdomen.

Let's follow the process when a mammal ventilates its lungs (**Figure 48.10**). At the start of a breath, the diaphragm contracts, pulling downward and enlarging the thoracic cavity. Simultaneously, the external intercostal muscles contract, moving the chest upward and outward, which also helps to enlarge the thoracic cavity. Recall that the pleural sacs adhere the lungs to the chest wall, so as the chest expands, the lungs expand with it. According to Boyle's law, as the volume of the lungs increases, the pressures of the gases within them must decrease. In other words, the pressure in the lungs becomes negative with respect to the outside air. Air, therefore, flows down its pressure gradient from outside the mouth and nose, into the lungs.

Once the lungs are inflated with air, the chest muscles and diaphragm relax and recoil back to their original positions as an animal exhales. This compresses the lungs and forces air out of the airways. Whereas inhaling requires the expenditure of significant amounts of energy, exhaling is mostly passive and does not require much energy. During exercise, however, exhalation is assisted by contraction of another set of rib muscles, called the internal intercostal muscles. The contraction of these muscles depresses the rib cage and provides a more forceful exhalation; this increases the rate at which ventilation occurs and helps empty the larger volume of air from the lungs.

Mammals Breathe by Tidal Ventilation

When mammals exhale, air leaves via the same route that it entered during inhalation, and no new oxygen is delivered to the airways at that time. As mentioned, this type of breathing is called **tidal ventilation** (think of air flowing in and out of the lungs like the tides of an ocean). Tidal ventilation is less efficient than the unidirectional, flow-through system of fishes in which gills are always exposed to oxygenated water during all phases of the respiratory cycle.

As you know from experience, the lungs are neither fully inflated nor deflated at rest. For example, you could easily take a larger breath than normal if you wished, or exhale more than the usual amount of air. The volume of air that is normally breathed in and out at rest is the **tidal volume**, about 0.5 L in an average-sized human. Tidal volume is proportional to body size both among humans and between species. A 6-foot-tall adult, for example, has a larger tidal volume than a 4-foot-tall child because the adult has larger lungs. Similarly, horses have larger tidal volumes than humans, and humans have larger tidal volumes than dogs (**Table 48.1**). During exertion, the lungs can be inflated further than the resting tidal volume to provide additional oxygen. Likewise, the lungs can be deflated beyond their normal limits at rest, by exerting a strong effort during exhalation. The lungs never fully deflate, however, partly because they are held open by their adherence to the chest wall. This is important for a simple reason. Think again of our analogy of a

1 The external intercostal muscles contract during inhalation to increase chest volume.

2 The diaphragm lowers during inhalation.

3 Air flows in during inhalation because air pressure in the lungs is lower than that outside.

4 Air flows out when muscles relax.

Intercostal muscles

Diaphragm

KEY
----- Inhalation
----- Exhalation

(a) Action of muscles during ventilation

After inhaling

After exhaling

(b) Change in lung volume during ventilation

Figure 48.10 Ventilation of the mammalian lung by negative pressure filling. (a) The external intercostal muscles contract, which expands the chest cavity by moving the ribs up and out. The diaphragm also contracts, causing it to pull downwards, further expanding the cavity. The muscular efforts require energy, whereas the return to the resting state by exhaling is primarily by recoil. (b) X-ray images of the chest of an adult man after inhaling and exhaling. The volume of the lungs after exhaling is superimposed on the left, using dashed lines to illustrate the relative change in lung volume after inhaling.

Table 48.1	Respiratory Characteristics of Different Mammals		
Animal	**Body mass**	**Breaths/min**	**Tidal volume**
Shrew	0.0024 kg	700	30 µL
Dog	25 kg	20	0.27 L
Human	75 kg	12	0.50 L
Horse	465 kg	9	6.50 L

Note: All values are averages from resting animals. Tidal volume is the volume of air breathed in with each breath. Note that tidal volume increases as an animal's mass and lung size increase. By contrast, breathing rate is higher in smaller animals. Similar relationships occur in other vertebrates, notably birds.

balloon. It is much easier to fill a balloon that is already partly inflated than it is to inflate a completely empty balloon. The same is true of the lungs. The most difficult breath is the very first one that a newborn mammal takes—the only time its lungs are ever completely empty of air.

Surfactant Facilitates Lung Inflation by Reducing Surface Tension

Like all cells, those that make up the lining of the alveoli are surrounded by extracellular fluid. This fluid layer is where gases dissolve. Unlike other internal body cells, however, alveolar cells come into contact with air, creating an air/liquid interface along the inner surface of the alveoli. This results in surface tension (see Chapter 2) within the alveoli. Surface tension results from the attractive forces between water molecules at the air/liquid interface and partly explains why droplets of water form beads. It also produces a force that makes alveoli tend to collapse as water molecules lining its surfaces are attracted to

each other. If many or all of the alveoli collapsed, however, the amount of surface area available for gas exchange in the lungs would be greatly reduced. What prevents them from collapsing? The Type II cells of the alveoli produce **surfactant**, a mixture of proteins and amphipathic lipids (that is, lipids with both polar and nonpolar regions), and secrete it into the alveolar lumen. There, it forms a barrier between the air and the fluid layer inside the alveoli. This barrier reduces surface tension in the alveolar walls, allowing them to remain open.

Surface tension is particularly important in the transition from fetal to postnatal life in mammals. Most mammalian fetuses are encased in fluid within the uterus. Consequently, their lungs do not have an air/liquid interface, and they do not start producing surfactant until the final stages of pregnancy. In humans, surfactant production begins around week 26 of gestation but does not increase to final levels until after week 33 (normal pregnancy length is about 40 weeks). If a human baby is born prematurely (defined as prior to week 37 of gestation), sufficient surfactant may not be available, and consequently, many alveoli may collapse after birth. This condition, known as **respiratory distress syndrome of the newborn**, can be partially alleviated by inserting a tube in the trachea and injecting synthetic surfactant. Each year in the U.S., approximately 450,000 babies are born prematurely; of those, roughly 40,000 will be born sufficiently premature as to be diagnosed with respiratory distress syndrome of the newborn.

The Avian Respiratory System Is a Flow-Through System

The avian respiratory system provides a stark contrast to that of mammals, despite the similarity that negative pressure is used to fill the lungs in both cases. Unlike mammals, ventilation in birds is a flow-through system. The avian respiratory system is unique among vertebrates in that it is supplemented with numerous **air sacs** (Figure 48.11a), which for simplicity we can group together as those at the anterior or posterior regions of the body. The air sacs—not the lungs—expand when a bird inhales and shrink when it exhales. They expand by movements of the chest muscles, not by a diaphragm.

The air sacs, however, do not participate in gas exchange, as first demonstrated in an experiment by French physiologist J. Soum in 1896. Soum surgically plugged the openings of one of a pigeon's air sacs and then injected poisonous carbon monoxide (CO) gas into the sac. Birds are very sensitive to the toxic effects of CO, but in this experiment, the pigeon showed no signs of being poisoned. Soum concluded that gases do not diffuse from the air sacs into the blood.

When a bird inhales, air enters its trachea and from there moves into two bronchi. Instead of branching into bronchioles and then alveoli, however, avian bronchi branch into a series of parallel air tubes called **parabronchi** that make up the lungs (Figure 48.11b). The parabronchi are the regions of gas exchange. They have enormous surface area and an extensive network of blood capillaries. Blood flows through the lungs in a crosscurrent direction with respect to movement of air. This is less efficient in extracting oxygen than the countercurrent flow arrangement in fish gills, because air and blood do not move in exactly opposite directions along the length of the capillaries; it is more efficient than the tidal ventilation system of mammalian lungs, however (Figure 48.12). Precisely how air moves through avian lungs and the function of the air sacs remained a mystery to physiologists for many years, however, until the ingenious work of American physiologist Knut Schmidt-Nielsen, as described next.

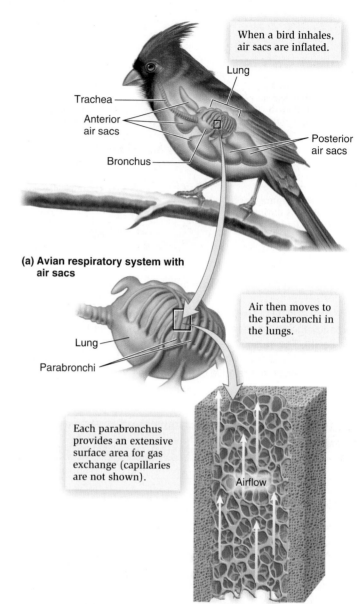

When a bird inhales, air sacs are inflated.

Lung
Trachea
Anterior air sacs
Posterior air sacs
Bronchus

(a) Avian respiratory system with air sacs

Air then moves to the parabronchi in the lungs.

Lung
Parabronchi

Each parabronchus provides an extensive surface area for gas exchange (capillaries are not shown).

Airflow

(b) Lung showing parabronchi and cross section of a parabronchus

Figure 48.11 Respiratory system of a bird.

Fish gills

H₂O flows in one direction.

Blood flow is countercurrent to water flow (most efficient for obtaining O₂).

Capillaries

Avian lungs

Air flows in one direction.

Blood flow is crosscurrent to airflow (more efficient than tidal flow but less efficient than countercurrent flow).

Mammalian lungs

Airflow is tidal.

Blood flows past a uniform pool of air (least efficient for obtaining O₂).

Figure 48.12 Relationships between direction of flow of blood and water or air in different vertebrates.

Concept check: *What might you conclude about the relative amounts of oxygen available in water and air based on the differences described in this figure?*

FEATURE INVESTIGATION

Schmidt-Nielsen Mapped Airflow in the Avian Respiratory System

In 1967, American researcher Knut Schmidt-Nielsen and coworkers conducted a series of experiments designed to determine the mechanisms of ventilation and gas exchange in avian lungs, using ostriches because of their large lung volumes. The researchers used O_2- and CO_2-sensitive probes to determine the concentrations of these gases in the birds' air sacs. In their first experiment, described in **Figure 48.13**, the investigators allowed the birds to breathe ordinary room air, which contained 21% oxygen. They used special syringes to extract a gas sample from the large air sacs of the birds and analyzed the gas composition with the probes. They discovered that the O_2 content was high (21%) and the CO_2 content low in the posterior air sacs, with the reverse pattern in the anterior sacs. They concluded that fresh air must move through the trachea and into the poste-

rior air sacs, while "stale" air, which had previously given up much of its oxygen and picked up CO_2 in the lungs, must have entered the anterior sacs.

To test this hypothesis, they developed a method to track the movement of oxygen through the ostrich's respiratory system, as shown in the second experiment of Figure 48.13. In this experiment, the scientists fitted ostriches with face masks, through which the birds inhaled a single breath of pure (100%) oxygen. O_2-sensitive probes were surgically inserted into the posterior and anterior air sacs to monitor when the pure oxygen gas appeared in each sac. The researchers discovered that during the first inhalation, high levels of oxygen appeared in the posterior air sac but not the anterior sac. As the ostrich exhaled, some oxygen remained in the posterior air sac, but still no oxygen appeared in the anterior one. When the animal inhaled its next breath (this one of plain air, not pure oxygen), pure oxygen began appearing in the anterior sacs and disappearing from

Figure 48.13 Schmidt-Nielsen and coworkers demonstrated the route that air travels in the avian (ostrich) lung.

4 **THE DATA**

Results from Experiment 2: Pure O_2 appears first in the posterior sacs and does not appear in anterior sacs until the next breath.

*The O_2 shown in these graphs is from the pure O_2 inhaled on the first breath and is above the normal levels of O_2.

5 **CONCLUSION**

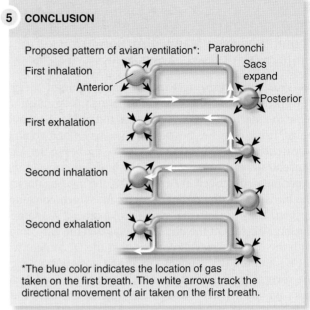

*The blue color indicates the location of gas taken on the first breath. The white arrows track the directional movement of air taken on the first breath.

6 **SOURCE** Schmidt-Nielsen, K. et al. 1969. Temperature regulation and respiration in the ostrich. *The Condor* 71:341–352.

the posterior sacs. Finally, as the ostrich exhaled a second time, the pure oxygen in the anterior sacs decreased.

Schmidt-Nielsen concluded from this study that air flows through the trachea, down the bronchi, and into the posterior air sacs during inhalation. As a bird exhales, air exits the posterior sacs and flows through the parabronchi in a posterior to anterior direction. During its passage through the lungs, some of the oxygen diffuses into the capillaries, while CO_2 diffuses out of the capillaries and into the lungs. During the next inhalation, air at the anterior part of the lungs flows through the remainder of the lungs and then into the anterior air sacs, which serve as a sort of holding chamber. Finally, a second exhalation causes this air from the previous inhalation to exit the anterior sacs and leave the animal's body through the trachea.

Therefore, it takes two complete breaths for air to move from the environment into the lungs and back out again to the environment. Air flows through the lungs during inhalation, as it is drawn by negative pressure through the parabronchi and into the expanding anterior sacs. Air is also forced through the parabronchi by positive pressure during exhalation, as the posterior air sacs are compressed. The efficiency of the avian flow-through system, together with its crosscurrent blood supply, is a major reason why many birds are able to fly at altitudes with extremely low atmospheric pressures (even over Mt. Everest—see Figure 48.2).

Experimental Questions

1. What was the purpose of these studies?

2. Considering the first experiment only, what were the results, and what conclusions did the researchers make based on these results?

3. Explain the purpose and procedure of the second experiment conducted by Schmidt-Nielsen and the conclusions he drew from his results.

48.4 Control of Ventilation in Mammalian Lungs

In the previous section, you saw that different vertebrates have adapted to air-breathing in very different ways. Now let's look at how the mechanisms of breathing are controlled, using mammals as our example.

Lungs are neither muscles nor electrically excitable tissue. Therefore, they cannot initiate or regulate their own expansion. Nonetheless, lungs require a mechanism to rhythmically expand and recoil, because animals cannot always consciously control breathing, such as during sleep. In this section, we examine the ways in which the nervous system and other structures control ventilation in mammals and how these control mechanisms are linked with metabolism.

The Nervous System Contains the Pacemaker for Ventilation

The pacemaker that initiates expansion of the lungs is a collection of nuclei in the nervous system. In mammals, these **respiratory centers** are located in the brainstem (Figure 48.14). Neurons within certain brainstem nuclei periodically generate action potentials. These electrical impulses then travel from the brainstem through two sets of nerves. The first set, called the intercostal nerves, stimulates the intercostal muscles, and the second set, the phrenic nerves, stimulates the diaphragm. When the lungs expand in response to the contraction of these muscles, stretch-sensitive neurons in the lungs and chest send signals to the respiratory centers, informing them that the lungs are inflated. This temporarily turns off the stimulating signal until the animal exhales, whereupon a new signal is sent to the breathing muscles.

Although the brainstem controls breathing automatically, it can be overridden. Animals that dive underwater—including humans when we go swimming—can hold their breath temporarily. We can also breathe faster than normal even while resting, at least for short periods of time. Normally, however, increased breathing occurs in response to physical activity. At such times, a variety of factors converge on the respiratory centers to increase the rate and strength of signals to the breathing muscles, resulting in faster and deeper breaths.

Chemoreceptors Modulate the Activity of the Respiratory Centers

The respiratory centers are influenced by the concentrations of certain substances in the blood, including the partial pressures of oxygen and carbon dioxide, and the concentration of hydrogen ions (in other words, blood pH). **Chemoreceptors**—special cells located in the aorta, carotid arteries, and the brainstem—detect the circulating levels of these substances and relay that information through nerves or interneurons to the respiratory centers.

Normally, the oxygen level is not the most important variable monitored by chemoreceptors. However, if the oxygen level falls far below normal, as might occur at high altitude or in certain respiratory diseases, the chemoreceptors signal the respiratory centers to increase the rate and depth of breathing to increase ventilation of the lungs. This brings in more oxygen. Similarly, a buildup of carbon dioxide in the blood, which would occur if an animal's ventilation were lower than normal (again, often the result of respiratory disease), signals the respiratory centers to stimulate breathing. Not only does increased ventilation bring in more oxygen, it also helps eliminate more carbon dioxide. Finally, an increased level of hydrogen ions in the blood activates chemoreceptors, which signal the brain that the blood is too acidic. This leads to an increase in the rate of breathing.

What is the link between ventilation rate and hydrogen ion concentration? The concentrations of carbon dioxide and hydrogen ions in the blood are related, because the concentration of hydrogen ions in the fluid bathing the brainstem chemoreceptors reflects the amount of carbon dioxide produced by cells during metabolism. This is because roughly two-thirds of the carbon dioxide produced during metabolism is converted into a less toxic and more soluble form, bicarbonate ions (HCO_3^-). In the process, hydrogen ions are formed according to the following reaction, where H_2CO_3 is a short-lived compound called carbonic acid that immediately dissociates to a hydrogen ion and a bicarbonate ion:

$$CO_2 + H_2O \leftrightarrow H_2CO_3 \leftrightarrow H^+ + HCO_3^-$$

This reaction is readily reversible. The first step is catalyzed in both directions by the enzyme carbonic anhydrase, which is present in high amounts in red blood cells. As you learned in Chapter 2, reactions of this type proceed in a net "direction" according to the concentrations of the reactants and products. For example, if CO_2 levels were to increase for any reason, the forward reaction (from left to right as illustrated here) would be favored, and the pH of the blood would decrease because hydrogen ions would increase. Conversely, if the levels of CO_2 decreased, the reactions would proceed from right to left, and the pH of the blood would increase as hydrogen ions were removed from the blood and combined with bicarbonate ions.

So what happens when an animal exercises? During exercise, lactic acid—a by-product of metabolism—is released into

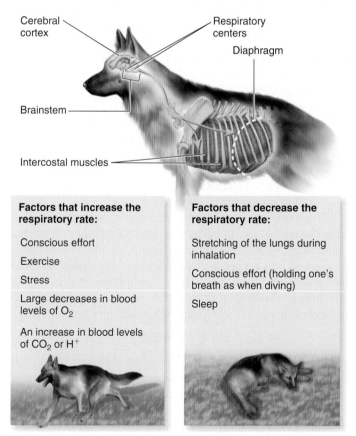

Factors that increase the respiratory rate:

Conscious effort

Exercise

Stress

Large decreases in blood levels of O_2

An increase in blood levels of CO_2 or H^+

Factors that decrease the respiratory rate:

Stretching of the lungs during inhalation

Conscious effort (holding one's breath as when diving)

Sleep

Figure 48.14 **The control of breathing via respiratory centers in the mammalian brain.** Neurons in the brainstem send action potentials along neurons in nerves that stimulate the intercostal muscles and diaphragm. The factors listed here can modulate the rate of action potential generation and therefore the respiratory rate.

the blood, and this lowers the pH of the blood. The increase in hydrogen ion concentration that results from lactic acid activates chemoreceptors to stimulate increased ventilation. This, in turn, helps prevent CO_2 levels from increasing in the blood despite the increased metabolism that is occurring.

In body tissues, CO_2 levels are high because metabolism is generating the gas. As stated earlier, the reactions proceed largely to the right, and CO_2 and H_2O are converted to H^+ and HCO_3^-. As blood flows through the lungs, however, dissolved CO_2 diffuses into the alveoli and is exhaled. Therefore, CO_2 levels in the lung capillaries decrease, which favors the reverse reactions (right to left), so HCO_3^- and H^+ are converted back to CO_2 and H_2O; the CO_2 is then exhaled. These reactions are remarkably fast, occurring within the space of time required for blood to move through a capillary, typically less than 1 second.

The chemoreceptors in the brain are very sensitive to certain drugs, such as ethanol, opiates, and barbiturates. These drugs decrease the sensitivity of chemoreceptors to CO_2, for example, and thereby reduce ventilation to potentially life-threatening levels. Particularly when taken together, these drugs may be a lethal combination partly because of their inhibitory effects on ventilation.

Respiratory Center Activity Varies Among Species with Different Metabolic Rates

The respiratory centers are sensitive to a variety of other factors besides gases and hydrogen ions, such as sleep, stress, hormones, and body size. Recall from Chapter 46 that small animals tend to have higher metabolic rates than larger animals. Thus, small animals would be expected to require more oxygen per unit mass than would larger animals. This is not achieved by having disproportionately large lungs in small animals. Instead, small animals have higher breathing rates than do large animals. The respiratory centers are set at a higher rate in smaller, active animals. For example, a typical adult human takes about 12 to 14 breaths per minute at rest, whereas hummingbirds, small rodents, and shrews have astonishingly high rates of breathing, up to many hundreds of breaths per minute (see Table 48.1)! Thus, the respiratory activity of small, highly active animals is matched with their metabolism by increasing the resting breathing rate. This is similar to the differences in cardiac function among species. Small animals have hearts that are roughly proportional to their body size, but with higher heart rates than are found in larger animals (refer back to Table 47.1). Parallel changes in the heart and breathing rates are adaptations that allow small animals to have small hearts and lungs yet achieve high metabolic rates.

48.5 Mechanisms of Oxygen Transport in Blood

Gases such as oxygen are soluble in the plasma component of blood, but as with all solutes, there are limits to how much of a gas can dissolve in a volume of water. The amount of gas dissolved in the body fluids is not sufficient to sustain life in most animals. In nearly all animals, therefore, the amount of oxygen in the body fluid compartments must be increased above that which can be physically dissolved. This is made possible because of the widespread occurrence in animals of oxygen-binding proteins, which increase the total amount of oxygen available to cells. In this section, we examine the nature, function, and evolution of these molecules.

Oxygen Binds to Respiratory Pigments

The oxygen-binding proteins that have evolved in animals are called **respiratory pigments**, because they have a color (blue or red). In vertebrates, the pigments are contained within red blood cells, whereas many invertebrates have these pigments in their hemolymph. Enclosing the pigments within blood cells is an efficient way to package them, because this keeps the pigments in close proximity to various cellular enzymes that assist in binding and unbinding the gases to the pigments.

Respiratory pigments are proteins containing one or more metal atoms in their cores; the metal atom binds to oxygen. Typically, the metal is iron (Fe^{2+}) in vertebrates and many marine invertebrates, but in decapod crustaceans, arachnids, and many mollusks (including cephalopods and some gastropods), it is copper (Cu^{2+}). The copper-containing pigment, **hemocyanin**, gives the blood or hemolymph a bluish tint. **Hemoglobin** is the major iron-containing pigment and gives blood a red color when oxygen is bound.

Hemoglobin gets its name because it is a globular protein—which refers to the shape and water solubility of a protein—and because it contains a chemical group called a heme in its core. An atom of iron is bound within the heme group. In vertebrates, hemoglobin consists of four protein subunits, each with its own heme group and iron atom to which a molecule of oxygen can bind (refer back to Figure 47.6). Thus, a single hemoglobin protein can bind up to four molecules of oxygen.

Respiratory pigments share certain characteristics that make them ideal for transporting oxygen. First, they all have a high affinity for binding oxygen. Second, the binding between the pigment and oxygen is noncovalent and reversible. It would not be very useful for a protein to bind oxygen if it could not later unload the oxygen to cells that need it. The formula for the reversible binding of oxygen to a respiratory pigment is:

$$Hb + O_2 \leftrightarrow HbO_2$$

where Hb is hemoglobin, HbO_2 is called oxyhemoglobin, and the double arrows indicate that the reaction is reversible.

The amount of pigment present in the blood is great enough to provide sufficient oxygen to meet all but the most strenuous exertion. In humans, for example, the presence of hemoglobin gives blood about 45 times more oxygen-carrying capacity than would plasma alone.

Differences exist among respiratory pigments, however, at the molecular level. For example, the amino acid sequences of hemoglobins from different species vary, in some cases considerably. Human hemoglobin is similar but not identical to that of other mammals, and mammalian hemoglobins have numerous amino acid differences from other vertebrate hemoglobins. The differences in amino acid sequences cause slight changes in the overall three-dimensional shape of hemoglobin, which alters its affinity for oxygen. As a general rule, animals with the highest metabolic rates tend to have hemoglobins with lower oxygen affinity, and animals with lower metabolic rates have higher affinities. That may sound paradoxical at first, but consider the preceding reactions again. Because the hemoglobin of small animals with high metabolic rates has lower oxygen affinity, the reverse reaction ($HbO_2 \rightarrow Hb + O_2$) will be more likely to occur when oxygen is diffusing from the blood to active cells. In other words, hemoglobin unloads its oxygen more readily in those animals whose cells use oxygen at the greatest rates.

The Amount of Oxygen Bound to Hemoglobin Depends on the P_{O_2} of Blood

Recall that the partial pressure of O_2 (P_{O_2}) is a measure of its dissolved concentration. When P_{O_2} is high, more O_2 binds to hemoglobin, whereas fewer O_2 molecules will be bound when P_{O_2} is low. **Figure 48.15** shows the relationship between O_2

Figure 48.15   The human oxygen-hemoglobin dissociation curve. Depending on the partial pressure of oxygen, oxygen is either loaded onto hemoglobin, as in the lungs, or unloaded from hemoglobin, as in the rest of the body tissues. When P_{O_2} is high, more Fe^{2+} atoms are bound to O_2, and the hemoglobin is more saturated with O_2.

Concept check: Members of the Channichthyidae family of Antarctic icefish are the only vertebrates that do not have red blood cells and hemoglobin. Can you propose some mechanisms or adaptations that allow them to survive without the oxygen reservoir provided by hemoglobin? (Hint: Think about the effect of temperature on oxygen solubility and on an animal's metabolic rate.)

binding and P_{O_2}, known as an **oxygen-hemoglobin dissociation curve**, for humans. At a P_{O_2} of 100 mmHg, which is found in the blood leaving the lungs, each hemoglobin protein binds four O_2 molecules. It is 100% saturated with oxygen. The P_{O_2} of blood leaving the tissue capillaries of other parts of the body is lower and depends on metabolic activity and exercise. At rest, the P_{O_2} of blood capillaries in these other parts of the body is typically around 40 mmHg. At this P_{O_2}, hemoglobin releases some O_2 molecules, becoming less saturated with O_2. At high levels of exercise, P_{O_2} in the capillaries drops even further (as low as 20 mmHg), and hemoglobin releases even more O_2. In this way, hemoglobin performs its role of oxygen delivery. In the lungs, it binds O_2, and in other tissues of the body, it releases some of the O_2 as needed.

You may have noticed that the curve in Figure 48.15 is not linear but S-shaped (sigmoidal). This is because the subunits of hemoglobin are said to cooperate with each other in binding oxygen. Once a molecule of oxygen binds to one subunit's iron atom, the shape of the entire hemoglobin protein changes, making it easier for a second oxygen to bind to the next subunit, and so on. Thus, the relationship between oxygen pressure and the amount of oxygen bound to hemoglobin becomes very steep in the midrange of the curve, which represents the pressures that occur in the tissue capillaries throughout the body except in the lungs. This steepness allows O_2 release to be very sensitive to changes in P_{O_2}.

One of the remarkable features of the oxygen-hemoglobin binding relationship is that it can be influenced by metabolic waste products such as CO_2 and H^+ and also by temperature. In addition to being transported as bicarbonate ions, about 25% of the CO_2 in blood is bound to hemoglobin. The remainder—about 7–10%—exists dissolved in solution in the plasma and within red blood cells. **Figure 48.16a** shows three curves, one obtained under normal resting conditions and the others in the presence of low or high levels of CO_2. Carbon dioxide binds to amino acids in the hemoglobin protein (not to the iron, like O_2), and when it does, the ability of hemoglobin to continue binding oxygen decreases. Note how increased CO_2 shifts the curve to the right, such that at any P_{O_2}, less O_2 is bound to hemoglobin. Another way of saying this is that at any P_{O_2}, more O_2 has been released from hemoglobin, thus becoming available to cells. A similar shift in the curve occurs in the presence of H^+, which can also bind to hemoglobin; these effects of CO_2 and H^+ on the oxygen-hemoglobin dissociation curve are known as the Bohr effect, or Bohr shift, after their discoverer, the Danish physiologist Christian Bohr. Elevated temperature will also reduce the affinity of hemoglobin for oxygen, resulting in a shifted curve. Cells generate each of these products—CO_2, H^+, and heat—when they are actively metabolizing fuel. The metabolic products enter the surrounding blood vessels, where they disrupt the normal shape of hemoglobin, causing it to release more of its oxygen than would normally occur at that oxygen pressure. This phenomenon is a way in which individual body tissues can obtain more oxygen from the blood to match their changing metabolic demands. Thus, when an animal exercises, the skeletal muscles generate more waste products and heat than does,

(a) Shifts in the oxygen-hemoglobin dissociation curve

Labels on graph:
- At the P_{O_2} in the lung, the shift is not as important.
- At the P_{O_2} in the tissue capillaries, the shift is most dramatic.
- Decreased CO_2, H^+, temperature
- Normal CO_2, H^+, temperature
- Increased CO_2, H^+, temperature

Axes: Percent saturation of hemoglobin with O_2 vs P_{O_2} (mmHg)

(b) Oxygen-hemoglobin dissociation curves of different animals

Labels on graph:
- Mouse Hb unloads more O_2 at any pressure.
- Elephant
- Human
- Mouse

Axes: Percent saturation of hemoglobin with O_2 vs P_{O_2} (mmHg)

Figure 48.16   Differences in oxygen-hemoglobin dissociation curves under different physiological conditions and among different species.   **(a)** Increasing or decreasing the amounts of carbon dioxide or hydrogen ions (acid), or the temperature of the blood, shifts the oxygen-hemoglobin dissociation curve. Actively metabolic tissues generate these products. The change in affinity of hemoglobin for oxygen allows different tissues to obtain oxygen in proportion to their metabolic requirements. **(b)** Oxygen-hemoglobin dissociation curves for three mammals with low (elephant), moderate (human), or high (mouse) relative metabolic rates. For any P_{O_2}, such as the one shown (40 mmHg), which is typical of the P_{O_2} of tissue capillaries, less oxygen is bound to mouse hemoglobin than to human or elephant hemoglobin, and less O_2 is bound to human hemoglobin than to elephant hemoglobin. Therefore, O_2 is unloaded from hemoglobin more readily in smaller animals.

Concept check: *What would happen to the position of the middle curve in part (a) following infusion of bicarbonate ions (HCO_3^-) into the blood of a resting, healthy individual?*

say, the skin. Therefore, the oxyhemoglobin in muscle blood vessels releases more oxygen to the cells where it is immediately required.

The shift in the oxygen-hemoglobin dissociation curve occurs in all classes of vertebrates (although not in all species), but it has different magnitudes in different species. Not surprisingly, perhaps, the most metabolically active animals, such as mice, show a greater shift for a given increase in CO_2, H^+, or heat than do less-active animals. Recall from Chapter 47 that these same waste products of metabolism also cause local vasodilation of arterioles. Thus, the more metabolically active a tissue is, the more blood flow it receives, which means more oxygen-bound hemoglobin. Moreover, the oxygen is unloaded from hemoglobin more readily due to the shift in the curve. This is another example of how adaptive changes in circulatory and respiratory functions often occur in parallel.

The hemoglobins of metabolically active animals also have a lower affinity for oxygen (**Figure 48.16b**). Small, active animals have curves that are displaced to the right of the human curve. In contrast, larger animals with slower metabolisms, such as the elephant, have curves shifted to the left of humans. At high oxygen pressures, such as those that occur in the lungs, these shifts have little relevance because nearly all the hemoglobin is bound to oxygen at those pressures. At lower oxygen pressures, however, such as those that would occur in the blood vessels of actively metabolic tissues, the difference in the curves becomes significant. For example, look at the three curves at an

oxygen pressure of 40 mmHg, a typical value found in tissues that are using oxygen at a resting rate. The mouse hemoglobin has less oxygen bound (its hemoglobin is less saturated) at that pressure than does the human hemoglobin, which, in turn, has less oxygen bound than does the elephant hemoglobin. In other words, the mouse hemoglobin has released more of its oxygen to the active tissues than have the other animal hemoglobins.

Oxygen-carrying molecules are among the most ancient proteins found in animals. They appear to have begun as simple, single-subunit proteins such as muscle myoglobin, which stores oxygen in muscle cells. Around 500 million years ago, the gene for this ancient molecule duplicated, resulting in one gene that coded for myoglobin and the other for an early form of hemoglobin (refer back to Figure 21.8). Thus, animals gained the ability to not only store oxygen in tissues but also to transport it throughout the body via the circulation. Let's take a closer look at the molecular structure and evolution of hemoglobin and how a simple mutation in the sequence of one hemoglobin gene can have devastating consequences in humans.

Genomes & Proteomes Connection

Hemoglobin Evolved Over 500 Million Years Ago

Modern hemoglobin in adult mammals contains four subunits, designated $\alpha1$, $\alpha2$, $\beta1$, and $\beta2$. The two alpha and two beta subunits are identical. Early in animal evolution, hemoglobin existed only as monomers (one subunit) or dimers (two subunits), which is still observed in lampreys. Later, gene duplication created the second set of subunits, leading to the four-subunit form, which first appeared in sharks and bony fishes. As we have seen in this chapter, the subunit structure of hemoglobin enables one hemoglobin protein to bind four oxygen molecules. Equally important, the subunits of hemoglobin cooperate with each other as shown by the sigmoidal shape of the oxygen-hemoglobin dissociation curve.

As noted earlier, modern hemoglobins differ among species within the same class of vertebrates and even more so between classes. This suggests that evolution of the hemoglobin genes continued even after the four-subunit structure appeared. The mutations that arose in the hemoglobin genes affected the structure of the hemoglobin protein. One consequence is that the affinity of hemoglobin for oxygen differs among animals, as discussed earlier. The differences in affinity may result, for example, in better oxygen-capturing ability of animals living in regions of low oxygen pressure, such as at high altitude. They also account, at least in part, for the ability of metabolically active animals to unload oxygen from hemoglobin more readily than less metabolically active organisms.

Despite the high frequency of mutations in the hemoglobin genes in vertebrates, certain regions of the molecule have remained well conserved. Not surprisingly, these are regions that are critically important for the function of the molecule. For instance, the amino acids that encase the heme group and iron atom of each subunit have changed little if at all during evolution. Similarly, the amino acid sequences associated with sites of subunit interactions have also resisted evolutionary change. Mutations that arise in these regions are typically not adaptive and may even be lethal.

Some mutations in the hemoglobin genes may appear to be detrimental because they negatively impact hemoglobin's function, and yet they provide a selective advantage under certain conditions. Consider the disease **sickle-cell disease** (also called sickle-cell anemia). In this case, substitution of a thymine for an adenine at one position in the hemoglobin gene results in a single amino acid substitution (refer back to Figure 14.1). This produces an abnormal form of hemoglobin that tends to form polymers and precipitate under low oxygen conditions, particularly in capillaries and veins. The protein forms long fibrous strands that may permanently deform the red blood cell, making it sickle-shaped. Sickled cells are less able to move smoothly through capillaries and can block blood flow, resulting in severe pain and cell death in the surrounding tissue.

The sickled red blood cells are also fragile and easily destroyed. The loss of those cells and their hemoglobin results in anemia, meaning that people with the disease have less oxygen-carrying capacity in their blood. Only individuals who are homozygous for this mutation show the dramatic phenotype of the disease. Heterozygotes are relatively unaffected.

The sickle-cell gene mutation is present in up to a third of individuals living in malaria-prone regions of Africa, to lesser extents in the Middle East and Eastern Europe, and in roughly 8% of African Americans (refer back to Figure 24.6). Why should such an obviously harmful mutation have persisted in the human population?

Malaria, a disease transmitted by bites from mosquitos of the genus *Anopheles*, is a global health threat that is responsible for over 300 million severe illnesses and 1 million deaths every year. When an infected mosquito bites, it injects a parasitic protozoan of the family Plasmodiidae (the most deadly being *Plasmodium falciparum*) into the circulation, where it spends part of its life cycle growing and multiplying within the host's red blood cells (refer back to Figure 28.29). The sickle-cell trait protects individuals from developing full-blown malaria, although the reason for this is uncertain. One favored hypothesis states that the parasite lowers the oxygen pressure in red blood cells by consuming oxygen to support its own metabolism. This may render the red blood cell susceptible to sickling even in heterozygous individuals. The damaged cells are destroyed and removed from the circulation, which also removes the parasite and limits its ability to multiply. Another hypothesis states that the heterozygotic condition results in a more efficient immune response of the body to the presence of the parasite. Thus, although both malaria and homozygous sickle-cell disease are life-threatening, heterozygotes for the sickle-cell trait develop neither pronounced anemia nor severe malaria and therefore have an advantage in areas where malaria is present. This heterozygote advantage explains why the sickle-cell mutation persists among certain human populations.

48.6 Adaptations to Extreme Conditions

Many animals are able to live permanently or temporarily in low-oxygen environments. Many humans, for instance, live in mountainous regions, and llamas and mountain goats spend most of their lives at extremely high altitudes. Other animals may only transiently encounter periods of oxygen deprivation, such as some reptiles, birds, and mammals that dive underwater to forage for food. As described in this section, several adaptations allow animals to exploit these environments, but all of them include changes in both cardiovascular and respiratory activities.

Life at High Altitudes Requires More Hemoglobin

Animals that live at high altitudes must have special adaptations that permit them to obtain the oxygen they need at such low atmospheric pressures. Llamas, such as the one shown in the chapter-opening photo, may live at altitudes up to 4,800 m, where the P_{O_2} is only about 85 mmHg (compared with 160 mmHg at sea level). Llama hemoglobin is quite different in amino acid sequence from that of other mammals, giving it an extraordinarily high affinity for binding oxygen even at very low atmospheric pressure. In other words, their hemoglobin curves are well to the left of a human's. In addition, llamas have larger hearts and lungs than would be predicted for their body size, and a higher number of red blood cells in a given volume of blood. These adaptations provide the oxygen-carrying capacity and cardiac output needed to deliver sufficient oxygen to the tissues, even at low partial pressures of oxygen.

When animals that normally inhabit lowland areas temporarily move to higher altitudes, they develop some of these same features. In humans, for example, the number of red blood cells increases from about 5.1×10^{12} cells/L to about 6.4×10^{12} cells/L. This is stimulated by the hormone erythropoietin, which is secreted by the kidneys when the arterial P_{O_2} is low. This hormone acts on bone marrow to stimulate maturation of new red blood cells. Moving to higher elevations also increases ventilation due to a higher rate of breathing. The number of capillaries in skeletal muscle increases at high altitudes, a modification that facilitates oxygen diffusion into muscle cells. Finally, myoglobin content increases in muscle cells, expanding the reservoir of oxygen in the cytosol of those cells. Thus, after several days or weeks at high altitude, the cardiovascular and respiratory systems adapt together to maximize oxygen uptake, diffusion of oxygen into the blood, and the oxygen-carrying capacity of blood.

Diving Animals Rely on Anaerobic Respiration Once Their Oxygen Is Depleted

Many birds, reptiles, and several species of mammals spend time under water, foraging for food or escaping predators. During that time, the animal cannot breathe. In short dives, this is not a problem. We are all familiar with our own ability to stay under water for a minute or so. After that time, the oxygen level in the blood decreases, the carbon dioxide level in the blood increases, and we must surface for air. Some marine mammals, however, have an astonishing ability to remain under water for very long periods—up to nearly 2 hours in beaked whales and elephant seals. During that time, they continue to be active and search for food. How do they do it?

Like high-altitude animals, many diving animals have unusually high numbers of red blood cells, allowing them to store more oxygen in their blood than nondiving animals. In some species, such as seals, the extra blood cells are stored in the spleen until they dive, at which time the spleen contracts like wringing a wet washcloth and ejects the blood cells into the circulation. When seals resurface, the blood cells are sequestered again in the spleen until needed. In addition to having more erythrocytes, diving mammals typically have larger blood volumes than comparably sized mammals that live exclusively on land.

The muscles of diving mammals usually contain large quantities of myoglobin and its bound oxygen. This means that the muscles do not need to consume the precious stores of oxygen circulating in blood. Instead, the blood and its oxygen can be routed to other critical structures that lack myoglobin, such as the eyes, certain glands, the brain, and the placenta if the animal is pregnant.

Eventually, even with these adaptations, the oxygen in muscles and blood becomes so depleted that the only way to prolong the dive is for cells to begin respiring anaerobically. Most of a long dive, in fact, occurs under anaerobic conditions.

This concludes our examination of the ways in which animals obtain oxygen and eliminate carbon dioxide. We now address the importance of healthy lung function in humans and the ways in which humans are affected by respiratory diseases.

48.7 Impact on Public Health

Respiratory diseases afflict over 10% of the U.S. population and cause an estimated 500,000 deaths per year, making lung disease among the top three causes of death in the U.S. As many as 25 million Americans have impaired lung function. The economic impact of respiratory disorders on the U.S. economy is staggering, with recent estimates of $40–150 billion per year in health-related costs and lost work time. Many of these diseases are chronic—once they appear, they last for the rest of a person's lifetime. Lifestyle factors, such as smoking tobacco and exposure to air pollution, cause some respiratory disorders or make existing conditions worse. In this section, we examine a few of the most common respiratory disorders, as well as some of their causes and treatments.

Asthma Is a Disease of Hyperreactive Bronchioles

You learned earlier that the bronchioles deliver fresh air to the alveoli. Bronchioles are thin tubes surrounded by smooth muscle cells that can contract in the presence of airborne pollutants or other potentially damaging substances. In the disease **asthma**, however, the muscles around the bronchioles are hyperexcitable and contract more than usual. Contraction of these muscles narrows the bronchioles, a process called bronchoconstriction. This makes it difficult to move air in and out of the lungs, because resistance to airflow increases when the diameter of the airways decreases. Often, the resistance to airflow can be so great that the movement of air creates a characteristic wheezing sound.

Asthma tends to run in families and therefore has a genetic basis. Several known triggers can elicit wheezing, including exercise, cold air, and allergic reactions. The latter is of interest because asthma is believed to be partly the result of an imbalance in the immune system (see Chapter 53), which controls inflammation and other allergic responses. During flare-ups of asthma, a viscous, mucus-like fluid may inhibit the flow of air in and out of the airways and make symptoms worse.

The symptoms of asthma can be alleviated by inhaling an aerosol mist containing **bronchodilators**, compounds that bind to receptors located on the plasma membranes of smooth muscles that comprise the outer part of bronchioles. These compounds, which are related to the neurotransmitter norepinephrine, cause bronchiolar smooth muscles to relax. This, in turn, allows the bronchioles to dilate (widen). To help reduce the inflammation of the lungs, patients may inhale a mist containing hormones with anti-inflammatory actions. Currently there is no cure for asthma, but with regular treatment and the avoidance of known triggers, most people with this disease can lead perfectly normal lives.

Tobacco Smoke Causes Respiratory Health Problems and Cancer

Smoking tobacco products is one of the leading global causes of death, contributing to about 430,000 deaths each year in the U.S. alone and over 5 million per year worldwide. According to the Centers for Disease Control and Prevention and the American Lung Association, people who smoke up to one pack of cigarettes each day live on average 7 years less than nonsmokers, and heavy smokers lose on average 15–25 years of life. Pregnant women who smoke run a high risk of their babies being born underweight, a potentially serious condition that may affect the newborn's long-term health.

Up to 85% of all new cases of lung cancer diagnosed each year are attributable to smoking, making lung cancer the leading cause of preventable death. Equally important, however, is that smoking is estimated to be responsible for nearly 30% of all cancers, including cancer of the mouth and throat, esophagus, bladder, pancreas, and ovaries. Smoking is also a leading cause of cardiovascular disease, high blood pressure, atherosclerosis, and stroke. Smoking as few as three to five cigarettes per day raises the risk of heart disease.

Because the products of tobacco smoke are inhaled directly into the lungs, the chemicals in smoke can do considerable damage to lung tissue. Even adolescents who have only recently started smoking have increased mucus (phlegm) production in their airways, shortness of breath, and reduced lung growth. Thousands of chemicals, including over 40 known cancer-causing compounds, have been identified in cigarette smoke. Some of these chemicals—such as formaldehyde—are toxic to all cells. Others, like the odorless gas carbon monoxide (CO), have harmful effects on lung function in particular. CO competes with oxygen for binding sites in hemoglobin, thereby reducing hemoglobin saturation. Heavy smokers who smoke more than a pack of cigarettes each day may have as much as 15% less oxygen-carrying capacity in their blood.

In addition to its effects on cancer, cardiovascular disease, and lung function, long-term smoking is the major cause of the serious and irreversible disease emphysema.

Emphysema Causes Permanent Lung Damage

Unlike asthma, in which the major problems are inflamed airways and hyperreactive bronchioles, **emphysema** involves extensive lung damage (Figure 48.17). The disease reduces the elastic quality of the lungs and the total surface area of the alveoli, which cuts the rate of oxygen diffusion from the lungs into the circulation. Consequently, one sign of emphysema is a lower than normal partial pressure of oxygen in the arteries. It is also physically harder to exhale because of the loss of

Figure 48.17 The effects of emphysema. These light micrographs compare a section of a normal lung (left) with that of a lung from a person who died of emphysema (right). The destruction of alveoli caused by this disease reduces the surface area for gas exchange in the lungs.

elasticity, and therefore, arterial CO_2 levels increase. Finally, the terminal ends of the bronchioles are often damaged, which increases resistance to airflow and creates asthma-like symptoms and shortness of breath.

Reduced blood oxygen and poor lung function limit the patient's ability to function, and in its late stages, emphysema results in a person being essentially bedridden. Oxygen therapy, in which the person breathes a mixture of air and pure oxygen from a portable gas tank, can provide some help. The extra oxygen increases the pressure gradient for oxygen from the alveoli to the lung capillaries, promoting oxygen diffusion into the blood.

Estimating how many people have emphysema is difficult because the symptoms appear gradually, but more than 3 million people have severe cases of the disease, and about 120,000 people die from it each year in the U.S. Emphysema is a progressive disease that does not go away. As the years go by, the disease worsens, although medical care can slow the rate at which this happens.

In some cases, emphysema results from an enzyme deficiency in the lungs that destroys the protein that provides the recoil during exhalation, or it may result from chronic exposure to air pollution. However, the overwhelming majority of cases, 85%, are due to smoking. Toxins in cigarettes and other tobacco products damage the lungs by stimulating white blood cells to release proteolytic enzymes that degrade lung tissue. The likelihood of developing emphysema is strongly correlated with the quantity of cigarettes smoked during a person's lifetime.

Summary of Key Concepts

48.1 Physical Properties of Gases

- Gas exchange is the process of moving oxygen and carbon dioxide in opposite directions between the environment, body fluids, and cells. The partial pressure of oxygen in the environment provides the driving force for its diffusion from air or water across a respiratory organ and into the blood. Atmospheric pressure decreases at higher elevations. (Figure 48.2)

- Three factors—the pressure of the gas, temperature of the water, and presence of any other solutes—are particularly important for affecting the solubility of a gas in water.

48.2 Types of Respiratory Systems

- Ventilation is the process of bringing oxygenated water or air into contact with a respiratory organ. All respiratory organs have moist surfaces in which gases can dissolve and diffuse, as well as other adaptations that increase the amount of surface area and blood flow. Water-breathing and air-breathing animals face different gas exchange challenges.

- The body surface is permeable to gases in some invertebrates and in amphibians, eels, and a few other species of fishes. (Figure 48.3)

- Water-breathing animals use external or internal gills for gas exchange. (Figures 48.4, 48.5, 48.6)

- Air-breathing animals have evolved two major mechanisms to exchange gas with the environment: tracheal systems in insects and lungs in terrestrial vertebrates. In insects, air moving down the tracheoles comes into contact with fluid at the tracheole tips. Oxygen from the air dissolves in this fluid and diffuses across the tracheole wall and into nearby cells. (Figure 48.7)

- Except for some rare amphibians, all air-breathing vertebrates use lungs to bring oxygen into the circulatory system and remove carbon dioxide. All lungs receive deoxygenated blood from the heart and return oxygenated blood to the heart.

- Frogs and most other amphibians ventilate their lungs with positive pressure filling, which uses the principles of Boyle's law. (Figure 48.8)

48.3 Structure and Function of the Mammalian and Avian Respiratory Systems

- The mammalian respiratory system includes the nose, mouth, airways, lungs, and muscles and connective tissues that encase these structures within the thoracic (chest) cavity. (Figure 48.9)

- Most reptiles and all birds and mammals ventilate their lungs by negative pressure filling. In mammals, the work is provided by the intercostal muscles and diaphragm.

- Mammals breathe by tidal ventilation. Tidal volume is proportional to body size within and between species. (Figure 48.10, Table 48.1)

- The avian respiratory system, a flow-through system, is unique among vertebrates because it is supplemented with numerous air sacs. Inhaled air is stored in air sacs before passing through the lungs. Blood flows through the lungs in a crosscurrent direction with respect to the movement of oxygen. This is less efficient in extracting oxygen than the countercurrent flow in fish gills, but more efficient than the tidal ventilation system of mammals. (Figures 48.11, 48.12, 48.13)

48.4 Control of Ventilation in Mammalian Lungs

- In mammals, respiratory centers in the brainstem initiate the rhythmic expansion of the lungs. (Figure 48.14)

- Chemoreceptors detect blood levels of hydrogen ions and the partial pressures of carbon dioxide and oxygen. They relay this information through nerves to the respiratory centers, which, in turn, affect the breathing rate. The respiratory centers are also sensitive to other factors, such as sleep, stress, hormones, and body size.

48.5 Mechanisms of Oxygen Transport in Blood

- There are limits to how much of a gas can dissolve in water. These limits are overcome in nearly all animals by either transporting a gas reversibly bound to a protein carrier or by transforming the gas into a more soluble form.

- Animals have evolved a way to carry a reservoir of oxygen on respiratory pigments, large proteins containing one or more metal atoms that bind to oxygen. Typically, the metal is iron in vertebrates and many marine invertebrates, and copper in decapod crustaceans, arachnids, and many mollusks, including cephalopods and some gastropods.

- The amount of oxygen bound to hemoglobin depends on the partial pressure of oxygen in the blood. Metabolic waste products can influence the oxygen-hemoglobin binding relationship. (Figures 48.15, 48.16)

- The evolution of the globin gene family has resulted in several specialized hemoglobin proteins. Heterozygotes for the sickle-cell allele are resistant to malaria.

48.6 Adaptations to Extreme Conditions

- Several adaptations in cardiovascular and respiratory activities allow animals to exploit low-oxygen environments. Animals that inhabit high altitudes have larger hearts and lungs and have hemoglobin with a high affinity for binding oxygen. Many diving animals have unusually high numbers of red blood cells and also muscles with large quantities of myoglobin.

48.7 Impact on Public Health

- In asthma, the muscles around the bronchioles contract more than usual, increasing resistance to airflow.

- Smoking tobacco products is one of the leading global causes of death. Smoking is strongly linked to cancer, cardiovascular disease, stroke, and emphysema. (Figure 48.17)

Assess and Discuss

Test Yourself

1. The driving force for diffusion of oxygen across the cells of a respiratory organ is
   a. the difference in the partial pressure of oxygen in the environment and in the blood.
   b. the humidity.
   c. the partial pressure of carbon dioxide in the blood.
   d. the air temperature.
   e. the partial pressure of carbon dioxide in the atmosphere.

2. Carbon dioxide is considered a harmful by-product of cellular respiration because it
   a. lowers the pH of the blood.
   b. lowers the hydrogen ion concentration in the blood.
   c. competes with oxygen for transport in the blood.
   d. does all of the above.
   e. a and b only

3. The process of bringing oxygenated water or air into contact with a gas-exchange surface is
   a. respiration.   c. ventilation.   e. exhalation.
   b. gas exchange.   d. gas transport.

4. The group of vertebrates with the greatest capacity for gas exchange across the skin is
   a. the fishes.   d. the birds.
   b. the reptiles.   e. the mammals.
   c. the amphibians.

5. The countercurrent exchange mechanism in fish gills
   a. maximizes oxygen diffusion into the bloodstream.
   b. is a less efficient mechanism for gas exchange compared to mammalian lungs.
   c. occurs because the flow of blood is in the same direction as water flowing across the gills.
   d. is the same phenomenon observed in birds' lungs.
   e. requires that the fish swallow water.

6. The tracheal system of insects
   a. consists of several tracheae that connect to multiple lungs within the different segments of the body.
   b. consists of extensively branching tubes that are in close contact with all the cells of the body.
   c. allows oxygen to diffuse across the thin exoskeleton of the insect to the bloodstream.
   d. cannot function without constant movement of the wings to move air into and out of the body.
   e. provides oxygen that is carried through the animal's body in hemolymph.

7. _____ is secreted by Type II alveolar cells in the mammalian lung to prevent the collapse of alveoli due to surface tension at the interface of air and extracellular fluid.
   a. Hemoglobin   c. Mucus   e. Surfactant
   b. Myoglobin   d. Water

8. In negative pressure filling, air moves into the lungs when
   a. the volume of the thoracic cavity increases.
   b. the pressure in the thoracic cavity decreases.
   c. air is forced down the trachea by muscular contractions of the mouth and pharynx.
   d. all of the above
   e. a and b only

9. Which of the following factors does not alter the rate of breathing by influencing the chemoreceptors?
   a. carbon dioxide partial pressures in the blood
   b. oxygen partial pressures in the blood
   c. blood pH
   d. blood glucose levels
   e. hydrogen ion concentration in the blood

10. With rare exceptions, the majority of oxygen is transported in the blood of vertebrates
    a. by binding to plasma proteins.
    b. by binding to hemoglobin in erythrocytes.
    c. as dissolved gas in the plasma.
    d. as dissolved gas in the cytoplasm in the erythrocytes.
    e. by binding to myoglobin.

Conceptual Questions

1. Define countercurrent exchange as it relates to gas exchange in fishes.

2. Explain the special features of the avian respiratory system that make birds so well-adapted to air-breathing.

3. Explain some of the special adaptations for life at high altitudes; why are such adaptations necessary?

Collaborative Questions

1. Discuss two ways in which animals exchange gases in an aqueous environment. What special adaptations facilitate this exchange?

2. Discuss the components of the mammalian respiratory system.

Online Resource

www.brookerbiology.com

Stay a step ahead in your studies with animations that bring concepts to life and practice tests to assess your understanding. Your instructor may also recommend the interactive ebook, individualized learning tools, and more.

Excretory Systems and Salt and Water Balance

49

The human kidneys could filter the volume of water of this pool in a few months.

I f you have ever noticed how quickly the water in an aquarium or a swimming pool becomes dirty if the filter is not functioning, you will have a good idea of the importance of filtering the wastes from the blood of an animal. The human kidneys, for example, are remarkable filtration devices. Although each one is only about the size of a computer mouse, the kidneys are able to filter blood at a rate of 150–200 liters/day. Considering that there are only five or so liters of blood in a typical adult, that is an astonishingly effective filtration mechanism. Despite their small size, our kidneys could filter the entire contents of a medium-sized swimming pool in a few months. By the time a person reaches 50 years of age, their kidneys have filtered roughly 3,000,000 liters of blood! Even more impressive is that the kidneys not only remove waste products of metabolism from that filtered blood, but recapture useful substances such as sodium ions and water that form part of the liquid being filtered. In this way, the kidneys and other excretory organs found in animals contribute to homeostasis.

Homeostasis has been a common theme of the previous several chapters. Animals maintain a variety of physiological processes—including energy intake and usage, blood pressure, body temperature, and blood oxygen levels—within normal and often narrow ranges. Homeostasis is also critical in the regulation of salt and water levels in body fluid compartments. As described in Chapter 2, the general term "salt" is used to refer to a compound formed from an attraction between a positively charged ion, such as Na^+, and a negatively charged ion, such as Cl^-. These ions are held together by ionic bonds, which are broken when the salt is dissolved in water. Changes in the concentrations of ions resulting from dissolved salts in the extracellular and intracellular fluids have the potential to disrupt proper cellular function; for example, they may alter the difference in electrical potential across plasma membranes in the heart and brain. Salt (ion) concentrations and water volumes in the different body fluid compartments are related to each other, because a major way in which water moves between compartments is by osmosis, which, in turn, depends on the numbers of dissolved solutes in water. Consequently, changes in salt concentrations in body fluids may also affect cell volume, which then can cause cell disruption or death.

As we have learned, homeostasis is an energy-requiring process. A significant portion of most animals' daily energy expenditures goes toward maintaining salt and water homeostasis. The ability to do so is complicated by many factors, such as the environment and climate in which an animal lives and its access to sufficient supplies of drinking water.

In this chapter, we examine why salt and water balance is vital for survival, how it is affected by the requirement to eliminate metabolic wastes, and how different excretory organs participate in these processes. We then highlight some of the major features of the vertebrate and mammalian kidney, explore challenges posed by an animal's life history and environment, examine how the kidney eliminates wastes and regulates salt and water balance, and consider how kidney disease affects human health.

49.1 Principles of Homeostasis of Internal Fluids

As we saw in Chapter 47, an animal's internal fluids exist in compartments within the body. In invertebrates, these fluids include the hemolymph and the intracellular fluid, whereas in vertebrates, it includes the plasma component of blood, the interstitial fluid surrounding cells, and the intracellular fluid. A description of the major salts found in animals' bodies and some of their chief functions was included in Chapter 45 (refer back to Table 45.3). As stated earlier, salts dissociate in solution to form charged ions. Because ions are electrically charged, such salts are referred to as electrolytes. In this section, we will examine why water and electrolyte homeostasis is so important and how an animal's electrolyte concentration and water volume are maintained within normal ranges.

The Balance of Water and Electrolytes Is Critical for Survival

Maintenance of normal body water levels is of great importance for all animals. Not only is water the major portion of an animal's body mass, it is also the solvent that permits dissolved solutes to participate in chemical reactions. As described in Chapters 2 and 3, water also participates in many important chemical reactions, notably hydrolysis reactions. In addition, water is the transport vehicle that brings oxygen and nutrients to cells and removes wastes generated by metabolism.

When an animal's water volume is reduced below the normal range, we say the animal is dehydrated. In terrestrial animals, dehydration may occur if sufficient drinking water is not available or when water is lost by evaporation (perspiring or panting). Dehydration can be a serious, potentially life-threatening condition. For example, because blood is roughly 50% plasma (water), blood volume may decrease in dehydrated animals. Reduced blood volume compromises the ability of the circulatory system to move nutrients and wastes throughout the body and to assist in the regulation of body temperature on hot days (see Chapters 46 and 47).

Electrolyte balance is also very important for animals. A change of only a few percent in the extracellular fluid concentrations of K^+, for example, can trigger changes in nerve, heart, and skeletal muscle function by altering membrane potentials (see Chapter 41). Other electrolytes, such as Ca^{2+}, Mg^{2+}, PO_4^{3-}, and SO_4^{2-}, also participate in various biological activities. These functions include serving as cofactors for enzyme activation, participating in bone formation, forming part of the extracellular matrices around cells, and activating cellular events such as exocytosis and muscle contraction. An imbalance in any of these ions can seriously disrupt cellular activities.

Water and Salts Move Between Fluid Compartments by Different Mechanisms

Water moves between adjacent body compartments by osmosis down an osmotic gradient (see Chapter 5). Changes in the salt concentration in one compartment will lead to changes in fluid distribution between the compartments. These changes can cause cells to shrink or swell. When, for example, the salt concentration of extracellular fluid increases, water moves by osmosis from inside cells to the extracellular fluid, causing the cells to shrink. Shrinking or swelling of cells in the brain, heart, and other vital organs can rupture plasma membranes, leading to cell death.

Whereas water moves between fluid compartments by osmosis, ions from dissolved salts such as sodium chloride move by different mechanisms. All ions have very limited ability to diffuse across plasma membranes, because of their high water solubility and low lipid solubility. As described in Chapter 5, ions may cross membranes through channels formed by proteins that create a pore in the membrane. Alternatively, ions may be actively transported across epithelial cells that line tube-like structures such as those in the kidneys. Still other epithelia, such as those in gills, can actively transport ions between the surrounding salt or fresh water and the animal's body fluids. As this type of transport is an active process, it requires energy stored in the chemical bonds of ATP. Animals that face exceptional challenges to maintaining salt balance, such as marine fishes, must expend a considerable share of their daily energy budget to transport ions across epithelial cells.

Salt and Water Balance Occurs Despite Obligatory Exchanges with the Environment

Many vital processes—eliminating nitrogenous wastes, obtaining oxygen and eliminating carbon dioxide, consuming and metabolizing food, and regulating body temperature—have the potential to disturb salt and water balance. Therefore, these processes require additional energy expenditure to minimize or reverse the disturbance. Exchanges of salt and water with the environment that occur as a consequence of such vital processes are called obligatory exchanges (because the animal is "obliged" to make them) (**Figure 49.1**). Next we will examine these and other obligatory exchanges and how they are related to an animal's environment and life history.

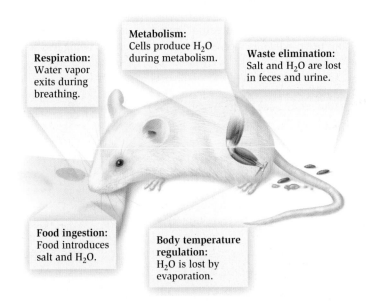

Figure 49.1 Types of obligatory salt and water exchanges in a terrestrial animal. Obligatory exchanges with the environment occur as the result of necessary life processes.

Concept check: Can animals completely avoid all the losses resulting from obligatory exchanges?

Elimination of Nitrogenous Wastes   When carbohydrates and fats are metabolized by animal cells, the major waste product is carbon dioxide gas, which is exhaled or diffuses across the body surface. By contrast, proteins and nucleic acids contain nitrogen; when these molecules are broken down and metabolized, nitrogenous wastes are generated. **Nitrogenous wastes** are molecules that include nitrogen from amino groups (NH_2). These wastes are toxic at high concentrations and must be eliminated from the body but, unlike carbon dioxide gas, cannot be eliminated by exhaling or diffusion. The elimination of nitrogenous wastes occurs via excretory organs such as the kidneys or other specialized structures, such as gills.

Nitrogenous wastes are usually found in three forms—ammonia (and ammonium ions), urea, or uric acid (Figure 49.2). Different animal groups produce a particular form of waste, depending on the species and the environment in which they live.

Ammonia (NH_3) and ammonium ions (NH_4^+) are the most toxic of the nitrogenous wastes because they disrupt pH, ion electrochemical gradients, and many chemical reactions that involve oxidations and reductions. Animals that excrete wastes in this form typically live in water. In marine invertebrates, ammonia and ammonium ions are continually excreted across the skin, whereas in freshwater and most saltwater fishes, these wastes are excreted via the gills and kidneys. Because ammonia is so toxic, aquatic animals excrete it as quickly as it is formed. However, some terrestrial snails and crustaceans can excrete ammonia in its gaseous form, which is less toxic and does not require water for elimination. The chief advantage of excreting nitrogenous wastes as NH_3 or NH_4^+ is that energy is not required for their conversion to a less toxic product, such as urea and uric acid, as described next.

All mammals, most amphibians, some marine fishes, some reptiles, and some terrestrial invertebrates convert ammonia into **urea**, which is then excreted. In addition to being less toxic than ammonia, urea does not require large volumes of water to be excreted. Animals can tolerate some accumulation of urea in their blood, tissues, and storage organs such as the urinary bladder. This conserves water, removes the necessity for constant excretion, and reduces the likelihood of toxicity. One drawback of producing urea is that the enzymatic conversion of ammonia into urea requires a moderate expenditure of ATP and thus uses up part of an animal's total daily energy budget.

Birds, insects, and most reptiles produce **uric acid** or other nitrogenous compounds called purines. Like urea, these compounds are less toxic than ammonia, but they are even more energetically costly to synthesize from ammonia. However, because they are poorly soluble in water, they are not excreted in a watery urine but instead are packaged with salts and other waste products into a semisolid, partly dried precipitate that is excreted. The energy investment required to produce uric acid, therefore, is balanced against the water conserved by excreting nitrogenous wastes in this form.

Respiration-Related Water and Electrolyte Exchanges   The requirements for both respiration and water and electrolyte balance present different challenges to air- and water-breathing animals. To ventilate its lungs, an air-breathing animal moves air in and out of its airways. Water in the form of water vapor in the mouth, nasal cavity, and upper airways exits the body with each exhalation. As an animal becomes more active, it requires more oxygen and produces more carbon dioxide. These changes are met by an increase in respiratory activity. Breathing becomes deeper and more rapid, which, in turn, increases the rate of water loss from the body. Therefore, respiration in animals with lungs is associated with significant water loss, as you can observe in cold weather when you can "see your breath."

As described in Chapter 46, small, active animals with high metabolic rates have higher breathing rates than do larger, less

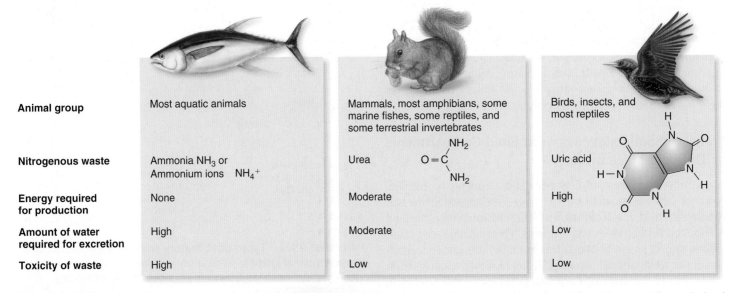

Animal group	Most aquatic animals	Mammals, most amphibians, some marine fishes, some reptiles, and some terrestrial invertebrates	Birds, insects, and most reptiles
Nitrogenous waste	Ammonia NH_3 or Ammonium ions NH_4^+	Urea	Uric acid
Energy required for production	None	Moderate	High
Amount of water required for excretion	High	Moderate	Low
Toxicity of waste	High	Low	Low

Figure 49.2   Nitrogenous wastes produced by different animal groups.   The three forms of nitrogenous wastes, which are derived from the breakdown of proteins or nucleic acids, have different properties.

active animals. Consequently, the potential for water loss due to respiration is considerably greater in small animals, particularly in endotherms. Hummingbirds, for example, may have 15 to 20 times the water loss per gram of body mass than would a large goose.

In water-breathing animals, the challenge of water and salt balance is more complex, because such animals move water, not air, over their respiratory organs (gills). Recall from Chapter 48 that gills, like all respiratory organs, are thin structures with large amounts of surface area and an extensive network of capillaries. Although these features make gills ideal for gas exchange by diffusion between the capillaries and the surrounding water, they also make them ideal for salt and water movement by diffusion and osmosis, respectively.

The solute concentration of a solution of water is known as the solution's **osmolarity**, expressed as milliosmoles/liter (mOsm/L). The number of dissolved solute particles determines a solution's osmolarity. For example, a 150 mM NaCl solution has an osmolarity of 300 mOsm/L, because each NaCl molecule dissociates into two ions, one Na^+ and one Cl^- (2 × 150 = 300).

When differences occur in salt concentration between a water-breathing animal's body fluids and the surrounding water, respiration via the gills has the potential to disrupt salt and water balance. The internal fluid osmolarity of most fishes is usually within the range of 225–400 mOsm/L, similar to that of most other vertebrates. Fishes or other water-breathing animals that live in fresh water and those that live in salt water face opposite challenges in maintaining this balance (**Figure 49.3**). Because freshwater lakes and rivers have very little salt content (usually <25 mOsm/L), a high concentration gradient for salts could promote the loss of salts from a fish's body into the fresh water. Likewise, a high osmotic gradient favors the movement of water into a freshwater fish. Freshwater fishes, therefore, gain water and lose salt when ventilating their gills (Figure 49.3a). If left uncorrected, this would cause a dangerous decrease in blood salt concentrations.

Freshwater fishes avoid this problem with two different mechanisms. First, their kidneys are adapted to producing copious amounts of dilute urine—up to 30% of their body mass per day (an amount that would be equivalent to about 25 L per day in an average-sized man!). Second, specialized gill epithelial cells actively transport Na^+ and Cl^- from the surrounding water into the fish's capillaries. Thus, these two important ions are recaptured from the water. As the preceding discussion suggests, freshwater fishes rarely, if ever, drink water, except for any that might be swallowed with food.

Other freshwater animals, such as frogs and other adult amphibians, have body surfaces that are permeable to water and are used in gas exchange. Like freshwater fishes, therefore, they tend to gain water by osmosis and compensate by excreting a copious dilute urine. Epithelial cells of the skin actively transport necessary electrolytes from the water into the blood.

Saltwater fishes have the opposite problem. They tend to gain salts and lose water across their gills, because seawater has a much higher osmolarity (about 1,000 mOsm/L) than that of their body fluids (Figure 49.3b). The gain of salts and the loss of water from the body are only partly offset by the kidneys, which in marine fishes produce very little urine so water can be retained in the body. The urine that is produced has a higher salt concentration than that of freshwater fishes.

To prevent dehydration from occurring, marine fishes must drink. However, the only water available to them is seawater, which has a very high salt content. Paradoxically, therefore, marine fish drink seawater to replenish the water lost by osmosis through their gills. This creates a new problem: What does the fish do with all of the salt it ingested?

The ingested salt must be eliminated, and this process is accomplished by gill epithelial cells. In contrast to the gills of freshwater fishes, which pump salt from the water into the fluids of the fish, the gills of marine fishes pump salt out of the fish and into the ocean. Thus, marine fishes drink seawater to replace the water lost through their gills by osmosis and then expend energy to transport the excess salt out of the body.

(a) Freshwater fish

(b) Saltwater fish

Figure 49.3 Salt and water balance in water-breathers. Water-breathing creates osmoregulatory challenges due to diffusion of salts and osmosis of water across gills. These challenges differ between **(a)** freshwater and **(b)** saltwater fishes and are met by drinking or not drinking water, by active transport of salts across the gills, and by alterations in urine output.

Ingestion-Related Fluid and Electrolyte Balance Because foods contain minerals and water, eating also involves an obligatory exchange of salt and water. Some plant products are over 95% water by weight, and other foods may contain high amounts of sodium or other minerals. Therefore, the type of diet an animal consumes determines how much salt and water it ingests.

Once food has been digested and absorbed, the unusable parts of food are excreted as solid wastes. Some salt and water are lost by this route in most animals, but exceptions exist. Desert-dwelling kangaroo rats such as *Dipodomys panamintensis* produce fecal pellets that are almost completely dry, which helps these animals conserve water.

As noted earlier, marine fishes drink seawater. Other animals besides marine fishes may also consume seawater, either because fresh water is unavailable or because they ingest some with the food they eat. Many marine reptiles and birds, for example, ingest seawater when consuming prey or, in some cases, when they spend prolonged periods at sea and have no access to fresh water for drinking. These animals have specialized epithelial cells that line structures called salt glands, located in groups around the nostrils, mouth, and eyes (**Figure 49.4**). Salts move from the blood into the interstitial fluid, from where they are actively transported by the epithelial cells of the salt glands into the tubules of the gland. The salts and a small amount of fluid then collect into a central duct and are excreted as highly concentrated solutions. In general, vertebrates without salt glands cannot survive by drinking seawater, because they have no means of creating and excreting such a highly concentrated salt solution. Some marine mammals have been observed to occasionally drink small amounts of seawater, but most appear to never drink at all. These animals get their water from the food they eat.

Regulation of Body Temperature with Water Endotherms use body water to cool off (see Chapter 46). For example, sweating and panting are used to cool the body. These behaviors use the evaporation of water to draw heat out of the body. In the process, however, the animal loses water and, in sweat, some salts. You know from tasting sweat that it is salty, but the saltiness of sweat and that of blood are not the same. Sweat is a hypoosmotic (more dilute) solution compared to blood, so the fluid left behind in the body after perspiration has both a lower volume and a higher salt concentration.

Other than perspiration and panting, very little water is gained or lost directly across the body surface of most terrestrial vertebrates, because their skin is impermeable to water. Exceptions include amphibians (and also some invertebrates). In invertebrates, the rate of water loss across the body surface depends on whether the animal is soft-bodied, like worms, or covered in a waxy, water-impermeable cuticle, like most insects.

Metabolism and Water Balance When food molecules are metabolized to provide energy that will be stored in the chemical bonds of ATP, oxygen captures electrons and combines with hydrogen ions, thereby making water. This water is sometimes called "metabolic water" to indicate its origin. Some animals—especially desert dwellers, which lack ready access to drinking

(a) Penguin with salt glands   **(b) Secretory tubules**

NaCl diffuses from blood to interstitial fluid.

Interstitial fluid (exaggerated size)

Lumen of salt gland tubule

NaCl is actively transported from interstitial fluid to the lumen of the salt gland tubule.

Blood flow   To central duct   Epithelial cell

(c) Collection of salt solution in the tubule

Figure 49.4 Salt glands as an adaptation for marine life. Many marine birds and reptiles have salt glands, which contain a network of secretory tubules that actively transport NaCl from the interstitial fluid into the tubule lumen. The viscous solution then moves through a central duct and to the outside environment through pores in the nose, around the eyes, and in other locations. The black arrows indicate direction of flow of blood or salt gland excretions.

Concept check: Why can't humans survive by drinking seawater?

water—depend on this water to provide all or nearly all of their water requirements. In other animals, the production of metabolic water may sometimes result in more water than is required at that time. This excess water is eliminated by the excretory organs or through other routes. Because metabolism is always ongoing and is required for survival, the excretion or retention of metabolic water can be considered a type of obligatory exchange.

As described next, the significance of obligatory exchanges and their effects on homeostasis was dramatically illustrated by a long-term investigation by a research team at the University of Florida. Their discovery would lead to a revolution in our understanding of exercise physiology in humans.

FEATURE INVESTIGATION

Cade and Colleagues Discovered Why Athletes' Performance Wanes on Hot Days

On a typically hot summer day in the mid-1960s in Gainesville, Florida, the University of Florida football team was practicing in full equipment. The players were rapidly becoming dehydrated and, unbeknownst to them, the osmolarity of their body fluids was increasing as their bodies produced copious amounts of dilute sweat in an effort to maintain body temperature. The athletes became aware of two things. First, they discovered that they did not need to urinate for long periods after a tough practice session, and second, their performance on the field suffered as they became increasingly fatigued and more susceptible to severe muscle cramps. Often, players would need to receive medical treatment or even hospitalization for their symptoms. In extreme cases, athletes exercising in these conditions have been known to occasionally develop seizures—uncontrolled activity of neurons in the brain. This situation did not escape the notice of the team physicians and, notably, university faculty member and kidney specialist Robert Cade.

Many of the symptoms experienced by the players could be readily explained. The fatigue was directly related to loss of water from the body, which put a strain on the circulatory system and reduced blood flow to muscles and other organs. It was worsened by a decrease in blood glucose levels that were not being replenished during the long practices or games. The muscle cramps and even occasional seizures arose from an imbalance in extracellular electrolytes—notably sodium and potassium ions—which are secreted outside the body by sweat glands in the process of perspiration. The resulting imbalance in extracellular electrolyte concentrations caused a change in the electrical potential across muscle and neuronal cell membranes, which triggered the spasms. Lastly, the decreased urine production is one of the body's mechanisms for reducing fluid loss when body water is decreasing.

The key question was, How can these effects of extreme exercise best be reversed or prevented? The answer was simple and clever. Cade and his colleagues rejected the prevailing view that drinking any fluids during heavy exercise somehow contributed to cramps and other problems. Instead, they hypothesized that the best way to maintain salt and water homeostasis in a profusely sweating person is to restore to the body exactly what was lost; that is, the person should drink a solution that resembles sweat!

The first thing Cade needed to do was analyze precisely how much sodium, potassium, and other ions are actually present in sweat. Fortunately, he had an abundance of human sweat at his disposal to analyze. Once the players left the field, their jerseys were wrung out, and the composition of the sweat was determined with an ion analyzer such as one called a flame spectrophotometer, as shown in **Figure 49.5**. The concentrations could then be compared with known values of ion concentrations in human blood. The composition of human sweat is now known to change under certain conditions and to vary among people, but Cade's results were typical. The sweat of the athletes contained mostly Na^+, K^+, and Cl^- at concentrations

Figure 49.5 Cade and colleagues discovered a way to improve athletic performance and prevent salt and water imbalance during strenuous exercise.

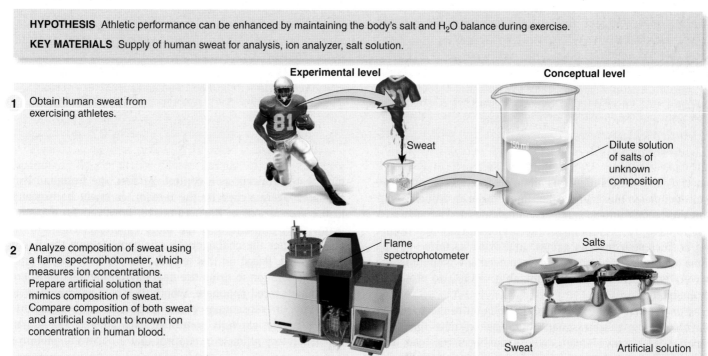

HYPOTHESIS Athletic performance can be enhanced by maintaining the body's salt and H_2O balance during exercise.

KEY MATERIALS Supply of human sweat for analysis, ion analyzer, salt solution.

Experimental level   Conceptual level

1 Obtain human sweat from exercising athletes.

Sweat

Dilute solution of salts of unknown composition

2 Analyze composition of sweat using a flame spectrophotometer, which measures ion concentrations. Prepare artificial solution that mimics composition of sweat. Compare composition of both sweat and artificial solution to known ion concentration in human blood.

Flame spectrophotometer

Salts

Sweat   Artificial solution

3 Add flavoring and sugar to artificial solution.

Sugar

Artificial solution

Sugar improves flavor and provides energy.

4 Provide freshman team with the artificial solution and varsity B-team with water. Hold scrimmage.

Freshman team — Artificial solution

Varsity B-team — Water

5 THE DATA

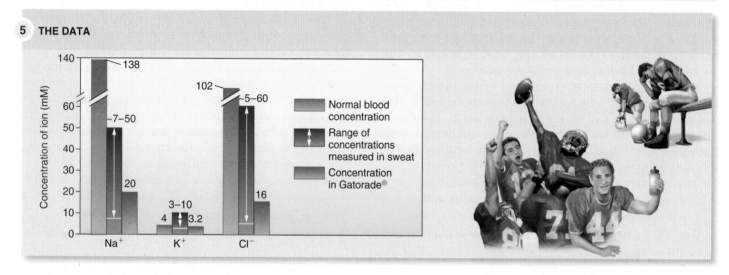

Concentration of ion (mM)

- 138
- ~7–50
- 20
- 4
- 3–10
- 3.2
- 102
- ~5–60
- 16

Na⁺ K⁺ Cl⁻

Legend:
- Normal blood concentration
- Range of concentrations measured in sweat
- Concentration in Gatorade®

6 CONCLUSION Replacement of fluid with solute concentrations similar to those found in human sweat improves athletic performance compared to water replacement alone.

7 SOURCE Most of the original studies described here were published in expanded form in a later report. See Cade, R. et al. 1972. Effect of fluid, electrolyte and glucose replacement during exercise on performance, body temperature, rate of sweat loss, and compositional changes of extracellular fluid. *Journal of Sports Medicine and Physical Fitness* 12:150–156.

that indicated the solution was dilute compared to blood. Once Cade completed this analysis, he simply made an artificial solution of a composition similar to human sweat. The next step was to have the players ingest the solution before and during the practice sessions and games. Improving its taste—adding some lemon flavoring and sugar—removed any inhibitions the players may have had about drinking it, while also providing an energy boost.

For the first trial, Cade gave the solution to the freshman players during an intrasquad scrimmage against the more experienced varsity B-team, whose members received only pure water to drink as a control. At first, the freshman team appeared overmatched by the B-team, as might be expected. In the second half of the scrimmage, however, the freshman team vastly outperformed the more experienced players and did not suffer the characteristic late-game fatigue the B-team experienced. Based on this test, the varsity A-team was given a similar solution to drink the next day during a game against a heavily favored opponent, whom they beat handily on a hot 39°C day. In subsequent years, Cade and other researchers would conduct carefully controlled experiments with humans and laboratory animals to confirm that a balanced solution of electrolytes similar to that present in sweat effectively improves exercise performance and reduces the possibility of dehydration and its consequences.

Because the solution was envisioned as "aid" for the team known as the University of Florida "Gators," the drink eventually came to be called Gatorade®. The year after its introduction, the Gators enjoyed their most successful season. In 1965, the Kansas City Chiefs of the former American Football League became the first professional sports team to try the drink, and shortly thereafter, the team enjoyed its greatest success. Nowadays, Gatorade® is used at sporting events around the world, and for good scientific reason.

The effectiveness of a solution like Gatorade® is due to its ability to restore the correct amounts of both water and ions lost during exercise. Importantly, it is very rapidly absorbed because its osmolarity is close to that of body fluids. Many of the other sports drinks subsequently invented contain additional solutes, such as vitamins and other minerals, and many contain higher amounts of sugar. Because of the presence of these other solutes, these drinks may be very hyperosmotic relative to body fluids. Therefore, when ingested, they initially tend to draw water out of the interstitial fluid and into the gut lumen by osmosis. This slows down the rate at which the water from the drink gets absorbed into the blood.

The story of Gatorade® is one of good common sense based on solid scientific principles of osmolarity and salt and water homeostasis. You can now understand why drinking a dilute salt solution during strenuous exercise is better than drinking water. Although drinking pure water prevents dehydration, if drunk in excess, it will actually reduce plasma salt concentrations to below normal. In other words, it will replace one type of salt imbalance with another.

Experimental Questions

1. What symptoms are sometimes seen in athletes after prolonged, strenuous exercise, particularly in hot weather? How are these symptoms related to water loss during exercise, and what did Cade and his colleagues hypothesize about this?

2. How did the researchers test their hypothesis?

3. What was the result of consuming the drink during exercise?

Animals Adapt to Osmotic Challenges by Regulating or Conforming

Animals adapt to osmotic challenges posed by the environment in one of two major ways. Some animals regulate their internal osmolarity at a very stable level, whereas others conform to the osmolarity of their environment (for example, the sea). Animals that maintain very stable internal salt concentrations and osmolarities, even when living in water with very different osmolarities than their body fluids or on land, are called **osmoregulators**. Such animals drink or excrete water and salt as necessary to maintain an internal osmolarity that is generally about 300 mOsm/L, or about one-third that of seawater and at least 10 times that of fresh water. All terrestrial animals are osmoregulators, as are all freshwater animals and many marine animals, including bony fishes and some crustaceans. Osmoregulators maintain stable cellular levels of ions and fluid, but this requires considerable expenditure of energy, primarily to pump ions into and out of epithelial cells.

Most marine invertebrates and some vertebrates—notably sharks—use a different means to control body fluid composition. In this case, the osmolarity of extracellular and intracellular fluids is matched with seawater. These animals are called **osmoconformers**, because their osmolarity conforms to that of their environment. The osmolarity of blood and other fluids of marine osmoconformers is like seawater, around 1,000 mOsm/L. An advantage of having body fluids conform to the osmolarity of the surrounding seawater is that there is much less tendency to gain or lose water by osmosis across the skin or gills. Thus, sharks and other osmoconformers expend less energy to compensate for water gain or loss than do other aquatic animals. However, osmoconformers are generally limited to the marine environment, whereas many species of osmoregulators such as salmon can migrate between waters of different salinities.

Vertebrate osmoconformers have a high concentration of uncharged molecules in their extracellular fluids. This allows the extracellular fluids and seawater to have similar osmolarities, but it prevents the excessive accumulation of ions in the body. The body fluids of sharks and other osmoconformers contain sugars, amino acids, and metabolic waste products—notably urea and an organic compound called trimethylamine oxide (TMAO). The total amount of salt and organic compounds in a shark's extracellular fluids produces an osmolarity very similar to that of seawater, even though the salt concentration is similar to that of osmoregulators.

Vertebrate osmoconformers cannot tolerate high ion concentrations in their body fluids any better than osmoregulators. One reason is because a proper ion balance is required for normal electrical signaling in their neurons and muscle cells. In addition, very high salt concentrations tend to disrupt the three-dimensional structure of many proteins, rendering them inactive. Consequently, the body fluids of vertebrate osmoconformers are less salty—that is, they have fewer ions—than seawater, as is also the case in all osmoregulators. Therefore, vertebrate osmoconformers tend to gain salt by diffusion across their gills. That excess salt is eliminated by the kidneys and a type of salt gland called the rectal gland.

Animals that cannot survive wide changes in the salinity of their surroundings—including most marine invertebrates and many types of fishes—are said to be stenohaline (from the Greek, meaning narrow salt). Other animals, such as migratory fishes and some species of shellfish (oysters and mussels), can tolerate certain changes in the salinity of their surroundings and are said to be euryhaline (from the Greek, meaning broad salt). One of the most dramatic examples of euryhaline animals is salmon, which migrate between fresh water and the sea without any ill effects. Although the mechanisms by which such animals adapt to changing salinities are not entirely understood, exposure to a sudden change in salinity is associated with large-scale changes in gene expression. A recent study in nematode worms, for example, reported that the expression of well over 100 genes was altered by changing the salt concentration of the medium on which the worms were living. Presumably, some of the genes are associated with such functions as ion transport in skin and other tissue.

Next we will examine the excretory organs found in different animals and how such organs not only function to eliminate wastes but also are involved in salt and water homeostasis.

49.2 Principles of Fluid Filtration and Waste Excretion

Most excretory organs operate by using one or more of the following processes: filtration, reabsorption, secretion, and excretion (Figure 49.6). In **filtration**, an organ acts like a sieve or filter, removing some of the water and its small solutes from the blood, interstitial fluid, or hemolymph, while retaining blood cells and large solutes such as proteins. A typical filtration system is that seen in the mammalian kidney, in which the plasma component of the blood is forced under pressure through leaky capillaries and into the kidney tubules. The material that passes through the filter and enters the excretory organ for either further processing or excretion is called a **filtrate**.

Some of the material in the filtrate can be recaptured and returned to the blood. This is an important feature of many excretory organs, because the formation of a filtrate is not selective, apart from the exclusion of proteins and blood cells. In other words, in order to filter the blood and remove soluble wastes, necessary molecules such as salts, sugars, and amino acids also get filtered in the process. Recapturing these useful solutes requires active transport pumps or other transport systems and is known as **reabsorption**. Much of the filtered water also gets reabsorbed along with useful solutes by osmosis.

In some cases, solutes may get excreted from the body in quantities greater than those found in the filtrate. How is this possible? Some solutes are actively transported from the interstitial fluid surrounding the epithelial cells of the tubules into the tubule lumens. This process, called **secretion**, supplements the amount of a solute that would normally be removed by filtration alone. This is often a way in which excretory organs eliminate particularly toxic compounds from an animal's body, and it can be very effective. Some marine fishes, for example, use secretion as the sole means of cleansing the blood. These animals do not form a filtrate at all.

Excretion is the process of expelling waste or harmful materials from the body. In animals that form a filtrate, the excreted part of the filtrate that remains after reabsorption has been completed is called **urine**. Now let's take a comparative look at some of the organ systems that perform these vital functions in different animals.

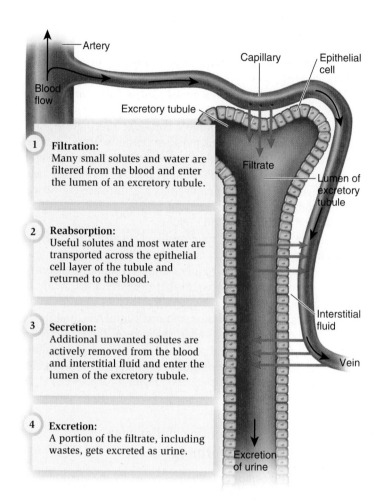

1 **Filtration:**
Many small solutes and water are filtered from the blood and enter the lumen of an excretory tubule.

2 **Reabsorption:**
Useful solutes and most water are transported across the epithelial cell layer of the tubule and returned to the blood.

3 **Secretion:**
Additional unwanted solutes are actively removed from the blood and interstitial fluid and enter the lumen of the excretory tubule.

4 **Excretion:**
A portion of the filtrate, including wastes, gets excreted as urine.

Figure 49.6 Basic features of the function of many excretory systems.

Concept check: *What is the benefit of secreting substances into the tubule?*

49.3 Comparative Excretory Systems

Although the mammalian kidney, and in particular that of humans, has been especially well studied, enough is known about other classes of animals to make general statements regarding the regulation of salt, water, and waste levels in the fluid compartments of an animal's body. This is an ancient and important function that arose early in evolution.

As described, animals make use of one or more different organs to rid themselves of metabolic wastes, excess water and salts, and toxins from their environment. Most excretory organs contain tubular structures lined with epithelial cells that have the capacity to actively transport ions across their membranes. Wastes are excreted out of the body by means of these tubes.

In some cases, animals may have considerable ability to regulate the rate at which waste is secreted and how much water is lost in the process. For example, even though a thirsty mammal on a hot, sunny day must continue to rid its body of soluble waste products, it must also conserve water. In this section, we consider invertebrate and vertebrate excretory organs.

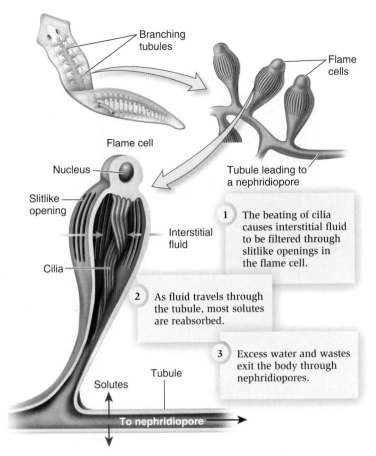

Figure 49.7 **The protonephridial filtration system of flatworms.** As the filtrate moves along the tubules, the composition of the filtrate can be changed by the action of the tubule epithelial cells. The final excreted fluid is typically hypoosmotic to body fluids.

Most Invertebrates Use a Filtration Mechanism to Cleanse the Blood

The simplest filtration mechanism in invertebrates is the **protonephridia** system of flatworms (**Figure 49.7**). A series of branching tubules filters fluids from the body cavity into the tubule lumens by means of ciliated cells that cap the ends of the tubule branches. The beating of the cilia bears some resemblance to a flickering flame, which is why these cells are known as flame cells. As fluid is drawn through slitlike openings of the flame cells and into the lumen, it percolates through the tubule, where most solutes are reabsorbed back into the interstitial fluid. Excess water and some wastes travel through the tubules and are emptied through tiny openings in the body wall called nephridiopores. Much of the nitrogenous waste in flatworms actually diffuses across the body surface into the surrounding water; therefore, the protonephridia are primarily osmoregulatory organs. The urine is generally hypoosmotic compared to the rest of the body fluids, an adaptation for life in fresh water.

Annelids use a different filtration mechanism, called a **metanephridial system** (**Figure 49.8**). In annelids, pairs of metanephridia are located in each body segment. They consist of a tubular network that begins with a funnel-like structure called a nephrostome. The nephrostomes are open to the body cavity and collect coelomic fluid, which contains nitrogenous wastes,

through tiny pores that exclude large solutes. Na⁺, Cl⁻, and other solutes are reabsorbed by active transport along the length of the tubules that extend from the nephrostomes, and from there diffuse into nearby capillaries. The nitrogenous wastes remain behind in the tubules and are excreted through nephridiopores in the body wall. Many annelids live in watery environments and thus, like flatworms, excrete a hypoosmotic urine.

Filtration mechanisms also operate in the metanephridia of many mollusks and in the crustacean excretory organs called antennal glands. As in worms, reabsorption of useful solutes occurs along the length of the excretory system. Research has clearly indicated that mollusks can reabsorb organic nutrients well enough that under most conditions, these nutrients do not appear at all in the urine. As in the protonephridial and metanephridial systems just discussed, urine is excreted via nephridiopores in mollusks and crustaceans.

Insects Excrete Wastes by Means of Secretory Organs Rather Than Filtration

The insect excretory system is quite different from other invertebrates, because it involves secretion rather than filtration of body fluids. In insects, a series of narrow, extensive tubes called **Malpighian tubules** arises from the midgut and extends into the surrounding hemolymph (**Figure 49.9**). The cells lining the tubules actively transport potassium ions, along with uric

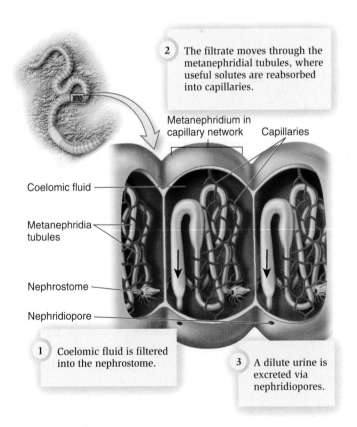

Figure 49.8 **The metanephridial filtration system of annelids.** Most internal body structures have been omitted for clarity. Only one of the two metanephridia in each segment is shown.

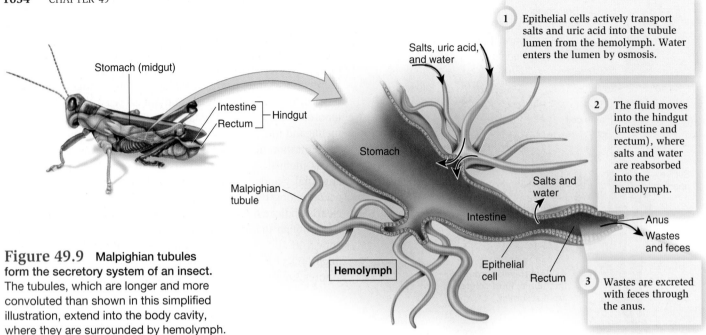

Figure 49.9 Malpighian tubules form the secretory system of an insect. The tubules, which are longer and more convoluted than shown in this simplified illustration, extend into the body cavity, where they are surrounded by hemolymph.

acid and other purines from the hemolymph, into the tubule lumen. This secretion process creates an osmotic gradient that draws water into the tubules. The fluid moves from the tubules into the hindgut—the intestine and rectum—where much of the useful salt and water is reabsorbed. The nitrogenous wastes and other waste compounds are excreted together with the feces through the anus.

Unlike other invertebrates, most terrestrial insects, apart from blood-sucking ones, excrete urine that is either isoosmotic or hyperosmotic to body fluids. This is a testament to the efficiency with which the insect hindgut reabsorbs water, and reflects the fact that life in dry environments is associated with a risk of dehydration.

The Kidney Is the Major Excretory and Filtration Organ in Vertebrates

The major excretory organ found in all vertebrates is the **kidney**. The kidneys of all vertebrates have many features in common. They typically contain specialized tubules composed of epithelial cells that participate in both salt and water homeostasis by promoting active transport of sodium, potassium, and other ions across their membranes. In addition, all kidneys participate in the excretion of wastes. In response to an animal's changing salt and water requirements, these processes can be controlled, that is, sped up or slowed down, by the actions of nerves and hormones. Most vertebrate kidneys are filtration kidneys, with the exception of purely secretory kidneys found in some marine fishes. Finally, filtration in the kidneys is controlled by mechanical forces, such as the hydrostatic pressure exerted by blood entering the capillaries of the kidneys. The mammalian kidney has been especially well studied and is examined in detail later in this chapter. Note that when we refer to kidneys, we use the adjective renal, meaning "pertaining to the kidneys." For example, we refer to renal physiology and renal functions.

The need to eliminate waste products while simultaneously maintaining salt and water homeostasis occurs in all animals. How animals achieve this balance in highly disparate environments, and with different life histories, is the subject of our next discussion.

49.4 Renal Function and Vertebrate Life History

As previously mentioned, not all animals have similar requirements for salt and water. **Table 49.1** summarizes some of the special requirements of different vertebrates and some major behavioral and physiological adaptations that allow them to maintain homeostasis.

The demands placed on an animal by its environment can often be predicted by examining the activities of that animal's kidneys. For example, the kidneys of freshwater fishes are specially adapted for rapid, large-scale filtering of the blood and producing a very dilute urine. Amphibians, because of their permeable skin, absorb fresh water from their environments, and consequently they have kidneys with an appearance and activity that resemble those of freshwater fishes. By contrast, the kidneys of marine fishes and desert mammals are adapted to produce urine that is more concentrated than that of freshwater fishes and amphibians. This is an important adaptation that helps conserve as much water as possible.

Some animals have a diet that is extreme in one way or another. This is often associated with differences in renal structure and function. Many mammals and insectivores, for example, subsist entirely or almost entirely on sporadic, high-protein meals. These animals consume so much protein at a single meal that the rate at which they generate nitrogenous waste is very high. Not surprisingly, perhaps, the rate at which blood is filtered through the kidneys of such animals tends also to be very high, an adaptation that helps eliminate the toxic waste.

At the other extreme, polar bears may go 4 to 5 months without feeding. During the nonfeeding periods, polar bears neither defecate nor urinate and subsist by breaking down stores of fat and protein in their bodies. Thus, even though they are not eating, they generate nitrogenous wastes such as urea. Despite not urinating during this time, the concentration of urea in their blood may actually decrease! The explanation for this surprising phenomenon appears to be that polar bears have the extraordinary ability to recycle nitrogenous waste into synthesis of new protein. This ability is of great interest to researchers who study the health consequences of kidney disease and consequent accumulation of urea in the bloodstream of humans.

We turn now to an examination of one of the best understood excretory organs: the mammalian kidney.

Table 49.1	A Comparison of the Vertebrate Mechanisms of Osmoregulation in Different Environments				
Animal group		Blood concentration relative to environment	Urine concentration relative to blood	Main nitrogenous waste	Osmoregulatory mechanisms and special features
Marine fish		Hypoosmotic	Isoosmotic or hyperosmotic	Ammonia	Drinks seawater; secretes salt from gills
Freshwater fish		Hyperosmotic	Strongly hypoosmotic	Ammonia	Drinks no water; transports salt across gills
Amphibian (fresh water)		Hyperosmotic	Strongly hypoosmotic	Most secrete urea	Absorbs water through skin
Marine reptile		Hypoosmotic	Isoosmotic	Ammonia, urea, uric acid	Drinks seawater; hyperosmotic salt-gland secretion
Terrestrial reptile		—	Isoosmotic	Uric acid	Drinks fresh water
Marine bird		Hypoosmotic	Weakly hyperosmotic	Uric acid	Drinks seawater; hyperosmotic salt-gland secretion
Terrestrial bird		—	Weakly hyperosmotic	Uric acid	Drinks fresh water
Desert mammal		—	Strongly hyperosmotic	Urea	Drinks no water; very long loops of Henle; depends on water generated by cellular metabolism
Marine mammal		Hypoosmotic	Strongly hyperosmotic	Urea	Drinks no water
Other mammals		—	Strongly hyperosmotic	Urea	Drink fresh water

49.5 Structure and Function of the Mammalian Kidney

All mammals have kidneys that are similar in structure and function, although some are more adapted to one type of life history and environment than another. Much of what we know today about the human kidney arose from observations made on the kidneys of rodents, dogs, and other mammals.

The two kidneys lie in the abdomen (**Figure 49.10a**). The urine formed in each kidney collects in a central area called the renal pelvis. From there it flows through tubes called the **ureters** into the **urinary bladder**. Urine is eliminated via the **urethra**, in a process called micturition, or urination. Collectively, the kidneys, ureters, urinary bladder, and urethra constitute the **urinary system** in humans.

Each kidney is composed of an outer portion called the renal cortex and an inner portion called the renal medulla, which, in turn, is composed of outer and inner regions (**Figure 49.10b**). The cortex is the primary site of blood filtration. In the medulla, the filtrate becomes concentrated by the reabsorption

of water back into the blood. In this section, we will examine the structural features of the kidney that allow it to function as a filtration system.

The Functional Units of the Kidney Are Called Nephrons

Depending on its size, the mammalian kidney contains as many as several million similar, single-cell–thick structures called **nephrons**. (We get the name of the medical specialty nephrology from the word nephron.) As shown in **Figure 49.10c**, each nephron consists of an initial filtering component called the **renal corpuscle** and a narrow tubule that extends out from the renal corpuscle. This tubule empties into a larger tubule called a **collecting duct**.

The renal corpuscle forms a filtrate from blood that is free of cells and proteins. This filtrate then leaves the renal corpuscle and enters the lumen of the tubule. As it flows through different regions along the length of the tubule, substances are reabsorbed from it or secreted into it. Ultimately the filtrate remaining at the end of each nephron combines in the collecting

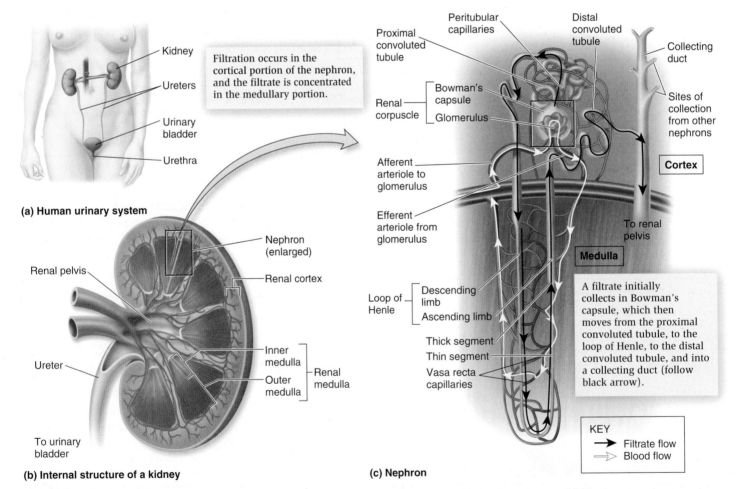

(a) Human urinary system

Filtration occurs in the cortical portion of the nephron, and the filtrate is concentrated in the medullary portion.

(b) Internal structure of a kidney

(c) Nephron

A filtrate initially collects in Bowman's capsule, which then moves from the proximal convoluted tubule, to the loop of Henle, to the distal convoluted tubule, and into a collecting duct (follow black arrow).

KEY
→ Filtrate flow
⇨ Blood flow

Figure 49.10 The mammalian urinary system, including the basic functional unit of the nephron. **(a)** The human urinary system in a woman. In the male, the urethra passes through the penis. **(b)** Enlarged view of a section through a kidney, showing the locations of the major internal structures and a single nephron (enlarged; at the scale of this illustration, nephrons would be microscopic). **(c)** Structure of a nephron. The nephron begins at Bowman's capsule and empties into a collecting duct. Many nephrons empty into a given collecting duct. Surrounding the nephron are capillaries, called peritubular capillaries in the cortex and vasa recta capillaries in the medulla.

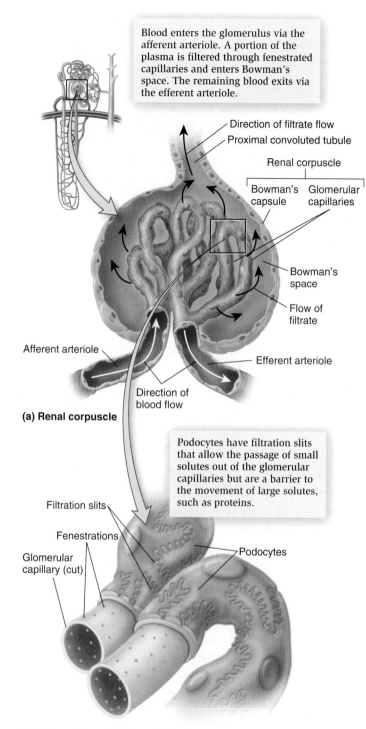

Blood enters the glomerulus via the afferent arteriole. A portion of the plasma is filtered through fenestrated capillaries and enters Bowman's space. The remaining blood exits via the efferent arteriole.

Direction of filtrate flow
Proximal convoluted tubule
Renal corpuscle
Bowman's capsule   Glomerular capillaries
Bowman's space
Flow of filtrate
Afferent arteriole
Efferent arteriole
Direction of blood flow

(a) Renal corpuscle

Podocytes have filtration slits that allow the passage of small solutes out of the glomerular capillaries but are a barrier to the movement of large solutes, such as proteins.

Filtration slits
Fenestrations
Glomerular capillary (cut)
Podocytes

(b) Glomerular capillaries with podocytes

Figure 49.11 The structure and function of the renal corpuscle.

ducts. Let's look more closely at each part of the nephron and its associated structures.

The Renal Corpuscle Each renal corpuscle contains a cluster of capillaries called the **glomerulus** (plural, glomeruli), or glomerular capillaries (**Figure 49.11a**). Each glomerulus is supplied with blood under pressure by an **afferent arteriole**, and blood exits the glomerulus via an **efferent arteriole**. The glomerulus protrudes into a fluid-filled space called Bowman's space, which lies within a capsule called **Bowman's capsule** (first identified

by the English physiologist William Bowman in 1841). The combination of a glomerulus and a Bowman's capsule constitutes a renal corpuscle. The glomerular capillaries contain fenestrations, tiny holes that permit rapid flow of plasma out of the capillaries. This filtration is modified by cells called podocytes that surround the capillaries and form filtration slits that allow the passage of small solutes, but are believed to help prevent proteins from entering the filtrate (**Figure 49.11b**).

The Tubular Part of the Nephron The tubule of the nephron, which is continuous with Bowman's capsule, is made of a single layer of epithelial cells resting on a basement membrane. The epithelial cells differ in structure and function along the tubule's length, and several distinct segments are recognized. The segment of the tubule that drains Bowman's capsule is the **proximal convoluted tubule** ("proximal" means adjacent to, with respect to the Bowman's capsule) (see Figure 49.10c). The next portion of the tubule is the **loop of Henle**, discovered by the German physician Friedrich Henle in 1861. It is a long, hairpin-like loop consisting of a descending limb coming from the proximal tubule and an ascending limb. The ascending limb has two segments, a thin and a thick segment. The thick segment leads to the next tubular segment, the **distal convoluted tubule** ("distal" means away from the Bowman's capsule). Fluid flows from the distal convoluted tubule into one of the many collecting ducts in the kidney.

Capillaries of the Nephron All along its length, each tubule is surrounded by capillaries. These include the **peritubular capillaries** in the cortex and the **vasa recta capillaries** in the medulla (see Figure 49.10c). Both sets of capillaries carry away reabsorbed solutes and water from the filtrate in the nephron and return them to the bloodstream. However, only the peritubular capillaries secrete solutes into the tubules.

Now that we have learned about the structural units of the mammalian kidney, we turn our attention to the mechanisms by which nephrons filter blood and produce urine for excretion and how these activities are controlled under different conditions.

Filtration: A Filtrate Is Produced in Bowman's Space

Filtration begins as blood flows through the glomerulus and a portion of the plasma leaves the glomerular capillaries and filters into Bowman's space. Only about 15–20% of the plasma is filtered from the blood as it circulates through the glomerulus. Most of the blood, therefore, exits the glomerulus by an efferent arteriole (see Figure 49.11a). Proteins and blood cells are prevented from leaving the glomerular capillaries because of the small diameter of the fenestrations and the presumed actions of the filtration slits mentioned earlier. The fluid that enters Bowman's space is called the glomerular filtrate. The rate at which the filtrate is formed is called the **glomerular filtration rate** (**GFR**). GFR can be increased by dilation (widening) of the afferent arteriole. When the afferent arteriole dilates, more blood enters the glomerulus, increasing the hydrostatic pressure in those capillaries and forcing more plasma through the fenestrations in the glomerular capillaries and into Bowman's space. This might happen, for example, when there is an excess of water in the body that must be excreted in the urine.

By contrast, constriction of the afferent arteriole would reduce the amount of blood entering the glomerular capillaries and therefore would reduce GFR. This might occur following a loss of blood due to a severe injury. In such a scenario, reducing GFR results in less urine production, which, in turn, minimizes how much water is lost from the body and helps compensate for the blood lost due to the injury.

Reabsorption and Secretion: Useful Solutes Are Reabsorbed from, and Toxic Solutes Are Secreted into, the Filtrate in the Proximal Tubule

The filtrate flows from the renal corpuscle to the proximal convoluted tubule. Anywhere from two-thirds to 100% of a particular useful solute is reabsorbed from the filtrate in the proximal tubule. This includes Na^+, K^+, Cl^-, HCO_3^-, and organic molecules such as glucose and amino acids. Some ions diffuse through channels in the membranes of the epithelial cells that form the single-cell–thick proximal tubule. Others are actively transported across the tubule. Organic molecules generally are reabsorbed by being coupled to transport of ions such as Na^+. The absorption of solutes and water is enhanced by microvilli that extend from the luminal surface of the epithelial cells of the proximal tubule. This anatomical adaptation, called a brush border, creates an enormous surface area for transporters and channels to be localized (**Figure 49.12**).

Most of the water in the filtrate leaves by osmosis as the ions and organic molecules are transported from the lumen of the proximal tubule to the interstitial fluid outside the tubules. The amount of water reabsorbed is proportional to the amount of solutes reabsorbed. From the interstitial fluid, the solutes and water enter peritubular capillaries to return to the blood.

Many of the reabsorptive mechanisms in the renal tubule have a limit to the amounts of material they can transport in a given amount of time—called the transport maximum (T_m). This occurs because the binding sites on the membrane-transport proteins become saturated with their ligands. Occasionally, certain molecules may reach such high concentrations in the blood that they exceed the T_m for that molecule in the renal tubule and consequently get excreted in the urine. For example, people who ingest very large quantities of water-soluble vitamins, such as vitamin C, have increased plasma concentrations of vitamin C. Eventually, the filtered load may exceed the tubular reabsorptive T_m for this vitamin, and any additional ingested vitamin C is excreted in the urine.

In addition to reabsorption, secretion also occurs in the proximal tubule. Some solutes that are either not required by an animal or that are potentially toxic at high concentrations are secreted by active transport mechanisms from the extracellular fluid into this tubule. Examples of such substances include foreign compounds (for example, penicillin), naturally occurring toxins, nucleoside metabolites, and ions such as K^+ and H^+. These solutes are excreted in the final urine.

Tubule lumen

Figure 49.12 Electron micrograph of cuboidal epithelial cells of the proximal convoluted tubule of a rat nephron. Note the extensive brush border microvilli on the luminal surface of the cells.

Concept check: *In what other organ have you learned about a brush border? Can you draw general conclusions about the function of such specialized epithelia?*

By the time the filtrate leaves the proximal tubule, its volume and composition have changed considerably. The amount of salts and water it contains is much reduced, and the organic molecules have all been removed. However, the final osmolarity of the filtrate is still about the same as blood.

Reabsorption: Water and Salt Continue to Be Reabsorbed from the Filtrate by a Countercurrent Osmotic Gradient Along the Loop of Henle

In the loop of Henle in the mammalian kidney, filtrate moves down the descending limb of the loop, makes a U-turn at the bottom, and then moves back up the ascending limb of the loop. The permeabilities and transport characteristics of the epithelial cells lining the loop change over its length as it descends from the cortex into the medulla and then ascends to the cortex again (**Figure 49.13**).

As we have seen, the filtrate that leaves the proximal convoluted tubule has had some solutes and water reabsorbed from it; the osmolarity of the initial filtrate is about 300 mOsm/L. This filtrate enters the descending limb of the loop of Henle, which is very permeable to water but not to solutes. Water leaves the filtrate by osmosis in this region because the surrounding interstitial fluid is hyperosmotic compared to the tubule contents. The hyperosmolarity of the interstitial fluid originates from three sources. First, the initial upturn of the thin segment of the ascending limb of the loop of Henle is very permeable to sodium and chloride but not to water. Therefore, these ions diffuse at high rates out of the loop into the interstitial fluid, significantly increasing the osmolarity of the inner medulla. Second, the epithelial cells of the thick segment of the ascending limb of the loop of Henle actively transport some of the remaining sodium and chloride ions out of the filtrate and into the interstitial fluid of the outer medulla. Third, although urea is a waste product that gets excreted, some of the urea that is present in the filtrate does not get excreted in the urine, but instead diffuses out of the lower ends of the collecting ducts and into the inner medulla interstitial fluid. Collectively, these solutes create the osmotic force that draws water out of the

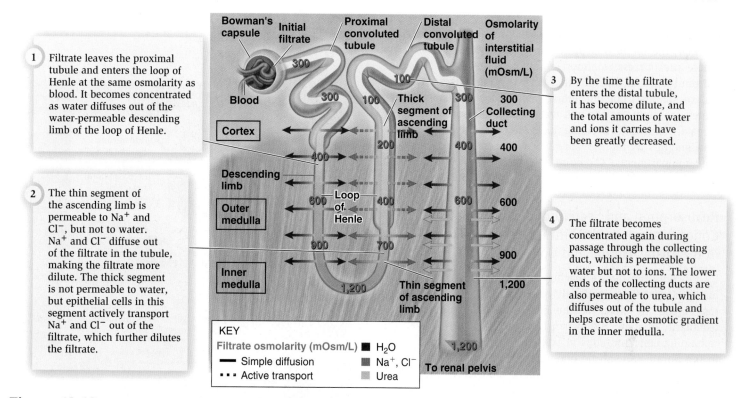

1 Filtrate leaves the proximal tubule and enters the loop of Henle at the same osmolarity as blood. It becomes concentrated as water diffuses out of the water-permeable descending limb of the loop of Henle.

2 The thin segment of the ascending limb is permeable to Na^+ and Cl^-, but not to water. Na^+ and Cl^- diffuse out of the filtrate in the tubule, making the filtrate more dilute. The thick segment is not permeable to water, but epithelial cells in this segment actively transport Na^+ and Cl^- out of the filtrate, which further dilutes the filtrate.

3 By the time the filtrate enters the distal tubule, it has become dilute, and the total amounts of water and ions it carries have been greatly decreased.

4 The filtrate becomes concentrated again during passage through the collecting duct, which is permeable to water but not to ions. The lower ends of the collecting ducts are also permeable to urea, which diffuses out of the tubule and helps create the osmotic gradient in the inner medulla.

KEY
Filtrate osmolarity (mOsm/L)   ■ H_2O
— Simple diffusion   ■ Na^+, Cl^-
••• Active transport   ■ Urea

Figure 49.13 Tubule permeabilities and concentrations of the filtrate in the nephron and collecting duct.

filtrate in the descending limb. The water then enters local capillaries and rejoins the blood circulation.

As water diffuses out of the filtrate in the descending limb of the loop of Henle, the osmolarity of the filtrate increases from 300 to about 1,200 (or higher in some species) mOsm/L. During its passage up the ascending limb, however, the osmolarity of the filtrate decreases to about 200 mOsm/L as ions diffuse out and are transported out of the tubule. The thin segment of the ascending limb is permeable to Na^+ and Cl^-, but not water, so Na^+ and Cl^- diffuse out of the filtrate in the tubule. This begins to dilute the filtrate. The thick segment is also not permeable to water, but epithelial cells in this segment actively transport Na^+ and Cl^- out of the filtrate, further diluting the filtrate.

As a consequence of ion movement out of the ascending limb, the osmolarity of the kidney interstitial fluid increases from cortex to inner medulla. This extracellular osmolarity gradient is what allows water to diffuse by osmosis from the descending limb of the loop of Henle all along its length. In other words, this is an example of a countercurrent exchange system, like those that operate in heat and gas exchange in some animals (see Chapters 46 and 48). A major difference between those countercurrent exchange systems and the one in the loop of Henle, however, is that the latter requires energy-dependent ion pumps to maintain the necessary gradient. Energy-requiring systems such as that in the kidney are referred to as countercurrent multiplication systems, because energy is used to increase—or multiply—the gradient. Energy is not used to establish a heat or oxygen gradient in those other countercurrent exchange systems.

The chief advantage to an animal provided by the loop of Henle is that the final volume of urine produced can be reduced and its contents concentrated by the recapture of water along the osmotic gradient. This is especially important in animals in which total body water stores are regularly in danger of being depleted. For example, desert mammals such as the kangaroo rat tend to have longer loops of Henle than other mammals. The extra length of the loop provides for a very large osmotic gradient in the medulla and, therefore, more water-reabsorbing capacity. At the other extreme, animals that generally must eliminate excess water have reduced or absent loops of Henle. Freshwater fishes, for example, have no loops of Henle in their nephrons. In their case, it is advantageous to excrete as much water as possible to compensate for the large amounts of water constantly entering the body by osmosis across the gills.

Reabsorption: The Vasa Recta Capillaries Help to Maintain the Medullary Osmotic Gradient and So Minimize the Loss of Useful Solutes from the Filtrate

Why does the blood flowing through the vasa recta capillaries of the medulla not eliminate the countercurrent osmotic gradient set up by the loops of Henle? One would think that as plasma, having the usual osmolarity of ~300 mOsm/L, enters the highly concentrated environment of the medulla, two types of massive net diffusion or osmosis would occur: of sodium and chloride into the capillaries and of water out of them. Thus, the interstitial gradient would be "washed away." However, this does not happen because the vasa recta form hairpin loops that run parallel to the loops of Henle and medullary collecting ducts (see Figure 49.10c). Near the top of the loop of Henle, blood in the vasa recta has a normal osmolarity. As the blood flows down the loop deeper and deeper into the inner medulla, sodium and chloride do indeed diffuse into, and water moves out of, the vasa recta. However, after the bend in the loop is reached, the blood then flows up the ascending vessel loop, where the process is almost completely reversed. Thus, the hairpin-loop structure of the vasa recta minimizes excessive loss of solutes from the interstitial fluid by diffusion. At the same time, both the salt and water being reabsorbed from the loops of Henle and collecting ducts are carried away in equivalent amounts by the movement of water and solutes between fluid compartments. Therefore, the steady-state countercurrent gradient set up by the loops of Henle is maintained.

Reabsorption and Secretion: Salt and Water Concentrations of the Filtrate Are Fine-Tuned in the Distal Tubule and Collecting Duct

By the time the filtrate reaches the distal convoluted tubule, most of the reabsorbed salt and water have already been restored to the blood, so the filtrate has been diluted to an osmolarity of about 100 mOsm/L (Figure 49.13). However, the remaining concentrations of sodium, chloride, and potassium in the filtrate can still be fine-tuned. This process is mediated by the actions of two hormones called aldosterone and antidiuretic hormone (ADH).

Aldosterone and Sodium Reabsorption and Potassium Secretion Aldosterone acts on the basolateral surface of epithelial cells of the distal tubule and upper part of the collecting duct. There it stimulates the active transport of three molecules of Na^+ out of the filtrate in the tubule into the interstitial fluid (reabsorption) for every two molecules of K^+ it pumps into cells from the interstitial fluid and ultimately into the filtrate (secretion) (Figure 49.14). Water from the filtrate follows the Na^+ by osmosis into the interstitial fluid and blood. This makes the filtrate that moves to the lower part of the collecting duct a bit more concentrated than it was before, with an osmolarity of about 400 mOsm/L. Aldosterone levels increase in the blood whenever the Na^+ concentration of the blood is lower than normal or the K^+ concentration is higher than normal. Through its actions on the nephron, aldosterone corrects such imbalances.

ADH and Water Reabsorption The lower part of the collecting duct is the final place where urine composition can be altered. The osmolarity of the filtrate increases further during passage along the collecting duct, which is permeable to water but not to ions, allowing water to diffuse out of the tubules by osmosis. As previously described, the cells of the collecting duct in the inner medulla are permeable to urea, which contributes to the osmotic gradient there.

In addition, however, the permeability of the epithelial cells of the collecting ducts to water (but not ions) can be regulated, depending on the body's requirement at that moment for retaining or excreting water. This happens under the influence of **anti<u>di</u>uretic <u>hormone</u>** (**ADH**). When ADH is present in the blood, it acts to increase the number of water channels in the membranes of the collecting duct cells. Because the collecting ducts travel through the hyperosmotic medulla of the kidney, an osmotic gradient draws water from the filtrate in the duct into

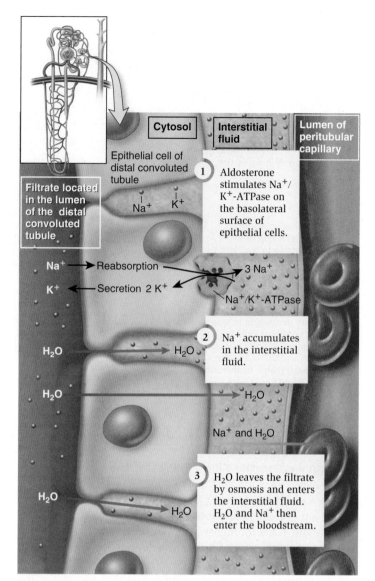

Figure 49.14 **Action of aldosterone on distal convoluted tubule epithelial cells.** Aldosterone stimulates the Na^+/ K^+-ATPase activity of distal tubule cells on the side facing the interstitial fluid (the basolateral membrane). This activity creates an osmotic gradient, as three Na^+ are reabsorbed from the filtrate for every two K^+ secreted into it. Water then leaves the tubule by osmosis. The net effect is reabsorption of Na^+ and water, and secretion of K^+, which gets excreted in the urine. Na^+ and water enter the peritubular capillaries and are carried away in the bloodstream.

Concept check: *What special property of epithelial cells is demonstrated by the action of aldosterone?*

the surrounding interstitial fluid, from where it enters the vasa recta capillaries and the blood. In this way, as the filtrate travels through the collecting ducts, it becomes greatly concentrated (hyperosmotic) compared to blood—as much as four to five times more concentrated in humans, for example. The ability of the kidneys to produce hyperosmotic urine is a major determinant of an animal's ability to survive in conditions where water availability is limited. When the body's stores of water are plentiful, the level of ADH in the blood decreases. This, in turn, decreases the water permeability of the collecting ducts. In such a case, urine increases in volume and becomes more dilute because less water diffuses out of the collecting ducts.

Biological membranes such as the plasma membranes of animal cells are composed of a lipid bilayer that inhibits the movement of water. The mystery of how water moves rapidly across membranes like those of the collecting ducts and loop of Henle was solved by the discovery in the early 1990s of a family of molecules that came to be known as aquaporins (from the Latin, meaning water pores), as described next.

Genomes & Proteomes Connection

Water Channels Called Aquaporins Comprise a Large Family of Proteins That Are Ubiquitous in Nature

In the early 1990s, the American physiologist Peter Agre and colleagues discovered the first of what would eventually be recognized as a new family of proteins, called aquaporins, a discovery that earned Agre the Nobel Prize in Chemistry in 2003 (see Chapter 5). Scientists now know that aquaporins are a subfamily of an even larger family of proteins with membrane transport capabilities, found in all kingdoms of living organisms. The functions of all the members of this family are not yet known, but some aquaporins can transport other small molecules, such as glycerol and urea. There are currently 11 known members of the aquaporin family in the human genome.

Aquaporins are proteins with six transmembrane domains and two short loops in the membrane (**Figure 49.15**). In animals, the two short loops come together to form the three-dimensional core of the water channel. The importance of the loops is reflected in the observation that these portions are the most highly conserved sequences of the protein among different species. Water must pass through a zone of constriction, created by the loops, that reduces the channel opening to about 30 picometers (30×10^{-12} meters), or just about the width of a water molecule. Scattered along the inner part of the channel are arginine amino acids, which are positively charged. The charged arginines participate in hydrogen bonding with water molecules, facilitating their single-file movement through each channel at rates that have been estimated to be up to billions of water molecules per second!

Within the various extracellular and intracellular domains of aquaporins are sites that can be modified by enzymes, such as kinases. This suggests that the opening and closing of these channels may be gated by stimuli, like the way ion channels are gated in neurons and other cells. In addition, the promoter region of certain aquaporin genes contains a site that is recognized by transcriptional activator proteins that are responsive to the presence of cAMP, a common intracellular signaling molecule and one that is generated by cells stimulated by ADH. Thus, one mechanism by which ADH promotes osmosis of water out of the renal collecting ducts appears to be by stimulating the transcription of one or more aquaporin genes.

Our understanding of aquaporin function has allowed us to explain the molecular basis of one form of an inherited human disease called hereditary nephrogenic diabetes insipidus. Patients with this disease are unable to produce a concen-

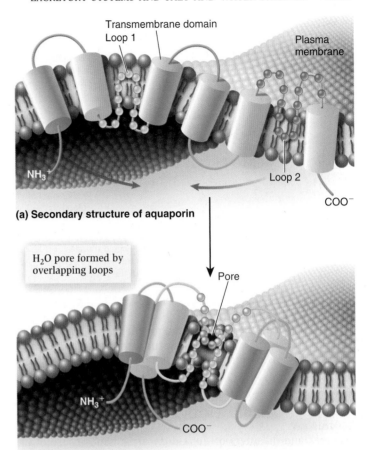

(a) Secondary structure of aquaporin

(b) Three-dimensional (tertiary) structure of aquaporin

Figure 49.15 **Detailed aquaporin structure.** All proteins of the aquaporin family share a similar structure, with six membrane-spanning domains (represented as cylinders) and two loops that come together to form a water channel. **(a)** Secondary structure of aquaporin. This highly schematic representation highlights the two separate regions of the molecule that come together to form the water pore. **(b)** Tertiary structure of aquaporin showing pore formation.

trated urine and consequently lose large amounts of water. A mutation in one aquaporin gene results in a form of the protein expressing any of several abnormalities: improper folding into its correct shape, impaired ability of the molecule to enter the plasma membrane, or impaired ability of the molecule to form a channel core. In addition, Agre's discovery may have widespread implications for other areas of biology and human health. For example, certain types of plant disease are associated with abnormal aquaporin expression in roots. In addition, some scientists are investigating the possibility that drugs that inhibit bacterial aquaporins may someday be useful antibiotics. The use of drugs to inhibit one class of aquaporins has even been suggested as a possible antiperspirant!

The mammalian kidney is a remarkably efficient organ serving many functions, most of them related to various aspects of homeostasis. Because of its central role in homeostasis, it should not be surprising that kidney damage or disease can have devastating consequences. In the final section, we consider kidney diseases in humans.

Figure 49.16 **Simplified diagram of hemodialysis.** The dialyzer is composed of many strands of very thin, sievelike tubing. In the dialyzer, blood within the dialysis tubing and the dialysis fluid bathing the tubing move in opposite directions (a countercurrent). The dialyzer provides a large surface area for diffusion of waste products out of the blood and into the dialysis fluid.

49.6 Impact on Public Health

Diseases and disorders of the kidney are a major cause of illness in the human population. According to statistics released by the Centers for Disease Control and Prevention, up to 4.5% of the U.S. population—or about 13 million people—suffers from one form of kidney disease or another, with approximately 15,000 individuals requiring kidney transplants each year. This section gives an overview of kidney diseases and disorders in humans and also discusses some of the available treatments for these conditions.

Kidney Damage and Disease Are Caused by Many Factors and Result in Disruption of Homeostasis

Many diseases affect the kidneys. Diabetes, bacterial infections, allergies, congenital defects, kidney stones (accumulation of mineral deposits in nephron tubules), tumors, and toxic chemicals are some possible sources of kidney damage or disease. A buildup of pressure due to obstruction of the urethra or a ureter may damage one or both kidneys and increase the likelihood of a bacterial infection.

The symptoms of renal malfunction are similar regardless of the cause of the disease, and all stem from the condition known as **uremia**—from the Greek, meaning urine in the blood. Assuming that a person with diseased kidneys continues to ingest a normal diet containing the usual quantities of nutrients and electrolytes, what problems might arise? Potentially toxic waste products that would normally enter the nephron tubules by filtration instead build up in the blood, because kidney damage significantly reduces the number of functioning nephrons. In addition, the excretion of potassium is impaired because too few nephrons remain capable of normal tubular secretion of

this ion. Increased K^+ in the blood is an extremely serious condition, because of the importance of stable extracellular concentrations of K^+ in the control of heart and nerve function.

The kidneys are still able to perform their homeostatic functions reasonably well as long as at least 10% of the nephrons are functioning normally. The remaining nephrons undergo alterations in function—filtration, reabsorption, and secretion—to compensate for the missing nephrons. For example, each remaining nephron increases its rate of potassium secretion so that the total amount of potassium excreted by the kidneys can be maintained at normal levels. The kidneys' regulatory abilities are limited, however. If, for example, someone with severe renal disease were to eat a diet high in potassium, the remaining nephrons might not be able to secrete enough potassium to prevent its concentrations from increasing in the extracellular fluid.

Kidney Disease May Be Treated with Hemodialysis and Transplantation

Diseased kidneys may eventually reach a point where they can no longer excrete and reabsorb water and ions at rates that maintain salt and water homeostasis, nor excrete waste products as fast as they are produced. Adjusting a person's diet can help reduce the severity of these problems; for example, lowering potassium intake reduces the amount of potassium to be excreted. However, such alterations may not eliminate the problems. In that case, doctors must use various procedures to artificially perform the kidneys' excretory functions.

The most important of these procedures is **hemodialysis**. The general term dialysis means to separate substances in solution using a porous membrane. An artificial kidney is an apparatus that removes excess substances from the blood by hemodialysis (**Figure 49.16**). During this procedure, blood from one of the patient's arteries flows into the dialysis machine, which is called a dialyzer. Within the dialyzer, blood flows through cellophane tubing that is surrounded by a special dialysis fluid. The tubing is highly permeable to most solutes but relatively impermeable to protein and completely impermeable to blood cells. These characteristics are designed to be quite similar to those of the body's own capillaries. The dialysis fluid has ion concentrations similar to those in normal plasma but contains no urea or other substances that are to be completely removed from the plasma.

As blood flows through the tubing in the dialyzer, small solutes diffuse out into the dialysis fluid until an equilibrium is reached. If, for example, the patient's plasma potassium concentration is above normal, potassium diffuses out of the blood across the cellophane tubing and into the dialysis fluid. Similarly, waste products and excess amounts of other substances also diffuse into the dialysis fluid and thus are eliminated from the body. Note in Figure 49.16 that blood and dialysis fluid flow in opposite directions through the dialyzer. This establishes an artificial countercurrent exchange system that increases the efficiency with which the blood is cleansed. The dialyzed, purified blood is then returned to one of the patient's veins through another type of tubing that leaves the dialyzer.

Some patients with reversible, temporary forms of kidney disease may require hemodialysis for only days or weeks. However, patients with chronic, irreversible kidney disease require treatment for the rest of their lives, unless they receive a kidney transplant. Such patients undergo hemodialysis several times a week. Each year nearly 300,000 Americans undergo some type of dialysis.

The treatment of choice for most patients with permanent kidney disease is kidney transplantation. Rejection of the transplanted kidney by the recipient's body is a potential problem with transplants, but great strides have been made in reducing the frequency of rejection. Many people who might benefit from a transplant, however, do not receive one, because the number of people needing a transplant exceeds the number of donors. Currently, the major source of kidneys for transplanting is from recently deceased persons. Improved public understanding may lead many more individuals to give permission to have their kidneys and other organs used following their death. Recently, donation from a living, related donor has become more common, particularly with improved methods for preventing rejection. As noted earlier in this chapter, the mammalian kidney can perform its functions with only a fraction of its nephrons intact. Due to this large safety factor, a person who donates one of his or her kidneys can function quite normally with only one kidney.

▌ Summary of Key Concepts

49.1 Principles of Homeostasis of Internal Fluids

- Exchanges of salt and water with the environment resulting from vital processes, such as respiration or the elimination of wastes, are called obligatory exchanges. (Figure 49.1)

- Among the important products of the breakdown of proteins and nucleic acids are nitrogenous wastes, molecules that include nitrogen from amino groups (NH_2). Most aquatic animals produce ammonia (NH_3) and ammonium ions (NH_4^+), which are the most highly toxic nitrogenous wastes but require no energy to produce. Many animals, including all mammals, convert ammonia into urea, which is less toxic than ammonia and requires moderate expenditures of water and energy. Birds, insects, and most reptiles produce uric acid or other nitrogenous compounds called purines. These nitrogenous wastes conserve water and are less toxic but energetically costlier than ammonia. (Figure 49.2)

- The solute concentration of a solution of water is known as the solution's osmolarity, expressed as milliosmoles/liter (mOsm/L). Fishes and other water-breathing animals that live in fresh water and those that live in salt water face opposite osmoregulatory challenges. (Figure 49.3)

- Salt glands are adaptations found in marine birds and reptiles that permit these animals to eliminate excess salt from the body. (Figure 49.4)

- Robert Cade and coworkers discovered that fluid replacement during exercise is particularly beneficial if the fluid contains solutes at concentrations resembling those in sweat. (Figure 49.5)

- Animals that maintain constant internal salt concentrations and osmolarities, even when living in water with very different osmolarities than their body fluids, are called osmoregulators. Animals in which internal osmolarity conforms to the osmolarity of the environment are called osmoconformers.

49.2 Principles of Fluid Filtration and Waste Excretion

- Most excretory organs operate by one or more of the following processes: (1) filtration, the removal of water and small solutes from the body fluids such as blood or hemolymph; (2) reabsorption, in which useful filtered solutes are returned to the body fluids via transport systems; (3) secretion, in which toxic or unnecessary solutes are actively transported from the blood or interstitial fluid and into the excretory tubule, and (4) excretion, in which wastes and unnecessary materials are passed out of the body. (Figure 49.6)

49.3 Comparative Excretory Systems

- In the protonephridial system of flatworms, a series of branching tubules filters fluids from the body cavity into the tubule lumens via the actions of ciliated flame cells. (Figure 49.7)

- In the metanephridial system of annelids, pairs of metanephridia located in each body segment filter interstitial fluid through funnel-like structures called nephrostomes. A dilute urine is excreted via nephridiopores in the body wall. (Figure 49.8)

- In insects, cells of Malpighian tubules secrete salts and uric acid from hemolymph into the tubule lumen; water follows by osmosis. After useful salts and water are reabsorbed into the hemolymph, wastes are excreted from the body. (Figure 49.9)

49.4 Renal Function and Vertebrate Life History

- The mechanisms by which vertebrates osmoregulate vary depending on the animal's environment. (Table 49.1)

49.5 Structure and Function of the Mammalian Kidney

- The urinary system in humans consists of the kidneys, ureters, urinary bladder, and urethra. Urine is excreted through the urethra. Each kidney is composed of an outer renal cortex and an inner renal medulla. (Figure 49.10)

- The functional units of the kidney, called nephrons, are composed of a filtering component, called the renal corpuscle, and a tubule that empties into a collecting duct. Each renal corpuscle contains a cluster of interconnected, fenestrated capillaries called the glomerulus within a structure called Bowman's capsule. Each glomerulus is supplied with blood under pressure by an afferent arteriole. Filtration occurs in the renal corpuscle. Each tubule of a nephron is composed of a proximal convoluted tubule, a loop of Henle, and a distal convoluted tubule. Tubules are surrounded by peritubular capillaries in the cortex and by the vasa recta capillaries in the medulla. (Figure 49.11)

- Different portions of the tubule have different permeabilities to solutes and water. Most reabsorption of useful solutes occurs in the proximal convoluted tubule. Water and ion reabsorption continues along the loop of Henle, using a medullary countercurrent osmotic gradient. The reabsorption of solutes by the vasa recta minimizes the loss of solutes from the renal medulla. Fine-tuning of urine composition by the hormone aldosterone occurs in the distal convoluted tubule and upper collecting duct, and the final concentration of urine is determined by ADH in the lower collecting duct. (Figures 49.12, 49.13, 49.14)

- Agre and coworkers discovered that water moves through plasma membranes through protein channels. Agre called these channels aquaporins. Aquaporins regulate water reabsorption in the kidneys. (Figure 49.15)

49.6 Impact on Public Health

- Regardless of the cause of renal malfunction, the symptoms are similar and stem from the condition known as uremia, which literally means "urine in the blood."

- One important treatment for kidney disease is hemodialysis, in which wastes in blood diffuse across a selectively permeable artificial membrane into a dialysis fluid. Hemodialysis uses a countercurrent exchange system. (Figure 49.16)

Assess and Discuss

Test Yourself

1. A change in electrolyte levels in the body may result in
   a. altered membrane potentials that disrupt normal cell function.
   b. disruption of certain biochemical processes that occur in the cell.
   c. cell death.
   d. a and b only.
   e. a, b, and c.

2. Nitrogenous wastes are the by-products of the metabolism of
   a. carbohydrates.   c. nucleic acids.   e. both c and d.
   b. lipids.   d. proteins.

3. Marine fishes avoid water balance problems by
   a. producing a large volume of dilute urine.
   b. not drinking water.
   c. having gill epithelial cells that recapture lost salts from the environment.
   d. producing small volumes of concentrated urine.
   e. a, b, and c.

4. Metabolic water is water
   a. necessary to stimulate the process of cellular respiration.
   b. found within cells.
   c. produced during cellular respiration.
   d. produced by sweat glands.
   e. used by cells during the uptake of glucose.

5. Animals that maintain a constant water balance despite changes in water concentrations in the environment are
   a. osmoregulators.   b. osmoconformers.

6. The excretory system found in insects is composed of
   a. protonephridia.   d. a filtration kidney.
   b. metanephridia.   e. a secretory kidney.
   c. Malpighian tubules.

7. In the mammalian kidney, filtration is driven by
   a. solute concentration in the tubular filtrate.
   b. solute concentration in the blood.
   c. water concentration in the blood.
   d. water concentration in the tubular filtrate.
   e. hydrostatic pressure in the blood vessels of the glomerulus.

8. In the mammalian urinary system, the urine formed in the kidneys is carried to the urinary bladder by
   a. the collecting duct.   d. the ureters.
   b. the renal tubule.   e. the urethra.
   c. the renal pelvis.

9. Which of the following causes an increase in sodium reabsorption in the distal convoluted tubule?
   a. an increase in aldosterone levels
   b. an increase in antidiuretic hormone levels
   c. a decrease in aldosterone levels
   d. a decrease in antidiuretic hormone levels
   e. none of the above

10. Aquaporins are
   a. ion channels.   d. small pores in the fenestrated
   b. water channels.   capillaries of the glomerulus.
   c. receptors for aldosterone.   e. both a and c.

Conceptual Questions

1. Define nitrogenous wastes, and list four types. What are some advantages and disadvantages of excreting different types of nitrogenous wastes?

2. Explain how salt glands are an adaptation for marine life.

3. List and define the three processes involved in urine production. Do all three processes occur for every substance that enters an excretory organ such as the kidney?

Collaborative Questions

1. Discuss two different types of filtration mechanisms found in invertebrates.

2. Briefly discuss the parts and functions of the nephron in the mammalian kidney.

Online Resource

www.brookerbiology.com

Stay a step ahead in your studies with animations that bring concepts to life and practice tests to assess your understanding. Your instructor may also recommend the interactive ebook, individualized learning tools, and more.

Endocrine Systems

50

A 22-year-old man was seen by his physician because of a complaint of hair loss and the appearance of severe acne over much of his face, neck, back, and shoulders. Upon examination, the man was found to have additional symptoms, including hypertension (high blood pressure), an elevated cholesterol level, an increased hematocrit (red blood cell count; see Chapter 47), and, alarmingly, shrunken testes. When questioned, the patient admitted that for 6 months he had been taking an oral form of the illegal drug stanozolol in an effort to improve his physique by building more muscle mass. Stanozolol is a synthetic version of a hormone known as an androgen (from the Greek, *andros*, meaning man, and *genein*, meaning to produce). A **hormone** is a chemical produced by cells in the body, which circulates throughout the body in the bloodstream and acts on one or more target tissues to effect a wide range of functions.

In males, androgens are normally produced by the testes, the endocrine organs that are also responsible for producing sperm. In normal amounts, androgens play a key role in the physiological events associated with puberty and reproduction (see Chapter 51), including the characteristic changes in the male physique that accompany puberty. Our patient hoped to accentuate those characteristics by consuming large amounts of androgens to build more muscle. At high concentrations, however, androgens promote increased red blood cell production, which, in turn, makes the heart work harder and can lead to hypertension. Likewise, high concentrations of androgens increase the cholesterol level in the blood, promote fluid retention, damage the liver, cause hair loss, and increase the activity of the sebaceous glands (the oil-producing glands in the skin that make acne worse). At such concentrations, androgens also strongly inhibit the activity of the testes. The latter phenomenon can be thought of for now as arising because the body senses that sufficient androgens are already available in the blood; therefore, the testes are not required to make more androgens, and the glands shrink in size.

Fortunately, this person was educated by his doctor about the consequences of misuse of these powerful hormones, and he discontinued the practice. Had he continued to take androgens, all of his symptoms would have worsened, and he would have run the risk of serious and irreversible damage to his heart, liver, and other organs, while increasing his risk of cancer and other diseases.

A section through a human brain, highlighting the pituitary gland and its connection to the hypothalamus (white). Both of these structures secrete numerous hormones. (Image is a three-dimensional MRI.)

Androgens are not unique to humans, or even to mammals, but are found in all vertebrates. Nor are hormones unique to vertebrates; most invertebrates require hormones to maintain homeostasis and to time when key events occur during their life history. Hormones are often produced in response to a homeostatic challenge, such as a change in an animal's blood pressure or body temperature. A chief function of hormones is to counter these challenges and maintain the body's homeostasis. Therefore, hormones affect a wide range of body functions, including gastrointestinal activity, blood pressure, cholesterol balance, fluid and mineral balance, and reproduction, among others.

Hormones are made by cells in nearly all the body's organs. In addition, hormone-producing cells are often found in specialized glands, called **endocrine glands**, whose primary function is hormone synthesis and secretion. An example is the pituitary gland highlighted in the chapter-opening photo. Collectively, all the endocrine glands and other organs with hormone-secreting cells constitute the **endocrine system**, as depicted for a human in **Figure 50.1**. Unlike some other organ systems such as the urinary or respiratory system, the endocrine system does not consist of a small number of physically connected tissues and organs. Rather, the endocrine system consists of glands and tissues that produce hormones and that are scattered throughout the body. Please take a moment to read through the information presented in Figure 50.1, in preparation for the rest of the chapter.

In this chapter, we will first learn about the chemical nature of hormones and their mechanisms of action, how the endocrine and nervous systems interact, and the ways in which hormones influence such diverse functions as metabolism, growth, and reproduction. We conclude with a discussion of how hormones impact the human condition, including how synthetic hormones are misused, as in the example just described.

Hypothalamus: Secretes several neurohormones that stimulate or inhibit anterior pituitary function. Synthesizes two neurohormones that are stored in and released from the posterior pituitary.

Anterior pituitary gland: Produces 6 hormones with diverse actions (ACTH, FSH, LH, GH, PRL, TSH).

Posterior pituitary gland: Secretes oxytocin, which stimulates uterine contractions during birth and milk secretion after birth; also secretes antidiuretic hormone, which increases water retention in the kidneys.

Heart: Makes atrial natriuretic peptide, which lowers blood sodium.

Pineal: Makes melatonin, which regulates daily rhythms and seasonal reproduction in some mammals.

Adrenal glands (medulla and cortex)

Medulla (not visible): Makes epinephrine and norepinephrine, which mediate the fight-or-flight response.

Parathyroids (attached to the back of the thyroid): Make parathyroid hormone, which increases blood calcium.

Cortex: Makes mineralocorticoids (such as aldosterone), which regulate sodium and potassium balance in the blood; makes glucocorticoids (such as cortisol), which regulate growth, metabolism, development, immune function, and the body's response to stress; makes androgens, which control reproduction.

Thyroid: Makes thyroxine (T_4) and triiodothyronine (T_3), which regulate metabolic rate, growth, and differentiation; makes calcitonin, which plays a role in Ca^{2+} homeostasis in some species.

Liver and kidneys: Secrete erythropoietin, which regulates production of red blood cells; also convert vitamin D to an active hormone.

Stomach and small intestine: Secrete gastrin and other hormones that facilitate digestion and regulate pancreatic activity.

Pancreas: Makes insulin, which lowers blood glucose, and glucagon, which raises blood glucose.

Ovaries (in females): Produce estrogens, such as estradiol and progesterone, which control female reproduction.

Adipose tissue: Produces hormones (for example, leptin), which regulate appetite and metabolic rate.

Testes (in males): Produce androgens, such as testosterone, which control male reproduction.

Figure 50.1 **Overview of the endocrine system in humans.** This figure shows many of the major endocrine glands and other structures that constitute the human endocrine system, as well as some major functions of the hormones produced by those glands.

50.1 Mechanisms of Hormone Action and Control

In this section, we will examine some of the general characteristics of hormones, how they act, and how they are controlled. Before proceeding, you might look at Figure 9.3 for a review of the different types of cell-to-cell communication mechanisms found in animals, of which hormones are only one.

Long-Distance Signaling Allows Remote Organs and Tissues to Communicate

Suppose an animal is injured and bleeding badly, which decreases the total amount of fluid in its body significantly. Therefore, the animal's heart would not fill adequately with blood. As a consequence, blood pressure would decrease, creating a potentially life-threatening situation. In vertebrates, the major organs capable of minimizing the loss of fluid from the body during such times are the kidneys. After a loss of blood, the heart communicates with the kidneys to improve their ability to retain water that would otherwise be lost in the urine. This helps prevent the fluid level of the body from dropping further. In this type of communication between organs, signaling molecules (hormones) must be released from the heart—which detects the loss of blood—into the bloodstream, where they can be carried to the kidneys to exert their effects.

Hormones Can Have Short-Term or Long-Term Effects

The effects of a given hormone may occur within seconds or require several hours to develop, and they may last for as short as a few minutes or as long as days. This is a key difference between the communication processes in the nervous and endocrine systems. In the nervous system, signals are transmitted from one cell to another within milliseconds, and the effect on the postsynaptic cell occurs immediately. In the endocrine system, one hormone may act on a cell quickly and for a very short time, while another may act very slowly and have a lingering action. The explanation for these differences lies in the mechanisms by which different hormones act. This principle is further illustrated by the fact that several hormones, including norepinephrine and dopamine, act as both hormones and neurotransmitters depending on where they are synthesized and how they are released. To understand this, we first examine the different chemical types of hormones, because it is the chemical nature of hormones that determines their mechanisms of action.

The Three Classes of Hormones Differ in Composition

Hormones fall into three broad classes: the amines, proteins/peptides, and steroid hormones (**Table 50.1**). The amines and the proteins/peptides generally share similar chemical properties and mechanisms of action, whereas the steroid hormones act very differently from the other two classes. Let's look in more detail at each of these three types of hormones.

The amine hormones are derived from an amino acid, either tyrosine or tryptophan. As shown in **Figure 50.2**, tyrosine is the precursor for the hormones epinephrine and norepinephrine, which are produced in the adrenal medulla and are important in the body's response to stress (the fight-or-flight reaction;

see Chapter 42), and for dopamine, a hormone made by the brain. Tyrosine is also the chemical backbone of the hormones made by the thyroid gland, which are important regulators of metabolic rate, growth, and differentiation. The major hormone derived from tryptophan is melatonin, which is produced within an endocrine gland called the pineal gland located in the brain. Melatonin is important for controlling seasonal reproduction in certain mammals such as sheep and hamsters.

The proteins and peptides (small polypeptides) class of hormones is the most abundant. These hormones participate in numerous body functions such as metabolism, mineral balance, growth, and reproduction. Examples include insulin and glucagon made by the pancreas, leptin made by adipose tissue, and many others.

Steroid hormones are synthesized from cholesterol, and thus all steroid hormones are lipids, unlike the other classes of hormones. Steroids are less soluble in water than are amines or protein/peptide hormones. Due to this limited solubility, steroids are usually bound to large, soluble proteins in the blood that serve as carriers. By combining with these proteins, steroids can reach high concentrations in the blood. The major steroid hormones found in vertebrates include aldosterone and cortisol made by the adrenal cortex, androgens made by the testes, and estrogens and progesterone made by the ovaries. In insects, the steroid hormone ecdysone is produced in the prothoracic gland located in the thorax. How do these different characteristics affect the function of hormones?

Hormones Act Through Membrane or Intracellular Receptors

The amine and protein/peptide hormones are generally water-soluble, and the steroid hormones are lipid-soluble. Therefore, the amines and proteins/peptides are not able to cross plasma membranes and must use receptors on the cell surface. Steroid hormones, however, being lipids, can diffuse across plasma membranes and access receptors in either the cytosol or nucleus. Hormones that work through membrane receptors tend to be those that elicit fast responses, whereas those that require a response from the genetic machinery may take longer. Let's look at each of these.

Water-Soluble Hormones and Their Membrane Receptors
All the amine and protein/peptide hormones, with one exception

Table 50.1	Chemical Classes of Hormones			
Class	Chemical properties	Location of target cell receptor	Mechanism of action	Examples
Amines	Derived from tyrosine or tryptophan; small, water-soluble	Plasma membrane (except thyroid hormones, which act via intracellular receptors)	Stimulate second messenger pathways (except thyroid hormones which act via changes in gene transcription)	Epinephrine, norepinephrine, dopamine, thyroid hormones, melatonin
Proteins/Peptides	Water-soluble	Plasma membrane	Stimulate second messenger pathways	Insulin, glucagon, leptin
Steroids	Derived from cholesterol, mostly lipid-soluble	Cytosol or nucleus	Usually stimulate gene transcription directly	Aldosterone, cortisol, testosterone, estradiol

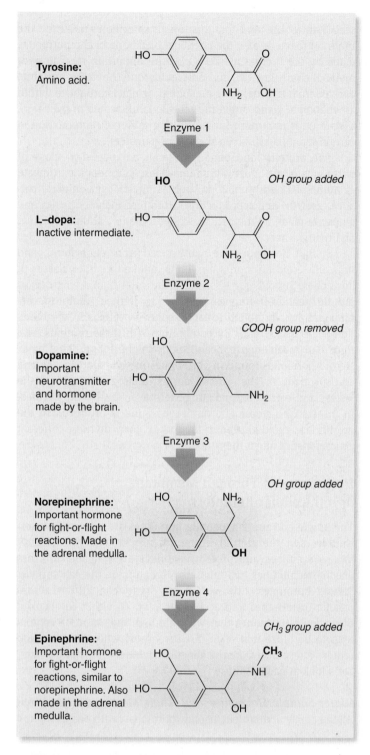

Figure 50.2 Synthesis of the amine hormones dopamine, norepinephrine, and epinephrine. The synthesis of thyroid hormones also begins with tyrosine but follows a different mechanism, as described later.

on their surfaces can respond to the hormone. Thus, although a hormone travels throughout the entire circulatory system, it activates only specific cells. The hormone interacts noncovalently and reversibly with the receptor. The reversibility of the binding between hormone and receptor is one way in which cells are prevented from being permanently stimulated.

Among cells throughout the body, different receptor proteins may recognize the same hormone. These different receptors, called subtypes or isoforms, may be the product of different genes or may be produced by alternative splicing, described in Chapter 13. By binding to receptor isoforms, the same hormone is able to elicit differing, sometimes even opposite, responses, depending on where it binds in the body. These isoforms may have different affinities for the hormone, such that low concentrations of a hormone may cause one type of effect by binding to high-affinity receptors on one cell, and higher concentrations may cause entirely different actions by binding to low-affinity receptors on another cell. In this way, most hormones are able to serve more than one function.

The binding of a water-soluble hormone to its plasma membrane receptor initiates intracellular signaling pathways that often involve second messengers (refer back to Figure 9.12). Three major signaling pathways that are activated by water-soluble hormones are the cyclic AMP pathway, the calcium/phosphatidylinositol pathway, and the tyrosine kinase pathway. These signal transduction processes may be rapid, occurring in some cases within seconds, and involve changing the activity of enzymes. Both of these features are important, because occasionally the rapidity of the cell response to a hormone can mean the difference between life and death, as may occur, for example, under conditions of extreme danger. The presence of enzymes in the signaling process ensures that the signal from a small amount of hormone is greatly amplified, because a single signaling molecule results in the production of many intracellular messengers. In this way, endocrine glands may synthesize and secrete relatively small amounts of hormone while still producing biological effects; this conserves energy and resources.

Activation of signaling pathways may also lead to changes in cellular activities that occur more slowly. These slower changes usually require activation or inhibition of genes in the nucleus, and are mediated by transcription factors (see Chapter 12) that are activated by the same signaling pathways described earlier. However, another means of activating gene transcription is through activation of receptors for lipid-soluble hormones.

Lipid-Soluble Hormones and Their Intracellular Receptors

All steroid hormones and the thyroid hormones are lipophilic and bind to intracellular receptors. The complex of a steroid or thyroid hormone bound to its intracellular receptor functions as a transcriptional activator (or less frequently as an inhibitor) by binding to enhancers of particular genes (refer back to Figure 9.9). Once bound, transcription of a gene is increased, which increases the amount of that gene's protein product. These proteins may be important in a variety of cellular activities, such as regulating the number of ion pumps in membranes, controlling cell differentiation, secretory activity, and growth.

(thyroid hormones, discussed later), act by binding to a receptor protein located in the plasma membrane (refer back to Figure 9.5). As these hormones diffuse from the blood into the interstitial fluid, they must bind to a specific receptor protein located on a cell surface. Only cells having the proper receptors

Steroid and thyroid hormones can influence several genes within a single cell or different cells. In this way, one hormone can exert a variety of actions throughout the body. The physical changes that accompany puberty in mammals result from the actions of two steroid hormones—androgens in males and estrogens in females—and are among the most striking and commonly recognized examples of this widespread action.

We have learned that hormones act in all cases by binding to receptor proteins located on the cell surface, in the cytosol, or in the nucleus. This raises an intriguing question: Which evolved first, a given hormone or its receptor? There would appear to be no selection pressure to evolve one without the other. However, recent research has generated intriguing new hypotheses regarding the evolution of these signaling systems, as described next.

Genomes & Proteomes Connection

Hormones and Receptors Evolved as Tightly Integrated Molecular Systems

All hormones act by binding to receptors. Without a receptor, a hormone has no function, and without a hormone, the receptor has no function. Hormones and their receptors function as tightly integrated molecular systems. Each molecule depends on the other for biological activity. How such systems could have evolved, or whether they evolved together, has long been a major puzzle.

American biologist and environmental scientist Joseph Thornton and colleagues recently addressed this puzzle by examining how two structurally related steroid hormones evolved separate activities and distinct receptors. **Aldosterone**, as noted, is a steroid hormone made by the adrenal cortex. Because it regulates the balance of two minerals (Na^+ and K^+) in the body, it is known as a **mineralocorticoid**. **Cortisol** is another steroid hormone made by the adrenal cortex and is known as a **glucocorticoid** because one of its major functions is to regulate glucose balance. The actions of these hormones are mediated by intracellular receptors known as the mineralo-corticoid receptor (MR) and the glucocorticoid receptor (GR), respectively. Phylogenetic analysis of the gene sequences that code for these two receptors suggests that they arose at least 450 million years ago by duplication of an ancient corticoid receptor (CR) gene.

The researchers analyzed the known sequences of the genes for the MR and GR of many vertebrate species, including the most ancient vertebrates, to deduce the theoretical sequence of an ancestral CR. They then synthesized this CR and tested its ability to bind aldosterone and cortisol. They discovered that the CR was capable of binding both hormones, particularly aldosterone. This was surprising, because aldosterone is only present in tetrapods, which arose long after the proposed gene duplication event that created the MR. Therefore, it appears that a receptor with high affinity for aldosterone was present long before animals acquired the capacity to synthesize aldosterone. The receptor seems to have evolved to bind other steroids. Around 450 million years ago, the gene for the CR duplicated, and then the two resulting genes seem to have evolved such that one receptor gained high affinity primarily for mineralocorticoids and the other for glucocorticoids.

These studies suggest that the ability of an animal to respond to aldosterone evolved long before aldosterone did, because a receptor with high affinity was already in place in animals. When aldosterone evolved in tetrapods, it was able to use the MR derivative of the ancestral CR. In this example, therefore, the answer to "which came first" appears to be that the receptor evolved first and the hormone later.

Hormone Levels in Blood Depend on Rates of Synthesis and Removal

When hormones are produced by endocrine cells, they are released into the interstitial space between the cell and adjacent blood vessels. The hormones then diffuse into the blood vessel. Although most hormones circulate in the blood at all times, their concentrations can increase or decrease dramatically when necessary. This can be accomplished in two major ways: by changing the rate of hormone production and release by an endocrine cell and by changing the rates at which hormones are removed from the blood or inactivated.

How hormones are synthesized and the rate of that synthesis differ for each of the three classes, amine, protein/peptide, and steroid. Synthesis of amine hormones is mediated by enzymatic conversion of either tyrosine or tryptophan into their respective hormones. These reactions require several enzymes that add or remove chemical side groups and achieve the final product (Figure 50.2). The synthesis of the hormone depends on the amounts and activities of the synthesizing enzymes. These enzymes are always present in the cell, but they can be greatly stimulated when additional hormone is required.

The protein and peptide hormones are synthesized at a steady rate in an unstimulated cell, until additional hormone is required. In that case, transcription factors within the cell direct the increased transcription of the gene coding for the hormone in question. Conversely, when less hormone is required, gene transcription is slowed or stopped. These hormones are too large and hydrophilic to diffuse across the plasma membrane and then into the extracellular fluid. Instead, they are packaged into secretory vesicles much the same way as neurotransmitters are packaged in neuron terminals. This packaging provides a ready means of secreting the hormones by exocytosis and a reservoir of stored hormone available for immediate release when required. When a cell is stimulated to secrete its stored protein or peptide hormones, it also is typically stimulated to synthesize new hormone molecules to replace them.

As noted earlier, the steroid hormones are derived from cholesterol, as depicted in **Figure 50.3**. To form steroid hormones, a cell must express enzymes capable of adding hydroxyl groups to specific carbons in the cholesterol skeleton. Other enzymes carry out reduction reactions, and still others can both hydroxylate a carbon and split carbon-carbon bonds. The activities and amounts of many of these enzymes change if more or less hormone is required. Steroid hormones, unlike water-soluble hormones, are not packaged into secretory vesicles because the steroid can diffuse out across the lipid membrane of the vesicle. Instead, steroid hormones are made on demand, and no significant storage exists for them.

Although hormones carry out vital functions, excessive stimulation of cells by hormones may produce detrimental effects. For example, hormones that are activated if an animal loses blood typically help restore blood volume and pressure. This cannot continue indefinitely, however, because once blood volume and pressure are restored, it would be harmful if they continued to increase until their levels were above normal. Therefore, once a hormone enters the blood and performs its functions, it is usually prevented from exerting its effects indefinitely. This is accomplished in one or more ways:

- Hormones that bind to plasma membrane receptors may be engulfed by endocytosis into a cell, where lysosomal enzymes degrade the hormones.
- Small, water-soluble hormones are excreted in the urine.
- The liver chemically modifies many hormones to render them inactive and more easily excretable via the kidneys.
- Negative feedback processes (see Chapter 40) turn off the signals that were responsible for stimulating the synthesis and secretion of the hormone.

Generally, these processes ensure that hormone levels in the blood remain within a normal range under most circumstances,

Figure 50.3 Synthesis and major functions of the chief steroid hormones in animals.

but have the capacity to be increased or decreased beyond that range if required. One of the ways in which changes in hormone levels are initiated is through sensory input to an animal's brain. As we see next, the nervous system and endocrine system are functionally linked in many animals, including all vertebrates.

50.2 Links Between the Endocrine and Nervous Systems

A key feature of the endocrine system in most animals is that the concentrations of many hormones in the extracellular fluid rise and fall with changes in an animal's environment. For this to happen, the endocrine system must somehow be informed of environmental changes. This often occurs when sensory input is received by an animal's nervous system, which, in turn, modulates the activity of one or more endocrine glands.

Sensory stimuli detected by the nervous system can activate the endocrine system. For example, when an antelope detects the presence of a nearby lioness, visual and olfactory sense information is relayed to the antelope's brain. The brain initiates responses in certain endocrine glands that release hormones to prepare the antelope for the possibility of an attack. As another example, the levels of certain hormones fluctuate in the blood of certain fishes as they migrate back and forth between feeding grounds in the sea and freshwater spawning sites. These hormones are activated by different salinities in the environment, which are detected by sensory cells of the fish's nervous system. The hormones act to prepare the gills of the fish to handle the large changes in salinity of the water.

The common feature of these examples and many others is that a sensory cue, such as a predator or the salinity of water, must be perceived by a sensory receptor and converted into an endocrine response. Electrical signals are transmitted from the sensory receptors to different parts of the brain, including the hypothalamus. In this section, we explore how the hypothalamus and pituitary gland play the major roles in linking the nervous and endocrine systems.

The Hypothalamus and Pituitary Gland Are Physically Connected

The **hypothalamus** is a collection of several nuclei that are located at the bottom surface of the vertebrate brain (**Figure 50.4a**). Neurons in these nuclei are connected to an endocrine gland sitting directly below the hypothalamus, called the **pituitary gland** (see the bottom part of the highlighted structure in the chapter-opening photo). The pituitary gland in humans is made up of two lobes, the anterior and posterior lobes. Some vertebrate species have an intermediate lobe that is believed to be vestigial in primates. The hypothalamus and pituitary are connected by a thin piece of tissue called the **infundibular stalk**. These two structures are also connected by a system of blood vessels called portal veins.

Portal veins differ from ordinary veins because not only do they collect blood from capillaries—like all veins do—but they

then also form another set of capillaries, as opposed to returning the blood directly to the heart like other veins. The portal veins extend through the length of the infundibular stalk before forming new capillaries. Within the anterior lobe of the pituitary gland—often simply called the anterior pituitary gland—the portal veins empty into a second set of capillaries. This arrangement of blood vessels bypasses the general circulation and allows the hypothalamus to communicate directly with the anterior pituitary gland. Let's now explore the nature of this communication.

The Hypothalamus and Anterior Pituitary Gland Have Integrated Functions

As described in Chapter 42, the hypothalamic nuclei are vital for such diverse functions as reproduction, bodily rhythms, appetite, metabolism, and responses to stress. The hypothalamus has such wide-ranging effects in part because it acts as a master control, signaling the pituitary gland when to produce and secrete its many hormones. However, the hypothalamus communicates differently with the anterior and posterior pituitary glands. Let's look at the interaction between the hypothalamus and anterior pituitary first.

Within the different nuclei of the hypothalamus are neurons that synthesize a class of hormones called neurohormones. A **neurohormone** is any hormone that is made in and secreted by neurons. All neurohormones are either amines or peptides. Although they are produced within neurons, these molecular signals are not referred to as neurotransmitters, because the endings of the neurons do not terminate in a synapse with another cell. Instead, the neuron terminals from the hypothalamus end near capillaries. Here, the neurons secrete their neurohormones into the capillaries, which, in turn, collect into the portal veins. This allows neurohormones to be delivered directly from the hypothalamus to the cells of the anterior pituitary gland in a quick, efficient manner (**Figure 50.4b**).

In response to the presence of these hypothalamic neurohormones, the anterior pituitary gland synthesizes several different hormones. **Table 50.2** lists the six major anterior pituitary hormones that have well-defined functions in vertebrates and the neurohormones that regulate them. As shown in the table, five of the anterior pituitary gland hormones are activated by neurohormones, one is inhibited by dopamine, and one is under both stimulatory and inhibitory control by the hypothalamus. Recent research suggests the possibility for dual control of some of the other pituitary gland hormones, but this remains uncertain. The stimulatory actions of several of the hypothalamic neurohormones has historically led to them also being known as "hypothalamic-releasing hormones," because they cause the release, or secretion, of other hormones from the anterior pituitary. The six hormones of the anterior pituitary gland are secreted into the general blood circulation, where they act on other glands or structures.

Adrenocorticotropic hormone (ACTH) stimulates the cells of the adrenal cortex to synthesize and release cortisol or related steroid hormones. Follicle-stimulating hormone (FSH) and luteinizing hormone (LH) act on the ovaries or testes to

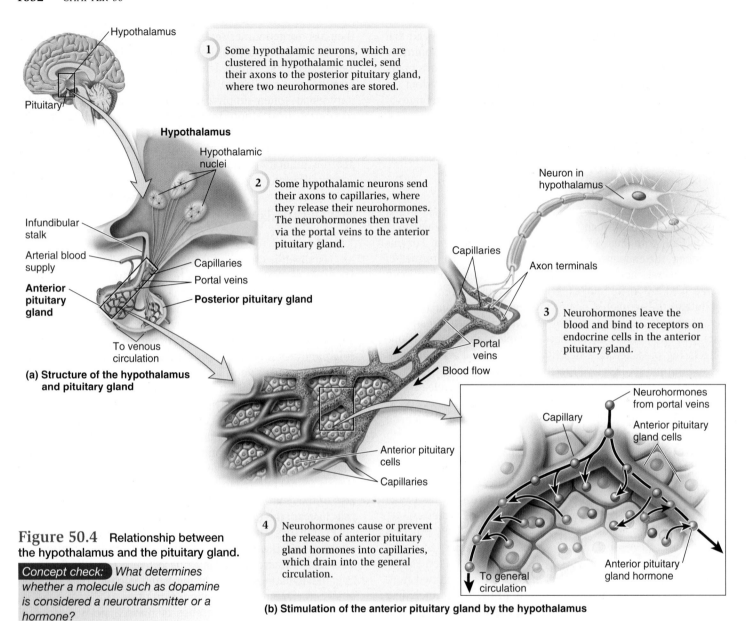

Hypothalamus

1 Some hypothalamic neurons, which are clustered in hypothalamic nuclei, send their axons to the posterior pituitary gland, where two neurohormones are stored.

2 Some hypothalamic neurons send their axons to capillaries, where they release their neurohormones. The neurohormones then travel via the portal veins to the anterior pituitary gland.

3 Neurohormones leave the blood and bind to receptors on endocrine cells in the anterior pituitary gland.

(a) Structure of the hypothalamus and pituitary gland

4 Neurohormones cause or prevent the release of anterior pituitary gland hormones into capillaries, which drain into the general circulation.

Figure 50.4 Relationship between the hypothalamus and the pituitary gland.

Concept check: What determines whether a molecule such as dopamine is considered a neurotransmitter or a hormone?

(b) Stimulation of the anterior pituitary gland by the hypothalamus

Table 50.2	Hormones of the Anterior Pituitary Gland and Hypothalamus		
Anterior pituitary gland hormone	**Stimulatory neurohormone from hypothalamus**	**Inhibitory neurohormone from hypothalamus**	**Major functions**
Adrenocorticotropic hormone (ACTH)	Corticotropin-releasing hormone (CRH)	None known	Stimulates adrenal cortex to make glucocorticoids
Follicle-stimulating hormone (FSH)	Gonadotropin-releasing hormone (GnRH)	None known	Stimulates germ cell development and sex steroid production in gonads
Luteinizing hormone (LH)	Gonadotropin-releasing hormone (GnRH)	None known	Stimulates release of eggs in females; stimulates sex steroid production from gonads
Growth hormone (GH)	Growth hormone-releasing hormone (GHRH)	Somatostatin	Promotes linear growth; regulates glucose and fatty acid balance in blood
Prolactin (PRL)	TRH and other factors have been suggested as stimulators of PRL, but this is not unequivocal	Dopamine	Stimulates milk formation in mammals; participates in mineral balance in other vertebrates
Thyroid-stimulating hormone (TSH)	Thyrotropin-releasing hormone (TRH)	None known	Stimulates thyroid gland to make thyroid hormones

stimulate egg or sperm development, as well as the synthesis and release of estrogens and androgens from these gonads. Growth hormone (GH) stimulates growth of immature animals and also participates in control of energy balance. The prolactin (PRL) level in the blood is increased in mammals during late pregnancy and during lactation. This hormone stimulates milk production in mammals but affects salt and water balance in other vertebrates. Finally, thyroid-stimulating hormone (TSH) acts on cells of the thyroid gland to stimulate growth of the gland and production of thyroid hormones.

Understanding the function of the hormones of the anterior pituitary gland is an ongoing and active area of research. Several other peptides from the anterior pituitary have been characterized and suggested to be hormones, but their functions have not yet been clearly defined in any vertebrate species. One example is β-endorphin, which is released into the blood during times of stress. Because it is related in structure to the opiate-type molecules such as morphine, it has been suggested that β-endorphin may act in some species as a painkiller during times of stress, but this is uncertain.

The Posterior Pituitary Gland Contains Axon Terminals from Hypothalamic Neurons That Store and Secrete Oxytocin and Antidiuretic Hormone

The posterior pituitary gland has a blood supply, but, in contrast to the anterior pituitary gland, it is not connected to the hypothalamus by portal veins and does not respond to neurohormones from the hypothalamus. Instead, the posterior pituitary gland is an extension from the hypothalamus that lies in close contact with the anterior pituitary gland (Figure 50.4). Axons from neurons extend from the hypothalamus into the posterior pituitary. In mammals, the axon terminals in the posterior pituitary store one of two hormones, oxytocin or antidiuretic hormone (ADH), which are produced by the cell bodies of those neurons. When the hypothalamus receives information that these hormones are required, they are released directly from the neuron axon terminals into the bloodstream.

Oxytocin concentrations increase in the blood of pregnant mammals just prior to birth. It stimulates contractions of the smooth muscles in the uterus, which facilitates the birth process. After the offspring are born, oxytocin becomes important in the let-down of milk. When the mother's nipples are stimulated by the suckling of a newborn (or the thought, sight, sound, or smell of a newborn), neurons transmit a signal from the breast to the mother's hypothalamus, which stimulates the cells that produce oxytocin. Oxytocin is released from the posterior pituitary gland into the blood where it travels to smooth muscle cells surrounding the milk glands and ducts of the breast. This stimulates the release of milk. Thus, oxytocin has two important and different functions, one during birth and one during lactation. In both cases, the hormone stimulates contraction of muscle cells.

Antidiuretic hormone (ADH) gets its name because it acts on kidney cells to decrease urine production—a process known as antidiuresis. Diuresis is an increased loss of water in the

urine, as happens, for example, when you drink large amounts of fluids. If the fluid content of the body is low, such as during dehydration or after blood loss, ADH is secreted into the blood from the posterior pituitary gland. It acts to increase the number of water-channel proteins called aquaporins present in the membranes of kidney tubule cells; water is recaptured from the forming urine through these aquaporins (see Chapter 49). Minimizing the volume of water used to form urine is an adaptation that conserves body water when necessary.

At high concentrations, ADH also increases blood pressure by stimulating vasoconstriction of blood vessels. Like oxytocin, therefore, ADH has more than one function. Both of its major functions are related in that they contribute to maintaining blood pressure and fluid levels in the body. The two actions of ADH are mediated by different receptor isoforms, a feature that has proven to be of medical value in humans. Researchers have synthesized versions of ADH that have been chemically modified to activate ADH receptors in the kidney but not those in blood vessels. One such compound is sometimes used to treat **nocturnal enuresis** (bed wetting) in children by decreasing urine production without the unwanted side effect of increasing blood pressure.

Oxytocin and ADH are well-studied examples of the evolution of hormones. These two hormones are found only in mammals. However, many invertebrates and all nonmammalian vertebrates secrete one or more peptides that share chemical similarities with oxytocin and ADH but are not identical to the mammalian hormones. One of these is **vasotocin**, which combines some of the chemical structure of both oxytocin and ADH. Research indicates that an ancestral vasotocin gene duplicated at some point, and then the two genes evolved into the oxytocin and ADH genes found in mammals. Because only mammals lactate, the role of vasotocin must be different in birds, fishes, and other vertebrates than that of oxytocin in mammals. Research has shown that vasotocin is responsible for regulating salt and water balance in the blood of nonmammalian vertebrates.

The observation that members of the vasotocin/oxytocin/ADH gene family arose early in animal evolution also suggests that oxytocin and ADH may have additional, unrecognized actions that have been retained in mammals. For example, human males have oxytocin in their blood, but its role cannot be the same as in females because, of course, only females give birth and lactate. The key roles of oxytocin in male humans and other animals have not been unequivocally identified.

50.3 Hormonal Control of Metabolism and Energy Balance

An important function of the endocrine system is to control metabolic rate and energy balance. Hormones are partly responsible for regulating energy use by modulating appetite, digestion, absorption of nutrients, and the levels of energy sources like glucose in the blood and its transport into cells. Although many hormones are involved in these processes, those from the thyroid gland, pancreas, adipose tissue, and the adrenal glands play particularly important roles, as described in this section.

Thyroid Hormones Contain Iodine and Regulate Metabolic Rate

The thyroid gland lies within the neck of vertebrates. In mammals, it straddles the trachea just below the larynx (Figure 50.5a). It consists of many small, spherical structures called follicles that consist of a shell of epithelial cells called follicular cells and a core of a gel-like substance called the **colloid**. The colloid consists primarily of large amounts of the protein **thyroglobulin (TG)**, which plays a major role in the synthesis of thyroid hormones.

Thyroid hormones are produced when the hypothalamic neurohormone thyrotropin-releasing hormone (TRH; see Table 50.2) stimulates the anterior pituitary gland to secrete thyroid-stimulating hormone (TSH) into the blood. TSH stimulates the follicular cells of the thyroid gland to begin the process of making thyroid hormones. The thyroid hormones have a negative feedback effect on the hypothalamus and anterior pituitary, preventing them from producing too much TRH and TSH.

Figure 50.5b shows the pathway leading to thyroid hormone synthesis. First, iodide—converted by the gut from dietary iodine—diffuses from the bloodstream into the interstitial fluid, from where it is transported across the basolateral membrane of the thyroid follicular cells. The iodide then diffuses through the apical membrane and enters the colloid, where it is oxidized and bonds to tyrosine side chains in thyroglobulin. When the thyroid follicular cells are stimulated by TSH, the apical membranes undergo endocytosis (see Chapter 5), bringing colloid with its iodinated thyroglobulin into the cell. The endocytotic vesicles fuse with lysosomes; there, lysosomal enzymes cleave the iodinated tyrosines from thyroglobulin to form two thyroid hormones, called **thyroxine** and **triiodothyronine**. These molecules are unique among hormones in that they contain iodine molecules: four in the case of thyroxine, also called T_4, and three in triiodothyronine, or T_3. Both T_4 and T_3 then diffuse out of the follicular cells across the basolateral membrane into the bloodstream. These hormones are carried throughout the body and diffuse into cells. Inside cells, most T_4 is converted by enzymes that remove one iodine, forming T_3. It is T_3 that actually binds to cellular receptor proteins. T_4 can be considered, therefore, as a circulating precursor of the active hormone T_3.

A major action of thyroid hormones in adult animals is to stimulate energy consumption by many different cell types. This occurs in large part by increasing the number and activity of the Na^+/K^+-ATPase pumps in plasma membranes. As these pumps hydrolyze ATP, the cellular level of ATP decreases. This decrease is compensated for by increasing the cell's metabolism of glucose. Whenever metabolism is increased, heat production is increased. Consequently, a person with a hyperactive thyroid gland (**hyperthyroidism**; from the Greek *hyper*, meaning over or above) generally feels warm, whereas the opposite condition (**hypothyroidism**; from the Greek *hypo*, meaning under or below) results in a sensation of coldness. Scientists have

estimated that up to 70% of the heat produced by some homeotherms is attributable solely to the actions of thyroid hormones on metabolic rate.

The observation that thyroid hormones cannot be made without iodine presents some interesting and unique features of these hormones. The availability of iodine in the diet of most animals is variable. As a consequence, the ability to store large amounts of thyroglobulin in the colloid of the thyroid was an important evolutionary adaptation. In this way, during times when iodine ingestion is high, many thyroglobulin molecules have their tyrosine amino acids bound to iodines, one of the first steps in forming T_4 and T_3. During times of low iodine availability, this reservoir of iodinated tyrosines in thyroglobulin molecules can be tapped. Humans, for example, have at least a 2-month supply of thyroid hormones even if dietary iodine were to become unavailable. In most industrialized countries, iodine deficiency is rarely a problem since the introduction of iodized salt in the mid-20th century. However, in some regions of the world, this is still a major health problem.

The left side of Figure 50.6a shows the pattern of T_4 and T_3 production from a normal thyroid gland when iodine ingestion is adequate. The right side of the figure shows the consequences when thyroid hormones are not produced in normal amounts, for example, due to a lack of iodine in the diet. In such a case, the decreased T_4 and T_3 levels provide less negative feedback on the hypothalamus and anterior pituitary gland, resulting in elevated TRH and, consequently, TSH levels. The thyroid gland responds to the increased TSH by increasing the cellular machinery needed to produce more and more thyroglobulin, even though in the absence of iodine, no additional T_4 or T_3 will be synthesized. What results is an overgrown gland that still lacks the resources to make thyroid hormone. This condition is known as an **iodine-deficient goiter** (Figure 50.6b). In humans, the problem can be alleviated either by adding iodine to the diet or by taking T_4 pills. Goiters are not unique to humans. Iodine deficiency is relatively common among vertebrates, and goiters are found frequently in reptiles and birds, particularly those that subsist on all-seed diets, which are generally low in iodine.

Hormones of the Pancreas and Adrenal Glands Regulate Fuel Levels in the Blood

Thyroid hormones regulate an animal's metabolism. For metabolism to proceed normally, however, body cells must have adequate sources of energy available, usually in the form of glucose and fatty acids. The brain, in particular, must have a constant supply of glucose because brain cells have relatively limited storage capacity for fuel. Regulation of energy availability to cells is in large part the job of the hormones of the pancreas and the adrenal glands.

The pancreas is a complex organ consisting of two parts, exocrine and endocrine cells (Figure 50.7). Most of the mass of the pancreas comprises exocrine glands. An **exocrine gland** is one in which epithelial cells secrete chemicals into a duct, which carries those molecules directly to another structure or to the outside surface of the body. Examples of exocrine glands

- Larynx
- Thyroid gland
- Trachea

(a) Location and structure of the thyroid gland

Figure 50.5   **The thyroid gland and synthesis of thyroid hormones.**   **(a)** Location and structure of the gland in a human. Vertebrates contain thyroid tissue in the neck region, but it is not always consolidated into a single structure as shown here. **(b)** The steps involved in production of thyroid hormones. Shown is a cross section through a small part of a single follicle in a thyroid gland, with a nearby capillary (not to scale). The blood delivers iodide to the gland and picks up thyroid hormones secreted by the gland.

Capillary   **Interstitial fluid**   **Follicular cell of thyroid**   **Colloid**

- I^- — Iodide ion
- Basolateral membrane

2 Iodides bond to tyrosine residues in thyroglobulin.

- Apical membrane

1 Iodide is transported across the follicular cell.

- Lysosome

4 The endocytotic vesicle fuses with lysosomes, which cleave off T_3 and T_4.

Thyroglobulin (TG)

Iodinated TG

- Endocytotic vesicle

T_3

T_4

3 When the cell is stimulated by TSH, iodinated TG enters the cell by endocytosis.

$$HO-\overset{\overset{O}{\parallel}}{C}-\underset{NH_2}{CH}-CH_2-$$

Triiodothyronine (T_3)

5 T_3 and T_4 diffuse out of the cell and into the blood.

$$HO-\overset{\overset{O}{\parallel}}{C}-\underset{NH_2}{CH}-CH_2-$$

Thyroxine (T_4)

(b) Synthesis of thyroid hormones

include the sweat glands, salivary glands, liver, and pancreas. The secretions of the exocrine pancreas empty into the small intestine, where they aid digestion. The non-exocrine portion of the pancreas consists of endocrine cells that produce peptide hormones, notably insulin and glucagon.

Spherical clusters of endocrine cells called **islets of Langerhans** are scattered in great numbers throughout the pancreas. Within the islets are alpha cells, which make **glucagon**, and beta cells, which make **insulin**. The actions of these two

hormones are antagonistic with respect to each other—insulin lowers and glucagon raises blood glucose concentrations.

Maintaining normal glucose and other nutrient levels in the blood is a vital process that keeps cells functioning optimally. When an animal has not fed for some time, its energy stores become depleted, and the blood glucose level falls. Under these conditions, glucagon is secreted into the blood, where it acts on the liver (**Figure 50.8**, right side). The liver contains a limited supply of glucose in the form of stored glycogen. Glucagon

Normal iodine intake

Normally, T$_4$ and T$_3$ prevent overstimulation of the thyroid gland by TSH.

Hypothalamus

TRH   Excess TRH

Anterior pituitary gland

TSH   Excess TSH

Thyroid

Normal thyroid

5.1 cm

Normal negative feedback

Normal levels of T$_4$ and T$_3$

Reduced iodine intake

In iodine deficiency, T$_4$ and T$_3$ levels decrease, allowing TSH to increase and overstimulate the thyroid.

Enlarged thyroid

5.1 cm

Low T$_4$ and T$_3$ due to lack of iodine

Reduced negative feedback

(a) Thyroid function with and without normal iodine intake

(b) Woman with iodine-deficient goiter

Figure 50.6 **Consequences of normal and inadequate iodine in the diet.** **(a)** With normal iodine intake, as shown on the left, T$_4$ and T$_3$ levels inhibit TSH secretion. Without enough iodine, as shown on the right, less T$_4$ and T$_3$ are synthesized, and the TSH level increases. This leads to the enlargement of the thyroid gland. **(b)** An extreme example of an enlarged thyroid gland, or goiter, due to iodine deficiency.

Small intestine   Pancreas

Pancreatic duct (carries exocrine secretions to small intestine)

Pancreatic exocrine cells

Pancreatic islets of Langerhans

Alpha cells produce glucagon

Beta cells produce insulin

stimulates the breakdown of glycogen into many molecules of glucose, which are then secreted into the blood. This process, known as glycogenolysis (see Chapter 46), is stimulated within seconds by glucagon. A second action of glucagon on the liver is important for responses to prolonged fasting. In that case, glucagon stimulates the process of gluconeogenesis (see Chapter 46), by which noncarbohydrates are converted into glucose, which is then released into the blood.

The adrenal glands, which sit atop the kidneys, also play a role in glucose metabolism by producing amine and steroid hormones. When blood glucose levels are lower than normal, neurons from the sympathetic nervous system activate the secretion of epinephrine by the adrenal medulla. Like glucagon, epinephrine also stimulates glycogenolysis and gluconeogenesis. In addition, cortisol is released from the adrenal cortex. Once in the blood, cortisol also activates gluconeogenesis in the liver. These processes are vital to long-term survival in the absence of food. Because of the combined short-term and long-term actions of glucagon, epinephrine, and cortisol, the blood glucose level rarely falls or remains significantly below normal except in extreme circumstances.

In contrast to fasting, after an animal eats a meal, the levels of glucose and other nutrients in the blood become elevated.

Figure 50.7 **Location, appearance, and internal structure of the mammalian pancreas.** Amidst the exocrine pancreas are scattered islets of Langerhans, which are endocrine tissue. Only the exocrine products are secreted into the intestine; the hormones from the islets of Langerhans are secreted into the blood (not shown).

Concept check: *The pancreas contains both exocrine and endocrine tissue. Is this property ever observed in other organs?*

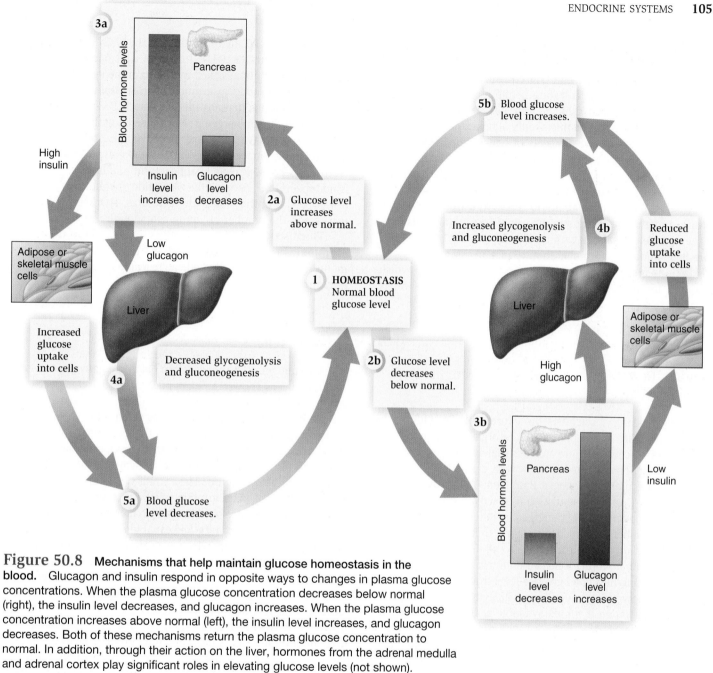

Figure 50.8 Mechanisms that help maintain glucose homeostasis in the blood. Glucagon and insulin respond in opposite ways to changes in plasma glucose concentrations. When the plasma glucose concentration decreases below normal (right), the insulin level decreases, and glucagon increases. When the plasma glucose concentration increases above normal (left), the insulin level increases, and glucagon decreases. Both of these mechanisms return the plasma glucose concentration to normal. In addition, through their action on the liver, hormones from the adrenal medulla and adrenal cortex play significant roles in elevating glucose levels (not shown).

Restoring the normal blood concentrations of glucose, fats, and amino acids is almost exclusively under the control of insulin, one of the few hormones that are absolutely essential for survival in animals. The secretion of insulin is directly stimulated by an increased concentration of glucose in the blood (Figure 50.8, left side). Once in the blood, insulin acts on plasma membrane receptors located primarily in cells of adipose and skeletal and cardiac muscle tissues to facilitate the transport of glucose across the plasma membrane into the cytosol. Once in the cell, glucose is used for cellular functions or converted to stored energy forms such as fat. As discussed in Chapter 46, the process of glucose transport involves the actions of proteins called glucose transporters (refer back to Figure 46.3). These proteins exist in cytosolic membrane vesicles. The major function of insulin is to stimulate movement of the vesicles with their glucose transporters to the plasma membrane. Once the vesicle fuses with the plasma membrane, the transporters begin transporting glucose from the extracellular fluid into the cell.

When the blood glucose level returns to normal, the stimulus for insulin secretion disappears, and the insulin level in the blood decreases. This results in a decrease in the number of glucose transporters in plasma membranes, because the glucose transporters are subjected to endocytosis and thereby return to membrane vesicles in the cytosol. The actions of insulin are not limited to glucose transport, however, because the hormone also stimulates transport of amino acids into cells and promotes fat deposition in adipose tissue. In other words, the broader role of insulin in the body is to facilitate the transfer of energy from the extracellular fluid into storage sites primarily in muscle and fat.

In the absence of sufficient insulin, as occurs in the disease **type 1 diabetes mellitus (T1DM)**, less extracellular glucose can cross plasma membranes, and consequently, glucose accumulates to a very high concentration in the blood. T1DM occurs in many mammals and other vertebrates. It is caused when the body's immune system mistakenly attacks and destroys the insulin-producing cells of the islets of Langerhans. One consequence of the disease is that muscle and fat cells cannot receive their normal amount of glucose to provide the ATP they require. In addition, the unusually large amount of glucose in the blood overwhelms the kidneys' ability to reabsorb it from the kidney filtrate, and glucose appears in the urine. Fortunately, this form of the disease is treatable with regular monitoring of blood glucose level and daily administration of insulin.

In humans, the more common form of diabetes, which accounts for roughly 90–95% of all cases (estimates range between 15 and 20 million people in the U.S. alone), is **type 2 diabetes mellitus (T2DM)**. In T2DM, the pancreas functions normally and is not attacked by the immune system. However, the cells of the body lose much of their ability to respond to insulin for reasons that are still unclear. T2DM is linked to obesity and can often be prevented or reversed with weight control. Drugs are also available that improve the ability of cells to respond to insulin. Formerly, T2DM was known as "adult-onset" diabetes, because it usually appeared in middle age. This name is no longer used, because the rising tide of childhood obesity in developed countries like the U.S. has sparked a dramatic increase in the number of young people with the disease.

The discovery of insulin and its application to the human condition was one of the greatest and most influential achievements in the history of medical research. Next, we describe how this occurred.

FEATURE INVESTIGATION

Banting, Best, MacLeod, and Collip Were the First to Isolate Active Insulin

In the 19th century, scientists discovered that in addition to the exocrine part of the pancreas, the organ contains spherical clusters of cells that are not associated with its exocrine digestive functions. These clusters were called islets by their discoverer, German physiologist Paul Langerhans. By the early 20th century, German scientists had discovered that in dogs, complete removal of the pancreas resulted immediately in T1DM. Researchers assumed, therefore, that cells of the islets of Langerhans produced a factor of some kind that prevented diabetes—in other words, it helped maintain glucose homeostasis by preventing the blood glucose level from getting too high. The factor, however, proved impossible to isolate using the chemical purification methods available at the time. Researchers assumed that during the process of grinding up a pancreas to produce an extract, the factor was destroyed by the digestive enzymes of the exocrine pancreas.

This problem was eventually solved in 1921 by a team of Canadian scientists working at the University of Toronto: surgeon Frederick Banting, medical student Charles Best, and biochemist James Bertram Collip. The work was performed in the laboratory of John MacLeod, a renowned expert in carbohydrate metabolism at the time. Banting had read a paper in a medical journal that described a deceased patient in whom the pancreatic duct, which carries digestive juices to the small intestine, had become clogged due to calcium deposits. The closed duct caused pressure to build up behind the blockage, which eventually caused the exocrine part of the pancreas to atrophy and die. The islets of Langerhans, however, survived intact. Banting hypothesized that if he were to tie off, or ligate, the pancreatic ducts of an animal, and then wait a sufficient time for the exocrine pancreas to die, he could more easily obtain an active glucose-lowering factor from the remaining islets without the problem of contamination by digestive enzymes.

Banting and Best proceeded to ligate the pancreatic ducts of several dogs as shown in **Figure 50.9**. After waiting 7 weeks, an amount of time they previously had determined was sufficient for the exocrine part of the pancreas to atrophy, they prepared extracts of the remaining parts of the pancreas, including its islets of Langerhans. This was done by removing the atrophied pancreas from each dog and grinding it up with a mortar and pestle in an acid solution. The extract was then injected into a second group of dogs in which diabetes had previously been induced by surgically removing the pancreas. The researchers discovered that the extract was capable of keeping the diabetic dogs alive for a brief time. However, they were unable to isolate sufficient quantities of the active factor in the extract to keep the experiments going longer than a day or so, nor were they able to obtain sufficiently pure factor to prevent side effects such as infection and fever.

This is where Collip's expertise came in. Collip developed a method to precipitate contaminating proteins from the extract by adding alcohol to the acid (step 3, Figure 50.9). Different proteins will precipitate from a solution in response to different concentrations of alcohol. Using a concentration too low to cause the active factor to precipitate, Collip was able to remove most of the contaminating proteins, leaving behind a much more purified and safe extract. The extract was then filtered to remove debris and further extracted to remove lipids. In this step, a hydrophobic solvent was added to the partially purified extract; the lipids in the extract dissolved in the solvent, which could then be decanted and removed. This was done because the researchers believed that the factor was either a peptide or a small protein and by removing lipids, they would obtain a more purified preparation. Finally, the purified extract was concentrated by evaporating the alcohol, which increased its potency. When increasing amounts of this purified factor were injected into diabetic dogs, their blood glucose levels decreased and even returned to normal, and there was no longer any glucose in their urine. Diabetic dogs that received only a control solution that did not contain the factor continued to show a high glucose level in the blood and urine, compared to healthy controls.

In an ironic twist, scientists later recognized that the ligation procedure was not necessary to obtain the active factor. In fact, previous attempts to obtain the factor probably failed not because of enzymatic actions of the exocrine pancreas, but rather because the purification steps were too crude to yield a sufficient amount of purified compound. In later experiments, two subsequent innovations enabled the researchers to obtain larger amounts of the factor and more accurately assess its potency without using the ligation procedure. First, the researchers chose the very large pancreases of cows, obtained at a local slaughterhouse, for the starting material from which to prepare the extracts. Second, Collip developed a highly sensitive assay

to more precisely measure the concentration of glucose in the blood of an animal before and after injection of the purified factor. The combination of the improved chemical purification steps, large amounts of starting material, and improved assays for testing the extract proved to be the keys that enabled the team to test the factor, which they eventually named insulin, on human patients. The first successful test came in 1922 on a 14-year-old boy in Toronto, who was seriously ill from T1DM. In 1982, recombinant human insulin first became available for widespread use.

The success of the team in rapidly isolating insulin and proving its effectiveness was so significant that Banting and MacLeod were awarded the Nobel Prize the following year, in 1923. The Nobel committee felt that Banting and MacLeod were the leaders of the project and that it was they who deserved

Figure 50.9   The isolation of insulin by Banting, Best, MacLeod, and Collip.

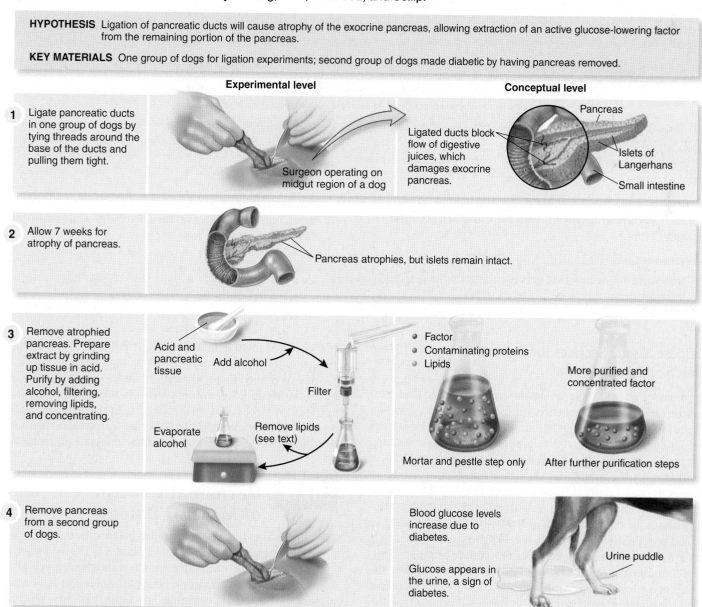

HYPOTHESIS Ligation of pancreatic ducts will cause atrophy of the exocrine pancreas, allowing extraction of an active glucose-lowering factor from the remaining portion of the pancreas.

KEY MATERIALS One group of dogs for ligation experiments; second group of dogs made diabetic by having pancreas removed.

Experimental level   Conceptual level

1 Ligate pancreatic ducts in one group of dogs by tying threads around the base of the ducts and pulling them tight.

Surgeon operating on midgut region of a dog

Ligated ducts block flow of digestive juices, which damages exocrine pancreas.

Pancreas
Islets of Langerhans
Small intestine

2 Allow 7 weeks for atrophy of pancreas.

Pancreas atrophies, but islets remain intact.

3 Remove atrophied pancreas. Prepare extract by grinding up tissue in acid. Purify by adding alcohol, filtering, removing lipids, and concentrating.

Acid and pancreatic tissue
Add alcohol
Filter
Evaporate alcohol
Remove lipids (see text)

• Factor
• Contaminating proteins
• Lipids

Mortar and pestle step only

More purified and concentrated factor

After further purification steps

4 Remove pancreas from a second group of dogs.

Blood glucose levels increase due to diabetes.

Glucose appears in the urine, a sign of diabetes.

Urine puddle

5 Inject either purified extract or control solution into the diabetic dogs. Determine the levels of glucose in the blood and in the urine.

Diabetic dog

6 THE DATA

7 CONCLUSION Atrophy of the exocrine portion of the pancreas eliminated digestive enzymes that would have degraded the glucose-lowering factor (insulin) during its purification from the pancreas. Along with improved chemical purification procedures, this atrophy step allowed researchers to obtain a highly purified and fully active glucose-lowering factor.

8 SOURCE Banting, F.G., and Best, C.H. 1922. The internal secretion of the pancreas. *Journal of Laboratory and Clinical Medicine* 7:256–271.

the prize, but the two scientists disagreed. Banting shared the monetary portion of his award with Best, and MacLeod shared his with Collip.

Experimental Questions

1. How did Banting and Best propose to obtain the glucose-lowering factor produced by the pancreas?

2. What was Collip's contribution to the isolation of insulin?

3. What subsequent innovations led to large-scale production of insulin for human patients?

Adipose Tissue Secretes the Hormone Leptin That Regulates Appetite

Hormones also contribute to metabolism by exerting effects on appetite and, as a consequence, on food consumption. As you might imagine from its importance as the body's major storage site for energy-yielding fat, a chief source of appetite-regulating hormones is adipose tissue. One such hormone introduced in Chapter 46 is the protein leptin, which has been observed in species from all vertebrate classes. Leptin is released by adipose cells into the blood in direct proportion to the amount of adipose tissue in the body, and it acts on the hypothalamus to inhibit appetite (refer back to Figure 46.9). When adipose stores are low, however, leptin secretion from adipose cells decreases, resulting in a reduced blood level of leptin. This removes the inhibitory effect of leptin on appetite, resulting in increased food consumption by the animal. In these ways, the amount of

energy stored in an animal's body in the form of fat is communicated to the brain to regulate how hungry an animal feels.

We have seen that hormones influence energy homeostasis by regulating appetite, nutrient levels in the blood, transport of nutrients into cells, and metabolic rate. In the next section, we consider how hormones control another feature of homeostasis, that of regulating concentrations of key minerals in the blood.

50.4 Hormonal Control of Mineral Balance

All animals must maintain a proper balance in their cells and fluids of minerals such as Ca^{2+}, Na^+, and K^+. These ions participate in numerous functions common throughout much of the animal kingdom. For example, the partitioning of ions across plasma membranes determines in large part the electrical properties of cells like neurons. In this section, we will see how maintaining a homeostatic balance of these ions in an animal's body is coordinated in large part by hormones.

Vitamin D and Parathyroid Hormone Regulate the Calcium Ion Level in Blood

Calcium ions play critical roles in neuronal transmission, heart function, muscle contraction, and numerous other events. Therefore, the concentration of Ca^{2+} in the blood is among the most tightly regulated variables in an animal's body.

Calcium is obtained from the diet and absorbed through the small intestine. Like all charged molecules, Ca^{2+} cannot readily cross plasma membranes, including those of the epithelial cells of the small intestine, and thus its transport must be facilitated. This process is controlled by the action of a derivative of **vitamin D**. In most mammals that receive regular exposure to the sun, skin cells produce vitamin D from a precursor called 7-dehydrocholesterol, a reaction that requires the energy of ultraviolet light (**Figure 50.10**). In addition, many animals obtain vitamin D from food or, in people today, from milk supplemented with the vitamin. This is especially important for people who live at latitudes that receive little sunlight for much or part of the year.

Before it can act, vitamin D must first be modified by two enzymes that both add one hydroxyl group to specific carbon atoms, first in the liver and then in the kidney. The final active product is called 1,25-dihydroxyvitamin D. This molecule is a hormone because it is secreted into the extracellular fluid by one organ—the kidneys—and then circulates in the blood and acts on a distant target tissue—the small intestine. The major function of 1,25-dihydroxyvitamin D in the intestine is to activate a Ca^{2+} transporter in the intestinal epithelium, thereby stimulating the absorption of Ca^{2+}. The calcium ions then enter the blood, from where they are delivered to tissues for such activities as building bone and maintaining nerve, muscle, and heart functions. If the active form of vitamin D is not present in the blood, the bones lose calcium and become weakened. In children, this results in the weight-bearing bones of the legs becoming deformed, a condition called rickets.

Even when calcium is not present in the diet, or when 1,25-dihydroxyvitamin D is not formed in normal amounts because of insufficient exposure to sunlight, the blood level of Ca^{2+} does not normally decrease dramatically, because of a hormone called **parathyroid hormone** (**PTH**). This hormone is secreted from several small glands called parathyroid glands, which in humans are attached to the back of the thyroid gland (**Figure 50.11a**). PTH acts on bone to stimulate the activity of cells that dissolve the mineral part of bone. This releases Ca^{2+}, which then enters the blood (**Figure 50.11b**). Typically, only a very small fraction of the total Ca^{2+} in bone is removed in this way. Therefore, besides providing a skeletal framework for the vertebrate body, bone also serves as an important reservoir of Ca^{2+}. PTH also acts to increase reabsorption of Ca^{2+} in the kidneys, such that less Ca^{2+} is excreted in the urine. Without PTH, calcium homeostasis is not possible. The total absence of PTH is fatal in humans and other mammals.

Figure 50.10 Synthesis of the active hormone formed from vitamin D or its precursor in the skin.

Concept check: *Would you predict that all mammals synthesize the active form of vitamin D using the energy of sunlight?*

In addition to 1,25-dihydroxyvitamin D and PTH, another hormone called **calcitonin** plays a role in Ca^{2+} homeostasis in some vertebrates, notably fishes and possibly some mammals, including young children (but not adult humans). Calcitonin is produced in and secreted from cells in the thyroid gland. Its function is the opposite in many respects to that of PTH. Calcitonin promotes excretion of Ca^{2+} via the kidneys and deposition of Ca^{2+} into bone, thereby lowering blood Ca^{2+} levels. This is especially important in marine fishes, because of the high Ca^{2+} content of seawater and the entry of this ion into the body fluids of the animals.

Several Hormones Regulate Sodium and Potassium Ions in Vertebrates

Like calcium, concentrations of sodium and potassium ions in the body fluids of most animals are tightly regulated, because

Pharynx (view from the back)

Thyroid gland

Parathyroid glands

Esophagus

Trachea

(a) Location of the parathyroid glands

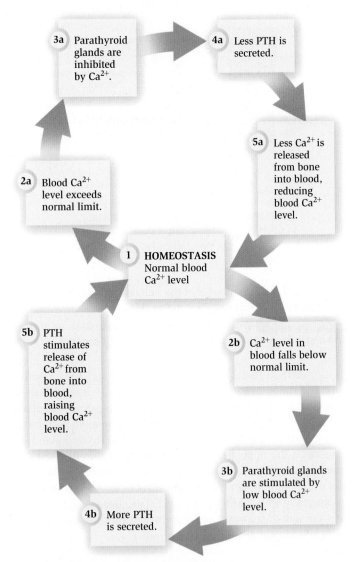

3a Parathyroid glands are inhibited by Ca^{2+}.

4a Less PTH is secreted.

2a Blood Ca^{2+} level exceeds normal limit.

5a Less Ca^{2+} is released from bone into blood, reducing blood Ca^{2+} level.

1 HOMEOSTASIS Normal blood Ca^{2+} level

5b PTH stimulates release of Ca^{2+} from bone into blood, raising blood Ca^{2+} level.

2b Ca^{2+} level in blood falls below normal limit.

3b Parathyroid glands are stimulated by low blood Ca^{2+} level.

4b More PTH is secreted.

(b) Homeostatic control of blood Ca^{2+} level

Figure 50.11 Parathyroid glands and the role of parathyroid hormone in calcium homeostasis. (a) The four small parathyroid glands are located behind the thyroid. (b) The action of PTH: Steps 2a–5a occur when Ca^{2+} is in excess; steps 2b–5b occur when Ca^{2+} is below normal.

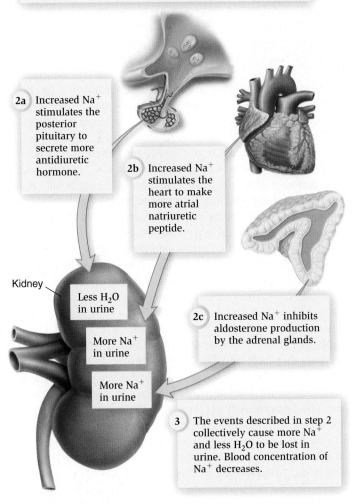

1 NaCl (table salt) is ingested. NaCl dissolves in the gut and eventually is absorbed into the blood.

2a Increased Na^+ stimulates the posterior pituitary to secrete more antidiuretic hormone.

2b Increased Na^+ stimulates the heart to make more atrial natriuretic peptide.

Kidney

Less H_2O in urine

More Na^+ in urine

More Na^+ in urine

2c Increased Na^+ inhibits aldosterone production by the adrenal glands.

3 The events described in step 2 collectively cause more Na^+ and less H_2O to be lost in urine. Blood concentration of Na^+ decreases.

Figure 50.12 An example of sodium balance achieved by the coordinated actions of three hormones.

Concept check: Why is there more than one hormone that regulates sodium and potassium balance?

these ions play crucial roles in membrane potential formation and action potential generation (see Chapter 41), among other functions. Like calcium, sodium and potassium are ingested in the diet and excreted in the urine in vertebrates. Unlike calcium, however, no large reservoirs exist in the body for sodium and potassium. One of the key mechanisms that regulates the concentrations of these ions in the blood is altering the rate of sodium, potassium, and water reabsorption from the urine as it is being formed in the kidneys. This is accomplished in large part by the actions of three hormones: ADH, aldosterone, and atrial natriuretic peptide (**Figure 50.12**). All three of these hormones exert their effects on the kidneys.

The vertebrate kidney normally reabsorbs most sodium and potassium from the fluid filtered through the kidney glomerulus (see Chapter 49). However, dietary and other changes can alter the concentrations of these ions in the blood. When this happens, the kidney works to restore the ions to their normal concentrations. For example, if the blood sodium level increases above normal, the osmolarity of the blood will increase. Osmoreceptors in the brain detect this and stimulate ADH secretion from the posterior pituitary gland. Recall from Chapter 49 that ADH acts on the kidney to reabsorb water from the forming urine. By increasing the amount of ADH available to act on the kidneys, less water is excreted in the urine. In addition, the synthesis and secretion of the adrenal steroid hormone aldosterone is directly inhibited in response to an increase in Na^+ concentration. Normally, aldosterone increases sodium reabsorption in the kidney, and therefore, its absence results in more sodium excretion in the urine. Finally, **atrial natriuretic peptide** (**ANP**) is secreted from the atria of the heart whenever the blood level of sodium increases. ANP causes a natriuresis (a loss of sodium in the urine; from the Latin *natrium*, meaning sodium) by decreasing sodium reabsorption. Thus, ANP and aldosterone have opposite effects on sodium balance in the body, which is why their concentrations in the blood tend to rise and fall in opposite directions. The combined effect of decreasing water loss in the urine and reabsorbing less sodium is to decrease the concentration of sodium in blood and other body fluids, returning its concentration to normal.

50.5 Hormonal Control of Growth and Differentiation

Hormones play a nearly universal role in controlling growth and differentiation in animals. Growth can occur slowly over long periods or in brief spurts, with some animals exhibiting both types of growth. Many mammals, for example, grow slowly but steadily until puberty, then experience a period of rapid growth, followed by slower rates of growth, and finally cessation of growth. Many insects, however, grow and develop in spurts during molting periods. Although growth is determined by many factors, notably adequate nutrition, it is regulated in large part by the endocrine system.

Growth is distinguished from differentiation, which is the process by which cells form tissues with specific functions, and tissues develop into larger and more complex structures. In this section, we will examine how both processes depend in part on the endocrine system.

Vertebrates Require a Balance of Several Hormones for Normal Growth

Several hormones stimulate growth in vertebrates. The anterior pituitary gland produces **growth hormone** (**GH**), which is under the control of the hypothalamus (see Table 50.2). GH acts on the liver to produce another hormone, called **insulin-like growth factor-1** (**IGF-1**). In mammals, IGF-1 stimulates the elongation of bones, especially during puberty, when mammals become reproductively mature. This growth is further accelerated by the steroid hormones of the gonads, leading to the rapid pubertal growth spurt. Eventually, however, the gonadal hormones cause the growth regions of bone to seal, preventing any further bone elongation.

GH, IGF-1, and gonadal steroid hormones continue to be produced in adult vertebrates—including humans—even though growth has ceased. In adulthood, GH serves metabolic functions, such as helping to regulate the levels of glucose and fatty acids in the blood. The gonadal steroids are important for reproduction in adults (see Chapter 51).

In rare cases, a tumor of the GH-secreting cells of the anterior pituitary gland produces excess GH. If this occurs during childhood, a person with this disorder can grow very tall and is known as a **pituitary giant**. Soon after puberty, growth stops. If a tumor causes a high GH level after puberty, the excess GH causes many bones, such as those of the hands and feet, to thicken and enlarge, a condition known as **acromegaly** (**Figure 50.13**). People with acromegaly are generally treated with a synthetic form of somatostatin, which as you may recall from Table 50.2 is the inhibitor of GH made by the hypothalamus. In many cases, however, the tumor must be surgically removed.

By contrast, if the pituitary fails to make adequate amounts of GH during childhood, the concentrations of GH and IGF-1 in the blood will be lower than normal. In such cases, growth is stunted, resulting in **short stature** (formerly called **pituitary dwarfism**). Individuals with this condition can be treated with injections of recombinant human GH and will grow to relatively normal height, as long as treatment begins before puberty ends so that the bones may still elongate.

Hormones are also required during fetal life, not for growth but for development of the brain, lungs, and other organs into fully mature, functional structures. For example, cortisol from the fetal adrenal gland is vital for proper lung formation. Premature babies born before the adrenal glands have matured have lungs that are not capable of expanding normally; therefore such babies require hospitalization to survive.

As another example, thyroid hormones impact differentiation among all vertebrates. In amphibians, thyroid hormones play a critical role in metamorphosis, notably in tadpoles, where they promote the resorption of the tail and development of the legs (**Figure 50.14**). This effect can be dramatically demonstrated by experimentally decreasing or increasing the tadpole's thyroid hormone level, which results respectively in a tadpole that does not transform into a frog or in a tadpole in which metamorphosis occurs sooner than normal resulting in a tiny frog. In fishes, thyroid hormones play an equally critical role in differentiation. For example, among species of flatfish such as flounder that live on the ocean floor, thyroid hormones are responsible for the characteristic change in appearance that occurs in these species as they settle into a sedentary existence on the ocean bottom. The fins and gill covers migrate to the dorsal surface facing the water; the dorsal body surface becomes pigmented; and most remarkably, the eyes migrate to the same side of the head such that one eye is not unused on the side facing the ocean floor. Each of these metamorphic events is under the direct control of thyroid hormones.

Invertebrates Grow and Develop in Spurts Under the Control of Three Major Hormones

Like vertebrate endocrine pathways, hormonal control systems in insects and other invertebrates often involve multiple glands and neural structures acting in concert. In insects, for example,

Figure 50.13 Acromegaly in one individual from a pair of identical twins. This disease is caused by a high level of growth hormone, which leads to the enlarged bones in the hands and feet.

Concept check: *Did the individual on the left develop a growth hormone disorder before or after puberty?*

the endocrine system is critical for the growth and development of larvae and their eventual differentiation into pupae (**Figure 50.15**). In the case of larval growth and metamorphosis, specialized neurosecretory cells periodically secrete a hormone called **prothoracicotropic hormone** (**PTTH**), which then stimulates a pair of endocrine glands called the prothoracic glands. These glands, located in the thorax, synthesize and secrete into the hemolymph a steroid hormone called **20-hydroxyecdysone** (not found in vertebrates). This hormone is secreted only periodically. In response to each burst of secretion, the larva undergoes a rapid differentiation and molts (sheds its cuticle) and begins a new developmental period until it molts again in response to another episode of secretion. The molting process is known as ecdysis, from which the name of the hormone is derived.

Throughout larval development, paired neurosecretory structures behind the brain, called the corpus allata, secrete

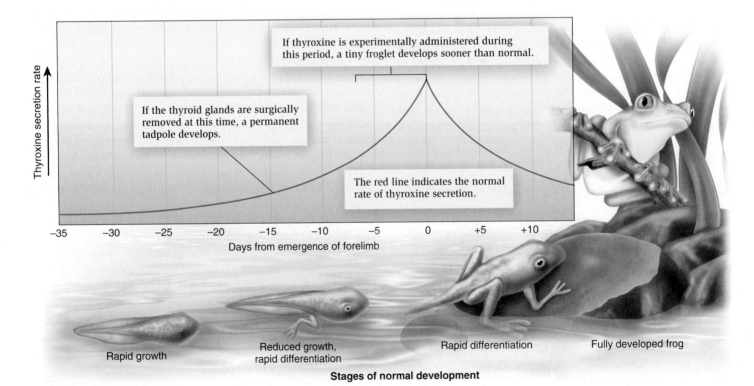

If thyroxine is experimentally administered during this period, a tiny froglet develops sooner than normal.

If the thyroid glands are surgically removed at this time, a permanent tadpole develops.

The red line indicates the normal rate of thyroxine secretion.

Thyroxine secretion rate

−35 −30 −25 −20 −15 −10 −5 0 +5 +10
Days from emergence of forelimb

Rapid growth

Reduced growth, rapid differentiation

Rapid differentiation

Fully developed frog

Stages of normal development

Figure 50.14 The effect of thyroid hormones on tadpole development. Experimental manipulation of thyroid hormone levels can retard or accelerate this development.

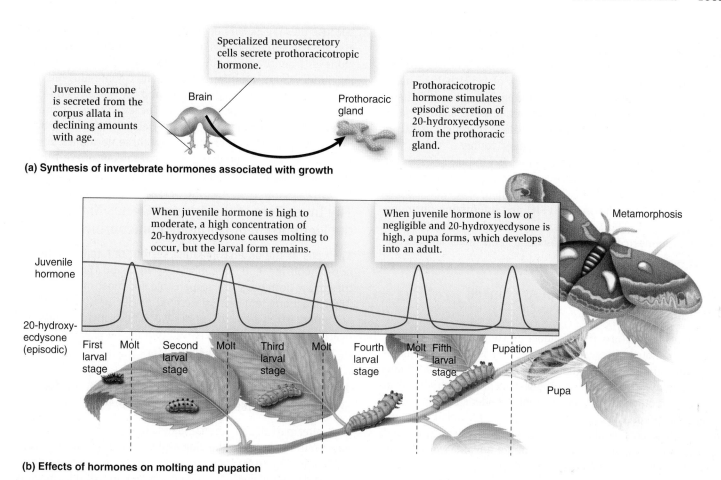

(a) Synthesis of invertebrate hormones associated with growth

(b) Effects of hormones on molting and pupation

Figure 50.15 **Hormonal control of insect development.** Development of insects requires the coordinated actions of three hormones. Note: The relative levels and patterns of 20-hydroxyecdysone and juvenile hormone in this figure are schematic and not representative of all insects.

Concept check: *In what part of a cell do you predict the receptor for the steroid 20-hydroxyecdysone would be located?*

another hormone, a protein called **juvenile hormone** (**JH**). JH determines the character of the molt that is induced by 20-hydroxyecdysone, by acting to prevent metamorphosis into an adult (hence the name "juvenile," which reflects that fact that JH fosters the larval stage). As the larva ages, however, the amount of JH it produces gradually declines until its concentration is nearly zero. During this time, 20-hydroxyecdysone continues to be periodically secreted. The decline in JH below a certain concentration results in the transition from larva to pupa in response to a burst of 20-hydroxyecdysone; the near absence of JH is a prerequisite for the final step of metamorphosis into an adult.

50.6 Hormonal Control of Reproduction

The topic of reproduction is covered in Chapter 51, but it is worth noting here that in all vertebrates and probably most invertebrates, reproduction is closely linked with endocrine function. In this section, we explore the most common reproductive hormones and their actions in male and female animals.

The Gonads Secrete Sex Steroids That Influence Most Aspects of Reproduction

Hormones produced by the gonads of animals play vital roles in nearly all aspects of reproduction, from reproductive behaviors to the ability to produce offspring. In males, the gonads are called testes. They produce and house the gametes, called sperm cells. In addition, the testes produce several related steroid hormones collectively termed **androgens**. In many vertebrates, including humans, the primary androgen is **testosterone**. The gonads of females are called the ovaries, which contain the female gametes, or egg cells. Other cells within the ovaries secrete two major steroid hormones, **progesterone** and a family of related hormones called **estrogens**. The major estrogen in many animals, including humans, is **estradiol**. Collectively, the male and female gonadal steroid hormones are referred to as the sex steroids.

The sex steroids are responsible for male- or female-specific reproductive behaviors associated with courtship and mating. In vertebrates, androgens from the developing testes are required for development of the male phenotype. Exposure of fetuses to abnormal levels of male or female hormones can lead to ambiguous phenotypes and sexual behavior after birth. Sex steroids are also chiefly responsible for development of secondary sex characteristics for each sex, which in humans include such things as growth of the appropriate external genitals, growth of the breasts in women, development of facial hair in men, and distribution and amount of fat and muscle in the body (see Chapter 51). Finally, sex steroids are required for maturation of gametes, the transition of young animals to reproductive maturity, and the ability of females to produce young.

The ability of the testes and ovaries to produce sex steroids depends on the presence of the gonadotropins secreted by the anterior pituitary, which are the same in both sexes and include **follicle-stimulating hormone** (**FSH**) and **luteinizing hormone** (**LH**) (see Table 50.2). Therefore, identical anterior pituitary gland hormones control the gonads in both male and female animals.

Nutrition and Reproduction Are Linked Through Hormones

An interesting feature of the endocrine control of reproduction is the long-noted observation that puberty is delayed in mammals, including humans, that are very undernourished. Similarly, fertility—the ability to produce offspring—is reduced in women and other adult female mammals under such conditions. This makes sense, because supporting the nutritional demands of a growing fetus would be difficult without good nutrition and energy stores. In such a case, it is more advantageous for an animal to delay reproduction until sufficient food is available.

How does the brain of a female mammal determine when sufficient energy is stored in her body to support pregnancy? The answer appears to be partly the result of hormonal signals. For example, if the amount of fat in a woman's body decreases, so does the concentration of leptin in her blood, as described earlier. Leptin has been demonstrated to stimulate synthesis and secretion of reproductive hormones such as FSH and LH. Consequently, undernourishment that causes a decrease in leptin results in reduced production of reproductive hormones, thereby contributing to a loss of fertility. Therefore, leptin acts as a link between adipose tissue and the reproductive system.

50.7 Hormonal Responses to Stress

Stress in animals is often defined as any real or perceived threat to survival. This can take many forms depending on the species. To a crab, a passing shadow may indicate a hungry shark passing overhead, but to a human, a shadow may simply mean that a cloud has passed in front of the sun. Similarly, a severe storm can be extremely stressful for birds in exposed nests, but not to animals that live in sheltered environments. Crowding is adaptive for penguins trying to keep warm on ice floes, but it is stressful to mice and rats.

In the 1920s, Canadian physiologist Hans Selye made the remarkable discovery that regardless of the nature of the stress in any given species, a mammal's adaptive responses to stress were highly similar. Later this was determined to be true

because these responses depend on hormones from the adrenal glands. The adrenal glands, so named because they sit atop the kidneys (from the Latin *ad*, meaning toward, and *renis*, meaning kidney), are multifunctional glands containing an inner region called the adrenal medulla and an outer region called the adrenal cortex (**Figure 50.16a**). These two regions produce different hormones that affect the vertebrate response to stress. In this section, we will examine how these hormones function and how they act to prepare an animal to confront or escape a challenge—the fight-or-flight response.

The Inner Core of the Adrenal Glands Makes the Fight-or-Flight Hormone Epinephrine

The cells of the adrenal medulla secrete the amine hormones norepinephrine and epinephrine (**Figure 50.16b**). Together, these two hormones are responsible for most of the fight-or-flight reactions that were described earlier in Chapter 42 and are summarized here in **Table 50.3**. These include improved heart and lung function, increased production of energy sources by the liver, and increased alertness.

The Outer Regions of the Adrenal Glands Make Steroid Hormones in Response to Stress

The outer part of the adrenal gland, the cortex, is itself subdivided into three zones: the glomerulosa, the fasciculata, and the reticularis (Figure 50.16b). The outer zone, the glomerulosa, is the region that makes aldosterone. As mentioned, this mineralocorticoid hormone acts to maintain mineral balance. The innermost cortical zone is known as the reticularis, which functions in humans to make certain androgens, but whose function in other animals is not as clear. The bulk of the cortex is the middle zone, the fasciculata. The glucocorticoid hormones are made here, among them cortisol and other structurally related steroids.

Table 50.4 summarizes some of the major actions of glucocorticoids in vertebrates. Glucocorticoids are catabolic hormones; that is, they promote the breakdown of molecules and macromolecules. For example, they act on bone, immune, muscle, and fat tissue to break down proteins and lipids to provide energy for the body's cells. This is important for an animal that is facing an acute stress, because feeding and digestion both stop during fight-or-flight reactions, leaving internal stores as the only source of energy. These adrenal steroids are called glucocorticoids because one of their major actions is to promote gluconeogenesis in the liver during times of stress, thus providing glucose to the blood.

Responding to acute stress is an important function of glucocorticoids, but excessive production of these steroids can create serious problems. Such a situation might arise if an animal were chronically stressed for some reason, for example, due to habitat destruction, intraspecies competition and aggression, or chronic infection. Glucocorticoids, together with amine hormones such as epinephrine, increase the activity of the heart and can raise blood pressure. Therefore, prolonged stress can lead to disorders of the heart and blood vessels. In addition, a chronic increase in the glucocorticoid level can suppress the

(a) The location and structure of the adrenal glands

(b) Cellular organization and hormones of the adrenal cortex and medulla

Figure 50.16 Location, cellular organization, and function of the adrenal glands.

Table 50.3	The Fight-or-Flight Response to Stress
System	**Responses**
Cardiovascular	Increases heart rate and strength of heart contractions to maximize pumping of blood to all parts of the body; dilates blood vessels entering tissues requiring more oxygen—such as skeletal muscle—and constricts blood vessels to regions of less immediate importance—such as the gut and kidney
Respiratory	Dilates small airways (bronchioles) to reduce resistance to airflow in mammals; increases rate and depth of breathing to maximize oxygen intake and carbon dioxide elimination
Metabolic	Increases glycogenolysis in muscle to provide glucose for muscle cells and in the liver to provide glucose to the blood, where it can reach all body cells; increases breakdown of adipose triglycerides into usable fuel (fatty acids) that can then enter the bloodstream; stimulates glucagon secretion, which acts on the liver to promote gluconeogenesis
Nervous	Increases arousal and alertness; inhibits nonessential functions such as appetite

Table 50.4	The Major Actions of Glucocorticoids in Stress
Site	**Actions**
Liver	Stimulate gluconeogenesis, thus providing glucose to the blood
Adipose tissue	Stimulate breakdown of triglycerides into fatty acids and glycerol for fuel
Muscle and adipose tissue	Inhibit sensitivity to insulin, making more glucose available to brain cells, which do not require insulin to move glucose across their plasma membranes
Bone	Inhibit bone growth and formation, because such processes require large amounts of nutrients that could be used to combat stress instead
Lungs	Stimulate lung maturation in the fetus
Immune	Suppress immune system function and reduce inflammation
Other	Regulate sodium and chloride balance in migratory fishes; stimulate nervous system development in most vertebrates; stimulate protein breakdown to provide amino acids to the liver for gluconeogenesis; inhibit reproduction

immune system because of the catabolic actions of glucocorticoids on immune tissue. This may compromise an animal's ability to fight infection. For example, the prolonged stress a salmon experiences when it makes its exhausting swim upriver to its spawning grounds causes massive suppression of its immune system, and many mature salmon that die after spawning are found to be riddled with infections. People who are chronically stressed also tend to be susceptible to infections or to having dormant viruses suddenly flare up, such as the type of herpesvirus that causes cold sores.

Even an animal's ability to reproduce and grow is negatively affected by chronic stress, as noted most dramatically in medical situations involving humans. In children, for example, chronic exposure to a high level of glucocorticoids may stunt growth, at least temporarily. In women, chronic stress may result in the loss of monthly reproductive cycles. The explanation for this may be understood from an evolutionary perspective. Chronic stress generally suggests that an animal's fitness or survival is in jeopardy. Therefore, it is advantageous to have all energetically demanding activities—such as growth, reproduction, and even immune activity—come to a halt if these activities are not immediately required for staying alive.

The preceding discussion illustrates only a few ways in which the endocrine system can affect health and fitness in many animals, including humans. In the last section of this chapter, we highlight some aspects of hormones that are exerting a large impact on public health in humans.

50.8   Impact on Public Health

The list of endocrine-related diseases in the human population is a long one, ranging from relatively common disorders such as diabetes and thyroid disease to rare ones that may affect only 1 in every 100,000 or more individuals. Hormones can be used to treat many of these diseases, but they can also be misused in ways that cause health problems. Also, the widespread industrial use of chemicals with hormone-like actions and their possible environmental consequences have become very much a part of the news. Let's consider how hormones are used therapeutically as well as misused, and then look at an example of how factors that alter the function of the human endocrine system can have important health implications.

Hormones Are Used to Treat Millions of People

Since the advent of synthetic hormone production, doctors have been able to provide individuals with hormones that their own bodies fail to produce. For example, millions of people self-administer insulin each day or take thyroxine to supplement their insufficient levels of thyroid hormones. Many women undergo hormone therapy to help induce pregnancy or control the symptoms of menopause. A growing number of men, too, are administered gonadal hormones if the levels produced by their bodies decline significantly in later life. Other examples include recombinant human growth hormone to treat abnormally short stature in prepubertal children, epinephrine inhalers to treat asthma, and glucocorticoids to treat inflammation, lung disease, and skin disorders, to name just a few.

Hormone Misuse Can Have Disastrous Consequences

Some individuals, typically those in competitive sports, self-administer hormones such as androgens. This practice may increase muscle mass and improve athletic performance. However, the price is steep. First, androgens exert negative feedback actions on FSH and LH secreted by the anterior pituitary gland. As a consequence, the anterior pituitary gland stops producing and secreting FSH and LH while the user is taking supplements containing androgens. In males, this causes the testes to shrink, as they no longer are making sperm, and the man becomes infertile; this is what happened to our case subject described at the beginning of the chapter. Androgen administration has also been linked to extreme aggressive behavior (**"roid" rage**), cardiovascular disease and heart attacks, skin problems, and certain cancers. Women using androgens run similar health risks as men but, in addition, develop masculinizing traits, including growth of body hair and thinning scalp hair.

Another example of hormone misuse is when they are used to boost the number of red blood cells in the circulation to increase the oxygen-carrying capacity of the blood (often called **blood-doping**). This has become common among athletes participating in long-distance aerobic activities, such as cycling and cross-country skiing. It has replaced the former practice of actually transfusing suspensions of red blood cells into a person's circulation. The hormone used is **erythropoietin (EPO)**, which acts by stimulating the maturation of red blood cells in the bone marrow and their release into the blood. EPO is normally made by the liver and kidneys in response to any situation where additional blood cells are required, such as following blood loss or when a person lives at high altitudes, where the oxygen pressure is low. When used abusively, however, the concentration of red blood cells reaches such a high level that the blood becomes much thicker than normal. This puts a serious strain on the heart, which must work harder to pump the thickened blood. Since the 1990s, the international cycling community has been rocked by an alarming trend of world-class European cyclists who died of heart attacks in the prime of their lives. These individuals had been using EPO to gain an unfair and, as it turns out, unwise advantage over their peers. Testing continues to detect injected EPO in the blood of some cyclists and other endurance athletes.

Synthetic Compounds May Act as Endocrine Disruptors

A recent phenomenon—disturbing because of its potential to impact reproduction in animals and people—is the growing amount of so-called **endocrine disruptors** found in lakes, streams, ocean water, and soil exposed to pollution runoff. These chemicals are derived in many cases from industrial waste and have molecular structures that in some cases resemble estrogen sufficiently to bind to estrogen receptors. If these compounds make their way into drinking water or food, they can exert estrogen-like actions or inhibit the actions of the body's own estrogen. This can lead to dramatic consequences on fertility and on the development of embryos and fetuses. The extent of the risk from these **xenoestrogens** (from the Greek *xeno*, meaning foreign) is hotly debated, but the number of mature, functional germ cells produced in animals as diverse as mollusks and human males has declined dramatically during the past 50 years in the U.S. In addition, researchers throughout the world have noted feminization of freshwater fishes downstream of wastewater facilities. For example, male fishes that were exposed to such conditions during development show increased production of proteins normally made by females bearing eggs. They also show changes in gonadal structures that resemble the female appearance. Further research is urgently needed to provide more information on the consequences of these contaminants to animal endocrine systems.

Summary of Key Concepts

50.1 Mechanisms of Hormone Action and Control

- The endocrine glands and other organs with hormone-secreting cells constitute the endocrine system.

- Endocrine glands contain epithelial cells that secrete hormones into the bloodstream, where they circulate throughout the body. Although slower than the electrical signaling of nervous systems, chemical signaling complements nervous system regulation through its varying actions in multiple locations across widely ranging time frames.

- The effects of a hormone may occur within seconds or require several hours to develop, and it may last for as short as a few minutes or as long as days.

- Hormones fall into three broad classes that include the amines, proteins/peptides, and steroids. The amines and the proteins/peptides generally share similar chemical properties and modes of action, whereas the steroid hormones act very differently from the other classes. (Table 50.1)

- Water-soluble hormones (amines and proteins/peptides) act on receptor molecules located in the plasma membrane, whereas lipid-soluble hormones (steroids) act on intracellular receptors.

- Blood levels of hormones can increase or decrease through two mechanisms: changing the rate of hormone synthesis by an endocrine cell and changing the rates at which hormones are removed or inactivated. (Figures 50.2, 50.3)

50.2 Links Between the Endocrine and Nervous Systems

- Sensory input from an animal's nervous system modulates the activity of certain endocrine glands and influences blood levels of many hormones.

- The hypothalamus is physically connected to the pituitary gland; the two lobes of the pituitary are the anterior and posterior pituitary glands. (Figure 50.4)

- Within the hypothalamus are numerous neurons that synthesize neurohormones and stimulate the anterior pituitary.

- The anterior pituitary gland synthesizes six different hormones that respond to the presence of hypothalamic neurohormones. They are adrenocorticotropic hormone, follicle-stimulating hormone, luteinizing hormone, growth hormone, prolactin, and thyroid-stimulating hormone. (Table 50.2)

- In mammals, the neuron terminals in the posterior pituitary gland store and secrete one of two hormones, oxytocin or antidiuretic hormone (ADH).

50.3 Hormonal Control of Metabolism and Energy Balance

- Hormones are partly responsible for regulating energy use by cells, that is, for modulating appetite, digestion, absorption of nutrients, and blood levels of energy sources such as glucose. Although many hormones are involved in these processes, two from the thyroid gland (thyroxine and triiodothyronine), two from the pancreas (insulin and glucagon), and one from adipose tissue (leptin) play especially important roles.

- The thyroid gland makes thyroid hormones, which contain iodide. A major action of thyroid hormones in adult animals is to stimulate energy consumption by many different cell types. (Figures 50.5, 50.6)

- The endocrine pancreas produces the peptide hormones insulin and glucagon, which have opposite effects on the blood glucose level. The adrenal glands produce steroid hormones known as glucocorticoids, which increase the blood glucose level. (Figure 50.7)

- Maintaining normal glucose and other nutrient levels in the blood is a vital process that keeps cells functioning optimally.

The combined short-term and long-term actions of insulin, glucagon, epinephrine, and cortisol help maintain normal blood glucose levels during fasting. (Figure 50.8)

- Groundbreaking research by Banting, Best, MacLeod, and Collip isolated insulin for therapeutic use in treating diabetes mellitus. (Figure 50.9)

- Adipose tissue is an important source of appetite-regulating hormones, including the protein leptin, which acts on the hypothalamus to inhibit appetite. Adipose cells release leptin into the blood in direct proportion to the amount of adipose tissue in the body.

50.4 Hormonal Control of Mineral Balance

- Because of the important roles that calcium plays in neuronal transmission, heart function, muscle contraction, and numerous other events, the concentration of Ca^{2+} in the blood is among the most tightly regulated variables in an animal's body. Vitamin D and parathyroid hormone, produced by the parathyroid glands, regulate the blood concentration of Ca^{2+}. (Figures 50.10, 50.11)

- Sodium and potassium play crucial roles in membrane potential formation, action potential generation, and other functions. A key mechanism that regulates blood concentrations of these ions is to alter the rate of sodium, potassium, and water reabsorption from the urine as it is being formed in the kidneys. This is accomplished in large part by the actions of ADH, aldosterone, and atrial natriuretic peptide (ANP). (Figure 50.12)

50.5 Hormonal Control of Growth and Differentiation

- Hormones play a crucial role in regulating growth and differentiation. In vertebrates, normal growth depends on a balance between growth hormone, insulin-like growth factor-1 (IGF-1), and gonadal hormones. Thyroid hormones affect differentiation among all vertebrates. (Figures 50.13, 50.14)

- In insects, prothoracicotropic hormone, 20-hydroxyecdysone, and juvenile hormone control the growth of larvae and their differentiation into pupae. (Figure 50.15)

50.6 Hormonal Control of Reproduction

- Hormones produced by the gonads play vital roles in nearly all aspects of reproduction. The ability of the testes and ovaries to produce the sex steroids depends on the gonadotropins, which are the same in both sexes and include follicle-stimulating hormone (FSH) and luteinizing hormone (LH).

- Leptin is required for fertility in mammals.

50.7 Hormonal Responses to Stress

- Regardless of the nature of the stress, animals' bodies respond to stress in similar ways. In vertebrates, hormones produced by the adrenal glands are the common denominator of the response to stress. Norepinephrine and epinephrine, produced by the adrenal medulla, are responsible for most fight-or-flight reactions. (Figure 50.16, Table 50.3)

- Glucocorticoids promote breakdown of storage compounds to provide energy for cells during stress. (Table 50.4)

50.8 Impact on Public Health

- Hormones are used therapeutically to treat a variety of human disorders, such as infertility and growth disorders.
- Androgen misuse can disrupt normal hormone levels and cause health risks such as aggression; cardiovascular disease; skin problems; cancer; and, in women, masculinizing traits. Blood-doping with erythropoietin can make blood dangerously thick.
- Endocrine disruptors, such as chemicals derived from industrial waste, may bind to estrogen receptors in animals' bodies. They may exert estrogen-like actions or inhibit the actions of the body's own estrogen.

Assess and Discuss

Test Yourself

1. Which is the defining feature of hormones?
   a. They are only produced in endocrine glands.
   b. They are secreted by one type of cell into the blood, where they may simultaneously reach many distant target cells and thereby alter cell function throughout the body.
   c. They are released only by neurons.
   d. They are never released by neurons.
   e. They are secreted into ducts, where they diffuse to another nearby gland or other structure.

2. Steroid hormones are synthesized from _____ and bind _____.
   a. proteins; membrane receptors
   b. fatty acids; membrane receptors
   c. tyrosine; intracellular receptors
   d. proteins; intracellular receptors
   e. cholesterol; intracellular receptors

3. Which of the following is not true about protein and peptide hormones?
   a. Most of them bind to receptors located on the cell membrane.
   b. Most of them are lipophilic.
   c. They are the most abundant class of hormones.
   d. They normally activate second messengers and intracellular signaling pathways.
   e. They bind noncovalently to receptors.

4. Which is not correct about the control of hormones?
   a. Many are regulated by negative feedback.
   b. In some cases, they may be controlled by changes in blood levels of certain nutrients such as glucose.
   c. In some cases, they may be controlled by changes in blood levels of certain minerals such as Ca^{2+}.
   d. Their blood levels are controlled by their rates of synthesis and degradation.
   e. The rate of degradation of a hormone may go up or down, but synthesis is always constant.

5. The hypothalamus and the pituitary gland are physically connected by
   a. arteries.
   b. the infundibular stalk and portal veins.
   c. the adrenal medulla.
   d. the spinal cord.
   e. the intermediate lobe.

6. Antidiuretic hormone
   a. increases water reabsorption in the kidneys.
   b. regulates blood pressure by constricting arterioles.
   c. decreases the volume of urine produced by the kidneys.
   d. increases blood pressure during times of high blood loss.
   e. does all of the above.

7. Which of the following pairs of hormones are involved in the regulation of blood calcium ion level in vertebrates?
   a. aldosterone and ANP
   b. insulin and glucagon
   c. parathyroid hormone and 1,25-dihydroxyvitamin D
   d. prolactin and oxytocin
   e. thyroxine and TSH

8. In invertebrates, molting of larvae is stimulated by
   a. growth hormone.   d. 20-hydroxyecdysone.
   b. cortisol.   e. aldosterone.
   c. juvenile hormone.

9. Which of the following is not true of glucocorticoids?
   a. They stimulate maturation of the fetal lungs.
   b. They promote a decrease in the blood glucose level.
   c. They inhibit the sensitivity of cells to insulin.
   d. They are lipophilic (lipid soluble).
   e. They reduce inflammation.

10. Endocrine disruptors are
    a. chemicals released by the nervous system to override the endocrine system.
    b. chemicals released by the male of a species to decrease the fertility of other males.
    c. drugs used to treat overactive endocrine structures.
    d. chemicals derived from industrial waste that may alter endocrine function.
    e. all of the above.

Conceptual Questions

1. What is the function of leptin, and what is the benefit of an adipose-derived signaling molecule in this context? Why does an animal have an appetite?

2. Describe the major functions of insulin and glucagon. When are they released into the blood? What might happen to a nonfasting mammal that was injected with a high dose of glucagon?

3. Distinguish between type 1 and type 2 diabetes mellitus.

Collaborative Questions

1. Discuss the role of hormones in insect development.

2. Discuss the role of the different steroid hormones and where they are produced in the human body.

Online Resource

www.brookerbiology.com

Stay a step ahead in your studies with animations that bring concepts to life and practice tests to assess your understanding. Your instructor may also recommend the interactive ebook, individualized learning tools, and more.

Animal Reproduction

51

Most animals reproduce by sexual reproduction. These mud daubers, shown here mating, reproduce by sexual reproduction, which favors genetic variation in species.

O ver a span of 40 years in the early 18th century, a woman known to history only as the "wife of Feodor Vassilyev," a Russian peasant, reportedly gave birth 27 times to a total of 69 children! If accurate, it is safe to say that this is about the uppermost realistic limit of human fertility, considering a pregnancy length of about 40 weeks, the rarity of a woman producing multiple births such as twins, and the reproductive life span of an adult woman. Impressive as such a number is, however, it pales in comparison to the numbers of offspring produced by many other animals. For example, many invertebrate species, such as the mud daubers shown in the chapter-opening photo, lay anywhere from several to dozens of eggs in 1 day. As an extreme example, a single female cod fish can lay several million eggs during one spawning season (although not all the eggs will result in offspring).

Indeed, when we examine all forms of life, we discover that individuals of every species are part of an unbroken cycle of life and death. Perpetuation of all life requires **reproduction**—the processes by which organisms replicate themselves and multiply. Because reproduction is the only way in which a species can perpetuate itself, enormous evolutionary pressure has been exerted on the processes of reproduction, and demands made on organisms by their environments have influenced how reproduction occurs.

The biological mechanisms that favor successful reproduction in the animal kingdom are extraordinarily diverse. Many of the observable differences in animal behavior and anatomy are the result of adaptations to environmental pressures that increase an animal's chances of reproducing. Both the behaviors and the anatomical specializations that promote reproduction are under the control of a variety of factors, particularly hormones.

In this chapter, we will initially focus on the diverse means of reproduction that occur throughout the animal kingdom, including asexual and sexual reproduction. Then we will highlight some of the anatomical, hormonal, and behavioral aspects of reproduction in mammals, including humans. We will conclude with a discussion of some key issues related to fertility (the ability to reproduce) in the human population today. We begin by examining the different ways animals reproduce.

51.1 Asexual and Sexual Reproduction

Asexual reproduction occurs when offspring are produced from a single parent, without the fusion of genetic material from two parents. The offspring are therefore clones of the parent. **Sexual reproduction** is the production of a new individual by the joining of two haploid reproductive cells called **gametes**, one from each parent. This produces offspring that are genetically different from both parents. In this section, we will consider the processes of asexual and sexual reproduction, as well as their advantages and disadvantages.

Asexual Reproduction

Asexual reproduction occurs in some invertebrates and in a small number of vertebrate species. Animals use one of three major forms of asexual reproduction: budding, regeneration, or parthenogenesis. **Budding**, which is seen in cnidarians, occurs

(a) Budding   (b) Regeneration

Figure 51.1 Examples of asexual reproduction. (a) A hydra with a bud on its left side. (b) A sea star regenerating a complete new body from a single arm.

when a portion of the parent organism pinches off to form a complete new individual (**Figure 51.1a**). In this process, cells from the parent undergo mitosis and differentiate into specific types of structures before the new individual breaks away from the parent. At any one time, a parent organism may have one, two, or multiple buds forming simultaneously. Budding continues throughout such an animal's lifetime.

Some animals, including certain species of sponges, echinoderms, and worms, reproduce by the **regeneration** of a complete organism from small fragments of their body. In the sea star, for example, an arm removed by injury or predation can grow into an entirely new individual (**Figure 51.1b**). Similarly, a flatworm bisected into two pieces will regenerate into two new individuals.

The asexual process called **parthenogenesis** (from the Greek, meaning virgin birth) is the development of offspring from an unfertilized egg. It occurs in several invertebrate classes and in a few species of fishes and reptiles. Animals produced by parthenogenesis are usually haploid. Some species—such as rotifers, many social insects, and the freshwater crustacean *Daphnia*—may reproduce either parthenogenically or sexually, depending on the time of year or environmental conditions. In honeybees, ants, and wasps, for instance, haploid males (drones) are produced by parthenogenesis, and diploid females (workers and queens) by sexual reproduction. Production of drones, whose major function is to mate with a queen, usually occurs in late spring. The queen stores sperm cells from several drones for up to 2 years, during which time she may lay hundreds of thousands of eggs, only some of which will be fertilized. As stored sperm runs out, the unfertilized eggs become new drones.

Sexual Reproduction

Sexual reproduction occurs in most animal species, invertebrate and vertebrate. As mentioned, sexual reproduction involves the joining of two gametes. The gametes are spermatozoa (usually shortened to **sperm cell**) from the male and **egg cells**, or **ova**,

(singular, **ovum**) from the female. When a sperm unites with an egg—a process called **fertilization**—each haploid gamete contributes its set of chromosomes to produce a diploid cell called a fertilized egg, or **zygote**. As the zygote undergoes cell divisions and begins to develop, it is called an **embryo**.

Advantages and Disadvantages of Asexual and Sexual Reproduction

Asexual reproduction provides a relatively simple way for an organism to produce many copies of itself, whereas sexual reproduction requires two individuals to produce offspring. What are the advantages and disadvantages of each?

Asexual reproduction has numerous advantages over sexual reproduction. First, an animal can reproduce asexually even if it is isolated from others of its own species, either because the animal is sessile (nonmotile) or rarely encounters another member of its species. Another advantage is that individuals can reproduce rapidly because they need not seek out, attract, and mate with an opposite sex. Asexual reproduction, therefore, is an effective way of generating large numbers of offspring. Although many kinds of animals reproduce asexually, it is more prevalent in species that live in very stable environments, with little selection pressure for genetic diversity in a population.

Compared with asexual reproduction, sexual reproduction is associated with unique costs. Two types of gametes (sperm and eggs) must be made, males and females require specialized body parts to mate with each other, and the two sexes must be able to find each other. Yet given that most eukaryotic species reproduce sexually, a question has intrigued biologists since the time of Darwin: What is the advantage of sexual reproduction for the perpetuation of a species?

In the context of species survival, the major difference between asexual and sexual reproduction is that sexual reproduction allows for greater genetic variation due to genetic recombination. Only certain alleles from each parent are passed on, and when a set of genes from one parent mix with a different set from the other parent, the offspring are never exactly like either of their parents. Thus, a hallmark of sexual reproduction is increased genetic variation between successive generations. One prevalent hypothesis about why sex evolved is that sexual reproduction allows more rapid adaptation to environmental changes than does asexual reproduction.

In particular, sexual reproduction allows alleles within a species to be redistributed via crossing over and independent assortment across many generations. As a result, some offspring carry combinations of alleles that promote survival and reproduction, while other offspring may carry less favorable combinations. As described in Chapter 24, natural selection can favor those combinations that promote greater reproductive success while eliminating offspring with lower fitness. By comparison, the alleles of asexual organisms are not reassorted from generation to generation. As a result, it is more difficult to accumulate potentially beneficial alleles within individuals of these species. Also, as described next, sexual reproduction may facilitate elimination of harmful alleles from a population.

FEATURE INVESTIGATION

Paland and Lynch Provided Evidence That Sexual Reproduction May Promote the Elimination of Harmful Mutations in Populations

Evolutionary biologists have suggested that the inability of asexual species to reassort alleles may be a key disadvantage compared to sexually reproducing species. To investigate this question, American researchers Susanne Paland and Michael Lynch recently studied the persistence of mutations in populations of *Daphnia pulex*, a freshwater organism commonly known as the water flea. The researchers chose this organism because some natural populations reproduce asexually, and others reproduce sexually.

In their experiment, shown in **Figure 51.2**, Paland and Lynch studied the sequences of several mitochondrial genes in 14 sexually reproducing and 14 asexually reproducing populations of *D. pulex*. The researchers hypothesized that asexual populations would be less able to eliminate harmful mutations.

As discussed in Chapter 24, random gene mutations that change the amino acid sequence of the encoded protein are much more likely to be harmful than beneficial. As we discussed, the alleles of sexually reproducing populations can be reassorted from generation to generation, thereby producing succeeding generations in which the detrimental alleles are lost from the population. As you can see in the data, the researchers discovered that both the sexual and asexual populations could eliminate highly deleterious mutations. Organisms harboring such mutations probably died rather easily. In addition, the sexual and asexual populations both retained mildly deleterious and neutral mutations. However, moderately deleterious mutations were eliminated from the sexual populations but not from the asexual ones. One interpretation of these data is that sexual reproduction allowed for the reassortment of beneficial and detrimental alleles, making it easier for sexual populations to eliminate those mutations that are moderately detrimental.

Figure 51.2   Paland and Lynch demonstrated the importance of sexual reproduction in reducing the frequency of maladaptive genetic mutations.

HYPOTHESIS   Sexual reproduction allows for greater mixing of alleles of different genes and thereby may prevent the accumulation of detrimental alleles in a population.

KEY MATERIALS   The researchers collected samples of *Daphnia pulex* from many natural populations. A total of 14 sexual populations and 14 asexual populations were studied.

Experimental level   Conceptual level

1   Isolate mitochondrial DNA from members of 28 populations of *D. pulex*. This involves breaking open cells and extracting the DNA (refer back to Figure 21.1).

Daphnia   DNA   Segments of mitochondrial DNA

2   Amplify regions of mitochondrial genes, using PCR. Subject the regions to DNA sequencing. The techniques of PCR and DNA sequencing are described in Chapter 20.

PCR (refer back to Figure 20.6)   DNA sequencing (refer back to Figure 20.8)

3   Using computer technology, align the sequences (refer back to Figure 21.12a) and determine the number of DNA changes that would cause amino acid substitutions. These amino acid changes were categorized as those that would be highly deleterious, moderately deleterious, mildly deleterious, or neutral for protein function. Compare the sexual and asexual populations for the persistence of these types of changes.

GGCACCTCACCC

GGCACCTAACCC

Stop codon

This change would be highly detrimental because it would put a stop codon into the gene.

4 **THE DATA**

Results from step 3:

Types of amino acid substitutions (The amino acid substitutions were due to rare mutations that occurred in the natural populations of *D. pulex*.)	% of total amino acid substitutions	Allowed to persist in	
		Sexual populations	Asexual populations
Highly deleterious	73.2	No	No
Moderately deleterious	13.3	No	Yes
Mildly deleterious	4.4	Yes	Yes
Neutral	9.1	Yes	Yes

5 **CONCLUSION** Moderately deleterious mutations are less likely to persist in populations of animals that reproduce sexually.

6 **SOURCE** Paland, S., and Lynch, M. 2006. Transition to asexuality results in excess amino acid substitutions. *Science* 311:990–992.

Experimental Questions

1. How did Paland and Lynch propose to test the hypothesis that sexual reproduction allowed for the reduction in deleterious mutations?

2. What did they discover?

3. What is the proposed evolutionary benefit of sexual reproduction?

51.2 Gametogenesis and Fertilization

This section covers how and where gametes are formed, and how two gametes join to form a new organism.

Sperm and Eggs Are Produced During the Process of Gametogenesis

Male and female gametes are formed within the **gonads**—the **testes** (singular, **testis**) in males and the **ovaries** (singular, **ovary**) in females. Some similarities are found in the ways gametes develop in the testes and the ovaries, as well as some differences.

Gametogenesis—the formation of gametes—begins with primordial cells called germ cells, which multiply by mitosis, resulting in diploid cells (carrying two copies of each gene; denoted as $2n$) called **spermatogonia** (singular, **spermatogonium**) in males and **oogonia** (singular, **oogonium**) in females (Figure 51.3). Some of these cells become **primary spermatocytes** or **primary oocytes** that may begin the process of meiosis. Until this point, the development of sperm and eggs is similar. From then on, gametogenesis differs between the two types of gamete.

Spermatogenesis The formation of haploid sperm from the original diploid germ cell is called **spermatogenesis**. As shown in Figure 51.3a, primary spermatocytes begin this process by undergoing the first of two meiotic divisions (meiosis I). (For a review of the processes of mitosis and meiosis, see Chapter

15.) Meiosis I in the primary spermatocyte produces two haploid (n) cells called **secondary spermatocytes**. These cells also undergo meiosis (meiosis II), producing four haploid **spermatids** that eventually differentiate into mature sperm cells. Gametogenesis in males, therefore, results in four gametes from each spermatogonium.

The most striking change in each spermatocyte as it differentiates into a sperm is the formation of a flagellum, also called the tail (Figure 51.3b). The movements of the tail require cellular energy and make the sperm motile. The sperm also has a head, which contains the nucleus that carries the chromosomes. The head and tail are separated by a midpiece containing one or more mitochondria, depending on the species, that produce the ATP required for tail movements. At the tip of the head is a special structure called the **acrosome** that contains proteolytic enzymes that help break down the protective outer layers surrounding the ovum.

Oogenesis Whereas spermatogenesis produces four gametes from each primary spermatocyte, gametogenesis in the female, called **oogenesis**, results in the production of a single gamete from each primary oocyte (Figure 51.3c). Each meiotic division in oogenesis results in one large cell—either a **secondary oocyte** (meiosis I) or (because meiosis II is not complete until fertilization) a fertilized egg (meiosis II)—plus a smaller cell, called a polar body, that eventually degenerates. Only the larger of the two daughter cells resulting from meiosis II (that is, the fertilized egg) contains the cellular components needed for development after fertilization.

(a) Spermatogenesis

(c) Oogenesis

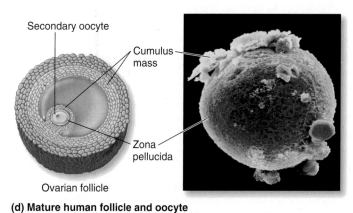

(b) Mature human sperm

(d) Mature human follicle and oocyte

Figure 51.3 **Gametogenesis and gametes in males and females.** **(a)** In the process of spermatogenesis, male diploid ($2n$) germ cells undergo two meiotic divisions to produce mature haploid (n) sperm. **(b)** The characteristic head, midpiece, and tail (flagellum) of a mature human sperm, as seen in a drawing and the accompanying scanning electron micrograph (SEM). **(c)** The process of oogenesis in females, which produces a haploid secondary oocyte that enters but does not complete meiosis II until it is fertilized. **(d)** Mature follicle and oocyte. The drawing depicts a secondary oocyte within its follicle; the SEM shows an isolated human oocyte covered by its zona pellucida and remnants of the cumulus mass.

Depending on the species, one or many oocytes can develop at a time. Within the ovaries, each oocyte undergoes growth and development within a structure called a **follicle** before it leaves the ovary in the process of **ovulation**. Once ovulated, a secondary oocyte can become fertilized if sperm are available. In mammals, the secondary oocyte is surrounded by two layers: an inner layer composed of glycoproteins called the **zona**

pellucida, which surrounds the surface of the secondary oocyte, and an outer layer of cells called the cumulus mass (Figure 51.3d). These two layers play important roles in the fertilization of the secondary oocyte.

In mammals, oogenesis begins in the female fetus before birth: A cohort of germ cells develop into primary oocytes and enter meiosis I, which is arrested partway through the process.

Meiosis I does not resume until puberty, the time when a mammal first becomes capable of reproducing. In selected primary oocytes, meiosis I is completed, producing haploid secondary oocytes. These cells then begin but do not complete meiosis II. A secondary oocyte will complete meiosis II and become a haploid egg if it is fertilized by a sperm. Once the haploid egg nucleus fuses with a haploid sperm nucleus, a diploid zygote is produced.

In a Given Species, Fertilization Occurs Either Outside or Inside the Female

For sperm to fertilize eggs, the two gametes must physically come into contact. This can occur either outside or inside the female's body. When this occurs outside of the female, the process is called **external fertilization**. This type of fertilization occurs in aquatic environments, when eggs and sperm are released into the water in close enough proximity for fertilization to occur. The aqueous environment protects the gametes from drying out.

Animals that reproduce by external fertilization show species-specific behaviors that bring the eggs and sperm together. For instance, very soon after a female fish lays her eggs, a male deposits his sperm in the water such that they spread over the clump of eggs. When frogs mate, the clasping behavior of the male stimulates the female to release her eggs into the water (**Figure 51.4**); the male then releases sperm onto the eggs. The fertilized eggs then develop outside the mother's body.

Although the aqueous environment protects against desiccation, fertilized eggs can be eaten by predators, washed downstream by currents, or subjected to potentially lethal changes in water temperature. Such environmental challenges have led to selection for some species, including many aquatic or amphibious animals, to release very large numbers of eggs at once, as noted previously for cod fish.

In contrast to external fertilization, most terrestrial animals and some aquatic animals use **internal fertilization**, in which sperm are deposited within the reproductive tract of the female during the act called **copulation**, as seen in the mud daubers in the chapter-opening photo. Internal fertilization protects the delicate gametes from environmental hazards and predation and also guarantees that sperm are placed and remain in very close proximity to eggs. Once fertilization occurs within the female, the zygotes then develop into offspring.

The behaviors and anatomical structures involved in achieving internal fertilization are extremely varied among species. Typically, mating involves accessory sex organs, which are reproductive structures other than the gonads. The external accessory sex organs involved in copulation are the genitalia (for example, the **penis** and the **vagina**), which are used to physically join the male and female so that sperm can be deposited directly into the female's reproductive tract. A penis or analogous structure is present in most insects, reptiles, some species of birds (rattites), and all mammals. However, males of other vertebrate species—such as most birds—that reproduce by internal fertilization lack a structure that can be inserted into the female, so they deposit sperm in the female by cloacal

Figure 51.4 **An example of external fertilization.** The male frog (*Rana palustris*) clasps the female, which stimulates her to release her eggs. He then releases his sperm (not visible here) over the eggs. This process occurs in aquatic environments, which protect gametes from drying out.

Concept check: *What kinds of aquatic environments are most suitable for external fertilization?*

contact. The cloaca is a common opening for the reproductive, digestive, and excretory systems in these animals.

Another form of internal fertilization involves an indirect means of depositing sperm. In this case, males produce small packets of sperm, called spermatophores, that are deposited externally and then inserted into the female's reproductive tract by either the male or the female. During copulation in cephalopods, the male uses a tentacle to transfer a spermatophore into the mantle cavity of the female. In spiders, the male places a droplet of sperm on a web and then uses a foreleg to insert the droplet into the female's reproductive tract. Subsequently the eggs are fertilized internally and deposited on the web or elsewhere in the environment.

Some Hermaphroditic Species Can Fertilize Themselves or Each Other

In some species, individuals have both male and female reproductive systems. This is called **hermaphroditism** (after the male and female Greek gods, Hermes and Aphrodite). In some hermaphroditic species, individuals can fertilize their own eggs with their own sperm, but in most hermaphroditic species, individuals exchange sperm with another individual. The latter situation has the selective advantage of creating additional genetic diversity in the population. An advantage of hermaphroditism is that all individuals of a sexually reproducing species can produce offspring.

There are two main types of hermaphroditism, in which animals are simultaneously both male and female or alternately male or female. This first type of hermaphroditism is known as synchronous hermaphroditism. Such hermaphrodites are often sessile animals or burrowing animals such as earthworms, which may live for long periods without encountering sexual partners. A single earthworm can fertilize its own eggs, or two earthworms can join together for several hours during which sperm from each worm fertilize the eggs of the other. In the latter case, individual worms act both as females (receiving sperm to fertilize their eggs) and males (giving sperm to fertilize another worm's eggs), so the offspring carry genes from both individuals.

The second type of hermaphroditism, called sequential hermaphroditism, involves sex reversal, in which a female may change into a male, or vice versa. In such animals, individuals express specific genes and therefore develop and maintain reproductive structures of either a male or a female, but not at the same time. This kind of sex reversal occurs in some species of animals with strong social hierarchies. In some reef-dwelling species of fishes, for example, a single dominant male defends a harem of several females within a specific territory. When that dominant male dies, the largest of the females reverses sex and becomes a male. Thus, these fishes are protogynous—that is, female first but capable of becoming males later during their life cycle. Oysters and clownfish are also sequential hermaphrodites, but they are protandrous; that is, they are males first and only later become females. An advantage of protandrous hermaphroditism is that males change into females when they are older and larger—and so likely to be capable of producing a greater number of eggs.

Fertilization Involves the Union of Sperm with Egg

Fertilization is a complex series of events by which the haploid male and female gametes unite and become a diploid zygote. Several important cellular and molecular processes must occur before the nuclei of the gametes can fuse. The mechanism by which the egg and sperm make contact has been studied extensively in sea urchins, and some evidence suggests that a similar process occurs in humans and other mammals. When chemical attractant molecules emitted by sea urchin eggs bind to nearby sperm, cellular respiration (that is, the breakdown of nutrients to synthesize ATP) within the sperm increases, which helps increase sperm motility. The sperm then swim toward the egg by following the attractant's concentration gradient. The attractants are species-specific; that is, sperm will respond to the attractants produced by eggs of only their own species.

For a sperm to physically contact the egg, it must first penetrate the layers surrounding the egg's plasma membrane. In mammals, when the head of a sperm contacts the layers surrounding a secondary oocyte, chemicals in the cumulus mass stimulate the breakdown of the membrane covering the acrosome of the sperm. Proteolytic enzymes released from the acrosome digest a local area of the zona pellucida, allowing the sperm to contact the plasma membrane of the secondary oocyte. The plasma membrane of the secondary oocyte then fuses with the sperm head, facilitating its movement into the secondary oocyte's cytoplasm.

Once a sperm fuses with the secondary oocyte, other sperm must be prevented from also fusing with the fertilized egg to avoid the formation of a polyploid zygote, a condition that few zygotes survive. In most mammals, the penetration of one sperm induces metabolic changes within the egg that prevent additional sperm from penetrating the zona pellucida and entering the egg (see Chapter 52).

Offspring Are Born Live or Hatch from Eggs That Are Laid or Retained Within the Mother

As we have seen, internal fertilization occurs within the female, and the resulting zygotes then continue their development into offspring. This occurs by three main processes. First, when most of embryonic development occurs within the mother and the animal is born alive, as in most mammals, the process is called **viviparity**. Second, if all or most of embryonic development occurs outside the mother and the embryo depends exclusively on yolk from an egg for nourishment, the process is called **oviparity**. Oviparity is the rule in avian species and is common in reptiles, fishes, amphibians, and insects. Among mammals, only the echidna and the platypus are oviparous. In some cases, embryos of oviparous animals grow within a protective covering, such as a shell, from which they hatch. Terrestrial animals lay eggs with shells that protect the future offspring from desiccation. Such eggs may have leathery shells, as in most reptiles and insects, or hard shells containing calcium carbonate, as in birds and many turtles. These eggs contain all the nutrients necessary for the development of the embryo. Oxygen enters, and carbon dioxide exits, through tiny pores in the shell. The eggs of aquatic species lack shells so that external fertilization may occur. Such eggs are protected from drying out by their environment.

Although some animals lay only a single egg at one time, some animals lay many more. For example, snakes and turtles lay between several and hundreds of eggs at a time, frogs and fishes can lay thousands or even millions of eggs, and honeybee queens produce as many as 250,000 eggs during a nearly continuous laying cycle over 1–2 years!

Although oviparity reduces the female's metabolic investment in the young, it increases the incidence of predation. In many amphibian and reptile species, the large number of eggs laid increases the chances of some young surviving even if predation occurs. The energetic cost to the female of producing numerous eggs is high, but usually little or no parental care is involved, freeing the parent to devote energy to other activities. In birds, some fish, and some reptiles (crocodilians), however, incubation of the eggs after laying requires to varying degrees additional energy expenditure in the form of parental care, which typically continues after the eggs have hatched, and thus relatively few eggs are laid.

Third and finally, some animals develop by a process called **ovoviviparity**, which features aspects of the first two modes of development. In this case, fertilized eggs covered with a shell that is little more than a thin sheath hatch inside the mother's body, but the offspring receive no nourishment from the mother. Ovoviviparity occurs in sharks, lizards, some snakes, and some invertebrates.

51.3 Mammalian Reproductive Structure and Function

We turn now to a detailed look at the mammalian reproductive system. For both sexes, we will begin with a description of the anatomy of the reproductive system, including the gonads and the accessory sex structures. We will then examine the hormones that control the production of the gametes and the preparation for and establishment of pregnancy.

The Male Reproductive Tract Is Specialized for Production and Ejaculation of Sperm

The external structures of the male reproductive tract—the genitalia—consist of the penis and the scrotum, the sac that contains the testes and holds them outside the body cavity (**Figure 51.5**). The testes develop within the body cavity, and just before birth in human males, they descend into the scrotum, where the temperature is approximately 2°C lower than core body temperature. The lower temperature is optimal for spermatogenesis.

Each testis is composed of tightly packed **seminiferous tubules**, encased in connective tissue (**Figure 51.6**). Surrounding the tubules are Leydig cells—scattered cells that secrete the steroid hormone testosterone. Spermatogenesis begins at puberty and continues throughout life. It occurs all along the walls of the seminiferous tubules. Cells at the earliest stages of spermatogenesis, the spermatogonia, are located nearest the wall. Cells of more advanced stages are located progressively inward, such that the sperm are released into the tubule lumen. Cells within the seminiferous tubules are continuously developing from spermatogonia into spermatocytes and eventually to

sperm, so at any one time, all types of cells are present along the seminiferous tubule. Support cells, called Sertoli cells, surround the developing spermatogonia and spermatocytes, providing them with nutrients and protection and playing a role in their maturation into sperm.

Sperm moving out of the seminiferous tubules are emptied into the **epididymis**, a coiled, tubular structure located on the surface of the testis (Figure 51.6). The epididymis is very long—approximately 6 m in humans—and in humans, it takes about 20 days for new sperm to reach its end. Here the sperm complete their differentiation by becoming motile and gaining the capacity to fertilize ova.

Sperm leave the epididymis through the **vas deferens** (or ductus deferens), a muscular tube leading to the **ejaculatory duct**, which then connects to the urethra (see Figure 51.5). As noted in Chapter 49, the urethra originates at the bladder and extends to the end of the penis. In males, the urethra not only conducts urine but also carries **semen**, a mixture containing fluid and sperm, that is released during **ejaculation**—the movement of semen through the urethra by contraction of muscles at the base of the penis. These contractions during ejaculation contribute to the pleasurable sensation of orgasm.

The liquid components of semen are important for the survival and movement of sperm through the female reproductive tract. This liquid is formed by three paired accessory glands that secrete substances into the urethra to mix with the sperm.

Figure 51.5 Male reproductive structures in humans (nonreproductive structures are identified in parentheses for orientation purposes).

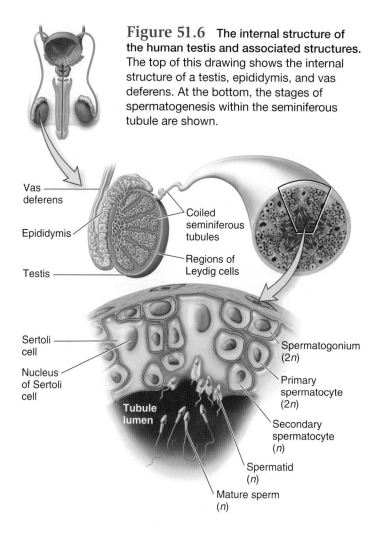

Figure 51.6 The internal structure of the human testis and associated structures. The top of this drawing shows the internal structure of a testis, epididymis, and vas deferens. At the bottom, the stages of spermatogenesis within the seminiferous tubule are shown.

Vas deferens

Epididymis

Testis

Coiled seminiferous tubules

Regions of Leydig cells

Sertoli cell

Nucleus of Sertoli cell

Tubule lumen

Spermatogonium (2*n*)

Primary spermatocyte (2*n*)

Secondary spermatocyte (*n*)

Spermatid (*n*)

Mature sperm (*n*)

The **seminal vesicles** secrete the monosaccharide fructose, the main nutrient for sperm, as well as other factors that enhance sperm motility and survival. The **prostate gland** secretes a thin, alkaline fluid that protects sperm from acidic fluids in the urethra and within the female reproductive tract. An additional small amount of alkaline fluid is secreted by the **bulbourethral glands**. Secretions constitute about 95% of semen, while sperm make up only about 5% of total semen volume. The volume of semen released at ejaculation in humans is about 2–5 ml and contains 20–130 million sperm per milliliter. Although this seems like a huge excess of sperm to fertilize a single ovum, relatively few sperm actually reach the egg.

Introduction of sperm into the female reproductive system during copulation is made possible by erection of the penis. Erection occurs when blood fills spongy erectile tissue located along the length of the penis (see Figure 51.5). In a number of mammals (excluding humans, but for example, in whales), the firmness of the erection is aided by the presence of a bone in the penis called a baculum.

Sexual arousal stimulates release of the gaseous neurotransmitter nitric oxide (NO) in the penis, causing vasodilation of arteries. The pressure of the blood flowing into the penis constricts nearby veins, causing a reduction in venous drainage from the penis, engorging it with blood. After ejaculation, NO

release is reduced, causing a reversal of the vascular changes responsible for erection. Both physiological and psychological factors can result in an inability to achieve an erection, a medical condition known as erectile dysfunction (also called impotence). Orally administered drugs are available that increase the occurrence of erections by stimulating the same intracellular signaling events as NO in the smooth muscle cells of penile blood vessels.

Male Reproductive Function Requires the Actions of Testosterone

The role of hormones in male reproductive function is diagrammed in **Figure 51.7**. Recall from Chapter 50 that the hypothalamus is a structure at the base of the brain containing neurosecretory cells (see Table 50.2). One of the neurohormones produced by some of these cells is gonadotropin-releasing hormone (GnRH). GnRH stimulates the anterior pituitary gland to release two gonadotropic hormones: luteinizing hormone (LH) and follicle-stimulating hormone (FSH).

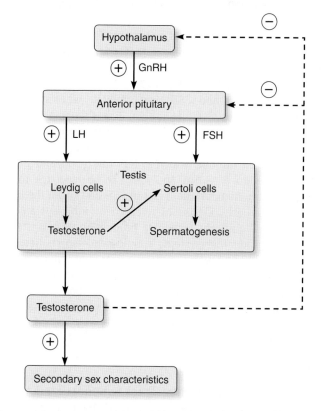

Figure 51.7 The endocrine control of spermatogenesis. In response to LH, Leydig cells in the testes secrete testosterone, which, along with FSH, acts on Sertoli cells to facilitate spermatogenesis. Negative signs indicate inhibitory effects via negative feedback, and plus signs indicate stimulatory effects. Not shown: Sertoli cells also secrete the protein hormone inhibin, which inhibits secretion of GnRH and FSH.

Concept check: What would happen to spermatogenesis in a man taking testosterone to increase muscle mass or improve athletic performance?

LH stimulates the Leydig cells of the testes to produce androgens, particularly testosterone. In turn, testosterone acts on Sertoli cells and germ cells to stimulate spermatogenesis. Testosterone also stimulates the growth of the male reproductive tract and the genitalia during development and puberty and the development of male secondary sex characteristics. In humans, these include facial hair growth, increased muscle size, and deepening of the voice. In other male mammals, they include such things as growth of the horns of a bull, enlargement of the nose of proboscis monkeys, and formation of cheek pads in some apes. Testosterone-dependent secondary sex characteristics are not unique to mammals, however. One familiar example is the bright coloration of extensive plumage in male birds such as the peacock.

The other pituitary gonadotropin, FSH, functions along with testosterone to stimulate spermatogenesis. FSH does this by stimulating the activity of the Sertoli cells within the seminiferous tubules. The Sertoli cells provide the nutritional and structural support necessary for development of the sperm. They also respond to testosterone produced by the Leydig cells, by stimulating mitosis and meiosis of the germ cells associated with them in the tubules.

Production of sperm and testosterone is kept in check by negative feedback mechanisms that control the amount of gonadotropins produced (Figure 51.7). Testosterone in high concentrations inhibits the secretion of GnRH from the hypothalamus, so both LH and FSH are inhibited when the blood concentration of testosterone is high. Testosterone also directly inhibits LH secretion by the anterior pituitary gland. In addition, when Sertoli cells are activated by FSH, they secrete a protein hormone called inhibin, which enters the blood and inhibits further secretion of GnRH and FSH. These feedback mechanisms maintain homeostatic levels of FSH and LH in the blood.

Before puberty, LH is not released in sufficient amounts to stimulate significant testicular production of testosterone, and the reproductive system is quiescent—spermatogenesis does not occur. Although the mechanisms that initiate puberty in mammals are still not completely understood, research has shown that increased GnRH production at that time initiates increased LH and FSH secretion from the pituitary. The testosterone induced by LH stimulates development of adult male characteristics. Testosterone is also responsible for an increased sex drive (libido) at this time.

The Female Reproductive Tract Is Specialized for Production and Fertilization of the Egg and Development of the Embryo

The female genitalia differentiate from the same embryonic tissues as the male genitalia. The female genitalia are composed of the larger, hair-covered outer folds called the **labia majora** (from the Latin, meaning major lips), which surround the external opening of the reproductive tract, plus the smaller, inner folds, called the **labia minora** (from the Latin, meaning minor lips; Figure 51.8). The labia majora originate from the same embryonic tissue as the scrotum of the male, whereas the labia minora

originate from urethral primordial tissue. At the anterior part of the labia minora is the **clitoris**, which is erectile tissue of the same origin as the penis. Like the penis, the clitoris becomes engorged with blood during sexual arousal (as do the labia minora) and is very sensitive to sexual stimulation. Unlike males, however, the openings of the reproductive tract and the urethra are separate in females. The opening of the urethra is located between the clitoris and the opening of the reproductive tract.

In mammals alone, the external opening of the reproductive tract leads to the vagina, a tubular, smooth muscle structure into which sperm are deposited. At the end of the vagina is the **cervix**, the opening to the **uterus**, which in humans is about the size and shape of an inverted pear. Sperm pass through the cervix into the uterus, which is specialized for carrying the developing fetus. It consists of an inner lining of glandular and secretory cells, called the endometrium, and a thick muscular layer, called the myometrium. We will discuss the functions of the uterus later in the chapter.

Oocytes develop within one of the two bilateral ovaries (Figure 51.8), which are suspended within the abdominal cavity by connective tissue. In humans, each ovary is typically a little larger than an almond. Usually, one secondary oocyte leaves the ovary and is quickly drawn into a thin tube, the **oviduct** (also called the fallopian tube), by the actions of undulating fimbriae (fingers) of the oviduct that extend out to the ovary.

The secondary oocyte is moved down the length of the oviduct by cilia on the oviduct's inner surface. For fertilization to take place, sperm must travel through the cervix and uterus and then into the oviduct where fertilization typically occurs. Upon contact with a sperm, the secondary oocyte completes meiosis II, and the union of sperm and secondary oocyte creates a fertilized egg, or zygote. The zygote undergoes several cell divisions to become a **blastocyst**, a ball of approximately 32 to 150 cells that enters the uterus, where it will develop into an embryo (the details of this development are covered in Chapter 52).

Gametogenesis in Females Is a Cyclical Process Within the Ovaries

The process of oogenesis occurs within the ovaries (**Figure 51.9**). In contrast to spermatogenesis, which continues throughout postpubertal life in the testes in males, most female mammals appear to be born with all the primary oocytes they will ever have, although limited recent evidence in mice suggests that new oocytes may form later in life. At birth, each ovary in a human female has about 1 million primary oocytes, which are arrested in prophase of meiosis I. Most of these degenerate before the onset of puberty, when each ovary contains about 200,000 primary oocytes. Other than this degeneration, the ovaries are quiescent until puberty, when they begin to show cyclical activity.

The major estrogen produced by mammalian ovaries is **estradiol**, which plays a critical role in ovulation and influences the secondary sex characteristics of females. The secondary sex characteristics, which begin to develop at puberty, include development of breasts, widening of the pelvis (an adaptation for giving birth), and a characteristic pattern of fat deposition.

Figure 51.8 **Female reproductive structure and function in humans.** Side and front views of the female reproductive system (nonreproductive structures are identified in parentheses for orientation purposes). An oocyte moves from the ovary into the oviduct (also called the fallopian tube), where it may be fertilized and develop into a blastocyst. Subsequently, the blastocyst enters the uterus, where it may implant in the endometrium, the inner lining of the uterus.

In an adult female, the process of producing mature oocytes occurs as a cycle that may be as brief as a few days in small rodents or as lengthy as 15–16 weeks in elephants. In humans, a typical cycle lasts approximately 28 days, during which time several oocytes in each ovary begin to mature. However, all but one of these oocytes usually degenerates, and only a single secondary oocyte fully matures and gets released (ovulated) from the ovary each cycle. Thus, only 300 to 500 secondary oocytes are ovulated over a woman's 30- to 40-year reproductive lifetime.

Ovulation and degeneration of additional primary oocytes continue throughout adulthood. Eventually, the oocytes become nearly depleted, and a woman stops having ovarian cycles, an event called **menopause**. The average onset of menopause in the U.S. is approximately 51 years of age. After menopause, a woman is no longer capable of becoming pregnant. Until 1900, the average life expectancy of a woman in the U.S. was no higher than 47 years; therefore, menopause was once much less common than it is today. At one time, scientists thought menopause was unique to humans, but it now appears that other female mammals, if they survive long enough, also become incapable of ovulation at some point.

The **ovarian cycle** involves the development of an ovarian follicle, the release of a secondary oocyte, and the formation and subsequent regression of a corpus luteum (Figure 51.9). During the first week of the ovarian cycle in humans, several primary oocytes begin to mature, each within a follicle. At the beginning of the second week, all but one of the growing follicles and its primary oocyte degenerate, and the single remaining follicle continues to develop and enlarge. During that time, the primary oocyte of that follicle completes meiosis I, becomes a secondary oocyte, and begins meiosis II. The developing

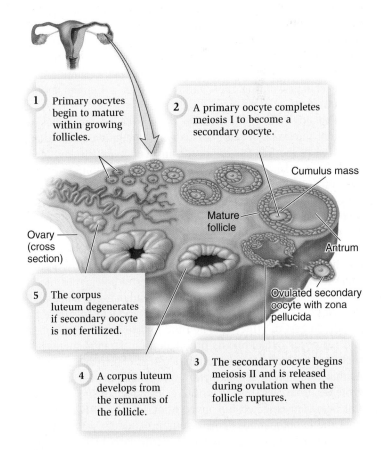

1 Primary oocytes begin to mature within growing follicles.

2 A primary oocyte completes meiosis I to become a secondary oocyte.

Cumulus mass

Mature follicle

Antrum

Ovary (cross section)

Ovulated secondary oocyte with zona pellucida

5 The corpus luteum degenerates if secondary oocyte is not fertilized.

4 A corpus luteum develops from the remnants of the follicle.

3 The secondary oocyte begins meiosis II and is released during ovulation when the follicle ruptures.

Figure 51.9 **Follicle and oocyte development.**
Development of an oocyte and corpus luteum within the ovary, illustrating the events that occur during a single ovarian cycle.

secondary oocyte is surrounded by cells of the cumulus mass, which both protect and nurture it and which secrete estradiol. The estradiol is secreted into the blood, where it functions to control the secretion of LH and FSH from the anterior pituitary gland. Some estradiol is also secreted into the follicle, where it stimulates fluid secretion into the inner core of the follicle, called the antrum. As the follicle grows in response to continued stimulation by LH and FSH, the fluid pressure inside the antrum increases, until the follicle begins to form a bulge. Eventually, ovulation occurs as the follicle ruptures, and the secondary oocyte, zona pellucida, and some surrounding supportive cells of the cumulus mass are released from the ovary.

Cells in the empty follicle subsequently proliferate and develop into a structure called the **corpus luteum**. In humans, the corpus luteum is active for approximately the second half of the ovarian cycle. It is responsible for secreting hormones that stimulate the development of the uterus needed for sustaining the embryo in the event of a pregnancy. If pregnancy does not occur, the corpus luteum degenerates, and a new group of follicles with their primary oocytes develops.

The Ovarian Cycle Results from Changes in Hormone Secretion

We saw that in males, testosterone produced in the testes exerts a negative feedback on secretion of GnRH and LH. In females, however, the situation is more complicated. Although GnRH also stimulates release of LH and FSH, the resulting estradiol can have both negative and positive feedback effects on the gonadotropic hormones. To understand this, let's examine hormone changes during the ovarian cycle in a human female (Figure 51.10).

The first half of the ovarian cycle is called the follicular phase of the cycle, because this is when the growth and differentiation of the follicles are occurring. The relatively low level of LH that exists during follicular development stimulates the cells of the follicle to make estradiol. The estradiol that is produced is important for enlargement and growth of the oocytes and it also is secreted into the blood, where it can influence the secretion of LH and FSH.

As the follicles develop, estradiol (and to a lesser extent, another steroid hormone called progesterone) production continues, and consequently, the level of estradiol in the blood slowly but steadily increases. Initially, estradiol exerts a negative feedback action on LH and FSH secretion, inducing death of all but the largest of follicles. When the follicle is fully developed and ready for ovulation, its production of estradiol increases, such that the blood concentration of estradiol increases sharply. At that time, the feedback action of estradiol on LH and FSH switches from negative to positive, by mechanisms that are still being investigated but involve increased GnRH secretion from the hypothalamus. This results in a sudden, sharp surge in gonadotropin levels in the blood, particularly LH.

The LH released from the pituitary as a result of positive feedback by estradiol induces rupture of the follicle and ovulation. This type of ovulation is known as spontaneous ovulation,

Figure 51.10 **The ovarian and uterine cycles in a human female.** The ovarian cycle is divided into the follicular and luteal phases. The uterine cycle is divided into menstruation and the proliferative and secretory phases.

Concept check: *Would similar surges in the levels of FSH and LH be expected to occur in human males?*

because it happens regularly on a cyclical basis without requiring any external stimulus. Some mammals, including rabbits, cats, and camels, undergo ovarian cycles that turn off unless mating occurs. If mating occurs at a time in the ovarian cycle when oocytes are ready, the mating act itself triggers ovulation. This mechanism, called induced ovulation, helps ensure that oocytes are not wasted by being ovulated when the female has not mated with a male.

Ovulation marks the end of the follicular phase and the beginning of the luteal phase of the ovarian cycle, named after the corpus luteum. Estradiol production decreases, and the continuing high LH level initiates development of the corpus luteum. The corpus luteum secretes progesterone, the dominant ovarian hormone of the luteal phase, plus some estradiol. Progesterone inhibits LH and FSH secretion, and it further prepares the uterus for receiving and nourishing the embryo. If fertilization of the secondary oocyte does not occur, the corpus luteum degenerates after 2 weeks, allowing LH and FSH to initiate development of a new set of oocytes. However, if fertilization does occur, the blastocyst develops a surrounding layer of cells that secrete an LH-like hormone, called **chorionic gonado-tropin**, which maintains the corpus luteum and its ability to secrete progesterone. Chorionic gonadotropin is the hormone that is tested for by home pregnancy tests.

Maternal Hormones Prepare the Uterus to Accept the Embryo

In humans, the ovarian cycle occurs in parallel with changes in the lining of the uterus called the **uterine cycle**, or **menstrual cycle**. The hormones produced by the ovarian follicle influence the development of the endometrium, the glandular inner layer of the uterus. As depicted at the bottom left of Figure 51.10, a period of bleeding called **menstruation** (from the Latin *mensis*, meaning month) marks the beginning of the uterine cycle and the follicular phase of the ovarian cycle. During menstruation, a portion of the well-developed uterine wall, including the blood vessels that grew during the previous cycle, is sloughed off and released from the body.

Menstrual cycles are found in many primates, including humans. Other mammals also have uterine cycles but without the bleeding associated with menstruation. These cycles are called estrous cycles and are usually associated with a period of sexual receptivity in females that is timed to coincide with the preovulatory period. Cyclical changes in female sexual receptivity may be present in some primates with menstrual cycles but have never been documented to occur in humans.

By about the end of the first week of the menstrual cycle in humans, the endometrium is ready to grow again in response to the newly increasing level of estrogen secreted by a developing follicle. This phase of the menstrual cycle, which corresponds to the latter part of the ovarian follicular phase, is called the proliferative phase (Figure 51.10). During this time, the endometrium becomes thicker and more vascularized. During the subsequent luteal phase of the ovarian cycle, progesterone from the corpus luteum initiates further endometrial growth, including the development of glands that secrete nutritive substances that sustain the embryo during its first 2 weeks in the uterus. This part of the menstrual cycle is called the secretory phase. If fertilization does not occur, degeneration of the corpus luteum and the associated decrease in progesterone and estrogen levels initiate menstruation and the beginning of the next uterine cycle. If fertilization does occur, however, and the blastocyst becomes embedded in the endometrium, pregnancy begins, as described in the next section.

51.4 Pregnancy and Birth in Mammals

Pregnancy, or gestation, is the time during which a developing embryo and fetus grows within the uterus of the mother. Physiologically, pregnancy is considered to begin not at fertilization but when the embryo is established in the uterine lining. This occurs within days of fertilization in animals with short gestation lengths but may take weeks in large animals with long gestations.

In mammals, gestation length varies widely and is roughly related to the size of adults in a particular species. Small animals such as hamsters and mice have gestation periods of 16–21 days, canines have longer pregnancies of about 60–65 days, humans average about 268 days, and the Asian elephant carries its fetus up to 660 days. The advantages of prolonging prenatal development are twofold: The embryo is protected while it is developing in the uterus, and the offspring can be more fully developed at birth. This is especially important for animals whose survival depends on mobility shortly after birth, such as horses and ruminants.

Gestation length is influenced not only by adult body size but also by the number of offspring in a single pregnancy. Rats, for example, which bear up to 12 or so offspring per litter, have a short gestation period and produce young that are relatively undeveloped at birth and are totally dependent on the mother. Horses, by contrast, have a long gestation period and typically give birth to a single, highly developed offspring. In this section, we will examine how mammals have evolved to retain their young in the uterus for extended periods, the structure and function of the nourishing placenta, and the role of hormones in pregnancy and birth.

Most Mammals Retain Their Young in the Uterus and Nourish Them via a Placenta

Three types of pregnancies are found in mammals, which correspond to the three subclasses of mammals (see Chapter 34). Monotremes or prototherians, such as the platypus, are the only mammals that lay fertilized eggs. In marsupials or metatherians such as the kangaroo, the young are born while still extremely undeveloped. They then crawl up the mother's abdomen to her pouch, where they attach to a nipple to suckle and obtain nourishment. They mature within the pouch. Compared to marsupials, humans and other eutherian mammals retain their young within the uterus for a longer period of time, and nourish them via transfer of nutrients and gases through a structure called the **placenta**.

During pregnancy, many physiological changes occur in both the embryo and the mother. The first event of pregnancy in eutherian (placental) mammals is **implantation**, when the blastocyst embeds within the uterine endometrium, which occurs in humans within 1–2 weeks, typically around 8–10 days after fertilization. Initially, the implanted blastocyst receives nutrients directly from the endometrial glands. However, shortly after implantation, newly developing embryonic tissues merge with the endometrium to form the placenta (**Figure 51.11**), which remains in place and grows larger as the embryo matures into a **fetus**. (In humans, an embryo is called a fetus after the eighth week of gestation.) The placenta, therefore, has a maternal portion and a fetal portion.

The placenta is rich in blood vessels from both the mother and the fetus. The maternal and fetal sets of vessels lie in close proximity. The fetal portion of the placenta, called the chorion, contains convoluted structures called chorionic villi that provide a large surface area containing capillaries for exchange of

(a) Location of the placenta in the uterus

(b) Detailed view of the placenta

Figure 51.11 **The structure of the placenta.** In all mammals, the placenta is composed of both fetal and maternal tissues. **(a)** Overview of placental structure in the human. **(b)** Enlarged view of the placenta showing the relationship between fetal and maternal structures. Note that in humans, blood in the fetal and maternal circulations does not mix.

Concept check: Note that the umbilical arteries are shown in blue to signify that they carry deoxygenated blood, and the umbilical vein is shown in red to signify it carries oxygenated blood. Normally, arteries carry oxygenated blood, and veins carry deoxygenated blood, except in the pulmonary circulation (see Chapter 48). Why is the circulation through the placenta like that of the lungs?

nutrients, gases, and other solutes. Nutrients and oxygen from the mother are carried through maternal arteries, where her blood pools in large areas of the fetal placenta surrounding the fetal capillaries. Solutes diffuse from the maternal blood into fetal capillaries, and from there flow into the umbilical vein, a portion of the fetal circulation. In turn, carbon dioxide and other waste products from the fetus are carried through the umbilical artery to the placenta, where they diffuse into the mother's circulation, from which they can be excreted. Because of this placental organization, the blood of the mother and the fetus do not mix together.

Prenatal development in humans is generally divided into three trimesters, each of which lasts about 3 months (**Figure 51.12**). During the first 2 months of pregnancy, the organs of

(a) First-trimester human embryo (6 weeks)

(b) Second-trimester human fetus (16 weeks)

(c) Early third-trimester human fetus (24 weeks)

Figure 51.12 Prenatal development in humans.

the embryo develop. At the end of the first trimester, the rudiments of the organs are present, and the developing fetus is about an inch long. The second trimester is an extremely rapid phase of growth. During the third trimester, the lungs of the fetus mature so that they are ready to function as gas-exchange organs.

Genomes & Proteomes Connection

The Evolution of the Globin Gene Family Has Been Important for Internal Gestation in Mammals

As discussed in Chapter 21, genes can become duplicated to create gene families. Gene families have been important in the evolution of complex traits because the various members of a gene family can enable the expression of complex, specialized forms and functions.

An interesting example is the globin gene family in animals. Globin genes encode polypeptides that are subunits of proteins that function in oxygen binding. Hemoglobin, which is made in erythrocytes, carries oxygen throughout the body in all vertebrates and many invertebrates, delivering oxygen to all of the body's cells. In humans, the globin gene family is composed of several homologous genes that were originally derived from a single ancestral globin gene (refer back to Figure 21.8).

All of the globin polypeptides are subunits of proteins that play a role in oxygen binding, but the various family members tend to have specialized functions. For example, certain globin genes are expressed only during particular stages of embryonic development. This has particular importance in placental mammals, because the oxygen demands of a growing embryo and fetus are quite different from the demands of its mother. These different demands are met by the differential expression of hemoglobin genes during prenatal development.

The hemoglobin protein of adult mammals is composed of four globin polypeptides—two encoded by the α-globin gene and two encoded by the β-globin gene. Altogether, five globin genes, designated α, β, γ, ε, and ζ, encode the major subunits that are found in hemoglobin proteins at different developmental stages. During embryonic development, the ε-globin and ζ-globin genes are turned on, resulting in embryonic hemoglobin with a very high affinity for oxygen (**Table 51.1**). At the fetal stage, these genes are turned off, and the α-globin and γ-globin genes are turned on, producing fetal hemoglobin with slightly less (but still high) affinity for oxygen. Finally, just before birth, expression of the γ-globin gene decreases, and the β-globin gene is turned on, resulting in adult hemoglobin, which has a lower affinity for oxygen than either the embryonic or fetal forms. The higher affinities of embryonic and fetal hemoglobins enable the embryo and fetus to remove oxygen from the mother's bloodstream and use that oxygen to meet their own metabolic demands. Therefore, the expression of different globin genes at particular stages of development enables placental mammals to

Table 51.1	Globin Gene Expression During Mammalian Development		
Stage of development	Globin genes expressed	Hemoglobin composition	Oxygen affinity (P_{50})*
Embryo	ε-globin and ζ-globin	Two ε-globin and two ζ-globin subunits	5–13.5 mmHg
Fetus	γ-globin and α-globin	Two γ-globin and two α-globin subunits	19.5 mmHg
Birth to adult	β-globin and α-globin	Two β-globin and two α-globin subunits	26.5 mmHg

*P_{50} values represent the partial pressure of oxygen required to half-saturate hemoglobin (see Chapter 48). A lower P_{50} indicates a higher affinity of hemoglobin for oxygen. The value for embryos is an estimate based on in vitro experiments. All values are for human hemoglobins.

develop in the uterus without either breathing on their own or being continually exposed to atmospheric oxygen (as occurs for vertebrate embryos that develop externally within eggs).

Birth Is Dependent on Hormones That Elicit a Positive Feedback Loop

Birth—also called **parturition**—is initiated by the actions of several hormones and other factors secreted by the mother and the placenta (**Figure 51.13**). In humans and many other mammals, the placenta begins to secrete large amounts of estrogens such as estradiol into the maternal circulation toward the end of the third trimester of pregnancy. Estrogens have at least two major effects on uterine tissue at this time. First, they promote gap junction formation between uterine smooth muscle cells, which enables coordinated uterine contractions. Second, estrogens enhance uterine sensitivity to oxytocin.

Recall from Chapter 50 that oxytocin is a posterior pituitary gland hormone that stimulates contraction of uterine muscle. The high level of estradiol in the mother's blood near the end of pregnancy stimulates the production of oxytocin receptors within the cells of the smooth muscle layer of the uterus, making the uterus more sensitive to oxytocin. At the same time, the fetus usually positions itself with its head above the uterine cervix in preparation for birth. The pressure of the fetus's head pressing on the cervix stretches the smooth muscle of the uterus and cervix. This stretch is detected by neurons in these structures. Signals from the stretch-sensitive neurons are sent to the mother's hypothalamus, triggering the release of still more oxytocin from the posterior pituitary gland.

Binding of oxytocin to its receptors initiates the strong uterine muscle contractions that are the hallmark of **labor**. In addition to its direct action on uterine muscle, oxytocin stimulates uterine secretion of prostaglandins that act with oxytocin to increase the strength of the muscle contractions. The stronger contractions elicit more oxytocin release from the mother's

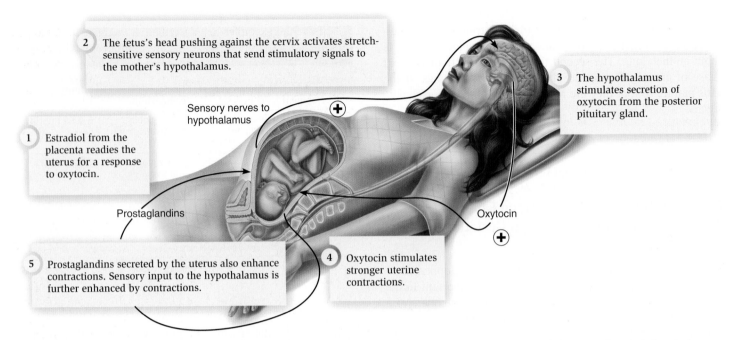

2 The fetus's head pushing against the cervix activates stretch-sensitive sensory neurons that send stimulatory signals to the mother's hypothalamus.

3 The hypothalamus stimulates secretion of oxytocin from the posterior pituitary gland.

Sensory nerves to hypothalamus

1 Estradiol from the placenta readies the uterus for a response to oxytocin.

Prostaglandins

Oxytocin

5 Prostaglandins secreted by the uterus also enhance contractions. Sensory input to the hypothalamus is further enhanced by contractions.

4 Oxytocin stimulates stronger uterine contractions.

Figure 51.13 **Hormonal control of parturition.** Birth relies on maternal hormones that act on the uterus, and neural signals from the uterus. In response to sensory neural input arising from the push of the fetus on the cervix, the maternal posterior pituitary gland releases oxytocin, which stimulates uterine smooth muscle contractions. The secretion of prostaglandins by the uterus also increases the strength of the contractions. Sensory receptors in the uterus detect the more forceful contractions and signal the mother's posterior pituitary gland to secrete more oxytocin, thus completing a positive feedback loop that further strengthens the contractions.

pituitary, which causes yet stronger contractions, setting up a positive feedback loop that continues until the baby is born.

Labor occurs in three stages (**Figure 51.14**). The initial stage induces dilation and thinning of the cervix to allow passage of the fetus out of the uterus. As the uterine contractions get stronger and more frequent toward the end of labor, the fetus is pushed, usually headfirst, through the cervix and the vagina and out into the world; this is the second stage of labor. In the third and final stage, the contractions continue for a short while. Blood vessels within the placenta and umbilical cord contract and block further blood flow, making the newborn independent from the mother. The placenta detaches from the uterine wall and is delivered a few minutes after the birth of the baby.

In mammals, the young are nurtured for a period after birth by milk produced within the mother's mammary glands and secreted via the nipples. The monotremes, in which the young of some species use their pliable bills to attach to the breast of the mother and draw milk directly through the skin, are an exception. The production of milk is called **lactation**. During pregnancy, an elevated blood progesterone level suppresses the secretion of the anterior pituitary hormone prolactin, which is required for milk production. In humans, lactation begins shortly after birth in response to a declining blood concentration of progesterone previously provided by the placenta. Whereas the formation of milk is dependent on prolactin, the actual release of milk from the breast is dependent on activation of smooth muscle cells surrounding secretory ducts in the breast. These cells are stimulated to contract by the presence of oxytocin. Oxytocin, therefore, plays a key role in two major processes, birth and nourishment of the young.

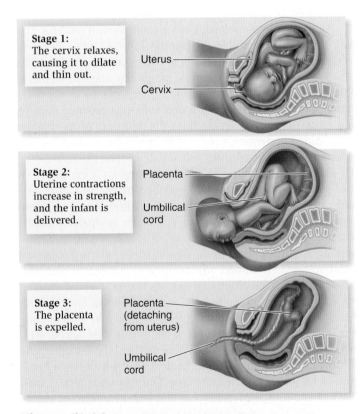

Stage 1: The cervix relaxes, causing it to dilate and thin out.

Uterus

Cervix

Stage 2: Uterine contractions increase in strength, and the infant is delivered.

Placenta

Umbilical cord

Stage 3: The placenta is expelled.

Placenta (detaching from uterus)

Umbilical cord

Figure 51.14 **The three stages of labor.**

Concept check: Many female mammals consume the placenta after giving birth. What is the benefit of this?

51.5 Timing of Reproduction with Favorable Times of Year

A species' survival depends on successful production of offspring, and the likelihood that those offspring will survive to adulthood and reproduce is influenced to a large degree by the environment into which they are born. For that reason, reproductive cycles often occur at times of the year when the likelihood of reproductive success is greatest—when the young will have sufficient nutrients to sustain them during their period of rapid growth. In general, animals that live under more uniformly favorable conditions—stable temperatures, ample rainfall, and abundant food—have less dramatic cycles of reproductive activity. Many tropical species, for example, reproduce several times each year. By contrast, many temperate zone animals have seasonal reproductive cycles that reflect large fluctuations in environmental conditions. For example, insects such as the mayfly live as underwater larvae for a year before emerging into the adult (reproductively active) stage, but the emergence occurs only during the favorable summer months.

Animals synchronize the production of offspring with favorable environmental conditions by several means. In **sperm storage**, females store and nourish sperm in their reproductive tract for long periods of time, as long as 2 years in honeybees and 4 years in some turtles. Certain insectivorous bats such as *Myotis lucifugus* use a different strategy. These bats mate in the fall, but the ovarian cycle in females is halted before ovulation, and sperm are stored and nourished in the female's uterus over the winter. Upon arousal from hibernation in the spring, the female ovulates one or more oocytes, which are fertilized by the stored sperm. This type of reproductive cycle, called **delayed ovulation**, ensures that such bats mate when they are in prime condition and that young are born when temperatures and food supplies are optimal.

Other animals have **delayed implantation**, in which a fertilized egg reaches the uterus but does not implant until environmental conditions are more favorable for the newly produced young. This type of reproduction is common among carnivores, notably the mustelids (weasels) and some bears.

Finally, many animals (cats, some hamsters, sheep, and many birds) have seasonal periods of mating followed immediately by implantation and pregnancy. Such seasonal breeding results from neuroendocrine changes in the hypothalamus in response to changes in the length of daylight. In some seasonal breeders with short gestation periods, such as hamsters, these neuroendocrine changes occur when day length increases in spring. The young are born in the late spring or in summer when conditions are favorable. Other seasonal breeders with long gestation times—sheep, for example—reproduce in response to shorter day length. Ewes are impregnated by rams in the fall, as the days grow shorter, and they carry the pregnancy through the winter. Lambs are born in the spring, providing a relatively long growth period before onset of the next winter. If sheep were to mate in summer, lambs would be born in late fall or winter, when conditions are much less favorable.

Unlike the examples just described, humans neither have a seasonal breeding cycle nor alter the timing of implantation or ovulation. Moreover, sperm can survive in a woman's reproductive tract for only a short time—2 or 3 days. Human reproduction, therefore, is far less responsive to environmental changes, day length, or other factors. However, human reproductive success can be seriously curtailed by sickness and other factors, as we see next.

51.6 Impact on Public Health

Human reproduction can be affected by many factors, some voluntary and some not. Approximately 5–10% of U.S. adults are not fertile; that is, they cannot reproduce. In men and women alike, fertility can be compromised by a variety of factors. In this section, we discuss some of the common causes of **infertility**—the inability of a man to produce sufficient numbers or quality of sperm to impregnate a woman, or the inability of a woman to become pregnant or maintain a pregnancy. We then conclude by examining the methods in use today to prevent pregnancy in fertile women.

Infertility May Result from Disease, Developmental Disorders, Inadequate Nutrition, and Stress

As many as 75% of infertility cases have some identifiable cause, and among the more prominent causes is disease. Primary among the diseases that affect fertility are sexually transmitted diseases (STDs). For example, some STDs may cause blockage in the ducts of the testes, preventing normal sperm transport.

Developmental disorders are conditions that are either present at birth or arise during childhood and adolescence. In some developmental disorders that affect fertility, inherited mutations of genes that code for enzymes involved in the biosynthesis of reproductive hormones cause abnormal expression of those genes. The result is either too much or too little of one or more of these hormones, notably estradiol or testosterone. Other developmental disorders that compromise fertility include malformations of the cervix or oviducts.

Adequate nutrition is required for normal growth and development of all parts of the body, including the reproductive system. Because the reproductive system is not essential for an individual's survival, it often becomes inactive when nutrients are chronically scarce, such as during starvation. In this way, precious stores of energy in the body are preserved for vital functions, such as the operation of the brain and heart.

Nutrition can also affect reproduction before adulthood. Undernourished children may enter puberty several years later than normal. The brains of mammals contain a center that monitors the body's fat stores. One of the triggers that initiate puberty in girls may be a signal—such as the hormone leptin—from adipose tissue to the brain. Very low fat stores in undernourished girls signal the brain that the body does not contain sufficient fuel to support the energetic demands of pregnancy; consequently, puberty is delayed.

Starvation or poor nutrition is considered a type of stress, defined as any real or perceived threat to an animal's homeostasis. Physical and psychological stress can and do affect fertility in humans. In the short term, stress can produce hormonal changes that are adaptive in meeting a crisis. However, long-term stress is damaging to many aspects of health, including reproductive health. Many nonessential functions—including the maintenance of menstrual cycles in women—are suppressed by chronic stress. The reproductive consequences of stress, including starvation, appear to be much greater in females than in males, perhaps because only females bear the energetic cost of pregnancy. Interestingly, the human body responds to long-term strenuous exercise in a way that is similar to its response to long-term stress. This is why many young ballerinas and gymnasts experience delayed puberty and why in female marathon runners, menstrual cycles may be abnormal or absent.

When the causes of infertility cannot be determined, a variety of factors come under suspicion. Among these possible causes are ingestion of toxins (for example, certain heavy metals such as cadmium), tobacco smoking, marijuana use, injuries to the gonads, and aging. Recall that as women age, they experience a loss of fertility, an event called menopause. Although reproductive function declines with age in men, they do not experience complete cessation of gamete production even at very advanced ages.

Among several currently available treatments for increasing the likelihood of pregnancy in infertile couples are hormone therapy for the woman to increase egg production and a collection of procedures known as assisted reproductive technologies (ART). In the most common ART procedure, called in vitro fertilization, sperm and eggs collected from a man and a woman are placed together in culture dishes. Once the sperm have fertilized the eggs, and the resulting zygotes have undergone several cell divisions, one or more embryos are inserted into a woman's uterus in the hope that one will implant. When this procedure was first used in 1978, the children born as a result came to be known as "test-tube babies." Since then, over 200,000 children have been born using this technology, which is typically effective about 30–35% of the time.

We turn now to a discussion of the other side of fertility issues—namely, the ways in which pregnancy is prevented.

Contraception Usually Prevents Pregnancy

The voluntary use of procedures to suppress fertility is commonly termed birth control. The use of such procedures to prevent fertilization or the implantation of a fertilized egg is termed **contraception**. By contrast, procedures or circumstances that cause the death of an embryo or fetus after implantation produce an **abortion**. Abortions can occur either spontaneously—usually when the embryo is in some way defective—or they can be induced by substances or surgical procedures. A substance that induces an abortion is called an abortifacient.

Methods of contraception can be either permanent or temporary. The permanent forms of contraception surgically prevent the transport of gametes through the reproductive tract (**Figure 51.15a**). **Vasectomy** is a surgical procedure in men that severs the vas deferens, thereby preventing the release of sperm at ejaculation (however, semen is still released). In women, **tubal ligation** involves the cutting and sealing of the oviducts. This procedure prevents the movement of the egg from the oviduct into the uterus. These procedures are considered permanent, because it is difficult—sometimes impossible—to reverse the surgery.

Temporary methods of preventing fertilization can be barrier methods, which prevent sperm from reaching an egg (**Figure 51.15b**). Barrier methods include **vaginal diaphragms**, which are placed in the upper part of the vagina just prior to intercourse and block movement of sperm to the cervix, and **condoms**, which are sheathlike membranes worn over the penis that collect the ejaculate. In addition to their contraceptive function, condoms significantly reduce the risk of STDs such as HIV infection, syphilis, gonorrhea, chlamydia, and herpes. Other types of contraception do not reduce this risk.

Another temporary form of contraception involves synthetic hormones. Oral contraceptives (birth control pills) are synthetic forms of estradiol and progesterone, taken by mouth, that prevent ovulation in women by inhibiting pituitary LH and FSH release. The hormones in these pills also affect the composition of cervical mucus such that sperm cannot easily pass through it into the uterus. In addition to the oral route, hormones can be administered by injections and skin patches.

Although no chemical contraceptive agents are widely used by males, a recent experimental formulation of testosterone and progesterone given by injection has proven effective in suppressing sperm production in men. In another approach, American researcher Michael O'Rand and colleagues demonstrated in 2004 that long-term administration of antibodies against a protein expressed specifically in the testis and epididymis of male primates completely suppressed fertility in 100% of the monkeys tested. Whether or not this novel approach to controlling fertility will be effective in humans remains to be determined.

A final temporary method of contraception involves placement in the uterus of an **intrauterine device** (**IUD**), a small object that interferes with the endometrial preparation required for acceptance of the blastocyst. Unlike the other forms of contraception described here, an IUD works after fertilization—by preventing implantation (although some IUDs also inhibit sperm movement and survival in the uterus). Although not as widely used as other contraceptives in the U.S., IUDs are the most commonly used means of contraception by women worldwide because of their effectiveness and simplicity of use.

In addition to the contraceptive methods used before or during intercourse, within 72 hours after intercourse, women can take a variety of drugs that typically interfere with ovulation or implantation. Approaches include a high dose of estrogen, or two large doses (12 hours apart) of a combined estrogen-progestin oral contraceptive. However, the drug RU486 (mifepristone) is more effective and produces fewer side effects; this drug antagonizes progesterone's effects on the endometrium, causing it to erode. RU486 can also be used as an abortifacient if pregnancy has occurred.

Used prior to the advent of modern contraception, and still in use today by individuals who prefer not to use contraceptives,

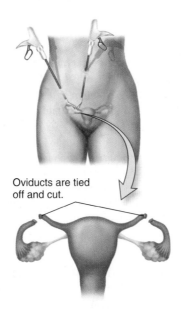

Oviducts are tied off and cut.

Each vas deferens is tied off and cut.

Vasectomy (<1.0%)

(a) Permanent methods

Tubal ligation (<1.0%)

Diaphragm (5–20%)

Condoms (male) (2–15%)

Oral contraceptive (1–2%)

(b) Temporary methods

Intrauterine device (IUD) (1–2%)

Figure 51.15 Examples of contraceptive methods. These methods may be used by men or women to **(a)** permanently or **(b)** temporarily prevent pregnancy. The estimated first year failure rates for each method are given in parentheses (collected from data published by the U.S. Food and Drug Administration and other organizations). A failure rate of 10% means that 10 of every 100 women using that method of contraception will become pregnant in the first year of use. The large range for use of condoms and diaphragms is due to improper use of these devices by many people. Female condoms are also available and have a failure rate of approximately 20%.

is the rhythm method, which involves abstaining from sexual intercourse near the time of ovulation. Its main drawback is the difficulty in precisely pinpointing the time of ovulation, even using laboratory techniques. One problem with predicting the time of ovulation is that several of the detectable changes characteristic of the midpoint of the ovarian cycle—including a small rise in body temperature and changes in the cellular characteristics of the vaginal epithelium—occur only after ovulation. This problem, plus the fact that ovulation can occur any time between days 5 and 15 of the 28-day cycle, explains why the rhythm method has a relatively high failure rate.

Summary of Key Concepts

51.1 Asexual and Sexual Reproduction

- Asexual reproduction occurs when offspring are produced from a single parent, without the fusion of genetic material from two parents. Budding occurs when part of the parent organism pinches off to form a complete new individual. Some animals reproduce by the regeneration of a complete organism from small body fragments. Parthenogenesis is the development of offspring from an unfertilized egg. (Figure 51.1)

- Sexual reproduction is the production of a new individual by the joining of two haploid gametes: a sperm from the father and an egg from the mother. The union of a sperm and an egg—fertilization—produces a diploid zygote, which develops into an embryo.

- Sexual reproduction was shown by Paland and Lynch to be more effective in eliminating deleterious mutations from a population of *Daphnia* than asexual reproduction. (Figure 51.2)

51.2 Gametogenesis and Fertilization

- Male and female gametes are formed within the gonads—the testes in males and the ovaries in females.

- Gametogenesis—the formation of gametes—begins with diploid primordial cells called germ cells. In spermatogenesis, one spermatogonium becomes a primary spermatocyte. This gives rise by meiosis to two haploid secondary spermatocytes, which yield four spermatids that mature into four sperm cells. In oogenesis, one oogonium becomes a primary oocyte. This produces by meiosis a haploid secondary oocyte that can yield a fertilized egg. (Figure 51.3)

- At the tip of a sperm's head is a special structure called the acrosome, which contains proteolytic enzymes that help break down the outer layers of the secondary oocyte.

- Within the ovaries, each oocyte undergoes growth and development within a structure called a follicle before it leaves the ovary in a process called ovulation.

- In external fertilization, sperm and eggs are released into an aquatic environment where they unite and avoid desiccation. (Figure 51.4)

- Terrestrial animals use internal fertilization, in which sperm are deposited within the reproductive tract of the female during the act called copulation.

- External accessory sex organs, such as the penis, physically join the male and female during copulation so that sperm can be deposited directly into the female's reproductive tract.

- In hermaphroditism, individuals can fertilize their own eggs with their own sperm, but in most hermaphroditic species, individuals exchange sperm with another individual.

- When an embryo develops within the mother and the mother gives birth to live young, the process is called viviparity. Development of an embryo that occurs primarily outside the mother, sometimes within a protective shell, is called oviparity. In ovoviviparity, eggs covered with a thin shell are produced and hatch inside the mother's body, but the offspring receive no nourishment from the mother.

51.3 Mammalian Reproductive Structure and Function

- Sperm produced within each testis move out of the seminiferous tubules and into the epididymis, which leads into the vas deferens, a muscular tube leading to the ejaculatory duct. The urethra conducts semen, a mixture containing fluid and sperm, during ejaculation. The fluid components of semen are produced in the seminal vesicles, the bulbourethral glands, and the prostate gland. (Figures 51.5, 51.6, 51.7)

- The female genitalia are composed of large outer folds called the labia majora, plus smaller, inner folds called the labia minora. At the anterior part of the labia minora is the clitoris, which is erectile tissue. In mammals, the external opening of the reproductive tract leads to the vagina, a tubular structure into which sperm are deposited. Sperm then move through a fibrous structure called the cervix, which forms the opening to the conical-shaped uterus. Tubes called oviducts extend between the uterus and each of two ovaries. In the ovary, primary oocytes develop into secondary oocytes (the ovarian cycle). Sperm typically fertilize secondary oocytes in the oviduct. The fertilized egg undergoes several cell divisions to become a blastocyst, a ball of cells that enters the uterus, where it will develop into an embryo. (Figures 51.8, 51.9)

- In females, changes in hormone secretion produce the ovarian cycle and also control the uterine cycle or menstrual cycle. The latter name refers to menstruation, a period of bleeding at the beginning of the uterine cycle. (Figure 51.10)

- Following the release of a secondary oocyte during ovulation, cells in the empty follicle proliferate and develop into a structure called the corpus luteum, which degenerates if fertilization does not occur. If fertilization does occur, the blastocyst develops a surrounding layer of cells that secrete an LH-like hormone, called chorionic gonadotropin (CG), which maintains the corpus luteum.

- The event during which a woman stops having ovarian cycles is called menopause.

51.4 Pregnancy and Birth in Mammals

- The time during which a developing embryo and fetus grow within the uterus of the mother is termed pregnancy, or gestation. Humans and other eutherian mammals retain and nourish their young within the uterus via transfer of nutrients and gases through a structure called the placenta. (Figure 51.11)

- The first event of pregnancy is implantation, when the blastocyst imbeds within the uterine endometrium. Shortly after implantation, newly developing embryonic tissues merge with the endometrium to form the placenta, which remains in place and grows larger as the embryo matures into a fetus. There are three trimesters to human pregnancy. (Figure 51.12)

- The evolution of the globin gene family contributed to the ability of placental mammals to develop inside the mother's uterus. (Table 51.1)

- Birth, or parturition, is initiated by the actions of hormones produced by the placenta and by the mother's endocrine system. The hormone oxytocin stimulates the strong uterine muscle contractions that are the hallmark of the three-stage process called labor. (Figures 51.13, 51.14)

- In mammals, the young are nurtured for a period after birth by milk produced in the process called lactation.

51.5 Timing of Reproduction with Favorable Times of Year

- Animals synchronize the production of offspring with favorable environmental conditions by several means. In sperm storage, females store and nourish sperm in their reproductive tract for long periods of time. In delayed ovulation, the ovarian cycle in females is halted before ovulation, and sperm are stored and nourished in the female's uterus until a more favorable time for birth of the young. In delayed implantation, a fertilized egg reaches the uterus but does not implant until later, when environmental conditions are more favorable for the newly produced young. In addition, seasonal mating is practiced by many species.

51.6 Impact on Public Health

- Infertility is the inability of a man to produce sufficient numbers or quality of sperm to impregnate a woman, or the inability of a woman to become pregnant or maintain a pregnancy. A primary cause is STDs.

- The use of procedures to prevent fertilization or implantation of a fertilized egg is termed contraception. Procedures or circumstances that cause the death of an embryo or fetus after implantation produce an abortion.

- Vasectomy is a form of contraception in which transport of gametes through the male reproductive tract is prevented by surgically severing the vas deferens. Tubal ligation involves cutting and sealing of the oviducts. Intrauterine devices (IUDs) prevent implantation of a blastocyst. (Figure 51.15)

- Two types of barrier contraceptives are vaginal diaphragms, which are placed in the upper part of the vagina just prior to intercourse and block movement of sperm to the cervix, and condoms, which are sheathlike membranes worn over the penis that collect the ejaculate. Oral contraceptives are used to prevent ovulation.

▌ Assess and Discuss

Test Yourself

1. The development of offspring from unfertilized eggs is
   a. budding.   d. parthenogenesis.
   b. cloning.   e. implantation.
   c. fragmentation.

2. Which is considered an advantage of sexual reproduction?
   a. necessity to locate a mate
   b. increased energy expenditure in producing gametes that may not be used in reproduction
   c. increased genetic variation
   d. decreased genetic variation
   e. both a and b

3. Spermatogonia
   a. are germ cells.
   b. are diploid cells.
   c. are male gametes.
   d. have flagella.
   e. a and b only

4. Compared to external fertilization, in internal fertilization,
   a. male gametes have a higher chance to come into close proximity to female gametes.
   b. gametes are less protected against predation or other harmful environmental factors.
   c. there is a decreased likelihood of desiccation of gametes.
   d. gametes come into contact only outside the mother's reproductive tract.
   e. b and c only

5. Which of the following is an example of ovoviviparity?
   a. Honeybees lay soft eggs within the hive.
   b. Sharks hatch from shell-covered fertilized eggs within the female's body.
   c. Birds hatch from shell-covered fertilized eggs laid within a nest.
   d. Fetal mammals obtain nourishment from the mother through a placenta.
   e. Amphibian eggs are released into the water column, where they may be fertilized.

6. The fructose in semen is secreted by
   a. the epididymis.
   b. the seminiferous tubules.
   c. the seminal vesicles.
   d. the prostate gland.
   e. the bulbourethral glands.

7. A major function of FSH is to
   a. stimulate the development of the gonads during early development.
   b. stimulate spermatogenesis in males and oocyte maturation in females.
   c. increase the secretion of testosterone by the testes.
   d. regulate the secretion of the bulbourethral glands.
   e. inhibit the activity of Sertoli cells in the testes.

8. During the human ovarian cycle, ovulation is stimulated by
   a. a decrease in FSH secretion.
   b. an increase in progesterone secretion.
   c. an increase in LH secretion.
   d. the presence of semen in the vagina.
   e. a decrease in estradiol level in the bloodstream.

9. During the secretory phase of the menstrual cycle, endometrial glands secrete
   a. hormones that increase the likelihood of pregnancy.
   b. nutritive substances that sustain an embryo during the first two weeks of development.
   c. hormones that prevent ovulation.
   d. waste products into the lumen of the uterus.
   e. both a and c.

10. During the _____ stage of labor in mammals, the placenta is expelled from the uterus.
    a. first
    b. second
    c. third
    d. It is not expelled; the placenta is reabsorbed.

Conceptual Questions

1. Distinguish between viviparity, ovoparity, and ovoviviparity, and give examples of animals for each type. Which type characterizes humans? What is an advantage of viviparity?

2. What are some of the costs associated with sexual reproduction? What outweighs those costs and accounts for the observation that most animals reproduce sexually?

3. How does the hypothalamus influence vertebrate reproduction?

Collaborative Questions

1. Define asexual reproduction and give three examples.

2. Compare and contrast internal fertilization and external fertilization.

Online Resource

www.brookerbiology.com

Stay a step ahead in your studies with animations that bring concepts to life and practice tests to assess your understanding. Your instructor may also recommend the interactive ebook, individualized learning tools, and more.

Animal Development

52

How embryonic development can go wrong—a normal zebrafish embryo (top) and one exposed to ethanol (bottom).

During the late 1950s and early 1960s in Europe, a drug called thalidomide was developed for use as an anticonvulsive treatment for epilepsy; it was also prescribed as an antihistamine for allergies. Although further testing revealed no beneficial effects on these conditions, it was noted that thalidomide seemed to calm nausea and help people sleep. In 1961, thalidomide was the most widely used sleeping medication in Europe, where it was also prescribed to combat morning sickness (nausea) in pregnant women.

Soon afterward, however, physicians in Europe and Australia noticed that babies born to mothers who had taken thalidomide sometimes had severely malformed limbs, with hands and feet often emanating directly from the body. It became clear that certain cells in human embryos, such as those that give rise to arms and legs, are selectively and adversely affected by this toxic compound. Ultimately, thousands of babies worldwide were deformed due to thalidomide. The drug was quickly eliminated from pharmaceutical markets, and the U.S. was largely spared from its effects due to prompt legislative action. Such tragedies underscore the importance

of continued research into the mechanisms that regulate the normal development of an embryo. Environmentally or drug-induced developmental abnormalities occur throughout the animal kingdom. The chapter-opening photo, for example, shows the effect of ethanol on development of a zebrafish embryo. (An **embryo** is an early stage of a multicellular organism during which the organization of the organism is largely formed.) Such models are helping researchers understand the effects of alcohol consumption and other environmental factors on human embryo development.

In Chapter 51, we learned how animals reproduce. Chapter 19 covered how the sequential actions of genes provide a program for the development of an organism from a fertilized egg to an adult, discussing both animals and plants. In this chapter, we will learn about the cellular and molecular processes that lead to the formation of an animal embryo. We will first briefly review some basic principles of animal development from Chapter 19. Then we will consider the five general events of embryonic development. Next, we will examine the cellular and molecular mechanisms that control development. We conclude with an overview of how abnormal development impacts the human condition.

52.1 Principles of Animal Development

The biological information that controls embryonic development resides in both the organism's genetic material—its DNA—and in the cytoplasm of the egg. A fertilized egg (that is, a **zygote**) first becomes transformed into a cluster of cells without specialized functions, and ultimately into a complex organism containing organs with specific and evolutionarily conserved functions. In biology, the **morphology** of an organism refers to its organization and structure. **Morphogenesis** is the process that creates morphology. These events occur during a period of time known as an animal's development. The process by which a fertilized egg is transformed into an organism with distinct physiological systems and body parts is called **embryonic development**. Development is often accompanied by growth, but they are different processes. Growth produces more or larger cells, whereas development produces organisms with a defined set of characteristics. Let's begin by reviewing some of the fundamental processes that underlie this development, first introduced in Chapter 19.

As animals develop, cells arrange themselves in coordinated ways that lead to the establishment of a **body plan** or **pattern**. The final, adult body pattern is organized along three axes: the **dorsoventral axis**, the **anteroposterior axis**, and the **right-left axis** (refer back to Figure 19.2a). Along these axes are often separate sections, or body segments, each containing specific body parts such as a wing or leg.

To establish the correct body plan, each cell in a developing animal must "know" where it is within the body, where it should move to, whether or not it should divide (or die), and what types of functions it will ultimately perform. This is possible because each cell receives positional information from its neighboring cells. This information is provided by cellular secretions into the extracellular fluid and by cell membrane proteins that interact with proteins on the membranes of other cells.

In addition to being stimulated to divide or to die (by apoptosis), positional information may cause the migration of a cell or group of cells from one region of the embryo to another (refer back to Figure 19.3). It may also cause **cellular differentiation**—the process by which different cells within a developing organism acquire specialized forms and functions, due to the expression of cell-specific genes. (In the thalidomide cases we discussed in the introduction, for example, cells that should have differentiated into those capable of developing into limbs were destroyed or damaged.) Now let's look at the general events that occur during embryonic development.

52.2 General Events of Embryonic Development

Even though the adult forms of animals vary immensely in size and morphology, embryonic development follows a similar pattern in most animals. As described in Chapter 32, most modern animals are triploblasts; that is, they develop from embryos with three germ cell layers. Such triploblasts include vertebrates, arthropods, echinoderms, and mollusks. Development in these animals can be categorized into five general events: fertilization, cleavage, gastrulation, neurulation, and organogenesis (**Figure 52.1**). In this section, we will examine the key aspects of each of the five general events of animal development. However, it should be noted that many species also go through an event called metamorphosis, which is a transition from a feeding larval form to an adult (refer back to Figure 33.28). Metamorphosis occurs after organogenesis and facilitates the rapid growth of young organisms into mature ones. Examples of metamorphosis include the transformation of a caterpillar into a butterfly and that of a tadpole into a frog.

Event 1: Fertilization Involves a Union Between Sperm and Egg to Create a Zygote-Stage Embryo

In all species of triploblasts, the events in **fertilization** are quite similar. The description that follows summarizes some of the hallmark events following external fertilization in sea urchins, a well-studied model organism.

In sea urchins, as in many animals, the sperm must penetrate a jellylike layer consisting of glycoproteins and

Event 1: Fertilization initiates the process of embryonic development.

Sperm
Egg

Event 2: Cleavage produces a mass of smaller cells in the embryo.

Event 3: Gastrulation produces 3 cell layers, called ectoderm, mesoderm, and endoderm.

Ectoderm
Mesoderm
Endoderm

Adult

Stages of Development

Future nervous system
Future body segment

Tadpole

Development and growth continue as the embryo matures through larval stage to adulthood.

Event 4: Neurulation produces the future nervous system; during this time, a segmented body plan also develops.

Event 5: Organogenesis results in tissues becoming organized into functional organs composed of differentiated cells.

Figure 52.1 Overview of events of embryonic development. This figure shows the general events that all vertebrate embryos go through, using a frog as an example.

Concept check: *What is the final process called by which a tadpole develops into an adult frog, and is this unique to frogs?*

polysaccharides before contacting the plasma membrane of the egg. The sperm is able to do this because of the **acrosomal reaction**, in which proteases and other hydrolytic enzymes are released from the acrosome in the tip of its head onto the jelly coat of the egg (**Figure 52.2a**). These enzymes dissolve a localized region of the jelly coat, allowing the sperm head to bind to proteins in the egg's plasma membrane. Binding is followed by fusion of the sperm head membrane with the egg membrane, and shortly thereafter by penetration of the sperm head with its nucleus into the egg.

Additional sperm are prevented from fusing with the egg because the fusion of sperm and egg depolarizes the egg. This depolarization blocks other sperm from binding to egg membrane proteins and is known as the **fast block to polyspermy**. Without this block, a single egg could receive chromosomes from two or more sperm, resulting in zygotes that fail to develop

normally or at all. Polyspermy does occur in some animals, notably urodeles (newts and salamanders), but in such cases, the nucleus from only one sperm fuses with the nucleus of the egg; the remaining nuclei are degraded inside the egg.

The acrosomal reaction is followed by the **cortical reaction** (**Figure 52.2b**). Normally, the cytosolic calcium concentration in eggs, as in most cells, is kept low by several mechanisms, including one in which Ca^{2+} is transported by an ATP-dependent pump out of the cytosol and into the endoplasmic reticulum. When the sperm binds to the egg, inositol trisphosphate (IP_3) (refer back to Figure 9.16) is released from the region of the plasma membrane nearest to the sperm entry point. IP_3 then binds to nearby sites on the endoplasmic reticulum and opens Ca^{2+} channels. Within 10 seconds after a sperm cell binds to an egg, Ca^{2+} is released from the lumen of the endoplasmic reticulum and into the cytosol. This signal is propagated across

(a) Acrosomal reaction

1. When a sperm cell contacts an egg, the acrosome releases hydrolytic enzymes that dissolve the jelly coat.

2. This exposes sperm-binding proteins on the egg cell plasma membrane that bind to the sperm.

3. The sperm and egg plasma membranes fuse. The sperm nucleus will then enter the egg.

Sperm head
Sperm nucleus
Acrosome
Sperm-binding proteins
Jelly coat
Vitelline layer
Hydrolytic enzymes
Cortical granules
Egg plasma membrane
Egg cell cytoplasm

(b) Cortical reaction

1. IP_3 is released from the plasma membrane near the site of sperm fusion with the egg.

2. IP_3 stimulates Ca^{2+} release from the endoplasmic reticulum.

3. Ca^{2+} stimulates exocytosis of cortical granules.

4. The contents of the cortical granules destroy the sperm-binding proteins and cause the vitelline layer and plasma membrane to separate. The vitelline layer of the egg hardens. This prevents polyspermy.

Egg cell cytoplasm
IP_3
Ca^{2+}
Endoplasmic reticulum

(c) The Ca^{2+} wave of the cortical reaction in a sea urchin egg

Site of sperm entry

Time after sperm entry

15 seconds   25 seconds   31 seconds   36 seconds

Figure 52.2 Fertilization of an egg by a sperm. (a) Acrosomal reaction. The contact of a sperm with an egg initiates a series of events that permits the head of the sperm to bind to the plasma membrane of the egg. This depolarizes the egg and blocks other sperm from entering; this is known as the fast block to polyspermy. (b) Cortical reaction. Sperm fusion leads to an increased level of cytosolic Ca^{2+} that ultimately causes the vitelline layer of the egg to harden, creating the slow block to polyspermy. (c) Calcium wave in a sea urchin egg. The increase in cytosolic Ca^{2+} begins near the site of sperm entry and propagates throughout the egg. Green and blue represent regions of low cytosolic Ca^{2+}; yellow and red represent regions of high cytosolic Ca^{2+}.

the entire endoplasmic reticulum, resulting in the transmission of a **calcium wave** across the egg over a period of about 30 seconds. This calcium wave can be visualized by injecting the cytosol of an unfertilized egg with a calcium-sensitive fluorescent dye that becomes highly fluorescent when Ca^{2+} is released from the endoplasmic reticulum (**Figure 52.2c**).

The release of Ca^{2+} in the cortical reaction has several important effects. First, membrane-bound vesicles in the egg's cytosol, called cortical granules, release enzymes and other substances that inactivate the sperm-binding proteins on the plasma membrane. In addition, the outer coating of the egg cell, known as the vitelline layer in sea urchins, or the zona pellucida in vertebrates, becomes hardened and begins to separate from the plasma membrane. These events create another barrier to more sperm fusing with the egg, a process called the **slow block to polyspermy**. Additionally, the burst of cytosolic Ca^{2+} both leads to the activation of molecular signaling pathways that initiate the first cell cycle and triggers an increase in protein synthesis and metabolism within the egg cell.

Shortly afterwards, the nucleus of the sperm fuses with the nucleus of the egg, creating a diploid zygote. The first cell division of the zygote occurs approximately 90 minutes after fertilization in sea urchins and amphibians, but it can take up to 24 hours in mammals.

Event 2: Cell Divisions Without Cell Growth Create a Cleavage-Stage Embryo

The initial cell cycles of embryos are unique because they involve repeated cell divisions without cell growth. The process by which these cell cycles occur is called **cleavage**. The embryonic cells repeatedly split in two, resulting in several generations of daughter cells that are roughly half the size of the cells that gave rise to them. These early cell cycles that lack cell growth are characterized as "biphasic," because they alternate only between the mitotic (M) phase and DNA synthesis (S) phase of the cell cycle—neither the G_1 nor G_2 phase occurs (see Chapter 9 for a discussion of the cell cycle).

In most species in which development occurs outside the mother, where eggs can be eaten by predators, cell division during cleavage represents some of the fastest cell cycles found in nature. The cell cycle during cleavage in amphibians, for example, requires only 20 minutes. During each 20-minute cell cycle, complete genome replication, mitosis, and duplication of the nuclear envelope are followed by cytokinesis. In eutherian (placental) mammals, however, in which development occurs within the protective environment of the mother's body, biphasic cell divisions during cleavage are relatively slow, requiring about 12 hours to complete.

The two half-size daughter cells produced by each cell division during cleavage are known as **blastomeres**. Individual blastomeres are bound together, and the outer single-cell layer of blastomeres forms a sheet of epithelial cells that separates the embryo from its environment. After formation of the outer epithelial layer, the embryos of many animals take up water and form a cavity called a **blastocoel**. The embryo at this stage

is called a **blastula**. The blastocoel provides a space into which cells will migrate to form the digestive tract and other structures of the embryo, as described later in Event 3.

Incomplete Cleavage: Birds and Fishes   Among triploblasts, cleavage-stage embryos can vary dramatically in size and appearance. This variation is in part related to whether or not the egg contains yolk and, if so, the location and amount of yolk that was deposited in the egg. Yolk is a nutrient-rich food store that will be used by the developing embryo. The eggs of birds, some fishes, and some other vertebrates have large amounts of yolk. In the eggs of these species, yolk is most concentrated toward one end—or pole—of the egg, called the **vegetal pole**. Much less yolk, and much more cytoplasm, is concentrated near the opposite pole, called the **animal pole** (**Figure 52.3**). These poles form the apices of the vegetal and animal hemispheres, which determine in part the future anteroposterior (head-tail) and dorsoventral (back-front or top-bottom, depending on the species) axes of the embryo.

In some but not all species that exhibit animal and vegetal poles, cleavage of the zygote is called incomplete because only the region of the zygote and embryo containing the animal hemisphere undergoes cell division. This process is also called **meroblastic cleavage** (**Figure 52.4**). Instead of forming a ball of cells (a blastula), in this type of cleavage, a flattened disc of blastomeres known as a **blastoderm** develops on top of the yolk mass.

Complete Cleavage: Amphibians and Mammals   In animals whose eggs have smaller amounts of yolk, cleavage during the first cell division is complete and bisects the entire zygote into two equal-sized blastomeres. Such **holoblastic cleavage** occurs in amphibians and mammals (Figure 52.4). In amphibians, cleavage-stage embryos form a blastula, as previously noted. In mammals, however, cleavage-stage embryos undergo a process called compaction, in which the amount of physical contact between cells is maximized. At this stage, the embryo in these species is called a **morula**. The resulting blastomeres in mammals then proceed to form a **blastocyst**, the mammalian counterpart of a blastula.

Cleavage and Implantation in Mammals   In mammals, the events of fertilization and cleavage occur in the oviduct

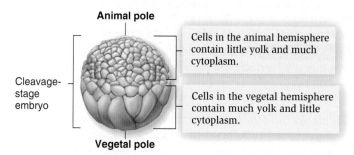

Figure 52.3 Polarity in an amphibian cleavage-stage embryo. Cells of the animal hemisphere contain little yolk and much cytoplasm. Cells of the vegetal hemisphere contain much yolk and little cytoplasm.

Figure 52.4 **Meroblastic and holoblastic cleavage.** As seen in these electron micrographs, early embryos of birds and many fishes undergo incomplete (meroblastic) cleavage, whereas most amphibian and mammalian embryos undergo complete (holoblastic) cleavage. The amount of yolk in the egg (not visible in these images) contributes to many of these morphological differences observed in various species. Source (top and middle row): © Dr. Richard Kessel & Dr. Gene Shih/Visuals Unlimited.

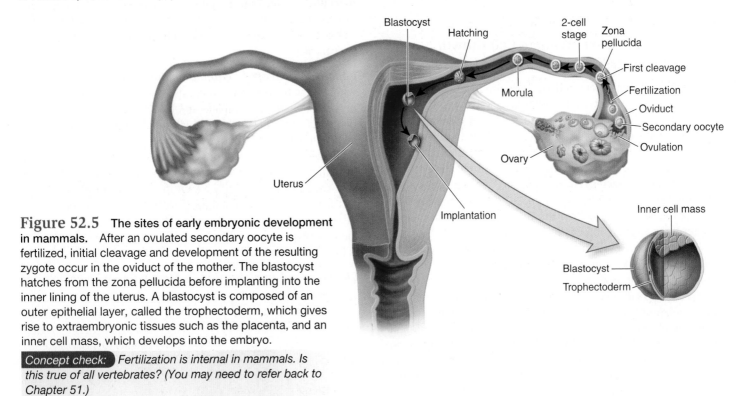

Figure 52.5 **The sites of early embryonic development in mammals.** After an ovulated secondary oocyte is fertilized, initial cleavage and development of the resulting zygote occur in the oviduct of the mother. The blastocyst hatches from the zona pellucida before implanting into the inner lining of the uterus. A blastocyst is composed of an outer epithelial layer, called the trophectoderm, which gives rise to extraembryonic tissues such as the placenta, and an inner cell mass, which develops into the embryo.

Concept check: *Fertilization is internal in mammals. Is this true of all vertebrates? (You may need to refer back to Chapter 51.)*

(**Figure 52.5**). The blastocyst has a different morphological appearance than the blastula or blastoderm embryos in non-mammalian species, and no animal-vegetal polarity that is analogous to that of other chordates exists. The blastocyst consists of an outer epithelial layer called the trophectoderm, which gives rise to the placenta, and an inner layer called the inner cell mass, which develops into the embryo. After the blastocyst forms, the embryo hatches from the zona pellucida, a layer of glycoproteins that surrounds the secondary oocyte and is retained up to this time. The embryo then becomes embedded in the endometrial wall of the mother's uterus, a process known as implantation (Figure 52.5; also see Chapter 51). This entire process takes 4 days in mice and about 8–10 days in humans.

Toward the end of cleavage, cell cycles become less synchronous, and the embryo begins to express its own genes. The embryo's shift from existing exclusively on maternal factors to developing in response to products derived from its own genome begins 6–24 hours after fertilization in vertebrates. This is followed by the next general event of development, called gastrulation.

Event 3: Gastrulation Establishes the Three Germ Layers in the Embryo

Following cleavage is **gastrulation**, one of the most dramatic events of embryonic development in animals because of the major cell movements that occur. During gastrulation, the hollow ball of cells that makes up a blastula or blastocyst is developed into a highly organized structure called a **gastrula** (refer back to Figure 32.3). In the gastrula-stage embryo, the three germ layers—**ectoderm**, **mesoderm**, and **endoderm**—become clearly established. These distinct germ layers are partially differentiated tissues that are easily recognized by their appearance under a light microscope. The layers occupy discrete regions of the embryo, with an outer ectoderm, a middle mesoderm, and an inner endoderm layer. Each type of germ layer eventually gives rise to different structures. The organization that emerges during gastrulation is most evident by the clear establishment of the digestive tube and body axes. Gastrulation is the first time when both the anteroposterior and dorsoventral body axes are clearly evident in the embryo.

Scientists have established the ultimate fate of the three germ layers with an experimental procedure called **fate**

Figure 52.7 Examples of cell types derived from ectoderm, mesoderm, and endoderm.

mapping. In this technique, a single cell or a small population of cells within an embryo is specifically labeled with a harmless dye, and the fate of these labeled cells is followed to a later stage of embryonic development (**Figure 52.6**).

Use of this technique in a variety of vertebrates has shown that the ectoderm in the gastrula forms the epidermis and nervous system in the later embryo (**Figure 52.7**). The mesoderm gives rise to muscles, kidneys, blood, heart, limbs, connective tissues, and the notochord, which is a key feature of all chordates, described later in this section of the chapter. The endoderm becomes the epithelial lining of the pancreas, thyroid, lungs, gut, liver, and urinary bladder.

Gastrulation occurs in all vertebrates. Many of the genes responsible for specifying gastrulation processes are conserved from fishes to humans. Although the developmental steps vary among vertebrate species, gastrulation usually is initiated by changes in a small number of cells within an external epithelial cell layer. This leads to a morphologically distinct structure that clearly defines the anteroposterior axis of the animal.

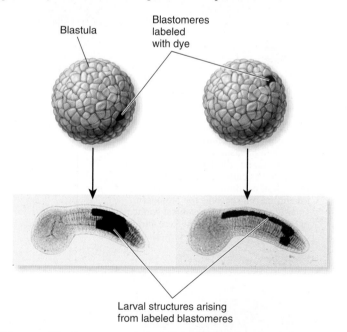

Figure 52.6 Fate mapping during embryonic development. In fate mapping, an individual embryonic cell (a blastomere) is labeled with a dye. The fate of that cell can then be followed, showing that different cells in the embryo (in this case, a urochordate blastula) form different structures in the tadpole-like larva, as shown by the dark regions.

Some of our most detailed descriptions of the events in gastrulation come from the study of frog embryos. Three features of amphibians make them ideal for analyzing embryonic development using nongenetic mechanisms. First, fertilization and embryonic development occur outside the mother, making it easy to observe and manipulate embryos as they develop. Second, eggs are abundant—adult females carry thousands of eggs and can be induced to lay them by injecting a specific hormone—and are easily fertilized, usually by simply placing a male in an aquarium with an ovulating female. Third, amphibian embryos develop rapidly. In frogs, for example, development from fertilized egg to tadpole formation takes only about 35 hours. The major events in gastrulation as described in amphibians are depicted in **Figure 52.8** and described next.

Invagination and Involution: Formation of Germ Layers and Archenteron   Prior to gastrulation, the blastula is enclosed in a simple, spherical epithelial cell layer. Gastrulation begins when a band of tissue extending perpendicularly to the animal-vegetal axis at the widest part of the embryo **invaginates** (pinches in), pushing cells from the outside of the embryo to the inside (Figure 52.8, step 1). This process creates a small opening called the **blastopore**, which defines the anteroposterior axis of the animal. Invagination begins when a few epithelial cells located at the vegetal hemisphere of the blastula—called bottle cells—undergo a drastic change in their morphology, causing them to elongate toward their basal end and forcing the cells toward the interior of the embryo. The initiating site of invagination becomes what is called the dorsal lip of the blastopore. This change in morphology of only a few key cells in the embryo initiates the gastrulation process in amphibians.

Once the bottle cells change their shape and push into the interior of the embryo, other cell movements occur, and together these orchestrated movements establish the mesoderm and endoderm of the organism, including its future digestive tract. Just before invagination begins near the "equator" or midline of the embryo, cells of the animal hemisphere spread out and move downward. The bottle cells then form and move in at the blastopore. Animal hemisphere cells continue to migrate downward. When they arrive at the blastopore, they too enter the opening and subsequently migrate upward along the roof of the blastocoel, toward the animal pole of the embryo. This folding back of sheets of surface cells into the interior of the embryo is called **involution** (Figure 52.8, step 2). Both endodermal and mesodermal cells involute from the blastopore toward the opposite end of the embryo. After involution, dorsal mesodermal cells migrate toward the animal pole by crawling along the roof of the blastocoel, with endoderm following closely behind.

As the opening from the blastopore extends into the embryo, a new cylindrically shaped cavity called the **archenteron** displaces the existing blastocoel (Figure 52.8, steps 2 and 3). The archenteron will become the organism's digestive tract. The blastopore opening remains sealed with a yolk-rich piece of tissue called the yolk plug until later in development.

1. **Formation of the blastopore by invagination of bottle cells.** Gastrulation is initiated by the invagination of bottle cells, which forms a blastopore. Invagination of bottle cells forces cells behind them to involute toward the future anterior end of the embryo. The curved arrows indicate directions of cell movements.

2. **Formation of the archenteron by invagination and involution.** The cavity that begins at the blastopore expands to form the archenteron (future digestive tract). Ectoderm spreads over the embryo.

3. **Completion of gastrulation with the beginning of notochord formation.** By the end of gastrulation, the archenteron has displaced the blastocoel and becomes closed by a yolk plug. Involution continues; some of the involuting cells become the mesoderm layer of the gastrula. The dorsal surface of the gastrula begins to thicken, and the dorsal mesoderm begins to form the notochord.

Figure 52.8   **The events of gastrulation in amphibians.**

(a) Apical constriction   **(b) Convergent extension**

Figure 52.9 Two mechanisms that affect cell shape and movement.

In chordates and echinoderms, the opening formed by the blastopore ultimately becomes the anus of the organism (refer back to Figure 32.5). Meanwhile, during involution, surface cells spread from the animal hemisphere to surround the entire vegetal hemisphere to become the future ectoderm. The result of these cellular rearrangements is an embryo with three distinct germ layers.

Mechanisms for Changes in Cell Shape and Position How do cells change shape and position during the process of embryonic development? Several processes are at work. For example, bottle cells initiate gastrulation through a process called **apical constriction** (**Figure 52.9a**). As discussed in Chapter 10, epithelial cells may have a ring of actin filaments that underlie anchoring junctions between neighboring cells (refer back to Figure 10.18). In apical constriction, the actin rings connected to the adherens junctions constrict, elongating the cells.

The spreading of ectoderm in the animal hemisphere toward the vegetal hemisphere is mediated by a cellular process called **convergent extension**. During this process, two rows of cells merge to form a single elongated layer (**Figure 52.9b**). Convergent extension produces the movement of sheets of cells.

Notochord Formation A distinguishing anatomical feature that begins forming at the end of gastrulation in all chordates is the **notochord**—a mesodermal structure that provides rigidity along the anteroposterior axis in the dorsal side of the gastrula (Figure 52.8, step 3). The presence of a notochord defines the phylum Chordates, which also includes the urochordates.

Interestingly, urochordates such as the tunicates form a tadpole-like larva but develop into adults that lack vertebrae. The similarities between urochordate and vertebrate larva—including the formation of a notochord—have led many to believe that urochordates may be quite similar to the ancestor species of current-day vertebrates. The notochord persists in the trunk and tail of fishes and amphibians; in birds and mammals, the notochord disappears by the time vertebrae have formed.

In the amphibian embryo, after formation of the archenteron, the dorsal surface of the gastrula begins to thicken, and the dorsal mesoderm forms the notochord. The notochord elongates through convergent extension. By the time the notochord has formed, the dorsal ectoderm overlying the notochord begins to thicken, which initiates the next general event in development, called neurulation. Before we discuss that event, however, we consider another important event that occurs during gastrulation: the establishment of germ cells.

Primordial Germ Cells During gastrulation, a specialized group of cells arise called **primordial germ cells** (**PGCs**). The PGCs often arise independently of the three germ layers in the embryo. This cell lineage has two primary functions: (1) to protect and propagate the genetic content of the species and (2) to undergo meiosis and differentiate into gametes—sperm or eggs—in the adult organism. PGCs are stem cells that can divide through mitosis to make copies of themselves. Some of the resulting daughter cells can later undergo meiosis and differentiate into gametes.

The eggs and resulting embryos of some species contain certain cytoplasmic determinants—called the **germ plasm** (**Figure 52.10a**)—that help define and specify the PGCs in the gastrula stage. In amphibian eggs, for example, the germ plasm occupies a small region of cytoplasm around the vegetal pole and contains a specific subset of maternal mRNAs. These cytoplasmic determinants are inherited by a subpopulation of blastomeres during cleavage. At the beginning of gastrulation, blastomeres that inherit these cytoplasmic determinants differentiate into PGCs. At a later stage of development, the PGCs migrate to primordial gonads that eventually will form testes

Germ plasm around the vegetal pole of an amphibian embryo

Pole cells at the posterior end of an early fly embryo

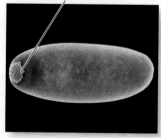

(a) Germ plasm   **(b) Pole cells**

Figure 52.10 Germ plasm and primordial germ cells (PGCs) during gastrulation. (a) Germ plasm in an amphibian embryo visualized by labeling an mRNA that is specifically expressed in PGCs. (b) In fly embryos, the PGCs are called pole cells.

or ovaries. In flies, by contrast, PGCs are the first cells to form at the posterior end of the embryo and are called pole cells (**Figure 52.10b**).

In mammals, neither germ plasm nor pole cells have been identified. Instead, a few mesoderm-like cells begin to express PGC-specific genes early during gastrulation. These cells migrate along the hindgut during gastrulation and, as in amphibians, eventually intermix with gonad primordial cells later in development. Thus, the establishment of PGCs occurs during the earliest phases of animal development and signifies the broad importance of this cell lineage for the propagation of species.

Event 4: Neurulation Involves Formation of the Central Nervous System and Segmentation of the Body

By studying development in several different vertebrate species, researchers are beginning to understand some of the fundamental steps in the formation of the central nervous system (CNS)—the brain and spinal cord—in vertebrates. The multistep embryological process responsible for initiating CNS formation is called **neurulation** (**Figure 52.11**). Neurulation occurs just after gastrulation and involves the formation of the **neural tube**

from ectoderm located dorsal to the notochord. All neurons and their supporting cells in the CNS originate from neural precursor cells derived from the neural tube. During neurulation, the embryo also develops segmented structures. Next, we discuss the processes of neurulation and body segmentation.

Neural Tube Formation   Neurulation in vertebrates occurs in four major steps, as shown in Figure 52.11. In the first step, ectoderm overlaying the notochord thickens by the elongation of cells in the dorsal region to form the neural plate, with adjacent regions that will eventually form a structure called the neural crest (discussed shortly) and the epidermis. The neural plate then elongates by convergent extension, resulting in the formation of a single, dorsal elongated epithelial cell layer that is aligned with the animal's anteroposterior axis.

Next, the neural plate forms the neural tube through a series of apical constrictions (steps 2 and 3, Figure 52.11). In the second step, a column of cells along the midline of the neural plate—the medial hinge point—undergoes apical constriction. This initiates the folding phase of neurulation and leads to the formation of the neural groove. After folding, in the third step, bilateral columns of cells in the dorsal lateral hinge points then undergo apical constriction, leading

1   Thickening and elongation:
   Ectoderm over the notochord thickens to form the neural plate, which elongates. Adjacent regions will form a neural crest and epidermis.

2   Folding:
   Cells along the medial hinge point undergo apical constriction, causing the formation of a neural groove.

3   Convergence:
   Two dorsal lateral hinge points undergo apical constriction to produce a tubelike structure.

4   Fusion:
   The dorsal-most cells on each side of the neural tube begin to fuse. Epidermis also fuses dorsal to the neural tube.

5   Neural crest cells migrate away as fusion is completed.

Figure 52.11   Neurulation and the beginning of neural crest formation in vertebrates.   The four major steps of neurulation include thickening and elongation; folding, which creates the neural groove; convergence, in which the neural tube begins to take shape; and fusion, in which the neural tube is completed. In a later event, cells migrate away from the neural tube to form several other structures, including the neural crest.

to convergence of the two sides of the neural groove and generation of a tubelike structure that is not yet sealed on the dorsal side.

In the fourth step of neurulation, called fusion, the dorsal-most cells on either side of the neural tube are released from adjacent ectoderm and fuse with each other, culminating in the closure of the neural tube. At the same time, ectoderm on either side of the neural tube moves toward the centerline, then up and over the neural tube, where it fuses and forms the dorsal epidermis of the embryo. At this time, the next step in the process involves the neural crest cells.

Neural Crest Formation

Neural Crest Formation   Another important cell lineage that arises during neurulation is the **neural crest**, which is unique to vertebrates. It consists of cells that originate from the ectoderm overlaying the dorsal side of the newly formed neural tube and that migrate to other regions of the embryo (Figure 52.11, step 5). Once these cells reach their final destination in the embryo, they differentiate into a variety of cell types different from those that arise from the rest of the ectoderm. All neurons and supporting cells of the peripheral nervous system in vertebrates are derived from neural crest cells. In addition, the neural crest gives rise to skeletal and cartilaginous structures in the head and face, melanocytes (specialized cells that provide pigmentation to the skin of vertebrates), the medulla (the inner region) of the adrenal glands, and connective tissue in numerous organs, notably the heart.

Segmentation and the Formation of Somites

Segmentation and the Formation of Somites   As embryos develop, distinct tissues acquire recognizable shapes and patterns, and the body often becomes segmented along the anteroposterior axis of the embryo. **Segmentation** allows individual body segments to have more specialized functions. In some animal taxa, such as insects, body segments are found in the adult animal. In vertebrates, the segmentation process helps define repeated structures such as vertebrae and ribs, which form later during development. Segmentation of the body plan along the anteroposterior axis becomes apparent during neurulation.

During neurulation, the mesoderm becomes segmented from the anterior end first, giving rise to blocklike structures of mesoderm called **somites** (Figure 52.12). The segmentation of somitic mesoderm continues toward the posterior end of the animal and into the tail, as can be seen in Figure 52.12b. The number of somite pairs that forms can vary considerably among species (50 in chicks, 65 in mice, and 500 in some snakes). Because the rate of embryonic development within a given species often varies with temperature and other environmental factors, stages of embryo development are often standardized according to the number of somites that have formed.

Somites arise from a group of loosely packed mesodermal cell aggregates that condense into epithelial somites, which consist of a hollow ball or vesicle enclosed by an epithelial layer. The scanning electron micrograph in Figure 52.12a shows a dorsal view of a chick embryo undergoing somite formation. These epithelial somites are located toward the anterior end,

Toward anterior end

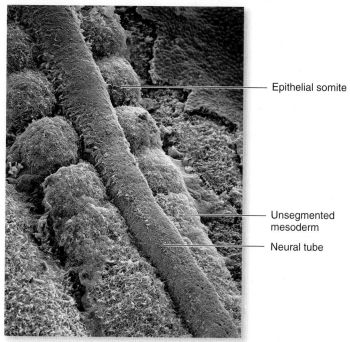

Epithelial somite

Unsegmented mesoderm

Neural tube

Toward posterior end

(a) Segmentation in an early chick embryo as seen with a scanning electron microscope

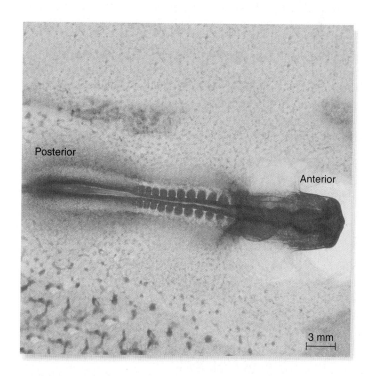

Posterior

Anterior

3 mm

(b) Segmentation in an early chick embryo as seen with a light microscope

Figure 52.12   Segmentation of mesoderm into somites during neurulation.   **(a)** This scanning electron micrograph shows early events in the segmentation process: the formation of epithelial somites toward the anterior end of a chick embryo. **(b)** Light-microscopic image of a chick embryo, showing body segments along the anteroposterior axis.

and unsegmented mesoderm can be seen extending toward the posterior end. The epithelial somites are transient structures, and the cells within them soon transform into two morphologically distinct structures that eventually form ribs/skeletal muscle and vertebrae, respectively.

Event 5: Organogenesis Is the Process of Organ Formation

As described in Chapter 40, organs are specialized structures that consist of arrangements of two or more tissue types. Most organs, such as the kidney, contain all four tissue types: nervous, muscle, epithelial, and connective tissue (see Chapters 10 and 40). The developmental event in which cells and tissues form organs is called **organogenesis**. Each germ layer gives rise to particular types of cells found within different organs (see Figure 52.7).

Many organs begin to form during or just after neurulation. However, these organs become functional at different times during development. For example, the lungs of mammals do not acquire the ability to function until shortly before birth. By contrast, the heart is the first functional organ to form in the vertebrate embryo. It begins to beat and pump blood before all the embryo's somites have formed (by 2.5 days after fertilization in chicks, 9 days in mice, and about 22 days in humans).

As we saw in Chapter 40, the development of different organs in animals is controlled by genes in the embryo, notably the *Hox* genes. *Hox* genes are important for establishing structures along the anteroposterior axis. Many of the genes controlling the processes of gastrulation, neurulation, and organogenesis encode secreted proteins or growth factors that induce cells in their local vicinity to differentiate along a specific developmental pathway. For example, the notochord produces many signaling proteins that help establish tissue patterns in the embryo. Proteins produced within it induce segment-specific expression of the *Hox* genes in subsequent stages of development. Next, we will see how such growth factors can be studied during embryonic development.

52.3 Control of Cell Differentiation and Morphogenesis During Animal Development

Thus far, we have examined the five general events of embryonic development. In this section, we will turn our attention to certain molecular mechanisms that are crucial for cell differentiation and morphogenesis.

Positional Information Is Conveyed Internally and Externally to Each Cell

The process of embryonic development requires that cells divide, move to specific sites, and acquire distinct functional properties (that is, differentiate). As we discussed, cells receive positional information that determines how they differentiate. There are two primary mechanisms of conveying such positional information. The first mechanism is an internal one, which involves the unequal acquisition by daughter cells of various cytoplasmic factors during cell division, a process known as **autonomous specification**. The second is a variety of external cell-to-cell signaling mechanisms, a process called **conditional specification**. These two mechanisms provide embryonic cells with a continuously changing internal and external environment, in which cells ultimately fulfill their unique spatial and functional fates. This differentiation process involves alterations of gene expression in which specific cells express a unique set of genes required for a particular function.

There are two main molecular mechanisms that mediate these forms of internal and external communication: morphogens and cell-to-cell contacts (refer back to Figure 19.5). **Morphogens** are molecules that impart positional information and promote developmental changes at the cellular level. Morphogens are the cytoplasmic factors involved in autonomous specification. They can also be used in conditional specification as cell-to-cell signals.

The concentration of a morphogen determines its activity. For example, some morphogens are only effective above a certain threshold concentration. Others may direct cells along different developmental pathways at different concentrations.

The concentrations of morphogens may vary within or between cells in an embryo. Some morphogenic gradients may be established in the cytoplasm of the oocyte. This is a form of autonomous specification (**Figure 52.13a**). In this process, the morphogen gradient results in daughter cells that have unequal amounts of the morphogen in their cytoplasm. The presence or lack of the morphogen at a threshold concentration will then determine how the cell differentiates. Second, a morphogenic gradient can be established in the embryo by secretion into extracellular fluids. This is a type of conditional specification (**Figure 52.13b**). For example, one or more cells may secrete a morphogen into the surrounding extracellular fluid at a specific stage of development. After secretion, the morphogen contacts neighboring cells, generating an intracellular signal such as a second messenger. This signal may influence the developmental fate of those cells. This process, whereby one or more cells governs the developmental fate of neighboring cells, is known as **induction**.

In contrast to morphogen secretion, the mechanism of direct cell-to-cell contacts used in conditional specification involves proteins that are present in the cell membrane and that can interact with other proteins in the cell membranes of other cells. This interaction generates intracellular signals. Specific cell-to-cell contacts play a major role in determining the final positioning of individual cells within different regions of an embryo.

Throughout development, mechanisms of autonomous and conditional specification function together so that distinct cells respond appropriately. We will focus in the rest of this section

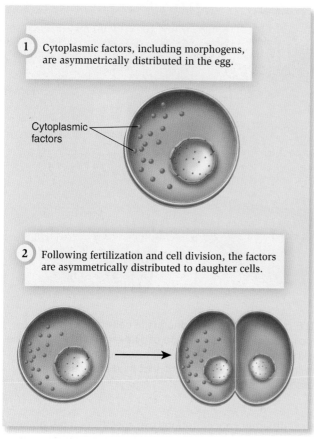

1. Cytoplasmic factors, including morphogens, are asymmetrically distributed in the egg.

Cytoplasmic factors

2. Following fertilization and cell division, the factors are asymmetrically distributed to daughter cells.

(a) Autonomous specification

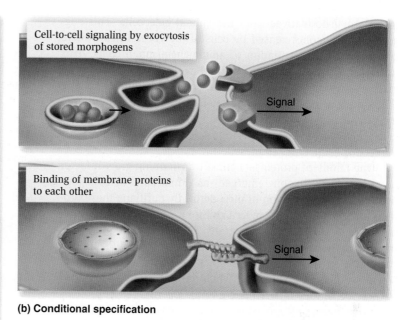

Cell-to-cell signaling by exocytosis of stored morphogens

Signal

Binding of membrane proteins to each other

Signal

(b) Conditional specification

Figure 52.13 **Mechanisms that convey positional information during embryonic development.** (a) In autonomous specification, cell fate is determined through the unequal segregation of subcellular components during mitosis. (b) In conditional specification, cells respond to signals generated by neighboring embryonic cells or to membrane-bound proteins.

on how these mechanisms lead cells to differentiate into each of the numerous diverse cell types with unique functions in the developing organism. Such differentiation involves changes in both gene expression and subcellular organization. Identifying the precise autonomous and conditional signals that specify each cell lineage in the embryo is one of the biggest challenges facing developmental biologists in the 21st century. Let's look at some examples of these mechanisms now.

Autonomous Specification: Asymmetric Distribution of Intracellular Morphogens in the Oocyte

An example of autonomous specification can be found during the cleavage events of embryonic development. During early cleavage, cell cycles can be extremely rapid, such that little or no gene transcription occurs during early cleavage. Consequently, nearly all cellular division and differentiation processes during cleavage are regulated by cytoplasmic factors that resided in the oocyte prior to fertilization. Subcellular components that were synthesized prior to fertilization are called maternal factors. These factors include many mRNAs and morphogens that are stockpiled in the maturing oocyte during oogenesis to facilitate cleavage in the absence of transcription.

Conditional Specification: Concentration Gradients of Extracellular Morphogens Control Cell Differentiation in Vertebrate Embryos

Conditional specification often involves cellular induction, the process by which a cell or group of cells induces a response in a neighboring group of cells in the embryo. The idea of cellular induction during embryonic development arose from experiments in which cells isolated from one part of an amphibian embryo were analyzed in the presence or absence of cells isolated from a different region of the embryo.

An example of cellular induction was observed in experiments performed by Dutch developmental biologist Pieter Nieuwkoop during the 1950s. Nieuwkoop isolated cells from a region near either the animal pole or the vegetal pole of late blastula stage (that is, during cleavage) amphibian embryos. When cells from around the animal pole, a region called the animal cap, were cultured in an appropriate growth medium, they formed undifferentiated spherical structures composed primarily of ectodermal cells. When cells isolated from around the vegetal pole were cultured, undifferentiated clumps of endodermal cells resulted. However, when animal and vegetal cells were cultured together, Nieuwkoop observed the formation of mesodermal derivatives from the animal pole cells. This suggested that factors released by cells of the vegetal hemisphere could induce differentiation of animal hemisphere cells. This type of experiment, called an **animal cap assay**, has been used extensively to identify morphogens secreted by embryonic cells that induce cells in the animal hemisphere to differentiate into mesoderm. Nieuwkoop went on to show that different vegetal cells isolated from various positions along the dorsoventral axis induced specific types of mesoderm.

One family of proteins—named transforming growth factor betas (TGF-βs)—soon emerged as important morphogens involved in the induction of mesoderm. By purifying these proteins and then adding them to a culture medium containing animal caps, researchers demonstrated that TGF-β proteins have mesoderm-inducing activity. In the absence of TGF-βs, animal caps grow in culture to form clumps of ectodermal tissue that resemble skin cells. However, when purified TGF-βs are added to these cells, they differentiate into mesoderm. TGF-β proteins bind to a specific receptor molecule expressed on the surface of animal cap cells, inducing them to differentiate into mesoderm.

A key question was whether different TGF-β proteins specify different types of mesodermal cells, or whether a single

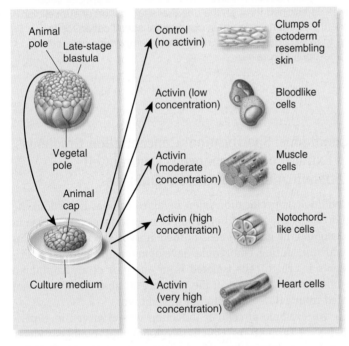

Figure 52.14 Use of an animal cap assay to demonstrate morphogen activity in conditional specification. Activin, a member of the TGF-β protein family, functions as a morphogen in an animal cap assay by inducing different types of mesoderm at different concentrations. In the absence of activin, no mesodermal derivatives develop, and the animal caps form cells resembling skin cells. Low concentrations of activin induce the formation of mesodermal derivatives resembling blood cells, whereas moderate concentrations produce muscle cells, and higher concentrations produce such mesodermal derivatives as notochord and heart cells.

Concept check: *How might it be possible for a single molecule to exert different effects on cells at different concentrations?*

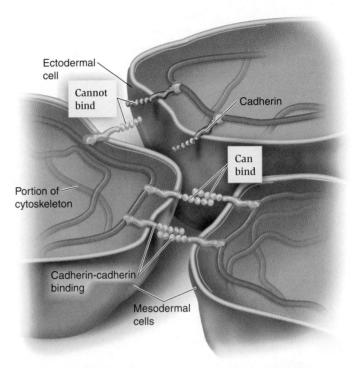

Figure 52.15 Role of cadherins in cell-to-cell contact. Developing cells express one type of cadherin. Similar cadherins bind with each other to promote cell-to-cell contacts. Cells expressing dissimilar cadherins cannot bind and thus do not form associations.

protein at different concentrations can specify all types of mesodermal cells. The morphogenic activity of one TGF-β protein—activin—in an animal cap assay is shown in **Figure 52.14**. At low concentrations, activin induces the ectoderm to differentiate into mesodermal, bloodlike cells. However, at incrementally higher concentrations, muscle cells, notochord, and fully differentiated heart cells are produced. These results show that certain TGF-β proteins exert different effects depending on their concentrations.

Cell-to-Cell Contact Involves Binding of Membrane Proteins

During the major cell movements of embryonic development, how do cells receive information about where they are supposed to be? One way is that the different classes of cells are held together by cell-to-cell contact. Cells within ectoderm, mesoderm, and endoderm express different genes that encode distinct cadherin proteins, which are cell adhesion molecules (described in Chapter 10). Only cadherins of the same type can bind to one another (**Figure 52.15**). As a result, for example, mesodermal cells tend to adhere to each other but not to endodermal or ectodermal cells.

This property of cells within distinct germ layers is best illustrated by cell dissociation and mixing experiments. When the cells of a gastrula are completely dissociated and then mixed with each other and allowed to reassociate, a fairly well-organized ball of cells forms. When the ectoderm, mesoderm, and endoderm cells within this ball are identified biochemically, the cells are observed to sort themselves out. The cells usually bind only to cells of the same type, and mesodermal cells tend to reside between the ectodermal and endodermal cells. This experiment indicates that cells within each germ layer can find each other and self-associate via cell-to-cell interactions.

Genomes & Proteomes Connection

Groups of Embryonic Cells Can Produce Specific Body Structures Even When Transplanted into Different Animals

One of the early ideas to emerge from the study of developmental biology was the concept of the morphogenetic field, a group of embryonic cells that ultimately produce a specific body structure. Long before genes and their encoded proteins were identified, embryologists discovered that particular groups of embryonic cells in amphibians have a striking characteristic: They form complete body structures when transplanted to another site in the embryo. This has been observed for cells that form the eyes, limbs, and heart in vertebrates and the eyes, antennae, legs, and wings in insects. In insects, these morphogenetic fields are called imaginal discs.

One of the first morphogenetic fields postulated in vertebrates was the limb field. Between 1910 and 1930, two important observations were made concerning the limb field. First, removing this group of cells from either side of an early developing embryo led to an embryo that lacked a limb at the corresponding position. Second, transplanting a limb field to a new location within pre-limb embryos led to the development of an additional limb at the new location. Interestingly, fate-mapping studies showed that certain cells within the field give rise to specific regions of the limb. These types of experiments suggested that cells within morphogenetic fields were uniquely specified to become particular embryonic structures before any physical evidence for the structure itself could be observed in the embryo.

In the 1920s, the German zoologists Hans Spemann and Hilde Mangold discovered an extremely important field with amazing properties in an early gastrula. This region of the gastrula is now known as **Spemann's organizer**. Their experiments involved dissecting a small piece of tissue from the dorsal lip of the blastopore in an early gastrula of a newt, and transplanting it to the opposite side of another gastrula (**Figure 52.16**). The result was the formation of a second notochord and neural tube during gastrulation and neurulation in the host embryo, and ultimately the formation of an entirely new body axis—the resulting embryo developed two bodies!

In their experiments, the researchers used two very closely related species of newts. *Triton taeniatus*, which is pigmented, served as the tissue donor, and *Triton cristatus*, which is nonpigmented, served as the host embryo. By transplanting a pigmented Spemann's organizer from the donor embryo into a nonpigmented host embryo, Spemann and Mangold could visually track the origin of the newly developed tissue. The results showed that the secondary notochord and neural tube were composed in large part of host (nonpigmented) cells, with some pigmented cells remaining from the transplanted tissue. This indicated that the transplanted tissue—which they named the "organizer"—had induced cells in the host to differentiate into neural tissue on the transplanted side of the embryo. More recent work has shown that the organizer secretes morphogens—unknown at the time of Spemann's experiments—responsible for inducing the formation of a new embryonic axis. In 1935, Spemann was acknowledged for his important discovery of cellular induction by receiving the first Nobel Prize for studies in developmental biology.

In the past two decades, scientists have identified several genes expressed specifically in Spemann's organizer. Strikingly, many of these genes, which were first discovered in amphibians, are conserved in all vertebrates. These organizer-specific genes are expressed in very small regions of early gastrula embryos, allowing researchers to identify the equivalent of Spemann's

Figure 52.16 **Experiment on newt embryos that led to the discovery of Spemann's organizer.** Note that the secondary body axis formed by the transplanted organizer is composed of host (nonpigmented) tissue, indicating that the organizer (donor) tissue induced the host tissue to differentiate.

organizer in these other species. The names given the organizer in different species vary. In chicks, it is called Hensen's node, whereas in mice, it is simply referred to as the node.

With advances in molecular genetics during the 1960s and 1970s, molecular biologists began to re-examine the concept of a morphogenetic field. An important goal of this new generation of molecular biologists was to identify the specific genes transcribed within Spemann's organizer that give this region

the ability to form an entirely new body structure. They thought that identifying these genes would provide key insights into the genetic control systems that govern embryonic patterning during gastrulation in vertebrates. However, identifying unknown genes and their protein products was a daunting challenge. Eventually, in 1992, the first secreted morphogen expressed specifically in the organizer was isolated, as we see next.

FEATURE INVESTIGATION

Richard Harland and Coworkers Identified Genes Expressed Specifically in the Organizer

In the early 1990s, Richard Harland and his colleagues hypothesized that morphogens expressed in the organizer should promote the formation of dorsal structures such as those found in the head of embryos. To search for genes that promote the formation of dorsal structures and that are expressed in the organizer during gastrulation, Harland's group took advantage of a key property of unfertilized eggs of the frog *Xenopus laevis*. When these eggs are exposed to UV light at their vegetal hemisphere, they fail to form tadpoles after the eggs are fertilized. However, embryos derived from UV-treated eggs do

form primitive mesoderm, but only ventral mesoderm forms. Dorsal mesoderm, which normally gives rise to the notochord and somites, does not form in UV-treated eggs, and the resulting "ventralized" embryos soon die. The embryos appear to be lacking one or more morphogens needed for proper dorsal development.

To identify genes in the organizer that encode morphogens that promote the formation of structures such as the notochord, Harland and colleagues used a strategy called expression cloning, as shown in **Figure 52.17**. First, after dissecting a frog embryo, they isolated and purified mRNA from dorsal-lip tissue containing the organizer. Second, they constructed a cDNA library (see Chapter 20) from the purified mRNA. This resulted

Figure 52.17 Harland and colleagues discovered the morphogen noggin using expression cloning.

HYPOTHESIS Cells within Spemann's organizer express specific genes that encode proteins that regulate the development of dorsal structures during gastrulation.

KEY MATERIALS *Xenopus laevis* gastrulas and eggs.

Experimental level	Conceptual level

1 Isolate mRNA from dorsal lip tissue, which contains the organizer.

Surgical scissors / Dorsal lip / Gastrula / Blastopore / Extract and purify mRNA. / mRNA / Contains many different mRNAs expressed in the organizer (~15,000 different mRNAs)

2 Create cDNA library. Use cDNA library to make 15,000 different mRNAs. (See Chapter 20 for a description of cDNA libraries.)

Plasmid / cDNA ligated into vector / Bacterium / Each bacterium takes up 1 plasmid. / Clones of bacteria, each containing a different cDNA / DNA is isolated and transcribed into mRNA in vitro.

3 Expose unfertilized *Xenopus laevis* eggs to UV light, then inject with an mRNA. Fertilize the eggs to determine which, if any, develop. Note: Each egg was injected with a single type of mRNA.

UV / Vegetal pole / *Xenopus laevis* eggs / mRNA injected into egg / UV light inactivates a morphogen and prevents dorsal mesoderm formation; the embryos usually die.

Most fertilized eggs formed only ventral mesoderm and died. One, however, developed into a tadpole (rescue).

4. Inject with UV-treated eggs increasing amounts of the mRNA that rescued the embryo in step 3.

Rescue mRNA from step 3

Egg

Fertilize

Fertilized egg   mRNA   Protein product of mRNA

Few copies of mRNA → Low number of proteins formed.

Moderate number of copies of mRNA → Moderate number of proteins formed.

Many copies of mRNA → Many proteins formed.

5 THE DATA

Embryo from UV-treated egg | Partial rescue | Normal tadpole | Abnormal tadpoles with excess dorsal and anterior development

Increasing amounts of dorsal mesoderm-promoting mRNA and its protein product

6 CONCLUSION Cells in Spemann's organizer secrete a morphogen—termed noggin—that directs normal dorsal development in *Xenopus laevis* embryos.

7 SOURCE Smith, W.C., and Harland, R.M. 1992. Expression cloning of noggin, a new dorsalizing factor localized to the Spemann organizer in *Xenopus* embryos. *Cell* 70:829–840.

in approximately 15,000 different cDNAs, each of which corresponded to a single gene that had been expressed in the organizer. Their next task was to transcribe each of these cDNAs back into mRNA in vitro. In a third step, the researchers injected these different mRNAs into thousands of unfertilized UV-treated eggs, which they then fertilized. Each egg was injected with a single type of mRNA. Their idea was that any mRNA that produced a protein with dorsal mesoderm-promoting activity might "rescue" the embryos and result in free-swimming tadpoles.

Using this procedure, the researchers identified one mRNA whose protein product rescued the embryo (step 3). They then injected increasing amounts of this mRNA into UV-treated eggs, fertilized the eggs, and examined the morphology of the resultant tadpoles. They discovered that the protein product from this mRNA acted as a morphogen because it induced different embryonic structures at different concentrations. At low concentrations, it partially rescued the embryos. At moderate concentrations, it resulted in the formation of normal embryos. However, at very high concentrations, the embryos developed too much dorsal mesoderm and not enough ventral mesoderm. These abnormal embryos developed large heads and extremely small trunks. Further studies identified the gene that coded for this mRNA and verified that it was transcribed in Spemann's organizer. They named the gene *noggin* (from a slang term for brain or head), because it caused embryos to develop large heads when present in abnormally high levels (see right side of the data).

Later work revealed that noggin protein promotes dorsal development by inhibiting ventral development. Noggin protein inhibits at least two other morphogens that are known to induce ventral, but not dorsal, mesoderm. These studies showed that antagonistically acting proteins expressed and secreted from certain cells in the organizer can specify precise structures in a concentration-dependent fashion. Further work in mice revealed that *noggin* is expressed in the node of mammals. Specific deletion of the *noggin* gene in mice leads to defects in dorsal structures. This indicates that *noggin* plays a fundamental role during gastrulation in mammals as well as amphibians.

The discovery of *noggin* and its mechanism of action revealed that the normal development of embryos requires a balance between stimulatory and inhibitory factors. The inhibition of developmental processes—such as the suppression of ventral mesoderm by *noggin*—appears to be a normal part of the mechanism by which development of an organism occurs.

Experimental Questions

1. Why were scientists interested in identifying genes expressed exclusively in Spemann's organizer?

2. What hypothesis did Harland and colleagues test?

3. How did the scientists identify possible genes that are important for the development of dorsal structures, and how did they test the gene products to determine their activity?

52.4   Impact on Public Health

Given the complexity and precision of the five general developmental events we have described in this chapter, it may not be surprising that on occasion these processes fail to occur properly. When this happens in humans, its impact can be devastating. Many health problems in humans are caused by genetic or environmental factors that disrupt embryonic development. As we have seen, the early development of an embryo is among the most complex processes found in nature. Because embryonic development is affected by genetic as well as environmental factors, a number of diseases and disorders in humans stem from problems that arise during this process.

An example of a condition involving defective embryonic development is spina bifida, caused by the failure of the neural tube to close at either the anterior or posterior end during neurulation. Signs and symptoms of spina bifida, which occurs in about 7 of every 10,000 births in the U.S., vary depending on the degree to which the neural tube fails to close. In minor forms, gaps may occur between a few vertebrae at the bottom of the back, and impairment of motor or sensory function may not be apparent at birth. However, neurological deterioration often becomes evident during childhood or adulthood. Surgical procedures soon after birth can improve the long-term quality of life for patients with spina bifida, but prevention of this condition—by increasing the amount of folic acid the embryo gets during pregnancy—is much more effective. Up to 75% of spina bifida cases may be prevented if women in the first trimester of pregnancy increase their dietary intake of folic acid, a common B vitamin needed during the rapid growth that occurs during embryonic development.

Whereas spina bifida appears to be related to deficient folic acid levels during neurulation, other embryonic defects are caused by foreign chemicals introduced during pregnancy. Given the deleterious effects of certain chemical compounds on specific phases of embryonic development in humans, it is not surprising that ingested or inhaled compounds such as tobacco, alcohol, and other drugs can also have severe and devastating consequences on a developing fetus. Indeed, the leading overall cause of mental retardation worldwide is fetal alcohol syndrome (FAS), which occurs in babies whose mothers drink excessive amounts of alcohol during pregnancy. In addition to cognitive disorders, children born with FAS also show malformed facial features and joints and altered overall growth characteristics (also seen in other vertebrate embryos exposed experimentally to ethanol; see the chapter-opening photo). These morphological features are likely caused by the generally detrimental effects that ethanol has on cell division. The cognitive deficits are thought to result from the death of developing CNS neurons, which are particularly susceptible to the toxic effects of ethanol.

Unlike the previous conditions, which are caused by environmental conditions surrounding the embryo, other embryonic defects are caused by genetic defects. The most common genetic disorder affecting humans is Down syndrome (formerly called Down's syndrome), which arises from an egg containing two copies of chromosome 21. The effects in an embryo that ultimately acquires three copies of chromosome 21—two from the egg and one from the sperm—are complex, such that multiple physical and neurological disorders are associated with this syndrome.

Finally, cleft lip or palate (or both combined) are relatively common (roughly 1:1,000 births in the U.S.) developmental disorders that appear to share both a genetic and an environmental basis. During early development in humans, the tissues of the lip and palate exist as two structures that merge around weeks 5–7 of embryonic development. In cleft lip, the tissues of the upper lip fail to fuse, whereas in cleft palate, those of the palate fail to fuse; in some cases, these two deformities occur together. In the latter case, the result is a gap or groove that extends from the mouth into the nasal cavity. Cleft lip/palate is a disfiguring condition that leads to speech impediments, difficulty eating and drinking (for example, an affected infant cannot readily produce the suction needed to drink from a bottle or nipple), and repeated infections. The only treatment for cleft lip/palate is surgical repair of the affected tissues, which is often quite successful but which usually entails numerous surgical procedures over several years (**Figure 52.18**). Despite intensive research, the causes of cleft lip/palate are still uncertain. However, certain environmental agents have been linked with an increased risk of cleft lip/palate, notably maternal ingestion of alcohol, lead, cocaine, certain antiseizure drugs, or possibly vitamin deficiencies during pregnancy. Recently, researchers have identified several genes in which loss-of-function mutations affect the way in which the palate develops, contributing to a cleft.

The World Health Organization has compiled the number of birth defects around the world by geographic region in order

Figure 52.18   Child with cleft lip/palate before and at two different ages after surgeries.

to assess the extent of the impact of developmental disorders on public health. The resulting document, entitled *The World Atlas of Birth Defects*, indicates that regional differences in the incidences of a given birth defect can be dramatic. For example, the rate of spina bifida varies from 1 in every 10,000 births in France to about 15–30 in every 10,000 births in Mexico and Ireland. The occurrence of Down syndrome, by contrast, varies from about 7 cases per 10,000 births in Cuba to 22 per 10,000 births in Chile. Together, 27 different birth defects affect 2–3 % of all babies born in the world today, an astonishing number when one considers the world's population of roughly 6.5 billion, with annual worldwide births of nearly 150 million.

 # Summary of Key Concepts

52.1 Principles of Animal Development

- The process by which a fertilized egg is transformed into an organism with distinct physiological systems and body parts is called embryonic development.

- The process by which different cells within a developing organism acquire specialized forms and functions, due to the expression of cell-specific genes, is called cellular differentiation.

- Four responses to positional information are cell division, cell migration, cell differentiation, and apoptosis.

52.2 General Events of Embryonic Development

- Development in many animals, including vertebrates, involves five general events: fertilization, cleavage, gastrulation, neurulation, and organogenesis. (Figure 52.1)

- Major events in fertilization include the acrosomal reaction and the cortical reaction. During the acrosomal reaction, the binding of a sperm with the egg membrane triggers a series of events producing the fast block to polyspermy, which prevents other sperm from binding to the egg. During the cortical reaction, events produce a calcium wave that leads to additional barriers to more sperm, a process called the slow block to polyspermy. (Figure 52.2)

- During cleavage, which involves cell divisions without cell growth, the daughter cells are called blastomeres. When the embryo forms an outer epithelial layer and an inner cavity, it is called a blastula. The mammalian counterpart of a blastula is called a blastocyst.

- Cleavage-stage embryos in triploblast organisms have animal and vegetal hemispheres. The hemispheres determine in part the future anteroposterior and dorsoventral axes of the embryo. (Figure 52.3)

- Incomplete or meroblastic cleavage occurs in birds, some fishes, and some other vertebrates whose eggs contain large amounts of yolk. Complete or holoblastic cleavage occurs in amphibians and mammals, whose eggs have smaller amounts of yolk. In mammals, cleavage occurs in the oviduct and the embryo implants in the uterine wall. (Figures 52.4, 52.5)

- In fate mapping, a small population of embryonic cells is specifically labeled with a harmless dye, and the fate of these labeled cells is followed to a later stage of embryonic development. (Figure 52.6)

- During gastrulation, the hollow ball of cells that makes up the blastula or blastocyst is converted into a highly organized structure in which the three germ layers—endoderm, mesoderm, and ectoderm—and the primordial germ cells are established. Each germ layer gives rise to specific structures. An embryo that is undergoing gastrulation is called a gastrula. (Figure 52.7)

- Gastrulation begins when a band of tissue invaginates (pinches in), creating a small opening called the blastopore. During gastrulation, the embryo forms a cavity called the archenteron, which will become the organism's digestive tract. (Figure 52.8)

- During gastrulation, two cellular processes are crucial to development: apical constriction, in which a reduction in the diameter of the actin rings connected to the adherens junctions causes the cells to elongate toward their basal end, and convergent extension, in which two rows of cells merge to form a single elongated layer. (Figure 52.9)

- A distinguishing anatomical feature that forms at the end of gastrulation in all chordates is the notochord.

- During gastrulation, primordial germ cells (PGCs) become established. Amphibian eggs contain certain cytoplasmic determinants, called the germ plasm, that help define and specify the PGCs in the gastrula stage. In flies, PGCs are the first cells to form at the posterior end of the embryo and are called pole cells. (Figure 52.10)

- Neurulation is the multistep embryological process responsible for initiating CNS formation. Neurulation occurs just after gastrulation and involves the formation of the neural tube from ectoderm located dorsal to the notochord. (Figure 52.11)

- Another important cell lineage that arises during neurulation is the neural crest, which gives rise to all neurons and supporting cells of the peripheral nervous system in vertebrates.

- During neurulation, the mesoderm becomes segmented from the anterior end first, giving rise to blocklike structures of mesoderm called somites. (Figure 52.12)

- The developmental event during which cells and tissues form organs is called organogenesis.

52.3 Control of Cell Differentiation and Morphogenesis During Animal Development

- Two mechanisms convey positional information during embryonic development: autonomous specification, which involves internal factors, and conditional specification, which involves external signals. (Figure 52.13)

- Morphogens and cell-to-cell contacts are the molecular mechanisms behind autonomous and conditional specification.

- Morphogens can elicit different cellular responses at distinct concentrations in an embryo.

- Autonomous specification involves the asymmetric distribution of intracellular morphogens and other factors in the oocyte.

- A type of experiment called an animal cap assay has been used extensively to identify morphogens secreted by embryonic cells that induce cells in the animal hemisphere to differentiate into mesoderm. This is a type of conditional specification. (Figure 52.14)

- In another type of cell-to-cell interactions, cells use membrane proteins called cadherins to bind to other like cells, ensuring that the cells will move together during development. (Figure 52.15)

- An extremely important region in the early gastrula is known as Spemann's organizer. The organizer secretes morphogens responsible for inducing the formation of a new embryonic axis. (Figure 52.16)

- Richard Harland and coworkers discovered the *noggin* gene in frog embryos and demonstrated that it acted to promote dorsal development by inhibiting ventral development. (Figure 52.17)

52.4 Impact on Public Health

- An important example of a condition involving defective embryonic development is spina bifida, caused by the failure of the neural tube to close completely during neurulation.

- The leading overall cause of mental retardation worldwide is fetal alcohol syndrome (FAS), which occurs in babies whose mothers drink excessive amounts of alcohol during pregnancy.

- In Down syndrome, the most common genetic disorder affecting humans, the embryo has three copies of chromosome 21. The effects are complex and include multiple physical and neurological disorders.

- Cleft lip or palate is a developmental disorder that occurs in 1:1,000 births in the U.S.; it has both genetic and environmental causes. (Figure 52.18)

Assess and Discuss

Test Yourself

1. The acrosomal and cortical reactions occur during which event of development?
   a. fertilization
   b. cleavage
   c. gastrulation
   d. neurulation
   e. organogenesis

2. Cadherins are
   a. adhesion protein molecules that attach cells to extracellular material.
   b. genes necessary for proper germ layer formation.
   c. adhesion protein molecules that allow cells of a given germ layer to adhere to each other.
   d. germ layers.
   e. proteins that provide structural support to the cytoplasm of cells.

3. Cell differentiation that results from cell-to-cell signaling is
   a. autonomous specification.
   b. conditional specification.
   c. communicative specification.
   d. gastrulation.
   e. embryogenesis.

4. The three germ layers of triploblasts are established during
   a. fertilization.
   b. cleavage.
   c. blastula formation.
   d. gastrulation.
   e. neurulation.

5. The limbs of vertebrates are formed from
   a. the endoderm.
   b. the mesoderm.
   c. the ectoderm.
   d. all of the above.

6. In vertebrates, the digestive tract forms from
   a. the blastopore.
   b. the dorsal lip.
   c. the archenteron.
   d. the mesoderm.
   e. both a and d.

7. The cells that give rise to gametes
   a. are derived from the mesoderm.
   b. often arise independently from the three germ layers.
   c. originate directly from the tissue that will develop into the gonads.
   d. are some of the last cells of the embryo to differentiate.
   e. are formed the same way in all animals.

8. Cells of the neural crest
   a. give rise to the central nervous system.
   b. originate from the ectoderm.
   c. migrate to different areas of the body and differentiate into a variety of cells, including neurons of the peripheral nervous system.
   d. all of the above
   e. b and c only

9. Which is true of Spemann's organizer?
   a. It is unique to amphibian embryos.
   b. It arises early in cleavage-stage embryos.
   c. It secretes morphogens, including the protein called noggin.
   d. It is found in the dorsal lip region of early gastrula-stage embryos.
   e. c and d are both correct.

10. Morphogens
    a. are proteins that stimulate responses in the cells of an embryo.
    b. are important signaling proteins in the process of embryonic development.
    c. can elicit different cell responses at different concentrations.
    d. contribute to autonomous specification.
    e. all of the above

Conceptual Questions

1. Distinguish between embryonic development, growth, and differentiation.

2. During organogenesis in vertebrates, some organs develop and become functional sooner than others. Why is it important, for example, that the heart of a terrestrial vertebrate develop sooner than its lungs?

3. Distinguish between autonomous and conditional specification.

Collaborative Questions

1. Discuss the process of gastrulation.

2. Discuss the process of neurulation.

Online Resource

www.brookerbiology.com

Stay a step ahead in your studies with animations that bring concepts to life and practice tests to assess your understanding. Your instructor may also recommend the interactive ebook, individualized learning tools, and more.

Immune Systems

53

In an immune system response, a macrophage engulfs numerous bacteria (false-color SEM).

All animals must defend themselves against environmental factors that threaten their survival, including predation, intraspecific fighting, accidents, and hazardous substances. In addition, animals must contend with various threats within their internal environment, including the invasion of potentially harmful microorganisms such as bacteria, the presence of foreign molecules such as the products of microorganisms, and the presence of abnormal cells such as cancerous cells.

The ability of an animal to ward off these internal threats—an animal's **immunity**, or immune defenses—is the subject of this chapter. The cells and organs within an animal's body that contribute to immune defenses collectively constitute an animal's **immune system**. The study of immunity is called immunology. Immunologists examine the processes by which the immune system protects an animal from foreign matter, whether living or nonliving. In these processes, the body's immune defenses recognize the body's own molecules as "self" and attack anything that is foreign, or "nonself."

Immune defenses are often divided into two types: innate and acquired. In **innate immunity**, the body's defenses are present at birth and act against foreign materials in much the same way regardless of the specific identity of the invading material. Thus, the innate immunity of animals is also known as "nonspecific immunity." Innate immunity includes the body's external barriers (skin and mucous membranes), plus a set of cellular and chemical defenses that oppose substances that breach those barriers. An example of an innate immune response is seen in the chapter-opening photo, in which a cell known as a macrophage is engulfing several bacteria. All animals have innate immune defenses. In contrast, **acquired immunity** develops only after the body is exposed to foreign substances. This type of immunity is characterized by the ability of certain cells of the immune system to recognize a foreign substance and initiate a response that targets that substance specifically. For this reason, acquired immunity is also known as "specific immunity." Another feature distinguishing acquired immunity is that repeated exposure to a foreign substance elicits greater and greater defense responses, unlike the situation for nonspecific immunity, where each exposure to the foreign material elicits the same defense responses. Thus, acquired immunity is also known by a third name, "adaptive immunity." Specific immune defenses have been identified in all vertebrates except for the jawless fishes, but they have not been unequivocally identified in invertebrates.

We begin the chapter with a brief overview of the different microorganisms that cause disease in animals. We then consider the mechanisms that provide animals with innate and acquired defenses against harmful microorganisms. We conclude with a discussion of the public health implications of some selected immunity-related conditions in humans.

53.1 Types of Pathogens

As just noted, the body's immune defenses must protect against a variety of foreign materials, but most important among them are disease-causing microorganisms, or **pathogens**. Pathogens exist in nearly every possible ecological niche on Earth. Both terrestrial and aquatic animals, including invertebrates and vertebrates, encounter each of the three major types of pathogens: certain bacteria, viruses, and eukaryotic parasites. A fourth type of infectious agent—proteins called prions—were considered in Chapter 18 and will not be discussed here.

As discussed in Chapter 27, bacteria are single-celled prokaryotic organisms that lack a true nucleus. Bacteria can either damage tissues at open wound sites or release toxins that enter the bloodstream and disrupt functions in other parts of the body. Bacteria are responsible for many diseases and infections, including typhoid fever, strep throat, skin infections, middle ear infections, and food poisoning. The major ways in which bacteria gain entry into an animal's body are through direct bodily contact, open wounds, inhalation through the respiratory tract, and ingestion via fecal contamination of food or water. The latter situation may arise because many infectious bacteria enter the intestines and are excreted in feces, which may be deposited near food or water sources used by some animals.

Viruses are nucleic acids enclosed within a protein coat (see Chapter 18). Unlike bacteria, viruses lack the metabolic machinery to synthesize the proteins they require to replicate themselves. Instead, they must infect a host cell and use its biochemical and genetic machinery, including nucleotides and energy sources, to make more viruses. The viral nucleic acid directs the host cell to synthesize the proteins required for viral replication. After entering a cell, some viruses, such as the common cold virus, multiply rapidly, kill the cell, and then infect other cells. Other viruses can lie dormant within host cells before suddenly undergoing rapid replication, which causes cell damage or death. Finally, certain viruses can transform their host cells into cancerous cells. Viruses are responsible for a great variety of illnesses, including some sexually transmitted diseases (refer back to Figure 18.2). Like bacterial infections, viral infections can spread rapidly among animals and can be lethal. Viruses typically enter the body through the respiratory tract or through open wounds.

Eukaryotic parasites—whether protists, fungi, or multicellular organisms such as worms—damage a host by using the host's nutrients for their own growth and reproduction or by secreting toxic chemicals. In humans, parasites account for an enormous number of cases of disease annually. For example, several hundred million people are infected each year with one of the mosquito-borne protists of the genus *Plasmodium* that causes malaria. Parasitic infections may enter a host through the bite of an infected insect, as in malaria; by ingestion of food or water containing parasitic organisms, such as roundworms; or in some cases, by penetrating the skin, as with blood flukes. What are the first defenses that such pathogens encounter?

53.2 Innate Immunity

Innate (nonspecific) immune defenses protect against foreign cells or substances without having to recognize the invaders' specific identities. This type of defense mechanism is called innate because animals inherit the ability to perform these protective functions and because this type of immunity does not require prior exposure to invaders. Instead of distinguishing among foreign materials, nonspecific defenses recognize some general, conserved property marking the invader as foreign, such as particular classes of carbohydrates or lipids present in the cell walls of many different kinds of microbes.

In this section, we will consider the innate immune defenses. These include defenses at the body surfaces, the actions of phagocytic cells, the response to injury known as inflammation,

and various proteins secreted by cells of the immune system that facilitate the destruction of pathogens.

The Body Surface Is an Initial Line of Defense

An animal's initial defenses against pathogens are the barriers provided by surfaces exposed to the external environment. Very few microorganisms can penetrate the intact skin or body surface of most animals, particularly the tough, thick, or scaly skin characteristic of many vertebrates or the rigid exoskeleton of many arthropods. In addition, glands in the body surfaces of both invertebrates and vertebrates secrete a variety of antimicrobial molecules, including mild acids and enzymes such as lysozyme that destroy bacterial cell walls.

The mucus secreted by cells in the mucous membranes lining the respiratory and upper gastrointestinal tracts of vertebrates also contains antimicrobial molecules. More importantly, mucus is sticky—microbes that become stuck in it are prevented from penetrating the mucous membrane barrier. They are either swept up by cilia into the pharynx and then swallowed, or they are engulfed by cells that are present in both tracts. Pathogens ingested with food are often destroyed by the acidic environment of an animal's midgut.

If a pathogen is able to penetrate a barrier and gain entry into an animal's internal tissues and fluids, other nonspecific defense mechanisms are activated. These mechanisms are mediated by several types of cells that reside in the body fluids and tissues, as described next.

Phagocytic Cells Provide Innate Defense Against Pathogens That Enter the Body

Several different types of cells in vertebrates play key roles in innate immunity (**Figure 53.1**). Many of these cells are **phagocytes**—cells capable of **phagocytosis**. Phagocytosis is a type of endocytosis in which the cell engulfs particulate matter, which usually is then destroyed by proteases or oxidizing compounds such as hydrogen peroxide. Phagocytes are found in the body fluids, such as hemolymph and blood, and also within various tissues and organs. Phagocytes are present in all classes of animals and are among the most fundamental and ancestral forms of immune defenses.

In vertebrates, most phagocytes belong to the type of blood cells called white blood cells, or **leukocytes** (Figure 53.1). All leukocytes are derived from a common type of stem cell (see Chapter 19), which in mammals and birds is found in the bone marrow. These stem cells give rise to several types of leukocytes and other cells that have specialized functions. The leukocytes involved in innate immunity include neutrophils, eosinophils, monocytes, macrophages, basophils, and natural killer cells.

Neutrophils are phagocytes and the most abundant leukocytes. They are found in blood and enter tissues during inflammation. After neutrophils engulf bacteria by phagocytosis, the bacteria are destroyed within endocytotic vacuoles by proteases, oxidizing compounds, and antibacterial proteins called defensins. The production and release of neutrophils from bone

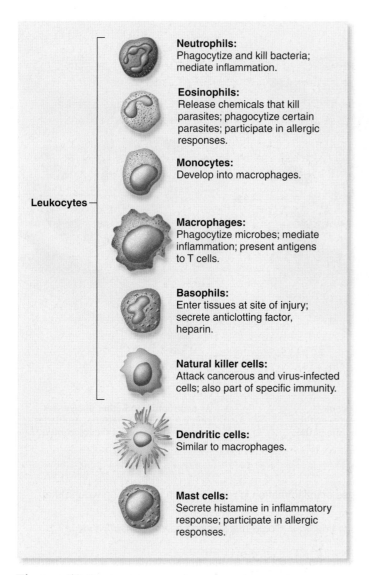

Neutrophils:
Phagocytize and kill bacteria; mediate inflammation.

Eosinophils:
Release chemicals that kill parasites; phagocytize certain parasites; participate in allergic responses.

Monocytes:
Develop into macrophages.

Leukocytes

Macrophages:
Phagocytize microbes; mediate inflammation; present antigens to T cells.

Basophils:
Enter tissues at site of injury; secrete anticlotting factor, heparin.

Natural killer cells:
Attack cancerous and virus-infected cells; also part of specific immunity.

Dendritic cells:
Similar to macrophages.

Mast cells:
Secrete histamine in inflammatory response; participate in allergic responses.

Figure 53.1   Cells involved in nonspecific immunity in vertebrates.   Note that six of these types of cells are leukocytes.

marrow is greatly stimulated during the course of an infection. **Eosinophils** are found in the blood, and mucosal surfaces lining the gastrointestinal, respiratory, and urinary tracts, where they fight off parasitic invasions. In some cases, eosinophils act by releasing toxic chemicals that kill parasites, and in other cases by phagocytosis. **Monocytes** are phagocytes that circulate in the blood for a short time, after which they migrate into tissues and organs and develop into **macrophages**. Macrophages are strategically located where they will encounter invaders, including epithelia in contact with the external environment, such as skin and the linings of respiratory and digestive tracts. Macrophages are large phagocytes capable of engulfing viruses and bacteria, as shown in the chapter-opening photo.

In contrast to these other leukocytes, **basophils** are secretory cells. They secrete an anticlotting factor called heparin at the site of an infection, which helps the circulation flush out the infected site. Basophils also secrete histamine, which attracts infection-fighting cells and proteins to the site.

Natural killer (NK) cells are another kind of leukocyte called lymphocytes, of which there are several types. We will learn more about the other types of lymphocytes later, because they play the major role in acquired immunity. NK cells, however, participate in both innate and acquired immunity. These cells are part of the body's innate defenses because they recognize general features on the surface of cancerous cells or any virus-infected cells. NK cells arise in the bone marrow, but their life history remains unclear. They act by releasing chemicals into the vicinity of cancerous or virus-infected cells, thereby killing those cells.

In addition to leukocytes, two other types of cells derived from bone marrow stem cells play important roles in nonspecific immunity. **Dendritic cells** are scattered throughout most tissues, where they perform various macrophage-like functions. **Mast cells** are found throughout connective tissues, particularly beneath the epithelial surfaces of the body. Mast cells secrete many locally acting molecules, including histamine. Histamine and other substances are involved in inflammation, an important component of the nonspecific defense mechanism to which we now turn.

Inflammation Is an Innate Response to Infection or Injury

Inflammation is an innate local response to infection or injury. The functions of inflammation are to destroy or inactivate foreign invaders, to clear the infected region of dead cells and other debris, and to set the stage for tissue repair. The key cellular components of this process are phagocytes, primarily neutrophils, macrophages, and dendritic cells, as well as mast cells.

The events of inflammation are induced and regulated by chemical mediators. These include a family of proteins called **cytokines** that function in both innate and acquired immune defenses. Cytokines provide a chemical communication network that synchronizes the components of the immune response. Most cytokines are secreted by more than one type of immune system cell and also by nonimmune cells such as endothelial cells and fibroblasts.

The sequence of local events in a typical inflammatory response to a bacterial infection is summarized in **Figure 53.2**. A tissue injury such as that caused by the splinter shown here begins the inflammation process, which causes the familiar signs and symptoms of local redness, swelling, heat, and pain.

Substances secreted into the extracellular fluid from mast cells, injured tissue cells, and neutrophils contribute to the inflammatory response. For example, histamine from mast cells and nitric oxide from endothelial cells (Figure 53.2, step 1) cause dilation of the small blood vessels in the infected and damaged area, and the vessels become leaky (step 2).

These vascular changes provide two benefits. First, the increased blood flow to the inflamed area, which accounts for the redness and heat, increases the delivery of beneficial proteins and leukocytes and increases local metabolism to facilitate healing. Second, the increased vascular permeability ensures that the plasma proteins that participate in inflammation can gain entry to the interstitial fluid. The swelling in an inflamed area also results from this increased leakiness of blood vessels.

Once neutrophils and other leukocytes arrive at the site of the injury, they begin the process of phagocytizing and

1. An injury introduces bacteria beneath the skin. Mast cells secrete histamine, and endothelial cells secrete nitric oxide.

2. Capillaries dilate and become leaky. Fluid and leukocytes exit the capillaries and enter the site of the wound.

3. As neutrophils and other phagocytes engulf and destroy bacteria, capillaries return to normal, and the infection is brought under control.

Figure 53.2   **The events in inflammation.**   Shown are the initial stages of inflammation in response to a penetrating wound that introduces bacteria beneath the skin.

Concept check: *Inflammation is often associated with swelling of the inflamed area. Could this swelling have an adaptive value?*

destroying invading microbes (Figure 53.2, step 3). The initial step in phagocytosis involves the interaction of phagocyte surface receptors with certain carbohydrates or lipids in the microbial cell walls. Subsequently, the neutrophils and other phagocytes also release antimicrobial substances into the extracellular fluid that can destroy microbes even before phagocytosis occurs. Other secreted substances such as nitric oxide function as inflammatory mediators. The result is positive feedback: Once phagocytes enter the area and encounter microbes, they release inflammatory mediators that bring even more phagocytes into the area.

Inflammation sets the stage for tissue repair. Nearby connective tissue cells called fibroblasts divide rapidly and secrete large quantities of collagen, the major component of extracellular matrices, and nearby blood vessel cells proliferate to restore blood supply. Ultimately, tissue repair may be imperfect, leaving a scar largely composed of fibroblasts, collagen, and other proteins.

Inflammation—indeed any stage of an illness—may be accompanied by **fever**, an increase in an animal's body temperature that results from an infection. This is distinguished from hyperthermia, an increase in body temperature resulting from any number of causes, such as overexertion. In humans, a temperature greater than 38°C (100.4°F) is considered to be a fever (normal is about 37°C [98.6°F]). The precise mechanisms by which a fever arises are not completely understood. It is

clear, however, that in mammals, cytokines released into the circulation by activated macrophages act within the hypothalamus to raise the body's set point for temperature. The result is a decrease in heat dissipation and an increase in heat generation. In ancient times, a fever was considered beneficial in helping fight off an infection, and this may indeed be true. While a sustained high fever requires medical attention because of the damaging effects of high temperature on membrane function, enzyme activity, and other processes (see Chapter 46), a low or moderate increase in temperature appears to stimulate leukocyte activity and proliferation and provides a less hospitable environment for at least some types of pathogens.

Antimicrobial Proteins Include Interferons and Complement Proteins

In addition to phagocytes and the inflammatory response, there are at least two other types of innate defenses against invading pathogens used by the body. These defenses allow for extracellular killing of microbes without prior exposure to those microbes. The first is used against viruses. **Interferons** are proteins that generally inhibit viral replication inside host cells. In response to viral infection, most cell types produce interferons and secrete them into the extracellular fluid. When the interferons bind to plasma membrane receptors on the secreting cell and on other cells, each cell synthesizes a variety of proteins

Primary lymphoid organs   **Secondary lymphoid organs**

- Tonsil
- Lymph nodes
- Thymus
- Spleen
- Bone marrow
- Lymphocyte accumulation in small intestine
- Lymphatic vessels

(a) Lymphatic system

- Lymph
- Lymphatic vessels
- Lymphatic vessel
- Valve
- To venous circulation
- Clusters of lymphocytes
- Blood-filled cavities (blood vessels not shown for clarity)

(b) Lymph node

Figure 53.3   **The lymphatic system in humans.**   (a) The major components of the human lymphatic system. Primary lymphoid organs are shown in red, and secondary lymphoid organs are shown in green. In adult humans, the primary lymphoid organs in bone are found in the sternum, ribs, parts of the skull, small regions of the femur and humerus, and, as shown here, the hip bones. (b) The structure of a lymph node. Lymph nodes occur along the course of lymphatic vessels, which drain interstitial fluid from tissues and return it to the venous circulation. Within a lymph node, lymph percolates through open cavities containing clusters of lymphocytes.

Concept check: *From where does lymph arise?*

that interfere with the ability of the viruses to replicate. Interferons are not specific. Many kinds of viruses induce interferon synthesis, and the same interferons, in turn, can inhibit the multiplication of many different kinds of viruses.

The second type of nonspecific defense is provided by the family of plasma proteins known as **complement**. Inactive complement proteins normally circulate in the blood at all times. When they contact the surface of microbes, a cascade of events results in the activation of complement proteins. Among their many actions, including those in specific immunity, complement proteins stimulate the release of histamine from mast cells, thereby increasing permeability of local blood vessels, as described earlier. Five of the active proteins generated in the complement cascade form a multiunit protein called the **membrane attack complex** (**MAC**), which, by embedding itself in the microbial plasma membrane, creates porelike channels. Water and salts enter the microbe through the channels, and the microbe bursts. We turn now to the type of immunity that is distinguished by an ability to recognize great numbers of specific foreign molecules.

53.3   Acquired Immunity

In acquired, or specific immune defense mechanisms, cells of the immune system first encounter and later recognize a specific foreign cell or protein to be attacked, as opposed to recognizing some general feature of pathogens. Any molecule that can trigger a specific immune response is called an **antigen**. An antigen is any molecule that the host does not recognize as self. Most antigens are either proteins or very large polysaccharides. Antigens include the protein coats of viruses, bacterial surface proteins, specific macromolecules on pollens and other allergens, cancerous cells, transplanted cells, and toxins.

This type of immunity is found in all classes of vertebrates except jawless fishes and was once thought to be absent in invertebrates. Recently, however, scientists who study the evolution of immune systems have uncovered several interesting features of invertebrate immune function that suggest some invertebrates have a limited ability to adapt immune activity to a specific invader. Despite this intriguing finding, however, we will consider acquired immune responses in the context in which they are best understood, the jawed vertebrates.

The Immune System Consists of Lymphoid Organs, Tissues, and Cells

The cells of the immune system that are responsible for acquired immunity are a type of leukocyte called **lymphocytes**. Like all leukocytes, lymphocytes circulate in the blood, but most of them reside in a group of organs and tissues that constitute the **lymphatic system** (Figure 53.3a). The system is composed primarily of a network of lymphatic vessels. These vessels drain the fluid known as lymph, which filters out of capillaries into the interstitial spaces. The lymph is eventually returned to the circulatory system. Various lymphoid organs and tissues are located throughout the lymphatic system. They are grouped into primary and secondary lymphoid organs.

The primary lymphoid organs are the structures in which lymphocytes differentiate into mature immune cells. These are the bone marrow in birds and mammals and the thymus gland in all vertebrates. In animals without extensive bone marrow, specialized regions of other organs such as the kidney and liver serve as primary lymphoid organs.

The primary lymphoid organs supply mature lymphocytes to secondary lymphoid tissues and organs, where the lymphocytes multiply and function. These include the lymph nodes of mammals (Figure 53.3b), the spleen (found in all jawed vertebrates and the largest secondary lymphoid structure), the tonsils (small, rounded lymphoid organs in the pharyngeal region of mammals), and scattered lymphocyte accumulations in the linings of the intestinal, respiratory, genital, and urinary tracts. With some exceptions (for example, adult humans, in which the thymus gland is no longer very active), destruction of or damage to a primary lymphoid organ results in a severe inability to fight off infections. The loss of any of the secondary lymphoid organs, while not as serious, nevertheless raises the risk of local or systemic infections throughout an animal's life. For example, in humans, the spleen must occasionally be surgically removed due to injury or disease. Such individuals must be monitored carefully for the rest of their lives because of their increased vulnerability to infection.

After leaving the bone marrow or thymus gland, lymphocytes circulate between the secondary lymphoid organs, blood, lymph, and all the tissues of the body. Lymphocytes from all the secondary lymphoid structures continually leave those structures and are carried to the bloodstream. Simultaneously, some circulating lymphocytes leave venules all over the body to enter the interstitial fluid. From there, they reenter lymphatic vessels and are carried back to secondary lymphoid organs. This constant recirculation of lymphocytes increases the likelihood that any given lymphocyte will encounter an antigen it is specifically programmed to recognize.

Lymphocytes Provide Specific Immunity Against Antigens in Humoral and Cell-Mediated Immunity

There are different kinds of lymphocytes that participate in coordinated specific immune system responses (Figure 53.4). In addition to NK cells described earlier, the two major types of lymphocytes are **B cells** and **T cells**. B cells were first observed to mature in an avian organ called the bursa of Fabricius—thus the name B cells. In mammals, B cells mature within bone marrow. Some B cells differentiate further into **plasma cells**, which synthesize and secrete antibodies, proteins that bind to and help destroy foreign molecules, as described later. T cells are so named because they mature within the thymus gland.

T cells may directly kill infected, mutated, or transplanted cells. T cells include two distinct types of lymphocytes: cytotoxic T cells and helper T cells. **Cytotoxic T cells** travel to the location of their targets, bind to these targets by recognizing an antigen, and directly kill those targets via secreted chemicals. In

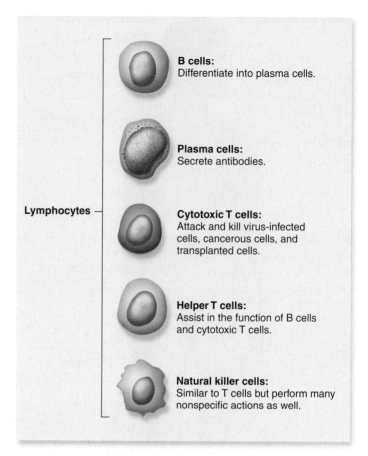

Figure 53.4 Cells involved in acquired immunity in vertebrates. Some important functions of each type of lymphocyte are included. The shape and color conventions shown in this figure will be used throughout this chapter.

addition, responses mediated by cytotoxic T cells are directed against body cells that have become cancerous or infected by pathogens. Natural killer cells also destroy such cells by secreting toxic chemicals.

As their name implies, **helper T cells** do not themselves function as "attack" cells. Instead, they assist in the activation and function of B cells and cytotoxic T cells. With only a few exceptions, B cells and cytotoxic T cells cannot function adequately unless they are stimulated by cytokines secreted from helper T cells.

Lymphocytes carry out their role by recognizing antigens such as those found on viruses, bacteria, and the surface of cancerous cells. The ability of lymphocytes to distinguish one antigen from another plays a central role in acquired immunity. Immunologists recognize two types of acquired immunity. In **humoral immunity**, plasma cells secrete antibodies that bind to antigens. In **cell-mediated immunity**, cytotoxic T cells directly encounter and destroy infected body cells, cancerous cells, or transplanted cells.

Acquired Immune Responses Occur in Three Stages

Thus far, we have considered the types of lymphocytes involved in an acquired immune response and how they circulate through the body. An acquired immune response can usually be divided into three stages (Figure 53.5): recognition of antigen, activation of lymphocytes, and attack against antigen. In the first stage, lymphocytes encounter and recognize an antigen. In the second stage, lymphocytes are activated and then proliferate and differentiate to produce effector cells and memory cells. **Effector cells** are plasma cells and cytotoxic T cells, which carry out the attack response; **memory cells** remain poised to recognize the antigen if it returns in the future. In the third stage, an attack is launched against the recognized antigen by effector cells and/or their secretions. Let's look at each of these stages more closely.

Stage 1: Recognition of Antigen During its development, each lymphocyte synthesizes a type of membrane receptor that can bind to a specific antigen. If subsequently the lymphocyte encounters that antigen, the antigen becomes bound to the receptor. This specific binding is the meaning of the word "recognize" in immunology. Antigens that bind to a lymphocyte receptor are said to be recognized by the lymphocyte. The ability of lymphocytes to distinguish one antigen from another, therefore, is determined by the nature of their plasma membrane receptors. Each lymphocyte is specific for just one type of antigen.

Stage 2: Activation of Lymphocytes The binding of an antigen to a receptor on a lymphocyte activates that lymphocyte. Upon activation, the lymphocyte undergoes multiple cycles of cell division. The result is the formation of many identical plasma cells called clones that express the same receptor as the receptor that first recognized the antigen. The chemical nature of the antigen determines which individual lymphocytes will be activated to form clones. This process requires the function of helper T cells, which divide when activated and then secrete the cytokines that promote cell division. Some of the cloned lymphocytes become plasma cells, and others function as memory cells.

In a typical person, the size of the lymphocyte population is staggering. Over 100 million different lymphocytes, each with the ability to recognize a unique antigen, are found in a person's immune system. This vast population explains why our bodies are able to recognize so many different antigens as foreign and eventually destroy them.

Stage 3: Attack Against Antigen The effector plasma cells and cytotoxic T cells attack all antigens of the kind that initiated the immune response. Plasma cells carry out a humoral

Figure 53.5 **The three stages of a specific (acquired) immune response.** All three cell types in step 1 recognize the same antigen. Helper T cells secrete cytokines that activate B cells and cytotoxic T cells, as indicated by the + symbols. Both B cells and T cells undergo cell division to form clones when activated, and in both cases, a portion of the cells are set aside as memory cells to fight off a future infection of the same type.

response by secreting antibodies into the blood. These antibodies then recruit and guide other molecules and cells that perform the actual attack. Activated cytotoxic T cells, by contrast, carry out cell-mediated immunity. They directly attack and kill the cells bearing the antigens.

Once the attack is successfully completed, the great majority of B cells, plasma cells, and cytotoxic T cells that participated in it die by apoptosis. The timely death of these effector cells prevents the immune defense from becoming excessive and possibly destroying the body's own tissues. However, memory cells persist even after the immune response has been successfully completed, so they can recognize and fight off any future infection with the same type of antigen. Let's look at each of these three stages in more detail, first for humoral immunity and then cell-mediated immunity.

In Humoral Immunity, B Cells Produce Immunoglobulins That Serve as Receptors or Antibodies

In stage 1 of humoral immunity (Figure 53.5), as mentioned, B cells recognize antigens with the help of B-cell receptors. When B cells are activated, they proliferate and differentiate into plasma cells, which secrete **antibodies**. These are proteins that travel all over the body to reach antigens identical to those that stimulated their production. In the extracellular body fluids, antibodies combine with these antigens and guide an attack that eliminates the antigens or the cells bearing them, a process we will discuss in more detail later. Such antibody-mediated responses are also called humoral immune responses, the adjective "humoral" denoting communication by way of

soluble chemical messengers (the old word "humors" was once used to refer to bodily fluids). Antibody-mediated responses are the major defense against bacteria, viruses, and other microbes in the extracellular fluid, and against toxin molecules.

B-cell receptors and antibodies share many structural and functional similarities (**Figure 53.6**). They are both members of a family of proteins called **immunoglobulins** (Ig). However, there are some differences between them. B-cell receptors have a transmembrane domain that anchors them in the plasma membrane of the B cell. These immunoglobulins function as receptors for antigens (Figure 53.6a). Antibodies are soluble proteins that are secreted from plasma cells (Figure 53.6b).

B-Cell Receptors   Prior to their activation, B cells express B-cell receptors that recognize specific antigens (Figure 53.6a). Interestingly, B-cell receptors and the antibodies made by plasma cells are encoded by the same genes. In plasma cells, the pre-mRNA is alternatively spliced, a phenomenon described in Chapter 13, so the transmembrane domain is not present in the protein. For this reason, the B-cell receptor in a particular B cell and secreted antibodies from the resulting plasma cells recognize the exact same antigen.

Immunoglobulin Structure   Each immunoglobulin molecule is composed of four interlinked polypeptides: two long **heavy chains** and two short **light chains** (Figure 53.6b). A hinge region that provides the molecule with flexibility separates the light chains and upper parts of the heavy chains from the lower parts of the heavy chains. A key feature of immunoglobulins is their **variable region**, which gets its name because it varies among different B cells. The variable region is the site that specifically recognizes a particular antigen.

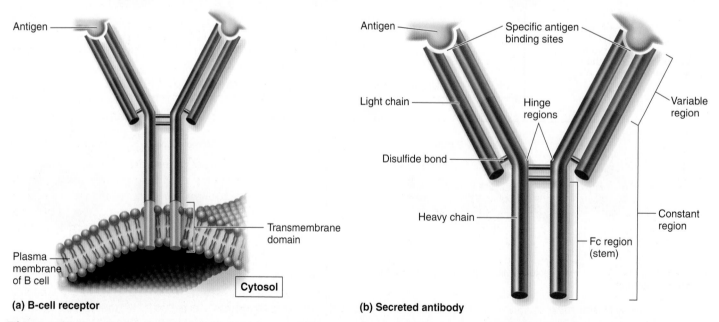

(a) B-cell receptor   (b) Secreted antibody

Figure 53.6   Immunoglobulins. (a) B-cell receptor and (b) secreted antibody.   Immunoglobulins are composed of two heavy chains and two light chains. Disulfide bonds hold the chains together. Within each immunoglobulin class, the constant regions, including the Fc regions (stems), have identical amino acid sequences. In contrast, the antigen-binding sites formed by the light- and heavy-chain variable regions have unique amino acid sequences and give each antibody its specificity for a particular antigen.

IgM
(pentamer)

Mammals have five classes of immunoglobulins, designated IgM, IgG, IgE, IgA, and IgD. All vertebrates have IgM molecules. These pentamers (made of five Ig molecules connected by disulfide bridges and other linkages) are the first Ig class produced after antigen exposure, but their blood concentration declines afterward. Some vertebrates have only some of the other classes and also express unique immunoglobulins not found in mammals. By contrast, invertebrates lack immunoglobulins. However, they have proteins containing regions called Ig-domains (or Ig-folds) with sequences that are similar to those of immunoglobulins. These proteins may be ancestral to immunoglobulins and in some cases have been shown to carry out immune activities.

IgG
(monomer)

The most abundant immunoglobulins in mammals are IgM and IgG. IgG, commonly called gamma globulin, is the most abundant Ig class in terms of blood concentration. Together these two immunoglobulin classes provide the bulk of specific immunity against bacteria and viruses in the extracellular fluid.

IgE
(monomer)

IgE antibodies are monomers that participate in defenses against multicellular eukaryotic parasites and also mediate allergic responses. Although present in blood at low concentrations, they also exist in mast cell membranes. When mast cell IgE molecules bind antigen, the mast cell secretes its histamine into the extracellular fluid, causing vasodilation and contributing to the allergic response. In people who are particularly sensitive to allergens, this response is easily demonstrated by a pinprick injection of an antigen, such as proteins associated with hay fever, into a small region under the skin. The resultant local inflammation and reddening of the skin are mediated in large part by IgE molecules.

IgA
(dimer)

IgA antibodies exist as dimers and are secreted by plasma cells in the linings of the gastrointestinal, respiratory, and genitourinary tracts and in tear ducts and salivary glands. They act locally in the linings of these structures or on their surfaces, as they are present in their secretions. For example, IgA molecules secreted into saliva help keep animals' mouths relatively free of pathogens. IgA molecules are also secreted by the mammary glands of mammals shortly after birth and therefore are the major antibodies in milk.

IgD
(monomer)

The functions of IgD are still unclear. However, IgD molecules are present both in blood and on the surface of B cells, and they are known to bind antigen on B cells, thus possibly contributing to B-cell activation.

The two heavy chains of an immunoglobulin have a stem called the Fc region (Figure 53.6b), which contains two or three domains depending on the class of immunoglobulin. The amino acid sequences of the Fc domains are identical for all immunoglobulins of a given class and are therefore known as **constant regions**.

Part of each heavy and light chain contains a constant region. In addition, as mentioned, they contain a variable region that serves as the antigen-binding site. In contrast to the constant regions of the heavy and light chains, the amino acid sequences of the variable regions vary widely from immunoglobulin to immunoglobulin in a given Ig class. The enormous number of variable sequences results in countless unique structures of immunoglobulins within each class. Thus, each of the five classes of antibodies contains up to millions of unique immunoglobulins, each capable of combining with only one specific antigen or, in some cases, with several antigens whose structures are very similar. The genetic basis for this remarkable array of immunoglobulins was first identified in the 1970s, as we see next.

Genomes & Proteomes Connection

Recombination and Hypermutation Produce an Enormous Number of Different Immunoglobulin Proteins

The human genome contains about 200 genes that encode immunoglobulins. This raises an intriguing question. How can the body produce millions of different immunoglobulin proteins if there are only 200 immunoglobulin genes? The answer is that the 200 genes undergo a unique process involving gene rearrangements. This phenomenon was discovered by Susumu Tonegawa and others in the 1970s.

Along the length of a typical immunoglobulin gene are numerous gene segments that code for a piece of the final immunoglobulin protein (Figure 53.7). In light chains, these gene segments are of three types, called variable, joining, and constant segments. A total of about 300 variable (V) segments code for the N-terminal region of the antigen-binding site. These are next to four joining (J) segments and a single constant (C) segment. Each segment along the length of the gene is associated with recognition sequences that bind two enzymes, called RAG-1 and RAG-2 (for recombination-activating gene). These enzymes, which are expressed only in developing lymphocytes, cut randomly at the end of a V segment and at the beginning of a J segment. The intervening region is lost, and then other enzymes paste the V and J segments together. The result is a new, permanent immunoglobulin gene for that B cell. Because any V segment can be linked with any J segment, the number of possible final genes among different B cells is huge. Additionally, heavy chains have multiple segments that are spliced together in this way, except that they have yet another segment (designated D) and more V segments, yielding an even greater number of possible heavy-chain genes. Because any heavy chain can combine with any light chain in a given B cell, the number of possible combinations of immunoglobulins is immense.

The number of possible immunoglobulins is increased even further in two more important ways within B cells. First, the joining process of the V, D, and J segments is not always

Organization of gene segments in a light-chain gene

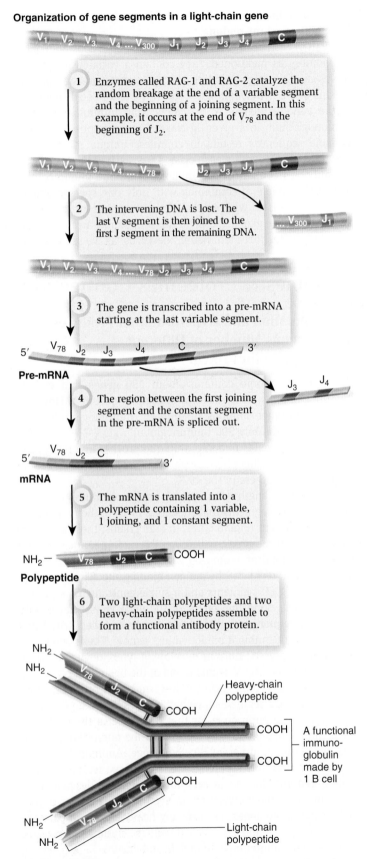

1. Enzymes called RAG-1 and RAG-2 catalyze the random breakage at the end of a variable segment and the beginning of a joining segment. In this example, it occurs at the end of V_{78} and the beginning of J_2.

2. The intervening DNA is lost. The last V segment is then joined to the first J segment in the remaining DNA.

3. The gene is transcribed into a pre-mRNA starting at the last variable segment.

Pre-mRNA

4. The region between the first joining segment and the constant segment in the pre-mRNA is spliced out.

mRNA

5. The mRNA is translated into a polypeptide containing 1 variable, 1 joining, and 1 constant segment.

Polypeptide

6. Two light-chain polypeptides and two heavy-chain polypeptides assemble to form a functional antibody protein.

Heavy-chain polypeptide

A functional immunoglobulin made by 1 B cell

Light-chain polypeptide

Figure 53.7 **The mechanism of immunoglobulin diversity.** Events similar to those depicted here also occur in the heavy chains.

precise. Occasionally, a few nucleotides may be lost at a joining end, resulting in a different amino acid sequence in the immunoglobulin protein. Second, in a subset of activated B cells, the DNA coding for the variable antigen-binding sites of immunoglobulins undergoes a process known as **hypermutation**, which primarily produces point mutations. During this process, cytosines are deaminated into uracils. After DNA replication, this leads to numerous C to T mutations. In addition, the presence of numerous uracils within the DNA recruits lesion-replicating DNA polymerases (see Chapter 11). These DNA polymerases are error prone, which leads to several additional types of mutations. The result is a hypervariable region of the light and heavy chains of all immunoglobulins. Hypermutation in lymphocytes appears to arise from lymphocyte-specific expression of a novel enzyme capable of deaminating cytosine.

The three processes of gene recombination, imprecise joining of gene segments, and hypermutation cause each lymphocyte within an individual's body to produce a unique type of immunoglobulin. The immune system can produce an incredibly diverse array of antibodies capable of recognizing many different antigens because the body makes hundreds of millions of different lymphocytes. Nearly any foreign antigen that is taken into the body will be recognized by some lymphocytes in this large population.

Activated B Cells Produce Plasma Cells Whose Antibodies Attack Pathogens

In stages 2 and 3 of the humoral immune response, B cells are activated and divide into plasma and memory cells. The plasma cells then secrete antibodies that attack the antigen detected (see Figure 53.5). In this section, we will take a closer look at these processes.

Clonal Selection B cells are activated by a specific antigen, with the aid of a helper T cell (a process which we will discuss later). When an antigen-stimulated lymphocyte divides and replicates itself, the progeny of this lymphocyte—all of which express the same receptor—are clones. The process by which these clones are formed is called **clonal selection** (Figure 53.8). This term emphasizes that lymphocyte proliferation is selected by exposure to an antigen.

Antibody Attack via Opsonization The antibodies secreted from the plasma cells circulate through the lymphatic system and the bloodstream. Eventually, the antibodies combine with the antigen that initiated the immune response. These antibodies then direct the attack against the pathogen to which they are now bound. Thus, immunoglobulins play two distinct roles in humoral immune responses. First, during antigen recognition, immunoglobulins (B-cell receptors) on the surface of B cells bind to antigen brought to them. Second, immunoglobulins (antibodies) secreted by the resulting plasma cells bind to pathogens bearing the same antigens, marking them as the targets to be attacked.

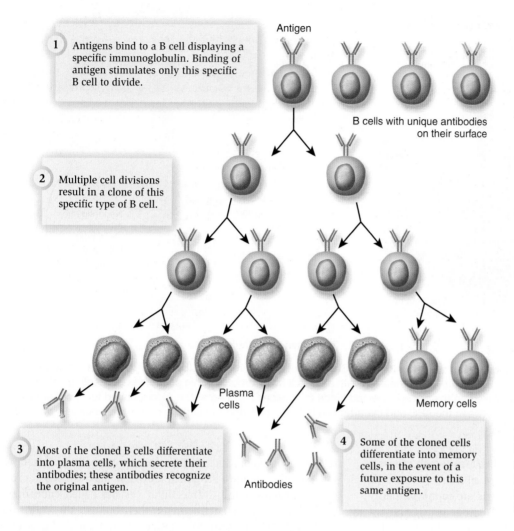

1 Antigens bind to a B cell displaying a specific immunoglobulin. Binding of antigen stimulates only this specific B cell to divide.

Antigen

B cells with unique antibodies on their surface

2 Multiple cell divisions result in a clone of this specific type of B cell.

Plasma cells

Memory cells

3 Most of the cloned B cells differentiate into plasma cells, which secrete their antibodies; these antibodies recognize the original antigen.

Antibodies

4 Some of the cloned cells differentiate into memory cells, in the event of a future exposure to this same antigen.

Figure 53.8 Clonal selection. In this example, a B cell with a specific surface immunoglobulin recognizes an antigen and is stimulated to divide into a clone of identical cells.

Instead of directly killing the pathogens, antibodies bound to antigen on the pathogen surface inactivate the pathogens in various ways. Antibodies may physically link the pathogens to phagocytes (neutrophils and macrophages), complement proteins, or NK cells. This linkage—called **opsonization**—triggers the attack mechanism and ensures that only the pathogens, and not nearby body cells, are killed.

In a second mechanism, antibodies directed against toxins produced by bacterial pathogens in the extracellular fluid bind to the toxins, thereby preventing them from harming susceptible body cells. The antibody-antigen complexes that are formed are then destroyed by phagocytes.

In a similar way, antibodies produced against certain viral surface proteins bind to the viruses in the extracellular fluid, preventing them from attaching to the plasma membranes of potential host cells. As with bacterial toxins, the antibody-virus complexes that are formed are subsequently phagocytized.

In Cell-Mediated Immunity, T Cells Recognize Antigens Complexed with Self Proteins

We now turn to stages 1 and 2 of cell-mediated immunity, in which T cells recognize and are activated by antigen (see

Figure 53.5). T-cell receptors for antigens have specific regions that differ from one T cell to another. As shown in **Figure 53.9a**, they are composed of two polypeptides, each with a variable and constant region, along with a transmembrane domain. The variable regions recognize an antigen. As in B-cell development, multiple DNA rearrangements occur during T-cell maturation, leading to millions of distinct types of T cells, each with a receptor of unique specificity. For T cells, this maturation occurs as they develop in the thymus.

In addition to their general structural differences, B-cell and T-cell receptors differ in a much more important way. T-cell receptors cannot combine with antigen unless the antigen is first complexed with certain of the body's own plasma membrane proteins. The T-cell receptor then recognizes and interacts with the entire self protein–antigen complex.

The plasma membrane self proteins that must be complexed with the antigen in order for T-cell recognition to occur are encoded by a gene family known as the **major histocompatibility complex** (**MHC**), and thus the proteins are called MHC proteins. Other than identical twins, no two individuals have the same sets of MHC genes, so no two individuals have the same MHC proteins on the plasma membranes of their cells. MHC proteins are cellular "identity tags" that serve as genetic markers of self.

(a) T-cell receptor

(b) Antigen presentation

Figure 53.9 **The T-cell receptor and antigen presentation.** (a) Structure of a T-cell receptor. (b) Antigen presentation. In the initial events in helper T-cell activation, antigen fragments are complexed with a class II MHC protein within an antigen-presenting cell such as a macrophage. The complex is then displayed on the cell surface and binds to a helper T-cell receptor. Also required for T-cell activation are the binding of nonantigenic proteins between the APC and the attached helper T cell, and the actions of the cytokines IL-1 and TNF.

Two major classes of MHC proteins are known. Class I MHC proteins are found on the surface of all human body cells except erythrocytes. Class II MHC proteins are found primarily on the surface of macrophages, B cells, and dendritic cells.

The two different types of T cells have different MHC requirements. Cytotoxic T cells require antigen to be associated with class I MHC proteins, whereas helper T cells require class II MHC proteins. One reason for this difference stems from the presence of different proteins on their surfaces; helper T cells can be identified by a unique membrane protein called CD4, and cytotoxic T cells are identified by a membrane protein known as CD8. CD4 binds to class II MHC proteins, whereas CD8 binds to class I MHC proteins.

How do antigens, which are foreign, end up complexed with MHC proteins on the surface of the body's own cells? The answer involves the mechanism known as antigen presentation, described next.

Antigen Presentation to Helper T Cells As previously noted, helper T cells can bind antigen only when the antigen appears on the plasma membrane of a host cell complexed with the cell's class II MHC proteins. Cells bearing these complexes, therefore, function as **antigen-presenting cells** (**APCs**). Because only macrophages, B cells, and dendritic cells express class II MHC proteins, only these cells can function as APCs for helper T cells.

Let's consider the function of macrophages as APCs for helper T cells (**Figure 53.9b**). After a microbe or noncellular antigen has been phagocytized by a macrophage in a nonspecific

response, antigens, such as proteins, are partially broken down into smaller peptide fragments by the macrophage's proteolytic enzymes within intracellular vesicles called endosomes. The resulting digested fragments then bind in the endosome to class II MHC proteins synthesized by the macrophage. Each fragment-MHC complex is then transported to the cell surface, where it is displayed in the plasma membrane. A specific helper T-cell receptor then binds this entire complex on the cell surface of the macrophage. The CD4 protein helps link the two cells. What is complexed to MHC proteins and presented to the helper T cells is not the intact antigen but instead a peptide fragment of the antigen—called an antigenic determinant, or **epitope**. Even so, it is customary to call this antigen presentation rather than epitope presentation.

B cells process antigen and present it to helper T cells in essentially the same way as macrophages do. The ability of B cells to present antigen to helper T cells is a second function of B cells in response to antigenic stimulation, in addition to their differentiation into antibody-secreting plasma cells.

The binding between the helper T-cell receptor and antigen bound to class II MHC proteins on an APC is the essential antigen-specific event in helper T-cell activation. However, by itself this specific binding will not result in T-cell activation. In addition, interactions occur between nonantigenic proteins on the surfaces of the attached helper T cell and the APC. These interactions provide a necessary costimulus for T-cell activation (Figure 53.9b).

Finally, the antigenic binding of the APC to the T cell plus the costimulus induce the APC to secrete large amounts of two cytokines—interleukin 1 (IL-1) and tumor necrosis factor (TNF). These molecules also stimulate the attached helper T cell.

Thus, the APC participates in the activation of a helper T cell in three ways: (1) antigen presentation, (2) provision of a costimulus, and (3) secretion of cytokines. As discussed, activated helper T cells then secrete cytokines that stimulate B cells and cytotoxic T cells.

Helper T Cells and B-Cell Activation

Now we can go back to B cells and understand how they are activated by the actions of helper T cells. This process begins when a helper T cell specific for a particular antigen binds to a complex of that antigen and a class II MHC protein on an APC, activating the helper T cell. Along with other signals, this binding induces the activated helper T cell to divide. Some of the resulting activated helper T cells then bind to B cells that display the same antigen on their surfaces. This binding, along with additional cytokines, stimulates the B cell to go through the process of clonal selection. Thus, helper T cells are so named because their secretions help activate B cells that have bound antigen, in addition to their participation in activation of cytotoxic T cells and antigen presentation, our next subject.

Antigen Presentation to Cytotoxic T Cells

Unlike helper T cells, cytotoxic T cells require class I MHC proteins for activation. Class I MHC proteins are synthesized by all nucleated cells. This distinction helps explain the major function of cytotoxic T cells—destruction of any of the body's own altered cells that have become cancerous or infected with viruses. The crucial point is that the antigens that complex with class I MHC proteins typically arise within body cells. They are endogenous antigens, synthesized by a body cell.

How do such antigens arise? In viral infections, once a virus has entered a host cell, the viral nucleic acid instructs the host cell to manufacture viral proteins, which are foreign to the cell. In cancerous cells, one or more of the cell's genes have become altered by chemicals, radiation, or other factors. The altered genes, called oncogenes, code for proteins that are not normally found in the body. Such abnormal proteins act as antigens.

In both virus-infected and cancerous cells, cytosolic enzymes hydrolyze some of the endogenously produced antigenic proteins into peptide fragments, which are transported into the endoplasmic reticulum. There the fragments are complexed with the host cell's class I MHC proteins and then shuttled by exocytosis to the plasma membrane, where a cytotoxic T cell specific for the antigen/MHC protein complex can bind to it. Once binding occurs, cytotoxic T cells release chemicals that kill the infected or cancerous cell, as discussed next.

Activated Cytotoxic T Cells Kill Infected or Cancerous Cells

The previous sections described how immune responses provide long-term defenses against bacteria, viruses, and individual foreign molecules that enter the body's extracellular fluid. We now examine how the body's own cells that have become infected by viruses or transformed into cancerous cells are destroyed by stage 3 cell-mediated immune responses (see Figure 53.5).

What is the value of destroying virus-infected host cells? First, such destruction prevents cells from making more viruses. Second, for cells that already are making mature viruses, it results in the release of the viruses into the extracellular fluid, where they can be neutralized by circulating antibody.

Role of Cytotoxic T Cells

A typical cytotoxic T-cell response triggered by viral infection of a vertebrate's body cells is summarized in **Figure 53.10**. The response triggered by a cancerous cell would be similar. In this discussion, we go back to stages 1 and 2 of the cell-mediated response to see how they lead up to the antigen attack in stage 3. A virus-infected cell produces foreign proteins, viral antigens that are processed and presented on the plasma membrane of the cell complexed with class I MHC proteins. Cytotoxic T cells specific for the particular antigen bind to the complex (Figure 53.10, step 1). As with B cells, binding to antigen alone does not cause activation of the cytotoxic T cell. Cytokines from nearby activated helper T cells are also required.

Macrophages phagocytize extracellular viruses (or, in the case of cancer, antigens released from the surface of cancerous cells) and then process and present antigen, in association with class II MHC proteins, to the helper T cells (step 2). In addition, the macrophages provide a costimulus and also secrete IL-1 and TNF. The activated helper T cell releases IL-2 and other cytokines, which stimulate proliferation of the helper T cell.

IL-2 and other cytokines also act on the cytotoxic T cell bound to the surface of the virus-infected or cancerous cell, stimulating this attack cell to proliferate. Why is proliferation important if a cytotoxic T cell has already located and bound to its target? The answer is that there is rarely just one virus-infected or cancerous cell. By expanding the population of cytotoxic T cells capable of recognizing the particular antigen, the likelihood is greater that the other virus-infected or cancerous cells will be encountered by an appropriate cytotoxic T cell.

The cytotoxic T cells specific for that virus then find and bind to other virus-infected cells (step 3). The cytotoxic T cell releases the contents of its secretory vesicles directly into the extracellular space between itself and the target cell to which it is bound (thereby ensuring that other nearby host cells will not be killed). These vesicles contain proteases, and a protein called perforin, which is similar in structure to the proteins of the complement system's membrane attack complex. Perforin is believed to insert into the target cell's membrane and form channels ("perforations") through the membrane (step 4). In this manner, it causes the attacked cell to take in the proteases secreted by the cytotoxic T cell, which is believed to induce apoptosis. The pores also cause the attached cell to take in water and burst. The cytotoxic T cell is not harmed by this process and can then continue to kill other virus-infected cells (step 5). Target cell killing by activated cytotoxic T cells occurs by several mechanisms, but this is one of the most important.

This shows that the body can eliminate viruses in two ways: through the humoral actions of antibodies in body fluids

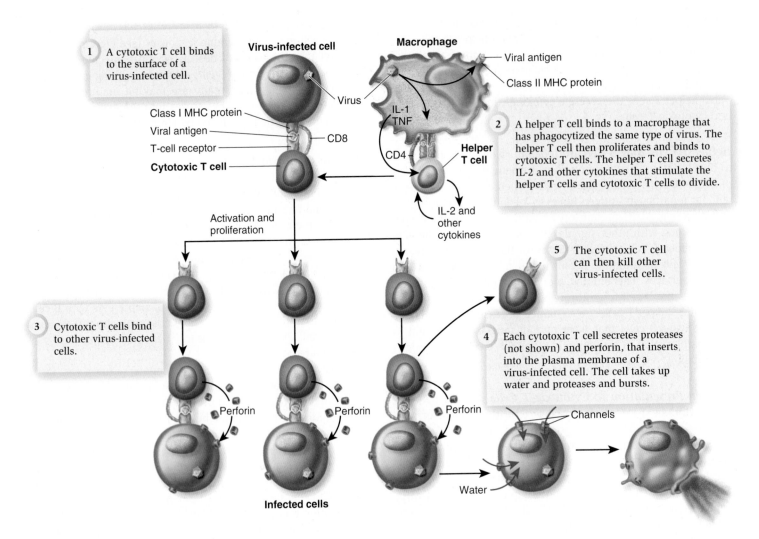

1 A cytotoxic T cell binds to the surface of a virus-infected cell.

Virus-infected cell

Virus

Macrophage

Viral antigen

Class II MHC protein

Class I MHC protein

Viral antigen

T-cell receptor

Cytotoxic T cell

CD8

IL-1
TNF

CD4

**Helper
T cell**

2 A helper T cell binds to a macrophage that has phagocytized the same type of virus. The helper T cell then proliferates and binds to cytotoxic T cells. The helper T cell secretes IL-2 and other cytokines that stimulate the helper T cells and cytotoxic T cells to divide.

IL-2 and other cytokines

Activation and proliferation

5 The cytotoxic T cell can then kill other virus-infected cells.

3 Cytotoxic T cells bind to other virus-infected cells.

4 Each cytotoxic T cell secretes proteases (not shown) and perforin, that inserts into the plasma membrane of a virus-infected cell. The cell takes up water and proteases and bursts.

Perforin

Perforin

Perforin

Channels

Water

Infected cells

Figure 53.10 Summary of events in the killing of virus-infected cells by cytotoxic T cells. The sequence is similar for cancerous cells attacked by a cytotoxic T cell.

and through the cell-mediated killing of virus-infected cells by cytotoxic T cells. Although cytotoxic T cells play an important role in the attack against such cells, they are not the only ones.

Role of Natural Killer (NK) Cells NK cells also destroy virus-infected and cancerous cells by secreting toxic chemicals. As mentioned earlier, NK cells can recognize general features on the surface of such cells and participate in innate immunity. In addition, in a cell-mediated immune response, NK cells can be linked to such target cells by antibodies and then can destroy them by release of toxic molecules.

Summary: Example of an Acquired Immune Response

Let's bring together our discussion of the humoral immune system by looking in detail at one example. One classic humoral immune response results in the destruction of bacteria. The sequence of events, which is quite similar to the humoral

response to a virus in the extracellular fluid, is summarized in **Figure 53.11**. For this example, we consider the response in mammals, in which lymph nodes are present. Many features of the response, however, are similar in other vertebrates.

This process starts the same way as for nonspecific responses, with the bacteria penetrating one of the body's linings through an injury and entering the interstitial fluid (Figure 53.11, step 1). The bacteria then move with lymph into the lymphatic system and are carried to lymph nodes (step 2). Within the lymph node, a macrophage and a B cell recognize one of the bacteria as a foreign substance and bind to it.

As we have discussed, the process of B-cell activation usually requires the activation of helper T cells. The helper T cell binds to a complex of processed antigen and class II MHC protein on an APC (step 3). In this case, the APC is a macrophage that has phagocytized the bacterium, hydrolyzed its proteins into peptide fragments, complexed the fragments with class II MHC proteins, and displayed the complexes on its surface. Once a helper T cell specific for the complex binds to it, the helper

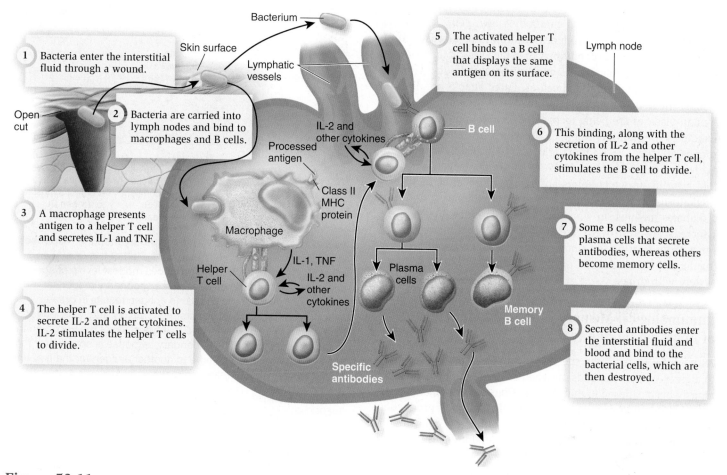

Figure 53.11 Summary of events in a typical humoral immune response. Most of the events depicted occur within a lymph node.

T cell becomes activated. The macrophage helps this process in two other ways: It provides a costimulus, and it secretes the cytokines IL-1 and TNF.

IL-1 and TNF stimulate the helper T cell to secrete another cytokine, IL-2. IL-2 stimulates the activated helper T cell to divide. This leads eventually to the formation of a clone of activated helper T cells (step 4), which bind to B cells and also secrete IL-2 and other cytokines (step 5). Certain of these cytokines provide the additional signals that are usually required to activate nearby antigen-bound B cells to proliferate (step 6). These cells differentiate into memory cells, which help ward off possible future attacks by the same antigen, and plasma cells, which then secrete specific antibodies (step 7). The antibodies enter the bloodstream and bind to bacterial cells, which are then destroyed (step 8).

B Cells and T Cells That Recognize Self Molecules Must Be Killed or Inhibited

As we have seen, the lymphocytes responsible for the specific immune response in vertebrates are very capable killers of pathogens—so capable, in fact, that it raises a question: Why don't these cells attack and kill normal self cells? In other words, how does the body distinguish between self and nonself components and develop what is called **immune tolerance**, or tolerance of your own antigens?

Recall that the huge diversity of lymphocyte receptors is ultimately the result of multiple random DNA cutting/recombination processes. It is virtually certain, therefore, that every animal possessing specific immune defenses would have clones of lymphocytes with receptors that could bind to that individual's own proteins. The continued existence and functioning of such lymphocytes would be disastrous, because such binding would launch an immune attack against all body cells expressing these proteins.

At least two mechanisms explain why individuals normally lack active lymphocytes that respond to self components. First, during early development in vertebrates, T cells are exposed to a wide mix of self proteins in the thymus. Those T cells with receptors capable of binding self proteins are destroyed by apoptosis in a process termed **clonal deletion**. The second process, termed **clonal inactivation**, occurs outside the thymus and causes potentially self-reacting T cells to become nonresponsive. B cells undergo similar processes. The mechanisms by which these two events occur are still under investigation.

Occasionally, however, these mechanisms fail, and the body's immune cells attack the body's own cells. When this happens in humans, it produces **autoimmune disease**. Auto-immune diseases are conditions in which the body's normal state of immune tolerance somehow breaks down, with the result that both humoral and cell-mediated attacks are directed against the body's own cells and tissues. A growing number of human diseases are being recognized as autoimmune in origin. Examples are multiple sclerosis, in which myelin around neurons is attacked; myasthenia gravis, in which the receptors for acetylcholine on skeletal muscle cells are the targets; rheumatoid arthritis, in which joints are damaged; systemic lupus erythematosus, in which numerous organs are damaged; and type 1 diabetes mellitus, in which the insulin-producing cells of the pancreas are destroyed. Treatments for autoimmune disease range from treating the symptoms (for example, administering insulin to individuals with diabetes) to suppressing the immune system with drugs.

Immunological Memory Is an Important Feature of Acquired Immunity

As we have learned, the specific or acquired immune response to a given antigen depends on whether or not the body has previously been exposed to that antigen. Consider, for example, the humoral immune response. In mammals, antibody production in response to the first contact with an antigen occurs slowly, over a few weeks. This response to an initial antigen exposure is termed a **primary immune response** (Figure 53.12). Any subsequent infection by the same pathogen elicits an immediate and heightened production of additional specific antibodies against that particular antigen, a reaction termed a **secondary immune response**.

In the case of humoral immunity, this secondary response occurs more quickly, is stronger, and lasts longer because memory B cells that were produced in response to the initial antigen exposure are quickly stimulated to multiply and differentiate into thousands of plasma cells. These cells then produce large amounts of specific antibodies. The immune system's ability to produce this secondary response is called **immunological memory**.

Immunological memory explains why we and other animals are able to fight off many illnesses to which we have been previously exposed, such as many common childhood diseases. The acquired response to exposure to any type of antigen is known as **active immunity**. Active immunity not only results from natural exposure to antigens, but it is also the basis for the artificial exposures to antigen that occur in vaccinations. In **vaccinations**, small quantities of living, dead, or altered microbes, small quantities of toxins, or harmless antigenic molecules derived from a microorganism or its toxin are injected into the body, resulting in a primary immune response, including the production of memory cells. Subsequent natural exposure to the immunizing antigen results in a rapid, effective response that can prevent or reduce the severity of disease.

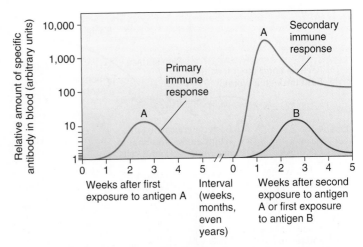

Figure 53.12   Primary and secondary immune responses. In a primary response, as shown on the left of this graph, an initial exposure to an antigen produces modest levels of specific antibody over a period of weeks. In a secondary response, subsequent exposure to the same antigen results in greater antibody production that occurs more rapidly and lasts longer than a primary response. (Note that the scale of the y-axis on the graph is logarithmic.) The secondary response is specific for that antigen. Exposure at that time to another antigen for the first time produces the usual primary response.

Concept check: *What is the advantage of a secondary immune response?*

In contrast to active immunity, another type of acquired immunity, called **passive immunity**, confers protection against disease through the direct transfer of antibodies from one individual to another. Passive immunity can occur naturally, as when IgG molecules cross the mammalian placenta to protect a fetus from various pathogens, or when a newborn mammal receives antibodies from breast milk. It can also occur artificially, as when a human patient is given an injection of IgG molecules shortly after exposure to hepatitis viruses. Recent advances in the creation of highly specific and pure antibodies, called **monoclonal antibodies** because they are derived from a single clone of cells prepared in a laboratory, have paved the way for the use of passive immunity to combat certain types of cancer. Because antibodies are proteins with a limited life span, the protection afforded by the transfer of antibodies in passive immunity is relatively short-lived, usually lasting only a few weeks or months.

Immunizations are very important in preventing disease among humans and domesticated animals. This is especially important when humans or other animals live in dense populations. A good example is seen in certain invertebrates such as social insects, which live in populations with enormous numbers of individuals in confined areas. Do such populations have any naturally occurring protection against widespread disease that could otherwise ravage an entire colony of animals? This question has been explored recently by biologists studying social insects such as ants and termites, as described next.

FEATURE INVESTIGATION

Traniello and Colleagues Demonstrated That Social Insects May Develop "Social Immunity"

Biologists have questioned how highly social animals living in dense colonies protect themselves from disease transmission. Such social animals include wood-eating termites that exist in very large numbers in complex nests. Termites may encounter a range of pathogens because they nest in moist, decaying wood, which provides a favorable environment for bacteria, fungi, and other parasites. A single infected termite could rapidly spread disease from insect to insect in a densely populated colony, but rarely are entire colonies wiped out by infection. In part, this occurs because termites kill microbes by producing antimicrobial secretions that are used to line their nests. In addition, behavioral adaptations, such as grooming to remove pathogens from nest mates' cuticles and the removal of sick or dead animals from the nest, help reduce disease transmission.

Another possible mechanism of protection from disease, called "social immunity," involves the transfer of infection resistance from immunized to susceptible nest mates. This phenomenon was recently identified by James Traniello and coworkers while studying the termite *Zootermopsis angusticollis*. The researchers discovered that these termites were able to survive a pathogen challenge better when they are living in a group than when they are isolated as individuals. This indicates that an individual's immunity is enhanced when they are surrounded by nest mates.

To determine how this may occur, Traniello and his colleagues tested the susceptibility of the termites to parasitic fungi (Figure 53.13). They divided the animals into two groups. To distinguish the groups from each other, one group was fed paper that contained a dye that marked the animals with a visible color. The investigators then exposed this group of dyed termites to a sublethal dose of fungal spores to immunize them

Figure 53.13   Traniello and colleagues demonstrated that social insects may develop "social immunity."

Concept check: What other animals besides social insects live in extremely dense colonies and require protection against devastating illnesses sweeping through the entire population?

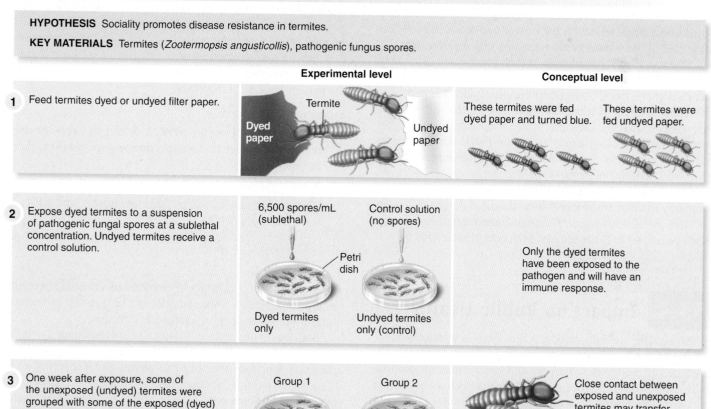

HYPOTHESIS Sociality promotes disease resistance in termites.

KEY MATERIALS Termites (*Zootermopsis angusticollis*), pathogenic fungus spores.

Experimental level

Conceptual level

1 Feed termites dyed or undyed filter paper.

Termite
Dyed paper
Undyed paper

These termites were fed dyed paper and turned blue.   These termites were fed undyed paper.

2 Expose dyed termites to a suspension of pathogenic fungal spores at a sublethal concentration. Undyed termites receive a control solution.

6,500 spores/mL (sublethal)   Control solution (no spores)

Petri dish

Dyed termites only   Undyed termites only (control)

Only the dyed termites have been exposed to the pathogen and will have an immune response.

3 One week after exposure, some of the unexposed (undyed) termites were grouped with some of the exposed (dyed) termites. This is called group 1. Group 2 consists of only undyed termites.

Group 1   Group 2

Allowed to coexist for 1 week.

Close contact between exposed and unexposed termites may transfer immunity.

4 Expose group 1 and group 2 to a potentially lethal concentration of fungal spores. Measure the percentage of undyed termites that survive after 6 days.

65,000 spores/mL

If immunity is transferred, a higher percentage of undyed animals in group 1 should survive exposure to the spores.

5 THE DATA

Results from step 4:

6 CONCLUSION Immunity against deadly pathogens can be transferred from exposed termites to unexposed termites.

7 SOURCE Traniello, J.F., Rosengaus, R.B., and Savoie, K. 2002. The development of immunity in a social insect: evidence for the group facilitation of disease resistance. *Proceedings of the National Academy of Sciences USA* 99:6838–6842.

against the fungus. The other (undyed) group was exposed only to a control solution. One week following exposure to spores or control solutions, half of the control group of animals was combined into a single nest with the immunized animals, while the other half of the control group remained in a separate nest.

One week later, the animals in each group were exposed to a potentially lethal dose of fungal spores, and their survival was recorded after 6 days. Undyed termites that had not experienced an initial exposure to spores but had later nested with animals that were exposed and thus immunized had a significantly higher survival rate following exposure to the lethal dose of spores compared to animals that did not nest with immunized animals. Thus, the investigators concluded that social contact between nonimmunized and immunized animals resulted in a transfer of immunity to the more-susceptible members of the group.

The mechanism of social immunization is not known but could be related to either the oral exchange of secretions or con-

tact with inactivated spores remaining on the bodies of immunized animals. Whatever the mechanism may prove to be, the consequence is striking: The close physical contact that is characteristic of social insect behavior helps propagate colonywide immunization against potentially deadly pathogens. Moreover, it is not unique to *Zootermopsis angusticollis*. Researchers in Denmark have recently demonstrated a similar phenomenon in the garden ant *Lasius neglectus*.

Experimental Questions

1. What features of termite physiology and/or behavior reduce the chance that an infection will spread throughout the colony?

2. What hypothesis was tested by Traniello and colleagues?

3. How did the researchers test this hypothesis, and what were the results of the study?

53.4 Impact on Public Health

In this section, we will consider a few ways in which the functioning of a person's immune system is affected by lifestyle, medical interventions, allergies, and destruction of immune cells. Collectively, the effects of these problems have an almost immeasurable impact on society in terms of worker productivity, health-care resources, and the economy.

Lifestyle Has an Important Influence on Immunity

Protein-calorie malnutrition is the single greatest contributor to decreased resistance to infection worldwide. When inadequate

amino acids are available to synthesize essential proteins, immune function is impaired. Deficits of specific nonprotein nutrients can also lower resistance to infection.

Both stress and state of mind can affect resistance to infection and to cancer. The immune system can alter neural and endocrine function, and, in turn, neural and endocrine activity modifies immune function. Lymphoid tissue is connected to nerves, and immune cells have receptors for many neurotransmitters and hormones. Conversely, immune cells release cytokines that have important effects on the brain and endocrine system. Moreover, lymphocytes secrete several hormones that are also produced by endocrine glands. The multiple "mind-body" interactions that affect disease resistance are the subject of a field called psychoneuroimmunology.

Of the hormones associated with stress, the adrenal hormone cortisol has received the most attention due to its powerful suppressive activity on inflammation and specific immunity. Among other things, cortisol inhibits production of inflammatory mediators, reduces capillary permeability in injured areas, and suppresses the growth and activity of certain types of leukocytes. In this way, it acts as a sort of brake on the immune system. During chronic stress or when cortisol is used to treat certain illnesses for long periods of time, it may cause immunosuppression. This is a key link between stress and health. Chronic stress may lead to high cortisol levels that, by suppressing the body's immune responses, lowers resistance to infection.

Another feature of a person's lifestyle that appears to affect immune function is exercise. The influence of physical exercise on the body's resistance to infection and cancer has been debated for decades. Present evidence indicates that the intensity, duration, regularity, and psychological stress of the exercise all have important influences, both negative and positive, on a variety of immune functions, such as the numbers of circulating NK cells. Despite these complexities, most experts currently consider moderate exercise and physical conditioning to have net beneficial effects on the immune system and on disease resistance. A 2005 study suggested that exercise may be particularly beneficial in warding off the onset of breast cancer, one of the most common types of cancer in women.

Organ Transplants and Blood Transfusions Are Medical Procedures That Can Cause Serious Immune Reactions

Both organ transplants and blood transfusions have saved numerous lives. However, both also carry the possibility of provoking immune reactions that can threaten the life of the recipient. Since the mid-20th century, organ transplants from a healthy or recently deceased donor to a recipient have become widespread. The United Network for Organ Sharing reports that approximately 27,000 organs are transplanted in the U.S. each year, with kidney (16,000), liver (6,000), heart (2,000), and lung (1,200) accounting for most of the transplants. The major obstacle to successful transplantation of tissues and organs is a reaction called graft rejection, in which the immune system recognizes the transplants (also called grafts) as foreign and attacks them as it would any other foreign cells. Although B cells and macrophages play some role in graft rejection, cytotoxic T cells and helper T cells are mainly responsible. To minimize this possibility, patients are given drugs that suppress immune function.

Except for grafts from identical twins, the class I MHC proteins on graft cells differ from those on the recipient's cells, as do the class II MHC proteins present on macrophages in the graft cells. Consequently, the recipient's T cells recognize the MHC proteins in the graft as foreign, and cytotoxic T cells (with the aid of helper T cells) destroy the graft cells.

Blood transfusion reactions, in which erythrocytes are destroyed after the transfusion, are a special example of tissue rejection in that antibodies rather than cytotoxic T cells are the major factor in rejection. Although erythrocytes lack MHC proteins, they have plasma membrane components that can function as antigens when transfused into another person's blood. Of the more than 400 erythrocyte antigens, the ABO system (see Chapter 16) of carbohydrates is the most important for transfusion reactions.

The four possible blood types—A, B, AB, and O—are genetically determined. Individuals with type A blood have erythrocytes bearing the A antigen. Those whose erythrocytes have the B antigen are type B, those that produce both antigens are type AB, and those that produce neither antigen are considered type O. Genes cannot code for the carbohydrates that function as antigens; instead, the genes code for the particular enzymes that catalyze the attachment of the carbohydrates onto cell surface proteins. Additionally, the plasma of type A individuals contains anti-B antibodies. Type B individuals have anti-A antibodies, AB individuals have neither antibody, and type O individuals have both. Scientists do not yet understand how these antibodies occur without prior exposure to the appropriate antigen.

What would happen if a person with type A blood were given type B blood (something that would not be done in actual practice)? There are two incompatibilities. The recipient's anti-B antibodies would bind to the transfused erythrocytes, and the anti-A antibodies in the transfused plasma would attack the recipient's erythrocytes. Therefore, a type A person cannot receive blood containing B antigens but can receive blood of type A.

By similar logic, type B individuals can receive blood of type B without an immune response occuring. Because AB individuals have neither anti-A nor anti-B antibodies, they can receive blood of any type and are considered "universal recipients." By contrast, type O individuals, who have both anti-A and anti-B antibodies, can receive only type O blood. Because type O blood has neither A nor B antigens, however, it can be transfused into individuals of all types without being destroyed by the recipient's immune system, and thus type O people are considered "universal donors." Nonetheless, type O blood is not an ideal match for a person with type A, B, or AB blood. Although the transfused (type O) blood is safe from immune attack, the recipient's blood may still be affected by the anti-A and anti-B antibodies in type O blood.

Allergies Affect the Quality of Life of Millions of People

An allergy (also known as hypersensitivity) is a condition in which immune responses to environmental antigens cause inflammation and damage to body cells. Antigens that induce allergic reactions are called allergens. Common examples of allergens include ragweed pollen and animal dander. Most allergens themselves are relatively or completely harmless. It is the immune responses to them that cause the damage. In essence, then, allergy is immunity gone awry, for the response is of inappropriate strength and duration for the stimulus. In the U.S. alone, as many as 40 million people (about 13% of the population) suffer from allergies.

For any allergy to develop, a genetically predisposed person must first be exposed to the allergen—a process called sensitization. Subsequent exposures elicit the damaging immune responses we recognize as an allergy.

Hypersensitivities can be broadly classified according to the speed of the reaction. Allergies that take up to several days to develop are considered delayed hypersensitivities. The skin rash that appears after contact with poison ivy is an example. More common are reactions considered immediate hypersensitivities, which can develop in minutes or up to a few hours. These allergies are also called IgE-mediated hypersensitivities because they involve IgE antibodies.

In immediate hypersensitivity, sensitization to the allergen leads to the production of specific antibodies and a clone of memory B cells. In individuals who are genetically susceptible to allergies, antigens that elicit immediate hypersensitivity reactions stimulate the production of IgE antibodies. Upon their release from plasma cells, these IgE molecules circulate throughout the body and become attached to mast cells in connective tissue. When the same antigen subsequently enters the body at some future time and binds with IgE that is bound to mast cells, the mast cell is stimulated to secrete many inflammatory mediators, including histamine, that then initiate an inflammatory response.

The signs and symptoms of IgE-mediated hypersensitivity reflect both the effects of inflammatory mediators and the body site in which the antigen–IgE–mast cell binding occurs. When, for example, a previously sensitized person inhales ragweed pollen, the antigen combines with IgE on mast cells in the airways. The mast cells release their contents, which induce increased mucous secretion, increased blood flow, swelling of the epithelial lining, and contraction of the smooth muscle surrounding airways. These effects produce the congestion, runny nose, sneezing, and difficulty in breathing characteristic of hay fever. Antihistamines are drugs taken by people to block the action of histamine that is released during allergic responses. These drugs prevent histamine from binding to its receptor protein on its target cells, thereby preventing or relieving some of the symptoms of allergy.

Acquired Immune Deficiency Syndrome (AIDS) Is a Growing Pandemic

Acquired immune deficiency syndrome (**AIDS**) is caused by the **human immunodeficiency virus** (**HIV**), which incapacitates the immune response by preferentially infecting helper T cells. HIV is a retrovirus; such viruses have a nucleic acid core of RNA rather than DNA. Once inside a helper T cell, HIV uses the enzyme reverse transcriptase to transcribe the virus's RNA into DNA, which is then integrated into the T cell's chromosomes. Viral replication within the T cell results in the death of the cell.

HIV infects helper T cells because the CD4 protein in their plasma membranes acts as a receptor for an HIV capsid protein. However, binding to CD4 is not sufficient to enable HIV to enter the helper T cell. Another T-cell surface protein, which normally acts as a receptor for certain cytokines, must serve as a coreceptor. Interestingly, individuals possessing a mutation in this cytokine receptor are highly resistant to HIV infection, so much research is now focused on the possible therapeutic use of chemicals that can bind to and block this coreceptor.

HIV not only directly kills helper T cells, but it also indirectly causes additional helper T-cell death by inducing cytotoxic T cells to kill HIV-infected helper T cells. In addition, by still poorly understood mechanisms, HIV causes the death of many uninfected helper T cells by apoptosis. Without adequate numbers of helper T cells, neither B cells nor cytotoxic T cells can function normally. Both humoral and cell-mediated immunity are compromised. Many individuals with AIDS die from infections and cancers that ordinarily would be readily handled by a fully functional immune system.

AIDS, first described in 1981, has since reached pandemic proportions. About 40 million people worldwide are infected with HIV, and an estimated 15,000 new infections occur each day. The major routes of HIV transmission are (1) unprotected sexual intercourse with an infected partner; (2) transfer of contaminated blood or blood products between individuals, such as during a blood transfusion or the sharing of needles among intravenous drug users; (3) transfer from an infected mother to her child across the placenta or during delivery; or (4) transfer via breast milk during nursing.

The great majority of individuals now infected with HIV show no signs of AIDS. Their infections are diagnosed by the presence of anti-HIV antibodies or HIV RNA in the blood. However, if left untreated, HIV infection commonly develops into AIDS in about 10 years. During the first 5 years, killed helper T cells are typically replaced by new cells, so T-cell levels remain normal, and the individual remains asymptomatic. Over the next 5 years, T-cell levels begin to decline, until at some point, AIDS reveals itself in the form of opportunistic viral, bacterial, and fungal infections. Certain unusual cancers, such as Kaposi's sarcoma, also occur with high frequency. In untreated individuals, death usually occurs within 2 years after the onset of AIDS symptoms.

Treatment for HIV-infected individuals has two components: one directed against the virus itself to delay progression of the disease, and one to prevent or treat the opportunistic infections and cancers that ultimately cause death. One current antiviral approach involves administering a "cocktail" of four drugs, known as HAART (highly active antiretroviral therapy). Two of the drugs inhibit the action of reverse transcriptase in converting viral RNA into DNA within the host cell, a third drug inhibits an HIV enzyme required for assembling new viruses, and a recently developed fourth class of drugs called fusion inhibitors prevents the virus from entering T cells. These treatments have been demonstrated to be effective in slowing the rate at which infection with HIV leads to AIDS. Unfortunately, however, the HAART regimen is associated with numerous side effects, including nausea, vomiting, diarrhea, metabolic disturbances resulting in part from insulin resistance, and liver damage. Much research is under way to find better treatments and ultimately to cure this scourge.

Summary of Key Concepts

- An animal's cells and organs that collectively contribute to its immune defenses, or immunity, constitute the animal's immune system.

- In innate (nonspecific) immunity, the body's defenses are present at birth and act against foreign materials in much the same way regardless of the specific identity of the invading material. Acquired (specific) immunity develops only after the body is exposed to foreign substances and targets those foreign substances specifically.

53.1 Types of Pathogens

- Pathogens are of three major types: bacteria, viruses, and eukaryotic parasites.

53.2 Innate Immunity

- An important innate defense is composed of phagocytes—cells capable of phagocytosis. In vertebrates, most phagocytes belong to the blood cells called leukocytes. The leukocytes involved in innate immunity include neutrophils, eosinophils, monocytes, macrophages, basophils, and natural killer (NK) cells. (Figure 53.1)

- Two other types of cells play important roles in innate immunity: dendritic cells and mast cells.

- Inflammation is an innate local response to infection or injury characterized by local redness, swelling, heat, and pain. The events of inflammation are induced and regulated by chemical mediators called cytokines. (Figure 53.2)

- Antimicrobial proteins include interferons, which inhibit viral replication, and complement proteins, which kill microbes without prior phagocytosis. Activation of the complement proteins results in the formation of a membrane attack complex (MAC), which creates water channels in the microbial plasma membrane and causes the microbe to swell and burst.

53.3 Acquired Immunity

- A foreign molecule that the host does not recognize as self and that triggers an acquired immune response is an antigen.

- Leukocytes called lymphocytes are responsible for acquired immune responses. Although lymphocytes circulate in the blood, most of them reside in a group of organs and tissues that constitute the lymphatic system. (Figure 53.3)

- Lymphocytes responsible for acquired immunity are B and T cells. B cells differentiate into antibody-producing cells called plasma cells. T cells include cytotoxic T cells, which directly kill target cells, and helper T cells, which assist in the activation and function of B cells and cytotoxic T cells. (Figure 53.4)

- Immunologists recognize two types of acquired immunity. In humoral immunity, plasma cells secrete antibodies that bind to antigens. In cell-mediated immunity, cytotoxic T cells directly attack and destroy abnormal body cells.

- Acquired immune responses occur in three stages. The first stage is recognition of antigen; the second is activation of lymphocytes; and the third is attack against antigen. When a lymphocyte is stimulated by an antigen, it divides and produces clones of itself, all of which express the same receptor. After activation, some of the cloned lymphocytes function as effector cells, which carry out the attack response; other cells function as memory cells, which remain poised to recognize the antigen if it returns in the future. (Figure 53.5)

- In humoral immunity, B cells recognize antigens with B-cell receptors. When B cells are activated, they proliferate and differentiate into plasma cells, which secrete antibodies.

- Both B-cell receptors and antibodies belong to a family of proteins called immunoglobulins. Immunoglobulins have two heavy chains and two light chains. The lower halves of the heavy chains contain constant regions, which are identical for all immunoglobulins of a given Ig class. At the other end of the Ig molecule, the two chains form a variable region that serves as the antigen binding site. (Figures 53.6, 53.7)

- The process known as hypermutation, which primarily involves numerous C to T point mutations, is crucial to enabling plasma cells to produce a diverse array of antibodies capable of recognizing many different antigens.

- B cells that are activated by an antigen differentiate into plasma cells by a process called clonal selection. (Figure 53.8)

- Antibodies combine with the antigen that activated the B cell and guide an attack that eliminates the antigen or the cells bearing it.

- In cell-mediated immunity, T cells can recognize antigens only when they are complexed with self proteins.

- Major histocompatibility complex (MHC) proteins are cellular "identity tags" that serve as genetic markers of self. Class I MHC proteins are found on the surface of all human body cells except erythrocytes. Class II MHC proteins are found only on the surface of macrophages, B cells, and dendritic cells.

- Antigen-presenting cells (APCs) are cells bearing fragments of antigen, called antigenic determinants or epitopes, complexed with the cell's MHC proteins.

- The binding between a helper T-cell receptor and an antigen bound to class II MHC proteins on an APC is essential to helper T-cell activation. Once activated, helper T cells can help to activate both B cells and cytotoxic T cells. (Figure 53.9)

- Cell-mediated immune responses are mediated by cytotoxic T cells, which directly kill virus-infected and cancerous cells via secreted chemicals. Humoral immune responses are mediated by B cells and plasma cells. In both types of responses, helper T cells are required. (Figures 53.10, 53.11)

- The process by which the body distinguishes between self and nonself components is called immune tolerance. Individuals normally lack active lymphocytes that respond to self components because of two mechanisms. T cells with receptors capable of binding self proteins are destroyed by apoptosis in a process termed clonal deletion. Clonal inactivation causes potentially self-reacting lymphocytes to become nonresponsive.

- When the body's immune cells attack the body's own cells, the result is autoimmune disease.

- Upon initial exposure to an antigen, the body produces a primary immune response. Any subsequent exposure to the same antigen elicits an immediate and heightened response termed a secondary immune response. The immune system's ability to produce this secondary response is called immunological memory. (Figure 53.12)

- The acquired response to exposure to any type of antigen is known as active immunity. The artificial exposures to antigen that occur in vaccinations and immunizations also induce active immunity. In contrast, passive immunity confers protection against disease through the direct transfer of antibodies from one individual to another. Monoclonal antibodies may pave the way for the use of passive immunity to combat certain types of cancer.

- Some animals, such as social insects, can confer immunity on each other through social contact, called "social immunity." (Figure 53.13)

53.4 Impact on Public Health

- Factors that affect the body's defense mechanisms include lifestyle; organ transplants and blood transfusions; allergies; and AIDS, caused by the human immunodeficiency virus (HIV). AIDS reduces the body's immunity by killing helper T cells.

Assess and Discuss

Test Yourself

1. Which of the following is <u>not</u> an example of a barrier defense in animals?
   a. skin
   b. secretions from skin glands
   c. exoskeleton
   d. mucus
   e. antibodies

2. The leukocytes that are found in mucosal surfaces and that play a role in defending the body against parasitic infections are
   a. neutrophils.
   b. eosinophils.
   c. basophils.
   d. monocytes.
   e. NK cells.

3. The vascular changes of inflammation
   a. lead to an increase in bacterial cells at the injury site.
   b. decrease the number of leukocytes at the injury site.
   c. allow plasma proteins to move easily from the bloodstream to the injury site.
   d. increase the number of antibodies at the injury site.
   e. activate lymphocytes.

4. Which is correct regarding acquired immunity?
   a. Acquired immunity only requires the presence of helper T cells to function properly.
   b. Acquired immunity does not require exposure to a foreign substance.
   c. Acquired immunity is triggered by contact with a particular antigen.
   d. Acquired immunity includes inflammation.
   e. All of the above are correct.

5. Memory B cells are
   a. cloned lymphocytes that are active in subsequent infections.
   b. cloned lymphocytes that are active during a primary infection.
   c. NK cells that recognize cancer cells and destroy them.
   d. cells that produce antibodies.
   e. macrophages that have recognized self antigens.

6. The immunoglobulin that is passed from mother to fetus across the placenta is
   a. IgA.
   b. IgD.
   c. IgE.
   d. IgG.
   e. IgM.

7. The region of an antibody that is the antigen binding site is
   a. the constant region.
   b. the variable region.
   c. the heavy chain.
   d. the light chain.
   e. the hinge region.

8. A major difference between the activation of B cells and T cells is that
   a. T cells must interact with antigens bound to plasma membranes.
   b. B cells interact only with free antigens.
   c. B cells are not regulated by helper T cells.
   d. T cells produce antibodies.
   e. none of the above

9. Cells that process foreign proteins and complex them with their MHC proteins are called
   a. cytotoxic T cells.
   b. plasma cells.
   c. NK cells.
   d. antigen presenting cells.
   e. helper T cells.

10. HIV causes immune deficiency because the virus
    a. destroys all the cytotoxic T cells.
    b. preferentially destroys helper T cells that regulate the immune system.
    c. directly inactivates plasma cells.
    d. causes mutations that lead to autoimmune diseases.
    e. does all of the above.

Conceptual Questions

1. Distinguish between innate immunity and acquired immunity.
2. Explain the function of cytotoxic T cells.
3. Describe the basic structure of an immunoglobulin.

Collaborative Questions

1. Discuss three types of pathogens that affect the health of animals.
2. Describe the functions of helper T cells.

Online Resource

www.brookerbiology.com

Stay a step ahead in your studies with animations that bring concepts to life and practice tests to assess your understanding. Your instructor may also recommend the interactive ebook, individualized learning tools, and more.

FEATURE INVESTIGATION

Callaway and Aschehoug's Experiments Showed That the Secretion of Chemicals Gives Invasive Plants a Competitive Edge Over Native Species

One important topic in the area of population ecology concerns **introduced** or **exotic species**, species moved from a native location to another location, usually by humans. Such species sometimes spread so aggressively that they crowd out native organisms, in which case they are considered **invasive species**. Of the 300 most invasive plants in the U.S., over half were brought in for gardening, horticulture, or landscaping purposes. Invasive species have traditionally been thought to succeed because they have escaped their natural enemies, primarily insects that remained in the country of origin and were not transported to the new locale. One way of controlling these species,

therefore, has been to import the plant's natural enemies. This is known as **biological control**. However, new research on the population ecology of diffuse knapweed (*Centaurea diffusa*), an invasive Eurasian plant that has established itself in many areas of North America, suggests a different reason for their success.

Researchers Ragan Callaway and Erik Aschehoug hypothesized that the roots of this particular species secrete powerful toxins, called **allelochemicals,** that kill the roots of other species, allowing *Centaurea* to proliferate. To test their hypothesis, Callaway and Aschehoug collected seeds of three native Montana grasses, *Koeleria cristata*, *Festuca idahoensis*, and *Agropyron spicata*, and grew each of them with or without the exotic *Centaurea* species (**Figure 54.3**). As hypothesized, *Centaurea* depressed the biomass of the native grasses. When the experiments were repeated with grasses native to Eurasia, *Koeleria*

Figure 54.3 Experimental evidence of the effect of allelochemicals on plant production.

HYPOTHESIS Exotic plants from Eurasia outcompete native Montana grasses by secreting allelochemicals from their roots.

STARTING LOCATION *Centaurea diffusa*, a Eurasian plant, is invading Montana grasslands because it outcompetes 3 native grasses: *Koeleria cristata*, *Festuca idahoensis*, and *Agropyron spicata*.

Experimental level   **Conceptual level**

1 Collect seeds of native Montana grasses and plant with and without seeds of invasive *C. diffusa* from Eurasia. Three months after sowing seeds, the plants are harvested, dried, and weighed.

C. diffusa significantly reduces biomass of native Montana grasses.

2 Collect seeds of grasses from Eurasia of the same three genera as the Montana grasses and plant with and without *C. diffusa*. Three months after sowing seeds, the plants are harvested, dried, and weighed.

C. diffusa doesn't depress the biomass of grasses native to Eurasia as much.

3 **THE DATA***

*The biomass is that of the genus noted at the top of each graph.

4   **CONCLUSION** *Centaurea diffusa*, a Eurasian grass, is invasive in the U.S. because it secretes allelochemicals, which inhibit the growth of native plants.

5   **SOURCE** Callaway, R.M., and Aschehoug, E.T. 2000. Invasive plants versus their old and new neighbors. *Science* 290:521–523.

laerssenii, Festuca ovina, and *Agropyron cristatum,* the growth of each species was inhibited, but to a significantly lesser degree than the growth of the Montana species.

In other experiments not described in Figure 54.3, Callaway and Aschehoug added activated carbon to the soil, which absorbs the chemical excreted by the *Centaurea* roots. With activated carbon added, the Montana grass species increased in biomass compared to the previous experiments. The researchers concluded that *C. diffusa* outcompetes Montana grasses by secreting an allelochemical and that Eurasian grasses are not as susceptible to the chemical's effect because they coevolved with it. If the reason for the success of invasive plants can be attributed to the chemicals they secrete, this calls into question the effectiveness of biological control of invasive plants by

importation of their natural enemies. This study on the population biology of an invasive plant has changed the way we think about why such species succeed and could affect the way we attempt to control them in the future.

Experimental Questions

1. Prior to Callaway and Aschehoug's study, what was the prevailing hypothesis of why invasive species succeed in new environments?

2. Briefly describe the evidence collected to support the allelochemical hypothesis.

3. What was the function of the activated carbon used in a subsequent test of the hypothesis?

Community Ecology Focuses on What Factors Influence the Number of Species in a Given Area

We've just seen how a population of an invasive plant may produce allelochemicals that give it a growth advantage over populations of other species. On a larger scale, **community ecology** studies how populations of species interact and form functional communities. In a forest, there are many populations of trees, herbs, shrubs, grasses, the herbivores that eat them, and the carnivores that prey on the herbivores. Community ecology focuses on why certain areas have high numbers of species (that is, are species rich), while other areas have low numbers of species (that is, are species poor).

Although ecologists are interested in species richness for its own sake, a link also exists between species richness and community function. Ecologists generally believe that species-rich communities perform better than species-poor communities. It has also been proposed that more species make a community more stable, that is, more resistant to disturbances such as introduced species. Community ecology also considers how species composition and community structure change over time and, in particular, after a disturbance, a process called succession.

Ecosystem Ecology Describes the Passage of Energy and Nutrients Through Communities

An **ecosystem** is an interacting system of a community of organisms and the physical environment in which they live. **Ecosystem ecology** deals with the flow of energy and cycling of chemical elements among organisms within a community and between organisms and the environment. Following this

flow of energy and chemicals necessitates an understanding of feeding relationships between species, called food chains. In food chains, each level is called a trophic level, and many food chains interconnect to form complex food webs.

As we learned in Chapter 6, the second law of thermodynamics states that in every energy transformation, free energy is reduced because heat energy is lost in the process, and the entropy of the system increases. There is, therefore, a unidirectional flow of energy through an ecosystem, with energy dissipated at every step. An ecosystem needs a recurring input of energy from an external source—in most cases, the sun—to sustain itself. In contrast, chemicals such as nitrogen do not dissipate and constantly cycle between abiotic and biotic components of the environment, often becoming more concentrated in organisms in higher trophic levels.

54.2 Ecological Methods Focus on Observation and Experimentation

How do ecologists go about studying their subject? In this section, we will examine the value of different ecological methods such as observation and experimentation. Let's suppose you are employed by the United Nations' Food and Agricultural Organization (FAO), which operates internationally out of Rome, Italy. As an ecologist, you are charged with finding out what causes outbreaks of locusts, a type of grasshopper whose population periodically erupts in Africa and other parts of the world, destroying crops and causing food shortages.

First of all, you might draw up a possible web of interaction between the factors that could affect locust population size

An Introduction to Ecology and Biomes

54

In 2006, a study led by J. Alan Pounds of the Monteverde Cloud Forest Preserve in Costa Rica reported that fully two-thirds of the 110 species of harlequin frogs in mountainous areas of Central and South America had become extinct over the previous 20 years. The researchers noted that populations of other species, such as the Panamanian golden frog (*Atelopus zeteki*), had been greatly reduced (**Figure 54.1**). The question was why. The culprit was identified as a disease-causing fungus, *Batrachochytrium dendrobatidis*, but Pounds's study implicated global warming—a gradual increase in the average temperature of the Earth's atmosphere—as the agent causing outbreaks of the fungus. One effect of global warming is to increase the cloud cover, which reduces daytime temperatures and raises nighttime temperatures. Researchers believe that this combination has created favorable conditions for the spread of *B. dendrobatidis* and other diseases, which thrive in cooler daytime temperatures. Pounds, the team's lead researcher and an ecologist, was quoted as saying, "Disease is the bullet killing frogs, but climate change is pulling the trigger."

Ecology is the study of interactions among organisms and between organisms and their environments. Interactions among organisms are called **biotic** interactions, and those between organisms and their nonliving environment are termed **abiotic** interactions. These interactions, in turn, govern the numbers of species in an area and their population densities. In this first chapter of the ecology unit, we will introduce you to the four broad areas of ecology: organismal, population, community, and ecosystems ecology. Next, we will explore how ecologists approach and conduct their work. We will then turn our focus to abiotic interactions and examine the effects of factors such as temperature, water, light, pH, and salt concentrations on the distributions of organisms. We conclude with a consideration of climate and its large influence on biomes, the major types of habitats where organisms are found.

In other chapters in this unit, we'll also learn how plants compete with one another, how herbivores affect plant abundance, and how natural enemies impact prey populations. We'll examine the effects of humans on the environment, including pollution, global warming, and the introduction of exotic species of plants and animals.

What controls the densities of these flowering plants at Glacier National Park, Montana? Is it temperature or rainfall? Or is it the availability of pollinators, herbivory by insects or vertebrates, or competition with other plant species for resources? Ecology seeks to answer questions such as these.

Figure 54.1 Diminishing and disappearing populations. Population sizes of the Panamanian golden frog (*Atelopus zeteki*) have diminished greatly over the past 20 years, and populations of many other species of harlequin frogs have disappeared entirely. Ecologists are investigating the reasons for this decline.

Ecologists are among the best-equipped scientists to study these phenomena. Before 1960, the field of ecology was dominated by taxonomy, natural history, and speculation about observed patterns. An ecologist's tools of the trade might have included sweep nets, quadrats (small, measured plots of land used to sample living things), and specimen jars. Since that time, an explosion in the number of ecological studies has occurred, and ecologists have become active in investigating environmental change on local, regional, and global scales. Ecologists have embraced experimentation and adapted concepts and methods derived from agriculture, physiology, biochemistry, genetics, physics, chemistry, and mathematics. Their tools have kept pace with technological innovations. Now an ecologist's equipment is just as likely to include portable computers, satellite-generated images, and chemical autoanalyzers.

Ecological studies have important implications in the real world, as will be amply illustrated by examples discussed throughout the unit. However, there is a distinction between ecology and **environmental science**, the application of ecology to real-world problems. To use an analogy: Ecology is to environmental science as physics is to engineering. Both physics and ecology provide the theoretical framework on which to pursue more applied studies. Engineers rely on the principles of physics to build bridges. Environmental scientists rely on the principles of ecology to solve environmental problems.

54.1 The Scale of Ecology

Ecology ranges in scale from the study of an individual organism through the study of populations to the study of communities and ecosystems (**Figure 54.2**). In this section, we introduce each of the broad areas of organismal, population, community, and ecosystem ecology and provide an investigation that helps illuminate the field of population ecology.

Organismal Ecology Investigates How Adaptations and Choices by Individuals Affect Their Reproduction and Survival

Organismal ecology can be divided into two subdisciplines. The first, **physiological ecology**, investigates how organisms are physiologically adapted to their environment and how the environment impacts the distribution of species. Much of this chapter discusses physiological ecology. The second area, **behavioral ecology**, focuses on how the behavior of an individual organism contributes to its survival and reproductive success, which, in turn, eventually affects the population density of the species.

Population Ecology Describes How Populations Grow and Interact with Other Species

Population ecology focuses on groups of interbreeding individuals, called populations. A primary goal is to understand the factors that affect a population's growth and determine its size and density. Although the attention of a population ecologist may be aimed at studying the population of a particular species, the relative abundance of that species is often influenced by its interactions with other species. Thus, population ecology includes the study of **species interactions**, such as predation, competition, and parasitism. Knowing what factors impact populations can help us lessen species endangerment, stop extinctions, and control invasive species.

(a) A single organism

(b) A population of zebras

(c) An African grassland community

(d) Nutrient flow in an African grassland community

Figure 54.2 **The scale of ecology.** (a) Organismal ecology. What is the temperature tolerance of this zebra? (b) Population ecology. What factors influence the growth of zebra populations in Africa? (c) Community ecology. What factors influence the number of species in African grassland communities? (d) Ecosystems ecology. How do water, energy, and nutrients flow among plants, zebras, and other herbivores and carnivores in African grassland communities?

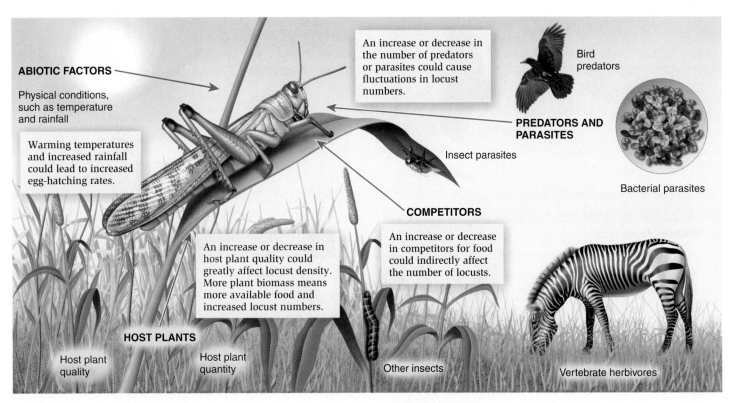

ABIOTIC FACTORS

Physical conditions, such as temperature and rainfall

Warming temperatures and increased rainfall could lead to increased egg-hatching rates.

An increase or decrease in the number of predators or parasites could cause fluctuations in locust numbers.

Bird predators

PREDATORS AND PARASITES

Insect parasites

Bacterial parasites

COMPETITORS

An increase or decrease in competitors for food could indirectly affect the number of locusts.

An increase or decrease in host plant quality could greatly affect locust density. More plant biomass means more available food and increased locust numbers.

HOST PLANTS

Host plant quality

Host plant quantity

Other insects

Vertebrate herbivores

Figure 54.4 Interaction web of factors that might influence locust population size.

(**Figure 54.4**). These interactions are many and varied, and they include

- abiotic factors, such as temperature, rainfall, wind, and soil pH;
- natural enemies, including bird predators, insect parasites, and bacterial parasites;
- competitors, including other insects and larger vertebrate grazers;
- host plants, including increases or decreases in either the quality or quantity of the plants.

With such a vast array of factors to be investigated, where is the best place to start? As discussed in Chapter 1, hypothesis testing involves a five-stage process: (1) observations, (2) hypothesis formation, (3) experimentation, (4) data analysis, and (5) acceptance or rejection of the hypothesis. In our study of locusts, we begin by careful observation of the organism in its native environment. We can analyze the fluctuations of locusts and determine if the populations vary with fluctuations in the other phenomena, such as levels of parasitism, numbers of predators, or food supply. Imagine we found that locust numbers are affected by bird predation levels and that an inverse relationship exists between predation levels and locust numbers. As predation levels increase, locust numbers decrease. If we plotted this relationship graphically, the resulting graph would look like that depicted in **Figure 54.5a**. This result would give us some confidence that predation levels determined locust numbers, and this would be our hypothesis.

In fact, we would have so much confidence that we could create a statistically determined "line of best fit" to represent a summary of the relationship between these two variables, which is shown in the graph.

However, if the points were not highly clustered, as in **Figure 54.5b**, we would have little confidence that predation affects locust density. Many statistical tests are used to determine whether or not two variables are significantly correlated. In the studies in this unit, unless otherwise stated, most graphs like Figure 54.5a imply that a meaningful relationship exists between the two variables. We call this type of relationship a significant **correlation**. In this graph, locust density shows a negative linear relationship with predation; therefore, we say that locust density is negatively correlated with predation.

We have to be cautious when forming conclusions based on correlations. For example, large numbers of locusts could be associated with large, dense plants. We might conclude from this that food availability controls locust density. However, an alternative conclusion would be that large plants provide locusts refuge from bird predators, which cannot attack them in the dense interior. While it would appear that biomass affects locust density by providing abundant food, in actuality, predation would still be the most important factor affecting locust density. Thus, correlation does not always mean causation. For this reason, after conducting observations, ecologists usually turn to experiments to test their hypotheses.

In our example, an experiment might involve removing predators from an area inhabited by locusts. If predators are having a significant effect, then removing them should cause

(a) Strong relationship

Line of best fit

Locust numbers

(b) No relationship

Level of predation

Figure 54.5   **How locust numbers might be correlated with predation.**   In this case, higher locust numbers are found in nature where predation levels are lowest. We can draw a line of best fit **(a)** to represent this relationship. In **(b)**, the relationship between locust numbers and predation levels might be so weak that we would not have much confidence in a linear relationship between the variables.

Concept check: What would it mean if the line of best fit sloped in the opposite direction?

an increase in locust numbers. We would have two groups: a group of locusts with predators removed (the experimental group) and a group of locusts with predators still present (the control group), with equal numbers of locusts in both groups at the start of the experiment. Any differences in locust population density would be due solely to differences in predation. Reduced predation might be achieved by putting a cage made of chicken wire over and around bushes containing locusts, so that birds are denied access. Experiments often have a defined time frame, and in our example, we could look at locust survivorship over the course of one generation of locust and predator.

Performing the experiments several times is called **replication**. We might replicate the experiment 5 times, 10 times, or even more. We would add up the total number of surviving locusts from each replicate and calculate the mean, which is the sum divided by the number of values. In the experimental

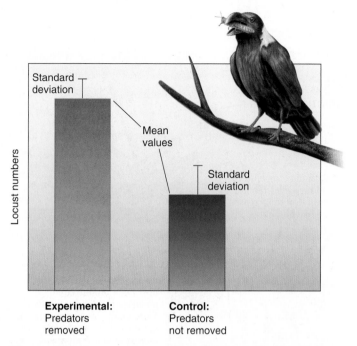

Standard deviation

Mean values

Standard deviation

Locust numbers

Experimental: Predators removed

Control: Predators not removed

Figure 54.6   Graphic display of hypothetical results of a predator removal experiment.   The two bars represent the average number of locusts where predators are removed (experimental group) and where predators are not removed (control group). The vertical lines (the standard deviations or standard errors) give an indication of how tightly the individual results are clustered around the mean. The shorter the lines, the tighter the cluster, and the more confidence we have in the result.

group, let's suppose that the numbers of surviving locusts in each replicate are 5, 4, 7, 8, 12, 15, 13, 6, 8, and 10; the mean number surviving would be 8.8. In the control group, which still allows predator access, the numbers surviving might be 2, 4, 7, 5, 3, 6, 11, 4, 1, and 3, with a mean of 4.6. Without predators, the mean number of surviving locusts would therefore be almost double the average number surviving with predators. Our data analysis would give us confidence that predators were indeed the cause of our changes in locust numbers. The results of such experiments can be illustrated graphically by a bar graph (**Figure 54.6**). Ecologists can use a variety of tests to see if the differences between the control data and the experimental data are statistically significant, which means that the differences are not likely to have occurred as a result of random chance. We won't look at the mechanics of these tests, but in this unit, when experimental and control data are presented as differing, these are considered to be statistically significant differences unless stated otherwise.

By the way, it turns out that predation is not the primary factor that controls locust populations. The results we have been discussing were hypothetical. Weather, in particular, rain, is the most important feature governing locust population size. Moist soil allows eggs to hatch and provides water for germinating plants, allowing a ready source of food for the hatchling locusts. In fact, physical or abiotic factors such as amount of moisture usually have powerful effects in most ecological systems. In the next part of the chapter, we turn our attention to an examination of the effects of the physical environment on the distribution patterns of organisms.

54.3   The Environment's Impact on the Distribution of Organisms

Both the distribution patterns of organisms and their abundance are limited by physical features of the environment such as temperature, wind, availability of water and light, salinity, and pH (**Table 54.1**). In this section, we will examine these features of the environment.

Temperature Has an Important Effect on the Distribution of Plants and Animals

Temperature is perhaps the most important factor in the distribution of organisms because of its effect on biological processes and because of the inability of most organisms to regulate their body temperature precisely. For example, the organisms that form coral reefs secrete a calcium carbonate shell. Shell formation and coral deposition are accelerated at high temperatures but are suppressed in cold water. Coral reefs are therefore abundant only in warm water, and a close correspondence is observed between the 20°C isotherm for the average daily temperature during the coldest month of the year and the limits of the distribution of coral reefs (**Figure 54.7**). An isotherm is a line on a map connecting points of equal temperature. Coral reefs are located between the two 20°C isotherm lines that are formed above and below the equator.

Frost is probably the single most important factor limiting the geographic distribution of tropical and subtropical plants. In plants, cold temperature can be lethal because cells may rupture if the water they contain freezes. In the Sonoran Desert in Arizona, saguaro cacti can easily withstand frost for one night as long as temperatures rise above freezing the following day, but they are killed when temperatures remain below freezing for more than 36 hours. This means that the cactus's

Table 54.1	Selected Abiotic Factors and Their Effects on Organisms
Factor	**Effect**
Temperature	Low temperatures freeze many plants; high temperatures denature proteins. Some plants require fire for germination.
Wind	Wind amplifies effects of cool temperatures (wind chill) and water loss; creates pounding waves.
Water	Insufficient water limits plant growth and animal abundance; excess water drowns plants and other organisms.
Light	Insufficient light limits plant growth, particularly in aquatic environments.
Salinity	High salinity generally reduces plant growth in terrestrial habitats; affects osmosis in marine and freshwater environments.
pH	Variations in pH affect decomposition and nutrient availability in terrestrial systems; directly influences mortality in both aquatic and terrestrial habitats.

distribution is limited to places where the temperature does not remain below freezing for more than one night (**Figure 54.8**).

The geographic range limits of endothermic animals are also affected by temperature. For example, the eastern phoebe (*Sayornis phoebe*), a small bird, has a northern winter range that coincides with an average minimum January temperature of above 4°C. Such limits are probably related to the energy demands associated with cold temperatures. Cold temperatures mean higher metabolic costs, which are, in turn, dependent on high feeding rates. Below 4°C, the eastern phoebe cannot feed fast enough or, more likely, find enough food to keep warm.

High temperatures are also limiting for many plants and animals because relatively few species can survive internal temperatures more than a few degrees above their metabolic optimum. We have discussed how corals are sensitive to low temperatures; however, they are sensitive to high temperatures

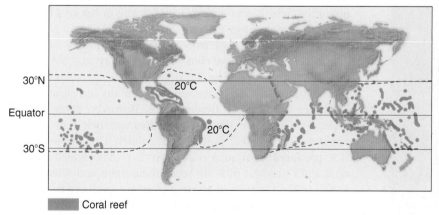

(a) Worldwide distribution of coral reefs

◻ Coral reef

(b) A coral reef

Figure 54.7   Locations of coral reefs.   (a) Coral reef formation is limited to waters bounded by the 20°C isotherm (dashed line), a line where the average daily temperature is 20°C during the coldest month of the year. (b) Coral reef from the Pacific Ocean.

Concept check: *Why are coral reefs limited to warm water?*

Arizona

- - - Boundary of saguaro cactus range

● Temperatures remain below freezing for 1 or more days/year

● Temperatures remain below freezing for less than 0.5 days/year

● No days below freezing on record

Figure 54.8   **Saguaro cacti in freeze-free zones.**   A close correspondence is seen between the range of the saguaro cactus (dark green area) and the area in which temperatures do not drop below freezing (0°C).

Figure 54.9   **Coral bleaching.**   Mantanani Island, Malaysia.

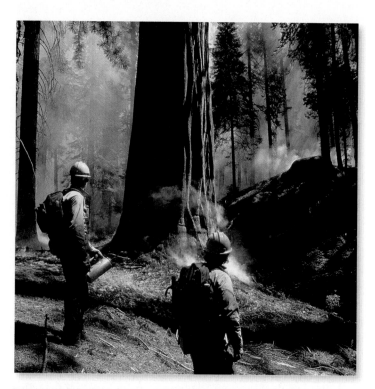

Figure 54.10   **Giant sequoia.**   A park ranger uses a drip torch to ignite a fire at Sequoia National Park in California. Periodic, controlled human-made fires mimic the sporadic wildfires that normally burn natural areas. Such fires are vital to the health of giant sequoia populations, because they serve to open the pine cones and release the seeds.

Concept check: *Why are some fires very destructive to natural systems?*

as well. When temperatures are too high, the symbiotic algae that live within coral die and are expelled, causing a phenomenon known as coral bleaching. Once bleaching occurs, the coral tissue loses its color and turns a pale white (**Figure 54.9**). El Niño is a weather phenomenon characterized by a major increase in the water temperature of the equatorial Pacific Ocean. In the winter of 1982–1983, an influx of warm water from the eastern Pacific raised temperatures just 2–3°C for 6 months, which was enough to kill many of the reef-building corals on the coast of Panama. By May 1983, just a few individuals of one species, *Millepora intricata*, were alive. Scientists are concerned that future climate changes may increase the frequency of coral bleaching.

The ultimate high temperatures that many terrestrial organisms face are brought about by fire. However, some species depend on frequent low-intensity fires for their reproductive success. The longleaf pine (*Pinus palustris*) of the southeast U.S. produces serotinous cones, which remain sealed by pine resin until the heat of a fire melts them open and releases the seeds. In the west, giant sequoia trees are similarly dependent on periodic low-intensity fires for germination of their seeds. Such fires both enhance the release of seeds and clear out competing vegetation at the base of the tree so that seeds can germinate and grow (**Figure 54.10**). Fire-suppression practices that attempt to protect forests from fires can actually have undesirable results by preventing the regeneration of fire-dependent species. Furthermore, fire prevention can result in an accumulation of vegetation beneath the canopy (the understory) that may later fuel hotter and more damaging fires.

Keep in mind that it is not average temperatures that usually limit the range of species but rather the frequency of occasional extremes, such as freezes for the saguaro cacti. Farmers know this only too well. The frequency and strength of periodic freezes limit the northern distribution of oranges in Florida and the southern distribution of coffee in Brazil, not average temperatures for the coldest months. Experimentally moving organisms outside their normal range and monitoring survivorship is a useful way to establish what factors control the natural limits of species. Of course, such movements must be done in a carefully controlled manner to ensure that there is no risk of species becoming invasive in their new habitats. Again, physical extremes, such as fires or severe freezes, may be apparent only in isolated years, so ecologists often have to wait many years to determine whether extremes are limiting the distribution and abundance of plants and animals in the field.

Despite the obvious relationships between species distributions and temperature, we need to be cautious about solely relating the two. The temperatures measured for constructing isotherm maps such as that in Figure 54.7a are not always the temperatures that the organisms experience. For example, in nature, an organism may choose to lie in the sun or hide in the shade, both of which affect the temperatures it experiences. Such local variations of the climate within a given area, or **microclimate**, can be important for a particular species. In the spring, temperate forests turn green from the ground upwards. Herbaceous species leaf out first, followed by shrubs, small trees, and large trees, because the air is warmer closer to the ground. Dark-colored surfaces such as soil absorb more light and produce more heat, which also helps to warm the ground cover first.

Because so many species are limited in their distribution patterns by global temperatures, ecologists are concerned that if global temperatures rise, as they are predicted to, many species will be driven to extinction or that their geographic ranges will shrink and the location of centers of agriculture and forestry will be altered. The increase in the average temperature of Earth's atmosphere and oceans is called global warming, and it is caused by a process known as the greenhouse effect.

The Greenhouse Effect
The Earth is warmed by the **greenhouse effect**. In a greenhouse, sunlight penetrates the glass and raises temperatures, with the glass acting to trap the resultant heat inside. Similarly, solar radiation in the form of short-wave energy passes through the atmosphere to heat the surface of the Earth. At night, this energy is radiated from the Earth's warmed surface back into the atmosphere, but in the form of long-wave infrared radiation. Instead of letting it escape back into space, however, atmospheric gases absorb much of this infrared energy and radiate it back to the Earth's surface, causing its temperature to rise further (**Figure 54.11**). The greenhouse effect is a naturally occurring process that is responsible for keeping the earth warm enough to sustain life. Without some type of greenhouse effect, global temperatures would be much

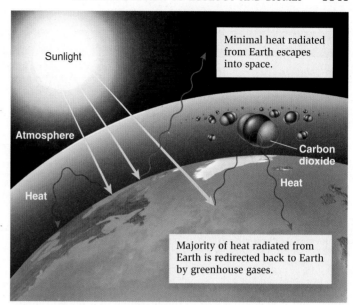

Figure 54.11 **The greenhouse effect.** Solar radiation, in the form of short-wave energy, passes through the atmosphere to heat the Earth's surface. Long-wave infrared energy is radiated back into the atmosphere. Much infrared energy is absorbed by atmospheric gasses, including carbon dioxide molecules, and reflected back to Earth, causing global temperatures to rise.

lower than they are, perhaps averaging only −17°C compared to the existing average of +15°C.

The greenhouse effect is caused by a group of atmospheric gases that together make up less than 1% of the total volume of the atmosphere. These gases—primarily water vapor, carbon dioxide, methane, nitrous oxide, and chlorofluorocarbons—are referred to as greenhouse gases (**Table 54.2**).

Global Warming Ecologists are concerned that human activities are increasing the greenhouse effect and causing **global warming**, a gradual elevation of the Earth's surface temperature. According to the Intergovernmental Panel on Climate Change 2007 report, warming of the climate is unequivocal, as is now evident from observations of increases in global average air and ocean temperatures, widespread melting of snow and ice, and rising average global sea level. Most greenhouse gases have increased in atmospheric concentration since industrial times. Of those increasing, the most important is carbon dioxide (CO_2). As Table 54.2 shows, although CO_2 has a lower global warming potential per unit of gas (relative absorption) than any of the other major greenhouse gases, its concentration in the atmosphere is much higher. Annual emissions of CO_2 have increased by about 80% between 1970 and 2004.

To predict the effect of global warming, most scientists focus on a future point, about 2100, when the concentration of atmospheric CO_2 will have doubled—that is, increased to about 700 ppm compared with the late-20th-century level of 350 ppm. Scientists estimate that at that time, average global temperatures will be somewhere in the range of 1–6°C (about 2–10°F) warmer than present and will increase an additional 0.5°C each decade. This increase in heat might not seem like

Table 54.2 Selected Greenhouse Gases and Their Contribution to Global Warming*

	Carbon dioxide (CO₂)	Methane (CH₄)	Nitrous oxide (N₂O)	Chlorofluorocarbons (CFCs)
Relative absorption per ppm of increase**	1	21	310	10,000
Atmospheric concentration ppm†	385	1.75	0.315	0.0005
Contribution to global warming	73%	7%	19%	1%
Percent from natural sources; type of source	20–30%; volcanoes	70–90%; swamps, gas from termites and ruminants	90–100%; soils	0%
Major human-made sources	Fossil fuel use, deforestation	Rice paddies, landfills, biomass burning, coal and gas exploitation	Cultivated soil, fossil-fuel use, automobiles, industry	Previously manufactured products (for example, aerosol propellants) but now banned in the U.S. and the E.U.

*Water vapor is not included in this table.
**Relative absorption is the warming potential per unit of gas.
†ppm = parts per million

much, but it is comparable to the warming that ended the last ice age. Future consequences would include a further contraction of snow cover and a decrease in sea ice extent, heat waves and drought in dry areas, heavier precipitation in moister areas, and an increase in tropical cyclone intensity.

Assuming this scenario of gradual global warming is accurate, we need to consider what the consequences might be for plant and animal life. At the beginning of the chapter, we saw how global warming is believed to be contributing to the decline and extinction of some amphibian species. Although many species can adapt to slight changes in their environment, the anticipated changes in global climate are expected to occur too rapidly to be compensated for by normal evolutionary processes such as natural selection. Plant species cannot simply disperse and move north or south into the newly created climatic regions that will be suitable for them. Many tree species take hundreds, even thousands, of years for seed dispersal. Paleobotanist Margaret Davis predicted that in the event of a CO₂ doubling, the sugar maple (*Acer saccharum*), which is presently distributed throughout the midwestern and northeastern U.S. and southeastern Canada, would die back in all areas except in northern Maine, northern New Brunswick, and southern Quebec (**Figure 54.12**). Of course, this contraction in the tree's distribution could be offset by the creation of new favorable habitats in central Quebec. However, most scientists believe that the climatic zones would shift toward the poles faster than trees could migrate via seed dispersal; therefore, extinctions would occur. Interestingly, scientists are beginning to be able to genetically modify organisms to change their temperature tolerances, as described next.

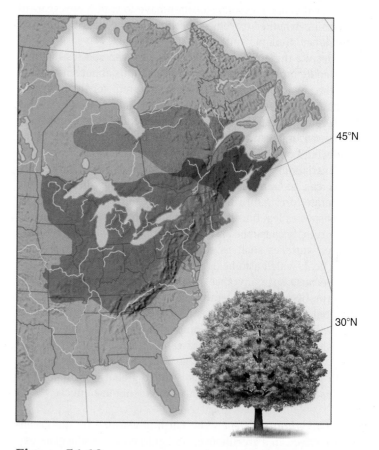

Figure 54.12 Possible changes in the range of sugar maples due to global warming. The present geographic range of the sugar maple (blue shading) and its potential range under doubled CO₂ levels (red shading) in North America. Purple shading indicates the region of overlap, which is the only area where the sugar maple would be found before it spread into its new potential range.

Genomes & Proteomes Connection

Temperature Tolerance May Be Manipulated by Genetic Engineering

Below-freezing temperatures can be very damaging to plant tissue, either killing the plant or greatly reducing its productivity. Frost injury causes losses to agriculture of more than $1 billion annually in the U.S. Frost has been considered an unavoidable result of subfreezing temperatures, but genetic engineering is beginning to change this view.

Between 0°C and −40°C, pure water will be a liquid unless provided with an ice nucleus or template on which an ice crystal can be built. Researchers discovered that some bacteria commonly found on leaf surfaces act as ice nuclei, triggering the formation of ice crystals and eventually causing frost damage. The genes that confer ice nucleation have been identified, isolated, and deactivated in an engineered strain of the bacteria *Pseudomonas syringae*. When this strain is allowed to colonize strawberries, frost damage is greatly reduced, and plants can withstand an additional 5°C drop in temperature before frost forms. The promise of this technique for increasing agricultural yields and altering normal plant-distribution patterns is staggering.

At the other end of the temperature spectrum, heat shock proteins (HSPs) help organisms cope with the stress of high temperatures. At high temperatures, proteins may denature, that is, either unfold or bind to other proteins to form misfolded protein aggregations. HSPs act as molecular chaperones, proteins that help in the proper folding of other proteins, to prevent these types of events from taking place. HSPs normally constitute only about 2% of the cell's soluble protein content, but this can increase to 20% when a cell is stressed, whether by heat, cold, drought, or other condition. The genes that encode HSPs are extremely common and are found in the genomes of all organisms, from bacteria to plants and animals.

In the tropics, high temperatures can substantially decrease the growth rates and productivity of many crop species. There is now substantial interest in identifying crop strains with naturally high HSP levels for use in crop-breeding programs. Given the projected continuation of global warming, such research seems particularly timely.

Wind Can Amplify the Effects of Temperature

Wind is created by temperature gradients. As air heats up, it becomes less dense and rises. As hot air rises, cooler air rushes in to take its place. For example, hot air rising in the tropics is replaced by cooler air flowing in from more temperate regions, thereby creating northerly or southerly winds.

Wind affects living organisms in a variety of ways. It increases the rate of heat loss by convection, the transfer of heat by the movement of air next to the body (the wind chill factor). Wind also contributes to water loss in organisms by increasing the rate of evaporation in animals and transpiration in plants. For example, the tree line in alpine areas is often determined by a combination of low temperatures and high winds, an environmental condition in which transpiration exceeds water uptake.

Winds can also intensify oceanic wave action, with resulting effects for aquatic organisms. On the ocean's rocky shore, seaweeds survive heavy surf by a combination of holdfasts and flexible structures. The animals of this zone have powerful organic glues and muscular feet to hold them in place (**Figure 54.13**).

The Availability of Water Has Important Effects on the Abundance of Organisms

Water has an important effect on the distribution of organisms. Cytoplasm is 85–90% water, and without moisture, there can

(a) Brown alga with a holdfast

(b) A mussel with byssal threads

Figure 54.13 Animals and plants of the intertidal zone adhering to their rocky surface. **(a)** The brown alga (*Laminaria digitata*) has a holdfast that enables it to cling to the rock surface. **(b)** The mussel (*Mytilus edulis*) attaches to the surface of a rock by proteinaceous threads (byssal threads) that extend from the animal's muscular foot.

be no life. As noted in Chapter 2, water performs crucial functions in all living organisms. It acts as a solvent for chemical reactions, takes part in hydrolysis and dehydration reactions, is the means by which animals eliminate wastes, and is used for support in plants and in some invertebrates as part of a hydrostatic skeleton.

The distribution patterns of many plants are limited by available water. Some plants, such as the water tupelo tree (*Nyssa aquatica*) in the southeast U.S., do best when completely flooded and are thus found predominantly in swamps. In contrast, coastal plants that grow on sand dunes experience very little fresh water. Their roots penetrate deep into the sand to extract moisture. In cold climates, water can be present but locked up as permafrost and, therefore, unavailable—this is termed frost drought. Alpine trees can be affected by frost drought. The trees stop growing at a point on the mountainside where they cannot take up enough moisture to offset transpiration losses. This point, known as the timberline, is readily apparent on many mountainsides. Not surprisingly, the density of many plants is limited by the availability of water. For example, a significant correlation is observed between rainfall and creosote bush density in the Mojave Desert.

Animals face problems of water balance, too, and their distribution and population density can be strongly affected by water availability. Because most animals depend ultimately on plants as food, their distribution is intrinsically linked to those of their food sources. Such a phenomenon regulates the number of buffalo (*Syncerus caffer*) in the Serengeti area of Africa. In this area, grass productivity is related to the amount of rainfall in the previous month. Buffalo density is governed by grass availability, so a significant correlation is found between buffalo density and rainfall (**Figure 54.14**). The only exception occurs in the vicinity of Lake Manyara, where groundwater promotes plant growth.

The importance of water in limiting animal population density was underscored by an extraordinary event caused by El Niño. As we noted in Chapter 23, seed abundance on some Galápagos Islands is strongly linked to rainfall. During the 1982–1983 rainy season, rainfall on Isla Genovesa in the Galápagos Islands increased from its normal 100–150 mm to 2,400 mm. Plants responded with prodigious growth, and certain finches (*Geospiza* spp.) bred up to eight times rather than their normal maximum of three, probably because of increased abundance of fruits and seeds. The population densities of finches were exceptionally high by the end of the breeding season.

Light Can Be a Limiting Resource for Plants and Algae

Because light is necessary for photosynthesis, it can be a limiting resource for plants. However, what may be sufficient light to support the growth of one plant species may be insufficient for another. Many plant species grow best in shady conditions, such as eastern hemlock (*Tsuga canadensis*). Its saplings grow in the understory below the forest canopy, reaching maximal photosynthesis at one-quarter of full sunlight. Other plants,

such as sugarcane (*Saccharum officinarum*) or the desert shrub *Larrea*, continue to increase their photosynthetic rate as light intensity increases.

In aquatic environments, light may be an even more limiting factor because water absorbs light, preventing photosynthesis at depths greater than 100 m. Most aquatic plants and algae are limited to a fairly narrow zone close to the surface, where light is sufficient to allow photosynthesis to occur. This zone is known as the **photic zone**. In marine environments, seaweeds at greater depths have wider thalli (leaflike light-gathering structures) than those nearer the surface, because wide thalli can collect more light. In addition, in aquatic environments, plant color changes with depth. At the surface, plants and algae appear green, as they are in terrestrial conditions, because they absorb red and blue light, but not green. At greater depths, red light is mostly absorbed by water, leaving predominantly blue-green light. Red algae occur in deeper water because they possess pigments that enable them to utilize blue-green light efficiently (**Figure 54.15**).

(a) Green algae at the ocean surface

(b) Red algae at a greater depth

Figure 54.14 **The relationship between the amount of rainfall and the density of buffalo.** In the Serengeti area of Africa, buffalo density is very much dependent on grass availability, which itself is dependent on annual rainfall. The main exception is where there is permanent water, such as Lake Manyara, where greater water availability leads to greater grass growth and buffalo densities.

Figure 54.15 **Algae growing at different ocean depths.** (a) In the eastern Pacific Ocean, off the coast of California, these giant kelp floating at the ocean surface are green, just like terrestrial plants. (b) In contrast, at 75-m depth, in the McGrail Bank off of the Gulf of Mexico, most seaweeds are pink and red because the pigments can absorb the blue-green light that reaches such depths.

The Concentration of Salts in Soil or Water Can Be Critical

Salt concentrations vary widely in aquatic environments and have a great impact on osmotic balance in animals. Oceans contain considerably more dissolved minerals than rivers because oceans continually receive the nutrient-rich waters of rivers, and the sun evaporates pure water from ocean surfaces, making concentrations of minerals such as salts even higher.

The phenomenon of osmosis influences how living organisms cope with different environments. Freshwater fishes cannot live in salt water, and saltwater fishes cannot live in fresh water. Each employs different mechanisms to maintain an osmotic balance with their environment. Freshwater fishes are hyperosmotic (having a higher concentration of ions) to their environment and tend to gain water by osmosis as it diffuses through the thin tissue of the gills and mouth. To counter this, the fish continually eliminate water in the urine. However, to avoid losing all dissolved ions, many ions are reabsorbed into the bloodstream at the kidneys. Many marine fishes are hypoosmotic (having a lower concentration of ions) to their environment and tend to lose water as seawater passes over the mouth and gills. They drink water to compensate for this loss, but the water contains a higher concentration of salt, which must then be excreted at the gills and kidneys (refer back to Chapter 49).

Salt in the soil also affects the growth of plants. In arid terrestrial regions, salt accumulates in soil where water settles and then evaporates. This can also be of great significance in agriculture, where continued watering in arid environments, together with the addition of salt-based fertilizers, greatly increases salt concentration in soil and reduces crop yields. A few terrestrial plants are adapted to live in saline soil along seacoasts. Here the vegetation consists largely of **halophytes**, species that can tolerate higher salt concentrations in their cell sap than regular plants. Species such as mangroves and *Spartina* grasses have salt glands that excrete salt to the surface of the leaves, where it forms tiny white salt crystals (**Figure 54.16**).

The pH of Soil or Water Can Limit the Distribution of Organisms

As discussed in Chapter 2, the pH of water can be acidic, alkaline, or neutral. Variation in pH can have a major impact on the distribution of organisms. Normal rainwater has a pH of about 5.6, which is slightly acidic because the absorption of atmospheric CO_2 and SO_2 into rain droplets forms carbonic and sulfuric acids. However, most plants grow best at a soil water pH of about 6.5, a value at which soil nutrients are most readily available to plants. Only a few genera, such as rhododendrons (*Rhododendron*) and azaleas (*Azalea*), can live in soils with a pH of 4.0 or less. Furthermore, at a pH of 5.2 or less, nitrifying bacteria do not function properly, which prevents organic matter from decomposing. In general, alkaline soils containing chalk and limestone have a higher pH and sustain a much richer flora (and associated fauna) than do acidic soils (**Figure 54.17**).

Generally, the number of fishes and other species also decreases in acidic waters. The optimal pH for most freshwater fishes and bottom-dwelling invertebrates is between 6.0 and 9.0. Acidity in lakes increases the amount of toxic metals, such as mercury, aluminum, and lead, which can leach into the water from surrounding soil and rock. Both too much mercury and too much aluminum can interfere with gill function, causing fishes to suffocate. In the U.S. northeast, acid rain has caused high mercury levels, and some states have advised that children and pregnant women should not eat freshwater fishes.

(a) Rich flora on alkaline soil   **(b) Sparse flora on acidic soil**

Figure 54.17   Species-rich floras of chalk grassland compared to species-poor floras of acid soils.   (a) At Mount Caburn, in the lime-rich chalk hills of Sussex County, England, there is a much greater variety of plant and animal species than at (b) a heathland site in England. Heathlands are a product of thousands of years of human clearance of natural forest areas and are characterized by acidic, nutrient-poor soils.

Concept check: *Why do acidic soils support fewer species of plants and animals than lime-rich soils?*

Figure 54.16   Plant adaptations for salty conditions. Special salt glands in the leaves of *Spartina* exude salt, enabling this grass to exist in saline intertidal conditions.

The susceptibility of both aquatic and terrestrial organisms to changes in pH explains why ecologists are so concerned about **acid rain**, precipitation with a pH of less than 5.6. Acid rain results from the burning of fossil fuels such as coal, oil, and natural gas, which releases sulfur dioxide and nitrogen oxide into the atmosphere. These react with oxygen in the air to form sulfuric acid and nitric acid, which falls to the Earth's surface in rain or snow. When this precipitation falls on rivers and especially lakes, it can turn them more acidic, and they lose their ability to sustain fishes and other aquatic life. For example, lake trout disappear from lakes in Ontario and the eastern U.S. when the pH dips below about 5.2. Although this low pH does not affect survival of the adult fish, it affects the survival of juveniles.

Acid rain is important in terrestrial systems, too. For example, acid rain can directly affect forests by killing leaves or pine needles, as has happened on some of the higher mountaintops in the Great Smoky Mountains. It can also greatly lower soil pH, which can result in a loss of essential nutrients such as calcium and nitrogen. Low soil calcium results in calcium deficiencies in plants, in the snails that consume the plants, and in the birds that eat the snails, ultimately causing weak eggshells that break before hatching. Decreased soil pH also kills certain soil microorganisms, preventing decomposition and recycling of nitrogen in the soil. Decreases in soil calcium and nitrogen weaken trees and other plants and may make them more susceptible to insect attack.

Acid rain is a common problem in the northeastern U.S. and Scandinavia, where sulfur-rich air drifts over from the Midwest and the industrial areas of Britain, respectively, causing the deposition of highly acidic rain (look ahead to Figure 59.23). The problem was particularly acute during the 1960s and 1970s, but decreased manufacturing and the use of low-sulfur coal and the introduction of sulfur-absorbing scrubbers on the smokestacks of coal-burning power plants have somewhat reduced the problem in recent years. Acid rain is clearly a problem with a wide-ranging impact on ecological systems.

54.4 Climate and Its Relationship to Biological Communities

Temperature, wind, precipitation, and light are components of **climate**, the prevailing weather pattern in a given region. As we have seen, the distribution and abundance of organisms are influenced by these factors. Therefore, to understand the patterns of abundance of life on Earth, ecologists need to study the global climate. In this section, we examine global climate patterns, focusing on how temperature variation drives atmospheric circulation and how features such as elevation and landmass can alter these patterns. We will see how climate largely determines the occurrence of different **biomes**, the major community types on Earth such as tropical forests and hot deserts.

Atmospheric Circulation Is Driven by Global Temperature Differentials

Substantial differences in temperature occur over the Earth, mainly due to latitudinal variations in the incoming solar radiation. In higher latitudes, such as northern Canada and Russia, the sun's rays hit the Earth obliquely and are spread out over more of the planet's surface than they are in equatorial areas (**Figure 54.18**). More heat is also lost in the atmosphere of higher latitudes because the sun's rays travel a greater distance through the atmosphere, allowing more heat to be dissipated by cloud cover. The result is that 40% less solar energy strikes polar latitudes than equatorial areas. Generally, temperatures increase as the amount of solar radiation increases (**Figure 54.19**). However, at the tropics, both cloudiness and rain reduce average temperature, so temperatures do not continue to increase toward the equator.

Global patterns of atmospheric circulation and precipitation are influenced by solar energy. In 1735, English meteorologist George Hadley made the initial contribution to a model of general atmospheric circulation. In his model, high temperatures at the equator cause the surface equatorial air to heat up and rise vertically into the atmosphere. As the warm air rises away from its source of heat, it cools and becomes less buoyant, but the cool air does not sink back to the surface because of the warm air behind it. Instead, the rising air spreads north and south away from the equator, eventually returning to the surface at the poles. From there it flows back toward the equator to close the circulation loop. Hadley suggested that on a nonrotating Earth, this air movement would take the form of one large convection cell in each hemisphere.

When the effect of the Earth's rotation is added, however, the surface flow is deflected to the right in the Northern

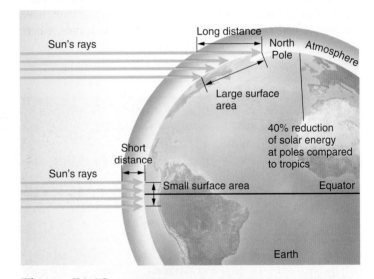

Figure 54.18 The intensity of solar radiation at different latitudes. In polar areas, the sun's rays strike the Earth at an oblique angle and deliver less energy than at tropical locations. In tropical areas, the energy is concentrated over a smaller surface and travels a shorter distance through the atmosphere.

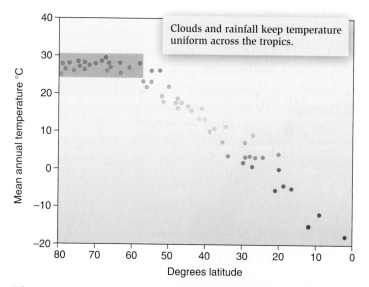

Figure 54.19 Variation of the Earth's temperature. The temperatures shown in this figure were measured at moderately moist continental locations of low elevation.

Concept check: Why is there a wide band of similar temperatures at the tropics?

Hemisphere and to the left in the Southern Hemisphere. This consequence is known as the **Coriolis effect**. Hadley's one-cell circulation has since been modified to account for the Coriolis effect and other newer data. In the 1920s, a three-cell circulation in each hemisphere was proposed to fit the Earth's heat balance (**Figure 54.20**). The contribution of George Hadley is still recognized, in that the cell nearest the equator is called the **Hadley cell**. In the Hadley cell, the warm air rising near the equator forms towers of cumulus clouds that provide rainfall, which, in turn, maintains the lush vegetation of the equatorial rain forests. As the upper flow in this cell moves toward the poles, it begins to subside, or fall back to Earth, at about 30° north and south of the equator. These **subsidence zones** are areas of high pressure and are the sites of the world's tropical deserts, because the subsiding air is relatively dry, having released all of its moisture over the equator. Winds are generally weak and variable near the center of this zone of descending air. Subsidence zones have popularly been called the horse latitudes. The name is said to have been coined by Spanish sailors crossing the Atlantic, whose ships were sometimes rendered motionless in these waters and who reportedly were forced to throw horses overboard as they could no longer water or feed them.

From the center of the subsidence zones, the surface flow splits into the westerlies, which flow toward the poles, and the equatorial flow, which is deflected by the Coriolis effect and forms trade winds. In the Northern Hemisphere, the trades are from the northeast, the direction from which they provided the sail power to explore the New World; in the Southern Hemisphere, the trades are from the southeast. The trade winds from both hemispheres meet near the equator in a region called the intertropical convergence zone (ITCZ), also known as the doldrums. Here the light winds and humid conditions provide the monotonous weather that may be the basis for the expression "in the doldrums."

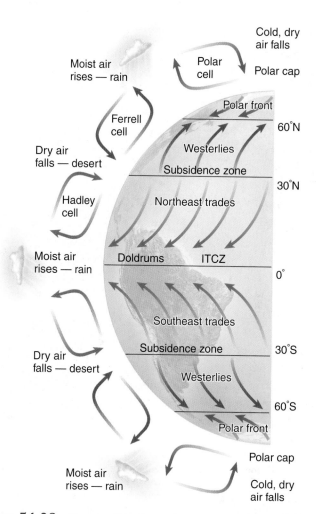

Figure 54.20 Three-cell pattern of atmospheric circulation. Three-cell model of the atmospheric circulation on a uniform, rotating Earth heated at the equator and cooled at the poles.

In the three-cell model, the circulation between 30° and 60° latitude, called the **Ferrell cell**, is opposite that of the Hadley cell. The net surface flow is poleward, and because of the Coriolis effect, the winds have a strong westerly (flowing from west to east) component. These prevailing westerlies were known to Benjamin Franklin, perhaps the first American weather forecaster, who noted that storms migrated eastward across the colonies. The final circulation cell is known as the **polar cell**. At the poles, the air has cooled and descends, but it has little moisture left, explaining why many high-latitude regions are actually desert-like in condition.

The three-cell model provides a good understanding of global circulation but is still oversimplified. The Hadley and polar cells are strong, but the Ferrell cell is much weaker, with passing high and low pressure systems, depending on the seasons. The sun does not remain constantly over the equator but moves annually between 23.5°N and 23.5°S and back again, creating seasonal changes. There is also differential heating between land and water. These effects modify the position of the three cells of circulation (**Figure 54.21**). In fact, the secondary

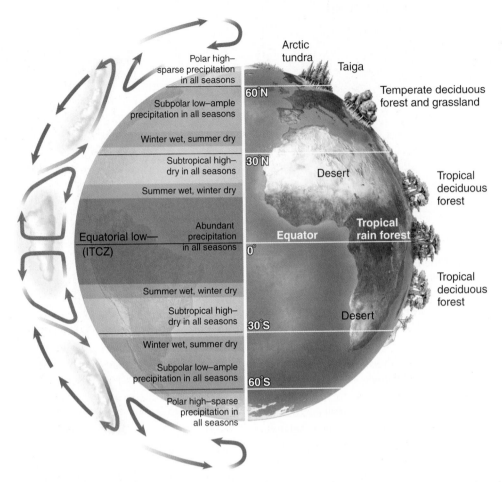

Figure 54.21 **Global circulation based on a modified three-cell model.** Tropical forests exist mainly in a band around the equator, where it is hot and rainy. At around 30° north and south, the air is hot and dry, and deserts exist. A secondary zone of precipitation exists at around 45° to 55° north and south, where temperate forests are located. The polar regions are generally cold and dry.

zones of high precipitation can come anywhere from about 35° to 65°, with between 45° and 55° being most common.

The distributions of the major biomes are largely determined by temperature differences and the wind patterns they generate. Hot, tropical forest blankets the tropics, where rainfall is high. At about 30° latitude, the air cools and descends, but it is without moisture, so the hot deserts occur around that latitude. The middle cell of the circulation model shows us that at about 35° to 65° latitude, the air has warmed and gained moisture, so it ascends, dropping rainfall over the wet, temperate forests of the Pacific Northwest and Western Europe in the Northern Hemisphere and New Zealand and Chile in the Southern Hemisphere.

Elevation and Other Features of a Landmass Can Also Affect Climate

Thus far, we have considered how global temperatures and wind patterns affect climate. The geographic features of a landmass can also have an important impact. For example, the elevation of a region greatly influences its temperature range. On mountains, temperatures decrease with increasing elevation. This decrease is a result of a process known as **adiabatic cooling**, in which increasing elevation leads to a decrease in air pressure. When air is blown across the Earth's surface and up over mountains, it expands because of the reduced pressure. As it expands, it cools at a rate of about 10°C for every 1,000 m in elevation, as long as no water vapor or cloud formation occurs. (Adiabatic cooling is also the principle behind the function of a refrigerator, in which refrigerant gas cools as it expands coming out of the compressor.) A vertical ascent of 600 m produces a temperature change roughly equivalent to that brought about by an increase in latitude of 1,000 km. This explains why mountaintop vegetation, even in tropical areas, can have the characteristics of a colder biome.

Mountains can also influence patterns of precipitation. For example, when warm, moist air encounters the windward side of a mountain, it flows upward and cools, releasing precipitation in the form of rain or snow. On the side of the mountain sheltered from the wind (the leeward side), drier air descends, producing what is called a **rain shadow**, an area where precipitation is noticeably less (**Figure 54.22a**). In this way, the western side of the Cascade Range in Washington State receives more than 500 cm of annual precipitation, whereas the eastern side receives only 50 cm.

As moist air blows across the windward side of a mountain, it rises and cools, and precipitation falls as rain or snow.

Moist air

Dry air

On the leeward side of a mountain, the cooler air descends and becomes warmer; little precipitation occurs.

(a) Rain shadow

Cool air

Warm air

During the day, as warm air rises, cooler air rushes in from the ocean to replace it.

(b) Sea breezes

Figure 54.22   The influence of elevation and proximity to water on climate.

The proximity of a landmass to a large body of water can affect climate because land heats and cools more quickly than the sea does. The specific heat of the land is much lower than that of the water, allowing the land to warm more quickly during the day. The warmed air rises and cooler air flows in to replace it. This pattern creates the familiar onshore sea breezes in coastal areas (**Figure 54.22b**). At night, the land cools quicker than the sea, and so the pattern is reversed, creating offshore breezes. The sea, therefore, has a moderating effect on the temperatures of coastal regions and especially islands. The climates of coastal regions may differ markedly from those of their climatic zones. Many never experience frost, and fog is often evident. Thus, along coastal areas, different vegetation patterns may occur

compared to those in areas farther inland. In fact, some areas of the U.S. would be deserts were it not for the warm water of the sea and the moisture-laden clouds that form above them.

Together with the rotation of the Earth, winds also create ocean currents. The major ocean currents act as "pinwheels" between continents, running clockwise in the ocean basins of the Northern Hemisphere and counterclockwise in those of the Southern Hemisphere (**Figure 54.23**). The Gulf Stream, equivalent in flow to 50 times the world's major rivers combined, brings warm water from the Caribbean and the U.S. coasts across to Europe, the climate of which is correspondingly moderated. The Humboldt Current brings cool conditions almost to the equator along the western coast of South America.

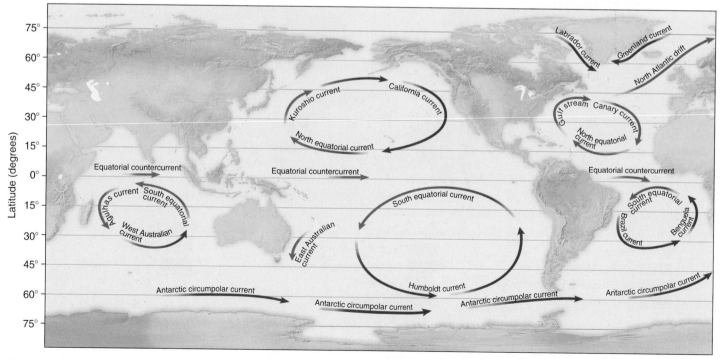

Figure 54.23   Ocean currents of the world.   The red arrows represent warm water; the blue arrows, cold water.

54.5 Biome Types Are Determined by Climate Patterns and Other Physical Variables

Differences in climate on Earth help to define its different terrestrial biomes. Many types of classification schemes are used for mapping the geographic extent of terrestrial biomes, but one of the most useful was developed by the American ecologist Robert Whittaker, who classified biomes according to the physical factors of average annual precipitation and temperature (**Figure 54.24**). In this scheme, we recognize 10 terrestrial biomes (**Figure 54.25**). Aquatic biomes are generally differentiated by water salinity or current strength. In this section, we explore the main characteristics of Earth's major terrestrial and aquatic biomes.

In this chapter we have discussed how the physical environment profoundly influences the distribution and abundance of life on Earth. In subsequent chapters, we will explore the impact of other factors on the distribution of plants and animals. On a smaller scale, the presence of predators, parasites, or competitors can also control where organisms are found. Part of an organism's fundamental niche may be occupied by a competitively superior species. We will examine the influences of such factors in Chapter 57. For animals, social interactions

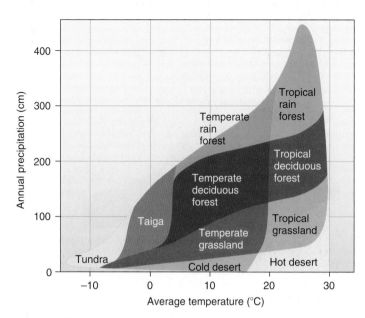

Figure 54.24 The relationship between the world's terrestrial biome types and temperature and precipitation patterns.

Concept check: What other factors may influence biome types?

with other members of the same species, such as fights over territory or mates, can also have an effect on population distributions. To discuss these issues, Chapter 55 examines animal behavior.

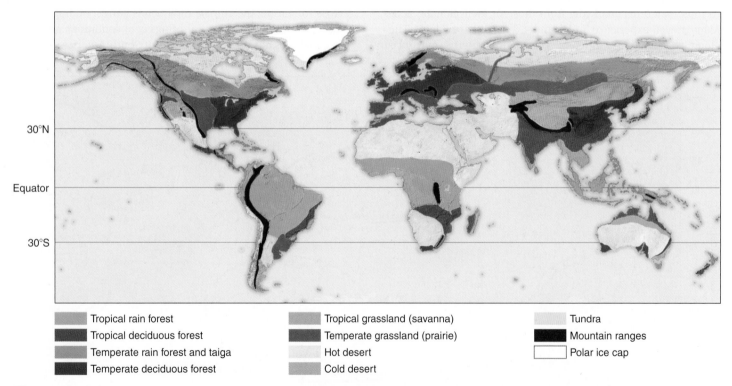

Tropical rain forest	Tropical grassland (savanna)	Tundra
Tropical deciduous forest	Temperate grassland (prairie)	Mountain ranges
Temperate rain forest and taiga	Hot desert	Polar ice cap
Temperate deciduous forest	Cold desert	

Figure 54.25 Geographic location of terrestrial biomes. The distribution patterns of taiga and temperate rain forest are combined because of their similarity in tree species and because temperate rain forest is actually limited to a very small area.

Figure 54.26a–j illustrates the 10 terrestrial biomes and identifies their main characteristics.

Although these broad terrestrial biomes are a useful way of defining the main types of communities on Earth, ecologists acknowledge that not all communities fit neatly into one of these 10 major biome types. Also, one biome type often grades into another, as seen on mountain ranges (Figure 54.26k). Soil conditions can also influence biome type. In California, serpentine soils, which are dry and nutrient-poor, support only sparse vegetation. In the eastern U.S., most of New Jersey's coastal plain, called the Pine Barrens, consists of sandy, nutrient-poor soil that cannot support the surrounding deciduous forest and instead contains grasses and low shrubs growing among open stands of pygmy pitch pine and oak trees.

Tropical Rain Forest

Figure 54.26a

Tropical rain forest in Fiji

Physical Environment: Rainfall exceeds 230 cm per year, and the temperature is hot year round, averaging 25–29°C. Soils are often shallow and nutrient-poor.

Location: This biome is found in equatorial regions. Tropical forests cover much of northern South America, Central America, western and central Africa, Southeast Asia, and various islands in the Indian and Pacific oceans.

Plant Life: The numbers of plant species found in tropical forests can be staggering, often reaching as many as 100 tree species per square kilometer. Leaves often narrow to "drip-tips" at the apex so that rainwater drains quickly. Many trees have large buttresses that help support their shallow root systems. Little light penetrates the **canopy**, the uppermost layer of tree foliage, and the ground cover is often sparse. Vines and epiphytes, plants that live perched on trees and are not rooted in the ground, are common.

Animal Life: Animal life in the tropical rain forests is diverse; insects, reptiles, amphibians, and mammals are well represented. Large mammals, however, are not common. Because many of the plant species are widely scattered in tropical forests, it is risky for plants to rely on wind for pollination or to disperse their seed. This means that animals are important in pollinating flowers and dispersing fruits and seeds. Mimicry and bright protective coloration, warning of bad taste or the existence of toxins, are common.

Effects of Humans: Humans are impacting tropical forests greatly by logging and by clearing the land for agriculture. Many South American tropical forests are cleared to create grasslands for cattle.

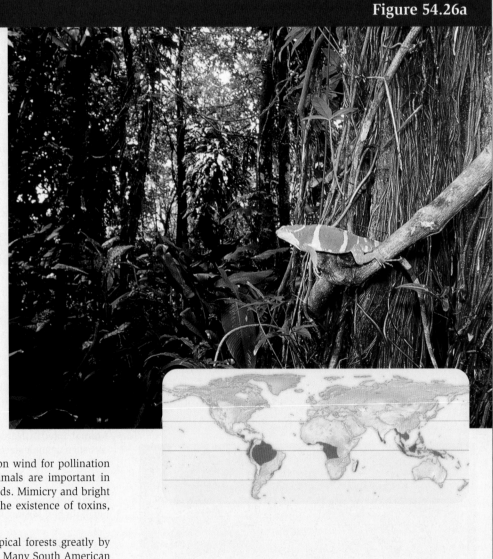

Tropical Deciduous Forest

Figure 54.26b

Tropical deciduous forest in Bandhavgarh National Park, India

Physical Environment: Rainfall is substantial, at around 130–280 cm a year, and temperatures are hot year round, averaging 25–39°C. This biome experiences a distinct dry season that is often 2 to 3 months or longer. Soil water shortages can occur in the dry season.

Location: This biome exists in equatorial regions where rainfall is more seasonal than in tropical rain forests. Much of India consists of tropical deciduous forest, containing teak trees. Brazil, Thailand, and Mexico also contain tropical deciduous forest. At the wet edges of this biome, it may grade into tropical rain forests; at the dry end, it may grade into tropical grasslands or savannas.

Plant Life: Because of the biome's distinct dry season, many of the trees in tropical deciduous forests shed their leaves, just as they do in temperate forests, and an understory of herbs and grasses may grow during this time. Indeed, because the canopy is often more open than in the tropical rain forest and more sunlight reaches the ground, a denser closed forest—what we might think of as a "tropical jungle"—exists at the forest floor. Where the dry season is 6 to 7 months long, tropical deciduous forests may contain shorter, thorny plants such as acacia trees, and the forest is referred to as a tropical thorn forest.

Animal Life: The diversity of animal life is high, and species such as monkeys, antelopes, wild pigs, and tigers are present. However, as with plant diversity, animal diversity is less than that of tropical rain forests. Tropical thorn forests may contain more browsing mammals; hence, the development of plant thorns as a defense.

Effects of Humans: The soil of tropical deciduous forests is more fertile than that of tropical rain forests. Land is increasingly being logged and cleared for agriculture and a growing human population.

Temperate Rain Forest

Figure 54.26c

Hoh Rain Forest in Olympic National Park, Washington

Physical Environment: There is abundant rainfall, usually exceeding 200 cm a year. The condensation of water from dense coastal fogs augments the normal rainfall. Temperatures seldom drop below freezing in the winter, and summer temperatures rarely exceed 27°C.

Location: The area of this biome type is small, consisting of a thin strip along the northwest coast of North America from northern California through Washington State, British Columbia, and into southeast Alaska (where it is called tongass). It also exists in southwestern South America along the Chilean coast. Indeed, it is found only in coastal locales because of the moderating influence of the ocean on air temperature.

Plant Life: The dominant vegetation type, especially in North America, consists of large evergreen trees such as western hemlock, Douglas fir, and Sitka spruce. The high moisture content allows epiphytes to thrive. Cool temperatures slow the activity of decomposers, so the litter layer is thick and spongy.

Animal Life: In North America, the temperate rain forest is rich in species such as mule deer, elk, squirrels, and numerous birds such as jays and nuthatches. Because of the abundant moisture and moderate temperatures, reptiles and amphibians are also common.

Effects of Humans: This biome is a prolific producer of wood and supplies much timber; logging threatens the survival of the forest in some areas.

Temperate Deciduous Forest

Figure 54.26d

Temperate deciduous forest in Minnesota

Physical Environment: Annual rainfall is generally between 75 and 200 cm. Temperatures fall below freezing each winter but not usually below −12°C.

Location: Large tracts of temperate deciduous forest are evident in the eastern U.S., eastern Asia, and western Europe. In the Southern Hemisphere, eucalyptus forests occur in Australia, and stands of southern beech are found in southern South America, New Zealand, and Australia.

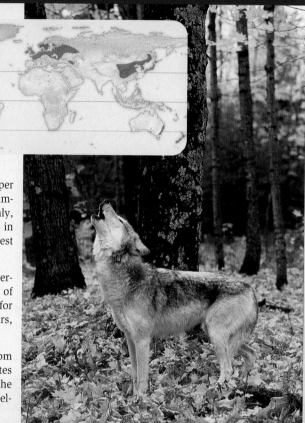

Plant Life: Species diversity is much lower in temperate deciduous forests than in the tropical forests, with about only 3 to 4 tree species per square kilometer, and several tree genera may be dominant in a given locality—for example, oaks, hickories, and maples are usually dominant in the eastern U.S. Commonly, leaves are shed in the fall and reappear in the spring. Many herbaceous plants flower in spring before the trees leaf out and block the light. Even in the summer, though, the forest is not as dense as in tropical forests, so there is abundant ground cover.

Animal Life: Animals are adapted to the vagaries of the climate; many mammals hibernate during the cold months, birds migrate, and insects enter diapause, a condition of dormancy passed usually as a pupa. Reptiles, which are dependent on solar radiation for heat, are relatively uncommon. Mammals include squirrels, wolves, bobcats, foxes, bears, and mountain lions.

Effects of Humans: Logging has eliminated much of the temperate deciduous forest from populated portions of Europe and North America. Because the annual leaf drop promotes high soil nutrient levels, soils are rich and easily converted to agriculture. Much of the human population lives in the temperate deciduous forest, and both agriculture and development are the threats to the biome.

Temperate Coniferous Forest (Taiga)

Figure 54.26e

Temperate coniferous forest in Canada

Physical Environment: Precipitation is generally between 30 and 70 cm and often occurs in the form of snow. Temperatures are very cold, often below freezing for long periods of time.

Location: The biome of coniferous forests, known commonly by its Russian name, taiga, lies north of the temperate-zone forests and grasslands. Vast tracts of taiga exist in North America and Russia, and mountain taiga exists on mountainous areas. In the Southern Hemisphere, little land area occurs at latitudes at which one would expect extensive taiga to exist.

Plant Life: Most of the trees are evergreens or conifers with tough needles, hence its similarity to temperate rain forest. In this biome, spruces, firs, and pines generally dominate, and the number of tree species is relatively low. Many of the conifers have conical shapes to reduce bough breakage from heavy loads of snow. As in tropical forests, the understory is sparse because the dense year-round canopies prevent sunlight from penetrating. Soils are poor because the fallen needles decay so slowly in the cold temperatures that a layer of needles builds up and acidifies the soil, reducing the numbers of understory species.

Animal Life: Reptiles and amphibians are rare because of the low temperatures. Insects are strongly periodic but may often reach outbreak proportions in times of warm temperatures. Mammals that inhabit this biome, such as bears, lynxes, moose, beavers, and squirrels, are heavily furred.

Effects of Humans: Humans have not extensively settled these areas, but they have been quite heavily logged. Exploration and development of oil and natural gas reserves are also a threat.

Tropical Grassland (Savanna)

Figure 54.26f

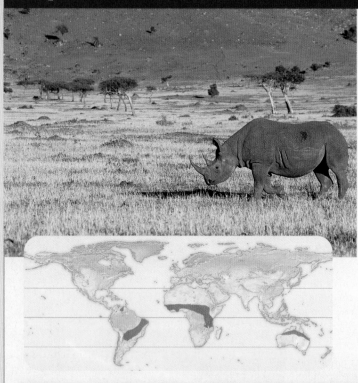

Tropical grassland of the Masai Mara Game Reserve in Kenya

Physical Environment: This biome includes hot, tropical areas, with a low or seasonal rainfall between 50 and 130 cm per year. There is often an extensive dry season. Temperatures average 24–29°C.

Location: Extensive savannas occur in Africa, South America, and northern Australia.

Plant Life: Wide expanses of grasses dominate savannas, but occasional thorny trees, such as acacias, may occur. Fire is prevalent in this biome, so most plants have well-developed root systems that enable them to resprout quickly after a fire.

Animal Life: The world's greatest assemblages of large mammals occur in the savanna biome. Herds of antelope, zebra, and wildebeest are found, together with their associated predators: cheetah, lion, leopard, and hyena. Termite mounds dot the landscape in some areas. The extensive herbivory of large grazers, together with frequent fires, may help maintain savannas and prevent their development into forests.

Effects of Humans: Savanna soils are often poor because the occasional rain leaches nutrients out. Nevertheless, conversion of this biome to agricultural land is rampant, especially in Africa. Overstocking of land for domestic animals can greatly reduce grass coverage through overgrazing, turning the area desert-like. This process is known as **desertification**.

Temperate Grassland (Prairie)

Figure 54.26g

Temperate grassland in Wyoming State

Physical Environment: Annual rainfall is generally between 25 and 100 cm, too low to support a forest but higher than that in deserts. Temperatures in the winter often fall below −10°C, while summers may be very hot, approaching 30°C.

Location: Temperate grasslands include the prairies of North America, the steppes of Russia, the pampas of Argentina, and the veldt of South Africa. In addition to the limiting amounts of rain, fire and grazing animals may also prevent the establishment of trees in the temperate grasslands. Where temperatures rarely fall below freezing and most of the rain falls in the winter, chaparral, a fire-adapted community featuring shrubs and small trees, occurs. Chaparral is seen at around 30° latitude, where cool ocean waters moderate the climate, as along the coasts of California, South Africa, Chile, and southwest Australia and in countries surrounding the Mediterranean Sea. Some ecologists recognize chaparral as a distinct biome type.

Plant Life: From east to west in North America and from north to south in Asia, grasslands show differentiation along moisture gradients. In Illinois, with an annual rainfall of 80 cm, tall prairie grasses such as big bluestem and switchgrass grow to about 2 m high. Along the eastern base of the Rockies, 1,300 km to the west, where rainfall is only 40 cm, prairie grasses such as buffalo grass and blue grama rarely exceed 0.5 m in height. Similar gradients occur in South Africa and Argentina.

Animal Life: Where the grasslands remain, large mammals are the most prominent members of the fauna: bison and pronghorn in North America, wild horses in Eurasia, and large kangaroos in Australia. Burrowing animals such as North American gophers and African mole rats are also common.

Effects of Humans: Prairie soil is among the richest in the world, having 12 times the humus layer of a typical forest soil. Worldwide, most prairies have been converted to agriculture, and original grassland habitats are among the rarest biomes in the world.

Hot Desert

Figure 54.26h

The Namib Desert, Namibia

Physical Environment: Rainfall is less than 30 cm per year. Temperatures are variable, from below freezing at night to as much as 50°C in the day.

Location: Hot deserts are found around latitudes of 30°

north and south. Prominent deserts include the Sahara of North Africa, the Kalahari and Namib of southern Africa, the Atacama of Chile, the Sonoran of northern Mexico and the southwest U.S., and the Simpson of Australia.

Plant Life: Three forms of plant life are adapted to deserts: annuals, succulents, and desert shrubs. Annuals circumvent drought by growing only when there is rain. Succulents, such as the saguaro cactus and other barrel cacti of the southwestern deserts, store water. Desert shrubs, such as the spraylike ocotillo, have short trunks, numerous branches, and small, thick leaves that can be shed in prolonged dry periods. In many plants, spines or volatile chemical compounds serve as a defense against water-seeking herbivores.

Animal Life: To conserve water, desert plants produce many small seeds, and animals that eat those seeds, such as ants, birds, and rodents, are

common. Reptiles are numerous, because high temperatures permit these ectothermic animals to maintain a warm body temperature. Lizards and snakes are important predators of seed-eating mammals.

Effects of Humans: Ambitious irrigation schemes and the prolific use of underground water have allowed humans to develop deserts and grow crops there. Salinization, a buildup in the salt content of the soil that results from irrigation in areas of low rainfall, is prevalent. Off-road vehicles can disturb the fragile desert communities.

Cold Desert

Figure 54.26i

The Gobi Desert of Mongolia

Physical Environment: Precipitation is less than 25 cm a year and is often in the form of snow. Rainfall usually comes in the spring. In the daytime, temperatures can be high in the summer, 21–26°C, but average around freezing, −2 to 4°C, in the winter.

Location: Cold deserts are found in dry regions at middle to high latitudes, especially in the interiors of continents and in the rain

shadows of mountains. Cold deserts are found in North America (the Great Basin Desert), in eastern Argentina (the Patagonian Desert), and in central Asia (the Gobi Desert).

Plant Life: Cold deserts are relatively poor in terms of numbers of plant species. Most plants are small in stature, being only between 15 and 120 cm tall. Many species are deciduous and spiny. The Great Basin Desert in Nevada, Utah, and bordering states is a cold desert dominated by sagebrush.

Animal Life: As in hot deserts, large numbers of plants produce small seeds on which numerous ants, birds, and rodents feed. Many species live in burrows to escape cold. In the Great Basin Desert, pocket mice, jackrabbits, kit fox, and coyote are common.

Effects of Humans: Agriculture is hampered because of low temperatures and low rainfall, and human populations are not extensive. If the top layer of soil is disturbed by human intrusions such as off-road vehicles, erosion occurs rapidly and even less vegetation is able to exist.

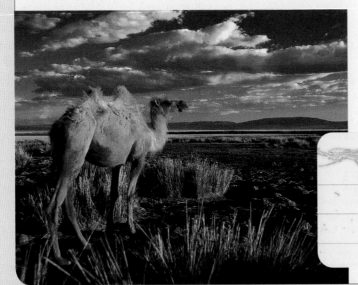

Terrestrial Biomes (continued)

Tundra

Figure 54.26j

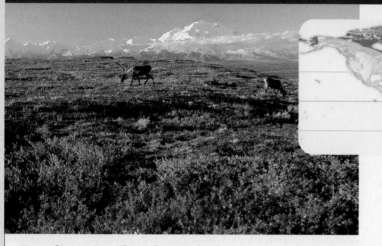

Denali National Park in Alaska

Physical Environment: Precipitation is generally less than 25 cm per year and is often locked up as snow and unavailable for plants. Deeper water can be locked away for a large part of the year in **permafrost**, a layer of permanently frozen soil. The growing season is short, only 50–60 days. Summer temperatures are only 3–12°C, and even during the long summer days, the ground thaws to less than 1 m in depth. Midwinter temperatures average −32°C.

Location: Tundra (from the Finnish *tunturia*, meaning treeless plain) exists mainly in the Northern Hemisphere, north of temperate coniferous forest, because there is very little land area in the Southern Hemisphere at the latitude where tundra would occur.

Plant Life: With so little available water, trees cannot grow. Vegetation occurs in the form of fragile, slow-growing lichens, mosses, grasses, sedges, and occasional shrubs, which grow close to the ground. Plant diversity is very low. In some places, desert conditions prevail because so little moisture falls.

Animal Life: Animals of the arctic tundra have adapted to the cold by having good insulation. Many birds, especially shorebirds and waterfowl, migrate. The fauna is much richer in summer than in winter. Many insects spend the winter at immature stages of growth, which are more resistant to cold than the adult forms. Larger animals include such herbivores as musk oxen and caribou in North America, called reindeer in Europe and Asia. Smaller animals include hares and lemmings. Common predators include arctic fox, wolves, and snowy owls, and polar bears near the coast.

Effects of Humans: Though this area is sparsely populated, mineral extraction, especially of oil, has the potential to significantly impact this biome. Ecosystem recovery from such damage would be very slow.

Mountain Ranges

Figure 54.26k

Rocky Mountains of Colorado

Physical Environment: Mountain ranges must be viewed differently than other biomes. Biome type relies predominantly on climate. On mountains, temperature decreases with increasing elevation through adiabatic cooling, as discussed previously. Thus, precipitation and temperature may change dramatically, depending on elevation and whether the mountainside is on the windward or leeward side.

Location: Mountain ranges exist in many areas of the world, but among the largest are the Himalayas in Asia, the Rockies in North America, and the Andes in South America.

Plant Life: A variety of biomes can be found on a single mountain range. Biome type may change from temperate forest through taiga and into tundra on an elevation gradient in the Rocky Mountains, and even from tropical forest to tundra on the highest peaks of the Andes in tropical South America. In tropical regions, daylight averages 12 hours per day throughout the year. Instead of a period of intense productivity, seen in arctic tundra, vegetation in the tropical alpine tundra exhibits slow but steady rates of photosynthesis and growth all year.

Animal Life: The animals of this biome are as varied as the number of habitats they contain. Generally, more species of plants and animals are found at lower elevations than at higher ones. At higher elevations, animals such as bighorn sheep and mountain goats have to be sure-footed to climb the craggy slopes and have skidproof pads on their hooves. Despite the often-strong winds, birds of prey, such as eagles, are frequent predators of the furry rodents found at higher elevations, including guinea pigs and marmots.

Effects of Humans: Logging and agriculture at lower elevations can cause habitat degradation. Because of the steep slopes, mountain soils are often well drained, thin, and especially susceptible to erosion following agriculture.

Aquatic Biomes Consist of Marine and Freshwater Regions

Within aquatic environments, several different biome types are also recognized, including marine aquatic biomes (intertidal zone, coral reef, and open ocean) and freshwater habitats (lakes, rivers, and wetlands). These biomes are distinguished primarily by differences in salinity, oxygen content, depth, current strength, and availability of light (Figures 54.27a–f). Freshwater habitats are traditionally divided into **lentic**, or standing-water habitats (from the Latin *lenis*, meaning calm), and **lotic**, or running-water habitats (from the Latin *lotus*, meaning washed).

Intertidal Zone

Figure 54.27a

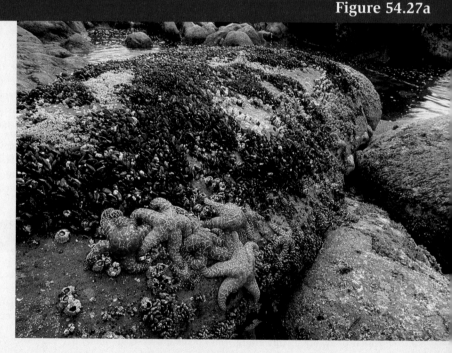

Olympic Coast National Marine Sanctuary in Washington State

Physical Environment: The **intertidal zone**, the area where the land meets the sea, is alternately submerged and exposed by the daily cycle of tides. The resident organisms are subject to huge daily variations in temperature, light intensity, and availability of seawater.

Location: Throughout the world, the area where the land meets the sea consists of sandy shore, mudflats, or rocky shore. Three broad zones occur in a vertical direction, which is most evident on rocky shores. The upper littoral zone is submerged only during the highest tides. The midlittoral zone is submerged during the highest regular tide and exposed during the lowest tide each day. The lower littoral zone is exposed only during the lowest tide.

Plant Life: Plant life may be quite limited because the sand or mud is constantly shifted by the tide. Mangroves may colonize mudflats in tropical areas, and salt marsh grasses may colonize mudflats in temperate locations. On the rocky shore, green algae and seaweeds predominate.

Animal Life: Animal life may be quite diverse. On the rocky shore, sea anemones, snails, hermit crabs, and small fishes live in tide pools. On the rock face, there may be a variety of limpets, mussels, sea stars, sea urchins, snails, sponges, tube worms, whelks, isopods, and chitons. At low tides, organisms may be dry and vulnerable to predation by a variety of animals, including birds and mammals. High tides bring predatory fishes. Sandy or muddy shores may contain burrowing marine worms, crabs, and small isopods.

Effects of Humans: Urban development has greatly reduced the beach area available to shorebirds and breeding turtles. Oil spills have greatly impacted some rocky intertidal areas.

Coral Reef

Figure 54.27b

Caribbean coral reef

Physical Environment: Corals need warm water of at least 20°C but less than 30°C. They are also limited to the photic zone, where light penetrates. Sunlight is important because many corals harbor symbiotic algae, or dinoflagellates, that contribute nutrients to the animals and that require light to live.

Location: Coral reefs exist in warm tropical waters where there are solid substrates for attachment and water clarity is good. The largest coral reef in the world is the Great Barrier Reef off the Australian coastline, but other coral reefs are found throughout the Caribbean Sea, the Red Sea, and the Pacific and Indian Oceans.

Plant Life: Dinoflagellate algae live within the coral tissue, and a variety of red and green algae live on the coral reef surface.

Animal Life: An immense variety of microorganisms, invertebrates, and fishes live among the coral, making the coral reef one of the most interesting and species-rich biomes on Earth. Probably 30–40% of all fish species on Earth are found on coral reefs. Prominent herbivores include snails, sea urchins, and fishes. These are consumed by octopuses, sea stars, and carnivorous fishes. Many species are brightly colored, warning predators of their toxic nature.

Effects of Humans: Collectors have removed many corals and fishes for the aquarium trade, and marine pollution threatens water clarity in some areas. Perhaps the greatest threat to coral reefs is from global warming. Water temperatures that are too high (over 30°C) and high pH caused by elevated CO_2 levels both contribute to coral bleaching.

The Open Ocean

Figure 54.27c

Manta ray in the open ocean

Physical Environment: In the open ocean, sometimes called the **pelagic zone**, water depth averages 4,000 m. Nutrient concentrations are typically low, though the waters may be periodically enriched by ocean **upwelling**, the circulation of cold, mineral-rich nutrients from deeper water to the surface. Pelagic waters are mostly cold, only warming near the surface.

Location: Across the globe, covering 70% of the Earth's surface.

Plant Life: In the photic zone, many microscopic photosynthetic organisms (**phytoplankton**) grow and reproduce. Phytoplankton account for nearly half the photosynthetic activity on Earth and produce much of the world's oxygen.

Animal Life: Open-ocean organisms include **zooplankton**, minute animal organisms consisting of some worms, copepods (tiny shrimplike creatures), small jellyfish, and small invertebrate and fish larvae that graze on the phytoplankton. The open ocean also includes free-swimming animals collectively called **nekton**, which can swim against the currents to locate food. Nekton includes large squids, fishes, sea turtles, and marine mammals. Only a few of these organisms live at any great depth. In some areas, a unique assemblage of animals is associated with deep-sea hydrothermal vents that spew hot (350°C) water rich in hydrogen sulfide. Large polychaete worms and other chemoautotrophic organisms exist together in this dark, oxygen-poor environment (refer back to Figure 22.3).

Effects of Humans: Oil spills and a long history of garbage disposal have polluted the ocean floors of many areas. Overfishing has caused many fish populations to crash, and the whaling industry has greatly reduced the numbers of most species of whales.

Lentic Habitats

Figure 54.27d

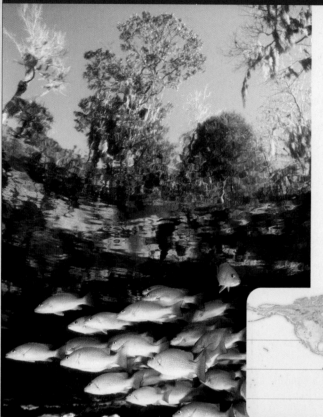

Everglades National Park, Florida

Physical Environment: The lentic habitat consists of still, often deep water. Its physical characteristics depend greatly on the surrounding land, which dictates what nutrients collect in the lake. Young lakes often start off clear and with little plant life. Such lakes are called **oligotrophic**. With age, the lake becomes richer in dissolved nutrients from erosion and runoff from surrounding land, with the result that cyanobacteria and algae spread, reducing the water clarity. Such lakes are termed **eutrophic**. The process of eutrophication occurs naturally but can be sped up by human activities (see Chapter 59).

Location: Throughout all the continents of the world.

Plant Life: In addition to free-floating cyanobacteria and algae, lentic habitats may have rooted vegetation, which often extends above the water surface (emergent vegetation), such as cattails, plus deeper-dwelling aquatic plants and algae.

Animal Life: Animals include fishes, frogs, turtles, crayfish, insect larvae, and many species of insects. In tropical and subtropical lakes, alligators and crocodiles are common.

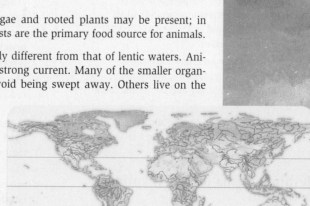

Effects of Humans: Agricultural runoff, including fertilizers and sewage, can greatly increase lake nutrient levels and speed up the process of eutrophication, resulting in algal blooms and fish kills. In some areas, invasive species of invertebrates and fishes are outcompeting native species.

Lotic Habitats

Figure 54.27e

Fast-flowing river in the Pacific Northwest

Physical Environment: In lotic habitats, flowing water prevents nutrient accumulations and phytoplankton blooms. The current also mixes water thoroughly, providing a well-aerated habitat of relatively uniform temperature. The current, oxygen level, and clarity are greater at the source of a stream (its headwaters) than in the lower reaches of rivers. Nutrient levels are generally less in headwaters.

Location: On all continents except Antarctica.

Plant Life: In slow-moving streams and rivers, algae and rooted plants may be present; in swifter-moving rivers, leaves from surrounding forests are the primary food source for animals.

Animal Life: Lotic habitats have a fauna completely different from that of lentic waters. Animals are adapted to stay in place despite an often-strong current. Many of the smaller organisms are flat and attach themselves to rocks to avoid being swept away. Others live on the underside of large boulders, where the current is much reduced. Fish such as trout may be present in rivers with cool temperatures, high oxygen, and clear water. In warmer, murkier waters, catfish and carp may be abundant.

Effects of Humans: Animals of lotic systems are not well adapted for low-oxygen environments and thus are particularly susceptible to oxygen-reducing pollutants such as sewage. Dams across rivers have prevented the passage of migratory species such as salmon.

Wetlands

Figure 54.27f

Yellow Waters River, Kakadu National Park, Northern Territory, Australia

Physical Environment: At the margins of both lentic and lotic habitats, wetlands may develop. Wetlands are areas regularly saturated by surface water or groundwater. They range from marshes, treeless areas where herbaceous species predominate; to swamps, wet areas dominated by trees; and bogs, depressions dominated by marshes. Many wetlands are seasonally flooded when rivers overflow their banks or lake levels rise. Some wetlands also develop along estuaries, where rivers merge with the ocean, and high tides can flood the land. Because of generally high nutrient levels, oxygen levels are fairly low. Temperatures vary substantially with location.

Location: Worldwide, except in Antarctica.

Plant Life: Wetlands are among the most productive and species-rich areas in the world. In North America, floating plants such as lilies and rooted species such as sedges, cattails, cypress, and gum trees predominate.

Animal Life: Most wetlands are rich in animal species. Wetlands are a prime habitat for wading and diving birds. In addition, they are home to a profusion of insects, from mosquitoes to dragonflies. Vertebrate predators include many amphibians, reptiles, otters, and alligators.

Effects of Humans: Long mistakenly regarded as wasteland by humans, many wetlands have been drained and developed for housing and industry. Wetlands play a valuable role in protecting coastal communities from hurricanes, and the loss of wetlands in Louisiana contributed to the severity of effects from hurricane Katrina in 2005.

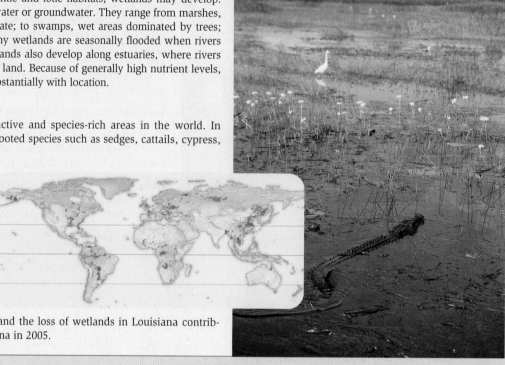

Summary of Key Concepts

54.1 The Scale of Ecology

- Ecologists study the interactions among organisms and between organisms and their environments. (Figure 54.1)
- The field of ecology can be subdivided into broad areas of organismal, population, community, and ecosystem ecology. (Figure 54.2)
- Organismal ecology considers how individuals are adapted to their environment and how the behavior of an individual organism contributes to its survival and reproductive success and the population density of the species.
- Population ecology explores those factors that influence a population's growth, size, and density. Community ecology studies how populations of species interact and form functional communities. (Figure 54.3)
- Ecosystem ecology examines the flow of energy and cycling of nutrients among organisms within a community and between organisms and the environment.

54.2 Ecological Methods Focus on Observation and Experimentation

- Ecological methods focus on observation and experimentation. Initially, ecologists may construct a possible web of interactions between species. (Figure 54.4)

- Interactions among species are often observed and analyzed graphically, and a hypothesis is formed. (Figure 54.5)
- Ecologists often test their hypotheses using well-replicated experiments. The results are often presented graphically and analyzed via a variety of statistical tests. (Figure 54.6)

54.3 The Environment's Impact on the Distribution of Organisms

- Abiotic factors such as temperature, wind, water, light, salinity, and pH can have powerful effects on ecological systems. (Table 54.1)
- Temperature exerts important effects on the distribution of organisms because of its effect on biological processes and the inability of most organisms to regulate their body temperature. (Figures 54.7, 54.8, 54.9)
- The greenhouse effect is the process in which short-wave solar radiation passes through the atmosphere to warm the Earth and is radiated back into the atmosphere as long-wave infrared radiation. Much of this radiation is absorbed by atmospheric gases and radiated back to the Earth's surface, causing its temperature to rise. (Figure 54.10)
- The major atmospheric gases causing the greenhouse effect are water vapor, carbon dioxide, methane, nitrous oxide, and chlorofluorocarbons. (Table 54.2)
- An increase in atmospheric gases is increasing the greenhouse effect, causing global warming, a gradual elevation of the Earth's surface temperature. Ecologists expect that global

warming will have a large effect on the distribution of the world's organisms. (Figure 54.11)

- Wind can amplify the effects of temperature and modify wave action. (Figure 54.12)

- The availability of water has an important effect on the abundance of organisms. (Figure 54.13)

- Light can be a limiting resource for plants in both terrestrial and aquatic environments. (Figure 54.14)

- The concentration of salts and the pH of soil and water can limit the distribution of organisms. (Figures 54.15, 54.16, 54.17)

54.4 Climate and Its Relationship to Biological Communities

- Global temperature differentials are caused by variations in incoming solar radiation and patterns of atmospheric circulation. (Figures 54.18, 54.19, 54.20, 54.21)

- Elevation and the proximity between a landmass and large bodies of water can similarly affect climate. (Figures 54.22, 54.23)

54.5 Biome Types Are Determined by Climate Patterns and Other Physical Variables

- Climate has a large effect on biomes, major types of habitats characterized by distinctive plant and animal life. (Figures 54.24, 54.25)

- Terrestrial biomes are generally named for their climate and vegetation type and include tropical rain forest, tropical deciduous forest, temperate rain forest, temperate deciduous forest, temperate coniferous forest (taiga), tropical grassland (savanna), temperate grassland (prairie), hot and cold deserts, and tundra. In mountain ranges, biome type may change on an elevation gradient. (Figure 54.26)

- Within aquatic environments, biomes include marine aquatic biomes (the intertidal zone, coral reef, and open ocean) and freshwater lakes, rivers, and wetlands. These are distinguished by differences in salinity, oxygen content, depth, current strength (lentic versus lotic), and availability of light. (Figure 54.27)

Assess and Discuss

Test Yourself

1. Which of the following is probably the most important factor in the distribution of organisms in the environment?
   a. light
   b. temperature
   c. salinity
   d. water availability
   e. pH

2. The greenhouse effect is
   a. a new phenomenon resulting from industrialization.
   b. due to the absorption of solar radiation by atmospheric gases.
   c. responsible for the natural warming of the Earth.
   d. all of the above.
   e. b and c only.

3. An examination of the temperature tolerances of locusts would best be described by which ecological subdiscipline?
   a. organismal ecology
   b. population ecology
   c. community ecology
   d. ecosystem ecology
   e. both a and b

4. Physics is to engineering as ecology is to
   a. biology.
   b. environmental science.
   c. chemistry.
   d. mathematics.
   e. statistics.

5. The most common biome type, by area occupied, is the
   a. open ocean.
   b. tropical rainforest.
   c. tundra.
   d. hot desert.
   e. lentic habitats.

6. What is the driving force that determines the circulation of the atmospheric air?
   a. temperature differences of the Earth
   b. winds
   c. ocean currents
   d. mountain ridges
   e. all of the above

7. In this biome, rainfall is between 25 cm and 100 cm and temperatures vary between −10°C in winter and 30°C in summer. Where are you?
   a. tropical rainforest
   b. tropical deciduous forest
   c. savanna
   d. prairie
   e. temperate deciduous forest

8. What characteristics are commonly used to identify the biomes of the Earth?
   a. temperature
   b. precipitation
   c. vegetation
   d. all of the above
   e. a and b only

9. Young lakes are often clear and with little plant life. Such lakes are called
   a. oligotrophic.
   b. eutrophic.
   c. lotic.
   d. lentic.
   e. pelagic.

10. Which gas contributes most to human-caused global warming?
    a. carbon dioxide
    b. nitrous oxide
    c. sulfur dioxide
    d. methane
    e. chlorofluorocarbons

Conceptual Questions

1. If mountains are closer to the sun than valleys, why aren't they hotter?

2. Explain the greenhouse effect.

3. In most locations on Earth, at about 30° latitude, air cools and descends, and hot deserts occur. Florida is situated between 31°N and 24°N. Why does it not support a desert biome?

Collaborative Questions

1. The so-called Telegraph fire, near Yosemite National Park, in 2008, was one of the worst in California that year. What could be done to prevent such a catastrophic fire in the park itself?

2. Based on your knowledge of biomes, identify the biome in which you live. In your discussion, list and describe the organisms that you have observed in your biome.

Online Resource

www.brookerbiology.com

Stay a step ahead in your studies with animations that bring concepts to life and practice tests to assess your understanding. Your instructor may also recommend the interactive ebook, individualized learning tools, and more.

Behavioral Ecology

55

A chimpanzee (*Pan troglodytes*) using a stick to catch food.

After their young hatch, nesting black-headed gulls (*Larus ridibundus*) pick up the empty eggshells and carry them away from the nest. One might think that they are being neat and tidy or are minimizing the risk of bacterial infection to the chicks, but there is more to the behavior than this. The chicks and unhatched eggs are well camouflaged in the nest, but the white color of the empty eggshell quickly attracts the attention of predators such as crows that would kill and eat the chicks or remaining eggs. By removing the old eggshells, the gull parents are increasing the chances that their offspring—and thus their genes—will survive. Although this behavior is likely to be an instinctive activity of the gulls, it is promoted because birds that performed this activity had a higher rate of chick survival than those that didn't, ensuring that the genes that code for this behavior are passed on.

Behavior is the observable response of organisms to external or internal stimuli. In this chapter, we focus our attention on the field of **behavioral ecology**, the study of how behavior contributes to the differential survival and reproduction of organisms. Contemporary behavioral ecology builds upon earlier work that focused primarily on how organisms behave. In the early 20th century, scientific studies of animal behavior, termed **ethology** (from the Greek *ethos*, meaning habit or manner), focused on the specific genetic

and physiological mechanisms of behavior. These factors are called **proximate causes**. For example, we could hypothesize that male deer rut or fight with other males in the fall because a change in day length stimulates the eyes, brain, and pituitary gland and triggers hormonal changes in their bodies. The founders of ethology, Karl von Frisch, Konrad Lorenz, and Niko Tinbergen, shared the 1973 Nobel Prize in Physiology or Medicine for their pioneering discoveries concerning the proximate causes of behavior.

However, we could also hypothesize that male deer fight to determine which deer get to mate with a female deer and pass on their genes. This hypothesis leads to a different answer than the one that is concerned with changes in day length. This answer focuses on the adaptive significance of fighting to the deer, that is, on the effect of a particular behavior on reproductive success. These factors are called **ultimate causes** of behavior. Since the 1970s, behavioral ecologists have focused more on understanding the ultimate causes of behavior.

In this chapter, we will explore the role of both proximate and ultimate causes of behavior. We begin the chapter by investigating how behavior is achieved, examining the roles of both genetics and the environment. In doing so, we will examine the important contributions of ethologists von Frisch, Lorenz, and Tinbergen. We consider how different behaviors are involved in movement, gathering food, and communication. Later, we investigate how organisms interact in groups, whether an organism can truly behave in a way that benefits others at a cost to itself, and how behavior shapes different mating systems. The chapter focuses on animal behavior, because the behavior of other organisms is more limited and less well understood.

55.1 The Impact of Genetics and Learning on Behavior

Behavior is controlled by both genetics and the environment, and in this chapter, we will discuss the influence of both. Determining to what degree a behavior is influenced by genes versus the environment will depend on the particular genes and environment examined. However, in a few cases, changes in behavior may be caused by variation in just one gene. Even if a given behavior is influenced by many genes, if one gene is altered, it is possible that the entire behavior can change. To use the analogy of baking a cake, a change in one ingredient of the recipe may change the whole taste of the cake, but that does not mean that the one ingredient is responsible for the entire cake.

In this section, we begin by examining how genes can affect behavior and consider several examples of simple, genetically programmed behaviors. Later, we explore several types of learned behavior, including classical and operant conditioning and cognitive learning, and conclude the section by exploring an example of the interaction of genetics and learning on behavior.

Genomes & Proteomes Connection

Some Behavior Results from Simple Genetic Influences

An example of the effect of genes on complex behavior was demonstrated in W. C. Rothenbuhler's 1964 work on honeybees. Some strains of bees are termed hygienic; that is, they detect and remove diseased larvae from the nest. This behavior involves two distinct maneuvers: uncapping the wax cells and then discarding the dead larvae. Other strains are not hygienic and do not exhibit such behavior. Using genetic crosses, Rothenbuhler demonstrated that one gene (*u*) controlled cell uncapping and another gene (*r*) controlled larval removal. Double recessives (*uurr*) were hygienic strains, and double dominants (*UURR*) were nonhygienic strains. When the two strains were crossed, all the F₁ hybrids were nonhygienic (*UuRr*). When the F₁ hybrids were crossed with the pure hygienic strain (*uurr*), four different genotypes were produced, as Mendel's law of independent assortment predicts (Chapter 16): one-quarter of the offspring were hygienic (*uurr*), one-quarter were nonhygienic and showed neither behavior (*UuRr*), one-quarter uncapped the cells but failed to remove the larvae (*uuRr*), and one-quarter removed the larvae but only if the cells were uncapped for them (*Uurr*).

As a second example, Jennifer Brown and her colleagues showed how a single gene, *fosB*, controls nurturing behavior in mice. Normal mice, which carry the gene, clean their newborn pups, nurse them, and crouch over them to keep them warm. In 1996, Brown's group created mutant mice lacking the *fosB* gene. Despite fully functioning mammary glands, mutant mothers paid little attention to their newborn pups, which remained scattered around the cage and died within a few days. If the pups of the *fosB* mutants were placed with normal females, they thrived, but if normal pups were given to mutant mothers, they too were neglected.

Genes for behavior act on the development of the nervous system and musculature, physical traits that evolve through natural selection. Many genes are needed for the proper development and function of the nervous system and musculature. Even so, as described by Rothenbuhler's and Brown's work, variation in a single gene can have a dramatic impact on behavior.

Fixed Action Patterns Are Genetically Programmed

Behaviors that seem to be genetically programmed are referred to as **innate** (also called instinctual). Although we recognize that the expression of genes varies, often in response to environmental stimuli, some behavior patterns evidently are genetically quite fixed. Most individuals will exhibit the same behavior regardless of the environment. A spider will spin a specific web without ever seeing a member of its own species build one. The courtship behaviors of many bird species are so stereotyped as to be virtually identical.

A classic example of innate behavior is the egg-rolling response in geese (**Figure 55.1**). If an incubating goose notices an egg out of the nest, she will extend her neck toward the egg, get up, and then roll the egg back to the nest using her beak. Such behavior functions to improve fitness because it increases the survival of offspring. Eggs that roll out of the nest get cold and fail to hatch. Geese that fail to exhibit the egg-rolling response would pass on fewer of their genes to future generations.

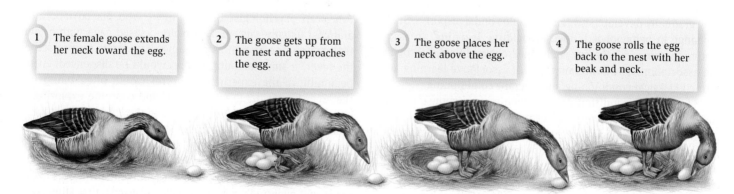

| 1 | The female goose extends her neck toward the egg. | 2 | The goose gets up from the nest and approaches the egg. | 3 | The goose places her neck above the egg. | 4 | The goose rolls the egg back to the nest with her beak and neck. |

Figure 55.1   A fixed action pattern as an example of innate behavior.   Female geese retrieve eggs that have rolled outside the nest through a set sequence of movements. The goose will complete this entire sequence even if a researcher takes the egg away before the goose has rolled it back to the nest.

Egg-rolling behavior is an example of what ethologists term a **fixed action pattern** (**FAP**), a behavior that, once initiated, will continue until completed. For example, if the egg is removed while the goose is in the process of rolling it back toward the nest, the goose still completes the FAP, as though she were rolling back the now-absent egg to the nest. The stimulus to initiate this behavior is obviously a strong one, which ethologists term a **sign stimulus**. The sign stimulus for the goose is that an egg had rolled out of the nest. According to ethologists, this stimulus acts on the goose's central nervous system, which provides a neural stimulus to initiate the motor program or FAP. Interestingly, any round object will elicit the egg-rolling response, from a wooden egg to a volleyball. Although sign stimuli usually have certain key components, they are not necessarily very specific.

Niko Tinbergen's study of male stickleback fish provides another classic example of an FAP. Male sticklebacks, which have a characteristic red belly, will attack other male sticklebacks that invade their territory. Tinbergen found that sticklebacks attacked small, unrealistic model fish having a red ventral surface (the sign stimulus), while ignoring a realistic male stickleback model that lacked a red underside (Figure 55.2).

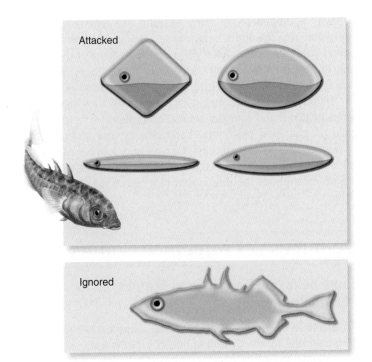

Figure 55.2 A fixed action pattern elicited by a sign stimulus. The sign stimulus for male sticklebacks to attack other males entering their territory is a red ventral surface. In experiments, male sticklebacks attacked all models that had a red underside, while ignoring a realistic model of a stickleback that lacked the red belly.

Conditioning Occurs When a Relationship Between a Stimulus and a Response Is Learned

Although many of the behavioral patterns exhibited by animals are largely innate, sometimes animals can make modifications to their behavior based on previous experience, a process that involves learning. Perhaps the simplest form of learning is **habituation**, in which an organism learns to ignore a repeated stimulus. For example, animals in African safari parks become habituated to the presence of vehicles containing tourists; these vehicles are neither a threat nor a benefit to them. After a while, birds can become habituated to the presence of a scarecrow, resulting in damage to crops. Habituation can be a problem at airports, where birds eventually ignore the alarm calls designed to scare them away from the runways.

Habituation is a form of nonassociative learning, a change in response to a repeated stimulus without association with a positive or negative reinforcement. Alternatively, an association may gradually develop between a stimulus and a response. Such a change in behavior is termed **associative learning**. In associative learning, a behavior is changed or conditioned through the association. The two main types of associative learning are termed classical conditioning and operant conditioning.

In **classical conditioning**, an involuntary response comes to be associated positively or negatively with a stimulus that did not originally elicit the response. This type of learning is generally associated with the Russian psychologist Ivan Pavlov. In his original experiments in the 1920s, Pavlov restrained a hungry dog in a harness and presented small portions of food at regular intervals. The dog would salivate whenever it smelled the food. Pavlov then began to sound a metronome when presenting the food. Eventually the dog would salivate at the sound of the metronome, whether or not the food was present. Pavlov termed the food the **unconditioned stimulus** for salivation, and the metronome was the **conditioned stimulus**. Likewise, he termed salivation in response to food as an **unconditioned response** and salivation in response to the metronome as a **conditioned response**. Classical conditioning is pervasive in human society and is widely observed in other animals. For example, many insects quickly learn to associate certain flower odors with nectar rewards and other flower odors with no rewards.

In **operant conditioning**, an animal's behavior is reinforced by a consequence, either a reward or a punishment. The classic example of operant conditioning is associated with the American psychologist B. F. Skinner, who placed laboratory animals, usually rats, in a specially devised cage with a lever that came to be known as a Skinner box. If the rat pressed on the lever, a small amount of food would be dispensed. At the beginning of the experiment, the rat would often bump into the lever by accident, eat the food, and continue exploring its cage. Later, it would learn to associate the lever with obtaining food. Eventually, if it was hungry, the rat would almost continually press the lever. Operant conditioning, also called trial-and-error learning, is common in animals. Often it is associated with negative rather than positive reinforcement. For example, toads will eventually refuse to strike at insects that sting, such as wasps and bees, and birds will learn to avoid bad-tasting butterflies (Figure 55.3).

(a) Blue jay eating monarch   **(b) Vomiting reaction**

Figure 55.3   Operant conditioning, also known as trial-and-error learning.   (a) A young blue jay will eat a monarch butterfly, not knowing that it is noxious. (b) After the first experience of vomiting after eating a monarch, a blue jay will avoid the insects in the future.

Concept check: *What's the difference between operant conditioning and classical conditioning?*

Cognitive Learning Involves Conscious Thought

Cognitive learning refers to the ability to solve problems with conscious thought and includes activities such as perception, analysis, judgment, recollection, and imagining. In the 1920s, psychologist Wolfgang Kohler conducted a series of classic experiments with chimpanzees that suggested animals could exhibit cognitive learning. In the experiments, a chimpanzee was left in a room with bananas hanging from the ceiling and out of reach (**Figure 55.4**). Also present in the room were several wooden boxes. At first, the chimp tried in vain to jump up and grab the bananas. After a while, however, it began to arrange the boxes one on top of another underneath the fruit. Eventually, the chimp climbed the boxes and retrieved the fruit. This clearly looks like an example of problem solving that involves cognitive learning.

Many other examples of such behavior have been observed. Chimps strip leaves off twigs and use the twigs to poke into ant nests, withdrawing the twig and licking the ants off (see chapter-opening photo). Captive ravens have been shown to retrieve meat suspended from a branch by a string, even though they have never encountered the problem before. They pull up on the string, step on it, and then pull up on the string again, repeating the process until the meat is within reach.

Both Genetics and Learning Influence Most Behaviors

Much of the behavior we have discussed so far has been presented as either innate or learned, but the behavior we observe in nature is a mixture of both. Bird songs present a good example. Many birds learn their songs as juveniles, when they hear their parents sing. If juvenile white-crowned sparrows are raised in isolation, their adult songs do not resemble the typical species-specific song (**Figure 55.5**). If they hear only the song of a different species, such as the song sparrow, they again sing a poorly developed adult song. However, if they hear the song of the white-crowned sparrow, they will learn to sing a fully developed white-crowned sparrow song. The birds are genetically programmed to learn, but they will sing the correct song only if the appropriate instructive program is in place to guide learning.

Another example of how innate behavior interacts with learning can occur during a limited time period of development, called a **critical period**. At this time, many animals develop irreversible species-specific behavior patterns. This process is called **imprinting**. One of the best examples of imprinting was demonstrated by the Austrian ethologist Konrad Lorenz in the 1930s. Lorenz noted that young birds of some species imprint

Figure 55.4   **Cognitive behavior involving problem-solving ability.**   This chimp has devised a solution to the problem of retrieving bananas that were initially out of its reach.

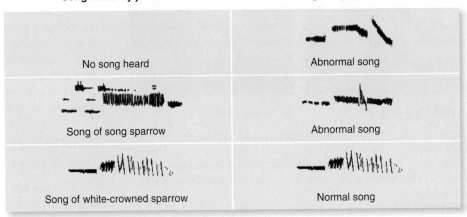

Song heard by juvenile   **Song sung by juvenile**

No song heard	Abnormal song
Song of song sparrow	Abnormal song
Song of white-crowned sparrow	Normal song

Figure 55.5 **The interaction between genetics and learning.**   The lines represent the different sound frequencies produced by the birds over a short time interval. The juvenile white-crowned sparrow will sing an abnormal song if it is kept in isolation or hears only the song of a different species. However, the juvenile will sing the normal white-crowned sparrow song if exposed to it.

Concept check: *Cuckoos lay their eggs in other birds' nests, so their young are reared by parent birds of a different species. However, unlike the white-crowned sparrow, adult cuckoos always sing their own distinctive song, not that of the host species they hear as juveniles. How is this possible?*

on their mother during a critical period that is usually within a few hours after hatching. This behavior serves them well, because in many species of ducks and geese, it would be hard for the mother to keep track of all her offspring as they walk or swim. After imprinting takes place, the offspring keep track of the mother.

The survival of the young ducks requires that they quickly learn to follow their mother's movements. Lorenz raised grey-lag geese from eggs, and soon after they hatched, he used himself as the model for imprinting. As a result, the young goslings

imprinted on Lorenz and followed him around (**Figure 55.6**). For the rest of their life, they preferred the company of Lorenz and other humans to geese. Studies have shown that even an object as foreign as a black box, watering can, or flashing light will be imprinted on if it is the first moving object the chick sees during the critical period. In nature, if young geese are not provided with any stimulus during this period, they will fail to imprint on anything, and without parental care, they will almost certainly die.

Other animals imprint in different ways. Newborn shrews imprint on the scent of their mother. Mothers also can imprint on their own young within a few hours. For example, if sheep mothers are kept apart from their offspring for only a few hours after birth, they will reject them. In these situations, the innate behavior is the ability to imprint soon after birth, and the factors in the environment are the stimulus to which the imprinting is directed.

Finally, innate behavior can interact with learning during animal migration. Inexperienced juvenile birds will migrate in a particular direction but will fail to correct for deviations if they are blown off course. Experienced adult birds, on the other hand, can often correct for storm-induced displacement, indicating they have more complex navigational skills. There are many complex behaviors involved in movement and migration, as we will explore in the next section.

Figure 55.6 **Konrad Lorenz being followed by his imprinted geese.**   Newborn geese follow the first object they see after hatching and later will follow that particular object only. They normally will follow their mother but can be induced to imprint on humans. The first thing these young geese saw after hatching was ethologist Konrad Lorenz.

55.2 Local Movement and Long-Range Migration

Organisms need to find their way, both locally and over what can be extremely long distances. Locally, organisms continually need to locate sources of food, water, mates, and perhaps nesting sites. Migration involves the longer-distance seasonal movement of animals, usually between overwintering areas and summer breeding sites; these are often hundreds or even thousands of kilometers apart. Several different types of behavior may be involved in these movements.

In this section, we begin by exploring local movement and, in particular, how one species uses landmarks to guide its movements. We then consider migration and examine the possible mechanisms used by migrating animals to find their way.

Local Movement Can Involve Kinesis, Taxis, and Memory

The simplest forms of movement are mere responses to stimuli. A **kinesis** is a movement in response to a stimulus, but one that is not directed toward or away from the source of the stimulus. A simple experiment often done in classrooms is to observe the activity levels of woodlice, sometimes called sow bugs or pill bugs, in dry areas and moist areas. The woodlice move faster in drier areas, and they slow down when they reach moist environments. This behavior tends to keep them in damper areas, which they prefer in order to avoid desiccation.

A **taxis** is a more directed type of response either toward or away from an external stimulus. Cockroaches exhibit negative phototaxis, meaning they tend to move away from light. Under low-light conditions, the photosynthetic unicellular flagellate *Euglena gracilis* shows positive phototaxis and moves toward a light source. Sea turtle hatchings are also strongly attracted to light. On emerging from their nests, they crawl toward the brightest location, traditionally the reflected moonlight on the ocean's surface. Lighted houses on the shore can disorient the hatchlings, however, and lead them to wander away from the ocean and succumb to dehydration, exhaustion, and predation. Male silk moths orient themselves in relation to wind direction (anemotaxis). If the air current carries the scent of a female moth, they will move upwind to locate it. Some freshwater fishes orient themselves to the currents of streams. Many fishes exhibit positive rheotaxis (from the Greek *rheos*, meaning current), in that they swim against the water current to prevent being washed downstream.

Sometimes memory and landmarks may be used to aid in local movements. Dutch-born ethologist Niko Tinbergen showed how the female digger wasp uses landmarks to relocate her nests, as described next.

FEATURE INVESTIGATION

Tinbergen's Experiments Show That Digger Wasps Use Landmarks to Find Their Nests

In the sandy, dry soils of Europe, the solitary female digger wasp (*Philanthus triangulum*) digs four to five nests in which to lay her eggs. Each nest stretches obliquely down into the ground for 40–80 cm. The wasp, also called a bee wolf, follows this by performing a sequence of apparently genetically programmed events. She catches and stings a honeybee, which paralyzes it; returns to the nest; drags the bee into the nest; and lays an egg on it. The egg hatches into a larva, which feeds on the paralyzed bee. However, the larva needs to ingest five to six bees before it is fully developed. This means the wasp must

catch and sting four to five more bees for each larva. She can carry only one bee at a time. After each visit, the wasp must seal the nest with soil, find a new bee, relocate the nest, open it, and add the bee. How does the wasp relocate the nest after spending considerable time away? Niko Tinbergen observed the wasps hover and fly around the nest each time they took off. He hypothesized that they were learning the nest position by creating a mental map of the landmarks in the area.

To test his hypothesis, Tinbergen experimentally adjusted the landmarks around the burrow that the wasps might be using as cues (Figure 55.7). First, he put a ring of pinecones around the nest entrance to train the wasp to associate the pinecones with the nest. Then, when the wasp was out hunting, he

Figure 55.7 How Niko Tinbergen discovered the digger wasp's nest-locating behavior.

Concept check: *How would you test what type of spatial landmarks are used by female digger wasps?*

HYPOTHESIS Digger wasps (*Philanthus triangulum*) use visual landmarks to locate their nests.

STARTING LOCATION The female digger wasp excavates an underground nest, to which she returns daily, bringing food to the larvae located inside.

Experimental level

1 Place a ring of pinecones around the nest to train the wasp to associate pinecones with the nest.

Pinecones

Digger wasp

2 After the wasp leaves the nest to hunt, move the pinecones 30 cm from the real nest. The wasp returns and flies to the center of the pinecone circle instead of the real nest. Repeated experiments yield similar results (see data), indicating that the wasp uses landmarks as visual cues.

Move pinecones 30 cm from the nest.

3 To test whether it is the shape or the smell of the pinecones that elicits the response, perform the same experiment as above, except use pinecones with no scent and add 2 small pieces of cardboard coated with pine oil.

Pine oil

Cardboard

4 After the wasp leaves the nest, move the pinecones 30 cm from the nest, but leave the scented cardboard at the nest. The wasps again fly to the pinecone nest (see data), indicating that it is the arrangement of cones, not their smell, that elicits the learning.

Move pinecones 30 cm from the nest.

5 THE DATA*

Results from steps 1 and 2:

Wasp #	Number of return visits per wasp to real nest without pinecones	Number of return visits per wasp to sham nest with pinecones
1–17	0	~9

Results from steps 3 and 4:

Wasp #	Number of return visits per wasp to real nest with scented cardboard	Number of return visits per wasp to sham nest with pinecones
18–22	0	~6

*Seventeen wasps, numbered 1–17, were studied as described in steps 1 and 2. Five wasps, numbered 18–22, were studied as described in steps 3 and 4.

6 CONCLUSION Digger wasps remember the positions of visual landmarks and use them as aids in local movements.

7 SOURCE Tinbergen, N. 1951. The study of instinct. Clarendon Press, Oxford.

moved the circle of pinecones a distance from the real nest and constructed a sham nest, making a slight depression in the sand and mimicking the covered entrance of the burrow. On returning, the wasp flew straight to the sham nest and tried to locate the entrance. Tinbergen chased it away. When it returned, it again flew to the sham nest. Tinbergen repeated this nine times, and every time the wasp chose the sham nest. Tinbergen got the same result with 16 other wasps, and not once did they choose the real nest.

Next Tinbergen experimented with the type of stimulus that might be eliciting the learning. He hypothesized that the wasps could be responding to the distinctive scent of the pinecones rather than their appearance. He trained the wasps by placing a circle of pinecones that had no scent and two small pieces of cardboard coated in pine oil around the real nest. He then moved the cones to surround a sham nest and left the scented cardboard around the real nest. The returning wasps again ignored the real nest with the scented cardboard and flew to the sham. He concluded that for the wasps, sight was apparently more important than smell in determining landmarks.

Experimental Questions

1. What observations were important for the development of Niko Tinbergen's hypothesis explaining how digger wasps located their nests?

2. How did Tinbergen test the hypothesis that the wasps were using landmarks to relocate the nest? What were the results?

3. Did the Tinbergen experiment rule out any other cue the wasps may have been using besides the sight of pinecones?

Migration Involves Long-Range Movement and More Complex Spatial Navigation

As well as learning to navigate over short-range distances, many animal species undergo **migration**, long-range seasonal movement. Migrations usually involve a movement away from a birth area to feed and a return to the birth area to breed, with the movement generally being linked to seasonal availability of food. For example, nearly half the birds of North America migrate to South America to escape the cold winters and feed, returning to North America in the spring to breed. Arctic terns that breed in Arctic Canada and Asia in summer migrate to the Antarctic to feed in the winter and then return to breed. This staggering journey involves up to a 40,000-km (25,000-mi) round-trip, most of it over the open ocean, during which the birds must stay airborne for days at a time!

Many mammals, including wildebeest and caribou, make migrations that track the appearance of new vegetation on which they feed. The monarch butterfly of North America migrates to overwinter in California, Mexico, and possibly south Florida and Cuba (**Figure 55.8**). An interesting point about the northward journey of the monarch is that it involves several generations of butterflies to complete. On their way back to the northern U.S. and Canada, the butterflies lay eggs and die. The caterpillars develop on milkweed plants, and the resultant adults continue to journey farther north. This cycle happens several times in the course of the return journey. The northward and southward migrations are unique in that none of the individuals has ever been to the destinations before; therefore, the ability to migrate must be an innate behavior.

How do migrating animals find their way? Three mechanisms may be involved: piloting, orientation, and navigation. In **piloting**, an animal moves from one familiar landmark to the next. For example, many whale species migrate between summer feeding areas and winter calving grounds. Gray whales migrate between the Bering Sea near Alaska to coastal areas of Mexico. Features of the coastline, including mountain ranges, and rivers, may aid in navigation. In **orientation**, animals have the ability to follow a compass bearing and travel in a straight line. **Navigation** involves the ability not only to follow a compass bearing but also to set or adjust it.

An experiment with starlings helps illuminate the difference between orientation and navigation (**Figure 55.9**). European starlings breed in Scandinavia and northeastern Europe and migrate in a southwest direction toward coastal France and southern England to spend the winter. Migrating starlings were captured and tagged in the Netherlands and then transported south to

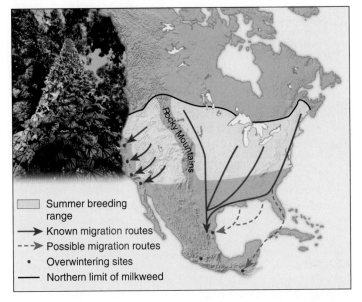

	Summer breeding range
→	Known migration routes
--→	Possible migration routes
•	Overwintering sites
—	Northern limit of milkweed

Figure 55.8 Monarch butterfly migration. Many monarch butterflies east of the Rocky Mountains migrate to a small area in Mexico to avoid the cold northern weather. Here they roost together in large numbers in fir trees (inset). Some butterflies may stay in Florida and Cuba. Butterflies west of the Rockies overwinter in mild coastal California locations.

Concept check: *Why is this an unusual example of migration?*

Switzerland and released. Juvenile birds, which had never made the trip before, flew southwest in their migration and were later recaptured in Spain. Adult birds, with more experience, returned to their normal wintering range by adjusting their course by approximately 90°. This implies that the adult birds can actually navigate, whereas the juveniles rely on orientation.

Many species use a combination of navigational reference points, including the position of the sun, the stars (for nighttime travel), and Earth's magnetic field. Homing pigeons have magnetite in their beaks that acts as a compass to tell direction (refer back to Section 43.4, Electromagnetic Sensing). Navigation by the sun or the stars also requires the use of a timing device to compensate for the ever-changing position of these reference points. Many migrants, therefore, possess the equivalent of an internal clock. Pigeons integrate their internal clock with the position of the sun. Researchers have altered the internal clock of pigeons by keeping them under artificial lights for certain periods of time. When the pigeons are released, they display predictable deviations in their flight. For every hour that their internal clock is shifted, the orientation of the birds shifts about 15°.

KEY
- Area of capture
- Breeding range
- Moved birds recovered
- Normal winter range

Scandinavia

Navigation: Mature birds corrected their course and flew to France and England.

Netherlands

Breeding range

Normal migration route

Area of capture

Normal winter range

Birds moved to Switzerland.

France

Switzerland

Spain

Moved birds recovered

Orientation: Young starlings did not fly northwest to their traditional grounds but continued on the same southwestern route, taking them to Spain.

Figure 55.9 Orientation versus navigation. Starlings normally migrate from breeding grounds in Scandinavia and northeastern Europe through the Netherlands and northern Germany to overwintering sites in France and England. This involves a southwest flight. When juveniles were captured in the Netherlands and moved to Switzerland, they continued on in a southwestern direction and ended up in Spain. When adult birds were captured and moved, they changed course and flew to the normal overwintering areas.

Not all examples of animal migration are well understood. Green sea turtles feed off the coast of Brazil yet swim east for 2,300 km (1,429 mi) to lay their eggs on Ascension Island, an 8-km-wide island in the center of the Atlantic Ocean between Brazil and Africa. It is not known why the turtles lay their eggs on this speck of an island or how they succeed in finding it. Perhaps fewer predators exist on Ascension than on other beaches. A combination of magnetic orientation and chemical cues may help them find it. Thus, while scientists have made many discoveries about animal navigation, much remains to be learned about how animals acquire a map sense.

To a large extent, local and long-distance movement involves searching for food. In the next section, we will investigate how such foraging decisions are made.

55.3 Foraging Behavior

Food gathering, or foraging, often involves decisions about whether to remain at a resource patch and look for more food or look for a completely new patch. The analysis of these decisions is often performed in terms of **optimality theory**, which predicts that an animal should behave in a way that maximizes the benefits of a behavior minus its costs. In this case, the benefits are the nutritional or caloric value of the food items, and the costs are the energetic or caloric costs of movement. When the difference between the energetic benefits of food gathering and the energetic costs of food gathering is maximized, an organism is said to be optimizing its foraging behavior. Optimality theory can also be used to investigate other behavioral issues such as how large a territory to defend. Too small a territory would contain insufficient resources, such as food and mates, and too large a territory would be too energetically costly to defend. Theoretically, then, there is an optimal territory size for a given individual.

Optimal Foraging Entails Maximizing the Benefits and Minimizing the Costs of Food Gathering

Optimal foraging proposes that in a given circumstance, an animal seeks to obtain the most energy possible with the least expenditure of energy. The underlying assumption of optimal foraging is that natural selection favors animals that are maximally efficient at propagating their genes and at performing all other functions that serve this purpose. In this model, the more net energy an individual gains in a limited time, the greater the reproductive success.

Shore crabs (*Carcinus maenas*) will eat many different-sized mussels but tend to feed preferentially on intermediate-sized mussels, which give them the highest rate of energy return (**Figure 55.10**). Very large mussels yield more energy,

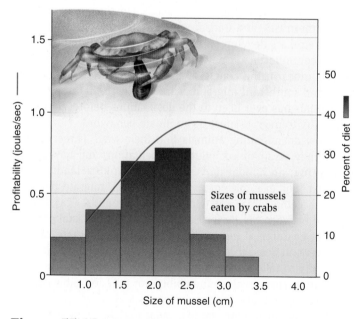

Figure 55.10 Optimal foraging behavior in shore crabs. In an aquarium setting, when offered a choice of equal numbers of each size mussel, shore crabs (*Carcinus maenas*) prefer intermediate-sized mussels that provide the highest rate of energy return. Profitability is the energy yield (joules) per second of time used in breaking open the shell.

(b) Cheetah

(c) Nesting gannets

Figure 55.11   Differing territory sizes among animals.   (a) The golden-winged sunbird of East Africa (*Nectarinia reichenowi*) has a medium territory size that is dependent on the number of flowers it can obtain resources from and defend. (b) Cheetahs (*Acinonyx jubatus*) hunt over large areas and can have extensive territories. This male is urine-marking part of his territory in the southern Serengeti, near Ndutu, Tanzania. (c) Nesting gannets (*Morus bassanus*) have much smaller territories, in which each bird is just beyond the pecking range of its neighbor.

(a) Golden-winged sunbird

but they take so long for the crab to open that they are actually less profitable, in terms of energy yield per unit time spent, than smaller sizes. Very small mussels are easy to crack open but contain so little flesh that they are not worth the effort. This leaves intermediate-sized mussels as the preferred size. Of course, the intermediate-sized mussels may take a longer time to locate, because more crabs are looking for them, so crabs eat some less profitable but more frequently encountered sizes of mussels. The result is that the diet consists of mussels in a range of sizes around the preferred optimal size.

In some cases, animals do not forage optimally. For example, animals seek not only to maximize food intake but also to minimize the risk of predation. Some species may only dart out to take food from time to time. The risk of predation thus has an influence on foraging behavior. Many animals also maintain territories to minimize competition with other individuals and control resources, whether food, mates, or nesting sites. As we will see, defending these territories also has an energetic cost.

Defending Territories Has Costs and Benefits

Many animals or groups of animals, such as a pride of lions, actively defend a **territory**, a fixed area in which an individual or group excludes other members of its own species, and sometimes other species, by aggressive behavior or territory marking. Optimality theory predicts that territory owners tend to optimize territory size according to the costs and benefits involved. The primary benefit of a territory is that it provides exclusive access to a particular resource whether it be food, mates, or sheltered nesting sites. Large territories provide more of a resource or resources but may be costly to defend, while small territories that are less costly to defend may not provide enough.

In studies of the territorial behavior of the golden-winged sunbird (*Nectarinia reichenowi*) in East Africa, researchers Frank Gill and Larry Wolf measured the energy content of nectar as the benefit of maintaining a territory and compared it

to the energy costs of activities such as perching, flying, and fighting (**Figure 55.11a**). Defending the territory ensured that other sunbirds did not take nectar from available flowers, thus increasing the amount of nectar in each flower. In defending a territory, the sunbird gained 780 calories a day in extra nectar content. However, the sunbird also spent 728 calories in defense of the territory, yielding a net gain of 52 calories a day and making territorial defense advantageous.

Territory size differs considerably among species. Because cheetahs need large areas to be able to hunt successfully, they establish large territories relative to their size (**Figure 55.11b**). In contrast, territories set up solely to defend areas for mating or nesting are often relatively small. For example, male sea lions defend small areas of beach. The preferred areas contain the largest amount of females and are controlled by the largest breeding bulls. The size of the territory of some nesting birds, such as gannets, is determined by how far the bird can reach to peck its neighbor without leaving its nest (**Figure 55.11c**).

Territories may be held for a season, a year, or the entire lifetime of the individual. Ownership of a territory needs to be periodically proclaimed; thus, communication between individuals is necessary for territory owners. This may involve various types of signaling, which we discuss next.

55.4 Communication

Communication is the use of specially designed signals or displays to modify the behavior of others. It may be used for many purposes, including defining territories, maintaining contact with offspring, courtship, and contests between males. The use of different forms of communication between organisms depends on the environment in which they live. For example, visual communication plays little role in the signals of nocturnal animals. Similarly, for animals in dense forests, sounds are of prime importance. Sound, however, is a temporary signal.

Scent can last longer and is often used to mark the large territories of some mammals. In this section, we outline the various types of communication—chemical, auditory, visual, and tactile—that occur among animals.

Chemical Communication Is Often Used to Mark Territories or Attract Mates

The chemical marking of territories is common among animals, especially among members of the canine and feline families (see Figure 55.11b). Scent trails are often used by social insects to recruit workers to help bring prey to the nest. Fire ants (genus *Solenopsis*) attack large, living prey, and many ants are needed to drag the prey back to the nest. The scout that finds the prey lays down a scent trail from the prey back to the nest. The scent excites other workers, which follow the trail to the prey. The scent marker is very volatile, and the trail effectively disappears in a few minutes to avoid mass confusion over old trails.

Animals frequently use chemicals to attract mates. Female moths attract males by powerful chemical attractants called **pheromones**. Male moths have receptors that can detect as little as a single molecule. Among social organisms, some individuals use pheromones to manipulate the behavior of others. For example, a queen bee releases pheromones that suppress the reproductive system of workers, which ensures that she is the only reproductive female in the hive.

Auditory Communication Is Often Used to Attract Mates and to Deter Competitors

Many organisms communicate by making sound. Because the Earth itself can absorb sound waves, sound travels farther in the air, which is why many birds and insects perch on branches or leaves when singing. Air is on average 14 times less turbulent at dawn and dusk than during the rest of the day, so sound carries farther then, which helps explain the preference of most animals for calling at these times. Some insects utilize the very plants on which they feed as a medium of song transmission. Many male leafhopper and planthopper insects vibrate their abdomens on leaves and create species-specific courtship songs that are transmitted by adjacent vegetation and are picked up by nearby females of the same species.

While many males use auditory communication to attract females, some females use calls to attract the attention of males. Female elephant seals use this behavior to their advantage. When a nondominant male attempts to mate with a female, she screams loudly, attracting the attention of the dominant male, which drives the nondominant male away. In this way, she is guaranteed a mating with the strongest male. Sound production can attract predators as well as mates. Some bats listen for the mating calls of male frogs to find their prey. Parasitic flies detect and locate chirping male crickets and then deposit larvae on or near them. The larvae latch onto and penetrate the cricket and eventually kill it. Sound may also be used by males during competition over females. In many animals, lower-pitched sounds come from larger males, so by calling to one another, males can gauge the size of their opponents and save energy that would be used in fighting.

Visual Communication Is Often Used in Courtship and Aggressive Displays

In courtship, animals use a vast number of visual signals to identify and select potential mates. Competition among males for the most impressive displays to attract females has led to elaborate coloration and extensive ornamentation in some species. For example, peacocks and males of many bird species have developed elaborate plumage to attract females.

Male fireflies have developed light flashes that are species specific with regard to number and duration of flashes (**Figure 55.12**). Females respond with a flash of their own. Such bright flashes are also bound to attract predators. Some female fireflies use mimicry to their advantage. Female *Photuris versicolor* fireflies mimic the flashing responses normally given by females of other species, such as *Photinus tanytoxus*, in order to lure the males of those species close enough to eat them.

Visual signals are also used to resolve disputes over territories or mates. Deer and antelope have antlers or horns that

(a) Firefly flashing

(b) Multiple firefly flashes

(c) Use of mimicry to lure prey

Figure 55.12 **Visual communication in fireflies.** (a) Communication between fireflies is conducted by species-specific light flashes emitted by organs located on the underside of the abdomen. (b) At dusk, many different light flashes can be seen. (c) Sometimes, females, such as this large *Photuris versicolor*, mimic the displays of other species, luring an unwitting male, such as this *Photinus tanytoxus*, and then eating him.

(a) Bees clustering around a recently returned worker, shown on the right

(b) Round dance

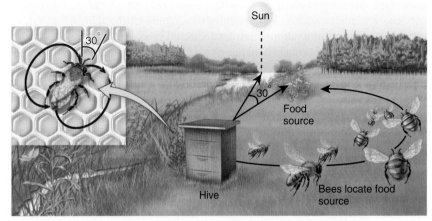

(c) Waggle dance: The angle of the waggle to the vertical orientation of the honeycomb corresponds to the angle of the food source from the sun.

Figure 55.13 Tactile communication among honeybees regarding food sources. (a) Bees gather around a newly returned scout to receive information about nearby food sources. (b) If the food is less than 50 m away, the scout performs a round dance. (c) If the food is more than 50 m away, the scout performs a waggle dance, which conveys information about its location. If the dance is performed at a 30° angle to the right of the hive's vertical plane, then the food source is located at a 30° angle to the right of the sun.

they use to display and spar over territory and females. Most of these matches never develop into outright fights, because the males gauge their opponent's strength by the size of these ornaments. Among insects, the "horns" of rhinoceros beetles and the eye stems of stalk-eyed flies send similar signals.

Tactile Communication Is Used to Strengthen Social Bonds and to Convey Information About Food

Animals often use tactile communication to establish bonds between group members. Primates frequently groom one another, and canines and felines may nuzzle and lick each other. Many insects use tactile communication to convey information on the whereabouts of food. Members of the ant genus *Leptothorax* feed on immobile prey such as dead insects. When a scouting ant encounters such prey, it usually needs an additional worker to help bring it back to the nest. Rather than laying a scent trail, which is energetically costly, the scout ant recruits a helper and physically leads it to the food source. The helper runs in tandem with the scout, its antennae touching the scout's abdomen.

Perhaps the most fascinating example of tactile communication among animals is the dance of the honeybee, elegantly studied by German ethologist Karl von Frisch in the 1940s. Bees commonly live in large hives; in the case of the European honeybee (*Apis mellifera*), the hive consists of 30,000–40,000 individuals. The flowering plants on which the bees forage can be located miles from the hive and are distributed in a patchy manner, with any given patch usually containing many flowers that store more nectar and pollen than an individual bee can carry back to the nest. The scout bee that locates the resource patch returns to the hive and recruits more workers to join it (**Figure 55.13a**). Because it is dark inside the hive, the bee uses a tactile signal. The scout dances on the vertical side of a honeycomb, and the dance is monitored by other bees, which follow

and touch her to interpret the message. If the food is relatively close to the hive, less than 50 m away, the scout performs a round dance, rapidly moving in a circle, first in one direction and then the other. The other bees know the food is relatively close at hand, and the smell of the scout tells them what flower species to look for (**Figure 55.13b**).

If the food is more than 50 m away, the scout will perform a different type of dance, called a "waggle dance." In this dance, the scout traces a figure 8, in the middle of which she waggles her abdomen and produces bursts of sound. Again, the other bees maintain contact with her. Occasionally, the scout will regurgitate a small sample of nectar so the bees know the type of food source they are looking for. The truly amazing part of the waggle dance is that the angle at which the central part of the figure 8 deviates from the vertical direction of the comb represents the same angle at which the food source deviates from the point at which the sun hits the horizon (**Figure 55.13c**). The direction is always up-to-date, because the bee adjusts the dance as the sun moves across the sky.

As we have seen, much communication occurs not only to defend territories but also to communicate information to other individuals in the population, including potential mates. Although living on your own and maintaining a territory has advantages, living in a group also has its benefits, including ready availability of mates and increased protection from predators. In the following section, we examine group living and the behavior it engenders.

55.5 Living in Groups

As we have seen, much of animal behavior is directed at other animals. Some of the more complex behavior occurs when animals live together in groups such as flocks or herds. If a central concern of ecology is to explain the distribution patterns of organisms, then one of our most important tasks is to understand why there is such variation in the degree of sociality. Although congregations promote competition for food and increased disease transmission, there are benefits of group living that compensate for the costs involved. Many of these benefits relate to locating food sources, assistance in rearing offspring, mate access, and group defense against predators. Group living can reduce predator success in at least two ways: through increased vigilance and through protection in numbers.

Living in Large Groups May Reduce the Risk of Predation Because of Increased Vigilance

For many predators, success depends on surprise. If an individual is alerted to an attack, the predator's chance of success is lowered. A woodpigeon (*Columba palumbus*) in a flock will take to the air when it spots a goshawk (*Accipiter gentilis*). Once one pigeon takes flight, the other members of the flock are alerted and follow suit. If each individual in a group occasionally scans the environment for predators, the larger the group, the less time an individual forager needs to devote to vigilance and the more time it can spend feeding. This is referred to as the **many-eyes hypothesis** (Figure 55.14). Of course, cheating is a possibility, because some birds might never look up, relying on others to keep watch while they keep feeding. However, the individual that happens to be scanning when a predator approaches is most likely to escape, a fact that tends to discourage cheating.

Figure 55.14 Living in groups and the many-eyes hypothesis. The larger the number of woodpigeons, the less likely an attack will be successful.

Concept check: What other advantages are there to large groups of individuals when being attacked by a predator?

Living in Groups Offers Protection by the "Selfish Herd"

Group living also provides protection in sheer numbers. Typically, predators take one prey animal per attack. In any given attack, an individual antelope in a herd of 100 has a 1 in 100 chance of being selected, whereas a single individual has a 1 in 1 chance. Large herds may be attacked more frequently than a solitary individual, but a herd is unlikely to attract 100 times more attacks than an individual, often because of the territorial nature of predators. Furthermore, large numbers of prey are able to defend themselves better than single individuals, which usually choose to flee. For example, groups of nesting black-headed gulls will mob a crow relentlessly, thereby reducing the crow's ability to steal the gulls' eggs.

Research has shown that within a group, each individual can minimize the danger to itself by choosing the location that is as close to the center of the group as possible. This was the subject of a famous paper, "The Geometry of the Selfish Herd," by the British evolutionary biologist W. D. Hamilton. The explanation of this type of defense is that predators are likely to attack prey on the periphery because they are easier to isolate visually. Many animals in herds tend to bunch close together when they are under attack, making it physically difficult for the predator to get to the center of the herd.

Overall, group size may be the result of a trade-off between the costs and benefits of group living. Although much group behavior serves to reduce predation, other complex behavior occurs in groups, including grooming behavior and behavior that appears to benefit the group at the expense of the individual. For example, a honeybee will sting a potential hive predator to discourage it. The bee's stinger is barbed, and once it has penetrated the predator's skin, the bee cannot withdraw it. The bee's only means of escape is to tear away part of its abdomen, leaving the stinger behind and dying in the process. In the next section, we explore the reasons for altruistic behavior, in which an individual incurs costs to itself for the benefit of others.

55.6 Altruism

In Chapter 23, we learned that a primary goal of an organism is to pass on its genes, yet we see many instances in which some individuals forego reproducing altogether, apparently to benefit the group. How do ecologists explain **altruism**, a behavior that appears to benefit others at a cost to oneself? In this section, we begin by discussing whether such behavior evolved for the good of the group or for the good of the individual. As we will see, most altruistic acts serve to benefit the individual's close relatives. We explore the concept of kin selection, which argues that acts of self-sacrifice indirectly promote the spread of an organism's genes, and see how this plays out in an extreme form in the genetics of social insect colonies. We conclude by examining reciprocal altruism as an attempt to explain the evolution of altruism among nonkin.

In Nature, Individual Selfish Behavior Is More Likely Than Altruism

One of the first attempts to explain the existence of altruism was called **group selection**, the premise that natural selection produces outcomes beneficial for the whole group or species. In 1962, the British ecologist V. C. Wynne-Edwards argued that a group containing altruists, each willing to subordinate its interests for the good of the group, would have a survival advantage over a group composed of selfish individuals. In concept, the idea of group selection seemed straightforward and logical: a group that consisted of selfish individuals would overexploit its resources and die out, while the fitness of a group with altruists would be enhanced.

In the late 1960s, the idea of group selection came under severe attack. Leading the charge was the biologist G. C. Williams, who argued that evolution acts through **individual selection**, which proposes that adaptive traits generally are selected for because they benefit the survival and reproduction of the individual rather than the group. Williams' arguments against group selection follow.

Mutation Mutant individuals that readily use resources for themselves or their offspring will have an advantage in a population where individuals limit their resource use. Consider a species of bird in which a pair lays only two eggs; that is, it has a clutch size of two, and the resources are not overexploited for the good of the group. Two eggs would ensure a replacement of the parent birds but would prevent a population explosion. Imagine a mutant bird arises that lays three eggs. If the population is not overexploiting its resources, sufficient food may be available for all three young to survive. If this happens, the three-egg genotype will eventually become more common than the two-egg genotype.

Immigration Even in a population in which all pairs laid two eggs and no mutations occurred to increase clutch size, selfish individuals that laid more could still immigrate from other areas. In nature, populations are rarely sufficiently isolated to prevent immigration of selfish mutants from other populations.

Individual Selection For group selection to work, some groups must die out faster than others. In practice, groups do not become extinct very frequently. Individuals die off more frequently than groups, so individual selection will be the more powerful evolutionary force.

Resource Prediction Group selection assumes that individuals are able to assess and predict future food availability and population density within their own habitat. There is little evidence that they can. For example, it is difficult to imagine that songbirds would be able to predict the future supply of the caterpillars that they feed to their young and adjust their clutch size accordingly.

Most ecologists accept individual gain as a more plausible result of natural selection than group selection. Population size

Figure 55.15 Infanticide as selfish behavior. Male Hanuman langurs (*Semnopithecus entellus*) can act aggressively toward the young of another male, even killing them, hastening the day the females come into estrus and the time when the males can father their own offspring. Note that the mother is running with the infant.

is more often controlled by competition in which individuals strive to command as much of a resource as they can. Such selfishness can cause some seemingly surprising behaviors. For example, male Hanuman langurs (*Semnopithecus entellus*) kill infants when they take over groups of females from other males (**Figure 55.15**). The reason for the behavior is that when they are not nursing their young, females become sexually receptive much sooner, hastening the day when the male can father his own offspring. Infanticide ensures that the male will father more offspring, and the genes governing this tendency spread by natural selection.

If individual selfishness is more common than group selection, how do we account for what appear to be examples of altruism in nature?

Apparent Altruistic Behavior in Nature Is Often Associated with Kin Selection

Because all offspring have copies of their parents' genes, parents taking care of their young are actually caring for copies of their own genes. Genes for altruism toward one's young are favored by natural selection and will become more numerous in the next generation, because offspring have copies of those same genes.

The probability that any two individuals will share a copy of a particular gene is a quantity, *r*, called the **coefficient of relatedness**. During meiosis in a diploid species, any given copy of a gene has a 50% chance of segregating into an egg or sperm. A mother and father are on average related to their children by an amount $r = 0.5$, because half of a child's genes come from its mother and half from its father. By similar reasoning, brothers or sisters are related by an amount $r = 0.5$ (they share half their mother's genes and half their father's); grandchildren and grandparents, by 0.25; and cousins, by 0.125 (**Figure 55.16**). In 1964, ecologist W. D. Hamilton realized the

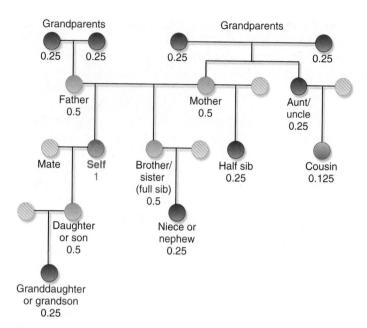

Figure 55.16   Degree of genetic relatedness to self in a diploid organism.   Pink hatched circles represent completely unrelated individuals.

Concept check: In theory, should you sacrifice your life to save two sisters or nine cousins?

implication of the coefficient of relatedness for the evolution of altruism. An organism not only can pass on its genes through having offspring, but also can pass them on through ensuring the survival of siblings, nieces, nephews, and cousins. This means an organism has a vested interest in protecting its brothers and sisters, and even their offspring.

The term **inclusive fitness** is used to designate the total number of copies of genes passed on through one's relatives, as well as one's own reproductive output. Selection for behavior that lowers an individual's own fitness but enhances the reproductive success of a relative is known as **kin selection**. Hamilton proposed that an altruistic gene will be favored by natural selection when

$$rB > C$$

where r is the coefficient of relatedness of donor (the altruist) to the recipient, B is the benefit received by the recipient of the altruism, and C is the cost incurred by the donor. This is known as **Hamilton's rule**.

Imagine two sisters who are not yet mothers. One has a rare kidney disease and needs a transplant from her sister. Let's assume both sisters will have two children of their own. The risk of the transplant to the donor involves a 1% chance of dying, but the benefit to the recipient involves a 90% chance of living and having children. In this example, $r = 0.5$, $B = 0.9 \times 2 = 1.8$, and $C = 0.01 \times 2 = 0.02$. Because the genetic benefit (rB) of 0.9 is much greater than the genetic cost (C) of 0.02, it makes evolutionary sense to proceed with the transplant. While humans are unlikely to do this type of calculation before deciding whether to risk their lives to save their siblings from a

life-threatening event, this example shows how such behavior could arise and spread in nature.

Let's examine a situation involving altruism within a group of animals. Many insect larvae, especially caterpillars, are soft-bodied creatures. They rely on possessing a bad taste or toxin to deter predators and advertise this condition with bright warning colors. For example, noxious *Datana ministra* caterpillars, which feed on oaks and other trees, have bright red and yellow stripes and adopt a specific posture with head and tail ends upturned when threatened (**Figure 55.17**). Unless it is born with an innate avoidance of this prey type, a predator has to kill and eat one of the caterpillars in order to learn to avoid similar individuals in the future. It is of no personal use to the unlucky caterpillar to be killed. However, animals with warning colors often aggregate in kin groups because they hatch from the same egg mass. In this case, the death of one individual is likely to benefit its siblings, which are less likely to be attacked in the future, and thus its genes will be preserved. This explains why the genes for bright color and a warning posture are successfully passed on from generation to generation. In a case where $r = 0.5$, B might be 50, and $C = 1$, the benefit of 25 is greater than 1, so the genes for this behavior will be favored by natural selection.

A common example of altruism in social animals occurs when a sentry raises an alarm call in the presence of a predator. This behavior has been observed in Belding's ground squirrels (*Spermophilus beldingi*). The squirrels feed in groups, with certain individuals acting as sentries and watching for predators. As a predator approaches, the sentry typically gives an alarm call, and the group members retreat into their burrows. Similar behavior occurs in prairie dogs (*Cynomys* spp.) (**Figure 55.18**). In drawing attention to itself, the caller is at a higher risk of being attacked by the predator. However, in many groups, those closest to the sentry are most likely to be offspring or brothers or sisters; thus, the altruistic act of alarm calling is reasoned to be favored by kin selection. Supporting this is the observation that most alarm calling is done by females, because they are more likely to stay in the colony where they were born

Figure 55.17   Altruistic behavior or kin selection?   *Datana ministra* caterpillars exhibit a bright, striped warning pattern to advertise their bad taste to predators.

Concept check: Why do these caterpillars congregate in clusters?

Figure 55.18 Alarm calling, a possible example of kin selection. This prairie dog sentry is emitting an alarm call to warn other individuals, which are often close kin, of the presence of a predator. It is believed that by doing so, the sentry draws attention away from the others but becomes an easier target itself.

and have kin nearby, whereas the males are more apt to disperse far from the colony.

Altruism in Social Insects Arises Partly from Genetics and Partly from Lifestyle

Perhaps the most extreme form of altruism is the evolution of sterile castes in social insects, in which the vast majority of females, known as workers, rarely reproduce themselves but instead help one reproductive female (the queen) to raise offspring, a phenomenon called **eusociality**. The explanation of eusociality lies partly in the particular genetics of most social insect reproduction. Females develop from fertilized eggs and are diploid, the product of fertilization of an egg by a sperm. Males develop from unfertilized eggs and are haploid.

Such a system of sex determination is called the **haplodiploid system** (refer back to Figure 16.14). If they have the same parents, each daughter receives an identical set of genes from her haploid father. The other half of a female's genes comes from her diploid mother, so the coefficient of relatedness (*r*) of sisters is 0.50 (from father) + 0.25 (from mother) = 0.75. The result is that females are more related to their sisters (0.75) than they would be to their own offspring (0.50). This suggests it is evolutionarily advantageous for females to stay in the nest or hive and care for other female offspring of the queen, which are their full sisters.

Elegant though these types of explanations are, they do not provide the whole picture. Large eusocial colonies of termites exist, but termites are diploid, not haplodiploid. In this case, how do we account for the existence of eusociality?

In the 1970s, Richard Alexander suggested it was the particular lifestyle of these animals, rather than genetics, that promoted eusociality. He argued that in a normal diploid organism, females are related to their daughters by 0.50 and to their sisters by 0.50, so it should matter little to them whether they rear siblings or daughters of their own. He predicted, well before eusociality was discovered to occur in mammals, that a eusocial mammalian species could exist when certain conditions were met, including that the nests or burrows be enclosed and subterranean, in order to house a large colony, and that the colony have a food supply such as large tubers and roots. In addition, the soil would need to be hard, dry clay to keep the colony safe from digging predators. He proposed that the colony would be defended by a few members of the colony willing to give their lives in defense of others, and he posited the existence of mechanisms by which a queen could manipulate other individuals.

At the time, Alexander had no idea that a mammal with such characteristics existed. Surprisingly, subsequent discoveries confirmed the existence of a eusocial mammal that satisfied all of the predictions of Alexander's model: the naked mole rat (*Heterocephalus glaber*). Naked mole rats are diploid species that live in arid areas of Africa in large underground colonies where only one female, the queen, produces offspring (Figure 55.19). A renewable food supply is present in the form of tubers of the plant *Pyrenacantha kaurabassana*. These weigh up to 50 kg and can provide food for a whole colony, though the food would be insufficient if all the mole rats reproduced. Because the burrows are hard packed, there are few ways to attack them, and a heroic effort by a mole rat blocking the entrance can effectively stop a predator (commonly a rufous-beaked snake). The queen mole rat does indeed manipulate the colony members; she suppresses reproduction in other females by producing a pheromone in her urine that is passed around the colony by grooming. Hence, the mole rats seem to have evolved the appropriate behavior to exploit this ecological niche. As Alexander argued, lifestyle characteristics can provide an explanation for the evolution of eusociality in species such as termites and naked mole rats, in which both sexes are diploid.

Figure 55.19 A naked mole rat colony (*Heterocephalus glaber*). In this mammalian species, most females do not reproduce; only the queen (shown resting on workers) has offspring.

Unrelated Individuals May Engage in Altruistic Acts If the Altruism Is Likely to Be Reciprocated

Even though we have argued that kin selection can explain instances of apparent altruism, cases of altruism are known to exist between unrelated individuals. What drives this type of behavior appears to be a "You scratch my back, I'll scratch yours" type of reciprocal altruism, in which the cost to the animal of behaving altruistically is offset by the likelihood of a return benefit. This occurs in nature, for example, when unrelated chimps groom each other.

Researcher Gerald Wilkinson has noted that female vampire bats exhibit reciprocal altruism via food sharing. Vampire bats can die after 60 hours without a blood meal, because they can no longer maintain their correct body temperature. Adult females will share their food with their young, the young of other females, and other unrelated females that have not fed. The females and their dependent young roost together in groups of 8 to 12. A hungry female will solicit food from another female by approaching and grooming her. The female being groomed then regurgitates part of her blood meal for the other. The roles of blood donor and recipient are often reversed, and Wilkinson showed that unrelated females are more likely to share with those that had recently shared with them. The probability of a female getting a free lunch is decreased because the roost consists of individuals that remain associated with each other for long periods of time.

55.7 Mating Systems

In many species, the majority of males seem superfluous because one male mates with many females in a local area. If one male can mate with many females, why in most species does the sex ratio remain at approximately 1 to 1? The answer lies with natural selection. Let's consider a hypothetical population that contains 10 females to every male; each male mates, on average, with 10 females. A parent whose children were exclusively sons could expect to have 10 times the number of grandchildren compared to a parent with the same number of daughters. Under such conditions, natural selection would favor the spread of genes for male-producing tendencies, and males would become prevalent in the population. If the population were mainly males, females would be at a premium, and natural selection would favor the spread of genes for female-producing tendencies. Such constraints operate on the numbers of both male and female offspring, keeping the sex ratio at about 1:1. This idea was developed in 1930 by the geneticist Ronald Fisher and has come to be known as Fisher's principle.

Even though the sex ratio is fairly even in most species, that doesn't mean that one female always mates with one male, or vice versa. In many species, mating is **promiscuous**, with each female mating with a different male every year or breeding season. In other species, more lasting pair bonds develop between individuals. In this section, we examine the characteristics of different types of mating systems that occur among

Figure 55.20 Female choice of males based on nuptial gifts. A male hangingfly, on the left, has presented a nuptial gift, a small moth, to a female, and now mates with her while she consumes the meal.

animals. Then we explore the role of sexual selection, a type of natural selection in which competition for mates drives the evolution of certain traits.

Sexual Selection Involves Mate Choice and Mate Competition

As we learned in Chapter 24, **sexual selection** promotes traits that will improve an organism's mating success. You will recall that sexual selection can take two forms. In intersexual selection, members of one sex, usually females, choose mates based on particular characteristics, such as the color of plumage or the sound of a courtship song. In intrasexual selection, members of one sex, usually males, compete over partners, and the winner performs most of the matings (refer back to Figure 24.7). Let's explore each of these in a little more detail.

Intersexual Selection Females have many different ways to choose their prospective mates. Female hangingflies (genus *Hylobittacus*) demand a nuptial gift of a food package, an insect prey item that the male has caught (**Figure 55.20**). Such a nutrient-rich gift may permit females to produce more eggs. The bigger the gift, the longer it takes the female to eat it and the longer the male can copulate with her. Females will not mate with males that do not offer such a package. Female spiders and mantids will sometimes eat their mate during or after copulation, with the male's body constituting the ultimate nuptial gift.

Males may also have parenting skills that females desire. Among 15-spined sticklebacks (*Spinachia spinachia*), males perform cleaning, guarding, and fanning the offspring. Males display their parental skills through body shakes during courtship, and females prefer to mate with males that shake their bodies the most energetically, apparently using this cue to assess the quality of the male as a potential father.

Often, females choose mates without the offering of obvious material benefits and make their choices based on plumage color or courtship display. The male African long-tailed widowbird (*Euplectes progne*) has long tail feathers that he displays to females via aerial flights. Researcher Malte Andersson

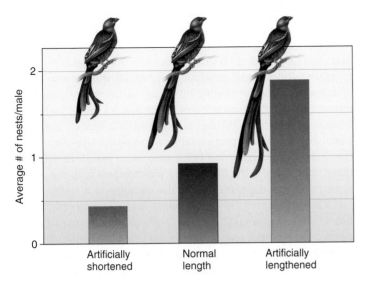

Figure 55.21 Female choice based on male appearance. Males with artificially lengthened tails mate with more females and, therefore, have more nests than males with a normal or artificially shortened tail.

Concept check: Why do you think it is rarer for female birds or mammals to have more colorful plumage or elaborate adornments than males of the same species?

experimentally shortened the tails of some birds by clipping their tail feathers, and he lengthened the tails of others by taking the clippings and sticking them onto other birds with superglue. Males with experimentally lengthened tails attracted four times as many females as males with shortened tails, and they fathered more clutches of eggs (**Figure 55.21**).

Some researchers have suggested ornaments such as excessively long tail feathers function as a sign of an individual's genetic quality, in that the bearer must be very healthy in order to afford this energetically costly trait. This hypothesis is called the handicap principle. However, in some species of birds,

other important benefits may be associated with plumage quality. Bright colors are often caused by pigments called carotenoids that help stimulate the immune system to fight diseases. In zebra finches and red jungle fowl, colorful plumage has been associated with heightened resistance to disease, suggesting that females that choose such males are choosing genetically healthier mates.

On rare occasions, a sex role reversal occurs, and the male discriminates among females. In the Mormon cricket (*Anabrus simplex*), males mate only once because they provide a nutrient-rich nuptial gift of a spermatophore to females, which is energetically costly to produce. Here, males choose heavier females to mate with because these females have more eggs, and the males can father more offspring.

Intrasexual Selection In many species, females do not actively choose their preferred mate; instead, they mate with competitively superior males. In such cases, dominance is determined by fighting or by ritualized sparring. Outcomes may be dictated by the size of weapons, such as antlers or horns, or by body size. In the southern elephant seal (*Mirounga leonina*), females haul up onto the beach to give birth and gain safe haven for their pups from marine predators. Following birth, they are ready to mate. In this situation, dominant males are able to command a substantial group of females and constantly lumber across the beach to fight other males and defend their harem. Over the course of many generations, such competition results in an increased body size. Thus, in species with male–male competition, males are often substantially larger than females (**Figure 55.22**).

Large body size does not always guarantee paternity. Smaller male elephant seals may intercept females in the ocean and attempt to mate with them there, rather than on the beach,

(a) Relationship between body size and mating success

(b) Male competition for mates

Figure 55.22 Large male size and mating success. (a) In elephant seals, the larger the male, in relation to female size, the greater the number of females that can be mated with and monopolized. (b) These male elephant seals are fighting to maintain control of their female harems.

Concept check: During fights between males, some pups are crushed as the males lumber across the beach. Why aren't the males careful to avoid the pups?

where the competitively dominant males patrol. Such "satellite" males, which are unable to acquire and defend territories, move around the edge of the mating arena. For example, small male frogs hang around ponds waiting to intercept females headed toward the call of dominant males. Thus, even though competitively dominant males father most offspring, smaller males can have reproductive success.

In Monogamous Mating Systems, Males and Females Are Paired for at Least One Reproductive Season

Mating success is also dependent on the type of mating system involved. In **monogamy**, each individual mates exclusively with one partner over at least a single breeding cycle and sometimes for longer. Males and females do not exhibit much **sexual dimorphism**, a pronounced difference in the morphologies of the two sexes within a species, and are generally similar in body size and structure (**Figure 55.23a**). Several hypotheses explain the existence of monogamy. The first is the **mate-guarding hypothesis**, which suggests that males stay with a female to protect her from being fertilized by other males. Such a strategy may be advantageous when receptive females are widely scattered and difficult to find.

The **male-assistance hypothesis** maintains that males remain with females to help them rear their offspring. Monogamy is common among birds, about 70% of which are socially monogamous; that is, the pairings remain intact during at least one breeding season. According to the male-assistance hypothesis, monogamy is prevalent in birds because eggs and chicks take a considerable amount of parental care. Most eggs need to be incubated continuously if they are to hatch, and chicks require almost continual feeding. It is therefore in the male's best interest to help raise his young, because he would have few surviving offspring if he did not.

The **female-enforced monogamy hypothesis** suggests that females stop their male partners from being polygynous. Male

and female burying beetles (*Nicrophorus defodiens*) work together to bury small, dead animals, which will provide a food resource for their developing offspring. Males will release pheromones to attract other females to the site. However, while an additional female might increase the male's fitness, the additional developing offspring might compete with the offspring of the first female, decreasing her fitness. As a result, on smelling these pheromones, the first female will interfere with the male's attempts at signaling, preserving the monogamous relationship.

Recent research by Larry Young and Elizabeth Hammock has shown that social behavior such as fidelity may have a genetic basis. These researchers found that fidelity of male voles depends on the length of microsatellites, short repeating genetic sequences, in a gene that codes for a key hormone receptor. Adult male voles with the long version of the microsatellite were more apt to form pair bonds with female partners and nurture their offspring than were voles with the short version.

In Polygynous Mating Systems, One Male Mates with Many Females

In **polygyny** (Greek, meaning many females), one male mates with more than one female in a single breeding season. Physiological constraints often dictate that female organisms must care for the young. Because of these constraints, at least in many organisms with internal fertilization, such as mammals and some fishes, males are able to desert and mate with more females. Polygynous systems are therefore associated with uniparental care of young, with males contributing little. Sexual dimorphism is typical in polygynous mating systems, particularly when males engage in competition over mates (**Figure 55.23b**). Sexual maturity is often delayed in males that fight because of the considerable time it takes to reach a sufficiently large size to compete for females.

Polygyny is influenced by the spatial or temporal distribution of breeding females. In cases where all females are sexually receptive within the same narrow period of time, little

(a) Monogamous species

(b) Polygynous species

(c) Polyandrous species

Figure 55.23 **Sexual dimorphism in body size and mating system.** **(a)** In monogamous species, such as these Manchurian cranes, *Grus japonensis*, males and females do not exhibit pronounced sexual dimorphism and appear very similar. **(b)** In polygynous species, such as white-tailed deer, *Odocoileus virginianus*, males are bigger than females and have large horns with which they engage in combat over females. **(c)** In polyandrous species, females are usually bigger, as with these golden silk spiders, *Nephila clavipes*.

opportunity exists for a male to garner all the females for himself. Where female reproductive receptivity is spread out over weeks or months, there is much more opportunity for males to mate with more than one female. For example, females of the common toad (*Bufo bufo*) all lay their eggs within a week, and males generally have time to mate with only one female. In contrast, female bullfrogs (*Rana catesbeiana*) have a breeding season of several weeks, and males may mate with as many as six females in a season.

Resource-Based Polygyny Where some critical resource is patchily distributed and in short supply, certain males may dominate the resource and breed with more than one visiting female. In the lark bunting (*Calamospiza melanocorys*), which mates in North American grasslands, males arrive at the grasslands first, compete for territories, and then display with special courtship flight patterns and songs to attract females. The major source of nestling death in this species is overheating from too much exposure to the sun. Prime territories are therefore those with abundant shade, and some males with shaded territories attract two females, even though the second female can expect no help from the male in the process of rearing young. Males in some exposed territories remain bachelors for the season. From the dominant male's point of view, resource-based polygyny is advantageous; from the female's point of view, there may be costs. Although by choosing dominant males, a female may be gaining access to good resources, she will have to share these resources with other females.

Harem Mating Structures Sometimes males defend a group of females without commanding a resource-based territory. This pattern is more common when females naturally congregate in groups or herds, perhaps to avoid predation, as with southern elephant seals (see Figure 55.22). Usually the largest and strongest males command most of the matings, but being a harem master is usually so exhausting that males may only manage to remain the dominant male for a year or two.

Communal Courting Polygynous mating can occur where neither resources nor harems are defended. In some instances, particularly in birds and mammals, males display in designated communal courting areas called **leks** (Figure 55.24). Females come to these areas specifically to find a mate, and they choose a prospective mate after the males have performed elaborate displays. Most females seek to mate with the best male, so a few of the flashiest males perform the vast majority of the matings. At a lek of the white-bearded manakin (*Manacus manacus*) of South America, one male accounted for 75% of the 438 matings where there were as many as 10 males. A second male mated 56 times (13% of matings), while six others mated only a total of 10 times.

In Polyandrous Mating Systems, One Female Mates with Many Males

In most systems in which one individual mates with more than one individual of the opposite sex, the polygamous sex is the

Figure 55.24 **Male birds at a lek.** Black grouse (*Tetrao tetrix*) congregate at a moorland lek in Scotland in April. Females visit the leks, and males display to them.

male. The opposite condition, **polyandry** (Greek, meaning many males), in which one female mates with several males, is more rare. Nevertheless, it occurs in some species of birds, fishes, and insects. Sexual dimorphism is present, with the females being the larger of the sexes (see Figure 55.23c). In the Arctic tundra, the summer season is short but very productive, providing a bonanza of insect food for 2 months. The productivity of the breeding grounds of the spotted sandpiper (*Actitis macularia*) is so high that the female becomes rather like an egg factory, laying up to five clutches of four eggs each in 40 days. Her reproductive success is limited not by food but by the number of males she can find to incubate the eggs, and females compete for males, defending territories where the males sit.

Polyandry is also seen in some species where egg predation is high and males are needed to guard the nests. For example, in the pipefish (*Syngnathus typhle*), males have brood pouches that provide eggs with safety and a supply of oxygen- and nutrient-rich water. Females produce enough eggs to fill the brood pouches of two males and may mate with more than one male.

Ultimately, as we have seen, most behaviors have evolved to maximize an individual's reproductive output. In a successful group of individuals, this leads to population growth. But we are not knee-deep in sandpipers or elephant seals, so there must be some constraints on reproductive output. In Chapter 56, we next turn to the realm of population ecology to explore how populations grow and what factors limit their growth.

Summary of Key Concepts

55.1 The Impact of Genetics and Learning on Behavior

- Behavior is usually due to the interaction of an organism's genes and the environment.

- Genetically programmed behaviors are termed innate and often involve a sign stimulus that initiates a fixed action pattern. (Figures 55.1, 55.2)

- Organisms can often make modifications to their behavior based on previous experience, a process called learning. Some forms of learning include habituation, classical conditioning, operant conditioning, and cognitive learning. (Figures 55.3, 55.4)

- Much behavior is a mixture of innate and learned behaviors. A good example of this occurs in a process called imprinting, in which animals develop strong attachments that influence subsequent behavior. (Figures 55.5, 55.6)

55.2 Local Movement and Long-Range Migration

- The simplest forms of local movement involve kinesis, taxis, and memory. (Figure 55.7)

- Many animals undergo long-range seasonal movement called migration in order to feed or breed. They do this using three proposed mechanisms: piloting, the ability to move from one landmark to the next; orientation, the ability to follow a compass bearing; and navigation, the ability to set, follow, and adjust a compass bearing. (Figures 55.8, 55.9)

55.3 Foraging Behavior

- Animals use complex behavior in food gathering or foraging. Optimality theory views foraging behavior as a compromise between the costs and benefits involved.

- The theory of optimal foraging assumes that animals modify their behavior to keep the ratio of their energy uptake to energy expenditure high. (Figure 55.10)

- The size of a territory, a fixed area in which an individual or group excludes other members of its own species, tends to be optimized according to the costs and benefits involved. (Figure 55.11)

55.4 Communication

- Communication is a form of behavior. The use of different forms of communication between organisms depends on the environment in which they live.

- Chemical communication often involves marking territories; auditory and visual forms of communication are often used to attract mates. A fascinating form of tactile communication involves the dance of the honeybee. (Figures 55.12, 55.13)

55.5 Living in Groups

- Many benefits of group living relate to defense against predators, offering protection through sheer numbers and through what is called the many-eyes hypothesis or the geometry of the selfish herd. (Figure 55.14)

55.6 Altruism

- Infanticide is an example of selfish behavior. (Figure 55.15)

- Altruism is behavior that benefits others at a cost to oneself. One of the first hypotheses to explain altruism, called group selection, suggested that natural selection produced outcomes beneficial for the group. Biologists now believe that most apparently altruistic acts are often associated with outcomes beneficial to those most closely related to the individual, a concept termed kin selection. (Figures 55.16, 55.17, 55.18)

- Altruism among eusocial animals may arise partly from the unique genetics of the animals and partly from lifestyle. (Figure 55.19)

- Altruism is known to exist among nonrelated individuals that live in close proximity for long periods of time.

55.7 Mating Systems

- Sexual selection takes two forms: intersexual selection, in which the female chooses a mate based on particular characteristics, or intrasexual selection, in which males compete with one another for the opportunity to mate with a female. (Figures 55.20, 55.21, 55.22)

- Several types of mating systems are found among animals, including monogamy, polygyny, and polyandry. (Figure 55.23)

- Polygynous mating can often occur in situations where males dominate a resource, defend groups of females (harems), or display in common courting areas called leks. (Figure 55.24)

Assess and Discuss

Test Yourself

1. What is the proximate cause of male deer fighting over females?
   a. to determine their supremacy over other males
   b. to injure other males so that these other males cannot mate with females
   c. to maximize the number of genes they pass on
   d. because changes in day length stimulate this behavior
   e. because fighting helps rid the herd of weaker individuals

2. Geotaxis is a response to the force of gravity. Fruit flies placed in a vial will move to the top of the vial. This is an example of _____ geotaxis.
   a. positive   c. innate   e. learned
   b. neutral   d. negative

3. Certain behaviors seem to have very little environmental influence. Such behaviors are the same in all individuals regardless of the environment and are referred to as _____ behaviors.
   a. genetically programmed   d. all of the above
   b. instinctual   e. b and c only
   c. innate

4. Patrick has decided to teach his puppy a few new tricks. Each time the puppy responds correctly to Patrick's command, the puppy is given a treat. This is an example of
   a. habituation.
   b. classical conditioning.
   c. operant conditioning.
   d. imprinting.
   e. orientation.

5. Whales have magnetite in their retinas, which aids in navigation during migration by
   a. piloting.
   b. locating the position of the sun.
   c. use of the Earth's magnetic fields.
   d. locating the positions of the stars.
   e. none of the above.

6. For group living to evolve, the benefits of living in a group must be greater than the costs of group living. Which of the following is an example of a benefit of living in a group?
   a. reduced spread of disease and/or parasites
   b. increased food availability
   c. reduced competition for mates
   d. decreased risk of predation
   e. all of the above

7. The modification of behavior based on prior experience is called
   a. a fixed action pattern.
   b. learning.
   c. navigation.
   d. adjustment behavior.
   e. innate.

8. When an individual behaves in a way that reduces its own fitness but increases the fitness of others, the organism is exhibiting
   a. kin selection.
   b. group selection.
   c. altruism.
   d. selfishness.
   e. ignorance.

9. In ants, which employ a haplodiploid mating system, fathers are related to sons by $r =$
   a. 0
   b. 0.125
   c. 0.25
   d. 0.5
   e. 0.75

10. In a polygynous mating system,
    a. one male mates with one female.
    b. one female mates with many different males.
    c. one male mates with many different females.
    d. many different females mate with many different males.

Conceptual Questions

1. Define ethology.

2. Why does male parental care occur in only 7% of fishes and amphibian families with internal fertilization but in 69% of families with external fertilization?

3. Describe the distinguishing features of the different mating systems.

Collaborative Questions

1. Whooping cranes (*Grus americana*) are an endangered species that are bred in captivity to increase their numbers. One problem is that these cranes are migratory. In the absence of other cranes, can you think of an innovative way human researchers might have used crane behavior to ensure their safe passage to overwintering sites?

2. Discuss several ways in which organisms communicate with each other.

Online Resource

www.brookerbiology.com

Stay a step ahead in your studies with animations that bring concepts to life and practice tests to assess your understanding. Your instructor may also recommend the interactive ebook, individualized learning tools, and more.

Population Ecology

56

A population of black-footed ferrets in Meeteetse, Wyoming.

The last known population of black-footed ferrets, *Mustela nigripes*, was discovered in 1981 near Meeteetse, Wyoming. Shortly thereafter, all but 18 of the 100 known ferrets in Meeteetse died of canine distemper. The remainder were captured between 1985 and 1987, inoculated against distemper, and bred in captivity. Seven females were used as genetic founders for the captive population, with the intent of reestablishing the population in the wild later on. Since then, populations have been established in Arizona, Colorado, Montana, South Dakota, Utah, Wyoming, and Chihuahua, Mexico.

In Wyoming, an area called Shirley Basin was one of those targeted for reintroductions of captive-born animals. During 1991 to 1994, Shirley Basin received 228 ferrets, but distemper again triggered a decline in the population size. By 1997, only 5 ferrets were found. Extinction seemed imminent. Monitoring efforts, which might disturb the animals, decreased. Surprisingly, by 2003, a total of 52 animals were found, and by 2006, 223 were present. How did the population increase this fast?

A **population** can be defined as a group of interbreeding individuals occupying the same area at the same time. In this way, we can think of a population of water lilies in a particular lake, the lion population in the Ngorongoro crater in Africa, or the human

population of New York City. The boundaries of a population can be a little difficult to define, though they may correspond to geographic features such as the boundaries of a lake or forest or be contained within a mountain valley or a certain island. Individuals may enter or leave a population, such as the human population of New York City or the deer population in North Carolina. Thus, populations are often fluid entities, with individuals moving into (immigrating) or out of (emigrating) an area.

This chapter explores **population ecology**, the study of what factors affect population size and how these factors change over space and time. To study populations, we need to employ some of the tools of **demography**, the study of birth rates, death rates, age distributions, and the sizes of populations. We begin our discussion by examining the ways that ecologists measure and categorize populations. We will explore characteristics of populations and how growth rates are determined by the number of reproductive individuals in the population and their fertility rate. These data are used to construct simple mathematical models that allow us to analyze and predict population growth. We will also look at the factors that limit the growth of populations and conclude the chapter by using the population concepts and models to explore the growth of human populations.

56.1 Understanding Populations

Within their areas of distribution, organisms occur in varying numbers. We recognize this pattern by saying a plant or animal is "rare" in one place and "common" in another. For more precision, ecologists quantify distribution further and talk in terms of population **density**, the numbers of organisms in a given unit area or volume. Population growth affects population density, and knowledge of both can help us make decisions about the management of species. How long will it take for a population of an endangered species to recover to a healthy level if we protect it from its most serious threats? For example, how quickly will the ferret populations increase in Wyoming? A knowledge of population growth rates and population densities would allow us to predict future ferret population sizes. Since 1994, several large parts of Georges Bank, an area of the sea floor in the North Atlantic that was once one of the world's richest fishing grounds, have been closed to commercial fishing because of overfishing. How many fishes can we reasonably trawl from the sea and still ensure that an adequate population will exist for future use? Such information is vital in making determinations of size limits, catch quotas, and length of season for fisheries to ensure an adequate future population size.

In this section, we will examine density and other characteristics of populations within their habitats. We will also discuss the different reproductive strategies organisms use and how ecologists assign individuals to different groups called age classes.

Ecologists Use Many Different Methods to Quantify Population Density

The simplest method to measure population density is to visually count the number of organisms in a given area. We can reasonably do this only if the area is small and the organisms are relatively large. For example, we can readily determine the number of gumbo limbo trees (*Bursera simaruba*) on a small island in the Florida Keys. Normally, however, population ecologists calculate the density of plants or animals in a small area and use this figure to estimate the total abundance over a larger area.

For plants, algae, or other sessile organisms such as intertidal animals, it is fairly easy to count numbers of individuals per square meter or, for larger organisms such as trees, numbers per hectare (an area of land equivalent to 2.471 acres). However, many plant individuals are clonal; that is, they grow in patches of genetically identical individuals, so that rather than count individuals, we can also use the amount of ground covered by plants as an estimate of vegetation density.

Plant ecologists use a sampling device called a **quadrat**, a square frame that often, but not always, measures 50 x 50 cm and encloses an area of 0.25 m² (**Figure 56.1a**). They then count the numbers of plants of a given species inside the quadrat to obtain a density estimate per m². For example, if you counted densities of 20, 35, 30, and 15 plants in four quadrats, you could reliably say that the density of this species was 25 individuals per 0.25 m², or 100 individuals per m². For larger plants, such as trees, a quadrat would be ineffective. To count such organisms, many ecologists perform a **line transect**, in which a long piece of string is stretched out and any tree along its length is counted. For example, to count tree species on larger islands in the Florida Keys, we could lay out a 100-m line transect and count all the trees within 1 m on either side of the transect. In effect, this transect is little more than a long thin quadrat encompassing 200 m². By performing five such transects, we could obtain estimates of tree density per 1,000 m² and then extrapolate that to a number per hectare or per island.

Several different sampling methods exist for quantifying the density of animals, which are more mobile than plants. Suction traps, like giant aerial vacuum cleaners, can suck flying insects from the sky. Pitfall traps set into the ground can catch species such as spiders, lizards, or beetles wandering over the surface, (**Figure 56.1b**). Sweep nets can be passed over vegetation to dislodge and capture the insects feeding there. Mist nets, consisting of very fine netting spread between trees, can entangle flying birds and bats (**Figure 56.1c**). Baited snap traps, such as mouse traps, or live traps can snare terrestrial animals (**Figure 56.1d**). Population density can thus be estimated as the number of animals caught per trap or per unit area where a given number of traps are set, for example, 10 traps per 100 m² of habitat.

Sometimes population biologists will capture animals and then tag and release them (**Figure 56.2**). The rationale behind the **mark-recapture technique** is that after the tagged animals are released, they mix freely with unmarked individuals and within a short time are randomly mixed within the population. The population is resampled, and the numbers of marked and unmarked individuals are recorded. We assume that the ratio of marked to unmarked individuals in the second sample is the

(a) Quadrat

(b) Pitfall trap

(c) Mist net

(d) Live mammal trap

Figure 56.1 **Sampling techniques.** (a) Quadrats are frequently used to count the number of plants per unit area. (b) Pitfall traps set into the ground catch wandering species such as beetles and spiders. (c) Mist nets consisting of very fine mesh entangle birds or bats. (d) Baited live traps catch terrestrial animals, including Komodo dragons, as here on Rinca Island, Indonesia.

Figure 56.2 **The mark-recapture technique to estimate population size.** An ear tag identifies this Rocky Mountain goat (*Oreamnos americanus*) in Olympic National Park, Washington. Recapture of such marked animals permits estimates of population size.

Concept check: If we mark 110 Rocky Mountain goats and recapture 100 goats, 20 of which have ear tags, what is the estimate of the total population size?

same as the ratio of marked individuals in the first sample to the total population size. Thus,

$$\frac{\text{Number of individuals marked in first catch}}{\text{Total population size, } N} = \frac{\text{Number of marked recaptures in second catch}}{\text{Total number of second catch}}$$

Let's say we catch 50 largemouth bass in a lake and mark them with colored fin tags. A week later, we return to the lake and catch 40 fish and 5 of them are previously tagged fish. If we assume no immigration or emigration has occurred, which is quite likely in a closed system like a lake, and we assume there have been no births or deaths of fish, then the total population size is given by rearranging the equation:

$$\text{Total population size, } N = \frac{\text{Number of marked individuals in first catch} \times \text{Total number of second catch}}{\text{Number of marked recaptures in second catch}}$$

Using our data,

$$N = \frac{50 \times 40}{5} = \frac{2,000}{5} = 400$$

From this equation, we estimate that the lake has a total population size of 400 largemouth bass. This could be useful information for game and fish personnel who wish to know the total size of a fish population in order to set catch limits.

However, the mark-recapture technique can have drawbacks. Some animals that have been marked may learn to avoid the traps. Recapture rates will then be low, resulting in an overestimate of population size. Imagine that instead of 5 tagged fish out of 40 recaptured fish, we get only 2 tagged fish. Now our population size estimate is 2,000/2 = 1,000, a dramatic increase in our population size estimate. On the other hand, some animals can become trap-happy, particularly if the traps are baited with food. This would result in an underestimate of the population size.

Because of the limitations of the mark-recapture technique, ecologists also use other, more novel methods to estimate population density. For some larger terrestrial or marine species, captured animals can be fitted with radio-collars and followed telemetrically, using an antennal tracking device. Their home ranges can be determined and population estimates developed based on the area of available habitat. For many species with valuable pelts, we can track population densities through time by examining pelt records taken from trading stations. We can also estimate relative population density by examining catch per unit effort, which is especially valuable in commercial fisheries. We can't easily expect to count the number of fishes in an area of ocean, but we can count the number caught, say, per 100 hours of trawling. For some species that leave easily recognizable fecal pellets, like rabbits or deer, we can count pellet numbers, and, if we know the pellet production per individual and how long pellets last in the environment, we can estimate population size. For frogs or birds, we can count chorusing or singing individuals. We can also count leaf scars or chewed leaves and, if we know the responsible herbivores and the rates of herbivory, use these as an estimate of the density of the animals that damage them.

Patterns of Spacing Individuals within a population can show different patterns of **dispersion**; that is, they can be clustered together or spread out to varying degrees. The three basic kinds of dispersion patterns are clumped, uniform, and random. We can visualize these patterns by imagining people in a meeting room. If some people know each other, they get together in small groups, creating a clumped pattern. If people do not know each other and are perhaps even wary of each other, they might maintain a certain minimum personal distance between themselves to produce a uniform dispersion. If nobody thinks or cares about their position relative to anyone else, we would get a random dispersion.

The type of dispersion observed in nature can tell us a lot about what processes shape group structure. The most common dispersion pattern is **clumped**, because resources in nature tend to be clustered. For example, certain plants may do better in moist conditions, and moisture is greater in low-lying areas (**Figure 56.3a**). Social behavior between animals that aggregate into flocks or herds reflects a clumped pattern.

On the other hand, competition may cause a **uniform** dispersion pattern between individuals, as between trees in a

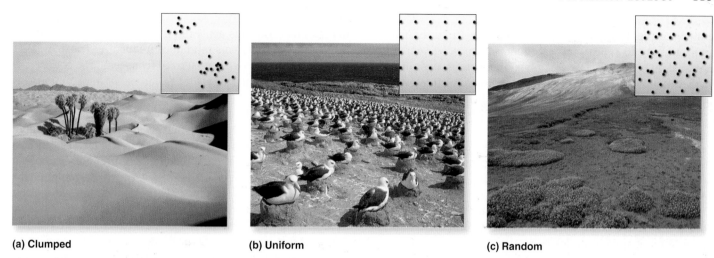

(a) Clumped   **(b) Uniform**   **(c) Random**

Figure 56.3   **Three types of dispersion.**   **(a)** A clumped distribution pattern, as in these plants clustered around an oasis, often results from the uneven distribution of a resource, in this case, water. **(b)** A uniform distribution pattern, as in these nesting black-browed albatrosses (*Diomedea melanophris*) on the Falkland Islands, may be a result of competition or social interactions. **(c)** A random distribution pattern, as in these bushes at Leirhnjukur Volcano in Iceland, is the least common form of spacing.

Concept check: *What is the distribution pattern of students in a half-empty classroom?*

forest. At first, the pattern of trees and seedlings may appear random as seedlings develop from seeds dropped at random, but competition between roots may cause some trees to be outcompeted by others, causing a thinning out and resulting in a relatively uniform distribution. Thus, the dispersion pattern starts out random but ends up uniform. Uniform dispersions may also result from social interactions, as between some nesting birds, which tend to keep an even distance from each other (**Figure 56.3b**).

Perhaps the rarest dispersion pattern is **random**, because resources in nature are rarely randomly spaced. Where resources are common and abundant, as in moist, fertile soil, the dispersion patterns of plants may lack a pattern (**Figure 56.3c**).

Reproductive Strategies   To better understand how populations grow in size, let's consider their reproductive strategies. Some organisms produce all of their offspring in a single reproductive event. This pattern, called **semelparity** (from the Latin *semel*, meaning once, and *parere*, meaning to bear), is common in insects and invertebrates and also occurs in organisms such as salmon, bamboo grasses, and agave plants (**Figure 56.4a**). These individuals reproduce once only and die. Semelparous organisms may live for many years before reproducing, like agaves, or they may be annual plants that develop from seed, flower, and drop their own seed within a year.

Other organisms reproduce in successive years or breeding seasons. The pattern of repeated reproduction at intervals

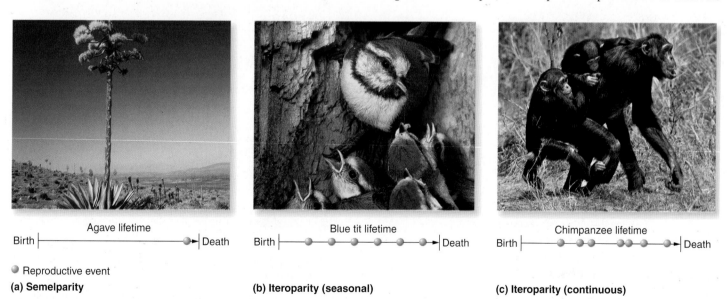

| Agave lifetime | Blue tit lifetime | Chimpanzee lifetime |

Birth ———————————●┤ Death   Birth ├—●—●—●—●—●—●—●►┤ Death   Birth ├——●—●●——●●—●—●►┤ Death

● Reproductive event

(a) Semelparity   **(b) Iteroparity (seasonal)**   **(c) Iteroparity (continuous)**

Figure 56.4   **Differences in reproductive strategies.**   Species such as **(a)** agave plants (*Agave shawii*) are semelparous, meaning they breed once in their lifetime and then die. This contrasts with **(b)** blue tits (*Parus caeruleus*) and **(c)** chimpanzees (*Pan troglodytes*), which are iteroparous and breed more than once in their lifetime.

throughout the life cycle is called **iteroparity** (from the Latin *itero*, meaning to repeat). It is common in most vertebrates, perennial plants, and trees. Among iteroparous organisms, much variation occurs in the number of reproductive events and in the number of offspring per event. Many species, such as birds or trees in temperate areas, have distinct breeding seasons (seasonal iteroparity) that lead to distinct generations (**Figure 56.4b**). For a few species, individuals reproduce repeatedly and at any time of the year. This is termed continuous iteroparity and is exhibited by some tropical species, many parasites, and many primates (**Figure 56.4c**).

Why do species reproduce in a semelparous or iteroparous mode? The answer may lie in part in environmental uncertainty. If survival of juveniles is very poor and unpredictable, then selection favors repeated reproduction and long reproductive life to increase the chance that juveniles will survive in at least some years. This is often referred to as bet-hedging. If the environment is stable, then selection favors a single act of reproduction, because the organism can devote all its energy to making offspring, not maintaining its own body. Under favorable circumstances, annual plants produce more seeds per unit biomass than trees, which have to invest a lot of energy in maintenance. However, when the environment becomes stressful, annuals run the risk of their seeds not germinating. They must rely on some seeds successfully lying dormant and germinating after the environmental stress has ended.

Age Classes The reproductive strategy employed by an organism has a strong effect on the subsequent age classes of a population. Semelparous organisms often produce groups of same-aged young called **cohorts** that grow at similar rates.

Iteroparous organisms generally have many young of different ages because the parents reproduce frequently. The age classes of populations can be characterized by specific categories, such as years in mammals, stages (eggs, larvae, or pupae) in insects, or size in plants.

We expect that a population increasing in size should have a large number of young, whereas a decreasing population should have few young. An imbalance in age classes can have a profound influence on a population's future. For example, in an overexploited fish population, the larger, older reproductive age classes are often removed. If the population experiences reproductive failure for 1 or 2 years, there will be no young fish to move into the reproductive age class to replace the removed fish, and the population may collapse. Other populations experience removal of younger-age classes. In the Florida Keys, where populations of Key deer are high, they overgraze the vegetation and eat many young Gumbo limbo trees, leaving older trees, whose foliage is too tall for them to reach (**Figure 56.5**). This can have disastrous effects on the future population of trees, for while the forest might consist of healthy mature trees, when these die, there will be no replacements. To accurately examine how populations grow, we need to examine and understand the demography of the population.

56.2 Demography

One way to determine how a population will change is to examine a cohort of individuals from birth to death. For most animals and plants, this involves marking a group of individuals in a population as soon as they are born or germinate and following

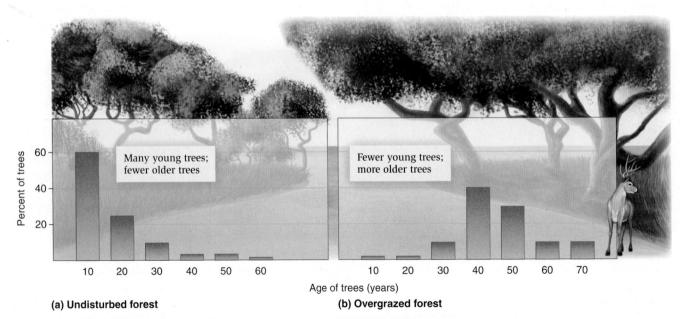

(a) Undisturbed forest   (b) Overgrazed forest

Figure 56.5 Theoretical age distribution of two populations of Gumbo limbo trees in the Florida Keys. (a) Age distribution of an island with no Key deer with numerous young trees, many of which die as the trees age and compete with one another for resources, leaving relatively few big, older trees. (b) Age distribution of a forest where overgrazing by Key deer has reduced the abundance of young trees, leaving only trees in the older age classes.

their fate through their lifetime. For some long-lived organisms such as tortoises, elephants, or trees, this is impractical, so a snapshot approach is used, in which researchers examine the age structure of a population at one point in time. Recording the presence of juveniles and mature individuals, researchers use this information to construct a **life table**, a table providing data on the number of individuals alive in each particular age class. Age classes can be created for any time period, but they often represent 1 year. Males are not usually included in these tables, because they are typically not the limiting factor in population growth. **Demography** is the study of how births and deaths change population sizes over time. In this section, we will determine how to construct life tables and plot survivorship curves, which show at a glance the general pattern of population survival over time. With a knowledge of life tables and additional information on birth rates, we can predict how populations will grow.

Life Tables and Survivorship Curves Summarize Survival Patterns

Let's examine a life table for the North American beaver (*Castor canadensis*). Prized for their pelts, by the mid-19th century, these animals had been hunted and trapped to near extinction. Beavers began to be protected by laws in the 20th century, and populations recovered in many areas, often growing to what some considered to be nuisance status. In Newfoundland, Canada, legislation supported trapping as a management technique. From 1964 to 1971, trappers provided mandibles from which teeth were extracted for age classification. If many mandibles were obtained from, say, 1-year-old beavers, then such animals

were probably common in the population. If the number of mandibles from 2-year-old beavers was low, then we know there was high mortality for the 1-year-old age class. From the mandible data, researchers constructed a life table (**Table 56.1**). The number of individuals alive at the start of the time period (in this case, a year) is referred to as n_x, where n is the number, and x refers to the particular age class. By subtracting the value of n_x from the number alive at the start of the previous year, we can calculate the number dying in a given age class or year, d_x. Thus $d_x = n_x - n_{x+1}$. For example, in Table 56.1, 273 beavers were alive at the start of their sixth year (n_5), and only 205 were alive at the start of the seventh year (n_6); thus, 68 died during the sixth year: $d_5 = n_5 - n_6$, or $d_5 = 273 - 205 = 68$.

A simple but informative exercise is to plot numbers of surviving individuals at each age, creating a **survivorship curve** (**Figure 56.6**). The value of n_x, the number of individuals, is typically expressed on a log scale. Ecologists use a log scale to examine rates of change with time, not change in absolute numbers. Although we could accomplish the same thing with a linear scale, the use of logs makes it easier to examine a wide range of population sizes. For example, if we start with 1,000 individuals and 500 are lost in year 1, the log of the decrease is

$$\log_{10}1,000 - \log_{10}500 = 3.0 - 2.7 = 0.3 \text{ per year}$$

If we start with 100 individuals and 50 are lost, the log of the decrease is similarly

$$\log_{10}100 - \log_{10}50 = 2.0 - 1.7 = 0.3 \text{ per year}$$

In both cases, the rates of change are identical, even though the absolute numbers are different. Plotting the n_x data on a log scale ensures that regardless of the size of the starting

Table 56.1	Life Table for the Beaver (*Castor canadensis*) in Newfoundland, Canada				
Age (years), x	**Number alive at start of year, n_x**	**Number dying during year, d_x**	**Proportion alive at start of year, l_x**	**Age-specific fertility, m_x**	**$l_x m_x$**
0–1	3,695	1,995	1.000	0.000	0
1–2	1,700	684	0.460	0.315	0.145
2–3	1,016	359	0.275	0.400	0.110
3–4	657	286	0.178	0.895	0.159
4–5	371	98	0.100	1.244	0.124
5–6	273	68	0.074	1.440	0.107
6–7	205	40	0.055	1.282	0.071
7–8	165	38	0.045	1.280	0.058
8–9	127	14	0.034	1.387	0.047
9–10	113	26	0.031	1.080	0.033
10–11	87	37	0.024	1.800	0.043
11–12	50	4	0.014	1.080	0.015
12–13	46	17	0.012	1.440	0.017
13–14	29	7	0.007	0.720	0.005
14+	22	22	0.006	0.720	0.004

Net reproductive rate, $\Sigma l_x m_x = 0.938$

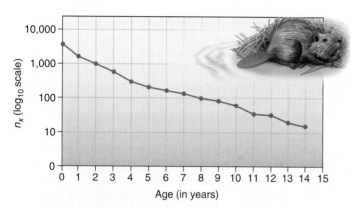

Figure 56.6 Survivorship curve for the North American beaver. The survivorship curve is generated by plotting the number of surviving individuals, n_x, from any given cohort of young, usually measured on a log scale, against age. This survivorship curve shows a fairly uniform rate of decline through time.

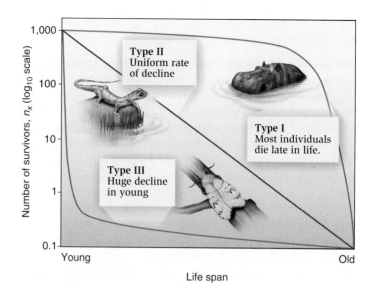

Figure 56.7 Idealized survivorship curves.

Concept check: Which type of survivorship curve would you expect in (a) mussels and (b) turtles?

population, the rate of change of one survivorship curve can easily be compared to that of another species.

Survivorship curves generally fall into one of three patterns (**Figure 56.7**). In a type I curve, the rate of loss for juveniles is relatively low, and most individuals are lost later in life, as they become older and more prone to sickness and predators (see Feature Investigation that follows). Organisms that exhibit type I survivorship have relatively few offspring but invest much time and resources in raising their young. Many large mammals, including humans, exhibit type I curves. At the other end of the scale is a type III curve, in which the rate of loss for juveniles is relatively high, and the survivorship curve flattens out for those organisms that have avoided early death. Many

fishes and marine invertebrates fit this pattern. Most of the juveniles die or are eaten, but a few reach a favorable habitat and thrive. For example, once they find a suitable rock face on which to attach themselves, barnacles grow and survive very well. Many insects and plants also fit the type III survivorship curve, because they lay many eggs or release hundreds of seeds, respectively. Type II curves represent a middle ground, with fairly uniform death rates over time. Species with type II survivorship curves include many birds, small mammals, reptiles, and some annual plants. The North American beaver population exhibits this survivorship curve. Keep in mind, however, that these are generalized curves and that few populations fit them exactly.

FEATURE INVESTIGATION

Murie's Collections of Dall Mountain Sheep Skulls Permitted Accurate Life Tables to Be Constructed

The Dall mountain sheep (*Ovis dalli*) lives in mountainous regions, including the Arctic and sub-Arctic regions of Alaska. In the late 1930s, the U.S. National Park Service was bombarded with public concerns that wolves were responsible for a sharp decline in the population of Dall mountain sheep in Denali National Park (then Mt. McKinley National Park). Shooting the wolves was advocated as a way of increasing the number of sheep. Because meaningful data on sheep mortality were nonexistent, the Park Service enlisted biologist Adolph Murie to collect relevant information. In addition to spending many hours observing interactions between wolves and sheep, Murie also collected sheep skulls, determining the sheep's age at death by counting annual growth rings on the horns.

In 1947, Edward Deevey put Murie's data in the form of a life table that listed each age class and the number of skulls in it (**Figure 56.8**). Although Murie had collected 608 skulls, Deevey expressed the data per 1,000 individuals to allow for comparison with other life tables. From the data, Deevey constructed a survivorship curve. For the Dall mountain sheep in Denali National Park, there was a slight initial decline in survivorship as young lambs were lost; then the survivorship curve flattened out, indicating that the sheep survived well through about age 7 or 8. Then the number of sheep declined rapidly as they aged. These data underlined what Murie had previously observed, which was that wolves preyed primarily on the most vulnerable members of the sheep population—the youngest and the oldest. Such predation would not be expected to dramatically reduce the sheep population. The Park Service ultimately ended a limited wolf-control program that had been in effect since 1929.

Figure 56.8 Examining the survivorship curve of a Dall mountain sheep population reveals information on the cause of death.

HYPOTHESIS Culling the wolf population would protect reproductively active adults in the Dall mountain sheep population.

STARTING LOCATION Denali National Park (formerly known as Mt. McKinley National Park) in Alaska, where wolf predation of sheep is common.

Experimental level   **Conceptual level**

1 Collect sheep skulls lying on the ground.

Only skulls with horns are collected in this sampling technique.

2 Determine the age of the skulls by counting their growth rings.

Annuli are the annual growth rings used to estimate a horned animal's age.

3 Organize the data into a life table (see step 4) and construct a survivorship curve using the data.

Survivorship curve for the Dall mountain sheep shows the number of sheep alive in each age class on a log scale, plotted against age in years.

4 **THE DATA**

Results used in step 3:

Age class	Number alive, n_x	$\log_{10} n_x$	Age class	Number alive, n_x	$\log_{10} n_x$
0–1	1,000	3.00	7–8	640	2.81
1–2	801	2.90	8–9	571	2.76
2–3	789	2.90	9–10	439	2.64
3–4	776	2.89	10–11	252	2.40
4–5	764	2.88	11–12	96	1.98
5–6	734	2.86	12–13	6	0.78
6–7	688	2.84	13–14	3	0.48

Experimental Questions

1. What problem led to the study conducted by Murie on the Dall mountain sheep population of Denali National Park?

2. Describe the survivorship curve developed by Deevey based on Murie's data.

3. How did the Murie and Deevey data affect the decision of the Park Service on the control of the wolf population?

Age-Specific Fertility Data Can Help to Predict Population Growth

To calculate how a population grows, we need information on birth rates as well as mortality and survivorship rates. For any given age, we can determine the proportion of female offspring that are born to females of reproductive age. Using these data, we can determine an **age-specific fertility rate**, called m_x. For example, if 100 females of a given age produce 75 female offspring, $m_x = 0.75$. With this additional information, we can calculate the growth rate of a population.

First, we use the survivorship data to find the proportion of individuals alive at the start of any given age class. This age-specific survivorship rate, termed l_x, equals n_x/n_0, where n_0 is the number alive at time 0, the start of the study, and n_x is the number alive at the beginning of age class x. Let's return to the beaver life table in Table 56.1. The proportion of the original beaver population still alive at the start of the sixth age class, l_5, equals $n_5/n_0 = 273/3,695$, or 0.074. This means that 7.4% of the original beaver population survived to age 5. Next we multiply the data in the two columns, l_x and m_x, for each row, to give us a column $l_x m_x$, an average number of offspring per female. This column represents the contribution of each age class to the overall population growth rate. An examination of the beaver age-specific fertility rates illustrates a couple of general points. First, for this beaver population in particular, and for many organisms in general, there are no babies born to young females. As females mature sexually, age-specific fertility goes up, and it remains fairly high until later in life, when females reach postreproductive age.

The number of offspring born to females of any given age class depends on two things: the number of females in that age class and their age-specific fertility rate. Thus, although fertility of young beavers is very low, there are so many females in the age class that $l_x m_x$ for 1-year-olds is quite high. Age-specific fertility for older beavers is much higher, but the relatively few females in these age classes cause $l_x m_x$ to be low. Maximum values of $l_x m_x$ occur for females of an intermediate age, 3–4 years old in the case of the beaver. The overall growth rate per generation is the number of offspring born to all females of all ages, where a generation is defined as the mean period between birth of females and birth of their offspring. Therefore, to calculate the generational growth rate, we sum all the values of $l_x m_x$, that is, $\Sigma l_x m_x$, where the Σ symbol means "sum of." This summed value, R_0, is called the **net reproductive rate**.

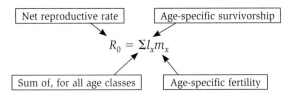

To calculate the future size of a population, we simply multiply the number of individuals in the population by the net reproductive rate. Thus, the population size in the next generation, N_{t+1}, is determined by the number in the population now, at time t, which is given by N_t, multiplied by R_0.

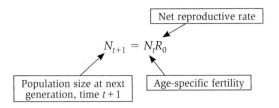

Let's consider an example in which the number of beavers alive now, N_t, is 1,000, and $R_0 = 1.1$. This means the beaver population is reproducing at a rate that is 10% greater than simply replacing itself. The size of the population next generation, N_{t+1}, is given by

$$N_{t+1} = N_t R_0$$
$$N_{t+1} = 1,000 \times 1.1$$
$$= 1,100$$

Therefore, the number of beavers in the next generation is 1,100, and the population will have grown larger.

In determining population growth, much depends on the value of R_0. If $R_0 > 1$, then the population will grow. If $R_0 < 1$, the population is in decline. If $R_0 = 1$, the population size stays the same, and we say it is at **equilibrium**. In the case of the beavers, Table 56.1 reveals that $R_0 = 0.938$, which is less than 1, and, therefore, the population is declining. This is valuable information, because it tells us that at that time, the beaver population in Newfoundland needed more protection (perhaps in the form of bans on trapping and hunting) in order to attain a population level at equilibrium.

Because of the effort involved in calculating R_0, the net reproductive rate, ecologists sometimes use a shortcut to predict population growth. Imagine a bird species that breeds annually. To measure population growth, ecologists count the number of birds in the population, N_0. Let's say $N_0 = 100$. The next year, ecologists count 110 birds in the same population, so $N_1 = 110$. The **finite rate of increase**, λ, is the ratio of the population size from one year to the next, calculated as

$$\lambda = N_1/N_0$$

In this case, $\lambda = 1.10$. λ is often given as percent annual growth, and t is a number of years. Let's consider a population of birds growing at a rate of 5% per year. After 5 years, how many birds would there be?

$$N_t = 100, \lambda = 1.05, \text{ and } t = 5$$
$$\text{therefore, } N_{t+5} = 100 \, (1.05)^5 = 127.6$$

What's the difference between R_0 and λ? R_0 represents the net reproductive rate per generation. λ represents the finite rate of population change over some time interval, often a year. Where species are annual breeders that live 1 year, such as annual plants, $R_0 = \lambda$. For species that breed for multiple years, $R_0 \neq \lambda$. Just as

$$N_t = N_0 R_0{}^t, \text{ where } t = \text{a number of generations}$$
$$\text{so } N_t = N_0 \lambda^t, \text{ where } t = \text{a number of time intervals}$$

Populations grow when R_0 or $\lambda > 1$; populations decline when R_0 or $\lambda < 1$; and they are at equilibrium when R_0 or $\lambda = 1$.

56.3 How Populations Grow

Life tables can provide us with accurate information about how populations can grow from generation to generation. However, other population growth models can provide us with valuable insights into how populations grow over shorter time periods. The most simple of these assumes that populations grow if, for any given time interval, the number of births is greater than the number of deaths. In this section, we will examine two different types of these simple models. The first assumes resources are not limiting, and it results in prodigious growth. The second, and perhaps more biologically realistic, assumes resources are limiting, and it results in limits to growth and eventual stable population sizes. We then consider how other factors might limit population growth, such as natural enemies, and discuss the overall life history strategies exhibited by different species that enable them to exist on Earth.

Knowing the Per Capita Growth Rate Helps Predict How Populations Will Grow

The change in population size over any time period can be written as the number of births per unit time interval minus the number of deaths per unit time interval.

For example, if in a population of 1,000 rabbits, there were 100 births and 50 deaths over the course of 1 year, then the population would grow in size to 1,050 the next year. We can write this formula mathematically as

$$\frac{\text{Change in numbers}}{\text{Change in time}} = \text{Births} - \text{Deaths}$$

or

$$\frac{\Delta N}{\Delta t} = B - D$$

The Greek letter delta, Δ, indicates change, so that ΔN is the change in number, and Δt is the change in time; B is the number of births per time unit; and D is the number of deaths per time unit.

Often, the numbers of births and deaths are expressed per individual in the population, so the birth of 100 rabbits to a population of 1,000 would represent a per capita birth rate, b, of 100/1,000, or 0.10. Similarly, the death of 50 rabbits in a population of 1,000 would be a per capita death rate, d, of 50/1,000, or 0.05. Now we can rewrite our equation giving the rate of change in a population.

$$\frac{\Delta N}{\Delta t} = bN - dN$$

For our rabbit example,

$$\frac{\Delta N}{\Delta t} = 0.10 \times 1,000 - 0.05 \times 1,000 = 50$$

so if $\Delta t = 1$ year, the rabbit population would increase by 50 individuals in a year.

Ecologists often simplify this formula by representing $b - d$ as r, the **per capita growth rate**. Thus, $bN - dN$ can be written as rN. Because ecologists are also interested in population growth rates over very short time intervals, so-called instantaneous growth rates, instead of writing

$$\frac{\Delta N}{\Delta t}$$

they write

$$\frac{dN}{dt}$$

which is the notation of differential calculus. The equations essentially mean the same thing, except that dN/dt reflects very short time intervals. Thus,

$$\frac{dN}{dt} = rN = (0.10 - 0.05)N = 50$$

Exponential Growth Occurs When the Per Capita Growth Rate Remains Above Zero

How do populations grow? Clearly, much depends on the value of the per capita growth rate, r. When $r < 0$, the population decreases; when $r = 0$, the population remains constant; and when $r > 0$, the population increases. When $r = 0$, the population is often referred to as being at equilibrium, where no changes in population size will occur and there is **zero population growth**.

Even if r is only fractionally above 0, population increase is rapid, and when plotted graphically, a characteristic J-shaped curve results (**Figure 56.9**). We refer to this type of population growth as geometric, or **exponential growth**. When conditions are optimal for the population, r is at its maximum rate and is called the **intrinsic rate of increase** (denoted r_{max}). Thus, the rate of population growth under optimal conditions is $dN/dt = r_{max}N$. Again, the larger the value of r_{max}, the steeper the slope of the curve. Because population growth depends on the value of N as well as the value of r, the population increase is even greater as time passes.

How do field data fit this simple model for exponential growth? Population growth cannot go on forever, as envisioned under exponential growth. But initially at least, in a new and expanding population where resources are not limited, exponential growth is often observed. Let's look at a few examples. Tule elk (*Cervus elaphus nannodes*) is a subspecies of elk that is native to California. Hunted nearly to extinction in the 19th century, less than a dozen individuals survived on a private ranch. In the 20th century, reintroductions resulted in the recovery of tule elk to around 3,500 individuals. One reintroduction was made in March 1978 at Point Reyes National Seashore in California, where 10 animals—2 males and 8 females—were released. By 1993, the herd had reached 214 individuals, and it continued to grow in an exponential fashion until 1998, when the herd size

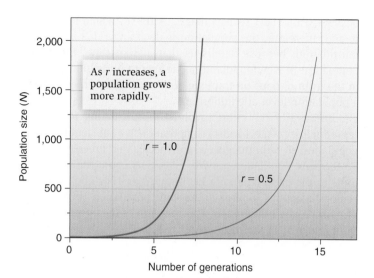

Figure 56.9 **Exponential population growth.** As the value of r increases, the slope of the curve gets steeper. In theory, a population with unlimited resources could grow indefinitely.

stood at 549 (**Figure 56.10a**). This was deemed an excessive number for the size of the available habitat, and animals were removed to begin herds in other locations. Since then, herd size at Point Reyes has been maintained at around 350.

The growth of the recovering black-foot ferret population in Wyoming that we mentioned at the beginning of the chapter also fit the exponential growth pattern (**Figure 56.10b**). The ferrets had been reintroduced in 1991, but the population declined and languished for many years. However, from 2000 to 2006, the population grew in an exponential fashion. A value of $r = 0.47$ was calculated for the increase in population size of the ferrets during those years. Ecologists have noted that in some cases, populations seem to languish at low levels before conditions become favorable for population growth.

(a) Tule elk

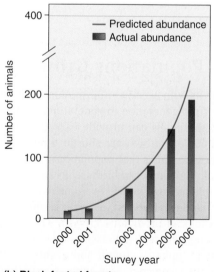

(b) Black-footed ferrets

Figure 56.10 **Exponential growth following reintroduction of a population into a habitat.** (a) A population of tule elk (*Cervus elaphus nannodes*) reintroduced to Point Reyes National Seashore in 1978 fits a pattern of exponential growth. (b) Black-footed ferrets reintroduced to Shirley Basin, Wyoming, since 2000.

The growth of some introduced species also seems to fit the pattern of exponential growth. The rapid expansion of rabbits after their introduction into South Australia in the late 19th century is a case in point. In 1859, Thomas Austin received two dozen European rabbits from England. Rabbit gestation lasts a mere 31 days, and in South Australia, each doe could produce up to 10 litters of at least six young each year. The rabbits had essentially no enemies and ate the grass used by sheep and other grazing animals. Even when two-thirds of the population was shot for sport, which was the purpose of the initial introduction, the population grew into the millions within a few short years. By 1875, rabbits were reported on the west coast of Australia, having moved over 1,760 km across the continent despite the deployment of huge, thousand-kilometer-long fences ("rabbit-proof fences") meant to contain them.

Finally, one of the most prominent examples of exponential growth is the growth of the global human population, which, because of its great importance, we will examine separately in Section 56.4.

Logistic Growth Occurs in Populations in Which Resources Are Limited

Despite its applicability to rapidly growing populations, the exponential growth model is not appropriate in many situations. The model assumes unlimited resources, which is not typically the case in the real world. For most species, resources become limiting as populations grow, and the per capita growth rate decreases. The upper boundary for the population size is known as the **carrying capacity** (**K**). A more realistic equation to explain population growth, one that takes into account the amount of available resources, is

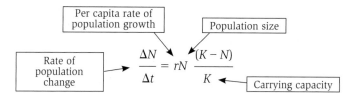

where $(K - N)/K$ represents the proportion of the carrying capacity that is unused by the population. This equation is called the **logistic equation**.

As the population size, N, grows, it moves closer to the carrying capacity, K, with fewer available resources for population growth. At large values of N, the value of $(K - N)/K$ becomes small, and population growth is small. If $K = 1,000$, $N = 900$, and $r = 0.1$, then

$$\frac{dN}{dt} = (0.1)(900) \times \frac{(1,000 - 900)}{1,000}$$

$$\frac{dN}{dt} = 9$$

In this instance, population growth is 9 individuals per unit of time.

Let's consider how an ecologist would use the logistic equation. First, the value of K would come from intense field and laboratory work where researchers would determine the amount of resources, such as food, needed by each individual and then determine the amount of available food in the wild. Field censuses would determine N, and field censuses of births and deaths per unit time would provide r. When this type of population growth is plotted over time, an S-shaped growth curve results (**Figure 56.11**). This pattern, in which the growth of a population slows down as it approaches K, is called **logistic growth**.

Does the logistic growth model provide a better fit to growth patterns of plants and animals in the wild than the exponential model, which is also shown in Figure 56.11? In some instances, such as laboratory cultures of bacteria and yeasts, the logistic growth model provides a very good fit (**Figure 56.12**). In nature, however, variations in temperature, rainfall, or resources cause changes in carrying capacity and thus in population size. The uniform conditions of temperature, moisture, and resource levels of the laboratory do not usually exist. Therefore, there are relatively few examples of logistic growth outside of the laboratory.

Is the logistic model of little value because it fails to describe population growth accurately? Not really. It is a useful starting point for thinking about how populations grow, and it seems intuitively correct. However, the carrying capacity is a difficult feature of the environment to identify for most species, and it also varies temporally, according to local climate patterns. For these reasons, logistic growth is difficult to measure accurately.

Also, as we will discover, populations are affected by interactions with other species. In Chapter 57, we will examine how

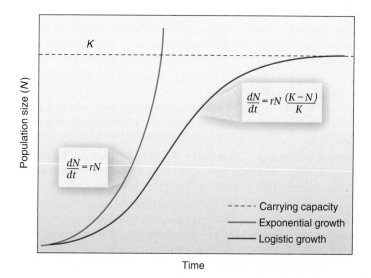

Figure 56.11 **Exponential versus logistic growth.** Exponential (J-shaped) growth occurs in an environment with unlimited resources, whereas logistic (S-shaped) growth occurs in an environment with limited resources.

Concept check: *What is the population growth per unit of time when r = 0.1, N = 200, and K = 500?*

Figure 56.12 **Logistic growth of yeast cells in culture.** Early tests of the logistic growth curve were validated by growth of yeast cells in laboratory cultures. These populations showed the typical S-shaped growth curve.

predators, parasites, and competitors affect population densities and explore situations in which species interactions commonly limit population growth. As described next, such population reductions are often influenced by a process known as density dependence.

Density-Dependent Factors May Regulate Population Sizes

A **density-dependent factor** is a mortality factor whose influence increases with the density of the population. Parasitism, predation, and competition are some of the many density-dependent factors that may reduce the population densities of living organisms and stabilize them at equilibrium levels. Such factors can be density dependent in that their impact depends on the density of the population; they kill relatively more of a population when densities are higher and less of a population when densities are lower. For example, many predators develop a visual search image for a particular prey. When a prey is rare, predators tend to ignore it and kill relatively few. When a prey is common, predators key in on it and kill relatively more. In England, for example, predatory shrews kill proportionately more moth pupae in leaf litter when the pupae are common compared to when they are rare. Density-dependent mortality may also occur as population densities increase and competition for scarce resources increases, reducing offspring production or survival. Parasitism may also act in a density-dependent manner. Parasites are able to pass from host to host more easily as the host's densities increase. Disease is often more common in denser populations.

Density dependence can be detected by plotting mortality, expressed as a percentage, against population density (**Figure 56.13**). If a positive slope results and mortality increases with density, the factor tends to have a greater effect on dense populations than on sparse ones and is clearly acting in a density-dependent manner.

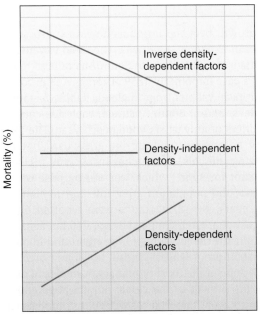

Figure 56.13 **Three ways that factors affect mortality in response to changes in population density.** For a density-dependent factor, mortality increases with population density; for a density-independent factor, mortality remains unchanged. For an inverse density-dependent factor, mortality decreases as a population increases in size.

Concept check: *Which types of factors tend to stabilize populations at equilibrium levels?*

A **density-independent factor** is a mortality factor whose influence is not affected by changes in population size or density. When mortality is plotted against density, a flat line results. In general, density-independent factors are physical factors, including weather, drought, freezes, floods, and disturbances such as fire. For example, in hard freezes, the same proportion of organisms such as birds or plants are usually killed, no matter how large the population size. However, even physical factors such as weather can act in a density-dependent manner. For example, in an environment where there were many beavers and a limited number of rivers to dam, some individuals would not have a lodge. In such a situation, a cold winter could kill a high percentage of beavers. If, on the other hand, there were few beavers, most would have a lodge to provide them with protection from a hard freeze. In this case, the cold would kill a lower percentage.

Determining which factors act in a density-dependent fashion has large practical implications. Foresters, game managers, and conservation biologists alike are interested in learning how to maintain populations at equilibrium levels. For example, if a specific disease were to act in a density-dependent manner on white-tailed deer, there wouldn't be much point in game managers attempting to kill off predators such as mountain lions to increase herd sizes for hunters, because proportionately more deer would be killed by disease.

Finally, a mortality factor that decreases with increasing population size is considered an **inverse density-dependent factor.**

In this case, a negative slope results when mortality is plotted against density. For example, if a territorial predator such as a lion always killed the same number of wildebeest prey, regardless of wildebeest density, it would be acting in an inverse density-dependent manner, because it is taking a smaller proportion of the population at higher density. Some mammalian predators, being highly territorial, often act in this manner on herbivore density.

Thus, natural enemies can act in a density-dependent manner and limit the sizes of prey populations, or they can act in a density-independent or even inverse density-dependent manner and not regulate them. To make generalizations about which factors control populations in nature, we need to know the prevalence of density-dependent factors in natural systems.

Which factors tend to act in a density-dependent manner? In the 1980s, a broad review of many research studies, which considered 51 populations of insects, 82 populations of large mammals, and 36 populations of small mammals and birds, showed a wide variety of density-dependent factors. No single process such as parasitism, competition, or predation could be regarded as a regulatory factor of overriding importance. Even for individual taxa, such as insects, density dependence varied from parasitism and predation to competition and abiotic factors. This finding is disconcerting to ecologists interested in population management, because it means that generalizations about which factors are likely to act in a density-dependent manner are not easily made. Each population has to be analyzed on a case-by-case basis.

Life History Strategies Incorporate Traits Relating to Survival and Competitive Ability

The population parameters we have discussed—including iteroparity versus semelparity, exponential versus logistic growth, and density-dependent versus density-independent factors—have important implications for how populations grow and indeed for the reproductive success of populations and species. These reproductive strategies can be viewed in the context of a much larger picture of life history strategies, sets of physiological and behavioral features that incorporate not only reproductive traits but also survivorship and length of life characteristics, habitat type, and competitive ability.

When comparing many different species, life history strategies follow a continuum. At the one end are species, termed **r-selected species**, that have a high rate of per capita population growth, r, but poor competitive ability (**Figure 56.14a**). An example is a weed, such as a dandelion, that produces huge numbers of tiny seeds and therefore has a high value of r. Weeds exist in disturbed habitats such as gaps in a forest canopy where trees have blown down, allowing light to penetrate to the forest floor. An r-selected species such as a weed grows quickly and reaches reproductive age early, devoting much energy to producing a large number of seeds that disperse widely. These weed species generally remain small, and individuals do not live long. Populations pass a few generations in the light gap before it closes.

At the other end are species, termed **K-selected species**, that have more or less stable populations adapted to exist at or near the carrying capacity, K, of the environment (**Figure 56.14b**). An example is an oak tree that exists in a mature forest. Oak trees grow slowly and reach reproductive age late, having to devote much energy to growth and maintenance. A K-selected species like a tree grows large and shades out r-selected species like weeds, eventually outcompeting them. Such trees live a long time and produce seeds repeatedly every year when mature. These seeds are bigger than those of r-selected species, but do not disperse widely. Acorns contain a large food reserve that helps them grow, whereas dandelion seeds must rely on whatever nutrients they can gather from the soil.

- Small size
- Rapid growth
- Short life span

- Many small seeds
- Good seed dispersal

(a) *r*-selected species

- Large size
- Slow growth
- Long life span

- Few large seeds
- Poor seed dispersal

(b) *K*-selected species

Figure 56.14 Life history strategies. Differences in traits of a dandelion **(a)** and an oak tree **(b)** illustrate some of the differences between *r*- and *K*-selected species.

Table 56.2	Characteristics of *r*- and *K*-Selected Species	
Life history feature	*r*-selected species	*K*-selected species
Development	Rapid	Slow
Reproductive rate	High	Low
Reproductive age	Early	Late
Body size	Small	Large
Length of life	Short	Long
Competitive ability	Weak	Strong
Survivorship	High mortality of young	Low mortality of young
Population size	Variable	Fairly constant
Dispersal ability	Good	Poor
Habitat type	Disturbed	Not disturbed
Parental care	Low	High

Although the weed–tree example is a useful way to think about the *r*- and *K*-selection continuum, other organisms can be *r*- or *K*-selected, too. Insects are mostly *r*-selected species that produce many young and have short life cycles. Mammals, such as elephants, that grow slowly, have few young, and reach large sizes are typical of *K*-selected species. Table 56.2 compares the general characteristics of *r*- and *K*-selected species.

In a human-dominated world, almost every life history attribute of a *K*-selected species sets it at risk of extinction. First, *K*-selected species tend to be larger, so they need more habitat in which to live. Florida panthers need huge tracts of land to establish their territories and hunt for deer. There is room for only about 22 panthers on publicly owned land in South Florida. Privately owned land currently supports another 50 panthers. As the amount of land shrinks through development, so does the number of panthers. *K*-selected species tend to have fewer offspring, so their populations cannot recover as fast from disturbances such as fire or overhunting. California condors, for example, produce only a single chick every other year. *K*-selected species breed at a later age, and their generation time and time to grow from a small population to a larger population is long. Gestation time in elephants is 22 months, and elephants take at least 7 years to become sexually mature. Large trees, such as the giant sequoia; large terrestrial mammals, such as elephants, rhinoceroses, and grizzly bears; and large marine mammals, such as blue whales and sperm whales, all run the risk of extinction. Interestingly, the coast redwood seems to be an exception, a fact perhaps attributable to its unusual genome (see the following Genomes & Proteomes Connection).

What are the advantages to being a *K*-selected species? In a world not disturbed by humans, *K*-selected species would fare well. However, in a human-dominated world, many *K*-selected species are selectively logged or hunted, or their habitat is altered, and the resulting small population sizes make extinction a real possibility.

Genomes & Proteomes Connection

Hexaploidy Increases the Growth of Coast Redwood Trees

Besides having the world's most massive tree, the giant sequoia (*Sequoiadendron giganteum*), California is also home to the world's tallest tree, the coast redwood (*Sequoia sempervirens*), a towering giant that can grow to over 90 m and can live for up to 2,000 years (Figure 56.15). These trees are currently confined to a relatively small 700-km strip along the Pacific coast from California to southern Oregon, an area characterized by year-long moderate temperatures, heavy winter rains, and dense summer fog. Interestingly, because this climate was far more common in an earlier era, these trees were once dispersed throughout the Northern Hemisphere.

How is this huge species different from other tree species? In 1948, researchers made the startling discovery that the tree is a hexaploid; that is, each of its cells contains six sets of chromosomes, with 66 chromosomes in total. (Keep in mind that humans have two sets of chromosomes in every cell.) While hexaploidy is not unknown in grasses and shrubs, it is unusual in trees and particularly gymnosperms. Of all the conifers on Earth, the coast redwood is the only known hexaploid. Having this quality means each tree may have several different alleles for any given gene, which leads to a very genetically diverse population. Molecular biologist Chris Brinegar has found that hardly any two trees have exactly the same genetic constitution. Such genetic diversity allows greater adaptation to environmental conditions and more adaptations against insect or fungal pests. Indeed, living redwoods have no known lethal diseases, and pests do not cause significant damage. What's more, with six sets of genes, trees also have the potential for great variety in their gene products, the proteins, which may help explain their prodigious growth. It grows faster than any conifer on Earth, and this is why it is an exception to most *K*-selected species.

56.4 Human Population Growth

In 2006, the world's population was estimated to be increasing at the rate of 146 people every minute: 2 in developed nations and 144 in less-developed nations. The United Nations' 2006 projections pointed to a world population stabilizing at around 10 billion near the year 2150, as would happen with logistic growth. However, until now, human population growth has better fit an exponential growth pattern than a logistic one. In this section, we examine human population growth trends in more detail and discuss how knowledge of the human population's age structure can help predict its future growth. We then investigate the carrying capacity of the Earth for humans and explore how the concept of an ecological footprint, which measures human resource use, can help us determine this carrying capacity.

Figure 56.15 The coast redwood (*Sequoia sempervirens*)—a hexaploid conifer. The coast redwood can grow to over 90 m, and the oldest living trees are over 2,000 years old. Their great genetic variation may help explain their incredible growth and longevity.

Human Population Growth Fits an Exponential Pattern

Until the beginning of agriculture and the domestication of animals, about 10,000 B.C.E., the average rate of population growth was very low. With the establishment of agriculture, the world's population grew to about 300 million by 1 C.E. and to 800 million by the year 1750. Between 1750 and 1998, a relatively tiny period of human history, the world's human population surged from 800 million to 6 billion (**Figure 56.16**). In 2009, the number of humans was estimated at 6.7 billion, with more than two people added to the world's population every second. Considering this phenomenal increase in growth, the two biggest questions are when and at what level will the human population level off?

Human populations can exist at equilibrium densities in one of two ways:

1. *High birth and high death rates.* Before 1750, this was often the case, with high birth rates offset by deaths from wars, famines, and epidemics.
2. *Low birth and low death rates.* In Western Europe, beginning in the 18th century, better health and living conditions reduced the death rate. Eventually, social changes such as increasing education for women and marriage at a later age reduced the birth rate.

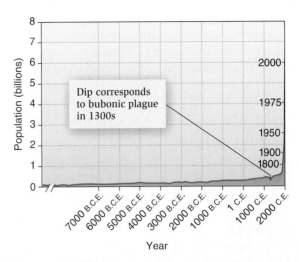

Figure 56.16 The growth pattern of the human population through history. If, and when, human population growth will level off are issues of considerable debate.

The shift in birth and death rates with development is known as the **demographic transition** (**Figure 56.17**). In the first stage of the transition, birth and death rates are both high, and the population is in equilibrium. In the second stage of this transition, which first occurred in Western Europe beginning in the late 18th century, the death rate declines first, while the birth rate remains high. High rates of population growth result. In the third stage, the birth rates drop and death rates stabilize, so population growth rates become lower. In the fourth stage, both birth and death rates are low, and the population is again at equilibrium.

The pace of the demographic transition between countries differs, depending on culture, economics, politics, and religion. This is illustrated by comparing the demographic transition in Sweden and Mexico (**Figure 56.18**). In Mexico, the demographic transition occurred more recently and was typified by a faster decline in the death rate, reflecting rapid improvements in public health. A relatively longer lag occurred between the decline

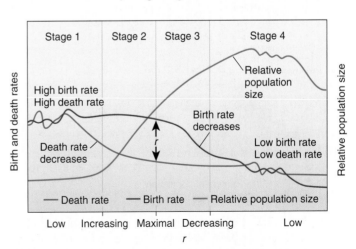

Figure 56.17 The classic stages of the demographic transition. The difference between the birth rate and the death rate determines the rate of population increase or decrease.

in the death rate and the decline in the birth rate, however, with the result that Mexico's population growth rate is still well above Sweden's, perhaps reflecting differences in culture or the fact that in Mexico, the demographic transition is not yet complete.

Knowledge of a Population's Age Structure Can Help Predict Its Future Growth

Changes in the age structure of a population also characterize the demographic transition. In all populations, **age structure** refers to the relative numbers of individuals of each defined age group. This information is commonly displayed as a population pyramid (**Figure 56.19**). In West Africa, for example, children under the age of 15 make up nearly half of the population, creating a pyramid with a wide base and narrow top. Even if fertility rates decline, there will still be a huge increase in the population as these young people move into childbearing age. The age structure of Western Europe is much more balanced. Even if the fertility rate of young women in Western Europe increased to a level higher than that of their mothers, the annual numbers of births would still be relatively low because of the low number of women of childbearing age.

Human Population Fertility Rates Vary Widely Around the World

Most estimates propose that the human population will grow to between 10 and 11 billion people by the middle of the 22nd century. Global population growth can be examined by looking

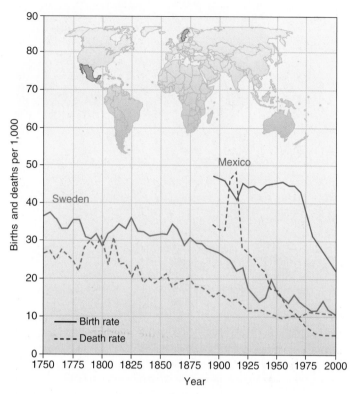

Figure 56.18 **The demographic transition in Sweden and Mexico.** Although the demographic transition began earlier in Sweden than it did in Mexico, the transition was more rapid in Mexico, and the overall rate of population increase remains higher. (The spike in the death rate in Mexico prior to 1920 is attributed to the turbulence surrounding the Mexican Revolution.)

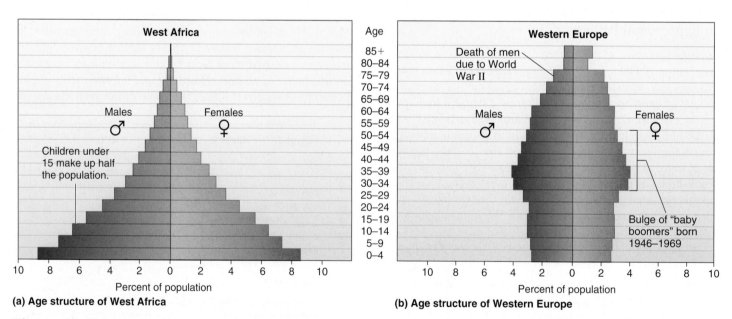

(a) Age structure of West Africa

(b) Age structure of Western Europe

Figure 56.19 **The age structure of human populations in West Africa and Western Europe, as of 2000.** **(a)** In developing areas of the world such as West Africa, there are far more children than any other age group. Population growth is rapid. **(b)** In the developed countries of Western Europe, the age structure is more evenly distributed. The bulge represents those born in the post-World War II baby boom, when birth rates climbed due to stabilization of political and economic conditions. Population growth is close to zero.

Concept check: *If the population pyramid in (a) was inverted, what would you conclude about the age structure of the population?*

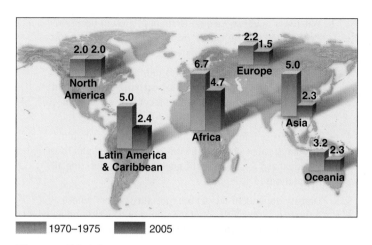

Figure 56.20 Total fertility rates (TFRs) among major regions of the world. Data refer to the average number of children born to a woman during her lifetime.

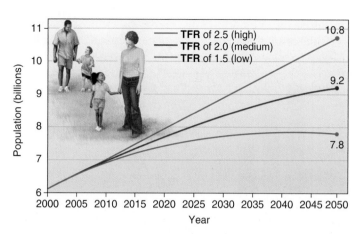

Figure 56.21 Population predictions for 2000–2050, using three different total fertility rates (TFRs).

at the **total fertility rate** (**TFR**), the average number of live births a woman has during her lifetime (**Figure 56.20**). The total fertility rate differs considerably between geographic areas. In Africa, the total fertility rate of 4.67 in 2007 has declined substantially since the 1970s, when it was around 6.7 children per woman. In Latin America and Southeast Asia, the rates have declined considerably from the 1970s and are now at around 2.37 and 2.34, respectively. Most countries in Europe and North America have a TFR of less than 2.0; in Russia, fertility rates have dropped to 1.34. In China, while the TFR is only 1.7, the population there will still continue to increase until at least 2025 because of the large number of women of reproductive age. Although the global TFR has declined from 4.47 in the 1970s to 2.59 in 2007, this is still greater than the average of 2.3 needed for zero population growth. The replacement rate is slightly higher than 2.0, to replace mother and father, due to natural mortality prior to reproduction. The replacement rate varies globally, from 2.1 in developed countries to between 2.5 and 3.3 in developing countries.

The wide variation in fertility rates makes it difficult to predict future population growth. The 2006 United Nations report shows world population projections to the year 2050 for three different growth scenarios: low, medium, and high (**Figure 56.21**). The three scenarios are based on three different assumptions about fertility rate. Using a low fertility rate estimate of only 1.5 children per woman, the population would reach a maximum of about 7.8 billion people by 2050. A more realistic assumption may be to use the fertility rate estimate of 2.0 or even 2.5, in which case the population would continue to rise to 9.2 or 10.8 billion, respectively.

The Concept of an Ecological Footprint Helps Estimate Carrying Capacity

What is the Earth's carrying capacity for the human population, and when will it be reached? Estimates have been quite varied. Much of the speculation on the upper boundary of the world's population size centers on lifestyle. To use a simplistic example, if everyone on the planet ate meat extensively and drove large cars, then the carrying capacity would be a lot less than if people were vegetarians and used bicycles as their main means of transportation.

In the 1990s, researcher Mathis Wackernagel and his coworkers calculated how much land is needed for the support of each person on Earth. Everybody has an impact on the Earth, because they consume the land's resources, including crops, wood, fossil fuels, minerals, and so on. Thus, each person has an **ecological footprint**, the aggregate total of productive land needed for survival in a sustainable world. The average footprint size for everyone on the planet is about 3 hectares (1 ha = 10,000 m^2), but a wide variation is found around the globe (**Figure 56.22**). The ecological footprint of the average Canadian is 7.5 hectares versus about 10 hectares for the average American.

In most developed countries, the largest component of land is for energy, followed by food and then forestry. Much of the land needed for energy serves to absorb the CO_2 emitted by the use of fossil fuels. If everyone required 10 hectares, as the average American does, we would need three Earths to provide us with the needed resources. Many people in less-developed countries are much more frugal in their use of resources. However, globally we are already beyond the Earth's carrying capacity for humans if we were to live in a sustainable manner. This is possible because many people currently live in an unsustainable manner, using supplies of nonrenewable resources, such as groundwater and fossil fuels.

What's your personal ecological footprint? Several different calculations are available on the Internet that you can use to find out. A rapidly growing human population combined with an increasingly large per capita ecological footprint makes it increasingly difficult to preserve other species on the planet, a subject we will examine further in our discussion of conservation biology (Chapter 60).

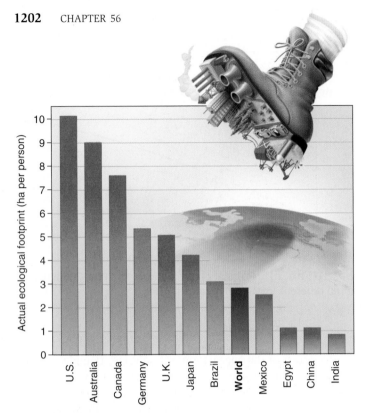

Figure 56.22 Ecological footprints of different countries. The term ecological footprint refers to the amount of productive land needed to support the average individual of that country.

Concept check: What is your ecological footprint?

Summary of Key Concepts

56.1 Understanding Populations

- Population ecology studies how populations grow and what factors promote and limit growth.

- Ecologists measure population density, the numbers of organisms in a given unit area, in many ways, including the mark-capture technique. (Figures 56.1, 56.2)

- Individuals within populations show different patterns of dispersion, including clumped (the most common), uniform, and random. Individuals also exhibit different reproductive strategies, and populations have different age classes. (Figures 56.3, 56.4, 56.5)

56.2 Demography

- Life tables summarize the survival pattern of a population. (Table 56.1)

- Survivorship curves illustrate life tables by plotting the numbers of surviving individuals at different ages. (Figures 56.6, 56.7, 56.8)

- Age-specific fertility and survivorship data help determine the overall growth rate per generation, or the net reproductive rate (R_0).

56.3 How Populations Grow

- The per capita growth rate (r) helps determine how populations grow over any time period.

- When r is > 0, exponential (J-shaped) growth occurs. Exponential growth can be observed in an environment where resources are not limited. (Figures 56.9, 56.10)

- Logistic (S-shaped) growth takes into account the upper boundary for a population, called carrying capacity, and occurs in an environment where resources are limited. (Figures 56.11, 56.12)

- Variations in temperature, rainfall, or resource quantity or quality cause changes in carrying capacities, and thus the idealized logistic growth model does not describe all populations.

- Density-dependent factors are mortality factors whose influence varies with population density. Density-independent factors are those whose influence does not vary with density. (Figure 56.13)

- Life history strategies are a set of features including reproductive traits, survivorship and length of life characteristics, and competitive ability.

- Life history strategies can be viewed as a continuum, with r-selected species (those with a high rate of population growth but poor competitive ability) at one end and K-selected species (those with a lower rate of population growth but better competitive ability) at the other. (Figures 56.14, 56.15, Table 56.2)

56.4 Human Population Growth

- Up to the present, human population growth has fit an exponential growth pattern. (Figure 56.16)

- Human populations have been moving from states of high birth and death rates to low birth and death rates, a shift called the demographic transition. (Figures 56.17, 56.18)

- Differences in the age structure of a population, the numbers of individuals in each age group, are also characteristic of the demographic transition. (Figure 56.19)

- Although they have been declining worldwide, total fertility rates (TFRs) differ markedly in less-developed and more-developed countries. Predicting the growth of the human population depends on the total fertility rate that is projected. (Figures 56.20, 56.21)

- The ecological footprint refers to the amount of productive land needed to support each person on Earth. Because people in many countries live in a nonsustainable manner, globally we are already in an ecological deficit. (Figure 56.22)

Assess and Discuss

Test Yourself

1. A student decides to conduct a mark-recapture experiment to estimate the population size of mosquitofish in a small pond near his home. In the first catch, he marked 45 individuals. Two weeks later, he captured 62 individuals, of which 8 were marked. What is the estimated size of the population based on these data?

   a. 134   c. 558   e. 22,320
   b. 349   d. 1,016

Questions 2–4 refer to the table that follows:

Age	n_x	d_x	l_x	m_x	$l_x m_x$
0	100	35	1.00	0	0
1	65	?	0.65	0	0
2	45	15	?	3	1.35
3	30	20	0.30	1	?
4	10	10	0.10	1	0.10
5	0	0	0.00	1	0.0

2. How many individuals die between their first and second birthday?
   a. 65   c. 35   e. 20
   b. 45   d. 25

3. What proportion of newborns survive to age 2?
   a. 0.55   c. 0.35   e. 0.15
   b. 0.45   d. 0.20

4. What is the net reproductive rate?
   a. 5   c. 1.75   e. 0.80
   b. 2.5   d. 1.45

5. _____ survivorship curves are usually associated with organisms that have high mortality rates in the early stages of life.
   a. Type I   c. Type III   e. Types II and III
   b. Type II   d. Types I and II

6. If the net reproductive rate (R_0) is equal to 0.5, what assumptions can we make about the population?
   a. This population is essentially not changing in numbers.
   b. This population is in decline.
   c. This population is growing.
   d. This population is in equilibrium.
   e. none of the above

7. The maximum number of individuals a certain area can sustain is known as
   a. the intrinsic rate of growth.   d. the logistic equation.
   b. the resource limit.   e. the equilibrium size.
   c. the carrying capacity.

Questions 8 and 9 refer to the following generalized growth patterns as plotted on arithmetic scales. Match the following descriptions with the patterns indicated below.

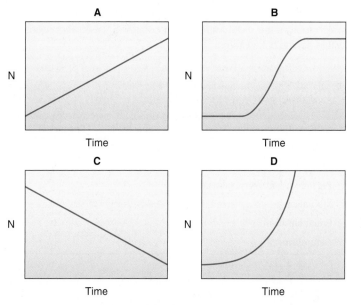

Each pattern may be used once, more than once, or not at all.

8. Which pattern is found where a population exhibits a constant per capita growth rate?
   a. A   d. D
   b. B   e. none of the above
   c. C

9. Which pattern is found where a population is heading toward extinction?
   a. A   d. D
   b. B   e. none of the above
   c. C

10. The amount of land necessary for survival for each person in a sustainable world is known as
    a. the sustainability level.   d. survival needs.
    b. an ecological impact.   e. all of the above.
    c. an ecological footprint.

Conceptual Questions

1. Define population and population ecology.

2. Describe and list the assumptions of the mark-recapture technique.

3. Using the logistic equation, calculate population growth when $K = 1,000$, $N = 500$, and $r = 0.1$ and when $K = 1,000$, $N = 100$, and $r = 0.1$. Compare the results to those shown in Section 56.3, where $K = 1,000$, $N = 900$, and $r = 0.1$.

Collaborative Questions

1. Discuss the two main extremes of life history strategies.

2. Describe where students on campus might show each type of dispersion pattern, and explain why this might occur.

Online Resource

www.brookerbiology.com

Stay a step ahead in your studies with animations that bring concepts to life and practice tests to assess your understanding. Your instructor may also recommend the interactive ebook, individualized learning tools, and more.

Species Interactions

57

An example of species interactions. This shark is feeding on a ray, which, in turn, feeds on bay scallops.

I n 2007, Ransom Myers and his colleagues showed that overfishing severely depleted the numbers of 11 shark species that occurred along the eastern seaboard of the U.S. Several shark species had declined by over 99% since the 1950s. Because of strong interactions between the sharks and other marine species, this drastic reduction of the shark population had at least two other effects. First, there was a large increase in the main prey species of the sharks, rays and skates. Second, the increase in rays and skates reduced the densities of their prey, bay scallops (*Argopecten irradians*). Such losses contributed to the closure of the bay scallop industry in North Carolina.

In this chapter, we turn from considering populations on their own to investigating how they interact with populations of other species that live in the same locality. Such species interactions can take a variety of forms (**Table 57.1**). **Competition** is an interaction that affects both species negatively (−/−), as both species compete over food or other resources. Sometimes this interaction is quite one-sided, where it is detrimental to one species and neutral to the other, an interaction called **amensalism** (−/0). **Predation**, **herbivory**, and **parasitism** all have a positive effect on one species and a negative

effect on the other (+/−). However, although predators always kill their prey, the hosts of parasites and herbivores often survive their attacks. **Mutualism** is an interaction in which both species benefit (+/+), whereas **commensalism** benefits one species and leaves the other unaffected (+/0). Last is the interaction, or rather lack of interaction, termed **neutralism**, when two species occur together but do not interact in any measurable way (0/0). Neutralism may be quite common, but few researchers have quantified its occurrence.

To illustrate how species interact in nature, let's consider a rabbit population in a woodland community (**Figure 57.1**). To determine what factors influence the size and density of the rabbit population, we need to understand each of its possible species interactions. For example, the rabbit population could be limited by the quality of available food. It is also likely that other species, such as deer, use the same resource and thus compete with the rabbits. The rabbit population could be limited by predation from foxes or by the virus that causes the disease myxomatosis, which is usually spread by fleas and mosquitoes. It is also possible that other associations, such as mutualism or commensalism, may occur.

This chapter examines each of these species interactions in turn, beginning with competition, an important interaction among species. We conclude with a discussion of conceptual models of species interactions that ecologists use when trying to determine which factors are most important in controlling population densities within ecological systems.

Table 57.1	Summary of the Types of Species Interactions	
Nature of interaction	**Species 1***	**Species 2***
Competition	−	−
Amensalism	−	0
Predation, herbivory, parasitism	+	−
Mutualism	+	+
Commensalism	+	0
Neutralism	0	0

*+ = positive effect; − = negative effect; 0 = no effect.

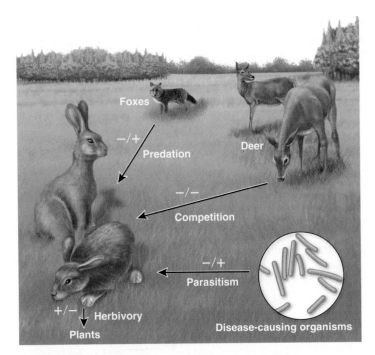

Figure 57.1 **Species interactions.** These rabbits can interact with a variety of species, experiencing predation by foxes, competition with deer for food, and parasitism from various disease-causing organisms. The species interaction known as herbivory also occurs when rabbits feed on various plants. The effects of each species on the other are shown by the terms assigned to the arrows, as discussed in the text.

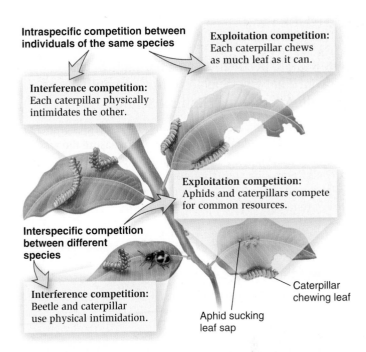

Figure 57.2 **The different types of competition in nature.**

Concept check: How would you classify competition between vultures feeding on roadkill?

57.1 Competition

In this section, we will see how ecologists have studied different types of competition and how they have shown that the competitive effects of one species on another can change as the environment changes or as different predators or parasites are present. Although species may compete for resources, we will also learn how sufficient differences in lifestyle or morphology can exist that reduce the overlap in their ecological niches, thus allowing them to coexist.

Several Different Types of Competition Occur in Nature

Several different types of competition are found in nature (Figure 57.2). Competition may be **intraspecific**, between individuals of the same species, or **interspecific**, between individuals of different species. Competition can also be characterized as exploitation competition or as interference competition. In **exploitation competition**, organisms compete indirectly through the consumption of a limited resource, with each obtaining as much as it can. For example, when fly maggots compete in a mouse carcass, not all the individuals can command enough of the resource to survive and become adult flies. In **interference competition**, individuals interact directly with one another by physical force or intimidation. Often this force is ritualized into aggressive behavior associated with territoriality (as discussed in Chapter 55). In these cases, strong individuals survive and take the best resource, and weaker ones perish or, at best, survive under suboptimal conditions.

Competition between species is not always equal. In many cases, one species has a detrimental effect on another but is unaffected itself. Such interactions are called amensalism. An extreme asymmetric competition can be observed between plants, where one species secretes and produces chemicals from its roots that inhibit the growth of another species. In Chapter 54's Feature Investigation, we saw how diffuse knapweed, an introduced species, secretes root chemicals called allelochemicals into the surrounding environment that kill the roots of native grass species. This phenomenon is termed **allelopathy**.

Researchers have established that one of the best methods of studying competition between two species is to temporarily remove one of them and examine the effect on the other species. A now-classic example of this method involved a study of the interactions between two species of barnacles conducted on the west coast of Scotland, as described next.

FEATURE INVESTIGATION

Connell's Experiments with Barnacle Species Showed That One Species Can Competitively Exclude Another in a Natural Setting

The most direct method of assessing the effect of competition is to remove individuals of species A and measure the response of species B. Often, however, such manipulations are difficult to conduct outside the laboratory. If individuals of species A are removed, what is to stop them from migrating back into the area of removal?

In 1954, ecologist Joseph Connell conducted an experiment that overcame this problem. *Chthamalus stellatus* and *Semibalanus balanoides* (formerly known as *Balanus balanoides*) are two species of barnacles that dominate the Scottish coastline. Each organism's ecological niche on the intertidal zone was well defined. *Chthamalus* occurs in the upper intertidal zone, and *Semibalanus* is restricted to the lower intertidal zone. Connell sought to determine what the range of *Chthamalus* adults might be in the absence of competition from *Semibalanus* (Figure 57.3).

Figure 57.3   Connell's experimental manipulation of species indicated the presence of competition.

HYPOTHESIS Adult *Chthamalus stellatus* were being competitively excluded from the lower intertidal zone by the species *Semibalanus balanoides*.

STARTING LOCATION The intertidal zone of the rocky shores of the Scottish coast, where the two species of barnacles occur.

Experimental level

1   Transfer rocks containing young and mature *Chthamalus* from the upper intertidal zone to the lower intertidal zone, and fasten them down in the new location with screws.

2   Allow *Semibalanus* to colonize the rocks.

3   After the colonization period is over, remove *Semibalanus* from half of each rock with a needle (leaving the other half undisturbed). Return the rocks to the lower intertidal zone, and fasten them down once again.

4   Monitor the survival of *Chthamalus* on both sides of the rocks.

Chthamalus grows on the side where *Semibalanus* has been removed, indicating that *Semibalanus* may exclude *Chthamalus* from certain habitats.

5 **THE DATA**

Rock No.	Side of rock	% *Chthamalus* mortality over 1 year	
		Young barnacles	Mature barnacles
13b	*Semibalanus* removed	35	0
	Semibalanus not removed	90	31
12a	*Semibalanus* removed	44	37
	Semibalanus not removed	95	71
14a	*Semibalanus* removed	40	36
	Semibalanus not removed	86	75

6 **CONCLUSION** The data from this study indicate that *Semibalanus* is found on the lower rock face because it outcompetes *Chthamalus*. Other studies indicate that *Chthamalus* occupies the upper rock face because it is more resistant to desiccation.

7 **SOURCE** Connell, J.H. 1961. The influence of interspecific competition and other factors on the distribution of the barnacle *Chthamalus stellatus*. *Ecology* 42:710–732.

To do this, Connell obtained rocks from high on the rock face, just below the high-tide level, where only *Chthamalus* grew. These rocks already contained young and mature *Chthamalus*. He then moved the rocks into the *Semibalanus* zone, fastened them down with screws, and allowed *Semibalanus* to also colonize them. Once *Semibalanus* had colonized these rocks, he took the rocks out, removed all the *Semibalanus* organisms from one side of the rocks with a needle, and then returned the rocks to the lower intertidal zone, screwing them down once again. As seen in the data, the mortality of *Chthamalus* on rock halves with *Semibalanus* was fairly high. On the *Semibalanus*-free halves, however, *Chthamalus* survived well.

In other studies, Connell also monitored survival of natural patches of both barnacle species where both occurred on the intertidal zone at the upper margin of the *Semibalanus* distribution. In a period of unusually low tides and warm weather, when no water reached any barnacles for several days, desiccation became a real threat to the barnacles' survival. During this time, young *Semibalanus* suffered a 92% mortality rate, and older individuals, a 51% mortality rate. At the same time, young *Chthamalus* experienced a 62% mortality rate compared with a rate of only 2% for more-resistant older individuals.

Clearly, *Semibalanus* is not as resistant to desiccation as *Chthamalus* and could not survive in the upper intertidal zone where *Chthamalus* occurs. *Chthamalus* is more resistant to desiccation than *Semibalanus* and can be found higher in the intertidal zone. Thus, while the lower limit of *Chthamalus* was set by competition with *Semibalanus*, the upper limit was controlled by desiccation.

Although the potential distribution, the fundamental niche, of *Chthamalus* extends over the entire intertidal zone, its actual distribution, the realized niche, is restricted to the upper zone. Connell's experiments were among the first to show that in a natural environment, one species can actually outcompete another, affecting its distribution within a habitat.

Experimental Questions

1. Describe the realized niches for the two species of barnacles used in Connell's experiment.

2. Outline the procedure that Connell used in the experiments.

3. How did Connell explain the presence of *Chthamalus* in the upper intertidal zone if *Semibalanus* was shown to outcompete the species in the first experiment?

The Outcome of Competition Can Vary with Changes in the Biotic and Abiotic Environments

Using experiments to temporarily remove individuals of one species and examine the results on the remaining species, as Connell did, is often the most direct method to investigate the effects of competition. It is especially valuable to do such manipulations in the field, because organisms can then also interact with all other organisms in their environment, or, as ecologists say, natural variation can be factored in.

Alternatively, in laboratory experiments, the investigator can often control and vary important factors systematically. In the late 1940s, biologist Thomas Park began a series of experiments examining competition between two flour beetles, *Tribolium castaneum* and *Tribolium confusum*, in which he systematically varied temperature, moisture, and a variety of other factors. These beetles were well suited to study in the laboratory, because large colonies could be grown in relatively small containers containing dry food medium. Thus, many replicates of each experiment were possible to confirm that results were consistent.

Park conducted the experiments by putting the same number of beetles of both species into a container and counting the number of each type that were still alive after a given time interval. *T. confusum* usually won, but in initial experiments, the beetle cultures were infested with a protozoan parasite called *Adelina triboli* that killed some beetles and preferentially killed *T. castaneum* individuals, influencing the outcome. In these experiments, *T. confusum* won in 66 out of 74 replicates (89%) because it was more resistant to the parasite. In subsequent experiments, the parasite was removed, and *T. castaneum* won in 12 out of 18 replicates (67%). Two things were evident from this experiment. First, the presence of a parasite (a biotic factor) was shown to alter the outcome of competition. Second, with or without the parasite, there was no absolute victor. For example, even with the parasite present, sometimes *T. castaneum* won. Thus, some random variation, which we call stochasticity, was evident when these species competed.

Park then began to vary the abiotic environment and found that competitive ability was greatly influenced by environmental factors (**Figure 57.4**). Generally, *T. confusum* did better in dry conditions, and *T. castaneum*, in moist conditions. However, *T. confusum* also won in cold-wet conditions. Once again, some stochasticity occurred, and victory was not absolute. Later, it was found that the mechanism of competition was largely predation of eggs and pupae by larvae and adults, which, as they ate the flour medium, would also bite the stationary eggs and pupae, killing them. In general, the species were mutually antagonistic; that is, they bit more eggs and pupae of the other species than they did of their own.

Park's important series of experiments illustrated that the results of competition could vary as a function of at least three factors: parasites, temperature, and moisture. The experiments also showed how much stochasticity occurred, even in controlled laboratory conditions.

Field Studies Show Competition Occurs Frequently in Nature

By reviewing studies that have investigated competition in nature, we can see how frequently it occurs and in what particular circumstances it is most important. In a 1983 review of field studies by Joseph Connell, competition was found in 55% of 215 species surveyed, demonstrating that it is indeed frequent in nature. Generally in studies of single pairs of species utilizing the same resource, competition is almost always reported (90%), whereas in studies involving more species, the frequency of competition drops to 50%. Why should this be the case? Imagine a resource such as a series of different-sized grains with four species—ants, beetles, mice, and birds—feeding on it (**Figure 57.5a**). The ants feed on the smallest grain, the beetles and mice on the intermediate sizes, and the birds, on the largest. If only adjacent species competed with each other, competition would be expected only between the ant–beetle, beetle–mouse, and mouse–bird. Thus, competition would be found in only three out of the six possible species pairs (50%). Naturally, the percent would vary according to the number of

Figure 57.4 **Influence of abiotic factors on competition between *T. castaneum* and *T. confusum*.** Results of competition between the flour beetles *Tribolium castaneum* and *Tribolium confusum* show that each species usually performs better in a given habitat; for example, *T. confusum* does better in dry conditions.

species on the resource spectrum. If there were only three species along the spectrum, we would expect competition in two of the three pairs (67%). If there were just two species utilizing the resource spectrum, however, we would expect competition in almost 100% of the cases (**Figure 57.5b**).

Some other general patterns were evident from Connell's review. Plants showed a high degree of competition, perhaps because they are rooted in the ground and cannot easily escape or perhaps because they are competing for the same set of limiting nutrients—water, light, and minerals. Marine organisms tended to compete more than terrestrial ones, perhaps because many of the species studied lived in the intertidal zone and were attached to the rock face, in a manner similar to that of plants. Because the area of the rock face is limited, competition for space is quite important, as we saw with Connell's work with barnacles in Scotland.

Species May Coexist If They Do Not Occupy Identical Niches

Although competition is common, researchers have proposed several mechanisms by which two competing species can coexist. One states that similar species can coexist if they occupy different niches. In 1934, the Russian microbiologist Georgyi Gause began to study competition between three protist species, *Paramecium aurelia*, *Paramecium bursaria*, and *Paramecium caudatum*, all of which fed on bacteria and yeast, which, in turn, fed on an oatmeal medium in a culture tube in the laboratory. The bacteria occurred more in the oxygen-rich upper part of the culture tube, and the yeast in the oxygen-poor lower part of the tube. Because each species was a slightly different size, Gause calculated population growth as a combination of numbers of individuals per milliliter of solution multiplied by

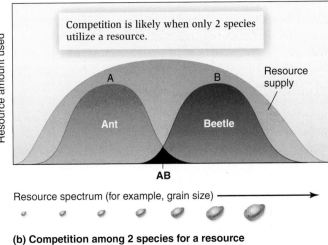

(a) Competition among 4 species for a resource

(b) Competition among 2 species for a resource

Figure 57.5   The frequency of competition according to the number of species involved.   **(a)** Resource supply and utilization curves of four species, A, B, C, and D, along the spectrum of a hypothetical resource such as grain size. If competition occurs only between species with adjacent resource utilization curves, competition would be expected between three of the six possible pairings: A and B; B and C; and C and D. **(b)** When only two species utilize a resource set, competition would nearly always be expected between them.

Concept check: *If five species utilized the resource set in part (a), what percent of the interactions would be competitive?*

their unit volume to give a population volume for each species. When grown separately, population volume of all three *Paramecium* species followed a logistic growth pattern (**Figure 57.6a**). When Gause cultured *P. caudatum* and *P. aurelia* together, *P. caudatum* went extinct (**Figure 57.6b**). Both species utilized bacteria as food, but *P. aurelia* grew at a rate six times faster than *P. caudatum* and was better able to convert the food into offspring.

However, when Gause cultured *P. caudatum* and *P. bursaria* together, neither went extinct (**Figure 57.6c**). The population volume of each was much less compared to when they were grown alone, because some competition occurred between them. Gause discovered, however, that *P. bursaria* was better able to utilize the yeast in the lower part of the culture tubes. From these experiments, Gause concluded that two species with exactly the same requirements cannot live together in the same place and use the same resources, that is, occupy the same niche. His conclusion was later termed the **competitive exclusion principle**.

If complete competitors drive one species to local extinction, at least in the laboratory, how different do two species have to be to coexist, and in what features do they usually differ? To address such questions, in 1958, ecologist Robert Mac-Arthur examined coexistence between five species of warblers feeding within spruce trees in New England. All belonged to the genus *Dendroica*, so one would expect these closely related bird species to compete strongly, possibly sufficiently strongly to cause extinctions. MacArthur found that the species occupied different heights and portions in the tree, and therefore, each probably fed on a different range of insects (**Figure 57.7**). In addition, the Cape May warbler fed on flying insects and tended to remain on the outside of the trees.

The term **resource partitioning** describes the differentiation of niches, both in space and time, that enables similar species to coexist in a community, just as the five species of warblers feeding in different parts of a spruce tree. We can think of resource partitioning as reflecting the results of past competition, in which competition leads the inferior competitor to eventually occupy a different niche. British ornithologist David Lack examined competition and coexistence among about 40 species of British passerines, or perching birds (**Figure 57.8**). As a group, these perching birds had fairly similar lifestyles. Most segregated according to some resource factor, with habitat being the most common one. For example, although all the passerines fed on insects, some would feed exclusively in grasslands, others in forests, some low to the ground, and others high in trees, where the insects present would likely be different. Birds also segregated by size—so bigger species would take different-sized food than smaller species—and by feeding habit—with some feeding on insects on tree branches, others on tree trunks, and so on. Some species also fed in different winter ranges, while others occurred in different parts of the country (separation by geography). About 15% of bird species showed no segregation at all.

What about the species that do not appear to live in different habitats or have different food habits? To answer this question, researchers have looked at differences in physical form, or morphology, between coexisting species.

Morphological Differences May Allow Species to Coexist

Although the competitive exclusion principle acknowledges that complete competitors cannot coexist, some partial level of competition may exist that is not severe enough to drive one of the competitors to extinction or to a different niche. In 1959, biologist G. Evelyn Hutchinson examined the sizes of mouthparts or other body parts important in feeding and compared their sizes

(a) Each *Paramecium* species grown alone

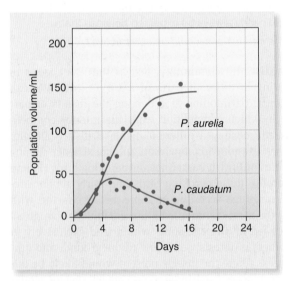

(b) Competition between *P. aurelia* and *P. caudatum*

(c) Competition between *P. caudatum* and *P. bursaria*

Figure 57.6 **Competition among *Paramecium* species.** (a) When grown alone, each of three species, *Paramecium aurelia*, *Paramecium bursaria*, and *Paramecium caudatum*, grows according to the logistic model. (b) When *P. aurelia* is grown with *P. caudatum*, the density of *P. aurelia* is lowered compared to when grown alone, and *P. caudatum* goes extinct. (c) When *P. caudatum* is grown with *P. bursaria*, the population densities of both are lowered, but they coexist.

between species when they were **sympatric** (occurring in the same geographic area) and **allopatric** (occurring in different geographic areas). Hutchinson's hypothesis was that when species were sympatric, each species tended to specialize on different types of food. This was reflected by differences in the size of body parts associated with feeding, also called feeding characters. The tendency for two species to diverge in morphology and thus resource use because of competition is called **character displacement**. Alternatively, in areas where species were allopatric, there was no need to specialize on a particular prey type, so the size of the feeding character did not evolve to become larger or smaller; rather it retained a "middle of the road" size that allowed species to exploit the largest range of prey size distribution.

One of the classic cases of character displacement involves a study of Galápagos finches, several closely related species of finches Charles Darwin discovered on the Galápagos Islands (refer back to the Feature Investigation in Chapter 23). When two species, *Geospiza fortis* and *Geospiza fuliginosa*, are sympatric, their beak sizes (bill depths) are different: *G. fortis* has a larger bill depth, which enables it to feed on bigger seeds, whereas *G. fuliginosa* has a smaller bill depth, which enables it to crack small seeds more efficiently. However, when both species are allopatric, that is, existing on different islands, their bills are more similar in depth. Researchers studying *Geospiza* concluded that the bill depth differences evolved in ways that minimized competition.

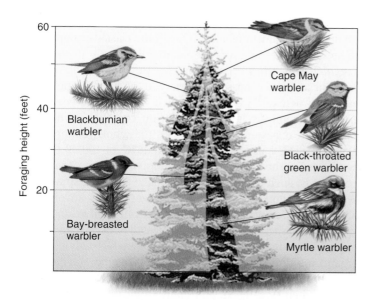

Figure 57.7 **Resource partitioning.** Among five species of warblers feeding in North American spruce trees, each species prefers to feed at a different height and portion of the tree, thus reducing competition.

How great do differences between characters have to be in order to permit coexistence? Hutchinson noted that the ratio between feeding characters when species were sympatric (and thus competed) averaged about 1.3 (**Table 57.2**). In contrast, the ratio between feeding characters when species were allopatric (and did not compete) was closer to 1.0. Hutchinson proposed that the value of 1.3, a roughly 30% difference, could be used as an indication of the amount of difference necessary to permit two species to coexist. One problem with Hutchinson's ratio is that some differences of 1.3 between similar species might have evolved for reasons other than competition. Some ecologists have argued that we should not conclude that a 30% difference is a strong indicator of coexistence. Nevertheless, while the actual ratio may be disputed, Hutchinson's findings have shown that competition in nature can cause character displacement.

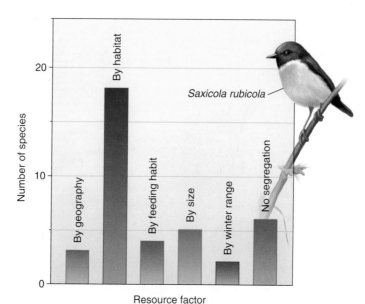

Figure 57.8 **Segregation according to resource factor.** Among 40 species of passerine birds, most segregation is by habitat, followed by size, feeding habit, geography, and type of winter range they forage in. In about 15% of cases, no obvious segregation was observed. More than half of all bird species, including *Saxicola rubicola*, are passerines, also known as perching birds.

Concept check: Do you think these results for passerine birds are typical for most other species? For example, do most other species segregate by habitat?

57.2 Predation, Herbivory, and Parasitism

Predation, herbivory, and parasitism are interactions that have a positive effect for one species and a negative effect for the other. These categories of species interactions can be classified according to how lethal they are for the prey and the length of association between the consumer and prey (**Figure 57.9**).

Table 57.2	Comparison of Feeding Characters of Sympatric and Allopatric Species				
		Measurement (mm) when		**Ratio* when**	
Animal (character)	**Species**	**Sympatric**	**Allopatric**	**Sympatric**	**Allopatric**
Weasels (skull)	*Mustela erminea*	50.4	46.0	1.28	1.07
	Mustela ivalis	39.3	42.9		
Mice (skull)	*Apodemus flavicollis*	27.0	26.7	1.09	1.04
	Apodemus sylvaticus	24.8	25.6		
Nuthatches (beak)	*Sitta tephronota*	29.0	25.5	1.23	1.02
	Sitta neumayer	23.5	26.0		
Galápagos finches (beak)	*Geospiza fortis*	12.0	10.5	1.43	1.13
	Geospiza fuliginosa	8.4	9.3		

*Ratio of the larger to smaller character.

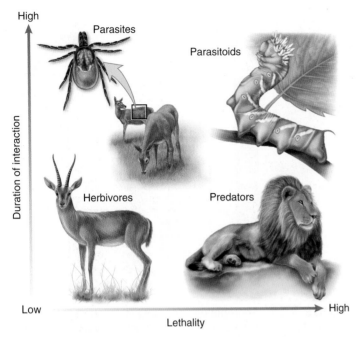

High

Parasites

Parasitoids

Duration of interaction

Herbivores

Predators

Low

High

Lethality

Figure 57.9 Possible interactions between populations. Lethality represents the probability that an interaction results in the death of the prey. Duration represents the length of the interaction between the consumer and the prey.

Concept check: Where might omnivores fit in this figure?

Each has particular characteristics that set it apart. Herbivory usually involves nonlethal predation on plants, whereas predation generally results in the death of the prey. Parasitism, like herbivory, is typically nonlethal and differs from predation in that the adult parasite typically lives and reproduces for long periods in or on the living host, as in Chinese liver flukes (refer back to Figure 33.9). Parasitoids, insects that lay eggs in living hosts, have features in common with both predators and parasites. They always kill their prey, as predators do, but unlike predators, which immediately kill their prey, parasitoids kill the host more slowly. Parasitoids are common in the insect world and include parasitic wasps and flies that feed on many other insects.

In this section, we begin by looking at antipredator strategies and how, despite such strategies, predation remains a factor affecting the density of prey. We survey the strategies plants use to deter herbivores and how, in turn, herbivores overcome host plant defenses. Finally, we investigate parasitism, which may be the predominant lifestyle on Earth, and explore the growing role of genomics in the fight against parasites.

Animals Have Evolved Many Antipredator Strategies

The variety of strategies that animals have evolved to avoid being eaten suggests that predation is a strong selective force. Common strategies include chemical defense; forms of camouflage and mimicry; displays of intimidation; agility; armor; and altering of reproductive patterns, as in masting.

Chemical Defense A great many species have evolved chemical defenses against predation. One of the classic examples of a chemical defense involves the bombardier beetle (*Stenaptinus insignis*), which has been studied by Tom Eisner and coworkers. These beetles possess a reservoir of hydroquinone and hydrogen peroxide in their abdomen. When threatened, they eject the chemicals into an "explosion chamber," where the subsequent release of oxygen causes the whole mixture to be violently ejected as a hot spray (about 88°C, or 190°F) that can be directed at the beetle's attackers (**Figure 57.10a**). Many other arthropods, such as millipedes, also have chemical sprays, and the phenomenon is also found in vertebrates, as anyone who has had a close encounter with a skunk can testify.

Often associated with a chemical defense is an **aposematic coloration**, or warning coloration, which advertises an organism's unpalatable taste. For instance, the ladybird beetle's bright red color warns of the toxic defensive chemicals it exudes when threatened, and many tropical frogs have bright warning coloration that calls attention to their skin's lethality (**Figure 57.10b**). In the 1960s, entomologist Lincoln Brower and coworkers showed that after inexperienced blue jays ate a monarch butterfly and suffered a violent vomiting reaction, they learned to associate the striking orange-and-black appearance of the butterfly with a noxious reaction (refer back to Figure 55.3). Monarch butterfly caterpillars feed exclusively on milkweed, which contains toxic chemicals called cardiac glycosides that pass into the caterpillars. Other animals, such as rattlesnakes, synthesize toxins using their own metabolic processes.

Cryptic Coloration **Cryptic coloration** is an aspect of camouflage, the blending of an organism with the background of its habitat. Cryptic coloration is a common method of avoiding detection by predators. For example, many grasshoppers are green and blend in with the foliage on which they feed. Stick insects mimic branches and twigs with their long, slender bodies. In most cases, these animals stay perfectly still when threatened, because movement alerts a predator. Maintenance of a fixed body posture is referred to as catalepsis. Cryptic coloration is prevalent in the vertebrate world, too. Many sea horses adopt a body shape and color pattern that are similar to the environment in which they are found (**Figure 57.10c**).

Mimicry **Mimicry**, the resemblance of a species (the mimic) to another species (the model), also secures protection from predators. There are two major types of mimicry. In **Müllerian mimicry**, two or more toxic species converge to look the same, thus reinforcing the basic distasteful design. One example is the black and yellow striped bands of several different types of bees and wasps. Müllerian mimicry is also found among noxious Amazonian butterflies. The viceroy butterfly (*Limenitis archippus*) and the monarch butterfly (*Danaus plexippus*) are examples of Müllerian mimicry. Both species are unpalatable and look similar, but the viceroy can be distinguished from the monarch by a black line that crosses its wings.

Batesian mimicry is the mimicry of an unpalatable species (the model) by a palatable one (the mimic). Some of the

(a) As it is held by a tether attached to its back, this bombardier beetle (*Stenaptinus insignis*) directs its hot, stinging spray at a forceps "attacker."

(b) Aposematic coloration advertises the poisonous nature of this blue poison arrow frog (*Dendrobates azureus*) from South America.

(c) Cryptic coloration allows this Pygmy sea horse (*Hippocampus bargibanti*) from Bali to blend in with its background.

(d) In this example of Batesian mimicry, an innocuous scarlet king snake (*Lampropeltis triangulum*) (left) mimics the poisonous coral snake (*Micrurus nigrocinctus*) (right).

(e) In a display of intimidation, this porcupine fish (*Diodon hystrix*) puffs itself up to look threatening to its predators.

Figure 57.10   Antipredator adaptations.

Concept check: *According to the classification of species interactions in Table 57.1, how would you classify Batesian and Müllerian mimicry?*

best examples involve flies, especially hoverflies of the family Syrphidae, which are striped black and yellow and resemble stinging bees and wasps but are themselves harmless. Among vertebrates, the nonvenomous scarlet king snake (*Lampropeltis triangulum*) mimics the venomous coral snake (*Micrurus nigrocinctus*), thereby gaining protection from would-be predators (**Figure 57.10d**).

Displays of Intimidation   Some animals put on displays of intimidation in an attempt to discourage predators. For example, a cat arches its back to make itself appear larger, frilled lizards extend their collars when frightened to create this same effect, and porcupine fish inflate themselves to large proportions when threatened (**Figure 57.10e**). All of these animals use displays to deceive potential predators about the ease with which they can be eaten.

Fighting   Though many animals developed horns and antlers for sexual selection, they can also be used in defense against predators (refer back to Figure 34.25). Invertebrate species often have powerful claws, pincers, or, in the case of scorpions, venomous stingers that can be used in defense as well as offense.

Agility   Some groups of insects, such as grasshoppers, have a powerful jumping ability to escape the clutches of predators. Many frogs are prodigious jumpers, and flying fish can glide above the water to escape their pursuers.

Armor   The shells of tortoises and turtles are a strong means of defense against most predators, as are the quills of porcupines (refer back to Figure 34.23c). Many beetles have a tough exoskeleton that protects them from attack from other arthropod predators such as spiders.

Masting   Masting is the synchronous production of many progeny by all individuals in a population to satiate predators and thereby allow some progeny to survive. Masting is more commonly discussed in relation to seed predation in trees, which tend to have years of unusually high seed production that reduces predation. However, a similar phenomenon is exhibited by the emergence of 13-year and 17-year periodical cicadas (genus *Magicicada*). These insects are termed periodical because the emergence of adults is highly synchronized to occur once every 13 or 17 years. Adult cicadas live for only a few weeks, during which time females mate and deposit eggs on the twigs of trees. The eggs hatch 6 to 10 weeks later, and the nymphs drop to the ground and begin a long subterranean development, feeding on the contents of the xylem of roots. Because the xylem is low in nutrients, it takes many years for nymphs to develop, though there appears to be no physiological reason why some cicadas couldn't emerge after, say, 12 years of feeding, and others after 14. Their synchrony of emergence is thought to maximize predator satiation. Worth noting in this context is the fact that both 13 and 17 are prime numbers, and thus predators on a shorter multiannual cycle cannot repeatedly utilize this resource. For example, a predator that bred every 3 years could not rely on cicadas always being present as a food supply.

How common is each of these defense types? No one has done an extensive survey of the entire animal kingdom. However, in 1989, Brian Witz surveyed studies that documented antipredator mechanisms in arthropods, mainly insects. By far the most common antipredator mechanisms were chemical defenses and associated aposematic coloration, noted in 51% of the examples. All other types of defense mechanisms occurred with considerably less frequency.

Despite the Impressive Array of Defenses, Predators Can Still Affect Prey Densities

The importance of predation on prey populations may be dependent on whether the system is donor controlled or predator controlled. In a donor-controlled system, prey supply is determined by factors other than predation, such as food supply, so that removal of predators has no significant effect on prey density. Examples include predators that feed on intertidal communities in which space is the limiting factor that controls prey populations.

In a predator-controlled system, the action of predator feeding eventually reduces the supply of prey. Therefore, the removal of predators would probably result in large increases in prey abundance. Research studies have shown that predators can have a significant effect on prey populations. Considerable data exist on the interaction of the Canada lynx (*Lynx canadensis*) and its prey, snowshoe hares (*Lepus americanus*), because of the value of the pelts of both animals. In 1942, British ecologist Charles Elton analyzed the records of furs traded by trappers to the Hudson's Bay Company in Canada over a 100-year period. Analysis of the records showed that a dramatic 9- to 11-year cycle existed for as long as records had been kept (**Figure 57.11**). As hare density increases, there is an increase in density of the lynx, which then depresses hare numbers. This is followed by a decline in the number of lynx, and the cycle begins again. By tracking individual hares by using radio collars, researchers were able to determine that 90% of individuals died of predation. However, more recent research has shown that hare densities may increase in times of food surpluses. When researchers added food supplements to large experimental areas containing both hares and lynxes, the hare densities increased threefold but still continued to cycle.

Invasive species provide striking examples of the effects of predators. The brown tree snake (*Boiga irregularis*) was

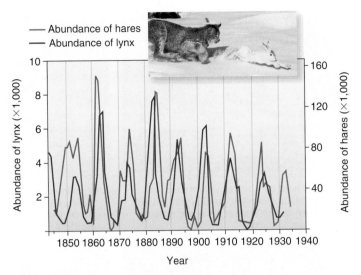

Figure 57.11 **Effect of predator on prey populations.** The 9- to 11-year oscillation in the abundance of the snowshoe hare (*Lepus americanus*) and the Canada lynx (*Lynx canadensis*) was revealed from pelt trading records of the Hudson's Bay Company.

inadvertently introduced by humans to the island of Guam, in Micronesia, shortly after World War II. The growth and spread of its population over the next 40 years closely coincided with a precipitous decline in the island's forest birds. On Guam, the snake had no natural predators to control it. Because the birds on Guam did not evolve with the snake, they had no defenses against it. Eight of the island's 11 native species of forest birds went extinct by the 1980s, such as the Guam rail and Micronesian kingfisher (**Figure 57.12**).

One of the biggest reductions in prey in response to predation has been the systematic decline of various whale species as a result of the human whaling industry. The history of whaling has been characterized by a progression from larger, more valuable or easily caught species to smaller, less valuable or easily caught ones, as numbers of the original targets have been depleted (**Figure 57.13**). Belatedly, the International Whaling Commission enacted a moratorium on all commercial whaling in 1986. Following the moratorium, the populations of some whales have increased. Blue whales are thought to have quadrupled their numbers off the California coast during the 1980s, and numbers of the California gray whales have recovered to prewhaling levels, showing the impact that an absence of a predator can have.

Ecologists have found that in nearly 1,500 predator-prey studies, over two-thirds (72%) showed a large depression of prey density by predators. Thus, we can conclude that in the majority of cases, predators influence the abundance of their prey in their native environment. The variety of antipredator mechanisms discussed earlier also shows how predation is important enough to select for the evolution of chemical defenses, camouflage, and mimicry in prey. Taken together, these data indicate that predation is a powerful force in nature.

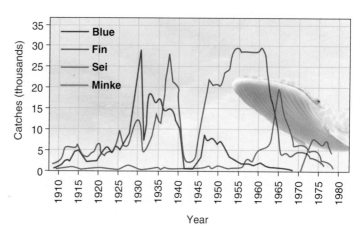

Figure 57.13 Sequential decline of whale catches in the Antarctic due to human predation. Whale catches are believed to be directly related to whale population sizes. The catches of the blue whale, the first species to be strongly affected by human predation, started a precipitous decline in the 1940s, as the whale was hunted to very low levels. Humans then began hunting more-abundant fin whales, and then sei and minke whales, as each species became depleted.

Figure 57.12 Predation by an invasive species. Population trends for two native Guam birds, as indicated by 100-km roadside surveys conducted from 1976 to 1998, show a precipitous decline because of predation by the introduced brown tree snake.

Concept check: Why can invasive predators have such strong effects on native prey?

Plants and Herbivores May Be Engaged in an Evolutionary Arms Race

Herbivory involves the predation of plants or similar life forms such as algae. Such predation can be lethal to the plants, especially for small species, but often it is nonlethal, because many plant species, particularly larger ones, can regrow. We can distinguish two types of herbivores: generalist herbivores and specialist herbivores. Generalist herbivores can feed on many different plant species and are usually mammals. Specialist herbivores are often restricted to one or two species of host plants and are typically insects. There are, however, exceptions. Pandas are specialists because they feed only on bamboo, and koalas specialize on eucalyptus trees. On the other hand, grasshoppers are generalists, because they feed on a wide variety of plant species, including agricultural crops.

Plants present a luscious green world to any organism versatile enough to use it, so why don't herbivores eat more of the food available to them? After all, unlike most animals, plants cannot move to escape being eaten. Two hypotheses have been proposed to answer the question of why more plant material is not eaten. First, predators and parasites may keep herbivore numbers low, thereby sparing the plants. The many examples of the strength of predation provide evidence for this view. Second, the plant world is not as helpless as it appears. The sea of green is armed with defensive spines, tough cuticles, noxious chemicals, and more. Let's take a closer look at plant defenses against herbivores and the ways that herbivores attempt to overcome them.

Plant Defenses Against Herbivores As described in Chapter 7, an array of unusual and powerful chemicals is present in plants, including alkaloids (nicotine in tobacco, morphine in poppies, cocaine in coca, and caffeine in coffee), phenolics (lignin in wood and tannin in leaves), and terpenoids (in peppermint) (**Figure 57.14a–c**, also refer back to Figure 7.17). Such compounds are not part of the primary metabolic pathway that plants use to obtain energy. They are therefore referred to as **secondary metabolites**. Most of these chemicals are bitter tasting or toxic, and they deter herbivores from feeding. The staggering variety of secondary metabolites in plants, over 25,000, may be testament to the large number of organisms that feed on plants. In an interesting twist, many of these compounds have medicinal properties that are beneficial to humans (see Chapter 7).

In addition to containing chemical compounds, many plants have an array of mechanical defenses, such as thorns and spines (**Figure 57.14d**). In other cases, an organism may provide a plant physical protection against herbivores in return for resources such as light, water, nutrients, or nesting sites (as we will see in Section 57.3).

An understanding of plant defenses is of great use to agriculturalists. The better that crops are defended against pests,

(a) Alkaloids in tobacco

(b) Phenolics in tea

(c) Terpenoids in peppermint

(d) Thorns on rose stems

Figure 57.14 **Defenses against herbivory.** Plants possess an array of unusual and powerful chemicals, including **(a)** alkaloids, such as nicotine in tobacco, **(b)** phenolics, such as tannins in tea leaves (near Mount Fuji, Japan), and **(c)** terpenoids in peppermint leaves. **(d)** Mechanical defenses include plant spines and thorns, as on this shrub, *Rosa multiflora*.

Concept check: *Of the defenses shown here, which type of defense would be most effective in deterring invertebrate herbivores?*

the higher the crop yield. The ability of plants to prevent herbivory via either chemical or mechanical defenses is also known as **host plant resistance**. One serious problem associated with commercial development of host plant resistance is that it may take a long time to breed into plants—between 10 and 15 years. This is because of the long time it takes to identify the responsible chemicals and develop the resistant genetic lines. Also, resistance to one pest may come at the cost of increasing susceptibility to other pests. Finally, some pest strains can overcome the plant's mechanisms of resistance.

Despite these problems, host plant resistance is a good tactic for the farmer. After the initial development of resistant varieties, the cost is minimal. Perhaps more importantly, host plant resistance reduces the need for chemical insecticides and is less environmentally harmful, generally having few side effects on other species in the community. About 75% of cropland in the U.S. utilizes pest-resistant plant varieties, most of these being

resistant to plant pathogens. Bt corn is a variety of corn that has been genetically modified to incorporate a gene from the soil bacterium *Bacillus thuringiensis* (Bt) that encodes a protein, Bt toxin, that is toxic to some insects. Genetic engineers have also produced Bt cotton, Bt tomato, and genetically modified varieties of many other crop species.

Overcoming Plant Resistance Herbivores can often overcome plant defenses. They can detoxify many poisons, mainly by two chemical pathways: oxidation and conjugation. Oxidation, the most important of these mechanisms, occurs in the liver of mammals and in the midgut of insects. It involves catalysis of the secondary metabolite to a corresponding alcohol by a group of enzymes known as mixed-function oxidases (MFOs). Conjugation, often the next step in detoxification, occurs by uniting the harmful compound or its oxidation product with another molecule to create an inactive and readily excreted product.

In addition, certain chemicals that are toxic to generalist herbivores actually increase the growth rates of adapted specialist species, which put the chemicals to good use in their own metabolic pathways. The Brassicaceae, the plant family that includes mustard, cabbage, and other species, contains acrid-smelling mustard oils called glucosinolates, the most important one of which is sinigrin. Large white butterflies (*Pieris brassicae*) preferentially feed on cabbage over other plants. They are able to detoxify even high levels of sinigrin. If newly hatched larvae are fed an artificial diet, they do much better when sinigrin is added to it. When larvae are fed cabbage leaves on hatching from eggs and are later switched to an artificial diet without sinigrin, they starve rather than eat. In this case, the secondary metabolite has become an essential feeding stimulant.

A good method to estimate the effects of herbivory on plant populations is to remove the herbivores and examine subsequent growth and reproductive output. Analyses of hundreds of such experiments have yielded several interesting generalizations. First, herbivory in aquatic systems is more extensive than in terrestrial ones. Aquatic systems contain species, such as algae, that are especially susceptible to herbivory, presumably because these organisms are the least sophisticated in terms of their ability to manufacture complex secondary metabolites. In terms of terrestrial systems, grasses and shrubs are significantly affected by herbivores, but woody plants such as trees are less so. Large and long-lived trees can draw on large resource reserves to buffer the impact of herbivores.

Second, invertebrate herbivores such as insects have a stronger effect on plants than vertebrate herbivores such as mammals, at least in terrestrial systems. Thus, while one might consider large grazers like bison in North America or antelopes in Africa to be of huge importance in grasslands, it is more likely that grasshoppers are the more significant herbivores because of their sheer weight of numbers. In forests, invertebrate grazers such as caterpillars have greater access to canopy leaves than vertebrates and are also likely to have a greater effect.

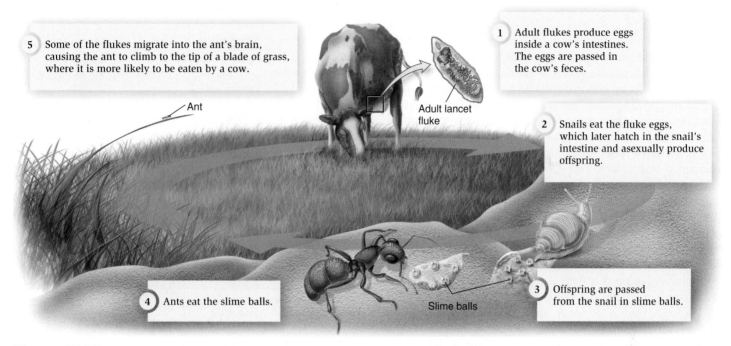

5 Some of the flukes migrate into the ant's brain, causing the ant to climb to the tip of a blade of grass, where it is more likely to be eaten by a cow.

Ant

Adult lancet fluke

1 Adult flukes produce eggs inside a cow's intestines. The eggs are passed in the cow's feces.

2 Snails eat the fluke eggs, which later hatch in the snail's intestine and asexually produce offspring.

4 Ants eat the slime balls.

Slime balls

3 Offspring are passed from the snail in slime balls.

Figure 57.15 **The life cycle of the lancet fluke.** The lancet fluke (*Dicrocoelium dendriticum*) causes behavioral changes in ants (one of its three hosts) that increase its transmission rate.

Parasitism Might Be the Predominant Lifestyle on Earth

When one organism feeds on another but does not normally kill it outright, the organism is termed a **parasite**, and the prey a **host**. Some parasites remain attached to their hosts for most of their life. For example, tapeworms spend their entire adult life inside the host's alimentary canal and even reproduce within their host. Others, such as the lancet fluke, have more complex life cycles that require multiple hosts (**Figure 57.15**). Some, such as ticks and leeches, drop off their hosts after prolonged periods of feeding. Others, like mosquitoes, remain attached for relatively short periods.

Some flowering plants are parasitic on other plants. **Holoparasites** lack chlorophyll and are totally dependent on the host plant for their water and nutrients. One famous holoparasite is the tropical *Rafflesia arnoldii*, which lives most of its life within the body of its host, a *Tetrastigma* vine, which grows in rain forests (**Figure 57.16**). Only the *Rafflesia* flower develops externally. It is a massive flower, 1 m in diameter and the largest known in the world. **Hemiparasites** generally do photosynthesize, but they depend on their hosts for water and mineral nutrients. Mistletoe (*Viscum album*) is a hemiparasite that grows on the stems of trees. Hemiparasites usually have a broader range of hosts than do holoparasites, which may be confined to a single or a few host species.

Parasites that feed on one species or just a few closely related hosts are termed **monophagous**. By contrast, **polyphagous** species can feed on many different host species, often from more than one family. We can also distinguish parasites

Figure 57.16 **A holoparasite.** *Rafflesia arnoldii*, the world's biggest flower, lives as a holoparasite in Indonesian rain forests.

as **microparasites** (for example, pathogenic bacteria), which multiply within their hosts, sometimes within the cells, and **macroparasites** (such as schistosomes), which live in the host but release infective juvenile stages outside the host's body. Usually, the host has a strong immunological response to microparasitic infections. For macroparasitic infections, however, the immunological response is short-lived. Such infections tend to be persistent, and the hosts are subject to continual reinfection.

Last, we can distinguish **ectoparasites**, such as ticks and fleas, which live outside of the host's body, from **endoparasites**, such as pathogenic bacteria and tapeworms, which live inside the host's body. Problems of definition arise with regard to plant parasites, which seem to straddle both camps. For example, some parasitic plants, such as dodder (*Cuscuta pentagona*), an orange, stringlike plant, exist partly outside of the host's body and partly inside (refer back to Figure 37.22). Outgrowths called haustoria penetrate inside the host plant to tap into nutrients. Being endoparasitic on a host seems to require greater specialization than ectoparasitism. Therefore, ectoparasitic animals such as leeches feed on a wider variety of hosts than do endoparasites such as liver flukes.

As we have seen throughout this textbook, parasitism is a common way of life. There are vast numbers of species of parasites, including bacteria, protozoa, flatworms (flukes and tapeworms), nematodes, and various arthropods (ticks, mites, and fleas). Parasites may outnumber free-living species by four to one. Most plant and animal species harbor many parasites. For example, leopard frogs have nematodes in their ears and veins, and flukes in their bladders, kidneys, and intestines. A free-living organism that does not harbor parasitic individuals of a number of species is a rarity.

As with studies of other species interactions, a direct method to determine the effect of parasites on their host population is to remove the parasites and to reexamine the population. However, this is difficult to do, primarily because of the small size and unusual life histories of many parasites, which makes them difficult to remove from a host completely. The few cases of experimental removal confirm that parasites can reduce host population densities. The nests of birds such as blue tits are often infested with parasitic blowfly larvae that feed on the blood of nestlings. In 1997, Sylvie Hurtrez-Boussès and colleagues experimentally reduced blowfly larval parasites of young blue tits in nests in Corsica. Parasite removal was cleverly achieved by taking the nests from 145 nest boxes, removing the young, microwaving the nests to kill the parasites, and then returning the nests and chicks to the wild. The success of chicks in microwaved nests was compared to that in nonmicrowaved (control) nests. The parasite-free blue tit chicks had greater body mass at fledging, the time when feathers first grow (**Figure 57.17**). Perhaps more important was that complete nest failure, that is, death of all chicks, was much higher in control nests than in treated nests.

Because parasite removal studies are difficult to do, ecologists have also examined the strength of parasitism as a mortality factor by studying introduced parasite species. Evidence from natural populations suggests that introduced parasites have substantial effects on their hosts. For example, chestnut blight, a fungus from Asia, was accidentally introduced to New York around 1904. By the 1950s, the airborne fungus had significantly reduced the density of American chestnut trees (*Castanea dentata*) in North Carolina (**Figure 57.18**). Eventually, it eliminated nearly all chestnut trees across North America. In Europe and North America, Dutch elm disease has similarly devastated populations of elms. The disease wiped out 25 million of Britain's original 30 million elm trees between the 1960s and the 1990s. In Italy, canker has had similar severe effects on cypress. The creation of transgenic plants through recombinant DNA methods is a recent development in the fight against plant diseases.

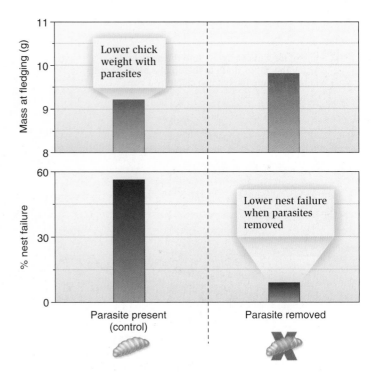

Figure 57.17   **Parasite removal experiments.**   The left side shows the results when blowfly larva were present in the nests of young blue tits. The right side shows the results when these parasites were removed.

Figure 57.18   **Effects of introduced parasites on American chestnut trees.**   The reduction in density of American chestnut trees in North Carolina following the 1904 introduction of chestnut blight disease from Asia shows the severe effect that parasites can have on their hosts. By the 1950s, this once-prevalent species was virtually eliminated.

Genomes & Proteomes Connection

Transgenic Plants May Be Used in the Fight Against Plant Diseases

Many important native forest trees, which are also grown in urban landscapes, have been almost entirely wiped out by diseases spread by the importation of exotic plants. Sudden oak death is a recently recognized disease that is killing tens of thousands of oak trees and other plant species in California. The symptoms vary between species but include leaf spots, oozing of a dark sap through the bark, and twig dieback (**Figure 57.19**). Although sudden oak death is a forest disease, the organism causing this disease is known to infect many woody ornamental plants, such as rhododendrons, that are commonly sold by nurseries. In March 2004, a California nursery was found to have unknowingly shipped plants infected with sudden oak death to all 50 states. Following this discovery, California nurseries halted shipments of trees to other states in an attempt to stop the spread of the disease, originally thought to have been imported on rhododendrons. In 2004, scientists mapped out the genome sequence of the disease-carrying fungus, *Phytophthora ramorum*. Identifying the genes and their proteins may help scientists develop specific diagnostic tests to quickly detect the presence of sudden oak death in trees, which is currently impossible to detect until a year or more after the tree is infected.

Scientists hope for much from the field of genomics in their fight against plant parasites that cause diseases. Many scientists have suggested limited, cautious transfer of resistance genes from the original host species in the source regions of the disease to newly threatened species. Original host species have usually evolved over millions of years of exposure to these diseases and have acquired genes that provide resistance. In the regions of recent introduction of parasites, there has been no selection for resistance, so the host plants are often killed en masse. Transgenic trees that have received pathogen resistance-enhancing genes could be produced and then be replanted in forests or urban areas. An advantage of this technique over traditional cross-breeding strategies involving two different species is that transgenic methods involve the introduction of fewer genes to the native species. Also, fewer tree generations would be required to develop resistance. For example, using traditional breeding technology, Asian chestnut trees (*Castanea mollissima*) are being bred with American chestnuts (*Castanea dentata*) to reduce the susceptibility of the latter to chestnut blight. The resultant hybrid is often significantly altered in appearance from the traditional American chestnut, and the process takes more than a decade to produce trees that are ready to plant. Transgenic technology could minimize these drawbacks. William Powell and colleagues are working to enhance the American chestnut's resistance by inserting a gene taken from wheat. The gene, which encodes an enzyme called oxalate oxidase, destroys a toxin produced by the fungus that causes chestnut blight.

57.3 Mutualism and Commensalism

In this section, we will examine the major types of mutualism and commensalism, interactions that are beneficial to at least one of the species involved. In mutualism, both species gain from the interaction. For example, in mutualistic pollination systems, the plant benefits by the transfer of pollen, and the pollinator typically gains a nectar meal. In commensalism, one species benefits, and the other remains unaffected. For example, in some forms of seed dispersal, barbed seeds are transported to new germination sites in the fur of mammals. The seeds benefit, but the mammals are generally unaffected.

It is interesting to note that humans have entered into mutualistic relationships with many species. For example, the association of humans with plants has resulted in some of the most far-reaching ecological changes on Earth. Humans have planted huge areas of the Earth with crops, allowing these plant populations to reach densities they never would on their own. In return, the crops have led to expanded human populations because of the increased amounts of food they provide.

Mutualism Is an Association Between Two Species That Benefits Both Species

Many close associations are known between species in which both species benefit. For example, leaf-cutting ants of the group

Figure 57.19   Sudden oak death disease.

Attini, of which there are about 210 species, enter into a mutualistic relationship with a fungus. A typical colony of about 9 million ants has the collective biomass of a cow and harvests the equivalent of a cow's daily requirement of fresh vegetation. Instead of consuming the leaves directly, the ants chew them into a pulp, which they store underground as a substrate on which the fungus grows. The ants shelter and tend the fungus, helping it reproduce and grow and weeding out competing fungi. In turn, the fungus produces specialized structures known as gongylidia, which serve as food for the ants. In this way, the ants circumvent the chemical defenses of the leaves, which are digested by the fungus.

The ant-fungus mutualism, which permits both species to live in close association, utilizing a common resource, is known as a **trophic mutualism**, where both species receive a benefit of resources. Other different types of mutualisms occur in nature. Some of these are **defensive mutualisms**, often involving an animal defending a plant or an herbivore. **Dispersive mutualisms** include plants and pollinators that disperse their pollen, and plants and fruit eaters that disperse the plant's seeds.

Defensive Mutualism One of the most commonly observed mutualisms occurs between ants and aphids. Aphids are fairly defenseless creatures and are easy prey for most predators. The aphids feed on plant sap and have to process a significant amount of it to get their required nutrients. In doing so, they excrete a lot of fluid, and some of the sugars still remain in the excreted fluid, which is called honeydew. The ants drink the honeydew and, in return, protect the aphids from an array of predators, such as ladybird beetle larvae, by driving the predators away. In some cases, the ants herd the aphids like cattle, moving them from one area to another (**Figure 57.20a**).

In other cases, ants enter into a mutualistic relationship with a plant itself. One of the most famous cases involves acacia trees in Central America, whose large thorns provide food and nesting sites for ants (**Figure 57.20b**). In return, the ants bite and discourage both insect and vertebrate herbivores from feeding on the trees. They also trim away foliage from competing plants and kill neighboring plant shoots, ensuring more light, water, and nutrient supplies for the acacias. In this case, neither species can live without the other, a concept called **obligatory mutualism**. This contrasts with **facultative mutualism**, in which the interaction is beneficial but not essential to the survival and reproduction of either species. For example, ant-aphid mutualisms are generally facultative. Both species benefit from the association, but each could live without the other.

Dispersive Mutualism Many examples of plant-animal mutualisms involve pollination and seed dispersal. From the plant's perspective, an ideal pollinator would be a specialist, moving quickly among individuals but retaining a high fidelity to a plant species. Two ways that plant species in an area promote the pollinator's species fidelity is by synchronized flowering within a species and by sequential flowering of different species through the year. The plant should provide just enough nectar

(a) An ant defending aphids in exchange for food

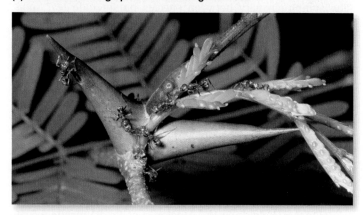

(b) Ants defending an acacia plant in exchange for food and shelter

Figure 57.20 **Defensive mutualism.** This form of mutualism involves species that receive food or shelter in return for providing protection. **(a)** This red carpenter ant, *Camponotus pennsylvanicus*, tends aphids feeding on a twig. The ants receive sugar-rich honeydew produced by the aphids, and, in return, they protect the aphids from predators. **(b)** Ants, usually *Pseudomyrmex ferruginea*, make nests inside the large, hornlike thorns of the bull's horn acacia and defend the plant against insects and mammals. In return, the acacia (*Acacia collinsii*) provides two forms of food to the ants: protein-rich granules called Beltian bodies and nectar from extrafloral nectaries (nectar-producing glands that are physically apart from the flower).

Concept check: *Is the relationship between red carpenter ants and aphids an example of facultative or obligatory mutualism?*

to attract a pollinator's visit. From the pollinator's perspective, it would be best to be a generalist and obtain nectar and pollen from as many flowers as possible in a small area, thus minimizing the energy spent on flight between patches. This suggests that although mutualisms are beneficial to both species, their optimal needs are quite different.

Mutualistic interactions are also highly prevalent in the seed-dispersal systems of plants. Fruits provide a balanced diet of proteins, fats, and vitamins. In return for this juicy meal, animals unwittingly disperse the enclosed seeds, which pass through the digestive tract unharmed. Fruits eaten by birds and mammals often have attractive colors (**Figure 57.21**); those dispersed by nocturnal bats are not brightly colored but instead give off a pungent odor that attracts the bats.

Figure 57.21 **Dispersive mutalism.** This blackbird (*Turdus merula*) is an effective seed disperser.

(a) An orchid without nectar mimicking a female bee

(b) Seed dispersal via hooked seeds

Figure 57.22 **Commensalisms.** (a) Bee orchids (*Ophrys apifera*) mimic the shape of a female bee. Male bees copulate with the flowers, transferring pollen but getting no nectar reward. (b) Hooked seeds of burdock (*Arctium minus*) have lodged in the fur of a white-footed mouse (*Peromyscus leucopus*). The plant benefits from the relationship by the dispersal of its seeds, and the animal is not affected.

In Commensalism, One Partner Receives a Benefit While the Other Is Unaffected

Commensalism is an interaction between species in which one benefits and the other is neither helped nor harmed. Such is the case when orchids or other epiphytes grow in forks of tropical trees. The tree is unaffected, but the orchid gains support and increased exposure to sunlight and rain. Cattle egrets feed in pastures and fields among cattle, whose movements stir up insect prey for the birds. The egrets benefit from the association, but the cattle generally do not. One of the best examples of commensalism involves **phoresy**, in which one organism uses a second organism for transportation. Hummingbird flower mites feed on the pollen of flowers and travel between flowers in the nostrils (nares) of hummingbirds. The flowers the mites inhabit live only a short while before dying, so the mite relocates by scuttling into the nares of visiting hummingbirds and hitching a ride to the next flower. When the hummingbird visits a new flower, the mite disembarks. Presumably, the hummingbirds are unaffected.

Some commensalisms involve one species "cheating" on the other without harming it. In the bogs of Maine, the grass-pink orchid (*Calopogon pulchellus*) produces no nectar, but it mimics the nectar-producing rose pogonia (*Pogonia ophioglossoides*) and is therefore still visited by bees. Another example involves bee orchids (*Ophrys apifera*) that mimic the appearance and scent of female bees. Males pick up and transfer pollen while trying to copulate with the flowers (**Figure 57.22a**). The stimuli of flowers of the bee orchid are so effective that male bees prefer to mate with them even in the presence of actual female bees! Many plants have essentially cheated their potential mutualistic seed-dispersal agents out of a meal by developing seeds with barbs or hooks that lodge in the animals' fur or feathers rather than their stomachs (**Figure 57.22b**). In these cases, the plants receive free seed dispersal, and the animals receive nothing, except perhaps minor annoyance. This type of relationship is fairly common; most hikers and dogs have at some time gathered spiny or sticky seeds as they wandered through woods or fields.

57.4 Conceptual Models

In this chapter, we have seen that interactions between species, such as competition, predation, and parasitism, are important in nature. How can we determine which of these factors, along with abiotic factors such as temperature and moisture, are the most important in influencing population size? The question is one asked by many applied biologists, such as foresters, marine biologists, and conservation biologists, who are interested in maximizing a population's size, as well as ecologists in general.

Some ecologists stress the importance of so-called bottom-up factors, such as plant or prey quality and abundance in controlling herbivores and the predators that feed on them. Others stress the importance of top-down factors, such as predators and parasites, acting to control their animal or plant prey (**Figure 57.23**). For example, in the beginning of the chapter, we noted how a decline in the size of shark populations along the east coast had led to an increase in their main prey, rays and skates, and a decrease in bay scallops, the prey of rays and skates. This is a top-down effect known as a trophic cascade, because its effects cascade down to all feeding levels of the system. In this section, we will briefly discuss some of the evidence for the existence of bottom-up versus top-down control and examine conceptual models where both mechanisms are deemed important.

Bottom-Up Control Suggests Food Limitation Influences Population Densities

At least two lines of evidence suggest that bottom-up effects are important in limiting population sizes. First, we know there is

(a) Bottom-up control

(b) Top-down control

Figure 57.23 Bottom-up control versus top-down control. **(a)** Bottom-up control proposes that host plant quantity or quality limits the density of herbivores, such as grasshoppers, which, in turn, sets limits on the abundance of predators, such as spiders. Taken together, this means that high quantity and quality of host plants would result in increased numbers of predators because of higher densities of the herbivores they prey on. **(b)** Top-down control proposes that predators limit the number of herbivores, which, in turn, increases host plant density. Taken together, this means that high levels of predation would result in high densities of host plants because there would be fewer herbivores.

Concept check: You add fertilizer to a bush, and this increases spider density on the bush. What is this an example of?

a progressive lessening of available energy passing from plants through herbivores to carnivores and to secondary carnivores (carnivores that eat other carnivores). This line of evidence, based on the thermodynamic properties of energy transfer, suggests that the quantity and quality of plants regulates the population size of all other species that rely on them.

Second, much evidence supports the **nitrogen-limitation hypothesis** that organisms select food in terms of the nitrogen content of the tissue. This is largely due to the different proportions of nitrogen in plants and animals. Animal tissue generally contains about 10 times as much nitrogen as plant tissue. For this reason, animals favor high-nitrogen plants. Fertilization has repeatedly been shown to benefit herbivores. Nearly 60% of 186

studies investigating the effects of fertilization on herbivores reported that increasing a plant's tissue nitrogen concentration through fertilization had strong positive effects on herbivore population sizes, survivorship, growth, and fecundity.

Top-Down Control Suggests Natural Enemies Influence Population Densities

Top-down models suggest that predators control populations of their prey (ultimately, herbivores) and that these herbivores control plant populations. Supporting evidence comes from the world of biological control, where natural enemies are released to control agricultural pests such as weeds. Many weeds are invaders that were accidentally introduced to an area from a different country, as seeds that lodge in ships' ballasts or in agricultural shipments. Over 50% of the 190 major weeds in the U.S. are invaders from outside the country. Many of these weeds have become separated from their native natural enemies, which is one reason the weeds become so prolific. Because chemical control is expensive and may have unwanted environmental side effects, many land managers have reverted to biological control, in which the invading weed is reunited with its native natural enemy.

Ecologists have seen many successes in the biological control of weeds. St. John's wort (*Hypericum perforatum*), a pest in California pastures, was controlled by two beetles from its homeland in Europe. Likewise, alligator weed has been controlled in Florida's rivers by the alligatorweed flea beetle (*Agasicles hygrophila*) from South America. The prickly pear cactus (*Opuntia stricta*) provides a prime example of effective biological control of a weed. The cactus was imported into Australia in the 19th century and quickly established itself as a major pest of rangeland. The small cactus moth (*Cactoblastis cactorum*) was introduced in the 1920s and, within a short time, successfully saved hundreds of thousands of acres of valuable rangeland from being overrun by the cacti (**Figure 57.24**). The numerous examples showing that pest populations are controlled when

(a) Before biological control

(b) After biological control

Figure 57.24 Successful biological control of prickly pear cactus. The prickly pear cactus (*Opuntia stricta*) in Chinchilla, Australia, **(a)** before and **(b)** after control by the cactus moth (*Cactoblastis cactorum*).

Table 57.3 Primary Mortality Factors According to the Availability of Plant Biomass

	Low ⟶ Plant biomass ⟶ High			
Taxa	Plants only	Plants and herbivores	Plants, herbivores, and carnivores	Plants, herbivores, carnivores, and secondary carnivores
Plants	Competition	Herbivory	Competition	Herbivory
Herbivores		Competition	Predation	Competition
Carnivores			Competition	Predation
Secondary carnivores				Competition

reunited with their natural enemies provide strong evidence of top-down control in nature.

Modern Models Suggest Both Top-Down and Bottom-Up Effects Are Important

More recently, different models have been proposed that take into account the effects of both natural enemies and limited resources on species. In 1981, Laurie Oksanen and coworkers suggested that the strength of mortality factors varies with availability of plant biomass involved, a model they termed the ecosystem exploitation hypothesis (**Table 57.3**). Thus, for very simple systems where mainly plants exist, such as Arctic tundra, not enough plant material is available to support herbivores, and plants must be resource limited (that is, limited by competition with each other). As plant biomass increases, some herbivores can be supported, but there are too few herbivores to support carnivores. In the absence of carnivores, levels of herbivory can be quite high. Plant abundance becomes limited primarily by herbivory, not competition. The abundance of herbivores, in the absence of carnivores, is limited by competition for plant resources.

As plant biomass increases still further and herbivores become common, carnivores become abundant and reduce the number of herbivores, which, in turn, increases plant abundance. Plants, being abundant, endure severe competition for resources, but herbivores, suffering high rates of mortality from natural enemies, are not abundant enough to compete. The predators, being limited only by the availability of their prey, are also common and thus experience competition. With the availability of yet more plant biomass, additional taxa such as secondary carnivores can be supported, which changes the relative importance of mortality factors among all taxa. Ecologists are currently examining the degree to which such a model holds true in nature.

Species interactions can clearly be very important in influencing both the growth of both individual populations and the structure of communities, groups of species living in a particular area. What factors determine the numbers of species in a given area? What factors influence the stability of a community, and what are the effects of disturbances on community structure? In the next chapter, on community ecology, we will explore these and other questions.

▶ Summary of Key Concepts

57.1 Competition

- Species interactions can take a variety of forms that differ based on their effect on the species involved. (Figure 57.1, Table 57.1)
- Competition can be categorized as intraspecific (between individuals of the same species) or interspecific (between individuals of different species). Competition can also be categorized as exploitation competition or interference competition. (Figure 57.2)
- One species can exclude the other in a natural environment, affecting its distribution within a habitat. (Figure 57.3)
- Laboratory and field experiments show that competition occurs frequently in nature and varies as a function of both biotic and abiotic factors. (Figures 57.4, 57.5)
- The competitive exclusion hypothesis states that two species with the same resource requirements cannot occupy the same niche. (Figure 57.6)
- Resource partitioning and morphological differences between species allow them to coexist in a community. (Figures 57.7, 57.8, Table 57.2)

57.2 Predation, Herbivory, and Parasitism

- The many antipredator strategies that animals have evolved suggest that predation is a strong selective force. Of these strategies, chemical defense and aposematic coloration are the most common. (Figures 57.9, 57.10)
- Despite these defenses, oscillations in predator-prey cycles, the effect of introduced species, and examples of human predation illustrate that predators can have a large effect on prey densities. (Figures 57.11, 57.12, 57.13)
- Plants have also evolved an array of defenses against herbivores, termed host plant resistance, which includes chemical defenses, such as secondary metabolites, and mechanical defenses, such as thorns and spines. (Figure 57.14)
- Parasitism is a common lifestyle on Earth, and some parasites have complex life cycles involving multiple hosts. (Figures 57.15, 57.16)
- Evidence from experimental removal of parasites and from the study of introduced plant and animal parasites confirms that parasites can greatly reduce prey densities. (Figures 57.17, 57.18, 57.19)

57.3 Mutualism and Commensalism

- Mutualism is an association between two species that benefits both. Defensive mutualisms typically involve an animal defending either a plant or herbivore; dispersive mutualisms involve plants and pollinators that disperse their pollen, and plants and fruit eaters that disperse the plant's seeds. (Figures 57.20, 57.21)

- In commensal relationships, one partner receives a benefit while the other is not affected. (Figure 57.22)

57.4 Conceptual Models

- Conceptual models of species interactions describe the importance of factors such as competition and predation. Bottom-up models propose that plant quality or quantity regulates the abundance of all herbivore and predator species; top-down models propose that the abundance of predators controls herbivore and plant densities. (Figures 57.23, 57.24)

- Modern models, which incorporate components of both bottom-up and top-down models, propose that mortality varies according to productivity of plant biomass. (Table 57.3)

Assess and Discuss

Test Yourself

1. A species interaction in which one species benefits but the other species is unharmed is called
   a. mutualism.   c. parasitism.   e. mimicry.
   b. amensalism.   d. commensalism.

2. Two species of birds feed on similar types of insects and nest in the same tree species. This is an example of
   a. intraspecific competition.   d. mutualism.
   b. interference competition.   e. none of the above.
   c. exploitation competition.

3. According to the competitive exclusion hypothesis,
   a. two species that use the exact same resource show very little competition.
   b. two species with the same niche cannot coexist.
   c. one species that competes with several different species for resources will be excluded from the community.
   d. all competition between species results in the extinction of at least one of the species.
   e. none of the above is correct.

4. In Lack's study of British passerine birds, different species seem to segregate based on resource factors, such as location of prey items. This differentiation among the niches of these passerine birds is known as
   a. competitive exclusion.   d. resource partitioning.
   b. intraspecific competition. e. allelopathy.
   c. character displacement.

5. Divergence in morphology that is a result of competition is termed
   a. competitive exclusion.   d. amensalism.
   b. resource partitioning.   e. mutualism.
   c. character displacement.

6. Tapeworms have
   a. low lethality and low duration of interaction.
   b. low lethality and high duration of interaction.
   c. high lethality and low duration of interaction.
   d. high lethality and high duration of interaction.
   e. none of the above.

7. Ticks are regarded as
   a. monophagous endoparasites.
   b. monophagous ectoparasites.
   c. polyphagous endoparasites.
   d. polyphagous ectoparasites.
   e. none of the above.

8. Batesian mimicry differs from Müllerian mimicry in that
   a. in Batesian mimicry, both species possess the chemical defense.
   b. in Batesian mimicry, one species possesses the chemical defense.
   c. in Müllerian mimicry, one species has several different mimics.
   d. in Müllerian mimicry, one species has several different chemical defenses.
   e. in Batesian mimicry, cryptic coloration is always found.

9. Deadly nightshade is protected from herbivores by
   a. the alkaloid capsaicin.
   b. the alkaloid atropine.
   c. the phenolic anthocyanin.
   d. the phenolic tannin.
   e. the terpenoid β-carotene.

10. Parasitic plants that rely solely on their host for nutrients are called
    a. hemiparasites.
    b. fungi.
    c. holoparasites.
    d. monophagous.
    e. polyphagous.

Conceptual Questions

1. Define the competitive exclusion hypothesis.

2. Distinguish between Müllerian and Batesian mimicry.

3. Why don't we see more evidence of plants being eaten by herbivores?

Collaborative Questions

1. Discuss how the reintroduction of wolves in Yellowstone National Park might be beneficial.

2. Discuss several antipredator strategies that animals have evolved.

Online Resource

www.brookerbiology.com

Stay a step ahead in your studies with animations that bring concepts to life and practice tests to assess your understanding. Your instructor may also recommend the interactive ebook, individualized learning tools, and more.

Community Ecology

58

A massive volcanic explosion in 1883 on the island of Krakatau in Indonesia destroyed two-thirds of the island, originally 11-km long and covered in tropical rain forest. Life on the remaining part was eradicated, suffocated by tens of meters of red-hot ash. Nine months after the eruption, the first reported colonist was a spider spinning its web. By 1896, there were 11 species of ferns and 15 species of flowering plants, mainly grasses. Most plant species at that time had been wind or sea dispersed. By the 1920s, 40 years after the original eruption, birds had become abundant and had dispersed additional plant species via their seeds. Ecologists have been examining the return of different species to the island ever since.

So far in this unit, we have examined ecology in terms of the behavior of individual organisms, the growth of populations, and interactions between pairs of species. Most populations, however, exist not on their own, but together with populations of many other species. This assemblage of many populations that live in the same place at the same time is known as a **community**. For example, a tropical forest community consists of not only tree species, vines, and other vegetation, but also the insects that pollinate them, the herbivores that feed upon the plants, and the predators and parasites of the herbivores. Communities can occur on a wide range of scales, and one community can be nested within another. For example, the tropical forest community also encompasses smaller communities, such as the water-filled recesses of bromeliads, which form a microhabitat for different species of insects and their larvae. Both of these entities—the tropical forest and the bromeliad tank—are viable communities, depending on one's frame of reference with regard to scale.

Community ecology is the study of how groups of species interact and form functional communities. In Chapter 57, we considered the interactions between individual species. In this chapter, we widen our focus to explore the factors that influence the number and abundance of species in a community. We begin by examining the nature of ecological communities. Are communities loose assemblages of species that happen to live in the same place at the same time, or are they more tightly organized groups of mutually dependent species? We explore why, on a global scale, the number of species is usually greatest in the tropics and declines toward the poles.

Krakatau and neighboring islands.   Many of the species formerly present on Krakatau have returned following the 1883 eruption that covered the island in volcanic dust.

Community ecology also addresses what factors act to stabilize the number and abundance of species in a community. However, ecologists recognize that communities may change, for example, following a disturbance such as a fire or a volcanic eruption. This recovery tends to occur in a predictable way, which ecologists have termed succession. In certain situations—for example, on islands recovering from physical disturbance—the structure of the community tends toward an equilibrium determined by the balance between the rates of immigration and extinction.

58.1 Differing Views of Communities

Ecologists have long held differing views on the nature of a community and its structure and functions. Some of the initial work in the field of community ecology considered a community to be equivalent to a superorganism, in much the same way that the body of an animal is more than just a collection of organs. In this view, individuals, populations, and communities have a stable relationship to each other that resembles the associations found between cells, tissues, and organs. American botanist Frederic Clements, the champion of this viewpoint, suggested in 1905 that ecology was to the study of communities what physiology was to the study of individual organisms. This view of community, with predictable and integrated associations of species separated by sharp boundaries, is termed the **organismic model**, which depicts the ecological community as a superorganism.

Clements's ideas were challenged in 1926 by botanist Henry Allan Gleason. Gleason proposed an **individualistic model**, which described a community as an assemblage of species coexisting primarily because of similarities in their physiological requirements and tolerances. Although acknowledging that some assemblages of species were fairly uniform and stable over a given region, Gleason suggested that distinctly structured ecological communities usually do not exist. Instead, communities are loose assemblages of species distributed independently along an environmental gradient. Viewed in this way, communities do not necessarily have sharp boundaries, and associations of species are much less predictable and integrated than in Clements's organismic model.

By the 1950s, many ecologists had abandoned Clements's view in favor of Gleason's. In particular, Robert Whittaker's studies proposed the **principle of species individuality**, which stated that each species is distributed according to its physiological needs and population dynamics; that most communities intergrade, or merge into one another gradually; and that competition does not create distinct vegetational zones. For example, let's consider an environmental gradient such as a moisture gradient on an uninterrupted slope of a mountain. Whittaker proposed that four hypotheses could explain the distribution patterns of plants and animals on the gradient (**Figure 58.1**):

1. Competing species, including dominant plants, exclude one another along sharp boundaries. Other species evolve toward a close, perhaps mutually beneficial association with the dominant species. Communities thus develop along the gradient, each zone containing its own group of interacting species giving way at a sharp boundary to another assemblage of species. This corresponds to Clements's organismic model.
2. Competing species exclude one another along sharp boundaries but do not become organized into groups of species with parallel distributions.
3. Competition does not usually result in sharp boundaries between species. However, the adaptation of species to similar physical variables will result in the appearance of groups of species with similar distributions.
4. Competition does not usually produce sharp boundaries between species, and the adaptation of species to similar physical variables does not produce well-defined groups of species with similar distributions. The centers and boundaries of species populations are scattered along the environmental gradient. This corresponds to Gleason's individualistic model.

To test these possibilities, Whittaker examined the vegetation on various mountain ranges in the western U.S. He

Figure 58.1 Four hypotheses for the distribution patterns of plants and animals along an environmental gradient. Each curve in each part of the figure represents one species and the way its population might be distributed along an environmental gradient.

sampled plant populations along an elevation gradient from the tops of the mountains to the bases and collected data on physical variables, such as soil moisture.

The results supported the fourth hypothesis, that competition does not produce sharp boundaries between species and that adaptation to physical variables does not result in defined groups of species. Whittaker concluded that his observations agreed with Gleason's predictions that (1) each species is distributed in its own way, according to its genetic, physiological, and life cycle characteristics; and (2) most communities grade into each other continuously rather than form distinct, clearly separated groups. The composition of species at any one point in an environmental gradient was largely determined by abiotic factors such as temperature, water, light, pH, and salt concentrations (features discussed in Chapter 54).

Even though most communities intergrade along environmental gradients such as a mountain slope, ecologists recognize distinct differences between communities. The community at the top of a mountain will be quite different from that at the bottom, so distinguishing between communities on a broad scale is useful. Also, some sharp boundaries between groups of species sometimes do exist, especially related to physical differences such as water quality and soil type that cause distinct communities to develop. For example, serpentine soils are rich

Figure 58.2 An example of a sharp boundary between two communities. In New Zealand's Dun Mountain area, the sparse vegetation on the serpentine soil on the left contrasts with that of the beech forest on the nonserpentine soil on the right.

in metals, including magnesium, iron, and nickel, but poor in plant nutrients. The species that have adapted to these harsh conditions form a unique community restricted to this area (**Figure 58.2**). Such sharp boundaries are not common between neighboring communities.

58.2 Patterns of Species Richness

One method to analyze communities is to determine the number of species in each community, or **species richness**. The number of species of most taxa varies according to latitudinal gradient, generally increasing from polar areas to temperate areas and reaching a maximum in the tropics. For example, the species richness of North American birds increases from Arctic Canada to Panama (**Figure 58.3**). A similar pattern exists for mammals and reptiles. Species richness is also influenced by topographical variation. More mountains mean more hilltops, valleys, and differing habitats; thus, there is an increased number of birds in the mountainous western U.S. In contrast, species richness is reduced by the peninsular effect, in which the number of species decreases as a function of distance from the main body of land.

Many hypotheses for the latitudinal gradient of species richness have been advanced. We will consider four of the most important, which propose that communities diversify with increased time, area, productivity, respectively, and that diversity is maximized in areas of intermediate disturbance. Although they are treated separately here, these hypotheses are not mutually exclusive. All four can contribute to patterns of species richness.

The Time Hypothesis Suggests Communities Diversify with Age

Many ecologists argue that communities diversify, or gain species, with time. Therefore temperate regions have less rich communities than tropical ones because they are younger and have only more recently (relatively speaking) recovered from recent

Figure 58.3 Species richness of birds in North America. The values indicate the numbers of different species in a given area. Contour lines show equal numbers of bird species, with colors indicating incremental changes. Note the pronounced latitudinal gradient toward the tropics and the high diversity in California and northern Mexico, regions of considerable topographical variation and habitat diversity.

glaciations and severe climatic disruption. The time hypothesis proposes that resident species of the temperate zone have not yet evolved new forms to exploit vacant niches. In addition, it suggests that species that could possibly live in temperate regions have not migrated back from the unglaciated areas into which the ice ages drove them.

In support of the time hypothesis, ecologists compared the species richness of bottom-dwelling invertebrates, such as worms, in historically glaciated (covered with ice) and unglaciated lakes in the Northern Hemisphere that occur at similar latitudes. Lake Baikal in Siberia is an ancient, unglaciated temperate lake and contains a very diverse fauna. For example, 580 species of invertebrates are found in the bottom zone. Great Slave Lake, a comparable-sized lake in northern Canada that was once glaciated, contains only four species in the same zone.

However, ecologists recognize drawbacks to the time hypothesis. For example, the time hypothesis may help explain variations in the species richness of terrestrial organisms, but it has limited applicability to marine organisms. Although we might not expect terrestrial species, particularly plants, to redistribute themselves quickly following a glaciation—especially if there is a physical barrier like the English Channel to overcome—there seems to be no reason that marine organisms couldn't relatively easily shift their distribution patterns during glaciations, yet the polar-equatorial gradient of species richness still exists in marine habitats.

The Area Hypothesis Suggests Large Areas Support More Species

The **area hypothesis** proposes that larger areas contain more species than smaller areas because they can support larger populations and a greater range of habitats. Much evidence supports the area hypothesis. For example, in 1974, Donald Strong showed that insect species richness on tree species in Britain was better correlated with the area over which a tree species could be found than with time of habitation since the last ice age (**Figure 58.4**). The relationship between the amount of available area and the number of species present is called the **species-area effect**. Some introduced tree species, such as apple and lime, were relatively new to Britain, but they bore many different insect species, a fact Strong argued did not support the time hypothesis. This means that, on average, an individual willow tree standing next to an individual maple tree in the same location would support an order of magnitude more insect species because of its greater abundance in that habitat.

The large, climatically similar area of the tropics has been proposed as a reason why the tropics have high species richness. However, the area hypothesis seems unable to explain why, if increased richness is linked to increased area, more species are not found in certain regions such as the vast contiguous landmass of Asia. Furthermore, although tundra may be the world's largest land biome, it has low species richness. Finally, the largest marine system, the open ocean, which has the greatest volume of any habitat, has fewer species than tropical nearshore waters, which have a relatively small volume.

The Productivity Hypothesis Suggests That More Energy Permits the Existence of More Species

The **productivity hypothesis** proposes that greater production by plants results in greater overall species richness. An increase in plant productivity, the total weight of plant material produced over time, leads to an increase in the number of herbivores and hence an increase in the number of predator, parasite, and scavenger species. Production itself is influenced by factors such as temperature and rainfall, because many plants grow better where it is warm and wet. For example, in 1987, David Currie and colleagues showed that the species richness of trees in North America is best predicted by the **evapotranspiration rate**, the rate at which water moves into the atmosphere through the processes of evaporation from the soil and transpiration of plants, both of which are influenced by the amount of solar energy (**Figure 58.5**).

Once again, however, there are exceptions to this rule. Some tropical seas, such as the southeast Pacific off of Colombia and Ecuador, have low productivity but high species richness. On the other hand, the sub-Antarctic Ocean has a high productivity but low species richness. Estuarine areas, where rivers empty into the sea, are similarly very productive yet low in species, presumably because they represent stressful environments for many organisms that are alternately inundated by fresh water and salt water with daily changes in the tide. Some lakes that are polluted with fertilizers also have high productivity but low species richness.

In 1993, Robert Latham and Robert Ricklefs showed that although patterns of tree richness in North America support the productivity hypothesis, the pattern does not hold for broad comparisons between continents. For example, the temperate forests of eastern Asia support substantially higher numbers of tree species (729) than do climatically similar areas of North

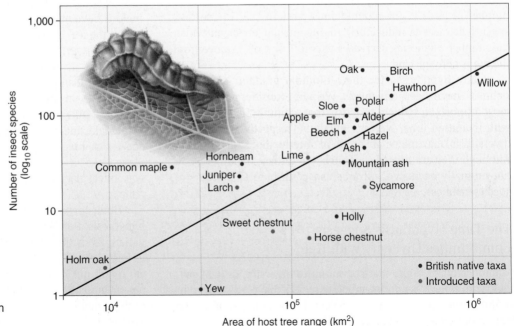

Figure 58.4 **Relationship between insect species richness on British host trees and area.** A positive correlation is found between insect species richness and the host tree's present range (in km²). Here, the range represents an area in Britain known to contain trees.

Figure 58.5 Tree species richness in North America.
Contour lines show equal numbers of tree species, with colors indicating incremental changes. Tree species richness and evapotranspiration rates are highest in the southeast.

Concept check: *Why doesn't the species richness of trees increase in mountainous areas of the West, as it does for birds?*

America (253) or Europe (124). These three areas have different evolutionary histories and different neighboring areas from which species might have invaded.

The Intermediate-Disturbance Hypothesis Proposes That Moderately Disturbed Communities Contain More Species

Ecologist Joseph Connell has argued that the highest numbers of species are maintained in communities with intermediate levels of disturbance, a concept called the **intermediate-disturbance hypothesis** (Figure 58.6a). Disturbance in communities may be brought about by many different phenomena such as droughts, fires, floods, and hurricanes or by species interactions such as herbivory, predation, or parasitism. Recall from Chapter 56 that some species, termed *r*-selected species, are better dispersers than other species and that *K*-selected species are better competitors. Connell reasoned that at high levels of disturbance, only colonists that were *r*-selected species would survive, giving rise to low species richness. This is because these species would be the only ones able to disperse quickly to a highly disturbed area. At low rates of disturbance, competitively dominant *K*-selected species would outcompete all other species, which would also yield low species richness. The most species rich communities would lie somewhere in between.

Connell argued that natural communities fit into this model fairly well. Tropical rain forests and coral reefs are both examples of communities with high species richness. Coral reefs exhibit highest species richness in areas disturbed by hurricanes, and the richest tropical forests occur where disturbance by storms causes landslides and tree falls. The fall of a tree creates a hole in the rain forest canopy known as a light gap, where direct sunlight is able to reach the rain forest floor. The light gap is rapidly colonized by *r*-selected species, such as small herbaceous plants, which are well adapted for rapid growth. Although these pioneering species grow rapidly, they are eventually overtaken by *K*-selected species, such as mature trees, which fill in the gap in the canopy (**Figure 58.6b**). Although events such as hurricanes and tree falls are fairly

(a) Relationship between species richness and disturbances

(b) Light gap in a tropical rain forest

Figure 58.6 The intermediate-disturbance hypothesis of community organization. **(a)** This hypothesis proposes that species richness is highest at intermediate levels of disturbances caused by events such as fires or windstorms. **(b)** A light gap in a tropical rain forest in Costa Rica promotes the growth of small herbaceous species until trees colonize the light gap and gradually grow over and outcompete the smaller species.

Concept check: *According to the intermediate-disturbance hypothesis, why are there so many species in the tropics?*

frequent events in these communities, their occurrence in any one area is usually of intermediate frequency.

In 1979, Wayne Sousa provided an elegant experimental verification of the intermediate-disturbance hypothesis in the marine intertidal zone. He found that small boulders, which were easily disturbed by waves, carried a mean of 1.7 sessile plant and animal species. These frequently moving boulders crushed or dislodged most colonizing species. Large boulders, which were rarely moved by waves, had a mean of 2.5 species. On these boulders, competitively dominant species supplanted many other species. Sousa found that intermediate-sized boulders had the most species, an average of 3.7 species per boulder, because they contained a mix of r- and K-selected species. To test the hypothesis, Sousa cemented small boulders to the shoreline and obtained an increase in species richness to near the value for large boulders, showing that the resulting number of species was a result of rock stability, not rock size.

When comparing the time, area, productivity, and intermediate disturbance hypotheses, each has some evidence to support it and some to contradict it. Different processes may occur over different scales. On a regional scale, we know that the time since the last glaciation has the potential to change patterns of species richness. On a more local scale, area, productivity of available habitat, and disturbance level may be important. At any given point on the globe, species richness may be affected by the interaction of these different factors.

58.3 Calculating Species Diversity

So far, we have discussed communities in terms of variations in species richness. However, to measure species diversity, ecologists need to take into account not only the number of species in a community but also their frequency of occurrence, or **relative abundance**. For example, consider two hypothetical communities, A and B, both with two species and 100 total individuals.

	Number of individuals of species 1	Number of individuals of species 2
Community A	99	1
Community B	50	50

The species richness of community B equals that of community A, because they both contain two species. However, community B is considered more diverse because the distribution of individuals between species is more even. One would be much more likely to encounter both species in community B than in community A, where one species dominates. **Species diversity** can be considered the community measure that incorporates both species number and relative abundance.

To measure the species diversity of a community, ecologists calculate what is known as a diversity index. Although many different indices are available, the most widely used is the **Shannon diversity index** (H_S), which is calculated as

$$H_S = -\Sigma p_i \ln p_i$$

where p_i is the proportion of individuals belonging to species i in a community, ln is the natural logarithm, and the Σ is a summation sign. For example, for a species in which there are 50 individuals out of a total of 100 in the community, p_i is 50/100, or 0.5. The natural log of 0.5 is -0.693. For this species, $p_i \ln p_i$ is then $0.5 \times -0.693 = -0.347$. For a hypothetical community with 5 species and 100 total individuals, the Shannon diversity index would be calculated as follows:

Species	Abundance	p_i	$p_i \ln p_i$
1	50	0.5	-0.347
2	30	0.3	-0.361
3	10	0.1	-0.230
4	9	0.09	-0.217
5	1	0.01	-0.046
Total	5	100	1.00	$\Sigma p_i \ln p_i$	-1.201

In this example, even the rarest species, species 5, contributes some value to the index. If a community had many rare species, their contributions would accumulate. This makes the Shannon diversity index very valuable to conservation biologists, who often study rare species and their importance to the community. Remember, too, that in the equation, the negative sign in front of the summation changes these values to positive, so the index actually becomes 1.201, not -1.201.

Values of the Shannon diversity index for real communities often fall between 1.5 and 3.5, with the higher the value, the greater the diversity. **Table 58.1** calculates the diversity of two bird communities in Indonesia with similar species richness but differing species abundance. The bird communities were surveyed in a pristine unlogged forest or in a selectively logged lowland forest. To document diversity, biologist Stuart Marsden established census stations in the two types of forest and recorded the type and number of all bird species for a number of 10-minute periods. Although a greater number of individual birds was seen in the logged areas (2,358) compared to unlogged ones (1,824), a high proportion of the individuals in the logged areas (0.386) belonged to just one species, *Nectarinia jugularis*. Although only one more bird species was found in the unlogged area than in the logged area, calculation of the Shannon diversity index showed a higher diversity of birds in the unlogged area, 2.284 versus 2.037, which is a considerable difference, considering the logarithmic nature of the index.

An accurate determination of community diversity depends on detailed knowledge of which and how many of each species are present. This is relatively easy to determine for communities of vertebrates and some invertebrates, but it is much more difficult for microbial communities. Yet knowledge of microbial communities is of great importance, because microbes carry out vital functions such as nitrogen fixation and decomposition. As described next, with the advent of modern molecular tools, our knowledge of the diversity of microbial communities is beginning to expand.

Table 58.1	Shannon Diversity Index of Bird Species on Logged and Unlogged Sites in Indonesia					
	Unlogged			Logged		
Species	N	p_i	$p_i \ln p_i$	N	p_i	$p_i \ln p_i$
Nectarinia jugularis, Olive-backed sunbird	410	0.225	−0.336	910	0.386	−0.367
Ducula bicolor, Pied imperial pigeon	230	0.126	−0.261	220	0.093	−0.221
Philemon subcorniculatus, Grey-necked friarbird	210	0.115	−0.249	240	0.102	−0.233
Nectarinia aspasia, Black sunbird	190	0.104	−0.235	120	0.051	−0.152
Dicaeum vulneratum, Ashy flowerpecker	185	0.101	−0.232	280	0.119	−0.253
Ducula perspicillata, White-eyed imperial pigeon	170	0.093	−0.221	180	0.076	−0.196
Phylloscopus borealis, Arctic warbler	160	0.088	−0.214	140	0.059	−0.167
Eos bornea, Red lory	88	0.048	−0.146	73	0.031	−0.108
Ixos affinis, Golden bulbul	76	0.042	−0.133	31	0.013	−0.056
Geoffroyus geoffroyi, Red-cheeked parrot	44	0.024	−0.089	54	0.023	−0.087
Rhyticeros plicatus, Papuan hornbill	24	0.013	−0.056	27	0.011	−0.050
Cacatua moluccensis, Moluccan cockatoo	12	0.007	−0.035	1	0.001	−0.007
Tanygnathus megalorynchos, Great-billed parrot	9	0.005	−0.026	11	0.005	−0.026
Electus roratus, Electus parrot	7	0.004	−0.022	0	0	0
Macropygia amboinensis, Brown cuckoo-dove	6	0.003	−0.017	7	0.003	−0.017
Cacomantis sepulcralis, Ruby-breasted cuckoo	3	0.002	−0.012	0	0	0
Trichoglossus haematodus, Rainbow lorikeet	0	0	0	64	0.027	−0.097
Total	**1,824**	**1.0**		**2,358**	**1.0**	
Shannon diversity index			**2.284**			**2.037**

Genomes & Proteomes Connection

Metagenomics May Be Used to Measure Community Diversity

Bacteria are abundant members of all communities and are vital to their functioning. They serve as food sources for other organisms and participate in the decomposition process. However, most microorganisms are taxonomically unknown, mainly because they cannot be cultivated on known culture media. The field of **metagenomics** seeks to identify and analyze the collective microbial genomes contained in a community of organisms, including those that are not easily cultured in the laboratory. This technique has been in existence only since the early 1990s, but significant progress has already been made in providing data that have advanced our understanding of which bacteria are present in various communities and how they function.

The process involves four main steps (**Figure 58.7**). First, an environmental sample containing an unknown number of bacterial species is collected, and its DNA is isolated from the cells using chemical or physical methods. Because the genomic DNA of each species is relatively large, it is cut up into fragments with restriction enzymes (see Chapter 20). Second, the fragments are combined with vectors, small units of DNA that can be inserted into a model laboratory organism, usually a bacterium. The third step begins with transformation in which the

DNA from step 2 is taken up into bacterial cells. Individual bacteria are then grown on a selective media so that only the transformed cells survive. Each cell will grow into a colony of cloned cells. A collection of thousands of clones, each containing a different piece of microbial DNA, is called a metagenomic library. Lastly, the DNA from the metagenomic library is analyzed. In some cases, expression of the new DNA results in the synthesis of a new protein that changes the phenotype of the host, for example, a new enzyme that is detected by a chemical technique or an unusual color or shape in the model organism.

In 2004, Jill Banfield and colleagues used metagenomics techniques to identify the five dominant species of bacteria living at temperatures of 42°C (107°F) and pH 0.8 in the acidic wastewater (the same pH as battery acid) from a mine in California. They detected 2,033 proteins from these species, including proteins from the most abundant bacterial species, a *Leptospirillum* group II bacteria. This represented the first large-scale proteomics-level expression of a natural microbial community. One of the proteins, a cytochrome, oxidizes iron and probably influences the rate of breakdown of acid mine drainage products. Many other proteins appear responsible for defending against free radicals, suggesting that this is an important metabolic trait for persistence in the acidic environment. The hope is that the team can now identify enzymes and metabolic pathways that will help in the cleanup of this and other environmentally contaminated sites in the future.

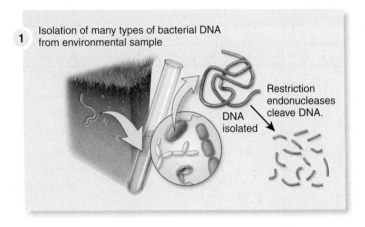

1 Isolation of many types of bacterial DNA from environmental sample

DNA isolated

Restriction endonucleases cleave DNA.

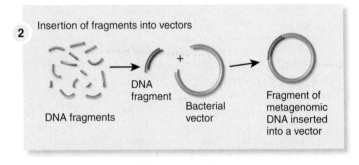

2 Insertion of fragments into vectors

DNA fragments

DNA fragment

Bacterial vector

+

Fragment of metagenomic DNA inserted into a vector

3 Construction of metagenomic library

Each colony contains a different piece of microbial DNA.

Cloned fragment of metagenomic DNA

Transform into host bacterial cell (*E. coli*)

A group of *E. coli* colonies (a library)

4 Analysis of metagenomic library

ATTGAACCAT
GGATTAACG
CTGGTTGA
CCTGATGT
TGAACCA
GATTAAC

Possible methods of analysis:

- Screen for particular sequences by PCR or hybridization

- Random DNA sequencing

- Screen for expression of particular phenotypes

Figure 58.7 **The standard protocol of a metagenomics experiment.** (1) Isolation and fragmentation of DNA from the sample; (2) insertion of fragments into bacterial vectors; (3) insertion of cloned DNA into host cell, and culturing in selective growth media to create a library; and (4) analysis of DNA sequences and protein expression.

58.4 Species Diversity and Community Stability

A community is often seen as stable when little to no change can be detected in the number of species and their abundance over a given time period. The community may then be said to be in equilibrium. The frame of reference for detecting change may encompass a study of a few years or, preferably, several decades. For example, long-term data from Bookham Common, England, revealed that the number of species of birds remained stable for nearly 30 years. Community stability is an important consideration in conservation biology. A decrease in the stability of a community over time may alert biologists to a possible problem. In the 1970s, the populations of many bird species in the U.S. and Europe declined precipitously. Raptor species such as peregrine falcons, bald eagles, and osprey were particularly hard hit. Eventually, the decline was traced to use of the pesticide DDT (dichlorodiphenyltrichloroethane), which caused eggshells to become thin and break before the birds could hatch. After DDT was banned later in the decade, raptor species began to recover.

In this section, we consider the relationship between species richness and community stability. We begin by exploring the question of whether communities with more species are more stable than communities with fewer species. We then examine the link between diversity and stability, using evidence from the field. Finally, we look at the relationship from a different angle and consider whether or not stable communities are more species rich than communities that have been disturbed.

The Diversity-Stability Hypothesis States That Species-Rich Communities Are More Stable Than Those with Fewer Species

Community stability may be viewed in several different ways. Some communities, such as extreme deserts, are considered stable because they are resistant to change by anything other than water. Other communities, such as river communities, are considered stable because they can recover quickly after a disturbance, such as pollution, being cleansed by the rapid flow of fresh water. Lake communities, on the other hand, may be seen as less stable because there is no drainage outlet and pollutants can accumulate quickly.

Because maintaining community stability is seen as important, much research has gone into understanding the factors that enhance it. This work has produced the prevailing idea that species-rich communities are more stable than species-poor communities. Even so, ecologists debate the effects of species richness. For example, are species-rich communities more resistant to invasion by introduced species, such as weeds, than species-poor communities?

The link between species richness and stability was first explicitly proposed by the English ecologist Charles Elton in the 1950s. He suggested that a disturbance in a species-rich community would be cushioned by large numbers of interacting species and would not produce as drastic an effect as it would on a species-poor community. Thus, an introduced predator or parasite could cause extinctions in a species-poor system but possibly not in a more diverse system, where its effects would be buffered by interactions with more species in the community. Elton argued that outbreaks of pests are often found on cultivated land or land disturbed by humans, both of which are species-poor communities with few naturally occurring species. His argument became known as the **diversity-stability hypothesis**.

However, some ecologists began to challenge Elton's association of diversity with stability. Mathematical models showed that communities with higher diversity tended to be less, not more, stable. Ecologists pointed out many examples of introduced species that have assumed pest proportions in species-rich areas, including rabbits in Australia and pigs in North America. They noted that disturbed or cultivated land may suffer from pest outbreaks not because of its simple nature but because individual species, including introduced species, often have no natural enemies in the new environment, in contrast to the long associations between native species and their natural enemies. For example, in Europe, coevolved predators such as foxes prevent rabbit populations from increasing to pest proportions. What was needed was research to determine if a link existed between diversity and stability.

In 1996, ecologist David Tilman reported the relationship between species richness and stability from an 11-year study of 207 grassland plots in Minnesota that varied in their species richness. He measured the biomass of every species of plant, in each plot, at the end of every year and obtained the average species biomass. He then calculated how much this biomass varied from year to year through a statistical measure called the coefficient of variation. Less variation in biomass signified community stability. Year-to-year variation in plant community biomass was significantly lower in plots with greater plant species richness (**Figure 58.8**). The results showed that greater diversity enhances community stability.

Tilman suggested that diversity stabilizes communities because they are more likely to contain disturbance-resistant species that, in the event of a disturbance, could grow and compensate for the loss of disturbance-sensitive species. For example, when a change in climate such as drought decreased the abundance of competitively dominant species that thrived in normal conditions, unharmed drought-resistant species increased in mass and replaced them. Such declines in the number of susceptible species and compensatory increases in other species acted to stabilize total community biomass.

58.5 Succession: Community Change

At 8:32 a.m. on May 18, 1980, Mount St. Helens, a previously little-studied peak in the Washington Cascades, erupted. The blast felled trees over a 600-km² area, and the landslide that followed—the largest in recorded history—destroyed everything in

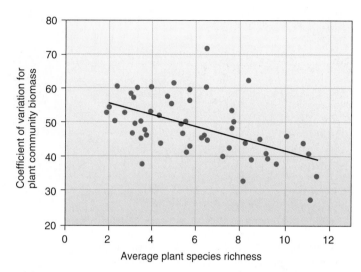

Figure 58.8 Biomass variation and species richness. Tilman's 11-year study of grassland plots in Minnesota revealed that year-to-year variability in community biomass was lower in species-rich plots. Each dot represents an individual plot. Only the plots from one field are graphed.

its path, killing nearly 60 people and millions of animals. However, as with Krakatau, noted at the beginning of this chapter, much of the area has experienced a relatively rapid recovery of plant and animal communities (**Figure 58.9**).

Ecologists have developed several terms to describe how community change occurs. The term **succession** describes the gradual and continuous change in species composition and community structure over time. **Primary succession** refers to succession on a newly exposed site that has no biological legacy in terms of plants, animals, or microbes, such as bare ground caused by a volcanic eruption or the sediment created by the retreat of glaciers. In primary succession on land, the plants must often build up the soil, and thus a long time—even hundreds of years—may be required for the process. Only a tiny proportion of the Earth's surface is currently undergoing primary succession, for example, around Mount St. Helens and the volcanoes in Hawaii and off the coast of Iceland, and behind retreating glaciers in Alaska and Canada.

Secondary succession refers to succession on a site that has already supported life but has undergone a disturbance, such as a fire, tornado, hurricane, or flood (as in the 2004 tsunami in Indonesia). In terrestrial areas, soil is already present. Clearing a natural forest and farming the land for several years is an example of a severe forest disturbance that does not kill all native species. Some plants and many soil bacteria, nematodes, and insects are still present. Secondary succession will occur if farming is ended. The secondary succession in abandoned farmlands (also called old fields) can lead to a pattern of vegetation quite different from one that develops after primary succession following glacial retreat. For example, the plowing and added fertilizers, herbicides, and pesticides may have caused substantial changes in the soil of an old field, allowing species that require a lot of nitrogen to colonize. These species would not be present for many years in newly created glacial soils.

(a) 1980

(b) 1997

Figure 58.9 Succession on Mount St. Helens. (a) The initial blast occurred on May 18, 1980. (b) By 1997, 17 years later, many of the areas initially flattened by the blast and covered in ash developed low-lying vegetation, and new trees sprouted up between the old dead tree trunks.

Frederic Clements is often viewed as the founder of successional theory as well as the organismic model of ecological communities. His work in the early 20th century emphasized succession as proceeding to a distinct end point, or **climax community**. Each phase of succession is called a **sere**, or seral stage. The initial sere is known as the pioneer seral stage. Although disturbance can return a community from a later seral stage to an earlier seral stage, generally the community headed toward climax.

Three mechanisms of succession are discussed here. While Clements's depiction of succession focused on a process termed facilitation, two other mechanisms of succession—inhibition and tolerance—have since been described. Let's examine the evidence for each of them.

Facilitation Assumes Each Invading Species Creates a More Favorable Habitat for Succeeding Species

A key assumption of Clements is that each colonizing species makes the environment a little different—a little shadier or a little richer in soil nitrogen—so that it becomes more suitable for other species, which then invade and outcompete the earlier residents. This process, known as **facilitation**, continues until the most competitively dominant species has colonized, when the community is at climax. The composition of the climax community for any given region is thought to be determined by climate and soil conditions.

Succession following the gradual retreat of Alaskan glaciers is often used as a specific example of facilitation as a mechanism of succession. Over the past 200 years, the glaciers in Glacier Bay have undergone a dramatic retreat of nearly 100 km (Figure 58.10). Succession in Glacier Bay follows a distinct pattern of vegetation. As glaciers retreat, they leave moraines, deposits of stones, pulverized rock, and debris that serve as soil. In Alaska, the bare soil has a low nitrogen content and scant organic matter. In the pioneer stages, the soil is first colonized by a black crust of cyanobacteria, mosses, lichens, horsetails (*Equisetum variegatum*), and the occasional river beauty (*Epilobium*

(a) Glacier Bay, Alaska

Degree of glacier retreat

(b) Glacial retreat

Figure 58.10 The degree of glacier retreat at Glacier Bay, Alaska, since 1794. (a) Primary succession begins on the bare rock and soil evident at the edges of the retreating glacier. (b) The lines reflect the position of the glacier in 1794 and its subsequent retreat northward.

Concept check: Why do ecologists sometimes view walking through Glacier Bay as the equivalent of being in a time machine?

latifolium) (**Figure 58.11a**). Because the cyanobacteria are nitrogen fixers, the soil nitrogen increases a little, but soil depth and litterfall (fallen leaves, twigs, and other plant material) are still minimal. At this stage, there may be a few seeds and seedlings of dwarf shrubs of the rose family commonly called mountain avens (*Dryas drummondii*), alders (*Alnus sinuata*), and spruce, but they are rare. After about 40 years, mountain avens dominates the landscape (**Figure 58.11b**). Soil nitrogen increases, as does soil depth and litterfall, and alder trees begin to invade.

At about 60 years, alders form dense, close thickets (**Figure 58.11c**). Alders have nitrogen-fixing bacteria that live mutualistically in their roots and convert nitrogen from the air into a biologically useful form. Soil nitrogen dramatically increases, as does litterfall. Spruce trees (*Picea sitchensis*) begin to invade at about this time. After about 75 to 100 years, the spruce trees begin to overtop the alders, shading them out. The litterfall is still high, and the large volume of needles turns the soil acidic. The shade causes competitive exclusion of many of the original understory species, including alder, and only mosses carpet the ground. At this stage, seedlings of western hemlock (*Tsuga heterophylla*) and mountain hemlock (*Tsuga mertensiana*) may also occur. After 200 years, a mixed spruce-hemlock climax forest results (**Figure 58.11d**).

What other evidence is there of facilitation? Experimental studies of early primary succession on Mount St. Helens, which show that decomposition of fungi allows mosses and other fungi to colonize the soil, provide evidence of facilitation. In New England salt marshes, *Spartina* grass facilitates the establishment of beach plant communities by stabilizing the rocky substrate and reducing water velocity, which enables other seedlings to emerge. Succession on sand dunes also supports the facilitation model, in that pioneer plant species stabilize the sand dunes and facilitate the establishment of subsequent plant species. The foredunes, those nearest the shoreline, are the most frequently disturbed and are maintained in a state of early succession, while more stable communities develop farther away from the shoreline.

Succession also occurs in aquatic communities. Although soils do not develop in marine environments, facilitation may still be encountered when one species enhances the quality of settling and establishment sites for another species. When experimental test plates used to measure settling rates of marine organisms were placed in the Delaware Bay, researchers discovered that certain cnidarians enhanced the attachment of tunicates, and both facilitated the attachment of mussels, which were the dominant species in the community. In this

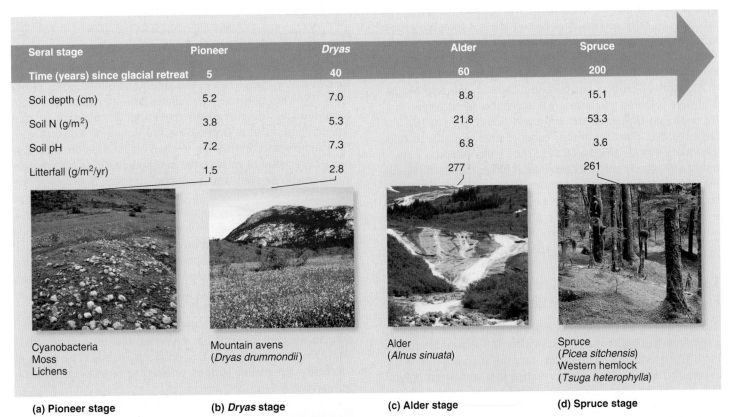

Seral stage	Pioneer	*Dryas*	Alder	Spruce
Time (years) since glacial retreat	5	40	60	200
Soil depth (cm)	5.2	7.0	8.8	15.1
Soil N (g/m^2)	3.8	5.3	21.8	53.3
Soil pH	7.2	7.3	6.8	3.6
Litterfall (g/m^2/yr)	1.5	2.8	277	261

Cyanobacteria
Moss
Lichens

Mountain avens
(*Dryas drummondii*)

Alder
(*Alnus sinuata*)

Spruce
(*Picea sitchensis*)
Western hemlock
(*Tsuga heterophylla*)

(a) Pioneer stage **(b) *Dryas* stage** **(c) Alder stage** **(d) Spruce stage**

Figure 58.11 **The pattern of primary succession at Glacier Bay, Alaska.** (a) The first species to colonize the bare Earth following retreat of the glaciers are small species such as cyanobacteria, moss, and lichens. (b) Mountain avens (*Dryas drummondii*) is a flower common in the *Dryas* seral stage. (c) Soil nitrogen and litterfall increase rapidly as alder (*Alnus sinuata*) invade. Note also the appearance of a few spruce trees higher up the valley. (d) Spruce (*Picea sitchensis*) and hemlock (*Tsuga heterophylla*) trees comprise a climax spruce-hemlock forest at Glacier Bay, with moss carpeting the ground. Two hundred years ago, glaciers occupied this spot.

Concept check: *Is facilitation the only mechanism fueling succession at Glacier Bay?*

experiment, the smooth surface of the test plates prevented many species from colonizing, but once the surface became rougher, because of the presence of the cnidarians, many other species were able to colonize. In a similar fashion, early colonizing bacteria, which create biofilms on rock surfaces, can facilitate succession of other organisms.

Inhibition Implies That Early Colonists Prevent Later Arrivals from Replacing Them

Although data on succession in some communities fit the facilitation model, researchers have proposed alternative hypotheses of how succession may operate. In the process known as **inhibition**, early colonists prevent colonization by other species. For example, removing the litter of *Setaria faberi*, an early successional plant species in New Jersey old fields, causes an increase in the biomass of a later species, *Erigeron annuus*. The release of toxic compounds from decomposing *Setaria* litter or physical obstruction by the litter itself blocks the establishment of *Erigeron*. Without the litter present, *Erigeron* dominates and reduces the biomass of *Setaria*. Plant species that grow in dense thickets, such as some grasses, ferns, vines, pine trees, and bamboo, can inhibit succession, as can many introduced plant species.

Inhibition has been seen as the primary method of succession in the marine intertidal zone, where space is limited. In this habitat, early successional species are at a great advantage in maintaining possession of valuable space. In 1974, ecologist Wayne Sousa created an environment for testing how succession works in the intertidal zone by scraping rock faces clean of all algae or putting out fresh boulders or concrete blocks. The first colonists of these areas were the green algae *Ulva*. By removing *Ulva* from the substrate, Sousa showed that the large red alga *Chondracanthus canaliculatus* was able to colonize more quickly (**Figure 58.12**). The results of Sousa's study indicate that early colonists can inhibit rather than facilitate the invasion of subsequent colonists. Succession may eventually occur because early colonizing species, such as *Ulva*, are more susceptible to the rigors of the physical environment and to attacks by herbivores, such as crabs (*Pachygrapsus crassipes*), than later successional species, such as *Chondracanthus*.

Tolerance Suggests That Early Colonists Neither Facilitate nor Inhibit Later Colonists

In 1977, researchers Joseph Connell and Ralph Slatyer proposed a third mechanism of succession, which they termed **tolerance**. In this process, any species can start the succession, but the eventual climax community is reached in a somewhat orderly fashion. The species that establish and remain do not change the environment in ways that either facilitate or inhibit subsequent colonists. Species have differing tolerances to the intensity of competition that results as more species accumulate. Relatively competitive-intolerant species are more successful early in succession when the intensity of competition is low and resources are abundant. Relatively competition-tolerant species appear later in succession and at climax. Connell and Slatyer found the best evidence for the tolerance model in Frank Egler's

Figure 58.12 Inhibition as a primary method of succession in the marine intertidal zone. Removing *Ulva* from intertidal rock faces allowed colonization by *Chondracanthus*. The inset shows *Chondracanthus* on a rock face with the striped shore crab *Pachygrapsus crassipes*, a herbivore.

earlier work on floral succession. In the 1950s, Egler showed that succession in plant communities is determined largely by species that already exist in the ground as buried seeds or old roots. Whichever species germinates first or regenerates from roots initiates the succession sequence. Germination or root regeneration, in turn, depends on the timing of a disturbance. For example, an early-season tree fall would promote early germinating species to grow in the subsequent light gap, whereas a late-season tree fall would promote the growth of late germinating species. As succession proceeds, earlier germinating or regenerating species may be outcompeted by different species.

The key distinction between the three models is in the manner in which succession proceeds. In the facilitation model, species replacement is facilitated by previous colonists; in the inhibition model, it is inhibited by the action of previous colonists; and in the tolerance model, species may be affected by previous colonists, but they do not require them (**Figure 58.13**). The type of model followed depends on the community in question. Subsequent research has suggested that other factors may also influence succession, especially on islands. The study of succession on islands is often referred to as island biogeography, which is described next.

58.6 Island Biogeography

In the 1960s, two eminent ecologists, Robert MacArthur and E. O. Wilson, developed a comprehensive model to explain the process of succession on new islands, where a gradual buildup of species proceeds from a sterile beginning. Their model, termed the **equilibrium model of island biogeography**, holds

Figure 58.13 **Three models of succession.** A, B, C, and D represent four stages or seres. D represents the climax community. An arrow indicates "is replaced by," and + = facilitation, − = inhibition, and 0 = no effect. The facilitation model is the classic model of succession. In the inhibition model, much depends on which species gets there first. The tolerance model is similar to the facilitation model, in that later species may be facilitated by earlier species, but they can also invade in their absence. The colored arrows show that succession may bypass some stages in the tolerance model.

Concept check: *Inhibition implies competition between species with early-arriving species tending to outcompete later arrivals, at least for a while. Does competition or mutualism feature more prominently in facilitation?*

that the number of species on an island tends toward an equilibrium number that is determined by the balance between two factors: immigration rates and extinction rates. In this section, we explore island biogeography and how well the model's predictions are supported by experimental data.

The Island Biogeography Model Suggests That During Succession, Gains in Immigration Are Balanced by Losses from Extinction

MacArthur and Wilson's model of island biogeography suggests that species repeatedly arrive on an island and either thrive or become extinct. The rate of immigration of new species is highest when no species are present on the island. As the number of species accumulates, the immigration rate decreases, since subsequent immigrants are more likely to represent species already present on the island. The rate of extinction is low at the time of first colonization, because few species are present and many have large populations. With the addition of new species, the populations of some species diminish, so the probability of extinction by chance alone increases. Over time, the number of species tends toward an equilibrium, $\hat{S}$, in which the rates of immigration and extinction are equal. Species may continue to arrive and go extinct, but the number of species on the island remains approximately the same.

MacArthur and Wilson reasoned that when plotted graphically, both the immigration and extinction lines would be curved, for several reasons (**Figure 58.14a**). First, species arrive

on islands at different rates. Some organisms, including plants with seed dispersal mechanisms and winged animals, are more mobile than others and will arrive quickly. Other organisms will arrive more slowly. This pattern causes the immigration curve to start off steep but get progressively shallower. On the other hand, extinctions rise at accelerating rates, because as later species arrive, competition increases and more species are likely to go extinct. As noted previously, earlier-arriving species tend to be *r*-selected species, which are better dispersers, whereas later-arriving species are generally *K*-selected species, which are better competitors. Later-arriving species usually outcompete earlier-arriving ones, causing an increase in extinctions.

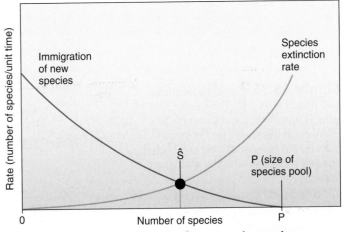

(a) Effects of immigration and extinction on species number

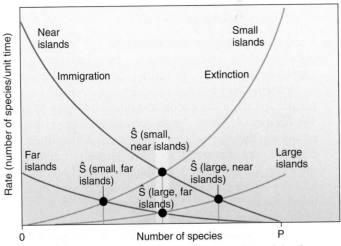

(b) Added effects of island size and proximity to the mainland

Figure 58.14 **MacArthur and Wilson's equilibrium model of island biogeography.** **(a)** The interaction of immigration rate and extinction rate produces an equilibrium number of species on an island, $\hat{S}$. $\hat{S}$ can vary from 0 species to P species, the total number of species available to colonize. **(b)** $\hat{S}$ varies according to the island's size and distance from the mainland. An increase in distance (near to far) lowers the immigration rate. An increase in island area (small to large) lowers the extinction rate.

Concept check: *Can you think of a scenario where there would be large numbers of species on a small island?*

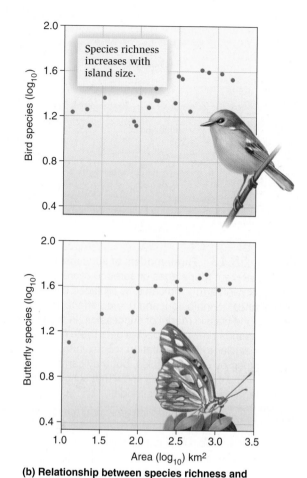

(a) Lesser Antilles Islands

(b) Relationship between species richness and island size

Figure 58.15 **Species richness and island size.** **(a)** The Lesser Antilles extend from Anguilla in the north to Grenada in the south. **(b)** On these islands, the number of bird and butterfly species increases with the area of an island.

Concept check: How large is the change in bird species richness across islands in the Lesser Antilles?

The strength of the island biogeography model was that it generated several testable predictions:

1. The number of species should increase with increasing island size. This concept is also known as the species-area effect (see Figure 58.4). Extinction rates would be greater on smaller islands because population sizes would be smaller and more susceptible to extinction (**Figure 58.14b**).

2. The number of species should decrease with increasing distance of the island from the mainland, or the **source pool**, the pool of potential species available to colonize the island. Immigration rates would be greater on islands near the source pool because species do not have as far to travel (Figure 58.14b).

3. The turnover of species should be considerable. The number of species on an island might remain relatively constant, but the composition of the species should vary over time as new species colonize the island and others become extinct.

Let's examine these predictions one by one and see how well the data support them.

Species-Area Relationships

The West Indies has traditionally been a key location for ecologists studying island biogeography. This is because the physical geography and the plant and animal life of the islands are well known. Furthermore, the Lesser Antilles, from Anguilla in the north to Grenada in the south, enjoy a similar climate and are surrounded by deep water (**Figure 58.15a**). In 1999, Robert Ricklefs and Irby Lovette summarized the available data on the richness of species of four groups of animals—birds, bats, reptiles and amphibians, and butterflies—over 19 islands that varied in area over two orders of magnitude (13 km^2 to 1,510 km^2). In each case, a positive correlation occurred between area and species richness (**Figure 58.15b**). Note that these relationships are traditionally plotted on a double logarithmic scale, a so-called log-log plot, in which the horizontal axis is the logarithm to the base 10 of the area and the vertical axis is the logarithm to the base 10 of the number of species. A linear plot of the area versus the number of species would be difficult to produce, because of the wide range of area and richness of species involved. Logarithmic scales condense this variation to manageable limits.

Species-Distance Relationships

MacArthur and Wilson provided evidence for the effect of the distance of an island from a source pool of colonists, usually the mainland. In studies of the numbers of lowland forest bird species in Polynesia, they found that the number of species decreased with the distance from the source pool of New Guinea (**Figure 58.16**). They expressed the richness of bird

(a) New Guinea and neighboring islands

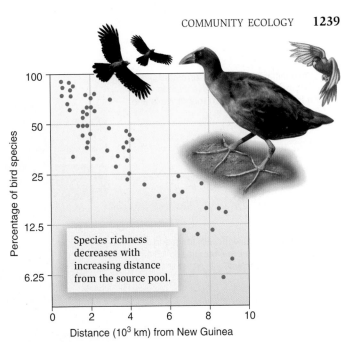

Species richness
decreases with
increasing distance
from the source pool.

(b) Relationship between species richness and distance from source

Figure 58.16 Species richness and distance from the source pool. **(a)** Map of Australia, New Guinea, and these Polynesian Islands: New Caledonia, Fiji Islands, Cook Islands, Marquesas Islands, Pitcairn, and Easter Island. **(b)** The numbers of bird species on the islands decreases with increasing distance from the source pool, New Guinea. The species richness is expressed as the percentage of bird species on New Guinea.

species on the islands as a percentage of the number of bird species found on New Guinea. A significant decline in this percentage was observed with increasing distance. More-distant islands contained lower numbers of species than nearer islands. This research substantiated the prediction of species richness declining with increasing distance from the source pool.

Species Turnover

Studies involving species turnover on islands are difficult to perform because detailed and complete species lists are needed over long periods of time, usually many years and often decades.

The lists that do exist are often compiled in a casual way and are not usually suitable for comparison with more modern data. In 1980, British researcher Francis Gilbert reviewed 25 investigations carried out to demonstrate turnover and found a lack of this type of rigor in nearly all of them. Furthermore, most of the observed turnover in these studies, usually less than 1% per year, or less than one species per year, appeared to be due to immigrants that never became established, not due to the extinction of well-established species. More recent studies have revealed similar findings. Most studies suggest that the rates of turnover are low, giving little conclusive support to the third prediction of the equilibrium model of island biogeography.

FEATURE INVESTIGATION

Simberloff and Wilson's Experiments Tested the Predictions of the Equilibrium Model of Island Biogeography

In the 1960s, Daniel Simberloff and E. O. Wilson conducted possibly the best test of the equilibrium model of island biogeography ever performed, using islands in the Florida Keys. They surveyed small red mangrove (*Rhizophora mangle*) islands, 11–25 m in diameter, for all terrestrial arthropods. They then enclosed each island with a plastic tent and had the islands fumigated with methyl bromide, a short-acting insecticide, to

remove all arthropods on them. The tents were removed, and periodically thereafter Wilson and Simberloff surveyed the islands to examine recolonization rates. At each survey, they counted all the species present, noting any species not there at the previous census and the absence of others that were previously there but had presumably gone extinct (results for four of the islands are shown in **Figure 58.17**). In this way, they estimated turnover of species on islands.

After 250 days, all but one of the islands had a similar number of arthropod species as before fumigation, even though population densities were still low. The data indicated that

Figure 58.17 Simberloff and Wilson's experiments on the equilibrium model of biogeography.

HYPOTHESIS Island biogeography model predicts higher species richness for islands closer to the mainland and significant turnover of species on islands.

STARTING LOCATION Mangrove islands in the Florida Keys.

Experimental level

Conceptual level

1 Take initial census of all terrestrial arthropods on 4 mangrove islands. Erect a framework over each mangrove island.

Each mangrove island is isolated.

Distant

Very near

Mainland

2 Cover the framework with tents and fumigate with methyl bromide to kill all arthropod species.

Methyl bromide is a low-persistent insecticide that at low levels will not kill plant life.

Distant

Mainland

Very near

3 Remove the tents and conduct censuses every month to monitor recolonization of arthropods and to determine extinction rates.

Mangrove islands are recolonized.

Distant

Mainland

Very near

4 **THE DATA** Island E2 was closest to the mainland and supported the highest number of species both before and after fumigation. E3 and ST2 were at an intermediate distance from the mainland, and E1 was the most distant.

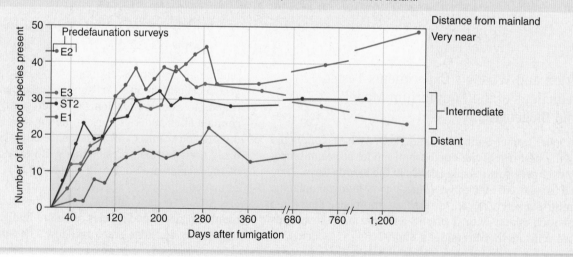

Distance from mainland

Very near

Intermediate

Distant

5 **CONCLUSION** Island distance from the mainland influences species richness on mangrove islands in the Florida Keys. However, species turnover is minimal, and species richness changes little following initial recolonization.

6 **SOURCE** Simberloff, D.S. 1978. Colonization of islands by insects: immigration, extinction and diversity. pp. 139–153 in L.A. Mound and N. Waloff (eds.). Diversity of insect faunas. *Blackwell Scientific Publications*, Oxford, U.K.

recolonization rates were higher on islands nearer to the mainland than on far islands—as the island biogeography model predicts. However, the data, which consisted of lists of species on islands before and after extinctions, provided little support for the prediction of substantial turnover. Rates of turnover were low, only 1.5 extinctions per year, compared to the 15 to 40 species found on the islands within a year. Wilson and Simberloff concluded that turnover probably involves only a small subset of transient or unimportant species, with the more-important species remaining permanent after colonization.

Experimental Questions

1. What was the purpose of Simberloff and Wilson's study?

2. Why did the researchers conduct a thorough species survey of arthropods before experimental removal of all the arthropod species?

3. What did the researchers conclude about the relationship between island proximity to the mainland and species richness and turnover?

MacArthur and Wilson's equilibrium model of island biogeography has stimulated much research that confirms the strong effects of area and distance on species richness. However, species turnover appears to be low rather than considerable, which suggests that succession on most islands is a fairly orderly process. This means that colonization is not a random process and that the same species seem to colonize first and other species gradually appear in the same order.

It is also important to note that the principles of island biogeography have been applied to wildlife preserves, which are essentially islands in a sea of developed land, either agricultural fields or urban sprawl. Conservationists have therefore utilized island biogeography modeling in the design of nature preserves, a topic we will return to in Chapter 60.

In this chapter, we have seen how important species richness is to community function. It affects not only stability but, as we shall see, many other features of a community and its surrounding environment, such as nutrient uptake, biomass, and productivity. In order to know how diversity affects these properties, we need to understand the basic processes of energy flow and chemical cycling through a community and its environment, and for that we next turn to a discussion of ecosystems ecology in Chapter 59.

▌ Summary of Key Concepts

58.1 Differing Views of Communities

- Community ecology studies how groups of species interact and form functional communities. Ecologists have differing views on the nature of a community. In one view, communities are tightly organized groups of mutually dependent species, and in another, they are loose assemblies of species that happen to live in the same place at the same time. (Figure 58.1)

- Although many observations support the idea that communities are loose assemblages of species, sharp boundaries between groups of species do exist, especially related to physical differences that cause distinct communities to develop. (Figure 58.2)

58.2 Patterns of Species Richness

- The number of species of most taxa varies according to geographic location, generally increasing from polar areas to tropical areas. (Figure 58.3)

- Varying hypotheses for the polar-equatorial gradient have been advanced, including the time hypothesis, the area hypothesis, the productivity hypothesis, and the intermediate disturbance hypothesis. (Figures 58.4, 58.5, 58.6)

58.3 Calculating Species Diversity

- The most widely used measure of the diversity of a community, called the Shannon diversity index, takes into account both species richness and species abundance. (Table 58.1)

- The field of metagenomics seeks to identify and analyze the genomes contained in a community of microorganisms. (Figure 58.7)

58.4 Species Diversity and Community Stability

- Community stability is an important consideration in ecology. The diversity-stability hypothesis maintains that species-rich communities are more stable than communities with fewer species.

- Tilman's field experiments, which showed that year-to-year variation in plant biomass decreased with increasing species diversity, established a link between diversity and stability. (Figure 58.8)

58.5 Succession: Community Change

- Succession describes the gradual and continuous change in community structure over time. Primary succession refers to succession on a newly exposed site with no prior biological legacy; secondary succession refers to succession on a site that has already supported life but has undergone a disturbance. (Figures 58.9, 58.10)

- Three mechanisms have been proposed for succession. In facilitation, each species facilitates or makes the environment more suitable for subsequent species. In inhibition, initial species inhibit later colonists. In tolerance, any species can start the succession, and species replacement is unaffected by previous colonists. (Figures 58.11, 58.12, 58.13)

58.6 Island Biogeography

- In the equilibrium model of island biogeography, the number of species on an island tends toward an equilibrium number determined by the balance between immigration rates and extinction rates. (Figure 58.14)

- The model predicts that the number of species increases with increasing island size; that the number of species decreases with distance from the source pool; and that turnover is high. (Figures 58.15, 58.16)

- Simberloff and Wilson's experiments on mangrove islands in the Florida Keys provided support for the first tenet of the island biogeography model but refuted the third tenet. (Figure 58.17)

Assess and Discuss

Test Yourself

1. A community with many individuals but few different species would exhibit
   a. low abundance and high species complexity.
   b. high stability.
   c. low species richness and high abundance.
   d. high species diversity.
   e. high abundance and high species richness.

2. Which of the following statements best represents the productivity hypothesis regarding species richness?
   a. The larger the area, the greater the number of species that will be found there.
   b. Temperate regions have a lower species richness due to the lack of time available for migration after the last ice age.
   c. The number of species in a particular community is directly related to the amount of available energy.
   d. As invertebrate productivity increases, species richness will increase.
   e. Species richness is not related to primary productivity.

3. Ecologists began to question Elton's link of increased stability to increased diversity because
   a. mathematical models showed communities with high diversity had high stability.
   b. cultivated land undergoes few outbreaks of pests.
   c. highly disturbed areas have high numbers of species.
   d. pest outbreaks are caused by lack of long associations with natural enemies, not because they occur in simple systems.
   e. all of the above.

4. Metagenomics is a field of study that
   a. determines the similarities of the genomes of all species in a community.
   b. focuses on the microbial genomes contained in a community.
   c. compares the genomes of similar species in different communities.
   d. none of the above
   e. both a and b

5. Extreme fluctuations in species abundance
   a. lead to more diverse communities.
   b. are usually seen in early stages of community development.
   c. may increase the likelihood of extinction.
   d. have very little effect on species richness.
   e. are characteristic of stable communities.

6. Which of the following statements best represents the relationship between species diversity and community disturbance?
   a. Species diversity and community stability have no relationship.
   b. Communities with high levels of disturbance are more diverse.
   c. Communities with low levels of disturbance are more diverse.
   d. Communities with intermediate levels of disturbance are more diverse.
   e. Communities with intermediate levels of disturbance are less diverse.

7. The process of primary succession occurs
   a. around a recently erupted volcano.
   b. on a newly plowed field.
   c. on a hillside that has suffered a mudslide.
   d. on a recently flooded riverbank.
   e. on none of the above.

8. Early colonizers excluding subsequent colonists from moving into a community is referred to as
   a. facilitation.   d. inhibition.
   b. competitive exclusion.   e. natural selection.
   c. secondary succession.

9. A tree falls in a forest in spring and flowers germinate in the light gap. Following a tree fall in autumn, different species of flowers germinate in the light gaps. This illustrates the principle of
   a. facilitation.   d. primary succession.
   b. tolerance.   e. climax communities.
   c. inhibition.

10. On which types of island would you expect species richness to be greatest?
    a. small, near mainland
    b. small, distant from mainland
    c. large, near mainland
    d. large, distant from mainland
    e. Species richness is equal on all these types of islands.

Conceptual Questions

1. Define a community and community ecology.

2. List some possible ecological disturbances, their likely frequency in natural communities, and the severity of their effects.

3. When is a community in equilibrium?

Collaborative Questions

1. Distinguish between the time hypothesis, area hypothesis, and productivity hypothesis as explanations for the latitudinal gradient in species richness.

2. Calculate the species diversity of the following four communities. Which community has the highest diversity? What is the maximum diversity each community could have?

Community	Relative abundance of species			H_S	Maximum possible diversity
	Species 1	Species 2	Species 3		
1	90	10	—		
2	50	50	—		
3	80	10	10		
4	33.3	33.3	33.3		

Online Resource

www.brookerbiology.com

Stay a step ahead in your studies with animations that bring concepts to life and practice tests to assess your understanding. Your instructor may also recommend the interactive ebook, individualized learning tools, and more.

an ecosystem. In examining energy flow, our main task will be to document the complex networks of feeding relationships, called food webs, and to measure the efficiency of energy transfer between organisms in an ecosystem.

Next, we will focus on the measurement of **biomass**, a quantitative estimate of the total mass of living matter in a given area, usually measured in grams or kilograms per square meter. We will examine the amount of biomass produced through photosynthesis, termed primary production, and the amount of biomass produced by the organisms that are the consumers of primary production.

The functioning of an ecosystem can sometimes be most limited by the availability of a scarce chemical or mineral. In the last section, we will examine the movement of chemicals through ecosystems, called **biogeochemical cycles**, and explore the cycling of elements, such as phosphorus, carbon, nitrogen, and sulfur, and the effects that human activities are having on these ecosystem-wide processes.

59.1 Food Webs and Energy Flow

Most organisms either make their own food using energy from sunlight or feed on other organisms. Simple feeding relationships between organisms can be characterized by an unbranched **food chain**, a linear depiction of energy flow, with each organism feeding on and deriving energy from the preceding organism. Each feeding level in the chain is called a **trophic level** (from the Greek *trophos*, meaning feeder), and different species feed at different levels. In a food-chain diagram, an arrow connects each trophic level with the one above it (**Figure 59.2**).

In this section, we will consider trophic relationships and the flow of energy in a food chain and also examine a food web, a more complex model of interconnected food chains. We will then explore two of the most important features of food webs—chain length and the pyramid of numbers—and learn how the passage of nutrients through food webs can result in the accumulation of harmful chemicals in the tissues of organisms at higher trophic levels.

The Main Trophic Levels Within Food Chains Consist of Primary Producers, Primary Consumers, and Secondary Consumers

Food chains typically consist of organisms that obtain energy in different ways. **Autotrophs** harvest light or chemical energy and store that energy in carbon compounds. Most autotrophs, including plants, algae, and photosynthetic bacteria, use sunlight for this process. These organisms, called **primary producers**, form the base of the food chain. They produce the energy-rich organic molecules upon which nearly all other organisms depend. However, not all primary producers utilize sunlight; some organisms, called chemoautotrophs, obtain their energy by oxidizing inorganic compounds such as sulfides.

Organisms in trophic levels above the primary producers are termed **heterotrophs**. These organisms receive their nutrition by eating other organisms. Organisms that obtain their food by consuming primary producers are termed **primary consumers** and include most protists, all animals, and even some plants such as mistletoe, which is parasitic on other plants. Animals that eat plants are also called **herbivores**. Organisms that eat primary consumers are **secondary consumers**. Animals that eat other animals are also called **carnivores** (from the Latin *carn*, meaning flesh). Organisms that feed on secondary consumers are **tertiary consumers**, and so on. Thus, energy enters a food chain through primary producers, via photosynthesis, and is passed up the food chain to primary, secondary, and tertiary consumers (**Figure 59.2**).

At each trophic level, many organisms die before they are eaten. Much energy from the first trophic level, such as the plants, goes unconsumed by herbivores. Instead, unconsumed plants die and decompose in place. This material, along with dead remains of animals and waste products, is called **detritus**. Consumers that get their energy from detritus, called **detritivores**, or **decomposers**, break down dead organisms from all trophic levels (**Figure 59.3**). In terrestrial systems, detritivores probably carry out 80–90% of the consumption of plant matter, with different species working in concert to extract most of

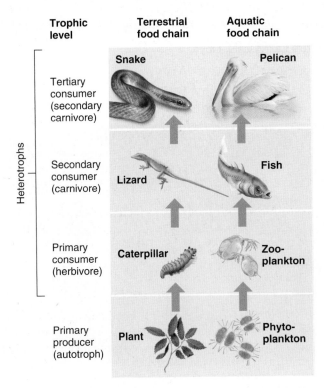

Trophic level	Terrestrial food chain	Aquatic food chain
Tertiary consumer (secondary carnivore)	Snake	Pelican
Secondary consumer (carnivore)	Lizard	Fish
Primary consumer (herbivore)	Caterpillar	Zooplankton
Primary producer (autotroph)	Plant	Phytoplankton

Heterotrophs

Figure 59.2 Food chains. Two examples of the flow of food energy up the trophic levels: a terrestrial food chain and an aquatic food chain.

Ecosystem Ecology

59

Mandara Lake Oasis, Libya.

F amiliar backyard earthworms are known for their ability to convert organic matter such as dead leaves into rich humus, improving soil fertility. However, earthworms are not native everywhere. North American glaciations exterminated earthworms from hardwood forests in Wisconsin and Minnesota some 11,000 to 14,000 years ago. In the absence of earthworms, fungi and bacteria dominated decomposition, a thick forest floor formed, and carbon built up in the soil. Exotic earthworms from Europe and Asia initially were introduced into Wisconsin and Minnesota by European settlers and have continued to be transported to the area through a range of human activities, including the dumping of fishing bait. Organisms that are beneficial in one location can be destructive when introduced to another, and earthworms are no exception. In northern forests, the worms accelerate the cycling of nutrients through the soil, drastically altering the structure of the forest floor soil and releasing soil carbon to the atmosphere. According to Cindy Hale, a biologist who has been studying the effect of earthworms on northern hardwood forests, "They have a cascading effect on plants, animals, and soil organisms. And we know they're causing significant damage to some forests. Their effect could be really profound."

The term **ecosystem** was coined in 1935 by the British plant ecologist A. G. Tansley to include not only the biotic community of organisms in an area but also the abiotic environment affecting that community. **Ecosystem ecology** is concerned with the movement of energy and materials through organisms and their communities. Just like the concept of a community, the ecosystem concept can be applied at any scale. A small pond inhabited by protozoa and insect larvae is an ecosystem, and an oasis with its plants, frogs, fishes, and birds constitutes another. Most ecosystems cannot be regarded as having definite boundaries. Even in a clearly defined pond ecosystem, species may be moving in and out (**Figure 59.1**). Nevertheless, studying ecosystem ecology allows us to use the common currency of energy and chemicals, or nutrients, to compare the functions between and within ecosystems.

In investigating the different processes of an ecosystem, at least three major constituents can be measured: energy flow, biomass production, and biogeochemical cycling. We begin the chapter by exploring **energy flow**, the movement of energy through

Figure 59.1 A small ecosystem. Even in this pond ecosystem, frogs or other species such as birds may move in and out, importing or exporting nutrients and energy with them.

the energy. Detritivores may, in turn, support a community of predators that feed on them. As we noted at the beginning of the chapter, changes in the decomposer community can lead to important changes in nutrient cycling.

The consumption of species between trophic levels is quite varied. For example, many different herbivore species may feed on the same plant species. Also, each species of herbivore may feed on several different plant species. Such branching of food chains also occurs at other trophic levels. For instance, on the African savanna, cheetahs, lions, and hyenas all eat a variety of prey, including wildebeest, impala, and Thompson's gazelle. These, in turn, eat a variety of trees and grasses. It is more correct, then, to draw relationships between these plants and animals not as a simple chain but as a more elaborate interwoven **food web**, in which there are multiple links between species (Figure 59.4).

In Most Food Webs, Chain Lengths Are Short

Let's examine some of the characteristics of food webs in more detail. The concept of chain length refers to the number of links between the trophic levels involved. For example, if a lion feeds on a zebra, and a zebra feeds on grass, the chain length would be two. In many food webs, chain lengths tend to be short, usually less than six levels, even including parasites and detritivores. The main reason why they are short comes from the well-established laws of physics and chemistry that we discussed in Chapter 2. The second law of thermodynamics states that energy conversions are not 100% efficient. In any transfer process, some useful energy, which can do work, is lost, often in the form of heat. This decreases the amount of available energy at higher trophic levels. We can construct energy budgets for food webs that trace energy flow from green plants to tertiary consumers (and if needed beyond) (Figure 59.5). In each trophic level, some energy is lost to maintenance, for example, to maintain body temperature. Because energy transfer between trophic levels is not 100% efficient, energy is lost in the passage from one trophic level to another.

As described next, two ways that ecologists evaluate the efficiency of consumers as energy transformers are production efficiency and trophic-level transfer efficiency.

Production Efficiency **Production efficiency** is defined as the percentage of energy assimilated by an organism that becomes incorporated into new biomass.

$$\text{Production efficiency} = \frac{\text{Net productivity}}{\text{Assimilation}} \times 100$$

Here, net productivity is the energy, stored in biomass, that has accumulated over a given time span, and assimilation is the total amount of energy taken in by an organism over the same time span. Invertebrates generally have high production efficiencies that average about 10–40% (Figure 59.6a). Microorganisms also have relatively high production efficiencies. Vertebrates tend to have lower production efficiencies than invertebrates, because they devote more energy to sustaining

Figure 59.3 Decomposers (detritivores) feeding on dead plant and animal matter.   Many dead plants and animals are eaten by a variety of decomposers. Here, bacteria and fungi feed on dead animals and rotting plant material. Dead animals, or carrion, may also be fed upon by blowfly larvae, carrion beetles, and other scavengers. These may, in turn, support a variety of predators, including centipedes or larger predators such as mammals and birds, which will also feed on the animal carcass.

Concept check: At which trophic level do decomposers feed?

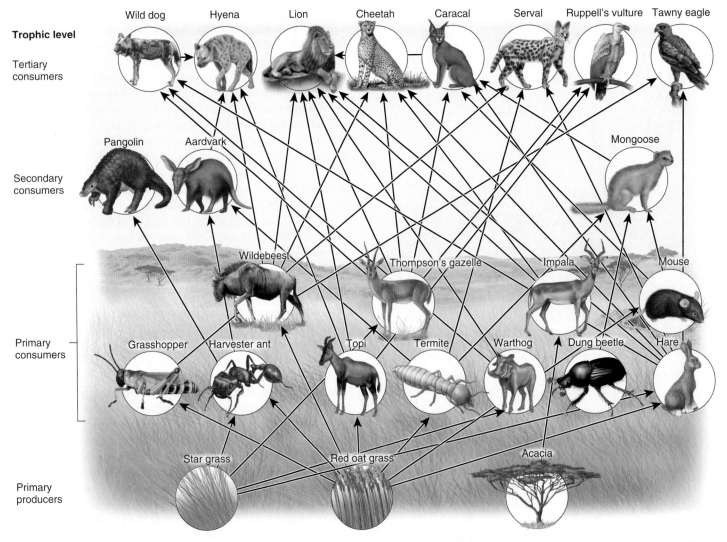

Trophic level

Tertiary consumers

Wild dog Hyena Lion Cheetah Caracal Serval Ruppell's vulture Tawny eagle

Secondary consumers

Pangolin Aardvark Mongoose

Primary consumers

Wildebeest Thompson's gazelle Impala Mouse

Grasshopper Harvester ant Topi Termite Warthog Dung beetle Hare

Primary producers

Star grass Red oat grass Acacia

Figure 59.4 **A simplified food web from an African savanna ecosystem.** Each trophic level is occupied by different species. Generally, each species feeds on, or is fed upon by, more than one species.

their metabolism than to new biomass production. Even within vertebrates, much variation occurs. Fishes, which are ectotherms, typically have production efficiencies of around 10%, and birds and mammals, which are endotherms, have production efficiencies in the range of 1–2% (**Figure 59.6b**). In large part, the difference reflects the energy cost of maintaining a constant body temperature. Production efficiencies are higher in young animals, which are rapidly accruing biomass, than in older animals, which are not. This is the main reason behind the practice of harvesting young animals for meat, at about the time when they first attain adult mass.

One consequence of differing production efficiencies is that sparsely vegetated deserts can support healthy populations of snakes and lizards, whereas mammals might easily starve. The largest living lizard known, the Komodo dragon, eats the equivalent of its own weight every 2 months, whereas a cheetah consumes approximately four times its own weight in the same period.

Trophic-Level Transfer Efficiency The second measure of efficiency of consumers as energy transformers is **trophic-level transfer efficiency**, which is the amount of energy at one trophic level that is acquired by the trophic level above and incorporated into biomass. This provides a way to examine energy flow between trophic levels, not just in an individual species. Trophic level transfer efficiency is calculated as follows:

$$\text{Trophic-level transfer efficiency} = \frac{\text{Production at trophic level } n}{\text{Production at trophic level } n-1} \times 100$$

For example, if there were 14 g/m² of zooplankton in a lake (trophic level n) and 100 g/m² of phytoplankton production (trophic level $n - 1$), the trophic level efficiency would be 14%. Trophic-level transfer efficiency appears to average around 10%, though there is much variation. In some marine food chains, for example, it can exceed 30%.

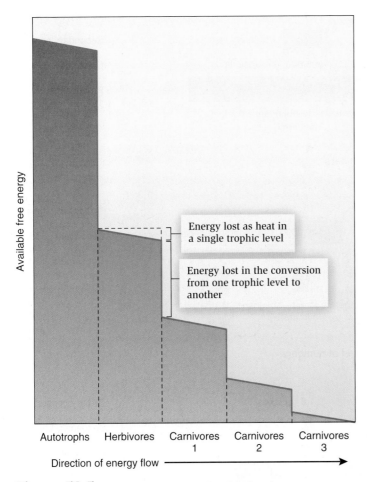

Figure 59.5   Energy flow through a food web.

Trophic-level transfer efficiency is generally low for two reasons. First, many organisms cannot digest all their prey. They take only the easily digestible plant leaves or animal tissue such as muscles and guts, leaving the hard wood or energy-rich bones behind. Second, much of the energy assimilated by animals is used in maintenance, so most energy is lost from the system as heat. The 10% average transfer rate of energy from one trophic level to another also necessitates short food webs of no more than four or five levels. Relatively little energy is available for the higher levels.

Eltonian Pyramids Describe the Distribution of Numbers, Biomass, or Energy Between Trophic Levels

Trophic-level transfer efficiencies can be expressed in a graphical form called an Eltonian pyramid, named after the British ecologist Charles Elton. The best-known pyramid, and the one described by Elton in 1927, is the **pyramid of numbers**, in which the number of individuals decreases at each trophic level, with a large number of individuals at the base and fewer individuals at the top. Elton's example was that of a small pond, where the numbers of protozoa may run into the millions and those of *Daphnia*, their predators, number in the hundreds of

(a) High production efficiency of an invertebrate

(b) Low production efficiency of a vertebrate

Figure 59.6   Production efficiencies.   **(a)** This caterpillar, an invertebrate, chews leaves to obtain its energy. If a mouthful of food contains 1,000 joules (J) of energy, about 320 J is used in cellular respiration to fuel metabolic processes (32%), and 500 J (50%) is lost in feces. This leaves about 180 J of the 500g assimilated to be converted into insect biomass, a production efficiency of 36%. **(b)** The production efficiency of this squirrel, a mammal, is much lower.

Concept check: *What is the production efficiency of the squirrel, using the numbers in the figure?*

thousands. Hundreds of beetle larvae may feed on *Daphnia*, and tens of fishes feed on the beetles. Many other examples of this type of pyramid are known. For example, in a grassland, there may be hundreds of individual plants per square meter, dozens of insects that feed on the plants, a few spiders feeding on the insects, and birds that feed on the spiders (Figure 59.7a).

Ecologists have, however, discovered many exceptions to this pyramid. One single producer such as an oak tree can support hundreds of herbivorous beetles, caterpillars, and other primary consumers, which, in turn, may support thousands of predators. This is called an inverted pyramid of numbers (Figure 59.7b).

One way to reconcile this apparent exception is to weigh the organisms in each trophic level, creating a **pyramid of biomass**. For example, an oak tree weighs more than all its herbivores and predators combined. Looking at the biomass at each trophic level rather than at numbers of organisms shows an upright pyramid. In 1957, Howard Odum measured the pyramid of biomass for a freshwater ecosystem, Silver Springs, Florida (**Figure 59.7c**). Beds of eelgrass (genus *Sagittaria*) and attached algae make up most of the producers. Insects, snails, herbivorous fishes, and turtles eat the producers. Other fishes form the secondary and tertiary consumers. Odum also noted the presence of fungi and bacteria, which were involved in decomposition on all trophic levels.

Even when biomass is used as the measure, inverted pyramids can still occur, albeit rarely. In some marine and lake systems, the biomass of phytoplankton supports a higher biomass of zooplankton (**Figure 59.7d**). This is possible because the rate of production of phytoplankton biomass is much higher than that of zooplankton, and the small phytoplankton **standing crop** (the total biomass in an ecosystem at any one point in time) processes large amounts of energy.

However, by expressing the pyramid in terms of production rate, it is no longer inverted. The **pyramid of energy**, which shows the rate of energy production rather than standing crop, is never inverted (**Figure 59.7e**). The laws of thermodynamics ensure that the highest amounts of free energy are found at the lowest trophic levels. Howard Odum's energy pyramid for Silver Springs also shows that large amounts of energy pass through decomposers, despite their relatively small biomass.

Figure 59.7 **Ecological pyramids in food webs.** **(a)** In this pyramid of numbers, the abundance of species decreases with increasing trophic level. **(b)** In an inverted pyramid of numbers, the abundance of species increases with increasing trophic level. **(c)** When the amount of biological material decreases with increasing trophic level, the pyramid is termed a pyramid of biomass. Note the presence of decomposers that decompose material at all trophic levels. **(d)** In an inverted pyramid of biomass, biomass increases with increasing trophic level. **(e)** In a pyramid of energy, energy production decreases with increasing tropic level. Note the large energy production of decomposers, despite their small biomass. (After Odum 1971.)

(a) Pyramid of numbers

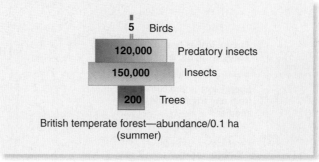

(b) Inverted pyramid of numbers

(c) Pyramid of biomass

(d) Inverted pyramid of biomass

(e) Pyramid of energy

Biomagnification Can Occur in Higher Trophic Levels

Thus far, we have considered how available biomass and energy flow can influence the properties of food chains and webs. An issue that faces organisms is the tendency of certain chemicals to concentrate in higher tropic levels in food chains, a process called **biomagnification**. The passage of DDT in food chains provides a startling example.

Dichlorodiphenyltrichloroethane (DDT) was first synthesized by chemists in 1874. In 1939, its insecticidal properties were recognized by Paul Müller, a Swiss scientist who won a Nobel Prize in 1948 for his discovery and subsequent research on the uses of the chemical. The first important application of DDT was in human health programs during and after World War II, particularly to control mosquito-borne malaria; at that time, its use in agriculture also began. The global production of DDT peaked in 1970, when 175 million kg of the insecticide were manufactured.

DDT has several chemical and physical properties that profoundly influence the nature of its ecological impact. First, DDT is persistent in the environment. It is not rapidly degraded to other, less toxic chemicals by microorganisms or by physical agents such as light and heat. The typical persistence in soil of DDT is about 10 years, which is two to three times longer than the persistence of most other insecticides. Another important characteristic of DDT is its low solubility in water and its high solubility in fats or lipids. In the environment, most lipids are present in living tissue. Therefore, because of its high lipid solubility, DDT tends to concentrate in biological tissues.

Because biomagnification occurs at each step of the food chain, organisms at higher tropic levels can amass especially high concentrations of DDT in their lipids. A typical pattern of biomagnification is illustrated in **Figure 59.8**, which shows the relative amounts of DDT found in a Lake Michigan food chain. The highest concentration of the insecticide was found in gulls, tertiary consumers that feed on fishes that, in turn, eat small insects. An unanticipated effect of DDT on bird species was its interference with the metabolic process of eggshell formation. The result was thin-shelled eggs that often broke under the weight of incubating birds (**Figure 59.9**). DDT was responsible for a dramatic decrease in the populations of many birds due to failed reproduction. Relatively high levels of the chemical were also found to be present in some game fishes, which became unfit for human consumption.

Because of growing awareness of the adverse effects of DDT, most industrialized countries, including the U.S., had banned the use of the chemical by the early 1970s. The good news is that following the outlawing of DDT, populations of the most severely affected bird species have recovered. However, had scientists initially possessed a more thorough knowledge of how DDT accumulated in food chains, some of the damage to the bird populations might have been prevented. DDT is still used in some developing countries to kill malaria-carrying mosquitoes, although there has been significant movement toward the use of alternative pest-control technologies to control the disease.

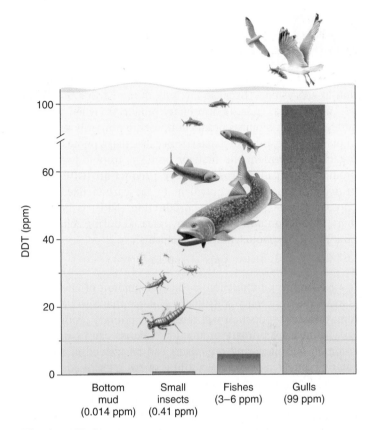

Figure 59.8   Biomagnification in a Lake Michigan food chain.   The DDT tissue concentration in gulls, a tertiary consumer, was about 240 times that in the small insects sharing the same environment. The biomagnification of DDT in lipids causes its concentration to increase at each successive link in the food chain.

DDT (dichlorodiphenyltrichloroethane)
• Persists in environment
• High solubility in lipids
• Found in high concentrations at higher trophic levels

Figure 59.9   Thinning of eggshells caused by DDT.
These ibis eggs are thin-shelled and have been crushed by the incubating adult.

59.2 Biomass Production in Ecosystems

In this section, we will take a closer look at biomass production in ecosystems. Because the bulk of the Earth's biosphere, 99.9% by mass, consists of primary producers, when we measure ecosystem biomass production, we are primarily interested in plants, algae, or cyanobacteria. Because these photosynthetic organisms represent the first, or primary, trophic level, their production is called **gross primary production** (GPP). Gross primary production is equivalent to the carbon fixed during photosynthesis. **Net primary production** (NPP) is gross primary production minus the energy released during cellular respiration (R) of photosynthetic organisms.

$$NPP = GPP - R$$

Net primary production is thus the amount of energy that is available to primary consumers. Unless otherwise noted, the term primary production will refer to net primary production.

To measure primary production, calories can be used as a common currency, and organisms can be viewed in terms of caloric content. For example, 100 g of rye grass seeds (*Secale cereale*) has a calorific content of about 380 calories, whereas 100 g of lead tree leaves (*Leucaena leucocephala*) has a calorific content of only 68 calories. Energy content is generally measured using dry biomass. Dry weight is used because the bulk of living matter in most species is water, and water content fluctuates widely, often according to wet or dry seasons. Of the dry weight, 95% is made up of carbon compounds, so measuring energy flow in ecosystems is in many ways equivalent to examining the carbon cycle (see Section 59.3). Ecologists often measure NPP in terms of carbon fixed per square meter or per hectare.

Understanding what factors limit primary production is of vital importance if we are to examine ecosystems as energy transformers. Furthermore, by determining these factors, we can understand how primary production varies globally. We can also examine the effects of primary production on **secondary production**, the gain in the biomass of heterotrophs and decomposers.

Primary Production Is Influenced in Terrestrial Ecosystems by Water, Temperature, and Nutrient Availability

In terrestrial systems, water is a major determinant of primary production, and primary production shows an almost linear increase with annual precipitation, at least in arid regions. Likewise, temperature, which affects production primarily by slowing or accelerating plant metabolic rates, is also important. Ecologist Michael Rosenzweig noted that, on a logarithmic scale, evapotranspiration rate could predict the aboveground primary production with good accuracy in North America (**Figure 59.10**). Recall from Chapter 58 that the evapotranspiration rate measures the amount of water entering the atmosphere

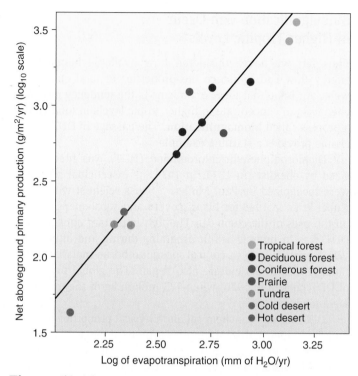

Figure 59.10   Relationship between primary production and the evapotranspiration rate.   A positive correlation is observed. Warm, humid environments are ideal for plant growth. Dots represent different ecosystems.

from the ground through the process of evaporation from the soil and transpiration of plants, so it is a measure of both temperature and available water. For example, a desert will have a low evapotranspiration rate because water availability is low despite high temperature. Rates of evapotranspiration are maximized when both temperature and moisture are at high levels, as in tropical rain forests.

A lack of **nutrients**, key elements in usable form, particularly nitrogen and phosphorus, can also limit primary production in terrestrial ecosystems, as farmers know only too well. Fertilizers are commonly used to boost the production of annual crops.

In 1984, Susan Cargill and Robert Jefferies showed how a lack of both nitrogen and phosphorus was limiting to salt marsh sedges and grasses in subarctic conditions in Hudson Bay, Canada (**Figure 59.11**). Of the two nutrients, nitrogen was the most limiting; without it, the addition of phosphorus did not increase production. However, once nitrogen was added, phosphorus became the **limiting factor**, that is, the one in shortest supply for growth. Once nitrogen was added and was no longer limiting, the addition of phosphorus increased production. The addition of nitrogen and phosphorus together increased production the most. This result supports a principle known as **Liebig's law of the minimum**, named for Justus von Liebig, a 19th-century German chemist, which states that species biomass or abundance is limited by the scarcest factor. This factor

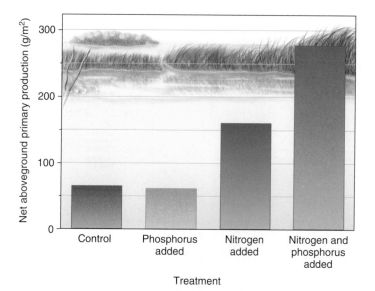

Figure 59.11 Limitation of primary production by nitrogen and phosphorus. Net aboveground primary production of a salt marsh sedge (*Carex subspathacea*) in response to nutrient addition. Nitrogen is the limiting factor. After nitrogen is added, phosphorus becomes the limiting factor.

can change, as the Hudson Bay experiment showed. When sufficient nitrogen is available, phosphorus becomes the limiting factor. Once phosphorus becomes abundant, then productivity will be limited by another nutrient.

Primary Production in Aquatic Ecosystems Is Limited Mainly by Light and Nutrient Availability

Of the factors limiting primary production in aquatic ecosystems, the most important are available light and available nutrients. Light is particularly likely to be in short supply because water readily absorbs light. At a depth of 1 m, more than half the solar radiation has been absorbed. By 20 m, only 5–10% of the radiation is left. The decrease in light is what limits the depth of algal growth.

The most important nutrients affecting primary production in aquatic systems are nitrogen and phosphorus, because they occur in very low concentrations. Whereas soil contains about 0.5% nitrogen, seawater contains only 0.00005% nitrogen. Enrichment of the aquatic environment by the addition of nitrogen and phosphorus can result in large, unchecked growths of algae called algal blooms. Such enrichment occurs naturally in areas of upwellings, where cold, deep, nutrient-rich water containing sediment from the ocean floor is brought to the surface by strong currents, resulting in very productive ecosystems and plentiful fishes. Some of the largest areas of upwelling occur in the Antarctic and along the coasts of Peru and California.

Primary Production Is Greatest in Areas of Abundant Warmth and Moisture

Knowing which factors limit primary production helps ecologists understand why the mean net primary production varies across the different biomes on Earth (**Table 59.1**). In general, primary production is highest in tropical rain forests and decreases

progressively toward the poles (**Figure 59.12**). As we saw in Chapter 58, the productivity hypothesis suggests that greater production by plants results in the latitudinal gradient of species richness. This primary productivity gradient occurs because temperatures decrease toward the poles, and, as we have just learned, temperatures affect primary production greatly. Wetlands also tend to be extremely productive, primarily because water is not limiting and nutrient levels are high. Productivity of the open ocean is very low, falling somewhere between the productivity of deserts and that of the Arctic tundra. Marine production is high on coastal shelves, particularly in upwelling zones. However, the greatest marine production occurs on algal beds and coral reefs, where temperatures are high and water levels are not so deep that light becomes limiting.

Secondary Production Is Generally Limited by Available Primary Production

What factors control secondary production—the productivity of herbivores, carnivores, and decomposers? This is a complex question, but it is generally thought to be limited largely by available primary production. A strong relationship exists between primary production in a variety of ecosystems and the biomass of herbivores (**Figure 59.13**). This means that more plant biomass, and thus more primary production, leads to an increased biomass of consumers. This is not such a trivial answer as might be assumed; for example, secondary production could be limited by the availability of a particular nutrient or by the presence of natural enemies.

Table 59.1	Net Primary Production for Earth's Ecosystems
Ecosystem type	**Mean net primary production (g/m²/yr)**
Terrestrial	
Tropical rain forest	2,500
Tropical deciduous forest	1,600
Temperate deciduous forest	1,550
Savanna	1,080
Prairie	750
Cultivated land	610
Taiga	380
Tundra	140
Hot desert	90
Aquatic	
Algal beds and coral reefs	2,500
Wetlands	2,000
Estuaries	1,500
Upwelling zones	500
Continental shelf	360
Lake and stream	250
Open ocean	125

Figure 59.12   **Annual net primary productivity on Earth.**   Primary productivity generally increases from the poles to the equator.

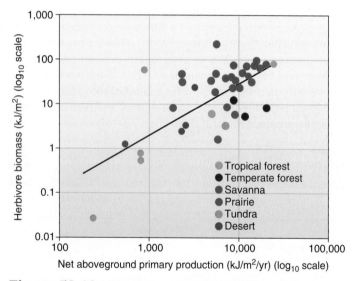

Figure 59.13   **A positive correlation between herbivore biomass and net aboveground primary production.**   These data are taken from a variety of case studies from different biomes. Herbivore biomass can be considered a surrogate for secondary production.

Concept check: *What does this relationship imply about the effects of plant secondary metabolites, many of which taste bad and some of which are toxic, on secondary production?*

As we have noted before, trophic-level transfer efficiency averages about 10%. Thus, after one link in the food web, only 1/10 of the energy captured by plants is transferred to herbivores, and after two links in the food web, only 1/100 of the energy fixed by plants goes to carnivores. Thus, secondary production is much smaller than that of primary production. We can see this in the work of John Teal. In 1962, he examined energy flow in a Georgia salt marsh (**Figure 59.14**). Salt marshes

are among the most productive habitats on Earth in terms of the amount of vegetation they produce. In salt marshes, most of the energy from the sun goes to two types of organisms: *Spartina* plants and marine algae. The *Spartina* plants are rooted in the ground, whereas the algae float on the water surface or live on the mud or on *Spartina* leaves at low tide. These photosynthetic organisms absorb about 6% of the sunlight. Most of the plant energy, 77.6%, is used in plant and algal cellular respiration. Of the energy that is accumulated in plant biomass, 22.4%, most dies in place and rots on the muddy ground, to be consumed by bacteria. Bacteria are the major decomposers in this system, followed distantly by nematodes and crabs, which feed on tiny food particles as they sift through the mud. Some of this dead material is also removed from the system (exported) by the tide. The herbivores take very little of the plant production, around 0.8%, eating only a small proportion of the *Spartina* and none of the algae. A fraction of herbivore biomass is then consumed by spiders. Overall, if we view the species in ecosystems as transformers of energy, then plants and algae are by far the most important organisms on the planet, bacteria are next, and animals are a distant third.

59.3   Biogeochemical Cycles

As we have seen, a unit of energy moves through an ecosystem only once, passing through the trophic levels of a food web from producer to consumer and dissipating as heat. In contrast, chemical elements such as carbon or nitrogen are recycled, moving from the physical environment to organisms and back to the environment, where the cycle begins again. Whereas an ecosystem constantly receives energy in the form of light, chemical elements are available in limited amounts and are continually recycled. Because the movements of chemicals through ecosystems involve biological, geological, and chemical

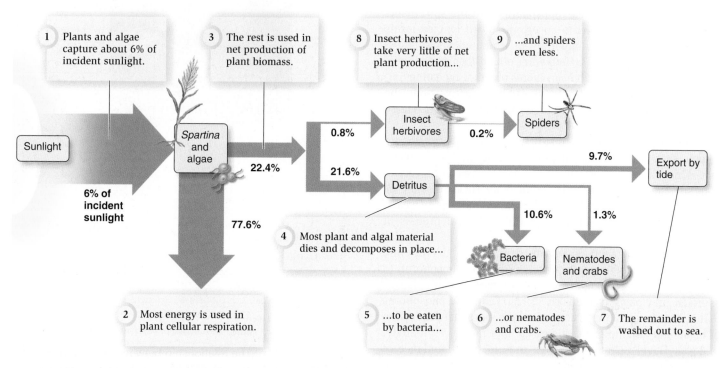

Figure 59.14 **Energy-flow diagram for a Georgia salt marsh.** Numbers represent the percentage of gross primary production that flows into different trophic levels or is used in plant respiration.

transport mechanisms, they are termed biogeochemical cycles. Biological mechanisms involve the absorption of chemicals by living organisms and their subsequent release back into the environment. Geological mechanisms include weathering and erosion of rocks, and elements transported by surface and subsurface drainage. Chemical transport mechanisms include dissolved matter in rain and snow, atmospheric gases, and dust blown by the wind.

In addition to the basic building blocks of hydrogen, oxygen, and carbon, which we discussed in Chapter 2, the elements required in the greatest amounts by living organisms are phosphorus, nitrogen, and sulfur. In this section, we take a detailed look at the cycles of these nutrients. These cycles can be divided into two broad types: local cycles, such as the phosphorus cycle, which involve elements with no atmospheric mechanism for long-distance transfer; and global cycles, which involve an interchange between the atmosphere and the ecosystem. Global nutrient cycles, such as the carbon, nitrogen, and sulfur cycles, unite the Earth and its living organisms into one giant interconnected ecosystem called the **biosphere**. In our discussion of biogeochemical cycles, we will take a particular interest in the alteration of these cycles through human activities that increase the input of chemical elements, such as the burning of fossil fuels.

Phosphorus Cycles Locally Between Geological and Biological Components of Ecosystems

All living organisms require phosphorus, which becomes incorporated into ATP, the compound that provides energy for most metabolic processes. Phosphorus is a key component of other biological molecules such as DNA and RNA, and it is also an essential mineral that in many animals helps maintain a strong, healthy skeleton.

The phosphorus cycle is a relatively simple cycle (**Figure 59.15**). Phosphorus has no gaseous phase and thus no atmospheric component; that is, it is not moved by wind or rain. As a result, phosphorus tends to cycle only locally. The Earth's crust is the main storehouse for this element. Weathering

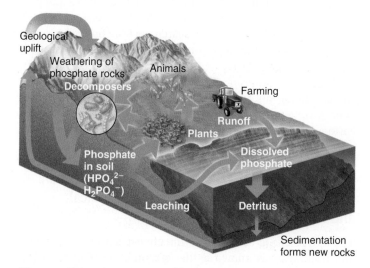

Figure 59.15 **The phosphorus cycle.** Unlike other major biogeochemical cycles, the phosphorus cycle does not have an atmospheric component and thus cycles only locally. The widths of the lines indicate the relative importance of each process.

and erosion of rocks release phosphorus into the soil. Plants have the metabolic means to absorb dissolved ionized forms of phosphorus, the most important of which occurs as phosphate (HPO_4^{2-} or $H_2PO_4^-$). Herbivores obtain their phosphorus only from eating plants, and carnivores obtain it by eating herbivores. When plants and animals excrete wastes or die, the phosphorus becomes available to decomposers, which release it back to the soil.

Leaching and runoff eventually wash much phosphate into aquatic systems, where plants and algae utilize it. Phosphate that is not taken up into the food chain settles to the ocean floor or lake bottom, forming sedimentary rock. Phosphorus can remain locked in sedimentary rock for millions of years, becoming available again through the geological process of uplift.

Plants can take up phosphate so rapidly and efficiently that they often reduce soil concentrations of phosphorus to extremely low levels, so phosphorus becomes a limiting factor, as noted previously (see Figure 59.11). The more phosphorus that is added to an aquatic ecosystem, the more that production of algae and aquatic plants increases. In a pivotal 1974 study, biologist David Schindler showed that an overabundance of phosphorus caused the rapid growth of algae and plants in an experimental lake in Canada (**Figure 59.16**). What is the consequence of the rapid growth of algae? When the algae die, they sink to the bottom, where bacteria decompose them and consume the dissolved oxygen in the water. Dissolved oxygen concentrations can drop too low for fishes to breathe, causing fish kills. The process by which elevated nutrient levels lead to an overgrowth of algae and the subsequent depletion of water oxygen levels is known as **eutrophication**. Cultural eutrophication refers to the enrichment of water with nutrients derived from human activities, such as fertilizer use and sewage dumping.

Lake Erie became eutrophic in the 1960s due to the fertilizer runoff from farms rich in phosphorus and to the industrial and domestic pollutants released from the many cities along its shores. Fish species such as white fish and lake trout became severely depleted. Based on research such as Schindler's that showed the dramatic effect of phosphorus on a lake system, the U.S. and Canada teamed together to reduce the levels of discharge by 80%, primarily through eliminating phosphorus in laundry detergents and maintaining strict controls on the phosphorus content of wastewater from sewage treatment plants. Fortunately, lake systems have great potential for recovery after phosphorous inputs are reduced, and Lake Erie has experienced fewer algal blooms, clearer water, and more fishes.

Carbon Cycles Among Biological, Geological, and Atmospheric Pools

The movement of carbon from the atmosphere into organisms and back again is known as the carbon cycle (**Figure 59.17**). Carbon dioxide is present in the atmosphere at a level of about 380 parts per million (ppm), or about 0.04%. Autotrophs, primarily plants, algae, and cyanobacteria, acquire carbon dioxide from the atmosphere and incorporate it into the organic matter

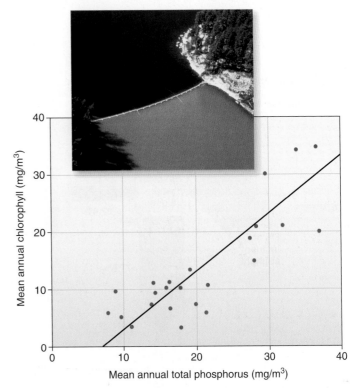

Figure 59.16 **The relationship between primary production and total phosphorus concentration.** As shown in this graph, primary production (measured by chlorophyll concentration) increases linearly with an increase in phosphorus. Each dot represents a different lake. The aerial photograph shows the contrast in water quality of two basins of an experimental lake in Canada separated by a plastic curtain. Carbon and nitrogen were added to the upper basin, and carbon, nitrogen, and phosphorus were added to the lower basin. The bright green color is from a surface film of algae that resulted from the added phosphorus.

of their own biomass via photosynthesis. Each year, plants, algae, and cyanobacteria remove approximately one-seventh of the CO_2 from the atmosphere. At the same time, respiration and the decomposition of plants recycle a similar amount of carbon back into the atmosphere as CO_2. Much material from primary producers is also transformed into deposits of coal, gas, and oil, which are known as **fossil fuels**. Herbivores can return some carbon dioxide to the atmosphere, eating plants and breathing out CO_2, but the amount flowing through this part of the cycle is minimal. Chemical processes such as diffusion and absorption of CO_2 in and out of oceans also contribute to changes in atmospheric CO_2.

Over time, much carbon is also incorporated into the shells of marine organisms, which eventually form huge limestone deposits on the ocean floor or in terrestrial rocks, where turnover is extremely low. As a result, rocks and fossil fuels contain the largest reserves of carbon. Natural sources of CO_2 such as volcanoes, hot springs, and fires release large amounts of CO_2. In addition, human activities, primarily deforestation and the burning of fossil fuels, are increasingly causing large amounts of CO_2 to enter the atmosphere together with large volumes of particulate matter.

Direct measurements over the past five decades show a steady rise in atmospheric CO_2 (**Figure 59.18**), a pattern that

Figure 59.17 **The carbon cycle.** Each year, plants and algae remove about one-seventh of the CO_2 in the atmosphere. Animal respiration is so small it is not represented. The width of the arrows indicates the relative importance of each process.

Concept check: *Where are the greatest stores of global carbon?*

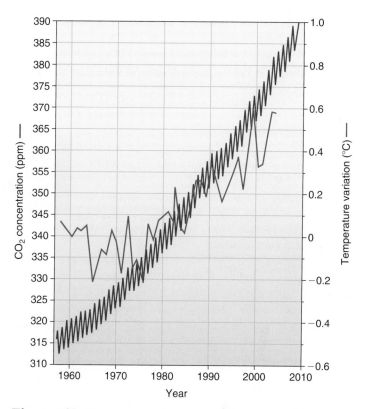

Figure 59.18 **The increase in atmospheric CO_2 levels and temperatures due to the burning of fossil fuels.** From 1958 to 2008, atmospheric CO_2 shows an increase of nearly 20%. In addition, the graph shows a seasonal variation in CO_2. Temperatures are annual deviations from the 1961–1990 average. Measurements were recorded at Mauna Loa Observatory in Hawaii.

Concept check: *Why does the amount of CO_2 fluctuate seasonally in the graph?*

shows no sign of slowing. Because of its increasing concentration in the atmosphere, CO_2 is the most troubling of the greenhouse gases, which are a primary cause of global warming (see Chapter 54). Elevated atmospheric CO_2 has other dramatic environmental effects, boosting plant growth but lowering the amount of herbivory (see Feature Investigation).

The amount of carbon dioxide in the atmosphere shows a seasonal variation, as can be seen in Figure 59.18. Concentrations of atmospheric carbon dioxide are lowest during the Northern Hemisphere's summer and highest during the winter, when photosynthesis is minimal. This phenomenon occurs asynchronously in both of the Earth's hemispheres. Because there is more land in the Northern Hemisphere than in the Southern Hemisphere, and therefore more vegetation, concentrations

of atmospheric carbon dioxide are lowest during the Northern Hemisphere's summer. The vegetation has a maximum photosynthetic activity during the summer, reducing the global amount of carbon dioxide. During the Northern Hemisphere's winter, photosynthesis is low, and decomposition is relatively high, causing a global increase in the gas.

FEATURE INVESTIGATION

Stiling and Drake's Experiments with Elevated CO_2 Showed an Increase in Plant Growth but a Decrease in Herbivory

How will forests of the future respond to elevated CO_2? To begin to answer such a question, ecologists ideally would enclose large areas of forests with chambers, increase the CO_2 content within the chambers, and measure the responses. This has proven to be difficult for two reasons. First, it is hard to enclose large trees in chambers, and second, it is expensive

to increase CO_2 levels over such a large area. However, in a discovery-based investigation, ecologists Peter Stiling and Bert Drake were able to increase CO_2 levels around small patches of forest at the Kennedy Space Center in Cape Canaveral, Florida. In much of Florida's forests, trees are small, only 3–5 m when mature, because frequent lightning-initiated fires prevent the growth of larger trees. Stiling and Drake teamed up with NASA engineers to create 16 circular, open-topped chambers (**Figure 59.19**). In eight of these they increased atmospheric CO_2 to double their ambient levels, from 360 ppm to 720 ppm, the

Figure 59.19 The effects of elevated atmospheric CO_2 on insect herbivory.

GOAL To determine the effects of elevated CO_2 on a forest ecosystem; effects on herbivory are highlighted here.

STUDY LOCATION Patches of forest at the Kennedy Space Center in Cape Canaveral, Florida.

	Experimental level	Conceptual level

1 Erect 16 open-top chambers around native vegetation. Increase CO_2 levels from 360 ppm to 720 ppm in half of them.

Expected atmospheric CO_2 level is 720 ppm by end of the 21st century. Open-top chambers allow movement of herbivores in and out of chambers.

2 Conduct a yearly count of numbers of insect herbivores per 200 leaves in each chamber.

3 Count number of herbivores that died due to nutritional inadequacy. Monitor attack rates on insect herbivores by natural enemies such as predators and parasitoids.

Elevated CO_2 reduces foliar nitrogen, inhibits normal insect development, and prolongs the feeding time of herbivores, allowing natural enemies greater opportunities to attack them.

4 THE DATA

Source of mortality*	Elevated CO_2 (% mortality)	Control (% mortality)
Nutritional inadequacy	10.2	5.0
Predators	2.4	2.0
Parasitoids	10.0	3.2

*Data refer only to mortality of larvae within leaves and do not sum to 100%. Mortality of eggs on leaves, pupae in the soil, and flying adults is unknown.

5 CONCLUSION Elevated CO_2 decreases insect herbivory in a Florida forest.

6 SOURCE Stiling, P., and Cornelissen, T. 2007. How does elevated carbon dioxide (CO_2) affect plant-herbivore interactions? A field experiment and meta-analysis of CO_2-mediated changes on plant chemistry and herbivore performance. *Global Change Biology* 13:1823–1842.

latter of which is the atmospheric level predicted by the end of the 21st century. The experiments were initiated in 1996 and lasted until 2007. Plants produced more biomass in elevated CO_2, because carbon dioxide is limiting to plant growth, but the data revealed much more.

Because the chambers were open-topped, insect herbivores could come and go. Insect herbivores cause the largest amount of herbivory in North American forests, because most vertebrate herbivores cannot access the high foliage. Censuses were conducted of all damaged leaves, but focused on leaves damaged by leaf miners, the most common type of herbivore at this site. Leaf miners are small moths whose larvae are small enough to burrow between the surfaces of plant leaves and mine tunnels through the leaves.

Densities of damaged leaves, including those damaged by leaf miners, were lower in elevated CO_2 in every year studied. Part of the reason for the decline was that even though plants increased in mass, the existing soil nitrogen was diluted over a greater volume of plant material, so the nitrogen level in leaves decreased. This increased insect mortality by two means. First, poorer leaf quality directly increased insect death because leaf nitrogen levels may have been too low to support the normal development of the leaf miners. Second, lower leaf quality increased the length of time insects had to feed to gain suffi-

cient nitrogen. Increased feeding times, in turn, led to increased exposure to natural enemies, such as predatory spiders and ants, and parasitoids, and top-down mortality also increased (see the data of Figure 59.19). Thus, in a world of elevated CO_2, plant growth may increase, and herbivory could decrease.

Experimental Questions

1. What was the hypothesis of the Stiling and Drake experiment?

2. What was the purpose of increasing the carbon dioxide levels in only half of the chambers in the experiment and not all of the chambers?

3. What were the results of the experiment?

The Nitrogen Cycle Is Strongly Influenced by Biological Processes That Transform Nitrogen into Usable Forms

Nitrogen is an essential component of proteins, nucleic acids, and chlorophyll. Because 78% of the Earth's atmosphere consists of nitrogen gas (N_2), it may seem that nitrogen should not be in short supply for organisms. However, nitrogen is often a limiting factor in ecosystems because N_2 molecules must be broken apart before nitrogen atoms are available to combine with other elements. Because of its triple bond, nitrogen gas is very stable, and only certain bacteria can break it apart into usable forms such as ammonia (NH_3). This process, called nitrogen fixation, is a critical component of the five-part nitrogen cycle (**Figure 59.20**):

1. A few species of bacteria can accomplish **nitrogen fixation**, that is, convert atmospheric nitrogen (N_2) to forms usable by other organisms. The bacteria that fix nitrogen are fulfilling their own metabolic needs, but in the process, they release excess ammonia (NH_3) or ammonium (NH_4^+), which can be used by some plants. An important group of nitrogen-fixing bacteria are known as rhizobia, which live in nodules on the roots of legumes, including peas, beans, lentils, and peanuts and some woody plants. In more natural systems, such as forests and savannas, nitrogen-fixing bacteria such as *Frankia* form a symbiosis with actinorhizal plants. Cyanobacteria are important nitrogen fixers in aquatic systems.

2. In the process of **nitrification**, soil bacteria convert NH_3 or NH_4^+ to nitrate (NO_3^-), a form of nitrogen commonly used by plants. The bacteria *Nitrosomonas* and *Nitrococcus* first oxidize the forms of ammonia to nitrite (NO_2^-), after which the bacteria *Nitrobacter* converts NO_2^- to NO_3^-.

3. **Assimilation** is the process by which inorganic substances are incorporated into organic molecules. In the nitrogen cycle, organisms assimilate nitrogen by taking up NH_3, NH_4^+, and NO_3^- formed through nitrogen fixation and nitrification and incorporating them into other molecules. Plant roots take up these forms of nitrogen through their roots, and animals assimilate nitrogen from plant tissue.

Figure 59.20 **The nitrogen cycle.** The five main parts of the nitrogen cycle are (1) nitrogen fixation, (2) nitrification, (3) assimilation, (4) ammonification, and (5) denitrification. The recycling of nitrogen from dead plants and animals into the soil and then back into plants is of paramount importance because this is the main pathway for nitrogen to enter the soil. The width of the arrows indicates the relative importance of each process.

4. Ammonia can also be formed in the soil through the decomposition of plants and animals and the release of animal waste. **Ammonification** is the conversion of organic nitrogen to NH_3 and NH_4^+. This process is carried out by bacteria and fungi. Most soils are slightly acidic and, because of an excess of H^+, the NH_3 rapidly gains an additional H^+ to form NH_4^+. Because many soils lack nitrifying bacteria, ammonification is the most common pathway for nitrogen to enter the soil.

5. **Denitrification** is the reduction of nitrate (NO_3^-) to gaseous nitrogen (N_2). Denitrifying bacteria, which are anaerobic and use NO_3^- in their metabolism instead

of oxygen, perform the reverse of their nitrogen-fixing counterparts by delivering nitrogen to the atmosphere. This process delivers only a relatively small amount of nitrogen to the atmosphere.

Human alterations of the nitrogen cycle have approximately doubled the rate of nitrogen input to the cycle. Industrial fixation of nitrogen for the production of fertilizer makes a significant contribution to the pool of nitrogen-containing material in the soils and waters of agricultural regions. One problem is that fertilizer runoff can cause eutrophication of rivers and lakes, and, as the resultant algae die, decomposition by bacteria depletes the oxygen level of the water, resulting in fish die-offs. Excess nitrates in surface or groundwater systems used for drinking water are also a health hazard, particularly for infants. In the body, nitrate is converted to nitrite, which then combines with hemoglobin to form methemoglobin, a type of hemoglobin that does not carry oxygen. In infants, the production of large amounts of nitrites can cause methemoglobinemia, a dangerous condition in which the level of oxygen carried through the body decreases.

Finally, burning fossil fuels releases not only carbon but also nitrogen in the form of nitrogen oxides, which can contribute to air pollution. Nitrous oxides (N_2O) can then react with rainwater to form nitric acid (HNO_3), which contributes to acid rain, decreasing the pH of lakes and streams and increasing fish mortality (see Chapter 54). Although much of the acid rain problem can be traced to the sulfur cycle, nitrogen oxides are also partially to blame.

The dramatic effects of human activities on nutrient cycles in general and the nitrogen cycle in particular were illustrated by a famous long-term study by ecosystem ecologists Gene Likens, Herbert Bormann, and their colleagues at Hubbard Brook Experimental Forest in New Hampshire in the 1960s. Hubbard Brook is a 3,160-hectare reserve that consists of six catchments along a mountain ridge. A catchment is an area of land where all water eventually drains to a single outlet. In Hubbard Brook, each outlet is fitted with a permanent concrete dam that enables researchers to monitor the outflow of water and nutrients (**Figure 59.21a**). In this large-scale experiment, researchers felled all of the trees in one of the Hubbard Brook catchments (**Figure 59.21b**). The catchment was then sprayed with herbicides for 3 years to prevent regrowth of vegetation. An untreated catchment was used as a control.

Researchers monitored the concentrations of key nutrients in the flow of water exiting the two catchments for over 3 years. Their results revealed that the overall export of nutrients from the disturbed catchment rose to many times the normal rate (**Figure 59.21c**). The researchers determined that two phenomena were responsible. First, the enormous reduction in plants reduced water uptake by vegetation and led to 40% more runoff discharged to the streams. This increased outflow caused greater rates of chemical leaching and rock and soil weathering. Second, and more significantly, in the absence of nutrient uptake in spring, when the deciduous trees would have resumed photosynthesis, the inorganic nutrients released by decomposer activity were simply leached in the drainage water. Similar processes operate in the majority of terrestrial ecosystems where clearance of forests is significant.

The Sulfur Cycle Is Heavily Influenced by Human Activities

Most naturally produced sulfur in the atmosphere comes from the gas hydrogen sulfide (H_2S), which is released from volcanic

(a) Hubbard Brook dam and weir

(b) Hubbard Brook Experimental Forest, New Hampshire

Figure 59.21 **The effects of deforestation on nutrient concentrations.** (a) Concrete dam and weir, which concentrates water flow and permits accurate measurement of discharge rate, used to monitor nutrient flow from a Hubbard Brook catchment. (b) Deforested catchment at Hubbard Brook. (c) Nutrient concentrations in stream water from the experimentally deforested catchment and a control catchment at Hubbard Brook. The timing of deforestation is indicated by arrows.

(c) Nutrient concentrations in deforested and control catchments

Figure 59.22   The sulfur cycle. This cycle can be dramatically affected by human outputs through the burning of fossil fuels. The width of the arrows indicates the relative importance of each process.

eruptions and during decomposition, especially in wetland environments, where sulfur is very common (**Figure 59.22**). The H_2S quickly oxidizes into sulfur dioxide (SO_2). Because SO_2 is soluble in water, it returns to Earth as weak sulfuric acid (H_2SO_4), or natural acid rain. H_2SO_4 dissociates into $2\ H^+ + SO_4^{2-}$, making the pH of natural rainwater slightly acidic, about 5.6 (see also Chapter 54). The sulfate ions, SO_4^{2-}, enter the soil, where sulfate-reducing bacteria may release sulfur as H_2S, or the sulfate may be incorporated by plants into their tissue.

In the presence of iron, sulfur can precipitate as ferrous sulfide, FeS_2, and be incorporated into pyritic rocks. The weathering of rocks and the decomposition of organic matter therefore releases sulfur to solution, which runs through rivers to the sea. Because such rocks commonly overlay coal deposits, mining exposes them to the air and water, resulting in a discharge of sulfuric acid and other sulfur-containing compounds into aquatic ecosystems. Mining has polluted hundreds of kilometers of streams and rivers in mid-Atlantic states such as West Virginia, Kentucky, and Pennsylvania in this way.

Interestingly, certain marine algae and a few salt marsh plants produce relatively large amounts of the sulfurous gas dimethyl sulfide (CH_3SCH_3), commonly abbreviated as DMS. Small DMS particles that diffuse to the atmosphere often form the nuclei around which water vapor can condense and form the water droplets making up clouds. Because of the sheer global extent of the oceans, changes in algal abundance and thus global DMS levels have the potential to alter cloud cover and thus climate. Because of its ability to cool the climate, some researchers are investigating how DMS production might potentially offset global warming.

Human activity involving the combustion of fossil fuels has altered the sulfur cycle more than any of the other nutrient cycles. The burning of coal and oil to provide energy for heating or to fuel electric power stations produces huge amounts of sulfur dioxide (SO_2). This reacts with rain or snow to make human-produced acid rain (see Chapter 54). One of the main differences between human-produced acid rain and natural acid rain is its relative pH. In North America, for example, natural acid rain has a pH of about 5.6, whereas measurements of rain falling in southern Ontario, Canada, in the 1980s showed pH values in the range of 4.1–4.5. Huge areas of the industrial midwest U.S. and Europe were affected by acid rain in the 1950s through the 1980s, but a reduction in the use of high-sulfur coal and the use of scrubbers to prevent sulfur dioxide from passing through smokestacks has reduced the problem in more recent times (**Figure 59.23**). The combustion of fossil fuel also results in the release of particulate matter into the air. Larger particles are visible as smoke or soot. Recently such particulate matter has been shown to increase heritable mutations (see Genomes & Proteomes).

Genomes & Proteomes Connection

Pollution Can Cause Heritable Mutations

Sulfur dioxide, nitrous oxide, and carbon dioxide are pollutants emitted from the burning of fossil fuels. In addition to these atmospheric gases, the combustion of fossil fuels also results in the release of particulate matter, tiny solid and liquid particles, into the air. Larger particles are visible as smoke or soot. The small particles, when inhaled, can penetrate deeply into the lungs and become distributed throughout the body. Such exposure, whether from transportation sources, factories, power plants, or tobacco smoke, can lead to the formation of lung tumors and induce mutations in somatic cells.

1990

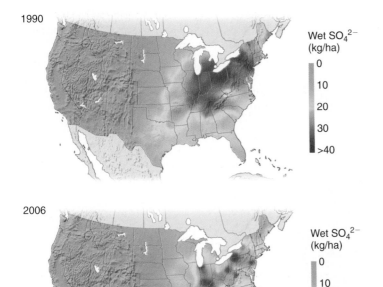

Wet SO_4^{2-}
(kg/ha)

0
10
20
30
>40

2006

Wet SO_4^{2-}
(kg/ha)

0
10
20
30
>40

Figure 59.23 The decrease in acid rain in the U.S. from 1989 to 2006. Sulfate dissolved in atmospheric water produces acid rain. The data show wet sulfate deposition in kilograms per hectare.

Concept check: Why did relatively high levels of wet sulfate deposition occur in upstate New York in the 1980s when this was not an industrial area?

A study by biologist James Quinn and colleagues in 2004 showed that polluted air can also induce genetic changes in mouse sperm. In the study, the researchers put two groups of mice into separate sheds that were downwind of steel mills in Hamilton, an urban-industrial center of Ontario, Canada. One shed was fitted with <u>h</u>igh-<u>e</u>fficiency <u>p</u>articulate <u>a</u>ir (HEPA) filters, which removed particulate matter from the air, and the other was not. The researchers discovered changes in the size of noncoding DNA sequences in the offspring of the mice in the filterless shed. They concluded that the increased rate of mutations was paternally derived. Offspring of mice from the unfiltered shed inherited twice as many mutations as offspring of mice in the filtered shed. The researchers concluded that the pollutant particles, or some chemical compound associated with them, were responsible for the observed, heritable DNA changes.

What is the mechanism behind these genetic changes? Quinn and colleagues hypothesized that the inhaled pollutants are transported to the liver and metabolized to DNA-reactive compounds. Then they are transported to the testes and finally to the sperm stem cells, where the mutations occurred. The damaged DNA would then be transmitted to first-generation offspring.

The Water Cycle Is Largely a Physical Process of Evaporation and Precipitation

The water cycle, also called the hydrological cycle, differs from the cycles of other nutrients in that very little of the water that cycles through ecosystems is chemically changed by any of the cycle's components (**Figure 59.24**). It is a physical process, fueled by the sun's energy, rather than a chemical one, because it consists of essentially two phenomena: evaporation and precipitation. Even so, the water cycle has important biological components. Over land, 90% of the water that reaches the atmosphere is moisture that has passed through plants and exited from the leaves via evapotranspiration. Only about 2% of the total volume of Earth's water is found in the bodies of organisms or is held frozen or in the soil. The rest cycles between bodies of water, the atmosphere, and the land.

As we noted in Chapter 54, water is limiting to the abundance of many organisms, including humans. It takes 228 L of water to produce a pound of dry wheat, and 9,500 L of water to support the necessary vegetation to produce a pound of meat. Industry is also a heavy user of water, with goods such as oil, iron, and steel requiring up to 20,000 L of water per ton of product. To increase the amount of available water, humans have interrupted the hydrological cycle in many ways, most prominently through the use of dams to create reservoirs. Such dams, such as those on the Columbia River in Washington State, can greatly interfere with the migration of fishes such as salmon and affect their ability to reproduce and survive. Other activities, such as tapping into underground water supplies, or **aquifers**, for drinking water removes more water than is put back by rainfall and can cause shallow ponds and lakes to dry up and sinkholes to develop.

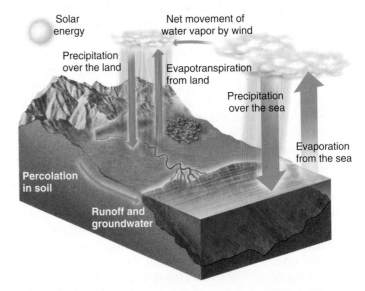

Figure 59.24 The water cycle. This cycle is primarily a physical process, not a chemical one. Solar energy drives the water cycle, causing evaporation of water from the ocean and evapotranspiration from the land. This is followed by condensation of water vapor into clouds and precipitation. The width of the arrows indicates the relative importance of each step.

Figure 59.25 Severe erosion following deforestation in Madagascar. After clearing and a few years of farming, the shallow soil can no longer support crops and is susceptible to erosion by rainfall.

Deforestation can also significantly alter the water cycle. When forests are cut down, less moisture is transpired into the atmosphere. This reduces cloud cover and diminishes precipitation, subjecting the area to drought. Reestablishing the forests, which calls for increased water, then becomes nearly impossible. Such a problem has occurred on the island of Madagascar, located off the east coast of Africa. In this country, clearing of the forests for cash crops such as cotton, coffee, and tobacco has been so rapid and extensive that areas devoid of vegetation have appeared (**Figure 59.25**). In Madagascar, as in so many other areas on Earth, deforestation and other environmental degradations are having an impact on much of the natural habitat. Appropriately, in the following chapter, Chapter 60, we finish our study of ecology in particular, and biology in general, with a discussion of what we gain from ecosystem diversity and how best to conserve the diversity of ecosystems in the future.

◼ Summary of Key Concepts

59.1 Food Webs and Energy Flow

- Ecosystem ecology concerns the movement of energy and materials through organisms and their communities. (Figure 59.1)

- Simple feeding relationships between organisms can be characterized by an unbranched food chain, and each feeding level in the chain is called a trophic level. (Figure 59.2)

- Organisms that obtain energy from light or chemicals are called autotrophs and are the primary producers. Organisms that feed on other organisms are called heterotrophs. Those organisms that feed on primary producers are called primary consumers. Animals that eat plants are also called herbivores. Organisms that feed on primary consumers are called secondary consumers. Animals that eat other animals are called carnivores. Consumers that get their energy from the remains and waste products of organisms are called detritivores or decomposers. (Figure 59.3)

- Food webs are a more complex model of interconnected food chains in which there are multiple links between species. (Figure 59.4)

- Food webs tend to have five or fewer levels. Energy conversions are not 100% efficient, and usable energy is lost within each tropic level and from one trophic level to the next. (Figure 59.5)

- Production efficiency measures the percentage of energy assimilated that becomes incorporated into new biomass. (Figure 59.6)

- Trophic-level transfer efficiency measures the energy available at one trophic level that is acquired by the level above. These efficiencies can be expressed in the form of ecological pyramids, the best known being the pyramid of numbers. (Figure 59.7)

- The increase in the concentration of a substance in living organisms, called biomagnification, can occur at each trophic level of the food web. (Figures 59.8, 59.9)

59.2 Biomass Production in Ecosystems

- Plant production can be measured as gross primary production. Net primary production is gross primary production minus the energy released during respiration via photosynthetic organisms.

- Net primary production in terrestrial ecosystems is limited primarily by temperature and the availability of water and nutrients. Net primary production in aquatic ecosystems is limited mainly by availability of light and nutrients. (Figures 59.10, 59.11)

- Ecosystems differ in their net primary production. (Table 59.1, Figure 59.12)

- Secondary production is limited by available primary production. (Figures 59.13, 59.14)

59.3 Biogeochemical Cycles

- Elements such as phosphorus, carbon, nitrogen, and sulfur recycle from the physical environment to organisms and back in what are called biogeochemical cycles.

- The phosphorus cycle lacks an atmospheric component and thus is a local cycle. An overabundance of phosphorus can cause the overgrowth of algae and subsequent depletion of oxygen levels, called eutrophication. (Figures 59.15, 59.16)

- In the carbon cycle, autotrophs incorporate carbon dioxide from the atmosphere into their biomass; decomposition of plants and respiration recycles most of this CO_2 back to the atmosphere. Human activities, primarily the burning of fossil fuels, are causing increased amounts of CO_2 to enter the atmosphere. Air pollutants may cause mutations. (Figures 59.17, 59.18)

• Experiments have shown that elevated levels of carbon dioxide result in an increase in plant growth but a decrease in herbivory. (Figure 59.19)

• The nitrogen cycle has five parts: nitrogen fixation, nitrification, assimilation, ammonification, and denitrification. In the nitrogen cycle, atmospheric nitrogen is unavailable for use by most organisms and must be converted to usable forms by certain bacteria. (Figure 59.20)

• The activities of humans, including fertilizer use, fossil fuel use, and deforestation, have dramatically altered the nitrogen cycle. (Figure 59.21)

• Sulfur enters the atmosphere through both natural sources (such as volcanoes and decomposition) and human sources, including the combustion of fossil fuels. Atmospheric sulfur dioxide returns to Earth as weak sulfuric acid (H_2SO_4), commonly called natural acid rain. (Figure 59.22)

• Human-produced acid rain results primarily from the combustion of fossil fuels. (Figure 59.23)

• The water cycle is a physical rather than a chemical process, because it consists of essentially two phenomena: evaporation and precipitation. (Figure 59.24)

• Alteration of the water cycle by deforestation can result in regional climatic changes because a reduction in transpiration causes a decrease in cloud cover and precipitation. (Figure 59.25)

Assess and Discuss

Test Yourself

1. The amount of energy that is fixed during photosynthesis is known as
   a. net primary production.
   d. gross primary production.
   b. biomagnification.
   e. production efficiency.
   c. trophic-level transfer efficiency.

2. Chemoautotrophic bacteria are
   a. primary consumers.
   d. primary producers.
   b. secondary consumers.
   e. decomposers.
   c. tertiary consumers.

3. When considering the average food chain, which of the following statements is true?
   a. Secondary consumers are the most abundant organisms in an ecosystem.
   b. The more lengths in the food chain, the more stable the ecosystem.
   c. Biomass decreases as you move up the food chain.
   d. The trophic level with the highest species abundance is usually the primary producers.
   e. All of the above are true.

4. Which organisms are the most important consumers of energy in a Georgia salt marsh?
   a. *Spartina* grass and algae
   d. crabs
   b. insects
   e. bacteria
   c. spiders

5. Primary production in aquatic systems is limited mainly by
   a. temperature and moisture.
   d. light and nutrients.
   b. temperature and light.
   e. light and moisture.
   c. temperature and nutrients.

6. The net primary production of cultivated land is closest to that of
   a. prairies.
   d. temperate deciduous forest.
   b. tropical rain forest.
   e. lakes and streams.
   c. wetlands.

7. The evapotranspiration rate
   a. can be used as a predictor for primary production.
   b. is increased when temperature decreases.
   c. is not affected by temperature.
   d. is highest in deserts.
   e. can be predicted by measuring only the water content of the soil.

8. Eutrophication is
   a. caused by an overabundance of nitrogen, which leads to an increase in bacteria populations.
   b. caused by an overabundance of nutrients, which leads to an increase in algal populations.
   c. the normal breakdown of algal plants following a pollution event.
   d. normally seen in dry, hot regions of the world.
   e. none of the above.

9. Terrestrial primary producers acquire the carbon necessary for photosynthesis from
   a. decomposing plant material.
   b. carbon monoxide released from the burning of fossil fuels.
   c. carbon dioxide in the atmosphere.
   d. carbon sources in the soil.
   e. both a and d.

10. Nitrogen fixation is the process
    a. that converts organic nitrogen to ammonia.
    b. by which plants and animals take up nitrates.
    c. by which bacteria convert nitrate to gaseous nitrogen.
    d. by which atmospheric nitrogen is converted to ammonia or ammonium ions.
    e. all of the above

Conceptual Questions

1. Define autotrophs and heterotrophs.

2. Explain why chain lengths are short in food webs.

3. What is the major difference among the phosphorous, carbon, and nitrogen cycles?

Collaborative Questions

1. Outline the main trophic levels within a food chain.

2. What might the atmospheric concentration of carbon dioxide be in 2100? What effects might this have on the environment?

Online Resource

www.brookerbiology.com

Stay a step ahead in your studies with animations that bring concepts to life and practice tests to assess your understanding. Your instructor may also recommend the interactive ebook, individualized learning tools, and more.

Biodiversity and Conservation Biology

60

Biological control, the control of pests by natural enemies, is seen as a preferable option to chemical control of pests, which leaves chemical residues in the environment. A 2006 study by William Snyder and colleagues showed that an increase in predator biodiversity resulted in greater suppression of insect herbivore populations and a subsequent increase in plant growth. This was important because it was previously thought that strong interspecific competition, predators feeding on other predators, could lead to weaker herbivore control. Snyder's data showed how biological control improved when the natural enemy community included multiple species.

Biological diversity, or **biodiversity**, encompasses the genetic diversity of species, the variety of different species, and the different ecosystems that they form. The field of **conservation biology** uses principles and knowledge from molecular biology, genetics, and ecology to protect the biological diversity of life at each of these three levels. Because it draws from nearly all chapters of this textbook, a discussion of conservation biology is an apt way to conclude our study of biology. In this chapter, we begin by examining the question of why biodiversity should be conserved, and explore how much diversity is needed for ecosystems to function properly. We then survey the main threats to the world's biodiversity. For many species, there are multiple threats from human activities, ranging from habitat loss, overexploitation, and the effects of introduced species to climate change and pollution. Even if species are not exterminated, many may exist only at very small population sizes. We will see how small populations face special problems such as inbreeding, genetic drift, and limited mating, emphasizing the importance of genetics in conservation biology.

Last, we consider what is being done to help conserve the world's endangered biota. This includes identifying global areas rich in species and establishing parks and refuges of the appropriate size, number, and connectivity. We also discuss conservation of particularly important types of species and outline how ecologists have been active in restoring damaged habitats to a more natural condition. We then examine how captive breeding programs have been useful in building up populations of rare species prior to their release back into the wild. Some programs have also used modern genetic techniques such as cloning to help breed and perhaps eventually increase populations of endangered species.

Increased natural enemy diversity increases the suppression of prey. These aphids are attacked by two species of natural enemy, a ladybird beetle larva and a parasitoid, which turns parasitized individuals copper colored and swollen.

60.1 What Is Biodiversity?

As noted previously, biodiversity can be examined at three levels: genetic diversity, species diversity, and ecosystem diversity. Each level of biodiversity provides valuable benefits to humanity. Genetic diversity consists of the amount of genetic variation that occurs within and between populations. Maintaining genetic variation in the wild relatives of crops may be vital to the continued success of crop-breeding programs. For example, in 1977, Rafael Guzman, a Mexican biologist, discovered a previously unknown wild relative of corn, *Zea diploperennis*, that is resistant to many of the viral diseases that infect domestic corn, *Zea mays*. Genetic engineers believe that this relative has valuable genes that can improve current corn crops. Because corn is the third-largest crop on Earth, the discovery of *Z. diploperennis* may well turn out to be critical to the global food supply.

Figure 60.1 Ecosystem biodiversity. This small cemetery in Bureau County, Illinois, contains the remains of a natural prairie ecosystem. Most of the prairie has been plowed under for agriculture.

The second level of biodiversity concerns species diversity, the number and relative abundance of species in a community. Species diversity is an area on which much public attention is focused. In 1973, the U.S. Endangered Species Act was enacted, which was designed to protect both endangered and threatened species. **Endangered species** are those species that are in danger of extinction throughout all or a significant portion of their range. **Threatened species** are those likely to become endangered in the future. Many species are currently threatened. According to the International Union for Conservation of Nature and Natural Resources (IUCN), more than 25% of the fish species that live on coral reefs and 22% of all mammals, 12% of birds, and 31% of amphibians are threatened with extinction.

The last level of biodiversity is ecosystem diversity, the diversity of structure and function within an ecosystem. Conservation at the level of species diversity has largely focused attention on species-rich ecosystems such as tropical forests. Some scientists have argued that other relatively species-poor ecosystems are highly threatened and need to be conserved. In North America, many of the native prairies have been converted to agricultural use, especially in Midwestern states such as Illinois and Iowa. In some counties, remnants of prairie exist only inside cemetery plots, which have been spared from the plow (**Figure 60.1**).

60.2 Why Conserve Biodiversity?

Biologists Paul Ehrlich and E. O. Wilson have suggested that the loss of biodiversity should be an area of great concern for at least three reasons. First, humans depend on plants, animals, and microorganisms for a wide range of food, medicine, and industrial products. The second reason to preserve biodiversity focuses on preserving the array of essential services provided by ecosystems, such as clean air and water. Finally, they proposed that we have an ethical responsibility to protect what are our only known living companions in the universe. In this section, we examine some of the primary reasons why preserving

biodiversity matters, and we explore the link between biodiversity and ecosystem functioning.

The Preservation of Biological Diversity Can Be Justified Based on the Economic and Ecological Values of Diversity as Well as on Ethical Grounds

During the latter half of the 20th century, the reduction of the Earth's biological diversity emerged as a critical issue, one with implications for public policy. A major concern was that loss of plant and animal resources would impair future development of important products and processes in agriculture, medicine, and industry. For example, as previously noted, *Z. diploperennis*, the wild relative of corn discovered in Mexico, is resistant to many corn viruses. Its genes are currently being used to develop virus-resistant types of corn. However, *Z. diploperennis* occurs naturally in only a few small areas of Mexico and could easily have been destroyed by development or cultivation of the land. If we allow such species to go extinct, we may unknowingly threaten the food supply on which much of the world depends.

The pharmaceutical industry is heavily dependent on products made by plants. About 25% of the prescription drugs in the U.S. alone are derived from plants, and the 2006 market value of such drugs was estimated to be $19 billion. Many medicines come from plants found only in tropical rain forests. These include quinine, a drug from the bark of the Cinchona tree (*Cinchona officinalis*) (**Figure 60.2**), which is used for malaria, and vincristine, a drug derived from rosy periwinkle (*Catharanthus*

Figure 60.2 The value of biodiversity. Bark of the Cinchona tree, *Cinchona officinalis*, found only in tropical rain forests, is used to produce quinine, an effective drug against malaria.

roseus), which is a treatment for leukemia and Hodgkin disease. Many plant chemicals of therapeutic importance are likely to be found in the numerous rain forest plant species that have not yet been fully analyzed. The continued destruction of rain forests thus could mean the loss of potential life-saving medical treatments.

On a smaller scale, individual species are valuable for research purposes. The blood of the horseshoe crab (*Limulus polyphemus*) clots when exposed to toxins produced by some bacteria. Pharmaceutical industries use the blood enzyme responsible for this clotting to ensure that their products are free of bacterial contamination. Desert pupfishes, in the genus *Cyprinodon*, found in isolated desert springs in the U.S. Southwest, tolerate salinity twice that of seawater and are valuable models for research on human kidney diseases.

Beyond this, humans benefit enormously from the processes that natural ecosystems provide (**Table 60.1**). Forests soak up carbon dioxide, maintain soil fertility, and retain water, preventing floods; estuaries provide water filtration and protect rivers and coastal shores from excessive erosion. The loss of biodiversity could disrupt an ecosystem's ability to carry out such functions. Other ecosystem functions include the maintenance of populations of natural predators to regulate pest outbreaks and reservoirs of pollinators to pollinate crops and other plants. As discussed in the chapter opening, scientists continue to discover the benefits of biodiversity to the functioning of communities.

A 1997 paper in the journal *Nature* by economist Robert Constanza and colleagues made an attempt to calculate the monetary value of ecosystems to various economies. They came to the conclusion that, at the time, the world's ecosystems were worth more than $33 trillion a year, nearly twice the gross national product of the world's economies combined ($19 trillion) (**Table 60.2**). Due to its massive size, open ocean has the greatest total global value of all ecosystems. Another way to view ecosystem value is its dollar value per hectare. From this perspective, shallow aquatic ecosystems, such as estuaries and swamps, are extremely valuable because of their role in nutrient cycling, water supply, and disturbance regulation. They also serve as nurseries for aquatic life. These habitats, once thought of as useless wastelands, are among the ecosystems most endangered by pollution and development.

Arguments can also be made against the loss of biodiversity on ethical grounds. As only one of many species, it has been argued that humans have no right to destroy other species and the environment around us. Philosopher Tom Regan suggests that animals should be treated with respect because they have a life of their own and therefore have value apart from anyone else's interests. Law professor Christopher Stone,

Table 60.1 Examples of the World's Ecosystem Services

Service	Example
Atmospheric gas supply	Regulation of carbon dioxide, ozone, and oxygen levels
Climate regulation	Regulation of carbon dioxide, nitrogen dioxide, and methane levels
Water supply	Irrigation; water for industry
Pollination	Pollination of crops
Biological control	Pest population regulation
Wilderness and refuges	Habitat for wildlife
Food production	Crops; livestock
Raw materials	Fossil fuels; timber
Genetic resources	Medicines; genes for plant resistance
Recreation	Ecotourism; outdoor recreation
Cultural	Aesthetic and educational value
Disturbance regulation	Storm protection; flood control
Waste treatment	Sewage purification
Soil erosion control	Retention of topsoil; reduction of accumulation of sediments in lakes
Nutrient cycling	Nitrogen, phosphorus, carbon, and sulfur cycles

Table 60.2 Valuation of the World's Ecosystem Services

Ecosystem type	Total global value* ($ trillion)	Total value (per ha) ($)	Main ecosystem service
Open ocean	8,381	252	Nutrient cycling
Coastal shelf	4,283	1,610	Nutrient cycling
Estuaries	4,100	22,832	Nutrient cycling
Tropical forest	3,813	2,007	Nutrient cycling/ raw materials
Seagrass and algal beds	3,801	19,004	Nutrient cycling
Swamps	3,231	19,580	Water supply/ disturbance regulation
Lakes and rivers	1,700	8,498	Water supply
Tidal marsh	1,648	9,990	Waste treatment/ disturbance regulation
Grasslands	906	232	Waste treatment/ food production
Temperate forest	894	302	Climate regulation/ waste treatment/ lumber
Coral reefs	375	6,075	Recreational/ disturbance regulation
Cropland	128	92	Food production
Desert	0	0	
Ice and rock	0	0	
Tundra	0	0	
Urban	0	0	
Total	33,260		

*In 1997 values

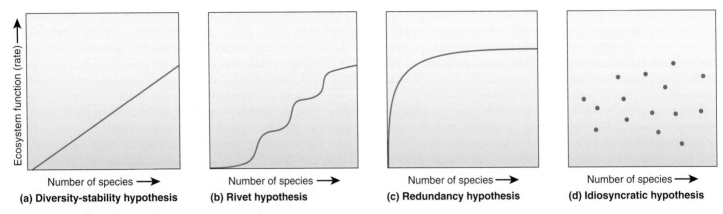

Figure 60.3 Four models that describe the relationship between ecosystem function and biodiversity. The relationship is strongest in **(a)** and weakest in **(d)**.

in an influential article titled "Should Trees Have Standing?" has argued that entities such as nonhuman natural objects like trees or lakes should be given legal rights just as corporations are treated as persons for certain purposes. As E. O. Wilson proposed in a 1984 concept known as biophilia, humans have innate attachment with species and natural habitats because of our close association for over millions of years.

How Much Diversity Is Needed for Ecosystems to Function Properly?

Because biodiversity has an impact on the health of ecosystems, ecologists have explored the question of how much diversity is needed for ecosystems to function properly. In doing so, they have described several possible relationships between biodiversity and ecosystem function. Recall that in the 1950s, ecologist Charles Elton proposed in the diversity-stability hypothesis that species-rich communities are more stable than those with fewer species (see Chapter 58). If we use stability as a measure of ecosystem function, Elton's hypothesis suggests a linear correlation between diversity and ecosystem function (Figure 60.3a).

In 1981, ecologists Paul and Anne Ehrlich proposed an alternative called the **rivet hypothesis** (Figure 60.3b). In this model, species are like the rivets on an airplane, with each species playing a small but critical role in keeping the plane (the ecosystem) airborne. The loss of a rivet weakens the plane and

causes it to lose a little airworthiness. The loss of a few rivets could probably be tolerated, but the loss of more rivets would prove critical to the airplane's function.

A decade later, Australian ecologist Brian Walker proposed an alternative to this idea, termed the **redundancy hypothesis** (or passenger hypothesis) (Figure 60.3c). According to this hypothesis, most species are more like passengers on a plane—they take up space but do not add to the airworthiness. The species are said to be redundant because they could simply be eliminated or replaced by others with no loss in function. Airworthiness is primarily affected by the activity of a few crucial species, in this case, the pilot or copilot, which are called keystone species (a concept we will discuss in detail later in Section 60.4). In this scenario, species can be lost without affecting ecosystem function as long as species of critical importance remain.

In another alternative, British ecologist John Lawton included the possibility that ecosystem function changes as the number of species increases or decreases but that the amount and direction of change is unpredictable. He called this the idiosyncratic hypothesis (Figure 60.3d).

Determining which model is most correct is very important, as our understanding of the effect of species loss on ecosystem function can greatly affect the way we manage our environment. As we will discuss next, experimental studies have provided data showing that reduced biodiversity does lead to reduced ecosystem functioning.

FEATURE INVESTIGATION

Ecotron Experiments Showed the Relationship Between Biodiversity and Ecosystem Function

In the early 1990s, Shahid Naeem and colleagues used a series of 14 environmental chambers in a facility termed the Ecotron, at Silwood Park, England, to determine how biodiversity affects ecosystem functioning. These chambers contained terrestrial communities that differed only in their level of biodiversity (Figure 60.4). The number of species in each chamber was

manipulated to create high-, medium-, and low-diversity ecosystems, each with four trophic levels. The trophic levels consisted of primary producers (annual plants), primary consumers (insects, snails, and slugs), secondary consumers (parasitoids that fed on the herbivores), and decomposers (earthworms and soil insects). The experiment ran for just over 6 months, and species were added only after the trophic level below them was established. For example, parasitoids were not added until herbivores were abundant.

Researchers monitored and analyzed a range of measures of ecosystem function, including community respiration, decomposition, nutrient retention rates, and community productivity. The data of Figure 60.4 focuses only on community productivity. The result was that community productivity, expressed as percent change in vegetation cover (the amount of ground covered by leaves of plants), increased as species richness increased. This occurred because of a greater variety of plant growth forms that could utilize light at different levels of the plant canopy. A larger ground cover also meant a larger plant biomass and greater community productivity, and

increased decomposition and nutrient uptake rates. For the first time, ecologists had provided an experimental demonstration that the loss of biodiversity can alter or impair the functioning of an ecosystem.

Experimental Questions

1. What was the goal of Shahid Naeem and colleagues in their experiment at Silwood Park, England?
2. What was the hypothesis tested by the researchers?
3. How did the researchers test for ecosystem functioning?

Figure 60.4 Ecotron experiments comparing species diversity and ecosystem function.

Concept check: What is one of the dangers in interpreting these results?

HYPOTHESIS Reduced biodiversity can lead to reduced ecosystem functioning.

STARTING LOCATION Ecotron, a controlled environment facility at the Natural Environment Research Council (NERC) Centre for Population Biology, in Silwood Park, England.

Experimental level | **Conceptual level**

1. Construct 14 identical experimental chambers.

Air exhaust, Irrigation lance, Cooling air for lights, Fans, Moisture and temperature sensors, Air input

Temperature- and humidity-controlled chambers are used to control environmental conditions and allow identical starting conditions in all chambers.

2. Add different combinations of species to the 14 chambers. The species added were based on 3 types of model communities (food webs), each with 4 trophic levels but with varying degrees of species richness.

2° consumers, 1° consumers, 1° producers, Decomposers

| Biodiversity | High | Medium | Low |
| Number of chambers analyzed | 6 | 4 | 4 |

- Species present in all 3 systems.
- Species present in 2 systems.
- Species present in most diverse system only.

The three diagrams to the left each represent a single chamber. Circles represent species, and lines connecting them represent biotic interactions among the species. Note that each lower-biodiversity community is a subset of its higher-diversity counterpart and that all community types have 4 trophic levels.

3. Measure and analyze a range of processes, including vegetation cover and nutrient uptake.

Measurements help determine how each different type of community functions.

4 **THE DATA**

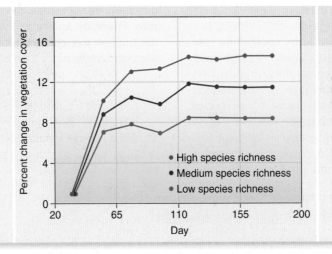

Plant productivity is linked to community diversity as measured by the percent change in vegetation cover from initial conditions.

Data reveal that low-diversity communities have lower vegetation cover and thus are less productive than high-diversity communities.

5 **CONCLUSION** Increases in biodiversity lead to increases in ecosystem function. In this case, increased plant diversity results in greater vegetation cover.

6 **SOURCE** Naeem, S. et al. 1994. Declining biodiversity can alter the performance of ecosystems. *Nature* 368:734–737.

Field Experiments Also Suggest That Biodiversity Is Important for Ecosystem Function

In the mid-1990s, David Tilman and colleagues performed experiments in the field to determine how much biodiversity was necessary for proper ecosystem functioning. Tilman's previous experiments had suggested that species-rich grasslands were more stable; that is, they were more resistant to the ravages of drought and recovered from drought more quickly than species-poor grasslands (refer back to Figure 58.8). In the newer experiments, Tilman's group sowed plots, each 3 m by 3 m and on comparable soils, with seeds of 1, 2, 4, 6, 8, 12, or 24 species of prairie plants. Exactly which species were sown into each plot was determined randomly from a pool of 24 native species. The treatments were replicated 21 times, for a total of 147 plots. The results showed that more-diverse plots had increased productivity and used more nutrients, such as nitrate (NO_3), than less-diverse plots (**Figure 60.5a,b**). Furthermore, the frequency of invasive plant species (species not originally planted in the plots) decreased with increased plant species richness (**Figure 60.5c**). In a separate experiment, where plots were planted with 1, 2, 4, 8, or 16 species, Tilman and colleagues showed that increased diversity also reduced the severity of attack by foliar fungal diseases (**Figure 60.5d**).

Although Tilman's experiments show a relationship between diversity and ecosystem function, they also suggest that most of the advantages of increasing diversity come with the first 5 to 10 species, beyond which adding more species appears to have little to no impact. This supports the redundancy hypothesis (see Figure 60.3c). For example, uptake of nitrogen remains relatively unchanged as the number of species

increases beyond 6. This is also observed on a larger scale. The productivity of temperate forests in different continents is roughly the same despite different numbers of tree species present—729 in East Asia, 253 in North America, and 124 in Europe. The presence of more tree species may ensure a supply of "backups" should some of the most-productive species die off from insect attack or disease. This can happen, as was seen in the demise of the American chestnut and elm trees. Diseases devastated both of these species, and their presence in forests dramatically decreased by the mid-20th century (refer back to Figure 57.18). The forests filled in with other species and continued to function as before in terms of nutrient cycling and gas exchange. However, although the forests continued to function without these species, some important changes occurred. For example, the loss of chestnuts deprived bears and other animals of an important source of food and may have affected their reproductive capacity and hence the size of their populations.

60.3 The Causes of Extinction and Loss of Biodiversity

In light of research showing that the loss of species influences ecosystem function, the importance of understanding and preventing species loss takes on particular urgency. Throughout the history of life on Earth, **extinction**—the process by which species die out—has been a natural phenomenon. As we saw in Chapter 22, five major mass extinctions have occurred over the past 500 million years. The average time span of a typical animal or plant species in the fossil record is about 4 million

(a) Plant cover increased with more species.

(b) Available nitrate decreased with more species.

(c) Invasive species decreased with more species.

(d) Fungal disease decreased with more species.

Figure 60.5 Increased species richness improves community function.

extinction rate. However, it can be argued that the fossil record is heavily biased toward successful, often geographically wide-ranging species, which undoubtedly have a longer than average persistence time. The fossil record is also biased toward vertebrates and marine mollusks, both of which fossilize well because of their hard body parts. If background extinction rates were, say, 10 times higher than the rates perceived from the fossil record, then extinctions among the living mammals today would be expected to occur at a rate of about 1 every 100 years. Because there are twice as many bird species as mammals, the background extinction rate for birds would be 2 species every 100 years.

However, the extinction rate for species in recent times has been far higher than this. In the past 100 years, approximately 20 species of mammals and over 40 species of birds went extinct (**Figure 60.6**). The rates of species extinctions on islands in the past confirm the dramatic effects of human activity. The Polynesians, who colonized Hawaii in the 4th and 5th centuries, appear to have been responsible for the extinction of half of the 100 or so species of endemic land birds in the period between their arrival and that of the Europeans in the late 18th century. A similar impact was felt in New Zealand, which was colonized by settlers some 500 years later than Hawaii. In New Zealand, an entire avian megafauna, consisting of huge land birds, was exterminated over the course of a century, probably through

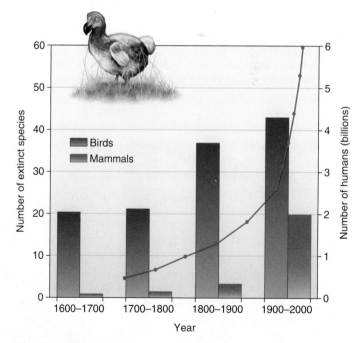

Figure 60.6 Animal extinctions since the 17th century in relation to human population growth. Increasing numbers of known extinctions in birds and mammals are concurrent with exponential increases in the global human population. These data suggest that as human numbers increase, more and more species will go extinct.

Concept check: Why might the increasing human population result in an increase in the extinction rate of other species?

years. To calculate the current extinction rate, we could take the total number of species estimated to be alive on Earth at present, around 10 million, and divide it by 4 million, giving an average extinction rate of 2.5 species each year. For the 5,500 species of living mammals, using the same average life span of around 4 million years, we would expect about 1 species to go extinct every 1,000 years. This is termed the background a combination of hunting and large-scale habitat destruction through burning.

The term **biodiversity crisis** is often used to describe this elevated loss of species. Many scientists believe that the rate of loss is higher now than during most of geological history, and most suggest that the growth in the human population has led to the increase in the number of extinctions of other species. Most scientists believe that we are in the middle of a sixth mass extinction.

To understand the process of extinction in more modern times, ecologists need to examine the role of human activities and their environmental consequences. In this section, we examine why species have gone extinct in the past and look at the factors that are currently threatening species with extinction.

The Main Threats to Species Are Human-Induced

Although all causes of extinctions are not known, introduced species, direct exploitation, and habitat destruction have been identified as the most important human-induced threats. In addition, climate change is increasingly being viewed as a significant threat to species.

Introduced Species **Introduced species** are those species moved by humans from a native location to another location. Most often the species are transported for agricultural purposes or as sources of timber, meat, or wool. Others, such as plants, insects, or marine organisms, are unintentionally transported in the cargo of ships or planes. Regardless of their method of introduction, some introduced species become **invasive species**, spreading and outcompeting native species for space and resources.

We can categorize the interactions between introduced and native species into competition, predation, and disease. Competition can eliminate local populations and cause huge reductions in the densities of native species, but it has not yet been clearly shown to exterminate entire species. On the other hand, many recorded cases of extinction have been due to predation. Introduced predators such as rats, cats, and mongooses have accounted for at least 43% of recorded extinctions of birds on islands. Lighthouse keepers' cats have annihilated populations of ground-nesting birds on small islands around the world. The brown tree snake, which was accidentally introduced onto the island of Guam, has decimated the country's native bird populations (refer back to Figure 57.12). Parasitism and disease carried by introduced organisms have also been important in causing extinctions. Avian malaria in Hawaii, spread by introduced mosquito species, is believed to have contributed to the demise of up to 50% of native Hawaiian birds (**Figure 60.7a**).

Direct Exploitation Direct exploitation, particularly the hunting of animals, has been the cause of many extinctions in the past. Two remarkable North American bird species, the passenger pigeon and the Carolina parakeet, were hunted to extinction by the early 20th century. The passenger pigeon (*Ectopistes migratorius*) was once the most common bird in North America, probably accounting for over 40% of the entire bird population (**Figure 60.7b**). Their total population size was estimated to be over 1 billion birds. It may seem improbable that the most common bird on the continent could be hunted to extinction for its meat, but that is just what happened! The flocking behavior of the birds made them relatively easy targets for hunters, who used special firearms to harvest the birds in quantity. In 1876, in Michigan alone, over 1.6 million birds were killed and sent to markets in the eastern U.S. The Carolina parakeet (*Conuropsis*

(a) Introduced species

(b) Direct exploitation

(c) Habitat destruction

Figure 60.7 **Causes of extinction.** **(a)** Many Hawaiian honeycreepers such as this Ou, *Psittirostra psittacea,* were exterminated by avian malaria from introduced mosquito species. **(b)** The passenger pigeon, which may have once been among the most abundant bird species on Earth, was hunted to extinction for its meat. **(c)** The ivory-billed woodpecker, the third-largest woodpecker in the world, was long thought to be extinct in the southeastern U.S. because of habitat destruction, but a possible sighting occurred in 2004. This nestling was photographed in Louisiana in 1938.

carolinensis), the only species of parrot native to the eastern U.S., was similarly hunted to extinction by the early 1900s.

Many whale species were driven to the brink of extinction prior to a moratorium on commercial whaling issued in 1988 (refer back to Figure 57.13). Steller's sea cow (*Hydrodamalis gigas*), a 9-meter-long manatee-like mammal, was hunted to extinction in the Bering Strait only 27 years after its discovery by humans in 1740. A poignant example of human excess in hunting was the dodo (*Raphus cucullatus*), a flightless bird native only to the island of Mauritius that had no known predators. A combination of overexploitation and introduced species led to its extinction within 200 years of the arrival of humans. Sailors hunted it for its meat, and the rats, pigs, and monkeys brought to the island by humans destroyed the dodo's eggs and chicks in their ground nests.

Habitat Destruction Habitat destruction through **deforestation**, the conversion of forested areas to nonforested land, has historically been a prime cause of the extinction of species. About one-third of the world's land surface is covered with forests, and much of this area is at risk of deforestation through human activities such as development, farming, animal grazing, or logging. Although tropical forests are probably the most threatened forest type, with rates of deforestation in Africa, South America, and Asia varying between 0.6% and 0.9% per year, the destruction of forests is a global phenomenon. The ivory-billed woodpecker (*Campephilus principalis*), the largest woodpecker in North America and an inhabitant of wetlands and forests of the southeastern U.S., was widely assumed to have gone extinct in the 1950s due to destruction of its habitat by heavy logging (**Figure 60.7c**). In 2004, the woodpecker was purportedly sighted in the Big Woods area of eastern Arkansas, though this has not been confirmed despite concerted efforts.

Deforestation is not the only form of habitat destruction. The scouring of land to plant agricultural crops can create soil erosion, increased flooding, declining soil fertility, silting of the rivers, and desertification. The average area of land under cultivation worldwide averages about 11%, with an additional 24% given over to rangeland; however, this amount varies tremendously between regions. For example, Europe uses 28% of its land for crops and pasturelands, with the result that many of its native species went extinct long ago. Wetlands also have been drained for agricultural purposes. Others have been filled in for urban or industrial development. In the U.S., as much as 90% of the freshwater marshes and 50% of the estuarine marshes have disappeared. Urbanization, the development of cities on previously natural or agricultural areas, is the most human-dominated and fastest-growing type of land use worldwide and devastates the land more severely than practically any other form of habitat degradation.

Climate Change As mentioned at the beginning of Chapter 54, human-induced climate change, or global warming, has been implicated in the dramatic decrease in the population sizes of frog species in Central and South America. We also noted that the distribution of trees would also change as climate zones

shifted. Unfortunately, such shifts would occur faster than many plants could migrate via seed dispersal. Indeed, a recent study of six biodiversity-rich regions employed computer models to simulate the movement of species' ranges in response to changing climate conditions. The models predicted that unless greenhouse gas emissions are cut drastically, climate change will cause 15–37% of the species in those regions to go extinct by the year 2050.

Other ecological properties of species, not just range limits, may change with global warming, including population densities and the timing of events, called phenology, such as flowering, egg laying, or migration. In 2003, Terry Root and colleagues examined the phenologies of 694 species over the past 30 years. Over an average decade, the estimated mean number of days changed in spring phenology was 5.1 days earlier. For example, the North American common murre (*Uria aalge*) bred, on average, 24 days earlier per decade, and Fowler's toad (*Bufo fowleri*) bred 6.3 days earlier. Potentially more critical than the absolute change in phenology is the possible disruption of timing between associated species such as herbivores and host plants or predators and prey. All is well if the phenologies of all species are sped up by global warming. However, limited evidence suggests that this is not the case. Only 11 systems have been examined in detail, but in 7 of them, interacting species responded differently enough to temperature changes to put them more out of synchrony than they were earlier. For example, in the Colorado Rockies, yellow-bellied marmots (*Marmota flaviventris*) now emerge from hibernation 23 days earlier than they did in 1975, changing the relative phenology of the marmots with their food plants.

Small Populations Are Threatened by the Loss of Genetic Diversity

Even if habitats are not destroyed, many become fragmented, leading to the development of small, isolated populations. Such populations are more vulnerable to the loss of genetic diversity resulting from three factors: inbreeding, genetic drift, and limited mating.

Inbreeding **Inbreeding**, which is mating among genetically related relatives, is more likely to take place in nature when population size becomes very small and there are a limited number of potential mates to choose from (see Chapter 24). In many species, the health and survival of offspring declines as populations become more inbred. This phenomenon was shown as long ago as the 19th century. Litter size in inbred rats declined from 7.5 in 1887 to 3.2 in 1892. Furthermore, during the same time span, the level of nonproductive matings, those where no offspring were born, rose from 2% to 50%. Generally, the more inbred the population, the more severe these types of problems become.

One of the most striking examples of the effects of inbreeding in conservation biology involves the greater prairie chicken (*Tympanuchus cupido*). The male birds have a spectacular mating display that involves inflating the bright orange air sacs on their throat, stomping their feet, and spreading their tail feathers. The prairies of the Midwest were once home to millions of these birds, but as the prairies were converted to farmland, the range and population sizes of the bird shrank dramatically. The population of prairie chickens in Illinois decreased from 25,000 in 1933 to less than 50 in 1989. At that point, according to studies by Ronald Westemeier and colleagues, only 10 to 12 males existed. Because of the decreasing numbers of males, inbreeding in the population had increased. This was reflected in the steady reduction in the hatching success of eggs (**Figure 60.8**). The prairie chicken population had entered a downward spiral toward extinction from which it could not naturally recover, a phenomenon called an **extinction vortex**. In the early 1990s, conservation biologists began trapping prairie chickens in Kansas and Nebraska, where populations remained larger and more genetically diverse, and moved them to Illinois, bringing an infusion of new genetic material into the population. This transfer resulted in a rebounding of the egg-hatching success rate to over 90% by 1993.

Genetic Drift   In small populations, there is a greater chance that some individuals will fail to mate successfully purely by chance. For example, finding a mate may be increasingly difficult as population size decreases. The decline in the reproduction and survival of individuals in small populations is known as the **Allee effect**, after ecologist W. C. Allee, who first described it. If an individual that fails to mate possesses a rare gene, that genetic information will not be passed on to the next generation, resulting in a loss of genetic diversity from the population. As noted in Chapter 24, **genetic drift** refers to the random change in allele frequencies attributable to chance. Because the likelihood of an allele being represented in just one or a few individuals is higher in small populations than in large populations, small, isolated populations are particularly vulnerable to this type of reduction in genetic diversity. Such isolated populations will lose a percentage of their original diversity over time, approximately at the rate of $1/(2N)$ per generation, where N = population size. This has a greater effect in smaller versus larger populations:

If $N = 500$, then $\dfrac{1}{2N}$ = 1/1,000 = 0.001, or 0.1% genetic diversity lost per generation

If $N = 50$, then $\dfrac{1}{2N}$ = 1/100 = 0.01, or 1.0% genetic diversity lost per generation

Due to genetic drift, a population of 500 will lose only 0.1% of its genetic diversity in a generation, whereas a population of 50 will lose 1%. Such losses become magnified over many generations. After 20 generations, the population of 500 will lose 2% of its original genetic variation, but the population of 50 will lose about 20%! For organisms that breed annually, this would mean a substantial loss in genetic variation over 20 years. Once again, this effect becomes more severe as the population size decreases.

Figure 60.8   **Changes in the abundance and egg-hatching success rate of prairie chickens.**   As the number of males decreased, inbreeding increased, resulting in a decrease in fertility, as indicated by a reduced egg-hatching rate. An influx of males in the early 1990s increased the egg-hatching success rate dramatically.

Concept check: *Is the fitness of all organisms decreased by inbreeding?*

As with inbreeding, the effects of genetic drift can be countered by immigration of individuals into a population. Even relatively low immigration rates of about one immigrant per generation (or one individual moved from one population to another) can be sufficient to counter genetic drift in a population of 100 individuals.

Limited Mating   In many populations, the **effective population size**, the number of individuals that contribute genes to future populations, may be smaller than the number of individuals in the population, particularly in animals with a harem mating structure in which only a few dominant males breed. For example, dominant elephant seal bulls control harems of females, and a few males command all the matings (refer back to Figure 55.22). If a population consists of breeding males and breeding females, the effective population size is given by

Effective population size → $N_e = \dfrac{4N_m N_f}{N_m + N_f}$ ← Number of breeding males / Number of breeding females

In a population of 500, a 50:50 sex ratio, and all individuals breeding, $N_e = (4 \times 250 \times 250)/(250 + 250) = 500$, or 100% of the actual population size. However, if 10 males breed with 250 females, $N_e = (4 \times 10 \times 250)/(10 + 250) = 38.5$, or 8% of the actual population size.

Knowledge of effective population size is vital to ensuring the success of conservation projects. One notable project in the U.S. involved planning the sizes of reserves designed to protect grizzly bear populations in the contiguous 48 states. The grizzly bear (*Ursus arctos*) has declined in numbers from an estimated 100,000 in 1800 to less than 1,000 at present. The range of the species is now less than 1% of its historical range and is restricted to six separate populations in four states (**Figure 60.9**). Research by biologist Fred Allendorf has indicated that the effective population size of grizzly populations is generally only about 25% of the actual population size because not all bears breed. Thus, even fairly large, isolated populations, such as the 200 bears in Yellowstone National Park, are vulnerable to the harmful effects of loss of genetic variation because the effective population size may be as small as 50 individuals.

Figure 60.9 **Past and current ranges of the grizzly bear.** The range of the grizzly bear is currently less than 1% of its historical range. The current range in the continental U.S. has contracted to just six populations in four states, as the population size has shrunk from 100,000 before the West was settled to about 1,000 today.

Concept check: If 500 male and 500 female grizzlies exist today, but only 25% of the males breed, what is the effective population size?

Allendorf and his colleagues proposed that an exchange of grizzly bears between populations or zoo collections would help tremendously in promoting genetic variation. Even an exchange of two bears per generation between populations would greatly reduce the loss of genetic variation.

60.4 Conservation Strategies

In their efforts to maintain the diversity of life on Earth, conservation biologists are currently active on many fronts and employ many strategies. We begin this section by discussing how conservation biologists identify the global habitats richest in species. Next, we explore the concept of nature reserves and consider questions such as how large conservation areas should be and how far apart they should be situated. These questions are within the realm of landscape ecology, which studies the spatial arrangement of communities and ecosystems in a geographic area. Next we discuss how conservation efforts often focus on certain species that can have a disproportionate impact on their ecosystem. We will also examine the field of restoration ecology, focusing on how wildlife habitats can be established from degraded areas and how captive breeding programs have been used to reestablish populations of threatened species in the wild. We conclude by returning to the theme of genomes and proteomes to show how modern molecular techniques of cloning can contribute to the fight to save critically endangered species.

Habitat Conservation Focuses on Identifying Countries Rich in Species, Areas Rich in Endemics, or Representative Habitats

Conservation biologists often must make decisions regarding which habitats should be protected. Many conservation efforts have focused on saving habitats in so-called megadiversity countries, because they often have the greatest number of species. However, more recent strategies have promoted preservation of certain key areas with the highest levels of unique species or the preservation of representative areas of all types of habitat, even relatively species-poor areas.

Megadiversity Countries One method of targeting areas for conservation is to identify those countries with the greatest numbers of species, the **megadiversity countries**. Using the number of plants, vertebrates, and selected groups of insects as criteria, biologist Russell Mittermeier and colleagues have determined that just 17 countries are home to nearly 70% of all known species. Brazil, Indonesia, and Colombia top the list, followed by Australia, Peru, Mexico, Madagascar, China, and nine other countries. The megadiversity country approach suggests that conservation efforts should be focused on the most biologically rich countries. However, although megadiversity areas may contain the most species, they do not necessarily contain the most unique species. The mammal species list for

Peru is 344, and for Ecuador, it is 271; of these, however, 208 species are common to both.

Areas Rich in Endemic Species

Another method of setting conservation priorities, one adopted by the organization Conservation International, takes into account the number of species that are **endemic**, or found only in a particular place or region and nowhere else. This approach suggests that conservationists focus their efforts on geographic **hot spots**. To qualify as a hot spot, a region must meet two criteria: It must contain at least 1,500 species of vascular plants as endemics and have lost at least 70% of its original habitat. Vascular plants were chosen as the primary group of organisms to determine whether or not an area qualifies as a hot spot, mainly because most other terrestrial organisms depend on them to some extent.

Conservationists Norman Myers, Russell Mittermeier, and colleagues identified 34 hot spots that together occupy a mere 2.3% of the Earth's surface but contain 150,000 endemic plant species, or 50% of the world's total (**Figure 60.10**). Of these areas, the Tropical Andes and Sundaland (the region including Malaysia, Indonesia, and surrounding islands) have the most endemic plant species (**Table 60.3**). This approach proposes that protecting geographic hot spots will prevent the extinction of a larger number of endemic species than would protecting areas of a similar size elsewhere. The main argument against using hot spots as the criterion for targeting conservation efforts is that the areas richest in endemics—tropical rain forests—would receive the majority of attention and funding, perhaps at the expense of protecting other areas.

Representative Habitats

In a third approach to prioritizing areas for conservation, scientists have recently argued that we need to conserve representatives of all major habitats. An example is the Pampas region of South America, which is arguably the most threatened habitat on the continent because of conversion of its natural grasslands to ranch land and agriculture. The Pampas does not compare well in richness or endemics to the rain forests, but it is a unique area that without preservation could disappear (**Figure 60.11**). By selecting habitats that are most distinct from those already preserved, many areas that are threatened but not biologically rich may be preserved in addition to the less immediately threatened, but richer, tropical forests.

Ecologists are divided as to which is the best way to identify areas for habitat conservation. Perhaps the best approach might be one that creates a "portfolio" of areas to conserve, containing some areas of high species richness, others with large numbers of endemic species, and some with various habitat types.

The Theory and Practice of Reserve Design Incorporate Principles of Island Biogeography and Landscape Ecology

After identifying areas to preserve, conservationists must determine the size, arrangement, and management of the protected land. Among the questions conservationists address is whether one large reserve is preferable to an equivalent area composed of smaller reserves. Ecologists also need to determine whether parks should be close together or far apart and whether they

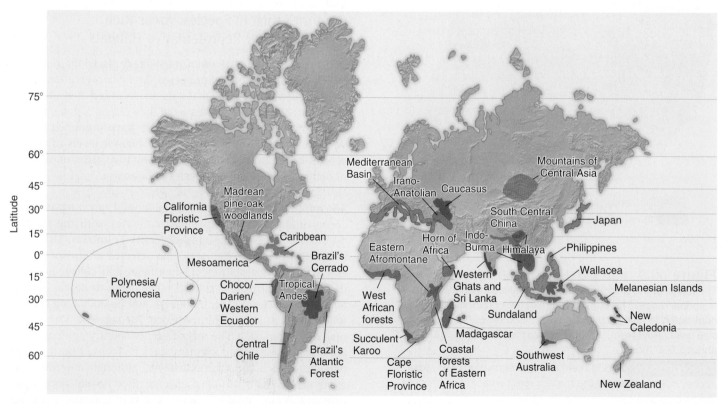

Figure 60.10 Location of major biodiversity hot spots around the world. Hot spots have high numbers of endemic species.

Table 60.3	Numbers of Endemic Species Present in the Top 10 Hot Spots of the World, Ranked by the Numbers of Endemic Plants					
Rank	Hot spot	Plants	Birds	Mammals	Reptiles	Amphibians	Freshwater fishes
1	Tropical Andes	15,000	584	75	275	664	131
2	Sundaland	15,000	146	173	244	172	350
3	Mediterranean Basin	11,700	32	25	77	27	63
4	Madagascar	11,600	183	144	367	226	97
5	Brazil's Atlantic Forest	8,000	148	71	94	286	133
6	Indo-Burma	7,000	73	73	204	139	553
7	Caribbean	6,550	167	41	468	164	65
8	Cape Floristic Province	6,210	6	4	22	16	14
9	Philippines	6,091	185	102	160	74	67
10	Brazil's Cerrado	4,400	16	14	33	26	200

should be connected by strips of suitable habitat to allow the movement of plants and animals between them. Conservationists also need to consider that park design is often contingent on economic factors. Let's examine some of the many issues that conservationists address in the creation and management of protected land.

The Role of Island Biogeography In our exploration of the equilibrium model of island biogeography (see Chapter 58), we noted that wildlife reserves and sanctuaries are, in essence, islands in a sea of human-altered land. Seen this way, the tenets of the equilibrium model of island biogeography can be applied not only to a body of land surrounded by water but also to nature reserves. One question for conservationists is how large a protected area should be (**Figure 60.12a**). According to island biogeography, the number of species should increase with increasing area (the species-area effect). Thus, the larger the area, the greater the number of species would be protected. In addition, larger parks have other benefits. For example, they are beneficial for organisms that require large spaces, including migrating species and species with extensive territories, such as lions and tigers.

A related question is whether it is preferable to protect one single, large reserve or several smaller ones (**Figure 60.12b**). This is called the **SLOSS debate** (for single large or several small). Proponents of the single, large reserve claim that a larger reserve is better able to preserve more and larger populations than an equal area divided into small areas. According to island biogeography, a larger block of habitat should support more species than several smaller blocks.

However, many empirical studies suggest that multiple small sites of equivalent area will contain more species, because a series of small sites is more likely to contain a broader variety of habitats than one large site. Looking at a variety of sites, researchers Jim Quinn and Susan Harrison concluded that animal life was richer in collections of small parks than in a smaller number of larger parks. In their study, having more habitat types outweighed the effect of area on biodiversity. In addition, another benefit of a series of smaller parks is a reduction of extinction risk by a single event such as a wildfire or the spread of disease.

Landscape Ecology Landscape ecology is a subdiscipline of ecology that examines the spatial arrangement of communities and ecosystems in a geographic area. In the design of nature reserves, one question that needs to be addressed is how close to situate reserves to each other, such as whether to place three or four small reserves close to each other or farther apart. A similar question is whether to have a linear or a cluster arrangement of small reserves. Island biogeography suggests that if an area must be fragmented, the sites should be as close as possible

Figure 60.11 **The pampas, Argentina.** This habitat is not rich in species but is threatened due to conversion to ranch land and agricultural land.

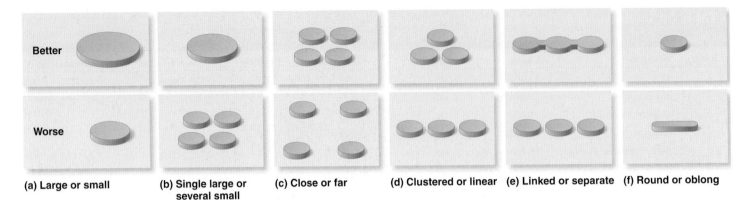

Better					
Worse					
(a) Large or small	**(b)** Single large or several small	**(c)** Close or far	**(d)** Clustered or linear	**(e)** Linked or separate	**(f)** Round or oblong

Figure 60.12 **The theoretical design of nature reserves.** **(a)** A larger reserve will hold more species and have low extinction rates. **(b)** A given area should be fragmented into as few pieces as possible. **(c)** If an area must be fragmented, the pieces should be as close as possible to permit dispersal. **(d)** To enhance dispersal, a cluster of fragments is preferable to a linear arrangement. **(e)** Maintaining or creating corridors between fragments may also enhance dispersal. **(f)** Circular-shaped areas will minimize the amount of edge effects.

Concept check: *What are some of the potential risks in connecting areas via habitat corridors?*

to permit dispersal (Figure 60.12c,d). In practice, however, having small sites far apart may preserve more species than having them close together, because once again, distant sites are likely to incorporate slightly different habitats and species.

Landscape ecologists have also suggested that small reserves should be linked together by biotic corridors, or **movement corridors**, which are thin strips of land that may permit the movement of species between patches (Figure 60.12e). Such corridors ideally facilitate movements of organisms that are vulnerable to predation outside of their natural habitat or that have poor powers of dispersal between habitat patches. In this way, if a disaster befalls a population in one small reserve, immigrants from neighboring populations can more easily recolonize it. This avoids the need for humans to physically move new plants or animals into an area.

Several types of habitat function as corridors, including hedgerows in Europe, which facilitate movement and dispersal of species between forest fragments (Figure 60.13). In China, corridors of habitat have been established to link small, adjacent populations of giant pandas. Riparian habitats, vegetated corridors bordering watercourses, are thought to help facilitate movement of species between habitats. In Florida, debate continues about whether to establish new populations of the Florida panther in refuges in both central and north Florida and whether to have these populations connect with the current population in south Florida via riparian movement corridors. In theory, this would lessen the threat of extinction of the Florida panther via inbreeding or a catastrophic event, such as a hurricane. However, disadvantages are associated with movement corridors. First, corridors also can facilitate the spread of disease, invasive species, and fire between small reserves. Second, it is not yet clear if species, such as female Florida panthers, would actually use such corridors.

Finally, parks are often designed to minimize **edge effects**, the special physical conditions that exist at the boundaries or edges of ecosystems. Habitat edges, particularly those between natural habitats such as forests and developed land, are often different in physical characteristics from the habitat core. For

Figure 60.13 Habitat corridors.

Concept check: *Why would these European hedgerows act as habitat corridors?*

example, the center of a forest is shaded by trees and has less wind and light than the forest edge, which is unprotected. Many forest-adapted species therefore shy away from forest edges and prefer forest centers. Because the amount of edge is minimized, circular parks are generally preferable to oblong parks (Figure 60.12f).

Economic Considerations in Conservation Although the principles of island biogeography theory and landscape ecology are useful in illuminating conservation issues, in reality, there

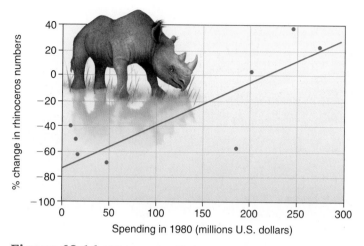

Figure 60.14 **The economics of conservation.** A positive relationship is seen between the percent change in the number of black rhinoceros and conservation spending in various African countries between 1980 and 1984.

is often little choice as to the location, size, shape, and extent of nature reserves. Management practicalities, such as costs of acquisition and maintenance, and politics often override ecological considerations, especially in developing countries, where costs for large reserves may be relatively high. Economic considerations often enter into the choice of which areas to preserve. Typically, many countries protect areas in those regions that are the least economically valuable rather than choosing areas to ensure a balanced representation of the country's biota. In the U.S., most national parks have been chosen for their scenic beauty, not because they preserve the richest habitat for wildlife.

When designing nature reserves, countries should also consider how to finance their management. Interestingly, the amount of money spent to protect nature reserves may better determine species extinction rates than reserve size. According to island biogeography theory, large areas minimize the risk of extinctions because they contain sizable populations. In Africa, several parks, such as Serengeti and Selous in Tanzania, Tsavo in Kenya, and Luangwa in Zambia, are large enough to fulfill this theoretical ideal. However, in the 1980s, populations of black rhinoceroses and elephants declined dramatically within these areas because

of poaching. A larger park is also more difficult to patrol. Research has shown that the rates of decline of rhinos and elephants, largely a result of poaching, have been related directly to conservation efforts and spending (Figure 60.14). Populations of the remaining black rhinos, lowland gorillas, and pygmy chimpanzees in Africa, and the vicuna, a llama-like animal in South America, have all shown the greatest stabilization in areas that have been heavily patrolled and where economic resources have been concentrated.

The Single-Species Approach Focuses Conservation Efforts on Particular Types of Species

Much public awareness of the biodiversity crisis results from efforts to preserve individual species that are at risk of extinction. The single-species approach to conservation focuses on saving species that are deemed particularly important. As with habitat conservation, there are different approaches to identifying ecologically important species to focus effort on, such as indicator, umbrella, flagship, or keystone species, as well as phylogenetically distinct taxa.

Indicator Species   Some conservation biologists have suggested that certain organisms can be used as **indicator species**, those species whose status provides information on the overall health of an ecosystem. Corals are good indicators of marine processes such as siltation, the accumulation of sediments transported by water. Because siltation reduces the availability of light, the abundance of many marine organisms decreases in such situations, with corals among the first to display a decline in health. Coral bleaching is also an indicator of climate change. A proliferation of the dark variety of the peppered moth (*Biston betularia*) has been shown to be a good indicator of air pollution. The darker-colored moths flourish because predators are less able to detect them on trees darkened by soot. Polar bears (*Ursus maritimus*) are thought to be an indicator species for global climate change (Figure 60.15a). Scientists believe that global warming is causing the ice in the Arctic to melt earlier in the spring than in the past. Because polar bears rely on the

(a) Indicator species: Polar bear

(b) Umbrella species: Northern spotted owl

(c) Flagship species: Florida panther

Figure 60.15 **Indicator, umbrella, and flagship species.** (a) Polar bears have been called an indicator species of global climate change. (b) The Northern spotted owl is considered an umbrella species for the old-growth forest in the Pacific Northwest. (c) The Florida panther has become a flagship species for Florida.

ice to hunt for seals, the earlier breakup of the ice is leaving the bears less time to feed and build the fat that enables them to sustain themselves and their young. A U.S. Geological survey concluded that future reduction of arctic ice could result in a loss of two-thirds of the world's polar bear population within 50 years. In May 2008, polar bears were listed as a threatened species under the U.S. Endangered Species Act (ESA).

Umbrella Species **Umbrella species** are species whose habitat requirements are so large that protecting them would protect many other species existing in the same habitat. The Northern spotted owl (*Strix occidentalis*) of the Pacific Northwest is considered to be an important umbrella species (**Figure 60.15b**). A pair of birds needs at least 800 hectares of old-growth forest for survival and reproduction, so maintaining healthy owl populations is thought to help ensure survival of many other forest-dwelling species. In the southeast area of the U.S., the red-cockaded woodpecker (*Picoides borealis*) is often seen as the equivalent of the spotted owl, because it requires large tracks of old-growth long-leaf pine (*Pinus palustris*), including old diseased trees in which it can excavate its nests. Some species, such as *Zea diploperennis*, which we discussed at the beginning of the chapter, fall into this category. For example, to protect *Z. diploperennis* in its natural environment, the state of Jalisco in Mexico bought the land where it grows and established a nature reserve and research facility there.

Flagship Species In the past, conservation resources were often allocated to a **flagship species**, a single large or instantly recognizable species. Such species were typically chosen because they were attractive and thus more readily engendered support from the public for their conservation. The concept of the flagship species, typically a charismatic vertebrate such as the American buffalo (*Bison bison*), has often been used to raise awareness for conservation in general. The giant panda (*Ailuropoda melanoleuca*) is the World Wildlife Fund's emblem for endangered species, and the Florida panther (*Puma concolor*) has become a symbol of the state's conservation campaign (**Figure 60.15c**).

Keystone Species A different, perhaps more effective conservation strategy focuses on **keystone species**, species within a community that have a role out of proportion to their abundance or biomass. The beaver, a relatively small animal, can completely alter the composition of a community by building a dam and flooding an entire river valley (**Figure 60.16**). The resultant lake may become a home to fish species, wildfowl, and aquatic vegetation. A decline in the number of beavers could have serious ramifications for the remaining community members, promoting fish die-offs, waterfowl loss, and the death of vegetation adapted to waterlogged soil.

In the southeastern U.S., gopher tortoises can be regarded as keystone species because the burrows they create provide homes for an array of other animals, including mice, opossums, frogs, snakes, and insects. Many of these creatures depend on the gopher tortoise burrows and would be unable to survive

Figure 60.16 **Keystone species.** The American beaver creates large dams across streams, and the resultant lakes provide habitats for a great diversity of species.

without them. Gopher tortoises and some other keystone species, including beavers, are also called **ecosystem engineers**, because they create, modify, and maintain habitats. African elephants act as ecosystem engineers through their browsing activity, destroying small trees and shrubs and changing woodland habitats into grasslands.

Tropical ecologist John Terborgh considers palm nuts and figs to be keystone species because they produce fruit during otherwise fruitless times of the year and are thus critical resources for tropical forest fruit-eating animals, including primates, rodents, and many birds. Together, these fruit eaters account for as much as three-quarters of the tropical forest animal biomass. Without the fruit trees, wholesale extinction of these animals could occur. Note that a keystone species is not the same as a **dominant species**, one that has a large effect in a community because of its abundance or high biomass. For example, *Spartina* cordgrass is a dominant species in a salt marsh because of its large biomass, but it is not a keystone species.

Few studies have analyzed the community importance of keystone species, and no set criteria have been established for designating a keystone species. Nevertheless, such species do seem to affect species diversity. The conservation community is eager to identify keystone species, because by managing a keystone species, they may also ensure the survival of many other species in the ecosystem. This supports the concept of the redundancy hypothesis that we talked about earlier in Section 60.2.

Restoration Ecology Attempts to Rehabilitate Degraded Ecosystems and Populations

Restoration ecology is the full or partial repair or replacement of biological habitats and/or their populations that have been degraded or destroyed. It can focus on restoring or rehabilitating a habitat, or it can involve attempting to return species to the wild following captive breeding. Following opencast mining for coal or phosphate, huge tracts of disturbed land must be replenished with topsoil, and a large number of species such as grasses, shrubs, and trees must be replanted. Aquatic habitats can be restored by reducing human impacts and replanting vegetation. In Florida, seagrass beds damaged by boat propellers are closed off to motorboats and the area replanted.

Habitat Restoration   The three basic approaches to habitat restoration are complete restoration, rehabilitation, and ecosystem replacement. In complete restoration, conservationists attempt to return a habitat to its condition prior to the disturbance. Under the leadership of Aldo Leopold, the University of Wisconsin pioneered the restoration of prairie habitats as early as 1935, converting agricultural land back to species-rich prairies (**Figure 60.17a**). The second approach aims to return the habitat to something similar to, but a little less than, full restoration, a goal called rehabilitation. In Florida, phosphate mining involves removing a layer of topsoil or "overburden," mining the phosphate-rich layers, returning the overburden, and replanting the area. Exotic species such as cogongrass (*Imperata cylindrica*), an invasive Southeast Asian species, often invade these disturbed areas, and the biodiversity of the restored habitat is usually not comparable to that of unmined areas (**Figure 60.17b**). The third approach, termed replacement, makes no attempt to restore what was originally present but instead replaces the original ecosystem with a different one. The replacement could be an ecosystem that is simpler but more productive, as when deciduous forest is replaced after mining by grassland to be used for public recreation.

Although any of these approaches can be employed in the habitat restoration process, complete restoration is not always the desired endpoint. In some cases, full restoration is appropriate, but there are also many cases where restoration is so difficult or expensive as to be impractical. Ecosystem replacement is particularly sensible for land that has been significantly damaged by past activities. It would be nearly impossible to re-create the original landscape of an area that was mined for stone or gravel. In these situations, however, wetlands or lakes may be created in the open pits (**Figure 60.17c**).

Bioremediation   Restoration can also involve **bioremediation**, the use of living organisms, usually microbes or plants, to detoxify polluted habitats such as dump sites or oil spills. Some bacteria can detoxify contaminants, while certain plants can accumulate toxins in their tissues and are then harvested, removing the poison from the system. As we noted in Chapter 20, biologists have used microbes in the degradation of sewage for over 100 years. More recently, as we saw in Chapter 27, the bacteria *Geobacter sulfurreducens* has been used to treat borrow pits contaminated with uranium.

In 1975, a leak from a military fuel storage facility released 80,000 gallons of jet fuel into the sandy soil at Hanahan, South Carolina. Soon the groundwater contained toxic chemicals such as benzene. By the 1980s, it was found that naturally occurring microorganisms in the soil were actively consuming many of these toxic compounds and converting them into carbon dioxide. In 1992, nutrients were delivered in pipes to the contaminated soils to speed up the action of the natural microbial community. By the end of 1993, contamination had been reduced by 75%. The increasing interest in bacterial genomes is providing opportunities for understanding the genetic and molecular bases of degradation of organic pollutants. Many novel biochemical reactions have been discovered.

Heavy metals such as cadmium or lead are not readily absorbed by microorganisms. Phytoremediation, the treatment of environmental problems through the use of plants, is valuable in these cases. Plants absorb contaminants in the root system and store them in root biomass or transport them to stems and leaves. After harvest, a lower level of soil contamination will remain. Several growth/harvest cycles may be needed to achieve cleanup. Sunflower (*Helianthus annuus*) may be used to extract arsenic and uranium from soils. Pennycress (*Thalsphi caenilescens*) is an accumulator of zinc and cadmium, and lead may be removed by Indian mustard (*Brassica juncea*) and ragweed (*Ambrosia artemisifolia*). Some polychlorinated biphenyls (PCBs) have been removed by transgenic plants containing genes for bacterial enzymes.

Captive Breeding   Captive breeding, the propagation of animals and plants outside their natural habitat to produce stock for subsequent release into the wild, has proved valuable in

(a) **Complete restoration**

(b) **Rehabilitation**

(c) **Ecosystem replacement**

Figure 60.17   Habitat restoration.   (a) The University of Wisconsin pioneered the practice of complete restoration of agricultural land to native prairies. **(b)** In Florida, phosphate mines are so degraded that complete restoration is not possible. After topsoil is replaced, some exotic species such as cogongrass often invade, allowing only habitat rehabilitation. **(c)** These old open-pit mines in Middlesex, England, have been converted to valuable freshwater habitats, replacing the wooded area that was originally present.

Figure 60.18 Captive breeding programs. The California condor, the largest bird in the U.S., with a wingspan of nearly 3 m, has been bred in captivity in California. **(a)** A researcher at the San Diego Wild Animal Park feeds a chick with a puppet so that the birds will not become habituated to the presence of humans. **(b)** This captive-bred condor soars over the Grand Canyon. Note the tag on the underside of its wing.

(a) A condor chick being fed using a puppet   (b) A released captive-bred condor

reestablishing breeding populations following extinction or near extinction. Zoos, aquariums, and botanical gardens often play a key role in captive breeding, propagating species that are highly threatened in the wild. They also play an important role in public education about the loss of biodiversity and the use of restoration programs.

Several classic programs illustrate the value of captive breeding and reintroduction. The peregrine falcon (*Falco peregrinus*) became extinct in nearly all of the eastern U.S. by the mid-1960s, a decline that was linked to the effects of DDT (refer back to Figure 59.9). In 1970, Tom Cade gathered falcons from other parts of the country to start a captive breeding program at Cornell University. Since then, the program has released thousands of birds into the wild, and in 1999, the peregrine falcon was removed from the Endangered Species List. A captive breeding program is also helping save the California condor (*Gymnogyps californicus*) from extinction. In the 1980s, there were only 22 known birds, some in captivity and some in the wild. Scientists made the decision to capture the remaining wild birds in order to protect and breed them (**Figure 60.18a**). By 2008, the captive population numbered 152 individuals, and 147 birds were living in the wild, 80 in California and 67 in Arizona (**Figure 60.18b**). A milestone was reached in 2002, when a pair of captive-reared California condors bred in the wild.

Because the number of individuals in any captive breeding program is initially small, care must be exercised to avoid inbreeding. Matings are usually carefully arranged to maximize resultant genetic variation in offspring. The use of genetic engineering to clone endangered species is a new area that may eventually help bolster populations of captive-bred species.

Genomes & Proteomes Connection

Can Cloning Save Endangered Species?

In 1997, Ian Wilmut and colleagues at Scotland's Roslin Institute announced to the world that they had cloned a now-famous sheep, Dolly, from mammary cells of an adult ewe (see Chapter 20). Since then, interest has arisen among conservation biologists about whether the same technology might be used to save species on the verge of extinction. Scientists were encouraged

that in January 2001, an Iowa farm cow called Bessie gave birth to a cloned Asian gaur (*Bos gaurus*), an endangered species. The gaur, an oxlike animal native to the jungles of India and Burma, was cloned from a single skin cell taken from a dead animal. To clone the gaur, scientists removed the nucleus from a cow's egg and replaced it with a nucleus from the gaur's cell. The treated egg was then placed into the cow's womb. Unfortunately, the gaur died from dysentery 2 days after birth, although scientists believe this was unrelated to the cloning procedure. In 2003, another type of endangered wild cattle, the Javan banteng (*Bos javanicus*), was successfully cloned (**Figure 60.19**). In 2005, clones of the African wildcat (*Felis libyca*) successfully produced wildcat kittens. This is the first time that clones of a wild species have bred.

Despite the promise of cloning, a number of issues remain unresolved:

1. Scientists would have to develop an intimate knowledge of different species' reproductive cycles. For sheep and cows, this was routine, based on the vast experience in breeding these species, but eggs of different species, even if they could be harvested, often require different nutritive media in laboratory cultures.

2. Because it is desirable to leave natural mothers available for breeding, scientists will have to identify surrogate females of similar but more common species that can carry the fetus to term.

3. Some argue that cloning does not address the root causes of species loss, such as habitat fragmentation or poaching, and that resources would be better spent elsewhere, for example, in preserving the remaining habitat of endangered species.

4. Cloning might not be able to increase the genetic variability of the population. However, if it were possible to use cells from deceased animals, for example, from their hair or feathers, these clones could theoretically reintroduce lost genes back into the population.

Many biologists believe that while cloning may have a role in conservation, it is only part of the solution and that we should address what made the species go extinct before attempting to restore it.

Figure 60.19 Cloning an endangered species. In 2004, this 8-month-old cloned Javan banteng made its public debut at the San Diego Zoo.

Conservation is clearly a matter of great importance, and a failure to value and protect our natural resources adequately could be a grave mistake. Some authors, most recently the ecologist and geographer Jared Diamond, have investigated why many societies of the past—including Angkor Wat, Easter Island, and the Mayans—collapsed or vanished, leaving behind monumental ruins. Diamond has concluded that the collapse of these societies occurred partly because people inadvertently destroyed the ecological resources on which their societies depended. Modern nations such as Rwanda face similar issues. The country's population density is the highest in Africa, and it has a limited amount of land that can be used for growing crops. By the late 1980s, the need to feed a growing population led to the wholesale clearing of Rwanda's forests and wetlands, with the result that little additional land was available to farm. Increased population pressure, along with food shortages fueled by environmental scarcity, were likely contributing factors in igniting the genocide of 1994.

As we've seen throughout this textbook, an understanding of biology is vital to comprehend and help solve many of society's problems. Within this large field, genomics and proteomics may have a huge potential for improving people's lives and society at large. These disciplines offer the opportunity to unlock new diagnoses and treatments for diseases, to improve nutrition and food production, and even to help us restore biological diversity.

▌Summary of Key Concepts

60.1 What Is Biodiversity?

- Biodiversity represents diversity at three levels: genetic diversity, species diversity, and ecosystem diversity.
- Conservation biology uses knowledge from molecular biology, genetics, and ecology to protect the biological diversity of life. (Figure 60.1)

60.2 Why Conserve Biodiversity?

- The preservation of biodiversity has been justified because of its economic value, because of the value of ecosystem services, and on ethical grounds. (Tables 60.1, 60.2, Figure 60.2)
- Four models exist that describe the relationship between biodiversity and ecosystem function: diversity-stability, rivet, redundancy, and idiosyncratic. (Figure 60.3)
- Experiments both in the laboratory and in the field have shown that increased biodiversity results in increased ecosystem function. (Figures 60.4, 60.5)

60.3 The Causes of Extinction and Loss of Biodiversity

- Extinction—the process by which species die out—has been a natural phenomenon throughout the history of life on Earth. Extinction rates in recent times, however, have been much higher than in the past, a phenomenon called the biodiversity crisis. (Figure 60.6)
- The main causes of extinctions have been and continue to be introduced species, direct exploitation, and habitat destruction. (Figure 60.7)
- Reduced population size can lead to a reduction of genetic diversity through inbreeding, genetic drift, and limited mating, which reduces effective population size.
- Inbreeding, mating among genetically related relatives, can lead to a reduction in fertility. (Figure 60.8)
- Knowledge of a species' effective population size is vital to ensure the success of conservation projects. (Figure 60.9)

60.4 Conservation Strategies

- Habitat conservation strategies commonly target megadiversity countries, countries with the largest number of species; biodiversity hot spots, areas with the largest number of endemic species, those unique to the area; and representative habitats, areas that represent the major habitats. (Figures 60.10, 60.11, Table 60.3)
- Conservation biologists employ many strategies in protecting biodiversity. Principles of the equilibrium model of island biogeography and landscape ecology are used in the theory and practice of park reserve design to determine, for example, whether the park should take the form of one single or several small reserves. (Figures 60.12, 60.13)
- Economic considerations also play an important role in reserve creation, and it has been shown that conservation spending is positively related to population size. (Figure 60.14)
- The single-species approach focuses conservation efforts on indicator species, umbrella species, flagship species, and keystone species. Indicator species are species whose status can provide information on the overall health of an ecosystem. Umbrella species are species whose habitat requirements are so large that preserving them would also preserve many other species. Flagship species are usually charismatic or instantly recognizable species. Keystone species have an effect out of proportion to their abundance. (Figures 60.15, 60.16)
- Restoration ecology seeks to repair or replace populations and their habitats. Three basic approaches to habitat restoration

are complete restoration, rehabilitation, and ecosystem replacement. (Figure 60.17)

- Captive breeding is the propagation of animals outside their natural habitat. Several programs have illustrated the success of captive breeding and reintroduction to the wild, including those for California condors. (Figure 60.18)

- Cloning of endangered species has been accomplished on a very small scale and despite its limitations may have a role in conservation biology. (Figure 60.19)

Assess and Discuss

Test Yourself

1. Which of the following statements best describes an endangered species?
   a. a species that is likely to become extinct in a portion of its range
   b. a species that has disappeared in a particular community but is present in other natural environments
   c. a species that is extinct
   d. a species that is in danger of becoming extinct throughout all or a significant portion of its range
   e. both b and d

2. Biological diversity is important and should be preserved because
   a. food, medicines, and industrial products are all benefits of biodiversity.
   b. ecosystems provide valuable services to us in many ways.
   c. many species can be used as valuable research tools.
   d. we have an ethical responsibility to protect our environment.
   e. all of the above are correct.

3. The research conducted by Tilman and colleagues demonstrated that
   a. as diversity increases, productivity increases.
   b. as diversity decreases, productivity increases.
   c. areas with higher diversity demonstrate less efficient use of nutrients.
   d. species-richness increases lead to an increase in invasive species.
   e. increased diversity results in increased susceptibility to disease.

4. Approximately what percentage of genetic variation remains in a population of 25 individuals after three generations?
   a. 98   c. 94   e. 84
   b. 96   d. 92

5. What is the effective population size of an island population of parrots of 30 males and 30 females, where only 10 of the males breed?
   a. 10   c. 30   e. 60
   b. 20   d. 40

6. Saving endangered habitats, such as the Argentine pampas, focuses on
   a. saving genetic diversity.
   b. saving keystone species.
   c. conservation in a megadiversity country.
   d. preserving an area rich in endemic species.
   e. preservation of a representative habitat.

7. Geographic hotspots are those areas rich in
   a. species.   d. biodiversity.
   b. habitats.   e. endemic species.
   c. rare species.

8. A new canine distemper pathogen that decimates a population of black-footed ferrets is known as a(n):
   a. keystone species.   d. umbrella species.
   b. dominant species.   e. flagship species.
   c. indicator species.

9. Small strips of land that connect and allow organisms to move between small patches of natural habitat are called
   a. biological conduits.   d. migration pathways.
   b. edge effects.   e. landscape breaks.
   c. movement corridors.

10. Bioremediation is
    a. a process that restores a disturbed habitat to its original state.
    b. a process that uses microbes or plants to detoxify contaminated habitats.
    c. the legislation requiring rehabilitation of a disturbed habitat.
    d. a process of capturing all of the living individuals of a species for breeding purposes.
    e. the process of removing tissue from a dead organism in the hopes of cloning it.

Conceptual Questions

1. What are the three levels at which biodiversity can be examined?

2. Which types of species are most vulnerable to extinction?

3. Distinguish between an umbrella species, a flagship species, and a keystone species.

Collaborative Questions

1. Discuss several causes of species extinction.

2. You are called upon to design a park to maximize biodiversity in a tropical country. What are your recommendations?

Online Resource

www.brookerbiology.com

Stay a step ahead in your studies with animations that bring concepts to life and practice tests to assess your understanding. Your instructor may also recommend the interactive ebook, individualized learning tools, and more.

Periodic Table of the Elements

MAIN–GROUP ELEMENTS

Legend:
- Metals (main-group)
- Metals (transition)
- Metals (inner transition)
- Metalloids
- Nonmetals

MAIN–GROUP ELEMENTS

TRANSITION ELEMENTS

1A (1)																	8A (18)
1 **H** 1.008	2A (2)											3A (13)	4A (14)	5A (15)	6A (16)	7A (17)	2 **He** 4.003
3 **Li** 6.941	4 **Be** 9.012											5 **B** 10.81	6 **C** 12.01	7 **N** 14.01	8 **O** 16.00	9 **F** 19.00	10 **Ne** 20.18
11 **Na** 22.99	12 **Mg** 24.31	3B (3)	4B (4)	5B (5)	6B (6)	7B (7)	8B (8)	8B (9)	8B (10)	1B (11)	2B (12)	13 **Al** 26.98	14 **Si** 28.09	15 **P** 30.97	16 **S** 32.07	17 **Cl** 35.45	18 **Ar** 39.95
19 **K** 39.10	20 **Ca** 40.08	21 **Sc** 44.96	22 **Ti** 47.88	23 **V** 50.94	24 **Cr** 52.00	25 **Mn** 54.94	26 **Fe** 55.85	27 **Co** 58.93	28 **Ni** 58.69	29 **Cu** 63.55	30 **Zn** 65.41	31 **Ga** 69.72	32 **Ge** 72.61	33 **As** 74.92	34 **Se** 78.96	35 **Br** 79.90	36 **Kr** 83.80
37 **Rb** 85.47	38 **Sr** 87.62	39 **Y** 88.91	40 **Zr** 91.22	41 **Nb** 92.91	42 **Mo** 95.94	43 **Tc** (98)	44 **Ru** 101.1	45 **Rh** 102.9	46 **Pd** 106.4	47 **Ag** 107.9	48 **Cd** 112.4	49 **In** 114.8	50 **Sn** 118.7	51 **Sb** 121.8	52 **Te** 127.6	53 **I** 126.9	54 **Xe** 131.3
55 **Cs** 132.9	56 **Ba** 137.3	57 **La** 138.9	72 **Hf** 178.5	73 **Ta** 180.9	74 **W** 183.9	75 **Re** 186.2	76 **Os** 190.2	77 **Ir** 192.2	78 **Pt** 195.1	79 **Au** 197.0	80 **Hg** 200.6	81 **Tl** 204.4	82 **Pb** 207.2	83 **Bi** 209.0	84 **Po** (209)	85 **At** (210)	86 **Rn** (222)
87 **Fr** (223)	88 **Ra** (226)	89 **Ac** (227)	104 **Rf** (263)	105 **Db** (262)	106 **Sg** (266)	107 **Bh** (267)	108 **Hs** (277)	109 **Mt** (268)	110 **Ds** (281)	111 **Rg** (272)	112 **Uub** (285)	113 **Uut** (?)	114 **Uuq** (291?)	115 **Uup** (288)	116 **Uuh** (289?)	117 **Uus** (291)	118 **Uuo** (?)

Period: 1–7

INNER TRANSITION ELEMENTS

		58 **Ce** 140.1	59 **Pr** 140.9	60 **Nd** 144.2	61 **Pm** (145)	62 **Sm** 150.4	63 **Eu** 152.0	64 **Gd** 157.3	65 **Tb** 158.9	66 **Dy** 162.5	67 **Ho** 164.9	68 **Er** 167.3	69 **Tm** 168.9	70 **Yb** 173.0	71 **Lu** 175.0
6	Lanthanides														
7	Actinides	90 **Th** 232.0	91 **Pa** (231)	92 **U** 238.0	93 **Np** (237)	94 **Pu** (242)	95 **Am** (243)	96 **Cm** (247)	97 **Bk** (247)	98 **Cf** (251)	99 **Es** (252)	100 **Fm** (257)	101 **Md** (258)	102 **No** (259)	103 **Lr** (260)

The complete Periodic Table of the Elements. Group numbers are different in some cases from those presented in Figure 2.5, because of the inclusion of transition elements. In some cases, the average atomic mass has been rounded to one or two decimal places, and in others only an estimate is given in parentheses due to the short-lived nature or rarity of those elements. The symbols and names of elements 112–118 are temporary until the chemical characteristics of these elements become better defined. Element 117 is currently not confirmed as a true element, and little is known about element 118. The International Union of Pure and Applied Chemistry (IUPAC) has recently proposed adopting the name copernicium (Cp) for element 112 in honor of scientist and astronomer Nicolaus Copernicus.

APPENDIX B

Answer Key

Answers to Collaborative Questions can be found on the website.

Chapter 1

Concept Checks

Figure 1.4 It benefits humans in many different ways. Discoveries in biology are important in the fields of medicine, research, agriculture, biotechnology, and many other disciplines. Most of the medicines we take were developed by biologists.

Figure 1.6 It would be at the population level.

Figure 1.8 In monkeys, the tail has been modified to grasp onto things, such as tree branches. In skunks, the tail is modified with a bright stripe; the tail can stick up and act as a warning signal to potential predators. In cattle, the tail has long hairs and is used to swat insects. Many more examples are possible.

Figure 1.9 Natural selection is the process that causes evolution to happen.

Figure 1.11 A tree of life suggests that all living organisms evolved from a single ancestor by vertical evolution with mutation. A web of life assumes that both vertical evolution with mutation and horizontal gene transfer were important mechanisms in the evolution of new species.

Figure 1.13 Taxonomy helps us to appreciate the unity and diversity of life. Organisms that are closely related evolutionarily are placed in smaller groups.

Figure 1.14 The genome stores the information to make an organism's proteins. In and of itself, the genome is merely DNA. The traits of cells and organisms are largely determined by the structures and functions of the hundreds or thousands of different proteins they make.

Figure 1.17 Discovery-based science does not require a preconceived hypothesis in order to carry out an experiment.

Figure 1.18 A researcher can compare the results in the experimental group and control group to determine if a single variable is causing a particular outcome in the experimental group.

Figure 1.19 After the *CF* gene was identified by discovery-based science, researchers realized that the *CF* gene was similar to other genes that encoded proteins that were already known to be transporters. This provided an important clue that the *CF* gene also encodes a transporter protein.

Feature Investigation Questions

1. In discovery-based science, a researcher does not need to have a preconceived hypothesis. Experimentation is conducted in the hope that it may have practical applications or may provide new information that will lead to a hypothesis. By comparison, hypothesis testing occurs when a researcher forms a hypothesis that makes certain predictions. Experiments are conducted to see if those predictions are correct. In this way, the hypothesis may be accepted or rejected.

2. This strategy may be described as a five-stage process:
   1. Observations are made regarding natural phenomena.
   2. These observations lead to a hypothesis that tries to explain the phenomena. A useful hypothesis is one that is testable because it makes specific predictions.
   3. Experimentation is conducted to determine if the predictions are correct.
   4. The data from the experiment are analyzed.
   5. The hypothesis is accepted or rejected.

3. In an ideal experiment, the control and experimental groups differ by only one factor. Biologists apply statistical analyses to their data to determine if the control and experimental groups are likely to be different from each other because of the single variable that is different between the two groups. This provides an objective way to accept or reject a hypothesis.

Test Yourself

1. d   2. a   3. c   4. c   5. d   6. b   7. d   8. d   9. a   10. b

Conceptual Questions

1. Cells and organization—All living organisms consist of cells; organisms maintain an internal order that is separated from the environment.
   - Energy use and metabolism—All living organisms acquire energy from the environment and use that energy to maintain their internal order. An organism's chemical reactions are collectively known as metabolism.
   - Response to environmental changes—Living organisms respond to environmental changes. These responses are adaptations.
   - Regulation and homeostasis—Living organisms maintain relatively stable internal conditions, homeostasis.
   - Growth and development—Growth produces more or larger cells; development produces organisms with a defined set of characteristics.
   - Reproduction—To sustain life, organisms must produce others like themselves; the genetic material maintains continuity over the generations.
   - Biological evolution—Populations of organisms change over the course of many generations.

2. The unity among different species occurs because modern species have evolved from a group of related ancestors. Some of the traits in those ancestors are also found in modern species and thereby unites them. The diversity is due to the variety of environments on the Earth. Each species has evolved to occupy its own unique environment. For every species, many traits are evolutionary adaptations to survival in a specific environment. For this reason, evolution also promotes diversity.

3. Domains and kingdoms are very diverse, being composed of hundreds of thousands or even millions of different species. The least-diverse groups are genera and species. A species is composed of just one type of organism, and a genus is typically composed of several or several dozen different species, though some genera are quite large.

Chapter 2

Concept Checks

Figure 2.4 An energy shell is a region outside the nucleus of an atom occupied by electrons of a given energy level. More than one orbital can be found within an electron shell. An orbital may be spherical or dumbbell-shaped and contains up to two electrons.

Figure 2.9 The octet rule states that atoms are stable when they have eight electrons in their outermost shell. Oxygen has six electrons in its outer shell. When two oxygen atoms share two pairs of electrons, each atom has eight electrons in its outer shell, at least part of the time.

Figure 2.11 Strand separation requires energy, because the DNA strands are held together by a large number of hydrogen bonds. Although each hydrogen bond is weak, collectively the vast number of such bonds in a molecule of DNA adds up to a considerable strength.

Figure 2.17 The oil would be in the center of the soap micelles.

Figure 2.19 Due to the colligative properties of water, the solutes in blood lower its freezing point. Human blood, for example, freezes at a temperature that is about half a degree Celsius lower than that of pure water.

Figure 2.21   It is 10^{-6} M. Because [H$^+$][OH$^-$] always equals 10^{-14} M, if [H$^+$] = 10^{-8} M (i.e., pH 8.0), then [OH$^-$] must be 10^{-6} M.

Feature Investigation Questions

1. Scientists were aware that atoms contained charged particles. Many believed that the positive charges and mass were evenly distributed throughout the atom.

2. Rutherford was testing the hypothesis that atoms are composed of positive charges evenly distributed throughout the atom. Based on this model of the structure of the atom, alpha particles, which are positively charged nuclei of helium atoms, should be deflected as they pass through the foil, due to the presence of positive charges spread throughout the gold foil.

3. Instead of detecting slight deflection of most alpha particles as they passed through the gold foil, the majority, 98%, of the alpha particles passed directly through the gold foil without deflection. A much smaller percentage either deflected or bounced back from the gold foil. Rutherford suggested that since most of the alpha particles passed unimpeded through the gold foil, most of the volume of atoms is empty space. Rutherford also proposed that the bouncing back of some of the alpha particles indicated that most of the positively charged particles were concentrated in a compact area. These results were counter to the hypothesized model.

Test Yourself

1. b   2. b   3. b   4. d   5. e   6. e   7. e   8. c   9. e   10. b

Conceptual Questions

1. Covalent bonds are bonds in which atoms share electrons. A hydrogen bond is a weak polar covalent bond that forms when a hydrogen atom from one polar molecule becomes electrically attracted to an electronegative atom. A nonpolar covalent bond is one between two atoms of similar electronegativities, such as two carbon atoms. The van der Waal forces are temporary, weak bonds, resulting from random electrical forces generated by the changing distributions of electrons in the outer shells of nearby atoms. The strong attraction between two oppositely charged atoms forms an ionic bond.

2. Hydrophobic: "Water-fearing"—molecules that are not attracted to water molecules. Hydrophilic: "Water-loving"—generally, ions and molecules that contain polar covalent bonds will dissolve in water and are said to be hydrophilic.

3. Within limits, bonds within molecules can rotate and thereby change the shape of a molecule. This is important because it is the shape of a molecule that determines, in part, the ability of that molecule to interact with other molecules. Also, when two molecules do interact through such forces as hydrogen bonds, the shape of one or both molecules may change as a consequence. The change in shape is often part of the mechanism by which signals are sent within and between cells.

Chapter 3
Concept Checks

Figure 3.1   Due to the fact that he had earlier purified urea from urine and then formed urea crystals, he already knew what urea crystals looked like. As seen in this figure, they are quite distinctive looking. Therefore, when he reacted ammonia and cyanic acid and got a compound that formed crystals, the distinctive look of the crystals made him realize that he had synthesized urea.

Figure 3.6   One reason is that the binding of a molecule to an enzyme depends on the spatial arrangements of the atoms in that molecule. Enantiomers have different spatial relationships that are mirror images of each other. Therefore, one may bind very tightly to an enzyme while the other may not be recognized at all.

Figure 3.7   Recall from Figure 3.5 that the reverse of a dehydration reaction is called a hydrolysis reaction, in which a molecule of water is added to the molecule being broken down, resulting in the formation of monomers.

Figure 3.11   Hydrogenation is adding hydrogens to double-bonded carbon atoms, changing them from unsaturated to saturated. This causes them to be solid at room temperature.

Figure 3.12   The phospholipids would be oriented such that their polar regions dissolved in the water layer and the nonpolar regions dissolved in the

oil. Thus, the phospholipids would form a layer at the interface between the water and oil.

Figure 3.15   71; one less than the number of amino acids in the polypeptide

Figure 3.19   If the primary structure of Protein 1 were altered in some way, this would, in turn, most likely alter the secondary and tertiary structures of Protein 1. Therefore, it is possible that the precise fit between Proteins 1 and 2 would be lost and that the two proteins would lose the ability to interact.

Figure 3.24   Yes. The opposite strand must be the mirror image of the first strand, because pairs can form only between A and T, and G and C. For instance, if a portion of the first strand is AATGCA, the opposite strand along that region would be TTACGT.

Feature Investigation Questions

1. Many scientists assumed that protein folding was directed by some cellular factor, meaning some other molecule in the cytoplasm, and therefore, protein folding could not occur spontaneously. Others assumed that protein folding was determined somehow by the ribosome, because this organelle is primarily responsible for synthesizing proteins.

2. Anfinsen was testing the hypothesis that the information necessary for determining the three-dimensional shape of a protein is contained within the protein itself. In other words, the chemical characteristics of the amino acids that make up a protein will determine the three-dimensional shape.

3. The urea disrupts hydrogen bonds and ionic interactions that are necessary for protein folding. The mercaptoethanol disrupted the S—S bonds that also form between certain amino acids of the same polypeptide chains. Both substances essentially allow the polypeptide chain to unfold, disrupting the three-dimensional shape. Anfinsen removed the urea and mercaptoethanol from the protein solution by size-exclusion chromatography. After removing the urea and mercaptoethanol, Anfinsen discovered that the protein refolded into its proper three-dimensional shape and became functional again. This was important because the solution contained only the protein and lacked any other cellular material that could possibly assist in protein folding. This demonstrated that the protein could refold into the functional conformation.

Test Yourself

1. b   2. b   3. e   4. b   5. c   6. b   7. b   8. d   9. b   10. b

Conceptual Questions

1. Isomers are two structures with an identical molecular formula but with different structures and arrangements of atoms within the molecule. There are two major types of isomers: structural and stereoisomers. Because many chemical reactions in biology depend upon the actions of enzymes, which are often highly specific for the spatial arrangement of atoms in a molecule, one isomer of a pair may have biological functions, and the other may not.

2. a. Carbohydrates—energy storage and structural support

   b. Lipids—energy storage and components of cellular membranes

   c. Proteins—many functions, including enzymes, defense, transport, structure, contraction

   d. Nucleic acids—information storage, gene expression

3. Saturated fatty acids are saturated with hydrogen and have only single (C—C) bonds, whereas unsaturated fatty acids have one or more double (C=C) bonds. The double bonds in unsaturated fatty acids alters their shape, resulting in a kink in the structure. Saturated fatty acids are unkinked and are better able to stack tightly together. Fats containing saturated fatty acids have a higher melting point than those containing mostly unsaturated fatty acids; consequently, saturated fats tend to be solids at room temperatures, and unsaturated fatty acids are usually liquids at room temperature.

Chapter 4
Concept Checks

Figure 4.1   You would use transmission electron microscopy. The other methods do not have good enough resolution.

Figure 4.3   The primary advantage is that it gives an image of the 3-D surface of a material.

Figure 4.6   They have different proteomes.

Figure 4.7   Centrioles: Not found in plant cells; their role is not entirely clear, but they are found in the centrosome, which is where microtubules are anchored.

Chloroplasts: Not found in animal cells; function in photosynthesis.

Cell wall: Not found in animal cells; important in cell shape.

Figure 4.12   Both dynein and microtubules are anchored in place. Using ATP as a source of energy, dynein tugs on microtubules. Because the microtubules are anchored, they bend in response to the force exerted by dynein.

Figure 4.15   The nuclear lamina organizes the nuclear envelope and also helps to organize/anchor the chromosomes. The nuclear matrix is inside the nucleus and helps to organize the chromosomes into chromosome territories.

Figure 4.18   The protein begins its synthesis in the cytosol and then is further synthesized into the ER. It travels via vesicles to the *cis-*, *medial-*, and *trans*-Golgi and then is secreted.

Figure 4.22   Of these three choices, membrane transport is probably the most important because it regulates which molecules can enter the cell and participate in metabolism and which products of metabolism are exported from the cell. Cell signaling may affect metabolism, but its overall effect is probably less important than membrane transport.

Figure 4.24   It increases the surface area where ATP synthesis takes place and thereby makes it possible to increase the amount of ATP synthesis.

Figure 4.27   Bacteria and mitochondria are similar in size; they both have circular chromosomes; they both divide by binary fission; and they both make ATP. Bacterial chromosomes are larger, and they make all of their cellular proteins. Mitochondria chromosomes are smaller, and they import most of their proteins from the cytosol.

Figure 4.29   The signal sequence of an ER protein is recognized by SRP, which halts translation. The emerging protein and its ribosome are then transferred to the ER membrane, where translation resumes.

Figure 4.31   If chaperone proteins were not found in the cytosol, the mitochondrial matrix protein would start to fold, which might prevent it from being able to pass through the channel in the outer and inner mitochondrial membrane. Normally, a protein is threaded through this channel in an unfolded state.

Feature Investigation Questions

1. In a pulse-chase experiment, radioactive material is provided to cells. This is referred to as the pulse, or single administration of the radioactive material to the cells. After a few minutes, a large amount of nonradioactive material is provided to the cells to remove or "chase away" any of the remaining radioactive material.

   The researchers were attempting to determine the movement of proteins through the different compartments of a cell. Radioactive amino acids were used to label the proteins and enable the researchers to visualize where the proteins were at different times.

2. Pancreatic cells produce large numbers of proteins that are secreted from the cell. The number and final location of the proteins would allow the researchers an ideal system for studying protein movement through the cell.

3. Using electron microscopy, the researchers found that the proteins, indicated by radioactivity, were first found in the endoplasmic reticulum of the cells. Later the radioactivity moved to the Golgi and then into vesicles near the plasma membrane.

   The researchers were able to conclude that secreted proteins moved through several cellular compartments before they were secreted from the cell. Also, the movement of proteins through these compartments was not random but followed a particular pathway: ER, Golgi, secretory vesicles, plasma membrane, and, finally, secreted.

Test Yourself

1. d   2. d   3. b   4. c   5. e   6. a   7. e   8. e   9. a It would go there first, because targeting to the ER occurs cotranslationally.   10. c It is true they carry out metabolism, but so do eukaryotic cells.

Conceptual Questions

1. There are a lot of possibilities. The interactions between a motor protein (dynein) and cytoskeletal filaments (microtubules) cause a flagellum to

bend. The interaction between v-snares and t-snares causes a vesicle to fuse with the correct target membrane.

2. If the motor is bound to a cargo and the motor can walk along a filament that is fixed in place, this will cause the movement of the cargo when the motor is activated. If the motor is fixed in place and the filament is free to move, this will cause the filament to move when the motor is activated. If both the motor and filament are fixed in place, the activation of the motor will cause the filament to bend.

3. The Golgi apparatus performs three overlapping functions: (1) processing of proteins and lipids, (2) protein sorting, and (3) secretion.

Chapter 5
Concept Checks

Figure 5.3   More double bonds and shorter fatty acyl tails make the membrane more fluid. Changing the cholesterol concentration can also affect fluidity, but that depends on the starting level of cholesterol. If cholesterol was at a level that maximized stability, increasing the cholesterol concentration would probably increase fluidity.

Figure 5.4   The low temperature prevents lateral diffusion of membrane proteins. Therefore, after fusion, all of the mouse proteins would stay on one side of the fused cell, and all of the human proteins would remain on the other.

Figure 5.6   In animal cells, the glycocalyx primarily plays a protective function. It protects proteins in the plasma membrane.

Figure 5.7   Probably not. The hydrophobic tails of both leaflets touch each other, so the heavy metal would probably show a single, thick dark line. Osmium tetroxide shows two parallel lines because it labels the polar head groups, which are separated by the hydrophobic interior of the membrane.

Figure 5.8   Lipids are transferred to the other leaflet of the ER via enzymes called flippases.

Figure 5.9   The most common way for a transmembrane segment to form is that it contains a stretch (about 20) of amino acids that mostly have hydrophobic (nonpolar) side chains.

Figure 5.12   Leucine would more readily cross a membrane because its side chain is nonpolar. The side chain of lysine is positively charged.

Figure 5.13   Although both of these molecules penetrate the bilayer fairly quickly, methanol has a polar —OH group and therefore crosses a bilayer more slowly than methane.

Figure 5.16   Water will move from outside to inside, from the hypotonic medium into the hypertonic medium.

Figure 5.18   Freshwater protists live in a hypotonic environment, which causes them to continually take in water. To avoid bursting, they use contractile vacuoles that regularly expel the water. Note that such protists lack a cell wall.

Figure 5.21   The purpose of gating is to regulate the function of channels.

Figure 5.25   The Na$^+$/K$^+$-ATPase could reach the point where the protein was covalently phosphorylated and Na$^+$ was released on the outside. At that stage, the reaction would stop, because it needs K$^+$ to proceed through the rest of the cycle.

Figure 5.26   The protein coat is needed for the membrane to bud from its site and form a vesicle.

Feature Investigation Questions

1. Most cells allow movement of water across the cell membrane by passive diffusion. However, it was noted that certain cell types had a much higher rate of water movement, indicating that something different was occurring in these cells.

2. The researchers identified water channels by characterizing proteins that are present in red blood cells and kidney cells but not other types of cells. Red blood cells and kidney cells have a faster rate of water movement across the membrane than other cell types. These cells are more likely to have water channels. By identifying proteins that are found in both of these types of cells but not in other cells, the researchers were identifying possible candidate proteins that function as water channels. In addition, CHIP28 had a structure that resembled other known channel proteins.

   Agre and his associates experimentally created multiple copies of the gene that produces the CHIP28 protein and then artificially transcribed the genes to produce many mRNAs. The mRNAs were injected into frog oocytes where they could be translated to make the CHIP28 proteins. After altering the frog oocytes by introducing the CHIP28 mRNAs, they compared the rate of water transport in the altered oocytes versus normal frog oocytes. This procedure allowed them to introduce the candidate protein to a cell type that normally does not have the protein present.

3. After artificially introducing the candidate protein into the frog oocytes, the researchers found that the experimental oocytes took up water at a much faster rate in a hypotonic solution as compared to the control oocytes. The results indicated that the presence of the CHIP28 protein did increase water transport into cells.

Test Yourself

1. c 2. c 3. b 4. d 5. b 6. e 7. d 8. e 9. e 10. c

Conceptual Questions

1. See Figure 5.1 for the type of drawing you should have made. The membrane is considered a mosaic of lipid, protein, and carbohydrate molecules. The membrane exhibits properties that resemble a fluid because lipids and proteins can move relative to each other within the membrane.

2. Integral membrane proteins can contain transmembrane segments that cross the membrane, or they may contain lipid anchors. Peripheral membrane proteins are noncovalently bound to integral membrane proteins or to the polar heads of phospholipids.

3. A solute would touch the tails of the phospholipids in the process of diffusion. For facilitated diffusion and active transport, it would avoid touching the tails because it would be moving through a protein. For exocytosis and endocytosis, the membrane vesicles fuse with the plasma membrane, and the solute never really crosses through a membrane.

Chapter 6

Concept Checks

Figure 6.2 The solution of dissolved Na^+ and Cl^- has more entropy. A salt crystal is very ordered, whereas the ions in solution are much more disordered.

Figure 6.3 The negative ΔG value tells that the direction will proceed from reactants to products. It does not tell us anything about the rate.

Figure 6.4 It speeds up the rate. When the activation energy is lower, it takes less time for reactants to reach a transition state where a chemical reaction can occur. It does not affect the direction of a reaction.

Figure 6.5 The activation energy is lowered during the second step when the substrates undergo induced fit.

Figure 6.6 At a substrate concentration of 0.5 mM, enzyme A would have a higher velocity. Enzyme A would be very near its V_{max}, whereas enzyme B would be well below its V_{max}.

Figure 6.11 The oxidized form is NAD^+.

Figure 6.12 The breakdown of ADP would have an adverse effect. Usually, ATP is made by the covalent attachment of pre-existing phosphate and ADP. If there was insufficient ADP, a cell would have to make more ADP to make ATP. This would take a lot of energy.

Figure 6.13 The metabolic pathway would not be controlled by feedback inhibition. This may result in an overaccumulation of the product of the pathway.

Figure 6.15 Protein degradation eliminates proteins that are worn out, misfolded, or no longer needed by the cell. Such proteins could interfere with normal cell function. In addition, the recycling of amino acids saves the cell energy.

Feature Investigation Questions

1. RNase P has both a protein and RNA subunit. To determine which subunit has catalytic function, it was necessary to purify them individually and then see which one is able to cleave ptRNA.

2. The experimental strategy was to incubate RNase P or subunits of RNase P with ptRNA and then run a gel to determine if ptRNA had been cleaved to a mature tRNA and a 5′ fragment. The control without protein

was to determine if the RNA alone could catalyze the cleavage. The control without RNA was to determine if some other factor in the experiment (e.g., Mg^{2+} or protein) was able to cleave the ptRNA.

3. The critical results occurred when the researchers incubated the purified RNA subunit at high Mg^{2+} concentrations with the ptRNA. Under these conditions, the ptRNA was cleaved. These results indicate that the RNA subunit has catalytic activity. A high Mg^{2+} concentration is needed to keep it catalytically active in the absence of a protein subunit.

Test Yourself

1. d 2. e 3. b 4. d 5. a 6. c 7. b 8. c 9. e 10. a

Conceptual Questions

1. Exergonic reactions are spontaneous. They proceed in a particular direction. An exergonic reaction could be slow or fast. By comparison, an endergonic reaction is not spontaneous. It will not proceed in a particular direction unless free energy is supplied. An endergonic reaction can be fast or slow.

2. During feedback inhibition, the product of a metabolic pathway binds to an allosteric site on an enzyme that acts earlier in the pathway. The product inhibits this enzyme and thereby prevents the overaccumulation of the product.

3. Recycling of amino acids and nucleotides is important because it conserves a great deal of energy. Cells don't have to remake these building blocks, which would require a large amount of energy. Eukaryotes primarily use the proteasome to recycle proteins.

Chapter 7

Concept Checks

Figure 7.1 Most of the NADH is oxidized to make ATP during oxidative phosphorylation.

Figure 7.2 The first phase is named the energy investment phase because some ATP is used up. The second phase is called the cleavage phase because a 6-carbon molecule is broken down into two 3-carbon molecules. The energy liberation phase is so named because NADH and ATP are made.

Figure 7.3 The molecules that donate phosphates are 1,3-bisphosphoglycerate and phosphoenolpyruvate.

Figure 7.5 For each acetyl group that is oxidized, the main products are 2 CO_2, 3 NADH, 1 $FADH_2$, and 1 GTP.

Figure 7.7 It is called cytochrome oxidase because it removes electrons from (oxidizes) cytochrome c. Another possible name would be oxygen reductase because it reduces oxygen.

Figure 7.9 No. The role of the electron transport chain is to make an H^+ electrochemical gradient. It is the H^+ electrochemical gradient that drives ATP synthase. If the H^+ electrochemical is made another way, such as by bacteriorhodopsin, the ATP synthase still makes ATP.

Figure 7.10 The gamma subunit turns clockwise, when viewed from the intermembrane space. The β subunit in the back right is in conformation 3 and the one on the left is in conformation 1.

Figure 7.12 Tumors often become hypoxic and therefore have trouble making ATP via oxidative phosphorylation. Being able to carry out a higher level of glycolysis, which doesn't require oxygen, allows them to make ATP even if they're hypoxic.

Figure 7.13 The advantage is that the cell can use the same enzymes to metabolize different kinds of organic molecules. This saves the cell energy because it is costly to make a lot of different enzymes, which are composed of proteins.

Figure 7.16 Depending on the species, almost any part of a plant may contain anthocyanins. They are most common in fruits, vegetables, and flowers, but they may also be found in stems and even in leaves. For example, the red color found in many leaves in the autumn is due, in part, to anthocyanins.

Figure 7.17 Animals that eat this plant and ingest atropine become very sick and may die. Any animal that eats deadly nightshade and survives would be unlikely to eat it a second time.

Figure 7.19 The advantage of making streptomycin is that it kills other bacterial species in the soil that might compete with *S. griseus* for the same resources.

Feature Investigation Questions

1. The researchers attached an actin filament to the γ subunit of ATP synthase. The actin filament was fluorescently labeled so the researchers could determine if the actin filament moved when viewed under the fluorescence microscope.

2. When functioning in the hydrolysis of ATP, the actin filament was seen to rotate. The actin filament was attached to the γ subunit of ATP synthase. The rotational movement of the filament was the result of the rotational movement of the enzyme. In the control experiment, no ATP was added to stimulate enzyme activity. In the absence of ATP, no movement was observed.

3. No, the observation of counterclockwise rotation is the opposite of what would be expected inside the mitochondria. During the experiment, the enzyme was not functioning in ATP synthesis but instead was running backwards and hydrolyzing ATP.

Test Yourself

1. a 2. d 3. b 4. b 5. b 6. d 7. d 8. b 9. e 10. e

Conceptual Questions

1. The purpose of the electron transport chain is to pump H^+ across the inner mitochondrial membrane to establish a H^+ electrochemical gradient. When the H^+ flows back across the membrane through ATP synthase, ATP is synthesized.

2. The movement of H^+ through the c subunits causes the γ subunit to rotate. As it rotates, it sequentially alters the conformation of the subunits, where ATP is made. This causes: (1) ADP and P_i to bind with moderate affinity, (2) ADP and P_i to bind very tightly such that ATP is made, and (3) ATP to be released.

3. There are two main reasons. First, NADH can be toxic at high levels because it haphazardly donates electrons and may generate free radicals. Second, the oxidation of NADH to NAD^+ is necessary to keep glycolysis running so that ATP can be made.

Chapter 8

Concept Checks

Figure 8.1   Both heterotrophs and autotrophs carry out cellular respiration.

Figure 8.3   The Calvin cycle can occur in the dark as long as there is sufficient CO_2, ATP, and NADPH.

Figure 8.4   Gamma rays have higher energy than radio waves.

Figure 8.5   To drop down to a lower orbital at a lower energy level, an electron could release energy in the form of heat, release energy in the form of light, or transfer energy to another electron by resonance energy transfer.

Figure 8.7   By having different pigment molecules, plants can absorb a wider range of wavelengths of light.

Figure 8.8   ATP and NADPH are made in the stroma. O_2 is made in the thylakoid lumen.

Figure 8.9   Noncyclic electron flow produces equal amounts of ATP and NADPH. However, plants usually need more ATP than NADPH. Cyclic photophosphorylation allows plants to make just ATP and thereby increases the relative amount of ATP.

Figure 8.10   Because these two proteins are homologous, this means that the genes that encode them were derived from the same ancestral gene. Therefore, the amino acid sequences of these two proteins are expected to be very similar, though not identical to each other. Because the amino acid sequence of a protein determines its structure, two proteins with similar amino acid sequences would be expected to have similar structures.

Figure 8.12   The answer is five. It catalyzes the following electron transfers.

1. H_2O to tyrosine (Tyr).
2. Tyr to P680.
3. P680 to pheophytin (Pp).
4. Pp to Q_A.
5. Q_A to Q_B.

Figure 8.13   No. The enhancement effect occurs because the flashes activate both photosystem II and photosystem I. Light at 700 nm is needed to activate P700 in photosystem I.

Figure 8.14   An electron has its highest amount of energy just after it has been boosted by light in PSI.

Figure 8.15   NADPH reduces organic molecules and makes them more able to form C—C and C—H bonds.

Figure 8.18   The arrangement of cells in C_4 plants makes the level of CO_2 high and the level of O_2 low in the bundle sheath cells.

Figure 8.19   When there is plenty of moisture and it is not too hot, C_3 plants are more efficient. However, under hot and dry conditions, C_4 and CAM plants have the advantage because they lose less water and avoid photorespiration.

Feature Investigation Questions

1. The researchers were attempting to determine the biochemical pathway of the process of carbohydrate synthesis in plants. The researchers wanted to identify different molecules produced in plants over time to determine the steps of the biochemical pathway.

2. The purpose for using ^{14}C was to label the different carbon molecules produced during the biochemical pathway. The researchers could "follow" the carbon molecules from CO_2 that were incorporated into the organic molecules during photosynthesis. The radioactive isotope provided the researchers with a method of labeling the different molecules.

   The purpose of the experiment was to determine the steps in the biochemical pathway of photosynthesis. By examining samples from different times after the introduction of the labeled carbon source, the researchers would be able to determine which molecules were produced first and, thus, products of the earlier steps of the pathway versus products of later steps of the pathway.

   The researchers used two-dimensional paper chromatography to separate the different molecules from each other. Afterwards, the different molecules were identified by different chemical methods. The text describes the method of comparing two-dimensional paper chromatography results of unknown molecules to known molecules and identifying the unknown with the known molecule it matched.

3. The researchers were able to determine the biochemical process that plants use to incorporate CO_2 into organic molecules. The researchers were able to identify the biochemical steps and the molecules produced at these steps in what is now called the Calvin cycle.

Test Yourself

1. c 2. c 3. c 4. a 5. b 6. b 7. c 8. b 9. e 10. c

Conceptual Questions

1. The two stages of photosynthesis are the light reactions and the Calvin cycle. The key products of the light reactions are ATP, NADPH, and O_2. The key product of the Calvin cycle is carbohydrate. The initial product is G3P, which is used to make sugars and other organic molecules.

2. NADPH is used during the reduction phase of the Calvin cycle. It donates its electrons to 1,3 BPG.

3. After one pigment molecule absorbs energy, resonance energy transfer allows that energy to be transferred among many pigment molecules, eventually reaching the reaction center. If resonance energy transfer did not occur, the light energy would have to be absorbed directly by the pigment in the reaction center, which is either P680 or P700. In contrast, the light-harvesting complex is composed of many pigment molecules of different types (chlorophylls and carotenoids), which can absorb light at different wavelengths and transfer that energy to the reaction center. The light-harvesting complex makes it easier for plants to absorb light energy.

Chapter 9

Concept Checks

Figure 9.1   It is glucose.

Figure 9.2   Auxin causes cells to elongate. Therefore, cells on the nonilluminated side grow faster, causing the plant to bend toward the light.

Figure 9.3   Endocrine signals are more likely to exist for a longer period of time. This is necessary because endocrine signals called hormones travel relatively long distances to reach their target cells. Therefore, the hormone must exist long enough to reach its target cells.

Figure 9.4 The effect of a signaling molecule is to cause a cellular response. Most signaling molecules do not enter the cell. Therefore, to exert an effect, they must alter the conformation of a receptor protein, which, in turn, stimulates an intracellular signal transduction pathway that leads to a cellular response.

Figure 9.6 Phosphorylation of a protein via a kinase involves ATP hydrolysis, which is an exergonic reaction. The energy from this reaction usually alters the conformation of the phosphorylated protein and thereby influences its function. Phosphorylation is used to regulate protein function.

Figure 9.7 The α subunit has to hydrolyze its GTP to GDP. This changes the conformation of the α subunit so that it can reassociate with the β/γ subunits.

Figure 9.10 The GTP-bound form of Ras is active and promotes cell division. To turn the pathway off, Ras hydrolyzes GTP to GDP. If this cannot occur due to a mutation, the pathway will be continuously on, and uncontrolled cell division will result.

Figure 9.12 The signal transduction pathway begins with the G protein and ends with protein kinase A being activated. The cellular response involves the phosphorylation of target proteins. The phosphorylation of target proteins will change their function in some way, which is how the cell is responding.

Figure 9.13 Depending on the protein involved, phosphorylation can activate or inhibit protein function. Phosphorylation of phosphorylase kinase and glycogen phosphorylase activates their function, whereas it inhibits glycogen synthase.

Figure 9.14 Signal amplification allows a single signaling molecule to affect many proteins within a cell and thereby amplify a cellular response.

Figure 9.20 The initiator caspase is part of the death-inducing signaling complex. It is directly activated when a cell receives a death signal. The initiator caspase then activates the executioner caspases, which degrade various cellular proteins and thereby cause the destruction of the cell.

Feature Investigation Questions

1. Compared to control rats, those injected with prednisolone alone would be expected to have a decrease in the number of cells because it suppresses ACTH synthesis. Therefore, apoptosis would be higher. By comparison, prednisolone + ACTH would have a normal number of cells because the addition of ACTH would compensate for effects of prednisolone. ACTH alone would be expected to have a greater number of cells; apoptosis would be inhibited.

2. Yes, when injected with ACTH, prednisolone probably inhibited the ability of the rats to make their own ACTH. Even so, they were given ACTH by injection, so they didn't need to make their own ACTH to prevent apoptosis.

3. The lowest level of apoptosis would occur in the ACTH alone group, because they could make their own ACTH plus they were given ACTH. With such high levels of ACTH, they probably had the lowest level of apoptosis; it was already known that ACTH promotes cell division.

Test Yourself

1. d   2. c   3. d   4. e   5. a   6. d   7. e   8. e   9. e   10. b

Conceptual Questions

1. Cells need to respond to a changing environment, and cells need to communicate with each other.

2. In the first stage, a signaling molecule binds to a receptor, causing receptor activation. In the second stage, one type of signal is transduced or converted to a different signal inside the cell. In the third stage, the cell responds in some way to the signal, possibly by altering the activity of enzymes, structural proteins, or transcription factors. When the estrogen receptor is activated, the second stage, signal transduction, is not needed because the estrogen receptor is an intracellular receptor that directly activates the transcription of genes to elicit a cellular response.

3. If apoptosis did not occur, embryonic development would not occur properly, and adults would not maintain a correct number of cells. Also, cells that are worn out, infected by viruses or intracellular bacteria, or have the potential to cause cancer, would not be eliminated. In mammals, the immune system would not function properly because apoptosis is needed to eliminate B and T cells that are ineffective or potentially damaging to the body.

Chapter 10
Concept Checks

Figure 10.1 The four functions of the ECM in animals are strength, structural support, organization, and cell signaling.

Figure 10.2 The extension sequences of procollagen prevent fibers from forming intracellularly.

Figure 10.3 The proteins would become more linear, and the fiber would come apart.

Figure 10.4 GAGs are highly negatively charged molecules that tend to attract positively charged ions and water. Their high water content gives GAGs a gel-like character, which makes them difficult to compress.

Figure 10.5 Because the secondary cell wall is usually rigid, it prevents cell growth. If it were made too soon, it might prevent a cell from attaining its proper size.

Figure 10.7 Adherens junctions and desmosomes are cell-to-cell junctions, whereas hemidesomosomes and focal adhesions are cell-to-ECM junctions.

Figure 10.9 Tight junctions in your skin prevent harmful things like toxins and viruses from entering your body. They also prevent materials like nutrients from leaking out of your body.

Figure 10.10 As opposed to the results shown in Figure 10.10, the dye would be in the side of the cell layer facing the intestinal tract. You would see dye up to the tight junction on this side of the cells, but not on the side of the tight junction facing the blood.

Figure 10.13 Middle lamellae are similar to anchoring junctions and desmosomes in that they all function in cell-to-cell adhesion. However, their structures are quite different. Middle lamellae are composed primarily of carbohydrates that involve linkages between negatively charged carbohydrates and divalent cations. By comparison, anchoring junctions and desmosomes hold cells together via proteins such as cadherins and integrins.

Figure 10.15 Connective tissue would have the most extensive ECM.

Figure 10.16 Dermal tissue would be found on the surfaces of leaves, stems, and roots.

Figure 10.19 Simple epithelium and epidermis are one cell layer thick, whereas stratified epithelium and periderm are several cell layers thick.

Figure 10.21 Both ground tissue in plants and connective tissue in animals are important in supporting the organism. These tissues have a large amount of ECM that provides structural support.

Feature Investigation Questions

1. The purpose of this study was to determine the sizes of molecules that can move through gap junctions from one cell to another.

2. The researchers used fluorescent dyes to visibly monitor the movement of material from one cell to an adjacent cell through the gap junctions. First, single layers of rat liver cells were cultured. Next, fluorescent dyes with molecules of various masses were injected into particular cells. The researchers then used fluorescence microscopy to determine whether or not the dyes were transferred from one cell to the next.

3. The researchers found that molecules of masses less than 1,000 daltons could pass through the gap junction channels. Molecules of masses larger than 1,000 daltons could not pass through the gap junctions. Further experimentation revealed variation in gap junction channel size of different cell types. However, the upper limit of the gap junction channel size was determined to usually be around 1,000 daltons.

Test Yourself

1. e   2. c   3. b   4. e   5. e   6. d   7. e   8. d   9. e   10. a

Conceptual Questions

1. The primary cell wall is synthesized first between the two newly made daughter cells. It is relatively thin and allows cells to expand and grow. The secondary cell wall is made in layers by the deposition of cellulose fibrils and other components. In many cell types, it is relatively thick.

2. Cadherins and integrins are both membrane proteins that function as cell adhesion molecules. They also can function in cell signaling. Cadherins bind one cell to another cell, whereas integrins bind a cell to

the extracellular matrix. Cadherins require calcium ions to function, but integrins do not.

3. To create tissues and organs, cells must undergo six basic processes that influence their shape, arrangement, and number:

- Cell division—Many cells are needed to make tissues and organs. These arise via cell division.
- Cell growth—After a cell divides, it needs to grow to reach its correct size.
- Differentiation—Due to the expression of different genes, cells can differentiate into specialized cells based on what cells are needed at that particular time in the tissue or organ.
- Migration—During embryonic development in animals, cells migrate to their appropriate position within the body. This event doesn't occur in plants.
- Apoptosis—Apoptosis is also known as programmed cell death. During development, organs and tissues need to be shaped to form their correct structure. For example, during the development of fingers and toes, the cells between the digits must be removed. This cell removal is done through the process of apoptosis.
- Cell connections—For tissues and organs to work properly, the cells must be held together in a specific arrangement. This is achieved through the different types of cell-to-cell connections.

Chapter 11

Concept Checks

Figure 11.1 Yes, they would be resistant to killing. These bacteria are derived from the type R bacteria, so they would have this trait. In addition, they have become type S because they obtained genetic material from the dead type S bacteria.

Figure 11.4 ^{35}S was used to label phage proteins, whereas ^{32}P was used to label phage DNA.

Figure 11.6 Cytosine is found in both DNA and RNA.

Figure 11.7 The phosphate is attached to the number 5′ carbon in a single nucleotide. In a DNA strand, it is attached to both the 5′ carbon and 3′ carbon.

Figure 11.8 A phosphoester bond is a single covalent bond between a phosphorus atom and an oxygen atom. A phosphodiester linkage involves two phosphoester bonds. This linkage occurs along the backbone of DNA and RNA strands.

Figure 11.10 Because it is antiparallel and obeys the AT/GC rule, it would be 3′–CTAAGCAAG–5′.

Figure 11.13 It would be 1/8 half-heavy and 7/8 light.

Figure 11.17 The oxygen in a new phosphoester bond comes from the sugar.

Figure 11.19 The lagging strand is made discontinuously in the direction opposite to the movement of the replication fork.

Figure 11.20 When primase is synthesizing a primer in the lagging strand, it moves from left to right in this figure. After it is done making a primer, it needs to hop to the opening of the replication fork to make a new primer. This movement is from right to left in this figure.

Figure 11.22 Telomerase uses a short strand of RNA as template to make the DNA repeat sequence.

Figure 11.25 Proteins hold the bottoms of the loops in place.

Figure 11.26 Proteins that compact the radial loop domains are primarily responsible for the X shape.

Feature Investigation Questions

1. Previous studies had indicated that mixing different strains could lead to transformation or the changing of a strain into a different one. Griffith had shown that mixing heat-killed type S with living type R would result in the transformation of the type R to type S. Though mutations could cause the changing of the identity of certain strains, the type R to type S transformation was not due to mutation but was more likely due to the transmission of a biochemical substance between the two strains. Griffith recognized this and referred to the biochemical substance as the "transformation principle." If Avery, MacCleod, and McCarty could determine the biochemical identity of this "transformation principle," they could identify the genetic material for this organism.

2. A DNA extract contains DNA that has been purified from a sample of cells.

3. The researchers could not verify that the DNA extract was completely pure and did not have small amounts of contaminating molecules, such as proteins and RNA. The researchers were able to treat the extract with enzymes to remove proteins (using protease), RNA (using RNase), or DNA (using DNase). Removing the proteins or RNA did not alter the transformation of the type R to type S strains. Only the enzymatic removal of DNA disrupted the transformation, indicating that DNA is the genetic material.

Test Yourself

1. a 2. b 3. d 4. d 5. b 6. c 7. b 8. d 9. d 10. c

Conceptual Questions

1. The genetic material must contain the information necessary to construct an entire organism. The genetic material must be accurately copied and transmitted from parent to offspring and from cell to cell during cell division in multicellular organisms. The genetic material must contain variation that can account for the known variation within each species and among different species.

   Griffith discovered something called the transformation principle, and his experiments showed the existence of biochemical genetic information. In addition, he showed that this genetic information can move from one individual to another of the same species. In his experiments, Griffith took heat-killed type S bacteria and mixed them with living type R bacteria and injected them into a live mouse, which died after the injection. By themselves, these two strains would not kill the mouse, but when they were put together, the genetic information from the heat-killed type S bacteria was transferred into the living type R bacteria, thus transforming the type R bacteria into type S.

2. In the case of the Hershey and Chase experiment, a radioactive isotope of sulfur was used to label the protein in the viral protein coat. The DNA was labeled using a radioactive isotope of phosphorus. This was an ideal way of labeling the different components, because sulfur is found in proteins but not DNA, and phosphorus is found in DNA but not proteins. By labeling the two candidate molecules with the radioactive isotopes, Hershey and Chase could determine the genetic material by seeing which isotope entered the bacterial cells.

3. Two long chains of nucleotides are coiled around a central axis forming a helix. The two chains run in opposite directions or are antiparallel. Hydrogen bonds between the bases in opposite strands stabilize the structure. Adenine always pairs with thymine, and cytosine always pairs with guanine. The width of the double helix is relatively constant. One complete turn of the double helix is composed of 10 base pairs.

Chapter 12

Concept Checks

Figure 12.1 A person with two defective copies of phenylalanine hydroxylase would have phenylketonuria.

Figure 12.2 The ability to convert ornithine into citrulline is missing.

Figure 12.3 The usual direction of flow of genetic information is from DNA to RNA to protein, though exceptions occur.

Figure 12.4 If a terminator was removed, transcription would occur beyond the normal stopping point. Eventually, RNA polymerase would encounter a terminator from an adjacent gene, and transcription would end.

Figure 12.10 The ends of structural genes do not have a poly T region that acts as a template for the synthesis of a poly A tail. Instead, the poly A tail is added after the pre-mRNA is made by an enzyme that attaches many adenine nucleotides in a row.

Figure 12.11 A structural gene would still be transcribed into RNA if the start codon was missing. However, it would not be translated properly into a polypeptide.

Figure 12.12 It would bind to a 5′–UGG–3′ codon, and it would carry tryptophan.

Figure 12.14 The function of the anticodon in tRNA is to recognize a codon in an mRNA.

Figure 12.15 The attachment of an amino acid to a tRNA is an endergonic reaction. ATP provides the energy to catalyze this reaction.

Figure 12.17   Each mammal is closely related to the other mammals, and *E. coli* and *S. marcescens* are also closely related. The mammals are relatively distantly related to the bacterial species.

Figure 12.19   A region near the 5′ end of the mRNA is complementary to a region of rRNA in the small subunit. These complementary regions hydrogen-bond with each other to promote the binding of the mRNA to the small ribosomal subunit.

Feature Investigation Questions

1. A triplet mimics mRNA because it can cause a specific tRNA to bind to the ribosome. This was useful to Nirenberg and Leder because it allowed them to correlate the binding of a tRNA carrying a specific amino acid with a triplet sequence.

2. The researchers were attempting to match codons with appropriate amino acids. By labeling one amino acid in each of the 20 tubes for each codon, the researchers were able to identify the correct relationship by detecting which tube resulted in radioactivity on the filter.

3. The AUG triplet would have shown radioactivity in the methionine test tube. Even though AUG acts as the start codon, it also codes for the amino acid methionine. The other three codons act as stop codons and do not code for an amino acid. In these cases, the researchers would not have found radioactivity trapped on filters.

Test Yourself

1. b   2. d   3. d   4. b   5. c   6. e   7. d   8. d   9. d   10. b

Conceptual Questions

1. Beadle and Tatum had the insight from their studies that a single gene controlled the synthesis of a single enzyme. In later years, it became apparent that genes code for all proteins and that some proteins consist of more than one polypeptide chain. So the modern statement is one gene codes for each polypeptide.

   Confirmation of their hypothesis came from studies involving arginine biosynthesis. Biochemists had already established that particular enzymes are involved in a pathway to produce arginine. Intermediates in this pathway are ornithine and citrulline. Mutants in single genes disrupted the ability of cells to catalyze just one reaction in this pathway, thereby suggesting that a single gene encodes a single enzyme.

2. Each of these 20 enzymes catalyzes the attachment of a specific amino acid to a specific tRNA molecule.

3. It would be complementary to the 5′–CCA–3′ codon. This tRNA should carry proline.

Chapter 13

Concept Checks

Figure 13.1   The advantage is energy savings. The bacterium saves energy by making these proteins only when they are needed for lactose uptake and breakdown.

Figure 13.2   Gene regulation causes each cell type to express its own unique set of proteins, which, in turn, are largely responsible for the morphologies and functions of cells.

Figure 13.3   The embryonic and fetal forms of hemoglobin have a higher affinity for oxygen. This allows the embryo and fetus to obtain oxygen from their mother's bloodstream.

Figure 13.6   The *lacZ*, *lacY*, and *lacA* genes are under the control of the *lac* promoter.

Figure 13.7   Negative control refers to the action of a repressor protein, which inhibits transcription when it binds to the DNA. Inducible refers to the action of a small effector molecule. When it is present, it promotes transcription.

Figure 13.11   In this case, the repressor keeps the *lac* operon turned off unless lactose is present in the environment. The activator allows the bacterium to choose between glucose and lactose.

Figure 13.12   Both proteins are similar in that they repress transcription. They prevent RNA polymerase from transcribing the operons. They are different with regard to the effects of their small effector molecules. For the lac repressor, the binding of allolactose causes a conformational change that

prevents the repressor from binding to its operator site. In contrast, the binding of tryptophan to the trp repressor allows it to bind to its operator site. Another difference is that the lac repressor binds to the DNA sequence found in the *lac* operator site, whereas the trp repressor recognizes a different DNA sequence that is found in the *trp* operator site.

Figure 13.16   When an activator interacts with mediator, it causes RNA polymerase to proceed to the elongation phase of transcription.

Figure 13.17   The common state of euchromatin, the 30-nm fiber, is too compact to be transcribed. This level of compaction must be loosened up so that transcription can occur.

Figure 13.18   Some histone modifications may promote a loosening of chromatin structure, whereas others cause the chromatin to become more compact.

Figure 13.19   The gene could not be activated in the presence of glucocorticoid hormone.

Figure 13.20   The advantage of alternative splicing is that it allows a single gene to encode two or more polypeptides. This enables organisms to have smaller genomes, which is more efficient and easier to package into a cell.

Figure 13.22   When iron levels rise in the cell, the iron binds to IRP and removes it from the mRNA that encodes ferritin. This results in the rapid translation of ferritin protein, which can store excess iron. Unfortunately, ferritin storage does have limits, so iron poisoning can occur if too much is ingested.

Feature Investigation Questions

1. The first observation was the identification of rare bacterial strains that had constitutive expression of the *lac* operon. Normally, the genes are expressed only when lactose is present. These mutant strains expressed the genes all the time. The researchers also observed that some of these strains had mutations in the *lacI* gene. These two observations were key to the development of hypotheses explaining the relationship between the *lacI* gene and the regulation of the *lac* operon.

2. The correct hypothesis is that the *lacI* gene encodes a repressor protein that inhibits the operon.

3. The researchers used an F′ factor to introduce the wild-type *lacI* gene into the cell. In this case, the cells that contained the F′ factor had both a mutant copy of the gene and a normal copy of the gene. By creating a merozygote with an F′ factor with a normal copy of the *lacI* gene, regulation of the *lac* operon was restored. The researchers concluded that the normal *lacI* gene produced adequate amounts of a diffusible protein that could interact with the operator on the chromosomal DNA as well as the F′ factor DNA and regulate transcription.

Test Yourself

1. d   2. b   3. c   4. c   5. c   6. d   7. c   8. c   9. d   10. c

Conceptual Questions

1. In an inducible operon, the presence of a small effector molecule causes transcription to occur. In repressible operons, a small effector molecule inhibits transcription. The effects of these small molecules are mediated through regulatory proteins that bind to the DNA. Repressible operons usually encode anabolic enzymes, and inducible operons encode catabolic enzymes.

2. a. regulatory protein; b. small effector molecule; c. segment of DNA; d. small effector molecule; and e. regulatory protein

3. The addition of methyl groups to CpG islands, especially near the promoters of eukaryotic genes, may prevent an activator from binding to an enhancer element or may convert chromatin from an open to a closed conformation.

Chapter 14

Concept Checks

Figure 14.1   At neutral pH, glutamic acid is negatively charged. Perhaps the negative charges repel each other and prevent hemoglobin proteins from aggregating into fiber-like structures.

Figure 14.3   Only germ-line cells give rise to gametes (sperm or egg cells). A somatic cell cannot give rise to a gamete and therefore cannot be passed to offspring.

Figure 14.4 This is a mutation in a somatic cell, so it cannot be transmitted to an offspring.

Figure 14.6 A thymine dimer is harmful because it can cause errors in DNA replication.

Figure 14.7 If we divide 44 by 2 million, the rate is 2.2×10^{-5}.

Figure 14.8 UvrC and UvrD are responsible for removing the damaged DNA. UvrC makes cuts on both sides of the damage, and then UvrD removes the damaged region.

Figure 14.9 The sun has UV rays and other harmful radiation that could damage the DNA. This person has a defect in the nucleotide excision repair pathway. Therefore, his DNA is more likely to suffer mutations, which cause growths on the skin.

Figure 14.11 Growth factors turn on a signaling pathway that ultimately leads to cell division.

Figure 14.14 The type of cancer associated with this fusion is leukemia, which is a cancer of blood cells. The gene fusion produces a chimeric gene that is expressed in blood cells because it has the *bcr* promoter. The abnormal fusion protein promotes cancer in these cells.

Figure 14.15 Checkpoints prevent cell division if a genetic abnormality is detected. This helps to properly maintain the genome, thereby minimizing the possibility that a cell harboring a mutation will divide to produce two daughter cells.

Figure 14.16 Cancer would not occur if both copies of the *Rb* gene and both copies of the *E2F* gene were rendered inactive due to mutations. An active copy of the *E2F* gene is needed to promote cell division.

Figure 14.18 Translocations could create oncogenes as in the *bcr/abl* fusion example found in the Philadelphia chromosome. Chromosome loss could result in the loss of tumor suppressor genes. Increases in chromosome number could contribute to cancer if the extra chromosomes carry proto-oncogenes.

Feature Investigation Questions

1. Some individuals believed that heritable traits may be altered by physiological events. This suggests that mutations may be stimulated by certain needs of the organism. Others believed that mutations were random. If a mutation had a beneficial effect that improved survival and/or reproductive success, these mutations would be maintained in the population through natural selection.

2. The Lederbergs were testing the hypothesis that mutations are random events. By subjecting the bacteria to some type of environmental stress, the bacteriophage, the researchers would be able to see if the stress induced mutations or if mutations occurred randomly.

3. When looking at the number and location of colonies that were resistant to viral infection, the pattern was consistent among the secondary plates. This indicates that the mutation that allowed the colonies to be resistant to viral infection occurred on the master plate. The secondary plates introduced the selective agent that allowed the resistant bacteria colonies to survive and reproduce while the other colonies were destroyed. Thus, mutations occurred randomly in the absence of any selective agent.

Test Yourself

1. c 2. d 3. d 4. e 5. d 6. b 7. b 8. b 9. c 10. e

Conceptual Questions

1. Random mutations are more likely to be harmful than beneficial. The genes within each species have evolved to work properly. They have functional promoters, coding sequences, terminators, and so on, that allow the genes to be expressed. Mutations are more likely to disrupt these sequences. For example, mutations within the coding sequence may produce early stop codons, frameshift mutations, and missense mutations that result in a nonfunctional polypeptide. On rare occasions, however, mutations are beneficial; they may produce a gene that is expressed better than the original gene or produce a polypeptide that functions better.

2. A spontaneous mutation originates within a living cell. It may be due to spontaneous changes in nucleotide structure, errors in DNA replication, or products of normal metabolism that may alter the structure of DNA. The causes of induced mutations originate from outside the cell. They may be physical agents, such as UV light or X-rays, or chemicals that

act as mutagens. Both spontaneous and induced mutations may cause a harmful phenotype such as a cancer. In many cases, induced mutations are avoidable if the individual can prevent exposure to the environmental agent that acts as a mutagen.

3. The effects of mutations are cumulative. If one mutation occurs in a cell, this mutation will be passed to the daughter cells. If a mutation occurs in the daughter cell, now there will be two mutations. These two mutations will be passed to the next generation of daughter cells, and so forth. The accumulation of many mutations eventually kills the cells. That is why mutagens are more effective at killing dividing cells compared to nondividing cells. It is because the number of mutations accumulates to a lethal level.

There are two main side effects to this treatment. First, some normal (noncancerous) cells of the body, particularly skin cells and intestinal cells, are actively dividing. These cells are also killed by chemotherapy and radiation therapy. Second, it is possible that the therapy may produce mutations that will cause noncancerous cells to become cancerous. For these reasons, there is a maximal dose of chemotherapy or radiation therapy that is recommended.

Chapter 15
Concept Checks

Figure 15.1 Chromosomes are readily seen when they are compacted in a dividing cell. By adding such a drug, you increase the percentage of cells that are actively dividing.

Figure 15.2 Interphase consists of the G_1, S, and G_2 phases of the cell cycle.

Figure 15.3 You would find 92 chromatids in a human cell at metaphase. These are attached to form 46 pairs of sister chromatids.

Figure 15.4 Checkpoint proteins monitor conditions and prevent cell division if the conditions are not appropriate. They also sense if there is DNA damage, incompletely replicated chromosomes, and chromosomes not attached to the spindle. If such abnormalities are detected, cell division is halted. This maintains the integrity of the genome.

Figure 15.7 As shown in the inset, each object is a pair of sister chromatids.

Figure 15.8 The astral microtubules, which extend away from the chromosomes, are important for positioning the spindle apparatus within the cell. The polar microtubules project into the region between the two poles. Polar microtubules that overlap with each other play a role in the separation of the two poles. Kinetochore microtubules are attached to kinetochores at the centromeres and are needed to sort the chromosomes.

Figure 15.9 The mother cell (in G_1 phase) and the daughter cells have the same chromosome composition. They are genetically identical.

Figure 15.10 Micrographs showing cytokinesis in animal and plant cells. Cytokinesis in both animal and plant cells separates a mother cell into two daughter cells. In animal cells, cytokinesis involves the formation of a cleavage furrow, which constricts like a drawstring to separate the cells. In plants, the two daughter cells are separated by the formation of a cell plate, which forms a cell wall between the two daughter cells.

Figure 15.14 The mother cell is diploid with two sets of chromosomes, whereas the four resulting cells are haploid with one set of chromosomes.

Figure 15.15 The reason for meiosis in animals is to produce gametes. These gametes combine during fertilization to produce a diploid organism. Following fertilization, the purpose of mitosis is to produce a multicellular organism.

Figure 15.17 Inversions and the translocations shown here do not affect the total amount of genetic material.

Feature Investigation Questions

1. Researchers had demonstrated that the binding of progesterone to receptors in oocytes caused the cells to progress from the G_2 phase of the cell cycle to mitosis. It appeared that progesterone acted as a signaling molecule for the progression through the cell cycle.

2. The researchers proposed that progesterone acted as a signaling molecule that led to the synthesis of molecules that cause the cell to progress through the cell cycle. These changes led to the maturation of the oocyte.

To test their hypothesis, donor eggs were exposed to progesterone for either 2 or 12 hours. Control donor oocytes were not exposed to progesterone. Cytosol from each treatment was then transferred to recipient oocytes. The researchers recorded whether or not the recipient oocytes underwent maturation.

3. The oocytes that were exposed to the progesterone for only 2 hours did not induce maturation in the recipient oocytes, whereas the oocytes that were exposed to progesterone for 12 hours did induce maturation in the recipient oocytes. The researchers suggested that a time span greater than 2 hours is needed to accumulate the proteins that are necessary to promote maturation.

Test Yourself

1. b   2. e   3. b   4. e   5. c   6. a   7. c   8. d   9. b   10. c

Conceptual Questions

1. In diploid species, chromosomes are present in pairs, one from each parent, and contain similar gene arrangements. Such chromosomes are homologous. When DNA is replicated, two identical copies are created, and these are sister chromatids.

2. There are four copies. A karyotype shows homologous chromosomes that come in pairs. Each member of the pair has replicated to form a pair of sister chromatids. Therefore, four copies of each gene are present. See the inset to Figure 15.1.

3. Two cells would have five chromosomes (one copy of each chromosome plus an extra copy of chromosome 3), and the other two cells would have only three chromosomes because they would be missing a chromosome (chromosome 3).

Chapter 16

Concept Checks

Figure 16.2   Having blue eyes is a variant (also called a trait). A character is a more general term, which in this case would refer to eye color.

Figure 16.4   In this procedure, stamens are removed from the purple flower to prevent self-fertilization.

Figure 16.5   The reason why offspring of the F_1 generation exhibit only one variant of each character is because one trait is dominant over the other.

Figure 16.6   The ratio of alleles (T to t) is 1:1. The reason why the phenotypic ratio is 3:1 is because T is dominant to t.

Figure 16.7   It was Pp. To produce white offspring, which are pp, the original plant had to have at least one copy of the p allele. Because it had purple flowers, it also had to have one copy of the P allele. So, its genotype must be Pp.

Figure 16.8   If the linked hypothesis had been correct, the ratio would have been 3 round, yellow : 1 wrinkled, green.

Figure 16.10   The word segregate means that alleles are separated into different places. In this case, the alleles are segregated into different cells during the process of meiosis. Alleles are located on chromosomes. A diploid cell has two copies of each allele. During meiosis, a diploid cell divides twice to produce four haploid cells that each have only one copy of an allele.

Figure 16.11   There would be four possible ways of aligning the chromosomes, and eight different types of gametes (ABC, abc, ABc, abC, Abc, aBC, AbC, aBc) could be produced.

Figure 16.12   No. If two parents are affected with the disease, they must be homozygous for the mutant allele if it's recessive. Two homozygous parents would have to produce all affected offspring. If they don't, then the inheritance pattern is not recessive.

Figure 16.13   When all affected offspring have at least one affected parent, this suggests a dominant pattern of inheritance.

Figure 16.14   The person would be a female. In mammals, the presence of the Y chromosome causes maleness. Therefore, without a Y chromosome, a person with a single X chromosome would develop into a female.

Figure 16.18   The mother is Bb, and the father is bb. Both of these parents would not be bald. The son is Bb and is bald because B is dominant in males.

Figure 16.19   No. You need a genetically homogenous population to study the norm of reaction. A wild population of squirrels is not genetically homogenous, so it could not be used.

Feature Investigation Questions

1. Morgan was testing the hypothesis of use and disuse. This hypothesis suggests that if a structure is not used, over time, it will diminish and/or disappear. In Morgan's experiments, originally he was testing to see if flies reared in the dark would lose some level of eye development.

2. When the F_1 individuals were crossed, only male F_2 offspring expressed the white eye color. At this time, Morgan was aware of sex chromosome differences between male and female flies. He realized that since males only possess one copy of X-linked genes, this would explain why only F_2 males exhibited the recessive trait.

3. In a cross between a white-eyed male and a female that is heterozygous for the white and red alleles, 1/2 of the female offspring would have white eyes. Also, a cross between a white-eyed male and a white-eyed female would yield all offspring with white eyes.

Test Yourself

1. c   2. b Mendel's law of segregation refers to the separation of the two alleles into separate cells. Meiosis is the cellular division process that produces haploid cells. During the first meiotic division, a diploid cell divides to produce haploid cells. This is the phase in which the two alleles segregate, or separate, from each other.   3. d   4. c
5. e   6. d   7. d   8. d   9. d   10. c

Conceptual Questions

1. If two affected parents had an unaffected offspring, that would rule out recessive inheritance. If two unaffected parents had an affected offspring, that would rule out dominant inheritance. However, it should be noted that this answer assumes that no new mutations are happening. In rare cases, a new mutation could cause or alter these results. For recessive inheritance, two affected parents could have an unaffected offspring if the offspring had a new mutation that converted the recessive allele to the dominant allele. Similarly for dominant inheritance, two unaffected parents could have an affected offspring if the offspring inherited a new mutation that was dominant. Note: New mutations are expected to be relatively rare.

2. The individual probabilities are as follows: $AA = 0.25$; $bb = 0.5$; $CC = 0.5$; and $Dd = 0.5$. These are determined by making small Punnett squares. We use the product rule to calculate the probability of $AAbbCCDd = (0.25)(0.5)(0.5)(0.5) = 0.03125$, or 3.125%.

3. Because males are hemizygous, they may display a recessive trait that is masked by the dominant allele in a heterozygous female. It only takes one recessive allele for the male to display the trait.

Chapter 17

Concept Checks

Figure 17.1   The recessive allele is the result of a loss-of-function mutation. In a $Ccpp$ individual, the enzyme encoded by the P gene is defective.

Figure 17.2   The pigmentation phenotype displays a continuous distribution because it is polygenic and because environmental variation has a big impact. This creates genotypes that result in a range of phenotypes that overlap.

Figure 17.4   Crossing over occurred during oogenesis in the heterozygous female of the F_1 generation to produce the recombinant offspring of the F_2 generation.

Figure 17.5   One strategy would be to begin with two true-breeding parental strains: $alal\ dpdp$ and $al^+al^+\ dp^+dp^+$ and cross them together to get F_1 heterozygotes $al^+al\ dp^+dp$. Then testcross female F_1 heterozygotes to male $alal\ dpdp$ homozygotes. In the F_2 generation, the recombinant offspring would be $al^+al\ dpdp$ and $alal\ dp^+dp$, and the nonrecombinants would be $al^+al\ dp^+dp$, and $alal\ dpdp$.

Figure 17.6   The evolutionary origin of these organelles is an ancient endosymbiotic relationship. Mitochondria are derived from purple bacteria, and chloroplasts are derived from cyanobacteria.

Figure 17.7   The gene is located in the chloroplast DNA. In this species, chloroplasts are transmitted from parent to offspring via eggs but not via sperm.

Figure 17.9   The Barr body is much more compact than the other X chromosome in the cell. This compaction prevents most of the genes on the Barr body from being expressed.

Figure 17.10 On rare occasions, a cat can be XXY. Because it carries a Y chromosome, it will be a male. If it is heterozygous for the orange and black coat-color alleles that are X-linked, it will be a calico male.

Figure 17.11 Only the male genes are transcriptionally active in the offspring. In this case, half the offspring would be normal, and half would be dwarf. The dwarf offspring would have inherited the *Igf-2* allele from their father.

Figure 17.14 Its mother must be *dd*. The genotype of the mother determines the phenotype of the offspring. A *dd* mother produces offspring that coil to the left.

Figure 17.15 A haploid egg that carries a *D* allele cannot produce an offspring that coils to the left. Such an egg would have to come from a mother that carried at least one *D* allele. The nurse cells would supply this egg with the *D* gene product, and the offspring would coil to the right because the *D* gene product is dominant. On the other hand, an egg that carries the *d* allele could possibly produce an offspring that coils to the right. Such an egg could be from a mother that is *Dd*. The nurse cells from such a mother would supply the egg with both gene products. Because the *D* gene product is dominant, such an offspring would coil to the right.

Feature Investigation Questions

1. Bateson and Punnett were testing the hypothesis that the gene pairs that influence flower color and pollen shape would assort independently of each other. The two traits were expected to show a pattern consistent with Mendel's law of independent assortment.

2. The expected results were a phenotypic ratio of 9:3:3:1. The researchers expected 9/16 of the offspring would have purple flowers and long pollen, 3/16 of the offspring would have purple flowers and round pollen, 3/16 of the offspring would have red flowers and long pollen, and 1/16 of the offspring would have red flowers and round pollen.

3. Though all four of the expected phenotype groups were seen, they were not in the predicted ratio of 9:3:3:1. The number of individuals with the phenotypes found in the parental generation (purple flowers and long pollen or red flowers and round pollen) was much higher than expected. Bateson and Punnett suggested that the gene controlling flower color was somehow coupled with the gene that controls pollen shape. This would explain why these traits did not always assort independently.

Test Yourself

1. d 2. c 3. d 4. c 5. d 6. b 7. a 8. d 9. a 10. a

Conceptual Questions

1. The correct answer is 6%. Individuals that are *Aabb* are recombinants that occurred as a result of crossing over. Because the genes are 12 map units apart, we expect that 12% will be recombinants. However, there are two types of recombinants: *Aabb* and *aaBb,* which would occur in equal amounts. Therefore, we expect 6% to be *Aabb.*

2. This may happen due to X inactivation. As a matter of bad luck, a female embryo may preferentially inactivate the X chromosome carrying the normal allele in the embryonic cells that will give rise to the eyes. If the X chromosome carrying the color-blind allele is preferentially expressed, one or both eyes may show color blindness to some degree.

3. The mother is *Nn*; the father and offspring could be any genotype. The mother has a small head, so her mother must have been *nn*. Because the mother with the small head also had offspring with normal heads, she also must carry the *N* allele. Therefore, the mother must be *Nn*.

Chapter 18

Concept Checks

Figure 18.3 Viruses vary with regard to their structure and their genomes. Genome variation is described in Table 18.1.

Figure 18.5 The advantage of the lytic cycle is that the virus can make many copies of itself and proliferate. However, sometimes the growth conditions may not be favorable to make new viruses. The advantage of the lysogenic cycle is that the virus can remain latent until conditions become favorable to make new viruses.

Figure 18.8 Here are four possible effects of a drug: 1. A drug could specifically recognize the PrP^{Sc} conformation and prevent it from binding to PrP^{C}.

2. It could convert PrP^{Sc} back to PrP^{C}. 3. It could recognize PrP^{C} and prevent the binding of PrP^{Sc}. 4. It could bind to PrP^{C} and stabilize its conformation, thereby preventing PrP^{Sc} from changing it to the abnormal conformation. You may also think of other interesting possibilities.

Figure 18.9 There appears to be three nucleoids in the bacterial cell to the far right.

Figure 18.11 The loop domains are held in place by proteins that bind to the DNA at the bases of the loops. The proteins also bind to each other.

Figure 18.12 Bacterial chromosomes and plasmids are similar in that they typically contain circular DNA molecules. However, bacterial chromosomes are usually much longer than plasmids and carry many more genes. Also, bacterial chromosomes tend to be more compacted due to the formation of loop domains and supercoiling.

Figure 18.13 16 hours is the same as 32 doublings. So, $2^{32} = 4,294,967,296$. (The actual number would be much less because the cells would deplete the growth media and grow more slowly than the maximal rate.)

Figure 18.16 Yes. The two strains would have mixed together, allowing them to conjugate. Therefore, there would have been colonies on the plates.

Figure 18.17 During conjugation, only one strand of the DNA from an F factor is transferred from the donor to the recipient cell. The single-stranded DNA in both cells is then used as a template to create double-stranded F factor DNA in both cells.

Figure 18.19 Transduction is not a normal part of the phage life cycle. It is a mistake in which a piece of the bacterial chromosome is packaged into a phage coat and is then transferred to another bacterial cell.

Feature Investigation Questions

1. Lederberg and Tatum were testing the hypothesis that genetic material could be transferred from one bacterial strain to another.

2. The experimental growth medium lacked particular amino acids and biotin. The mutant strains were unable to synthesize these particular amino acids or biotin. Therefore, they were unable to grow due to the lack of the necessary nutrients. The two strains used in the experiment each lacked the ability to make two essential nutrients necessary for growth. The appearance of colonies growing on the experimental growth medium indicated that some bacterial cells had acquired the normal genes for the two mutations they carried. By acquiring these normal genes, the ability to synthesize the essential nutrients was restored.

3. Bernard Davis placed samples of the two bacterial strains in different arms of a U-tube. A filter allowed the free movement of the liquid in which the bacterial cells were suspended, but prevented the actual contact between the bacterial cells. After incubating the strains in this environment, Davis found that genetic transfer did not take place. He concluded that physical contact between cells of the two strains was required for genetic transfer.

Test Yourself

1. c 2. e 3. c 4. b 5. e 6. a 7. d 8. d 9. b 10. c

Conceptual Questions

1. Viruses are similar to living cells in that they contain a genetic material that provides a blueprint to make new viruses. However, viruses are not composed of cells, and by themselves, they do not carry out metabolism, use energy, maintain homeostasis, or even reproduce. A virus or its genetic material must be taken up by a living cell to replicate.

2. Conjugation—The process involves a direct physical contact between two bacterial cells in which a donor cell transfers a strand of DNA to a recipient cell.

   - Transformation—This occurs when a living bacteria takes up genetic information that has been released from a dead bacteria.
   - Transduction—When a virus infects a donor cell, it incorporates a fragment of bacterial chromosomal DNA into a newly made virus particle. The virus then transfers this fragment of DNA to a recipient cell.
   - Horizontal gene transfer is the transfer of genes from another organism without being the offspring of that organism. These acquired genes sometimes increase survival and therefore may have an evolutionary advantage. Such genes may even promote the formation of new species. From a medical perspective, an important example of horizontal gene transfer is when one bacterium acquires

antibiotic resistance from another bacterium and then itself becomes resistant to that antibiotic. This phenomenon is making it increasingly difficult to treat a wide variety of bacterial diseases.

3. If neither cell has a selective growth advantage, we would expect that the F^+ cells would eventually overrun the population. This is because a mating starts with an F^+ and F^- cell and ends with two F^+ cells. Therefore, F^+ cells can convert F^- cells into F^+ cells, but the opposite cannot occur.

Chapter 19

Concept Checks

Figure 19.1 Model organisms are studied by many different researchers so they can compare their results and determine scientific principles that apply more broadly to other species.

Figure 19.3 Cell division and cell migration are common in the earliest stages of development, whereas cell differentiation and apoptosis are more common as tissues and organs start to form.

Figure 19.4 If apoptosis did not occur, the fingers would be webbed.

Figure 19.5 During development, positional information may a cause a cell to respond in one of four ways: cell division, cell migration, cell differentiation, and cell death.

Figure 19.6 Cell division and migration would be the most prevalent in the early phases of development, such as phase 1 and 2.

Figure 19.8 The larva would have anterior structures at both ends and would lack posterior structures such as a spiracle.

Figure 19.9 The Bicoid protein functions as a transcription factor. Its function is highest in the anterior end of the zygote.

Figure 19.11 There are 15 pink stripes in this embryo. Each stripe corresponds to a portion of the 15 segments in the embryo.

Figure 19.14 The last abdominal segment would have legs!

Figure 19.18 Stem cells can divide, and they can differentiate into specific cell types.

Figure 19.20 Hematopoietic stem cells are multipotent.

Figure 19.23 Most stem cells in plants are found in meristems, which are located at the tips of roots and shoots.

Figure 19.24 The pattern would be sepal, petal, stamen, stamen.

Feature Investigation Questions

1. The researchers were interested in the factors that cause cells to differentiate. For this particular study, the researchers were attempting to identify genes involved in the differentiation of muscle cells.

2. Using genetic technology, the researcher compared the gene expression in cells that could differentiate into muscle cells to the gene expression in cells that could not differentiate into muscle cells. Though many genes were expressed in both, the researchers were able to isolate three genes that were expressed in muscle cell lines that were not expressed in the nonmuscle cell lines.

3. Again, using genetic technology, each of the candidate genes was introduced into a cell that normally did not give rise to skeletal muscle. This procedure was used to test whether or not these genes played a key role in muscle cell differentiation. If the genetically engineered cell gave rise to muscle cells, the researchers would have evidence that a particular candidate gene was involved in muscle cell differentiation. Of the three candidate genes, only one was shown to be involved in muscle cell differentiation. When the *MyoD* gene was expressed in fibroblasts, these cells differentiated into skeletal muscle cells.

Test Yourself

1. c 2. d 3. e 4. b 5. c 6. e 7. a 8. a 9. b 10. d

Conceptual Questions

1. a. This would be consistent with a defect in a segmentation gene, such as a gap gene.

   b. This would be consistent with a mutation in a homeotic gene because the characteristics of a particular segment have been changed.

2. Both types of genes encode transcription factors that bind to the DNA and regulate the expression of other genes. The effects of *Hox* genes are to determine the characteristics of certain regions of the body, whereas the *myoD* gene is cell specific—it causes a cell to become a skeletal muscle cell.

3. In plants, there are organized groups of cells that are very active with respect to cell division and producing stem cells. These areas are called meristems. At these areas, cells divide and retain the ability to differentiate into several different types of cells. In plants, there are two major places where these meristems are found. First, meristem tissue found in the roots is called the root meristem, which gives rise to the roots and tissue associated with the roots. The second is the shoot meristem, which produces all of the aerial parts of the plant, such as stems, leaves and flowers.

Chapter 20

Concept Checks

Figure 20.2 No. A recombinant vector has been made, but it has not been cloned. In other words, many copies of the recombinant vector have not been made yet.

Figure 20.3 The insertion of chromosomal DNA into the vector disrupts the *lacZ* gene and thereby prevents the expression of β-galactosidase. The functionality of *lacZ* can be determined by providing the growth medium with a colorless compound, X-Gal, which is cleaved by β-galactosidase into a blue dye. Bacterial colonies containing recircularized vectors will form blue colonies, whereas colonies containing recombinant vectors carrying a segment of chromosomal DNA will be white.

Figure 20.5 The 600 bp piece would be closer to the bottom. Smaller pieces travel faster through the gel.

Figure 20.6 The primers are complementary to sequences at each end of the DNA region to be amplified.

Figure 20.7 It means that part of their inserts are exactly the same, but other regions are not.

Figure 20.8 If a dideoxynucleotide ddNTP is added to a growing DNA strand, the strand can no longer grow because the 3′—OH group, the site of attachment for the next nucleotide, is missing.

Figure 20.9 A fluorescent spot identifies a cDNA that is complementary to a particular DNA sequence. Because the cDNA was generated from mRNA, this technique identifies a gene that has been transcribed in a particular cell type under a given set of conditions.

Figure 20.10 The reason why the A and B chains are made as fusion proteins is because the A and B chains are rapidly degraded when expressed in bacterial cells by themselves. The fusion proteins, however, are not.

Figure 20.12 Only the T DNA within the Ti plasmid is transferred to a plant cell.

Figure 20.14 Not all of Dolly's DNA came from a mammary cell. Her mitochondrial DNA came from the oocyte donor.

Figure 20.15 The bands match suspect 2.

Feature Investigation Questions

1. Gene therapy is the introduction of cloned genes into living cells to correct genetic mutations. The hope is that the cloned genes will correct or restore the normal gene function and thereby eliminate the clinical effects of the disease.

   ADA deficiency is a recessive genetic disorder in which an enzyme, adenosine deaminase, is not functional. The absence of this enzyme causes a buildup of deoxyadenosine, which is toxic to lymphocytes. When lymphocytes are destroyed, a person's immune system begins to fail, leading to a severe combined immunodeficiency disease, or SCID.

2. The researchers introduced normal copies of the ADA gene into lymphocytes, restoring normal cell metabolism. The researchers isolated lymphocytes from the patient and used a viral vector to introduce the gene into the lymphocytes. These lymphocytes were then reintroduced back into the patient.

3. Following several rounds of treatment with gene therapy, researchers were able to document continued production of the correct enzyme by the lymphocytes over the course of 4 years. However, because the

patients were also receiving other forms of treatment, it was not possible to determine if the gene therapy reduced the negative effects of the genetic disease.

Test Yourself

1. e   2. d   3. b   4. b   5. b   6. c   7. d   8. c   9. e   10. e

Conceptual Questions

1. The restriction enzyme cuts the plasmid at a specific site, leaving sticky ends. The gene of interest, cut with the same enzyme, will have complementary sticky ends that allow hydrogen bonding between the gene of interest and the plasmid. The connections are then made permanent, using DNA ligase that connects the DNA backbones.

2. A ddNTP is missing an oxygen at the 3′ position. These prevents the further growth of a DNA strand and thereby causes chain termination.

3. A mouse model is a strain of mice that carries a mutation that is analogous to a mutation that causes a human disease. Such mice exhibit disease symptoms resembling those found in humans. These mice can be used as model organisms to study a human disease. Such mice models have also been used to test the effects of various therapies in the treatment of human diseases.

Chapter 21

Concept Checks

Figure 21.2   One reason is that more complex species tend to have more genes. A second reason is that species vary with regard to the amount of repetitive DNA that is found in their genome.

Figure 21.3   The answer is a matter of opinion. Many people find it surprising that the vast majority of DNA is not part of genes or the coding region of genes.

Figure 21.5   For DNA transposons, inverted repeats are recognized by transposase, which cleaves the DNA and inserts the transposon into a new location.

Figure 21.6   Retroelements. A single retroelement can be transcribed into multiple copies of RNA, which can be converted to DNA by reverse transcriptase, and inserted into multiple sites in the genome.

Figure 21.8   The overall advantage is specialization. When multiple copies of a gene are found in the genome, each copy can become specialized to suit the needs of particular cell types or particular stages of development.

Figure 21.9   Differential gene regulation. Genes that encode metabolic enzymes are highly expressed in liver cells, whereas those same genes are expressed in lower amounts in muscle cells. Conversely, genes that encode cytoskeletal and motor proteins are highly expressed in muscle cells, but less so in liver cells.

Figure 21.10   Reversible post-translation covalent modifications provide a way to modulate protein function. Certain types can turn off protein function, whereas others can turn on protein function. These modifications provide a rapid way for a cell to control protein function.

Figure 21.11   The two main advantages of having a computer program translate a genetic sequence is that it's faster and probably more accurate.

Figure 21.12   It is possible for orthologs to have exactly the same DNA sequence if neither of them has accumulated any new mutations that would cause their sequences to become different. This is likely only for closely related species that have diverged relatively recently from each other.

Feature Investigation Questions

1. The goal of the experiment was to sequence the entire genome of *Haemophilus influenzae*. By conducting this experiment, the researchers would have information about genome size and the types of genes the bacterium has.

2. One strategy requires mapping the genome prior to sequencing. After mapping is completed, each region of the genome is then sequenced. The shotgun approach does not require mapping of the genome prior to sequencing. Instead, many fragments are randomly sequenced.

   The advantage of the shotgun approach is the speed at which the sequencing can be conducted because the researchers do not have to spend time mapping the genome first. The disadvantage is that because the researchers are sequencing random fragments, some fragments may be sequenced more than necessary.

3. The researchers were successful in sequencing the entire genome of the bacterium. The genome size was determined to be 1,830,137 base pairs, with a predicted 1,743 structural genes. The researchers were also able to predict the function of many of these genes. More importantly, the results were the first complete genomic sequence of a living organism.

Test Yourself

1. c   2. e   3. a   4. c   5. e   6. b   7. b   8. d   9. c   10. c

Conceptual Questions

1. a. yes

   b. No, it's only one chromosome in the nuclear genome.

   c. yes

   d. yes

2. The two main reasons why the proteomes of eukaryote species are usually much larger than their genomes are alternative splicing and post-translational covalent modifications. During alternative splicing, a pre-mRNA is spliced in two or more different ways to yield two or more different polypeptides. Post-translational covalent modifications can affect protein structure in a variety of ways, including proteolytic processing; disulfide bond formation; the attachment of prosthetic group, sugars, or lipids; phosphorylation; acetylation; and methylation.

3. Because they are derived from the same ancestral gene, homologous genes usually carry out similar or identical functions. In many cases, the first way to identify the function of a newly determined gene sequence is to find a homologous gene whose function is already known.

Chapter 22

Concept Checks

Figure 22.2   Organic molecules form the chemical foundation for the structure and function of living organisms. Modern organisms can synthesize organic molecules. However, to explain how life got started, biologists need to explain how organic molecules were made prior to the existence of living cells.

Figure 22.3   These vents release hot gaseous substances from the interior of the Earth. Organic molecules can form in the temperature gradient between the extremely hot vent water and the cold water that surrounds the vent.

Figure 22.4   A liposome is more similar to real cells, which are surrounded by a membrane that is composed of a phospholipid bilayer.

Figure 22.5   Certain chemicals, such as RNA molecules, may have properties that provide advantages and therefore cause them to increase in number compared to other molecules.

Figure 22.7   In a sedimentary rock formation, the layer at the bottom is usually the oldest.

Figure 22.8   For this time frame, you would analyze the relative amounts of the rubidium-87 and strontium-87 isotopes.

Figure 22.12   In part (a) at the top, the two cells are associating with each other, but one cell is not inside of the other cell, as in part (b).

Figure 22.13   The number of cells increases, and parts (c) and (d) have two cell types (somatic and reproductive). When comparing (c) and (d), the number of somatic cells increases relative to the number of reproductive cells.

Figure 22.14   Most animal species exhibit bilateral symmetry, including fruit flies, fishes, and humans.

Feature Investigation Questions

1. Chemical selection occurs when a particular chemical in a mixture has advantageous properties that allow it to increase in number compared to the other chemicals in the mixture. Bartel and Szostak hypothesized that variation in the catalytic abilities of RNA molecules would allow for chemical selection in the laboratory. Bartel and Szostak proposed to select for RNA molecules with higher catalytic abilities.

2. The short RNA molecules allowed the researchers to physically separate the mixture of longer RNA molecules based on catalytic properties. Long RNA molecules with catalytic abilities would covalently bond

with the short RNA molecules. The short RNA molecules had a specific region that caused them to be attracted to column beads in the experimental apparatus. The long RNA molecules that did not have catalytic abilities passed through the column and therefore could be separated from the ones that had catalytic activity and became bound to the column beads.

3. The researchers found that with each round of selection, the enzymatic activity of the selected pool of RNA molecules increased. These results provided evidence that chemical selection could improve the functional characteristics of a group of molecules. Much of the explanation of the evolution of life on Earth is theoretical, meaning it is based on scientific principles but has not been experimentally verified. Researchers are attempting to develop laboratory experiments that test the explanations of the evolution of life. The experiment conducted by Bartel and Szostak provided experimental data to support the hypothesis of chemical selection as a possible mechanism for the early evolutionary process that led to living cells.

Test Yourself

1. b 2. e 3. b 4. c 5. e 6. a 7. d 8. c 9. b 10. d

Conceptual Questions

1. Nucleotides and amino acids were produced prior to the existence of cells.

   Nucleotides and amino acids became polymerized to form DNA, RNA, and proteins.

   Polymers became enclosed in membranes.

   Polymers enclosed in membranes evolved cellular properties.

2. The relative ages of fossils can be determined by the locations in sedimentary rock formation. Older fossils are found in lower layers. A common way to determine the ages of fossils is via radioisotope dating, which is often conducted on igneous rock in the vicinity of the fossil. A radioisotope is an unstable isotope of an element that decays spontaneously, releasing radiation at a constant rate. The half-life is the length of time required for a radioisotope to decay to exactly one-half of its initial value. To determine the age of a rock (and that of a nearby fossil), scientists can measure the amount of a given radioisotope as well as the amount of the decay product.

3. The Cambrian period lasted from 543 million years ago to 490 million years ago. The Cambrian explosion was an abrupt increase on a geological scale in the diversity of animal species. During the Cambrian period, all of the existing major types of marine invertebrates arose as well as many other major animal groups. The basic body types of all modern animal species are based on body types that arose during the Cambrian.

Chapter 23

Concept Checks

Figure 23.2 A single organism does not evolve. Populations may evolve from one generation to the next.

Figure 23.7 Due to a changing global climate, the island fox became isolated from the mainland species. Over time, natural selection resulted in adaptations for the population on the island and eventually resulted in a new species with characteristics that are somewhat different from the mainland species.

Figure 23.8 Many answers are possible. One example is the wing of a bird and the wing of a bat.

Figure 23.11 The relative sizes of traits are changing. For example, in dogs, the lengths of legs, body size, etc., are quite different. Artificial selection is often aimed at changing the relative sizes of body parts.

Figure 23.13 Rhesus and green monkeys = 0, Congo puffer fish and European flounder = 2, and Rhesus monkey and Congo puffer fish = 10. Pairs that are closely related evolutionarily have fewer differences than do pairs that are more distantly related.

Figure 23.14 Orthologs have similar gene sequences because they are derived from the same ancestral gene. The sequences are not identical because after the species diverged, each one accumulated different random mutations that changed their sequences.

Figure 23.15 It creates multifunctional proteins that may have new properties that can be acted upon by natural selection.

Figure 23.17 Humans have one large chromosome 2, but this chromosome is divided into two separate chromosomes in the other three species. In chromosome 3, the banding patterns among humans, chimpanzees, and gorillas are very similar, but the orangutan has a large inversion that flips the arrangement of bands in the centromeric region.

Feature Investigation Questions

1. The island has a moderate level of isolation but is located near enough to the mainland to have some migrants. The island is an undisturbed habitat, so the researchers would not have to consider the effects of human activity on the study. Finally, the island had an existing population of ground finches that would serve as the study organism over many generations.

2. First, the researchers were able to show that beak depth is a genetic trait that has variation in the population. Second, the depth of the beak is an indicator of the types of seeds the birds can eat. The birds with larger beaks can eat larger and drier seeds; therefore, changes in the types of seeds available could act as a selective force on the bird population.

   During the study period, there were annual changes in rainfall. This had an impact on the seed sizes produced by the plants on the island. In the drier year, fewer small seeds were produced, so the birds would have to eat larger, drier seeds.

3. The researchers found that following the drought in 1978, the average beak depth in the finch population increased. This indicated that birds with larger beaks were better able to adapt to the environmental changes due to the drought and produce more offspring. This is direct evidence of the phenomenon of natural selection.

Test Yourself

1. d 2. d 3. b 4. b 5. b 6. d 7. c 8. b 9. d 10. e

Conceptual Questions

1. Some random mutations result in a phenotype with greater reproductive success. If so, natural selection results in a greater proportion of such individuals in succeeding generations. These individuals are more likely to survive and reproduce, which means they have evolved to be better adapted to their environment.

2. The process of convergent evolution produces two different species from different lineages that show similar characteristics because they occupy similar environments. An example is the long snout and tongue of both the giant anteater, found in South America, and the echidna, found in Australia. This enables these animals to feed on ants, but the two structures evolved independently. These observations support the idea that evolution results in adaptations to particular environments.

3. Homologous structures are two or more structures that are similar because they are derived from a common ancestor. An example is the same set of bones that is found in the human arm, turtle arm, bat wing, and whale flipper. The forearms in these species have been modified to perform different functions. This supports the idea that all of these animals evolved from a common ancestor.

Chapter 24

Concept Checks

Figure 24.2 If C^R is 0.4, then C^W must be 0.6, because the allele frequencies add up to 1.0. The heterozygote ($2pq$) equals $2(0.4)(0.6)$, which equals 0.48, or 48%.

Figure 24.3 Over the short run, alleles that confer better fitness would be favored and increase in frequency, perhaps enhancing diversity. Over the long run, however, an allele that confers high fitness in the homozygous state may become monomorphic, thereby reducing genetic diversity.

Figure 24.4 Stabilizing selection eliminates alleles that give phenotypes that deviate significantly from the average phenotype. For this reason, it tends to decrease genetic diversity.

Figure 24.6 If malaria was eradicated, there would be no selective advantage for the heterozygote. The H^S allele would eventually be eliminated because the H^SH^S homozygote has a lower fitness. Directional selection would occur.

Figure 24.7   This is likely to be a form of intersexual selection. Such traits are likely to be involved in mate choice.

Figure 24.11   The bottleneck effect decreases genetic diversity. This may eliminate adaptations that promote survival and reproductive success. Therefore, the bottleneck effect makes it more difficult for a population to survive.

Figure 24.12   There are several possibilities. Two examples would be changing the DNA sequence within an intron or changing the DNA sequencing in a chromosomal region where a gene is not found.

Figure 24.13   Gene flow tends to make the allele frequencies in neighboring populations more similar to each other. It also promotes genetic diversity by introducing new alleles into populations.

Figure 24.14   Inbreeding favors homozygotes. If a homozygote has a higher Darwinian fitness, inbreeding would accelerate the prevalence of the allele in the population. Alternatively, if a homozygote has a lower fitness, inbreeding would accelerate the elimination of the allele from the population.

Feature Investigation Questions

1. The two species of cichlids used in the experiment are distinguishable by coloration, and the researchers were testing the hypothesis that the females make mate choices based on this variable.

2. Individual females were placed in tanks that contained one male from each species. The males were held in small glass tanks to limit their movement but allowed the female to see each of the males. The researchers recorded the courtship behavior between the female and males and the number of positive encounters between the female and each of the different males. This procedure was conducted under normal lighting and under monochromatic lighting that obscured the coloration differences between the two species. Comparing the behavior of the females under normal light conditions and monochromatic light conditions allowed the researchers to determine the importance of coloration in mate choice.

3. The researchers found that the female was more likely to select a mate from her own species in normal light conditions. However, under monochromatic light conditions, the species-specific mate choice was not observed. Females were as likely to choose males of the other species as they were males of their own species. This indicated that coloration is an important factor in mate choice in these species of fish.

Test Yourself

1. d   2. c   3. c   4. e   5. b   6. c   7. b   8. d   9. b   10. a

Conceptual Questions

1. The frequency of the disease is a genotype frequency because it represents individuals with the disease. If we let q^2 represent the genotype frequency, then q equals the square root of 0.04, which is 0.2. If $q = 0.2$, then $p = 1 - q$, which is 0.8. The frequency of heterozygous carriers is $2pq$, which is $2(0.8)(0.2) = 0.32$, or 32%.

2. Directional selection—This is when natural selection favors an extreme phenotype that makes the organism better suited to survive and reproduce in its environment. As a result, the extreme phenotype will become predominant in the population. This can occur either through new mutation or through a prolonged environmental change. In addition to selecting for a certain phenotype, the opposite end of the extreme is removed from the gene pool.

   - Stabilizing selection—In this type of selection, natural selection favors individuals with intermediate phenotypes, whereas organisms with extreme phenotypes are less likely to reproduce. This selection tends to prevent major changes in the phenotypes of populations.
   - Disruptive selection—This type of selection favors both extremes and removes the intermediate phenotype. It is also known as diversifying selection.
   - Balancing selection—This type of selection results in a balanced polymorphism in which two or more alleles are stably maintained in a population. Examples include heterozygote advantage, as in the sickle-cell allele, and negative frequency-dependent selection, as in certain prey.
   - Sexual selection—This is a type of natural selection that is directly aimed at reproductive success. It can occur by any of the previous four mechanisms. Male coloration in African cichlids is an example.

3. The founder effect occurs when a small group of individuals separates from a larger population and establishes a colony in a new location. Genetic drift can occur for two reasons. First, the founding population, which is relatively small, may have less genetic variation and different allele frequencies than the larger, original population from which it was derived. Second, as a matter of chance, the allele frequencies in the founding population will drift substantially while the population is small.

Chapter 25

Concept Checks

Figure 25.1   There are a lot of possibilities. Certain grass species look quite similar. Elephant species look very similar. And so on.

Figure 25.3   Temporal isolation is an example of a prezygotic isolating mechanism. Because the species breed at different times of the year, hybrid zygotes are not formed between the two species.

Figure 25.5   Hybrid sterility is a type of postzygotic isolating mechanism. A hybrid forms between the two species, but it is sterile.

Figure 25.7   The Hawaiian Islands have many different ecological niches that can be occupied by birds. The first founding bird inhabitants evolved to occupy those niches and thereby evolved into many different species.

Figure 25.11   The offspring would inherit 16 chromosomes from *G. tetrahit*, and from the hybrid, it would inherit anywhere from 8 to 16. So the answer is 24 to 32. The hybrid parent would always pass the 8 chromosomes that are found in pairs. With regard to the 8 chromosomes not found in pairs, it could pass 0 to 8 of them.

Figure 25.12   The insects on different host plants would tend to breed with each other, and natural selection would favor the development of traits that are an advantage for feeding on that host. Over time, the accumulation of genetic changes may lead to reproductive isolation between the populations of insects.

Figure 25.14   If the *Gremlin* gene was underexpressed, this would cause less Gremlin protein to be produced. Because Gremlin protein inhibits apoptosis, more cell death would occur, and the result would probably be smaller feet, and maybe they would not be webbed.

Figure 25.15   By comparing the number of *Hox* genes in many different animal species, a general trend is observed that animals with more complex body structures have a greater number of *Hox* genes.

Figure 25.17   The tip of the mouse's tail would have a mouse eye!

Feature Investigation Questions

1. Podos hypothesized that the morphological changes in the beak would also affect the birds' songs. A bird's song is an important component for mate choice. If changes in the beak alter the song of the bird, this would impact reproductive ability. Podos suggested that changes in the beak morphology could thus lead to reproductive isolation among the birds.

2. Podos first caught male birds in the field and collected data on beak size. The birds were banded for identification and released. Later, the banded birds' songs were recorded and analyzed for range of frequencies and trill rates. The results were then compared to similar data from other species of birds to determine if beak size constrained the frequency range and trill rate of the song.

3. The results of the study did indicate that natural selection on beak size due to changes in diet could lead to changes in song. Considering the importance of bird song to mate choice, the changes in the song could also lead to reproductive isolation.

   The phrase "by-product of adaptation" refers to changes in the phenotype that are not directly acted on by natural selection. In the case of the Galápagos finches, the changes in beak size were directly related to diet; however, as a consequence of that selection, the song pattern was also altered. The change in song pattern was a by-product.

Test Yourself

1. b   2. b   3. e   4. d   5. c   6. a   7. b   8. d   9. c   10. c

Conceptual Questions

1. Prezygotic isolating mechanisms prevent the formation of the zygote. An example is mechanical isolation, the incompatibility of genitalia. Postzygotic isolating mechanisms act after the formation of the zygote.

An example is inviability of the hybrid that is formed. (Other examples shown in Figure 25.2 would also be correct.) Postzygotic mechanisms are more costly because some energy is spent in the formation of a zygote and its subsequent growth.

2. The concept of gradualism suggests that each new species evolves continuously over long spans of time (Figure 25.13a). The principal idea is that large phenotypic differences that produce new species are due to the accumulation of many small genetic changes. According to the punctuated equilibrium model, species exist relatively unchanged for many generations. During this period, the species is in equilibrium with its environment. These long periods of equilibrium are punctuated by relatively short periods during which evolution occurs at a far more rapid rate. This rapid evolution is caused by relatively few genetic changes.

3. One example involves the *Hox* genes, which control morphological features along the anteroposterior axis in animals. An increase in the number of *Hox* genes during evolution is associated with an increase in body complexity and may have spawned many different animal species.

Chapter 26

Concept Checks

Figure 26.2   A phylum is broader than a family.

Figure 26.3   Yes. They can have many common ancestors, depending on how far back you go in the tree. For example, dogs and cats have a common ancestor that gave rise to mammals, and an older common ancestor that gave rise to vertebrates. The most recent common ancestor is the point at which two species diverged from each other.

Figure 26.4   An order is a smaller taxa that would have a more recent common ancestor.

Figure 26.9   A hinged jaw is the character common to the salmon, lizard, and rabbit, but not to the lamprey.

Figure 26.10   Changing the second G to an A is common to species A, B, and C, but not to species G.

Figure 26.12   The pair with a 3% difference would have an older common ancestor. A higher difference (3% compared to 1%) has occurred because the common ancestor is older and there has been more time to accumulate mutations.

Figure 26.13   Gorillas and humans would be expected to have fewer genetic differences because their common ancestor (named C) is more recent than that of orangutans and gorillas, which is ancestor B.

Figure 26.15   The kiwis are found in New Zealand. Even so, the kiwis are more closely related to Australian and African flightless birds than they are to the moas, which were found in New Zealand.

Figure 26.16   Monophyletic groups are based on the concept that a particular group of species descended from a common ancestor. When horizontal gene transfer occurs, not all of the genes in a species were inherited from the common ancestor, so this muddles the concept of monophyletic groups.

Feature Investigation Questions

1. Molecular paleontology is the sequencing and analysis of DNA obtained from extinct species. Tissue samples from specimens of extinct species may contain DNA molecules that can be extracted, amplified, and sequenced. The DNA sequences can then be compared to living species to study evolutionary relationships between modern and extinct species.

   The researchers extracted DNA from tissue samples of moas, extinct flightless birds that lived in New Zealand. The DNA sequences from the moas were compared to the DNA sequences of modern species of flightless birds to determine the evolutionary relationships of this particular group of organisms.

2. The researchers compared the DNA sequences of the extinct moas and modern kiwis of New Zealand to the emu and cassowary of Australia and New Guinea, the ostrich of Africa, and rheas of South America. All of the birds are flightless birds. With the birds selected, the researchers could look for similarities between birds over a large geographic area.

3. The sequences were very similar among the different species of flightless birds. Interestingly, the sequences of the kiwis of New Zealand were more similar to those of the modern species of flightless birds found on other land masses than they were to those of the moas found in New Zealand.

The researchers constructed a new evolutionary tree that suggests that kiwis are more closely related to the emu, cassowary, and ostrich. Also, based on the results of this study, the researchers suggested that New Zealand was colonized twice by ancestors of flightless birds. The first ancestor gave rise to the now-extinct moas. The second ancestor gave rise to the kiwis.

Test Yourself

1. c   2. d   3. e   4. d   5. b   6. d   7. b   8. b   9. c   10. e

Conceptual Questions

1. The scientific name of every species has two parts, which are the genus name and the species epithet. The genus name is always capitalized, whereas the species name is not. Both names are italicized. An example is *Canis lupus.*

2. Morphological analysis focuses on morphological features of extinct and modern species. Many traits are analyzed to obtain a comprehensive picture of two species' relatedness. Convergent evolution leads to similar traits that arise independently in different species as they adapt to similar environments. Convergent evolution can, therefore, cause errors if a researcher assumes that a particular trait arose only once and that all species having the trait have the same common ancestor.

3. If neutral mutations occur at a relatively constant rate, they act as a molecular clock on which to measure evolutionary time. Genetic diversity between species that is due to neutral mutation gives an estimate of the time elapsed since the last common ancestor. A molecular clock can provide a timescale to a phylogenetic tree.

Chapter 27

Concept Checks

Figure 27.2   The red color arises from rhodopsins, pigment-protein complexes located in the cell membranes of the halophytic archaea growing abundantly in these salty ponds.

Figure 27.5   The cell will tend to float because it is full of intact gas vesicles.

Figure 27.11   The fluorescence staining process may be faster but requires the use of a fluorescence microscope, which is more expensive and complex than a simple compound microscope. The Gram stain can be assessed with a simple compound microscope of the type often found in biology class laboratories.

Figure 27.12   The motion of the stiff filament of a prokaryotic flagellum is more like that of a propeller shaft than the flexible arms of a human swimmer.

Figure 27.14   Cells having pili tend to move with a twitching or gliding motion.

Figure 27.15   When DNA sequencing studies show that samples contain many uncultured bacterial species, the fluorescence method is preferred, though it requires the use of a fluorescence microscope. Under such conditions, the culture method will give an underestimate of bacterial numbers. But when the goal is to estimate numbers of bacteria whose culture preferences are known, the culture method may provide good estimates.

Figure 27.16   Endospores allow bacterial cells to survive treatments and environmental conditions that would kill ordinary cells.

Figure 27.18   The transition point for both genes occurs around dusk.

Figure 27.19   Structural similarities to bacterial flagella and pili indicate that these types of attack systems evolved from these structures.

Feature Investigation Questions

1. *Deinococcus radiodurans* has the ability to survive levels of ionizing radiation that would kill most other organisms, and for periods of time that are not tolerated by other organisms. Investigators wanted to know why *D. radiodurans* was so resistant to radiation.

2. Daly and associates suggested that cellular levels of manganese ion might play a role in radiation resistance.

3. Bacteria that had been grown on media containing higher levels of manganese ion were better able to resist radiation than those grown on media containing lower levels of manganese ion.

Test Yourself

1. c   2. b   3. c   4. e   5. a   6. a   7. b   8. e   9. d   10. d

Conceptual Questions

1. Small cell size and simple division processes allow many bacteria to divide much more rapidly than eukaryotes. This helps to explain why food can spoil so quickly and why infections can spread very rapidly within the body. Other factors also influence these rates.

2. Pathogen populations naturally display genetic variation in their susceptibility to antibiotics. When such populations are exposed to antibiotics, even if initially only a few cells are resistant, the numbers of resistant cells will eventually increase and could come to dominate natural populations.

3. Humans. Only when humans pollute natural waters with high levels of fertilizers originating from sewage effluent or crop field runoff are cyanobacterial populations able to grow large enough to produce harmful blooms.

Chapter 28

Concept Checks

Figure 28.1 Diatoms are algal protists, meaning that they are typically photosynthetic and produce organic compounds. These organic compounds can be digested and respired as food by heterotrophic protists.

Figure 28.6 After particles are ingested via feeding grooves, particles are enclosed by membrane vesicles and then digested by enzymes.

Figure 28.7 The intestinal parasite *Giardia lamblia* is transmitted from one person to another via fecal wastes, whereas the urogenital parasite *Trichomonas vaginalis* can be transmitted by sexual activity.

Figure 28.14 The nucleomorph encodes some of the materials needed for plastid function.

Figure 28.16 The sail-like portions of the dinoflagellate cell wall help keep the cells afloat.

Figure 28.17 Flagellar hairs function like oars, helping to pull cells through the water.

Figure 28.18 Kelps are harvested for the production of industrially useful materials. In addition, they nurture fishes and other wildlife of economic importance.

Figure 28.21 Genes that encode cell adhesion and extracellular matrix proteins are likely essential to modern choanoflagellates' ability to attach to surfaces, where they feed. Similar proteins are involved in the formation of multicellular tissues in animals. Evolutionary biologists would say that ancient choanoflagellates were preadapted for the later evolution of multicellular tissues in early animals.

Figure 28.24 Cysts allow protists to survive conditions that are not suitable for growth. One such condition would be the dry or cold environment outside a parasitic protist's warm, moist host tissues.

Figure 28.29 Gametes of *Plasmodium falciparum* undergo fusion to produce zygotes while in the mosquito host.

Feature Investigation Questions

1. One strain had earlier been reported to be toxic to fishes, while the other had been reported to be nontoxic, a difference that could be attributed to differing experimental conditions. The investigators wanted to determine the degree of toxicity of the two strains when grown under the same conditions.

2. Producing toxins requires considerable ATP and other resources, so many organisms produce such compounds only when needed. In the case of *Pfiesteria shumwayae*, this might be when a major food source, fish, was present, but not when they fed primarily upon algal cells. The investigators needed to know if this dinoflagellate produces toxin even when feeding on algae alone (which would not require toxin production) or only when exposed to fishes.

3. The team knew that fishermen and scientists had suffered amnesia and other neurological impairments when they were near water containing large populations of the genus *Pfiesteria*. These observations suggested that the toxin was volatile or suspended in water droplets that people could inhale. As a precaution, they used the biohazard containment system to avoid personal harm. The use of biohazard containment systems is generally recommended for scientists who work with hazardous or potentially hazardous biological materials.

Test Yourself

1. c   2. a   3. b   4. b   5. e   6. b   7. e   8. d   9. b   10. c

Conceptual Questions

1. Protists are amazingly diverse, reflecting the occurrence of extensive adaptive radiation after the origin of eukaryotic cells, widespread occurrence of endosymbiosis, and adaptation to many types of moist habitats, including the tissues of animals and plants. As a result of this extensive diversity, protists cannot be classified into a single kingdom or phylum.

2. Several protists, including the apicomplexans *Cryptosporidium parvum* and *Plasmodium falciparum* and the kinetoplastids *Leishmania major* and *Trypanosoma brucei*, cause many cases of illness around the world, but few treatments are available, and organisms often evolve drug resistance. Genomic data allow researchers to identify metabolic features of these parasites that are not present in humans and are therefore good targets for development of new drugs. An example is provided by metabolic pathways of the apicoplast, a reduced plastid that is present in cells of the genus *Plasmodium*. Because the apicoplast plays essential metabolic roles in the protist but is absent from humans, drugs that disable apicoplast metabolism would kill the parasite without harming the human host.

3. Most protist cells cannot survive outside moist environments, but cysts have tough walls and dormant cytoplasm that allow them to persist in habitats that are unfavorable for growth. While cysts play important roles in the asexual reproduction and survival of many protists, they also allow protist parasites such as *Entamoeba histolytica* (the cause of amoebic dysentery) to spread to human hosts who consume food or water that have been contaminated with cysts. Widespread contamination can sicken thousands of people at a time.

Chapter 29

Concept Checks

Figure 29.3 Liverworts grow very close to surfaces such as soil or tree trunks. Raising their sporophytes off the surface helps to disperse spores into air currents.

Figure 29.4 Wind speed varies, so if the moss released all the spores at the same time into a weak air current, the spores would not travel very far and might have to compete with the parent plant for scarce resources. By releasing spores gradually, some spores may enter strong gusts of wind that carry them long distances, reducing competition with the parent.

Figure 29.8 Larger sporophytes are able to capture more resources for use in producing larger numbers of progeny and therefore have greater fitness than do smaller sporophytes.

Figure 29.14 The capacity to produce both wood and seeds are key features of lignophytes.

Figure 29.16 The polyester cutin found in cuticle, sporopollenin on spore walls, and lignin on water-conducting tracheids of vascular tissues are resistant to decay and thus help plants fossilize.

Figure 29.17 During the Carboniferous (Coal Age), atmospheric oxygen levels reached historic high levels that were able to supply the large needs of giant insects, which obtain oxygen by diffusion.

Figure 29.20 Increase in the surface area of placental transfer cells provides room for increased numbers of transport proteins. The greater the number of transport proteins per cell, the more rapidly solutes can be transferred from one cell to another.

Figure 29.22 Because the veins of fern leaves reflect the vascular systems of branched stem systems, you might infer that leaves evolved from more highly branched stem systems would be more-densely veined, that is, have more veins per unit area than fern leaves.

Figure 29.23 Although some angiosperm seeds, such as those of corn and coconut, contain abundant endosperm, many angiosperm embryos consume most or all of the nutritive endosperm during their development.

Figure 29.25 Because the lacy integument of *Runcaria* does not completely enclose the megasporangium, it probably did not function to protect the megasporangium before fertilization nor as an effective seed coat after fertilization, as do the integuments of modern seed plants. However, the lacy integument of *Runcaria* might have retained the megasporangium on the parent sporophyte during the period of time when nutrients flowed from parent to developing ovule and seed. That function would prevent megasporangia from dropping off the parent plant before fertilization occurred, allow the parent plant to provide nutrients needed during embryo development, and allow seeds time to absorb and store more nutrients from the parent. Such a function would illustrate how one mutation having a positive reproductive benefit can lay the foundation for subsequent mutations that confer additional fitness. *Runcaria* illustrates a first step in the multistage evolutionary process that gave rise to modern seeds.

Feature Investigation Questions

1. The experimental goals were to determine the rate at which organic molecules produced by gametophyte photosynthesis were able to move into sporophytes and to investigate the effect of sporophyte size on the amount of organic molecules transferred from the gametophyte.

2. The investigators shaded sporophytes with black glass covers to ensure that all of the radioactive organic molecules detected in sporophytes at the end of the experiment came originally from the gametophyte.

3. The investigators measured the amount of radioactivity in gametophytes and sporophytes, and in sporophytes of different sizes. These measurements indicated the relative amounts of labeled organic compounds that were present in different plant tissues.

Test Yourself

1. c   2. d   3. d   4. e   5. b   6. a   7. c   8. e   9. c   10. b

Conceptual Questions

1. Charophycean algae, particularly the complex genera *Chara* and *Coleochaete,* share many features of structure, reproduction, and biochemistry with land plants. Examples include cell division similarities and plasmodesmata and sexual reproduction by means of flagellate sperm and eggs.

2. Bryophytes are well adapted for sexual reproduction when water is available for fertilization. Their green gametophytes efficiently transfer nutrients to developing embryos, enhancing their growth into sporophytes. Their sporophytes are able to produce many genetically diverse spores as the result of meiosis and effectively disperse these spores by means of wind.

3. Vascular tissues allow tracheophytes to effectively conduct water from roots to stems and to leaves. Waxy cuticle helps prevent loss of water by evaporation through plant surfaces. Stomata allow plants to achieve gas exchange under moist conditions and help them avoid losing excess water under arid conditions.

Chapter 30

Concept Checks

Figure 30.3   People should not eat cycad seeds and other plant parts because they typically contain toxic compounds.

Figure 30.4   The nitrogen-fixing cyanobacteria that often occur within the coralloid roots of cycads are photosynthetic organisms that require light. If coralloid roots occurred underground, symbiotic cyanobacteria would not receive enough light to survive.

Figure 30.9   In the dry or cold conditions in which conifers live, air bubbles may form in tracheids; these air bubbles can interfere with water flow if they spread from one tracheid to another. When such bubbles form, the torus seals off connections to other tracheids, thereby helping to maintain water flow.

Figure 30.10   Ways in which conifer leaves are adapted to resist water loss include low surface area/volume needle- or scale-shape, thick surface coating of waxy cuticle, and stomata that are sunken into the leaf and are therefore less exposed to drying winds.

Figure 30.12   Wide vessels are commonly present in the water transport tissues of angiosperms and much less commonly in other plants. The vessels occasionally found in nonangiosperms are thought to have evolved independently from those of angiosperms.

Figure 30.20   A large, showy perianth would not be useful to grass plants because they are wind pollinated; such a perianth would interfere with

pollination in grasses. By not producing a showy perianth, grasses increase the chances of successful pollination and save resources that would otherwise be consumed during perianth development.

Figure 30.24   The flower characteristics of *Brighhamia insignis* shown in this figure (white color and deep, narrow nectar tubes) are consistent with pollination by a moth (see Table 30.1).

Figure 30.26   Importantly, ears of modern *Zea mays* do not readily shatter when the fruits are mature, as do those of teosinte. This feature fosters human ability to harvest the fruits.

Feature Investigation Questions

1. The investigators obtained many samples from around the world because they wanted to increase their chances of finding as many species as possible.

2. The researchers grew plants in a greenhouse under consistent environmental conditions because they wanted to reduce possible impact of environmental variation on the ratio of cannabinoids produced.

3. Although cannabinoids are produced in glandular hairs that cover the plant surface, these compounds are most abundant on leaves near the flowers. Collecting such leaves reduces the chances that compounds might be missed by the analysis.

Test Yourself

1. d   2. a   3. e   4. e   5. b   6. d   7. e   8. c   9. d   10. e

Conceptual Questions

1. Humans should generally not consume food products made from cycads because these plants typically produce toxins that protect against herbivory. At least some cycads harbor cyanobacterial symbionts that are known to produce an unusual amino acid (BMAA) that is associated with dementia in humans.

2. Apple, strawberry, and cherry plants coevolved with animals that use the fleshy, sweet portion of the fruits as food and excrete the seeds, thereby dispersing them. Humans have sensory systems similar to those of the target animals and likewise are attracted by the same colors, odors, and tastes.

3. A sunflower is not a single flower, but rather is an inflorescence, a group of flowers.

Chapter 31

Concept Checks

Figure 31.3   Fungal hyphae growing into a substrate having much higher solute concentration will tend to lose cell water to the substrate, a process that could inhibit fungal growth. This process explains how salting or drying foods helps to protect them from fungal degradation and thus are common preservation techniques.

Figure 31.6   You might filter the air entering the patient's room and limit the entry of visitors and materials that could introduce fungal spores from the outside environment.

Figure 31.8   Early-diverging fungi that are adapted to live and reproduce in the water would not display the reproductive features shown.

Figure 31.11   The toxin amanitin could kill eukaryotic cells by interfering with transcription, thus stopping gene expression.

Figure 31.16   Compounds that interfere with the function of histidine kinase would not harm humans (because we lack such proteins) but would prevent dimorphic fungi from producing the yeast form that spreads so readily in the body by means of budding.

Figure 31.18   Modern AM (arbuscular mycorrhizal fungi), also known as Glomeromycota, do not occur separately from plant hosts, as far as is known.

Figure 31.19   Ectomycorrhizal fungi provide their plant partners with water and minerals absorbed from a much larger area of soil than plant roots can exploit on their own.

Figure 31.22   Lichens do not necessarily contain the same algae as the soredia from which the lichens develop, because lichens often switch algal partners.

Feature Investigation Questions

1. Plants growing on soils up to 65°C would be expected to have fungal endophytes that aid in heat stress tolerance.

2. The investigators cured some of their *C. protuberata* cultures of an associated virus; then they compared the survival of plants infected with fungal endophytes that had virus versus endophytes lacking virus under conditions of heat stress. Only plants having fungal endophytes that possessed the virus were able to survive growth on soils of high temperature.

3. The fungus *C. protuberata* might be used to confer heat stress tolerance to crop plants, as the investigators demonstrated in tomato.

Test Yourself

1. c   2. b   3. e   4. b   5. a   6. d   7. e   8. b   9. e   10. a

Conceptual Questions

1. Fungi are like animals in being heterotrophic, having absorptive nutrition, and storing surplus organic compounds in their cells as glycogen. Fungi are like plants in having rigid cell walls and reproducing by means of walled spores that are dispersed by wind, water, or animals.

2. Toxic or hallucinogenic compounds likely help to protect the fungi from organisms that would consume them.

3. Some fungi partner with algae or cyanobacteria to form lichens. Some fungi associate with plant roots to form mycorrhizae. Some fungi grow as endophytes within the bodies of plants. In all cases, the heterotrophic fungi receive photosynthetic products from the autotrophic partner.

Chapter 32

Concept Checks

Figure 32.2   Simple choanoflagellates are single-celled organisms. Only later, when such organisms become colonial and groups of cells acquire specialized functions, as in sponges, can we consider them early animals.

Figure 32.5   The coelom functions as a hydrostatic skeleton, which aids in movement. This feature permitted increased burrowing activity and contributed to the development of a profusion of wormlike body shapes.

Figure 32.10   Molecular analysis splits the protostomes into two distinct clades—the Lophotrochozoa and the Ecdysozoa—whereas the traditional phylogeny does not.

Figure 32.11   The sister group to the deuterostomes is the protostomes, but in the body plan phylogeny, the sister group is a more loosely defined part of the protostomes, the coelomate protostomes, including mollusks, annelids, and arthropods.

Figure 32.12   The main members of the Ecdysozoa are the arthropods (insects, spiders, and crustaceans) and the nematodes.

Feature Investigation Questions

1. The researchers sequenced the complete gene that encodes small subunit rRNA from a variety of representative taxa of animals to determine their phylogenetic relationships, particularly the relationships of arthropods to other animal taxa.

2. The results indicated a monophyletic clade containing arthropods and nematodes, plus several other smaller phyla. This clade was called the Ecdysozoa. The results of this study indicated that nematodes were more closely related to the arthropods than previously believed.

3. The fruit fly, *Drosophila melanogaster,* and the nematode, *Caenorhabditis elegans,* have been widely studied to understand early development. Under the traditional phylogeny, these two species were not considered to be closely related, so similarities in development were assumed to have arisen early in animal evolution. With the closer relationship indicated by this study, these similarities may have evolved after the divergence of the Ecdysozoan clade. This puts into question the applicability of studies of these organisms to the understanding of human biology.

Test Yourself

1. b   2. c   3. e   4. c   5. c   6. e   7. d   8. d   9. b   10. e

Conceptual Questions

1. (1) Absence or existence of different tissue types. (2) Type of body symmetry. (3) Presence or absence of a true body cavity. (4) Patterns of embryonic development.

2. Radially symmetric animals can be divided equally by a longitudinal plane passing through the central axis. Bilaterally symmetric animals can be divided along a vertical plane at the midline to create two halves; thus, a bilateral animal has a left side and a right side, which are mirror images.

3. Sea urchins are deuterostomes. Deuterostomes have indeterminate cleavage, and all cells have the ability to develop into a complete embryo. Humans are also deuterostomes, so sea urchin embryos can be used as a model for human development.

Chapter 33

Concept Checks

Figure 33.2   Sponges aren't eaten by other organisms because they produce toxic chemicals and contain needle-like silica spicules that are hard to digest.

Figure 33.4   The dominant life stages are jellyfish: medusa; sea anemone: polyp; Portuguese man-of-war: polyp (in a large floating colony).

Figure 33.7   Having no specialized respiratory or circulatory system, flatworms obtain oxygen by diffusion. A flattened shape ensures no cells are too far from the body surface.

Figure 33.11   (1) A ciliary feeding device, and (2) a respiratory device are the two main functions of the lophophore.

Figure 33.12   Technically, most mollusks pump hemolymph into vessels and then into tissues. The hemolymph collects in open, fluid-filled cavities called sinuses, which flow into the gills and then back to the heart. This is known as an open circulatory system. Only closed circulatory systems pump blood, as occurs in the cephalopods.

Figure 33.17   Some advantages of segmentation are organ duplication, minimization of body distortion during movement, and specialization of some segments.

Figure 33.20   Other parasitic nematodes in humans are roundworms, *Ascaris lumbricoides;* hookworms, *Necator americanus;* and pinworms, *Enterobius vermicularis.*

Figure 33.25   All arachnids have a body consisting of two tagmata: a cephalothorax and an abdomen. Insects have three tagmata: a head, thorax, and abdomen.

Figure 33.27   Two key insect adaptations are the development of wings and an exoskeleton that reduced water loss and aided in the colonization of land.

Figure 33.33   In embryonic development, deuterostomes have radial cleavage, indeterminate cleavage, and the blastopore becomes the anus. (In protostomes, cleavage is spiral and determinate, and the blastopore becomes the mouth.)

Figure 33.34   Two unique features of an echinoderm are an internal skeleton of calcified plates and a water vascular system.

Feature Investigation Questions

1. The researchers tested the hypothesis that an octopus can learn by observing the behavior of another octopus.

2. The results indicated that the observer learned by watching the training of the other octopus. The observer was much more likely to choose the same color ball that the demonstrator was trained to attack. These results seem to support the hypothesis that octopuses can learn by observing the behavior of others.

3. The untrained octopuses had no prior exposure to the demonstrators. The results indicated that these octopuses were as likely to attack the white ball as the red ball. No preference for either color was indicated. The untrained octopuses acted as a control. This is an important factor to ensure the results from the trials using observers indicate response to learning and not an existing preference for a certain color.

Test Yourself

1. b   2. d   3. d   4. d   5. b   6. c   7. b   8. a   9. c   10. a

Conceptual Questions

1. The five main feeding methods used by animals are (1) suspension feeding, (2) decomposition, (3) herbivory, (4) predation, and (5) parasitism. Suspension feeding is usually used to filter out food particles from the water column. A great many phyla are filter feeders, including sponges, rotifers, lophophorates, some mollusks and echinoderms and tunicates. Decomposers usually feed on dead material such as animal carcasses or dead leaves. For example, many fly and beetle larvae feed on dead animals, and earthworms consume dead leaves from the surface of the Earth. Earthworms and crabs also sift through soil or mud, eating the substrate and digesting the soil-dwelling bacteria, protists, and dead organic material. Herbivores eat plants or algae and are especially common in the arthropoda. Adult moths and butterflies also consume nectar. Snails are also common plant feeders. Predators feed on other animals, killing their prey, and may be active hunters or sit-and-wait predators. Many scorpions and spiders actively pursue their prey, whereas web-spinning spiders ambush their prey using webs. Parasites also feed on other animals but do not normally kill their hosts. Endoparasites live inside their hosts and include flukes, tapeworms, and nematodes. Ectoparasites live on the outside of their hosts and include ticks and lice.

2. The nematocyst is a powerful capsule with an inverted coiled and barbed thread that functions to immobilize small prey so they can be passed to the mouth and ingested. It is a unique and characteristic feature of the cnidarians.

3. Complete metamorphosis has four stages: egg, larva, pupa, and adult. The larval stage is often spent in an entirely different habitat from that of the adult, and larval and adult forms utilize different food sources. Incomplete metamorphosis has only three stages: egg, nymph, and adult. Young insects, called nymphs, look like miniature adults when they hatch from their eggs.

Chapter 34

Concept Checks

Figure 34.1   Vertebrates (but not invertebrates) usually possess a (1) notochord; (2) dorsal hollow nerve chord; (3) pharyngeal slits; (4) postanal tail, exhibited by all chordates; (5) cranium; (6) neural crest, exhibited by all craniates; (7) vertebral column; (8) endoskeleton of cartilage or bone; and (9) diversity of internal organs.

Figure 34.2   The hagfish is not a true fish because it does not possess vertebrae.

Figure 34.7   Ray-finned fishes (but not sharks) have a (1) bony skeleton; (2) mucus-covered skin; (3) swim bladder; and (4) operculum covering the gills.

Figure 34.11   The advantages to animals that moved onto land included an oxygen-rich environment and a bonanza of food in the form of terrestrial plants and the insects that fed on them.

Figure 34.14   No. Caecilians and some salamanders give birth to live young.

Figure 34.15   Besides the amniotic egg, other critical innovations in amniotes are thoracic breathing; internal fertilization; a thicker, less permeable skin; and more efficient kidneys.

Figure 34.18   Both classes have four-chambered hearts and care for their young.

Figure 34.21   Adaptations in birds to reduce body weight for flight include a lightweight skull; reduction of organ size; and a reduction of organs outside of breeding season. Also female birds have one ovary and relatively few eggs, and no urinary bladder.

Figure 34.28   Defining features of primates are grasping hands; eyes situated on the front of the head to facilitate binocular vision; a large brain; and digits with flat nails instead of claws.

Feature Investigation Questions

1. The researchers were interested in determining the method in which *Hox* genes controlled limb development.

2. The researchers bred mice that were homozygous for certain mutations in specific *Hox* genes. This allowed the researchers to determine the function of individual genes.

3. The researchers found that homozygous mutants would develop limbs of shorter lengths compared to the wild-type mice. The reduced length was due to the lack of development of particular bones in the limb, specifically, the radius, ulna, and some carpels. These results indicated that simple mutations in a few genes could lead to dramatic changes in limb development.

Test Yourself

1. e   2. d   3. a   4. d   5. d   6. c   7. c   8. a   9. c   10. d

Conceptual Questions

1. Both taxa have external limbs that move when the attached muscles contract or relax. The difference is that arthropods have external skeletons with the muscles attached internally, whereas vertebrates have internal skeletons with the muscles attached externally.

2. The sensors of the lateral line pick up pressure waves and send nervous signals to the brain. The operculum is a protective flap that covers the gills.

3. 1. The amnion, the innermost membrane, protects the developing embryo in a fluid-filled sac.

   2. The yolk sac provides a stockpile of nutrients.

   3. The allantois functions as a disposal sac for metabolic wastes.

   4. The chorion allows gas exchange between the embryo and the surrounding air.

Chapter 35

Concept Checks

Figure 35.3   Because organ systems are defined as structures that are composed of more than one organ, roots lack organ systems.

Figure 35.5   As in the case of shoots, the capacity to divide the root into two equal pieces by means of a line drawn from the circular edges through the center would indicate that a root has superficial radial symmetry. In order to determine that an organ has radial symmetry at the cellular level, you would have to compare the microscopic views of randomly chosen, wedge-shaped pieces of cross-slices. If the structure of the wedges is similar, the organ has radial symmetry at the microscopic level.

Figure 35.8   Locating stomata on the darker and cooler lower leaf surface helps reduce water loss from the leaf.

Figure 35.12   A twig having five sets of bud scale scars is likely to be approximately 6 years old.

Figure 35.17   Cactus stems are green and photosynthetic, playing the role served by the leaves of most plants.

Figure 35.22   A woody stem builds up a thicker layer of wood than inner bark in part because older tracheids and vessel element walls are not lost during shedding of bark, which is the case for secondary phloem. In addition, plants typically produce a greater volume of xylem than phloem tissue per year, in part because vessel elements are relatively wide. A large volume of water-conducting tissue helps plants maintain a large amount of internal water.

Figure 35.26   Lateral roots are produced from internal meristematic tissue because roots do not produce axillary buds like those from which shoot branches develop. Internal production of branch roots helps to prevent them from shearing off as the root tip grows through abrasive soil.

Feature Investigation Questions

1. The advantages of using natural plants include the opportunity to avoid influencing plants with unnatural environmental factors, such as artificial light, and the exposure of all experimental plants to similar growth conditions. In addition, the investigators studied the leaves of some large trees, which would be hard to accommodate in a greenhouse.

2. Pinnately veined leaves were splinted to prevent their breaking, since they were cut at the single main vein, which has both support and conducting functions.

3. Sack and associates measured leaf water conduction at two or more places on each leaf because the effect of cutting a vein might have affected some portions of leaves more than others.

Test Yourself

1. d   2. c   3. b   4. a   5. c   6. a   7. d   8. e   9. 6   10. d

Conceptual Questions

1. If overall plant architecture were bilaterally symmetrical, plants would be shaped like higher animals, with a distinct front (ventral surface) and back (dorsal surface). By comparison to radially symmetrical organisms, bilaterally symmetrical plants would have reduced ability to deploy branches and leaves in a way that would fill available lighted space and would thus not be able to take optimal photosynthetic advantage of their habitats.

2. If leaves were generally radially symmetrical (shaped like spheres or cylinders), leaves would not have maximal ability to absorb sunlight, and they would not be able to optimally disperse excess heat from their surfaces.

3. Although tall herbaceous plants exist (palms and bamboo are examples), the additional support and water-conducting capacity that are provided by secondary xylem allow woody plants to grow tall.

Chapter 36

Concept Checks

Figure 36.4   Auxin efflux carriers could be located on the upper sides of root cells, thereby allowing auxin to move upward in roots.

Figure 36.6   Once a callus has been established from a single plant having desirable characteristics using plant tissue culture, the callus can be divided into many small calluses. A grower could transfer these to separate containers having the appropriate hormone mixtures to induce root and shoot growth, then transplant the young plant clones to soil. This would allow the grower to produce many identical plants.

Figure 36.9   The triple response that dicot seedlings show in response to internally produced ethylene allows them to protect the delicate apical meristem from damage as the seedling emerges through the soil.

Figure 36.10   The active conformation of phytochrome absorbs far-red light. Such absorption causes the active conformation of phytochrome to change

to the inactive conformation and to move out of the nucleus and into the cytosol.

Figure 36.11   The inactive conformation of phytochrome would absorb the red portion of sunlight, thereby converting phytochrome into the active conformation.

Figure 36.12   Exposing plants to brief periods of darkness during the daytime will have no effect on flowering because flowering is determined by night length.

Figure 36.14   Yes, just as shoots exhibit negative gravitropism in upward growth, roots are capable of using negative phototropism to grow downward, because light decreases with depth in the soil.

Figure 36.16   In some plants, aerenchyma development is genetically determined and occurs even in the absence of flooding. In other cases, aerenchyma develops only under flooding conditions as a result of controlled cell death.

Figure 36.17   Predators are more likely to be able to find their prey if the latter are concentrated and exposed while feeding on plants. Plants benefit when predation removes herbivores, a process that lessens damage to plants.

Figure 36.19   Similar suites of protective plant hormones, such as jasmonic acid, are used in both types of defenses.

Feature Investigation Questions

1. The following experiment by Briggs falsified the hypothesis that light destroys auxin:

2. Hypothetically, auxin enhances the rate at which cell membrane proton pumps acidify the plant cell wall, thereby allowing cells to extend. Although the evidence for acid effects on cell wall extension is strong, the molecular basis of possible auxin effects on proton pumps is not as yet clear.

3. A small number of seedling tips could display atypical responses for a variety of reasons. The investigators actually performed the experiment with many replicate seedling tips (coleoptiles), in order to gain confidence that the responses are general.

Test Yourself

1. c   2. c   3. a   4. e   5. d   6. d   7. c   8. d   9. d   10. d

(b) Briggs experiment 1

HYPOTHESIS   Light destroys auxin on lit side of shoot tips, causing unequal auxin distribution. Unlit side should grow more than lit side.

STARTING MATERIALS   Corn seedlings.

	Experimental level	Conceptual level
1 Collect auxin into agar blocks from: A dark-grown tips B tips grown with directional light	Dark-grown tip — Auxin diffusion — Directional light-grown tip A B	If light destroys auxin on one side, less auxin will enter the block.
2 Place agar blocks on right side of decapitated shoots.	Agar block — Dark-grown Shoot — Light-grown A B	If the block on the right side has less auxin, it will cause less bending.

3 **THE DATA**	4 **CONCLUSION**
Dark-grown 26° Light-grown 26° A B	Similar bending demonstrates that light did not destroy auxin in the directionally lit shoot tip. If it had, less auxin would have been present in the agar block in B, and the degree of bending would have been less. The hypothesis described under conceptual level (above) is incorrect.

Conceptual Questions

1. Behavior is defined as the responses of living things to a stimulus. Therefore, because plants display many kinds of responses to diverse stimuli, they display behavior.

2. Many kinds of disease-causing bacteria and fungi occur in nature, and these organisms evolve very quickly, producing diverse elicitors. Thus, plants must maintain a stock of resistance genes, each having many alleles.

3. Talking implies a conversation with "listeners" who detect a message and respond to it. Thus, plants that exude volatile compounds that attract enemies of herbivores could be interpreted as "talking" to those enemies. The message is "Hey, you guys, there's food for you over here." In addition, research has revealed that some plants near those under attack respond to volatile compounds by building up defenses. "Talking" to other plants does not enhance the "talker's" fitness. But the ability to "listen" enhances the "listener's" fitness, because it can take preemptive actions to prevent attack.

Chapter 37
Concept Checks

Figure 37.4   Plants adapted to deep shade may have green and photosynthetic plastids in their epidermal cells, whereas the epidermal cells of most plants have plastids that lack chlorophyll.

Figure 37.5   Plastids that occur in a cluster near the nucleus would have more rubisco than plastids at the periphery.

Figure 37.6   Chlorosis is not always a sign of iron deficiency; it can be a deficiency symptom for several mineral nutrients, including zinc in corn.

Figure 37.10   Mineral leaching occurs more readily from sandy soils than from clay soils.

Figure 37.11   Tropical soils are often acidic, as are those of higher latitude regions that are impacted by acid rain resulting from air pollution.

Figure 37.12   The list could include chlorophyll, carotenoids, ATP, NADP, rubisco, and many other enzymes.

Figure 37.13   Soil crusts containing nitrogen-fixing cyanobacteria increase soil fertility, fostering the growth of larger plants that stabilize soils against erosion and provide forage for animals.

Figure 37.14   Oxygen, which makes up 21% of Earth's present atmosphere, can bind to the active site of nitrogenase, thereby inactivating it.

Figure 37.16   Acidic conditions cause soil aluminum ions to become more mobile and thus to enter plants more easily. Hydrangea stores aluminum ions in the cell vacuole, a protective adaptation that prevents aluminum from damaging other cell components. The aluminum causes pigments that are also located in the vacuoles of sepals to change from pink to blue.

Feature Investigation Questions

1. During this period of time, the amount of phosphorus in plant tissues had significantly decreased, but plant growth had not yet been affected. Thus, a monitoring system based on gene expression changes occurring during this time would allow farmers time to apply fertilizer in order to prevent crop losses resulting from nutrient deficiency.

2. *SDQ1* expression is induced by phosphorus deficiency. This gene fosters replacement of plastid phospholipids with sulfur-containing lipids, thereby reducing the plant's phosphorus requirement.

3. They used genetic engineering techniques to place a reporter gene under the control of the *SQD1* promoter, so that when *SQD1* was expressed, the reporter gene was expressed also. After growing plants in nutrient solutions containing various levels of phosphorus, they removed sample leaves and treated them with a compound that turns blue when the reporter gene is expressed. When they saw blue leaves, the investigators could infer (1) that the plants from which those leaves had been taken were beginning to experience phosphorus deficiency, and (2) that application of fertilizer at this point could prevent damage to the plants.

Test Yourself

1. e   2. a   3. c   4. b   5. e   6. c   7. d   8. d   9. a   10. b

Conceptual Questions

1. Agricultural experts are concerned that adding excess fertilizer to crop fields increases the costs of crop production. Ecologists are concerned that excess fertilizers will wash from crop fields into natural waters and cause harmful overgrowths of cyanobacteria, algae, and aquatic plants. Methods for closely monitoring crop nutrient needs so that only the appropriate amount of fertilizer is applied would help to allay both groups' concerns.

2. Boron deficiency induces the expression of a gene that encodes a membrane transporter protein that moves boron from living root cells into root xylem for transport throughout the plant. This action helps to reduce boron deficiency in plant tissues. When plants are exposed to too much boron, the transporter protein is removed from root cell membranes, with the result that boron cannot move into the xylem and the rest of the plant. This action protects the plant from boron toxicity.

3. Use Figure 37.10 as a reference. A first arrow could be drawn from a root to rhizobia in the soil, and the arrow labeled "flavonoids." A second arrow could be drawn from rhizobia to roots and labeled "Nod factors." A third arrow from rhizobia to roots could be labeled "infection proteins." A fourth arrow from roots to rhizobia could be labeled "nodulins" and the resulting nodule environmental conditions, which influence the formation of bacteroids. A fifth arrow could represent the flow of fixed nitrogen from bacterioids to plant. A sixth arrow could represent the flow of organic compounds from plant to bacteroids.

Chapter 38
Concept Checks

Figure 38.5   When placed in pure water, a turgid cell having a water potential of 1.0 will lose water, because 1 is greater than 0. When placed in pure water, a plasmolyzed cell having a water potential of -1.0 MPa will gain water. When placed in pure water, a flaccid cell having a water potential of -0.5 MPa will gain water. This is because water moves from a region of higher water potential to a region of lower water potential, and 0 is greater than -0.5.

Figure 38.8   In most plants, most Al^{3+} will remain in the roots, because the ions can move through outer root tissues but cannot cross the endodermis. However, some aluminum may be able to get into root vascular tissues by seeping into root tips where the endodermis has not yet formed or at places where the endodermis has been broken by the emergence of branch roots. These processes, together with the fact that aluminum ions are very abundant in soils, explain why plant tissues may contain some aluminum even though it is not a plant nutrient.

Figure 38.11   You would likely see stained rings or helical ribbons extending up the insides of the long walls of extensible tracheids. You would not see staining at the ends of tracheids, where they connect to form cell files.

Figure 38.12   The large perforations in vessel element end walls allow an air bubble to extend from one element to another, thereby clogging vessels and preventing water flow through them. In contrast, the much smaller pores in the end walls of tracheids do not allow water to flow as efficiently as it does through vessels, but these smaller pores also retard the movement of air bubbles. As a result, air bubbles are confined to a single tracheid where they do little harm.

Figure 38.13   Root pressure can help to reverse embolism, thereby aiding water flow through xylem.

Figure 38.16   The evaporation of water has a powerful cooling effect because it disperses heat so effectively. Water has the highest heat of vaporization of any known liquid.

Figure 38.17   You could model a stomatal guard cell with an elongate balloon by partially inflating it, then attaching thick tape along one side to represent thickened inner walls and circles of string or thin tape to represent radial cellulose, then adding more air to the balloon. The balloon should curve as it expands, just as a guard cell does when the stomatal pore opens. Two such balloons could be used to model both guard cells and the stomatal pore.

Figure 38.19   In its desert habitat, times of drought and contrasting availability of water sufficient to support the development and photosynthetic function of leaves do not occur at predictable times, as is the case for temperate forests. For this reason, ocotillo leaf abscission is not amenable to the evolution of genetic mechanisms that allow leaf drop to be precisely timed in anticipation of the onset of drought.

Figure 38.25 You could note the relatively few genes that are plotted along the middle left side of the triangle, then try to localize the encoded proteins within the tissues of very young stem tissue, using microscopy.

Feature Investigation Questions

1. This design allowed investigators to compensate for variation among plants, which might have influenced the results had they used separate plants for experiments and controls.

2. Transpiration! Water evaporating from the surfaces of leaves exerted a tension on the water column of the xylem, pulling sap and water through it.

3. The effects of ions on sap flow rates did not directly depend on a biological process, so xylem sap of the same ionic concentration moved through dead plants at the same rate as in living plants.

Test Yourself

1. c 2. b 3. e 4. d 5. a 6. d 7. d 8. e 9. a 10. c

Conceptual Questions

1. In the case of plant fertilizers, more is not better, because the ion concentration of overfertilized soil may become so high as to draw water from plant cells. In this case, the cells would be bathed in a hypertonic solution and would likely lose water to the solution. If plant cells lose too much water, they will die.

2. When the natural vegetation is removed, transpiration stops, so water is not transported from the ground to the atmosphere, where it may be an important contributor to local rainfall. Extensive removal of plants actually changes local climates in ways that reduce agricultural productivity and human survival.

3. You cannot assume that an ocotillo plant lacking leaves is dead, because this plant responds to drought by shedding its leaves, and living plants can produce new leaves when the drought stress is relieved. However, if the ocotillo plants do not produce new leaves after normal rainstorms, you might suspect that they have died.

Chapter 39

Concept Checks

Figure 39.2 Because gametophytes are haploid, they lack the potential for allele variation at each gene locus that is present in diploid sporophytes. Hence, gametophytes are more vulnerable to environmental stresses. By living within the diploid tissues of flowers, flowering plant gametophytes are protected to some extent, and the plant does not lose its gamete-producing life cycle stage.

Figure 39.3 Some flowers lack some of the major flower parts.

Figure 39.4 By clustering its stamens around the pistil, the hibiscus flower increases the chance that a pollinator will both pick up pollen and deliver pollen from another hibiscus flower on the same trip.

Figure 39.7 The absence of showy petals often correlates with wind pollination, because large petals would interfere with the shedding of pollen in the wind.

Figure 39.10 The rim flowers of *Gerbera* inflorescences have bilateral symmetry, conferred by expression of a *CYCLOIDEA*-like gene. Rim flowers also possess showy petals that attract pollinators, but lack pollen-producing stamens. By contrast, central flowers display radial symmetry, lack showy petals, and possess pollen-producing stamens.

Figure 39.11 The maximum number of cells in a mature male gametophyte of a flowering plant is three: a tube cell and two sperm cells.

Figure 39.12 Female gametophytes are not photosynthetic and cannot produce their own food. Enclosed within ovules, female gametophytes lack direct access to the outside environment. Carpels contain veins of vascular tissue that bring nutrients from sporophytic tissue to ovules.

Figure 39.17 An embryo in which the TOPLESS genes were nonfunctional would have two roots and no shoots.

Figure 39.18 During their maturation, the cotyledons of eudicot seeds absorb the nutrients originally present in endosperm.

Feature Investigation Questions

1. Plant gametes, particularly egg cells, are produced by microscopic gametophytes enclosed by sporophytic flower tissues and were thus difficult to isolate.

2. The investigators used a pulse of electricity to stimulate cell fusion. This process is similar to electroporation, the use of electrical discharges to cause small pores to appear in the membranes of cells prior to their transformation during genetic engineering.

3. Investigators obtained egg cells and sperm cells from parents that differed in the color of stigmas and stigma hairs. They demonstrated that progeny plants were hybrids that had stigmas that were colored the same as one parent and stigma hairs that were colored the same as the other parent.

Test Yourself

1. a 2. b 3. d 4. b 5. d 6. e 7. b 8. c 9. d 10. c

Conceptual Questions

1. Pollen grains are vulnerable to mechanical damage and microbial attack during the journey through the air from the anthers of a flower to a stigma. Sporopollenin is an extremely tough polymer that helps to protect pollen cells from these dangers. The function of the beautiful sculptured patterns of sporopollenin on pollen surfaces is unclear.

2. The embryos within seeds are vulnerable to mechanical damage and microbial attack after they are dispersed. Seed coats protect embryos from these dangers and also help to prevent seeds from germinating until conditions are favorable for seedling survival and growth.

3. Flower diversity is an evolutionary response to diverse pollination circumstances. For example, plants such as oak and corn that are wind-pollinated produce flowers having a poorly developed perianth. If such wind-pollinated flowers had large, showy perianths, they would get in the way of pollen dispersal or acquisition. On the other hand, flowers that are pollinated by animals often have diverse shapes and attractive petals of differing colors or fragrances that have coevolved with different types of animal pollinators.

Chapter 40

Concept Checks

Figure 40.2 Locomotion is the movement of an animal's body from one place to another. This is achieved by the actions of skeletal muscle. However, smooth muscle contraction promotes movement of internal structures, like those of the digestive system, and contraction of cardiac muscle causes movement (beating) of the heart.

Figure 40.6 No. The brain, for example, does not contain muscle tissue (although the blood vessels supplying the brain do contain smooth muscle).

Figure 40.7 Blood, including plasma and blood cells, would leak out of the blood vessel into the interstitial space. The fluid level of the bloodstream would decrease, and that of the interstitial space near the site of the injury would increase. Eventually the blood that entered the interstitial space would be degraded by enzymes, resulting in the characteristic skin appearance of a bruise. If the injury were very severe, the fluid level in the blood could decrease to a point where the various tissues and organs of the body would not receive sufficient nutrients and oxygen to function normally.

Figure 40.8 A decrease in intracellular fluid volume, like that shown in the cell in this figure, would result in an increase in intracellular solute concentration (likewise, an increase in intracellular fluid volume would decrease intracellular solute concentrations). This may have drastic consequences on cell function. For example, some solutes, like Ca^{2+} and certain other ions, are toxic to cells at high concentrations.

Figure 40.10 Surface area is important to any living organism that needs to exchange materials with the environment. A good example of a high surface area/volume ratio is that of most tree leaves. This makes leaves ideally suited for such processes as light absorption (required for photosynthesis; see Chapter 8) and the exchange of gases and water with the environment.

Figure 40.12 No, not necessarily. Body temperature, for example, is maintained at different set points in birds and mammals. Other vertebrates and most invertebrates do not have temperature set points; their body temperature simply conforms close to that of the environment. As another example, a giraffe has a set point for blood pressure that is higher than that of a human being, because a giraffe's circulatory system must generate enough pressure to pump blood up its long neck.

Figure 40.15 In nature, an animal such as a horse would have the same type of responses shown here if threatened by a predator. Upon sensing the presence of the predator, the horse's respiratory and circulatory systems would begin increasing their activities in preparation for the possibility that the horse might have to flee or defend itself. This would occur even before the horse began to flee.

Feature Investigation Questions

1. Pavlov studied feedforward regulation of saliva production that occurs in hungry dogs even before they receive food. He hypothesized that the feedforward response could be conditioned to other, nonrelevant stimuli such as sounds, as long as the sounds were presented simultaneously with food.

2. Pavlov remained outside the room where the dog was housed when the conditioning stimulus—a metronome—was started. In addition, the room was carefully sealed to prevent any other stimuli, including smells, sights, and sounds, from interfering with the conditioning response.

3. He measured the amount of saliva secreted by salivary glands in the dog's mouth by collecting the saliva through a tube and funnel, and then recording the number of drops. He discovered that once a dog had become conditioned to hearing the sound of the metronome whenever presented with food, the sound itself was sufficient to stimulate the feedforward response of salivation. This experiment revealed that feedforward processes could be modulated by experience and learning.

Test Yourself

1. d 2. c 3. d 4. b 5. c 6. c 7. d 8. b 9. d 10. e

Conceptual Questions

1. Anatomy is the study of structure (form), and physiology is the study of function. For example, an anatomist might be interested in the arrangement of cells and tissues in an organ such as the heart, while a physiologist might be interested in the mechanisms by which heart cells contract to produce a heartbeat.

2. Structure and function are related in that the function of a given organ, for example, depends in part on the organ's size, shape, and cellular and tissue arrangement. Clues about a physical structure's function can often be obtained by examining the structure's form. For example, the extensive surface area of a moth's antennae suggests that the antennae are important in detecting the presence of airborne chemicals. Likewise, any structure that contains a large surface area for its volume is likely involved in some aspect of signal detection, cell-cell communication, or transport of materials within the animal or between the animal and the environment. Surface area increases by a power of 2, and volume increases by a power of 3 as an object enlarges; this means that in order to greatly increase surface area of a structure such as an antenna, without occupying enormous volumes, specializations must be present (such as folds) to package the structure in a small space.

3. Homeostasis is the ability of animals to maintain a stable internal environment by adjusting physiological processes, despite changes in the external environment. Examples include maintenance of salt and water balance, pH of body fluids, and body temperature. Some animals conform to their external environment to achieve homeostasis, but others regulate their internal environment themselves.

Chapter 41

Concept Checks

Figure 41.1 This is an example of a feedforward response, most famously demonstrated by the conditioning experiments of Ivan Pavlov. In this case, the peripheral and central nervous systems interact to prepare the hyena for feeding.

Figure 41.4 Many reflexes, such as the knee-jerk reflex, cannot be prevented once started. Others, however, can be controlled to an extent. Open your eyes widely and gently touch your eyelashes. A reflex that protects your eye will tend to make you close your eyelid. However, you can overcome this reflex with a bit of difficulty if you need to, for example, when you are putting in contact lenses.

Figure 41.5 The squid axon is not coated in myelin sheaths. This is ano[ther] feature of the squid giant axon that makes it a convenient model for condu[ct]ing in vitro experiments such as the one depicted in this figure.

Figure 41.7 Yes, the flow of K^+ down its chemical gradient does create a[n] electrical gradient because K^+ is electrically charged. The net flow of K^+ wi[ll] stop when the chemical gradient balances the electrical gradient. This occur[s] at the equilibrium potential.

Figure 41.10 When the K^+ channels open (at 1 msec), the Na^+ channels would still be opened, so the part of the curve that slopes downward would not occur as rapidly, and perhaps the cell would not be able to restore its resting potential.

Figure 41.12 The action potential can move faster down an axon. This is especially important for long axons, such as those that carry signals from the spinal cord to distant muscles.

Figure 41.14 In the absence of such enzymes, neurotransmitters would remain in the synapse for too long, and the postsynaptic cell could become overstimulated. In addition, the ability of the postsynaptic cell to respond to multiple, discrete inputs from the presynaptic cell would be compromised.

Feature Investigation Questions

1. Loewi was aware that electrical stimulation of the vagus nerve associated with heart muscle would slow down the rate of heart contractions in a frog. Also, he knew that electrical stimulation of other nerves associated with the frog heart produced opposite results. If the effects of the different nerves on heart muscle were mediated directly by electrical activity only, the heart muscle cells would have no way to distinguish between stimulatory and inhibitory signals. Loewi hypothesized that nerves released chemicals onto heart muscle cells and that it was these different chemicals that produced the varied effects on the heart.

2. Loewi placed two hearts in separate chambers, one heart with its vagus nerve intact and the other with its vagus nerve removed. He electrically stimulated the vagus nerve of the first heart, then removed some of the saline solution surrounding the heart and transferred it to the second heart. He then observed whether or not the second heart responded as if its vagus nerve had been intact and had been stimulated.

3. When fluid from the saline solution around the stimulated heart was added to the saline solution of the second, unstimulated heart, the rate of contraction in the second heart was decreased just as if its own vagus nerve had been intact and was stimulated. This suggested that chemicals were released into the saline solution of the first heart following the electrical stimulation of its vagus nerve, and that it was these chemicals that caused the cardiac muscle to slow its rate of contraction. The results did support Loewi's hypothesis.

Test Yourself

1. c 2. d 3. e 4. e 5. b 6. d 7. b 8. e 9. a 10. e

Conceptual Questions

1. Neurons are highly specialized, electrically excitable cells that communicate with another cell of its kind and with other types of cells by electrical or chemical signals.

   Glia are cells that surround the neurons; they are a major class of cells in nervous systems that perform various functions.

   Schwann cells are glial cells that form myelin around axons in the peripheral nervous system.

2. In a graded potential, a weak stimulus causes a small change in the membrane potential while a strong stimulus produces a greater change. Graded potentials occur along the dendrites and cell body. If a graded potential reaches the threshold potential at the axon hillock, this results in an action potential. This is a change in the membrane potential that is of a constant value and is propagated from the axon hillock to the axon terminal.

3. An increase in extracellular Na^+ concentration would slightly depolarize neurons, thereby changing the resting membrane potential. This effect would be minimal, however, because the resting membrane is not very permeable to Na^+. However, the shape of the action potentials in such neurons would be a little steeper, and the peak a little higher, because the electrochemical gradient favoring Na^+ entry into the cell through voltage-gated channels would be greater.

Chapter 42

Concept Checks

Figure 42.4   Not necessarily. Brain mass is not the sole determinant of intelligence or the ability to perform complex tasks. The degree of folding of the cerebral cortex is also important.

Figure 42.6   A spinal nerve is composed of both afferent and efferent neurons.

Figure 42.7   The major symptom experienced by patients undergoing a lumbar puncture is headache, in part because the brain is no longer cushioned adequately by CSF. Within 24–48 hours, however, the CSF is replenished to normal levels.

Figure 42.9   Damage to the cerebellum would result in loss of balance and a lack of fine motor control, such as picking up fine objects or making graceful, smooth movements.

Figure 42.12   It was in her right hand.

Figure 42.15   Thinking requires energy! Even daydreaming requires energy; imagine how much energy the brain uses when you concentrate for 60 minutes on a difficult exam. In fact, you just expended energy thinking about this question!

Feature Investigation Questions

1. Gaser and Schlaug hypothesized that repeated exposure to musical training would increase the size of certain areas of the brain associated with motor, auditory, and visual skills. All three skills are commonly used in reading and performing musical pieces.

2. The researchers used MRI to examine the areas of the brain associated with motor, auditory, and visual skills in three groups of individuals: professional musicians, amateur musicians, and nonmusicians. The researchers found that certain areas of the brain were larger in the professional musicians compared to the other groups, and larger in the amateur musicians compared to the nonmusicians.

3. Schmithorst and Holland found that, when exposed to music, certain regions of the brains of musicians were activated differently compared to nonmusicians. This study supports the hypothesis that there is a difference in the brains of musicians compared to nonmusicians.

   The experiment conducted by Gaser and Schlaug compared the size of certain regions of the brain among professional musicians, amateur musicians, and nonmusicians. Schmithorst and Holland, however, were also able to detect functional differences between musicians and nonmusicians.

Test Yourself

1. b   2. a   3. e   4. c   5. c   6. b   7. e   8. b   9. a   10. d

Conceptual Questions

1. All animals with nervous systems have reflexes, which allow rapid behavioral responses to changes in the environment. When a cnidarian senses a tactile stimulus, its nerve net responds immediately and the animal reflexively contracts nearly all of its muscles, making the animal a smaller target. This behavior protects the animal from predators. When you hear a loud, unexpected, and frightening sound (such as a firecracker), you hunch your shoulders and slightly lower your head; this reflex protects you from danger by minimizing exposure of your neck and head to danger. Dilation of the pupils of the eyes in darkness, and constriction of the pupils in bright light, are reflexes that help us see in the dark and protect our retinas in bright light. Reflexes are particularly adaptive because they occur rapidly, typically with very few synapses involved, and without the need for conscious thought.

2. White matter consists of the myelinated axons that are bundled together in large tracts in the central nervous system and which connect different CNS regions. The lipid-rich myelin gives the tracts a whitish appearance. It is distinguished from gray matter, which are the cell bodies, dendrites, and some unmyelinated axons of neurons in the CNS.

3. The sympathetic division is responsible for rapidly activating "fight-or-flight" systems that provide beneficial responses to stress or danger. For example, activation of the sympathetic division results in increased heart rate, increased breathing rate, and increased energy production.

   The parasympathetic system is involved in "rest-or-digest" processes that stimulate digestion of food, slow the heart rate, and decrease energy production.

Chapter 43

Concept Checks

Figure 43.1   The term sensory receptor refers to a type of cell that can respond to a specific type of stimulus. The term membrane receptor refers to a protein within a cell membrane that binds a ligand and thereby generates signals that initiate a cellular response.

Figure 43.3   To think about what types of touch you are aware of, let's take the example of sitting in a chair reading this textbook while holding it on your lap. You are aware of the constant weight of the book, the brush of the pages on your fingertips as you turn a page, a gentle breeze that may be circulating in your environment, the deep pressure from regularly adjusting your posture in your chair, an itch you may have on your skin, and the heat or cold of the room. Even a simple exercise such as this one is filled with stimuli of numerous types and durations.

Figure 43.11   This orientation permits animals to detect circular or angular movement of the head in three different planes. The fluid in a canal that is oriented in the same plane as the plane of movement will respond maximally to the movement. For example, the canal that is oriented horizontally would respond greatest to horizontal movements, while the other two canals would not. Overall, by comparing the signals from the three canals, the brain can interpret the motion in three dimensions.

Figure 43.18   The discs of photoreceptors are an example of increasing surface area without greatly expanding the volume of a structure. The greater surface area allows for more rhodopsin molecules per photoreceptor and thus a greater likelihood of capturing even low levels of light.

Figure 43.19   Because red-green color blindness is a sex-linked recessive gene, males require only a single defective allele on an X chromosome, whereas females require two defective alleles, one on each X chromosome.

Figure 43.27   Salt is a vital nutrient needed to maintain plasma membrane potentials and fluid balance in animals' bodies. Sugar provides glucose and other monosaccharides, important energy-yielding compounds. Sour (acidic) foods, like citrus fruits, provide nutrients and important antioxidants (vitamin C, for example) that protect against disease. Bitter substances are often toxic, and their bad taste discourages animals from eating them.

Feature Investigation Questions

1. One possibility is that many different types of odor molecules might bind to one or just a few types of receptor proteins, with the brain responding differently depending on the number or distribution of the activated receptors. The second hypothesis is that organisms can make a large number of receptor proteins, each type binding a particular odor molecule or group of odor molecules. According to this hypothesis, it is the *type* of receptor protein, and not the number or distribution of receptors, that is important for olfactory sensing.

   The researchers extracted RNA molecules from the olfactory receptor cells of the nasal epithelium. They then used this RNA to identify genes that encoded G-protein-coupled receptor proteins.

2. In their study, they identified 18 different genes that encoded different G-protein-coupled receptor proteins.

3. The results of the experiment conducted by Buck and Axel support the hypothesis that animals discriminate between different odors based on having a variety of receptor proteins that recognize different odor molecules. Current research suggests that each olfactory receptor cell has a single type of receptor protein that is specific to particular odor molecules. Because most odors are due to multiple chemicals that activate many different types of odor receptor proteins, the brain detects odors based on the combination of the activated receptor proteins. Odor seems to be discriminated by many olfactory receptor proteins, which are in the membrane of separate olfactory receptor cells.

Test Yourself

1. d   2. d   3. a   4. c   5. e   6. d   7. b   8. b   9. d   10. b

Conceptual Questions

1. Sensory transduction—The process by which incoming stimuli are converted into neural signals. An example would be the signals generated in the retina when a photon of light strikes a photoreceptor.

   Perception—An awareness of the sensations that are experienced. An example would be an awareness of what a particular visual image is.

2. The organ of Corti contains the hair cells and sensory neurons that initiate signaling. The hair cells sit on top of the basilar membrane, and their stereocilia are embedded in the tectorial membrane at the top of the organ of Corti. Pressure waves of different frequencies cause the basilar membrane to vibrate at particular sites. This bends the stereocilia of hair cells back and forth, sending oscillating signals to the sensory neurons. Consequently, the sensory neurons send intermittent action potentials to the CNS via the auditory nerve. Hair cells at the end of the basilar membrane closest to the oval window respond to high-pitched sounds, and lower-pitched sounds trigger hair cell movement further along the basilar membrane.

3. Animals that have both eyes located at the front of the head facing forward have a large degree of binocular vision, because the overlapping images coming into both eyes are processed together to form one perception. Binocular vision provides excellent depth perception. Predators depend on binocular vision to locate prey precisely. In contrast, animals with eyes toward the side of the head have reduced binocular vision, but a wide visual field, which allows them to see a wide area at one time. Most prey animals benefit from a wide visual field, which allows them to detect the presence of predators even, in some cases, if a predator is behind the prey.

Chapter 44

Concept Checks

Figure 44.1   Yes. In addition to not having a requirement to shed their skeletons periodically, animals with endoskeletons can use their skin as an efficient means of heat transfer (and, to an extent in amphibians, water transfer). In addition, the body surface of such animals is often a highly sensitive sensory organ.

Figure 44.3   If a tendon is torn, its ability to link a muscle to bone is reduced or lost. Therefore, when a muscle such as the one shown in this illustration contracts, it will not be able to move the bone to which the tendon has become dislodged.

Figure 44.8   The ATP concentration in cells becomes depleted after death, because oxygen and nutrients are not being provided to cells. Consequently, the cross-bridge cycle becomes locked before Step 3. Without ATP, the cross-bridges cannot dissociate until many hours later, when the muscle tissue sufficiently decomposes.

Figure 44.11 Na^+ enters the muscle cell because all cells have an electrochemical gradient for Na^+ that favors diffusion of Na^+ from extracellular to intracellular fluid (see Chapter 41). This is because cells have a negative membrane potential and because Na^+ concentrations are higher in the extracellular fluid. The acetylcholine receptor on skeletal muscle cells is also a ligand-gated ion channel; when acetylcholine binds the receptor, it induces a shape change that opens the channel. This allows the entry of Na^+ into the cell.

Figure 44.14   No. The data are expressed as "per kg"; this means that when normalized to a standard body mass (1 kg), the amount of energy expended for any type of locomotion by a small animal tends to be greater than that of a larger animal. However, these are *relative* values. For example, the *absolute* amount of energy expended by a tiny minnow is much less than that of a large tuna over any given distance.

Feature Investigation Questions

1. PPAR-δ is a nuclear receptor that regulates the expression of genes that enable cells to more efficiently burn fat instead of glucose for energy.

2. Evans suggested that if PPAR-δ were highly activated in mice, the mice would lose weight because of the high level of fat metabolism.

3. They developed transgenic mice with highly activated PPAR-δ. Then they fed the transgenic mice and a strain of normal mice high-fat diets. They then compared the weights of the two strains of mice to determine if the change in PPAR-δ activity affected weight. The weights of the transgenic mice were considerably lower than those of the normal mice. These results supported the hypothesis that highly activated PPAR-δ would lead to lower weight gain due to fat metabolism. Interestingly, the researchers also discovered that the transgenic mice could perform prolonged exercise for a much longer time than the normal mice. The muscle tissue of the transgenic mice was more specialized for long-term exercise.

Test Yourself

1. c   2. e   3. d   4. c   5. e   6. c   7. a   8. e   9. c   10. e

Conceptual Questions

1. Locomotion is the ability of an animal to move from place to place. Examples include swimming, walking or running, jumping, flying, crawling, and sliding.

2. Exoskeletons are on the outside of an animal's body, and endoskeletons are inside the body. Both function in support and protection, but only exoskeletons protect an animal's outer surface. Exoskeletons must be shed when an animal grows, whereas endoskeletons grow with an animal.

3. a. The cycle begins with the binding of an energized myosin cross-bridge to an actin molecule on a thin filament.

    b. The cross-bridge moves, and the thin filaments slide past the thick filaments.

    c. The ATP binds to myosin, causing the cross-bridge to detach.

    d. The ATP bound to myosin is hydrolyzed by ATPase, re-forming the energized state of myosin.

Chapter 45

Concept Checks

Figure 45.4   After a large blood meal, the body mass of a flying blood-sucking animal increases sufficiently as to make it nearly impossible to fly. The problem is solved, however, by a unique adaptation that allows such animals to concentrate the nutrients from blood and excrete most of the water portion of blood as soon as they begin eating. By the time the meal is finished, much of the water they consumed has already been excreted.

Figure 45.7   Sauropod dinosaurs were herbivores that probably contained a gizzard-type stomach in which stones helped to grind coarse vegetation. Such stones would have become smooth after months or even years of rumbling around in the gizzard. Some of these sauropods are known to have lacked the sort of grinding teeth characteristic of modern mammalian herbivores, and thus a gizzard would have aided in their digestion much as it does in modern birds.

Figure 45.10   By having bile stored in a gallbladder, bile can be released precisely when needed in response to a meal; this is particularly useful for animals that consume large or infrequent meals. In the absence of a gallbladder, bile flows into the intestine continuously and cannot be increased to match the amount or timing of food intake.

Figure 45.11   Secondary active transport requires energy provided by ATP. Thus, absorption of nutrients by this mechanism is an energy-requiring event, and some portion of an animal's regular nutrient consumption is used to provide the energy required to absorb the nutrients.

Figure 45.13   CCK inhibits stomach activity. This is an example of negative feedback. The arrival of chyme in the small intestine stimulates CCK, which promotes digestion as shown in the figure. At the same time, CCK inhibits contraction of the smooth muscles of the stomach so that the entry of chyme into the small intestine is slowed down. This allows time for controlled digestion and absorption of nutrients in the intestine, without the intestine becoming overfilled with chyme. Simultaneously, CCK inhibits acid production by the stomach so that the pH of the intestine does not become dangerously low before bicarbonate ions are able to neutralize it.

Feature Investigation Questions

1. The researchers severed the nerves that connected to the small intestine in a dog. Following the removal of the nerves, the researchers introduced an acidic solution directly into the intestine of the dog. The introduction of the acid into the intestine caused pancreatic secretion. This suggested that non-neural factors must have mediated communication between the digestive tract and the secretory cells of the pancreas.

2. Other researchers were not convinced that all the nerves were dissected from the intestine, because of the technical difficulty in performing such a procedure. To provide more conclusive evidence of other regulatory factors that were produced by the intestine, the researchers conducted a second experiment. First, they dissected a portion of a small intestine from a dog, treated it with acid, ground it up to produce a mash, and then filtered the mash to obtain an extract. The extract—which was expected to contain any secretions of the intestine that occurred following acid exposure—was then injected into the circulatory system of a second dog. The results indicated that the second dog had pancreatic secretion following the injection.

3. The results suggested that factors were secreted by the small intestine following exposure of the intestine to acid, as would occur when chyme enters the intestine from the stomach. These factors probably reached the pancreas through the bloodstream. The researchers called these factors hormones. Thus, the digestive system was regulated not only by the nervous system, but also by chemical secretions, and different parts of the digestive system were able to communicate with each other via hormones.

Test Yourself

1. d 2. e 3. e 4. b 5. c 6. b 7. d 8. c 9. a 10. e

Conceptual Questions

1. Digestion is the breakdown of large molecules into smaller ones by the action of enzymes and acid. Absorption is the transport of digested molecules and small molecules that do not require digestion, across the epithelial cells of the alimentary canal and from there into the extracellular fluid of an animal.

2. Carnivores eat live animal flesh and/or fluids or may scavenge dead animals. Carnivores' teeth are adapted for seizing, grasping, piercing, biting, slicing, tearing, or holding prey; they generally do not chew their food extensively, but may chew to facilitate swallowing. Herbivores have powerful jaw muscles and large, broad molars for grinding tough, fibrous plant material; they may also have incisors adapted for nipping grass or other vegetation. Simply examining the type of teeth an animal has is often sufficient to determine whether that animal eats vegetation, animals, or both.

3. The crop is a dilation of the esophagus, which stores and softens food. The gizzard contains swallowed pebbles that help pulverize food. Both of these functions are adaptations that assist digestion in birds, who do not have teeth and therefore do not chew food. Humans, like many animals, can chew food before swallowing.

Chapter 46

Concept Checks

Figure 46.3 The time required for the vesicles to move to the plasma membrane and fuse with it is much shorter than the time required for new GLUTs to be synthesized by activation of GLUT genes. Thus, the action of insulin on cells is very quick, because the GLUTs are already synthesized.

Figure 46.4 The glycerol and fatty acids used to make glucose are the breakdown products of triglycerides that were stored in adipose tissue during the absorptive period. The amino acids used to make glucose are derived from the breakdown of protein in muscle and other tissue.

Figure 46.6 Even though the goose was resting, sampling the air from the mask would be only a rough estimate of BMR. That is because the artificial setting and the placement of the mask would be enough of a stimulus to affect the activity and behavior of the goose and thereby increase its metabolism.

Figure 46.7 As shown in Figure 46.7, for humans exercise is a voluntary activity. In nature, however, "exercise" is often a component of the fight-or-flight reaction, such as when an animal attempts to escape danger. During such times, digestion and absorption of food are less important than providing as much blood flow, oxygen, and nutrients as possible to skeletal muscle. The gut, therefore, temporarily reduces its activity and requires less blood flow.

Figure 46.8 Nearly all animals today show a similar relationship between body mass and metabolic rate, and there is no reason why it should not always have been true. Thus, the tiny 1-foot-tall ancestral horse *Eohippus* most likely had a higher BMR than do today's larger horses.

Figure 46.11 Humans are homeothermic endotherms. We maintain our body temperature within a very narrow range, and we supply our own body heat.

Feature Investigation Questions

1. Scientists were interested in knowing why animals seemed to regulate their body mass around a particular level, even though many animals experience changes in food supply throughout the year. This seemed to indicate that a mechanism existed within the body that monitored when fuel stores were higher or lower than normal, and that initiated changes in behavior and metabolism to compensate.

2. Coleman hypothesized that communication regarding energy status must take place between the brain and the rest of the body. He suggested that chemical signals were transported through the blood from outside the brain to feeding or satiety centers within the brain, where they regulated appetite and thus regulated body weight. He tested this by linking the blood circulations of normal mice and genetically obese mice and then monitoring the mice for changes in body weight.

3. In most cases, the obese mice lost weight and ate less during the experimental procedure. This confirmed that something in the bloodstream of the wild-type mice was regulating body weight but was missing in the obese mice. When the unknown factor crossed into the bloodstream of the obese mice, it caused them to lose weight. In another group of parabiosed mice, however, the wild-type mice lost weight, but the obese mice did not. Coleman concluded that these obese mice were not able to respond to the chemical signal that regulates body weight, even though they made the signal themselves and it was active in their parabiosed wild-type partners.

Test Yourself

1. c 2. a 3. d 4. c 5. a 6. e 7. e 8. c 9. c 10. e

Conceptual Questions

1. Insulin acts on adipose and skeletal muscle cells to facilitate the diffusion of glucose from extracellular fluid into the cell cytosol. This is accomplished by increasing the translocation of glucose-transporter (GLUT) proteins from the cytosol to sites within the plasma membrane of insulin-sensitive cells. Insulin also inhibits glycogenolysis and gluconeogenesis in the liver, which decreases the amount of glucose secreted into the blood by the liver. Insulin is required for glucose transport because like many other polar molecules, glucose cannot move across the lipid bilayer of a plasma membrane by simple diffusion. The inhibitory effects of insulin on liver function help to ensure that liver glycogen stores will be spared for the postabsorptive period.

2. Appetite is controlled by a satiety center in the brain that receives signals from the stretched stomach and intestines after a meal. When digestion and absorption are complete, the stomach and intestines return to their original size, and the brain no longer senses that an animal feels "full." In addition, appetite is controlled by leptin, a hormone secreted by adipose cells in direct proportion to the amount of fat stored in an animal's body. When leptin levels in the blood are high, appetite is suppressed. When leptin levels are low, as occurs when an animal is losing weight, appetite is increased. The presence of a hormone that is released into the blood in proportion to fat mass in the body allows the brain to monitor the amount of energy stored in the body. A decrease in the concentration of leptin in the blood, for example, is the mechanism that communicates to the brain that fat stores are lower than normal. This initiates the sensation of hunger, which encourages an animal to seek food.

3. Countercurrent heat exchange is a mechanism for retaining body heat. As warm blood travels through arteries down a bird's leg, for example, heat moves by conduction from the artery to adjacent veins carrying cooler blood in the other direction, toward the heart. By the time the arterial blood reaches the tip of the leg, its temperature has dropped considerably, reducing the amount of heat loss to the environment, while the heat is returned to the body's core via the warmed veins.

Chapter 47

Concept Checks

Figure 47.2 Open circulatory systems evolved prior to closed systems. However, this does not mean that open systems are in some way inferior to closed circulatory systems. It is better to think of open systems as being ideally suited to the needs of those animals that have them. Arthropods are an incredibly successful order of animals, with the greatest number of species, and inhabiting virtually every ecological niche on the planet. Clearly, their type of circulatory system has not prevented arthropods from achieving their great success.

Figure 47.4 Keeping oxygenated and deoxygenated blood fully separate allows the arterial blood of birds and mammals to provide the maximum amount of oxygen to tissues. This means that those tissues can achieve higher metabolisms and be more active at all times.

Figure 47.6 Each hemoglobin molecule contains four subunits, each of which has an iron atom at its core. Each iron binds one oxygen molecule (O_2); therefore, a total of eight oxygen atoms can bind to one hemoglobin molecule.

Figure 47.11 Body fluids, both extracellular and intracellular, contain large amounts of charged ions, which are capable of conducting electricity. The slight electric currents generated by the beating heart muscle cells are conducted through the surrounding body fluids by the movements of ions in those fluids. This is recorded by the surface electrodes and amplified by the recording machine.

Figure 47.14 When the animal is active, the arterioles of its leg muscle would dilate, bringing more blood and, consequently, nutrients and oxygen to the active muscle tissue.

Figure 47.17 The valves open toward the heart. When the head is upright, the valves are open, and blood drains from the head to the right atrium by gravity. When the giraffe lowers its head to drink, however, gravity would prevent the venous blood from reaching the heart; instead, blood would pool in the head and could raise pressure in the head and brain. The valves in the neck veins work the same way as those in the legs of other animals, helping to propel blood against gravity to the heart.

Figure 47.20 Baroreceptors are mechanoreceptors. Like all mechanoreceptors (e.g., those in distensible or deformable structures such as the urinary bladder and stomach), their ion channels are opened by physical deformation or stretching of the plasma membrane. They are, therefore, mechanically gated ion channels.

Feature Investigation Questions

1. Furchgott noted that acetylcholine had different effects on the rabbit aorta depending on the manner in which the aorta was isolated and prepared. When applied to flattened strips of the aorta, acetylcholine caused contraction of the aorta smooth muscle; however, when applied to circular rings of the aorta, acetylcholine caused relaxation. Furchgott suggested that the difference was due to the absence of the endothelial layer of tissue in the flattened strips of aorta.

2. Furchgott hypothesized that acetylcholine stimulated the endothelial cells to secrete a substance that functioned as a vasodilator, causing the muscle layer to relax. Furchgott performed several experiments to test his hypothesis. He compared the effects of acetylcholine on circular rings of aorta that either had the endothelial layer intact or experimentally removed. The results of this experiment demonstrated that when the endothelial layer was present, relaxation occurred in the presence of acetylcholine. Removal of the endothelial layer, however, resulted in contraction of muscle in the presence of acetylcholine. In a second experiment, a strip of the aorta with the endothelial layer removed was put in contact with a strip of aorta with an intact endothelial layer. When this "sandwiched" treatment was exposed to acetylcholine, both muscle layers relaxed.

3. Furchgott concluded that the endothelial layer produced a vasodilator in the presence of acetylcholine. The vasodilator diffused from the intact strip of muscle to the denuded strip and caused the muscle layer to relax.

Test Yourself

1. b 2. a 3. c 4. b 5. a 6. d 7. c 8. d 9. c 10. d

Conceptual Questions

1. The three main components of a circulatory system are (1) blood or hemolymph, an internal body fluid containing dissolved solutes; (2) blood vessels, a system of hollow tubes within the body through which blood travels; and (3) one or more hearts, muscular structures that pump blood through the blood vessels.

2. *Closed circulatory system*—In a closed circulatory system, the blood and interstitial fluid are contained within tubes called blood vessels and are transported by a pump called the heart. All of the nutrients and oxygen that tissues require are delivered directly to them by the blood vessels. Advantages of closed circulatory systems are that different parts of an animal's body can receive blood flow in proportion to that body part's metabolic requirements at any given time. Due to its efficiency, a closed circulatory system allows organisms to become larger.

Open circulatory system—In an open circulatory system, the organs are bathed in hemolymph that ebbs and flows into and out of the heart(s) and body cavity, rather than blood being directed to all cells. Like a closed circulatory system, there are a pump and blood vessels, but these two structures are less developed and less complex compared to a closed circulatory system. Partly as a result, organisms such as mollusks and arthropods are generally limited to being relatively small, although exceptions do exist.

3. The cardiac cycle can be divided into two phases. The first is diastole, during which the ventricles fill with blood coming from the atria through the open AV valves. This is followed by systole, the contraction of the ventricles that ejects blood through the open semilunar valves. Valves must be one-directional because if, for example, the AV valves opened in both directions, then blood would flow from the ventricles into the atria whenever the ventricles contracted. This would reduce the amount of blood flowing into the arteries, and consequently, cardiac output and blood pressure would decrease to dangerously low levels.

Chapter 48
Concept Checks

Figure 48.2 Regardless of whether the atmosphere is measured on Mt. Everest or at sea level, the percentage of gas molecules that are oxygen remains close to 21%. However, the pressure exerted by those gas molecules decreases as one ascends in elevation.

Figure 48.3 If a lungless salamander were to dry out, its capacity for gas exchange would be greatly reduced. Gases diffuse into and out of the body of the salamander by dissolving in the moist fluid layer on the skin.

Figure 48.5 Imagine holding several thin sheets of a wet substance, such as paper. If you wave them in the air, what happens? The sheets stick to one another because of surface tension and other properties of moist surfaces. This is what happens to the lamellae in gills when they are in air. When the lamellae stick to each other, this reduces the surface area available for gas exchange, and the fish suffocates.

Figure 48.7 There are probably several factors that limit insect body size, but the respiratory system most likely is one such factor. If an insect grew to the size of a human, for example, the trachea and tracheoles would be so large and extensive that there would be little room for any other internal organs in the body! Also, the mass of the animal's body and the forces generated during locomotion would probably collapse the tracheoles. Finally, diffusion of oxygen from the surface of the body to the deepest regions of a human-sized insect would take far too long to support the metabolic demands of internal structures.

Figure 48.12 Because fishes have the most efficient means of extracting oxygen from their environment, one might conclude that this is an adaptation to cope with low environmental oxygen. Based on that logic, you would conclude that the oxygen content of water was less than that of air, which is indeed correct.

Figure 48.15 The waters off the coast of Antarctica are extremely cold, rarely warmer than 0.30°C. As we saw earlier in this chapter, more oxygen dissolves in cold water than in warm water, and therefore icefish have the potential to obtain more oxygen across their gills. Cold temperatures also decrease the metabolic rate of the animals, because all chemical reactions slow down at low temperature. Thus, the oxygen demands of icefish are lower than those of warm-water fish. Several other adaptations have evolved to enable these animals to live without hemoglobin. Large gills with exceptionally high surface area facilitate diffusion of oxygen into the animal's blood. In addition, cardiovascular adaptations evolved to help increase the total amount of oxygen in the blood and its ability to be pumped to all body tissues. For example, icefish have larger blood volumes and a larger heart than warm-water fish of a similar size. Also, the absence of red blood cells makes the blood less viscous (makes it "thinner") and therefore easier to pump through the body.

Figure 48.16 An increase in the blood concentration of HCO_3^- would favor the reaction $HCO_3^- + H^+ \rightarrow H_2CO_3 \rightarrow CO_2 + H_2O$. This would reduce the hydrogen ion concentration of the blood, thereby raising the pH; the CO_2 formed as a result would be exhaled. These changes would shift the hemoglobin curve to the left of the usual position.

Feature Investigation Questions

1. The study conducted by Schmidt-Neilsen intended to determine the route of air through the avian respiratory system. This would provide a better understanding of the functions of the air sacs and the process of gas exchange in birds.

2. The first experiment by Schmidt-Neilsen compared the composition of air between the posterior and anterior air sacs. Oxygen content was high in the posterior sacs but low in the anterior sacs. Carbon dioxide levels, however, were low in the posterior sacs but high in the anterior sacs. The researchers concluded that when inhaled, the air moves first to the posterior sacs; then to the lungs where oxygen diffuses into the blood and carbon dioxide diffuses into the lungs; and, finally, to the anterior sacs before being exhaled.

3. The second experiment by Schmidt-Neilsen was conducted to verify the pathway of air through the respiratory system of the bird. In this experiment, the researcher monitored oxygen levels by surgically implanting oxygen probes in the anterior and posterior air sacs. The bird was fitted with a face mask and allowed to take one breath of pure oxygen. The researcher was then able to track the movement of this oxygen through the respiratory tract. Schmidt-Neilsen concluded that it takes two complete breaths for air to move from the environment through the lungs and back out again to the environment. The two breaths are required to move the air from the posterior air sacs through the lungs and, finally, to the anterior air sacs before exiting the body.

Test Yourself

1. a 2. e 3. c 4. c 5. a 6. b 7. e 8. e 9. d 10. b

Conceptual Questions

1. Countercurrent exchange maximizes the amount of oxygen that can be obtained from the water in fishes. Oxygenated water flows across the lamellae of a fish gill in the opposite direction in which deoxygenated blood flows through the capillaries of the lamellae. In this way, a diffusion gradient for oxygen is maintained along the entire length of the lamellae, facilitating diffusion of oxygen even when much of it has already entered the blood.

2. The avian respiratory system is unique among vertebrates in that it is supplemented with air sacs, which do not participate in gas exchange. They do, however, create a unidirectional flow of air through the respiratory system. Air enters the trachea and flows into the two bronchi and then into a series of sacs and parallel tubes called parabronchi, which comprise the avian lungs. Air moves through the system in two cycles. In the first inhalation, air flows into the posterior air sacs. On exhalation, air exits the posterior air sacs and flows through the parabronchi from the back to the front of the lungs. During the next inhalation, air flows from the anterior area of the lungs into the anterior air sacs, which serve as a holding chamber, while fresh air enters the posterior sacs again. The efficiency of this avian flow-through system is a major reason why birds can fly at altitudes with extremely low atmospheric pressure and, consequently, low partial pressures of oxygen.

3. Animals that live at high altitudes face the special challenge of obtaining oxygen where the atmospheric pressure is low. When atmospheric pressure is low, the partial pressure of oxygen in the air is also low. This means that there is less of a driving force for the diffusion of oxygen from the air into the body of the animal. Several adaptations have arisen that help animals cope with such habitats. For example, many high-altitude animals have more red blood cells and have hemoglobin with a higher affinity for oxygen than that of sea-level animals. This means their hemoglobin can bind oxygen even at the low partial pressures of high altitudes and thereby saturate their blood with oxygen. In addition, such animals generally have larger hearts and lungs for their body size than animals that live at lower altitudes. Animals that move to high altitude show increases in the number of red blood cells in their circulation and in respiratory rates. The number of capillaries in skeletal muscle increases to facilitate oxygen diffusion into the muscle cells. Myoglobin content of muscle cells also increases, expanding the reservoir of oxygen in the cytosol.

Chapter 49

Concept Checks

Figure 49.1 No, obligatory exchanges must always occur, but animals can minimize obligatory losses through modifications in behavior. For example, terrestrial animals that seek shade on a hot, sunny day reduce evaporative water loss. As another example, reducing activity minimizes water loss due to respiration.

Figure 49.4 Humans cannot survive by drinking seawater because we do not possess specialized salt glands to rid ourselves of the excess sodium and other ions ingested with seawater. The human kidneys cannot eliminate that much salt. The high blood levels of sodium and other ions would cause changes in cellular membrane potentials, disrupting vital functions of electrically excitable tissue such as cardiac muscle and nerve tissue.

Figure 49.6 Secretion of substances into excretory organ tubules is advantageous because it increases the amount of a substance that gets removed from the body by the excretory organs. This is important, because many substances that get secreted are potentially toxic. Filtration, though efficient, is limited by the volume of fluid that can leave the capillaries and enter the excretory tubule.

Figure 49.12 A brush border composed of microvilli is also present along the epithelial cell layer of the vertebrate small intestine (see Chapter 45). In the intestine, the brush border serves to increase the absorption of nutrients. In both the intestine and the proximal tubule of nephrons, therefore, a brush border provides extensive surface area for the transport of substances between a lumen and the epithelial cells (and from there to extracellular fluid).

Figure 49.14 Epithelial cells like those in the kidney tubules can distribute proteins between the luminal and basolateral sides of the plasma membrane. In this way, the Na^+/K^+-ATPase pumps that are stimulated by aldosterone are present and active only on one side of the cell, the basolateral surfaces. If the pumps were activated on the luminal surface of the cell, aldosterone would not be able to promote reabsorption of Na^+ and water, because Na^+ would also be transported from the cell into the tubule lumen.

Feature Investigation Questions

1. Symptoms of prolonged, heavy exercise include fatigue, muscle cramps, and even occasionally seizures. Fatigue results from the reduction in blood flow to muscles and other organs. Muscle cramps and seizures are the results of imbalances in plasma electrolyte levels. Cade and his colleagues hypothesized that maintaining proper water and electrolyte levels would prevent these problems, and that if water and electrolyte levels were maintained, athletic performance should not decrease as rapidly with prolonged exercise.

2. To test their hypothesis, the researchers created a drink that would restore the correct proportions of lost water and electrolytes within the athletes. If the athletes consumed the drink during exercise, they should not experience as much fatigue or muscle cramping, and thus their performance should be enhanced compared to a control group of athletes that drank only water.

3. The performance of a group of exercising athletes given the electrolyte-containing drink was better than that of the control group that drank only water during exercise. This could be attributed to the replacement of normal electrolyte levels by the drink.

Test Yourself

1. e 2. e 3. d 4. c 5. a 6. c 7. e 8. d 9. a 10. b

Conceptual Questions

1. Nitrogenous wastes are the breakdown products of the metabolism of proteins and nucleic acids. They consist of ammonia, ammonium ions, urea, and uric acid. The predominant type of waste excreted depends in part on an animal's environment. For example, aquatic animals typically excrete ammonia and ammonium ions, whereas many terrestrial animals excrete primarily urea and uric acid. Urea and uric acid are less toxic than the other types but require energy to be synthesized. Urea and uric acid also result in less water excreted, an adaptation that is especially useful for organisms that must conserve water, such as many terrestrial species.

2. Salt glands contain a network of secretory tubules that actively transport NaCl from the extracellular fluid into the tubule lumen. This solution then moves through a central duct and to the outside environment through pores in the nose, around the eyes, and in other locations. The ability to remove salt from body fluids is an adaptation for many marine reptiles and birds, which do not have ready access to fresh water and would otherwise run the risk of having very high levels of salts in their blood and other body fluids.

3. In filtration, an organ acts like a sieve or filter, removing some of the water and its small solutes from the blood, interstitial fluid, or

hemolymph, while retaining blood cells and large solutes such as proteins. Reabsorption is the process whereby epithelial cells of an excretory organ recapture useful solutes that were filtered. Secretion is the process whereby epithelial cells of an excretory organ transport unneeded or harmful solutes from the blood to the excretory tubules for elimination. Some substances such as glucose and amino acids are reabsorbed but not secreted, while some other substances such as toxic compounds are not reabsorbed and are secreted. Still other substances, namely proteins, are not filtered at all.

Chapter 50

Concept Checks

Figure 50.4   When dopamine is secreted from an axon terminal into a synapse where it diffuses to a postsynaptic cell, it is considered a neurotransmitter. When it is secreted from an axon terminal into the extracellular fluid, from where it diffuses into the blood, it is considered a hormone.

Figure 50.7   In addition to the pancreas, certain other organs in an animal's body may contain both exocrine and endocrine tissue or cells. For example, you learned in Chapters 45 and 46 that the vertebrate alimentary canal is composed of several types of secretory cells. Some of these cells release hormones into the blood that regulate the activities of the pancreas and other structures, such as the gallbladder. Other cells of the alimentary canal secrete exocrine products such as acids or mucus into the gut lumen that directly aid in digestion or act as a protective coating, respectively.

Figure 50.10   Not all mammals use the energy of sunlight to synthesize vitamin D. There are many animals, such as those that inhabit caves or that are strictly nocturnal, that rarely are exposed to sunlight. Some of these animals get their vitamin D from dietary sources. How others maintain calcium balance without dietary or sunlight-derived active vitamin D remains uncertain.

Figure 50.12   Sodium and potassium ion balance is of vital importance for most animals because of the critical role these ions play in nervous system and muscle function. It is more the rule than the exception that such important physiological variables are under multiple layers of control. This grants a high degree of fine-tuning capability such that these ions—and other similarly important molecules—rarely exceed or fall below the normal range of concentrations for a given animal.

Figure 50.13   The great height of the twin on the left in Figure 50.13 clearly indicates that his condition arose prior to puberty. The enlarged bones further suggest that the disease continued for a time after puberty, when further linear bone growth was no longer possible.

Figure 50.15   Because 20-hydroxyecdysone is a steroid hormone, you would predict that its receptor would be intracellular. All steroid hormones interact with receptors located either in the cytosol or, more commonly, in the nucleus. The hormone : receptor complex then acts to promote or inhibit transcription of one or more genes. The receptor for 20-hydroxyecdysone is indeed found in cell nuclei.

Feature Investigation Questions

1. Banting and Best based their procedure on a medical condition that results when pancreatic ducts are blocked. The exocrine cells will deteriorate in a pancreas that has obstructed ducts; however, the islet cells are not affected. The researchers proposed to experimentally replicate the condition to isolate the cells suspected of secreting the glucose-lowering factor. From these cells, they assumed they would be able to extract the substance of interest without contamination or degradation due to exocrine products.

2. The extracts obtained by Banting and Best did contain insulin, the glucose-lowering factor, but were of low strength and purity. Collip developed a procedure to obtain a more purified extract with higher concentrations of insulin.

3. The researchers chose to use bovine pancreases as their starting material for preparing the extracts. Because of the large size of these animals and their availability at local slaughterhouses, the researchers were able to obtain great yields of insulin. Second, Collip developed a highly sensitive assay for monitoring changes in blood glucose levels after injection of insulin. This allowed the researchers to better estimate how much insulin was in a preparation and how much was necessary to give to a patient.

Test Yourself

1. b   2. e   3. b   4. e   5. b   6. e   7. c   8. d   9. b   10. d

Conceptual Questions

1. Leptin acts in the hypothalamus to reduce appetite and increase metabolic rate. Because adipose tissue is typically the most important and abundant source of stored energy in an animal's body, the ability to relay information to the appetite and metabolism centers of the brain about the amount of available adipose tissue is a major benefit. In this way, the brain's centers can indirectly monitor the minute-to-minute energy status in the body. A decrease in leptin, for example, would indicate that a decrease in adipose tissue existed—as might occur during a fast. Removal of the leptin signal would cause appetite to increase and metabolism to decrease, thereby conserving energy. The presence of an appetite and the subjective sensations associated with hunger is a motivation that drives an animal to seek food at the expense of other activities, such as seeking shelter, finding a mate, and so on.

2. Insulin acts to lower blood glucose concentrations, for example, after a meal, whereas glucagon elevates blood glucose, for example, during fasting. Insulin acts by stimulating the insertion of glucose transporter proteins into the cell membrane of muscle and fat cells. Glucagon acts by stimulating glycogenolysis in the liver. If a high dose of glucagon were injected into an animal, including humans, the blood concentration of glucagon would increase rapidly. This would stimulate increased glycogenolysis, resulting in blood glucose concentrations that were above normal.

3. Type 1 DM is characterized by insufficient production of insulin due to the immune system destroying the insulin-producing cells of the pancreas. In type 2 DM, insulin is still produced by the pancreas, but adipose and muscle cells do not respond normally to insulin.

Chapter 51

Concept Checks

Figure 51.4   Aquatic environments in which the water is stagnant or only gently moving, as shown in this figure, are generally best for external fertilization. Fast-moving bodies of water reduce the likelihood of a sperm contacting an egg and increase the chances that gametes will be washed away in the current. Many river-dwelling fishes lay eggs in gently moving streams, and many marine fishes do so in relatively shallow waters.

Figure 51.7   The elevated testosterone levels would inhibit LH and FSH production through negative feedback. This would result in reduced spermatogenesis and possibly even infertility (an inability to produce sufficient sperm to cause a pregnancy).

Figure 51.10   FSH and LH concentrations do not surge in males, but instead remain fairly steady, because the testes do not show cyclical activity. Sperm production in males is constant throughout life after puberty.

Figure 51.11   In addition to its other functions, the placenta must serve the function of the lungs for the fetus, because its real lungs are not breathing air during this time. Arteries always carry blood away from the heart; veins carry blood to the heart. Consequently, blood leaving the heart of the fetus and traveling through arteries to the placenta is deoxygenated. As blood leaves the placenta and returns to the heart, the blood has become oxygenated as oxygen diffuses from the maternal blood into fetal blood. That oxygenated blood then gets pumped from the fetal heart through other arteries to the rest of the fetus' body.

Figure 51.14   Pregnancy and subsequent lactation require considerable energy and, therefore, nutrient ingestion. Consuming the placenta provides the female with a rich source of protein and other important nutrients.

Feature Investigation Questions

1. Using *Daphnia*, Paland and Lynch compared the accumulation of mitochondrial mutations between sexually reproducing populations and asexually reproducing populations.

2. The results—that sexually reproducing populations had a lower rate of deleterious mutations compared to asexually reproducing populations—indicate that sexual reproduction does decrease the accumulation of deleterious mutations, at least in this species.

3. Sexual reproduction allows for mixing of the different alleles of genes with each generation, thereby increasing genetic variation within the population. This could prevent the accumulation of deleterious alleles in the population.

Test Yourself

1. d   2. c   3. e   4. a   5. b   6. c   7. b   8. c   9. b   10. c

Conceptual Questions

1. In viviparity, most of embryonic development occurs within the mother, and the animal is born alive, as occurs in most mammals. If all or most of embryonic development occurs outside the mother and the embryo depends exclusively on yolk from an egg for nourishment, the process is called oviparity; this occurs in most vertebrates and in insects. In ovoviviparity, which occurs in some reptiles, sharks and some invertebrates, fertilized eggs covered with a very thin shell hatch inside the mother's body, but the offspring receive no nourishment from the mother. Humans are viviparous. An advantage of viviparity is that the embryo and fetus develop in a protected environment.

2. Sexual reproduction requires that males and females of a species produce different gametes and that these gametes come into contact with each other. This requires males and females to expend energy to locate mates. It also may require specialized organs for copulation and in some cases requires the production of very large numbers of gametes to increase the likelihood that the eggs are fertilized. These costs are outweighed by the genetic diversity afforded by sexual reproduction.

3. Cells of the hypothalamus produce two important hormones that regulate reproduction. GnRH stimulates the anterior pituitary gland to release two gonadotropic hormones, LH and FSH. These two hormones regulate the production of gonadal hormones and development of gametes in both sexes. In addition, increased secretion of GnRH contributes to the initiation of puberty. The hypothalamus also produces oxytocin, a hormone that is stored in the posterior pituitary gland and that acts to stimulate milk release during lactation. Finally, changes in neuroendocrine activity in the hypothalamus are linked to seasonal changes in day length and therefore contribute to seasonal breeding in certain mammals.

Chapter 52

Concept Checks

Figure 52.1   The process by which a tadpole develops into an adult frog is called metamorphosis. This process is widespread in animals and occurs in many arthropods, certain fishes, numerous marine invertebrates such as gastropods, and amphibia.

Figure 52.5   No, all vertebrates do not use internal fertilization. External fertilization is common in fishes and amphibia; these animals lay unfertilized eggs, over which males deposit sperm (see Chapter 51).

Figure 52.14   Different concentrations of a signaling protein can exert different effects on cells when, for example, different cells express different isoforms of a plasma membrane receptor for the protein. If one cell expresses a high-affinity receptor and another cell a low-affinity receptor, the two cells would respond to the signaling protein at different concentrations. Likewise, the different receptors may be linked with different second messenger molecules generated within the cell. These messengers, such as camp and Ca^{2+}, may have different effects on cell function.

Feature Investigation Questions

1. Knowing the genes expressed in this region of a developing embryo would provide important information about the control of the patterning of embryonic tissues and structures.

2. Harland and colleagues tested the hypothesis that cells within the Spemann organizer expressed certain genes important in the development of dorsal structures, such as the notochord.

3. The scientists used a procedure called expression cloning. In this process, they isolated the various mRNAs that were present in the tissue of the dorsal lip of the embryo. After purifying these mRNAs, they produced a cDNA library. This library contained all the genes expressed in the particular tissue at that particular time of development. The scientists then transcribed the different genes in the cDNA library into

mRNAs and injected these into UV-damaged eggs, which were subsequently fertilized. UV-damaged fertilized eggs fail to develop dorsal structures. The scientists were interested in any mRNA that "rescued" the developing embryo and restored some level of normal development. One protein, noggin, was found to rescue the embryos and acted as a morphogen.

Test Yourself

1. a   2. c   3. c   4. d   5. b   6. c   7. b   8. e   9. e   10. e

Conceptual Questions

1. Embryonic development is the process by which a fertilized egg is transformed into an organism with distinct physiological systems and body parts. Cell differentiation is the process by which different cells within a developing organism acquire specialized forms and functions, due to the expression of cell-specific genes. Growth is the enlargement of an embryo, as cells divide and/or enlarge.

2. The timing of the final development of an embryo's organs is typically linked with the requirement for that organ's function. In mammals, for example, the heart is required early in development to pump blood through the embryonic and fetal circulation, thereby delivering nutrients and removing wastes. Fully functional lungs, however, are not required until the animal is born and begins breathing air for the first time.

3. Autonomous specification results from the asymmetrical distribution of intracellular proteins and mRNAs during the cleavage events of embryonic development. The resulting daughter cells will contain different amounts of these cytoplasmic determinants, and this will direct these cells into different developmental fates. Conditional specification results from the interactions of proteins on the extracellular surface of the cell membranes of different cells or from proteins secreted from one cell and acting on another cell. This type of specification determines where a cell ends up within the embryo and what type of cell develops.

Chapter 53

Concept Checks

Figure 53.2   Although swelling is one of the most obvious manifestations of inflammation, it has no significant adaptive value of its own. It is a consequence of fluid leaking out of blood vessels into the interstitial space. It can, however, contribute to pain sensations, because the buildup of fluid may cause distortion of connective tissue structures such as tendons and ligaments. Pain, while obviously unpleasant, is an important signal that alerts many animals to the injury and serves as a reminder to protect the injured site.

Figure 53.3   Recall from Chapter 47 that as blood circulates, a portion of the plasma—the fluid part of blood—exits venules and capillaries and enters the interstitial fluid. Most of the plasma is reabsorbed back into the capillaries, but a portion gets left behind. That excess fluid is drained away by lymph vessels and becomes lymph. Without lymph vessels, fluid would accumulate outside of the blood, in the interstitial fluid.

Figure 53.12   Because an animal may encounter the same type of pathogen many times during its life, having a secondary immune response means that future infections will be fought off much more efficiently.

Figure 53.13   Although social insects live in colonies that may reach enormous numbers of individuals, many other insect species inhabit densely populated colonies, as do many species of vertebrates. The free-tailed bat *Tamarind brasiliensis,* for example, lives in caves that contain as many as 10–20 million individuals! It is highly likely that all such species have evolved immune defense mechanisms that enable them to ward off infections that might devastate the entire population.

Feature Investigation Questions

1. Termites produce and line their nests with antimicrobial secretions that reduce the probability that an infection can spread. Termites also practice social grooming that removes foreign objects and possible pathogens from nest mates. Sick or dead termites are also removed from the nest, thus decreasing the likelihood of disease transmission.

2. The researchers tested the hypothesis that social interaction among termites promotes disease resistance. That is, termites that had contact with nest mates that had previously been exposed to a pathogen were

subsequently less likely to become infected with that pathogen than termites that lacked prior contact with nest mates. The researchers regarded this as social immunity.

3. The researchers formed two groups of termites. The experimental group was exposed to a microbial pathogen, whereas the control group was not. After 1 week, the control group was divided into two groups. One control subgroup was introduced to the group of termites previously exposed to the pathogen, whereas the other control subgroup was kept isolated from the exposed termites. Finally, the researchers challenged both groups with a lethal concentration of the same microbial pathogen. The researchers then evaluated the effect of social interaction on survival. The results indicated that the termites that previously had been introduced to the exposed termites had higher survival rates compared to the control group that was not introduced to exposed termites. The researchers concluded that the interaction with the exposed termites allowed the control termites to acquire immunity to the microbial pathogen.

Test Yourself

1. e   2. b   3. c   4. c   5. a   6. d   7. b   8. a   9. d   10. b

Conceptual Questions

1. Nonspecific immunity is present at birth and is found in all animals. These defenses recognize general, conserved features common to a wide array of pathogens and include external barriers, such as the skin, and internal defenses involving phagocytes and other cells. Specific immunity develops *after* an animal has been exposed to a *particular* antigen. The responses include humoral and cell-mediated defenses. Specific immunity appears to be largely restricted to vertebrates. Unlike nonspecific immunity, in specific immunity, the response to an antigen is greatly increased if an animal is exposed to that antigen again at some future time.

2. Cytotoxic T cells are "attack" cells that are responsible for cell-mediated immunity. Once activated, they migrate to the location of their targets, bind to the targets by combining with an antigen on them, and directly kill the targets via secreted chemicals.

3. Each immunoglobulin molecule is Y shaped and composed of four interlinked polypeptide chains: two long heavy chains and two short light chains. The chains within the Y have complex looplike structures due to disulfide bonds within and between the chains. This accounts for the characteristic tertiary (three-dimensional) structure of immunoglobulins. At one region of immunoglobulins, the amino acid sequence is highly variable; this accounts for the specificity of binding of immunoglobulins to particular antigens.

Chapter 54

Concept Checks

Figure 54.5   Higher predation would occur where locust numbers are highest. This means that predators would be responding to an increase in prey density by eating more individuals.

Figure 54.7   Cold water suppresses the ability of the coral-building organisms to secrete their calcium carbonate shell.

Figure 54.10   In some areas when fire is prevented, fuel, in the form of old leaves and branches, can accumulate. When a fire eventually occurs, it can be so large and hot that it destroys everything in its path, even reaching high into the tree canopy.

Figure 54.17   Acid soils are low in essential plant and animal nutrients such as calcium and nitrogen and are lethal to some soil microorganisms that are important in decomposition and nutrient cycling.

Figure 54.19   This occurs because increasing cloudiness and rain at the tropics maintain fairly constant temperatures across a wide latitudinal range.

Figure 54.24   Soil conditions can also influence biome type. Nutrient-poor soils, for example, may support vegetation different than that of the surrounding area.

Feature Investigation Questions

1. Most believe that invasive species succeed in new environments due to the lack of natural enemies and that diseases and predators present in

the original environment controlled the growth of the population. When these organisms are introduced into a novel environment, the natural enemies are usually absent. This allows for an unchecked increase in the population of the invasive species.

2. Callaway and Aschehoug were able to demonstrate through a controlled experiment that the presence of *Centaurea*, an invasive species, reduced the biomass of three other native species of grasses by releasing allelochemicals. Similar experiments using species of grasses that are found in the native region of *Centaurea* indicate that these species have evolved defenses against the allelochemicals.

3. The activated charcoal helps to remove the allelochemical from the soil. The researchers conducted this experiment to provide further evidence that the chemical released by the *Centaurea* was reducing the biomass of the native Montana grasses. With the removal of the chemicals by the addition of the charcoal, the researchers showed an increase in biomass of the native Montana grasses compared to the experiments lacking the charcoal.

Test Yourself

1. b   2. e   3. a   4. b   5. a   6. a   7. d   8. d   9. a   10. a

Conceptual Questions

1. Mountains are cooler than valleys because of adiabatic cooling. Air at higher altitudes expands because of decreased pressure. As it expands, air cools, at a rate of 10°C for every 1,000 m in elevation. As a result, mountain tops can be much cooler than the plains or valleys that surround them.

2. Solar radiation in the form of short-wave energy passes through the atmosphere to heat the Earth's surface. This energy is then radiated back to the atmosphere in the form of infrared radiation. Much of this radiation is absorbed by atmospheric gases and reradiated toward Earth, causing an increase in temperature.

3. Florida is a peninsula that is surrounded by the Atlantic Ocean and the Gulf of Mexico. Differential heating between the land and the sea creates onshore sea breezes on both the east and west coasts. These breezes often drift across the whole peninsula, bringing heavy rain.

Chapter 55

Concept Checks

Figure 55.3   In classical conditioning, an involuntary response comes to be associated with a stimulus that did not originally elicit the response, as with Pavlov's dogs salivating at the sound of a metronome.

Figure 55.5   The ability to sing the same distinctive song must be considered innate behavior because the cuckoo has had no opportunity to learn its song from its parents.

Figure 55.7   Tinbergen manipulated pinecones, but not all digger wasp nests are surrounded by pinecones. You could manipulate branches, twigs, stones, and leaves to determine the necessary size and dimensions of objects that digger wasps use as landmarks.

Figure 55.8   This is an unusual example because the return trip involves several different generations to complete: One generation overwinters in Mexico, but these individuals lay eggs and die on the return journey, and their offspring continue the return trip.

Figure 55.14   The individuals in the center of the group are less likely to be attacked than those on the edge of the group. This is referred to as the geometry of the selfish herd.

Figure 55.16   Because of the genetic benefit, the answer is nine cousins. Consider Hamilton's rule, expressed in the formula $rB > C$. Using cousins, $B = 9$, $r = 0.125$, and $C = 1$, and $1.125 > 1$. Using sisters, $B = 2$, $r = 0.5$, and $C = 1$. Because rB would not be greater than C, there would be no net genetic benefit in self-sacrifice.

Figure 55.17   All the larvae in the group are likely to be the progeny of one egg mass from one adult female moth. The death of the one caterpillar teaches a predator to avoid the pattern and benefits the caterpillar's close kin.

Figure 55.21   Because sperm are cheaper to produce than eggs, males try to maximize their fitness through attracting multiple females, whereas female fitness is maximized by choosing a mate with good genetic quality and parenting skills. Colorful plumage and elaborate adornments are thus signals of the male's overall health.

Figure 55.22 The males aren't careful because it is likely the pups were fathered in the previous year by a different male. Being a harem master is demanding, and males may often only perform this role for a year or two.

Feature Investigation Questions

1. Tinbergen observed the activity of digger wasps as they prepared to leave the nest. Each time, the wasp hovered and flew around the nest for a period of time before leaving. Tinbergen suggested that during this time, the wasp was making a mental map of the nest site. He hypothesized that the wasp was using characteristics of the nest site, particularly landmarks, to help relocate it.

2. Tinbergen placed pinecones around the nest of the wasps. When the wasps left the nest, he removed the pinecones from the nest site and set them up in the same pattern a distance away, constructing a sham nest. For each trial, the wasps would go directly to the sham nest, which had the pinecones around it. This indicated to Tinbergen that the wasps identified the nest based on the pinecone landmarks.

3. No. Tinbergen also conducted an experiment to determine if the wasps were responding to the visual cue of the pinecones or the chemical cue of the pinecone scent. The results of this experiment indicated that the wasps responded to the visual cue of the pinecones and not their scent.

Test Yourself

1. d   2. d   3. d   4. c   5. c   6. d   7. b   8. c   9. a   10. c

Conceptual Questions

1. Ethology is the scientific study of animal behavior that focuses on the specific genetic and physiological mechanisms of behavior.

2. Certainty of paternity influences degree of parental care. With internal fertilization, certainty of paternity is relatively low. With external fertilization, eggs and sperm are deposited together, and paternity is more certain. This explains why males of some species, such as mouth-breeding cichlid fish, are more likely to engage in parental care.

3. In monogamy, each individual mates exclusively with one partner for at least one breeding cycle. In polygamy, individuals mate with more than one individual during a breeding cycle. There are two types of polygamy. In polygyny, one male mates with more than one female, and in polyandry, females mate with more than one male.

Chapter 56

Concept Checks

Figure 56.2 The total population size, N, would be estimated to be 110 × 100/20, or 550.

Figure 56.3 In a half-empty classroom, the distribution is often clumped because friends sit together.

Figure 56.7 (a) type III, (b) type II

Figure 56.11 $dN/dt = 0.1 \times 200 (300)/500 = 12.$

Figure 56.13 Only density-dependent factors operate in this way.

Figure 56.19 There were very few juveniles in the population and many mature adults. The population would be in decline.

Figure 56.22 Many different ecological footprint calculators are available on the Web. Does altering inputs such as type of transportation, amount of meat eaten, or amount of waste generated make a difference?

Feature Investigation Questions

1. It became apparent that the sheep population was declining. Some individuals felt that the decline in the population was due to increased wolf predation having a negative effect on population growth. This led to the suggestion of culling the wolf population to reduce the level of predation on the sheep population.

2. The survivorship curve is very similar to a typical type I survivorship curve. This suggests that survival is high among young and reproductively active members of the population and that mortality rates are higher for older members of the population. One difference between the actual survivorship curve and a typical type I curve is that the mortality

rate of very young sheep was higher in the actual curve, and then it leveled off after the second year. This suggests that very young and older sheep are more at risk for predation.

3. It was concluded that wolf predation was not the primary reason for the drop in the sheep population. It appeared that wolves prey on the vulnerable members of the population and not on the healthy, reproductively active members. The Park Service determined that several cold winters may have had a more important impact on the sheep population than wolf predation did. Based on these conclusions, the Park Service ended a wolf population control program.

Test Yourself

1. b   2. e   3. b   4. c   5. c   6. b   7. c   8. d   9. c   10. c

Conceptual Questions

1. A population can be defined as a group of interbreeding individuals occupying the same habitat at the same time. Population ecology is the study of how populations grow and what promotes and limits growth.

2. Population biologists first capture and tag the animals. The tagged animals are freed and assumed to mix freely with unmarked individuals in the population. When the population is resampled, the numbers of marked and unmarked individuals are recorded. The assumption is that the ratio of marked to unmarked individuals in the second sample is the same as the ratio of marked to unmarked individuals in the first sample, and in this way, an estimate of population size is obtained.

3. In promiscuity each female mates with a different male every year or breeding cycle. At medium values of N, (K − N)/K is closer to a value of 1, and population growth is relatively large. If K = 1,000, N = 500, and r = 0.1, then

$$\frac{dN}{dt} = (0.1)(500) \times \frac{(1,000 - 500)}{1,000}$$

$$\frac{dN}{dt} = 25$$

However, if population sizes are low (N = 100), (K − N)/K is so small that growth is low.

$$\frac{dN}{dt} = (0.1)(100) \times \frac{(1,000 - 100)}{1,000}$$

$$\frac{dN}{dt} = 9$$

By comparing these two examples with that shown in Section 56.3, we see that growth is small at high and low values of N and is greatest at immediate values of N. Growth is greatest when N = K/2. However, when expressed as a percentage, growth is greatest at low population sizes. Where N = 100, percentage growth = 9/100 = 9%. Where N = 500, percentage growth = 25/500 = 5%, and where N = 900, percentage growth = 9/100 = 1%.

Chapter 57

Concept Checks

Figure 57.2 Individual vultures often fight one another over small carcasses. These interactions would constitute intraspecific interference competition.

Figure 57.5 There would be 10 possible pairings (AB, AC, AD, AE, BC, BD, BC, CD, CE, DE), of which only neighboring species (AB, BC, CD, DE) competed. Therefore, competition would be expected in 4/10 pairings, or 40% of the cases.

Figure 57.8 A 1974 review by Tom Schoener examined segregation in a more wide-ranging literature review of over 80 species, including slime molds, mollusks, and insects, as well as birds. He found segregation by habitat occurred in the majority of the examples, 55%. The other most common form of segregation was by food type, 40%.

Figure 57.9 Omnivores, such as bears, can feed on both plant material, such as berries, and animals, such as salmon. As such, omnivores may act as both predators and herbivores depending on what they are feeding on.

Figure 57.10 Batesian mimicry has a positive effect for the mimic, and the model is unaffected, so it is a +/0 relationship, like commensalism. Müllerian mimicry has a positive effect on both species, so it is a +/+ relationship, like mutualism.

Figure 57.12 Because there is no evolutionary history between invasive predators and native prey, the native prey often have no defenses against these predators and are very easily caught and eaten.

Figure 57.14 Invertebrate herbivores can eat around mechanical defenses; therefore, chemical defenses are probably most effective against invertebrate herbivores.

Figure 57.20 It's an example of facultative mutualism, because in this case, both species can live without the other.

Figure 57.23 Fertilizer increases plant quality and hence herbivore density, which, in turn, increases the density of spiders. This is a bottom-up effect.

Feature Investigation Questions

1. The two species of barnacles can be found in the same intertidal zone, but there is a distinct difference in niche of each species. *Chthamalus stellatus* is found only in the upper intertidal zone. *Semibalanus balanoides* is found only in the lower tidal zone.

2. Connell moved rocks with young *Chthamalus* from the upper intertidal zone into the lower intertidal zone to allow *Semibalanus* to colonize the rocks. After the rocks were colonized by *Semibalanus*, he removed *Semibalanus* from one side of each rock and returned the rocks to the lower intertidal zone. This allowed Connell to observe the growth of *Chthamalus* in the presence and the absence of *Semibalanus*.

3. Connell observed that *Chthamalus* was more resistant to desiccation compared to *Semibalanus*. Though *Semibalanus* was the better competitor in the lower intertidal zone, the species was at a disadvantage in the upper intertidal zone when water levels were low. This allowed *Chthamalus* to flourish and outcompete *Semibalanus* in a different region of the intertidal zone.

Test Yourself

1. d   2. c   3. b   4. d   5. c   6. b   7. d   8. b   9. b   10. c

Conceptual Questions

1. The hypothesis states that two species with exactly the same requirements cannot live together in the same place and use the same resources, that is, occupy the same niche.

2. In Müllerian mimicry, many unpalatable species evolve to look the same, reinforcing the distasteful design; in Batesian mimicry, a palatable species mimics an unpalatable one.

3. There are at least three reasons why we don't see more herbivory in nature. First, plants possess an array of defensive chemicals, including alkaloids, phenolics, and terpenes. Second, many herbivore populations are reduced by the action of natural enemies. We see evidence for this in the world of biological control. Third, the low nutritive value of plants ensures herbivore populations remain low and unlikely to impact plant populations.

Chapter 58

Concept Checks

Figure 58.5 Species richness of trees doesn't increase because rainfall in the western United States is low compared to that in the east.

Figure 58.6 Hurricanes, tropical storms, heavy rainfall, and mudslides are disturbances that maintain a mosaic of disturbed and undisturbed habitats, favoring high species richness in the tropics.

Figure 58.10 As we walk forward from the edge of the glacier to the mouth of the inlet, we are walking backward in ecological time to communities that originated hundreds of years ago.

Figure 58.11 No, competition is also important. For example, the shade from later-arriving species, such as spruce trees, causes competitive exclusion of some of the original understory species.

Figure 58.13 Competition features more prominently. Although early colonists tend to make the habitat more favorable for later colonists, it is the later colonists who outcompete the earlier ones, and this fuels species change.

Figure 58.14 If a small island was extremely close to the mainland, it could continually receive migrating species from the source pool. Even though these species could not complete their life cycle on such a small island, extinctions would rarely be recorded because of this continual immigration.

Figure 58.15 At first glance, the change looks small, but the data are plotted on a log scale. On this scale, an increase in bird richness from 1.2 to 1.6 equals an increase from 16 to 40 species, a change of over 100%.

Feature Investigation Questions

1. Simberloff and Wilson were testing the three predictions of the theory of island biogeography. One prediction suggested that the number of species should increase with increasing island size. Another prediction suggested that the number of species should decrease with increasing distance of the island from the source pool. Finally, the researchers were testing the prediction that the turnover of species on islands should be considerable.

2. Simberloff and Wilson used the information gathered from the species survey to determine whether the same types of species recolonized the islands or if colonizing species were random.

3. The data suggested that species richness did increase with island size. Also, the researchers found that in all but one of the islands, the number of species was similar to the number of species before fumigation.

Test Yourself

1. c   2. c   3. d   4. b   5. c   6. d   7. a   8. d   9. b   10. c

Conceptual Questions

1. A community is an assemblage of many populations living in the same place at the same time. Community ecology studies how groups of species interact and form functional communities.

2.

Disturbance	Frequency	Severity of effects
Forest fire	Low to high, depending on lightning frequency	High to low, depending on frequency
Hurricane	Low	Severe
Tornado	Very low	Severe
Floods	Medium to high in riparian areas	Fairly low; many communities can recover quickly
Disease epidemics	Low	High; may cause catastrophic losses of species
Droughts	Low	Potentially severe
High winds	High	May kill large trees and create light gaps
Hard freezes	Low	May cause deaths to tropical species, such as mangroves

3. A community is in equilibrium when no change can be detected in the number of species and their population sizes over a given period of time.

Chapter 59

Concept Checks

Figure 59.3 It depends on the trophic level of their food, whether dead vegetation or dead animals. Many decomposers feed at multiple trophic levels.

Figure 59.6 The production efficiency is $(16/823) \times 100$, or 1.9%.

Figure 59.13 On a population level, plant secondary metabolites can deter herbivores from feeding. However, on an ecosystem level, these effects are not as important because higher primary production tends to result in higher secondary production.

Figure 59.17 The greatest stores are in rocks and fossil fuels.

Figure 59.18 It fluctuates because less CO_2 is emitted from vegetation in the summer and more is emitted in the winter. This pattern is driven by the large land masses of the Northern Hemisphere relative to the smaller land masses of the Southern Hemisphere.

Figure 59.23   High levels occurred because prevailing westerly winds carried acid rain from the industrial areas of the Midwest, where it was produced, to areas of the U.S. northeast.

Feature Investigation Questions

1. The researchers were testing the effects of increased carbon dioxide levels on the forest ecosystem. The researchers were testing the effects of increased carbon dioxide levels on primary production as well as other trophic levels in the ecosystem.

2. By increasing the carbon dioxide levels in only half of the chambers, the researchers were maintaining the control treatments necessary for all scientific studies. By maintaining equal numbers of control and experimental treatments, the researchers could compare data to determine what effects the experimental treatment had on the ecosystem.

3. The increased carbon dioxide levels led to an increase in primary productivity, as expected. Since photosynthetic rate is limited by carbon dioxide levels, increases in the available carbon dioxide should increase photosynthetic rates. Interestingly, though, the increase in primary productivity did not lead to an increase in herbivory. The results indicated that herbivory actually decreased with increased carbon dioxide levels.

Test Yourself

1. d   2. d   3. d   4. a   5. d   6. a   7. a   8. b   9. c   10. d

Conceptual Questions

1. Autotrophs harvest light or chemical energy and store that energy in carbon compounds. Heterotrophs obtain their nutrition by eating other organisms.

2. Chain lengths are short in food webs because there is low production efficiency and only a 10% rate of energy transfer from one level to another, so only a few links can be supported.

3. The phosphorous cycle is a local cycle with no capacity for long-distance transport. Both the carbon and nitrogen cycles are global cycles with biological, geological, and atmospheric pools.

Chapter 60

Concept Checks

Figure 60.4   It is possible that the results are driven by what is known as a sampling effect. As the numbers of species in the community increase, so does the likelihood of including a "superspecies," a species with exceptionally large individuals that would use up resources. In communities with higher diversity, care has to be taken that increased diversity is driving the results, not the increased likelihood of including a superspecies.

Figure 60.6   The extinction rate could increase because an increasing human population requires more space to live, work, and grow food, resulting in less available habitat and resources for other species.

Figure 60.8   No, some species, such as self-fertilizing flowers, appear to be less affected by inbreeding.

Figure 60.9   The effective population size (N_e) would be = $(4 \times 125 \times 500) / (125 + 500)$, or 400.

Figure 60.12   Corridors might also promote the movement of invasive species or the spread of fire between areas.

Figure 60.13   They act as habitat corridors because they permit movement of species between forest fragments.

Feature Investigation Questions

1. The researchers hoped to replicate terrestrial communities that differed only in their level of biodiversity. This would allow the researchers to determine the relationship between biodiversity and ecological function.

2. The hypothesis was that ecological function was directly related to biodiversity. If biodiversity increased, the hypothesis suggested that ecological function should increase.

3. The researchers tested for ecosystem function by monitoring community respiration, decomposition, nutrient retention rates, and productivity. All of these indicate the efficiency of nutrient production and use in the ecosystem.

Test Yourself

1. d   2. e   3. a   4. c   5. c   6. e   7. e   8. a   9. c   10. b

Conceptual Questions

1. The first level is genetic diversity; the second is species diversity; and the third is ecosystem diversity.

2. The most vulnerable are those with small population sizes, low rates of population growth, *K*-selected (Chapter 56), with inbreeding and possible harem mating structure, tame and unafraid of humans, possibly limited to islands, flightless, possibly valuable to humans as timber, a source of meat or fur, or desirable by collectors (Chapter 60).

3. An umbrella species is a species whose habitat requirements are so large that protecting it would also protect many other species existing in the same habitat. A flagship species is a single large or instantly recognizable species typically chosen because it is attractive and thus more readily engenders support for its conservation. A keystone species is a species within the community that has a role out of proportion to its abundance.

Glossary

A

A band A wide, dark band in a myofibril produced by the orderly parallel arrangement of the thick filaments in the middle of each sarcomere.

abiotic The term used to describe interactions between organisms and their nonliving environment.

abortion A procedure or circumstance that causes the death of an embryo or fetus after implantation.

abscisic acid One of several plant hormones that help a plant cope with environmental stress.

absolute refractory period The period during an action potential when the inactivation gate of the voltage-gated sodium channel is closed; during this time, it is impossible to generate another action potential.

absorption The process in which digested nutrients are transported from the digestive cavity into an animal's circulatory system.

absorption spectrum A diagram that depicts the wavelengths of electromagnetic radiation that are absorbed by a pigment.

absorptive nutrition The process whereby an organism uses enzymes to digest organic materials and absorbs the resulting small food molecules into its cells.

absorptive state One of two alternating phases in the utilization of nutrients; occurs when ingested nutrients enter the blood from the gastrointestinal tract. The other phase is the postabsorptive state.

acclimatization A long-term and persistent physiological adaptation to an extreme environment.

accommodation In the vertebrate eye, the process in which contraction and relaxation of the ciliary muscles adjust the lens according to the angle at which light enters the eye.

acetylcholinesterase An enzyme located on membranes of muscle fibers in a neuromuscular junction; breaks down excess acetylcholine released into the synaptic cleft.

acid A molecule that releases hydrogen ions in solution.

acid hydrolase A hydrolytic enzyme found in lysosomes that functions at acidic pH and uses a molecule of water to break a covalent bond.

acid rain Precipitation with a pH of less than 5.6; results from the burning of fossil fuels.

acidic A solution that has a pH below 7.

acoelomate An animal that lacks a fluid-filled body cavity.

acquired antibiotic resistance The common phenomenon of a previously susceptible strain of bacteria becoming resistant to a specific antibiotic.

acquired immune deficiency syndrome (AIDS) A disease caused by the human immunodeficiency virus (HIV) that leads to a defect in the immune system of infected individuals.

acrocentric A chromosome in which the centromere is near one end.

acromegaly A condition in which a person's growth hormone level is abnormally elevated after puberty, causing many bones to thicken and enlarge.

acrosomal reaction An event in fertilization in which enzymes released from a sperm's acrosome break down the outer layers of an egg cell, allowing the entry of the sperm cell's nucleus into the egg cell.

acrosome A special structure at the tip of a sperm's head containing proteolytic enzymes that break down the protective outer layers of the egg cell at fertilization.

actin A cytoskeletal protein.

actin filament A thin type of protein filament composed of actin proteins that forms part of the cytoskeleton and supports the plasma membrane; plays a key role in cell strength, shape, and movement.

action potential An electrical signal along a cell's plasma membrane; occurs in animal neuron axons muscle cells and some plant cells.

action spectrum The rate of photosynthesis plotted as a function of different wavelengths of light.

activation energy An initial input of energy in a chemical reaction that allows the molecules to get close enough to cause a rearrangement of bonds.

activator A transcription factor that binds to DNA and increases the rate of transcription.

active immunity An animal's ability to fight off a pathogen to which it has been previously exposed. Active immunity can develop as a result of natural infection or artificial immunization.

active site The location in an enzyme where a chemical reaction takes place.

active transport The transport of a solute across a membrane against its gradient (from a region of low concentration to a region of higher concentration). Active transport requires an input of energy.

adaptations The processes and structures by which organisms adjust to changes in their environment.

adaptive radiation The process whereby a single ancestral species evolves into a wide array of descendant species that differ greatly in their habitat, form, or behavior.

adenine (A) A purine base found in DNA and RNA.

adenosine triphosphate (ATP) A molecule that is a common energy source for all cells.

adenylyl cyclase An enzyme in the plasma membrane that synthesizes cAMP from ATP.

adherens junction A mechanically strong cell junction between animal cells that typically occurs in bands. The cells are connected to each other via cadherins, and the cadherins are linked to actin filaments on the inside of the cells.

adhesion The ability of two different substances to cling to each other; the ability of water to be attracted to, and thereby adhere to, a surface that is not electrically neutral.

adiabatic cooling The process in which increasing elevation leads to a decrease in air temperature.

adventitious root A root that is produced on the surfaces of stems (and sometimes leaves) of vascular plants; also, roots that develop at the bases of stem cuttings.

aerenchyma Spongy plant tissue with large air spaces.

aerobic respiration A type of cellular respiration in which O_2 is consumed and CO_2 is released.

aerotolerant anaerobe A microorganism that does not use oxygen but is not poisoned by it either.

afferent arterioles Blood vessels that provide a pathway for blood into the glomeruli of the vertebrate kidney.

affinity The degree of attraction between an enzyme and its substrate.

aflatoxins Fungal toxins that cause liver cancer and are a major health concern worldwide.

age-specific fertility rate The rate of offspring production for females of a certain age; used to calculate how a population grows.

age structure The relative numbers of individuals of each defined age group in a population.

AIDS *See* acquired immune deficiency syndrome.

air sac A component of the avian respiratory system; air sacs—not lungs—expand when a bird inhales and shrink when it exhales. They do not participate in gas exchange, but help direct air through the lungs.

akinete A thick-walled, food-filled cell produced by certain bacteria or protists that enables them to survive unfavorable conditions in a dormant state.

aldosterone A steroid hormone made by the adrenal glands that regulates salt and water balance in vertebrates.

algae (singular, **alga**) A term that applies to about 10 phyla of protists that include both photosynthetic and nonphotosynthetic species; often also includes cyanobacteria.

alimentary canal In animals, the single elongated tube of a digestive system, with an opening at either end through which food and eventually wastes pass from one end to the other.

alkaline A solution that has a pH above 7.

alkaloids A group of secondary metabolites that contain nitrogen and usually have a cyclic, ringlike structure. Examples include caffeine and nicotine.

allantois One of the four extraembryonic membranes in the amniotic egg. It serves as a disposal sac for metabolic wastes.

Allee effect The phenomenon that some individuals will fail to mate successfully purely by chance, for example, because of the failure to find a mate.

allele A variant form of a gene.

allele frequency The number of copies of a particular allele in a population divided by the total number of alleles in that population.

allelochemical A powerful plant chemical, often a root exudate, that kills other plant species.

allelopathy The suppressed growth of one species due to the release of toxic chemicals by another species.

allopatric The term used to describe species occurring in different geographic areas.

allopatric speciation A form of speciation that occurs when a population becomes geographically isolated from other populations and evolves into one or more new species.

alloploid An organism having at least one set of chromosomes from two or more different species.

allosteric site A site on an enzyme where a molecule can bind noncovalently and affect the function of the active site.

alpha (α) helix A type of protein secondary structure in which a polypeptide forms a repeating helical structure stabilized by hydrogen bonds.

alternation of generations The phenomenon that occurs in plants and some protists in which the life cycle alternates between multicellular diploid organisms, called sporophytes, and multicellular haploid organisms, called gametophytes.

alternative splicing The splicing of pre-mRNA in more than one way to create two or more different polypeptides.

altruism Behavior that appears to benefit others at a cost to oneself.

alveoli 1. Saclike structures in the lungs where gas exchange occurs. 2. Saclike cellular features of the protists known as alveolates.

Alzheimer disease (AD) The leading worldwide cause of dementia; characterized by a loss of memory and intellectual and emotional function (formerly called Alzheimer's disease).

AM fungi A phylum of fungi that forms mycorrhizal associations with plants.

amensalism One-sided competition between species, where the interaction is detrimental to one species but not to the other.

Ames test A test that helps ascertain whether or not an agent is a mutagen by using a strain of a bacterium, *Salmonella typhimurium*.

amino acid The building block of proteins. Amino acids have a common structure in which a carbon atom, called the α-carbon, is linked to an amino group (NH$_2$) and a carboxyl group (COOH). The α-carbon also is linked to a hydrogen atom and a particular side chain.

aminoacyl site (A site) One of three sites for tRNA binding in the ribosome during translation; the other two are the peptidyl site (P site) and the exit site (E site). The A site is where incoming tRNA molecules bind to the mRNA (except for the initiator tRNA).

aminoacyl tRNA *See* charged tRNA.

aminoacyl-tRNA synthetase An enzyme that catalyzes the attachment of amino acids to tRNA molecules.

amino terminus *See* N-terminus.

ammonia (NH$_3$) A highly toxic nitrogenous waste typically produced by many aquatic animal species.

ammonification The conversion of organic nitrogen to NH$_3$ and NH$_4^+$ during the nitrogen cycle.

amnion The innermost of the four extraembryonic membranes in the amniotic egg. It protects the developing embryo in a fluid-filled sac called the amniotic cavity.

amniotes A group of tetrapods with amniotic eggs that includes turtles, lizards, snakes, crocodiles, birds, and mammals.

amniotic egg A type of egg produced by amniotic animals that contains the developing embryo and the four separate extraembryonic membranes that it produces: the amnion, the yolk sac, the allantois, and the chorion.

amoeba (plural, **amoebae**) A protist that moves by pseudopodia, which involves extending cytoplasm into filaments or lobes.

amoebocyte A mobile cell within a sponge's mesophyl that absorbs food from choanocytes, digests it, and carries the nutrients to other cells.

amphibian An ectothermic, vertebrate animal that metamorphoses from a water-breathing to an air-breathing form but must return to the water to reproduce.

amphipathic Molecules containing a hydrophobic (water-fearing) region and a hydrophilic (water-loving) region.

ampulla (plural, **ampullae**) 1. A muscular sac at the base of each tube foot of a echinoderm; used to store water. 2. A bulge in the walls of the semicircular canals of the mammalian inner

ear; important for sensing circular motions of the head.

amygdala An area of the vertebrate forebrain known to be critical for understanding and remembering emotional situations.

amylase A digestive enzyme in saliva and the pancreas involved in the digestion of starch.

anabolic reaction A metabolic pathway that involves the synthesis of larger molecules from smaller precursor molecules. Such reactions usually require an input of energy.

anabolism A metabolic pathway that results in the synthesis of cellular molecules and macromolecules; requires an input of energy.

anaerobic Refers to a process that occurs in the absence of oxygen; a form of metabolism that does not require oxygen.

anaerobic respiration The breakdown of organic molecules in the absence of oxygen.

anagenesis The pattern of speciation in which a single species is transformed into a different species over the course of many generations.

analogous structure A structure that is the result of convergent evolution. Such structures have arisen independently, two or more times, because species have occupied similar types of environments on Earth.

anaphase The phase of mitosis during which the sister chromatids separate from each other and move to opposite poles; the poles themselves also move farther apart.

anatomy The study of the morphology of organisms, such as plants and animals.

anchoring junction A type of junction between animal cells that attaches cells to each other and to the extracellular matrix (ECM).

androecium The aggregate of stamens that forms the third whorl of a flower.

androgens Steroid hormones produced by the male testes (and, to a lesser extent, the adrenal glands) that affect most aspects of male reproduction.

anemia A condition characterized by lower than normal levels of hemoglobin, which reduces the amount of oxygen that can be stored in the blood.

aneuploidy An alteration in the number of particular chromosomes so that the total number of chromosomes is not an exact multiple of a set.

angina pectoris Chest pain during exertion due to the heart being deprived of oxygen.

angiosperm A flowering plant. The term means enclosed seed, which reflects the presence of seeds within fruits.

animal cap assay A type of experiment used to identify proteins secreted by embryonic cells that induce cells in the animal pole to differentiate into mesoderm.

animal pole In triploblast organisms, the pole of the egg with less yolk and more cytoplasm.

Animalia A eukaryotic kingdom of the domain Eukarya.

animals Multicellular heterotrophs with cells that lack cell walls. Most animals have nerves, muscles, the capacity to move at some point in their life cycle, and the ability to reproduce sexually, with sperm fusing directly with eggs.

anion An ion that has a net negative charge.

annual A plant that dies after producing seed during its first year of life.

antagonist A muscle or group of muscles that produces oppositely directed movements at a joint.

antenna complex *See* light-harvesting complex.

anterior Refers to the end of an animal where the head is found.

anteroposterior axis In bilateral animals, one of the three axes along which the adult body pattern is organized; the others are the dorsoventral axis and the right-left axis.

anther The uppermost part of a flower stamen, consisting of a cluster of microsporangia that produce and release pollen.

antheridia Round or elongate gametangia that produce sperm in plants.

anthropoidea A member of a group of primates that includes the monkeys and the hominoidea; these species are larger-brained and diurnal.

antibiotic A chemical, usually made by microorganisms, that inhibits the growth of certain other microorganisms.

antibody A protein secreted by plasma cells that is part of the immune response; antibodies travel all over the body to reach antigens identical to those that stimulated their production, combine with these antigens, then guide an attack that eliminates the antigens or the cells bearing them.

anticodon A three-nucleotide sequence in tRNA that is complementary to a codon in mRNA.

antidiuretic hormone (ADH) A hormone secreted by the posterior pituitary gland that acts on kidney cells to decrease urine production.

antigen Any foreign molecule that the host does not recognize as self and that triggers a specific immune response.

antigen-presenting cells (APCs) Cells bearing fragments of antigen, called antigenic determinants or epitopes, complexed with the cell's major histocompatibility complex (MHC) proteins.

antiparallel The arrangement in DNA where one strand runs in the 5′ to 3′ direction while the other strand is oriented in the 3′ to 5′ direction.

antiporter A type of transporter that binds two or more ions or molecules and transports them in opposite directions across a membrane.

anus In mammals, the final portion of the rectum through which feces are expelled.

aorta In vertebrates, a large blood vessel that exits a ventricle of the heart and leads to the systemic circulation.

apical-basal-patterning genes A category of genes that are important in early stages of plant development during which the apical and basal axes are formed.

apical-basal polarity An architectural feature of plants in which they display an upper, apical pole and a lower, basal pole; shoot apical meristem occurs at the apical pole, and root apical meristem occurs at the basal pole.

apical constriction A cellular process during gastrulation that occurs in bottle cells, where a reduction in the diameter of the actin rings connected to the adherens junctions causes the cells to elongate toward their basal end.

apical meristem In plants, a group of actively dividing cells at a growing tip.

apical region The region of a plant seedling that produces the leaves and flowers.

apomixis A natural asexual reproductive process in which plant fruits and seeds are produced in the absence of fertilization.

apoplast The continuum of water-soaked cell walls and intercellular spaces in a plant.

apoplastic transport The movement of solutes through cell walls and the spaces between cells.

apoptosis Programmed cell death.

aposematic coloration Warning coloration that advertises an organism's unpalatable taste.

aquaporin A transport protein in the form of a channel that allows the rapid diffusion of water across the cell membrane.

aqueous humor A thin liquid in the anterior cavity behind the cornea of the vertebrate eye.

aqueous solution A solution made with water.

aquifer An underground water supply.

arbuscular mycorrhizae Symbiotic associations between AM fungi and the roots of vascular plants.

Archaea One of the three domains of life; the other two are Bacteria and Eukarya.

archaea When not capitalized, refers to a cell or species within the domain Archaea.

archegonia Flask-shaped plant gametangia that enclose an egg cell.

archenteron A cavity formed in an animal embryo during gastrulation that will become the organism's digestive tract.

area hypothesis The proposal that larger areas contain more species than smaller areas because they can support larger populations and a greater range of habitats.

arteriole A single-celled layer of endothelium surrounded by one or two layers of smooth muscle and connective tissue that delivers blood to the capillaries and distributes blood to regions of the body in proportion to metabolic demands.

artery A blood vessel that carries blood away from the heart.

artificial selection *See* selective breeding.

asci Fungal sporangia shaped like sacs that produce and release sexual ascospores.

ascocarp The type of fruiting body produced by ascomycete fungi.

ascomycetes A phylum of fungi that produce sexual spores in saclike asci located at the surfaces of fruiting bodies known as ascocarps.

ascospore The type of sexual spore produced by the ascomycete fungi.

aseptate The condition of not being partitioned into smaller cells; usually refers to fungal cells.

asexual reproduction A reproductive strategy that occurs when offspring are produced from a single parent, without the fusion of gametes from two parents. The offspring are therefore clones of the parent.

A site *See* aminoacyl site.

assimilation During the nitrogen cycle, the process by which plants and animals incorporate the ammonia and NO_3^- formed through nitrogen fixation and nitrification.

associative learning A change in behavior due to an association between a stimulus and a response.

asthma A disease in which the smooth muscles around the bronchioles contract more than usual, decreasing airflow in the lungs.

AT/GC rule Refers to the phenomenon that an A in one DNA strand always hydrogen-bonds with a T in the opposite strand, and a G in one strand always bonds with a C.

atherosclerosis The condition in which large plaques may occlude (block) the lumen of an artery.

atmospheric (barometric) pressure The pressure exerted by the gases in air on the body surfaces of animals.

atom The smallest functional unit of matter that forms all chemical substances and cannot be further broken down into other substances by ordinary chemical or physical means.

atomic mass An atom's mass relative to the mass of other atoms. By convention, the most common form of carbon, which has six protons and six neutrons, is assigned an atomic mass of exactly 12.

atomic nucleus The center of an atom; contains protons and neutrons.

atomic number The number of protons in an atom.

ATP *See* adenosine triphosphate.

ATP-dependent chromatin remodeling enzyme An enzyme that catalyzes a change in the positions of nucleosomes.

ATP synthase An enzyme that utilizes the energy stored in a H^+ electrochemical gradient for the synthesis of ATP via chemiosmosis.

atrial natriuretic peptide (ANP) A peptide secreted from the atria of the heart whenever blood levels of sodium increase; ANP causes a natriuresis by decreasing sodium reabsorption in the kidney tubules.

atrioventricular (AV) node Specialized cardiac cells in most vertebrates that sit near the junction of the atria and ventricles and conduct the electrical events from the atria to the ventricles.

atrioventricular (AV) valve A one-way valve into the ventricles of the vertebrate heart through which blood moves from the atria.

atrium In the heart, a chamber to collect blood from the tissues.

atrophy A reduction in the size of a structure, such as a muscle.

audition The ability to detect and interpret sound waves; present in vertebrates and arthropods.

autoimmune disease In humans and many other vertebrates, a disorder in which the body's normal state of immune tolerance breaks down, with the result that attacks are directed against the body's own cells and tissues.

autonomic nervous system The division of the peripheral nervous system that regulates homeostasis and organ function.

autonomous specification The unequal acquisition of cytoplasmic factors during cell division in a developing vertebrate embryo.

autophagosome A double-membrane structure enclosing cellular material destined to be degraded; produced by the process of autophagy.

autophagy A process whereby cellular material, such as a worn-out organelle, becomes enclosed in a double membrane and is degraded.

autosomes All of the chromosomes found in the cell nucleus of eukaryotes except for the sex chromosomes.

autotomy In echinoderms, the ability to detach a body part, such as a limb, that will later regenerate.

autotroph An organism that has metabolic pathways that use energy from either inorganic molecules or light to make organic molecules.

auxin One of several types of hormones considered to be "master" plant hormones because they influence plant structure, development, and behavior in many ways.

auxin efflux carrier One of several types of PIN proteins, which transport auxin out of plant cells.

auxin influx carrier A plasma membrane protein that transports auxin into plant cells.

auxin-response genes Plant genes that are regulated by the hormone auxin.

avirulence gene (*Avr* gene) A gene in a plant pathogen that encodes a virulence-enhancing elicitor, which causes plant disease.

Avogadro's number As first described by Italian physicist Amedeo Avogadro, 1 mole of any element contains the same number of atoms—6.022×10^{23}.

axillary bud A bud that occurs in the axil, the upper angle where a twig or leaf emerges from a stem.

axillary meristem A meristem produced in the axil, the upper angle where a twig or leaf emerges from a stem. Axillary meristems generate axillary buds, which can produce flowers or branches.

axon An extension of the plasma membrane of a neuron that is involved in sending signals to neighboring cells.

axon hillock The part of the axon closest to the cell body; typically where an action potential begins.

axon terminal The end of the axon that sends electrical or chemical messages to other cells.

axoneme The internal structure of eukaryotic flagella and cilia consisting of microtubules, the motor protein dynein, and linking proteins.

B

bacilli Rod-shaped prokaryotic cells.

backbone The linear arrangement of phosphates and sugar molecules in a DNA or RNA strand.

Bacteria One of the three domains of life; the other two are Archaea and Eukarya.

bacteria (singular, bacterium) When not capitalized, refers to a cell or species within the domain Bacteria.

bacterial artificial chromosome (BAC) A cloning vector derived from F factors that can contain large DNA inserts.

bacterial colony A clone of genetically identical cells formed from a single cell.

bacteriophage A virus that infects bacteria.

bacteroid A modified bacterial cell of the type known as rhizobia present in mature root nodules of some plants.

balanced polymorphism The phenomenon in which two or more alleles are kept in balance and maintained in a population over the course of many generations.

balancing selection A type of natural selection that maintains genetic diversity in a population.

balloon angioplasty A common treatment to restore blood flow through a blood vessel. A thin tube with a tiny, inflatable balloon at its tip is threaded through the artery to the diseased area; inflating the balloon compresses the plaque against the arterial wall, widening the lumen.

barometric pressure *See* atmospheric pressure.

baroreceptor A pressure-sensitive region within the walls of certain arteries that contains the endings of nerve cells; these regions sense and help to maintain blood pressure in the normal range for an animal.

Barr body A highly condensed X chromosome present in female mammals.

basal body A site at the base of flagella or cilia from which microtubules grow. Basal bodies are anchored on the cytosolic side of the plasma membrane.

basal metabolic rate (BMR) The metabolic rate of an animal under resting conditions, in a postabsorptive state, and at a standard temperature.

basal nuclei Clusters of neuronal cell bodies in the vertebrate forebrain that surround the thalamus and lie beneath the cerebral cortex; involved in planning and learning movements.

basal region The region of a plant seedling that produces the roots.

basal transcription   A low level of transcription resulting from just the core promoter.

basal transcription apparatus   In a eukaryotic structural gene, refers to the complex of RNA polymerase II, GTFs, and a DNA sequence containing a TATA box.

base   1. A molecule that when dissolved in water lowers the H^+ concentration. 2. A component of nucleotides that is a single or double ring of carbon and nitrogen atoms.

base pair   The structure in which two bases in opposite strands of DNA hydrogen-bond with each other.

base substitution   A mutation that involves the substitution of a single base in the DNA for another base.

basic local alignment search tool (BLAST)   *See* BLAST.

basidia   Club-shaped cells that produce sexual spores in basidiomycete fungi.

basidiocarp   The type of fruiting body produced by basidiomycete fungi.

basidiomycetes   A phylum of fungi whose sexual spores are produced on the surfaces of club-shaped structures (basidia).

basidiospore   A sexual spore of the basidiomycete fungi.

basilar membrane   A component of the mammalian ear that vibrates back and forth in response to sound and bends the stereocilia in one direction and then the other.

basophil   A type of leukocyte that secretes the anticlotting factor heparin at the site of an infection, which helps flush out the infected site; basophils also secrete histamine, which attracts infection-fighting cells and proteins.

Batesian mimicry   The mimicry of an unpalatable species (the model) by a palatable one (the mimic).

Bayesion method   One method used to evaluate a phylogenetic tree based on an evolutionary model.

B cell   A type of lymphocyte responsible for specific immunity.

behavior   The observable response of organisms to external or internal stimuli.

behavioral ecology   A subdiscipline of organismal ecology that focuses on how the behavior of an individual organism contributes to its survival and reproductive success, which, in turn, eventually affects the population density of the species.

benign tumor   A precancerous mass of abnormal cells.

beta (β) pleated sheet   A type of protein secondary structure in which regions of a polypeptide lie parallel to each other and are held together by hydrogen bonds to form a repeating zigzag shape.

bidirectional replication   The process in which DNA replication proceeds outward from the origin in opposite directions.

biennial   A plant that does not reproduce during the first year of life but may reproduce within the following year.

bilateral symmetry   An architectural feature in which the body or organ of an organism can be divided along a vertical plane at the midline to create two halves.

Bilateria   Bilaterally symmetric animals.

bile   A substance produced by the liver that contains bicarbonate ions, cholesterol, phospholipids, a number of organic wastes, and a group of substances collectively termed bile salts.

bile salts   A group of substances produced in the liver that solubilize dietary fat and increase its accessibility to digestive enzymes.

binary fission   The process of cell division in bacteria and archaea in which one cell divides into two cells.

binocular vision   A type of vision in animals having two eyes located at the front of the head; the overlapping images coming into both eyes are processed together in the brain to form one perception.

binomial nomenclature   The standard method for naming species. Each species has a genus name and species epithet.

biochemistry   The study of the chemistry of living organisms.

biodiversity   The diversity of life forms in a given location.

biodiversity crisis   The idea that there is currently an elevated loss of species on Earth, far beyond the normal extinction rate of species.

biofilm   An aggregation of microorganisms that secrete adhesive mucilage, thereby gluing themselves to surfaces.

biogeochemical cycle   The continuous movement of nutrients such as nitrogen, carbon, sulfur, and phosphorus from the physical environment to organisms and back.

biogeography   The study of the geographic distribution of extinct and modern species.

bioinformatics   A field of study that uses computers to study biological information.

biological control   The use of an introduced species' natural enemies to control its proliferation.

biological nitrogen fixation   Nitrogen fixation that is performed in nature by certain prokaryotes.

biological species concept   An approach used to distinguish species, which states that a species is a group of individuals whose members have the potential to interbreed with one another in nature to produce viable, fertile offspring but cannot successfully interbreed with members of other species.

biology   The study of life.

bioluminescence   A phenomenon in living organisms in which chemical reactions give off light rather than heat.

biomagnification   The increase in the concentration of a substance in living organisms from lower to higher trophic levels in a food web.

biomass   A quantitative estimate of the total mass of living matter in a given area, usually measured in grams or kilograms per square meter.

biome   A major type of habitat characterized by distinctive plant and animal life.

bioremediation   The use of living organisms, usually microbes or plants, to detoxify polluted habitats such as dump sites or oil spills.

biosphere   The regions on the surface of the Earth and in the atmosphere where living organisms exist.

biosynthetic reaction   Also called an anabolic reaction; a chemical reaction in which small molecules are used to synthesize larger molecules.

biotechnology   The use of living organisms or the products of living organisms for human benefit.

biotic   The term used to describe interactions among organisms.

biparental inheritance   An inheritance pattern in which both the male and female gametes contribute organellar genes to the offspring.

bipedal   Having the ability to walk on two feet.

bipolar cells   Cells in the vertebrate eye that make synapses with photoreceptors and relay responses to the ganglion cells.

bipolar disorder   A neurological disorder characterized by alternating periods of euphoria and depression, resulting from an imbalance in neurotransmitter levels in the brain.

bivalent   Homologous pairs of sister chromatids associated with each other, lying side by side.

blade   The flattened portion of a leaf.

BLAST   A computer program that can identify homologous genes that are found in a database.

blastocoel   A cavity formed in a cleavage-stage vertebrate embryo (blastula); provides a space into which cells of the future digestive tract will migrate.

blastocyst   The mammalian counterpart of a blastula.

blastoderm   A flattened disc of dividing cells in the embryo of animals that undergo incomplete cleavage; occurs in birds and some fishes.

blastomeres   The two half-size daughter cells produced by each cell division during cleavage.

blastopore   A small opening created when a band of tissue invaginates during gastrulation. It forms the primary opening of the archenteron to the outside.

blastula   An animal embryo at the stage when it forms an outer epithelial layer and an inner cavity.

blending inheritance   An early hypothesis of inheritance that stated that the genetic material that dictates hereditary traits blends together from generation to generation, and the blended traits are then passed to the next generation.

blood   A fluid connective tissue in animals consisting of cells and (in mammals) cell fragments suspended in a solution of water containing dissolved nutrients, proteins, gases, and other molecules.

blood-doping   An example of hormone misuse in which the number of red blood cells in the circulation is boosted to increase the oxygen-carrying capacity of the blood.

blood pressure   The force exerted by blood on the walls of blood vessels; blood pressure is responsible for moving blood through the vessels.

body mass index (BMI)   A method of assessing body fat and health risk that involves calculating the ratio of weight compared to height; weight in kilograms is divided by the square of the height in meters.

body plan   The organization of cells, tissues, and organs within a multicellular organism; also known as a body pattern.

bone   A relatively hard component of the vertebrate skeleton; a living, dynamic tissue composed of organic molecules and minerals.

bottleneck effect   A situation in which a population size is dramatically reduced and then rebounds. While the population is small, genetic drift may rapidly reduce the genetic diversity of the population.

Bowman's capsule   A saclike structure that houses the glomerulus at the beginning of the tubular component of a nephron in the mammalian kidney.

brain   Organ of the central nervous system of animals that functions to process and integrate information.

brainstem   The part of the vertebrate brain composed of the medulla oblongata, the pons, and the midbrain.

brassinosteroid   One of several plant hormones that help a plant to cope with environmental stress.

bronchi (singular, **bronchus**)   Tubes branching from the trachea and leading into the lungs.

bronchiole   A thin-walled, small tube branching from the bronchi and leading to the alveoli in mammalian lungs.

bronchodilator   A compound that binds to the muscles of the bronchioles of the lung and causes them to relax, thereby widening the bronchioles.

brown adipose tissue   A specialized tissue in small mammals such as hibernating bats, small rodents living in cold environments, and many newborn mammals, including humans, that can help to generate heat and maintain body temperature.

brush border   The collective name for the microvilli in the vertebrate small intestine.

bryophytes   Liverworts, mosses, and hornworts, the modern nonvascular land plants.

buccal pumping   A form of breathing in which animals take in water or air into their mouths, then raise the floor of the mouth, creating a positive pressure that pumps water or air across the gills or into the lungs; found in fishes and amphibians.

bud   A miniature plant shoot having a dormant shoot apical meristem.

budding   A form of asexual reproduction in which a portion of the parent organism pinches off to form a complete new individual.

buffer   A compound that acts to minimize pH fluctuations in the fluids of living organisms. Buffer systems can raise or lower pH as needed.

bulbourethral glands   Paired accessory glands in the human male reproductive system that secrete an alkaline mucus that protects sperm by neutralizing the acidity in the urethra.

bulk feeders   Animals that eat food in large pieces.

bulk flow   The mass movement of liquid in a plant caused by pressure, gravity, or both.

C

C_3 plant   A plant that incorporates CO_2 into organic molecules via RuBP to make 3PG, a three-carbon molecule.

C_4 plant   A plant that uses PEP carboxylase to initially fix CO_2 into a four-carbon molecule and later uses rubisco to fix CO_2 into simple sugars; an adaptation to hot, dry environments.

cadherin   A cell adhesion molecule found in animal cells that promotes cell-to-cell adhesion.

calcitonin   A hormone that plays a role in Ca^{2+} homeostasis in some vertebrates.

calcium wave   A brief increase in cytosolic Ca^{2+} concentrations in an egg that has been penetrated by a sperm cell; the change in Ca^{2+} moves through the cell and contributes to the slow block to polyspermy.

callose   A carbohydrate that plays crucial roles in plant development and plugging wounds in plant phloem.

calorie   The amount of heat required to raise the temperature of 1 gram of water 1 degree Celsius.

Calvin cycle   The second stage in the process of photosynthesis. During this cycle, ATP is used as a source of energy, and NADPH is used as a source of high-energy electrons so that CO_2 can be incorporated into carbohydrate.

calyx   The sepals that form the outermost whorl of a flower.

Cambrian explosion   An event during the Cambrian period (543 to 490 mya) in which there was an abrupt increase (on a geological scale) in the diversity of animal species.

cAMP   *See* cyclic adenosine monophosphate.

CAM plants C_4 plants that open their stomata at night to take up carbon dioxide.

cancer   A disease caused by gene mutations that lead to uncontrolled cell growth.

canopy   The uppermost layer of tree foliage.

capillary   A tiny thin-walled vessel that is the site of gas and nutrient exchange between the blood and interstitial fluid.

capping   The process in which a 7-methylguanosine is covalently attached at the 5′ end of mature mRNAs of eukaryotes.

capsid   A protein coat enclosing a virus's genome.

CAP site   One of two regulatory sites near the *lac* promoter; this site is a DNA sequence recognized by the catabolite activator protein (CAP).

capsule   A very thick, gelatinous glycocalyx produced by certain strains of bacteria that may help them avoid being destroyed by an animal's immune (defense) system.

carapace   The hard protective cuticle covering the cephalothorax of a crustacean.

carbohydrate   An organic molecule often with the general formula, $C(H_2O)$; a carbon-containing compound that includes starches, sugars, and cellulose.

carbon fixation   A process in which carbon from inorganic CO_2 is incorporated into an organic molecule such as a carbohydrate.

carboxyl terminus   *See* C-terminus.

carcinogen   An agent that increases the likelihood of developing cancer, usually a mutagen.

carcinoma   A cancer of epithelial cells.

cardiac cycle   The events that produce a single heartbeat, which can be divided into two phases, diastole and systole.

cardiac muscle   A type of muscle tissue found only in hearts in which physical and electrical connections between individual cells enable many of the cells to contract simultaneously.

cardiac output (CO)   The amount of blood the heart pumps per unit time, usually expressed in units of L/min.

cardiovascular disease   Diseases affecting the heart and blood vessels.

cardiovascular system   A system containing three components: blood or hemolymph, blood vessels, and one or more hearts.

carnivore   An animal that consumes animal flesh or fluids.

carotenoid   A type of photosynthetic or protective pigment found in plastids that imparts a color that ranges from yellow to orange to red.

carpel   A flower shoot organ that produces ovules that contain female gametophytes.

carrier   *See* transporter.

carrying capacity (K)   The upper boundary for a population size.

Casparian strips   Suberin ribbons on the walls of endodermal cells of plant roots; prevent apoplastic transport of ions into vascular tissues.

caspase   An enzyme that is activated during apoptosis.

catabolic reaction   A metabolic pathway in which a molecule is broken down into smaller components, usually releasing energy.

catabolism   A metabolic pathway that results in the breakdown of larger molecules into smaller molecules. Such reactions are often exergonic.

catabolite activator protein (CAP)   An activator protein for the *lac* operon.

catabolite repression   In bacteria, a process whereby transcriptional regulation is influenced by the presence of glucose.

catalase   An enzyme within peroxisomes that breaks down hydrogen peroxide to water and oxygen gas.

catalyst   An agent that speeds up the rate of a chemical reaction without being consumed during the reaction.

cataract   An accumulation of protein in the lens of the eye; causes blurring and poor night vision.

cation   An ion that has a net positive charge.

cation exchange   With regard to soil, the process in which hydrogen ions are able to replace mineral cations on the surfaces of humus or clay particles.

cDNA   *See* complementary DNA.

cDNA library   A type of DNA library in which the inserts are derived from cDNA.

cecum   The first portion of a vertebrate's large intestine.

cell   The simplest unit of a living organism.

cell adhesion   A vital function of the cell membrane that allows cells to bind to each other. Cell adhesion is critical in the formation of multicellular organisms and provides a way to convey positional information between neighboring cells.

cell adhesion molecule (CAM)   A membrane protein found in animal cells that promotes cell adhesion.

cell biology   The study of individual cells and their interactions with each other.

cell body   A part of a neuron that contains the cell nucleus and other organelles.

cell coat   Also called the glycocalyx, the carbohydrate-rich zone on the surface of animal cells that shields the cell from mechanical and physical damage.

cell communication   The process through which cells can detect and respond to signals in their extracellular environment. In multicellular organisms, cell communication is also needed to coordinate cellular activities within the whole organism.

cell cycle   The series of phases a eukaryotic cell progresses through from its origin until it divides by mitosis.

cell differentiation   The phenomenon by which cells become specialized into particular cell types.

cell division   The process in which one cell divides into two cells.

cell doctrine   *See* cell theory.

cell junctions   Specialized structures that adhere cells to each other and to the ECM.

cell-mediated immunity   A type of specific immunity in which cytotoxic T cells directly attack and destroy infected body cells, cancer cells, or transplanted cells.

cell nucleus   The membrane-bound area of a eukaryotic cell in which the genetic material is found.

cell plate   In plant cells, a structure that forms a cell wall between the two daughter cells during cytokinesis.

cell signaling   A vital function of the plasma membrane that involves cells sensing changes in their environment and communicating with each other.

cell surface receptor   A receptor found in the plasma membrane that enables a cell to respond to different kinds of signaling molecules.

cell theory   A theory that states that all organisms are made of cells, cells are the smallest units of living organisms, and new cells come from pre-existing cells by cell division.

cell-to-cell communication A form of cell communication that occurs between two different cells.

cellular differentiation The process by which different cells within a developing organism acquire specialized forms and functions due to the expression of cell-specific genes.

cellular respiration A process by which living cells obtain energy from organic molecules and release waste products.

cellular response Adaptation at the cellular level that involves a cell responding to signals in its environment.

cellulose The main macromolecule of the primary cell wall of plants and many algae; a polymer made of repeating molecules of glucose attached end to end.

cell wall A relatively rigid, porous structure located outside the plasma membrane of prokaryotic, plant, fungal, and certain protist cells; provides support and protection.

centiMorgan (cM) *See* map unit (mu).

central cell In the female gametophyte of a flowering plant, a large cell that contains two nuclei; after double fertilization, it forms the first cell of the nutritive endosperm tissue.

central dogma Refers to the steps of gene expression at the molecular level. DNA is transcribed into mRNA, and mRNA is translated into a polypeptide.

central nervous system (CNS) In vertebrates, the brain and spinal cord.

central region The region of a plant seedling that produces stem tissue.

central vacuole An organelle that often occupies 80% or more of the cell volume of plant cells and stores a large amount of water, enzymes, and inorganic ions.

central zone The area of a plant shoot meristem where undifferentiated stem cells are maintained.

centrioles A pair of structures within the centrosome of animal cells. Most plant cells and many protists lack centrioles.

centromere The region where the two sister chromatids are tightly associated; the centromere is an attachment site for kinetochore proteins.

centrosome A single structure often near the cell nucleus of eukaryotic cells that forms a nucleating site for the growth of microtubules; also called the microtubule-organizing center.

cephalization The localization of a brain and sensory structures at the anterior end of the body of animals.

cephalothorax The fused head and thorax structure in species of the class Arachnida and Crustacea.

cerebellum The part of the vertebrate hindbrain, along with the pons, responsible for monitoring and coordinating body movements.

cerebral cortex The surface layer of gray matter that forms the outer part of the cerebrum of the vertebrate brain.

cerebral ganglia A paired structure in the head of invertebrates that receives input from sensory cells and controls motor output.

cerebrospinal fluid Fluid that exists in ventricles within the central nervous system and surrounds the exterior of the brain and spinal cord; it absorbs physical shocks to the brain resulting from sudden movements or blows to the head.

cerebrum A region of the vertebrate forebrain that is responsible for the higher functions of conscious thought, planning, and emotion, as well as control of motor function.

cervix A fibrous structure at the end of the female vagina that forms the opening to the uterus.

channel A transmembrane protein that forms an open passageway for the direct diffusion of ions or molecules across a membrane.

chaperone A protein that keeps other proteins in an unfolded state during the process of post-translational sorting.

character A characteristic of an organism, such as the appearance of seeds, pods, flowers, or stems.

character displacement The tendency for two species to diverge in morphology and thus resource use because of competition.

character state A particular variant of a given trait.

charged tRNA A tRNA with its attached amino acid; also called aminoacyl tRNA.

charophyceans The lineages of freshwater green algae that are most closely related to the land plants.

checkpoint One of three critical regulatory points found in the cell cycle of eukaryotic cells. At these checkpoints, a variety of proteins act as sensors to determine if a cell is in the proper condition to divide.

checkpoint protein A protein that senses if a cell is in the proper condition to divide and prevents a cell from progressing through the cell cycle if it is not.

chemical energy The potential energy contained within covalent bonds in molecules.

chemical equilibrium A state in a chemical reaction in which the rate of formation of products equals the rate of formation of reactants.

chemical mutagen A chemical that causes mutations.

chemical reaction The formation and breaking of chemical bonds, resulting in a change in the composition of substances.

chemical selection Occurs when a chemical within a mixture has special properties or advantages that cause it to increase in amount. May have played a key role in the formation of an RNA world.

chemical synapse A synapse in which a chemical called a neurotransmitter is released from the axon terminal of a neuron and acts as a signal from the presynaptic to the postsynaptic cell.

chemiosmosis A process for making ATP in which energy stored in an ion electrochemical gradient is used to make ATP from ADP and P_i.

chemoautotroph An organism able to use energy obtained by chemical modifications of inorganic compounds to synthesize organic compounds.

chemoheterotroph An organism that must obtain organic molecules both for energy and as a carbon source.

chemoreceptor A sensory receptor in animals that responds to specific chemical compounds.

chiasma The connection at a crossover site of two chromosomes.

chimeric gene A gene formed from the fusion of two gene fragments to each other.

chitin A tough, nitrogen-containing polysaccharide that forms the external skeleton of many insects and the cell walls of fungi.

chlorophyll A photosynthetic green pigment found in the chloroplasts of plants, algae, and some bacteria.

chlorophyll *a* A type of chlorophyll pigment found in plants, algae, and cyanobacteria.

chlorophyll *b* A type of chlorophyll pigment found in plants, green algae, and some other photosynthetic organisms.

chloroplast A semiautonomous organelle found in plant and algal cells that carries out photosynthesis.

chloroplast genome The chromosome found in chloroplasts.

chlorosis The yellowing of plant leaves caused by various types of mineral deficiencies.

choanocyte A specialized cell of sponges that functions to trap and eat small particles.

chondrichthyans Members of the class Chondrichthyes, including sharks, skates, and rays.

chordate An organism that has or at some point in its life has had a notochord and a hollow dorsal nerve cord; includes all vertebrates and some invertebrates.

chorion One of the four extraembryonic membranes in the amniotic egg. It exchanges gases between the embryo and the surrounding air.

chorionic gonadotropin (CG) An LH-like hormone made by the blastocyst that maintains the corpus luteum.

chromatin Refers to the biochemical composition of chromosomes, which contain DNA and many types of proteins.

chromosome A discrete unit of genetic material composed of DNA and associated proteins. Eukaryotes have chromosomes in their cell nuclei and in plastids and mitochondria.

chromosome territory A distinct, nonoverlapping area where each chromosome is located within the cell nucleus of eukaryotic cells.

chromosome theory of inheritance An explanation of how the steps of meiosis account for the inheritance patterns observed by Mendel.

chylomicrons Large fat droplets coated with amphipathic proteins that perform an emulsifying function similar to that of bile salts; chylomicrons are formed in intestinal epithelial cells from absorbed fats in the diet.

chyme A solution of water and partially digested food particles in the stomach and small intestine.

chymotrypsin A protease involved in the breakdown of proteins in the small intestine.

chytrids Simple, early-diverging phyla of fungi; commonly found in aquatic habitats and moist soil, where they produce flagellate reproductive cells.

cilia (singular, **cilium**) Cell appendages that have the same internal structure as flagella and function like flagella to facilitate cell movement; cilia are shorter and more numerous on cells than are flagella.

ciliate A protist that moves by means of cilia, which are tiny hairlike extensions that occur on the outsides of cells and have the same internal structure as flagella.

circulatory system A system that transports necessary materials to all cells of an animal's body and transports waste products away from cells. Three basic types are gastrovascular cavities, open systems, and closed systems.

***cis*-acting element** *See cis*-effect.

***cis*-effect** A DNA segment that must be adjacent to the gene(s) that it regulates. The *lac* operator site is an example of a *cis*-acting element.

cisternae Flattened, fluid-filled tubules of the endoplasmic reticulum.

***cis/trans* isomers** Organic molecules with the same chemical composition but existing in two different configurations determined by the positions of hydrogen atoms on the two carbons of a C=C double bond. When the hydrogen atoms are on the same side of the double bond, it is called a *cis* isomer; when on the opposite sides of the double bond, it is a *trans* isomer.

citric acid cycle   A cycle that results in the breakdown of carbohydrates to carbon dioxide; also known as the Krebs cycle.

clade   A group of species derived from a single common ancestor.

cladistic approach   An approach used to construct a phylogenetic tree by comparing primitive and shared derived characters.

cladogenesis   A pattern of speciation in which a species is divided into two or more species.

cladogram   A phylogenetic tree constructed by using a cladistic approach.

clamp connection   In basidiomycete fungi, a structure that helps distribute nuclei during cell division.

clasper   An extension of the pelvic fin of a chondrichthyan, used by the male to transfer sperm to the female.

class   In taxonomy, a subdivision of a phylum.

classical conditioning   A type of associative learning in which an involuntary response comes to be associated positively or negatively with a stimulus that did not originally elicit the response.

cleavage   A succession of rapid cell divisions with no significant growth that produces a hollow sphere of cells called a blastula.

cleavage furrow   In animal cells, an area that constricts like a drawstring to separate the cells during cytokinesis.

climate   The prevailing weather pattern of a given region.

climax community   A distinct end point of succession.

clitoris   Located at the anterior part of the labia minora, erectile tissue that becomes engorged with blood during sexual arousal and is very sensitive to sexual stimulation.

clonal deletion   One of two mechanisms that explain why normal individuals lack active lymphocytes that respond to self components; T cells with receptors capable of binding self proteins are destroyed by apoptosis.

clonal inactivation   One of two mechanisms that explain why normal individuals lack active lymphocytes that respond to self components; the process occurs outside the thymus and causes potentially self-reacting T cells to become nonresponsive.

clonal selection   The process by which an antigen-stimulated lymphocyte divides and forms a clone of cells, each of which recognizes that particular antigen.

cloning   Making many copies of something such as a DNA molecule.

closed circulatory system   A circulatory system in which blood flows throughout an animal entirely within a series of vessels and is kept separate from the interstitial fluid.

closed conformation   Tightly packed chromatin that cannot be transcribed into RNA.

clumped   The most common pattern of dispersion within a population, in which individuals are gathered in small groups.

cnidocil   On the surface of a cnidocyte, a hairlike trigger that detects stimuli.

cnidocyte   A characteristic feature of cnidarians; a stinging cell that functions in defense or the capture of prey.

coacervates   Droplets that form spontaneously from the association of charged polymers such as proteins, carbohydrates, or nucleic acids surrounded by water.

coactivator   A protein that increases the rate of transcription but does not directly bind to the DNA itself.

coat protein   A protein that surrounds a membrane vesicle and facilitates vesicle formation.

cocci   Sphere-shaped prokaryotic cells.

cochlea   A coiled structure in the inner ear of mammals that contains the auditory receptors (organ of Corti).

coding sequence   The region of a gene or a DNA molecule that encodes the information for the amino acid sequence of a polypeptide.

coding strand   The DNA strand opposite to the template (or noncoding strand).

codominance   The phenomenon in which a single individual expresses two alleles.

codon   A sequence of three nucleotide bases that specifies a particular amino acid or a stop codon; codons function during translation.

coefficient of relatedness (r)   The probability that any two individuals will share a copy of a particular gene.

coelom   A fluid-filled body cavity in an animal.

coelomate   An animal with a true coelom.

coenzyme   An organic molecule that participates in a chemical reaction with an enzyme but is left unchanged after the reaction is completed.

coevolution   The process by which two or more species of organisms influence each other's evolutionary pathway.

cofactor   Usually an inorganic ion that temporarily binds to the surface of an enzyme and promotes a chemical reaction.

cognitive learning   The ability to solve problems with conscious thought and without direct environmental feedback.

cohesion   The ability of like molecules to noncovalently bind to each other; the attraction of water molecules for each other.

cohesion-tension theory   The explanation for long-distance water transport as the combined effect of the cohesive forces of water and evaporative tension.

cohort   A group of organisms of the same age.

coleoptile   A protective sheath that encloses the first bud of the epicotyl in a mature monocot embryo.

coleorhiza   A protective envelope that encloses the young root of a monocot.

colinearity rule   The phenomenon whereby the order of homeotic genes along the chromosome correlates with their expression along the anteroposterior axis of the body.

collagen   A protein secreted from animal cells that forms large fibers in the extracellular matrix.

collecting duct   A tubule in the mammalian kidney that collects urine from nephrons.

collenchyma cells   Flexible cells that make up collenchyma tissue.

collenchyma tissue   A plant ground tissue that provides support to plant organs.

colligative property   A property of a solution that depends only on the concentration of solute molecules.

colloid   A gel-like substance in the follicles of the thyroid gland.

colon   A part of a vertebrate's large intestine consisting of three relatively straight segments—the ascending, transverse, and descending portions. The terminal portion of the descending colon is S-shaped, forming the sigmoid colon, which empties into the rectum.

colony hybridization   A method that uses a labeled probe to identify bacterial colonies that contain a desired gene.

combinatorial control   The phenomenon whereby a combination of many factors determines the expression of any given gene.

commensalism   An interaction that benefits one species and leaves the other unaffected.

communication   The use of specially designed visual, chemical, auditory, or tactile signals to modify the behavior of others.

community   An assemblage of populations of different species that live in the same place at the same time.

community ecology   The study of how populations of species interact and form functional communities.

compartmentalization   A characteristic of eukaryotic cells in which many organelles separate the cell into different regions. Cellular compartmentalization allows a cell to carry out specialized chemical reactions in different places.

competent   The term used to describe bacterial strains that have the ability to take up DNA from the environment.

competition   An interaction that affects two or more species negatively, as they compete over food or other resources.

competitive exclusion hypothesis   The proposal that two species with the same resource requirements cannot occupy the same niche.

competitive inhibitor   A molecule that binds to the active site of an enzyme and inhibits the ability of the substrate to bind.

complement   The family of plasma proteins that provides a means for extracellular killing of microbes without prior phagocytosis.

complementary   Describes the specific base pairing that occurs between strands of nucleic acids; A pairs only with T (in DNA) or U (in RNA), and G pairs only with C.

complementary DNA (cDNA)   DNA molecules that are made from mRNA as a starting material.

complete flower   A flower that possesses all four types of flower organs.

complete metamorphosis   During development in the majority of insects, a dramatic change in body form from larva to a very different looking adult.

compound   A molecule composed of two or more different elements.

compound eyes   Image-forming eyes in arthropods and some annelids consisting of several hundred to several thousand light detectors called ommatidia.

computational molecular biology   An area of study that uses computers to characterize the molecular components of living things.

concentration   The amount of a solute dissolved in a unit volume of solution.

condensation reaction   A chemical reaction in which two or more molecules are combined into one larger molecule by covalent bonding, with the loss of a small molecule.

conditional specification   The acquisition by cells of specific properties through a variety of cell-to-cell signaling mechanisms in a developing vertebrate embryo.

conditioned response   The learned response that is elicited by a newly conditioned stimulus.

conditioned stimulus   A new stimulus that is delivered at the same time as an old stimulus, and that over time, is sufficient to elicit the same response.

condom A sheathlike membrane worn over the penis; in addition to their contraceptive function, condoms significantly reduce the risk of sexually transmitted diseases.

conduction The process in which the body surface loses or gains heat through direct contact with cooler or warmer substances.

cone pigments The several types of visual pigments found in the cones of the vertebrate eye.

cones 1. Photoreceptors found in the vertebrate eye; they are less sensitive to low levels of light but can detect color. 2. The reproductive structures of coniferous plants.

congenital hypothyroidism A condition characterized by poor differentiation of the central nervous system due to a failure of neurons to become myelinated in fetal development; results in profound mental defects.

congestive heart failure The condition resulting from the failure of the heart to pump blood normally; results in fluid buildup in the lungs (congestion).

conidia A type of asexual reproductive cell produced by many fungi.

conifers A phylum of gymnosperm plants, Coniferophyta.

conjugation A type of genetic transfer between bacteria that involves a direct physical interaction between two bacterial cells.

connective tissue Clusters of cells that connect, anchor, and support the structures of an animal's body; includes blood, adipose (fat-storing) tissue, bone, cartilage, loose connective tissue, and dense connective tissue.

connexon A channel that forms gap junctions consisting of six connexin proteins in one cell aligned with six connexin proteins in an adjacent cell.

conservation biology The study that uses principles and knowledge from molecular biology, genetics, and ecology to protect the biological diversity of life.

conservative mechanism In this incorrect model for DNA replication, both parental strands of DNA remain together (are conserved) following DNA replication. The two newly made daughter strands also occur together.

consortia A community of many microbial species.

constant regions The portions of amino acid sequences in the heavy and light chains that are identical for all immunoglobulins of a given class.

constitutive gene An unregulated gene that has constant levels of expression in all conditions over time.

contig A series of clones that contain overlapping pieces of chromosomal DNA.

continental drift The process by which, over the course of billions of years, the major landmasses, known as the continents, have shifted their positions, changed their shapes, and, in some cases, have become separated from each other.

contraception The use of birth control procedures to prevent fertilization or implantation of a fertilized egg.

contractile vacuole A small, membrane enclosed, water-filled compartment that eliminates excess liquid from the cells of certain protists.

contrast In microscopy, relative differences in the lightness, darkness, or color between adjacent regions in a sample.

control group The sample in an experiment that is treated just like an experimental group

except that it is not subjected to one particular variable.

convection The transfer of heat by the movement of air or water next to the body.

convergent evolution The process whereby two different species from different lineages show similar characteristics because they occupy similar environments.

convergent extension A cellular process during gastrulation that is crucial to development; two rows of cells merge to form a single elongated layer.

convergent trait *See* analogous structure.

coprophagy The practice of certain birds and mammals in which feces are consumed to maximize absorption of water and nutrients.

copulation The process of sperm being deposited within the reproductive tract of the female.

coral reef A type of aquatic biome found in warm, marine environments.

corepressor A small effector molecule that binds to a repressor protein to inhibit transcription.

core promoter Refers to the TATA box and the transcriptional start site of a eukaryotic structural gene.

Coriolis effect The effect of the Earth's rotation on the surface flow of wind.

cork cambium A secondary meristem in a plant that produces cork tissue.

cornea A thin, clear layer on the front of the vertebrate eye.

corolla The petals of a flower, which occur in the whorl to the inside of the calyx and the outside of the stamens.

corona The ciliated crown of members of the phylum Rotifera.

coronary artery An artery that carries oxygen and nutrients to the heart muscle.

coronary artery bypass A common treatment to restore blood flow through a coronary artery. A small piece of healthy blood vessel is removed from one part of the body and surgically grafted onto the coronary circulation in order to bypass the diseased artery.

coronary artery disease A condition that occurs when plaques form in the coronary arteries.

corpus callosum The major tract that connects the two hemispheres of the cerebrum.

corpus luteum A structure that develops from a ruptured follicle following ovulation; it is responsible for secreting hormones that stimulate the development of the uterus during pregnancy.

correlation A meaningful relationship between two variables.

cortex The area of a plant stem or root beneath the epidermis that is largely composed of parenchyma tissue.

cortical reaction An event in fertilization in which IP_3 and calcium signaling produces barriers to more than one sperm cell binding to and uniting with an egg; called the slow block to polyspermy.

cortisol A steroid hormone made by the adrenal cortex.

cotranslational sorting The sorting process in which the synthesis of certain eukaryotic proteins begins in the cytosol and then halts temporarily until the ribosome has become bound to the ER membrane.

cotransporter *See* symporter.

cotyledon An embryonic seed leaf.

countercurrent exchange mechanism An arrangement of water and blood flow in which water enters a fish's mouth and flows between

the lamellae of the gills in the opposite direction to blood flowing through the lamellar capillaries.

countercurrent heat exchange A method of regulating heat loss to the environment; many animals conserve heat by returning it to the body's core and keeping the core much warmer than the extremities.

covalent bond A chemical bond in which two atoms share a pair of electrons.

CpG island A cluster of CpG sites. CG refers to the nucleotides of C and G in DNA, and p refers to a phosphodiester linkage.

cranial nerve A nerve in the peripheral nervous system that is directly connected to the brain.

craniate A chordate that has a brain encased in a skull and possesses a neural crest.

cranium A protective bony or cartilaginous housing that encases the brain of a craniate.

crenation The process of cell shrinkage that occurs if animal cells are placed in a hypertonic medium.

cristae Projections of the highly invaginated inner membrane of a mitochondrion.

critical innovations New features that foster the diversification of phyla.

critical period A limited period of time in which many animals develop species-specific patterns of behavior.

crop A storage organ that is a dilation of the lower esophagus; found in most birds and many invertebrates, including insects and some worms.

cross-bridge A region of myosin molecules that extend from the surface of the thick filaments toward the thin filaments in skeletal muscle.

cross-bridge cycle During muscle contraction, the sequence of events that occurs between the time when a cross-bridge binds to a thin filament and when it is set to repeat the process.

cross-fertilization Fertilization that involves the union of a female gamete and a male gamete from different individuals.

crossing over The exchange of genetic material between homologous chromosomes during meiosis; allows for increased variation in the genetic information that each parent may pass to the offspring.

cross-pollination The process in which a stigma receives pollen from a different plant of the same species.

cryptic coloration The blending of an organism with the background color of its habitat; also known as camouflage.

cryptochrome A type of blue-light receptor in plants and protists.

C-terminus The location of the last amino acid in a polypeptide; also known as the carboxyl terminus.

CT scan Computerized tomography, which is an X-ray technique used to examine the structure and activity level of the brain without anesthesia or surgery.

cupula A gelatinous structure within the lateral line organ of fishes that detects changes in water movement.

cuticle A coating of wax and cutin that helps to reduce water loss from plant surfaces. Also, a nonliving covering that serves to both support and protect an animal.

cycads A phylum of gymnosperm plants, Cycadophyta.

cyclic adenosine monophosphate (cAMP) A small effector molecule that acts as a second messenger and is produced from ATP.

cyclic AMP (cAMP)   *See* cyclic adenosine monophosphate.

cyclic electron flow   *See* cyclic photophosphorylation.

cyclic photophosphorylation   During photosynthesis, a pattern of electron flow in the thylakoid membrane that is cyclic and generates ATP alone.

cyclin   A protein responsible for advancing a cell through the phases of the cell cycle by binding to a cyclin-dependent kinase.

cyclin-dependent kinase (cdk)   A protein responsible for advancing a cell through the phases of the cell cycle. Its function is dependent on the binding of a cyclin.

cyst   A one-to-few celled structure that often has a thick, protective wall and can remain dormant through periods of unfavorable climate or low food availability.

cytogenetics   The field of genetics that involves the microscopic examination of chromosomes.

cytokines   A family of proteins that function in both nonspecific and specific immune defenses by providing a chemical communication network that synchronizes the components of the immune response.

cytokinesis   The division of the cytoplasm to produce two distinct daughter cells.

cytokinin   A type of plant hormone that promotes cell division.

cytoplasm   The region of the cell that is contained within the plasma membrane.

cytoplasmic inheritance   *See* extranuclear inheritance.

cytosine (C)   A pyrimidine base found in DNA and RNA.

cytoskeleton   In eukaryotes, a network of three different types of protein filaments in the cytosol called microtubules, intermediate filaments, and actin filaments.

cytosol   The region of a eukaryotic cell that is inside the plasma membrane and outside the organelles.

cytotoxic T cell   A type of lymphocyte that travels to the location of its target, binds to the target by combining with an antigen on it, and directly kills the target via secreted chemicals.

D

dalton (Da)   A measure of atomic mass. One dalton equals one-twelfth the mass of a carbon atom.

data mining   The extraction of useful information and often previously unknown relationships from sequence files and large databases.

database   A large number of computer data files that are collected, stored in a single location, and organized for rapid search and retrieval.

daughter strand   The newly made strand in DNA replication.

day-neutral plant   A plant that flowers regardless of the night length, as long as day length meets the minimal requirements for plant growth.

deafness   Hearing loss, usually caused by damage to the hair cells within the cochlea.

death-inducing signaling complex (DISC)   A complex consisting of death receptors, FADD, and procaspase that initiates apoptosis via the extrinsic pathway.

death receptor   A type of receptor found in the plasma membrane of eukaryotic cells that can promote apoptosis when it becomes activated.

decomposer   A consumer that gets its energy from the remains and waste products of other organisms.

defecation   The expulsion of feces that occurs through the anus of an animal's digestive canal.

defensive mutualism   A mutually beneficial interaction often involving an animal defending a plant or herbivore in return for food or shelter.

deforestation   The conversion of forested areas by humans to nonforested land.

degenerate   In the genetic code, the observation that more than one codon can specify the same amino acid.

dehydration   A reduction in the amount of water in the body.

dehydration reaction   A type of condensation reaction in which a molecule of water is lost.

delayed implantation   A reproductive cycle in which a fertilized egg reaches the uterus but does not implant until later, when environmental conditions are more favorable for the newly produced young.

delayed ovulation   A reproductive cycle in which the ovarian cycle in females is halted before ovulation and sperm are stored and nourished in the female's uterus over the winter. Upon arousal from hibernation in the spring, the female ovulates one or more eggs, which are fertilized by the stored sperm.

deletion   A type of mutation in which a segment of genetic material is missing.

demographic transition   The shift in birth and death rates accompanying human societal development.

demography   The study of birth rates, death rates, age distributions, and the sizes of populations.

dendrite   A treelike extension of the plasma membrane of a neuron that receives electrical signals from other neurons.

dendritic cell   A type of cell derived from bone marrow stem cells that plays an important role in nonspecific immunity; these cells are scattered throughout most tissues, where they perform various macrophage functions.

denitrification   The reduction of nitrate to gaseous nitrogen.

density   In the context of populations, the numbers of organisms in a given unit area.

density-dependent factor   A mortality factor whose influence increases with the density of the population.

density-independent factor   A mortality factor whose influence is not affected by changes in population density.

deoxynucleoside triphosphates   Individual nucleotides with three phosphate groups.

deoxyribonucleic acid (DNA)   One of two classes of nucleic acids; the other is ribonucleic acid (RNA). A DNA molecule consists of two strands of nucleotides coiled around each other to form a double helix, held together by hydrogen bonds according to the AT/GC rule.

deoxyribose   A five-carbon sugar found in DNA.

depolarization   The change in the membrane potential that occurs when a cell becomes less polarized, that is, less negative relative to the surrounding fluid.

dermal tissue   The covering on various parts of a plant.

descent with modification   Darwin's theory that existing life-forms on our planet are the product of the modification of pre-existing life-forms.

desertification   The overstocking of land with domestic animals that can greatly reduce grass coverage through overgrazing, turning the area more desert-like.

desmosome   A mechanically strong cell junction between animal cells that typically occurs in spotlike rivets.

determinate cleavage   In animals, a characteristic of protostome development in which the fate of each embryonic cell is determined very early.

determinate growth   A type of growth in plants that is of limited duration, such as the growth of flowers.

determined   The term used to describe a cell that is destined to differentiate into a particular cell type.

detritivore   *See* decomposer.

detritus   Unconsumed plants that die and decompose, along with the dead remains of animals and animal waste products.

deuterostome   An animal whose development exhibits radial, indeterminate cleavage and in which the blastopore becomes the anus; includes echinoderms and vertebrates.

development   In biology, a series of changes in the state of a cell, tissue, organ, or organism; the underlying process that gives rise to the structure and function of living organisms.

developmental genetics   A field of study aimed at understanding how gene expression controls the process of development.

diaphragm   A large muscle that subdivides the thoracic cavity from the abdomen in mammals; contraction of the diaphragm enlarges the thoracic cavity during inhalation.

diastole   The phase of the cardiac cycle in which the ventricles fill with blood coming from the atria through the open AV valves.

diazotroph   A bacterium that fixes nitrogen.

dideoxy chain-termination method   The most common method of DNA sequencing; utilizes dideoxynucleotides as a reagent.

differential gene regulation   The phenomenon in which the expression of genes differs under various environmental conditions and in specialized cell types.

diffusion   In a solution, the process that occurs when a solute moves from a region of high concentration to a region of lower concentration.

digestion   The process of breaking down nutrients in food into smaller molecules that can be absorbed across the intestinal epithelia and directly used by cells.

digestive system   In animals, the long tube through which food is processed. In a vertebrate, this system consists of the alimentary canal plus several associated structures.

dihybrid   An offspring that is a hybrid with respect to two traits.

dihybrid cross   A cross in which the inheritance of two different traits is followed.

dikaryotic   The occurrence of two genetically distinct nuclei in the cells of fungal hyphae after mating has occurred.

dimorphic fungi   Fungi that can exist in two different morphological forms.

dinosaur   A term, meaning "terrible lizard," used to describe some of the extinct fossil reptiles.

dioecious   The term to describe plants that produce staminate and carpellate flowers on separate plants.

diploblastic   Having two distinct germ layers— ectoderm and endoderm but not mesoderm.

diploid   Refers to cells containing two sets of chromosomes; designated as $2n$.

diploid-dominant species   Species in which the diploid organism is the prevalent organism in the life cycle. Animals are an example.

direct calorimetry   A method of determining basal metabolic rate that involves quantifying the amount of heat generated by the animal.

direct repair   Refers to a DNA repair system in which an enzyme finds an incorrect structure in the DNA and directly converts it back to the correct structure.

directionality   In a DNA or RNA strand, refers to the orientation of the sugar molecules within that strand. Can be 5′ to 3′ or 3′ to 5′.

directional selection   A pattern of natural selection that favors individuals at one extreme of a phenotypic distribution.

disaccharide   A carbohydrate composed of two monosaccharides.

discovery-based science   The collection and analysis of data without the need for a preconceived hypothesis; also called discovery science.

discrete trait   A trait with clearly defined phenotypic variants.

dispersion   A pattern of spacing in which individuals in a population are clustered together or spread out to varying degrees.

dispersive mechanism   In this incorrect model for DNA replication, segments of parental DNA and newly made DNA are interspersed in both strands following the replication process.

dispersive mutualism   A mutually beneficial interaction often involving plants and pollinators that disperse their pollen, and plants and fruit eaters that disperse the plant's seeds.

dissociation constant   An equilibrium constant between a ligand and a protein, such as a receptor or an enzyme.

distal convoluted tubule   The segment of the tubule of the nephron through which fluid flows into one of the many collecting ducts in the kidney.

diversifying selection   A pattern of natural selection that favors the survival of two or more different genotypes that produce different phenotypes.

diversity-stability hypothesis   The proposal that species-rich communities are more stable than those with fewer species.

DNA (deoxyribonucleic acid)   The genetic material that provides a blueprint for the organization, development, and function of living things.

DNA fingerprinting   A technology that identifies particular individuals using properties of their DNA.

DNA helicase   An enzyme that uses ATP to separate DNA strands during DNA replication.

DNA library   A collection of recombinant vectors, each containing a particular fragment of chromosomal DNA (cDNA).

DNA ligase   An enzyme that catalyzes the formation of a covalent bond between nucleotides in adjacent DNA fragments to complete the replication process.

DNA methylase   An enzyme that attaches methyl groups to bases in DNA.

DNA methylation   A process in which methyl groups are attached to bases in DNA.

DNA microarray   A technology used to monitor the expression of thousands of genes simultaneously.

DNA polymerase   An enzyme responsible for covalently linking nucleotides together during DNA replication.

DNA primase   An enzyme that synthesizes a primer for DNA replication.

DNA repair systems   One of several systems to reverse DNA damage before a permanent mutation can occur.

DNA replication   The process by which DNA is copied.

DNase   An enzyme that digests DNA.

DNA sequencing   A method to determine the base sequence of DNA.

DNA supercoiling   A method of compacting chromosomes through the formation of additional coils around the long, thin DNA molecule.

DNA topoisomerase   An enzyme that alleviates DNA supercoiling during DNA replication.

DNA transposon   A type of transposable element that moves as a DNA molecule.

domain   1. A defined region of a protein with a distinct structure and function. 2. One of the three major categories of life: Bacteria, Archaea, and Eukarya.

domestication   A process that involves artificial selection of plants or animals for traits desirable to humans.

dominant   A term that describes the displayed trait in a heterozygote.

dominant species   A species that has a large effect in a community because of its high abundance or high biomass.

dormancy   A phase of metabolic slowdown in a plant.

dorsal   Refers to the upper side of an animal.

dorsoventral axis   In bilateral animals, one of the three axes along which the adult body pattern is organized; the others are the anteroposterior axis and the right-left axis.

dosage compensation   The phenomenon that gene dosage is compensated between males and females. In mammals, the inactivation of one X chromosome in the female reduces the number of expressed copies (doses) of X-linked genes from two to one.

double bond   A bond that occurs when the atoms of a molecule share two pairs of electrons.

double fertilization   In angiosperms, the process in which two different fertilization events occur, producing both a zygote and the first cell of a nutritive endosperm tissue.

double helix   Two strands of DNA hydrogen-bonded with each other. In a DNA double helix, two DNA strands are twisted together to form a structure that resembles a spiral staircase.

Down syndrome   A human disorder caused by the inheritance of three copies of chromosome 21.

duplication   A type of mutation in which a section of a chromosome occurs two or more times.

dynamic instability   The oscillation of a single microtubule between growing and shortening phases; important in many cellular activities, including the sorting of chromosomes during cell division.

E

Ecdysozoa   A clade of molting animals that encompasses primarily the arthropods and nematodes.

echolocation   The phenomenon in which certain species listen for echoes of high-frequency sound waves in order to determine the distance and location of an object.

ECM   *See* extracellular matrix.

ecological footprint   The amount of productive land needed to support each person on Earth.

ecological species concept   An approach used to distinguish species; considers a species within its native environment and states that each species occupies its own ecological niche.

ecology   The study of interactions among organisms and between organisms and their environments.

ecosystem   The biotic community of organisms in an area as well as the abiotic environment affecting that community.

ecosystem engineer   A keystone species that creates, modifies, and maintains habitats.

ecosystems ecology   The study of the flow of energy and cycling of nutrients among organisms within a community and between organisms and the environment.

ecotypes   Genetically distinct populations adapted to their local environments.

ectoderm   In animals, the outermost layer of cells formed during gastrulation that covers the surface of the embryo and differentiates into the epidermis and nervous system.

ectomycorrhizae   Beneficial interactions between temperate forest trees and soil fungi.

ectoparasite   A parasite that lives on the outside of the host's body.

ectotherm   An animal whose body temperature changes with the environmental temperature.

edge effect   A special physical condition that exists at the boundary or edge of an area.

effective population size   The number of individuals that contribute genes to future populations, often smaller than the actual population size.

effector   A molecule that directly influences cellular responses.

effector cell   A cloned lymphocyte that carries out the attack response during specific immunity.

efferent arteriole   A blood vessel that carries blood away from a glomerulus of the vertebrate kidney.

egg cell   The female gamete; also called an ovum.

ejaculation   The movement of semen through the urethra by contraction of muscles at the base of the penis.

ejaculatory duct   The structure in the male reproductive system through which sperm leave the vas deferens and enter the urethra.

elastin   A protein that makes up elastic fibers in the extracellular matrix of animals.

electrical synapse   A synapse that directly passes electric current from the presynaptic to the postsynaptic cell via gap junctions.

electrocardiogram (ECG or EKG)   A record of the electrical impulses generated by the cells of the heart during the cardiac cycle.

electrochemical gradient   The combined effect of both an electrical and chemical gradient across a membrane; determines the direction that an ion will move.

electrogenic pump   A pump that generates an electrical gradient across a membrane.

electromagnetic receptor   A sensory receptor in animals that detects radiation within a wide range of the electromagnetic spectrum, including visible, ultraviolet, and infrared light, as well as electrical and magnetic fields in some animals.

electromagnetic spectrum   All possible wavelengths of electromagnetic radiation, from relatively short wavelengths (gamma rays) to much longer wavelengths (radio waves).

electron   A negatively charged particle found in orbitals around an atomic nucleus.

electron microscope   A microscope that uses an electron beam for illumination.

electron transport chain (ETC)   A group of protein complexes and small organic molecules within the inner membranes of mitochondria and chloroplasts and the plasma membrane of prokaryotes. The components accept and donate electrons to each other in a linear manner and produce a H$^+$ electrochemical gradient.

electronegativity A measure of an atom's ability to attract electrons to its outer shell from another atom.

element A substance composed of specific types of atoms that cannot be further broken down by ordinary chemical or physical means.

elicitor A compound produced by bacterial and fungal pathogens that promotes virulence.

elimination In animals, the process of undigested material passing out of the body.

elongation factor A protein that is needed for the growth of a polypeptide during translation.

elongation stage The second step in transcription or translation where RNA strands or polypeptides are made, respectively.

embryo The early stages of development in a multicellular organism during which the organization of the organism is largely formed.

embryogenesis The process by which embryos develop from single-celled zygotes by mitotic divisions.

embryonic development The process by which a fertilized egg is transformed into an organism with distinct physiological systems and body parts.

embryonic germ cell (EG cell) A cell in the early mammalian embryo that later gives rise to sperm or egg cells. These cells are pluripotent.

embryonic stem cell (ES cell) A cell in the early mammalian embryo that can differentiate into almost every cell type of the body. These cells are pluripotent.

embryophyte A synonym for the land plants.

emerging virus A newly arising virus.

emphysema A progressive disease characterized by a loss of elastic recoil ability of the lungs, usually resulting from chronic tobacco smoking.

empirical thought Thought that relies on observation to form an idea or hypothesis, rather than trying to understand life from a nonphysical or spiritual point of view.

emulsification A process during digestion that disrupts large lipid droplets into many tiny droplets, thereby increasing their total surface area and exposure to lipase action.

enantiomer One of a pair of stereoisomers that exist as mirror images.

endangered species Those species that are in danger of extinction throughout all or a significant portion of their range.

endemic The term to describe organisms that are naturally found only in a particular location.

endergonic Refers to chemical reactions that require an addition of free energy and do not proceed spontaneously.

endocrine disruptor A chemical found in polluted water and soil that resembles a natural hormone; a common example are chemicals that resemble estrogen and can bind to estrogen receptors in animals.

endocrine gland A structure that contains epithelial cells that secrete hormone molecules into the bloodstream, where they circulate throughout the body.

endocrine system All the endocrine glands and other organs containing hormone-secreting cells.

endocytosis A process in which the plasma membrane invaginates, or folds inward, to form a vesicle that brings substances into the cell.

endoderm In animals, the innermost layer of cells formed during gastrulation; lines the gut and gives rise to many internal organs.

endodermis In vascular plants, a thin cylinder of root tissue that forms a barrier between the root cortex and the central core of vascular tissue.

endomembrane system A network of membranes that includes the nuclear envelope, the endoplasmic reticulum, Golgi apparatus, lysosomes, vacuoles, and plasma membrane.

endomycorrhizae Partnerships between plants and fungi in which the fungal hyphae grow into the spaces between root cell walls and plasma membranes.

endoparasite A parasite that lives inside the host's body.

endophyte A mutualistic fungus that lives compatibly within the tissues of various types of plants.

endoplasmic reticulum (ER) A convoluted network of membranes in a cell's cytoplasm that forms flattened, fluid-filled tubules or cisternae.

endoskeleton An internal hard skeleton covered by soft tissue; present in echinoderms and vertebrates.

endosperm A nutritive tissue that increases the efficiency with which food is stored and used in the seeds of flowering plants.

endospore A cell with a tough coat that is produced in certain bacteria and then released when the enclosing bacterial cell dies and breaks down.

endosporic gametophyte A plant gametophyte that grows within the confines of microspore or megaspore walls.

endosymbiont A smaller species that lives within a larger species in a symbiotic relationship.

endosymbiosis A symbiotic relationship in which the smaller species—the symbiont—lives inside the larger species.

endosymbiosis theory A theory that mitochondria and chloroplasts originated from bacteria that took up residence within a primordial eukaryotic cell.

endosymbiotic Describes a relationship in which one organism lives inside the other.

endothelium The single-celled inner layer of a blood vessel; forms a smooth lining in contact with the blood.

endotherm An animal that generates its own internal heat and maintains a relatively stable body temperature independent of the environment.

endothermic A term to describe the ability of an organism to generate and retain body heat through its metabolism.

energy The ability to promote change or to do work.

energy expenditure The amount of energy an animal uses in a given period of time to power all of its metabolic requirements.

energy flow The movement of energy through an ecosystem.

energy intermediate A molecule such as ATP or NADH that stores energy and is used to drive endergonic reactions in cells.

energy shell In an atom, an energy level of electrons occupied by one or more orbitals; each energy level is a characteristic distance from the nucleus, with outer shells having more energy than inner shells.

enhancement effect The phenomenon in which maximal activation of the pigments in photosystems I and II is achieved when organisms are exposed to two wavelengths of light.

enhancer A response element in eukaryotes that increases the rate of transcription.

enthalpy (*H*) The total energy of a system.

entomology The study of insects.

entropy The degree of disorder of a system.

environmental science The application of ecology to real-world problems.

enzyme A protein that acts as a catalyst to speed up a chemical reaction in a cell.

enzyme-linked receptor A receptor found in all living species that typically has two important domains: an extracellular domain, which binds a signaling molecule, and an intracellular domain, which has a catalytic function.

enzyme-substrate complex The binding between an enzyme and its substrate.

eosinophil A type of phagocyte found in large numbers in mucosal surfaces lining the gastrointestinal, respiratory, and urinary tracts, where they fight off parasitic infections.

epicotyl The portion of an embryonic plant stem with two tiny leaves in a first bud; located above the point of attachment of the cotyledons.

epidermis A layer of dermal tissue that helps protect a plant from damage.

epididymis A coiled, tubular structure located on the surface of the testis in which sperm complete their differentiation.

epigenetic inheritance An inheritance pattern in which modification of a gene or chromosome during egg formation, sperm formation, or early stages of embryonic growth alters gene expression in a way that is fixed during an individual's lifetime.

epinephrine A hormone secreted by the adrenal glands; also known as adrenaline.

episome A plasmid that can integrate into a bacterial chromosome.

epistasis A gene interaction in which the alleles of one gene mask the expression of the alleles of another gene.

epithalamus A region of the vertebrate forebrain that includes the pineal gland.

epithelial tissue In animals, a sheet of densely packed cells that covers the body, covers individual organs, and lines the walls of various cavities inside the body.

epitopes Antigenic determinants; the peptide fragments of an antigen that are complexed to MHC proteins and presented to a helper T cell.

equilibrium 1. In a chemical reaction, occurs when the rate of the forward reaction is balanced by the rate of the reverse reaction. 2. In a population, the situation in which the population size stays the same.

equilibrium model of island biogeography A model to explain the process of succession on new islands; states that the number of species on an island tends toward an equilibrium number that is determined by the balance between immigration rates and extinction rates.

equilibrium potential In membrane physiology, the membrane potential at which the flow of an ion is at equilibrium, with no net movement in either direction.

ER lumen A single compartment enclosed by the ER membrane.

ER signal sequence A sorting signal in a polypeptide usually located near the amino terminus that is recognized by SRP (signal recognition particle) and directs the polypeptide to the ER membrane.

erythrocyte A cell that serves the critical function of transporting oxygen throughout an animal's body; also known as a red blood cell.

erythropoietin (EPO)   A hormone made by the liver and kidneys in response to any situation where additional red blood cells are required.

E site   *See* exit site.

esophagus   In animals, the tubular structure that forms a pathway from the throat to the stomach.

essential amino acids   Those amino acids that are required in the diet of particular organisms.

essential fatty acid   A polyunsaturated fatty acid, such as linoleic acid, that cannot be synthesized by animal cells and must therefore be consumed in the diet.

essential nutrient   In animals, a compound that cannot be synthesized from any ingested or stored precursor molecule and so must be obtained in the diet in its complete form. In plants, those substances needed to complete reproduction while avoiding the symptoms of nutrient deficiency.

estradiol   The major estrogen in many vertebrates, including humans.

estrogens   Steroid hormones produced by the ovaries that affect most aspects of female reproduction.

ethology   Scientific studies of animal behavior.

ethylene   A plant hormone that is particularly important in coordinating plant developmental and stress responses.

euchromatin   The less condensed regions of a chromosome; areas that are capable of gene transcription.

eudicots   One of the two largest lineages of flowering plants in which the embryo possesses two seed leaves.

Eukarya   One of the three domains of life; the other two are Bacteria and Archaea.

eukaryote   One of the two categories into which all forms of life can be placed. The distinguishing feature of eukaryotes is cell compartmentalization, including a cell nucleus; includes protists, fungi, plants, and animals.

eukaryotic   Refers to organisms having cells with internal compartments that serve various functions; includes all members of the domain Eukarya.

Eumetazoa   A subgroup of animals having more than one type of tissue and, for the most part, different types of organs.

euphyll   A leaf with branched veins.

euphyllophytes   The clade that includes pteridophytes and seed plants.

euploid   An organism that has a chromosome number that is a multiple of a chromosome set ($1n$, $2n$, $3n$, etc.).

eusociality   An extreme form of altruism in social insects in which the vast majority of females, known as workers, do not reproduce. Instead, they help one reproductive female (the queen) raise offspring.

Eustachian tube   In mammals, a connection from the middle ear to the pharynx; maintains the pressure in the middle ear at atmospheric pressure.

eustele   In plants, a ring of vascular tissue arranged around a central pith of nonvascular tissue; typical of progymnosperms, gymnosperms, and angiosperms.

eutherian   A placental mammal and member of the subclass Eutheria.

eutrophic   Waters that contain relatively high levels of nutrients such as phosphate or nitrogen and typically exhibit high levels of primary productivity and low levels of biodiversity.

eutrophication   The process by which elevated nutrient levels in a body of water lead to an overgrowth of algae or aquatic plants and a subsequent depletion of water oxygen levels when these photosynthesizers decay.

evaporation   The transformation of water from the liquid to the gaseous state at normal temperatures. Animals use evaporation as a means of losing excess body heat.

evapotranspiration rate   The rate at which water moves into the atmosphere through the processes of evaporation from the soil and transpiration of plants.

evolution   The phenomenon that populations of organisms change over the course of many generations. As a result, some organisms become more successful at survival and reproduction.

evolutionarily conserved   The term used to describe homologous DNA sequences that are very similar or identical between different species.

evolutionary developmental biology (evo-devo)   A field of biology that compares the development of different organisms in an attempt to understand ancestral relationships between organisms and the developmental mechanisms that bring about evolutionary change.

evolutionary lineage concept   An approach used to distinguish species; states that a species is derived from a single distinct lineage and has its own evolutionary tendencies and historical fate.

excitable cell   The term used to describe neurons and muscle cells because they have the capacity to generate electrical signals.

excitation-contraction coupling   The sequence of events by which an action potential in the plasma membrane of a muscle fiber leads to cross-bridge activity.

excitatory postsynaptic potential (EPSP)   The response from an excitatory neurotransmitter that depolarizes the postsynaptic membrane; the depolarization brings the membrane potential closer to the threshold potential that would trigger an action potential.

excretion   In animals, the process of expelling waste or harmful materials from the body.

exercise   Any physical activity that increases an animal's metabolic rate.

exergonic   Refers to chemical reactions that release free energy and occur spontaneously.

exit site (E site)   One of three sites for tRNA binding in the ribosome during translation; the other two are the peptidyl site (P site) and the aminoacyl site (A site). The uncharged tRNA exits from the E site.

exocrine gland   A gland in which epithelial cells secrete chemicals into a duct, which carries those molecules directly to another structure or to the outside surface of the body.

exocytosis   A process in which material inside a cell is packaged into vesicles and excreted into the extracellular medium.

exon   A portion of RNA that is found in the mature mRNA molecule after splicing is finished.

exon shuffling   A form of mutation in which exons and their flanking introns are inserted into genes and thereby create proteins with additional functional domains.

exonuclease   An enzyme that cleaves off nucleotides, one at a time, from the end of a DNA or RNA molecule.

exoskeleton   An external skeleton made of chitin and protein that surrounds and protects most of the body surface of animals such as insects.

exosome   A multiprotein complex that degrades mRNA.

expansin   A protein that occurs in the plant cell wall and fosters cell enlargement.

experimental group   The sample in an experiment that is subjected to some type of variation that does not occur for the control group.

exploitation competition   Competition in which organisms compete indirectly through the consumption of a limited resource.

exponential growth   Rapid population growth that occurs when the per capita growth rate remains above zero.

extensor   A muscle that straightens a limb at a joint.

external fertilization   Fertilization that occurs in aquatic environments, when eggs and sperm are released into the water in close enough proximity for fertilization to occur.

external intercostal muscles   Muscles of the rib cage that contract during inhalation, thereby expanding the chest.

extinction   The end of the existence of a species or a group of species.

extinction vortex   A downward spiral toward extinction from which a species cannot naturally recover.

extracellular fluid   The fluid in an organism's body that is outside of the cells.

extracellular matrix (ECM)   A network of material that is secreted from animal cells and forms a complex meshwork outside of cells. The ECM provides strength, support, and organization.

extranuclear inheritance   In eukaryotes, the transmission of genes that are located outside the cell nucleus.

extremophile   An organism that occurs primarily in extreme habitats.

eye   The visual organ in animals that detects light and sends signals to the brain.

eyecup   An eye in planaria that detects light and its direction but which does not form an image.

F

facilitated diffusion   A method of passive transport that involves the aid of a transport protein.

facilitation   A mechanism for succession in which a species facilitates or makes the environment more suitable for subsequent species.

facultative anaerobe   A microorganism that can use oxygen in aerobic respiration, obtain energy via anaerobic fermentation, or use inorganic chemical reactions to obtain energy.

facultative mutualism   An interaction between mutualistic species that is beneficial but not essential to the survival and reproduction of either species.

family   In taxonomy, a subdivision of an order.

fast block to polyspermy   A depolarization of the egg that blocks other sperm from binding to the egg membrane proteins.

fast fiber   A skeletal muscle fiber containing myosin with a high rate of ATP hydrolysis.

fast-glycolytic fiber   A skeletal muscle fiber that has high myosin ATPase activity but cannot make as much ATP as oxidative fibers because its source of ATP is glycolysis; best suited for rapid, intense actions.

fast-oxidative fiber   A skeletal muscle fiber that has high myosin ATPase activity and can make large amounts of ATP; used for long-term activities.

fate   The ultimate morphological features that a cell or a group of cells will adopt.

fate mapping   A technique in which a small population of cells within an embryo is specifically labeled with a harmless dye, and the fate of these labeled cells is followed to a later stage of embryonic development.

feedback inhibition   A form of regulation in which the product of a metabolic pathway inhibits an enzyme that acts early in the pathway, thus preventing the overaccumulation of the product.

feedforward regulation   The process by which an animal's body begins preparing for a change in some variable before it even occurs.

female-enforced monogamy hypothesis   The hypothesis that a male is monogamous due to various actions employed by his female mate.

female gametophyte   A haploid multicellular plant generation that produces one or more eggs but does not produce sperm cells.

fermentation   The breakdown of organic molecules to produce energy without any net oxidation of an organic molecule.

Ferrell cell   The middle cell in the three-cell model of atmospheric circulation.

fertilization   The union of two gametes, such as an egg cell with a sperm cell, to form a zygote.

fertilizer   A soil addition that enhances plant growth by providing essential elements.

fetus   The maturing embryo after the eighth week of gestation in humans.

fever   An increase in an animal's temperature due to infection.

F factor   A type of bacterial plasmid called a fertility factor that plays a role in bacterial conjugation.

F_1 generation   The first filial generation in a genetic cross.

F_2 generation   The second filial generation in a genetic cross.

fiber   A type of tough-walled plant cell that provides support.

fibrin   A protein that forms a meshwork of threadlike fibers that wrap around and between platelets and blood cells, enlarging and thickening a blood clot.

fibrous root system   The root system of monocots, which consists of multiple adventitious roots that grow from the stem base.

fight-or-flight   The response of vertebrates to real or perceived danger; associated with increased activity of the sympathetic branch of the autonomic nervous system.

filament   1. The elongate portion of a flower's stamen; contains vascular tissue that delivers nutrients from parental sporophytes to anthers. 2. In fishes, a part of the gills.

filtrate   In the process of filtration in an excretory system, the material that passes through the filter and enters the excretory organ for either further processing or excretion.

filtration   The passive removal of water and small solutes from the blood during the production of urine.

finite rate of increase   In ecology, the ratio of a population size from one year to the next.

first law of thermodynamics   States that energy cannot be created or destroyed; also called the law of conservation of energy.

fitness   The relative likelihood that a genotype will contribute to the gene pool of the next generation as compared with other genotypes.

5′ cap   The 7-methylguanosine cap structure at the 5′ end of most mature mRNAs in eukaryotes.

fixed action pattern (FAP)   An animal behavior that, once initiated, will continue until completed.

fixed nitrogen   Atmospheric nitrogen that has been combined with other elements into a form of nitrogen that can be used by plants. An example is ammonia, NH_3.

flagella (singular, **flagellum**)   Relatively long cell appendages that facilitate cellular movement or the movement of extracellular fluids.

flagellate   A protist that uses one or more flagella to move in water or cause water motions useful in feeding.

flagship species   A single large or instantly recognizable species.

flame cell   A cell that exists primarily to maintain osmotic balance between an organism's body and surrounding fluids; present in flatworms.

flavonoid   A type of phenolic secondary metabolite that provides plants with protection from UV damage or imparts color to flowers.

flexor   A muscle that bends a limb at a joint.

florigen   The hypothesized flowering hormone, now identified as the FT (flowering time) protein that moves from leaves, where it is produced, into the shoot apex.

flower   A reproductive shoot; a short stem that produces reproductive organs instead of leaves.

flowering plants   The angiosperms, which produce ovules within the protective ovaries of flowers. The ovules develop into seeds, and the ovaries develop into fruits, which function in seed dispersal.

flow-through system   The method of ventilation in fishes in which water moves unidirectionally such that the gills are constantly in contact with fresh, oxygenated water. Buccal pumping and ram ventilation are examples.

fluid-feeder   An animal that licks or sucks fluid from plants or animals and does not need teeth except to puncture an animal's skin.

fluidity   A property of biomembranes in which individual molecules remain in close association yet have the ability to move rotationally or laterally within the plane of the membrane. Membranes are semifluid.

fluid-mosaic model   The accepted model of the plasma membrane; its basic framework is the semifluid phospholipid bilayer with a mosaic of proteins. Carbohydrates may be attached to the lipids or proteins.

fMRI   *See* functional magnetic resonance imaging.

focal adhesion   A mechanically strong cell junction that connects an animal cell to the extracellular matrix (ECM).

follicle   A structure within an animal ovary where each ovum undergoes growth and development before it is released.

follicle-stimulating hormone (FSH)   A gonadotropin that stimulates follicle development.

food chain   A linear depiction of energy flow between organisms, with each organism feeding on and deriving energy from the preceding organism.

food-induced thermogenesis   A rise in metabolic rate for a few hours after eating that produces heat.

food vacuole   *See* phagocytic vacuole.

food web   A complex model of interconnected food chains in which there are multiple links between species.

foot   In mollusks, a muscular structure usually used for movement.

forebrain   One of three major divisions of the vertebrate brain; the other two divisions are the midbrain and hindbrain.

fossil   Recognizable preserved remains of past life on Earth.

fossil fuel   A fuel formed in the Earth from protist, plant, or animal remains, such as coal, petroleum, and natural gas.

founder effect   Genetic drift that occurs when a small group of individuals separates from a larger population and establishes a colony in a new location.

fovea   A small area on the retina directly behind the lens, where an image is most sharply focused.

frameshift mutation   A mutation that involves the addition or deletion of a number of nucleotides that are not in multiples of three.

free energy (G)   In living organisms, the amount of available energy that can be used to do work.

free radical   A molecule containing an atom with a single, unpaired electron in its outer shell. A free radical is unstable and interacts with other molecules by removing electrons from their atoms.

frequency   In regard to sound, the number of complete wavelengths that occur in 1 second, measured in hertz (Hz).

frontal lobe   One of four lobes of the cerebral cortex of the human brain; important in a variety of functions, including judgment and conscious thought.

fruit   A structure that develops from flower organs, encloses seeds, and fosters seed dispersal in the environment.

fruiting bodies   The visible fungal reproductive structures that are composed of densely packed hyphae that typically grow out of the substrate.

functional genomics   Genomic methods aimed at studying the expression of a genome.

functional group   A group of atoms with chemical features that are functionally important. Each functional group exhibits the same properties in all molecules in which it occurs.

functional magnetic resonance imaging (fMRI)   A technique used to determine changes in brain activity while a person is performing specific tasks.

Fungi   A eukaryotic kingdom of the domain Eukarya.

fungus-like protists   Heterotrophic protists that often resemble true fungi in having threadlike, filamentous bodies and absorbing nutrients from their environment.

G

G_0   A phase in which cells exit the cell cycle and postpone making the decision to divide.

G_1   The first gap phase of the cell cycle.

G_2   The second gap phase of the cell cycle.

gallbladder   In many vertebrates, a small sac underneath the liver that is a storage site for bile; allows the release of large amounts of bile to be precisely timed to the consumption of fats.

gametangia   Specialized structures produced by many land plants in which developing gametes are protected by a jacket of tissue.

gamete   A haploid cell that is involved with sexual reproduction, such as a sperm or egg cell.

gametic life cycle   A type of life cycle where all cells except the gametes are diploid, and gametes are produced by meiosis.

gametogenesis   The formation of gametes.

gametophyte   In plants and many multicellular protists, the haploid stage that produces gametes by mitosis.

ganglion   A group of neuronal cell bodies in the peripheral nervous system that are involved in a similar function.

GLOSSARY

ganglion cells Cells in the vertebrate eye that send their axons into the optic nerve.

gap gene A type of segmentation gene; a mutation in this type of gene may cause several adjacent segments to be missing in the larva.

gap junction A type of junction between animal cells that provides a passageway for intercellular transport.

gas exchange The process of moving oxygen and carbon dioxide in opposite directions between the environment and blood and between blood and cells.

gas vesicle A cytoplasmic structure used to adjust buoyancy in cyanobacteria and certain other bacteria that live in aquatic habitats.

gastrovascular cavity In certain invertebrates such as cnidarians, a body cavity with a single opening to the outside; it functions as both a digestive system and circulatory system.

gastrula A stage of an animal embryo that is the result of gastrulation and has three cellular layers: the ectoderm, endoderm, and mesoderm.

gastrulation In animals, a process in which an area in the blastula invaginates and folds inward, creating different embryonic cell layers called germ layers.

gated A property of many channels that allows them to open and close to control the diffusion of solutes through a membrane.

gel electrophoresis A technique used to separate macromolecules by using an electric field that causes them to pass through a gel matrix.

gene A unit of heredity that contributes to the characteristics or traits of an organism. At the molecular level, a gene is composed of organized sequences of DNA.

gene addition The insertion of a cloned gene into the genome of an organism.

gene amplification An increase in the copy number of a gene.

gene cloning The process of making multiple copies of a gene of interest.

gene expression Gene function both at the level of traits and at the molecular level.

gene family A group of homologous genes within a single species.

gene flow Occurs when individuals migrate between different populations and results in changes in the genetic composition of the resulting populations.

gene interaction A situation in which a single trait is controlled by two or more genes.

gene knockout An organism in which both copies of a functional gene have been replaced with nonfunctional copies. Experimentally, this can occur via gene replacement.

gene mutation A relatively small change in DNA structure that alters a particular gene.

gene pool All of the genes in a population.

genera (singular, **genus**) In taxonomy, a subdivision of a family.

general lineage concept A widely accepted approach used to distinguish species; states that each species is a population of an independently evolving lineage.

general transcription factors (GTFs) Five different proteins that play a role in initiating transcription at the core promoter of structural genes in eukaryotes.

generative cell In a seed plant, one of the cells resulting from the division of a microspore; a generative cell divides to produce two sperm cells.

gene regulation The ability of cells to control their level of gene expression.

gene replacement The phenomenon in which a cloned gene recombines with the normal gene on a chromosome and replaces it.

gene therapy The introduction of cloned genes into living cells in an attempt to cure disease.

genetic code A code that specifies the relationship between the sequence of nucleotides in the codons found in mRNA and the sequence of amino acids in a polypeptide.

genetic drift The random change in a population's allele frequencies from one generation to the next that is attributable to chance. It occurs more quickly in small populations.

genetic engineering The direct manipulation of genes for practical purposes.

genetic map A chart that shows the linear arrangement of genes along a chromosome.

genetic mapping The use of genetic crosses to determine the linear order of genes that are linked to each other along the same chromosome.

genetically modified organisms (GMOs) *See* transgenic.

gene transfer The process by which genetic material is transferred from one bacterial cell to another.

genome The complete genetic composition of a cell or a species.

genomic imprinting A phenomenon in which a segment of DNA is imprinted, or marked, in a way that affects gene expression throughout the life of the individual who inherits that DNA.

genomic library A type of DNA library in which the inserts are derived from chromosomal DNA.

genomics Techniques that are used in the molecular analysis of the entire genome of a species.

genotype The genetic composition of an individual.

genotype frequency In a population, the number of individuals with a given genotype divided by the total number of individuals.

geological timescale A time line of the Earth's history from its origin about 4.55 billion years ago to the present.

germination In plants, the process in which an embryo absorbs water, becomes metabolically active, and grows out of the seed coat, producing a seedling.

germ layer An embryonic cell layer such as ectoderm, mesoderm, or endoderm.

germ line Cells that give rise to gametes such as egg and sperm cells.

germ plasm Cytoplasmic determinants that help define and specify the primordial germ cells in the gastrula stage of animal development.

gestation *See* pregnancy.

giant axon A very large axon in certain species such as squids that facilitates high-speed neuronal conduction and rapid responses to stimuli.

gibberellic acid A type of gibberellin.

gibberellin A plant hormone that stimulates both cell division and cell elongation.

gills Specialized filamentous organs in aquatic animals that are used to obtain oxygen and eliminate carbon dioxide.

ginkgos A phylum of gymnosperms; Ginkgophyta.

gizzard The muscular portion of the stomach of birds and some reptiles that is capable of grinding food into smaller fragments.

glaucoma A condition in which drainage of aqueous humor in the eye becomes blocked and the pressure inside the eye increases. If untreated, this pressure damages cells in the retina and leads to irreversible loss of vision.

glia Cells that surround the neurons; a major class of cells in nervous systems that perform various functions.

global warming A gradual elevation of the Earth's surface temperature caused by an increasing greenhouse effect.

glomerular filtration rate (GFR) The rate at which a filtrate of plasma is formed in all the glomeruli of the vertebrate kidneys.

glomerulus A cluster of interconnected, fenestrated capillaries in the renal corpuscle of the kidney; the site of filtration in the kidney.

glucagon A hormone found in animals that stimulates the processes of glycogenolysis, gluconeogenesis, and the synthesis of ketones in the liver.

glucocorticoid A steroid hormone that regulates glucose balance and helps prepare the body for stress situations.

gluconeogenesis A mechanism for maintaining blood glucose level; enzymes in the liver convert noncarbohydrate precursors into glucose, which is then secreted into the blood.

glucose sparing A metabolic adjustment that reserves the glucose produced by the liver for use by the nervous system.

glycocalyx 1. An outer viscous covering surrounding a bacterium that traps water and helps protect bacteria from drying out. 2. A carbohydrate-rich zone on the surface of animal cells; also called a cell coat.

glycogen A polysaccharide found in animal cells (especially liver and skeletal muscle) and sometimes called animal starch; also, the major carbohydrate storage of fungi.

glycogenolysis A mechanism for maintaining blood glucose level; stored glycogen can be broken down into molecules of glucose which are then secreted into the blood.

glycolipid A lipid that has carbohydrate attached to it.

glycolysis A metabolic pathway that breaks down glucose to pyruvate.

glycolytic fiber A skeletal muscle fiber that has few mitochondria but possesses both a high concentration of glycolytic enzymes and large stores of glycogen.

glycoprotein A protein that has carbohydrate attached to it.

glycosaminoglycan (GAG) The most abundant type of polysaccharide in the extracellular matrix (ECM) of animals, consisting of repeating disaccharide units that give a gel-like character to the ECM of animals.

glycosidic bond A bond formed between two sugar molecules.

glycosylation The attachment of carbohydrate to a protein or lipid, producing a glycoprotein or glycolipid.

glyoxysome A specialized organelle within plant seeds that contains enzymes needed to convert fats to sugars.

gnathostomes All vertebrate species that possess jaws.

gnetophytes A phylum of gymnosperms; Gnetophyta.

Golgi apparatus A stack of flattened, membrane-bound compartments that performs three overlapping functions: secretion, processing, and protein sorting.

gonadotropins Hormones secreted by the anterior pituitary gland that are the same in both sexes; gonadotropins influence the ability of the testes and ovaries to produce the sex steroids.

gonads The testes in males and the ovaries in females, where the gametes are formed.

G protein An intracellular protein that binds guanosine triphosphate (GTP) and guanosine diphosphate (GDP) and participates in intracellular signaling pathways.

G-protein-coupled receptors (GPCRs) A common type of receptor found in the cells of eukaryotic species that interacts with G proteins to initiate a cellular response.

graded potential A depolarization or hyperpolarization in a neuron that varies with the strength of a stimulus.

gradualism A concept suggesting that species evolve continuously over long spans of time.

grain The characteristic single-seeded fruit of cereal grasses such as rice, corn, barley, and wheat.

Gram stain A staining process that can help to identify bacteria and predict their responses to antibiotics.

granum A structure composed of stacked membrane-bound thylakoids within a chloroplast.

gravitropism Plant growth in response to the force of gravity.

gray matter Brain tissue that consists of neuronal cell bodies, dendrites, and some unmyelinated axons.

greenhouse effect The process in which short-wave solar radiation passes through the atmosphere to warm the Earth but is radiated back to space as long-wave infrared radiation. Much of this radiation is reflected by atmospheric gases back to Earth's surface, causing its temperature to rise.

groove In the DNA double helix, an indentation where the atoms of the bases make contact with the surrounding water.

gross primary production (GPP) The measure of biomass production by photosynthetic organisms; equivalent to the carbon fixed during photosynthesis.

ground meristem In plants, a type of primary plant tissue meristem that gives rise to ground tissue.

ground tissue Most of the body of a plant, which has a variety of functions, including photosynthesis, storage of carbohydrates, and support. Ground tissue can be subdivided into three types: parenchyma, collenchyma, and sclerenchyma.

group selection A premise that attempts to explain altruism. States that natural selection produces outcomes beneficial for the whole group or species rather than for individuals.

growth An increase in weight or size.

growth factors Proteins in animals that stimulate certain cells to grow and divide.

growth hormone (GH) A hormone produced in vertebrates by the anterior pituitary gland; GH acts on the liver to produce insulin-like growth factor-1 (IGF-1).

guanine (G) A purine base found in DNA and RNA.

guard cell A specialized plant cell that allows epidermal pores (stomata) to close when conditions are too dry and to open under moist conditions, allowing the entry of CO_2 needed for photosynthesis.

gustation The sense of taste.

gut The gastrointestinal (GI) tract of an animal.

guttation Droplets of water at the edges of leaves that are the result of root pressure.

gymnosperm A plant that produces seeds that are exposed rather than seeds enclosed in fruits.

gynoecium The aggregate of carpels that forms the innermost whorl of a flower.

H

habituation The form of nonassociative learning in which an organism learns to ignore a repeated stimulus.

Hadley cell The most prominent of the three cells in the three-cell model of atmospheric circulation; the cell nearest to the equator.

hair cell A mechanoreceptor in animals that is a specialized epithelial cell with deformable stereocilia.

half-life 1. In the case of organic molecules in a cell, refers to the time it takes for 50% of the molecules to be broken down and recycled. 2. In the case of radioisotopes, the time it takes for 50% of the molecules to decay and emit radiation.

halophile A bacterium or archaeon that can live in an extremely salty environment.

halophyte A plant that can tolerate higher than normal salt concentrations and can occupy coastal salt marshes or saline deserts.

Hamilton's rule The proposal that an altruistic gene will be favored by natural selection when $rB > C$, where r is the coefficient of relatedness of the donor (the altruist) to the recipient, B is the benefit received by the recipient, and C is the cost incurred by the donor.

haplodiploid system A genetic system in which females develop from fertilized eggs and are diploid but males develop from unfertilized eggs and are haploid.

haploid Containing one set of chromosomes; designated as $1n$.

haploid-dominant species Species in which the haploid organism is the prevalent organism in the life cycle. Examples include fungi and some protists.

haplorrhini Larger-brained diurnal species of primates; includes monkeys, gibbons, orangutans, gorillas, chimpanzees, and humans.

Hardy-Weinberg equation An equation ($p^2 + 2pq + q^2 = 1$) that relates allele and genotype frequencies; the equation predicts an equilibrium if no new mutations are formed, no natural selection occurs, the population size is very large, the population does not migrate, and mating is random.

heart A muscular structure that pumps blood through blood vessels.

heart attack *See* myocardial infarction (MI).

heat of fusion The amount of heat energy that must be withdrawn or released from a substance to cause it to change from the liquid to the solid state.

heat of vaporization The heat required to vaporize 1 mole of any substance at its boiling point under standard pressure.

heavy chain A part of an immunoglobulin molecule.

H$^+$ electrochemical gradient A transmembrane gradient for H$^+$ composed of both a membrane potential and a concentration difference for H$^+$ across a membrane.

helper T cell A type of lymphocyte that assists in the activation and function of B cells and cytotoxic T cells.

hematocrit The volume of blood that is composed of red blood cells, usually between 40 and 65% in vertebrates.

hemidesmosome A mechanically strong cell junction that connects an animal cell to the extracellular matrix (ECM).

hemiparasite A parasitic organism that photosynthesizes, but lacks a root system to draw water and thus depends on its host for that function.

hemispheres The two halves of the cerebrum.

hemizygous The term used to describe the single copy of an X-linked gene in a male.

hemocyanin A copper-containing pigment that binds oxygen and gives blood or hemolymph a bluish tint.

hemodialysis A medical procedure used to artificially perform the kidneys' function.

hemoglobin An iron-containing protein that binds oxygen and is found within the cytosol of red blood cells.

hemolymph Blood and interstitial fluid combined in one fluid compartment; present in many invertebrates.

hemophilia An inherited disorder characterized by the deficiency of a specific blood clotting factor.

hemorrhage A loss of blood from a ruptured blood vessel.

herbaceous plant A plant that produces little or no wood and is composed mostly of primary vascular tissues.

herbivore An animal that eats only plants.

herbivory Refers to herbivores feeding on plants.

hermaphrodite In animals, an individual that can produce both sperm and eggs.

hermaphroditism A form of sexual reproduction in which individuals have both male and female reproductive systems.

heterochromatin The highly compacted regions of chromosomes that are usually transcriptionally inactive because of their tight conformation.

heterochrony Evolutionary changes in the rate or timing of developmental events.

heterocyst A specialized cell of some cyanobacteria in which nitrogen fixation occurs.

heterospory In plants, the formation of two different types of spores: microspores and megaspores; microspores produce male gametophytes, and megaspores produce female gametophytes.

heterotherm An animal that has a body temperature that is not constant; both exotherms and endotherms may be heterotherms.

heterotroph Organisms that cannot produce their own organic molecules and thus must obtain organic food from other organisms.

heterotrophic Requiring organic food from the environment.

heterozygote advantage A phenomenon in which a heterozygote has a higher Darwinian fitness than either corresponding homozygote.

heterozygous An individual with two different alleles of the same gene.

hibernation The state of torpor in an animal that can last for months.

highly repetitive sequence A DNA sequence found tens of thousands or even millions of times throughout a genome.

hindbrain One of three major divisions of the vertebrate brain; the other two divisions are the midbrain and forebrain.

hippocampus The area of the vertebrate forebrain that functions in establishing memories for spatial locations, facts, and the sequence of events.

histone acetyltransferase An enzyme that loosens the compaction of chromatin by attaching acetyl groups to histone proteins.

histone code hypothesis Refers to the pattern of histone modification recognized by particular proteins. The pattern of covalent modifications of amino terminus tails provides binding sites for proteins that subsequently affect the degree of chromatin compaction.

histones   A group of proteins involved in the formation of nucleosomes that aid in the compaction of eukaryotic DNA.

HIV   *See* human immunodeficiency virus.

holoblastic cleavage   A complete type of cell cleavage in certain animals in which the entire zygote is bisected into two equal-sized blastomeres.

holoparasite   A parasitic organism that lacks chlorophyll and is totally dependent on a host plant for its water and nutrients.

homeobox   A 180-bp sequence within the coding sequence of homeotic genes.

homeodomain   A region of a homeotic protein that functions in binding to the DNA.

homeostasis   The process whereby living organisms regulate their cells and bodies to maintain relatively stable internal conditions.

homeostatic control system   A system designed to regulate particular variables in an animal's body, such as body temperature; consists of a set point, sensor, integrator, and effectors.

homeotherm   An animal that maintains its body temperature within a narrow range.

homeotic gene   A gene that controls the developmental fate of particular segments or regions of an animal's body.

hominin   Either an extinct or modern species of humans.

hominoidea (hominoid)   A member of a group of primates that includes gibbons, orangutans, gorillas, chimpanzees, and humans.

homologous genes   Genes derived from the same ancestral gene that have accumulated random mutations that make their sequences slightly different.

homologous structures   Structures that are similar to each other because they are derived from the same ancestral structure.

homologue   A member of a pair of chromosomes in a diploid organism.

homology   A fundamental similarity that occurs due to descent from a common ancestor.

homozygous   An individual with two identical copies of an allele.

horizontal gene transfer   A process in which an organism incorporates genetic material from another organism without being the offspring of that organism.

hormone   A chemical messenger that is produced in a gland or other structure and acts on distant target cells in one or more parts of an animal or plant.

hornworts   A phylum of bryophytes; Anthocerophyta.

host   The prey organism in a parasitic association.

host cell   1. A cell that is infected by a virus, fungus, or a bacterium. 2. A eukaryotic cell that contains photosynthetic or nonphotosynthetic endosymbionts.

host plant resistance   The ability of plants to prevent herbivory.

host range   The number of species and cell types that a virus or bacterium can infect.

hot spot   A human-impacted geographic area with a large number of endemic species. To qualify as a hot spot, a region must contain at least 1,500 species of endemic vascular plants and have lost at least 70% of its original habitat.

Hox genes   In animals, a class of genes involved in pattern formation in early embryos.

Human Genome Project   A 13-year international effort coordinated by the U.S. Department of Energy and the National Institutes of Health that characterized and sequenced the entire human genome.

human immunodeficiency virus (HIV)   A retrovirus that is the causative agent of acquired immune deficiency syndrome (AIDS).

humoral immunity   A type of specific immunity in which plasma cells secrete antibodies that bind to antigens.

humus   A collective term for the organic constituents of soils.

hybridization   A situation in which two individuals with different characteristics are mated or crossed to each other; the offspring are referred to as hybrids.

hybrid zone   An area where two populations can interbreed.

hydrocarbon   Molecules with predominantly hydrogen–carbon bonds.

hydrogen bond   A weak chemical attraction between a partially positive hydrogen atom of a polar molecule and a partially negative atom of another polar molecule.

hydrolysis reaction   A chemical reaction that utilizes water to break apart molecules.

hydrophilic   Refers to ions and molecules that contain polar covalent bonds and will dissolve in water.

hydrophobic   Refers to molecules that do not have partial charges and therefore are not attracted to water molecules. Such molecules are composed predominantly of carbon and hydrogen and are relatively insoluble in water.

hydrostatic skeleton   A fluid-filled body cavity in certain soft-bodied invertebrates that is surrounded by muscles and provides support and shape.

hydroxide ion   An anion with the formula, OH^-.

hypermutation   A process that primarily involves numerous C to T point mutations that are crucial to enabling lymphocytes to produce a diverse array of immunoglobulins capable of recognizing many different antigens.

hyperpolarization   The change in the membrane potential that occurs when the cell becomes more polarized.

hypersensitive response (HR)   A plant's local defensive response to pathogen attack.

hyperthermophile   An organism that thrives in extremely hot temperatures.

hyperthyroidism   A medical condition resulting from a hyperactive thyroid gland.

hypertonic   Any solution that causes a cell to shrink due to osmosis of water out of the cell.

hypha   A microscopic, branched filament of the body of a fungus.

hypocotyl   The portion of an embryonic plant stem located below the point of attachment of the cotyledons.

hypothalamus   A part of the vertebrate brain located below the thalamus; it controls functions of the gastrointestinal and reproductive systems, among others, and regulates many basic behaviors such as eating and drinking.

hypothesis   In biology, a proposed explanation for a natural phenomenon based on previous observations or experimental studies.

hypothesis testing   Also known as the scientific method, a strategy for testing the validity of a hypothesis.

hypothyroidism   A medical condition resulting from an underactive thyroid gland.

hypotonic   Any solution that causes a cell to swell when placed in that solution.

H zone   In a myofibril, a narrow, light region in the center of the A band that corresponds to the space between the two sets of thin filaments in each sarcomere.

I

I band   In a myofibril, a light band that lies between the A bands of two adjacent sarcomeres.

immune system   The cells and organs within an animal's body that contribute to immune defenses.

immune tolerance   The process by which the body distinguishes between self and nonself components.

immunity   The ability of an animal to ward off internal threats, including harmful microorganisms, foreign molecules, and abnormal cells such as cancer cells.

immunoglobulin   A Y-shaped protein with two heavy chains and two light chains that provides immunity to foreign substances; antibodies are a type of immunoglobulin.

immunological memory   The immune system's ability to produce a secondary immune response.

imperfect flower   A flower that lacks either stamens or carpels.

implantation   The first event of pregnancy, when the blastocyst embeds within the uterine endometrium.

imprinting   1. The development of a species-specific pattern of behavior that occurs during a critical period; a form of learning, with a large innate component. 2. In genetics, the marking of DNA that occurs differently between males and females.

inactivation gate   A string of amino acids that juts out from a channel protein into the cytosol and blocks the movement of ions through the channel.

inborn error of metabolism   A genetic defect in the ability to metabolize certain compounds.

inbreeding   Mating among genetically related relatives.

inbreeding depression   The phenomenon whereby inbreeding produces homozygotes that are less fit, thereby decreasing the reproductive success of a population.

inclusive fitness   The term used to designate the total number of copies of genes passed on through one's relatives, as well as one's own reproductive output.

incomplete dominance   The phenomenon in which a heterozygote that carries two different alleles exhibits a phenotype that is intermediate between the corresponding homozygous individuals.

incomplete flower   A flower that lacks one or more of the four flower organ types.

incomplete metamorphosis   During development in some insects, a gradual change in body form from a nymph into an adult.

incurrent siphon   A structure in a tunicate used to draw water through the mouth.

indeterminate cleavage   In animals, a characteristic of deuterostome development in which each cell produced by early cleavage retains the ability to develop into a complete embryo.

indeterminate growth   Growth in which plant shoot apical meristems continuously produce new stem tissues and leaves, as long as conditions remain favorable.

indicator species   A species whose status provides information on the overall health of an ecosystem.

indirect calorimetry   A method of determining basal metabolic rate in which the rate at which an animal uses oxygen is measured.

individualistic model   A view of the nature of a community that considers it to be an assemblage of species coexisting primarily because of similarities in their physiological requirements and tolerances.

individual selection The proposal that adaptive traits generally are selected for because they benefit the survival and reproduction of the individual rather than the group.

induced fit Occurs when a substrate(s) binds to an enzyme and the enzyme undergoes a conformational change that causes the substrate(s) to bind more tightly to the enzyme.

induced mutation A mutation brought about by environmental agents that enter the cell and then alter the structure of DNA.

inducer In transcription, a small effector molecule that increases the rate of transcription.

inducible operon In this type of operon, the presence of a small effector molecule causes transcription to occur.

induction 1. In development, the process by which a cell or group of cells governs the developmental fate of neighboring cells. 2. In molecular genetics, refers to the process by which transcription has been turned on by the presence of a small effector molecule.

industrial nitrogen fixation The human activity of producing nitrogen fertilizers.

infertility The inability to produce viable offspring.

inflammation An innate local response to infection or injury characterized by local redness, swelling, heat, and pain.

inflorescence A cluster of flowers on a plant.

infundibular stalk The structure that physically connects the hypothalamus to the pituitary gland.

ingestion In animals, the act of taking food into the body.

ingroup In a cladogram, a group of interest.

inheritance The acquisition of traits by their transmission from parent to offspring.

inheritance of acquired characteristics Jean-Baptiste Lamarck's incorrect hypothesis that species change over the course of many generations by adapting to new environments.

inhibition A mechanism for succession in which early colonists exclude subsequent colonists.

inhibitory postsynaptic potential (IPSP) The response from an inhibitory neurotransmitter that hyperpolarizes the postsynaptic membrane; this hyperpolarization reduces the likelihood of an action potential.

initiation factor A protein that facilitates the interactions between mRNA, the first tRNA, and the ribosomal subunits during the initiation stage of translation.

initiation stage The first step in the process of transcription or translation.

initiator tRNA A specific tRNA that recognizes the start codon AUG in mRNA and binds to it, initiating translation.

innate The term used to describe behaviors that seem to be genetically programmed.

inner bark The thin layer of secondary phloem that carries out most of the sugar transport in a woody stem.

inner ear One of the three main compartments of the mammalian ear. The inner ear is composed of the bony cochlea and the vestibular system, which plays a role in balance.

inner segment The part of the vertebrate photoreceptors (rods and cones) that contains the cell nucleus and cytoplasmic organelles.

inorganic chemistry The study of the nature of atoms and molecules, with the exception of those that contain rings or chains of carbon.

insulin A hormone found in animals that regulates metabolism in several ways, primarily by regulating the blood glucose concentration.

insulin-like growth factor-1 (IGF-1) A hormone in mammals that stimulates the elongation of bones, especially during puberty.

integral membrane protein A protein that cannot be released from the membrane unless it is dissolved with an organic solvent or detergent. Includes transmembrane proteins and lipid-anchored proteins.

integrase An enzyme, sometimes encoded by viruses, that catalyzes the integration of the viral genome into a host-cell chromosome.

integrin A cell adhesion molecule found in animal cells that connects cells to the extracellular matrix.

integument In plants, a structure that encloses the megasporangium to form an ovule.

interference competition Competition in which organisms interact directly with one another by physical force or intimidation.

interferon A protein that generally inhibits viral replication inside host cells.

intermediate-disturbance hypothesis The proposal that moderately disturbed communities are more diverse than undisturbed or highly disturbed communities.

intermediate filament A type of protein filament of the cytoskeleton that helps maintain cell shape and rigidity.

internal fertilization Fertilization that occurs in terrestrial animals in which sperm are deposited within the reproductive tract of the female during copulation.

interneuron A type of neuron that forms interconnections between other neurons.

internode The region of a plant stem between adjacent nodes.

interphase The G_1, S, and G_2 phases of the cell cycle. It is the portion of the cell cycle during which the chromosomes are decondensed and found in the nucleus.

intersexual selection Sexual selection between members of the opposite sex.

interspecies hybrid The offspring resulting from the mating of two different species.

interspecific competition Competition between individuals of different species.

interstitial fluid The fluid that surrounds cells.

intertidal zone The area where the land meets the sea, which is alternately submerged and exposed by the daily cycle of tides.

intracellular fluid The fluid inside cells.

intranuclear spindle A spindle that forms within an intact nuclear envelope during nuclear division in fungi and some protists.

intrasexual selection Sexual selection between members of the same sex.

intraspecific competition Competition between individuals of the same species.

intrauterine device (IUD) A small object that is placed in the uterus and interferes with the endometrial preparation required for acceptance of the blastocyst; used as a form of contraception.

intrinsic rate of increase The situation in which conditions are optimal for a population and the per capita growth rate is at its maximum rate.

introduced species A species moved by humans from a native location to another location.

intron Intervening DNA sequences that are found in between the coding sequences of genes.

invagination The act of pinching inward, as during early embryonic development in animals.

invasive cell A cancer cell that can invade healthy tissues.

invasive species Introduced species that spread on their own, often outcompeting native species for space and resources.

inverse density-dependent factor A mortality factor whose influence decreases as population size or density increases.

inversion A type of mutation that involves a change in the direction of the genetic material along a single chromosome.

invertebrate An animal that lacks vertebrae.

in vitro Meaning, "in glass." An alternative to studying a process in living cells that involves isolating and purifying cellular components and studying their functions outside the cell.

in vivo Meaning, "in life." Studying a process in living cells or organisms.

involution During embryogenesis, the folding back of sheets of surface cells into the interior of an embryo.

iodine-deficient goiter An overgrown thyroid gland that is incapable of making thyroid hormone due to a lack of dietary iodine.

ion An atom or molecule that gains or loses one or more electrons and acquires a net electric charge.

ion electrochemical gradient A dual gradient for an ion that is composed of both an electrical gradient and a chemical gradient for that ion.

ionic bond The bond that occurs when a cation binds to an anion.

ionotropic receptor One of two types of postsynaptic receptors, the other being a metabotropic receptor. Consists of a ligand-gated ion channel that opens in response to binding of a neurotransmitter.

iris The circle of pigmented smooth muscle and connective tissue that is responsible for eye color.

iron regulatory element (IRE) A response element within the ferritin mRNA to which the iron regulatory protein binds.

iron regulatory protein (IRP) An RNA-binding protein that regulates the translation of the mRNA that encodes ferritin.

islets of Langerhans Spherical clusters of endocrine cells that are scattered throughout the pancreas; the cells secrete insulin or glucagon, among other hormones.

isomers Two structures with an identical molecular formula but different structures and characteristics.

isotonic Condition in which the solute concentrations on both sides of a plasma membrane are equal, which does not cause a cell to shrink or swell.

isotope An element that exists in multiple forms that differ in the number of neutrons they contain.

iteroparity The pattern of repeated reproduction at intervals throughout an organism's life cycle.

J

joint The juncture where two or more bones of a vertebrate endoskeleton come together.

juvenile hormone A hormone made in arthropods that inhibits maturation from a larva into a pupa.

K

karyogamy The process of nuclear fusion.

karyotype A photographic representation of the chromosomes in an actively dividing cell.

K_d The dissociation constant between a ligand and its receptor.

ketones Small compounds generated from fatty acids. Ketones are made in the liver and released into the blood to provide an important energy source during prolonged fasting for many tissues, including the brain.

keystone species A species within a community that has a role out of proportion to its abundance.

kidney The major excretory organ found in all vertebrates.

kilocalorie (kcal) One thousand calories; the amount of heat energy required to raise the temperature of 1 kg of water by 1 degree Celsius.

kinesis A movement in response to a stimulus, but one that is not directed toward or away from the source of the stimulus.

kinetic energy Energy associated with movement.

kinetic skull A characteristic of lizards and snakes in which the joints between various parts of the skull are extremely mobile.

kinetochore A group of proteins that bind to a centromere and are necessary for sorting each chromosome.

kingdom A taxonomic group; the second largest division after domain.

kin selection Selection for behavior that lowers an individual's own fitness but enhances the reproductive success of a relative.

K_M The substrate concentration at which an enzyme-catalyzed reaction is half of its maximal value.

knowledge The awareness and understanding of information.

Koch's postulates A series of steps used to determine whether a particular organism causes a specific disease.

K-selected species A type of life history strategy where species have a low rate of per capita population growth but good competitive ability.

K/T event An ancient cataclysm that involved at least one large meteorite or comet that crashed into the Earth near the present-day Yucatán Peninsula in Mexico about 65 million years ago.

L

labia majora In the female genitalia, large outer folds that surround the external opening of the reproductive tract.

labia minora In the female genitalia, smaller, inner folds near the external opening of the reproductive tract.

labor The strong rhythmic contractions of the uterus that serve to deliver a fetus during childbirth.

lac **operon** An operon in the genome of *E. coli* that contains the genes for the enzymes that allow it to metabolize lactose.

lac repressor A repressor protein that regulates the *lac* operon.

lactation In mammals, a period after birth in which the young are nurtured by milk produced by the mother.

lacteal A lymphatic vessel in the center of each intestinal villus; lipids are absorbed by the lacteals, which eventually empty into the circulatory system.

lagging strand During DNA replication, a DNA strand made as a series of small Okazaki fragments that are eventually connected to each other to form a continuous strand.

lamellae Platelike structures in the internal gills of fishes that branch from structures called filaments; gas exchange occurs here.

larva A free-living organism that is morphologically very different from the embryo and adult.

larynx The segment of the respiratory tract that contains the vocal cords.

latent The term used to describe a prophage or provirus that remains inactive for a long time.

lateral line system Microscopic sensory organs in fishes and some toads that allows them to detect movement in surrounding water.

lateral meristem *See* secondary meristem.

law of independent assortment States that the alleles of different genes assort independently of each other during gamete formation.

law of segregation States that two copies of a gene segregate from each other during gamete formation and during transmission from parent to offspring.

leaching The dissolution and removal of inorganic ions as water percolates through materials such as soil.

leading strand During DNA replication, a DNA strand made in the same direction that the replication fork is moving. The strand is synthesized as one long continuous molecule.

leaf abscission The process by which a leaf drops after the formation of an abscission zone.

leaflet 1. Half of a phospholipid bilayer. 2. A portion of a compound leaf.

leaf primordia Small outgrowths that occur at the sides of a shoot apical meristem and develop into young leaves.

leaf vein In plants, a bundle of vascular tissue in a leaf.

learning The ability of an animal to make modifications to a behavior based on previous experience; the process by which new information is acquired.

leaves Flattened plant organs that emerge from stems and function in photosynthesis.

leghemoglobin A protein found in legume plants that helps to regulate local oxygen concentrations around rhizobial bacteroids in root nodules.

legume A member of the pea (bean) family; also their distinctive fruits.

lek A designated communal courting area.

lens 1. A structure of the eye that focuses light. 2. The glass components of a light microscope or the electromagnetic parts of an electron microscope that allow the production of magnified images of microscopic structures.

lentic Refers to a freshwater habitat characterized by standing water.

leptin A hormone produced by adipose cells in proportion to fat mass; controls appetite and metabolic rate.

leukocyte A cell that develops from the marrow of certain bones of vertebrates; all leukocytes (also known as white blood cells) perform vital functions that defend the body against infection and disease.

lichens The mutualistic association between particular fungi and certain photosynthetic green algae or cyanobacteria. This association results in a body form distinctive from that of either partner alone.

Liebig's law of the minimum States that species' biomass or abundance is limited by the scarcest factor.

life cycle The sequence of events that characterize the steps of development of the individuals of a given species.

life table A table that provides data on the number of living individuals in a population in particular age classes.

ligand An ion or molecule that binds to a protein, such as an enzyme or a receptor.

ligand-gated ion channel A type of cell surface receptor that binds a ligand and functions as an ion channel. Ligand binding either opens or closes a channel.

ligand•receptor complex The structure formed when a ligand and its receptor noncovalently bind.

light chain 1. A part of an immunoglobulin molecule. 2. Two of the polypeptides that comprise each myosin molecule.

light-harvesting complex A component of photosystem II and photosystem I composed of several dozen pigment molecules that are anchored to proteins in the thylakoid membranes of a chloroplast. The role of these complexes is to absorb photons of light.

light microscope A microscope that utilizes light for illumination.

light reactions The first of two stages in the process of photosynthesis. During the light reactions, photosystem II and photosystem I absorb light energy and produce ATP, NADPH, and O_2.

lignin A tough polymer that adds strength and decay resistance to cell walls of tracheids, vessel elements, and other cells of plants.

lignophytes Modern and fossil seed plants and seedless ancestors that produced wood.

limbic system In the vertebrate forebrain, the areas involved in the formation and expression of emotions; also plays a role in learning, memory, and the perception of smells.

limiting factor A factor whose amount or concentration limits the rate of a biological process or a chemical reaction.

lineage A progression of changes in a series of ancestors.

line transect A sampling technique used by plant ecologists in which the number of plants located along a length of string are counted.

linkage The phenomenon that two genes close together on the same chromosome are transmitted as a unit.

linkage group A group of genes that usually stay together during meiosis.

lipase The major fat-digesting enzyme from the pancreas.

lipid A molecule composed predominantly of hydrogen and carbon atoms. Lipids are nonpolar and therefore very insoluble in water. They include fats (triglycerides), phospholipids, and steroids.

lipid-anchored protein A type of integral membrane protein that is attached to the membrane via a lipid molecule.

lipid-exchange protein A protein that extracts a lipid from one membrane, diffuses through the cell, and inserts the lipid into another membrane.

lipid raft In a membrane, a group of lipids and proteins that float together as a unit in a larger sea of lipids.

lipopolysaccharides Lipids with covalently bound carbohydrates; prevalent in the thin, outer envelope that encloses the cell walls of Gram-negative bacteria.

liposome A vesicle surrounded by a lipid bilayer.

liver An organ in vertebrates that performs diverse metabolic functions and is the site of bile production.

liverworts A phylum of bryophytes; formally called Hepatophyta.

lobe fins The Actinistia (coelacanths), Dipnoi (lungfishes), and tetrapods; also called Sarcopterygii.

lobe-finned fishes Fishes in which the fins are part of the body; the fins are supported by skeletal extensions of the pectoral and pelvic areas.

locomotion   The movement of an animal from place to place.

locus   The physical location of a gene on a chromosome.

logistic growth   The pattern in which the growth of a population typically slows down as it approaches the carrying capacity.

long-day plant   A plant that flowers in spring or early summer, when the night period is shorter (and thus the day length is longer) than a defined period.

long-term potentiation (LTP)   The long-lasting strengthening of the connection between neurons that is believed to be part of the mechanism of learning and memory.

loop domain   In bacteria, a chromosomal segment that is folded into loops by the attachment to proteins; a method of compacting bacterial chromosomes.

loop of Henle   A segment of the tubule of the nephron of the kidney containing a sharp hairpin-like loop that contributes to reabsorption of ions and water. It consists of a descending limb coming from the proximal tubule and an ascending limb leading to the distal tubule.

lophophore   A horseshoe-shaped crown of tentacles used for feeding in several invertebrate species.

Lophotrochozoa   A clade of animals that encompasses the mollusks, annelids, and several other phyla; they are distinguished by two morphological features—the lophophore, a crown of tentacles used for feeding, and the trochophore larva, a distinct larval stage.

lotic   Refers to a freshwater habitat characterized by running water.

lumen   The internal space of an organelle.

lungfishes   The Dipnoi; fish with primitive lungs that live in oxygen-poor freshwater swamps and ponds.

lungs   In terrestrial vertebrates, internal paired structures used to bring oxygen into the circulatory system and remove carbon dioxide.

luteinizing hormone (LH)   A gonadotropin that controls the production of sex steroids in both males and females.

lycophyll   A relatively small leaf having a single unbranched vein; produced by lycophytes.

lycophytes   Members of a phylum of vascular land plants whose leaves are lycophylls; Lycopodiophyta.

lymphatic system   A system of vessels along with a group of organs and tissues where most leukocytes reside. The lymphatic vessels collect excess interstitial fluid and return it to the blood.

lymphocytes   A type of leukocyte that is responsible for specific immunity; the two types are B cells and T cells.

lysogenic cycle   The growth cycle of a bacteriophage consisting of integration, prophage replication, and excision.

lysosome   A small organelle found in animal cells that contains acid hydrolases that degrade macromolecules.

lytic cycle   The growth cycle of a bacteriophage in which the production and release of new viruses lyses the host cell.

M

macroalgae   Photosynthetic protists that can be seen with the unaided eye; also known as seaweeds.

macroevolution   Evolutionary changes that create new species and groups of species.

macromolecule   Many molecules bonded together to form a polymer. Carbohydrates, proteins, and nucleic acids (for example, DNA and RNA) are important macromolecules found in living organisms.

macronutrient   An element required by plants in amounts of at least 1 g/kg of plant dry matter.

macroparasite   A parasite that lives in a host but releases infective juvenile stages outside the host's body.

macrophage   A type of phagocyte capable of engulfing viruses and bacteria.

macular degeneration   An eye condition in which photoreceptor cells in and around the fovea of the retina are lost; one of the leading causes of blindness in the U.S.

madreporite   A sievelike plate on the surface of an echinoderm where water enters the water vascular system.

magnetic resonance imaging (MRI)   An imaging method that relies on the use of magnetic fields and radio waves to visualize the internal structure of an organism's body.

magnification   The ratio between the size of an image produced by a microscope and a sample's actual size.

major depressive disorder   A neurological disorder characterized by feelings of despair and sadness, resulting from an imbalance in neurotransmitter levels in the brain.

major groove   A groove that spirals around the DNA double helix; provides a location where a protein can bind to a particular sequence of bases and affect the expression of a gene.

major histocompatibility complex (MHC)   A gene family that encodes the plasma membrane self proteins that must be complexed with an antigen for T-cell recognition to occur.

male-assistance hypothesis   A hypothesis to explain the existence of monogamy that maintains that males remain with females to help them rear their offspring.

male gametophyte   A haploid multicellular plant life cycle stage that produces sperm.

malignant tumor   A growth of cells that has progressed to the cancerous stage.

Malpighian tubules   Delicate projections from the digestive tract of insects and some other taxa that function as an excretory organ.

mammal   A vertebrate that is a member of the class Mammalia that nourishes its young with milk secreted by mammary glands. Another distinguishing feature is hair.

mammary gland   A gland in female mammals that secretes milk.

manganese cluster   A site where the oxidation of water occurs in photosystem II during photosynthesis.

mantle   In mollusks, a fold of skin draped over the visceral mass that secretes a shell in those species that form shells.

mantle cavity   The chamber in a mollusk mantle that houses delicate gills.

many-eyes hypothesis   The idea that increased group size decreases predators' success because of increased detection of predators.

map distance   The distance between genes along chromosomes, which is calculated as the number of recombinant offspring divided by the total number of offspring times 100.

mapping   The process of determining the relative locations of genes or other DNA segments along a chromosome.

map unit (mu)   A unit of distance on a chromosome equivalent to a 1% recombination frequency.

mark-recapture technique   The capture and tagging of animals so they can be released and recaptured, allowing an estimate of population size.

marsupial   A member of a group of seven mammalian orders and about 280 species found in the subclass Metatheria.

mass extinction   When many species become extinct at the same time.

mass-specific BMR   The amount of energy expended per gram of body mass.

mastax   The circular muscular pharynx in the mouth of rotifers.

mast cell   A type of cell derived from bone marrow stem cells that plays an important role in nonspecific immunity.

masting   The synchronous production of many progeny by all individuals in a population; serves to satiate predators and thereby allow some progeny to survive.

mate-guarding hypothesis   The hypothesis that a male is monogamous to prevent his mate from being fertilized by other males.

maternal effect   An inheritance pattern in which the genotype of the mother determines the phenotype of her offspring.

maternal effect gene   A gene that follows a maternal effect inheritance pattern.

maternal inheritance   A phenomenon in which offspring inherit particular genes only from the female parent (through the egg).

matrotrophy   In plants, the phenomenon in which zygotes remain enclosed within gametophyte tissues, where they are sheltered and fed.

matter   Anything that has mass and takes up space.

maturation promoting factor (MPF)   A factor, now known to be a complex of cyclin and cyclin-dependent kinase, important in the division of all types of eukaryotic cells.

mature mRNA   In eukaryotes, transcription produces a long RNA, pre-mRNA, which undergoes certain processing events before it exits the nucleus; mature mRNA is the final functional product.

maximum likelihood   One method used to evaluate a phylogenetic tree based on an evolutionary model.

mean fitness of the population   The average reproductive success of members of a population.

mechanoreceptor   A sensory receptor in animals that transduces mechanical energy such as pressure, touch, stretch, movement, and sound.

mediator   A large protein complex that plays a role in initiating transcription at the core promoter of structural genes in eukaryotes.

medulla oblongata   The part of the vertebrate hindbrain that coordinates many basic reflexes and bodily functions, such as breathing.

medusa   A type of cnidarian body form that is motile and usually floats mouth down.

megadiversity country   Those countries with the greatest numbers of species; used in targeting areas for conservation.

megaspore   In seed plants and some seedless plants, a large spore that produces a female gametophyte within the spore wall.

meiosis   The process by which haploid cells are produced from a cell that was originally diploid.

meiosis I   The first division of meiosis in which the homologues are separated into different cells.

meiosis II   The second division of meiosis in which sister chromatids are separated into different cells.

Meissner's corpuscles   Structures that sense touch and light pressure and lie just beneath the skin surface of an animal.

melatonin   A hormone produced by the pineal gland of vertebrates; plays a role in light-dependent behaviors such as seasonal reproduction and daily rhythms.

membrane attack complex (MAC)   A multiunit protein formed by the activation of complement proteins; the complex creates water channels in the microbial plasma membrane and causes the microbe to swell and burst.

membrane potential   The difference between the electric charges outside and inside a cell; also called a potential difference (or voltage).

membrane transport   The movement of ions or molecules across a cell membrane.

memory   The retention of information over time.

memory cells   A cloned lymphocyte that remains poised to recognize a returning antigen; a component of specific immunity.

Mendelian inheritance   The inheritance patterns of genes that segregate and assort independently.

meninges   Three layers of sheathlike membranes that cover and protect the brain and spinal cord.

meningitis   A potentially life-threatening infectious disease in which the meninges become inflamed.

menopause   The event during which a woman permanently stops having ovarian cycles.

menstrual cycle   The cyclical changes that occur in the uterus in parallel with the ovarian cycle in a female mammal. Also called the uterine cycle.

menstruation   A period of bleeding at the beginning of the menstrual cycle in a female mammal.

meristem   In plants, an organized tissue that includes actively dividing cells and a reservoir of stem cells.

meroblastic cleavage   An incomplete type of cell cleavage in which only the region of the egg containing cytoplasm at the animal pole undergoes cell division. Occurs in birds and some fishes.

merozygote   A strain of bacteria containing an F′ factor.

mesoderm   In animals, a layer of cells formed during gastrulation that develops between the ectoderm and endoderm; gives rise to the skeleton, muscles, and much of the circulatory system.

mesoglea   A gelatinous substance between the epidermis and the gastrodermis in the Radiata.

mesohyl   A gelatinous, protein-rich matrix in between the choanocytes and the epithelial cells of a sponge.

mesophyll   The internal tissue of a plant leaf; the site of photosynthesis.

messenger RNA (mRNA)   RNA that contains the information to specify a polypeptide with a particular amino acid sequence.

metabolic cycle   A biochemical cycle in which particular molecules enter while others leave; the process is cyclical because it involves a series of organic molecules that are regenerated with each turn of the cycle.

metabolic pathway   In living cells, a series of chemical reactions in which each step is catalyzed by a specific enzyme.

metabolic rate   The total energy expenditure of an organism per unit of time.

metabolism   The sum total of all chemical reactions that occur within an organism. Also, a specific set of chemical reactions occurring at the cellular level.

metabotropic receptor   A G-protein-coupled receptor that initiates a signaling pathway in response to a neurotransmitter. One of two types of postsynaptic receptors, the other being an ionotropic receptor.

metacentric   A chromosome in which the centromere is near the middle.

metagenomics   A field of study that seeks to identify and analyze the collective microbial genomes contained in a community of organisms, including those not easily cultured in the laboratory.

metamorphosis   The process in which a pupal or juvenile organism changes into a mature adult with very different characteristics.

metanephridia   Excretory organs found in a variety of invertebrates.

metanephridial system   The filtration system used by annelids to filter out wastes and excess water.

metaphase   The phase of mitosis during which the chromosomes are aligned along the metaphase plate.

metaphase plate   A plane halfway between the poles of the spindle apparatus on which the sister chromatids align during the metaphase stage of mitosis.

metastasis   The process by which cancer cells spread from their original location to distant parts of the body.

Metazoa   The collective term for animals.

methanogens   Several groups of anaerobic archaea that convert CO_2, methyl groups, or acetate to methane, and release it from their cells.

methanotroph   An aerobic bacterium that consumes methane.

methyl-CpG-binding protein   A protein that binds methylated sequences and inhibits transcription.

micelle   The sphere formed by long amphipathic molecules when they are mixed with water. In animals, micelles aid in the absorption of poorly soluble products during digestion.

microbiome   All of the microorganisms in a particular environment.

microclimate   Local variations of the climate within a given area.

microevolution   Changes in a population's gene pool from generation to generation.

microfilament   *See* actin filament.

micrograph   An image taken with the aid of a microscope.

micronutrient   An element required by plants in amounts at, or less than, 0.1 g/kg of plant dry matter; also known as a trace element.

microparasite   A parasite that multiplies within its host, usually within the cells.

micropyle   A small opening in the integument of a seed plant ovule through which a pollen tube grows.

microRNAs (miRNAs)   Small RNA molecules, typically 22 nucleotides in length, that silence the expression of specific mRNAs by inhibiting translation.

microscope   A magnification tool that enables researchers to study very small structures such as cells.

microspore   In seed plants and some seedless plants, a relatively small spore that produces a male gametophyte within the spore wall.

microtubule   A type of hollow protein filament composed of tubulin proteins that is part of the cytoskeleton and is important for cell shape, organization, and movement.

microtubule-organizing center   *See* centrosome.

microvilli   Small projections in the surface membranes of epithelial cells in the small intestine and many other absorptive cells.

midbrain   One of three major divisions of the vertebrate brain; the other two divisions are the hindbrain and the forebrain.

middle ear   One of the three main compartments of the mammalian ear; contains three small bones

called ossicles that connect the eardrum with the oval window.

middle lamella   An extracellular layer in plants composed primarily of carbohydrate; cements adjacent plant cell walls together.

migration   Long-range seasonal movement among animals in order to feed or breed.

mimicry   The resemblance of an organism (the mimic) to another organism (the model).

mineral   An inorganic ion or inorganic molecule required by a living organism.

mineralization   The general process by which phosphorus, nitrogen, CO_2, and other minerals are released from organic compounds.

mineralocorticoid   A steroid hormone such as aldosterone that regulates the balance of sodium and potassium ions in the body.

minor groove   A smaller groove that spirals around the DNA double helix.

miRNA   *See* microRNAs.

missense mutation   A base substitution that changes a single amino acid in a polypeptide sequence.

mitochondrial genome   The chromosome found in mitochondria.

mitochondrial matrix   A compartment inside the inner membrane of a mitochondrion.

mitochondrion   A semiautonomous organelle found in eukaryotic cells that supplies most of a cell's ATP.

mitogen-activated protein kinase (MAP kinase)   A type of protein kinase that is involved with promoting cell division.

mitosis   In eukaryotes, the process in which nuclear division results in two nuclei, each of which receives the same complement of chromosomes.

mitotic cell division   A process whereby a eukaryotic cell divides to produce two new cells that are genetically identical to the original cell.

mitotic spindle   The structure responsible for organizing and sorting the chromosomes during mitosis; also called the mitotic spindle apparatus.

mixotroph   An organism that is able to use autotrophy as well as phagotrophy or osmotrophy to obtain organic nutrients.

M line   In a myofibril, a narrow, dark band in the center of the H zone where proteins link the central regions of adjacent thick filaments.

model organism   An organism studied by many different researchers so they can compare their results and determine scientific principles that apply more broadly to other species.

moderately repetitive sequence   A DNA sequence found a few hundred to several thousand times in a genome.

molar   A term used to describe a solution's molarity; a 1 molar solution contains 1 mole of a solute in 1 L of water.

molarity   The number of moles of a solute dissolved in 1 L of water.

mole   The amount of any substance that contains the same number of particles as there are atoms in exactly 12 g of carbon.

molecular biology   A field of study spawned largely by genetic technology that looks at the structure and function of the molecules of life.

molecular clock   A method for estimating evolutionary time; based on the observation that neutral mutations occur at a relatively constant rate.

molecular evolution   The molecular changes in genetic material that underlie the phenotypic changes associated with evolution.

molecular formula   A representation of a molecule that consists of the chemical symbols for all of the atoms present and subscripts that indicate how many of those atoms are present.

molecular homologies   Similarities at the molecular level that indicate that living species evolved from a common ancestor or interrelated group of common ancestors.

molecular mass   The sum of the atomic masses of all the atoms in a molecule.

molecular pharming   An avenue of research that involves the production of medically important proteins in agricultural crops or animals.

molecular systematics   A field of study that involves the analysis of genetic data, such as DNA sequences, to identify and study genetic homology and construct phylogenetic trees.

molecule   Two or more atoms that are connected by chemical bonds.

monoclonal antibodies   Antibodies of a specific type that are derived from a single clone of cells.

monocots   One of the two largest lineages of flowering plants in which the embryo produces a single seed leaf.

monocular vision   A type of vision in animals that have eyes on the sides of the head; the animal sees a wide area at one time, though depth perception is reduced.

monocyte   A type of phagocyte that circulates in the blood for only a few days, after which it takes up permanent residence in various organs as a macrophage.

monoecious   The term to describe plants that produce carpellate and staminate flowers on the same plant.

monogamy   A mating system in which one male mates with one female, and most individuals have mates.

monohybrid   The F_1 offspring, also called single-trait hybrids, of true-breeding parents that differ with regard to a single trait.

monohybrid cross   A cross in which the inheritance of only one trait is followed.

monomer   An organic molecule that can be used to form larger molecules (polymers) consisting of many repeating units of the monomer.

monomorphic gene   A gene that exists predominantly as a single allele in a population.

monophagous   The term used to describe parasites that feed on one or a few closely related species.

monophyletic group   A group of species, a taxon, consisting of the most recent common ancestor and all of its descendants.

monosaccharide   A simple sugar.

monosomic   An aneuploid organism that has one too few chromosomes.

monotreme   A member of the mammalian order Monotremata, which consists of three species found in Australia and New Guinea: the duck-billed platypus and two species of echidna.

morphogen   A molecule that imparts positional information and promotes developmental changes at the cellular level.

morphogenesis   The process that creates morphology.

morphology   The structure or form of a body part or an entire organism.

morula   An early stage in a mammalian embryo in which physical contact between cells is maximized by compaction.

mosaic   An individual with somatic cells that are genetically different from each other.

mosses   A phylum of bryophytes; Bryophyta.

motor end plate   The region of a skeletal muscle cell that lies beneath an axon terminal at the neuromuscular junction.

motor neuron   A neuron that sends signals away from the central nervous system and elicits some type of response.

motor protein   A category of cellular proteins that uses ATP as a source of energy to promote movement; consists of three domains called the head, hinge, and tail.

movement corridor   Thin strips of habitat that may permit the movement of individuals between larger habitat patches.

M phase   The sequential events of mitosis and cytokinesis.

mRNA   *See* messenger RNA.

Müllerian mimicry   A type of mimicry in which many noxious species converge to look the same, thus reinforcing the basic distasteful design.

multicellular   Describes an organism consisting of more than one cell, particularly when cell-to-cell adherence and signaling processes and cellular specialization can be demonstrated.

multimeric protein   A protein with more than one polypeptide chain; also said to have a quarternary structure.

multiple alleles   Refers to the occurrence of a gene that exists as three or more alleles in a population.

multiple sclerosis (MS)   A disease in which the patient's own body attacks and destroys myelin as if it were a foreign substance; impairs the function of myelinated neurons that control movement, speech, memory, and emotion.

multipotent   A term used to describe a stem cell that can differentiate into several cell types, but far fewer than pluripotent cells.

muscle   A grouping of muscle cells (fibers) bound together by a succession of connective tissue layers.

muscle fibers   Individual cells within a muscle.

muscle tissue   Bundles of muscle fibers that are specialized to contract when stimulated.

muscular dystrophy   A group of diseases associated with progressive degeneration of skeletal and cardiac muscle fibers.

mutagen   An agent known to cause mutation.

mutant allele   An allele that has been altered by mutation.

mutation   A heritable change in the genetic material of an organism.

mutualism   A symbiotic interaction in which both species benefit.

myasthenia gravis   A disease characterized by loss of ACh receptors on skeletal muscle, due to the body's own immune system destroying the receptors.

mycelium   A fungal body composed of microscopic branched filaments known as hyphae.

mycorrhizae   Associations between the hyphae of certain fungi and the roots of plants.

myelin sheath   In the nervous system, an insulating layer made up of specialized glial cells wrapped around the axons.

myocardial infarction (MI)   The death of cardiac muscle cells, which can occur if a region of the heart is deprived of blood for an extended time.

myofibrils   Individual muscle cells within a muscle, each of which contains thick and thin filaments.

myogenic heart   A heart in which the signaling mechanism that initiates contraction resides within the cardiac muscle itself.

myoglobin   An oxygen-binding protein that provides an intracellular reservoir of oxygen for muscle fibers.

myosin   A motor protein found abundantly in muscle cells and also in other cell types.

N

NAD⁺   Nicotinamide adenine dinucleotide; a dinucleotide that functions as an energy intermediate molecule. It combines with two electrons and H^+ to form NADH.

NADPH   Nicotinamide adenine dinucleotide phosphate; an energy intermediate that provides the energy and electrons to drive the Calvin cycle during photosynthesis.

natural killer (NK) cells   A type of leukocyte that participates in both nonspecific and specific immunity; recognizes general features on the surface of cancer cells or any virus-infected cells.

natural selection   The process that eliminates those individuals that are less likely to survive and reproduce in a particular environment, while allowing other individuals with traits that confer greater reproductive success to increase in numbers.

nauplius   The first larval stage in a crustacean.

navigation   A mechanism of migration that involves the ability not only to follow a compass bearing but also to set or adjust it.

negative control   Transcriptional regulation by repressor proteins.

negative feedback loop   A homeostatic system in animals in which a change in the variable being regulated brings about responses that move the variable in the opposite direction.

negative frequency-dependent selection   A pattern of natural selection in which the fitness of a genotype decreases when its frequency becomes higher; the result is a balanced polymorphism.

negative pressure filling   The mechanism by which reptiles, birds, and mammals ventilate their lungs.

nekton   Free-swimming animals in the open ocean that can swim against currents to locate food.

nematocyst   In a cnidarian, a powerful capsule with an inverted coiled and barbed thread that functions to immobilize small prey.

neocortex   The layer of the brain that evolved most recently in mammals.

nephron   One of several million single-cell-thick tubules that are the functional units of the mammalian kidney.

Nernst equation   The formula that gives the equilibrium potential for an ion at any given concentration gradient.

nerve   A structure found in the peripheral nervous system that is composed of multiple myelinated neurons bound by connective tissue; carries information to or from the central nervous system.

nerve cord   In many invertebrates, a ventral structure that extends from the anterior end of the animal to the tail; a dorsal nerve cord is found in chordates.

nerve net   Interconnected neurons with no central control organ.

nervous system   Groups of cells that sense internal and environmental changes and transmit signals that enable an animal to respond in an appropriate way.

nervous tissue   Clusters of cells that initiate and conduct electrical signals from one part of an animal's body to another part.

net primary production (NPP)   Gross primary production minus the energy lost in plant cellular respiration.

net reproductive rate   The population growth rate per generation.

neural crest   In vertebrates, a group of embryonic cells derived from ectoderm that disperse throughout the embryo and contribute to the development of the skeleton and other structures, including peripheral nerves.

neural tube In chordates, a structure formed from ectoderm located dorsal to the notochord; all neurons and their supporting cells in the central nervous system originate from neural precursor cells derived from the neural tube.

neurogenesis The production of new neurons by cell division.

neurogenic heart A heart that will not beat unless it receives regular electrical impulses from the nervous system.

neurohormone A hormone made in and secreted by neurons whose cell bodies are in the hypothalamus.

neuromodulator Another term for a neuropeptide, which is a neurotransmitter that can alter or modulate the response of a postsynaptic neuron to other neurotransmitters.

neuromuscular junction The junction between a motor neuron's axon and a skeletal or cardiac muscle fiber.

neuron A highly specialized cell found in nervous systems of animals that communicates with other cells by electrical or chemical signals.

neuroscience The scientific study of nervous systems.

neurotransmitter A small signaling molecule that is released from an axon terminal and diffuses to a postsynaptic cell where it elicits a response.

neurulation The embryological process responsible for initiating central nervous system formation.

neutral theory of evolution States that most genetic variation is due to the accumulation of neutral mutations that have attained high frequencies in a population via genetic drift.

neutral variation Genetic variation in which natural selection does not favor any particular genotype.

neutralism The phenomenon in which two species occur together but do not interact in any measurable way.

neutron A neutral particle found in the center of an atom.

neutrophil A type of phagocyte and the most abundant type of leukocyte. Neutrophils engulf bacteria by endocytosis.

nitrification The conversion by soil bacteria of NH_3 or NH_4^+ to nitrate (NO_3^-), a form of nitrogen commonly used by plants.

nitrogen fixation A specialized metabolic process in which certain prokaryotes use the enzyme nitrogenase to convert inert atmospheric nitrogen gas into ammonia; also, the industrial process by which humans produce ammonia fertilizer from nitrogen gas.

nitrogenase An enzyme used in the biological process of fixing nitrogen.

nitrogen-limitation hypothesis The proposal that organisms select food based on its nitrogen content.

nitrogenous wastes Degradation products of proteins and nucleic acids that are toxic at high concentrations and must be eliminated from the body.

nociceptor A sensory receptor in animals that responds to extreme heat, cold, and pressure, as well as to certain molecules such as acids; also known as a pain receptor.

nocturnal enuresis Bed-wetting.

node The region of a plant stem from which one or more leaves, branches, or buds emerge.

nodes of Ranvier Exposed areas in the axons of myelinated neurons that contain many voltage-gated Na^+ channels and are the sites of regeneration of action potentials.

Nod factor Nodulation factor; a substance produced by nitrogen-fixing bacteria in response to flavonoids secreted from the roots of potential host plants. Nod factors bind to receptors in plant root membranes, starting a process that allows the bacteria to invade roots.

nodule A small swelling on a plant root that contains nitrogen-fixing bacteria.

nodulin One of several plant proteins that foster root nodule development.

noncoding strand *See* template strand.

noncompetitive inhibitor A molecule that binds to an enzyme at a location that is outside the active site and inhibits the enzyme's function.

noncyclic electron flow The combined action of photosystem II and photosystem I in which electrons flow in a linear manner to produce NADPH.

non-Darwinian evolution The idea that much of the modern variation in gene sequences is explained by neutral variation rather than adaptive variation.

nondisjunction An event in which the chromosomes do not sort properly during cell division.

nonpolar covalent bond A strong bond formed between two atoms of similar electronegativities in which electrons are shared between the atoms.

nonpolar molecule A molecule composed predominantly of nonpolar bonds.

nonrandom mating The phenomenon that individuals choose their mates based on their genotypes or phenotypes.

nonrecombinant An offspring whose combination of traits has not changed from the parental generation.

nonsense codon *See* stop codon.

nonsense mutation A mutation that changes a normal codon into a stop codon; this causes translation to be terminated earlier than expected, producing a truncated polypeptide.

nonshivering thermogenesis An increase in an animal's metabolic rate that is not due to increased muscle activity; occurs primarily in brown adipose tissue.

nonspecific (innate) immunity The body's defenses that are present at birth and act against foreign materials in much the same way regardless of their specific identity; includes the skin and mucous membranes, plus various cellular and chemical defenses.

nonvascular plant A plant that does not produce lignified vascular tissue; includes the bryophytes.

norepinephrine A type of neurotransmitter; also known as noradrenaline.

norm of reaction A description of how a trait may change depending on environmental conditions.

notochord A defining characteristic of all chordate embryos; consists of a flexible rod that lies between the digestive tract and the nerve cord.

N-terminus The location of the first amino acid in a polypeptide; also known as the amino terminus.

nuclear envelope A double-membrane structure that encloses the cell's nucleus.

nuclear genome The chromosomes found in the nucleus of a eukaryotic cell.

nuclear lamina A collection of filamentous proteins that line the inner nuclear membrane; part of the nuclear matrix.

nuclear matrix A filamentous network of proteins that is found inside the nucleus and lines the inner nuclear membrane. The nuclear matrix serves to organize the chromosomes.

nuclear pore A passageway for the movement of molecules and macromolecules into and out of the nucleus; formed where the inner and outer nuclear membranes make contact with each other.

nucleic acid An organic molecule composed of nucleotides. The two types of nucleic acids are deoxyribonucleic acid (DNA) and ribonucleic acid (RNA).

nucleoid region A site in a bacterial cell where the genetic material (DNA) is located.

nucleolus A prominent region in the nucleus of nondividing cells where ribosome assembly occurs.

nucleosome A structural unit of eukaryotic chromosomes composed of an octamer of histones (eight histone proteins) wrapped with DNA.

nucleotide An organic molecule having three components: one or more phosphate groups, a five-carbon sugar (either deoxyribose or ribose), and a single or double ring of carbon and nitrogen atoms known as a base.

nucleotide excision repair (NER) A common type of DNA repair system that removes (excises) and repairs a region of the DNA where damage has occurred.

nucleus (plural, **nuclei**) 1. In cell biology, an organelle found in eukaryotic cells that contains most of the cell's genetic material. 2. In chemistry, the region of an atom that contains protons and neutrons. 3. In neurobiology, a group of neuronal cell bodies in the brain that are devoted to a particular function.

nutrient Any substance taken up by a living organism that is needed for survival, growth, development, repair, or reproduction.

O

obese According to current National Institutes of Health guidelines, a person having a body mass index (BMI) of 30 kg/m² or more.

obligate aerobes Microorganisms that require oxygen.

obligate anaerobes Microorganisms that are poisoned by oxygen.

obligatory mutualism An interaction in which two mutualistic species cannot live without each other.

occipital lobe One of four lobes of the cerebral cortex of the human brain; controls aspects of vision and color recognition.

ocelli Photosensitive organs in some animal species.

octet rule The phenomenon that some atoms are most stable when their outer shell is full with eight electrons.

Okazaki fragments Short segments of DNA synthesized in the lagging strand during DNA replication.

olfaction The sense of smell.

olfactory bulbs Part of the limbic system of the forebrain of vertebrates; the olfactory bulbs carry information about odors to the brain.

oligodendrocytes Glial cells that produce the myelin sheath around neurons in the central nervous system.

oligotrophic The term used to describe aquatic systems that are low in nutrients such as phosphate and combined nitrogen and are consequently low in primary productivity and biomass, but typically high in species diversity.

ommatidium An independent visual unit in the eye of insects that functions as a separate photoreceptor capable of forming an independent image.

omnivore   An animal that has the ability to eat and survive on both plant and animal products.

oncogene   A type of mutant gene derived from a proto-oncogene. An oncogene is overactive, thus contributing to uncontrolled cell growth and promoting cancer.

one gene–one enzyme hypothesis   An early hypothesis by Beadle and Tatum that suggested that one gene encodes one enzyme. It was later modified to the one gene–one polypeptide theory.

one gene–one polypeptide theory   The concept that one structural gene codes for one polypeptide.

oogenesis   Gametogenesis in a female animal resulting in the production of an egg cell.

oogonia (singular, **oogonium**)   In animals, diploid germ cells that give rise to the female gametes, the eggs.

open circulatory system   In animals, a circulatory system in which hemolymph, which is not different than the interstitial fluid, flows throughout the body and is not confined to special vessels.

open complex   Also called the transcription bubble; a small bubble-like structure between two DNA strands that occurs during transcription.

open conformation   Loosely packed chromatin that can be transcribed into RNA.

operant conditioning   A form of behavior modification; a type of associative learning in which an animal's behavior is reinforced by a consequence, either a reward or a punishment.

operator   A DNA sequence in bacteria that is recognized by activator or repressor proteins that regulate the level of gene transcription.

operculum   A protective flap that covers the gills of a bony fish.

operon   An arrangement of two or more genes in bacteria that are under the transcriptional control of a single promoter.

opposable thumb   A thumb that can be placed opposite the fingers of the same hand; gives animals a precision grip that enables the manipulation of small objects.

opsin   A protein that is a component of visual pigments in the vertebrate eye.

opsonization   The process by which an antibody binds to a pathogen and provides a means to link the pathogen with a phagocyte.

optic disc   In vertebrates, the point on the retina where the optic nerve leaves the eye.

optic nerve   A structure of the vertebrate eye that carries electrical signals to the brain.

optimal foraging   The concept that in a given circumstance, an animal seeks to obtain the most energy possible with the least expenditure of energy.

optimality theory   The theory that predicts an animal should behave in a way that maximizes the benefits of a behavior minus its costs.

orbital   The region surrounding the nucleus of an atom where the probability is high of finding a particular electron.

order   In taxonomy, a subdivision of a class.

organ   Two or more types of tissue combined to perform a common function.

organelle   A subcellular structure or membrane-bound compartment with its own unique structure and function.

organic chemistry   The study of carbon-containing molecules.

organic farming   The production of crops without the use of commercial inorganic fertilizers, growth substances, and pesticides.

organic molecule   A carbon-containing molecule, so named because they are found in living organisms.

organism   A living thing that maintains an internal order that is separated from the environment.

organismal ecology   The investigation of how adaptations and choices by individuals affect their reproduction and survival.

organismic model   A view of the nature of a community that considers it to be equivalent to a superorganism; individuals, populations, and communities have a relationship to each other that resembles the associations found between cells, tissues, and organs.

organizing center   A group of cells in a plant shoot meristem that ensures the proper organization of the meristem and preserves the correct number of actively dividing stem cells.

organogenesis   The developmental stage during which cells and tissues form organs in animal embryos.

organ system   Different organs that work together to perform an overall function in an organism.

orientation   A mechanism of migration in which animals have the ability to follow a compass bearing and travel in a straight line.

origin of replication   A site within a chromosome that serves as a starting point for DNA replication.

ortholog   A homologous gene in different species.

osmoconformer   An animal whose osmolarity conforms to that of its environment.

osmolarity   The solute concentration of a solution of water, expressed as milliosmoles/liter (mOsm/L).

osmoregulator   An animal that maintains stable internal salt concentrations and osmolarities, even when living in water with very different osmolarities than its body fluids.

osmosis   The movement of water across membranes to balance solute concentrations. Water diffuses from a solution that is hypotonic (lower solute concentration) into a solution that is hypertonic (higher solute concentration).

osmotic adjustment   The process by which a plant modifies the solute concentration of its cytosol.

osmotic lysis   Occurs when a cell in a hypotonic environment takes up so much water that it ruptures.

osmotic pressure   The hydrostatic pressure required to stop the net flow of water across a membrane due to osmosis.

osmotroph   An organism that relies on osmotrophy (uptake of small organic molecules) as a form of nutrition.

osteomalacia   Bone deformation in adults due to inadequate mineral intake or absorption from the intestines.

osteoporosis   A disease in which the mineral and organic components of bone are reduced.

otoliths   Granules of calcium carbonate found in the gelatinous substance that embeds hair cells in the vertebrate ear.

outer bark   Protective layers of mostly dead cork cells that cover the outside of woody stems and roots.

outer ear   One of the three main compartments of the mammalian ear; consists of the external ear, or pinna, and the auditory canal.

outer segment   The highly convoluted plasma membranes found in the rods and cones of the eye.

outgroup   In a cladogram, a species or group of species that does not exhibit one or more shared derived characters found in the ingroup.

ovarian cycle   The events beginning with the development of an ovarian follicle, followed by release of a secondary oocyte, and concluding with formation and subsequent degeneration of a corpus luteum.

ovaries (singular, **ovary**)   1. In animals, the female gonads where eggs are formed. 2. In plants, the lowermost portion of the pistil that encloses and protects the ovules.

overweight   According to current National Institutes of Health guidelines, a person having a body mass index (BMI) of 25 kg/m² or more.

oviduct   A thin tube with undulating fimbriae (fingerlike structures) that is connected to the uterus and extends out to the ovary; also called the fallopian tube.

oviparity   Development of an embryo outside the mother, usually in a protective shell or other structure from which the young hatch.

oviparous   An animal whose young hatch from eggs laid outside the mother's body.

ovoviparous   An animal that retains fertilized eggs covered by a protective sheath or other structure within the body, where the young hatch.

ovoviviparity   Development of an embryo involving aspects of both viviparity and oviparity; fertilized eggs covered with a protective sheath are produced and hatch inside the mother's body, but the offspring receive no nourishment from the mother.

ovulation   The process by which a mature oocyte is released from an ovary.

ovule   In a seed plant, a megaspore-producing megasporangium and enclosing tissues known as integuments.

ovum (plural, **ova**)   See egg.

oxidation   A process that involves the removal of electrons; occurs during the breakdown of small organic molecules.

oxidative fiber   A skeletal muscle fiber that contains numerous mitochondria and has a high capacity for oxidative phosphorylation.

oxidative phosphorylation   A process during which NADH and $FADH_2$ are oxidized to make more ATP via the phosphorylation of ADP.

oxygen-hemoglobin dissociation curve   A curve that represents the relationship between the partial pressure of oxygen and the binding of oxygen to hemoglobin proteins.

oxytocin   A hormone secreted by the posterior pituitary gland that stimulates contractions of the smooth muscles in the uterus of a pregnant mammal, facilitating the birth process; after birth, it is important in milk secretion.

P

pacemaker   See sinoatrial (SA) node.

Pacinian corpuscles   Structures located deep beneath the surface of an animal's skin that respond to deep pressure or vibration.

paedomorphosis   The retention of juvenile traits in an adult organism.

pair-rule gene   A type of segmentation gene; a mutation in this gene may cause alternating segments or parts of segments to be deleted.

paleontologist   A scientist who studies fossils.

palisade parenchyma   Photosynthetic ground tissue of the plant leaf mesophyll that consists of closely packed, elongate cells adapted to efficiently absorb sunlight.

palmate   A type of leaf vein pattern in which veins radiate outward, resembling an open hand.

pancreas   In vertebrates, an elongated gland located behind the stomach that secretes digestive enzymes and a fluid rich in bicarbonate ions.

parabronchi   In birds, a series of parallel air tubes that make up the lungs and are the regions of gas exchange.

paracrine   Refers to a type of cellular communication in which molecules are released into the interstitial fluid and act on nearby cells.

paralogs   Homologous genes within a single species.

paraphyletic group   A group of organisms that contains a common ancestor and some, but not all, of its descendants.

parapodia   Fleshy, footlike structures in the polychaetes that are pushed into the substrate to provide traction during movement.

parasite   A predatory organism that feeds off another organism but does not normally kill it.

parasitism   A symbiotic association in which one organism feeds off another but does not normally kill it.

parasympathetic division   The division of the autonomic nervous system that is involved in maintaining and restoring body functions.

parathyroid hormone (PTH)   A hormone that acts on bone to stimulate the activity of cells that dissolve the mineral part of bone.

Parazoa   A subgroup of animals lacking specialized tissue types or organs, although they may have several distinct types of cells; the one phylum in this group is the Porifera (sponges).

parenchyma cell   A type of plant cell that is thin-walled and alive at maturity.

parenchyma tissue   A plant ground tissue that is composed of parenchyma cells.

parental strand   The original strand in DNA replication.

parietal lobe   One of four lobes of the cerebral cortex of the human brain; receives and interprets sensory input from visual and somatic pathways.

parthenogenesis   An asexual process in which an offspring develops from an unfertilized egg.

partial pressure   The individual pressure of each gas in the air; the sum of these pressures is known as atmospheric pressure.

particulate inheritance   The idea that the determinants of hereditary traits are transmitted intact from one generation to the next.

parturition   The birth of an organism.

passive diffusion   Diffusion through a membrane without the aid of a transport protein.

passive immunity   A type of acquired immunity that confers protection against disease through the direct transfer of antibodies from one individual to another.

passive transport   The diffusion of a solute across a membrane in a process that is energetically favorable and does not require an input of energy.

paternal inheritance   A pattern in which only the male gamete contributes particular genes to the offspring.

pathogen   A microorganism that causes disease symptoms in its host.

pattern formation   The process that gives rise to a plant or animal with a particular body structure.

pedal glands   Glands in the foot of a rotifer that secrete a sticky substance that aids in attachment to the substrate.

pedicel   1. A flower stalk. 2. A narrow, waistlike point of attachment between the body parts of spiders and some insects.

pedigree analysis   An examination of the inheritance of human traits in families.

pedipalps   In spiders, a pair of appendages that have various sensory, predatory, or reproductive functions.

peer-review process   A procedure in which experts in a particular area evaluate papers submitted to scientific journals.

pelagic zone   The open ocean, where the water depth averages 4,000 m and nutrient concentrations are typically low.

penis   A male external accessory sex organ found in many animals that is involved in copulation.

pentadactyl limb   A limb ending in five digits.

PEP carboxylase   An enzyme in C_4 plants that adds CO_2 to phosphoenolpyruvate (PEP) to produce the four-carbon compound oxaloacetate.

pepsin   An active enzyme in the stomach that begins the digestion of protein.

peptide bond   The covalent bond that links amino acids in a polypeptide.

peptidoglycan   A polymer composed of carbohydrates cross-linked with peptides that is an important component of the cell walls of most bacteria.

peptidyl site (P site)   One of three sites for tRNA binding in the ribosome during translation; the other two are the aminoacyl site (A site) and the exit site (E site). The P site holds the tRNA carrying the growing polypeptide chain.

peptidyl transfer reaction   During translation, the transfer of the polypeptide from the tRNA in the P site to the amino acid at the A site.

per capita growth rate   The per capita birth rate minus the per capita death rate; the rate that determines how populations grow over any time period.

perception   An awareness of the sensations that are experienced.

perennial   A plant that lives for more than 2 years, often producing seeds each year after it reaches reproductive maturity.

perfect flower   A flower that has both stamens and carpels.

perianth   The term that refers to flower petals and sepals collectively.

pericarp   The wall of a plant's fruit.

pericycle   A cylinder of plant tissue having cell division (meristematic) capacity that encloses the root vascular tissue.

peripheral membrane protein   A protein that is noncovalently bound to regions of integral membrane proteins that project out from the membrane, or they are noncovalently bound to the polar head groups of phospholipids.

peripheral nervous system (PNS)   In vertebrates, all nerves and ganglia outside the brain and spinal cord.

peripheral zone   The area of a plant shoot meristem that contains dividing cells that will eventually differentiate into plant structures.

periphyton   Communities of microorganisms that are attached by mucilage to underwater surfaces such as rocks, sand, and plants.

peristalsis   In animals, the rhythmic, spontaneous waves of muscle contractions that propel food through the digestive system.

peritubular capillaries   Capillaries near the junction of the cortex and medulla that surround the nephron of the mammalian kidney.

permafrost   A layer of permanently frozen soil found in tundra.

peroxisome   A relatively small organelle found in all eukaryotic cells that catalyzes detoxifying reactions.

personalized medicine   A medical practice in which information about a patient's genotype is used to individualize their medical care.

petal   A flower organ that usually serves to attract insects or other animals for pollen transport.

petiole   A stalk that connects a leaf to the stem of a plant.

P generation   The parental generation in a genetic cross.

pH   The mathematical expression of a solution's hydrogen ion concentration, defined as the negative logarithm to the base 10 of the H^+ concentration.

phage   See bacteriophage.

phagocyte   A cell capable of phagocytosis; phagocytes provide nonspecific defense against pathogens that enter the body.

phagocytic vacuole   A vacuole that functions in the degradation of food particles or bacteria; also called a food vacuole.

phagocytosis   A form of endocytosis that involves the formation of a membrane vesicle, called a phagocytic vacuole, which engulfs a particle such as a bacterium.

phagotroph   An organism that specializes in phagotrophy (particle feeding) by means of phagocytosis as a form of nutrition.

pharyngeal slit   A defining characteristic of all chordate embryos. In early-diverging chordates, pharyngeal slits develop into a filter-feeding device, and in some advanced chordates, they form gills.

pharynx   A portion of the vertebrate alimentary canal; also known as the throat.

phenolics   A group of secondary metabolites that contain a benzene ring covalently linked to a single hydroxyl group. Includes tannins, lignins, and flavonoids.

phenotype   The characteristics of an organism that are the result of the expression of its genes.

pheromone   A powerful chemical attractant used to manipulate the behavior of others.

phloem   A specialized conducting tissue in a plant's stem.

phloem loading   The process of conveying sugars to sieve-tube elements for long-distance transport.

phoresy   A form of commensalism in which individuals of one species use individuals of a second species for transportation.

phosphodiesterase   An enzyme that breaks down cAMP into AMP.

phosphodiester linkage   Refers to a double linkage (two phosphoester bonds) that holds together adjacent nucleotides in DNA and RNA strands.

phospholipid   A class of lipids that are similar in structure to triglycerides, but the third hydroxyl group of glycerol is linked to a phosphate group instead of a fatty acid; a key component of biological membranes.

phospholipid bilayer   The basic framework of the cellular membrane, consisting of two layers of lipids.

phosphorylation   The attachment of a phosphate to a molecule.

photic zone   A fairly narrow zone close to the surface of an aquatic environment, where light is sufficient to allow photosynthesis to exceed respiration.

photoautotroph   An organism that uses the energy from light to make organic molecules from inorganic sources.

photoheterotroph   An organism that is able to use light energy to generate ATP but must take in organic compounds from the environment.

photon A discrete particle that makes up light. A photon is massless and travels in a wavelike pattern.

photoperiodism A plant's ability to measure and respond to amounts of light; used as a way of detecting seasonal change.

photoreceptor A specialized cell in an animal that responds to visible light energy.

photorespiration The metabolic process occurring in C_3 plants that occurs when the enzyme rubisco combines with oxygen instead of carbon dioxide and produces only one molecule of 3PG instead of two, thereby reducing photosynthetic efficiency.

photosynthesis The process whereby light energy is captured by plant, algal, or bacterial cells and is used to synthesize organic molecules from CO_2 and H_2O (or H_2S).

photosystem I (PSI) A distinct complex of proteins and pigment molecules in chloroplasts that absorbs light during the light reactions of photosynthesis.

photosystem II (PSII) A distinct complex of proteins and pigment molecules in chloroplasts that generates oxygen from water during the light reactions of photosynthesis.

phototropin The main blue-light sensor involved in phototropism in plants.

phototropism The tendency of a plant to grow toward a light source.

phylogenetic tree A diagram that describes a phylogeny; such a tree is a hypothesis of the evolutionary relationships among various species, based on the information available to and gathered by systematists.

phylogeny The evolutionary history of a species or group of species.

phylum (plural, **phyla**) In taxonomy, a subdivision of a kingdom.

physical mutagen A physical agent, such as UV light, that causes mutations.

physiological ecology A subdiscipline of organismal ecology that investigates how organisms are physiologically adapted to their environment and how the environment impacts the distribution of species.

physiology The study of the functions of cells and body parts of living organisms.

phytochrome A red and far-red-light receptor in plants.

phytoplankton Microscopic photosynthetic protists that float in the water column or actively move through water.

phytoremediation The process of removing harmful metals from soils by growing hyperaccumulator plants on metal-contaminated soils, then harvesting and burning the plants to ashes for disposal and/or metal recovery.

pigment A molecule that can absorb light energy.

pili (singular, **pilus**) Threadlike surface appendages that allow prokaryotes to attach to each other during mating or to move across surfaces.

piloting A mechanism of migration in which an animal moves from one familiar landmark to the next.

pinnate A type of leaf vein pattern in which veins appear feather-like.

pinocytosis A form of endocytosis that involves the formation of membrane vesicles from the plasma membrane as a way for cells to internalize the extracellular fluid.

pistil A flower structure that may consist of a single carpel or multiple, fused carpels and is differentiated into stigma, style, and ovary.

pit A thin-walled circular area in a plant cell wall where secondary wall materials such as lignin are absent and through which water moves.

pitch The tone of a sound wave that depends on its length and frequency.

pituitary dwarfism A condition in which a person's anterior pituitary gland fails to make adequate amounts of GH during childhood; results in stunted growth. The currently accepted name is short stature.

pituitary giant A person who has a tumor of the GH-secreting cells of the anterior pituitary gland and thus produces excess GH during childhood and, if untreated, during adulthood; the person can grow very tall before growth ceases after puberty.

pituitary gland A multilobed endocrine gland sitting directly below the hypothalamus of the brain.

placenta A structure through which humans and other eutherian mammals retain and nourish their young within the uterus via the transfer of nutrients and gases.

placental transfer tissue In plants, a nutritive tissue that aids in the transfer of nutrients from maternal parent to embryo.

plant A multicellular eukaryotic organism that is photosynthetic, generally lives on land, and is adapted in many ways to cope with the environmental stresses of life on land.

Plantae A eukaryotic kingdom of the domain Eukarya.

plant tissue culture A laboratory process to produce thousands of identical plants having the same desirable characteristics.

plaque 1. A deposit of lipids, fibrous tissue, and smooth muscle cells that may develop inside arterial walls. 2. A bacterial biofilm that may form on the surfaces of teeth.

plasma The fluid part of blood that contains water and dissolved solutes.

plasma cell A cell that synthesizes and secretes antibodies.

plasma membrane The biomembrane that separates the internal contents of a cell from its external environment.

plasmid A small circular piece of DNA found naturally in many strains of bacteria and occasionally in eukaryotic cells; can be used as a vector in cloning experiments.

plasmodesma (plural, **plasmodesmata**) A membrane-lined, ER-containing channel that connects the cytoplasm of adjacent plant cells.

plasmogamy The fusion of the cytoplasm between two gametes.

plasmolysis The shrinkage of algal or plant cytoplasm that occurs when water leaves the cell by osmosis, with the result that the plasma membrane no longer presses on the cell wall.

plastid A general name given to organelles found in plant and algal cells that are bound by two membranes and contain DNA and large amounts of either chlorophyll (in chloroplasts), carotenoids (in chromoplasts), or starch (in amyloplasts).

platelets Cell fragments in the blood of mammals that play a crucial role in the formation of blood clots.

pleiotropy The phenomenon in which a mutation in a single gene can have multiple effects on an individual's phenotype.

pleural sac A double layer of moist sheathlike membranes that encases each lung.

pluripotent Refers to the ability of embryonic stem cells to differentiate into almost every cell type of the body.

point mutation A mutation that affects only a single base pair within DNA or that involves the addition or deletion of a single base pair to a DNA sequence.

polar cell The highest latitude cell in the three-cell model of atmospheric circulation.

polar covalent bond A covalent bond between two atoms that have different electronegativities; the shared electrons are closer to the atom of higher electronegativity than the atom of lower electronegativity. This distribution of electrons around the atoms creates a polarity, or difference in electric charge, across the molecule.

polarized 1. In cell biology, refers to cells that have different sides, such as the apical and basal sides of epithelial cells. 2. In neuroscience, refers to the electrical gradient across a neuron's plasma membrane.

polar molecule A molecule containing significant numbers of polar bonds.

polar transport The process whereby auxin flows primarily downward in shoots.

pole A structure of the spindle apparatus defined by each centrosome.

pollen In seed plants, tiny male gametophytes enclosed by sporopollenin-containing microspore walls.

pollen coat A layer of material that covers the sporopollenin-rich pollen wall.

pollen grain The immature male gametophyte of a seed plant.

pollen tube In seed plants, a long, thin tube produced by a pollen grain that delivers sperm to the ovule.

pollen wall A tough, sporopollenin wall at the surface of a pollen grain.

pollination The process in which pollen grains are transported to an angiosperm flower or a gymnosperm cone primarily by means of wind or animal pollinators.

pollination syndromes The pattern of coevolved traits between particular types of flowers and their specific pollinators.

pollinator An animal that carries pollen between angiosperm flowers or cones of gymnosperms.

polyandry A mating system in which one female mates with several males, but males mate with only one female.

poly A tail A string of adenine nucleotides at the 3′ end of most mature mRNAs in eukaryotes.

polycistronic mRNA An mRNA that contains the coding sequences for two or more structural genes.

polycythemia A condition of increased hemoglobin due to increased hematocrit.

polygenic A trait in which several or many genes contribute to the outcome of the trait.

polygyny A mating system in which one male mates with several females in a single breeding season, but females mate with only one male.

polyketides A group of secondary metabolites produced by diverse organisms. Examples include streptomycin, erythromycin, and tetracycline.

polymer A large molecule formed by linking many smaller molecules called monomers.

polymerase chain reaction (PCR) A technique to make many copies of a gene in vitro; primers are used that flank the region of DNA to be amplified.

polymorphic gene A gene that commonly exists as two or more alleles in a population.

polymorphism The phenomenon that many traits or genes may display variation within a population.

polyp   A type of cnidarian body form that is sessile and occurs mouth up.

polypeptide   A linear sequence of amino acids; the term denotes structure.

polyphagous   Parasites that feed on many host species.

polyphyletic group   A group of organisms that consists of members of several evolutionary lines and does not include the most recent common ancestor of the included lineages.

polyploid   An organism that has three or more sets of chromosomes.

polyploidy   In an organism, the state of having three or more sets of chromosomes.

polysaccharide   Many monosaccharides linked to form long polymers.

pons   The part of the vertebrate hindbrain, along with the cerebellum, responsible for monitoring and coordinating body movements.

population   A group of individuals of the same species that occupy the same environment and can interbreed with one another.

population ecology   The study of how populations grow and what factors promote or limit growth.

population genetics   The study of genes and genotypes in a population.

portal vein   A vein that not only collects blood from capillaries—like all veins—but also forms another set of capillaries, as opposed to returning the blood directly to the heart.

positional information   Molecules that are provided to a cell that allow it to determine its position relative to other cells.

positive control   Transcriptional regulation by activator proteins.

positive feedback loop   In animals, the acceleration of a process, leading to what is sometimes called an explosive system.

positive pressure filling   The method by which amphibians ventilate their lungs. The animals gulp air and force it under pressure into the lungs, as if inflating a balloon.

postabsorptive state   One of two alternating phases in the utilization of nutrients; occurs when the gastrointestinal tract is empty of nutrients and the body's own stores must supply energy. The other phase is the absorptive state.

postanal tail   A defining characteristic of all chordate embryos; consists of a tail of variable length that extends posterior to the anal opening.

posterior   Refers to the rear (tail-end) of an animal.

postsynaptic cell   The cell that receives the electrical or chemical signal sent from a neuron.

post-translational covalent modification   A process of changing the structure of a protein, usually by covalently attaching functional groups; this process greatly increases the diversity of the proteome.

post-translational sorting   The uptake of proteins into the nucleus, mitochondria, chloroplasts, or peroxisomes that occurs after the protein is completely made in the cytosol (that is, completely translated).

postzygotic isolating mechanism   A mechanism that prevents interbreeding by blocking the development of a viable and fertile individual after fertilization has taken place.

potential energy   The stored energy that a substance possesses due to its structure or location.

power stroke   In muscle, a conformation change in the myosin cross-bridge that results in binding between myosin and actin and the movement of the actin filament.

P protein   Phloem protein; the proteinaceous material used by plant phloem as a response to wounding.

prebiotic soup   The medium formed by the slow accumulation of organic molecules in the early oceans over a long period of time prior to the existence of life.

predation   An interaction in which the action of a predator results in the death of its prey.

predator   An animal that kills its prey.

prediction   An expected outcome based on a hypothesis that can be shown to be correct or incorrect through observation or experimentation.

pregnancy   The time during which a developing embryo and fetus grows within the uterus of the mother. The period of pregnancy is also known as gestation.

preinitiation complex   The structure of the completed assembly of RNA polymerase II and GTFs at the TATA box prior to transcription of eukaryotic structural genes.

pre-mRNA   In eukaryotes, the mRNA transcript prior to any processing.

pressure-flow hypothesis   Explains sugar translocation in plants as a process driven by differences in turgor pressure between cells of a sugar source, where sugar is produced, and cells of a sugar sink, where sugar is consumed.

pressure potential (P)   The component of water potential due to hydrostatic pressure.

presynaptic cell   The neuron that sends an electrical or chemical signal to another cell.

prezygotic isolating mechanism   A mechanism that stops interbreeding by preventing the formation of a zygote.

primary active transport   A type of transport that involves pumps that directly use energy to transport a solute against a gradient.

primary cell wall   In plants, a relatively thin and flexible cell wall that is synthesized first between two newly made daughter cells.

primary consumer   An organism that obtains its food by eating primary producers; also called a herbivore.

primary electron acceptor   The molecule to which a high-energy electron from an excited pigment molecule such as P680* is transferred during photosynthesis.

primary endosymbiosis   The process by which a eukaryotic host cell acquires prokaryotic endosymbionts. Mitochondria and the plastids of green and red algae are examples of organelles that originated with primary endosymbiosis.

primary growth   Plant growth that occurs from primary meristems and produces primary tissues and organs of diverse types.

primary immune response   The response to an initial exposure to an antigen.

primary meristem   A meristematic tissue that increases plant length and produces new organs.

primary metabolism   The synthesis and breakdown of molecules and macromolecules that are found in all forms of life and are essential for cell structure and function.

primary oocytes   In animals, cells that undergo meiosis to begin the process of egg production.

primary plastid   A plastid that originated from a prokaryote as the result of primary endosymbiosis.

primary producer   An autotroph, which typically harvests light energy from the sun; located at the base of the food chain.

primary spermatocytes   In animals, cells that undergo meiosis to begin the process of sperm production.

primary structure   The linear sequence of amino acids of a polypeptide; one of four levels of protein structure.

primary succession   Succession on newly exposed sites that were not previously occupied by soil and vegetation.

primary vascular tissue   Plant tissue composed of primary xylem and phloem, which is the conducting tissue of nonwoody plants.

primer   A short segment of RNA, typically 10 to 12 nucleotides in length, that is needed to begin DNA replication.

primordial germ cells (PGCs)   In animals, the embryonic cells that eventually give rise to gametes.

principle of parsimony   The concept that the preferred hypothesis is the one that is the simplest.

principle of species individuality   A view of the nature of a community in which each species is distributed according to its physiological needs and population dynamics; most communities intergrade continuously, and competition does not create distinct vegetational zones.

prion   An infectious protein that causes disease by inducing the abnormal folding of other protein molecules.

probability   The chance that an event will have a particular outcome.

proboscis   The coiled tongue of a butterfly or moth, which can be uncoiled, enabling it to drink nectar from flowers.

procambium   In plants, a type of primary tissue meristem that produces vascular tissue.

producer   An organism that synthesizes the organic compounds used by other organisms for food.

product   The end result of a chemical reaction.

production efficiency   The percentage of energy assimilated by an organism that becomes incorporated into new biomass.

productivity hypothesis   The proposal that greater production by plants results in greater overall species richness.

product rule   The probability that two or more independent events will occur is equal to the product of their individual probabilities.

progesterone   A hormone secreted by the female ovaries that plays a key role in pregnancy.

progymnosperms   An extinct group of plants having wood but not seeds, which evolved before the gymnosperms.

prokaryote   One of the two categories into which all forms of life can be placed. Prokaryotes lack a nucleus and include archaea.

prokaryotic   Refers to organisms having cells lacking a membrane-enclosed nucleus and cell compartmentalization; includes all members of the domains Bacteria and Archaea.

prometaphase   The phase of mitosis during which the mitotic spindle is completely formed.

promiscuous   In ecology, a term for animals that have different sexual mates every year or breeding season.

promoter   The site in the DNA where transcription begins.

proofreading   The ability of DNA polymerase to identify a mismatched nucleotide and remove it from the daughter strand.

prophage   Refers to the DNA of a phage that has become integrated into a bacterial chromosome.

prophase   The phase of mitosis during which the chromosomes condense and the nuclear membrane begins to vesiculate.

proplastid   Unspecialized structures that form plastids.

prostate gland   A structure in the male reproductive system that secretes a thin fluid that protects sperm once they are deposited within the female reproductive tract.

prosthetic group   Small molecules that are permanently attached to the surface of an enzyme and aid in catalysis.

protease   An enzyme that cuts proteins into smaller polypeptides.

proteasome   A molecular machine that is the primary pathway for protein degradation in archaea and eukaryotic cells.

protein   A functional unit composed of one or more polypeptides. Each polypeptide is composed of a linear sequence of amino acids.

protein kinase   An enzyme that transfers phosphate groups from ATP to a protein.

protein kinase cascade   The sequential activation of multiple protein kinases.

protein phosphatase   An enzyme responsible for removing phosphate groups from proteins.

protein-protein interactions   The specific interactions between proteins that occur during many critical cellular processes.

protein subunit   An individual polypeptide within a functional protein; most functional proteins are composed of two or more polypeptides.

proteoglycan   A glycosaminoglycan in the extracellular matrix linked to a core protein.

proteolysis   A processing event within a cell in which enzymes called proteases cut proteins into smaller polypeptides.

proteome   The complete complement of proteins that a cell or organism can make.

proteomics   Techniques used to identify and study groups of proteins.

prothoracicotropic hormone (PTTH)   A hormone produced in certain invertebrates that stimulates a pair of endocrine glands called the prothoracic glands.

protist   A eukaryotic organism that is not a member of the animal, plant, or fungal kingdoms; lives in moist habitats and is typically microscopic in size.

Protista   Formerly a eukaryotic kingdom. Most protists can be placed into seven eukaryotic supergroups.

protobiont   The term used to describe the first nonliving structures that evolved into living cells.

protoderm   In plants, a type of primary tissue meristem that generates the outermost dermal tissue.

proton   A positively charged particle found in the nucleus of an atom. The number of protons in an atom is called the atomic number and defines each type of element.

protonephridia   Simple excretory organs found in flatworms that are used to filter out wastes and excess water.

proton-motive force   *See* H^+ electrochemical gradient.

proto-oncogene   A normal gene that, if mutated, can become an oncogene.

protostome   An animal whose development exhibits spiral determinate cleavage and in which the blastopore becomes the mouth; includes mollusks, annelid worms, and arthropods.

protozoa   A term commonly used to describe diverse heterotrophic protists.

proventriculus   The glandular portion of the stomach of a bird.

provirus   Refers to viral DNA that has become incorporated into a eukaryotic chromosome.

proximal convoluted tubule   The segment of the tubule of the nephron in the kidney that drains Bowman's capsule.

proximate cause   A specific genetic and physiological mechanism of behavior.

pseudocoelomate   An animal with a pseudocoelom.

P site   *See* peptidyl site.

pteridophytes   A phylum of vascular plants having euphylls, but not seeds; Pteridophyta.

pulmocutaneous circulation   The routing of blood from the heart to the gas exchange organs (lungs and skin) of frogs and some other amphibians.

pulmonary circulation   The pumping of blood from the right side of the heart to the lungs to pick up oxygen from the atmosphere and release carbon dioxide.

pulmonary hypertension   A condition that usually results from a diseased or damaged left ventricle that fails to pump out the usual amount of blood with each beat of the heart. This causes blood to back up in the pulmonary vessels, raising their pressure.

pulse-chase experiment   A procedure in which researchers administer a pulse of radioactively labeled materials to cells so that they make radioactive products. This is followed by the addition of nonlabeled materials called a chase.

pump   A transporter that directly couples its conformational changes to an energy source, such as ATP hydrolysis.

punctuated equilibrium   A concept that suggests that the tempo of evolution is more sporadic than gradual. Species rapidly evolve into new species followed by long periods of equilibrium with little evolutionary change.

Punnett square   A common method for predicting the outcome of simple genetic crosses.

pupa   A developmental stage in some insects that undergo metamorphosis; occurs between the larval and adult stages.

pupil   A small opening in the eye of a vertebrate that transmits different patterns of light emitted from images in the animal's field of view.

purine   The bases adenine (A) and guanine (G), with double rings of carbon and nitrogen atoms.

pyramid of biomass   A measure of trophic-level transfer efficiency in which the organisms at each trophic level are weighed.

pyramid of energy   A measure of trophic-level transfer efficiency in which rates of energy production are used rather than biomass.

pyramid of numbers   An expression of trophic-level transfer efficiency in which the number of individuals decreases at each trophic level, with a huge number of individuals at the base and fewer individuals at the top.

pyrimidine   The bases thymine (T), cytosine (C), and uracil (U) with a single ring of carbon and nitrogen atoms.

Q

quadrat   A sampling device used by plant ecologists consisting of a square frame that often encloses 0.25 m^2.

quantitative trait   A trait that shows continuous variation over a range of phenotypes.

quaternary structure   The association of two or more polypeptides to form a protein; one of four levels of protein structure.

quorum sensing   A mechanism by which prokaryotic cells are able to communicate by chemical means when they reach a critical population size.

R

radial cleavage   A mechanism of animal development in which the cleavage planes are either parallel or perpendicular to the vertical axis of the embryo.

radial loop domain   A loop of chromatin, often 25,000 to 200,000 base pairs in size, that is anchored to the nuclear matrix.

radial pattern   A characteristic of the body pattern of plants.

radial symmetry   1. In plants, an architectural feature in which embryos display a cylindrical shape, which is retained in the stems and roots of seedlings and mature plants. In addition, new leaves or flower parts are produced in circular whorls, or spirals, around shoot tips. 2. In animals, an architectural feature in which the body can be divided into symmetrical halves by many different longitudinal planes along a central axis.

Radiata   Radially symmetric animals; includes cnidarians and ctenophores.

radiation   The emission of electromagnetic waves by the surfaces of objects; a method of heat exchange in animals.

radicle   An embryonic root, which extends from the plant hypocotyl.

radioisotope   An isotope found in nature that is inherently unstable and usually does not exist for long periods of time. Such isotopes decay and emit energy in the form of radiation.

radioisotope dating   A common way to estimate the age of a fossil by analyzing the elemental isotopes within the accompanying rock.

radula   A unique, protrusible, tonguelike organ in a mollusk that has many teeth and is used to eat plants, scrape food particles off of rocks, or bore into shells of other species.

rain shadow   An area on the side of a mountain that is sheltered from the wind and experiences less precipitation.

ram ventilation   A mechanism used by fishes to ventilate their gills; fishes swim or face upstream with their mouths open, allowing water to enter into their buccal cavity and across their gills.

random   The rarest pattern of dispersion within a population, in which the location of individuals lacks a pattern.

random sampling error   The deviation between the observed and the expected outcomes.

rate-limiting step   The slowest step in a pathway.

ray-finned fishes   The Actinopterygii, which includes all bony fishes except the coelacanths and lungfishes.

reabsorption   In the production of urine, the process in which useful solutes in the filtrate are recaptured and transported back into the body fluids of an animal.

reactant   A substance that participates in a chemical reaction and becomes changed by that reaction.

reading frame   Refers to the way in which codons are read during translation, in groups of three bases beginning with the start codon.

receptacle   The enlarged region at the tip of a flower peduncle to which flower parts are attached.

receptor   1. A cellular protein that recognizes a signaling molecule. 2. A structure capable of detecting changes in the environment of an animal, such as a touch receptor.

receptor-mediated endocytosis   A common form of endocytosis in which a receptor is specific for a given cargo.

receptor potential The membrane potential in a sensory receptor cell of an animal.

receptor tyrosine kinase A type of enzyme-linked receptor found in animal cells that can attach phosphate groups onto tyrosines that are found in the receptor itself or in other cellular proteins.

recessive A term that describes a trait that is masked by the presence of a dominant trait in a heterozygote.

reciprocal translocation A type of mutation in which two different types of chromosomes exchange pieces, thereby producing two abnormal chromosomes carrying translocations.

recombinant An offspring that has a different combination of traits from the parental generation.

recombinant DNA technology The use of laboratory techniques to isolate and manipulate fragments of DNA.

recombinant vector A vector containing a piece of chromosomal DNA.

recombination frequency The frequency of crossing over between two genes.

red blood cell *See* erythrocyte.

redox reaction A type of reaction in which an electron that is removed during the oxidation of an atom or molecule is transferred to another atom or molecule, which becomes reduced; short for a reduction-oxidation reaction.

reduction A process that involves the addition of electrons to an atom or molecule.

reductionism An approach that involves reducing complex systems to simpler components as a way to understand how the system works. In biology, reductionists study the parts of a cell or organism as individual units.

redundancy hypothesis A biodiversity proposal that is an alternative to the rivet hypothesis. In this model, most species are said to be redundant because they could simply be eliminated or replaced by others with no effect.

reflex arc A simple circuit that allows an organism to respond rapidly to inputs from sensory neurons and consists of only a few neurons.

regeneration A form of asexual reproduction in which a complete organism forms from small fragments of its body.

regulatory element In eukaryotes, a DNA sequence that is recognized by regulatory transcription factors and regulates the expression of genes.

regulatory gene A gene whose function is to regulate the expression of other genes.

regulatory sequence In the regulation of transcription, a DNA sequence that functions as a binding site for genetic regulatory proteins. Regulatory sequences control whether a gene is turned on or off.

regulatory transcription factor A protein that binds to DNA in the vicinity of a promoter and affects the rate of transcription of one or more nearby genes.

relative abundance The frequency of occurrence of species in a community.

relative refractory period The period near the end of an action potential when voltage-gated potassium channels are still open; during this time a new action potential can be generated if a stimulus is sufficiently strong to raise the membrane potential to threshold.

relative water content (RWC) The property often used to gauge the water content of a plant organ or entire plant; RWC integrates the water potential of all cells within an organ or plant and is thus a measure of relative turgidity.

release factor A protein that recognizes a stop codon in the termination stage of translation and promotes the termination of translation.

renal corpuscle A filtering component in the nephron of the kidney.

repetitive sequence Short DNA sequences that are present in many copies in a genome.

replica plating A technique in which a replica of bacterial colonies is transferred from one petri plate to a new petri plate.

replication 1. The copying of DNA strands. 2. The performing of experiments several or many times.

replication fork The area where two DNA strands have separated and new strands are being synthesized.

repressible operon In this type of operon, a small effector molecule inhibits transcription.

repressor A transcription factor that binds to DNA and inhibits transcription.

reproduce To generate offspring by sexual or asexual means.

reproductive cloning The cloning of a multicellular organism, such as a plant or animal.

reproductive isolating mechanisms Mechanisms that prevent interbreeding between different species.

reproductive isolation Refers to the concept that a species cannot successfully interbreed with other species.

reproductive success The likelihood of contributing fertile offspring to the next generation.

resistance (R) The tendency of blood vessels to slow down the flow of blood through their lumens.

resistance gene (R gene) A plant gene that has evolved as part of a defense system in response to pathogen attack.

resolution In microscopy, the ability to observe two adjacent objects as distinct from one another; a measure of the clarity of an image.

resonance energy transfer The process by which energy (not an electron itself) can be transferred to adjacent pigment molecules during photosynthesis.

resource partitioning The differentiation of niches, both in space and time, that enables similar species to coexist in a community.

respiration Metabolic reactions that a cell uses to get energy from food molecules and release waste products.

respiratory centers Several regions of the brainstem in vertebrates that initiate expansion of the lungs.

respiratory chain *See* electron transport chain.

respiratory distress syndrome of the newborn The situation in which a human baby is born prematurely, before sufficient surfactant is produced in the lungs, causing the collapse of many alveoli.

respiratory pigment A large protein that contains one or more metal atoms that bind to oxygen.

respiratory system All components of the body that contribute to the exchange of gas between the external environment and the blood; in mammals, includes the nose, mouth, airways, and lungs and the muscles and connective tissues that encase these structures within the thoracic (chest) cavity.

resting potential The difference in charges across the plasma membrane in an unstimulated neuron.

restoration ecology The full or partial repair or replacement of biological habitats and/or their populations that have been damaged.

rest-or-digest The response of vertebrates to situations associated with nonstressful states, such as feeding; mediated by the parasympathetic branch of the autonomic nervous system.

restriction enzyme An enzyme that recognizes particular DNA sequences and cleaves the DNA backbone at two sites.

restriction point A point in the cell cycle in which a cell has become committed to divide.

restriction sites The base sequences recognized by restriction enzymes.

reticular formation An array of nuclei in the brainstem of vertebrates that plays a major role in controlling states such as sleep and arousal.

retina A sheetlike layer of photoreceptors at the back of the vertebrate eye.

retinal A derivative of vitamin A that is capable of absorbing light energy; a component of visual pigments in the vertebrate eye.

retroelement A type of transposable element that moves via an RNA intermediate.

retrovirus An RNA virus that utilizes reverse transcription to produce viral DNA that can be integrated into the host cell genome.

reverse transcriptase A viral enzyme that catalyzes the synthesis of viral DNA starting with viral RNA as a template.

rhizobia The collective term for proteobacteria involved in nitrogen-fixation symbioses with plants.

rhodopsin The visual pigment in the rods of the vertebrate eye.

ribonucleic acid (RNA) One of two classes of nucleic acids; the other is deoxyribonucleic acid (DNA). RNA consists of a single strand of nucleotides.

ribonucleoprotein A complex between an RNA molecule and a protein.

ribose A five-carbon sugar found in RNA.

ribosomal RNA (rRNA) An RNA that forms part of ribosomes, which provide the site where translation occurs.

ribosome A structure composed of proteins and rRNA that provides the site where polypeptide synthesis occurs.

ribozyme A biological catalyst that is an RNA molecule.

rickets A condition in children characterized by bone deformations due to inadequate mineral intake or malabsorption in the intestines.

right-left axis In bilateral animals, one of the three axes along which the adult body pattern is organized; the others are the dorsoventral axis and the anteroposterior axis.

ring canal A central disc in the water vascular system of echinoderms.

rivet hypothesis An alternative to the diversity-stability hypothesis of biodiversity. In this model, species are like the rivets on an airplane, with each species playing a small but critical role in keeping the plane (the ecosystem) airborne.

RNA *See* ribonucleic acid.

RNA-induced silencing complex (RISC) A complex consisting of miRNA or siRNA and proteins; mediates RNA interference.

RNA interference (RNAi) Refers to a type of mRNA silencing; miRNA or siRNA interferes with the proper expression of an mRNA.

RNA polymerase The enzyme that synthesizes strands of RNA during gene transcription.

RNA processing A step in gene expression between transcription and translation in eukaryotes; the RNA transcript, termed pre-mRNA, is modified in ways that make it a functionally active mRNA.

RNase An enzyme that digests RNA.

RNA world A hypothetical period on primitive Earth when both the information needed for life and the enzymatic activity of living cells were contained solely in RNA molecules.

rods Photoreceptors found in the vertebrate eye; they are very sensitive to low-intensity light but do not readily discriminate different colors. Rods are utilized mostly at night, and they send signals to the brain that generate a black-and-white visual image.

"roid" rage Extremely aggressive behavior brought about by androgen administration.

root A plant organ that provides anchorage in the soil and also fosters efficient uptake of water and minerals.

root apical meristem (RAM) The region of rapidly dividing cells at plant root tips.

root hair A specialized, long, thin root epidermal cell that functions to absorb water and minerals, usually from soil.

root meristem The collection of cells at the root tip that generate all of the tissues of a plant root.

root pressure Osmotic pressure within roots that causes water to rise for some distance through a plant stem, under conditions of high soil moisture or low transpiration.

root-shoot axis The general body pattern of plants in which the root grows downward and the shoot grows upward.

root system The collection of roots and root branches produced by root apical meristems.

rough endoplasmic reticulum (rough ER) The part of the ER that is studded with ribosomes; this region plays a key role in the initial synthesis and sorting of proteins that are destined for the ER, Golgi apparatus, lysosomes, vacuoles, plasma membrane, or outside of the cell.

rRNA *See* ribosomal RNA.

***r*-selected species** A type of life history strategy, where species have a high rate of per capita population growth but poor competitive ability.

rubisco The enzyme that catalyzes the first step in the Calvin cycle in which CO_2 is incorporated into an organic molecule.

Ruffini corpuscle Tactile (touch) receptors in the skin of mammals that respond to deep pressure and vibration.

ruminants Animals such as sheep, goats, llamas, and cows that have complex stomachs consisting of several chambers.

S

saltatory conduction The conduction of an action potential along an axon in which the action potential is regenerated at each node of Ranvier instead of along the entire length of the axon.

sarcoma A tumor of connective tissue such as bone or cartilage.

sarcomere One complete unit of the repeating pattern of thick and thin filaments within a myofibril.

sarcoplasmic reticulum A cellular organelle that provides a muscle fiber's source of the calcium involved in muscle contraction; a specialized form of the endoplasmic reticulum.

satiety A feeling of fullness.

satiety signal A response to eating that removes the sensation of hunger and sets the time period before hunger returns again.

saturated fatty acid A fatty acid in which all the carbons are linked by single covalent bonds.

scanning electron microscopy (SEM) A type of microscopy that utilizes an electron beam to produce an image of the three-dimensional surface of biological samples.

scavenger An animal that eats the remains of dead animals.

Schwann cells The glial cells that form myelin on axons that travel outside the brain and spinal cord.

science In biology, the observation, identification, experimental investigation, and theoretical explanation of natural phenomena.

scientific method A series of steps to test the validity of a hypothesis. This approach often involves a comparison between control and experimental groups.

sclera The white of the vertebrate eye; a strong outer sheath that in the front is continuous with a thin, clear layer known as the cornea.

sclereid Star- or stone-shaped plant cells having tough, lignified cell walls.

sclerenchyma tissue A rigid plant ground tissue composed of tough-walled fibers and sclereids.

secondary active transport A type of membrane transport that involves the utilization of a pre-existing gradient to drive the active transport of another solute.

secondary cell wall A thick rigid plant cell wall that is synthesized and deposited between the plasma membrane and the primary cell wall after a plant cell matures and has stopped increasing in size.

secondary consumer An organism that eats primary consumers; also called a carnivore.

secondary endosymbiosis A process that occurs when a eukaryotic host cell acquires a eukaryotic endosymbiont having a primary plastid.

secondary growth Plant growth that occurs from secondary meristems and increases the girth of woody plant stems and roots.

secondary immune response An immediate and heightened production of additional specific antibodies against the particular antigen that previously elicited a primary immune response.

secondary meristem A meristem in woody plants forming a ring of actively dividing cells that encircle the stem.

secondary metabolism The synthesis of chemicals that are not essential for cell structure and growth and are usually not required for cell survival but are advantageous to the organism.

secondary metabolite Molecules that are produced by secondary metabolism.

secondary oocyte In animals, the large haploid cell that is produced when a primary oocyte undergoes meiosis I during oogenesis.

secondary phloem The inner bark of a woody plant.

secondary plastid A plastid that has originated by the endosymbiotic incorporation of a eukaryotic cell containing a primary plastid into a eukaryotic host cell.

secondary production The measure of production of heterotrophs and decomposers.

secondary spermatocytes In animals, the haploid cells produced when a primary spermatocyte undergoes meiosis I during spermatogenesis.

secondary structure The bending or twisting of proteins into α helices or β sheets; one of four levels of protein structure.

secondary succession Succession on a site that has previously supported life but has undergone a disturbance.

secondary xylem In plants, a type of secondary vascular tissue that is also known as wood.

second law of thermodynamics States that the transfer of energy or the transformation of energy from one form to another increases the entropy, or degree of disorder, of a system.

second messengers Small molecules or ions that relay signals inside the cell.

secretion 1. The export of a substance from a cell. 2. In the production of urine, the process in which some solutes are actively transported into the tubules of the excretory organ; this supplements the amount of a solute that would normally be removed by filtration alone.

secretory pathway A pathway for the movement of larger substances, such as carbohydrates and proteins, out of a cell.

secretory vesicle A membrane vesicle carrying different types of materials that fuses with the cell's plasma membrane to release the contents extracellularly.

seed A reproductive structure having specialized tissues that enclose plant embryos; produced by gymnosperms and flowering plants, usually as the result of sexual reproduction.

seed coat A hard and tough covering that develops from the ovule's integuments and protects a plant embryo.

seed plant The informal name for gymnosperms and angiosperms.

segmentation The division of an animal's body into clearly defined regions.

segmentation gene A gene that controls the segmentation pattern of an animal embryo.

segment-polarity gene A type of segmentation gene; a mutation in this gene causes portions of segments to be missing either an anterior or a posterior region and for adjacent regions to become mirror images of each other.

segregate To separate, as in chromosomes during mitosis.

selectable marker A gene whose presence can allow organisms (such as bacteria) to grow under a certain set of conditions. For example, an antibiotic-resistance gene is a selectable marker that allows bacteria to grow in the presence of the antibiotic.

selective breeding Programs and procedures designed to modify traits in domesticated species.

selectively permeable The property of membranes that allows the passage of certain ions or molecules but not others.

selective serotonin reuptake inhibitors Drugs used to treat major depressive disorder that act by increasing concentrations of serotonin in the brain.

self-fertilization Fertilization that involves the union of a female gamete and male gamete from the same individual.

self-incompatibility (SI) Rejection of pollen that is genetically too similar to the pistil of a plant.

selfish DNA hypothesis The hypothesis that transposable elements exist because they have the characteristics that allow them to insert themselves into the host cell DNA but do not provide any advantage.

self-pollination The process in which pollen from the anthers of a flower is transferred to the stigma of the same flower or between flowers of the same plant.

self-splicing The phenomenon that RNA itself can catalyze the removal of its own intron(s); occurs in rRNA and tRNA.

SEM *See* scanning electron microscopy.

semelparity A reproductive pattern in which organisms produce all of their offspring in a single reproductive event.

semen A mixture containing fluid and sperm that is released during ejaculation.

semicircular canals Structures of the vertebrate ear that can detect circular motions of the head.

semiconservative mechanism The correct model for DNA replication; double-stranded DNA is half conserved following replication, resulting in new double-stranded DNA containing one parental strand and one daughter strand.

semifluid A quality of motion within biomembranes; considered two-dimensional because movement occurs only within the plane of the membrane.

semilunar valves One-way valves into the systemic and pulmonary arteries through which blood is pumped from the ventricles.

seminal vesicles Paired accessory glands in the male reproductive system that secrete fructose, the main nutrient for sperm, into the urethra.

seminiferous tubule A tightly packed tubule in the testis, where spermatogenesis takes place.

senescent Cells that have doubled many times and have reached a point where they have lost the capacity to divide any further.

sense A system in an animal that consists of sensory cells that respond to a specific type of chemical or physical stimulus and send signals to the central nervous system, where the signals are received and interpreted.

sensory neuron A neuron that detects or senses information from the outside world, such as light, sound, touch, and heat; sensory neurons also detect internal body conditions such as blood pressure and body temperature.

sensory receptor In animals, a specialized cell whose function is to receive sensory inputs.

sensory transduction The process by which incoming stimuli are converted into neural signals.

sepal A flower organ that occurs in a whorl located outside whorls of petals of eudicot plants.

septum (plural, **septa**) A cross wall; examples include the cross walls that divide the hyphae of most fungi into many small cells and the structure that separates the old and new chambers of a nautilus.

sere Each phase of succession in a community; also called a seral stage.

setae Chitinous bristles in the integument of many invertebrates.

set point The normal value for a controlled variable, such as blood pressure, in an animal.

sex chromosomes A distinctive pair of chromosomes that are different in males and females.

sex-influenced inheritance The phenomenon in which an allele is dominant in one sex but recessive in the other.

sex linked Refers to genes that are found on one sex chromosome but not on the other.

sex pili Hairlike structures made by bacterial F^+ cells that bind specifically to other F^- cells.

sexual dimorphism A pronounced difference in the morphologies of the two sexes within a species.

sexual reproduction A process that requires a fertilization event in which two gametes unite to produce a cell called a zygote.

sexual selection A type of natural selection that is directed at certain traits of sexually reproducing species that make it more likely for individuals to find or choose a mate and/or engage in successful mating.

Shannon diversity index (H_S) A means of measuring the diversity of a community; $H_S = -\Sigma p_i \ln p_i$.

shared derived character A trait that is shared by a group of organisms but not by a distant common ancestor.

shared primitive character A trait shared with a distant ancestor.

shattering The process by which ears of wild grain crops break apart and disperse seeds.

shell A tough, protective covering on an amniotic egg that is impermeable to water and prevents the embryo from drying out.

shivering thermogenesis Rapid muscle contractions in an animal, without any locomotion, in order to raise body temperature.

shoot The portion of a plant comprised of stems and leaves.

shoot apical meristem (SAM) The region of rapidly dividing plant cells at plant shoot apices.

shoot meristem The tissue that produces all aerial parts of the plant, which include the stems as well as lateral structures such as leaves and flowers.

shoot system The collection of plant organs produced by shoot apical meristems.

short-day plant A plant that flowers only when the night length is longer than a defined period.

short stature A condition characterized by stunted growth; formerly called pituitary dwarfism.

short tandem repeat sequences (STRs) Short sequences repeated many times in a row and found in multiple sites in the genome of humans and other species; often vary in length among different individuals.

shotgun DNA sequencing A strategy for sequencing an entire genome by randomly sequencing many different DNA fragments.

sickle-cell disease A disease due to a genetic mutation in a hemoglobin gene in which sickle-shaped red blood cells are less able to move smoothly through capillaries and can block blood flow, resulting in severe pain and cell death of the surrounding tissue.

sieve plate The perforated end wall of a mature sieve-tube element.

sieve plate pore One of many perforations in a plant's sieve plate.

sieve-tube elements A component of the phloem tissues of flowering plants; thin-walled cells arranged end to end to form transport pipes.

sigma factor A protein that plays a key role in bacterial promoter recognition and recruits RNA polymerase to the promoter.

signal Regarding cell communication, an incoming or outgoing agent that influences the properties of cells.

signal recognition particle (SRP) A protein/RNA complex that recognizes the ER signal sequence of a polypeptide, pauses translation, and directs the ribosome to the ER to complete translation.

signal transduction pathway A group of proteins that convert an initial signal to a different signal inside a cell.

sign stimulus In animals, a trigger that initiates a fixed-action pattern of behavior.

silencer A regulatory element in eukaryotes that prevents transcription of a given gene.

silencing RNAs (siRNAs) Small RNA molecules, typically 22 nucleotides in length, that silence the expression of specific mRNAs by promoting their degradation.

silent mutation A gene mutation that does not alter the amino acid sequence of the polypeptide, even though the nucleotide sequence has changed.

simple Mendelian inheritance The inheritance pattern of traits affected by a single gene that is found in two variants, one of which is completely dominant over the other.

simple translocation A type of mutation in which a single piece of chromosome is attached to another chromosome.

single-factor cross *See* monohybrid cross.

single nucleotide polymorphism (SNP) A type of genetic variation in a population in which a particular gene sequence varies at a single nucleotide.

single-strand binding protein A protein that binds to both of the single strands of parental DNA and prevents them from re-forming a double helix during DNA replication.

sinoatrial (SA) node A collection of modified cardiac cells in the right atrium of most vertebrates that spontaneously and rhythmically generates action potentials that spread across the entire atria; also known as the pacemaker of the heart.

siRNAS *See* silencing RNAs.

sister chromatids The two duplicated chromatids that are still joined to each other after DNA replication.

skeletal muscle A type of muscle tissue that is attached by tendons to bones in vertebrates and to the exoskeleton of invertebrates.

skeleton A structure or structures that serve one or more functions related to support, protection, and locomotion.

sliding filament mechanism The way in which a muscle fiber shortens during muscle contraction.

SLOSS debate In conservation biology, the debate over whether it is preferable to protect one single, large reserve or several smaller ones.

slow block to polyspermy Events initiated by the release of Ca^{2+} that produce barriers to more sperm penetrating an already fertilized egg.

slow fiber A skeletal muscle fiber containing myosin with a low rate of ATP hydrolysis.

slow-oxidative fiber A skeletal muscle fiber that has a low rate of myosin ATP hydrolysis but has the ability to make large amounts of ATP; used for prolonged, regular activity.

small effector molecule With regard to transcription, refers to a molecule that exerts its effects by binding to a regulatory transcription factor, causing a conformational change in the protein.

small intestine In vertebrates, a tube that leads from the stomach to the large intestine where nearly all digestion of food and absorption of food nutrients and water occur.

smooth endoplasmic reticulum (smooth ER) The part of the ER that is not studded with ribosomes. This region is continuous with the rough ER and functions in diverse metabolic processes such as detoxification, carbohydrate metabolism, accumulation of calcium ions, and synthesis and modification of lipids.

smooth muscle A type of muscle tissue that surrounds hollow tubes and cavities inside the body's organs; it is not under conscious control.

soil horizon Layers of soil, ranging from topsoil to bedrock.

solute A substance dissolved in a liquid.

solute potential (S) The component of water potential due to the presence of solute molecules.

solution A liquid that contains one or more dissolved solutes.

solvent The liquid in which a solute is dissolved.

soma *See* cell body.

somatic cell The type of cell that constitutes all cells of an animal or plant body except those that give rise to gametes.

somatic embryogenesis The production of plant embryos from body (somatic) cells.

somatic nervous system The division of the peripheral nervous system that senses the external environmental conditions and controls skeletal muscles.

somites   Blocklike structures resulting from the segmentation of mesoderm during neurulation.

soredia   An asexual reproductive structure produced by lichens consisting of small clumps of hyphae surrounding a few algal cells that can disperse in wind currents.

sorting signal   A short amino acid sequence in a protein that directs the protein to its correct location; also known as a traffic signal.

source pool   The pool of species on the mainland that is available to colonize an island.

spatial summation   Occurs when two or more postsynaptic potentials are generated at one time along different regions of the dendrites and their depolarizations and hyperpolarizations sum together.

speciation   The formation of new species.

species   A group of related organisms that share a distinctive form in nature and (for sexually reproducing species) are capable of interbreeding.

species-area effect   The relationship between the amount of available area and the number of species present.

species concepts   Different approaches for distinguishing species.

species diversity   A measure of biological diversity that incorporates both the number of species in an area and the relative distribution of individuals among species.

species interactions   A part of the study of population ecology that focuses on interactions such as predation, competition, parasitism, mutualism, and commensalism.

species richness   The numbers of species in a community.

specific heat   The amount of energy required to raise the temperature of 1 gram of a substance by 1°C.

specific (acquired) immunity   An immunity defense that develops only after the body is exposed to foreign substances; believed to be unique to vertebrates.

specificity   Refers to the concept that enzymes recognize specific substrates.

Spemann's organizer   An extremely important morphogenetic field in the early gastrula; the organizer secretes morphogens responsible for inducing the formation of a new embryonic axis.

spermatids   In animals, the haploid cells produced when the secondary spermatocytes undergo meiosis II; these cells eventually differentiate into sperm cells.

spermatogenesis   Gametogenesis in a male animal resulting in the production of sperm.

spermatogonia (singular, **spermatogonium**)   In animals, diploid germ cells that give rise to the male gametes, the spermatozoa.

spermatophytes   All of the living and fossil seed plant phyla.

sperm cell   Refers to a male gamete that is generally smaller than the female gamete (egg); also called a sperm.

sperm storage   A method of synchronizing the production of offspring with favorable environmental conditions in which female animals store and nourish sperm in their reproductive tract for long periods of time.

S phase   The DNA synthesis phase of the cell cycle.

spicules   Needle-like structures that are usually made of silica and form lattice-like skeletons in sponges, possibly helping to reduce predation.

spinal cord   In chordates, the structure that connects the brain to all areas of the body and together with the brain constitutes the central nervous system.

spinal nerve   A nerve that connects the peripheral nervous system and the spinal cord.

spiracle   A pore on the body surface of insects that leads to the trachea.

spiral cleavage   A mechanism of animal development in which the planes of cell cleavage are oblique to the axis of the embryo.

spirilli   Rigid, spiral-shaped prokaryotic cells.

spirochaetes   Flexible, spiral-shaped prokaryotic cells.

spliceosome   A complex of several subunits known as snRNPs that removes introns from eukaryotic pre-mRNA.

splicing   The process whereby introns are removed from RNA and the remaining exons are connected to each other.

spongin   A tough protein that lends skeletal support to a sponge.

spongocoel   A central cavity in the body of a sponge.

spongy parenchyma   Photosynthetic ground tissue of the plant leaf mesophyll that contains round cells separated by abundant air spaces.

spontaneous mutation   A mutation resulting from abnormalities in biological processes.

sporangia   Structures that produce and disperse the spores of plants, fungi, or protists.

spore   A haploid, typically single-celled reproductive structure of fungi and plants that is dispersed into the environment and is able to grow into a new fungal mycelium or plant gametophyte in a suitable habitat.

sporic life cycle   *See* alternation of generations.

sporophyte   The diploid generation of plants or multicellular protists that have a sporic life cycle; this generation produces haploid spores by the process of meiosis.

sporopollenin   The tough material that composes much of the walls of plant spores and helps to prevent cellular damage during transport in air.

stabilizing selection   A pattern of natural selection that favors the survival of individuals with intermediate phenotypes.

stamen   A flower organ that produces the male gametophyte, pollen.

standard metabolic rate (SMR)   The metabolic rate of ectotherms measured at a standard temperature for each species—one that approximates the average temperature that a species normally encounters.

standing crop   The total biomass in an ecosystem at any one point in time.

starch   A polysaccharide composed of repeating glucose units that is produced by the cells of plants and some algal protists.

start codon   A three-base sequence—usually AUG—that specifies the first amino acid in a polypeptide.

statocyst   An organ of equilibrium found in many invertebrate species.

statoliths   1. Tiny granules of sand or other dense objects located in a statocyst that aid equilibrium in many invertebrates. 2. In plants, a starch-heavy plastid that allows both roots and shoots to detect gravity.

stem   A plant organ that produces buds, leaves, branches, and reproductive structures.

stem cell   A cell that divides so that one daughter cell remains a stem cell and the other can differentiate into a specialized cell type. Stem cells construct the bodies of all animals and plants.

stereocilia   Deformable projections from epithelial cells called hair cells that are bent by movements of fluid or other stimuli.

stereoisomers   Isomers with identical bonding relationships, but different spatial positioning of their atoms.

sternum   The breastbone of a vertebrate.

steroid   A lipid containing four interconnected rings of carbon atoms; functions as a hormone in animals and plants.

steroid receptor   A transcription factor that recognizes a steroid hormone and usually functions as a transcriptional activator.

sticky ends   Single-stranded ends of DNA fragments that will hydrogen-bond to each other due to their complementary sequences.

stigma   In a flower, the topmost portion of the pistil, which receives and recognizes pollen of the appropriate species or genotype.

stomach   A saclike organ in some animals that most likely evolved as a means of storing food; it partially digests some of the macromolecules in food and regulates the rate at which the contents empty into the small intestine.

stomata   Surface pores on plant surfaces that can be closed to retain water or open to allow the entry of CO_2 needed for photosynthesis and the exit of oxygen and water vapor.

stop codon   One of three three-base sequences—UAA, UAG, and UGA—that signals the end of translation; also called termination codon or nonsense codon.

strain   Within a given species, a lineage that has genetic differences compared to another lineage.

strand   A structure of DNA (or RNA) formed by the covalent linkage of nucleotides in a linear manner.

strepsirrhini   Smaller species of primates; includes bush babies, lemurs, and pottos.

stretch receptor   A type of mechanoreceptor found widely in an animal's organs and muscle tissues that can be distended.

striated muscle   Skeletal and cardiac muscle with a series of light and dark bands perpendicular to the muscle's long axis.

stroke   The condition that occurs when blood flow to part of the brain is disrupted.

stroke volume (SV)   The amount of blood ejected with each beat, or stroke, of the heart.

stroma   The fluid-filled region of the chloroplast between the thylakoid membrane and the inner membrane.

stromatolite   A layered calcium carbonate structure in an aquatic environment generally produced by cyanobacteria.

strong acid   An acid that completely ionizes in solution.

structural gene   Refers to most genes, which produce an mRNA molecule that contains the information to specify a polypeptide with a particular amino acid sequence.

structural isomers   Isomers that contain the same atoms but in different bonding relationships.

style   In a flower, the elongate portion of the pistil through which the pollen tube grows.

stylet   A sharp, piercing organ in the mouth of nematodes and some insects.

submetacentric   A chromosome in which the centromere is off center.

subsidence zones   Areas of high pressure that are the sites of the world's tropical deserts because the subsiding air is relatively dry, having released all of its moisture over the equator.

subspecies   A subdivision of a species; this designation is used when two or more geographically restricted groups of the same species differ, but not enough to warrant their placement into separate species.

substrate   1. The reactant molecules and/or ions that bind to an enzyme at the active site and participate in a chemical reaction. 2. The organic compounds such as soil or rotting wood that fungi use as food.

substrate-level phosphorylation A method of synthesizing ATP that occurs when an enzyme directly transfers a phosphate from an organic molecule to ADP.

succession The gradual and continuous change in species composition and community structure over time.

sugar sink The plant tissues or organs in which more sugar is consumed than is produced by photosynthesis.

sugar source The plant tissues or organs that produce more sugar than they consume in respiration.

sum rule The probability that one of two or more mutually exclusive outcomes will occur is the sum of the probabilities of the possible outcomes.

supergroup One of the seven subdivisions of the domain Eukarya.

surface area/volume (SA/V) ratio The ratio between a structure's surface area and the volume in which the structure is contained.

surface tension A measure of how difficult it is to break the interface between a liquid and air.

surfactant A mixture of proteins and amphipathic lipids produced in certain alveolar cells that prevents the collapse of alveoli by reducing surface tension in the lungs.

survivorship curve A graphical plot of the numbers of surviving individuals at each age in a population.

suspension feeder An aquatic animal that sifts water, filtering out the organic matter and expelling the rest.

suspensor A short chain of cells at the base of an early angiosperm embryo that provides anchorage and nutrients.

swim bladder A gas-filled, balloon-like structure that helps a fish to remain buoyant in the water even when the fish is completely stationary.

symbiosis An intimate association between two or more organisms of different species.

symbiotic Describes a relationship in which two or more different species live in direct contact with each other.

sympathetic division The division of the autonomic nervous system that is responsible for rapidly activating body systems to provide immediate energy in response to danger or stress.

sympatric The term used to describe species occurring in the same geographic area.

sympatric speciation A form of speciation that occurs when members of a species that initially occupy the same habitat within the same range diverge into two or more different species.

symplast All of a plant's protoplasts (the cell contents without the cell walls) and plasmodesmata.

symplastic transport The movement of a substance from the cytosol of one cell to the cytosol of an adjacent cell via membrane-lined channels called plasmodesmata.

symplesiomorphy See shared primitive character.

symporter A type of transporter that binds two or more ions or molecules and transports them in the same direction across a membrane; also called a cotransporter.

synapomorphy See shared derived character.

synapse A junction where a nerve terminal meets a target neuron, muscle cell, or gland and through which an electrical or chemical signal passes.

synapsis The process of forming a bivalent.

synaptic cleft The extracellular space between a neuron and its target cell.

synaptic plasticity The formation of additional synaptic connections that occurs as a result of learning.

synaptic signaling A specialized form of paracrine signaling that occurs in the nervous system of animals.

synergids In the female gametophyte of a flowering plant, the two cells adjacent to the egg cell that help to import nutrients from maternal sporophyte tissues.

syntrophy The phenomenon in which one species lives off the products of another species.

systematics The study of biological diversity and evolutionary relationships among organisms, both extinct and modern.

systemic acquired resistance (SAR) A whole-plant defensive response to pathogenic microorganisms.

systemic circulation The pumping of blood from the left side of an animal's heart to the body to drop off oxygen and nutrients and pick up carbon dioxide and wastes. The blood then returns to the right side of the heart.

systemic hypertension An arterial blood pressure above normal; in humans, normal blood pressure ranges from systolic/diastolic pressures of about 90/60 to 120/80 mmHg; often called hypertension or high blood pressure.

systems biology A field of study in which researchers investigate living organisms in terms of their underlying networks—groups of structural and functional connections—rather than their individual molecular components.

systole The second phase of the cardiac cycle, in which the ventricles contract and eject the blood through the open semilunar valves.

T

tagmata The fusion of body segments into functional units.

taproot system The root system of eudicots, consisting of one main root with many branch roots.

taste buds Structures located in the mouth and tongue of vertebrates that contain the sensory cells, supporting cells, and associated neuronal endings that contribute to taste sensation.

TATA box One of three features found in most eukaryotic promoters; the others are the transcriptional start site and regulatory elements.

taxis A directed type of response to a stimulus that is either toward or away from the stimulus.

taxon A group of species that are evolutionarily related to each other. In taxonomy, each species is placed into several taxons that form a hierarchy from large (domain) to small (genus).

taxonomy The field of biology that is concerned with the theory, practice, and rules of classifying living and extinct organisms and viruses.

T cell A type of lymphocyte that directly kills infected, mutated, or transplanted cells.

telocentric A chromosome in which the centromere is at the end.

telomerase An enzyme that catalyzes the replication of the telomere.

telomere A region at the ends of eukaryotic chromosomes where a specialized form of DNA replication occurs.

telophase The phase of mitosis during which the chromosomes decondense and the nuclear membrane re-forms.

TEM See transmission electron microscopy.

temperate phage A bacteriophage that may spend some of its time in the lysogenic cycle.

template strand The DNA strand that is used as a template for RNA synthesis or DNA replication.

temporal lobe One of four lobes of the cerebral cortex of human brain; necessary for language, hearing, and some types of memory.

temporal summation Occurs when two or more postsynaptic potentials arrive at the same location in a dendrite in quick succession and their depolarizations and hyperpolarizations sum together.

tepal A flower perianth part that cannot be distinguished by appearance as a petal or a sepal.

termination codon See stop codon.

termination stage The final stage of transcription or translation in which the process ends.

terminator A sequence that specifies the end of transcription.

terpenoids A group of secondary metabolites synthesized from five-carbon isoprene units. An example is β-carotene, which gives carrots their orange color.

territory A fixed area in which an individual or group excludes other members of its own species, and sometimes other species, by aggressive behavior or territory marking.

tertiary consumer An organism that feeds on secondary consumers.

tertiary endosymbiosis The acquisition by eukaryotic protist host cells of plastids from cells that possess secondary plastids.

tertiary plastid A plastid acquired by the incorporation into a host cell of an endosymbiont having a secondary plastid.

tertiary structure The three-dimensional shape of a single polypeptide; one of four levels of protein structure.

testcross A cross to determine if an individual with a dominant phenotype is a homozygote or a heterozygote. Also, a cross to determine if two different genes are linked.

testes (singular, **testis**) In animals, the male gonads, where sperm are produced.

testosterone The primary androgen in many vertebrates, including humans.

tetrad See bivalent.

tetraploid An organism or cell that has four sets of chromosomes.

tetrapod A vertebrate animal having four legs or leglike appendages.

thalamus A region of the vertebrate forebrain that plays a major role in relaying sensory information to appropriate parts of the cerebrum and, in turn, sending outputs from the cerebrum to other parts of the brain.

theory In biology, a broad explanation of some aspect of the natural world that is substantiated by a large body of evidence. Biological theories incorporate observations, hypothesis testing, and the laws of other disciplines such as chemistry and physics. A theory makes valid predictions.

thermodynamics The study of energy interconversions.

thermoreceptor A sensory receptor in animals that responds to cold and heat.

theropods A group of bipedal saurischian dinosaurs.

thick filament A section of the repeating pattern in a myofibril composed almost entirely of the motor protein myosin.

thigmotropism Touch responses in plants.

thin filament A section of the repeating pattern in a myofibril that contains the cytoskeletal protein actin, as well as two other proteins—troponin and tropomyosin—that play important roles in regulating contraction.

30-nm fiber Nucleosome units organized into a more compact structure that is 30 nm in diameter.

thoracic breathing Breathing in which coordinated contractions of muscles expand the rib cage, creating a negative pressure to suck air in and then forcing it out later; found in amniotes.

threatened species Those species that are likely to become endangered in the future.

threshold concentration The concentration above which a morphogen will exert its effects but below which it is ineffective.

threshold potential The membrane potential, typically around -50mV, which is sufficient to trigger an action potential in an electrically excitable cell such as a neuron.

thrifty genes Genes that boosted our ancestors' ability to store fat from each feast in order to sustain them through the next famine.

thrombocytes Intact cells in the blood of vertebrates other than mammals that play a crucial role in the formation of blood clots; in mammals, cell fragments called platelets serve this function.

thylakoid A flattened, platelike membranous region found in cyanobacterial cells and the chloroplasts of photosynthetic protists and plants; the location of the light reactions of photosynthesis.

thylakoid lumen The fluid-filled compartment within the thylakoid.

thylakoid membrane A membrane within the chloroplast that forms many flattened, fluid-filled tubules that enclose a single, convoluted compartment. It contains chlorophyll and is the site where the light-dependent reactions of photosynthesis occurs.

thymine (T) A pyrimidine base found in DNA.

thymine dimer In DNA, a type of pyrimidine dimer that can cause a mutation; a site where two adjacent thymine bases become covalently cross-linked to each other.

thyroglobulin A protein found in the colloid of the thyroid gland that is involved in the formation of thyroid hormones.

thyroxine (T_4) A weakly active thyroid hormone that contains iodine and helps regulate metabolic rate; it is converted by cells into the more active triiodothyronine (T_3).

tidal ventilation A type of breathing in mammals in which the lungs are inflated with air and then the chest muscles and diaphragm relax and recoil back to their original positions as an animal exhales. During exhalation, air leaves via the same route that it entered during inhalation, and no new oxygen is delivered to the airways at that time.

tidal volume The volume of air that is normally breathed in and out at rest.

tight junction A type of junction between animal cells that forms a tight seal between adjacent epithelial cells and thereby prevents molecules from leaking between cells; also called an occluding junction.

Ti plasmid Tumor-inducing plasmid found in *Agrobacterium tumefaciens*; it is used as a cloning vector to transfer genes into plant cells.

tissue The association of many cells of the same type, for example, muscle tissue.

tolerance A mechanism for succession in which any species can start the succession, but the eventual climax community is reached in a somewhat orderly fashion; early species neither facilitate nor inhibit subsequent colonists.

tonoplast The membrane of the central vacuole in a plant or algal cell.

torpor The strategy in endotherms of lowering internal body temperature to just a few degrees above that of the environment in order to conserve energy.

torus The nonporous, flexible central region of a conifer pit that functions like a valve.

total fertility rate The average number of live births a female has during her lifetime.

total peripheral resistance (TPR) The sum of all the resistance in all arterioles.

totipotent The ability of a fertilized egg to produce all of the cell types in the adult organism; also the ability of unspecialized plant cells to regenerate an adult plant.

toxins Compounds that have adverse effects in living organisms; often produced by various protist and plant species.

trace element An element that is essential for normal function in living organisms but is required in extremely small quantities.

trachea 1. A sturdy tube arising from the spiracles of an insect's body; involved in respiration. 2. The name of the tube leading to the lungs of air-breathing vertebrates.

tracheal system The respiratory system of insects consisting of a series of finely branched air tubes called tracheae; air enters and exits the tracheae through spiracles, which are pores on the body surface.

tracheary elements Water-conducting cells in plants that, when mature, are always dead and empty of cytosol; include tracheids and vessel elements.

tracheid A type of dead, lignified plant cell in xylem that conducts water, along with dissolved minerals; also provides structural support.

tracheophytes A term used to describe vascular plants.

tract A parallel bundle of myelinated axons in the central nervous system.

traffic signal *See* sorting signal.

trait An identifiable characteristic; usually refers to a variant.

transcription The use of a gene sequence to make a copy of RNA.

transcriptional start site The site in a eukaryotic promoter where transcription begins.

transcription factor A protein that influences the ability of RNA polymerase to transcribe genes.

transduction A type of genetic transfer between bacteria in which a virus infects a bacterial cell and then subsequently transfers some of that cell's DNA to another bacterium.

***trans*-effect** In both prokaryotes and eukaryotes, a form of genetic regulation that can occur even though two DNA segments are not physically adjacent. The action of the lac repressor on the *lac* operon is a *trans*-effect.

transepithelial transport The process of moving solutes across an epithelium, such as in the gut of animals.

transfer RNA (tRNA) An RNA that carries amino acids and is used to translate mRNA into polypeptides.

transformation A type of genetic transfer between bacteria in which a segment of DNA from the environment is taken up by a competent cell and incorporated into the bacterial chromosome.

transgenic The term used to describe an organism that carries genes that were introduced using molecular techniques such as gene cloning.

transitional form An organism that provides a link between earlier and later forms in evolution.

transition state In a chemical reaction, a state in which the original bonds have stretched to their limit; once this state is reached, the reaction can proceed to the formation of products.

translation The process of synthesizing a specific polypeptide on a ribosome.

translocation 1. A type of mutation in which one segment of a chromosome becomes attached to a different chromosome. 2. A process in plants in which phloem transports substances from a source to a sink.

transmembrane gradient A situation in which the concentration of a solute is higher on one side of a membrane than on the other.

transmembrane protein A protein that has one or more regions that are physically embedded in the hydrophobic region of a cell membrane's phospholipid bilayer.

transmembrane segment A region of a membrane protein that is a stretch of nonpolar amino acids that spans or traverses the membrane from one leaflet to the other.

transmembrane transport The export of material from one cell into the intercellular space and then into an adjacent cell.

transmission electron microscopy (TEM) A type of microscopy in which a beam of electrons is transmitted through a biological sample to form an image on a photographic plate or screen.

transpiration The evaporative loss of water from plant surfaces into sun-heated air.

transporter A membrane protein that binds a solute and undergoes a conformational change to allow the movement of the solute across a membrane; also called a carrier.

transport protein Proteins embedded within the phospholipid bilayer that allow plasma membranes to be selectively permeable by providing a passageway for the movement of some but not all substances across the membrane.

transposable element (TE) A segment of DNA that can move from one site to another.

transposase An enzyme that facilitates transposition.

transposition The process in which a short segment of DNA moves within a cell from its original site to a new site in the genome.

transverse tubules (T-tubules) Invaginations of the plasma membrane of skeletal muscle cells that open to the extracellular fluid and conduct action potentials from the outer surface to the myofibrils.

triacylglycerol *See* triglyceride.

trichome A projection, often hairlike, from the epidermal tissue of a plant that offers protection from excessive light, ultraviolet radiation, extreme air temperature, or attack by herbivores.

triglyceride A molecule composed of three fatty acids linked by ester bonds to a molecule of glycerol; also known as a triacylglycerol.

triiodothyronine (T_3) A thyroid hormone that contains iodine and helps regulate metabolic rate.

triplet A group of three bases that function as a codon.

triploblastic Having three distinct germ layers—endoderm, ectoderm, and mesoderm.

triploid An organism or cell that has three sets of chromosomes.

trisomic An aneuploid organism that has one too many chromosomes.

tRNA *See* transfer RNA.

trochophore larva A distinct larval stage of many invertebrate phyla.

trophic level Each feeding level in a food chain.

trophic-level transfer efficiency The amount of energy at a trophic level that is acquired by the trophic level above and incorporated into biomass.

trophic mutualism A mutually beneficial interaction between two species in which both species receive the benefit of resources.

tropism In plants, a growth response that is dependent on a stimulus that occurs in a particular direction.

tropomyosin A rod-shaped protein that plays an important role in regulating muscle contraction.

troponin A small globular-shaped protein that plays an important role in regulating muscle contraction through its ability to bind Ca^{2+}.

trp **operon** An operon of *E. coli* that encodes enzymes required to make the amino acid tryptophan, a building block of cellular proteins.

true-breeding line A strain that continues to exhibit the same trait after several generations of self-fertilization or inbreeding.

trypsin A protease involved in the breakdown of proteins in the small intestine.

t-snare A protein in a target membrane that recognizes a v-snare in a membrane vesicle.

tubal ligation A means of contraception that involves the cutting and sealing of the fallopian tubes in a woman, thereby preventing movement of a fertilized egg into the uterus.

tube cell In a seed plant, one of the cells resulting from the division of a microspore; stores proteins and forms the pollen tube.

tube feet Echinoderm structures that function in movement, gas exchange, feeding, and excretion.

tumor An abnormal overgrowth of cells.

tumor-suppressor gene A gene that when normal (that is, not mutant) encodes a protein that prevents cancer; however, when a mutation eliminates its function, cancer may occur.

tunic A nonliving structure that encloses a tunicate, made of protein and a cellulose-like material called tunicin.

turgid The term used to describe a plant cell whose cytosol is so full of water that the plasma membrane presses right up against the cell wall; as a result, turgid cells are firm or swollen.

turgor pressure *See* osmotic pressure.

20-hydroxyecdysone A hormone produced by the prothoracic glands of arthropods that stimulates molting.

two-factor cross *See* dihybrid cross.

type 1 diabetes mellitus (T1DM) A disease in which the pancreas does not produce sufficient insulin; as a result, extracellular glucose cannot cross plasma membranes, and glucose accumulates to very high concentrations in the blood.

type 2 diabetes mellitus (T2DM) A disease in which the pancreas produces sufficient insulin, but the cells of the body lose much of their ability to respond to insulin.

U

ubiquitin A small protein in eukaryotic cells that directs unwanted proteins to a proteasome by its covalent attachment.

ultimate cause The reason a particular behavior evolved, in terms of its effect on reproductive success.

umbrella species A species whose habitat requirements are so large that protecting them would protect many other species existing in the same habitat.

unconditioned response An action that is elicited by an unconditioned stimulus.

unconditioned stimulus A trigger that elicits an original response.

uniform A pattern of dispersion within a population in which individuals maintain a certain minimum distance between themselves to produce an evenly spaced distribution.

uniporter A type of transporter that binds a single ion or molecule and transports it across a membrane.

unipotent A term used to describe a stem cell found in the adult that can produce daughter cells that differentiate into only one cell type.

unsaturated The quality of a lipid containing one or more C=C double bonds.

unsaturated fatty acid A fatty acid that contains one or more C=C double bonds.

upwelling In the ocean, a process that carries mineral nutrients from the bottom waters to the surface.

uracil (U) A pyrimidine base found in RNA.

urea A nitrogenous waste commonly produced in many terrestrial species, including mammals.

uremia A condition characterized by the presence of nitrogenous wastes, such as urea, in the blood; typically results from kidney disease.

ureter A structure in the mammalian urinary system through which urine flows from the kidney into the urinary bladder.

urethra The structure in the mammalian urinary system through which urine is eliminated from the body.

uric acid A nitrogenous waste produced by birds, insects, and reptiles.

urinary bladder The structure in the mammalian urinary system that collects urine before it is eliminated.

urinary system The structures that collectively act to filter blood or hemolymph and excrete wastes, while recapturing useful compounds.

urine The part of the filtrate formed in the kidney that remains after all reabsorption of solutes and water is complete.

uterine cycle *See* menstrual cycle.

uterus A small, pear-shaped organ capable of enlarging and specialized for carrying a developing fetus in female mammals.

V

vaccination The injection into the body of small quantities of weakened or dead pathogens, resulting in the development of immunity to those pathogens without causing disease.

vacuole Specialized compartments found in eukaryotic cells that function in storage, the regulation of cell volume, and degradation.

vagina The birth canal of female mammals; also functions to receive sperm during copulation.

vaginal diaphragm A barrier method of preventing fertilization in which a diaphragm is placed in the upper part of the vagina just prior to intercourse; blocks movement of sperm to the cervix.

valence electron An electron in the outer shell of an atom that is available to combine with other atoms. Such electrons allow atoms to form chemical bonds with each other.

van der Waals forces Attractive forces between molecules in close proximity to each other, caused by the variations in the distribution of electron density around individual atoms.

variable region A unique domain within an immunoglobulin that serves as the antigen-binding site.

vasa recta capillaries Capillaries in the medulla in the nephron of the kidney.

vascular bundle Primary plant vascular tissues that occur in a cluster.

vascular cambium A secondary meristematic tissue of plants that produces both wood and inner bark.

vascular plant A plant that contains vascular tissue. Includes all modern plant species except liverworts, hornworts, and mosses.

vascular tissue Plant tissue that provides both structural support and conduction of water, minerals, and organic compounds.

vas deferens A muscular tube through which sperm leave the epididymis.

vasectomy A surgical procedure in men that severs the vas deferens, thereby preventing the release of sperm at ejaculation.

vasoconstriction A decrease in blood vessel radius; an important mechanism for directing blood flow away from specific regions of the body.

vasodilation An increase in blood vessel radius; an important mechanism for directing blood flow to specific regions of the body.

vasotocin A peptide hormone that is responsible for regulating salt and water balance in the blood of nonmammalian vertebrates.

vector A type of DNA that acts as a carrier of a DNA segment that is to be cloned.

vegetal pole In triploblast organisms, the pole of the egg where the yolk is most concentrated.

vegetative growth The production of new nonreproductive tissues by the shoot apical meristem and root apical meristem during seedling development and growth of mature plants.

vein 1. In animals, a blood vessel that returns blood to the heart. 2. In plants, a bundle of vascular tissue in a leaf.

veliger In mollusks, a free-swimming larva that has a rudimentary foot, shell, and mantle.

ventilation The process of bringing oxygenated water or air into contact with a respiratory surface such as gills or lungs.

ventral Refers to the lower side of an animal.

ventricle In the heart, a chamber that pumps blood out of the heart.

venule A small, thin-walled extension of a capillary that empties into larger vessels called veins that return blood to the heart for another trip around the circulation.

vertebra A bony or cartilaginous column of interlocking structures that provides support and also protects the nerve cord, which lies within its tubelike structure.

vertebrate An organism with a backbone.

vertical evolution A process in which species evolve from pre-existing species by the accumulation of mutations.

vesicle A small membrane-enclosed sac within a cell.

vessel In a plant, a pipeline-like file of dead, water-conducting vessel elements.

vessel element A type of plant cell in xylem that conducts water, along with dissolved minerals and certain organic compounds.

vestibular system The organ of balance in vertebrates, located in the inner ear next to the cochlea.

vestigial structure An anatomical feature that has no apparent function but resembles a structure of a presumed ancestor.

vibrios Comma-shaped prokaryotic cells.

villi Finger-like projections extending from the luminal surface into the lumen of the small intestine; these are specializations that aid in digestion and absorption.

viral envelope A structure enclosing a viral capsid that consists of a membrane derived from the plasma membrane of the host cell; is embedded with virally encoded spike glycoproteins.

viral genome The genetic material of a virus.

viral reproductive cycle The series of steps that result in the production of new viruses during a viral infection.

viral vector A type of vector used in cloning experiments that is derived from a virus.

viroid An RNA particle that infects plant cells.

virulence The ability of a microorganism to cause disease.

virulent phage A phage that follows only the lytic cycle.

virus A small infectious particle that consists of nucleic acid enclosed in a protein coat.

visceral mass In mollusks, a structure that rests atop the foot and contains the internal organs.

vitamin An organic nutrient that serves as a coenzyme for metabolic and biosynthetic reactions.

vitamin D A vitamin that is converted into a hormone in the body; regulates the calcium level in the blood through an effect on intestinal transport of calcium ions.

vitreous humor A thick liquid in the large posterior cavity of the vertebrate eye, which helps maintain the shape of the eye.

viviparity Development of an embryo within the mother, resulting in a live birth.

viviparous The term used to describe an animal whose embryos develop within the uterus, receiving nourishment from the mother via a placenta.

V_{max} The maximal velocity of an enzyme-catalyzed reaction.

volt A unit of measurement of potential difference in charge (electrical force) such as the difference between the interior and exterior of a cell.

voltage-gated ion channels Ion channels that open and close in response to changes in the amount of electric charge across a membrane.

v-snare A protein incorporated into a vesicle membrane during vesicle formation that is recognized by a t-snare in a target membrane.

W

water potential The potential energy of water.

water vascular system A network of canals powered by water pressure generated by the contraction of muscles; enables extension and contraction of the tube feet, allowing echinoderms to move slowly.

wavelength The distance from the peak of one sound wave or light wave to the next.

waxy cuticle A protective, waterproof layer of polyester and wax present on most surfaces of vascular plant sporophytes.

weak acid An acid that only partially ionizes in solution.

weathering The physical and chemical breakdown of rock.

white blood cell *See* leukocyte.

white matter Brain tissue that consists of myelinated axons that are bundled together in large numbers to form tracts.

whorls In a flower, concentric rings of sepals and petals (or tepals), stamens, and carpels.

wild-type allele One or more prevalent alleles in a population.

wood A secondary plant tissue composed of numerous pipelike arrays of dead, empty, water-conducting cells whose walls are strengthened by an exceptionally tough secondary metabolite known as lignin.

woody plant A type of plant that produces both primary and secondary vascular tissues.

X

xenoestrogen A synthetic compound that exerts estrogen-like actions or, in some cases, inhibits the actions of the body's own estrogen.

X inactivation The phenomenon in which one X chromosome in the somatic cells of female mammals is inactivated, meaning that its genes are not expressed.

X inactivation center (Xic) A short region on the X chromosome known to play a critical role in X inactivation.

X-linked gene A gene found on the X chromosome but not on the Y.

X-linked inheritance The pattern displayed by pairs of dominant and recessive alleles located on X chromosomes.

X-ray crystallography A technique in which researchers purify molecules and cause them to form a crystal. When a crystal is exposed to X-rays, the resulting pattern can be analyzed mathematically to determine the three-dimensional structure of the crystal's components.

xylem A specialized conducting tissue in plants that transports water, minerals, and some organic compounds.

xylem loading The process by which root xylem parenchyma cells transport ions and water across their membranes into the long-distance conducting cells of the xylem, which include the vessel elements and tracheids.

Y

yeast A fungus that can occur as a single cell and that reproduces by budding.

yolk sac One of the four extraembryonic membranes in the amniotic egg. The yolk sac encloses a stockpile of nutrients, in the form of yolk, for the developing embryo.

Z

zero population growth The situation in which no changes in population size occur.

Z line A network of proteins in a myofibril that anchors thin filaments at the ends of each sarcomere.

zona pellucida The glycoprotein covering that surrounds a mature oocyte.

zone of elongation The area above the root apical meristem of a plant where cells extend by water uptake, thereby dramatically increasing root length.

zone of maturation The area above the zone of elongation in a plant root where root cell differentiation and tissue specialization occur.

zooplankton Aquatic organisms drifting in the open ocean or fresh water; includes minute animals consisting of some worms, copepods, tiny jellyfish, and the small larvae of invertebrates and fishes.

Z scheme A model depicting the series of energy changes of an electron during the light reactions of photosynthesis. The electron absorbs light energy twice, resulting in an energy curve with a zigzag shape.

zygomycete A phylum of fungi that produces distinctive, large zygospores as the result of sexual reproduction.

zygospore A dark-pigmented, thick-walled spore that matures within the zygosporangium of zygomycete fungi during sexual reproduction.

zygote A diploid cell formed by the fusion of two haploid gametes.

zygotic life cycle The type of life cycle of most unicellular protists in which haploid cells develop into gametes. Two gametes then fuse to produce a diploid zygote.

GLOSSARY

Photo Credits

Chapter 1: Opener © Gary Kavanagh; 1.1: © Photo W. Wüster, courtesy Instituto Butantan; 1.2: © blickwinkel/Alamy; 1.3: © SciMAT/Photo Researchers; 1.4: © Robert Winslow/Animals Animals; 1.5a: © Biophoto Associates/Photo Researchers; 1.5b: © Alexis Rosenfeld/Photo Researchers; 1.5c: © Cathlyn Melloan/Stone/Getty Images; 1.5d: © Adam Jones/Visuals Unlimited; 1.5e: © Patti Murray/Animals Animals; 1.5f: © Paul Hanna/Reuters/Corbis; 1.5g: © AP Photo/Mehgan Murphy, National Zoo; 1.7: © Michael L. Smith/Photo Researchers; 1.12a: © Dr. David M. Phillips/Visuals Unlimited; 1.12b: © B. Boonyaratanakornkit & D.S. Clark, G. Vrdoljak/EM Lab, U of C Berkeley/Visuals Unlimited; 1.12c(1): © Dr. Dennis Kunkel/Visuals Unlimited; 1.12c(2): © Kent Foster/Photo Researchers; 1.12c(3): © Carl Schmidt-Luchs/Photo Researchers; 1.12c(4): © Fritz Polking/Visuals Unlimited; 1.15a: © Sierra Blakely; 1.15b: © Erkang Ai and Ahna Skop, 2009; 1.16a: © Fred Bavendam/Minden Pictures; 1.16b: © Eastcott/Momatiuk/Animals Animals; 1.16c-e: © Andrew Brookes/Corbis; 1.16e(inset): © Alfred Pasieka/Photo Researchers; 1.17a: © Age fotostock/SuperStock; 1.17b: © Laurent/Photo Researchers; 1.20: © AP Photo/Dita Alangkara.

Chapter 2: Opener © Kevin Schafer/Peter Arnold; 2.6: © The McGraw-Hill Companies, Inc./Al Tesler, photographer; 2.12b: © Charles D. Winters/Photo Researchers; 2.17: © Jeremy Burgess/Photo Researchers; 2.19a: © Norbert Wu/Peter Arnold; 2.19b: © Jerome Wexler/Visuals Unlimited; 2.20b: © Aaron Haupt/Photo Researchers; 2.20d: © Rainer Drexel/Bilderberg/Peter Arnold; 2.20e: © Ethel Davies/Imagestate; 2.20f: © Anthony Bannister/Gallo Images/Corbis; 2.20g: © OSF/Photolibrary.

Chapter 3: Opener © de Vos, A. M., Ultsch, M., Steitz, A. A., (1992). "Human growth hormone and extracellular domain of its receptor: crystal structure of the complex. *Science*. 255 (306). Image Courtesy of Daniel Gage, University of Connecticut; 3.1: © The McGraw-Hill Companies, Inc./Al Tesler, photographer; 3.11a-b: © Tom Pantages; 3.11b: © Felicia Martinez/PhotoEdit; 3.13b: © Adam Jones/Photo Researchers; 3.13c: © Adam Jones/Photo Researchers.

Chapter 4: Opener © Biophoto Associates/Photo Researchers; 4.2a-b: © Images courtesy of Molecular Expressions; 4.3a(inset): © Dr. Donald Fawcett & L. Zamboni/Visuals Unlimited; 4.3b-4b: © Dr. Dennis Kunkel Microscopy/Visuals Unlimited; 4.6a: © Ed Reschke/Peter Arnold; 4.6b: © Eye of Science/Photo Researchers; p. 74 (left): © Thomas Deerinck/Getty Images; p. 74 (middle, left): © Dr. Gopal Murti/Visuals Unlimited; 4.12a: © Charles Brokaw/California Institute of Technology; 4.12b: © Brian P. Piasecki; 4.12c: © SPL/Photo Researchers; 4.13a: © Aaron Bell; 4.13b-c: © Dr. William Dentler/University of Kansas; 4.15(inset 1): © Dr. Don W. Fawcett/Visuals Unlimited; 4.16: Reprinted by permission from Macmillan Publishers Ltd. Cremer, T. & Cremer, C. Chromosome territories, nuclear architecture and gene regulation in mammalian cells. *Nature Reviews/Genetics*, Vol. 2(4), Figure 2, 292–301, 2001; 4.17(right): © Dennis Kunkel Microscopy, Inc/Phototake; 4.19: Reproduced from *Journal of Cell Biology*, 1964, 20:473-495. Copyright 1964 Rockefeller University Press; 4.20a: © E. H. Newcomb & S. E. Frederick/Biological Photo Service; 4.20b: © Courtesy Dr. Peter Luykx, Biology, University of Miami; 4.20c: © Dr. David Patterson/Photo Researchers, Inc.; 4.21: © The McGraw-Hill Companies, Inc./Al Tesler, photographer; 4.24: © Dr. Don W. Fawcett/Visuals Unlimited; 4.25: © Dr. Jeremy Burgess/Photo Researcher; 4.26: © T. Kanaseki and Donald Fawcett/Visuals Unlimited.

Chapter 5: Opener © Tom Pantages; 5.6: © Biochemistry: The Molecular Basis Of Life. 3rd Ed. by McKee and McKee. McGraw Hill. Courtesy of Audrey M. Glauert and G. M. W. Cook; 5.7a-b: © The McGraw-Hill Companies, Inc./Al Tesler, photographer; 5.17a-b: © James Strawser/Grant Heilman Photography; 5.18a: © Carolina Biological Supply/Visuals Unlimited; 5.18b: © Carolina Biological Supply/Visuals Unlimited; 5.19(4): © Courtesy Dr. Peter Agre. From Preston GM, Carroll TP, Guggino WP, Agre P (1992). Appearance of water channels in Xenopus oocytes expressing red cell CHIP28 protein. *Science*. 256:385.

Chapter 6: Opener © Patrick Olear/Photo Edit; 6.1a: © Bob Daemmrich/The Image Works; 6.1b: © amanaimages/Corbis RF; 6.9(5): © Altman, S., (1990). Nobel Lecture: Enzymatic Cleavage of RNA by RNA. *Bioscience Reports*, 10, 317–337; 6.14: © Liu, Q., Greimann, J. C., and Lima, C. D., (2006). Reconstitution, activities, and structure of the eukaryotic exosome. *Cell*, 127, 1223-1237. Graphic generated using DeLano, W. L. (2002). The PyMOL Molecular Graphics System (San Carlos, CA, USA, DeLano Scientific).

Chapter 7: Opener © PCN Photography/Duomo/Corbis; 7.11(5): © Reprinted by permission from Macmillan Publishers Ltd. Noji, H., Yasuda, R., Yoshida, M., Kinosita K., (1997). Direct Observation of the Rotation of F1-ATPase. *Nature*, 386, (6622), 299-302; 7.12: © Custom Medical Stock Photo; 7.13: © Ernie Friedlander/Cole Group/Getty Images RF; 7.15a(left): © Bill Aron/Photo Edit; 7.15b: © Jeff Greenberg/The Image Works; 7.16a: © Chris Hellier/SPL/Photo Researchers; 7.16b: © Michael P. Gadomski/Photo Researchers; 7.17: © Naturfoto Honal/Corbis; 7.18: © James Randklev/Photographer's Choice/Getty Images; 7.19: © SciMAT/Photo Researchers.

Chapter 8: Opener © Wolfgang Kaehler/Alamy; 8.2(top): © Norman Owen Tomalin/Bruce Coleman Inc./PhotoShot; 8.2(middle): © J. Michael Eichelberger/Visuals Unlimited; 8.2(bottom): © Visuals Unlimited; 8.12b: © Reproduction of Fig 1A from Ferreira, K. N., Iverson, T. M., Maghlaoui, K., Barber, J. and Iwata, S. (2004). Architecture of the photosynthetic oxygen evolving center. *Science*. 303, 1831-1838 with permission; 8.16 (step 6): © Elin Sowle. Reprinted from Chemistry 1942-1962, M. Calvin, "The Path of Carbon in Photosynthesis." *Nobel Lectures*, Elsevier Publishing Company. Copyright (1961); 8.17a: © David Norton Photography/Alamy; 8.17b: © David Sieren/Visuals Unlimited; 8.19(left): © Walter H. Hodge/Peter Arnold; 8.19(right): © John Foxx/Alamy RF.

Chapter 9: Opener © David McCathy/SPL/Photo Researchers; 9.15(both): © Courtesy of Brian J. Bacskai, from Bacskai et al., *Science*. 260:222-226, 1993. Reprinted with permission from AAAS; 9.18(all): © Prof. Guy Whitely/Reproductive and Cardiovascular Disease Research Group at St George's, University of London; 9.19: © Wyllie, A. H., Kerr J. F. R., Currie A. R., (1973). Adrenocortical cell deletion: The role of ACTH. *Journal of Pathology*. 111, 85–94.

Chapter 10: Opener © Altrendo Panoramic/Getty Images; 10.1a: © Courtesy of Dr. Joseph Buckwalter/University of Iowa; 10.1b: © J. Gross/Biozentrum, University of Basel/Photo Researchers RF; 10.6(inset): © Dr. Dennis Kunkel Microscopy/Visuals Unlimited; 10.10: © Dr. Daniel Friend; 10.11: © Courtesy Dr. Dan Goodenough/Harvard Medical School; 10.13: © Lee W. Wilcox; 10.14: © E. H. Newcomb & W. P. Wergin/Biological Photo Service; 10.19a: © Carolina Biological Supply Company/Phototake; 10.19b: © Biodisc/Visuals Unlimited; 10.20: © Robert Brons/Biological Photo Service; 10.21: © J. N. A. Lott, McMaster Univ./Biological Photo Service.

Chapter 11: Opener © Jean Claude Revy/ISM/Phototake; 11.3b: © Eye of Science/Photo Researchers; 11.13: © Meselson M. Stahl, F. The Replication of DNA in Escherichia Coli. *PNAS*. Vol. 44, 1958. f. 4a, p. 673; 11.24: © Photo courtesy of Dr. Barbara Hamkalo; 11.26a: © Dr. Gopal Murti/Visuals Unlimited; 11.26b: © Ada L. Olins and Donald E. Olins/Biological Photo Service; 11.26c: © Courtesy Dr. Jerome B. Rattner, Cell Biology and Anatomy, University of Calgary; 11.26d: © Courtesy of Paulson, J. R. & Laemmli, U.K. This article was published in *Cell*. Vol. 12, James R. Paulson and U.K. Laemmli, "The Structure of Histone-Depleted Metaphase Chromosomes," pp. 817-28. Copyright Elsevier 1977; 11.26e-f: © Peter Engelhardt/Department of Virology, Haartman Institute.

Chapter 12: Opener © Kiseleva and Donald Fawcett/Visuals Unlimited; 12.6: © Darst, S., (2001). Bacterial RNA polymerase. *Current Opinion in Structural Biology*. 11 (2), 155-162.

Chapter 13: Opener © Daniel Gage, University of Connecticut. Appeared in Schultz, S. C., Shields, G. C. and T. A. Steitz. (1991). "Crystal structure of a CAP-DNA complex: the DNA is bent by 90 degrees," *Science*. 253:1001; 17.9b: © Courtesy of I. Solovei, University of Munich (LMU); 13.2a: © Ed Reschke/Peter Arnold; 13.2b: © Triarch/Visuals Unlimited; 13.2c: © SIU BioMed/Custom Medical Stock Photo; 13.10: © Thomas Steitz, Howard Hughes Medical Institution, Yale.

Chapter 14: Opener © Yvette Cardozo/Workbook Stock/Getty Images; 14.1a-b: © Stan Flegler/Visuals Unlimited; 14.1c: © Courtesy of Thomas Wellems and Robert Josephs. Electron microscopy and Image Processing Laboratory, University of Chicago; 14.4: © Scott Aitken, www.scottpix.com; 14.9: © Dr. Kenneth Greer/Visuals Unlimited; 14.10b: © St. Bartholomew's Hospital/Photo Researchers; 14.17(all): © Dr. Oscar Auerbach, reproduced with permission; 14.18(both): © Courtesy Dr. Ruhong Li, Molecular and Cell Biology, University of California at Berkeley.

Chapter 15: Opener © Biophoto Associates/Photo Researchers; 15.1(4): © Burger/Photo Researchers;

PHOTO CREDITS

15.1(5): © Courtesy of the Genomic Centre for Cancer Research and Diagnosis, CancerCare Manitoba, University of Manitoba, Winnipeg, Manitoba, Canada; 15.3a: © Don W. Fawcett/Photo Researchers; 15.9a-f: © Dr. Conly L. Rieder/Wadsworth Center, Albany, N.Y.; 15.10a: © Visuals Unlimited; 15.10b: © Carolina Biological Supply Company/Phototake; 15.19a-b: © A. B. Sheldon; 15.20a: © Sylvan Wittwer/Visuals Unlimited; 15.20b: © B. Runk/S. Schoenberger/Grant Heilman Photography.

Chapter 16: Opener © Reuters/Radu Sigheti; 16.1: © SPL/Photo Researchers; 16.3: © Nigel Cattlin/Photo Researchers; 16.18: © Chabruken/Photographer's Choice/Getty Images; 16.20: © March of Dimes.

Chapter 17: Opener © Dave King/Dorling Kindersley/Getty Images; 17.9a: © Courtesy of I. Solovei, University of Munich (LMU); 17.11: © Courtesy of Dr. Argiris Efstratiadis; 17.13: John Mendenhall, Judy Lundelius, and Gary Freeman.

Chapter 18: Opener © CAMR/A. Barry Dowsett/Photo Researchers; 18.1: © Norm Thomas/Photo Researchers; 18.3a: © Robley C. Williams/Biological Photo Service; 18.3b: © Courtesy of R. C. Valentine and H. G. Pereira. Reprinted from *Journal of Molecular Biology*, Vol. 13, No. 2, R. C. Valentine and H. G. Pereira, "Antigens and Structure of the Adenovirus," pages 71-83, 1965, with permission from Elsevier; 18.3c: © Chris Bjornberg/Photo Researchers; 18.3d: © Omikron/Photo Researchers; 18.6: © CDC/Peter Arnold; 18.7: © Photograph by J. W. Randles, at Albany Research Center, Philippines; 18.8: © Eye of Science/Photo Researchers; 18.9: © Carl Robinow & Eduard Kellengerger. The Bacterial Nucleoid Revisited. *Microbiological Reviews*, 1994 June, 58(2):211-32. Fig. 4. © American Society of Microbiology; 18.12: © Stanley Cohen/Photo Researchers; 18.13: © Dr. Jeremy Burgess/SPL/Photo Researchers; 18.17a: © Dr. L. Caro/SPL/Photo Researchers.

Chapter 19: Opener © Medical-on-Line/Alamy; 19.1a: © Herman Eisenbeiss/Photo Researchers; 19.1b: © Sinclair Stammers/Photo Researchers; 19.1c: © J-M. Labat/Photo Researchers; 19.1d: © Mark Smith/Photo Researchers; 19.1e: © Nigel Cattlin/Photo Researchers; 19.4a(all): © Courtesy of the National Museum of Health and Medicine, Washington, D.C.; 19.6(top): © Courtesy of the National Museum of Health and Medicine, Washington, D.C.; 19.6(center): © Congenital Anomaly Research Center of Kyoto University; 19.6(bottom): © Courtesy of the National Museum of Health and Medicine, Washington, D.C.; 19.9b-c: © Christiane Nusslein-Volhard, Development, Supplement 1, 1991; 19.11(top center): © Christiane Nusslein-Volhard, Development, Supplement 1, 1991; 19.11(top right, bottom): © Jim Langeland, Steve Paddock and Sean Carroll/University of Wisconsin–Madison; 19.12a-b: © Courtesy of the Archives, California Institute of Technology; 19.14a: © Juergen Berger/Photo Researchers; 19.14b: © F. R. Turner, Indiana University; 19.24a: © Darwin Dale/Photo Researchers; 19.24b-c: © John Bowman.

Chapter 20: Opener © Corbis; 20.13a: © Bill Barksdale/Agestockusa/Age fotostock; 20.14(6): © Roslin Institute/Phototake; 20.15a: © Leonard Lessin/Peter Arnold.

Chapter 21: Opener © Coston Stock/Alamy; 21.2b: © The Picture Store/SPL/Photo Researchers;

21.2c: © Photo by Michael Beckmann, Institute of Geobotany and Botanical Garden, Halle, Germany; 21.4a: © Topham/The Image Works; 21.4b: © Jerome Wexler/Visuals Unlimited.

Chapter 22: Opener © George Bernard/SPL/Photo Researchers; 22.1a: © Stanley M. Awramik/Biological Photo Service; 22.1b: © Michael Abbey/Visuals Unlimited; 22.3b: © Woods Hole Oceanographic Institution/Visuals Unlimited; 22.4a: © A. I. Oparin. From the *Origin of Life*, New York: Dover, 1952; 22.4b: © Mary Kraft; 22.7: © Simon Fraser/SPL/Photo Researchers; 22.11a: © Dirk Wiersma/SPL/Photo Researchers; 22.11b: © Roger Garwood & Trish Ainslie/Corbis; 22.13a: © Dr. Cristian A. Solari, Department of Ecology and Evolutionary Biology, University of Arizona; 22.13b: © Carolina Biological Supply Company/Phototake; 22.13c-d: © Dr. Cristian A. Solari, Department of Ecology and Evolutionary Biology, University of Arizona; 22.14: © We thank Prof. Jun-yuan Chen for permission to use this image; 22.15a: © Newman & Flowers/Photo Researchers; 22.15b: © Courtesy of the Smithsonian Institution, Photograph by Chip Clark. Marrella Splendens, most commonly occurring fossil from the Burgess Shale. Collected by Charles D. Walcott for the Smithsonian Research Collections; 22.16a: © Francois Gohier/Photo Researchers; 22.16b: © B. Runk/S. Schoenberger/Grant Heilman Photography; 22.17(right): © Ken Lucas/Visuals Unlimited; 22.19: © Jason Edwards/Getty Images RF.

Chapter 23: Opener © Mark Dadswell/Getty Images; 23.1a: © Rainer Heubeck/Peter Arnold; 23.3a: © Worldwide Picture Library/Alamy; 23.3b: © Gerald & Buff Corsi/Visuals Unlimited; 23.7a: © Wm. Grenfell/Visuals Unlimited; 23.7b: © Stephen J. Krasemann/Photo Researchers; 23.8a(Left): © Tom Brakefield/Corbis; 23.8a(Right): © Tom McHugh/Photo Researchers; 23.8b(Left): © David Sieren/Visuals Unlimited; 23.8b(Right): © 2003 Steve Baskauf - bioimages.vanderbilt.edu; 23.8c(Left): © Herb Segars/Animals Animals; 23.8c(Right): © David Wrobel/SeaPics.com; 23.9a: © Robert Smith/Grant Heilman; 23.9b: © Martin Rugner/Age fotostock; 23.9c: © Gerard Lacz/Peter Arnold.

Chapter 24: Opener © Dynamic Graphics Group/IT StockFree/Alamy RF; 24.1: © Paul Harcourt Davies/SPL/Photo Researchers; 24.5a: © Courtesy Mark McNair/University of Exeter; 24.7a: © Gerald Cubitt; 24.7b: © Topham/The Image Works; 24.7c: © Courtesy Darren Croft/University of Exeter; 24.8a-b: © Ole Seehausen.

Chapter 25: Opener © Frederic B. Siskind; 25.1a(Left): © Mark Smith/Photo Researchers; 25.1a(Right): © Pascal Goetgheluck/ardea.com; 25.1b(Left): © Gary Meszaros/Visuals Unlimited; 25.1b(Right): © R. Andrew Odum/Peter Arnold; 25.3a: © Bryan E. Reynolds; 25.3b: © C. Allan Morgan/Peter Arnold; 25.4a: © Rod Planck/Photo Researchers; 25.4b: © Ron Austing/Photo Researchers; 25.5(bottom): © Stephen L. Saks/Photo Researchers; 25.5(top left): © Mark Boulton/Photo Researchers; 25.5(top right): © Carolina Biological Society/Visuals Unlimited; 25.6(Left): © Hal Beral/V&W/imagequestmarine.com; 25.6(Right): © Guillen Photography/Alamy; 25.7(Left): © FLPA/Alamy; 25.7b(Bottom Center): © Jim Denny; 25.7b(all): © Jack Jeffrey Photography; 25.12: © Dr. Sara Via, Department of Biology and Department of Entomology, University of Maryland; 25.14a(all): © Courtesy Ed Laufer; 25.14b-c(all): © Courtesy of

Dr. J. M. Hurle. Originally published in *Development*. 1999 Dec; 126(23):5515-22; 25.16b: © David Cavagnaro/Visuals Unlimited; 25.17a-b: © Prof. Walter J. Gehring, University of Basel.

Chapter 26: Opener © Peter Weimann/Animals Animals.

Chapter 27: Opener © Dr. Jeremy Burgess/SPL/Photo Researchers; p. 548 top: © Dr. David M. Phillips/Visuals Unlimited; p. 548 middle: © Michael J. Daly, Uniformed Services University of the Health Sciences, Bethesda, MD, U.S.A.; p. 548 middle: © Wim van Egmond/Visuals Unlimited/Getty Images; p. 548 (bottom): © Prof. Dr. Hans Reichenbach/GBF; 27.2: © saturno dona'/Alamy; 27.2: © David Parker/SPL/Photo Researchers; T27.1d: © Prof. Dr. Hans Reichenbach/GBF; 27.3: © B. Runk/S. Schoenberger/Grant Heilman Photography; 27.4a: © The McGraw-Hill Companies, Inc./Linda Graham, photographer; 27.4b: © Wim van Egmond/Visuals Unlimited; 27.4c: © Michael Abbey/Visuals Unlimited; 27.4d: © Lee W. Wilcox; 27.5: © H. Stuart Pankratz/Biological Photo Service; 27.6: © Dr. Richard P. Blakemore, University of New Hampshire; 27.7a: © SciMAT/Photo Researchers; 27.7b-d: © Dennis Kunkel Microscopy; 27.8a: © Science Photo Library/Alamy RF; 27.10a: © CNRI/Photo Researchers, Inc.; 27.10b: © Lee W. Wilcox; 27.11: © Fig. 1 from "A Fluorescent Gram Stain for Flow Cytometry and Epifluorescence Microscopy" *Appl Environ Microbiol*, July 1998, pp. 2681-2685, Vol. 64, No. 7. American Society for Microbiology; 27.13a: © Dennis Kunkel Microscopy/Phototake; 27.13b: © Scientifica/Visuals Unlimited; 27.14: © Dr. T. J. Beveridge/Visuals Unlimited; 27.15a: © Scientifica/RML/Visuals Unlimited; 27.15b: © Cytographics/Visuals Unlimited; 27.15c: © Lee W. Wilcox; 27.16a: © Lee W. Wilcox; 27.16b: © Dr. Kari Lounatmaa/Photo Researchers; 27.17: © Roger Ressmeyer/Corbis.

Chapter 28: Opener 28.1: © Photographs by H. Cantor-Lund reproduced with permission of the copyright holder J. W. G. Lund; 28.2a: © Brian P. Piasecki; 28.2b: © Roland Birke/Phototake; 28.2c-d: © The McGraw-Hill Companies, Inc./Linda Graham, photographer; 28.2e: © The McGraw-Hill Companies, Inc./Claudia Lipke, photographer; 28.3: © Dr. Dennis Microscopy/Visuals Unlimited; 28.4: © Dr. Stanley Flegler/Visuals Unlimited; 28.7a: © David Phillips/Mediscan; 28.7b: © Jerome Paulin/Visuals Unlimited; 28.8a: © Wim van Egmond/Visuals Unlimited; 28.8b: © Ross Waller, University of Melbourne, Australia; 28.8c: © Eye of Science/Photo Researchers; 28.10a: © Dr. Brandon D. Cole/Corbis; 28.10b: © Walter H. Hodge/Peter Arnold, Inc.; 28.11: © Joe Scott, Department of Biology, College of William and Mary, Williamsburg, VA 23187; 28.13a: © Dennis Kunkel Microscopy; 28.13b: © Steve Gschmeissner/Photo Researchers; 28.13c: © Bob Krist/Corbis; 28.14: © Lee W. Wilcox; 28.16a-b: © Lee W. Wilcox; 28.16c: © Claude Taylor, III; 28.18a: © Dennis Kunkel Microscopy/Visuals Unlimited; 28.18b: © Walter H. Hodge/Peter Arnold; 28.19a: © Claude Nuridsany & Marie Perennou/SPL/Photo Researchers; 28.19b: © Manfred Kage/Peter Arnold; 28.21: © Stephen Fairclough, King Lab, University of California at Berkeley; 28.22: © Lee W. Wilcox; 28.24: © The McGraw-Hill Companies, Inc./Linda Graham, photographer.

Chapter 29: Opener © Craig Tuttle/Corbis; 29.1a: © Roland Birke/Phototake; 29.1b: © The McGraw-Hill Companies, Inc./Linda Graham, photographer;

29.1c-e: © Lee W. Wilcox; 29.1f: © B. Runk/S. Schoenberger/Grant Heilman Photography; 29.1g: © Ed Reschke/Peter Arnold; 29.1h: © Patrick Johns/CORBIS; 29.1i: © Bob Evans/Peter Arnold; 29.1j: © Wolfgang Kaehler/Corbis; 29.1k: © Fred Bruemmer/Peter Arnold; 29.1l: © Gallo Images/Corbis; 29.2a(chara): © Lee W. Wilcox; 29.2a(Coleochaete): © The McGraw-Hill Companies, Inc./Linda Graham, photographer; 29.2b(Mesostigma): © Lee W. Wilcox; 29.2b(Micrasterias): © The McGraw-Hill Companies, Inc./Linda Graham, photographer; 29.3a: © Dr. Jeremy Burgess/SPL/Photo Researchers; 29.3b-c: © Lee W. Wilcox; 29.4(mniumSEM): © Eye of Science/Photo Researchers; 29.5: © Lee W. Wilcox; 29.7(male): © The McGraw-Hill Companies, Inc./Claudia Lipke, photographer; 29.7(mature): © L. West/Photo Researchers; 29.7(young): © The McGraw-Hill Companies, Inc./Linda Graham, photographer; 29.9-29.10a: © Lee W. Wilcox; 29.10b: © Walter H. Hodge/Peter Arnold; 29.10c: © Patrick Johns/Corbis; 29.10d: © Rich Reid/Animals Animals; 29.11a: © The McGraw-Hill Companies, Inc./Linda Graham, photographer; 29.11b: © Lee W. Wilcox; 29.12(antheridia): © Lee W. Wilcox; 29.12(blue gametophyte): © Lee W. Wilcox; 29.12(gametangium): © Lee W. Wilcox; 29.12(gametophyte): © Lee W. Wilcox; 29.12(sori): © Carolina Biological Supply Company/Phototake; 29.12(sporangium, spore): © The McGraw-Hill Companies, Inc./Linda Graham, photographer; 29.12(sporophyte): © Barrett & MacKay Photography; 29.12(youngsporophyte): © Dr. Richard Kessel & Dr. Gene Shih/Visuals Unlimited; 29.13: © Mitsuyasu Hasabe, National Institute for Basic Biology, Japan; 29.14: © Stephen P. Parker/Photo Researchers; 29.15: © Brand X Pictures/PunchStock RF; 29.16: © Photo by Steven R. Manchester, University of Florida. Courtesy Botanical Society of America, St. Louis, MO., www.botany.org; 29.18(left): © Lee W. Wilcox; 29.18(middle): © Charles McRae/Visuals Unlimited; 29.18(right): © David R. Frazier/The Image Works; 29.19: © Marjorie C. Leggitt; 29.20: © Courtesy Prof. Roberto Ligrone. Fig. 6 in Ligrone et al., *Protoplasma* (1982) 154:414-425; 29.23c: © Lee W. Wilcox.

Chapter 30: Opener © Gallo Images/Corbis; 30.3a: © Walter H. Hodge/Peter Arnold; 30.3b: © Ed Reschke/Peter Arnold; 30.4a-b: © Lee W. Wilcox; 30.5a: © Karlene V. Schwartz; 30.5b: © Wolfgang Kaehler/Corbis; 30.5c: © B. Runk/S. Schoenberger/Grant Heilman Photography; 30.6a: © Lee W. Wilcox; 30.6b: © Bryan Pickering/Eye Ubiquitous/Corbis; 30.8a: © Zach Holmes Photography; 30.8b: © Duncan McEwan/Naturepl.com; 30.8c: © Ed Reschke/Peter Arnold; 30.10a: © Unknown Photographer/Grant Heilman Photography; 30.10b: © Ken Wagner/Phototake; 30.10c: © Lee W. Wilcox; 30.11a: © Robert & Linda Mitchell; 30.11b: © Walter H. Hodge/Peter Arnold; 30.11c: © Michael & Patricia Fogden/Corbis; 30.12: © Bill Ross/Corbis RF; 30.17: © Sangtae Kim, Ph.D.; 30.18a: © Medioimages/PunchStock RF; 30.18b: © Ed Reschke/Peter Arnold; 30.20a: © James Strawser/Grant Heilman Photography; 30.20b: © Corbis RF; 30.20c: © Image released under GFDL license. Photographer Florence Devouard, France 2003; 30.21a-e: © Lee W. Wilcox; 30.21f: © Michel Viard/Peter Arnold; 30.21g: © Biophoto/Hazan Muriel/Peter Arnold; 30.21h: © Jerome Wexler/Visuals Unlimited; 30.22a: © Eddi Boehnke/zefa/Corbis; 30.22b: © Jonathan Buckley/GAP Photo/Getty Images; 30.22c: © John Colwell/Grant Heilman Photography; 30.23(Cannabis): © Phil Schermeister/Peter Arnold; 30.23(Cannabis close): © Joao Luiz

Bulcao; 30.24: © Jack Jeffrey/Photo Resource Hawaii; 30.25: © Beng & Lundberg/naturepl.com.

Chapter 31: Opener © Brian Lightfoot/naturepl.com; 31.3a: © Fig. 16, Kaminskyj, S. G. W., and Heath, I. B. (1996). "Studies on Saprolegnia ferax suggest the general importance if the cytoplasm in determining hyphal morphology." *Mycologia* 88: 20-37. The New York Botanical Garden, Bronx, NY; Allen Press, Lawrence Kansas; 31.5a: © Agriculture and Agri-Food Canada, Southern Crop Protection and Food Research Centre, London ON; 31.5b: © CDC; 31.6: © Dr. Dennis Kunkel Microscopy/Visuals Unlimited; 31.7: © David Scharf/Peter Arnold, Inc.; 31.9a: © Felix Labhardt/Taxi/Getty Images; 31.9b: © Bob Gibbons/ardea.com; 31.10: © Rob Casey/Alamy RF; 31.11: © Gary Meszaros/Visuals Unlimited/Getty Images; 31.12: © Hans Pfletschinger/Peter Arnold; 31.13: © N. Allin & G. L. Barron, University of Guelph/BPS; 31.14a: © Nigel Cattlin/Photo Researchers; 31.14b: © Herve Conge/ISM/Phototake; 31.15: © Dr. Dennis Kunkel Microscopy/Phototake; 31.16: © Courtesy Bruce Klein. Reprinted with permission. *J Exp Med.* 1999; 189: cover illustration; 31.17: © Photo credit Dr. D. P. Donelley and Dr. J. R. Leake, University of Sheffield Department of Animal & Plant Sciences; 31.18a: © Mark Brundrett; 31.19a: © Jacques Landry, Mycoquebec.org; 31.19b: © Courtesy of Larry Peterson and Hugues Massicotte; 31.21a: © Joe McDonald/Corbis; 31.21b: © Lee W. Wilcox; 31.21c: © Ed Reschke/Peter Arnold; 31.21d: © Lee W. Wilcox; 31.22: © The McGraw-Hill Companies, Inc./Linda Graham, photographer; 31.23: Thomas Kuster, USDA, FS, Forest Products Laboratory; 31.24: © Photographs by H. Cantor-Lund reproduced with permission of the copyright holder J. W. G. Lund; 31.25a: © Peres/Custom Medical Stock Photo; 31.25b: © William E. Schadel/Biological Photo Service; 31.26: Yolande Dalpé, Agriculture and Agri-Food Canada; 31.27a: Micrograph courtesy of Timothy M Bourett, DuPont Crop Genetics, Wilmington, DE USA; 31.27b: © Charles Mims; 31.28: © Peter Arnold/Alamy; 31.29: © Clemson University - USDA Cooperative Extension Slide Series, ipmimages.org; 31.30a: © Darlyne A. Murawski/Peter Arnold; 31.30b: © Mark Turner/Botanica/Getty Images; 31.31(left): © Biophoto Associates/Photo Researchers; 31.31(right): © Dr Jeremy Burgess/Photo Researchers.

Chapter 32: Opener © Visuals Unlimited/Corbis; 32.2a: © Tom Stack/Tom Stack & Associates; 32.2b: © Doug Perrine/SeaPics.com; 32.2c: © Tui De Roy/Minden Pictures; 32.12: © Dwight Kuhn.

Chapter 33: Opener © Georgie HollandAge fotostock; 33.2a: © Jand Grushow/Grant Heilman Photography; 33.5b: © Peter Parks/OSF/Animals Animals; 33.6: © Matthew J. D'Avella/SeaPics.com; 33.8: © Biophoto Associates/Photo Researchers; 33.11a: © Daniel L. Geiger/SNAP/Alamy; 33.11b: © Wim van Egmond/Visuals Unlimited; 33.11c: © Fred Bavendam/Peter Arnold; 33.13a: © Breck P. Kent/Animals Animals; 33.13b: © Kjell Sandved/Visuals Unlimited; 33.13c: © W. Gregory Brown/Animals Animals; 33.13d: © Alex Kerstitch/Visuals Unlimited; 33.14: © Sinclair Stammers/Photo Researchers; 33.15b: © Jonathan Blair/Corbis; 33.18: © B. Runk/S. Schoenberger/Grant Heilman Photography; 33.19: © Biophoto Associates/Photo Researchers; 33.20: © Johnathan Smith; Cordaiy Photo Library Ltd./Corbis; 33.22: © John Cancalosi/Peter Arnold; 33.24a-c: © NASA/SPL/Photo Researchers; 33.25a: © Duncan Usher/ardea.

com; 33.25b: © Paul Freed/Animals Animals; 33.25c: © Dr. Dennis Kunkel Microscopy/Visuals Unlimited; 33.25d: © Roger De LaHarpe; Gallo Images/Corbis; 33.26a: © David Aubrey/Corbis; 33.26b: © Larry Miller/Photo Researchers; 33.29a: © Alex Wild/myrmecos.net; 33.29b: © Christian Ziegler; 33.31: © Wim van Egmond/Visuals Unlimited; 33.32a: © Kjell Sandved/Visuals Unlimited; 33.32b: © Richard Walters/Visuals Unlimited; 33.32c: © Franklin Viola/Animals Animals; 33.34: © Leslie Newman & Andrew Flowers/Photo Researchers; 33.37a: © OSF/Photolibrary; 33.38c: © Reinhard Dirscherl/Visuals Unlimited/Getty Images.

Chapter 34: Opener © Steve Bloom/stevebloom.com; 34.2: © Pat Morris/ardea.com; 34.3a: © Breck P. Kent/Animals Animals; 34.3b: © Jacana/Photo Researchers; 34.5a: © Valerie & Ron Taylor/ardea.com; 34.5b: © Jeff Rotman/naturepl.com; 34.5c: © OSF/Photolibrary; 34.5d: © Bill Curtsinger/National Geographic/Getty Images; 34.7a: © Reinhard Dirscherl/Visuals Unlimited; 34.7b: © Andrew Dawson/Age fotostock; 34.7c: © Brandon Cole Marine Photography/Alamy; 34.8: © Peter Scoones/SPL/Photo Researchers; 34.9: © D. R. Schrichte/SeaPics.com; 34.13a: © Gary Meszaros/Visuals Unlimited; 34.13b: © Dwight Kuhn; 34.13c: © Michael Redmer/Visuals Unlimited; 34.14a: © Gregory G. Dimijian/Photo Researchers; 34.14b: © Juan Manuel Renjifo/Animals Animals Enterprises; 34.14c: © Gary Meszaros/Photo Researchers; 34.16a: © Pat Morris/ardea.com; 34.16b: © David M. Dennis/Animals Animals; 34.16c: © Jim Merli/Visuals Unlimited; 34.17: © Michael & Patricia Fogden/Minden Pictures; 34.18a: © Royalty-Free/Corbis RF; 34.18b: © J. & C. Sohns/Animals Animals; 34.21c: © Gilbert S. Grant/Photo Researchers; 34.22a: © B. G. Thomson/Photo Researchers; 34.22b: © Jean-Claude Canton/Bruce Colemen/Photoshot; 34.22c: © Sylvain Condier/Peter Arnold; 34.22d: © Brand X Pictures/Jupiter Images RF; 34.22e: © Rick & Nora Bowers/Visuals Unlimited; 34.22f: Mervyn Rees/Alamy; 34.23a: © Kelvin Aitken/Peter Arnold; 34.23b: © Charles Krebs/Corbis; 34.23c: © Anthony Bannister/Photo Researchers; 34.24a: © Royalty-Free/Corbis RF; 34.24b: © Joe McDonald/Corbis; 34.24c: © S. J Krasemann/Peter Arnold; 34.24d: © Michel Renavdeau/Age fotostock; 34.24e: © Ken Lucas/Visuals Unlimited; 34.25a: © Martin Harvey/Peter Arnold; 34.25b: © John Shaw/Photo Researchers; 34.25c: © Paul A. Souders/Corbis; 34.26a: © Dave Watts/naturepl.com; 34.26b: © Theo Allofs/Visuals Unlimited; 34.26c: Jeffrey Oonk/Foto Naturaa/Minden Pictures; 34.28a: © David Haring/Photolibrary; 34.28b: © Lacz, Gerard/Animals Animals; 34.28c: © Martin Harvey/Corbis; 34.29a: © Joe McDonald/Corbis; 34.29b: © Creatas/PunchStock RF; 34.29c: © Royalty-Free/Tetra Images/Getty Images RF.

Chapter 35: Opener © Jeff Carroll/Age fotostock; 35.4: © Prof. Dr. Gerd Jürgens/Universität Tübingen. Image Courtesy Hanno Wolters; 35.5: © Figure adapted from (Jackson and Hake, 1999); 35.7a-b: © Lee W. Wilcox; 35.7c: © Dr. Dennis Drenner/Visuals Unlimited/Getty Images; 35.7d: © Lee W. Wilcox; 35.8a: © Lee W. Wilcox; 35.8c: © Lee W. Wilcox; 35.13a-d: © Lee W. Wilcox; 35.15: © Eye of Science/Photo Researchers; 35.16: © Figure adapted from (Jackson and Hake, 1999); 35.17: © Ed Reschke/Peter Arnold; 35.17b: © John Farmar; Cordaiy Photo Library Ltd/Corbis; 35.17c: © Steve Terrill/Corbis; 35.17d: © John Gerlach/Animals Animals; 35.18-35.19: © Lee W. Wilcox; 35.22: © Photo: Bruce Iverson; 35.23a: © Stephen Ingram/

Animals Animals; 35.23b: © Peter E. Smith, Natural Sciences Image Library; 35.26: © Lee W. Wilcox.

Chapter 36: Opener © Gerald & Buff Corsi/Visuals Unlimited; 36.1a: © Digital Photography by Ash Kaushesh, University of Central Arkansas, Conway, Arkansas 72035 USA; Image courtesy Botanical Society of America, www.botany.org St. Louis, MO 63110; 36.1b(all): © Lee W. Wilcox; 36.11(all): © Prof. and Mrs. M. B. Wilkins/University of Glasgow; 36.12a(bottom): © Pat O'Hara/Corbis; 36.12a(middle): © Robert Maier/Animals Animals; 36.12a(top): © Pat O'Hara/Corbis; 36.12b(top and bottom): © Henry Schleichkorn/Custom Medical Stock Photo; 36.12b(middle): © Michael Gadomski/Animals Animals; 36.13: © Lee W. Wilcox; 36.16: © Courtesy Dr. Malcolm Drew, Texas A&M University; 36.18: © G. R. 'Dick' Roberts/Natural Sciences Image Library.

Chapter 37: Opener © Dwight Kuhn; 37.2: © Walter H. Hodge/Peter Arnold; 37.3a: © Raymon Donahue; 37.3b: © Raymon Donahue; 37.4: © Royal Botanical Gardens Kew/Wellcome Trust Medical Photographic Library; 37.5: © Confocal fluorescence micrograph courtesy Simon D. X. Chuong; 37.6: © Geoff Kidd/SPL/Photo Researchers; 37.7b: © Courtesy of C. A. Stiles, University of Wisconsin; 37.7c: © Ruddy Gold/Age fotostock; 37.8: © USDA Photo; 37.13: © Photo courtesy of Jayne Belnap, U.S. Geological Survey; 37.16: © Eberhard Grames/Bilderberg/Peter Arnold; 37.18: © J. Hyvönen; 37.18(Inset): © Photo by Birgitta Bergman, Department of Botany, Stockholm University (Sweden); 37.19: © M. Kalab/Custom Medical Stock Photo; 37.21a: © Dani/Jeske/Animals Animals; 37.21b: © Breck P. Kent/Animals Animals; 37.22: © B. Runk/S. Schoenberger/Grant Heilman Photography.

Chapter 38: Opener © Barry Mason/Alamy RF; 38.8: © James S. Busse; 38.10: © Jerry Cooke, Inc./Animals Animals; 38.11b: © Astrid & Hanns - Frieder Michler/Photo Researchers; 38.12b: © John N. A. Lott/Biological Photo Service; 38.12c: © John N. A. Lott/Biological Photo Service; 38.13: © David Cavagnaro/Peter Arnold; 38.16: Adalberto Rios Szalav/Sexto Sol/Photodisc Blue/Getty Images RF; 38.17: © Andrew Syred/Photo Researchers; 38.19a: © B. Runk/S. Schoenberger/Grant Heilman Photography; 38.19b: © Dr. William J. Weber/Visuals Unlimited; 38.20a-38.22: © Lee W. Wilcox.

Chapter 39: Opener © E. R. Degginger/Animals Animals; 39.3b: © Lee W. Wilcox; 39.5a: © Natural Sciences Image Library; 39.5b: © Lee W. Wilcox; 39.7a: © Photo: Bruce Iverson; 39.7b: © Kate Land Photography, iamkate.com; 39.8a: © Richard Shiell/Animals Animals; 39.8b: © Richard Jorgensen; 39.10: Burke/Triolo Productions/Getty Images RF; 39.9a-b: © Courtesy of John Innes Centre; 39.11b: © RMF/Scientifica/Visuals Unlimited; 39.12(Top): Ed Reschke/Peter Arnold; 39.14: © Courtesy J. M. Escobar-Restrepo and A. J. Johnston, University of Zurich, Institute of Plant Biology. From "Bernasconi et al. *Science*, Feb 13, 2004. p. 973, Fig. 2; 39.19a-b: © Lee W. Wilcox; 39.22: © Lee W. Wilcox.

Chapter 40: Opener © John Rowley/Digital Vision/Getty Images RF; 40.2(left): Michael Abbey/Photo Researchers, Inc.; 40.2(middle): © Sinclair Stammers/Photo Researchers, Inc.; 40.2(right): © Dr. Richard Kessel/Visuals Unlimited; 40.5(inset a): © Dennis Kunke Microscopy/Phototake; 40.5(inset b): © Ed

Reshke/Peter Arnold; 40.5(inset c): © Innerspace Imaging/Photo Researchers, Inc.; 40.5(inset d): © Dr. John D. Cunningham/Visuals Unlimited; 40.5e-f: © The McGraw-Hill Companies, Inc./Al Tesler, photographer; 40.8(left): © Dr. Stanley Flegler/Visuals Unlimited; 40.8(middle): © Dr. Stanley Flegler/Visuals Unlimited; 40.8(right): © Dr. David M. Phillips/Visuals Unlimited; 40.10a: © Biophoto Associates/Photo Researchers; 40.10b: © Pete Oxford/Minden Pictures; 40.10c: © Thomas Deerinck/Visuals Unlimited; 40.10d: © Anthony Bannister/Photo Researchers; 40.15(inset): © Mitch Wojnarowicz/The Image Works.

Chapter 41: Opener © James Cavallini/Photo Researchers; 41.2a: © James Cavallini/BSIP/Phototake; 41.5a: Courtesy of Dr. R. F. Rakowski, Ohio University, Athens, OH; 41.14b: © The McGraw-Hill Companies, Inc./Al Tesler, photographer.

Chapter 42: Opener © Courtesy Ann-Shyn Chiang, Tsing Hua Chair Professor/Brain Research Center & Institute of Biotechnology/National Tsing Hua University; 42.6b: © Jan Leestma, M.D./Custom Medical Stock Photo; 42.13a: © David Wrobel/Visuals Unlimited; 42.15(both): © Dr. Scott T. Grafton/Visuals Unlimited; 42.17: © The McGraw-Hill Companies, Inc./Al Tesler, photographer.

Chapter 43: Opener © Jane Burton/Bruce Coleman/Photoshot; 43.9: © Gary Meszaros/Photo Researchers; 43.12: © Daniel Heuclin/Photo Researchers; 43.14a: © Eye of Science/Photo Researchers; 43.16: © Omikron/Photo Researchers; 43.17(right): © Dr. Donald Fawcett/Visuals Unlimited; 43.19b: © Royalty-Free/Corbis RF; 43.23a: © Corbis/SuperStock RF; 43.23b: © Cal Vornberger/Peter Arnold; 43.24b: © Courtesy of Russell Jurenka, Iowa State University; 43.24c(left): © Anthony Bannister/NHPA/Photoshot; 43.24c(right): © Courtesy of Louisa Howard, Dartmouth College; 43.28a-c: © The Royal Society for the Blind of South Australia Inc.

Chapter 44: Opener © Stephen Dalton/Photo Researchers; 44.1a(inset): © Dwight Kuhn; 44.1b: © Michele Hall/SeaPics.com; 44.1c: © Georgette Douwma/Photo Researchers; 44.1c(inset): © The Natural History Museum London; 44.7: © Reprinted by permission from Macmillan Publishers Ltd: Stedman, Hansell H. et al. Myosin gene mutation correlates with anatomical changes in the human lineage, *Nature*, Vol. 428, Issue Number 6981, pages 415-418, 2004; 44.11a: © Don W. Fawcett/Photo Researchers; 44.12(inset a): © Mark Richards/PhotoEdit; 44.17a: © Dr. LR/Photo Researchers; 44.17b(both): © James C. Weaver and Paul K. Hansma.

Chapter 45: Opener © Richard Hutchings/PhotoEdit; 45.2a: © Daniel Hornbach and Tony Deneka, Courtesy of Daniel Hornbach, Macalester College; 45.2b: © 2006 James Watt/www.norbertwu.com; 45.3a: © David A. Northcott/Corbis; 45.3b: © John Giustina/Photolibrary RF; 45.3c: © Kim Karpeles/Age fotostock; 45.3d: © Reinhard Dirscherl/WaterFrame - Underwater Images/Photolibrary; 45.3e: © CFranke/F1Online RF/Photolibrary RF; 45.4a: © Royalty-Free/Corbis RF; 45.4b: © Pascal Goetgheluck/Photo Researchers.

Chapter 46: Opener © AP Photo/The Rockefeller University; 46.9a(inset): © Darren Bennett/OSF/Animals Animals; 46.9b(inset): © William S. Clark;

Frank Lane Picture Agency/Corbis; 46.13: © Frank Oberle/Stone/Getty Images; 46.14: © Nutscode/T Service/Photo Researchers; 46.17(both): © Courtesy of Uffe Midtgård, University of Copenhagen.

Chapter 47: Opener © SPL/Photo Researchers; 47.5: © Dr. David Phillips/Visuals Unlimited/Getty Images; 47.7a (inset): © Dennis Kunkel Microscopy; 47.11(left): © Laurent/BSIP; 47.13(inset): © Carolina Biological Supply/Visuals Unlimited; 47.15: © Ed Reschke/Peter Arnold; 47.21: © Biophoto Associates/SPL/Photo Researchers; 47.22(inset): © Sovereign/TSM/Phototake

Chapter 48: Opener © Xela/Alamy RF; 48.3a: © Ken Lucas/Visuals Unlimited; 48.3b: © Courtesy Alan Cressier; 48.4a: © Hal Beral/V+W/The Image Works; 48.4b: © Jane Burton/Bruce Coleman/Photoshot; 48.5a: © B. Runk/S. Schoenberger/Grant Heilman Photography; 48.5c: © Electron Microscopy Unit, Royal Holloway University of London; 48.7(inset): © Ed Reschke/Peter Arnold; 48.9: © Ed Reschke/Peter Arnold; 48.10b(both): © Pr. M. Brauner/Photo Researchers, Inc.; 48.17(left): © Astrid & Hanns-Frieder Michler/SPL/Photo Researchers; 48.17(right): © Dr. John D. Cunningham/Visuals Unlimited.

Chapter 49: Opener © Guido Alberto Rossi/Photolibrary; 49.12: Steve Gschmeissner/SPL/Photo Researchers.

Chapter 50: Opener © Sovereign/ISM/Phototake; 50.6a: © S. Goodwin & Dr. Max Hincke, Division of Clinical and Functional Anatomy, University of Ottawa; 50.6b: © Bob Daemmrich/The Image Works; 50.7: © Carolina Biological Supply/Visuals Unlimited; 50.13(left, right): © 2007 Massachusetts Medical Society. All rights reserved. *New England Journal of Med.* Vol. 340, p. 524, 1999; 50.13(middle): © Bart's Medical Library/Phototake.

Chapter 51: Opener © David Liebman/Pink Guppy; 51.1a: © Clouds Hill Imaging Ltd./Corbis; 51.1b: © Norbert Wu/Peter Arnold; 51.3b: © Dr. Stanley Flegler/Visuals Unlimited; 51.3d: © P. M. Motta & G. Familiari/Univ. La Sapienza/Photo Researchers; 51.4: © Rannels/Grant Heilman Photography; 51.12a: © Claude Edelmann/Photo Reseachers; 51.12b-c: © Photo Lennart Nilsson/Albert Bonniers Forlag AB, *A Child Is Born*, Dell Publishing Company; 51.15(all) © The McGraw-Hill Companies, Inc./Jill Braaten, Photographer

Chapter 52: Opener © Matsui et al. Ethanol and Vision Development, *IOVS*, October 2006, Vol. 47, No. 10, p. 4591. http://www.iovs.org/cgi/reprint/47/10/4589.pdf. Image Courtesy of Jonathan Matsui; 52.2c(all): © Courtesy Prof. M. Whitaker, University of Newcastle upon Tyne; 52.4(bottom): © Tom Fleming; 52.6: © Courtesy Hiroki Nishida, Biological Sciences, Osaka University; 52.10a: © Courtesy Dr. Laurence D. Etkin, University of Texas M.D. Anderson Cancer Center. With permission of Malgorzata Kloc. This article was published in *Mechanisms of Development* Vol. 75, No. 1-2, Malgorzata Kloca, Carolyn Larabellb, Agnes Pui-Yee Chana and Laurence D. Etkin "Contribution of METRO pathway localized molecules to the organization of the germ cell lineage," pages 81-93. Elsevier Science Ireland Ltd. July 1998; 52.10b: © F. R.Turner/Indiana University; 52.11a-d: © Courtesy Kathryn Tosney; 52.12b: © Ed Reschke/Peter Arnold; 52.17(5): © Richard Harland, U. C. Berkeley; 52.18(all): © Courtesy Edward M. King.

Chapter 53: Opener © SPL/Photo Researchers.

Chapter 54: Opener © David Muench/Corbis; 54.1: © David M. Dennis/Animals Animals; 54.2a: © Brand X Pictures/PunchStock RF; 54.2b: © Biosphoto/Pons Alain/Peter Arnold; 54.2c: © Brand X Pictures/PunchStock RF; 54.2d: © Art Wolfe/Photo Researchers; 54.7b: © Michael McCoy/Photo Researchers, Inc.; 54.10: © Raymond Gehman/Corbis; 54.9: Jonathan Bird/Peter Arnold; 54.13a: © Heather Angel/Natural Visions; 54.13b: © Biophoto Associates/Photo Researchers; 54.15a: © Gregory Ochocki/Photo Researchers; 54.15b: © Image courtesy of FGBNMS/UNCW-NURC/NOAA; 54.16: © Virginia P. Weinland/Photo Researchers; 54.17a: © Peter Wakely/English Nature; 54.17b: © G. A. Matthews/SPL/Photo Researchers; 54.26a: © Jean-Paul Ferrero/ardea.com; 54.26b: © Theo Allofs/theoallofs.com; 54.26c: © Scott W. Smith/Animals Animals; 54.26d: © altrendo nature/Altrendo/Getty Images; 54.26e: © Tom and Pat Leeson; 54.26f: © Joe McDonald/Visuals Unlimited; 54.26g: © D. Robert & Lorri Franz/Corbis; 54.26h: © Joe McDonald/Visuals Unlimited/Getty Images; 54.26i: © Art Wolfe/Getty Images; 55.6: © Nina Leen/Time Life Pictures/Getty Images; 54.26j: © Michio Hoshino/Minden Pictures; 54.26k: © Howie Garber/Animals Animals; 54.27a: © Nature Picture Library/Alamy; 54.27b: © Stephen Frink/Corbis; 54.27c: © Jeffrey L. Rotman/Corbis; 54.27d: © Phillip Colla/OceanLight/Bruce Coleman/PhotoShot; 54.27e: © Tom and Pat Leeson; 54.27f: Larry Mulvehill/Photo Researchers.

Chapter 55: Opener © James Balog/Getty Images; 55.3a-b: © L. P. Brower; 55.4(all): © Lilo Hess/Time Life Pictures/Getty Images; 55.5: © Joe McDonald/Corbis; 55.8(inset): © Frans Lanting/Minden Pictures; 55.11a: © Tony Camacho/Photo Researchers; 55.11b: © Gregory G. Dimijian/Photo Researchers; 55.11c: © Wolfgang Kaehler/Corbis; 55.12a: © Phil Degginger/Alamy; 55.12b: © Kuribayashi Satoshi/OSF/Photolibrary; 55.12c: © J. E. Lloyd, UF Entomology; 55.13a: © Mark Moffett/Minden Pictures; 55.15: © BIOS/Peter Arnold; 55.17: © The McGraw-Hill Companies, Inc./

Peter Stiling, photographer; 55.18: Danita Delimont/Alamy; 55.19: © Raymond Mendez/Animals Animals; 55.20: © Stephen A. Marshall; 55.22b: © Frans Lanting/Minden Pictures; 55.23a: © Masahiro Iijima/ardea.com; 55.23b: © Bill Terry/Grant Heilman Photography; 55.23c: © Millard H. Sharp/Photo Researchers; 55.24: © Chris Knights/ardea.com.

Chapter 56: Opener © Mike Lockhart; 56.1a: © Paul Glendeli/Peter Arnold; 56.1b: © Nigel Cattlin/Photo Researchers; 56.1c: © AP Photo/The Baxter Bulletin, Kevin Pieper; 56.1d: © Cyril Ruoso/JH Editorial/Minden Pictures; 56.2: © W. Wayne Lockwood, M.D./Corbis; 56.3a: © Phil Banko/Corbis; 56.3b: © Fritz Polking; Frank Lane Picture Agency/Corbis; 56.3c: © Bob Krist/Corbis; 56.4a: © A. Paul Jenkin/Animals Animals; 56.4b: © John Foxx/ImageState RF/Agefotostock RF; 56.4c: © M. Watson/ardea.com; 56.15: © Jenny Hager/The Image Works.

Chapter 57: Opener © Stephen Wong & Takako Uno; 57.10a: © Dr. Thomas Eisner/Visuals Unlimited; 57.10b: © Hans D. Dossenbach/ardea.com; 57.10c: © Thomas Aichinger/V&W/The Image Works; 57.10d(left): © Suzanne L. & Joseph T. Collins/Photo Researchers; 57.10d(right): © Michael Fogden/OSF/Animals Animals; 57.10e: © Paul Springett/Alamy RF; 57.11: © Tom & Pat Leeson/Photo Researchers; 57.14a: © Jim Tuten/Animals Animals; 57.14b: Toyofumi Mori/The Image Bank/Getty Images; 57.14c: © Gilbert S. Grant/Photo Researchers; 57.14d: © Donald Specker/Animals Animals; 57.16: © Compost/Peter Arnold; 57.19: © Dan Honda/Contra Costa Times/Newscom; 57.20a: © Dynamic Graphics Group/Creatas/Alamy RF; 57.20b: © Michael & Patricia Fogden/Corbis; 57.21: © Mike Wilkes/naturepl.com; 57.22a: © E. A. Janes/Age fotostock/Maxximages.com; 57.22b: © Dwight Kuhn; 57.24a-b: © From Alan P. Dodds *The Biological Campaign against Prickly Pear*, Published under the Authority of the Commonwealth Prickly Pear Board. Brisbane, Queensland. 30th October, 1940.

Chapter 58: Opener © Wolfgang Kaehler/Corbis; 58.2: © G. R. "Dick" Roberts/Natural Sciences Image

Library; 58.6b: © Plowes Proteapix; 58.10a: © Charles D. Winters/Photo Researchers; 58.9a: © AP Photo/Gary Stewart; 58.9b: © David M. Dennis/Animals Animals; 58.11a: © Tom Bean; 58.11b: © Courtesy of Lawrence R. Walker; 58.11c: © Howie Garber/Accent Alaska.com; 58.11d: © Tom Bean; 58.12: © Wayne Sousa/University of California, Berkeley; 58.17(all): © Courtesy Dr. D. Simberloff, University of Tennessee.

Chapter 59: Opener © Juan Carlos Muñoz/Age fotostock; ; 59.1: © Stephen Dalton/Photo Researchers; 59.9: © R. T. Smith/ardea.com; 59.16: © Courtesy of Experimental Lakes Area, Fisheries and Oceans Canada. Reproduced with the permission of the Minister of Public Works and Government Services Canada, 2009; 59.19(all): © The McGraw-Hill Companies, Inc./Peter Stiling, photographer; 59.21a: © Science VU/Visuals Unlimited; 59.21b: © Provided by the Northeastern Forest Experiment Station, Forest Service, USDA; 59.25: © Frans Lanting/Minden Pictures.

Chapter 60: Opener © David Cappaert, Michigan State University, Bugwood.org; 60.1: © Willard Clay Photography; 60.2: © Heather Angel/Natural Visions; 60.4: © Pete Manning, Ecotron Facility, NERC Centre for Population Biology; 60.7a: © The Bridgeman Art Library; 60.7b: © Topham/The Image Works; 60.7c: © James T. Tanner/Photo Researchers; 60.8: © Bruce Coleman/Photoshot; 60.11: © Vicki Fisher/Alamy; 60.13: © Andrew Parker/Alamy; 60.15a: © Robert E. Barber/Alamy; 60.15b: © Rick A. Brown/rick@moosephoto.com; 60.15c: © T Kitchin & V Hurst/NHPA/Photoshot; 60.16: © Johnny Johnson/Animals Animals - Earth Scenes; 60.17a: © University of Wisconsin-Madison Arboretum; 60.17b: © Courtesy of DL Rockwood, School of Forest Resources and Conservation, University of Florida, Gainesville, FL; 60.17c: © Sally A. Morgan; Ecoscene/Corbis; 60.18a: © Corbis; 60.18b: © Tom Brownold 2006; 60.19: AFP/Getty Images.

Index

Page numbers followed by *f* denote figures; those followed by *t* denote tables.

A

ABA (abscisic acid), 754*t*, 761, 803, 804*f*, 826
A band, 922, 922*f*, 923*f*
ABC model for flower development, 408, 409*f*
Abelson leukemia virus, 296*t*
abiotic interactions, 1133, 1137, 1137*f*, 1139*t*
abl gene, 295–296, 295*f*, 296*f*
ABO blood groups
  inheritance of, 343*t*, 344–345
  transfusion incompatibilities and, 1129
abortion, 324, 1088
abscisic acid (ABA), 754*t*, 761, 803, 804*f*, 826
abscission, leaf, 803–806, 805*f*
absolute refractory period, 859, 859*f*
absorption
  definition of, 937, 937*f*
  in large intestine, 950
  mechanisms of, 945
  of nutrients
    amino acids, 961*f*, 962
    carbohydrates, 961, 961*f*
    triglycerides, 961, 961*f*
absorption spectrum, 161, 162*f*
absorptive nutrition, of fungi, 631–632
absorptive state
  nutrient absorption, use, and storage, 960, 961–962, 961*f*, 963*t*
  regulation of, 963–966
acacia trees, mutualistic relationship with ants, 1220, 1220*f*
Acanthodii, 703–704
Acanthoplegma, 575*f*
acceptor stem, tRNA, 252, 252*f*
acclimatization, 976
accommodation, 904, 904*f*
ACE (angiotensin-converting enzyme) inhibitors, 1*f*
Acetabularia, 567*f*
acetaminophen, 100
acetic acid, 44
acetylation, of histones, 276–277, 276*f*
acetylcholine
  discovery of, 865, 866*f*, 867
  functions of, 864, 864*t*
  receptors for, 190–191, 927, 928*f*, 934
  release at neuromuscular junctions, 927, 928*f*
acetylcholinesterase, 927, 928*f*, 934
acetyl CoA, 140–141, 141*f*
acetyl group, production from pyruvate, 138–139, 140–141, 141*f*
acetylsalicylic acid, 1, 1*f*, 119
achondroplasia, 343
acid
  amino acids, 53–54, 55*f*
  definition of, 39
  strong, 39
  weak, 39
acid-base buffers, 40
acid hydrolases, 83
acidic solution, 39
acid rain, 1145–1146, 1258, 1259–1260, 1260*f*
acoelomates, 655, 656*f*, 663
acquired immune deficiency syndrome (AIDS), 377, 1130
acquired immunity. *See* immunity, acquired
acrocentric, centromere location, 321, 321*f*
acromegaly, 1063, 1064*f*
acrosomal reaction, 1094, 1094*f*

acrosome, 1074, 1075*f*
ACTH (adrenocorticotropic hormone), 191–193, 1051, 1052*t*
actin filaments
  anchoring junctions and, 203, 203*f*, 204, 204*f*
  cytokinesis, role in, 314
  description, 74*t*, 75
  fluorescently-labeled, 147, 148*f*
  in skeletal muscle structure and function, 922–926, 922*f*, 923*f*, 925*f*, 926*f*
Actinistia, 700*f*, 702*t*, 703, 705, 706
actinobacteria, in woody plant symbioses, 784
Actinopterygii, 10, 10*f*, 700*f*, 702*t*, 703, 705–706, 706*f*
action potentials
  conduction of, 860–861, 860*f*, 861*f*
  generation of, 858–860, 859*f*
  in muscle, 926, 927, 927*f*, 928*f*
  in sensory receptors, 893, 894*f*
action spectrum, of photosynthesis, 161–162, 162*f*
activation energy, 123, 123*f*
activator
  combinatorial control, 273
  description of action, 264, 265*f*
  eukaryotic, 273, 275, 275*f*, 276*f*, 277, 277*f*
  lac operon, 271, 271*f*
active site, enzyme, 123–125, 124*f*, 125*f*
active transport, 791*f*, 792, 792*f*, 839, 1025, 1038, 1040
  definition of, 106, 113
  ion pumps, 113–115, 114*f*
  primary, 113, 114*f*
  secondary, 113, 114*f*
activin, morphogenic activity of, 1104, 1104*f*
adaptations. *See also specific characteristics*
  cellular response, 178
  definition of, 494
  to local environments, 519
adaptive radiation, 514*f*, 515
adenine, 61–62, 62*f*, 220, 221*f*
adenosine deaminase (ADA) deficiency, 427–428, 428*f*
adenosine triphosphate (ATP)
  hydrolysis of
    cellular processes driven by, 130–131, 130*t*, 131*f*
    chemical equilibrium for, 122
    description of, 121, 121*f*
    exergonic reaction, 122
    ion pumps and, 114–115, 114*f*
  production of
    in anaerobic respiration, 151, 151*f*
    by chemiosmosis, 129
    in citric acid cycle, 139, 142, 142*f*, 143*f*
    in glycolysis, 129, 138, 138*f*, 139–140, 140*f*–141*f*
    in mitochondria, 86
    by oxidative phosphorylation, 139, 142, 144*f*, 145–147, 146*f*
    in photosynthesis light reactions, 162–164, 163*f*, 164*f*, 168
    by substrate-level phosphorylation, 129, 138, 138*f*, 139, 140, 142
  use of
    in Calvin cycle, 168, 169*f*, 170
    in charging tRNAs, 253, 253*f*
    in cross-bridge cycle, 924–925, 925*f*
    in endergonic reactions, 122
    as energy intermediate, 129
    by proteins as energy source, 130–131, 130*t*
adenovirus, 370*t*, 371, 372*f*
adenylyl cyclase, 186, 186*f*, 187*f*
ADH. *See* antidiuretic hormone (ADH)
adherens junctions, 203, 203*f*
adhesion, 38

adhesive proteins, in extracellular matrix, 198–199, 199*t*
adiabatic cooling, 1148
adipose tissue
  brown, 976
  as connective tissue, 835*f*
  endocrine functions of, 1046*f*, 1060
adrenal cortex, 1066, 1067*f*
  cortisol release from, 1056
  endocrine functions of, 1046*f*
adrenal glands
  anatomy of, 1067*f*
  endocrine functions of, 1046*f*, 1056, 1066–1067
adrenal medulla, 1046*f*, 1056, 1066, 1067*f*
adrenocorticotropic hormone (ACTH), 191–193, 1051, 1052*t*
adventitious roots, 733*t*, 736
aequorin, 11
aerenchyma, 766, 766*f*
aerobe, obligate, 557
aerobic respiration, 137–150
afferent arteriole, of glomerulus, 1037, 1037*f*
affinity, enzyme-substrate, 124
aflatoxins, 637
African forest elephant (*Loxodonta cyclotis*), 528, 528*f*, 531
African long-tailed widowbird (*Euplectes progne*), 1178–1179
African wildcat (*Felis libyca*), 1280
agar, 756
age classes, population, 1188, 1188*f*
age-specific fertility rate, 1192
age structure, population, 1200, 1200*f*
aging, telomere shortening and, 233
Agre, Peter, 110–112, 1041
agriculture
  angiosperm domestication and, 628–629, 629*f*
  Bt crops, 424, 425*f*
  habitat destruction and, 1271
  selective breeding, 481–482, 482*f*, 483*f*
Agrobacterium tumefaciens, 423–424, 424*f*
  chromosomes of, 432
  gall formation by, 549, 549*f*
  type IV secretion system of, 560, 560*f*
Agrostis tenuis, 496, 497*f*
Aguinaldo, Anna Marie, 660, 661*f*
AIDS (acquired immune deficiency syndrome), 377, 1130
Aintegumenta gene, 408, 408*t*
air sacs, 716–717, 1012–1014, 1012*f*, 1013*f*–1014*f*
akinetes, 556, 556*f*
alanine, 55*f*
alarm call, 1176–1177, 1177*f*
albinism, 327, 327*f*
alcohol
  effects on neurotransmission and mood, 869*t*
  fetal alcohol syndrome (FAS), 1108
  metabolism of, 79
aldehyde (functional group), 45*t*
aldosterone
  action of, 1040, 1040*f*, 1049, 1062*f*, 1063
  synthesis of, 1050*f*
aleurone, 827
Alexander, Richard, 1177
algae
  brown, 574, 574*f*, 1143*f*
  contractile vacuoles in, 83*f*, 84
  cryptomonads, 572, 572*f*
  diatoms, 566*f*
  distribution, effect of light on, 1144, 1144*f*
  food storage molecules of, 577
  green (*See* green algae)